Refrigeration, Air Conditioning, and Cold Storage

REFRIGERATION, AIR CONDITIONING, and COLD STORAGE

PRINCIPLES AND APPLICATIONS

FIRST BRITISH EDITION REVISED AND ENLARGED

RAYMOND C. GUNTHER

Colonel, USAR Ret.; Registered Professional Engineer; Formerly, Officer in Charge, Refrigeration School, Army Service Forces Training Centre

BAILEY BROTHERS AND SWINFEN LTD

FOLKESTONE

Published by Bailey Bros. & Swinfen Ltd.
1973
First British edition revised.

Second revised and enlarged edition.

SBN: 561–00138–3

Printed in England by Kingprint Ltd, Richmond, Surrey.

Dedicated to

My Mother and Father

Annie Gilmore Gunther and Frederick A. Gunther

Preface

This revised edition includes all the material contained in the original volume, in addition to a number of new subjects whose techniques have been developed since the date of the first publication. Retaining the original material is feasible because the work is based on fundamental principles, which never change but are often modified, expanded, or substituted by other techniques to attain a desired result.

The subject matter remains in its original chronological order related to the three phases of refrigeration, i.e., (1) development, (2) production, and (3) application. Nomenclature has been changed, where appropriate, to meet the trend posed by modern usage toward more logical definition, or truer description. While many changes have been made in outward appearance, most of the equipment continues to successfully operate on time-proven principles but with modification in some areas.

New equipment has been developed to broaden the industry. Included in this category are screw compressors, cold-gas refrigerators, cryogenic freezers, thermoelectric coolers, magnetic refrigerators, vortex-tube refrigerators, single-stage centrifugal compressors, two-stage reciprocating compressors, and cryogenically cooled containers and transportation vehicles. New type controls, such as transducers, thermistors and other solid state devices are coming into more general use. All of the foregoing are described in the revised work.

The continuing development of space equipment, containerization and more rapid product freezing methods promises greater changes in the future. More efficient and effective engineering materials, semi-conductors and lubricants are some of the keys to future progress.

An expression of appreciation to all those individuals and organizations who so kindly cooperated in amassing the material, both written and portrayed within this book and acknowledged in the first edition, is repeated here. This also includes the author's wife, Almira Doak Gunther, who typed the original and revised manuscripts and continued to serve as grammatical critic and counselor throughout both texts.

RAYMOND C. GUNTHER

Westgate Hills,
Havertown, Pennsylvania

New equipment has been developed to broaden the industry. Included in this category are screw compressors, [illegible] rotary [illegible], thermoelectric [illegible], vortex tube [illegible] single-stage [illegible] centrifugal compressors, two-stage [illegible] compressors, [illegible] and [illegible] heat exchangers, thermoelectric and other solid state devices [illegible]. All of the foregoing are detailed in the revised work.

The recent development of [illegible] equipment, [illegible] construction and [illegible] progress [illegible] greater changes in the [illegible] and electronic [illegible] are some of the keys to future progress.

[illegible]

[illegible], Pennsylvania

Contents

Refrigeration, Air Conditioning, and Cold Storage

STUDY OUTLINE FOR CHAPTER 1

Definition of refrigeration:
A process of reducing temperature and removing heat from any substance.

Purpose of refrigeration:
(*a*) Reduce the temperature of a substance,
(*b*) Transform a substance from one state to another, as by condensation or freezing,
(*c*) Maintain substances in a desired state.

Methods of refrigeration development:
(*a*) Simple temperature difference,
(*b*) Evaporation,
(*c*) Combination of (*a*) and (*b*).

Classification of systems:
Domestic (household)
Commercial
Industrial
Air conditioning
Marine

Principle of mechanical refrigeration:
Transformation of a liquid to a gas and reconversion of that gas to its liquid phase.

Sequence of functions:
First step, Evaporation at reduced pressure
Second step, Compression, act of increasing pressure
Third step, Condensation at elevated pressure
Fourth step, Pressure reduction

Basic operational equipment:
Evaporator
Compressor
Condenser
Throttling device (expansion valve)

Operational stages:
Low pressure
High pressure

Mechanical refrigeration cycle:
Completion of all four functions
Evaporation
Compression
Condensing
Expansion

Adaptation:
(*a*) Direct expansion, in which the cooling element (evaporator) is directly exposed to substances to be cooled,
(*b*) Indirect expansion, in which the cooling element is exposed to a secondary medium (air, water, etc.), which, in turn, is exposed to the substance to be cooled.

I

Introduction to Refrigeration

The ice cream we buy at the corner drugstore, the frozen vegetables we have for dinner, the refreshing water we drink at the office water cooler are things we take for granted which are all dependent on the science of refrigeration. Without it there would be no air conditioning in offices, homes, or factories; no refrigerators or home freezers to keep food; no ice cubes for cold drinks on hot summer days. The outer space programs and many other scientific and commercial activities would not succeed if environmental conditions could not be controlled by refrigeration processes.

While refrigeration in its specialized forms is a comparatively modern development, it has been practiced for generations. The springhouse on the farm might be considered an example of nature's refrigeration technique. The porous clay jugs used in hot desert countries to cool water employ an important refrigeration principle. The icebox, widely used to preserve food from spoiling, utilizes another principle.

The purpose of the author in this book is to explain the various principles involved in the science of refrigeration, and their application. The study outline and the summary of each chapter will aid in assimilation of the material.

GENERAL DEFINITION AND DEVELOPMENT

Refrigeration may be defined as the process of removing heat from a substance. The American Society of Refrigerating Engineers defines refrigeration as "the science of providing and maintaining temperatures below that of the surrounding atmosphere." This implies the development of temperature differentials rather than the establishment of a given temperature level.

The science of refrigeration utilizes several methods of providing a temperature differential. They vary from the simplicity of the springhouse, where cool water removes heat from warm fresh milk, to the complex machinery required for the manufacture of ice cream or the production of properly conditioned air for theaters, auditoriums, industrial plants, aircraft, space capsules, laboratories, to name a few.

The simplest way of producing refrigeration is to expose a warm substance to a colder one, thereby setting up a flow of heat energy. Since heat always flows from a warmer object to a colder one, an exchange of heat takes place and the warmer body becomes colder and the colder one warmer. This is the process which takes place in the springhouse where containers of milk, butter and other foods are placed in cold water. It is the *direct* heat transfer method of refrigeration.

An example of the *indirect* method of heat transferral is the icebox, which introduces a third substance. Ice placed in the upper compartment of the box absorbs heat from the trapped air and reduces its temperature. As the air becomes colder, it grows heavier and flows to the bottom compartment of the box where food is stored. There it forces out the lighter, warmer air, establishing a circulatory movement within the icebox. The colder air flowing from the ice compartment above absorbs heat from the warmer food in the lower compartment. As its temperature increases, the air becomes lighter and again rises to the ice compartment, where it transfers heat to the ice. This circulatory action continues until all the ice melts. An initial heat transfer occurs during exposure of the warm food to the cold air. A secondary transfer of heat takes place when the warmed air from the lower compartment comes in contact with the ice. Heat from the food thus is transferred to the ice by means of the air.

If the ice in the chest were replaced by a coil through which flowed cold water, cooled by mechanical means, an additional refrigerating medium would be introduced. The heat absorbed by the air from the food would be transferred to the cold water, which, in turn, would be cooled by a fluid in the mechanical refrigeration system. Such use of more than one medium of heat transfer involves a complex secondary method of refrigeration.

Refrigeration is accomplished also by evaporation of liquids. The liquids are exposed directly or indirectly to warmer substances, from which they absorb heat. A common method of cooling water in hot desert countries is to place it in porous clay jars. The water gradually seeps through the pores to the outer surface, where it evaporates into the dry desert atmosphere. Heat is absorbed during the evaporation process and carried away by the resultant vapor. In the case of the porous jar, the heat of evaporation is absorbed from the jar and its contents, thus lowering their temperature. The same principle applies when a cooling sensation is experienced by waving wet hands vigorously in the air, or allowing a breeze to sweep over perspiring parts of the body.

Refrigeration may be accomplished either by temperature differential or by evaporation. It may be brought about also by using a combination of both methods of removing heat from a substance.

PURPOSE

The purposes of refrigeration are threefold: to reduce the temperature of a substance (the act of cooling); to transform a substance from one state to another (as water into ice); and to maintain substances in a desired state (as in the storage of ice or the preservation of food).

Its purpose can be understood better by describing its effects.

If the temperature of a gas is lowered sufficiently, it will be transformed to its liquid state. For example, when exposed to a cooling medium, such as cold air or water, steam will liquefy and return to its natural state, water. Water can be transformed into a solid (ice) if its temperature and heat content are reduced sufficiently. One of the effects of refrigeration is to change a gas into a liquid, or a liquid into a solid. When refrigeration is applied to water, the solid, ice, will be produced. Production of ice is one purpose of refrigeration.

Many micro-organisms, such as bacteria, yeasts, and molds, cause spoilage of foods—meats, dairy products, eggs, fruits, and vegetables. At normal room temperatures, these micro-organisms are active, causing food decomposition in varying degrees, the rate depending on the type and the condition of the food product. If the temperature of these foods is reduced sufficiently, the action of these micro-organisms is retarded greatly, making it possible to keep foods longer. The preservation of foods is therefore another highly important use for refrigeration.

The scope of the science of refrigeration is indicated by this list of its uses:

Production of ice and ice storage
Air conditioning
Cold storage of food products, flowers, furs, and medicines
Quick freezing of food products
Brewing
Ice cream manufacturing and storage
Water coolers
Display cabinets—meat, candy, etc.
Household refrigeration
Skating rinks
Chemical manufacturing
Rubber products manufacturing
Candy making
Industrial testing laboratories
Creamery operation
Pharmaceutical manufacturing
Metallurgical control
Wine making
Machine tool; e.g., metal grinding fluids
Oil refining—dewaxing certain lubricating oils
Construction
Space activities

Aircraft
Concentrated fruit juice industry
Freeze drying foods, drugs, etc. (sublimation)

Refrigeration systems are classified as domestic, commercial, industrial, air conditioning, or marine.

Mechanical refrigeration commonly is termed artificial refrigeration. Ice produced by it is often, but erroneously, called artificial ice to distinguish it from ice produced by nature's method of freezing open bodies of water.

MECHANICAL REFRIGERATION PRINCIPLES

The repeated transformation of liquid to gas and gas back to its liquid state is the basis on which mechanical refrigeration operates. This gas or liquid is known as refrigerant, and is the medium from which the cooling effect on a substance is obtained.

Liquid refrigerant leaving a container, called a liquid receiver, under pressure, passes through a throttling device that can regulate its flow to any desired amount. Throttling devices vary in design according to the type of service, and include the expansion valve, a capillary tube, and a liquid float valve. The purpose of the throttling device is simply to regulate the amount of liquid allowed to pass through it on its way to the evaporator or cooling coil. From the evaporator, it is drawn into a compressor cylinder on the suction side, through a suction valve. Entrance of the refrigerant into the compressor cylinder is caused by the vacuum created by the action of the piston on its downward or suction stroke. The refrigerant enters the cylinder as a gas.

Refrigerants in their liquid state have a definite boiling point, and *the boiling point of any liquid varies with the pressure exerted on it,* differing with the type of refrigerant. The pressure within the evaporator is lowered because of the action of the compressor piston during its suction stroke, causing a differential in pressures between the receiver (or high-pressure side) and the evaporator (or low-pressure side) of the system. The result is a movement of the liquid refrigerant from the receiver through the throttling device into the evaporator. Upon entering the low-pressure area, the boiling point of the refrigerant is reduced and the heat absorbed by the refrigerant from the substance raises its temperature above the new boiling point *which is governed by the evaporator pressure,* causing it to boil or evaporate. The heat absorbed during evaporation of the liquid refrigerant causes the temperature to drop within the substance to be cooled.

Entering the compressor, the refrigerant in its gaseous form is started on the third step in its journey around the system. Upon reaching the bottom limits of its stroke, the piston immediately starts upward and the suction valve closes, trapping the gas within the cylinder. As the piston moves upward, *compression of the gas occurs, raising the pressure, which, in turn, increases its temperature and boiling point.* The pressure to which the gas is

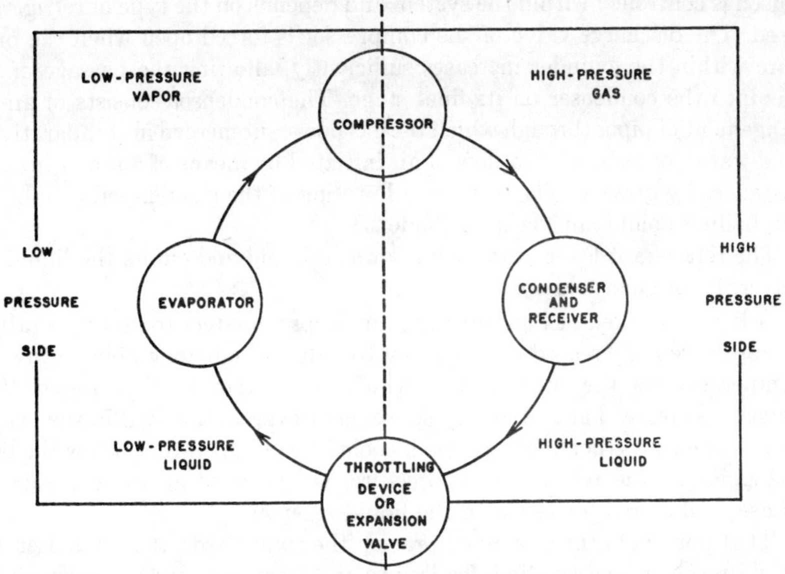

Fig. 1. Schematic diagram of a simple refrigeration system.

Fig. 2. A well-kept and efficiently maintained refrigeration compressor room. Note both horizontal and vertical type compressors. (Courtesy, Hershey Chocolate Corporation.)

raised is controlled within the system and depends on the type of refrigerant used. The discharge valve of the compressor is forced open when the pressure within the cylinder increases sufficiently, allowing the passage of the gas into the condenser on its final stage. The condenser consists of an arrangement of pipes through which the gas flows, submerged in comparatively cold water, or exposed to a flow of air initiated by means of a fan or in some instances by gravity. The result is a lowering of the gas temperature below the boiling point, causing it to condense.

The refrigerant leaves the condenser as a liquid and enters the liquid receiver, its original location.

A liquid is taken at high pressure and approximately room temperature. Its pressure is decreased considerably to reduce the boiling point below the temperature of the substance to which the evaporator is exposed. This causes its conversion to a gas by subsequent evaporation within the evaporator. The gas is then recompressed, cooled to a temperature below the boiling point at the recompressed pressure, transformed again to the liquid phase, and stored for re-use in the liquid receiver.

That portion of the system embracing the compressor, the condenser, the liquid receiver, and the lines leading to and from each one is known as the high-pressure side, and the pressure maintained within this side is called the head or discharge pressure. The evaporator, and the lines leading to it from the throttling device and from it to the compressor suction valve, is called the low-pressure side. The pressure within this side is known as the suction pressure.

From the time the refrigerant leaves the throttling device to enter the evaporator until it returns in its liquid form to the receiver, it will have completed each phase of its activity once. This is known as a mechanical refrigeration cycle.

PRIMARY AND SECONDARY SYSTEMS

Systems wherein heat removal is accomplished by the direct exposure of the evaporator to the space or the substance being cooled are referred to as primary or direct expansion systems. Household refrigerators, ice cream hardening cabinets, and small cold-storage rooms employ the direct expansion method of heat removal.

Systems employing a secondary medium such as brine, air, or water, which is circulated to remove heat from the space or the substance to be cooled which heat in turn is removed by an evaporator from the space or the substance to be cooled, are known as secondary or indirect cooling systems. Ice manufacture, large cold-storage plants, and true air conditioning are examples of heat removal by the indirect or secondary method of refrigeration.

SUMMARY OF CHAPTER 1

Refrigeration may be defined as the process of removing heat from any substance to:

Render colder—Reduce temperature
Change its state—As freezing water to make ice
Maintain its state—As in preservation of foods or storage of ice

Refrigeration is accomplished by:

Establishing temperature differentials
Evaporation of liquids
Or—a combination of both methods

Refrigeration systems may be divided into five classifications;

Domestic:

Household refrigerators
Unit air conditioners (home)
Small water coolers
Domestic freezer units—small

Commercial:

Cold-storage rooms—butchers, florists, restaurants, food lockers, etc.
Display cabinets—delicatessens, candy, meats, frozen foods
Skating rinks

Industrial:

Where refrigeration is part of manufacturing processes or necessary to manufacturing control, such as rubber, rayon, petroleum, and candy
Cold-storage warehouses, packing plants, ice making

Air conditioning:

Air-conditioning systems for stores, offices, theaters, laboratories, ships, space capsules, aircraft and enclosed arenas

Marine:

All cooling and freezing functions aboard ships, where safety and space economy are important factors

In addition to the conventional compression-expansion process other refrigeration systems include absorption, cryogenic, magnetic, vortex tube and thermoelectric processes.

The three basic factors involved in the refrigeration process are:

Heat exchange
Pressure control
Liquid-gas relationship

The function of mechanical refrigeration is based on repeated transformation of a liquid to gas and gas back to its liquid form.

The fluid employed in the refrigeration process is called a refrigerant. There are many kinds, but only a few are used extensively.

The mechanical refrigeration system is comprised of four major fundamental functions:

(*a*) Evaporation of liquid—to form gas—*function of evaporator;*
(*b*) Compression of gas—to increase pressure—to raise boiling point—*function of compressor;*
(*c*) Condensation of gas—to reconvert to liquid at higher pressure—*function of condenser;*
(*d*) Pressure reduction of liquid—to support evaporation by lowering boiling point—*function of throttling device.*

The action of the mechanical refrigeration system in completing each of its functions once is known as a refrigeration cycle.

The mechanical refrigeration cycle is divided into two stages:

(*a*) High-pressure stage, in which the condenser and the receiver are located;
(*b*) Low-pressure stage, in which the evaporator is located.

Pressure increase is effected by the compressor located in the path of the low-pressure gas as it passes from evaporator to condenser.

Pressure reduction is effected by the throttling device located in the path of the high-pressure liquid as it passes from the receiver to the evaporator.

Absorption of heat from the substance to be cooled is the function of the evaporator.

Elimination from the refrigerant of heat absorbed in the evaporator and the compressor is the function of the condenser.

A direct expansion system of refrigeration is one in which the substance to be cooled is exposed directly to the evaporator.

An indirect expansion system of refrigeration is one in which the heat of the substance to be cooled is absorbed, first, by a secondary medium—such as air, water, or brine—and conveyed by that medium to the evaporator.

STUDY OUTLINE FOR CHAPTER 2

Thermodynamics
- Definition
- Laws

Energy
- Definition
- Conservation, universal law
- Capacity to do work
- Forms of energy

State of energy
- Kinetic energy
- Fixed energy
- Potential energy

Thermal energy in transition—heat

Thermal energy—availability

Thermodynamic laws—first and second laws

Energy—conversion to other forms

Work—energy in the process of being expended

Heat—thermal analogy to mechanical motion

Power—rate of doing work
- Mechanical processes
- Electrical processes
- Thermodynamic processes

Thermal energy—molecular theory
- Random molecular translatory motion
- Rotational molecular motion about their own centers of mass
- Relative motion of atoms comprising the molecules, to each other

Thermal energy—measurement
- Intensity—temperature
  - Fahrenheit scale
  - Centigrade scale
  - Absolute scale
    - (°Rankine)
    - (°Kelvin)
- Quantity
  - British thermal unit
  - Calorie

Thermal energy—intensity measuring devices
- Thermometers
- Pyrometers

Temperature—absolute
- Theory of diminishing energy
- Degree Kelvin—°K
- Degree Rankine—°R

British thermal unit—(Btu)
- Method of determination—definition

Specific heat—heat absorbing capacity of substances as related to their temperature change
- Standard substance—water having a specific heat value of 1.0

Latent heat—involving change of state
- Of fusion—from solid to liquid or vice versa
- Of evaporation—from liquid to gas or vice versa

Total internal energy—energy stored within a substance at a given state point

Potential thermal energy—thermal energy responsible for a change in state and/or volume of a substance

Thermal kinetic energy—thermal energy responsible for a change in temperature only of a substance

Enthalpy—function of internal energy, specific volume, absolute pressure, and Joules heat equivalent of mechanical energy
- Constant pressure process
- Steady-flow systems

Entropy—term denoting unavailable energy proportionality under given state conditions of a fluid
- Unavailable energy
- Available energy
- Wasted energy
- Efficiency of a system
- Shaft-work
- State path
- Reversible and irreversible processes

Temperature differential

Joule equivalent—mechanical equivalent of thermal energy

Expansion and contraction of substances due to the changes in temperature

Heat transfer
- Conduction—transfer of heat within, or between, solid bodies in contact
- Radiation—transfer of heat between bodies through a substance by waves whose length varies with intensity
- Convection—transfer of heat within a fluid mass due to density changes

2

Thermal Energy

The division of science which deals with *thermal energy* of substances is called *thermodynamics,* from Greek *therme* = heat and *dynamic* = power. *Substance* in this text includes any material that forms a body whether the body is solid, liquid or gaseous. Other words such as body, body-mass, matter, mass and material may be accepted as being synonymous with substance. They all describe the physical *medium* employed to perform the thermodynamic changes by reason of its capability to absorb, store and, in turn, discharge thermal energy. It is the quantity of such substance involved that constitutes the thermodynamic *system.* The thermodynamic changes that occur within the system are called the *process.* Except for the system, i.e., the quantity of substance necessary to carry out the thermodynamic changes demanded of the process, everything else is called the *surroundings.* Upon the completion of a process during which the system performs its intended functional thermal changes and the medium reverts to its original *state,* a *cycle* is concluded.

When such thermal energy changes occur within the substance, the *state* of the substance also changes. A change in state involves a change in the *properties* of the substance. Working systems deal with liquids, vapors and gases, although solids may be included in certain circumstances as the thermodynamic medium. While a change in state may mean a conversion of a solid to a liquid or a liquid to a vapor, or the reverse in each case, it also involves changes in certain fluid properties, such as degree of intensity or *temperature, pressure, specific volume, specific weight, internal energy, enthalpy* and *entropy.*

Thermodynamics, or more specifically here, engineering thermodynamics embrace all laws governing thermal energy within a system and its influence on the surroundings of the system. This influence may be the effect it produces to cause changes in other thermal systems, as in the production of ice, or to convert thermal energy to other forms, such as mechanical, electrical or chemical energy, or vice versa. Included in the laws governing thermal energy are those dealing with its measurement, characteristics and potentialities.

Refrigeration is built primarily on thermodynamic laws. This chapter deals with such laws, as well as the measurement theory and practical behavior of thermal energy and its relation to other forms of energy.

ENERGY

Energy is the capacity of a body to do work. It is *the* impelling force universally distributed and present in some form in all substances. Energy is apparent by the changes it causes, either in the position or in the state of a body. Its presence enables all substances to move or be moved from one location to another, or to effect a transformation in their physical or chemical nature.

The universal law governing the conservation of energy states in effect that, *while energy may be converted in form, it cannot be created or destroyed.*

Energy is the same thing regardless of its source or existing form. Its variation lies only in identification, and that by association with the system from which its productive use is derived, and to which it finally will be converted. By determining the source from which energy may be drawn, a suitable system may be devised for its utilization in producing a desired result. Energy may be mechanical, electrical, chemical, nuclear, or thermal in nature.

THE STATE OF ENERGY

Energy usually is associated with bodies in motion rather than with those at rest.

A moving locomotive hauling a long train of freight cars exemplifies a condition wherein energy is displayed with great power and animation. Powerful effort is evidenced by every stroke of the driving rods or revolution of the wheels. The energy responsible for this action originates in the fuel fed to the firebox or the diesel cylinders of the locomotive. The fuel, inanimate as it lies in the tender or oil storage tank compared with the dynamic power displayed by the locomotive in action, indicates that energy exists in more than one state.

Locomotion in this case results from a transfer of force from the driving rods or gears to the drive wheels. It follows that one state of energy must exist due to motion. Since all motion would cease if the fuel supply suddenly were exhausted, the energy displayed by locomotion originates from within the fuel, in a fixed or inanimate state.

Energy thus exists in two states: (1) Energy of motion, or kinetic energy, and (2) fixed energy.

Kinetic Energy

Energy of motion, also called kinetic energy, exists wherever action occurs. Falling bodies, projectiles, flowing water, rotating wheels, sliding bodies—

all possess kinetic energy. Boiling water, escaping gases, and burning fuels also represent conditions wherein kinetic energy exists. The first set of conditions indicates changes in location, while the second set denotes changes in physical or chemical state. As action occurs in both instances, the presence of kinetic energy is apparent.

Regardless of the kind of change taking place, the action is caused by an effort outside the apparent action itself. For example, the pull of gravity influences the action of falling bodies. As a result of this action, some effect on other bodies occurs in the form of either motion or a change of state. Flowing water may rotate a wheel which, in turn, will generate electricity. Boiling water produces steam which is capable of driving steam turbines.

Kinetic energy exists only in a transitory state. Its form may be mechanical, chemical, thermal, electrical, magnetic and nuclear. The transition may be merely one of transfer involving only one form of energy, or it may be accomplished by the conversion of one form of energy to another.

Born of a source outside its own action and expiring as the result of its effect on other bodies, kinetic energy is a phase through which any form of energy must pass to accomplish its purpose. *The measure of kinetic energy is the measure of energy about to be expended in its intensity.*

Fixed Energy

Substances—such as coal, fuel oil, explosives, and chemicals—are composed of complex particles called molecules which are unstable under certain conditions. Coal oxidizes rapidly when exposed to high temperature, and will continue the process aided by its own temperature once the process is begun. Explosives, such as dynamite, react violently when subjected to shock. Chemicals tend to readjust themselves in their individual manner when placed in contact with other substances with which they have a tendency to unite. The reaction in any case is due to awakened energy stored within such substances, which would remain dormant as long as conditions tending to disturb it did not appear. The energy contained in such substances is called fixed energy, a form of potential energy.

Potential Energy

If a heavy weight is carried to the top of a building and placed on the edge of the roof, that weight will possess energy by reason of its position above the ground. The energy possessed by the weight is classified as fixed energy because it will remain unchanged as long as the weight keeps its position. Because the actions involved in placing the weight on the roof and its subsequent motion when falling are mechanical in nature, the energy possessed by the weight is also called potential energy. Potential energy is fixed energy by reason of position. It may also be called "stored" energy.

The water of a river whose flow is arrested by a dam, a coil spring held under tension, or a hammer poised in the air—all are illustrative of potential energy conditions. Immediately upon the release of fixed energy, kinetic energy is born, and the amount of the latter is proportional to the amount of fixed energy expended.

The amount of potential energy possessed by the weight on the roof is governed by the weight of the object and the height of the building. An increase in either weight or height would result in an increase of potential energy.

In its static condition, the weight does not possess kinetic energy, but, the moment it is dislodged and starts to fall, kinetic energy develops. Because, in falling, the distance between the weight and the ground diminishes, potential energy decreases. With every increment of potential energy lost,

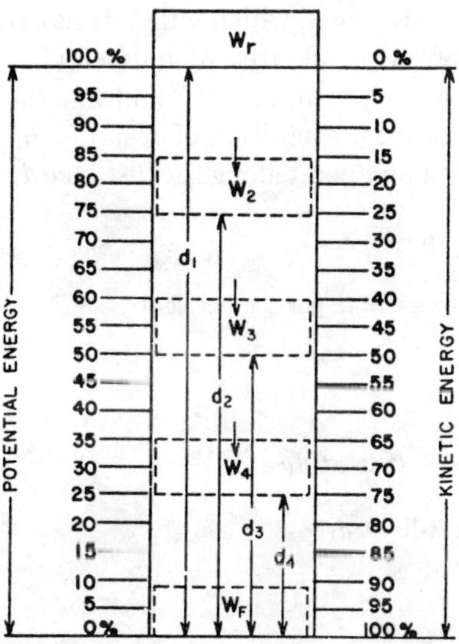

Fig. 3. (*Top*) Graph showing the relative values of potential and kinetic energy.

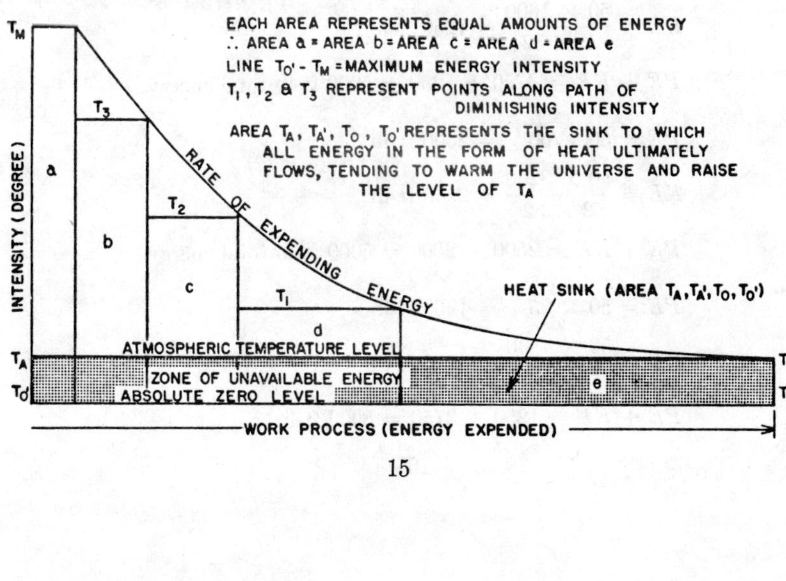

Fig. 4. (*Bottom*) Graph showing decrease in available energy as the temperature approaches absolute zero.

there is a corresponding increase in kinetic energy. This exchange progresses continually until the instant of impact with the ground, when all the potential energy is converted to kinetic energy.

Graphically illustrated (Fig. 3), the total energy possessed by the weight when it started to fall would be indicated as potential energy, and kinetic energy would be zero (0). At the instant of impact, the reverse would be true; i.e., the total energy would be kinetic and the potential energy zero.

Upon striking the ground, the energy passed from the weight to the ground in the form of (1) mechanical energy, as evidenced by the resulting dent, and (2) kinetic thermal energy, apparent by the warming of the earth in the immediate area of impact. While the potential energy of the weight as it left the roof was expended through kinetic energy developed as a result of motion, there was no loss of energy, except that which is used to overcome friction created by air resistances, which will be ignored here. All energy in this case was transferred to the area of impact.

The graph in Fig. 3 shows relative values of potential (*PE*) and kinetic (*KE*) energy of a falling weight W starting from a position of rest W_r, in which the sum of both energies is constant at any point along the distance d_1, where

$$PE = W \times d_1.\ W \times d_2,\ W \times d_3 \text{ and } W \times d_4$$

and

$$KE = \frac{W \times v^2}{2g}, \text{ where } v^2 = 2gd \quad (g = 32 \text{ ft per sec per sec})$$

If

$$W = 50 \text{ lb}$$

$$d_1 = 100 \text{ ft}$$

Then

$$PE = 50 \times 100 = 5000 \text{ ft-lb at } d_1$$

and

$$KE = \frac{50 \times 0}{2g} = 0 \text{ ft-lb at } d_1$$

At d_2,

$$PE = 50 \times 75 = 3750 \text{ ft-lb}$$

$$KE = \frac{50 \times 1600}{2 \times 32} = 1250 \text{ ft-lb}$$

$$PE + KE = 3750 + 1250 = 5000 \text{ ft-lb total energy}$$

At d_3,

$$PE = 50 \times 50 = 2500 \text{ ft-lb}$$

$$KE = \frac{50 \times 3200}{2 \times 32} = 2500 \text{ ft-lb}$$

$$PE + KE = 2500 + 2500 = 5000 \text{ ft-lb total energy}$$

At d_4.

$$PE = 50 \times 25 = 1250 \text{ ft-lb}$$

$$KE = \frac{50 \times 4800}{2 \times 32} = 3750 \text{ ft-lb}$$

$$PE + KE = 1250 + 3750 = 5000 \text{ ft-lb}$$

At W_f,

$$PE = 50 \times 0 = 0 \text{ ft-lb}$$

$$KE = \frac{50 \times 6400}{2 \times 32} = 5000 \text{ ft-lb}$$

$$PE + KE = 0 + 5000 = 5000 \text{ ft-lb (at the instant weight strikes the ground)}$$

HEAT

Thermal energy may be *stored* in a substance comprising a thermodynamic system. Where the temperature of the system is above its surroundings, a flow of thermal energy will be initiated from the system to its surroundings. The thermal energy in the act of flowing is said to be in transit. Mechanical energy in transit is defined as *work*. Thermal energy in transition is defined as *heat*. Work has been defined as mechanical energy in the act of being expended. Heat then may be defined as thermal energy in the act of being expended. Therefore, both terms, work and heat, are analogous, but only in regard to their own peculiar system of energy transfer.

It is important to distinguish between the term heat and stored or *internal energy*. Heat is thermal energy in transition across the boundaries of a thermodynamic system and can exist only in an environment of temperature differential. Internal energy is energy stored within a system by reason of having heat transferred to the medium from some external source.

Consider again the graph shown in Fig. 3 in which W_r represents the body W at rest atop the building. In this position, the body possessed potential energy only because of the absence of motion. Its potential energy is mechanical in form and originated during the process of being physically elevated to the roof. The thermo-mechanical energy system analogy here is that the weight represents the system, mechanical potential energy represents stored energy, and the act of elevating the weight to the roof represents a part of the work process; also, that the body while at rest atop the building exists in an environment of equal forces represented by the weight of the body as opposed to the upward thrust of the building structure. Again in the thermo-mechanical energy system analogy, the equal and upward thrust of the building structure of the mechanical system represents a thermal condition wherein no temperature differential exists. The mechanical body W under the conditions described so far is analogous to a thermodynamic medium (system), the temperature of which is equal to the temperature of its surroundings. As long as the condition exists no work is accomplished because no kinetic energy has been developed.

If the weight is dislodged from its position of rest and allowed to fall toward the earth, kinetic energy is created at the instant of initiating motion. This mechanical action introduces work to the system and thus removing the structural force from under the weight has its analogy in the thermodynamic system of creating a lower temperature in its surroundings relative to the system. By doing so, a flow of thermal energy from the warm system to the colder surroundings is initiated, thus transforming part of the stored or internal energy within the system to kinetic energy. Whereas in the mechanical system, work is initiated simultaneously with motion, in the thermodynamic system heat is initiated simultaneously with the flow of thermal energy when an environment of temperature differential is created. Heat then is thermal energy in transition, and during the thermodynamic process the ratios of potential energy to kinetic energy within the system may be compared with those depicted on the graph (Fig. 3) for a mechanical process with certain reservations, as discussed below. It may be mentioned here that the flow of thermal energy reversed direction during the mechanical refrigeration process, where heat enters the system in the region of cooling or low pressure side and emits heat in the condensing region or high pressure side. The medium that constitutes the system is the refrigerant, a fluid whose selection is based on certain inherent characteristics as required by the very nature of the process. The compression process in the refrigeration cycle is analogous to raising the weight to the top of the building.

ENERGY IS RELATIVE

In the graph shown in Fig. 3 both potential and kinetic energies are given definite percentage values ranging from the top to the bottom or vice versa. These values are only relative to the top and bottom extremities of the graph. The bottom line represents the ground in our discussion of potential energy. Therefore, the bottom line represents the base line or *datum* for the percentage values shown on the graph. If the energy values as computed for potential energy and kinetic energy in foot-pounds were substituted in the graph for percentage values, the portrait of energy behavior during the work process would become more realistic. It would show a diminishing value of potential energy in ft-lbs from the elevation W_r to ground level W_f. However, if an excavation were made alongside the building on which the weight rested and in line of the falling path of the object, the energy values as computed would no longer be true. The answer to this lies in the increase in distances as represented by an elongation of d_l. Also, the relative values would change because the base line (W_f) or datum becomes lower relative to W_r.

If for some technical reason it were desirable to devise a universal system of plotting the energy values of a given weight falling along a vertical path, some base line or datum would necessarily need to be established at which relative potential energy would diminish to zero or kinetic energy reach its maximum relative value. Were the distance of fall to become infinite the values would become infinite, creating complexities that would preclude reasonable standard interpretation. The system of computation would become unwieldy.

Accepting ground level to be the recognized datum for the conditions prevailing in the graph (Fig. 3), it must be realized that the body retains its original weight as it comes to rest on the ground. (Ignored here is the increase in the pull of gravity as it approaches the center of the earth.) It may be said then that the body still retains potential energy in relation to the earth's center. This is evident by its tendency to remain stationary only because its vertical force (weight) is countered by an equal and upward force of ground thrust.

Because of its position on the ground, the weight becomes useless as a work system as related to ground level. Its kinetic energy, relative to the datum, is spent. Further, the value of its kinetic energy becomes increasingly negative (−) as the distance from ground level to any point below the earth's surface increases. Kinetic energy is energy in transition within a system, be it mechanical, thermal or whatever. Kinetic energy is a relative quantity and is indicated by its intensity (degree of animation, heat, etc.) as related to the intensity of another body, system or region, or its datum. Potential energy is energy due to position, head, or internal forces contained within a body. It is measured by its capacity to perform work in returning a body to a state or position relative to a given datum or base-line. The amount of internal potential energy governs the state existence of a substance. In a thermodynamic system, the solid, liquid, vaporous or gaseous state of the medium is a function of the quantity of its potential energy. The unit measurement of all classes of thermal energy is expressed as the British thermal unit or Btu.

As in the mechanical work process, the amount of thermal kinetic energy available in a thermodynamic medium is relative to a standard reference point or datum. This reference point is analogous to ground level as described for the hypothetical condition portrayed in Fig. 3. Mechanical effort or work is the product of force × distance. In the case of a falling body, force is spontaneous so that the only variable is distance to the datum, as shown in computations related to the graph (Fig. 3).

In the thermodynamic system, the quantity of potential energy contained in the medium is measured from a point of zero thermal intensity at which all energy ceases to exist. This means that under certain absolute conditions molecular motion is absent and the magnitude of all classes of thermal energy equals zero both as to quantity and intensity.

This point represents the lowest level of measurement indicated by any system of temperature determination, i.e., 0° Absolute. Refer to the graphic development of this phenomenon in Fig. 8, where the temperature represented on both the Kelvin (°K) (Absolute Centigrade temperature) and the Rankine (°R) (Absolute Fahrenheit temperature) scales indicate 0° Absolute.

Considering the magnitude of potential energies, the temperature level of 0° Absolute would be the thermal analogy of mechanical potential energy existing at the earth's center as described in the above discussion.

For practical reasons, the thermal level of 0° Absolute (Kelvin or Rankine) cannot be adopted as the datum for thermal kinetic energy just as it is not practical to adopt the earth's center for the datum of mechanical kinetic energy. The explanation for this is that practical work systems deal with energy *changes* occurring well above the center of the earth, as in the mechanical process, or 0° Absolute in the thermal process. When work or heat is unleashed from one body, it must immediately be transferred to another body. In thermodynamic processes, such transfer of energy results in either a warming effect or change in state of the recipient body, and the temperatures of both bodies tend toward equalization, or become equalized in some cases. As energy is neither created nor destroyed, the energy given up by the warmer body is equal to the amount of energy absorbed by the recipient or colder body. Because a certain amount of stored or internal energy always remains in the medium from which the energy was taken and some energy was already present in the recipient body before the transfer of heat occurred, providing it exists above the region of Absolute Zero, it becomes obvious that the important consideration in the entire transaction is the *change in energy* throughout the process. This follows the laws of cause and effect.

In order to determine any change in energy, it is necessary to know the original state of the substances involved in the transaction. In the case of fluids as in the mechanical refrigeration process, the properties governing the state of the medium are pressure, specific volume, temperature, internal energy, enthalpy and entropy (the repetition of property nomenclature here is made purposely). Property values of the various fluids, including water, ammonia, etc., are published in appropriate tables known as "Tables of Thermodynamic Properties." They are also included on charts (curves) exemplified by the Mollier Chart of Properties of Ammonia (U.S. Bureau of Standards) found in Chapter 7. It will be noted from the tables and charts that with each change in one property, such as temperature, a corresponding change occurs in all other properties. The condition of the fluid at any given value is called the *state point* for that condition. Therefore, the tables and charts for the various fluids have been developed for a series of state points.*

The energy values shown in the property tables for the various fluids are based on a datum of -40 F. At this temperature, the fluid in its liquid phase has an energy value of zero. The arbitrary designation of -40 F is the thermal analogy of the base-line or datum represented by ground level in the mechanical energy system, as represented in the graph (Fig. 3). Below the datum value of -40 F, the kinetic energy value becomes negative. Note also that the entropy value at -40 F for liquid refrigerants is zero.

The selection of -40 F or °C was made "partly because $-40°$ represents the same temperature on both the Fahrenheit and Centigrade scales and partly because negative values of enthalpy or entropy would not be encountered in the usual working range of tables."**

The expansion process of the refrigeration cycle is analogous to the act of the weight in its fall from the top of the building to the ground level.

THERMAL ENERGY—AVAILABILITY

Since all other forms of energy may be converted to thermal energy, it is considered to be energy in its lowest form. Although energy cannot be created or destroyed, the theory of diminishing availability remains and thermal energy will become the form to which all other forms are ultimately reduced. This theory states that *universally the availability of all energy diminishes as its conversion to thermal energy at a lower intensity level continues.* Consider the thermal losses encountered with the simplest machine through friction, or as a result of conduction and radiation of heat in steam and internal combustion engines.

The theory of diminishing energy availability is based on the fact that, *where intensity (temperature) differentials exist, there is a decrease of thermal energy in the body with higher temperature as heat flows to a body of lower intensity.* The phenomenon of diminishing energy availability is the basis of the theory of *entropy,* one of the properties of fluids.

*In the earlier tables and charts, energy content in Btu is designated as "Heat Content," both for the liquid and the vapor. This designation is a misnomer made obvious by our definition of heat as "energy in transition." For reasons discussed under the subject, the energy content at each of the state points should be designated as *Enthalpy*—Btu per lb. This new concept appears on more recent tables and charts as in the Freon refrigerants tables included in the Appendix, pp. 1342-44.

**Quoted from the Circular of the Bureau of Standards, No. 142 (First Edition), April 16, 1923, entitled "Tables of Thermodynamic Properties of Ammonia."

The factor of diminishing availability of energy in any system of work is illustrated by the rectangles shown in Fig. 4. While the amount of energy in any form, as represented by the area, is equal in all figures, the intensity of the effort is reduced progressively as work is done, as represented by the height of the rectangles. Rectangle *a* is representative of maximum effort as based on a given level of intensity. Rectangle *e* is indicative of a static condition when, upon ultimate conversion to thermal energy, its intensity finally becomes equal to that of the surrounding atmosphere. Rectangles *b*, *c*, and *d* represent intermediate steps in the gradual reduction of available energy during the work process.

Thermodynamic Laws

The fact that all forms of energy can be converted to thermal energy without any loss is borne out by the universal law of energy conservation.

This is the fundamental principle underlying the relationship of thermal energy to mechanical energy, as expressed in the First Law of Thermodynamics. It says:

When thermal energy is transferred to mechanical energy, a definite amount of thermal energy disappears to emerge in like amount as mechanical energy; and, conversely, when mechanical energy is transformed to thermal energy, a definite amount of mechanical energy disappears to emerge in like amount as thermal energy.

That work can be performed thermodynamically only when a temperature differential exists is consistent with the principle that heat always flows from a level of high intensity. When the temperatures of two bodies are equal, there is no flow of heat and no thermal energy is available for performing work relative to the two bodies.

It is this latter principle upon which the Second Law of Thermodynamics, as stated by Clausius, is based. It reads:

It is impossible for a self-acting machine, unaided by any external agency, to convey heat from a body of low temperature to a body of higher temperature. Heat cannot of itself pass from a cold to a warmer body.

A practical illustration of energy conversion from one form to another was well described by Dr. Julius Robert Mayer, a German physician, in 1842. He wrote:

> Our locomotives may be compared to distilling apparatus: the heat beneath the boilers passes into the motion of the train and is again deposited as heat in the axles and wheels.

This establishes one concept of how energy may be converted in form to perform work compatible with that form. This transformation enables man to obtain motion or a change of state of a substance from energy sources already present, but it requires the construction of a suitable system whereby its conversion is made possible.

Here are six examples of kinds of energy sources and their converted forms as utilized by man's industry:

(*a*) **Chemical energy to mechanical energy:**
Coal burns to give off heat, forming steam from water, to operate a steam engine. Thus chemical energy is converted to thermal energy, which in turn is converted to mechanical energy.

(*b*) Mechanical energy to electrical energy:
The steam engine drives an electrical generator that produces an electric current.

(*c*) Electrical energy to thermal energy:
An electric current flowing through a wire possessing high resistance to electrical flow emits heat.

(*d*) Thermal energy to chemical energy:
The heat emitted from an electrical element may be utilized in ovens for processing products of all kinds.

(*e*) Electrical energy to chemical energy:
An electric current passing through an electrolyte via electrodes causes metal plating.

(*f*) Chemical energy to electrical energy:
A storage battery connected to a starting motor of an automobile causes it to start.

WORK—ENERGY EXPENDED

When energy is released, some physical or chemical change occurs either in the form of motion or in the state of the body. Energy is the cause and change the effect in any interchange or conversion of energy regardless of its form. When conditions are applicable, useful change occurs, producing effects which can be utilized to the benefit of man. Energy, the cause, is expended in such a manner as to perform useful work, the effect.

Work, therefore, is the effect of applied energy or effort, and implies change.

Work denotes activity. To be real, activity suggests the occurrence of some form of energy transformation. This depends on the kind of energy source and the system employed for its conversion and utilization.

A horse pulling a loaded vehicle over a road, a pump lifting water from a well and delivering it to a tank suspended aboveground, a water wheel rotated by the force of water flowing downstream, or a man driving a spike into a railroad tie with a sledge hammer—all are examples of work being performed. It is not difficult to determine the cause in any of these cases because motion accompanies the effort throughout the entire action.

Where tangible results are obtained through an effort accompanied by visible motion, the action is considered to be wholly mechanical in nature. Under such conditions, the work performed is defined as mechanical work. In the examples above, energy was transferred from one body to another, and not primarily converted in form. (Some conversion did occur in the action in all cases which may be considered as loss through friction, a subject discussed later in the text.)

Mechanical work is a function of force and distance, expressed by the formula:

$$\text{Mechanical work} = \text{force} \times \text{distance}$$

or

$$W = f \times d$$

where f = force expressed in pounds (lb)
d = distance expressed in feet (ft) or inches (in.)

The results are expressed as either foot-pounds (ft-lb) or inch-pounds (in.-lb).

EXAMPLE: If a force of 500 lb is required to pull a loaded vehicle over a road through a distance of 100 ft, the amount of work performed will be:

$$\text{Work} = \text{force} \times \text{distance}$$

or

$$W = f \times d$$

$$W = 500 \text{ lb} \times 100 \text{ ft}$$

$$W = 50{,}000 \text{ ft-lb}$$

or

$$W = 500 \times 100 \times 12$$

$$W = 600{,}000 \text{ in.-lb}$$

While indicating the effort involved, such a measure as foot-pounds is not sufficiently complete to enable man to design a continuous productive work system. Speed also must be considered. Time now enters the picture, and with it the term power.

POWER—RATE OF DOING WORK

Power may be defined as the ability of a body to do work in a given period of time, or, as the rate of doing work.

The conventional method of expressing power is by ratio of work done in foot-pounds per minute, as compared with an arbitrary value adopted as a standard. One arbitrary value is known as a horsepower (hp), 1 hp being equivalent to 33,000 ft-lb of work performed in 1 minute (min).

To determine the power of any system in terms of horsepower, the following formula is employed:

$$\text{hp} = \frac{\text{force in pounds} \times \text{distance in feet moved in 1 min}}{33{,}000}$$

or

$$\text{hp} = \frac{f \times d}{33{,}000\, t}$$

where t = time in minutes

If 50,000 ft-lb of work were to be performed in 1 min, it would require $50{,}000 \div 33{,}000 \times 1$, or 1.51 hp to accomplish the task.

While the transfer of mechanical energy is always accompanied by motion of some kind, this does not apply to electrical energy. This does not preclude the possibility of its being measured in terms of power. The factors may be likened to those present in a mechanical system. The analogy is portrayed better by comparing the flow of electrical energy with the flow of water in a hydraulic system.

The power of a hydraulic system is determined by the amount of water handled per minute relative to the resistance through which it is forced. The resistance in a hydraulic system is the result of friction in the pipes and the pressure against which the pump must act, as a result of either restricted flow or the height to which the water must be lifted, or both. The quantity of water handled by the hydraulic system usually is expressed in gallons (gal), which may be converted to pounds for the purpose of determining power requirements in terms of horsepower. No expression has been coined for rate of flow of water, so it is necessary to employ such terms as "gallons per minute" (gpm), or "pounds per minute," or "cubic feet per second" (cfs). The power requirements of a hydraulic system are based on the quantity of water in pounds per minute (lb per min) handled *against a force expressed in pounds per square inch* (psi).

The flow of electrical energy or current also is expressed, as in hydraulics, in quantity per unit of time, or coulombs per second. The amount of current flowing through an electrical system is expressed in amperes (amp), 1 amp being the quantity of electricity flowing through a circuit at the rate of 1 coulomb per second.

The pressure, in pounds per square inch, in the hydraulic system finds its analogy in the term *volt* (v) of the electrical circuit. If a term existed to indicate the frictional resistance in the hydraulic system, it would be equivalent to the *ohm* in the electrical circuit. The volt is an expression of force (or, more exactly, electromotive force), whose value is based on the amount of electrical pressure necessary to cause 1 amp to flow through a resistance of 1 ohm.

Therefore,

Amperes indicate quantity per second
Volts indicate force or pressure required

As horsepower is indicative of mechanical power, electrical power has a similar expression in the kilowatt (kw). While horsepower and kilowatt do not represent equivalent power values, they are determined in the same manner; i.e., both are functions of work and time.

The basic electrical power formula becomes:

$$\text{Watts} = \text{volts} \times \text{amperes}$$

or

$$\text{Kilowatts} = \frac{\text{volts} \times \text{amperes}}{1000}$$

(As watts are relatively small units of power, it is more convenient to use the term denoting higher value, or kilowatt, which is equal to 1000 watts.)

Another factor in any work system involves the *continuous expenditure of power;* i.e., energy expended at a specific rate for extended periods of time.

Mechanically, expended energy is determined by computing horsepower against time in hours (hr). If a machine delivering power rated at 5 hp were allowed to operate at full load for 10 hr, the total energy expended would be determined by multiplying the power output by the number of hours of operation, or

$$\text{hp} \times \text{number of hours operated} = \text{horsepower hours (hp-hr)}$$

or

$$5 \text{ hp} \times 10 \text{ hr} = 50 \text{ hp-hr}$$

If allowed to operate ½ hr, the total power expended would be

$$5 \times 0.5 = 2.5 \text{ hp-hr}$$

Electrically, total power expended is determined by multiplying the kilowatt load by time, or

$$\text{kw} \times \text{hr} = \text{kilowatthours (kwhr)}$$

Substituting kw for horsepower in the example above:

$$5 \text{ kw} \times 10 \text{ hr} = 50 \text{ kwhr}$$

$$5 \text{ kw} \times 0.5 \text{ hr} = 2.5 \text{ kwhr}$$

Because such expressions as horsepower and kilowatt are indicative of a specific amount of work done, an equivalent relationship exists between them. However, their equivalent values cannot be determined by any arithmetical formula, but must be derived through experiment. This is done by first converting both forms of energy to thermal energy, and then calculating their values from certain results obtained.

A study of the basic technology of thermal energy is essential before delving into equivalent values for the various forms of energy. Such a study embraces measurement of both heat intensity and quantity of heat transferred, the effect of heat on various substances, and the capacity of all substances to absorb and retain thermal energy. Thermal energy also may be rated in terms of power, as were mechanical and electrical energy, but in a manner peculiar to thermal characteristics.

THERMAL ENERGY—MOLECULAR THEORY

Like any other form of energy, thermal energy can be measured because it has quantity and intensity. Heat, the transitional form of thermal energy, is not visible, but manifests itself in its effect on various substances either by changing their state of being or by creating relative degrees of sensation as warmness or coldness (intensity) when in contact with the human body.

All matter, whether solid, liquid, or gas, is made up of minute particles called molecules. Molecules vary in size and weight according to their type, which, in turn, determines the nature of the substances they form. It is this variance in the types of molecules that creates and establishes the nature of different substances, such as iron, copper, water, sugar, gasoline, ammonia, and oil, to name only a few. All matter contains thermal energy in some quantity. Some substances are capable of retaining more than one form of energy at the same time, such as a steel magnet, which contains both thermal and electrical energy.

Thermal energy causes a vibration or molecular motion to occur. Because all substances contain thermal energy, some degree of molecular motion is carried on within them at all times. The rate of molecular activity depends on the amount of heat applied and increases in activity in proportion to the quantity added. Most substances expand when heated, probably resulting from the tendency of the molecules to separate with increased activity upon the application of heat. The state of the substance also governs the rate of molecular motion within it. This rate is lowest in solids, greater in liquids, and reaches its highest rate of activity in gases. The state of a substance depends on its molecular activity. A relationship exists between the activity of the molecules of a substance and its internal kinetic energy. Molecular activity is motion as represented by (1) random translatory motion or the tendency of the individual molecules to separate from others to leave the mass, (2) the motion of each molecule in rotating about their individual centers of mass, and (3) rotating motion of the atoms that constitute the molecule in respect to each other. If heat is added to a body in sufficient quantity, its molecular activity increases to such a rate as to change its state. When enough heat is applied, a solid such as ice changes to its liquid state, water, and water in turn to its gaseous state, steam. The state of any substance depends, therefore, on its intensity, which increases with the amount of heat added.

Thermal energy is a form of molecular energy, capable of being measured as to its intensity and amount. When applied in sufficient quantities, it will change the state of matter.

MEASUREMENT OF THERMAL ENERGY

A large stone lying on the ground has no available mechanical energy relative to the ground. Suspended in the air by some mechanical means, that same stone would possess potential energy, and in an amount relative to its weight and height. As the weight of the stone remains the same, any change in energy value is a function of height alone.

If two stones of different weight were suspended in the air at the same height, the heavier of the two would possess the greater amount of potential energy. The difference in energy value here is due to weight only.

In either case, a greater amount of energy becomes available as the height increases.

With two stones of different weights, it would be necessary to raise the lighter one a greater height proportional to the difference in weight in order to gain the equivalent amount of energy possessed by the heavier stone. While the weight of the lighter stone remains the same, to increase its energy value it is necessary to increase its height or its "mechanical intensity."

In any mechanical system of work, two variables may be present, mass and position. Where mass is limited or fixed, energy availability depends entirely on position or mechanical intensity. Mass is only as effective in doing work as its measure of mechanical intensity.

What is true of mechanical energy is also true of thermal energy. Where an analogy is created between the two forms of energy, mass becomes the equivalent of thermal energy quantity and position or mechanical intensity is represented by intensity of heat or temperature. The availability of heat energy depends on the relative intensity, or its value in relation to the intensity of surrounding bodies, just as height in the case of the stones governs the availability of mechanical energy. In the thermodynamics system availability of thermal energy depends, then, on temperature differential.

Intensity

Thermal energy has intensity and quantity, both of which can be measured. The thermal intensity of a substance is its relative degree of warmth or coldness as manifested by normal reactions of the human body. Thus ice is accepted as being cold, boiling water as hot, and molten iron as extremely hot. These approximate thermal intensity levels, though somewhat indicative of intensity, are not sufficiently descriptive or accurate for scientific purposes.

To develop a more specific means of measurement, a comparative standard is required. Consider the freezing point of water as a standard for the low level, and boiling water as the high level in a scale of thermal intensity. Place an arbitrary value of zero (0) and one hundred (100), respectively, to these points and a basis for determining intermediate intensity levels is established.

Elements, such as iron, mercury, and hydrogen, expand when heat is applied to them and contract when their thermal energy content is reduced. Mercury, having a quick reaction and a high expansion characteristic, has been adopted as one of the most practical mediums for determining the intermediate points in the scale of intensity levels encountered in refrigeration practice.

If mercury is placed in a long hollow glass tube closed at the bottom to prevent leakage, and the closed end of this tube is inserted in a block of melting ice near the surface, the column of mercury will contract and finally settle at some definite height from the bottom of the tube. Marking the tube at this point fixes the zero level. Next, if the tube is placed in an open pan of boiling water, the mercury will expand and rise to a definite height above the zero mark and remain constant. Marking the tube at this point, the high level of 100 is determined. The next step is to divide the portion of the tube between the two marks into 100 equal spaces, marking each space. An instrument known as a thermometer is thus provided for the accurate measuring of thermal intensity.

Each of the marks on the tube or thermometer is called a degree, and the level at which the thermal intensity in any body is found is its *temperature.* Melting ice on the thermometer, has a temperature of 0°, or zero degree, whereas boiling water (at atmospheric pressure, 14.7 psi) has a temperature of 100°.

If divisions equal to those placed between 0 and 100 on the glass tube are extended below the 0° point or above the 100° point, a greater range of temperature measurements can be obtained. The highest point on the scale will be governed by the temperature limitations of the glass tube. The lowest point will be governed by the nature of the expanding fluid employed.

There are several types of thermometers, but those used most extensively in refrigeration work employ either the Centigrade or the Fahrenheit scale. The Centigrade scale is divided into 100 equal parts or degrees, starting at 0° for melting ice to 100° for boiling water. The Fahrenheit scale is divided into 180 equal parts or degrees, starting at 32° for melting ice to 212° for boiling water. Each degree rise on the Centigrade scale corresponds with a 1.8° rise on the Fahrenheit scale. This ratio cannot be used directly as a factor in converting one scale to the other due to the differences in their designation of water freezing points.

To convert from Centigrade to Fahrenheit, and vice versa, the following formulas are used:

$$^\circ\text{ Fahrenheit} = \frac{9^\circ\text{C}}{5} + 32$$

$$^\circ\text{ Centigrade} = \frac{5}{9}(^\circ\text{F} - 32)$$

The mercury thermometer is employed most generally for temperatures ranging from −40 F to 500 F. Usually the maximum temperature range for mercury-filled glass thermometers does not exceed 800 F, although 1000 F temperatures are possible, using special glass. Thermometers used in high-temperature work (500 F to 1000 F) contain an inert gas, usually nitrogen, placed in the column above the mercury. The presence of nitrogen raises the boiling point of the mercury because of the increased pressure resulting from compression of the gas when the mercury expands within the column. Thermometers containing alcohol in place of the mercury are employed for low-temperature work of −40 F and below. The latter are not adaptable

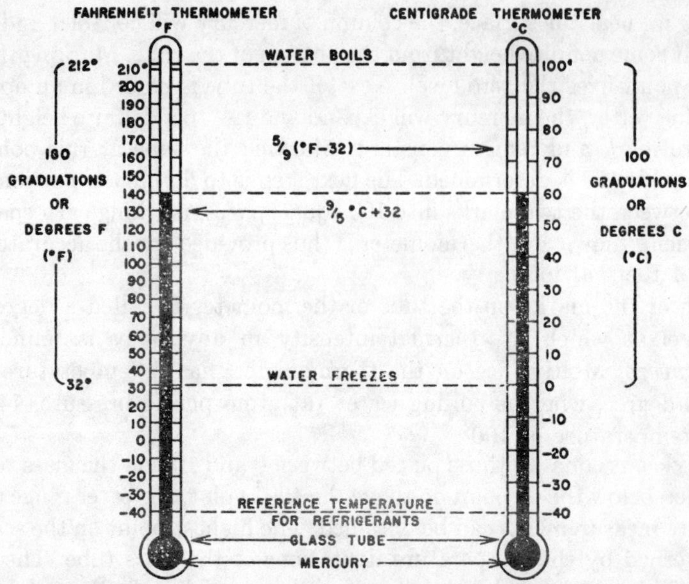

Fig. 5. Temperature relationship between Fahrenheit and Centigrade scales.

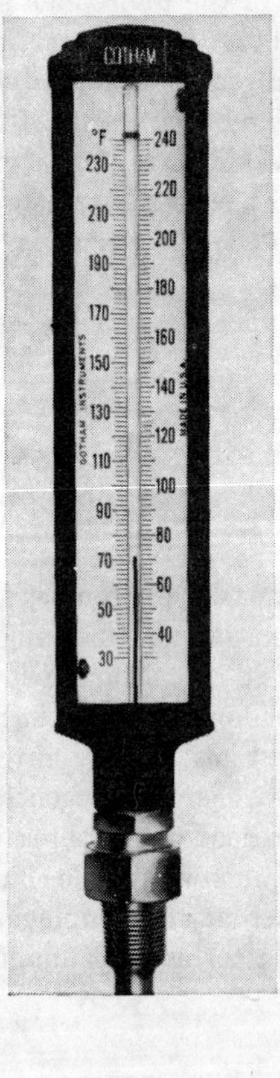

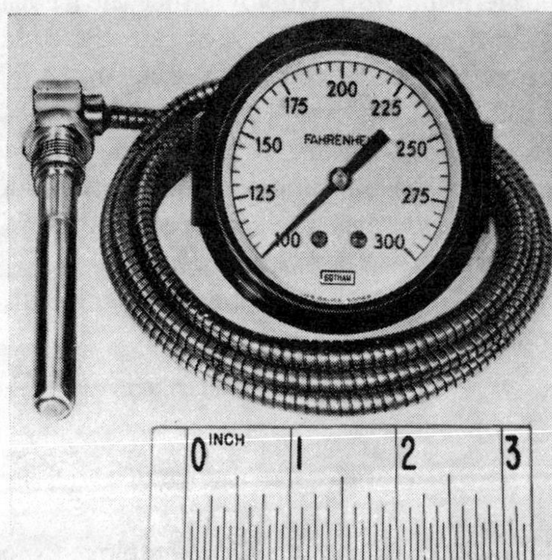

Fig. 6. Thermometers. (*Left*) Mercury column, glass tube. (*Right*) Mercury bulb and capillary type for remote reading. (Courtesy, United States Gauge, Division of American Machine and Metals, Inc.)

for temperatures above 172 F, the boiling point of alcohol. To increase visibility of colorless alcohol, a red dye is added.

A change in temperature of any body represents a change in translatory kinetic energy of that body.

Pyrometry

For measuring temperatures above the limits of the mercury thermometer, a group of instruments called pyrometers are used. Pyrometers employ different principles according to their intended use and temperature range. There are three basic types:

Thermocouples. This type of pyrometer measures the temperatures of bodies through physical contact. Thermocouples consist of a circuit made

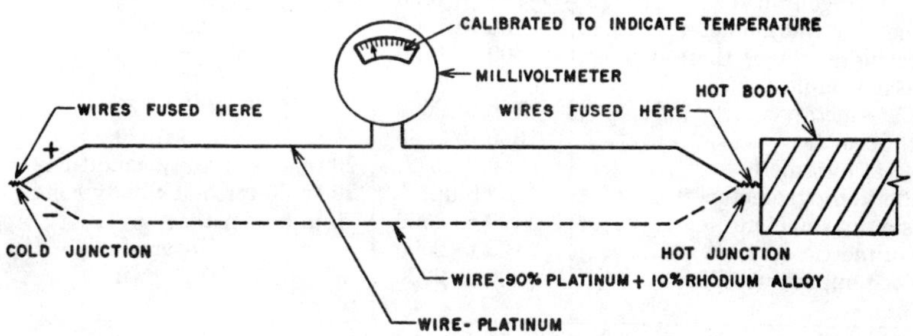

Fig. 7. Thermoelectric pyrometer or thermocouple.

up of two wires of dissimilar metals fused together at both ends and encased in a suitable protecting tube. When the temperatures of the junction points differ, an electrical potential is developed between them. The magnitude of the voltage so developed, which varies in proportion to the difference in temperature between the cold junction and the hot junction, can be measured by means of a millivoltmeter or potentiometer placed in the circuit. For convenience, the voltage indicator is calibrated to register directly the temperature (°F) of the hot junction which is in contact with the surface to be measured.

Various combinations of metals are used to make up the wire circuit of thermocouples. Their selection is based on the temperature-range requirements of the instrument and the accuracy demanded. Metal combinations include the following for the maximum temperatures shown:

Alloy of platinum (90%) and rhodium (10%)—platinum	Up to 2700 F
Chromel (alloy of chromium and nickel)—Alumel (alloy of aluminum and nickel)	Up to 2000 F
Nichrome—constantan (copper-nickel alloy)	Up to 1600 F
Iron—constantan	Up to 1200 F
Copper—constantan	Up to 500 F

TABLE 1. RANGE AND ACCURACY OF THERMOMETERS AND PYROMETERS*

Type	Range, °F	Accuracy, °F
Glass thermometers:		
Mercury-filled	−38 to 760	0.5 to 2
Mercury-and-nitrogen-filled	−38 to 1000	0 to 10
Alcohol-filled	−95 to 150	1 to 2
Pentane-filled	−300 to 70	1 to 2
Pressure-gage thermometers:		
Vapor-pressure type:		
Alcohol-filled	200 to 400	2 to 10
Ether-filled	100 to 300	2 to 10
Sulfur-dioxide-filled	20 to 250	2 to 10
Liquid-filled or gas-filled:		
Alcohol-filled	−50 to 300	2 to 10
Mercury-filled	−38 to 1000	2 to 10
Nitrogen-filled	−200 to 1000	2 to 10
Bimetallic thermometer	−100 to 1000	0.5 to 25
Electric resistance thermometer	−400 to 1800	0.005 to 5
Thermocouples:		
Base-metal	−300 to 2000	0.5 to 20
Rare-metal	−300 to 2800	0.5 to 20
Optical pyrometers	1400 up	20 for black body conditions
Radiation pyrometers	1000 up	20 to 30 for black body conditions
Fusion pyrometers	1100 to 3600	20 in best makes
Colorimetric devices	100 to 2500	Uncertain
Color-temperature charts	800 to 2900	Uncertain

Radiation Pyrometers. The radiation pyrometer is made up of a thermocouple whose hot junction is located so that the radiant heat rays from a hot body passing through a lens will focus on it. The electrical potential developed by the thermocouple is a function of emissivity or the ratio of heat radiated by a body to that radiated by a black body at the same temperature. Calibrated for black body conditions, the voltage indicator is calibrated to register temperatures directly in °F. Radiation pyrometers are employed generally for temperatures above 1000 F, but they can be modified to measure temperatures considerably lower.

Optical Pyrometers. Comparative brightness of a hot body with a standard lamp of variable light intensity is the working principle of the optical pyrometer. As the name of the instrument implies, comparisons are made visually. By varying the intensity of the lamp, through filament-current adjustment, it can be equalized with the intensity of the hot body. Intensity equalization is accomplished when the brightness of the hot body and the lamp filament are the same, as indicated by the disappearance of the filament within the incandescent background. The filament current is measured by means of a milliammeter, which usually is calibrated so that direct

* By permission from *Mechanical Engineers Handbook,* Fifth Edition, edited by Lionel S. Marks, copyright 1951, McGraw-Hill Book Co., Inc.

temperature readings can be made. Another type of optical pyrometer employs a photoelectric cell to determine temperature.

Optical pyrometers are employed to determine temperatures above 3000 F.

Other Methods for Measuring Temperature

In addition to the glass thermometers and the pyrometers already described, the following instruments also are employed for measuring temperature:

Bimetal strip or coil—acts on the principle of the compound bar described elsewhere in this text (p. 477). The bending motion of the bimetal strip that occurs with a change in temperature operates a pointer through suitable linkage. The pointer is mounted on a dial calibrated to indicate temperature.

Pressure-gage thermometer—consists of a liquid-, a vapor-, or a gas-filled bulb connected to a Bourdon spring (see Bourdon gage, Chap. 3) through a capillary tube. Expansion of the fluid occurs with any rise in temperature, causing the Bourdon spring to move outward (straighten) to operate a temperature dial-indicating pointer with which it is linked. The bulb, which is charged with various fluids, is exposed to the temperature-measured area.

Electrical-resistance thermometer—employs the principle that the electrical resistance of a fine wire coil, usually platinum or nickel, will vary according to temperature. Unknown temperatures are determined by the amount of change in resistance, as measured by a Wheatstone bridge, as compared with the standard resistance of the instrument, predetermined at a given temperature. The resistance coil is the sensitive element exposed to the temperature being measured.

TABLE 2. MELTING-POINT TEMPERATURES OF FUSION CONES (SEGER CONES)

Cone Number	*Melting Point °F*
022	1110
016	1380
010	1740
02	2030
7	2320
15	2615
20	2790
26	2880
30	3040
35	3220
39	2420
42	3600

Fusion pyrometers—are a series of pyramid-shaped cones having graduated melting points. Prepared from a mixture of different ratios of various oxides and glass, individual melting points range from 1100 F to 3600 F through intervals of approximately 20 F to 70 F. Temperature is determined by interpolation, noting the highest melting-point cone that slumps and the lowest melting-point cone that remains unchanged of a series of cones placed in a space of unknown temperature. Also called Seger cones, each cone is numbered to identify its melting point, as shown in Table 2. Time of exposure is an important factor to be considered when using Seger cones.

Color-temperature charts—based on a color-temperature relationship of high-temperature bodies, are not accurate since results depend on human estimate of color. When color is determined, temperature is read from a chart.

Magnetic thermometer—for measuring temperatures below 1°R (−458.6°F or −272°C). Involves the use of paramagnetic salts.

Vapor-pressure thermometers—for cryogenic temperatures using liquid cryogenes such as helium, hydrogen, neon, nitrogen and oxygen.

Absolute Temperature

All matter contains thermal energy, depending on the type, the weight, the physical state, and the temperature of the substance. Any reduction in temperature of a body results in a decrease in its thermal energy. Theoretically, if the temperature of a body continually were reduced, a point finally would be reached where all the thermal energy would be extracted and molecular motion cease to exist. This point is called Absolute Zero, and is the lowest temperature theoretically obtainable.

By means of graphs plotted from data secured as a result of experiment, using a manometer to measure gas pressure at descending temperature levels, absolute zero on the Centigrade temperature scale was found to be 273° below the freezing point of water. On the Fahrenheit scale, absolute zero corresponds to 459.6° below 0°, or 491.6° below the freezing point of water. The scale using zero as the point at which thermal energy ceases to exist is called the *absolute-temperature scale*.

When the temperature according to either the Centigrade or the Fahrenheit scales is known, the absolute temperature may be determined as follows:

$$\text{Absolute Centigrade temperature} = {}^\circ C + 273^\circ$$

$$\text{Absolute Fahrenheit temperature} = {}^\circ F + 459.6^\circ$$

Thus, the absolute Fahrenheit temperature of freezing water is equal to 32 F + 459.6, or 491.6 A. Of boiling water, it is 212 F + 459.6, or 671.6 A. On the Centigrade scale, the freezing point of water is 273 A and its boiling point is 273° plus 100 C, or 373 A.

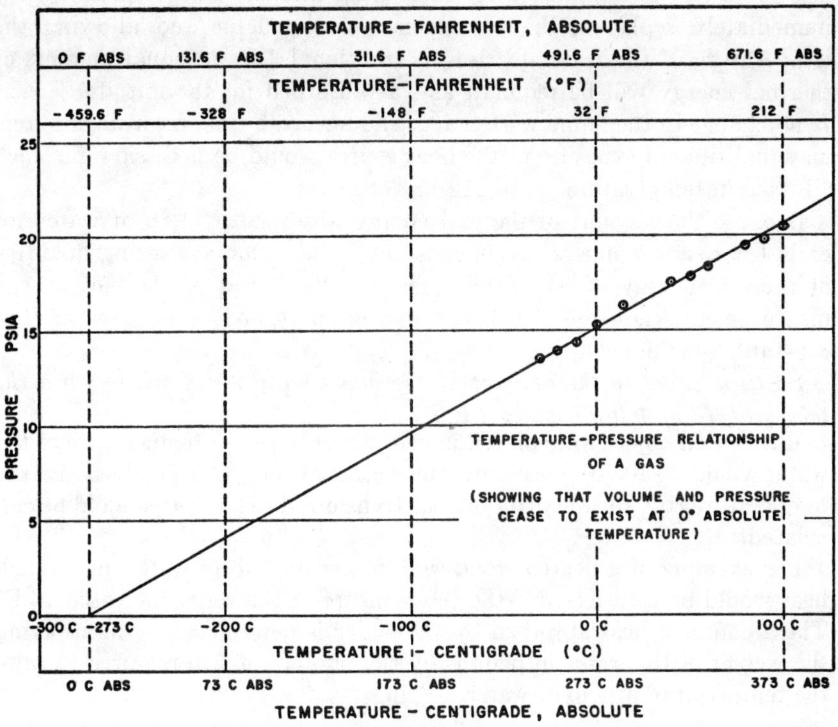

Fig. 8. Temperature-pressure relationship of air.

Absolute Centigrade temperature is referred to also as degrees Kelvin or °K, and absolute Fahrenheit temperature as degrees Rankine or °R.

Quantity

Since thermal energy has neither weight, nor shape, nor form, its quantity cannot be measured by any direct method such as we employ in the case of solids, liquids, or gases. Hence, it cannot be expressed in such tangible terms as quarts, pounds, or cubic feet. Since in itself it does not exert pressure or cause movement, as do mechanical systems, a direct determination of thermal quantity cannot be made through the use of gages, scales, or other forms of measuring instruments.

If a container of water is placed over a bunsen burner, a temperature rise may be noted by means of a thermometer immersed in water. If left exposed to the flame for a definite period of time, the temperature will rise to a level whose value depends on two factors: (1) the intensity of the flame and (2) the amount of water in the container.

By placing a measured amount of water in the container, say 1 lb, and allowing the temperature to rise 1 F before removing from the flame, a definite procedure is established for measuring thermal energy quantity. By immediately replacing the container over the flame and allowing the temperature of the water to rise an additional 1 F, an equal amount of thermal energy will be required as was absorbed for the initial 1 F rise. If no change in the flame were made, and accurate time recordings taken, an equal time of exposure to the heat source would be necessary for each 1 F rise in temperature up to 212 F.

Because the amount of thermal energy absorbed by 1 lb of water for each 1 F rise in temperature is constant, a basis for measuring quantity of thermal energy is established. The standard unit of thermal energy measurement, known as the British thermal unit, or Btu, is based on that constant. It is defined thus:

1 Btu is equal to the amount of thermal energy necessary to raise the temperature of 1 lb of water 1 F.

If a heated object were placed in cold water, a rise in temperature of the water would result. By measuring the weight of the water involved and the temperature rise, the quantity of heat transferred to the water could be calculated.

For example, if a heated stone were placed in 1 lb of water, how much heat would be transferred if the temperature of the water increased 10 F? The amount of heat absorbed by the water is determined by multiplying the weight of the water in pounds by the number of Btu required to raise the temperature of 1 lb of water 1 F, or

$$\text{Btu} = 1 \text{ lb} \times 1 \text{ Btu} \times 10 \text{ F}$$

$$= 10 \text{ Btu absorbed by the water}$$

EXAMPLE: 150 lb of water at 70 F is placed in a boiler and heated to 180 F. What quantity of heat is absorbed by the water?

(*a*) 1 Btu is required to raise the temperature of 1 lb of water 1 F
(*b*) Weight of water equals 150 lb (W)
(*c*) Temperature rise equals 180 F − 70 F or 110 F (t_d) (temperature difference, F)

Applying the equation

$$\text{Btu} = W \times 1 \text{ Btu} \times t_d$$

$$= 150 \times 1 \times 110$$

$$= 16{,}500 \text{ Btu absorbed by the water}$$

SPECIFIC HEAT

For all practical purposes, any reasonably clean water would require an equal amount of heat per pound per 1 F temperature rise, regardless of its initial temperature. Heat requirements would be essentially the same whether a given amount of clean water were raised 1 F from 32 F or 1 F from 211 F. The standard value of 1 Btu is based on a temperature rise of water from 62 F to 63 F. However, the amount of heat required to raise 1 lb of water from 32 F (melting temperature) to 212 F (boiling temperature), or through a range of 180 F, is 180 Btu, an *average* value of 1 Btu per lb per 1 F temperature rise.

Actually, the amount of heat required to raise the temperature of 1 lb of water 1 F varies slightly with each increment of temperature rise.

Stated another way, water has an inherent thermal energy capacity of 1 Btu per lb per 1 F. Different substances vary in their capacity to absorb heat. That is, the ability of some substances to rise in temperature under equal heat applications is greater than others. Therefore, the inherent thermal energy capacity of all substances is not the same. Their heat-absorbing qualities may be evaluated by comparison with that of water, thus establishing a basis for determining the quantity of heat required to raise the temperature of any substance 1 F.

If two vessels were placed side by side on a heater, one containing water and the other an equal amount, by weight, of petroleum oil, it would be found that the temperature of the oil would increase at a greater rate than that of the water. Petroleum oil would be proved to have a greater thermal energy susceptibility than water, which means a lower inherent thermal energy capacity. This method of observing temperature reaction could be pursued by substituting for the oil all the known elements and the element combinations. From test results obtained, a table indicating the respective inherent thermal energy capacity of each substance could be compiled, and a relationship of each to the other could be ascertained. Carried farther, a numerical factor could be established that would express the inherent thermal energy capacity of any substance based on a comparative value with that of water.

For example, if, in heating the oil, it is found that the rate of temperature rise is approximately twice that of water, the conclusion is reached that it requires only one half the amount of heat to raise its temperature 1 F than that required to raise the temperature of water an equal amount. According to this experiment, the numerical factor for petroleum oil, based on a value of 1 for water, could be established at ½ or 0.5, and the inherent thermal energy capacity of oil would be equal to one half that of water. If ice were applied to the vessel instead of heat, the temperature of the oil would drop at a correspondingly increased rate as compared with water. The ability of a substance to retain its thermal energy content is related directly to its heat-absorbing qualities. In this respect, the old adage of "easy come, easy go" applies.

This comparative thermal energy susceptibility value is called the specific heat (sp ht) of a substance, and may be defined as follows:

The specific heat of any substance is the amount of heat required to raise its temperature 1 F compared with the amount necessary to raise an equal weight of water 1 F.

The specific heat of petroleum oil as determined under laboratory test conditions is 0.46. Thus, it requires approximately 46/100 of the amount of heat to raise the temperature of petroleum oil 1 F as it does to raise an equal amount of water 1 F. Numerically, the specific heat of a substance is the number of Btu required to raise the temperature of 1 lb 1 F.

The specific heat of any substance is not a constant factor, but is subject to change with a change in physical state or even at different temperature

TABLE 3. SPECIFIC HEAT OF VARIOUS SOLIDS, LIQUIDS, AND GASES*

Solids	*Specific Heat Btu per lb per °F*
Aluminum	0.21
Antimony	0.05
Asbestos	0.20
Bismuth	0.03
Boron	0.31
Brass	0.086
Bronze	0.086
Calcium	0.15
Carbon	0.17
Concrete	0.156
Copper	0.0951
Cork	0.485
Glass	0.18 (av)
Glass wool	0.16
Gypsum	0.26
Ice @ −40 F	0.504
Iron, cast	0.110
Iron, wrought	0.114
Lead	0.030
Leather	0.360
Magnesium	0.25
Manganese	0.11
Marble	0.210
Mercury	0.03
Nickel	0.11
Porcelain	0.26
Pyrex (glass)	0.20
Rubber	0.481
Silver	0.055
Sodium	0.30
Steel	0.119
Tin	0.053
Vanadium	0.12
Wood	0.40 (av)
Zinc	0.09

* Specific heat values of most substances vary with temperature, either slightly or substantially according to the item. Except where otherwise noted, average values are shown. All values are approximate.

TABLE 3. SPECIFIC HEAT OF VARIOUS SOLIDS, LIQUIDS, AND GASES—(*Continued*)

Liquids	*Specific Heat Btu per lb per °F*
Alcohol @ 70 F	0.60
Ammonia @ 32 F	1.12
Ammonia @ 212 F	1.50
Calcium chloride brine (20% solution)	0.73
Carbon dioxide	0.766
Gasoline	0.53
Glycerin	0.58
Mercury	0.033
Methylene chloride	0.33
Oil, castor	0.43
Oil, lubricating	0.46 (av)
Oil, olive	0.47
Sodium chloride brine (20% solution)	0.810
Sulfur dioxide	0.346
Turpentine	0.41
Water (32 F to 212 F)	1.0 (av)

Gases	*Specific Heat Btu per lb per °F Constant Pressure*	*Specific Heat Btu per lb per °F Constant Volume*
Air (dry)	0.241	0.103 to 0.170
Ammonia	0.54	0.402
Carbon dioxide	0.205	0.1599
Chlorine	0.112	0.082
Ethyl chloride	0.274	0.23
Freon-11 @ 110 F	0.1374	0.1213
Freon-12 @ 86 F	0.1476	0.1297
Freon-113 @ 160 F	0.1633	0.1516
Freon-114 @ 110 F	0.1629	0.1502
Hydrogen	3.41	
Methane	0.59	0.446
Methyl chloride	0.24	0.20
Methyl formate	0.20	0.18
Methylene chloride	0.154	0.313
Nitrogen	0.24	0.170
Nitrous oxide	0.21	0.166
Oxygen	0.22	
Steam @ 212 F	0.47	0.316
Sulfur dioxide	0.154	0.119

ranges. Tables compiled to show the specific heat for various substances usually indicate the temperature range within which the value given is an average for that range. Under these circumstances, the value assumes no change in state.

The specific heat value varies according to the state of the substance. Water is an excellent example. As a liquid, its specific heat is 1, while that of its solid, ice, is 0.504. In its gaseous state, steam, its specific heat is 0.480 under constant pressure conditions. Gases of the same substance also vary in their specific heat value, depending on whether the value is based on constant volume or constant pressure. (See Table 3, Specific Heat of Various Solids, Liquids, and Gases.)

The practical use of specific heat values of various substances is shown in the following:

EXAMPLE: What quantity of heat is required to raise the temperature of an aluminum vessel weighing 30 lb from 75 F to 200 F? Assume no losses.

(sp ht of aluminum = 0.21)

Applying the formula previously stated,

$$\text{Btu} = \text{weight} \times \text{Btu per lb per 1 F} \times \text{temperature difference}$$

Substituting specific heat value, the formula is now written

$$\text{Btu} = W \times \text{sp ht} \times (t_2 - t_1)$$

$$\text{Btu} = 30 \times 0.21 \times (200 - 75)$$

$$= 6.30 \times 125$$

$$= 787.5 \text{ Btu required}$$

EXAMPLE: How much heat is required to raise the temperature from 50 F to 185 F of an iron vessel weighing 300 lb containing 500 lb of water, in which has been immersed 60 lb of copper pellets?

In solving problems involving more than one substance, it is necessary to determine first the heat required by each one individually and then add to obtain the total amount necessary. (sp ht of iron = 0.11)

(sp ht of copper = 0.0951)

Applying the formula,

$$\text{Btu} = \text{weight} \times \text{sp ht} \times (t_2 - t_1)$$

$$\text{Btu} = 300 \times 0.11 \times 135 = 4455 \text{ Btu—iron vessel}$$

$$= 500 \times 1 \times 135 = 67{,}500 \text{ Btu—water}$$

$$= 60 \times 0.0951 \times 135 = 77.03 \text{ Btu—copper}$$

Total amount of heat required by all substances = 72,032.03 Btu

KINETIC THERMAL ENERGY

Thermal energy in the act of being transferred by reason of a temperature differential is referred to as *heat*. Heat, then, is analogous to *work* which denotes the act of transferring mechanical energy initiated in an environment of unbalanced forces. Heat causes one or more of the following:

(*a*) Change in temperature
(*b*) Change in state
(*c*) Change in volume

A change in temperature indicates a transfer of *kinetic thermal energy* from the body of higher temperature to the lower temperature body or recipient. The transfer of kinetic thermal energy denotes an increase or decrease of internal energy within both bodies, depending on whether it is the source or recipient under consideration. Kinetic thermal energy, then, describes the energy absorbed or rejected by a body causing a change in its temperature. It will be so interpreted throughout this text. Kinetic thermal energy was formerly described in some texts as "sensible heat."

The equation employed to determine thermal energy requirements in the discussion of specific heat is the same for determining the kinetic thermal energy necessary to change the temperature of a body. The amount of kinetic thermal energy involved in any heat transfer situation dealing only with temperature changes is found by applying the same formula as shown in the discussion of specific heat, viz:

$$\text{Kinetic thermal energy} = \text{weight} \times \text{sp ht} \times \text{temperature differential}$$

or

$$\text{Btu} = W \times \text{sp ht} \times (t_2 - t_1)$$

The amount of kinetic thermal energy added to each pound of water to raise the temperature from 40 F to 140 F would be

$$\begin{aligned} \text{Btu} &= w \times \text{sp ht} \times t_d \\ &= 1 \text{ lb} \times 1 \times (140 \text{ F} - 40 \text{ F}) \\ &= 1 \times 1 \times 100 \text{ F} \\ &= 100 \text{ Btu, kinetic thermal energy absorbed per lb of water} \end{aligned}$$

(Note that 100 Btu per lb of water represents only the amount of thermal energy *added* to 40 F water. It does not represent all the thermal energy contained in the water at the final temperature of 140 F because the water already possessed some thermal energy at its initial temperature of 40 F. This factor is a highly important one, and should be understood thoroughly to avoid confusion in calculating refrigeration loads.)

EXAMPLE: How much kinetic thermal energy per lb is required to bring the temperature of cast iron to 1000 F from a room temperature of 70 F?

$$\begin{aligned} \text{Btu} &= w \times \text{sp ht} \times t_d \\ &= 1 \times 0.11 \times (1000 - 70) \\ &= 102.3 \text{ Btu, kinetic thermal energy surrendered by the copper} \end{aligned}$$

Changes in state and/or volume indicate a change in potential thermal energy as exemplified by a spring in the state of being compressed or a bowstring being drawn in preparation for projecting an arrow.

LATENT HEAT

Latent heat differs from kinetic thermal energy in that it does not cause a temperature rise within a body, but does effect a change in its state of being.

There are two phases of latent heat: (1) latent heat of vaporization and (2) latent heat of fusion.

If a thermometer is placed in an open vessel of water and a constant flow of heat is applied to the vessel, a gradual rise in temperature takes place. This rise will continue until the temperature of the water reaches 212 F (under atmospheric pressure of 14.7 psi), or 100 C, depending on the type of thermometer used. Upon reaching this point, the temperature of the water will cease to rise, although the application of heat continues. The water continues to absorb this heat, and, when a sufficient amount has been taken up, vaporization occurs, transforming the water to a vapor or steam. The heat absorbed by the water from the time the temperature rise ceased at 212 F until vaporization occurred is called the latent heat of vaporization. During this period, it appears that the heat absorbed by the water is unproductive of any work, and for this reason is called *latent* heat. Actually, it is this latent heat that provides the necessary energy to transform a liquid to a gas.

If further thermal observation is made, it will be found that the resulting vapor or steam will have the same temperature as the boiling water from which it emanated at the time the temperature ceased to rise at 212 F. *Latent heat of vaporization is that heat necessary to change a liquid to a gas without a change in temperature. Conversely, it is the amount of heat extracted from any gas upon liquefaction without a change in temperature.*

Latent heat of fusion has characteristics similar to those of latent heat of vaporization, whereby heat is absorbed by a body without a change in temperature. It differs in that the transformation is made from a solid to a liquid, or a liquid to a solid. *Latent heat of fusion is that amount of heat necessary to change a solid to a liquid without a change in temperature or the amount of heat extracted from a liquid upon its solidification without changing its temperature.*

The quantity of heat for both types of latent heat is expressed in Btu per lb. Heat of vaporization for water is 970 Btu per lb. This means that it requires 970 Btu to change 1 lb of water at 212 F to steam at 212 F (under atmospheric pressure of 14.7 psi). Heat of fusion for water is 144 Btu per lb, which means that 144 Btu are required to change 1 lb of ice at 32 F to water at 32 F.

TOTAL INTERNAL ENERGY

The total thermal energy contained within a body is the sum of the kinetic thermal energy and the latent heat applied to the mass. It is calculated from a given point or temperature where it is assumed that there is an absence of all energy. To determine the total internal energy value of any substance, calculations must be based on a definite starting temperature, known as a reference point or datum. The total internal energy in any body can be determined if the reference point or temperature indicated as a starting temperature is given.

As an example, to determine the amount of internal energy in 1 lb of saturated steam at 212 F and atmospheric pressure, starting with ice at 0 F, the first step is to find the internal energy content of the ice. Table 3, Specific Heat of Various Solids, Liquids, and Gases, shows that ice has a specific heat of 0.504; i.e., it requires 0.504 Btu to raise the temperature

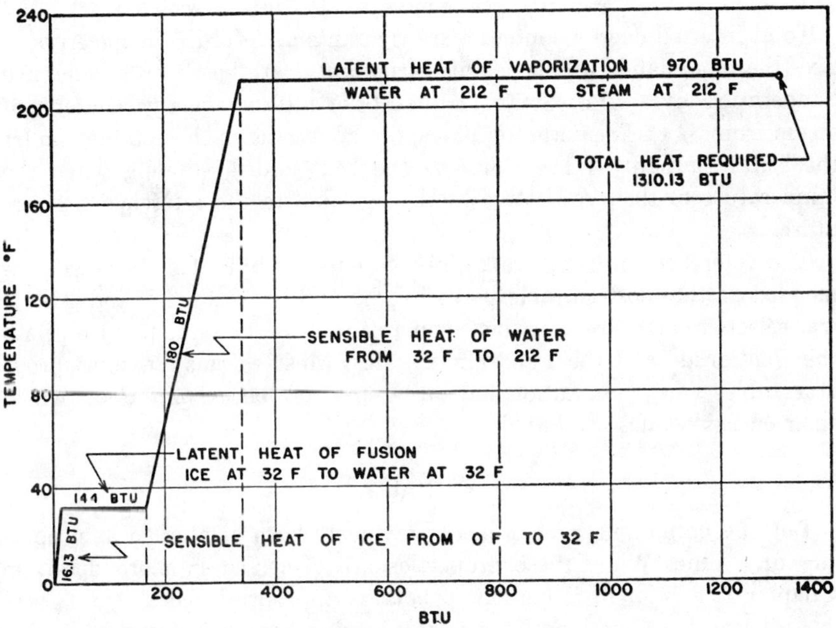

Fig. 9. Total internal thermal energy chart. Conversion of 0 F ice to 212 F steam.

of 1 lb of ice 1 F. Since the melting point of ice is 32 F, this gives a temperature differential of 0 − 32 or 32°. In 1 lb of ice, the kinetic thermal energy content is 32×0.504, or 16.13 Btu.

Secondly, the latent heat of fusion of ice must be known. By tests, it is found that 144 Btu are required to change 1 lb of ice at 32 F to water at 32 F.

The next step is to determine the heat required to raise the temperature of the water from 32 F to 212 F, or through a temperature differential of 180 F (212 − 32). As it requires 1 Btu to raise the temperature of 1 lb of water 1 F, the kinetic thermal energy content would be 180 × 1, or 180 Btu.

The latent heat of vaporization for water is known to be 970 Btu per lb, or the amount required to change 1 lb of water at 212 F to steam at 212 F, and at atmospheric pressure.

The final step is to add these values together:

16.13	Btu,	Internal energy of the ice
144	"	Latent heat of fusion
180	"	Internal energy of the liquid
970	"	Latent heat of vaporization
1310.13	"	Total heat required to be applied to the substance starting with 0 F ice to produce steam at 212 F and atmospheric pressure (14.7 psi) above 0 F

Total internal energy content of refrigerants as recorded in the Appendix (thermodynamic properties of refrigerants) are based on a reference temperature of −40 F. Steam Tables base internal energy content of steam from 32 F. The enthalpy of vapors appearing in these tables under the Total Enthalpy (H) column is the heat added, starting when the temperature of their respective liquids is equal to the reference temperature.

The reference temperature of refrigerants, −40 F, has been recommended as the most suitable point by the U.S. Bureau of Standards. It was selected partly because this temperature is at the same point on both the Centigrade and the Fahrenheit scales. Most thermodynamics property tables carry a notation indicating the reference point upon which their energy values are based.

ENTHALPY

Both *enthalpy* and *entropy* have previously been mentioned as properties of a fluid. While these properties are discussed in more detail in Chapter 7, a brief introduction to both is appropriate here.

Enthalpy is a function of the state of a thermodynamic system. It is a composite property in that it represents the heat transferred to the system, plus the energy involved in the change of state, affecting pressure and volume. The word enthalpy is often misconstrued as representing total internal energy of a fluid. This is not so as shown in the following energy equation in which the mechanical equivalent of thermal energy is represented

$$h = u + \frac{pv}{J}$$

where h = specific enthalpy, Btu per lb
u = internal energy, Btu per lb
v = specific volume, cu ft per lb
p = absolute pressure, lb per sq ft
J = Joules equivalent = 778.26 ft lb per Btu

The development of enthalpy in this text deals with practical work processes in accordance with engineering thermodynamics. Such work processes as the steam turbine and mechanical refrigeration fall in this category. In both of these processes heat is received into the system at constant pressure and emitted from the system at *constant pressure.* Pressures are not necessarily the same in both functions. To maintain constant pressure in a fluid system *volume* must increase or decrease as heat is respectively received or rejected. To maintain constant pressure under variable volume conditions the fluid must not be restricted or confined within an enclosed chamber but allowed to flow continuously at a rate consistent with the amount of heat absorbed or rejected at any given increment of time during which the system is in operation. As heat is absorbed or rejected by the fluid, work against an external force, or from an external force, must be exerted by the system or on the system to effect the appropriate change in volume. This work required to either increase or decrease volume must be considered along with the thermal energy added to or subtracted from the system during the thermodynamic process. Under the condition described above, the system is said to be one of *continuous* or *steady flow.* While the conditions in some types of machines, such as the steam engine and internal combustion engine, tend more toward *intermittent flow,* as within the cylinder, the conditions throughout the rest of the system have a steadying influence on the whole process. Enthalpy values, then, include not only the thermal energy values resulting from heat transfer in or out of the system but also reflect the work gained by or lost to the system in the change of volume under constant pressure prevailing in steady-flow processes.

ENTROPY

Engineering thermodynamics deals with continuous change in the state of the system or medium. This involves steady-flow processes in which heat passes into and out of the system. Practical work systems encounter friction or temperature differential or both. Energy utilized in performing useful work is called *shaft-work.* Energy that is capable of doing work, but is lost due to the design inefficiencies of the process is considered *wasted energy.* Internal energy that remains in the system, in which a temperature differential no longer exists, becomes unavailable to perform additional work as related to a given system. This last is regarded as *unavailable energy. Entropy* is the term used in measuring the proportion of unavailable thermal energy within a thermodynamic system. Its value increases as the temperature differential between the system and its surroundings diminishes. The total internal energy of a system is the energy *available* to perform work, plus unavailable energy.

Thermal energy transferred from a system to its surroundings because of a temperature difference, or converted to mechanical or other forms of energy, can never be fully retrieved in order to return the system to its original state. In a steady-flow process, the original state of a system is maintained by heat from an external source, as from the boilers of a power plant, or from the substances being reduced in temperature, as in a mechanical refrigeration system. A process experiencing a loss of energy as a result of friction (mechanical work) or temperature differential (thermal) cannot retrieve such energy for its return along its original state path to its original state condition. All such energy finally reverts to thermal energy at a lower level of intensity. Processes in which energy finally reverts to a lower level of thermal energy are said to be *irreversible*.

The theoretical process in which the loss of energy is precluded and whose *state path* during its original state change is duplicated exactly on its return to its original state is called a *reversible process*. In the reversible process, there is no change in entropy. (See discussion of the development of the reversible process as related to the Carnot Cycle in Chapter 7.) In the irreversible process, entropy changes with changes in temperature, increasing as the temperature differential between state conditions and datum increases.

In the case of wasted heat mentioned above, a definite improvement can be made in decreasing this loss by intelligent system design and utilization of various types of economizers. (See the discussion of centrifugal refrigerant compressors in Chapter 8.)

EXPANSION OF SUBSTANCES

All solids, liquids, and gases expand when heated and contract upon cooling. Temperature variations tend to cause dimensional changes in all directions so that substances are subject to linear, areal, and cubical expansion or contraction.

The rate of expansion or contraction of the same substance is nearly uniform for each degree temperature change, but varies with different substances. The rate of dimensional change per unit of measurement (inch, foot, etc.) per degree temperature variation for any specific substance is called the coefficient of expansion. The coefficient of expansion of any substance represents the average dimensional increase per unit of measurement per degree temperature rise within the range of temperatures between the freezing point (32 F) and the boiling point (212 F) of water. Coefficients are established experimentally and determined by applying the following equations:

$$\text{Coefficient of linear expansion} = \frac{\text{increase in length}}{\text{original length} \times {}^\circ\ \text{temperature rise}}$$

$$\text{Coefficient of areal expansion} = \frac{\text{increase in area}}{\text{original area} \times {}^\circ\ \text{temperature rise}}$$

$$\text{Coefficient of cubical expansion} = \frac{\text{increase in volume}}{\text{original volume} \times {}^\circ\ \text{temperature rise}}$$

To determine dimensional changes of any substance when the measurements at a given temperature and temperature change are known, the following general rules may be applied:

(*a*) Coefficient of linear expansion as found in appropriate tables is employed to compute changes in length, breadth, or thickness;
(*b*) Coefficient of areal expansion is twice the coefficient of linear expansion;
(*c*) Coefficient of cubical expansion is three times the coefficient of linear expansion.

Coefficients of expansion values are based on a temperature change of 1° and vary according to the type of thermometer (Centigrade or Fahrenheit) used. As there are 180 divisions on the Fahrenheit scale as compared with 100 divisions on the Centigrade scale, coefficients of expansion based on the former are equal to

$$\tfrac{100}{180} \text{ or } \tfrac{5}{9}$$

of the latter. To determine equivalent coefficient of expansion values from tables based on either the Centigrade or the Fahrenheit scale, proceed as follows:

(*a*) From values based on the Centigrade scale to Fahrenheit equivalent, multiply Centigrade coefficient of expansion by $\frac{5}{9}$;
(*b*) From values based on the Fahrenheit scale to Centigrade equivalent, multiply Fahrenheit coefficient of expansion by $\frac{9}{5}$.

The coefficient of expansion values for the various substances are listed in Table 4.

The coefficient of expansion is not based on any particular unit of measurement. The result of any calculation, therefore, will be expressed in terms of the unit employed.

EXAMPLE: What will be the final length of a copper wire 600 ft long at 50 F when the temperature rises to 95 F?

As indicated in Table 4, the coefficient of linear expansion of copper is 0.00000933. This means that, for every inch of length, copper will expand 0.00000933 in. for each 1 F temperature rise; or that, for every foot of cop-

Table 4. Coefficient of Linear Expansion for Various Solids

	Per 1 C	$\times \frac{5}{9}$ =	Per 1 F
Aluminum	0.000023	"	0.0000127
Brass	0.000019	"	0.00001052
Copper	0.00017	"	0.00000933
Glass (average)	0.000009	"	0.000005
Iron, cast	0.000011	"	0.0000059
Iron, wrought	0.000012	"	0.0000067
Lead	0.00003	"	0.0000170
Nickel	0.000010	"	0.0000056
Platinum	0.000009	"	0.000005
Rock salt	0.00000846	"	0.0000047
Silver	0.000019	"	0.0000105
Silver, German	0.0000184	"	0.0000102
Steel (average)	0.0000123	"	0.00000685
Tin	0.000021	"	0.000012
Zinc	0.000029	"	0.000016

Coefficients of Cubical Expansion for Various Liquids/°F

Calcium chloride		Petroleum oil	0.00054
6% solution	0.000140	Sodium chloride	
40% solution	0.000253	20% solution	0.00023
Glycerin	0.000208	Sulfur	0.00127
Ice	0.00069	Turpentine	0.00054
Mercury	0.000101	Water	0.000105
Olive oil	0.00040		

per, an increase of 0.00000933 ft per 1 F temperature rise will occur, etc. As the problem is based on feet, the results of the calculation will be in feet.

Data:

Initial length	= 600 ft
Initial temperature	= 50 F
Final length	= Unknown
Final temperature	= 95 F
Coefficient of expansion of copper	= 0.00000933

Solution:

Final length = original length + (original length × coefficient of expansion) × temperature rise

$$= 600 + (600 \times 0.00000933) \times (95 - 50)$$
$$= 600 + (0.005598 \times 45)$$
$$= 600 + 0.25191$$
$$= 600.252 \text{ ft—final length}$$

(*Note:* The plus sign of expansion will change to minus when contraction occurs.)

Example: A steel bridge is 2500 ft long when completed on a summer day when the temperature registers 105 F. Assuming that the coldest day of winter will approach −10 F, how much total overlap must be allowed on the two expansion joints on the day of completion? What will be the final length of the bridge steel at −10 F?

(Complete value is 0.00000685)

(Coefficient of expansion for steel is 0.00000685.)

Applying the formula,

$$\begin{aligned}
\text{Final length} &= \text{original length} - (\text{original length} \times \text{coefficient of expansion}) \times \text{temperature difference} \\
&= 2500 - (2500 \times 0.00000685) \times (105 - (-10)) \\
&= 2500 - 1.97 \\
&= 2498.03 \text{ ft—final length}
\end{aligned}$$

Considering that all bridge joints, except the extreme ends, are free, the overlap would be found by dividing the total amount of expansion by 4. (Because of overlap, the total expansion at each joint must be divided by 2. As there are two expansion joints, the denominator becomes 2 × 2, or 4.) Therefore,

$$\frac{1.97}{4} = 0.4925 \text{ ft}$$

$$= 5.91 \text{ in. overlap}$$

Example: What will be the final area of a cast-iron manhole cover whose diameter on a 0 F day measures 24 in. if the temperature on a summer day registers 110 F? What will be its new diameter?

Solution:

Coefficient of area expansion is equal to twice the coefficient of linear expansion.

Coefficient of linear expansion for cast iron = 0.0000059

$$\begin{aligned}
\text{Area on 0 F day} &= \frac{\pi d^2}{4} \\
&= \frac{3.1416 \times 24^2}{4} \\
&= 452.39 \text{ sq in.}
\end{aligned}$$

$$\begin{aligned}
\text{Area on 110 F day} &= \text{original area} + (\text{original area} \times 2 \times \text{coefficient of linear expansion} \times t_d) \\
&= 452.39 + (452.39 \times 2 \times 0.0000059 \times 110) \\
&= 452.978 \text{ sq in.}
\end{aligned}$$

$$\text{Diameter on 110 F day} = \text{original diameter} + (\text{original diameter} \times \text{coefficient of linear expansion} \times t_d)$$

$$= 24 + (24 \times 0.0000059 \times 110)$$
$$= 24.0156 \text{ in.}$$

THERMAL ENERGY—WORK

Thermal energy is capable of imparting the capacity to do work to a body. In a work system employing mechanical or electrical energy, the capacity to perform work is expressed in horsepower or kilowatts, respectively. The work capacity of a system utilizing thermal energy is expressed in Btu per minute or Btu per hour.

Energy, being the same thing regardless of its origin, allows the capacity of any work system to be expressed in terms equivalent to all other work systems. The amount of mechanical work done as expressed in horsepower is representative of an equivalent amount of electrical work done as expressed in kilowatts, or, an equivalent amount of thermal work done as expressed in Btu per minute.

Before developing the means by which the work values of one form of energy may be converted to equivalent values of all other forms, a summary of heat characteristics is given.

SUMMARY—THERMAL ENERGY

1. Thermal energy is another form of energy.
2. It may be measured as to its intensity and quantity.
3. The presence of thermal energy causes translatory, rotating and vibratory motion of the molecules of all substances within an orbit whose range is governed by the state of the substance, and at a velocity determined by its temperature. Velocity of the molecules increases as the temperature of the substance increases.
4. As a result of increased molecular motion, expansion occurs in all substances. When added in sufficient quantities, heat disturbs the molecular bond to such an extent that separation is encouraged, resulting in melting of solids and evaporation in liquids. There is a definite relationship between the rate of molecular velocity and the ability of any substance to maintain its state.
5. Since thermal energy is present in all substances, molecular motion is sustained perpetually. The tendency to transfer thermal energy to a body whose molecular velocity is less rapid continues wherever the relative molecular velocities of two bodies differ. This bears out the law that "wherever temperature differentials exist, a tendency of thermal energy to flow to the lower one also exists." Heat flow will continue until the energy contained in both bodies reaches an equilibrium.
6. Because available thermal energy is present only where a tempera-

ture differential exists, it is always found in a transitory state. Therefore, heat is a form of *kinetic energy*.

7. As kinetic energy, heat imparts to all substances the capacity of doing work.

Confined gases do not suppress their rate of molecular velocity by reason of their confinement when heat is added; only their range of movement is limited. Their tendency to expand with increased temperature remains intact, although they are prevented from doing so by their confining walls. The force built up finds outlet in the impact against the walls as a result of high velocity and the combined effect of all molecules generates a pressure directly proportional to the temperature of the gas.

If a gas is confined in a fully enclosed cylinder having as one of its retaining walls a movable, weighted piston, a thermal-work system is created. By introducing a high-temperature gas to the chamber when the position of the piston is such that only a small amount of the cylinder volume remains, subsequent expansion of the hot gas will move the piston, causing the weight to move with it. The total force capable of being exerted by the gas depends on (1) its temperature, which controls pressures, and (2) the area of the piston against which pressure is exerted. The action is similar to that which is carried on in a steam-engine cylinder and is the way of utilizing thermal energy to perform useful work.

If a relatively cold gas instead of steam were injected into the cylinder and heat applied after injection was complete, either by a hot spot or a spark to promote combustion, a similar movement of the piston would result because of the resulting expansion of the gas. This is the principle of the internal combustion engine and represents another widely used method of employing thermal energy.

Work can be accomplished also by applying heat to a solid and exploiting the movement resulting from its expansion. In using solids, tremendous forces become available that are not present when a similar amount of heat is applied to gases. The reason for this lies in the compressibility of gases as compared with the noncompressibility of true solids. Gases possess the advantage of a greater range of movement due to their indefinite expansion characteristics. Liquids also expand in much the same way as solids, and, as they are only slightly compressible, exert great pressures when expansion occurs due to increased temperature. Expansion qualities of liquids are limited by their boiling points, above which they act as gases.

To determine the relationship between various units of energy, it is necessary to perform certain experiments. J. P. Joule, an Englishman, performed a series of such experiments during the nineteenth century to determine the equivalent values of various work systems. By the use of weights, cords, and pulleys, he was able to rotate paddles within an insulated chamber or calorimeter containing water. During the stirring process, heat was

TABLE 5. WORK EQUIVALENTS

1 Btu =	778 ft-lb	
	252 calories (cal)	
	0.000293 kwhr	
	0.000393 hp-hr	
	0.001036 lb of water evaporated from 212 F to steam at 212 F	
	1.055 watt-sec	
1 ft-lb =	0.001285 Btu	
1 kw =	56.9	Btu per min
	3412	Btu per hr
	0.948	Btu per sec
	1000	watts
	1.34	hp
	737.3	ft-lb per sec
	44,240	ft-lb per min
	2,654,200	ft-lb per hr
	3.52	lb of water evaporated per hr from 212 F to steam at 212 F
1 hp =	42.4	Btu per min
	2545	Btu per hr
	0.707	Btu per sec
	33,000	ft-lb
	0.746	kw
	746	watts
	2.64	lb of water evaporated per hr from 212 F to steam at 212 F
1 watt =	3.412	Btu per hr

developed which raised the temperature of the water in the calorimeter. Thus he was able to determine the numerical relationship between mechanical and thermal energy, known variously as the mechanical equivalent of thermal energy or Joule's equivalent.

By such experimental methods as were carried on by Joule and others, the work equivalents shown in Table 5 were established.

Practical uses of the equivalents listed in Table 5 are shown in the following examples:

EXAMPLE: Convert 75 hp to Btu per min and Btu per hr.

From the table, 1 hp is equivalent to 42.4 Btu per min or 2545 Btu per hr.

$$\text{hp} \times \text{Btu per min equivalent} = \text{total thermal units}$$

$$75 \times 42.4 = 3180 \text{ Btu per min}$$

or

$$75 \times 2545 = 190{,}875 \text{ Btu per hr}$$

EXAMPLE: How much heat is given off by three 60-watt electric light bulbs and a 25-hp electric motor running at full load?

From the table, 1 watt is equivalent to 3.412 Btu per hr

1 hp is equivalent to 2545 Btu per hr

$$\therefore 3 \times 60 \times 3.412 + 25 \times 2545 = \text{Btu per hr total}$$

$$615.16 + 63{,}625 = 64240.16 \text{ Btu per hr}$$

EXAMPLE: What is the thermal energy equivalent for the work performed by a machine handling 60,000 lb per min through a distance of 15 ft? How much horsepower is developed? How many kilowatts?

$$60{,}000 \text{ lb} \times 15 \text{ ft} = 900{,}000 \text{ ft-lb per min}$$

$$\frac{900{,}000}{33{,}000} = 27.27 \text{ hp}$$

$$27.27 \text{ hp} \times 0.746 = 20.34 \text{ kw}$$

$$27.27 \text{ hp} \times 42.4 \text{ Btu} = 1156.24 \text{ Btu per min}$$

HEAT TRANSFER

Heat seeks its own level, and will flow from a body of high temperature to one having a relatively lower temperature. The rate of heat flow depends on the temperature differential of the two bodies, increasing correspondingly as the difference in their temperature becomes greater. This flow of heat, called heat transfer, may be accomplished by means of conduction, convection, or radiation, or by any combination of the three methods.

Conduction

The transfer of heat from one part of a body to another part of the same body, or from one body to another in direct physical contact, without apparent motion of the mass, is known as conduction. Conduction usually is associated with solid bodies, and develops as a result of increased molecular activity within the mass. The rate of conduction varies with the nature of the material used as the carrier or conductor of heat. Different materials offer more or less resistance to heat flow, which classifies them as good or poor conductors. Extremely poor heat-conducting materials are called insulators.

A metal rod, one end of which is placed in a flame, will conduct heat from the source of heat to the other end. An iron pipe carrying cold water through a mass of warm air will conduct heat from the air to the water. In neither of these examples of heat transfer was there any apparent external disturbance of the mass acting as the conductor.

Metals are good conductors of heat, some being better than others. Nonferrous metals, such as copper, bronze, and brass, are excellent heat conductors, while iron and steel are considered good. Both copper and iron are used extensively in heat-transfer equipment. Their individual selection is subject to considerations of strength and corrosive tendencies.

Cork, sawdust, hair felt, asbestos, magnesium, and seaweed are extremely poor heat conductors and are employed as insulators. Insulators are used to retard the entrance of conducted heat through walls of refrigerated compartments, brine tanks, and air-conditioned spaces. They are employed also to reduce thermal losses from boilers and steam or hot-water pipes.

Heat-conduction values of various materials are listed in Table 72. Methods of calculating heat-conduction losses through various thicknesses and combinations of materials are described in Chapter 12, "Heat Transfer."

Convection

The density, or weight per unit volume, of a fluid decreases with any increase in its temperature. As a result, warmer layers tend to rise above colder layers within the same fluid mass. When heat is applied to the bottom of a fluid body, a continuous displacement of the colder upper layers by the warmer lower layers occurs. The current so established carries and distributes stored internal thermal energy within the fluid mass and tends to establish a temperature level throughout the entire body. Activity within the fluid is initiated by a difference in densities between the several regions of the mass (such as the bottom, intermediate and top areas, when diffusion is the result of gravity). This principle of transporting stored energy from one region to another within the same mass is known

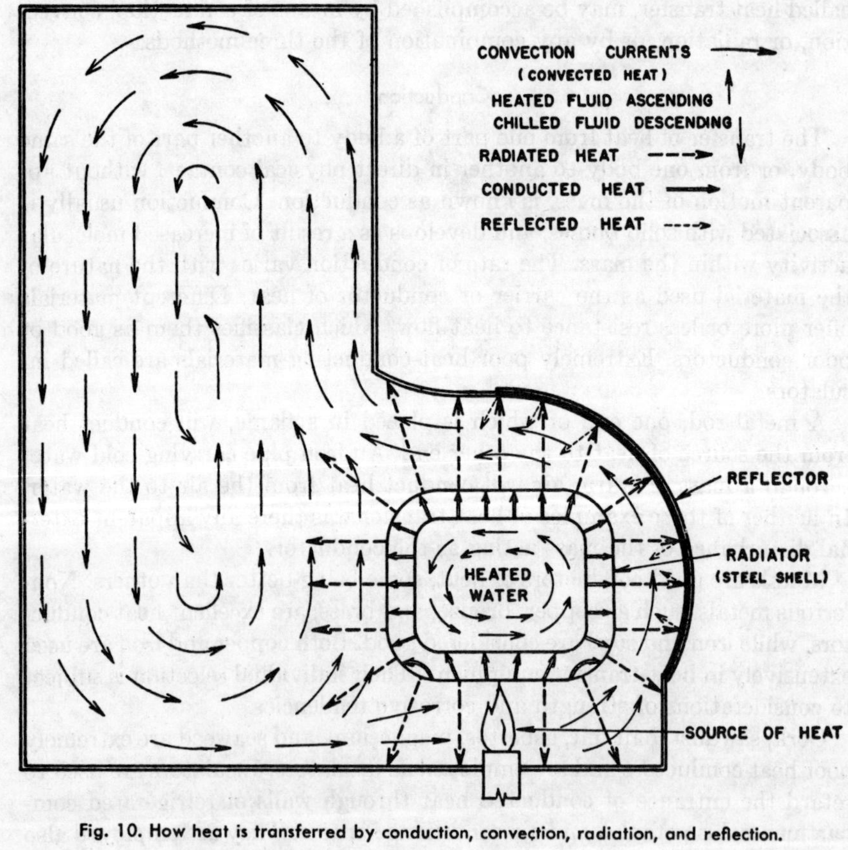

Fig. 10. How heat is transferred by conduction, convection, radiation, and reflection.

as *convection.* It must be realized that it is not the thermal energy that is being transferred from and to the various density zones in the same manner as occurs in a conduction system. Rather it is the fluid itself that responds to the changes in density carrying its stored thermal energy with it. *Thus, in a convection system thermal energy is transported or carried by its host fluid where each portion of the fluid continually changes its location within the mass as a result of density differentials.* While this principle of convection is included under the classification of heat transfer methods, along with radiation and conduction, any transitional flow of thermal energy *within* the fluid mass is of minor importance compared to the major phenomenon of current (flow) formation within the mass initiated by density changes within the same mass. Heat, according to its definition (thermal energy in transit) is theoretically not involved when convection is considered by itself, only when it is being transferred to or from the fluid mass in which convection occurs.

The principle of convection is employed in cold-storage rooms. The air heated by contact with warm food products rises to displace the heavier air already cooled by the refrigerating coils that purposely are installed near the ceiling. The domestic electric refrigerator and the icebox employ the principle of convection.

Where a fan or a pump is used to gain greater current velocity throughout a fluid, the system becomes one of forced convection. Natural convection systems are those that depend solely on gravity-actuated currents.

Radiation

Heat transferred by waves from a warm body to a colder one is called radiant heat. Radiant heat waves travel in a straight line between bodies of different temperatures and do not require a carrier for their transfer as does conducted heat. The intensity of radiation depends on the temperature of its source and the distance between the source and the surface of impact. The length of radiant waves is the distance between the crest of one wave to the crest of its neighboring wave. Wave length is constant for a given temperature of heat source, but varies, as does its amplitude, according to any change in that temperature. As temperatures increase, radiant waves diminish in length but become more numerous. Dispersed in all directions, radiant waves impinge on all surfaces within their sphere of intensity.

The rays of the sun or the heat from an electric light bulb or a hot stove are examples of radiation.

Surface character and color of bodies subject to the impact of radiant waves govern the amount of energy absorbed by such bodies. Polished surfaces, such as mirrors and chromium plate, reflect the rays of radiant heat. Dull and dark surfaces absorb them. They will pass through a vacuum without loss of energy and through gases with very little reduction in their intensity. Radiant heat penetrates transparent substances, such as clear glass, but will be absorbed by those that are opaque.

Some provision should be made to shade or reflect the sun's rays to combat the effect of radiant heat from that source on cold-storage plants, air-conditioned buildings, etc. Coating the exterior surfaces with white or aluminum paint will reflect radiation.

SUMMARY OF CHAPTER 2

Thermodynamics is a physical science dealing with the laws of thermal energy.

Thermodynamic laws:

1st Law—When thermal energy is converted to mechanical energy, a definite amount of thermal energy disappears to emerge in like amount as mechanical energy and, conversely, when mechanical energy is converted to thermal energy, a definite amount of mechanical energy disappears to emerge in like amount as thermal energy.

2nd Law—It is impossible for a self-acting machine, unaided by an external source, to convey heat continuously from a body of relatively lower temperature to a body of higher temperature. Heat cannot of itself pass from a cold to a warmer body.

Energy is the impelling force that creates change in a body physically, chemically, etc. It is the capacity of a body to do work.

Conservation of energy—A universal law stating that energy is never created or destroyed.

Energy exists in various forms, including mechanical, electrical, chemical, thermal, magnetic and nuclear energy.

Energy may be converted from one form to another. All forms may be converted to thermal energy. Thermal energy is energy in its lowest form.

Energy may exist as:

Kinetic energy—energy of motion or translation. Released by temperature differential in thermodynamic systems.

Fixed energy—latent energy existing in substances, such as fuels, explosives, chemicals, etc. Released by change in environment, e.g., combustion, shock, reaction or atom bombardment.

Potential energy—latent energy existing by reason of position as in mechanical systems.

Thermodynamic medium—A working substance acting as the agent to perform a given thermodynamic change, or series of changes, appropriate to the process under consideration. Water, steam, air, oil, refrigerants or other substances either alone or combined represent various types of thermodynamic media. The medium is the substance that receives, stores, conducts and transfers thermal energy.

Thermodynamic system—includes all or that part of the medium upon which the appropriate thermodynamic changes are imposed to perform the thermal process under consideration. It does not include the equipment employed to house the medium, or the thermal energy source or sink.

Surroundings—are all substances outside the thermodynamic system and are the recipients, or sources of heat in transit across the boundaries of the system either by design or otherwise, i.e., losses, etc.

Sink—the region to which thermal energy flows. It includes all recipient bodies whose temperature is equal to or below that of the datum for the particular process systems, internal energy stored in a substance as in thermodynamic systems, etc. The level of potential thermal energy increases by an ingress of kinetic energy from a higher temperature source and decreases by egress as kinetic energy to a sink having a lower temperature.

Thermal energy can be measured as to its intensity and quantity.

Thermal intensity is identified as the temperature of a substance as measured by a thermometer, pyrometer, etc., and expressed in degrees according to the scale selected, i.e., Fahrenheit (F), Centigrade (C), etc.

Thermal energy quantity is expressed as British thermal units (Btu), which is the amount of heat required to raise the temperature of 1 lb of water 1 F.

Heat is thermal energy in its kinetic state, i.e., thermal energy in transit across the boundaries of a thermodynamic system by reason of the existence of a temperature differential.

Specific heat is the amount of thermal energy required to raise the temperature of 1 lb of any substance 1 F as compared with the amount required to raise the temperature of 1 lb of water 1 F. Specific heat of water is 1.0, which defines the British thermal unit.

Latent heat is the amount of thermal energy required to change the state of a substance without a change in temperature. It is identified with the fusion or evaporation of substances.

Latent heat of fusion is the thermal energy absorbed or rejected by a substance during its conversion, respectively, from a solid to a liquid, as from ice to water, or from a liquid to a solid, as from water to ice.

Latent heat of vaporization is the thermal energy absorbed or rejected by a fluid during its conversion, respectively, from a liquid to a gas, as from water to steam, or from a gas to a liquid, as from steam to water.

Temperature differential is the difference in temperature between two bodies, such as between the thermodynamic system and its surroundings.

Work is energy being expended and represents kinetic energy, i.e., mechanical energy being transmitted from one body to another in the form of motion. Motion involves distance so that

Work = force (energy intensity) × distance (magnitude of motion)

or $W = f \times d$

Work is measured in terms of foot-pounds.

Power is the rate of doing work or the amount of work performed in a given time. It is expressed mechanically in terms of horsepower and represented by the equation

$$\text{horsepower} = \frac{\text{force in lbs} \times \text{distance in feet}}{33{,}000}$$

Power is expressed electrically as kilowatts where

$$\text{kilowatts} = \frac{\text{volts} \times \text{amperes}}{1000}$$

Molecular theory of thermal energy is based on the behavior of the molecules that make up a substance due to the internal energy present at any given time, or during periods when heat is being absorbed or rejected by the substance. Heat absorbed by a substance tends to intensify molecular activity even to the point of liquefying solids or vaporizing liquids. Heat rejected by a substance tends to slow down molecular activity even to the point of freezing or solidifying liquid or condensing gases. Molecular activity is created by

(*a*) random translatory motion of the molecules;
(*b*) rotational motion of the molecules about their own centers of mass;
(*c*) motion of the atoms comprising the molecules relative to each other.

Absorbed heat lengthens the distance traveled by the molecules tending to separate each from the other within the mass. Rejection of thermal energy in the form of heat has the opposite effect on the molecules within the mass.

Kinetic thermal energy—thermal energy absorbed or surrendered by a substance resulting in a change in temperature of that substance. Formerly described as sensible heat, it is sometimes referred to as internal kinetic energy.

Potential thermal energy—thermal energy absorbed or surrendered by a substance resulting in: (*a*) a change in the state of that substance (e.g., water to steam), (*b*) a change in its volume, (*c*) or both.

Total internal energy is intrinsic thermal energy stored within a substance at a given state point. It is expressed in Btu of a substance and represented by the symbol U. Specific internal energy is expressed in Btu per lb of substance and represented by the symbol u. U and u equal zero at the datum.

Joule's equivalent is the mechanical work equivalent of heat usually identified by the symbol J.

$$J = 778 \text{ ft-lb per 1 Btu}$$

Thus the conversion factor of 778 ft-lb, the mechanical equivalent of 1 Btu thermal energy, is the mathematical demonstration of the First Law of Thermodynamics that energy is never created or destroyed, but is mutually convertible to other forms. See Table 5 for work equivalents for the various forms of energy as related to each other.

Enthalpy—a property of a substance, the value of which is fixed by the state of the substance. It is given the symbol of H for total enthalpy, Btu, and h for specific enthalpy, Btu per lb. It is defined as the sum of the total internal energy, or the specific internal energy of a substance, plus a function of the ratio of pressure, psia and total or specific volume, cu ft or cu ft per lb at the state point under consideration and the mechanical equivalent of thermal energy 778 ft-lb per 1 Btu, e.g.:

$$H = U + \frac{pV}{J} = \text{Btu total enthalpy}$$

or

$$h = u + \frac{pv}{J} = \text{Btu per lb, specific enthalpy}$$

See Chapter 7 for practical mathematical demonstrations of enthalpy determinations for several state points.

Entropy is a property of the thermodynamic working medium usually represented by the symbol S for total entropy Btu per °F Abs. and s for specific entropy, Btu per lb per 1 F Abs. It is usually referred to as the appropriate number of units of entropy which eliminates the reference to the unity value of lb-degree dimension. Entropy (according to Clausius) is the measure of unavailability of thermal energy of a system. It is demonstrated by the mathematical definition

$$ds = \frac{dq}{T}$$

where ds is change in entropy, dq is change in the quantity of thermal energy within the medium, Btu per lb, and T is the temperature, °Absolute. According to the calculus, the change in specific entropy is the definite integral of the change in specific entropy (ds) between initial and final state points 1 and 2 ($s_2 - s_1$) as being equal to the definite integral of the ratio of the change in internal energy (dq) to the absolute temperature (T), for reversible processes. In engineering thermodynamics, it is a change in entropy rather than the total entropy that is important. Entropy increases with a loss of available energy from any thermodynamic process.

Expansion or contraction in the physical dimension of a substance is the result of an increase or decrease in temperature, respectively. Changes occur in linear, areal and cubical dimension.

Coefficient of expansion is the numerical rate of dimensional change of a substance per degree change in temperature. Each substance possesses its individual rate of dimensional change or coefficient of expansion or contraction.

Heat transfer from one body to another may be accomplished by

Conduction—via solids

Convection—via fluids

Radiation—via heat rays

Some substances are better conductors of heat than others. Radiated heat is reflected by a smooth, light-colored surface, but is absorbed by rough, dark surfaces.

STUDY OUTLINE FOR CHAPTER 3

Pressure:
- Definition
- Cause—compression, expansion
- Effectiveness
- Sources—mechanical, hydraulic, pneumatic

Atmospheric pressure:
- Source, intensity, and measurement

Gage pressures

Absolute pressure

Vacuum

Pascal's law

Liquid pressures

Effect of pressure:
- On boiling point of liquids
- On freezing point
- On thermal qualities of fluids

Vaporization and evaporation:
- Vapor pressure

Condensation

Pressure-evaporation relationship

Saturated liquid

Saturated vapor

Superheated gases

Tables of thermal properties of fluids

3

Pressure

In dealing with pressure, consideration must be given to the state of the substance from which a force comes, or through which it is transmitted. Pressure is simply the effect of force or energy, in either a kinetic or a potential state. When not opposed by a like force, it is capable of doing work.

While pressure is the same thing regardless of its conveyor, the form of utilization depends entirely on that conveyor. This may involve one or more of the three states of matter—solids, liquids, or gases. Therefore, any consideration of pressure must be associated with at least one of the following:

1. Perfect solids
2. Perfect liquids:
   (*a*) Free surface
   (*b*) Confined
3. Gases:
   (*a*) Constant volume
   (*b*) Variable volume (constant pressure)

Solids. Perfect solids may be defined as any solids free of air spaces and therefore incompressible. As they are capable of expanding to a limited degree with increased temperature, solids are able to exert tremendous pressure in all directions. They are not elastic. Some types of controls utilize the expansion characteristics of solids, particularly metals, for their actuation. Solids are the basic state of most substances.

Liquids. Perfect liquids may be defined as any liquids free of air or other gases and therefore practically incompressible. Like solids, liquids expand slightly with increased temperature. Liquids are the intermediate state between a solid and the gaseous phase of most substances. Their advantage lies in their vaporizing quality, and their usefulness in this respect is contingent on several factors, including boiling-point temperatures. All refrigerants must possess a practical pressure–boiling-point relationship to be so considered.

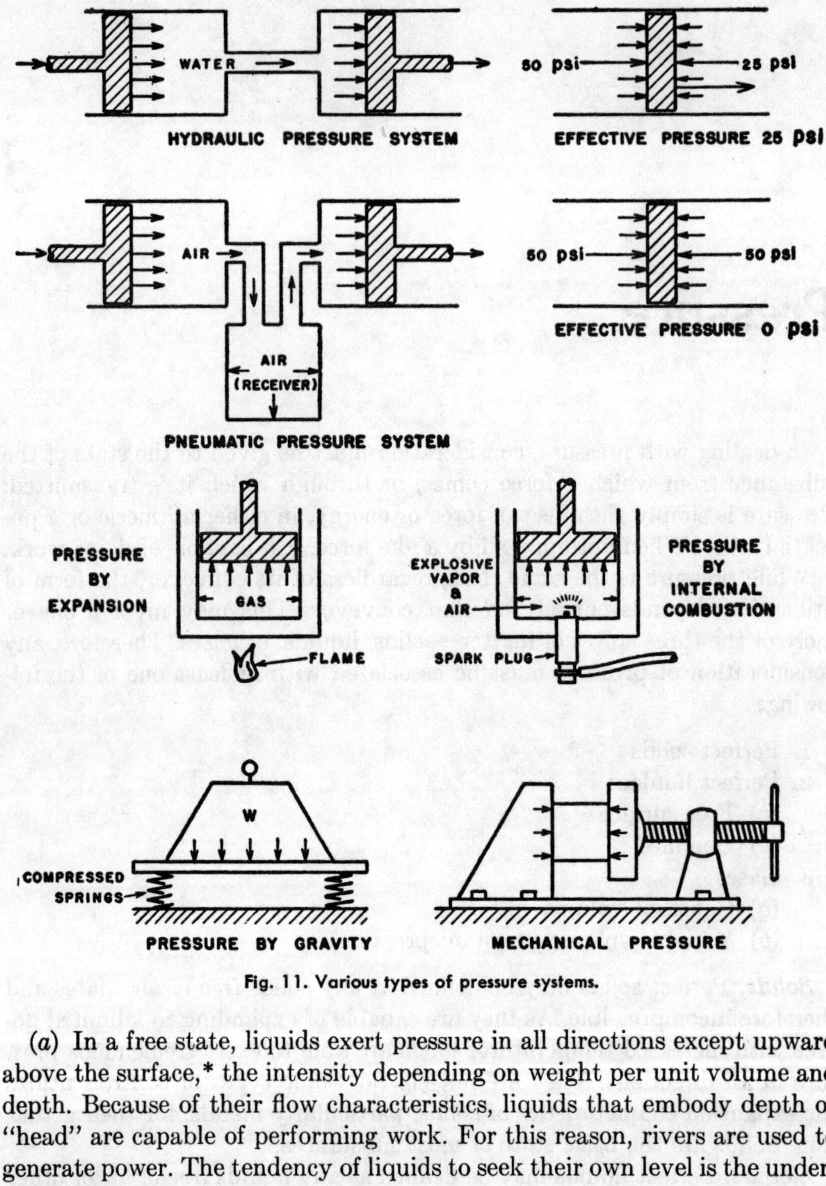

Fig. 11. Various types of pressure systems.

(*a*) In a free state, liquids exert pressure in all directions except upward above the surface,* the intensity depending on weight per unit volume and depth. Because of their flow characteristics, liquids that embody depth or "head" are capable of performing work. For this reason, rivers are used to generate power. The tendency of liquids to seek their own level is the underlying principle.

(*b*) When confined to restricted space, liquids are capable of transmitting energy with equal intensity in *all* directions. When employed for this purpose, its effect (called hydraulic pressure), together with its cause, is me-

* Free liquids do exert surface pressure in the form of vapor pressure, which varies with their temperature and atmospheric conditions. The sense in which liquids are treated here allows such pressures to be disregarded.

chanical in nature and depends on such a source of energy to perform work. The use of liquids in this manner constitutes the basic principle of all hydraulic machines.

Gases. Gases usually are derived from liquids. Because of their elastic qualities, they are compressible, which also accounts for their indefinite expansibility even to the point of complete diffusion. They will condense or return to their liquid phase when their temperature is reduced sufficiently. When restricted, their tendency to expand upon exposure to heat develops a pressure against their restraining walls with equal intensity in all directions. Unlike pneumatic pressure systems, whose source of energy is basically mechanical by reason of prior compression, this discussion of gases assumes a thermal system whereby heat is transferred to a liquid to produce gas by vaporization and possible expansion of the gas as additional heat is applied. Therefore, a major part of this refrigeration study embraces pressure in its relationship to the expanding qualities of gases that are generally in intimate contact with their liquid phase. Hence it becomes necessary to consider the expansion qualities of refrigerants under conditions of both constant volume and constant pressure. When in contact with their liquid phase, gases generally are spoken of as vapors.*

(*a*) Constant volume maintained by a gas while absorbing heat suggests that expansion is restricted within a definite space while the *tendency* to expand farther continues. Because the amount of force exerted on fluids governs the orbit within which molecular activity is concentrated, the increased velocity of the molecules resulting from higher temperature finds outlet in the greater impact force with which they impinge against their restraining walls, thus substituting such force for greater orbit or range of travel. The latter is called distance in the work formula, $W = f \times d$, which also applies to the work done by any one molecule. The combined impact force of all molecules equals the total force exerted against the walls of the restriction chamber. As a gas, the force exerted on any specific area is equal in intensity to that exerted on all other areas within the chamber having similar dimensions. Therefore, the force exerted on any area measuring, say, 1 sq in., is equal to the force exerted on any other square inch of area located at all other points within the chamber. The measurement of such force on any similar area would be expressed as pressure in pounds per square inch (psi).

Continuing heat application increases the molecular velocity, thereby providing greater impact force of each molecule. As a result, the total force increases, raising the unit area pressures proportionately. This action can be continued until the gas reaches a temperature above which it cannot be liquefied by any amount of compression. This point is called the critical temperature, and the pressure corresponding to this temperature is called the critical pressure. Both vary with different gases.

*** In this text, gas and vapor are synonymous, except for superheated conditions, at which time the term gas appears to be correct.**

(*b*) If the gas in the restricted chamber above were allowed to expand at a rate proportional to its temperature rise, by some piston-cylinder arrangement, a condition would exist whereby the pressure would remain constant while the volume increased. Under such conditions, the force of molecular impact would find outlet by performing work in moving the piston. By removing the permanent restriction to molecular activity, its orbit widens, thereby increasing the travel distance of each molecule and reducing impact force. Provided that the same amount of heat were added to the gas in case (*b*) and in case (*a*), then the amount of work in each case would be the same, and $W = f \times d =$ constant, holds true. However, under constant pressure conditions, the value of f decreases while the value of d increases proportionately at a rate governed by the resistance met in moving the piston. While the amount of work capable of being performed remains constant where applied heat energy is constant, greater *available* energy would be obtained if the value of f were allowed to increase. As f is the measure of pressure, it follows that energy becomes more available as pressures increase.

As the expanding gas moves the piston, thermal energy is converted to mechanical energy, lowering the temperature of the gas. Continued to its ultimate conclusion, a sufficient quantity of thermal energy becomes lost to the gas, so as to decrease molecular velocity to a point where it is no longer able to maintain an orbit of vaporous proportions, thereby reverting to its liquid phase. Returned to a heating chamber or boiler, this liquid would vaporize again and the process would be repeated. This is actually what occurs in steam-engine and steam-turbine operation.

Pressure, then, is a uniformly distributed force exerted on the surface of a body. When applied through a confined fluid medium, an equal force is exerted in all directions. Usually expressed in terms of psi, the total pressure on a surface of any size may be determined by multiplying the total number of square inches by the pressure in psi.

The latter is well demonstrated in determining indicated horsepower ratings of any thermal engine, as shown in the formula:

$$\text{Indicated horsepower} = \frac{PLAN}{33{,}000}$$

where P = pressure in pounds per square inch
L = length of piston stroke in feet
A = area of piston in square inches
N = number of strokes per minute

By employing unit pressures, the total pressure may be determined by multiplying the value of P by the total number of square inches of the piston, as represented by A.

Pressure is created by and its intensity depends on the amount of external energy. The amount of pressure is relative in that its effectiveness depends on surrounding pressure conditions. It can be equalized by the application of a similar counterpressure. Where pressure differentials exist, there is a tendency toward equalization. If accomplished, it results in a decrease in the higher pressure and a corresponding increase in the lower one.

ATMOSPHERIC PRESSURE

Atmospheric pressure is the amount of gravitational force exerted by free air surrounding the earth's surface, created by its weight. It decreases correspondingly with increased altitude above sea level. Normal atmospheric pressure at sea level is 14.7 psi, varying slightly with changes in air currents.

Atmospheric pressure is determined by means of an instrument called a barometer (Fig. 12), which expresses the intensity in inches of mercury. The barometer consists of a glass tube, approximately 32 in. long, closed at one end, and filled with mercury. During the filling process, all air is excluded from the tube. The tube is then inverted and the open end placed in a mercury reservoir, the top of which is open to the atmosphere. The weight of the mercury in the tube will cause its level to drop, creating a vacuum in the top of the tube above the column. When the upward surface pressure of the mercury in the reservoir is equal to the downward pressure of the atmosphere, the column will cease falling, with its level coming to rest at a definite height above the reservoir. Measuring from the surface of the mercury in the reservoir to the top of the mercury column in the tube, the number of inches of mercury capable of being sustained by the pressure of the atmosphere is determined. Any increase in atmospheric pressure causes the mercury column to rise and any decrease to fall. The pressure of the atmosphere is capable of sustaining a mercury column 0.491 in. high for each pound per square inch exerted on the surface of the mercury in the reservoir. Under normal conditions at sea level, atmospheric pressure will sustain a mercury column nearly 30 in. high. Due to the expansion characteristics of mercury, corrections in height must be made according to the temperature at the time of reading.

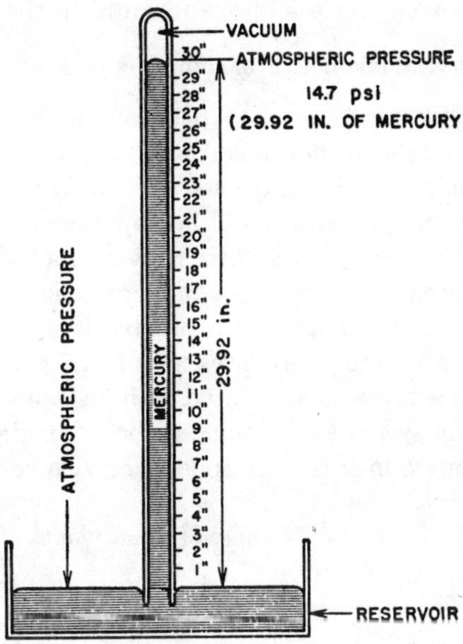

Fig. 12. Simple barometer.

The number of inches of mercury usually referred to as barometric pressure is used generally in connection with atmospheric pressure or pressures below atmospheric, whereas pressures above atmospheric are expressed in pounds per square inch. In order to convert psi to inches of mercury or vice versa, the following formula is employed:

$$\text{Inches of mercury (in. Hg)} \times 0.491 = \text{pressure, psi}$$

or

$$\text{Inches of mercury (in. Hg)} = \frac{\text{pressure, psi}}{0.491}$$

Applying the above formula to the normal pressure of the atmosphere at sea level, the barometric pressure would read $\frac{14.691}{0.491}$ or 29.92 in. of mercury. As pressure decreases with an increase in altitude, the barometer would also show a corresponding decrease at a rate of approximately 1 in. for each 900 ft above sea level. Thus, at an altitude of 2700 ft, the barometric pressure would be approximately 27 in. of mercury, and at 5400 ft, 24 in. of mercury. The same rule applies to pressures below the surface of the earth, except that an increase in pressure results as the depth increases. Depths of mines are estimated in this way.

An atmospheric pressure of 14.7 psi is equivalent to 1 atmosphere (atm), an expression associated with high pressures, such as are found in refrigeration systems employing carbon dioxide. To convert pressure in pounds per square inch to atmospheres or vice versa, this formula is used:

$$\text{Number of atmospheres} = \frac{\text{psi}}{14.7}$$

or

$$\text{psi pressure} = \text{number of atm} \times 14.7$$

Atmospheric pressure can be determined also by using water under an arrangement similar to that described for mercury. 1 atm, or 14.7 psi, will sustain a column of water 33.947 ft high when all the air in the tube above the water column is exhausted. However, such an instrument would be most cumbersome to employ generally. Torricelli found that, by using mercury, barometric pressure could be measured more conveniently because of the smaller instrument made possible by the greater weight of mercury over water, the ratio being 13.6 to 1. This meant that the mercury barometer could be made $\frac{1}{13.6}$ the length of the instrument employing water as the medium, or approximately 30 in. long. To provide some flexibility for expansion and handling, the barometer tube is made several inches longer, the amount of which is not important as long as all the air is exhausted above the mercury column.

GAGE PRESSURES

The pressure of any gas or liquid within a chamber may be determined by instruments called pressure gages, which register the force of pressure above that of the atmosphere. Their zero point is set at atmospheric pres-

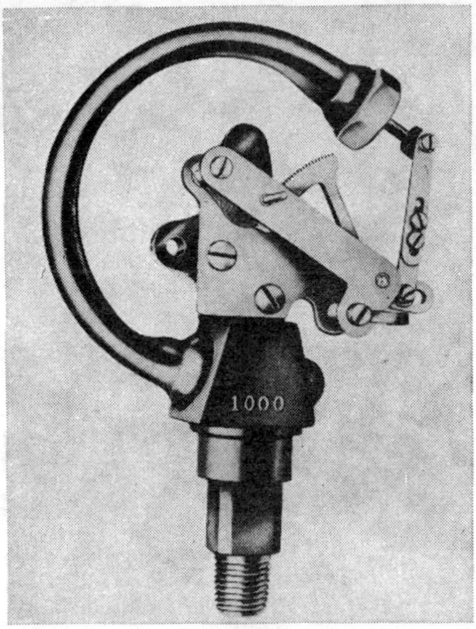

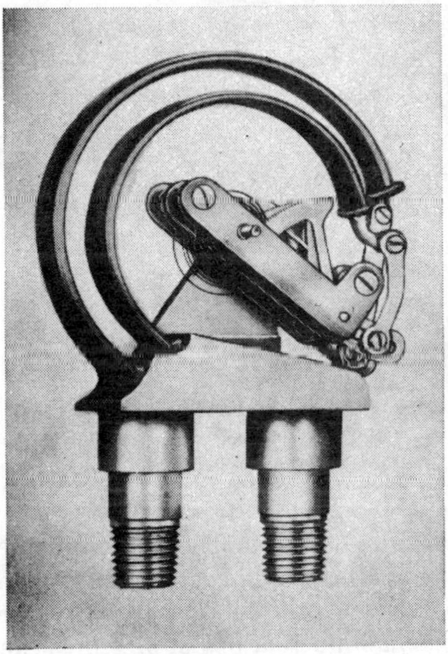

Fig. 13. Bourdon pressure gage. (*Top, left*) Internal view of single-tube Bourdon gage. (*Top, right*) External view of gage on left. (*Bottom, left*) Internal view of double-tube (duplex) Bourdon gage. (*Bottom, right*) External view of gage on left. Note double pointers. (Courtesy, United States Gauge, Division of American Machine and Metals, Inc.)

sure, or 14.7 psi. When a reading of 10 psi is indicated on the gage dial, it means that a pressure of 10 psi above the pressure of the atmosphere is being exerted within the chamber to which the gage is attached.

There are several types of pressure gages in use, the most common types being the diaphragm and the Bourdon gages.

The diaphragm gage, employed most widely where low pressures are prevalent, consists of two corrugated disks joined together to form a simple

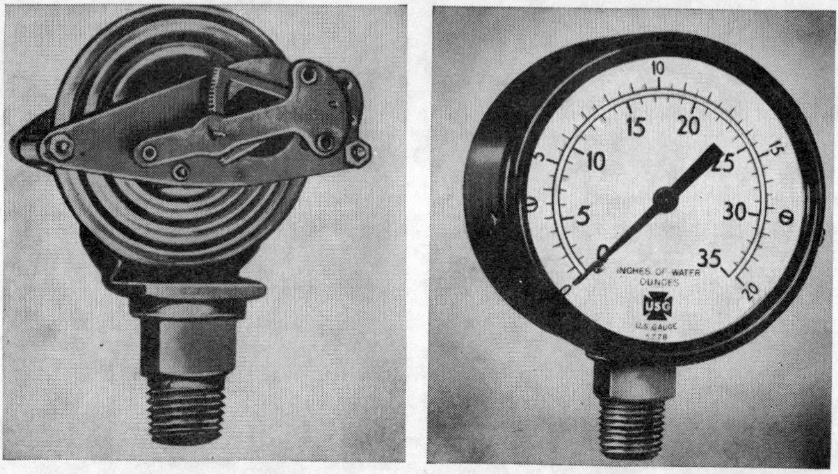

Fig. 14. (*Top*) Diaphragm pressure gage, internal and external views. (Courtesy, United States Gauge, Division of American Machine and Metals, Inc.)

Fig. 15. (*Bottom*) Compound pressure gage. (Courtesy, United States Gauge, Division of American Machine and Metals, Inc.)

bellows, which flexes in or out with pressure changes. The motion of the diaphragm is transmitted to a pointer by mechanical linkage, which in turn revolves about a dial indicating the pressure within the chamber. Because of the accuracy of the diaphragm gage, small fluctuations in pressures can be detected. The aneroid barometer operates on the principle of a diaphragm gage.

The Bourdon pressure gage also operates a pointer through a series of linkages in much the same way as does the diaphragm type of gage, but employs a flattened circular tube called a Bourdon spring instead of the bellows. As the pressure increases, there is a tendency for the flattened cross-section to become circular, thereby straightening the tube, creating a motion that in turn is transmitted to the dial pointer. The Bourdon pressure gage generally is employed for relatively high pressures, but is also adaptable for measuring pressures below atmospheric. In the latter case, the circular tube is turned in the opposite direction in order that the pointer will revolve in a clockwise direction with a decrease in pressure.

Gages that are capable of indicating pressures above and below atmospheric are called compound pressure gages, and operate on the principle of the Bourdon gage.

It is advisable to use gages that are calibrated at double the maximum working pressure to which they will be exposed so as to insure their safety against any abnormal rise above regular operating pressure. Figures 13, 14, and 15 show various types of pressure gages.

ABSOLUTE PRESSURE

The atmosphere, because of its weight, creates a gravitational force on the earth's surface and on all objects that exist within its sphere of influence. Such a force at sea level has an average intensity of 14.7 psi, exerted in all directions.

If it were possible to eliminate entirely the gases that make up the atmosphere from around the earth's surface, a void would be created, resulting in the absence of weight. This, in turn, would reduce the force or pressure created by the weight of the atmosphere to a zero value. This condition would then be considered as the absolute minimum to which pressures could be reduced, and would be termed the absolute zero of pressure. Any pressure created above and relative to the absolute zero point would be identified as an absolute force. Atmospheric pressure when measured in pounds per square inch is actually 14.7 psi above the absolute zero point of pressure and is termed 14.7 psi absolute pressure, or 14.7 psia.

Pressure which is created either pneumatically or hydraulically above that of the atmosphere is added to the pressure of the atmosphere in order to determine absolute pressures. Thus, if a pressure gage registers 10 psi, the absolute pressure would be 10 + 14.7, or 24.7 psia.

Absolute pressure may be described as the total pressure within a chamber, and includes the pressure exerted by the atmosphere.

Thermodynamic property tables covering steam and refrigerants are all based on absolute pressures, and usually are listed accordingly.

VACUUM

Space that is void of any pressure is referred to as a perfect vacuum. Any space that contains a gas, the pressure of which is less than that of the atmosphere, is considered to be under partial vacuum. Therefore, any pressure below a normal atmospheric pressure of 14.7 psi, and above an absolute vacuum, is a partial vacuum and is expressed in inches of mercury or in psi, absolute.

The range of pressure between the point of zero absolute and that of the atmosphere may be considered to be a zone of negative pressures wherein a pressure differential exists that creates a tendency on the part of the atmosphere to bring about a state of equilibrium. As the negative pressure approaches zero absolute, the greater becomes the pressure differential between that of the vacuum chamber and the atmosphere, which, in turn, increases atmospheric influence. This influence is apparent when measured by means of a mercury column somewhat similar to that described for the barometer. In the latter case, a U-tube, or manometer, is used in place of the barometer reservoir.

Referring to Fig. 16, the mercury in the top U-tube is equal in height in both stems when the openings at both ends are exposed to the atmosphere. When one end is connected to a chamber from which a part of the air has been withdrawn by means of an air pump, thus creating a pressure differential, atmospheric pressure exerted on the mercury column in the open stem forces that column to fall and the column in the closed stem to rise correspondingly. A new level for both columns will then be established, as shown in Fig. 16 (*center*), and their difference in heights will depend on the amount of air withdrawn from the chamber. As additional air is removed, the influence of the atmosphere becomes greater, until complete exhaustion of air in the chamber results in suspension of a column of mercury in the closed stem of the U-tube 29.92 in. above that in the open stem. Under the final conditions, a perfect vacuum will be established in the chamber, whereas only a partial vacuum exists at any time prior to reaching that point. Figure 16 (*bottom*) shows the mercury level as it would appear if the closed end of the U-tube were subjected to pressures slightly above atmospheric.

The mercury reading determined by measuring the difference in heights (h) in inches of the mercury columns in both stems of the U-tube is not a direct indication of the pressure within the chamber, but represents the differential in force between atmospheric and zero pressure absolute.

If an amount of air were withdrawn from the chamber so as to establish a differential of 28 in. in levels of both columns, a 28-in. vacuum would exist. Actually, a differential in pressures between the atmosphere and that of the chamber has been established in an amount equal to the height of mercury, 28 in. in this case, multiplied by the height of mercury equivalent to a pressure of 1 psi, or 0.491 in. The difference, therefore, would be 28 in. $\times$ 0.491 in., or 13.75 psi. Assuming a barometer reading of 29.92 in. of mercury as the prevailing atmospheric pressure which corresponds to 14.69 psi, the pressure in the chamber would be 14.69 $-$ 13.75, or 0.94 psia.

Some confusion results when an attempt is made to portray various degrees of pressure within a series of pressure chambers. This is especially true when the pressures involved are below atmospheric. For this reason, it is well to consider pressures according to their absolute rather than their gage values.

Figure 17 shows a series of air-tight compartments. Each compartment is divided into two sections of equal volume, connected by a by-pass line in

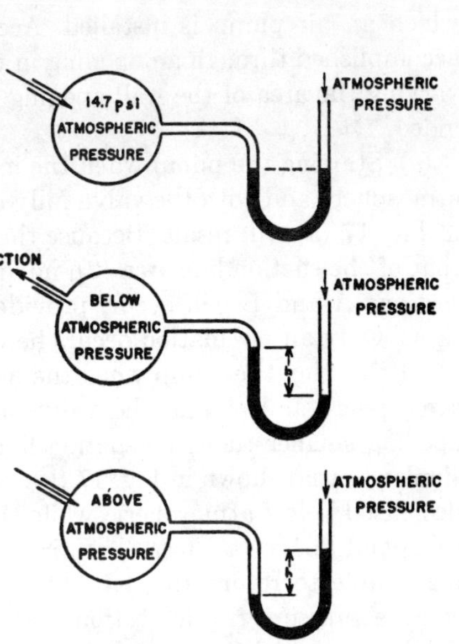

Fig. 16. (*Top*) **Effects of various pressure differences as measured by a manometer.**

Fig. 17. (*Bottom*) **Pressure differential development.**

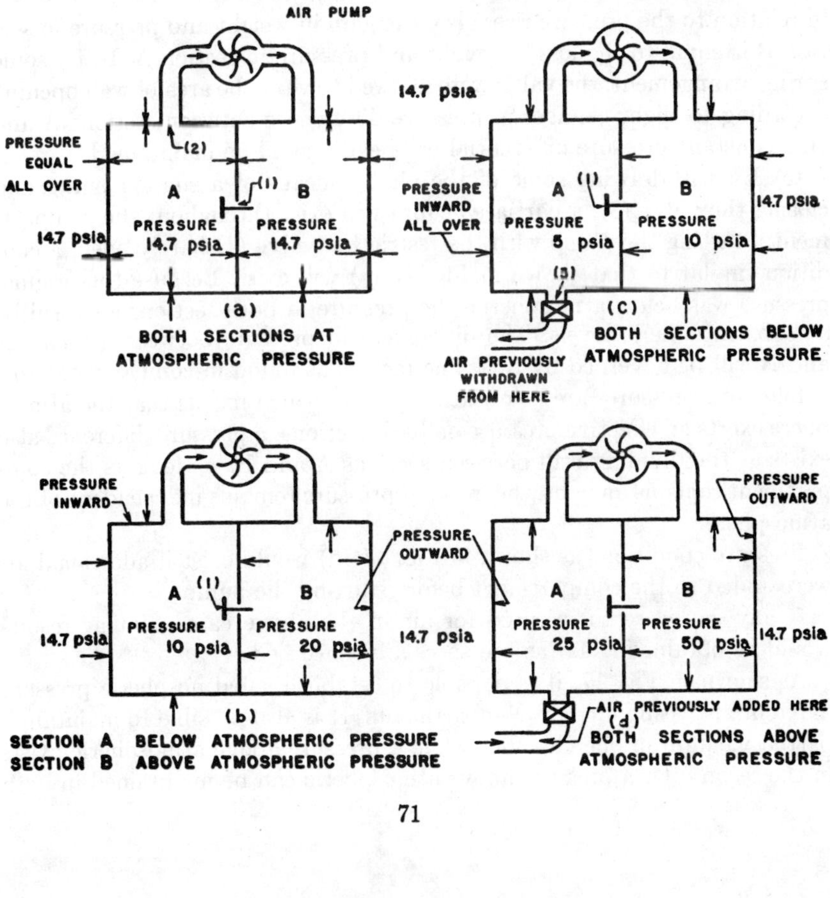

which an air pump is installed. Access from one section to the other is accomplished through an opening in their dividing wall. A valve capable of varying the area of the wall opening by moving to or from the wall is provided.

(*a*) Starting the pump when the internal pressure is equal to that of the atmosphere and with the valve fully opened, a pressure condition as shown in Fig. 17 (a) will result. Because the valve opening (1) is equal in area to that of the suction line area (2), no appreciable change in pressure between sections A and B will occur, provided that the temperature remains the same. Only air circulation occurs here.

(*b*) Starting the pump when the internal pressure is equal to that of the atmosphere and closing the valve sufficiently to render the area of wall opening smaller than the suction line area results in a pressure condition similar to that shown in Fig. 17 (b). Note that the pressure in section A has decreased below atmospheric while the pressure in section B has increased to a point above 14.7 psia. The air pumped from section A to section B, being unable to return through the wall opening fast enough, builds up to cause a pressure rise in section B. Here we have a pressure difference involving a partial vacuum and a positive pressure. An important point to consider here is that the weight per unit volume of air in section B increased in relation to the pressure rise; also, the gain in weight and pressure in section B is equal to the loss in weight and pressure in section A. If, by some spring arrangement, the valve were allowed to vary the area of wall opening according to slight changes in pressure differences between sections A and B, a constant pressure differential or "head" could be maintained.

(*c*) By withdrawing some of the air by means of a suction pump and closing the valve (3), a partial vacuum will exist throughout the compartment. Starting the pump with the restriction valve (1) barely open, a condition similar to that shown in Fig. 17 (c) will exist. Because the original pressure was below atmospheric, the pressure in both sections also will be at subatmospheric levels. The differences in pressure between sections A and B will be governed by the same factors as found in compartment (b). While both pressures are below atmospheric, which means that the atmosphere exerts an effective pressure on both sections, a pressure difference also exists at the dividing wall between sections A and B. As long as the compartment remains airtight, the internal pressure remains independent of the atmosphere.

Pressure conditions as shown in Fig. 17 (d) would exist if additional air were added to the compartment before starting the pump.

If any gas were substituted for air in the above cases, similar results would be obtained as far as pressures are concerned.

As shown in Fig. 17, it is possible to establish subatmospheric pressure differentials within airtight compartments. It is also possible to maintain a partial vacuum in one chamber while a pressure above atmospheric exists in the second. Or a pressure above atmospheric can be maintained in both chambers. In general practice, it is usual to express the value of pressure at or below atmospheric in terms of inches of mercury, and in pounds per square inch gage pressure where pressures are found to be above that of the atmosphere.

Refrigeration systems are operated under both conditions, and in some instances a combination of the two is employed, depending on the refrigerant used and the temperatures involved.

Draft Gages

Partial vacuums are measured also by means of water gages, the results of which are expressed in terms of inches or ounces of water. Here, again, the values express a negative rather than a positive pressure. Therefore, a measurement given as 1 in., or 1 oz of water, does not indicate absolute pressure directly, but merely the intensity of the pressure below that of the atmosphere.

One atmosphere (atm) is equivalent to 14.7 psi, or 33.947 ft of water. If a vacuum is created so as to reduce the water column 1 in., it is evident that the pressure differential between it and the atmosphere is slight. However, it also indicates that readings of small variations in pressure differentials are possible, and, because the coefficient of expansion of water is greatly below that of mercury, its use is advantageous under certain conditions. One of the foremost uses for water gages is in measuring chimney drafts, where only slight pressure differentials are prevalent.

Instead of using an instrument requiring a height of nearly 34 ft, as would be the case if a water barometer were employed, a much simpler arrangement in the form of a manometer, generally called a draft gage, is used. The same principle applies to the draft gage as for the U-tube, the chimney substituting for the vacuum chamber.

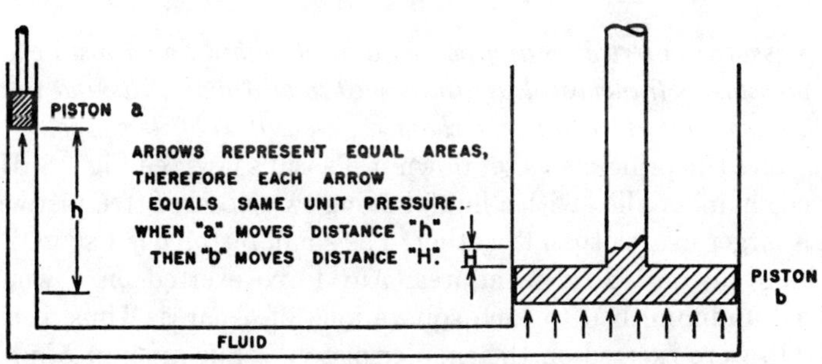

Fig. 18. Pressure distribution within an enclosed system.

Simple Hydraulic System

Where w = load applied on small piston
W = load transmitted by large piston
a = area of small piston
A = area of large piston

h = distance moved by small piston
H = distance moved by large piston

Rule: Load varies directly as areas
Distance varies inversely as areas

Then $w:W::a:A$ or $wA = Wa$ (Law of direct proportion)

And $h:H::A:a$ or $ha = HA$ (Law of inverse proportion)

If a, A, and w are known,

Then $W = \frac{Aw}{a}$ Pascal's law

If a, A, and h are known,

Then $H = \frac{ah}{A}$ Ratio of liquid displaced

Assume $a = 2$ sq in., $A = 8$ sq in.
$w = 150$ lb, $h = 6$ in.

1st case: Then $W = \frac{Aw}{a} = \frac{8 \times 150}{2} = 600$ lb transmitted by large piston

2nd case: And $H = \frac{ah}{A} = \frac{2 \times 6}{8} = 1.5$ in. distance large piston moved

Work performed by both pistons is also equal:

$$W \times H = w \times h$$

$$600 \times 1.5 = 150 \times 6$$

$$900 \text{ in.-lb} = 900 \text{ in.-lb}$$

PASCAL'S LAW

When pressure is exerted on any part of a fluid within an enclosed chamber, an equal pressure is transmitted undiminished to all equal areas and in all directions through the fluid within the chamber, regardless of its shape.

To illustrate the principle as set down in Pascal's law, see Fig. 18. Here is an enclosed chamber filled with a fluid, having two closely fitted pistons, one of which is larger in area than the other. The small piston *a* is 1 sq in. in area and the large piston *b* 8 sq in. in area. Any force exerted on *a* would be transmitted undiminished to each square inch of *b*'s area. Thus, if a pressure of 10 lb were exerted on the smaller piston, a like amount of pressure would be transmitted to every square inch of the larger piston's internal surface. To hold piston *b* in equilibrium with piston *a* at the time such force is applied at *a*, a counterweight equal to 10 psi $\times$ 8 sq in., or a total force of 80 lb, would be required at *b*.

An equal force of 10 lb would be exerted to each square inch of area throughout the internal surface, including the top, the bottom, and all sides of the chamber.

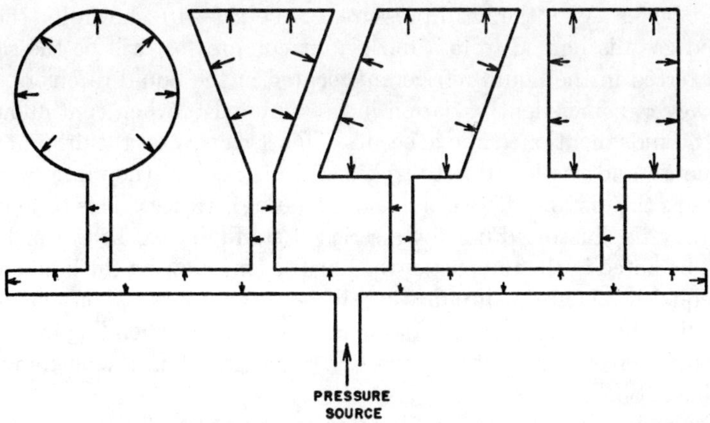

Fig. 19. Pressure is independent of the shape of a vessel. Unit pressure is equal in all vessels regardless of shape. Pressure is transmitted undiminished to all equal areas and in all directions. Pressure is exerted perpendicular to the retaining surface.

According to Pascal's principle, every square inch of internal surface area within the high-pressure side of a refrigeration system, including the condenser and the receiver, is subjected to the same pressure, the intensity—or head pressure—of which would be equivalent to that indicated on the pressure gage. The same law applies to the internal surface area of parts of the low-pressure side, including the evaporator. However, in the latter unit, pressures would be much lower than those prevailing in the high-pressure side.

TABLE 6. PRESSURE AND WEIGHT EQUIVALENTS

1 psi = 144 lb per sq ft (psf) 2.04 in. of mercury 2.31 ft of water 27.7 in. of water	**1 oz per sq in.** = 1.73 in. of water 0.128 in. of mercury
1 atm = 14.7 psi 33.947 ft of water 29.92 in. of mercury	**1 in. of mercury** = 0. 491 psi 13.61 in. of water 1.134 ft of water
1 in. of water = 0.036 psi 0.58 oz per sq in. 0.0735 in. of mercury	**1 ft of water** = 62.4 psf 0.433 psi 0.882 in. of mercury

LIQUID PRESSURES

An enclosed system containing liquid and gas will exert the same pressure on both. Therefore, the steam pressure in a boiler is also the water pressure in that boiler. Also, the head pressure, or the pressure shown for the compressed gas discharged from a refrigerant compressor, will be the same as that exerted in the liquid refrigerant located in the liquid receiver.

However, some slight variation in pressure exists when continuous flow of both fluids mentioned above occurs. This is caused by the drop in potential energy and the loss due to frictional resistance of the pipes, which increases as the distance between inlet and outlet increases. This total drop in head may be illustrated best by a series of standpipes as shown in Fig. 20. It will be noted in the top part of the illustration, showing the liquid at rest, that equal levels are maintained in all four standpipes. Upon opening the gate valve, however, all four standpipes show levels descending in height as the distance from the inlet becomes greater, as shown in the bottom part of the illustration.

Systems in which liquids are handled are subject to fluctuations in pressure when the pump alone is depended on to maintain constant head. This is especially true with reciprocating pumps, even though continuous flow is maintained. And where any pump, including centrifugal types, is used, some means of pressure control must be employed in order to provide sufficient head when a valve is opened after a period of interrupted flow, during which the pump ceases operation. The reason for this lies in the incompressibility of liquids which prevents any head being built up within it, as is possible in the case of gases. This incompressible characteristic of liquids also requires that the pump cease operating when flow ceases, otherwise the subsequent pressure might burst the pipes or cause the pump to labor and destroy itself.

Where pressure demands vary, as they do with hydraulic operation, either a variable displacement pump or a suction-valve control may be employed. Both methods of liquid volume control allow the pump to continue operating even during those periods when the demand for pressure does not exist. The suction-valve control holds the suction valves open, through which the liquid entering the cylinder during the suction stroke is discharged back into the suction line. The variable displacement pumps, of which there are several types, are designed to deliver zero to maximum volume upon gradual changes in discharge capacity. With plunger pumps, this discharge capacity change is made by varying the length of piston stroke. Capacity control of centrifugal pumps is obtained by throttling the suction to reduce the amount of liquid entering the pump.

Reciprocating pumps usually are equipped with an air dome installed on the discharge side. One purpose of the air dome is to obtain smooth flow of liquid, the pressure of which otherwise would fluctuate when the piston is in the process of reversing its stroke. It also provides a cushion against the force of liquid discharge. The air dome also provides a slight potential head in the discharge line facilitating immediate flow upon opening of the shut-off valve.

The air dome is a chamber in which trapped air is compressed as a result

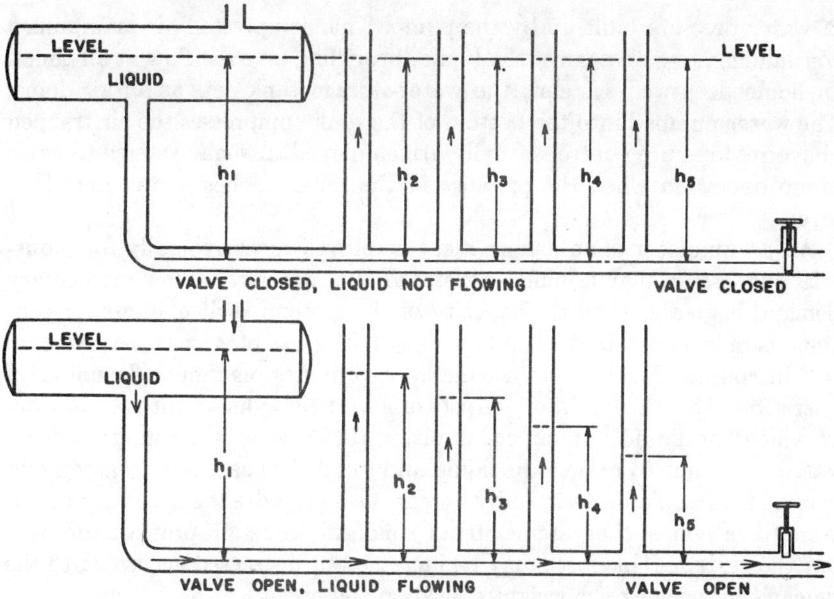

Fig. 20. Effect of friction on pressure in pipes.

Fig. 21. Air dome. Shown on discharge of steam-driven pump. (Courtesy, Wagener Steam Pump Co., of Wagener-Schrader Pump Division, Canton Stoker Corporation.)

of water pressure built up by the pump. This compressed air maintains a continuous pressure against the liquid, providing an even flow at all times. In home deep-well systems, the water-storage tank acts as an air dome. The water pumped into the bottom of the tank compresses the air trapped above it. Pressure controls usually are employed in such systems to cease pump operation when the pressure in the tank reaches a predetermined level.

An accumulator is an arrangement employed most generally for industrial purposes in maintaining a constant liquid pressure when a momentary demand is greater than the capacity of the system. It also allows for continuous pump operation when flow ceases. The use of a suction-valve control in conjunction with an accumulator provides maximum flexibility of operation. This accumulator consists of a vertical cylinder into the bottom of which the liquid is pumped. A piston, inserted at the top, provides a variable volume to receive the liquid and to fill the space left by any liquid leaving it. The piston assisted by weights also provides the necessary downward force against the water column to maintain constant pressure throughout the system. This force may be increased or decreased according to the demand by varying the weight placed on the piston.

Compressed-air systems also require some arrangement for maintaining constant pressure even though air, being a gas, is compressible. For

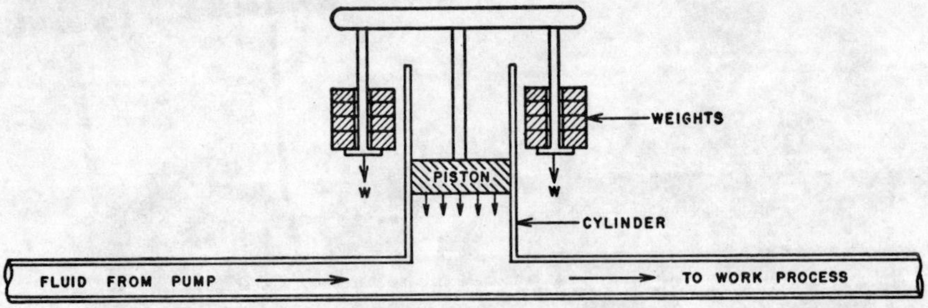

Fig. 22. Accumulator.

the purpose of providing a greater amount of air for short periods than the compressor can deliver, and to maintain constant pressure, an air reservoir, called a receiver, forms an integral part of most compressed-air systems.

Pressure in a tank increases as the depth of a liquid increases, but the pressure at any given depth is the same in all directions. Where no external pressure is applied to the body of a liquid, the force at any point within the tank is equal to the weight of the liquid above that point. Where external pressure is applied, the total force at any point will be equal to the weight of the liquid above it, plus the force of the external pressure.

Water weighs 62.4 lb per cu ft. This means that for every square foot of water 1 ft in depth, a pressure of 62.4 lb is exerted on the bottom of a vessel containing it, or

$$\frac{62.4 \text{ lb}}{144 \text{ sq in.}} = 0.433 \text{ psi}$$

A flat-bottomed tank filled with water to a depth of 10 ft, then, would have a pressure of 10 × 0.433, or 4.33 psi exerted on every square inch of its bottom. It would mean also that the seams of the tank would be required to withstand a pressure of 4.33 psi because fluid pressure is the same in all directions at any given point. Halfway between the top surface of the water and the bottom of the tank, the pressure would be only one half that exerted on the bottom, or 5 feet × 0.433, or 2.165 psi in all directions.

If air pressure of 100 psi were applied to the tank, then the total pressure at the half mark, and the bottom, would be 102.165 and 104.33 psi, respectively.

To obtain similar values for tanks with uneven or graded bottoms, an average depth must be considered in computing total force on the bottom. Intermediate levels, of course, still will be dependent on the weight of the liquid above them.

In the discussion above, water was used as the liquid in illustrating the subject. However, the principles apply to all liquids, but the pressure values for each will vary according to their weight per cubic foot, or *density*. The density of water is 62.4 lb per cu ft. The pressure exerted by any liquid at any depth, then, would be found by applying the formula:

$$\text{Unit pressure} = \text{density} \times \text{depth}$$

In order to compute pressure in psi, divide by 144 when density is given in lb per cu ft.

Boiling Points of Liquids

The boiling point of any liquid is the highest temperature at which it will remain in its liquid form at any given pressure. As was stated, the rate of molecular motion within any substance increases in proportion to the amount of heat applied. The energy thus transferred increases molecular velocity that causes a correspondingly greater impact force against the adjoining substance, whether it be the walls of a container or the atmosphere. The force of molecular impact, then, is the pressure of the substance created internally. It is known as vapor pressure. When this internal or vapor pressure becomes great enough, it tends to separate the molecules, thereby destroying the bond with which they are held together as a liquid. This pressure, known as vapor pressure, exists in all liquids and vapors at an intensity dependent on their temperature and thermal energy content. When sufficient heat is added to a liquid to raise its temperature to a level at which the vapor pressure equals the external pressure exerted on its surface, its boiling point has been reached.

Table 7. Boiling Points of Liquids at Atmospheric Pressure
(14.7 psia)

Liquid	*°F*
Ammonia	−28 F
Carbon dioxide	−110 F
Ethyl chloride	54 F
Freon-11	74.7 F
Freon-12	21.5 F
Freon-21	48 F
Freon-22	41.44 F
Freon-113	117.6 F
Freon-114	38.4 F
Methyl chloride	−11 F
Methylene chloride	103.7 F
Sulfur dioxide	14 F
Water	212 F

However, the characteristics of various liquids are such that their boiling points differ at similar pressures. Table 7 lists the boiling points of some liquids at atmospheric pressure.

The boiling point of any liquid is the temperature at which the force of molecular impact within the liquid is equal to the force of molecular impact of the surrounding substance. According to Table 7, liquids vary in their ability to maintain a molecular bond when exposed to similar external pressure conditions.

Effect of Pressure on Boiling Point

Because the boiling point of any liquid is that temperature at which its saturated vapor pressure is equal to the surrounding pressure, it would necessitate an increase in its vapor pressure to induce boiling if the surrounding pressure were increased. Therefore, any increase in the surrounding pressure exerted on the surface of a liquid results in an increase in its boiling-point temperature. Conversely, any decrease in pressure exerted on a liquid decreases its boiling-point temperature.

This principle may be illustrated if water is heated at atmospheric pressure (14.7 psia) until its temperature reaches 212 F. It will boil under these conditions. However, if the pressure is increased to 5 psig (pounds per square inch gage), its temperature must be raised to 227 F before boiling occurs; whereas, at pressures below atmospheric, a temperature of less than 212 F will be required. Every liquid has a definite pressure-temperature relationship that is constant under similar conditions and usually may be determined by means of thermodynamic property tables covering the various liquids in common use.

The law can thus be stated, that *an increase or a decrease in pressure on any liquid results in the raising or the lowering of its boiling point.* Therefore, one of the effects of pressure in relation to liquids is that it governs their boiling points.

TABLE 8. EFFECT OF PRESSURE ON BOILING POINT OF WATER

Pressure	*Boiling Point*
28 in. Hg (Vacuum)	101.8 F
0 psig (14.7 abs)	212.0 F
70 psig	316.0 F
100 psig	338.0 F

Effect of Pressure on Freezing Point

Pressure affects the freezing point of substances in that a lower temperature is required with an increase in pressure. However, the effect is slight and for all practical purposes can be disregarded.

Effect of Pressure on Thermal Qualities of Fluids

Because of the additional heat required to raise the boiling point of liquids to meet the increased pressure requirements, an increase in thermal energy results. If reference is made to the tables of thermodynamic properties for a given liquid, it will be noted that, as the pressure increases, the internal energy also increases.

However, the opposite is true of the latent heat of vaporization characteristics of fluids. With each increment of pressure increase exerted on a liquid, a corresponding decrease in latent heat of vaporization results. The reason for this lies in the amount of thermal energy contained in the liquid. At atmospheric pressure, for example, only 970 Btu are required to transform 1 lb of water at 212 F to vapor at 212 F. However, the internal energy of the liquid above 32 F is 180 Btu under similar pressure conditions. To boil water at 1 psia, it requires 1034.7 Btu to transform it from a liquid at 101.8 F to vapor at 101.8 F. The energy content of the liquid under the latter conditions is only 69 Btu per lb, resulting in a lower total internal energy of the vapor as compared with the amount in the vapor at the higher pressure. Therefore, the total internal energy of the vapor is greater at the higher pressure by reason of a decrease in energy of the liquid by a corresponding amount that is necessary to promote vaporization. The existence of higher pressures will always demand a greater total internal energy content to initiate the transformation of water to vapor, because of the increased kinetic energy requirements necessary to overcome the additional force.

Thus, it is evident that pressure affects the thermal qualities of a fluid.

VAPORIZATION AND EVAPORATION

The process of converting a liquid to its vaporous form is known as vaporization. Evaporation usually is considered to be slow vaporization, and will take place when the vapor pressure of the surrounding air is below the vapor pressure of the liquid even when the temperature of the liquid is below the boiling point. The technical details of this phenomenon are discussed more fully in Chapter 4. A liquid boils when it contains sufficient thermal energy to cause its vapor pressure to equal the surrounding pressure. If, however, additional heat over that required to promote boiling is absorbed by a liquid, its vapor pressure becomes greater than the surrounding pressure, and, as a result, molecular separation occurs. These molecules leave the liquid mass and become suspended independently at a violent rate of motion. They will remain in such suspension as long as the temperature remains at or above the boiling point. Vaporization, therefore, can be described as the resulting action whereby the vapor pressure of a liquid is increased above the pressure exerted on its surface. And for the same reason, as is true in the case of liquid boiling points, vaporization temperatures are governed by the surrounding pressure.

Table 9 shows the relationship of vapor pressures of various refrigerants under similar temperature conditions of 86 F. Consider the pressures shown as those exerted as the result of the combined molecular impact force being

TABLE 9. VAPOR PRESSURE OF REFRIGERANTS AT 86 F

Ammonia (NH_3)	169 psia
Carbon dioxide (CO_2)	1039 "
Dichlorodifluoromethane (F-12) (CCl_2F_2) (Freon-12)	108 "
Ethyl chloride (C_2H_5Cl)	27 "
Methyl chloride (CH_3Cl)	95 "
Methylene chloride (CH_2Cl_2)	10 "
Sulfur dioxide (SO_2)	66 "

concentrated over an area of 1 sq in. of their confining walls. Therefore, to maintain them in a liquid form, it is necessary to exert a similar counterforce equal to the pressure shown.

CONDENSATION

Condensation is actually the reverse process of evaporation, or the conversion of a vapor to its liquid form. It is caused by decreasing the temperature of the vapor below the boiling point of the liquid from which it originated. Condensation also occurs when the volume is decreased by increasing the pressure above the vapor pressure of the fluid.

The relationship of latent heat to evaporation and condensation should be noted. Evaporation occurs when additional heat is absorbed *after* the liquid reaches its boiling point. This additional heat is the latent heat of vaporization, and in the case of water, at atmospheric pressure, 970 Btu are required to evaporate 1 lb. When condensation occurs at a similar pressure, an equivalent amount of thermal energy is lost by the vapor during its conversion to a liquid. Therefore, steam upon condensing loses 970 Btu for each pound.

PRESSURE-EVAPORATION RELATIONSHIP

The ability of a liquid to maintain its state of equilibrium—that is, to remain a liquid—depends on its vapor pressure as compared with the vapor pressure of the surrounding atmosphere. When any change occurs causing the surrounding pressure to fall below that of the liquid, evaporation takes place, and at a rate directly proportional to the pressure difference. Such a change may be brought about either by increasing the temperature content of the liquid to provide greater kinetic energy, or by reducing the pressure exerted on the surface of the liquid. In either case, a differential in vapor pressures or kinetic "head" is established.

If it is possible to curb vaporization by applying external pressure, it must be true that the process is reversed when pressure is relieved.

Consider the several conditions under which the amount of pressure is governed as it affects any liquid-gas relationship: (1) pressures above atmospheric, (2) atmospheric pressure, (3) pressure below atmospheric (vacuum).

In the first case, external energy must be provided over that already possessed by the atmosphere under normal conditions. In the second case, it may be assumed that the amount of energy contained by the surrounding atmosphere is natural and remains constant within a reasonable degree. The third case presupposes a deliberate reduction in kinetic energy content of the surrounding atmosphere. Therefore, while pressures above atmospheric depend entirely on some means of compression for their support, a partial vacuum is formed by evacuating some quantity of vapor from the space immediately exposed to the substance under consideration.

According to the laws of kinetic energy, a state of equilibrium exists when the rate of evaporation and the rate of condensation are equal for any vapor-liquid relationship. Equilibrium is thus upset when, for any reason, either of these functions is disturbed, which occurs with pressure changes.

Therefore, to maintain a force favorable to condensation, the pressure on a liquid must be increased, and to maintain a force favorable to evaporation, the pressure must be decreased.

The subject is of greater interest when applied to liquids whose boiling-point temperatures range from slightly below to considerably above normal atmospheric conditions. Such a range would include all refrigerants whose boiling temperatures fall between that of ammonia (−28 F) to that of water (212 F) under atmospheric pressure.

In order to accelerate evaporation of ammonia so that ultra-low temperatures approaching −50 F to −90 F may be obtained, it is necessary to operate under partial vacuum, as borne out by the "Thermal Properties of Saturated Ammonia Vapor" tables (see Appendix). Also, to create a substantial refrigerating effect through the use of water, it becomes necessary to reduce pressures considerably below atmospheric so that rapid evaporation may be induced. This is the principle employed in water-vapor refrigeration and steam jet processes.

From the above, it may be stated that *liquid evaporation proceeds at a rate relative to the existing kinetic energy differential, which may be created either by adding heat energy to the liquid under constant pressure or by reducing kinetic energy from the surrounding vapor by maintaining a vacuum.*

SATURATED LIQUID

A liquid is said to be saturated when sufficient heat has been added to raise its temperature to a boiling level at the pressure existing. Any increase in heat would result in vaporization, whereas a decrease in temperature only brings about a reduction in kinetic thermal energy.

When liquid saturation exists, the maximum amount of kinetic thermal energy in relation to the pressure under which saturation occurs is said to be held by the liquid. Thermal energy of the liquid, therefore, reaches its maximum value when liquid saturation exists.

In the tables of "Properties of Saturated Vapors," a column headed "Enthalpy of Liquid" will be found. The values listed in that column for any kind of vapor (such as steam, ammonia, etc.) represent the amount of enthalpy in Btu contained by the liquid at the temperature of boiling corresponding to the pressure existing. This means that the liquid contains the maximum amount of thermal energy possible and still maintains its liquid form under the specific pressure exerted for the condition selected. The value so listed for any vapor is called the *enthalpy of the liquid,* and any liquid containing that amount of thermal energy prior to vaporization is said to be a *saturated liquid.*

EXAMPLE: How many Btu will a liquid contain under saturated conditions if the absolute pressure exerted is 50 psi? (*a*) Calculate this value for steam, (*b*) for ammonia.

(*a*) In the steam tables, in "Properties of Saturated Vapor," column 1 represents absolute pressure. Following this column down, a value of 50 psia is found. Now find column 5, headed "Enthalpy of the Liquid." The value of 250.09 is found at the intersection of the 50 psi pressure line and the "Enthalpy of the Liquid" of the Liquid column. This value, then, is the number of Btu required to bring 1 lb of water to its boiling temperature of 281 F from its reference point of 32 F. The value of 250.4 Btu per lb did not in any way enter into the process of changing the water at 281 F to steam at 281 F. Furthermore, 250.5 Btu per lb is the maximum amount of heat that water under 50 psia pressure can absorb without vaporization occurring.

It may be stated that water under a pressure of 50 psia is a saturated liquid when it contains 250.4 Btu per lb. The "Enthalpy of Liquid" column, therefore, justifiably could be called "Enthalpy of *Saturated* Liquid."

(*b*) Following the same procedure, using the table "Properties of Saturated Ammonia Vapor," it is found that "Enthalpy of Liquid"* corresponding to 50.36 psia is 66.9 Btu per lb. This means that 66.9 Btu are required to raise the temperature of 1 lb of ammonia to its boiling temperature of 22 F at 50.36 psia from its reference point of −40 F. It also means that liquid ammonia is a saturated liquid if it contains 66.9 Btu per lb when its temperature is 22 F, corresponding to a pressure of 50.36 psia. As in the case of steam, any additional heat added to liquid ammonia under the above stated conditions would promote vaporization.

The problem stipulated a pressure of 50 psia. To determine the enthalpy of the liquid for this pressure, interpolation between values found at the next lowest pressure listed and that corresponding to 50.36 psia must be accomplished. In this case, the next lowest pressure listed is 49.28 psia. Reading across, under "Enthalpy of the Liquid" column, there is a value of 65.8 Btu per lb of ammonia. The difference between the two is 66.9 − 65.8 = 1.1 Btu. There is a difference of 50.36 − 49.28 = 1.08 psi pressure, of which 50 − 49.28 = 0.72 is under the pressure specified. Therefore, the Btu reading at 49.28 psi is $\frac{0.72}{1.08}$ of the total Btu difference below that required for 50 psi ammonia, or 1.1, or

$$\frac{0.72}{1.08} \times 1.1 = 0.73 \text{ Btu}$$

The Btu value at 49.28 psi of 65.8 + 0.73 Btu = 66.53 Btu contained in saturated liquid ammonia at a pressure of 50 psia.

SATURATED VAPOR

Vapor whose temperature equals that of its boiling liquid is said to be saturated. Any increment of decrease in the temperature of a saturated vapor results in condensation. Saturated vapors are usually in contact with their liquid phase. Any extraction of heat will cause some of the vapors to condense, reducing their total volume and therefore their total weight.

A saturated vapor must contain only that energy required to raise the temperature of the liquid from its reference point, plus its latent heat of vaporization corresponding to the pressure employed. At atmospheric pressure, saturated steam would contain 180 Btu (212° − 32° × sp ht of 1) as kinetic thermal energy, plus 970 Btu as latent heat of vaporization, or a total of 1150 Btu.

The tables of "Properties of Saturated Vapor" include a column wherein the total enthalpy of vapors corresponding to the various pres-

* Table 1, page 1316, uses the heading 'Heat Content'. See the 2nd paragraph on Heat on page 17, and the last paragraph on Thermodynamic Property Tables on page 212, for explanation of change in nomenclature.

sures is listed under the heading "Total Enthalpy." As stated above, the values shown in this column are the sum of the values shown in the "Enthalpy of Liquid" and the "Enthalpy of Vaporization" columns, and represent the total enthalpy contained in 1 lb of saturated vapor.

EXAMPLE: What is the total enthalpy of a saturated vapor at a pressure of 120 psia? (*a*) Determine for steam, (*b*) for ammonia.

Referring to the steam tables, the pressure of 120 psia is found in column 1. Reading across until the 120 psia line intersects the "Total Enthalpy" column (3), a reading of 1189.2 is found. This means that 1 lb of saturated steam at 120 psia contains a total of 1189.2 Btu per lb.

Another method of determining the total enthalpy quantity of this steam is to continue the 120 psia line across to the "Enthalpy of Liquid" column (6), where the value of 312.3 is listed. Continue on to the next column headed "Enthalpy of Vaporization," where the figure 876.9 is found Adding the two, or $312.3 + 876.9 = 1189.2$, or the same amount of enthalpy as listed in column (3). Hence, saturated steam at 120 psia contains 1189.2 Btu per lb.

In the ammonia tables, a value of 628.4 Btu is found under the "Total Enthalpy" column on the line corresponding to 120 psia. This is found also to be the sum of values under "Enthalpy of Liquid" and "Enthalpy of Vaporization" columns and on the 120 psia line, or $116 + 512.4$, respectively, which is equal to 628.4 Btu per lb.

Saturated ammonia vapor at 120 psia, then, contains 628.4 Btu per lb.

SUPERHEAT

Superheated gases are gases whose temperatures are above the boiling

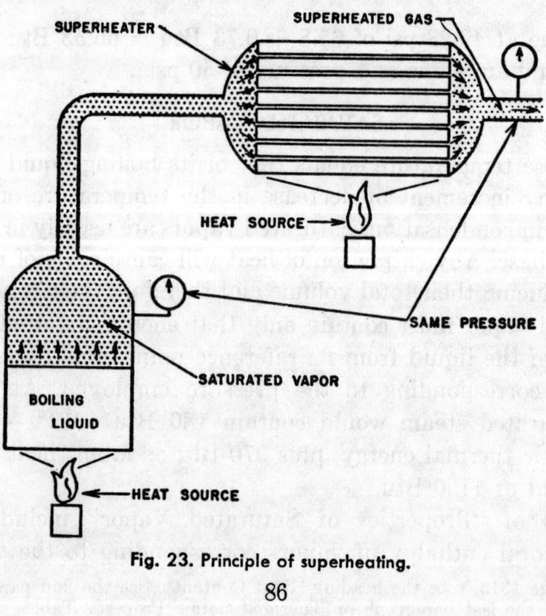

Fig. 23. Principle of superheating.

point of their liquids corresponding to the existing pressure. To become superheated, vapors are separated from their liquid phase.

By applying additional heat to a saturated vapor and maintaining constant pressure, the vapor becomes superheated. Steam generated in a boiler leaves its liquid phase and enters a separate chamber, where additional heat is applied. When it left the boiler, the steam was a saturated vapor, but upon absorbing more heat in the second chamber, its temperature is raised above the temperature of saturation. Because the pressure remains constant throughout the entire process, the additional heat is considered to be in excess of that required for saturation at the existing pressure, and the vapor is said to be superheated.

Superheat is kinetic thermal energy, the quantity of which is computed in a fashion similar to that used for determining thermal energy content of liquids or solids, viz.:

$$\text{Superheat} = \text{wt of vapor} \times \text{sp ht} \times \text{superheat °F}$$

EXAMPLE: (*a*) How much superheat is contained in 1000 lb of steam, the pressure of which remains constant at 150 psia when its temperature is increased 100 F above that required for saturation? (*b*) What is the total enthalpy per pound of steam?

(sp ht of 450 F steam = 0.57)

Solution: When the temperature of a vapor is raised to some definite degree above the temperature of saturation, it contains a definite degree of superheat. In this example, the steam contains 100 F of superheat.

(*a*) Applying the formula:

$$\begin{aligned} \text{Btu} &= \text{wt} \times \text{sp ht} \times td \\ &= 1000 \times 0.57 \times 100 \\ &= 57{,}000 \text{ Btu superheat} \end{aligned}$$

(*b*) It is found in the steam tables that 150 psia steam contains a total enthalpy of 1193 Btu when saturated. Adding to this amount

$$1 \times 0.57 \times 100 = 57$$

or from the above $\dfrac{57{,}000}{1{,}000} = 57$ Btu of superheat, the total amount of heat contained in 100 F superheated 150 psia steam is $1193 + 57 = 1250$ Btu per lb, or a total of 1000×1250, or 1,250,000 Btu in 1000 lb of 150 psia steam, 100 F superheated. The total enthalpy per pound of superheated steam may be found also in the superheated steam tables under the given pressure and temperature conditions.

TABLES OF THERMAL PROPERTIES OF FLUIDS

It was stated previously that pressure exerted on fluids affects their thermal properties, such as boiling point, kinetic energy, and latent heat of vaporization. Because pressure affects all three properties, it also changes the total enthalpy as measured above the reference point.

For the sake of comparison, standard boiling points for liquids usually are expressed in values corresponding to atmospheric pressure of 14.7 psi. This is also true in expressing other factors, such as the latent heat of vaporization, or the kinetic thermal energy content of liquids above their reference points. However, because such liquids as water (when converted to steam to perform work as in a power plant) or ammonia (when used as a refrigerant) perform their functions at pressures other than atmospheric, it becomes essential to know their properties at various pressure levels.

The thermal qualities of any liquid depend on the internal vapor tensions, as compared with external pressure under which it exists at any specific time, and the ability of different fluids to maintain their liquid phase depends on the amount of vapor tension they are capable of withstanding without molecular separation. The boiling point of liquids at a definite pressure is an excellent criterion of the molecular cohesion possessed by various fluids. A relative value can be obtained when this property is made the standard of comparison.

Water, whose boiling point is 212 F at atmospheric pressure, obviously possesses greater molecular cohesiveness than does ammonia, whose boiling point is −28 F at a similar pressure. Carrying this farther, it is found that the effect of vapor pressure within the fluid is less effective in ammonia than in carbon dioxide, whose boiling point of −110 F denotes extreme molecular separability.

The effect of such characteristics may be realized if a quantity of each of these three fluids is exposed openly to the atmosphere under normal room temperatures of, say, 70 F. It will be found that the carbon dioxide flashes rapidly into vaporous form, as will ammonia but at a much reduced rate, while water will not evidence any change in state whatever. By exerting pressure to the surface of the carbon dioxide and ammonia, a counteraction to the effect of internal vapor pressure is affected and the tendency toward molecular separation decreases, assuming no rise in temperature. As increased pressure is applied, the tendency toward increased molecular separation diminishes until a force exerted on the fluids becomes equal to or above the intensity of the vapor pressure, at which time molecular separation ceases. It will be found that a total pressure of 853.4 psia will be required to bring about stabilization within the carbon dioxide, and 128.8 psia for ammonia, provided that a temperature of 70 F prevailed in both cases. At this point, it will be found also that the latent heat of vaporization of the carbon dioxide and ammonia will be changed, showing decreased values in both instances. The kinetic thermal energy content of the liquids will be increased. The increase in energy of the liquid content and the decrease in latent heat of vaporization are the result of gradual changes through each increment of pressure increase whereby each increment will be found to have its own specific set of values. This is true throughout the range from 14.7 psi to the final stabilizing pressure in each case, which, as stated above for carbon dioxide and ammonia, will be 853.4 and 128.8 psia, respectively.

In order to determine heat-load capacities of various fluids, the thermal property values at any pressure level within the working range of the different fluids must be known. Such values can be determined by referring to the tables of thermal properties covering the respective fluids, values which have been calculated by the use of thermal formulas based on experimentation and using the Btu as the standard of thermal energy measurement. Tables of thermal properties have been organized for a wide range of fluids, including steam and refrigerants commonly used.

"Thermal Property Tables" for all fluids are similar in their content in that they include the values of the more important properties according to the pressure involved. In their usual form, such tables list, in the order named, temperature, pressure (both absolute and gage), volume of vapor in cubic feet per pound, density of vapor in pounds per cubic foot, enthalpy of the liquid, and latent heat of the vapor. Some forms list the gage pressure rather than the temperature in the first column. The tables are drawn up also to cover fluids in their several forms, being designated as "Properties of Liquid and Saturated Vapor" and "Properties of Superheat Vapor." While some use of the tables was experienced in the foregoing paragraphs, it is necessary for the understanding of refrigeration to be familiar with all phases of their content in order to calculate load factors necessary in the design and size estimation of cooling systems.

As an example showing the practical use of such tables, reference is made to ammonia vapor, saturated at 5 F. Referring to Table I, Thermal Properties of Saturated Ammonia, the temperature of 5 F is found in column 1. Reading across the table in a line corresponding to the temperature of 5 F, an absolute pressure of 34.27 psi is listed in column 2, or a gage pressure of 19.6 psi, as shown in column 3, is required to control the boiling point of ammonia at that temperature level. Saturated ammonia vapor will require a space whose volume is equal to 8.15 cu ft for each pound, according to column 4, the integral of which corresponds to 0.1227 lb of vapor for each cubic foot of space, as listed in column 5. For the same set of temperature-pressure conditions, the thermal properties of saturated ammonia vapor will be found in columns 6, 7, and 8, respectively, which shows the enthalpy of the liquid to be 48.3 Btu, a total enthalpy content of 613.3 Btu, and the latent heat of vaporization 565.0 Btu, column 7 being the sum of columns 6 and 8.

Carrying out a similar procedure for saturated ammonia vapor having a temperature of 86 F, the following values are noted: absolute pressure 169.2 psi, gage pressure 154.5 psi, volume per pound 1.772 cu ft per lb, density 0.5643 lb per cu ft, enthalpy of liquid 138.9 Btu per lb, enthalpy of vapor 631.5 Btu per lb, and latent heat of vaporization 492.6 Btu per lb.

A practical demonstration of the use of the information found in the "Thermal Property Tables" can be made by determining the amount of heat that must be extracted from 1 lb of ammonia vapor at 86 F to produce an equal amount of liquid at 70 F at the same pressure.

Again referring to Table I, Properties of Saturated Ammonia Vapor, it is found that the latent heat of 86 F saturated ammonia vapor is 492.6 Btu per lb. Upon the extraction of that amount of heat from the vapor, conversion to its liquid form is accomplished, resulting in liquid ammonia at 86 F and 169.2 psia pressure. The enthalpy of the liquid content for 86 F is 138.9 Btu per lb above the reference point of −40 F, as shown in column 6.

On the line corresponding to the temperature of 70 F of the same table, column 6 lists enthalpy of the liquid content of 120.5 Btu per lb. To reduce the temperature of 86 F liquid ammonia to 70 F, it is necessary to extract an amount of thermal energy equal to the difference in the two enthalpy of liquid values. This difference is found to be 138.9 Btu − 120.5 Btu, or 18.4 Btu per lb.

To find the total amount of thermal energy to be extracted from 86 F saturated ammonia vapor in order to produce 70 F liquid ammonia, add the latent heat of vaporization of the 86 F vapor to the differential in enthalpy of the liquids at the two temperatures. The total amount of thermal energy to be extracted is 492.6 Btu plus 18.4 Btu, or 511.0 Btu per lb.

This method of determining the amount of thermal energy to be extracted from a saturated vapor to produce a liquid at a lower temperature shows the relationship of the values in the several enthalpy columns. A much simpler method of determining the same result would be to subtract the enthalpy of the liquid of 70 F ammonia from the total enthalpy of saturated ammonia vapor. Referring again to the table, these values are found to be:

Total enthalpy of vapor at 86 F . . .	631.5 Btu per lb	(column 7)
Enthalpy of the liquid at 70 F . . .	−120.5 Btu per lb	(column 6)
Thermal energy to be extracted . . .	511.0 Btu per lb	

Properties of Superheated Vapor Tables

Following the tables covering properties of saturated ammonia will be found Table IV, Properties of Superheated Ammonia Vapor. In dealing with the data as listed in these tables, as well as tables covering all other vapors, it must be remembered that superheat does not affect the pressure of the vapor, but merely increases the thermal energy content above that of saturation. However, because of this additional heat, the temperature of the vapor rises and causes expansion to occur, lowering the density as the volume increases. Because of the additional energy involved under conditions of superheat, it can be expected that the (*a*) pressure will remain constant, (*b*) temperature of saturation (vapor) will remain the same for the corresponding pressures, (*c*) temperature of the vapor due to superheating will increase, (*d*) enthalpy will increase, (*e*) specific volume will increase, and (*f*) density will decrease.

The tables of superheated vapors include the absolute pressures maintained during evaporation and superheating. These values are the numerals located in the space that heads the data column. Under the absolute pressure values, in italics, the temperature of saturation at that pressure is shown. Referring to the properties of *superheated* ammonia vapor under absolute pressure reading of 6 psi, the temperature value of −57.64 F is found. Referring to the tables covering *saturated* ammonia vapor, it will be found that the boiling temperature of ammonia corresponding to 6 psia is −57.64 F, or the temperature of saturation.

The temperatures listed in the left column are the final temperatures of the vapor and *not the degree of superheat.* To find the degree of superheat, it is necessary to subtract the temperature of saturation from the temperature of the superheated vapor. At a pressure of 6 psi and a final temperature of −40 F, the degree of superheat in ammonia vapor would be −57.64 minus −40, or 17.64 F. Therefore, under the above conditions, the ammonia vapor would be 17.64 F *above* the temperature of saturation at 6 psia. This is important in view of the universal custom of specifying superheat in terms of "degree of superheat" rather than in final temperatures. As will be learned later in "dry vapor systems" of refrigeration, a 10 F superheat is considered to be ideal in preventing liquid refrigerant from being drawn into the compressor during the suction stroke. This means that the vapor is allowed to rise 10 F above its temperature of saturation, corresponding to the pressure, in the evaporator.

The data columns are headed V, H, and S. The values in the respective columns represent (V) volume of superheated vapor in cubic feet per pound, (H) enthalpy of the vapor, including superheat, and (S) entropy. The latter values will be discussed in detail later in the text.

Again referring to the columns under an absolute pressure of 6 psi and reading across the line corresponding to a final temperature of 50 F, the value of 53.26 is listed under V. That is the volume in cubic feet per pound of saturated vapor *after* it has been superheated to a final temperature of 50 F. In allowing a rise of +50 F minus −57.64 F or 107.64 F superheat to occur, the volume increased from 41.59 cu ft per lb (shown in italics on the saturation line at the top of the column V) to 53.26 cu ft per lb.

For the same reason, the enthalpy content increased to 645.2 Btu per lb, or 645.2 − 590.6 = 54.6 Btu above that of saturated vapor. Pursuing the calculations here, it will be found that, by dividing the superheat value of 54.6 Btu by the temperature increase due to superheat, the quotient corresponds to the specific heat of ammonia vapor (at constant pressure). Therefore,

$$\frac{54.6}{107.64} = 0.508$$

specific heat of ammonia vapor at constant pressure. This value changes slightly with temperature so that the specific heat for ammonia, as listed in various authoritative handbooks, may range from a mean of 0.508 to 0.52, depending on the temperature range covered and the source of the data listed in the thermal tables. This is based on table's compiled by the U.S. Bureau of Standards.

SUMMARY OF CHAPTER 3

Pressure is a uniformly distributed force exerted on the surface of a body. When transmitted by means of a fluid, the intensity is equal in all directions, regardless of the size or the shape of the confining vessel.

Pressure is measured and expressed in terms of force per unit area, such as pounds per square inch, tons per square foot, etc.

Pneumatic pressure is a force exerted by a gas acting as an energy-transmission agent.

Hydraulic pressure is a force exerted by a liquid acting as an energy-transmission agent.

Mechanical pressure is a force exerted by a solid acting as an energy-transmission agent.

Vapor pressure of gases may be increased by means of compression, employing mechanical energy, *or by maintaining constant volume while increasing the tendency to expand as a result of adding* thermal energy.

The intensity of pressure is relative in that its effectiveness depends on surrounding pressures and may be equalized by providing a similar counterpressure.

Atmospheric pressure is the amount of force exerted by the weight of the air mass surrounding the earth. It is expressed in terms of pounds per square inch (psi), which is the weight of a column of air 1 sq in. in area and extending upward from the earth to the outer surface of the air mass. Atmospheric pressure is the gravitational forces of the air exerted on or above the earth's surface.

Atmospheric pressure averages 14.7 psi at sea level, but decreases correspondingly in intensity with each increment of rise in altitude. *The barometer is an instrument employed for measuring atmospheric pressure.* Its principle is based on the height of a mercury column capable of being sustained by atmospheric pressure. For each 1 lb of pressure exerted, the column rises 0.491 in. *A standard atmosphere, which is equivalent to 14.7 psi, sustains a mercury column 29.92 in. in height,* or a water column 33.946 ft high.

Gage pressure is the pressure exerted by a fluid, within an enclosed chamber, above that of the atmosphere. The gage, of either the Bourdon or the diaphragm type, is set to register zero at atmospheric pressure. Compound gages register pressures both above and below atmospheric pressure.

Absolute pressure is any pressure that includes atmospheric pressure. It is equal to the gage pressure plus 14.7 psi.

A vacuum is a space void of any pressure.

The term is used commonly to denote partial vacuum, which is any pressure below that of atmospheric and expressed in terms of absolute pressure or inches of mercury (in. Hg).

Pascal's law states that, when a pressure is exerted on any part of a fluid within an enclosed chamber, an equal pressure is transmitted undiminished to all equal areas and in all directions through the fluid within the chamber, regardless of shape or size.

There are three states of matter:

1. *Solids,* which have both shape and volume.
2. *Liquids,* which have definite volume, but no regular shape.

3. *Gases,* which have neither definite volume nor shape.

The boiling point of any liquid is the highest temperature at which it can remain a liquid under existing pressure. It varies with different liquids.

The *effects of increased pressure on fluids* are that it:

(*a*) *Raises their boiling point,*
(*b*) *Increases kinetic thermal energy content,*
(*c*) *Decreases latent heat of vaporization*

The process of converting a liquid to its vaporous phase is known as *vaporization.*

The process of converting a vapor to its liquid phase is known as *condensation.* It is vaporization in reverse.

A state of equilibrium exists when the rate of evaporation and the rate of condensation are equal in any vapor-liquid relationship.

A saturated liquid is one whose boiling-point temperature corresponding to existing pressure has been reached.

A saturated vapor is one whose temperature is equal to the boiling point of its liquid phase at the same pressure.

Superheated gases are gases whose temperatures have been raised above that of the saturated vapor while maintaining constant pressure. They usually are separated from their liquid phase.

Thermal properties of gases are listed in tables compiled for individual fluids. They include boiling-point temperature at various pressures and corresponding enthalpy values for both liquid and vapor and total enthalpy content, as well as density and specific volume. Such tables are essential in calculating load factors of thermodynamic systems, such as refrigeration and power-generating plants.

Internal energy content of fluids is based on a reference point temperature, usually 32 F for steam tables and −40 F for refrigerant tables.

STUDY OUTLINE FOR CHAPTER 4

Definition of a gas

Origin

Characteristics:
- Pressure-temperature relationship
- Gas-liquid relationship

Density:
- Weight and volume
- Avogadro's law

Critical temperature and pressure

Gas laws:
- Boyle's law
- Charles' law
- Gay-Lussac's law
- Multiple factor changes

Gas constant:
- Ratio $\frac{C_p}{C_v}$

Dalton's law of partial pressures

Specific heat of gases:
- At constant volume C_v
- At constant pressure C_p

Isothermal expansion and compression

Adiabatic expansion and compression

Use of pressure-volume diagrams

4

Gases

One of the three states of matter, gases are elastic fluids having indefinite expansion characteristics when not confined. They are compressible to an extent dependent on their liquefying temperature-pressure relationship below what is known as the *critical point.* They are derived in most cases by the process of vaporization, either from liquids whose temperatures have exceeded their boiling points or by slow evaporation of the same liquids at lower temperatures. In some instances, gases are formed as a result of evaporation directly from their solid phase without first entering the liquid state. This action is known as *sublimation,* and occurs more readily with some substances than with others. Carbon dioxide is an excellent example of a substance that converts to a vapor form from its solid phase, and for this reason is called "dry ice." Sublimation is brought about by the rapid absorption of surrounding heat that is utilized in the vaporizing process, causing the temperature of the solid mass to remain below its melting point.

All gases will liquefy at some definite temperature, depending on the substance from which they emanate, and prevailing pressure. The temperature range of boiling points for the various gases is extremely wide, and for this reason allows ample potentiality on the part of users to select a type most suitable for their requirements. This is especially true where such selections involve substances to be used as refrigerants, the boiling points of which dictate size and type of equipment and temperature limitations. In many instances, the boiling point determines the service for which the refrigeration system may be employed.

Gases present a uniform rate of expansion for equal increments of temperature increase, provided that pressures remain constant. Because they are compressible, their volume diminishes with equal uniformity for each pound of pressure exerted, while the temperature of the mass increases proportionately to the amount of work expended during compression. Such physical characteristics on the part of all gases have been the subject of scientific experiment and investigation, resulting in the formulation of laws governing such action. These laws are of great importance in the study of refrigeration, and are presented here under headings identifying the individuals credited with their formulation.

DENSITY

Because gases emanate from all liquids, their weight varies accordingly and is expressed in pounds per cubic foot (lb per cu ft). To specify the weight of any gas merely in lb per cu ft, however, would be an incomplete description, since its expansion characteristics require that the pressure and the temperature also be known. When expressed in its entirety, such a specification is referred to as the density of the gas. The density of a gas is the weight in lb per cu ft of volume at any given temperature and the pressure in psi. Decreasing the volume of a gas by compression, thereby increasing its pressure, does not change the weight but does increase its density. For example, a cylinder holding 1 lb of gas when the piston begins its compression stroke still contains the same weight of gas when the piston reaches any point along its line of travel. However, its density changes with each increment of distance the piston moves in compressing the gas, so that, at the halfway point, it is double its original value because the same weight of gas is contained in one half the space.

Actually, the density of any gas is an indication of the relative position of its molecules to each other, and also their velocity. As density increases, the molecules are forced closer together. If the force exerted in reducing gas volume is continually increased, the sphere of molecular movement becomes more limited, until a point is reached whereby their reconversion to a liquid phase occurs, assuming no increase in temperature. Avogadro's law states that *equal volumes of all gases contain the same number of molecules when similar conditions of temperature and pressure exist.*

The specific volume of a gas is the volume in cubic feet of 1 lb of any gas existing under specified conditions of temperature and pressure. Numerically, it equals the reciprocal of density, or

$$\text{Specific volume} = \frac{1}{\text{density}}$$

Written simply

$$V_{sp} = \frac{1}{D}$$

where V_{sp} = specific volume of any gas in cu ft per lb
D = density of any gas in lb per cu ft

Saturated ammonia vapor at a temperature of 86 F and 169.2 psia, according to the "Properties of Saturated Ammonia Tables," has a density of 0.5643 lb per cu ft. This corresponds to a specific volume of $\frac{1}{0.5643}$, or 1.772 cu ft per lb. When the temperature remains constant, (*a*) the pressure of a gas is proportional to its density, or (*b*) the weight of a gas in a given volume is directly proportional to its pressure.

CRITICAL TEMPERATURES AND PRESSURES OF GASES

As was stated, the boiling point of any substance is dependent on its pressure-temperature relationship, and any increase in pressure is accompanied by a corresponding increase in the boiling point. A gas will liquefy when sufficient pressure is exerted to raise its boiling point above the temperature of the gas at the time of initial compression. True of any gas, this phenomenon is utilized in obtaining and controlling temperatures within the refrigeration systems.

Every gas, whether it be an element or a compound, has a temperature above which it cannot exist in a liquid state, and, regardless of the pressure exerted, it will not liquefy. This temperature is called the critical temperature, and the corresponding pressure for dry saturated vapor at the critical temperature is known as the critical pressure. Table 10 shows the critical temperature and the critical pressure for some of the important refrigerants.

TABLE 10. CRITICAL TEMPERATURES AND PRESSURES OF SOME GASES

Refrigerant	*Critical Temperature °F*	*Critical Pressure psia*
*Air	−221.0	573.3
*Ammonia	266.0	1690.5
*Carbon dioxide	88.16	1073.1
Ethyl chloride	368.7	764.4
†Freon-11	388.4	635.0
†Freon-12	232.7	582.0
†Freon-21	353.3	750.0
†Freon-22	204.8	716.0
†Freon-113	417.4	495.0
†Freon-114	294.3	474.0
Methyl chloride	289.4	967.3
Methylene chloride	421.0	670.0
*Sulfur dioxide	312.0	1161.3
Water vapor	705.2	3187.3

* Smithsonian Tables (Converted).
† E. I. duPont de Nemours, Inc.

At temperatures above 266 F, ammonia will not liquefy, regardless of the pressure exerted, as that is the critical temperature for this substance.

It is necessary to reduce the temperature of air to −221 F in order to liquefy it at a pressure of 573.3 psia. For temperatures below the critical temperature of air, a correspondingly lower pressure would be required to obtain liquefaction.

Pressure, temperature, and volume of any (perfect) gas form a definite relationship in that a change in any one causes a change in either one of the others.

Boyle's law states that, *with constant temperature, the volume of a perfect gas varies inversely as the pressure applied to it.* If a body of gas remaining at constant temperature is allowed to increase to three times its original volume, its pressure will decrease to one third its original pressure. Conversely, the volume of a gas will decrease to one third its original volume if the pressure applied to it is increased three times, provided that the temperature remains constant.

When gas is expanded or compressed according to Boyle's law—i.e., at constant temperature—it is said to be isothermal. During isothermal expansion or compression, the product of the absolute pressure times the volume at any point along the line of volumetric change is constant and can be stated:

$$P_1V_1 = P_2V_2, \text{ etc.}$$

or

$$P \times V = \text{constant}$$

If P_2 is unknown, then

$$P_2 = \frac{P_1V_1}{V_2}$$

where P_1 = original absolute pressure at V_1
P_2 = final absolute pressure at V_2
V_1 = original volume
V_2 = final volume

Example: What is the final pressure of a gas the initial pressure of which is 15 psig? Assume a reduction in volume from 10 cu ft to 2 cu ft. Compression is isothermal.

Applying the formula

$$P_1V_1 = P_2V_2$$

$$P_2 = \frac{P_1V_1}{V_2}$$

$$P_2 = \frac{29.7 \times 10}{2}$$

P_2 = 148.5 psia Pressure

Data:

$P_1 = 15 + 14.7 = 29.7$ psia

P_2 = unknown

V_1 = 10 cu ft

V_2 = 2 cu ft

Example: What would be the final pressure if the gas above were further reduced in volume to 1 cu ft? Final pressure here is indicated at P_3.

$$P_3 = \frac{P_2V_2}{V_3}$$

$$P_3 = \frac{148.5 \times 2}{1}$$

$$P_3 = 297 \text{ psia}$$

Thus, the law, that the volume of a gas varies inversely with the absolute pressure at constant temperature, is borne out in the above examples when it is considered that, if the first case,

(1) A reduction in volume to one fifth the original volume resulted in a final pressure (P_2) five times greater than the original pressure (P_1), viz.:

$$29.7 \times 5 = 148.5 \text{ psia,}$$

and in the second case,

(2) A reduction in volume to one half the volume at P_2 resulted in a final pressure (P_3) double that for P_2, viz.:

$$148.5 \times 2 = 297 \text{ psia}$$

That

$$P_1V_1 = P_2V_2 = P_3V_3$$

or

$$P_2V_2 = \text{constant}$$

is also true in that

$$P_1V_1 = \text{constant, or } 29.7 \times 10 = 297$$

$$P_2V_2 = \text{constant, or } 148.5 \times 2 = 297$$

and

$$P_3V_3 = \text{constant, or } 297 \times 1 = 297$$

Therefore,

$$(P_1V_1) = 297 = (P_2V_2) = 297 = (P_3V_3) = 297, \text{ etc.}$$

It may be noted here that the various volumes in cubic feet in these calculations—namely, 1, 2, and 10—may be regarded also as percentages of total volume of the cylinder in which the gas is compressed. Volume percentages are used to develop the various curves to demonstrate the reactions of gases obeying the several thermodynamic laws, of which isothermal law is one.

CHARLES' LAW OF GASES

Whereas Boyle's law treated the action of a gas at constant temperature, Charles' law deals with gas at constant volume. This law states that *the pressure of a specific volume of gas is directly proportional to its absolute temperature.* Upon an increase in temperature of a constant volume of gas, a corresponding increase in pressure will result. Conversely, a corresponding decrease in pressure will result when its temperature is lowered. This law being true, the resulting pressure of any body of gas can be determined at any temperature if the original pressure and temperature are known. Its proportion can be expressed as follows:

$$\frac{P_1}{P_2} = \frac{T_1}{T_2} \text{ when volume is constant}$$

where P_1 = original absolute pressure
P_2 = final absolute pressure
T_1 = original absolute temperature
T_2 = final absolute temperature

EXAMPLE: Find the final pressure exerted by a specific volume of gas at 60 F and 100 psia, if its temperature is increased to 150 F. Applying the formula:

$$\frac{P_1}{P_2} = \frac{T_1}{T_2}$$

$$P_1T_2 = P_2T_1$$

$$P_2 = \frac{P_1T_2}{T_1}$$

Then

$$P_2 = \frac{100 \times 609.6}{519.6}$$

$$P_2 = 117.3 \text{ psia final pressure}$$

Data:

P_1 = 100 psia

P_2 = unknown

$T_1 = 60 + 459.6 = 519.6$ F abs

$T_2 = 150 + 459.6 = 609.6$ F abs

Charles' law of gases also may express pressure changes in terms of its coefficient of expansion. In doing so, note the uniform rate at which the pressure of any gas will change for each equal rate of temperature change when the volume remains constant. The coefficient of expansion is based on the pressure of the gas at the temperature equal to the freezing point of water for the respective scales employed, which is 0 C, or 32 F. This is equivalent to $\frac{1}{273}$ of its pressure at 0 C for 1 C change or $\frac{1}{492}$ for 1 F change in temperature. Converting to a decimal, pressure coefficient of expansion for all gases for both Centigrade and Fahrenheit scales is $1 \div 273$, or 0.00366, and $1 \div 492$, or 0.00203, respectively.

Restating the law of Charles, according to the pressure coefficient of expansion of gases, it says:

All gases have the same pressure coefficient of expansion equivalent to the reciprocal of the freezing point of water expressed as absolute temperature, being $\frac{1}{273}$ (0.00366) on the Centigrade scale and $\frac{1}{492}$ (0.00203) on the Fahrenheit scale.

The formula for the restated law of Charles for the calculation of pressure changes on gases due to temperature differentials, as measured on the Centigrade scale, may be expressed as follows:

$$P_2 = P_0 \left(1 + \frac{1}{273} t\right)$$

where P_2 = pressure at t°C
P_0 = pressure at 0 C
t = temperature above 0 F

and, for Fahrenheit temperatures,

$$P_2 = P_0\left(1 + \frac{1}{492} [t - 32]^*\right)$$

In employing the formula

$$P_2 = P_0(1 + \tfrac{1}{493}[t - 32]),$$

P_0 must first be determined in order to solve for P_2. This can best be demonstrated by applying the formula to the same conditions as in the foregoing example:

First step: $P_2 = P_0(1 + \frac{1}{492}[t - 32])$

Solve for P_0: $100 = P_0(1 + \frac{1}{492}[60 - 32])$

$$P_0 = \frac{100}{1 + 0.057}$$

$P_0 = 94.6$ psi of gas at 32 F

Data:

$P_2 = 100$ psia

$P_0 =$ unknown

$t = 60$ F

Second step: $P_2 = P_0(1 + \frac{1}{492}[t - 32])$

Solve for P_2: $P_2 = 94.6(1 + \frac{1}{492}[150 - 32])$

$P_2 = 94.6 \times 1.24$

$= 117.3$ psia final pressure

Data:

$P_2 =$ unknown

$P_0 = 94.6$ psia

$t = 150$ F

According to Charles' law, a decrease in absolute temperature results in a corresponding decrease in pressure when the volume of gas remains constant. Theoretically, if the temperature of a body of gas were decreased to 0° absolute, there would be a total absence of pressure. This being true, it would prove that the rate of molecular motion within a body depends on its temperature. As pressure is the result of an increased rate of molecular motion induced by heat, the absence of thermal energy, which is the case at 0° absolute, would bring about an absence of molecular motion and hence an absence of pressure. This is indicated by the pressure-temperature relationship shown on the curve. Actually, all gases would liquefy before reaching 0° absolute.

GAY-LUSSAC'S LAW OF GASES

Having treated gases at constant volume and constant temperature under laws proved by Boyle and Charles, their behavior when maintained at constant pressure can be determined from Gay-Lussac's law of gases. This states that *the volume of a specific weight of gas varies in direct proportion to its absolute temperature at constant pressure.* This law varies from Charles' law in the substitution of volume for pressure as the constant; with this one exception, the proportion may be expressed similarly, viz.:

$$\frac{V_1}{V_2} = \frac{T_1}{T_2} \text{ when pressure is constant}$$

where V_1 = original volume
V_2 = final volume
T_1 = original absolute temperature
T_2 = final absolute temperature

*** 32 F is subtracted from *t* because the pressure coefficient of expansion is based on gas pressure at the freezing point of water.**

EXAMPLE: Find the final volume of 10 cu ft of gas when its temperature is raised from 70 F to 250 F. Applying the formula:

$$\frac{V_1}{V_2} = \frac{T_1}{T_2}$$

Data:

$V_1 = 10$ cu ft

$V_2 =$ unknown

$T_1 = 70\text{ F} + 459.6 = 529.6°$ abs

$T_2 = 250\text{ F} + 459.6 = 709.6°$ abs

Then

$$V_1T_2 = V_2T_1$$

$$V_2 = \frac{V_1T_2}{T_1}$$

$$V_2 = \frac{10 \times 709.6}{529.6}$$

$V_2 = 13.39$ cu ft, final volume

Gay-Lussac's law also may be expressed in terms of coefficient of expansion as related to gas-volume changes due to temperature variations when the pressure remains constant. Any gas will expand or contract uniformly with each equal increment of temperature change. This means that the pressure will change if volume remains constant. Also, the volume will change uniformly with each degree of temperature change when the pressure remains constant, as provided in the law of Gay-Lussac.

Restated, Gay-Lussac's law says:

All gases have the same volume coefficient of expansion equivalent to the reciprocal of the freezing point of water expressed as absolute temperature, being $\frac{1}{273}$ *(0.00366) on the Centigrade scale and* $\frac{1}{492}$ *(0.00203) on the Fahrenheit scale.*

Reducing the definition to formula, we have, using the Centigrade scale:

$$V_2 = V_0\left(1 + \frac{1}{273}t\right)$$

or

$$V_2 = V_0\,(1 + 0.00366\,t)$$

where $V_2 =$ final volume
$V_0 =$ volume at 0 C
$t =$ temperature above 0 C

and for Fahrenheit temperatures:

$$V_2 = V_0\left(1 + \frac{1}{492}[t - 32]\right)$$

or

$$V_2 = V_0\,(1 + 0.00203\,[t - 32])$$

MULTIPLE FACTOR CHANGES

Under the three laws of gases—Boyle's, Charles', and Gay-Lussac's—their treatment at constant volume, pressure, and temperature has been described. In each case, the relationship of the remaining factors has been determined. If the proportions of changes as proved by all the above laws are consolidated in one formula, final results can be determined when a change in all factors takes place.

Boyle's law: $\frac{P_1}{P_2} = \frac{V_2}{V_1}$ at constant temperature

Charles' law: $\frac{P_1}{P_2} = \frac{T_1}{T_2}$ at constant volume

Gay-Lussac's law: $\frac{V_1}{V_2} = \frac{T_1}{T_2}$ at constant pressure

Consolidated, we have:

$$\frac{P_1V_1}{P_2V_2} = \frac{T_1}{T_2} \text{ change in all factors}$$

or

$$\frac{P_1V_1}{T_1} = \frac{P_2V_2}{T_2}$$

and

$$\frac{P_1V_1}{T_1} = \text{constant}$$

$$\therefore \frac{P_1V_1}{T_1} = \frac{P_2V_2}{T_2} = \frac{P_3V_3}{T_3}, \text{ etc.}$$

DENSITY-PRESSURE-TEMPERATURE RELATIONSHIP

It was stated previously that the density is the weight of a specific volume of any gas, usually expressed as lb per cu ft. It is determined by dividing the total weight of the mass by the total volume, or

$$\text{Density} = \frac{\text{weight in pounds of any volume of gas}}{\text{total volume, in cubic feet}}$$

(*a*) Density varies as the pressure, or

$$D \text{ varies as } P$$

(*b*) Density varies inversely as the absolute temperature, or

$$D \text{ varies as } \frac{1}{T}$$

According to the laws of direct variation,

(*a*) then $\frac{D_1}{D_2} = \frac{P_1}{P_2}$

and the law of inverse variation,

(*b*) then $\frac{D_1}{D_2} = \frac{T_2}{T_1}$

Consolidating, we have

$$\frac{D_1}{D_2} = \frac{P_1T_2}{P_2T_1}$$

Therefore,

$$\frac{D_1T_1}{P_1} = \frac{D_2T_2}{P_2} = \frac{D_3T_3}{P_3} = \text{etc.}$$

GAS CONSTANTS

In finding the solution to gas problems, many steps may be saved by incorporating known constants in various formulas. In doing so, the employment of complicated fractions is eliminated. This will be understood readily in computing gas equations involving weight as well as volume. Selecting air for an example in demonstrating the use of constants, it is first necessary to determine the constant.

From the study of multiple factor changes, it is known that

$$\frac{PV}{T} = \text{constant}$$

In determining the constant for any gas, it should be noted that volume is based on the number of cubic feet contained in 1 lb at atmospheric pressure of 14.7 psi and a temperature of 32 F. Therefore,

$$\frac{PV}{T} = \frac{14.7 \times V \text{ (of 1 lb of air)}}{460 + 32} = \text{constant}$$

The volume of 1 lb of air at 14.7 psi and 32 F, according to the tables, is 12.39 cu ft. Remembering the second method of determining changes in volume, according to Gay-Lussac's law—i.e., $V_2 = V_0(1+[\frac{1}{492}t - 32])$—it is known that the volume coefficient expansion is equivalent to $\frac{1}{492} \times$ volume at freezing (32 F). The formula therefore could be written

$$\frac{PV}{T} = \frac{14.7 \times 1 \times \text{volume of 1 lb gas (cu ft)}}{492} = \text{constant}$$

Substituting the value of cubic feet per pound of air for unity in the formula,

$$\frac{PV}{T} = \frac{14.7 \times 12.39}{492} = 0.37 \text{ constant for air}$$

Designating R as the symbol for any gas constant, the formula can be written

$$\frac{PV}{T} = R, \text{ or } PV = RT$$

and for air, the formula for 1 lb is

$$PV = 0.37\, T$$

For any weight of air in pounds is

$$PV = 0.37\, MT$$

where M = the number of pounds of air.

As noted above, 0.37 as the constant for air is based on pressure in psi. To convert to pressure per sq ft, it is necessary to multiply the constant by the number of square inches in 1 sq ft, or 144. This is equal to

$$0.37 \times 144 = 53.3$$

For any gas, the fundamental thermodynamic equation as based on the above description becomes

$$PV = MRT$$

where P = pressure of gas in psia
V = total volume, cu ft
M = weight of gas in lb
R = gas constant
T = absolute temperature

EXAMPLE: Air at the rate of 50 cu ft per min is pumped into a tank from the outside of a building where the prevailing temperature is 55 F. What is the weight of air handled?

$$PV = MRT$$

$$M = \frac{PV}{RT} = \frac{14.7 \times 50}{0.37 \times (55 + 460)}$$

$$M = 3.86 \text{ lb air}$$

EXAMPLE: Determine the volume of 6 lb of air compressed from 14.7 psia to 50 psig at a final temperature of 150 F. (R = 53.3 for air)

$$V = \frac{MRT}{P} = \frac{6 \times 53.3 \times (150 + 459.6)}{144 \times (50 + 14.7)}$$

$$= 20.9 \text{ cu ft}$$

EXAMPLE: If the air in the above example is pumped into a 50-cu ft tank where its temperature is raised to 210 F, what is its final pressure?

Applying the formula of multiple factor change, we have

$$\frac{P_1V_1}{P_2V_2} = \frac{T_1}{T_2}$$

Incorporating the constants,

$$\frac{P_1V_1}{P_2V_2} = \frac{MRT_1}{MRT_2}$$

Because the weight of air handled is similar in both equations, M is canceled out. R being the constant for air is also canceled out for the same reason. $V_1 = V_2$ is also canceled, leaving pressure and temperature the only variables. Therefore,

$$\frac{P_1\cancel{V_1}}{P_2\cancel{V_2}} = \frac{\cancel{M}\cancel{R}T_1}{\cancel{M}\cancel{R}T_2}, \text{ or } \frac{P_1}{P_2} = \frac{T_1}{T_2}$$

$$P_2 = \frac{P_1T_2}{T_1}$$

$$P_2 = \frac{14.7 \times 670}{515}$$

$$P_2 = 191.2 \text{ psia}$$

If the same conditions existed in the above examples except that the tank had a capacity of 100 cu ft, then

$$\frac{P_1V_1}{P_2V_2} = \frac{\not{M}\not{R}T_1}{\not{M}\not{R}T_2}$$

$$\frac{P_1V_1}{P_2V_2} = \frac{T_1}{T_2}$$

$$P_2 = \frac{P_1V_1T_2}{V_2T_1}$$

$$P_2 = \frac{14.7 \times 50 \times 670}{100 \times 515}$$

$$P_2 = 95.6 \text{ psia}$$

DALTON'S LAW OF PARTIAL PRESSURES

This law states that *total pressure exerted by a mixture of gases is equal to the sum of the pressures exerted by the individual gases.*

The atmosphere surrounding the earth is made up of several gases (namely, oxygen, nitrogen, argon, and water vapor), the combined weight of which exerts a force expressed as barometric pressure.

According to Dalton's law, the barometric pressure is equal to the sum of the pressures exerted by the above gases individually, and may be expressed as follows:

Barometric pressure = oxygen pressure + nitrogen pressure + argon pressure + water vapor pressure

The law also may be stated that each individual gas exerts the same pressure when combined with other gases to form a mixture, as it would if it existed alone at the same temperature as the mixture.

Assume that a given volume of gas when expanded to 10 cu ft exerted a pressure of 10 psi. Assume another gas contained in a cylinder, the volume of which is 10 cu ft, at a pressure of 5 psi. The pressure resulting in mixing both gases within the 10-cu ft cylinder would be the sum of the pressures of both gases, viz.: 10 psi, pressure of first gas, +5 psi, pressure of second gas = 15 psi, total pressure within the cylinder.

Dalton's law can be stated as the formula: Total pressure = $P_1 + P_2 + P_3 + P_4$, etc., for any number of gases where P_1, P_2, etc., equals the pressure of each gas.

In calculating total pressures as determined by the sum of the separate pressures, corrections must be made for temperature differences in each case. It is also necessary that pressures for the individual gases represent the level occurring when compressed or expanded to the volume at which mixture takes place.

If the initial volumes for all gases involved in the mixture are equal and their temperatures are also equal, then the total pressure of the mixture is the sum of the individual pressures.

EXAMPLE: 5 cu ft of gas "A" having a pressure of 30 psi when mixed with 5 cu ft of gas "B," at 20 psi, in a cylinder having a volume of 5 cu ft, results in a total pressure of 30 + 20 or 50 psi, assuming the temperature of both gases to be the same.

If the initial volumes and pressures of the individual gases vary, then the pressure of the mixture is found by converting the pressure of each gas to that level relating to the volume of the mixture and adding. If the temperature remains the same,

$$P_m V_m = P_1 V_1 + P_2 V_2 + P_3 V_3, \text{ etc.}$$

where P_m and V_m are the pressure and the volume of the gas mixture and $P_1 V_1$, $P_2 V_2$, $P_3 V_3$ are the original pressure and volume of the individual gases.

EXAMPLE: If two cylinders of gas, one of which contained 10 cu ft at 32 psi, and the other 5 cu ft at 20 psi, are mixed together in a cylinder, the volume of which is 7 cu ft, what is the total pressure of the mixture? Assume the temperature of both gases to be the same.

Applying the formula

$$P_m V_m = P_1 V_1 + P_2 V_2$$

$$P_m \times 7 = (32 \times 10) + (20 \times 5)$$

$$P_m = \frac{320 + 100}{7}$$

$$P_m = 60 \text{ psi, pressure of mixture}$$

If the initial volumes, pressures, and temperatures of the individual gases vary, then the total pressure of the mixture is found by converting the pressure and the temperature of each gas to the level related to the temperature and the volume of the mixture and adding. If temperature, pressure, and volume of the individual gases vary, then

$$\frac{P_m V_m}{T_m} = \frac{P_1 V_1}{T_1} + \frac{P_2 V_2}{T_2} + \frac{P_3 V_3}{T_3}, \text{ etc.}$$

EXAMPLE: Two gases are to be mixed together in a cylinder having a volume of 10 cu ft, the final temperature to be 150 F. If the individual gases have the following initial volumes, temperatures, and pressures, what will be the total pressure of the mixture?

	First Gas	*Second Gas*
Volume..............	3 cu ft	15 cu ft
Pressure............	25 psi	20 psi
Temperature.........	60 F	200 F

Applying the formula

$$\frac{P_m V_m}{T_m} = \frac{P_1 V_1}{T_1} + \frac{P_2 V_2}{T_2}$$

then

$$\frac{P_m \times 10}{150\text{ F} + 460} = \frac{25 \times 3}{60 + 460} + \frac{20 \times 15}{200 + 460}$$

$$P_m = \frac{610}{10}\left(\frac{75}{520} + \frac{300}{660}\right)$$

$$P_m = 61\,(0.144 + 0.454)$$

$$P_m = 36.5 \text{ psi, total pressure of mixture}$$

Dalton's law of partial pressures is important in the study of absorption refrigeration where hydrogen is used to balance the pressure between the low- and high-pressure sides of the system.

SPECIFIC HEAT OF GASES

That all substances contain thermal energy was discussed earlier. It was stated also that various substances are capable of absorbing such energy in amounts compatible with their inherent thermal capacity. The relative capacity of each substance, as well as the state of any particular substance, is established by comparing their specific heat values. Although the specific heat of solids and liquids varies according to temperature levels for any particular substance, the variations are slight enough to be disregarded for practical consideration. Therefore, the specific heat of any solid or liquid substance, which represents an average, is accepted as being constant.

This statement applies also to gases when their volume remains constant or when their pressure remains constant, but the specific heat values are not similar under both conditions. Therefore a gas has two specific heats: (1) at constant volume and (2) at constant pressure.

(1) The specific heat of gas at constant volume is the amount of kinetic thermal energy required to raise the temperature of a specific volume 1 degree, and usually is expressed as Btu per lb per °F. Under constant volume conditions, no volume increase or decrease occurs, but a corresponding change in pressure takes place due to a change in temperature when thermal energy is added or subtracted.

(2) The specific heat of gas at constant pressure is the amount of thermal energy required to raise the temperature of a specific volume 1 degree, and usually is expressed as Btu per lb per °F. Under constant pressure conditions, no pressure increase or decrease occurs, but a corresponding change in volume takes place due to a change in temperature.

When thermal energy is added to a gas at constant volume, an increase in both temperature and pressure occurs. As a result, the gas retains all its energy as no work is performed. Therefore, the internal energy of the gas per unit volume increases in proportion to the amount of heat added.

When heat is added to a gas at constant pressure, an increase in both temperature and volume occurs. As a result, some energy is expended in increasing the volume against the pressure surrounding the cylinder in which the gas is contained. Therefore, any heat transferred to the mass is divided into two forms of energy: (1) that necessary to raise the temperature of gas and retained by the gas as internal energy and (2) that necessary to perform the work of increasing its volume.

It follows, then, that more thermal energy is necessary to raise the temperature of a specific volume of gas at constant pressure than it does to raise the temperature of a specific volume of gas at constant volume. Because this is true, the specific heat of gas at constant pressure is higher in value than that at constant volume. (See Table 11.)

ISOTHERMAL EXPANSION AND COMPRESSION

According to Boyle's law, the volume of a perfect gas varies inversely with the pressure when there is no change in temperature. This is true during either gas compression or gas expansion. It is recognized that the kinetic thermal energy necessary to perform work during expansion is provided from within the gas, but its replacement from an outside source in order to obey Boyle's law must be assumed. Likewise, during compression, external energy is expended which is transformed into internal energy that appears in the gas, tending to raise its temperature. Again, in order to act in accordance with Boyle's law, it must be assumed that the thermal energy which would accumulate normally during the performance of the work of compression is eliminated by some means of cooling.

Therefore, it may be stated that changes in volume and pressure of any gas, the temperature of which remains the same, act in accordance with Boyle's law, and the work performed is said to be isothermal.

Isothermal expansion and compression, then, is expansion and compression that is performed according to the formula

$$P \times V = \text{constant}$$

Referring to the isothermal curve shown in Fig. 24, it will be found that the product of pressure × volume anywhere along the curve is always identical. This would not be true if the temperatures were allowed to vary, a statement borne out by Charles' law, which may be applied by adopting any point along the isothermal curve as constant volume. If the temperature were allowed to change as a result of the work performed during expansion or compression under conditions not isothermal, a pressure variation would occur that would render $PV =$ constant no longer true. This would result in a curve whose pressure-volume relationship would not be constant at any point, but would more nearly approach the adiabatic line also shown in Fig. 24.

ADIABATIC EXPANSION AND COMPRESSION

In performing work of any kind, energy undergoes a transformation. Thus, in compressing a gas, mechanical energy is transformed to thermal energy, which is taken up by the gas. The transformation is reversed during expansion, whereby the thermal energy leaves the gas and reverts to mechanical energy. The air compressor in the first case and the internal combustion engine in the second case are examples of machines that operate according to such established laws of energy transformation.

If it were possible to construct a cylinder from a nonheat-conducting material and design a frictionless leakproof piston to fit, any gas within it would embrace or reject only that amount of enthalpy necessary to perform the actual work required to compress it to any predetermined volume, or expand it against any lower counterpressure to any specific volume. As there would be no loss of thermal energy in this hypothetical cylinder, the enthalpy of the gas would be increased in an amount equal to the mechanical equivalent required to compress it to any specific pressure and volume. A similar amount of enthalpy would be given up upon allowing the compressed air to expand and return to its original volume and pressure. Work performed in accordance with the above conditions is said to be *adiabatic*.

Adiabatic expansion, then, is expansion of a gas with no energy being gained or lost, except that which is given up in performing external work. Adiabatic compression follows a similar rule, except that the procedure is reversed in that the gas neither gains nor rejects energy other than that received as internal energy required to compress it.

It follows that the thermal energy gained or rejected by the gas as a result of energy transformation during compression and expansion changes the temperature of the gas accordingly. Where similar conditions of volume and pressure exist, an equal amount of energy is involved, but the temperature change is contingent on the kind of gas employed. This is true by reason of variations in specific heat characteristics of different substances, and because the temperature of any substance is a function of its thermal energy content and specific heat. Since the specific heat of a gas at constant volume differs from that at constant pressure, the change in temperature of any gas in the cylinder during adiabatic compression or expansion must be determined as a function of the ratio of both values.

The amount of temperature change will vary with each kind of gas, as will its pressure. Under such conditions brought about adiabatically, changes in temperature, volume, and pressure do not conform to Boyle's, Charles', and Gay-Lussac's laws. Instead, changes are experienced that occur due to increased or decreased temperatures for every increment of travel of the piston. These changes, in turn, vary according to the specific heat of the different gases. Under adiabatic conditions, $P \times V$ is no longer constant.

Referring to the adiabatic curve in Fig. 24, it will be noted that its rise is more rapid along the line of pressure with each increment of volume decrease than is the isothermal curve whose pressure varies in direct propor-

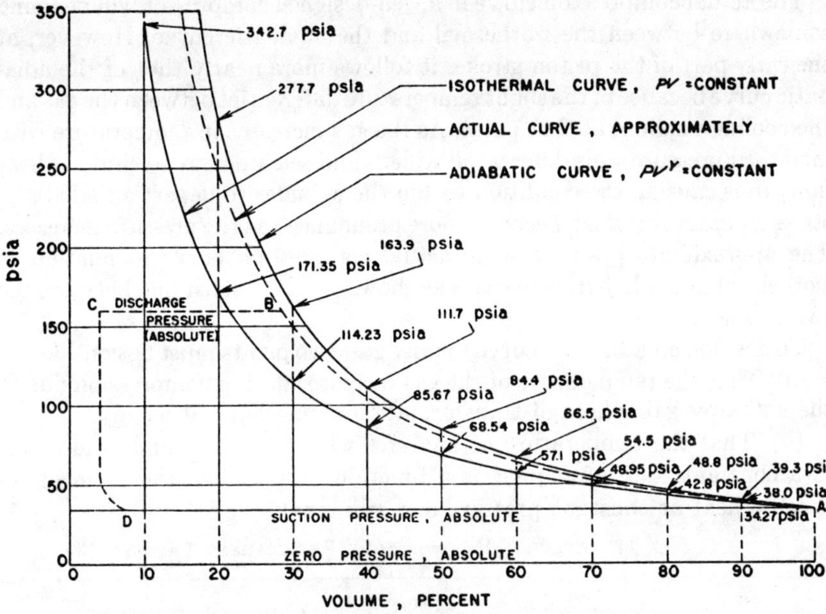

Fig. 24. Pressure-volume diagram.

tion to its volume. Here are shown graphically the effects of increased temperature occurring due to adiabatic law, whereby the work energy transformed to enthalpy within the gas during compression develops an inhibited tendency on the part of the gas to expand. As all gases possess expansion tendencies when their temperature is increased, the rate of expansion will depend on their specific heat values. Although the compressive force of the piston prevents any expansion within the cylinder, the energy developed nevertheless remains, being manifested as work in the form of pressure. It is this additional energy developed as a result of adiabatic conditions, in excess of that developed isothermally, which causes the piston to do more work in order to reduce the gas to a similar volume in each case. This is borne out by comparing the areas embraced by both curves.

For actual operations, adiabatic conditions are never realized because of the heat-conduction characteristics of materials used in compressor construction. Nor would such a condition be desirable because of the additional amount of work required to compress gases according to adiabatic law. On the other hand, isothermal compression, although an ideal condition for which to strive, eludes every attempt to eliminate all the thermal energy gained by the gas due to the work of compression because of the ineffectiveness of any cooling system so far devised. Under ideal conditions, thermal energy dissipation would have to occur simultaneously with its development.

The actual compression curve in a well-designed compressor will be found somewhere between the isothermal and the adiabatic curve. However, at the early part of the piston stroke, it follows more nearly that of the adiabatic curve because of the slight temperature differential between the gas and the cooling medium at that point. As the gas increases in temperature with increased pressure, this differential widens and sets up a greater rate of heat flow, thus causing the condition within the cylinder to depart an adiabatic state, a departure which becomes more pronounced as the pressure increases. The approximate position of an actual compression curve in relation to isothermal and adiabatic curves is as shown by the dotted line between the two in Fig. 24.

To develop an adiabatic curve for any gas, two points must be considered:

(*a*) That the temperature of the gas increases during compression due to the enthalpy gain through transformation of mechanical energy.

(*b*) That the temperature of any gas, whose pressure and volume are in a constant state of change, is a function of enthalpy and ratio of the specific heat at constant pressure and volume.

TABLE 11. SPECIFIC HEAT OF SOME REFRIGERANT VAPORS

Vapor	*Pressure*	*Temperature °F*	*Specific Heat, Constant Pressure* (C_p)	*Specific Heat, Constant Volume* (C_v)	*Ratio* C_p/C_v (y)
*Air	Atmospheric	60	0.24	0.169	1.4
*Ammonia	"	60	0.58	0.39	1.3
*Carbon dioxide	"	60	0.203	0.156	1.3
*Ethyl chloride	"	60	0.275	0.231	1.13–1.19
†Freon-11	"	110	0.1374	0.1213	1.133
†Freon-12	"	86	0.1476	0.1297	1.138
†Freon-21	"	60	0.1364	0.1156	1.18
†Freon-22	"	60–212	0.148–0.1702	0.125–0.148‡	1.18–1.15
†Freon-113	"	160	0.1633	0.1516	1.077
†Freon-114	"	110	0.1629	0.1502	1.085
*Methyl chloride	"	60	0.24	0.20	1.20–1.28
*Methylene chloride	"	60	..	..	1.18–1.224
*Sulfur dioxide	"	60	0.154	0.123	1.25–1.29
*Water vapor	"	212°	0.48	0.36	1.27–1.33

* From *Compressed Air Data*—Ingersoll-Rand Company.
† From Kinetic Chemicals Company, div. of E. I. duPont de Nemours & Co.
‡ Calculated.

Identifying the specific heat ratio with the letter y for the sake of simplicity, the formulas for the ratio may be written, viz.:

$$y = \frac{\text{specific heat at constant pressure}}{\text{specific heat at constant volume}}$$

or

$$y = \frac{C_p}{C_v}$$

Referring to Table 11, Specific Heat of Some Refrigerant Vapors, it will be found that the value of C_p for air is 0.238, and for C_v, the value is 0.169.

The value of y for air, then, is

$$y = \frac{C_p}{C_v} = \frac{0.238}{0.169} = 1.4$$

During adiabatic expansion and compression, the pressure is affected to the extent of the tendency of the gas to increase its volume raised to a power of y. The formula for adiabatic compression and expansion then becomes, with y as an exponent of V,

$$\frac{P_1}{P_2} = \frac{V_2^y}{V_1^y}$$

or

$$P \times V^y = \text{constant}$$

For air,

$$P_1 \times V^{1.4} = \text{constant for air}$$

For ammonia,

$$C_p = 0.508$$

$$C_v = 0.391$$

$$y = \frac{0.508}{391} = 1.3$$

$$P_1 \times V^{1.3} = \text{constant for ammonia}$$

For carbon dioxide,

$$C_p = 0.203$$

$$C_v = 0.156$$

$$y = \frac{0.203}{0.156} = 1.30 \text{ for } CO_2$$

$$P_1 V^{1.30} = \text{constant for } CO_2$$

The adiabatic curve, then, conforms to

$$P_1V_1^y = P_2V_2^y = P_3V_3^y = P_4V_4^y, \text{ etc.}$$

ISOTHERMAL AND ADIABATIC CURVES

To demonstrate the use of the formulas described above and to provide a graphic study of the effect of both isothermal and adiabatic laws as related to the compression and the expansion of gases, the curves shown in Fig. 21 are constructed. These curves represent the changes in pressure-volume relationship with each increment of piston travel as it moves within the cylinder during the compression or expansion stroke.

Curves are constructed on cross-section paper, the vertical lines of which are called the *ordinates* or Y axis, and the horizontal lines identified as the *abscissas* or X axis. Any point plotted on the graph in the construction of a curve is located by relating its distance from the *origin* along the ordinate line and thence along the abscissa line. The plotted point will then be found at the intersection of the ordinate and the abscissa whose values are represented by the distances specified along each line from the point of origin, which may be located on either side of the graph. Distances along abscissa and ordinate lines may be represented in weight, pressure, volume, temperature, heat content, or any other value under consideration. In this discussion, only pressure and volume are involved, pressure being represented as the abscissas and volume as the ordinates.

The Isothermal Curve

In order to construct a curve, it is necessary that at least one set of values be known, as well as the law governing the action. In the first case here, the law will be that of isothermal compression and expansion, and the curve will be developed according to the formula PV = constant.

Dividing the length of the graph into 10 equal sections along the abscissa, starting from the point of origin, indicated by the letter O, each section will represent 10 percent of the total volume of the cylinder, forming a basis for the value of V in the formula. Assuming the initial pressure of the gas to be 34.27 psia, as it enters the cylinder, the constant in the formula is established, viz.:

$$PV = \text{constant}$$

$$34.27 \text{ psi} \times 100\% \text{ volume} = \text{constant}$$

$$34.27 \times 100 = 3427, \text{ a constant}$$

The first point on the graph in constructing the isothermal curve may now be plotted at the intersection of the abscissa, or X axis, whose value is 100, and the ordinate, or Y axis, representing a value of 34.27 psia.

As the piston moves forward on its journey toward the discharge end of the cylinder, the volume decreases with each increment of travel. For the sake of convenience, however, the next point on the graph will be plotted according to conditions found when the piston has moved 10 percent of the total length of the cylinder or a point representing 90 percent of total volume. For the second point, the pressure must be determined for the 90 percent volume. Again employing the isothermal formula,

$$PV = \text{constant}$$

$$P \times 90 = 3427$$

$$P = \tfrac{3427}{90}$$

$$P = 38.0 \text{ psia at } 90\% \text{ volume}$$

The second point on the graph in constructing the curve may now be plotted at the intersection of the abscissa, or X axis, whose value is 90, and the ordinate, or Y axis, representing a value of 38.0 psia.

Continuing to apply the isothermal formula for each 10 percent decrease in volume, the pressure for each respective point will be established. Assuming that the discharge valve opens when the volume is reduced to 10 percent of its original value, the following results are obtained:

$$80\% \text{ volume} \qquad P_3 = \frac{\text{constant}}{V_3} = \frac{3427}{80} = 42.8 \text{ psia}$$

$$70\% \text{ volume} \quad P_4 = \frac{\text{constant}}{V_4} = \frac{3427}{70} = 48.95 \text{ psia}$$

$$60\% \text{ volume} \quad P_5 = \frac{\text{constant}}{V_5} = \frac{3427}{60} = 57.1 \text{ psia}$$

$$50\% \text{ volume} \quad P_6 = \frac{\text{constant}}{V_6} = \frac{3427}{50} = 68.54 \text{ psia}$$

$$40\% \text{ volume} \quad P_7 = \frac{\text{constant}}{V_7} = \frac{3427}{40} = 85.67 \text{ psia}$$

$$30\% \text{ volume} \quad P_8 = \frac{\text{constant}}{V_8} = \frac{3427}{30} = 114.23 \text{ psia}$$

$$20\% \text{ volume} \quad P_9 = \frac{\text{constant}}{V_9} = \frac{3427}{20} = 171.35 \text{ psia}$$

$$10\% \text{ volume} \quad P_{10} = \frac{\text{constant}}{V_{10}} = \frac{3427}{10} = 342.7 \text{ psia}$$

Although the isothermal curve as constructed from the number of points selected, as represented in the calculations above, is accurate enough for all practical purposes, scientific investigation would demand more exacting details in plotting the outline. This demand would be met by determining the pressure values for points representative of the lesser volume changes than were employed here. That is, volume changes of 5 percent instead of 10 percent increase the accuracy in plotting the curve, as twice the number of points would be provided in determining the outline. Plotting pressure points for each 2 percent of volume change would tend toward still greater accuracy.

Dividing the pressure values as found at the several positions on the curve representing volume reductions (i.e., 90 percent V, 80 percent V, 70 percent V, etc.) by the suction pressure, a ratio is established for each position that may be employed for determining the pressures at comparable points on isothermal curves having any suction pressure. By identifying the suction pressure as P and the progressive pressures as determined for 90 percent V, 80 percent V, 70 percent V, etc., as P_1, P_2, P_3, the ratios are established in accordance with $\frac{P_1}{P_1}, \frac{P_2}{P_1}, \frac{P_3}{P_1}$, etc., for any position along the isothermal line. These ratios for positions represented by progressive reductions of 10 percent volume for the entire travel of the piston are as follows, and are true for comparable positions on any isothermal curve:

$$\text{Ratio } \frac{P_2}{P_1} = \frac{38.0}{34.27} = 1.11 \text{ at } 90\% \text{ volume}$$

$$\text{“} \quad \frac{P_3}{P_1} = \frac{42.8}{34.27} = 1.25 \text{ at } 80\% \text{ volume}$$

$$\text{Ratio } \frac{P_4}{P_1} = \frac{48.95}{34.27} = 1.43 \text{ at } 70\% \text{ volume}$$

$$\text{" } \frac{P_5}{P_1} = \frac{57.1}{34.27} = 1.66 \text{ at } 60\% \text{ volume}$$

$$\text{" } \frac{P_6}{P_1} = \frac{68.54}{34.27} = 2.00 \text{ at } 50\% \text{ volume}$$

$$\text{" } \frac{P_7}{P_1} = \frac{85.67}{34.27} = 2.50 \text{ at } 40\% \text{ volume}$$

$$\text{" } \frac{P_8}{P_1} = \frac{114.23}{34.27} = 3.33 \text{ at } 30\% \text{ volume}$$

$$\text{" } \frac{P_9}{P_1} = \frac{117.35}{34.27} = 5.00 \text{ at } 20\% \text{ volume}$$

$$\text{" } \frac{P_{10}}{P_1} = \frac{342.7}{34.27} = 10.00 \text{ at } 10\% \text{ volume}$$

EXAMPLE: Gas entering a cylinder under 65 psia would have a discharge pressure of 650 psia, assuming isothermal conditions and discharge valve opening when volume is reduced to 10 percent of its total volume. This was found by multiplying the 10 percent volume ratio, $\frac{P_{10}}{P_1}$, or 10.0 by the suction pressure of 65 psia.

The Adiabatic Curve

The adiabatic curve is developed in a manner similar to that employed for the isothermal curve. However, instead of following the action as dictated by Boyle's law, PV = constant, the adiabatic law must be abided by: $P\ (V^y)$ = constant. Considering ammonia as the gas in constructing the adiabatic curve, shown in Fig. 24, the formula becomes $P\ (V^{1.3})$ = constant.

Employing the same suction pressure as used for the isothermal curve, the first point for the adiabatic curve will be plotted at the intersection of the abscissa and the ordinate lines represented by the values 100 and 34.27, respectively.

The second point will be determined by solving for pressure prevailing at the 90 percent volume position.

Applying the formula,

$$P\ (V^{1.3}) = \text{constant}$$

then

$$P_1\ (V_1{}^{1.3}) = P_2\ (V_2{}^{1.3})$$

or

$$P_2 = \frac{P_1\ (V_1{}^{1.3})}{V_2{}^{1.3}}$$

$$\therefore P_2 = \frac{34.27 \times 100^{1.3}}{90^{1.3}} = \frac{13643.2}{90^{1.3}} = 39.31$$

$$P_2 = 39.31 \text{ psi at } 90\% \text{ volume}$$

At a constant pressure of 34.27 psi, the constant for the adiabatic curve is

$$P\,(V^{1.3}) = 13643.2$$

And the second point for the curve may now be plotted at the intersection of the abscissa and the ordinate lines represented by the values 90 and 39.31, respectively.

The remaining pressure-volume points are determined as follows:

$$80\% \text{ volume} \quad P_3 = \frac{13643.2}{80^{1.3}} = 48.8$$

$$70\% \text{ volume} \quad P_4 = \frac{13643.2}{70^{1.3}} = 54.5$$

$$60\% \text{ volume} \quad P_5 = \frac{13643.2}{60^{1.3}} = 66.5$$

$$50\% \text{ volume} \quad P_6 = \frac{13643.2}{50^{1.3}} = 84.4$$

$$40\% \text{ volume} \quad P_7 = \frac{13643.2}{40^{1.3}} = 111.7$$

$$30\% \text{ volume} \quad P_8 = \frac{13643.2}{30^{1.3}} = 163.9$$

$$20\% \text{ volume} \quad P_9 = \frac{13643.2}{20^{1.3}} = 277.7$$

$$10\% \text{ volume} \quad P_{10} = \frac{13643.2}{10^{1.3}} = 682.2$$

Here again, as with the isothermal curve, ratios can be established based on the pressure at any point along the line of piston travel and the suction pressure. Employing the pressure values found for the above volume points and carried out to two decimal places, the ratios for an adiabatic curve are as follows.

$$\frac{P_1}{P_1} = \frac{34.27}{34.27} = 1.0 \quad 100\% \text{ volume}$$

$$\frac{P_2}{P_1} = \frac{39.3}{34.27} = 1.14 \quad 90\% \text{ volume}$$

$$\frac{P_3}{P_1} = \frac{45.8}{34.27} = 1.34 \quad 80\% \text{ volume}$$

$$\frac{P_4}{P_1} = \frac{54.5}{34.27} = 1.59 \quad 70\% \text{ volume}$$

$$\frac{P_5}{P_1} = \frac{66.5}{34.27} = 1.94 \quad 60\% \text{ volume}$$

$$\frac{P_6}{P_1} = \frac{84.4}{34.27} = 2.46 \qquad 50\% \text{ volume}$$

$$\frac{P_7}{P_1} = \frac{111.7}{34.27} = 3.26 \qquad 40\% \text{ volume}$$

$$\frac{P_8}{P_1} = \frac{163.9}{34.27} = 4.78 \qquad 30\% \text{ volume}$$

$$\frac{P_9}{P_1} = \frac{277.7}{34.27} = 8.1 \qquad 20\% \text{ volume}$$

$$\frac{P_{10}}{P_1} = \frac{682.2}{34.27} = 20 \text{ (approx.)} \quad 10\% \text{ volume}$$

USE OF PRESSURE-VOLUME DIAGRAMS

Only those curves have been constructed that show pressure-volume relationship during the compression (or expansion) stroke of a piston as it moves within a compressor cylinder, according to isothermal and adiabatic laws. These curves are, of course, mathematical predictions of what would happen in a cylinder if it were possible to abide by the two thermal laws exactly. However, in actual operation, compressors should nearly approach the adiabatic curve.

The diagram shown in Fig. 24 not only embraces the compression (or expansion) curve, but also includes an outline of the pressure-volume relationship of a gas through a complete cycle of the piston; i.e., one revolu-

Fig. 25. Indicator. (Courtesy, Crosby Steam Gage & Valve Company.)

tion of the flywheel or two strokes of the piston. However, this diagram portrays an ideal condition and merely provides a basis for improvement in the design and the operation of compressors. Compressor performance should nearly coincide with the outline of the diagram as indicated by *A*, *B*, *C*, *D*, *A*, assuming a 4 percent clearance between the piston at the end of the compression stroke and the cylinder head. There are numerous reasons that dictate otherwise, such as worn piston rings and cylinders, carbon deposits, inefficient cooling, and improperly adjusted valves, to name a few.

To obtain a volume-pressure diagram similar to *A*, *B*, *C*, *D*, *A* shown in Fig. 24, an instrument called an *indicator* is used. The indicator is provided with a piston, closely fitted into a cylinder. The cylinder is open at the bottom and designed so that it may be attached readily and exposed to the interior of the compressor cylinder. Because the bottom of the indicator cylinder is open to the compressor cylinder, the indicator piston is subjected to the pressure changes that occur during operation, and moves vertically upward against a calibrated spring as pressure increases, and downward as pressure decreases. The piston is connected to a pencil, through a series of linkages, which reciprocates vertically along the surface of a vertical drum to which a paper card, commonly referred to as an indicator card, is attached. This pencil records the distance of indicator piston travel, which is governed by the pressures in the compressor cylinders. The vertical drum is connected by means of a cord to the crosshead of the compressor, which, with the action of an internal spring, causes an alternating horizontal motion of the drum coincident with and in constant ratio to the travel of the compressor piston.

As the compressor piston moves forward in compressing a gas, the increase in pressure causes the indicator piston to rise, the pencil tracing its motion on the indicator card. Simultaneously with the movement of the piston, the crosshead, also being in motion, causes the drum to revolve via the cord arrangement, thereby allowing the pencil to record a horizontal as well as a vertical motion. In completing one cycle, a diagram such as that shown in Fig. 24 results. Note the space between the zero pressure line or base line and the line *DA* forming the bottom of the diagram *A*, *B*, *C*, *D*, *A*. This represents the pressure exerted by the atmosphere, and the line *DA* is known as the atmospheric line.

In addition to providing a means for determining mechanical defects within the compression chamber of heat engines in actual operation, the indicator diagram may be used also to establish the mean effective pressure, usually designated mep. When the mean effective pressure is known, the indicated horsepower (ihp) may then be computed by application of the standard formula

$$\text{ihp} = \frac{(\text{mep}) \times L \times A \times N}{33{,}000}$$

where mep = mean effective pressure in psi
L = length of stroke in feet
A = area of piston in square inches
N = number of revolutions per minute

To determine the mean effective pressure of an engine from the pressure-volume diagram, an instrument called a planimeter is used. This instrument is equipped with a roller wheel, a pivot vernier, and a trace point. By outlining the diagram with the tracer point, the area is established by the subsequent turning of the roller wheel which is connected to a counting disk. Upon completion of travel by the tracer point around the outline once, the total area of the diagram is found from the figures on the roller wheel, the vernier, and the counting disk. If the planimeter is designed or adjusted to read directly in square inches, then that is the unit in which the readings will appear, and, by dividing the result by the length of the diagram in inches, the average height will be obtained. Multiplying the average height of the diagram by the scale of the spring employed in taking the diagram determines the mean effective pressure.

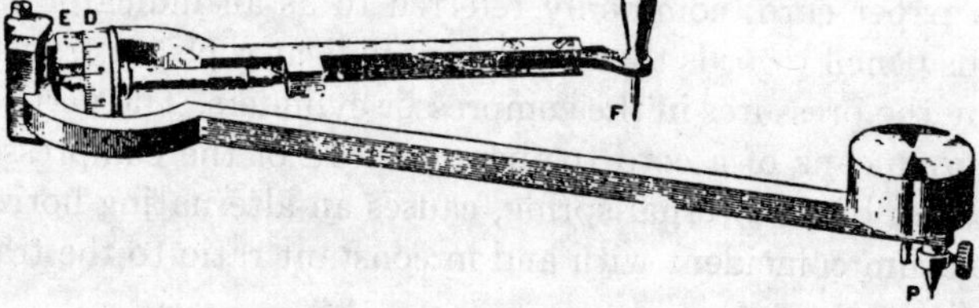

Fig. 26. Planimeter. (Courtesy, Crosby Steam Gage & Valve Company.)

While it is not the purpose of the author to provide in this text a study of compressor troubles as determined by the use of indicators, it is desirable to become acquainted with such instruments and the operating peculiarities that their diagrams reveal.

SUMMARY OF CHAPTER 4

Gases:

Definition:

An elastic fluid having indefinite expansion characteristics.

Compressible to the extent of their temperature-pressure relationship below their critical point.

Origin:

Emanate generally from liquids as a result of evaporation or vaporization at their boiling temperature corresponding to existing pressure.

Characteristics:

The existence of a gas depends on the temperature at the existing pressure. Expansion occurs uniformly for equal increments of temperature increase. The converse is true for contractions.

Density:

Weight:

The weight of a gas per cubic foot varies as temperature and pressure

$$D \propto \frac{I}{T}, \text{ and } D \propto P$$

Volume:

Measured as cubic feet per pound at a given temperature and pressure. Specific volume is equal to the reciprocal of density, or

$$V = \frac{1}{D}$$

Avogadro's law:

Equal volumes of all gases have an equal number of molecules.

Critical Point:

Critical temperature:

The temperature above which a gas cannot be liquefied regardless of pressure applied.

Critical pressure:

The pressure corresponding to the critical temperature.

Gas Laws:

Boyle's law:

Volume of a gas varies inversely as the pressure at constant temperature.

$$V \propto \frac{1}{P} \qquad T = \text{constant}$$

Charles' law:

Pressure of a gas varies directly in proportion to the absolute temperature at constant volume.

$$P \propto T \qquad V = \text{constant}$$

Gay-Lussac's law:

Volume of a gas varies directly in proportion to the absolute temperature at constant pressure.

$$V \propto T \qquad P = \text{constant}$$

Multiple factor changes:

Consolidation of Boyle's, Charles', and Gay-Lussac's laws when changes in all factors occur

$$\frac{PV}{T} = \text{constant}$$

$$\frac{P_1V_1}{T_1} = \frac{P_2V_2}{T_2} = \frac{P_3V_3}{T_3}, \text{ etc.}$$

Dalton's law of partial pressures:

The total pressure of any gas mixture is equal to the sum of the individual pressures of all gases in the mixture

$$P_1 + P_2 + P_3, \text{ etc.} = \text{total pressure}$$

Specific Heat of Gases:

Constant volume (C_v):

When work is being performed, the gas retains all its energy. Therefore, the internal heat energy increases proportionately as heat is added.

Constant pressure (C_p):

Some work is performed in increasing volume. The thermal energy retained by the gas is equal to the amount added less the amount expended in expanding. More heat is required to raise the temperature of a gas at constant pressure; consequently, its specific heat value is higher than when constant volume is maintained.

Ratio C_p/C_v or y:

The temperature of any gas whose pressure-volume relationship is in a state of constant change is a function of its specific heat at both constant pressure and constant volume, and of its enthalpy at any instant of change. The specific heat value, then, at any time during change, is the ratio of specific heat at constant pressure (C_p) to the specific heat at constant volume (C_v), $\frac{C_p}{C_v} = y$, a constant for any gas.

Expansion and Compression of Gases:

Isothermally:

The compression or the expansion of a gas during which changes in both pressure and volume occur without any change in temperature. Internal energy is neither gained nor lost.

$$P \times V = \text{constant}$$

Adiabatically:

The compression or the expansion of a gas during which changes in both pressure and volume occur and whose temperature changes as a result of retaining or rejecting only that amount of energy necessary to perform the work of expansion or compression.

$$P \times V \text{ is no longer constant}$$

but

$$P \times V^y = \text{constant}$$

where y is the ratio of C_p/C_v for any gas. Therefore, the temperature and the pressure are higher under adiabatic conditions at any point along the line of piston travel than they would be for isothermal conditions, everything else being equal.

STUDY OUTLINE FOR CHAPTER 5

Definition of a refrigerant:
- General
- As applicable to the mechanical refrigeration system

Factors governing the ideal refrigerant:
- (*a*) Chemical,
- (*b*) Thermal,
- (*c*) Physical,
- (*d*) Physiological,
- (*e*) Economic.

Factors governing the selection of a refrigerant based on:
1. Operating conditions:
   - (*a*) Temperatures,
   - (*b*) Pressures,
   - (*c*) Safety,
   - (*d*) Load capacity.
2. Utility:
   - (*a*) Domestic,
   - (*b*) Commerical,
   - (*c*) Industrial,
   - (*d*) Air conditioning,
   - (*e*) Marine.
3. Type of compressor:
   - (*a*) Reciprocating,
   - (*b*) Centrifugal,
   - (*c*) Rotary.

Refrigerant classification:
- Group I. Ammonia
- " II. Carbon dioxide
- " III. Sulfur dioxide
- " IV. Hydrocarbons
- " V. Halogenated group:
  - (*a*) Chlorinated hydrocarbon family
  - (*b*) Fluorinated hydrocarbon family (methane series)
  - (*c*) Fluorinated hydrocarbon family (ethane series)
  - (*d*) Bromines
- " VI. Azeotropes
- " VII. Miscellaneous
- " **VIII. Cryogenic Fluids**

5

Refrigerants

While a refrigerant may be any substance capable of absorbing kinetic thermal energy from another substance, such as ice, water, air, brine, etc., only those fluids adaptable to the mechanical refrigeration system will be discussed here. Because the mechanical system is based on the evaporation and the subsequent condensation of a fluid for absorption and dissipation of thermal energy, it follows that refrigerants must possess such physical characteristics as will enable them to repeat continuously a liquid-to-gas-to-liquid transformation. They are required also to perform such repeated transformations at temperatures demanded by the service for which they are selected and at pressures that are both convenient and appropriate to economy of design, construction, and operation.

In addition to physical characteristics, other factors concerning their thermodynamic, chemical, safety, and economic qualifications also must be considered in selecting fluids for refrigeration purposes. Refrigerant selection, then, must be approached on the basis of its compatibility with the conditions prevailing at the specific location of intended use or with the circumstances under which it will be generally adopted. Any fluid selected approaches the status of an ideal refrigerant only to the extent that it meets the combined qualifications of all factors involved.

REFRIGERATING EFFECT

The refrigerating effect of a fluid is measured by the amount of thermal energy it is capable of absorbing from the time it enters the evaporator as a liquid and leaves as a vapor to enter the suction line leading to the compressor. As the greatest amount of thermal energy is absorbed during the evaporation of the liquid, it follows that fluids having high latent heat of vaporization values tend to possess the greater refrigerating effect. This is true only when the advantage of high latent heat of vaporization is accompanied by reasonably high vapor density, and to some extent by low specific heat of the liquid. Everything else being equal, the ideal refrigerant should possess a high latent heat of vaporization.

The total refrigerating effect of a fluid is determined by calculating the difference between the enthalpy of the entering liquid at the entering temperature and the enthalpy of the vapor leaving the evaporator. Therefore, both the latent heat of vaporization and the kinetic thermal energy of the liquid and the vapor must be considered in determining the total refrigerating effect. Under standard ton conditions, such values are determined for 86 F liquid and 5 F saturated vapor.

For example, liquid ammonia entering the evaporator at a temperature of 86 F and leaving as a saturated vapor at 5 F possesses a refrigerating effect of 474.4 Btu per lb, determined as follows:

Referring to the tables of saturated ammonia vapor, it is found that, at 5 F and 34.27 psia,

Enthalpy of the liquid	= 48.3 Btu per lb
Latent heat	= 565.0 Btu per lb
Total enthalpy of vapor	= 613.3 Btu per lb

And, at 86 F, liquid ammonia

Enthalpy	= 138.9 Btu per lb
Refrigerating effect = 613.3 − 138.9	= 474.4 Btu per lb

By allowing the vapor to absorb additional thermal energy after saturation is complete, superheating occurs, thereby increasing the total refrigerating effect. If a 10 F superheat were allowed, the total energy absorbed, assuming that the above conditions prevailed, would be found in the following manner:

Enthalpy of 5 F liquid	= 48.3 Btu per lb
Latent heat	= 565.0 Btu per lb
Superheat—Btu = wt × sp ht × $(t_2 - t_1)$	
= 1 × 0.510 × 10 F	= 5.1 Btu per lb
Total enthalpy of vapor at 15 F and 34.27 psia	= 618.4 Btu per lb
Enthalpy of 86 F liquid	= 138.9 Btu per lb
Refrigerating effect = 618.4 − 138.9	= 479.5 Btu per lb

The total enthalpy of the superheated vapor may be determined also from the tables "Properties of Superheated Ammonia Vapor" (Appendix). Using the 10 F superheat, as in the above example, the temperature of the vapor leaving the evaporator would rise to 15 F at a constant pressure of 34.27 psia. Referring to the tables, it is found that, at 34 psia and 15 F (interpolation), the enthalpy of the vapor is equal to 619.4 Btu per lb. At 34.27 psia, again by interpolation, using the next highest recorded pressure of 35 psia, the total enthalpy of the superheated vapor is found to be approximately 618.5 Btu per lb, which checks closely with the calculated results above.

By aftercooling the liquid to 76 F—i.e., cooling the liquid from the condensing temperature of 86 F prior to entering the evaporator—the refrigerating effect is increased further, viz.:

Enthalpy of 76 F liquid = 127.4 Btu per lb
Refrigerating effect when superheating to 15 F = 618.4 − 127.4 = 491.0 Btu per lb

It is readily seen that greater economies may be secured by precooling the liquid below condenser temperatures than by allowing a similar degree of superheating to occur. In order to avoid liquid refrigerant entering the compressor, it is usual practice to allow at least a 10 F superheating in the suction side. Both actions are resorted to in commercial systems generally.

BOILING POINT

The boiling temperature of a refrigerant at atmospheric pressure is indicative of the type of equipment required for its use, and, to some extent, the type of service for which it is best suited. While it is not always true that extremely low-boiling-point fluids demand extra heavy equipment, the high pressures that accompany the temperature control of those refrigerants most commonly employed make this specification essential.

Carbon dioxide (CO_2), for example, having a boiling point at atmospheric pressure of −109.4 F, requires a condenser pressure of 1045.7 psia for its liquefaction at the standard temperature of 86 F, and an evaporator pressure of 331 psia in order to control vaporization at 5 F. At the other extreme, methylene chloride (CH_2Cl_2), whose boiling point at atmospheric pressure is 103.6 F, evaporates at 5 F and condenses at 86 F under partial vacuum conditions of 27.4 in. of mercury and 8.4 in. of mercury, respectively. Between these, and considered moderate in its requirements, is ammonia (NH_3), with a boiling point of −28 F at 1 atm. With the latter refrigerant, evaporator and condenser pressures, under standard ton conditions of 5 F and 86 F, correspond to 34.27 psia and 169.2 psia, respectively.

For general purposes, such as ice making, cold storage, and domestic refrigerators, refrigerants whose boiling-point temperatures at atmospheric pressure are within the range of −30 F to +14 F have proved satisfactory. Equipment properly designed and constructed to create and maintain low absolute pressures (below 1 atm) successfully employs refrigerants of high boiling points (bp), such as Refrigerant 113 (117.6 F bp). However, the advantage of preventing the entrance of air, moisture, and other contaminants into the system lies with the refrigerant capable of maintaining desired evaporator pressures slightly above that of the atmosphere.

Temperature requirements govern the boiling point, and hence determine the individual refrigerant to be selected. Such requirements may be divided into four temperature categories based on evaporator temperatures; i.e., ultra low temperatures (−65 F and below), low (−65 F to −20 F), intermediate (−20 F to 20 F), and high (above 20 F). As a matter of equipment design and construction convenience, individual refrigerants suitable for any one of these ranges should have a boiling point at atmospheric pressure low enough to evaporate at temperatures within its respective range at a pressure only slightly above or below atmospheric.

For example, ammonia boils at −28 F under normal atmospheric pressure of 14.7 psia. Its boiling point increases to 5 F if pressure is raised to 34.27 psia, or about 20 psi above atmospheric. This is within the range of intermediate temperatures and no difficult problem exists in evaporator design to withstand such a relatively low pressure, which in this case is exerted in an outward direction. If, instead of 5 F, an evaporator temperature of −40 F is required, it would become necessary to draw a partial vacuum of 8.7 in. of mercury, or 10.41 psia, in order to obtain it with ammonia as the refrigerant. This situation presents all the disadvantages of a negative-pressure system. By employing another refrigerant whose boiling-point temperature at atmospheric pressure is slightly below the desired evaporator temperature, a positive-pressure system can be attained. Freon-22 has a boiling point of −41.4 F at atmospheric pressure. It will evaporate at −40 F if a pressure of 15.31 psia, or 0.61 psi above atmospheric pressure, exists in the evaporator. If a −40 F temperature is to be normal for the system, the selection of Refrigerant 22 on the basis of boiling point alone would be the better choice of the equipment designer over ammonia in this case. A design advantage still would favor Refrigerant 22 if an evaporator temperature of −60 F were desired, for, while negative pressures would be required for both refrigerants, a greater pressure difference relative to the atmosphere would exist by using ammonia than with Refrigerant 22, being 5.55 psia and 8.86 psia, respectively.

CONDENSING TEMPERATURE AND PRESSURE

Refrigerants vary as to the pressure required to promote condensation or maintain their liquid form under similar temperature conditions. It follows that condensing pressures of any refrigeration system depend on the refrigerant employed, and will vary with each kind of fluid. (See Table 12.)

From Table 12, several rules may be deduced which generally hold true for any list of fluids and may be expressed as follows:

(*a*) *The boiling temperature at atmospheric pressure of a fluid is not always indicative of condenser pressures required.* This is borne out by comparing Refrigerant 12 with ammonia. With only a 6.4 F difference in boiling point at atmospheric pressure, a 61.8 psi greater condenser pressure at 86 F is required when ammonia is employed. This factor is more pronounced in the case of ethane versus carbon dioxide. Here the atmospheric boiling temperature is much lower for ethane (−127.5 F as com-

TABLE 12. PRESSURE–BOILING-POINT CHARACTERISTICS OF SOME REFRIGERANTS

Refrigerant	*Boiling Point Atmospheric Pressure °F*	*Evaporating Pressure @ 5 F psia*	*Condensing Pressure @ 86 F psia*	*Compression Ratio 5 F to 86 F psia cond psia evap*	*Pressure–Boiling-Point Factor*
Ammonia	−28.0	34.27	169.2	4.94	(1.67)
Carbon dioxide	−109.0	332.0	1043.0	3.15	(8.8)
Ethane	−127.5	236.0	675.8	2.87	(5.43)
Refrigerant 12	−21.6	26.51	107.4	4.08	(1)
Refrigerant 21	48.0	5.24	31.23	5.97	(0.32)
Refrigerant 22	−41.4	43.0	174.5	4.06	(1.6)
Refrigerant 113	117.6	0.982	7.856	8.02	(0.085)
Methyl chloride	−11.0	21.0	95.5	4.48	(0.92)
Sulfur dioxide	14.0	11.81	66.4	5.63	(0.67)

pared with −109 F for CO_2), and yet ethane requires a condensing pressure of only 675.8 psia at 86 F as compared with 1043 psia for carbon dioxide.

(*b*) *The boiling temperature-pressure relationship is not an equal numerical ratio for all fluids.* By developing a factor showing the difference in pressure requirements for each degree rise in boiling-point temperature, the characteristic difference in various fluids in this respect may be found readily. For example, ammonia requires a pressure of 34.27 psia to control its boiling point at 5 F, whereas it must maintain a pressure of 169.2 psia to control its boiling point at 86 F. This means an increase in pressure of 169.2 − 34.27, or 134.9 psi to raise its boiling temperature 86 − 5, or 81 F. For each degree rise in temperature, a

$$\frac{134.9}{81} = 1.67 \text{ psi}$$

rise in pressure is required to maintain ammonia in its liquid form. With Refrigerant 12, this factor becomes

$$\frac{81.4}{81} = 1 \text{ psi}$$

per 1 F temperature rise. Carrying out this procedure for each of the refrigerants listed in Table 12, a ready means of determining comparative condensing pressure-temperature relationships is established for various fluids. This factor is represented in parentheses in the table.

(*c*) *The condenser pressure depends on the refrigerant employed and under similar temperature conditions varies with each kind of fluid.* Comparing the liquefying pressure of CO_2 of 1043 psia, corresponding to a condenser temperature of 86 F, with that of 169.2 psia for ammonia and again with 107.4 psia for Refrigerant 12 under similar temperature conditions, is indicative of considerable differences in that quality for the various fluids. It is obvious that the strength of materials used in high-pressure side construction must meet the service demands of the refrigerant selected. By simple comparison of the three fluids mentioned above, materials employed in their high side construction would be classified as follows:

Carbon dioxide:	extra heavy construction materials
Ammonia:	heavy construction materials
Refrigerant 12:	medium construction materials
Sulfur dioxide:	light construction materials

By similar analysis, all refrigerants may be so classified.

COMPRESSION RATIO

The condenser pressure divided by the evaporator pressure is equal to the compression ratio. As the difference between these two pressures increases —i.e., with higher compression ratios—the power demand to operate the compressor becomes greater. The reason for this is twofold: (1) increased pressure range through which the piston must act and (2) losses in volumetric efficiency.*

Difference between the low and the high sides of the refrigeration system is governed by their respective temperatures and the boiling-point characteristics of the refrigerant employed. Where temperatures are fixed, as in the evaporator, or depend on surrounding atmospheric and condensing water temperatures prevalent in condenser operation, then the only means of controlling the compression ratio lies in the selection of the refrigerant. This assumes that single-stage operation is desirable.

Where all other factors are equally suitable for the service demands, the refrigerant having the lowest compression ratio should be selected. (See Table 12.)

COEFFICIENT OF PERFORMANCE

The coefficient of performance value of a refrigerant is a measure of its efficiency in utilizing the energy expended during compression in ratio to energy absorbed in the evaporator, and is one of the most important properties considered in selecting a refrigerant. The less energy that need be expended by the compressor, the higher will be the coefficient of performance of the system.

As the coefficient of performance (cop) value depends on the combined thermal and physical properties of a refrigerant, including refrigerating effect, specific volume, specific heat of the liquid and vapor, and compression ratio, it becomes the significant factor in predicting performance results of any fluid. Where other factors are equal, the refrigerant having the highest cop should be the one selected.

DENSITY

The resistance of a fluid to flowing through a tube or an orifice is an

* See Chapter 11, "Refrigeration Calculations."

indication of its density. This is true whether the fluid is in liquid or vapor form, the rule being that resistance to flow becomes greater as density increases. Thus, with refrigerants of high density, a more pronounced pressure drop occurs between two terminal points, using pipe of the same size and equal initial pressure, than would take place with fluids of lower density. In order to overcome this disadvantage, a larger size pipe is necessary to convey the heavier fluid.

In selecting refrigerants, however, the added expense of providing larger size pipe may often be counterbalanced by the greater refrigerating effect of some high-density fluids. It may be desirable, in certain instances, to select a refrigerant on the basis of its refrigerating effect, even though it may possess a higher density than some others. Where no thermal or physical advantage exists between any two fluids, the one of lower density should be selected in order to eliminate the necessity of providing larger piping in cases of equal load demands.

The drop in pressure caused by high-density refrigerants may be great enough in some instances to cause premature evaporation in the liquid line, thus interfering with the function of the throttling device by reducing its capacity to handle the required amount of liquid.

The resistance to flow on the part of a fluid through a pipe is measured in terms of pressure drop in psi. This value usually is expressed as pressure drop in psi per 100 ft of straight pipe or *equivalent*, meaning that a bend in the pipe, or a valve, or a series of valves and bends, will add to the resistance to flow equal to some definite number of feet of pipe length.

SPECIFIC HEAT OF LIQUID

The refrigerating effect (R E) of a fluid is equal to its latent heat of vaporization, less the thermal energy lost in cooling the liquid from its temperature upon entering the evaporator to the temperature corresponding to the pressure maintained in the latter unit. This loss in energy is determined by the fundamental heat equation,

$$\text{Btu} = \text{weight} \times \text{specific heat} \times (t_1 - t_2)$$

where t_1 = temperature of liquid entering evaporator
t_2 = temperature of evaporation corresponding to pressure in evaporator

Thus, the equation for the refrigerating effect of a fluid may be expressed as

$$\text{R E} = (\text{weight} \times \text{latent heat of vaporization}) - [(\text{weight} \times \text{specific heat}) \times (t_1 - t_2)]$$

It is obvious, from this equation, that the refrigerating effect varies as the specific heat, increasing in value as the latter decreases. For reasons of thermal economy, the specific heat of liquid refrigerants should be low.

SPECIFIC HEAT OF VAPORS

It is usual practice to allow superheating of vapors (about 10 F) to occur in the evaporator in order to prevent liquid refrigerant from entering the compressor. This reduces the amount of refrigerant handled by the compressor, effecting some volumetric loss as a result. Some of this loss is offset by the gain in kinetic thermal energy absorbed by the vapor during the process of superheating, the amount of which is governed by specific heat of the vapor. A thermal advantage is gained by employing refrigerants having high specific heat values in their vaporous form.

CRITICAL TEMPERATURE AND PRESSURE

The fluid selected as a refrigerant should be one whose critical temperature is safely above the highest temperature occurring within the condenser. Otherwise, liquefaction becomes impossible regardless of the pressure exerted. With the exception of carbon dioxide, the critical temperatures of all widely used refrigerants are well above the maximum temperatures found in normally operated systems.

Critical pressures of refrigerants also should be well above the maximum condenser pressure expected in order to assure liquefaction of the vapors at the condensing pressures. The amount of the condensing pressure is governed by the quantity and the temperature of the cooling medium.

Carbon dioxide again is an exception to all other refrigerants in that its critical pressure at the critical temperature is only slightly above the required condenser pressure of normally operated systems. The use of air, unless precooled, would be impractical for use as a cooling medium in condensing carbon dioxide vapors. This is especially true during summer months when the thermometer on numerous occasions registers several degrees above the critical temperature of that fluid, thereby making liquefaction impossible.

FREEZING POINT

While all refrigerants in wide use have freezing points well below normal evaporator temperatures of, say, −20 F to 10 F and above, ultra-low temperature requirements demand that greater attention be paid this thermal specification. The freezing point of a refrigerant should be safely below the lowest expected evaporator temperature.

CHEMICAL STABILITY AND EFFECT OF MOISTURE

Refrigerants should be of such chemical nature as to remain indefinitely unchanged by constant exposure to the highest discharge temperatures or the lowest evaporator temperatures prevailing within the system for which they are selected. They should also resist any chemical deterioration as a result of contamination, air, water, and oil being the greatest offenders.

The ideal refrigerant is one that is compatible with all metals considered practical for the construction of refrigeration equipment. While the majority of widely used refrigerants, in their pure state, have no deleterious effect on such metals, it requires but a small amount of moisture present to convert most of them into highly corrosive fluids.

For example, ammonia in its dry (anhydrous) state is not injurious to nonferrous metals such as brass and copper. When moisture is present, however, the resultant ammonia hydroxide formed is highly corrosive to such metals and to gamble with the combination would be to preclude all chances of moisture gaining entrance to the system, a risk not apt to prove very successful.

In dealing with refrigerants, it should be borne in mind that chemical reactions are always possible when the right combination of substances is present. This is because many of the important ones are made up of such elements as sulfur, chlorine, and fluorine, whose acids possess great affinity for all engineering metals with which they form salts. When contamination is slight, such reactions would remain negligible until such time as the opposing chemical forces neutralized each other. Due to the catalytic effect of some metals, even a slight reaction once initiated may be accelerated and expanded to a dangerous degree. This is particularly true where copper and lead are present.

Copper plating of cylinder walls, pistons, valves, and bearings is prevalent in those systems employing certain refrigerants in conjunction with evaporator and condenser coils made of that material. Copper plating is reduced considerably when properly refined oils are used for compressor lubrication accompanied by every precaution to eliminate moisture from the system.

REFRIGERANT-OIL RELATIONSHIP

The necessity of lubricating the frictional surfaces of the compressor, including bearings, pistons, cylinder walls, and valves, is the only reason for the presence of oil in the refrigeration system. Because the refrigerant and the oil ultimately become exposed to each other by either design or accident, it is essential that they be compatible, both chemically and physically, in order to avoid deterioration of either.

The ideal refrigerant is one that remains chemically stable in the presence of lubricating oil while it influences no chemical change within the oil.

An important physical quality affecting refrigerant-oil relationship, and one that differs with various refrigerants, is that of miscibility, or the ability of a refrigerant to mix with oil. Many refrigerants are miscible with lubricating oil in all proportions. A few mix with oil to a slight degree, while others do not mix at all. Several kinds such as Refrigerants 22 and 114 are miscible with oil under condenser pressure, but will separate when normal evaporator conditions are reached. This characteristic influences not only the operation of the refrigeration systems, but the design of the equipment as well.

Refrigerants that are considered immiscible or miscible to a slight degree with oil include ammonia, carbon dioxide, and sulfur dioxide. Those that are miscible are methyl chloride, ethyl chloride, methylene chloride, and members of the hydrocarbon and halogen groups. The degree of miscibility on the part of some of the latter group of refrigerants depends on the pressure and the temperature conditions becoming greater as pressure increases and temperature decreases.

The effect of miscibility is to decrease the viscosity of the oil; i.e., to render the lubricating oil "thinner" in body by dilution. Therefore, to counteract the effect of such dilution, it is necessary to employ heavier-bodied oils for the lubrication of compressor parts in those systems using miscible refrigerants than in the case of immiscible refrigerants. Miscibility also lowers the temperature at which an oil will congeal.

The question arises as to the advantages and the disadvantages of both types of refrigerants. This is best answered by first considering the effect of oil in the refrigeration system. Any oil in the system acts only to lower the operating efficiency, and must be regarded in the light of a "necessary evil."

Because oil has a tendency to congeal at lower temperatures, it forms a coating over the heat-transfer surfaces of the evaporator when carried to that area by the refrigerant. This coating acts as an insulation material, thereby reducing the heat-transfer capacity of that unit, which, in turn, decreases the load capacity of the entire system. Where this action occurs, it becomes necessary to provide a greater amount of cooling surface in the original design in order to counterbalance the loss in load capacity due to coating.

The presence of oil vapors in the system must displace some of the refrigerant vapors as it enters the compressor from the evaporator. For the compressor to handle sufficient refrigerant to maintain full load demands, it is necessary that the capacity of this unit be made larger than theoretically required to overcome the deficiency brought about by the presence of oil and/or its vapor.

When oil leaves the compressor to enter the other phases of the cycle, the refrigerant must carry it back where it may again perform its lubricating function. Otherwise, it devolves upon the designer to provide some means for removal of the oil before entering the evaporator. Here, again, it depends on the tendency of the refrigerant to mix with the oil as to which system the designer will select.

The lubrication of valves usually depends on the ability of the refrigerant to carry sufficient oil to their rubbing surfaces. If continuously operated dry, undue wear occurs, resulting in the necessity for constant replacement.

Separators and chillers are depended on to remove oil from the system in which immiscible refrigerants are employed. Such equipment acts to separate oil from the refrigerant by mechanical methods, through either centrifugal force, gravity, or precongealing, the latter being effected prior to the entrance of the refrigerant to the condenser. Even with the most efficient separation in use, some oil is carried to the evaporators, where its removal is accomplished by periodically allowing hot gases to flow directly from the compressor discharge via a by-pass line installed for the purpose. This "flushing" with hot refrigerant vapors also acts to defrost the cooling coils simultaneously with the removal of congealed oil within. This method of defrosting and oil removal is employed most generally in larger industrial installations, as it is in such systems that the use of immiscible refrigerants such as ammonia and carbon dioxide predominate. On smaller commercial and domestic units employing sulfur dioxide, some means of returning the oil to the crankcase from the entrance to the evaporator (float valve), where it floats on the liquid refrigerant, usually is provided. While sulfur dioxide is immiscible with oil, the description is more or less relative in respect to the same quality possessed by such highly miscible refrigerants as the Freon group. Actually, sulfur dioxide is partially miscible, and, as a result, compressor valves receive sufficient lubrication as a consequence of some oil carry-over in addition to that provided by the natural lubricating qualities of this refrigerant itself. However, the amount of miscibility of sulfur dioxide under ordinary temperatures does not materially reduce the viscosity of the oil.

With miscible refrigerants, lubrication is provided by the mixture of refrigerant and oil. While the mixture of oil with the refrigerant does not materially affect the boiling point of the latter, a reduction in the congealing temperature (pour point) of the oil occurs, an advantage not gained with immiscible refrigerants.

Another advantage of using miscible refrigerants lies in the elimination of oil deposits on evaporator surfaces, a condition brought about by the consequent foaming of the oil during refrigerant vaporization, a condition which also is responsible for the return of the oil to the compressor. Because of this latter action, it is not always necessary to provide elaborate oil return lines or oil separating devices, as required with systems employing immiscible fluids.

The problem of adequate valve lubrication is greatly overcome with the use of miscible refrigerants. The loss in volume of refrigerant handled by the compressor due to oil-vapor displacement remains one of the disadvantages presented by miscible refrigerants, a condition not so prevalent with immiscible types.

TOXICITY

Because of the possibility of leakage occurring in any refrigeration system, the effect on human life as a result of exposure to the escaping gases becomes an important factor to be considered when selecting a refrigerant. That property of fluids denoting their lethal tendencies is identified by the general term "toxicity," which implies that injury or death may be brought about by any one of several causes, including suffocation and poisoning.

TABLE 13. SOME REFRIGERANTS AND THEIR DEGREE OF TOXICITY (AS INDICATED BY THE NATIONAL FIRE UNDERWRITERS' GROUP NUMBERS)

Refrigerant	*Relative Toxicity: N.F.U. Group No.*	*Quantity and Time Required to Produce Serious or Fatal Results*	
		% by Volume	Time
Sulfur dioxide.......	1	0.7	5 min
Ammonia..........	2	0.5	30 min
Methyl formate.....	3	2.0–2.5	1 hr
Methyl chloride.....	4	2.5	2 hr
Ethyl chloride......	4	4	1 hr
Refrigerant 113......	4	5	1 hr
Dichloroethylene ...	4		
Refrigerant 21.......	Between 4 & 5	10	30 min
Methylene chloride..	4A	5	30 min
Carbon dioxide......	5	30	30 min–1 hr
Butane.............	5		
Propane............	5		
Refrigerant 11.......	5	10	2 hr
Refrigerant 22.......	5A	..	
Refrigerant 12.......	6	30	2 hr
Refrigerant 114......	6	21	2 hr
Refrigerant 503......	6		
Refrigerant 13.......	6		
Refrigerant 23.......	6		
Refrigerant 500......	5A		
Refrigerant 13B1....	6		

As all fluids with the exception of air are toxic, varying in degree according to the kind of substance, the term is a relative one, significant only when specified as to the amount required and the time necessary for exposure to impose permanent or fatal injury.

Such refrigerants as are incapable of inflicting serious injury except when present in the air in large quantities and for extended periods of exposure are rated as being mildly toxic or, for most practical purposes, "nontoxic." Others, that cause permanent injury or death in a matter of minutes with only small amounts present in the air, by similar standards must be considered highly toxic or lethal. Those in between the two extremes are rated according to their relative toxic effect by comparison.

Sulfur dioxide, for example, imposes serious injury on or causes the death of persons exposed for 5 min to air containing only 0.7 percent by volume of its vapor, thereby placing this refrigerant in the category of extreme toxicity. At the other extreme, and considered as being relatively nontoxic, are such refrigerants as Refrigerants 114 and 12, the latter requiring an exposure of 2 hr to concentrations of 30 percent and over in order to effect serious results.

If tests were made on all refrigerants to determine their relative toxicity and if then the refrigerants were listed according to their effect on human beings, as measured by the quantity and the time required to impose serious results, a basis for individual comparison would be established. Such tests have been made by the National Fire Underwriters and the results published on the more widely used refrigerants. Based on their degree of toxicity, the individual refrigerants are separated into groups, and each group is identified by numbers starting with group 1 through groups 2, 3, 4, 5, and 6, representing refrigerants in the order of descending toxic values. Under this arrangement, group 1 includes those refrigerants most highly toxic, and group 6 is the category of extremely mild or nontoxic fluids.

The degree of toxicity is a governing factor in the selection of a refrigerant, particularly for such fields as air conditioning, marine, hospitals, direct expansion cold-storage plants, or industrial systems. The reason lies in the hazardous condition brought about by exposing large groups of people to escaping toxic vapors in the event of leakage, or even a small number of persons confined to limited areas, either because of physical incapacity, as in hospitals, or due to the limited possibilities of escape, as aboard ship. As toxic refrigerants usually affect the palatable qualities of many items of food, their use in cold-storage systems should be limited to indirect cooling methods in which a secondary medium such as brine is employed.

Some compromise may be made by weighing the size of the refrigeration plant against the space and the conditions surrounding the installation. Sulfur dioxide presents minimum hazard when employed in small quantities, such as the amount required for domestic units or individual water coolers. However, it does impose a potentially dangerous condition if used in larger systems.

Carbon dioxide and Refrigerant 12 are particularly adaptable to marine service because of their relatively nontoxic and nonirritating properties, taking precedence over ammonia in this respect, even though space, an important consideration in ship design, could be saved because of the thermal advantage possessed by ammonia.

FLAMMABILITY

From the standpoint of safety, the ideal refrigerant should be nonflammable and nonexplosive. Here, again, the danger lies in the possibility of leaks occurring in the refrigeration system and causing a critical concentration of vapor in the surrounding air. A hazardous condition exists when flammable refrigerants are involved.

With the majority of commonly employed refrigerants, the danger of fires and explosions does not exist due to their nonflammability. Included in this group are the halogen group, sulfur dioxide, and carbon dioxide. Within certain limits of concentration, ammonia, methyl chloride, and ethyl chloride will burn, with only slight danger of their forming an explosive mixture.

Fluids possessing highly flammable and explosive characteristics are those belonging to the hydrocarbon family, such as methane, ethane, propane, and butane. Because of the extremely low boiling point of some hydrocarbons under atmospheric conditions, they are employed in systems of the ultra-low temperature class (evaporator temperatures of −125 F and below), and introduce a hazardous note in modern design.

Where highly explosive mixtures may occur, every precaution against leakage must be maintained. It is also necessary to provide well-ventilated space for equipment employing highly flammable fluids, even to the extent of installing exhaust fans to disperse vapors in case of serious leaks.

LEAK DETECTION

The ideal refrigerant should possess some characteristic that facilitates leak detection.

Leakage in and out of a refrigeration system is due to openings in the joints or flaws in construction materials, such as sand holes (improperly made castings), ruptures, and, in some instances, corrosion. They may occur anywhere in the system, but for obvious reasons are more prevalent in the high-pressure side. However, it is not unusual for leakage to occur in the evaporator. When this unit is submerged in brine, as is often the case, sometimes leak detection and location are difficult.

Some refrigerants have a tendency to leak more than others, which means that they are able to penetrate smaller openings. A dense fluid is less likely to leak than one of lower density. When a combination of high-discharge pressures and low-density refrigerants exists, the possibility of leakage is greatly increased. The same is true where evaporation pressures are considerably below atmospheric, a condition that favors inward rather than outward leaks due to partial vacuum.

Leakage of refrigerants must be guarded against continually. In the first place, sufficient loss of fluid decreases the load-carrying capacity of the system. Secondly, many of them are flammable or toxic when mixed with air—two characteristics that present a hazardous condition when gases are allowed to escape to the surrounding atmosphere. Lastly, the loss of refrigerant is costly and adds to the expense of operation.

Despite efforts to prevent leakage, it cannot be eliminated entirely. In the case of domestic units. hermetically sealed units have decreased the tendency, to a large extent, and are given rigid pressure tests before leaving the factory. A well-constructed industrial or commercial system is also tested thoroughly before the initial charge of refrigerant is installed. These tests are made largely by building up air pressure throughout the system and allowing it to set for at least 24 hours to note any pressure drop. Nevertheless, fluctuations in ambient temperature, internal pressure, and the like cause strains in the various pipe connections which ultimately cause openings to occur, with subsequent leakage.

Leak detection, therefore, becomes an important subject in the study of refrigeration, and is somewhat complex due not only to the various pressure conditions that exist in different type systems but also to the dissimilar nature of individual refrigerants. Hence, the same method cannot be applied to all refrigerants to determine whether, and if so, where, leakage is occurring.

Slow leakage inward continually tends to occur in subatmospheric pressure operation, as represented by the centrifugal-compressor type system. This condition requires the use of an evacuator or a pump to eliminate the air, which, if not removed, would considerably reduce the efficiency of the

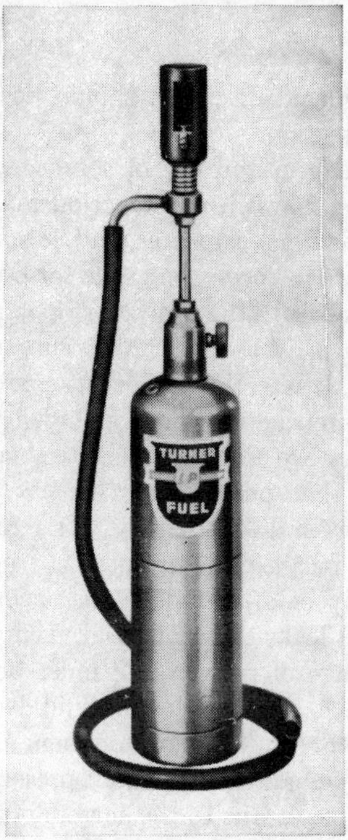

Fig. 27. Halide lamp for detecting leaks in Refrigerant 12 and other noncombustible halide refrigerant systems. (Courtesy, The Turner Brass Works.)

system. To localize the exact spot of any aggravated leakage, the use of oil or soapy water solution is necessary. When applied to a defective joint, the force of suction created by the partial vacuum within the system will draw such liquids inward and provide evidence of its existence. On positive or outward pressure systems, such leakage will be evidenced by bubbling of the applied solution or oil.

Another principle used in leak detection is chemical in nature, in which the refrigerant itself is exploited to cause a reaction with another compound. Burning sulfur will react with ammonia to form ammonium sulfite, a heavy white gas. Such a gas gives evidence of where the leak in an ammonia system is occurring when suspected connections are probed with a burning sul-

TABLE 14. METHODS OF DETECTING LEAKS IN REFRIGERATION SYSTEMS

Refrigerant	*Leak Detection Method*	*Reaction*	*Reference No. Table 27*
Refrigerants 11, 12, 13, 14, 21, 114; Kulene-131, ammonia, Carrene-7, sulfur dioxide, methyl chloride, ethyl chloride, iso-butane, butane, ethylamine, methylamine	Place system under high pressure using dry air (or inert gas) and submerge in lighted tank of clear water or kerosene (Small units)	Bubbles will emanate from leaking areas	1
Methane, ethylene, ethane, nitrous oxide, carbon dioxide, propylene, propane, methyl formate, methylene chloride, Freon-113, dichloroethylene, trichloroethylene, water	Place system under definite pressure using CO_2 or nitrogen (Large units)	Loss of gage pressure in 24 hr indicates leaks. Allow for changes in surrounding temperature	2
Refrigerants 11, 12, 13, 14, 21, 114; Kulene-131, ammonia, methyl chloride, ethyl chloride, iso-butane, sulfur dioxide, methylamine, butane.	Place system under low vacuum by vacuum pump. Shut pump down	Rise in vacuum gage after period of shutdown indicates leakage	3
Carbon dioxide, ethane, methane, propylene, propane, methyl chloride, ethyl chloride, nitrous oxide, ethylene	Soap and water solution (or heavy oil) swabbed over joints or other suspected areas	Bubbles form in leaky areas when system is under pressure of refrigerant or air. Solution is drawn into system under vacuum	4
Refrigerants 11, 12, 13, 14, 21, 114; methyl formate, Kulene-131, Carrene-7	Halide torch (using methyl alcohol)	Green flame	5
Ammonia—in air.	Sulfur candle	Forms ammonia sulfite, a white cloud	6
	Phenolphthalein paper	Paper turns pink	. .
	Odor	Pungent	. .
	See 1 and 3 also		. .
Ammonia—in brine or water. . .	Nessler's solution	Ammonia turns mixture yellow	. .
Sulfur dioxide.	A cloth soaked with 25% ammonia and water solution. See 1 and 3 also	Forms ammonia sulfite, a heavy white cloud	6
Methyl chloride.	1% acrolein added to refrigerant. See 1, 3, 4 also	Odor	. .

fur candle. The same principle is reversed for sulfur dioxide systems by probing the various parts with a cloth saturated with a 25 percent ammonia and water solution. Again, a white cloud will appear when the refrigerant and the saturated cloth are exposed to each other.

A hot copper element in the presence of a flame will cause the flame to turn a bright green to greenish purple if exposed to halogenated gases. This is the principle underlying the halide torch employed to detect leaks in systems using halide fluids. Flame testing should never be used around systems employing highly flammable refrigerants, such as ethylene, propane, and iso-butane.

Odor also serves as a warning that leaks are occurring, but, unfortunately, many of the refrigerants are practically odorless when mixed with air. About 1 percent of acrolein, an irritant, is sometimes added to methyl chloride, an anesthetic refrigerant, to serve as a leak-warning agent.

Phenolphthalein paper may be used in leak detection for ammonia systems. In the presence of ammonia or alkali, this paper, which is normally white, will turn pink. Nessler's solution is used to detect the presence of ammonia in water or brine after proper treatment of such liquids. Ammonia causes the mixture to turn yellow.

In addition to the above, there are several new proprietary leak detection methods available. One is an electric device capable of detecting the presence of various foreign gases in an air sample. It is based on variations in electronic resistance to current flow when such gases that are present become exposed to the sensor element within the instrument. Changes in current flow are indicated on a dial (milliammeter) or by the sounding of an alarm (bell).

Another method of a different type uses liquid chemicals that react with certain refrigerants and other types of gases to create a foam. The detector material (Leak Tec) is contained in a polyethylene dispenser that emits a small stream of detector fluid toward the leak-suspected area when squeezed by hand. There are several formulae available, each of which tests for a specific refrigerant or a specific group of related fluids, e.g., one for ammonia, and another for the halogenated group, such as the Freons and Genetrons. The manufacturers of these foam types of detectors list a number of physical advantages for their material, including non-toxic, non-residual, non-flammable, non-injurious, and non-effective with rubber or metals. They also guarantee the product's ability to detect leaks of such minute magnitude as 0.00001 standard cubic centimeter per second or 0.013 cubic foot per year.

Methods by which leaks may be detected for various refrigerants are shown in Table 14.

ODOR

Here, again, the possibility of leaks governs the selection of a refrigerant, making it necessary to consider the effect of odor on surrounding conditions. While the absence of odor would add to the dangers of a highly toxic refrigerant by making leak detection difficult, the presence of pungent vapors is also undesirable.

Among the widely used refrigerants, sulfur dioxide follows its toxic tendency by being the worst offender in respect to odor. When present in only small amount, it will cause severe irritation to mucous membranes. Ammonia is also an irritant having an offensive pungent odor when present in relatively small quantities.

Methyl chloride and methylene chloride have a slight sweetish odor not unlike that of chloroform, while the Freons are practically odorless when mixed with air but have a smell similar to that of carbon tetrachloride in concentrated form. Carbon dioxide is odorless, and, being reasonably harmless, does not present the danger to human life so prevalent with some odoriferous fluids whose only safety aspects lie in the warning the obnoxiousness provides.

Refrigerant odors affect the flavor of some foods, rendering many items unpalatable. This is particularly true where such odors are sharp and penetrating and leaks continue undiscovered for extended periods. Foods sensitive to odors, such as butter and meats, are especially affected as to taste when exposed to high concentrations of ammonia and sulfur dioxide. However, sulfur dioxide, not unlike the pure sulfur used in dried fruits, has certain food-preserving qualities, while ammonia causes rotting of meats and produce after long exposure.

Everything considered, refrigerants go from one extreme to the other where odor is concerned. Either they are practically odorless in air and not of sufficient strength to warn of their presence, or, like ammonia and sulfur dioxide, they are so strong as to be irritating. There seems to be no compromise.

COST AND AVAILABILITY

In small units, such as those found in commercial installations, the cost of refrigerant is not too important because of the small amount occasionally required for make-up in the event of leaks. This factor becomes magnified, of course, in the case of manufacturers producing a considerable number of such units where any appreciable difference in cost per pound between one refrigerant and another may mean an expenditure that directly affects profits. The cost of refrigerants also affects the expenses involved in operating large industrial systems where the leakage problem is greater by reason of size and number of pipe connections involved in its assembly.

It should be realized that cost cannot be based primarily on price per pound, but must be considered from the standpoint of over-all thermal properties and purity. Unless other factors enter into the selection of a refrigerant, the one that will absorb and reject heat at the lowest cost per Btu will prove to be the most economical, regardless of price per pound.

The refrigerant selected should be one immediately available in the event of a serious leak to avoid prolonged shutdown.

REFRIGERANT CHARACTERISTICS

There are a number of fluids available that possess some of the necessary qualifications to establish their adaptability as refrigerants. Because many of them present one or more undesirable characteristics that offset their better qualities, they must be eliminated for consideration as general-purpose refrigerants. The field of selection is thereby reduced to comparatively few fluids whose desirable characteristics are sufficiently predominant as to render their status as being the most important group of refrigerants available to the refrigeration industry.

Fluids in general used as refrigerants are representative of certain families of chemical compounds whose variations lie in the nature of the chemical elements of which they are composed. These elements differ for each family of refrigerants, resulting in their physical, chemical, and thermal property variations. Such variations in these properties also exist between refrigerants within the same family due to differences in the relative amount of each element constituting the fluid, or, in some instances, by the presence of an additional element.

Separated in family groups, the important refrigerants are listed below:

I. Ammonia
II. Carbon dioxide
III. Sulfur dioxide
IV. Hydrocarbon group:
  Ethane
  Ethylene
  Propane
  Iso-butane
  Butane
V. Halogenated or halocarbon group
  (*a*) Chlorinated hydrocarbon family:
    Methyl chloride
    Ethyl chloride
    Methylene chloride
    Dichloroethylene
    Trichloroethylene
  (*b*) Fluorinated hydrocarbon family (methane series):
    R-11—Trichloromonofluoromethane
    R-12—Dichlorodifluoromethane
    R-13—Monochlorotrifluoromethane
    R-14—Tetrafluoromethane
    R-21—Dichloromonofluoromethane
    R-22—Monochlorodifluoromethane
  (*c*) Fluorinated hydrocarbon family (ethane series):
    R-113—Trichlorotrifluoroethane
    R-114—Dichlorotetrafluoroethane
  (*d*) Bromines:
    Kulene-131—Bromotrifluoromethane
    R-13B1
VI. Azeotropes:
  R-500
  R-502
VII. Miscellaneous:
  Water vapor

Air
Nitrous oxide
Methyl formate
Ethylamine
Methylamine

VIII. Cryogenes

REFRIGERANT IDENTIFICATION

Evolution processes leave behind certain nomenclature, phrases and other identifying marks that continue to express specific definitions, describe certain materials or become synonyms for a given item or product. Classic results of this carry-over of words that have become synonymous with either an activity or a trademark, even though its reference may not be directed to its originator, include the following:

"Simonizing" is based on a proprietary name for a product used to polish the bodies of automobiles, etc., and is a verb still used to express that activity regardless of who manufactures the product.

"600-W" is a proprietary name of a specific petroleum cylinder oil and one of the oldest copywritten trademarks that identifies the product. The name is used today for the promiscuous identification of similar products regardless of their origin of manufacture.

There are two pertinent observations to be made regarding any such colloquialisms stated above:

First, that such colloquialisms must stand as a compliment to the originators of the product, and, secondly, that it does not preclude the possibility that those dealing in the same commodity to which the synonym is attached have failed to put out a product equal in quality to that of the originator.

This led to a similar situation in the field of refrigerants, particularly as it applies to the family of halogen hydrocarbons. For years, such refrigerants were universally identified as Freons, a proprietary tradename with individual fluids labeled by number, according to their properties, with the prefix F, such as F-12, F-22, etc. This was done primarily to avoid the necessity of employing their real names, which to say the least are long and cumbersome, e.g., dichlorodifluoromethane (F-12), monochlorodifluoromethane (F-22), etc. Because some of these halogen hydrocarbon refrigerants are produced by other manufacturers under their own brand names, such as Genetron, some universal means of identification became necessary. Therefore, all refrigerants, regardless of their basic chemistry were assigned numbers prefaced by the word "Refrigerant," or briefly by the letter "R." Reference is now made to the various refrigerants as Refrigerant 12, or R-12, Refrigerant 22, or R-22, etc. Where appropriate, their actual names are used as with ammonia, ethylene, propane, etc.

All the important refrigerants are listed in Table 15 showing their identifying numbers and their chemical symbols.‡

TABLE 15. STANDARD DESIGNATION OF REFRIGERANTS ADOPTED BY THE AMERICAN SOCIETY OF HEATING, REFRIGERATION AND AIR CONDITIONING ENGINEERS

Refrigerant Number	*Chemical Name*	*Chemical Formula*
Halocarbon Compounds		
10	Carbontetrachloride	CCl_4
11	Trichloromonofluoromethane	CCl_3F
12	Dichlorodifluoromethane	CCl_2F_2
13	Monochlorotrifluoromethane	$CClF_3$
13B1	Monobromotrifluoromethane	$CBrF_3$
14	Carbontetrafluoride	CF_4
20	Chloroform	$CHCl_3$
21	Dichloromonofluoromethane	$CHCl_2F$
22	Monochlorodifluoromethane	$CHClF_2$
23	Trifluoromethane	CHF_3
30	Methylene Chloride	CH_2Cl_2
31	Monochloromonofluoromethane	CH_2ClF
32	Methylene Fluoride	CH_2F_2
40	Methyl Chloride	CH_3Cl
41	Methyl Fluoride	CH_3F
50*	Methane	CH_4
110	Hexachloroethane	CCl_3CCl_3
111	Pentachloromonofluoroethane	CCl_3CCl_2F
112	Tetrachlorodifluoroethane	CCl_2FCCl_2F
112a	Tetrachlorodifluoroethane	CCl_3CClF_2
113	Trichlorotrifluoroethane	CCl_2FCClF_2
113a	Trichlorotrifluoroethane	CCl_3CF_3
114	Dichlorotetrafluoroethane	$CClF_2CClF_2$
114a	Dichlorotetrafluoroethane	CCl_2FCF_3
114B2	Dibromotetrafluoroethane	$CBrF_2CBrF_2$
115	Monochloropentafluoroethane	$CClF_2CF_3$
116	Hexafluoroethane	CF_3CF_3
120	Pentachloroethane	$CHCl_2CCl_3$
123	Dichlorotrifluoroethane	$CHCl_2CF_3$
124	Monochlorotetrafluoroethane	$CHClFCF_3$
124a	Monochlorotetrafluoroethane	CHF_2CClF_2
125	Pentafluoroethane	CHF_2CF_3
133a	Monochlorotrifluoroethane	CH_2ClCF_3
140a	Trichloroethane	CH_3CCl_3
142b	Monochlorodifluoroethane	CH_3CClF_2

‡ It may be added here that such tables and charts compiled using specifications of a proprietary brand of refrigerant will carry the appropriate identifying brand name. Such data are published only with the permission of the originators in most instances and the courtesy is appreciated in all cases by the author.

TABLE 15 (*Continued*)

Refrigerant Number	*Chemical Name*	*Chemical Formula*
143a	Trifluoroethane	CH_3CF_3
150a	Dichloroethane	CH_3CHCl_2
152a	Difluoroethane	CH_3CHF_2
160	Ethyl Chloride	CH_3CH_2Cl
170*	Ethane	CH_3CH_3
218	Octafluoropropane	$CF_3CF_2CF_3$
290*	Propane	$CH_3CH_2CH_3$
Cyclic Organic Compounds		
C316	Dichlorohexafluorocyclobutane	$C_4Cl_2F_6$
C317	Monochloroheptafluorocyclobutane	C_4ClF_7
C318	Octafluorocyclobutane	C_4F_8
Azeotropes		
500	Refrigerants 12/152a 73.8/26.2 wt %†	CCl_2F_2/CH_3CHF_2
501	Refrigerants 22/12 75/25 wt %	$CHClF_2/CCl_2F_2$
502	Refrigerants 22/115 48.8/51.2 wt %	$CHClF_2/CClF_2CF_3$
Miscellaneous Organic Compounds		
Hydrocarbons		
50	Methane	CH_4
170	Ethane	CH_3CH_3
290	Propane	$CH_3CH_2CH_3$
600	Butane	$CH_3CH_2CH_2CH_3$
601	Isobutane	$CH(CH_3)_3$
1150**	Ethylene	$CH_2{=}CH_2$
1270**	Propylene	$CH_3CH{=}CH_2$
Oxygen Compounds		
610	Ethyl Ether	$C_2H_5OC_2H_5$
611	Methyl Formate	$HCOOCH_3$
Nitrogen Compounds		
630	Methyl Amine	CH_3NH_2
631	Ethyl Amine	$C_2H_5NH_2$

* Methane, ethane, and propane appear in the Halocarbon section in their proper numerical order, but these compounds are not halocarbons.

** Ethylene and propylene appear in the Hydrocarbon section to indicate that these compounds are hydrocarbons, but are properly identified in the section Unsaturated Organic Compounds.

† Carrier Corporation Document 2-D-127, p. 1.

TABLE 15 (*Continued*)

Refrigerant Number	*Chemical Name*	*Chemical Formula*
Inorganic Compounds (Cryogenic)		
702	Hydrogen (Normal and Para)	H_2
704	Helium	He
720	Neon	Ne
728	Nitrogen	N
729	Air	$.21O_2$, $.78N_2$, .01A
732	Oxygen	O_2
740	Argon	A
Inorganic Compounds (Noncryogenic)		
717	Ammonia	MH_3
718	Water	H_2O
744	Carbon Dioxide	CO_2
744A	Nitrous Oxide	N_2O
764	Sulfur Dioxide	SO_2
Unsaturated Organic Compounds		
1112a	Dichlorodifluoroethylene	$CCl{=}CF_2$
1113	Monochlorotrifluoroethylene	$CClF{=}CF_2$
1114	Tetrafluoroethylene	$CF_2{=}CF_2$
1120	Trichloroethylene	$CHCl{=}CCl_2$
1130	Dichloroethylene	$CHCl{=}CHCl$
1132a	Vinylidene Fluoride	$CH_2{=}CF_2$
1140	Vinyl Chloride	$CH_2{=}CHCl$
1141	Vinyl Fluoride	$CH_2{=}CHF$
1150	Ethylene	$CH_2{=}CH_2$
1270	Propylene	$CH_3CH{=}CH_2$

STATUS OF REFRIGERANTS

While all the more important refrigerants are either described or listed herein, they are not all in use at the present time. Some of them have been eliminated because of their deleterious physical properties, such as toxicity, flammability, etc., or have been replaced by more efficient and desirable fluids.

Compression type fluids *in use* include the following:

Ammonia (717)
Refrigerants 12, 22, 500, 502, 11, 113, 114, 13, 13B1

Compression type fluids *in limited use* include the following hydrocarbon type fluids:

Propane (290), Methane (50),
Ethylene (1150), Propylene (1270),
Ethane (170), R C 318

Fluids used for special purposes, such as brine, diluents, degressing agents, etc., are:

Methylene chloride (30), Refrigerant 12, Trichloroethylene (1120)

Cryogenic fluids in use for various refrigeration purposes, i.e., cooling, metallurgical testing and processes, experimental work, etc., include:

Nitrogen (728), Hydrogen (702), Helium (704),
Neon (720), Oxygen (732), Argon (740),
Air (729), Krypton and Xenon

Ammonia

Ammonia possesses many of the advantages expected in the ideal refrigerant, and for this reason finds popularity with a large number of plant operators and equipment manufacturers. One of its widest uses is found in the cold-storage and the ice-manufacturing industries, where its excellent properties prevail in maintaining low production and maintenance costs. It is the oldest refrigerant now in common use.

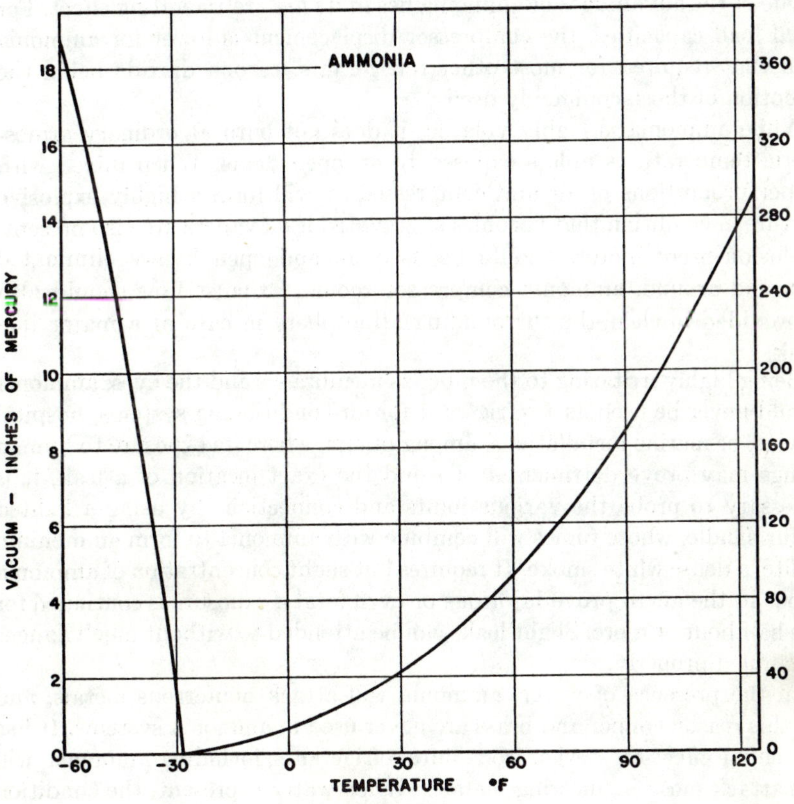

Fig. 28. Ammonia pressure-temperature chart.

The boiling point of ammonia under atmospheric pressure, being −28 F, enables evaporation and condensation to occur at moderately low pressures when normal refrigerating temperatures are maintained, thus eliminating the necessity of providing extra-heavy equipment. Its discharge temperatures are relatively high, requiring water cooling of the compressor cylinder head. Its gas is colorless and approximately one half as heavy as air.

Ammonia is chemically stable, and, except for leakage, an original charge will remain in good condition indefinitely. Anhydrous (dry) ammonia will not affect lubricating oil chemically, or reduce its viscosity, as the two fluids are practically immiscible. Where moisture is present, oil and ammonia will form an emulsion that will cause operating difficulties, the extent of which depends on the amount of water and the quality of the lubricating oil. Only anhydrous (below 0.5 percent moisture) ammonia must be used for refrigeration purposes, and systems in which it is employed should be maintained free of any water. As delivered, commercial anhydrous ammonia is limited to 0.01 to 0.02 percent water.

The high specific heat of liquid ammonia is offset to some extent by its high specific heat of vapor.

One of the advantages of ammonia lies in its high refrigerating effect. For equal load capacities, the compressor displacement is lower for ammonia than that required for most other refrigerants, carbon dioxide being the exception of those commonly used.

While ammonia is highly volatile, it does not burn at ordinary atmospheric temperatures unless exposed to an open flame. When mixed with proper proportions of air and compressed, it will form a highly explosive mixture, a condition that becomes aggravated if oil vapors are also present. Explosion-proof motors should be used in, and open flames eliminated from and around, ammonia compressor rooms. Exhaust fans should also be provided to clear the surrounding atmosphere in case of a major line break.

Being highly irritating to the mucous membranes and the eyes, ammonia should never be used as a refrigerant for air-conditioning systems, hospital service, or marine installations, among others, where its exposure to human beings may prove detrimental. To find the exact location of a leak, it is necessary to probe the various joints and connections by using a lighted sulfur candle, whose fumes will combine with ammonia to form ammonium sulfite, a dense white smoke. It requires but slight concentration of ammonia vapor in the air to prove injurious or even fatal if exposure is continued for one half hour or more. Slight leaks can be attended to without much danger if handled properly.

In the presence of water, ammonia will attack nonferrous metals, and for this reason copper and brass are never used in ammonia systems. It has but slight effect on steel or iron. Pure refrigerants, including ammonia, will not attack most engineering metals. When water is present, the condition is changed by reason of new chemicals formed due to the association. In the case of ammonia, the water forms ammonia hydroxide, a strong alkali.

The practical low limits for evaporator temperatures using ammonia lie between −80 F and −90 F, which correspond to pressures of 2.74 psia and 1.86 psia, respectively.

Carbon Dioxide

The use of carbon dioxide as a refrigerant demands extra-heavy piping and equipment by reason of the high pressures encountered, in both the evaporator and the condenser, which, under standard operating conditions, average 331 psia and 1043 psia, respectively. It also requires more power per ton of refrigeration to operate the compressor as compared with all other refrigerants, hence larger prime movers must be provided for equal load capacities.

Theoretically, the critical temperature of carbon dioxide is 87.8 F. This means that low-temperature condenser water should be provided in order to liquefy it, especially when surrounding atmospheric temperatures are above the critical point of this fluid. However, cooling water temperature of 90 F has been used successfully with carbon dioxide systems.

Because of its low boiling-point characteristics, carbon dioxide proves excellent as an extremely low-temperature refrigerant. A study of its properties in the tables, "Properties of Liquid and Saturated Carbon Dioxide," will show that an evaporation temperature of −110 F may be obtained at a pressure approximately that of the atmosphere, or 14.67 psia. It will not remain a liquid when pressures are below its triple-point pressure of 75.1 psia. This is due to evaporation occurring so rapidly as to reduce the temperature of liquid below that of freezing, forming what is virtually snow, or solid carbon dioxide—a product popularly known as "dry ice." Under atmospheric pressure, solid carbon dioxide sublimes, or passes directly to a vapor without first liquefying. This occurs due to the considerable drop in temperature within the solid as a result of the heat absorbed in the evaporation process under the reduced pressure. The temperature of solid carbon dioxide or "dry ice" reaches temperatures of −110 F to −114 F under atmospheric conditions, depending on prevailing pressures, air movement, etc.

Carbon dioxide is 1.53 times heavier than air, tending to displace the latter rather than rise above it. It dissolves in water, increasing in solubility as pressures become greater. It forms a weak carbonic acid, an unstable compound that decomposes when pressure is released. Soda and carbonated water are made by forcing carbon dioxide gas through water, to which it imparts a slight acid taste and a "lively" appearance. Being odorless and colorless, leaks are difficult to detect. Except for the use of soap-and-water solutions, which produce bubbles when applied to the defective spot in the system, there are no other simple means of leak detection.

This refrigerant is nontoxic in the sense that it is nonpoisonous, but will cause death by suffocation when present in large concentrations. Because it is also nonflammable, it possesses those advantages that have made it widely acceptable for marine and air-conditioning systems. Neither does it support combustion, but rather will extinguish fires by forming a blanket because of its weight advantage relative to air, preventing the access of oxygen to the flames. For this reason it is used in some types of fire extinguishers.

Carbon dioxide is chemically stable under all conditions of temperature and pressure changes occurring within the refrigeration system. While remaining unaffected by the presence of water, air, or numerous other foreign substances, these impurities should be eliminated from the system in the interest of efficient operation. Being immiscible with lubricating oils, with which it is also compatible, no allowance need be made in oil viscosity by reason of dilution. It does not affect any of the metals used in constructing refrigeration equipment, copper, brass, and iron included.

While the refrigerating effect per pound appears to be low, the small specific volume of carbon dioxide requires relatively smaller cylinder displacement than other refrigerants for similar load capacities.

Sulfur Dioxide

Sulfur dioxide is a colorless, suffocating fluid having a relatively higher boiling point than either ammonia or carbon dioxide, and therefore is liquefied more readily. It is over twice the weight of air, which it tends to displace.

Vapors of sulfur dioxide are highly soluble in water, with which it forms sulfurous acid, a compound that changes to sulfuric acid when exposed to oxygen in the air. This acid is highly corrosive to most metals.

Sulfur dioxide is nonflammable and will not support combustion at ordinary temperatures. Leakage is not detrimental to most food products, but will act as a bleach and preservative when applied to dried fruits during their processing.

Since sulfur dioxide is only partially miscible with lubricating oil, it does not affect its viscosity, and for all practical purposes may be regarded as being immiscible.

It is not compatible with all types of lubricating oil, however, acting as a solvent in many instances to precipitate sludge that tends to obstruct valves and passages of refrigeration systems.

Compared with other refrigerants, sulfur dioxide has a low refrigerating effect, and, coupled with a high specific volume, requires a relatively large piston displacement for equal refrigeration capacities, being 2.3 times greater than is necessary for ammonia and about 8.5 times that for CO_2 compressors.

Its high toxicity and irritating effect on mucous membranes and eyes, even when present in small quantities, eliminate all consideration for its use in any but small domestic-type systems, a service in which it remains chemically stable throughout the range of pressures and temperatures prevailing.

A partial vacuum (5.9 in. mercury) must be maintained to produce a standard evaporator temperature of 5 F with sulfur dioxide. Its advantage as a refrigerant for small household units lies in its low condenser pressure of 66.5 psia at 86 F. It has been replaced by halocarbon refrigerants such as R-12 in modern domestic refrigerators.

Leaks become evident immediately by reason of its strong pungent odor. Defective spots are located by applying a swab saturated with aqueous ammonia until a dense white fog, ammonium sulfite, appears as a result of contact with sulfur dioxide.

The use of sulfur dioxide as a refrigerant demands that the system be kept free of moisture to prevent the formation of acids (generally sulfurous acid) that creates a harmful effect on metal parts. Free of moisture, sulfur dioxide does not affect steel, iron, copper, brass, or any of the bearing metals. It is adaptable to air-cooled condensers.

Hydrocarbon Group

The hydrocarbons are a group of colorless fluids, normally gaseous, made up in varying numerical combinations of the two elements, hydrogen and carbon. They are derived from several sources, the most important of which are petroleum or natural gases. While some of the hydrocarbon gases listed below are not employed commonly for general refrigeration purposes, they are mentioned here as a matter of completeness and to indicate their refrigerant possibilities.

Of those listed, iso-butane has been considered of some importance for small domestic refrigerators; it is known to the trade as Freezol. Butane, iso-butane, propane, and ethane are used to a limited extent, the latter being adaptable for extremely low temperatures, a purpose for which ethylene is also suitable. Ethylene has been adopted to act as the low temperature refrigerant of cascade systems* employed for the storage of liquid natural gas.

The hydrocarbon group of fluids are flammable and highly explosive when mixed with air. They are not poisonous, but are anesthetic in varying degrees, from ethane and propane, which are mild in this respect, to ethylene, a gas noted for its highly anesthetic characteristics. They will not affect metals and are miscible with lubricating oil in all proportions. Leaks from systems employing any one of the hydrocarbon series are detected by the bubble method, using soap-and-water solutions. They all affect rubber.

Refrigerants as represented by the hydrocarbon group may be divided further according to their chemical structure and their behavior with other elements. Ethylene and propylene are the first and the second gases of a

* See Chapter 16.

series of hydrocarbons known as olefins, a group of unsaturated compounds that are chemically reactive. Methane, ethane, propane, and butane are saturated compounds not readily affected by reagents, belonging to the paraffin series of hydrocarbons, being the first, second, third, and fourth members, respectively, of that series. While some of the properties of the different gases are influenced by the series of hydrocarbons to which they belong, their boiling points are not.

The boiling point of each gas is governed by the number of hydrogen and carbon atoms that compose one of its molecules, and, as a result, this property is related directly to the molecular weight of the individual fluid. It is also related directly to the specific weight of the vapor as borne out in

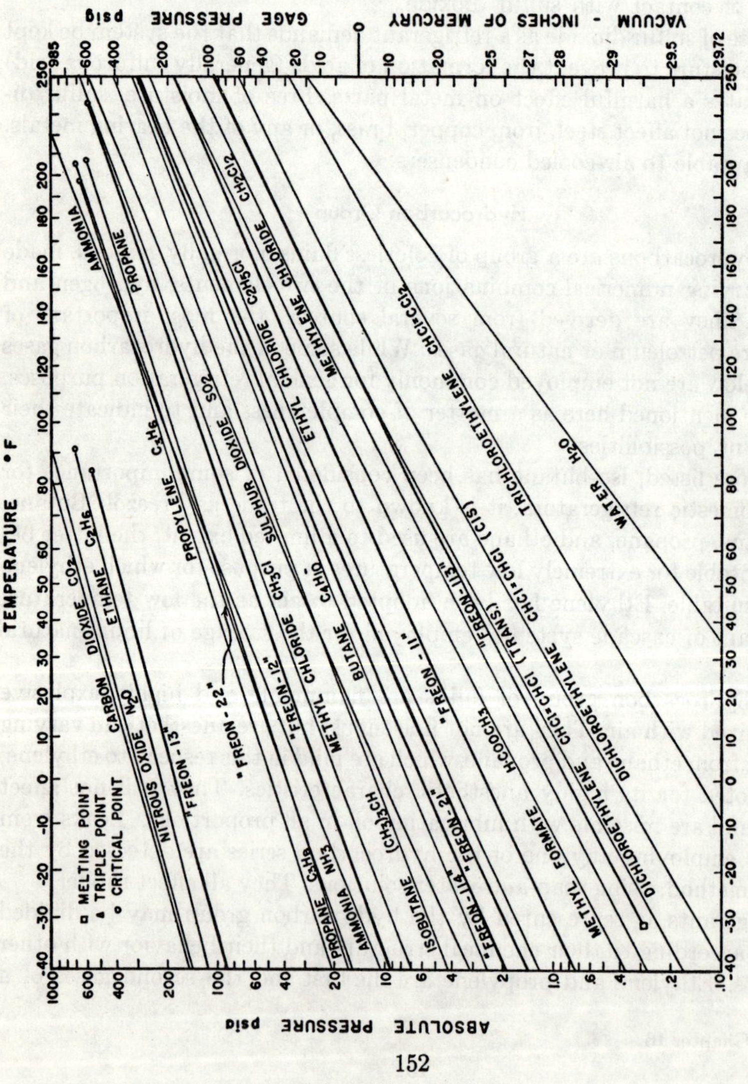

Fig. 29. Pressure-temperature relationships of various refrigerants. (Courtesy, E. I. duPont de Nemours & Company, Inc. Reprinted by permission.)

Table 16. Some Characteristics and Chemical Structure of Hydrocarbon Refrigerants

Refrigerants		*Chemical Formula*	*Molecular Constituents*	*Molecular Weight*	*Specific Gravity of Vapor Air = 1*	*Boiling Point °F 1 atm*
Paraffinic series:	**Olefinic series:**					
Methane		CH_4	1 Carbon atom 4 Hydrogen atoms	16.03	0.544	−259
	Ethylene	C_2H_4	2 Carbon atoms 4 Hydrogen atoms	28.03	0.9749	−155
Ethane		C_2H_6	2 Carbon atoms 6 Hydrogen atoms	30.04	1.05	−128
	Propylene	C_3H_6	3 Carbon atoms 6 Hydrogen atoms	42.05	..	−54
Propane		C_3H_8	3 Carbon atoms 8 Hydrogen atoms	44.06	1.562	−44
Iso-butane		C_4H_{10} $(CH_3)_3CH$	4 Carbon atoms 10 Hydrogen atoms	58.12	2.067	10.3
	Butylene	C_4H_8	4 Carbon atoms 8 Hydrogen atoms	..	..	20
Butane		C_4H_{10}	4 Carbon atoms 10 Hydrogen atoms	58.12	2.085	31.3

Table 16, in which the various members of this group are listed according to their ascending boiling point. It also will be noted that listing hydrocarbons according to their ascending boiling points automatically places the gas of lightest weight first and the heaviest one last.

While butane and iso-butane have the same chemical formula, C_4H_{10} (i.e., a similar number of carbon and hydrogen atoms in one molecule of both gases), their boiling points differ. This is due to a difference in the way the carbon and the hydrogen atoms are bonded together (chemical structure, as shown in parentheses for iso-butane) in the two fluids that accounts for their dissimilar properties.

As stated previously, natural gas is the greatest source of hydrocarbon refrigerants, and found in that product in the following proportions according to a typical analysis:

Methane—79%

Ethane —13%

Propane — 5%

Butane — 2%

The remaining 1 percent is made up of a higher boiling-point fluid that is ordinarily a liquid under atmospheric conditions and is identified as pentane. To distinguish the difference between various fluids of the hydrocarbon family, the boiling-point characteristics of each must be understood. Fluids, such as methane, ethane, propane, and butane, as refrigerants must be considered as "pure" products containing only those constituents in the correct proportions that identify them as such. Therefore, methane is not "pure" if it contains any ethane, nor is propane if it contains butane, etc. One of the results of such a mixture would be found in its boiling point, which would be representative of a range of temperatures rather than a specific temperature, as shown in Table 16 for the individual fluids. This range, having an initial boiling point governed by the most volatile fluid and an end point or final boiling temperature influenced by the least volatile fluid of the mixture, is representative of what is expected in some hydrocarbon liquids, such as gasoline, kerosine, and fuel oils. In these products, a boiling-point range is advantageous and purposely established by blending several fluids having different boiling points, such as pentane, hexane, heptane, octane, etc., to list a few in their descending order of volatility. However, such blends cannot be used in refrigeration service.

These same requirements apply to the olefins, particularly ethylene, the more important one of this series from the standpoint of refrigeration. Ethylene is produced as a result of decomposing high boiling-point hydrocarbons through several stages of petroleum processing, known as "cracking."

Ethane

As a member of the hydrocarbon group, ethane is highly flammable and explosive when mixed with air. It has a pungent odor and will attack rubber seals or gaskets, although it does not affect metals. Because of its low boiling point (−127.3 F at 1 atm), it is employed as an ultra-low temperature refrigerant, usually in the lower stage of cascade systems.

Its coefficient of performance is low, however, possessing only approximately half the thermal qualities of ammonia but approaching those of carbon dioxide. The specific heat of liquid ethane exceeds that of both ammonia and carbon dioxide, being 1.140 as compared with 1.119 and 0.765 for the latter two, respectively.

Because of the close relationship of ethane and lubricating oil, it is often necessary to employ a mixture of glycol and graphite for compressor lubrication in order to assure protection against wear of the moving parts.

Ethylene

This member of the hydrocarbon group of refrigerants is highly anesthetic. Like ethane, it also possesses an extremely low boiling point (−154.7 F at 1 atm), and therefore employed for ultra-low temperature requirements, usually in conjunction with other refrigerants of higher boiling points, in cascade systems.

The problem of lubrication with this fluid is similar to that prevalent with all members of the hydrocarbon group, caused by their great affinity for oil.

In some instances, ethylene absorbs the protective film of oil before it has an opportunity to perform its function. For this reason, glycol mixed with graphite has been the lubricant for ethylene compressors.

With the exception of Freon-14, it requires higher evaporator and condenser pressures corresponding to any given temperature than all other refrigerants. To reach a temperature of −165 F, it is usual practice to employ ethylene in the lowest pressure stage with Freon-12 and Freon-22, or equivalent fluids, in the second and the third stages, respectively, of a three-stage cascade system. Ethylene is also used as the low temperature refrigerant in cascade systems along with propane in maintaining natural gas in its liquid state during storage. It has a 5 to 1 compression ratio, and a coefficient of performance of 2.57.

Ethylene is highly flammable and forms an explosive mixture with air.

Propane

The third member of the paraffinic series of the hydrocarbon group, propane, has a boiling point of −44 F, higher than ethane and propylene and lower than butane or iso-butane. It has a lower compression ratio than ammonia and its normal operating pressures in both evaporator and condenser are not radically different from those of ammonia, being 155.3 psia and 42.1 psia, respectively, under standard temperature conditions. The refrigerating effect per cubic foot of piston displacement, being 48.7 Btu, is higher for propane than for many of the more commonly used refrigerants, including sulfur dioxide, methyl chloride, and Freon-12.

Propane has an affinity for lubricating oil, and the same care must be taken with cylinders, pistons, and rings of compressors handling this fluid as with those in which other members of the hydrocarbon group are employed. It is used as a selective solvent in the solvent refining of some lubricating oils.

Its use as a booster for manufactured illuminating gas in maintaining a constant heat value per cubic foot serves as an excellent example of the flammability and the explosive nature of propane. For this reason, its utilization as a refrigerant is not extensive, being limited to smaller units, or to a special purpose such as the storage of natural gas mentioned above.

It attacks rubber, but does not affect metals. Propane is anesthetic and practically odorless.

Iso-butane

Butane and iso-butane are made up of the same number of carbon and hydrogen atoms, but have different chemical structures. While both compounds have the general formula, C_4H_{10} (i.e., four carbon atoms and ten hydrogen atoms in each molecule), the structural formula of iso-butane is written $(CH_3)_3CH$. Chemically, iso-butane is defined as an *isomer*, or a compound containing the same elements and in the same proportion as another compound, but whose chemical and physical properties differ due to a change in their chemical structure.

Iso-butane as a refrigerant has several advantages over butane, including a lower and more suitable boiling point at atmospheric pressure, lower compression ratio, and greater vapor density, which permits an increase in refrigerating effect per cubic foot of piston displacement.

Under a standard condition of 5 F evaporator temperature, a partial vacuum, 13.1 psia, is necessary, but, because of its relatively low condenser pressure, 59.5 pisa, liquefaction may be accomplished with air cooling.

Iso-butane is flammable when mixed with air, is miscible with oil, is anesthetic, and has a slight sweetish odor. It does not affect metals, but does attack rubber, a common trait of all hydrocarbon refrigerants. It is suitable for small domestic refrigerating units and is known to the trade for this purpose under the name of Freezol.

Butane

This refrigerant has the highest boiling point of any member of the hydrocarbon refrigerant group, being 31.1 F at atmospheric pressure. Butane is suitable for domestic service from the standpoint of evaporator and condenser pressures. At 5 F evaporator temperature, it is necessary to operate under partial vacuum, 8.2 psia, which corresponds to 13.2 in. of mercury below atmospheric pressure. A relatively low condenser pressure of 41.6 psia is required for butane under the standard temperature of 86 F. It may be cooled with air to liquefy at that temperature and pressure.

Although butane possesses some of the desirable qualities for a refrigerant suitable for normal temperature requirements, its highly flammable and explosive nature prohibit its use except in limited circumstance. Butane is anesthetic, a characteristic that is made more serious because it is practically odorless.

Butane is often identified by the letter "N" and written butane-N, meaning normal butane. This is to distinguish it from iso-butane, sometimes written butane-iso.

Halogenated Group

Chlorinated Hydrocarbon Family

The chlorine group of halogenated hydrocarbon refrigerants is composed of those fluids made up of the elements carbon and hydrogen, in addition to one or more atoms of chlorine. Compared with the straight hydrocarbon group, the properties of the chlorinated fluids differ, tending toward an improvement in several respects.

Of the five in this group, only two are used widely; namely, methyl chloride and methylene chloride. The remaining three have been of more or less importance, especially in the field of air conditioning, and for that reason are included in the discussion here.

In the order of their ascending boiling points, those refrigerants included in the chlorinated group are shown in Table 17. Note that the evaporator pressures for the refrigerants listed in Table 17, with the exception of methyl chloride, are below that of the atmosphere; also that methylene chloride, dichloroethylene, and trichloroethylene function under a partial vacuum in both the evaporator and the condenser, making them particularly suitable for use in centrifugal compressors.

TABLE 17. CHLORINATED HYDROCARBONS

Refrigerant	*Boiling Point °F 1 atm*	*Condenser Pressure 86 F psia*	*Evaporator Pressure 5 F psia*
Methyl chloride.......	−11.3	95.5	21.0
Ethyl chloride.........	53.9	27.1	4.76
Methylene chloride....	104.6	10.05	1.18
Dichloroethylene......	122.4	6.97	0.827
Trichloroethylene......	187.0	1.82	0.156

Methyl Chloride. A colorless anesthetic, methyl chloride has a slight odor not unlike that of chloroform and is hardly distinguishable when mixed with air. For that reason, an irritating substance, such as acrolein (1 percent), is added to give an odoriferous warning in case of leaks. It is flammable, and, when present in the air at concentrations of 8.2 percent to 17 percent by volume, becomes explosive.

Methyl chloride will not attack commonly used engineering metals if kept free of water, but, when moisture is present, hydrochloric acid may form and cause corrosion of both ferrous and nonferrous metals. Even in the presence of small amounts of moisture, aluminum, zinc, and magnesium are attacked and should never be employed in methyl chloride systems. It dissolves rubber.

The lubrication of compressors handling methyl chloride presents two difficult problems. While it is miscible with oil and serves to decrease the viscosity of the latter substance, a factor common with several refrigerants, it also acts as a selective solvent, causing the precipitation of paraffinic (waxy) compounds, which often clog the expansion valves. The second difficulty lies in the copper-plating action that occurs, not only in methyl chloride, but also in other halogenated hydrocarbon systems as well. The answer to both problems lies in the use of properly refined oils.

Methyl chloride meets a number of the thermal and physical requirements demanded of an ideal refrigerant, whatever its shortcomings otherwise. Having a boiling point of −10.8 F under atmospheric pressure, it has a compression ratio of 4.55, and for standard ton conditions of 5 F and 86 F, a corresponding pressure of 21 psia in the evaporator and 95.5 psia in the condenser is required. This means an outward or positive pressure on both sides of the system. Air-cooled condensers can be used with methyl chloride.

Compared with other widely used low-pressure refrigerants, such as Refrigerant 12 and sulfur dioxide, the refrigerating effect of methyl chloride is slightly higher at 148.7 Btu per lb. It has a coefficient of performance of 4.72, which is equal to the first-named fluid and slightly below the second. A piston displacement of 6.09 cu ft per min per ton for methyl chloride compares with 5.81 cu ft for Refrigerant 12 and 9.08 cu ft for sulfur dioxide.

Ethyl Chloride. Ethyl chloride possesses certain physical disadvantages that account for its limited use as a refrigerant.

It is a colorless liquid having a relatively high boiling point of 54 F at atmospheric pressure, which necessitates negative-pressure operation in the evaporator side of the system. As a low-pressure refrigerant, coupled with its anesthetic flammable characteristics, and requiring large piston displacements in cubic feet per minute per ton of refrigeration, its counterparts free of most such disadvantages are readily available.

Ethyl chloride presents lubrication difficulties similar to those of methyl chloride, including the problem of copper plating. Systems using this fluid must be maintained in a dry condition in order to prevent the formation of acids that become highly corrosive to metals. In its dry state, ethyl chloride will not affect metals, but will attack and dissolve rubber.

Under standard ton conditions of 5 F and 86 F, a negative evaporator pressure of 4.76 psia and a positive condenser pressure of 27.1 psia are required.

Methylene Chloride. Variously referred to as dichloromethane, or Carrene-1, methylene chloride is a liquid under atmospheric pressure having a normal boiling point of 103.6 F. It is practically nonflammable (smothering its own flame) and nontoxic. Water has no effect on it at ordinary temperatures.

Being a low-pressure refrigerant, as indicated by its boiling point, methylene chloride is employed in conjunction with centrifugal compressors. Because of its safety features, this refrigerant has been used extensively in the air-conditioning field, being particularly adaptable to large installations such as theaters, auditoriums, and office buildings.

Under standard ton conditions, both the low- and the high-pressure sides of the methylene chloride system operate under partial vacuum, being 10.05 psia (8.4 in. of mercury) in the condenser (86 F), and 1.18 psia (27.4 in. of mercury) in the evaporator (5 F). It has a fairly high coefficient of performance at 4.87 and a compression ratio of 8.5, which is not unusually high for low-pressure type refrigerants. It requires a piston displacement of 74.3 cu ft per min per ton of refrigeration.

Methylene chloride does not affect metals, but will dissolve rubber. That it is miscible with oil is of little importance generally as there is no contact between the two in well-sealed centrifugal compressors, the lubricant not ordinarily circulating with the refrigerant.

Dichloroethylene. Known also as dielene, dichloroethylene is, like ethylene chloride, a colorless, low-pressure refrigerant. However, unlike its counterpart, dichloroethylene is flammable, and hence limited in its use as a refrigerant, especially since the centrifugal type of compressor in which it must be handled generally is adapted to large industrial and air-conditioning systems. At atmospheric pressure, it has a boiling point of 118 F.

The dichloroethylene system operates under negative pressure in both the evaporator and the condenser, being 0.827 psia and 6.97 psia, respectively, at standard ton conditions. It has a coefficient of performance of 4.83 and a compression ratio of 8.42. It requires a piston displacement of 111.2 cu ft per min per ton of refrigeration.

Dichloroethylene is miscible in oil, but lubrication is not affected by the refrigerant for the same reason cited in the preceding section relating to ethylene chloride. It is mildly corrosive to nonferrous metals, such as copper and aluminum, but does not affect iron or steel when dry. In the presence of water, dichloroethylene attacks most metals.

This refrigerant was replaced by methylene chloride for most refrigerant purposes, which in turn is giving way to R-11 and R-113.

Trichloroethylene. Trielene, as this refrigerant is often called, has the highest boiling point corresponding to any given pressure of all other refrigerants, water excepted. Under atmospheric pressure, its boiling point is 187 F. For standard ton conditions, pressures in both evaporator and condenser must be maintained at an exceptionally low level, amounting to 0.157 psia (29.6 in. of mercury) in the former, and 1.82 psia (26.2 in. of mercury) in the latter. This means that refrigeration systems using trielene must be kept perfectly tight on both high- and low-pressure sides in order to prevent the entrance of air and moisture. As these systems are usually comparatively large ones of the industrial type, constant care in maintaining seals is necessary. This applies to all systems operating under negative pressure on one or both sides, whether the refrigerant is trielene or any of the other low-pressure fluids. Evacuation of air is generally required before starting such systems, during which a partial vacuum is pulled. The vacuum is slightly above the boiling-point pressure of the refrigerant, corresponding to the surrounding temperature.

Trichloroethylene is a heavy, colorless liquid at atmospheric pressure, nonflammable and nonexplosive. It has no effect on metals, and, as it is not readily soluble in water, the presence of moisture is of little concern except for operating reasons.

Its ratio of compression is higher at 11.65 than that of any other refrigerant, including those of low-pressure classification. Its required piston displacement is extremely high, necessitating the handling of 502.0 cu ft per min per ton, having a low vapor density at 5 F of 0.0044 lb per cu ft. More efficient fluids have gradually replaced trielene as a refrigerant.

Fluorinated Hydrocarbon Family

Methane Series. This group of halogen compounds is the result of an effort to produce fluids that more nearly meet the demands of the ideal refrigerant. This family of refrigerants also supplies a product for each level throughout the entire temperature range, from ultra-low to high boiling-point requirements.

The halocarbons are normally nontoxic, nonirritating, and nonflammable, and, when employed under reasonably dry conditions, are noncorrosive. Because they are colorless and practically odorless, leaks are difficult to detect or locate, requiring a halide lamp for the purpose. All members of this family are chemically stable under temperature conditions far beyond those encountered within normally operated refrigeration systems. They do not affect nor are they affected, chemically, by clean and highly refined lubricating oil, although they vary as to their degree of miscibility with the latter. The degree of oil miscibility depends on the individual fluid and the conditions of temperature and pressure under which they operate.

Halocarbons are divided into two series: (1) the methane series and (2) the ethane series. Properties of the individual fluids in both series are controlled by the replacement of hydrogen atoms with halogens, either chlorine or fluorine, or both.

For example, the hydrocarbon compound methane is made up of one carbon atom and four hydrogen atoms, chemically expressed as CH_4. By substituting one chlorine atom for one hydrogen atom, a new compound is formed having one carbon atom, three hydrogen atoms, and one chlorine atom, expressed as CH_3Cl. By continuing the process of hydrogen replacement with chlorine, a new group of compounds is formed having a progressively descending hydrogen content and a corresponding increase in chlorine content, respectively expressed as CH_2Cl_2, and finally CCl_4.

If one atom of fluorine were added as a replacement for either a hydrogen or a chlorine atom in the foregoing compounds, an additional group would be formed; i.e., CH_4 would change to CH_3F, CH_2Cl_2 to CH_2ClF, $CHCl_3$ to $CHCl_2F$, and CCl_4 to CCl_3F. By adding two atoms of fluorine, the compounds would again change respectively to CH_2F_2, $CHClF_2$, and CCl_2F_2. With the substitution of three fluorine atoms, two additional compounds would be formed as CHF_3 and $CClF_3$, and, by replacing all hydrogen and chlorine atoms with fluorine, the result would be CF_4.

As the replacement of hydrogen and/or chlorine continues as indicated, the properties of the resultant fluids also change. Methane, the basic fluid, has an extremely low boiling point (−260 F), but, with each additional

* See Refrigerant Identification, Chap. 5.

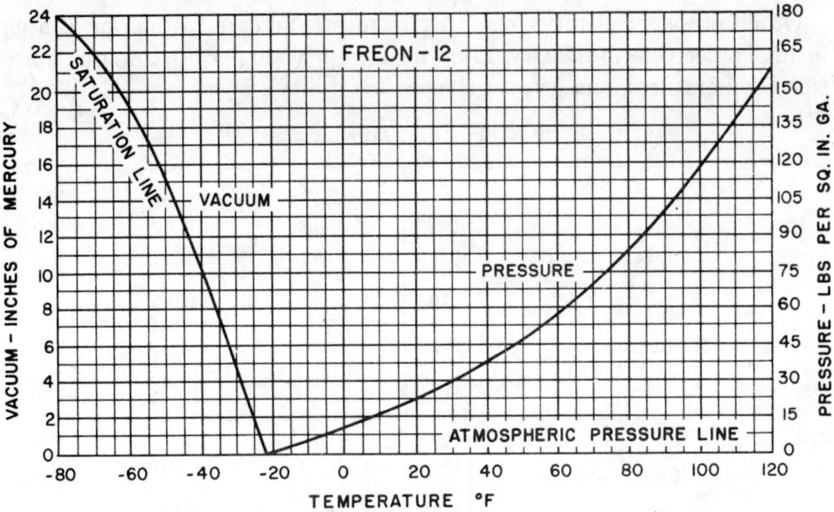

Fig. 30. Freon-12 pressure-temperature chart. (Courtesy, Worthington Corporation.)

atom of chlorine, the boiling point of the resultant compound increases until it reaches a maximum with CCl_4 of 170 F in which all the hydrogen atoms have been replaced. Decreasing flammability also follows an increase in hydrogen content.

The replacement of hydrogen and/or chlorine with fluorine also causes changes in fluid properties. Again using the boiling point as an example, it will be noted in Table 22 that the temperature of boiling decreases correspondingly with each additional atom of fluorine, reaching a minimum of −198 F for CF_4.

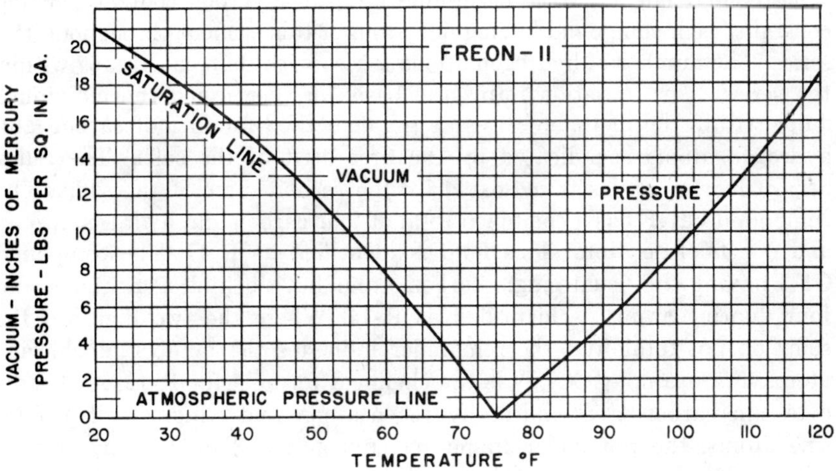

Fig. 31. Freon-11 pressure-temperature chart. (Courtesy, Worthington Corporation.)

Other property changes resulting from an increase in fluorine content include a decrease in toxicity, lower heat value (Btu per lb), lower compression temperatures, and greater chemical stability.

TABLE 18. METHANE SERIES*

	Freons	*Chemical Formula*	*Boiling Point °F*
No hydrogen	**10**	CCl_4	170
	F-11	CCl_3F	74.7
	F-12	CCl_2F_2	−21.6
	F-13	$CClF_3$	−114.5
	F-14	CF_4	−198.2
1 Hydrogen atom	**20**	$CHCl_3$	142
	F-21	$CHCl_2F$	48.0
	F-22	$CHClF_2$	−41.4
	F-23	CHF_3	−119.9
2 Hydrogen atoms	**30**	CH_2Cl_2	104
	F-31	CH_2ClF	16
	F-32	CH_2F_2	−61
3 Hydrogen atoms	**40**	CH_3Cl	−11
	F-41	CH_3F	−109
Methane	**50**	CH_4	−260

* Courtesy, Thompson, R. J.: "The 'Freon' Family." By permission of *Air Conditioning and Refrigeration News.*

It will be noted that there are a total of 5 atoms in each of the methane series compounds as in the basic fluid CH_4. The carbon content remains constant, each compound having one atom of this element throughout the series. The number of hydrogen atoms varies with different groups within the series. This variation forms the basis for identifying the individual fluids within the methane series as the combined number of carbon and hydrogen atoms is indicated by the first digit of the suffix. The number of fluorine atoms is indicated by the second digit of the suffix. The compound CH_3F has a combined total of four carbon and hydrogen atoms and one fluorine atom. This fluid is identified as R-41. The compound CF_4, having the usual single carbon atom but no hydrogen atoms and four fluorine atoms, is identified as R-14. Where the two digits in the suffix do not total five, the remainder indicates the number of chlorine atoms. For example, CCl_2F_2 is the chemical formula for R-12. As 1 indicates one carbon atom and no hydrogen atoms and 2 indicates two fluorine atoms, the remainder found by subtracting 3 or $(1 + 2)$ from 5 indicates the number of chlorine atoms, or two.

Table 18 shows the chemical construction of the methane series of halocarbons as well as their boiling points. The adaptation of the letter F to signify the proprietary name Freon, a standard procedure, is made in the table.

Of the methane series R-11, R-12, R-13, R-14, R-21, and R-22 are the more widely used. Methane is also used to a limited extent, but is a member of the hydrocarbon group of refrigerants.

TABLE 19. HALO-FLUORO DERIVATIVES OF METHANE*

Series No. *Formula* *Boiling Point*	*Freon Number* *Formula* *Boiling Point—1 atm*			
10 CCl_4 170 F	Freon-11 CCl_3F 74.7 F	Freon-12 CCl_2F_2 −21.6 F	Freon-13 $CClF_3$ −114.5 F	Freon-14 CF_4 −198.2 F
20 $CHCl_2$ 142 F	Freon-21 $CHCl_2F$ 48.0 F	Freon-22 $CHClF_2$ −41.4 F	Freon-23 CHF_2 −119.9 F	X
30 CH_2Cl_2 104 F	Freon-31 CH_2ClF 16F	Freon-32 CH_2F_2 −61F	X	X
40 CH_3Cl −11 F	Freon-41 CH_3F −109 F	X	X	X
50 CH_4 −260 F	X	X	X	X

* Courtesy, Thompson, R. J.: "The 'Freon' Family." By permission of *Air Conditioning and Refrigeration News.*

Ethane Series. The second series of halocarbons are halogenated compounds having ethane as the basic fluid, and carry three digits as their identifying suffix. R-113 and R-114 are the commercially more important members of this group.

Ethane is made up of two carbon atoms and six hydrogen atoms, expressed as C_2H_6, a member of the hydrocarbon family, as is methane. The Freons in this group continue to carry two carbon atoms throughout the entire series, but, as in the methane series, chlorine or fluorine, or both, replace hydrogen in various amounts to form the many saturated compounds possible with this combination.

The chemical formula of Freons within the ethane series may also be identified by their numerical suffix, in which the last digit signifies the number

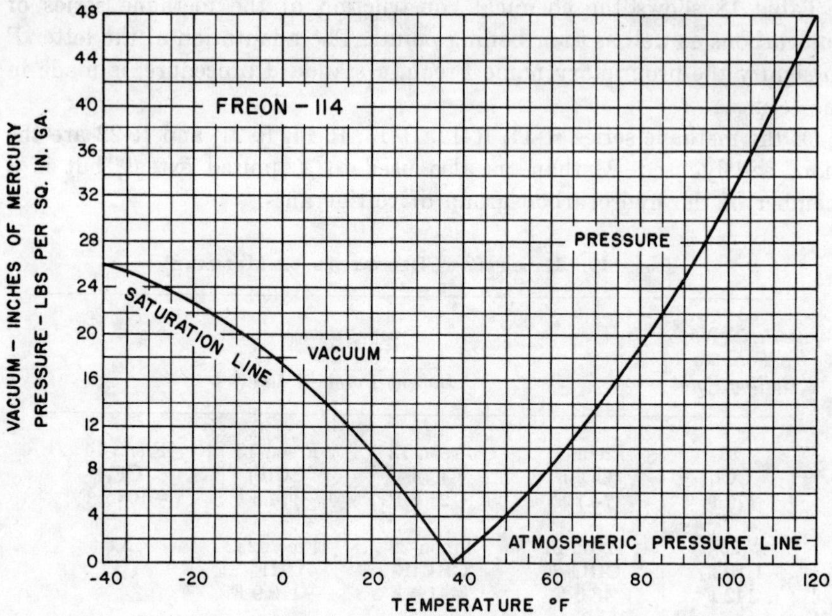

Fig. 32. Freon-114 pressure-temperature chart. (Courtesy, Worthington Corporation.)

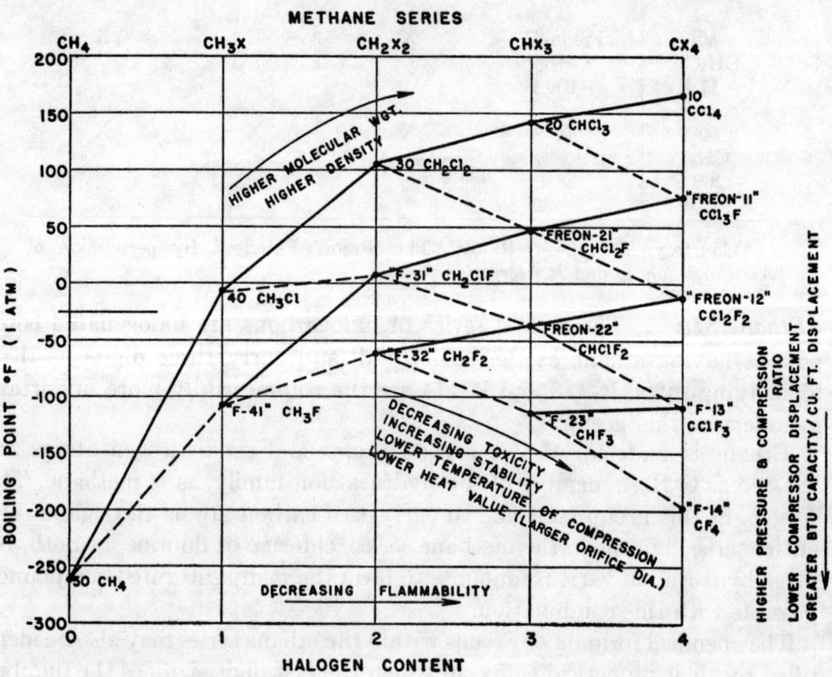

Fig. 33. Development of methane series. (Courtesy, *Air Conditioning and Refrigeration News.*)

TABLE 20. ETHANE SERIES*

Freons	*Chemical Formula*	*Boiling Point °F*
110	C_2Cl_6	Sublimes
F-111	$C_2Cl_5F_1$	279
F-112	$C_2Cl_4F_2$	199
F-113	$C_2Cl_3F_3$	117.6
F-114	$C_2Cl_2F_4$	38.4
F-115	C_2ClF_5	−36
F-116	C_2F_6	−109
120	C_2HCl_5	324
F-121	C_2HCl_4F	242
F-122	$C_2HCl_3F_2$	161
F-123b	$C_2HCl_2F_3$	82
F-124a	C_2HClF_4	14
F-125	C_2HF_5	−55
130c	$C_2H_2Cl_4$	295
F-131c	$C_2H_2Cl_3F$	217
F-132c	$C_2H_2Cl_2F_2$	140
F-133c	$C_2H_2ClF_3$	63
F-134c	$C_2H_2F_4$	−9
140a	$C_2H_3Cl_3$	236
F-141a	$C_2H_3Cl_2F$	165
F-142a	$C_2H_3ClF_2$	95
F-143a	$C_2H_3F_3$	41
150a	$C_2H_4Cl_2$	183
F-151a	C_2H_4ClF	95
F-152a	$C_2H_4F_2$	−13
160	C_2H_5Cl	56
F-161	C_2H_5F	−35
170	C_2H_6	−128

* Courtesy, Thompson, R. J.: "The 'Freon' Family." By permission of *Air Conditioning and Refrigeration News.*

of fluorine atoms in the molecule of the various fluids. Table 21 shows the Freon derivatives of ethane according to series numbers, giving their chemical formulas and boiling points. Note that the hydrogen content, while it remains constant for fluids within the same group, increases for each group according to descending series numbers.

Table 20 lists 21 of the possible 45 saturated compounds that may be derived from ethane by the substitution of chlorine or fluorine, or both, for hydrogen. Additional saturated compounds may be formed by making fluorine substitutions for hydrogen or chlorine in the first or the second radi-

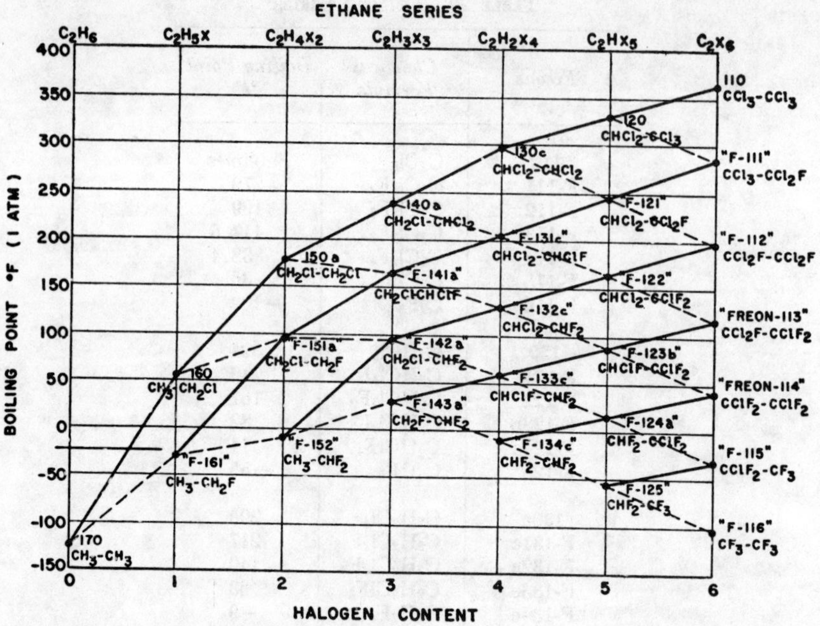

Fig. 34. Development of ethane series. (Courtesy, *Air Conditioning and Refrigeration News.*)

TABLE 21. HALO-FLUORO DERIVATIVES OF ETHANE*

Series No. *Formula* *Boiling Point*	*Freon Number* *Formula* *Boiling Point—1 atm*					
110 C_2Cl_6 Sublimes	Freon-111 C_2Cl_5F 279 F	Freon-112 $C_2Cl_4F_2$ 199 F	Freon-113 $C_2Cl_3F_3$ 117.6 F	Freon-114 $C_2Cl_2F_4$ 38.4 F	Freon-115 C_2ClF_5 −36 F	Freon-116 C_2F_6 −109 F
120 C_2HCl_5 324 F	Freon-121 C_2HCl_4F 242 F	Freon-122 $C_2HCl_3F_2$ 162 F	Freon-123b $C_2HCl_2F_3$ 82 F	Freon-124a C_2HClF_4 14 F	Freon-125 C_2HF_5 −55 F	X
130c $C_2H_2Cl_4$ 295 F	Freon-131c $C_2H_2Cl_3F$ 217 F	Freon-132c $C_2H_2Cl_2F_2$ 140 F	Freon-133c $C_2H_2ClF_3$ 63 F	Freon-134c $C_2H_2F_4$ −9 F	X	X
140a $C_2H_3Cl_3$ 236 F	Freon-141a $C_2H_3Cl_2F$ 165 F	Freon-142a $C_2H_3ClF_2$ 95 F	Freon-143a $C_2H_3F_3$ 41 F	X	X	X
150a $C_2H_4Cl_2$ 183 F	Freon-155a C_2H_4ClF 95 F	Freon-152 $C_2H_4F_2$ −13 F	X	X	X	X
160 C_2H_5Cl 56 F	Freon-161 C_2H_5F −35 F	X	X	X	X	X
170 C_2H_6 −128 F	X	X	X	X	X	X

* Courtesy, Thompson, R. J.: "The 'Freon' Family." By permission of *Air Conditioning and Refrigeration News.*

cals to produce symmetrical or asymmetrical arrangements. This is shown below for R-132, which also explains the suffix letters in the table.

R-132a: $CCl_2F—CH_2F$

R-132b: $CClF_2—CH_2Cl$

R-132c: $CHF_2—CHCl_2$

R-132d: $CHClF—CHClF$

While all four fluids may be expressed generally as $C_2H_2Cl_2F_2$, their properties may vary in comparison with each other, as is the rule for all isomers.

Because refrigerants in this group possess cumbersome chemical names, they are identified universally according to their Refrigerant number.

TABLE 22. SERVICE CLASSIFICATION OF FREON REFRIGERANTS

Refrigerant	*Chemical Formula*	*Boiling Point °F 1 atm*	*Temperature Service Range*
Tetrafluoromethane (Freon-14)—F-14	CF_4	−198.2	Ultra-low
Monochlorotrifluoromethane (Freon-13)—F-13	$CClF_3$	−114.5	Ultra-low
Monochlorodifluoromethane (Freon-22)—F-22	$CHClF_2$	−41.44	Low
Dichlorodifluoromethane (Freon-12)—F-12	CCl_2F_2	−21.6	Low to int.
Dichlorotetrafluoroethane (Freon-114)—F-114	$C_2Cl_2F_4$	38.4	Int.
Dichloromonofluoromethane (Freon-21)—F-21	$CHCl_2F$	48.0	Int.
Trichloromonofluoromethane (Freon-11)—F-11	CCl_3F	74.7	High
Trichlorotrifluoroethane (Freon-113)—F-113	$C_2Cl_3F_3$	117.6	High

Table 22 lists the more commonly used halocarbon refrigerants, showing their chemical names, identifying numbers, chemical formulas, boiling points, and the temperature service range to which they are adaptable. They are listed in the order of their boiling points rather than according to identifying numbers.

Bromines. Another group of halogen refrigerants is the bromines, of which Kulene-131, a proprietary name for bromotrifluoromethane, CF_3Br, is the more widely known.

Kulene-131 is ethereal in odor, nontoxic, nonflammable, and, under dry operating conditions, noncorrosive. It is slightly soluble in lubricating oil and can be used in systems constructed of any of the standard engineering

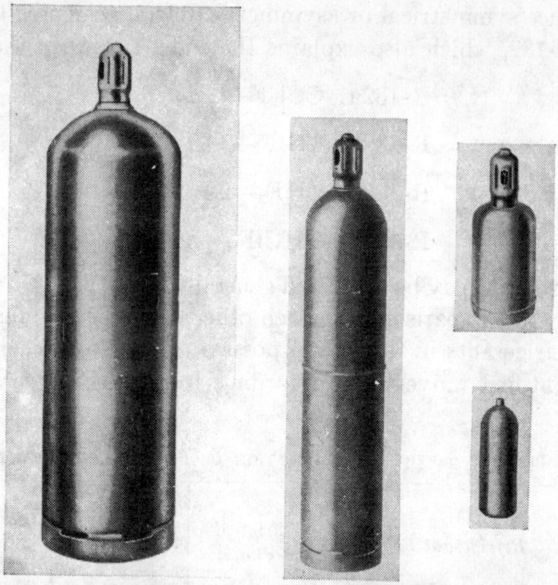

Fig. 35. Refrigerant containers. (Courtesy, Pressed Steel Tank Company.)

metals, including copper. Because of its basic characteristics and low boiling point, Kulene-131 is suited for small-size low-temperature equipment.

While Kulene-131 has a comparatively lower compression ratio (3.36 under standard ton conditions), its coefficient of performance is lower (4.07) than other refrigerants in the same operating category. Kulene-131 also has a relatively lower refrigerating effect, which for small units is not disadvantageous because of the closer control that may be effected. This is due to the slower response in temperature changes in relation to the quantity of liquid flowing through the throttling device. Its power requirements are greater at 1.160 hp per standard ton than other comparable fluids.

Monobromotrifluoromethane—(*R-13B1*). Refrigerant 13B1 (R-13B1) is another bromine type fluid having the chemical symbol $CBrF_2$. It has a boiling point of -72 F at atmospheric pressure 14.7 psig. It is nonflammable, rated in the Underwriters Laboratories Group Classification 6 so that it is similar to R-12 and R-114 in this respect. Under standard ton conditions of 5 F evaporator temperature and 86 F condenser temperature, its corresponding pressure is 63.2 psig and 247.1 psig, respectively. R-13B1 has a net refrigerating effect of 29.3 Btu per lb when circulated at a rate of 6.86 lbs per minute.

Its compression ratio of 3.36 and required horsepower of 1.030 compare favorably with other refrigerants. R-13B1 is recommended for use at temperatures between those at which R-22 and R-13 are used. All values are expressed for standard ton conditions.

Chloropentofluoroethane—(*R-115*). One of the halohydrocarbon family of fluids of the ethane series R-115 has the chemical symbol $CClF_2CF_3$. It has a boiling point of −37.7 F at atmospheric pressure. R-115 is combined with R-22 to form R-502, an azeotrope.

Azeotropes

Azeotropic fluids are formed by the mixture of two or more different liquids which, in combination, act as one compound. They cannot be separated by distillation. A distinguishing feature of this type of fluid is its ability to maintain permanent composition and uniform boiling point during evaporation—essential properties of a refrigerant. The thermal properties of azeotropes differ from either of the individual liquids used in their composition. In some cases, this difference results in an improvement in thermal qualities. If the liquid improved upon is a refrigerant that normally would be used under a given set of conditions, it becomes advantageous to employ the azeotropic fluid which it helps form.

As an example, the refrigerant R-500, an azeotropic refrigerant made up of 73.8% of R-12 and 26.2% of an asymmetrical difluoroethane,* R-152a has a lower boiling point (−28 F) than R-12 (21.6 F), and a theoretical refrigeration capacity 17 percent to 18 percent greater than the latter fluid for the same displacement and temperature range. This means that more capacity within the limits of 17 percent to 18 percent can be obtained with the same equipment without increasing speeds by substituting R-500 for R-12.

This capacity increase in thermal qualities of an azeotrope over its refrigerant constituent leads to other possibilities of gaining capacity in an otherwise static operating situation. In addition to increasing the capacity of existing installations using R-12, it meets other conditions, such as (1) compressor sizes, (2) speeds, (3) motor speed reductions by reason of lower electrical frequencies, and (4) boiling points.

1. Where estimated loads fall between the capacity ratings of two compressors, the larger of which is rated one step higher than the other, the smaller of the two may be employed by substituting R-500 for R-12. Thus, a gap between standard compressor sizes is filled without the necessity of using the next larger unit. Considering that an 18 percent increase in capacity is possible by using R-500, a 50-hp compressor may be increased to

$$\frac{50}{100 - 18} = 60.98 \text{ hp}$$

2. Compressors driven by directly connected electric motors lack speed flexibility, and, in this respect, are unlike belt-driven units whose speeds can be varied according to pulley size ratios to meet the power of the motor.

* Genetron-100, manufactured by General Chemical Div., Allied Chemical Co.

TABLE 23. PROPERTIES AND PERFORMANCE DATA
R-500, R-502 AND R-13B1

Properties	*R-500 (Azeotrope)*	*R-502 (Azeotrope)*	*R-13B1*
Chemical symbol	CCl_2F_2/CH_3CHF_2	$CHClF_2/CClF_2CF_2$	$CBrF_2$
Molecular weight	90.29	111.64	148.9
Boiling Point °F at 1 atm	−28	−50.1	−72
Freezing Point at 1 atm	−254	—	−270
Critical temperature	221.1	194.1	152.6
Critical pressure	631.0	618.7	575.0
Evaporation pressure 5 T	16.4	36.0	63.2
Condenser pressure 86 F	113.4	175.1	247.1
Ratio of compression $\frac{\text{psia}}{\text{psia}}$ at 86 F and 5 F	4.12	3.75	3.36
Net refrigerating effect, Btu/lb	61.1	45.7	29.3
Refrigerant circulated/ton, lbs/min	3.27	4.38	6.86
Specific volume, cu ft/lb	1.52	.82	.38
Compressor displacement/ton, cfm	4.97	3.61	2.63
Horsepower/ton	1.022	1.079	1.030
COP	4.61	4.37	4.25
Compressor discharge, temp. °F	105	99	124.0
Compressor type	Recip. & Rotary	Recip.	Recip.
Underwriters Lab. Class. No.	5a	—	6
Relative toxicity	Non-toxic	Non-toxic	Non-toxic

R-500 supplies the additional capacity to meet inflexible speed conditions in the former type of unit.

3. The speed of alternating-current motors depends on the number of poles in the field winding and the frequency of voltage supplied to those windings, or

$$\text{rpm} = \frac{Kf}{P}$$

where K = constant. (The value of K is 120 where frequency is expressed in cycles per second and speed in revolutions per minute)
f = frequency in cycles per second
P = number of poles

For a four-pole motor operating with a 60-cycle power supply, the speed would be

$$\text{rpm} = \frac{120 \times 60}{4} = 1800 \text{ rpm}$$

If the same motor operated under a frequency of 50 cycles, its speed would be reduced to 1500 rpm, a reduction of 16.7 percent. The additional capacity of R-500 over R-12 would compensate for the speed deficiencies of the 50-cycle power supply.

4. A lower evaporator temperature can be obtained with R-500 than with R-12 without changes in condensing temperature or power requirements.

R-500 does not support combustion and has toxic characteristics similar to those of the halocarbons generally. R-500 exerts a pressure of 31.1 psia, corresponding to a temperature of 5 F and 128.12 psia at 86 F. Refrigerant R-500 was formerly described as Carrene 7.

Azeotrope 22/15—(R-502). Another azeotropic fluid composed of 48.8% of R-22 and 51.2% of R-115, by weight. It has a boiling point of −50.1 F at atmospheric pressure. Under standard ton conditions of 5 F evaporator temperature and 86 F condenser temperature, its corresponding pressures are 36.0 psig and 175.1 psig, respectively. R-502 has a refrigerating effect of 45.7 Btu per lb when recirculating at the rate of 4.38 lbs per minute. It requires 1.079 horsepower per ton and a compression ratio of 3.75, all values expressed for standard ton conditions.

The capacity of R-502 is higher than R-22 with discharge temperatures comparable to R-12 in compressors of the reciprocating type. This latter is particularly true at temperatures down to −20 F in single stage compression systems. R-502 has the chemical symbol $CHClF_2/CClF_2CF_2$.

Genetron 23/13 Azeotrope. Given the designation R-503, Genetron 23/13 Azeotrope is a relatively new refrigerant composed of 40.1% R-23 and 59.9% R-13, by weight, and is useful for temperatures in the −100 F range. It has a boiling point of −127.6 F at 14.7 psia and a latent heat of vaporization of 74.2 Btu/lb. The specific heat of the liquid is .29 Btu/lb/F and a liquid density of 79.2 lbs/cu ft, both of which are reported for −30 F. Because of its low boiling point, it can be employed for low operating temperatures without the necessity of utilizing negative suction pressures. Compression ratios of R-503 are lower than either R-13 or R-23 under corresponding suction and condensing temperatures.

The manufacturers of Genetron 23/13 Azeotrope claim that with the relatively lower compression ratio and improved volumetric efficiency attained with this refrigerant, it will extend the capabilities of two-stage cascade systems to lower temperatures than have heretofore been practical, providing evaporator and condenser temperature remain constant. It follows that for any fixed compression ratio and condenser temperature, a lower evaporator temperature can be attained. Genetron 23/13 Azeotrope provides faster temperature reduction (pull-down), will handle larger loads and allow greater reduction in equipment size as compared to either R-23 or R-13, the refrigerants usually selected for similar service.

A cascade system comprising R-12 or R-22 in the upper stage and R-503 in the lower stage offers an efficient combination. Such a system is applicable to processes such as metallurgical, high vacuum (sublimation), food and beverage, pharmaceutical (penicillin, test units, etc.).

Some of the property values and other pertinent data relative to R-503, and comparative values of R-13 and R-23, are listed below.

TABLE 24. PROPERTIES OF R-503, R-13 AND R-23

Properties	*Genetron 23/13 Azeotrope (R-503)*	*Genetron 13 (R-13)*	*Genetron 23 (R-23)*
Chemical formula	$CClF_3/CHF_3$	$CClF_3$	CHF_3
Molecular weight	87.5	104.5	70
Boiling point °F at 14.7 psia	−127.6	−114.6	−115.7
Critical temperature °F	67.1	83.9	78.1
Critical pressure, psia	607	561	701.4
Latent heat of vaporization, Btu/lb	74.2	63.9	103.1
Liquid density at −30 F/lbs/cu ft	79.2	82.4	77.8
Specific heat of liquid at −30 F, Btu/lb/F	0.29	0.24	0.34
Comparative performance at −30 F Condensing/−120 F Evaporating Saturated Cycle Condition			
Evaporating pressure	3.75 psig	4.51″Hg	3.94″Hg
Condensing pressure, psig	131.6	90.9	111.3
Compression ratio	7.93	8.46	9.90
Volume of suction vapor, cu ft/lb	2.19	2.68	3.93
Coefficient of performance	2.96	2.89	2.86
Horsepower per ton	1.59	1.63	1.65
Refrigerant circulated per ton, lbs/min	3.75	4.50	2.63
Compression displacement per ton, cu ft/min	8.20	12.10	10.32
Liquid circulated per ton, cu in./min	81.5	94.2	58.3
Discharge temperature °F	28	14	65
Underwriters Lab classification (toxicity)	6	6	6
Flammability and explosivity	None	None	None

Properties of R-503 at saturation are shown in Table 25.

A new fluorocarbon refrigerant Genetron 23/13 is suitable for use in the low stage of cascade systems has been developed by Allied Chemical Corporation's General Chemical Division. This new refrigerant is useful for temperatures in the −100 F range. There is a growing demand for equipment at these temperatures, such as in environmental chambers, metal treating units, laboratory storage and test units, chemical and pharmaceutical processing equipment, high vacuum equipment, etc. Cascade systems normally use the common refrigerants, such as Genetron 12®

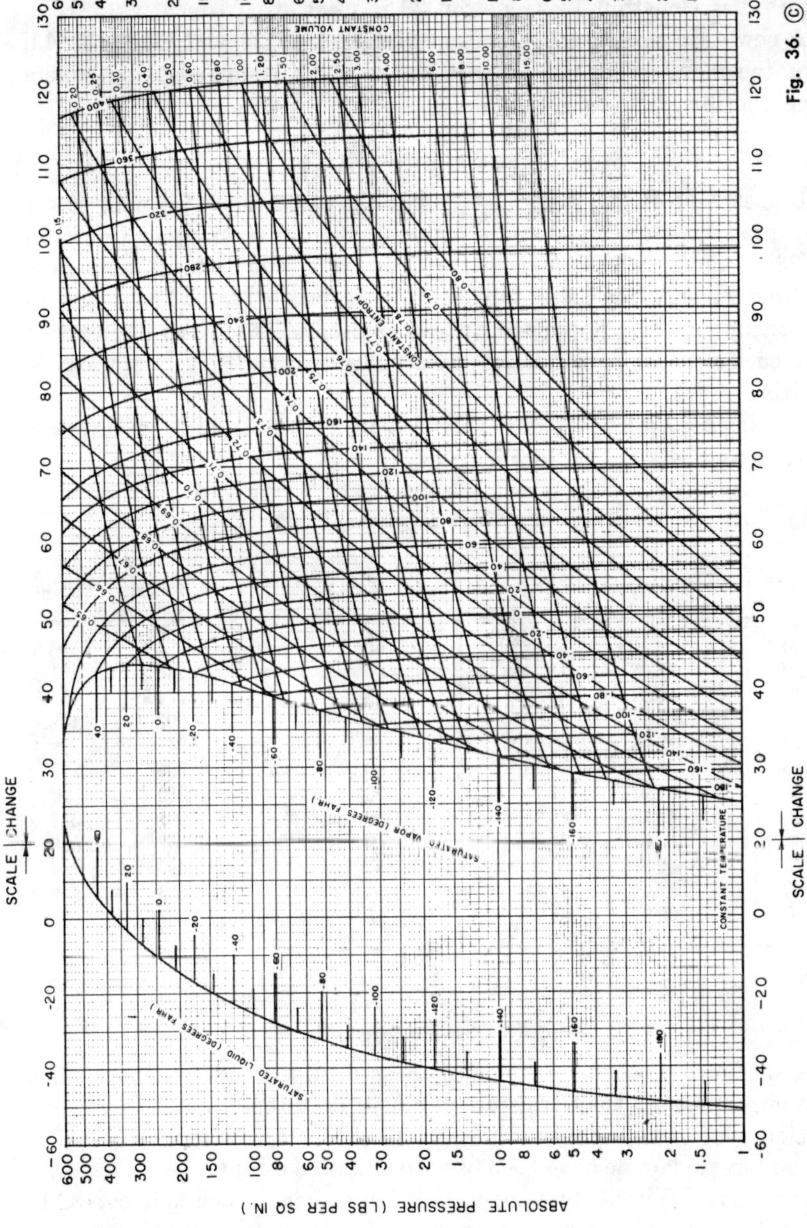

Fig. 36. © by General Chemical Division, Allied Chemical Corporation.

or Genetron 22 in their upper stage. The most widespread low-stage refrigerant is Genetron 13. Genetron 23 is used less frequently.

The new refrigerant, an azeotrope of Genetron 23 and Genetron 13, retains the many beneficial characteristics of its components but offers superior performance to either alone.

Properties of R-503

The azeotrope R-503 has a lower boiling point, operates with lower compression ratios, and generates higher capacities than either Genetron 13 or 23. It offers the potential for reaching lower temperatures and at the same time will provide faster pull-down, will handle larger loads, or will allow reduction in equipment size. Preliminary data also indicate a higher coefficient of performance (less power consumed per unit of refrigeration).

The following tables list its vapor pressure characteristics, its physical properties and selected performance data.

TABLE 25. VAPOR PRESSURE, R-503

Temp. (°F.)	*Psia*	*Psig*	*Temp. (°F.)*	*Psia*	*Psig*
−160	4.93	19.88*	−40	121.5	106.8
−150	7.08	15.51*	−30	146.3	131.6
−140	9.94	9.69*	−20	174.8	160.1
−130	13.67	2.09*	−10	207	192.4
−120	18.45	3.75	0	244	229
−110	24.48	9.78	10	285	270
−100	31.97	17.27	20	330	316
−90	41.15	26.5	30	381	367
−80	52.3	37.6	40	437	423
−70	65.6	50.9	50	499	484
−60	81.4	66.7	60	566	552
−50	99.9	85.2	67.1 (critical)	607	592

* Inches Hg vacuum.

Miscellaneous Refrigerants

In addition to those fluids previously described, there are others of more or less importance to the refrigeration industry. Of these, water vapor and air are of greatest importance, while nitrous oxide, methyl formate, ethylamine, and methylamine have been used to a limited extent.

Water Vapor. Within the temperature limitations, which is above 32 F (the freezing point of water), water vapor is an ideal refrigerant. Being nontoxic and nonflammable, it is the safest fluid for air-conditioning purposes. It has a high refrigerating effect, being 1025.3 Btu per lb at 86 F condensing temperature and 40 F evaporator temperature. Its compression ratio is favorable at 5.06, as is its power requirement of 1.125 hp per ton, both of which are based on the temperatures above. It has a comparable coefficient of performance of 4.20.

Its greatest disadvantage lies in the volume of vapor that must be handled, which amounts to 476.6 cu ft per min per ton of refrigeration. This is approximately 4.7 times greater than the volume necessary with R-113, its nearest counterpart of the more commonly used refrigerants.

Because of the high volumetric requirements of water vapor, it is employed in either centrifugal or steam jet compression systems.

Air. As a refrigerant, air has an extremely low coefficient of performance, 1.67 under standard ton conditions, and a relatively high power requirement of 2.82 hp per ton of refrigeration. Usually it is employed in connection with an air expander in which its temperature is further reduced after an intermediate cooling to remove heat of compression. Air, as a refrigerant, is becoming more important in the aviation industry for the removal of frictional heat generated by high plane speeds. In this connection, both compression and expansion are accomplished by rotary blowers capable of handling large volumes of air, with much of the heat of compression dissipated in driving the expansion unit. The advantages of air lie in its safety, availability, and lack of cost.

Nitrous Oxide (N_2O). Nitrous oxide is a colorless gas, is soluble in water, and has a faint pleasant odor. While it is nonflammable, nitrous oxide supports combustion almost as well as oxygen. It is highly anesthetic, being known widely as "laughing gas."

It possesses a low boiling point of −127 F at atmospheric pressure, requiring high condenser (915.3 psig at 86 F) and evaporator (318.3 psig at 5 F) pressures. Thermal qualities are fairly good with a 3.59 coefficient of performance and a refrigerating effect of 85.2 Btu per lb, both of which are based on standard ton conditions.

Methyl Formate ($C_2H_4O_2$). A low-pressure refrigerant, methyl formate is employed in rotary or centrifugal compressor systems. Having a relatively high boiling point (89.2 F under atmospheric pressure), it operates under a partial vacuum in both the condenser (26.3 in. Hg at 86 F) and the evaporator (1.8 in. Hg at 5 F). Methyl formate is slightly flammable, and forms an explosive mixture when present in air at concentrations of 4.5 percent to 20 percent by volume. It is miscible with lubricating oil. The system in which this refrigerant is employed must be maintained in a dry condition as it forms a corrosive compound with water. Under standard ton conditions, it has a refrigerating effect of 189.2 Btu per lb and requires only 1.06 lb per ton circulated.

Methylamine and Ethylamine. Methylamine and ethylamine are ammonia type compounds having some basic characteristics that classify them as refrigerants. However, they are not employed to any great extent in the refrigerating field. They derive their names from the combinations *methyl* + *amine* and *ethyl* + *amine*. Amines are a general group of compounds formed

TABLE 26. REFRIGERANT CHARACTERISTICS (PERFORMANCE BASED ON 5 F EVAPORATION AND 86 F CONDENSATION)[1]

Refrigerant	Chemical symbol	Molecular weight	Boiling temperature at 0 psig	Freezing temperature at 0 psig	Critical temperature	Critical pressure	Evaporator pressure at 5 F	Condensing pressure at 86 F	Ratio of compression at 86 F, 5 F (psia/psia)	Net refrigerating effect of liquid 86 F, 5 F	Refrigerant circulated per ton	Liquid circulated per ton 86 F, 5 F	Specific volume of vapor 5 F	Compressor displacement per ton 86 F, 5 F	Horsepower per ton 86 F, 5 F	Coefficient of performance 86 F, 5 F	Temperature of compressor discharge
			F	F	F	psia	psig	psig	psia	Btu/lb	lb/min	cu in./min	cu ft/lb	cfm	hp		F
Air		28.95	−318.0		−221.0	547.0	0	58.8	5.00	28.5	7.02		11.7	82.3	2.82	1.67	
Methane	CH_4	16.03	−258.9	−297.0	−115.8	673.0	‡										
Freon-14	CF_4	88.00	−198.2	−312.0	− 49.9	542.4	‡										
Ethylene	C_2H_4	28.03	−155.0	−272.0	48.8	731.8	399.8	‡									
Ethane	C_2H_6	30.04	−127.5	−278.0	90.1	708.3	221.3	661.1	2.86	58.6	3.41	342.9	0.53	1.82	1.953	2.41	122
Nitrous Oxide	N_2O	46.01	−127.0	−128.0	96.5	1050.0	294.3	922.3	3.03	85.2	2.35	71.2	0.28	0.66	1.310	3.60	
Freon-13	$CClF_3$	104.40	−114.5	−296.0	83.9	561.3	177.1	‡									
Carbon Dioxide	CO_2	44.00	−109.3	− 69.9	87.8	1071.1	317.5	1031.0	3.15	55.5	3.60	167.1	0.27	0.96	1.840	2.56	151
Kulene-131	CF_3Br	148.93	− 73.6	−226.0	153.5	587.0	63.2	247.1	3.36	29.3	6.82	123.8	0.38	2.63	1.030	4.25	124
Propylene	C_3H_6	42.08	− 53.7	−301.0	196.5	667.2	37.0	167.0	3.51	173.0	1.16	61.5	2.61	3.03	1.046	4.51	108
Propane	C_3H_8	44.06	− 44.2	−309.8	202.0	661.5	27.2	140.5	3.70	121.0	1.65	94.0	2.48	4.09	1.030	4.58	97
Freon-22	$CHClF_2$	86.48	− 41.4	−256.0	204.8	716.0	28.3	159.8	4.06	69.3	2.89	68.0	1.25	3.60	1.011	4.66	131
Ammonia	NH_3	17.03	− 28.0	−107.9	271.4	1657.0	19.6	154.5	4.94	474.4	0.422	19.6	8.15	3.44	0.989	4.76	210
Carrene-7	†	99.29	− 28.0	−254.0	221.1	631.0	16.4	113.4	4.12	61.1	3.27	79.3	1.52	4.97	1.022	4.61	105
Freon-12	CCl_2F_2	120.9	− 21.6	−252.0	232.7	582.0	11 8	93.2	4.07	51.1	3.92	83.9	1.49	5.81	1.002	4.70	100
Methyl Chloride	CH_3Cl	50.48	− 10.8	−144.0	289.4	968.7	6.5	80.0	4.48	150.2	1.33	40.9	4.47	5.95	0.962	4.90	172
Isobutane	C_4H_{10}	58.12	10.3	−229.0	272.7	537 0	3 3*	44.8	4.54	111.5	1.79	91.0	6.41	11.50	1.083	4.36	80
Sulfur Dioxide	SO_2	64.06	14.0	−103.9	314.8	1141.5	5.9*	51.8	5.63	141.4	1.41	26.6	6.42	9.09	0.968	4.87	191
Methylamine	CH_5N	31.05	20.3	−134.5	314.0	1082.0	9.9*	46.8	6.13	304.0	0.66	28.2	15.54	10.23	0.978	4.81	
Butane	C_4H_{10}	58.12	31.3	−211.0	306.0	550.1	13.2*	26.9	5.07	128.6	1.56	75.9	9.98	15.52	0.953	4.95	88
Freon-114	$C_2Cl_2F_4$	170.93	38.4	−137.0	294.3	474.0	16.1*	22.0	5.42	43.1	4.64	89.2	4.22	19.59	1.015	4.64	86
Freon-21	$CHCl_2F$	102.93	48.0	−211.0	353.3	750.0	19.2*	16.5	5.96	89.4	2.24	45.7	9.13	20.43	0.943	5.05	142
Ethyl Chloride	C_2H_5Cl	64.52	54.0	−217.7	369.0	764.0	20.5*	12.4	5.83	142.3	1.45	45.8	17.06	24.82	0.906	5.21	106
Ethylamine	C_2H_7N	45.06	61.8	−115.0	362.0	815.0	23.1*	10.0	7.40	225.5	0.89	349.0	32.32	28.67	0.855	5.52	
Freon-11	CCl_3F	137.38	74.7	−168.0	388.4	635.0	24.0*	3.6	6.24	67.5	2.96	56.0	12.27	36.32	0.927	5.09	112
Methyl Formate	$C_2H_4O_2$	60.04	89.2	−147.5	418.0	607.0	26.3*	1.6*	7.74	189.2	1.06	29.9	48.25	51.00			
Ethyl Ether	$C_4H_{10}O$	74.08	94.3	−177.3	522.1	380.8	26.9*	4.9*	8.20	126.3	1.58	62.9	35.00	55.40	0.822	5.74	
Methylene Chloride	CH_2Cl_2	84.93	105.2	−142.0	480.0	670.0	27.6*	9.5*	8.60	134.6	1.49	30.9	49.90	74.30	0.963	4.90	205
Freon-113	$C_2Cl_3F_3$	187.39	117.6	− 31.0	417.4	495.0	27.9*	13.9*	8.02	53.7	3.73	66.5	27.04	100.76	0.960	4.92	86
Dichloroethylene	$C_2H_2Cl_2$	96.93	118.0	− 70.0	470.0	795.0	28.3*	15.8*	8.42	114.3	1.75	38.3	63.60	111.20	0.973	4.83	
Trichloroethylene	C_2HCl_3	131.37	187.0	−124.3	520.0	728.0	29.6*	26.2*	11.65	91.7	2.18	41.6	229.40	502.00	0.980	4.82	
Water (40 F and 86 F	H_2O	18.02	212.0	32.0	706.1	3226.0	29.7*	28.6*	5.06	1025.3	0.195	5.51	2444.40	476.70	1.125	4.10	282

* Inches of Mercury Vacuum.
† Azeotropic Mixture of Genetron-100 (CH_3CHF_2) and Freon-12 (CCl_2F_2).
‡ Above Critical.

[1] Reprinted from the "Air Conditioning-Refrigerating Data Book, 9th Edition—1955-56 Design Volume," by courtesy of The American Society of Refrigerating Engineers.

Table 27. Applications, Leak Methods, and Oil Miscibility*

	Application for which the refrigerant is used or is suitable							Applicable leak test methods[4,5]	Miscibility with oil	Dielectric strength of vapor[6]
	Industrial with attendance[1]	Commercial up to 25 hp[1]	Air conditioning[1]	Household[1]	Evaporator temperature range[2]	Type of compressor[3]	Air or water cooled			
Methane	Ltd.	None	None	None	Ult.	Recip.	[9]	2, 4	Yes	1.00
Freon-14	Ltd.	Ltd.	None	None	Ult.	Recip.	[9]	1, 3, 5	No	
Ethylene	Ltd.	Few	None	None	Ult.	Recip.	[9]	2, 4	Yes	1.21
Nitrous oxide	Ltd.	Few	Suit.	None	Ult.	Recip.	Water	2, 4	No	1.41
Freon-13	Ltd.	Ltd.	None	None	Ult.	Recip	Water	1, 3, 5	No	
Ethane	Ltd.	Few	None	None	Ult.	Recip.	Water	2, 4	Yes	
Carbon dioxide	Ext.	Few	Ltd.	None	Low & Med.	Recip.	Water	2, 4	No	.88
Kulene-131	Ltd.	Ltd.	None	None	Ult. & Low	Rec.-Rot	Water	1, 3, 5	[7]	
Propylene	Ltd.	Ltd.	None	None	Ult. & Low	Recip.	Water	2, 4	Yes	
Propane	Ltd.	Ltd.	None	None	Ult. & Low	Recip.	Water	2, 4	Yes	
Freon-22	Ltd.	Ext.	[8]	Suit.	Ult. to High	Recip.	Wat. or Air	1, 3, 5	[7]	1.31
Ammonia	Ext.	Ltd.	Ltd.	Ext.	Low to High	Recip.	Water	1, 3, 6	No	.82
Carrene-7	Suit.	Suit.	Suit.	Suit.	Med. & High	Rec.-Rot.	Wat. or Air	1, 3, 5	Yes	
Freon-12	Ext.	Ext.	Ext.	Ext.	Med. & High	Rec.-Rot.	Wat. or Air	1, 3, 5	Yes	2.40
Methyl chloride	Ltd.	Ext.	Ltd.	Ext.	Med. & High	Rec.-Rot.	Wat. or Air	1, 3, 5	Yes	1.06
Isobutane	Few	Ltd.	None	Ltd.	Med. & High	Rec.-Rot.	Wat. or Air	1, 3	Yes	
Sulfur dioxide	Few	Ext.	Few	Ext.	Med. & High	Rec.-Rot.	Wat. or Air	1, 3, 6	No	1.90
Methylamine	Few	Suit.	None	Suit.	Med. & High	Rec.-Rot.	Wat. or Air	1, 3	Yes	
Butane	Few	Suit.	None	Ltd.	Med. & High	Rec.-Rot.	Wat. or Air	1, 3	Yes	
Freon-114	None	None	None	Ext.	Med. & High	Rotary	Air	1, 3, 5	[7]	2.80
Freon-21	Few	None	None	None	Med. & High	Rotary	Air	1, 3, 5	Yes	1.33
Ethyl chloride	None	Few	None	Ltd.	Med. & High	Rotary	Air	1, 3, 5	Yes	1.00
Ethylamine	Suit.	None	None	None	High	Rot.-Cen.	Air	1	Yes	
Freon-11	Suit.	Few	Ext.	Ext.	High	Rot.-Cen.	Water	1, 5	Yes	3.00
Methyl formate	Suit	None	None	None	High	Centrif.	Water	2, 5	Yes	
Ethyl ether	Suit.	None	None	None	High	Centrif.	Water	2	Yes	
Methylene chloride	Suit.	None	Ext.	None	High	Centrif.	Water	2, 5	Yes	1.11
Freon-113	Suit.	None	Ext.	None	High	Centrif.	Water	2	Yes	2.60
Dichlorethylene	None	None	None	None	High	Centrif.	Water	2, 5	Yes	
Trichlorethylene	None	None	None	None	High	Centrif.	Water	2, 5	Yes	
Water	Ltd.	None	Ext.	None	High	Cen.-Jet.	Water	2	No	

[1] Ext.—Extensive; Ltd.—Limited; Few, if any; Suit.—Suitable but few if any.

[2] Ult.—Ultra-Low, −130 to −75 F; Low— −75 to 0 F; Med.—Medium, 0 to 30 F; High—30 to 50 F

[3] Rec.—Reciprocating; Cen.—Centrifugal; Rot.—Rotary.

[4] Numbers refer to those described on page 139.

[5] When using the halide torch on equipment charged with refrigerants listed in section 5, Table 14 (page 139), make certain that good ventilation is provided.

[6] Ethyl chloride as unity.

[7] Separate into two phases, depending on pressure, temperature and type of oil.

[8] Limited in central systems, but extensive in self-contained air conditioners.

[9] Brine or refrigerant cooled; especially suitable for low stage in cascade systems.

* Reprinted from the "Air Conditioning-Refrigerating Data Book, 9th Edition—1955 56 Design Volume," by courtesy of The American Society of Refrigerating Engineers.

when one or more of the hydrogen atoms of ammonia are replaced by an alcohol or a dicarbon radical.

Methylamine is formulated when one hydrogen atom of ammonia (HN_3) is replaced by the group CH_3 of methyl alcohol (CH_3OH). Methyl alcohol is also called wood alcohol. The group CH_3 is called *methyl* and acts the same as a radical. Combining, then,

$$NH_3 + CH_3 \rightarrow CH_5N \text{ (Methylamine)}$$

The boiling point of methylamine at atmospheric pressure (14.7 psia) is 20.3 F. Under standard ton conditions of 5 F evaporator and 86 F condenser temperatures, pressures of 9.9 in. Hg (vacuum) corresponding to 9.83 psia, and 46.8 psig (61.5 psia), respectively, are required. Thus, methylamine is within the medium- to high-evaporator temperature range and capable of being handled by a reciprocating or rotary compressor. It has a relatively high refrigerating effect of 304 Btu per lb under standard ton conditions. Either air or water may be used to condense it. It is miscible with oil. It is suitable for industrial, commercial, and domestic applications. Leaks in methylamine systems may be detected by bubbles resulting when soapy water is placed over the leak area of systems placed under pressure.

Ethylamine is formulated when one hydrogen atom of ammonia is replaced by the (hypothetical) radical C_2H_5, the base of ethyl alcohol (C_2H_5OH). Ethyl alcohol is grain alcohol. The base radical C_2H_5 is called *ethyl*. Combining as above,

$$NH_3 + C_2H_5 \rightarrow C_2H_7N \text{ (Ethylamine)}$$

Having a boiling point of 61.8 F at atmospheric pressure places ethylamine in the high-temperature evaporation range. Under standard ton conditions of 5 F evaporator and 86 F condenser temperatures, pressures of 23.1 in. Hg (vacuum) corresponding to 2.95 psia, and 10 psig (24.7 psia), respectively, are required. Ethylamine also has a relatively high refrigerating effect, having 225.5 Btu per lb under standard ton conditions. Ethylamine may be handled by rotary or centrifugal compressors while air-cooled condensers are adequate for conversion to its liquid form. The method of detecting leaks in an ethylamine system is similar to that used for systems employing methylamine. Ethylamine is miscible with oil. It is suitable for industrial applications.

SUMMARY OF CHAPTER 5

A refrigerant generally may be any substance capable of absorbing thermal energy from another substance.

A refrigerant adaptable to the mechanical refrigeration system is a fluid that will absorb thermal energy during its evaporation at a desired temperature corresponding to a convenient pressure, and subsequently condense at a higher pressure corresponding to available cooling water, air, or other condensing medium temperatures, during which it rejects the latent heat of evaporation in preparation for re-use.

The ideal refrigerant is a fluid that possesses certain thermal, physical, chemical, physiological, and economic characteristics that provide for the efficient removal of thermal energy safely, conveniently, and economically.

The refrigerant used in mechanical refrigeration systems should possess the following characteristics:

High refrigerating effect
High latent heat of vaporization
Low specific heat of the liquid
High specific heat of the vapor
Low specific volume
High weight per cubic foot—density
Boiling point as required at atmospheric pressure
Low condensing pressure corresponding to available cooling medium temperature
Low compression ratio
Extremely low freezing point
High critical pressure and temperature

Chemically stable at maximum operating conditions
No effect on lubricating oil
Stable in the presence of moisture
Noncorrosive to metals
Nonflammable
Nontoxic
A faint inoffensive warning odor
Low viscosity of liquid and vapor
Low cost
Readily available
By its chemical nature affords some method of leak detection

There are two ways of obtaining low evaporator temperatures. One is by employing a refrigerant whose atmospheric boiling point is higher than the temperature required and operating under partial vacuum; or by maintaining a positive suction pressure on a refrigerant whose boiling point at atmospheric pressure is below the desired temperature.

A positive pressure is said to exist when the pressure of the refrigerant within the system is exerted outward or above atmospheric.

A negative pressure is said to exist when a partial vacuum or pressure below atmospheric is maintained within the system.

A low-pressure refrigerant is one that functions under partial vacuum in both the low- and the high-pressure sides of the system. Usually used for high temperatures (above 32 F).

An intermediate-pressure refrigerant is one that functions under relatively low positive pressure in the condensing unit, but may evaporate under high partial vacuum or low positive pressure, depending on the temperature desired, usually just below freezing.

A high-pressure refrigerant is one that functions under high condensing pressures and relatively high to relatively low evaporator pressures, depending on the temperature desired, usually below 10 F.

An extremely high-pressure refrigerant is one that functions under extremely high condenser pressures, but may evaporate at high to low negative pressures, depending on the temperature desired, usually ultra-low or below −100 F.

Utility classification of refrigerants is governed primarily by their boiling points at atmospheric pressure and condensing pressures at available cooling medium temperatures. According to this definition, the refrigerants in use are classified as follows:

LOW-PRESSURE–HIGH-TEMPERATURE REFRIGERANTS:

Utility: Air conditioning —Theaters
Office buildings
Auditoriums
Legislative chambers
Factories
Cold-water supply—Equipment cooling
(industrial) Gas absorption
Distillation
Heat exchangers

Refrigerants: R-113
R-11
Methylene chloride (Carrene-1)
Dichloroethylene (Dielene)
Trichloroethylene (Trielene)
Water

INTERMEDIATE-PRESSURE–INTERMEDIATE-TEMPERATURE REFRIGERANTS:

Utility: Domestic —Household refrigerators
Water coolers (small)
Unit air conditioners
Commercial—Display cases
Walk-in coolers
Refrigerated carriers
Beverage coolers
Unit air conditioners
Automobile air conditioners
Motor truck refrigeration
Industrial —Cold-water supply
Chemical storage
Cutting oil coolers

Refrigerants: R-12
R-114
Methyl chloride
Sulfur dioxide
Iso-butane
R-500

HIGH-PRESSURE–LOW-TEMPERATURE REFRIGERANTS:

Utility: Industrial—Cold-storage plants
Ice cream manufacturers
Breweries
Skating rinks
Creameries
Freeze-drying (sublimation)
Domestic and Commercial—Freezers

Refrigerants: Ammonia
Carbon dioxide
R-22
R-12
Propane
Kulene-131

EXTREMELY HIGH-PRESSURE–ULTRA-LOW TEMPERATURE REFRIGERANTS:

Utility: Industrial—Aerodynamic wind tunnels
Metallurgy
Gas liquefaction

Refrigerants: Ethylene
Ethane
R-13, R-14 and R-503

Leakage of refrigerant from a system may reduce load capacity and increase cost of operation. In many instances, it is dangerous to life and property. Leak-detection methods for all refrigerants should be based on a simple procedure. Unfortunately, leakage of some refrigerants cannot be determined until load capacity losses are noticeable. Methods of leak detection for various refrigeration according to Group Classification are as follows:

Ammonia —Sulfur candle
Sulfur dioxide —Ammonia water solution
Freons, Carrene-7 and bromines—Halide torch

Halogenated hydrocarbons and methyl formate	—Halide torch
All others—Hydrocarbons, nitrous oxide, etc.	—Soap-and-water solution

Leakage of air into the system is evident by higher than normal condenser pressures and loss of capacity.

STUDY OUTLINE FOR CHAPTER 6

Mechanical refrigeration system:
- Basic equipment composition:
  - Evaporator
  - Compressor
  - Condenser
  - Throttling device

Compression cycle:
- *First step:*
  - Evaporation of liquid refrigerant:
    - Controlled pressure
    - Controlled rate
  - Equipment:
    - Evaporator or cooling coil
  - Equipment function:
    - Liquid-to-vapor conversion
    - Substance or space cooling
  - State of operation:
    - Low pressure
- *Second step:*
  - Compression of refrigerant vapor:
    - Low pressure to high pressure
  - Equipment:
    - Compressor
  - Equipment function:
    - Reclaim spent refrigerant
    - Increase boiling point of refrigerant
  - Thermal energy of compression:
    - Conversion of mechanical energy
  - Superheat:
    - Compression ratio
- *Third step:*
  - Condensing of refrigerant vapor:
    - Vapor temperature reduction
  - Equipment:
    - Condenser
  - Equipment function:
    - Removal of thermal energy from high-pressure vapor
    - Transfers thermal energy to cooling medium
    - Vapor to liquid conversion
  - Stage of operation:
    - High pressure
- *Fourth step:*
  - Pressure reduction:
    - Refrigerant flow throttle
    - High-pressure to low-pressure conversion
  - Equipment:
    - Throttling device (expansion valve)
  - Equipment function:
    - Refrigerant supply to evaporator
    - Decreases boiling point of refrigerant

6

The Refrigeration Cycle

The mechanical refrigeration system, also known as the vapor-compression system, is designed to provide a method for the continual recovering and re-use of an initial charge of refrigerant upon the completion of each cycle. A refrigerant in a refrigeration system completes its functions upon evaporation, during which it absorbs thermal energy in an amount equivalent to its refrigerating effect, or R E. This action takes place within the evaporator.

In a simple refrigeration system, a costly operation would result if recovery of the refrigerant for re-use were not attempted and the gas were allowed to escape into the atmosphere. This method of producing refrigeration would entail the continual replenishment of refrigerant in order to maintain required temperatures. It is the equipment required for storing and evaporating the liquid, recovering the gas after evaporation takes place, and preparing it for re-use, that constitutes the mechanical refrigeration system.

The refrigeration system is an enclosed, air-tight system of tubes and equipment so constructed as to allow a controlled quantity of fluid (refrigerant) to flow from one necessary step to another under definite and predetermined pressures, for the purpose of absorbing thermal energy in a zone of low pressure (evaporator) by means of liquid evaporation, and dissipating that energy in a zone of high pressure by means of condensation (condenser). It is so designed that the preparation of the fluid for repeated performance of this cycle is made possible.

From this definition, it is found that four fundamental steps are required to complete the mechanical refrigeration cycle: evaporation, compression, condensation, and pressure reduction.

The study of each of the fundamental steps of the refrigeration cycle involves the use and the understanding of all laws, principles, and terms described previously. To facilitate this study, a list of those laws, principles, and terms relative to each step is given under their respective headings:

STEP 1. EVAPORATION

Laws, Principles, and Terms Involved. (*a*) Evaporation, (*b*) kinetic thermal energy, (*c*) superheat, (*d*) latent heat of vaporization, (*e*) thermal energy measurement—intensity or degree, quantity or Btu, (*f*) boiling points of liquids, (*g*) pressures—gage and absolute, (*h*) pressure—boiling point relationship, (*i*) pressure—enthalpy relationship, (*j*) heat transfer—conduction, (*k*) enthalpy, (*l*) heat flow, (*m*) refrigerating effect, and (*n*) effect of pressure on thermal properties of fluids.

Equipment. The first step in the refrigeration cycle is the evaporation of the liquid refrigerant. It is during this phase that the actual work of refrigeration is performed. The equipment which forms a part of the complete system and in which evaporation takes place is called the evaporator or the cooling coil. It is the evaporator that is exposed to the area or the substance under refrigeration and that is the cooling element of the system.

Evaporator Functions. To effect a decrease in temperature, thermal energy must be removed from a body. The energy thus removed is not lost, but is absorbed by a surrounding body whose total internal energy increases in proportion to the amount absorbed. The rule that *energy lost equals energy gained* is thus established.

The cooling coil or the evaporator of a refrigeration system is designed to effect a heat flow from the area, or the substance to which it is exposed, to a fluid medium (refrigerant) forced through the coil. The medium employed should induce such a flow of heat by one of the following two methods, or a combination of both: (1) temperature differential, whereby the refrigerant is lower in temperature than the area or the substance to be cooled; (2) evaporation of the refrigerant, whereby the latent heat of vaporization is utilized to initiate heat flow from the area or the substance to be cooled.

Temperature Differential. If temperature reduction is effected by means of the first method alone, the heat thus transferred is that of kinetic thermal energy only. An example of this method of heat exchange is the case of a copper coil submerged in a body of warm substance with one end of the coil connected to a pump delivering cold water and the other end connected to a drain. Heat flow would then be induced from the warmer substance to the cold water flowing through the coil, resulting in a decrease in temperature of the warm substance and a corresponding increase in temperature of the cold water. (See Fig. 37.)

To determine the quantity of thermal energy removed from the body of warm substance, it is necessary that the weight of cold water passing through the coil be known as well as its initial and final temperatures.

EXAMPLE: Determine the amount of thermal energy removed per minute from a body of warm water if 50 lb of cold water per minute having

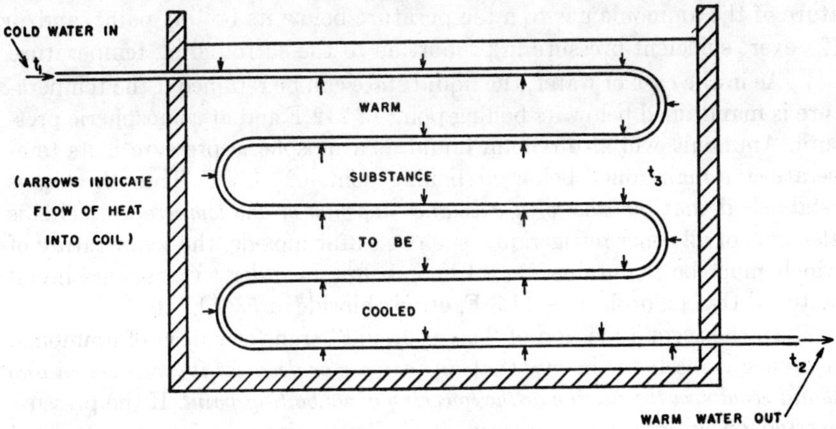

Fig. 37. Heat exchange by temperature differential.

an initial temperature of 50 F is pumped through a copper coil submerged within it and leaving the coil at a temperature of 76 F. Assume that no thermal energy is lost or gained because of other factors.

$$\text{Wt in lb} \times \text{sp ht} \times (T_2 - T_1) = \text{Btu absorbed}$$

$$50 \times 1 \times (76 - 50) = 1300 \text{ Btu absorbed by the cold water}$$

As heat loss equals heat gained, it can be stated also that 1300 Btu is the amount of heat removed from the body of warm water.

In the process of removing heat by utilizing only the temperature reducing capacity of fluids, it is necessary to circulate a large volume of cooling medium per unit of heat removed. This entails the employment of cooling coils of large surface areas in order to effect an appreciable rate of heat flow. Owing to the extremely low temperature conditions which some mechanical refrigeration systems are designed to create, the method of heat removal by temperature reduction alone is not always practical. This is due partly to the natural temperature limitations to which most liquids can be exposed without changing their state.

In secondary systems, such as the brine circulation system employed in ice manufacturing, the method of utilizing the flow of kinetic thermal energy does have its place and is used to excellent advantage.

Evaporation of Refrigerant. Because of the low boiling-point characteristics of many commercial refrigerants, they exist as a gas under normal atmospheric pressure–temperature conditions. Ammonia, for example, has a boiling point of −28 F at atmospheric pressure and, if exposed to such pressure at a room temperature of 86 F, it would evolve in its gaseous state. To transform the gaseous or vaporized ammonia to a liquid and to maintain it in that form, it is necessary that one of two steps, or a combination of both, be taken, namely (1) reduce the temperature of the ammonia gas to a temperature below its boiling point, and/or (2) exert sufficient pressure in proportion to the surrounding temperature.

1. As in the case of water, the liquid state can be retained if the temperature is maintained below its boiling point of 212 F and at atmospheric pressure. Ammonia will also remain liquid at atmospheric pressure if its temperature is maintained below *its* boiling point of −28 F. Thus the law is established that *the state of a substance depends on the temperature.* This is also true of all other refrigerants, such as sulfur dioxide, the temperature of which must be maintained at 14 F or below in order to retain its liquid state, methyl chloride at −11.3 F, ethyl chloride at 53.9 F, etc.

2. In the second method of "gas to liquid" transformation of ammonia, reference is made to the law that *an increase or decrease in pressure on any liquid results in the raising or the lowering of its boiling point.* If the pressure exerted on ammonia is increased, its boiling point also increases proportionately. If sufficient pressure is exerted, the boiling point of the ammonia can be raised above the room temperature of 86 F, and thus prevent its evaporation.

Referring to the ammonia temperature-pressure table, it is found that a pressure of 169.2 psia is required to retain liquid ammonia at the temperature of 86 F.

Conversely, a reduction in pressure exerted on any liquid decreases its boiling point. As has been stated, the boiling point of water is 212 F when exposed to normal atmospheric pressure (14.7 psia). According to the law of pressure–boiling-point relationship, the boiling point of water can be reduced to 193.21 F if the pressure exerted on it is reduced to 10 psia, and to 162.25 F at 5 psia, etc. Similarly with ammonia, whose boiling point becomes −41.34 F if exerted pressure is reduced below atmospheric to 10 psia.

The boiling point of a refrigerant definitely can be controlled by controlling the pressure exerted on it. In the evaporator of a refrigeration system, the pressure is controlled at a level influenced by the boiling point of the refrigerant used.

The significance of the pressure–boiling-point relationship of refrigerants lies in the fact that it is necessary to utilize the latent heat of vaporization characteristics of liquids to induce a high rate of heat flow into the evaporator or the cooling coil.

Figure 38 shows a container of liquid ammonia. As long as the temperature of the ammonia remains at 86 F and the valve is kept closed, a pressure of 169.2 psia will be maintained. If the valve is opened to the atmosphere, the resultant release of pressure will cause the liquid to boil. The pressure to which the liquid mass is reduced will depend on the amount of vapor that is allowed to escape. Allowing the valve to remain open, a continuous escape of vapor results, causing a gradual reduction in pressure within the container. As the pressure continues to decrease with each increment of vapor loss, the temperature of the liquid gradually becomes lower. When sufficient vapor escapes to reduce the pressure in the container to that of atmospheric (14.7 psia), the temperature of the boiling liquid will be −28 F. Upon reaching atmospheric pressure, the liquid ammonia will continue to boil. The temperature will not fall below −28 F while the pressure remains at 14.7 psia. By creating a partial vacuum within the container, as with some kind of suction device, a lower temperature can be attained.

Repeated measurements of both pressure and temperature taken during the evaporation process will show a definite relationship between the two values. However, the relationship between pressure and temperature will vary with different fluids. In the case of liquid ammonia, the relationship would be as represented by the curve in Fig. 28. Corresponding temperatures related to various pressures are given for the more important refrigerants in their respective thermodynamic property tables.

The reason for the drop in pressure and temperature within the container lies in the escape of resultant vapor during the boiling process. Replacement of liquid ammonia at a rate equal to the vapor loss would tend to equalize the pressure in the container at any level desired. For example, if a constant temperature of 5 F is desired, a corresponding ammonia pressure of 34.27 psia would have to be maintained. This is accomplished by constant replenishment of liquid ammonia at a rate equal to the quantity lost through evaporation. Replenishment in this case would be initiated when the pressure in the container reached 34.27 psia. To maintain a constant temperature of, say, 10 F, replenishment of liquid ammonia at an adequate rate would be initiated at 38.51 psia.

In this case, the surrounding atmosphere is the source of heat that promotes the boiling action of ammonia within the container. A temperature of 86 F surrounding the container induces a flow of heat by conduction through the container wall where it is absorbed by the ammonia and utilized as latent heat of vaporization. When the valve of the container is first opened, the temperature of the liquid ammonia is 86 F, having previously absorbed sufficient kinetic thermal energy from the atmosphere to reach that temperature. When the valve is opened, the ammonia, prompted by the presence of the energy already absorbed, and by its high internal vapor pressure incurred by a liquid temperature of 86 F, immediately starts to boil. The previously absorbed energy is quickly utilized by its conversion to latent heat of vaporization, and leaves the container with the vapor. As this kinetic thermal energy is converted and leaves the container, the temperature of the liquid body is reduced below its original level of 86 F. The temperature differential thus created causes heat to flow from the surrounding 86 F atmosphere to the now colder liquid ammonia. This, in turn, is converted to latent heat of vaporization, and thus the boiling process continues. As the temperature of the liquid ammonia

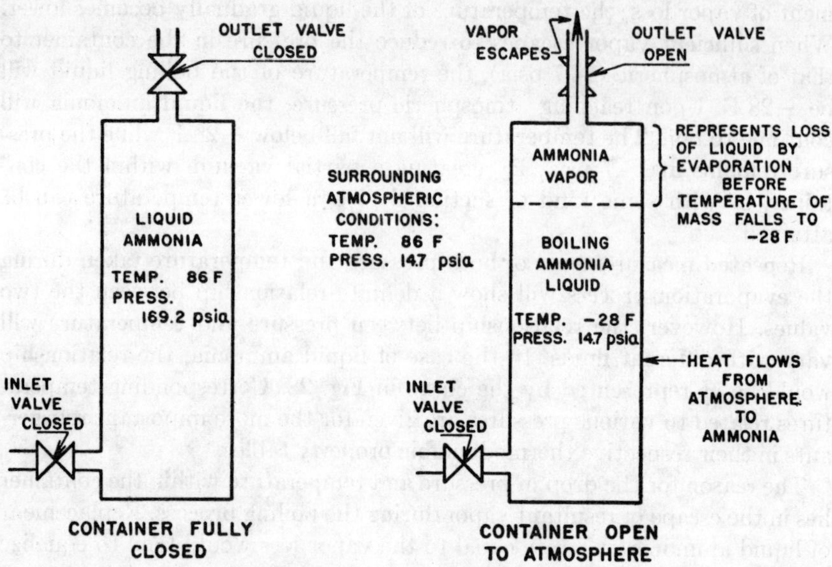

Fig. 38. Pressure-boiling point relationship of refrigerants.

is reduced, heat transfer becomes more rapid as a result of greater temperature difference. Evaporation progresses until the pressure in the container reaches any controlled level governed by the temperature desired. At this point, the rate of evaporation becomes constant, provided that the surrounding temperature remains the same and an adequate rate of liquid replenishment is accomplished.

The amount of liquid ammonia that will evaporate in a given time depends on the amount of thermal energy transferred from the atmosphere. The rate of heat transfer through a given container wall is governed by the temperature differential and the area of container surface exposed to the atmosphere or to any substance to be cooled. If the temperature differential becomes constant, such as would be the case under controlled liquid pressure conditions, the only means remaining for increasing the amount of liquid evaporation is to provide more exposed surface area.

The refrigeration capacity of any system is based on the quantity of fluid evaporated per unit of time. The refrigeration effect of ammonia under standard ton conditions of 5 F evaporation temperature and 86 F liquid temperature is 474.6 Btu per lb. The evaporation of 1 lb of liquid ammonia per minute is equivalent to a refrigeration capacity of $\frac{474.6}{200} = 2.373$ tons. A container having only enough surface area exposed to the substance to be cooled to allow $\frac{1}{2}$ lb liquid ammonia to evaporate per minute would have a capacity of $\frac{474.6}{2 \times 200} = 1.187$ tons. If the area of the container were increased to allow for the evaporation of 10 lb of ammonia per minute, then the refrigeration capacity would be $\frac{10 \times 474.6}{200} = 23.73$ tons, assuming standard ton conditions in all three cases.

It was stated above that liquid ammonia starts to boil immediately upon opening the container valve. This rapid boiling action is initiated by the internal energy contained within the liquid mass, which starts to convert to latent heat of vaporization the moment of pressure release. The presence of internal energy within the mass of 86 F liquid ammonia causes a loss of refrigeration effect. This loss is represented by the amount of liquid evaporated by the internal energy before the temperature of the

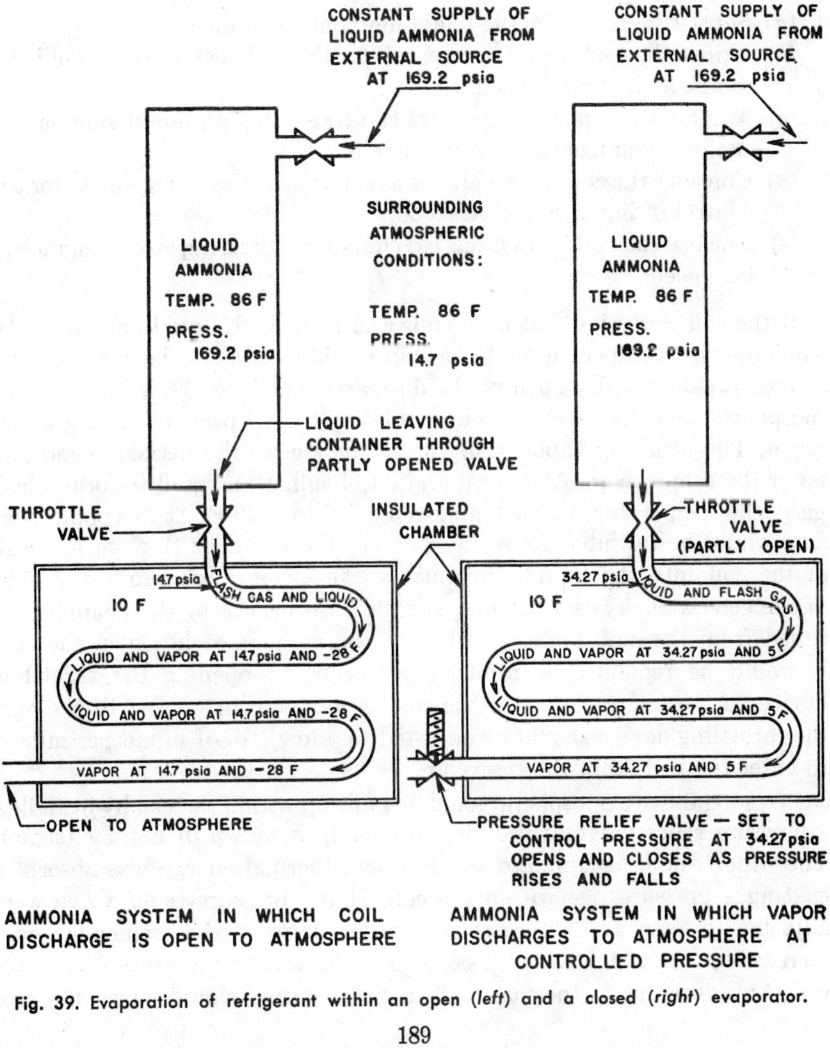

Fig. 39. Evaporation of refrigerant within an open (*left*) and a closed (*right*) evaporator.

mass is reduced to that corresponding to the controlled pressure level. To cool 86 F liquid ammonia to 5 F liquid ammonia by internal energy conversion requires the evaporation of 16 percent of the total mass. The vapor generated by internal energy is called "flash gas." The remaining 84 percent of the liquid mass, now at 5 F, is capable of absorbing external energy and thus perform work of refrigeration. To reduce the amount of "flash gas," thereby increasing refrigeration effect, it is necessary to cool the liquid ammonia below 86 F before releasing its pressure. This is the function of the heat exchangers, after-coolers, etc., described in later chapters of this text.

To convert the system described above to one similar to the evaporator of a mechanical refrigeration system, the following changes would be made:

(*a*) Substitute a coil for the liquid ammonia container,
(*b*) Utilize the container as a reservoir for the replenishment of liquid ammonia as it evaporates in the coil,
(*c*) Assume that constant pressure can be exerted on liquid ammonia in the reservoir from an external source,
(*d*) Consider the container valve as a metering or throttling device for the control of liquid flow to the coil,
(*e*) Enclose the coil in an insulated chamber containing any substance to be cooled.

If the coil were closed at both ends and filled with liquid ammonia, the same pressure-temperature relationship would exist as in the container described previously. By opening the discharge end (Fig. 39, *left*) to the atmosphere, the pressure in the coil would drop to 14.7 psia and boiling would begin. The resulting vapor would leave the coil at the discharge end and enter the atmosphere. After evaporation of sufficient liquid to form "flash gas," the temperature of the liquid would fall to −28 F. At this point, heat flow would be initiated from the space in the chamber through the wall of the coil into the liquid, to continue the evaporation process. As the vapor generated by the thermal energy absorbed from the chamber escaped from the coil, tending to reduce the amount of liquid in the coil, it would be replenished from the reservoir by opening the throttling device. If the loss of liquid in the coil by evaporation were 1 lb per min, the throttling device should be adjusted to allow 1 lb of liquid per minute to enter the coil from the reservoir.

Pressure above atmospheric could be built up within the coil by installing a pressure-relief valve at the discharge end, as shown in Fig. 39 (*right*). The vapor would accumulate as additional thermal energy was absorbed, causing a pressure rise within the coil. If the pressure-relief valve were adjusted to open at 34.27 psia, the boiling point of the ammonia would increase to 5 F, which would become the constant temperature within the coil. Any decrease in thermal load in the chamber would lessen the rate of evaporation, and the pressure within the coil would drop below 34.27 psia, at which point the relief valve would close. When enough vapor had accumulated, the pressure would again rise, opening the relief valve and bringing about stabilization at 34.27 psia. Adjustment of the throttling device would be required to meet load-changing conditions.

To reduce the boiling point of liquid ammonia below −28 F, a partial vacuum must exist in the coil. This is accomplished by connecting the discharge end of the coil to some kind of suction device, such as a piston and cylinder arrangement, as shown in Fig. 40. As the piston moves away from the evaporator, it draws vapor into the cylinder, reducing the pressure within the coil. Lowering the coil pressure below atmospheric, say, to 10.41 psia, reduces the boiling point of ammonia to −40 F, which becomes the new coil temperature.

In the mechanical refrigeration system, the suction device is the component called a compressor, of which there are several kinds. The compressor not only draws vapor from the cooling coil to effect a lower pressure, as in

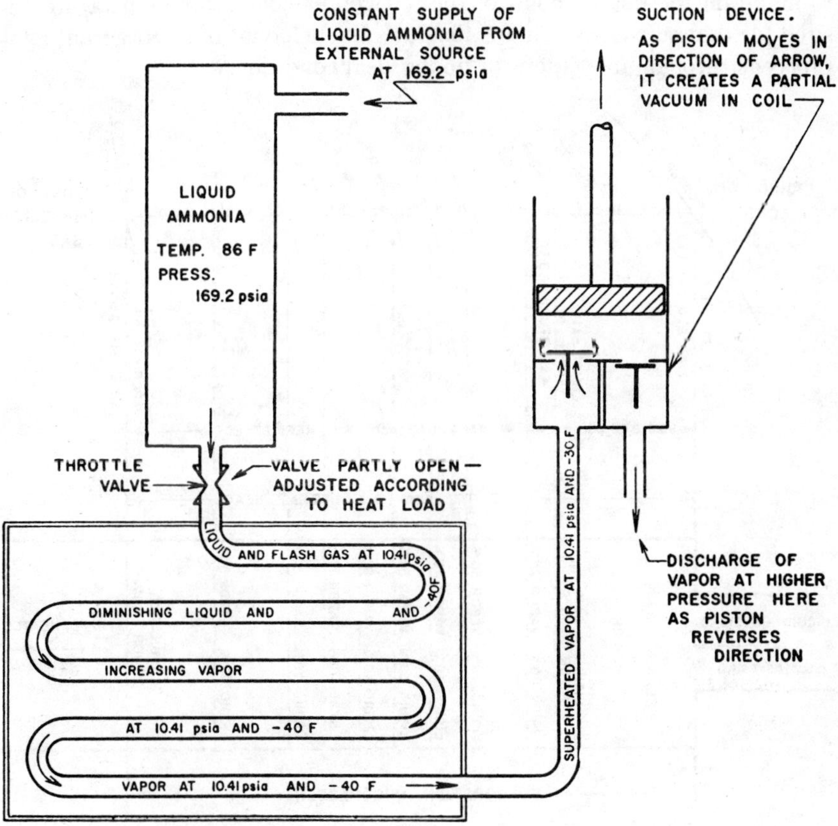

Fig. 40. Conditions within an evaporator operating under partial vacuum.

the above case, but also acts to reclaim the refrigerant vapor and compress it to higher pressure. The component described as a reservoir in the above discussion is referred to more commonly as a liquid receiver.

It is the balance provided by the adjustment of the throttling device in relation to the amount of suction established by the compressor that controls the pressure within the evaporator and the boiling point of the liquid. This balance is brought about by either automatic or hand control of the throttling device.

The evaporator or suction pressure, as it commonly is called, depends on various operating conditions, some of which are as follows: (*a*) boiling point of the refrigerant employed, (*b*) temperature maintained within the area or the substance being cooled, (*c*) heat-transfer capacity of the evaporator, (*d*) capacity of the compressor, (*e*) temperature differential desired between the evaporator and the substance to be cooled.

For reasons of power economy and certain thermal advantages, the suction pressure should be kept as high as possible, commensurate with the temperature requirement of the substance to be cooled. It should be low enough to maintain the boiling point of the refrigerant below the temperature desired for that substance. Figure 40 shows the behavior of a refrigerant as it proceeds through an evaporator under ideal conditions.

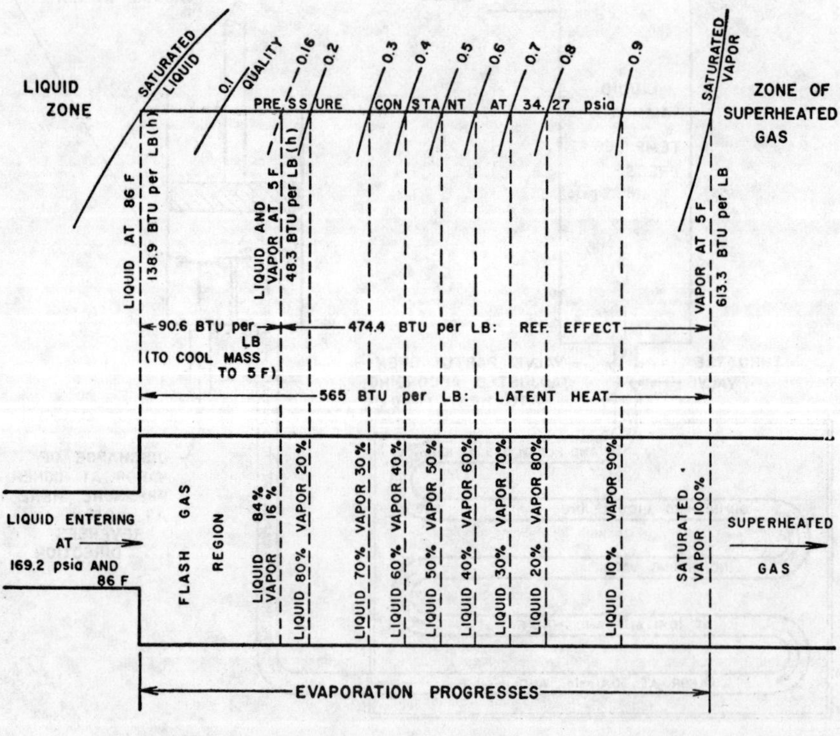

Fig. 41. Behavior of ammonia under standard ton conditions in evaporator.

Location and Operation of the Evaporator. The evaporator is installed within the low-pressure side of the refrigeration system. It is connected with a throttling device, usually called an expansion valve, through which it receives liquid refrigerant from the high-pressure side of the system. Upon evaporation of the liquid in the evaporator, the resulting vapor enters a suction device, or compressor, to which the discharge end is connected. The evaporator is located within the space or the substance to be cooled. It is during the evaporation process that heat flows from the refrigerated substance to the refrigerant.

Evaporators differ in their design and application. Shapes vary from hollow plates to intricate coils, depending on the purpose for which they are designed. Sometimes they are used as structures for the products refrigerated, as when they act as shelves, floors, and walls of domestic and commercial freezers, ice-cube compartments of domestic refrigerators, etc.

Some systems include more than one evaporator. Where multiple evaporators are installed, they may be connected in series or parallel. Parallel evaporators may be serviced by individual expansion valves or attended by one common to all. With proper controls, each evaporator of a multiple system may operate at a different temperature.

STEP 2. COMPRESSION

Laws, Principles, and Terms Involved: (*a*) Vacuum, (*b*) saturated gas, (*c*) pressures—gage and absolute, (*d*) Charles' law, (*e*) Boyle's law, (*f*) Dalton's law, (*g*) pressure–boiling-point relation, (*h*) compressibility of gases, (*i*) heat measurement—intensity, (*j*) pressure differential; high side —low side, (*k*) superheat, (*l*) heat of compression.

Equipment. As has been pointed out, the actual process of refrigeration takes place within the evaporator, during which the latent heat of vaporization is utilized to absorb the heat flowing from the area or the substance being cooled. If it were not for reasons of economy, additional equipment would not be necessary, as the refrigerant gas would be allowed to escape into the atmosphere and the refrigeration process continued by constant replenishment of liquid refrigerant.

However, it is necessary to reclaim the vapor and to prepare it for re-use. In order to do this, suitable equipment must be employed.

In the vapor-compression system, a compressor of either the reciprocating centrifugal, gear, or rotary type is employed for this purpose. As the reciprocating type compressor is used most generally, it is the one selected here for the purpose of describing the compression step of the refrigeration cycle.

The state of a substance depends on its temperature and pressure. One of the reasons for this is that, as a liquid, all substances have definite boiling points in relation to the pressure exerted on its surface. If the pressure varies, its boiling point varies likewise. In the study of evaporator functions, it was brought out that the pressure maintained within it was such that the boiling point of the refrigerant employed was below the temperature of the area

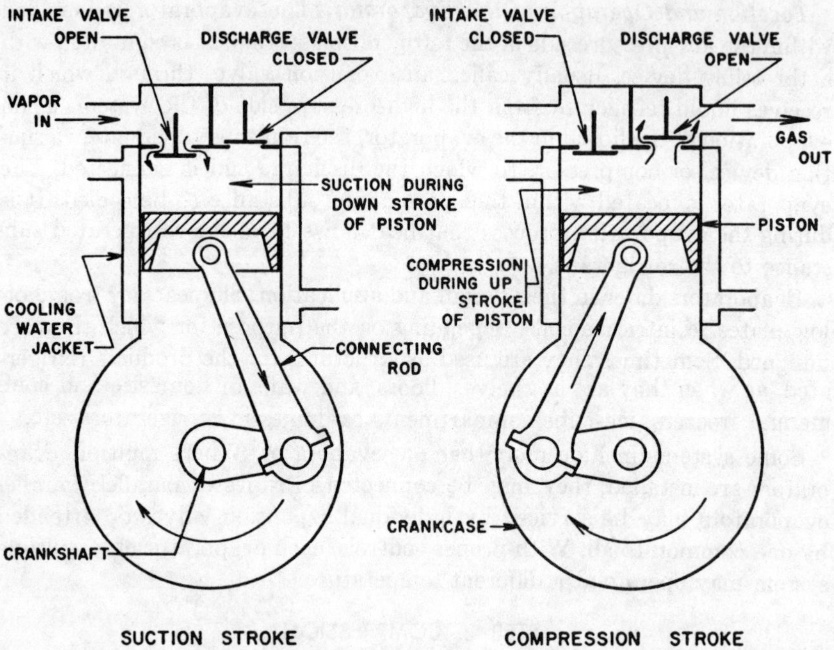

Fig. 42. Simple vertical compressor.

or the substance being cooled, which resulted in the evaporation of the liquid upon entering it.

In the compression phase of the refrigeration cycle, the objective is reversed. The function of the compressor is to recover the vaporized refrigerant by suction and to compress it to a sufficiently high pressure so as to raise the boiling point to a level suitable for conversion to its liquid form upon extraction of the heat absorbed in the evaporator, plus the internal energy developed as a result of converting mechanical energy during compression.

Figure 42 illustrates a simple vertical compressor. Embraced within a housing are the following:

(*a*) A *crankshaft* for transferring power from a prime mover, such as an electric motor, a gasoline engine, a steam engine, etc., to the connecting rod that moves the

(*b*) *piston*, that provides suction and compression by reciprocating within a

(*c*) *cylinder*, which is part of the housing, to the refrigerant vapor which is drawn in from the evaporator through a

(*d*) *suction valve*, located in the cylinder head or the piston crown which opens upon external pressure and closes when the pressure in the cylinder is equalized with that of the evaporator. The refrigerant vapor, upon reaching a pressure equal to or greater than the pressure maintained in the high-pressure side of the system, by compression, forces its way through a

(*e*) *discharge valve,* located in the cylinder head, which opens by internal pressure caused by compression of the vapor upon the upward stroke of the piston and closes when pressure in the cylinder is slightly below that in the discharge line.

Examining the activity taking place within the compressor cylinder through a complete cycle, or two strokes of the piston, reference is made to Fig. 43. The illustration portrays the travel of a piston through its upward (compression) and downward (suction) strokes, and a graph on which the pressure activity of the refrigerant vapor, drawn from the evaporator during the suction stroke, is plotted.

Starting the cycle at the moment the piston is at the extreme bottom limit of its stroke (*a*), the cylinder is completely filled with vapor which had been drawn from the evaporator through the suction valve (*s*) on the previous down stroke of the piston. At the moment the piston reaches position (*a*), the suction valve is open and the discharge valve (*d*) is closed. Theoretically, the pressure of the vapor in the cylinder is equal to that in the evaporator when the piston is in (*a*) position. As the piston starts upward, compression immediately takes place, closing the suction valve and causing the pressure within the cylinder to rise. As the piston moves upward, the pressure rises progressively according to Boyle's law of gases, as indicated by the curve AB in the graph. Actually, adiabatic compression and expansion more nearly occur. For descriptive purposes here, however, compression and expansion will be dealt with isothermally, or at constant temperature. Thus, it is found that $P_1V_1 = P_2V_2 = P_3V_3$, etc., or the product of the pressure times the volume is constant at any point along the line AB, and isothermal curve.

Compression continues as indicated by the pressure curve AB until the piston reaches a point in the cylinder corresponding to B on the graph. Because standard ton conditions are assumed for purposes here, the pressure at point B equals that corresponding to 86 F condenser temperature or 169.2 psia for ammonia.

Immediately after the piston reaches B, the pressure in the cylinder rises slightly above that maintained in the discharge line leading to the condenser —in this case, 169.2 psia. This slight pressure difference, as indicated by the small deviation on the graph extending line AB a little beyond B, is caused by the extra force above the discharge pressure required to open the discharge valve. At this point, the pressure is great enough to open the discharge valve, and, with continuing travel of the piston, the compressed gas is forced into the discharge line. During the period when the discharge valve is open and the gas flows through it, the pressure in the cylinder remains constant. This is indicated by the line BC of the graph. Point C on the graph corresponds with position (*b*) in the cylinder, which is the upward limit

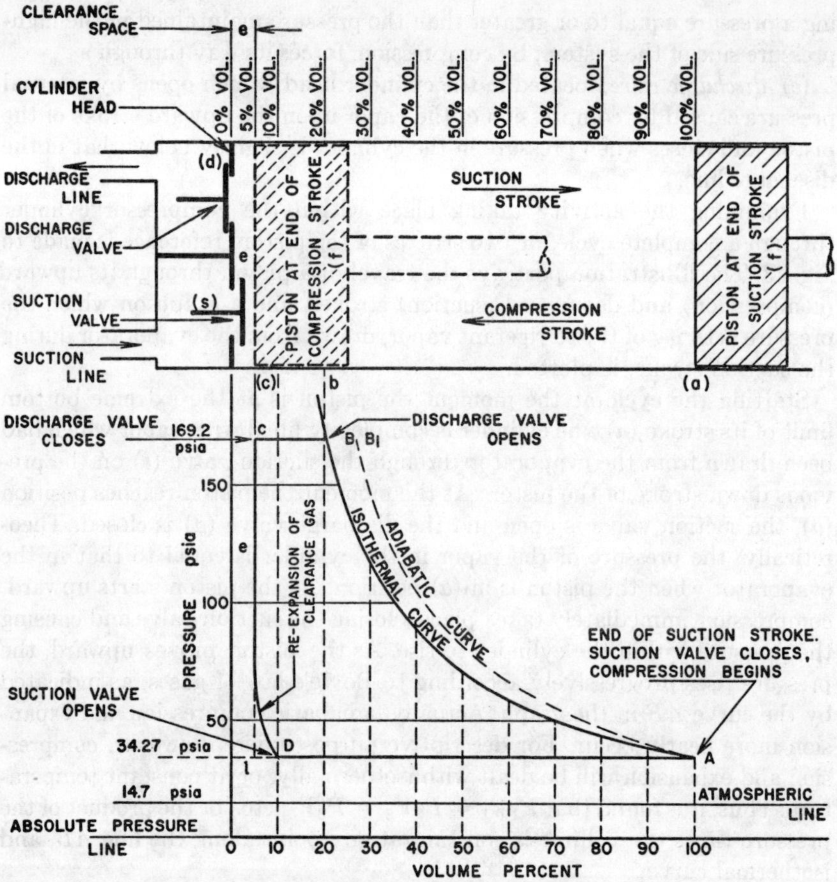

Fig. 43. Behavior of a gas under compression.

of piston travel or the end of the discharge stroke. The position of the piston at (*b*) usually is referred to as top-dead-center.

Immediately upon reaching top-dead-center (*b*), the piston momentarily rests in preparation of reversing direction. At the latter point, the pressure in the cylinder drops rapidly to a level equal to that in the evaporator or suction pressure. Under standard ton conditions, this means that a rapid drop in pressure from 169.2 psia to 34.27 psia occurs at the moment the piston starts on its suction stroke. This pressure drop is indicated on the graph by the line *CD*.

Because of the necessity of allowing clearance space between the top of the piston and the cylinder head to prevent metallic contact, a small amount of gas remains in the cylinder upon completion of its discharge stroke. The distance (*e*) represents the amount of clearance space, and therefore the volume of vapor remaining in the cylinder when the piston reaches top-dead-center (*b*). This vapor remains under compression during the hesitation period when the piston reverses direction. Because the amount of clearance gas remaining in the cylinder is small, its effective pressure is expended rapidly during the initial movement of the piston on the suction stroke. Its presence does cause a slight delay in allowing the pressure within the cylinder to equalize that in the evaporator. This causes a time lag in suction-valve opening, resulting in some volumetric loss. The time lag is indicated by the curvature of the line *CD*, the magnitude of which is represented by the distance *lD*.

Immediately after reaching point *D*, the pressure within the cylinder drops slightly below the evaporator pressure. This difference in pressure allows the suction valve to open and would be indicated on the graph by a slight deviation extending line *CD* below the evaporation pressure line *DA*. With the opening of the suction valve, vapor from the evaporator flows into the cylinder. This flow continues with the travel of the piston on its downward or suction stroke. During the suction stroke, the pressure of the vapor as it progressively fills the cylinder remains constant until the piston reaches the end of the suction stroke (*a*) or bottom-dead-center. This phase of the cycle is indicated by the line *DA* on the graph.

Upon reaching position (*a*), the piston momentarily pauses and immediately starts again on its compression stroke. At the instant of starting, compression takes place, causing the pressure in the cylinder to rise, closing the suction valve. Thus the cycle is repeated.

The number of such cycles performed by any compressor depends on the speed at which it is operated. The amount of refrigerant vapor capable of being handled by a compressor may be determined if the diameter of the cylinder, the length of the stroke, the number of the cylinders, the speed of the compressor, and the volumetric efficiency are known, using the formula

(Volume of cylinder in cu ft − clearance volume in cu ft)

× number of cylinders × speed in rpm × volumetric efficiency* ÷ 100

= cu ft of vapor per min handled at the low pressure

Heat of Compression. Applying the law that energy is never destroyed to the work performed in compressing a gas, it is found that, during compression, the temperature of the gas rises. This rise in temperature is due to the transfer of work energy required to increase pressure of the gas. From Charles' law, it is known that the pressure varies directly with temperature, and, as a result, an additional force is created. This, in turn, causes an additional increase in pressure over and above that which is the result of the actual compression. This is due to the inhibited tendency of the gas to expand. It results in an equalization of pressures within the cylinder and the discharge line (Fig. 43) prior to the piston reaching point *B*, as occurred under isothermal compression, causing the discharge

* See Volumetric Efficiency, Chapter 11, "Refrigeration Calculations."

valve to open at the point B_1 and *not* at point B. It will be noted that the energy transferred to the gas during compression (mechanical energy to thermal energy) must be added to the energy absorbed in the evaporator to determine the total amount of internal energy contained in the fluid as it leaves the discharge valve. To this must be added the external energy applied to the fluid in compressing it as described in the discussion of enthalpy. So instead of a theoretically dry saturated vapor, it becomes a superheated gas.

One of the reasons for allowing a slight amount of liquid to enter the compressor, as is practiced in many ammonia installations, is to counteract the thermal effect of compression through the evaporation of the liquid carried over. Thus, the energy of compression to some extent is absorbed by the refrigerant upon evaporation as latent heat. This practice is thought by many refrigeration engineers to be an ideal method of reducing the heat resulting from compression of the gas.

Consider the basis upon which the pressure of the vapor leaving the compressor is determined, and why it must be maintained at that level. In the study of the evaporator, it was learned that a pressure necessary to control the boiling point of the refrigerant at or slightly below the temperature in the area to be cooled was required to produce evaporation.

As an example, consider liquid ammonia, the boiling point of which is −28 F at atmospheric pressure. If a pressure of 34.27 psia, or 19.5 psig, is maintained in the evaporator, the new boiling point of the ammonia is 5 F. If the temperature of the area to be cooled is at 20 F, evaporation occurs, and there is a saturated vapor at 5 F (assuming no superheating) entering the compressor.

Since the purpose of the refrigeration equipment, aside from the evaporator, is to prepare the refrigerant for re-use, it is first necessary that the vapor leaving the evaporator be converted back to its liquid form. As the equipment beyond the evaporator usually is exposed to atmospheric temperatures, the preliminary action is to raise the boiling point of the refrigerant to a point above the prevailing surrounding temperature. That is accomplished in the compressor.

Consider an ammonia refrigeration system operating under a prevailing atmospheric temperature of 80 F. Establishing a boiling point of 86 F, or 6° above the room temperature, it is found upon referring to the tables of saturated ammonia (Appendix) that a pressure of 169.2 psia, or 154.5 psig, is necessary to maintain a boiling point of 86 F for ammonia.

Under the above-stated conditions, it is the requirement of the compressor to raise the pressure of the ammonia vapor from a suction pressure of 34.27 psia to 169.2 psia, and maintain the final pressure continuously in the discharge side of the system.

In this example, where ammonia is used, the action applied is similar for any other refrigerant employed. Because the boiling temperatures of dif-

TABLE 28. PRESSURES OF VARIOUS REFRIGERANTS CORRESPONDING TO BOILING POINTS OF 5 F IN EVAPORATOR AND 86 F IN CONDENSER

Refrigerant	*Evaporator Pressure 5 F*		*Condenser Pressure 86 F*	
	psig	psia	psig	psia
Ammonia (NH_3)	19.6	34.27	154.5	169.2
Carbon dioxide (CO_2)	316.8	331.5	1028.3	1043.0
Freon-12 (CCl_2F_2)	11.81	26.51	93.2	107.9
Sulfur dioxide (SO_2)	*5.88 in. Hg	11.81	51.75	66.45
Methyl chloride	6.455	21.15	80.00	94.70
Freon-113	*27.92 in. Hg	0.9802	*13.93 in. Hg	7.856
Freon-21	*19.25 in. Hg	5.243	16.53	31.23

* Refrigerants operating under partial vacuum—inches of mercury.

ferent refrigerants vary, their evaporator and condenser pressures corresponding to the same boiling-point temperature will also vary. Table 28 shows comparative pressures of the various commerical refrigerants corresponding to boiling points of 5 F in the evaporator and 86 F in the condenser. (To obtain absolute pressures, add 14.7 psi to positive pressures, or convert negative pressures to psi and subtract from 14.7 psi.)

Location and Operation. There are various kinds of compressors employed in mechanical refrigeration. In addition to the reciprocating type described previously, centrifugal and rotary compressors also are used extensively.

The primary function of the compressor is to recapture the refrigerant as it is converted to vapor in the evaporator and to prepare it for re use. Preparation of the vapor consists of raising its pressure, the function of the compressor, to a level corresponding to a temperature at which it can be condensed or returned to its liquid phase by utilizing an available cooling medium, such as water, air, or even another refrigerant.

The compressor is located in the mechanical refrigeration system between the evaporator and the condenser. The former is connected to the suction side while the latter is connected to the discharge side of the compressor. The compressor is the dividing component that joins the low-pressure side to the high-pressure side of the system, within the vapor zone.

STEP 3. CONDENSATION

Laws, Principles, and Terms Involved. (*a*) Pressure, (*b*) thermal energy measurement—intensity or degree, quantity or Btu, (*c*) condensation, (*d*) latent heat of vaporization, (*e*) kinetic thermal energy—of the liquid—superheat, (*f*) total internal energy, (*g*) pressure–boiling-point

Fig. 44. Small condensing unit showing drive motor, condenser, and air-cooled condenser. (Courtesy, Tecumseh Products Company.)

relationship, (*h*) boiling point of liquids, (*i*) heat transfer—conduction, (*j*) heat flow.

Equipment. The refrigerant, upon leaving the compressor as a superheated gas, enters a condenser through the discharge line. The purpose of the condenser is to provide a means by which the vapor may eject the heat absorbed during its passage through the evaporator, as well as the thermal energy of compression, and reconvert it to a liquid.

The condenser is an arrangement of pipes or tubes, possessing high heat conduction properties, and so designed as to allow for the entrance (inlet) and the exit (outlet) of the refrigerant. These pipes or tubes are exposed to a continuous flow of cooling medium, usually water or air, which absorbs and dissipates the heat from the refrigerant.

Connected to the outlet of the condenser, and in some systems forming an integral part of the condenser itself, is a tank or reservoir known as liquid receiver. The purpose of the latter is to receive the liquid refrigerant upon condensing and to hold it until such time as it is required by the evaporator. The liquid receiver should be large enough to hold the entire refrigerant charge of the system in a liquid state.

The compressor, the condenser, and the liquid receiver combination is referred to as the condensing unit because it requires the services of all three of these pieces of equipment to complete the functions of the high-pressure side of a vapor compression system.

Condenser Functions. The function of the condenser is to convert the superheated gas to its liquid form. It is a heat exchanger involving the principle of temperature differential. It extracts from the vapor the latent heat of vaporization and the kinetic thermal energy absorbed by the refrigerant in the evaporator, plus the superheat resulting from compression. This results in transforming the vapor to a liquid at the boiling-point temperature corresponding to the pressure exerted by the compressor. Usually the removal of additional thermal energy from the liquid is necessary to reduce its temperature sufficiently below the boiling point so that it will remain in its liquid state in the receiver under surrounding temperature conditions.

Consider first the condition in which the vapor enters the condenser:

(*a*) *Vapor state*—result of evaporation within the evaporator.
(*b*) *Containing latent heat of vaporization*—absorbed during evaporation, in the evaporator.
(*c*) *Under pressure*—as a result of compression to raise its boiling point above prevailing temperatures of the condenser cooling medium, air or water. The actual pressure depends on (1) the temperature of the surrounding atmosphere if air cooled, or of the available water supply if the latter is employed, and (2) the boiling point of the refrigerant at normal atmospheric pressure.
(*d*) *Superheated*—caused by energy of compression. An additional source of superheat originates within the evaporator when the supply of the liquid refrigerant is not great enough to meet load demands. This condition results from exposure of the vapor to the heat from the area being cooled after evaporation has taken place. Superheating is often allowed to occur purposely to prevent liquid being carried over to the compressor.

In view of these conditions, consider what action is necessary within the condenser to transform the vapor to a liquid at a temperature that will establish liquid stability under existing pressure.

1. *Eliminate the superheat contained within the vapor*, thus reducing its temperature to a level equal to its new boiling point. In so doing, heat is extracted in the form of kinetic thermal energy and in accordance with the principle of temperature differential.

2. *Extract the latent heat of vaporization.* Referring now to the definition of latent heat of vaporization, it is that energy required to change a liquid to a vapor without changing the temperature. Conversely, if this energy is extracted, the action is reversed and the vapor reverts to its liquid form. This is what actually takes place within the condenser when the latent heat of vaporization absorbed in the evaporator is extracted. This is known as condensation.

3. *Reduce the temperature of the refrigerant*, now in its liquid form somewhat below the new boiling point corresponding to the condenser pressure to assure its remaining in liquid form. This is accomplished by extracting energy in the form of kinetic thermal energy, again involving the principle of temperature differential.

EXAMPLE: Assume ammonia gas leaving the compressor, superheated at

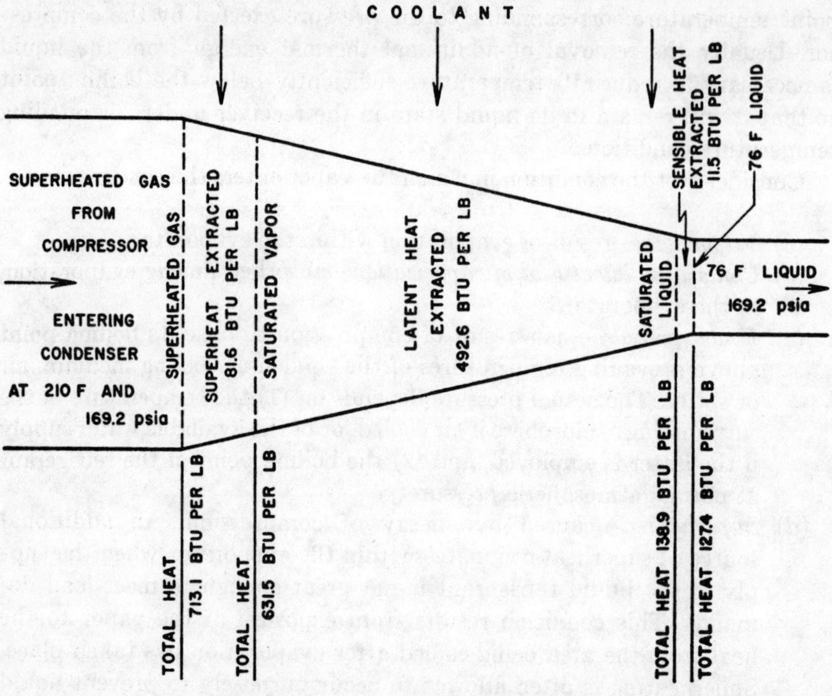

Fig. 45. Behavior of ammonia from condenser to receiver.

a temperature of 210 F. The system operates at 5 F suction temperature and 86 F condenser temperature. Allowing for an additional cooling of the liquid to 76 F by the condenser, determine the amount of thermal energy contained in each pound:

DATA:

Refrigerant	Ammonia
Superheat temperature	210 F
Condenser temperature	86 F
Condenser pressure	154.5 psig
Evaporator temperature	5 F
Evaporator pressure	19.6 psig
Reference point	−40

1. *Superheat*—Reference, "Properties of Superheated Ammonia Vapor" tables. At 154.5 psig pressure and 210 F superheat, ammonia vapor contains 713.1 Btu per lb. Saturated ammonia vapor at 154.5 psig contains 631 Btu per lb. The amount of superheat contained in each pound of vapor is 713.1 − 631.5, or 81.6 Btu per lb.

2. *Latent Heat of Vaporization*—Reference, "Saturated Ammonia Temperature" tables (Appendix). Heat of ammonia vapor at an 86 F boiling point and 154.5 psig has a latent heat of vaporization value of 492.6 Btu per lb.

3. *Enthalpy of Liquid*—Reference, "Properties of Liquid Ammonia," tables. Liquid ammonia at 86 F contains an enthalpy of 138.9 Btu per lb and at 76 F contains 127.4 Btu per lb. The difference equals the amount to be extracted in reducing the temperature of the liquid from 86 F to 76 F, or 138.9 − 127.4 = 11.5 Btu per lb.

To determine the amount of thermal energy to be extracted from each pound of ammonia entering the condenser, the above values are added:

$$81.6 + 492.6 + 11.5 = 585.7 \text{ Btu per lb}$$

Location and Operation. The condenser is located in the mechanical refrigeration system between the compressor and the throttling device. Actually, the refrigerant flows from the condenser to a liquid receiver, thence to the throttling device. For purposes here, the liquid receiver is considered as part of the condenser which in some systems it is.

There are a number of types of condensers. The more important types include surface, atmospheric, double pipe, evaporative and air-cooled condensers, all of which are described in Chapter 8, "Mechanical Equipment."

Ratio of Compression. The compression ratio is the quotient obtained by dividing the condenser pressure, absolute, by the evaporator pressure, absolute. For reasons of thermal and mechanical efficiency, as discussed in Chapter 11, "Refrigeration Calculations," the compression ratio should be kept as low as practical operation will permit. Therefore, the condenser pressure should be maintained at the lowest possible level as one step in the right direction.

STEP 4. PRESSURE REDUCTION

Laws, Principles, and Terms Involved: (*a*) Pressure differential, (*b*) high side—low side, (*c*) pressure, (*d*) pressure–boiling-point relationship, (*e*) temperature, (*f*) boiling point of liquids, (*g*) expansion, (*h*) flash gas.

Equipment. Up to this point in the refrigeration cycle, the refrigerant has been followed while it completed its function of cooling by evaporation, reclaimed, compressed to raise its boiling point, condensed to a liquid, and stored in a liquid receiver under pressure. At this point, it is ready for re-use and is able to again repeat its cooling function by evaporation when released in the evaporator.

As the refrigerant in its liquid state is under high pressure in the receiver and ready for re-use in the evaporator where a much lower pressure is maintained, some means of retaining the pressure difference is required. This is accomplished by placing a throttling device in the line between the receiver and the evaporator to restrict the flow of refrigerant to the evaporator. This restriction makes possible the maintaining of a high pressure on one side of it while allowing a lower pressure condition to exist on the other side, while flow of the refrigerant occurs.

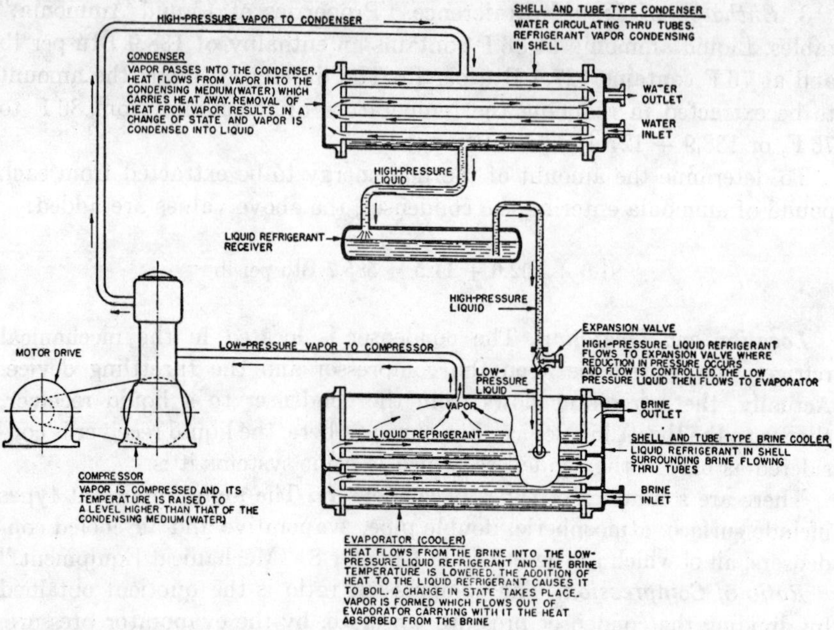

Fig. 46. Flow diagram of refrigeration system. (Courtesy, Worthington Corporation.)

The throttling device contains a small adjustable opening that limits the amount of refrigerant passing through it. There are various types of throttling devices in use today, differing in their restrictive methods but all serving the same purpose. These include the hand expansion valve, capillary tube, high-side and low-side float valves, and automatic and thermostatic expansion valves. (See Chapter 9, "Controls.")

Equipment Functions. To better understand the function of the throttling device or expansion valve, first consider the conditions that require its employment. Primarily, it is the dividing mechanism within the vapor zone between the high-pressure side and the low-pressure side of the refrigeration system. It is directly in line connecting the receiver containing the liquid refrigerant under high pressure and the evaporator in which a much lower pressure is maintained. If the expansion valve is closed, the refrigerant does not flow and consequently there is no activity.

Upon opening the valve slightly, the liquid in the receiver, being under pressure, is forced through the valve into the evaporator at a rate governed by the amount of valve opening. While under pressure in the receiver, the boiling point of the refrigerant is above that corresponding to the surrounding room temperature, and, as a result, retains its liquid form. Upon entering the evaporator, its boiling point is immediately reduced to a level corresponding to the lower pressure. The temperature of the space or of the substance to which the evaporator is exposed, being higher than the new boiling point, causes the liquid to boil. The refrigerant vapor is drawn out of the evaporator as boiling occurs, giving room for additional liquid to enter through the expansion valve. This continual evacuation of vapor prevents an increase in pressure within the evaporator and also helps to control superheating.

While the throttling device is often referred to as an expansion valve, its real purpose is to achieve a pressure reduction and consequently a lower boiling point of the liquid passing through it. That expansion occurs as a result of immediate evaporation beginning in the throttling device is an unfortunate circumstance that presents disadvantages involving both refrigerating effect and liquid flow.

As stated previously, a loss of refrigerating effect occurs as the liquid cools itself upon reduction of pressure. During the process, some volume is lost by evaporation as flash gas. This loss may be reduced by precooling the liquid before it enters the throttling device. To eliminate it entirely would necessitate precooling the liquid to a temperature equal to that correspond-

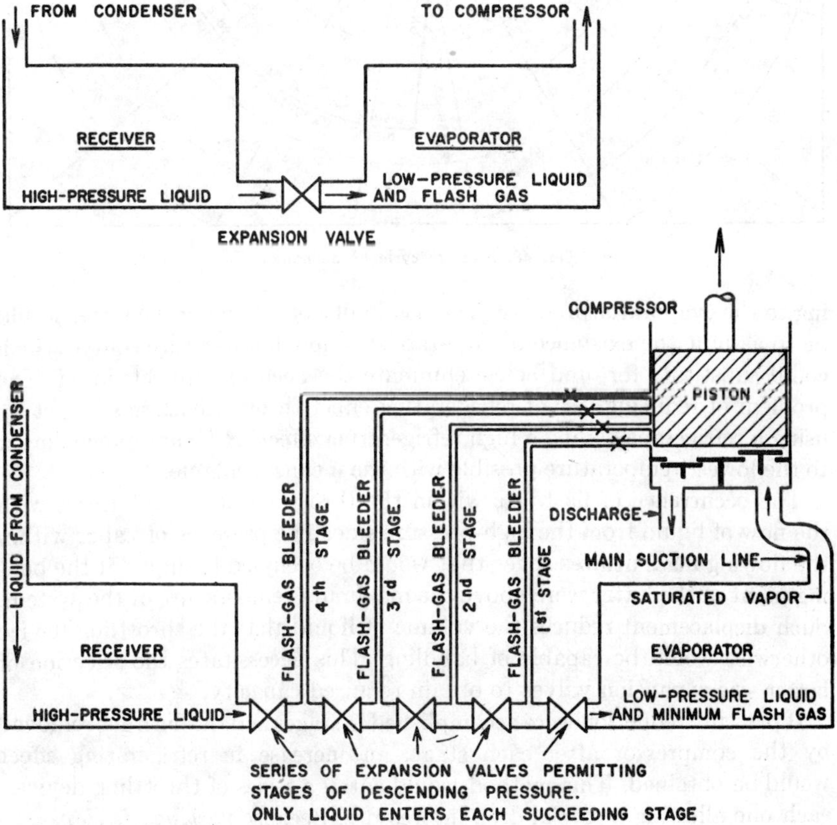

Fig. 47. Single vs. multi-expansion. (*Top*) Single expansion system—maximum flash gas developed in evaporator. (*Bottom*) Multi-expansion system (theoretical)—minimum flash gas developed in evaporator.

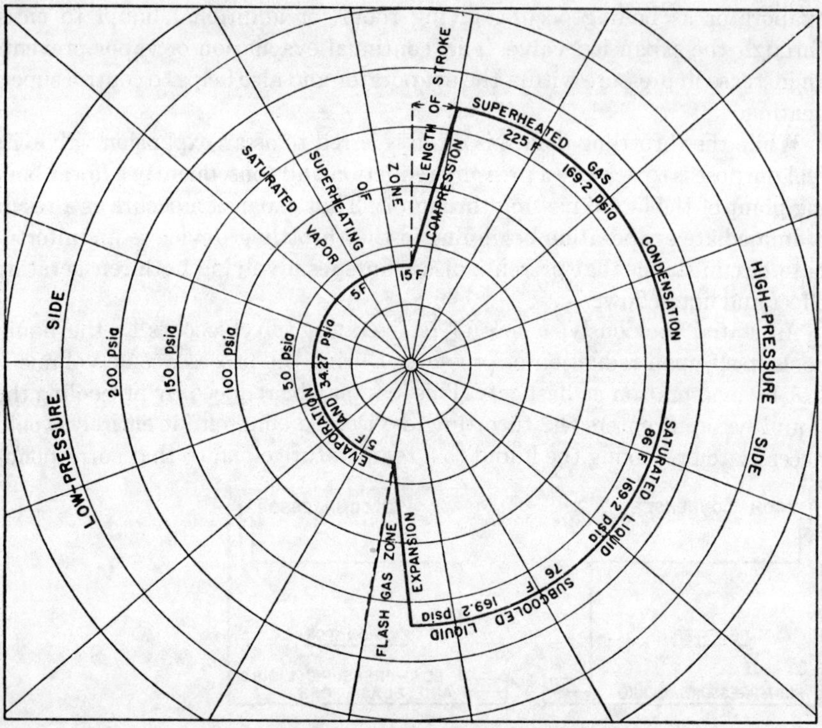

Fig. 48. Pressure cycle of ammonia.

ing to the evaporator pressure. The possibility of achieving the latter would be to assume the existence of an established low-temperature source which could substitute for, and hence eliminate the need of, subject liquid. The problem of reducing loss of refrigerating effect through flash gas is met by using a refrigerant having a high refrigerating effect (R E) and precooling it to the lowest temperature possible with the means available.

The occurrence of flash gas within the throttling device interferes with the flow of liquid from the high-pressure side. The presence of vapor within the flowing mass utilizes space that would be occupied by liquid if the boiling point of the latter were above the maximum temperature of the system. Such displacement reduces the volume of liquid that the throttling device otherwise would be capable of handling. This necessitates the selection of larger size expansion valves to obtain required capacity.

If pressure reduction were accomplished in stages and the flash gas removed by the compressor after each stage, an increase in refrigerating effect would be obtained. This method would entail a series of throttling devices, each one allowing a partial drop in pressure. Because *flash gas develops progressively with diminishing pressure*, it is possible to draw saturated vapor from each stage. Having served its usefulness in reducing liquid temperature to

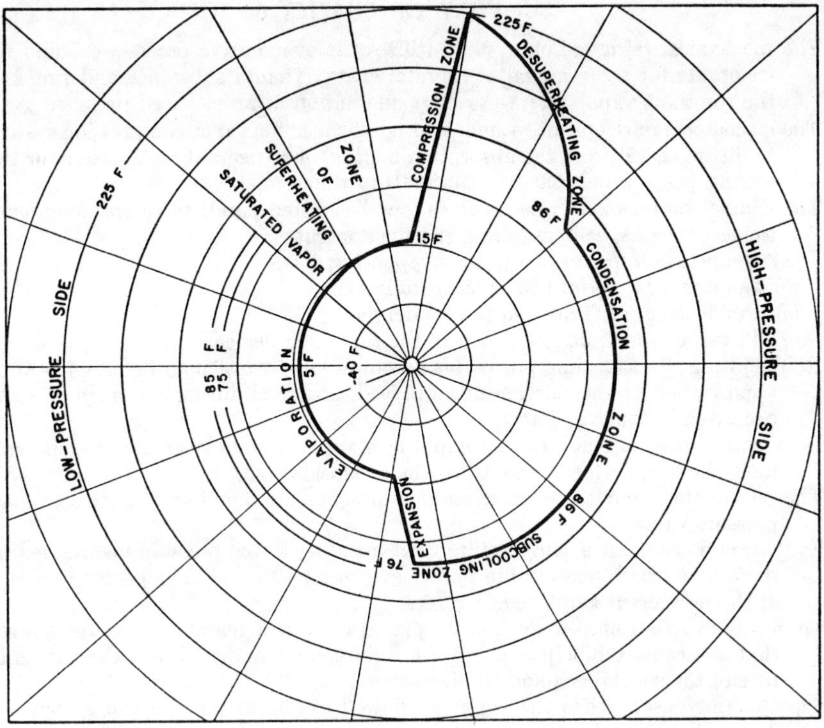

Fig. 49. Temperature cycle of ammonia.

the extent of the pressure drop in each stage, the drawn flash gas is spared undergoing a further useless pressure reduction. (See Fig. 47, *bottom.*)

Two advantages are gained through stage expansion: (1) It reduces the work of the compressor by an amount proportional to the accumulated useless pressure drops spared the several portions of flash gas at their respective expansion levels. (2) Only saturated liquid remains to proceed through each succeeding pressure level, resulting in a total increase in refrigerating effect.

The complications and the cost of installing and operating a multi-stage expansion system are obvious. Notwithstanding the thermal advantages, its general adoption is impractical.

Practical adaptation of stage expansion is found in centrifugal compression systems. The employment of an economizer as an intermediate step in reducing condenser pressure to evaporator pressure is described in Chapter 8, "Mechanical Equipment."

Maintaining a lower evaporator temperature than is necessary adds to refrigerating losses. The greater pressure drop increases the amount of flash gas resulting in a lower refrigerating effect. This is one reason that the difference in pressure between the high and the low sides of the system should be kept at a minimum.

SUMMARY OF CHAPTER 6

The mechanical refrigeration system utilizes the evaporative process of liquid refrigerants for the removal of thermal energy from a substance and prepares the resultant vapor for re-use by condensation at an elevated pressure.

The mechanical refrigeration system is also known as the vapor-compression system to distinguish it from the absorption method of refrigerant reclamation, or the cooling process induced by temperature differential.

The simple compression cycle involves four basic steps, and, therefore, four functional processes, each requiring specific apparatus:

(*a*) Evaporation, performed in the evaporator,
(*b*) Compression, performed in the compressor,
(*c*) Condensing, performed in the condenser,
(*d*) Pressure reduction, performed in the throttling device.

Refrigerating effect (cooling process) is accomplished by utilizing the latent heat of vaporization of the refrigerant employed, absorbed during a change of state occurring in the evaporator.

Because a change of state, from a liquid to a vapor, results in greater specific volume, the evaporator is also called the expansion coil.

The refrigerant enters the evaporator as a low-pressure liquid and leaves as a low-pressure vapor.

Evaporation occurs at a temperature corresponding to the pressure maintained in the evaporator, a relationship that is governed by the individual characteristics of the refrigerant employed.

To induce evaporation, the evaporator pressure must be maintained at some level that will cause the boiling point of the refrigerant to drop below the temperature of the substance under refrigeration.

Superheating occurs when the vapor is allowed to absorb thermal energy beyond its temperature of saturation at the evaporator pressure. A rise in temperature of the suction vapor increases as a result of superheating.

In order to produce refrigeration economically, the spent refrigerant leaving the evaporator, now a vapor, must be reclaimed.

One of the functions of the compressor is to reclaim the refrigerant vapor as it leaves the evaporator.

The second function of the compressor is to raise the pressure of the vapor to a level that will cause its boiling point to reach a temperature above, or at least equal to, that of the surrounding atmosphere.

The pressure to which the vapor must be raised by the compressor is governed by the temperature of the surrounding atmosphere and the pressure-temperature relationship of the refrigerant employed.

Evaporator pressure subtracted from the compressor discharge pressure is the pressure range through which work is done.

Compressor discharge pressure, also known as the head or condenser pressure, divided by the evaporator or suction pressure, is equal to the compression ratio, or

$$\frac{\text{Head pressure psia } (P_2)}{\text{Suction pressure psia } (P_1)} = \text{Compression ratio}$$

Higher compression ratios are accompanied by increased power requirements and lower operating efficiencies.

The mechanical work done during compression is transferred to the vapor in the form of thermal energy.

It is the energy of compression that causes the temperature of discharged vapor to be higher than the suction temperature, by an amount depending on the compression ratio, for a particular fluid.

Upon discharge from the compressor, the vapor enters the condenser in a superheated condition.

The refrigerant leaves the condenser as a liquid under high pressure and enters a receiver where it is stored for re-use.

The throttling device throttles the flow of refrigerant from the receiver to the evaporator to cause a drop in pressure.

Either in the form of an expansion valve, a float valve or a capillary tube, the throttling device separates the low- and the high-pressure sides of the refrigeration system and controls the evaporator pressure.

The purpose of the condenser is to allow the refrigerant to eject thermal energy and reconvert to its liquid form.

The condenser cooling medium may be air, water, brine, or another refrigerant.

The total amount of thermal energy ejected in the condenser by the refrigerant is equal to the sum of the energy absorbed in the evaporator, plus the energy of compression, or superheat.

Because the thermal energy absorbed in the evaporator is in the form of latent heat of vaporization, its ejection will cause the fluid to return to its liquid form.

For economy of operation, the evaporator pressure should be maintained as high as is practicable for the condition to be maintained.

Subcooling the liquid refrigerant prior to its entering the expansion valve reduces the amount of flash gas necessary to lower its own temperature to that being maintained in the evaporator.

Superheating of suction vapor within the evaporator increases refrigerating effect but reduces the amount of refrigerant handled by the compressor due to the subsequent decrease in density.

The ideal condition for suction vapors to enter the compressor is saturated at the evaporator temperature.

While entrained liquid refrigerant entering the compressor suction offsets to some extent the energy of compression, the possibility of causing damage by slugging is increased as a result of the practice.

STUDY OUTLINE FOR CHAPTER 7

Thermodynamic property tables

Enthalpy of a working fluid

Entropy

The temperature-entropy or *Ts* diagram

The pressure-enthalpy or *Ph* diagram

Power requirements of a refrigeration system as determined from the *Ph* diagram

Effect of liquid subcooling on the *Ph* diagram

Effect of heat exchanger on the *Ph* diagram

Coefficient of performance of a refrigeration system

Theoretical vs. practical considerations:
- Drop in pressure due to friction
- Rise in vapor temperature
- Volumetric efficiency
- Loss of heat in discharge lines
- Increase in head pressure

Effect of different compression ratios on the *Ph* diagram

The Mollier diagram

The Carnot cycle:
- Reversible vs. irreversible work processes

Relative efficiency of refrigeration systems

7

Thermodynamic Functions

The science of refrigeration is based on thermodynamic functions. Practical operation of any system involves factors that interfere with ideal results. Nevertheless, theoretical considerations are necessary for proper design, including equipment sizes, quantity of refrigerant handled, and the amount of cooling medium required. It is the difference between theoretical possibility and actual refrigeration production that determines the efficiency of operation. The sum of the efficiencies of the individual components is the over-all efficiency of the system. To obtain maximum efficiency of any system, every step must be taken to conserve energy during operation. What those steps should be can be determined only by ascertaining theoretical performance values. For example, the advantages or the disadvantages of adopting such steps as stage compression, liquid subcooling by internal heat exchange, superheating of vapor, higher or lower evaporator pressures, etc., may be decided previously by theoretical evaluation based on thermodynamic functions.

While the thermodynamic processes within the refrigeration cycle are few, they are complicated. In the study of the evaporator, the efficiency of stage expansion was shown. In a later chapter, the advantage of stage compression under certain operating conditions will be discussed. These are only two of the several refrigeration processes that may be evaluated through application of thermodynamic functions.

Thermodynamic functions involve pressure, volume, and temperature, all of which are physical properties of substances such as refrigerants. The thermodynamic properties of the same substances are functions of P, V, and T and deal with energy as related to refrigerants in a given state at any point in the work-flow process. Thermodynamic properties of refrigerants are (1) internal energy, (2) enthalpy, and (3) entropy.

The principal functions of refrigerants are those that associate physical and thermodynamic properties according to a given pattern of ascending or descending values. Such associations are accomplished primarily by graphic description. Hence the relationship of physical properties is shown on the pressure-volume curves discussed in Chapter 4 and the pressure-temperature curves in Chapter 5. This chapter deals with the association of physical properties to thermodynamic properties, the functions of which are developed graphically in the form of (*a*) temperature-entropy (*Ts*) diagrams and (*b*) pressure-enthalpy (*Ph*) diagrams.

The reversible and the irreversible work processes and their relationship to the Carnot cycle are also dealt with in the following discussion.

THERMODYNAMIC PROPERTY TABLES

The thermodynamic medium is the basic factor in the operation of certain work systems. For example, water is the required medium in steam power plant operation; ammonia, carbon dioxide, halogenated hydrocarbons, etc., are important media employed in refrigeration systems. Intelligent design of such work systems requires detailed knowledge of the capacity of the medium used to perform its work according to sound scientific prediction. Prediction in this case is based on accurate mathematical calculations. Calculations are dependent on known numerical values associated with the behavior of thermodynamic media under all physical conditions related to the work system. The tables of thermodynamic properties of fluids furnish the necessary data to enable the designer to accurately estimate the physical requirements of a system to meet a given work demand. Such tables are important reference tools for the thermodynamic engineer.

Thermodynamic Property Tables (see Appendix, Tables I, II, III) covering various working fluids consist of certain numerical values representing the various state points and thermal properties of individual media at any given state point. Such tables are usually organized according to the state of the fluid. Therefore, separate tables covering the properties of the *liquid, dry saturated vapor* and *superheat—vapor*, of a given fluid are published in appropriate literature.

The dry saturated vapor tables include the state points of temperature (°F), absolute pressure (psia) and specific volume (cu ft/lb). Some published tables also include density (lbs/cu ft) and specific heat (Btu/lb) for the state points listed. The thermal properties listed in the abovementioned tables include the enthalpy values of the liquid, evaporation process (latent heat) and dry saturated vapor, the latter being usually referred to as *total enthalpy* of the vapor. Formerly enthalpy of the various state conditions was called heat content of the liquid or vapor and so listed in older tables. The enthalpies for the several state conditions are designated by the letter symbols h_f for liquids, h_{fg} for latent heat and h_g for dry saturated vapor and stated in terms of their specific values in Btu per lb. Another thermal property included is entropy designated by the letter s, which is modified by an appropriate subscript for the various states, i.e., s_f for liquid, s_{fg} for evaporation and s_g for saturated vapor.

Values of entropy are stated in specific terms of Btu/lb/°F.

The Superheat Tables list specific volume (v) in cu ft per lb, total enthalpy (h) of the superheated gas in Btu per lb and entropy (s) in Btu per lb, °F. The values v, h and s are listed to correspond with absolute pressure (psia) and temperature (°F) of the superheated gas.

Tables covering the liquid phase of fluids either separately or combined with saturated vapor data usually include specific volume of the liquid at the various state points.

ENTHALPY

Consider all energy factors that make up a working system and which were described in Chapter 2. To demonstrate the first law of thermodynamics (*energy lost equals energy gained*), it is necessary to organize such factors to show an association with one another in which "energy in" and "energy out" of a working system are in balance.

The energy factors involved are given below along with their symbol, nature and term of expression:

J = Work—heat equivalent	Energy transformation	Energy conversion (position)	ft lbs/Btu
Z = Potential energy	Stored mechanical energy	Height above datum	ft lbs/lb
$\frac{V^2}{2g}$ = Kinetic energy	Stored mechanical energy	Due to motion	ft lbs/lb
pv = Flow energy	Stored mechanical energy	Maintain motion	ft lbs/lb
u = Internal energy	Stored thermal energy	Magnitude above datum	Btu/lb
W = Work	Mechanical energy in transit	Force × distance	ft lbs/lb
q = Heat	Thermal energy in transition	Temperature differential	Btu/lb

If Z_1, $V^2{}_1$, p_1v_1, u and q_c represent the input energies where q_c is the heat added or subtracted during the work flow interval, let Z_2, $V^2{}_2$, p_2v_2, u_2 and W_c represent the output energies where W is the work done on or by the working substance, the subscript c denoting change in both cases.

The association of the above energies when combined as an equation in which both sides are equal composes the *general energy equation,* viz.:

$$\frac{Z_1}{J} + \frac{V^2{}_1}{2gJ} + \frac{p_1v_1}{J} + u_1 + q_c = \frac{Z_2}{J} + \frac{V^2{}_2}{2gJ} + \frac{p_2v_2}{J} + u_2 + \frac{W_c}{J}$$

input = output

= Btu per lb

Note that the equation includes three properties of the thermodynamic medium, absolute pressure (p), specific volume (v) and internal energy (u) on both sides. It also includes the mechanical energy equivalent of thermal energy or the Joule equivalent having the symbol J and the numerical value of which is constant at 778 ft lbs per 1 Btu. All three properties of p, v and u are fixed according to the state of the fluid. The value of J being constant, the term $\frac{pv}{J}$ is fixed by the properties p and v. The property of u is also fixed by the state of the fluid. Because p, v and u are the only properties in the general energy equation and J is constant, the sum of $u + \frac{pv}{J}$ is also fixed by the state. Combining these terms a convenient new property appears which avoids the complex calculation involving all properties. By combining the properties as shown, the internal energy u and the flow work $\frac{pv}{J}$ can be determined by a single value. The sum of the two terms is called the *enthalpy* of the fluid at a given state. The new property specific enthalpy is indicated by the letter h, whose value is expressed in Btu per lb and included in published tables of thermodynamic properties of fluids. The following is a demonstration of enthalpy determination.

Supposing that 1 lb of liquid refrigerant is placed in an open vertical cylinder of extended length, but whose area measures 1 sq ft. Consider the refrigerant to be ammonia, whose temperature is -40 F, under which condition it would have a density of 43.08 lb per cu ft. This means that 1 lb would require a space in the cylinder 1 sq ft in area by $\frac{1}{43.08}$, or 0.02321 ft high. By applying enough heat to raise the temperature of the ammonia to -28 F, its volume would increase, thereby decreasing its density according to the equation $D = \frac{1}{V}$. The density of liquid ammonia at -28 F is 42.56 lb per cu ft, so that its new height in the cylinder would be $\frac{1}{42.56}$, or 0.02349 ft.

The cylinder, being open to the atmosphere, allows a pressure of 14.7 psi to be exerted on the surface of the ammonia. This means that a total of 14.7 psi × 144 sq in., or 2116.8 lb pressure, is exerted against the ammonia during its expansion from a height of 0.02321 ft to 0.02349 ft, or through an expanded distance of 0.00028 ft. The energy exerted by the liquid to overcome the resistance of atmospheric pressure, then, is 0.00028 ft × 2116.8 lb, or 0.593 ft-lb. 777.46 ft-lb is the mechanical equivalent of 1 Btu, so that $\frac{0.593}{777.46}$, or 0.00076 Btu, was required to effect expansion. This value, therefore, can be interpreted as being the energy required to perform *external work*.

While the energy required to expand liquid volume may be of negligible value, it does not remain so when a change of state occurs. This will be noted upon continuing the heating process in the cylinder to bring about evaporation.

Liquid ammonia has an average specific heat of 1.064 between the temperatures of −40 F (at which point the heat content of liquid is assumed to be 0.0 Btu per lb) and −28 F. The heat content of −28 F liquid ammonia would be 1.064 [(−40) − (−28)], or 12.768 Btu per lb, which corresponds with the figure 12.8 shown in the table. Part of the latter amount, or 0.00076 Btu, is energy resulting from external work as determined above.

By adding sufficient heat to the liquid ammonia to convert it to saturated vapor at −28 F, its volume will increase so that it will require a space within the cylinder of 18 cu ft. Because the cylinder has an area of 1 sq ft, the height of vapor will be 18 ft. Expansion here is effected against a total pressure of 2116.8 lb* so that the energy required to overcome atmospheric pressure during evaporation amounts to 2116.8 × (18 − 0.02349), or 38,060 ft-lb. The external energy, then, would be $\frac{38,060}{777.46} = 49$ Btu per lb.

From this, it appears logical that the Energy of Vaporization column in the tables may be divided into two subheadings; i.e., Internal Energy and External Work or Flow Energy—or, to further simplify the table, include the latter values in separate columns. If this procedure were followed, the columns in the property tables would be in the nature of those shown below, using the −28 F line as an example.

PROPERTIES OF SATURATED AMMONIA: TEMPERATURE TABLE
(REFERENCE POINT −40 F)

1 *Temperature*	2 *Pressure*		3 *Volume*	4 *Density*	5 *Energy of Liquid*	6 *Energy of Vapor*	7 *Energy of Vaporization*	8 *Internal Energy*	9 *External Energy*
°F	psia	psig	ft³/lb	lb/ft³	Btu/lb	Btu/lb	Btu/lb	Btu/lb	Btu/lb
t	p	p.g.	V	$1/V$	h	H	L	h_1	e
−28°	14.7	0.0	18.00	0.05555	12.8	602.1	589.3	540.3	49.0

The values shown in column 7 as "Energy of Vaporization" (L) are the latent heat of vaporization values of the refrigerant corresponding to prescribed temperature and pressure. Because the saturated vapor contains energy that was required in expanding to the corresponding volume, shown

* 14.7 psia × 144 sq in.

under V (column 3), and also required to maintain itself in that condition, H must be considered as being energy in its combination of forms. The same reasoning must be applied to column 5, "Energy of Liquid." Carried to its ultimate conclusion, "H," signifying energy of vapor listed in column 6, includes the kinetic thermal energy required, first, to raise the temperature of the liquid from the reference point of −40 F to the boiling point at the prescribed pressure, and secondly, to effect vapor conversion at constant pressure and temperature. It also contains the energy required to expand against the pressure under which the process occurs. Energy of the vapor "H," then, is the sum of the individual energies, viz.:

$$\text{SUM}\left\{\begin{array}{l}\left.\begin{array}{l}\text{Internal energy of liquid}\\ \text{External energy of liquid}\end{array}\right\} h\\ \text{Internal energy of vaporization } (h_1)\\ \text{External energy of vaporization } (e)\end{array}\right\} = \begin{array}{l}\text{total energy content}\\ \text{of vapor } (H)\end{array}$$

which, for saturated ammonia vapor under atmospheric pressure, is shown to be

$$h + h_1 + e = H$$

or

$$12.8 + 540.3 + 49 = 602.1 \text{ Btu per lb}$$

Column 6 is equal to the sum of columns 5 and 7, while column 7 is the sum of columns 8 and 9. Column 5, "Energy of Liquid," may also be divided to show internal and external energy values, but, because of negligible quantities involved for the latter, it would serve no useful purpose.

Vapor contains that energy required to overcome the pressure exerted during the process of expansion, or flow of work, in addition to the energy necessary to raise the temperature of the liquid and cause a change of state. It follows that all the energy cannot have originated as thermal energy even though such external work required to cause changes is expressed as Btu per lb. Therefore, it is illogical to brand the energy values as shown in either columns 5 or 6 as true thermal content, especially during the period of fluid flow which is continuous in a refrigeration cycle. To eliminate the erroneous implication, such values are described as the *enthalpy* of the fluid.

Enthalpy, then, is defined as the internal energy (u), plus the value representing flow energy of a working fluid mathematically expressed by the equation:

$$H = u + \frac{pv}{J}$$

where u = internal energy of the fluid, in Btu per lb, above the datum

and $\frac{pv}{J}$ = external work or flow energy in Btu per lb

in which p = pressure in psf, against which expansion occurs

v = specific volume, cu ft per lb

J = 777.46 ft-lb (mechanical energy equivalent of 1 Btu)

Substituting values already considered above for ammonia at −28 F and 14.7 psia,

$$\text{Enthalpy } H = u + \left(\frac{Pv}{J}\right)$$

$$= 540.3 + \left(\frac{14.7 \times 144 \times (18 - 0.02349)}{777.46}\right)$$

$$= 540.3 + 49$$

$$= 589.3 \text{ Btu per lb, energy of vaporization}$$

$$589.3 + 12.8 = 602.1 \text{ Btu, total enthalpy}$$

In the past, the practice of listing the internal and the external energy values was followed in compiling thermodynamic property tables. For some scientific purposes, such values serve a useful purpose. In solving practical refrigeration problems, very little if any use is found for such data, and, consequently, it has been eliminated from tables now published. If required, internal and external energy values are determined easily, as shown above. Of greater importance to the field are the values listed for L, latent energy of vaporization; h, energy of liquid; and H, total energy content of vapor.

ENTROPY

Practical work systems are *irreversible*. This means that the available energy contained *within* the system diminishes during the process of doing work. According to the law of energy conservation, energy is never destroyed, but converts to thermal energy at a lower level of availability as work processes continue. Unless replenished from a source external to the work system, the availability of internal energy decreases until it reaches a level of thermal equilibrium with the surrounding atmosphere. When thermal equilibrium is accomplished, energy availability as related to the conditions peculiar to the work system involved ceases to exist. When energy is no longer available, work ceases.

Heat flows from a warm body to a colder one. In actual work systems, heat flow continues until it reaches the lowest temperature prevailing for heat discard. If the lowest temperature is representative of the surrounding atmosphere, then the atmosphere is considered to be the heat sink whose temperature is T_o. In the refrigeration process, the water from melted ice or the condenser cooling medium may act as the sink, being the discard of the heat from substances being cooled.

According to Clausius, the German physicist, the universe tends toward maximum energy unavailability as thermal processes continue. The measure of energy unavailability, as expressed by Clausius, is a thermodynamic function known as entropy. Entropy is taken from the Greek to denote a "turning in" process. According to Clausius, then, entropy is the measure of the *unavailable* thermal energy within a substance for conversion into mechanical work. Obviously it is not a physical property of a substance in the same sense as pressure, volume, temperature, etc.

If the work system is to continue to function, the energy that becomes unavailable in the process must be replaced constantly from an external source. In order to continue the work process at a desired rate, it is necessary to determine the amount of spontaneous energy that is available at all times. Stated in another way, the amount of energy constantly becoming unavailable within the work system must be known in order to develop an equilibrium of input to output.

Entropy is an important factor in computing compressor power requirements. Its importance lies in the need for an instrument that serves as an index of energy availability under any specific set of conditions related to the refrigerating fluid. In this connection, it is the change in entropy rather than the absolute entropy which is of interest. Absolute entropy is based on a reference point of absolute zero in which area specific heat of substances is unknown. Entropy is indicative of the inherent ability of a fluid to transform itself from one state or condition to another state or condition spontaneously. Spontaneous transformation is always accompanied by a decrease in available heat energy. It follows that entropy also indicates the amount of heat energy required from external sources to return the fluid to its condition prior to spontaneous transformation, or to any desired condition above the reference point.

The total energy of a body is equal to the sum of its available energy plus its unavailable energy. Entropy is a numerical factor that indicates relationship of two kinds of energy during work processes when considered in conjunction with changes in thermodynamic properties of fluids.

Change in entropy occurs as a result of any one or all of the following: (*a*) energy conversion to friction during the work process, (*b*) introduction of heat from a source external to the work system, (*c*) loss of heat to an external source, (*d*) heat transfer between parts of the work system.

Under adiabatic conditions, wherein no gain or loss of thermal energy occurs, entropy remains constant. Adiabatic processes are said to be *isentropic*.

Because entropy is subject to change in accordance with changes in thermal conditions, it becomes a mathematical function of working substances. Entropy is related to any change in the properties of fluids such as pressure (p) psi, specific volume (v) cu ft per lb, temperature (T) °F absolute, enthalpy (h) Btu per lb, and internal energy (h_1) Btu per lb. Entropy is designated as s and expressed as Btu per lb °F. The properties h and h_1 as well as s are functions of p, v, and T.

Entropy determinations, as listed in the thermodynamic property tables, are based on the respective reference points of the various fluids. For steam, the reference point or datum is 32 F, while the datum for refrigerant fluids is -40 F. At the temperature of reference, the enthalpy of liquids (h) is assumed to be zero. Therefore, the enthalpy of the vapor under saturated conditions at the reference point is latent heat of vaporization.

Entropy of a substance is the sum of the separate entropies of the liquid, evaporation, and vapor, where

(*a*) entropy of the liquid (s) equals the sum of the increment changes in enthalpy divided by the average absolute temperature at which the enthalpy change occurs. Expressed as a mathematical equation,

$$\text{Entropy of liquid} = ds$$

where $ds = \dfrac{dh}{T}$

in which ds = change in entropy

dh = increment change in enthalpy

T = average absolute temperature during enthalpy change

(*b*) entropy of evaporation (S_{evap}) equals the quotient of latent heat of vaporization divided by the absolute temperature of saturation, or

$$S_{evap} = \frac{L}{T}$$

where S_{evap} = entropy of evaporation

L = latent heat of vaporization

T = absolute temperature of saturation

(*c*) entropy of the vapor (S) equals sum of the entropy of the liquid plus entropy of evaporation, or

$$S = \text{sum } ds + \frac{L}{T}$$

$$S = s + \frac{L}{T}$$

EXAMPLES:

1. Find the entropy of the vapor for steam saturated at 32 F.

$$\text{Entropy of the liquid at 32 F} = 0.0$$

$$\text{" " evaporation} = \frac{L}{T}$$

$$\text{" " "} = \frac{1073.4}{459.58 + 32}$$

$$\text{" " "} = 2.183$$

which at the reference point is also the entropy of the fluid.

2. Find the entropy of the vapor for ammonia saturated at −40 F.

$$\text{Entropy of liquid at } -40 \text{ F} = 0.0$$

$$\text{Entropy of evaporation} = \frac{L}{T}$$

$$S_{evap} = \frac{597.6}{459.6 + (-40)} = \frac{597.6}{419.58}$$

$$S_{evap} = 1.4242$$

which at the reference point is also the entropy of the vapor.

3. Determine the entropy of steam saturated at 33 F.

$$\text{Entropy of liquid } s = \text{sum } \frac{dh}{T}$$

$$ds = \frac{\text{specific heat of liquid}}{\text{average temperature change}}$$

For water,

$$ds = \frac{1}{459.58 + 32.5} = \frac{1}{492.08}$$

$$ds = 0.0020 \text{ change in entropy of liquid}$$

$$\text{Entropy of evaporation} = \frac{L}{T}$$

$$= \frac{1072.9}{459.58 + 33} = \frac{1072.9}{492.58}$$

$$= 2.178$$

$$\text{Entropy of the vapor} = \text{sum } ds + \frac{L}{T}$$

$$S = 0.0020 + 2.178$$

$$S = 2.180 \text{ entropy of the vapor}$$

4. Determine entropy of the vapor for ammonia saturated at −39 F.

$$\text{Entropy of the liquid } s = \text{sum } \frac{dh}{Tav}$$

$$ds = \frac{\text{specific heat of liquid}}{\text{average temperature change}}$$

$$ds = \frac{dh}{Tav}$$

$$\text{Average temperature} = 419.58 + \frac{(-40) - (-39)}{2}$$

$$= 419.58 + 0.5$$

$$= 420.08$$

$$\text{Specific heat of liquid ammonia at } -40 \text{ F} = 1.05$$

$$ds = \frac{1.05}{420.08}$$

$$= 0.0025 \text{ entropy of liquid}$$

$$\text{Entropy of evaporation} = \frac{L}{T}$$

$$= \frac{596.9}{420.58}$$

$$= 1.4192 \text{ entropy of evaporation}$$

$$\text{Entropy of vapor } S = \text{sum}\left(\frac{dh}{Tav}\right) + \left(\frac{L}{T}\right)$$

$$S = 0.0025 + 1.4192$$

$$S = 1.14217 \text{ entropy of } -39° \text{ ammonia vapor}$$

For any condition above the reference point, entropy of the vapor is found by adding the entropy of evaporation at the specific condition temperature to the sum of the separate entropies of the liquid as determined for each increment of average temperature rise.

5. Determine the entropy of ammonia vapor saturated under conditions of 150 psia and 78.81 F.

Enthalpy of the liquid (h) at 78.81 F = 130.6 Btu per lb

$$\text{Average temperature} = \frac{78.81 - (-40 \text{ F})}{2} + 459.58$$

$$= 19.4 + 459.58$$

$$= 478.98 \text{ F abs}$$

$$\text{Entropy of liquid } s = \frac{130.6}{478.98}$$

$$s = 0.2726$$

$$\text{Entropy of evaporation } S_{evap} = \frac{L}{T} = \frac{499.9}{459.58 + 78.81} = \frac{499.9}{538.39}$$

$$= 0.9285$$

$$\text{Entropy of vapor } S = S_{evap} + s$$

$$= 0.9285 + 0.2726$$

$$= 1.2010 \text{ entropy of vapor at 150 psia and 78.81 F}$$

6. What would be the entropy of the vapor in example 5 if it were superheated to 220 F?

According to the thermodynamic property tables for superheated ammonia, the total enthalpy of vapor at 150 psia and 220 F is 720.7 Btu per lb. The total enthalpy of saturated vapor is 630.5 Btu per lb as found in the thermodynamic tables for saturated ammonia. The heat added during superheating is equal to 720.7 − 630.5 = 90.2 Btu per lb. The average temperature from 78.81 F to 220 F $= \frac{220 + 78.81}{2} = 149.4$ F.

$$s = \frac{90.2}{459.58 + 149.4} + 1.2010 \text{ (from example 5)}$$

$$s = 0.148 + 1.2010$$

$$s = 1.3490 \text{ entropy of superheated vapor at 150 psia and 220 F}$$

EXAMPLE: Determine the entropy of superheated ammonia vapor whose temperature and pressure are 220 F and 150 pisa, respectively.

According to the thermodynamic property tables for superheated ammonia, the enthalpy at 220 F and 150 psia is 720.7 Btu per lb of vapor. By applying the formula sum $ds = \frac{dh}{T}$ for each increment of temperature change and adding to the entropy of saturated vapor, the same results as shown above would be obtained, viz.:

$$ds = \frac{dh \text{ (average specific heat of vapor at constant pressure)}}{T \text{ (average increment temperature rise)}}$$

$$s = \frac{0.508}{538.89} + \frac{0.508}{539.89} + \frac{0.508}{540.89} + \frac{0.508}{541.89}, \text{ etc., to } \frac{0.508}{678.89} = 0.148$$

$$s = 0.148 + 1.201$$

$$s = 1.3490 \text{ entropy of superheated vapor at 150 psia and 220 F}$$

THE TEMPERATURE-ENTROPY OR *Ts* DIAGRAM

Figure 50 is a diagram constructed by plotting temperature, the ordinate, against entropy, the abscissa. It is known as a temperature-entropy or *Ts* diagram. The *Ts* diagram shows the relationship of entropy to temperature for each phase of a complete refrigeration cycle. The diagram in Fig. 50 shows the temperature-entropy relationship within an ammonia-vapor refrigeration system operating dry compression. A dry compression or simple saturated system is one in which the vapor leaves the evaporator and enters the compressor in a dry saturated state—i.e., all the liquid is evaporated but the resulting vapor is not subject to superheating before it enters the compressor. Dry compression is an ideal condition for efficient operation. The values used in plotting the diagram are taken from the thermodynamic property tables of the refrigerant employed.

The complete cycle is represented by the lines connecting points 1, 2, 3, 4, 5, 1. The throttling process is indicated by 5–1. It is slanted toward higher entropy as a result of loss of refrigerating capacity which occurs in irreversible systems such as one employing a single expansion valve. The loss of refrigerating capacity is indicated by the shaded area below the evaporation line and is indicative of the greater amount of work performed in the

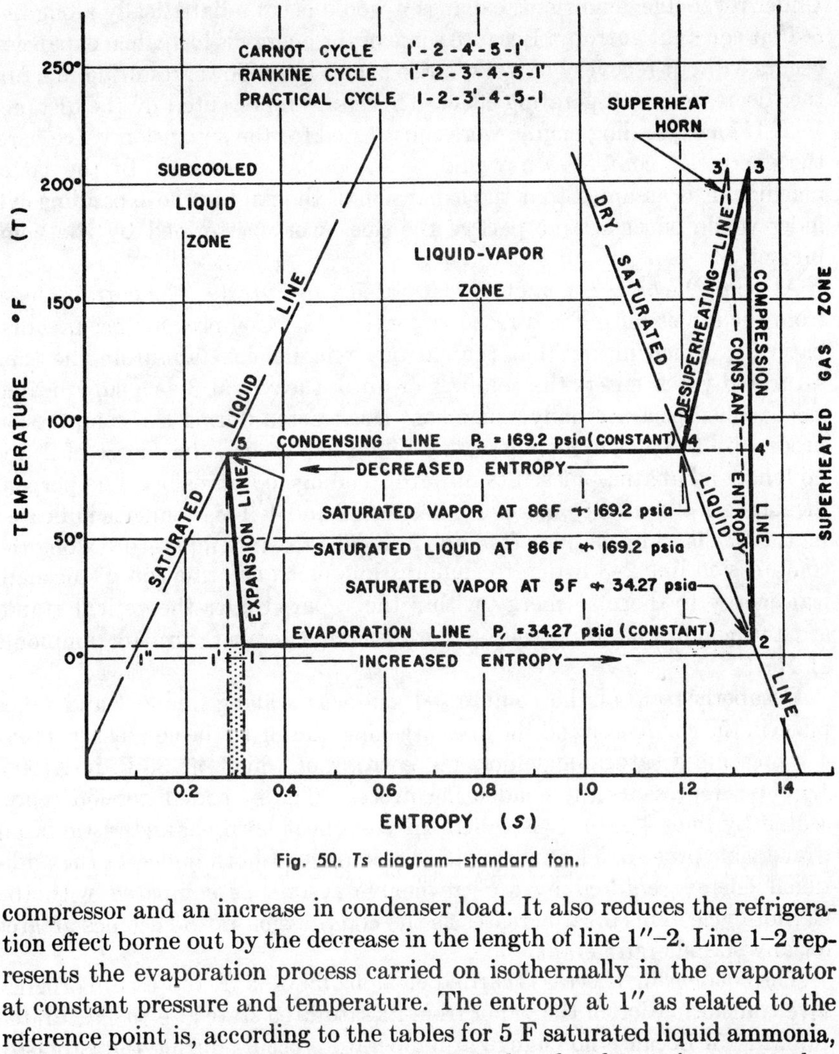

Fig. 50. Ts diagram—standard ton.

compressor and an increase in condenser load. It also reduces the refrigeration effect borne out by the decrease in the length of line 1″–2. Line 1–2 represents the evaporation process carried on isothermally in the evaporator at constant pressure and temperature. The entropy at 1″ as related to the reference point is, according to the tables for 5 F saturated liquid ammonia, 0.1092. The entropy at 1 is equal to 0.1092 plus the change in entropy between 1″–1 due to loss of refrigerating effect or

$$s = \frac{Q \times L}{T_2} + 0.1092$$

where s = entropy at point 1

Q = quality at point 1

L = latent heat of vaporization at 5 F

T_2 = absolute temperature at which vaporization occurs

$$s = \frac{0.1603 \times 565}{459.58 + 5} + 0.1092$$

$$s = 0.1941 + 0.1092$$

$$s = 0.3033 \text{ at point 1}$$

Under reversible conditions, expansion would occur adiabatically along line 5–1′ at constant entropy. Because some of the energy is lost when expansion occurs without recovery of work, a rise in entropy follows, resulting in a further decrease in refrigerating effect. This loss is represented by the distance 1′–1. If an expanding engine were substituted for the expansion valve here, the energy of compression would be somewhat recovered. In the latter machine, it is assumed that the expansion of the fluid in the expanding cylinder would offset a large part of the shaft work performed by the compressor.

At point 2, compression of the saturated vapor starts. The entropy here, from the tables for 5 F saturated vapor, is 1.325. Compression occurs adiabatically, which means that the entropy remains constant from the time saturated vapor enters the compressor until it leaves at 3 as a superheated gas at 169.2 psia. Actually, some heat is extracted during the compression process by the jacket-cooling water. When this occurs, the course of 2–3 is no longer adiabatic, but slants upward, tending to approach a line perpendicular to 1–2 drawn from a point between 2 and 4. This course is indicated on the *Ts* diagram as the dotted line 2–3′. The rise in temperature along the compression line 2–3 is due to compression, or transformation of mechanical energy to thermal energy within the vapor. Under theoretical standard ton conditions, this results in a discharge temperature for ammonia of 211 F.

Desuperheating of the compressed gas occurs along line 3–4 and takes place in the condenser and the discharge line leading to the condenser. Point 4 represents a saturated vapor, the entropy of which, at 86 F, is 1.1904. Line 4–5 represents the condensing process. The extended portion represented by lines 4′–3–4–4′ is called the superheat horn characteristic of an irreversible process. The area within the superheat horn indicates the additional energy required by an irreversible system as compared with the Carnot cycle. This area increases as the compression ratio becomes greater for any specific refrigerant.

The condensing process is carried on along line 4–5. It represents progressive transformation of the vapor from its saturated state at 4 to a saturated liquid at 5. Because no change in temperature occurs during the vapor-to-liquid transformation, the process is isothermal. Point 5 represents a saturated liquid at the condenser temperature and pressure, being 86 F and 169.2 psia in this case. The cycle is thus completed.

A similar *Ts* diagram is shown in Fig. 51. In this illustration, however, the cycle is designed to show changes in entropy during liquid precooling, line 5–5′, and vapor superheating after evaporation of liquid is complete, line 2–2′.

Note that line 5–5′ protrudes slightly into the liquid zone, indicating that subcooling of the liquid (below the temperature of saturated liquid) occurs.

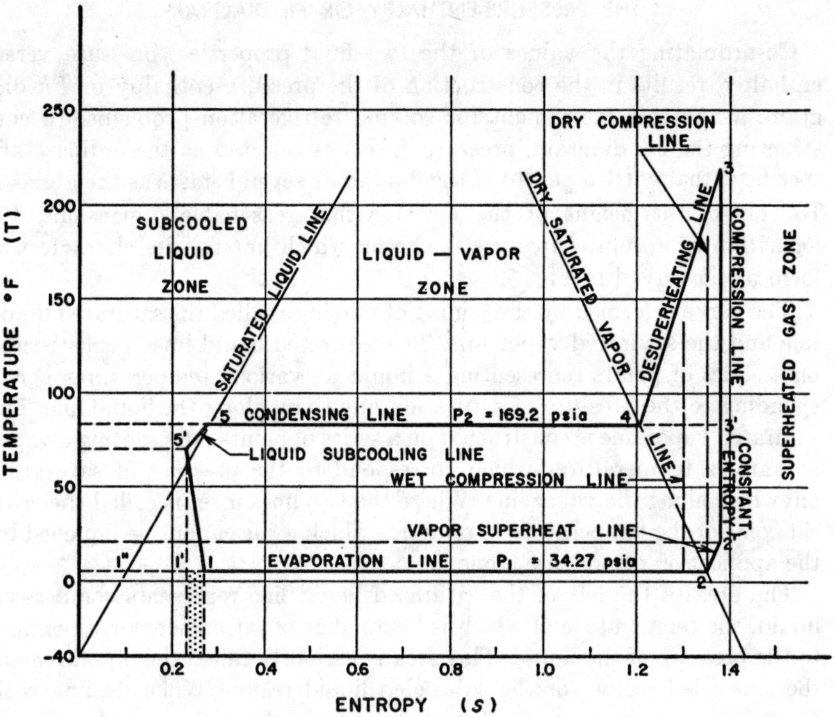

Fig. 51. *Ts* diagram subcooled and superheated.

Actually, the protrusion is exaggerated, as it is customary to show this line coincident with the saturated liquid line. The reason for this lies in the minute difference in entropy during the subcooling process.

Another significant difference in Fig. 51 from the dry compression diagram (Fig. 50) is the protrusion of the compression line farther into the zone of superheat by a distance 2–2′. The latter depicts a vapor that absorbs additional heat beyond that necessary for saturation before reaching the compression phase. This results in a higher vapor temperature and a subsequent increase in entropy. Because the compression ratio $\frac{P_2}{P_1}$ is the same in both Figs. 50 and 51, the superheating of the vapor results in a higher discharge temperature as represented by point 3 in Fig. 51. The area enclosed by lines 3′–3–4–3′ is also greater when superheating of the suction gas occurs before entering the compressor.

Relative locations of both dry compression and wet compression lines are shown. Wet compression indicates the entrance of vapor into the compressor before all liquid has been evaporated—i.e., the quality of the suction vapor is below 100 percent.

THE PRESSURE-ENTHALPY OR *Ph* DIAGRAM

Co-ordinating the values of the two fluid properties, pressure versus enthalpy, results in the construction of the pressure-enthalpy or *Ph* diagram, a valuable instrument for solving refrigeration problems. In constructing the *Ph* diagram, pressure (psia) is selected as the ordinate and specific enthalpy (Btu per lb) of the fluid in its several states as the abscissa. To control the height of the curve within reasonable dimensions, the logarithm of absolute pressure is chosen which imparts its characteristic form as portrayed in Fig. 52.

The curve is formed by the joining of two lines called the saturated liquid line and the saturated vapor line. The saturated liquid line is constructed on a series of points representing a liquid at various temperatures corresponding to the pressure of saturation anywhere along the liquid line. The saturated vapor line is constructed on a series of points representing a vapor at various temperatures which correspond to the pressure of saturation anywhere along the vapor line. Where the two lines meet is called the critical point or the temperature at or above which a gas cannot be liquefied by the application of pressure alone.

The area to the left of the saturated liquid line represents compressed liquid, the temperature of which is below that of saturation corresponding to the pressure of the liquid. This area is variously called the liquid region, the subcooled region, or the subcooled liquid region. When dealing with co-ordinates within that area left of the saturated liquid line, only a liquid is dealt with and vapors are not involved.

The area to the right of the saturated vapor line represents compressed gas, the temperature of which is above that of the saturated vapor corresponding to the pressure of the gas. This area is variously called the superheat region or the superheated gas region. When dealing with co-ordinates in that area to the right of the saturated vapor line, only a dry superheated gas is dealt with and liquid is not involved.

In the area between the saturated liquid line and the saturated vapor line is the region of (1) liquid-to-gas transformation or (2) gas-to-liquid transformation, depending on whether the chart is read from left to right as in the first instance, or right to left as in the second instance. Starting will *all* liquid at the saturated liquid line and ending with *all* vapor at the saturated vapor line, a mixture of liquid and vapor exists, the amount of each depending on the location of a given co-ordinate relative to either end of the region. Moving from left to right along a straight line or constant pressure in the direction of ascending enthalpy values, the amount of liquid existing in the mixture becomes less and the amount of vapor existing becomes greater. This area also may be called the evaporation region. Reversing the direction of movement starting from the saturated vapor line, the amount of vapor in the mixture becomes less while the amount of liquid existing becomes greater. It now becomes a region of condensation.

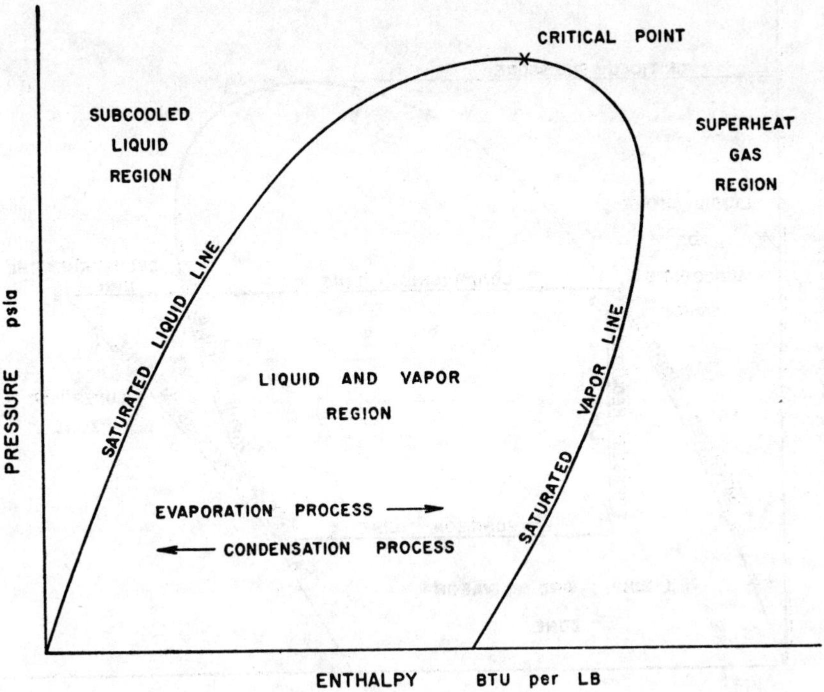

Fig. 52. *Ph* diagram, basic.

The degree to which liquid-to-vapor or vapor-to-liquid transformation has developed depends on the location of a given co-ordinate relative to the distance from both the saturated liquid line and the saturated vapor line. If a given co-ordinate as measured on regular graph is found to be equidistant from the two lines within the evaporation region, the mixture at that point will be composed of 50 percent liquid and 50 percent vapor. The degree of transformation is usually expressed in percent of vapor existing relative to the total amount of refrigerant involved. It is identified as the "quality of the mixture." The mixture at the halfway point between the two lines would be said to have a quality of 50 percent.

As the saturated vapor line is approached and the amount of vapor becomes greater, the quality of the mixture increases from 0 at the saturated liquid line until all the liquid evaporates, leaving only a dry saturated vapor whose quality is said to be 100 percent. Quality, then, denotes the amount of refrigerant that exists as a dry saturated vapor within a liquid-vapor mixture expressed in terms of percent of the total amount involved.

A vapor leaving an evaporator whose quality is said to be 98 percent still retains 2 percent of its total weight in liquid form and therefore has never been completely evaporated. It is referred to as "wet" saturated vapor. Vapor leaving an evaporator under 100 percent quality has forfeited part of its latent heat of vaporization by an amount equal to the difference cal-

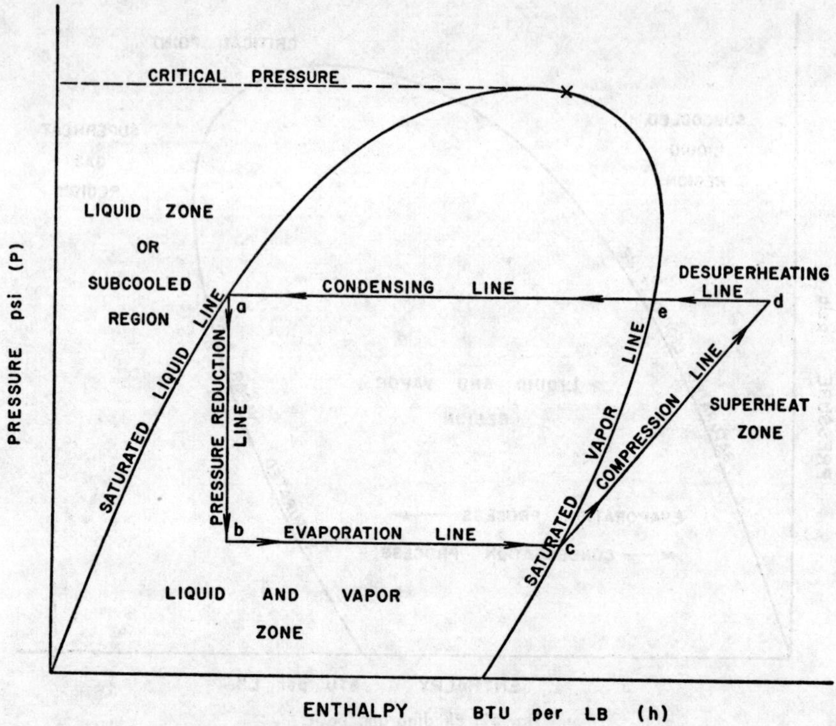

Fig. 53. *Ph* diagram with refrigeration cycle.

culated on a percentage basis, or, as in the case of the vapor cited, this would be equivalent to 2 percent of the latent heat per pound of refrigerant handled.

The four steps of the mechanical refrigeration cycle can be portrayed graphically on the *Ph* diagram by superimposed lines drawn from points that correspond to the operating pressures of any system, either real or hypothetical. This is accomplished diagrammatically in Fig. 53, and establishes a simple method of determining thermal behavior and distribution throughout any refrigeration system.

The cycle superimposed on the *Ph* diagram in Fig. 54 is that of a simple, dry compression refrigeration cycle using ammonia and operating under standard ton conditions of 5 F evaporator and 86 F condenser temperatures. In constructing the *Ph* chart, note that the absolute pressure, ranging from 0 psi to 175 psi, is selected as the ordinate while enthalpy from 0 Btu per lb to 800 Btu per lb establishes the abscissa. The values selected are sufficiently beyond those required to construct the cycle described above.

Employing a regular graph form—i.e., one whose lines, both vertical and horizontal, are of equal distance apart—and plotting a sufficient number of co-ordinate points, pressure versus enthalpy, the saturated liquid and the vapor lines are established. Enthalpy values that form the abscissa are found in the Properties of Saturated Ammonia tables on the lines of corresponding pressures. Both lines may be constructed simultaneously by reading and plotting enthalpy values for saturated liquid and saturated vapor against each selected pressure point at the same time. For example, read 34.3 Btu per lb for saturated liquid and 609.1 Btu per lb for saturated vapor and plot each of these values on the graph line corresponding to a pressure of 25 psia. Next, read 66.5 Btu per lb for saturated liquid and 618.2 Btu per lb for saturated vapor and plot on the graph line corresponding to a pressure of 50 psia. Continue this procedure for reasonably spaced pressure levels up to, say, 175 psia. Joining the points by a curve representative of intermediate values establishes both the saturated liquid and the saturated vapor lines. These lines are the *plane loci* of the infinite number of state points of both liquid and vapor whose temperature and pressure correspond to establish a condition of saturation.

The next step is to construct the thermal diagram for the ammonia refrigeration cycle, operating simple saturated, meaning (1) no superheating in evaporator or suction line and (2) no precooling of the liquid between the condenser (receiver) and the expansion valve.

First, the evaporation line is established. This is done by first determining the absolute pressure for 5 F saturated ammonia. Referring to the Properties of Saturated Ammonia tables, it is found that an absolute pressure of 34.27 psia corresponds to a 5 F boiling point. Drawing a line from the

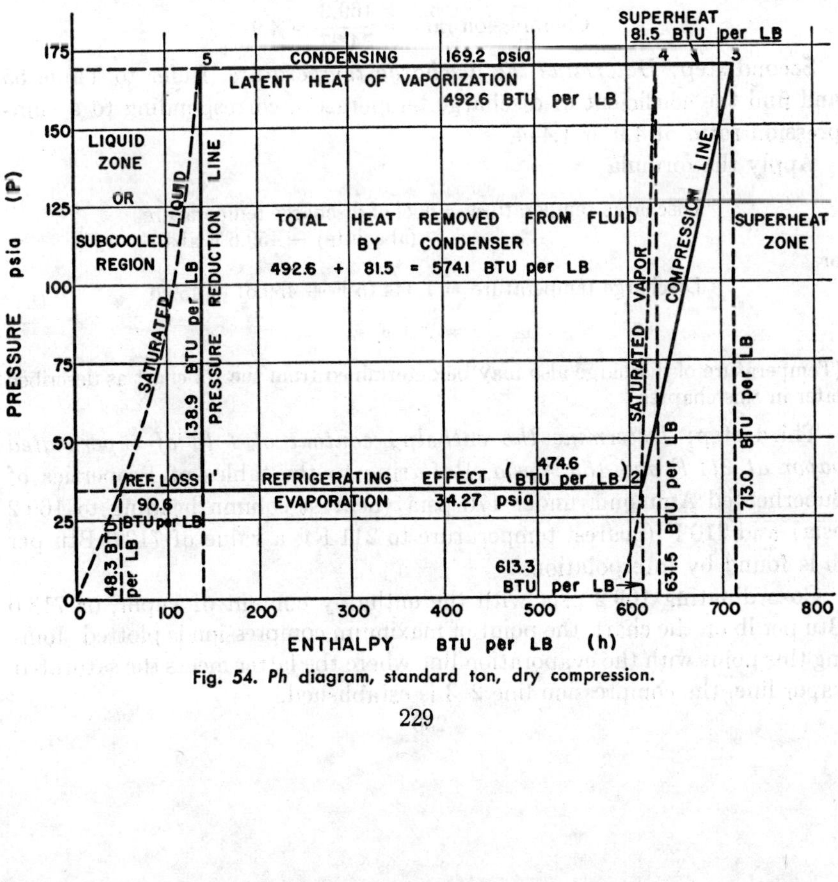

Fig. 54. Ph diagram, standard ton, dry compression.

saturated liquid line horizontally across the chart to the saturated vapor line corresponding to a pressure of 34.27 psia on the ordinate, the *evaporation line* is established. It is represented on the chart as line 1–2.

Now to establish the *condensing line*. Again referring to the tables, it is found that an absolute pressure of 169.2 psia corresponds to an 86 F ammonia boiling point. The condensing line is established by connecting the saturated liquid line with the saturated vapor line by a line drawn horizontally across the chart at a level corresponding to 169.2 psia on the ordinate. The condensing line is represented in Fig. 54 as 4–5.

By drawing a line from the intersection of the condensing line and the saturated liquid line perpendicular with and to the evaporation line 1–2, the pressure reduction or expansion line is established. This line becomes 5–1′.

To complete the diagrammatic cycle, it remains to establish the *compression line* 2–3, and to extend the condensing line to the point of maximum compression indicated as point 3.

Note that the compression line slopes. The path of this line is established as follows:

First step: *Determine the compression ratio of the system under consideration.* The compression ratio is equal to final pressure divided by initial pressure, or condensing pressure divided by evaporator pressure. Under standard conditions, assuming ammonia to be the refrigerant,

$$\text{Compression ratio} = \frac{169.2}{34.27} = 4.9$$

Second step: *Determine the discharge temperature.* Refer to Table 55 and find the coefficient of discharge temperature corresponding to a compression ratio of 4.9 or 1.444.

Apply the formula

$$\text{Discharge temperature} = \text{coef of discharge temperature (absolute)} - 459.6 \text{ F abs}$$

or

$$\text{Discharge temperature} = 1.444\,(5 \text{ F} + 459.6) - 459.6$$
$$= 211 \text{ F}$$

(Temperature of discharge also may be determined from the *Ph* chart as described later in this chapter.)

Third step: *Determine the enthalpy content of 1 lb of superheated vapor at 211 F and 169.2 psia.* Referring to the tables of Properties of Superheated Ammonia under 170 psia (nearest column heading to 169.2 psia) and 210 F (nearest temperature to 211 F), a value of 713.0 Btu per lb is found by interpolation.

Co-ordinating 169.2 psia with the enthalpy content of vapor, or 713.0 Btu per lb on the chart, the point of maximum compression is plotted. Joining this point with the evaporation line where the latter meets the saturated vapor line, the compression line 2–3 is established.

By extending the condensing line to the point of maximum compression to form 4–3, the desuperheating line is established and the diagrammatic cycle is completed.

In following the lines of the *Ph* diagram in the order in which they represent the four principal steps of the refrigeration cycle—i.e., expansion, evaporation, compression, and condensation—its use may be understood better, as well as its value in analyzing thermal distribution within a refrigeration system. This type of chart affords an excellent method of comparing the thermal efficiency of systems operating under various conditions of superheating, liquid subcooling, or combinations of both. For the purposes here, a simple ammonia system operating under standard conditions of 86 F condenser and 5 F evaporator temperatures in which the refrigerant enters the compressor as a saturated vapor (no superheating), shall be assumed. Proceeding in the order mentioned above, the cycle begins at point 5.

Point 5 is located on the saturated liquid line so that the refrigerant begins the cycle as a saturated liquid. This means that any increase in its temperature promotes vaporization. It is also on a line corresponding to a condenser pressure of 169.2 psia, which corresponds to a saturated temperature of 86 F. In the refrigeration cycle, assuming no external loss in heat or pressure, point 5 represents saturated liquid under high pressure located anywhere in the system between the entrance of the receiver to the entrance of the expansion valve.

If it were possible to extract heat from the refrigerant during its passage through the expansion valve, the path of the diagram could be made to follow the saturated liquid line to its intersection with the evaporation line at point 1.

As it is, this heat is taken up internally by the liquid, causing some evaporation in the form of flash gas to occur, which, in turn, causes a reduction in temperature—in this case, from 86 F to 5 F. Because energy is neither gained nor lost during the process of pressure reduction, the action is adiabatic. Since no work is performed, the process is said to occur at constant enthalpy. The only change occurring during the pressure reduction process involves the state of the refrigerant so that its enthalpy at the evaporation pressure is the same as it was when saturated at the condenser pressure. Therefore, this step in the *Ph* diagram must follow a path representing constant enthalpy, which in Fig. 54 is vertical and in a direction of descending pressure or downward to form the line 5–1′, known as the pressure reduction or expansion line. Line 5–1′ represents the flow of the liquid refrigerant through the expansion valve, during which it undergoes a pressure decrease under adiabatic conditions.

To cause pressure reduction to follow the saturated liquid line, it would be necessary to extract heat from the refrigerant during its passage through the expansion valve in an amount equal to that represented by the distance, point 1–1′, on the chart. This represents a loss in refrigerating effect because of internal or spontaneous cooling of the refrigerant from condensing to evaporation temperature. This loss becomes greater as the specific heat of the liquid refrigerant increases.

In the case of ammonia operating under standard ton conditions, this loss would be equal to 138.9 Btu per lb, enthalpy of the mixture (point 1′), minus 48.3 Btu per lb, enthalpy of the saturated liquid (point 1), or 90.6 Btu per lb. This value corresponds with that found by employing the formula introduced earlier, viz.:

$$\text{Btu} = \text{weight in lb} \times \text{sp ht}\ (t_1 - t_2)$$

$$= 1 \times 1.119\ (86\ \text{F} - 5\ \text{F})$$

$$= 1.119^* \times 81$$

$$= 90.6\ \text{Btu per lb lost}$$

Point 1′, then, represents a mixture of liquid and vaporous refrigerant, the ratio of which depends on (1) the difference in condenser and evaporator pressure and (2) the evaporator pressure. This ratio is indicated by the quality and is determined by the formula

$$\text{Quality} = \frac{\left\{\begin{array}{l}\text{Enthalpy of liquid at condenser temperature}\\ \quad - \text{enthalpy of liquid at evaporator temperature}\end{array}\right\}}{\left\{\begin{array}{l}\text{total enthalpy of vapor at evaporator temperature}\\ \quad - \text{enthalpy of liquid at evaporator temperature}\end{array}\right\}} \times 100$$

or

$$Q = \frac{h_c - h_e}{H_e - h_e} \times 100 \quad \left\{H_e - h_e = \text{latent heat of vaporization at evaporator temperature}\right\}$$

which, under standard ton conditions,

$$Q = \frac{h_{86\ \text{F}} - h_{5\ \text{F}}}{H_{5\ \text{F}} - h_{5\ \text{F}}} \times 100$$

and for ammonia under standard ton conditions,

$$Q = \frac{138.9 - 48.3}{613.3 - 48.3} \times 100$$

$$Q = 16.03\%$$

The mixture at point 1′, therefore, contains 83.97 percent liquid and 16.03 percent vaporous ammonia.

Loss due to cooling ammonia from 86 F to 5 F may be calculated, viz.:

$$\text{Btu} = \frac{Q\,(H_e - h_e)}{100}$$

$$= \frac{Q\,(613.3 - 48.3)}{100}$$

* 1.119 is the average specific heat of liquid ammonia for temperatures involved here.

$$= 0.1603 \times 565$$

$$= 90.6 \text{ Btu per lb}$$

Beginning at point 1′, the remaining liquid, which under the conditions used in the discussion here equals 83.97 percent of the amount that entered the expansion valve at point 5, starts to evaporate and continues the process under constant pressure conditions until its transformation to a saturated vapor is complete. This process is represented on the *Ph* diagram as line 1′–2. Because heat is absorbed during evaporation, the path of the diagram follows the direction of ascending enthalpy values, which is to the right and at a level corresponding to the evaporator pressure, or 34.27 psia.

Because the system under discussion is operated by allowing the vapor to leave the evaporator and enter the compressor saturated, the evaporation line terminates at the saturated vapor line, or point 2. If additional heat were absorbed by the refrigerant without any increase in pressure after complete evaporation occurred, as represented by point 2, the gas would become superheated, and this action would be indicated by continuing the line 1′–2 along the same pressure level into the superheated region to the right of the saturated vapor line. The distance to which this line would be extended beyond the saturated vapor line would depend on the amount of superheating allowed.

Under saturated (no superheating) conditions, the heat absorbed by the refrigerant is in the form of latent heat of vaporization. The amount absorbed is equal to the refrigerating effect and determined by computing the difference in enthalpy value between point 2 and point 5. This may be calculated as follows:

Refrigerating effect ($R\,E$) = latent heat of vaporization at evaporator temperature $\times$ (100 − % quality)

and, for ammonia under standard ton conditions,

$$R\,E = \frac{565\,(100 - 16.03)}{100}$$

$$R\,E = 474.4 \text{ Btu}$$

or it may be determined from the diagram,

$R\,E$ = total enthalpy at evaporator temperature − enthalpy of liquid at condenser temperature and pressure

$$R\,E = H_e - h_c$$

$$R\,E = 613.3 - 138.9$$

$$R\,E = 474.4 \text{ Btu per lb}$$

Because the line 5–1′ is adiabatic, the enthalpy of the refrigerant at point 1′ is the same as that for point 5, the latter being represented as h_c in the above formula. Therefore, the refrigerating effect may be determined also by subtracting the enthalpy of the refrigerant at point 1′ from that at point 2, the latter being represented by H_e in the formula, or

$$RE = 613.3 - 138.9 = 474.4 \text{ Btu per lb}$$

Line 1′–2 represents the heat absorbed in the evaporator and is equal to the refrigerating effect of the refrigerant when the vapor leaving the evaporator and entering the compressor is saturated at the evaporator pressure and temperature.

Still referring to Fig. 54, point 2 represents the vapor as it leaves the evaporator, and its location along the constant pressure line relative to the saturated vapor line depends on its thermal state. If it contains some liquid, which means its quality is less than 100 percent, point 2 will be located in the region of liquid and vapor, *or left of the saturated vapor line,* indicated as point 2″ in Fig. 57. If the refrigerant leaves as a dry saturated vapor, then point 2 will be located *on the saturated vapor line,* as shown in Fig. 54. It will be found in the region of superheat or *right of the saturated vapor line,* as indicated by point 2′ (Fig. 55) if superheating in the evaporator or suction line occurs.

Line 2–3 is the compression line and represents the vapor from the time it enters the compressor in its state at point 2 until it is discharged as a superheated gas compressed to the condenser pressures at point 3, where it intersects the condensing line 3–5.

Mechanical work done during compression is transferred to the gas as thermal energy. Assuming that no energy is lost or gained otherwise, the energy taken up by the gas will be just equal to the energy expended mechanically in compressing it, and the process therefore will be adiabatic. The amount of energy gained as a result of compression depends on the ratio of final pressure to initial pressure (compression ratio), and the kind of refrigerant employed. The internal energy of a perfect gas is a function of its temperature, so that any changes in the latter reflect the amount of change that has taken place due to thermal kinetic energy. This circumstance affords a basis upon which energy changes may be calculated when some other variables are known, such as initial and final pressures or volumes, initial temperature, and kind of gas.

It was stated in the study of gases (Chap. 4) that the temperature of any gas is a function of enthalpy and ratio of specific heat at constant pressure and constant volume (y). It was shown that the temperature is also a function of pressure, and that adiabatically their relationship is represented by the equation

$$\frac{T_2}{T_1} = \left(\frac{P_2}{P_1}\right)^{\frac{y-1}{y}}$$

where T_2 = final absolute temperature
T_1 = initial absolute temperature
P_2 = final absolute pressure
P_1 = initial absolute pressure

$$y = \frac{\text{sp ht at constant pressure}}{\text{sp ht at constant volume}} = \frac{c_p}{c_v}$$

For a standard ton ammonia system,

$$T_1 = 5° + 459.6 = 464.6 \text{ F abs}$$

$$P_1 = 34.27$$

$$P_2 = 169.2$$

and y for ammonia (Table 57) is 1.3.

To determine the enthalpy of gas at the final pressure, it is first necessary to establish the value of T_2. Transposing the factors in the above formula, then

$$T_2 = T_1 \frac{(P_2)}{P_1}^{\frac{y-1}{y}} \quad \text{where } \left(\frac{P_2}{P_1}\right) \text{ equals the compression ratio}$$

Under standard ton conditions,

$$T_2 = 464.6 \text{ F abs} \left(\frac{169.2 \text{ psia}}{34.27 \text{ psia}}\right)^{\frac{1.3-1}{1.3}} \quad \text{(compression ratio} = 4.93)$$

Following the method outlined in Chapter 11, "Refrigeration Calculations," for determining discharge temperatures, it is found that

$$T_2 = 211 \text{ F}$$

Referring to the properties of superheated ammonia tables under 169.2 psi and 210 F, the enthalpy of 1 lb of refrigerant is (by interpolation) 713.0 Btu.

Co-ordinating this value with 169.2 psia on the chart and connecting with point 2, the compression line 2–3 is established.

The compression line also may be established by following a path of constant entropy, the values of which are not shown on the *Ph* chart (Fig. 54) but which may be found on the Mollier diagram, described later in this chapter. This method of determining the discharge temperature is described in more detail in Chapter 11, "Refrigeration Calculations."

Point 3 represents the refrigerant as a superheated, high-pressure gas leaving the discharge valve of the compressor to enter the condenser where the heat absorbed in the evaporator and the thermal energy of compression are removed in order to effect its conversion to a liquid.

Line 4–5 is the condensing line. It is established along the condenser or high-pressure line, which, under standard ton conditions for ammonia of 86 F condensing temperature, corresponds to 169.2 psia. This pressure remains at the same level on the chart throughout its path of conversion.

Starting from point 3, the refrigerant comes in contact with the cooling medium—either air, water, or another refrigerant—and the process of heat transfer is initiated. Following line 3–4 from point 3, a continual decrease in enthalpy occurs, as indicated along the abscissa *h*. The thermal energy of compression, plus any additional superheat, is removed until the refrigerant reaches its temperature of saturation at the existing pressure. This latter condition occurs at point 4. Further cooling removes its latent heat, resulting in condensation of the vapor, a process that occurs along the line 4–5 until all the refrigerant has been converted to liquid at point 5, the starting point. By subtracting the enthalpy of the refrigerant at point 5 from that of point 3, the amount of thermal energy to be removed by the condenser is determined. Under standard ton conditions for ammonia, this is found to be, by referring to the diagram,

$$713 - 138.9 = 574.1 \text{ Btu per lb}$$

amount of thermal energy to be removed by condenser.

POWER REQUIREMENTS AS DETERMINED FROM THE *Ph* CHART

From the *Ph* chart, the theoretical horsepower requirements, as well as the amount of refrigerant necessary per ton of refrigeration, may be determined. To determine the amount of refrigerant necessary to absorb the thermal equivalent of 1 ton of refrigeration or 200 Btu per min, two factors must be known: (1) evaporator temperature and (2) refrigerating effect.

The refrigerating effect for an ammonia system operating under standard ton conditions, assuming no superheat or liquid subcooling, is equal to the enthalpy of the saturated vapor at 5 F evaporator temperature (point 2) or 613.3 Btu per lb minus the enthalpy of liquid at 86 F condenser temperature (point 5) or 138.9 Btu per lb equals 474.4 Btu per lb.

The amount of refrigerant, by weight, to absorb the thermal equivalent of 1 ton of refrigeration is

$$\text{Weight} = \frac{\text{thermal equivalent of 1 ton of refrigeration}}{\text{refrigerating effect}}$$

$$= \frac{200}{474.4} = 0.4216 \text{ lb of ammonia per min}$$

Multiplying the weight of refrigerant necessary per ton of refrigeration by the specific volume (cu ft per lb) of saturated vapor, the volume of refrigerant handled by the compressor for each ton of refrigeration is determined. Referring to the properties of saturated ammonia tables, it is found that the specific volume of 5 F ammonia vapor, saturated, is 8.15 cu ft per lb. The volume required per ton is:

$$V_T = \text{specific volume @ temperature of saturation} \times \text{weight per ton}$$

$$V_T = 8.15 \times 0.4216$$

$$V_T = 3.436 \text{ cu ft per min (cfm) per ton}$$

Knowing the weight of refrigerant necessary to absorb the thermal equivalent of 1 ton of refrigeration, it is now necessary to determine the thermal work performed by the compressor in order to solve the horsepower requirement. The work of compression is found by following the compression line 2–3 in the *Ph* chart. Entering the compressor as a 5 F saturated vapor (point 3), each pound of ammonia contains 613.3 Btu. Leaving the compressor as a high-pressure, high-temperature (210 F) gas, each pound of ammonia contains 713.0 Btu. Therefore,

$$\text{Thermal work of compression} = 713.0 - 613.3$$

$$= 99.7 \text{ Btu per lb per min}$$

In Chapter 2, "Thermal Energy," it was stated that 1 Btu is the thermal energy equivalent of 778 ft-lb of mechanical energy and that 33,000 ft-lb per min is equal to 1 hp. The theoretical horsepower requirements per ton of refrigeration may be determined for any refrigerant under any set of operating conditions by application of the formula

$$\text{hp per ton} = \frac{\text{thermal work of compression} \times \text{weight of refrigerant} \times 778}{33{,}000}$$

For ammonia under standard ton conditions,

$$\text{hp per ton} = \frac{99.7 \times 0.4216 \times 778}{33{,}000}$$

$$= 0.988 \text{ hp per ton of refrigeration}$$

THERMAL VARIATIONS IN SIMPLE SYSTEMS

There are three basic types of simple systems. They are (1) the simple saturated, (2) the superheated (in evaporator), and (3) the subcooled liquid systems. A fourth type is the system employing a heat exchanger wherein superheating occurs as a result of cooling the liquid. Another type commonly found embraces both evaporator superheating and liquid subcooling.

EFFECT OF SUPERHEATING ON *Ph* DIAGRAM

In Fig. 54, point 2 represents the condition of the vapor as it enters the suction side of the compressor. Because this latter point lies on the saturated vapor line, the suction gas enters the compressor as a dry saturated gas; i.e., its quality is 100 percent.

If, before leaving the evaporator, the refrigerant vapor absorbed additional heat above that necessary for dry saturation, a condition of saturation would no longer exist. Because this additional heat is absorbed without any change in suction line pressure, the refrigerant is now said to be superheated.

Assuming that the additional heat absorbed in the evaporator causes a rise of 15 F above evaporator temperature, then the vapor would be 15 F of superheating the suction vapor is shown in Fig. 55 by the extension of the constant evaporator pressure line beyond the saturated vapor line to a point corresponding to the enthalpy value of the superheated refrigerant vapor. This is indicated as point 2′.

As the initial temperature has been increased due to superheating, the discharge temperature is also expected to become greater than that for saturated conditions at the same pressures. The point representing the vapor when discharged from the compressor at the higher pressure would now be located at point 3′ instead of point 3, as was the case when operating under saturated conditions at point 2 as portrayed in Fig. 54.

By allowing superheating to occur, several functional and thermal values change as a result and affect the construction of the *Ph* diagram. Assuming that no other modifications are made within the refrigeration system, superheating of suction vapors brings about the following changes:

(*a*) Increases refrigerating effect by an amount governed by the length of line 2–2′, provided that superheating occurs in the evaporator;

(*b*) Causes a rise in discharge temperature in an amount determined by the formula $T_2 = T_1 \left(\frac{P_2}{P_1}\right)^{\frac{y-1}{y}}$, wherein it is obvious that any increase in T_1

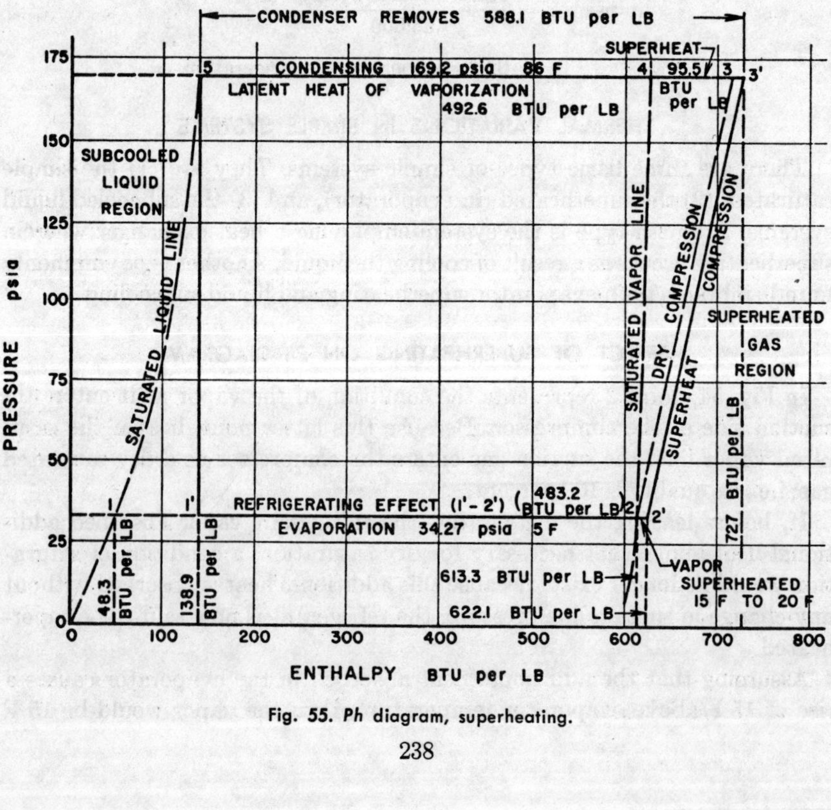

Fig. 55. *Ph* diagram, superheating.

will also raise the value of T_2. (See Chapter 11, "Refrigeration Calculations");

(*c*) Reduces the weight of refrigerant handled by the compressor per stroke due to the subsequent increase in specific volume as a result of higher temperature;

(*d*) Increases power requirements; i.e., horsepower per ton of refrigeration.

Comparing an ammonia system operated under saturated and standard ton conditions represented by the *Ph* diagram in Fig. 54 with one in which a 15 F superheating is allowed to occur, as represented by the *Ph* diagram in Fig. 55, the following numerical changes are found:

(*a*) *Refrigerating effect.* Referring to Table IV, Properties of Superheated Ammonia Vapor, it is found, by interpolation, that 1 lb of ammonia under 34.27 psia and superheated to 20 F has an enthalpy of 622.1 Btu. This value is represented by point 2′ (Fig. 55). Refrigerating effect is equal to point 2′ minus point 5, or 622.1 − 138.9 = 483.2. This is an increase of 483.2 − 474.4 = 8.8 Btu per lb over the refrigerating effect under saturated conditions.

(*b*) *Rise in discharge temperature.*

$\frac{P_2}{P_1}$ = 4.9 compression ratio, which is the same for both systems.

Coefficient of discharge temperature for 4.9 compression ratio equals 1.444.

$$\text{Discharge temperature} = 1.444\,(20\text{ F} + 459.6) - 459.6$$
$$= 233\text{ F for superheated conditions}$$

This compares with 211 F discharge temperature under saturated conditions.

(*c*) *Total enthalpy.* Referring to the tables of Properties of Superheated Ammonia under 169.2 psia and 233 F discharge temperature, it is found (by interpolation) that each pound of ammonia contains 727 Btu (point 3′). This represents an increase of 727 − 713.0 = 14.0 Btu per lb contained in the discharge vapor under 15 F superheated conditions or 20 F suction temperature. 727.0 − 138.9 = 588.1 Btu per lb to be removed by condenser. This, compared with 574.1 Btu per lb for saturated conditions, increases condenser load by 14 Btu per lb.

(*d*) *Reduces compressor capacity.*

Volume of vapor saturated 5 F	= 8.150 cu ft per lb
Volume of vapor superheated to 20 F (34.27 psia)	= 8.465 cu ft per lb

This means that the compressor would have to handle 8.465 − 8.150 = 0.315 cu ft more vapor under 15 F superheat than would be necessary under saturated conditions for each pound of refrigerant. The amount of refrigerant, by weight, handled per ton is equal to 200 divided by $R\,E$, or

$$\frac{200}{483.2} = 0.414 \text{ lb of ammonia per ton refrigeration}$$

$$0.414 \times 8.465 = 3.50 \text{ cu ft per ton refrigeration}$$

as compared with 3.436 cu ft per ton under 5° saturated conditions, or a difference of 3.50 − 3.436 = 0.064 cu ft.

(*e*) *Increased power requirements*. Thermal work of compression equals difference in enthalpy at points 3 and 4, or 727 − 622.1 = 104.9 Btu.

Weight of refrigerant handled was found to be 0.4139 lb of ammonia per ton.

Applying the formula

$$\text{hp} = \frac{H_c \times W \times 778}{33{,}000}$$

where H_c = thermal work of compression

W = weight in pounds of refrigerant per ton

then

$$\text{hp per ton} = \frac{104.9 \times 0.4139 \times 778}{33{,}000}$$

$$\text{hp per ton} = 1.023$$

This amounts to an increase of 1.023 − 0.988, or 0.035 hp per ton of refrigeration more when operating under 15 F superheat with ammonia as compared with a system whose vapors enter the compressor dry saturated.

EFFECT OF LIQUID SUBCOOLING ON THE *Ph* DIAGRAM

Cooling the liquid below the saturation temperature corresponding to condenser pressure increases the capacity of the evaporator. Liquid subcooling is accomplished between the receiver and the expansion valve. The principal purpose in subcooling the liquid before it enters the expansion valve is to reduce loss of refrigerating effect by decreasing the amount of flash gas resulting from pressure drop.

Liquid subcooling may be accomplished by means internal or external to the system. Internal subcooling is accomplished by (1) a heat exchanger whereby the flowing liquid contacts the cold suction line or (2) a by-passed liquid bled from the receiver that flashes into vapor within a low-pressure expansion chamber through which the liquid line passes. External subcooling is accomplished by exposure of the liquid line to a cooling medium whose refrigerating capacity originates at a source other than the immediate system. In the latter method of liquid cooling, water, air, or another refrigerant may be used.

The effect of external liquid cooling in an ammonia system operating at 5 F evaporator and 86 F condenser temperature is shown in the *Ph* diagram (Fig. 56). Note that the line 3–5 is extended to the point 5′. The line 5–5′ then, represents the amount of heat absorbed by the external cooling medium in reducing the temperature of the liquid—in this case, from 86 F to 76 F,

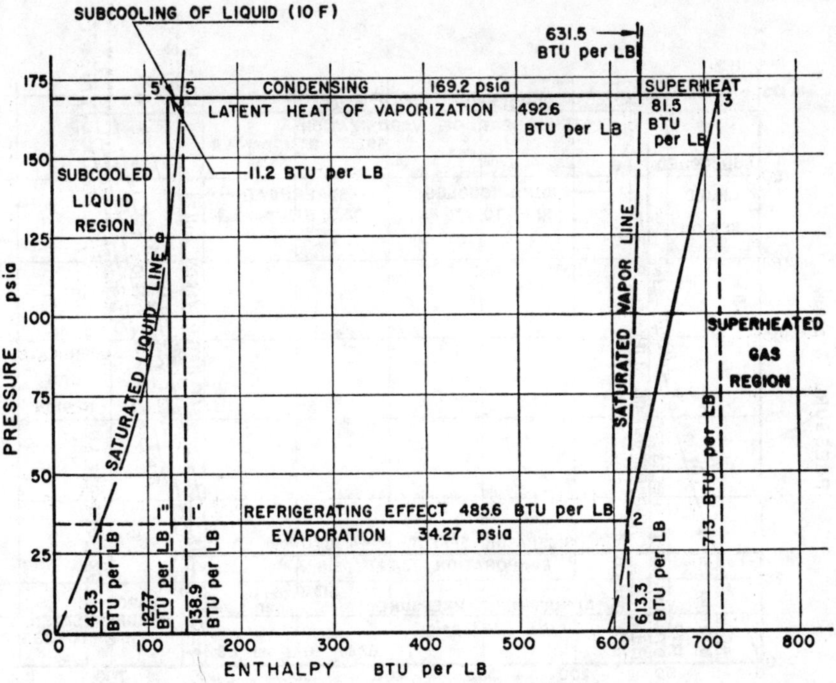

Fig. 56. *Ph* diagram, subcooling.

or 11.2 Btu per lb. This means that the loss in flash gas will be 11.2 Btu per lb less when pressure reduction occurs after cooling the liquid 10 F below the saturation temperature for ammonia at 169.2 psia. This is indicated at point 1″, the intersection of the pressure reduction line 5′–1″, and the evaporation line 1″–2. The line 1″–1′, being equal to line 5–5′, extends the evaporation line a similar distance beyond that of the system in which no subcooling occurs, or to point 1′. The refrigerating effect increases by an amount of thermal energy represented by the distance 1″–1′ for a total of 485.8 Btu per lb, provided that no superheating occurs. This compares with a refrigerating effect of 474.6 Btu per lb for the standard ton system represented in Fig. 54. By extracting additional heat from the liquid after it has left the condenser, it no longer enters the throttling device in a saturated condition but at some temperature below that of saturation corresponding to the condenser pressure. This is apparent on the diagram by the extension of the condensing line, which is now well into the liquid or subcooled region to the left of the saturated liquid line.

The new expansion line crosses the saturated liquid line at *a*, which is below the condenser pressure level of 169.2 psia. At point *a*, evaporation starts, but at a much reduced pressure than was the case when no subcooling was carried on, as represented in Fig. 54. The intersection of the pressure reduction line and the saturated liquid line represents the liquid at its new

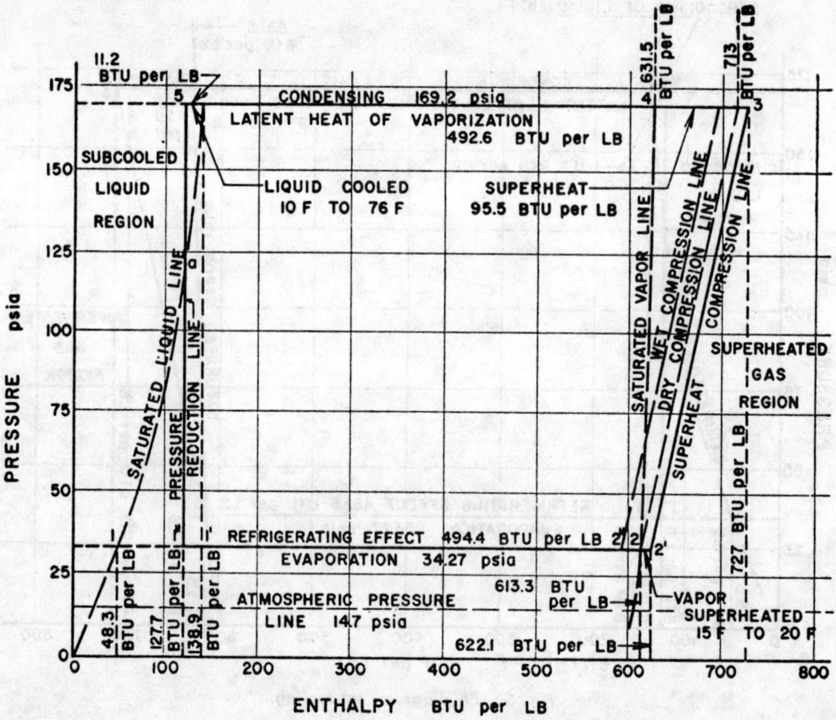

Fig. 57. *Ph* diagram, superheating and subcooling.

condition of saturation. The diagram in Fig. 56 shows the condition at *a* to be 145.4 psia, corresponding to a temperature of 76 F. Reducing the temperature of the liquid further would lower point *a* along the saturated liquid line.

Subcooling does not affect the thermal conditions during the compression or the condensing phases of the cycle. Neither does it always represent an over-all gain in system efficiency. However, it does increase refrigerating effect. If the cooling medium is obtained from a natural source such as well water, a definite gain may be accomplished through subcooling. If mechanical means must be provided to reduce the cooling medium to a suitable temperature, the advantages gained by subcooling may be offset entirely or reduced to a point that would make the procedure impractical. Where a pump or a blower is used to convey the cooling medium, power requirements must be considered in the total energy analysis.

EFFECT OF HEAT EXCHANGER ON THE *Ph* DIAGRAM

A simple heat exchanger is one in which the warm liquid from the receiver is exposed to the cold suction vapor line. Exposure may be accomplished by a double pipe arrangement in which the suction line is installed within the liquid line. Cooling the liquid in this manner results in a gain in refrigerating

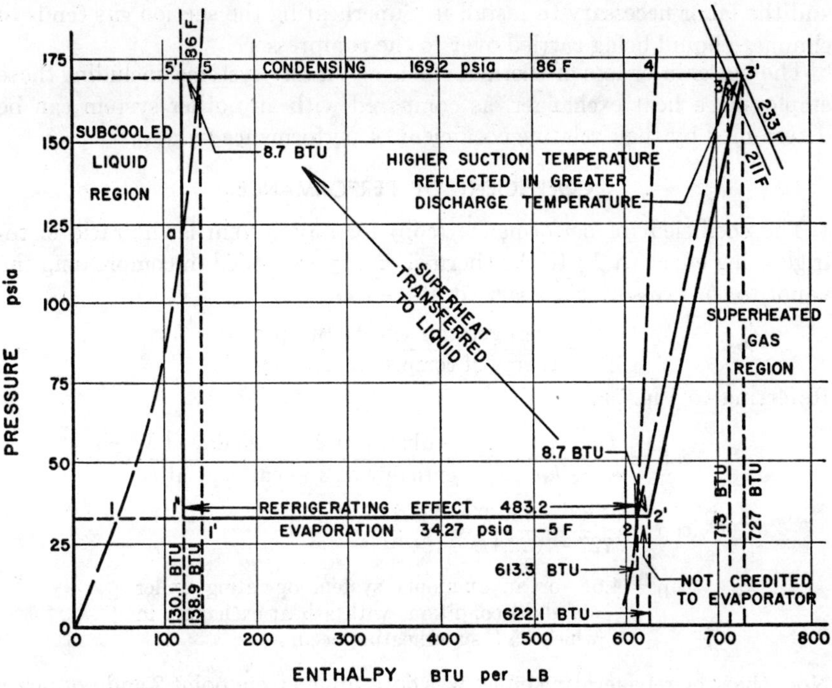

Fig. 58. *Ph* diagram, subcooling by superheating with heat exchanger (economizer).

effect by reducing the amount of flash gas, thereby increasing the capacity of the system. Superheating of the vapor by the warm liquid decreases the density of the former, reducing the weight of refrigerant handled by the compressor. Because superheating in this case occurs outside the evaporator, no gain in refrigerating effect from this source is attained. While the refrigerating effect line of the *Ph* diagram is extended to allow for increased enthalpy due to superheating and subsequent liquid subcooling, only that heat transferred to the refrigerant within the evaporator can be considered.

Whether the gain in refrigerating effect is offset by other losses may be determined by studying thermal behavior of the refrigerant in a system employing a heat exchanger. The gain in refrigerating effect due to liquid subcooling over the simple saturated system (Fig. 54) is obvious by the extension of line 1′–2 by an amount 5–5′ (or 1′–1″), as shown in Fig. 58. The length of line 5–5′, in this case, depends on the amount of superheating, so that the enthalpy represented by 5–5′ equals that represented by 2–2′, or $h_5 - h_{5'} = h_2 - h_{2'}$.

In addition to obtaining more capacity in the evaporator, a heat exchanger presents other advantages. When it is designed so that the vapor suction line is installed within the liquid line, atmospheric influences are avoided, thus preventing the absorption of heat from that source. Providing insulation in this manner eliminates the cost of regular insulating materials and the labor necessary to install it. Superheating the suction gas tends to eliminate liquid being carried over to the compressor.

The net gain or loss of thermal efficiency of any systems, including those employing a heat exchanger, as compared with any other system can be determined by their relative coefficient of performance values.

COEFFICIENT OF PERFORMANCE

The coefficient of performance (cop) for any system is the ratio of refrigerating effect ($R\ E$) to the thermal energy expended in compressing the vapor to the pressure necessary for its re-use.

$$\text{cop} = \frac{\text{refrigerating effect (Btu per lb)}}{\text{energy of compression (Btu per lb)}}$$

Referring to Fig. 58,

$$\text{cop} = \frac{h_2 - h_5}{h_{3'} - h_{2'}}, \quad \text{or} \quad \frac{\text{enthalpy at 2} - \text{enthalpy at } h_5}{\text{enthalpy at 3} - \text{enthalpy at } h_{2'}}$$

$$\text{cop} = \frac{613.3 - 130.1}{727 - 622.1} = \frac{483.2}{104.9}$$

cop = 4.61 for an ammonia system operating under standard conditions with a heat exchanger in which 15 F superheating occurs

Note that the refrigerating effect was determined from point 2 and not point 2′. This is because the enthalpy represented by line 2–2′ is superheat developed in the heat exchanger and not in the evaporator, as previously stated.

For a simple saturated ammonia system operating under standard ton conditions (Fig. 54):

$$\text{cop} = \frac{613.3 - 138.9}{713.0 - 613.3} = \frac{474.4}{99.7}$$

$$\text{cop} = 4.76$$

For an ammonia system whose liquid is subcooled from 86 F to 76 F by external means and whose vapor is superheated 15 F to a temperature of 20 F within the evaporator (Fig. 57):

$$\text{cop} = \frac{622.1 - 127.7}{727 - 622.1} = \frac{494.4}{104.9}$$

$$\text{cop} = 4.71$$

For an ammonia system whose vapor is superheated 15 F to a temperature of 20 F in the evaporator, no liquid subcooling (Fig. 55):

$$\text{cop} = \frac{622.1 - 138.9}{727 - 622.1} = \frac{483.2}{104.9}$$

$$\text{cop} = 4.61$$

The superheated system is the least efficient system of the three basic types.

For an ammonia system whose liquid is externally subcooled from 86 F to 76 F, no superheating

$$\text{cop} = \frac{613.3 - 127.7}{713 - 613.3} = \frac{485.6}{99.7}$$

$$\text{cop} = 4.87$$

The subcooled liquid system is the most efficient type, provided that the cooling medium is supplied economically.

Ammonia systems employing external liquid subcooling attain highest thermal efficiencies for the same operating conditions. This is apparent in Table 29 by comparing the coefficient of performance values of each type of operation. A definite decrease in cop is noted when ammonia is subjected to superheating. This is true whether superheating occurs in the evaporator or as the result of employing a heat exchanger. The advantage of superheating within the evaporator or heat exchanger lies in the resulting increase in refrigerating effect. This is indicated by the value given in column 6 of Table 29, determined by dividing the *R E* of any system by the *R E* of a simple saturated cycle having similar operating conditions.

The effect of superheating on the refrigeration capacity of a system varies in degree according to the refrigerant employed. For comparison with the foregoing examples treating with ammonia, a similar treatment of R-12 follows, the numerical results of which are also listed in Table 29.

For a simple saturated R-12 system operating under standard ton conditions of 86 F and 107.9 psia in the condenser and 5 F and 26.51 psia in the evaporator (Fig. 59):

$$\text{cop} = \frac{\text{refrigerating effect}}{\text{thermal work of compression}} = \frac{51.07}{10.91}$$

$$\text{cop} = \frac{78.79 - 27.72}{89.7 - 78.79} = \frac{51.07}{10.91}$$

$$\text{cop} = 4.69$$

For a R-12 system whose liquid is subcooled from 86 F to 76 F by external means, no superheating (Fig. 59):

$$\text{cop} = \frac{h_2 - h_{1''}}{h_3 - h_2}$$

Enthalpy of liquid at 76 F and 107.9 psia is determined by the equation

$$h_{5'} = h_{1'} - (t_d \times \text{sp ht})$$

$$h_{5'} = 27.72 - (10 \times 0.243^*)$$

$$h_{5'} = 25.29 \text{ Btu per lb}$$

$$\text{cop} = \frac{78.79 - 25.29}{89.7 - 78.79} = \frac{53.5}{10.91}$$

$$\text{cop} = 4.90$$

* Specific heat of liquid R-12.

Table 29. Thermal Variations of Ammonia and Refregerant 12 Systems Operating under Different Conditions of Evaporator Superheating and Liquid Subcooling*

Type of Thermal Operation	*Refrigerant*	cop $\frac{h_2 - h_5}{h_3 - h_2}$, etc.	*Discharge temperature °F (from Ph Diagram)*	*Thermal Work of Compression* Btu/lb $h_3 - h_2$	*Refrigerating Effect (R E)* $h_2 - h_5$	*Ratio* = $\frac{R\ E\ of\ any\ system}{R\ E\ of\ saturated\ system}$	*hp/ton* $\frac{(h_3 - h_2) \times lb/min/ton \times 778}{33{,}000}$ or $\frac{(h_3 - h_2) \times lb/min/ton}{42.42}$	*Lb refrigerant per min per ton or 200 ÷ R E*	*Volume—cu ft/min/ton* *lb/min/ton × sp vol at suction*	*Ratio* $\frac{hp/ton\ any\ system}{hp/ton\ any\ system\ saturated}$
Simple saturated, standard ton conditions. 5 F evap. and 86 F cond.	Ammonia	4.76	211	99.7	474.4	1.00	0.988	0.4216	3.436	1.00
	R-12	4.69	102	10.91	51.07	1.00	1.005	3.91	5.81	1.00
Evaporator superheating. From 5 F to 20 F. No liquid subcooling. 86° cond.	Ammonia	4.61	233	104.9	483.2	1.019 (+1.9%)	1.023	0.414	3.50	1.034 (+3.4%)
	R-12	4.42	122	12.05	53.23	1.042 (+4.2%)	1.066	3.757	5.77	1.060 (+6%)
Liquid subcooling, external source. From 86 F to 76 F. No superheating. 5 F evap.	Ammonia	4.87	211	99.7	485.6	1.024 (+2.4%)	0.931	0.412	3.36	0.942 (−5.8%)
	R-12	4.90	102	10.91	53.5	1.047 (+4.7%)	0.961	3.738	5.55	0.956 (−4.4%)
Evaporator superheating. From 5 F to 20 F. External liquid subcooling. From 86 F to 76 F	Ammonia	4.71	233	104.9	494.4	1.042 (+4.2%)	0.999	0.4045	3.422	1.011 (+1.1%)
	R-12	4.62	122	12.05	55.66	1.090 (+9.0%)	1.02	3.59	5.51	1.014 (+1.4%)
Liquid subcooling by superheating within heat exchanger. Assume 15 F superheat from 5 F to 20 F	Ammonia	4.61	233	104.9	483.2	1.019 (+1.9%)	1.023	0.414	3.50	1.034 (+3.4%)
	R-12	4.60	122	12.05	55.39	1.057 (+8.5%)	3.611	3.611	5.54	1.051 (+5.1%)

* Based on 34.27 psia evaporator and 169.2 psia condenser pressures for ammonia and 26.51 psia evaporator and 107.9 psia condenser pressures for Refrigerant 12.

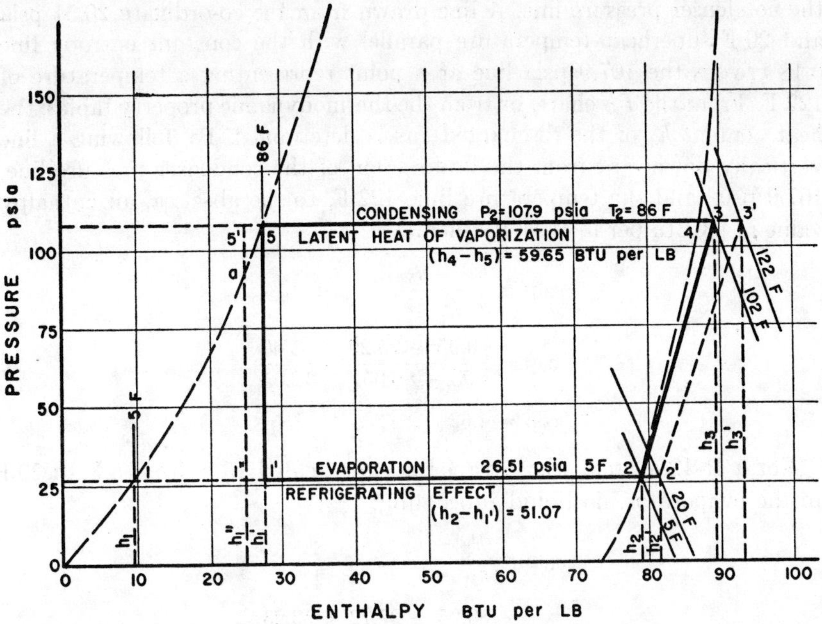

Fig. 59. *Ph* diagram representing various conditions.

For a R-12 system whose liquid is externally subcooled from 86 F to 76 F and whose vapor is superheated 15 F to a temperature of 20 F in the evaporator (Fig. 59):

$$\text{cop} = \frac{h_{2'} - h_{5'}}{h_{3'} - h_{2'}}$$

$$h_{5'} = h_5 - (t_d \times \text{sp ht})$$

$$h_{5'} = 27.72 - (10 \times 0.243)$$

$$h_{5'} = 25.29 \text{ Btu per lb liquid}$$

$$h_{2'} = h_2 + (t_d \times \text{sp ht})$$

$$h_{2'} = 78.79 + (15 \times 0.144^*)$$

$$h_{2'} = 78.79 + 2.16$$

$$h_{2'} = 80.95 \text{ Btu per lb superheated vapor}$$

$$h_{3'} = \text{enthalpy of compressed vapor}$$

at discharge temperature and condenser pressure. It may be determined by the method already shown or it can be found by determining the point on the *Ps* chart where a line of constant entropy drawn from the intersection of the evaporator pressure line and the superheat temperature line crosses

* Average specific heat of R-12 vapor at constant pressure.

the condenser pressure line. A line drawn from the co-ordinate 26.51 psia and 20 F superheat temperature parallel with the constant entropy line 0.18 crosses the 107.9 psia line at a point representing a temperature of 122 F. From the *Ph* chart, or from the thermodynamic property tables, the heat content h_3 of the discharged gas is determined. By following a line vertically downward from the intersection of the condenser pressure line, 107.9 psia, and the temperature line, 122 F, to the abscissa, an enthalpy value of 93 Btu per lb is established.

$$\text{cop} = \frac{h_{2'} - h_{5'}}{h_{3'} - h_{2'}}$$

$$\text{cop} = \frac{80.95 - 25.29}{93 - 80.95} = \frac{55.66}{12.05}$$

$$\text{cop} = 4.62$$

For a R-12 system whose vapor is superheated 15 F from 5 F to 20 F in the evaporator, no liquid subcooling:

$$\text{cop} = \frac{h_{2'} - h_5}{h_{3'} - h_{2'}}$$

$$\text{cop} = \frac{80.95 - 27.72}{93 - 80.95} = \frac{53.23}{12.05}$$

$$\text{cop} = 4.42$$

For a R-12 system employing a heat exchanger operating under otherwise standard conditions, assuming that the vapor is superheated (15 F) to 20 F before entering the compressor (Fig. 59):

$$\text{cop} = \frac{h_2 - h_{5'}}{h_{3'} - h_{2'}}$$

$$h_{1'} = h_1 - (t_s \times \text{sp ht})$$

$$h_{1'} = h_1 - (15 \times 0.144)$$

$$h_{1'} = 27.72 - 2.16$$

$h_{1'} = 25.56$ Btu per lb enthalpy of liquid

$h_{2'} = 80.95$ Btu per lb enthalpy of superheated gas entering compressor

$h_3 = 93$ Btu per min enthalpy of discharged gas

$$\text{cop} = \frac{80.95 - 25.56}{93.0 - 80.95} = \frac{55.39}{12.05}$$

$$\text{cop} = 4.60$$

The net effect of superheating or subcooling on the refrigeration capacity of any system may be determined by combining pertinent values and comparing them with similar values for the simple saturated system. This is accomplished as follows, using ammonia:

Gain or loss of refrigerating effect

Simple saturated	$R\,E = \frac{474.4}{474.4} = 1.00$
Superheating	$R\,E = \frac{483.2}{474.4} = 1.019$ or 1.9% increase per lb
Liquid subcooling	$R\,E = \frac{485.6}{474.4} = 1.024$ or 2.4% " "
Superheating and subcooling	$R\,E = \frac{494.4}{474.4} = 1.042$ or 4.2% " " "
Heat exchanger	$R\,E = \frac{483.2}{474.4} = 1.019$ or 1.9% " " "

The amount of refrigerant handled by the compressor depends on the specific volume of the gas at suction-line conditions and varies according to the thermal type of system. For example, while the *R E* increases with evaporator superheating, the specific volume diminishes so that the weight of refrigerant handled per minute is less.

Gain or loss due to change is specific volume:

Simple saturated	Ratio $\frac{3.436}{3.436} = 1.00$
Superheating	" $\frac{3.436}{3.50} = 0.956$
Liquid subcooling	" $\frac{3.436}{3.36} = 1.023$
Superheating and subcooling	" $\frac{3.436}{3.422} = 1.004$
Heat exchanger	" $\frac{3.436}{3.50} = 0.981$

Net increase or decrease in refrigeration capacity is equal to the product of all gains or losses. The net gain or loss as compared with a saturated system is a function of *R E* and specific volume of the gas entering the compressor. Therefore, it is expected that, under superheated conditions, any gain in *R E* will be offset in varying degree by a decrease in weight of refrigerant handled per unit of time, viz.:

Superheating:

$$\text{Gain or loss} = 1.019 \times 0.956 = 0.975$$
$$(1.00 - 0.975) \times 100 = 2.5\%\ \text{loss}$$

Liquid subcooling:

$$\text{Gain or loss} = 1.024 \times 1.023 = 1.04$$
$$(1.04 - 1.00) \times 100 = 4\%\ \text{gain}$$

Superheating and subcooling:

$$\text{Gain or loss} = 1.042 \times 1.004 = 1.046$$
$$(1.046 - 1.00) \times 100 = 4.6\% \text{ gain}$$

Heat exchanger:

$$\text{Gain or loss} = 1.019 \times 0.981 = 0.9996$$
$$(1.00 - 0.9996) \times 100 = 0.04\% \text{ loss}$$

Note that the gain in *R E* made with a heat exchanger is entirely offset by the loss due to the decrease in weight of refrigerant handled per minute. However, the convenience of the greater *R E* is often the deciding factor in adopting a heat exchanger, notwithstanding the negative influence of superheating. The latter, plus other operating advantages already mentioned, makes the heat exchanger a valuable auxiliary. This is especially true with commercial and railway installations, where every means toward self-contained apparatus and maximum *R E* must be adopted.

Industrial installations are more adaptable to external liquid cooling, and the foregoing data point to the advantages gained by this method for increasing refrigeration capacity as to warrant practical consideration.

As there is always the possibility of liquid carry-over to the compressor when a system is operated saturated, some superheating is advisable. At least 10 F above the saturation temperature is recommended. While some drop in performance is incurred as a result of superheating, the loss may be charged profitably to insurance against compressor damage.

THEORETICAL VS. PRACTICAL CONSIDERATIONS

The *Ph* diagrams referred to in the foregoing discussion are representative of refrigeration cycles based purely on theoretical considerations. This means that, aside from the energy required to complete the various phases of the cycle, no other heat is considered to be gained or lost due to external influences.

Theoretically, the following equation is always true for a simple system:

$$R\,E + \text{thermal work of compression} = \text{heat extracted in condenser}$$

Graphically, the equation may be written,

$$\text{Line } 1'\text{–}2 + \text{line } 2\text{–}3 = \text{line } 3\text{–}4 + \text{line } 4\text{–}5 \text{ (Fig. 54)}$$

where line 3–4 = superheat of vapor extracted in the condenser
line 4–5 = latent heat of vaporization extracted in the condenser

There are several operating factors that tend to offset theoretical values. These must be considered in estimating final performance results.

1. *Drop in pressure* in the evaporator from the point of entrance at the expansion valve to the entrance of compressor, due to internal (within refrigerant) and surface (along the pipe) friction. This varies with the (*a*) refrigerant employed, (*b*) density of the refrigerant which changes with evaporator temperature, (*c*) velocity of vapor (pipe size), (*d*) length of run, (*e*)

number of bends, and (*f*) internal pipe surface condition (rough or smooth). This is made up by reducing the suction pressure at the compressor end of the suction line, which in turn interferes with the thermal balance of the system.

2. *Rise in vapor temperature* in the suction line between the evaporator outlet and the compressor. If vapor leaves the evaporator saturated at the suction temperature, then the rise in temperature is caused by superheating and cannot be considered as useful refrigeration work. To reduce the amount of heat absorbed at this point in the system, the suction line should be insulated. The small amount of heat absorbed by the vapor along an insulated suction line may be sufficient to convert any entrained liquid to a vapor before entering the compressor.

3. *Volumetric efficiency* of compressor, the value of which depends on (*a*) drop in pressure across the valves due to friction, which in turn is governed by the velocity of the gas (area of opening and compressor speed), spring tension on the valves and directional changes of the gas, both in and out of the valve, (*b*) temperature of the compression chamber surfaces which tends to superheat the gas, subsequently decreasing the weight of refrigerant handled per stroke. The temperature of such surfaces and consequently the effect on the entering vapor is reduced in some instances by cooling the cylinder walls and the head, and by increasing the speed with which vapor is forced through the cylinder. This latter is based on the time element or the amount of exposure allowed for the vapor to absorb heat from the exposed piston, cylinder, and valve surfaces. Higher operating speeds (rpm) afford less time in which heat transfer can be accomplished. Hence high-speed compressors bring about conditions more nearly adiabatic. This tends to allow practical values to more closely approach theoretical values in modern design than was formerly the case with slow-speed compressors. This is particularly true with low-temperature refrigerants. (*c*) Loss of compression due to gas leakage past the piston and the valves. This loss increases with higher compression ratios and is aggravated further when valves and rings designed to prevent "blow-by" become worn or otherwise lose their sealing ability. (*d*) Recompression of vapors already discharged but which re-enter the cylinder during the suction stroke as a result of faulty or late closing discharge valves. This defect is referred to as "wire drawing." (*e*) Re-expansion of clearance gases, a condition that prevents suction valves from opening at the instant of piston reversal. The volume represented by the distance of piston movement on its suction stroke below top-dead-center during which the suction valve remains closed should be considered lost, and must be deducted from the theoretical displaced volume. Clearance volume of itself should not be deducted as it is offset by slightly increasing the length of the cylinder over theoretical requirement. The increase in cylinder length does not materially affect horsepower requirements per cubic foot of refrigerant vapor handled.

4. *Loss of heat in discharge line* to surrounding air prior to entering con-

denser is a desirable occurrence. Any dissipation of heat here reduces the condenser load an equal amount.

5. *Drop in pressure* in the liquid line due to friction within the liquid and along the pipe surfaces. The amount of pressure drop here becomes great enough in some instances to lower the boiling point of the refrigerant sufficiently to cause the occurrence of flash gas in the liquid line leading to the expansion valve. This is especially true where the evaporator is located at a higher level than the condenser, causing a critical difference in liquid pressure due to the weight of refrigerant in the riser. High liquid risers should be avoided.

6. *Increase in head pressure* due to the presence of noncondensible gases such as air. Purgers are employed to eliminate such gases.

Considering these factors, it is obvious that the theoretical results obtained from the *Ph* diagrams, as computed solely on the basis of thermal property tables of refrigerants, cannot be accepted entirely as accurately representing all practical values. Because of the erratic gains and losses of heat as a result of external influences that are felt in all phases of the operating cycle, it becomes necessary to depend more on computation wherein the effects of such outside factors are considered. Dependence also should be placed on related empirical values based on experiment or sound hypothesis, especially in those cases where more practical data cannot be provided. Theoretical values do represent those ideal conditions for which every designer should strive to build into each component part of the refrigeration system some improvement that will tend toward the defeat of those forces acting to reduce operating efficiency.

EFFECT OF DIFFERENT COMPRESSION RATIOS ON THE *Ph* DIAGRAM

Greater refrigeration capacity and more efficient operation results with an increase in suction pressure. As observed in the *Ph* diagram (Fig. 60), a decrease in compression ratio reduces the thermal losses incurred by flash gas. This increases the refrigerating effect, thus raising the capacity of the evaporator. It also allows for a greater flow of liquid through the throttling device. It is also obvious in the *Ph* diagram that the thermal work of compression decreases with any decrease in compression ratio. This latter results in lower horsepower per ton of refrigeration. Another advantage of maintaining highest practical suction pressures lies in the greater amount of refrigerant handled per minute by the compressor, thus increasing the capacity of the latter component.

The following shows comparative power and refrigerating capacity values for the same system operating at different compression ratios.

Assume an ammonia system operating simple saturated at a constant head pressure of 169.2 psia, whose evaporator pressure has been increased from 34.27 psia corresponding to 5 F (first case), to 53.73 psia corresponding to 25 F (second case), and again to 73.32 psia and 40 F (third case).

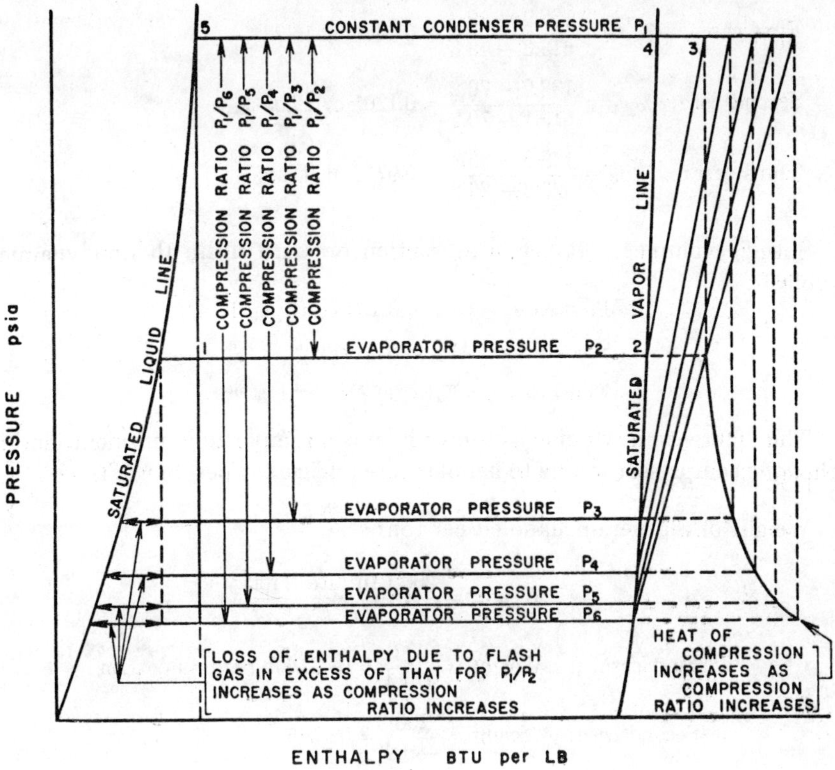

Fig. 60. Effect of different compression ratios on *Ph* diagram.

Refrigerating effect (from *Ph* chart or thermodynamic property tables):

$$RE = h_5 - h_2$$

First case: 5 F and 34.27 psia $RE = 613.3 - 138.9$
$= 474.4$ Btu per lb

Second case: 25 F and 57.73 psia $RE = 619.1 - 138.9$
$= 480.2$ Btu per lb, 1% increase

Third case: 40 F and 73.72 psia $RE = 623.0 - 138.9$
$= 484.1$ Btu per lb, 2% increase

an increase in capacity of approximately 1 percent for each 20 psia rise in suction pressure for an ammonia system.

Quality (% Q) (from *Ph* chart or thermodynamic property tables):

$$\% \, Q = \frac{h_5 - h_1}{h_2 - h_1} = \frac{\text{enthalpy of 86° liquid} - \text{enthalpy of 5 F liquid}}{\text{enthalpy of 5° vapor} - \text{enthalpy of 5 F liquid}}, \text{ etc.}$$

First case: $\% \ Q = \frac{138.9 - 48.3}{613.3 - 48.3} = 0.1603$ or 16.03%

Second case: $\% \ Q = \frac{138.9 - 70.2}{619.1 - 70.2} = 0.1205$ or 12.05%

Third case: $\% \ Q = \frac{138.9 - 86.8}{623.0 - 86.8} = 0.9712$ or 9.71%

Specific volume cu ft per lb at suction pressure (from thermodynamic tables):

First case: $V_s = 8.150$ cu ft per lb

Second case: $V_s = 5.334$ " " " "

Third case: $V_s = 3.971$ " " " "

The difference in volume as shown here is an important consideration in that the compressor is able to handle more refrigerant per unit of time.

Weight of refrigerant handled per ton:

$$\text{Weight} = \frac{200 \text{ Btu (1 ton ref)}}{R\ E}$$

First case: $\text{Weight} = \frac{200}{474.4} = 0.4216$ lb per min per ton

Second case: $\text{Weight} = \frac{200}{480.2} = 0.4165$ " " " " "

Third case: $\text{Weight} = \frac{200}{484.1} = 0.4131$ " " " " "

Horsepower per ton:

$$\text{hp} = \frac{(h_3 - h_2) \times \text{lb per min per ton} \times 777.46 \text{ Btu per hp}}{33{,}000}$$

or

$$\text{hp} = \frac{h_3 - h_2 \times \text{lb per min per ton}}{42.42}$$

First case: $\text{hp} = \frac{99.7 \times 0.4216}{42.42} = 0.989$ hp per ton

Second case: $\text{hp} = \frac{69.6 \times 0.4165}{42.42} = 0.683$ hp per ton

Third case: $\text{hp} = \frac{50.3 \times 0.4131}{42.42} = 0.489$ hp per ton

The saving in power as a result of a reduction in suction pressure amounts, in the second case, to

$$\frac{\frac{0.989 - 0.683}{0.683} \times 100}{53.73 - 34.27} = \frac{44.8\%}{19.46} = 2.3\% \text{ per psi}$$

and, in the third case,

$$\frac{\frac{0.989 - 0.489}{0.489} \times 100}{73.32 - 34.27} = \frac{102.3\%}{39.05} = 2.62\% \text{ per psi}$$

or approximately $2\frac{1}{2}$ percent savings in power for each psi increase in suction pressure. Similarly, for each psi decrease in condenser pressure, the power savings amounts to approximately $\frac{1}{2}$ percent. While the above represents comparative values for an ammonia system, the trend is the same for other refrigerants.

Volume or compressor displacement per ton (cfm):

V_d = lb per min × specific volume at suction pressure

First case: $V_d = 0.4216 \times 8.150 = 3.436$ cfm per ton

Second case: $V_d = 0.4165 \times 5.334 = 2.222$ " " "

Third case: $V_d = 0.4131 \times 3.971 = 1.640$ " " "

The above values show a rapid decrease in the volume of refrigerant handled by the compressor per ton per minute with any appreciable increase in suction pressure. Based on the results shown, the rate of volume decrease amounts to

$$100 - \left(\frac{1.640}{3.436}\right) \div (73.72 - 34.27) = 1.32\%$$

for each 1 psi increase in suction pressure. Thus, a smaller compressor or a larger compressor operated at reduced speed, depending on the flexibility required, can be considered when compression ratios are decreased.

THE MOLLIER DIAGRAM

The Mollier diagram is a chart on which the properties of a fluid are presented graphically. The co-ordinates are enthalpy (called heat content in older charts) in Btu per lb as the abscissa and the logarithm of pressure as the ordinate. These co-ordinates are the same as those selected for the *Ph* diagrams. The Mollier chart goes much farther than the simple *Ph* diagram in that the various state points represent values of enthalpy, temperature, volume, entropy, and quality of liquid-vapor mixtures. As for the regions of liquid, liquid-vapor, and superheated gas, the description given earlier for the *Ph* diagram also applies to the Mollier chart.

According to the Bureau of Standards Circular No. 142, Tables of Thermodynamic Properties of Ammonia

The use of the tables and chart is based on three properties of thermodynamic function called the heat content (now enthalpy) of a fluid. The three properties referred to are the following:

1. In any process which occurs at constant pressure the increase in heat content (enthalpy) is equal to the heat added.
2. In adiabatic compression—that is, at constant entropy—the thermal equivalent of the work done by the compressor is equal to the change in heat content (enthalpy) of the fluid between inlet and outlet.
3. In a throttling process, such as occurs at the expansion valve, the increase in heat content (enthalpy) is equal to the heat added. Therefore, when no heat is added, the heat content (enthalpy) is constant.*

By the application of these three propositions, the various heat and work involved in any ideal compression refrigerating cycle can be calculated from either the tables or chart.

Based on the above propositions, examples showing the use of the Mollier chart and the property tables are given in this text wherever apropos. Most of the basic examples given herein are based on standard ton conditions of 5 F evaporator and 86 F condenser temperatures as a matter of comparison and convenience. By means of either the chart or the tables, volume, enthalpy, temperature, entropy, and quality of the fluid at any point in the refrigerating cycle can be determined if conditions of saturation within the evaporator and the condenser are known. From the data compiled, the work done by the compressor (hp per ton), heat removed by the condenser per pound of refrigerant, refrigerating effect, weight in pounds of refrigerant circulated per minute per ton, compressor discharge temperature, and volume of refrigerant handled by the compressor per minute per ton of refrigeration, can be calculated.

REVERSIBLE WORK PROCESS, CARNOT CYCLE

A work system is said to be thermodynamically *reversible* when it (1) performs in either direction without change in available energy, (2) exactly retraces its original state path, and (3) reverts to its original state.

Because of certain operating and physical factors, practical work systems are *irreversible*. Friction immediately reduces available energy within any work system. Thermal energy losses due to temperature differentials between the working fluid and surrounding substances also create a condition in which reversible processes cannot exist. Such losses of available energy must be supplied externally if the working fluid is to be returned to its original state and the work process continued. Conditions present within a work system that tend to promote reversibility also tend to increase its efficiency and raise its capacity.

If a swinging weight were suspended by a chain from the ceiling of a compartment under vacuum, and the chain supported on the edge of a very thin steel blade, an approach to a reversible process would be made. The classical example of a reversible process is given in the following discussion.

* Words in parentheses added by author.

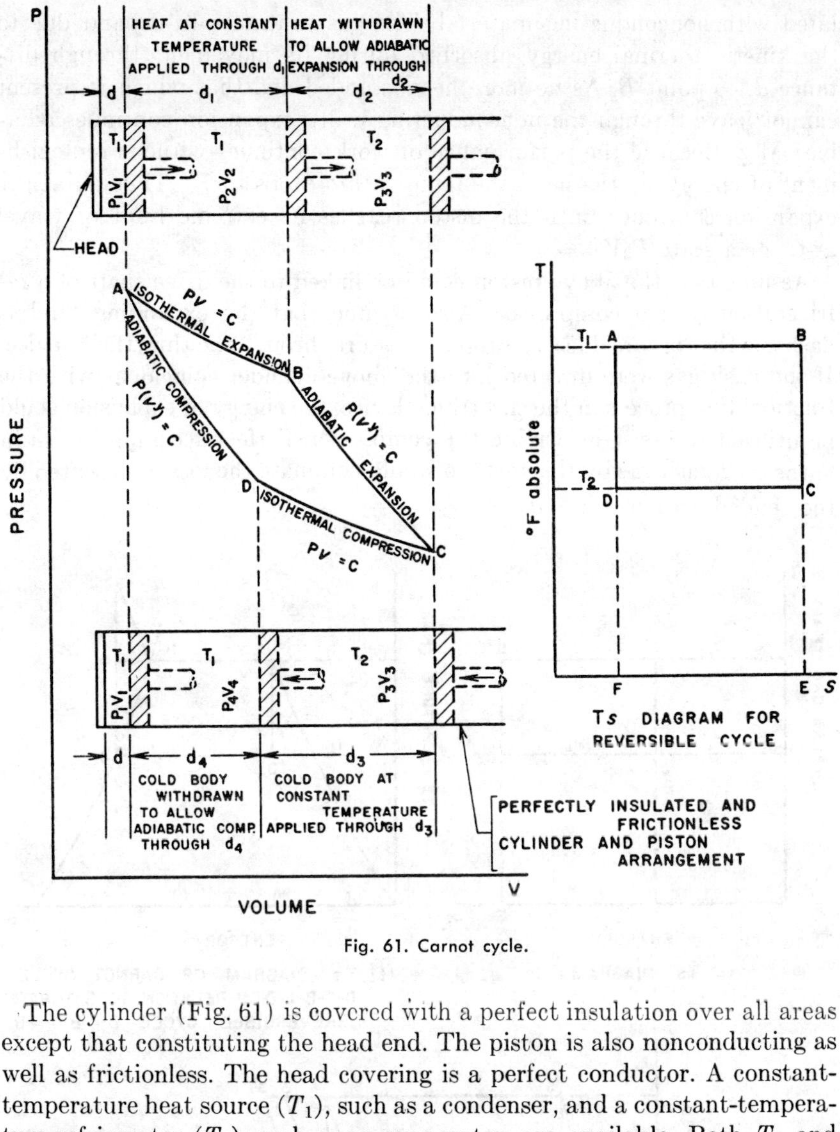

Fig. 61. Carnot cycle.

The cylinder (Fig. 61) is covered with a perfect insulation over all areas except that constituting the head end. The piston is also nonconducting as well as frictionless. The head covering is a perfect conductor. A constant-temperature heat source (T_1), such as a condenser, and a constant-temperature refrigerator (T_2), such as an evaporator, are available. Both T_1 and T_2 are absolute temperatures ($459.6 + t$ F).

A volume (V) of a perfect gas occupies the space d having a pressure P_1. Heat at constant temperature is applied to the head, causing the gas to expand. The internal energy supplied by the heat in this case acts to move the piston so that the energy is utilized in performing work, thus maintaining a constant temperature, T_1. Because expansion occurs at constant temperature, the process occurs isothermally. When the piston reaches point B or a state P_2V_2, the heat source is removed and the head insulated with nonconducting material. The gas continues to expand due to the kinetic thermal energy absorbed during its movement through distance d_1 to point B. As no more heat is added and that which is present cannot leave through the nonconducting walls, expansion continues adiabatically. Because the performance of work continues without replenishment of energy by the heat, the temperature drops to T_2. Thus adiabatic expansion continues until the piston reaches its extreme limit of travel at C, or a state P_3V_3.

Assume that the above piston could be linked to the drive shaft of a refrigeration-system compressor. Also assume that the expanding fluid is flash gas that occurs due to a drop in pressure through the throttling device. If the flash gas were directed into the above cylinder coincident with the fractionating process in the throttling device, the energy of expansion could be utilized to assist in driving the compressor. Perfect accomplishment in harnessing flash gas by this method would eliminate the loss represented by the shaded area in Fig. 50.

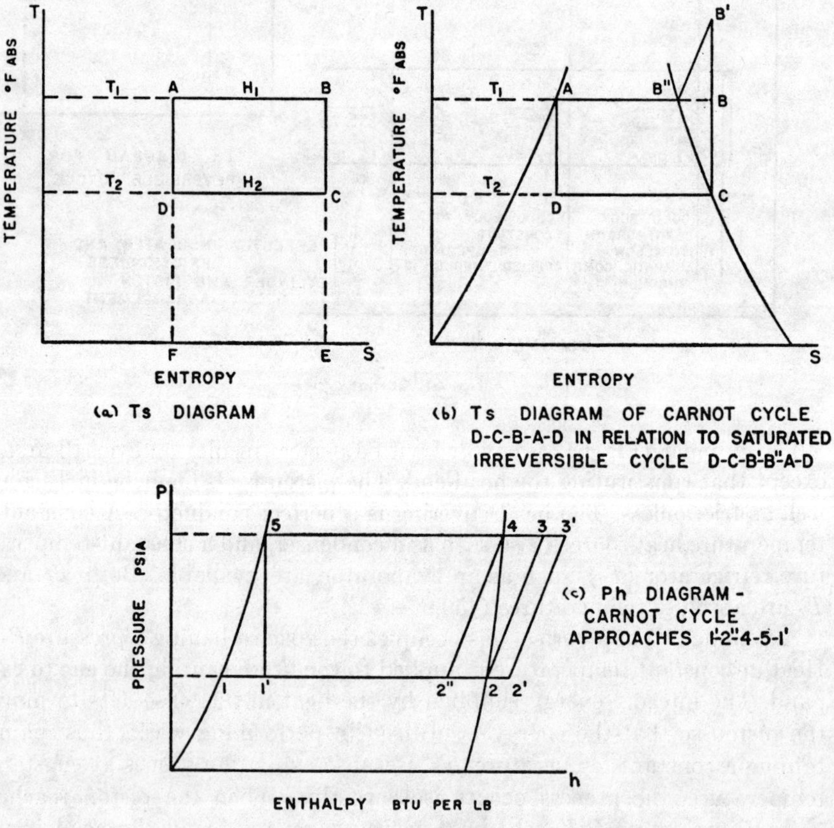

Fig. 62. Carnot cycle, *Ts* diagram.

Exposing the gas to a cold body applied to the head would cause a reduction in thermal energy. However, contraction or compression of the gas would occur counteracting the loss of energy so as to maintain constant temperature. Compression through d_3, therefore, would be isothermal. Removing the cold body and replacing it with the nonconducting head covering, compression would continue by reason of the reduction of thermal energy. The piston will continue to travel. From D to A isothermal compression occurs, through d_4, until the gas reaches its original state P_1V_1 and its temperature rises again to T_1.

Thus the cycle described above is composed of two reversible isothermal processes and two reversible adiabatic processes. It was first developed by the Frenchman, Sadi Carnot, in 1824, and therefore is known as the Carnot cycle.

The reversible cycle is the most efficient cycle. To assume otherwise would be to state that heat will flow from a cold body to a warmer body, a contradiction of the Second Law of Thermodynamics.

A temperature-entropy diagram for a reversible work system is shown in Fig. 62 (a). Lines AB and BC represent isothermal and adiabatic expansion of a perfect fluid according to the PV diagram (Fig. 61). The heat absorbed along AB is converted to work during the entire expansion process ABC. Lines CD and DA represent isothermal and adiabatic compression during which heat is rejected and the state of the working fluid returns to its original state.

Let H_1 equal the total heat absorbed by the system along AB and H_2 equal the heat rejected along CD. Then

$$H_1 = \text{area } FABE \text{ or available energy}$$

and

$$H_2 = \text{area } FDCE \text{ or unavailable energy}$$

The energy utilized during the work cycle is therefore equal to area $FABE$ – area $FDCE$ or area $ABCD$. Area $ABCD$ therefore equals $H_1 - H_2$.

From the definition of work efficiency—i.e., the ratio of energy utilized to heat supplied—then for a reversible engine

$$\text{Efficiency} = \frac{FABE - FDCE}{FABE} = \frac{ABCD}{FABE}$$

or

$$E = \frac{H_1 - H_2}{H_1}$$

If each unit of energy were equivalent to 1° of absolute temperature, then H_1 and H_2 would equal T_1 and T_2, respectively, and

$$E = \frac{T_1 - T_2}{T_1}$$

Instead of efficiency, the effectiveness of a refrigeration cycle is measured in terms of coefficient of performance. As defined earlier, cop is equal to the ratio of refrigerating effect ($R\ E$) to the thermal work of compression (h_c), or

$$\text{cop} = \frac{R\ E}{h_c}$$

Refrigeration effect is the heat absorbed by the refrigerant during the cooling phase of the cycle. Thermal work of compression is thermal energy absorbed by the refrigerant during the process of raising its pressure from that in the evaporator to that in the condenser.

The Carnot cycle for a refrigeration system differs from the heat engine cycle described above in that the process is reversed. This means that the cycle starts at point C of the Ts diagram and proceeds in a counterclockwise direction as follows:

Line CB adiabatic (or isentropic) compression (constant entropy)

Line BA isothermal compression (constant temperature)

Line AD adiabatic (or isentropic) expansion (constant entropy)

Line DC isothermal expansion (constant temperature)

Heat is absorbed by the fluid along the isothermal expansion line which is now DC and rejected along the isothermal compression line, now BA. Line BA therefore represents the condensing process at the higher temperature T_1, while DC represents the evaporation process at the lower temperature T_2.

The refrigerating effect is the amount of heat absorbed by the cold fluid in the evaporator, being equivalent to that represented by the change in entropy along DC. The amount of heat absorbed or $R\ E$ then is represented by the area $CDFE$, which shall be designated as H_2. The amount of work done in compressing the fluid isentropically along CB is represented by the area $CBAD$. The amount of heat rejected to the condenser is equivalent to that represented by the change in entropy along BA or by the area $CDFE$ + area $CBAD$. The amount of heat rejected along BA is equal to H_2, plus the thermal energy of compression (h_c), which sum shall be designated as H_1. The coefficient of performance for a refrigeration system operating under Carnot cycle conditions is therefore

$$\text{cop} = \frac{H_2}{H_1 - H_2}$$

where $H_1 - H_2$ equals the thermal work of compression h_c. Again assuming that

$$H_1 = T_1 \quad \text{and} \quad H_2 = T_2$$

then

$$\text{cop} = \frac{T_2}{T_1 - T_2}$$

As stated previously, no system has a higher coefficient of performance than one operating under Carnot cycle conditions. While the reversible cycle is purely theoretical (not practical), Carnot values may be used to determine relative efficiencies of any refrigeration system, viz.:

$$\text{Relative efficiency} = \frac{\text{cop of any system}}{\text{cop of Carnot cycle}}$$

Refrigeration systems are often employed as heating systems. This is accomplished by reversing the function of the condenser and the evaporator. This means that the original condenser now acting as an evaporator absorbs heat from some source such as the atmosphere or beneath the earth's surface. The resulting vapor is then compressed in the conventional manner, after which the warmed gas enters the original evaporator, now the condenser. One method of utilizing the thermal energy of the compressed gas is to force air over the condenser surface by means of a fan. The air acts as a cooling medium to condense the vapor as it passes through the condenser. The warmed air then acts to warm a space into which it is forced by the fan. Methods of converting refrigeration systems to heating systems are discussed more fully in a later chapter.

Referring to the *Ts* diagram (Fig. 62(*a*)) and the designations H_1 and H_2 in the previous discussion of the reversible (Carnot) refrigeration cycle, the cop for a heating cycle would be

$$\text{cop} = \frac{H_1}{H_1 - H_2}$$

or

$$\text{cop} = \frac{T_1}{T_1 - T_2}$$

COEFFICIENT OF PERFORMANCE, CARNOT CYCLE

It was stated earlier that any change in the availability of energy is accompanied by a change in entropy. It was also shown in Fig. 50 that any work system operating under Carnot cycle conditions gained an advantage of an increase in energy utilized over irreversible systems. For an ammonia refrigeration system operating under standard ton conditions, it was also shown in Fig. 50 that an increase in energy was gained when expansion is carried on adiabatically or isentropically (constant entropy). The increase in energy was represented by a decrease in entropy from an irreversible state or 0.3033 at point 1 to a saturated state, or 0.1092 (5 F).

As compression occurs isentropically and the maximum temperature of the working fluid at any time cannot be greater than T_1 or 86 F in this case, compression must originate at some point in the liquid-vapor zone. From the *Ph* chart for ammonia (Mollier diagram), it is found that the entropy of 86 F saturated vapor is 1.1904. Following the 1.1904 entropy line downward to the constant temperature line T_2 or 5 F in this case, it is found that the enthalpy for the liquid-vapor mixture is 552 Btu per lb. This establishes the location of 2″ of the *Ph* diagram (Fig. 62(*c*)), at which point compression starts.

The change in entropy during isothermal expansion *DC* (Fig. 62(*b*)), then, is

$$(C_s - D_s) = 1.1904 - 0.1092 = 1.0812$$

The heat absorbed by the fluid or refrigerating effect is equal to the latent heat between the point of introduction of the fluid at *D* and its exit at *C*. From the previous entropy equation for latent heat of vaporization,

$$ds = \frac{H}{T}$$

therefore

$$H = T \times ds$$

The heat absorbed along the line *DC* is

$$H_2 = T_2 (C_s - D_s)$$

and the thermal work of compression is

$$h_c = H_1 - H_2 = (T_1 - T_2) \times (C_s - D_s)$$

As stated previously,

$$\text{cop} = \frac{H_2}{H_1 - H_2}$$

Therefore

$$\text{cop} = \frac{T_2 (C_s - D_s)}{(T_1 - T_2) \times (C_s - D_s)}$$

Reducing the fraction to its simplest form by dividing both numerator and denominator by $(C_s - D_s)$, then

$$\text{cop} = \frac{T_2}{T_1 - T_2}$$

proving, as previously stated, that

$$\frac{H_2}{H_1 - H_2} = \frac{T_2}{T_1 - T_2}$$

Substituting numerical values in both equations,

$$\text{cop} = \frac{H_2}{H_1 - H_2} = \frac{464.6\ (1.1904 - 0.1092)}{(545.6 - 464.6)(1.1904 - 0.1092)} = \frac{502.32}{87.58} = 5.74$$

or

$$\text{cop} = \frac{T_2}{T_1 - T_2} = \frac{464.6}{545.6 - 464.6} = \frac{464.6}{81} = 5.74$$

The inference in the above equation

$$\frac{H_2}{H_1 - H_2}$$

appears to place a value of 501.77 Btu per lb as the refrigerating effect and 87.48 Btu per lb as the thermal work of compression. This would mean that the system would present an impractical situation in that more heat would be absorbed by the fluid than extracted in the condenser. Actually, the gain in capacity is reflected in a lower compressor demand as the flash gas in a reversible cycle is assumed to expand in the compressor cylinder and not in the expansion valve. Because, under standard ton conditions, the loss in weight of liquid by flash gas is 16.03 percent, it follows that, by directing a similar amount to the compressor suction, the thermal work will diminish proportionately. In directing the flash gas to the compressor, only 5 F liquid enters the evaporator. The expansion process now being isentropic, the loss represented by the shaded area in Fig. 50 is eliminated and the entropy of the fluid entering the evaporator is now 0.2875, and not 0.3033 as previously calculated for an irreversible cycle. The increase in capacity is represented by the difference in entropy or

$$T\,(s_1 - s_{1'})$$

For standard ton conditions,

$$\begin{aligned} ds &= (459.6 + 5) \times (0.3033 - 0.2875) \\ &= 464.6 \times 0.0158 \\ &= 7.3 \text{ Btu per lb} \end{aligned}$$

The refrigerating effect for the Carnot cycle is 474.4 + 7.3 = 481.7, or approximately 482 Btu per lb.

The thermal work of compression under irreversible standard ton conditions was found to be 99.7 Btu per lb as calculated previously. Under Carnot cycle conditions, this external load is reduced by an amount equivalent to that augmented by the expansion of the flash gas, or 16.03 percent in this case. The thermal work of compression is therefore

$$99.7 \times \left(\frac{100\% - 16.03\%}{100}\right) = 83.7 \text{ Btu per lb}$$

$$\text{cop} - \frac{482}{83.7} = 5.74 \text{ Carnot cycle}$$

The cop of 5.74 for an ammonia system operating under Carnot cycle conditions compares with 4.76 for a simple saturated ammonia system when both maintain 5 F evaporator and 86 F condenser temperatures. For comparisons with ammonia systems operating other than simple saturated, see Table 29.

The efficiency of a simple saturated ammonia system operating under conditions stated as related to the ideal is

$$\text{Relative efficiency} = \frac{\text{cop of any system}}{\text{cop of Carnot cycle}}$$

$$\text{Relative efficiency} = \frac{4.76}{5.74} = 82.92\%$$

SUMMARY OF CHAPTER 7

The efficiency of any process is the ratio of actual to theoretical performance.

The total efficiency of any work system is the product of the efficiencies of its several parts.

Refrigerating processes are based on the thermodynamic functions.

Thermodynamic functions involve the physical properties of fluid, viz.:

(a) Pressure (P),
(b) Volume (v),
(c) Temperature (T).

Functions of the physical properties of fluids are the thermodynamic properties:

(a) Internal energy,
(b) Enthalpy,
(c) Entropy.

A relationship exists between all properties of a fluid at any given state. Their relationship is shown on thermodynamic property diagrams for the various fluids.

Enthalpy is the measure of total energy of a working fluid in any given state, usually expressed in Btu per lb.

Enthalpy is determined by the equation

$$\text{Enthalpy} = H + \frac{(Pv)}{J}$$

where H = enthalpy of the vapor, Btu per lb

and $\frac{(Pv)}{J}$ = external work in Btu per lb

in which P = pressure psf against which expansion occurs
v = volume, cu ft per lb, expansion
J = 777.46 ft-lb (mechanical equivalent of 1 Btu)

Entropy is the measure of unavailable energy.

Energy in the form of heat constantly diminishes as the universe tends toward maximum energy unavailability. Theoretically, then, all substances ultimately will be reduced to a similar but lower temperature.

The total energy of a body is the sum of its available and unavailable energy.

Entropy is a numerical factor indicating the relationship of the two energy values. It is a function of P, v, and T.

The entropy of a substance is the sum of the separate entropies of the liquid, evaporation, and vapor.

Entropy of a liquid is expressed by the equation

$$ds = \frac{dh}{T}$$

of evaporation

$$S_{evap} = \frac{L}{T}$$

and of vapor

$$S = \text{sum } ds + \frac{L}{T}$$

where ds = change in entropy
dh = increment change in enthalpy
T_a = average absolute temperature during thermal energy change
S_{evap} = entropy of evaporation
S = entropy of vapor
L = latent heat of vaporization
T = absolute temperature of saturation

Coefficient of performance for any system is the ratio of refrigerating effect to thermal energy expended in compressing the vapor, or

$$\text{cop} = \frac{\text{refrigerating effect (Btu)}}{\text{energy of compression (Btu)}}$$

The Carnot cycle is composed of two reversible isothermal processes and two reversible adiabatic processes.

The most efficient cycle is a reversible one.

The relative efficiency of a given system is the ratio of coefficient of performance of that system to the coefficient of performance of a system operating under Carnot cycle conditions, or

$$\text{Relative efficiency} = \frac{\text{cop of a given system}}{\text{cop of Carnot cycle}}$$

STUDY OUTLINE FOR CHAPTER 8

- Receivers:
  - Use and design
- Evaporators:
  - Function
  - Flooded systems
  - Dry expansion systems
  - Frost back
  - Types:
    - Shell-and-tube
    - Shell-and-coil
    - Tank-and-coil
    - Double pipe
    - Baudelot
    - Coil
    - Plate
    - Raceway
  - Finned coils
  - Air coolers, forced circulation
    - Low velocity
    - Medium velocity
    - High velocity
- Compressors:
  - Types:
    - Positive displacement:
      - Reciprocating
      - Rotary
      - Screw
      - Gear
    - Kinetic displacement:
      - Centrifugal
  - Reciprocating compressors:
    - Single-acting, vertical enclosed:
      - Speeds
      - Pistons, types:
        - Plug, trunk, double trunk
      - Cylinders and cylinder heads
      - Valves, types:
        - Poppet
        - Ring plate
        - Reed
      - Crankshaft seals and packing
        - Stuffing box:
          - Bellows seals
  - Rotary compressors:
    - Multiple-blade
    - Single-blade
    - Screw (helical)
  - Gear compressors
  - Centrifugal compressors:
    - Theory and design
    - Operation
- Condensers:
  - Surface:
    - Shell-and-tube, horizontal closed type
    - Shell-and-tube, vertical open type
    - Shell-and-coil
  - Atmospheric:
    - Standard
    - Bleeder
  - Double pipe
  - Evaporative
  - Air-cooled
- Boosters
- Accumulators, surge tanks
- Intercoolers—stage compression:
  - Water
  - Flash
- Heat exchangers or economizers
- Purgers
- Dryers:
  - Adsorption type
  - Absorption type
  - Moisture indicators
- Strainers
- Filters
- Scale traps

8

Equipment

The mechanical refrigeration system is a composite arrangement of equipment so connected and interconnected as to absorb and dissipate heat through the use and the control of a fluid medium (refrigerant), continuously repeating that function as a result of cyclic liquid-to-gas-to-liquid transformations. The equipment involved may be of the type essential to simple operation or may include such auxiliary components installed for the purpose of increasing the efficiency and the safety of the system.

Separating the components of a mechanical refrigeration system according to their basic functions, the majority would be included in one or more of the following operational categories:

1. Basic equipment—that which is essential to simple operation, such as compressor, evaporators, condensers, and expansion valves. Piping.
2. Thermal economizers—installed for the purpose of increasing thermal efficiency. Includes heat exchangers and precoolers.
3. Power economizers—to reduce power requirements per ton of refrigeration and increase mechanical efficiency. Boosters and stage compressors are included here, also purgers.
4. Control, operational—temperature, pressure, and capacity controls. Starting switches.
5. Control, safety—high- and low-pressure controls, overloads, electrical controls, etc.
6. Maintenance equipment—by-pass lines, service valves, defrosting arrangements, oil separators, dryers, etc.
7. Water economizers—to reduce water consumption. Includes equipment necessary for recirculating condenser cooling water, such as cooling towers or spray ponds, pumps, and regulating valves.

While all systems must include certain basic equipment in order to function at all, the type of auxiliary equipment necessary is governed by local conditions, type of plant, and economic factors. For example, an industrial plant located on a river ordinarily would have little use for a cooling tower to cool condenser water. In most cases of this kind, an ample supply is available merely for the cost of pumping and allowing the water to drain back to the river after passing through the condenser once. On the other hand, an air-conditioned building in the center of a city would find it expensive, if not prohibited, to waste an ever-diminishing supply of water in this way. In the latter case, some method for cooling and recirculating their condenser water must be employed. To meet the water-conservation problem in metropolitan areas, the evaporative condenser is gaining wide use.

Heat exchangers increase the refrigerating effect of refrigerants and are used in many systems as a result. Intercoolers are required for multi-stage systems, but would have no place where a single-stage compressor is employed.

These are but a few of the many instances where the selection of auxiliary equipment must be weighed against circumstances. Regardless of the utility limits of many auxiliary components, it is necessary that all the more important ones be included in any complete study of refrigeration equipment.

To acquaint the student with the scope of the subject, and to provide the means for associating the various components with each other in their relative location within the system, a list of equipment is given here. In chronological order, the following list includes the various components divided according to their essential or auxiliary nature, beginning with the liquid receiver, the true starting point of the refrigerant.

The first column includes equipment essential to the complete simple system. The second column lists such auxiliary components as will either expand the capacity thermally and/or mechanically of the simple system or provide for its greater operational efficiency and safety.

MECHANICAL REFRIGERATION SYSTEM

Basic Components	*Auxiliary Components*
I. Receiver	
	Heat exchanger or Liquid cooler
	Dryer
II. Throttling device	
	Accumulator
III. Evaporator	
	Scale trap
(IV. Booster)*	
	(Intercooler)*
IV. Compressor	
	Oil separator
V. Condenser	
	Purger

* For multi-stage systems.

THE RECEIVER

The liquid receiver, as its name implies, receives the liquid refrigerant as it is converted from a vapor in the condenser and holds it until such time as required by the evaporator. The receiver is usually a cylindrical tank so constructed as to withstand the maximum pressures encountered with the refrigerant employed and large enough to hold the entire charge of liquid with some space to spare. In industrial systems, it is located usually below the condenser in order to gain the advantage of gravity flow.

Acting as a refrigerant storage tank, the receiver holds the excess liquid during operation of the system while continually feeding the evaporator with the amount necessary to maintain required temperatures through some means of manual or automatic control of the throttling device. Simultaneously, it continues to receive liquid from the condenser at approximately the same rate with which it feeds the evaporator, causing the liquid level to remain nearly constant during normal operation. To determine the liquid level, industrial type receivers usually are equipped with a gage glass called a sight level indicator. They are provided also with a drain pocket or a sump located at the bottom to catch any oil or sediment finding its way to this unit. These drain pockets are designed to form a leg equipped with a sump valve through which foreign substances may be removed periodically. On some installations, the sump is connected by piping to a settling tank where the oil is reclaimed for further service. To facilitate separation of the oil and the sediment within the receiver, baffles are often installed to reduce agitation. The drain pockets are effective only with refrigerants that are nonmiscible with and lighter (lower specific gravity) than lubricating oil, such as ammonia and carbon dioxide.

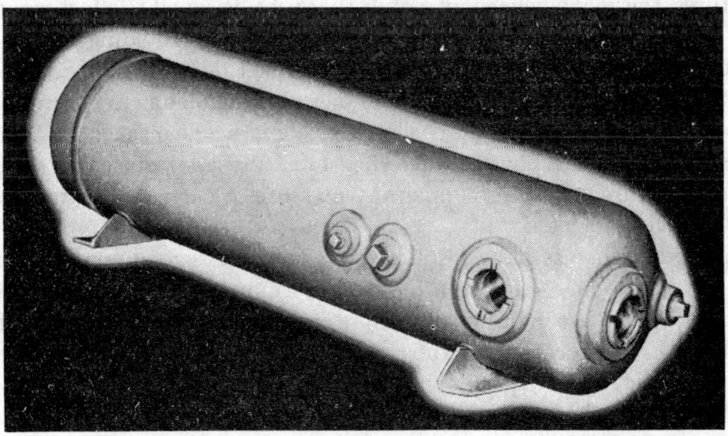

Fig. 63. Liquid receiver. (Courtesy, Pressed Steel Tank Company.)

Small domestic refrigeration units are often designed to utilize the lower part of the condenser as a receiver, thus saving much-needed space. This is possible because of the small amount of refrigerant involved.

Because liquid refrigerant should be maintained at the lowest possible temperature, the receiver usually is placed in a cool location—never in the vicinity of a steam boiler or other high-temperature equipment.

Receivers are natural traps for air entering the system. To eliminate air or any other noncondensible gases, a purge-line connection is made at the top of the receiver from which they are expelled by a purger that is also connected by a line from the top of the condenser. Fusible plugs are often used as a safety measure against excessive pressures in small receivers.

THE EVAPORATOR

That part of the refrigeration system within which the refrigerant is converted from a liquid to a vapor by the process of evaporation is called the evaporator. The high-pressure liquid enters the evaporator from the receiver through a throttling device whose function is to maintain a pressure differential between the two so as to initiate boiling within the evaporator at a temperature below the substance under refrigeration. The resulting vapor is evacuated from the evaporator through a *low-pressure* line and into the compressor by flow created by suction of the compressor piston during its intake stroke.

The evaporator is the element through which the heat load of the substance being cooled is transferred to the refrigerant and is absorbed by that fluid, principally as latent heat, during the evaporation process.

While the evaporator may be any enclosed vessel in which the pressure of the refrigerant can be controlled at a level corresponding to desired temperatures, thermal efficiency demands that it be designed to effect maximum heat transfer rate. This is accomplished partly by providing the greatest amount of exposed area between the contained fluid and the refrigerated substance. They should also be placed in such a manner as to permit highest mean temperature differential. They are constructed of such materials as will afford high heat conduction. They must possess sufficient strength to prevent rupture or collapse under any pressure or partial vacuum to which they may be subjected.

Such requirements have resulted in variously designed combinations of pipes or tubes enclosed within a vessel or compartment wherein the substances involved are maintained in close contact. Evaporators may be found in all shapes and forms, from the small compact units commonly associated with domestic refrigerators or the continuous-coil arrangement employed in water coolers to the long series of heavy pipe extended throughout numerous compartments of a storage cellar, such as are found in older breweries for aging beer. Its name does not denote shape, only a function.

There are several principles involved in obtaining the maximum thermal efficiency which concerns both the design and the method of evaporation employed. These principles provide the means of identifying the various evaporators as to type and classification. The following is a list of commonly employed types of evaporators, the selection of which is based primarily on (1) operational convenience and (2) highest thermal efficiency under different operating conditions: shell-and-tube, shell-and-coil, tank-and-coil, double pipe, Baudelot cooler, coil, plate, and raceway.

The evaporator element (piping) may be so arranged that the flow of refrigerant is continuous wherein the entire charge follows a common course. In such a system, all the refrigerant enters the evaporator element at the same point and passes through a continuous coil or a number of coils in series until it emerges in the suction line leading to the compressor. Another arrangement provides for the liquid refrigerant to be introduced to the element through a common header or series of headers from which it divides to flow through a number of tubes to emerge in another common or discharge header, and thence to the suction line. The latter is known as parallel flow.

Evaporators are classified as to the method used to effect and control evaporation of the liquid refrigerant, of which there are two types, the flooded system and the dry expansion system.

Flooded Systems

In this system, the evaporator element is filled (flooded) with liquid refrigerant to a level controlled, usually by means of a ball-float actuated valve. The evaporator, under flooded conditions, becomes a boiling mass of liquid whose vapors are drawn off at the top by the suction created by the compressor through a trap or accumulator. While evaporation tends to lower the level by reducing the amount of refrigerant in the evaporator, the action is offset by the simultaneous dropping of the ball float, allowing fresh liquid to enter the element. The primary function of the trap or accumulator is to separate entrained liquid carried over with the vapor and to return it to the evaporator for recirculation, thus allowing only saturated or slightly superheated gas to enter the compressor.

Flooded systems have the advantage of excellent heat transmission by reason of liquid-wetted surfaces as compared with the vapor-wetted surfaces prevalent in dry expansion systems. For this reason, smaller evaporators may be employed for equal capacities. They are also more flexible in their distribution of refrigerant, particularly where a number of coils are connected to a common header to form a multi-pass arrangement. Flooded systems also have the advantage of employing several evaporators in conjunction with one accumulator without using more than one throttling device.

Dry Expansion Systems

Employed long before the general adoption of the flooded system, dry expansion presupposes the entrance into the evaporator of only that amount

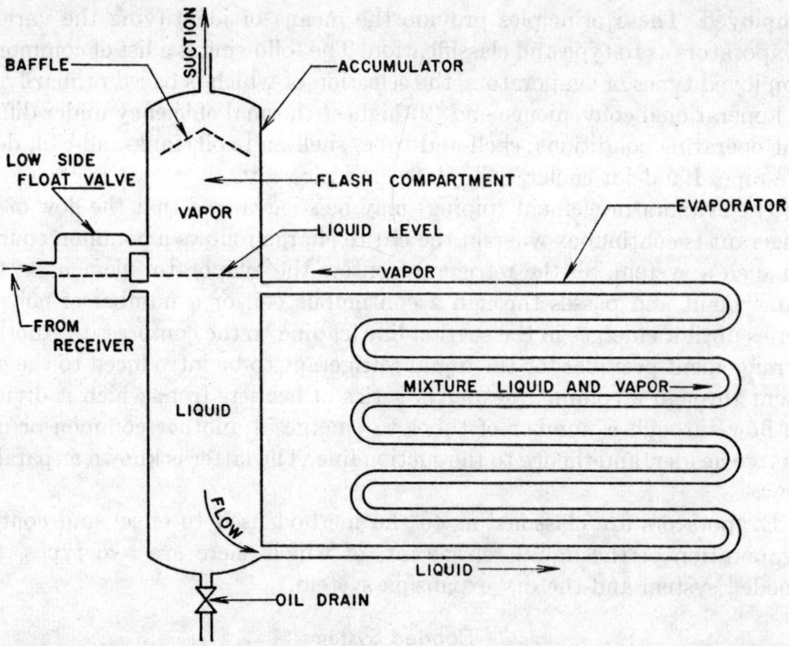

Fig. 64. Principle of the flooded system employing an accumulator. Difference in weight per unit volume between liquid in tank and mixture in coil causes fluid to flow by gravity as shown.

of refrigerant needed to maintain required temperature during the vaporization period. This means that the throttling device allows only enough liquid refrigerant to enter the evaporator as will leave in a saturated vapor state (or slightly superheated) upon absorbing the required amount of heat necessary to maintain desired temperatures in the chamber under refrigeration. Compared with the flooded system, the dry expansion system contains only a small amount of liquid at any given time, the quantity of which is controlled by the load demands. When loads are light, the quantity of liquid in the evaporator is small. With heavy loads, the evaporator will take on more nearly the condition of a flooded system in order to make available a larger quantity of liquid commensurate with the greater load demands and to avoid excessive superheating of the resultant vapor.

The amount of liquid entering the dry expansion type evaporator may be governed manually, as is done in some larger industrial systems, or automatically, using a thermostatic expansion valve. The dry expansion system is generally employed for continuous-coil evaporators. Flooded systems are used on split-run or header type evaporators.

Length of Run vs. Tube Diameter

Particular attention must be paid to the diameter of element tubes in relation to the length of run within the cooling space; i.e., the ratio of refrigerant volume within the tubes to the amount of evaporating surface should be balanced. If the tubes are too large in diameter, the refrigerant will not evaporate fast enough during its passage through the cooling area. This allows excessive liquid to be carried over into the suction lines and to enter the compressor, where its presence may prove damaging to the head or to the moving parts, an action known as "slugging." If the tubes are too small in diameter, the frictional resistance to fluid flow causes a pressure drop between the throttling device and the suction line great enough to reduce the efficiency of the system to an excessive degree. The diameter of the tube must be calculated on the basis of such fluid velocity as will minimize friction, hasten evaporation, and carry any oil entering the evaporator back to the compressor. This latter factor is especially important in the case of miscible refrigerants. Best results have been obtained with suction-line velocities of from 800 feet per minute (fpm) to upward of 2000 fpm, depending on the density of the refrigerant vapor. Such high-density vapors as carbon dioxide would be handled at reduced velocities of 800 fpm to 1000 fpm, depending on the temperatures involved.

Length of Run vs. Number of Expansion Valves

In dry expansion systems employing a continuous-coil arrangement, the length of run for each expansion valve is an important efficiency factor. If the run is too long per expansion valve, there is every possibility of excessive superheating. This would occur due to early evaporation of the liquid, leaving only vapor to absorb heat during the remainder of its travel to the suction line. Any demand in load above normal expectancy would further increase superheating. To avoid excessive superheating, only enough coil length should be connected to one expansion valve as will efficiently absorb

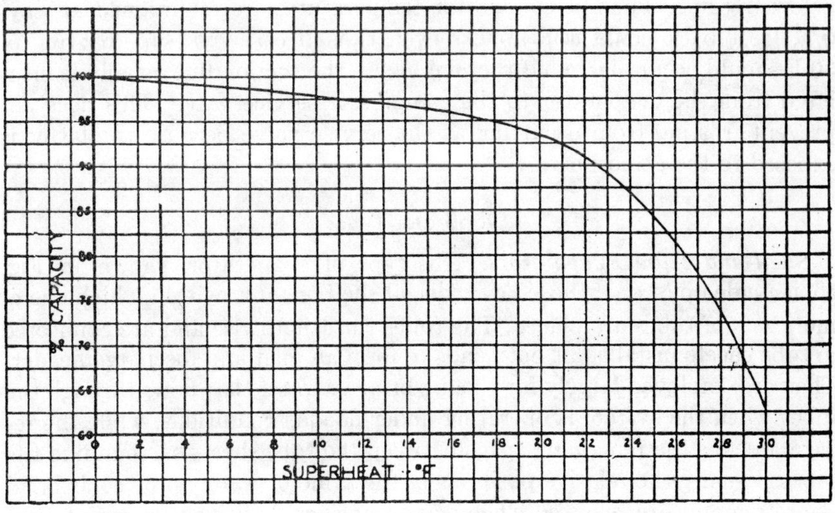

Fig. 65. Effect of superheating suction gas on refrigeration capacity of a Refrigerant 12 system. (Courtesy, Alco Valve Company.)

the heat load while maintaining a saturated or slightly superheated vapor condition in the suction line. Where load demands require greater heat-transfer area, additional coil length may be installed and fed from a separate expansion valve, or through a distributor.

Another disadvantage of long continuous coils lies in the pressure drop that occurs between the expansion valve and the compressor intake valve. In order to maintain proper pressures within the evaporator corresponding to the required temperatures, the suction pressure at the compressor intake must be reduced to offset the loss due to friction. The longer the run, the greater the drop in suction pressure. As operational efficiency decreases with increased compression ratio, long runs tend toward reduced economy. To obtain maximum efficiency through greater compressor capacity and reduced power requirements, the suction pressure must be maintained at the highest possible level. It is estimated that savings up to $2\frac{1}{2}$ percent in horsepower requirements may be obtained for each 1 psi increase in suction pressure.

Frost Back

This term is applied to the frost condition of the suction line which may often extend to the compressor. While the lack of frost might indicate excessive superheating in some instances, it should never serve as a criterion for good or bad operation. The reason for this lies in the fact that frosting will occur whenever the temperature of the refrigerant vapor falls below 32 F, the freezing point of water. Condensation of atmospheric water vapor occurs whenever it comes in contact with a surface whose temperature is below the prevailing dewpoint and will freeze to form frost if the temperature is below freezing. Thus, if evaporator temperatures are maintained at, say, 0 F, frost back could still appear even though excessive superheating of 30 F should occur. To avoid thermal losses, the suction line usually is insulated from the evaporator to the compressor service valve. This does not prevent frosting from occurring at the service valve because the latter is exposed to the atmosphere.

Evaporator Types

Shell-and-Tube Evaporators. This type of evaporator, shown in Fig. 66, is made up of an enclosed cylindrical shell or tank, within which one or more sets of tubes are placed. The tubes run lengthwise and are connected to tube sheets installed at both ends to form an inlet and discharge header. They are employed largely for cooling brine or water that flows through the tubes when the system is operating under flooded conditions, while the refrigerant fills the remaining space between the tube sheets surrounding the tubes. Usually the relative positions of the brine or water to be cooled and the refrigerant are reversed when dry expansion conditions are maintained within the shell-and-tube evaporator; i.e., the refrigerant flows through the

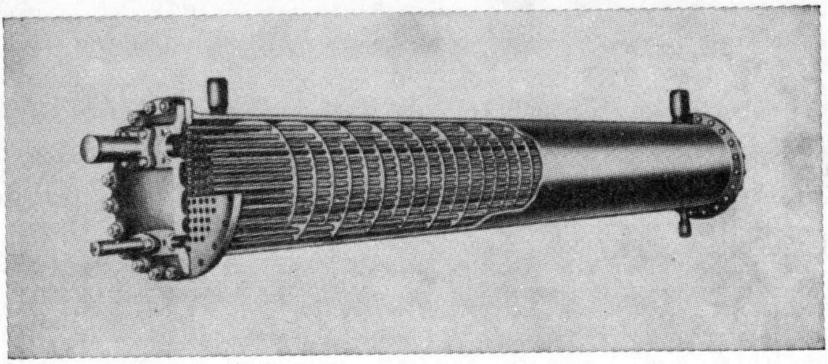

Fig. 66. Shell-and-tube evaporator, direct expansion type, showing tubes, tube sheets, baffles, and header. Refrigerant passes through tubes, water through shell. (Copyright, The Trane Company. Reproduced by permission.)

tubes. Where the refrigerant enters and leaves at one end of the shell, a baffled or manifolded head is used.

The brine or the water to be cooled enters the inlet header to be routed through two or more passes usually by means of baffles. When operated dry expansion, it is general practice to limit the number of refrigerant passes to two in order to minimize pressure drop.

A method of introducing refrigerant by spray is also used in connection with this type of cooler. In such installations, the liquid refrigerant is forced through spray nozzles located in headers or domes at the top of the shell. The brine or the water which flows through the tubes is cooled as the liquid refrigerant is sprayed and evaporated on the tube surfaces. The excess liquid gravitates to a sump at the bottom of the shell and by means of a pump is returned to the spray header and recirculated. The spray type shell-and-tube evaporators are highly efficient and do not require as much refrigerant to charge them as do flooded type coolers. They require a circulating pump, however.

Shell-and-tube evaporators sometimes are designed with an attached heat exchanger called a liquid cooler (Fig. 67). This auxiliary cooler is located on top of the shell and contains a set of superheater tubes through which liquid refrigerant from the receiver passes on its way to the throttling device. The liquid is thus cooled by the vapor leaving the shell, which, in turn, absorbs some superheat from the liquid. Reducing the temperature of the liquid in this way increases its refrigerating effect by decreasing the amount of flash gas necessary to bring its temperature down to that maintained in the evaporator. The saving so made in decreasing the temperature of the liquid prior to its passing through the throttling device amounts to approximately 1 percent in compressor load for each 5 F drop. While superheating the suction vapor reduces the weight of refrigerant handled by the compressor per minute to offset some of the thermal gain made in the evaporator by this method of liquid precooling, the savings remain considerable when superheating is limited to about 10 F. This slight superheating also

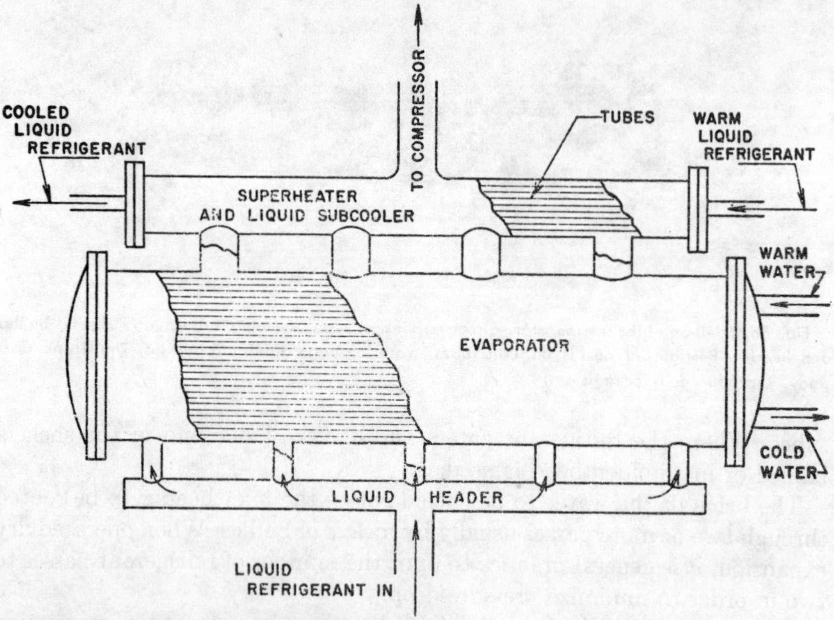

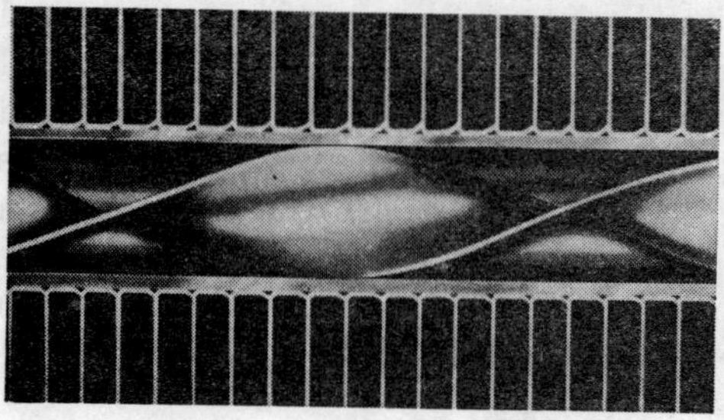

Fig. 67. (*Top*) Shell-and-tube evaporator, flooded type for Freon, with superheater to subcool liquid.

Fig. 68. (*Bottom*) Turbulator. Used in water tubes to increase heat transfer when water velocity falls below 4 fps. (Copyright, The Trane Company. Reproduced by permission.)

assures liquid-free gases entering the compressor. This type of evaporator-superheater arrangement is employed mainly with R-12 systems.

For refrigerants having corrosive tendencies toward nonferrous metals, such as ammonia, steel tubes and shell are used in constructing this type of evaporator. For R-12 and other refrigerants considered noncorrosive, copper tubes may be used to effect a greater rate of heat transfer.

To obtain a greater rate of heat flow, finned tubes generally are employed with refrigerants of low thermal conductance. Otherwise, large evaporators would be necessary for equal capacities using bare tubes. Fins are thin metal disks placed an equal distance apart along the length of a tube, or strips wound edgewise around a tube, forming a spiral from one end to the other. Their widths vary depending on the service to which they are applied. The rapid cooling of thin metal projecting into the cooling medium is the principle underlying the greater heat-transfer rate using finned tubes. Being of

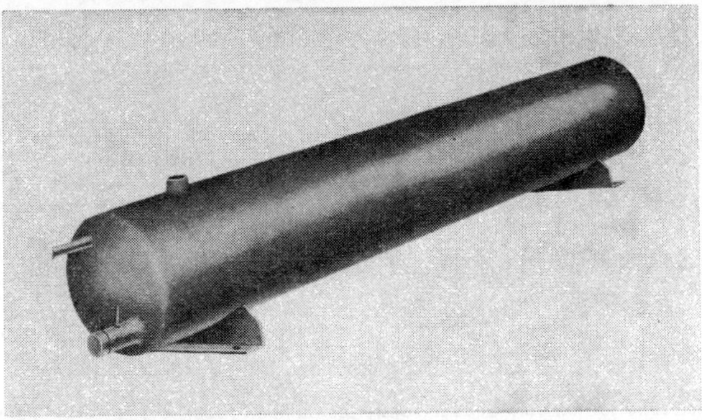

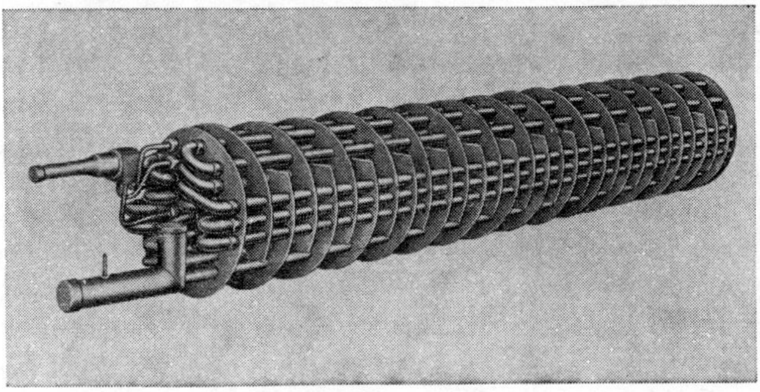

Fig. 69. (*Top*) Dry expansion evaporator—shell. (Copyright, The Trane Company. Reproduced by permission.)

Fig. 70. (*Bottom*) Dry expansion evaporator—coil and baffles. Inserted in shell shown in Fig. 69. (Copyright, The Trane Company. Reproduced by permission.)

small cross-sectional area, their ability to hold heat is reduced to a minimum. This tendency to cool rapidly creates a temperature differential between the warmer tube body and the end of the fin. Heat, having a tendency to seek its own level, continues to travel by conduction along the width of the fin just so long as the temperature differential is maintained at a rate proportional to the value of the latter. The fins adopted for this type of equipment are comparatively short, protruding from the tube approximately $\frac{1}{16}$ in. in most cases. Allowable fin protrusion remains small because maximum heat-transfer capacity is obtained with liquid-to-liquid contact and nothing is to be gained by extending fin width. In fact, low fins offer the advantage of reduced friction on liquids passing through the cooler.

The principle of fin cooling has found wide adaptation on air-cooled internal combustion engines and compressors, as well as on unit heaters and convection type radiators used for home heating.

Shell-and-tube type evaporators are often referred to in connection with the number of times the brine or the water to be cooled is forced to traverse the refrigerated area of flooded systems. If it enters the inlet header and

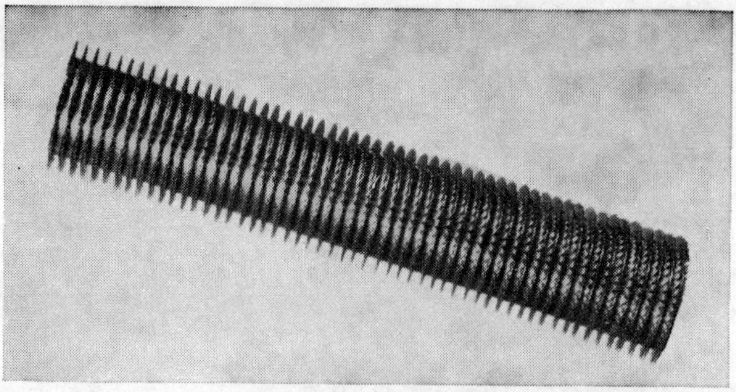

Fig. 71. Finned tube section used in heat-transfer equipment. (Courtesy, Edwards Engineering Corp.)

traverses the shell once to be immediately evacuated from the discharge header, the unit is considered to be a single-pass evaporator. Where the brine or water is routed to traverse the refrigerated area more than once, it is referred to as a double-pass, triple-pass, etc. (or more commonly as a multi-pass) evaporator or cooler. Usually they are designed to provide sufficient heat-transfer surface to effect a mean temperature difference (mtd) of between 8 F and 10 F, although numerous industrial systems operate at a much lower mtd.

Shell-and-Coil Evaporators. This type of cooler, shown in Fig. 72, is employed generally in connection with dry expansion. While it may be used to cool any liquid, its usual role is that of a water cooler. The coil is continuous and may be made up in either a single or a double spiral with refrigerant entrance and discharge located at the top.

The shell may be covered and sealed firmly to prevent the leakage of water circulated under pressure, or it may be open. In this case, the cooled water leaves by gravity over a weir in the discharge line.

Evaporators of this type are employed where temperature demands are such that the leaving water does not fall much below 40 F in order to maintain a fair safety margin to prevent freezing. The closed type is used most generally to cool drinking water, whereas the open type is employed for unit air conditioners where the water is recirculated and constant level maintained by means of a ball-float valve and standpipe to carry any overflow from dehumidification.

Double-Pipe Evaporators. Comprised of a pipe within a pipe, the double-pipe evaporator provides excellent contact between the fluids involved, resulting in a high heat-transfer rate. In this type of cooler, the refrigerant flows through the annular space around the inner pipe carrying the fluid under refrigeration. A stand of double pipe is formed by doubling one length back over the other, using return bends at each end to form a continuous coil. This arrangement, while allowing highest mean temperature difference, an important factor in heat transfer because of the counterflow principle employed, does require considerably more floor and head space than the

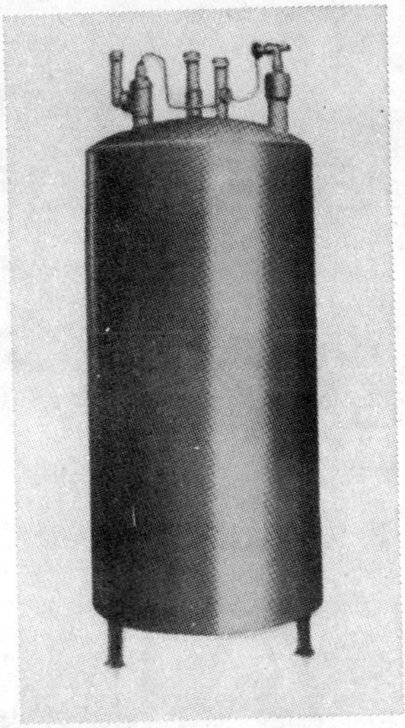

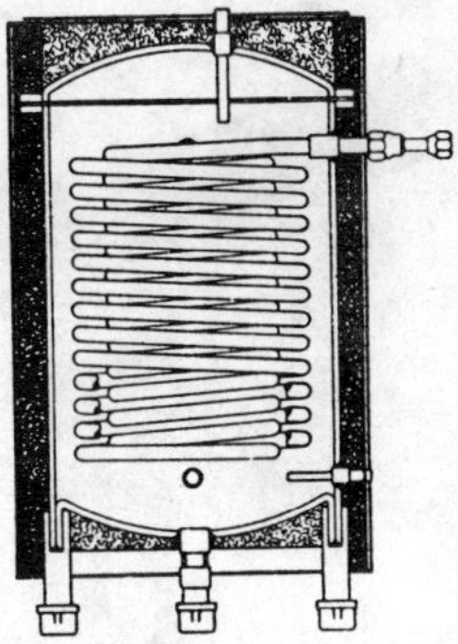

Fig. 72. Shell-and-coil evaporator, spiral coil type. Hi-Peak water cooler. (Courtesy, Acme Industries, Inc.)

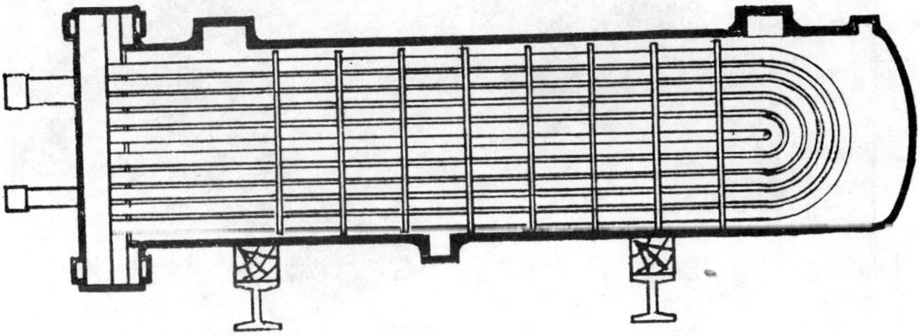

Fig. 73. Shell-and-coil evaporator, dry expansion cooler. (Courtesy, The Patterson-Kelley Co., Inc.)

shell-and-tube type evaporator. For that reason, it gradually is being eliminated in favor of the latter type, especially for brine cooling. The double-pipe principle is still used widely in condenser design, described later in this chapter. Usually it is employed under "flooded" conditions.

Baudelot Cooler. In cooling milk or water to temperatures approaching the freezing point, the danger of forming ice and rupturing the tubes or pipes is present where fully enclosed coolers, such as shell-and-tube or shell-and-coil types, are employed. This danger does not exist with the Baudelot cooler (Fig. 74) because the liquids under refrigeration flow or cascade over the external surface of the coil through which the refrigerant passes. Although ice does form occasionally, it can be detected visually and steps taken for its immediate control.

The Baudelot cooler consists of a series of horizontal pipes placed one over the other, through which the refrigerant passes, employing either the flooded or the dry expansion system of evaporation. The liquid to be re-

Fig. 74. Baudelot type cooler. Liquid to be cooled flows over external surface of coil or plate. Casing opened. (Courtesy, Mojonnier Bros. Co.)

frigerated cascades over these horizontal pipes from a trough located above the coil bank by way of holes drilled in the trough floor, and finally gravitates to a reservoir placed below the cooler.

This type of evaporator is used extensively in creameries for the rapid cooling of milk, where it also acts as an aerator. It is employed also for cooling water for dehumidification in older air-conditioning systems.

Coil Type Evaporators. Coils permit a wide flexibility of evaporator design, and for this reason remain one of the most important types used. While they are often employed to allow one continuous passage of the refrigerant from the expansion valve to the compressor intake valve, they are also connected in such a manner as to form a series of coils fed from a common header. Coil evaporators are made in various shapes, including oval, round, rectangular, square, or according to the layout of the compartment in which they are to be installed. They may be formed as a spiral, a double spiral, or bent back over each other to form a bank of parallel passes, either flat or boxlike. They often serve the double role of evaporator and shelving, such as used in ice cream hardening rooms. In cold-storage plants, where direct expansion refrigeration is carried on, the evaporator may be a series of pipes and bends connected to form an endless coil extending throughout several rooms, where it is attached to the ceiling or walls in bunkers.

The various forms of coils are shown in Figs. 76, 77, and 78. Coils may be made up of wrought iron, steel, or copper, the selection of which depends largely on the kind of refrigerant employed and the service for which they are intended. The term *pipe* refers to any kind of tubing employed for conveying fluids, but particularly to that which is formed from sheets of metal made cylindrical and having their edges bound together by butt—or lap—welding. If made from a solid piece of metal by drawing or forming, it is referred to as *tubing*, or, more specifically, as *seamless tubing*. Seamless tubing has the advantage over welded pipe in strength and in facility of bending without splitting at the seam.

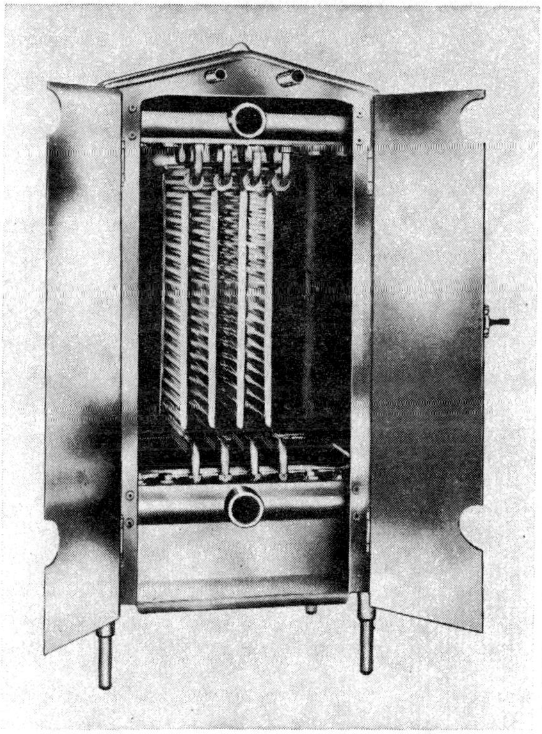

Fig. 75. Baudelot type cooler, end view of cooler shown in Fig. 74, showing refrigerant headers. (Courtesy, Mojonnier Bros. Co.)

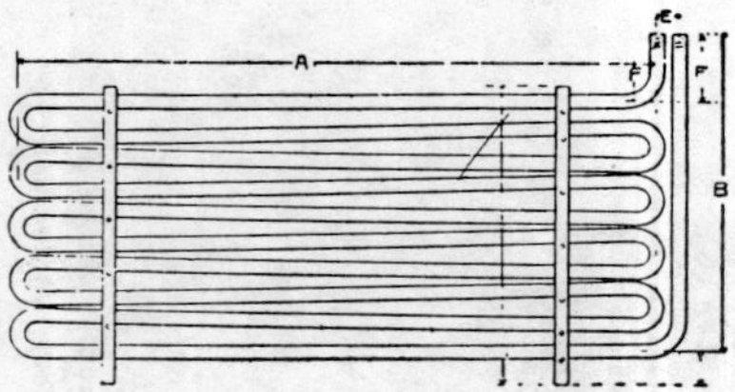

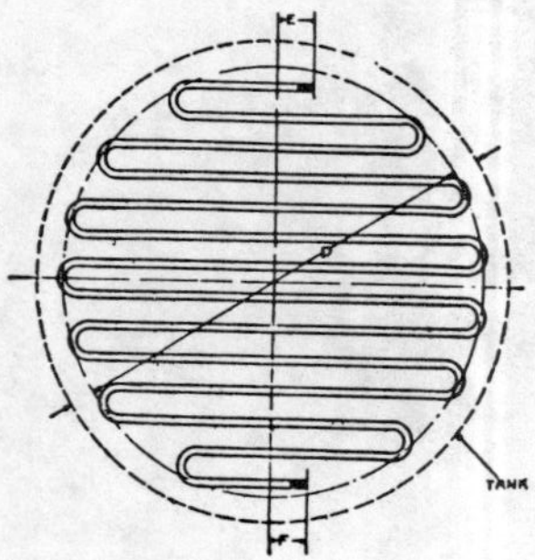

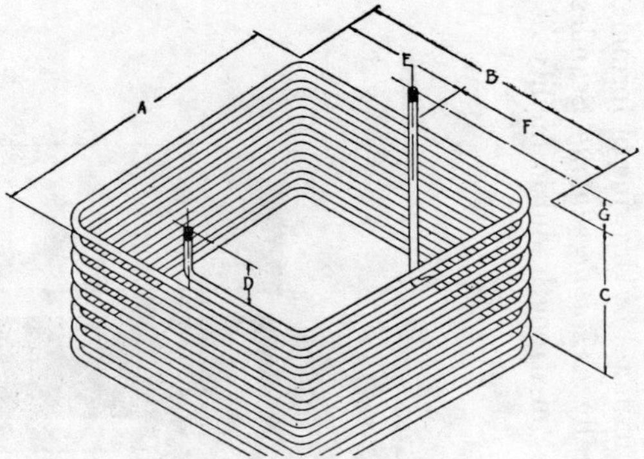

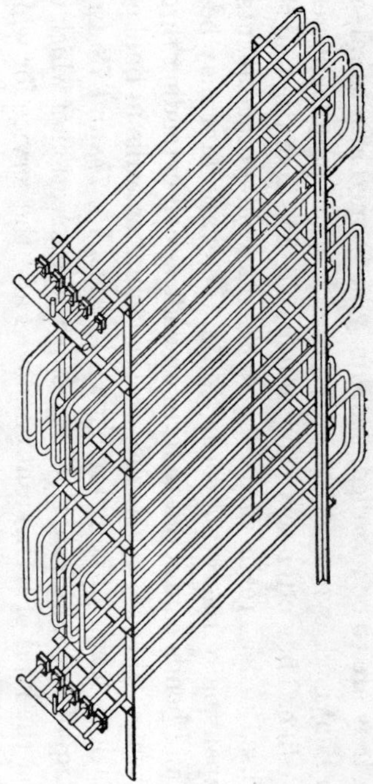

Fig. 76. (*Left*) Coil type evaporator. (*Top*) Rectangular flat. (*Bottom*) Round flat. (Courtesy, Acme Industries, Inc.)

Fig. 77. (*Center*) Coil type evaporator. Rectangular or "crate." (Courtesy, Acme Industries, Inc.)

Fig. 78. (*Right*) Coil type evaporator, stand for hardening room. Note location of headers that split refrigerant flow. (Courtesy, Acme Industries, Inc.)

There are normally three weights of iron piping, graded according to their weight per linear foot and known as standard, extra strong, and double extra strong. The size of pipe is usually listed in three dimensions; namely, nominal diameter, internal diameter, and external diameter. Nominal diameter is the internal dimension nearest to the standard division of the common foot rule. Below 1 in. their nominal sizes increase by $\frac{1}{8}$ in., such as $\frac{1}{8}$, $\frac{1}{4}$, $\frac{3}{8}$, $\frac{1}{2}$, etc., and are identified with such dimensions as being $\frac{1}{8}$-in. pipe or tubing, $\frac{1}{4}$-in. pipe, etc. Pipe sizes 1 in. and over up to 2 in. increase by $\frac{1}{4}$ in., the standard being 1 in., $1\frac{1}{4}$ in., $1\frac{1}{2}$ in., $1\frac{3}{4}$ in., and 2 in., respectively. Above that, the sizes increase by $\frac{1}{2}$ in. up to 4 in., from which the increment changes to 1-in. progressions. From the smallest size of $\frac{1}{8}$ in. to large pipe of 14 in., the listed dimension is indicative of its nominal internal (or inside) diameter, or I.D. Over 14 in., listed sizes indicate external (or outside) diameter, or O.D.

In order to maintain the same external diameter for all weights of piping, the internal diameter for standard, extra strong, and double extra strong pipe varies slightly. For example, standard size pipe having a nominal internal diameter of, say, 2 in., has, under standard specifications, an actual internal diameter of 2.067 in. For the same nominal dimension of extra strong pipe, the actual I.D. would be 1.939 in., and for double extra strong, it is 1.503 in., making the I.D. of the two heavier grades less than their nominal sizes. All three weights have the same external diameter of 2.375 in.

Because of the variation in thickness, and as their names imply, each weight classification has different weights per linear foot of pipe, and also different internal areas measured in square feet per linear foot. A standard 2-in. pipe has a nominal weight of 3.652 lb per foot and an internal area of 1 sq ft per 1.847 ft of length. For extra strong and double extra strong pipe, these values amount to 5.022 lb and 1.969 ft, and 9.02 lb and 2.561 ft, respectively. Tables of piping list the above information for all standard sizes and weights in addition to length of pipe containing 1 cu ft, circumferences, etc., information necessary in designing refrigeration systems for various load conditions. (See Table 61.)

To find the number of square feet of pipe surface required for any heat load, the following formula may be used:

$$\text{sq ft of pipe} = \frac{\text{heat load per day}}{U \times H \times (t_1 - t_2)}$$

where U = Btu transferred per hr per sq ft per °F t_d
H = hours of operation
$t_1 - t_2$ = temperature differential

Plate Evaporators. Because plate type evaporators may be installed to form walls, ceilings, or shelving, they are employed to good advantage where space is an important factor. They are particularly adaptable in refrigerated motor trucks, freezing lockers, and small coolers. They are made up of two

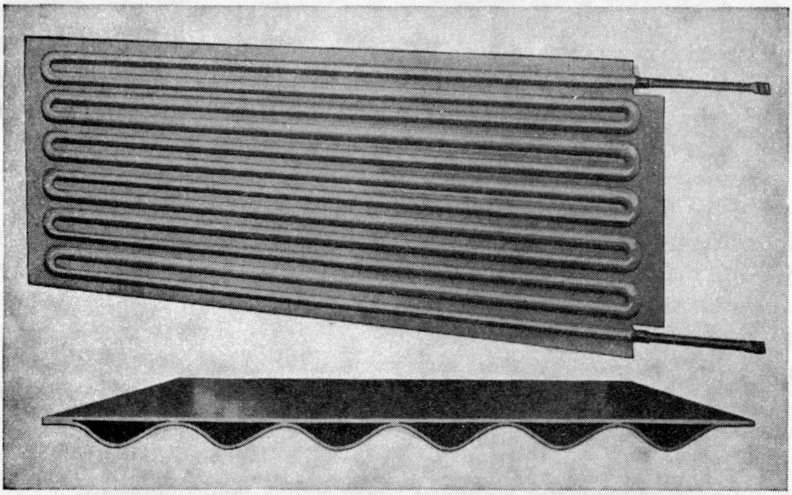

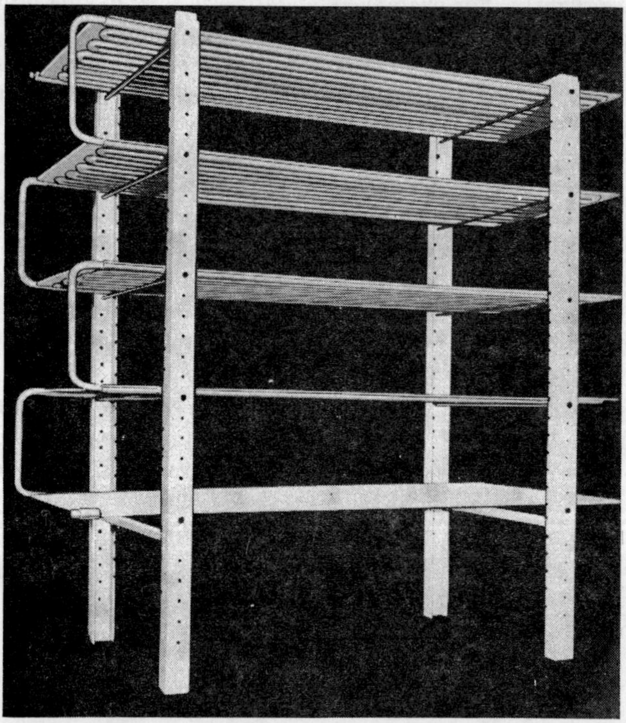

Fig. 79. (*Top*) Plate evaporator, serpentine, showing cross-section of refrigerant passages. (Courtesy, Kold-Hold, Division of Tranter Mfg., Inc.)

Fig. 80. (*Bottom*) Plate evaporator, stand as used in hardening rooms. (Courtesy, Kold-Hold, Division of Tranter Mfg., Inc.)

metal sheets having raised edges so that they form a flat compartment when placed together. One type contains tubing of small diameter through which the refrigerant passes. To increase heat transfer, the space between the sheets and around the tubes is held under partial vacuum, causing the

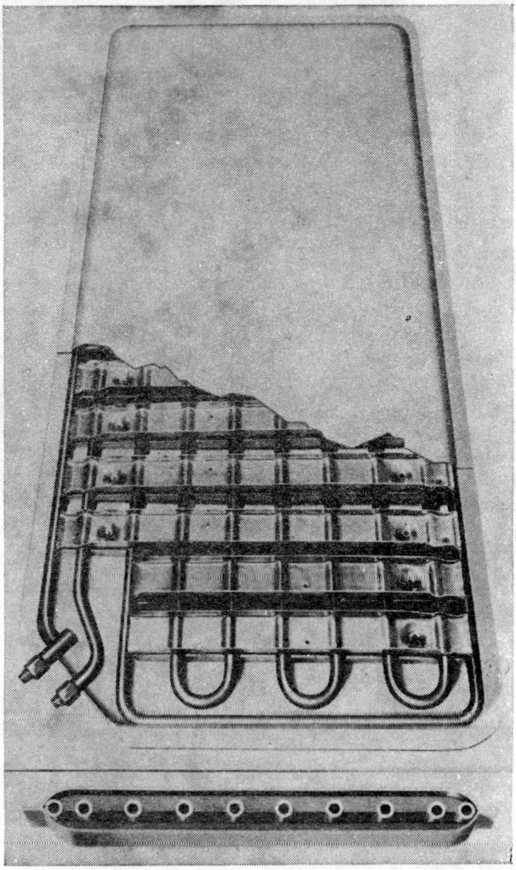

Fig. 81. Plate evaporator. Hold-over plate of the type used in refrigerated motor trucks. Note use of tubes that may be immersed in eutectic solution. (Courtesy, Kold-Hold, Division of Tranter Mfg., Inc.)

plates to be drawn tightly against the tubing. Another type requires no tubing but merely allows the refrigerant to enter the hollow space between the plates, where it evaporates.

The eutectic plate is so called because of the introduction of a brine or eutectic solution between the plates. Tubing is employed in eutectic plates. They are used to increase "hold-over" capacity, such as required for transportation units not having their own refrigeration equipment. The procedure under these circumstances is to "charge" or solidify the eutectic

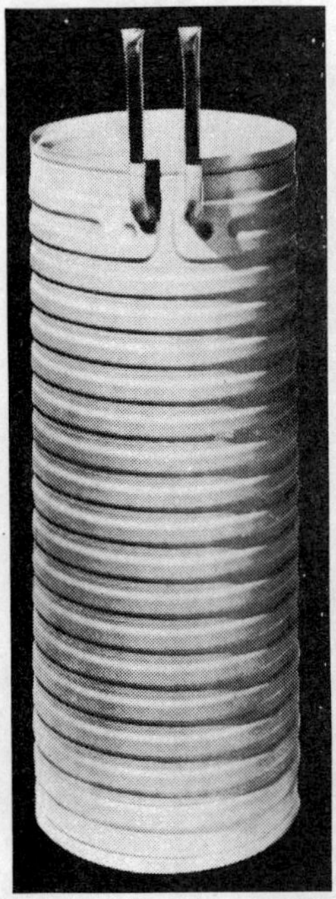

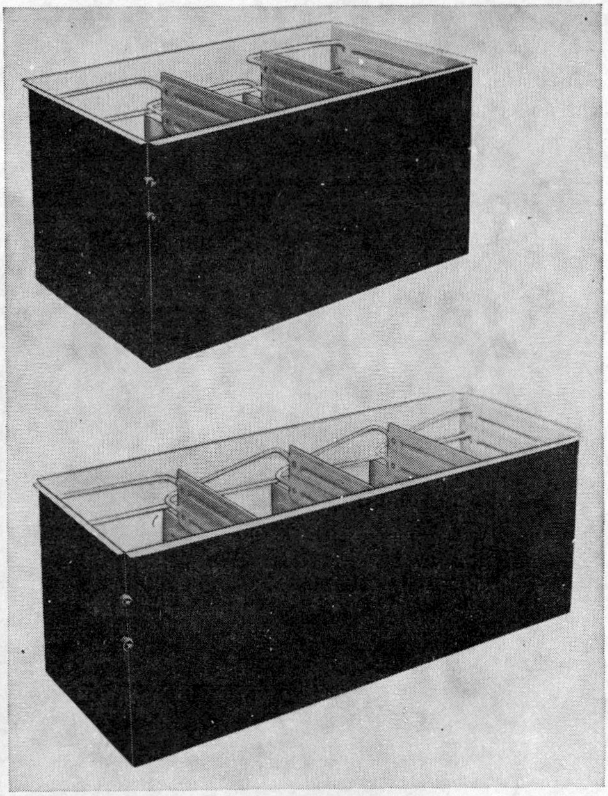

Fig. 82. (*Left*) Plate evaporator, cylindrical type. (Courtesy, Dean Products, Inc.)

Fig. 83. (*Right*) Plate evaporator. Soda fountain tank type. (Courtesy, Dean Products, Inc.)

solution by means of a stationary system located at the point of loading. They are used also on refrigerated trucks having their own condensing equipment, where they serve to lengthen the cycling periods.

The eutectic solution is a brine solution containing the correct proportions of salt and water to effect a lower freezing point than is possible with any other combination of these substances. The eutectic point of sodium chloride (common salt) brine, for example, is -6 F. To obtain this freezing temperature, the solution must contain 23 percent salt by weight. If a greater or lesser amount of salt is present, the freezing point of the solution will be higher. For example, a brine containing 24 percent salt will have a freezing temperature of $+3.8$ F, and with 22 percent salt the solution will freeze at -3 F. With calcium chloride brine, the eutectic point is -59.8 F. This is the lowest temperature at which this type of brine will remain liquid, regardless of how much more or less salt is added above or below the eutectic concentration of 29.6 percent by weight. With concentrations of 30 percent and 29 percent, the freezing temperature rises to −50.8 F and −49.4 F, respectively.

Plate evaporators are mounted in series or in parallel, using a header to distribute the liquid refrigerant. Using R-12 or methyl chloride, the amount of plate area exposed, including both sides of the plate, should not exceed 75 sq ft for each expansion valve. This figure may be doubled for ammonia. Plates usually are employed with dry expansion evapora-

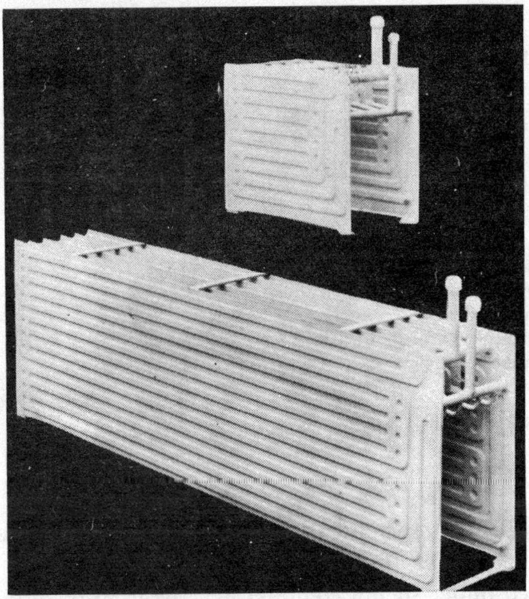

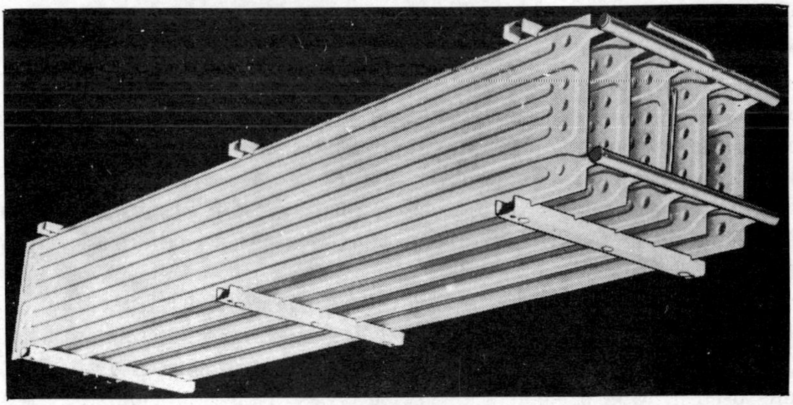

Fig. 84. (*Top*) Plate evaporator, tunnel freezer. (Courtesy, Dean Products, Inc.)

Fig. 85 (*Bottom*) Plate type evaporator, overhead bank. (Courtesy, Dean Products, Inc.)

tion, fed from either the top or the bottom, with the suction located at the top.

Another form of cold plate is made by placing two metal sheets together much the same as described above, except that one or both are embossed to provide evaporation space. This type of plate provides wide flexibility in the possible shapes that can be produced, including cylinders, tanks, cones, boxes, etc.

One manufacturer lists the following U factors for plate evaporators (see Table 30).

TABLE 30. HEAT TRANSFER VALUES FOR PLATE EVAPORATORS UNDER VARIOUS CONDITIONS*

Condition	*U Factor (Btu/sq ft/hr/°F)*
Still air above 20 F	2.5
Still air below 20 F	2.0
Agitated air	3 to 6
Water (still)	12
Brine (still)	12
Agitated water or brine	18
Direct contact	4 to 5

* Courtesy, Dean Products, Inc.

Finned Coils

This discussion relates to fins whose function it is to transfer heat from a gas (air) to a vaporizing refrigerant, and because the rate of heat transfer is much reduced under such a combination, the need for fins of greater exposed area increases. To obtain this additional area, fins are subject to two principal variations: they may be (1) extended in height or (2) more closely spaced on the tubing. While maximum heat-transfer efficiency definitely is related to spacing, as well as to the height of the fin in proportion to the tube diameter, the spacing of fins is also governed by the possibility of ice formation, which prevents the passage of air. This latter depends on the temperatures involved, and, since food storage is being dealt with as well as air conditioning, it may be assumed that ice remains an important problem which requires close control. Manual or automatic defrosting should be employed whenever finned coils are installed so that periodic frost removal may be accomplished.

Finned coils are employed extensively in commercial refrigeration, particularly as the cooling element of display cases for numerous perishable commodities. They are used also in connection with forced-draft cooling units used in walk-in coolers and some types of air-conditioning systems.

In addition to the type of coil employing spiral fins that makes each tube an individual unit, another type is used on which the fins act to strap a

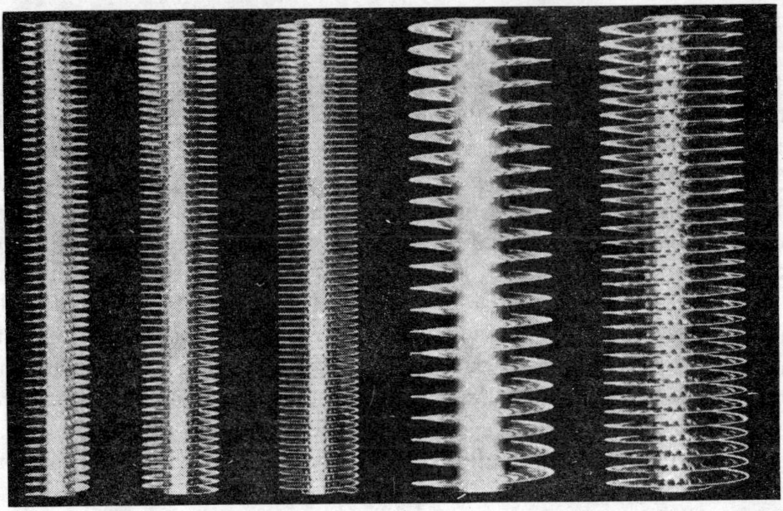

Fig. 86. Finned tubes as used in coil banks for air cooling. Refrigerant flows through tubes while air is forced over the coils. Banks are formed of one or more rows of tubes. (Courtesy, Aerofin Corporation.)

Fig. 87. (*Left*) Air-cooling coil with finned tubes for direct expansion. Note centrifugal distribution header. (Courtesy, Aerofin Corporation.)

Fig. 88. (*Right*) Air-cooling coil with finned tubes for direct expansion. Header is of the weir distributor type. (Courtesy, Aerofin Corporation.)

number of tubes together to form a nest. In this latter type, the tubes may be bent one over the other or variously staggered to form two or more rows of tubes for the purpose of increasing surface area exposed to the air passing through it. Such types, usually referred to as flat-finned evaporators, are made in either square or rectangular shapes and in different lengths.

Where natural convection is the only means of air circulation, a greater amount of fin and tube area is required than where forced draft is used. This is because heat-transfer rate becomes greater as the velocity of both the cooling and the refrigerated mediums increases up to a certain point. In those units depending entirely on natural convection, as is usual in display cases, the length of the coil should be at least 75 percent of the total length of the refrigerated space to insure uniform temperatures throughout. Where the heat load is not sufficient for the amount of coil necessary to run the required length using a specific size, then the number and the size of tubes should be reduced to allow for the greater length to obtain uniform temperatures.

Fins on either type of coil are spaced four to seven fins per inch depending on the temperatures involved and the amount of heat to be absorbed in relation to the length of coil required. Examples of typical finned coils are shown in Figs. 87 and 88.

Headers for finned coils are variously designed to effect uniform distribution of liquid refrigerant in order to avoid differences in temperature over the face of the coil. Note the method of distributing liquid refrigerant to a number of points in a coil by means of multi-outlet header in Fig. 87. A fairly uniform face temperature may be obtained with this distributing arrangement.

Air Coolers

Air coolers having forced circulation to establish rapid convection in coil type refrigerant evaporators provide a positive method of delivering cooled air into areas to be so conditioned. Such coolers are manufactured having either prime (bare) or extended (finned) surface tubing, and employ a fan to move the air to be cooled over the heat transfer elements.

Forced circulation air coolers vary in the velocity of the exit air according to the purpose for which they are selected. They are classified relative to their velocity capabilities as (*a*) low, (*b*) medium and (*c*) high velocity air coolers.

Low Velocity Units. These are contained within a maximum limit of *500 ft per min* exit velocity. Extended surfaces or fins are placed an average distance of $\frac{1}{4}$ inch apart or 3 to 5 fins per inch of tube length. They are recommended where minimum draft, low noise level and/or low evaporation rate in cooled spaces are desirable. Low velocity units are usually ceiling mounted and have dual exits so as to provide flow in angular or opposite directions. Both centrifugal and propeller type fans

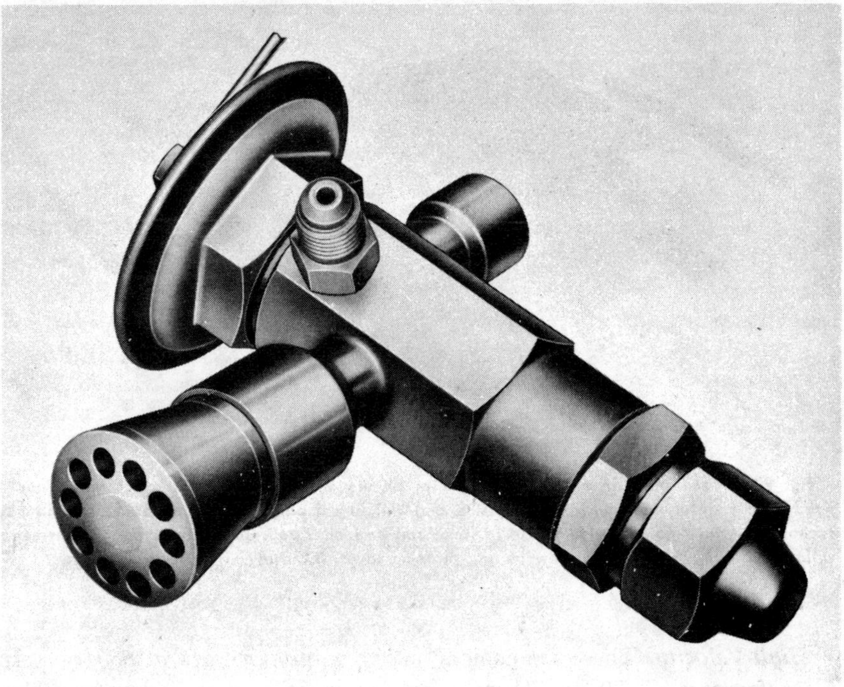

Fig. 89. Refrigerant distributor. Receives refrigerant from expansion valve and divides it to provide uniform feed to all phases of a direct expansion multi-circuit evaporator. Should be connected as close as possible to expansion valve. (Courtesy, Sporlan Valve Company.)

are used. They can be operated as a *dry coil* or *wet coil* unit and are capable of maintaining temperatures within a tolerance of 2 F. The dry coil air cooler is used in space conditioning where low velocity is essential to comfort, food shrinkage (produce storage, beef-aging, etc.). The wet coil is used in conjunction with brine, or water spray systems, in which the liquid is applied from a header over the coil surfaces. They are used to provide high humidity and as meat chillers where appropriate. Wet coil air coolers are usually, but not necessarily, floor mounted.

Medium Velocity. These air coolers provide velocities *over 500 fpm* but *not over 800 fpm.* They usually use propeller fans with coil tubes having fins spaced about three per inch of tube length, a greater number permissible with surface temperatures above freezing.

Medium velocity coolers are used where noise and draft are not a factor. Cold storage rooms where product is either frozen, chilled, or kept at constant temperature are usually equipped with air coolers in this category. Direction of air flow is outward above the product so that it will diffuse uniformly throughout the storage compartment.

Fig. 90. Brazed plate-fin exchanger. Made of aluminum, such components offer exceptionally good heat transfer with much reduced size and weight compared to conventional type fin-tube exchangers. Have gained wide usage in land, sea and air applications for cryogenic gas separation. Shown is an exploded view of one plate-fin layer. (Courtesy, The Trane Company.)

High Velocity. These air coolers have a *minimum rate of 800 fpm.* In blast freezers (−30 F), where they are mostly used, velocities often reach 1200 fpm. Centrifugal fans are predominant in this category because of their ability to develop a high static head and overcome resistance through the coil and matrix of product stacking. High velocity maintains fairly free ice coating conditions on the coils. They also provide rapid flow of cold air over the product, resulting in quick freezing and the development of a hard surface coating to trap fluids in the product substance.

THE COMPRESSOR

In the evaporator, the liquid refrigerant enters a space wherein the pressure is low enough to promote boiling under surrounding temperature conditions. As far as the function of refrigeration is concerned, the process is complete when the subsequent vapor containing the latent heat of vaporization absorbed from the substance being cooled is evacuated by some means from the evaporator. If the cost of refrigerant were low enough, evacuation could be accomplished by merely allowing the vapor to escape to the atmosphere through a pressure-relief valve and constantly replenished with new liquid refrigerant in order to continue the cooling process.

The continual loss of refrigerant in that manner after performing its cooling function once would prove costly. For reasons of economy, then, it becomes essential that some means of continuously recapturing the spent vapor and reconverting it to liquid form for re-use be established.

To do this, the vapor must first be drawn from the evaporator into an enclosure and compressed to a point at which its conversion to liquid may be accomplished during the process of extracting the latent heat of vaporization, and the heat imparted to it while being compressed. The pressure under which reconversion occurs should be high enough to maintain it in liquid form under prevailing atmospheric temperatures until it again enters the evaporator.

The process of recapturing and compressing the spent vapor is the function of the compressor, which becomes the next basic component following the evaporator in the simple mechanical refrigeration system.

Types of Compressors. Despite differences in design, principles, and methods of operation, the function of the compressor remains the same in any refrigeration systems—i.e., to recapture and raise the pressures of the spent vapor.

Basically, all compressors may be considered as belonging to one of two categories—i.e., their principles involve either that of true mechanical compression (positive displacement) or centrifugal compression.

Compressors utilizing true mechanical compression are so considered because the act of volumetric reduction is accomplished by means of a compressing element. The compression element may be in the form of a piston, blade, rotor, lobe or gear tooth, which in its particular motion entraps and displaces vapor within a suitably designed and fully enclosed housing. Motion may be circular, in which case the compressing element, which may be a rotor or impeller, rotates within its housing continuously in one direction. Motion may be reciprocating, during which the element in the form of a piston passes back and forth within dimensional limits over the same course within a cylinder in a straight line direction, either vertically, horizontally, or angularly.

Centrifugal compression is accomplished by centrifugal force exerted on an entrapped vapor during rotation of an impeller at high speed. They have no compressing element and depend on centrifugal force and high tangential velocity of the fluid at the periphery of the impeller to produce the desired head or discharge pressure.

Refrigeration compressors may be divided into types according to their category as follows:

Category I. Positive displacement types:
- (*a*) Reciprocating,
- (*b*) Rotary,
- (*c*) Gear,
- (*d*) Screw.

Category II. Kinetic displacement type:
- (*a*) Centrifugal.

Individual selection of any of the types of compressors listed above depends largely on the service for which it is intended and the refrigerant to be used. The more widely used types include reciprocating, rotary, and centrifugal compressors, all of which possess certain advantages in their own field of use. Screw compressors however are finding wide use in the refrigerating field.

The first consideration in the selection of compressor type must be the pressures to be maintained within the system involved. Such pressures would depend on the evaporator temperature that is to be obtained. A definite relationship must exist between the characteristics of the refrigerant selected on the basis of suitability under a given combination of service factors and the operating characteristics of the compressor selected if it is to perform efficiently.

If the refrigerant selected can maintain desired evaporator temperatures at corresponding pressures above 1 atm, a reciprocating compressor is generally the type preferred. Where suction pressures are below but condenser pressures are above 1 atm, it may be stated generally that the rotary type compressor would prove more satisfactory. If conditions are such as to require negative pressures in both evaporator and condenser, or slight positive condenser pressures, then the centrifugal compressor is the logical choice.

The following rules may be used as a guide in selecting the type of compressor:

(*a*) With refrigerants having a boiling point of 15 F or below, and when the discharge pressure is above 25 psi gage, use a reciprocating compressor, or screw compressor;

(*b*) With refrigerants having a boiling point between 10 F and 65 F, and when the suction pressure is below 1 atm, use a rotary compressor;

(*c*) With refrigerants having a boiling point of 65 F or above, and operating at low negative suction pressures and negative or low positive discharge pressures, use a centrifugal compressor.

There are many exceptions to the above general rules, one of which lies in the use of a rotary compressor as a booster in an ammonia system operating at low or ultra-low evaporator temperatures. Highly successful results are being obtained with rotary compressors in conjunction with Refrigerant 12 for air-conditioning systems and domestic units.

Reciprocating Compressors

Reciprocating compressors are employed in all fields of refrigeration. They vary in size from small $\frac{1}{8}$-hp domestic units to the large industrial types of high tonnage capacity. They are adapted best to systems operating at positive evaporator and condenser pressures, although they are used successfully in ultra-low and low-temperature work where, with refrigerants such as ammonia and Freon-12, evaporator pressures are considerably below 1 atm. Under the latter conditions, either stage compression is employed or a rotary compressor acting as a booster augments the main reciprocating compressor in handling the low-density vapor as it leaves the evaporator.

Restricted piston speeds, governed by allowable vapor velocity through the valves, tend to limit their economic employment in connection with low-pressure refrigerants requiring large volumetric displacement per ton of refrigeration. They are capable of developing extremely high head pressures and handle carbon dioxide requiring condenser pressures of 1043 psi abs with little difficulty when properly designed.

There are two basic types of reciprocating compressors in general use:

(*a*) Single-acting, of vertical, V-, W-, and VW types of enclosed compressors,

(*b*) Double-acting, horizontal compressors.

Single-Acting, Enclosed Compressors

The term single-acting is applied to those compressors in which vapor compression occurs only once during each revolution of the crankshaft, or every two strokes of the piston. Compression results when vapor, drawn into the cylinder during the downward or backward stroke of the piston when the suction valve is open, is trapped within the cylinder and reduced in volume during the upward or forward stroke of the piston when all valves are closed. Hence, in a single-acting compressor, compression always occurs in the head end of the cylinder, causing one side of the piston to perform all the work.

The principal parts of the single-acting enclosed compressor are as follows:

(*a*) One or more compressing elements, called *pistons*, that travel back and forth (reciprocate) within their own

(*b*) *cylinders*, that are fully enclosed when the piston is inserted in one end to act as a movable bulkhead and capped at the other end by a

(*c*) *cylinder head* or stationary bulkhead, in which a

(*d*) *discharge valve* (or valves) is located that opens when the pressure in the cylinder increases, as a result of piston travel toward the head, to a point slightly above the combined pressure in the discharge line and the force exerted by the

(*e*) *valve spring*, or reed, which causes rapid closing of the valve when the cylinder pressure drops due to reverse motion of the piston which is driven through a

(*f*) *connecting rod* and *wrist pin* arrangement from a rotating

(*g*) *crank or eccentric shaft*, located in a pressure-tight

(*h*) *crankcase*, which acts as a lubricating oil reservoir, and a

(*i*) *main bearing* support as well as a support for the

(*j*) *compressor seal*, installed to prevent the escape of oil and refrigerant through the opening in the crankcase that allows for passage of the crank shaft, to which is keyed a

(*k*) *flywheel*, installed for the purpose of providing the necessary inertia to maintain constant speed during every increment of piston travel, and which also acts as a

(*l*) *pulley* driven through a belt arrangement by a

(*m*) *prime mover*, which may be an electric motor, a steam engine, an internal combustion engine, that provides the rotating motion to the crank shaft.

(*n*) The compressor is equipped with a *suction valve* (or valves) that may be located either in the cylinder head or in the top of the piston and usually operates on the same principle as the discharge valve, but not always.

(*o*) The cylinder is provided with a cooling system either in the form of a water jacket that surrounds both the head and the head end of the cylinder or the same parts may be provided with fins to facilitate cooling with air when operating conditions permit the use of the latter. On some types of compressors having both suction and discharge valves in the head, a suction manifold surrounding the head and the top of the cylinder, through which cold suction vapors pass on their way to the intake, effects the necessary cooling.

The flexibility of single-acting reciprocating compressors as a result of numerous possible variations in design and operating features makes them adaptable to a wide range of temperature and load conditions. Some of the variable features that may exist in any combination within compressors of this type are as follows:

(*a*) Number of cylinders,
(*b*) Arrangement of cylinders,
(*c*) Diameter of cylinders (bore),
(*d*) Length of piston stroke,
(*e*) Speed of crankshaft—revolutions per minute,
(*f*) Speed of piston—feet per minute,
(*g*) Type of piston,
(*h*) Type of valves,
(*i*) Arrangement of valves,
(*j*) Passage of vapor,
(*k*) Capacity control,
(*l*) Method of cooling cylinders
(*m*) Enclosure of crankcase,
(*n*) Method of lubrication,
(*o*) Staging or cascading,
(*p*) Compressor arrangement—duplex, stage, etc.

The number of cylinders built into a single-acting compressor varies from 1 to 16, which may be arranged vertically, horizontally, or radially, or placed in line or at relative angles to each other as to form a V or W (double V). The diameter of cylinders depends on the capacity requirement of the unit, ranging from $1\frac{1}{4}$ in. for domestic types to large industrial ammonia compressors having bores upward of 18 in. While the length of stroke in many instances equals the cylinder diameter, for various reasons this rule does not hold true for all units and may vary with different manufacturers and even with different models of the same manufacturer. In single-acting compressors designed primarily for "booster" service, the size of bore is invariably greater than the length of stroke. Compressor sizes are usually specified in terms of cylinder diameter and length of stroke expressed as inches in the order named as bore and stroke. The number of cylinders is also given so that a complete specification would include this latter information in addition to the type of compressor, cylinder arrangement, bore diameter, and length of stroke—for example,

2 cylinders, vertical, single-acting $6'' \times 6''$ compressor

or

6 cylinders, V-type, single-acting $4\frac{3}{4}'' \times 3\frac{5}{8}''$ compressor

The speed of reciprocating compressors may be expressed either in terms of average piston velocity or rotational speed of the crankshaft. Piston velocity is given in feet per second or feet per minute. Crankshaft speeds are expressed as revolutions per minute. While modern single-acting compressors range from 250 rpm to 1750 rpm, and in some instances as high as 3600 rpm, their piston speeds are restricted in velocities below the recognized maximum of 750 fpm or $12\frac{1}{2}$ fps.

To compensate for the higher rotational speeds, shorter strokes are employed. For example, a $6'' \times 6''$ compressor operating at 720 rpm would have a piston velocity of 750 fpm, whereas an allowable rotational speed for a $3'' \times 3''$ compressor for the same piston velocity would be 1500 rpm. Following a similar rule for very high-speed compressors, the maximum length of stroke corresponding to 3600 rpm would be $1\frac{1}{4}$ in. It is good operating practice to keep compressor speeds well within the limits of maximum allowances in the interest of longer service life; for that reason, piston speeds are usually maintained at 600 fpm or below.

Piston speeds are governed principally by two important considerations: (1) cylinder valve efficiency as it affects vapor frictional losses and (2) cylinder lubrication.

Fig. 91. Enclosed reciprocating compressor, V-type. (Courtesy, Carrier Corporation.)

Vapor frictional loss is a function of vapor velocity and rate of acceleration from zero to maximum velocity. The effect of vapor velocity may be described best by drawing an analogy between the action that occurs in the compressor cylinder, at either the suction or the discharge valve, and the common toy balloon. If the air within an inflated balloon is allowed to escape slowly and nonrestricted, it would do so noiselessly and with little effort. If the neck of the balloon were held between two fingers so as to restrict the flow of air while pressure is applied gradually to the inflated body,

Fig. 92. Enclosed reciprocating compressor, nine-cylinder ammonia. Each piston connected on the crankshaft to obtain 40° angularity relative to each other. This results in exceptionally smooth compressor operation. (Courtesy, Frick Company.)

escape of air would be accompanied by a sound whose pitch would vary with the degree of restriction and the amount of pressure applied. This principle forms the basis of some wind instruments and whistles. In order to evacuate the air from the balloon rapidly enough to cause sound, more pressure is required either through extreme force or by greater inflation. The noise in either case is the result of friction between the internal surface of the balloon neck and the air passing through it acting to initiate a vibratory motion.

Fig. 93. Enclosed reciprocating compressor, radial (VW) type. Note safety springs in head. (Copyright, The Trane Company. Reproduced by permission.)

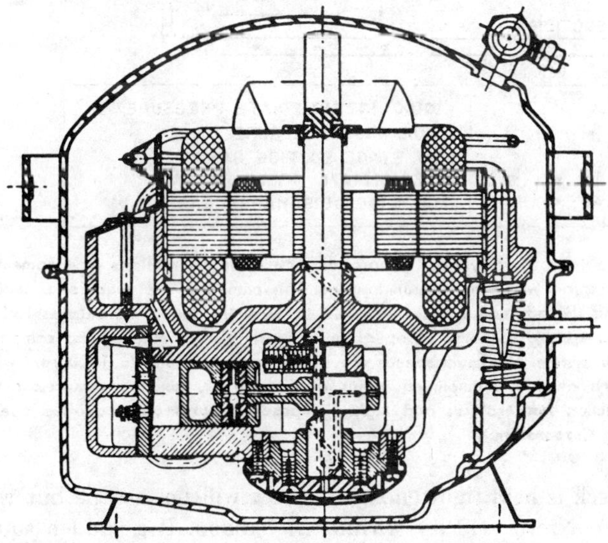

Fig. 94. Hermetically sealed unit with single cylinder reciprocating compressor. (Courtesy, Tecumseh Products Company.)

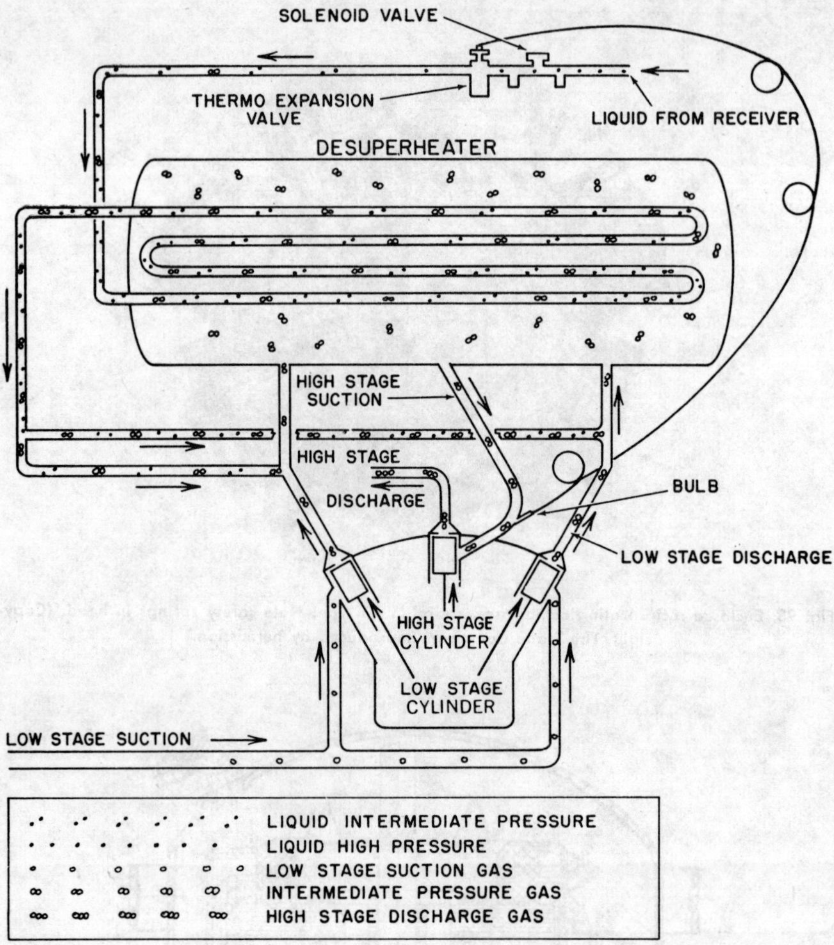

Fig. 95. Two-stage, multi-cylinder compressor. This compressor allows development of low and high stage pressures within a common housing. This eliminates the need for a separate booster compressor and the additional motor required to drive the latter. Advantages include compactness (less floor space), economy of operation, and simpler to maintain as compared with the two-compressor system. Maximum speeds are in the range of 1000 to 1200 rpm with 6 and 12 cylinder models available. Employed either as a main or auxiliary compressor for situations where low suction temperatures and high discharge pressures are requisites. (Courtesy, Vilter Manufacturing Corporation.)

If the neck is held tight enough, the air will not escape but will remain static or at a zero velocity within the balloon. If a sudden squeeze were applied simultaneously with release of the neck, the air would pass out of the balloon at a maximum velocity corresponding to the force of the squeeze. Because the force is applied so suddenly, the rate of acceleration from zero to maximum velocity is rapid enough to have eddy effects and fluttering (vibration) in and around the neck due to excessive friction within the air

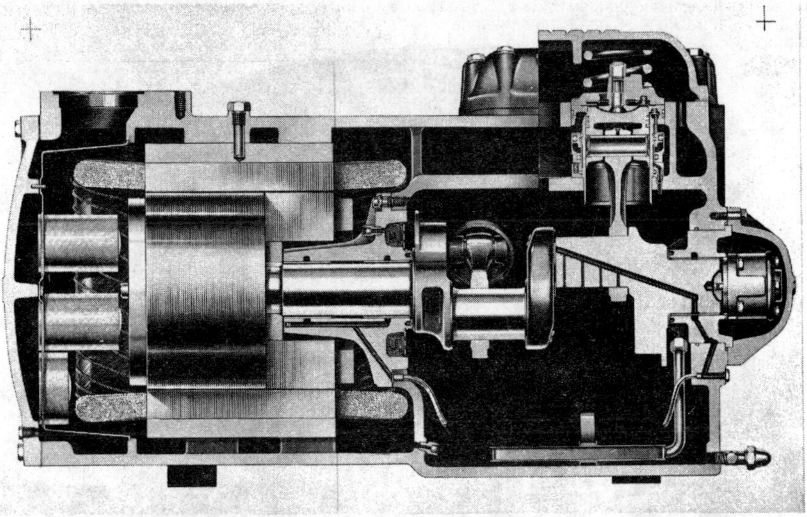

Fig. 96. Open and accessible hermetic reciprocating compressor. Sizes range from 10 to 100 tons and are equipped with multi-stage capacity control to meet wide fluctuation in loads prevalent with air conditioning systems. (Courtesy, The Trane Company.)

itself and also between the internal surfaces and the air. Actually, the small amount of time saved in emptying the balloon by applying sudden and great force over the time involved in accomplishing the same purpose with a slow, steady pressure usually does not warrant the extra energy required or the risk of destroying the balloon.

Applying these same conditions to the compressor cylinder, it is obvious that the vapor activity in the vicinity of both the suction and the discharge valves would be analogous to the activity of the air in the balloon. As piston speeds increase, the conditions within the cylinder become progressively aggravated and finally reach a point where loss of displacement offsets any volumetric gain resulting from the increased speed. Excessive piston speeds may be defined as that speed at or beyond which the pressure drop across the valves is great enough to cause (1) a decrease in volumetric efficiency initiated by

(*a*) high friction losses that follow subsequent increases in vapor velocity through the valves (wire drawing),
(*b*) increased vapor temperature as a result of wire drawing,
(*c*) eddying as a result of rapid velocity acceleration from zero to maximum during conversion from pressure head (static) to velocity head and vice versa,

and (2) abnormal power requirements to overcome the losses resulting from (*a*), (*b*), and (*c*) when the latter are excessive.

Individual compressors have an optimum speed at which their volumetric

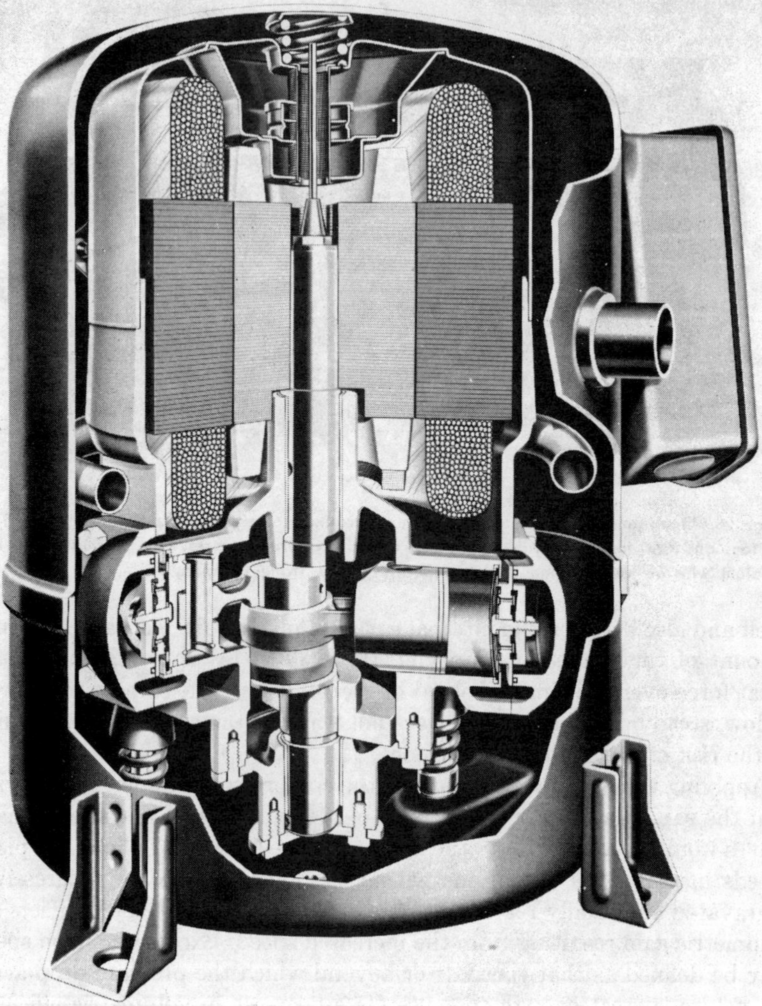

Fig. 97. Fully hermetic refrigeration compressor. This compact unit is designed for capacities of $2\frac{1}{2}$ to $7\frac{1}{2}$ tons. (Courtesy, The Trane Company.)

efficiency is greatest. Above that speed, volumetric efficiency falls off rapidly, accompanied by an abnormal increase in horsepower requirements per ton of refrigeration.

The allowable vapor velocity through an orifice (valve) is a function of fluid density and varies with different refrigerants under similar temperature conditions. For example, ammonia having a density of 0.1227 lb per cu ft at 5 F has a maximum allowable valve velocity of 12,000 to 15,000 fpm, while Refrigerant 12 with a density of 0.673 at a similar temperature is limited to approximately 6000 fpm. The use of the term "maximum allowable" here means that velocity at or below which (1) no apparent loss in volumetric efficiency or (2) no abnormal increase in horsepower per ton of refrigeration occurs.

Because vapor velocity is a function of density (viscosity), a definite relationship exists between different refrigerants and their maximum allowable valve velocities. This relationship may be expressed according to the inverse ratio

$$v \propto \frac{1}{\sqrt{d}}$$

where v = velocity in feet per minute
d = density in pounds per cubic foot

By adopting ammonia as a reference refrigerant and accepting 13,500 fpm as its maximum allowable valve velocity based on experience under average design conditions, the maximum allowable velocity for any refrigerant may be determined, viz.:

If

$$v \propto \frac{1}{\sqrt{d}}$$

then

$$v_2 = v_1 \frac{\sqrt{d_1}}{\sqrt{d_2}}$$

or

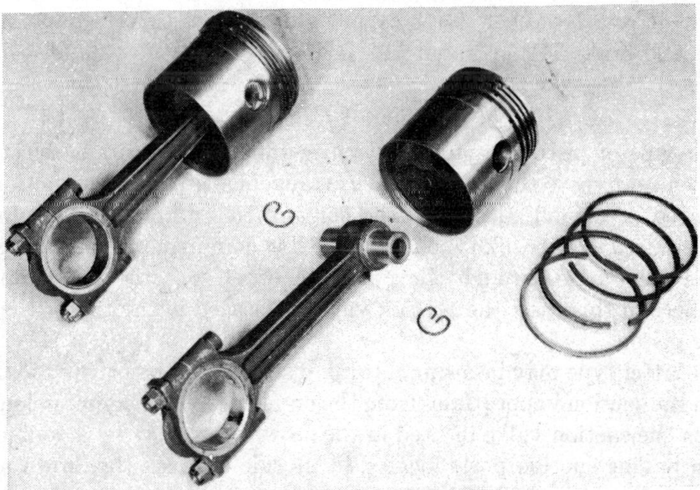

Fig. 98. Reciprocating compressor parts, showing automotive type piston, wrist-pin, connecting rod, connecting rod bearing, wrist-pin retainers, and piston rings (compression, top, and oil rings, *bottom*). Assembly is shown on left. Note opening in connecting rod bearing for oil passage to wrist pin. (Courtesy, Worthington Corporation.)

$$v_1 = v_2 \frac{\sqrt{d_2}}{\sqrt{d_1}}$$

Solving for Refrigerant 12 at 0.673 lb per cu ft density,

then

$$v_2 = 13{,}500 \frac{\sqrt{0.1227}}{\sqrt{0.673}}$$

$$v_2 = 5755 \text{ fpm}$$

Internally Compounded Compressors

Compressors that combine high pressure cylinders and booster (low pressure) cylinders in a common frame are called *internally compounded* compressors. In such an arrangement both stages can be powered by a single electric motor, their respective pistons being reciprocated by a common drive. Internally compounded compressors also possess the advantage of floor space economy. A schematic drawing of a two-stage multi-cylinder compressor is shown in Fig. 95. Compressors of this kind are available in 6 and 12 cylinder models and run at standard speeds of 1000 or 1200 rpm. They are made to handle ammonia, Refrigerants −12 and −22. To avoid excessive superheat or liquid carry-over to the first stage cylinders, the suction gas must be properly conditioned. Proprietary conditioners, such as the Vilter Vaporizer (R), are employed for the purpose.

Two-stage compressors may be direct-connected to their drive motor or belt-driven. However, for 6 cylinder units, belt drives cannot be used above 100 brake HP or about 125 BHP for 12 cylinder units.

Types of Pistons

The type of piston employed in refrigerant compressors is selected primarily according to the desired valve arrangement. Where the suction valve is located in the cylinder head, the conventional automotive or plug type piston is used. The trunk type is employed in compressors where the suction vapors must pass through the piston in order to enter the compression chamber. In this case, the suction valve is located in the top or the crown of the piston.

The latter type may be a single-trunk piston, an automotive type through which the suction vapor passes from the crankcase to the compression chamber via the suction valve located in the crown; or it may be a double-trunk piston having suction ports located in its side between the crown and the skirt to allow for the entrance of low-pressure vapor from the manifold through companion ports in the cylinder wall and thence into the compression chamber through suction valves in the crown.

The automotive type pistons are cylindrically grooved in the upper por-

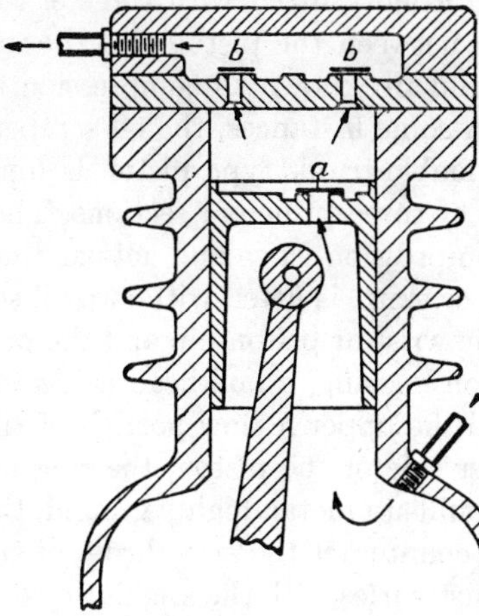

Fig. 99. (*Left*) Trunk type piston, showing side-thrust bearing surface, inlet ports, and ring grooves. (Courtesy, Frick Company.)

Fig. 100. (*Right*) Single-trunk piston of an air-cooled reciprocating compressor. Note path of refrigerant gas into crankcase and up through suction valves located in piston crown, and then through discharge valves in head. (Courtesy, Sandvik Steel, Inc.)

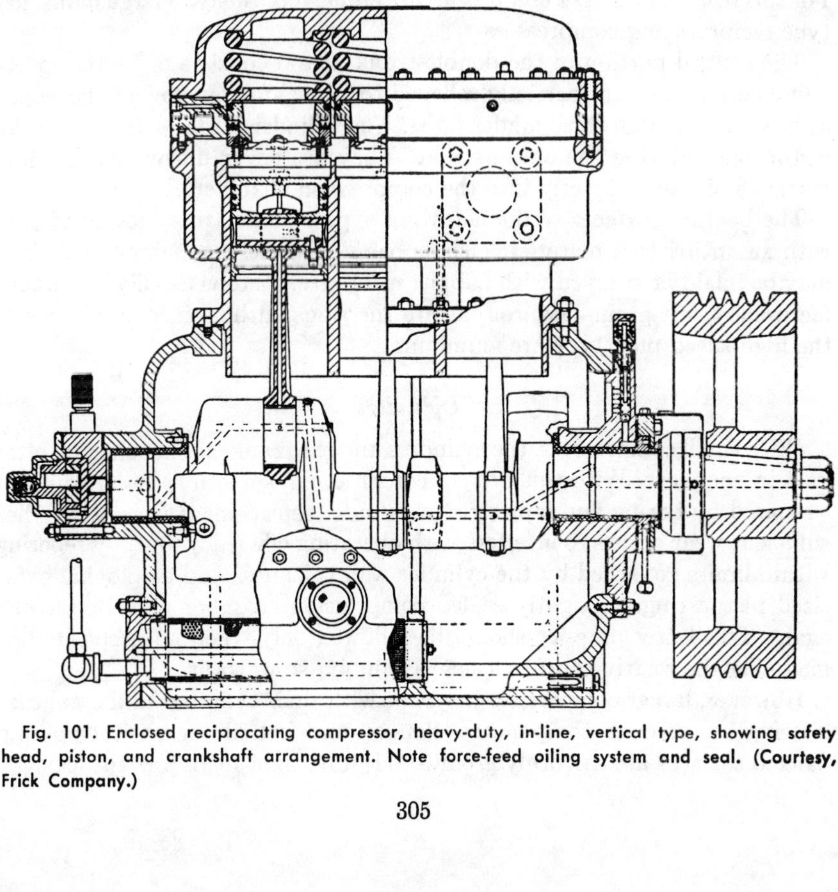

Fig. 101. Enclosed reciprocating compressor, heavy-duty, in-line, vertical type, showing safety head, piston, and crankshaft arrangement. Note force-feed oiling system and seal. (Courtesy, Frick Company.)

tion and usually fitted with three or four piston rings. Two act to prevent blow-by between the piston and the cylinder wall during the compression stroke, and are known as compression rings, while the third are oil scraper rings. In some instances, the oil scraper rings are in the skirt.

The double-trunk type piston is longer than the automotive type and consists of three principal sections. The top section is fitted with three or more compression rings and acts as the compressing element. The bottom section, or skirt, is fitted with two oil scraper rings. This lower section also holds the wrist or piston pin and the pin bushings, above which, and inside the piston housing, a bulkhead is located to effect separation of the crankcase and the upper hollow portion of the piston. Opposite the wrist pin on the outer side of the piston, the ring land is extended upward to form an area of built-up metal, highly ground, that acts as a slipper bearing or crosshead to counteract the side thrust of the connecting rod. The force of this side thrust varies with the angularity of the connecting rod and ranges from its maximum when the crank reaches a location of 90° with the center-line of piston movement to zero thrust when the crank is at top- or bottom-dead-center. As this angular thrust becomes greater with longer piston strokes, the need for extended bearing area at this part of the piston follows. For this reason, double-trunk pistons are employed widely in large industrial type reciprocating compressors.

The central portion of the double-trunk piston consists of suction ports and columns that are partially relieved to allow annular flow of the vapor as it enters through the cylinder ports. The cylinder ports register with the piston ports during the entire stroke. They are never uncovered to allow entrance of vapor directly into the compression chamber.

The bearing surfaces of double-trunk type pistons are sometimes inlaid with an antifriction material, such as babbitt, while the compression rings may be inlaid or sprayed with babbitt or bronze. Pistons usually are manufactured of fine-grain cast iron, as are the rings, although some pistons of the high-speed plug type are aluminum.

Cylinders

With small compressors, the cylinders and the crankcase housing are often cast in one piece. With this type of construction, smooth cylinder walls are prepared by boring and lapping. No cylinder replacement is possible when sufficient wear occurs to interfere with operating efficiency, but, by reboring within limits governed by the cylinder wall thickness and employing oversized piston rings, a tightly sealed compression chamber may be secured again. This latter practice should be followed only when all other factors, including the relative cost of a new assembly, so warrant.

However, larger compressors are built with their cylinders and crankcase housing as separate components, flanged and bolted together. Opposing flanged surfaces are smoothly ground to form a tight fit to prevent leakage.

In case of severe wear, the cylinder alone is replaceable without the added cost of a new crankcase.

Where removable cylinders are not employed, a liner or sleeve that can be inserted into the block to form a cylinder is generally used. When the liners become sufficiently worn, they may be replaced without the necessity of changing other parts, with the possible exception of the piston rings. This latter practice is far more preferable than reboring, which must be performed by specialists, usually at a greater cost than either a new cylinder or liner replacement and not always as satisfactory.

In some of the older compressors, especially of the large horizontal type not equipped with either removable cylinders or replaceable liners, the

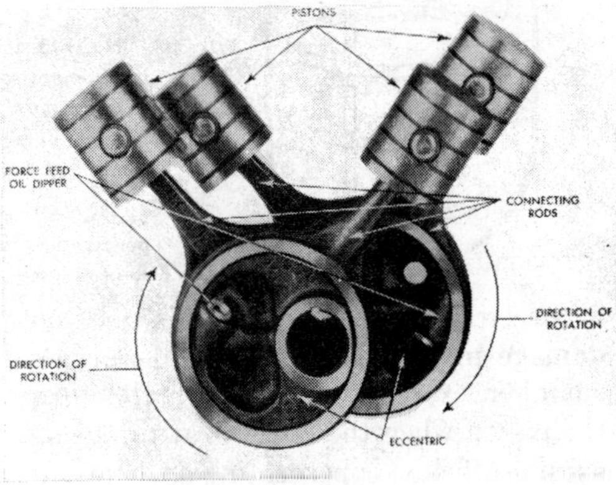

Fig. 102. Eccentric drive arrangement of a reciprocating compressor. Shown is a four-cylinder V-type unit. (Courtesy, Tecumseh Products Company.)

thickness of a worn cylinder wall should first be built up by hot metal spraying, a technique that has been highly developed in recent years, and then rebored to its original diameter. Thus an expensive cylinder may be reconditioned without weakening the wall or resorting to over-sized rings.

Cylinders and crankcase housings are constructed of fine-grained cast iron. When both the cylinder and the piston are of fine-grain cast iron, the diameter of the cylinder should be greater than the piston by an amount equal to one thousandth (0.001) of an inch per inch of piston diameter. This is called piston clearance.

Cylinder Heads

The cylinder head forms a bulkhead that permits entrapment of the vapor after it has been drawn into the cylinder and subjected to the compressive force of the piston. It may be bolted firmly to the housing that forms the cylinder wall (in which case it acts as a stationary head). When so designed, the head takes the form of an inverted cup in which the hollow space acts as the discharge chamber being separated from the compression chamber by a valve plate sandwiched between the head and the cylinder housing. Where both suction and discharge vapor pass through the head, the hollow chamber is provided with a partition to separate the two. The valve plate, as its name implies, holds the valves in place.

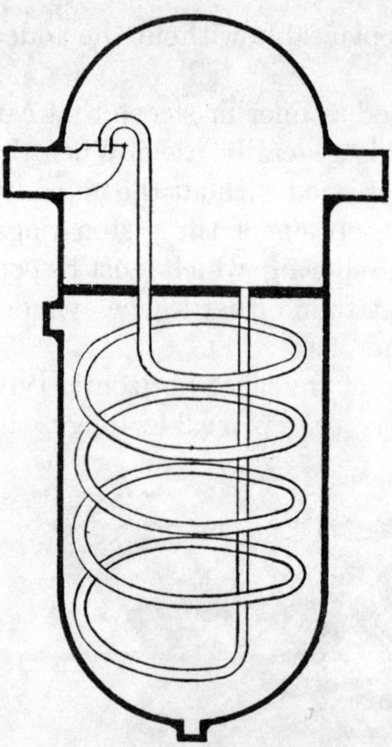

Fig. 103. (*Left*) Slug eliminator. Prevents liquid entering compressor and acts as an economizer to subcool liquid. Liquid in suction line evaporates in coil submerged in high-pressure liquid. (Courtesy, The Patterson-Kelley Co., Inc.)

Fig. 104. (*Right*) Safety head spring. Note relative position in Fig. 93. (Copyright, The Trane Company. Reproduced by permission.)

An arrangement such as this does not provide any protection against damage in case a slug of liquid refrigerant or oil enters the compression chamber, except by allowing greater clearance between the valve plate and the piston when the latter is at top-dead-center. Alternate practices are to assure sufficient superheating or to use a surge tank or accumulator to avoid slugging. Because compressor efficiencies are affected adversely by greater clearances, and the use of a surge tank or accumulator is not always feasible, systems employing stationary head units should be controlled so that only superheated vapor enters the suction manifold. (See Fig. 103.)

In larger compressors, a spring-loaded or safety head is employed to prevent damage from slugging. Here the head, which also acts as a valve plate, is supported by a powerful coil spring having as its counterthrust a cylinder head cap that encloses the discharge chamber. The spring acts to hold the safety head against shoulders formed by the top of the cylinder housing, and is strong enough to keep that head assembly in place under normal operating pressures. However, if slugging occurs, the forces developed by the hydraulic action of an incompressible fluid serve to lift the head when the piston is at top-dead-center, which returns to its original position when the piston starts on its suction stroke. Damage to the compressor is thus avoided. Where safety heads are employed, wasteful clearance space that otherwise would be needed to offset slugging is eliminated.

Arrangement of Valves

Suction valves of single-acting compressors may be located in either the cylinder head or the top of the piston, while the discharge valves are placed in the head.

Ammonia compressors, used most extensively in the industrial field, are usually two-cylinder, vertical, single-acting machines, having uniflow characteristics; i.e., the suction vapor is introduced through registering ports in the cylinder wall and trunk type piston and thence through the suction valve in the top of the piston to the compression chamber. Upon compression, the high-pressure gas is discharged through the discharge valve located in the head. However, multi-cylinder ammonia compressors having both their suction and discharge valves in-head are also employed, many of which are found in the ice cream and the dairy products industries, or as boosters.

The trend in the field of air conditioning, and also industrial water cooling, is toward higher speed compressors using Refrigerants 12, 22 or 500 as the refrigerant. These are generally multi-cylinder compressors having various cylinder arrangements, including in-line, V, W, VW, or rotary, the majority of which are valve-in-head units with automotive type pistons.

Regardless of their location, suction valves should be designed and inserted in such a manner as not to require additional clearance space within the cylinder when the piston is at top-dead-center. This also applies to discharge valves.

Typical compressors for domestic and small commercial systems are single-acting vertical units of uniflow design. However, instead of the refrigerant vapor entering the piston through cylinder ports, it is introduced to the crankcase first and then into the compression chamber via the suction valve in the piston. In some units of this kind, the flow of vapor from the crankcase to the compression chamber is augmented by an additional flow from the suction manifold to ports in the cylinder which are uncovered when the piston reaches bottom-dead-center.

Where the suction and discharge valves are located in the head, the pressure in the crankcase and the suction line are equalized by means of a relief passage between the two. In large compressors of this type, the relief port also acts as a housing for a foam breaker, the principle of which is described later in Chapter 16, "Lubrication."

Types of Valves

Valve design is based primarily on permissible velocities at which vapor enters or leaves the compressor cylinder. Velocity, in turn, is governed by the allowable pressure drop across the valve as a result of friction both within the vapor itself and between the vapor and the surface of the orifice and the connecting pipes. Because frictional resistance increases with higher fluid velocity, it follows that the power required to drive the compressor also increases. Since mechanical energy ultimately is transformed to heat energy within the vapor, a rise in temperature may well be expected which decreases the volumetric efficiency as compared with a frictionless machine. So, to approach as nearly as possible the ideal conditions of a frictionless compressor, it becomes necessary to minimize the amount of pressure drop across the valves, both suction and discharge. This can be accomplished largely by providing the largest possible valve area and allowing the valve to open with the least possible effort. These two factors, accompanied by straight-line vapor flow and the means to close the valve rapidly at the instant of reversed pressure, tend toward greater valve efficiency in general.

To accomplish their purpose efficiently, valves should be constructed of light but strong materials, operate with minimum lift, have uniflow characteristics, function quietly, operate automatically, and allow proper and tight seating. Their design should permit installation without necessitating extra clearance.

There are three basic types of valves employed in most modern compressors, all of which are self-actuating either through some form of spring arrangement or as a result of tension maintained by the restriction element. While there are numerous modifications in their design and construction, the three basic types are as follows: (*a*) the poppet valve, (*b*) the ring plate valve, and (*c*) the reed valve.

Individual compressors usually are designed to employ the same type of valve for both suction and discharge purposes. However, this practice is not always followed.

Poppet Valves. The earliest type still in use, the poppet valve is shaped like a mushroom having a head or disk whose edge is beveled to insure proper seating, and a stem that provides support and directional guidance for the head and acts as a mandrel for the valve springs. The beveled edge is called the valve face and is accurately ground to fit a valve seat inserted in the valve plate or cage. The face of the suction valve is beveled under on the stem side of the disk, whereas the discharge valve is beveled to face in the opposite direction.

The poppet valve is set in a cage which is attached to the valve plate and forms the seat, the stem guide, and the cavity for containing the valve

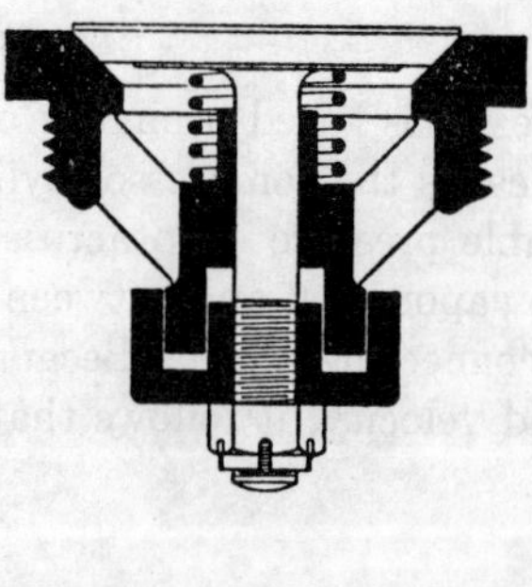

Fig. 105. Poppet valve in cage. (Courtesy, Mobil Oil Corp.)

springs. Because the actuating principles of the suction and the discharge valves differ somewhat, the mechanism held within their respective cages varies and must be described separately.

The suction valve must open at the start of the downward strokes of the piston, which means that movement takes place toward the compression chamber. To prevent it from leaving the cage entirely, it is equipped with a collar that is held on the stem by a nut and lock nut at the end opposite the head, which forms the upper thrust bearing for the actuating coil spring within the spring cavity and through which the valve stem passes. The opposite or lower end of the actuating spring rests against the shoulder formed by the bottom of the cavity and the stem guide opening. The actuating spring is compressed during the outward movement of the valve and therefore permits rapid closing of the valve at the end of the suction stroke. To limit the valve opening or lift, a second but shorter coil spring is placed on the stem. When the collar moves along the cavity during the period of valve opening, compressing the actuating spring in its travel, it is gradually halted upon contact with this second or buffer spring. Using a spring to control valve lift eliminates the shock that would accompany the employment of a more solid object, such as a shoulder.

Because the suction valve opens and closes when the piston is at minimum velocity, its actuation offers little difficulty from the standpoint of noise or excessive shock in either case. Correct actuating spring tension is important in suction valves of the poppet type. Insufficient tension results in valve lag upon closing and consequent loss of part of the charge when the piston starts on its compression stroke. Excessive tension, on the other hand, does not allow the valve to open soon enough, thereby causing some loss of the effective suction stroke. It also promotes rapid valve wear and noisy operation.

The discharge valve operates in the opposite direction from the suction valve. Because of the higher pressures involved, the problems attending their opening and closing without excessive shock become somewhat aggravated and require more effective means of cushioning than that of the suction valve.

Instant closing of the discharge valve is essential in preventing the return to the cylinder of vapor already compressed. The importance of this becomes apparent when it is realized that a pressure drop of approximately 170 psi in ammonia compressors, or 650 psi in carbon dioxide compressors, exists across the valve when the piston starts its suction stroke. As in the suction valve, closing is accomplished by means of a coil spring placed on the stem which must be strong enough to seat the valve at the instant when the piston is at top-dead-center. If closing is effected beyond that point, the valve will receive the assistance of the high-pressure vapor already discharged and the unbalanced pressure condition will drive the valve downward with excessive force. Repeated often enough, such performance eventually succeeds in destroying the valve seat and the face and makes it mandatory that actuating springs for poppet type discharge valves be selected carefully.

To prevent opening shock, the actuating spring alone, if properly selected, does not have sufficient tension. One of the oldest and most effective methods of arresting upward valve travel is to employ the collar described for the suction valve in conjunction with the spring cavity as a piston-cylinder arrangement. The collar, closely fitting in the spring cavity and attached to the stem, is located at the bottom of the cavity when the valve is closed. Refrigerant vapor trapped within the cavity during the period when the valve is closed is forced out through restricted openings in the upper part of the cavity wall during the upward motion of the valve. When the valve reaches a height equal to that of desired lift, the collar closes the small ports and prevents further escape of the entrapped vapor. Subsequent compression of the remaining vapor arrests valve travel without shock.

As the collar also acts as a movable thrust bearing for the actuating spring, the valve is instantly forced downward on its seat when the piston reaches top-dead-center. In doing so, the collar uncovers the cavity ports, allowing vapor to re-enter. The upper end of the actuating spring rests on the shoulder formed by the top of the spring cavity and the upper stem guide opening.

Sometimes a simple dashpot is used to provide cushioning of poppet valves in place of the more complicated bleeder type described above. Others employ only an adjustable spring, the tension of which may be varied by means of an adjusting nut on the valve stem.

The valve face and the seat of poppet valves, of course, must be parallel and highly ground to form a tight seal and prevent leakage. They are usually beveled at an angle of 45°.

Because of its nature, the poppet valve is necessarily a relatively heavy one. While the weight factor contributes an advantage in the form of inertia to assist in opening and closing the valve, the same factor may render them sluggish and limit their use to slow-speed compressors operating under high pressure, such as would be the case with ammonia and carbon dioxide. Another factor adding to their limited use is the low time interval during which they are fully opened, which is attained only at the peak of highest piston speeds of the suction stroke. During the remainder of the stroke they are opened only partially, which reduces the ability of the cylinder to take in its full capacity of vapor.

Having but one opening, the poppet valve necessitates relatively high lift in order to provide sufficient area. This is partially overcome by installing a number of poppet valves, particularly on larger compressors.

The design of the poppet valve makes them susceptible to sticking, caused chiefly by carbonaceous deposits between the stem and the valve guide. This is especially true of the discharge valve subjected to relatively high temperatures. The use of highly stable oil and periodic cleaning will often eliminate this difficulty.

Because the poppet valve is flush with the top of the piston or valve plate when closed, it requires minimum clearance when employed as either a suction or a discharge valve. Other advantages of the poppet valve include long life due to sturdiness and efficiency at slow speed as a result of its tight sealing characteristics.

Ring Plate Valves. The ring plate valve is replacing the poppet valve in many makes of compressors. This type valve consists of a valve seat, a ring plate, a valve spring, and a guard or retainer. The ring plate, of which there may be one or more, is held firmly on its seat and allowed to actuate by one or more valve springs whose opposite end rests against the guard. Both the seat and the guard contain comparable circular ports through which the vapor must pass. The entire assembly is held together by a machine screw.

Ring plate valves are employed for either suction or discharge. Some are so designed as to serve either purpose by reversing them. They lend themselves readily to either light or heavy construction, and therefore are suited

Fig. 106. Ring plate valve. (Courtesy, Frick Company.)

to high-speed as well as slow-speed operation, and where medium or high pressures are encountered.

One of the advantages of ring plate valves lies in their low lift requirements when ample port areas are provided. Such a combination affords sufficient area to become immediately available at the start of the stroke and to allow effective passage of the vapor during the greater portion of valve opening time and at minimum velocity. One authority, J. H. H. Voss,* gives comparisons of ring plate versus poppet valves under similar circumstances, and in the same compressor, as follows:

At $\frac{3}{8}''$ lift, which is 25% of the total, the heavy poppet valve has only 2.99 square inches effective area. It is true that the heavy poppet valve has 20 square inches

* Voss, J. H. H.: Automatic valves for ammonia compressors, *Refrigerating Engineering*, **17:** No. 5, May, 1929.

maximum area when fully opened, but this full opening is attained only for one moment at the point of highest piston speed. The average effective passage area of this particular poppet valve is only $9\frac{1}{2}$ square inches and these are effective only during 66% of the valve opening time.

Against this small area, the high speed light plate valve has an effective area of 16 square inches with only $\frac{3}{16}''$ lift and these 16 square inches are effective during 96% of the valve operating time.

Ring plates are usually made from high-grade steel, either nickel-chromium or stainless, while springs are of annealed chrome-vanadium or stainless steel. Guards and seats are of high-grade alloy steel bar stock or forgings, depending on size.

Reed Valves. Like all other types, the reed valve is made with numerous modifications and variations. A widely used type is the Feather valve, a proprietary design,† consisting of a seat, a series of flexible steel strips, and a guard. The seat is slotted to provide a number of straight, parallel ports that normally are covered with their respective steel strips which are held in place and rise against a curved guard plate. The guard is also slotted

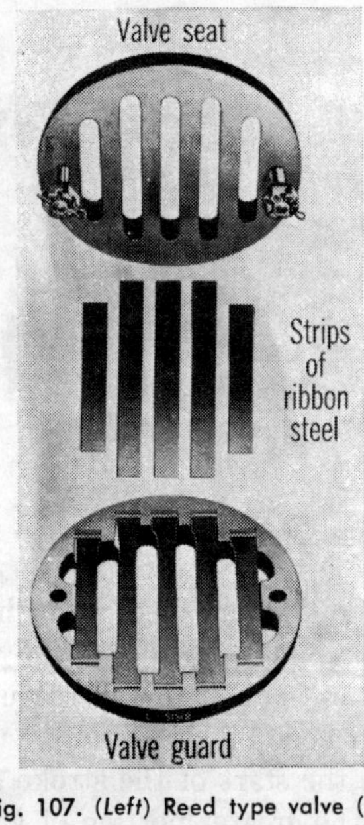

Fig. 107. (*Left*) Reed type valve (Feather valve). (Courtesy, Worthington Corporation.)

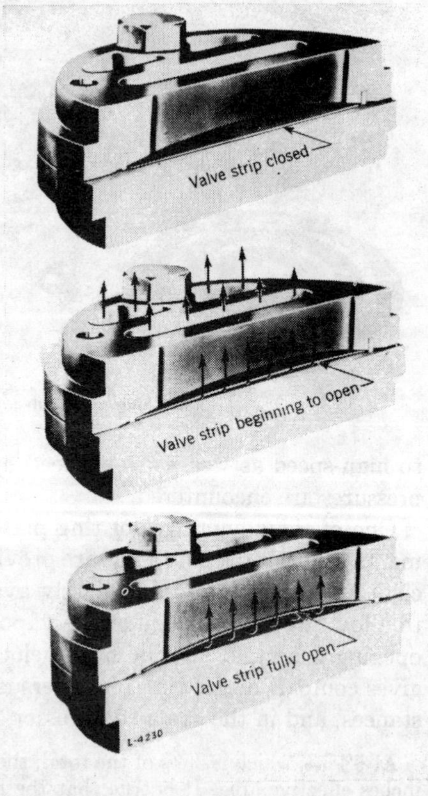

Fig. 108. (*Right*) Assembly and operating principle of Feather valve. (Courtesy, Worthington Corporation.)

† Worthington Pump and Machinery Corporation.

to form ports that are parallel with those in the seat but are offset to allow flow of vapor as it passes through the valve openings. The reeds or strips provide their own spring action and automatically cover their respective ports upon release of pressure.

The steel strips are not anchored firmly at any point, but are allowed to flex readily when pressure is applied on the seat side.

Because of the multiple-port arrangement and the large area provided when all strips are flexed, the reed valve allows vapor flow with minimum frictional loss by reason of relatively low velocity. Their light weight makes this type of valve especially suited to high-speed operation with low to medium pressures.

The clearance is necessarily greater in cylinders using either the ring plate or reed type valve due to the presence of the port spaces.

Miscellaneous Type Valves. The poppet, the ring plate, and the reed valves are all used in commercial and industrial compressors and selected for the purpose to which they are most adaptable. While the principles of the ring plate and the reed valves have been adopted for light commercial

Fig. 109. Beam valve. (Courtesy, Tecumseh Products Company.)

and domestic units, the need for extreme lightness without sacrifice of sturdiness resulted in their modification in numerous ways.

Modifications of the ring plate valve include the disk and the diaphragm valve. The disk valve consists of a round metal disk held down over a port in the valve plate by a valve spring set in a cap or retainer. The retainer, in turn, is held down by a larger coil spring whose opposite end rests against the cylinder cap and acts as a safety head. Under pressure of discharge, the disk is lifted, uncovering the port and allowing the vapor to pass through. The valve spring then snaps the disk over the port when the piston reaches top-dead-center.

The diaphragm valve is also a circular disk, but flexible in nature, held down by a bolt and nut passing through its center. The diaphragm covers a series of ports that form a circle near its outer edge. Under pressure, the disk flexes radially upward, allowing the vapor to pass through, snapping back automatically when pressure is released.

Modification of the reed type valve includes the flapper and the beam or leaf valves. The flapper valve consists of a thin, flat, flexible reed anchored

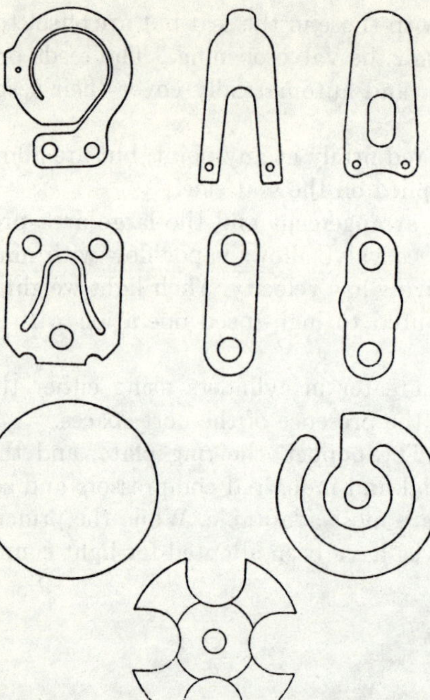

Fig. 110. Flapper valves, showing various shapes. (Courtesy, Sandvik Steel, Inc.)

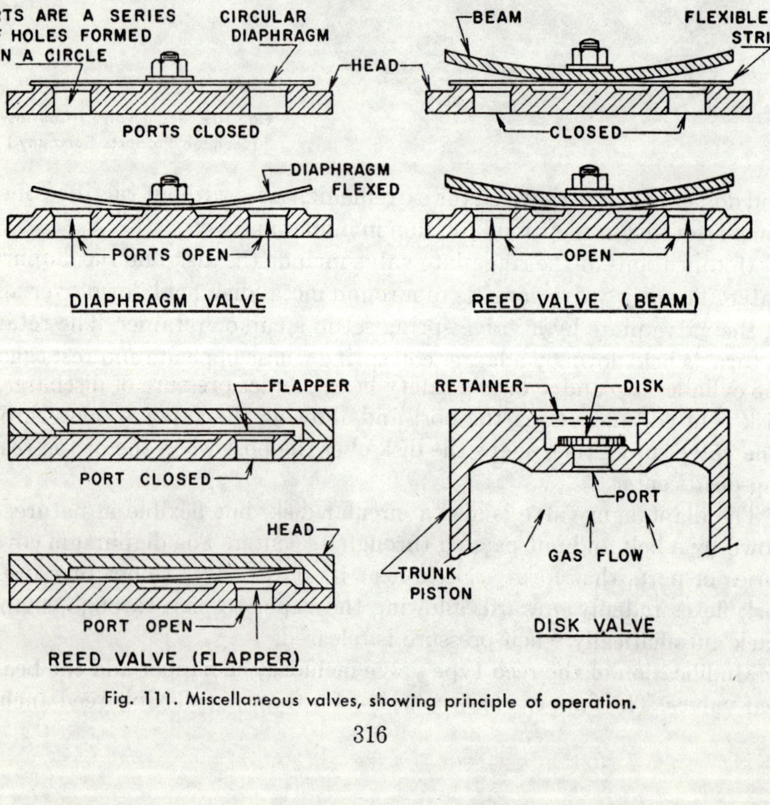

Fig. 111. Miscellaneous valves, showing principle of operation.

at one end and allowed to move or "flap" between the seat on which it rests, and in which the ports are located and a guard plate that controls the amount of lift.

The beam type has several variations. One includes a rigid beam or leaf, bent upward in the center and anchored at both ends to act as a guard for a flexible reed that normally covers the port in the valve plate or seat. The shoulder screws are equipped with coils or shoulder springs that hold the rigid beam in place but act to safeguard the valve in the event liquid refrigerant enters. Fig. 109 shows the safety spring feature of the beam type valve.

Another type of beam or leaf valve arrangement consists of a rigid leaf anchored in the middle and bent upward toward each end, which acts as a

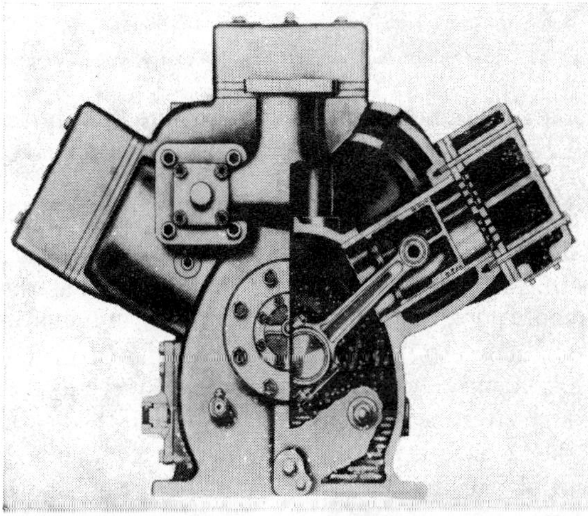

Fig. 112. Compressor showing installation of Feather valve. (Courtesy, Worthington Corporation.)

guard for a flexible reed that covers ports on opposite sides of the anchor screw in the valve plate.

Flapper valves are used for both suction and discharge, or, as in some compressors, in conjunction with the leaf type acting as discharge valve.

Crankshaft Seals and Packing

To transfer power from a prime mover, such as an electric motor, an internal combustion engine, etc., to a vertical compressor, an opening must be provided in the crankcase to allow passage of the crankshaft through which such power is transmitted. This opening presents a source of leakage in an otherwise pressure-tight system if not properly sealed and the seal carefully maintained. In the domestic field, this problem has been eliminated by the adoption of hermetically sealed units in which the electric motor and

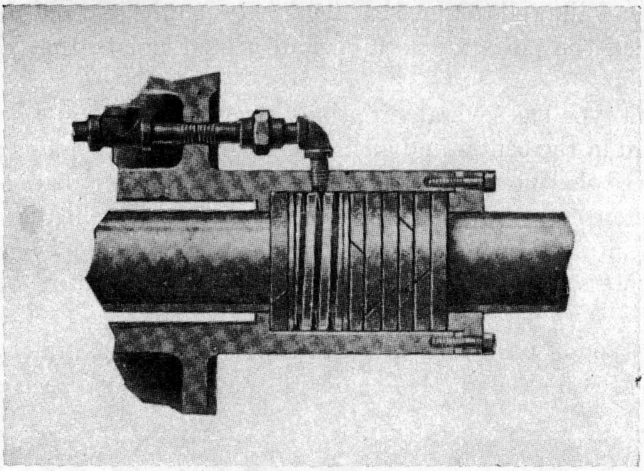

Fig. 113. Rotary shaft seal, stuffing box. (Courtesy, Frick Company.)

the compressor are contained together in a pressure-tight housing. However, in both the industrial and the commercial fields, the problem of shaft sealing remains.

Dealing here with vertical compressors only, seals capable of preventing leakage both inward or outward in the presence of a rotating shaft are of primary concern. As there are almost as many designs as there are manufacturers of compressors, many of which vary only in minor detail, this discussion will be confined to basic types.

The escape of oil and refrigerant vapor outward or entrance of air to the crankcase through clearances between the shaft and the bearing or between the shaft and the housing is prevented by means of either a stuffing box involving the use of packing materials, or a shaft seal, of which there are several basic types.

The stuffing box is a cylindrical housing of some length forming an integral part of the crankcase through which the crankshaft extends. The internal diameter of the stuffing box, being somewhat greater than that of the shaft, allows a series of packing rings, placed on the shaft, to be inserted which fill the annular space between the shaft and the housing. The rings are held in place at the compressor end by a shoulder formed by the shaft opening, and at the flywheel end by a gland secured by adjusting nuts.

The arrangement within a stuffing box usually consists of three sections. The first, nearest the crankcase, is made up of one or more rings followed by an oil lantern that forms the second. The third section of five or more rings fills out the remainder of the stuffing box, the outer one having contact with the gland. Various forms of oil lanterns are used, but, by employing a relatively heavy coil spring, some means of self-adjustment is provided to counteract wear and temperature changes.

The gland should not be drawn up so tight as to cause binding of the packing rings on the shaft, nor should it remain so loose as to preclude sufficient radial expansion to provide effective sealing. However, as packing wear occurs, a compensating adjustment must be made by tightening the gland screws. If leaking prevails after the gland has been drawn up close to the housing, it is an indication that the packing is worn and must be rebuilt or replaced.

To effect greater sealing and to provide some means for lubrication and cooling, oil is fed to the stuffing box from the compressor lubricating system by a line introduced at a point where the lantern is located.

Using split or sectionalized rings of the correct size and materials assures a close fit on the shaft when only slight pressure is applied at the gland if installed properly. They should also fit tightly to the wall of the stuffing

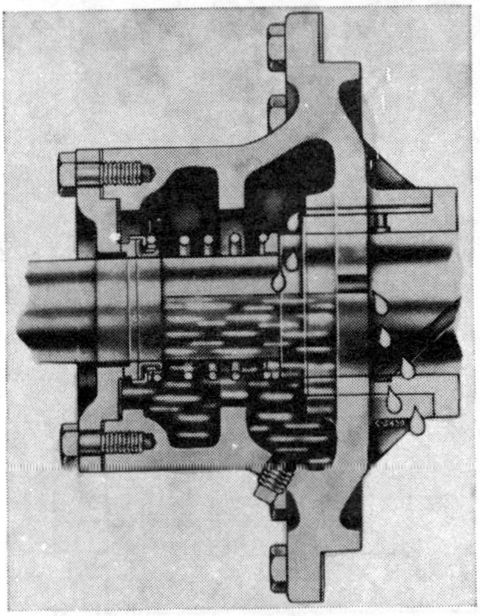

Fig. 114. Rotary shaft seal, spring type. (Courtesy, Worthington Corporation.)

box. The shaft should be smooth and centered in the stuffing box. Off-centered shafts may be caused by worn bearings. Materials considered for use as packing must possess definite characteristics to assure their preventing leaks. That they be nonporous is obvious. Because of temperature changes, they must be self-adjusting or resilient, especially if lanterns other than coil springs are employed. Their necessary contact with the shaft requires materials that are nonabrasive and possess low frictional qualities.

There are numerous types of packing, both metallic and nonmetallic. Metallic packings, usually some form of babbitt or lead foil, are made in such a manner as to offer resiliency through flexibility or collapsibility. The presence of graphite either on or in the packing assists in lubricating the shaft. Nonmetallic packings are usually rubber in combination with cotton or asbestos and processed to form a heterogeneous material. In some instances, both metallic and nonmetallic rings are used together, placed alternately on the shaft.

While the stuffing box is effective in preventing ammonia and carbon dioxide leaks, it is not successful in combating the difficulties posed by

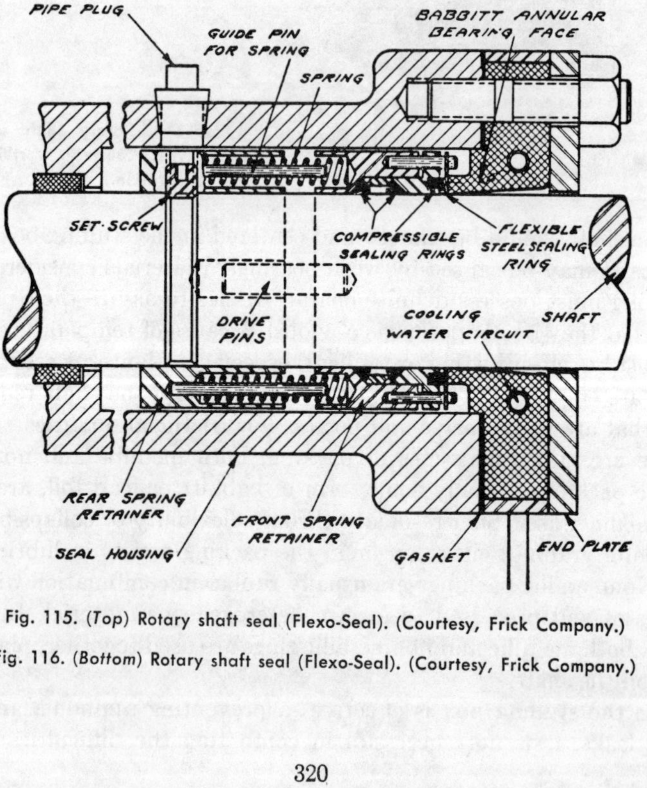

Fig. 115. (*Top*) Rotary shaft seal (Flexo-Seal). (Courtesy, Frick Company.)

Fig. 116. (*Bottom*) Rotary shaft seal (Flexo-Seal). (Courtesy, Frick Company.)

R-12 in this connection. As a result, they are being widely replaced by shaft seals on many makes of modern compressors, even those handling ammonia.

One of the more important shaft seals is the bellows type, of which there are two kinds distinguished as having either a rotating or a stationary nosepiece. Basically, both kinds consist of a sealing ring, a bellows, and a spring, all of which fit over the crankshaft and are encased within a housing and maintained in position by a cover plate.

The sealing ring bears against a shaft shoulder, the contacting surfaces of which are highly ground to form a tight seal. Attached to the sealing ring is a flexible, corrugated metal bellows, the other end of which is fastened securely between the end of the housing and a cover plate. A strong coil spring, which may be located within or outside the bellows, serves the double purpose of keeping the sealing ring tight against the shaft shoulder and expanding the bellows. In this type, the sealing ring acts as the nosepiece and remains stationary. A packed collar, that rotates with the shaft and against which the sealing ring bears, is a basic modification that distinguishes the two types mentioned above. All seals should be lubricated.

The bellows is necessary to compensate for wear on the bearings or the shaft. Any increase in leakage caused by such wear tends to further expand the bellows, which, in turn, exerts a correspondingly greater pressure at the sealing surfaces. An important consideration in this type seal concerns the relative position of (*a*) the center line of the convolutions of the bellows to (*b*) the center line of the sealing ring.

Because of the difficulty in combining sufficient strength with adequate flexibility, the size of bellows is limited. Therefore, the bellows type seal is not often used in compressors whose shafts exceed 2 in. in diameter.

Where bellows are not practical, sealing is accomplished by the use of one or more coil springs to maintain proper contact between a stationary ring and the rotating member. The usual material for sealing rings is bronze.

One form of shaft seal that functions without the bellows is the patented Flexo-Seal shown in Figs. 109 and 110 and used with various refrigerants, including Freon-12.

Double-Acting, Horizontal Compressors

The double-acting reciprocating compressor draws into its cylinder and compresses two charges of vapor for each revolution of the crankshaft. This means that both sides of the piston are alternately subjected to high and low pressures with each directional change in piston travel or stroke.

While there are some vertical double-acting compressors, the great majority are horizontal and have their driving mechanism remote of the cylinder, which means that the latter must be made gas tight at both ends. Many of them are built for two-stage operation. In this case, they have a low-pressure and a high-pressure cylinder with an intercooler installed between the two. In the latter, both pistons are driven from a common crankshaft.

Compared with vertical type industrial compressors, the horizontal unit requires more floor space but less head room. All parts are more accessible in the latter type compressor, which facilitates repairing and cleaning.

Plain pistons are employed in horizontal compressors to which a piston rod is rigidly attached. Because the rod enters the cylinder from one end, known as the tail end, it takes up space that ordinarily would be filled with vapor, as in the head end. Therefore, the volume displaced during the back stroke is less than during the forward stroke by an amount equal to the cross-sectional area of the rod times the length of stroke.

Valves are located in the cylinder wall at each end of the cylinder, with the suction valves on top and the discharge valves on the bottom. Poppet, ring plate, or reed type valves are all suitable for use in double-acting cylinders.

The driving mechanism of double-acting compressors departs considerably from that of single-acting types as a result of the straight-line motion of the piston rod as compared with the oscillating motion of the connecting rod. The piston of the former is driven by a piston rod connected to a crosshead, all of which have straight-line reciprocating motion. The crosshead, in turn, is driven through a connecting rod by a crankshaft which receives power from a prime mover, usually via a belt and pulley arrangement. A flywheel is hung on the end of the crankshaft opposite to that supporting the pulley. The crankshaft is counterbalanced for smooth operation.

The crosshead is supported by a plain bearing and held in position by means of a crosshead guide. Where the crankshaft is fully enclosed, the crosshead assembly usually is located within the same housing to facilitate its lubrication. In such units, it is necessary to install a breather or a vent to equalize internal and external pressures.

The necessity of preventing leakage in a double-acting compressor poses a more serious problem than in the enclosed vertical type. The opening through which the piston rod passes is directly exposed to the alternately high and low pressures occurring within the cylinder. Both the piston rod and the stuffing box are exposed to drastic temperature changes which range from that of compression to surrounding conditions when the compressor is not operating. Even the operating temperature of the rod is subject to change by reason of variations in suction pressure or as a result of liquid carry-over, so that much consideration need be given to the expansion and contraction of the rod and the probable response of the packing to compensate such changes.

To reduce leakage, an extra long stuffing box is used, partly because of the higher pressures involved and partly to counteract the pumping action of the piston rod which has a reciprocating rather than the rotary motion of the vertical machine. A typical stuffing box is so arranged as to include

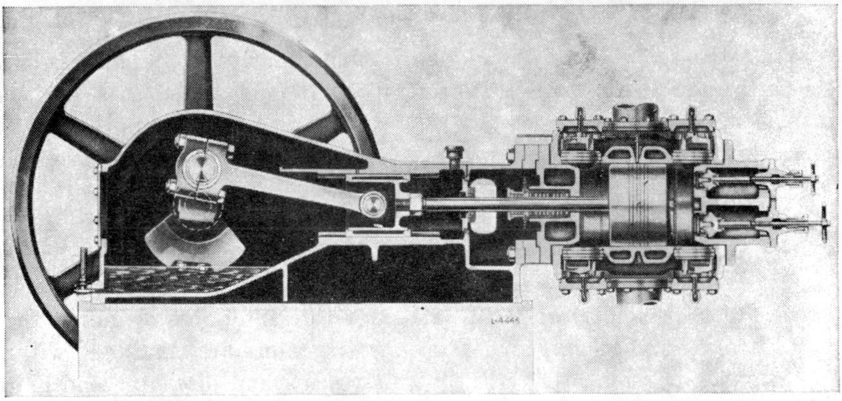

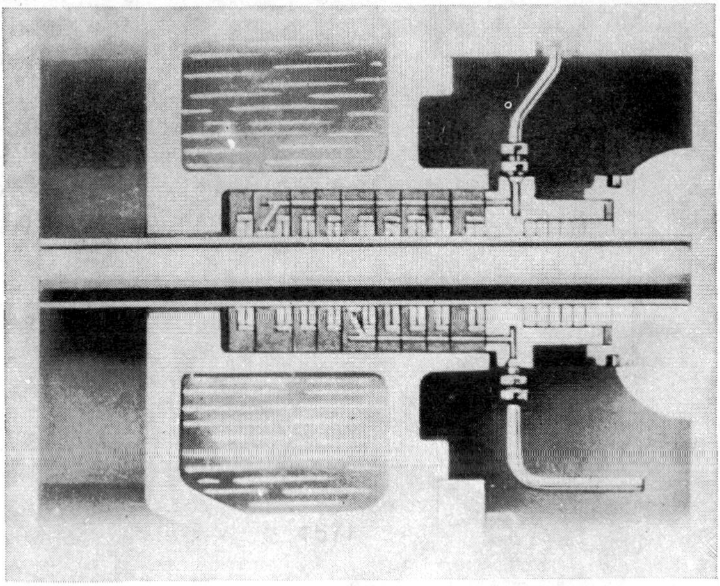

Fig. 117. (*Top*) Horizontal double-acting reciprocating compressor. (Courtesy, Worthington Corporation.)

Fig. 118. (*Bottom*) Stuffing box for piston rod of reciprocating double-acting compressor. Note method of oil distribution. (Courtesy, Worthington Corporation.)

a primary or main packing section nearest the cylinder. This is backed by an oil lantern, which, in turn, is followed by a secondary or auxiliary packing. The lantern is maintained full of oil, fed from the force feed lubricator, that also services the cylinder. In addition to lubricating the stuffing box, the lubricating oil forms a seal and in conjunction with the auxiliary packing prevents leakage during shutdown periods. The stuffing box should be jacketed and water cooled.

Screw Compressor

While the *screw compressor* was developed early in the 1930's, it is relatively new in the field of refrigeration. It is classified as a positive displacement compressor of the rotary type, achieving compression by direct volume reduction of a gas as opposed to dynamic compression, previously described for centrifugal machines.

As shown in Fig. 119, the screw compressor consists essentially of two mating helically grooved rotors, a male rotor having lobes (see Chapter 10, Fig. 344, to discern lobe design as applied to pumps or blowers, known as the Roots principle) and a female rotor with interlobes or gullies that provide meshing action during rotation. The rotors are contained within a stationary housing fitted with an inlet casing and discharge end plate which embody the suction and discharge ports, respectively. A hydraulically operated sliding valve forming the lower cusp (sharp protrusion formed by the overlapping of two cylinders), moved by a spindle parallel with the rotor journals, furnishes capacity control to the compressor, and thus the refrigeration system. There are two basic types of screw compressors which differ in their mechanical operation. One is classified as an *oil free* or "dry" high speed unit, and the other as an *oil flooded* or oil-injected type of screw compressor that operates at relatively lower

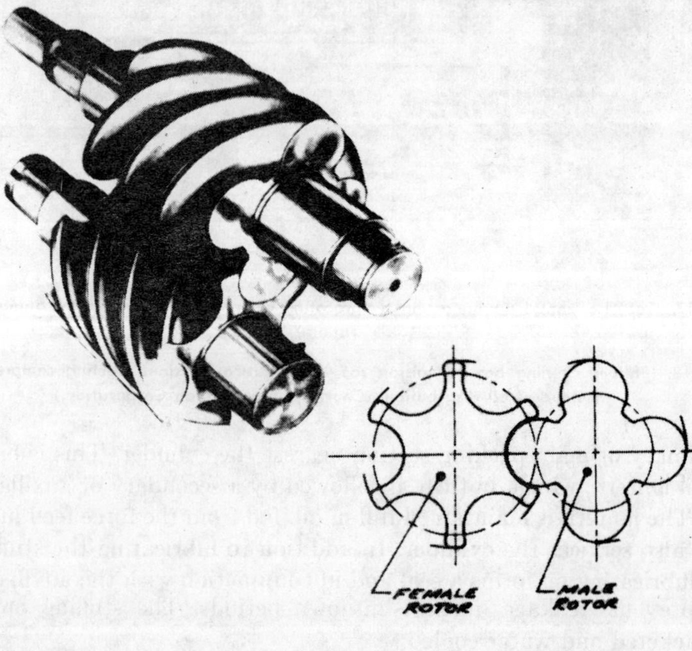

Fig. 119. Male and female rotors, screw compressor. (Courtesy, Dunham-Bush, Inc.)

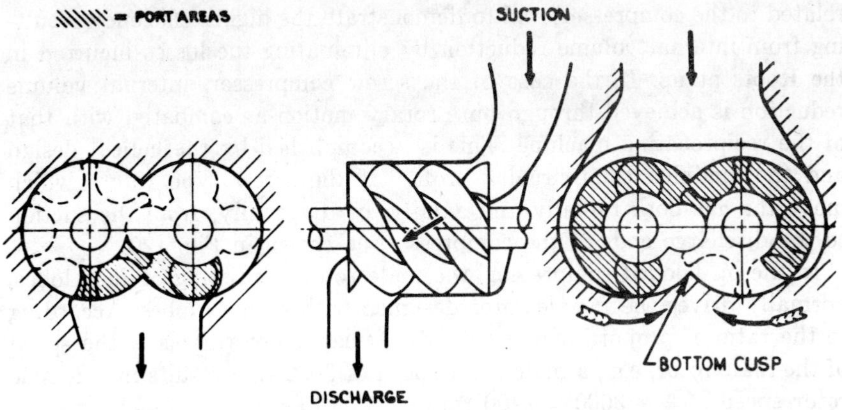

Fig. 120. Screw compressor showing radial and axial flow. (Courtesy, Dunham-Bush, Inc.)

speeds. As in all machines, oil is the necessary evil required to lubricate the various frictional parts and so must enter the discussion here.

In the *oil free screw compressor,* the power to drive the rotors is transmitted through timing gears so that it is unnecessary for the contour surfaces of the lobes and interlobes to have intimate contact (i.e., no metal-to-metal contact). Therefore, lubrication of the grooves and gullies is not a factor. However, a lubrication system is provided to protect the journal bearings and valve against frictional losses and wear. There are certain disadvantages associated with "dry" operation that will be discussed below in more detail, as will the oil flooded operation which is now replacing the older dry type of screw compressor. While the student should be acquainted with both types as a matter of complete knowledge, our discussion is based on the newer or oil flooded type.

The Roots principle is based on radial flow of fluid that involves certain compression losses due to an abrupt increase in pressure resulting from flow-back of gas through the discharge port from the higher pressure zone. As is obvious from the design of the Roots type lobe, volume reduction does not occur within the compressor, but relies on forcing or "crowding" additional weight of gas into the pressure area beyond the discharge port. Unlike the reciprocating compressor wherein pressure is built up gradually as the piston moves against entrapped gas within the cylinder, as portrayed by the pressure-volume diagram, Fig. 24, in which the overall compression ratio is the result of incremental changes in the value $P \times V^y$ for adiabatic processes, the straight lobe pump approaches a rectangular configuration as $P_{suction}$ $V^y{}_{suction}$ suddenly changes to $P_{discharge} \times V^y{}_{discharge}$ at the discharge port exit.

The point brought out here lies in the *method* of converting a low pressure gas to a higher pressure gas, either externally or internally, as related to the compressor, and to demonstrate the higher efficiency resulting from internal volume reduction by eliminating the losses incurred in the Roots pump. In the case of the screw compressor, internal volume reduction is achieved through pure rotary motion as compared with that of the reciprocating machine. This is accomplished by the helical design and the modified semi-circular profile of the screw type rotors which move the gas both radially and axially, or diagonally, from the suction to the discharge ends of the compressor, as shown in Fig. 120.

In the oil-flooded compressor, the male rotor, consisting of four lobes, normally drives the female rotor designed with six interlobes. According to the ratio of proportional speeds, the female rotor travels $\frac{2}{3}$ the speed of the male rotor, e.g., a male rotor speed of 3600 rpm results in a female rotor speed of $\frac{2}{3} \times 3600 = 2400$ rpm.

The four basic phases of operation are (refer to Fig. 121):

1. *Suction:* As a pair of lobes unmeshes on the inlet port side, gas flows in the increasing volume formed between the lobes and the housing (cylindrical side and faces) until the lobes are completely unmeshed.

2. *Transfer:* The trapped pocket now isolated from the inlet and outlet is moved around circumferentially at a constant suction pressure.

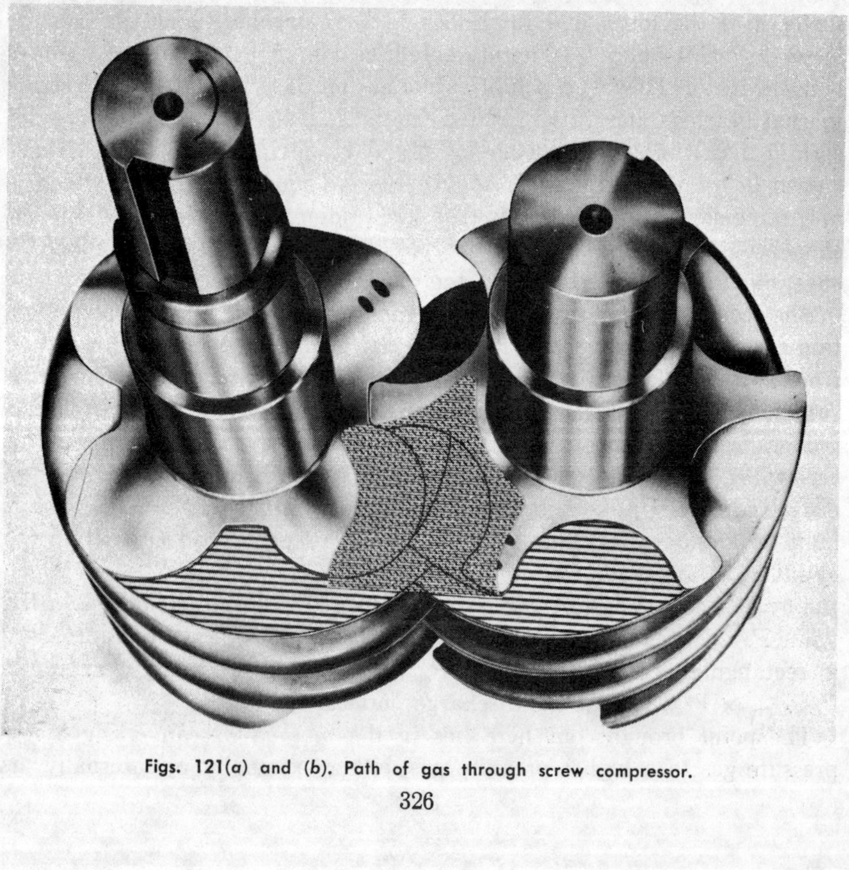

Figs. 121(*a*) and (*b*). Path of gas through screw compressor.

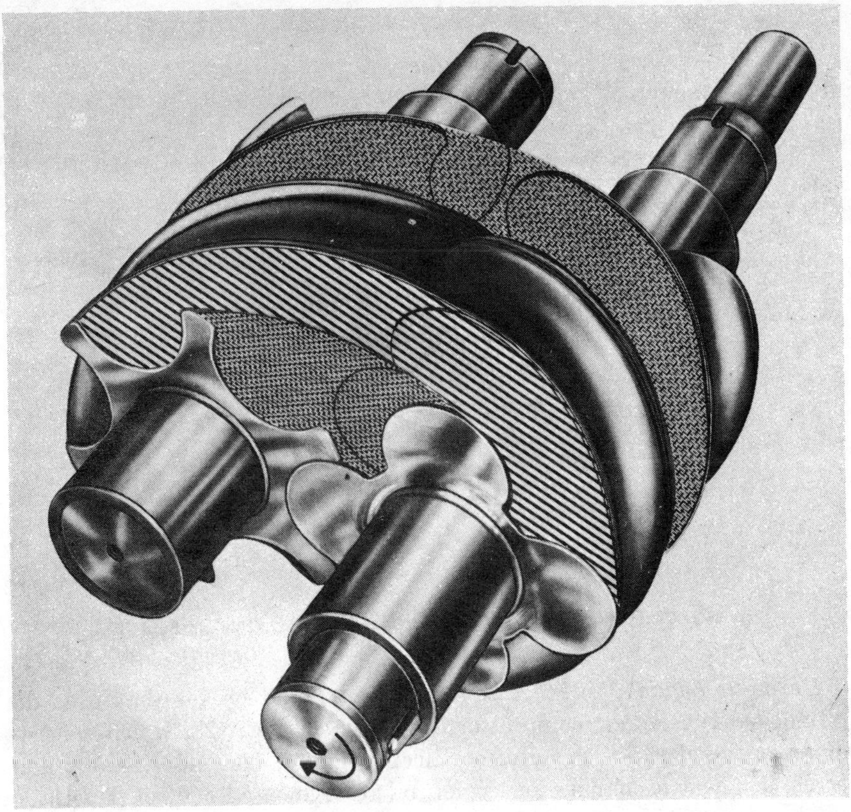

Fig. 121(*b*).

3. *Compression:* When remeshing starts at the inlet end, the trapped gas volume is reduced, the gas charge gradually compressed and moved toward the discharge end as the mesh point moves axially along the line of mesh. (Bottom cusp indicated in Fig. 120.)

4. *Discharge:* Starts when the compressed charge uncovers the radial and axial ports on the discharge end of the compressor (ports position determines the built-in volume reduction and corresponding design internal compression ratio) and continues until all the trapped gas is completely forced out. Thus, no residual gas volume to re-expand remains, as in the clearance space required in the reciprocating compressor cylinder.

Simultaneously with remeshing during phase 3 and 4 above, a fresh gas charge is drawn in on the opposite or inlet side of the meshing point. With four male lobes rotating four interlobes are filled and discharged per revolution and because of the two-thirds ratio of driver to driven rotor lobes and interlobes causes an overlap of several lobes, a smooth, continuous flow of gas is attained with uniform torque.

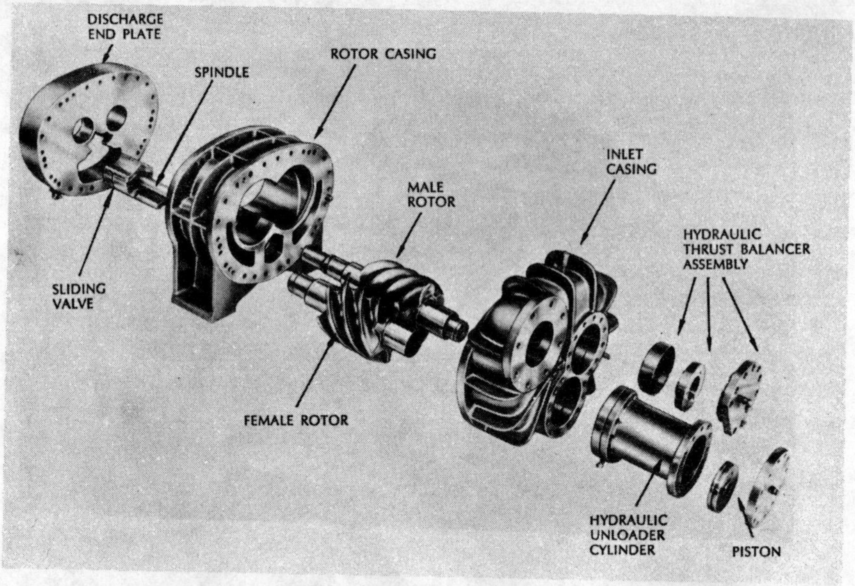

Fig. 122. Exploded view of screw compressor. (Courtesy, Dunham-Bush, Inc.)

Hermetic Liquid Chiller. A self-contained liquid chiller that uses the oil-flooded-type screw compressor, as shown in Fig. 123, which consists of an enclosed screw compressor, electric motor, condenser, evaporator or chiller, heat exchanger, oil cooler, oil reservoir and control panel.

The *evaporator* is the shell and tube design.

The *heat exchanger* is an extension (when installed) of the evaporator, that provides an interchange of heat to cool the liquid refrigerant by the suction vapor.

The *condenser* is of the cleanable shell and tube type.

Oil cooler is a water cooled shell and tube (bundle) type.

Electric motor is a 2 pole, squirrel cage 3600 rpm hermetic motor normally wound for star-delta or across-the-line starting.

Oil separator is of proprietary design having a unique flow pattern within the separator itself combined with an oversized mechanical demister. The result is highly effective in separating the oil from the discharge gas before it enters the condenser. The separated oil returns to the oil reservoir.

Oil reservoir that consists of a hermetically sealed gear type oil pump to supply oil to all frictional parts, and an oil heater for controlling lubricating oil temperature. The oil pump is controlled independently of the compressor to assure lubrication during start-up and coast-down (gradual stopping).

Control center containing all operating controls for automatic opera-

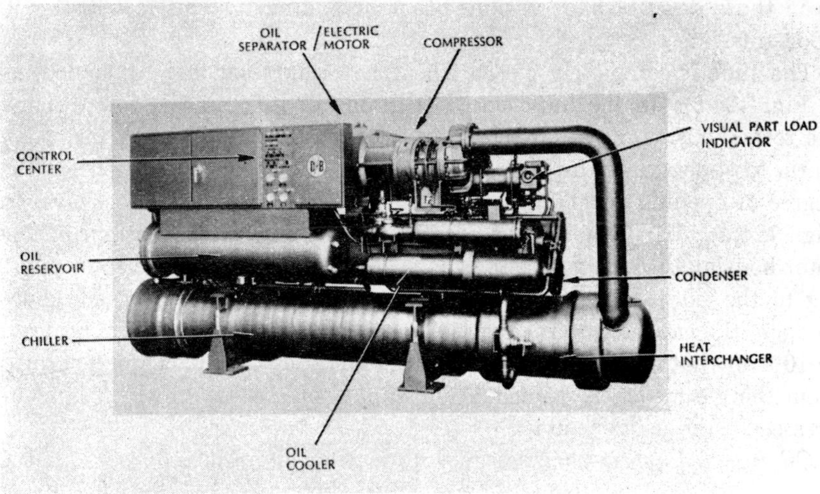

Fig. 123(a). Hermetic package chiller with screw compressor. (Courtesy, Dunham-Bush, Inc.)

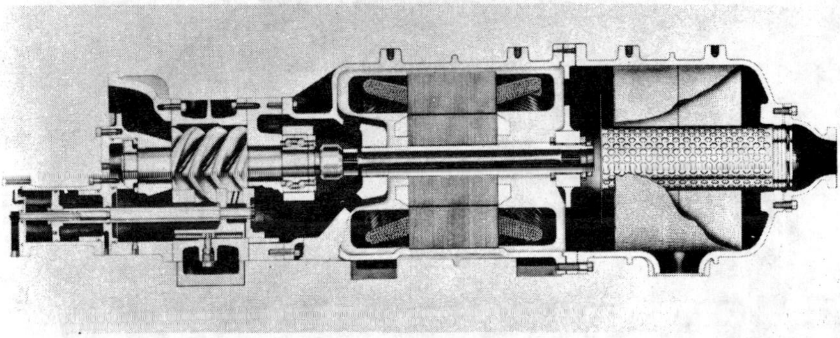

Fig. 123(b). Hermetic package chiller. Showing direct drive arrangement, hydraulic valve activator, screw compressor, hydraulic motor, and hermetic oil separator. (Courtesy, Dunham-Bush, Inc.)

tion, including load limiter control, anti-recycle timer, low pressure unloader control, oil pressure switch, oil temperature control, hi-lo pressure switch and operating and freeze protection thermostats.

Capacity Control. The slide valve within the rotor housing is the component that governs capacity control. It has a similar relationship to the screw compressor system as the programming procedure of cutting-in and cutting-out cylinders of multi-cylinder reciprocating compressor systems.

The slide valve moves axially to open or close as programmed by an electronic-pressure transducer through hydraulic actuation. The hydraulic piston-cylinder-spindle arrangement is shown in Fig. 123 (far left). The slide valve capacity control is sensitized by changes in evaporator load, as in those described for various other unloading systems in Chapter 9, "Controls."

The slide is completely closed when the compressor is fully loaded, as in Fig. 124(*b*). In the fully closed position, all the gas will flow through the rotor housing, as described above. As the load diminishes, the valve starts to slide axially away from the valve stop to an open position, the degree of opening depending on the magnitude of load drop, as shown in Fig. 124(*a*). Valve movement creates an opening in the bottom of the rotor housing through which the gas is shunted back from the rotor housing to the suction area before it is compressed. As the valve continues to open, the capacity of the compressor diminishes correspondingly down to 10% of that of full load. The reduced compressor capacity is obtained from the gas which is inside the inner part of the rotors which is compressed in the usual manner.

Oil injected screw compressor. Lubricating oil is introduced into the

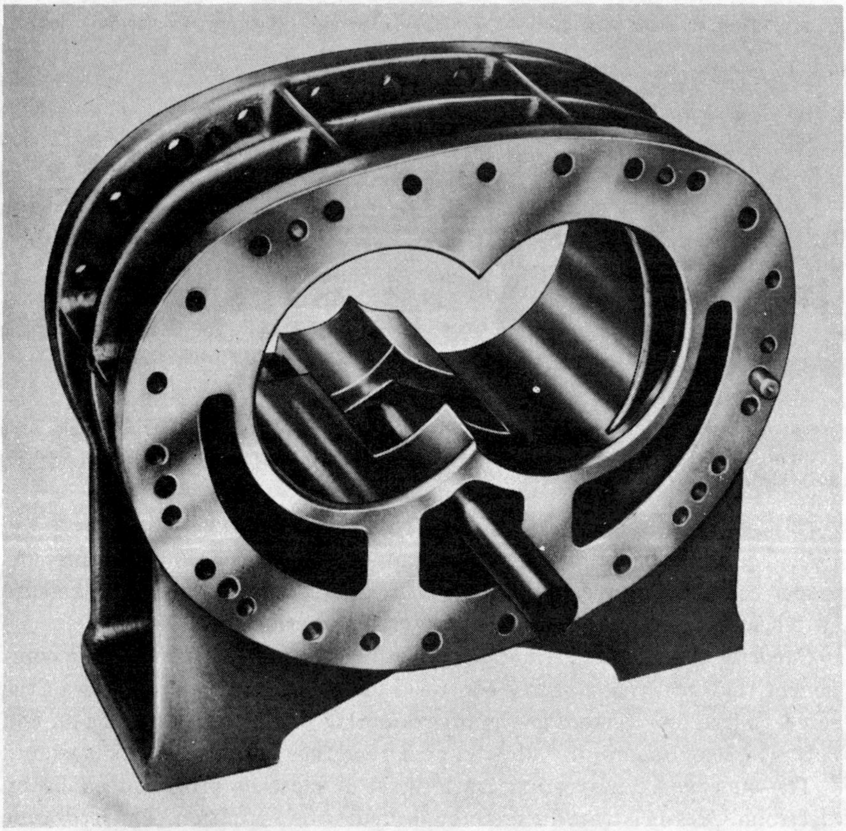

Fig. 124(a). Screw compressor housing showing capacity control valve open. (Courtesy, Dunham-Bush, Inc.)

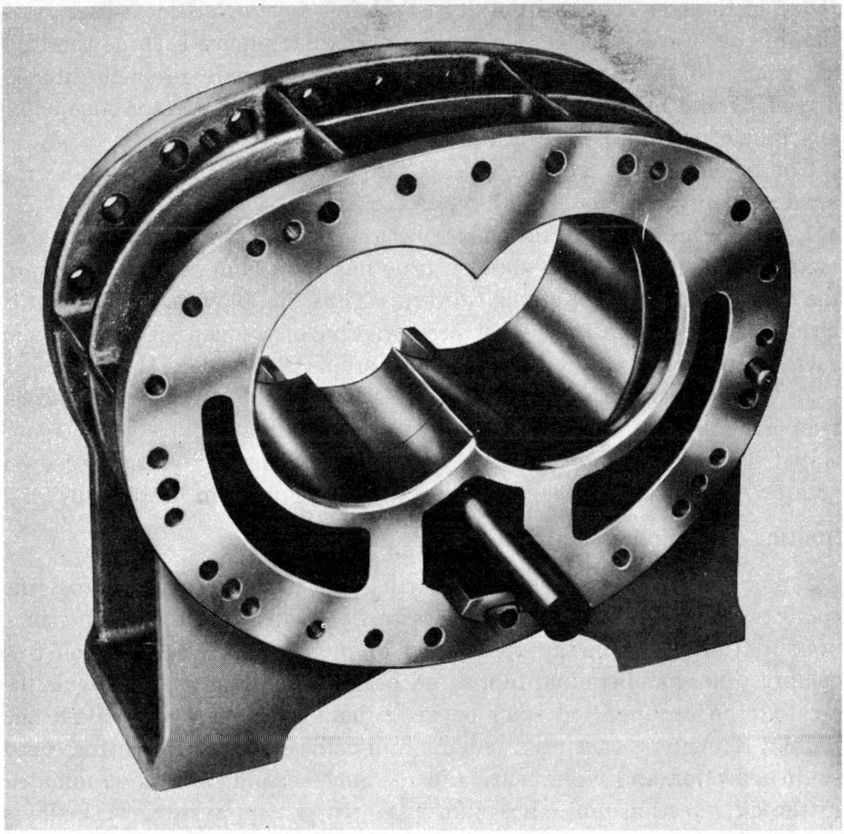

Fig. 124(b). Screw compressor housing showing capacity control valve closed. (Courtesy, Dunham-Bush, Inc.)

compressor at various points, depending on the frictional part. However, all the oil, including that introduced through the slide valve cusp, eventually passes through the hermetic motor compressor along with the refrigerant gas. Forced feed lubrication is employed to supply oil to all frictional parts, including the bearings and journals.

The rotor is lubricated by the oil entering through a small tube (main injection) in the slide valve cusp beyond the zone of inlet pressure, where the gas is partially compressed and superheated. While there is no provision for atomizing the oil, it is automatically dispersed by the high speed revolving action of the rotors to form a mist-like environment within the compressor housing. This mist, along with any carry-over of larger oil particles, supplies lubrication to the rubbing surfaces of the rotor lobes and interlobes.

This method of lubrication allows for intimate contact between the opposing surfaces of the lobe of the male rotor and the female interlobe.

It led to the conversion of the dry or oil-free type operation where rotation of the rotors is accomplished by an auxiliary drive such as the timing gears, to the wet type operation where the male rotor is directly driven by the motor, which in turn rotates the female rotor much like any pair of mating helical gears.

The oil-flooded or modern type of screw compressor has the following advantages over the older dry type:

1. The necessary clearance space between lobe and interlobes to avoid metal-to-metal contact in the dry type has been eliminated in favor of the no-clearance design of the wet type. Thus, the blow-by of gas in the direction away from the higher pressure end of the compressor through such clearances is reduced to minimum proportions.

2. Wet injection has also reduced discharge temperature and uneven temperatures, due to recompression, between mating components that resulted in thermal distortion, which had to be offset by providing even greater clearances, thus aggravating the basic problem of blow-by and limiting the compression ratio, i.e., reducing $\frac{P_d}{P_s}$.

3. For purely mechanical reasons, the pressure differential across the compressor ($\Delta P = P_d - P_s$) was necessarily limited in the dry type compressor. It resulted in the application of internal seals between oil-free compression chambers and lubricated bearings, gears, etc. Such seals increased the unsupported span between main bearings and reduced the journal diameters, causing a reduction in rotor stiffness, which increased shaft deflection and main bearing loads. Such seals have been eliminated in the oil injection unit, thus affording greater compression with all its resulting advantages. Summarizing, the important advantages of the oil injection unit are:

(*a*) Low degree of blow-by
(*b*) Greater compactness
(*c*) Reduction in shaft deflection
(*d*) Substantial reduction in vibration and noise level with the end results of
(*e*) high compression ratios
(*f*) high pressure differentials at lower compressor speeds, particularly
(*g*) 2 pole motor rpm in the range of somewhat below 100 tons to 800 tons of air conditioning loads, which means
(*h*) a simple direct drive without speed increasers

Lubrication System—Oil Injection Type. As mentioned previously, lubrication for the helically grooved rotors is provided by dispersing lubricating oil through the slide valve cusp into the hermetic area of the "wet" type screw compressor. A unique feature of this procedure of oil applica-

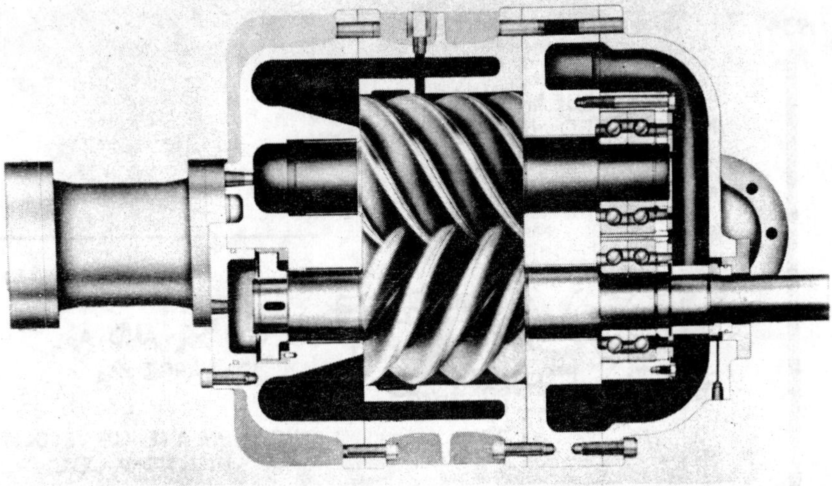

Fig. 125. Cross-section of screw compressor. (Courtesy, Dunham-Bush, Inc.)

tion is the cooling effect the mixture of oil vapor and hot discharge gas exerts on the electric motor that drives the rotors.

Referring to Fig. 125, the oil enters the system in the compression area of the rotor where it mixes with the cold refrigerant vapors and diffuses throughout the internal cavities of the compressor. As it spreads, contact is made with all frictional surfaces of the helical rotors. It is this method of applying lubrication that made possible the direct contact between the male and female rotors and eliminated timing gears.

The mixture of refrigerant gas and oil continues along the path of discharge enveloping the electric motor windings, after which it enters the hermetic oil separator where the gas is relieved of the oil before its discharge into the condenser. The manufacturer's (Dunham-Bush, Inc.) description of motor cooling under both full load and part load conditions is presented as follows:

Motor Cooling. Motor cooling is so effective that the hermetic motors are actually selected on the basis of torque requirements rather than maximum winding temperatures as in most modern hermetic gas-cooled reciprocating and liquid refrigerant-cooled centrifugal machines. Maximum winding temperatures ranging from 150 to 185 F (or 66 to 85 C) are appreciably lower than the set trip temperature (221 F = 105 C) of the solid-state protectors embedded in the "hot spot" of the winding, thus avoiding nuisance trip-outs even under extreme operating conditions. These winding temperatures are quite conservative for a Class B (266 F or 130 C) insulation, since the life of an insulation system roughly doubles for each 18 F reduction. The motor cooling method is rather unique

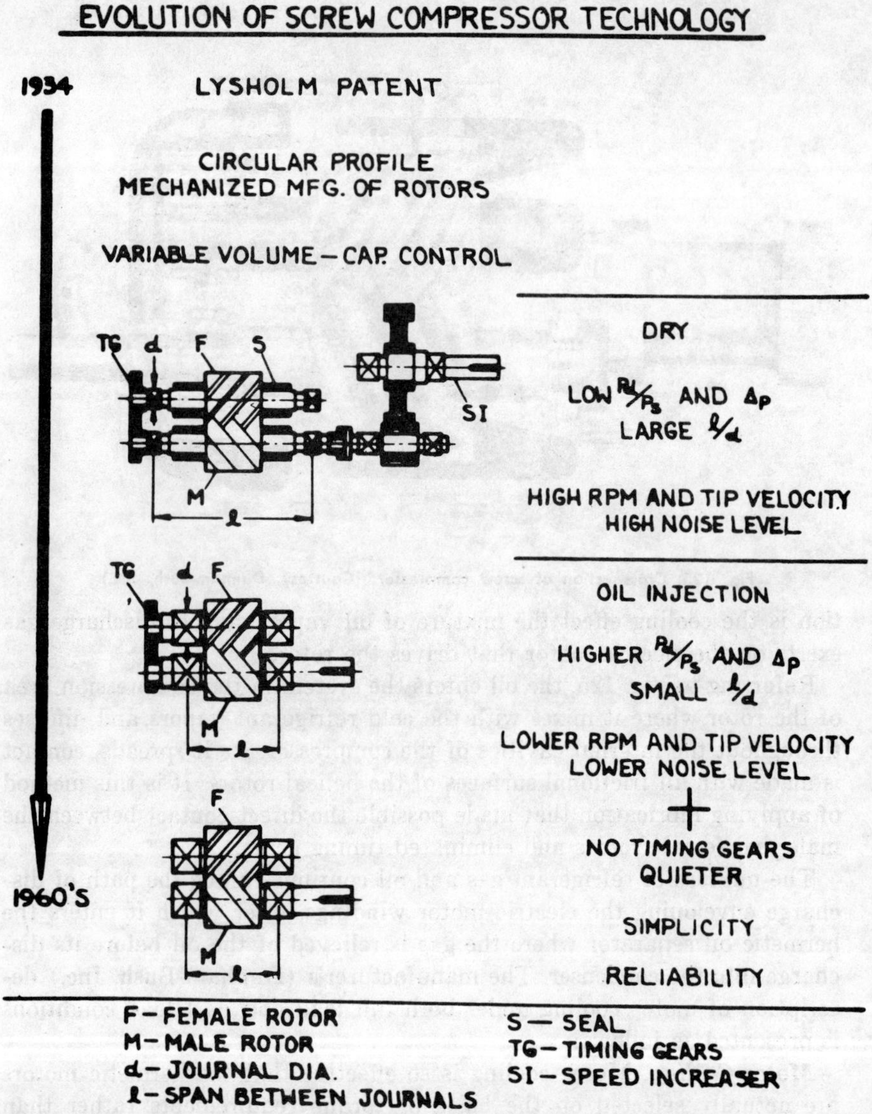

Fig. 126. Evolution of screw compressor technology. Note the absence of timing gears in lower unit. (Courtesy, Dunham-Bush, Inc.)

for large units of this kind and may deserve a few words of explanation. As shown in Fig. 125, the *dense high pressure gas-oil mist mixture* is directed from the compressor through the motor to uniformly blanket and cool the stator and rotor. As an important fringe benefit, *the motor serves as a first-stage oil separator* removing, by centrifugal action, the bulk (over 95%) of the entrained oil from the gas; the remaining fine oil mist is separated in a special demister section downstream of the motor. Although a similar motor cooling approach has been used successfully with sliding-vane rotary compressors for household or residential appliances, it may seem rather surprising at first to rely on the discharge mixture for cooling a large hermetic motor. The low winding temperatures actually measured in oil-injected screw compressors can be explained in this manner:

(*a*) *Full Load Conditions.* With large oil quantities, by weight ≅ 30% of full load refrigerant flow rates but less than 1% of suction cfm, the compressor discharge mixture ranges from 35–45 F above condensing temperature, i.e., from 125–155 F in the present maximum application range of these chillers, and increases by less than 5 F or to a maximum of 130–160 F as the mixture absorbs the heat equivalent of the motor losses. Furthermore, the heat transfer coefficient between the dense, high mass velocity gas-oil mixture and the motor is apparently so high that the winding never exceeds the leaving oil-gas mixture temperature by more than 20–25 F at full load, the maximum winding temperature thus ranging from 150–185 F.

(*b*) *Part Load Conditions.* The oil, i.e., the main coolant flow rate remains constant, whereas the motor kw and losses decrease and the winding temperature is *lower than at full* load. This is exactly the opposite of what occurs in conventional suction gas-cooled large hermetics that tend to overheat at part load or low suction pressures due to decreasing coolant flow rate. This unique characteristic is particularly desirable for low temperature applications of oil-injected hermetic, as well as open-type, screw compressors. (For instance, in laboratory tests a model 200-10 open-type machine could be run single-stage without ever exceeding 200 F discharge temperatures with R-22, at 90 F return gas temperatures down to −80 evaporating at 110 F condensing in fully loaded position, as well as down to −50 evaporating at less than 10% capacity.)

Since the oil itself is cooled by condenser water, the motor is ultimately cooled by the refrigerant cycle heat sink. Thermodynamically, this is more efficient than using liquid refrigerant evaporating at low side saturation temperature (about 33 F) and then compressing the vapor generated to discharge pressure without useful refrigeration effect (low pressure refrigerant liquid-cooling produces higher irreversible losses). Although the gains in efficiency, or horsepower per ton, attainable with high-side cooling, as compared to low-side liquid refrigerant cooling, are less significant at normal full load design points (of the order of 1 to 3%), they become more significant at part loads. In normal comfort cooling applications, the system operates at reduced load most of the time, and the advantage of high-side cooling may be more significant over a full operating season than would at first appear from an evaluation of the effect of motor cooling at design load only.

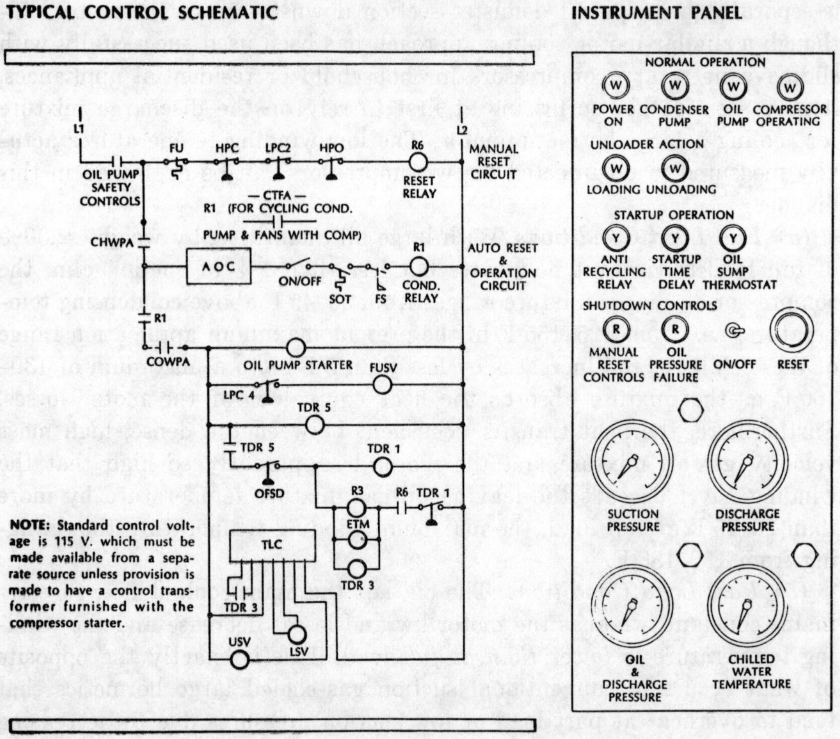

Fig. 127. Instrument panel for screw compressor showing schematic control arrangement. (Courtesy, Dunham-Bush, Inc.)

Control Components

HPO—High Pressure Oil Cut-Out Switch: Stops compressor if oil pressure exceeds a set point.

HPC—High Pressure Cut-Out Switch: Stops compressor motor if condensing pressure reaches excessively high set point.

CMT-3 and OPMT-3—Compressor Motor and Oil Pump Protectors (not shown on diagram): Inherent solid state motor winding protection to stop motors if temperature reaches a set point.

FU—Freeze-Up Switch: Pre-set safety switch shuts down compressor if chilled water temperature falls below set point.

FS—Flow Switch: Stops compressor if there is insufficient chilled water flow through evaporator. Supplied unmounted for field installation and electrical interconnection to control terminals.

SOT—Oil Sump Thermostat: Controls immersion type oil heater to maintain oil at proper temperature before start up and during prolonged off cycles.

TDR-1—Oil Failure Time Delay Relay: Prevents compressor from starting if oil pressure is inadequate and similarly prevents compressor shut down if momentary fluctuations in oil pressure are experienced.

OFSD—Low Oil Pressure Switch: Shuts down machine if oil pressure falls below pre-set minimum.

TDR-5—Compressor Starter Time Delay Relay: Delays compressor start until oil circulates through system for predetermined time period.

TLC—Temperature/Load Controller: Solid state controller for manual and automatic operation of capacity control system. Load limiter is manually adjustable control designed to limit maximum current draw to between 50% and 100% of full load.

CHWPA and COWPA—Chiller and Condenser Water Pump Auxiliary Contacts: Normally open contacts to prevent compressor from starting unless chilled water pump and condenser water pump have started.

CTFA—Cooling Tower Fan Auxiliary Contacts.

AOC—Automatic Operating Controls: Low water temperature switch; anti-recycle timer (senses pre-set low chilled water temperature condition that will initiate normal shut down and subsequently restart machine after pre-set recycle time has elapsed).

LPC-2—Compressor Low Pressure Control: Low pressure "safety" control.

LPC-4—Unloader Low Pressure Control: Maintains minimum operating suction pressure above pre-set minimum value by unloading compressor. Control overrides temperature controller and maintains normal evaporator pressure level during transient conditions.

FUSV—Fast Unload Solenoid Valve: Energized by the unloader low pressure control. Also unloads compressor during start up and shut down.

LSV and USV—Load and Unload Solenoid Valves: Energized by temperature/load controller to load or unload machine as chilled water temperature or compressor motor current conditions dictate.

ETM—Elapsed Time Meter: Indicates total accumulated time that unit has operated.

Screw Compressor—Summary

Some of the important factors associated with the design and operation of screw compressors follow:

(*a*) Positive displacement by rotary motion

(*b*) Helical lobes and interlobes are employed as compressing elements (Fig. 119)

(*c*) Movement of gas through housing is circumferential (Fig. 120)

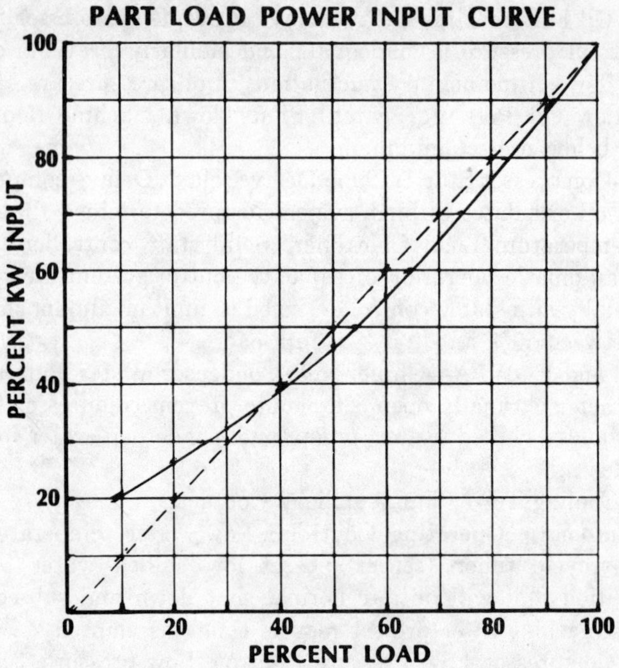

Fig. 128. Power input vs. load variation with screw compressor. (Courtesy, Dunham-Bush, Inc.)

(*d*) Compression is developed internally (Fig. 121)
(*e*) Uniform gas flow
(*f*) Even torque
(*g*) No clearance volume
(*h*) Speed range is 1500 to 24,000 rpm (depending on compressor size and duty)
(*i*) Load capacity is up to 800 tons and over
(*j*) Refrigerants used are R-12, R-22, R-502 and R-717 (NH_3)
(*k*) Lubricating oil is 300 SUS @ 100 F, having appropriate cold test to meet evaporator conditions
(*l*) Built-in automatic capacity control via axially sliding valve in flow-path (Fig. 124)
(*m*) Unloading limits are 100% at full load down to 10% of full load by modulation
(*n*) Suction throttling to 0% capacity at constant rpm
(*o*) Oil circulation remains constant regardless of capacity level (0% to 100%)
(*p*) Effective stroke of compressor (volume discharged to condenser) varies according to load via sliding valve
(*q*) Flexibility in condensing temperatures, both in low and high ranges

(*r*) Can be applied to remote evaporators and condensers, as compared to the limitations of dynamic compressors
(*s*) Applicable for use with internal or external speed increasers
(*t*) Application ranges from air conditioning to commercial freezing processes
(*u*) Two basic types—open and hermetically sealed. The newer open types of screw compressors are now available in the oil flooded version without timing gears. These are quieter than the older dry types with timing gears and also frequently with speed increasers, but not as quiet as the hermetic oil flooded version. In the open type machine, the major source of noise is from the electric motor
(*v*) Oil injection type is relatively free of vibration and has a low noise level
(*w*) Low compressor weight per ton of refrigeration produced
(*x*) Applicable both as boosters and high stage compressors and can be staged in series or cascaded as with other types of compressors
(*y*) Some screw compressors have been built as self-contained two-stage units for air and gas compressor service, which can also be applied to refrigeration processes, particularly for ultra-low temperature requirements, the latter in lieu of staging two different compressors

Rotary Compressors

Like the reciprocating type, the rotary compressor is a positive displacement unit. In place of a reciprocating piston, compression is accomplished by a radially bladed rotor that revolves within a cylinder of much larger diameter and with which it is eccentric. The blades, of which there are a number spaced equidistant around the rotor, are inserted in radial slots and are free to move in and out. As the rotor turns, centrifugal force acts on the blades and causes them to move radially outward against the cylinder wall. Because the rotor is eccentric with the cylinder but concentric with the shaft on which it is mounted and keyed, the blades move in and out of the slots as they follow the contour of the cylinder wall. The diameter of the rotor and its relative axis with that of the cylinder are such as to cause the rotor to very nearly touch the cylinder wall, usually at a point located at the bottom. At this point, the blades are forced back into their slots to their maximum depth, after which they emerge gradually until they reach their maximum allowable height at the top.

On each end of the rotor are closely fitted end plates or cylinder heads that also carry the rotor shaft. The suction vapor enters the compressor housing, which encloses the rotor-cylinder assembly, on one side and is discharged at a higher pressure at the opposite side. As the vapor enters the housing, it is drawn through a suction port in the cylinder wall, entrapped within the spaces between the blades, and compressed as the surfaces of the rotor and the cylinder wall converge to diminish the space volume. The compressed vapor is forced out through a discharge port located on the opposite side of the cylinder and thence into the high-pressure line through the housing exit. The vapor is prevented from being forced out the ends of the rotor by the cylinder heads.

To prevent leakage, stuffing boxes or rotary seals are provided and placed on both sides of the rotor. Any number of blades can be used, usually four or eight. They are made of softer material than that of the cylinder wall, so that any wear will occur on the more easily replaceable member. Larger rotary compressors of the industrial type are jacketed and water cooled.

In smaller rotary compressors, coil springs are placed in the slots behind the blades for the purpose of obtaining sufficient radial thrust to provide a tight seal between the end of the blade and the cylinder wall at all times.

Another type of rotary compressor having a single blade and employed principally in the domestic field is shown in Fig. 130. It consists of a steel ring mounted on a cam shaft and a steel cylinder, the diameter of which is greater than the ring. As the shaft on which the cam is located rotates concentrically, it causes the ring to revolve eccentrically about the same axis. Because the distance between the center of the concentric shaft and the perimeter of the ring at the point of greatest departure is equal to the radius of the cylinder, the two are in constant contact. The ring, being appreciably smaller in diameter than the cylinder, leaves a crescent-shaped space around the remaining perimeter of the ring between the two. The ring, being free on the cam, rotates as it rolls on the cylinder wall in a gyratory manner. The cylinder is located within a hermetically sealed housing partially filled with oil, which also houses the electric motor that drives the shaft. End plates are installed to enclose the cylinder at both ends.

The cylinder wall is drilled in two places in close proximity along its circumference to allow for the entrance of low-pressure vapor and its discharge at a higher pressure. Between the two drilled holes or ports a slot is cut across the entire width of the cylinder wall parallel with its axis. From the slot a single blade, backed by a coil spring, emerges to press against the ring. It is free to move in and out with the contour of the ring as the latter revolves. Sometimes the blade is backed by oil pressure.

When the suction port is uncovered, low-pressure vapor is drawn in the cylinder within that portion of the crescent-shaped space between the blade and the point where the ring and the cylinder wall are in contact. As the ring continues to revolve, it covers the suction port and traps the vapor within the diminishing space between the ever-changing contact point and the blade, forcing it out under higher pressure through the discharge port which is located now on the near side of the blade. At the instant the ring uncovers the suction port, low-pressure vapor enters the cylinder and continues to enter until the port is again covered. Similarly, at the instant the ring uncovers the discharge port, high-pressure vapor leaves the cylinder and continues to be forced out until the port is again covered.

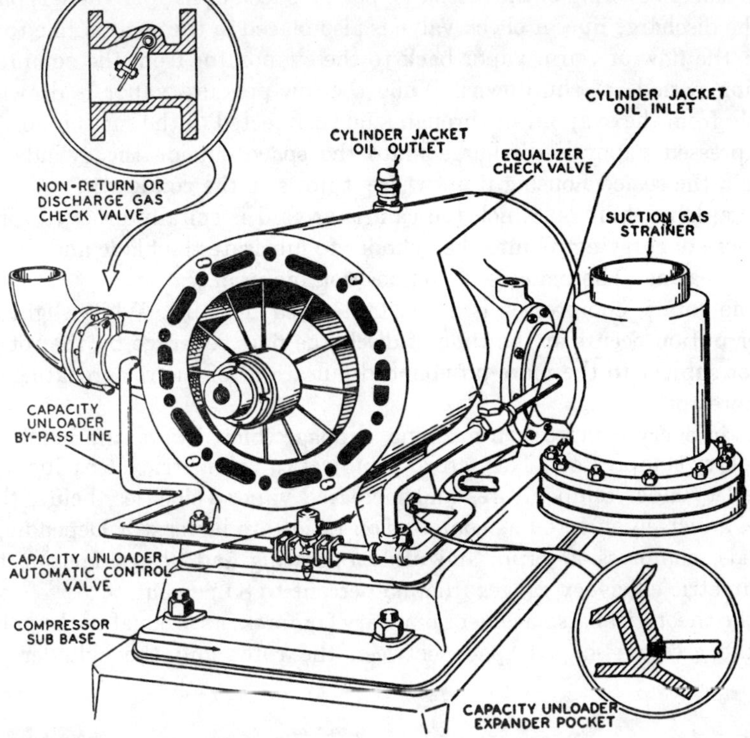

Fig. 129. Rotary compressor. Industrial type with vanes in rotor. Note position of non-return check valve and capacity control arrangement. End view with cylinder head removed. (Courtesy, Freezing Equipment Sales, Inc.)

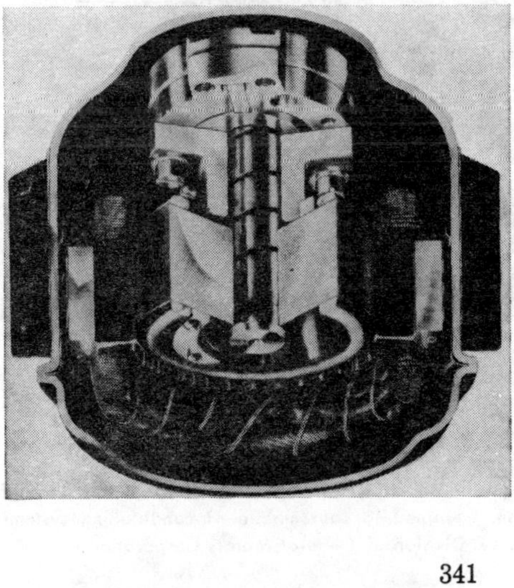

Fig. 130. Rotary compressor. Type used in domestic refrigerators and freezers. Blade in cylinder (Meter-Miser). (Courtesy, Frigidaire, Division of General Motors Corporation.)

To prevent the high-pressure vapor from flowing back into the cylinder with the uncovering of the discharge port, a check or flapper valve is placed in the discharge line. A check valve is also placed in the suction line to prevent the flow of warm vapor back to the evaporator from the compressor during periods of shutdowns. While the low-pressure vapor is drawn directly from the evaporator through a line connected to the suction port, the compressed vapor is discharged into the space outside the cylinder but within the sealed housing from where it flows tc the condenser.

The drive shaft on which the cam is ground is spirally grooved for the purpose of carrying oil into the cylinder to lubricate the blade and the cam shaft bearing. The blade is of composition material.

The rotary compressor is a constant-speed machine. While slight flow interruption occurs at the time of discharge port coverage by the rotor, it is not subject to the more pronounced pulsations of the reciprocating type compressor.

As in every compression machine, slippage (blow-by) occurs in all types of rotary compressors. They are also subject to volumetric losses due to elevated cylinder temperatures and increased vapor velocities, being, therefore, adversely affected as compression ratios are increased. Depending on speeds, compression ratio, and other operating and design factors, their volumetric efficiency ranges from 65 percent to 80 percent.

The theoretical displacement of rotary compressors is equal to the volume of the crescent-shaped space between the rotor and the cylinder wall.

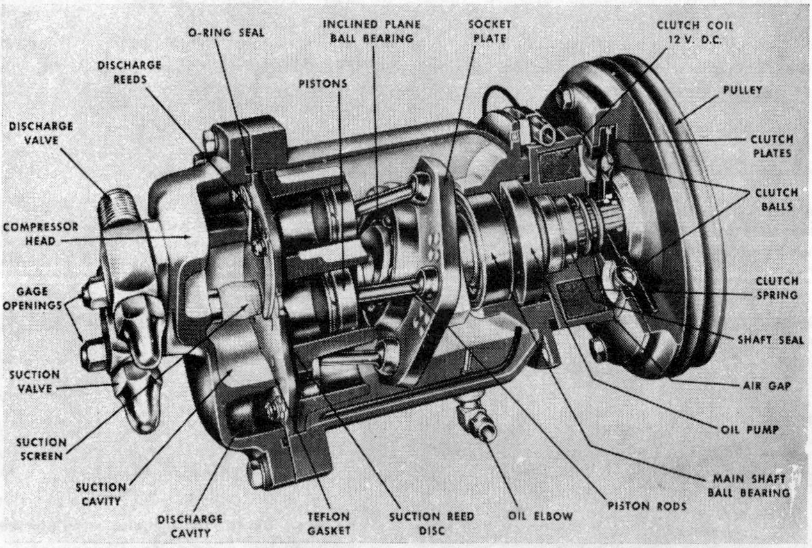

Fig. 131. Axial compressor. Pistons reciprocate in direction parallel to axis of drive shaft through inclined plane ball-bearing and socket plate. Designed for automobile air-conditioning systems. Belt driven from fan shaft. (Courtesy, Frigidaire, Division of General Motors Corporation.)

Therefore, it is equal to the difference in total cylinder volume and that of the rotor and shaft multiplied by the revolutions per minute of the rotor, as expressed by the following equation:

$$\text{Theoretical rotor displacement} = \text{rpm}\left(\frac{\pi D^2 \times W}{4} - \frac{\pi d^2 \times w}{4}\right)$$

where D = diameter of cylinder
d = diameter of rotor
W = depth of cylinder
w = depth of rotor

As the rotor depth is usually equal to the cylinder depth, then $W = w$.

Rotary compressors are employed in the industrial field as boosters for multi-stage systems where, because of the low temperatures involved, discharge pressures are correspondingly low. Their advantage over the reciprocating type lies in greater volume displacement capacity in handling low-density vapors. They are capable of maintaining discharge pressures above 80 psi, and therefore render efficient service in domestic refrigerators with such refrigerants as Freon-12, sulfur dioxide, and methyl chloride.

Gear Compressors

Not widely used, the gear compressor is the only positive displacement machine free of discharge pulsations. It consists of two meshing gears, mounted and keyed on their own shafts, which rotate within a gas-tight housing. Power is transmitted to one gear, which in turn drives the other. Both gears are fitted closely to the housing so that only slight clearance is maintained between their outermost surface (addendum circle), and the circular contour of the case through approximately 180°. A lubricating oil film completes the seal in that area.

The low-pressure vapor enters the case at the bottom and is carried between the gear teeth to the top of the case where it is forced out into the discharge line. The vapor is displaced from between the teeth by meshing teeth of the mating gear.

To maintain uniform center and eliminate end thrust, double-helical gears generally are used. The herringbone effect of such gears gives greater strength because of extended line contact between opposing teeth and allows additional teeth to mesh for greater load distribution. Such an arrangement also provides better sealing than would be possible with the usual type of spur gear. A check valve is placed in the suction line to prevent leakage back to the evaporator during periods of shutdown.

Like those of the rotary and reciprocating types, gear compressors can be arranged to operate as a duplex unit; i.e., two compressors placed on either side and directly connected to an electric motor. They are capable of handling large refrigeration tonnage.

Centrifugal Compressors

Centrifugal compressors function according to the basic laws of centrifugal force and the effect of that force on gases.

Centrifugal force may be defined as the amount of radial force generated within a rotating mass resulting from its tendency to travel in a straight line according to the law of inertia. To compel a mass to move in a circular path about a given axis requires a constant force acting toward the center, known as centripetal force, equal to the effort tending to maintain straight-line motion. The magnitude of centrifugal force is a function of several variables, viz.: It is (*a*) directly proportional to the mass, (*b*) inversely proportional to the radius of curvature, (*c*) directly proportional to the square of the velocity of the mass.

Fig. 132. Duplex arrangement. Two reciprocating compressors directly connected to and driven by a single motor. Rotary and gear compressors may be arranged similarly. (Courtesy, Carrier Corporation.)

It may be expressed mathematically by the fundamental equation, embracing the above variables, as follows:

$$\text{Centrifugal force} = \frac{\text{weight} \times \text{velocity}^2}{g \times \text{radius}} = \frac{WV^2}{gr} = \text{pounds}$$

where W = weight in pounds (mass)
V = velocity in feet per second
r = radius of curvature in feet
g = gravitational acceleration; 32.2 ft per sec²

Centrifugal machines may be divided into three groups based on their behavior in relation to their axis or center of rotation.

The first group is comprised of centrifugal machines whose rotating mass is permanently fixed to their axis and whose weight and radius remain constant. The only variable in the above equation for a given machine in this group is the velocity of the rotating mass. The flywheel, a pail of water whirled about in a circle, or a locomotive rounding a curve are all examples of centrifugal machines included in this group.

The second group consists of those machines whose radius is variable within the limits governed by mechanical design. With such a machine, any increase in velocity causes an outward movement of the weight, thereby increasing its radius also. Because any variation in centrifugal force is directly proportional to the square of the velocity and inversely proportional to the radius, the net effect is a gain in centrifugal force. Variables in the basic equation for machines of this group include velocity and radius. The ball governor and the mercury clutch fall in this category. Incidentally, mercury is used in certain type clutches because of its greater weight compared with other substances, creating a higher pressure in the rotating flexible element (diaphragm), causing that member to expand and contact its opposing number. Contact here is gradual because the mercury does not cause maximum expansion until the rotating element reaches full speed. Such progressive friction tends toward smoother clutch engagement.

The centrifugal pump and the centrifugal compressor comprise the third group in this discussion. The rotating member of either the pump or the compressor, which is in reality a gas pump, is an impeller consisting of a series of radial columns or vanes through which fluid, entering their axial inlet, is free to flow to their peripheral outlet, where it is discharged. The impeller is mounted on and rotated by a shaft to form an integral component which constitutes the only moving part of the centrifugal pump or compressor. As the impeller rotates, the fluid present in the space surrounding the shaft is literally scooped up by the blades whose axial extremities project a short distance into the cavity or "eye" formed by the shaft and the back disk or shroud of the impeller attached to the shaft. As the fluid enters the blades at low velocity, it is forced radially outward by centrifugal action and finally thrown from the peripheral outlet at high velocity.

The impeller is surrounded by diffuser vanes, which in turn are encased within a voluted or scroll-shaped housing that fully encloses it and collects the fluid as it emerges from the diffuser vanes where it is thrown from the peripheral outlets. The scroll is so designed as to maintain constant static-velocity head ratio as additional gas enters along the line of peripheral discharge.

Static pressure is developed within the impeller due to centrifugal action. The resulting force acts to compress the fluid much the same as the upper layers of a fluid column tend to compress the layers below. The fact that the compressive effort succeeds in the case of gases and does not with liquids spells the basic difference between the centrifugal pump and the centrifugal compressor.

Additional pressure is developed within the housing as a result of the high velocity of the fluid leaving the impeller. Therefore, the total pressure developed within a centrifugal compressor is equal to the sum of the static pressure and the velocity pressure.

To better describe the action taking place within a rotating gas column or vane, consider the water tank discussed earlier in this text. It was shown that the unit pressure exerted at the base of the tank depended on the height of the water level, and that, the higher the level, the greater becomes that pressure. This is in accordance with the liquid column equation that

$$\text{Pressure} = \text{height} \times \text{density}$$

It is true that the pressure at any point above the base would be less than the base pressure by an amount corresponding to the distance between that point and the base. A column of water 50 ft high having a base pressure of $\frac{50}{2.31} = 21.65$ psi would have a pressure 10 ft above the base of $21.65 - \frac{10}{2.31}$, or 17.32 psi. Water, being incompressible, would have a fairly constant density at any point in the column. This means that a cubic foot of water would weigh 62.4 lb measured either at the base, 10 ft above, or at the top. This would not be true if the column were gas instead of liquid, as will be shown.

First, an explanation concerning the expression "head."

In this discussion, the height of the water level in the tank was said to be 50 ft. The base pressure corresponding to a column of water 50 ft high and 1 in. in area was found to be 21.65 psi. It was also shown that a definite relationship exists between the water-column height and the pressure at any point in the column. For water, that relationship has a constant ratio of 2.31 ft to 1 psi. For mercury, the ratio would be constant at 2.04 in. to 1 psi. This rule holds true for any liquid, all of which have a definite height-pressure ratio corresponding to the density of the individual liquid. The height corresponding to unit pressure exerted by any fluid is known as the "head" in terms of feet of the individual fluid and the unit pressure corresponding to the height of any fluid is called the "head pressure" in terms of psi of the individual fluid. When the column is stationary, the head is referred to as being "static," and the pressure is termed "static pressure." The condition here is one having potential energy differently divided between energy of position and energy of pressure according to elevation.

If the uppermost layer of water in the tank were allowed to drop unrestricted to the base, assuming nonfrictional conditions, it would progressively gain speed at a rate of 32.2 fps per second through the entire falling distance, which in this case is 50 ft. It would have a definite velocity at any point in the course of travel corresponding to its relative position with respect to the top, obtaining a maximum speed at the instant of reaching the

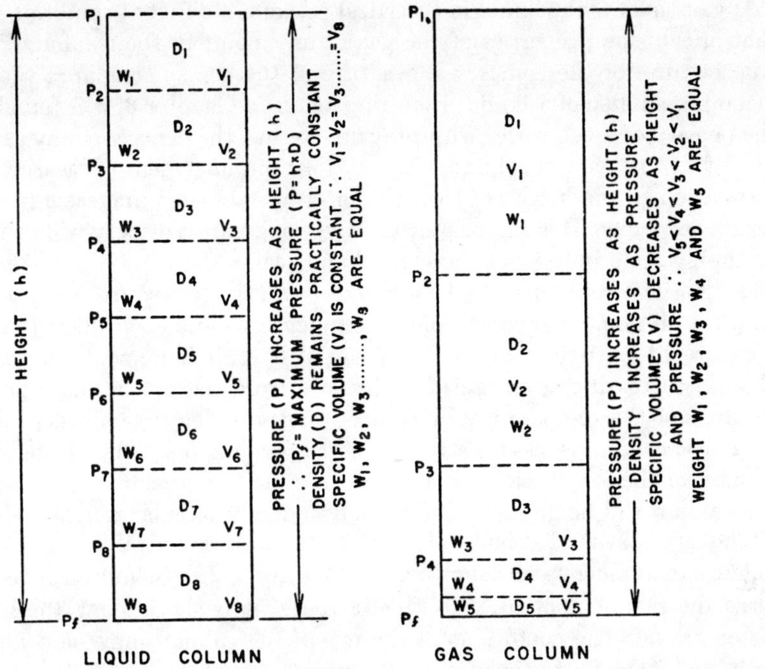

Fig. 133. Specific volume change in gas column vs. constant specific volume in liquid column. Each column division equals equal weight of substance for that column.

base. According to Galileo's law, "all bodies regardless of weight fall with equal acceleration and their final velocity varies only as the height from which they fell." This means that, regardless of the substance in the tank—whether water, mercury, gas, iron, or paper—disregarding friction, their final velocities would be the same. *The height in feet through which a body must fall to attain a given velocity is called the "velocity head," and is independent of weight.* The velocity head is equal to the static head causing that velocity. The sum of the static head and the velocity head are always constant in any flowing fluid.

The velocity head for any substance may be determined by the basic equation

$$h_v = \frac{V^2}{2g} = \frac{V^2}{2 \times 32.2} = 0.0155\ V^2$$

where h_v = velocity head in feet
V = velocity in feet per second

Because head may be converted to velocity or velocity to head, this equation may be transposed as follows and remain true, viz.:

$$V^2 = 2gh_v$$

or

$$V = \sqrt{2gh_v}$$

If gas replaces the liquid in a vertical column, a different condition exists that affects the properties of the gas at any point in the column. This is true because of the compressible nature of the gas as compared with the incompressibility of a liquid. Referring again to Chapter 3, it is found that the density of a gas varies with pressure so that the density of any gas employed in the vertical column will vary for each increment of measurement in relation to the height of the column of gas above it, increasing as that height increases. This means that the layer of greatest density will be found at the extreme bottom of the gas column, while the top layer will be the least dense. Where isothermal conditions exist, the temperature will be constant at all points along the column. Under adiabatic conditions, the temperature at each increment will vary with each increment of pressure change, increasing as the relative distance from the top becomes greater. Because of the temperature difference, then, the density anywhere in the adiabatic column is less than that at corresponding points in an isothermal column of the same height. It is also true that the pressure anywhere in a gas column will be greater with fluids of higher molecular weight.

The gravitational column, Fig. 133, as it is technically described, is somewhat analogous to conditions in a reciprocating compressor as concerns the ratio of potential to kinetic energy anywhere along the line of piston travel. The conditions at the top of the column may be compared with those at the initial point of compression stroke, and likewise for each point of downward progression where the compressive force developed by the weight of the gases above it is definitely related to some point along the line of piston travel during the compression stroke. It is obvious from the above statement that the changes in properties of a vapor under compression in a reciprocating machine will also occur in a gravitational column and with the same degree when the conditions of elevation causing such changes are similar. Therefore, the relationship of pressure, volume, density, and temperature remains the same as that in a positive displacement machine, both isothermally and adiabatically.

The ratio of compression in a gas column is found by dividing the pressure exerted on the bottom by the pressure at the top, and may be increased or reduced by lengthening or shortening the column. This means that the compression ratio varies directly as the difference in head, as in other compression machines.

By substituting a short column containing a liquid or a gas for the long vertical column above, and whirling it radially about in a circle, a pressure condition similar to that in the vertical column can be developed. It means simply that mechanical work through centrifugal force is replacing that work performed previously by gravity. To compensate for difference in height, it would be necessary merely to increase or decrease the speed of whirling. If some provision were made to continually feed gas or liquid into the axial opening of the radial column during the whirling action and allow the fluid to be discharged at the opposite end, an implement similar to the impeller of a centrifugal pump or compressor would be devised.

To develop a head comparable with that of a given gravitational column, it is necessary to rotate the radial column at a speed capable of imparting velocity to the discharge gas equal to the final velocity of a falling body through the vertical distance represented by the head. Therefore, the head is dependent solely on the angular velocity of the discharge outlet, which in a centrifugal compressor is the tip or peripheral speed measured in degrees or radians.

$$\text{Angular velocity} = \frac{\text{angle}}{\text{time}} = \frac{\text{degrees}}{\text{seconds}} \text{ or } \frac{\text{radians}}{\text{seconds}}$$

where 1 radian $= \dfrac{360°}{2\pi} = 57.3°$ (approx.)

The angular velocity multiplied by the radius equals the tangential velocity in feet per second when radius is given in feet. Applying the fundamental equation, then

$$h = \frac{V^2{}_h}{2g}$$

and

$$V_h = \sqrt{2gh}$$

The impeller of a centrifugal compressor, as stated above, develops both static and velocity head, the sum of which represents the total head developed. Because the static head is also a function of the peripheral speed, then

$$\text{Static head} = \frac{V^2}{2g}$$

The total head produced by the impeller becomes

$$\text{Total head} = \text{static head} + \text{velocity head}$$

$$h_t = h_s + h_v$$

Therefore, $h_t = \dfrac{V^2}{g}$ when the full radius of the impeller is considered.

Because of the necessity of allowing some area through which the suction gas must pass to enter the vanes, it is necessary that the column be made somewhat short of center. This means a loss of head represented by the velocity at the point where the vane commences. The power requirements remain the same as if the flow were initiated at the center, so that there is a net loss in the efficiency of the rotor by reason of the axial inlet or "eye."

The power required to produce a differential in head depends on the density of the gas handled, which also governs the pressure produced by centrifugal compressors. Work by definition being the effort required to accomplish physical change as expressed in foot-pounds, is determined for centrifugal machines on the basis of head and the weight of the fluid acting against a given head. Work, then, is related to head, as shown by the equation

$$\text{Work} = \text{weight} \times \text{head}$$

$$W = M \frac{V^2}{2g}$$

The head at any point in the gas column, which would also be the head at any point in the impeller vane, is a function of temperature (T), specific heat at constant pressure and constant volume of the fluid involved, and the pressure differential between the points in the column under consideration. It is determined by the adiabatic relationship of the foregoing factors according to the equation

$$h_2 - h_1 = \left(\frac{y}{y - 1}\right) RT_1 \left[\left(\frac{P_2}{P_1}\right)^{\frac{y-1}{y}} - 1\right]$$

where $h_2 - h_1$ = height differential
h_2 = final height
h_1 = initial height
R = gas constant
T_1 = initial temperature, abs
P_2 = final pressure, abs
P_1 = initial pressure, abs
$y = \frac{C_P}{C_V}$ = ratio of specific heats

and $R = \frac{PV}{T} = 778\,(C_p - C_v)$

The temperature at any point in the column is determined as in any compression system operating under adiabatic conditions, viz.:

$$\frac{T_2}{T_1} = \left(\frac{P_2}{P_1}\right)^{\frac{y-1}{y}}$$

where T_2 = final temperature, abs

From the above discussion, the following laws relating to centrifugal compression may be established:

(*a*) The total head produced by centrifugal action is the sum of the static head or pressure built up within the impeller and the velocity head, or pressure resulting from the tangential throw of the impeller.

(*b*) The head produced by a centrifugal compressor is independent of the impeller radius or its speed of rotation (angular velocity).

(*c*) The head produced by a centrifugal compressor is dependent on the peripheral velocity of the impeller.

(*d*) The head is independent of the fluid density. This means that the same amount of energy is required to impart the same velocity to a given weight of any substance and that 1 lb of liquid has the same amount of energy as 1 lb of gas, provided that their velocities are equal.

(*e*) Pressure developed within a centrifugal compressor is proportional to the fluid density.

(*f*) Total head produced is twice that produced by velocity alone in a radial-bladed (straight-column) impeller $\left(\text{viz., } h_t = \frac{V^2}{g}\right)$.

(*g*) Power requirements increase as the weight per unit volume of the fluid increases. This is borne out by the formula

$$\text{Horsepower (brake)} = \frac{\text{head, in feet of gas} \times \text{weight of gas handled per minute}}{33{,}000 \times \text{efficiency}}$$

(*h*) The power required per unit weight of vapor increases as the square of the velocity.

By following the action taking place at both the inlet and the outlet of the centrifugal gas column, or impeller blade, a means of establishing total head for any design becomes obvious. It will be shown that certain forces are developed by the rotating impeller, the magnitude of which depends on (1) the peripheral speed at both ends of the blade, (2) the angle of the inlet nozzle in respect to the direction of rotation, (3) the angle of the discharge nozzle in respect to the direction of rotation, and (4) the direction of discharge in respect to the direction of rotation.

In any mechanical system, the resultant force may be determined if the magnitude and the direction of all individual forces are known. One method is shown using a series of vectors to represent the direction of each force and their magnitude. For example, if two forces A and B act at point 1 with unequal magnitude, represented by the distances 1–2 and 1–4, and in different directions, as shown in Fig. 134, the resultant force would have a magnitude of C, represented by the distance 1–3 bisecting the parallelogram.

Substituting the forces shown in Fig. 134 with those developed at the outlet nozzle of an impeller blade, then

A = tangential velocity of the discharge nozzle = U_2

B = relative velocity of the gas leaving the blade in respect to the rotor = V_2

C = resultant velocity of the gas leaving the nozzle in respect to the casing = V_h

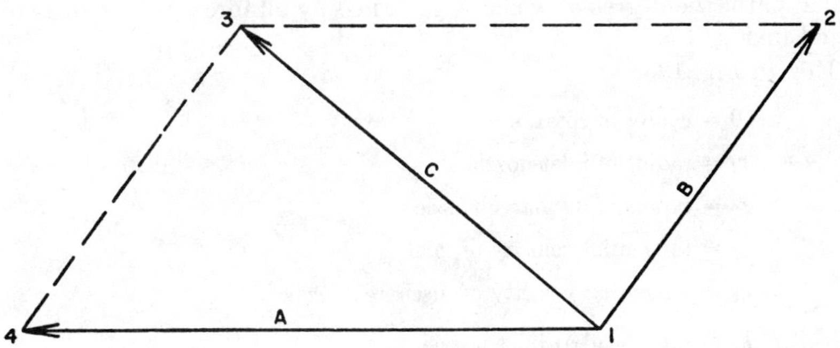

Fig. 134. Parallelogram of forces.

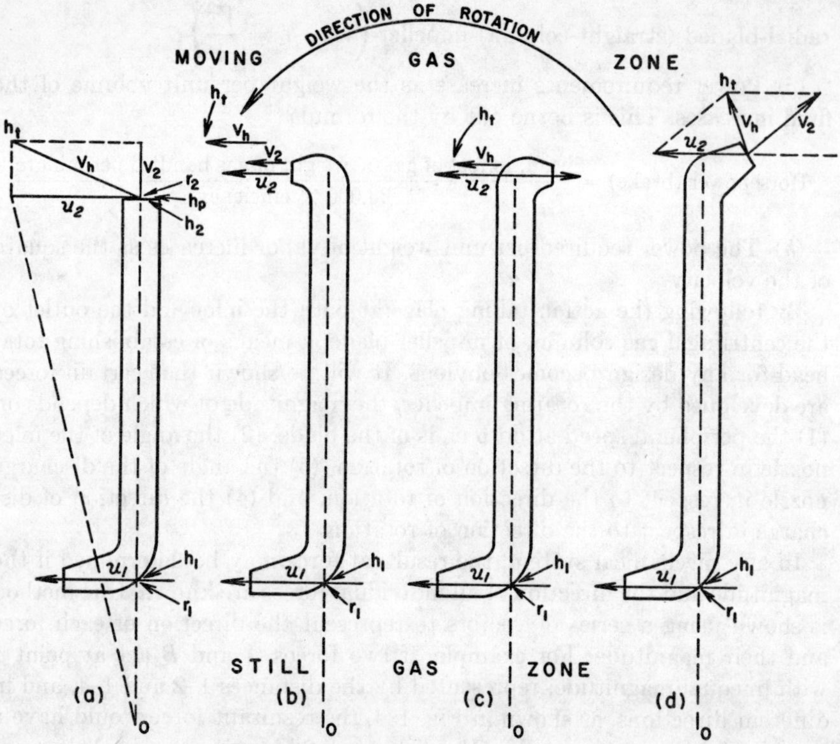

Fig. 135. Relative velocities of a gas leaving centrifugal compressor impeller, the blades of which are (*from left to right*) radial, forward; curved, backward; curved; angular. Parallelograms shown also apply to centrifugal fans, discusseed in Chapter 10.

While there are numerous combinations possible regarding the direction that V_2 might take as a result of nozzle or blade design, Fig. 135 (*d*) shows the conventional method of angular blading used in centrifugal compressors. Shown also are the resultant force V_h for (*a*) radial blades, (*b*) forward blades, and (*c*) backward blades. While the direction of a body leaving a rotating column appears to be radial, it actually travels at an angle to its radial path, the degree of which is governed by all forces involved, as indicated in Fig. 135.

Key to Fig. 135:

0 = center of rotation

r_1 = radius of inlet nozzle

r_2 = radius of discharge nozzle

u_1 = tangential velocity of inlet nozzle

u_2 = tangential velocity of discharge nozzle

h_1 = total head at inlet nozzle

Key to Fig. 135—(*Continued*):

h_2 = total head at outlet nozzle before discharge

V_2 = relative velocity of gas leaving nozzle in respect to impeller

V_h = relative velocity of gas leaving nozzle in respect to housing

h_t = total head of gas leaving nozzle

h_p = static head of gas leaving nozzle

In all cases, h_1 is equal to

$$\frac{(\text{tangential velocity of inlet nozzle})^2}{2g} = \frac{u^2{}_1}{2g}$$

In all cases, h_2 is equal to

$$\frac{(\text{tangential velocity of discharge nozzle})^2}{2g} = \frac{u^2{}_2}{2g}$$

However,

$$h_T = \text{velocity head} + \text{static (pressure) head.}$$

But static head $= h_2 - V_2$, which means the difference in total head at the nozzle before discharge minus the energy of velocity. And as

$$h_V = \frac{V^2{}_2}{2g}$$

therefore

$$h_p = \frac{u^2{}_2}{2g}$$

so that

$$h_T = \text{velocity head} + \frac{u^2{}_2}{2g} - \frac{V^2{}_2}{2g}$$

or

$$h_T = \frac{V^2{}_h}{2g} + \frac{U^2{}_2}{2g} - \frac{V^2{}_2}{2g}$$

However, V_h varies according to design, so that for a, b, c, and d in Fig. 135, h_T may be established as follows:

$$(a)\quad h_T = \frac{V^2{}_h}{2g} + h_p = \frac{u^2{}_2}{2g} + \frac{V^2{}_2}{2g} + \frac{u^2{}_2}{2g} - \frac{V^2{}_2}{2g} = \frac{u^2{}_2}{g}$$

$$(b)\quad h_T = \frac{V^2{}_h}{2g} + h_p = \frac{(u_2 + V_2)^2}{2g} + \frac{u^2{}_2}{2g} - \frac{V^2{}_2}{2g} = \frac{u^2{}_2 + u_2V_2}{g}$$

$$(c)\quad h_T = \frac{V^2{}_h}{2g} + h_p = \frac{(u_2 - V_2)^2}{2g} + \frac{u^2{}_2}{2g} - \frac{V^2{}_2}{2g} = \frac{u^2{}_2 - u_2V_2}{g}$$

$$(d)\quad h_T = \frac{V^2{}_h}{2g} + h_p = \frac{u^2{}_2 - u_2V_2 \cos \alpha}{g}$$

Disregarding friction, the shape of the blade does not affect the power required or the pressure produced as long as similar conditions exist at both inlet and outlet in all cases. While the discussion up to this point has considered only a single column in describing the action of centrifugal force on a compressible gas, the impeller is usually made up of a series of following blades. However, single stage compressors are available.

Where there is a limit to the speed with which the impeller may be rotated in any given design governed primarily by strength of materials, it

Fig. 136. Impellers and impeller shaft assembly of a centrifugal compressor. Note diminishing width of wheels from first to fifth stage to compensate for volume decrease (increased density). (Courtesy, Allis-Chalmers Manufacturing Company.)

follows that the employment of one impeller or wheel would restrict the pressure capabilities of the compressor within the maximum peripheral velocity of that one impeller. To increase the pressure range beyond the maximum produced by one wheel, it is necessary to equip the compressor with more than one, each of which is capable of receiving vapor at the pressure produced by a preceding wheel and imparting to that vapor additional pressure or kinetic energy. In this way, the total pressure increase between

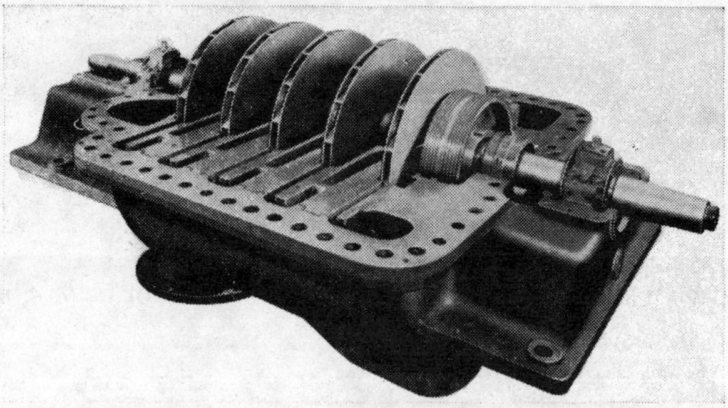

Fig. 138. Centrifugal compressor of the diffuser type with top of housing removed. (Courtesy, Allis-Chalmers Manufacturing Company.)

compressor suction and compressor discharge is imparted to the vapor in stages, the number of which is determined by the number of impellers. Therefore, a one-, two-, three-, five-, or six-stage compressor will have one, two, three, five, or six wheels, respectively.

The primary function of the centrifugal compressor is to effect a pressure increase in the vapor. Due to its operating characteristics, it imparts both static and velocity energy to the vapor leaving the impeller. It becomes necessary to convert a large part of the velocity energy to pressure without loss through impact. This is achieved by a series of stationary diffuser vanes installed around the periphery of the impeller. These vanes are curved in

Fig. 137. Centrifugal compressor impeller. (Courtesy, The Trane Company.)

the opposite direction to that of the discharged vapor and are so designed as to accomplish a gradual reduction in velocity pressure with a corresponding increase in static pressure without any appreciable loss in total energy. This gradual reduction of velocity pressure is achieved by a continuous increase in diffuser area leaving the impeller. In some types of centrifugal

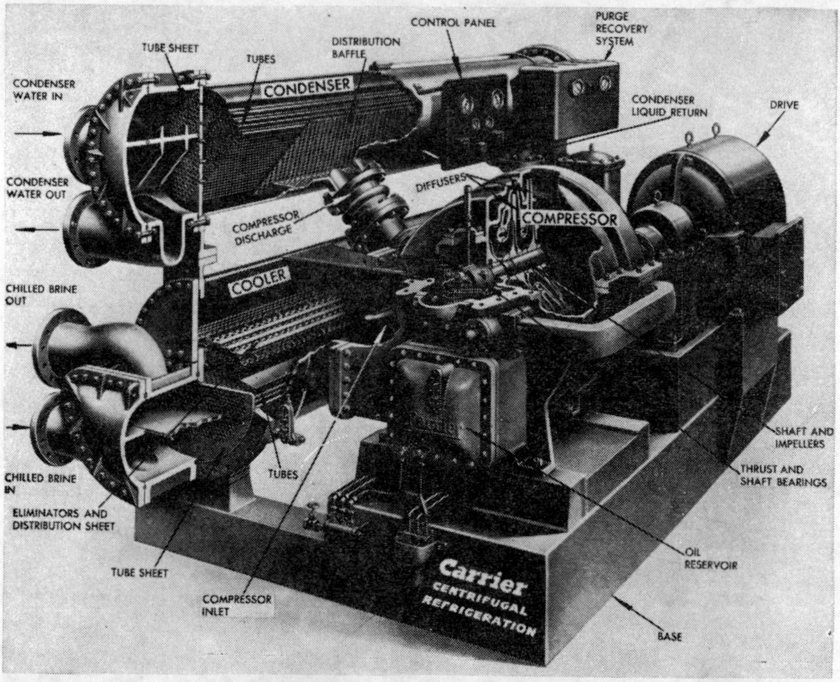

Fig. 139. Cut-away view of centrifugal compressor of the multi-stage diffuser design, showing its relative position to condenser, evaporator, and other parts. (Courtesy, Carrier Corporation.)

compressors, the gas discharges directly into one or more volutes to accomplish gradual velocity reduction. The latter are called volute type compressors. Still another type utilizes diffuser vanes followed by volutes.

A stationary diaphragm is employed to separate individual stages of multi-stage compressors. These diaphragms are augmented with a permanent type seal or packing to prevent leakage between stages. They also act to support vanes for guiding the vapor to the next higher stage in some types.

Fig. 138 shows some of the details of a multi-stage compressor with the rotor assembly mounted in its casing. Note that the wheels are made progressively smaller in the direction of vapor flow to compensate for the diminishing volume due to density increase resulting from the greater pressure head developed by each wheel.

The Centrifugal Compression System

The refrigeration system employing a centrifugal compressor operates according to similar laws and principles governing any other mechanical compression system. Due to certain peculiarities of the centrifugal compressor and the low pressures involved, some slight departures in refrigerant handling are necessary which cause some variation between the two.

The vapor leaving the evaporator is drawn through a suction line into the compressor, as in any other type of system. However, between the time it leaves the liquid body surrounding the tube bundle and enters the first-stage impeller, there are two major details involved peculiar to the centrifugal compression system.

First, to prevent entrained liquid from being carried over to the compressor, a set of baffles are installed in the shell-and-tube type evaporator immediately above and extending completely over the entire surface area of the liquid body (flooded system). Such baffle assemblies are known as *eliminators*.

Secondly, the suction line must be as short as possible and of large diameter in proportion to those employed in reciprocating systems. The reason for this lies in the larger volume of low-pressure refrigerant handled and the small allowable pressure drop between the evaporator and the compressor. Because the centrifugal system operates under high vacuum (low absolute pressure), even a small pressure drop substantially affects the suction temperature. Referring to R-11 tables, it is found that partial vacuum of 28.86 in. of mercury (0.520 psia) corresponds to a temperature of −50 F. At 29.19 in. (0.358 psia) of mercury, the corresponding temperature is −60 F. Therefore, there is a (−60) − (−50) or 10 F differential with only 0.520 psia − 0.358 psia = 0.162 psia pressure drop. Operating at higher suction temperatures, a small pressure drop still affects operating efficiency adversely to a considerable degree, as noted in the same table for higher pressures. For example, there is a 5 F difference with a pressure drop of

Fig. 140. Cut-away view of centrifugal compressor of the multi-stage volute design. (Courtesy, Worthington Corporation.)

2.931 psia − 2.555 psia = 0.376 psia at temperatures corresponding to 5 F and 0 F, respectively.

The vapor enters the compressor, is compressed to a higher pressure, and is discharged to the condenser, where it is reconverted to its liquid phase. The liquid leaves the condenser to enter the evaporator, but here, again, a variation in its handling exists that differs somewhat from the conventional reciprocating system. It applies only to multi-stage centrifugal compressors.

Upon leaving the condenser tubes, the liquid drains to a reservoir which acts as a trap for the warm refrigerant. When the liquid under higher pressure reaches a predetermined level, a ball float acts to allow its passage to an economizer chamber, where the pressure is lower than the condenser pressure, a portion of which flashes into vapor. The flash gas passes through eliminators located above the flash chamber and enters a pipe that carries it back to the compressor. However, this flash gas enters the compressor between the first- and the second-stage impellers, where it mixes with and reduces the temperature of the gas from the first stage. It is obvious from this action that the pressure in the economizer chamber is the same as that found between the first- and the second-stage impellers, which is below the condenser pressure but above the evaporator pressure. Its temperature, therefore, would correspond to that pressure.

The remaining liquid, cooled by the flash action, flows to a second float trap located below the condenser float chamber. From there it enters the evaporator via a conduit leading to a distribution sheet located below the tube bundle. Because a pressure reduction occurs here also, some additional flashing occurs reducing the temperature of the main liquid body. The remaining liquid acts as make-up for the main liquid body in the evaporator.

Through this method of interstage cooling of the liquid refrigerant, the capacity of the compressor is increased and the brake horsepower per ton is reduced.

Because the power required to operate centrifugal compressors rises with an increase in head, it is essential for purposes of economy to relieve the system of noncondensible gases that cause condenser pressures to go above that normally required for the operating temperature. This means that some method of noncondensible gas and air removal is essential.

Centrifugal compressor systems employ what is usually referred to as a purge-recovery system which not only allows for the removal of noncondensible gases but also performs that function with negligible loss of refrigerant. The purge-recovery system consists primarily of a small reciprocating compressor, an auxiliary condenser (air cooled), and an evacuator.

As in all systems employing refrigerants that are heavier than air and other noncondensible gases, the latter rises to the top of the condenser or receiver and, therefore, conveniently separates from the refrigerant as to permit their removal. In some systems, such gases collect in a purge drum and are vented to the atmosphere through a pressure-relief valve when the

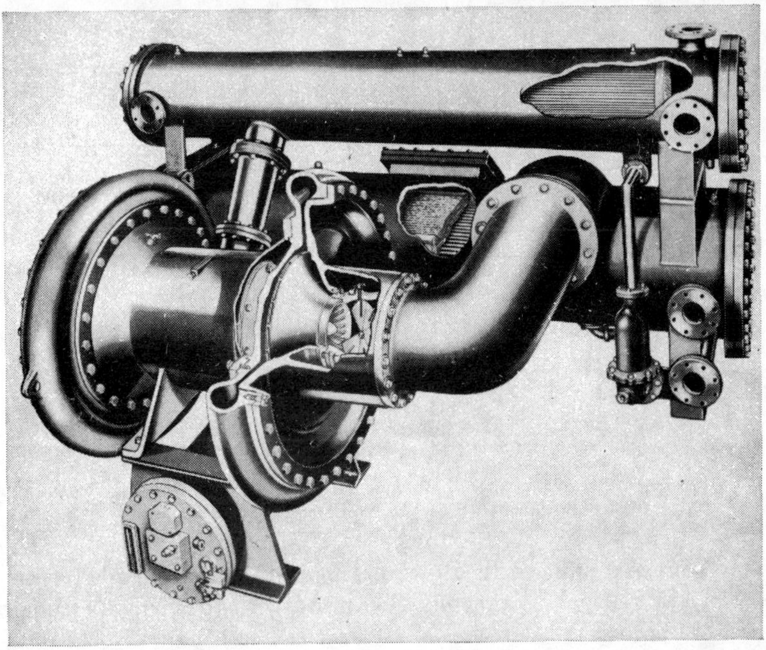

Fig. 141. Centrifugal compressor, two-stage volute design. Motor located between first and second stage impellers (CenTraVac). (Copyright, The Trane Company. Reproduced by permission.)

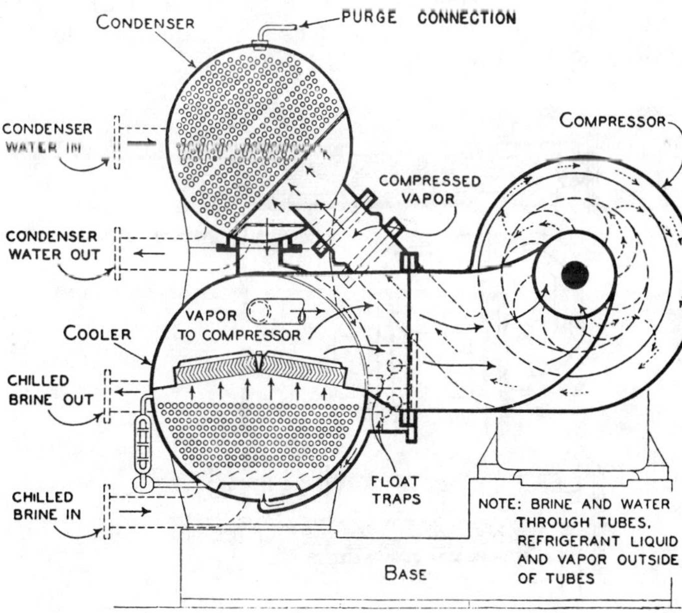

Fig. 142. Flow diagram of Carrier centrifugal refrigeration system. (Courtesy, Carrier Corporation.)

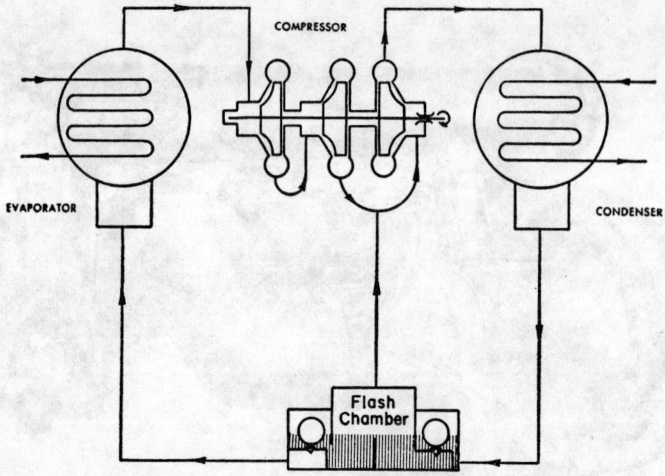

Fig. 143. Flow diagram of Worthington centrifugal refrigeration system, showing relative position of flash chamber. (Courtesy, Worthington Corporation.)

pressure increases sufficiently. However, no provisions are made for the recovery of any refrigerant that may be mixed with the foul gas using this method of purging.

By evacuating the noncondensible gases from the top of the condenser by means of a compressor and compressing to a higher pressure any refrigerant vapor accompanying, such gases subsequently are reconverted to their liquid phase in an air-cooled condenser immediately following compression. An oil separator is often installed between the auxiliary compressor

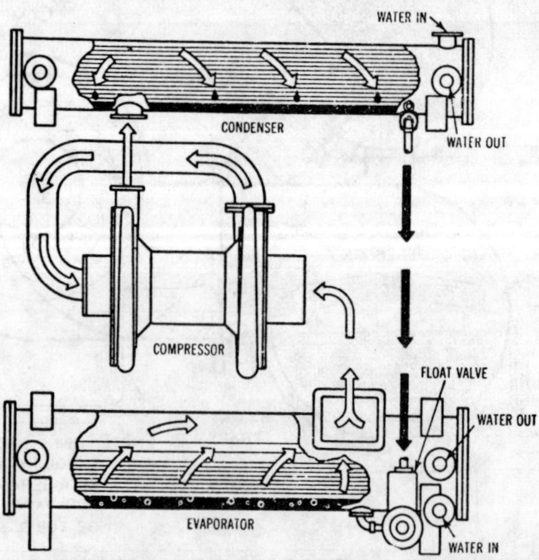

Fig. 144. Flow diagram of CenTraVac refrigeration system. Arrows indicate refrigerant path. (Copyright, The Trane Company. Reproduced by permission.)

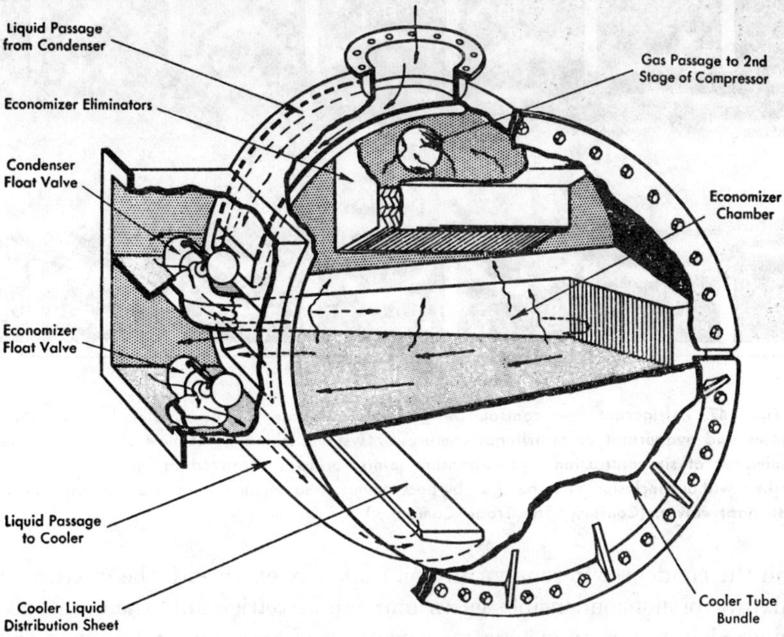

Fig. 145. Economizer (flash chamber) for system shown in Fig. 138. (Courtesy, Carrier Corporation.)

Fig. 146. Economizer (flash chamber) for system shown in Fig. 139. Note low-side float valve. (Courtesy, Worthington Corporation.)

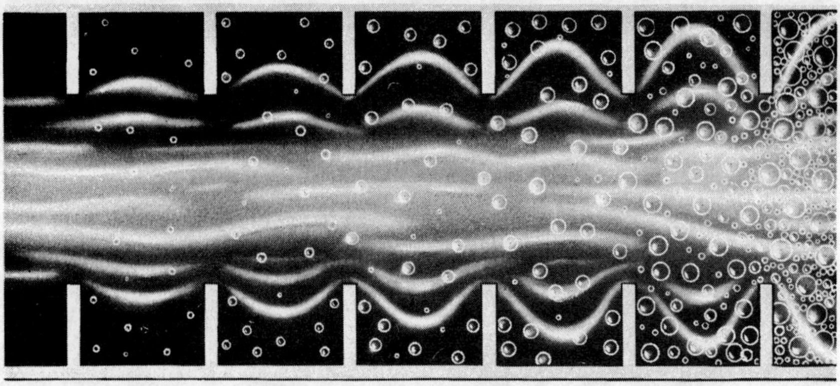

Fig. 147. Refrigerant flow control. Designed to eliminate the float valve between the condenser and evaporator of centrifugal compressor systems. This type of flow control provides the equivalent of six contraction and expansion joints (orifices) arranged in series. Flow is accomplished without inclusion of a hot gas by-pass device, and eliminates all moving parts prevalent with float valves. (Courtesy, The Trane Company.)

and the condenser to remove any oil that may enter with the mixture. The mixture of noncondensible gases and liquid refrigerant, then, enters the upper chamber of an evacuator, consisting of an upper and a lower chamber, connected by means of a trap. The liquid refrigerant separating from the gases is trapped in the lower chamber and returned to the main system through a float valve when a predetermined level is reached. The noncondensible gases remain in the top chamber, from which they are vented to the atmosphere through an automatic relief valve when the pressure rises sufficiently as a result of accumulation. Any water present floats on the liquid refrigerant and is removed either automatically or manually through a drain connection in the upper chamber, being forced out under pressure of the evacuator condenser when the level of the water reaches a given height.

While continuous operation of the purge system during running periods of the compressor is suggested, the recommendation is not always followed. To prevent reconversion of vapor to liquid in the compressor cylinder during those periods when the purge system is not in operation, a solenoid valve in the suction line automatically closes, isolating the system.

An important consideration in the design of centrifugal compressors is the prevention of leaks, especially through the shaft openings. Metallic seals lapped to provide smooth contact surfaces somewhat similar to those described for reciprocating compressors are often used. The surfaces are maintained in contact under oil pressure which also serves to provide for their lubrication and acts as a seal during operation. The oil pressure to the seal is closely controlled, and, in the case of any appreciable drop, the compressor is automatically shut down. A shaft seal reservoir located above the shaft provides sufficient oil head at the seal to prevent leakage during shut-down periods.

As previously mentioned, seals are also required to prevent leakage between stages as a result of pressure differentials. A labyrinth-type seal is used here. The labyrinths are made up of a number of narrow strips or blades placed concentrically with the shaft and close enough together to form a series of constricted spaces. Because of its tendency to expand, the gas is subjected to some throttling action as it attempts to pass from one space to another. This retards its flow sufficiently to prevent leakage, provided that enough concentric blades are installed.

One of the primary advantages of centrifugal compressors lies in their capacity to handle large volumes of low-pressure refrigerant vapor per unit of time to obtain maximum tonnage with a substantial saving in floor area

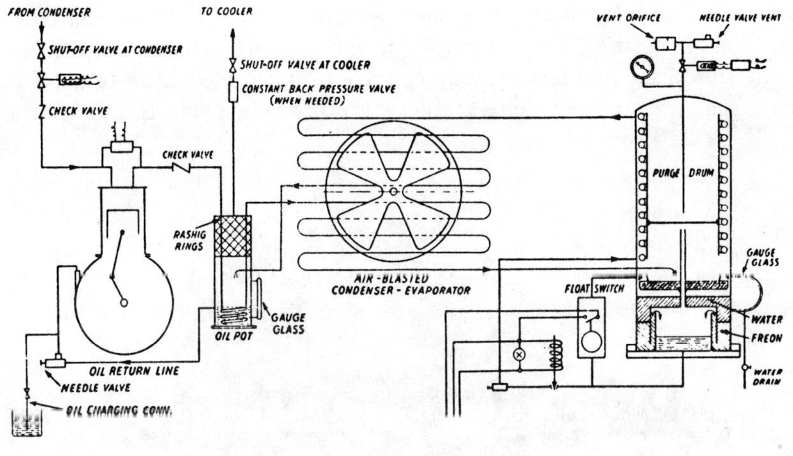

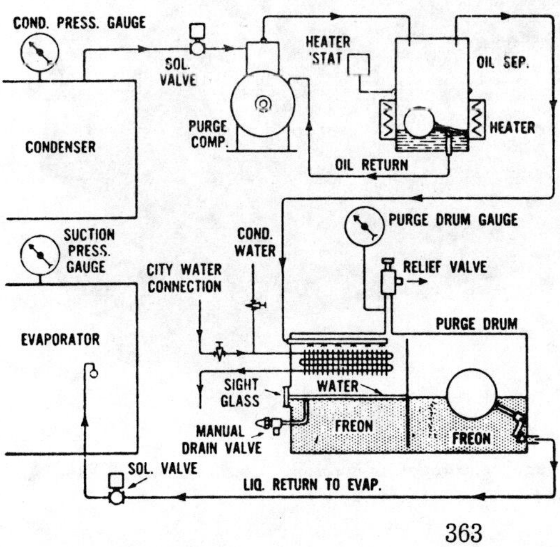

Fig. 148. (*Top*) Purge-recovery system for centrifugal refrigeration system. (Courtesy, Worthington Corporation.)

Fig. 149. (*Bottom*) Purge-recovery system for centrifugal refrigeration system. (Copyright, The Trane Company. Reproduced by permission.)

as compared with reciprocating compressors of equal capacity. Speeds depend on individual design based on the pressure differential between inlet and outlet in relation to the number of impellers (stages) provided. As the pressure produced by one impeller is proportional to the square of its peripheral velocity, it follows that high rotative speeds are necessary to obtain high pressure differentials with a relatively small number of impellers. The other alternatives are to enlarge the impeller diameter or provide additional impellers and keep speeds within the limits dictated by the strength of materials. Depending on the design combination employed, speeds of centrifugal compressors range anywhere from 2800 rpm to 8000 rpm and much higher in individual cases. Capacities range from 50 tons to 2600 tons.

The efficiencies of centrifugal compressors are high, running well over 80 percent in some instances. However, the efficiency of most industrial compressors ranges between 70 percent and 80 percent.

Because the centrifugal compressor has only one moving component, lubrication considerations are confined mainly to the impeller shaft bearings. This fact, in conjunction with proper sealing, eliminates to a large extent the usual problems of oil separation and fouled heat transfer elements that prevail in systems employing other type compressors. As there

Fig. 150. Hermetic chiller with single-stage centrifugal compressor. (Courtesy, Worthington Air Conditioning Company.)

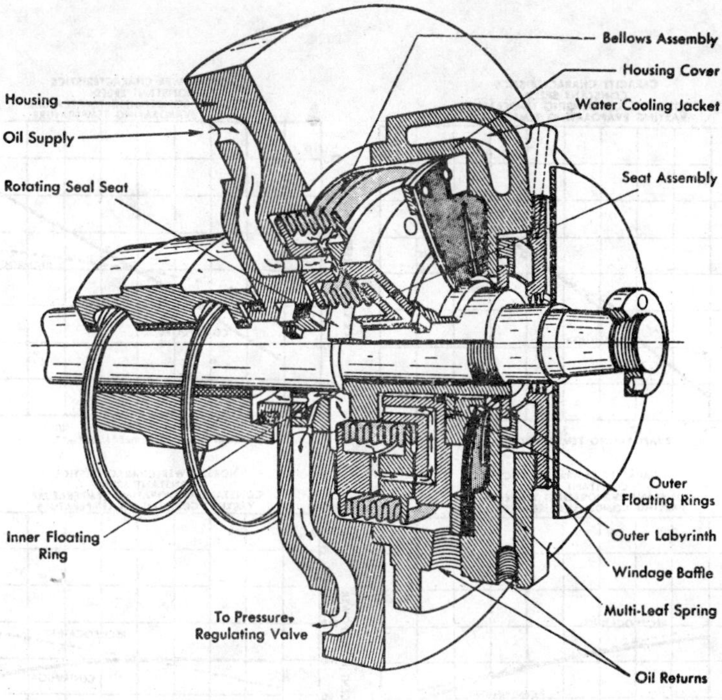

Fig. 151. Rotary shaft seal for centrifugal compressor. (Courtesy, Carrier Corporation.)

are no internal parts to wear, such as pistons, valves, etc., found in reciprocating compressors, their initial high efficiencies are sustained during the life of the machine.

Surging is an operating characteristic of all centrifugal compressors. It is caused when the amount of vapor leaving the evaporator is reduced as a result of decreased load to a point (surging point) usually below 50 percent of the handling capacity of the compressor. Surging may be noted by sound fluctuations that occur as the flow of gas momentarily stops and reverses as the compressor attempts to remove vapor from the evaporator faster than the load is capable of generating it. It can be eliminated by providing a by-pass arrangement accompanied by some reduction in efficiency or by means of a suction damper. This is discussed in greater detail in Chapter 9, "Controls." Surging of itself does not cause any serious loss in economy or affect the mechanical function of the compressor. The surging point is also variously called the pulsation point or the pumping limit.

Refrigerants employed in centrifugal compressors should be comparatively heavy in order to produce sufficient pressure ratio in any one stage. Because pressure produced is proportional to the density of the vapor, refrigerants such as R-11 and R-113 having high molecular weights are widely used in centrifugal systems. However, other refrigerants, such as carbon dioxide, ammonia, methyl chloride, sulfur dioxide, dichloroethyl-

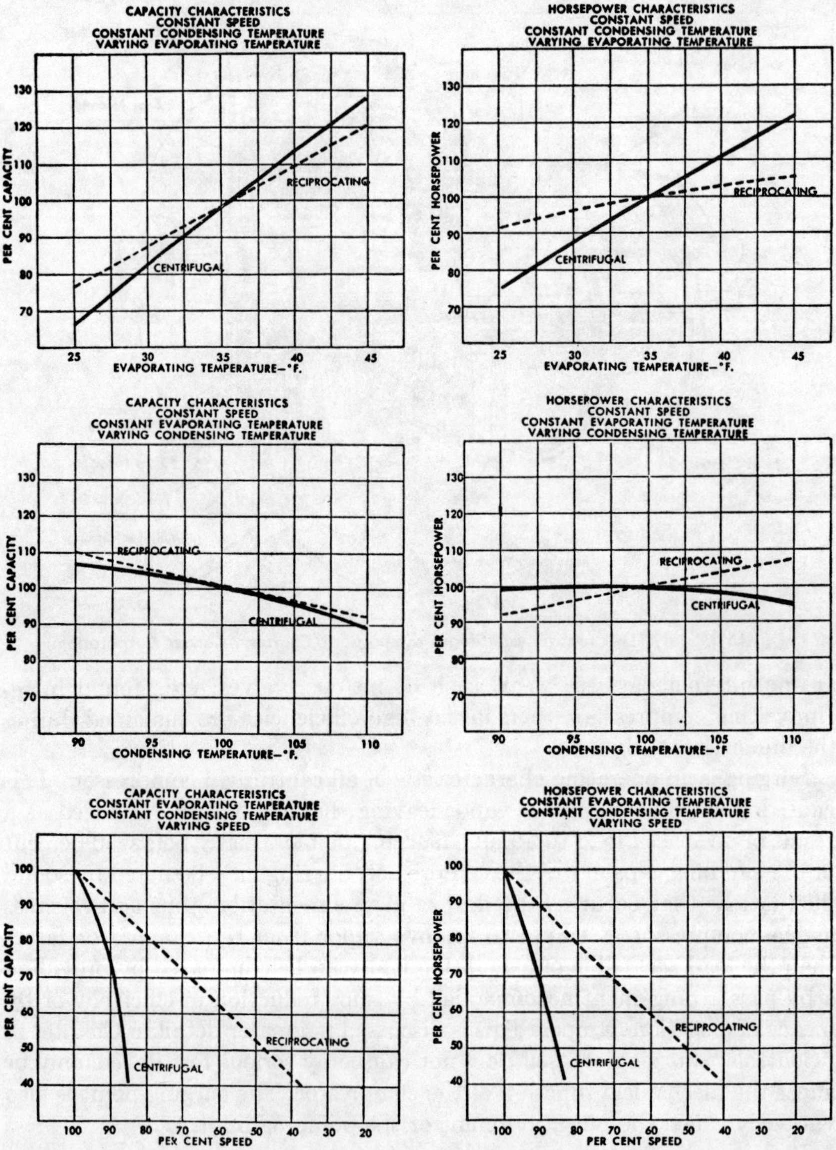

Fig. 152. Curves showing capacity and horsepower characteristics of centrifugal compressors for various operating conditions. (Courtesy, Worthington Corporation.)

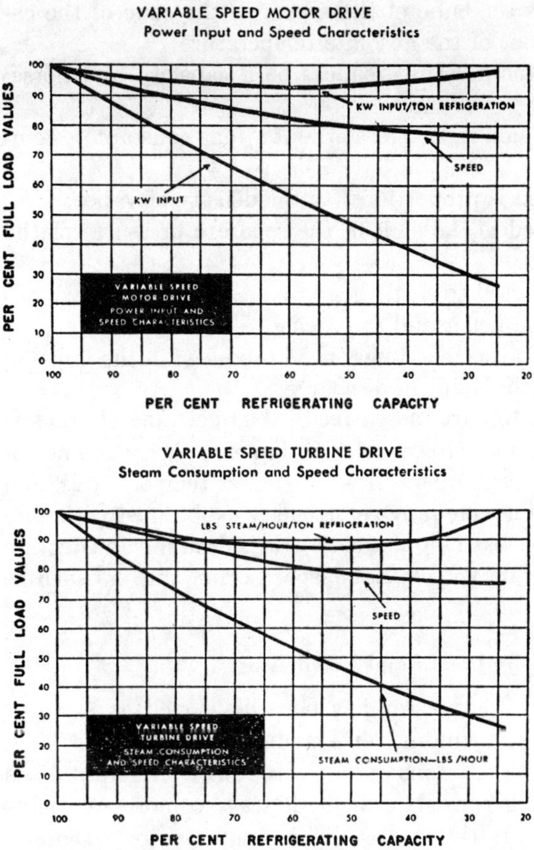

Fig. 153. Partial load characteristics of centrifugal compressors. (Courtesy, Worthington Corporation.)

ene, trichloroethylene, ethyl chloride, and water vapor, have been used with varying degrees of success.

Centrifugal compressors are adaptable to low-temperature operation. They are also used in conjunction with other types of compressors in obtaining very low suction temperatures either in compound arrangement, in which a common refrigerant is used, or by cascading, wherein a high-pressure system acts to condense vapors of the low-pressure system. (See Chap. 16.)

Some of the more important characteristics of the centrifugal compressor may be summarized as follows:

(*a*) They are primarily low-pressure machines;

(*b*) Capacity varies as speed and as the square root of the absolute temperature, making it sensitive to load changes;

(*c*) The discharge pressure varies as the square of the speed of a given machine under constant conditions;

(*d*) Horsepower varies as the cube of the speed, as the cube of the capacity, and as the square root of the absolute temperature;

(*e*) A rise in condenser temperature reduces both capacity and horsepower requirements;

(*f*) The ratio of compression increases as the RT (gas constant $\times$ temperature, abs) decreases;

(*g*) The pressure produced is proportional to the density of vapor;

(*h*) The pressure produced is the sum of the separate pressures of the individual impellers;

(*i*) Velocity head is converted to static head by means of stationary diffusers located between the impeller and the casing;

(*j*) Static head is equal to the centrifugal head produced in the impeller minus the velocity head at the point of discharge.

The curves shown in Fig. 152 are indicative of the operating characteristics of centrifugal compressors as to the relationship of capacity and horsepower to variable evaporating temperatures, condenser temperatures, and speeds. It will be noted that a wide range of capacity is obtained with only slight changes in rotative speeds. This characteristic is desirable in effecting capacity control in conjunction with variable-speed prime movers such as steam turbines.

Single-stage Centrifugal Compressors

A single-stage centrifugal compressor has been developed as a water chiller for various purposes, including air conditioning systems. They range in capacity from 130 to 637 tons of refrigeration. Refrigerant 11 is used in this machine. The unit consists of a single stage compressor (having only one impeller) and a refrigerant-cooled hermetic motor assembly, which is mounted on a nested chiller-condenser combination, together with necessary gears, controls and auxiliaries, including float valve and purge unit. A single stage, hermetic type centrifugal compressor is shown in Fig. 150.

Having smaller dimensions and being lighter in weight than multi-stage compressors, they are advantageously employed for air conditioning modern buildings having upper floor equipment locations. The effect of cooling motors by allowing the liquid to evaporate on the windings results in much lower winding temperature, approximately 100 F below those of water or refrigerant jacket or gas cooled units. The decrease in motor-insulation temperature results in an increased winding life.

The single-stage centrifugal compressor eliminates the need for an inter-stage flash chamber or economizer. It also avoids the possibility of unbalance due to the relationship between interstage pressure and the suction and discharge pressure which causes inefficiencies and unstable operation in some multi-stage machines. The absence of more than one stage is compensated by an appropriately higher speed depending on the load and desired evaporator temperature.

Capacity control is accomplished by suction inlet guide vanes driven by an electric motor. Capacities from 10% to 100% of full load may be attained by the latter control method.

Hermetic Compressor and Condensing Units

One of the characteristics of hermetic compressor operation is that it permits the suction gas completely to surround the cylinder walls. This arrangement insures maximum heat removal and eliminates the need for auxiliary water cooling. They are employed generally with refrigerants R-12 and R-22.

One type of hermetic unit is equipped with a centrifugal evacuating device which maintains an operating pressure below the suction pressure that develops a positive oil return to the crankcase.

THE CONDENSER

Upon discharge from the compressor, the warm gas enters the condenser in which the heat absorbed in the evaporator as well as the heat of compression is removed and it is converted to liquid state. Because the pressure of the discharge gas is higher than the suction pressure, the boiling point is raised so that conversion to liquid phase is accomplished at a much higher temperature, the boiling point and the condensing temperature being the same for any given pressure.

Like the evaporator, the condenser is a heat exchanger equipped with heat-transfer elements that enable opposing fluids to come in close contact and through which heat passes from the warmer to the colder fluid. This transfer of heat should be accomplished rapidly in order to minimize surface area requirements. Condensers should also be designed to allow the cooling medium to maintain effective temperature differential over all areas of the heat-transfer element.

There are a number of condenser types employed, selection of which is based on the operating conditions under which the system operates and the type, the source, and the comparative cost of available cooling medium. The refrigerant employed and the functional demands on the system most generally govern the kind of medium employed for condenser cooling, which in turn limits selection of condenser type. Air, water, brine, or another refrigerant all serve as cooling mediums, depending on their adaptability and the requirements of the refrigeration system.

The various types of condensers in general use are as follows:

Surface condensers:
- Shell-and-tube, horizontal closed type
- Shell-and-tube, vertical open type
- Shell-and-coil

Atmospheric:
Standard
Bleeder
Double pipe
Evaporative
Air-cooled

Surface Condensers

Shell-and-Tube Condensers—Horizontal Closed Type

This type of condenser, shown in Fig. 154, consists of a cylindrical shell in which are installed a number of parallel tubes, running lengthwise and connected to tube sheets at both ends. The tubes, either steel or copper, form a passage for the cooling medium from one end of the enclosed shell to the other opening into a water head or space between the tube sheet and

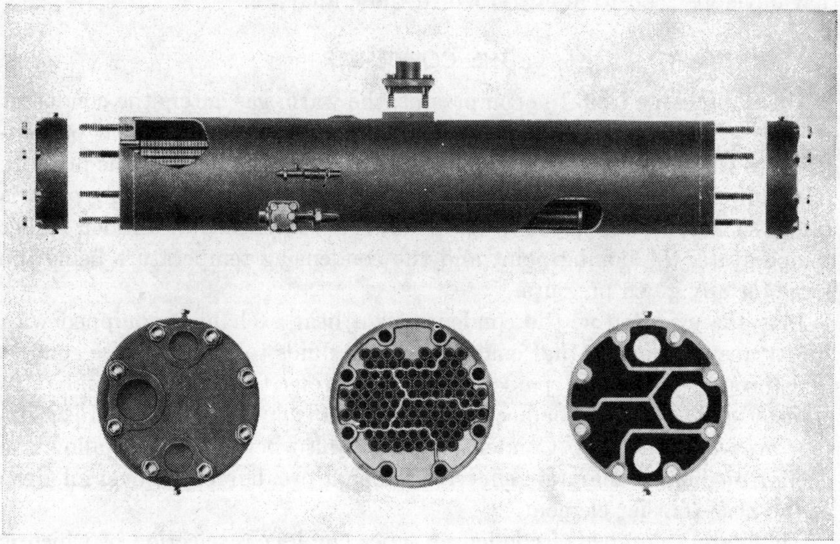

Fig. 154. Shell-and-tube condenser of the enclosed horizontal type. Note manifold head for reversing and diverting flow. (Courtesy, Carrier Corporation.)

the end plate of the condenser. The water heads are manifolded in such a manner as to receive cooling medium through one group of tubes and return it to the opposite end via a different group. The number of divisions formed by the manifold governs the number of passes made by the cooling medium during its journey through the condenser. Their design influences efficiency of heat transfer by reason of cooling medium distribution and its effect on temperature differential at all points.

The relative locations of cooling medium inlet and outlet vary, typical connections being shown in Fig. 155.

Horizontal closed shell-and-tube condensers provide a maximum amount

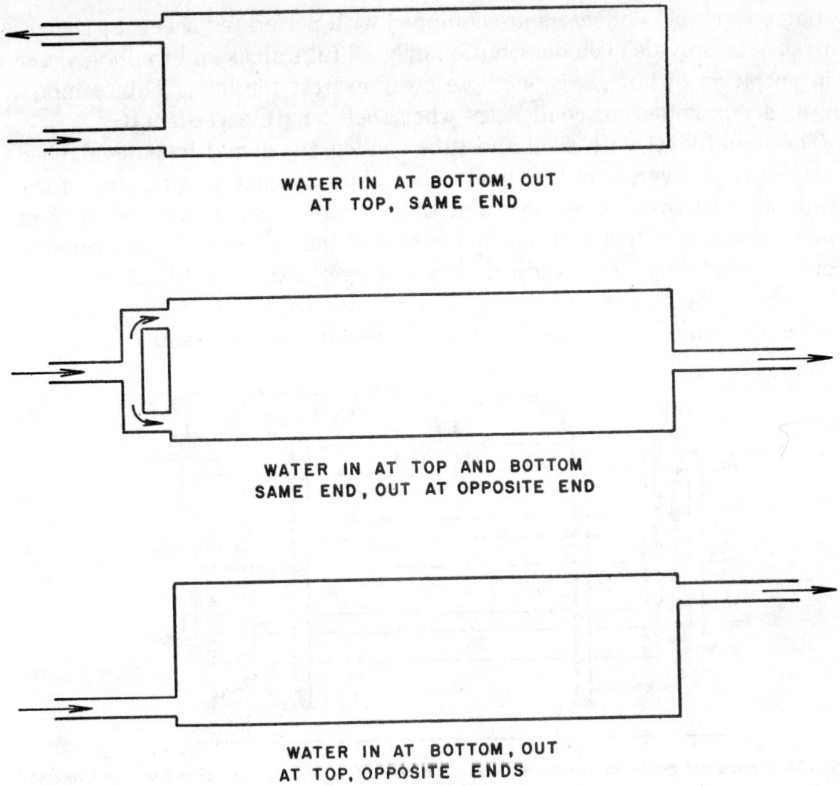

Fig. 155. Various water inlet and outlet arrangements for closed shell-and-tube condensers.

of heat-transfer surface in a relatively small space and may be installed anywhere within a building or outside, either on or off the floor or the ground. They can be employed singly or in multiple units of two or more, placed either side by side or one above the other.

The warm gas usually enters the condenser at the top, either directly or through a distribution header having two or more entrance points, condenses as it passes downward over the surface of the tubes, and leaves at the bottom as a liquid. Where multiple units are employed, the discharge gas is either proportionately divided between each unit by means of a common header, in which case the arrangement is said to be in parallel, or it enters the top unit and passes consecutively through the others, the liquid leaving to enter the receiver from the bottom unit. The latter is known as a series arrangement. In either case, the cooling medium enters each of the shells simultaneously and after completing the desired number of passes leaves through the individual outlets into a common discharge header. By providing control valves for each unit connected in parallel, repairs can be made on any one while continuing service in the others, thus establishing an advantage for this arrangement over those connected in series.

Some types of condensers are equipped with perforated sheets at the gas entrance to provide even distribution over all tube areas and to prevent the concentration of hot gases on those areas nearest the inlet. Tube support sheets are provided for condensers when their length exceeds 9 ft.

The liquid level with shell-and-tube condensers should be so controlled as to prevent covering of the lower tubes, thus eliminating a portion of the available heat-transfer surface and decreasing capacity. The number of tubes varies according to their diameter and the size of the shell, ranging from a low of seven to several hundred or over a thousand in large units. They should be staggered in such a manner as to offer a maximum amount of effective condensing surface to the refrigerant vapors passing over them.

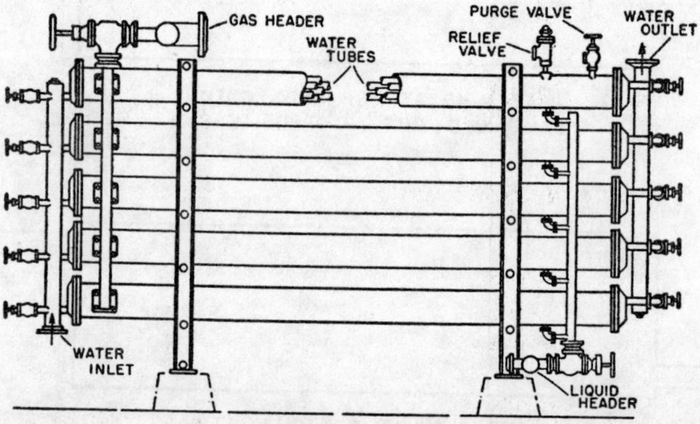

Fig. 156. Horizontal multi-pass shell-and-tube condenser. This one consists of a stack of five small individual condensers. (Courtesy, Mobil Oil Corp.)

Tube diameters also vary for different condensers with $\frac{3}{4}$ in., 1 in., $1\frac{1}{4}$ in., $1\frac{1}{2}$ in., and 2 in. O.D. seamless copper or steel tubing predominating. The length of tubes ranges from 5 ft to 18 ft, while shell diameters of $3\frac{1}{2}$ in. to 60 in. are the usual limits.

The number of passes, which is the number of times that a given volume of cooling medium passes from one end of the shell to the other, is controlled by the baffle arrangement or manifolding within the water heads and varies with different units, being anywhere from 2 to 20 passes. Where the number of passes is represented by even numbers, such as 12, 14, 18, or 20, it indicates that the cooling medium inlet and outlet are on the same end, while units having their inlet and outlet on opposite ends have an odd number of passes: 7, 15, 19, etc.

R-12 condensers usually differ from ammonia condensers in the type of tube employed, the former refrigerant requiring fins for greater surface area to compensate for its relatively lower film conductance as compared with the film conductance of water. The relatively higher film conductance of ammonia eliminates the requirement for fins, and consequently the tubes are not extended for the latter refrigerant. It holds true, however, that shorter lengths of cooling-medium travel accompany the use of fins as compared with bare tube, in accordance with the rule that the length of travel decreases for a given condition as the rate of heat transfer per lineal foot increases, regardless of the refrigerant employed.

The end plates of horizontal shell-and-tube condensers are removable to enable the tubes to be cleaned. While this type of condenser is suitable for most operating conditions, its use should be avoided where untreated water containing appreciable amounts of algae and silt, such as found in many fresh-water creeks, is to be used. Impurities of this kind act to foul the tube surfaces and finally restrict the flow of water resulting in high head pressures. The tubes should be maintained free of oil by installing a suitable trap in the discharge line between the compressor and the condenser.

As in all types of condensers, some arrangement should be provided for the removal of noncondensible gases. Usually this is accomplished by a connection from the top of the condenser to a purger located above, which also serves the receiver in a similar capacity. A drain should likewise be installed, located at the lowest point of the unit.

Shell-and-Tube Ammonia Condenser, Vertical Open Type

This type of condenser (Fig. 157) consists of a vertical shell enclosing a bundle of parallel tubes, also vertical, a water box located at the top, and a drain connection at the bottom. The tubes are locked in to tube sheets at both ends of the shell, opening out to the water box at the top and to a drain in the bottom. The refrigerant vapor enters the side of the shell about midway between the tube sheets to condense and flow down the outside surfaces of the tubes, leaving through a side opening at the bottom. Each tube is provided with a swirl or distributing ferrule at the top to lend a swirling action to the water as it starts its downward journey. This feature directs the water over the internal surfaces of the tubes rather than allowing it to drop relatively free to the bottom of the shell. It also prevents warm air from entering the tube.

The water box which receives the cold water is sometimes equipped with a perforated sheet steel distributing ring, which allows for even distribution of water to all tubes and also acts to prevent any solid material of size from entering and clogging the swirls or tubes.

While the shell is covered, the water box is not pressure tight, so that the water flows downward through the tubes by force of gravity only. When the water has completed its journey from the water box to the drain, it has completed its function and is conducted to either the sewer or the cooling tower, making this type condenser a single-pass unit. Purge connections for the removal of noncondensible gases are usually made immediately above the vapor-inlet level, but are found also near the top of the shell on some types. They are also equipped with safety valves as precaution against excessive head pressures. An oil drain is provided at the bottom of the shell.

Tube diameters range from $1\frac{1}{4}$ in. to $2\frac{1}{2}$ in. O.D., although the majority

Fig. 157. Vertical shell-and-tube open type ammonia condenser. Single pass. Water entering at top flows by gravity through vertical water tubes located within the shell. A swirling action is given the water as it leaves the header to effect greater heat-transfer capacity. (Courtesy, The Vilter Manufacturing Company.)

are 2 in., while their lengths vary from 10 ft to 16 ft. Shell diameters of from 12 in. to 60 in., having from 14 to several hundred tubes, give the vertical condenser wide capacity selection. Because this type of condenser is used in conjunction with ammonia, steel tubes or pipe are employed.

The vertical shell-and-tube condenser has several advantages, viz.:

(*a*) It requires comparatively small floor space,
(*b*) It may be installed inside or outside a building,
(*c*) It has a low installation cost,
(*d*) It requires simple water, refrigerant, and purge connections,
(*e*) It gives easy access to tubes which may be cleaned while the system is operating,
(*f*) It requires low water head even under maximum demand,

(*g*) It is adaptable to untreated water conditions because of easy tube cleaning.

To determine the water consumption of any size condenser, reference should be made to the method shown in Chapter 11, "Refrigeration Calculations." For a given temperature rise of the water through the condenser, an approximate consumption is given in Table 31.

TABLE 31. WATER CONSUMPTION PER DEGREE RISE IN CONDENSER

Water temperature rise °F (Difference in water temperature out and in).....	5	10	15	20	25	30	35
Gallons per minute per ton required...............	6	3	2	1.5	1.2	1.0	0.857
Pounds per minute per ton required (1 gal = 8.33 lb)	50	25	17	12.5	10	8.33	7.14

Shell-and-Coil Condensers

Instead of the longitudinal-tube bundle used in shell-and-tube condensers, this type employs one or more continuous coils. In some instances, the cooling medium flows through the coil, whereas older designs provided for

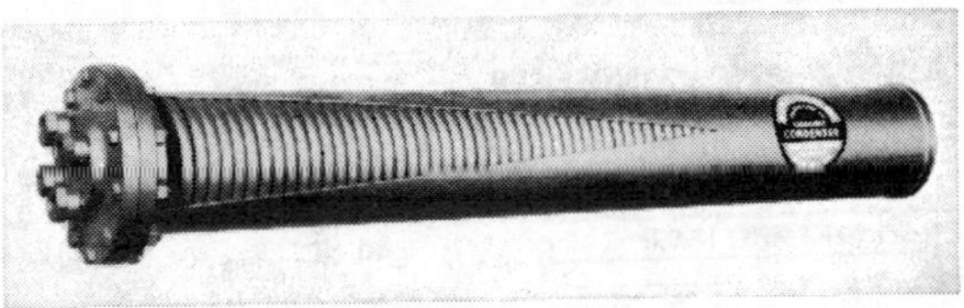

Fig. 158. Shell-and-coil condenser. (Courtesy, Metrex Valve Company.)

the cooling medium to surround the coil while the refrigerant passed through it. In the former type, clogging is serious because of the obvious difficulty in cleaning a coil so that only treated water can be used.

Atmospheric Condensers

This type of condenser consists of a number of vertical pipes, placed one over the other and connected at each end by return bends to form a continuous passage through which refrigerant vapors flow. Such an arrangement connects each pipe in series with all others in the stand. An atmospheric condenser may be made up of one or more stands, each of which numbers 8 or more pipes, fed from a common inlet header and discharging into a common outlet header.

A trough, located immediately above each stand, distributes water over the entire length of the top pipe, from which it cascades over the other pipes below and thence to a drain. Drip strips are sometimes provided between the pipes for the purpose of directing the flow of water and to reduce splash. It is standard practice to feed about 2 gal of water per minute (gpm) per foot of condenser length, but greater quantities may be used if drip strips are installed.

Cooling occurs as the result of temperature differential between the hot vapor and the cooling water so that over-all heat transfer increases as the quantity of water passing over the pipes increases. Cooling is also gained by evaporation of the water, the amount of which depends on the relative humidity of the surrounding air.

Standard atmospheric condensers provide for the entrance of the discharge gas at the top and the exit of the liquid at the bottom. This arrangement has the disadvantage of parallel flow, which means that the coldest

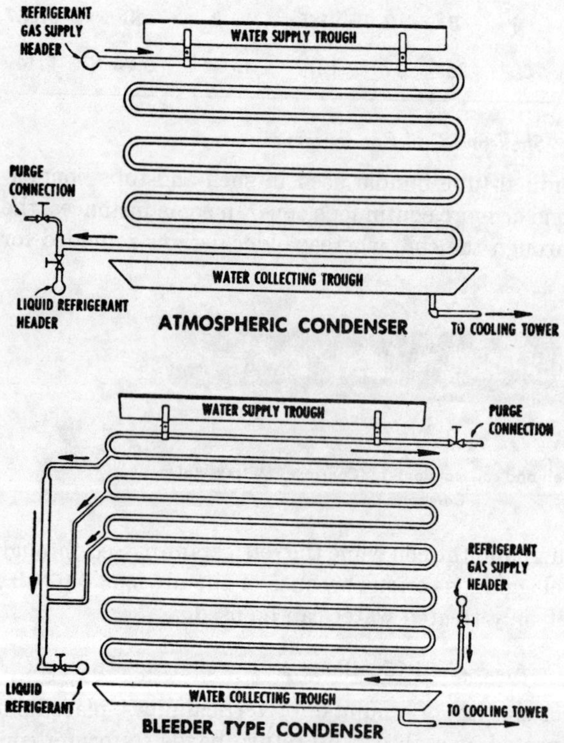

Fig. 159. Atmospheric condensers. (*Top*) Standard type; (*bottom*) bleeder type. (Courtesy, Armstrong Machine Works.)

water is in contact with the warmest surface and vice versa, which limits the terminal temperature of the liquid at some point above or equal to the final temperature of the water.

To overcome the disadvantage of parallel flow, the *bleeder type* of atmospheric condenser was developed. This type allows for the hot gases to enter at the bottom and pass upward against the downward flow of water, thereby gaining advantage of the counterflow principle. To prevent the liquid flowing back and settling in the lower pipes, traps are installed at intervals along the path of flow, usually at every other bend, from which bleeder lines conduct it to a liquid header and thence to the receiver.

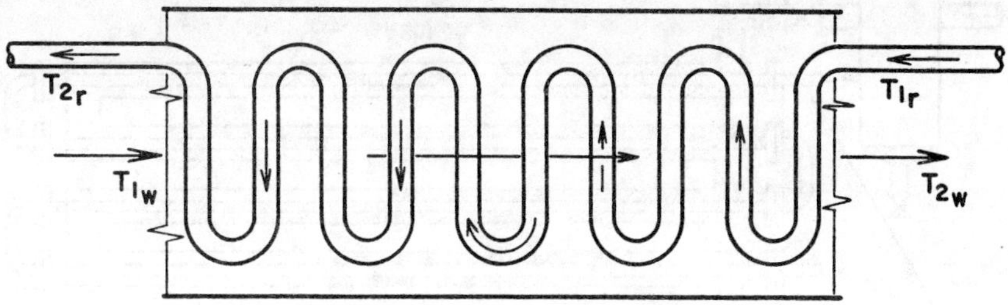

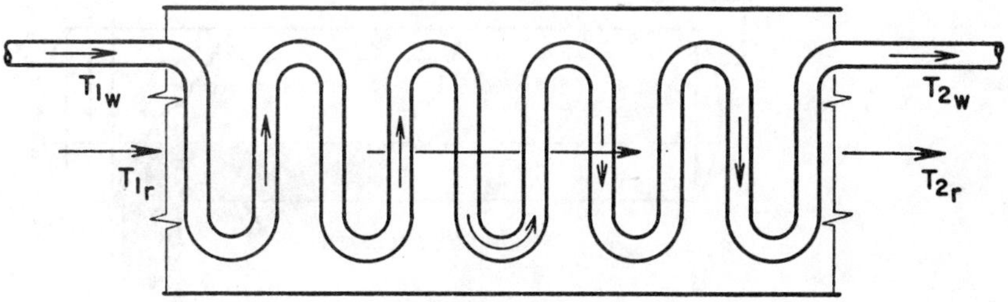

Fig. 160. Relative flow of water to refrigerant in condenser. (*Top*) Counter flow; (*bottom*) parallel flow.

Double-Pipe Condensers

The double-pipe condenser is similar in construction to that of the double pipe evaporator described previously. The refrigerant gas enters at the top while the cooling medium is fed from the bottom.

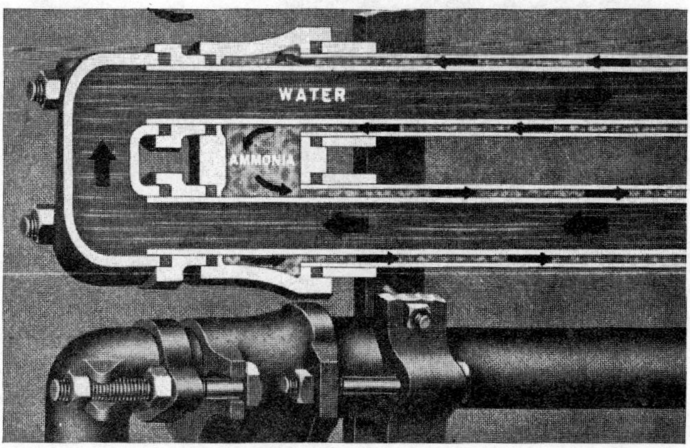

Fig. 161. Double pipe condenser, cross-sectional view. (Courtesy, Mobil Oil Corp.)

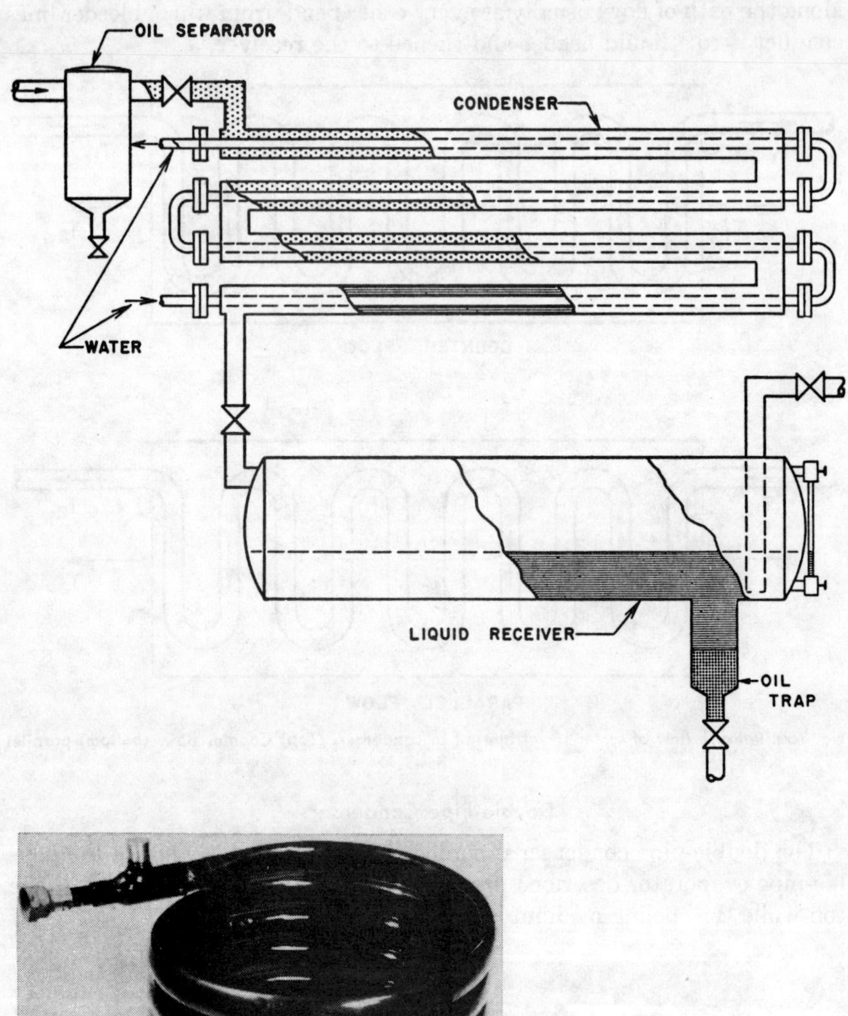

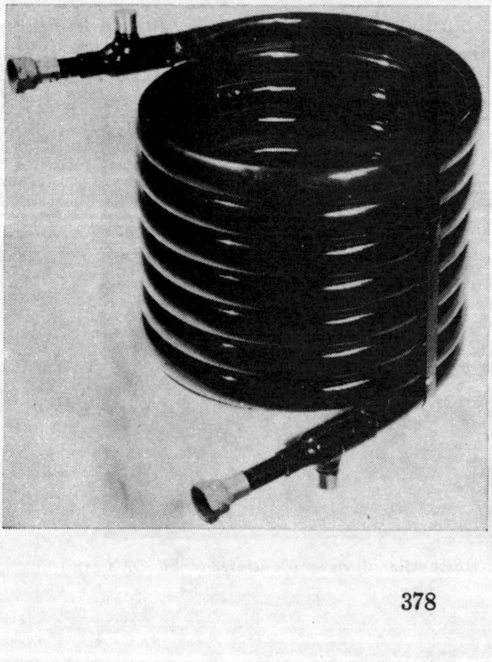

Fig. 162. (*Top*) Double pipe condenser—stand. (Courtesy, Mobil Oil Corp.)

Fig. 163. (*Bottom*) Double pipe condenser—helix. (Courtesy, Edwards Engineering Corp.)

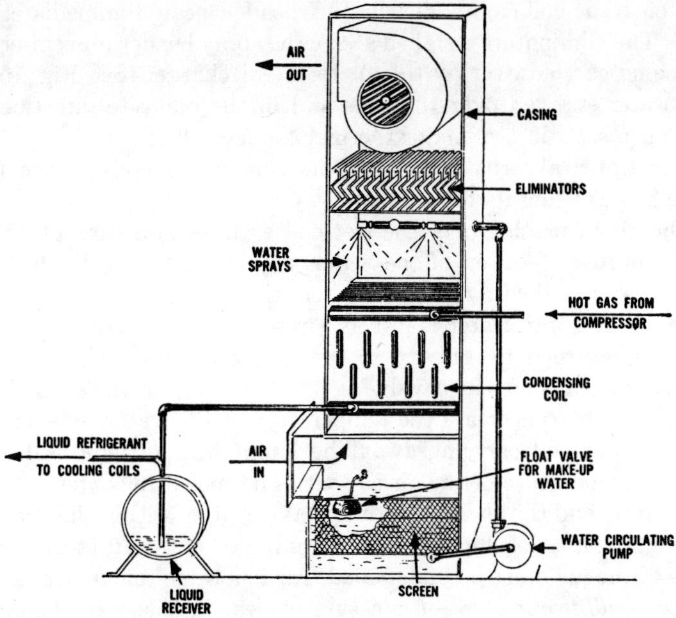

Fig. 164. Evaporative condenser. (Copyright, The Trane Company. Reproduced by permission.)

Usual construction consists of a $1\frac{1}{4}$-in. pipe within a 2-in. pipe, the refrigerant flowing in the annular space between the two, and the cooling medium in the inner pipe. By causing the two fluids to flow in opposite directions within each double pipe, the advantage of counterflow is gained.

An oil drain is placed at the lowest point in the condenser where this substance is removed by periodic blow-downs. Purge connections are made in the top pipe.

Evaporative Condensers

A comparatively new type, the evaporative condenser, is designed primarily to conserve water, an essential consideration in the operation of refrigerating plants serving large air-conditioning systems installed in metropolitan areas. Its underlying principle involves evaporative cooling, the rate of which depends on the wet-bulb temperature of the air introduced to the condenser. Approximately 1000 Btu* are removed per pound of water evaporated.

The various components of the evaporative condenser, consisting of a fan, a condensing coil, a water-spray header, and eliminators, are housed within a cabinet to form a self-contained unit. Air entering a grilled or hooded opening at the bottom is drawn by the fan through the condensing coil and past the spray header to be discharged from outlets above the fan. Water pumped from a reservoir at the bottom is sprayed against the air

* Latent heat of vaporization for water vapor is equal to $1091.6 - 0.56t$, approximately.

stream onto the coil from a distributing header located immediately above the coil. The eliminators installed above the spray header prevent entrained water being carried over by the air to the discharge. (See Fig. 164.)

The water sprayed over the coils and in the opposite direction to the flow of air results in two major thermal changes, viz.:

1. The water absorbs heat from the condensing coils in the form of sensible heat, raising its temperature.

2. The air in passing over the wet coil surfaces and through the spray area evaporates a portion of the water, carrying away as latent heat the sensible heat absorbed by the water from the coil.

Some of the spray water is lost as a result of evaporation while the remainder is returned by gravity to the reservoir below where the loss is made up by means of a float valve.

The air tends to decrease the temperature of the water by evaporation, thus absorbing and carrying away the latent heat of vaporization. This raises the temperature of the air as well as its moisture content.

To comprehend the thermal activity taking place within the evaporative condenser, the relationship of moisture contained in the air to the total heat content of the air must be understood. Air has three temperatures.

(*a*) *Dry-bulb temperature*—a measure of heat intensity of the air which decreases when moisture is added. This is the temperature determined by the conventional type thermometer called a dry-bulb thermometer.

(*b*) *Wet-bulb temperature*—by placing a water-wet wick around the mercury bulb of an ordinary (dry-bulb) thermometer and allowing air to move across its surface as by whirling, the water will evaporate, causing the dry-bulb temperature to drop. The reason for this drop lies in the absorption of latent heat by the air, cooling the water remaining in the wick. The rate of evaporation depends on the amount of moisture in the air passing over the wick so that a point is reached where the drop in temperature as recorded by the thermometer ceases and remains constant. This point is called the wet-bulb temperature. If the heat is neither added nor subtracted in the process, the wet-bulb temperature remains constant regardless of the amount of moisture added. If more moisture is added than the air can absorb at any given wet-bulb temperature, it will not evaporate but merely drop out as free water. *The wet-bulb temperature, then, is the measure of the sum of the sensible heat of the air plus the latent heat corresponding to the weight of moisture contained within it, or total heat of the air above 0 F.* When the air contains all the moisture it will absorb at any temperature, it is said to be saturated. Under saturated conditions, the wet-bulb and the dry-bulb temperatures are the same. The difference between the wet-bulb and the dry-bulb temperature is called the wet-bulb depression.

(*c*) *Dewpoint temperature*—the measure of the moisture content of the air. It is the point at which the air would become saturated if cooled to that temperature. If cooled further, condensation would occur. The dewpoint temperature remains constant if the moisture content remains constant regardless of the amount of heat added or subtracted above the limits of this temperature. It also governs the vapor pressure of the air. Under saturated conditions, the dewpoint, the wet-bulb and the dry-bulb temperatures are the same. The difference between the dewpoint and the dry-bulb temperatures is called the dewpoint depression.

The total heat of the air can be determined from the psychrometric chart if the wet-bulb temperature is known. The capacity of evaporative condensers is based on the maximum wet-bulb temperature of the region in which it is installed. A 20 F to 35 F difference between the entering wet-bulb and condenser temperature is recommended. For that reason, 75 F and 105 F, respectively, have been adopted for standard ton conditions. Other standards include atmospheric pressure of 29.92 in. Hg, entering and ambient air temperatures of 90 F, and an entering water temperature of 75 F.

To determine the amount of air required for a given condenser load, the following factors must be known or assumed:

(*a*) t_d = dry-bulb temperature of entering air
(*b*) t_w = wet-bulb temperature of entering air
(*c*) S_1 = weight of moisture per pound of entering air
(*d*) S_2 = weight of moisture per pound of leaving air
(*e*) h_{a1} = total heat of entering air – Btu per pound above 0 F
(*f*) h_{a2} = total heat of leaving air – Btu per pound above 0 F
(*g*) h_{w1} = total heat of entering water – Btu per pound above 32 F
(*h*) h_{w2} = total heat of leaving water – Btu per pound above 32 F
(*i*) W_w = weight of entering water, pounds per minute
(*j*) W_a = weight of entering air, pounds per minute

(1) S_1 and h_1 are determined from the psychrometric chart if t_d and t_w are known. (2) Assuming conditions of saturation for leaving air at a predetermined temperature somewhere between condensing temperature and t_w, then S_2 and h_2 may be determined from the psychrometric chart also. (3) The heat load is determined in the conventional manner, as described in Chapter 11, "Refrigeration Calculations," and used to find h_{w2}, assuming a terminal temperature differential between water in and water out. (4) h_{w1} is found by assuming an entering water temperature and subtracting 32 F. Both (3) and (4) are determined from the fundamental heat equation Btu = weight $\times$ sp ht $(t_2 - t_1)$, t_1 being 32 F in both cases here.

With the above factors known, the amount of air per minute passing through the unit may be determined from the equation

$$W_a = \frac{W_w (h_{w2} - h_{w1})}{h_{a2} - h_{a1} - (S_2 - S_1)h_{w2}}$$

EXAMPLE: Determine the volume of air required to pass through an evaporative condenser per ton of refrigeration, assuming the following conditions: refrigerant is Freon-12; water enters at 75 F and leaves at 90 F; air enters at 90 F db and 75 F wb, and leaves saturated at 90 F; suction and condensing temperatures 20 F and 105 F, respectively:

S_1 corresponding to 90 F db and 75 F wb (psychrometric chart)
= 0.0152 lb of moisture

S_2 corresponding to 90 F saturated air (psychrometric chart)
= 0.0301 lb of moisture

h_{a1} corresponding to 90 F db and 75 F wb (psychrometric chart)
= 37.6 Btu per lb

h_{a2} corresponding to 90 F saturated air (psychrometric chart) = 54.0 Btu per lb

Condenser load per ton — Freon-12 — 20 F suction and 105 F condensing
= 215 Btu per min

$$h_{w1} \text{ — entering at 75 F} = 1 \times 1\,(75 - 32) = 43 \text{ Btu per lb}$$

$$h_{w2} \text{ — leaving at 90 F} = 1 \times 1\,(90 - 32) = 58 \text{ Btu per lb}$$

$$W_w = \frac{\text{condensing load}}{h_{w2} - h_{w1}} = \frac{215}{58 - 43} = 14.33 \text{ lb of water}$$

Applying the equation,

$$W_a = \frac{W_w\,(h_{w2} - h_{w1})}{h_{a2} - h_{a1} - (S_2 - S_1)h_{w2}}$$

$$= \frac{14.33\,(58 - 43)}{54 - 37.6 - (0.0301 - 0.0152)58}$$

$$= \frac{215}{16.4 - 0.8642}$$

$$= 13.84 \text{ lb of air per min per ton}$$

$$= 13.84 \text{ lb} \times 14.2 \text{ cu ft per lb of air entering}$$

$$= 196.5 \text{ cu ft of air per min per ton}$$

The elements of time, intimacy of air and water contact, air velocity, length of spray chamber, and other factors affect the saturating efficiency of evaporative condensers. Complete saturation of air leaving is actually never attained, but will range from 50 percent to 75 percent.

Therefore, the amount of air in the above example should be increased for poorly designed condensers to

$$\frac{196.5}{0.50} = 393 \text{ cfm}$$

or for well-designed condensers to

$$\frac{196.5}{0.75} = 262 \text{ cfm}$$

Air velocities of 500 fpm to 600 fpm should be maintained for best results.

It will be noted in the equation that any decrease in h_{a1} would result in a decrease in the amount of air required (W_a). As condensers are operated at constant air flow, this means that an increase in capacity of the unit would be obtained with any reduction in the wet-bulb temperature of the air. Another way of increasing the temperature differential between condensing temperature and entering wet bulb to obtain greater unit capacity would be to raise the condensing pressure. However, this would also increase the power requirements of the compressor and reduce its volumetric efficiency.

Evaporative condensers are rated on the basis of Btu they are capable of absorbing at various combinations of suction and condensing temperatures for specific refrigerants. Thus, a unit might be said to have a refrigerating capacity of 85,000 Btu per hour for a R-12 system operating 40 F suction and 131.6 psia condensing pressure (100 F) when the entering air wet-bulb temperature is 75 F.

Coils are usually bare rather than finned tubing because of the difficulty in cleaning scale formed when other than treated (soft) water is used. Because of the continual evaporation occurring in this type of condenser, concentration of salts, introduced with the water, becomes a factor that must be attended to through periodic draining and cleaning.

Air-Cooled Condensers

Because the rate of heat transfer to a gaseous fluid is relatively low per unit area of surface, the use of air-cooled condensers is limited to smaller commercial and domestic type refrigerating units. For industrial service, the amount of heat transfer surface required would be so great as to render them impractical.

Air-cooled condensers usually consist of a copper or steel coil, fan and

Fig. 165. Air-cooled condenser, with motor shield. (Courtesy, Peerless of America, Inc.)

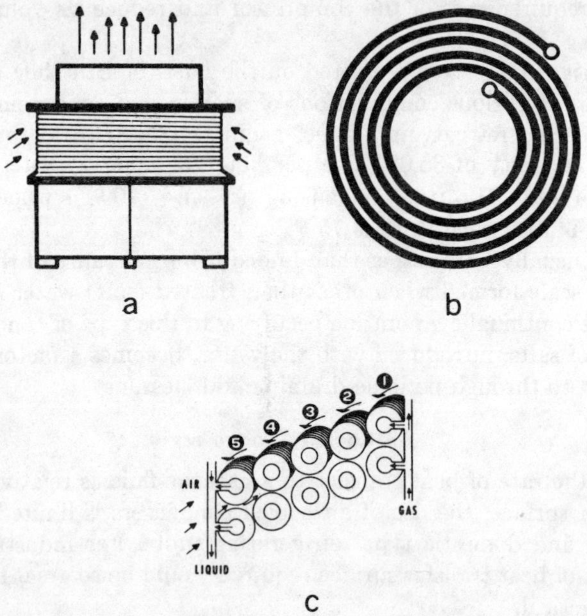

Fig. 166. Air cooled condenser designed for outside use. Single units have capacity up to 85 tons, up to 175 tons with duplex units. (*a*) Vertical air discharge; (*b*) circular coil from continuous tube; (*c*) finned coils. (Courtesy, Worthington Air Conditioning Company.)

housing, with the fan driven from the crankshaft of the compressor or an individual electric motor, depending on the circumstances of system design. Finned tubing is replacing the bare tubing, resulting in less bulky condensers. However, bare tubing continues in use for domestic service with excellent results due to the small loads.

Average air velocities range from 500 fpm to 1000 fpm with correspondingly less surface required as velocity increases within the limits of turbulent flow. Velocities are usually given according to face area, which is the area facing the tubes expressed as face velocity. At 500 fpm, about 150 sq ft of air-cooled finned condensing surface per ton is required. This compares with 6 to 10 sq ft of water-cooled surface.

As in all types, the condensing pressure varies according to the temperature of the cooling medium, which in this case is the surrounding air. For best results in preventing excessive head pressures, adequate ventilation should be provided.

Air-cooled condensers often act to supplement water-cooled condensers when the latter type is inadequate to maintain normal head pressures.

BOOSTERS

Boosters are employed to increase the performance efficiency of low-temperature refrigeration systems. They are compressors that draw and partially compress low-pressure suction gas immediately from the evaporator

Fig. 167. Booster compressor, reciprocating type. (Courtesy, The Vilter Manufacturing Company.)

and deliver it to another compressor for additional compression. Rotary, centrifugal, and reciprocating compressors are all employed to act as boosters, their selection depending on the type of system, the pressures involved, and the volume of refrigerant handled. When a booster is used, the process is known as stage compression.

The necessity for boosters, especially for low-temperature operation, becomes apparent when comparing the specific volume of any refrigerant at various suction pressures. Not only would a single-stage compressor have to operate at greater than maximum allowable speeds to handle sufficient volume but also would be subject to excessive compression ratios. It is obvious, then, that a booster is actually an initial stage compressor for multistage systems, capable of handling comparatively large volumes of low-density gases.

Ammonia, for example, has a specific volume of 33.08 cu ft per lb at −50 F and 7.67 psia suction pressure. At 5 F and 34.27 psia, the same refrigerant has a specific volume of 8.150 cu ft per lb. If only one single-stage compressor were used in a system operating under the first set of evaporator conditions, it would have to handle over four times as much volume as when operating under the second set of conditions. Assuming that the condenser temperature was the same in both instances, 86 F, their respective compression ratios would be 22.0 and 4.95 to 1. Not only would the power consumption be excessive in the first case, but also the temperatures involved and the volumetric loss would be so high as to make any such operation economically unsound. To handle any volume of gas at 33.08 cu ft per lb would necessitate either considerable speed or an exceptionally large reciprocating compressor, factors that would eliminate any such operation. By installing a rotary compressor capable of handling large volumes and raising

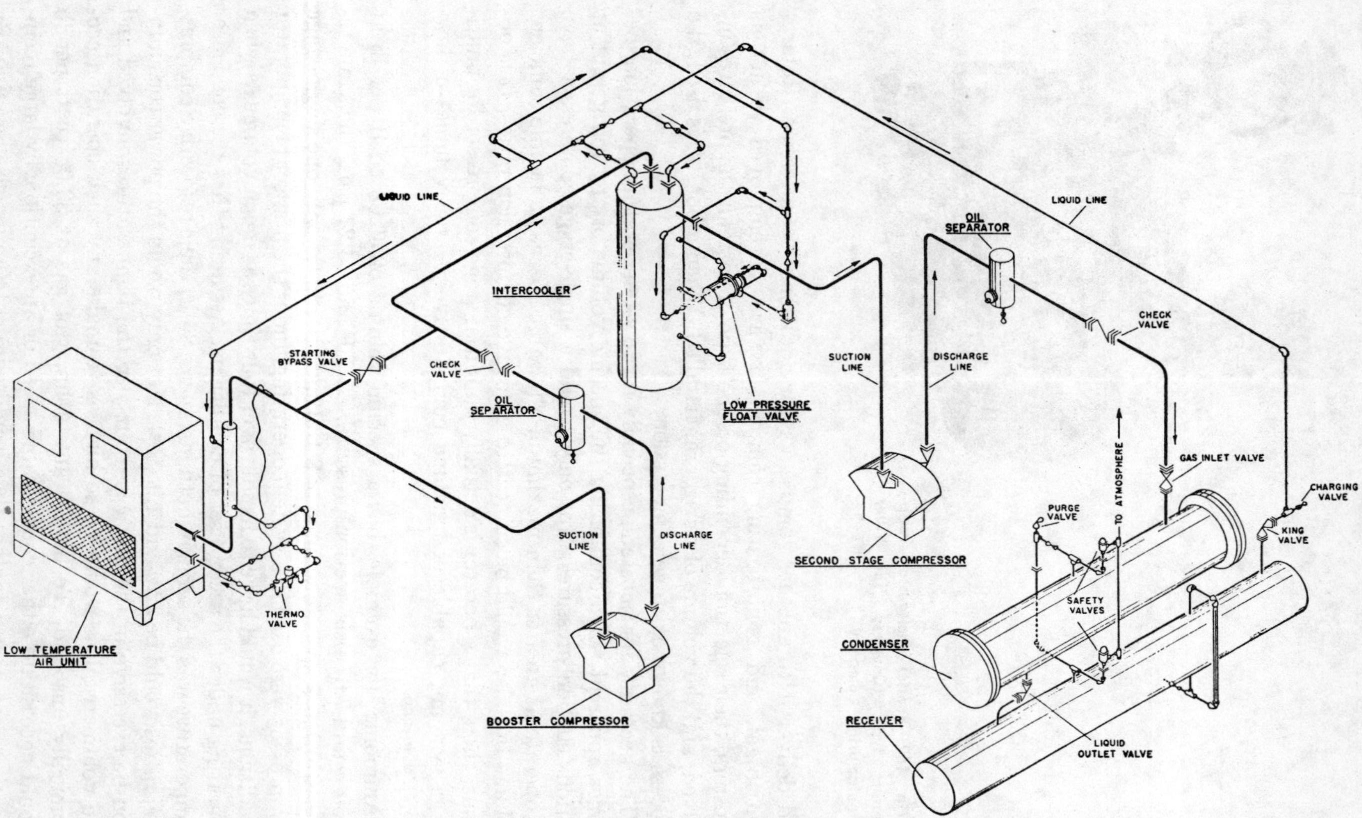

Fig. 168. Reciprocating booster compressor plant layout. Note relative location of ammonia intercooler. (Courtesy, The Vilter Manufacturing Company.)

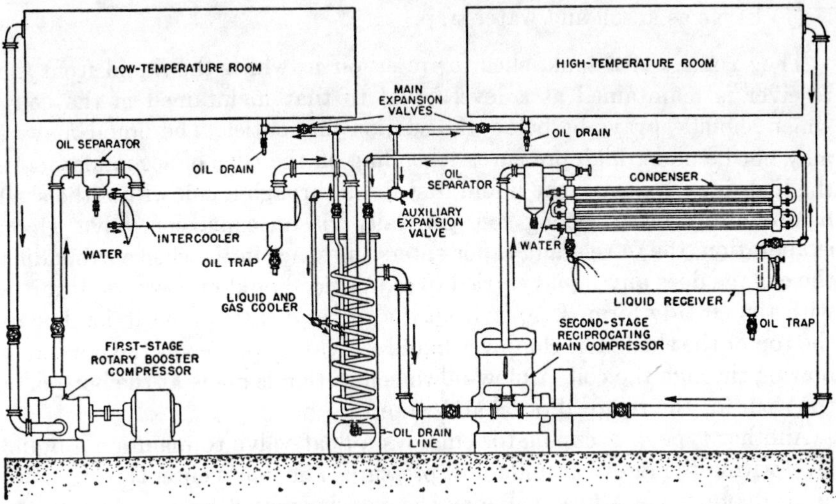

Fig. 169. Rotary booster plant layout. (Courtesy, Mobil Oil Corp., Inc.)

the pressure of the suction vapor 20 psi before it is delivered to the final-stage compressor, a more practical arrangement is obtained.

When a reciprocating compressor is employed as a booster for another of the same type, it must be capable of handling larger volumes than the second-stage compressor to compensate for the difference in densities. While the usual procedure is to install one of greater volumetric displacement, the differential in density can also be offset with higher speeds within practical ranges of suction temperature. Greater volumetric displacement is obtained in reciprocating compressors generally by increasing the bore diameter considerably beyond the length of stroke, such as $4\frac{3}{4}$ in. bore × $3\frac{5}{8}$ in. stroke, $8\frac{3}{4}$ in. bore × 6 in. stroke or $11\frac{1}{2}$ in. bore × 8 in. stroke, etc., and providing multi-cylinder units of the vertical, *V* or *W* types.

For ultra-low temperature conditions, favorable results are obtained with a rotary booster compressor of the type shown in Fig. 169.

ACCUMULATORS

As its name implies, the accumulator causes the refrigerant, both liquid and gas, to collect in a space for division to their ultimate destination; i.e., the liquid to the cooling coils and gas to the compressor suction line. Accumulators serve several purposes:

(*a*) Eliminate flash gas entering the cooling coils,
(*b*) Prevent liquid entering the suction line,
(*c*) Allow for recirculation of liquid not evaporated in the cooling coils and carried over by the gas (slop-over),
(*d*) Enable division of liquid to two or more evaporators,
(*e*) Act to subcool liquid refrigerant before entering evaporator, reducing flash gas thereby,

(*f*) Serve as an oil and water trap.

They consist of a tank, shell, or reservoir in which the liquid from the receiver is maintained at a level equal to that maintained in the coils, which should always be below the suction line outlet. The liquid may or may not be under high pressure, depending on the type of accumulator. In the high-pressure type, the liquid first flows through a coil within the shell before entering the cooling coil, generally via an expansion valve. Upon evaporation, the refrigerant vapor enters the space in the shell surrounding the coil, as does any liquid carried over. As this liquid evaporates, it mixes with the already formed vapor from the coil and enters the suction line at the top of the shell. The slop-over liquid in evaporating cools the hot liquid flowing through the coil. A blow-down connection is made at the bottom of the shell for the removal of oil and water.

Another type of accumulator employs a float valve to maintain a liquid level within the reservoir under low pressure. The liquid enters the accumulator chamber via a float valve and passes downward by gravity into the cooling coil. The flash gas passes upward and directly into the suction line outlet at the top. Upon evaporation, the refrigerant vapors are returned to the accumulator above the liquid level and pass from there to the suction-line outlet. Any slop-over entering with the vapor drops down to be recirculated with the liquid in the float chamber. Where a number of cooling coils are fed from a central accumulator, a pump is used to circulate the liquid from the reservoir through the system, using similar connections as described for the gravity type above. The pump is necessary for multiple-coil systems, as well as for evaporators having long runs, to overcome any back pressure built up during rapid vaporization. This would prevent the flow of liquid where gravity alone is depended on for circulation.

INTERCOOLERS

Multi-stage compression becomes necessary when the difference between evaporator and condenser pressures is great enough to result in excessive compression ratios. It may be accomplished within the same compressor having a low- and a high-pressure cylinder or by providing a booster in series with a single-stage compressor having higher pressure capacity. In some instances, three-stage compression is necessary, depending on the temperature conditions involved.

Multi-stage compression has several advantages over single-stage operation under low-temperature conditions, viz.:

(*a*) Lower final discharge temperature,
(*b*) Reduced power consumption per ton of refrigeration,
(*c*) Increased refrigeration capacity per cubic foot of refrigerant handled,
(*d*) Greater volumetric efficiency.

Vapors ordinarily leave the evaporator in either a saturated or a slightly superheated condition to enter the suction line where it is further superheated before entering the first compressor. At low or ultra-low suction temperatures, the vapor is extremely rare, denoting reduced refrigeration capacity per cubic foot. To counteract this condition and maintain required system capacity, a booster capable of handling the necessary volume of vapor is employed. The temperature of booster discharge, then, depends on the refrigerant, the compression ratio at this stage, and the suction temperature. If thermal conditions of the booster discharge vapor remain reasonably constant, it becomes the governing factor in the amount of refrigerant capable of being handled by the secondary or second-stage compressor.

One method of increasing the capacity of the second-stage compressor is to raise the weight of refrigerant per cubic foot handled, which means to increase the density of the primary discharge gas. This may be accomplished by removing the superheat absorbed in the evaporator suction line and the heat of compression added in the booster. This action is known as desuperheating, and results in the vapor entering the secondary compressor at a reduced volume per pound of refrigerant and in nearly a saturated condition at the intermediate pressure. An intercooler or desuperheater located between stages of multi-stage systems is employed for this purpose.

There are various types of intercoolers, the selection of which depends entirely on the refrigerant used and the temperature conditions under which the system operates. Those types most commonly employed are described below.

Water Intercoolers

This type of intercooler employs water as the desuperheating agent and is designed much the same as the shell-and-tube condenser described previously. Its employment is limited by the temperature conditions of the primary discharge gas as it is obvious that negative results will occur if the water temperature is too high. This is most likely when low or ultra-low evaporator temperatures prevail, using some kinds of refrigerants. The effectiveness of any water intercooler in raising system economy depends primarily on the temperature difference between the water available and that of the discharged gas, becoming greater as the cooling water temperature decreases. Some other means must be employed for intercooling where intermediate temperatures are below the freezing point of water (32 F).

Flash Intercoolers

The principle underlying the use of flash intercoolers involves the injection of some portion of the liquid refrigerant from the receiver into the vapor passing from the primary to the secondary compressor. The liquid is bled from the liquid line and enters an inter-stage chamber, through which the primary discharge gas must pass on its way to the secondary compressor. Upon entering the inter-stage chamber, the liquid flashes into vapor at the

Fig. 170. Intercooling, using water to cool vapor between first and second stage compressors. (Courtesy, Mobil Oil Corp., Inc.)

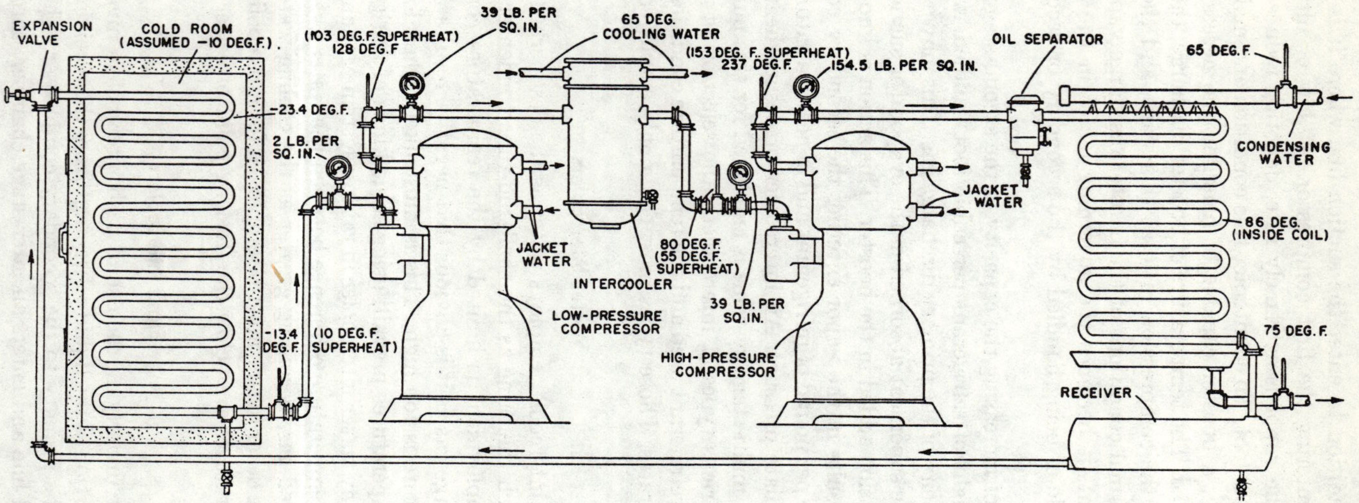

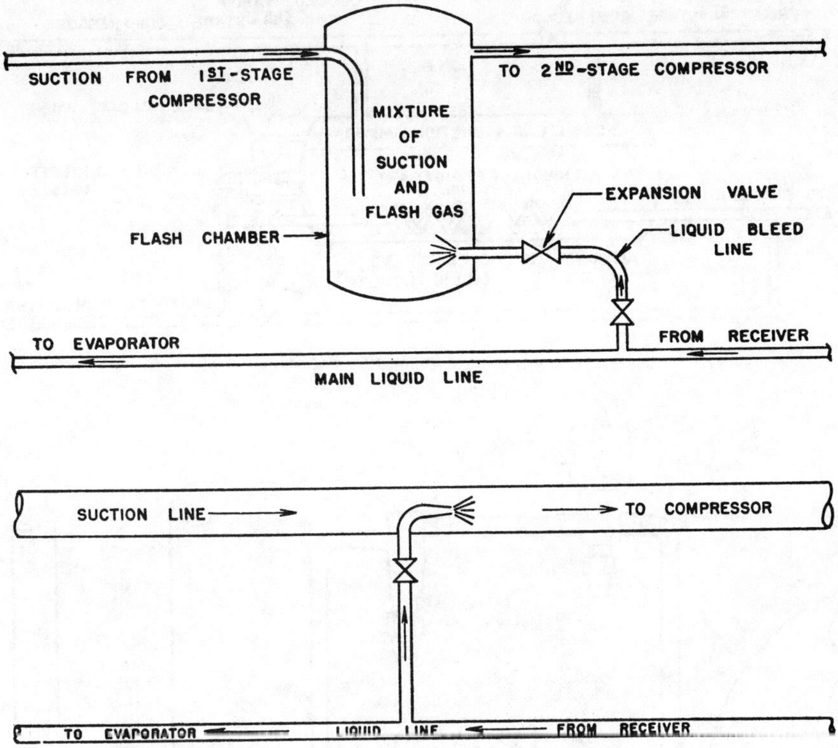

Fig. 171. Flash intercooler for ammonia system.

intermediate pressure, thereby reducing the temperature of the primary discharge gas as a result of latent heat absorption. The amount of inter-stage vapor cooling effected with this type of intercooling depends on the amount of liquid entering the chamber and the pressures involved.

The use of an inter-stage chamber is restricted to refrigerants that are nonmiscible with oil, such as ammonia. With oil-miscible refrigerants (R-12), much of the oil would settle out of the vapor in the intercooler due to velocity reduction, preventing its return to the secondary compressor. To eliminate this problem in systems employing oil-miscible refrigerants, the liquid is often fed directly into the suction line of the secondary compressor, where it flashes to effect necessary cooling of the primary discharge gas. A thermal expansion valve (see Chapter 9, "Controls") is employed to control the amount of liquid entering the suction line.

Liquid Intercooler—Flash Type

This type of intercooler combines the flash type intercooler with a heat exchanger for reducing the temperature of the main liquid body flowing to the evaporator. The main liquid line passes through the inter-stage chamber in the form of a coil. Heat transfer occurs between the warm liquid passing

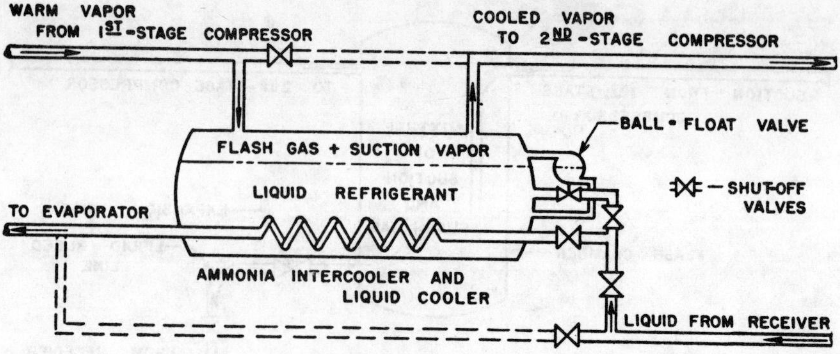

Fig. 172. Liquid intercooler, flash type.

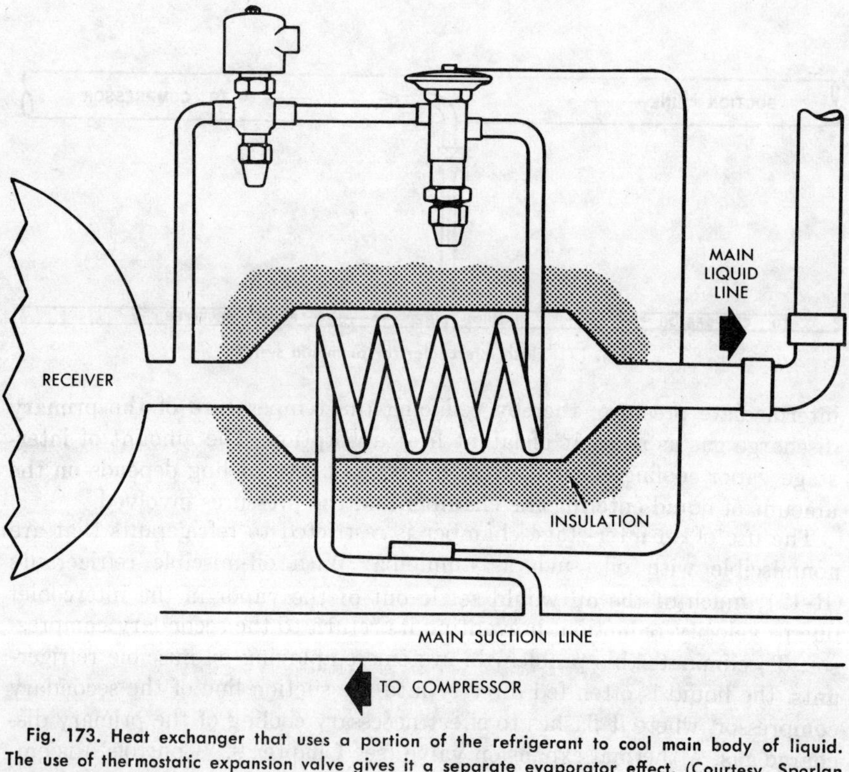

Fig. 173. Heat exchanger that uses a portion of the refrigerant to cool main body of liquid. The use of thermostatic expansion valve gives it a separate evaporator effect. (Courtesy, Sporlan Valve Company.)

through the coil and the primary discharge gas whose temperature is lowered by the flash gas.

While less liquid is available to the evaporator as a result of bleeding to the inter-stage chamber, the loss is more than compensated for by the thermal advantages of cooling both the vapor and the main liquid body. One of the most important factors to be considered in intercooler design of any kind is to keep suction-line pressure drop to a minimum. Otherwise, the subsequent drop in suction pressure at the secondary compressor will cause a loss in system capacity that may, in extreme instances, offset the benefit derived by intercooling. Another type, the shell-and-coil liquid cooler, is now used widely for cooling the liquid only. The liquid is bled from the liquid line and flashes in the shell and around the coil carrying the main liquid body. The flash gas then passes from the shell into the suction line of the secondary compressor. This system is employed mainly on Freon systems (see Fig. 172).

HEAT EXCHANGERS OR ECONOMIZERS

While the term "heat exchanger" may apply to any component of the refrigeration system responsible for the transfer of heat, it is used specifically in reference to liquid coolers placed in the suction line of single-stage systems. They are also called economizers. The purpose of the heat exchanger is to cool the liquid prior to its entrance to the evaporator by utilizing the cold suction gas leaving the evaporator to enter the compressor. The result is a gain in refrigeration effect with some kinds of refrigerants.

With some types of refrigerants, particularly R-12, a net gain in refrigeration capacity results from precooling the liquid even though the suction gas is superheated in doing so. While a net gain of 3 percent to 5 percent may be expected with a heat exchanger in conjunction with R-12, the high specific heat of liquid ammonia offsets the advantages gained by this method of precooling by causing excessive superheating of the suction vapor in systems employing the latter refrigerant. For this reason, vapor-liquid heat exchangers are not normally employed on ammonia systems. (See Chapter 7 for mathematical conclusions.)

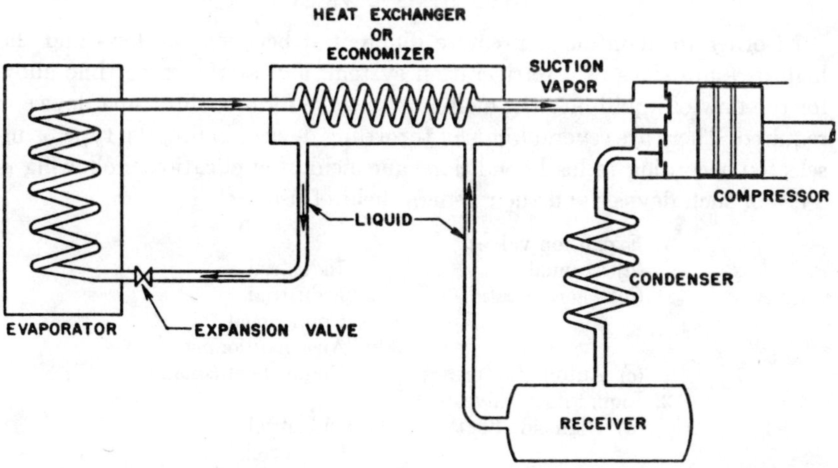

Fig. 174. Relative location of heat exchanger (economizer) in a Freon system.

A Freon heat exchanger and its relative position in a single-stage system are shown in Fig. 174.

Fig. 175. Heat exchanger or economizer, fin type. (Courtesy, Superior Valve & Fittings Co.)

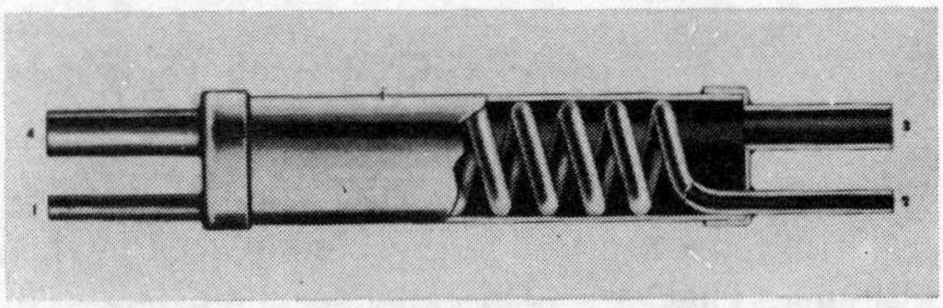

Fig. 176. Heat exchanger or economizer, coil type. (Courtesy, Superior Valve & Fittings Co.)

Liquid Precoolers

Any decrease in the temperature of the liquid refrigerant below that of condensing, prior to its entering the evaporator, increases its refrigerating effect. If precooling is accomplished by an external medium such as water, a gain in total refrigeration capacity follows regardless of the kind of refrigerant. Liquid precoolers are employed for this purpose where the temperature of available water is low enough to allow a reasonable differential between it and the liquid. Any heat removed from the liquid correspondingly reduces the flash loss in the evaporator. Counterflow double-pipe liquid precoolers are the most effective types because of their efficient heat-transfer characteristics.

THROTTLING DEVICES

In order to maintain a pressure differential between the low- and the high-pressure sides of a refrigeration system, and at the same time allow for the flow of liquid into the evaporator, some form of throttling device is required. There are several kinds of throttling devices in use, the type being selected according to load conditions and field of application. Following is a list of such devices and their general field of use:

1. Expansion valves:	
(*a*) Manual	Industrial
(*b*) Thermostatic	Industrial Commercial Air conditioning
(*c*) Automatic (pressure)	Commercial (small)
2. Liquid float valves:	
(*a*) High-side float	Industrial Commercial Domestic

(*b*) Low-side float	Industrial Commercial Domestic
(*c*) Magnetic	Commercial
(*d*) Electronic	Commercial
3. Capillary tube	Domestic

The purpose of any throttling device is to allow the liquid to flow from the high-pressure side of the system to the evaporator at a rate necessary to maintain a constant, predetermined pressure within the evaporator or low-pressure side. This means that the rate of liquid flow through the throttling device should be equal to the rate of evaporation as governed by the heat load. This obviously poses a problem of maintaining a correct balance between the two, notwithstanding fluctuations in heat load, a problem that may be solved by selecting the correct type of throttling device for the operating conditions involved. Such selection is based primarily on the size of the system, the load characteristics, the number of evaporators, and their classification (flooded or dry expansion), etc.

Because the subject of throttling devices is essentially one of control, the operating characteristics and the basis of selection of the various types are described in Chapter 9, "Controls."

PURGERS

The presence of noncondensible gases in the refrigeration system is one of the major causes of abnormally high head pressures. Their elimination is essential to efficient operation. While impurities in both the lubricating oil and the refrigerant provide limited sources for such gases, the leakage of air into the system remains the most serious factor in this regard. Except for hermetically sealed units, some provision must be made for the continual removal of noncondensible gases, notwithstanding all precautions adopted to prevent inward leakage, including the installation of properly fitted seals and gaskets.

When noncondensible gases are present in a system, they are pumped to the condenser and a portion of them finally settles in the receiver or travels along with the refrigerant to the low side. Because they are noncondensible, they tend to restrict the space intended for refrigerant, and, as the volume of such gases increases, so does the head pressure in accordance with Dalton's law of partial pressures. This condition results in unnecessarily greater compressor loads as well as decreasing the rate of heat transfer through condenser tubes by reason of the air film that collects around the tube surfaces to act as an insulator.

To continually eliminate air and other noncondensible gases from the system, a purging operation is necessary. While there are numerous methods by which this function is performed, the unit employed is called a purger.

The purging of a refrigeration system is rendered less difficult than would otherwise be the case if it were not that air is lighter in weight than the refrigerants commonly used. This allows the air to rise and settle in the top of the condenser and the receiver, from where it can be drawn off in various ways and discharged to the atmosphere.

The simplest method of purging is to provide a pipe connection from the top of the condenser and the receiver to some height above both. It is equipped with a purge cock which, when opened periodically, allows the noncondensible gases to escape to the atmosphere. This type cannot be used on condensers operating under pressures below atmospheric nor can it provide for the removal of all air without the loss of refrigerant under

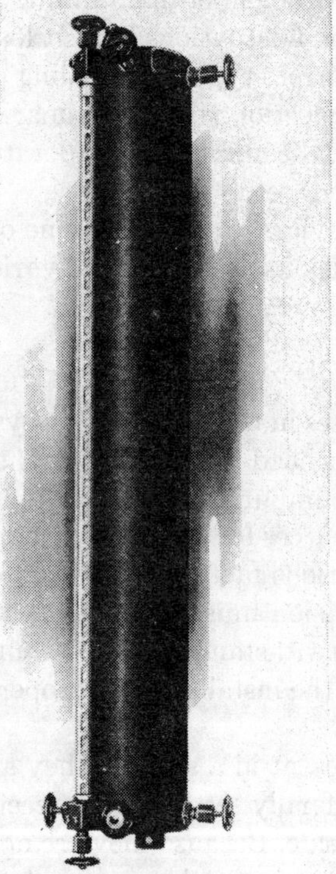

Fig. 177. Manually operated purger. (Courtesy, Frick Company.)

positive pressures. To purge systems operating under partial vacuum or to eliminate refrigerant losses in positive-pressure systems during the purging operation, certain fundamental steps must be taken, viz.:

1. Draw off the gases from the top of the condenser and the receiver into a sealed chamber that forms part of the purger and usually is isolated from the system. The gases thus drawn off are generally a mixture of noncondensibles and refrigerant vapor.

2. Chill the gases to condense the refrigerant and cause its separation from the noncondensibles.

3. Trap and discharge to the atmosphere the noncondensible gases that rise to the top of the purge chamber upon separation.

4. Trap and return the liquid refrigerant to the system.

Purgers may be operated manually or automatically.

On those systems whose condensers operate under partial vacuum, a small compressor is required to draw off the noncondensible gases. The compressed gases are then passed through an air-cooled condenser where the liquid refrigerant and the inert gases separate to be trapped individually and disposed of, as outlined above.

Where positive condenser pressures are maintained, the gas mixture is forced into the purge chamber under condenser pressure, where it is exposed to and chilled by a cooling coil fed through a liquid bleeder line and expansion valve from the receiver. The refrigerant vapor condenses out of the mixture and drains to the bottom of the chamber, while the inert gases rise to fill the space above the liquid. From this point on, the action of the purger depends on whether it is operated automatically or by hand.

If hand operated, the noncondensibles are released to the atmosphere through a discharge line leading from the top of the purge chamber when a hand-operated valve is cracked. The liquid level, depressed by the gas pressure, rises in the chamber by the force of the main liquid line as the volume of noncondensibles decreases. When the liquid in the chamber, as indicated by means of a sight level glass, reaches a maximum level, the discharge valve is closed, as is the expansion valve feeding the cooling coil.

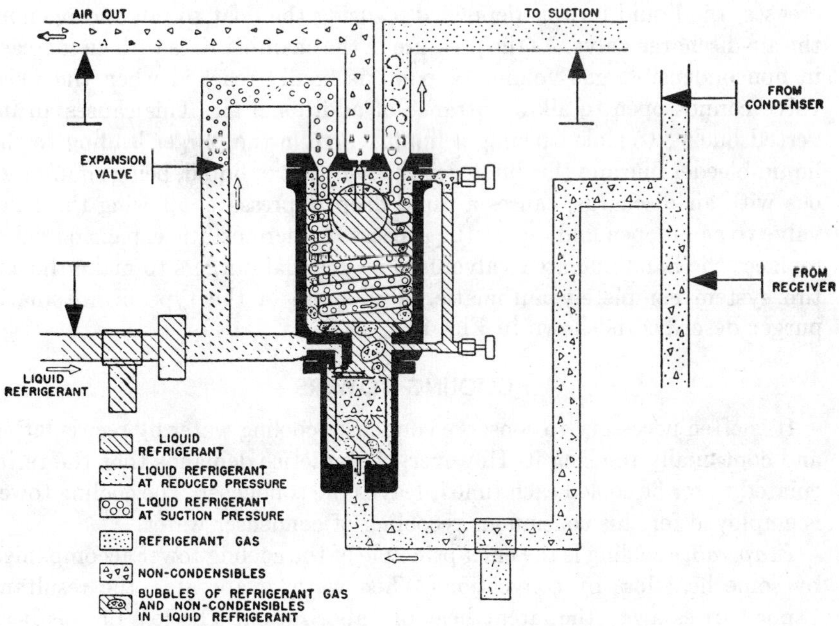

Fig. 178. Automatic purger, cross-sectional view. (Courtesy, Armstrong Machine Works.)

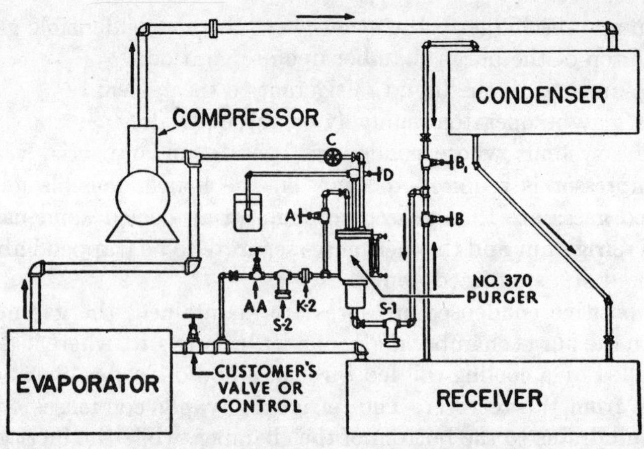

Fig. 179. Automatic purger hook-up. (A) Expansion valve, (B) receiver shut-off valve, (B_1) condenser shut-off valve, (AA) liquid supply valve, (C) compressor draw-off valve, (D) suction line valve, (S-1 and S-2) strainers, (K-2) differential valve (reducer). (Courtesy, Armstrong Machine Works.)

The purger becomes warm and the liquid refrigerant drains back to the system. All valves are then closed until purging is again required. A manually operated purge system is shown in Fig. 177.

The automatic purger works on the same principle except that the hand-operated purge valve and the gas-inlet valve are replaced by a liquid float and check valve, respectively. As the volume of noncondensible gas increases, the liquid level is depressed, causing the float to fall and opening the air-discharge valve. As the pressure in the chamber rises with an increase in noncondensible gas volume, a point is finally reached when the relief valve cannot open to allow entrance of additional air. This causes an inverted bucket to sink, opening a liquid valve in the purger leading to the liquid bleeder line and the differential valve. Exit of liquid, being simultaneous with air discharge, causes a rapid drop in pressure, allowing the relief valve to again open and repeat the process. A thermostatic expansion valve replaces the hand-operated valve used on manual purgers to make the entire system completely automatic. A diagram of the type of automatic purger described is shown in Fig. 179.

COOLING TOWERS

It is often necessary to conserve condenser cooling water by recirculating and continually re-using it. However, the practice demands that the recirculated water be cooled each time it leaves the condenser. The cooling tower is employed for this continuous recooling of condenser water.

Evaporative cooling is the chief principle of the cooling tower accompanied by some heat loss by convection. When water evaporates, the resultant vapor carries away the latent heat of vaporization. The loss of this heat causes a reduction in temperature of the body of water from whose surface evaporation occurs. The rate at which heat is removed depends on several factors: (1) the amount of water surface exposed, (2) the time of exposure, (3) the velocity of the air passing over the surface, (4) the relative humidity of the air or the relation of dry-bulb to wet-bulb temperature, (5) the relative direction of air flow relative to the water surface; i.e., parallel or transverse flow.

1. The amount of water exposed to the atmosphere may be increased either by (*a*) allowing it to spread thinly over a large area or (*b*) breaking it up into small droplets as by splashing or spraying.

2. The time of exposure may be increased or decreased by varying the dimensions of the space through which the water is allowed to flow. It may be controlled also by the design of the component or carrier that supports the flow.

3. The velocity of the air passing over the water surface is a function of winds and atmospheric conditions in natural-draft towers. It depends on the action of a fan in forced-draft towers.

4. The relative humidity as related to the outside atmosphere is a function of the weather and may vary from 30 percent to 100 percent. Design usually is based on a wet-bulb temperature that represents maximum for

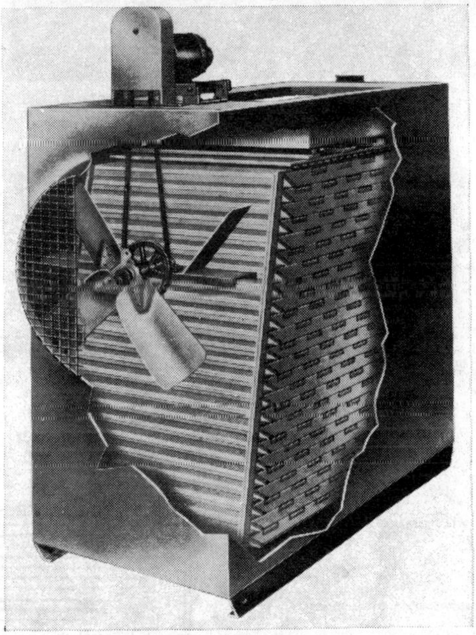

Fig. 180. Cooling tower, commercial type. Note staggered decks. (Courtesy, Halstead & Mitchell.)

about 95 percent of the time for the area involved. In the United States, the design maximum wet-bulb ranges from 64 F to 80 F, depending on the location.

5. The relative direction of air flow to water surface depends on the design of the carriers.

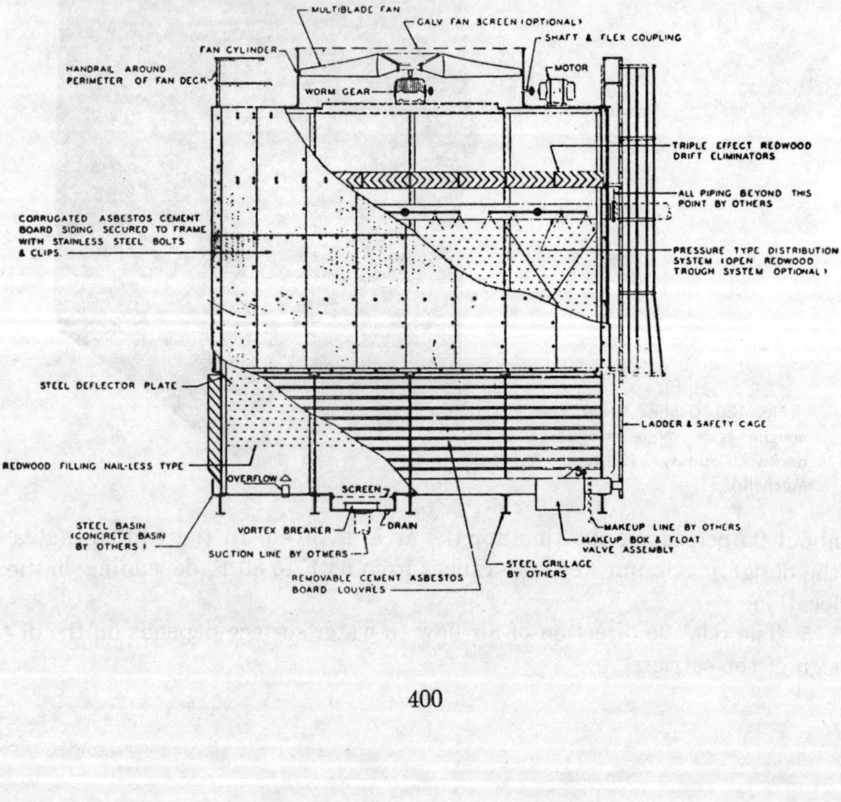

Fig. 181. Cooling tower, large industrial type. (*Top*) External and (*bottom*) cross-sectional views. (Courtesy, Phillips Cooling Tower Co., Inc.)

Cooling towers are built to exploit the flow of natural draft or to utilize the air velocity created by a fan known as forced draft. A combination of the two is also employed.

Natural-Draft Towers

The open type natural-draft tower is a boxlike structure with a bank of louvers placed fairly close together running horizontally across all four sides, supported by corner posts. The louvers are slanted downward (approximately 50°) toward the inside of the tower. The latter prevents the loss of water and helps direct flow to the reservoir in the bottom. The space between the louvers allows for the passage of air in any horizontal direction.

The warm condenser water is pumped to a trough or header located at the top of the tower, from which it falls or is sprayed downward. Spraying is one method of breaking the water up into droplets. Impact is another.

Generally, the reservoir is equipped with a constant-level float valve through which make-up water enters to replace that lost by evaporation

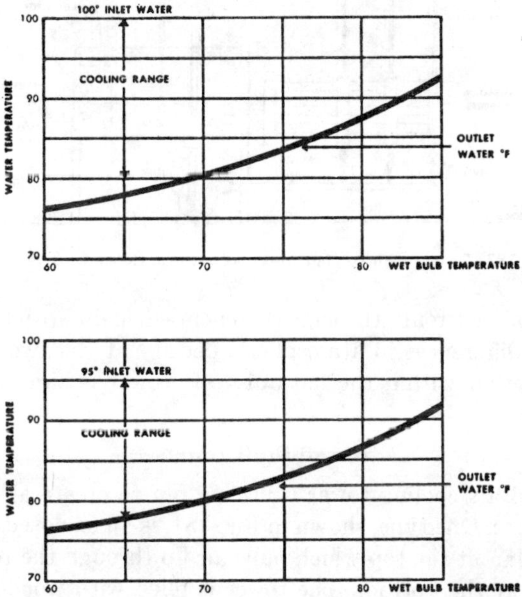

Fig. 182. Cooling tower temperature range as related to wet bulb temperature. Curves based on a fixed water flow of 3 gpm per ton and constant inlet water temperature. (Courtesy, Halstead & Mitchell.)

and drift. Drift is the entrained water, in the form of droplets, carried away by the air as it passes through the tower. The drift loss increases as air velocity becomes greater.

Sometimes a closed type natural-draft tower is used. Open at top and

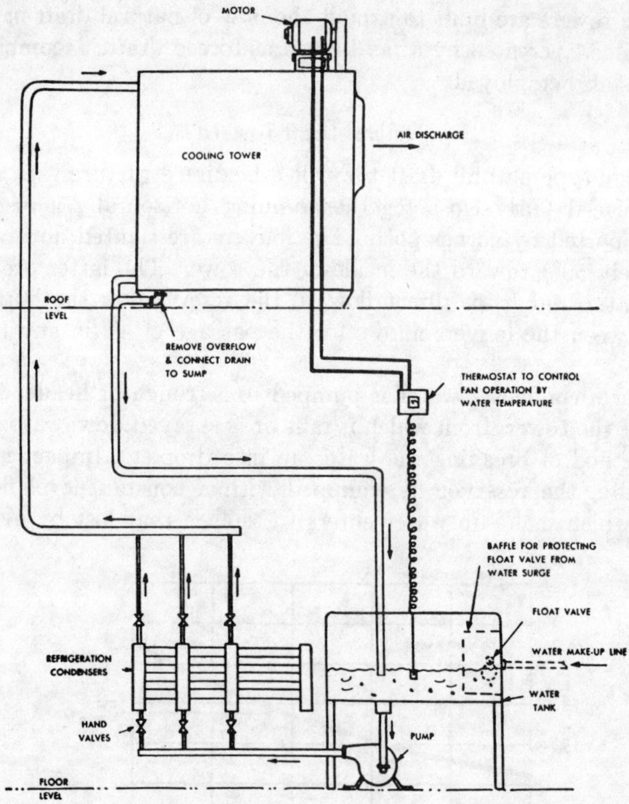

Fig. 183. System with cooling tower. (Courtesy, Halstead & Mitchell.)

bottom, only air currents through the enclosed main structure are set up as a result of differences in atmospheric density at the two levels. In this respect, the action within the natural-draft tower is similar to that of a chimney.

Forced-Draft Towers

The forced-draft cooling tower utilizes a fan to create air velocity over the water surface. One type, shown in Fig. 181, is an enclosed boxlike structure having a fan at the top which pulls air up through the tower from the louvered inlet at the bottom. The tower is filled with wooden splash bars called filling which retard the fall of the water and promote splashing. The splash bars are staggered to allow the water to fall and the air to pass upward.

The water enters the tower at the top into a distributing header or basin, from which it falls to the filling. As the water falls from one deck to the next, it splashes to form small droplets, thereby increasing the amount of surface exposed. It finally reaches a reservoir at the bottom of the tower, from where it again flows to the condenser. As in the natural-draft tower, the level in the reservoir is held constant by a float valve connected to a make-up water line. To prevent free water leaving with the air, baffles are placed between the water distributing header and fan.

The problem of freezing during winter operation is sometimes overcome by reversing air flow for short periods. Revolving the fans in the opposite direction sets up a down draft. The air warmed by the high-temperature entering water de-ices the louvers and other surfaces.

Forced-draft cooling towers are constructed also with the fan located on one side. The inlet louvers, being installed on the side opposite the fan, allow horizontal flow of air across the filling surfaces.

The heat load of a cooling tower is the number of Btu per unit of time (hours or minutes) absorbed by the air as latent heat of vaporization during its passage. The heat load of any tower is determined by the number of pounds of water circulated per minute multiplied by the cooling range. The cooling range is the difference in temperature between entering and leaving water.

For example, a tower through which 100 gal of water per min (gpm) is pumped and having a temperature range on a given day of 7 F would have a heat load of

$$\begin{aligned}\text{Btu} &= 100 \times 8.3 \times 7^\circ \\ &= 5810 \text{ Btu per min}\end{aligned}$$

Considering that condenser loads approximate 250 Btu per min per ton, then the tower in the example would be capable of serving a system of

$$\frac{5810}{250} = 23.24 \text{ tons}$$

In selecting natural-draft towers for a specific load, 1 sq ft of cross-sectional area should be provided for each 12 lb of water per minute to be cooled. For forced-draft towers, a lesser ratio is required, being about 1 sq ft for each 25 lb of water per minute or approximately half that of the natural-draft tower. The area referred to is the free area through the tower.

The weight of air required for a given load is determined by the equation

$$W_a = \frac{W_w \times t_d}{h_{a2} - h_{a1}}$$

and the free area across the tower relative to air flow will be

$$A_f = \frac{W_a \times V_s}{v_a}$$

where W_a = weight of entering air, lb
W_w = weight of water circulated per min
t_d = temp difference, leaving air (t_2) entering air (t_1)
h_{a2} = total heat in leaving air, Btu per lb
h_{a1} = total heat in entering air, Btu per lb
A_f = free area, sq ft

V_s = specific volume of entering air, cu ft per lb
v_a = velocity of air, fpm

To compensate for deficiencies in the tower, some modification in the total heat of leaving air should be made. This may be done by applying a tower *efficiency* factor, *e*, which may be assumed to be approximately 70 percent to 75 percent for forced-draft systems. This factor can be applied to the temperature range value as represented by t_d in the equation. If the maximum range is, say, 12 F between entering and leaving air, then h_{a2} is based on a leaving-air temperature of $(12 \times e) + t_1$. For an entering-air temperature of 70 F, then, the leaving air would be $(12 \times 0.70) + 70$ F = 78.4 F, based on a tower efficiency of 70 percent.

It cannot be expected that the water leaving the cooling tower will reach the atmospheric wet-bulb temperature. This means that a temperature difference will exist between the two which is referred to as approach. The temperature difference between the incoming water and that of the wet bulb, then, may be divided into two ranges: (1) the cooling range (°F water in − °F water out) and approach (°F water out − °F wet bulb).

The amount of make-up water depends on the loss due to evaporation, windage, and bleeding.

The rate of evaporation varies according to the difference in the vapor pressure of the water and that of the atmosphere moisture. It is determined for parallel flow by the empirical formula

$$W_w = \frac{95 + 0.425V}{r}(e_w - e_a)$$

or for transverse flow

$$W_w = \frac{201 + 0.88V}{r}(e_w - e_a)$$

where W_w = pounds of water evaporated per sq ft of water surface per hr
V = velocity of air, ft per min
r = latent heat of vaporization
e_w = vapor pressure in. of Hg corresponding to water temperature
e_a = vapor pressure of moisture in the atmosphere at wet-bulb temperature

Based on a latent heat value of 1000 Btu per lb and 250 Btu per ton per min, the ratio of evaporation to refrigeration produced is $\frac{250}{1000} = 0.25$ lb per min per ton evap. This is equivalent to about 2 gal of water per hr per ton.

The loss of water by windage (drift) is governed by the design of the tower. Drift loss of at least 1 percent of the water circulated should be considered for natural-draft units. This means a loss of 1 gpm for each 100 gal of water circulated per min. Where efficient baffles are installed, as in forced-draft towers, drift loss will be about 50 percent less than that for natural-draft units.

Water contains certain mineral salts such as the sulfates, the bicarbonates, the nitrates, and the chlorides of calcium, magnesium, and sodium.

It also contains silica, alumina, and compounds formed by iron and manganese. Many of these minerals cause water hardness, especially the carbonates and the sulfates of calcium and magnesia. The salts that cause water hardness are also responsible for the formation of scale that tends to deposit on the surfaces of pipes and reservoirs of water systems. The retardation of scale formation is accomplished by water-softening processes where it is feasible to do so. Water treatment is absolutely necessary in steam-boiler operation.

As evaporation of water continues, the scale-forming minerals (as well as all others) build up and become more concentrated in the system. If allowed to continue, the subsequent scale formed by the minerals eventually will clog the passages and attack (corrode) the metal parts of the system. The formation of scale will occur also in any evaporation system regardless of the relative hardness or softness of the water. For this reason, even where the water is treated, such as in boiler operation, the system must be subject to periodic "blow-down." One of the reasons for blowing down is to bring the system into better chemical balance. This is accomplished by the subsequent replacement of the concentrated salt solution with an equal amount of water having relatively low salt content.

Because the process of evaporation is continuous in the cooling tower, the circulating water is subject to an increasing concentration of mineral salts just as it is in the steam boiler. While this condition is relieved to some extent by make-up water replacing drift, it is not sufficient to keep the system in balance. It becomes necessary to bleed the system periodically or continuously. Bleeding rids the system of a specific amount of water whose mineral salt content has been built up somewhat by continuous evaporation. The addition of make-up water replacing that which is bled off acts to reduce the relative salt concentration of the remaining mass, keeping the system in balance. The bleed-off valve should be located in the warm water line connecting the condenser outlet to the water-cooling tower header. The amount of bleed-off depends on the chemical condition of the available water supply, being much heavier in localities where the mineral content is high. It may be determined by direct proportion, viz.:

$$\frac{\text{Amount of bleed-off water}}{\text{Amount of water evaporated}} :: \frac{\text{Amount of salts in make-up water}}{\text{Allowable salt in system} - \text{salt in make-up}}$$

therefore

$$\text{Bleed-off} = H_2O \text{ evaporated} \times \frac{\text{salt in make-up } H_2O}{\text{allowable salt in system} - \text{salt in make-up } H_2O}$$

The amount of make-up water for a cooling tower is equal to the sum of (*a*) loss by evaporation, (*b*) loss by drift, and (*c*) loss by bleeding.

Spray Ponds

Spray ponds are also used for cooling condenser water. The principle of cooling is the same as for the cooling tower, that of evaporation. The water is pumped from the condenser through spray nozzles that direct it vertically into the air above the pond. The draft caused by the spray plus the natural draft of the atmosphere supports evaporation. The nozzles are adjusted to control the particle size of the spray. The latter is important as water particles that are too fine will result in excessive evaporation and/or drift. A loss of 2 percent to 4 percent water may be expected.

One of the difficulties experienced with spray ponds, cooling towers, condensers, and other water circuits is the growth of microscopic plants called algae. Fresh-water species of algae consist of two varieties, blue-green and green. Highly productive, algae in time will choke drains and other water outlets. If they reach the condenser, algae will not only restrict and finally block the flow of water but also will reduce heat-transfer efficiency commensurate with the thickness of coating. The expression "fouling" is applied to the coating of water circuits by micro-organisms and scale. As fouling organisms, algae are included in the study of fouling control.

Of the organisms responsible for fouling, which include many types of bacteria, algae are considered one of the higher plants. The presence of chlorophyll in algae imparts a catalytic action responsible for the transformation of carbon dioxide and water into glucose with the help of energy supplied by the sun's rays and other light sources. It is the glucose that helps give algae-infected areas a slimy characteristic. Other forms of bacterial growth also support this condition. Surface fouling is aggravated by the intermingling of scale, mud, and other inorganic material with the slime to form a heterogeneous mass. Water and its problems are discussed further in Chapter 13.

DRYERS

Because of the harmful effect of moisture within a refrigeration system, every precaution should be taken to prevent its entrance or to eliminate it if already present. It may enter with the lubricating oil, which is hygroscopic, or with the refrigerant during initial charge. Moisture is also present in the atmosphere, which enters the system during periods when the various components are opened up for repairs. Faulty seals and joints also furnish avenues through which moisture-laden air gains entrance to the system.

Some of the more harmful effects of moisture are that it

(*a*) Combines with certain kinds of refrigerants to form corrosive compounds,
(*b*) Freezes and interferes with the function of expansion valves,
(*c*) Causes rapid deterioration of lubricating oil,
(*d*) Reduces system capacity by replacement of refrigerant,
(*e*) May enter into an oil-moisture-refrigerant relationship that influences copper plating. (See Chapter 16, "Lubrication.")

Fig. 184. Dehydrator, factory sealed. (Courtesy, The McIntire Company.)

The amount of moisture may be held to a minimum by taking the following steps:

(*a*) Completely drying the system and all parts before charging or recharging,
(*b*) Evacuating all air before charging, checking all seals and joints in the process,
(*c*) Avoiding opening any component after evacuation of air,
(*d*) Using dehydrated lubricating oil,
(*e*) Employing anhydrous refrigerants,
(*f*) Installing suitable dryers.

Dryers are cartridge-like units consisting of a demountable outer shell having tubing connections at both ends. A typical specimen shown in Fig. 184 contains a filtering element and a dehydrating material or desiccant usually contained in a bag or tube for easy replacement.

The preferred location of any dryer within the system with the exception of SO_2 units is in the liquid line between the receiver and the expansion valve and in the cold compartment if at all possible when either Freon-12 or methyl chloride is employed. The reasons for selecting this location are as follows:

1. The moisture is concentrated in a much smaller volume of fluid when refrigerant is in its liquid phase as compared with its vapor phase.

2. It affords greater protection to the expansion valve by removal of moisture before it reaches that component.

3. The refrigerant travels at its minimum velocity in the liquid line, allowing more time for contact with the drying agent, an important factor in drying efficiency.

4. Solubility diminishes as temperature decreases. As free moisture is removed more easily than when in solution with the refrigerant, some advantage is gained by placing the dryer in the cold compartment.

To prevent the freezing of moisture at the throttling device, methyl alcohol is sometimes added to the system in an amount equal to 1 cc per pound of refrigerant. This practice is based on the principle that methyl alcohol reduces the freezing point of water.

Sulfur dioxide has a high affinity for water, which it readily absorbs and which is difficult to remove when this refrigerant is in its liquid form. For this reason, dryers are located in the suction line of systems employing SO_2. Because water and SO_2 form an acid immediately upon contact with each other, it is highly important that the unit and all its parts, including the lubricating oil, be thoroughly dry when this refrigerant is introduced.

Regardless of the kind of refrigerant employed, if the dryer is placed in the condenser discharge line, one of larger size is required to compensate for the higher temperature.

Drying materials or desiccants used in refrigerant dryers may be divided into two general types, chemically referred to as (1) the adsorption type and (2) the absorption type.

Adsorption Type

The adsorption type of desiccant collects moisture mechanically by capillary attraction, whereby the fluid adheres to solid surfaces to form a thin film without apparent chemical change in either substance. Because the bond between the desiccant and the moisture is not permanent, reactivation of the material is possible, allowing for its continual re-use. Reactivation is accomplished by heating at mild temperatures to drive off the moisture collected in the system.

Having numerous microscopic pores or capillaries, adsorption type desiccants offer a considerable amount of drying surface per unit volume.* To retain their high capacity, such materials should be kept free of dirt and oil so as not to plug their capillaries and reduce their adsorption capacity.

Because no apparent chemical changes occur in adsorption materials as a result of moisture exposure, dryers in which they are employed can remain in the system without fear of damage.

The most commonly used adsorption-type materials are silica gel (pure silicon dioxide) and activated alumina (aluminum oxide). Granular in form, both of these desiccants vary in size, having considerably large areas of drying surface. Silica gel has a greater capacity than activated alumina, although the latter can produce a lower moisture level (dew point) than silica gel. Activated alumina can be used satisfactorily on R-11, R-12, methyl chloride, methylene chloride, and SO_2.

A proprietary material known as Mobilbead is a form of silica gel that has proved highly satisfactory as a refrigerant desiccant due to its high adsorption capacity, spherical shape, durability, uniform size, and chemical stability. Sovabead, used for charging refrigerant dryers, range in size from 4 to 8 mesh, which, coupled with their spherical shape, provide minimum resistance to the flow of refrigerant. Like other kinds of adsorption type desiccants, Sovabead may be left in the system indefinitely without harmful effects and can be reactivated repeatedly with no appreciable loss in adsorp-

* Makers of Mobilbead (Mobil Oil Corporation) estimate 2,000,000 sq ft per lb of their product.

tion capacity. Sovabead will not pack down in the cartridge to leave open spaces for moisture-wet refrigerant to pass through, or react with normally used refrigerants or metals of construction.

Absorption Type

The absorption type of desiccant has an affinity for moisture and will react with it to form another compound. This means that the relationship is chemical, rather than mechanical as with the type described above. Moisture *absorbed* by the desiccant of this type becomes an essential ingredient of the newly formed compound. Because of the chemical change that occurs, both the composition and the characteristics of the new substance differ from the original materials.

Some materials in this group disintegrate and form a powder or dust as a result of moisture reaction. Others deliquesce; i.e., they are dissolved by the moisture absorbed and finally liquefy. Those that dust may act to clog the system while some deliquescent materials become corrosive upon liquefying.

As chemical dryers cannot be reactivated, they must be replaced by a fresh charge. Because of their detrimental characteristics, they are installed only on a temporary basis and should never be left in a system more than 12 to 18 hr. The advantage of this type dryer lies in its ability to absorb moisture rapidly. If installed on a new unit or on one recently repaired, a chemical dryer will remove the major portion of moisture present in the system in a short time, after which it should be removed and replaced by a mechanical or adsorption type dryer. Considerable heating accompanies moisture absorption, such as that produced when water is added to quicklime (calcium oxide). However, heating alone is not a positive criterion that moisture is being absorbed nor will the lack of temperature rise serve to indicate a good or bad dryer in all cases.

Materials of the chemical type used as refrigerant dryers include the following:

Calcium oxide:	Tends to dust, cannot be used with SO_2.
Phosphorus pentoxide:	Very effective dryer, its powdery form causes refrigerant flow resistance.
Barium oxide:	Excessive powdering in the presence of moisture.
Calcium chloride:	Liquefies and becomes highly corrosive.
Calcium sulfate:	Tends to dust—granular form minimizes flow resistance.
Drierite:	Anhydrous calcium sulfate. Granular, also cast in sticks. Tends to dust somewhat, but does not cause trouble.

Moisture Indicators

To determine the moisture conditions within halocarbon and methyl chloride refrigeration systems, an indicator may be employed. Contained in a transparent cartridge, the indicator is made up of a material such as anhydrous calcium sulfate (Drierite) crystals treated with a chemical agent that will change color when wet. The time required for the crystals to change color indicates the temperature (dewpoint) at which refrigerant will release water. An indicator and cartridge dryer are shown mounted as a moisture-control unit in Fig. 185. The coloring agent used in the moisture-control unit shown is *green when dry* and *purple when wet*.

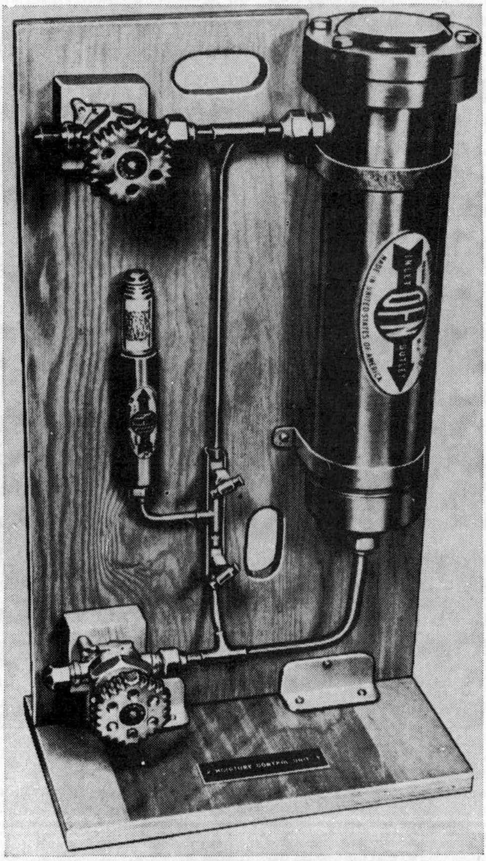

Fig. 185. Moisture-control unit. (Courtesy, The McIntire Company.)

When exposed to moisture in the refrigerant, the green indicator will change to purple. The manufacturer of the control unit shown gives the following data for systems operating at −10 F and above:

	Time in Minutes	*Refrigerant Dry to °F Dewpoint*
If green cartridge starts to show purple in	1	20°
	2	0
	4	−10

If the indicator is purple either by blowing on it or from prior exposure to moisture, then

	Time in Minutes	*Refrigerant Dry to °F Dewpoint*
If purple in cartridge starts to fade in	$\frac{1}{2}$	−60
	1	−50
	2	−40
	3	−30

When exposed for 5 min during which the color does not change either way, then the refrigerant will be dry to −15 F dewpoint.

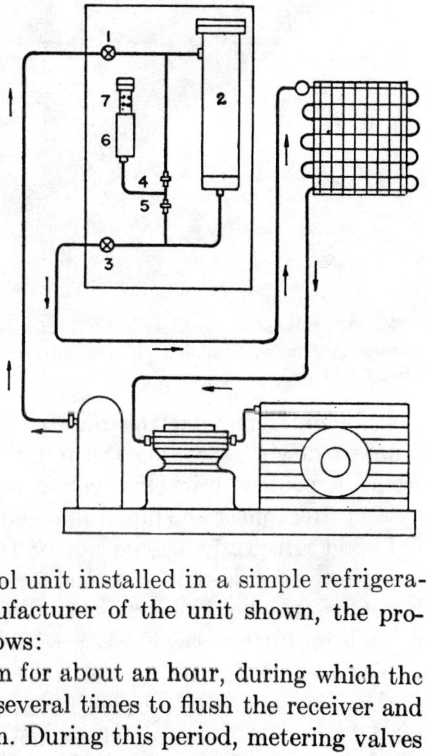

Fig. 186. Moisture-control unit hook-up. (Courtesy, The McIntire Company.)

Fig. 186 shows the moisture-control unit installed in a simple refrigeration system. According to the manufacturer of the unit shown, the procedure for drying a system is as follows:

1. Operate the refrigeration system for about an hour, during which the receiver valve is opened and closed several times to flush the receiver and remove any moisture trapped therein. During this period, metering valves *4* and *5* are closed and liquid passes through the dryer (*2*).

2. After 1 hr of operation, remove the caps from moisture indicator cartridge *7* and screw into the chamber (*6*). Crack valve *4* and test for gas flow by holding a wetted finger about $\frac{1}{2}$ in. above the cartridge opening. Note the time of opening valve *4*. Observe any color change in the indicator crystals and again note the time when such a change starts. Based on the time required for the crystals to change color, the moisture conditions can be determined from the time charts above.

3. To check the dryer (*2*), proceed as above with valve *4* closed and valve *5* open. If the indicator changes to purple in 5 min or less, the dryer cartridge should be changed

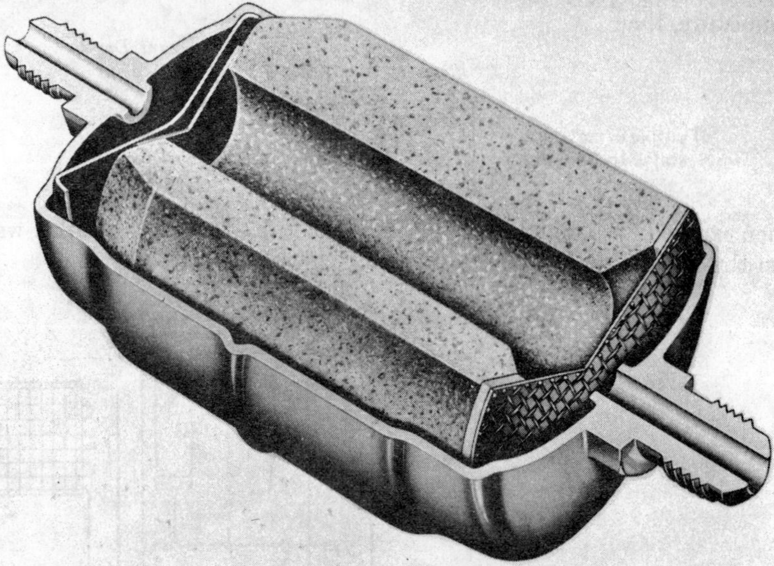

Fig. 187. Filter-Drier. Removes moisture and oil by absorption, acids by absorption, and solid particles by filtering (straining). It is called a Catch-All by the manufacturer. (Courtesy, Sporlan Valve Company.)

4. Before removing the moisture-control unit, the system should be pumped down by closing the receiver valve. After all liquid is removed from the control unit, close valves *1* and *3* and remove the unit from the system. Reconnect the liquid line and purge the system.

Liquid refrigerant should not be left in the moisture-control unit. The metering valve should be opened far enough to allow only gas to enter the indicator. Any liquid or oil getting into the indicator cartridge will ruin the crystals for further use. For best results, the system should be warmed sufficiently to melt any ice crystals that may be present in any part. If dried to the point where the purple color of the slightly moistened indicator fades out in $\frac{1}{2}$ min, the system will be moisture-free down to -60 F.

FILTERS AND STRAINERS

Any fluid flow system will accumulate a certain amount of foreign matter that eventually interferes with good operation. Gums, sludge, fibers, waxes, and abrasive particles occur in refrigeration systems as well as in water, oil, or any other circulating system. The source of such materials includes poor quality lubricating oil, packing, sealing compound, interior pipe surfaces (scale), and frictional surfaces (metallic wear), including compressor, pumps, and valves. In new equipment various residues of manufacture and construction are also found. Unless continually removed, improper valve seating, fouled control mechanism, clogged orifices, and aggravated wear of frictional surfaces will result.

In large industrial systems, periodic blow-down of evaporators, condensers, and oil traps will remove some of the foreign matter. Better results will be obtained by the use of filters, strainers, and scale traps placed at strategic locations in the refrigerant flow lines. Usually, each individual expansion valve of the self-actuating type is equipped with an individual strainer at the inlet. Where dryers are present, they are also equipped with a straining element.

The screening ability of any flow-line cleaner depends on the density and the depth of the filtering material, or the mesh size of the straining element. The amount of exposed area is also an important factor, as is the pressure drop across the cleaning assembly. Pressure drop should be maintained as low as possible not only because it interferes with efficient performance of the refrigeration system but also because more effective cleaning is obtained with gentle fluid flow.

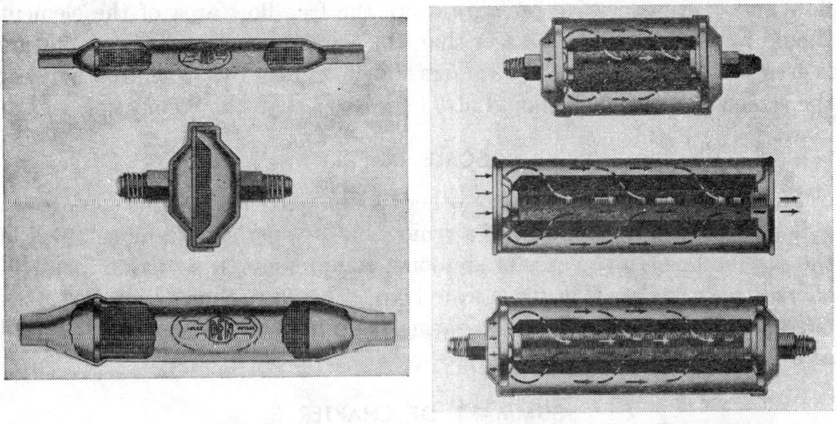

Fig. 188. (*Left*) Strainers and (*right*) filters. (Courtesy, The McIntire Company.)

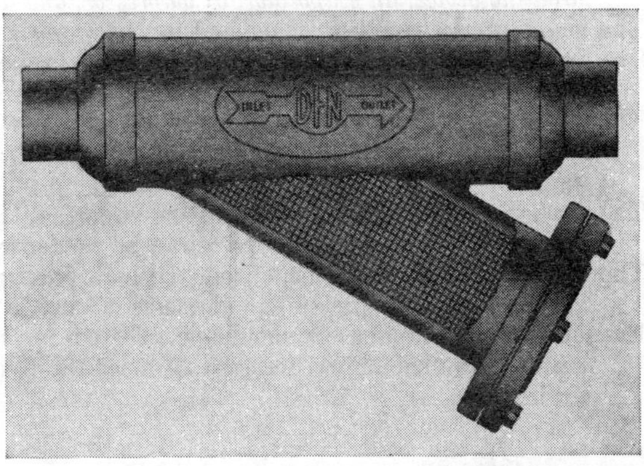

Fig. 189. Strainer, cleanable Y-type. (Courtesy, The McIntire Company.)

Strainers are effective in trapping foreign material of relatively small size such as would be capable of preventing proper control seating. Usually they are made up of screening material that resists corrosion under operating conditions. The screens of the strainers shown in Fig. 188 are of Monel metal. Mesh sizes of strainers may vary from 100 mesh to 60 mesh. They are designed for straight-line flow or in the form of a demountable trap such as that shown in Fig. 189. The small-mesh screen is often reinforced with a 10-mesh brass screen. A demountable angle type strainer is also used. Both the angle and the Wye (Y) types allow cleaning of their screens without removal of the shell from the line. Screens are placed in the liquid line to protect expansion valves and should also be present in the suction line to protect the compressor, especially those having plug type pistons.

Filters are capable of removing material of microscopic size. Equipped with graded density fiber filters, they will trap particles of sizes from 10 to 1 micron (0.001 millimeter). Filters are recommended for use in new systems particularly during the run-in period.

To compensate for the tendency of the cleaning elements to restrict fluid flow and cause excessive pressure drop, the free flow area of the element should be proportionately larger than the connection area. This proportion is greater than 5 to 1 in the strainers shown in Fig. 188. The area between the screen and the shell should also be greater than the connection area.

SCALE TRAPS

To avoid carry-over of scale or other foreign particles into the compressor cylinders, a trap in the form of a removable strainer should be installed in the suction line. In the case of ammonia compressors, the suction manifold offers a positive location for a scale trap; otherwise, the suction line itself offers a suitable channel for separating scale from the refrigerant if placed at a point approaching the compressor.

SUMMARY OF CHAPTER 8

The basic equipment of a simple refrigeration system consists of a *liquid receiver, a throttling device, an evaporator, a compressor, and a condenser.*

The *liquid receiver* receives, holds, and feeds the refrigerant charge.

The *throttling device* meters the liquid refrigerant to the cooling element or evaporator to control refrigerating temperature through proper pressure.

The *evaporator* is the cooling element of the system wherein the liquid evaporates under reduced pressure to absorb heat of vaporization from surrounding substances.

The *compressor* raises the pressure of the vapor as it comes from the evaporator, raising its boiling point so that it can be condensed and re-used.

The *condenser* extracts the heat from the high-pressure vapor, transforming it to liquid at the same pressure, and passes it back to the receiver.

Evaporators and condensers should be selected *on the basis of their heat-transfer ability.* They should be designed to meet the requirements of their intended service.

More *complex systems employ auxiliary equipment* to further operating economies and convenience. Auxiliary equipment includes boosters, accumulators, heat exchangers, purgers, liquid or precoolers, surge tanks, cooling towers, fans, and pumps.

Boosters are employed where low-density gas from ultra-low temperature evaporators requires compression by stages to offset volumetric losses. They are also called initial or intermediate-stage compressors, depending on the number of stages involved.

Accumulators are modified surge tanks used on flooded systems to

(*a*) Maintain required liquid level in evaporator,
(*b*) Allow recirculation of liquid returning from evaporator,
(*c*) Form a dome from which spent vapor can be returned to compressor,
(*d*) Act to prevent surging in suction line and evaporator.

Heat exchangers act to precool the liquid by transferring heat to the cold vapor as the latter comes from the evaporator. Some superheating of the vapor occurs in the process. Heat exchangers also prevent sweating or frosting of the suction line when suitably designed.

Liquid precoolers employ external cooling sources to reduce liquid temperatures.

Flash intercoolers exploit a portion of the liquid to reduce the temperature of the remaining liquid.

Purgers are employed to extract air and other foreign gases from the system.

Oil separators are used in systems employing immiscible refrigerants. They extract the oil from the compressor discharge gas. The oil can be either removed from the system by blowing down or returned to the compressor crankcase by some automatic trap arrangement.

Cooling towers used for the cooling of condenser water are designed to provide a large area over which a thin flow of water may pass. With or without fans, cooling is accomplished by partial evaporation of the water by air passing over its surface. *The process is called evaporative cooling.*

Spray ponds have the same purpose as cooling towers. *Cooling is accomplished by evaporation* as air passes through the spray. Spraying breaks up the liquid into small particles, thereby exposing a large surface area to the air.

Dryers are employed to reduce moisture in the refrigeration system. They are usually of the cartridge type, so may be replaced when saturated. There are two basic types: (*a*) adsorption (capillary) and (*b*) absorption (chemical).

Moisture indicators determine the moisture conditions of a refrigerant. They are based on color change of a chemically treated substance having moisture susceptibility.

Compressors are classified as to principle of operation, which, in turn, governs the refrigerant best handled by each type. The three principal types are: (*a*) reciprocating, (*b*) rotary, and (*c*) centrifugal.

Reciprocating and rotary compressors are *positive displacement types*. The centrifugal compressor is a *kinetic displacement type*.

The type of compressor used often depends on the kind of refrigerant employed. *Selection is based on the density of the vapor handled.* Density may be naturally low for some refrigerants while for some others it becomes low as a result of operation (low temperature).

Compressors may be classified also according to the density of the refrigerant handled, as follows:

Reciprocating and screw:	high-density fluids
Rotary:	medium-density fluids
Centrifugal:	low-density fluids

Reciprocating compressor cylinders (single-acting) may be arranged so they are in line or radially to form a V, W, VW.

Compressor valves vary in design, three types of which are commonly employed: poppet, ring plate, and reed valves. The type and the size of the valve employed are governed to a large extent by the velocity of the gas and the design of the compressor.

Compressor pistons are of two basic types: the *plug* or automotive type and the *trunk* type. The plug type is used where valves are in the cylinder head or some location other than in the piston. The trunk type allows for the vapor to enter the compression chamber through valves in the top of the piston.

Piston rings are usually the *compression* and *oil wiper* types. Sometimes they are backed up by ring expanders to obtain a tighter seal with the cylinder wall.

STUDY OUTLINE FOR CHAPTER 9

I. *Purpose of controls:*
  (*a*) Basic function requirements,
  (*b*) Capacity control,
  (*c*) Pressure control,
  (*d*) Temperature control,
  (*e*) Water regulation,
  (*f*) Safety and protection of equipment.

II. *Types of controls:*
  1. Basic function requirements—throttling devices:
    (*a*) Hand expansion valve,
    (*b*) Automatic expansion valve,
    (*c*) Thermostatic expansion valve,
    (*d*) Float valves:
      Low-side float,
      High-side float,
    (*e*) Capillary tube.
  2. Capacity controls:
    (A) Reciprocating compressors:
      (*a*) Variable-speed drives,
      (*b*) Cycling—start and stop
      (*c*) Suction-valve unloaders,
      (*d*) Cylinder by-pass,
      (*e*) Clearance pockets,
      (*f*) Multiple compressors.
    (B) Centrifugal compressors:
      (*a*) Variable-speed drives,
      (*b*) Suction-line throttling damper,
      (*c*) Condenser temperature control,
      (*d*) Variable-inlet guide vanes,
      (*e*) Discharge by-pass
      (*f*) Multiple compressors.
  3. Pressure controls:
    (*a*) Liquid pressure valves,
    (*b*) Evaporator pressure-regulating valves,
    (*c*) Suction pressure hold-back valves,
    (*d*) Dual pressure controllers:
      High-pressure cut-out,
      Low-pressure cut-out,
    (*e*) Check valves.
  4. Temperature controls:
    (*a*) Thermostats:
      Bi-metal type,
      Thermal bulb type,
      Fluid-filled element type,
    (*b*) Pressure switch.
  5. Water-flow regulation:
    (*a*) Condenser water regulators.
  6. Solenoid valves
    (*a*) Pilot solenoid valves
  7. Program control
  8. Safety controls:
    (*a*) Water-failure switch,
    (*b*) Overload cut-out,
    (*c*) Pressure-relief valves,
    (*d*) Fusible plugs.
  9. Stop valves
  10. Thermometers
  11. Pressure gages
  12. Flow gages
  13. Liquid level gages
  14 Thermistors
  15. Transducers
  16. Solid state controls

9

Controls

In order to govern the flow of refrigerant, maintain required suction pressures, and operate safely and economically, it is necessary that the refrigeration system be provided with suitable controls. While some types of controls such as throttling devices are essential to the operation of the basic system, others are employed to impart flexibility in meeting load fluctuations, to increase operating efficiency, and to promote safety.

Controls may be divided according to the medium through which they are actuated, and may be classified, in general, as follows: (*a*) manual, (*b*) self-contained, (*c*) mechanical (*external force*), (*d*) electrical, (*e*) pneumatic (*any gas*), (*f*) hydraulic (*any fluid*), and (*g*) electronic.

In numerous instances, a combination of two or more of the above mediums is utilized. From the description of the various controls covered in the following sections of this chapter, it will become obvious that self-contained and electric-pneumatic (vapor) actuated devices are in the majority as compared with other types used for refrigeration purposes. However, other mediums listed above find important places in the study of controls for refrigeration systems, such as liquid-pressure (hydraulic) and vapor-pressure (pneumatic) instruments as represented by the liquid-filled thermostat and the vapor-filled thermostatic expansion valve, respectively. While the pneumatic control system actuated by air pressure from an external source finds its widest use in the field of air conditioning, particularly as a means of liquid-valve and air-damper regulation, it is employed to a limited extent on the refrigeration system. To acquaint the student with the latter medium of control, a section devoted to the pneumatic suction damper regulator for centrifugal compressors is presented in this chapter.

The controls described in this chapter are some of the more important ones used in the refrigeration industry. Many of them may be found also in several fields of service. They are separated according to functional classification as follows:

1. Throttling devices,
2. Capacity controls,
3. Pressure controls,
4. Temperature controls,
5. Water regulator,
6. Solenoid valves,
7. Safety controls,
8. Auxiliary components:
   - Stop valves,
   - Thermometers,
   - Liquid level gages,
   - Flow gages,
   - Pressure gages,
   - Temperature indicators,
   - Pressure indicators.

Although some have dual purposes, they are included herein under the head of their more basic function.

THROTTLING DEVICES

To maintain constant temperature in the space under refrigeration, it is necessary that vaporization within the evaporator be carried on at a rate commensurate with the load and at a pressure corresponding to the required vaporizing temperature. To accomplish this, a balance must be maintained between the amount of vapor withdrawn by the compressor and the quantity of liquid entering the evaporator. This balance is maintained by a metering or throttling device which is so regulated as to feed liquid from the high-pressure side at a rate equal to that of vaporization. Because vaporization occurs at or above the temperature corresponding to the desired pressure, which in turn is governed by the compressor, it follows that the throttling device also acts to reduce the pressure of the refrigerant from a variable condenser pressure to a constant evaporator pressure. While these throttling devices commonly are referred to as "expansion valves," they are actually pressure-reducing valves. All expansion valves are located in the liquid line.

Hand Expansion Valves

The hand-operated expansion valve is employed to manually regulate the flow of liquid to the evaporator. However, its use is limited to systems operating under fairly constant loads for long periods of time, a condition that enables the operator to make necessary adjustments in liquid flow in response to any slight change in suction pressure as shown on the suction gage. Where loads are subject to repeated and rapid fluctuations, a condition that prevails in many industrial and commercial systems, the need for greater flexibility in liquid-flow control is required than can be obtained with hand adjustment. Therefore, while manually operated expansion valves are em-

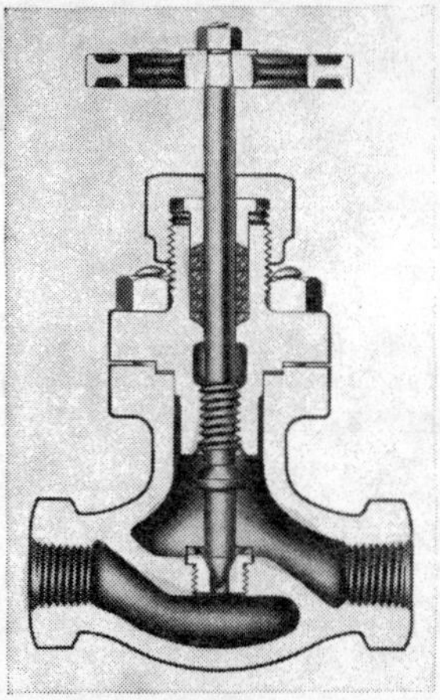

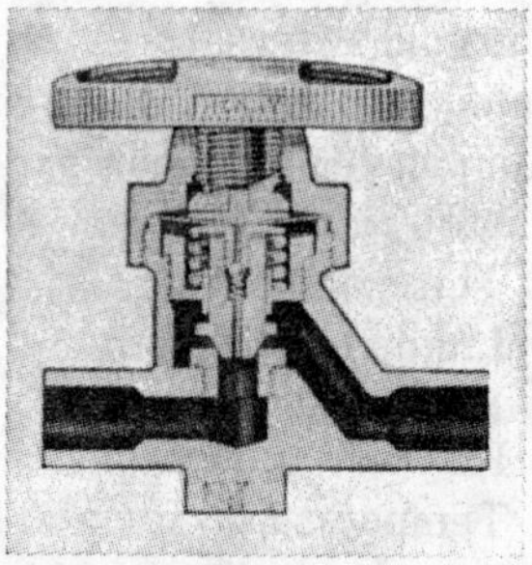

Fig. 190. (*Left*) **Expansion valve, hand-operated, packed. Bolted bonnet, rising stem type. Note needle-point stem. For ammonia systems. (Courtesy, Crane Co.)**

Fig. 191. (*Right*) **Expansion valve, hand-operated, packless. Diaphragm type for Freon. (Courtesy, Henry Valve Company.)**

ployed on by-pass lines and bleeder connections, they have been largely replaced by the more positive and quicker responding automatic types of throttling devices on main liquid lines.

Automatic Expansion Valves

The automatic expansion valve functions in response to pressure changes occurring within the evaporator to allow more or less liquid refrigerant to flow according to load demand. To distinguish this type from other automatically operated valves, it is often referred to as a constant-pressure expansion valve. Its use is limited to systems operated dry expansion under fairly constant loads and generally confined to small units of the domestic type. (See Fig. 192.)

The simple automatic expansion valve is divided in two parts: (*1*) power chamber and (*2*) refrigerant end, separated by a sensitive element in the form of a diaphragm or bellows.

The power chamber is under atmospheric pressure and contains an adjustable power spring (*a*) supported at one end by the sensitive element (*b*) and at the other end by the adjustment screw washer (*c*). This spring is adjusted to provide sufficient tension to maintain the sensitive element in equilibrium when opposed by a predetermined evaporator pressure. The refrigerant end of the valve consists of an inlet chamber (*d*) containing a strainer and an outlet chamber (*e*), a portion of whose restraining wall includes the opposite side of the sensitive element to that supporting the power spring. Some models also include a spring placed beneath the diaphragm exerting force against the latter and tending to close the valve. The only access to the outlet chamber from the inlet chamber is through a valve port (*f*) when that passage is not covered by a valve stem (*g*) or needle. The latter is connected to the sensitive element by a push rod or yoke (*h*) and

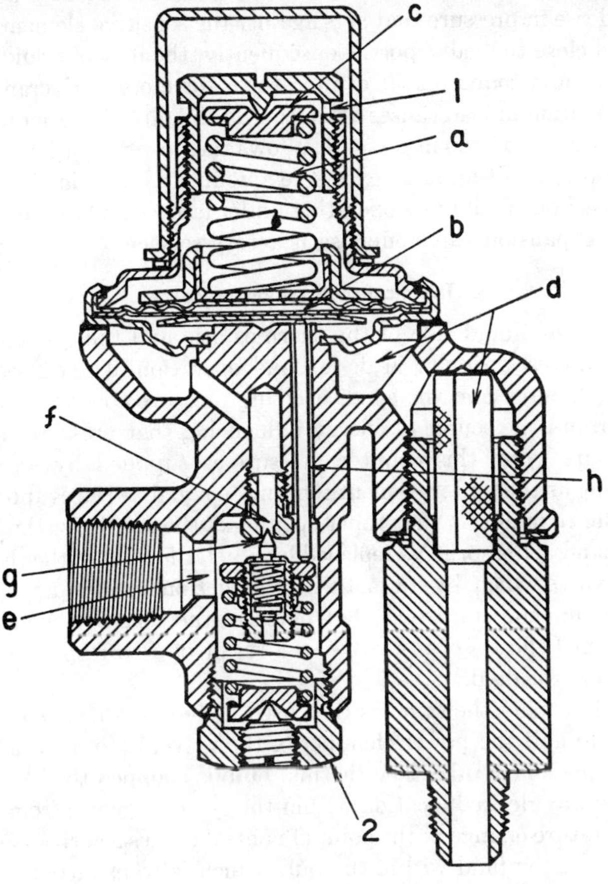

Fig. 192. Automatic or constant-pressure expansion valve. (Courtesy, Detroit Controls Corporation.)

actuated by that component when the diaphragm or bellows acting as a flexible bulkhead expands or contracts according to the relative force of the power spring as opposed to the pressure in the outlet chamber or evaporator and lower spring.

As the evaporator pressure decreases, the power spring acts against the sensitive element to force the needle away from its seat, allowing more liquid to enter the evaporator. As the evaporator pressure rises as a result of additional liquid, the sensitive element is forced to move against the power spring, tending to close the valve and retard or stop liquid flow. The action is one of modulation rather than that of abrupt (snap) opening or closing of the valve. Because the sensitive element is exposed to the evaporator, the automatic expansion valve has the advantage of direct action in maintaining constant pressure in the low side of the system.

The disadvantage of this type of valve lies in its inability to cope with sudden, appreciable load changes. An abrupt increase in load is accompanied by a rapid rise in pressure that acts against the sensitive element to seat the needle and close the valve port. Consequently, the flow of liquid ceases at a time when the demand on the evaporator is for more refrigerant. Again an abrupt reduction in load causes a decrease in the rate of evaporation, which, followed by a rapid drop in pressure, allows the power spring to expand and open the port, resulting in an excessive amount of liquid in the evaporator.

When used on small units operating under fairly constant conditions, the automatic expansion valve offers satisfactory service.

Thermostatic Expansion Valves

Basically, the liquid side of the thermostatic and the constant-pressure expansion valves is similar in design and operation. The difference lies in the power chamber and the agent exerting pressure counter to that in the evaporator and the source of energy influencing that pressure. In the constant-pressure type, this counter pressure is supplied by an adjustable spring acting as the power element tending to equalize the evaporator pressure in order to maintain the proper opening of the needle for the immediate liquid demand. This spring is replaced by a volatile fluid in the thermostatic expansion valve which, by contact with the suction line of the system, relays any change in temperature at that point to the sensitive member (either diaphragm or bellows) of the valve as a result of a corresponding pressure change within the fluid.

The fluid, usually the same as that used as refrigerant, is contained in a thermal bulb and the power chamber of the valve, both of which are connected by a capillary tube. The thermal bulb is clamped to the suction line or placed in any desired location within the system remote from the valve. Any temperature change at the point of control causes a corresponding pressure change in the fluid within the bulb which is transmitted through the capillary tube to the power side of the valve. This change in pressure actuates the sensitive element which, in turn, opens or closes the needle to allow more or less liquid to enter the evaporator.

Thermostatic expansion valves act to increase refrigerant flow with any rise in temperature at the control point and reduce flow when the temperature decreases.

While the constant-pressure valve is subject to only one variable, that of the evaporator pressure, the thermal valve must function as intermediary between two variables; i.e., evaporator pressure versus bulb fluid pressure, the latter being governed by the suction-line temperature. This means that the thermal valve must be designed to provide highly compensating characteristics in order to maintain a proper balance between the two since the pressure differential at the valve is subject to continual variation.

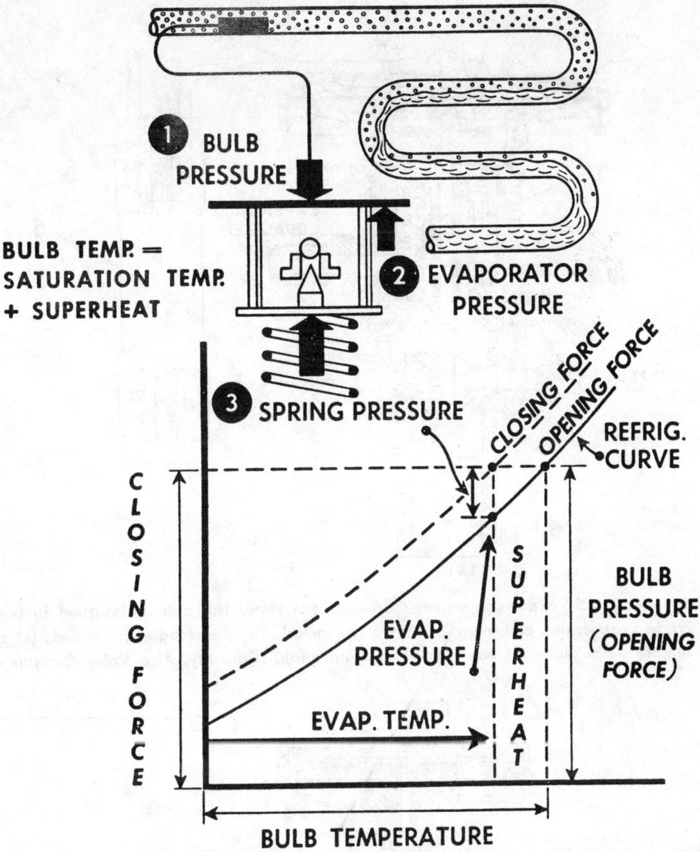

Fig. 193. Thermostatic expansion valve, principle of operation. (Courtesy, Sporlan Valve Company.)

The function of the thermal valve is to keep the evaporator fully charged with refrigerant while preventing liquid carry-over to the compressor. This is accomplished by controlling the degree of superheat usually at about 7 F to 10 F above saturation in the suction line. By clamping the thermal bulb tightly to the suction line at a point immediately beyond where it leaves the evaporator, any change in suction-vapor temperature will be transmitted to the fluid within the bulb. Because the fluid in the bulb is normally the same as that employed as refrigerant—i.e., when temperatures are within commercial range (above 0 F)—the pressure and temperature changes occurring in the suction line will be reflected by the sensitive element of the thermal valve to a corresponding degree. Therefore, if the suction-line temperature is controlled at some selected level above the saturated temperature corresponding to the evaporator pressure, a superheat condition will exist at that point. And as the thermal valve is adjusted to increase or decrease liquid flow according to suction-temperature fluctuations, only superheated vapor will enter the compressor.

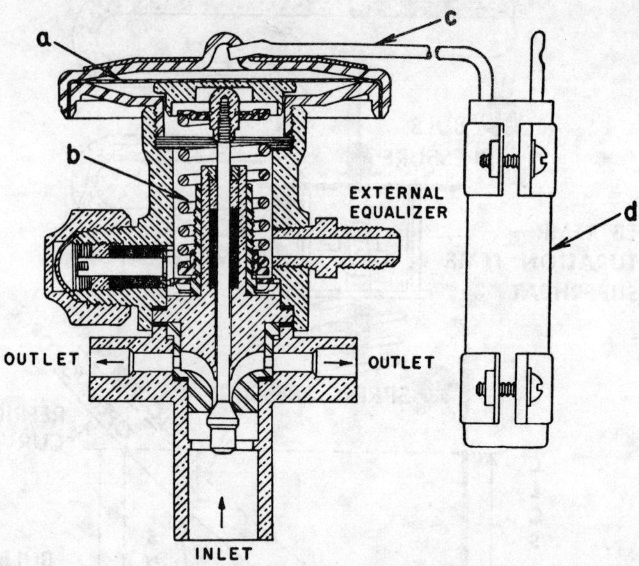

Fig. 194. Thermostatic expansion valve, cross-sectional view. This one is designed to feed more than one Freon evaporation coil (multi-outlet). Parts include (*a*) diaphragm, (*b*) spring, (*c*) capillary tube, (*d*) remote bulb. Note external equalizer connection. (Courtesy, Alco Valve Company.)

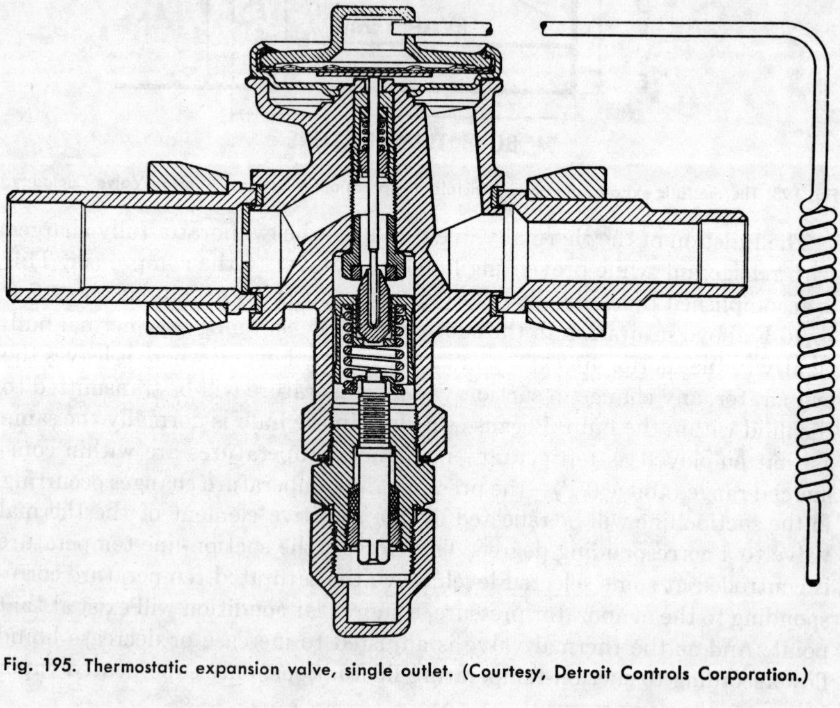

Fig. 195. Thermostatic expansion valve, single outlet. (Courtesy, Detroit Controls Corporation.)

Adjustments of thermal valves are accomplished by means of a spring located on the refrigerant side of the sensitive element so that any tension exerted will supplement the pressure in the evaporator. The tension of the valve spring, therefore, must be equal to the difference between evaporator pressure (inlet) and the pressure corresponding to the desired superheat temperature in the suction line. At that setting, the needle would be in equilibrium when the suction temperature is at the desired superheat level. Any increase in superheat would upset that equilibrium by the subsequent rise in the bulb fluid pressure and tend to force the needle away from the seat. The outward distance of needle movement would be proportional to the amount of superheat in excess of the desired level.

For example, if R-12 were employed as both the refrigerant and the bulb fluid on a system operating at 5 F evaporator temperature and 10 F superheat in the suction line, the equivalent valve spring tension would be found as follows:

$$\text{Evaporator pressure @ 5 F} = 11.81 \text{ psig}$$

$$\text{Pressure in bulb @ 5 F} + 10\text{ F} = 15\text{ F} = 17.74 \text{ psig}$$

Spring tension to maintain the needle at equilibrium and in a closed position would be equivalent to

$$17.74 - 11.81 = 5.93 \text{ psi}$$

A 1 F rise in superheat above the required 10 F would result in a suction-line temperature of 16 F. The bulb pressure corresponding to 16 F Freon-12 is 18.38 psig. This would result in an excessive pressure above equilibrium of 18.38 − 17.74 or 0.64 psi, enough to open the needle slightly and allow liquid to enter the evaporator, which upon vaporizing would reduce the suction temperature to the desired superheat level and again close the valve. Fig. 193 shows the relative influence of each of the three pressure elements; i.e., bulb fluid, evaporator, and superheat spring.

Considering the role of each element controlling the opening or closing of the valve, it is found that: (*a*) the pressure of the bulb fluid acts to open the valve, (*b*) the pressure of the refrigerant in the evaporator acts to close the valve, (*c*) the spring acts with the evaporator pressure to keep the valve closed in relation to the amount of superheat setting.

On systems having considerable length of run between expansion valve and suction line, a drop in pressure occurs due to friction in the coil. Because the thermostatic expansion valve depends primarily on the pressure differential between the evaporator pressure and the pressure of the bulb fluid as influenced by the desired superheat in the suction line, any drop in pressure between the two points will disturb the necessary balance and adversely affect control. To overcome this drop in pressure, an *external* equalizer is employed. This is made by connecting the suction line with the low-pressure chamber of the expansion valve by means of a small-diameter

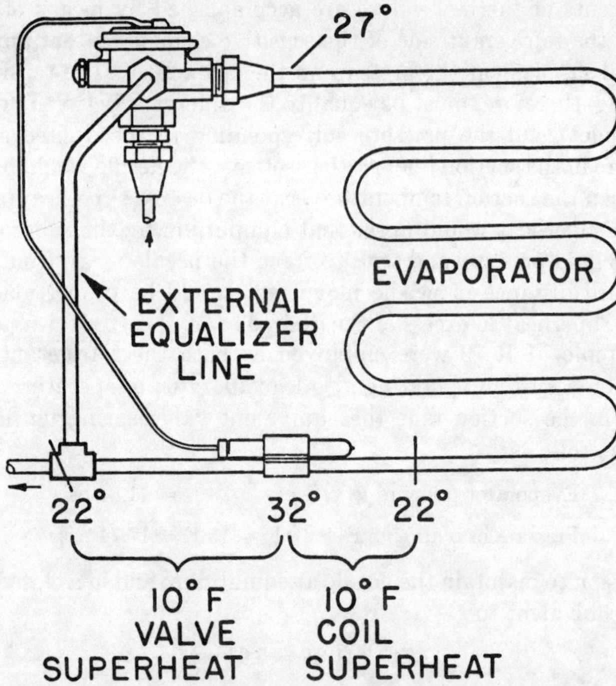

Fig. 196. System arrangement with thermostatic expansion valve and external equalizer. (Courtesy, Detroit Controls Corporation.)

tube. This equalizer tends to create an average evaporator pressure opposing the power element. (See Fig. 196.)

The pressure within the power chamber depends not only on the kind of fluid used in the sensitive element but also on the amount of fluid it contains. For example, if the remote bulb contains half liquid and half gas, the maximum operating pressure (MOP) would be greater than it would if less than half were liquid, assuming a similar temperature in each case. Where it is desirable to use the same fluid in the sensitive element as the refrigerant in the system on which it will be used, it follows that, for lower maximum operating pressures, correspondingly less fluid is required in the bulb. The reason for this is that all the liquid in the bulb evaporates at a relatively lower pressure. Any heat absorbed beyond the point of saturation superheats the gas, causing only a slight increase in pressure above that of saturation. Therefore, the maximum operating pressure may be controlled by the charge or amount of liquid introduced into the bulb.

Where the same fluid is used in the valve as in the evaporator, the pressure-temperature relationship of both remains constant. As long as liquid remains in the bulb, this relationship will follow a constant pattern over a wide range. This is shown by the curves in Fig. 197. The distance between the two curves represents the amount of superheat setting.

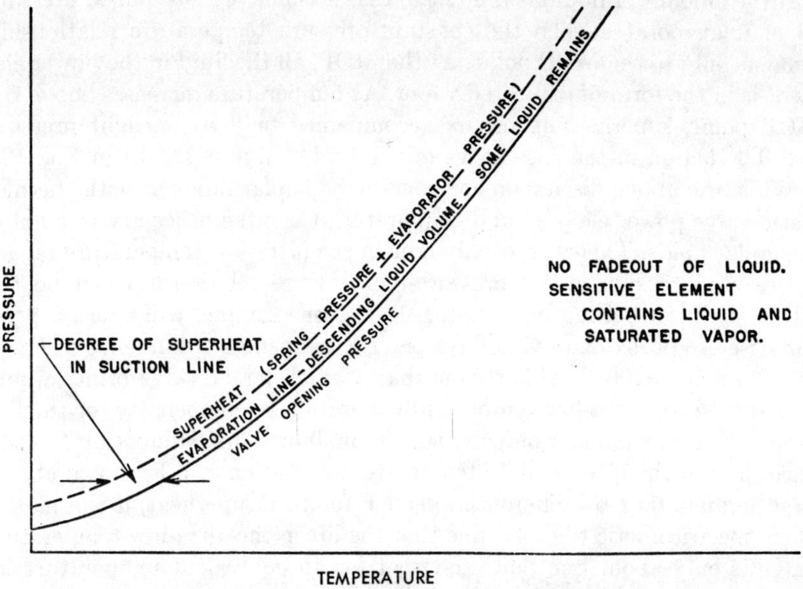

Fig. 197. Behavior of fluid in a liquid-charged thermostatic expansion-valve thermal element.

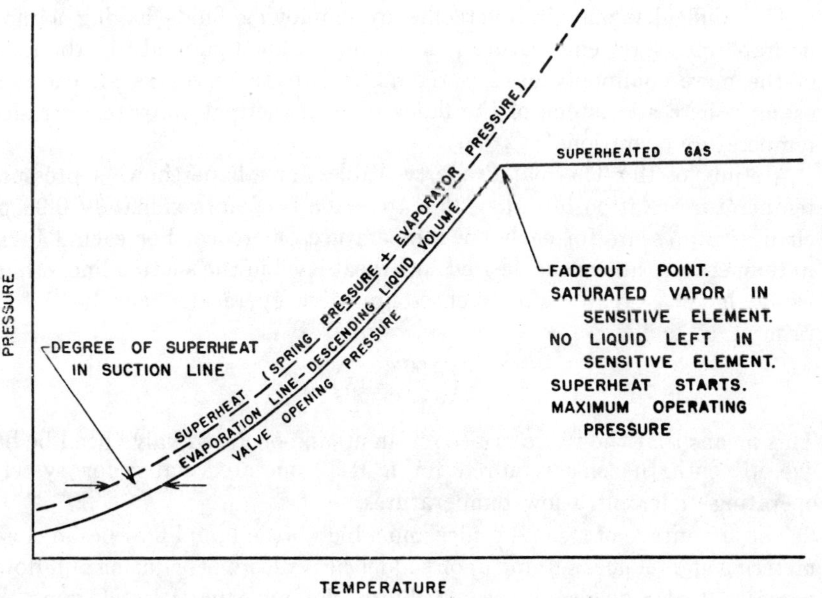

Fig. 198. Behavior of fluid in a limited (gas-) charged thermostatic expansion-valve thermal element.

If the amount of liquid is limited so that at some predetermined pressure all of it evaporates, then the constant-pressure–temperature relationship remains only to the MOP point. At the MOP, all the fluid in the power element is in the form of saturated vapor. As temperature increases above the MOP point, superheating occurs accompanied only by a slight pressure rise. The action in the case of a limited fluid charge is shown in Fig. 198.

While the above discussion considers only similar fluids in both thermostatic-valve power element and evaporator, it is often necessary to employ dissimilar fluids. This is especially true in the ultra-low temperature ranges where a more substantial pressure-temperature relationship can be obtained with low boiling point fluids. R-12, for example, will cause a pressure rise of approximately 0.06 psi per 1 F temperature difference at levels of -95 F to -100 F. This means that a much larger valve orifice would be required for ultra-low temperature operation to compensate for the low needle lift that would accompany such a small pressure differential. In comparison with the intermediate-temperature operation of 5 F, shown above, resulting in a 0.64 psi differential per 1 F for 16 F superheat, it is a matter of simple arithmetic to determine that the lift in the ultra-low temperature valve is only about one tenth as great for an equivalent temperature increase. In other words, the ultra-low temperature system would require a valve having an orifice 10 times larger than that of the system operating at 5 F, using R-12 in the bulb.

This difficulty may be overcome by employing fluids having a more desirable pressure-temperature relationship at low temperatures than any of the more commonly used refrigerants. Ethane serves as an excellent example here and is one of the fluids used in thermal bulbs for ultra-low temperature operation.

A study of the Thermal Property Tables for ethane shows a pressure-temperature relationship at -95 F to -105 F of approximately 0.90 psi change in pressure for each 1 F temperature difference. For each 1 F rise in temperature above the desired superheat level in the suction line, ethane would have a greater effectiveness on valve operation over R-12 by a ratio of 15 to 1,

$$\left(\frac{0.90}{0.06} = 15\right)$$

This means that the valve orifice for an ethane-operated valve need be but one fifteenth the size required for a R-12-operated valve for systems operating under ultra-low temperatures.

The advantage of a small orifice and a high lift in liquid flow lies in closer metering and reduces surging upon sudden heavy loading or during pull-down periods. It also decreases the possibility of foreign particles clogging the valve to prevent proper seating.

Using low-temperature fluids poses the problem of high pressures, which often reach levels beyond the ability of the sensitive element to withstand.

Ethane, for example, exerts a pressure of approximately 500 psia at 60 F. To use such a fluid necessitates limiting the charge by introducing it into the power element in small quantities, which in turn lowers the maximum operating pressure (MOP).

According to the foregoing discussion, thermostatic expansion valves may be classified as to their type of fluid charge, viz.:

(*a*) Liquid charge, using the same fluid in the power element as in the refrigeration system to which they will be attached. In this category, the volume of liquid is greater than the cubical capacity of both the capillary tube and the valve power chamber.

(*b*) Gas charge, using the same fluid in power element as in the refrigeration system to which it will be attached. The volume of liquid in the power element will be limited in amount as to be completely evaporated at a specific maximum operating pressure.

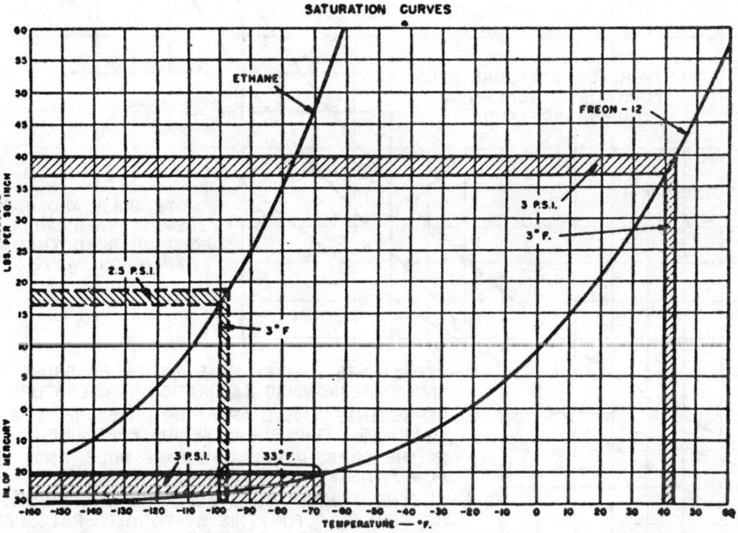

Fig. 199. Relative pressure changes of ethane vs. Freon-12 according to temperature. Note steep curve for ethane in ultra-low temperature range compared with slight slope for Freon-12 in same range. (Courtesy, Detroit Controls Corporation.)

(*c*) Cross charge, using a different fluid in the power element as in the refrigeration system to which it is attached. The pressure-temperature relationship of the valve fluid is such as to cause its curve to cross the pressure-temperature curve of the refrigerant. This relationship is shown in Fig. 200.

The liquid charge is more positive and will react immediately to create greater or less pressure on the sensitive element with any change in temperature above the boiling point of the fluid employed. While this characteristic is highly desirable when the system is already operating at required evaporator temperatures, it does have the disadvantage of causing flooded conditions when the system is started up after an extended period of idleness. This occurs because the high pressure in the power element under elevated temperatures holds the valve open. Flooding a warm evaporator places a greater than normal load on the compressor due to the high density of the refrigerant and in many instances increases the capacity to the extent of overloading the compressor motor.

When employing a fluid in the power element similar to the refrigerant under control, the changes in pressure are in direct relationship to changes in suction-line temperature. This means that the valve reaction will corre-

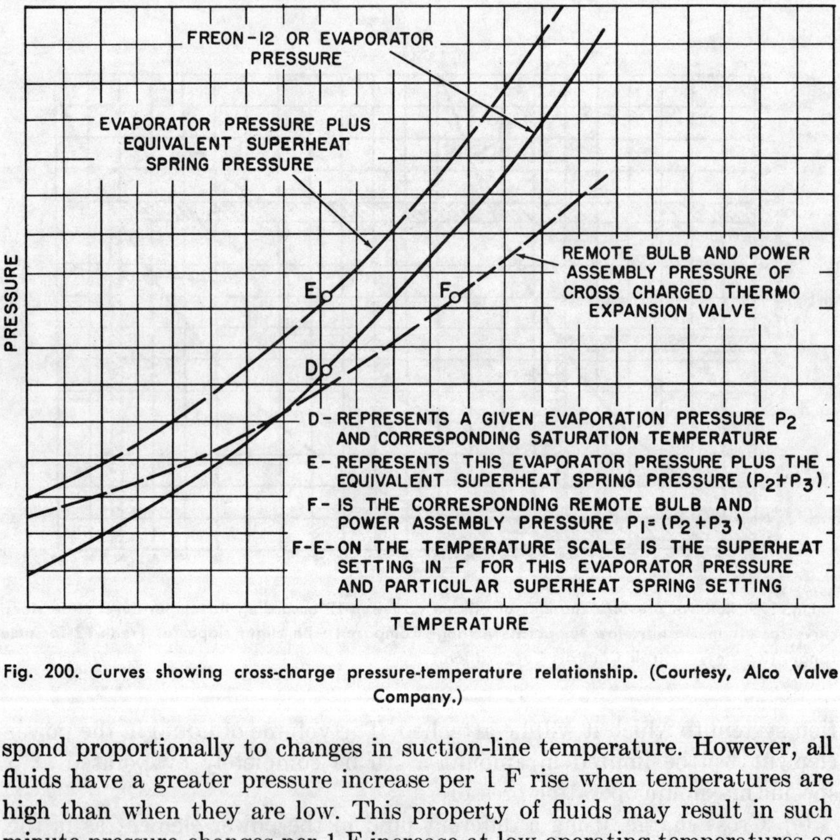

Fig. 200. Curves showing cross-charge pressure-temperature relationship. (Courtesy, Alco Valve Company.)

spond proportionally to changes in suction-line temperature. However, all fluids have a greater pressure increase per 1 F rise when temperatures are high than when they are low. This property of fluids may result in such minute pressure changes per 1 F increase at low operating temperatures as to restrict valve motion and limit the flow of liquid refrigerant to the extent of starving the evaporator. If this condition prevails with bulb fluids that are the same as the refrigerant by reason of low temperature operation, then it becomes necessary to employ another fluid having a lower boiling point and a more desirable pressure-temperature relationship at the lower temperature than the refrigerant. The dissimilar fluid having a greater pressure rise per

1 F temperature increase than the refrigerant under similar conditions causes sufficient valve motion for ample liquid flow.

The disadvantage of employing a liquid charge of low boiling point fluid, however, is the high saturation pressures prevailing under normal atmospheric temperatures. As pointed out in the case of ethane, the saturation pressure under the latter conditions becomes so high as to preclude the use of conventional type sensitive elements. Therefore, the use of limited or gas-charged power elements is resorted to as previously described.

Using a gas-charged power element eliminates the flooded condition that causes overloading prevalent in systems controlled by liquid-charged valves. This is because the suction pressure in the evaporator after a period of idleness is higher than the limited amount of fluid in the power element will allow its pressure to become at a similar temperature. Therefore, the valve remains closed until the compressor has had an opportunity to reduce the pressure in the evaporator. When the evaporator pressure reaches a point just below the bulb pressure, the valve opens slightly and feeds refrigerant slowly enough to prevent any abrupt rise in evaporator pressure. During subsequent cooling, a pressure balance is finally reached when the valve operates normally to maintain the required superheat in the suction line.

The pressure at which the limited charge forces the valve to open is governed by the amount and the kind of fluid employed in the power element. If the fluid is similar to the refrigerant, then the amount of charge can be so regulated as to cause valve opening at any given temperature above that corresponding to desired evaporator pressure. Conversely, the valve may be made to close completely at any temperature above that corresponding to desired evaporator pressure. Above that temperature, the fluid in the bulb vaporizes completely, causing the change in pressure per 1 F to become smaller in relation to a similar change in pressure in the evaporator resulting from the mixture of liquid and vapor in the latter component.

In selecting valves having limited charges, it is well to specify the maximum operating pressure above which the valve will close fully and below which the valve will commence to function. One manufacturer standardizes valves at the following maximum operating pressures for various refrigerants (see Table 32).

Fig. 201 shows the pull-down characteristics of gas-charged versus liquid-charged thermostatic-valve power elements.

In the low-temperature thermostatic expansion valve shown in Fig. 202, the refrigerant flow is governed by the pressure differential set up within two opposing power elements by reason of varying temperatures between the evaporator and the suction line acting on similar fluids. Thus temperature and not pressure is the controlling factor. One thermal bulb is attached to the evaporator while the other is in contact with the suction line. The valve mechanism is in equilibrium when the evaporator pressure plus spring

TABLE 32. MAXIMUM OPERATING PRESSURE STANDARDS FOR THERMOSTATIC EXPANSION VALVES*

Valve No.	Freon-12	Methyl Chloride	Sulfur Dioxide
Air Conditioning and Commercial Service			
573	45 lb	35 lb	..
673 and 674	55 lb	40 lb	20 lb
893 and 894	55 lb	40 lb	20 lb
896 and 897	55 lb	40 lb	20 lb
899	60 lb	50 lb	20 lb
786	60 lb	50 lb	20 lb
787	60 lb	50 lb	20 lb
788	60 lb	50 lb	20 lb
Low-Temperature Service			
573	10 lb	5 lb	..
673 and 674	15 lb	10 lb	5 lb
893 and 894	15 lb	10 lb	5 lb
896 and 897	15 lb	10 lb	5 lb

* Courtesy, Detroit Controls Corporation.

tension is equal to the opposing pressure in the suction-line bulb induced by the desired superheat setting. When the superheat in the suction line rises above the desired level, the pressure of the fluid in the superheat power element increases to disturb the balance between the two forces, whereupon the subsequent mechanical motion allows the valve to open. The low-temperature or differential thermostatic expansion valve fills the need for a positive-acting control for continuous-operating low-temperature systems.

On low- or ultra-low temperature systems, the employment of low-pressure refrigerants in the power element has the disadvantage of low-pressure change per 1 F rise in temperature. This must be overcome by the use of large valves or a high range of superheat in the suction line in order to provide sufficient liquid to meet the load. The alternative is to use either a high-pressure fluid, such as ethane, the charge of which must be extremely limited to avoid excessive pressures under normal atmospheric temperatures. While the advantage of better pressure-temperature relations at low temperatures is gained by the use of high-pressure fluid, it follows that the temperature at which the valve is allowed to open must be relatively low by reason of the extremely limited charge previously pointed out. The maximum temperature at which a limited charge of ethane (160 psi) will allow valve opening is −30 F. If the evaporator lacks an adequate supply of refrigerant during the pull-down period, the longer it takes to bring the system to the required temperature. Where this extension of time is inconvenient to the

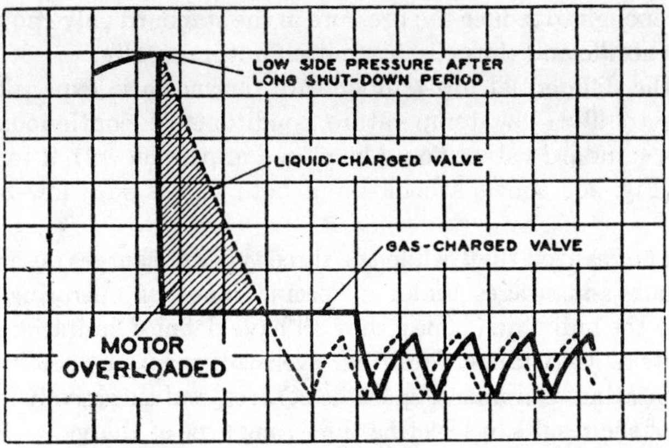

Fig. 201. Pull-down performance of gas-charged vs. liquid-charged thermostatic expansion-valve thermal elements. (Courtesy, Detroit Controls Corporation.)

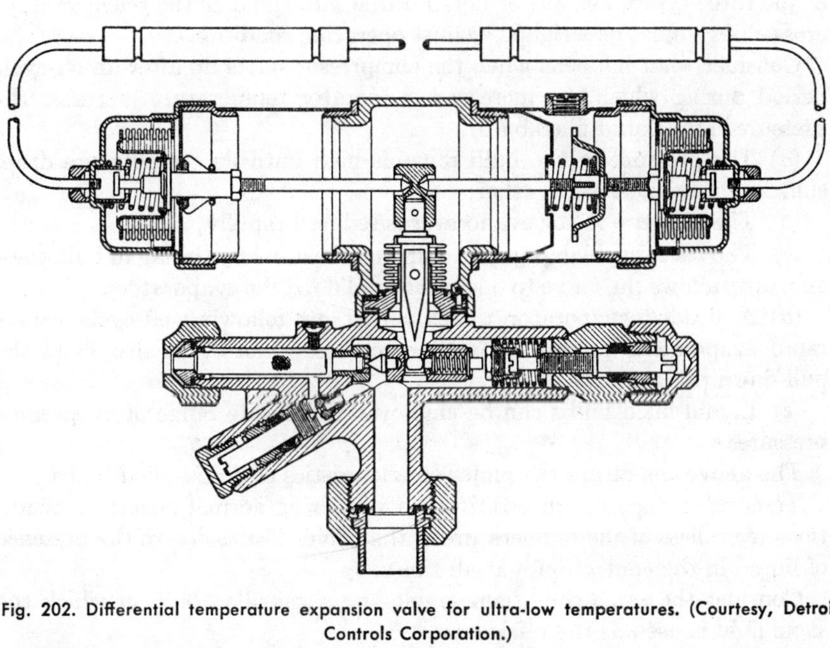

Fig. 202. Differential temperature expansion valve for ultra-low temperatures. (Courtesy, Detroit Controls Corporation.)

immediate process under refrigeration, the low-temperature valve employing the high-pressure fluid must be augmented by a standard valve using a low-pressure fluid and in parallel with it to allow liquid flow at higher temperature. By so overriding the period prior to the time the system reaches a pressure low enough to allow the low-temperature valve to open, the pull-down period is considerably reduced. When the system reaches the point at which the low-temperature valve cuts in, the temperature will become low enough to reduce the pressure in the standard valve power element to seat its needle and thus eliminate itself automatically.

While the differential low-temperature thermostatic expansion valve is adaptable to ultra-low temperature conditions of continuous operating systems, a standard valve should be placed in parallel with it for pull-down purposes. Fig. 203 shows a hook-up of both valves on a low-temperature system.

It must be realized that, while the three types of charges described above have certain advantages under a specific range of operating conditions peculiar to the individual types, they all have definite limitations when employed outside that range. Hunting, overload, reduced capacity, the need for compensating equipment, sporadic control, and excessively large valves are some of the results of selecting the wrong type of charge.

The service in which the control is to be used should govern the selection of the charge employed in the bulb. The advantages and the disadvantages of the three types; i.e., gas or liquid of the same fluid as the refrigerant, or cross charge must be weighed against operating conditions.

Consider what happens when the compressor starts up after an off-cycle period during which the increased evaporator temperature increases the pressure in the liquid-filled bulb:

(*a*) The pressure in the bulb remains high until its temperature drops sufficiently to cause some relief;

(*b*) The pressure in the evaporator is reduced rapidly;

(*c*) The resulting high pressure differential caused by the lag in bulb pressure drop allows the valve to open wide and flood the evaporator;

(*d*) A flooded evaporator under conditions following off-cycle causes rapid evaporation to overload the compressor motor. It also slows the pull-down process;

(*e*) Liquid-filled bulbs can be employed for a wide range of evaporator pressures.

The above constitute the major characteristics of liquid-filled bulbs.

Their advantage lies in positive control during normal operating conditions regardless of the temperature of the valve. This is due to the presence of liquid in the control bulb at all times.

Consider the same conditions employing a gas-filled bulb in which the same fluid is used as the refrigerant:

(*a*) The pressure in the bulb does not rise appreciably above that of vaporization upon off-cycle due to its limited charge. This means that the pressure tending to open the valve nearly becomes constant at the MOP;

(*b*) The pressure in the evaporator rises to the extent governed by the heat absorbed during the off-cycle period becoming greater than the bulb pressure;

(*c*) Assisted by the spring, the evaporator pressure tends to keep the valve closed;

(*d*) Until the compressor pulls down the pressure in the evaporator sufficiently upon starting up, the valve remains closed;

(*e*) In the meantime, the bulb pressure takes effect and starts to open the valve, allowing the gradual entrance of refrigerant. Normal modulating control follows until the next off-cycle period;

(*f*) In preventing the valve from suddenly opening wide, a flooded condition is avoided, placing only normal load on the compressor motor;

(*g*) Limited charges give best results on evaporators operating between 30 F and 50 F.

The above, then, are the major characteristics of gas-filled bulbs.

Their disadvantage lies in limited charge. Should the fluid condense in the valve because of lower temperature at the latter point, the bulb would lose control temporarily. The valve would close until a rise in temperature again vaporized the fluid. Therefore, the valve must always remain warmer than the remote bulb.

The effective pressure of fluids becomes correspondingly less with any drop in temperature, thereby reducing the needle lift when applied to thermal valves. This means that the tonnage capacity of an individual valve is correspondingly reduced as evaporator temperatures decrease. It follows that low-temperature operation requires a larger valve than a similarly loaded system operating at a higher evaporator temperature. Thus, a valve designed for 1-ton capacity at 40 F evaporator temperature may be reduced to 0.9-ton capacity at 20 F, or 0.6-ton capacity at −10 F, the rate of decrease depending on the kind of refrigerant, the temperature of the liquid, and the efficiency of the valve.

For similar evaporator temperatures, the capacity of an expansion valve increases with a rise in condenser temperature. This is true because of the greater pressure reduction across the valve at higher condenser temperatures, forcing the liquid through the port at increased velocity.

Capacities of expansion valves, then, are based on the orifice size in relation to the pressure drop across the valve and the evaporator temperatures. Because of differences in thermal characteristics of various refrigerants, as shown in Chapter 5, different size valves are required for many of them under equal load conditions. For example, the same valve will have nearly twice the capacity when handling methyl chloride as with R-12 or SO_2 under similar evaporator and condenser temperatures.

Table 33 shows a typical valve capacity chart, based on 3 F to 4 F superheat change, with orifice size shown for different refrigerants under various condenser and evaporator temperatures.

Some of the special features of thermostatic expansion valves include: (*a*) provision for connecting an external equalizer on some types, (*b*) discharge tubes for restricting expansion within the valve, (*c*) external superheat adjustments, (*d*) removable and cleanable strainers, (*e*) multi-outlet for distributing refrigerant to more than one evaporator carrying equal loads.

TABLE 33. CAPACITIES OF THERMOSTATIC EXPANSION VALVES BASED ON 3° TO 4° SUPERHEAT CHANGE*

(*Capacities Shown Between Heavy Lines Are Based on Recommendations of A.S.R.E.*)

Nominal Capacity	Suction Temp. °F.	TONS CAPACITY — Condenser Temperature °F.—Saturated Inlet Liquid												Connections
		Freon-12 & SO_2				Freon-22				Methyl Chloride				Inlet and Outlet
		80°	100°	120°	140°	80°	100°	120°	140°	80°	100°	120°	140°	
	+50	1.80	2.10	2.35	2.50	2.70	3.20	3.70	4.00	3.00	3.75	4.30	4.60	
	+40	1.65	2.00	2.10	2.25	2.60	3.00	3.40	3.70	2.90	3.50	4.10	4.50	
	+20	1.45	1.60	1.75	1.80	2.40	2.60	2.80	3.00	2.55	3.00	3.50	3.80	
1.6 Tons F-12	+ 5	1.20	1.35	1.45	1.50	2.00	2.20	2.30	2.40	2.20	2.50	2.80	3.00	
2.6 Tons F-22	−10	1.00	1.05	1.15	1.20	1.50	1.80	1.90	2.00	1.70	1.95	2.20	2.35	
3 Tons M.C.	−20	.80	.85	.85	.90	1.10	1.25	1.40	1.50	1.30	1.50	1.60	1.70	
	−30	.60	.65	.65	.65	.85	.90	1.05	1.10	.95	1.10	1.25	1.30	
	−40	.50	.55	.55	.55	.70	.85	.90	.95	.85	.95	1.05	1.10	
	+50	2.70	3.20	3.50	3.70	4.20	4.80	5.60	5.90	4.50	5.70	6.60	7.00	
	+40	2.45	3.00	3.20	3.40	4.00	4.50	5.00	5.30	4.30	5.30	6.20	6.70	
	+20	2.15	2.35	2.60	2.65	3.45	3.85	4.20	4.60	3.80	4.50	5.20	5.70	
3 Tons F-12	+ 5	1.80	2.00	2.15	2.20	2.95	3.30	3.50	3.70	3.25	3.70	4.10	4.45	
4.5 Tons F-22	−10	1.50	1.60	1.70	1.75	2.30	.270	2.80	2.90	2.55	2.95	3.30	3.50	
5.3 Tons M.C.	−20	1.20	1.25	1.30	1.35	1.70	1.90	2.10	2.20	2.00	2.25	2.50	2.60	3/8", 1/2", 5/8", 7/8", 1-1/8" O. D. Tube Sweat Tail Pipe with Union Nut
	−30	.90	1.00	1.00	1.00	1.25	1.40	1.60	1.70	1.50	1.65	1.85	2.00	
	−40	.75	.85	.85	.85	1.05	1.25	1.35	1.45	1.25	1.50	1.60	1.70	
	+50	4.00	4.75	5.25	5.60	6.40	7.50	8.70	9.20	7.00	8.50	10.00	11.00	
	+40	3.85	4.50	4.90	5.20	6.20	7.00	7.80	8.20	6.80	8.00	9.60	10.50	
	+20	3.30	3.70	4.10	4.20	5.40	6.00	6.50	6.90	5.90	7.00	8.10	9.00	
4.5 Tons F-12	+ 5	2.80	3.10	3.35	3.40	4.60	5.20	5.40	5.60	5.00	5.75	6.45	6.95	
7 Tons F-22	−10	2.35	2.50	2.65	2.75	3.70	4.20	4.40	4.60	4.00	4.60	5.10	5.50	
8 Tons M.C.	−20	1.80	1.90	2.00	2.05	2.60	3.00	3.20	3.50	3.00	3.35	3.70	4.00	
	− 30	1.40	1.50	1.55	1.55	1.95	2.20	2.45	2.60	2.25	2.50	2.80	3.00	
	−40	1.15	1.25	1.25	1.25	1.65	1.95	2.10	2.25	1.90	2.25	2.40	2.60	
	+50	5.35	6.35	7.00	7.45	8.40	10.00	10.70	12.70	9.20	11.80	13.70	14.70	
	+40	5.20	6.00	6.60	7.00	8.30	9.60	10.50	11.10	9.10	11.00	13.00	14.00	
	+20	4.50	5.00	5.50	5.60	7.30	8.10	8.80	9.30	8.00	9.50	11.00	12.00	
6 Tons F-12	+ 5	3.80	4.20	4.50	4.60	6.25	7.00	7.30	7.90	6.80	7.80	8.70	9.40	
9.6 Tons F-22	−10	3.20	3.40	3.60	3.70	5.30	5.70	5.90	6.10	5.40	6.20	6.90	7.40	
11 Tons M.C.	−20	2.40	2.50	2.65	2.70	3.55	4.00	4.40	4.70	4.05	4.60	5.00	5.40	
	−30	1.85	1.95	2.05	2.05	2.70	2.95	3.45	3.55	3.10	3.40	3.85	4.10	
	−40	1.45	1.65	1.65	1.65	2.30	2.70	2.85	3.05	2.65	3.10	3.30	3.50	

* Courtesy, Detroit Controls Corporation.

The position and the location of the thermal bulb are important considerations that often affect the degree of accuracy of thermostatic expansion-valve control. The bulb should be inserted in or strapped to the suction line at a point immediately beyond the evaporator outlet but not in close proximity to any bends or connections that tend to interfere with the even flow of superheated vapor. When present, equalizer connections should be made well beyond the bulb location. Never install the bulb on any part of the suction line that may possibly act as a liquid trap. Best results are obtained when bulbs are installed on a horizontal run of suction line with the valve body and the capillary tube located above. If installed on a vertical run, the capillary tube should leave at the top. The contacting surfaces of both bulb and suction line should be parallel, thoroughly clean, and close together to provide maximum heat conduction.

The bulb should be located in the space or the liquid being cooled, but never at a point where air blasts or other external disturbances will interfere to upset the thermal reaction to superheat changes in the suction line. Heat, as from a blow-torch, must never be applied to any part of the power element of thermostatic expansion valves.

With limited charges, the valve body should always remain warmer than the bulb at all times, otherwise the fluid will condense or expand in the valve chamber, causing the bulb to lose control. Gas-charged power elements should always be so located as to prevent the limited amount of liquid from leaving the bulb. When that element is installed above the valve body, suitable bends in the capillary tube should be provided to act as liquid traps by using wide sweeps when a change of direction of run is necessary.

Pressure-Limiting Thermostatic Expansion Valves

Pressure limiting of thermostatic expansion valves to prevent overloading is also accomplished mechanically. Placing a spring or a collapsible cartridge between the diaphragm (power element) and the needle-operating mechanism or push rod offers a limiting motion to the latter when the evaporator pressure becomes greater than the pressure in the cartridge or the tension of the spring.

In the collapsible cartridge type, the cartridge is filled with a noncondensible gas. While the pressure in the evaporator remains below the pressure

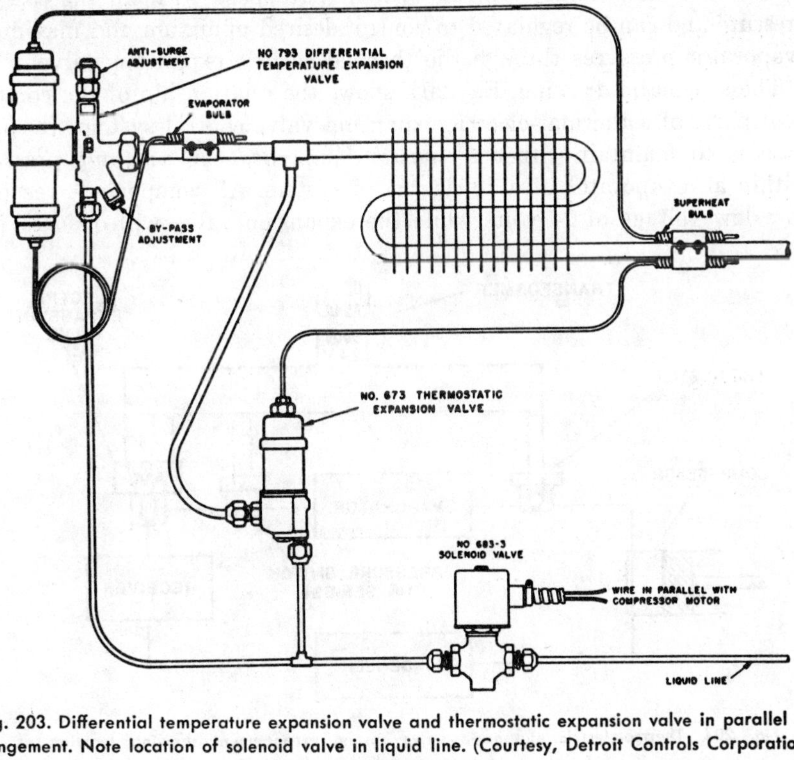

Fig. 203. Differential temperature expansion valve and thermostatic expansion valve in parallel arrangement. Note location of solenoid valve in liquid line. (Courtesy, Detroit Controls Corporation.)

in the cartridge, the latter acts as a solid link between the diaphragm and the push rod. As long as this pressure relationship continues, the motion of the push rod will respond to diaphragm pressure. When the evaporator pressure, which opposes diaphragm motion, exceeds the cartridge pressure, the cartridge collapses, allowing the superheat spring to throttle the valve. The latter condition remains until the evaporator pressure falls below that of the cartridge.

In the spring type pressure-limit valve, the spring will collapse when the pressure in the evaporator exceeds the tension for which it is set. There are a number of different types of spring-loaded pressure-limit valves, some of which are adjustable for various operating pressures.

Thermal-electric Expansion Valve

The thermal-electric expansion valve is a throttling device consisting of a simple heat motor operated needle valve motivated by a thermistor and/or a pressure switch. A thermistor is an electrical resistor composed of a material whose resistance varies in a definite manner with thermal energy changes. Thermistor resistance increases with a decrease in thermal energy and decreases as its thermal energy increases. The pressure switch is a slow make-or-break switch that opens with an increase in pressure and can be regulated to control desired minimum and maximum evaporator pressures through the thermal-electric expansion valve.

The schematic drawing, Fig. 204, shows the relationship of the component parts of a thermal-electric expansion valve control system. Its purpose is to maintain either a *constant superheat* or *constant pressure* within an evaporator of a refrigeration system. All components respond to a low voltage of 24 volts while the expansion valve will open to full

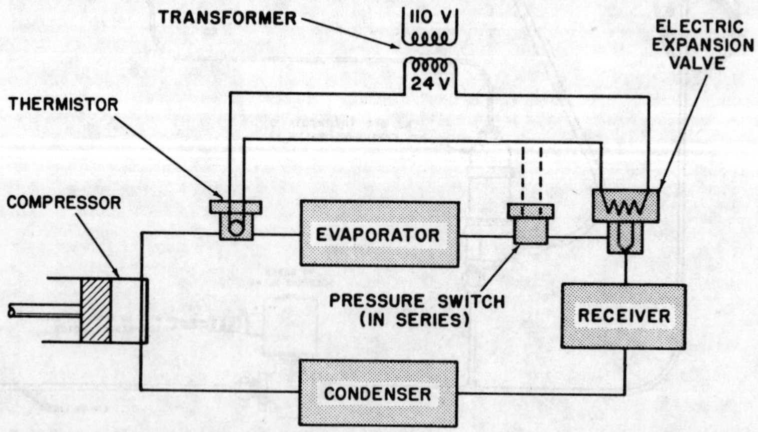

Fig. 204. Thermo-electric expansion valve system arrangement showing location of pressure switch if used.

capacity at 4.5 watts and closes when the power is zero. Approximately one minute is required for the valve, at ambient room temperature, to open to capacity upon starting the system. It will respond equally well with either a-c, or d-c electrical current.

Fig. 204 shows the thermistor inserted in the suction line and exposed to the refrigerant fluid. It is connected in electrical series with the pressure switch and expansion valve heat motor. An existing gaseous or superheated condition within the suction line causes the thermistor to self-heat, resulting in a decrease in electrical resistance which, according to Ohm's law, $W = \frac{E^2}{R}$ (see Chapter 10), increases the power to the expansion valve heater. As stated above, any increase in current tends to open the expansion valve and allow the flow of liquid refrigerant to increase. With increased flow the possibility exists of lowering the quality of the vapor (wet gas) or of allowing liquid refrigerant to reach the suction line, depending on load fluctuations. Either of the two fluid states exerts an immediate cooling effect on the thermistor to increase its resistance and initiate a tendency on the part of the expansion valve to close. The word "tend" in the foregoing discussion is used in a guarded sense. Its implication here is one of modulation rather than abrupt substantial degree of dimensional change in either direction of the valve needle. The thermistor material being extremely sensitive to refrigerant state changes succeeds in nearly approaching a flat modulating frequency so that the "hunting" procedure described for centrifugal compressor controls is greatly diminished. (See Chapter 10.) For this reason, the combination of thermistor and thermal-electric expansion valve can maintain a constant state of the refrigerant vapor at the location of the thermistor in the suction line. Beyond the thermistor toward the compressor suction inlet, slight superheating will occur to insure liquid-free gas reaching the compression chamber and so prevent damage in that area.

Because the thermal-electric expansion valve is free of any pressure-temperature influence by the passing fluid the same valve can be employed with any of the conventional refrigerants, except ammonia. For the same reason, refrigerant flow will occur in either direction and can be reversed at any time.

Refrigerant temperature level is not a controlling factor in the function of the thermistor. It does respond rapidly, however, if the state of the refrigerant changes from a vapor to liquid without a change in temperature. Therefore, the thermistor will function with the same positive response at any refrigerant temperature level.

The thermal-electric expansion valve acts as *constant pressure* control when its heat motor is connected in series with a simple make-and-break pressure switch exposed to evaporator pressure. Adjusted to close at a given minimum pressure, it will allow current to flow and energize the expansion valve heater when the pressure drops below the adjusted level. The valve opens upon being energized allowing a greater flow of refrigerant to raise evaporator pressure and open the switch. The expansion valve will then modulate toward the closed position once more. The characteristics of the valve heater cause the pressure to modulate at low frequency so as to maintain nearly constant pressure within the evaporator.

A limit pressure switch in conjunction with the thermistor will maintain constant pressure and temperature conditions within the evaporator during operation.

An arrangement involving a high pressure limit switch connected in electrical series with the expansion valve heater and a low pressure switch connected in parallel with the thermistor will maintain saturated vapor (100% quality—no superheat) between pressure limits governed by the respective adjustments of both switches. Both switches open with a pressure increase, but the low pressure limit switch closes and shunts the thermistor when the pressure in the evaporator falls below its adjusted setting. Thus, the pressure switches automatically transform the expansion valve from a superheat control to a constant pressure control.

The thermal-electric expansion valve method of refrigerant flow control with its close control tolerances makes it advantageous in several fields. One is in the field of air conditioning where constant evaporator control is a critical factor. Because of its bi-directional characteristics, it is adaptable to heat pump .(reverse cycle) applications in which a four-way valve is involved as described in Chapter 16.

Bleed Type Automatic Expansion Valves

Smaller systems using off-on cycle capacity control on which low starting torque motors, such as split-phase type, require some modification to prevent overloading upon on-cycle starting. The problem is overcome in systems having capillary tubes as throttling devices. In such systems, a continual bleeding of liquid refrigerant to the evaporator occurs during off-cycle, thereby providing a pressure balance between the high and low side upon initiation of start-up. The pressure balance conditions reduce starting torque to a minimum. The process is referred to as *system unloading.*

By providing an orifice either in the form of a drilled hole paralleling the main orifice of the automatic expansion valve, or by simply providing one or more nicks or slots in the valve orifice, succeeds in furnishing the bleed function. The size of the bleed orifice provided depends on the capacity of the system, i.e., as the capacity increases, the size of the bleed orifice is made necessarily larger. Orifice sizes are usually indicated in terms of Btu per hr under given condenser and evaporator temperatures. For example, under conditions of 100 F and 20 F condenser and evaporator temperatures, respectively, using R-12, the sizes range from about 1500 Btu per hr for small water coolers up to 14,000 Btu per hr for larger capacity systems. Above 6000 Btu per hr, the valve is usually drilled to provide an independent bleed orifice. During the on-cycle, the bleed orifice acts to supplement the main valve stem (or ball) opening, the latter of which will feed the necessary amount of liquid less the amount passing through the bleed opening, to the evaporator to meet load demands.

Float Valves

The float valve consists of a hollow, buoyant, metal float in the form of a ball, a cylinder, or a pan, attached to one end of an arm whose opposite end is connected to a pivot that allows the buoyant member to rise and fall according to the level of the liquid body on which it rests. A needle or plunger is linked to the pivot arm and moves back and forth to open or close a liquid port according to the rise and fall of the float. The float assembly is housed in a float chamber, accumulator, or evaporator, while the valve port provides the only link between the high- and the low-pressure sides of the system. Float valves are designed to respond to the liquid level in whatever side of the system the buoyant member is located, and to maintain a predetermined height of liquid at all times irrespective of pressure or load. If the buoyant member is located on the condenser side of the port, the assembly is known as a high-side float. If located on the evaporator side of the port, it is called a low-side float.

Low-Side Float Valves. The function of the low-side float valve is to maintain a constant liquid level in the evaporator when operating under flooded conditions, irrespective of pressure and load. As the heat load tends to raise the pressure in the evaporator by vaporizing the liquid, the compressor acts to maintain it at a constant level by continual evacuation of the resulting vapors.

It is the role of the low-side float valve continually and automatically to replace the liquid refrigerant at a rate equal to the amount lost by evaporation, and thus sustain an operating balance in conjunction with the heat load and the compressor. This it accomplishes by actuating the valve needle to open or close the valve as the need arises, allowing more or less refrigerant to enter the evaporator in direct response to the liquid level in that component. The action of the valve may conform directly with any variation in liquid level and thus provide modulated flow, or it may be designed to respond only to a maximum and minimum level at which point the port is fully open or fully closed as a result of trigger action of the needle.

While the former method of flow subjects the needle and the port to rapid wear and the latter has the obvious disadvantage caused by wide variations in liquid level, the remedy for both lies in the addition of a restriction cham-

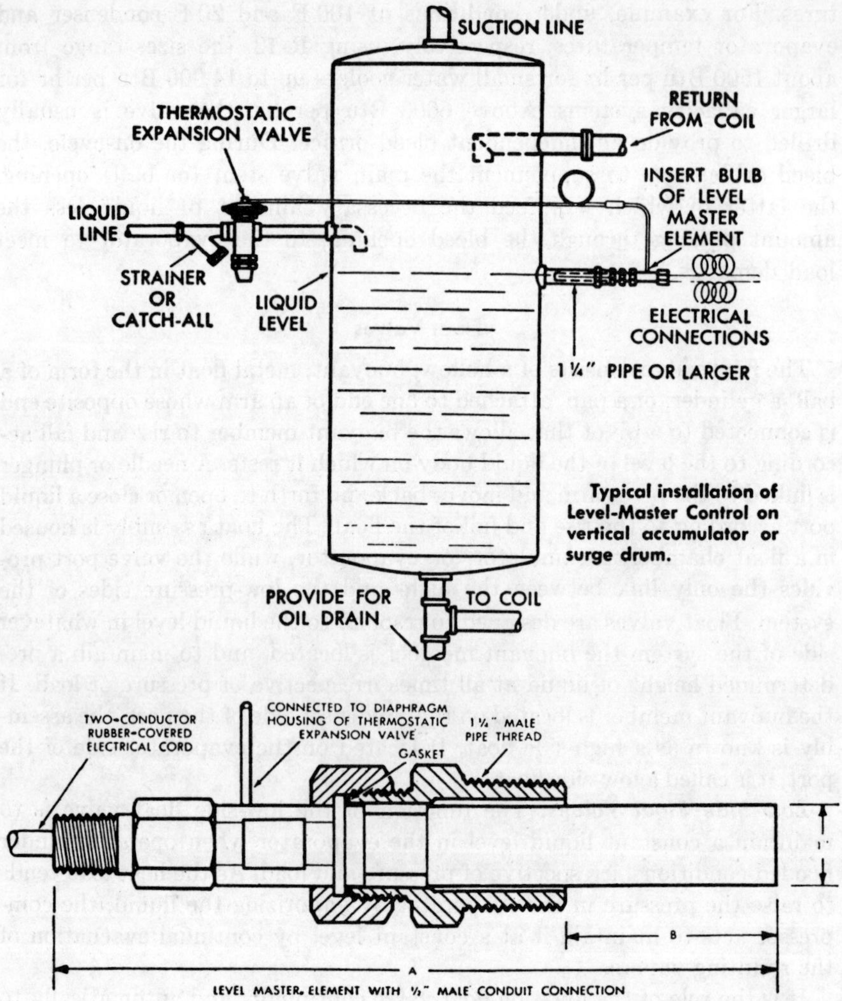

Fig. 205. Liquid-level control using a thermostatic expansion valve and electrical superheat element. (*Top*) Assembly arrangement. (*Bottom*) Detail of heating element. (Courtesy, Sporlan Valve Company.)

ber through which the liquid must flow before entering the evaporator or accumulator. Used in conjunction with modulated flow, the velocity of the entering liquid is reduced in two steps by providing a flow-restricting element beyond the valve port on the suction side. Thus, the pressure of the entering liquid is reduced only partially in passing through the valve to the restriction chamber. It is again reduced, this time to the suction pressure, as it enters the evaporator through the restricting port.

The low-side float valve can serve a number of evaporators. Whether located in an accumulator, in a float trap, or in one of the evaporators, the liquid level in all components served by the same float must be the same. If installed in the evaporator, sufficient space must be allowed between the liquid surface and the suction line exit to eliminate the possibility of liquid carry-over to the compressor. On large systems, a by-pass connection with a hand expansion valve should be provided in the event of float-valve failure.

When the compressor is stopped, the liquid in the evaporator continues to vaporize until the pressure reaches a point corresponding to the temperature. At that point, evaporation ceases and the valve closes to isolate the evaporator.

Liquid-Level Control—Using Thermostatic Expansion Valve. Another method of controlling the liquid level in low-side evaporators and accumulators employs a thermostatic expansion valve. Fig. 205 shows an arrangement for this method of control for flooded systems.

An electrical heating element provides artificial superheat to the thermostatic valve charge when the liquid falls below the desired level. This causes the pressure of the charge to rise and open the thermostatic expansion valve. As liquid level rises in the drum and contacts the bulb, the effect of

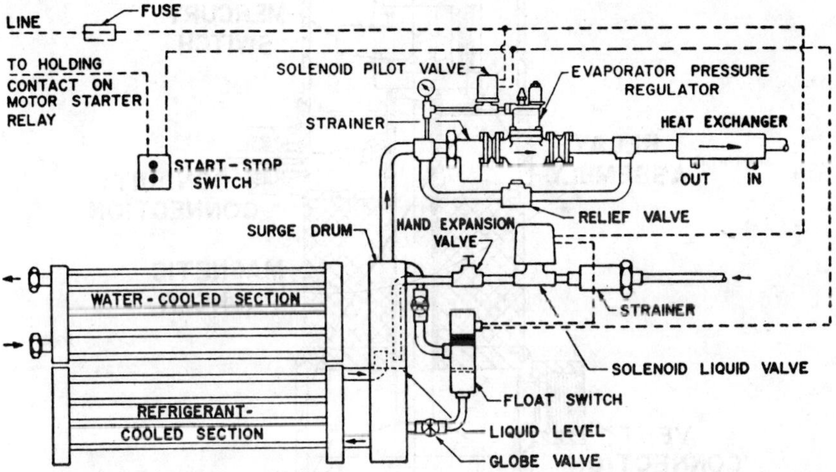

Fig. 206. Application of float switch controlling liquid valve (solenoid) to maintain liquid level in surge drum feeding Baudelot cooler. Note locaton of evaporator pressure regulator. (Courtesy, Alco Valve Company.)

the heater is nullified, thus causing a decrease in bulb pressure. The valve then closes or throttles to maintain the liquid at constant level in the evaporator or surge drum (accumulator).

The heat element is of low wattage, approximately 15 watts. The insert bulb must be in direct contact with the refrigerant and not inserted in a well that would act as a shield from the liquid. If inserted in the surge tank, the bulb should be protected from splash at those points where liquid enters from either the expansion valve or the coil return line.

Liquid-Level Control—Using Float Switch. The low-side liquid level of flooded systems may be controlled by the use of a float switch, as shown in Fig. 206. The liquid level in the float chamber is controlled by the level in the evaporator or accumulator. This is in accordance with the law that liquids seek their own level. The float ball is connected to a magnetic plunger which extends into the relay case through an enclosing tube as in Fig. 207. The relay consists of a permanent magnet and mercury switch. The mercury switch is mounted on a lever type magnetic armature. The relay assembly is positioned on the enclosing tube by spacers for proper operation.

When the float ball is below a given level governed by the desired height of liquid in the evaporator, the magnetic plunger tends to depart the magnetic field of the permanent magnet. When it descends far enough out of the field, the permanent magnet draws the lever armature in, closing the mag-

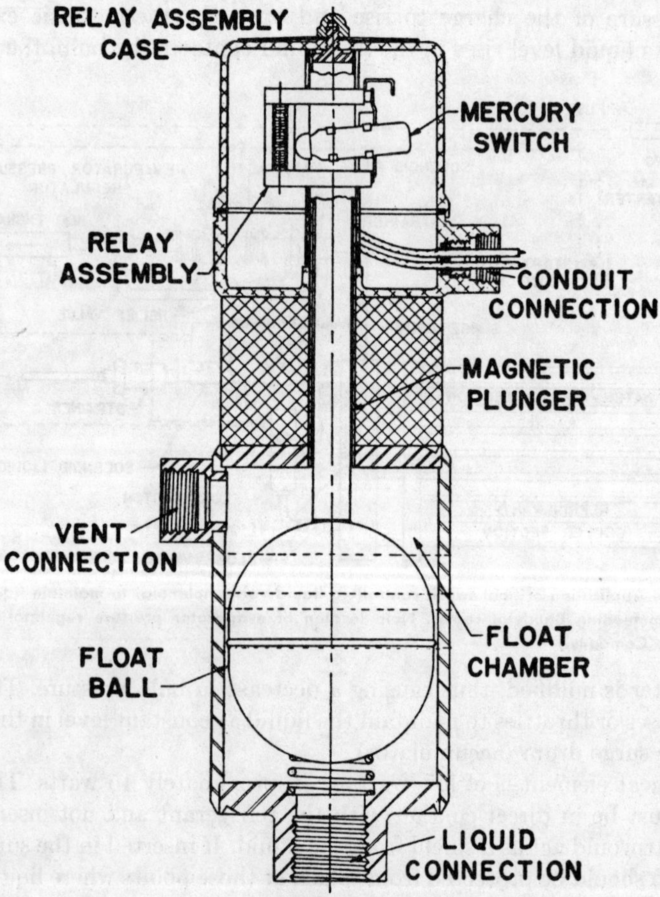

Fig. 207. Float switch. (Courtesy, Alco Valve Company.)

netic circuit. This action tilts the mercury switch, opening the electrical circuit to the solenoid. The solenoid becomes energized, opening the liquid valve to allow flow of refrigerant to the evaporator or accumulator.

When the liquid rises to the desired level, the float ball also rises. This causes the plunger to re-enter and short circuit the magnetic field. When the entire magnetic field has been short circuited, the lever armature and mercury switch are tilted by a spring to open the electrical circuit. The solenoid is de-energized and the liquid valve closes to stop the flow of liquid.

If it is desirable to open the electrical circuit upon fall, and to close it on rise, of liquid level, a reverse-acting relay may be used.

A hand expansion valve is placed in the liquid line between the solenoid valve and the evaporator. This is done to prevent surging caused by the sudden entry of liquid to the evaporator and to eliminate short-cycling of the solenoid valve.

High-Side Float Valves. This type of float valve allows only a minimum amount of liquid to remain in the high-pressure side of the system so that nearly all the refrigerant charge is in the evaporator. As the compressed vapor condenses, it flows into the float chamber from which it is forced under high pressure into the evaporator when a sufficient amount has collected to raise the float and open the valve. Because the float is set to open at a given level in the high-side float chamber or receiver, it follows that the liquid level in the evaporator is nearly constant at all times. And as almost the entire charge of refrigerant is in the evaporator, it is necessary that the system contain only enough liquid to maintain the desired level in the evaporator in order to prevent its carry-over to the compressor.

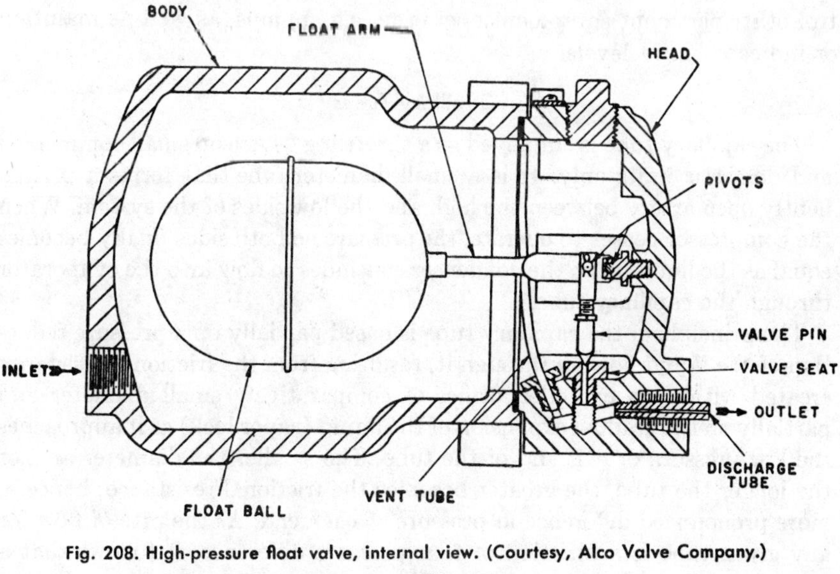

Fig. 208. High-pressure float valve, internal view. (Courtesy, Alco Valve Company.)

There are several differences between the high- and the low-side floats. In the former, the buoyant member is exposed to high pressure and must be elevated to open the valve and provide an exit for the liquid. The buoyant member of the low-side float is under low pressure and must fall to open the valve and allow liquid to enter the evaporator. While the maintenance of a constant liquid level in the evaporator is a direct function of the low-side float, it is controlled indirectly by the high-side float which balances the flow of liquid from the condensing side with the rate of evaporation in the evaporator.

To eliminate gas binding in the float chamber, a vent tube connecting the space above the liquid surface with a point immediately beyond the valve seat, as shown in Fig. 208, usually is installed.

The location of the high-side float valve is independent of the liquid level in the evaporator, another feature that departs from the requirements of the low-side float. It may be installed above or below that component. If located any distance from the evaporator, however, it becomes necessary to provide some means of maintaining sufficient pressure on the liquid to prevent premature boiling in the low-side liquid line. This is often accomplished by installing a liquid-pressure valve at the inlet of the evaporator.

The high-side float valve is limited to one evaporator, and its principle of operation precludes the use of more than one float valve in any one system. They are used principally on domestic type units.

Electronic Float Switch. In addition to the magnetic switch shown in Fig. 207 an electronic float switch is also employed. The actuating element as in the magnetic type is a pick-up coil governing the level of the ball. The coil is operated from an electronic controller that is substituted for the magnetic relay. Such switches are used to operate an alarm, control other pertinent equipment operation, e.g., pumps, as well as maintain or indicate liquid levels.

Capillary Tubes

The capillary tube is employed as a throttling device on small commercial and domestic units only. It is a small-diameter tube that forms a permanently open artery between the high and the low sides of the system. When the compressor ceases to operate, the pressure on both sides finally becomes equal as the liquid from the condenser continues to flow into the evaporator through the capillary tube.

The principle of the capillary tube is based partially on a pressure reduction of the liquid flowing through it, resulting from the frictional resistance created within an extended orifice of comparatively small diameter and partially on the gradual expansion of the liquid (vapor lock) as it approaches the low pressure or exit end of the tube. The smaller the diameter and/or the longer the tube, the greater becomes the frictional resistance; hence, a more pronounced difference in pressure at each end. As the rate of flow for any given tube is a function of that pressure difference, it follows that a greater supply of refrigerant is forthcoming when the summer load conditions demand it by reason of the accompanying effect of higher condensing temperatures with air-cooled condensers used on domestic units. The converse is also true, making the capillary tube self-compensating over a limited range of pressure differences.

With larger units, the need for more refrigerant grows due to greater heat loads. This means that a lesser frictional resistance is necessary to provide greater flow. As the rate of liquid flow depends on tube diameter and/or tube length, then the former must be enlarged and/or the latter shortened to compensate for any increased load capacity, assuming similar operating conditions in all cases. This relationship of load, tube diameter, and tube length, as shown in Table 34, bears out the latter statement.

The degree of tube bending also governs the resistance to flow.

Capillary tubes have been found satisfactory on domestic and small commercial units where loads are not subject to severe fluctuations. However, systems on which they are employed must be charged with the exact amount of refrigerant so that the length and the diameter of the tube must be determined for the specific installation. Their advantage lies in simple construction and operation, having no moving parts and creating low starting

TABLE 34. RELATION OF CAPILLARY TUBE BORE AND LENGTH TO REFRIGERATING CAPACITY*

Horsepower	*Load Btu per Hour*	*Suction Temperature*	*Tube Bore*	*Tube Length*
$\frac{1}{3}$	2800	26 F	0.040	3 ft
$\frac{1}{3}$	2800	26 F	0.050	5 ft
$\frac{1}{4}$	2125	26 F	0.040	4 ft
$\frac{1}{4}$	2125	26 F	0.050	$6\frac{1}{2}$ ft
$\frac{1}{5}$	1550	26 F	0.040	6 ft
$\frac{1}{5}$	1550	26 F	0.050	$9\frac{1}{2}$ ft

* Courtesy, Heath, D. P.: Data on capillarics, *Refrigerating Engineering,* **39**:304, May, 1940.

torque due to equalized pressures in evaporator and compressor on off-cycle.

Because the length and internal diameter, along with discrete bending, of the capillary tube governs its refrigerant flow capacity new designs are capable of some modification in this regard. By providing the means to increase or decrease its length by various telescoping arrangements and/or by simplifying, or compounding the bending design, capillary tubes may be adjusted to meet load changing conditions. Such adjustments may be the subject of future research to allow for their automatic modulation according to load demands, and where the operating condition warrants the cost of such control.

CAPACITY CONTROLS

Because refrigeration systems are subject to load fluctuations, it often becomes advantageous to provide some means by which the compressor may compensate for the subsequent changes in the weight of vapor handled. Continuous operation at full displacement for loads considerably below the capacity of the compressor results in a reduction in suction pressure accompanied by rarefaction of the vapor drawn from the evaporator. This results in increased power cost per ton of refrigeration. Compensating for the smaller amount of refrigerant vaporized under reduced loads entails the adoption of suitable automatic or manual controls for varying compressor capacity from maximum to a fractional part of maximum displacement or to zero when desirable and circumstances permit.

The various methods employed to vary compressor capacity are described below, divided according to the kind of compressor to which they apply; i.e., reciprocating or centrifugal machines.

Capacity Control of Reciprocating Compressors

Variable-Speed Drive. Speed is a direct function of compressor capacity. Therefore, any change in the number of strokes per unit of time varies the volume displacement by direct proportion for a similar period. Thus, if the normal operating speed of a reciprocating compressor is reduced 50 percent, its capacity is also reduced 50 percent, within certain limits.

Multi-speed induction motors are often used to obtain speed changes for the purpose of regulating compressor capacity, being available for 2-, 3- and 4-speeds. However, speed variations are not gradual with multi-speed induction motors but change abruptly from one speed to the other, an action that is not conducive to exact control for each increment of load fluctuation. The reason for this is that speed depends on the *frequency* of the power circuit and the number of poles and is determined by

$$\text{Speed (rpm)} = \frac{\text{alternations}}{\text{no. of pairs of poles}} \times 60$$

Therefore, as the number of poles increase, speed decreases. The simplest variation is made from maximum to 50 percent speed by merely cutting out half the poles. This can be accomplished externally by changing stator connections by means of a drum switch to which stator windings are connected. If more than two speeds are desired, or if the two speeds are not of 2 to 1 ratio, it becomes necessary to have more than one separate winding about the stator, each of which may be connected for two speeds.

Three-phase induction motors having but two speeds, the lower of which is either 25 percent or 50 percent below maximum, commonly are used. They may be controlled either manually or automatically.

When driven by a steam engine or an internal combustion engine, a less complex method for varying compressor speeds is obtained as a result of governor control. Because gradual speed changes can be made in such prime movers, it becomes possible to attain capacity modulation.

Stop and Start (Cycling). The method of controlling compressor capacity by stopping and starting is familiar through being the one used on domestic units. Volumetric displacement under such an arrangement is either maximum or zero, which presupposes temperature cycling rather than constant temperature through capacity modulation. Control here is generally vested in a thermostat placed in the refrigerated medium that acts to make or break the electric circuit leading to the motor when the temperature reaches predetermined minimum and maximum levels, respectively.

This method of capacity control requires an electric motor having exceptionally high starting torque characteristics (double squirrel-cage motor) when used on larger systems without benefit of compressor unloaders, or (capacitor start–induction run motors) on smaller systems not equipped with capillary tubes to equalize pressures during off-cycle periods. When stop-and-start capacity control is applied to commerical or industrial systems, a liquid stop valve actuated from the power circuit through a solenoid usually is installed in the liquid line. The valve closes whenever the compressor stops operating.

Capacity Control and Unloaded Starting System Using Oil-Pump and Crankcase Pressure.* Trane compressors from 10- to 100-tons are equipped with an integral and completely internal mechanism that provides automatic capacity control and permits the compressor to start unloaded. Loading and unloading of cylinders is actuated by suction pressure variations, but the unloader mechanism operates hydraulically on power from the oil pressure of the lubrication system.

Without oil pressure, the unloader mechanism holds suction valves open and individual cylinders operate unloaded. Under oil pressure, the unloader permits suction valves to function normally and cylinders to operate at full capacity. After shut down and without oil pressure, the compressor will start and operate unloaded until oil pressure is supplied to the unloader mechanism. Permanent, positive capacity control with substantial power savings are provided by this fool-proof, quiet mechanism.

Number of controlled cylinders and the capacity steps for various sizes are:

No.'s 10, 215, 25, and 240 BT & CT Units—Three cylinders unload with capacity steps of 100%, 75%, 50%, and 25%.

No.'s 15, 225, 40, and 260 BT & CT Units—Four cylinders unload with capacity steps of 100%, 66⅔%, 50%, and 33⅓%.

No.'s 20, 230, 50, and 275 BT & CT Units—Six cylinders unload with capacity steps of 100%, 75%, 50%, and 25%.

No.'s 60 BT & CT Units—Seven cylinders unload with capacity steps of 100%, 80%, 70%, 60%, 50%, 40%, and 30%.

No.'s 75 BT & CT Units—Eight cylinders unload with capacity steps of 100%, 83⅓%, 66⅔%, 58⅓%, 50%, 41⅔%, and 33⅓%.

No.'s 100 BT & CT Units—Twelve cylinders unload with capacity steps of 100%, 87½%, 75%, 62½%, 50%, 37½%, and 25%.

HOW THE UNLOADER OPERATES. To understand the complete operational cycle think of the mechanism as two distinct components.

*** Courtesy, The Trane Company, LaCrosse, Wisconsin.**

1. *Capacity Control Actuator*—one for each compressor. (See Fig. 209, *top*.)

2. *Cylinder Unloader Mechanism*—one for each controlled cylinder (Fig. 209, *bottom*).

(*Note:* Fig. 209 is a schematic drawing only and shows only the flow of operation.)

Capacity Control Actuator. The Capacity Control Actuator reacts to variations in refrigeration load requirements and transmits them to the Cylinder Unloader Mechanisms which act to load and unload the compressor cylinders. To perform this dual function, the Capacity Control Actuator consists of: (1) A pressure sensing device (Fig. 209, lower left quadrants) which is sensitive to variation in suction pressure. (2) A valving mechanism (Fig. 209, lower right quadrants) which regulates the oil pressure to the various cylinder unloader mechanisms.

Pressure Sensing Device. The pressure sensing device consists of a chamber *A*, connected to suction pressure through line *B*, and a bellows *C* the inside of which is connected to atmospheric pressure through vent *D*. The tendency of the pressure sensing device is to maintain as nearly as possible a predetermined suction pressure. This pressure is the maximum pressure required to satisfy the system and may range from 0 to 50 psi. The specific point is maintained by a balance of forces. Suction pressure balanced against a combination of atmospheric pressure and force from spring *E*. The amount of tension adjustable by set screw *F*. When the system requires less than full refrigeration load, the suction pressure will fall below the predetermined point, causing an unbalance within the device, and the unloading cycle will commence. The drop in suction pressure permits bellows *C* to expand, forcing plunger *G* against lever *H*, moving it downward. The downward movement of this lever opens the regulated orifice *I*. The opening and closing of this orifice controls the action of the valving mechanism.

Valving Mechanism. The function of the valving mechanism is to supply each of the cylinder unloaders with oil under pump pressure when full compressor capacity is required and to relieve this pressure when cylinders are to operate unloaded. This valving mechanism consists of a hydraulic cylinder containing an annularly grooved, floating piston *K*. The annular grooves are constantly connected to oil pump pressure through line *X*.

Above the piston is a chamber *L* which is vented to the crankcase through orifice *J*. Below the piston is another chamber *M* connected to the annular grooves in the piston by orifice *N* and connected to crankcase pressure through regulated orifice *I*. Located within the hydraulic cylinder is a spring *O* which tends to move the floating piston toward the lower chamber.

Under full capacity operation, as shown in Fig. 209, *left*, regulated orifice *I* is shut off; oil pressure in lower chamber *M* increases because oil under pump pressure is being supplied through line *X*. This pressure overcomes the force of spring *O* and floating piston *K* rises in the cylinder. As it rises, the annular grooves in the floating piston coincide in sequence with lines *P1*, *P2*, and *P3* to the cylinder unloaders, providing them with full oil pressure and permitting them to operate at full capacity. To make Fig. 209 as simple as possible, only line *P1* is connected to a cylinder unloader mechanism. Lines *P2* and *P3* are in reality connected to identical mechanisms, and while this discussion is concerned with only one unloader mechanism, it can be extended to cover them all.

When full compressor capacity is not required, regulated orifice *I* is opened through movement of lever *H*; oil bleeds through it and pressure within the lower chamber approaches crankcase pressure, as shown in Fig. 209, *right*. Under these circumstances, the force of spring *O* overcomes the pressure in the lower chamber and floating piston *K* is moved downward so that lines *P1*, *P2*, and *P3* become

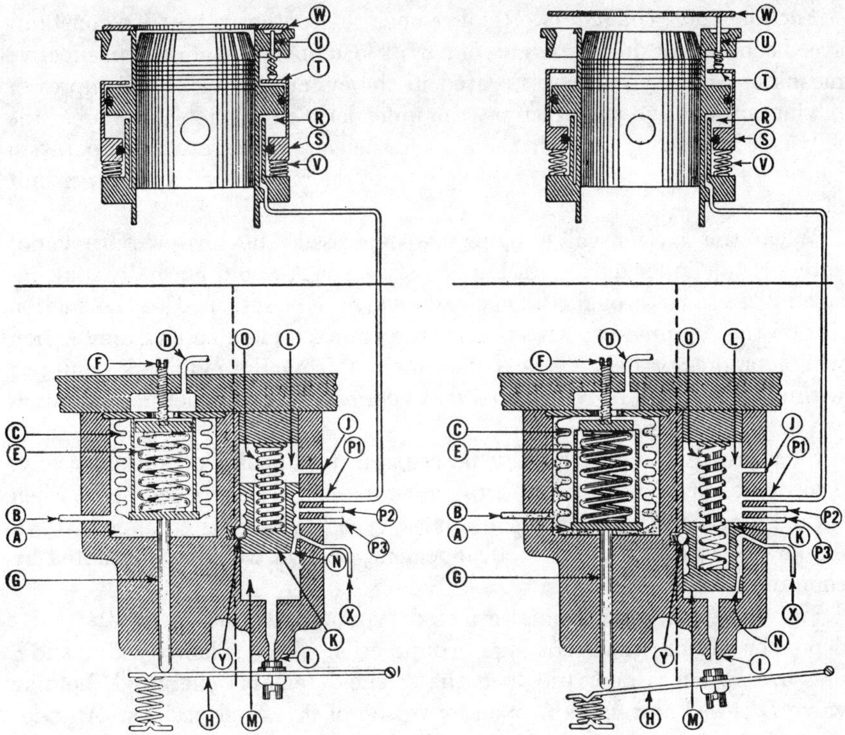

Fig. 209. Capacity control and unloaded starting system employing hydraulic cylinder and floating piston assembly using oil pump and crankcase pressure. (Copyright, The Trane Company. Reproduced by permission.)

connected in sequence to crankcase pressure through orifice *J*. The spring-loaded ball *Y* permits the piston to move only in distinct increments; one groove at a time.

In this manner, the valving mechanism supplies or withdraws from each cylinder unloader the oil pressure that operates the unloader mechanism.

THE UNLOADER MECHANISM. When oil from the forced feed lubricating system flows through line *P1*, from the valving mechanism to the cylinder unloader, it enters annular chamber *R*. The inner wall or unloader cylinder is firmly anchored to the cylinder liner; the unloader piston *S*, however, is free to move. The up and down movement of this unloader piston raises and lowers take-up ring *T* which raises and lowers suction valve lift pins *U*.

Under full capacity operation (Fig. 209, *left*) oil flows into annular chamber *R* under pressure sufficient to contract the unloader piston springs *V*. When oil pressure forces springs to contract, the unloader piston *S* moves down and take-up ring *T* and the suction valve lift pins *U* move with it. This permits the suction valve *W* to function normally and the cylinder operates at full capacity. When the compressor is to operate at less than full capacity (Fig. 209, *right*) oil line *P1* from the cylinder unloader mechanism is connected to crankcase pressure through orifice *J* which allows the pressure in the annular chamber *R* to dissipate; the cylinder unloader springs *V* expand, lifting the unloader piston *S*. This raises the take-up ring *T*, the valve lift pins *U* and holds the suction valve *W* open so that the controlled cylinder is operating in an unloaded condition.

Suction-Valve Unloaders. Holding open the suction valves of a constant-speed compressor during periods of partial load operation affords an effective means of capacity control. Located in the cylinder head of a compressor having automotive type pistons, the unloader element acts to depress the valve reeds or strips and hold them open when a reduced rate of evaporation causes a corresponding drop in weight of refrigerant to be handled per unit of time.

When the suction-valve strips are depressed, the low-pressure vapor enters the cylinder during the suction stroke as it would normally, but, instead of leaving through the discharge valve, it is returned to the suction line on the compression stroke. Thus the vapors undergo a churning action while compressor delivery to the condenser from the cylinders equipped with unloaders will be reduced to zero volume as long as the suction valves are held in open position.

Suction-valve unloaders may be controlled manually or automatically. The depressor of the manual type is actuated by a threaded stem when turned by a hand wheel. The automatic type depressor is equipped with a diaphragm or piston-cylinder arrangement, both of which are actuated by compressor discharge pressure.

Fig. 210 shows an automatic piston type cylinder unloader, the valve depressor withdrawn, operating in conjunction with a solenoid valve and a pressure switch. Under full load the solenoid remains energized, holding valve *D* closed and valve *C* open by reason of its raised position. Any decrease in load causes the suction pressure to fall off, which opens a pressure

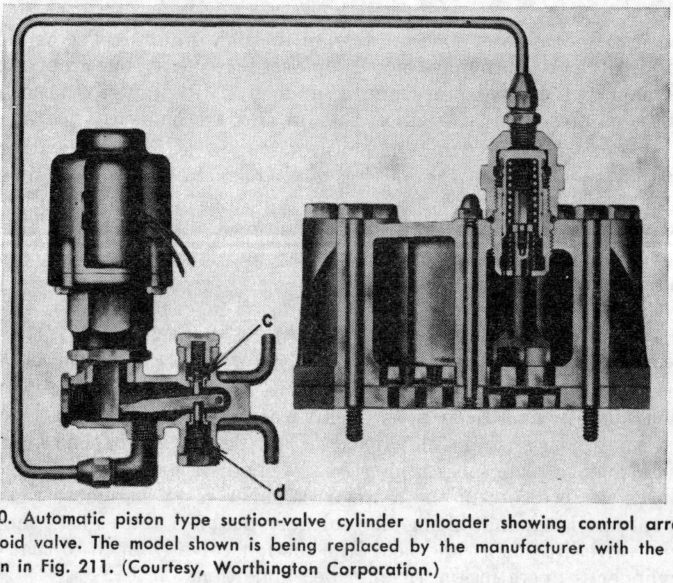

Fig. 210. Automatic piston type suction-valve cylinder unloader showing control arrangement with solenoid valve. The model shown is being replaced by the manufacturer with the improved type shown in Fig. 211. (Courtesy, Worthington Corporation.)

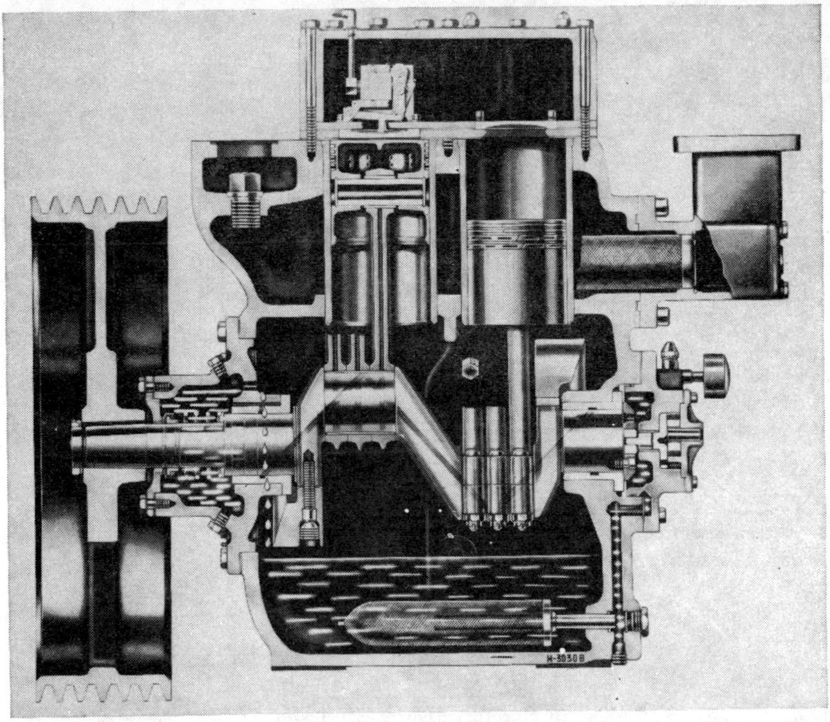

Fig. 211. Cross-sectional view of reciprocating compressor with improved type suction-valve unloader. (Courtesy, Worthington Corporation.)

switch to de-energize the solenoid. De-energized, the solenoid element falls, opening valve *D* and closing valve *C*. With valve *D* open, high-pressure gas from the discharge line enters the unloader, forcing the piston and the depressor fork down to open the suction valves.

A rise in suction pressure due to increased load closes the pressure switch to energize the solenoid. Upon rising, the solenoid closes valve *D* and opens valve *C*. With the release of discharge pressure from the depressor piston, that member is lifted by spring tension, allowing the suction valves to return to their normal operating position. No longer under high pressure, the gas is drawn into the suction line through valve *C*.

Unloaders may be installed on one or more cylinders of the same compressor, depending on the degree of flexibility required and the number of cylinders. A four-cylinder compressor having three of its cylinders equipped with unloaders may have a capacity reduction in steps of 25 percent of its rated volumetric displacement. This means that reductions of 25, 50, or 75 percent of rated capacity may be obtained with one, two, or three unloaders in depressed position. This form of capacity control cannot be used with trunk type pistons, for obvious reasons.

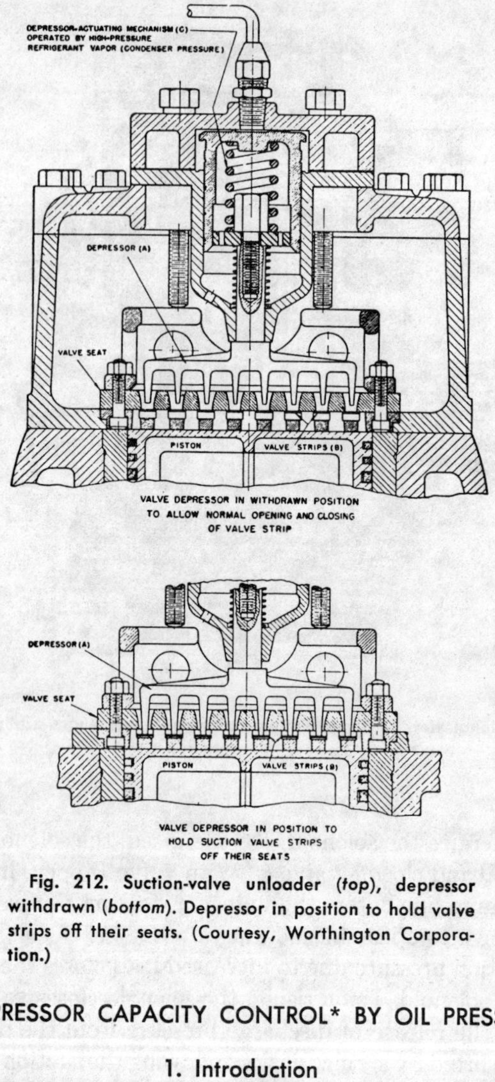

Fig. 212. Suction-valve unloader (*top*), depressor withdrawn (*bottom*). Depressor in position to hold valve strips off their seats. (Courtesy, Worthington Corporation.)

COMPRESSOR CAPACITY CONTROL* BY OIL PRESSURE

1. Introduction

The cylinder unloader and capacity control are the main components of the unloading mechanism. While usually referred to as the "unloading" mechanism, this component actually performs both loading and unloading functions, activating or de-activating cylinders as required by the increasing or decreasing demands of the system.

2. The Unloader

A. Function

When load on the system decreases, the unloader reduces the output of the

* Courtesy, Worthington Air Conditioning Co.

compressor by lifting, and maintaining the suction valve ring off the valve seat of the cylinder liner.

B. Method of Operation

Loading is accomplished when oil is fed to the unloader from the capacity control, through tubes (9, 10, or 11) and the connector (1). Pressure will then increase in the unloader chamber (4), causing the outer sleeve (3) to move downward. The inner sleeve (5), held by the retainer ring (7), stays in place. The return springs (2) are compressed by the outer sleeve. This permits the springs on the lifting pins (8) to push down the lifting ring (6), which, as it bears on the outer sleeve, follows its motion. This permits the suction valve ring to seat itself on the cylinder liner.

Unloading is accomplished when a drop in the cooling load causes oil in the unloader chamber (4) to return to the capacity control. The reduction in unloader chamber pressure permits the return springs (2) to raise the outer sleeve (3). This pushes the lifting ring (6) against the lifting pins (8), raising the suction valve ring. The suction ports, therefore, are kept open during the upstroke of the piston, and the refrigerant gas is returned to the suction manifold.

The action of the outer sleeve must be as smooth as possible. At the same time, an effective seal must be maintained to prevent the leakage of oil from the unloader chamber. Both of these requirements are effectively met through the use of Teflon®, in the form of O-rings.

Fig. 213. Reciprocating compressor (VC type) showing location of unloader and solenoid valve for Worthington external capacity control.

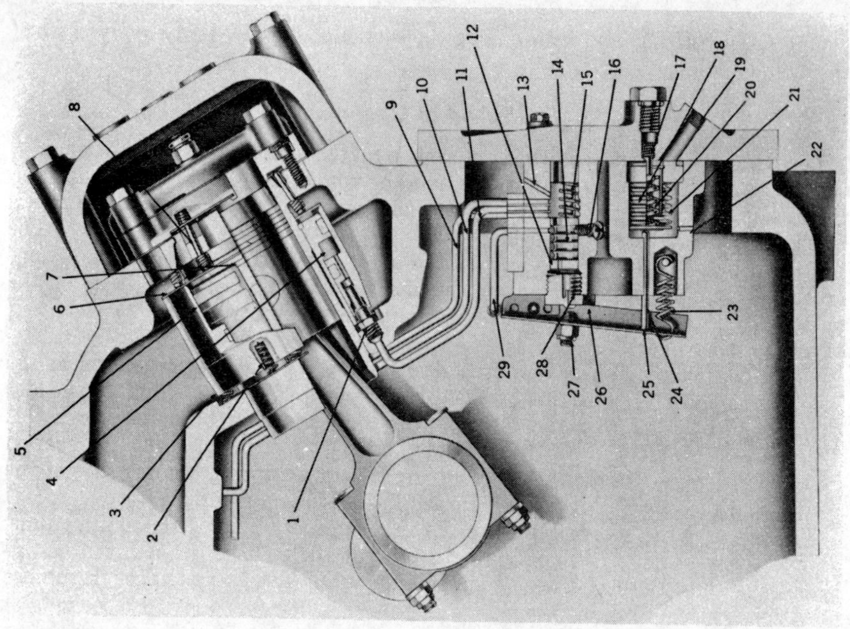

Fig. 214. Reciprocating compressor (VC type) showing location of components of Worthington internal pressure activated capacity control.

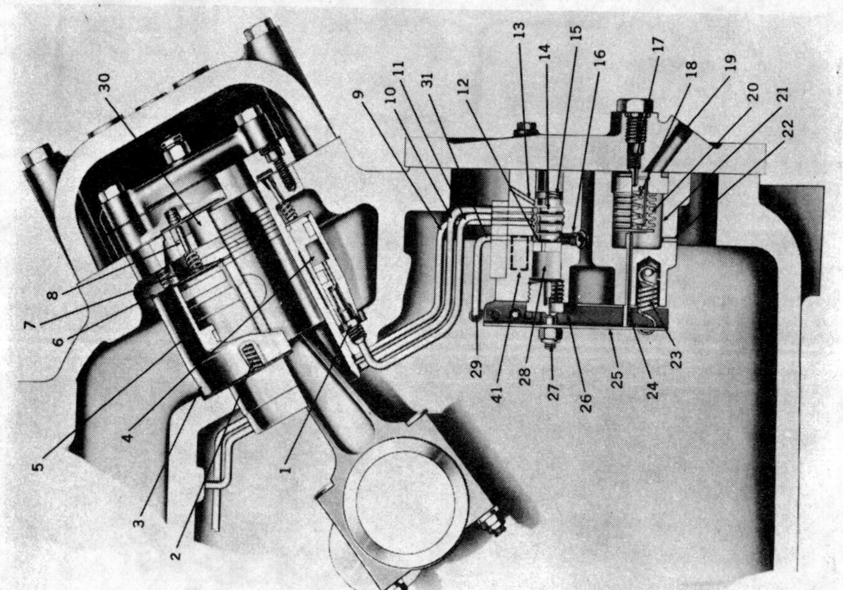

Fig. 215. Reciprocating compressor (VC type) showing location of components of Worthington external and internal capacity control.

3. The Capacity Control

A. Function

The capacity control is designed to automatically react to changes in cooling load by valving control oil to or from cylinder unloaders. A drop in load is accompanied by a drop in suction pressure. This device responds to this pressure drop by cutting off control oil supply to the correct number of cylinder unloaders and bleeding off oil pressure from them, decreasing compressor output and maintaining suction gas pressure at the proper level. On the other hand, with an increase in load it directs control oil to unloaders to match compressor output to system requirements.

B. Method of Operation

The capacity control works as it does because of the motion of the bellows (18). The bellows is at rest when compressor suction pressure, atmospheric pressure, the bellows compression spring (20), and the tension spring (23) counterbalance each other. With changes in any of these forces, the bellows (18) expands and contracts. This moves the push rod in and out, changing the position of the metering lever (25). As the metering lever pivots, the distance between the end of the metering screw (27) and the mouth of the orifice (26) changes.

This regulates amount of oil leaving the chamber (28). Oil flows through the tube (29) and a V-notch (12) in the floating piston (14) into the chamber (28) at a uniform rate. Therefore, the factor controlling chamber pressure is the change in outward flow of oil through the metering orifice.

The pressure differential between that acting on one face of the spring-loaded piston (15) (14) and that acting on the opposite face causes the piston to move. This differential must be sufficiently high to overcome the spring and, in addition, is in graduated steps due to the action of the spring-loaded detents (16) in the annular grooves of the piston, permitting oil to flow to the unloaders through tubes (9, 10, and 11), with all being fed simultaneously when the piston (14) is farthest from the orifice (26). When the differential drops as the chamber pressure decreases, the piston moves toward the orifice cutting off oil flow to each of the unloaders in turn.

Key to capacity control, Figs. 213, 214 and 215 for both *internal pressure activated* and *external electric solenoid activated.*

1. Connector
2. Return springs
3. Outer sleeve
4. Unloader chamber
5. Inner sleeve
6. Lifting ring
7. Snap ring
8. Lifting pins
9. Tube
10. Tube
11. Tube
12. V-notch
13. Vent port
14. Floating piston
15. Spring
16. Detent balls
17. Capacity control
18. Bellows
19. Pressure port and atmospheric vent
20. Compression spring
21. Chamber
22. Port
23. Tension spring
24. Push rod
25. Metering lever
26. Orifice
27. Metering screw
28. Chamber
29. Tube
30. Valve ring

31. Tube
32. Compression spring
33. Plunger
34. Upper valve seat
35. Lower valve seat
36. Ball seat
37. Solenoid valve body
38. Port
39. Tube
40. Tube
41. Manifold (shown in broken lines)
42. Tubes lead to manifold thence to unloader chamber through tubes 9, 10 and 12
43. Solenoid
44. Unloader

EXTERNAL COMPRESSOR CAPACITY CONTROL

The Worthington external compressor capacity control is a system which loads or unloads the compressor to adjust its capacity to meet load requirements. The system has two major components:

A. The cylinder unloader (44)

B. The solenoid valve (43)

Each major component is described below with regard to function and to method of operation.

1. The Cylinder Unloader

A. Function

The cylinder unloader is a power device that moves the suction valve ring (30) to start or stop the pumping of the cylinder whenever the control pressure is changed by the solenoid valve. The cylinder will deliver its normal output when the suction valve ring (30) is in its regular place, bearing on the cylinder liner seat. Unloading is accomplished when the suction valve ring (30) is lifted.

2. Method of Operation

A. Loading a Cylinder to Accommodate an Increase in System Load

Loading a system is accomplished when oil is fed to the unloader from the solenoid valve through the supply tube (12) and the connector (1). Pressure will then increase in the unloader chamber (4), causing the outer sleeve (3) to move downward. The inner sleeve (5), held by the retainer ring (7) stays in place. This permits the springs on the lifting pins (8) to push down the lifting ring (6) which, as it bears on the outer sleeve (3), follows its motion. This permits the suction valve ring (30) to seat itself on the cylinder liner. This cylinder can now pump refrigerant gas and has become a fully loaded working cylinder in the compressor.

The additional control tubes (10 and 11) are connected to other cylinders in the compressor and all three tubes (10), (11), (12) may be supplied by the same solenoid valve, or by individual solenoid valves so that loading may be accomplished in one, two, or three successive steps.

B. Unloading a Cylinder to Accommodate a Decrease in System Load

Unloading a cylinder is accomplished when the solenoid valve shuts off the supply of control pressure coming from the tube (31). The oil remaining in the unloader chamber (4) drains off slowly through the solenoid valve (37) and passage (38) into the compressor crankcase. The reduction in unloader chamber pressure permits the return springs (2) to raise the outer sleeve (3). This pushes the lifting ring (6) against the lifting pins (8), raising the suction ring valve (30).

The suction ports, therefore, are kept open during the upstroke of the piston, and the refrigerant gas is returned to the suction manifold. The action of the outer sleeve must be as smooth as possible. At the same time, an effective seal must be maintained to prevent the leakage of oil from the unloader chamber. Both of these requirements are effectively met through the use of Teflon® in the form of O-rings.

3. The Solenoid Valve

A. Function

The solenoid valve turns on, or off, the supply of control oil pressure from tube (31) into tubes (10), (11), (12) to load or unload the cylinder in response to an electric signal.

B. Method of Operation

a. Loading a Cylinder to Accommodate an Increase in System Load· When the solenoid valve (37) is energized the cylinder is loaded. In detail this is accomplished when the energized solenoid valve causes the plunger (33) to move upward compressing spring (32) changing the two-way valve (35) from closed to open condition by sealing the two-way valve (35) against the upper valve seat (34). The supply of oil pressure in tube (31), (38) passes through the open solenoid valve and is admitted to the unloader chamber (4) by tube (39), manifold (41), and tube (12). The solenoid valve (37) may be used to supply all three tubes (10), (11), (12) or each tube (10), (11), (12), may be connected to a separate solenoid valve, if loading is to be accomplished in one, two, or three successive steps.

b. Unloading a Cylinder to Accommodate a Decrease in System Load When the solenoid valve (37) is energized the cylinder is unloaded. In detail this is accomplished when the de-energized solenoid valve permits the compression spring (32) to push the plunger (33) downward changing the two-way valve from open to closed condition, by sealing the two-way valve (35) against the lower valve seat (36) thereby shutting off the supply of oil pressure from tube (31) (38). The oil remaining in the unloaded chamber (4) slowly drains out through tube (12) (39), solenoid valve (37) and passage (40) into the compressor crankcase. With the loss of supply oil pressure the cylinder becomes fully unloaded. As described above, this may be accomplished as a single step for all cylinders, or as two or three successive steps.

4. Suction Pressure Control

Because of the relative stability of the other forces acting on the bellows (18), the suction pressure can be considered, for practical purposes, the only force actuating the capacity control. How it does this is explained below.

A. Suction Pressure Decreases—the Compressor Unloads

The suction manifold is connected to the bellows chamber (21) through a port (22). A decrease in suction pressure lessens the load on the bellows (18). The bellows extends, forcing the push rod (24) against the metering lever (25), moving it outward. This moves metering screw (27) away from the orifice (26), permitting oil to flow from the chamber (28). This lowers chamber pressure so that the piston moves toward the orifice and cuts off the flow of oil to the unloaders, as explained above. The reduction in pressure causes the suction valve rings to be raised, and unloading takes place.

B. Suction Pressure Increases—the Compressor Loads

When cooling loads increase, with a corresponding increase in suction pressure, the bellows chamber (21) pressure increases, compressing the bellows (18). The tension spring (23) then pulls in the metering lever (25), to which the metering screw (27) is attached. This covers the orifice (26) and reduces the flow of oil from the chamber (28), causing the chamber pressure to rise. The piston (14) then moves away from the orifice and permits control oil under pressure to flow to the unloaders, causing the suction valve rings to seat against the cylinder liners and the compressor output to increase.

5. Low Starting Torque

Oil pressure does not build up until the crankshaft is turning over at rated speed. Since cylinder unloading is dependent upon oil pressure, this means the starting requirements of V-line compressor is low.

6. Adjusting the Compressor Capacity Control

The standard capacity control can be adjusted to balance compressor output with desired suction pressure at any point from approximately 25 psig to 95 psig. Compressors for low temperature or other special applications have a capacity control which is factory calibrated below the standard range. This range can be set from 8 inches vacuum to 38 psig.

The capacity control screw (17) is used to set the control point at which the capacity control must operate. The more it is screwed into the bellows, the higher the controlled suction pressure. Within the above range for R-22 or R-12, the sensing device will operate, with standard sensitivity setting, from maximum compressor capacity to completely unloaded condition or vice-versa with a total differential of 9 to 11 psi for R-22 and 6 to 7 psi for corresponding lower R-22 and R-12 pressures. Each step of loading or unloading is controlled to a differential of 3 to 4 psi at the higher psi range and 2 to 2½ psi at the corresponding lower range.

If pneumatic control is required, control pressure is introduced at the port next to the adjusting screw.

Cylinder By-Pass. Another means of varying the capacity of a reciprocating compressor operating at constant speed is to provide some means of by-passing some or all of the cylinder-entrapped vapor back to the suction from some point in the cylinder or the discharge manifold.

When entrapped vapor is returned from some point in the cylinder, the by-pass arrangement has the effect of reducing the length of compression stroke so that the volume actually delivered to the discharge line is less than rated capacity. The amount of reduction is equal to the space between the point where the piston starts its compression stroke and the point in the cylinder where the by-pass exit is located. Thus, if the by-pass valve port is installed immediately below the midpoint of the stroke, the vapor representing 50 percent of the volumetric displacement of the piston will be returned to the suction line when the valve is opened. The remaining 50 percent, trapped in the upper half of the cylinder when the piston covers

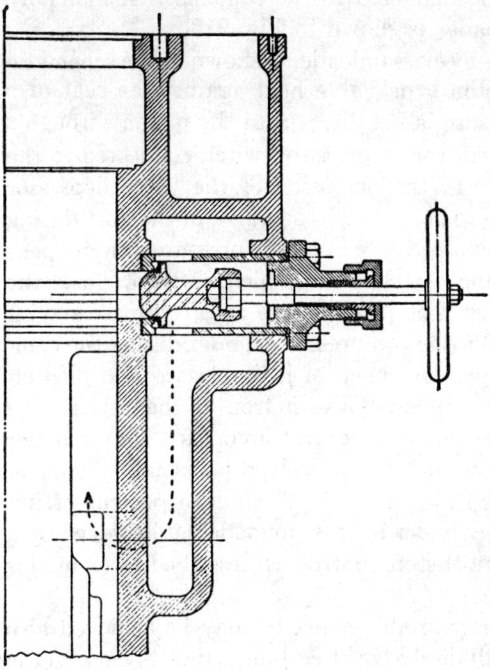

Fig. 216. (*Top*) Cylinder by-pass, manually operated. Opening valve provides passage of gas from middle of cylinder to suction port. (Courtesy, Frick Company.)

Fig. 217. (*Bottom*) Cylinder by-pass, automatically operated. Schematic diagram.

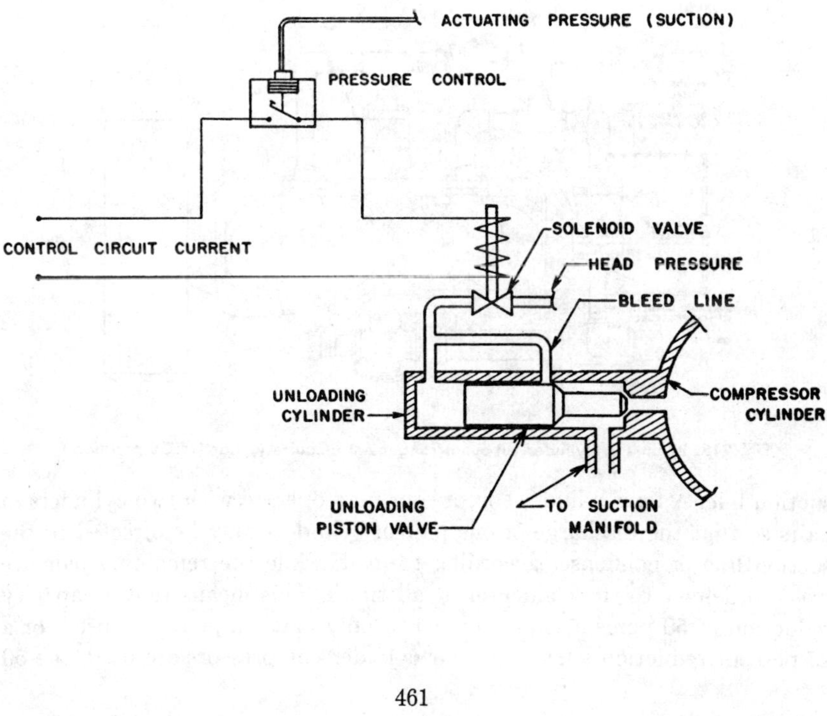

the port, will be compressed and delivered to the condenser. A manually operated cylinder by-pass unloader is shown in Fig. 216.

An automatic type of cylinder by-pass unloader is shown in the schematic arrangement in Fig. 217. A piston type valve held against the seat of a cylinder port by discharge pressure is fed directly to the piston through a solenoid valve which is controlled from a pressure switch connected to the suction line. However, contrary to the operation of the valve depressor type, the discharge pressure acts to hold the piston valve in the port during full load periods. When the suction pressure falls, the pressure switch opens the electrical circuit de-energizing the solenoid, which, in turn, closes the discharge line to the piston valve. The piston valve is forced back and off its seat by the pressure built up in the compressor cylinder during the compression stroke when the high pressure back of it is released and bled off through a small by-pass to the low-pressure area in front of the piston.

One arrangement of this type capacity control involves a four-cylinder compressor having all cylinders equipped with valved ports and capable of capacity variations in steps of 18.75, 37.5, 56.25, and 75 percent of full capacity. Operation of multiple-cylinder by-pass unloaders is achieved by a thermostat acting through a multi-step controller, described later in this chapter.

On multi-cylinder compressors, variable capacity may be achieved also by causing the by-pass of the full discharge of certain cylinders back to the

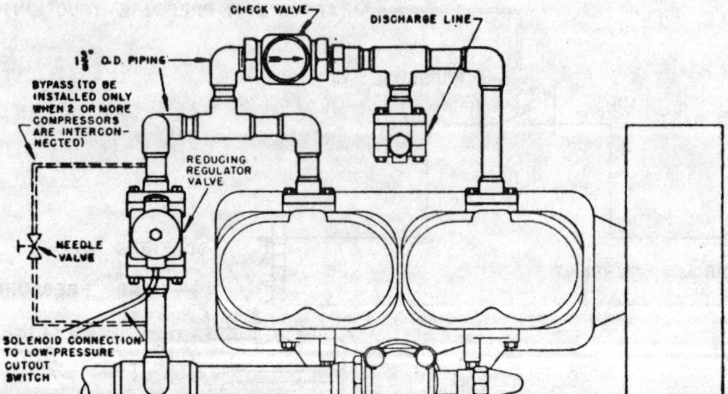

Fig. 218. By-pass or cylinder cut-out arrangement. (Courtesy, Carrier Corporation.)

suction line. A four-cylinder compressor may operate with two cylinders in pairs so that the discharge of one pair of cylinders may be directed to the suction line or condenser according to load, while the remaining pair are fixed to deliver to the condenser at all times. This means that a capacity reduction of 50 percent is possible when only one compressor is used, or a 25 percent reduction when two four-cylinder compressors are used, the 50 percent reduction being obtained in the latter case by cutting out (stopping) the second compressor.

The employment of this method of by-passing involves the use of a low-pressure cut-out switch whose closing energizes a solenoid that controls a reducing regulator valve. When energized, the solenoid opens a port to allow high-pressure gas to enter the regulator valve body, open the valve, and allow the compressor discharge to enter the suction line. As the suction pressure again increases to a predetermined level with added load, the cut-out switch opens to de-energize the solenoid, thereby closing the port in the reducing regulator valve. This causes that valve to close and direct the discharge gas to the condenser line between the by-pass tee and the condenser line tee that closes when the reducing regulating valve is open. (See Fig. 218.) To eliminate chattering of the check valve when two or more compressors are connected together, a $\frac{1}{4}$-in. by-pass line with a needle valve should be installed around the reducing regulator valve.

According to one manufacturer (Carrier Corporation) of this type control, the power required at 50 percent of full-load rating is reduced to 70 percent of full-load horsepower.

Clearance Pockets. Clearance in a compression cylinder has the tendency to maintain the suction valves in a closed position until the clearance gas has expanded sufficiently to allow equalization of pressure between the cylinder and that of the suction line. Any increase in clearance volume has

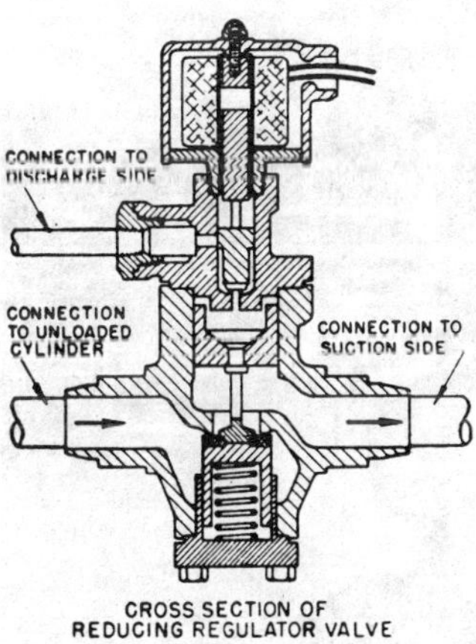

Fig. 219. (*Left*) Cross-section of reducing regulator valve. (Courtesy, Carrier Corporation.)

Fig. 220. (*Right*) By-pass connection (cross-over) for by-passing discharge to suction to relieve work of compressor during start-up. (Courtesy, Frick Company.)

the effect of shortening the suction stroke by an amount equal to that lost through clearance gas expansion. By providing a means whereby the amount of clearance can be varied, a method of controlling capacity is established.

This is accomplished by installing cylindrical clearance pockets at the compression end of the cylinder, the volume of which may be varied by an adjustable piston capable of clearance volume modulation, or by installing one or more fixed-volume clearance pockets. Both types lend themselves to manual or automatic operation. When two or more fixed-volume pockets are present on each cylinder, a greater flexibility in capacity is gained as the opening or closing of each pocket may be used as a step toward increasing or decreasing the effectiveness of each suction stroke below full volumetric rating.

It is obvious that the effect of clearance gas expansion will depend on both suction and discharge pressure, so that the capacity variation for a given compressor will not be the same under all operating conditions. To determine the capacity of a compressor having a known clearance pocket volume for any compression ratio, expressed in terms relative to full volumetric capacity, apply the following formula:

$$\text{Relative capacity} = \left[\frac{V_C}{V_D}\left(\frac{P_2}{P_1}\right)^{\frac{1}{y}} - 1\right]$$

where V_C = clearance volume
V_D = theoretical displacement of piston
P_1 = suction pressure, psia
P_2 = condenser pressure, psia
y = specific heat ratio C_P/C_V $\left(\frac{\text{sp ht of gas, constant pres.}}{\text{sp ht of gas, constant vol.}}\right)$

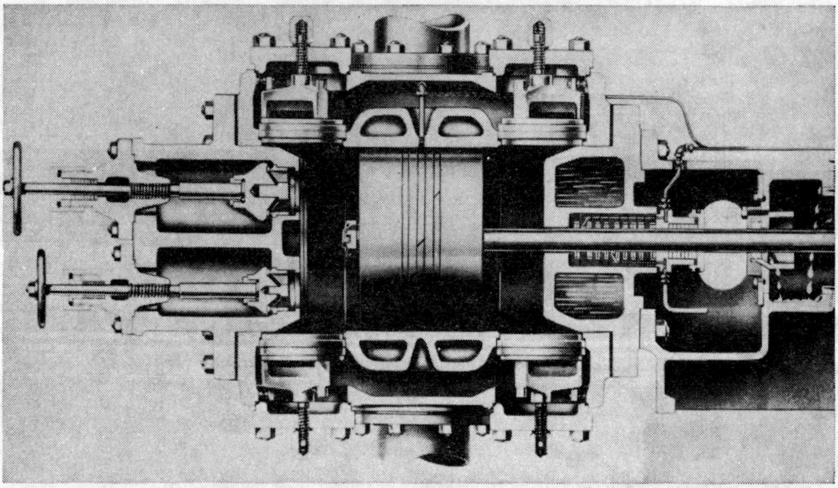

Fig. 221. Clearance pockets, fixed-volume type for double-acting compressor. (Courtesy, Worthington Corporation.)

A manually operated fixed-volume capacity regulator installed on a horizontal compressor is shown in Fig. 221. If the volume of each clearance pocket of a multiple-pocket capacity control differs so that their sizes progressively increase, a greater flexibility in compressor capacity will prevail

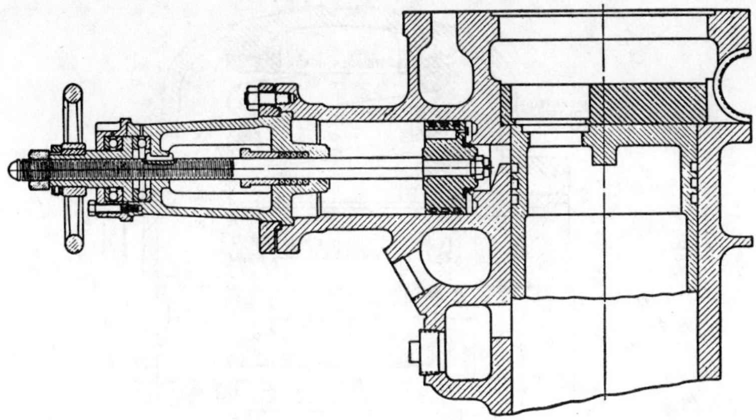

Fig. 222. Clearance pockets, variable or adjustable volume. Pocket is placed within limit of piston stroke. Delivery can be adjusted from 50 percent to 80 percent rated output by changing position of piston. Water jacket surrounding pockets prevents overheating. (Courtesy, Frick Company.)

due to the subsequent number of volumetric combinations possible. By installing three pockets having a volume ratio of 1, 2 to 1, and 3 to 1, it is possible to obtain six different capacities within the same cylinder between zero and maximum delivery.

Excessive superheating of compressed gases is one of the disadvantages of capacity control by means of clearance pockets and often poses the problem of injecting liquid refrigerant into the suction line when employing high-pressure fluids. This latter practice is becoming obsolete except where high condenser pressures prevail.

Slide Valve Capacity Control—Screw Compressors

A loading and unloading device that acts to control screw compressor system capacity is shown in the discussion relative to screw compressors. (See Chap. 8.) The axial motion of a sliding valve located in the lower cusp of the rotary cylinders controls the amount of vapor entering the compression area of the compressor. It acts in a similar manner as reciprocating compressor cylinders that cut-in and cut-out, returning any unnecessary vapor back to the suction area.

Multiple Compressors. Capacity control may be accomplished by installing two or more compressors whose total tonnage would equal the maximum load demand. As the load diminishes, a proportional number of compressors would be cut out. Flexibility here would depend on the number and the size of compressors so that for a maximum load of 100 tons, two 50-ton compressors would provide a capacity variation of 0, 50, and 100 percent, whereas one 50- and two 25-ton compressors would have 0, 25, 50, 75, and 100 percent variation in capacity. Despite the disadvantage of higher initial cost, the multiple compressor system is used widely to obtain capac-

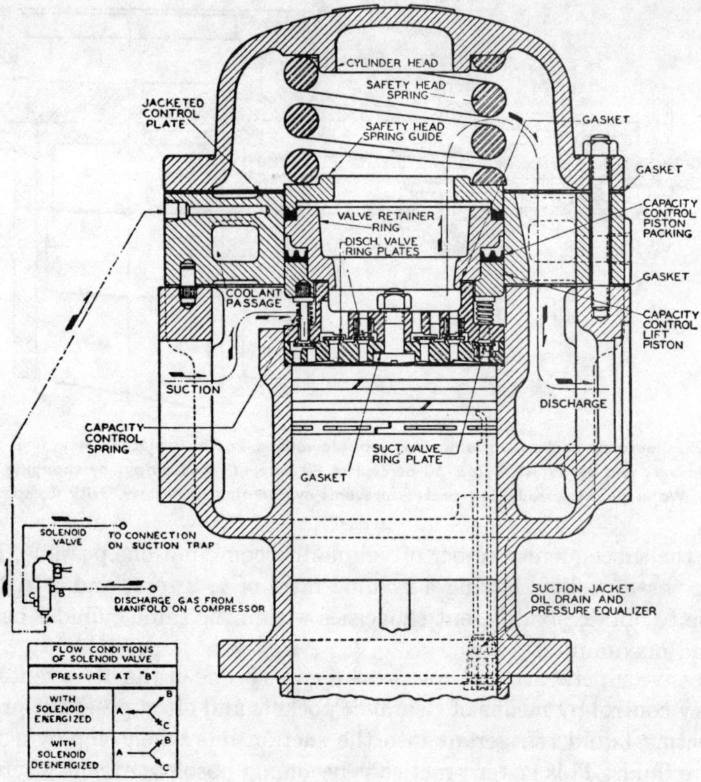

Fig. 223. Capacity control in compressor head. (Courtesy, Frick Company.)

ity control. In some instances, a cylinder cut-out on one or more compressors gives greater flexibility of capacity control.

Capacity Control of Centrifugal Compressors

Variable-Speed Drives. One of the characteristics of centrifugal compressors is that a relatively small reduction in speed considerably reduces the capacity, making the variable-speed drive the most preferable capacity control for most operating conditions. A 15 percent and 25 percent decrease in speed brings about an approximate capacity reduction of 50 percent and 75 percent, respectively. The steam turbine with governor speed control makes an ideal prime mover for centrifugal compressors, especially since increment speed variation can be obtained and its high speeds are conducive to increased operating efficiency.

Where centrifugal compressors are electrically driven, the wound-rotor slip-ring induction motor is the conventional prime mover. Speed variation with the slip-ring motor is achieved in steps with multiple resistances centralized for control purposes in a drum switch that may be operated by hand or automatically. If a constant-speed synchronous motor is used, it is necessary to employ either a hydraulic or an eddy current (magnetic slip) clutch to attain speed variation.

Suction-Line Throttling Damper. The suction-line throttling damper is employed to vary centrifugal-compressor capacity when driven with a constant-speed prime mover. By throttling the flow of vapor in the suction line by means of a butterfly arrangement, the pressure and the temperature between the damper and the compressor inlet are reduced below evaporator conditions. Thus, while the compressor handles an equal volume of vapor at all times, throttling acts to lower the density so that capacity variation lies in the lesser weight of refrigerant handled per unit of time. The effect of the damper is primarily one of increasing the pressure difference between inlet and outlet, which, in turn, reduces centrifugal compressor capacity. Positioning of the damper may be accomplished manually or automatically.

Capacity control by means of a suction-line throttling damper is not so economical as some other methods employed. This is particularly true when compared with variable-speed drives where the power requirements at similarly reduced capacities are much less than with the damper control.

Another effect of the damper control is to lower the surge point.

Condenser Temperature Control. The refrigerating capacity of centrifugal compressors operating at constant speed is reduced as the differential in suction and discharge pressures increase. Therefore, any increase in

Fig. 224. Impeller and runner units of hydraulic coupling. (Courtesy, Heldt, P. M.: "Torque Converters or Transmissions," p. 88, Philadelphia, Chilton Company.)

condenser temperature (thermodynamic head) will effect a capacity reduction which, under partial load conditions, may be accomplished by decreasing the supply of condenser cooling water.

Controlling capacity of centrifugal compressors by regulating condenser cooling has definite limitations as related to the surge or pulsation point of the compressor involved. This is true because the characteristics of these machines are such that unstable operation follows a certain capacity reduction for a given speed. Thus, if the load falls off sufficiently to require a thermodynamic head approaching the surge point at the operating speed, condenser temperature control should not be used unless supplemented with a step speed control. Any speed reduction has the effect of thrusting the surge point back toward lighter loads. Brake horsepower per ton of refrigeration is relatively higher with condenser water control than with some other methods, particularly speed variation, under fractional load conditions.

The water supply to the condenser is controlled from a thermostat whose element is located in either the substance being cooled (brine or water) or the liquid refrigerant so that the amount of condenser cooling is a function of load fluctuation.

Variable-Inlet Guide Vanes. The principle involved with the variable-inlet guide vane for the capacity control of centrifugal compressors differs

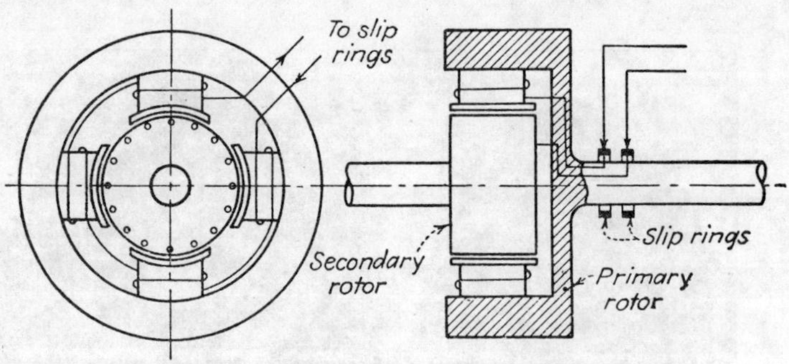

Fig. 225. (*Top*) Diagrammatic view of electric slip coupling. (Courtesy, Heldt, P. M.: "Torque Converters or Transmissions," p. 90, Philadelphia, Chilton Company.)

Fig. 226. (*Bottom*) Variable-inlet guide vanes. (Copyright, The Trane Company. Reproduced by permission.)

from that of the suction damper control. While the latter actually throttles the suction line as the result of butterfly positioning, the former accomplishes a volume reduction by changing the direction of vapor flow immediately prior to its entering the first wheel or impeller. This is accomplished by installing a number of variable guide vanes in the compressor housing directly ahead of the impeller inlet, as shown in Fig. 226. These impart a swirling action to the vapor as it passes through when the vanes are not entirely open or closed. This swirling action becomes more pronounced as the vanes approach their closed position. Because the vanes are made to throttle in the direction of impeller rotation, the column increases, effecting a change in the column-outlet velocity, pressure, and volume that is carried over to the following stages to bring about a capacity reduction within the entire unit. Some throttling action occurs immediately as the vanes start to close and increases as the free area decreases. However, not until the vanes are substantially near their closed position does throttling reach major proportions.

The individual vanes may move through a 90° arc and are positioned by means of small pinions keyed to the vane shafts and rotated by means of a ring gear actuated by a pneumatic control motor. The air pressure operating the control motor to open or close the vanes is governed by a temperature controller whose sensitive member (thermal bulb) is located in the substance under refrigeration.

Discharge By-Pass. By-passing the discharge gas to the suction line of centrifugal compressors is supplemental to other types of capacity controls. While it reduces the surge point normal to load and speed by substituting by-passed gas for the volume of vapor eliminated as a result of load reduction, the necessity of recompressing superheated gases considerably reduces operating efficiency.

PRESSURE CONTROLS

In order to maintain suitable pressures in various parts of the system, certain types of pressure-regulating and holding valves are required. Some of the more important types are described below:

Liquid Pressure Valves. This valve is used in conjunction with high-side float valves on domestic units when the latter is located below the entrance to the evaporator. Its purpose is to maintain sufficient pressure on the liquid to prevent vaporization after leaving the throttling device and before entering the evaporator. This valve is located on the low-pressure side between the float valve and the evaporator at the upper end of the vertical low-pressure liquid line. It consists of a weighted or spring-loaded needle that must be lifted from its seat by the pressure of the liquid before entrance to the evaporator can be gained. If it were not for the presence of this valve, the liquid under low pressure upon leaving the high-pressure float would commence boiling in the line leading to the evaporator.

The weight or spring tension is generally fixed to cause the needle to lift at a predetermined pressure differential across the valve orifice.

Evaporator Pressure-Regulating Valves (Back-Pressure Regulator). The purpose of this valve is to maintain constant pressure in the evaporator regardless of load changes. It is located in the suction line and can be adjusted to keep the evaporator pressure at a level corresponding to desired temperatures. Equipped with a spring-loaded diaphragm actuating a needle, suction vapor flow can be modulated according to the amount of pressure tending to build up in the evaporator. As pressure in the evaporator rises with increased load or tends to rise under normal load, it acts against the diaphragm which lifts the needle and allows the compressor to draw off the excess vapor. As the load falls off, evaporation diminishes followed by a drop in pressure which, in turn, allows the diaphragm, aided by spring tension, to close the valve. While the evaporator pressure-regulating valve will prevent the pressure in the evaporator from falling below the level for which

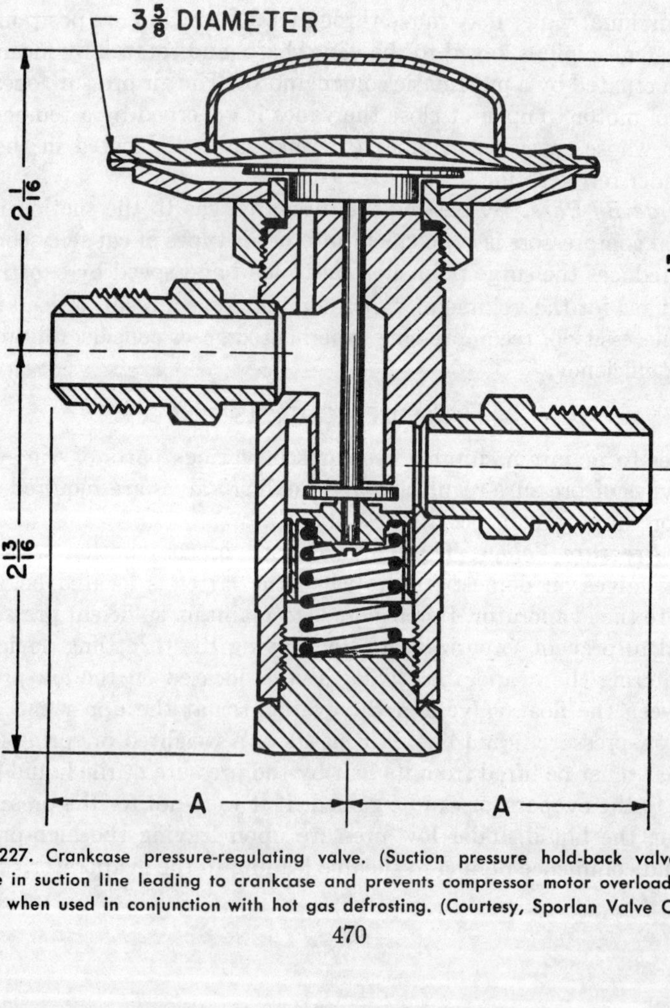

Fig. 227. Crankcase pressure-regulating valve. (Suction pressure hold-back valve.) Controls pressure in suction line leading to crankcase and prevents compressor motor overloading. Advantageous when used in conjunction with hot gas defrosting. (Courtesy, Sporlan Valve Company.)

it is set, it has no influence in regulating pressures when held wide open due to excessive load.

This type of valve can be used to control pressure in a single evaporator or in multiple evaporators having similar temperature requirements. They may be used also in multiple dual-temperature evaporator systems having but one compressor, to maintain minimum pressures in the high-temperature coils. In the latter instance, the valve is placed in the common suction line of the higher-temperature evaporators, maintaining constant conditions therein and independent of conditions in the low-temperature coils.

The evaporator pressure regulator finds wide use where minimum temperatures must be maintained in order to prevent freeze-ups or where a lower temperature might be detrimental to processes involved.

Suction Pressure Hold-Back Valves. The action of the suction pressure hold-back valve is the reverse of the regulating valve. Its purpose is to pre-

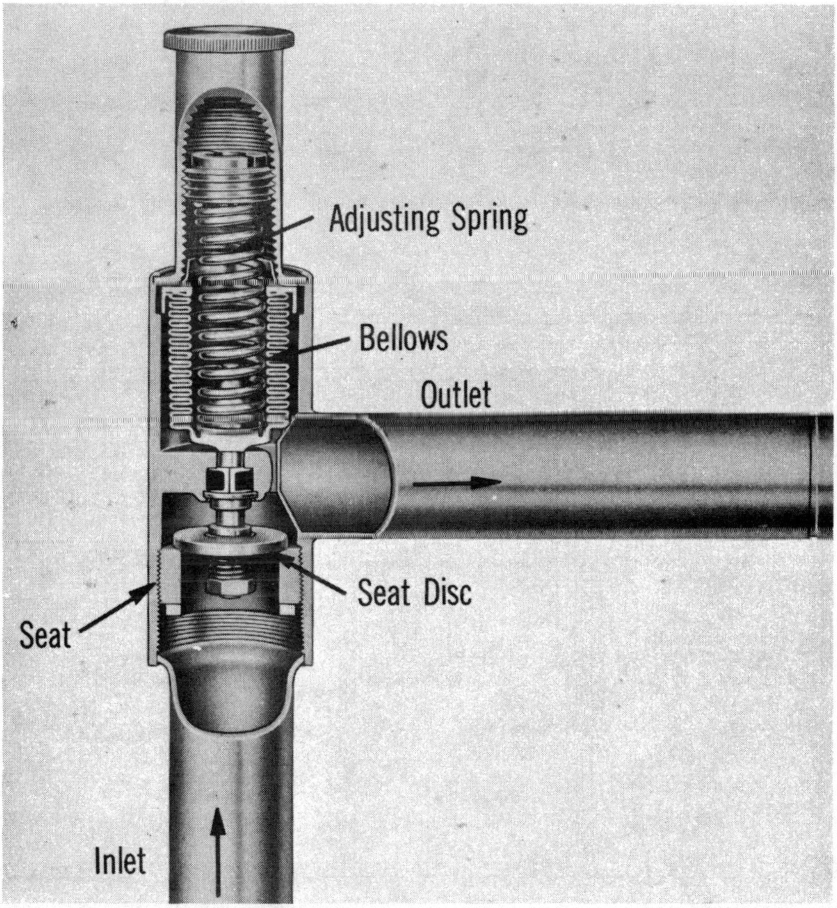

Fig. 228. Evaporating pressure regulating valve. (Courtesy, Sporlan Valve Company.)

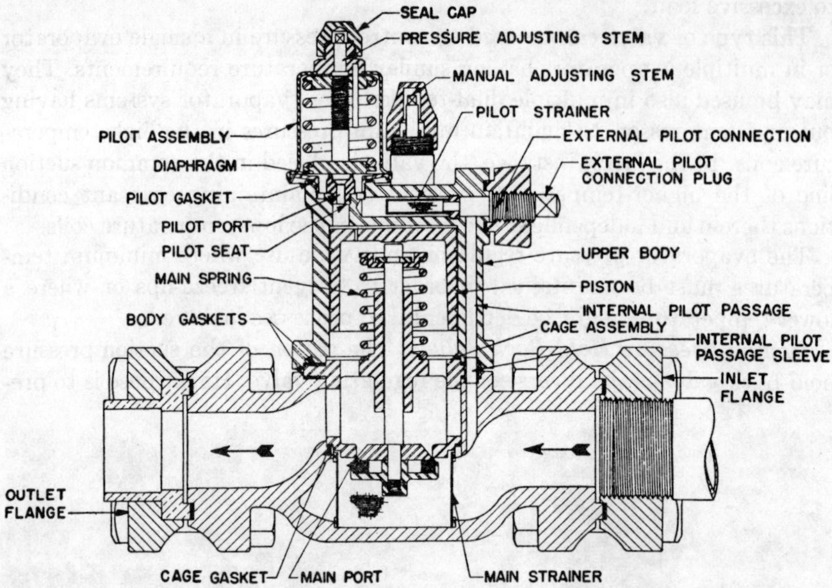

Fig. 229. Evaporator pressure regulator. Pilot operated. (Courtesy, Alco Valve Company.)

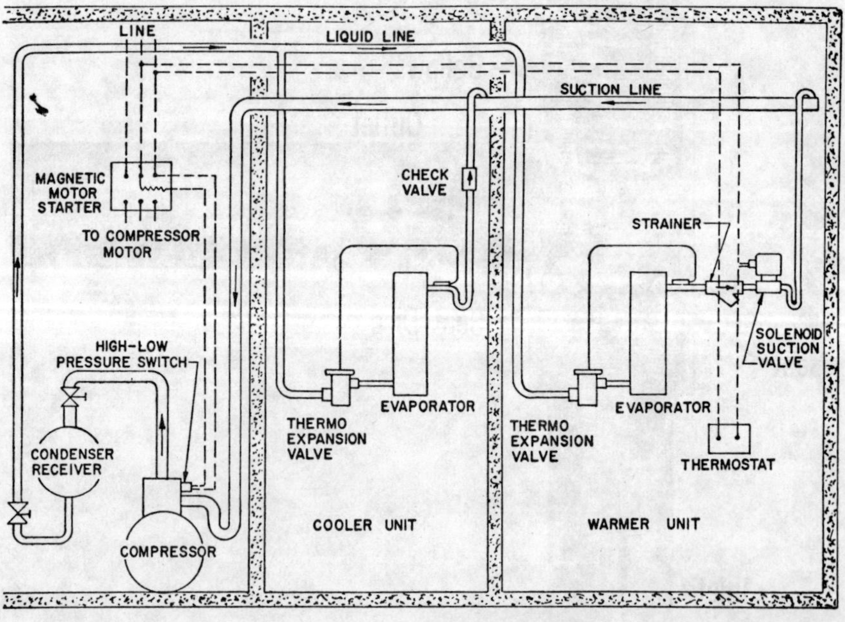

Fig. 230. Solenoid suction valve on two-temperature application. (Courtesy, Alco Valve Company.)

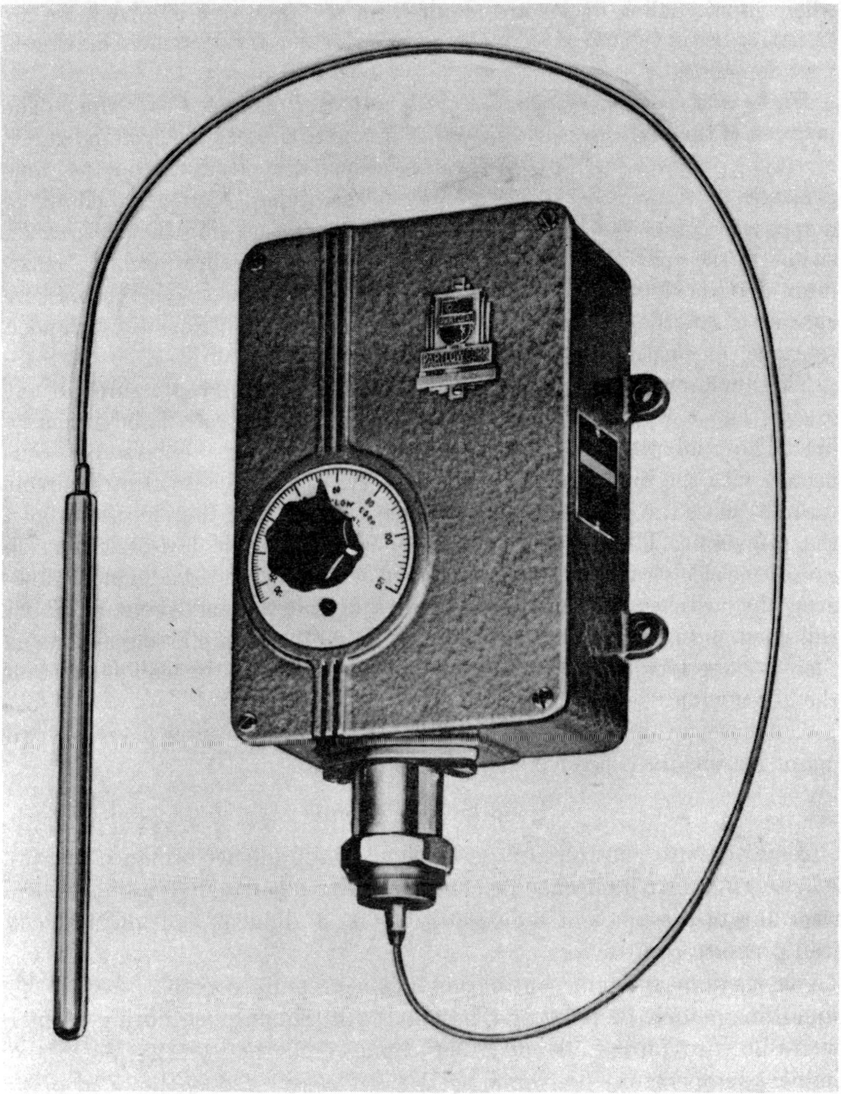

Fig. 231. Multi-stage control, non-indicating, for several stages of cooling. (Courtesy, The Partlow Corporation.)

vent compressor overloading during periods of excessive evaporation when high-temperature differentials exist between refrigerated space and cooling coils. With this type of valve, the pressure on the compressor side actuates the spring-loaded diaphragm to modulate the flow of vapor from the evaporator and cause the pressure in the latter to rise freely commensurate with the heat load. Because any rise in evaporator pressure causes a reduction in rate of heat transfer, the cooling time is extended, a definite disadvantage where rapid cooling cycles are required for the processes involved. An alternative lies in the use of an over-sized compressor motor to meet maximum load demands.

High- and Low-Pressure Cut-Outs (Dual Pressure Controller). The purpose of the high-pressure cut-out switch is to protect the system against excessive pressure build-up in the condensing side. It works on the same principle as some other types of pressure switches, having a bellows or diaphragm connected by tubing to the high-pressure side that actuates a switch in the control circuit. When the condenser pressure reaches a maximum safe working level, the predetermined point at which the high-pressure cut-out is set, the bellows or diaphragm has become sufficiently expanded to cause the electric circuit to break and stop the motor.

The low-pressure cut-out switch is similar to the high-pressure cut-out except that it works from the suction line pressure instead of the discharge line. The dual pressure controller, equipped with two bellows (one connected with the high-pressure side of the system), holds its electric circuit (switch) closed at all times except when the pressure therein rises above the safe limit. The other bellows connected with the low-pressure side causes the electric motor to start and stop according to load demands. However, the switches are connected in series so that the high-pressure switch will open and remain open and prevent the motor from starting as long as there is excessive pressure in the condenser regardless of the load demand on the low-pressure switch.

Condenser water failure or the presence of noncondensible gases are two major reasons for excessive condenser pressures.

Head Pressure Control

Head pressure controls are designed to maintain performance of systems serving air conditioned or chilled compartments. They are classified according to the ambient temperature as mild ambient and low ambient head pressure controls.

The medium ambient control employs a pressure control valve in the liquid line before the receiver. The valve will remain open during normal operation, providing full condenser capacity to the system. When the ambient temperature (surrounding the condenser) drops, the head pressure will start dropping until it reaches the predetermined setting of the valve, say 35 F. At this point the valve starts to close, flooding the condenser. This flooding neutralizes a proportional part of the condenser, which proportion is governed by the degree of valve closing. The flooding action reduces the capacity of the condenser. If the ambient continues to fall, the valve will further close until the ambient reaches, say 25 F, at which point the valve will close entirely and the condenser becomes 75% inoperative. The action of the pressure control valve again opens gradually when the ambient temperature starts to rise to allow the liquid to leave the condenser.

A similar control may be attained by linking the pressure control valve through a bellows arrangement to a switch that will open or close an electric circuit to stop and start the fan of an air cooled condenser.

The low ambient temperature operation, down to 0 F, consists of a specially sized vessel located in the liquid line before the receiver. When the head pressure is normal for the system, liquid is forced into the accumulator-like vessel. When the ambient temperature drops, the refrigerant pressure in the condenser coil also drops. With a decreased condenser pressure, liquid flows from the vessel into the condenser to reduce the condensing surface, resulting in reduced capacity.

Check Valves. The purpose of the check valves is to prevent fluids from reversing their direction of flow in pipes. Check valves remain open when fluids are flowing in their normal direction, but will close automatically when changing conditions within the refrigeration system tend to force a reversal of flow.

There are two basic types of check valves, swing and lift checks. In the swing check, a disk hinged at the top swings closed, aided by gravity, when the pressure on the flow side of the valve becomes equal with or greater than the up-stream pressure. The lift type may be spring loaded if operation is horizontal or operate by gravity when vertically positioned. The needle or disk, aided by the spring or gravity, closes when the up-stream pressure

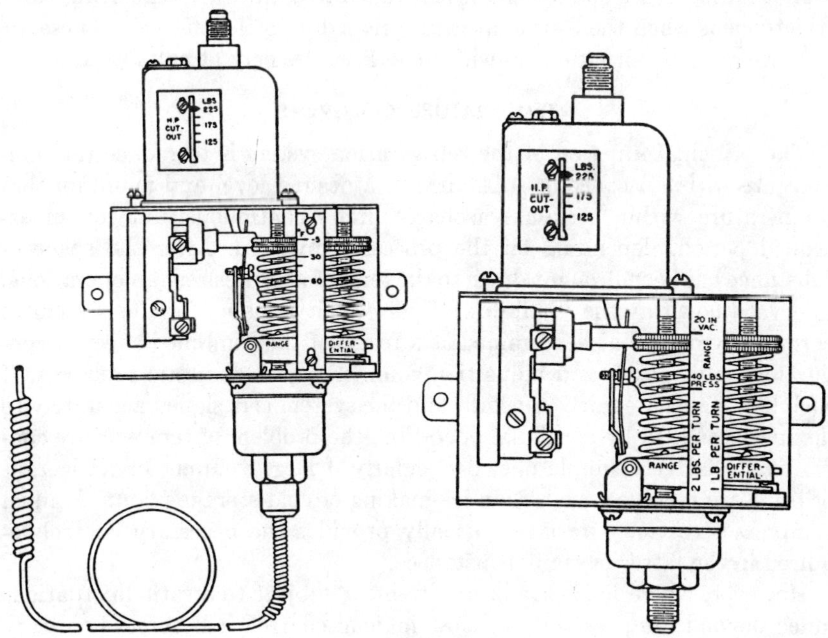

Fig. 232. (*Left*) Dual temperature control. (*Right*) Dual pressure control. (Courtesy, Detroit Controls Corporation.)

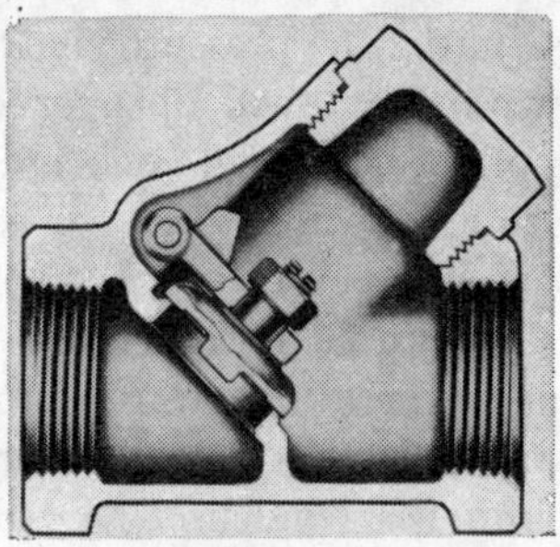 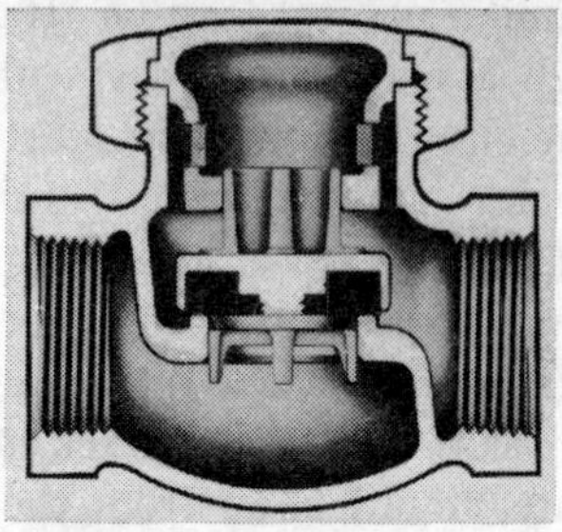

Fig. 233. (*Left*) Swing check valve. (Courtesy, Crane Co.)

Fig. 234. (*Right*) Lift check valve, gravity type. (Courtesy, Crane Co.)

falls below the tension of the spring in the first or below the down-stream pressure in the second type. The lift check valve causes a change in direction of flow whereas the swing check is a straight-through type.

A typical location for a check valve is in the suction line where it allows vapor to flow to the compressor but prevents its return to the evaporator when the compressor is idle. Check valves are also employed in dual-temperature systems where they are installed in the suction line of the low-temperature coil to prevent flow of vapor from the higher temperature coil when the compressor is shut down. In the latter system, the pressure in the higher-temperature coil usually is controlled by a pressure-regulating valve which opens when the coil temperature rises during the off-cycle. These are but two of many situations in which check valves may be employed.

TEMPERATURE CONTROLS

The principal function of the refrigeration system is to reduce the temperature within a space or substance to a desired level and maintain that temperature within certain reasonable limits continuously or for an extended period, depending on the processes involved. After the space or substance has been brought down to the temperature desired (an operational interval known as the "pull-down" period), it remains for the system to carry away only that heat gained as a result of the combined effect of conduction, infiltration, and industrial, commercial, or domestic processes. If such heat gains are fairly constant and the system is designed accurately to dissipate the load by continual operation, the problem of temperature control is a relatively simple one, particularly if large volumes of refrigerant or brine are involved, such as in ice-making or cold-storage plants. Manual compressor capacity variation usually provides the necessary control required in the latter or similar instances.

However, where loads are intermittent or subject to erratic fluctuations much beyond the possibility of close manual control, it becomes necessary to adopt full, or at least partial, automatic operation. The control of any system should be made as simple as possible, and, while modern automatic

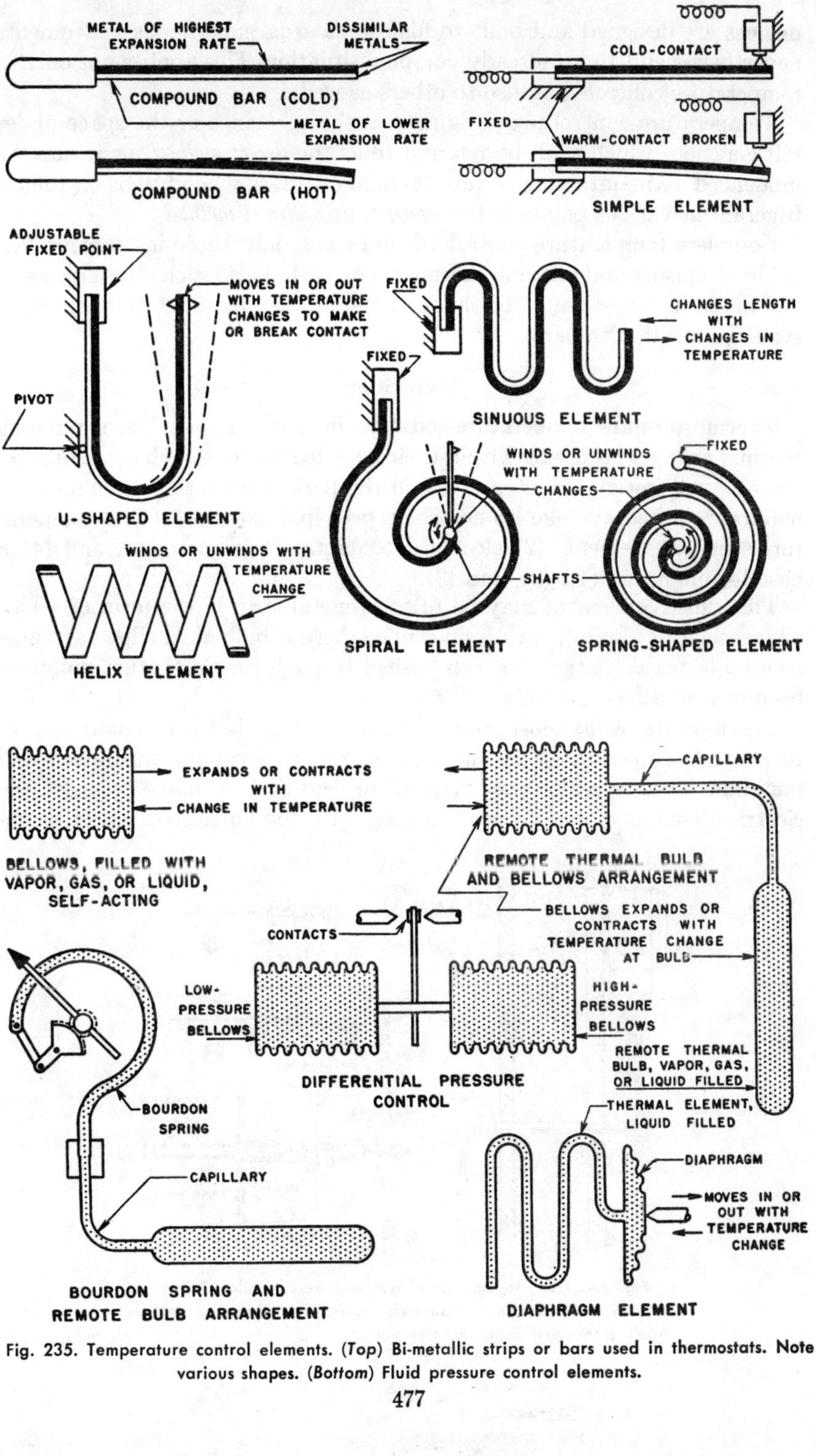

Fig. 235. Temperature control elements. (*Top*) Bi-metallic strips or bars used in thermostats. Note various shapes. (*Bottom*) Fluid pressure control elements.

devices are designed and built to high standards, superfluous instruments nevertheless add to an already complex situation. This applies not only to temperature controls but also to others as well.

Temperature control may originate in the substance or the space under refrigeration, which shall be referred to as the *direct method*, or it may be influenced by maintaining certain thermal or pressure conditions on the refrigerant at various points in the system, an *indirect method.*

Complete temperature control involves not only those instruments capable of sensing and relaying thermal changes, but also such mechanisms as will move to compensate for those changes in the required manner. Both types are dealt with here.

Thermostats

Thermostats are temperature-sensitive instruments capable of relaying thermal changes occurring within their installed space to such actuating devices as will initiate counteraction on the part of operating equipment to maintain desired average levels. Their principal parts are (1) a temperature-sensitive element, (2) electrical contacts, (3) adjustment, and (4) a base for mounting (1), (2), and (3).

The sensitive element may be (*a*) a bi-metal strip in the form of a bar, a horseshoe, or a spiral, or (*b*) a fluid-filled thermal bulb and bellows arrangement. Electrical contact is accomplished through simple contact points or by means of a mercury tube.

Thermostats, while sensitive to minute thermal changes, usually are set to control temperatures within a given range, involving maximum and minimum levels at which they "cut in" or "cut out" to make or break the electrical control circuit. If used to start and stop an electric motor to ob-

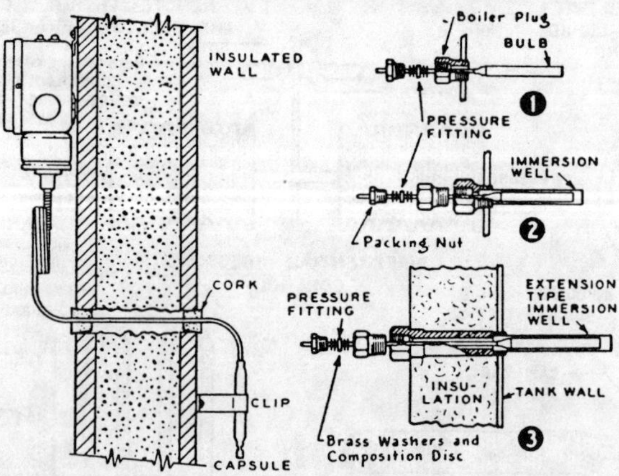

Fig. 236. (*Left*) Installation of temperature controller, fluid pressure type. (*Right*) Immersion well connections. (Courtesy, Minneapolis-Honeywell Regulator Company.)

tain capacity control, the maximum and the minimum levels are set wide enough apart to prevent short cycling but sufficiently close to maintan desired space or fluid temperatures within reasonable tolerances. To compensate for lag within the control system and to prevent temperature carryover beyond the maximum and the minimum levels desired, the thermostat should be set within limits narrower than those required for the refrigerated space. For example, if an average space temperature of 30 F is desired, a typical operating range would have a minimum of 25 F and a maximum of 35 F, or a 10 F operating differential. To obtain this condition, the instrument range warrants a lesser differential, which might be found by the "cut and try" method to be 6 F, wherein the maximum and the minimum instrument settings are set 2 F closer to the desired average temperature of 30 F in the refrigerated area.

The bi-metal sensitive element is made up of dissimilar metals bonded together. They may be invar (64 percent iron and 36 percent nickel) and brass, having different coefficients of expansion. Any change in temperature initiates a bending or straightening action, depending on the direction in which the thermal change occurs, caused by stress variations in the two metals. One end of the bi-metal element is fixed to the base of the thermostat while the other is free to move through a radius governed by the instrument range setting. When only one control circuit is involved, contact with the opposing electrode is made by movement of the bi-metal element in one direction along the radius. It is broken when movement occurs in the opposite direction. This type of thermostat is called a *single-pole, single-throw switch.* When motion in either direction opens one circuit and ultimately closes another, a thermostat described as a *single-pole, double-throw switch* is involved. The bi-metal sensitive element is used primarily in space or room thermostats.

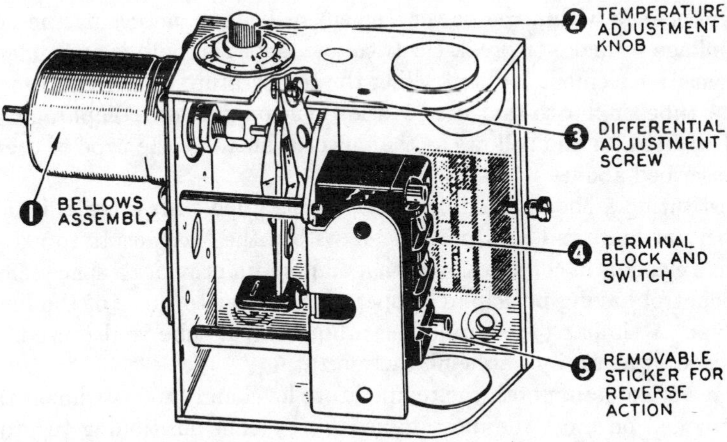

Fig. 237. Temperature controller actuated by changes in fluid pressure within capillary and remote bulb on bellows. (Courtesy, Minneapolis-Honeywell Regulator Company.)

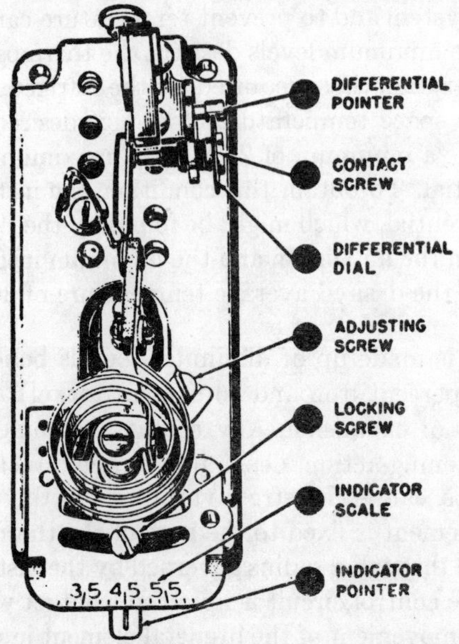

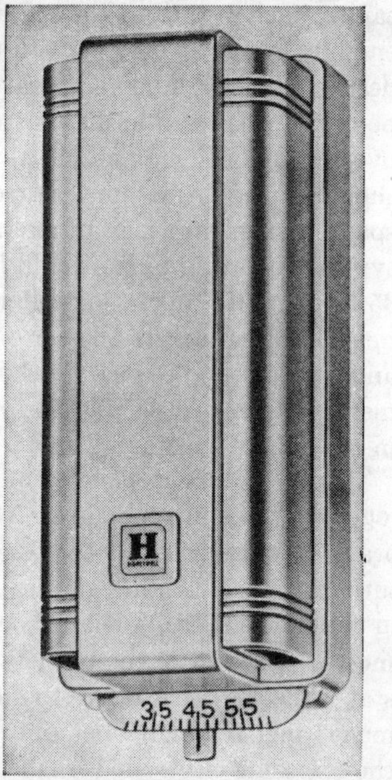

Fig. 238. Thermostat. (*Left*) Internal view, showing bimetallic element, contact points, and adjustment. For direct control of motors, starters, and relays. (Courtesy, Minneapolis-Honeywell Regulator Company.)

The second type of thermostat most generally used employs a **thermal bulb,** similar to that described for thermostatic expansion valves, connected to a bellows through a capillary tube. Charged with a fluid, either volatile liquid or vapor, temperature changes within the bulb reflect a corresponding pressure change within the power element or bellows, actuating the necessary linkage to make or break electrical contact. The bulb is an immersion type sensitive element for controlling the temperature of any substance. A circular tube similar to that used in the Bourdon gage or a diaphragm may be substituted for the bellows as the power element in the type of thermostat described above.

Substituting a short, small-diameter **tubing filled with volatile liquid** for the thermal bulb and incorporating it within the thermostat to act as a sensitive element in conjunction with a diaphragm provides a space temperature control having principles of operation similar to those of the immersion type. A similar type employing a liquid-filled tube is also used. This latter is known as a hydraulic fill thermostat.

Means of adjusting both the temperature level and the instrument range are provided on most thermostats usually by cam positioning to provide lost motion within the actuating mechanism in the first case and spring tension regulation in the second. The adjustments are made externally by means of a temperature-adjustment dial knob and a scale-adjustment screw.

A departure from the simple open-contact points is the mercury tube for opening and closing the electric control circuit. This switch consists of an enclosed glass tube having two electrodes inserted in one end and containing a small amount of mercury. As the sensitive element tilts the mercury tube in response to temperature change, the mercury flows from one end to the other. When the tilt position is such as to cause the mercury to flow to the end in which the electrodes are located, an electrical connection is made by reason of the latter's submersion.

Arcing and chattering at the electrodes when an electrical circuit is being opened or closed is undesirable for several reasons, not the least of which is the rapid destruction of the points. Another is the possibility of rapid and

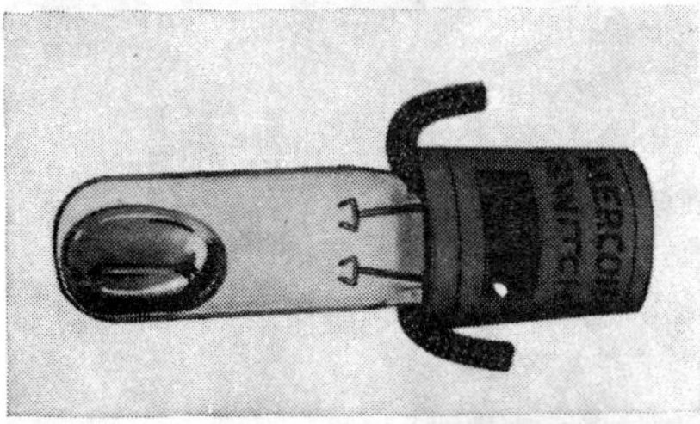

Fig. 239. Single-pole, single-throw mercury switch. (Courtesy, The Mercoid Corporation.)

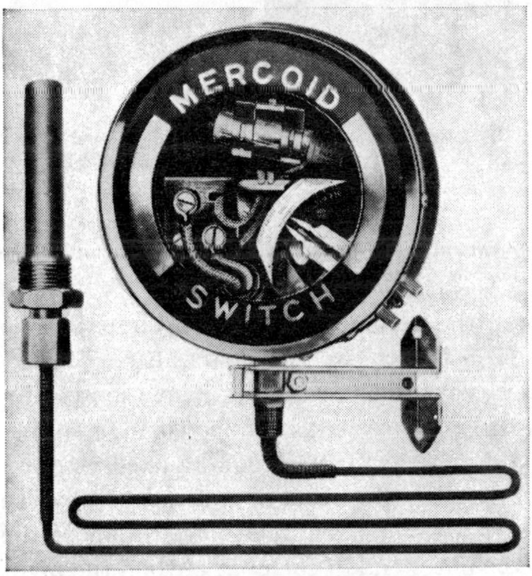

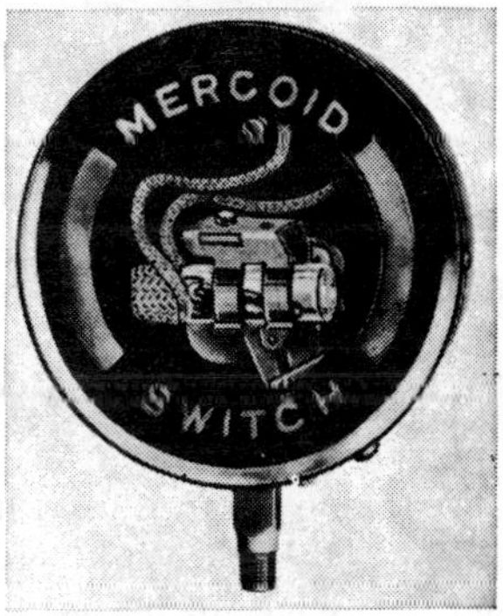

Fig. 240. (*Left*) Temperature and (*right*) pressure controllers with mercury switch. (Courtesy, The Mercoid Corporation.)

repeated starting and stopping of the electric motor during the period of chattering, which may prove detrimental to the motor. A permanent magnet located so as to impart snap action to the bi-metal element in the final stage of closing the gap between the two electrodes eliminates arcing and chattering during that phase of thermostatic operation. Snap action in opening the circuit is gained when the stresses within the bi-metal element develop sufficient force to overcome the attraction of the permanent magnet.

Fig. 241. Two-switch indicating temperature control. Provides two-point sequence control. (Courtesy, The Partlow Corporation.)

A toggle is often used in place of the permanent magnet to impart snap action to the switch. This device consists of two rigid bars, usually of unequal length, placed end to end, having their near ends joined loosely together to provide lost motion and their outer ends connected by a spring, with the whole forming a triangle. Each bar pivots around a fixed point on the base of the thermostat within an arc whose length is controlled by the extent of lost motion at their near ends and the relative position of the pivots. The spring is attached to the outer end of each bar beyond the pivot points. While always under tension, it reaches its greatest potential when an external force transforms the bars from their normal V-formation to an in-line formation. If the external force is continued so that motion extends beyond the in-line position of the bars to upset the equilibrium of the three toggle members, the spring acts to snap the outer ends closer together and reverse the formation to form an inverted V or Λ. The power element of the thermostat linked to either of the bars provides the necessary external force to upset the toggle in either direction and impart snap action to the make and break mechanism. In some instances, the bars are placed inside the spring but the same principle applies.

In the mercury tube, arcing is eliminated when the space not taken up by the mercury is filled with an inert gas.

Thermostats are employed either directly or in conjunction with other devices to control the temperature of the space or substance to which they are exposed. In fractional horsepower systems, not hermetically sealed, the thermostat may be placed in the main power circuit to start and stop the electric motor according to temperature demands. Split-phase induction

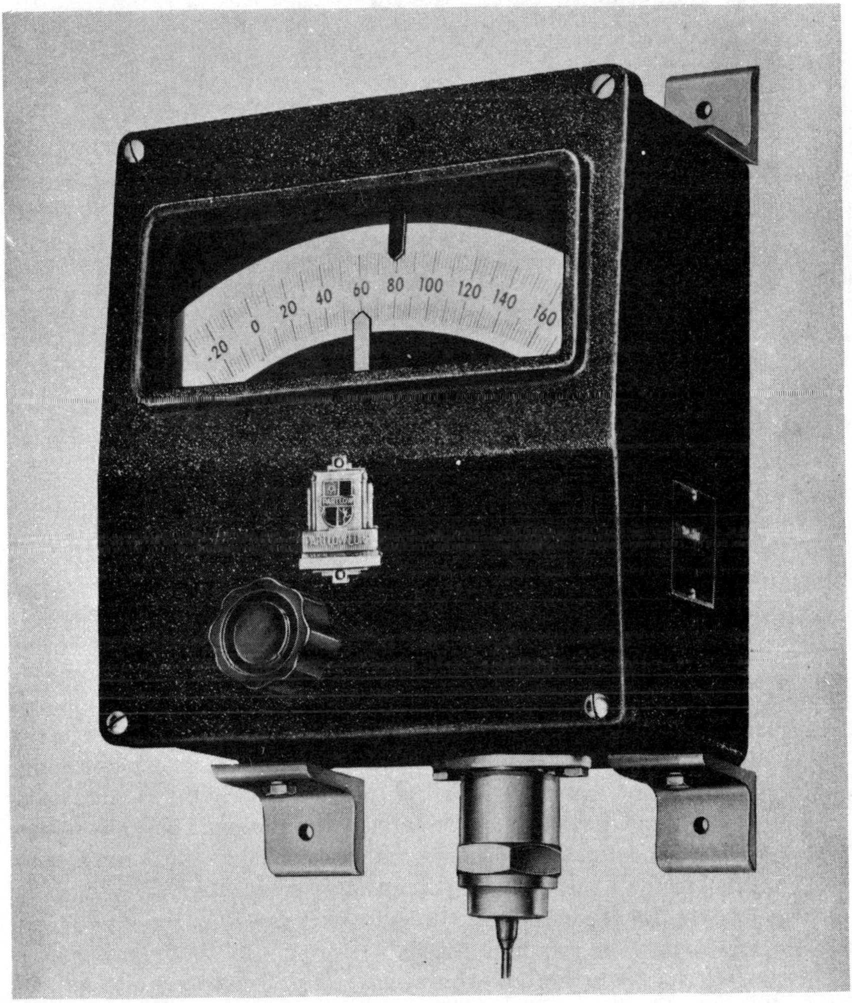

Fig. 242. Three-switch indicating temperature control. Provides regulation of three electrical circuits on rise or fall of temperature. (Courtesy, The Partlow Corporation.)

Fig. 243. Recording temperature control. Operates relays to start and stop cooling systems. (Courtesy, The Partlow Corporation.)

motors used in the latter instance are not self-starting and have separate starting and running windings. When up to full speed, split-phase motors develop greater torque with the running windings alone so that a centrifugally operated switch is installed in the motor to cut the starting winding out at some predetermined speed short of its maximum rating. Hermetically sealed units are not equipped with the centrifugal switch because of the presence of oil and refrigerant surrounding the motor. They have their starting windings cut out by means of an electromagnet or thermal relay located outside the housing. It is through this relay that the thermostat must act to control the electric motor.

Motors of over 1 hp require a magnetic starter. The thermostat here is connected to the power line to form a separate or control circuit and acts through a relay in the starter to actuate the switch to start and stop the motor.

Thermostats are often used to control the flow of electric current to a solenoid valve located in the liquid line. Start and stop control of the motor is the function of a pressure switch acting through a magnetic starter relay, while the thermostat causes the liquid valve to open or close according to the cycle.

Pressure Controllers

Temperature control is obtained by using a pressure-control switch in place of the thermostatic switch discussed previously. The pressure controller consists of a set of electrical contacts, an adjustment spring, and a power element in the form of bellows, diaphragm, or Bourdon tube. The power element is connected to the suction line through suitable tubing so that any change in pressure at that point will be reflected in the power element to make or break electrical contact.

This instrument works on the same principle as the thermostat (temperature controller) employing a thermal bulb as the sensitive element, except that power is supplied by the refrigerant directly from the system to a bellows, a diaphragm, or a Bourdon tube. A sufficient rise in suction pressure will cause the power element to overcome the opposing force of the adjustment spring to make electrical contact for starting the motor. As the pressure decreases, so will the force of the power element, and the spring tension bearing against the free contact arm ultimately will break the circuit. This type of temperature control cannot be used with an automatic expansion valve because the latter does not afford sufficient pressure change to actuate the power element.

CONDENSER WATER-REGULATING VALVES

The internal view of one type of condenser water-regulating valve is shown in Fig. 244. This valve is used to automatically control the amount of water flowing through the condenser and serves two purposes: (*a*) maintains constant condenser pressure when inlet water temperature is low enough, (*b*) conserves water, an economy factor if drawn from city supply.

In operation, the power chamber is connected by tubing to the high-pressure side of the system, thereby exposing the bellows to conditions similar to those found there, which tends to collapse it. The force of head pressure in the power chamber is opposed by the power spring supported at one end by the under side of the bellows and at the opposite end by the adjustment nut. The valve stem is attached to the bellows and passes through power spring, adjustment nut, and valve port. The valve disk is attached to the valve stem at the end projected through the valve port. The valve disk separates the water-inlet chamber from the outlet chamber when in a closed position.

When sufficient head pressure develops to overcome the tension of the power spring, the bellows partially collapse, moving the valve stem downward to unseat the disk. This is the normal position of the valve during

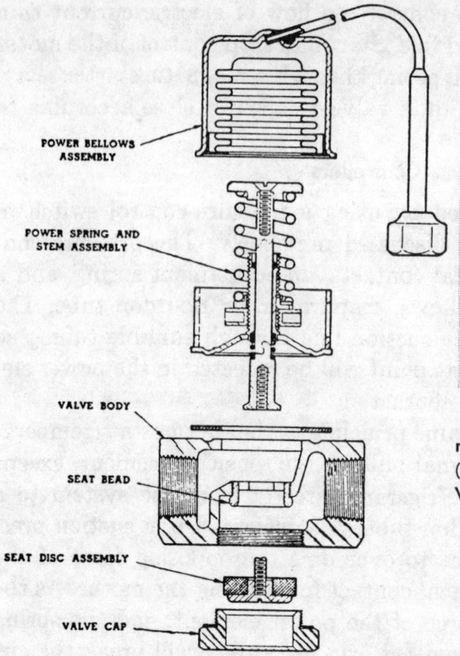

Fig. 244. Condensing water regulator, modulating. (*Left*) Internal mechanism. (*Right*) External view. (Courtesy, Metrex Valve Company.)

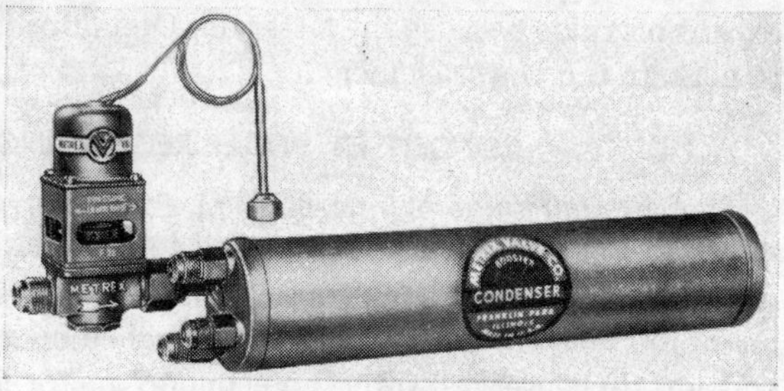

Fig. 245. Condensing water regulator connected to a booster condenser. The booster condenser supplements the air-cooled condenser when the latter cannot meet load conditions. (Courtesy, Metrex Valve Company.)

operation of the system. The rate of water flow is governed by the distance of disk opening allowed by the power spring, which, in turn, is controlled by the setting of the adjustment nut. If the head pressure rises above that for which the spring is adjusted, further collapse of the bellows occurs to increase the valve opening and allow a higher rate of water flow. As the amount of water increases to meet the higher condenser load, a drop in head pressure follows, allowing the bellows to re-expand and shorten the distance of valve lift and reduce the rate of water flow. The action of the valve is continuous during the operation of the system, and its control is one of modulation. When the system is shut down, the head pressure falls gradually, whereupon the tension of the power spring becomes the greater force to close the valve completely. Except for variations in mechanical arrangement, most condenser water-regulating valves operate according to the principles described above. Water leaks constituted the major fault with older models.

SOLENOID VALVES

Electromagnetic principles are involved in the operation of the solenoid, which is a coil made by superimposing layers of helically wound insulated copper wire to form a cylinder. The two ends of the wire are brought out to provide leads for connection to an electrical circuit.

If an electric current is passed through a conductor, lines of force or magnetic flux are set up about the conductor as shown in Fig. 291. The direction of the lines of force will be the same as the fingers of the right hand grasping the conductor when the thumb is pointing in the direction of the current. Forming the conductor into a helical coil and passing a current through it creates a magnetic field not only around the strand but also

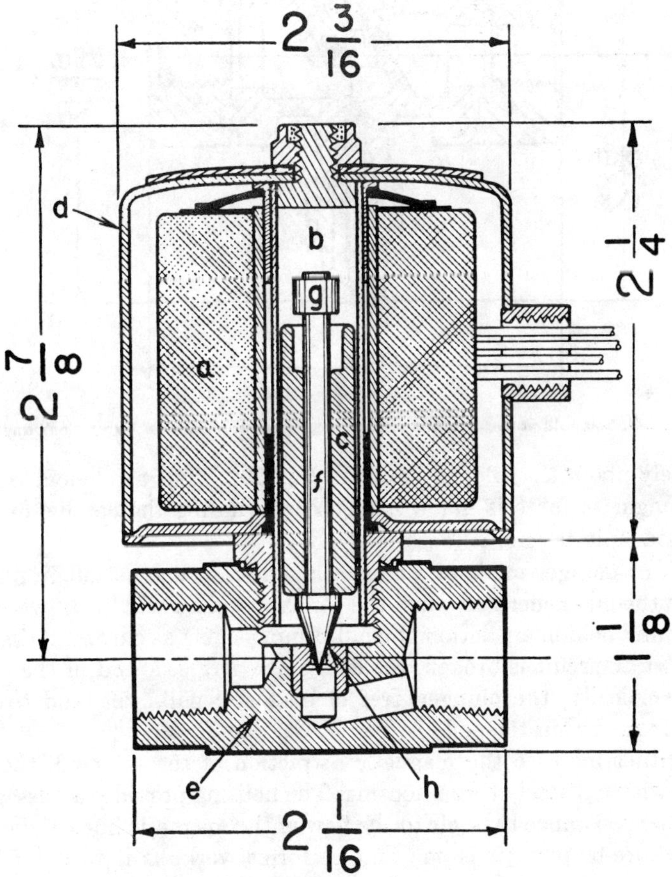

Fig. 246. Solenoid valve, direct acting type. (Courtesy, Sporlan Valve Company.)

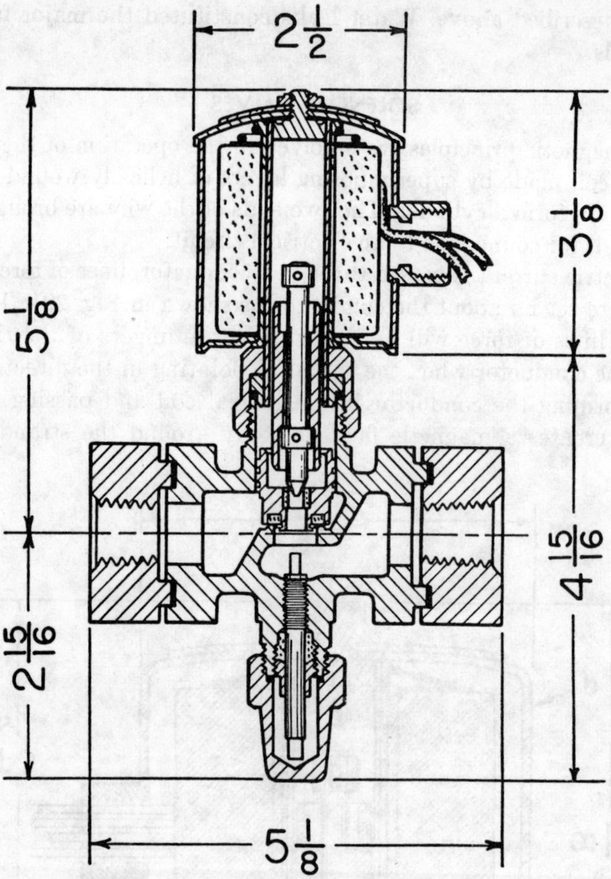

Fig. 247. Solenoid valve, floating piston type. (Courtesy, Sporlan Valve Company.)

completely about the pile, as shown in Fig. 292. Thus a solenoid is formed whose magnetic force is made stronger by winding the conductor closely and in multiple layers, as described above.

A core or plunger made of a magnetic material such as soft iron and exposed to the flux generated when the coil is energized will be drawn into the cylinder and held in a position of equilibrium as long as current flows. When the electrical circuit is broken, the coil is de-energized, and, if the cylinder is held vertically, the plunger, free of magnetic pull, will tend to fall by gravity from within the coil.

To further increase the magnetic attraction of the solenoid, the coil is encased within a steel or iron housing. The housing provides a passage having a lesser resistance than air to the flow of the magnetic lines of force.

If the core or plunger is one that performs work, as it would if it were made to move up and down to effect the opening and closing of a valve, the solenoid is considered *tractive* as compared with the hold or *portative* type.

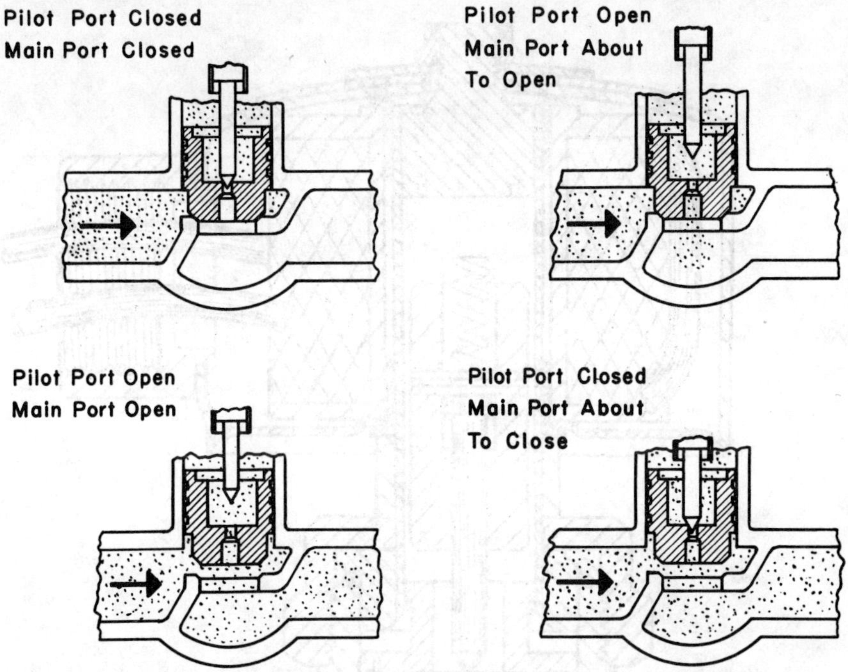

Fig. 248. Solenoid valve showing pilot operation. (Courtesy, Sporlan Valve Company.)

It is the tractive type only that is considered here and employed in the refrigeration control system for the automatic opening and closing of liquid valves.

The solenoid valve as shown in Fig. 246 is placed in an electrical control circuit which is made or broken by means of a suitable instrument such as a thermostat. It is composed of a coil (*a*), a guide tube (*b*), and a plunger (*c*), all contained in a coil housing (*d*), that is connected to a valve body (*e*). The valve stem (*f*) passes up through the plunger, having a collar (*g*) on the upper end and tapered at the lower end (the needle) for the purpose of sealing the port (*h*) when the coil becomes de-energized and allows it to drop. The valve-stem collar is fixed at a point beyond the upper end of the plunger when the valve is closed to provide lost motion and allow the plunger to gain momentum before attempting to unseat the needle when the coil becomes energized. Thus, if the needle is stuck to the valve seat for some reason, the slight shock of the moving plunger upward against the collar will break it loose.

On small systems, the needle opens or closes the valve port and thereby directly controls the flow of liquid through the valve. On valves of large capacity having ports greater than $\frac{1}{4}$ in. diameter, a floating piston arrangement is used to eliminate the use of extraordinarily large solenoids that would be required to operate the valve against high operating pressures. The floating piston acts to open and close the valve port while the needle,

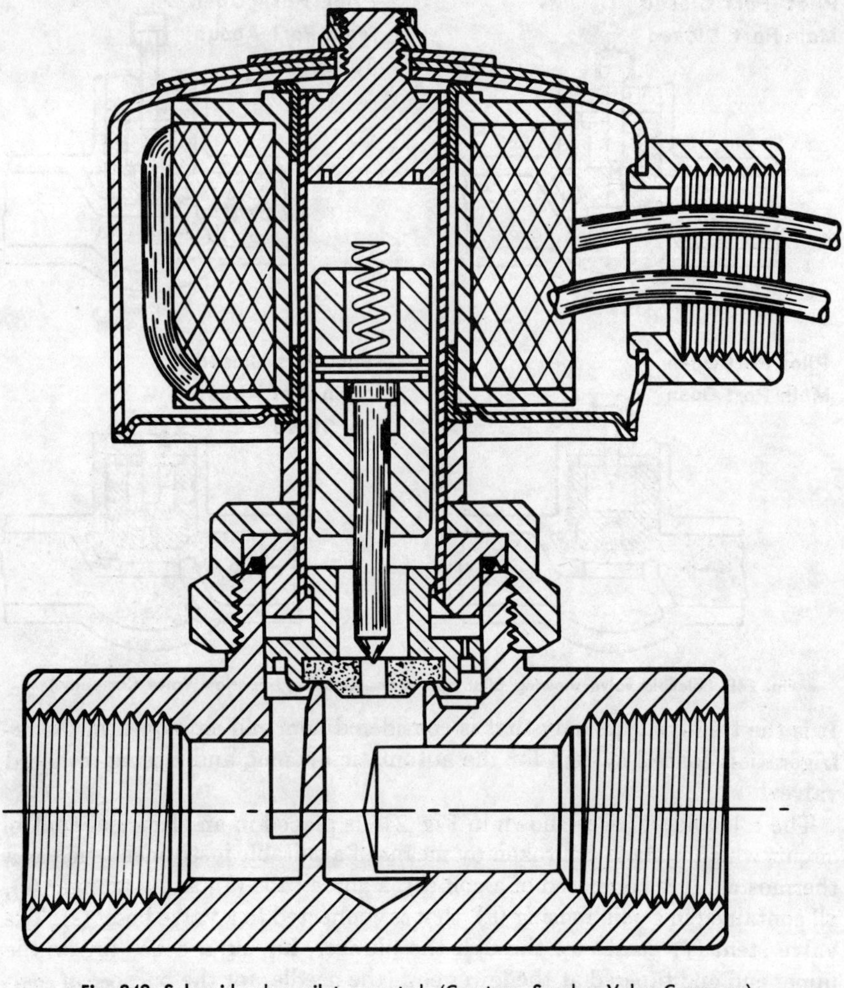

Fig. 249. Solenoid valve, pilot operated. (Courtesy, Sporlan Valve Company.)

actuated by the solenoid, opens and closes a pilot port drilled through the center of the floating piston. It operates in the following manner:

(*a*) When the solenoid is de-energized, the needle falls and closes the pilot port in the floating piston.

(*b*) With the pilot port closed, the liquid from the incoming line enters the space about the floating piston through the clearance between the floating piston and the valve body.

(*c*) This leakage causes the pressure above and below the floating piston to equalize and allow the piston to fall by gravity to close the valve port.

(*d*) As long as the solenoid is de-energized, the valve will remain completely closed.

(*e*) Energized, the solenoid lifts the needle off the pilot port.

(*f*) The area of the pilot port being greater than the total area of clearance between the floating piston and the valve body allows the trapped liquid to escape faster than it can accumulate, thus establishing a pressure differential to force the piston upward and unseat the main port.

(*g*) As long as the solenoid is energized, the valve will remain open *during liquid flow*.

(*h*) The small pressure differential across the valve (normal to any open valve through which liquid flows) will keep the piston afloat. The level, or distance above the main port at which it will float, varies according to the pressure drop across the valve, which, in turn, increases as the velocity of the liquid increases. Therefore, the piston will rise when the demand for liquid increases and drop when the demand decreases. The action of the floating piston, then, is one of modulation when the solenoid is energized to hold the pilot port open.

Solenoid valves are employed as liquid refrigerant stop valves located between the receiver and the expansion valve. Connected in series with the space or liquid thermostat, the solenoid valve acts to open or close the liquid line in response to temperature changes in the refrigerated area. This action affects the suction-line pressure, which actuates a pressure switch to start and stop the compressor.

Solenoid valves are also placed in the suction line leaving the high-temperature coil of a dual-temperature system. A thermostat placed in the high-temperature zone controls the flow of low-pressure vapor through the solenoid valve while compressor operation is a function of a pressure switch in the suction line common to both the low- and the high-temperature coils. Thus the flow of refrigerant to the two coils is controlled according to the demands of both. The pressure in the colder coil, being lower, requires that the limits of the pressure switch be set to meet the demands of the latter while the high-temperature control is self-sustaining by means of the solenoid and thermostatic expansion valve.

Sometimes it is necessary to allow the refrigerant to pass through the lines equipped with solenoid valves when the solenoid is de-energized and the valve is normally closed, such as during a pull-down period. For this reason, some arrangement should be made to lift the plunger manually or install a by-pass around the solenoid equipped with a hand stop valve.

The rating of solenoid valves is based on pressure drop across the valve, the standard being 2 psi. As the pressure drop increases, the valve capacity also increases, as shown for a given type in Table 35.

The strength or maximum pull of the solenoid coil governs the pressure against which the valve will operate. This means that systems operating under high pressures require larger coils to open the valve than required with lower pressures. As the magnetic pull varies with voltage changes, it is necessary to design their electrical circuit so as to avoid wide voltage fluctuations. A 10-v drop below their normal rating may in some instances decrease their pressure capacity by as much as 30 percent. On direct-current

TABLE 35. REFRIGERANT CAPACITY OF VALVE*

Refrigerant	*Tons Refrigerant at Various Pressure Drops*				
	1 lb	*2 lb*	*3 lb*	*4 lb*	*5 lb*
Refrigerant 12..........	1.9	3.0	3.4	3.8	4.2
Methyl chloride and sulfur dioxide...........	3.8	6.0	6.8	7.6	8.4

* Courtesy, Sporlan Valve Company.

circuits, a condenser should be provided to protect the coil against the high counter-voltage developed when the circuit is broken. Placed across the coil, the condenser acting as a surge protector will absorb the high voltage and prevent the coil from burning out. A similar arrangement is found in automobile ignition systems.

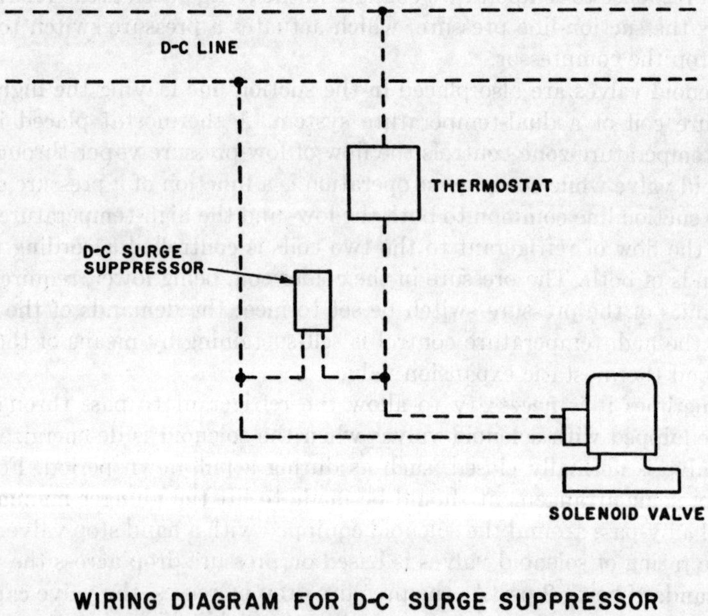

WIRING DIAGRAM FOR D-C SURGE SUPPRESSOR

Fig. 250. Surge protector hook-up for solenoid operating with direct-current electricity. (Courtesy, Sporlan Valve Company.)

Solenoid Pilot Control

Tighter shut-off of liquid flow to the evaporator at the thermostatic expansion valve is accomplished by a proprietary solenoid pilot control. It replaces the larger solenoid valve usually employed in the liquid line. This method utilizes the external equalizer line in which the pilot control is installed, as shown in Fig. 251.

When the solenoid coil is energized, the top port is closed and the bottom port is open. In this position, the pressure in the equalizer line is normal; i.e., average evaporator pressure. When the solenoid coil is de-energized, the top port is open and the bottom port is closed. In this position, the equalizer line is closed. The pressure in the liquid line (high pressure) is exerted on the under side of the thermostatic valve diaphragm through the pilot control. This pressure acts with the thermostatic expansion valve spring but against the existing bulb pressure to tightly close the port of the thermostatic expansion valve.

PROGRAM CONTROL

In the discussion of suction-valve unloaders, it was stated that volumetric displacement of multi-cylinder compressors could be varied in steps according to load demand. A four-cylinder compressor, for example, having three of its cylinders equipped with suction-valve unloaders, separately controlled, can be made to vary its capacity in progressive steps of 25 percent from 100 percent of its full rating to zero. This means that some method of control must be provided to unload the three cylinders separately and progressively for each 25 percent load reduction as the load diminishes from 100 percent to 25 percent with the last step from 25 percent to zero being accomplished by stopping the compressor. Such a control must also allow for the progressive relief of each unloader as the load steadily increases.

To accomplish sequence or program control, as it is commonly called, for the four-cylinder compressor mentioned above, the equivalent of four separate pressure switches must be used in conjunction with three 3-way solenoid valves (similar to that described under "Suction-Valve Unloaders") and an automatic electric motor starter. The four pressure switches are installed in the suction line of the compressor, three of them being electrically connected to their respective 3-way solenoid valve, while the fourth is connected to the motor starter relay. The solenoid valves are located in separate refrigerant vapor lines connecting the unloader cylinders with the compressor discharge and suction lines.

All four pressure switches are adjusted so that they cut-out at different pressures, starting with the first having the highest cut-out pressure adjusted to correspond with 75 percent full load, the second cutting out at a pressure corresponding to 50 percent full load, the third at 25 percent, and the fourth having the lowest setting for pressures indicating no load. Under full load, all four pressure switch circuits are closed because the suction pressure is above the limit for which the first switch is adjusted, so that the compressor delivers 100 percent rated capacity; i.e., all cylinders operate effectively. When the load falls to 75 percent of its full rating, the corresponding suction-pressure drop causes the first pressure switch to open its circuit, de-energizing its respective solenoid valve. With the solenoid de-energized, the three-

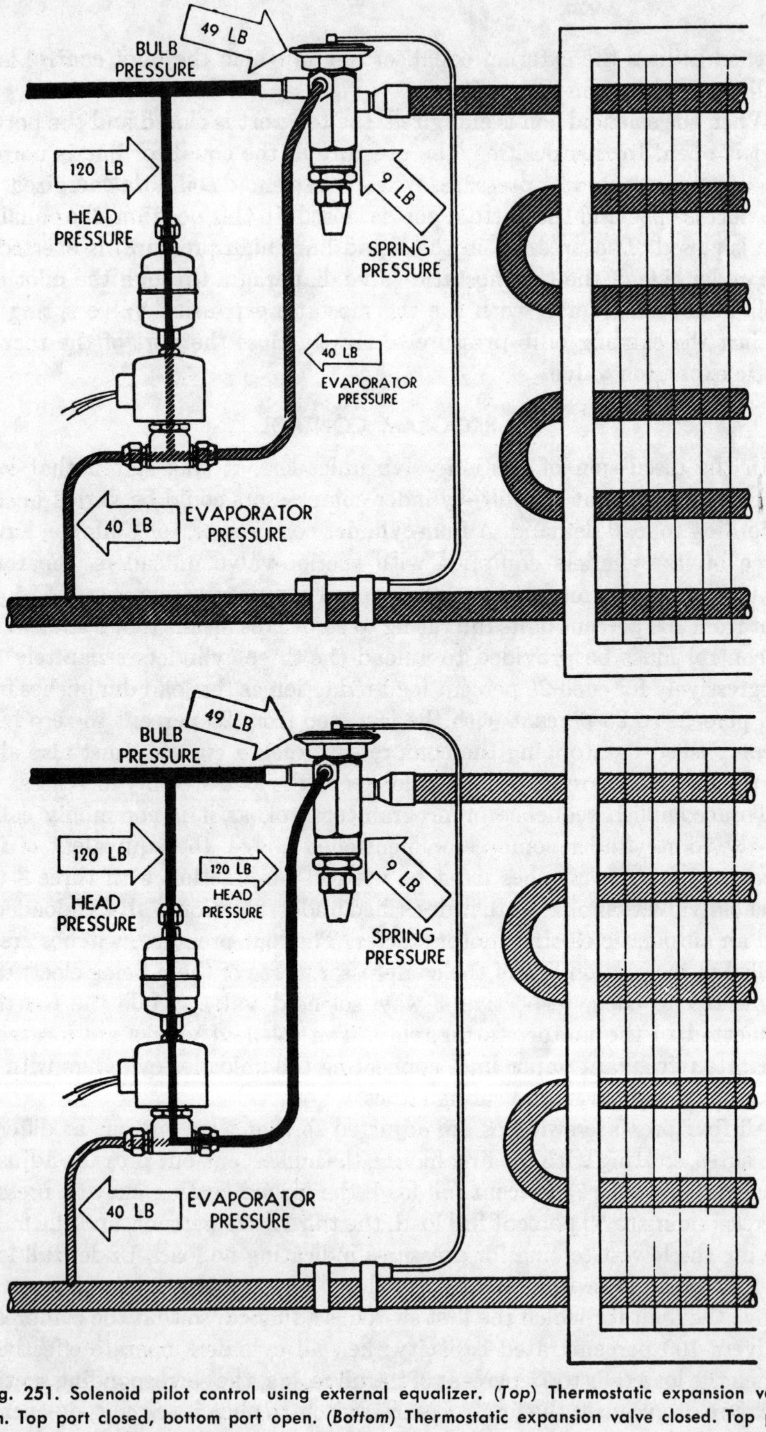

Fig. 251. Solenoid pilot control using external equalizer. (*Top*) Thermostatic expansion valve open. Top port closed, bottom port open. (*Bottom*) Thermostatic expansion valve closed. Top port open, bottom port closed. (Courtesy, Sporlan Valve Company.)

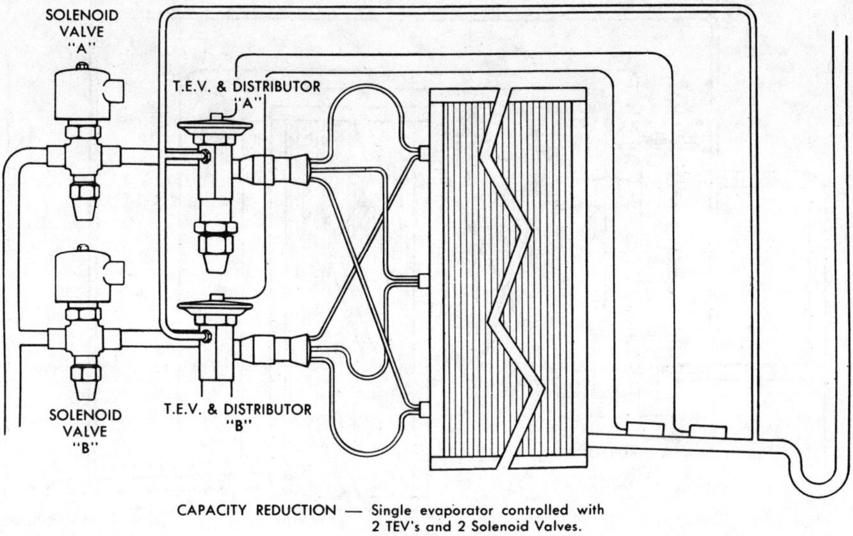

Fig. 252. Capacity reduction—2 or more evaporator sections handling the same load. (Courtesy, Sporlan Valve Company.)

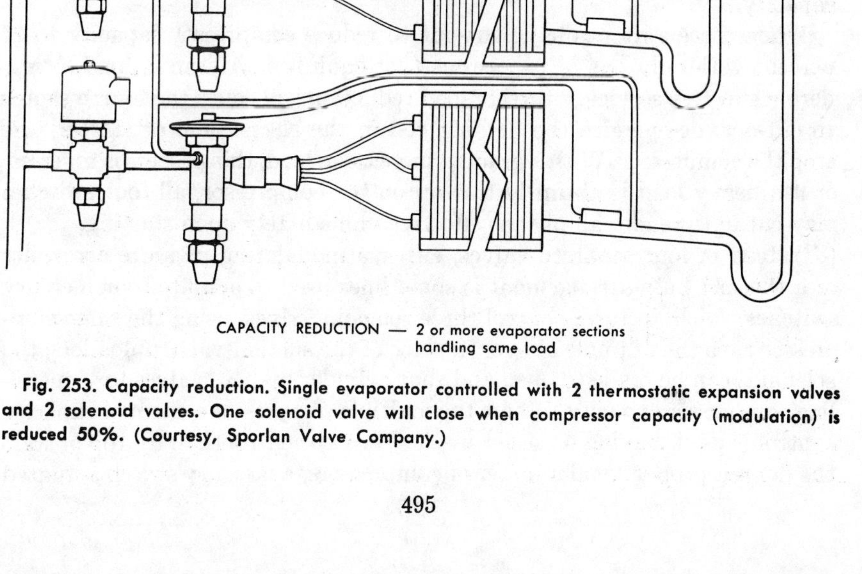

Fig. 253. Capacity reduction. Single evaporator controlled with 2 thermostatic expansion valves and 2 solenoid valves. One solenoid valve will close when compressor capacity (modulation) is reduced 50%. (Courtesy, Sporlan Valve Company.)

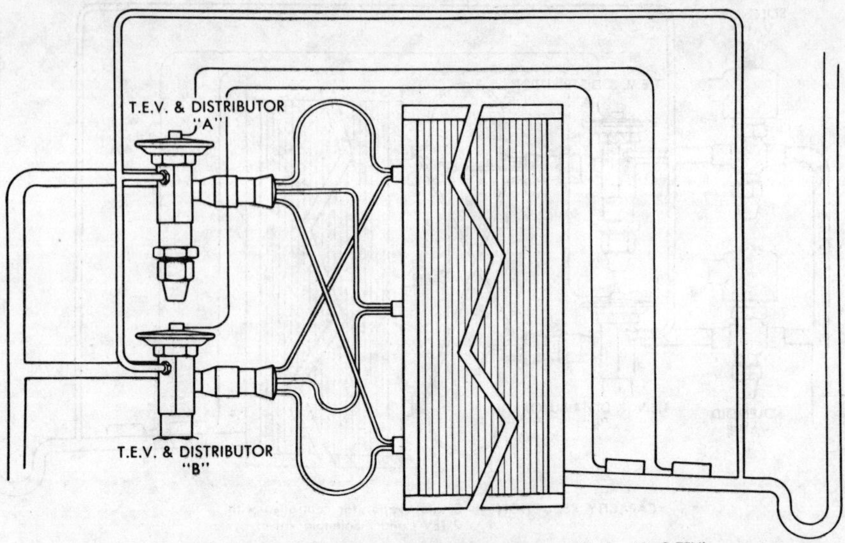

Fig. 254. Capacity reduction, single evaporator controlled with 2 thermostatic expansion valves having different superheat settings. (Courtesy, Sporlan Valve Company.)

way valve allows passage of the compressor discharge gas to the suction valve unloader of the cylinder to which it is connected, rendering that one compressor piston ineffective. The compressor now delivers only 75 percent of its rated volumetric capacity.

If the load continues to fall, say to 50 percent of full rating, the further drop in suction pressure causes the second pressure switch to cut-out, de-energizing its respective solenoid valve and rendering the second compressor piston ineffective. Now the compressor delivers only 50 percent of its rated capacity.

When three switches have cut-out to reduce compressor capacity to 25 percent with only one of its pistons (not equipped with an unloader) rendering effective service, a further load reduction will cause the fourth switch to cut-out, de-energize the holding coil in the electric motor starter, and stop the compressor. With a gradual increase in load, this process is reversed, or if a heavy load is abruptly thrown on the compressor, all four switches may cut in to render all pistons effective immediately upon starting.

Instead of four separate valves, either a modulating pressure controller or a thermal bulb arrangement is sometimes used to actuate four mercury switches which, in turn, control three solenoid valves, using the same compressor as in the example above. In place of the suction-valve unloaders, the solenoids can be made to open and shut cylinder by-pass valves for controlling compressor capacity. Where the multiple compressor system of capacity control is used, having no other method of step control than to stop or start the desired proportional number of compressors, a mercury switch actuated

Fig. 255. Step or program controller. (Courtesy, Minneapolis-Honeywell Regulator Company.)

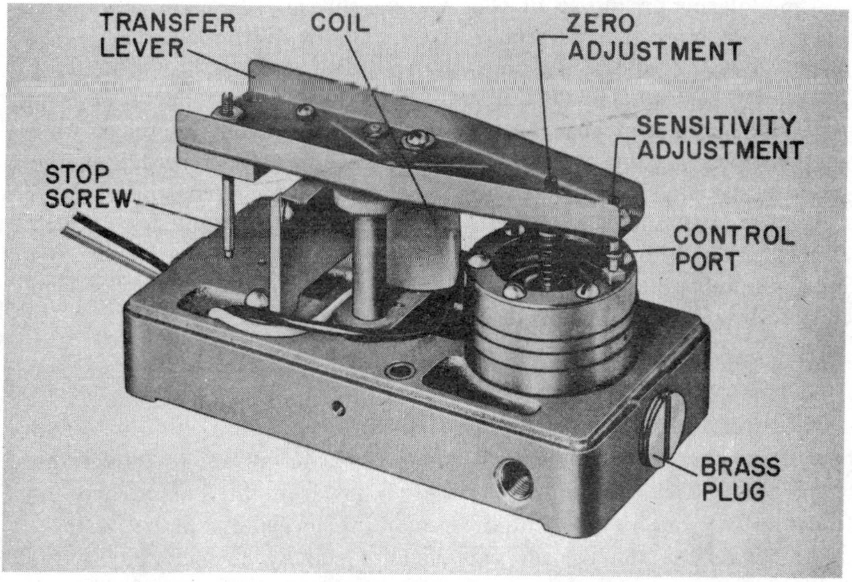

Fig. 256. Electro-pneumatic transducer. (Courtesy, Johnson Service Company.)

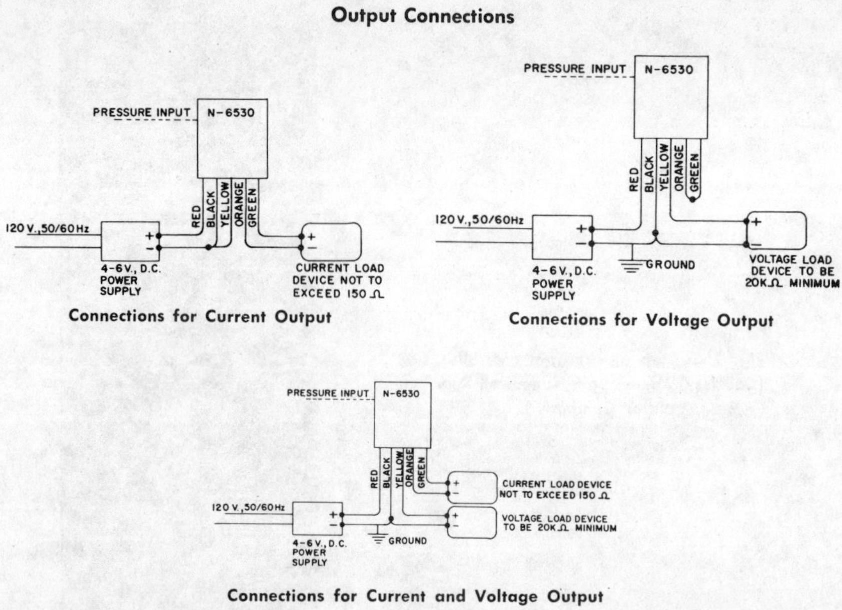

Fig. 257. Transducer connections, electro-pneumatic type. (Courtesy, Johnson Service Company.)

by a thermal bulb in the suction line is provided for each unit. In the latter case, the mercury switches control the electrical circuit of the motor starter holding coils. This method of control is employed mainly on large systems.

A modulating pressure controller working through a program or sequence controller is often used to replace the multiple valve and switch arrangements described above. The modulating pressure controller, sensitive to pressure changes, is installed in the suction line. Pressure fluctuations due to load changes are transmitted to the program controller motor on whose shaft a series of cams are located that operate switches according to their radial position, as governed by the modulating pressure controller.

While a four-cylinder compressor was used to simplify the description of step control methods, such instruments can be applied to compressors having any number of cylinders and more than one piston may be rendered ineffective through a common solenoid.

TRANSDUCERS

The transducer allows energy of one form of transmission system to flow to another form of energy system. Transducers act to convert mechanical, electrical, solar, or acoustical energy from any one of these four forms to any one or more of the remaining forms. Examples of transducer conversion include the photoelectric cell which converts solar or light energy to electrical energy, an electrocoustical device, such as the microphone which converts sound waves to electrical energy, the pressure-electric transducer that converts mechanical energy to electrical energy, etc.

Depending on the type of transducer, the activating elements may include such mechanical components as springs, bellows, bourdon tubes (C-shaped, twisted, or helical), cantilevers and various torque arrangements. Electrical and sound transducer converter elements include capacitors, resistors, inductors, ionizers, piezometers, potentiometers, vibrators, photocells, oscillators, transistors, semiconductors, vacuum tubes, etc.

The essential characteristics of any energy translation control system are:

1. Low sensitivity to
   (*a*) effects of temperature
   (*b*) shock
   (*c*) vibration
   (*d*) magnetic effects
2. Accurate response
   (*a*) within the temperature range limits of the system
   (*b*) within the pressure range limits of the system
3. Excellent linearity
4. Small volumetric motion
5. Temperature compensation
6. Excellent response to
   (*a*) static forces
   (*b*) dynamic forces
7. Efficient output (e.g., pressure input to electrical output [amps])
8. Applicable to required electrical specifications of current, voltage, or internal resistance
9. Easy calibration and adjustment
10. Definite and continuous revolution
11. Versatility, where required (e.g., both the production of voltage and/or current output)

Instrument transducers use various physical effects in their activity in either converting energy or controlling it. Conversion of energy from one form to another relies on some *active* principle to complete its function, while control measures depend more or less on *passive* principles to maintain continuity.

Conversion principles include electro-magnetic, magnetostrictive (change in magnetic induction), piezoelectric (electrical generation due to distortion of crystals by pressure of certain materials, such as rochelle salt), thermoelectric, etc. Controlling principles include resistance, inductance, capacitance, strain, thermo-resistance, piezo-resistance, etc.

The act of energy conversion carries with it the necessity of expressing the values of one form in equivalent terms of the other. To do so, an understanding of analogous terms is essential. If the conversion involves mechanical-electrical energies, then the proper analogies or expression relationship must be used for the various quantities.

Because the transducers used in the refrigeration field for both active and passive requirements are generally of the mechanical-electrical type, the analogies shown in previous discussions relative to energy (Chapter 2) apply here with certain additions; i.e., displacement (magnetic flux), momentum (charge), mass (capacitance), compliance (inductance) in which the electrical analogy is shown in parenthesis for the mechanical quantities mentioned.

One of the essential characteristics listed above is *linearity*, an important feature in the type of transducer used in the refrigeration control work particularly. Linearity, as its name implies relative to straight-line conversion, is the capability of a device to convert one form of energy to another without any change in quantity magnitude. Therefore, a linear transducer is one in which a given change in input energy produces a change in direct proportion in the output quantity. Linearity is capable of being interrupted as a result of certain physical influences that prevail in instrument circuits. Such influences include the effect of friction, current densities, heat conduction, imperfect response in some elements, such as springs, bellows, conductors, etc., and from other sources relative to thermal changes and their effect on dimensional changes, heating, etc.

Where a simple conversion is all that is required, an open-loop or straightforward transducer may be used which generally requires no appreciable compensating modifications to produce more or less linearity. In more articulate systems, a closed-loop or *feedback* arrangement is used. This provides for the return of a fraction of the energy from one point in the system to a preceding point as from the region of output to the region of input. Such an arrangement tends to balance the forces in order to maintain the straight-line flow of energy conversion and establish equilibrium between the input force generated, say, by pressure and a counterforce exerted by the output voltage or current. Feedback may be positive or negative and when correctly designed will ensure an electrical output that is linear and independent of wide variations in supply voltage in pressure-electric transducers.

Where transducer lead lines are placed in close proximity to power lines (e.g., d-c power vs. a-c power), proper shielding is necessary to avoid ripple pickups that may become apparent, especially on indicators, flow charts, etc.

Pressure-electric transducers are prevalent in refrigeration systems and may be used in various capacities. One type of pressure-electric transducer reacts to pressure changes by causing a transfer lever to vary compression on a set of carbon contacts, which reduces their electrical resistance with increased pressure and vice versa. Only a small change in pressure is required to produce a substantial change in contact resistance. The resultant change in current is sensitized in a torque motor having a high degree of (1) linearity and (2) high force output for relatively low input current. The torque power is applied to the transfer lever as either a positive or negative force feedback to balance the force of the input pressure. The carbon contacts are protected by a transistor and diode circuit (voltage regulator) that limits the voltage applied across them. Changes in ambient temperature causing variations in feed-back coil resistance do not affect the current output.

Instrumentation consists of a pressure input, pressure-electric transducer, transformer, and converter (a-c to d-c) where required, and either a current load, or voltage load device, or both depending on whether the output is in the form of current, voltage or both. Voltage or voltage-current output devices must be grounded. An example of energy values may be shown as typical for some control systems in the following table:

An electro-pneumatic transducer is shown in Fig. 256 on which its components are identified. It operates as follows:

A change in temperature, pressure, or humidity at the controller sensing element (remote from the transducer), produces a change in the voltage from the controller to the transducer electro-magnetic coil, establishing a force which changes the position of the transfer lever in relation to the control port. This actuates the pneumatic flapper-valve relay, and produces an output pressure which corresponds to the signal received from the controller. Pneumatic feed-back is incorporated to provide maximum accuracy and stability of this relationship between the input voltage and the output pressure. On a current failure the output of the transducer will be zero psi.

SOLID-STATE DEVICES

In contrast to transducers, solid-state devices have no moving parts, or more correctly, no physical displacement of intrinsic components. One of the older types of solid state systems is the electrical transformer used to step-up or step-down voltage and current without a change in frequency. Solid state systems also operate without heated filaments or vacuum gaps. Semiconductors, such as those used in transistors, and thermo-electric couples are elements of the solid state phenomena.

Controls of the solid state type are used in some refrigeration systems and are the ones used for the hermetic package chiller described for the screw compressor in Chapter 8.

The phenomena of energy transition in solid state systems is complex involving as it does the principles of lattice behavior, as described for thermoelectric processes covered in Chapter 17, and the thermistor discussed briefly in Chapter 9. It may be noted here that thermistors com-

TABLE 36. TRANSDUCER OUTPUT VS. INPUT FOR PRESSURE-ELECTRONIC INSTRUMENTS*

Type	*Model No.*	*Output for Input*		*E to P Int. Res. (ohms) P to E Load Res. (ohms)*	*I*	*V*
Electronic to Pneumatic	N-6810-1	3 psi 15 psi	4 ma 20 ma	600	X	
	N-6810-2	3 psi 15 psi	10 ma 50 ma	600	X	
	N-6810-3	3 psi 15 psi	1 ma 5 ma	2.5K	X	
	N-6810-4	3 psi 15 psi	1 ma 5 ma	12K	X	
	N-6810-5	3 psi 15 psi	4 ma 20 ma	180	X	
	N-6810-6	3 psi 15 psi	10 ma 50 ma	57	X	
	N-6820-1	3 psi 15 psi	0 16 VDC	1K		X
	N-6820-2	3 psi 15 psi	1 VDC 9 VDC	1.4K		X
Pneumatic to Electronic	N-6510-1	0 ma 20 ma	0 psi 20 psi	150 max.	X	
	N-6510-2	4 ma 20 ma	3 psi 15 psi	150 max.	X	
	N-6520-1	0 2 VDC	0 psi 20 psi	20K min.		X
	N-6520-2	0 200 mv	0 psi 20 psi	20K min.		X
	N-6530-1	4 ma @ 40 mv 20 ma @ 200 mv	3 psi 15 psi	I Load = 150 max. V Load = 20K min.	X	X

* Courtesy of Johnson Service Co.

pared to metal coil resistors have the characteristics of high value of non-linear resistance change and negative temperature coefficient. They are considered as being semiconductive. The transistor using the elements germanium and silica as semiconductors is another energy converting device belonging to this group.

SAFETY CONTROLS

The controls described in this section do not necessarily improve the flexibility or the efficiency of a refrigeration system, but are nevertheless highly important in their capacity as equipment and personnel protective devices.

Failure of Condenser Water Supply

Any failure of the condenser water supply causes a rapid rise in pressure. In addition to the high-pressure cut-out, some systems are equipped with a water-failure switch installed in the condenser water line. The pressure of the water flowing through the line keeps the motor-control circuit closed by means of a bellows or diaphragm connected to a switch. If the pressure drops below that considered normal for the operation or the water fails entirely, the circuit opens and stops the compressor.

Overload Cut-out

This is a device for protecting the electric motor against overheating due to overloads or too frequent starting, two of the major causes of overheating if proper ventilation is otherwise provided. The overload protector, as it is commonly called, is used principally on small single-phase motors. It consists of bi-metal strips or disks that are placed in such a manner on the motor frame as to respond to abnormal temperature change. Reacting in the same manner as the bi-metal strip described for space thermostats, any abnormal rise in motor temperature causes the power circuit to open and stop the motor.

Also included on some single-phase motors is a heater coil placed immediately beneath the bi-metal strip or disk, designed to carry normal starting and running currents. Any increase in current to the windings causes a temperature rise in the coil, whereupon the bi-metal element reacts to break the power circuit.

On larger motors, thermal overload relays are located in the starter to trip the contacts and stop the motor in case of any prolonged overload. *Low-voltage protection* or *low-voltage release* circuits are included in starters of this type. The former allows only for the manual starting of the motor after shutdown occurs, while with the latter the motor will start automatically when voltage returns to normal.

Relief Valves and Fusible Plugs

The principal function of pressure-relief valves and fusible plugs is to

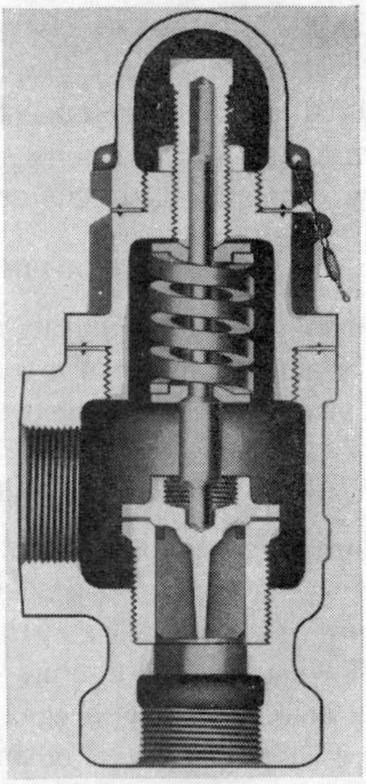

Fig. 258. Relief valve, ammonia type. (Courtesy, Crane Co.)

provide some means of relieving pressures within certain parts of the refrigeration system before they rise to levels that may cause equipment destruction or threaten the safety of personnel.

The pressure-relief valve is spring loaded and normally held in a closed position as shown in Fig. 258. One end of the comparatively heavy spring bears on the valve disk that holds the high-pressure fluid within its compartment, while the other end is supported by a threaded cap. The threaded

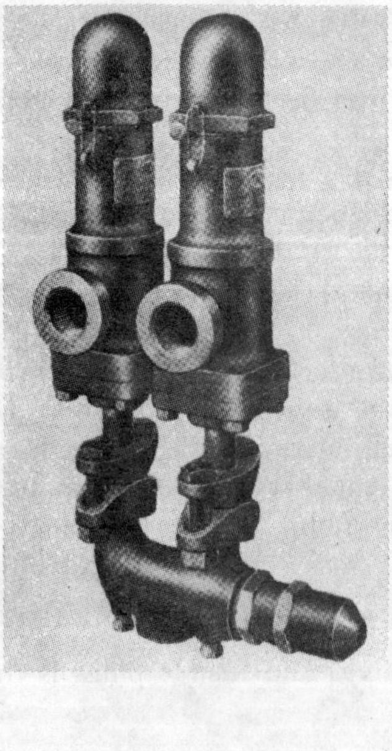

Fig. 259. Relief valves mounted on dual-outlet valves; permits adjustments to one safety valve while the other protects the system. (*Left*) Assembly. (*Right*) Cross-sectional view of outlet valves. (Courtesy, Frick Company.)

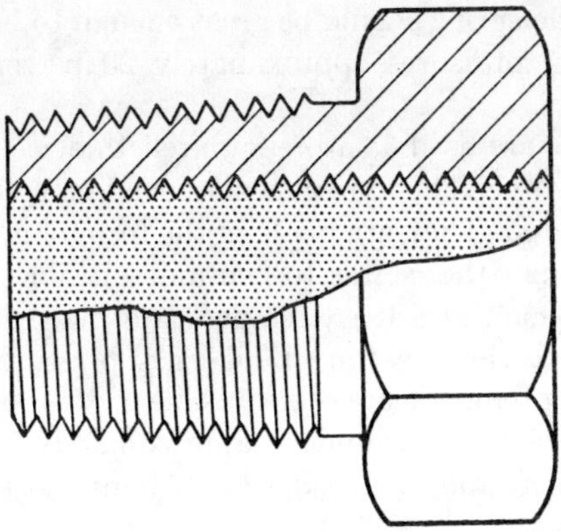

Fig. 260. Fusible plug.

cap acts to hold the spring in place and in some types provides a means for tension adjustment by turning it in or out.

The high-pressure side of the system must be provided with a relief valve installed on either the condenser or the receiver.

The atmospheric type allows the fluid to escape directly into the atmosphere immediately surrounding the valve, while another type provides for the attachment of a vent line leading either outside the building or to the low side of the system, a practice that should not be followed with all refrigerants.

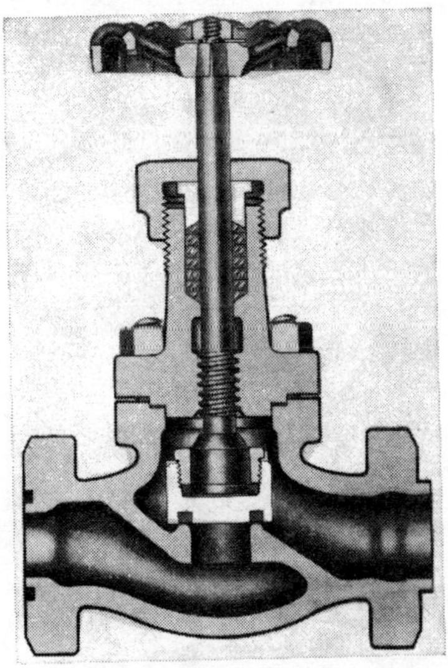

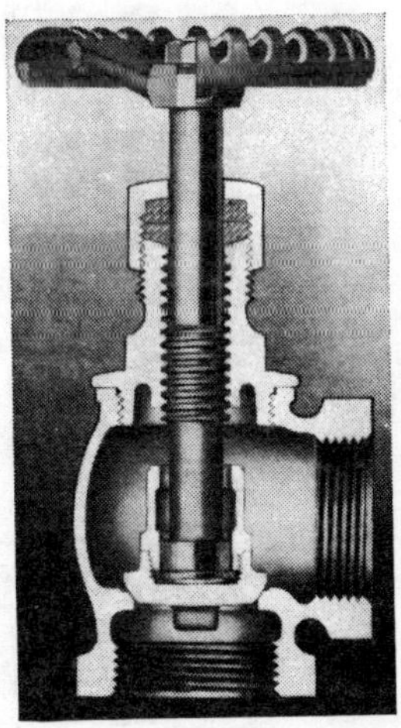

Fig. 261. (*Left*) Globe valve, straight through, packed. Flat face disk, flanged ends, bolted bonnet type. Note lead inserts in disk. (Courtesy, Crane Co.)

Fig. 262. (*Right*) Angle valve, screwed ends. (Courtesy, Crane Co.)

The tension on the spring should be great enough to hold the disk tightly on the seat against pressures approximately 20 percent in excess of expected maximum.

The fusible plugs are used as a pressure-relief device on smaller units in place of the relief valve. They are also employed on cylinders in which the refrigerant is delivered. They are made in the form of a threaded plug with a hole drilled through the center to contain a fusible metal. The fusible metal is alloyed to melt at a fixed temperature that will allow the hole to blow clear to relieve the subsequent high gas pressure. For refrigeration systems, the melting temperature of the fusible metal is usually about 250 F, and for refrigerant shipping cylinders, approximately 165 F, although the metal will become soft and be dislodged at a temperature somewhat below both in either case.

One advantage of the relief valve over the fusible plug lies in the saving of refrigerant that remains in the system after the pressure has been reduced sufficiently to allow the valve to close again.

STOP VALVES

To repair or replace any one component of the refrigeration system, it must be isolated from the remainder of the system by closing the leading and leaving lines by means of stop valves. In the simple system, stop valves are placed in the following locations: (*a*) compressor suction lines, (*b*) compressor discharge line, (*c*) receiver inlet, (*d*) receiver outlet (king valve). Where applicable, stop valves should also be installed between each evaporator and their liquid and suction headers, inlet and outlet of all auxiliary

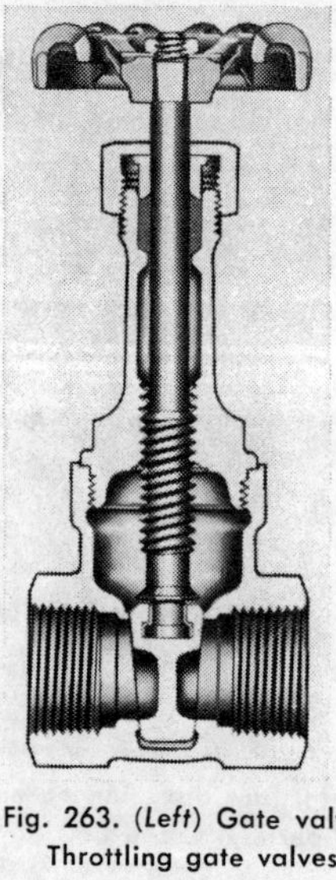

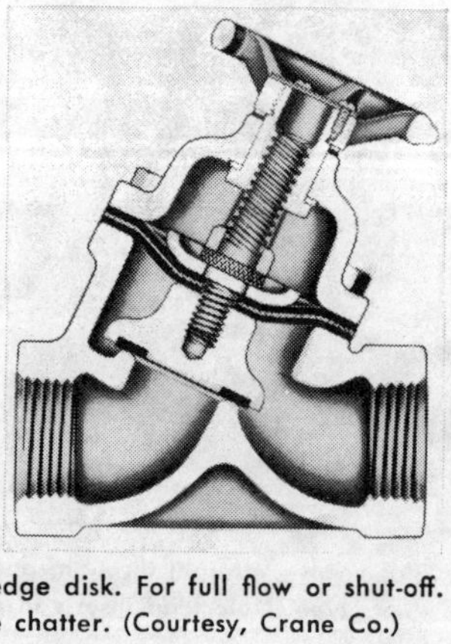

Fig. 263. (*Left*) Gate valve, wedge disk. For full flow or shut-off. Throttling gate valves cause chatter. (Courtesy, Crane Co.)

Fig. 264. (*Right*) Diaphragm valve, packless. (Courtesy, Crane Co.)

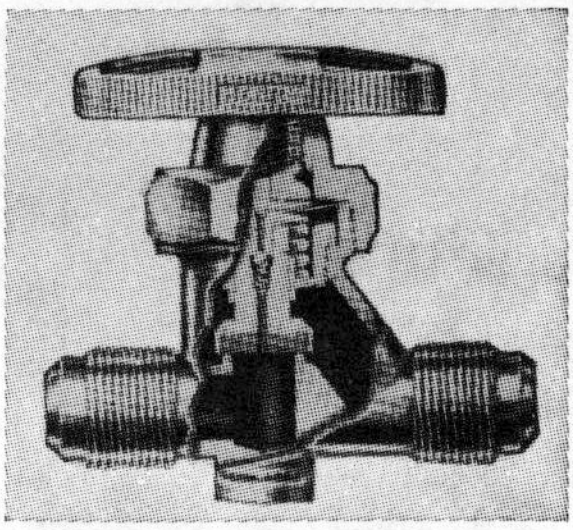

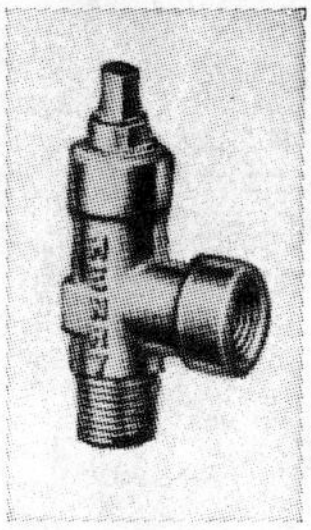

Fig. 265. (*Left*) Line shut-off valve, packless. (Courtesy, Henry Valve Company.)

Fig. 266. (*Right*) Auxiliary valve for purging purposes or mounting pressure gages. (Courtesy, Henry Valve Company.)

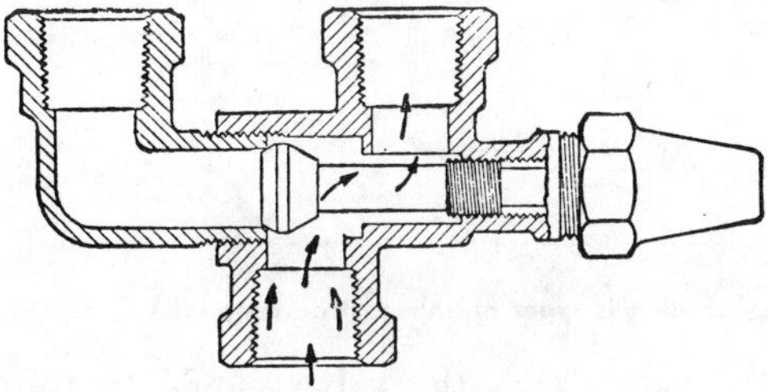

RIGHT OUTLET PORT OPEN
LEFT OUTLET PORT CLOSED

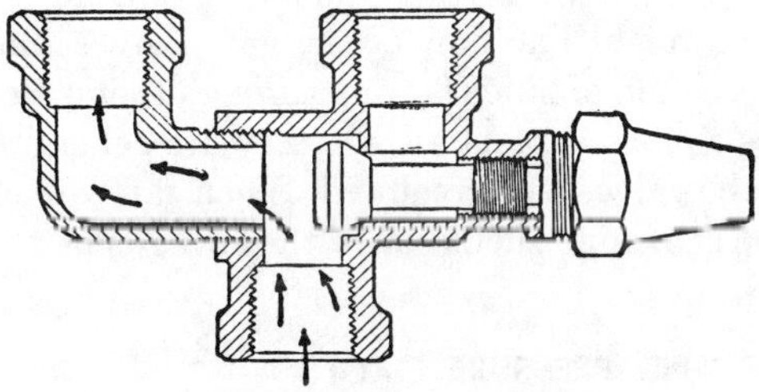

LEFT OUTLET PORT OPEN
RIGHT OUTLET PORT CLOSED

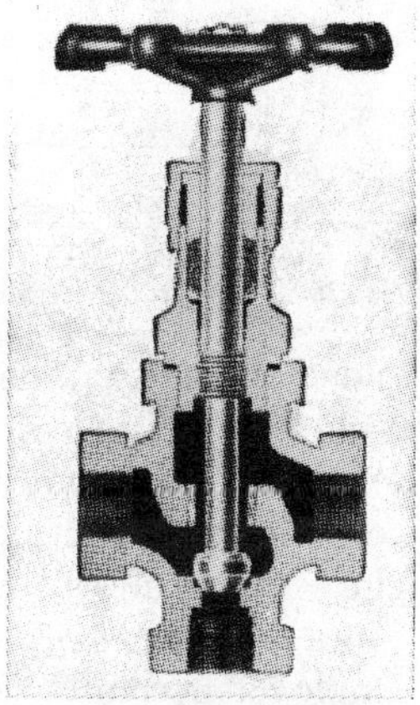

Fig. 267. (*Left*) Three-way dual shut-off valve. (*Right*) Three-way valve, tee type. (Courtesy, Henry Valve Company.)

components, compressor cross-overs, by-pass lines, purge and drain lines, and charging connections.

Because gate valves are not practical for throttling the flow of fluids, globe valves are generally employed for refrigeration purposes. Angle valves are also used extensively. The selection of valves as to their ferrous or non-ferrous construction should conform to the same principles involved when selecting pipe or tubing, which is generally based on the corrosive characteristics of the refrigerant. Gate valves and swing check valves are designed for straight-line flow and are, therefore, compatible when both are required on the same fluid line. Gate valves should be used only where conditions de-

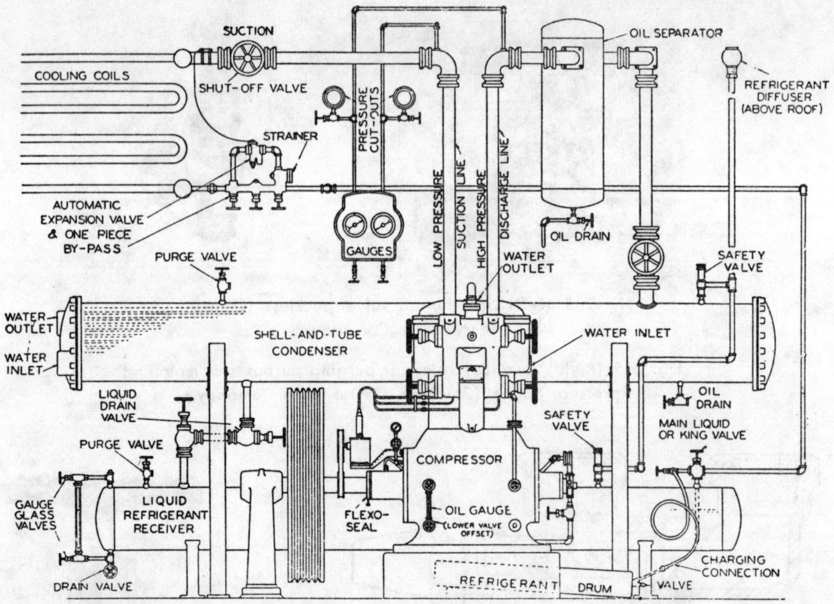

Fig. 268. Refrigeration system showing uses of various types of valves. (Courtesy, Frick Company.)

mand all-off or all-on flow, which means that the valve should be fully opened or fully closed. Globe valves, on the other hand, lend themselves well to partial flow or throttling and therefore are adaptable to refrigeration conditions. In contrast with the straight-line flow of the gate valves, the globe valve causes a change of direction of fluids flowing through them. Because lift type check valves have similar characteristics, they are generally employed in conjunction with globe valves when required. When gate valves are used, any check valves in the same line should be of the swing type because both give straight-line flow.

THERMOMETERS AND PRESSURE GAGES

To determine refrigerant conditions within the several parts of the system, thermometers and/or pressure gages should be installed. Among the more important locations for pressure gages are the suction and discharge lines of the compressor which enable the operator to determine the back and head pressures at all times. On large industrial systems, thermometers placed before and after the compressor serve added convenience in establishing conditions at those points. Other points include:

Liquid outlet of liquid refrigerant coolers,
Water inlet and outlet of surface condensers,
Air outlet of evaporative condensers,
Brine (or water) inlet and outlet of brine (or water) coolers,
Space under refrigeration (at strategic level),
Between the intercooler and the second-stage compressor.

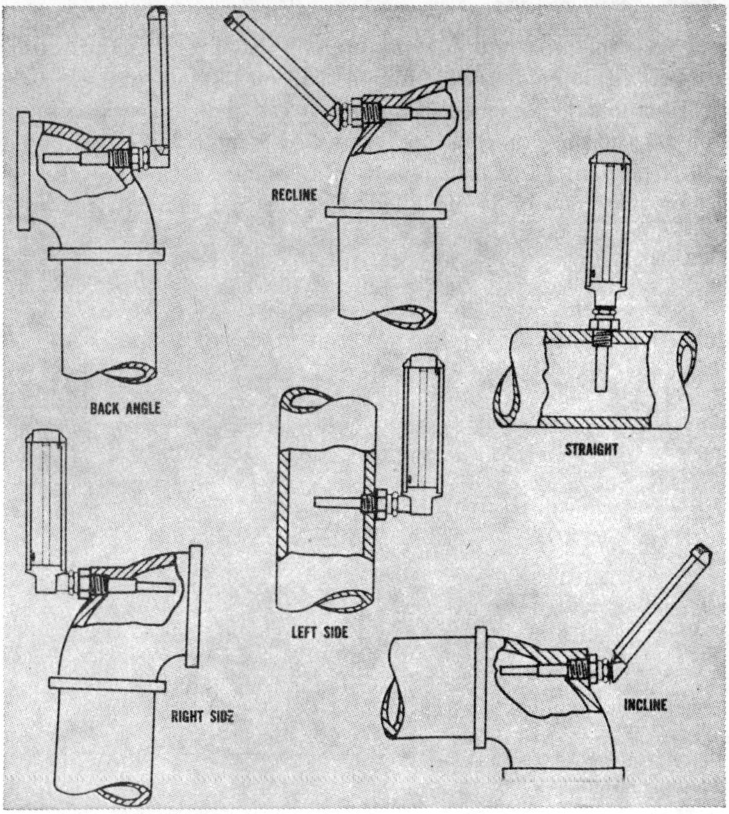

Fig. 269. Typical application for mercury column thermometers showing various fixed angles of stem. Multi-angle (adjustable) stems are also available. (Courtesy, United States Gauge, Division of American Machine and Metals, Inc.)

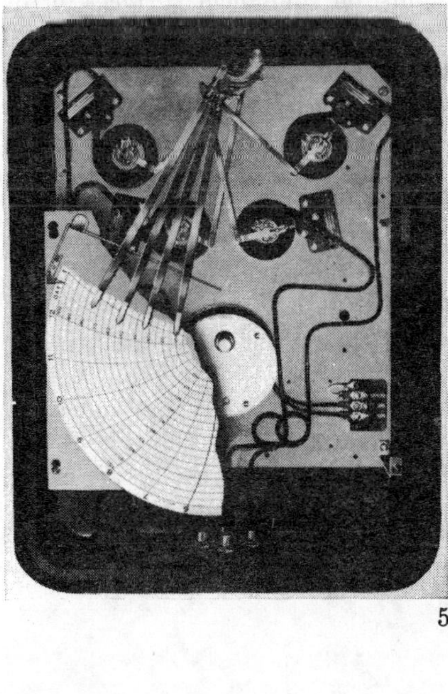

Fig. 270. Recorder, multi-pen. Pens are actuated by spiral Bourdon tubes which are connected to temperature-sensitive elements. Mercury-, vapor-, gas-, or liquid-filled elements may be used, depending on temperature and range. When used for pressure recordings, the spiral Bourdon tube may be replaced with bellows to obtain extreme sensitivity, or a C-type Bourdon tube for high pressures. Chart may be rotated by spring-wound or electric drives. The instrument shown will record conditions at four locations. (Courtesy, United States Gauge, Division of American Machine and Metals, Inc.)

Where close tolerance must be maintained on outlet brine or water temperatures or within cold-storage rooms, a recording indicator is employed. Such an instrument plots a continuous chart of temperatures at any point to which its thermocouple or mercury-filled remote bulb is exposed and serves as a permanent record for future reference. Remote bulbs may be mercury, vapor, gas, or liquid filled. For thermal element (remote bulb) characteristics, see Table 37.

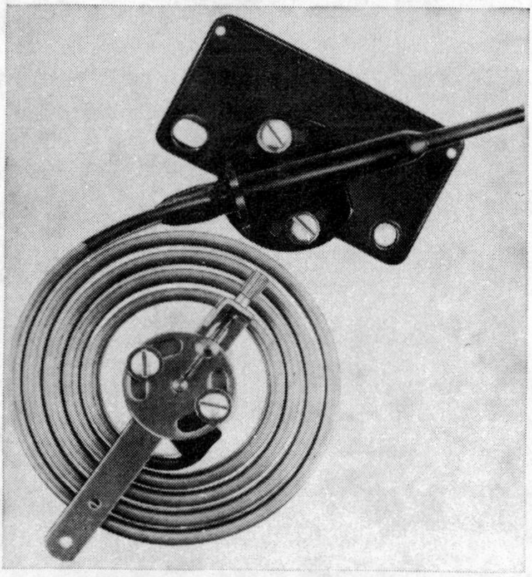

Fig. 271. Bourdon tube, spiral, for recorder shown in Fig. 232. (Courtesy, United States Gauge, Division of American Machine and Metals, Inc.)

Thermocouples

To determine the temperatures in fluids, on equipment surfaces, or of remote areas where it may be difficult to apply other measuring devices, thermocouples are employed. The thermocouple is based on the principle of Seebeck effect, as described in Chapter 17 and demonstrated in Chapter 2.

LIQUID FLOW GAGES

To determine the amount of liquid flowing through a pipe, liquid indicators are used. Such indicators are installed in the liquid line so that liquid passes directly through them. Liquid indicators are pipe fittings equipped with a sight-glass through which a view of the flowing liquid may be obtained. Partial flow of liquid refrigerant is indicated by a bubbling action not prevalent in full lines. Some advantage is gained by using an indicator having double ports which are on opposite sides of the pipe. This type allows the operator to view the liquid with more ease because of the lighter background furnished by the opposite port. The double port indicator is helpful where poor lighting prevails.

Another type is equipped with a sight-glass in which a cross is milled. A

TABLE 37. THERMAL ELEMENT CHARACTERISTICS*

	High Limit	*Low Limit*	*Min. Practical Span*	*Std. Bulb Dia.*	*Std. Length*	*Line Length Limit, Uncompensated*	*Line Length Limit, Compensated*
Mercury....	1200 F	−39 F	100 F	$\frac{3}{8}$	Depends on range	40 ft	200 ft
Vapor......	350 F	−225 F	..	$\frac{1}{2}$	$4\frac{3}{4}$ $7\frac{3}{4}$	125 ft	..
Gas........	1200 F	−350 F	200 F	$\frac{3}{4}$	$7\frac{3}{4}$ $14\frac{3}{4}$	200 ft	..
Liquid.....	400 F	−125 F	50 F	$\frac{1}{4}$ $\frac{3}{8}$	Depends on range	15 ft	125 ft

* Courtesy, United States Gauge, Division of American Machine and Metals, Inc.

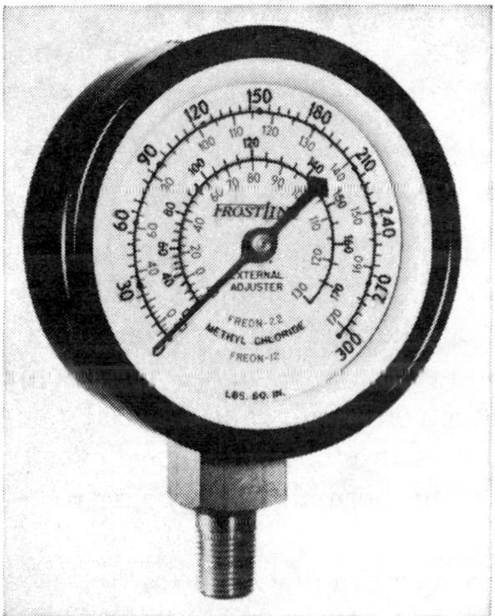

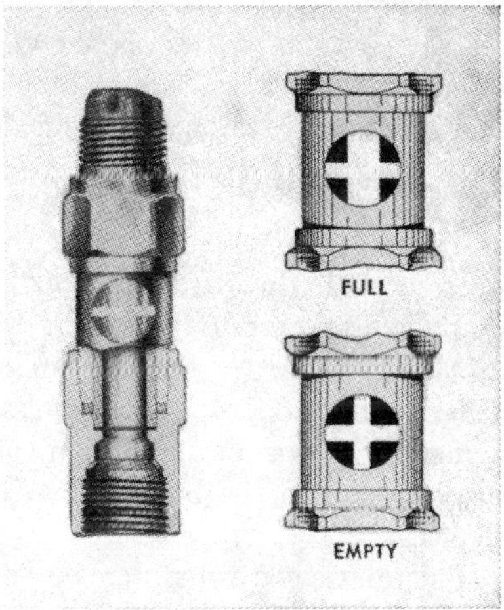

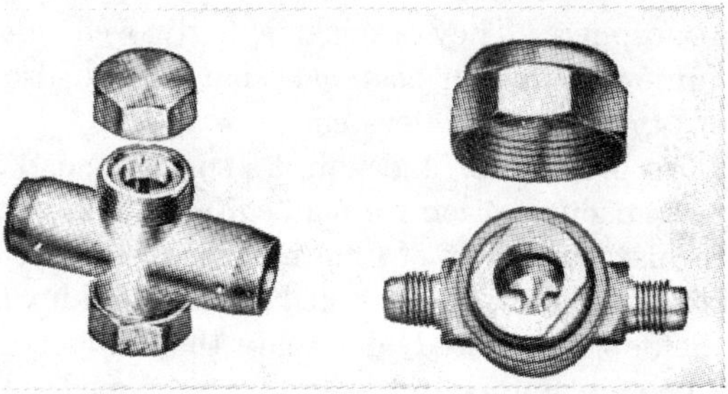

Fig. 272. (*Top, left*) Pressure gage designed for use on systems employing either Freon-12, methyl chloride, or Freon-22. (Courtesy, United States Gauge, Division of American Machine and Metals, Inc.)

Fig. 273. (*Bottom*) Liquid indicator, double port type. (Courtesy, Henry Valve Company.)

Fig. 274. (*Top, right*) Liquid indicator, cross-view type (Cross-Vu). (Courtesy, Henry Valve Company.)

full flow of liquid magnifies the line parallel to flow and extends its width. The line running across the direction of flow remains unchanged in width at all times. When the parallel line is not magnified, i.e., remains unchanged, which can be determined by comparison with the cross line, a partial or no flow of liquid is indicated.

LIQUID LEVEL GAGES

For determining the liquid level in reservoirs such as receivers, accumulators, purgers, oil traps, compressor crankcases, etc., liquid level gages or sight-glasses are employed. Basically, they consist of two angle valves connected by a glass tube having some convenient length. Installed in a vertical position on the side or the end of a vessel, the valves when open will allow any liquid contained therein to seek its own level in the sight-glass. Thus the liquid level in the sight-glass indicates the height of the liquid in the vessel.

To prevent the escape of liquid or gas in the event that the glass is broken, the valves are usually equipped with an automatic check arrangement, such as ball checks. In the case of breakage, the valves should be closed as soon as possible. Closing the valves also allows repairs to be made, such as replacing a broken sight-glass or repacking the valves.

Armored glass tubes, as shown in Fig. 275, vertical rods installed between the valves, screens, or other protective devices, are employed to safeguard against breakage of the gage glass.

PNEUMATIC CONTROLS (SUCTION DAMPER REGULATOR)

Pneumatic controls are used widely on air-conditioning systems in connection with the regulation of liquid valves and air dampers. They are employed to a limited extent in the field of mechanical refrigeration.

The equipment required includes the necessary control instruments (thermostat, etc.), an air compressor and receiver, piping, and especially designed air motors or diaphragm valves. Air pressures of 15 to 25 psig usually are required in the branch line to actuate the normally heavy controlled devices.

Pneumatic control systems do not require a return line as the air released for the purpose of reducing pressures at any given location is vented to the atmosphere. They are particularly well adapted to those situations requiring gradual or *floating* action but are also applied to the *positive* (off or on) type of control system.

Air pressure acting against a spring-loaded diaphragm (or bellows) serves to transmit motion through suitable linkage to the damper or valve which regulates the flow of fluid in their respective conduits. A diaphragm valve is used where the controlled fluid is liquid, whereas, in the case of a vapor, the device required to actuate the damper is referred to as an air motor. Some air motors and valves are equipped with weighted levers to increase tension on the diaphragm in order to retard motion. With the latter arrange-

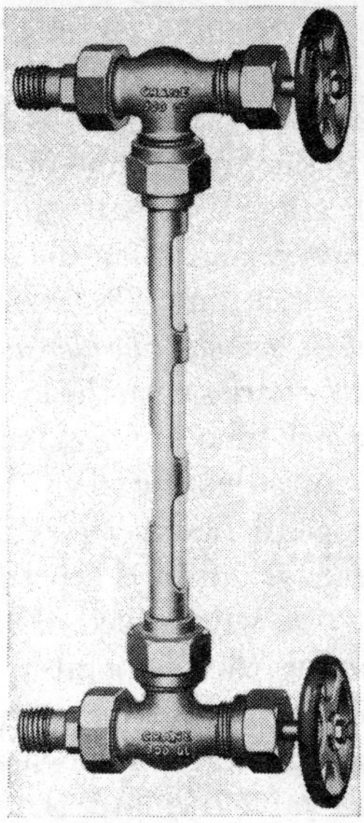

Fig. 275. (*Top*) Liquid level gage Armored to safeguard against breakage. (Courtesy, Crane Co.)

Fig. 276. (*Bottom*) Liquid level gage showing ball-check arrangement. (Courtesy, Henry Valve Company.)

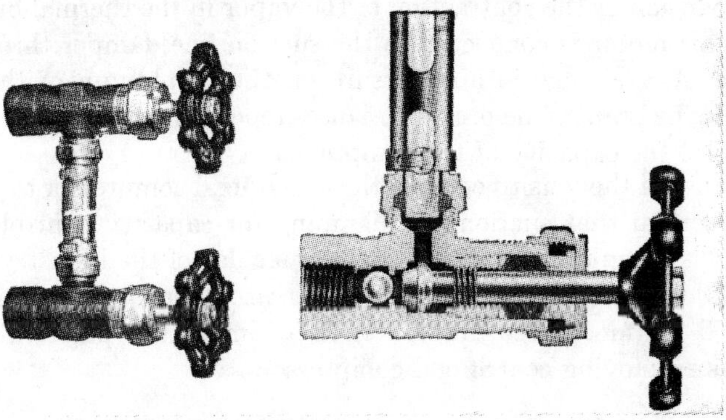

ment, tension of the spring may be varied according to the number of weights and their distance from the fulcrum.

Air-actuated diaphragm valves and motors are located in branch lines taken off the main header leading from the compressor receiver. The air entering these branch lines is controlled by an air valve, which is usually an integral part of a control instrument, such as a thermostat. As the control instrument calls for a change at the diaphragm valve or air motor, it causes its own air valve to partially open or close and allow more or less air to enter the branch line.

However, merely varying the valve port area would not affect the pressure in the branch line if, for any reason, it became equalized with the pressure in the header. One method of reducing branch line pressures is to provide an escape port in the controlled side of the valve whose area is less than the maximum opening of the main valve port. The escape port is allowed to remain permanently opened, which means that the branch line pressure drops to zero gage whenever the main valve port is completely closed. Its permanent opening also prevents the branch line pressure from ever reaching a level equal to that in the header. If the air valve is closed sufficiently as to cause its port area to become smaller than the escape port area, a drop in pressure will occur in the branch line. *The pressure in the branch line bears the same relationship to that in the main header as the escape port area does to the main port area.* As the latter varies with the load demand, it is obvious that the pressure in the branch line changes accordingly. Usually the escape port is equipped with an adjusting screw so that its area may be varied to meet any desired changes in operating conditions.

A typical suction damper regulator control system for a centrifugal compressor employs a thermostat actuated by a vapor-filled thermal bulb located in the brine or cold water line leaving the evaporator. Any load variation causes a corresponding change in bulb vapor pressure which is transmitted through a capillary tube to the diaphragm (or bellows) of the thermostat. The diaphragm linked to the air valve regulates the amount of air entering the branch line according to the degree of motion initiated by the expansion or the contraction of the vapor in the thermal bulb.

The air motor is connected to the suction line damper through suitable linkage. Any change in air pressure on the diaphragm of the air motor, which is the branch line pressure, causes repositioning of the damper, acting to control the capacity of the compressor.

Because of the sensitiveness of the centrifugal compressor to load changes, it is essential that suction line damping for capacity control purposes be accomplished gradually. To use the vernacular of the industry, the damper must "float" with the load. For this reason, pneumatic control systems designed for modulated air flow hold an important place in refrigeration systems employing centrifugal compressors.

TRANE PNEUMATIC INLET VANE CONTROL*

Capacity Control

The Trane CenTraVac is a hermetically sealed, self-contained water-chilling unit with an operating range of from maximum capacity down to 10 percent. To permit operation over this wide range, variable inlet control vanes are located ahead of each stage of the compressor. These vanes are operated by a pneumatic motor which is controlled in turn by a temperature controller located in the chilled water line leaving the machine.

As the temperature of the chilled water rises, due to increasing load on the system, the air pressure from the controller increases. This causes the damper motor to open the vanes, increasing the capacity of the machine. With decreasing load, the fall in the chilled water temperature causes lowering of the air pressure which closes the vanes and thereby reduces the capacity of the CenTraVac.

On start-ups, or when cooling requirements are abnormal, the CenTraVac will

* Courtesy, The Trane Company, La Crosse, Wisconsin.

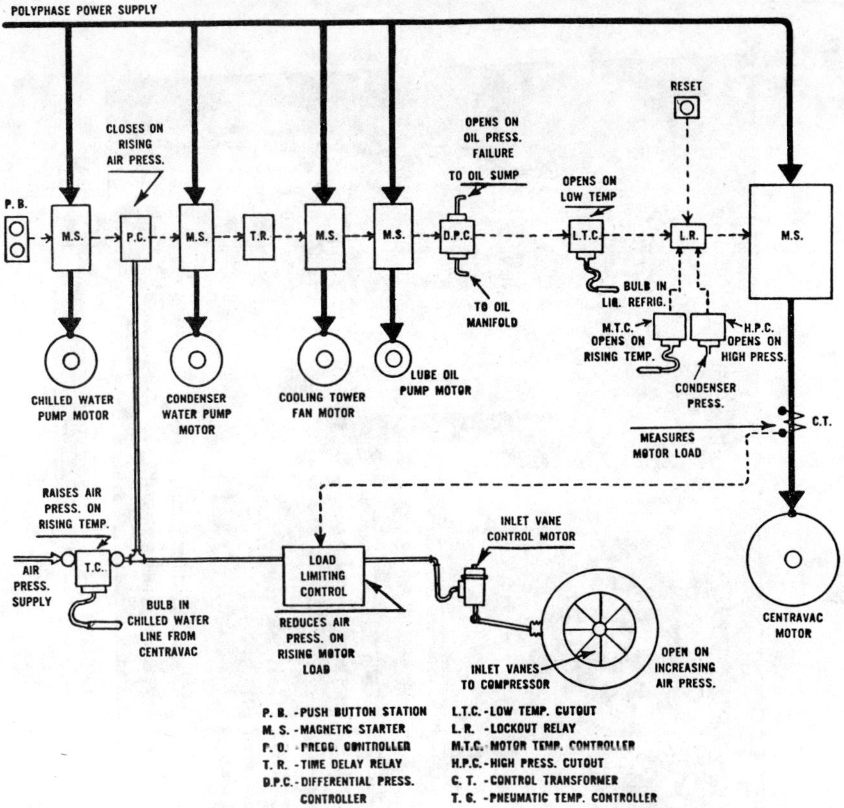

Fig. 277. Pneumatic control system. (Copyright, The Trane Company. Reproduced by permission.)

try to meet this demand. To prevent overload of the motor, the extent of inlet vane opening is restricted by a load limiting device. When motor current increases above normal, this device limits the amount of air pressure that the temperature controller can deliver to the damper motor, thereby limiting the extent of vane opening.

As soon as temperature conditions become normal, the temperature controller again takes over control of the machine capacity.

The load limiting device is generally set to permit 110 percent rated current as maximum load. The motor of the CenTraVac is therefore protected from unusual overloads.

Safety Control

The machine is protected from damage due to freezing by a low-limit thermostat located in the liquid refrigerant in the bottom of the evaporator. This thermostat is connected electrically to the compressor motor controller, and, if the liquid drops to a predetermined low temperature, the machine is cut off.

A high-pressure cutout shuts down the unit if the condenser pressure becomes excessive due to failure of condenser water supply. This cutout is connected to an interlock relay in a circuit to the motor controller which shuts down the motor. A push button on the instrument panel permits resetting of this control.

The motor driving the compressor is water cooled and is protected from over-

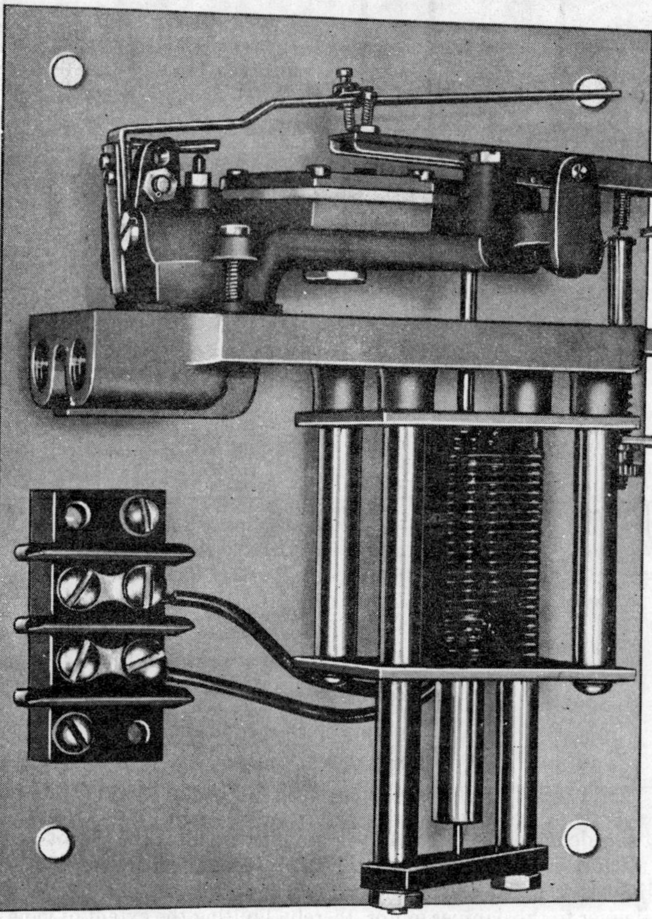

Fig. 278. Load limit control. This instrument is designed for use in pneumatic control systems of the CenTraVac (Trane centrifugal compressor) and provides a positive means of limiting the current drawn by the drive motor. It controls the position of the variable inlet guide vanes even to the point of closing them on both stages to assure no-load coast-down in case of a power failure in a peak load situation. (Courtesy, The Trane Company.)

heating due to failure of the cooling water supply by a thermostat located in the motor water jacket. This is connected into the same interlock circuit as the high-pressure cutout. It stops the motor when its temperature becomes excessive. This control locks out and is restored by the reset button on the instrument panel.

The compressor motor has force-feed lubricated bearings supplied by an independently driven oil pump. A differential pressure controller connected across the lubricating oil pump is interlocked to prevent operation of the compressor motor unless the oil pressure is adequate. A failure of the oil pressure stops the compressor and operation is automatically resumed upon restoration of the oil pressure.

Sequence of Control in a Single Machine Installation

With the control components discussed above, the sequence of control for a typical single unit installation is as follows:

Refer to the control diagram (Fig. 277). To put the system in operation, the chilled water pump is manually started by its push button. If the water temperature leaving the evaporator is above the control setting, air from the chilled water thermostat is supplied to pressure switch *P.C.* closing its contacts. This starts the condenser water pump, which, in turn, energizes the time delay relay *T.R.* At the end of a predetermined time interval the cooling tower fan is started. An interlock on this starter *M.S.* completes a circuit to the lube oil pump starting it. When the lube oil pump has established the oil pressure to the CenTraVac bearings, the contacts in DPC are closed.

Since the liquid Freon temperature in the evaporator is above the setting of low-temperature cutout *L.T.C.*, its contacts will be closed and the circuit is completed through the contacts of the lockout relay *L.R.* This relay is picked up by the reset button and maintained through the contacts of the motor temperature controller *T.C.* and the high-pressure cutout *H.P.C.* When this relay is energized, the circuit is completed to the starter *M.S.* of the CenTraVac.

This interlocking arrangement assures that the chilled water pump, the condenser water pump, the cooling tower fan, and the lube oil system are all operating normally before the CenTraVac can start. The unit will stop immediately upon failure of any of these pieces of equipment as well as through the action of the safety controls. Thus, the system is protected against mechanical failure and improper operating procedure.

As the chilled water temperature falls, the temperature controller *T.C.* causes the vanes to close gradually reducing machine capacity. *P.C.* is so adjusted that the system is shut off when the vanes are fully closed and the machine capacity is at the minimum point.

Sequence of Control in a Multiple Machine Installation

The control arrangement for a multiple machine installation will be essentially the same for each unit as it is for a single machine. There will, however, be only a single chilled water temperature controller for the entire group. This controller will then operate the several machines in sequence through the action of additional pneumatic devices.

In its simplest form, on a two-machine installation, the chilled water lines from the evaporators would be parallel; the units would be operated by a single temperature controller. This controller would start one machine unloaded and bring it up to full capacity and then start the second machine in a similar manner. On temperature drop, the procedure would be reversed, shutting down one machine and then the other.

To insure equal wear on both machines, assuming No. 1 machine has been starting first, the control system can be switched periodically so that the No. 2 machine starts first. When more than two machines are used, the control system can be expanded to give full control over all machines.

The foregoing discussion of the control is merely a basic description of the principles involved. The actual equipment and diagrams for any specific installation will vary quite widely with the requirements of the job and should be determined by The Trane Company, La Crosse, Wisconsin.

SUMMARY OF CHAPTER 9

Control	*Function*	*Method of Actuation*
Expansion Valves:		
Manual	Control flow of liquid to evaporator	Hand adjustment
Automatic	"	Directly by evaporator pressure at valve against adjustment spring
Thermostatic	"	By pressure of fluid in capillary tube from bulb in contact with suction line against spring or diaphragm
Thermal electric valve ...	"	Controls liquid flow by heat motor actuated by thermistor
Low-side float	"	Liquid level in low side of system raises or lowers ball float to open or close valve
High-side float	"	Liquid level in high side of system raises or lowers ball float to open or close valve
Capillary tube	"	Allows permanent communication between high and low sides of system and controls flow by gradual pressure drop to promote some vapor lock. Rate of flow governed by length, diameter, and shape of tube
Capacity Controls:		
Variable-speed	Control delivery capacity of compressor	Multi-speed induction motors—not modulating. By change of stator connections via drum switch cutting out or cutting in poles. Reciprocating compressors only
	"	Wound-rotor slip-ring induction motors. Modulating. By multiple resistance centralized in drum switch. Centrifugal compressors mostly
	"	Steam turbines—speed changes through modulating control in communication with speed governor to control steam flow. Centrifugal compressors
	"	Eddy current clutch—connects synchronous motor with centrifugal compressor. Slippage varies according to changes in magnetic field of clutch
Cycling	"	Stop and start of compressor. Displacement is zero or maximum. No modulation. Thermostat acts to break electrical circuit in relay
Suction valve unloaders...	"	Depressors hold suction valves open while compressor continues to operate. Flexibility depends on number of cylinders equipped with depressors. Not modulating. May be hand or automatically controlled, the latter by solenoid and pressure switch
Cylinder by-pass	"	By-passing discharge gas back to suction from either the cylinder or the discharge manifold. Controlled automatically by a solenoid and pressure switch to open valves in by-pass to suction. Can be applied also to centrifugal compressors, and screw compressors via slide valve
Clearance pockets	"	Pockets in cylinder at discharge end of cylinder act the same as shortening the piston stroke. Compressed gas enters pockets during compression and returns to cylinder on discharge. Flexibility is governed by number of pockets. Modulation can be accomplished by an adjustable piston where used. May be hand or automatically operated
Multiple compressors	"	Capacity controlled by cutting in or cutting out one or more compressors of any type
Program controls	"	Cutting in and cutting out cylinders of multi-cylinder compressors by means of solenoids and pressure switches, each of which is adjusted for graduated suction pressures. Used in connection with a holding coil for stopping the compressor. In centralized program controller, switches are operated by cams located on a common shaft
Throttling dampers	"	A damper placed at the suction intake of centrifugal compressors. Acts to reduce vapor density by increasing pressure difference between inlet and outlet of compressor, thereby decreasing weight of refrigerant handled

Control	*Function*	*Method of Actuation*
Condenser temperature control	Control delivery capacity of compressor	By increasing condenser temperature by decreasing cooling water supply. Raising the pressure differential (compression ratio) between inlet and outlet reduces capacity of compressor. Used on centrifugal compressors
Variable inlet guide vanes		Changes direction of flow of suction gas to coincide with impellar rotation. This reduces amount of gas picked up by impellar and lowers capacity of centrifugal compressors
Pressure Controls:		
Liquid pressure..........	Prevents evaporation in liquid line	Actuated by pressure of liquid against a spring-loaded needle. Tension of spring maintains proper pressure on liquid
Suction-pressure regulatting valve	Maintains constant evaporator pressure	Pressure in evaporator regulates opening of valve to allow more or less vapor flow to compressor. Spring-loaded diaphragm acts against evaporator pressure to modulate flow. Also called a dual-temperature valve. Snap action dual-pressure valves are also employed
Suction-pressure hold-back valve	Prevents compressor overloading	Acts in reverse to regulating valve. Pressure between hold-back valve and compressor acts against spring-loaded diaphragm to control flow of gas from evaporator
Dual pressure controller..	Prevents excessive condenser pressures and too low suction pressure	Pressure in discharge line acts against a diaphragm or pressure switch to break electrical circuit to stop motor. In series with low-pressure switch that also stops motor when evaporator pressure reaches a given level. One will not work until both pressures are in safe range
Heat Motor Valve	"	Controls valve stem opening by electrical resistance in heater coil
Head Pressure Control	"	Controls system operation by reducing or increasing condenser surface area by flooding
Condenser Water-Regulating Valves	Control flow of cooling water to condenser	Connected by tubing to condenser which pressure-regulates flow of water by action on diaphragm and spring to raise or lower valve stem
Check Valves	Prevent reverse flow of fluid in line	Fluid flowing in one direction forces a swing or lift gate open to continue flow. Any reversal of flow closes gate and prevents return of fluid
Thermostats	Temperature control	Actuated by effect of temperature changes on expansion qualities of solids, liquids, or gases. Sensitive element may be a single metal strip which shortens or lengthens, a bi-metal strip which distorts or bends, fluids which expand or contract, condense or vaporize, etc., with temperature changes. Changes in dimension cause electrical circuits to "make or break," influencing other control components
Thermocouples	Temperature control and/or measuring	Remote areas applying Seebeck effect
Solenoids	Actuate valves, dampers, etc.	Electric current through coil sets up magnetic field around central plunger drawing plunger into coil. De-energizing of coil allows plunger to drop. Plunger attached to valve stem allows opening and closing of valve. Used extensively on liquid lines
Transducers	Unlimited uses	Converting one form of energy to another for control, measurement, etc.
Thermistors	Various controls, i.e., liquid flow, etc.	Sensitive to temperature changes that vary electrical resistance
Water-Failure Switch	Prevents excessive condenser temperatures due to cooling water failure	Water pressure maintains positive pressure on diaphragm to keep electrical circuit to motor closed. Water failure causes pressure to drop and open circuit to stop motor
Overload Cut-out	Protects motor against overheating	Placed in motor or as relays in starters, the overload thermal element breaks electrical contact upon excessive temperature rise to stop motor
Relief Valves	Prevent damage to equipment due to excessive pressure	Pressure in vessels or lines acts against a spring to open valve and relieve pressure when predetermined level has been reached. They are set to open at pressure much below the safe working levels of the equipment

Control	*Function*	*Method of Actuation*
Fusible Plugs	Pressure relief	The purpose of the fusible plug is similar to that of the relief valve. They melt at certain temperatures, thus leaving an opening through which high-pressure gas escapes
Stop Valves..............	Close lines or vessels	Usually hand-operated stop valves are placed in various locations to control flow through the system. Globe valves are used mostly in refrigeration work. They are designed for straight through or angle flow
Damper Regulators.........	Control flow of gases or air	Pneumatic. Air pressure against diaphragm or air motors transmits motion through linkage to modulate damper
Float Switches.............	Control liquid level	Ball float controls solenoid liquid valve through magnetic plunger–armature arrangement. Also electronically operated via pick-up coil

STUDY OUTLINE FOR CHAPTER 10

Prime movers or driving components
- Steam turbines
  - Construction and principal parts
  - Types
  - Operating principles
    - Impulse forces
    - Velocity compounding
      - Axial re-entry
      - Radial re-entry
    - Multi-velocity compounding
      - Constant pressure
    - Pressure compounding
    - Reactive force
    - Newton's "Third Law of Motion"
    - Combined impulse and reactive type turbine

Electric motors
- Types

Fundamental electricity
- Definition of terms
  - Volts, amperes, ohms, watts
- Analogy of electrical terms
- Circuits
- Measurement
- Ohm's law
- Kirchhoff's laws
- Power

Fundamentals of magnetic field
- Analogy of magnetic to electrical circuit
- Definition of magnetic terms
- Measurement of magnetic forces
- Ohm's law as applied to magnetic circuits
- Permeability
- Hysteresis
- Magnetic circuit as applied to electric motors and generators
  - The sine curve

Electric generator
- Principle of producing an electric current

Electric currents
- Alternating current
- Direct current

Alternating-current circuits
- Definition of terms
  - Frequency, phase
- Inductance
  - Lenz's law
- Capacitance
  - Dielectrics
- Effect of inductance and reactance on alternating-current circuit
  - Reactance, inductive and capacitive
  - Impedance

Values of current and voltage in alternating-current systems
- Maximum, instantaneous, effective, and average

Relationship of voltage to current in inductance circuits—lagging current

Relationship of voltage to current in capacitance circuits—leading current

Trigonometric functions as applied to sine curves

Power in alternating-current circuits
- Apparent power
- True power

Power factor

Polyphase circuits

Relationship of magnetic field to induction motors
- Rotating magnetic field
- Torque and slip
- Synchronous speed
- Running torque
- Starting torque
- Breakdown torque
- Pull-up torque

Classification of alternating-current electric motors and their characteristics
- Single-phase induction motor
- Split-phase induction motor
- Capacitor-start induction run motor
- Repulsion-start induction run motor
- Shaded-pole motor
- Squirrel-cage induction motor
- Slip-ring or wound rotor induction motor
- Synchronous motor
  - Pull-in torque
  - Pull-out torque
- Multi-speed alternating-current motor

Direct-current motors
- Characteristics
- Types
  - Shunt wound
  - Series wound
  - Compound wound
    - Short shunt compound
    - Long shunt compound
      - Cumulative
      - Differential

Mechanical power transmission
- Direct drive
- Belt drive
- Gear drives
- Chain drives

Air-handling equipment
- Types
  - Fans
  - Blowers
  - Compressors

Fan laws, design, and performance characteristics
- Propeller
- Centrifugal
  - Forward-curved, radial (straight), and backward-curved blades
  - Volute and diffuser types

Pumps
- Types
  - Reciprocating
    - Direct-acting pumps
    - Power pumps
  - Rotary
  - Centrifugal
    - Radial
    - Mixed flow
    - Axial flow
  - Priming of centrifugal pumps
    - Manual priming
    - Self-priming
    - Automatic priming

Pump data
- Capacity
- Head
  - Suction lift—static and total
  - Suction head—static and total
  - Discharge head—static and total
  - Velocity head
  - Friction head
  - Total head
  - Entrance losses
- Water horsepower
- Brake horsepower
- Motor horsepower

Flexible couplings
- Purpose
- Types
- Principle of operation

10

Prime Movers, Fans, and Pumps

There are four principal types of prime movers employed to drive refrigeration compressors: (1) steam turbines, (2) steam engines, (3) internal combustion engines, and (4) electric motors. Of the four types, electric motors are used the most widely. Because of the important role of steam turbines as prime movers for centrifugal compressors, a description of their fundamental principles is included also.

STEAM TURBINES

The main function of steam turbines in refrigeration is to drive centrifugal compressors. They are ideal for this purpose because of their speed flexibility. Turbines are economical to operate, especially if the steam to drive them renders additional service before or after passing through the turbine. This last indicates somewhat whether low or moderate steam pressure will be available, a basic factor in selecting the type of turbine to be used.

Steam Turbine Construction

The steam turbine consists primarily of a cylinder called the casing, a rotor or spindle (quill), and a governor.

Casing. The casing forms the enclosure for the rotor and provides support for the stationary nozzles or blading, the governor, the rotor shaft bearings, the steam admission valves, and such steam extraction mechanism as may be involved. It also forms the cylinder through which highly energized steam flows from its point of entrance at one end to the area of discharge at the opposite end. It is during the flow process that static energy is converted to kinetic energy by transforming axial thrust to radial thrust.

Rotor. The rotor or spindle is the rotative member of the steam turbine. Mounted on outboard bearings placed at opposite ends of the casing, the spindle forms the axis of the rotating buckets (blades) and/or nozzles. It is through the rotor shaft that power is transmitted to the driven component, which, in conformity with the subject of this text, would be a centrifugal compressor.

Nozzles and Blades. Nozzles are stationary openings through which steam passes into the turbine casing. They also act as stationary openings through which steam passes and changes direction while traveling from one row of moving blades to each succeeding row, or re-entry point, as the case may be. Nozzles are of the converging-diverging type for high expansion ratios and convergent for low expansion ratios. Nozzles of impulse turbines are placed at intervals around the periphery of the casing to which they are attached. In the reaction turbine, nozzles are extended completely around the periphery of the casing.

Blades—or buckets, as they are called in impulse turbines—are revolving members upon which the steam acts in transforming its energy to mechanical work. Transformation of energy is accomplished by either impulse or reactive force, or, as is often the case, by a combination of both. Buckets of impulse turbines are placed around the periphery of a wheel attached to the shaft. Blades of reaction turbines are attached to a rotor from which they extend radially outward into the casing.

Governor. The speed governor controls the speed of the steam turbine. Any increase in load tends to decrease speed while a drop in load tends toward higher speed. To compensate for load fluctuations and to maintain constant speed, more or less steam must be admitted to the turbine. It is the function of the governor to modulate steam flow to meet such changes in load demands and to hold speeds at desired levels. Governors also act to stabilize performance in extraction type turbines where it is necessary to compensate for process steam bled from one or more stages. These last are known as pressure governors. Governors may be designed to vary the speed in response to external controls such as would be employed in centrifugal compressor operation.

The principle of governor operation is centrifugal force supplied by the rotation of the turbine shaft. The sensitive or centrifugal element varies in design according to the builder, but all are somewhat similar in their operation. The usual type of governor consists of weights or flyballs hung on arms attached to the turbine shaft or to an auxiliary shaft whose speed is definitely related to that of the turbine shaft. Another set of arms connect the weights to a sliding collar placed around the same shaft. Being pivoted, the arms allow the collar to move axially along the shaft. A spring placed on the shaft between the fixed arm connection and the sliding collar exerts force on the latter and tends to separate the two. The sliding collar is linked to the steam admission valve to control its opening or closing. As rotating speed of the shaft increases, the weights move outward by centrifugal force. This outward movement forces the sliding collar to move toward the fixed connection against the tension of the spring. Motion of the collar is transmitted through the linkage to operate the steam admission valve and decrease steam flow. Upon slowing down of the shaft, the weights move inward and with the help of the spring force the sliding collar away from the fixed connection, allowing an increase in steam flow. The distance through which the sliding collar moves is a function of spring tension, weight of the flyballs, diameter of the weight's circle, linkage factors, and speed ratios.

Mechanical speed governors have their weights connected directly to the turbine shaft. Other governors have flyballs mounted on an auxiliary shaft driven through a worm gear from the turbine shaft. Linkage may be mechanical or a combination of mechanical and hydraulic systems.

The pressure governor employs a bellows as the sensitive element which responds to pressure changes due to extraction of steam at one or more points. They are linked mechanically and hydraulically with steam extraction control valves and the speed governor. The speed and pressure governor elements act together to maintain speed in automatic-extraction turbines. Basic considerations in governor system design include their ability to (*a*) relieve work load on admission valves, (*b*) allows smooth flow of steam into turbine, (*c*) minimize loss of motion of any part, (*d*) obtain positive control, (*e*) create rapid response to operational changes, (*f*) reduce friction between governor parts, (*g*) prevent wide range of speed fluctuation, and (*h*) promote safety.

Thrust. Axial thrust is met by bearings designed to prevent movement of the shaft in a direction parallel to the casing. Thrust bearings vary in design, but usually are of the tapered-land or tilting pad types described in Chapter 15, "Lubrication."

Bearings. Radial load—i.e., the weight of the rotor and the blade assembly—is supported by two main outboard bearings. All bearings of larger turbines are lubricated and cooled by oil circulated through the system. Small turbines are equipped with ring oiled bearings which may or may not be water cooled.

Seals. Seals are installed to prevent steam leakage where the shaft passes through the casing.

Types of Steam Turbines

Steam turbines may be condensing or noncondensing. In the former, the exhaust steam enters a condenser where it is liquefied rapidly, creating a partial vacuum in the process. The steam from the noncondensing turbine is exhausted at some pressure above atmospheric and may be used again to drive a low-pressure turbine or for some other purpose.

Steam turbines may be divided further into types according to their steam-flow pattern. In this connection, steam may pass straight through under full throttle from inlet to exhaust, or part of the steam may be extracted at one or more stages. In the latter types, only part of the initial quantity of steam reaches the final outlet. Other types include the reheat, the mixed pressure, and the extraction-induction turbine. Where steam is extracted under automatic pressure control from one or two stages, these types are called single- or double-extraction or automatic-extraction turbines. The practice of extracting steam without pressure control from one or more stages is known as bleeding, and the unit so designed is called a bleeder or nonautomatic turbine. The steam extracted is used either for heating purposes, as in the bleeder type, or for industrial processes requiring closer pressure control—a purpose for which the extraction turbine is best suited.

Units from which all the steam is returned to the boiler to be reheated to its original temperature after passing through the high-pressure stage and returned to a lower pressure stage is called a reheat turbine. In large turbines, an intermediate stage may also be included. High-pressure turbines whose exhaust steam is used to power intermediate- or low-pressure turbines are known as topping units. Topping turbines are generally installed when it is advisable to convert to high-pressure steam but where the continued use of the older intermediate-pressure unit is practical. The mixed-pressure turbine is one in which low-pressure steam is introduced at an intermediate stage and mixes with the initial steam to power the low-pressure stage.

Turbines may be single-flow or double-flow. In the single-flow turbine, steam enters at one end and flows in one direction, ultimately to exhaust at the opposite end. The steam enters a double-flow turbine in the center and divides to flow in opposite directions. In essence, the double-flow unit is two turbines mounted on the same journal. The steam exhausts at both ends. The double-flow arrangement tends to equalize axial thrust.

There are numerous other turbine arrangements such as vertical double-flow, triple-compound, cross-compound, cross-compound double-flow, etc. The principal purpose in any arrangement, aside from convenience of equipment location and size, is that of obtaining maximum energy from each pound of steam generated. (See Fig. 279.)

The type of turbine selected for refrigeration purposes should be one that best suits operating conditions in the plant where it is to be installed. The most suitable arrangement appears to be one in which steam is used to power the turbine, after which it can be utilized for processing and/or heating purposes. In this arrangement, the turbine acts as a pressure-reducing valve through which high- or moderate-pressure steam is converted to low-pressure steam for use in industrial processes. However, the inefficiency of the pressure-reducing valve is eliminated in favor of the power-producing action of the turbine during such conversion. Economy also lies in the small additional cost of producing high-pressure steam with its greater performance potential.

An additional potential is gained by condensing the steam immediately following discharge from the turbine. The additional potential gained results from the partial vacuum developed in the condenser, thereby increasing the pressure drop through the turbine. The partial vacuum develops by

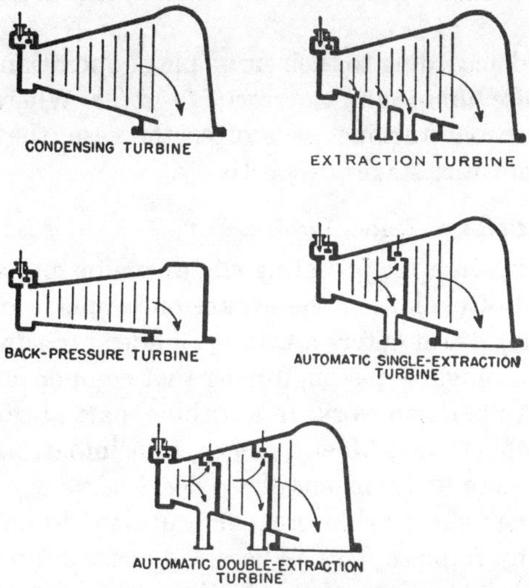

Fig. 279. (*Top*) Various turbine arrangments. (Courtesy, Socony Mobil Oil Company, Inc.)

Fig. 280. (*Bottom*) Turbine-generator unit. Noncondensing, it supplies power and steam for plant processes. (Courtesy, Allis-Chalmers Manufacturing Company.)

reason of the rapid reduction in volume upon extraction of heat from the steam, including its latent heat of vaporization, by the condenser cooling water.

In the noncondensing turbines, the back pressure is greater because the steam discharges against atmospheric or even higher processing pressures. Where power generation is the major function of the turbine, as in utility power plants, noncondensing operation would result in too great a loss in thermal efficiency. Therefore, most turbo-generators are of the condensing type.

Steam turbines may be classified according to their principle of operation. The two basic types are the *impulse turbine* and the *reaction turbine*. Where the two principles are combined in one turbine, as is often the case, they form the velocity stages and the pressure stages, respectively.

Operating Principles of Steam Turbines

The fundamental difference between a reciprocating steam engine and a steam turbine lies in the method of exploiting the expansion property of steam. In the reciprocating engine, steam enters a cylinder under pressure where it expands directly against a movable piston, forcing that component to move in a rectilinear direction to perform work. In a turbine, part of the work of expansion (enthalpy) is converted to kinetic energy, thus imparting velocity in the form of a moving steam jet by means of nozzles.

If the nozzles are stationary, the velocity thus created is utilized to impart motion to a rotative wheel by *impulse force*. Where the nozzles form part of the rotative wheel, then the force of reaction caused by expansion of the steam in the nozzle forces the wheel to rotate in a direction opposite to that of the leaving jet.

Thus it becomes obvious that the energy (enthalpy) of the steam is divided in order to accomplish two transformations, viz.: (*a*) to produce kinetic energy of motion or velocity of the steam, (*b*) to perform work on a rotative wheel.

Impulse Force. The impulse turbine is driven by impulse force of steam directed into a concave blade or bucket mounted on the periphery of a wheel which, in turn, is mounted on a shaft. The power developed by impulse force is transmitted through the shaft to the utilizing component.

Impulse force is not force alone. It is a function of force and time. To gain velocity in a moving body, force must be applied for a definite period in order to overcome the various resistances to motion. The time element must be of such magnitude as to develop a final velocity commensurate with circumstances surrounding the action. Time magnitude depends on the mass of the moving body relative to the propelling force, where

$$M = \frac{w}{g}$$

or mass equals weight divided by the gravitational constant, 32.2 ft per sec per sec. Since force is a function of mass and rate of velocity increase or acceleration (a), then the equation

$$F = \frac{w}{g} \times a$$

is significant, and for simplification $F = M \times a$ (force = mass $\times$ acceleration).

But impulse force is based on time, or, more definitely, it is the product of force and time—i.e., the time during which the force is exerted. Therefore, impulse force may be expressed as $F \times t$ or Ft. Multiplying through by t, then $F = Ma$ becomes $Ft = Mat$.

Final velocity is a function of rate of velocity increase and the time through which such an increase continues. Acceleration being the rate of velocity increase, then $Ft = Mat$ by substitution becomes $Ft = Mv$ where v is the velocity of the body.

Since the product of mass × velocity is significant of the *momentum* of a body, then the impulse of a force must represent a *change in momentum*. Representative of change, then, the equation for impulse force also may be written

$$Ft = M(v_2 - v_1)$$

For rotational bodies, the radius of the point at which force is applied must be considered along with the impulse force. The corresponding equation for rotation to that of translation ($Ft = Mv$) is $Frt = Iw$ where w is the angular velocity (radians per sec, etc.) and Iw is the angular momentum of the rotating body. Since I represents the moment of inertia, established by sum (mr^2), where m is particle mass and r the distance from axis of rotation, it involves a method of the calculus for determination and is beyond the scope of this text.

The force exerted on the blade or bucket of an impulse turbine is steady as a result of steady flow of steam against a succession of buckets on the wheel. If the steam jet impinged on only one blade, the efficiency would be low. This is because of the time function as represented by t in the equation $Ft = Mat$. By allowing the jet to impinge on a series of blades, i.e., one following the other in rapid succession, it would be the same as having the nozzle follow a single blade. This would bring about a continuation of impingement and allow a greater value of M (mass) by reason of the extended time factor. This is accomplished in steam turbines by providing each wheel with a row of blades.

In the above discussion, it was shown that impulse force is a function of mass and velocity and represents a *change* in momentum. Therefore, the impulse turbine is a high-speed machine depending on high-steam velocity at the point of applied force. Steam as stored in the steam chest contains potential energy that, upon release to a lower pressure zone, is converted to kinetic energy. In the case of the impulse, turbine kinetic energy is in the form of high velocity. To obtain high velocity, properly designed nozzles are employed. The type of nozzle used to gain high velocity in any flow of gases or vapors is dependent on the ratio of initial to discharge pressure. If the discharge pressure is equal to or less than the critical pressure,* the nozzle should be divergent at the discharge side, as shown in Fig. 281.

* See "Steam Jet Refrigeration," Chapter 16.

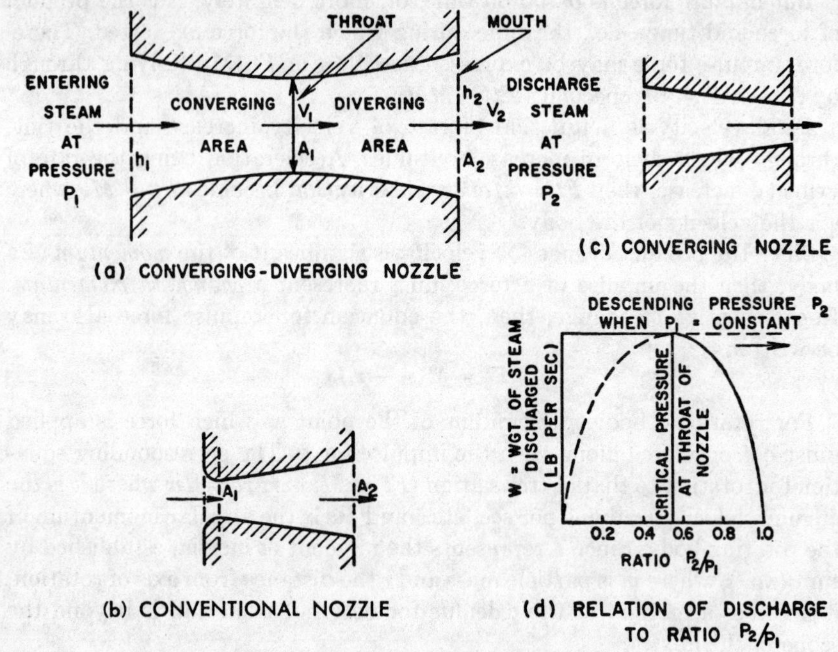

Fig. 281. Nozzle design and theory. Key follows:

V_1 = velocity of steam at A_1 (throat)
V_2 = velocity at discharge
A_1 = area of throat, sq in.
A_2 = area of mouth, sq in.
P_1 = initial steam pressure, psia

P_2 = discharge pressure, psia
P_c = critical pressure, psia
h_1 = enthalpy at P_1, Btu per lb
h_2 = enthalpy at P_2, Btu per lb
W = weight of steam flowing, lb per sec

Pressure drops as gas proceeds from A_1 to A_2
Specific volume and velocity increase from A_1 to A_2
Nozzle (a) for compressible fluid in which velocity first increases faster than specific volume until critical pressure is reached at A_1, when specific volume increases faster than velocity
Nozzle (b) in which specific volume increases faster than velocity from A_1 to A_2, for compressible fluids
Nozzle (c) for liquids where specific volume remains constant but velocity increases
Critical pressure P_c is reached
when $P_2 = 0.58P_1$ (saturated steam)
when $P_2 = 0.55P_1$ (steam superheated 100 F +), $P_2 = 0.53P_1$ (air)
Under which conditions A_2 should be = or < A_1

$$V_2 = 223.7\sqrt{h_1 - h_2}$$

$W = \frac{A_2 P_1}{70}$ (Napier's equation for approximate saturated steam flow). When P_2 is below P_c, discharge remains constant at maximum flow

$\frac{P_1}{P_2}$ = ratio of expansion

Ratio $\frac{A_2}{A_1}$ is function of $\frac{P_1}{P_2}$, A_2 becoming greater as $P_1 - P_2$ increases

Ratio $\frac{P_c}{P_1}$ = constant for a given fluid (gas)

Note in the curve (d) maximum flow is reached when $\frac{P_2}{P_1} = P_c$ for any gas or $0.58P_1$ for saturated steam.

Co-ordinates—solid line—wt vs. ratio $\frac{P_2}{P_1}$ (bottom)
dotted line—wt vs. pressure (top)

EXAMPLE: When initial saturated steam pressure = 100 psia, then

$$P_c = 0.58 \times P_1 = 0.58 \times 100$$
$$= 58 \text{ psia}$$

Actually to obtain high expansion ratios (rapid pressure drop) necessary for high velocity, a converging-diverging nozzle, as in the type shown, is used in impulse turbines.

Nozzles are stationary in an impulse turbine, a number of which are placed radially in the casing on a radius line corresponding to the rotating power elements or buckets at the periphery of the wheel. Usually nozzles do not extend completely around the wheel in impulse turbines.

The second step in turbine design is to supply a rotating element suitable for converting velocity of steam to shaft work by impulse force without shock. This is accomplished by allowing the high-velocity steam to enter a U-shaped blade or bucket where it follows the circular contour and acts to move the bucket while it simultaneously reverses direction of flow. Thus a continuous smooth force is applied to the rotating elements, a number of which are placed completely around the rim of the wheel, which, in turn, is fixed to the shaft.

When the steam enters the casing, its pressure drops immediately to that at which the turbine exhausts or at which the impulse section of combination turbines is controlled. This means that the pressure is equalized on both sides of the wheel so that, theoretically, no pressure drop occurs during passage of the steam through the impulse blades. In some instances, holes are provided in the wheel to facilitate pressure equalizing throughout the casing. For the same reason, shrouding of impulse wheels is not necessary.

Impulse turbines are high-speed machines. Maximum steam velocity occurs at the entrance to the blades. As it converts to shaft work, velocity decreases so that kinetic energy is given over to mechanical energy. For greatest thermal efficiency, steam velocity should drop from maximum to zero in passing through the blades. To obtain a proper balance between steam velocity and blade speed in order to obtain highest efficiency, and to keep peripheral speeds reduced to a safe level, some means of reducing the velocity in steps is necessary. This is done by velocity compounding, either by employing more than one row of blades, which is known as staging, or by some form of steam re-entry to the same row of blades. Both types of velocity compounding are described more fully later in this chapter.

To better understand the method of utilizing impulse force of steam, consider the concave object or blade shown in Fig. 282. The arrow represents a jet of steam as it leaves a nozzle and enters the circular pattern of the blade to follow the contour and reverse its direction of flow. The force of the jet tends to move the blade in the direction of flow or away from the nozzle. The effort turns the stream so that the exit direction is opposite that of entrance. It is apparent that a mathematical relationship exists between all moving bodies with respect to each other and also with respect to stationary bodies such as the turbine casing. This relationship deals with the principle of relative and absolute velocities.

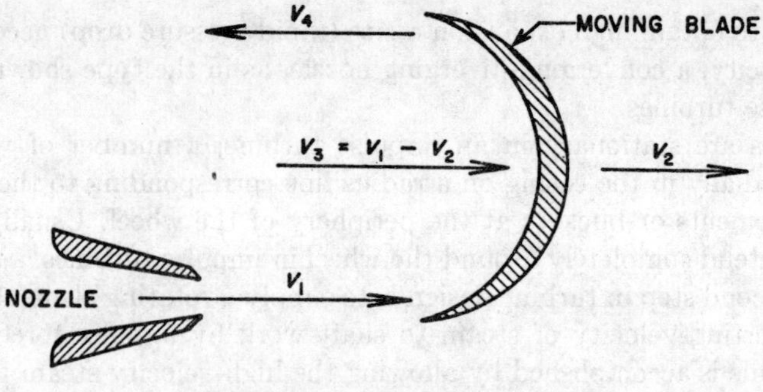

Fig. 282. Development of impulse force.

Absolute velocity is the velocity of a body relative to some stationary body, such as the earth or a turbine casing. Relative velocity is the velocity of a body relative to another moving body.

The relationship of the various components in a turbine with respect to velocity may be expressed as follows:

v_1 = (absolute) velocity of the entering jet relative to the earth or casing

v_2 = (absolute) velocity of the blade relative to the earth or casing

v_3 = (relative) velocity of the entering jet relative to the blade

v_4 = (absolute) velocity of the leaving jet relative to the earth or casing

Let v_1, v_2, v_3, and v_4 represent the various velocities as indicated above. Then

$$v_3 = v_1 - v_2$$

and

$$v_4 = (v_1 - v_2) - v_2$$

Note that where v_2 is zero, v_4 is constant.

EXAMPLES: A jet of steam leaves the nozzle to enter a frictionless blade at an absolute velocity of 3000 fps. Assume the blade to have an absolute velocity of 1000 fps. The relative velocity of the jet with respect to the blade is

$$v_3 = v_1 - v_2$$

$$v_3 = 3000 - 1000$$

$$v_3 = 2000 \text{ fps relative velocity}$$

The absolute velocity of steam leaving the blade is

$$v_4 = (v_1 - v_2) - v_2$$

$$v_4 = (3000 - 1000) - 1000$$

$$v_4 = 1000 \text{ fps absolute velocity}$$

When the absolute velocity of the blade is increased to 1500 fps, then

$$v_3 = 3000 - 1500$$

$$v_3 = 1500 \text{ fps relative velocity of the entering jet}$$

and

$$v_4 = (3000 - 1500) - 1500$$

$$v_4 = 0 \text{ absolute velocity of the leaving jet}$$

When the absolute velocity of the blade is reduced to zero, then

$$v_3 = 3000 - 0$$

$$v_3 = 3000 \text{ fps relative velocity of the entering jet}$$

and

$$v_4 = (3000 - 0) - 0$$

$$v_4 = 3000 \text{ fps absolute velocity of the leaving jet}$$

From the above examples, the following rules may be formulated:

1. When two bodies have some relative velocity to the earth, then the velocity of one relative to the other is the sum difference of their absolute velocities.
2. The velocity of a jet remains constant when direction of flow is reversed by a stationary frictionless blade.
3. The absolute velocity of a leaving jet is zero when the blade travels one half the speed of the entering jet.
4. It follows from (3) that all the kinetic energy of the steam has been imparted to the blade when the absolute velocity of the leaving jet is zero, and that such a condition is one of maximum efficiency.
5. For the impulse turbine, the most efficient operation is obtained when the blade travels one half the speed of the entering steam jet.
6. It is possible to obtain velocity energy from steam at diminishing expansion ratios by a series of pressure-reduction stages.
7. Velocity energy may be divided so that it may be reduced from maximum to zero in two or more steps (compounded).

Velocity Compounding and Pressure Compounding. Considering that steam under high pressure releases with tremendous velocity, the magnitude of which depends on the expansion ratio $\left(\frac{\text{discharge pressure}}{\text{initial pressure}}\right)$, it is realized that single-stage, single-wheel (one row of blades) turbines are terrifically high-speed machines. Following the rule that maximum efficiency is obtained when blade speed is one half the velocity of the steam jet, it is obvious that, with high-pressure steam, speeds would reach destructive proportions in single-row machines. To exploit maximum energy of steam and still maintain speeds within the bounds of safe turbine construction, some method of reducing velocity energy in steps is necessary. This is accomplished in several ways, viz.: velocity compounding and pressure compounding.

Velocity compounding is the method of dividing velocity reduction from maximum to zero in steps under constant pressure in the casing. Pressure compounding is the method of decreasing jet velocity by dividing the pressure drop into two or more steps within the casing, thereby maintaining lower expansion ratios between each stage.

Velocity compounding—an impulse force exerted by a steam jet on a single blade—was discussed previously. Consider now a steam jet that exerts an impulse force on one moving blade, reverses direction to enter a second moving blade. This is shown in Fig. 283.

Assume the steam jet velocity to be 2000 fps. Assume also that both blades are mounted on the same shaft and are moving at 500 fps.

1st blade:

Then

$$v_3 = v_1 - v_2$$
$$= 2000 - 500$$
$$= 1500 \text{ fps rel vel of entering steam jet } (a)$$

and

$$v_4 = (v_1 - v_2) - v_2$$
$$= (2000 - 500) - 500$$
$$= 1000 \text{ fps abs vel of leaving jet } (b)$$

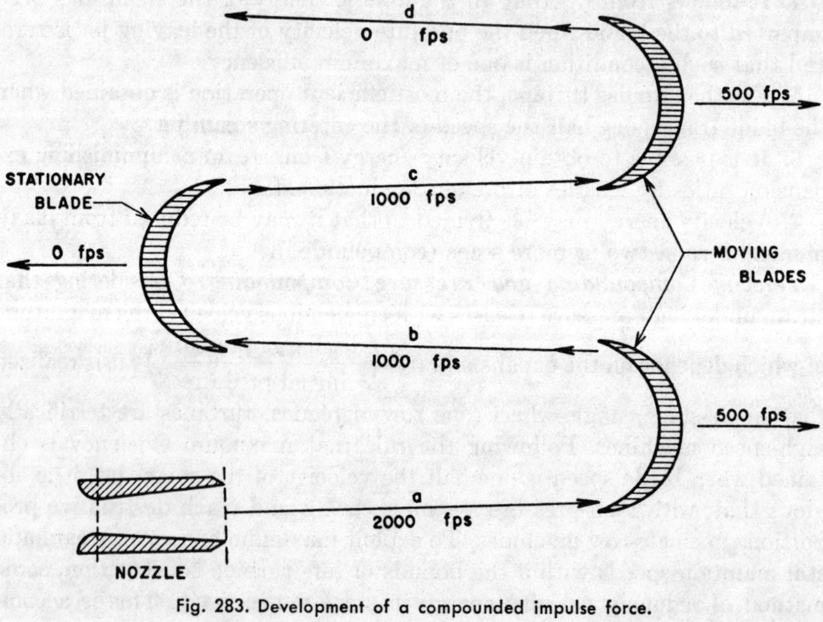

Fig. 283. Development of a compounded impulse force.

Stationary nozzle:

$$v_4 = (v_1 - v_2) - v_2$$
$$= (1000 - 0) - 0$$
$$= 1000 \text{ fps abs vel of leaving jet } (c)$$

2nd blade:

$$v_4 = (v_1 - v_2) - v_2$$
$$v_4 = (1000 - 500) - 500$$
$$v_4 = 0 \text{ fps abs vel of leaving jet } (d)$$

Thus the jet velocity of the steam of 2000 fps is reduced to zero in two steps by allowing it to exert energy on two rows of moving blades. If the second row of nozzles were replaced by a reversing chamber and the steam directed back to the first row of blades, thus eliminating the second row, the same result would be obtained.

Velocity compounding may be accomplished by (*a*) some method of re-entering the steam to a common row of blades or (*b*) employing two rows of blades and two rows of stationary nozzles. The re-entry method of velocity compounding may be divided into two types, viz.: (1) axial re-entry of steam, (2) radial re-entry of steam.

AXIAL RE-ENTRY. The re-entry method of velocity compounding in impulse turbines employs only one row of moving blades or buckets, i.e., a single wheel. The axial re-entry type (Fig. 284, *left*) is one in which the steam enters the blades in an axial direction from nozzles located at intervals around the circumference of the casing adjacent to the blades. The nozzles are of the high expansion ratio type (converging-diverging).

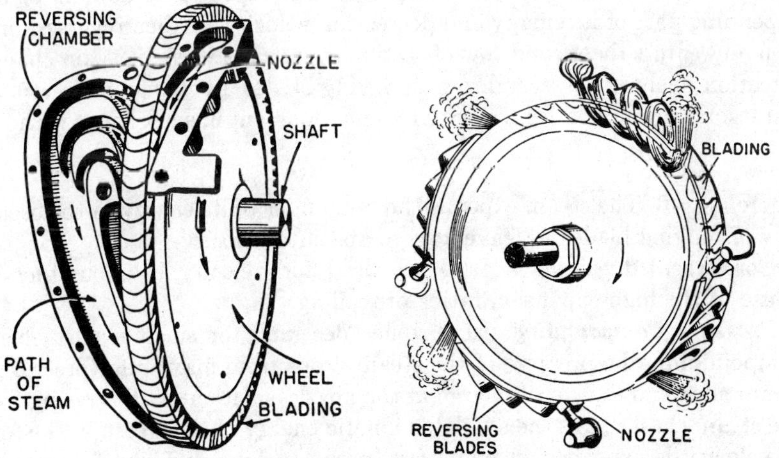

Fig. 284. Velocity compounding. (*Left*) Axial re-entry of steam. (*Right*) Radial re-entry of steam. (Courtesy, Mobil Oil Corporation.)

Steam enters the moving blades from the stationary nozzles. Instead of passing directly to discharge after leaving the blades, the steam enters a reversing chamber that forms part of the casing. Within the reversing chamber, vanes redirect the steam back to the moving blades at some point along the periphery of the wheel other than that of initial entry. The action of reversing and redirecting the steam back to the wheel may be repeated several times, depending on the state of the steam and the turbine design. Pressure is constant in the casing at discharge level during the entire action.

Radial Re-entry. In the radial-re-entry turbine, steam enters circular buckets built into the rim of the wheel. The nozzles located just outside the rim radius, and at intervals around the periphery of the wheel, direct steam into the buckets in a radial and nearly tangential direction. (See Fig. 284, *right.*) Following the contour of the buckets, the steam reverses direction to enter a reversing chamber, a series of which are located at each nozzle. Here the direction of the steam is again reversed to re-enter the buckets. This action is repeated several times until the velocity energy of the steam is spent. Here, again, pressure remains constant in the casing at discharge pressure during the entire action.

Multi-velocity Compounding—Constant Pressure. The third basic method of velocity compounding involves two rows of moving blades and two rows of stationary nozzles. The blades usually are located on a single wheel. It is this type of velocity compounding that furnishes the velocity stage of combination impulse-reaction turbines.

The velocity drop from maximum to zero is divided between the two moving blades, as previously demonstrated. The steam enters the casing through the first row of nozzles in which the pressure drops to that of discharge. From the nozzles the steam enters the first row of moving blades, expending part of its energy and decreasing velocity. It then reverses direction and enters the second row of stationary nozzles where it again changes direction to enter the second row of moving blades. In changing direction in the second row of nozzles, velocity remains constant in accordance with

$$v_4 = (v_1 - v_2) - v_2$$

where $v_2 = 0$. The steam expends the remainder of its energy in the second row of moving blades to leave at zero absolute velocity.

Convergent-divergent nozzles are used for velocity compounding because of the high expansion ratios prevailing.

Pressure Compounding. In turbines designed for single-stage velocity compounding, the pressure of the steam drops from maximum (pressure in steam chest) to that of discharge in the nozzles as it enters the casing from the steam chest. Thus the available kinetic energy of the steam in the form of velocity is expended in one step. In the turbine designed for pressure compounding, the drop in steam pressure is graduated so that a change from

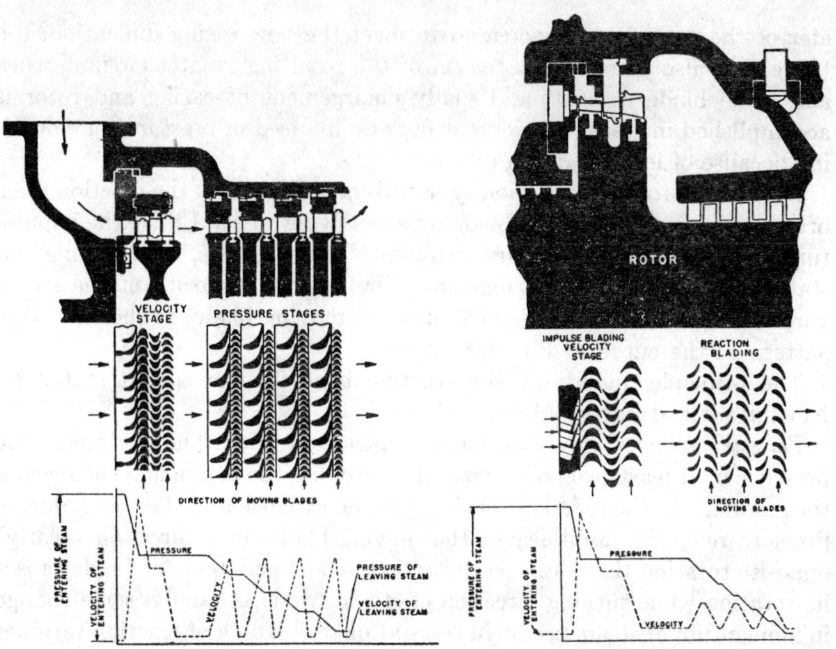

Fig. 285. (*Left*) Velocity and pressure characteristics of turbine having impulse type blading in which pressure compounding is carried on. (*Right*) Velocity and pressure characteristics of turbine having impulse and reaction stages. Note difference in velocity change from that of pure impulse in Fig. 202. (Courtesy, Mobil Oil Corporation.)

maximum to discharge pressure is effected through a series of stages. Each stage may be likened to a separate chamber having a specific pressure lower than its preceding one, but higher than its succeeding one. Each stage is comprised of a row of stationary nozzles and a row of moving blades. As the steam passes from a higher to the succeeding lower pressure stage, conversion to velocity energy occurs, the magnitude of which is commensurate with the difference in pressure between the two stages involved. The process of pressure reduction continues through a number of stages until it reaches the pressure of discharge in the last stage.

The pressure drop between stages is usually quite small. To control the pressure drop between the various stages, nozzles of the nonexpanding type are used. This type of nozzle is designed either convergent or with uniform area along its entire length. Some turbines are designed with an initial velocity stage followed by a series of pressure stages. Fig. 285 (*left*) shows an arrangement of velocity and pressure stage combination with characteristic curves.

To compensate for the increase in steam volume that accompanies pressure drop, the cross-sectional area of the casing is enlarged. While the diameter of the rotor is also increased to meet the new casing dimension, the blade area also increases by reason of the resulting greater circumference and some blade elongation. Usually enlargement of casing and rotor is accomplished in steps. Convergent nozzles are used in pressure compounding because of low expansion ratios.

Reactive Force. The reaction type turbine is driven by the reaction force of steam as it expands in the blades of a revolving wheel. Unlike the impulse turbine, a pressure drop occurs across each row of blades, both moving and stationary, of the reaction machine. The absolute velocity of the steam varies as it flows from stationary blade to moving blade, as shown by the pattern of the curve in Fig. 285 (*right*).

The principle underlying the reactive type turbine is that stated by Newton's Third Law of Motion, viz.:

For every action, there is an equal and opposite reaction. The law implies the presence of at least two bodies: one that supplies the action and the second, the reaction. In the reaction turbine it is the expansion of the steam during pressure reduction as it leaves the moving blades in a direction (nearly) opposite rotation that supplies the action. The reaction of the blade causes its rotation while turning direction of steam travel. Actually, some change in momentum of steam occurs in the stationary blading of reaction turbines so that impulse is also obtained.

Because some pressure drop occurs across each row of moving blades and nozzles (stationary blades), it becomes a gradual process from steam inlet to discharge in reaction turbines. Starting with high pressure at the inlet, the steam exerts every effort to reduce its pressure to that of discharge, which may be atmospheric or partial vacuum. Blocked by alternate rows of stationary nozzles and rotative bladed wheels from proceeding in a straight line, it must follow the turns and restrictions posed by each element. Between expansion and changing direction frequently, it expends its energy. Most of this energy is transferred to the moving elements by the impulse of turning the steam within the blades accompanied by the reaction of expansion during exit from them. As expansion progresses, so does volume. Therefore, the cross-sectional area of the turbine casing must be enlarged gradually to meet the ever-increasing volume of steam. This gives the characteristic conelike shape to reaction turbines.

To compensate for the increased volume and decreased pressure of steam along its path of travel through the turbine, some variation in blade design is necessary. Such variations include progressively greater length, tapering, and more pronounced twisting or warping, the latter to compensate for difference in speed between root and tip. Nozzles or stationary blades are made convergent because of low expansion ratios found in reaction turbines. Often blades of the last row are especially constructed to resist the erosive effect of moisture condensing from the steam at this point.

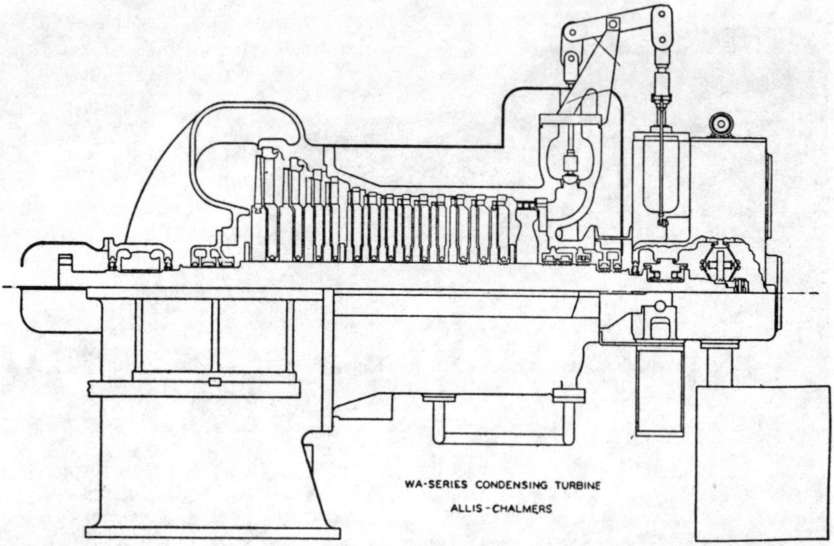

Fig. 286. Design of steam turbine having both velocity and pressure stages, showing governor linkage with steam admission valve. Condensing type. (Courtesy, Allis-Chalmers Manufacturing Company.)

The reaction turbine exerts an axial thrust which places a heavy load on the bearings installed to meet this condition. Such bearings are called thrust bearings and usually are designed to control axial thrust in either direction. This latter is particularly necessary in those uniflow turbines in which thrust reverses direction upon reaching, or exceeding, certain load levels peculiar to the individual turbine.

The use of the double-flow principle tends to eliminate thrust. This is simply a case of where steam traveling in one direction counteracts the thrust effect of steam traveling in the opposite direction while both perform work on the same shaft.

To keep the length of turbines at a minimum, impulse wheels usually are used in the first or velocity stage of an otherwise reaction type turbine.

Whereas the impulse turbine does not require sealing between stationary and moving parts, it is necessary in the reaction turbine. To prevent leakage from occurring at the tips of the moving blades, shrouds are provided. Seals are then installed between the shrouds and the casing. Seals are installed also between the ends of the nozzle diaphragms and the rotor to prevent short circuiting of steam in that area. Thus every precaution is taken to force a maximum amount of steam through the working elements and prevent its by-passing along nonproductive channels. Sealing between shaft and casing where the former passes through the latter is also necessary to prevent loss of steam pressure outward and to eliminate air leakage inward.

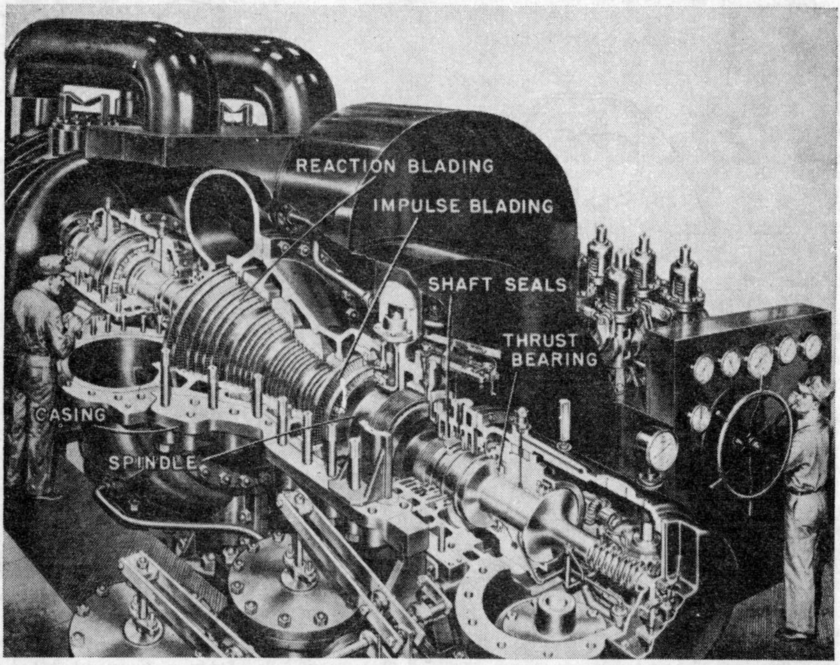

Fig. 287. Tandem-compound reaction turbine with an initial velocity-compounded impulse stage. (Courtesy, Mobil Oil Corporation.)

ELECTRIC MOTORS

Electric motors employed for refrigeration purposes, which include fans and pumps as well as compressors, may be divided into two categories: fractional-horsepower motors and motors of 1 hp and over.

Fractional-horsepower motors used to drive small compressors, centrifugal pumps, and fans include the following alternating-current single-phase types: (*a*) split-phase induction motor, (*b*) capacitor-start induction motor, (*c*) repulsion-start induction motor, (*d*) permanent-split capacitor motor, and (*e*) shaded-pole motor. Where alternating current is not available, (*f*) direct-current motors are used.

Motors of 1 hp and over used to drive larger compressors, fans, and pumps are usually alternating-current polyphase types, as follows:

(*a*) Polyphase-induction motors:
- (1) Squirrel cage—high torque,
- (2) Squirrel cage—normal torque,
- (3) Slip-ring,

(*b*) Synchronous motors.

For small units (1 to 25 hp), single-phase motors of the capacitor-start induction run and the repulsion induction types are also used. To grasp the significance of the nomenclature as given to electric motors and the operating principles of each, an understanding of fundamental electricity is required.

FUNDAMENTAL ELECTRICITY

Electricity is a form of energy. It has intensity and quantity, both of which can be measured. Like thermal energy, electrical energy measurements cannot be expressed in such terms as pounds, gallons, pounds per square inch, gallons per minute, feet, foot-pounds, etc.; however, it can be expressed in electrical terms that denote pressure, quantity, rate of flow, etc. In many ways, the electrical circuit or system is analogous to a closed hydraulic system. The terms denoting some equivalency between the two systems are given in Table 38.

TABLE 38. HYDRAULIC VS. ELECTRICAL TERMS

Data	*Hydraulic Term*	*Electrical Term*
Pressure	psi or feet-head	Voltage or electromotive force
Quantity	cu ft, lb or gal	Coulomb
Rate of flow	fps or gps	Coulombs per second or ampere
Force opposing flow	Drop in head	Resistance or ohm
Source of energy	Pump	Electrical generator (battery)
Carrier	Pipe	Conductor or wire
Work	horsepower	Watts or kilowatts

Ampere. The ampere is the measurement of electrical flow rate (quantity per unit of time) of a circuit. It corresponds to gallons per second in the hydraulic system. One ampere is equivalent to the amount of unvarying current that will deposit 0.001118 gram (g) of silver per second from a solution of pure nitrate of silver under certain specified conditions. It is indicated by the capital letter I.

Coulomb. The coulomb is the quantity measurement of electricity. It corresponds to the gallon or the cubic foot in the hydraulic system. One coulomb represents the amount of electricity that will pass through a given cross-section of wire or conductor when the current is 1 amp. Therefore, 1 coulomb per sec equals 1 amp. Because rate of flow is more important than quantity, the ampere and not the coulomb is of major significance.

Ohm. The ohm is the measure of resistance to electrical flow; it has no counterpart in the hydraulic system. However, the friction of pipes corresponds with such forces present in an electrical circuit tending to resist current flow. The ohm is the resistance offered to an unvarying current of electricity by a mercury column 14.4521 g in mass having a constant cross-section corresponding to a length of 106.3 cm when the temperature of the mercury is 0 C (32 F). A conductor has a resistance of 1 ohm when it requires 1 v to force a current of 1 amp through it. A copper wire 0.1 in. in diameter and 1000 ft long has a resistance of 1 ohm (approx). The ohm is indicated by the capital letter R.

Volt. The volt is the measure of electromotive force necessary to set up current flow in an electrical circuit. It corresponds to the pressure (psi) exerted by the pump of a hydraulic system to overcome the resistance to liquid flow or head. One volt is the amount of electromotive force required to set up a current of 1 amp through a conductor having a resistance of 1 ohm. Like pressure in a pipe, a voltage drop occurs along the length of a conductor and in an increasing amount as the length is extended. Volts are indicated by the capital letter E.

Watt, or Kilowatt. The watt is the rate of electrical work done, or the power of a circuit. It corresponds to ft-lb per min of a mechanical work system. Just as the rate of doing work or the power of a mechanical system is a function of force, distance, and time, i.e.,

$$\text{power} = \frac{\text{force} \times \text{distance}}{\text{time}} \text{ where } \frac{\text{distance}}{\text{time}} = \text{velocity},$$

electrical power is a function of volts and amperes. The relationship is shown in the equation,

$$\text{watts} = \text{volts} \times \text{amperes, or } w = EI$$

One watt is the power equivalent of thermal energy developed when a current of 1 amp passes through a resistance of 1 ohm. 1 watt = 3.412 Btu per hr.

Resistance. Resistance, being the force opposing current flow in an electrical circuit, varies according to conditions along the conductor. The four major factors on which the magnitude of resistance depends are:

1. CONDUCTING MATERIAL—metals offer less resistance than other materials. Silver and copper are excellent conductors, i.e., they offer less resistance to current flow than other engineering metals of equivalent length and cross-sectional area.

2. LENGTH OF CONDUCTOR—resistance increases as length increases.

3. CROSS-SECTIONAL AREA OF CONDUCTOR—resistance decreases with any increase in cross-sectional area.

4. TEMPERATURE OF CONDUCTOR—resistance increases as temperature increases. Generally, however, carbon has negative resistance.

The Electrical Circuit

The electrical circuit is similar in many respects to a pipe line to which water is fed by a pump at one end. As long as the pump operates and the pipe is open, say to the atmosphere, water will flow. If a valve were placed

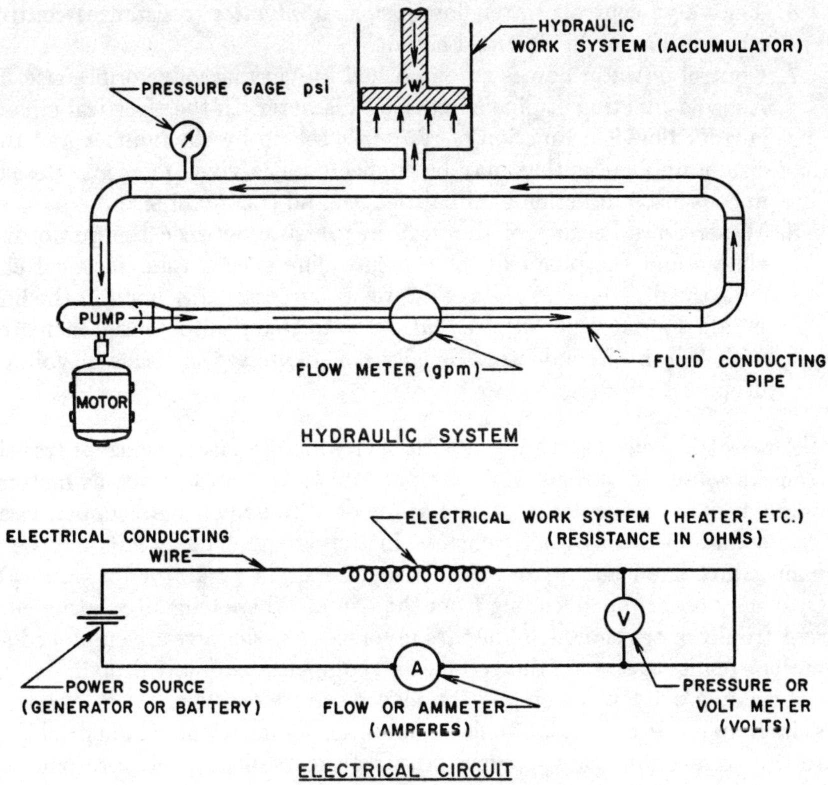

Fig. 288. Analogy of hydraulic system to electrical circuit.

at the open end of the pipe and partially closed, flow would still occur. The added resistance caused by the partially closed valve would increase pressure but reduce the amount of flow. Flow would cease entirely by complete closure of the valve. The pressure would remain if, by some method of by-pass control, the pump were allowed to continue in operation.

Applying the actions prevalent in the pipe line to an electrical circuit, the following analogies are found:

1. The pump acts as the *generator* to supply power.
2. The pipe acts as the *conductor* to distribute power.
3. Water acts as the *medium* to transmit power just as electricity in the electrical circuit.
4. Rate of water flow or water current is the substitute for *amperes* or electric current.
5. Pressure served to force flow of water as *voltage* (electromotive force) forces *current* flow.

6. The valve controls water flow just as *appliances* (resistance) control current flow in an electrical circuit.
7. Control of water flow is accomplished by varying valve orifice size. It is also a function of pipe length and diameter. In the electrical circuit current flow is a function of *resistance* set up by the number and the size of appliances that may be connected at a given moment. Resistance is also a function of circuit length and conductor size.
8. Water flows because of the drop in pressure between the pump discharge and the pipe exit, or pressure difference. In an electrical circuit, the difference in voltage between source and any point in the line is called a difference of *potential*. Drop in electrical potential increases with length of conductor. Usually it is expressed in terms of voltage drop.

An electrical circuit, then, is a path over which an electric current travels from its source to various work components or appliances, such as motors, lights, heaters, magnets, etc. The path is usually two or more copper wires that communicate with all elements in the circuit. The circuit may be a main trunk line leading from the source, usually a generator (or battery). Or it may be a branch leading from the trunk. Where long alternating-current trunks or transmission lines are involved, *transformers* are employed at various points to step up the voltage and compensate for any drop in potential occurring along the circuit. Transformers are employed also to reduce voltage to levels more adaptable to domestic, commercial, or industrial use. In this connection, the two types of transformers mentioned correspond to booster pumps or compressors and pressure-reducing valves, respectively, used in pipe lines transmitting fluids over long distances.

An *open* circuit is one in which a switch has broken connections to stop the flow of current just as a valve stops the flow of water when shut off. The circuit may still be alive between the switch and the source, which means that voltage is present just as pressure exists between the pump and the shut-off valve of the water system. A *closed* circuit is one in which the line forms an unbroken loop from the source to the load and back to the source. To complete the analogy of the above-described water system and an electrical circuit, it must be assumed that the water leaving the open end of the pipe enters a reservoir under pressure, from which it is again drawn by the pump during its suction stroke.

Electrical Measurements

Voltage is measured by a high-resistance instrument called a voltmeter. When determining voltage, the terminals of the voltmeter must be placed in parallel with the circuit, commonly expressed as *across the line*. Amperage is measured by a low-resistance instrument or ammeter placed *in-line* or in series with the circuit. The practical method of determining circuit resistance is by calculation based on measured voltage and amperage according to Ohm's law.

Ohm's Law

According to Ohm's law, a definite relation exists in an electrical circuit between electromotive force (E), current (I), and resistance (R). This relation may be stated as follows for a circuit having steady current:

1. The current is directly proportional to the total electromotive force of the circuit.

2. The current is inversely proportional to the total resistance of the circuit.

Ohm's law is expressed by the equation:

$$\text{Current} = \frac{\text{electromotive force}}{\text{resistance}}$$

or

$$I = \frac{E}{R}$$

and

$$R = \frac{E}{I}$$

and

$$E = IR$$

EXAMPLE:

1. What is the resistance of a simple circuit having a current of 12 amp when the voltage across the resistance is 120 v?

$$R = \frac{E}{I}$$

$$R = \frac{120}{12}$$

$$R = 10 \text{ ohms}$$

2. What electromotive force is required to force a current of 40 amp through a simple circuit having a resistance of 11 ohms?

$$E = IR = 40 \times 11 = 440 \text{ v}$$

3. Determine the number of amperes that can be put through a simple 110-v circuit having a resistance of 22 ohms.

$$I = \frac{E}{R} = \frac{110}{22} = 5 \text{ amp}$$

Basic Types of Circuits. The above examples apply to circuits in general. However, the values of all three electrical variables must be based on the

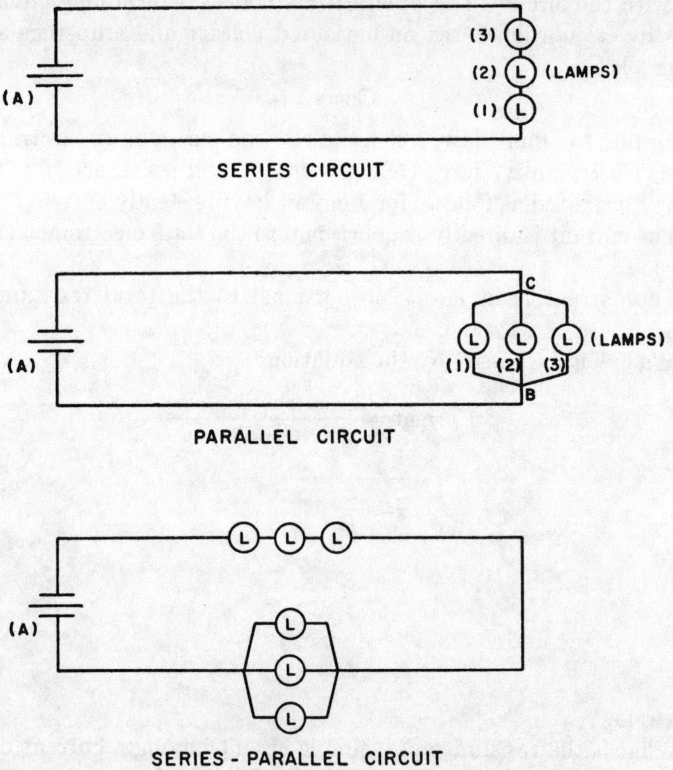

Fig. 289. Electrical circuits showing series, parallel, and series-parallel combinations.

factors existing in the separate parts of more complex circuits. The following rules apply to the three basic types of circuits, namely:

Series circuit—a circuit in which the various resistances are connected in tandem, such as that shown in Fig. 289 (*top*).

Parallel circuit—a circuit in which the various resistances are connected in multiple or side by side, as shown in Fig. 289 (*center*).

Series-parallel circuit—a circuit which combines series and parallel resistances, as shown in Fig. 289 (*bottom*).

Rules for Series Circuits: Voltage. Just as water in a pipe line reduces in pressure as it flows away from the pump through a series of various resistances, so does the pressure of electricity reduce in voltage as current passes through resistances in a series circuit. Therefore, *the voltage in a series circuit is equal to the sum of the voltages across the separate resistances.*

Example: Assume an emf of 210 v generated (A) and a current of 3 amp in the circuit in Fig. 289 (*top*). The voltage across the separate resistances (1), (2), and (3), according to Ohm's law, when R-5, 25 and 40 respectively would be

$$E = IR$$

(1) $E = 3 \times 5 = 15$ v

(2) $E = 3 \times 25 = 75$ v

(3) $E = 3 \times 40 = 120$ v

$$E_c = 15 + 75 + 120 = 210 \text{ v, total}$$

CURRENT. The amount of water leaving the pipe line is equal to that put into it by the pump so that the current is constant at every point along the circuit. This remains true for an electrical circuit, i.e., the amount of current in all parts is constant. *The number of amperes flowing through any one resistance is the same for all other resistances in a series circuit.*

EXAMPLE: Referring again to Fig. 289 (*top*), the circuit current when the emf is 210 v is

$$I = \frac{E}{R}$$

$$I_c = \frac{210}{5 + 25 + 40}$$

$$I_c = 3 \text{ amp}$$

RESISTANCE. The resistance of a series circuit is equal to the sum of the separate resistances. An example of this was shown in the previous example.

Rules for Parallel Circuits: VOLTAGE. Just as the pressure in each tube of a heat exchanger is equal to the pressure in the header and therefore equal to each other, the electrical pressure in each branch of a parallel circuit is the same and equal to that in the circuit. Therefore, the voltage in a parallel circuit is constant for each separate resistance. In Fig. 289 (*center*), the voltage across *BC* is the same regardless of which branch—(1), (2), or (3)—it is measured.

CURRENT. It is obvious that the liquid in the heat exchanger header represents the full current of the system. It is also true that each tube carries only its proportionate part of that current. Therefore, the current in the header is equal to the sum of the currents in the separate tubes. This rule also applies to the parallel electrical circuit—i.e., the current in the circuit equals the sum of the currents flowing through the separate resistances.

EXAMPLE: Referring to Fig. 289 (*center*), and assuming a generated emf of 120 v, the current in each resistance, then, is

$$I = \frac{E}{R}$$

(1) $I = \frac{120}{8} = 15$ amp

(2) $I = \frac{120}{6} = 20$ amp

(3) $I = \frac{120}{2} = 60$ amp

RESISTANCE. The physical law that flow resistance decreases as the number of parallel escape paths increases holds true for electrical circuits. *The total resistance of a parallel circuit is less than that of the lowest resistance offered by any one of its paths.*

EXAMPLE: The resistance across BC in Fig. 289 (*center*), when generated emf equals 120 v and total current is $15 + 20 + 60 = 95$ amp, as determined in previous example, is as follows:

$$R = \frac{E}{I} = \frac{120}{95} = 1.26 \text{ ohms}$$

It is interesting to note that the reciprocal of the sum of reciprocals of the separate resistances also gives a similar result, viz.:

$$R = \frac{1}{\frac{1}{2} + \frac{1}{6} + \frac{1}{8}} = \frac{1}{0.792} = 1.26$$

The reciprocal in the denominator in each case is the number of amperes per volt or *conductance* of each resistance. Thus the rule can be stated that the resistance in a parallel circuit is equal to the reciprocal of the sum of the separate conductances where conductance equals the reciprocal of the separate resistances (r), or

$$\text{total conductance} = \frac{1}{r_1} + \frac{1}{r_2} + \frac{1}{r_3} + \text{, etc.}$$

Kirchhoff's Laws

Ohm's law is covered more extensively by Kirchhoff's laws as applying to various electrical states (ohms, amperes, and voltage) distributed throughout a complete and possibly a more complex circuit. Kirchhoff's laws state in effect:

First law: *the algebraic sum of the currents meeting at a point in a branch circuit equals zero.*

Second law: *the algebraic sum of the separate potential drops around a complete circuit equals zero.*

For an example of Kirchhoff's laws, refer to the circuit in Fig. 290. To

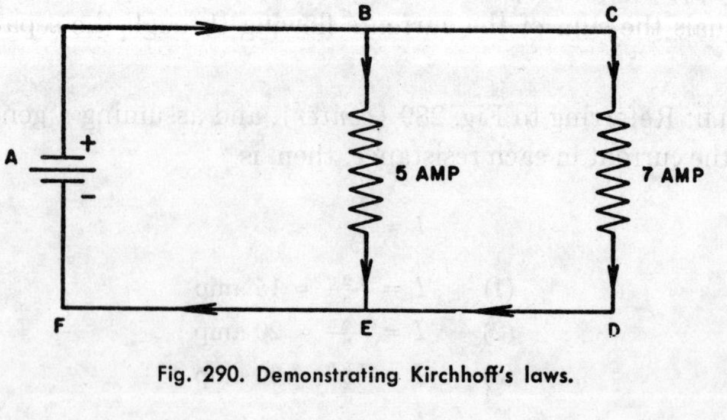

Fig. 290. Demonstrating Kirchhoff's laws.

satisfy the first law, consider point B. There are 5 amp flowing away from B through the path BE, and 7 amp flowing away from B through the path $BCDE$, or a total of 12 amp flowing away from B. If this is true, there must be 12 amp flowing toward B from the source along AB. Also, as there are $5 + 7 = 12$ amp flowing toward E, there must be 12 amp flowing away from it through EF. Thus,

$$\text{current in } BE + CD = \text{current in } AB$$

and

$$(BE + CD) - AB = 0$$

$$12 - 12 = 0$$

also

$$\text{current in } BE + CD = \text{current in } EF$$

and

$$(BE + CD) - EF = 0$$

$$12 - 12 = 0$$

To satisfy the second law, consider the electromotive force of the source (generator) as being E_t v. Consider also that IR drops along the separate paths as being E_1 along AB, E_2 along BC, E_3 along CD, E_4 along DE, E_5 along EF, and E_6 along FA (source IR drop). Therefore, the electromotive force of the source must equal the sum of the separate IR drops or

$$E_t = E_1 + E_2 + E_3 + E_4 + E_5 + E_6$$

Also consider that the IR drop across BE equals the sum of the IR drops $E_2 + E_3 + E_4$. This is obvious because (1) the path $BCDE$ contains no source of electromotive force except that which passes through point B, (2) the path BE runs counter to the circuit $BCDE$, and (3) an unbalance of emf at E otherwise would be supposed, an impractical supposition under the circumstances here. It can be said, then, that

$$E_t - (E_1 + E_2 + E_3 + E_4 + E_5 + E_6) = 0$$

Power

Power is the rate of doing work. In a mechanical system, power is defined as the number of foot-pounds of work done per minute. In a mechanical system, foot-pounds is determined by multiplying the force by distance. Power of an electrical system is based also on similar factors, i.e., electrical power is a function of electrical pressure or voltage and electrical flow rate or amperage. As these two terms of measurement embody all the necessary values of force, quantity, and time, their values are all that are required to determine the power of a circuit—i.e., watts = volts × amperes, or $W = EI$. Because both the voltage and the amperage are functions of resistance, the value of the latter may also be employed in power computations, viz.:

$$W = EI$$

as

$$E = IR$$

then

$$W = I \times (IR) = I^2R$$

as

$$I = \frac{E}{R}$$

then

$$W = \left(\frac{E}{R}\right) \times E = \frac{E^2}{R}$$

so that the power of a circuit in watts may be determined when any two unit values are known.

For large circuits, power is rated in units equivalent to 1000 watts, or kilowatts.

$$1 \text{ kw} = 1000 \text{ watts}$$

therefore

$$\text{kw} = \frac{\text{watts}}{1000}$$

$$1 \text{ kw} = 1.34 \text{ hp (equivalent)}$$

$$1 \text{ hp} = 0.746 \text{ kw or } 746 \text{ watts}$$

By multiplying the number of kilowatts consumed in any electrical system by the hours in operation, the amount of work can be determined.

EXAMPLE: If 50 kw were consumed steadily for a period of 6 hr, then

$$\text{kw} \times \text{hr} = \text{kwhr}$$

$$50 \times 6 = 300 \text{ kwhr}$$

THE MAGNETIC CIRCUIT

The operation of electric motors and generators is based on magnetic phenomena. The principles of a magnetic circuit are analogous to those of an electrical circuit. Therefore, if the reasoning applied to the latter is applied similarly to the former, no difficulty should be experienced in understanding magnetic current. In many instances, it is the change in nomenclature that tends toward some confusion.

An introduction to magnetic terms and their analogy to the electrical circuit is found in Table 39.

Magnetic Field. When an electric current passes through a conductor, a magnetic field is produced. This magnetic field is made up of an energy mass that may be likened to numerous concentric rings revolving around the conductor. When this occurs, the conductor is said to be energized. The energy mass or rings tend to grow by expanding either outward or axially along the conductor. (See Fig. 291.)

TABLE 39. MAGNETIC CIRCUIT TERMS AND THEIR ANALOGY TO THE ELECTRICAL CIRCUIT

Electrical Circuit				*Magnetic Circuit*				
Variable	Measure	Symbol	Ohm's Law	Variable	Measure (cgs)*	Measure (English Rational)	Symbol	Ohm's Law
Electromotive force	Volt	E	$E = IR$	Magnetomotive force	Gilbert	Ampere-turn (N.I.)	F	$F = \phi \mathcal{R}$
Current	Ampere	I	$I = \frac{E}{R}$	Flux	Maxwell	Line of force	ϕ	$\phi = \frac{F}{\mathcal{R}}$
Resistance	Ohm	R	$R = \frac{E}{I}$	Reluctance	oersted (cgs)	oersted (Eng. Rat.)	$\mathcal{R}$	$\mathcal{R} = \frac{F}{\phi}$
				Magnetizing force	gilberts per cm	ampere-turns/in.	H	
Ampere density	amp/sq in.			Flux density	gauss	Lines per sq cm	B	$B = \frac{\phi}{A}$†
Conduction	Specific resistance	K		Permeability	Lines per sq cm set up by 1 gilbert	Lines per sq in. set up by 1 amp turn	μ	

* cgs and English Rational system equivalents.
† (A = area in sq cm)

$$1 \text{ gilbert} = \frac{1 \text{ amp turn}}{1.26}$$
$$1 \text{ maxwell} = 1 \text{ line of force}$$
$$1 \text{ oersted (cgs)} = \frac{1 \text{ oersted (Eng. Rat.)}}{1.26}$$
$$1 \text{ gauss} = 1 \text{ line per sq cm} = \frac{1 \text{ line per sq in.}}{0.155}$$
$$1 \text{ gilbert per cm} = 2.02 \text{ amp turns per inch}$$
$$\mu \text{ (cgs)} = 3.19 \times \mu \text{ (Eng. Rat.)}$$

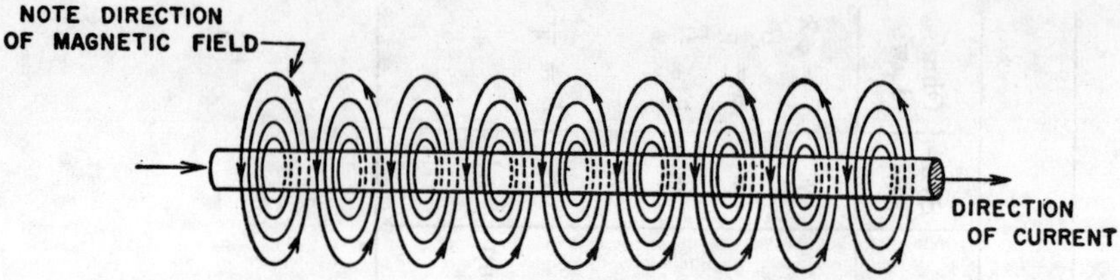

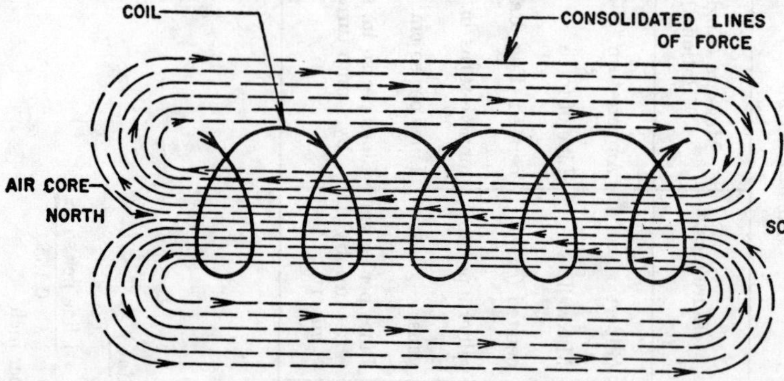

Fig. 291. (*Top*) Magnetic field around a conductor through which electric current flows.

Fig. 292. (*Bottom*) Solenoid. Showing consolidation of magnetic lines due to coiling about an air core.

The direction in which the rings revolve is always clockwise relative to the flow of electric current as observed from its source to destination. It can be determined readily by grasping the wire by the right hand with the fingers encircling the wire while the thumb is held parallel with the wire and pointing in the direction of current flow. The direction of the encircling fingers is the direction in which magnetic rings revolve around the wire. This method of determining direction of a magnetic field is known as the right-hand rule.

Winding a wire about an axis to form a coil produces a solenoid. An electric current passing through a solenoid tends to set up a magnetic field around each turn of wire. If the turns are close together, the individual fields will unite to create a consolidated force in the form of lines, as shown in Fig. 292. When energized, the solenoid takes on the characteristics of a bar magnet having a north and a south pole. When exposed to another similar coil, end to end, both coils obey the law of magnetism, i.e., like poles repel. It should be noted that the lines enter the south pole and exit at the north pole. The right-hand rule also holds true for solenoids. When the right hand grasps the coil so that the fingers are parallel with the windings in the direction of current flow, the thumb points in the direction of the magnetic field.

If the coil is wound around an iron core, the magnetic lines will follow the latter in the same manner as they do the air core shown in Fig. 292. An energized coil wound around an iron ring will set up a magnetic current in the ring as demonstrated by the broken lines in Fig. 293. If the ring

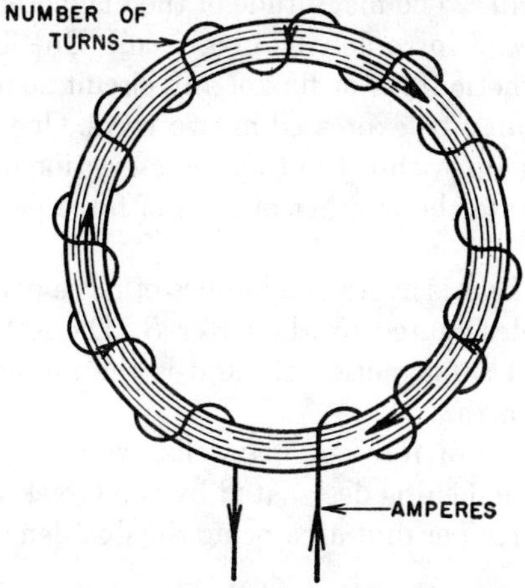

Fig. 293. Magnetic current set up in an iron ring.

formed an unbroken circle, it would have no north or south pole and would be an impractical machine. By removing a small segment to form an *air gap,* it becomes a practical apparatus.

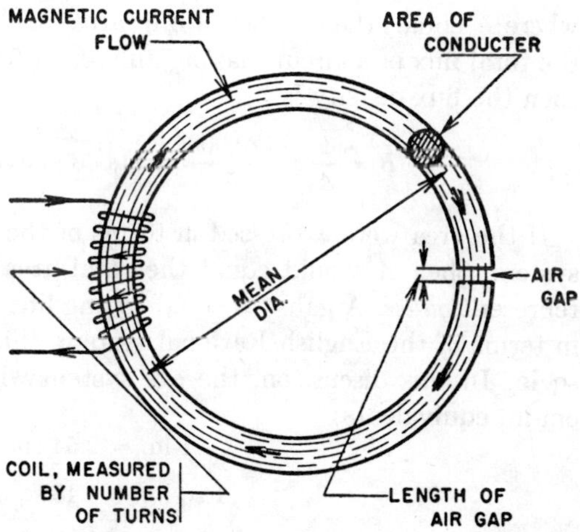

Fig. 294. Magnetic current set up in an iron ring with an air gap.

Magnetic Current. It was stated that a magnetic field is composed of expanding rings of energy emanating from and revolving around an energized conductor. The rings consolidate to become lines of energy when the conductor or the wire is wound about an axis to form a coil or solenoid. The direction of the magnetic lines through the core is toward the north pole. The strength of the magnetic field depends on the number of such lines set up in the core or ring.

Magnetic lines, represented by the broken lines following the core, are called *lines of force,* referred to as *flux.* The magnitude of the magnetic field is measured by the number of lines of force flowing in a circuit. One line of force is called a *maxwell.* The magnetic force or flux of any circuit as measured in lines of force or maxwells may be expressed in two ways. One is by indicating the number of lines set up within the total cross-sectional area of the core. The second is to indicate the number of lines of force per unit area of the core.

When the strength of magnetic field is expressed in lines of force per unit area, it is called *flux density* and designated by the letter B. The letter B, then, represents the flux density of a magnetic field and is the number of lines of force or maxwells per unit area.

Usually the total number of lines of force or maxwells, which may be described as *total flux,* in a magnetic field is designated by the Greek letter phi or ϕ. The number of lines of force per unit area being the flux density of a given circuit, then

$$\text{Flux density} = \frac{\text{total flux}}{\text{area}}$$

or

$$B = \frac{\phi}{A}$$

where A equals the number of unit areas such as sq in., sq cm, etc. Thus, if the total flux of a circuit having an area of 5 sq in. is 120,000 lines of force, then the flux density is

$$B = \frac{\phi}{A} = \frac{120{,}000}{5} = 24{,}000 \text{ lines of force per sq in.}$$

If the area were expressed in terms of the centimeter gram second (cgs) system, then A would equal the total area in centimeters and B would represent *gauss.* A gauss is equal to one line of force per sq cm. Expressed in terms of the English Rational system, then B would represent lines per sq in. In this discussion, the cgs system will be used. The following are cm-in. equivalents:

$$1 \text{ in.} = 2.54 \text{ cm}$$

$$1 \text{ sq in.} = 6.45 \text{ sq cm}$$

$$1 \text{ cm} = 0.39 \text{ in.}$$

$$1 \text{ sq cm} = 0.155 \text{ sq in.}$$

EXAMPLE: What is the flux density of a circuit having a cross-sectional area of 7.1 sq cm if a total of 2500 lines is set up in the circuit?

$$\text{Flux density} = \frac{\text{total number of lines in circuit}}{\text{area of circuit}}$$

$$B = \frac{\phi}{A}$$

$$B = \frac{2500}{7.1}$$

$$B = 352.1 \text{ lines per sq cm}$$

or

$$B = 352.1 \text{ gauss}$$

Magnetomotive Force. The magnetic pressure or magnetomotive force of a magnetic circuit is analogous to electromotive force of an electrical circuit. It is the force that causes magnetic current or flux to flow in a circuit. The magnitude of magnetomotive force created depends on (1) the number of turns of wire that form the solenoid (N) and (2) the amount of electric current in amperes flowing in the wire (I). The product of these two factors or amperes × number of turns in the coil is the measure of magnetomotive force in the magnetic circuit. In terms of the English Rational system, the product is expressed as *ampere-turns* of magnetomotive force. Ampere-turns usually are indicated by the letters NI. The measure of magnetomotive force as related to the cgs system is expressed in terms of *gilberts* and indicated by the letter F. Therefore, F equals magnetomotive force in number of gilberts. The relationship of gilbert to ampere-turn may be expressed by the ratio $\frac{1}{4\pi}$, or 1 gilbert equals $0.796NI$. Expressing the relationship another way, $1NI$ equals $\frac{1}{0.796}$, or 1.26 gilberts.

EXAMPLE: What is the magnetomotive force of a circuit having 520 turns of wire through which 4 amp are impressed?

$$NI = 520 \times 4 = 2080 \text{ ampere-turns}$$

$$F = 1.26 \times 2080 = 2620.8 \text{ gilberts}$$

Magnetizing Force. The magnetic circuit is rated on its magnetizing force. Magnetizing force is indicated by the letter H and is based on magnetomotive force per unit length of a given circuit. It is obtained by dividing the magnetomotive force by the length of the circuit, or

$$H = \frac{1.26NI}{L}$$

$$H = \frac{F}{L}$$

$$H = \text{gilberts per cm or ampere-turns per inch}$$

Hence the circuit in the previous example would have a magnetizing force, assuming the length of the circuit to be 25.12 cm or 9.9 in.

$$NI = 2080 \text{ ampere-turns}$$

also

$$F = 2620.8 \text{ gilberts}$$

as

$$H = \frac{F}{L}$$

then

$$H = \frac{2080}{9.9} = 210 \text{ ampere-turns per in.}$$

or

$$H = \frac{2620.8}{25.12} = 104.3 \text{ gilberts per cm}$$

The magnetic circuit is said to have a *field strength* of 104.3 gilberts per cm.

Magnetic Resistance. As in an electrical circuit, there is resistance to magnetic current flow in a magnetic circuit. However, in the magnetic circuit current-flow resistance is called *reluctance*. It is often expressed as the *reluctivity* of a magnetic circuit. The reluctance of a magnetic current to flow is governed by (1) the nature of the material used in the circuit and (2) the length of the circuit. Thus the material used should be selected with the same care as would be given the selection of a good conductor, say copper wire, for an electrical circuit.

As pertains to the magnetic circuit, materials may be classified as being magnetic or nonmagnetic just as materials may be classified as good or bad conductors of electric current. In this regard, (1) iron and steel are good conductors of magnetic current and are classed as magnetic materials, and (2) most all other materials, including air, are considered nonmagnetic.

Even with magnetic materials, current-carrying capacity is not constant but diminishes after reaching a certain degree of magnetism, a phenomenon to be discussed later. For nonmagnetic materials, reluctance may be assumed to be practically constant.

The measure of electrical resistance is expressed in terms of ohms. Similarly, the measure of magnetic reluctance is expressed in terms of *oersteds* and indicated by the script letter $\mathcal{R}$.

For magnetic circuits in which the core is made up of air, the reluctance is considered to be unity, i.e., $\mathcal{R} = 1$ oersted per cu cm of air. All other nonmagnetic materials are said to have a reluctance equivalent to air, or $\mathcal{R} = 1$ for any material except iron and steel.

The reluctance of any nonmagnetic circuits, then, is determined by the simple equation

$$\mathcal{R} = \frac{1 \times \text{length of circuit}}{\text{cross-sectional area of circuit}}$$

or

$$\mathcal{R} = \frac{1 \times L}{A}$$

where $\mathcal{R}$ = reluctance in oersteds
L = length of circuit in cm
A = area of circuit in sq cm
and 1 = reluctance of 1 cu cm of circuit in oersteds

To convert to English Rational system values, multiply oersteds (cgs) by 0.796 or divide by 1.26.

EXAMPLE: Determine the reluctance of a circuit made up, say, of a wooden ring, 25.12 cm long (mean circumference), having a cross-sectional area of 7.1 sq cm.

$$\mathcal{R} = \frac{1 \times L}{A} = \frac{1 \times 25.12}{7.1}$$

$$\mathcal{R} = 3.53 \text{ oersteds (cgs)}$$

or

$$\mathcal{R} = 0.796 \times 3.53 = 2.81 \text{ oersteds (English Rational system)}$$

Ohm's Law Applied to Magnetic Circuits

As in the electrical circuit, the same relationship exists in the magnetic circuit as pertains to pressure, current, and resistance. This relationship for the electrical circuit was stated to be:

$$\text{Current} = \frac{\text{electromotive force}}{\text{resistance}}$$

or

$$\text{amperes} = \frac{\text{volts}}{\text{ohms}}$$

or

$$I = \frac{E}{R}$$

Similarly, in the magnetic circuit

$$\text{Magnetic current} = \frac{\text{magnetomotive force}}{\text{reluctance}}$$

or

$$\text{flux} = \frac{\text{gilberts}}{\text{oersteds}}$$

or

$$\phi = \frac{F}{\mathcal{R}}$$

therefore

$$F = \phi\mathcal{R}$$

and

$$\mathcal{R} = \frac{F}{\phi}$$

so that *flux* is directly proportional to the total magnetomotive force of the circuit and inversely proportional to the total resistance of the circuit.

EXAMPLE: What flux is developed in a circuit having resistance of 1.2 oersteds and a total magnetomotive force of 1600 gilberts?

$$\phi = \frac{F}{\mathcal{R}} = \frac{1600}{1.2}$$

$$\phi = 1333.3 \text{ lines or maxwells}$$

EXAMPLE: Determine the total magnetomotive force required in a circuit of 0.75 oersted in order to develop a flux of 1200 maxwells.

$$F = \phi\mathcal{R} = 1200 \times 0.75$$

$$F = 900 \text{ gilberts}$$

EXAMPLE: For a circuit having a total magnetomotive force of 4200 gilberts and a flux of 1800 maxwells, compute the reluctance.

$$\mathcal{R} = \frac{F}{\phi} = \frac{4200}{1800}$$

$$\mathcal{R} = 2.33 \text{ oersteds}$$

The rules for parallel and series magnetic circuits are similar to those for electrical circuits, i.e., for *series circuits:*

(*a*) Magnetic flux is the same in all parts of the circuit. However, flux density will vary in the same circuit having various cross-sectional areas,

(*b*) Magnetomotive force is equal to the sum of the separate magnetomotive forces required by the separate parts (branches) of the circuit,

(*c*) The reluctance of a circuit is equal to the sum of the separate reluctances offered by the separate parts.

And for *parallel circuits:*

(*a*) The electromotive force in each branch of the circuit is the same,

(*b*) The magnetic flux in a circuit is equal to the sum of the separate flux of the separate resistances,

(*c*) The total reluctance is less than that of the lowest reluctance offered by any one of its parts.

Air Gap. The closed circuit in Fig. 293 was said to be impractical. By removing a segment of the ring, an area would be provided through which magnetic current could leave the material and set up an external field within the resulting gap (Fig. 294). This field would be set up by lines of force passing through the air gap from the now existing north pole to the now existing south pole. This air gap offers more resistance to magnetic flow than would any magnetic material in the ring body. Hence, an air gap increases the reluctance of a magnetic circuit.

Employing the equation for nonmagnetic materials, it is found that the reluctance offered by an air gap is

$$\mathcal{R} = \frac{L}{A}$$

so that an air gap formed by removing a 0.5-cm segment from a ring having a cross-sectional area of 32 sq cm will have a reluctance of

$$\mathcal{R} = \frac{0.5}{32} = 0.0156 \text{ oersted}$$

As this reluctance represents only part of the total reluctance of the system, it is necessary to determine the remaining unknown factors by the process of elimination to satisfy the reluctance offered by the ring material.

Permeability. The reluctance of nonmagnetic materials was said to be constant and assumed to have a value of 1. This is true according to Ohm's law and definition of terms, i.e., 1 gilbert is required to produce 1 maxwell in 1 cu cm of air, and therefore

$$\mathcal{R} = \frac{F}{\phi} = \frac{1}{1} = 1$$

The reluctance of magnetic materials is not nearly that value during the initial stages of the magnetizing process. As more lines of force are produced in iron and steel, the value of $\mathcal{R}$ tends to approach unity. Thus it may be said that iron and steel normally have less reluctance than nonmagnetic materials and therefore possess greater permeability. Permeability, as related to magnetic circuits, is the capacity of a material to develop a magnetizing force, i.e., readily passes a high magnetic current. Iron and steel possess this quality and are therefore magnetic.

Permeability, then, is directly contrary to reluctance, so that the latter becomes a reciprocal of the former, or

$$\text{permeability} = \frac{1}{\mathcal{R}}$$

The reluctivity being the ratio

$$\mathcal{R} = \frac{\text{gilberts per cu cm}}{\text{lines per cu cm}}$$

then permeability being the reciprocal of $\mathcal{R}$ must equal the inverse ratio

$$\frac{\text{lines per cu cm}}{\text{gilberts per cu cm}}$$

Represented by the letter μ, permeability may be expressed by the equation

$$\mu = \frac{B}{H}$$

It is obvious that reluctivity of a magnetic circuit is analogous to resistivity to an electrical or heat transfer circuit where

$$r = \frac{1}{\text{conductivity}}$$

Permeability being analogous to conductivity, then

$$\mu = \frac{1}{\mathcal{R}}$$

as shown above. Because reluctance varies directly as length and inversely as cross-sectional area, then $\mathcal{R}$ for magnetic materials may be expressed by the equation

$$\mathcal{R} = \frac{L}{A\mu}$$

It was stated that the reluctance of iron and steel approach unity with an increase in flux. It follows that they become less permeable as they become more magnetized. Therefore, permeability of iron and steel is not constant but diminishes as magnetic current increases. The relation of permeability to flux density may be described as follows:

1st phase: during the initial stage of magnetization the ratio of gilberts to gauss is low—i.e., it requires only a relatively small number of gilberts to produce a moderate flux.

2nd phase: as the flux density increases from moderate to a relatively higher level, the ratio of gilberts to gauss increases. During this stage, the relative rate of increase in gilberts is more rapid than the relative rate of flux increase.

3rd phase: a point is reached where the ratio of gilberts to lines produced is very high. At this point it requires an extremely large number of gilberts to cause a relatively small increase in flux.

Where the transition from low gilbert-flux ratio to extremely high gilbert-flux ratio occurs, the point of magnetic saturation is reached. The above bears out the rule that μ is variable and its value depends on the degree of magnetization and also on the kind and the purity of iron and steel.

To establish the relation of B to H, it is necessary to compile test data for individual specimens of each type and kind of magnetic material. The most convenient method of publishing such data is in the form of curves called magnetization or BH curves. As shown in Fig. 295, such curves are constructed by plotting flux density in gauss (B) against magnetizing force in gilberts per cm (H).

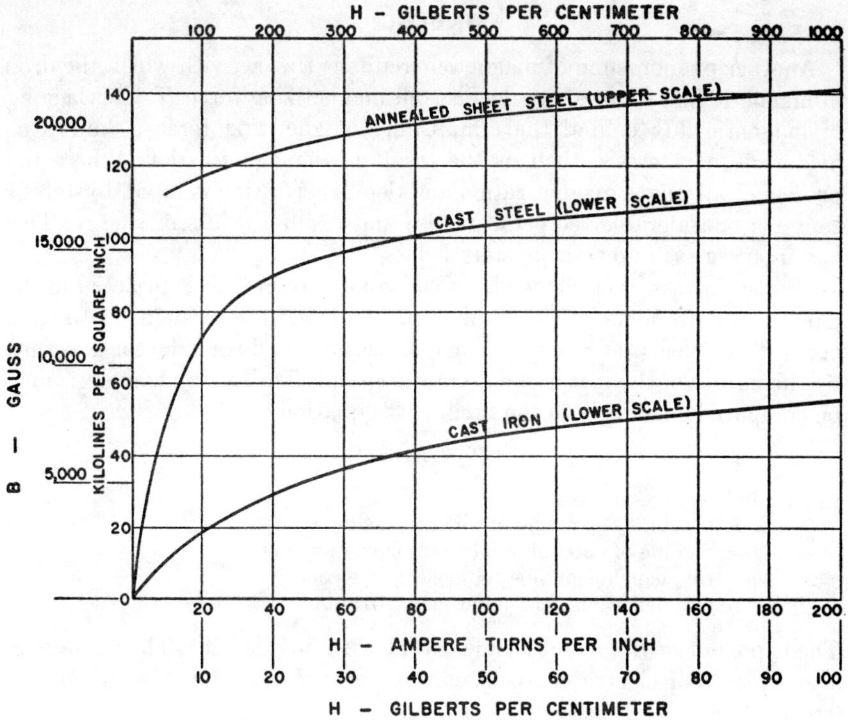

Fig. 295. Magnetization or *BH* curve for soft steel casting.

EXAMPLE: Assume the ring in Fig. 293 to be a soft steel casting 25.12 cm long and 7.1 sq cm in cross-sectional area. Determine the ampere-turns required to set up a flux of 113,600 lines.

$$B = \frac{\phi}{A} = \frac{113,600}{7.1} = 16,000 \text{ gauss (lines per sq cm)}$$

2nd step: Determine H (gilberts per cm).

From the *BH* curve it is found that it requires 40 gilberts per cm to produce 16,000 gauss.

3rd step: Determine F (gilberts in circuit).

$$F = L \times H = 25.12 \text{ cm} \times 40 \text{ gilberts per cm}$$

$$F = 1004.8 \text{ gilberts}$$

4th step: Determine NI.

$$NI = \frac{F}{1.26} = \frac{1004.8}{1.26}$$

$$NI = 797 \text{ ampere-turns (approx)}$$

Hysteresis

Another phenomenon of magnetic circuits is the lag with which the drop in magnetic flux follows any decrease in magnetizing force H. This lagging of magnetic flux behind the diminishing magnetizing force is known as hysteresis. Hysteresis, then, is the result of demagnetization. Where the process of alternate magnetization and demagnetization is rapid, hysteresis causes a considerable temperature rise and results in loss of energy. This loss in energy is known as hysteresis loss.

Hysteresis loss, then, depends on the volume of magnetic material in the circuit, the frequency with which magnetization and demagnetization occur, flux density at maximum magnetization, and some factor governed by the nature of the material used in the circuit. The loss by hysteresis may be computed according to the Steinmetz equation,

$$P_h = KvfB^{1.6}$$

where P_h = hysteresis loss in watts
K = constant depending on the core material
v = volume of core material in cubic centimeters
f = frequency or number of cycles per second
B = flux density at maximum magnetization

The exponent of B varies to some extent with flux density. The value of K also varies. For electric motor armature cores, K may be given a value of 0.003.

Magnetic Circuit as Applied to Electric Motors and Generators

Induced Electromotive Force. Consider the gap in the ring in Fig. 293 to be wide enough to allow a wire of any reasonable size to pass through it. Assume the ring to be energized so as to force a magnetic current to flow through the air gap from the north to the south pole. If a wire is placed at a right angle to the ring and moved up and down through the gap so as to cut the lines of force, an induced current will be set up in the wire. The direction of current flow in the wire will depend on whether it is moving upward or downward across the lines of force (Fig. 296). The Fleming three-finger rule governing the direction of an induced current in a wire moving across a magnetic field is as follows:

1. Hold the right hand outward so that the index finger points forward, the thumb points upward, and the middle finger points to the left and at a right angle to the index finger.

2. If the thumb is pointing in the direction in which the wire is moved, and the index finger points in the direction of the magnetic current, then the middle finger will be pointing in the direction of the induced current in the wire.

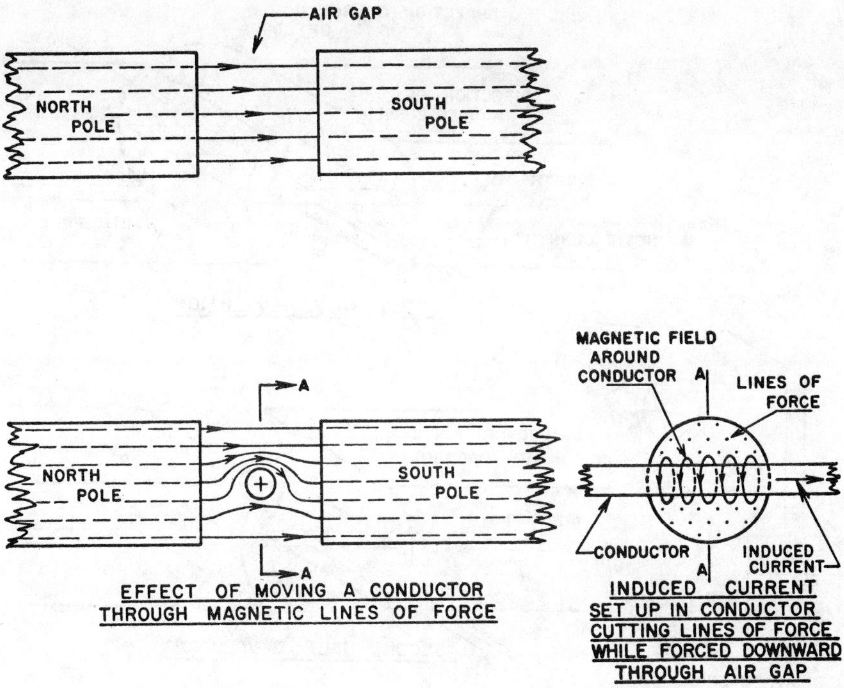

Fig. 296. Development of an induced current in a conductor passing through at right angles to a magnetic field.

The voltage developed in the wire is a function of the flux magnitude and the speed with which the wire cuts across the field. *One volt will be produced for each 100,000,000 lines of force cut by the wire in 1 sec.*

EXAMPLE: What is the induced electromotive force in a wire that cuts a magnetic field 30 times per second, assuming that the pole faces have a cross-sectional area of 400 sq cm and the flux density is 12,000 lines?

$$\phi = 400 \times 12{,}000 = 4{,}800{,}000$$

$$\text{lines cut per sec} = 4{,}800{,}000 \times 30 = 144{,}000{,}000$$

$$\text{induced emf} = \frac{144{,}000{,}000}{100{,}000{,}000} = 1.44 \text{ v}$$

Replace the single wire with a rectangular loop of wire the axis of which is located midway between and parallel with the pole faces. Rotating the loop around its axis during the first half of its revolution will cause one side (AB) to move downward while the other side (CD) moves upward across the field (see Fig. 297). This results in an induced emf to be produced in each

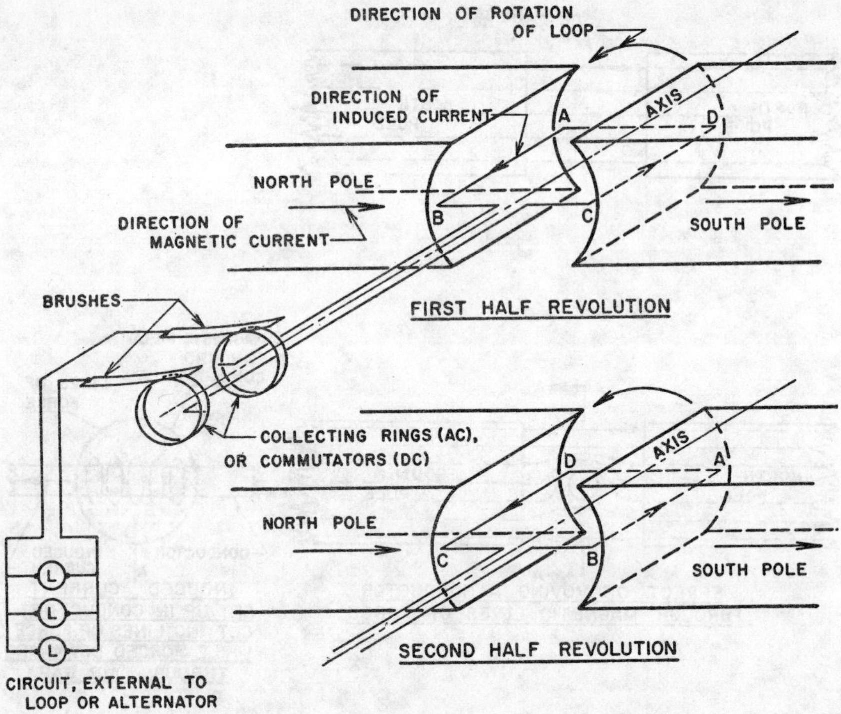

Fig. 297. Induced current developed in a rotating loop.

side of the loop. The effect, however, is to cause the current to flow through the loop in either a clockwise or a counterclockwise direction, depending on which way the loop is rotated. Continuing to rotate the loop and complete one revolution causes the side that previously moved upward now to move downward across the field. This action has the same effect as when a single wire is moved up and down across a magnetic field, i.e., the induced emf changes direction in the wire. With the loop, the emf changes direction in both sides so that flow will be clockwise during the second half revolution if it is counterclockwise during the first half revolution or vice versa. Thus, by continually rotating the loop, the direction of emf reverses itself twice per revolution, so that an *alternating* current is developed. Referring to Fig. 297, the direction of induced emf during the first half of revolution is along the path $ABCDA$ (*top*) and during the second half revolution it is along $ADCBA$ (*bottom*). It is only during the period when the loop wire is cutting the magnetic field that an emf is induced. This means that, when the loop is vertical relative to the horizontal plane of the poles, no emf is produced. This results in an emf that reaches maximum voltage when the loop lies along the center line of the poles, as shown in Fig. 297. The change in voltage from maximum to zero is not an abrupt one. The magnitude of emf varies as the angle of the loop changes in relation to the center line of the poles. This is best portrayed by use of a *sine curve* which graphically shows the gradual changes in emf as the loop turns through one full revolution. The significance of the name sine curve will be discussed later.

Sine Curve of Induced Voltage. Referring to Fig. 298, it will be noted that the curve originates on the center line at a point designated as zero voltage. This represents the emf produced when the loop is in a zero position or at a right angle to the pole center line. In this position, the sides of the loop are moving parallel with, and not cutting, the lines of force. Therefore, no emf is produced and the loop is said to be in zero position. As the loop revolves, its center line departs from its position at zero so that the sides of the loop cut an ever-increasing number of lines as they proceed downward or upward through the magnetic field. The rate at which the sides of the loop cut lines of force increases until a maximum is reached when the angle of departure from zero position becomes 90°. In this latter position, the center lines of the loop and the poles are congruent. Because the wires cut the maximum number of lines in the 90° position, the emf produced also

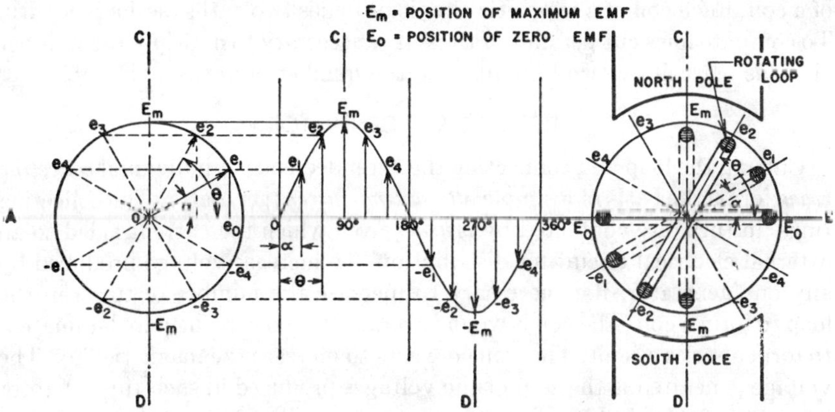

(a) e_1, e_2, e_3, ETC., REPRESENT LOOP POSITION AT SUCCEEDING INSTANCES OF 1 REVOLUTION. THEY ALSO REPRESENT EMF VALUE AT EACH POSITION.

(b) ANGLES α AND θ REPRESENT ANGULAR POSITION OF e_1, e_2, e_3, ETC., AT VARIOUS INSTANCES RELATIVE TO ZERO EMF ALONG LINE AB.

(c) TRIANGLE oe_1e_0 IS THE VECTOR DIAGRAM FOR EMF WHEN LOOP IS IN POSITION e_1 RELATIVE TO ZERO EMF.

(d) EMF IS AT MAXIMUM VALUE WHEN LOOP IS IN POSITION CD INDICATED AS E_m.

(e) VALUE OF EMF AS REPRESENTED BY e_1, e_2, e_3, ETC., EQUALS RATIO LINE e_1e_0 OVER LINE oe_1, ETC., TIMES MAXIMUM (E_m) EMF.

(f) ∴ EMF $e_1 = E_m \sin \alpha$, $e_2 = E_m \sin \theta$, ETC.

Fig. 298. Development of a sine curve showing vector method for determining instantaneous emf.

reaches a maximum value. Up to this point, the loop has made a one-quarter revolution.

As the loop continues to revolve, it passes through the 90° position. At the instant of leaving the state of center line congruency, the rate at which lines of force are cut starts to diminish and continues to do so until the loop again is at a right angle to the pole center line and the emf is again zero. However, the sides of the loop are now located opposite their starting point, or 180° from the original zero position, having completed half a revolution. The variation in emf produced during the first half revolution is shown by the curve above the zero voltage line extending from 0° to 180°. As the sides of the loop pass through the 180° position, the current reverses direction. However, the voltage varies during the second half revolution as the loop reaches the 270° position and finally returns to zero position, in the same manner as during the first half revolution. The positive and the negative (+ and −) signs indicate change of current direction only and have no arithmetical significance.

While the system described assumed a revolving loop and stationary poles, the same results are obtained by revolving the poles and keeping the loop stationary. As long as the lines of force are cut, it is not important which of the two components act as the rotor or stator. By looping a wire a number of times, a continuous or a series of paths is constructed in the form of a coil. Each coil of a magnetic circuit provides two paths per loop or turn. To compute lines cut per unit of time, it is necessary to multiply the number of turns per coil by two to determine the number of paths.

THE ELECTRIC GENERATOR

Cutting the loop and connecting the extended ends to individual *collecting rings* forms the basis of a simple *alternating-current* generator. The collecting rings, in turn, are contacted by *brushes* from which the current is led to an external electrical circuit for distribution. To increase voltage produced by any one generator, it is necessary to increase the number of turns in the loop to form a coil. The coil is wound around an iron core, usually laminated, to form a series circuit. The iron core acts to increase the magnetic flux. The voltage generated is the sum of the voltages produced in each turn. A rotor assembly of this kind is called an armature, usually made up of a number of coils.

The directional change of emf occurs in any revolving armature. Using collecting rings, which are unbroken circular conductors mounted on the shaft that rotates the armature, an alternating current is produced in the external electrical circuit. The collector rings are so called because they collect the individual voltage produced in each turn. They are sometimes called *slip rings* due to the sliding action of the brushes over their peripheral surface.

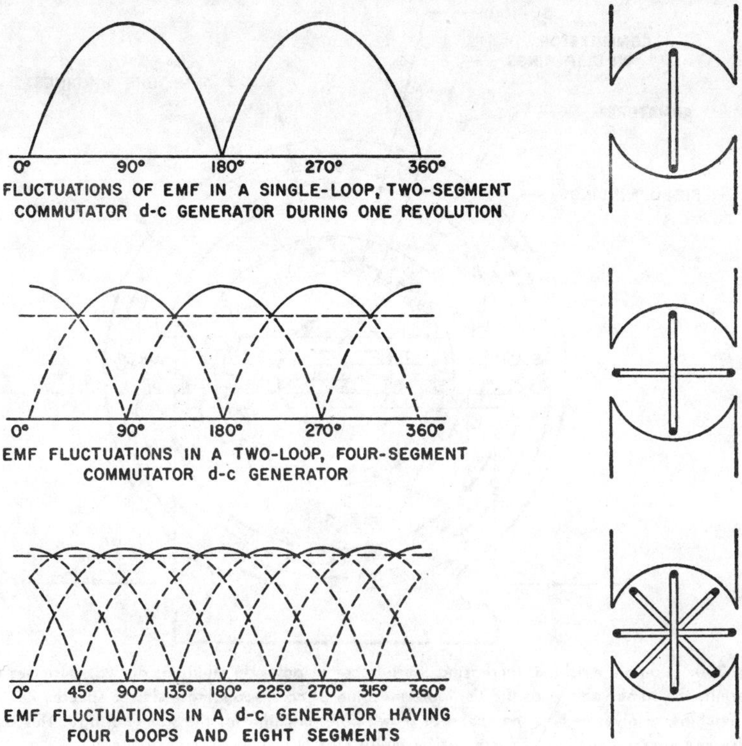

Fig. 299. Curves representing direct current and the effect of increasing number of coils on emf pulsations.

To convert the alternations of current to produce an external *direct current*—i.e., a current that always flows in one direction—it is necessary to employ a *commutator*. The commutator is not a continuous circular conductor as is the slip ring. It is a segmented ring whose segments are insulated against contact with each other. In the case of the single-loop armature, there will be two segments, each connected to different ends of the wire. As one side of the loop passes through the magnetic field to produce an emf in the direction of its own segment, the segment will be in contact with one of the brushes. This is the positive brush because it is the one toward which the induced emf flows and through which the current leaves the generator. As the same side reaches the 270° point (−0°) and the current reverses direction, its segment moves out of contact with the positive brush to make contact with the second brush. The second brush becomes the negative one because the current in the side of loop which it now contacts is going away from it. During this period of reversal the opposite side of the loop starts to cut lines of force so that its current, in turn, flows toward the positive brush.

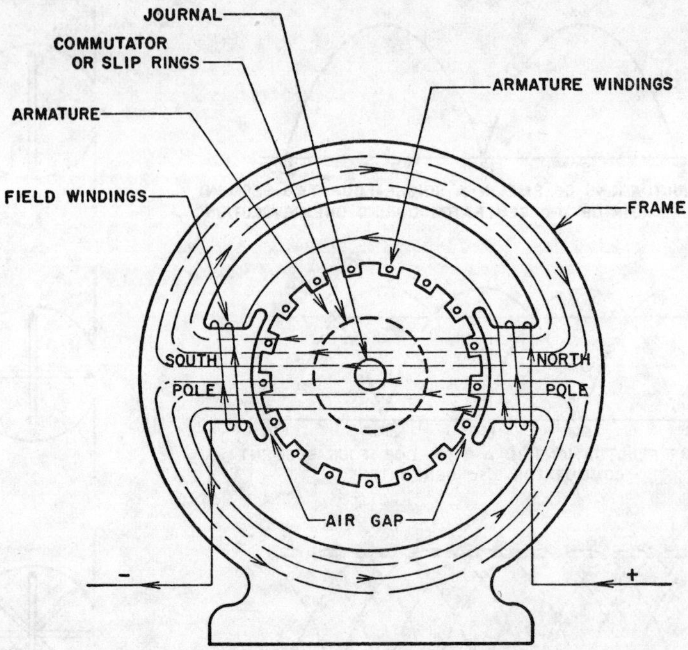

Fig. 300. Two-pole machine. Increasing the number of poles, in multiples of two, increases voltage or permits the same voltage as the two-pole machine at correspondingly slower speeds, i.e., a four-pole machine revolves at half the speed of a two-pole machine for the same voltage. This is due to the greater number of lines of force cut by multipolar machines.

The process is repeated every revolution of the armature so that the one brush is always positive while the other is always negative. As the current always leaves the same brush to enter an external circuit in the same direction, it is known as direct current. The generator producing such a current is called a *direct-current generator*. The curve shown in Fig. 299 represents changes in emf for a direct-current generator. Note that the curve is always above the line, indicating no reversal of direction after it leaves the positive terminal. Note also that pulsations level off as more coils are added to the armature.

Any generator, then, is made so as to provide a magnetic field winding and an induced emf winding having one or more coils. The ring in Fig. 255 used as a demonstrator in the foregoing discussion embodies all the principles involved in the practical *two-pole* machine with the loop (Fig. 297) representing the armature and the ring acting as the *field*. The only difference lies in the shape as seen in the two-pole generator (Fig. 300). More than two poles are often used for reasons given in the legend of Fig. 300. The voltage produced by a generator is that voltage induced in one armature coil or multiple loop—i.e., that circuit through the armature connecting the two brushes. The current produced by a generator is equal to the sum of the currents produced in each coil. Coils are designed to have the same resistance so that the current produced in each one is equal. The resistance in the armature from the positive brush to the negative brush is equal to the resistance of each *path* divided by the number of paths.

EXAMPLE: The armature of a two-pole generator is wound with 12 turns in each coil and travels at a speed of 3600 rpm. If the poles are wound to produce a flux of 6,000,000 lines per pole, what will be the average emf between brushes?

1st step: determine number of conductors

$$= 12 \text{ turns} \times 2 = 24$$

2nd step: determine revolutions per second

$$= \tfrac{3600}{60} = 60 \text{ rps}$$

3rd step: determine number of lines cut per conductor per revolution

$$= 6{,}000{,}000 \times 2 = 12{,}000{,}000 \text{ lines}$$

4th step: determine number of lines cut per second

$$24 \text{ conductors} \times 60 \text{ rps} \times 12{,}000{,}000 \text{ lines} = 17{,}280{,}000{,}000 \text{ lines}$$

5th step: determine voltage

$$1 \text{ v} = 100{,}000{,}000 \text{ lines cut per second}$$

$$\frac{17{,}280{,}000{,}000}{100{,}000{,}000} = 172.8 \text{ v}$$

ELECTRIC CURRENTS

From the previous discussion, it was learned that there are two kinds of electric current produced in the armature of a generator: (1) alternating current and (2) direct current.

Alternating-Current Electricity

Alternating current (ac) is one in which flow continually reverses itself to travel first in one direction and then in the opposite direction. The reason for these alternations in direction was explained earlier. The hydraulic analogy of alternating current is a valveless reciprocating pump forcing water through a by-pass line connecting the extreme ends of the cylinder, as shown in Fig. 301. As the piston moves back and forth, it forces the water to flow in the by-pass, first in one direction and then in the opposite direction. The water represents current in the electrical system while the pump acts as the generator. The pressure applied by the pump to the water represents voltage and the rate at which the water flows through the pipe represents amperes in the electrical circuit. The pipe friction, the

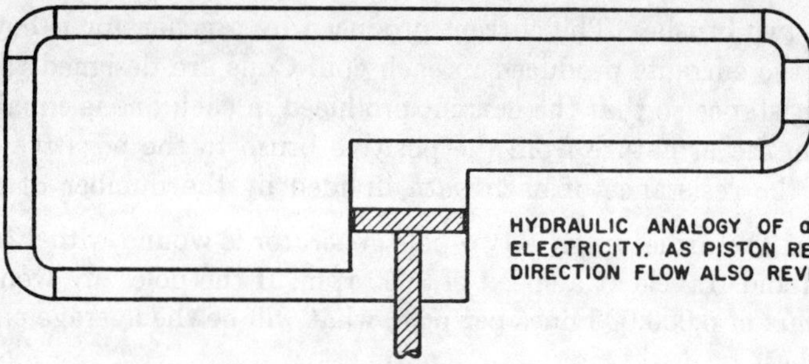

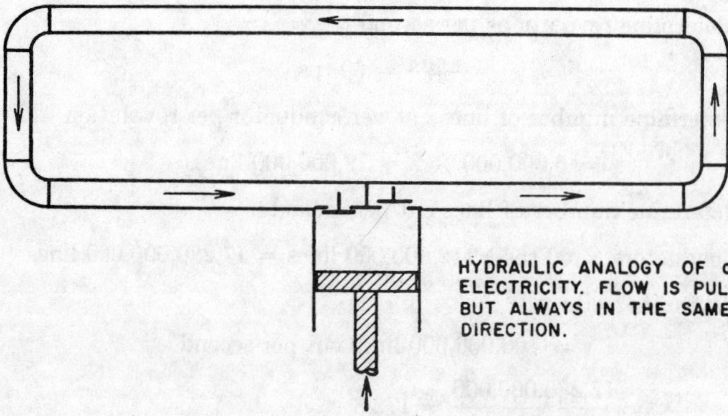

Fig. 301. Hydraulic analogy of a-c and d-c electrical circuits.

length of the pipe, and the bends of the by-pass are all factors of flow opposition and represent resistance in the electrical circuit.

The directional alternations of alternating-current electricity are represented by the sine curve shown in Fig. 298. The sine curve as shown in that figure represents the changing magnitude of emf during one revolution of a one-loop coil as it revolves on its axis between the faces of two energized electromagnets or poles. Thus the emf changes from zero to maximum twice during one revolution, considering that its return to zero at 360° is taken as the starting point of the next revolution. When a current changes from zero to maximum back to zero, reverses to again increase to maximum and return to 0, one cycle has been completed. If it completed 1 cycle in 1 sec, the frequency would be 1 cycle per sec. A frequency of 60 cycles is the completion of 60 cycles per sec. The radian line of the sine curve also represents time. The length of the radian line from 0° to 360° would represent 1 sec if the frequency is 1 cycle per sec and 1/60th of a second in the event that the frequency were 60 cycles.

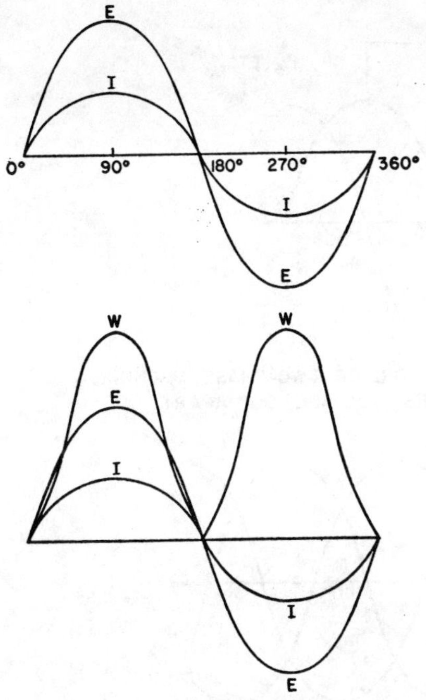

Fig. 302. Electromotive force (*E*), amperage (*I*), and power (*W*) relationship as shown on sine curve for a single-coil or single-phase machine.

While the sine curve in Fig. 298 was described as representing emf of an alternating current, the same kind of curve may represent changes in current magnitude or amperage. This is shown in Fig. 302 where the current produced is related directly to emf produced for the same circuit. Note the conformity of both curves to each other, a condition that can occur only in a circuit, external to the generator, having resistance only. This is explained later.

Still another curve representing power may be superimposed on the sine curve as shown in Fig. 302. Just as in mechanical work systems where power is the rate of doing work as expressed by the equation

$$\text{Work} = \frac{f \times d}{t},$$

the rate of doing electrical work is expressed by the equation, watts $= E \times I$. If the values of emf and current are computed for each increment distance along the sine curve, the power curve representing watts may be constructed as shown. Note that the power curve is completely above the line as it remains a positive value during the cycle. This is not always the case, as shall be learned later.

Up to this point, a two-pole generator having only one loop or one continuous coil has been considered. Adding an additional loop or coil placed at

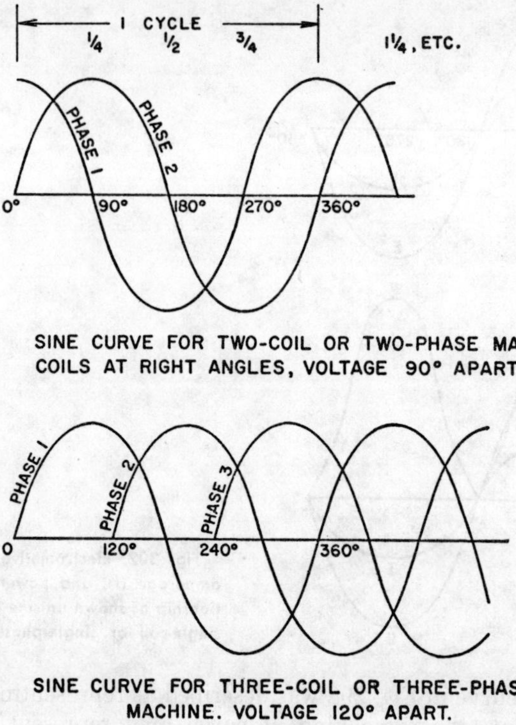

Fig. 303. Sine curves for two- and three-phase current.

right angles to each other would cause maximum cutting of lines of force by one while the other is at zero position. This means that two separate voltages are being produced one in each loop. Because the separate loops are at right angles to each other, the two voltages are said to be 90° apart. This results in two separate sine curves plotted on a common radian line, one of which is at zero while the other is at maximum emf, as shown in Fig. 303 (*top*). This is called a *two-phase* voltage curve. A two-phase system is one in which the voltage across each of the two phases is 90° apart. The loops, while rotating on the same shaft and turning through the same magnetic field, are connected to separate pairs of slip rings so that the effect is the same as having two alternators.

The addition of a third separate loop, so that each one is positioned at an angle of 120° to either of its neighbors, results in a *three-phase* alternator. The sine curve shown in Fig. 303 (*bottom*) shows the voltage variation and their relative locations for a three-phase alternator. Each loop could be connected to its individual pair of slip rings to establish the effect of three alternators. If the three loops are connected together within the generator, only three slip rings will be required as two external circuits may be served by each slip ring.

A three-phase system is one in which the voltages across each of three phases are 120° apart.

The single-phase and the three-phase systems of alternating current are the two most widely used.

Number of Poles in a Generator. The generator used in describing the foregoing principles consisted of only two poles. It is often necessary to have many more than two poles. The number must always be in multiples of two as it is necessary to have unlike poles opposing each other to create a magnetic field. The number of poles does not determine phase but does govern the number of cycles per revolution of the rotating member. This is shown by the equation in which the desired frequency, speed, and the number of poles are all related functions, viz.:

$$\text{Frequency} = \frac{\text{revolutions per min} \times \text{number of poles}}{120}$$

or

$$f = \frac{\text{rpm} \times P}{120}$$

Where a frequency of, say, 60 cycles is desirable and the only means of rotating the armature is by a relatively slow-moving prime mover such as a steam engine or a hydraulic turbine, it becomes necessary to compensate for the speed to satisfy the value of f in the equation. The only increasable variable here being the number of poles, it is obvious that P is the compensating factor.

Consider the problem of producing a 60-cycle current when rotating speeds are limited to, say, 300 rpm with any constancy.

Transposing, the equation becomes

$$P = \frac{120f}{\text{rpm}} = \frac{120 \times 60}{300}$$

$$P = 24 \text{ poles}$$

In contrast, let a steam turbine having a speed of 1800 rpm be available. Then

$$P = \frac{120f}{1800} = \frac{120 \times 60}{1800}$$

$$P = 4 \text{ poles}$$

Generators having two poles are used more often with steam turbines having a speed of 3600 rpm, to which they are directly connected. Because turbines used for power generation usually are operated at a constant speed of 3600 rpm, some compensation in speed is necessary for any given number of poles and frequency. In the latter case, the necessary speed for any given frequency may be obtained by using a speed changer such as a gear set having the proper reduction ratio. For example, to produce a 25-cycle current by such a turbine through a two-pole generator, the speed necessary is determined as follows:

$$\text{rpm} = \frac{f \times 120}{P} = \frac{25 \times 120}{2} = 1500 \text{ rpm}$$

The ratio $\frac{3600}{1500} = 2.4$, or the number of times the pinion on the turbine shaft must revolve for each revolution of the gear mounted on the generator shaft.

Inductance

Inductance in an electrical circuit is similar to inertia in a mechanical system. Inertia being that force opposing any state change, may be summarized in the statement that *bodies at rest tend to remain at rest and bodies in motion tend to remain in motion.*

Lenz's law covers the action within an electrical circuit called inductance by stating in effect that *an induced current always opposes the action that produces it.* Consider the flow direction of water in the by-pass system in Fig. 301 (*top*) during the forward stroke of the pump piston. Then picture the turbulence that results in the water line when the piston reverses direction of travel. It may be assumed safely that, due to the abruptness of directional change, the water will continue on in the same direction momentarily, even though the piston is exerting pressure in the opposite direction. In the valveless pump system, this momentary state is repeated each time the reciprocating piston reverses direction. It may be said that the force exerted by the pump on the water to send it in one direction momentarily opposes that force when the same pump reverses that force to send the water in the opposite direction. Thus Lenz's law is restated. It is obvious that the action creates an additional although momentary resistance to flow throughout the water circuit by reason of inertia.

To apply Lenz's law to the electrical circuit, consider two coils of wire of different diameters, the smaller of which is placed within the larger one but not connected in any way electrically. Let the smaller or primary coil be connected to an external electrical source and the larger or secondary coil have its terminals connected through a lamp or voltmeter. The secondary coil has no contact with an external electrical source. Closing the external circuit to allow current to flow through the primary coil in, say, a clockwise direction will create a flow of current in the secondary coil momentarily, as evident by the lighting of the lamp or movement of the voltmeter in a counterclockwise direction. However, the current in the secondary coil will disappear immediately the current in the primary coil becomes steady. Opening the circuit to stop the flow of current in the primary coil will again create a current to flow momentarily in the secondary coil, but this time in the same direction as the current now shut off in the primary coil. When the current in the primary coil completely disappears, so will the induced current in the secondary coil. In each case, the magnetic field or flux created in the secondary coil opposes the action of the impressed current—i.e., it opposed the initial flow of current when the circuit was closed by creating flow in the opposite direction, and then opposes the stopping of the impressed current by flowing in the same direction when the circuit is opened.

The induced current would not be of any significance in itself even though its direction is always contrary to the impressed current if it were not for the field created in the secondary coil. It is now known that the induced current is created by the magnetic field set up by the impressed current. But a magnetic field is created in any circuit whether the current is impressed or induced. Therefore, the resistance to change in either case is due to the conflictions of the opposing lines of force, both of which tend to expand outward from their respective axes. When these lines of force are in the process of spreading out in a common path during initial stages of flow, but traveling in opposite directions, which in effect tends to change *polarity*, opposition to change in the existing state is set up. Conversely, when the lines of force are in the process of spreading out in a common path, but in the same direction during the final stages of flow, opposition to change is again created by the tendency to continue flow.

It is only during the instant of closing and opening the electrical circuit that the lines of force in the primary coil spread out far enough to envelop the secondary coil. This action can be compared with the extra force necessary to start any heavy object in motion, or to stop it once it is in motion. After the impressed current settles down to steady flow, the lines of force contract beyond the reach of the secondary coil, preventing any further build-up of induced current.

Because opposition to change in the primary coil is created by an induced current in the secondary coil, the two coils have a relationship of *mutual inductance*. The magnitude of emf induced in the secondary coil is equal to the flux per turn per sec multiplied by the number of turns divided by 10^8. It is higher when the current in the primary coil diminishes upon being shut off. This is because the rate at which the impressed current diminishes is greater than when it is developing, giving more impetus to the lines of force to expand at the time the circuit is opened. The ignition coil used on internal combustion engines is based on the principles of mutual inductance.

If the secondary coil in Fig. 304 were eliminated, only a single coil connected to an external source of current would be left. Closing the circuit and allowing a current to flow through the coil would result in a gradual build-up of current. The maximum current would depend on the resistance of the coil and the impressed voltage according to Ohm's law. Note that it was stated that a gradual build-up of current would result. In contrast, the current in a straight-wire circuit would build up instantaneously. The reason

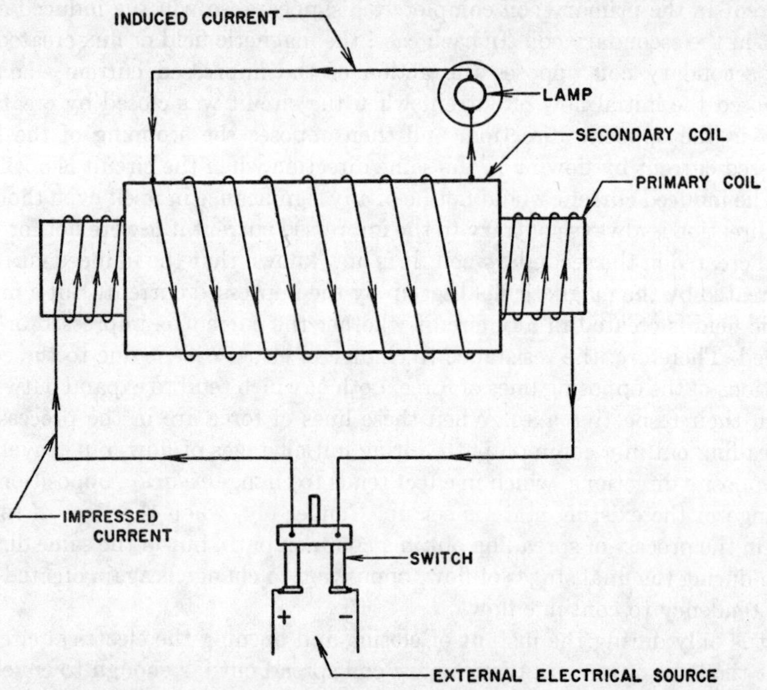

Fig. 304. Mutual inductance coil. Induced current in secondary coil is set up momentarily when switch in impressed current line to primary coil is closed, as shown. Direction of momentary induced current is reversed when switch is opened, or if polarity changes as it does in a-c circuits.

for the time difference between the straight-wire circuit and the one in which a coil is present lies in a property known as *self-inductance.*

A current flowing through any turn in a coil creates an expanding magnetic field around it, as previously described. The lines of this field will envelop the turn next to it, which is also carrying the same current in the same direction around the coil. It has already been established that a magnetic field in one coil will set up a counter flow in another. This is also true in the case of neighboring wires. As the current passes through the first turn in the coil, it instantly sets up a momentary field in the second turn. This creates an induced-current flow in the direction opposite to that of the impressed current, which must be overcome by the latter in traveling toward the opposite end of the coil. As this occurs in all turns, one after the other, while the impressed current continues to travel toward the last turn in the coil, it is obvious that a time lag results. The same time lag occurs when the current is shut off. Because the induced current in the neighboring turns reverse direction as it did in the double coil, the tendency is to maintain current flow toward the last turn. This, of course, opposes the diminishing action of the impressed current, making it again a gradual process.

Where changes in direction are repeated rapidly, as they are in an alternating-current circuit, the current lags considerably behind the voltage. The more rapid the alternations, the greater this lag becomes. In fact, the lag becomes so great in alternating-current circuits that the current is never able to reach its maximum as established by a direct current where the impressed current always flows in the same direction. *Inductance, then, always causes the current to lag behind the voltage in an alternating-current circuit.*

Inductance is not an individual property such as volts and amperes, but only a term applied to an algebraic expression which considers the design (such as shape, size, number of coils, permeability, etc.) of an electrical circuit.

The algebraic equation expressing inductance is as follows:

$$L = \frac{N\phi}{I \times 10^8}$$

or

$$L = \frac{1.26N^2A\mu}{10^8 \times l}$$

where L = self-inductance in *henries*
N = number of turns
I = current in amperes
A = area of core in sq cm
μ = permeability of core
l = length of core in cm

where 1 henry is the inductance of a circuit having a change in current at a rate of 1 amp per sec to induce an emf of 1 v.

To obtain the induced emf, then

$$E_{av} = L\frac{I_1 - I_2}{t}$$

where E_{av} = average induced emf (volts)
I_1 = maximum amperage
I_2 = steady amperage
L = henries
t = time of ampere change

EXAMPLE: A circuit consists of a coil having 600 turns about an iron core whose permeability is 2700 under the circumstances. (*a*) Determine the average inductance in henries if the core is 160 cm long and 225 sq cm in cross-sectional area. (*b*) What is the average emf if a current change of 6 amp occurs in $\frac{1}{2}$ sec?

1st step: determine inductance in henries.

$$L = \frac{1.26N^2\mu A}{10^8 l}$$

$$L = \frac{1.26 \times 600^2 \times 2700 \times 225}{100{,}000{,}000 \times 160}$$

$$L = 17.2 \text{ henries}$$

2nd step: determine emf.

$$E_{av} = L\frac{I_1 - I_2}{t}$$

$$= 17.2 \times \frac{6}{0.5}$$

$$= 206.4 \text{ v}$$

It was stated previously that the current in a circuit was a function of emf and resistance in accordance with Ohm's law. That is true for circuits in which the current is steady, such as would be the case for continuous flow of direct current. It is also quite true for alternating-current circuits in which there are no components tending to set up an induced emf and where only forces of resistance are present. The practical alternating-current circuit usually consists of components that create inductance, such as *induction motors*. Inductance, then, is a highly important property affecting the general laws of electrical circuits as they have been treated earlier in this discussion.

Capacitance

A simple condenser or capacitor is composed of two or more plates made of a conducting material separated by an insulating material. Insulating materials may be glass, air, paper, porcelain, etc. Conductors may be copper, silver, or any other good conducting material. An electrical condenser may be likened to a flexible diaphragm placed across a water line to block the flow. Starting a centrifugal pump will cause the water to flow momentarily on both sides of the diaphragm to the extent of elasticity of the separating membranous material. When the limit of elasticity is reached, flow will cease. The membranous material will assume its original position when the pump is stopped and some flow will occur in the opposite direction by reason thereof. If the line is a by-pass and the valveless pump reciprocates as shown in Fig. 301, then the diaphragm will flex back and forth while a corresponding momentary flow will again occur within the elastic dimensions of the membranous material in either direction. It could be said that the amount of flow occurring momentarily is the measure of diaphragm capacity.

The same principle applies to a condenser in an electrical circuit. When the circuit is closed initially, a current will flow for an instant on either side of the condenser. In case of an alternating current, instantaneous flow will occur each time the current reverses direction. If, by some means, the circuit is opened near the condenser, which would be analogous to closing a valve near the diaphragm of the water line, no return of current could take place. The condenser, then, would be under pressure and said to be charged with electricity.

When a condenser is charged, it may be considered as containing current "locked up" under pressure of an emf. The current does not flow when the

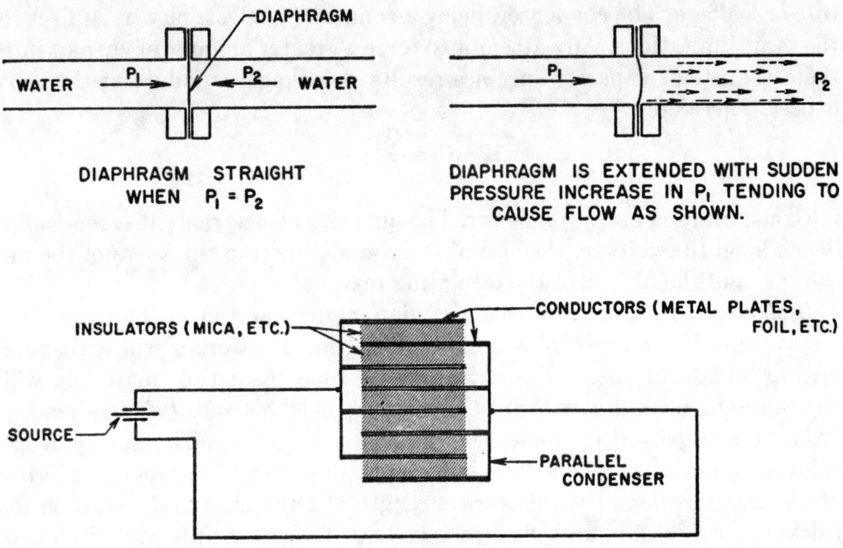

Fig. 305. Electrical condenser and its analogy to hydraulic system.

condenser is charged, so it cannot be measured in terms of flow rate or amperes. Because it is quantity of electricity being dealt with, measurement is in terms of *coulombs*. As the quantity of electricity is under pressure, just as a certain amount of water in the diaphragm is under pressure, a relationship of coulombs to emf exists. This relationship is called capacitance, or *electrostatic capacity*, of a condenser, and is expressed by the equation

$$C = \frac{Q}{E}$$

or

$$Q = CE$$

or

$$E = \frac{Q}{C}$$

where C = capacitance in *farads*
Q = coulombs (quantity of electricity)
E = volts

The size of a condenser is rated according to the charge of electricity that can be built up in it, as expressed in terms of capacitance or farads. *A condenser that can maintain a charge of 1 coulomb when a difference of potential across its terminals is 1 v is said to have a capacity of 1 farad.* As the farad is a large unit, condenser capacities are more often stated in *microfarads*, or one millionth of a farad.

Containers such as a 5-gal can or a 1-qt bottle can have a constant capacity. Any attempt to force more liquid in such containers results in rupture or leakage. The condenser, being a container of electricity, is subject to the same limitations. Any attempt to force a greater amount of current into them than their capacity specifies results in leakage or failure of the condenser. Therefore,

$$C = \frac{Q}{E}$$

is a constant for a given condenser. The quantity of electricity in a condenser depends on the voltage, the size of the conductors, the thickness of the insulator, and the nature of the insulating material.

Dielectrics. It is obvious that the diaphragm material in the water line will rupture if the pressure is great enough. Some materials will withstand greater pressures than others. Similarly, some insulating materials will withstand higher voltages than others when placed between two conductors. Insulators employed in condensers are called dielectrics. The ability of insulators to prevent voltage "break-through" or leakage is expressed in terms of *dielectric strength.* The dielectric strength of any material depends on its thickness and is the number of volts or kilovolts (1000 v) it will withstand per unit of thickness. Air has a dielectric strength of 3000 v per mm or 30,000 v per cm. Mica possesses greater ability in this regard, having a dielectric strength of 60,000 v per cm.

All condenserlike components in an electrical circuit are not parallel plates, so that the field strength is not uniform at all points. Consider the opposite situation of a round wire covered with an insulating material of some kind, transmitting current. In the latter case, the electrical field spreads out as it leaves the wire. The electrostatic field is more dense in the immediate vicinity of the wire, becoming less so as the distance from the wire increases. Hence the forces tending to rupture the insulation diminish with distance from the primary conductor. In a parallel plate arrangement, the electrostatic density is uniform throughout the entire dielectric thickness. The drop in potential per unit thickness of dielectric between plates may be determined by the equation

$$E_d = \frac{E}{t}$$

where E_d = drop in potential per unit thickness, or *potential gradient*
E = voltage across condenser terminals
t = thickness of dielectric

For parallel plates, E_d is constant. In nonuniform conductors, E_d diminishes as t becomes greater. At any point along the gradient where E_d exceeds the dielectric strength of the insulator, rupture will occur while the remainder of the material will not be disturbed. This often occurs in high-voltage transmission lines and is the reason for the bluish cast or corona around the wires

It was stated above that the nature of the dielectric is one factor governing the capacity of condensers. Each kind of material has a relative value in this regard and is based on the capacity of air. This relative value is called the *specific inductive capacity*. Just as water is assumed to have a specific heat of 1 upon which the specific heat of all other materials is related, air is assumed to have a specific conductive capacity of 1. This is also called the *dielectric constant* of air. The dielectric constant of any material, then, is its *condensive* capacity as compared with the condensive capacity of air under similar dimensional circumstances. The dielectric constant for any material, then, is equal to the ratio

$$K = \frac{C_{\text{material}}}{C_{\text{air}}},$$

where C in both instances represents farads and the ratio is represented by the letter K.

Table 40 shows the dielectric strength in kilovolts per centimeter as well as the dielectric constant for some materials.

TABLE 40. DIELECTRIC STRENGTH AND DIELECTRIC CONSTANTS OF SOME MATERIALS

Material	*Dielectric Constant K (av)*	*Dielectric Strength Kilovolts per cm*
Air	1	30
Glass	5.5–10	300–1500
Guttapercha	3.3–4.9	80–200
Rubber	2.75	120–200
Ebonite	2.72	300–1100
Mica	5.6–6.6	60
Paraffin	2.1	120
Glycerin	56.2	
Transformer oil	2.5	120
Paper, treated:		
beeswax	2.5	770
parafined	2.0	500
Lubricating oil (dehydrated)	2.1	

The capacity of simple condensers may be determined by the following formula:

$$C = \frac{KA}{4.452 \times d \times 10^6}$$

where C = capacity in microfarads
K = specific inductive capacity
A = area of plate (one side of one plate)
d = distance between plates
10^6 = conversion factor for microfarads (1,000,000)

Capacitance assists current flow, whereas inductance opposes it. By providing elasticity to an electrical circuit, capacitance provides a force tending to overcome inductance, thereby reducing the lag in current flow caused by the latter. In an alternating-current circuit, the property of capacitance is highly important, especially when *induction motors* are employed. The action here may be summarized by the following statement, viz.:

Inductance causes an alternating current to lag behind the voltage, whereas capacitance causes the current to lead the voltage.

The total capacitance of a circuit having more than one condenser in series is equal to the reciprocal of the sum of the reciprocals of the separate capacitances in the circuit. It is expressed by the equation:

$$C_s = \frac{1}{\frac{1}{C_1} + \frac{1}{C_2} + \frac{1}{C_3} + \text{etc.}}$$

where C_s = total capacitance in microfarads
and C_1, C_2, C_3, etc. = capacitance in microfarads of the separate condensers

The total capacitance of a circuit in which two or more condensers are in parallel is the sum of the separate capacitances, viz.:

$$C_s = C_1 + C_2 + C_3 + \text{etc.}$$

Reactance, Inductive and Capacitive

The sine curve shown in Fig. 302 portrays the current and the voltage of a circuit *in phase*. This means that the flow of current is at its maximum exactly the instant the voltage reaches its maximum. Both reach their zero value at the same time also. When the current and the voltage are in phase with each other, there is no lagging or leading on the part of either. This could happen only under certain circuit conditions. One condition assumes the absence of induction and capacitance, only resistance. This latter explains the use of the phrase, "a circuit having resistance only," in connection with Ohm's law earlier in this discussion. Another condition would be one in which the effects of inductive forces and capacitance forces are in equilibrium throughout the circuit.

It was stated earlier that induction in an alternating-current circuit tends to cause the current to lag behind the voltage for reasons already described. The force causing this lag is the reaction of self-induction coils in the circuit opposing any directional change in flow. This force is known as *inductive reactance*. Fig. 306 shows the effect on the sine curve when the current lags 90° behind the voltage, a condition that exists when a circuit contains inductance only.

Inductive reactance is measured in ohms and represented by X_L. Its value is determined by the equation:

$$X_L = 2\pi f L$$

where X_L = inductive reactance in ohms
f = frequency in cycles per second
L = inductance in henries
π = 3.1416

Thus, the reactance of a coil having an inductance of 0.16 henry in a 60-cycle circuit is

$$X_L = 2 \times 3.1416 \times 60 \times 0.16$$
$$= 60.32 \text{ ohms}$$

Ohm's law, then, may be restated in terms of inductive reactance by changing the equation from

$$\text{current in amps} = \frac{\text{emf in volts}}{\text{resistance in ohms}}$$

to

$$\text{current in amps} = \frac{\text{emf in volts}}{\text{inductive reactance in ohms}}$$

$$I = \frac{E}{X_L}$$

The value of E in the equation depends on whether the maximum, effective, average, or instantaneous voltage for the circuit is considered. In either case, the amperage will be identified with the voltage.

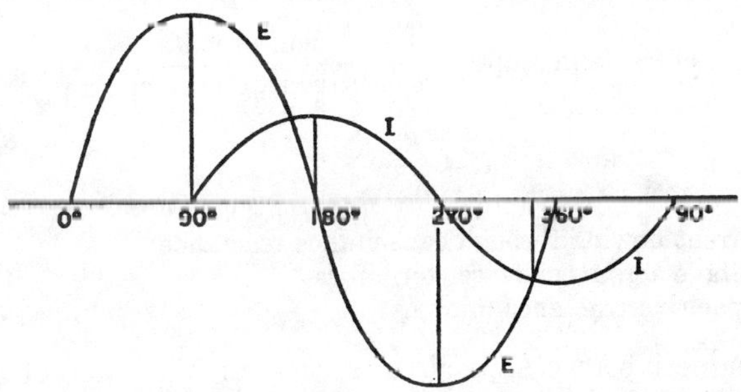

Fig. 306. Relationship of voltage and current for inductance circuit. Current lags 90° behind voltage.

Capacitance in a circuit assists any change in flow direction. Thus, it has the opposite reaction to the effect of inductance. The elastic force with which capacitance acts to assist the reversal of flow is called *capacitive reactance*. The difference between the two properties, inductance vs. capacitance, may be restated as follows:

(*a*) *A circuit having both inductance and resistance will cause the current to rise* gradually when a sudden steady voltage is applied.*

* Maximum current is never reached in the inductance coils of an alternating-current circuit.

(*b*) *A circuit having both capacitance and resistance will cause a maximum current gradually to diminish to zero when a sudden steady voltage is applied.*

Fig. 307 shows the effect of capacitance only on the sine curve of an alternating-current circuit. In this instance, the current leads the voltage by 90°. Note that the current flow is greatest at the 90° point when the voltage increases most rapidly at zero. Note also that the voltage is at its maximum when the current is zero and that the current again is at maximum flow when the voltage falls to zero at the 270° point. All through the cycle the current leads the voltages, in this case by 90°.

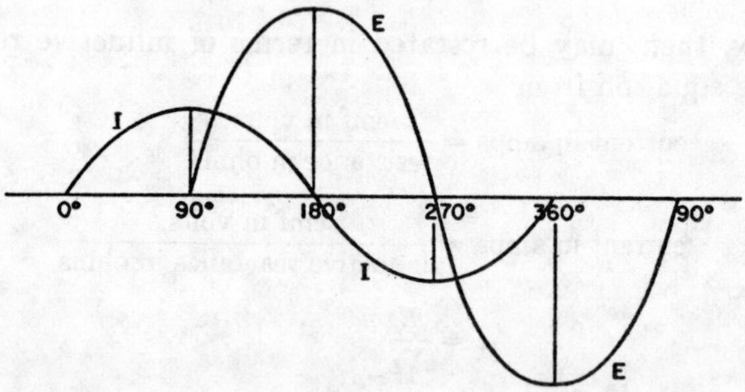

Fig. 307. Relationship of voltage and current for capacitance circuit. Current leads voltage by 90°.

Ohm's law restated in terms of capacitive reactance is as follows:

$$\text{currents in amps} = \frac{\text{emf in volts}}{\text{capacitive reactance in ohms}}$$

or

$$I = \frac{E}{X_c}$$

where I = current flowing through capacitance reactance
E = voltage across condenser terminals
X_c = capacitive reactance in ohms

As with inductive reactance circuits, the current value of a capacitive reactance circuit will depend on whether average, maximum, instantaneous, or effective voltage is considered.

The value of capacitive reactance may be determined by the equation:

$$X_c = \frac{1}{2\pi fc}$$

where X_c = capacitive reactance in ohms
π = 3.1416
f = frequency, cycles per second
c = capacitance in farads

(To convert to microfarads, divide c by 10^6 or multiply by 10^{-6})

Thus, an a-c circuit having a capacitance of 25 microfarads and a frequency of 60 cycles would have a capacitive reactance of

$$X_c = \frac{1}{2 \times 3.1416 \times 60 \times 25 \times 10^{-6}}$$

$$X_c = 106.3 \text{ ohms}$$

Where the circuit contains both inductive reactance and capacitive reactance, the one counteracts the other, tending to reduce reactance to zero. It is necessary, then, to subtract the value of X_c from the X_L, viz.:

$$X = X_L - X_c$$

where X = the reactance in ohms

A circuit having an inductive reactance of 22 ohms and a capacitive reactance of 7 ohms would have a reactance of 22 − 7 = 15 ohms.

Impedance

Impedance is the term used to indicate a circuit condition where both reactance and resistance are involved. It is expressed in ohms and represented by the letter Z.

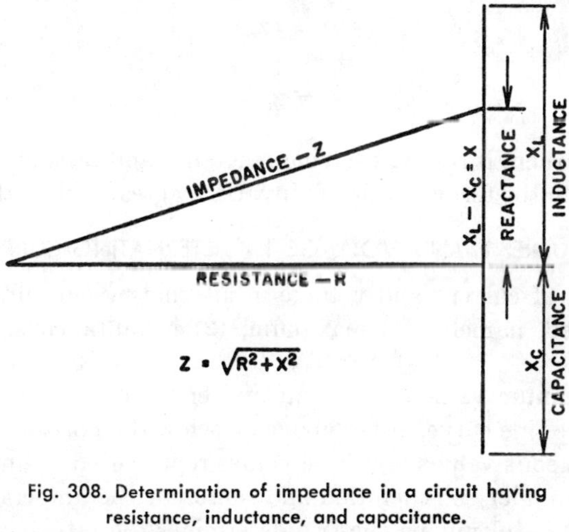

Fig. 308. Determination of impedance in a circuit having resistance, inductance, and capacitance.

Because the effect of reactance is always at 90° with resistance in a circuit, which means at right angles with each other, impedance cannot be determined by the simple process of adding, or subtracting the two as was done in the case of reactance alone. The rule here is that of the right-angle triangle in which the side opposite the right angle or hypotenuse, being the longest, is equal to the square root of the sum of the square of the base plus the square of the altitude.

Considering reactance as inductive, capacitive, or both in combination, in relation to resistance in the same circuit, then impedance is determined by the suitable equation of the three given below, viz.:

$$Z = \sqrt{R^2 + X^2} \quad \text{(combination)}$$

$$Z = \sqrt{R^2 + X_L^2} \quad \text{(inductive reactance only)}$$

$$Z = \sqrt{R^2 + X_c^2} \quad \text{(capacitive reactance only)}$$

where R = resistance of a circuit in ohms
X = resultant, i.e. $(X_L - X_c)$, reactance in ohms
X_L = inductive reactance in ohms
X_c = capacitive reactance in ohms
Z = impedance in ohms

Ohm's law applied to alternating-current circuits is modified by replacing impedance for resistance as follows:

$$\text{Current} = \frac{\text{voltage}}{\text{impedance}}$$

or

$$I = \frac{E}{Z}$$

$$E = IZ$$

$$Z = \frac{E}{I}$$

Impedance, then, is result of having reactance and resistance in the same circuit, in which the former is almost invariably greater than the latter.

VALUES OF CURRENT AND VOLTAGE IN ALTERNATING-CURRENT CIRCUITS

Four values of current and voltage in alternating-current circuits have been mentioned, namely: (1) maximum, (2) instantaneous, (3) effective, (4) average.

1. The **maximum values** of current and emf are found at the peaks of their respective sine curve, either above or below the horizontal line.

2. **Instantaneous values** are those values represented at any point along the sine curve under consideration and represent the amperage or voltage at that instant. Multiplying the two determines the rate at which work is done at that instant.

3. The **effective value** of alternating current and voltage is the more practical value. Upon this value lie the results that are expected through the frequent alternations of electrical energy from zero to maximum. In direct-current circuits, the determination of effective current and voltage values poses no problem because flow is steady and therefore can be measured directly. To determinine effective values for alternating current, some

method of establishing direct-current equivalency may be considered. This may be accomplished by comparing the heating effect of both types of current. When the heating effect of an alternating current and a direct current is the same, under identical conditions, an equivalent effective current exists for both. This method of obtaining equivalent values is resorted to because it is impossible to deposit silver electrolytically with an alternating current, the method used to establish the value of 1 amp of direct current.

The effective value of an alternating current may be determined also by the equation:

$$I = \sqrt{\text{average } (i_1^2 + i_2^2 + i_3^2 + \text{etc.})}$$

where i_1, i_2, i_3, etc., are instantaneous values of the current in amperes throughout a complete cycle. This complex method of determining the effective current results in a value called *the square root of the mean square* or the *root mean square* (rms) of the current.

If it is assumed that the alternations of a current follow a uniform sine curve, then

$$I_{eff} = 0.707 I_{max}$$

or

$$I_{eff} = \frac{I_{max}}{\sqrt{2}} = \frac{I_{max}}{1.41}$$

where I_{eff} = effective current in amperes
I_{max} = maximum current in amperes from the sine curve or by ammeter

The effective voltage may be determined in a similar manner as the same rules of alternating current apply equally to the emf as it does to the current. Therefore

$$E_{eff} = 0.707 E_{max}$$

or

$$E_{eff} = \frac{E_{max}}{\sqrt{2}} = \frac{E_{max}}{1.41}$$

4. The **average value** of current or emf of an alternating-current circuit is not too important. It may be determined by establishing the average for the complete cycle by adding a number of instantaneous values and dividing by the number of points represented. The average value is always less than the effective value and may be established more practically by multiplying the maximum current by 0.637, or

$$I_{av} = 0.637 E_{max}$$

The average voltage may be similarly determined as

$$E_{av} = 0.637 E_{max}$$

where I_{av} and E_{av} are average values in each case
and I_{max} and E_{max} are maximum values in each case

Thus, the average current of a circuit having a maximum current of 30 amp is $0.637 \times 30 = 19.11$ amp.

The Sine Curve. The sine curve, as previously described, graphically portrays the alternations of pressure, rate of flow, and power in an alternating-current circuit. It is based on the current or voltage changes in a single conductor passing through a magnetic field in the course of revolving about its axis. For every two passes through the field during which the direction of the current changes once, one cycle is completed. The number of cycles completed per unit of time identifies the frequency or the rate at which current reverses its direction. During the complete cycle, the current changes four times, viz.:

(*a*) Rises from zero to maximum in one direction,
(*b*) Drops from maximum to zero in the same direction,
(*c*) Rises from zero to maximum in the opposite direction,
(*d*) Drops from maximum to zero.

A circuit in which the four changes occur 60 times per second is said to have a frequency of 60 cycles.

If the voltage and the current change uniformly and simultaneously, they are in phase. This latter is the characteristic of a circuit having resistance only.

In a circuit containing inductive reactance only, the current will lag behind the voltage by 90° (i.e., 90° to the effect of resistance).

In a circuit containing capacitive reactance only, the current will lead the voltage by 90° (i.e., 90° to the effect of resistance).

In a circuit containing both, the degree of lead or lag will depend on the sum relationship of inductive reactance to capacitive reactance, i.e., $X = X_L - X_c$.

Impedance, the square root of the sum of resistance squared plus reactance squared, is the total opposition to alternating-current flow.

When inductive reactance and capacitive reactance are equal, the current will be in phase with the voltage. The circuit will then act as one having resistance only, and $W = EI$.

Trigonometric Functions. In most alternating-current systems, the effect of reactance causes the current to lag behind the voltage. This is due to the presence in the circuit of such inductive reactance equipment as induction motors, transformers, and electric furnaces. Because inductive reactance tends to oppose changes in direction of current flow, the actual line current will have some resultant value between that of the active and the reactive current. Since the reactive current lags the voltage by 90° (electrical), the resultant or actual line current must be out of phase with the voltage. The lag caused by inductance may be measured in degrees.

The determination of instantaneous values of an alternating-current circuit in which the current is out of phase with its voltage involves trigono-

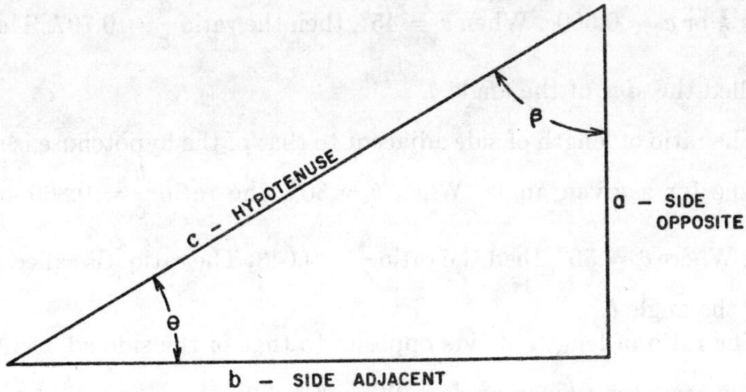

$$\text{SIN } \theta = \frac{a - \text{SIDE OPPOSITE}}{c - \text{HYPOTENUSE}}$$

$$\text{COS } \theta = \frac{b - \text{SIDE ADJACENT}}{c - \text{HYPOTENUSE}}$$

$$\text{TAN } \theta = \frac{a - \text{SIDE OPPOSITE}}{b - \text{SIDE ADJACENT}}$$

$$\text{COT } \theta = \frac{b - \text{SIDE ADJACENT}}{a - \text{SIDE OPPOSITE}}$$

Fig. 309. Trigonometric functions.

metric functions. To acquaint the student with simple trigonometry, the functions of the sine (sin), cosine (cos), and tangent (tan) are explained briefly here. All three are functions of a right-angle triangle.

In Fig. 309, a right angle triangle is shown having an acute angle θ and the three sides a, b, and c.

a is known as the side opposite the angle θ

b is known as the side adjacent the angle θ

and

c is known as the hypotenuse

Trigonometry deals with *functions of angles. A function of any quantity is a second quantity on which the value of the first quantity depends.* For example, the circumference of a circle is a function of its diameter, or, circumference = 3.1416 × diameter. In trigonometry, the dimensions of the sides of a right triangle are all functions of the acute angle—i.e., the relative lengths of a, b, and c depend on the amount of opening θ. The converse is also true.

There are several important principles upon which the laws of trigonometry are based. They are given below:

1. The sum of the three angles of a triangle is equal to 180°.
2. The ratio of length of side opposite to the length of the hypotenuse is always the same for a given angle θ. That is, if $\theta = 30°$, then the ratio $\frac{a}{c}$ is

always $\frac{1}{2}$ or $a = 0.500c$. When $\theta = 45°$, then the ratio $\frac{a}{c} = 0.707$. The ratio $\frac{a}{c}$ is called the *sine* of the angle θ.

3. The ratio of length of side adjacent to that of the hypotenuse is always the same for a given angle. When $\theta = 30°$, the ratio $\frac{b}{c} = 0.866$ or $b = 0.866c$. Where $\theta = 50°$, then the ratio $\frac{b}{c} = 0.643$. The ratio $\frac{b}{c}$ is called the *cosine* of the angle θ.

4. The ratio of length of side opposite to that of the side adjacent is always the same for a given angle θ. When $\theta = 30°$, then the ratio $\frac{a}{b} = 0.577$ or $a = 0.577b$. Where $\theta = 60°$, the ratio $\frac{a}{b} = 1.73$. The ratio $\frac{a}{b}$ is called the *tangent* of the angle θ.

5. One other function involves the ratio of the side adjacent to the side opposite, or $\frac{b}{a}$. The ratio $\frac{b}{a}$ is always the same for a given angle and is called the *cotangent* of the angle θ.

The laws for the above trigonometric functions may be expressed as follows:

Function	*Abbreviation*	*Ratio*
Sine of θ	$\sin\theta$	$= \frac{\text{side opposite}}{\text{hypotenuse}} = \frac{a}{c}$
Cosine of θ	$\cos\theta$	$= \frac{\text{side adjacent}}{\text{hypotenuse}} = \frac{b}{c}$
Tangent of θ	$\tan\theta$	$= \frac{\text{side opposite}}{\text{side adjacent}} = \frac{a}{b}$
Cotangent θ	$\cot\theta$	$= \frac{\text{side adjacent}}{\text{side opposite}} = \frac{b}{a}$

The ratios shown are true only for the angle θ in Fig. 309, i.e.,

$$\sin\theta = \frac{a}{c}, \quad \cos\theta = \frac{b}{c}, \text{ etc.}$$

If the angle β is the one under consideration, then the sides a, b, and c become related to that angle in the same relationship as they are to θ when the latter is considered; i.e., side b becomes the side opposite, a the side adjacent, etc.

The ratios of various functions being constant for any given angle, it is possible to compile a table that shows numerical values for each angle from 1° to 90°. Such a table is called a Table of Trigonometric Functions. The

Table 41. Trigonometric Functions

A	Sin	Cos	Tan	A	Sin	Cos	Tan
0	.000	1.000	.000				
1	.017	.999	.017	**46**	.719	.695	1.04
2	.035	.999	.035	**47**	.731	.682	1.07
3	.052	.999	.052	**48**	.743	.669	1.11
4	.070	.998	.070	**49**	.755	.656	1.15
5	.087	.996	.087	**50**	.766	.643	1.19
6	.105	.995	.105	**51**	.777	.629	1.23
7	.122	.993	.123	**52**	.788	.616	1.28
8	.139	.990	.141	**53**	.799	.602	1.33
9	.156	.988	.158	**54**	.809	.588	1.38
10	.174	.985	.176	**55**	.819	.574	1.43
11	.191	.982	.194	**56**	.829	.559	1.48
12	.208	.978	.213	**57**	.839	.545	1.54
13	.225	.974	.231	**58**	.848	.530	1.60
14	.242	.970	.249	**59**	.857	.515	1.66
15	.259	.966	.268	**60**	.866	.500	1.73
16	.276	.961	.287	**61**	.875	.485	1.80
17	.292	.956	.306	**62**	.883	.469	1.88
18	.309	.951	.325	**63**	.891	.454	1.96
19	.326	.946	.344	**64**	.899	.438	2.05
20	.342	.940	.364	**65**	.906	.423	2.14
21	.358	.934	.384	**66**	.914	.407	2.25
22	.375	.927	.404	**67**	.921	.391	2.36
23	.391	.921	.424	**68**	.927	.375	2.48
24	.407	.914	.445	**69**	.934	.358	2.61
25	.423	.906	.466	**70**	.940	.342	2.75
26	.438	.899	.488	**71**	.946	.326	2.90
27	.454	.891	.510	**72**	.951	.309	3.08
28	.469	.883	.532	**73**	.956	.292	3.27
29	.485	.875	.554	**74**	.961	.276	3.49
30	.500	.866	.577	**75**	.966	.259	3.73
31	.515	.857	.601	**76**	.970	.242	4.01
32	.530	.848	.625	**77**	.974	.225	4.33
33	.545	.839	.649	**78**	.978	.208	4.70
34	.559	.829	.675	**79**	.982	.191	5.14
35	.574	.819	.700	**80**	.985	.174	5.67
36	.588	.809	.727	**81**	.988	.156	6.31
37	.602	.799	.754	**82**	.990	.139	7.12
38	.616	.788	.781	**83**	.993	.122	8.14
39	.629	.777	.810	**84**	.995	.105	9.51
40	.643	.766	.839	**85**	.996	.087	11.43
41	.656	.755	.869	**86**	.998	.070	14.30
42	.669	.743	.900	**87**	.999	.052	19.08
43	.682	.731	.933	**88**	.999	.035	28.64
44	.695	.719	.966	**89**	.999	.017	57.29
45	.707	.707	1.000	**90**	1.000	.000	Infinity

values shown in the table (Table 41) are the ratios of lengths of sides respective of the function.

EXAMPLE: What is the sine of an 80° angle? What is the tangent?

Solution: Refer to Table 41 and find the line corresponding to an 80° angle. Under the sin A column is found the value of 0.985, the sine of 80°. On the same line find the tangent 5.67.

EXAMPLE: What is the length of the hypotenuse (c) in Fig. 309 if $\theta = 80°$ and the length of the side opposite (a) is 3 in.?

$$\sin \theta = \frac{a}{c}$$

Therefore

$$c = \frac{a}{\sin \theta} = \frac{3}{0.985}$$

$$c = 3.0456 \text{ in.}$$

EXAMPLE: What is the length of the side adjacent in the above example?

$$\tan \theta = \frac{a}{b}$$

$$b = \frac{a}{\tan \theta} = \frac{3}{5.67}$$

$$b = 0.529 \text{ in.}$$

Employing the table of trigonometric functions, any dimensional value may be determined if at least two other values are known.

INSTANTANEOUS VALUES OF VOLTAGE AND AMPERAGE. To determine instantaneous values of emf or current, three methods are offered. The first is by graphical representation of the sine curve. The second is by vector diagrams. The third method involves trigonometric functions and is in the form of an equation:

$$e = E_m \sin \theta$$

where e = instantaneous voltage
E_m = maximum voltage
θ = degree of angle of voltage for point of instantaneous voltage

1st Method: Sine Curve. 1st Step: draw a circle whose radius is scaled to equal maximum voltage or maximum current (Fig. 298).

2nd Step: divide the circle into convenient but equal segments along its circumference.

3rd Step: draw a horizontal line extending from the center line of the circle, the length of which is equal to the circumference of the circle.

4th Step: divide the horizontal line in equal parts whose dimensions are multiple to 360°, say 24, each of which represents 15°.

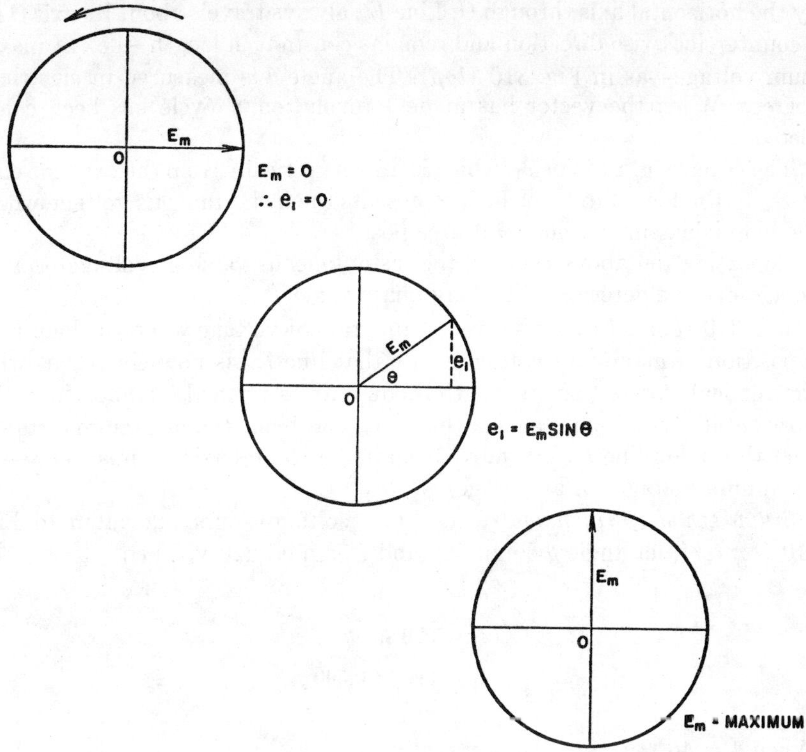

Fig. 310. Vector diagrams representing instantaneous voltage at any position of rotor loop (coil) during rotation. (Or instantaneous amperage by substituting E_m for I_m.)

5th Step: extend the points on the circle to meet the extension lines of the horizontal line points starting at the extreme right of the circle to form the extreme left intersection on the horizontal line.

6th Step: connect the successive points to form a sine curve. Scale length of the vertical line from the horizontal line to the sine curve intersection represents the instantaneous value at any point selected.

2nd Method: Graphically by Vector Diagram. Plotting a vector diagram as shown in Fig. 310 eliminates the necessity of the more complex sine curve in determining instantaneous values of voltage or amperage.

Fig. 310 (*top*) shows a vector diagram of voltage when the loop in Fig. 297 is in zero position. The length of line E_m represents maximum voltage, drawn to scale. Note that E_m is congruent with the vertical axis so that the phase angle is zero. Therefore, the instantaneous voltage is zero. (Refer to Fig. 298.)

Fig. 310 (*center*) shows a vector diagram of voltage when the loop in Fig. 297 is at some phase angle θ with the pole center line represented here by the horizontal axis through O. Line E_m always travels about its axis O in a counterclockwise direction and remains constant in length—i.e., at maximum voltage—as in Fig. 310 (*top*). The angle θ is measured in electrical degrees. When the vector has made 1 revolution, 1 cycle has been completed.

The scale length of line e_1, which is drawn vertically from the extreme end of E_m to the horizontal axis line, represents the instantaneous voltage when the loop is passing through θ degree position.

Repeating the above process, the instantaneous value of voltage or amperage may be determined for any phase angle.

Fig. 310 (*bottom*) shows a vector diagram of voltage when the loop is in a position of maximum voltage. Note that line E_m is now congruent with the vertical axis so that its length is equal to its vertical distance from the horizontal axis, which is at 0 in this case. This being the maximum distance that the end of line E_m can move from its horizontal axis, it must represent maximum voltage (or amperage).

3rd Method: Trigonometrically. (*a*) Voltage—referring again to Fig. 310 (*center*), let angle θ equal 30° and E_m equal 220 v. Then

$$e_1 = E_m \sin\theta$$

$$e_1 = 220 \sin\theta$$

$$e_1 = 220 \times 0.500$$

$$e_1 = 110 \text{ v}$$

When $\theta = 45°$

$$e_2 = 220 \sin 45°$$

$$e_2 = 220 \times 0.707$$

$$e_2 = 155.54 \text{ v}$$

(*b*) Amperage—let θ equal 30° and I_m equal 25 amp.

$$i_1 = I_m \sin\theta$$

$$i_1 = 25 \sin 30°$$

$$i_1 = 25 \times 0.500$$

$$i_1 = 12.5 \text{ amp}$$

POWER IN ALTERNATING-CURRENT CIRCUITS

In direct-current circuits, power in watts can be determined by multiplying volts by amperes, both of which may be measured directly by their respective meters. Power for d-c circuits, then, may be expressed as

$$\text{watts} = \text{volts} \times \text{amperes}$$

or

$$W = EI$$

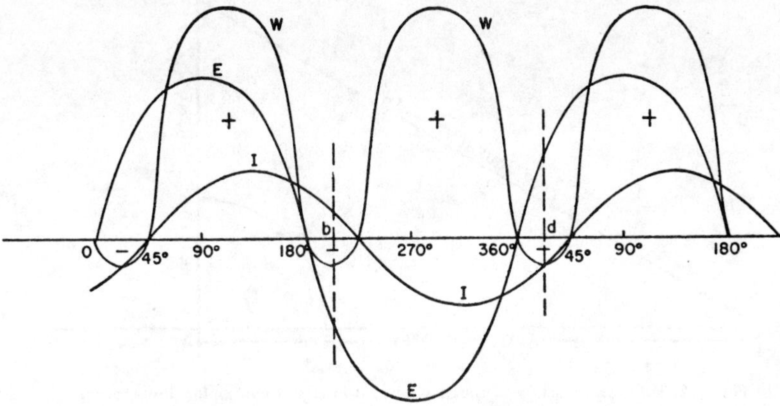

Fig. 311. Power curve for a circuit containing inductance, in which the current lags the voltage.

Power of alternating-current circuits in which the voltage and the amperage are in phase with each other, i.e., the circuit is free of inductance and capacitance (pure resistance), is the product of effective voltage times effective amperage, or

Real power = effective voltage × effective amperage

or

$$W = E_{eff} \times I_{eff}$$

Where inductance is present in the circuit, and usually is, the current will lag behind the voltage. In this case, the product of effective voltage and effective amperage will be greater than the real power. Therefore, $W = E_{eff} \times I_{eff}$ does not apply to circuits containing inductance.

As already stated, the power curve is plotted according to the product corresponding to volts × amperes for each instant along their sine curves. When the circuit is in phase, the entire power curve will be positive in value and will appear entirely above the axis. The reason for this is that the voltage and the amperage are always simultaneously positive or negative. (See Fig. 302.) Therefore, according to the algebraic law that *when the signs of the multiplicand and the multiplier are alike, the product is positive,* the power value remains positive throughout the cycle when the circuit is in phase.

When the current lags its voltage, they no longer remain simultaneously negative and positive throughout the cycle. As shown in Fig. 311, the voltage curve is below the axis at b and is therefore negative, while the current curve is above the axis at the same point and is therefore positive. Completing the foregoing algebraic law that *when the signs of the multiplicand and the multiplier are unlike, the product is negative,* it follows that the power value is negative at b. The power value is also negative at d where the signs of voltage and current are reversed; i.e., voltage is positive and current is

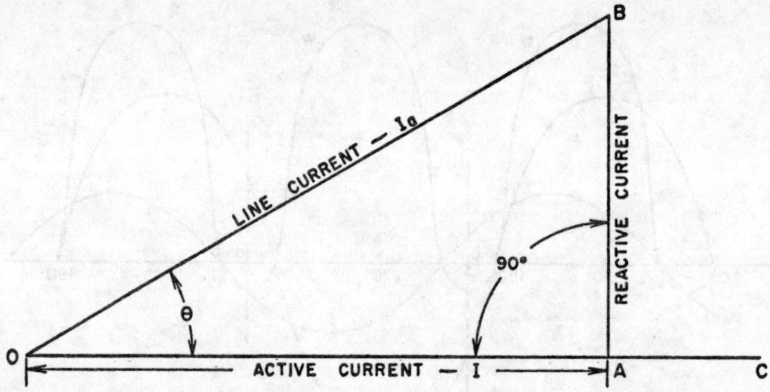

Fig. 312. Relation of active to reactive current in circuit containing inductance.

negative. During the remainder of the cycle, both voltage and current are simultaneously positive or negative so that power is positive except at *b* and *d*.

When the power is negative, energy is being returned to its source. It is the amount of power returned to the source that cannot be credited to that available in the circuit. Thus $W = EI$ is not true for inductive circuits.

In any inductive reactance circuit there are two currents. One is the active current which is in phase with the voltage. The second is the magnetizing or reactive current which opposes any change in direction of flow of current. This latter current lags 90° (electrical) behind the voltage. The action here may be compared with the effort of rowing a boat across a river having a strong downstream current. The distance the boat moves toward the opposite shore is the active effort and is in phase with the rowing effort. The distance downstream the boat is forced by the current is the reactive effort and is out of phase with the rowing effort.

The relationship of active to reactive current is shown by the vector diagram, Fig. 312.

Angle θ is the angle of lag or the angle at which the line current (measured by an ammeter) lags behind the voltage.

Line *OA* represents active current which is in phase with the voltage represented by *OC*.

Line *AB* represents the reactive current opposing change in direction of current flow.

If current and voltage were in phase as in a circuit having pure resistance, then *OB* and *OA* would be congruent.

The power in the circuit represented by the vector diagram is expressed by the equation:

$$W = EI_a \text{ (or } W = OC \times OA)$$

But

$$I_a = \imath \cos\theta \left(\begin{array}{l}\text{or } I_a = OB \times \text{cosine } \theta \\ \text{where cosine } \theta = \dfrac{OA}{OB}\end{array}\right)$$

Therefore

$$W \quad EI \cos\theta$$

Example: What power is consumed in a 220-v a-c circuit containing inductance if 22 amp are flowing, in which the current lags the voltage by 30°?

$$W = EI \cos\theta$$

$$W = 220 \times 22 \times \cos 30°$$

$$W = 220 \times 22 \times 0.866$$

$$W = 4191.4 \text{ watts, true power}$$

Where current is in phase with voltage in the example, then

$$W = EI \cos\theta \qquad \cos\theta = \frac{OA}{OB} = 1$$

$$W = 220 \times 22 \times 1$$

$$W = 4840 \text{ watts}$$

POWER FACTOR

Power supplied to a circuit is consumed in the resistance only. Inductance does not consume power but stores it during the period of increasing current and returns it to the generator during the period of decreasing current. A circuit having inductance and resistance then returns some power to the source. It follows that the true power of an inductance circuit is not the product of volts times amperes as measured by a voltmeter and an ammeter, respectively. It is less than $P = EI$ (power = volts × amperes) by the amount returned to the source. The amount of power not available for consumption by the circuit due to inductive reactance is represented on the sine curve by the negative areas discussed earlier (Fig. 311).

The power of an alternating current circuit when determined by the equation $P = EI$ where E and I are actual line measurements, is called the apparent power. Apparent power is expressed in terms of volt-amperes (va) or kilovolt-amperes (kva).

The apparent power less the power returned to the source by inductance is the *true* power of the circuit. The true power of the circuit is determined by a wattmeter designed to compensate for the power not consumed in the circuit. True power is expressed in terms of watts or kilowatts (kw = $\frac{\text{watts}}{1000}$). Thus an alternating-current circuit generally has two expressions of power: apparent power (expressed as kva) and true power (expressed in watts).

The ratio of true power to apparent power is called the *power factor*. It is expressed in one way by the equation

$$\text{Power factor} = \frac{\text{true power}}{\text{apparent power}}$$

For example, a circuit measuring 220 v and 30 amp by voltmeter and ammeter, which measures 5940 watts by wattmeter, would have the following power factor:

$$\text{Apparent power} = 220 \text{ v} \times 30 \text{ amp} = 6600 \text{ va}$$

$$\text{True power} = 5940 \text{ watts}$$

Then

$$\text{Power factor} = \tfrac{5940}{6600}$$

$$= 0.90$$

or

$$0.90 \times 100 = 90\% \text{ power factor}$$

Referring to the vector diagram (Fig. 312), it was found that $P = EI \cos\theta$, and that $\cos\theta = \dfrac{OA}{OB}$, where OA is current in phase with its voltage and OB the line current measured by an ammeter. OA, then, is the active current and OB is the active current plus the current influenced by inductive reactance. Therefore, OA is the true current and OB is the apparent current, so that OA times the voltage OC is the true power and OB times the voltage OC is the apparent power. It follows, then, that

$$\frac{OA}{OB} = \text{power factor}$$

but

$$\frac{OA}{OB} = \cos\theta$$

Therefore

$$\text{Power factor} = \cos\theta$$

If the

$$\text{Power factor} = \frac{\text{true power}}{\text{apparent power}}$$

then

$$\text{True power} = \text{apparent power} \times \text{power factor}$$

also

$$\text{True power} = \text{apparent power} \times \cos\theta$$

Where current is in phase with its voltage, then the true power equals apparent power and the ratio is unity (1). The cosine is therefore 1.00 so that the power factor for circuits in phase is 1.00 or 100%.

The power factor also equals the ratio of resistance in ohms to impedance in ohms. This is true whether inductive resistance, capacitance reactance, or both are present in the circuit. Therefore

$$\text{Power factor} = \frac{\text{resistance}}{\text{impedance}}$$

or

$$pf = \frac{R}{Z}$$

POLYPHASE CIRCUITS

Using one coil on the armature of an alternating-current generator provides a single-phase circuit. Where two or three coils are used, the circuits will be two-phase or three-phase, respectively. The latter are known as polyphase circuits. Single- and three-phase circuits are used most generally for refrigeration purposes. Single-phase circuits are made up of two wires while three-phase requires three- and sometimes four-wire circuits. The three-phase, three-wire circuit is the more common type.

In a true three-phase, three-wire system, each line or wire carries equal voltages which are 120 electrical degrees apart in phase. The equal voltages of a two-phase system are 90° apart in phase (Fig. 303). In order to use only three wires for a three-phase circuit instead of the six that would be necessary if the three generator coils were connected separately, a star or Y connection is used. Sometimes a delta (Δ) connection is employed. Conventional diagrams for the two types of connections are shown in Fig. 313.

The direction of current flow in a three-phase, three-wire circuit is as follows:

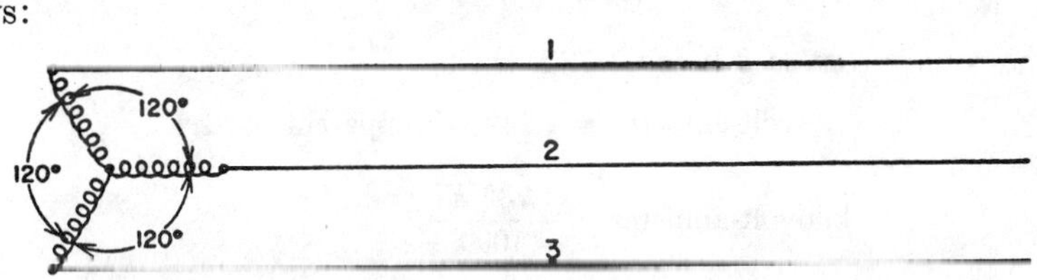

STAR OR WYE (Y) CONNECTION

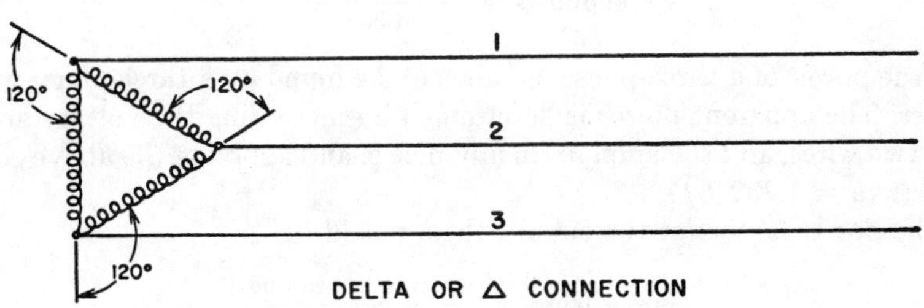

DELTA OR Δ CONNECTION

Fig. 313. Connection for three-phase three-wire a-c circuit.

Referring to Fig. 313:

(*a*) When current is leaving the generator through wires 1 and 2, current returns through wire 3.

(*b*) When current is leaving through wires 2 and 3, it is returning through wire 1.

(*c*) When current is leaving through wires 3 and 1, it is returning through wire 2.

The rule for three-phase, three-wire circuits is that each wire in turn carries current back to the generator as the two remaining wires feed the circuit.

Single-phase circuits may be obtained from a single-phase generator or from two wires of a three-wire, three-phase circuit. In the latter method, connections may be made from any two of the three wires. Where a number of single-phase connections are taken from a three-phase circuit, loads should be distributed evenly between the three wires in order to maintain a balance.

The apparent power of a three-phase circuit, when the currents in all three wires are equal and the voltage across any two wires is equal, is determined by the equation

$$\text{volt-amperes} = \sqrt{3} \times \text{volts} \times \text{amperes}$$

$$\text{va} = \sqrt{3}\, EI$$

The square root of 3 being 1.732, then

$$\text{volt-amperes} = 1.732\, EI \text{ (apparent power)}$$

or

$$\text{kilovolt-amperes} = \frac{1.732\, EI}{1000}$$

The true power of the same three-phase circuit is

$$\text{watts} = \sqrt{3}\, EI \cos\theta$$

$$\text{watts} = 1.732\, EI \cos\theta$$

or

$$\text{kilowatts} = \frac{1.732\, EI \cos\theta}{1000}$$

True power of a three-phase circuit may be found by a three-phase wattmeter. The apparent power is determined by measuring the voltage across any two wires, and the amperes in any phase, and applying the above equation (kva = 1.732 *EI*).

The power factor for the circuit, then, would be

$$\text{Power factor} = \frac{\text{wattmeter reading}}{1.732\, EI}$$

RELATIONSHIP OF A MAGNETIC FIELD TO AN INDUCTION MOTOR

Self-inductance is the principle of operation of an alternating-current motor called an *induction motor*. There are two major types of induction motors: (1) squirrel cage and (2) wound rotor. The squirrel-cage type will be discussed here.

The major parts of a squirrel-cage motor are (1) the stationary member or *stator* and (2) the rotating member or *rotor*. The cylindrical rotor revolves within a cylindrical stator but has no electrical connections, with either the stator or the power circuit.

The stator is made up of a laminated silicon steel core that lines the motor frame. Slots cut in the inner periphery of the stator core hold insulated coils that are grouped and connected to establish definite poles. The slots are partially closed to gain maximum pole area.

The rotor is made up of a number of copper or aluminum bars inserted axially in slots in the outer periphery of a laminated silicon steel core. These bars are not insulated but are joined together (short-circuited) at each end by end rings, the assembly of which resembles a squirrel cage—hence the name of this type motor. The core is mounted on a shaft or journal which, in turn, is supported by bearings fixed to end plates attached to the frame.

An air gap between the inner periphery of the stator core and the outer periphery of the rotor allows sufficient clearance between the two to allow the rotor to turn about its axis. The air gap is held to a minimum dimension, being approximately 0.15 in. for fractional horsepower motor to about 0.030 in. for larger motors.

The current from a power circuit flows around the stator windings setting up a magnetic field. Thus the stator becomes the primary circuit within the motor. The magnetic field produced by the stator causes an induced current to flow in the rotor circuit which, as stated above, has no other electrical contact. Therefore, the rotor becomes the secondary circuit in the motor, making the stator and the rotor the principal parts of a self-induced magnetic system.

It may be noted here that the induction motor, being similar to the self-inductance coil, causes the current in the power circuit to lag behind the voltage. Therefore, any circuit that includes induction motors tends to have a lagging current.

The simple act of energizing the stator windings does not by itself cause the rotor to revolve. It becomes necessary to direct the current through the stator windings in such a manner as to continually change polarity. This means that the strength of the magnetic field produced in the stator windings alternately changes from 0 to maximum in one phase of a two-phase circuit, for example, while changing from maximum to 0 in the other phase. These changes act to continually and progressively relocate the effective magnetic field in the air gap around the stator. Thus the resultant magnetic flux produced in the stator windings is relative in magnitude to current flow at any instant according to the sine curve of that current. This continual and progressive relocation of effective magnetic flux around the stator of a polyphase motor is called a *rotating magnetic field*. As the magnetic field rotates around the stator, it pulls the rotor after it.

Rotation of the rotor is due to the reaction of the magnetic field produced in the stator windings and the field set up by the induced current in the closed rotor circuit. The induced currents are at right angles to the magnetic lines of the stator current. The induced currents flow within the magnetic field. As the magnetic lines of force cross uniformly from the north to the south pole, they are distorted by the magnetic field of the induced current in the rotor conductors. The direction of distortion about the conductor depends on the direction in which induced current flows. The reason for this is that the lines of force of the magnetic field tend to follow the direction of the lines set up by induced current.

The lines of force tending to continue in a straight path exert a force on the conductor to push the latter out of the field. It is this force that causes the rotor to rotate. The action may be likened to that of a rubber band. If one end of the rubber band were held fast at the north pole, wound loosely over the nearest conductor and under the farthest conductor, to be held by the other end at the south pole, it would represent one line of force. By stretching the band suddenly, a force would be exerted downward on the first conductor and upward on the second conductor. If the conductors were free to move, they would be pushed out of the way by the rubber band in its effort to become straight during the stretching action.

The direction of rotation of the rotor can be determined best by applying Fleming's left-hand rule. By extending the thumb and the first two fingers of the left hand at right angles to each other, and letting the index finger point in the direction of the magnetic lines of force, the middle finger in the direction of the induced current, the thumb will point in the direction of the force or rotation.

Rotating Magnetic Field

Let the sine curve (Fig. 314, *top*), represent the currents in a two-phase circuit having a phase difference of 90°. Let Fig. 314 (*bottom*) represent the stator windings of a two-phase induction motor. Let *A* and *B* represent each phase of the two-phase circuit.

When the current in *A* is maximum at instant *I*, the current in *B* is zero. When the current in *B* is maximum at *II*, the current in *A* is zero. The case is similar at instants *III* and *IV*, respectively, except that direction of current flow is reversed.

When the cycle is at instant *I* on the sine curve, the flow of magnetic

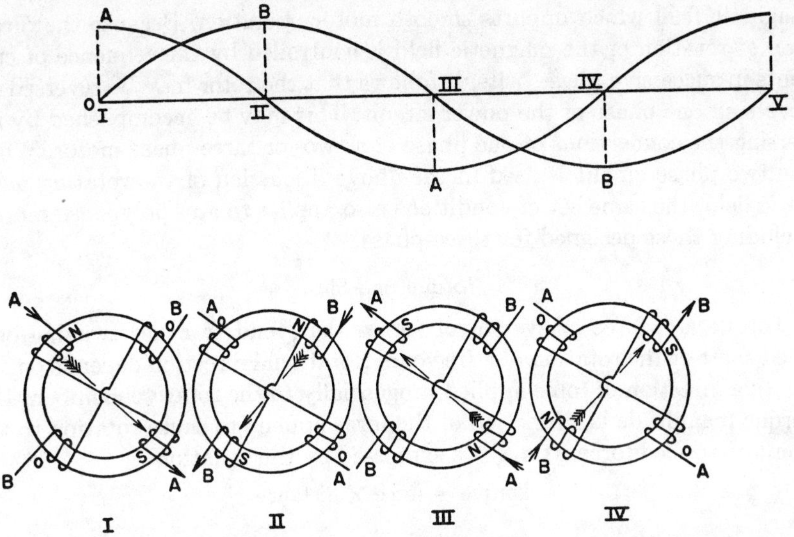

Fig. 314. Rotating magnetic field.

flux will be in the direction shown by the arrow in case *I* in Fig. 314 (*bottom*). Note that current flows into winding *A* at maximum value at the instant represented. The current in *B* winding is zero at the same instant. Polarity is as shown.

At instant *II*, no current flows in phase *A* but is at maximum in phase *B*. The flow of magnetic flux has now tilted, as shown by the new direction of the arrow.

At instant *III*, the current in *A* is reversed so that it flows into the windings as indicated to change polarity. Magnetic flux has again tilted to flow in the direction opposite to that at instant *I*. At instant *III*, current *B* is zero while *A* is maximum.

At instant *IV*, current *B* has reversed direction to cause the magnetic flux to flow as shown.

At instant *V*, the magnetic field will have made one revolution and is prepared to repeat the above performance, it being similar to instant *I* at this point.

The whole action is similar to that if a magnetic field produced by an external source is allowed to rotate mechanically around the rotor. Here the stator itself does not move (i.e., rotate), but the magnetic field does by reason of travel of magnetic flux around the air gap between the rotor and the stator.

While the description shows only four basic positions of magnetic field flow, many more could be plotted by considering midpoints along the sine curve. Actually, the induction motor is designed with a series of phase belts in which some overlapping of windings occurs. This results in a continuous magnetic field which imparts smooth motor operation. Because the direction of rotation of the magnetic field is controlled by the sequence of currents in successive phase belts, it follows that the rotor may be reversed by reversing one phase of the power circuit. This may be accomplished by reversing the connections of one phase of a two- or three-phase motor. While the two-phase circuit is used in the above discussion of the rotating magnetic field, the same set of conditions also applies to any polyphase motor, including those designed for three-phase.

Torque and Slip

Torque is a force or system of forces that tend to cause twisting of a fixed shaft or the rotation of a free one. Rotational direction depends on the relative direction of force applied tangentially to the rotative member. The torque magnitude is a function of the force applied to cause rotation to the point of applied force. It may be expressed by the equation

$$\text{Torque} = \text{force} \times \text{distance}$$

or

$$T = f \times d$$

where f in pounds $\times$ d in feet is the moment of force expressed in pound-feet.

The torque of an electric motor is dependent on the reaction of the rotating field flux and the induced current in the rotor. The lines of force must be cut by the rotor conductors to gain this reaction to obtain any torque. If the speed of the rotor were equal to that of the revolving field, no lines of force would be cut and there would be no induced rotor current, and no torque would be developed. It is necessary that the rotor of an induction motor rotate at some speed below that of the magnetic field to enable the rotor conductors to cut lines of force. This is made possible by the effect of any load, however small, placed on the shaft tending to reduce rotor speed. Under so-called no-load conditions, the electromotive force required to force no-load current through the rotor windings is small, and is obtained by a slight difference between the speed of the rotating magnetic field and rotor speed. This slight difference develops sufficient torque to overcome friction, windage, and other forces tending to resist motion and thus cause the rotor to turn. When load is applied, the rotating field cuts the rotor bars at a greater rate. This in turn develops higher induced currents in the rotor, resulting in greater torque.

The torque of an alternating-current motor is proportional to the square of the voltage impressed across the stator windings. It may be expressed approximately by the equation $T = E^2$. Modifying the product for certain design and other motor factors, then $T = E^2 \times k$, where k is a design constant.

Running torque is expressed as

$$\text{Torque (in ft-lb)} = 5250 \times \frac{\text{hp}}{\text{rpm}}$$

The speed of the magnetic field is in synchronization with the power frequency and is known as synchronous speed. The speed of the rotor under load is called load or running speed. The difference between synchronous and running speed is known as slip. Slip varies with load; i.e., slip increases as load increases and is expressed as percent of synchronous speed determined by the equation

$$\%\ \text{Slip} = \frac{\text{synchronous speed} - \text{running speed}}{\text{synchronous speed}} \times 100$$

Percentage of slip, being associated with load, is expressed usually in terms of full load.

Example: Determine the slip of an induction motor having four poles and operating at 60 cycles, when its load speed measures 1746 rpm.

$$\text{Synchronous speed} = \frac{\text{frequency, cycles per minute}}{\text{no. of pairs of poles}}$$

$$\text{rpm} = \frac{60 \times 60}{2}$$

$$\text{rpm} = 1800 \text{ synchronous speed}$$

$$\%\ \text{Slip} = \frac{\text{synchronous speed} - \text{load speed}}{\text{synchronous speed}} \times 100$$

$$= \frac{1800 - 1746}{1800} \times 100$$

$$= 3\%\ \text{slip}$$

The synchronous speeds for polyphase induction motors having different number of poles and operating under a frequency of 60 cycles are shown in Table 38.

Table 42. Synchronous Speeds of Alternating-Current Motors According to Number of Poles and Frequency

No. of Poles (Pairs)	*Speed (rpm)*		
	60 cycles	50 cycles	25 cycles
2 (1)	3600	3000	1500
4 (2)	1800	1500	750
6 (3)	1200	1000	500
8 (4)	900	750	375
10 (5)	720	600	300
12 (6)	600	500	250
24 (12)	300	250	125

Actual full-load speeds of induction motors will average 4 percent less than synchronous speed.

Starting Torque. The effort exerted by the motor to turn the rotor when full voltage is applied at the instant of starting is called the *starting* torque. Motors are designed to develop a normal starting torque or a high starting torque. The torque characteristics of an induction motor are a function of the design and the resistance of the rotor circuit—the greater the resistance, the higher the starting torque. The starting torque of any motor is based on its full-load torque and expressed in percent of the latter value. For example, usually a two-pole induction motor will have a minimum starting torque 1.5 times (150 percent) its full-load torque. This ratio diminishes as the number of poles increase so that an eight-pole motor should have a starting torque no less than 125 percent of full-load torque. While the above values represent normal starting-torque ratios, high starting-torque values reach 300 to 325 percent of full-load torque. To compensate for any decrease in running torque during switch-over periods, repulsion start motors are designed to produce starting torques up to 500 percent.

To increase the starting torque of squirrel-cage motors, end rings of high resistance are employed. However, there is a certain rotor resistance value at which torque reaches its maximum for a given polyphase motor. Above the optimum resistance value, torque decreases.

Breakdown Torque. The torque of an induction motor increases with slip until a maximum value is reached. Any further increase in slip beyond the point of maximum value results in a decrease in torque. The maximum torque value of any induction motor operating under rated voltage and frequency is called breakdown torque.

Pull-up Torque. The minimum torque developed by an induction motor during its acceleration from zero to full speed under rated conditions is called pull-up torque.

Pull-in Torque and Pull-out Torque. Torques designed as pull-in and pull-out will be discussed in the section on synchronous motors, the type with which they are associated.

CLASSIFICATION OF ELECTRIC MOTORS AND THEIR CHARACTERISTICS

Electric motors employed to drive fans, pumps, and compressors may be classified according to their horsepower rating, viz.:

(*a*) Fractional-horsepower motors—less than 1 hp

(*b*) Integral-horsepower motors—1 hp or more

Fractional-Horsepower Motors

Small fractional-horsepower motors are manufactured in sizes from $\frac{1}{10}$ to $\frac{3}{4}$ hp at 1800 rpm. They are employed to drive smaller types of fans, pumps, and compressors. Because they are usually single-phase, these motors are not primarily self-starting, hence they require special automatic devices to eliminate the necessity of starting them by hand. The method used to produce self-starting usually provides the basis of their nomenclature as follows:

Single-phase induction motor, manual starting
Split-phase induction motor
Capacitor-start induction run motor
Capacitor-start and run motor
Repulsion-start and run motor
Shaded-pole motor

Single-phase motors are designed for either 110 v or 220 v.

Single-Phase Induction Motor

The single-phase induction motor is similar to the polyphase squirrel-cage induction motor in its construction. Only one winding is employed, which does not cover the entire internal periphery of the stator as do the two or three windings of polyphase motors. The rotor is of the squirrel-cage design.

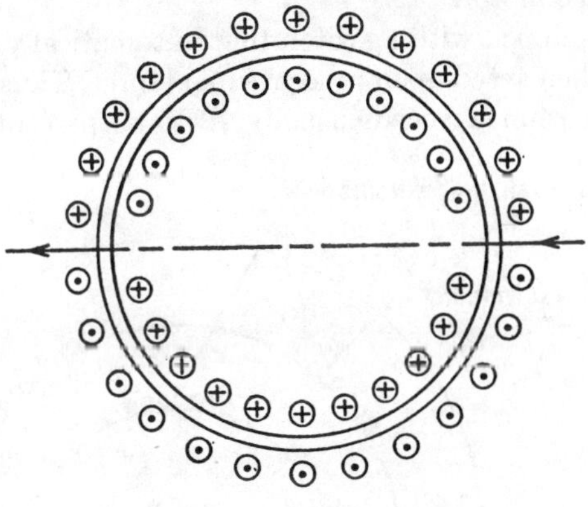

Fig. 315. Single-phase motor.

Because an alternating flux is produced along one axis only, as shown in Fig. 315, there is no rotating magnetic field. It is also evident from the polarity of the windings and by applying Fleming's rule that the current in the rotor flows in the opposite direction to that in the stator on both halves of the motor. Therefore, no torque will develop to start the motor. Such a motor must be started by turning the rotor by hand.

If the rotor of a single-phase motor is started by hand in either direction, lines of force will be cut and a rotating field will be produced to keep the rotor turning. Upon approaching synchronous speed, the single-phase motor operates in a manner similar to a polyphase induction motor. However, instead of the smooth progressive rotating field developed in the latter, the single-phase motor produces more of an oscillating field. As it is not self-starting, the simple single-phase induction motor is not used widely.

To develop starting torque and make the single-phase motor self-starting, the devices described for the following types of fractional-horsepower motors are employed.

Split-Phase Induction Motor

The split-phase induction motor is a single-phase machine equipped with a main or running winding and a starting winding. Both windings are located on the stator 90 electrical degrees apart, making it similar to a two-phase motor arrangement.

The starting winding is made up of fine wire to give it high resistance and low inductance. The running winding is more coarse and has low resistance and high inductance. The high inductance of the running winding gives it a choke-coil effect, causing the running current to lag behind the current in the starting winding. This lag plus the 90° winding displacement around the stator produces a twist on the rotor to effect a weak but sufficient rotating magnetic field, thus giving the single-phase motor the starting characteristics of a polyphase motor.

The rotor is equipped with a switch that automatically breaks the starting-coil circuit when acted upon by centrifugal force. Thus, when the motor reaches a predetermined speed, usually 70 to 80 percent of synchronous

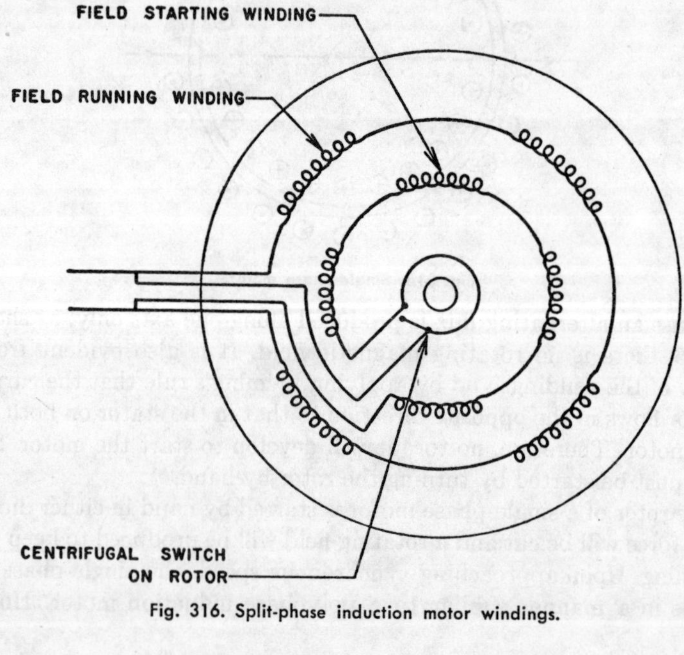

Fig. 316. Split-phase induction motor windings.

speed, the centrifugal switch acts to cut out the starting winding. When this occurs, the only circuit energized is the running winding which causes it to operate as a single-phase induction run motor.

The split-phase induction motor may be designed for normal or high torque; the former is the more desirable type because of the greater starting current required by the high-torque split-phase motor. By appropriately reversing the starting winding connections, the split-phase motor can be made to rotate in either direction. These motors are employed for loads up to $\frac{1}{3}$ hp with no-load starting. The hermetically sealed domestic unit equipped with a capillary tube in which pressures become equalized during off-cycle is a good example of where split-phase motors are suitable. When used on systems not designed for off-cycle pressure equalization, some method of compressor unloading must be employed.

Capacitor-Start Induction Run Motor

To obtain high starting torque without excessive starting current, the capacitor-start induction run motor is employed. This motor is similar to the split-phase motor except that it is equipped with a condenser or capacitor placed in series with the starting winding. Usually the capacitor is mounted on top of and outside the motor frame.

By placing a capacitor in series with the starting winding, the excessive current flow experienced in the split-phase motor is reduced appreciably. The analogy here would be the initial inrush of fluid into an empty hydraulic cylinder upon demand for power compared with the smoother reaction obtained when a surge of pressure is applied to the cylinder already filled with fluid. Thus the capacitor has the opposite effect to the inductance of the split-phase motor. Upon reversal of alternating-current flow, the capacitor discharges, as previously explained, and a flow of current is initiated in the starting coil circuit. When this occurs, the motor acts in a manner similar to that of the split-phase motor. Again the higher resistance of the starting winding causes its current to lag behind that in the running winding not withstanding their common origin in the single-phase power circuit.

When the rotor reaches a predetermined speed, an automatic centrifugal switch mounted on the rotor breaks the capacitor circuit, cutting out the condenser and starting winding. With the starter winding cut out, the motor operates on the running winding only, as did the split-phase motor.

It is noted here that motors requiring heavy current flow (inrush) to produce sufficient starting torque have several disadvantages. One is the effect on other electrical components connected in the same circuit, such as the temporary fading of lamps, radios, and television sets.

The capacitor-start induction run single-phase motor is also manufactured in sizes from fractional up to 5 hp. Having a starting torque of 300 percent or better, they are used for compressors starting loaded.

Capacitors are rated in microfarads and should always be sized according to the load demands of the motor, never undersized.

A capacitor-start capacitor run motor is one in which the capacitor remains in the circuit even when the motor is running at full speed.

Fig. 317. (*Top*) Capacitor-start induction motor. (Courtesy, General Electric Company.)

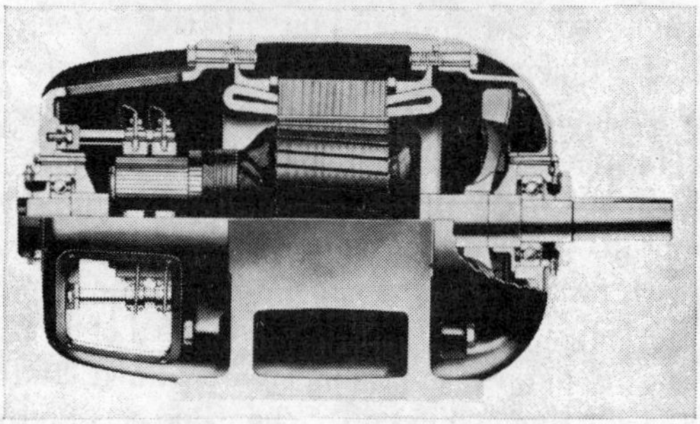

Fig. 318. (*Bottom*) Single-phase repulsion induction motor. (Courtesy, General Electric Company.)

Repulsion-Start Induction Run Motor

The single-phase repulsion-start induction run motor has one winding in the stator instead of two, as found in the split-phase and the capacitor types. In place of the axial bars used in squirrel-cage motors, the rotor is wire wound to become a wound armature. The armature windings are connected to a commutator composed of axial bars or rings encircling the motor shaft, each of which is insulated in respect to its neighboring bars or rings. In contact with the commutator is a set of stationary carbon brushes that ride the commutator as the latter revolves. The brushes are connected to each other but are not connected to the power line. Having no electrical contact with either the stator winding or the power circuit, the armature is excited by induction produced in the stator winding. The position of the brushes in respect to the poles governs the starting ability of this motor.

If the brushes were placed on the pole axis, no torque would develop because half the armature conductors under the pole would have their current flowing in the direction opposite to that flowing in the other half of the conductors. By placing the brushes at right angles to the pole axis, no current would flow in the armature because the induced voltage in each path would neutralize each other. There would be no voltage at the brushes under the latter condition. By placing the brushes between the two or at an angle to both axes, the direction of the current in the armature windings will be such as to produce a rotating magnetic field, cut lines of force, and develop torque. With the brushes in the latter position, the armature will turn when flux is produced in the stator winding.

When the rotor reaches full load speed, the commutator bars are automatically short circuited, thus forming a squirrel-cage circuit in the armature. When this occurs, the motor runs as a single-phase induction machine.

The repulsion-start induction run motor has a very high starting torque developed with relatively low current. Motors of this type are manufactured in both fractional and integral horsepower ratings. In some types of repulsion-start motors, the brushes are lifted away from the commutator by the same centrifugal device that short-circuits the bars. This is done to reduce brush and commutator wear. To distinguish between the two types, sometimes they are designated as either brush-riding or brush-lifting motors.

Shaded-Pole Motor

The shaded-pole motor is a small single-phase motor used for fans and blowers mounted on the rotor shaft. They are made in sizes from $\frac{1}{12}$ up to $\frac{1}{6}$ hp.

The stator of a shaded pole motor has a single winding that forms one coil around each pole. The coils are connected in series. In addition, each

Fig. 319. Integral horsepower, three-phase, squirrel-cage motor. (Courtesy, General Electric Company.)

pole is provided with a copper coil that is wound around one half of each pole piece. Sometimes a copper plate is inserted in one half of each pole piece in place of the copper coil. Either one acts as a shading device in which an induced current is produced by the flux from the main field windings. The latter flux also produces an induced current in the rotor.

As the impressed current flows through the pole coils, it sets up a magnetic field in the stator at each pole, which are 90 electrical degrees apart. At the same time, an induced current is produced in the shading pieces but at a slower rate than that set up by the free-pole areas. This causes a sufficient lag in flux between the shaded- and the free-pole areas to set up a rotating magnetic field to start the rotor turning. This action is maintained as the polarity of the poles changes with each alternation of the alternating current to continue rotation until full speed is reached. Upon approaching full speed, the action is similar to any other single-phase squirrel-cage motor.

It is evident from the slight lag in flux build up between two halves of a pole that the starting torque of a shaded-pole motor is very low. It is also subject to high slip if loaded much beyond forces incurred by overhung fans or blowers, i.e., fans or blowers having their bearing in common with the motor.

Integral-Horsepower Motors

When considering requirements of motors of larger than fractional horsepower sizes, several operating factors must be considered in addition to the normal features expected, such as proper size, quality, etc. Such operating factors include the following:

1. Phase requirements—single or polyphase,
2. Normal or high starting torque,
3. Constant or multi-speed,
4. Open, partially or totally enclosed—explosion-proof, drip-proof, moisture-proof, lint-proof, etc.,
5. Starting current limitations,
6. Available voltage,
7. Power factor,
8. Alternating or direct current,
9. Frequency—25, 50, or 60 cycles (or other),
10. Continuous or intermittent service.

While most larger motors are of the polyphase type—i.e., two- or three-phase—some of the smaller integral motors, 1 to 10 hp, are single phase. The latter requirement may be satisfied by the use of single-phase capacitor-start induction run or repulsion-start induction run motors previously described. Whether normal or high starting torque motors are used depends on the starting load conditions. Intermittent operation in which compressors are subject to frequent starting under load would demand a high starting torque motor. Continuous operation of compressors started unloaded requires a normal starting torque motor. The method of capacity control often governs the speed characteristics of a motor. Where such controls are based on a variable refrigerant displacement, say in one compressor, a multi-speed motor becomes necessary. Such a motor would not necessarily be required where constant loads prevail or where capacity changes are met through the use of multiple compressors, cylinder unloading devices, cycling, by-pass to suction, etc.

Motors are often manufactured having their windings fully enclosed. One reason for doing this is to prevent explosion that may result from sparking within an atmosphere containing explosive gases, dust, etc. Exclusion of foreign materials such as lint and water from the internal parts of the motor requires partially closed motors. Surrounding conditions then dictate whether a motor should be open or totally or partially enclosed. Because of the effect of high starting current on the main power supply, certain restrictions set by utility companies limit the amount of current required in starting motors connected to their lines. These same limitations apply to privately operated power plants. Motors possessing normal starting current characteristics should always be selected. In this connection, the National Electrical Manufacturers Association (NEMA) classifies squirrel-cage induction motors as follows:

Class A—Normal starting torque, normal starting current,
Class B—Normal starting torque, low starting current,
Class C—High starting torque, low starting current,
Class D—High slip.

Class B motors are the more widely used. As stated above, they are suitable for compressors started unloaded as well as for centrifugal pumps and fans. Class C motors are required for compressors started loaded.

Integral-horsepower motors are available for 220, 440, and 550 v. Motors of higher voltage are also available for special purposes. Both alternating-current and direct-current motors are used. However, except for those requirements necessitating direct-current machines, the alternating-current motor rapidly is replacing the latter in all phases of the refrigeration industry. Where alternating current is supplied for power, an important factor is its frequency. Motors are more efficient when operated at their design frequency or if the frequency variation is maintained within a 5 percent range above or below normal rating. Circuit frequency is fast becoming standardized at 60 cycles for general purposes. However, frequencies of 25, 40, and 50 cycles are also to be found for both general and special purposes.

Where a number of induction motors are in service within an industrial plant, the circuit power factor is lowered. A low power factor means a greater current demand by induction motors for a given power output. To correct a low power factor, consideration should be given the possible use of one or more synchronous motors for reasons discussed later in this chapter.

Usually motors are designed to give highest efficiency when loaded to their capacity rating or full load. For highest plant efficiency, then, the size of the motor selected should be not much greater than the maximum load it is expected to drive. The efficiency of a motor is expressed by the equation

$$\% \text{ Efficiency} = \frac{\text{output}}{\text{input}} \times 100$$

or

$$\% \text{ E} = \frac{\text{horsepower delivered}}{\text{kilowatts required}} \times 100$$

in which horsepower must first be converted to kilowatts. Efficiencies of motors used for general purposes range from 70 to 90 percent, depending on size, load, speed, and type.

A temperature rise occurs in motors being placed in operation. The degree of rise normal for a given motor is usually shown on the name plate along with other pertinent information such as horsepower rating, full-load current speed, frequency, etc., and indicates the temperature rise above ambient temperature. The standard temperature rating is 40 C or 72 F rise at full load. The temperature rise represents loss of energy converted to heat as a result of conductor resistance, core losses, friction (bearings and brushes), and windage. The power loss by conductor resistance is a major factor determined by the equation

$$\text{Power loss} = \text{resistance} \times \text{amperes}^2$$

$$P = RI^2$$

where P = power loss in watts
R = resistance in ohms
I = amperes

It is obvious that any increase in current demand by a motor for any reason raises its power loss and increases operating temperature.

To dissipate heat from a motor, some form of ventilation or radiation is necessary. Open type motors are equipped with fan blades attached to the rotor which forces a continual stream of room air through the motor. Fully enclosed motors are designed either fan cooled or with an extra large casing to provide sufficient radiating surface to control the temperature. It is often necessary to provide ventilation by an external fan if the location of the motor is such that proper cooling cannot be maintained otherwise. Overloading a motor will heat it beyond the normal temperature rise, even to the point of smoking and eventually burning the windings. If the current input of a motor, as measured by an ammeter, exceeds the full load current as marked on the name plate, then overloading exists. Steps should then be taken to correct the condition.

To protect the motor against overloading, fuses, circuit breakers, or overload relays are used. Overload relays are usually part of the motor starter while fuses and circuit breakers are located at the circuit switch. Overloads are expressed in terms of percent above full load rating.

Electric motor bearings may be a plain sleeve or of an anti-friction type. The latter includes ball or roller bearings. Usually the sleeve types are lubricated by means of a ring carrying oil up from a reservoir to the journal on which it rides. Anti-friction bearings are run in oil or grease lubricated. Application of grease may be by pressure gun or hand packed. Hand grease packed anti-friction bearings are usually sealed and one application should last about four years, depending on the quality and the adaptability of the grease.

There are two classes of alternating-current integral horsepower motors used in the refrigeration field. One is the induction motor, of which there are two general types: (1) squirrel cage and (2) slip-ring or wound rotor. The other is the synchronous motor.

Squirrel-Cage Induction Motor

The construction and the principle of operation of the squirrel-cage motor have already been discussed. This motor finds wider use than any other type. It is highly efficient when correctly applied. The speed of a squirrel-cage motor is nearly constant when operated at rated load and voltage. It is subject to slip, the average being 4 percent at normal load. Any induction motor lowers the power factor of its circuit and the squirrel cage is no exception.

The starting torque of a squirrel-cage motor increases with increased rotor resistance. High starting torque motors are motors having increased slip which adversely affects efficiency. To obtain one, some sacrifice is made in the other. Therefore, where normal torque motors can be used, high torque motors should not be substituted. Polyphase squirrel-cage motors are either two- or three-phase. They may have any number of poles multiple by two. Table 42 lists the synchronous speed of induction motors having various numbers of poles.

Low starting current motors are made so by incorporating within them high reactance. Motors classed as high slip motors have low starting current, but have a slip of 15 to 20 percent compared with a normal value of $3\frac{1}{4}$ to 4 percent for conventional Class B (NEMA) motors.

There is also a double squirrel-cage motor. This is one having two squirrel cages in the rotor: an inner one and an outer one relative to the periphery. The outer cage has high resistance while the inner one has low resistance. The high resistance of the outer cage produces high starting torque. The low-resistance inner cage allows most of the current to be carried in it during full-load speed periods.

The inductance of the outer cage is lower than that of the inner cage because of the better magnetic conditions of the latter, surrounded as it is on practically all sides with iron. During the starting period, the choke coil action of self-induction at line frequency greatly impedes current flow in the inner cage. This causes a greater current flow in the outer cage upon starting, creating a higher torque than would be obtained in a conventional squirrel-cage motor. As speed increases, the choke coil effect of the inner cage decreases as the frequency of the currents induced in it decreases. Because of the lower resistance of the inner cage, it carries the major percentage of the rotor current when the motor is running. The advantages of the double squirrel-cage motor are (1) fairly high starting torque with relatively low starting current, (2) better operating efficiency and speed control than the high-resistance single squirrel-cage (high torque) motor.

To obtain low starting current in any squirrel-cage motor, the rotor bars must be of such design and location relative to the periphery (of the rotor) as to produce high reactance. Thus the variously classed motors such as those listed by NEMA as Class A, B, C, and D will have their rotor bars designed and located somewhat as follows:

Class A—normal starting torque, normal starting current squirrel-cage motor—rotor bars close to the outer periphery of the rotor and rectangular in cross-section. Reactance is low.

Class B—normal starting torque, low starting current squirrel-cage motor—rotor bars are narrow and of some depth within the rotor. Reactance is high during starting.

Class C—high starting torque, low starting current squirrel-cage motor. Double squirrel-cage rotor is used. Reactance is high, slip at rated load is relatively low.

Class D—high slip squirrel-cage motor—rotor bars are rectangular and somewhat narrow, having high resistance. Slip of 15 to 20 percent limits starting current.

Because of the inductance of squirrel-cage motors, their power factor is low—between 0.70 and 0.80 according to the number of poles and the size of the motor. It becomes lower as loads decrease. The power factor is also affected adversely as the air gap becomes wider. Air gaps, therefore, should be made as small as possible within mechanical feasibility and in consideration of bearing wear.

Slip-Ring or Wound Rotor Induction Motor

The slip-ring or wound rotor induction motor is similar to the squirrel-cage motor except in the rotor winding. Instead of a squirrel cage, the slip-ring rotor is wound with insulated coils. The coils are grouped as a star-connected three-phase winding in such a manner as to have an equal number of poles as the stator. The terminals of the rotor windings are connected to slip rings that encircle the shaft. Stationary carbon brushes are in contact with the slip rings. Leads from the brushes are connected to an external variable resistance. There is no electrical connection between rotor and stator, and operation of slip-ring motors is the same as that of the squirrel-cage motor.

The external resistance may be adjusted by means of a drum controller. It was stated previously that the starting torque of an induction motor may be increased by increasing rotor resistance. With the external variable resistance in each phase of the rotor circuit, starting torque may be controlled. With any adjustment toward greater rotor resistance, higher starting torque can be developed. By gradually diminishing the rotor resistance as the motor develops speed until, at full load speed, all resistance is cut out, the rotor windings become short-circuited and the slip-ring motor operates with the low slip and the high efficiency characteristics of a squirrel-cage motor.

Another effect of resistance in a rotor circuit is to reduce speed. By using a drum controller in which resistance can be increased or decreased by bringing more or less resistors into the circuit, the speed of the motor can be

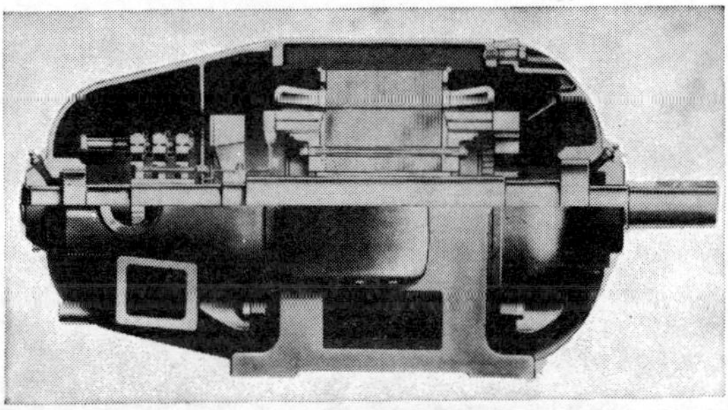

Fig. 320. (*Top*) Integral horsepower, wound rotor (slip ring) motor. (Courtesy, General Electric Company.)

Fig. 321. (*Bottom*) Motor-generator set (Amplidyne) consisting of four generators and 7½-hp motor. Note flexible couplings. (Courtesy, General Electric Company.)

controlled between normal full load speed (with all resistances cut out) and 50 percent of normal full load speed.

Characteristics of the slip-ring motor are: (*a*) high starting torque, (*b*) low starting current, (*c*) adjustable speed, (*d*) a higher power factor than a comparable squirrel-cage motor, (*e*) better starting conditions as to torque vs. current can be obtained by careful manipulation of the controller.

Because of its high starting torque, the slip-ring motor is used on reciprocating compressors that are started under load. Their speed-adjustment feature applied as a capacity control makes this type of motor adaptable for centrifugal compressor operation. To obtain maximum compressor speed, the compressor and the motor are connected through a speed-increasing gear set.

Synchronous Motors

The synchronous motor is a constant-speed alternating-current motor usually of the polyphase type. In the larger sizes, 100 hp and above, it is the most efficient electric motor exceeding the squirrel-cage induction motor in this regard. It consists of a revolving field attached to its journal and a stationary armature that surrounds the field and mounted inside the motor housing or frame. The armature is so wound as to have the same number of poles as the field.

Fig. 322. Synchronous motor, revolving field. Showing grids, and rings that connect field with exciter. (Courtesy, Allis-Chalmers Manufacturing Company.)

Fig. 323. Stationary armature of synchronous motor. (Courtesy, Allis-Chalmers Manufacturing Company.)

The armature is connected and draws its current from the alternating-current power line while the field is energized or excited by direct current from a d-c generator. The latter may be mounted on or connected by a belt to the same journal as the revolving field or located at some point remote from the motor. Connection is made through a brush and ring assembly.

The speed of a synchronous motor is fixed by the frequency of the alternator supplying current and the number of poles. The speed in revolutions per minute (rpm) may be determined by the equation

$$\text{rpm} = \frac{\text{frequency in cycles per sec} \times 60 \text{ sec per min}}{\text{no. of pairs of poles}}$$

Thus, for a four-pole motor operating at a frequency of 60 cycles, the speed would be

$$\text{rpm} = \frac{60 \times 60}{2}$$

$$\text{rpm} = 1800$$

Fig. 324. Synchronous motors driving ammonia compressors. (Courtesy, Hershey Chocolate Corporation.)

Its speed, therefore, is the same as the alternator from which it receives its a-c power, or some fixed multiple of that speed according to the number of poles.

The synchronous motor is not inherently a self-starting machine, although it develops a revolving magnetic field because of its polyphase characteristics. This is due to conflicting polarity by reason of basic design. In a simple synchronous motor, the turning force reverses itself each time a pole of the revolving magnetic field, set up in the armature, passes a nonrotating field pole located on the rotor. This action neutralizes the turning effort of the magnetic field so that the rotor remains motionless.

Synchronous motors are modified to make them self-starting, otherwise starting must be accomplished by using some external means, such as a squirrel-cage induction motor. Actually, the modification is carried out by placing squirrel-cage windings on the revolving field poles. Consisting of short-circuited copper bars placed in slots cut in the pole faces to form a grid, such windings enable the synchronous motor to start as a polyphase induction machine. Because the grid provides low resistance to induced current developed by the revolving magnetic flux, a torque is produced that starts the rotor to turn. At the time of starting, the a-c power circuit is closed, developing a revolving magnetic field in the stator. The d-c circuit is open so that only induced current flows in the grid.

The rotor does not attain synchronous speed when operating by induction only because of slippage involved even under no-load conditions. However, when maximum no-load induction speed is reached, say from 3 to 5 percent below synchronous speed, the direct-current line is closed to energize or excite the field. When this occurs, the rotor is pulled into synchronous speed and is "locked" at that speed by the force developed by the revolving magnetic flux and the excited field.

As long as the motor operates at no load, the field poles of the rotor and the poles of the revolving magnetic field of the armature tend to remain locked in respect to each other. However, even the slight load imposed by internal friction of the motor prevents the center line of the opposing poles from becoming congruent although they are rotating at the same speed. If this were not so, no torque would be developed to keep the rotor turning. Under no-load conditions, very little current is drawn from the power line. As load in the motor increases, the relative positions of the field poles shift and the angularity of the pole center lines increases correspondingly. This develops an ever-increasing torque. The rotor speed remains constant regardless of the pole center-line angularity within the torque limitations of the motor. However, when the pole center-line angularity exceeds its position of maximum torque, which is just beyond the half-way point between adjacent poles, synchronism no longer exists and the motor will stop. In other words, when the load is great enough to force the rotating poles of the field to lag the magnetic flux beyond the angle of maximum torque, the balance between motor-torque and load-torque requirements is broken. When the relationship of magnetic flux poles to rotor field poles is such that they are not locked, the motor is out of step with the alternations of a-c current and will cease running. Unless the power line is opened immediately at this point, the motor will burn out.

The maximum torque of a synchronous motor is called the "pull-out torque." The torque or load capable of being pulled into synchronism by the motor upon starting up is called the "pull-in torque." The starting torque is the torque developed when full voltage is impressed across the armature winding. Starting torque and pull-in torque usually are expressed in terms of percent of full-load torque.

The starting torque of a synchronous motor depends on the size and the resistance of the squirrel-cage grid. Incidentally, the grids are also called dampers because they damp out oscillations of the rotor when running at synchronous speed, an action better known as "hunting." They are known also as amortisseur windings. Starting or inrush current becomes greater with increases in starting torque.

In addition to driving such equipment as compressors, the synchronous motor may be operated so as to act as a condenser for controlling the power factor of the a-c circuit. This is highly advantageous where a number of induction motors are in service, especially where a plant generates its own power. Where synchronous motors are so used, they should be located at the end of the power transmission line and in parallel with induction motors.

Changes in power factor are brought about by varying the strength of the direct current applied to the field windings. It has been demonstrated previously that, when the current in an a-c circuit is in phase with its voltage, the power factor for the circuit is unity. In a synchronous motor when the field strength—i.e., the direct-current amperage—is changed, the angle of phase between the armature current and the armature voltage is also changed. There is a point, however, when the relation of the field current to load is such as to cause a unity power factor. When this condition exists, the current in the armature is lowest for a given load. Reducing the field strength below that corresponding to unit power factor for a given load, the armature current will lag behind its voltage. Therefore, a lagging power factor less than unity results, and a lagging power factor results when the field strength is weakened.

Increasing the field strength above that corresponding to unity power

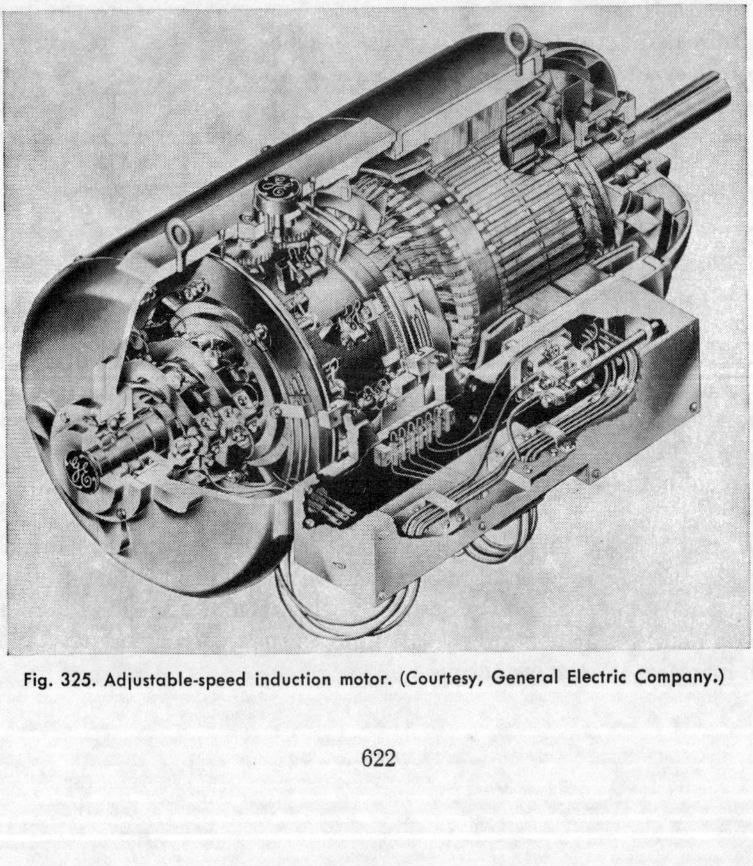

Fig. 325. Adjustable-speed induction motor. (Courtesy, General Electric Company.)

factor for a given load, the motor acts as a condenser and the armature current will lead its voltage.

When the field current is such as to bring armature current and voltage in phase and cause a unity power factor, it is said to be normal. The field is underexcited when its current is below normal, causing a lagging power factor. When the field current is greater than normal to cause a leading power factor, it is overexcited.

Maintaining a power factor at some value approaching unity, closer voltage regulation and higher circuit efficiency can be maintained.

Assume a circuit having a rated load capacity of 100 amp at 24,000 v, or 2400 kw, at unity power factor. If the circuit is reduced to a power factor of, say, 0.707, corresponding to a current lag of 45° behind its voltage, the true power would not be 2400 kw. Applying the equation:

$$\text{True power} = \text{apparent power} \times \text{power factor} (\cos\theta)$$

then

$$\begin{aligned}\text{True power} &= IE\cos\theta \\ &= 2400\cos 45^\circ \\ &= 2400 \times 0.707 \\ &= 1696.8 \text{ kw}\end{aligned}$$

Installing a synchronous motor and overexciting the field would tend to establish a leading power factor by counteracting the lagging effect of induction motors. This would bring the total power factor nearer to unity. At, say, 0.95 power factor, the true power for the above circuit would be:

$$\begin{aligned}\text{True power} &= 2400 \times 0.95 \\ &= 2280 \text{ kw}\end{aligned}$$

which represents a considerable savings for any power-generating plant.

Fig. 324 shows several ammonia compressors driven by synchronous motors.

A typical installation may consist of a 375-ton reciprocating compressor driven by a 500-hp, 300-rpm synchronous motor. In such a case, the electrical requirements of the armature may be 2300 v, 99.5 amp, unity power factor, three-phase, 60 cycles. The field would be excited normally with a direct current of 125 v and 41.5 amp when the load on the motor is 100 percent. Using a slower speed synchronous motor to drive the compressor, the field current normally would be higher. At 325 rpm it may reach 50 amp. Synchronous motors for refrigeration purposes are also made to operate on 220- and 440-v circuits. Larger type ammonia compressors driven by synchronous motors usually are run at speeds of 180, 225, or 300 rpm. They are started unloaded.

High-speed synchronous motors are used to drive centrifugal compres-

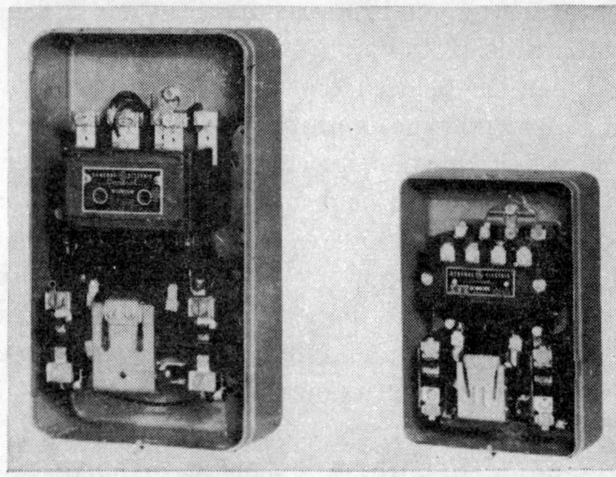

Fig. 326. Motor controls. (*Top, left*) Overload relay; (*top, right*) magnetic motor starter; (*center, left*) synchronous motor, magnetic, full-voltage starter; (*center, right*) primary-resistance motor starter; (*bottom, left*) multispeed controller; (*bottom, right*) drum switch. (Courtesy, General Electric Company.)

sors. To attain speeds greater than that of the motor, say, 6500 rpm, a speed increaser such as that shown in Fig. 333 is required. The speed increaser is placed between the motor and the compressor.

Capacity control by speed variation using a constant-speed machine, such as the synchronous motor, is attained by a variable-speed fluid coupling located between the motor and the speed increaser. A magnetic (eddy current) clutch is also used with synchronous motors to obtain speed variation.

Multi-Speed Alternating-Current Motors

As previously stated, the speed of a-c motors depends on the frequency of the power circuit in cycles per second and the number of poles. Therefore, the theoretical speed of any a-c motor in revolutions per minute is found by the equation shown for synchronous motors; i.e.,

$$\text{rpm} = \frac{\text{cycles per minute}}{\text{no. of pairs of poles}}$$

The frequency of a given power circuit is fixed, usually at 60 cycles per second. The only factor remaining in the equation that can be changed in order to vary speed, then, is the number of pole. Once a motor is built, it is impossible to add or subtract poles as such. Therefore, it becomes necessary to control the activity of a certain number of the poles built into the motor if speeds are to be changed. Because both a south and a north pole are required, poles must be dealt with in pairs. To increase or decrease the rpm value (i.e., the speed of the motor), the number of pairs of poles must be decreased or increased, respectively, within the limitation of the motor. Decreasing the number of pairs of poles by 50 percent doubles the speed of the motor and vice versa. Multi-speed motors are designed with other than 2 to 1 ratios in many cases.

The flexibility of speed control lies in the design of the stator windings. By furnishing extra external leads, half the poles may be cut in or cut out by a simple switch arrangement. Special changes less than or other than 2 to 1 ratio may be obtained by furnishing additional separate stator windings, each having two speed connections.

The drum switch is the principal type of control employed with multi-speed motors.

Multi-speed motors may be either (*a*) variable torque, (*b*) constant torque, or (*c*) constant horsepower machines. Their application depends on which of the three categories they belong to.

DIRECT-CURRENT MOTORS

While the majority of electric motors used in the United States are alternating-current types, a number of direct-current (d-c) motors also are employed.

TABLE 43. CLASSIFICATION OF ALTERNATING-CURRENT MOTORS*

	Type	Speed Characteristics	Full Voltage		HP Range	Application See Footnotes (a) to (e)
			Starting Torque	Starting Current		
			Constant Speed Drives			
Polyphase a-c	Squirrel-cage general purpose Design A	Constant	Normal 1-2.5 times	High 6-8 times	All	(a) Fans and (c) centrifugal pumps and centrifugal compressors
Polyphase a-c	Squirrel-cage Design B	Constant	Normal 1-2.5 times	Normal 5-6 times	Medium Small	(a) Fans and centrifugal pumps and centrifugal compressors
Polyphase a-c	Squirrel-cage Design C	Constant	High 2-2.5 times	Normal 5-6 times	Medium Small	(b) Reciprocating pumps (e) and compressors started loaded
Polyphase a-c	Wound rotor	Constant or variable	High 1-2.5 times (with secondary control)	Low 1-3 times	All	(a) Hoists (b) reciprocating pumps and compressors (c) and frequent (e) or hard start
Polyphase a-c	Synchronous high speed	Exactly constant	Normal 0.75-1.75 times	Normal 5-7 times	Medium Large	(a) Fans and centrifugal pumps and centrifugal compressors
Polyphase a-c	Synchronous low speed	Exactly constant	Low 0.3-0.4 times	Low 3-4 times	Medium Large	(a) Reciprocating compressors starting unloaded
Single phase a-c	Two value capacitor	Constant	High	Normal	Small	(b) Pumps and compressors
Single phase a-c	Permanent split capacitor	Constant	Low	Normal	Fractional	(a) Fans, Blowers
Single phase a-c	Capacitor start	Constant	Moderate	Normal	Small Fractional	(a) Fans and pumps
Single phase a-c	Repulsion Induction	Constant	High	Normal	Medium Small	(a) Fans (b) pumps and compressors
Single phase a-c	Split phase	Constant and adjustable	Normal	Normal	Fractional	(a) Fans (b) pumps and compressors (d) fans—direct
			Adjustable Speed Drives			
Polyphase a-c	Squirrel-cage high slip. Transformer adjustment	Variable	Normal	Normal	Medium Small	(a) Fans
Polyphase a-c	Squirrel-cage separate winding or regrouped poles	Constant multi-speed	Normal or high	Normal or low	All	(a) Fans (b) pumps and (c) compressors
Polyphase a-c	Wound rotor	Variable	High (with secondary control)	Low	All	(a) Fans (b) centrifugal pumps and compressors
Single phase a-c	Repulsion	Variable	High	Normal	Low and Fractional	(a) Fans, centrifugal pumps (b) compressors
Single phase a-c	Capacitor low torque tapped winding	Variable two speed	Low	Normal	Fractional	(d) Fans, direct
Single phase a-c	Capacitor low torque transformer adjustment	Variable	Low	Low	Fractional	(d) Fans
Single phase a-c	Split phase regrouped poles	Constant	Normal	Normal	Fractional	(d) Fans

a. Drives having medium or low starting torque and inertia (WR^2) such as fans and centrifugal pumps or reciprocating pumps and compressors started unloaded.

b. Drives having high starting torques, such as reciprocating pumps and compressors started loaded.

c. Similar to (*a*) except where frequent or hard starting (large WR^2) requires a higher starting and accelerating torque. *d*. Fans direct connected. *e*. Stoker drives.

* Reprinted by permission from "Heating Ventilating Air Conditioning Guide 1956," Chapter 40.

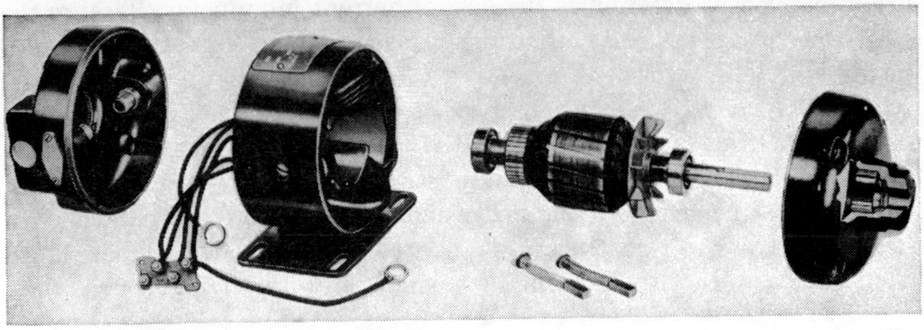

Fig. 327. Direct-current motor, exploded view, showing (*left to right*) end plate with brush-holders; field windings mounted in frame; rotor parts including bearing, commutator, armature winding, fan and bearing; end plate. Brushes with their tension springs shown in foreground. (Courtesy, General Electric Company.)

There are three general types of direct-current motors: shunt wound, series wound, and compound wound.

All three types consist of a stationary field winding located within the motor housing and surrounding a rotating armature attached to the motor journal. A commutator made up of individually insulated copper segments or bars that are clamped to and completely encircle the journal is an important part of a d-c motor. Brushes made up of carbon and graphite or graphite mixed with metallic powder, supported by brush holders, ride the commutator and provide the means of furnishing current to the armature conductors from the power circuit. The field windings receive current direct from the power circuit through terminals located on the motor. The commutator bars are connected to the terminals of the individual coils that make up the armature winding, which usually are wound around the core or drum wound. Each coil has its terminals connected to adjacent commutator bars.

The turning force necessary to operate a d-c motor is developed by the effect of a magnetic field on a coil carrying a current. In the a-c motor it was shown that the current in the coil winding of the rotor resulted from induction alone set up as within a solenoid. The current carried by a coil located on the armature of a d-c motor comes from the power circuit through the commutator and brush assembly. The basic principle of electric motor operation described earlier in this chapter applies also to d-c motors.

The effective voltage in an armature conductor, however, is not that of the power circuit. It is the difference between the impressed voltage and the counter electromotive force set up by the action of a conductor moving through a magnetic field. In an a-c motor, this action was called induction. The actual voltage in an armature conductor, then, is the difference between impressed electromotive force and induced electromotive force. It is the difference in opposing voltages divided by the resistance of the armature in ohms that determines the flow of current in amperes. Because the impressed voltage is greater than the induced voltage, current will flow in the direction of the former.

The difference between the three types of d-c motors lies in the manner in which their fields are connected. It is this difference that lends to each type its individual operating characteristics.

In a *shunt-wound* motor, the field winding is connected in parallel with the armature circuit, i.e., the field is shunted across the armature as shown in Fig. 328 (*top*). Only part of the current flows through the field.

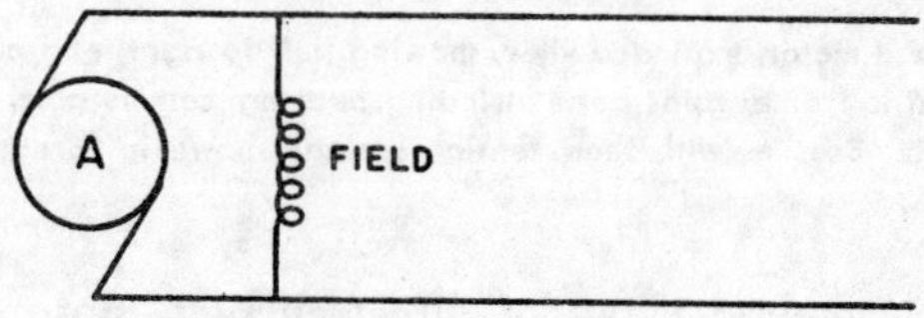

SHUNT MOTOR WINDING

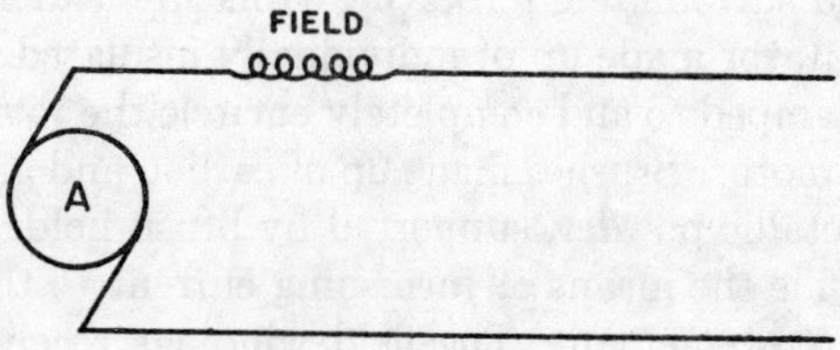

SERIES MOTOR WINDING

Fig. 328. Wiring diagrams for shunt and series direct-current motors.

The field winding of a *series-wound* motor is connected in series with the armature circuit. All the current flows through the field. (See Fig. 328, *bottom.*)

The *compound* motor has both a shunt and a series field winding connected. Where the series winding lies within the shunt field (i.e., the shunt field is connected to the motor terminals), it is known as a long shunt compound motor. (See Fig. 329, *top.*) A short shunt compound motor is one in which the series windings lie outside the shunt field terminals, as in Fig. 329 (*bottom*).

In modern motors, commutating poles are installed whose fields are connected in series with the series field. The purpose of the commutating poles, which are located between main poles, is to prevent sparking at the brushes when the load changes.

If the direction of current is reversed in either the armature or the field, by reversing terminal connections, the d-c motor will change its direction of rotation. However, there will be no change in direction of rotation if the current flow in both armature and field are reversed at the same time.

If the series and shunt windings of a long shunt compound motor are wound in such a manner as to increase the field strength (flux) upon an increase in load, a cumulative effect is obtained. This arrangement, shown in Fig. 330 (*top*), produces what is known as a cumulative compound motor. Where the series and shunt windings oppose each other as in Fig. 330 (*bottom*), a differential effect is set up. Such an arrangement is employed in

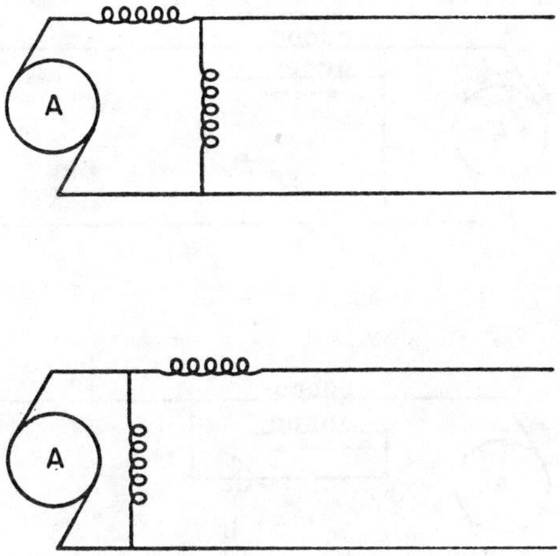

Fig. 329. Wiring diagram for (*top*) long shunt and (*bottom*) short shunt compound motors.

differential compound motors. The latter are seldom used because shunt motors having better speed regulation at the higher loads may be substituted effectively.

The circuit arrangements shown for the different types of d-c motors provide the key to their individual operating characteristics. For example, the series motor because of its peculiar circuit develops internal conditions that cause it to lose speed control at no load. The circuit of the shunt motor provides a stabilizing influence that allows this type of machine to remain at practically constant speed from no load to full load. The compound motor, being both series and shunt wound, tends to develop characteristics that are a combination of the series and shunt motors. In this respect, the compound motor (cumulative) decreases somewhat in speed as load increases (about 25 percent between no load and full load) drop, although the variation is not so great as in series motors. Actual speed change between no load and full load depends on the field strength. On the other hand, the compound motor will not "run away" at no load, because of the influence of the shunt winding. It has a given maximum no-load speed.

The same reasoning may be applied in considering torque of d-c motors as for speeds discussed above. The starting torque of a series motor is extremely high with heavy current, that is to say, heavy load. The starting torque of a shunt motor, having no series winding, is relatively low. The cumulative compound motor has a potential starting torque (depending on current) greater than that of a shunt motor but less than a series motor.

Speed of d-c motors is a function of armature voltage and/or strength of field. A proportionate reduction in speed will follow any decrease in arma-

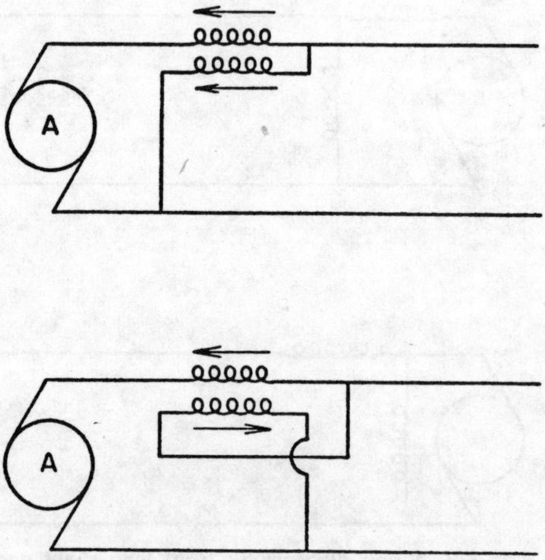

Fig. 330. Compound motor connections having (*top*) cumulative, and (*bottom*) differential series and shunt arrangements.

ture voltage. Control of speed, therefore, usually is exerted by placing a variable resistance in series with the armature.

Performance characteristics of the different types of d-c motors are given in Table 44.

DRIVES FOR TRANSMITTING POWER

There are four basic types of mechanical drives used in the operation of fans, pumps, compressors, and other equipment receiving its motive power from an external source: (*a*) direct drive, (*b*) belt drive, (*c*) gear drive, and (*d*) chain drive.

Direct Drive

Where a driven component is fixed to the same journal or whose journal is coupled to the journal of the driving component, the relationship is said to be one of direct drive. An overhung fan is an example of a direct drive in which the driven member is keyed to the driving shaft, usually that of an electric motor. In this latter case, the motor bearings also support the fan.

TABLE 44. PERFORMANCE CHARACTERISTICS OF DIRECT-CURRENT MOTORS

Type	*Speed Characteristics*	*Torque Characteristics*	*Refrigeration and Air-Conditioning Applications*
1. *Shunt wound*........	Slight drop in speed from no load to full load. Otherwise considered a constant speed machine. Adjustment from one constant speed level to another. Constant speed level can be obtained by use of adjustable field resistance in series with the field. One of the best adjustable speed motors of any kind or current requirements	Starting torque proportional (nearly) to the load and to the armature current. Full speed running torque equal to starting torque when full load current is applied. To increase starting torque above full load running torque requires correspondingly greater current	Centrifugal pumps, centrifugal fans, blowers, etc. In general, for moderate loads infrequently started and run at constant speed. Wide range of speeds can be obtained with field resistance control
2. *Series wound*........	Speed of series motors react according to load, decreasing as load increases and increasing as load decreases. At no load, series motors tend to "run away" and destroy themselves by centrifugal force exerted on the armature	Starting torque is nearly proportional to the square of the armature current (I^2_{arm}), therefore capable of developing high starting torque. Will start greater loads with less current than shunt motor and will pull extremely heavy loads without stalling	Should always be connected directly to load or through fixed gearing. In general, for starting heavy loads or where frequent starting and stopping of heavy loads prevail
3. *Compound wound*.... (cumulative type)	Speed varies as load within limitations controlled by field flux. A drop of 20 to 25% in speed between no load and full load is usual, but can be made greater or less by original design. Has a no-load speed limit by reason of shunt winding	With high current starting torque of compound motor is greater than that of shunt motor but less than series motor	Reciprocating compressors starting loaded or subject to frequent stops and starts. In general, for operating under heavy loads with frequent stops and starts

A driven member, the journal of which is supported by its own bearings, is fixed to the driving journal by a rigid or flexible coupling when directly connected. Direct drive indicates that the driving and the driven journals are in a straight line relative to each other and travel at the same speed. Direct drives are highly efficient. Speed flexibility depends on the characteristics of the driver.

Belt Drive

The belt drive consists of a continuous belt, either flat or V-shaped, that encompasses two or more wheels or pulleys, one of which is the driver. A simple industrial arrangement includes a driving pulley and a driven pulley mounted on separate shafts that may or may not be parallel. An idler wheel mounted independently of both pulleys so that it rides and exerts tension on the belt is often used. Depression of the belt by the idler maintains tautness and effects greater traction at the pulley faces. A pivoted base motor accomplishes the same purpose although it will not increase the arc of contact as does the idler.

Except where belt shifting is necessary to change speeds, the pulley face is crowned, i.e., tapered toward the middle. Pulleys for V-belts are grooved. They are grooved in such a manner as to have the belt flush with the top of the groove and a slight clearance between the root of the groove and the apex of the V. Two or more individual belts are used in multiple V-belt drives. Single belts are used on smaller units. For best results, the spacing between the centers of the driver and the driven shafts should be not less than the diameter of the largest pulley and not greater than the combined diameters of both pulleys. V-belt drives are highly efficient when installed properly (95 percent and over).

The basic purpose of most belt drives is to transmit power. They are also

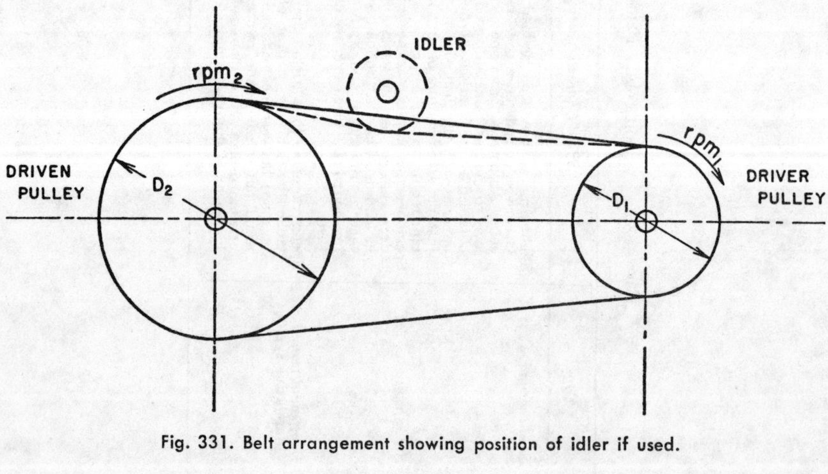

Fig. 331. Belt arrangement showing position of idler if used.

capable of increasing or decreasing the speed of the driven component as well as controlling its direction of rotation.

Reversing the rotation of the driven pulley is achieved by crossing the belt. Driving pulleys whose shafts are at right angles to the driver shaft, either vertically, horizontally, or angularly, may be accomplished by quarter turning the belt.

The relative pulley diameters govern the speed ratio of a belt drive. Where the driver and the driven pulleys have equal diameters, the speed of both will be the same, minus that lost by the driven as slippage. There are four factors in any speed-diameter equation, namely, (*a*) speed of driver (rpm_1), (*b*) speed of driven (rpm_2), (*c*) diameter of driver (D_1), (*d*) diameter of driven (D_2). The relation of each factor to the other is as follows:

$$\frac{D_1}{D_2} = \frac{rpm_2}{rpm_1}$$

If any three of the factors are known, the fourth may be determined, viz.:

1st—speed of driven unknown:

$$rpm_2 = \frac{rpm_1 \times D_1}{D_2}$$

2nd—speed of driver unknown:

$$rpm_1 = \frac{rpm_2 \times D_2}{D_1}$$

3rd—diameter of driven unknown:

$$D_2 = \frac{rpm_1 \times D_1}{rpm_2}$$

4th—diameter of driver unknown:

$$D_1 = \frac{rpm_2 \times D_2}{rpm_1}$$

EXAMPLES: What is the necessary diameter of a driven pulley to obtain a speed of 300 rpm when the diameter of the driver is 3 in. and has a speed of 1740 rpm?

$$D_2 = \frac{rpm_1 \times D_1}{rpm_2} = \frac{1740 \times 3}{300}$$

$$D_2 = 17.4 \text{ in.}$$

What must be the speed of a driver pulley having a diameter of 5 in. to obtain a driven speed of 625 rpm through a 14-in. diameter pulley?

$$rpm_1 = \frac{rpm_2 \times D_2}{D_1} = \frac{625 \times 14}{5}$$

$$rpm_1 = 1750$$

Gear Drive

Gears provide a highly positive method for transmitting power. In their various forms they are capable of transmitting power from motors, engines,

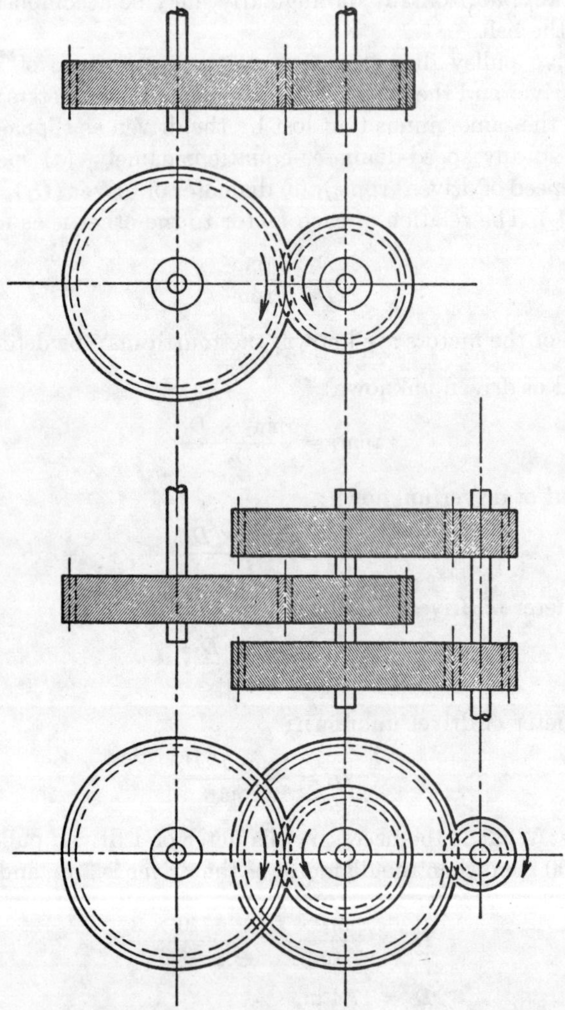

Fig. 332. Gear speed reducers. (*Top*) Single reduction, (*bottom*) double reduction.

and turbines in many directions at any speed ratio. They can develop tremendous mechanical advantage. Gears may be classified according to their pattern of tooth design, of which there are several. The more important types include spur, helical, and herringbone (double helical); bevel (straight tooth, spiral, zerol, and hypoid); and worm gears.

Gear sets are made up of two or more elements, a pinion (the smaller of the two), and a gear. They are designed to transmit mechanical power at speeds equal to or different from the speed of the driving shaft. When the drive gear is directly in mesh with its driven gear, the direction of rotation is reversed. Direction of the drive gear can be maintained for the driven gear by using an intermediate or idler gear between the two. Units accomplishing a reduction in speed are called speed reducers; step-up drives are called speed increasers. Gear drives are identified according to the number of steps involved in changing from initial (input) to final (take-off) speed: such terms of reference as single, double, and triple reduction gear sets indicate that their rated speed ratios are accomplished in one, two, or three steps, respectively.

The speed ratio of gear sets is expressed by a number based on the relationship of initial and final speed. When applied to a speed reducer, this number becomes a divisor, and a multiplier in the case of speed increasers.

A reducer having a speed ratio of 4.22:1 and an input speed of 1750 rpm will have a take-off speed of $\frac{1750}{4.22} = 414.7$ rpm.

A speed increaser having a ratio of 3.8:1 and an input speed of 1750 rpm will have a take-off speed of $1750 \times 3.8 = 6650$ rpm.

The speed ratio of any gear set is based on the relative number of teeth on each gear. For gears having an equal number of teeth, their speeds will be equal. If the driven gear has twice the number of teeth as the drive gear, the speed of the driven will be one half that of the driver. For any combination of gear sizes and/or speeds, the relationship is as follows:

$$\frac{N_1}{N_2} = \frac{\text{rpm}_2}{\text{rpm}_1}$$

in which rpm_1 and rpm_2 are the speed of the drive and the driven gears, respectively, and N_1 and N_2 represent the number of teeth on the drive and the driven gears, respectively. Therefore, if any three factors are known, the fourth factor may be determined by the following equations:

1st—Speed of driven gear unknown:

$$\text{rpm}_2 = \frac{\text{rpm}_1 \times N_1}{N_2}$$

2nd—Speed of drive gear unknown:

$$\text{rpm}_1 = \frac{\text{rpm}_2 \times N_2}{N_1}$$

3rd—Number of teeth on drive gear unknown:

$$N_1 = \frac{\text{rpm}_2 \times N_2}{\text{rpm}_1}$$

4th—Number of teeth on driven gear unknown:

$$N_2 = \frac{\text{rpm}_1 \times N_1}{\text{rpm}_2}$$

Fig. 333. Speed increaser with conventional herringbone gears. (Courtesy, Worthington Corporation.)

Where one gear has a greater number of teeth than the other, a difference in size must exist if they are to mesh properly. This means that gear sets capable of reducing or increasing speed have both a large and a small gear or pinion.

Gear drives hold an important place in the refrigeration field, particularly as speed increasers for centrifugal compressors. Herringbone or double helical gears are used in sets involving power transmission between parallel shafts, an illustration of which is given in Fig. 333.

Chain Drive

Chain drives are employed much in the same manner as belt drives, but are more positive. Their use varies over a wide range from driving small lubricating oil pumps to that of transmitting relatively high loads direct from electric motors. This type of drive consists of a continuous flexible chain that encompases two or more gearlike wheels or drums called sprockets. While there are a number of different types of chains used, the roller and leaf (silent) types are the most important.

Roller chains are made up of a number of rollers, each one connected to another by link plates securely fastened to pins on which the roller turns. The space between the adjacent rollers allows the sprocket teeth to project through. The silent chain consists of a number of flat leaves cut in such a manner as to provide teeth that mesh with the sprocket but not allowing the teeth of the latter to project through.

Chain drives are strong, are highly efficient (98 percent +), and do not allow slippage. The roller chain is made up in both single and multiple strands. The multiple strand type is made up of a number of parallel single strands whose rollers in the same row are mounted on a common pin. The widths of silent chains are varied in a like manner, mounting the required number of leaves in parallel on the same pin. Multiple type chains are employed where load requirements warrant additional strength.

Chains are identified according to their width and pitch, the latter being the distance in inches between the center lines of adjacent pins. Widths of roller chains vary from $\frac{1}{4}$ in. (roller) for single strands to an over-all width of 24 in. for multiple types. The same range of widths applies generally for silent chains. Pitch varies from $\frac{3}{8}$ in. to $2\frac{1}{2}$ in. for roller chains and to 2 in. for silent chains.

Chain drives perform best when installed in a horizontal position. Vertical drives require some form of tightening adjustment to take up slack resulting from wear. Idler sprockets are often used to avoid slack as well as to perform other functions involving chain travel, direction of rotation, motion of other machine parts, etc.

The sprocket pitch must conform to that of its chain. However, the sprocket diameters may vary greatly according to service requirements, the number of teeth corresponding to the pitch and the sprocket diameter. Because a large number of teeth tend to increase efficiency, allow greater speed, and reduce noise, the largest sprocket consistent with operating conditions should be used. The smaller sprocket should have not less than 17 teeth, although a higher number is preferable. An uneven number of teeth on both sprockets is desirable as it allows the chain to bear on a different surface with each revolution. This retards wear.

When the center line distance between sprockets is too great, the chain tends to whip. If too close together, a limitation may be placed on the number of teeth of the small sprocket in mesh with its chain. These and other factors govern the distance between sprocket center lines. While the minimum distance must be more than one half the sum of the diameters of both sprockets to permit clearance, good operation generally will be obtained by spacing them about 40 pitch lengths apart.

Maximum chain speed is a function of chain pitch and the number of teeth in the smaller sprocket. As speed increases, pitch should decrease somewhat correspondingly. The number of teeth should also be increased, which means also that sprocket diameter will increase with higher speed and load commensurate with pitch. For silent chains the relationship of pitch to speed of the smaller wheel may be in the magnitude of 5000 rpm for $\frac{3}{8}$ pitch, decreasing correspondingly with each increment of pitch increase to below 1000 rpm for 2 pitch. Roller chains may be considered similarly except that maximum speed for each increment of pitch increase is much less than for silent chains, i.e., about 3000 rpm for $\frac{3}{8}$ pitch diminishing to below 250 rpm for $2\frac{1}{2}$ pitch.

Chains operating at comparatively high speeds and load should be encased so as to run in an oil bath. Those not encased require frequent lubricant applications of an oil that is capable of reaching all frictional surfaces.

Single reductions should be limited to a speed ratio not greater than about 7:1. Above that ratio two or more reductions should be made. Speed ratios between driver and driven sprockets are determined in the way described for belt drives and gears, using the number of teeth (N) or diameter (D) of the two sprockets in the direct proportion equation: i.e., $\text{rpm}_1 \times N_1 = \text{rpm}_2 \times N_2$ or $\text{rpm}_1 \times D_1 = \text{rpm}_2 \times D_2$.

AIR-HANDLING EQUIPMENT

The movement of air from one place to another is a major consideration in some refrigeration processes. Air-conditioning systems require the handling of large quantities of air while air-cooled condensers, forced convection systems, cooling towers, evaporative condensers, pneumatic controls, and clear ice production all necessitate air displacement in more or less volume. Air handling also involves various pressure changes. It is the magnitude of pressure difference between air source and its final use that governs the type of air-handling machine selected.

The various kinds of air-handling machines are classified according to the amount of pressure change capable of being developed between their point of inlet or suction and point of outlet or discharge. The major kinds are: fans, blowers (centrifugal), and compressors. The primary factor separating the three major kinds, aside from design in some cases, is a matter of compressibility or density variation. This being true, some limiting range of compression values must be attached to each kind of air-handling machine to distinguish one from the other. Because compressibility or density variation is a function of pressure, identification is made according to their pressure-changing ability, i.e., between suction and discharge orifices.

The general rule is to identify the three kinds of air-handling machines according to their pressure-changing capabilities as follows:

Fans. In general, compressibility is limited to a change in density of 7 percent or less. This corresponds to a maximum pressure increase between suction and discharge of 1 psi or 27.7 in. of water. There are exceptions to this, as some fans are designed especially for pressure increases up to 3 psi. Higher pressures are possible by staging, i.e., by placing two or more fans in series within the same channel. In moving air with fans, the principal function is to overcome flow friction so that compressibility is relatively unimportant and usually disregarded.

Blowers. The terms blower and fan are often used synonymously. No error is incurred in doing so when referring to machines in which pressure changes are held within the range identified with fans. While blowers are actually low-pressure compressors, the Compressed Air Institute defines centrifugal blowers as machines in which air (or gas) is compressed by centrifugal force to a final pressure not exceeding 35 psi. On the other hand, some blowers are designed for pressure ratings as low as 20 in. of water.

Density becomes a factor associated with blower systems because of the higher pressures involved. As friction loss is directly proportional to air (or gas) density, the amount of compression becomes an important consideration.

Blowers, like centrifugal compressors, are built with any number of stages, according to desired pressure increase. Fig. 334 shows a multistage blower with five impellers or stages.

Fig. 334. Centrifugal blower. (Courtesy, United States Hoffman Machinery Corporation.)

One use for a blower similar to the type shown in Fig. 334 would be for agitating raw water in ice-freezing cans to produce clear, clean ice. Because of the relatively low compression ratios involved, blowers usually do not require water cooling.

Compressors. Air-handling equipment that develop pressures above 35 psi are called compressors. There are three major types: reciprocating, centrifugal, and rotary compressors. An additional type suitable only for handling large volumes of air (6000 cfm and over) is the axial-flow compressor. This latter type is built with rows of alternate fixed and moving blades such as those of a turbine. Flow is axial rather than radial, as in centrifugal compressors.

The laws governing reciprocating, centrifugal, and rotary refrigerant compressors previously discussed in this text apply similarly to air compressors and blowers. Therefore, no further comment regarding this type of equipment need be made here.

FAN DESIGN AND PERFORMANCE CHARACTERISTICS

Fans are divided into two types according to their air-flow pattern: axial-flow fans and radial-flow fans.

Axial-Flow Fans. This type of fan is designed to cause air to flow parallel to the fan axis. The two major types of axial-flow fans are the propeller and the disk fans.

The *propeller fan* is so named because of the characteristic twist given to each of its blades, which project radially outward from the hub. Desk, floor, and wall types are familiar examples of the propeller fan. The number of blades vary from two to six, depending on the design. The amount of blade twist is called the pitch angle. The pitch angle varies from slight or small pitch (about 6°) to steep pitch (about 40°). The capacity (cfm) of a fan depends on the blade pitch angle, increasing as the pitch becomes steeper. However, any increase in pitch angle is followed by a corresponding loss of efficiency and stability. Every fan has an optimum pitch angle at which maximum efficiency is obtained.

The capacity of a propeller fan is increased by increasing the number of blades. All other factors being equal, the increased volume based on a two-bladed fan in free air will be slightly above 10 percent, 20 percent, and 30 percent for three-, four-, and six-bladed fans, respectively. Capacity also varies with speed, increasing directly as speed increases. Other performance characteristics follow closely the speed laws of centrifugal fans described in the succeeding discussion. Propeller fans are used to move free air within a room or through channels having low frictional resistance.

Fan tests are based on standard atmospheric conditions, viz.: barometric pressure of 29.92 in. of mercury, density 0.075 lb per cu ft of air, and a

Fig. 335. Propeller fan. Four- and six-blade types with and without ring. (Courtesy, The Torrington Manufacturing Company.)

temperature of 68 F. Fans run either clockwise or counterclockwise, the direction of which is determined by facing the discharge area.

Propeller fans are either mounted on the drive motor shaft or belt driven. Their specifications usually include diameter, pitch, and number of blades. Some installations require that propeller fans be mounted in a ring that fully encircles the periphery of the blades. Wall or exhaust fans are of the latter design.

Disk type fans are modified propeller fans having the greater proportion of twist or pitch in the spider (that part holding the blade to the hub) rather than in the blade, which is relatively straight.

Tube-axial fans are propeller type fans mounted in a cylindrical housing of some length. They are used where some pressure head is prevalent, such as in a duct system.

Vane-axial fans are also mounted in a cylindrical housing. Their purpose is to meet still higher pressure heads beyond the capability of tube-axial fans. Guide vanes located either before or after the fan convert the tangential component of the velocity head into static head. The blades of vane-axial fans are designed to embrace the advantages of an airfoil or a blade surface so shaped as to gain some useful reaction from the air through which it moves.

A change in delivery volume from maximum, when the orifice is fully open (Fig. 336), to that of zero volume, when the orifice is entirely closed, is accompanied by a corresponding increase in horsepower to drive the fan. There is always an unstable region or break in the static pressure curve of a

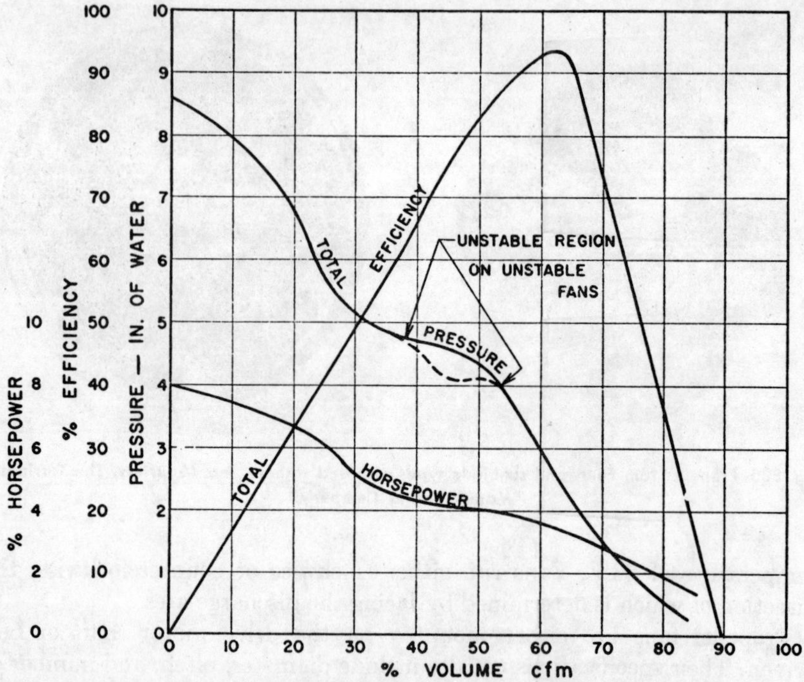

Fig. 336. Characteristics of propeller fans through operating range from zero to maximum air delivery.

propeller fan. This is due to the effect of head pressure on the lower part of the blades, which tends to prevent axial flow at that point. When speeds reach a definite level for the individual fan, the air tending to flow through the lower blade area is thrown outward by centrifugal force. It is at the region of activity when all forces of speed, pressure, and volume tend to convert axial flow to radial flow (centrifugal action) that the break in the curve occurs.

Radial-Flow Fans. Fans of the radial-flow type include the variously designed centrifugal fans. Centrifugal fans consist of an impeller mounted on a shaft and encased in a spiral or scroll housing having an axial inlet and a peripheral outlet. The air entering the fan in the area of its axis is thrown radially into the space between impeller periphery and housing formed by the scroll by centrifugal force imposed by the blades. The scroll design tends to permit uniform velocity of the air as it travels from the point of cut-off to the outlet, during which volume steadily increases. In this connection, the action is similar to that previously described for volute type centrifugal compressors. (See Chapter 8.)

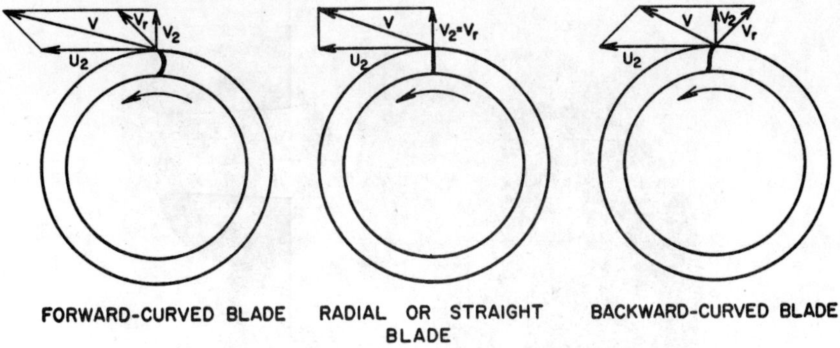

U_2 = LINEAL VELOCITY OF BLADE TIP

V_2 = RADIAL VELOCITY OF AIR LEAVING BLADE TIP

V = ACTUAL VELOCITY OF AIR RELATIVE TO FAN CASING

V_r = RELATIVE VELOCITY OF AIR

Fig. 337. Centrifugal fan blades and parallelogram of velocities of each type.

There are three general types of centrifugal fans whose difference lies in the curvature or position of their blades relative to the impeller radii. The three types of blades are:

Forward-curved, having their concave surfaces facing the direction of rotation. The number of blades varies from 22 to 66.

Straight or radial, having their surfaces parallel with impeller radii. Any number from 4 to 24 blades are used, usually 6 to 12.

Backward-curved, having their convex surfaces facing the direction of rotation. Blades number from 14 to 24.

Fig. 337 shows relative position and shape of the three types of blades and their comparative velocity parallelograms. Pressure is developed by centrifugal force, the total pressure being equal to the sum of the static and the velocity pressures developed between the columns formed by adjacent blades. Pressure development within a centrifugal fan follows the same pattern as that described for centrifugal compressors, and, like the latter machine, impellers having the greatest depth of blade will develop highest pressures, assuming equal conditions otherwise.

Comparing the absolute velocities as represented by V in the parallelograms (Fig. 337), it shows those of radial and backward-curved blades to be of less magnitude than that of comparable forward-curved blade fans. In order to obtain equal capacity of the latter under similar pressure conditions, the straight and backward-curved blade fans must be operated at higher tip speeds. The increase in tip speed will be found proportional to the difference in magnitude of V in both cases. (See Table 173.)

Fig. 338. Centrifugal fan. (*Right*) Fan rotor showing blade design. (Courtesy, Buffalo Forge Company.)

Centrifugal fans are subject to a number of variables involving both design and operation. Design variables include:

(*a*) Shape of blade,
(*b*) Number of blades,
(*c*) Depth of blade,
(*d*) Width of impeller (single, double, or fractional),
(*e*) Number of inlets (single or double),
(*f*) Size of inlet,
(*g*) Diameter of impeller,
(*h*) Scroll design,
(*i*) Use of diffusers (or not),
(*j*) Use of inlet guide vanes (or not).

All variables must be in proper balance to insure maximum fan performance. For example, a sensitive relationship exists between inlet size, wheel diameter, depth, and number of blades. It follows that individual fans have a definite inlet size that will impart maximum operational economy.

Centrifugal fans are designed with single or double inlets, i.e., an inlet on each side of the impeller. They are also built with single, double, or fractional width impellers. Double-width impellers necessitate double inlets. Stationary inlet guide vanes are often used. The latter is a series of vanes placed across the inlet area to lend prerotational action to the entering air, thus preventing the shock of directional change. Variable-inlet guide vanes somewhat similar to that shown in Fig. 226 are sometimes used to control air volume through the fan under constant speed conditions.

Fig. 339(a). Airfoil blade giving radial direction, having a configuration approaching ideal centrifugal fan design. (Courtesy, The Trane Company.)

Operating variables are as follows:

Speed, revolutions per minute—rpm
Volume delivered, cubic feet per minute—cfm
Static pressure, inches of water—press.
Size of fan, preferably diameter of wheel—size
Horsepower, input to fan—hp
Density, pounds per cubic foot of air handled, usually considered constant for fans—relative density—δ
Absolute temperature, of air handled—abs temp

Fig. 339(*b*). Mixed flow blade, having the compactness of a propeller fan with the centrifugal fan's ability to operate quietly against static pressure. (Courtesy, The Trane Company.)

The relationship of operating variables is stated by a number of fan performance laws. These laws (1 to 10) are given in Table 45. If weight of air is considered instead of volume, then for constant amount of weight and constant size the following three principal laws apply. When density of air varies:

11. cfm, rpm, and pressure vary inversely as density; i.e., inversely as barometric pressure and directly as the absolute temperature.

12. Horsepower varies inversely as the square of the density; i.e., inversely as the barometric pressure and directly as the square of the absolute temperature.

TABLE 45. FAN LAWS†

Fan Law Number		Ratio of	=	Variables* Ratio	×	Variables* Ratio		Normally Constant** Ratio
	a	cfm		size^3	×	rpm		1
1	b	press.	=	size^2	×	rpm^2	×	δ
	c	hp		size^5	×	rpm^3		δ
	a	cfm		size^2	×	press.$^{1/2}$		$1/\delta^{1/2}$
2	b	rpm	=	1/size	×	press.$^{1/2}$	×	$1/\delta^{1/2}$
	c	hp		size^2	×	press.$^{3/2}$		$1/\delta^{1/2}$
	a	rpm		1/size^3	×	cfm		1
3	b	press.	=	1/size^4	×	cfm^2	×	δ
	c	hp		1/size^4	×	cfm^3		δ
	a	cfm		size$^{4/3}$	×	hp$^{1/3}$		$1/\delta^{1/3}$
4	b	press.	=	1/size$^{4/3}$	×	hp$^{2/3}$	×	$\delta^{1/3}$
	c	rpm		1/size$^{5/3}$	×	hp$^{1/3}$		$1/\delta^{1/3}$
	a	size		cfm$^{1/2}$	×	1/press.$^{1/4}$		$\delta^{1/4}$
5	b	rpm	=	1/cfm$^{1/2}$	×	press.$^{3/4}$	×	$1/\delta^{3/4}$
	c	hp		cfm	×	press.		1
	a	size		cfm$^{1/3}$	×	1/rpm$^{1/3}$		1
6	b	press.	=	cfm$^{2/3}$	×	rpm$^{4/3}$	×	δ
	c	hp		cfm$^{5/3}$	×	rpm$^{4/3}$		δ
	a	size		press.$^{1/2}$	×	1/rpm		$1/\delta^{1/2}$
7	b	cfm	=	press.$^{3/2}$	×	1/rpm^2	×	$1/\delta^{3/2}$
	c	hp		press.$^{5/2}$	×	1/rpm^2		$1/\delta^{3/2}$
	a	size		1/hp$^{1/4}$	×	cfm$^{3/4}$		$\delta^{1/4}$
8	b	rpm	=	hp$^{3/4}$	×	1/cfm$^{5/4}$	×	$1/\delta^{3/4}$
	c	press.		hp	×	1/cfm		1
	a	size		hp$^{1/2}$	×	1/press.$^{3/4}$		$\delta^{1/4}$
9	b	rpm	=	1/hp$^{1/2}$	×	press.$^{5/4}$	×	$1/\delta^{3/4}$
	c	cfm		hp	×	1/press.		1
	a	size		hp$^{1/5}$	×	1/rpm$^{3/5}$		$1/\delta^{1/5}$
10	b	cfm	=	hp$^{3/5}$	×	1/rpm$^{4/5}$	×	$1/\delta^{3/5}$
	c	press.		hp$^{2/5}$	×	rpm$^{4/5}$		$\delta^{3/5}$

* Either one of these variables may be used independently, in which case the other must remain constant. Then the desired data will vary as power of the variable indicated. These laws are written as if both varied at the same time. In this case, the desired data will vary as the powers of the variables indicated.

** In the last column, the effect of relative density (δ) is shown. This is to apply an extra variable when needed, otherwise density is supposed to be constant.

† Courtesy, Madison, Richard D., ed.: "Fan Engineering," ed. 5, p. 242, Buffalo, N. Y., Buffalo Forge Company.

When both temperature and pressure vary,

13. cfm and rpm vary as $\sqrt{\text{press.} \times \text{abs temp}}$
    hp varies as $\sqrt{\text{press.}^3 \times \text{abs temp}}$

Note that consideration may be given density changes as shown in the last column of Table 45. Using a density ratio factor, the need for including actual density values in the equation is eliminated. The density ratio is determined by dividing the density of air at existing temperature and barometric pressure by the density of standard air; i.e., at 68 F and 29.92 in. of mercury or 0.07458 lb per cu ft or 0.075 approximately. Standard air, then, has a density ratio of

$$\delta = \frac{0.075}{0.075} = 1.0$$

and at 100 F and 29.92 in. mercury,

$$\delta = \frac{0.0692}{0.075} = 0.923$$

etc.

EXAMPLE: A centrifugal fan operates at 330 rpm against a static pressure resistance of 1.0 in delivering 8000 cfm. 2.2 hp are required. What will be the cfm, press., and hp if rpm is increased to 400 rpm? Density is constant.

Referring to Table 45:

$$cfm_2 = \frac{rpm_2}{rpm_1} \times cfm_1$$

$$cfm = \frac{400}{330} \times 8000$$

$$cfm = 9696 \text{ cfm, new volume}$$

$$press._2 = \left(\frac{rpm_2}{rpm_1}\right)^2 \times press._1$$

$$press._2 = \left(\frac{400}{330}\right)^2 \times 1.0$$

$$press._2 = 1.47 \text{ in. of water, new static pressure}$$

$$hp_2 = \left(\frac{rpm_2}{rpm_1}\right)^3 \times hp_1$$

$$hp_2 = \left(\frac{400}{330}\right)^3 \times 2.2$$

$$hp_2 = 3.8 \text{ hp, new power input}$$

The fans under discussion here are multi-blade types which are designed to give the appearance of a revolving squirrel cage. While some overlapping prevails, the following statements apply generally.

In the order mentioned, forward-curved, radial, and backward-curved blade fans are primarily low-, medium-, and high-speed machines. The radial-blade fan, then, is a compromise between the other two types. Large inlet areas signify large volume of air handled at moderate pressures. Small inlet fans having narrow blades signify moderate volume at higher pressures. Assuming constant speed, the volume of air delivered by a fan increases with increased impeller width and the pressure becomes greater with larger wheel diameter. The area of approach, i.e., the duct area, should be at least $1\frac{1}{2}$ times the fan inlet area.

The direction of rotation of centrifugal fans is designated as being clockwise or counterclockwise in respect to the driving side, or the side opposite the inlet of single-inlet fans. The direction of discharge for both clockwise and counterclockwise fans may be (1) top or bottom horizontal, (2) upblast or downblast, (3) top or bottom angular down, and (4) top or bottom angular up.

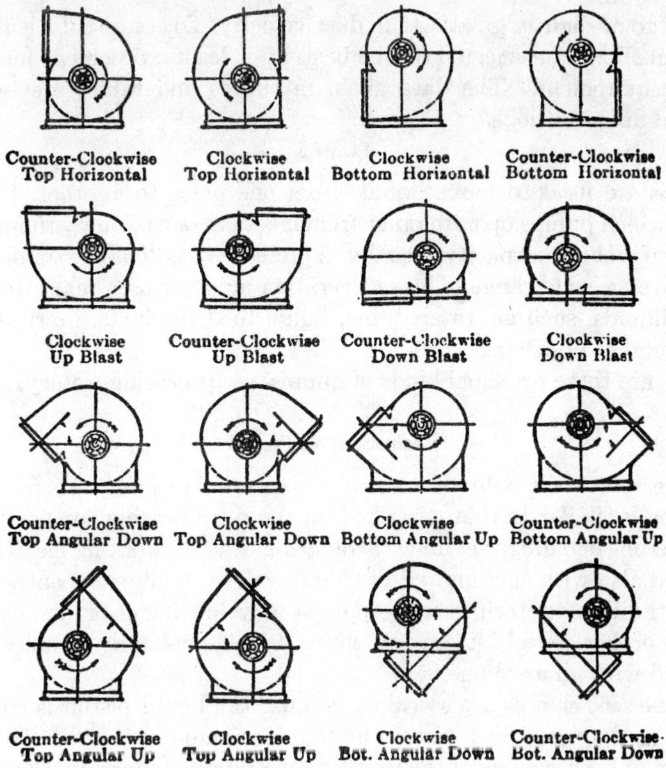

Fig. 340. Various discharge arrangements of centrifugal fans. (Courtesy, Richard D. Madison, ed.: "Fan Engineering," ed. 5, p. 654, Buffalo, N. Y., Buffalo Forge Company, 1949.)

The National Association of Fan Manufacturers (NAFM) classifies fans according to operational pressure limits, viz.:

Class I. Fans up to $3\frac{3}{4}$ in. water
Class II. Fans up to $6\frac{3}{4}$ in. water
Class III. Fans up to $12\frac{1}{4}$ in. water
Class IV. Fans over $12\frac{1}{4}$ in. water

Class I fans are the type most suitable for refrigeration and air-conditioning purposes.

Fan manufacturers publish tables or curves covering operating data for their various individual models. From these tables, or curves, corresponding values of capacity (cfm), speed (rpm), and horsepower (hp) can be determined for different static pressure heads. Outlet velocities are also included in some tables. Thus, by knowing the quantity of air required and the frictional resistance against which the fan must work, the size, the speed, and the horsepower input of the fan may be determined. Outlet velocity should be equal to or slightly greater than duct velocity. To assure sufficient capacity, a fan should be selected on the basis of at least a 5 percent increase in actual requirements. The data given in curves and tables are based on standard air conditions.

PUMPS

Pumps are used to move liquids from one place to another. Pressures against which pumps operate range from less than one to many thousands of psi. Their volume capacity, usually expressed in gallons per minute, also varies over a wide range. The refrigeration industry uses many pumps to handle liquids, such as water, brine, liquid food products, lubricating oil, and a number of others.

There are three principal kinds of pumps: reciprocating, rotary, and centrifugal.

Reciprocating Pumps

The reciprocating pump is a positive displacement pump. Its principle of operation is similar to that described for the reciprocating compressor having a piston- or plunger-cylinder arrangement as its working element. It is equipped also with suction and discharge valves to allow for entrance and exit of the liquid. Reciprocating pumps may be single- or double-acting, vertical or horizontal, steam- or power-driven, and may or may not be equipped with an air dome.

Direct-acting pumps are so called because the liquid piston is connected directly to a steam-driven piston through a common piston rod. The length of stroke of both pistons of the direct-acting pump is therefore the same, as is their speed. The direct-acting drive possesses the advantage of variable speed, thus providing a variable-capacity pump.

Power pumps are those whose power is derived from an external source, usually through speed-reduction gears and crankshaft arrangement. They are generally constant-speed machines, although speed variation may be obtained by suitable control methods. Power pumps are identified by the number of pistons or cranks, such as simplex (single piston), duplex (two pistons), triplex (three pistons), and quintuplex (five pistons). The throw of the crankshaft governs the length of stroke. Power pumps should be protected by a safety device, such as a relief valve, to avoid damage if subjected to excessive pressure for some reason.

Double-acting pumps have double the capacity of single-acting pumps of comparable size, less the volume taken up by the piston rod when in a position of maximum inward stroke. Fig. 21 shows a direct-acting reciprocating pump with an air dome.

Reciprocating pumps are designed to handle liquids having varying degrees of fluidity, from low viscosity solvents to viscous substances such as chocolate syrup, molasses, etc. They are capable of developing tremendous hydraulic pressures, such as required for large presses.

The discharge flow from reciprocating pumps is pulsating. This is due to the variable velocity of the piston as it moves through one stroke and the reversal of piston direction. The characteristics of pulsations vary according to the type of pump. If driven from a crankshaft, as in power pumps, the piston starts its stroke from zero velocity, accelerates gradually to maximum at mid-stroke, and then gradually diminishes again to zero velocity at the

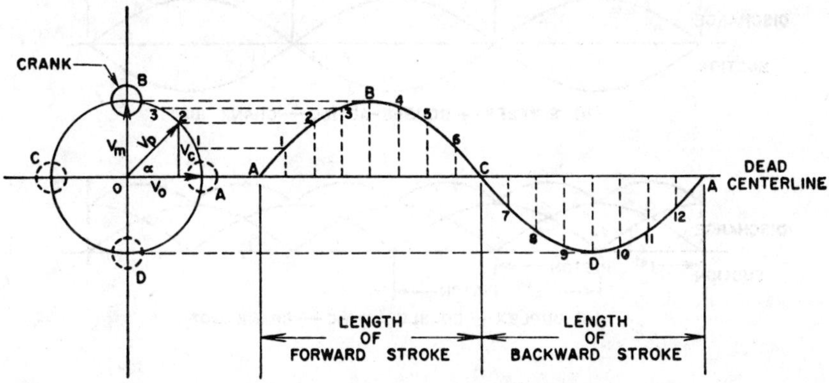

Fig. 341. Sine curve for power reciprocating pump.

end of the stroke. Gradual acceleration and deceleration of power pump pistons is due to continual change in angularity of the crank. As the crank revolves 360° to move the piston first in one direction and then in the reverse direction, the angle of the crank constantly changes in respect to the line of dead center. If the motion of the crankpin were plotted along a straight line, it would take the form of a sine curve similar to that shown for alternating-current frequencies. And the velocity of the piston at any point in its travel is a function of velocity of the crank and its angular relationship to the dead center line. Therefore, the piston velocity (V_p) is equal to the velocity of the crankpin (V_c) times the sine of the angle made by the crank with the dead center line at any point, or (see Fig. 341),

$$V_p = V_c \sin \alpha$$

Fig. 342 (a) shows the pattern of pulsating flow for a single-acting power pump. The frequency of pulsations would follow the magnitude of distance between center lines of succeeding positive-pressure curves. If the same

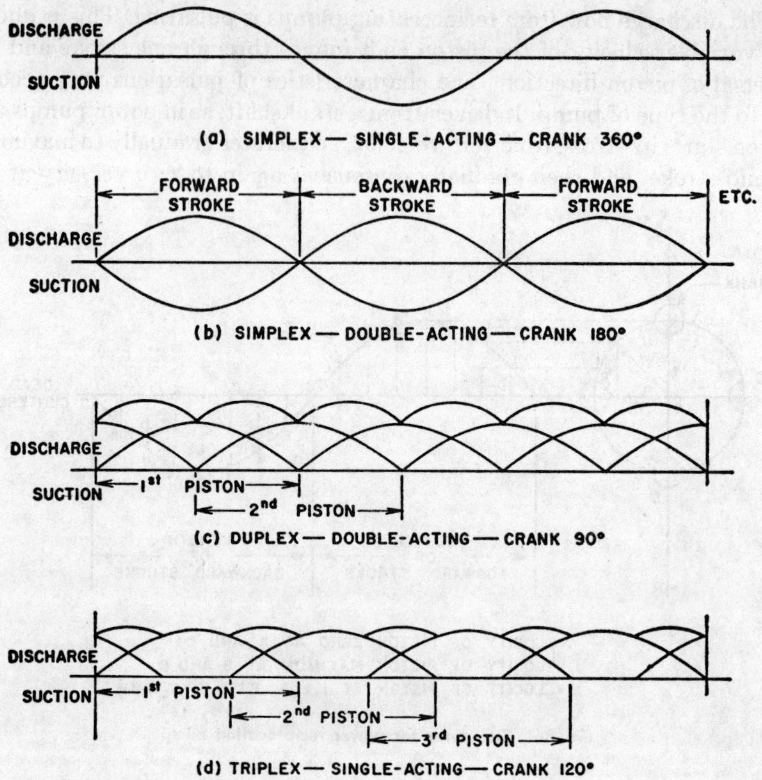

Fig. 342. Flow pattern of various types of reciprocating power pumps.

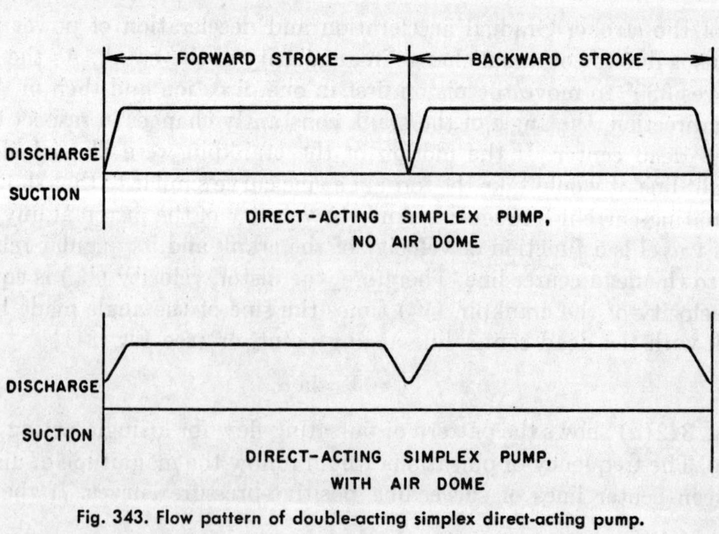

Fig. 343. Flow pattern of double-acting simplex direct-acting pump.

pump were made double-acting, the frequency of pulsation would be smaller, thus smoothing the flow, as shown in Fig. 342(*b*). Any arrangement of pistons, i.e., duplex, triplex, etc., single-acting and double-acting, has its own pulsating pattern, so that the one having the more constant discharge flow reduces pulsation.

Direct-acting pumps also are pulsating, but their flow pattern differs from power pumps because of the absence of crank effect. Starting from zero velocity, the piston accelerates rapidly to maximum velocity, which it maintains throughout the greater part of the stroke. Upon approach to end of stroke, piston velocity decelerates rapidly to zero. Fig. 343 shows the flow pattern of direct-acting pumps with characteristic flat curve. Note that Fig. 343 (*top*) shows the effect of no air dome by the sharp interruption in flow between strokes, while with an air dome, as in Fig. 343 (*bottom*), some flow continues, as represented by the rounded connections between reversed stroke patterns.

Rotary Pumps

Rotary pumps are also positive-displacement machines. Their principle of operation is similar to that previously described for rotary compressors. The liquid displacement element may be any one of a number of different kinds, including gear, lobe, screw, and vane types as shown in Fig. 344. The rotary pump is a constant-flow machine so that liquid discharge is not subject to the pulsating action experienced with reciprocating pumps. They are used widely in the food industry for handling milk and fruit and vegetable juices of all kinds.

The displacement of rotary pumps is found by determining the volume of the pumping element and multiplying this value with the rpm, as shown for rotary compressors. Some slip is experienced due to pump clearances. Slip loss does not include losses due to inlet conditions, such as the presence of air or other gases to which rotary pumps are sensitive. Wear of the frictional surfaces within a rotary pump is minimized if only those liquids capable of imparting a lubricating effect are handled.

Centrifugal Pumps

The operating principles of the centrifugal pump are similar to those described previously for centrifugal compressors and fans. Consisting of an impeller that carries the vanes, enclosed in a stationary casing, centrifugal pumps are of three general classes, viz.: radial flow, mixed flow, and axial flow.

Radial or centrifugal flow is one in which liquid is forced radially outward from the impeller axis by centrifugal force. The pump may be either single suction, in which the liquid enters only one side, or double suction, that allows for liquid entrance on both sides of the impeller.

Mixed flow pumps have their vanes inclined in such a manner as to

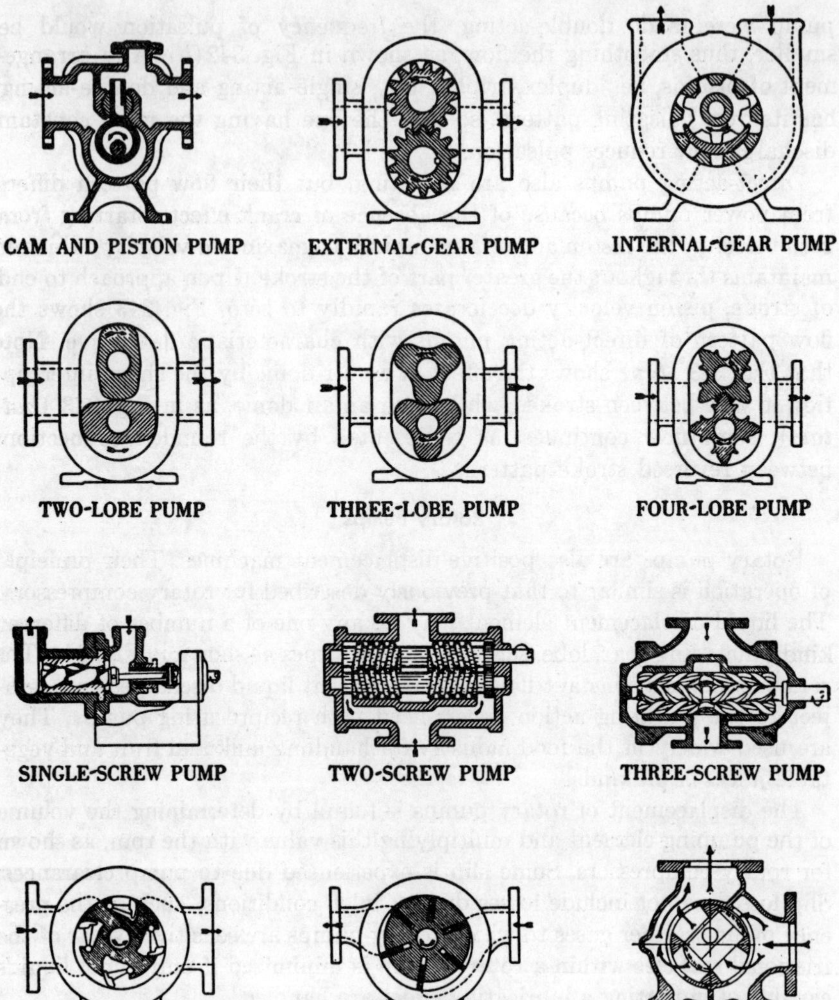

Fig. 344. Various types of rotary pumps. (Reprinted from the "Standards of the Hydraulic Institute," ed. 10. The Hydraulic Institute, 122 E. 42nd St., New York 17, N. Y.)

impart a lifting action to the liquid. Lift plus centrifugal force acts to create the velocity head developed by the pump and to discharge the liquid in an axial and radial direction.

Axial flow is similar to that given to air by a propeller fan, i.e., the liquid passes through the pump in a direction essentially parallel to the shaft or impeller axis.

Thus the three types of centrifugal pumps provide radial, axial-radial (inclined), or axial flow. They rotate in either a clockwise or a counterclock-

Fig. 345. Centrifugal pump. Showing cross-section of impeller, volute, and wear plates. Single inlet. (Courtesy of Ingersoll-Rand Company.)

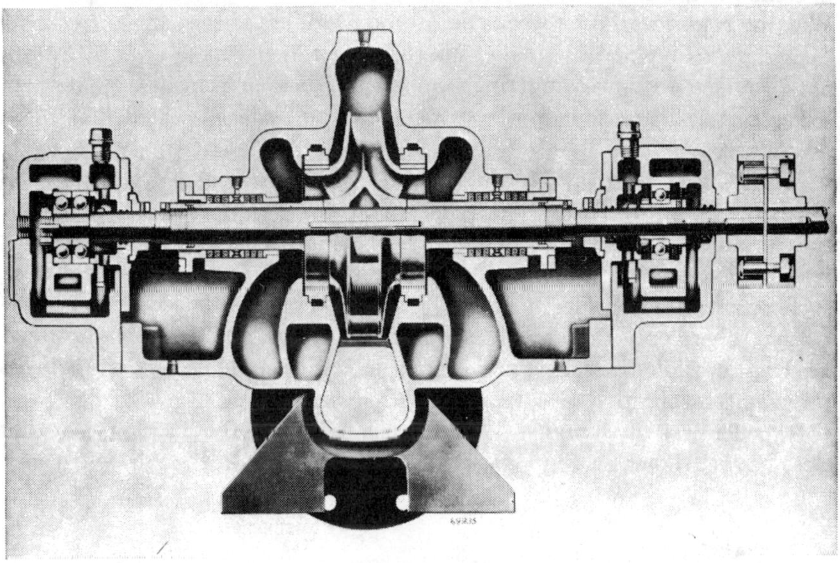

Fig. 346. Centrifugal pump. Double suction. Note packing, antifriction bearings, and coupling. (Courtesy of Ingersoll-Rand Company.)

wise direction, determined by facing the pump from the driver or motor. Like centrifugal compressors, pumps of this kind are made with one (single) or more (multi-stage) stages according to the desired pressure increase. The impellers of multi-stage pumps are mounted successively on the same shaft and enclosed in a common casing to impart a gradual rise in pressure to the liquid passing through.

The casing of centrifugal pumps may be circular, in which the cross section of the area surrounding the impeller is constant, or it may be of the volute design, as shown in Fig. 348. Diffuser (or guide) vanes are also used

Fig. 347. Centrifugal pump and electric motor mounted on common base. Note location of petcock. (Courtesy of Ingersoll-Rand Company.)

in some pumps for the same purpose described for centrifugal compressors. Wearing rings located between the rubbing surfaces of the impeller and the casing act as replaceable seals that reduce internal leakage (slip). Pumps not provided with wearing rings or seals have their impellers made of a softer material, usually bronze, which may be replaced when it becomes excessively worn.

Bronze, cast iron, and stainless steel are the principal materials of centrifugal pump construction. Such metals are used in suitable combination, or alone, depending on the nature of liquid to be handled. Thus, if all parts in contact with the liquid being handled are bronze, or stainless steel, such pumps are considered to be all bronze, or all stainless steel, respectively. Where only the impeller, the wearing rings, and probable bushings are bronze, while the casing is cast iron, in this case such a pump is called a bronze-fitted pump.

Impellers are of the open, the semi-enclosed, and the closed types. The closed type, shown in Fig. 345, is used more widely, although, when han-

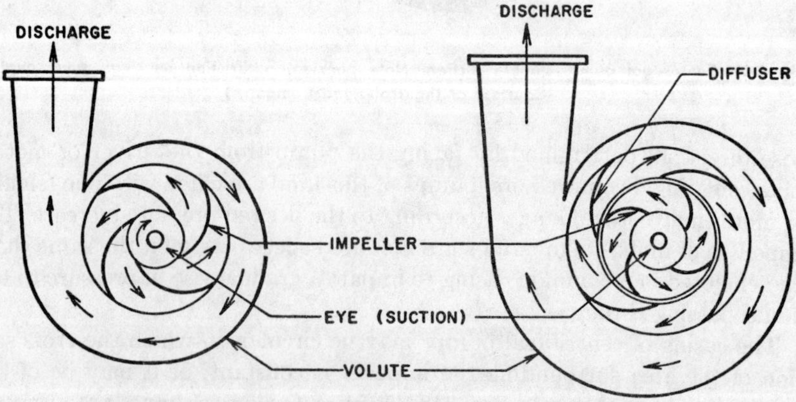

Fig. 348. (*Left*) Volute type, and (*right*) diffuser type centrifugal pump.

dling material that tends to clog, pumps having open or semi-enclosed impellers are employed.

Pump Priming

When a pump is shut down, the liquid left in the suction line drains back to the suction reservoir, allowing air to enter the space left void. In order to initiate flow when pump operation is resumed, the air in the suction line must be eliminated. Replacing suction-line air with liquid to be pumped is called priming. In the case of the hand pump on the farm, a bucket of water was kept on hand for the purpose. Pouring the water into the cylinder provided a seal for the plunger to draw a partial vacuum in the suction pipe, allowing the well water to rise sufficiently to be pumped. The time required and the vigorousness of working the pump handle depended on the depth of the well, the condition of the plunger, and the amount of water poured into the cylinder.

A self-priming pump is one capable of eliminating air from the suction line without external help. Such pumps are self-priming only if the total suction lift is held to definite limits corresponding to the nature of the liquid handled. For example, some positive-displacement types, such as reciprocating and certain rotary pumps, are self-priming if the distance between the pump and the water reservoir surface is not greater than 25 to 28 ft. Theoretically, the pressure of the atmosphere (14.7 psi) is capable of supporting a column of water 33.83 ft (approximately 34 ft) high. This does not hold true for practical suction-lift applications either for pump operation or for self-priming. Therefore, total suction lift for any one pump must be less than the theoretical value.

Centrifugal pumps are not self-priming, so some method of initiating flow must be provided after a period of shutdown. There are a number of methods used for priming centrifugal pumps.

Centrifugal pumps are provided with a petcock located at the top of the casing. If the pump is connected to city water supply, which is under pressure, or to a suction reservoir located above the pump, air can be forced out of both suction line and pump by opening the petcock and allowing the water to flood the system. Cracking the suction-line valve until water flows out the petcock usually frees the system of air. The old cliché that "easy does it" holds true here.

Where the suction reservoir is open (under atmospheric pressure only) and below the level of the pump, then an external flooding system can be used. Connect a small city water line, or any source of supply, to the top of the pump and provide a foot valve in the suction line (below the suction reservoir level). Open the petcock and the city water line valve to flood the system. Close the petcock, open the suction line valve, and start the pump. Purge several times after the pump is started by opening and closing the petcock.

Priming may be accomplished also by producing a vacuum in the suction line. This may be done with a steam, water, or air ejector connected in a pipe line leading from the top of the pump to a suitable drain. Or it may be accomplished by means of a vacuum pump. The installation of priming tanks placed above the pump level, submerging, circulating water through the suction line, and a number of other methods are used to make priming either a manual or an automatic operation.

PUMP DATA

The location of a pump in relation to its suction and discharge reservoirs, volume to be handled, total resistance against which it must operate, and a number of other details are important in selecting a suitable pump. Friction of fluid flow in pipes, an important associated factor, is treated in Chapter 13, "Secondary or Indirect Refrigeration."

The capacity of a pump is expressed generally in the gallons per minute it is capable of delivering. The resistance to flow against which a pump must work is expressed in terms of head or feet-head. Feet-head is the pressure equivalent exerted by a column of any open liquid. It is determined by the equation:

$$\text{head in feet} = \text{psi} \times \frac{144}{w}$$

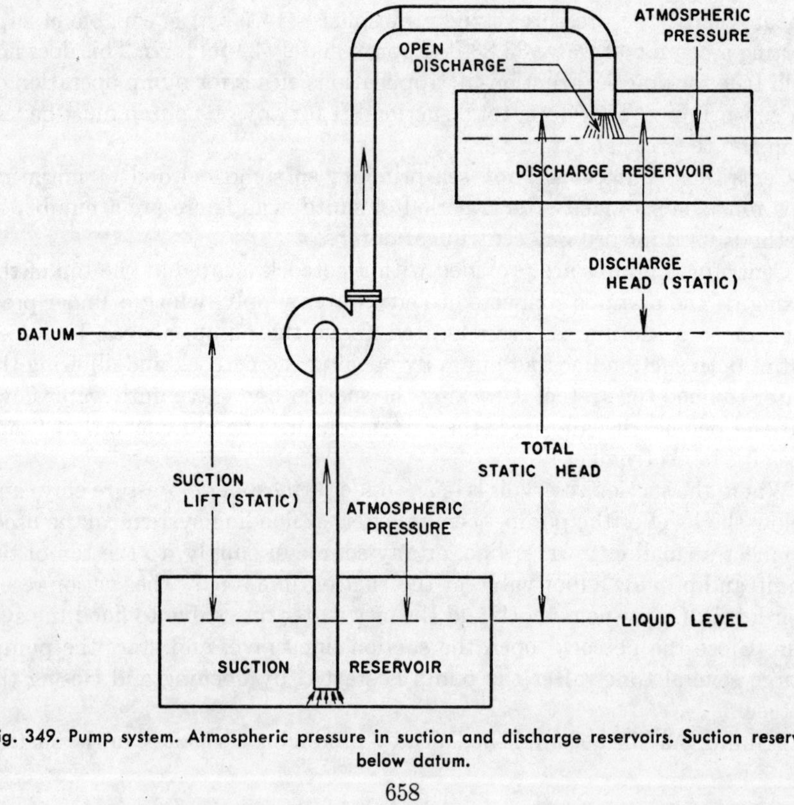

Fig. 349. Pump system. Atmospheric pressure in suction and discharge reservoirs. Suction reservoir below datum.

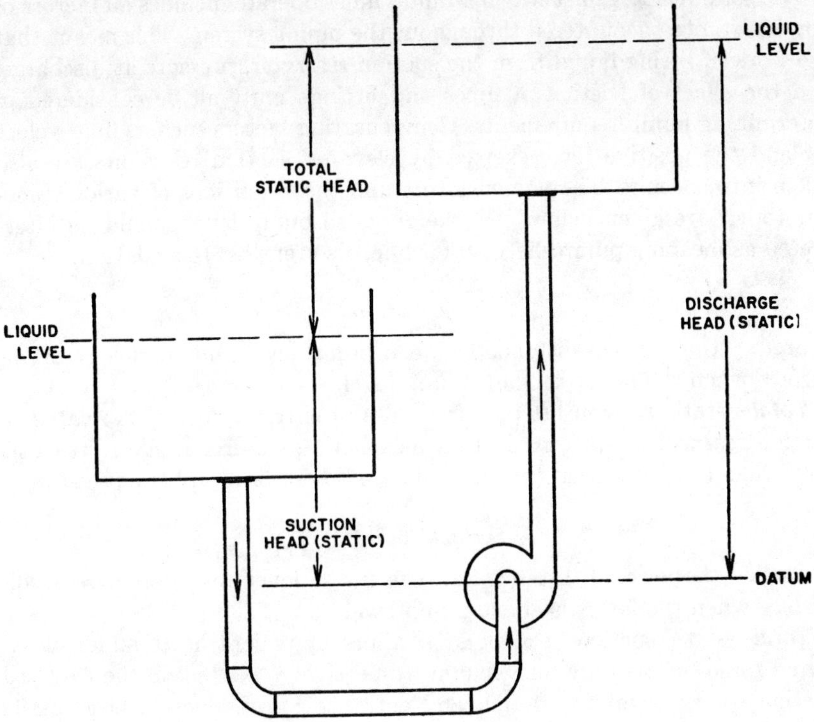

Fig. 350. Pump system. Suction reservoir above datum. Discharging against a head greater than atmospheric pressure at pipe exit.

where w = weight in lb per cu ft of any substance

For water,

$$\text{head in feet} = \text{psi} \times \frac{144}{62.3}$$
$$= \text{psi} \times 2.31$$

For a pressure of 125 psi, the equivalent head is

$$\text{head in feet} = 125 \times 2.31$$
$$= 288.75 \text{ ft-head}$$

To determine the pressure equivalent for feet-head of water,

$$\text{psi} = \frac{1}{2.31} \times \text{ft-head}$$
$$= 0.433 \times \text{ft-head}$$

For a column of water 210 ft high, then

$$\text{psi} = 0.433 \times 210$$
$$= 90.93$$

The total head against which a pump must operate includes all factors of flow resistance encountered throughout the piping system. This means that the work of lifting liquid from the suction reservoir, pressure at discharge, and the effect of friction in pipes and fittings must all be considered in determining pump requirements. Compensating factors such as fluid velocity and the positive force exerted by elevated suction reservoirs are also taken into account. The nomenclature and the definitions of various pipeline forces are given below. All references to pump level should be interpreted as meaning pump shaft center line, also referred to as datum.

Suction Lift (h_l)

Static—the vertical distance between pump level and suction reservoir surface when the latter is below pump level.

Total—static suction lift plus friction and entrance losses, as related to suction reservoirs subject only to atmospheric pressure. Suction lift exists only when the pressure at the pump nozzle is *below* atmospheric pressure.

Suction Head (h_s)

Static—the vertical distance between pump level and suction reservoir surface when the latter is above pump level.

Total—static suction head plus or minus equivalent head values of external forces of pressure, or vacuum, respectively, exerted on the elevated suction reservoir minus friction and entrance losses. Suction head exists only when the pressure at the pump nozzle is *greater* than atmospheric pressure.

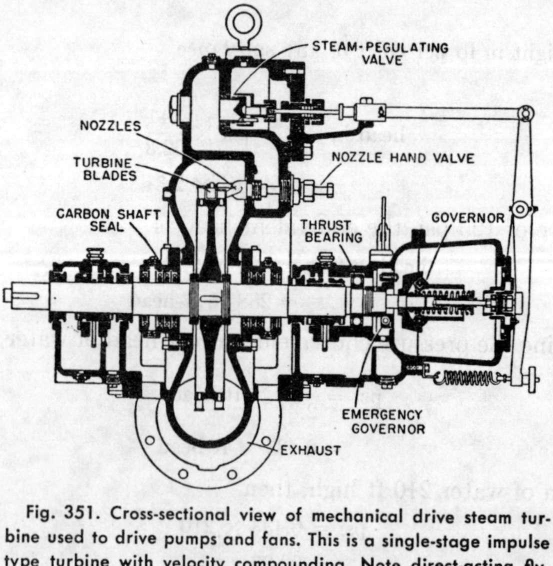

Fig. 351. Cross-sectional view of mechanical drive steam turbine used to drive pumps and fans. This is a single-stage impulse type turbine with velocity compounding. Note direct-acting flyball governor. (Courtesy, Mobil Oil Corporation.)

Discharge Head (h_d)

Static—the vertical distance between pump level and the point of free discharge.

Total—static discharge head plus equivalent head values of any external pressure exerted on the discharge reservoir, plus friction losses.

Velocity Head (h_v)

Velocity of the flowing fluid has some slight effect on the total pumping effort. The velocity head is the distance in feet that the liquid would have to fall in order to develop a given velocity. It may often be neglected because of its small value. Velocity head is expressed by the equation:

$$\text{Head due to velocity} = \frac{(\text{fluid velocity in pipe, fps})^2}{2g}$$

or

$$h_v = \frac{V^2}{2g}$$

The velocity head developed by a fluid flowing through a pipe at 8 fps is:

$$h_v = \frac{V^2}{2g} = \frac{8^2}{2 \times 32.2} = \frac{64}{64.4}$$

$$h_v = 1 \text{ ft (approx)}$$

Friction Head (h_f)

The loss due to friction in pipes and fittings expressed in terms of equivalent feet-head. To be added to total head in any case. (For determining h_f, see Chapter 13, "Secondary or Indirect Refrigeration.")

Total Head (H)

The algebraic sum of all positive and negative forces within a given pumping system:

$$H = h_d + h_l - h_s + h_f + (\text{any other losses})$$

Entrance Losses (h_e)

Losses due to the entrance of fluids into pipes depend on the shape and the condition of the entrance. Such losses are also expressed in feet-head (h_e) as determined by the equation:

$$h_e = k \frac{V^2}{2g}$$

in which V = fluid velocity in fps
and k = constant for various entrance shapes as follows:

	Value of k
Bell-mouthed	0.04
Slightly rounded	0.23
Sharp cornered	0.50
Inward projecting pipe	0.78

For a sharp-cornered pipe entrance, the losses in feet-head for a fluid flowing at 20 fps would be:

$$h_e = 0.50 \times \frac{20^2}{2 \times 32.2}$$

$$h_e = 0.50 \times \frac{400}{64.4}$$

$$h_e = 3.11 \text{ ft}$$

Other Losses

Losses also develop within a piping system due to sudden enlargements, sudden contractions, partially closed valves, and other causes. Loss of head in feet of liquid due to such conditions usually can be found in tables covering individual circumstances.

EXAMPLE: What is the total head for a water system having an open suction reservoir 16 ft below pump level (datum) and discharging into an accumulator reservoir 40 ft above pump level (datum), which is under 125 psi pressure? Assume a friction head of 9 ft and no entrance or discharge losses. Neglect velocity head.

Solution:

Static suction lift	=	16 ft
Static discharge head	=	40 ft
Dynamic or head due to external discharge pressure		
$= \text{psi} \times \frac{144}{62.3}$		
$= 125 \times 2.31$	=	288.75 ft
Friction head	=	9 ft
Total head	=	353.75 ft

Water Horsepower

To determine the water horsepower, apply the basic horsepower equation, i.e.,

$$\text{hp} = \frac{\text{force} \times \text{distance}}{33{,}000}$$

Force in a hydraulic system is the weight of fluid to be moved per minute. Distance is represented by the head against which the fluid is to be moved. Then

$$\text{Water hp} = \frac{\text{weight of fluid, lb per min} \times \text{heat in ft}}{33{,}000}$$

However, the quantity of fluid handled by a pump is expressed generally in gallons per minute (gpm) rather than weight. Therefore, it is necessary to convert pounds to gallons and correct for the specific gravity of the individual fluid. Then

$$\text{Water hp} = \frac{\text{gpm} \times \text{lb per gal} \times \text{head in ft (or } H)}{33{,}000}$$

For water

$$\text{Water hp} = \frac{\text{gpm} \times 8.33 \times H}{33{,}000}$$

Simplified,

$$\text{Water hp} = \frac{\text{gpm} \times H}{3962} \text{ (for water)}$$

$$\text{Water hp} = \frac{\text{gpm} \times H \times \text{sp gr}}{3962} \text{ (for any liquid)}$$

EXAMPLE: Determine the water hp for the above pumping system if 225 gpm of water is handled:

$$\text{Water hp} = \frac{\text{gpm} \times H \times \text{sp gr}}{3962}$$

$$\text{Water hp} = \frac{225 \times 353.75 \times 1}{3962}$$

$$\text{Water hp} = 20.0$$

Brake Horsepower and Motor Horsepower

Assume the above pump to be 65 percent efficient and driven by an electric motor having an efficiency of 89 percent. What is the brake hp of the pump and the size of the motor required?

$$\text{Brake hp} = \frac{\text{water hp}}{\text{pump efficiency } \%} \times 100$$

$$\text{Brake hp} = \tfrac{20}{65} \times 100$$

$$\text{Brake hp} = 30.77$$

$$\text{Motor hp} = \frac{\text{brake hp of pump}}{\text{motor efficiency } \%} \times 100$$

$$\text{Motor hp} = \frac{30.77}{89} \times 100$$

$$\text{Motor hp} = 34.5 \text{ or a 40-hp motor}$$

FLEXIBLE COUPLINGS

Some type of coupling is required to connect the shaft of a rotating machine, such as centrifugal pumps and compressors, rotary pumps, fans and blowers, to the shaft of its driving component. Couplings may be rigid, flexible, or have variable speed or slip characteristics found in hydraulic and magnetic types.

Torque transmission is often subject to some degree of misalignment brought about by machine foundation settling, bearing wear and other reasons. Misalignment imparts torsional stresses on the shaft, the bearings, and other parts of machines that are coupled together rigidly. Rigid coup-

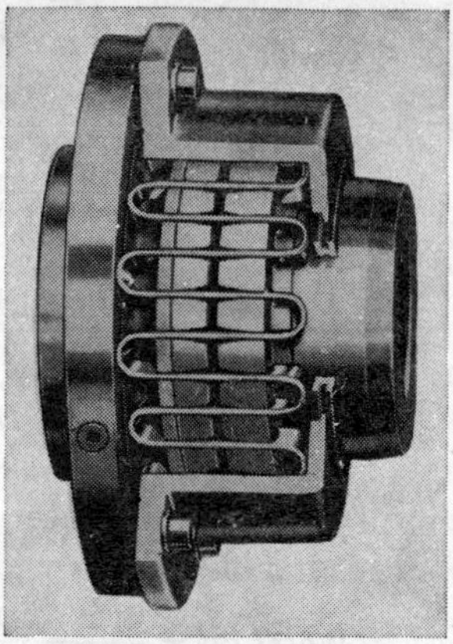

Fig. 352. Flexible coupling. The type shown is designed for high speeds as required for centrifugal compressors. (Steelflex coupling.) (Courtesy, The Falk Corporation.)

lings also transmit rotative shock and vibration from one machine to the other.

Where excessive misalignment exists, repositioning of at least one of the machines is essential. To relieve the stress due to slight misalignment and absorb rotative shock and vibration, flexible couplings are employed.

A rigid coupling consists essentially of two flangelike parts each of which is fixed to adjoining shafts and bolted tightly together. The corresponding parts of flexible couplings are not similarly attached to each other, but establish positive torsional contact through some suitable intermediate element or system of elements. It is the intermediate section of the flexible coupling that permits some slight degree of angular displacement.

There are a number of different elements employed to obtain angular flexibility in couplings. They include gears, chains (link, roller and silent), hardened steel rollers, crosses, springs, laminated pins, floating keys, double floating rings, disks, and a number of others. Some types of couplings are designed to protect machines against overload, such as friction plates that slip when definite limits of torsional strain are reached.

One type of flexible coupling in various operational positions is shown in Fig. 353.

SUMMARY OF CHAPTER 10

The four principal sources of power for refrigeration equipment are:
- Steam turbines
- Steam engines
- Internal combustion engines
- Electric motors

Steam turbines are of several types, including:
- Impulse
- Reaction
- Impulse-reaction
- Others

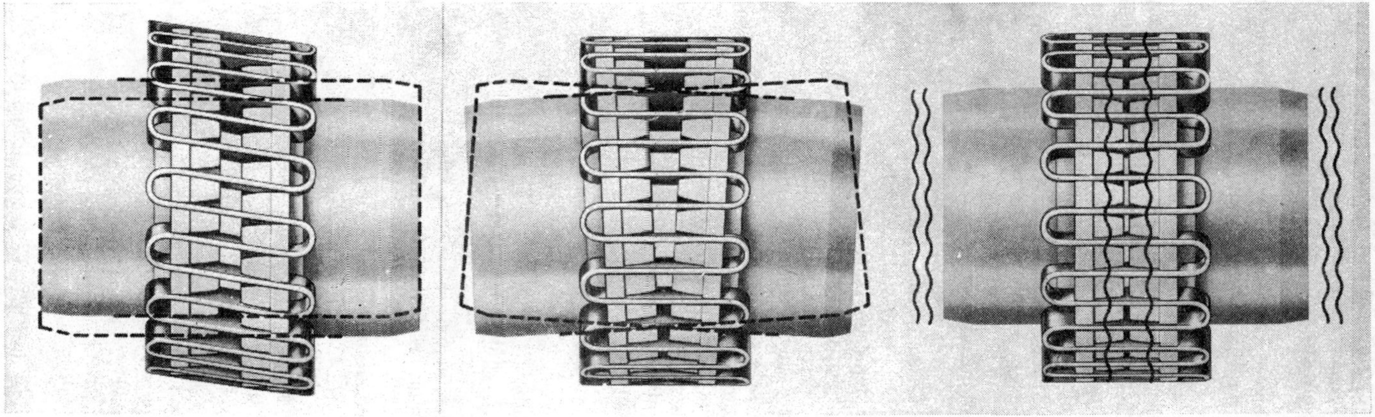

Fig. 353. Operating principle of flexible coupling showing response to different operating conditions: (*Left*) parallel misalignment; (*center*) angular misalignment; (*right*) free-end float. Conditions are exaggerated for the purpose of illustration. The type coupling shown is for standard motor speeds. (Courtesy, The Falk Corporation.)

Pure impulse single-row (of buckets) turbine is:

A high-speed machine. Velocity compounding is one means of reducing inherent speed. Also pressure staging.

Pure reaction turbine:

Requires a large number of blade rows for speed development. To overcome the latter and to shorten the rotor assembly, a velocity stage usually augments the reaction stages.

Steam turbines:

Are used to drive centrifugal compressors, fans, and centrifugal pumps. They have speed flexibility, an advantage in driving centrifugal compressors and fans.

Electric motors:

Represent the most widely employed driving component.

There are two types of electric current:

Alternating current
Direct current

Alternating-current motors include:

Fractional-horsepower:
- Split-phase induction
- Capacitor-start induction run
- Capacitor-start and run
- Repulsion-start and run
- Shaded-pole

Integral-horsepower:
- Squirrel-cage induction
- Slip-ring or wound rotor induction
- Synchronous

Fractional-horsepower motors:

Usually are of the single-phase type, but not always.

Integral-horsepower motors:

Usually are of the polyphase type, mostly three-phase.

Direct-current motors include:

Shunt wound
Series wound
Compound wound

The shunt-wound d-c motor:

Possesses characteristics required for most refrigeration purposes.

The synchronous motor:

Usually is operated at leading power factor to offset the lagging effect of induction motors. The latter improves voltage regulation and reduces power line losses.

Synchronous speed:

Is determined by the equation

$$\text{rpm} = \frac{\text{frequency per min}}{\text{no. of pairs of poles}}$$

Alternating current:

May have frequencies of 25, 40, or 60 cycles per sec. The latter is found more widely. Voltages may be 110, 208, 220, 440, 550, or 2300.

Electric motors:

Can be obtained enclosed variously, including:

Totally enclosed (explosion-proof)
Drip-proof (for damp areas)
Splash-proof
Open-frame (for normal use)

Power transmission:

Between driver and driven unit may be accomplished by:

Direct drives (centrifugal and rotary compressors, fans, pumps)

Belt drives (compressors, fans)

Gear drives ((*a*) speed increasers for centrifugal compressors, (*b*) speed reducers for pumps, processing machinery, etc.)

Chain drives (lubricating oil pumps, processing machinery, etc.)

Speed of drives (except direct connected):

$$rpm_2 = \frac{rpm_1 \times D_1}{D_2}$$

or

$$D_2 = \frac{rpm \times D_1}{rpm_2}$$

where rpm_1 and rpm_2 = speed of driver and driven (pulley or gear, respectively)
and D_1 and D_2 = diameter, or number of teeth, respectively

Air-handling equipment:

Is of three basic types:
- Fans (up to 1 psi)
- Blowers (up to 35 psi)
- Compressors (over 35 psi)

Fans:

Are axial-flow (propeller) or radial-flow (centrifugal)

Centrifugal fans:

May have their blades forward curved (low speed)
radial (to axis) (medium speed)
backward curved (high speed)

NAFM classification of fans as to pressure developed:

- Class I—Up to $3\frac{3}{4}$ in. water
- Class II—Up to $6\frac{3}{4}$ in. water
- Class III—Up to $12\frac{1}{4}$ in. water
- Class IV—Over $12\frac{1}{4}$ in. water

(Class I fans used most extensively in refrigeration and the air-conditioning field.)

Pumps are of three principal types:

Reciprocating (direct-acting or power)
Rotary
Centrifugal

Centrifugal pumps include the following types:

Radial flow (centrifugal)
Mixed flow (inclined)
Axial flow (propeller)

Priming:

Is necessary for centrifugal pumps, either manually or automatically. They are not self-priming.

The capacity of a pump:

Is expressed in terms of gallons per minute delivered.

Pumps are rated on their capacity:

In gallons per minute against a resistance expressed in feet-head or psi equivalent.

Head in feet:

Is equal to psi × wt of vol 1 sq in. × 1 ft high of liquid pumped.

Pressure:

In psi is equal to head in feet $\times \frac{1}{w}$

where w = wt, lb per vol 1 sq in. × 1 ft high of liquid pumps

or $$psi = \frac{\text{height in ft}}{w}$$

Total head (*H*):

Is the algebraic sum of all positive and negative forces within a pumping system.

Forces that make up total head include:

Discharge head (h_d)
Suction head, or lift (h_s) or (h_l)
Friction head (h_f)
Entrance losses (h_e)
Velocity head (h_v)
or $H = h_d + h_l - h_s + h_f + h_e + \text{etc.}$

where $h_e = K \dfrac{V^2}{2g}$ (*K* is a constant for shape of entrance)

and $\dfrac{V^2}{2g} = h_v$

Water horsepower:

$$\text{water hp} = \frac{\text{gpm} \times \text{wt} \times H\,(\text{ft})}{33{,}000}$$

where gpm = gal per min
and wt = lb per gal of pumped liquid

Flexible couplings:

Are employed to join the journals of a driver and a driven machine so that their axes are in a straight line relative to each other. Flexible couplings reduce torsional stresses due to misalignment. They also reduce transmission of rotative shock and vibration.

STUDY OUTLINE FOR CHAPTER 11

Heat balance factor in refrigeration calculations

Standard measurement of refrigeration loads

Temperature conversion:
- Fahrenheit to Centigrade
- Centigrade to Fahrenheit
- Réaumur scale

Heat-work equivalents:
- Mechanical energy
- Electrical energy

Specific heat factors:
- Practical use

Resultant temperature:
- Temperature result of mixing several substances having different temperature levels

Refrigerating effect:
- Determination for various refrigerants

Heat capacity of refrigerants

Condenser loads

Latent heat of fusion—food products

Infiltration loads

Human beings—heat load

Heat-transfer calculations—equipment

Mean temperature difference

Evaporation pressure determinations

Specific volume

Condenser pressure determinations

Compressor discharge temperatures

Specific volume—compressor discharge

Compression ratio

Refrigerant circulated per ton of refrigeration

Compressor work

Coefficient of performance—cop:
- Performance factor

Horsepower per ton of refrigeration

Mean effective pressure—mep:
- Coefficients of mep

Pressure drop due to liquid risers

Pressure drop due to friction in pipes

Drop in pressure due to bends, fittings, and valves

Pressure drop across an expansion valve

Horsepower—gas compression machines

Head-feet of gas of compression machines

Torque-horsepower relationship of electric motors

11

Refrigeration Calculations

Just as the student should understand the basic principles of the refrigeration cycle before attempting a study of the mechanical equipment involved, it also is necessary that he familiarize himself with mathematical formulas which will enable him to estimate refrigeration loads and select equipment of correct size and capacity to meet estimated demands. In this chapter it is the purpose of the author to present such formulas and to demonstrate their use by means of practical examples that will permit the student to apply them to the common problems arising in connection with refrigeration design and operation.

Many of the problems presented here may be solved by simple arithmetic; their only complexity lies in the intelligent application and selection of correct functional values. A few others base their solutions on logarithmic functions, and, while these problems necessarily are included, their conclusions may also be derived through the use of appropriate factors developed and demonstrated herein.

Before entering on a discussion of the various refrigeration calculations, it is essential to outline the fundamental laws that affect the results of any problems dealing with the exchange of thermal energy. Guided by these laws, the student should experience little difficulty in obtaining a thermal balance essential to the application of such mathematical equations as are presented throughout this text.

I

(*a*) Any gain in heat by one body must be accounted for by an equal loss of heat from another body.

(*b*) Any loss of heat from one body must be accounted for by an equal gain in heat by another body.

II

(*a*) The direction of heat flow must always be from a body of higher temperature to one of lower temperature.

(*b*) Heat cannot flow unaided from a body of lower temperature to one of higher temperature.

III

(*a*) Exchange or conversion of energy is always accompanied by some loss. Therefore, theoretical conclusions, while mathematically correct, are not always final but depend on the efficiency of the system.

(*b*) The efficiency of any system of energy exchange or conversion is the product of the separate efficiencies; i.e., thermal efficiency × mechanical efficiency × etc.

STANDARD MEASUREMENT OF REFRIGERATION LOAD

Refrigeration systems are rated on their capacity to remove heat from any space or substance for the purpose of either reducing temperature or maintaining a desired condition, or to change the existing state. It follows that the size of such a system varies according to the number of Btu that must be absorbed during a specific time interval in order to meet the load demand successfully.

If, for example, it is required to remove 1000 Btu per minute from a substance in order to accomplish a desired purpose (such as freezing meat or ice or to cool an office space 10 F below outside temperatures during the summer months), the size of the refrigeration system could be specified as 1000 Btu per minute system. Assuming that all systems are rated on minute time intervals, then, for the purpose of brevity, the foregoing load would demand a system having a 1000 Btu load capacity.

While small domestic refrigerators frequently are rated on such a basis, this method of specifying industrial loads would become quite confusing because of the greater load demands encountered in this field, resulting in the necessity of correlating extremely high Btu values. It would be analogous to specifying mechanical load requirements in foot-pounds per minute rather than in horsepower.

For the purpose of simplification, refrigeration systems are rated on the amount in tons of 32 F water they are capable of converting to 32 F ice during a 24-hr period. Thus, a system rated at, say, 10 tons capacity is of sufficient size to absorb an amount of heat equivalent to that rejected by 10 tons of 32 F water during its conversion to 32 F ice in a 24-hr period.

The latent heat of fusion of water is 144 Btu per pound, or 144 × 2000 lb = 288,000 Btu per ton. One ton of refrigeration capacity, therefore, is equivalent to 288,000 Btu per 24 hours. Reducing this value further,

$$1 \text{ standard ton of refrigeration} = \begin{cases} 2{,}000 \times 144 = 288{,}000 \text{ Btu per day} \\ 288{,}000 \div 24 = 12{,}000 \text{ Btu per hr} \\ 12{,}000 \div 60 = 200 \text{ Btu per min} \end{cases}$$

A refrigeration system rated at 1 ton has the capacity of absorbing heat at the rate of 200 Btu per minute, or 12,000 Btu per hour.

To determine the refrigeration tonnage for any requirement, it is only necessary to divide the load demand (total Btu) by the factor corresponding to the time on which the load is based.

EXAMPLE. Determine the amount of refrigeration in tons required to maintain a given condition if it is necessary to absorb 60,000 Btu per hour.

$$\frac{\text{Total load in Btu per hr}}{12{,}000} = \text{tons refrigeration}$$

$$\frac{60{,}000}{12{,}000} = 5 \text{ tons refrigeration}$$

EXAMPLE: What size refrigeration system is required to cool a building to the desired temperature if the heat load due to people, infiltration, electric lights, and heat transfer through walls is calculated to be (*a*) 564,000 Btu per hr? (*b*) 6600 Btu per min?

$$\frac{\text{Total load in Btu per hr}}{12{,}000} = \text{tons refrigeration}$$

$$(a) \quad \frac{564{,}000}{12{,}000} = 47 \text{ tons refrigeration}$$

$$(b) \quad \frac{6{,}600}{200} = 33 \text{ tons refrigeration}$$

Reference is often made to the *ice-making capacity* of refrigeration systems. This term should not be confused with the Standard Ton of Refrigeration defined above. The ice-making capacity of any system is based on the presupposition that the initial water temperature of 60 F or above must be reduced to 32 F before freezing begins, and, after conversion, the temperature of ice is further reduced to 26 F.

TEMPERATURE CONVERSIONS

Because of the wide use of several different types of thermometers, it is often necessary to convert the readings as determined by one type to correspond to the scale of another. While such conversions usually involve Fahrenheit-Centigrade relationship because of their wide use throughout the refrigeration industry, occasionally other types of scales are encountered. One of these includes the Réaumur scale employed where fermentation temperature control is important, such as in the brewing industry.

As described in Chapter 2, thermometer scales are based on the liquid range of water under standard atmospheric conditions. A comparison of the three temperature ranges based on the division of their scales between the freezing point and the boiling point of water follows:

Fahrenheit scale........	Freezing point of water......	32 F
	Boiling point of water.......	212 F
	Division of range...........	180°
Centigrade scale........	Freezing point of water......	0 C
	Boiling point of water.......	100 C
	Division of range...........	100°
Réaumur scale.........	Freezing point of water......	0 R
	Boiling point of water.......	80 R
	Division of range...........	80°

It will be noted that the temperatures of 32 F, 0 C, and 0 R represent the freezing points of water on their respective scales, and, therefore, are of equal intensity. The boiling points of water—212 F, 100 C, and 80 R—also represent an equal intensity level. Because the Centigrade and the Réaumur scales base their freezing-point values at zero, their readings at any level are mutually convertible by the simple ratio of 80 to 100, or vice versa. However, this is not true where Fahrenheit temperatures are involved, due to the designation of 32 F as the freezing point of water on the latter scale. This difference must be considered in the conversion equations involving Fahrenheit temperatures, as shown below:

(1) To convert Centigrade values to Fahrenheit:

$$°F = (\tfrac{180}{100} \times °C) + 32$$

$$°F = (\tfrac{9}{5} \times °C) + 32$$

(2) To convert Fahrenheit values to Centigrade:

$$°C = \tfrac{100}{180} \times (°F - 32)$$

$$°C = \tfrac{5}{9} \times (°F - 32)$$

(3) To convert Réaumur values to Centigrade:

$$°C = \tfrac{100}{80} \times °R$$

$$°C = \tfrac{5}{4} \times °R$$

(4) To convert Centigrade values to Réaumur:

$$°R = \tfrac{80}{100} \times °C$$

$$°R = \tfrac{4}{5} \times °C$$

(5) To convert Réaumur values to Fahrenheit:

$$°F = (\tfrac{180}{80} \times °R) + 32$$

$$°F = (\tfrac{9}{4} \times °R) + 32$$

(6) To convert Fahrenheit values to Réaumur:

$$°R = \tfrac{80}{180} \times (°F - 32)$$

$$°R = \tfrac{4}{9} \times (°F - 32)$$

Table 46. Table for Converting Centigrade Temperatures to Fahrenheit*

C.	F.	C.	F.	C.	F.	C.	F.	C.	F.
-30	-22.0	20	68.0	70	158.0	120	248.0	170	338.0
-29	-20.2	21	69.8	71	159.8	121	249.8	171	339.8
-28	-18.4	22	71.6	72	161.6	122	251.6	172	341.6
-27	-16.6	23	73.4	73	163.4	123	253.4	173	343.4
-26	-14.8	24	75.2	74	165.2	124	255.2	174	345.2
-25	-13.0	25	77.0	75	167.0	125	257.0	175	347.0
-24	-11.2	26	78.8	76	168.8	126	258.8	176	348.8
-23	-9.4	27	80.6	77	170.6	127	260.6	177	350.6
-22	-7.6	28	82.4	78	172.4	128	262.4	178	352.4
-21	-5.8	29	84.2	79	174.2	129	264.2	179	354.2
-20	-4.0	30	86.0	80	176.0	130	266.0	180	356.0
-19	-2.2	31	87.8	81	177.8	131	267.8	181	357.8
-18	-0.4	32	89.6	82	179.6	132	269.6	182	359.6
-17	+1.4	33	91.4	83	181.4	133	271.4	183	361.4
-16	3.2	34	93.2	84	183.2	134	273.2	184	363.2
-15	5.0	35	95.0	85	185.0	135	275.0	185	365.0
-14	6.8	36	96.8	86	186.8	136	276.8	186	366.8
-13	8.6	37	98.6	87	188.6	137	278.6	187	368.6
-12	10.4	38	100.4	88	190.4	138	280.4	188	370.4
-11	12.2	39	102.2	89	192.2	139	282.2	189	372.2
-10	14.0	40	104.0	90	194.0	140	284.0	190	374.0
-9	15.8	41	105.8	91	195.8	141	285.8	191	375.8
-8	17.6	42	107.6	92	197.6	142	287.6	192	377.6
-7	19.4	43	109.4	93	199.4	143	289.4	193	379.4
-6	21.2	44	111.2	94	201.2	144	291.2	194	381.2
-5	23.0	45	113.0	95	203.0	145	293.0	195	383.0
-4	24.8	46	114.8	96	204.8	146	294.8	196	384.8
-3	26.6	47	116.6	97	206.6	147	296.6	197	386.6
-2	28.4	48	118.4	98	208.4	148	298.4	198	388.4
-1	30.2	49	120.2	99	210.2	149	300.2	199	390.2
0	32.0	50	122.0	100	212.0	150	302.0	200	392.0
1	33.8	51	123.8	101	213.8	151	303.8	201	393.8
2	35.6	52	125.6	102	215.6	152	305.6	202	395.6
3	37.4	53	127.4	103	217.4	153	307.4	203	397.4
4	39.2	54	129.2	104	219.2	154	309.2	204	399.2
5	41.0	55	131.0	105	221.0	155	311.0	205	401.0
6	42.8	56	132.8	106	222.8	156	312.8	206	402.8
7	44.6	57	134.6	107	224.6	157	314.6	207	404.6
8	46.4	58	136.4	108	226.4	158	316.4	208	406.4
9	48.2	59	138.2	109	228.2	159	318.2	209	408.2
10	50.0	60	140.0	110	230.0	160	320.0	210	410.0
11	51.8	61	141.8	111	231.8	161	321.8	211	411.8
12	53.6	62	143.6	112	233.6	162	323.6	212	413.6
13	55.4	63	145.4	113	235.4	163	325.4	213	415.4
14	57.2	64	147.2	114	237.2	164	327.2	214	417.2
15	59.0	65	149.0	115	239.0	165	329.0	215	419.0
16	60.8	66	150.8	116	240.8	166	330.8	216	420.8
17	62.6	67	152.6	117	242.6	167	332.6	217	422.6
18	64.4	68	154.4	118	244.4	168	334.4	218	424.4
19	66.2	69	156.2	119	246.2	169	336.2	219	426.2

* Courtesy, "Trane Air Conditioning Manual," p. 291, copyright, 1941, by The Trane Company, La Crosse, Wisconsin. Reproduced by permission.

Examples:

(1) Convert 40 C to °F:

$$°F = (\tfrac{9}{5} \times °C) + 32$$
$$= (\tfrac{9}{5} \times 40) + 32$$
$$= 72 + 32$$
$$= 104 \text{ F}$$

(1) Convert −35 C to °F:

$$°F = (\tfrac{9}{5} \times -35) + 32$$

$$= -63 + (+32)$$

$$= -31\ F^{*}$$

(2) Convert 122 F to °C:

$$°C = \tfrac{5}{9} \times (°F - 32)$$

$$= \tfrac{5}{9} \times (122 - 32)$$

$$= \tfrac{5}{9} \times 90$$

$$= 50\ C$$

(2) Convert −40 F to °C:

$$\tfrac{5}{9} \times (-40 - (+32))^{\dagger} = \tfrac{5}{9} \times -72$$

$$= -40\ C^{\ddagger}$$

(3) Convert 6 R to °C:

$$°C = \tfrac{5}{4} \times °R$$

$$= \tfrac{5}{4} \times 6$$

$$= 7.5\ C$$

(4) Convert 36 C to °R:

$$°R = \tfrac{4}{5} \times °C$$

$$= \tfrac{4}{5} \times 36$$

$$= 28.8\ R$$

(6) Convert 60 F to °R:

$$°R = \tfrac{4}{9} \times (°F - 32)$$

$$= \tfrac{4}{9} \times (60 - 32)$$

$$= \tfrac{4}{9} \times 28$$

$$= 12.4\ R$$

HEAT-WORK EQUIVALENTS

Mechanical Energy

In reviewing the essential facts relative to energy and its conversion to various forms, the following points are listed:

1. Energy is never created or destroyed.
2. Energy may be converted from one form to another.

* When adding a negative number to a positive number, the rule is to subtract and impose the sign (+ or −) of the larger amount.

† When subtracting a positive number from a negative number, the rule is to add and impose a negative sign.

‡ This is the only temperature at which °F and °C coincide, and, partly for this reason, the U.S. Bureau of Standards has adopted −40 F as a reference point for Thermal Property Tables.

3. By reason of point 2, mechanical energy may be converted to thermal energy.

4. Because energy is the same thing regardless of form, it follows that equivalent values may be established for the various forms.

5. By experiment (Joule), it was found that 778 ft-lb of mechanical energy is the equivalent of 1 Btu of thermal energy.

6. As 33,000 ft-lb per min is equal to 1 hp, then the thermal energy equivalent for a like amount of mechanical energy may be found as follows:

$$1 \text{ hp} = 33{,}000 \text{ ft-lb per min}$$

and

$$1 \text{ Btu} = 778 \text{ ft-lb}$$

Therefore,

$$\frac{33{,}000}{778} = 42.4 \text{ Btu per min}$$

The thermal energy equivalent for 1 hp of mechanical energy is equal to 42.4 Btu per min.

EXAMPLE: What is the thermal equivalent of the mechanical energy produced by equipment having a total power capacity of 175 hp operating at full load? Applying the formula,

$$\text{hp} \times \text{thermal equivalent} = \text{Btu}$$

$$175 \times 42.4 = 7420 \text{ Btu per min total thermal equivalent}$$

EXAMPLE: Determine the amount of thermal energy converted from mechanical energy of equipment having a total power capacity of 325 hp operating at 75 percent load. Applying the formula,

$$\text{hp} \times \text{thermal energy equivalent} = \text{Btu}$$

$$\frac{75 \times 325 \times 42.4}{100} = 10{,}335 \text{ Btu per min total thermal energy equivalent}$$

EXAMPLE: Calculate the total heat load per minute resulting from the operation of the following mechanical equipment driven by electric motors under load conditions as shown:

1 Compressor	10 hp	Full load
1 Drill press	3 hp	50% load
1 Mechanical press	50 hp	75% load
1 Mechanical press	25 hp	Full load

$$10 + (3 \times 0.50) + (50 \times 0.75) + 25 = 74 \text{ hp}$$

$$74 \times 42.4 = 3137.6 \text{ Btu per min total heat load}$$

Electrical Energy

Electrical energy is indicated as watts or kilowatts, 746 watts being the electrical equivalent of 1 hp of mechanical energy.

$$1 \text{ kilowatt} = 1000 \text{ watts}$$

$$1 \text{ kilowatt} = \frac{1000 \text{ watts}}{746 \text{ watts}} = 1.34 \text{ hp}$$

$$1.34 \times 42.4 = 56.8 \text{ Btu per min thermal energy equivalent of 1 kw of electrical energy}$$

$$\frac{56.8}{1000} = 0.0568 \text{ Btu per min thermal energy equivalent of 1 watt of electrical energy}$$

or

$$\frac{56.8 \times \text{number of watts}}{1000} = \text{Btu per min thermal energy equivalent of any given number of watts}$$

EXAMPLE: Determine the thermal energy equivalent of six 50-watt electric light bulbs lighted for a period of 1 hr. Applying the formula,

$$\frac{56.8 \times \text{total wattage}}{1000} \times 60 \text{ min} = \text{Btu per hr total thermal energy equivalent}$$

$$\frac{56.8 \times 300}{1000} \times 60 = 1022.4 \text{ Btu per hr}$$

EXAMPLE: How many Btu per minute are given off by the following pieces of equipment when they are all in use?

Electric heater	300 watts
Electric light bulb	150 watts
Electric oven	2200 watts

Applying the formula,

$$\frac{56.8 \times \text{total wattage}}{1000} = \text{Btu per min}$$

$$\frac{56.8 \times 2650}{1000} = 150.52 \text{ Btu per min}$$

PRACTICAL USE OF SPECIFIC HEAT FACTORS

It has been pointed out that the specific heat of any substance may be expressed in two ways: (*a*) as the relative heat susceptibility of any substance to that of water (water accepted as unity) (*b*) as the number of Btu required to raise the temperature of 1 lb of a substance 1 F.

The thermal energy content of a substance above any given temperature may be determined if the weight and the final temperature are known. It may also be stated that any unknown factor can be determined if all other factors are known. In any problem of heat transfer involving substances of known specific heat, the resultant temperature, weight, or thermal energy content may be found by applying the basic formula

$$\text{Btu content} = \text{weight} \times \text{specific heat} \times \text{temperature difference}$$

or

$$\text{Btu} = W \times \text{sp ht} \times (t_2 - t_1)$$

Expressed as a formula, the number of pounds of a given substance multiplied by the number of Btu required to raise the temperature of each pound 1 F, multiplied by the number of degrees of temperature rise, is equal to total thermal energy required expressed as Btu.

EXAMPLE: How many Btu are required to raise the temperature of 300 lb of cottonseed oil from 68 F to 195 F?

(a) Unknown factor—Btu

(b) Specific heat of cottonseed oil = 0.47

$$\begin{aligned} \text{Btu} &= W \times \text{sp ht} \times (t_2 - t_1) \\ &= 300 \times 0.47 \times (195 - 68) \\ &= 300 \times 0.47 \times 127 \\ &= 17{,}907 \text{ Btu required} \end{aligned}$$

EXAMPLE: Determine the final temperature of 630 lb of ethyl alcohol at 140 F if 955 Btu are extracted.

(a) Unknown factor—final temperature (t_2)

(b) Specific heat of ethyl alcohol = 0.505

$$\text{Btu} = W \times \text{sp ht} \times (t_1 - t_2)$$

$$955 = 630 \times 0.505 \times (140 - t_2)$$

$$t_2 = 140 - \frac{955}{630 \times 0.505}$$

$$t_2 = 137 \text{ F final temperature}$$

EXAMPLE: What quantity of petroleum oil is required to extract 15,500 Btu from high-temperature steel submerged within it to reach a final temperature of 180 F? Initial temperature of oil is 60 F.

(a) Unknown factor—quantity (W)

(b) Specific heat of petroleum oil = 0.46

$$\text{Btu} = W \times \text{sp ht} \times (t_2 - t_1)$$

$$W = \frac{\text{Btu}}{\text{sp ht} \times (t_2 - t_1)}$$

$$W = \frac{15{,}500}{0.46 \times (180 - 60)}$$

$$W = 280.8 \text{ lb oil required}$$

RESULTANT TEMPERATURE CALCULATIONS

When two or more substances, each having a different temperature, are mixed together, there is a tendency to establish an intensity level common to all. Thus, if two or more substances or two or more bodies of a similar substance, having different temperatures, are combined, a change will take place in all. This results in a common temperature level at a point somewhere between the temperatures of all bodies, depending on their respective quantities and specific heat values.

EXAMPLE: What is the resultant temperature when 100 lb of water at 180 F is mixed with 75 lb of water at 40 F? Assume no other heat loss. Applying the formula,

$$\text{Heat lost} = \text{Heat gained}$$

$$W \times \text{sp ht} \times (t_1 - t_2) = W \times \text{sp ht} \times (t_2 - t_1)$$

$$100 \times 1 \times (180 - t_2) = 75 \times 1 \times (t_2 - 40)$$

$$175t_2 = 21{,}000$$

$$t_2 = 120 \text{ F}$$

Proof:

$$W \times \text{sp ht} \times (t_1 - t_2) = W \times \text{sp ht} \times (t_2 - t_1)$$

$$100 \times 1 \times (180 - 120) = 75 \times 1 \times (120 - 40)$$

$$100 \times 60 = 75 \times 80$$

$$6000 \text{ Btu lost} = 6000 \text{ Btu gained}$$

EXAMPLE: Determine the final temperature when 30 lb of air at 90 F is mixed with 160 lb of air at 42 F in an aluminum chamber weighing 300 lb at a room temperature of 80 F. Assume no other heat loss.

Data:

$$W_1 = \text{wt of 90 F air} = 30 \text{ lb}$$

$$W_2 = \text{wt of aluminum} = 300 \text{ lb}$$

$$W_3 = \text{wt of 42 F air} = 160 \text{ lb}$$

$$\text{Sp ht of air} = 0.24$$

$$\text{Sp ht of aluminum} = 0.22$$

$$t_1 \text{ values} = 90 \text{ F } W_1 \text{ air}$$

$$42 \text{ F } W_3 \text{ air}$$

$$t_2 \text{ value} = \text{unknown}$$

Applying the formula,

$$\text{Heat lost} = \text{heat gained}$$

$$W_1 \times \text{sp ht} \times (t_1 - t_2) + W_2 \times \text{sp ht} \times (t_1 - t_2) = W_3 \times \text{sp ht} \times (t_1 - t_2)$$

$$30 \times 0.24 \times (90 - t_2) + 300 \times 0.22 \times (80 - t_2) = 160 \times 0.24 \times (t_2 - 42)$$

$$648 - 7.2t_2 + 5280 - 66t_2 = 38.4t_2 - 1612.8$$

$$111.6t_2 = 7540.8$$

$$t_2 = 67.57 \text{ F resultant temperature}$$

Proof:

$$W_1 \times \text{sp ht} \times (t_1 - t_2) + W_2 \times \text{sp ht} \times (t_1 - t_2) = W_3 \times \text{sp ht} \times (t_2 - t_1)$$

$$30 \times 0.24 \times (90 - 67.57) + 300 \times 0.22 \times (80 - 67.57)$$

$$= 160 \times 0.24 \times (67.57 - 42)$$

$$982 \text{ Btu lost} = 982 \text{ Btu gained}$$

EXAMPLE: How many pounds of 50 F water are required to bring about a resultant temperature of 110 F when mixed with 200 lb of water at 190 F in a copper vat weighing 150 lb and having a room temperature of 85 F? Assume no other losses.

Data:

W_1 = wt of 190° water = 200 lb

W_2 = wt of 50° water = unknown

W_3 = wt of copper vat = 150 lb

Specific heat of water = 1

Specific heat of copper = 0.093

t_1 values = 190 F W_1 water

50 F W_2 water

85 F copper vat

t_2 value = 110 F

Applying the formula,

$$\text{Heat lost} = \text{heat gained}$$

$$W_1 \times \text{sp ht} \times (t_1 - t_2) = W_2 \times \text{sp ht} \times (t_2 - t_1) + W_3 \times \text{sp ht} \times (t_2 - t_1)$$

$$200 \times 1 \times (190 - 110) = W_2 \times 1 \times (110 - 50) + 150 \times 0.093 \times (110 - 85)$$

$$60\ W_2 = 16{,}000 - 348.75$$

$$W_2 = 260.85 \text{ lb water @ 50 F}$$

Proof:

$$W_1 \times \text{sp ht} \times (t_1 - t_2) = W_2 \times \text{sp ht} \times (t_2 - t_1) + W_3 \times \text{sp ht} \times (t_2 - t_1)$$

$$200 \times 80 = (260.85 \times 60) + 348.75$$

$$16{,}000 \text{ Btu lost} = 16{,}000 \text{ Btu gained}$$

REFRIGERATING EFFECT

In calculating the refrigerating effect of any refrigerant, allowances must be made for the cooling of the liquid as it enters the evaporator from the condenser temperature to the evaporator temperature. Under standard conditions, these temperatures are 86 F and 5 F, respectively, so it is necessary to base this cooling through a range of 86° − 5° or 81 F.

This can be accomplished by the standard formula

$$\text{Btu} = \text{wt} \times \text{sp ht} \times (t_1 - t_2)$$

or determined directly from the Thermodynamic Property Tables.

Using the first method, and ammonia as the refrigerant, the formula would read

$$\begin{aligned}\text{Btu} &= 1 \times 1.1185^* \times (86 - 5)\\ &= 1.1185 \times 81\\ &= 90.6 \text{ Btu}\end{aligned}$$

Thus, 90.6 Btu must be subtracted from the latent heat of vaporization of ammonia at 5 F to determine the actual refrigerating effect. Therefore,

$$565.0 \text{ Btu} - 90.6 = 474.4 \text{ Btu per min R E}$$

The results of this method correspond to the results obtained when the refrigerating effect is determined from the enthalpy contents of the entering liquid and the leaving vapor, as shown in the Thermodynamic Property Tables; that is,

Enthalpy of saturated ammonia vapor @ 5 F	= 613.3 Btu
Enthalpy of liquid ammonia @ 86 F	= 138.9 Btu
Refrigerating effect	= 474.4 Btu

If the liquid is subcooled in a heat exchanger, thus reducing the temperature of the liquid before it passes through the throttling device, a greater efficiency of operation may be obtained. This is demonstrated by comparing the refrigerating effect of liquid entering the evaporator cooled to 70 F.

Enthalpy of saturated ammonia vapor @ 5 F	= 613.3 Btu
Enthalpy of liquid ammonia @ 70 F	= 120.54 Btu
Refrigerating effect	= 492.76 Btu

or an increase in R E of 492.76 − 474.4 or 18.36 Btu per lb.

HEAT CAPACITY OF REFRIGERANTS

Problems of heat absorption capacities of refrigerants involve the use of refrigerant property tables. These tables are of utmost importance to stu-

* Average specific heat of liquid ammonia between the temperatures of 5 F and 86 F.

dents of refrigeration. Several examples illustrating their use are presented here.

As previously stated, a definite amount of heat absorption capacity on the part of the particular refrigerant is required in order to lower the temperature of the space or the substance from which the heat is to be extracted. In the study of refrigerant characteristics, it was found that this capacity varies according to the liquid employed. As a result, the quantity of refrigerant necessary to perform the same amount of cooling also varies. The amount of refrigerant to be evaporated can be determined if the amount of heat to be taken from the space or the substance to be cooled is known. It is also possible to determine the potential heat load capacity of a system if the amount of refrigerant allowed to flow through the evaporator is known, assuming, of course, that enough evaporator surface is provided to allow sufficient conduction of heat.

The amount of heat that may be extracted by a refrigerant within an evaporator depends on several factors, all of which must be considered in calculating the capacity of any system. They are as follows:

1. Refrigerant employed
2. Evaporator pressure
3. Condenser pressure
4. Quality of refrigerant leaving the evaporator (wet, dry [saturated], or superheated)

The quality of the refrigerant will determine its condition upon leaving the evaporator. That condition may be one of the following: (1) there will be a portion of the liquid that leaves the evaporator before it vaporizes (wet), in which case it has not utilized its full potential heat absorption

TABLE 47. WEIGHT OF REFRIGERANT REQUIRED PER TON ASSUMING STANDARD TON CONDITIONS

(5 F Evaporator and 86 F Condenser Temperature)

Refrigerant	*Lb per Ton*
Ammonia	0.422
Carbon dioxide	3.60
Carrene-7 (R-500)	3.27
Ethyl chloride	1.45
Refrigerant 12	3.92
Refrigerant 21	2.24
Refrigerant 22	2.89
Refrigerant 113	3.73
Methyl chloride	1.33
Sulfur dioxide	1.41

capacity as established by its latent heat. Systems employing this method of vapor return are known as wet compression systems. (2) The refrigerant will be completely vaporized (dry) and leave the evaporator at a temperature denoting saturation; the system is said to be dry compression. (3) The refrigerant will have absorbed heat in addition to the amount required for saturation and therefore become superheated, a superheated system. The usual practice is to superheat 10 F in order to avoid liquid refrigerant entering the compressor.

In these calculations, ideal conditions will be contemplated, and in all examples it will be assumed that the refrigerant leaves the evaporator as a saturated vapor.

EXAMPLE: Determine the weight of ammonia per minute required to absorb the heat load of a cold-storage room totaling 156,000 Btu per hour. Assume that the ammonia plant operates at a suction pressure of 22.9 psig and a condenser pressure of 138 psig. Ammonia leaves the evaporator as a saturated vapor. Assume no subcooling.

Data:

Heat load	156,000 Btu per hr
Refrigerant	Ammonia
Suction pressure	22.9 psig
Head pressure	138.3 psig
Amount of ammonia	Unknown

From Tables of Properties of Saturated Ammonia Vapor:

Suction temperature @ 22.9 psi	9 F
Condenser temperature @ 138.3 psi	80 F
Total enthalpy of 9 F vapor	614.6 Btu per lb
Enthalpy of 80 F liquid	132.0 Btu per lb
Refrigerating effect	482.6 Btu per lb

The refrigerating effect is the actual capacity of heat absorption of the ammonia from its liquid to vapor form. It is calculated from the temperature at which it enters the evaporator, and *not* from the evaporator temperature.

For each pound of ammonia employed under the stated conditions, 482.6 Btu can be absorbed. The number of pounds required to absorb the total heat load would be:

$$\frac{\text{Heat load per hour}}{\text{R E} \times 60} = \text{lb refrigerant per min}$$

$$\frac{156{,}000}{482.6 \times 60} = 5.39 \text{ lb ammonia per min circulated}$$

In order to compute the compressor size, it is necessary that the weight of refrigerant circulated be converted to specific volume, or cubic feet per minute passing through the evaporator. This value must be based on evaporator conditions and determined from the refrigerant property tables.

As it is assumed that ammonia leaves the evaporator in a dry saturated condition in the above example, the specific volume of 1 lb of ammonia is read from the line in the Properties of Saturated Ammonia Vapor corresponding to conditions prevailing in the evaporator, or 9 F and 22.9 psig.

Referring to the ammonia tables, it is found that for 9 F and 22.9 psi, gage pressure, the specific volume is equal to 7.464 cu ft per lb of ammonia.

Therefore, the compressor must be of sufficient size to handle

$$5.39 \text{ lb} \times 7.464 \text{ cu ft per lb} = 40.23 \text{ cu ft per min}$$

EXAMPLE: (*a*) What is the potential heat load capacity per hour of a system in which 240 lb of sulfur dioxide, leaving the evaporator as a saturated vapor, is circulated per hour and having a suction pressure of 4.61 in. of mercury and a condenser pressure of 59 psig? (*b*) per minute? (*c*) What is its rated capacity?

Data:

Heat load capacity...............	Unknown
Refrigerant.......................	Sulfur dioxide
Suction pressure..................	4.61 in. Hg
Head pressure.....................	59 psig

From Tables of Properties of Saturated SO_2 vapor:

Suction temperature	7 F
Discharge temperature	92 F
Total enthalpy of 7 F vapor	183.65 Btu per lb
Enthalpy of 92° liquid	44.19 Btu per lb
Refrigerating effect	139.46 Btu per lb

To determine the heat that may be absorbed by SO_2 under stated condition, apply the formula

$$\frac{\text{Heat load}}{\text{R E}} = \text{lb refrigerant}$$

$$\text{Heat load} = \text{R E} \times \text{lb refrigerant}$$

$$(a) \quad \text{Heat load} = 139.46 \times 240 = 33{,}470.4 \text{ Btu per hr}$$

or

$$(b) \quad \text{Heat load} = 557.8 \text{ Btu per min}$$

and

$$(c) \quad \frac{557.8}{200} = 2.79 \text{ tons (approx) rated capacity}$$

CONDENSER LOAD CALCULATIONS

The total amount of heat extracted from the refrigerant by the cooling medium circulated through the condenser is the sum of all the heat absorbed during evaporation and compression. Any superheat absorbed in the evaporator as well as heat resulting from temperature reduction of the refrigerant below the point of liquefaction, corresponding to the head pressure, must also be considered as part of the condenser load.

Under standard conditions of 5 F evaporator temperature and 86 F condenser temperature, and without superheating in the evaporator or after cooling, the total heat to be extracted in the condenser is determined as follows, using ammonia in the example:

1st Step: Superheat resulting from work of compression. At a head pressure of 169.4 psia and a discharge temperature of 210 F, each pound of refrigerant contains 713 Btu. (Reference: Properties of Superheated Ammonia Vapor tables.)

Saturated ammonia vapor at 86 F and 169.4 psia contains 631.5 Btu per lb. (Reference: Properties of Saturated Ammonia Vapor tables.)

The amount of superheat, therefore, is equal to 713 − 631.5 = 81.5 Btu per lb.

This means that 81.5 Btu must be extracted from each pound of ammonia vapor in order to reduce its temperature from 210 F to 86 F. A saturated vapor still exists, however.

2nd Step: To convert 86 F ammonia vapor to 86 F ammonia liquid, an amount of thermal energy equal to its latent heat of vaporization must be extracted. This value for ammonia at 86 F = 492.6. (Reference: Properties of Saturated Ammonia Vapor tables.)

3rd Step: To find the total heat to be extracted in the condenser, assuming standard conditions and no other sources of heat, add the heat of vaporization and the thermal energy (work) of compression, viz.:

Superheat − Btu per lb of 210 F ammonia and 169.4 psia = 81.5 Btu

Latent heat of vaporization of 86 F saturated ammonia vapor = 492.6 Btu

$$81.5 + 492.6 = 574.1 \text{ Btu}$$

that must be extracted from each pound of ammonia by the condenser.

In order to reduce the temperature of the liquid refrigerant below its point of liquefaction in the condenser, the kinetic thermal energy of the liquid through the desired temperature range must be considered as part of the condenser load. For example, to reduce 86 F saturated ammonia to 76 F in the above, it is necessary to extract an additional amount of heat corresponding to a 10 F temperature change in order to accomplish such a result.

1 lb ammonia liquid at 86 F contains.....	138.9 Btu
1 lb ammonia liquid at 76 F contains.....	127.4 Btu
or a difference of.....................	11.5 Btu

In addition to the 574.1 Btu that must be extracted to bring about a saturated liquid ammonia condition at 86 F, a further increase in condenser load of 11.5 Btu per lb is necessary to reduce the liquid temperature to 76 F.

$$574.1 + 11.5 = 585.6 \text{ Btu per lb ammonia, heat removed by condenser}$$

If the enthalpy of the liquid ammonia at 76 F, or 127.4 Btu per lb, is subtracted from the total enthalpy of the superheated discharge gas, or 713 Btu per lb, as in the above set of conditions, the result is the same as when each phase is calculated separately. Hence,

$$713.0 - 127.4 = 585.6 \text{ Btu per lb}$$

To determine the amount of condenser coolant necessary for the removal of heat contained in the refrigerant for the purpose of converting the superheated gas to a liquid at any temperature below the point of liquefaction, it is necessary only to apply the conventional thermal energy balance equation:

$$\text{Btu} = \text{weight of coolant} \times \text{sp ht} \times t_d$$

Assume that water is employed and that refrigerant conditions are as they appear in the example above. Also assume that the temperature of the cooling water enters the condenser at 58 F and leaves at a predetermined temperature of 68 F. Then

$$\text{Btu} = W \times 1 \times (t_2 - t_1)$$

$$W = \frac{\text{Btu}}{\text{sp ht} \times (t_2 - t_1)}$$

$$W = \frac{585.6}{1 \times 10}$$

$$W = 58.56 \text{ lb water per lb refrigerant}$$

Water quantities are given in gallons, and for convenience it is desirable to convert pounds to gallons.

1 gal water weighs 8.33 lb. Therefore,

$$\frac{58.56}{8.33} = 7.03 \text{ gal water per lb of ammonia}$$

LATENT HEAT OF FUSION—FOOD PRODUCTS

The latent heat of fusion of food products must be considered as part of the total refrigeration load for any plant in which the freezing of such substances may be involved. This factor represents the amount of heat to be extracted from the fluids that form part of food substances in order to bring about solidification.

The amount of fluid found in the various products is considered as water content and expressed in percent by weight. However, such fluids contain soluble compounds tending to change the thermal characteristics from those of pure water, not unlike the effect of chlorides in brine solutions. While these compounds reduce the freezing point of most food substances from 1 F to more than 30 F in some instances, below the freezing temperature of water (32 F), their effect on the latent heat of fusion of the water content appears to be negligible for all practical purposes. This statement is borne out by the close similarity in the results obtained when experimental values are compared with calculated values based on direct relationship between water content and latent heat of fusion for any individual item.

The freezing of food substance is gradual in that all the fluid does not solidify at any specific temperature. This is due to the rejection of soluble compounds by the already frozen fluid during its crystallization and subsequent concentration of these materials in that portion of the fluid still in liquid form. For this reason, complete solidification of many types of food products is reduced to −65 F and below.

Assuming that a direct relationship nearly exists between the percent of water content contained in the various food products and the latent heat of fusion of water, the latent heat of fusion for the individual item may be determined as follows:

$$\frac{\%\ \text{Water content} \times 144}{100} = \text{latent heat of fusion}$$

EXAMPLE:

(1) Asparagus contains 93 percent water by weight. Therefore,

$$\frac{93.0 \times 144}{100} = 134\ \text{Btu, latent heat of fusion}$$

(2) Peas contain 74.3 percent water by weight. Therefore,

$$\frac{74.3 \times 144}{100} = 107\ \text{Btu, latent heat of fusion}$$

(3) Pears contain 83.5 percent water by weight. Therefore

$$\frac{83.5 \times 144}{100} = 120\ \text{Btu, latent heat of fusion}$$

(4) Lean beef contains 72 percent water by weight. Therefore

$$\frac{72.0 \times 144}{100} = 104\ \text{Btu, latent heat of fusion}$$

Table 118 shows the latent heat of fusion for some food products, including fruits, vegetables, meat, and dairy products, based on calculated values.

INFILTRATION LOAD

All cold rooms or freezing compartments are subject to a certain amount of air leakage. This leakage acts as a source of heat which must be extracted by the evaporator as part of the refrigeration load. It is caused by the opening of doors incident to the storing and the removal of products, leakage around windows, and the loose construction which allows air to enter through the joints. The amount of leakage, of course, depends on the quality of construction and the length of time doors are permitted to remain open.

TABLE 48. AVERAGE AIR CHANGES OCCURRING IN COLD-STORAGE ROOMS PER 24 HOURS BASED ON INFILTRATION AND DOOR OPENING*

Volume of Room, cu ft	*Air Changes per 24 hr; Room Temperature Above 32 F*	*Air Changes per 24 hr; Room Temperature Below 32 F*
250	38.0	29.0
300	34.5	26.2
400	29.5	22.5
500	26.0	20.0
600	23.0	18.0
800	20.0	15.3
1,000	17.5	13.5
2,000	12.0	9.3
4,000	8.2	6.3
6,000	6.5	5.0
8,000	5.5	4.3
10,000	4.9	3.8
15,000	3.9	3.0
20,000	3.5	2.6
25,000	3.0	2.3
30,000	2.7	2.1
40,000	2.3	1.8
50,000	2.0	1.6
75,000	1.6	1.3
100,000	1.4	1.1

* Reprinted from the "Air Conditioning-Refrigerating Data Book 9th Edition—1955–1956 Design Volume," by courtesy of The American Society of Refrigerating Engineers.

For average conditions, the number of air changes per 24 hours for cold-storage rooms is shown in Table 48. The method of calculating infiltration heat loads, which involves both sensible and latent heat gain, is given in Chapter 18, "Principles of Air Conditioning."

HEAT LOAD—HUMAN BEINGS

The amount of heat given off by human beings depends on several factors, the more important of which concern the intensity of their physical activity and the temperature differential between that of the refrigerated space and normal body temperature. Men at work generate a greater amount of heat than when resting. Therefore, the load requirements are greater for each individual occupied in handling and storing producé within a cold-storage plant, for example, than it would be for each person seated in an air-conditioned theater or office.

The heat load of human beings must be considered as kinetic thermal energy in determining total refrigeration load requirements. Therefore, each person occupied within the compartment to be cooled represents a specific number of Btu, radiated from their bodies during a definite time interval, that tend to increase the ambient temperature above that desired. For this reason, the human heat load must be considered in estimating the total heat load factor, based on the number of persons that normally would be present.

In estimating refrigeration requirements for air-conditioning purposes, the latent heat of vaporization becomes a factor as a result of the moisture loss from human bodies by evaporation.

While authorities differ as to the allowances that should be made in coping with the heat load resulting from human occupancy, the values listed in Table 49 for both industrial and nonindustrial activity have been found by the author to give accurate results.

TABLE 49. TOTAL HEAT LOAD—HUMAN OCCUPANCY SENSIBLE AND LATENT HEAT

Temperature °F Refrigerated Space	*Industrial (Cold-Storage Plants, Air-Conditioned Factories, etc.)*		*Nonindustrial (Theaters Offices, etc.)*	
	Btu per Person per hr	Btu per Person per min	Btu per Person per hr	Btu per Person per min
70 F to 85 F........	480	8	180* to 300	3 to 5
32 F to 50 F........	720	12		
0 F to 32 F.........	1200	20		

* Seated in a theater.

EXAMPLE: What is the human heat load in an air-conditioned office occupied by 60 persons? (*a*) per hour? (*b*) per minute?

Solution:

Air-conditioned office temperatures = 70 F to 85 F

Heat factor (Table 45) = 300 Btu per person per hr

60 persons × 300 = 18,000 Btu per hr

$$\frac{18,000}{60} = 300 \text{ Btu per min}$$

EXAMPLE: If a maximum of eight men are employed in a cold-storage chill room 16 hours per day, (*a*) what is the total human heat load for any hour during their occupancy? (*b*) What is the average heat load per hour for a 24-hr operation?

Solution:

(*a*) Chill room temperatures = 32 F to 50 F

Heat factor (Table 45) = 720 Btu per person per hr

8 men × 720 = 5760 Btu per hr

(*b*) $\frac{8 \text{ men} \times 720 \times 16}{24} = 3840$ Btu per hr

HEAT TRANSFER EQUIPMENT CALCULATIONS

Heat transfer equipment as applied to the refrigeration system includes the evaporator, the condenser, the heat exchanger, the liquid precooler, and the intercooler. They usually consist of a series of pipes or tubing, termed the heat transfer element, submerged in the substance under refrigeration and through which the refrigerant passes, or vice versa. In some instances, the heat transfer element is in the form of a plate rather than piping, depending on the purpose for which it is intended.

The purpose of the heat transfer element is to provide a surface through which heat from the warmer substance can be transmitted to the colder substance. It is designed to accomplish heat transfer at a rate commensurate with the capacity requirements of the refrigeration system for which it is intended. Design consideration, therefore, must include such factors as materials employed in constructing the element, area of exposed surface, fluid velocities, fluid characteristics, and temperature differentials existing between the two substances involved.

Usually the element is constructed of iron, steel, monel metal, or copper, depending on the pressures maintained, the chemical stability of the metal when exposed to substances involved, and general operating conditions. While the heat transfer values of these metals differ in comparison with each other, the variations are slight enough to be considered negligible for practical purposes. Therefore, average heat transmission values may be adopted as being applicable for all standard size piping used as elements made from any of the materials mentioned above.

The rate of heat transfer is effected as follows:

(*a*) It is more rapid when both substances are liquid. (Water-to-water.)
(*b*) It is less rapid when one or both substances are gases. (Brine-to-air.)
(*c*) It decreases as the viscosity of one or both substances increases. (Water-to-oil.)

(*d*) Rust, scale, and oil films retard the flow of heat when deposited on element surfaces.
(*e*) It increases as the relative velocity of the substances increases.
(*f*) It increases as the temperature differential between the two substances increases.

Because surface deposits in the form of scale, rust, and oil film are variable and undeterminable factors, they nevertheless present a condition that must be considered in estimating heat transfer surface requirements. It is usual practice to allow a safety factor of 10 sq ft per ton of refrigeration to overcome the retarding action of such deposits. Where any program of cleanliness is maintained, this amount of added surface should prove sufficient for most normal operating conditions.

The advantageous effect of moving fluids on heat transmission over systems that depend on natural convection only is the reason for installing brine tank agitators and cooler room blowers in ice and cold-storage plants.

The rate at which heat is transmitted through a wall separating the substances involved is expressed in terms of Btu per hour per square foot of exposed element surface per degree F temperature differential. This value is called the over-all coefficient of heat transfer, and varies with any change occurring in the factors mentioned above. Designated by the letter "U," the over-all coefficient of heat transfer may be stated briefly as

$$U = \text{Btu per hr per sq ft per °F } t_d$$

The value of U for any material or condition cannot be determined except by experiment. Therefore, it is necessary to refer to tables listing experimental results in order to obtain such values of U as meet the conditions for which information is required. Because there are many variables that effect the transmission of heat through a heat transfer element, exact values based on constant operating conditions would be difficult to maintain. Therefore, an average value based on average operating conditions is justifiable.

Table 86 and Table 87 (see Chapter 12) list the average over-all coefficients of heat transfer for various combinations of factor changes. The values listed are applicable for any pipe metal commonly used as heat transmission elements, such as iron, copper, etc., whose thickness conforms to a standard ratio with its diameter as listed in tables of pipe sizes.

In order for a refrigeration system to extract the required amount of heat from the substance or the space for which it is designed in a definite interval of time, an evaporator of sufficient size must be provided. This means that the heat element through which the refrigerant passes during the cooling phase must possess the necessary amount of surface area exposed to the substance under refrigeration commensurate with the operating conditions or its coefficient of heat transfer.

Size, then, for any heat transfer equipment is designated by the area of heat transfer surface exposed to the substance from which heat is extracted. Usually it is expressed in terms of *square feet of surface area.* In refrigeration, this surface is referred to commonly as *cooling surface* in connection with the evaporator or *condensing surface* when applied to the condenser.

The area of exposed surface required for any given heat load may be determined by means of the conventional heat transfer formula,

$$H = U \times A \times t_d$$

where

H = total heat to be transmitted per hour
U = Over-all coefficient of heat transfer—Btu per hr per sq ft per °F t_d
A = area of transmission surface in square feet
t_d = temperature differential of substances

By transposing the factors in the formula and solving for A,

$$A = \frac{H}{U \times t_d}$$

EXAMPLE: Determine the length of $1\frac{1}{2}$-in. iron pipe required for an evaporator in maintaining a 28 F temperature in a cold-storage compartment, the heat load of which is equivalent to 3.5 tons of refrigeration. Suction temperature of refrigerant is 5 F using ammonia. Air circulation is provided by natural convection only.

$$\text{Heat load per hour} = 3\tfrac{1}{2} \text{ tons} \times 12{,}000 \text{ Btu} = 42{,}000 \text{ Btu}$$

$$\text{Temperature differential} = 28 \text{ F} - 5 \text{ F} = 23 \text{ F } t_d$$

$$U = 2.0 \text{ Btu per hr per sq ft per °F (see Table 87)}$$

Applying the formula

$$A = \frac{H}{U \times t_d} = \frac{42{,}000}{2.0 \times 23}$$

$$A = 913 \text{ sq ft of pipe surface}$$

The number of lineal feet of pipe may be determined by first finding the number of lineal feet of $1\frac{1}{2}$-in. pipe having an external area of 1 sq ft. This may be either calculated or found in tables giving pipe sizes. Multiplying the length of pipe having an area of 1 sq ft by the number of square feet of pipe surface required, the lineal feet of pipe is found.

For $1\frac{1}{2}$-in. pipe − 2.01 lineal feet equals 1 sq ft of external surface area

$$\therefore 2.01 \times 913 = 1835.13 \text{ lineal feet of } 1\tfrac{1}{2}\text{-in. pipe}$$

Allowing 10 sq ft per ton of refrigeration load for surface scale, rust, frosting, or oil film, an additional 10 × 3.5 or 35 sq ft is required. Multiplying 35 by 2.01 = 70.35 lineal feet of pipe.

$$1835.13 + 70.35 = 1905.48 \text{ ft of } 1\tfrac{1}{2}\text{-in. pipe}$$

MEAN TEMPERATURE DIFFERENCE

In the foregoing section, it was pointed out that the rate of heat transfer through a separating wall increases as the difference in temperature of the separated fluids becomes greater. Under conditions already described, it was assumed that temperatures of both fluids remained practically constant at their respective levels during the process of heat transmission, as would be the case where the quantity of heat extracted equals the amount gained. Such a condition is found in chill rooms, coolers, and brine tanks after desired temperatures are once established. In each case, the employment of arithmetically determined temperature differentials obtained by subtracting the lower temperature from the higher one is acceptable in deriving practical results. Nor are such results affected by fluid movement which occurs when cooler-room blowers or brine-tank agitators are employed because it is the same body of fluid being recirculated continuously.

Where the temperatures of either or both fluids are constantly changing by reason of heat transfer as a result of continuous replenishment and rejection of the fluids, temperature differential determination is no longer one of simple arithmetic. In this case, differential values are based on progressive temperature changes occurring in both fluids simultaneously, and therefore

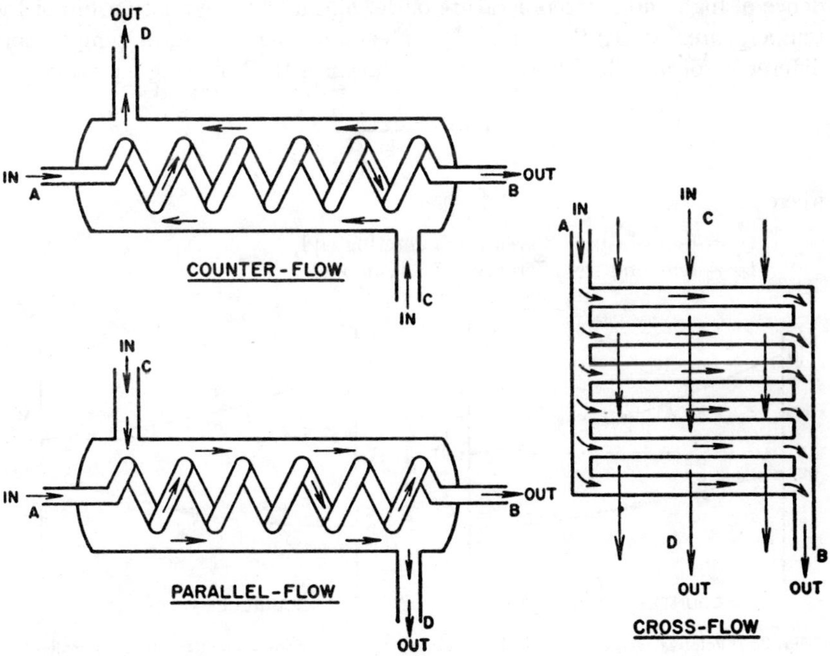

Fig. 354. Directional flow of fluids in heat exchangers.

become a mean value to all original and final temperatures. Such a value is known as the log mean temperature difference, or mtd. For similar inlet temperatures of both fluids, the log mean temperature difference will vary according to the relative direction of fluid flow. All other factors being equal, a lesser value results in parallel flow systems, i.e., fluids flowing in the same direction, than in systems employing counterflow principles in which the fluids flow in opposite directions. (See Fig. 354.)

If cold water is passed through a coil submerged in a bath of warmer circulated water, the result is a gradual temperature change in both fluids. At every point along the line of flow the temperature differential varies by reason of continuous heat transfer and a continual change in fluid temperature. This is true regardless of the flow principle employed.

Refer to Fig. 354, wherein two fluids are circulated through their respective elements as to effect close contact, separated only by the walls of the pipe carrying one of the fluids, say the warmer one. Then *A* becomes the inlet for the warm fluid and *B* the outlet of the warm fluid, *C* becomes the cold fluid inlet, and *D* its outlet. The actual temperatures must be known or prescribed, otherwise a definite ratio between *A*, *B*, *C*, and *D* cannot be established as the results are governed by the relationship of the highest temperature difference to the least temperature difference and not by coincidence of high and low temperature outlet and inlet. The relationship of the temperatures at *A*, *B*, *C*, and *D* for establishing the mean temperature difference for any fluid flow system is shown in the following equation

$$\text{mtd} = \frac{t_{d_1} - t_{d_2}}{\log_e \left(\frac{t_{d_1}}{t_{d_2}}\right)}$$

where

t_{d_1} = temperature difference (F) entering end
t_{d_2} = temperature difference (F) leaving end

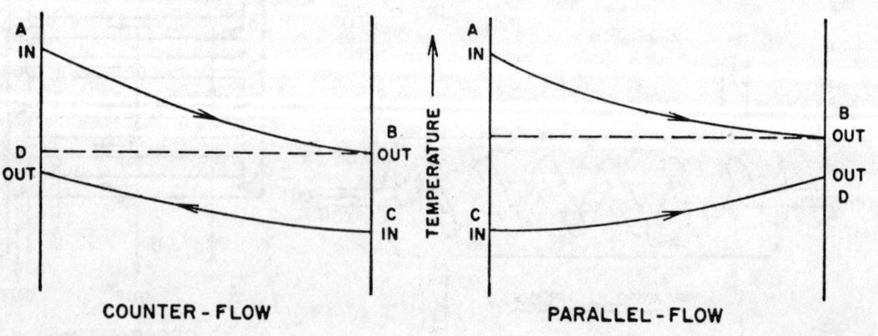

Fig. 355. Relative temperatures of fluids leaving heat exchanger. Counter-flow vs. parallel-flow. Note that counter-flow tends to bring temperatures closer together at outlets of fluids, allowing greater cooling effect.

The entering end is defined as the end at which the heating substance enters. The leaving end is the end at which the heating substance leaves. As an example, the compressed vapor from a compressor would represent the hottest fluid entering a condenser and so becomes the heating medium. The end at which such vapors enter the condenser then becomes the entering end. The leaving end would be the end at which the condensed fluid leaves. The reason for this distinction becomes apparent when the types of flow—i.e., counter- and parallel-flow—are observed in Fig. 355. Another example would be the case of an evaporator in which the refrigerant temperature remains constant or varies slightly due to superheat, and the substance to be cooled enters at a higher than leaving temperature. The substance to be cooled then becomes the heating medium and the end at which it enters the cooler is the entering end as interpreted for purposes of establishing log mean temperature difference.

Substituting the expression "greatest temperature difference" (gtd)* for t_{d_1} and "least temperature difference" (ltd) for t_{d_2}, the log mean temperature difference may be written as follows:

$$\text{mtd} = \frac{\text{gtd} - \text{ltd}}{\log_e \dfrac{\text{gtd}}{\text{ltd}}}$$

where

mtd = log mean temperature difference
gtd = greatest temperature difference
ltd = least temperature difference
$\log_e = 2.3 \times \log_{10}$

Converting the ratio in terms of $\log_{10}$, then

$$\text{mtd} = \frac{\text{gtd} - \text{ltd}}{2.3 \log_{10} \dfrac{\text{gtd}}{\text{ltd}}} = \frac{0.4343(\text{gtd} - \text{ltd})}{\log_{10} \left(\dfrac{\text{gtd}}{\text{ltd}}\right)}$$

By assigning temperature values to the points of inlet and outlet of both fluids in counter- and parallel-flow systems shown in Fig. 354, the mtd equation can best be demonstrated.

EXAMPLE:

Let $A = 210$ F $\quad C = 60$ F

$B = 86$ F $\quad D = 75$ F

For counter-flow, referring to Fig. 355 (*left*):

$$\text{gtd} = A - D = 210 - 75 = 135 \text{ F}$$

$$\text{ltd} = B - C = 86 - 60 = 26 \text{ F}$$

* gtd is variously expressed greatest, highest, or large temperature difference, while ltd may be described as either least or small temperature difference.

where A being the heating factor AD becomes the entering side

$$\text{mtd} = \frac{\text{gtd} - \text{ltd}}{2.3 \log_{10} \frac{\text{gtd}}{\text{ltd}}} = \frac{135 - 26}{2.3 \log_{10} \left(\frac{135}{26}\right)} = \frac{109}{2.3 \times .716}$$

$$\text{mtd} = 68.3 \text{ F}$$

For parallel flow, referring to Fig. 355 (*right*)

$$\text{gtd} = \text{A minus C} = 210 - 60 = 150 \text{ F}$$

$$\text{ltd} = \text{B minus D} = 86 - 75 = 11 \text{ F}$$

$$\text{mtd} = \frac{\text{gtd} - \text{ltd}}{2.3 \log_{10} \frac{\text{gtd}}{\text{ltd}}} = \frac{150 - 11}{2.3 \log_{10} \left(\frac{150}{11}\right)} = \frac{139}{2.3 \times 1.1345}$$

$$= 53.26 \text{ F}$$

The higher log mean temperature difference between counter- and parallel-flow of $68.3 - 53.26 = 15.04$ F bears out the statement made previously relative to the two systems. It becomes obvious that, to obtain the same cooling result, the parallel heat exchanger requires more cooling surface area than the counter-flow apparatus.

The mtd equation may be simplified further by reducing known values to a coefficient which, when multiplied by the greatest temperature difference, will result in the log mean temperature difference. The value of the coefficient depends on and corresponds to the ratio $\frac{\text{ltd}}{\text{gtd}}$ of a given set of conditions. Consider the ratio $\frac{\text{ltd}}{\text{gtd}} = \frac{4}{20} = 0.200$ in the above example. Then consider the ratio $\frac{\text{mtd}}{\text{gtd}} = \frac{9.95}{20} = 0.497$. It is true that $\frac{\text{mtd}}{\text{gtd}} = 0.497$ always corresponds to $\frac{\text{ltd}}{\text{gtd}} = 0.200$. A similar relationship may be established for any value of ratio $\frac{\text{mtd}}{\text{gtd}}$. For example, when the ratio $\frac{\text{mtd}}{\text{gtd}} = \frac{18.2}{30} = 0.6066$, it will always correspond to the ratio $\frac{\text{ltd}}{\text{gtd}} = 0.33$.

To demonstrate: Assume, in the above example, that the water enters the shell at 55 F and leaves at 45 F.

$$\text{mtd} = \frac{15 - 5}{2.3 \log_{10} \frac{15}{5}} = \frac{10}{1.097} = 9.1 \text{ F}$$

$$\text{Ratio} \frac{\text{ltd}}{\text{gtd}} = \frac{5}{15} = 0.33$$

$$\text{Ratio} \frac{\text{mtd}}{\text{gtd}} = \frac{9.1}{15} = 0.6066$$

Regardless of the actual values of ltd and gtd, as long as the ratio $\frac{\text{ltd}}{\text{gtd}} = 0.33$, the coefficient will equal 0.6066. For each ratio of $\frac{\text{ltd}}{\text{gtd}}$ there is a corresponding coefficient that is constant for that ratio. Carrying the computations through ratios $\frac{\text{ltd}}{\text{gtd}} = 0.01$ to, say, 0.99, a table of coefficients for any set of temperature conditions within a heat exchanger may be established. Table 50 lists the coefficients corresponding to the ratios $\frac{\text{ltd}}{\text{gtd}}$ for the range mentioned.

To determine the log mean temperature difference, using the coefficients in Table 50, then

$$\text{mtd} = \text{gtd} \times \text{coefficient corresponding to} \frac{\text{ltd}}{\text{gtd}}$$

EXAMPLE: Determine the mtd of a heat exchanger having an ltd of 12 F and a gtd of 38 F.

$$\text{Ratio} \frac{\text{ltd}}{\text{gtd}} = \frac{12}{38} = 0.316$$

Referring to column 1, Table 50, find $\frac{\text{ltd}}{\text{gtd}} = 0.316$. The coefficient corresponding to 0.316 as read in column 2 (by interpolation) is 0.594.

$$\begin{aligned} \text{mtd} &= \text{gtd} \times \text{coefficient} \\ &= 38 \times 0.594 \\ &= 22.57 \text{ F} \end{aligned}$$

The log mean temperature difference may also be determined from Table 51 if the values of both gtd and ltd are known. To use the table, locate ltd (small temperature difference) along the top horizontal line. Follow down its column until intersection is established with the horizontal row representing gtd (large temperature difference), whose value is shown in the extreme left or the extreme right columns. Read mtd where the two values intersect. To check the results obtained in the preceding example, it is found that the mtd = 22.55 F at the intersection of 12 F ltd and 38 F gtd.

TABLE 50. COEFFICIENTS OF LOG MEAN TEMPERATURE DIFFERENCE*

$\frac{\text{ltd}}{\text{gtd}}$	Coefficient or $\frac{\text{mtd}}{\text{gtd}}$	$\frac{\text{ltd}}{\text{gtd}}$	Coefficient or $\frac{\text{mtd}}{\text{gtd}}$
0.00	0.0	0.51	0.728
0.01	0.215	0.52	0.734
0.02	0.251	0.53	0.740
0.03	0.277	0.54	0.746
0.04	0.298	0.55	0.753
0.05	0.317	0.56	0.759
0.06	0.334	0.57	0.765
0.07	0.350	0.58	0.771
0.08	0.364	0.59	0.777
0.09	0.378	0.60	0.783
0.10	0.391	0.61	0.789
0.11	0.403	0.62	0.795
0.12	0.415	0.63	0.801
0.13	0.427	0.64	0.806
0.14	0.438	0.65	0.813
0.15	0.448	0.66	0.818
0.16	0.458	0.67	0.823
0.17	0.469	0.68	0.829
0.18	0.478	0.69	0.836
0.19	0.488	0.70	0.840
0.20	0.497	0.71	0.848
0.21	0.506	0.72	0.852
0.22	0.515	0.73	0.858
0.23	0.524	0.74	0.864
0.24	0.533	0.75	0.870
0.25	0.541	0.76	0.874
0.26	0.549	0.77	0.879
0.27	0.558	0.78	0.886
0.28	0.566	0.79	0.890
0.29	0.574	0.80	0.896
0.30	0.582	0.81	0.902
0.31	0.589	0.82	0.907
0.32	0.597	0.83	0.913
0.33	0.604	0.84	0.918
0.34	0.612	0.85	0.923
0.35	0.619	0.86	0.928
0.36	0.626	0.87	0.934
0.37	0.634	0.88	0.939
0.38	0.641	0.89	0.944
0.39	0.648	0.90	0.949
0.40	0.655	0.91	0.955
0.41	0.662	0.92	0.959
0.42	0.669	0.93	0.964
0.43	0.675	0.94	0.970
0.44	0.682	0.95	0.975
0.45	0.689	0.96	0.979
0.46	0.695	0.97	0.986
0.47	0.702	0.98	0.991
0.48	0.709	0.99	0.995
0.49	0.715	1.00	1.00
0.50	0.721		

* Courtesy, "Trane Air Conditioning Manual," copyright, 1941, by The Trane Company, La Crosse, Wisconsin. Reproduced by permission.

TABLE 51. MEAN TEMPERATURE DIFFERENCES*

LARGE TEMP. DIFF.	SMALL TEMPERATURE DIFFERENCE 1	2	3	4	5	6	7	8	9	10	11	12	13	14	15	16	17	18	19	20	LARGE TEMP. DIFF.
1	1.00																				1
2	1.44	2.00																			2
3	1.82	2.47	3.00																		3
4	2.16	2.89	3.48	4.00																	4
5	2.49	3.27	3.92	4.48	5.00																5
6	2.79	3.64	4.33	4.93	5.48	6.00															6
7	3.08	3.99	4.72	5.36	5.94	6.49	7.00														7
8	3.37	4.33	5.10	5.77	6.38	6.95	7.49	8.00													8
9	3.64	4.65	5.46	6.17	6.81	7.40	7.96	8.49	9.00												9
10	3.91	4.97	5.81	6.55	7.21	7.83	8.41	8.96	9.49	10.00											10
11	4.17	5.28	6.16	6.92	7.61	8.25	8.86	9.42	9.97	10.49	11.00										11
12	4.43	5.58	6.49	7.28	8.00	8.66	9.28	9.87	10.43	10.97	11.49	12.00									12
13	4.68	5.88	6.82	7.64	8.37	9.05	9.69	10.30	10.89	11.43	11.97	12.50	13.00								13
14	4.93	6.17	7.14	7.98	8.74	9.44	10.10	10.72	11.31	11.89	12.44	12.97	13.50	14.00							14
15	5.17	6.45	7.46	8.32	9.10	9.82	10.50	11.14	11.74	12.33	12.89	13.44	13.98	14.50	15.00						15
16	5.41	6.73	7.77	8.66	9.46	10.19	10.89	11.54	12.16	12.77	13.33	13.92	14.45	14.97	15.49	16.00					16
17	5.65	7.01	8.07	8.98	9.81	10.56	11.27	11.94	12.58	13.19	13.79	14.35	14.90	15.45	15.98	16.49	17.00				17
18	5.88	7.28	8.37	9.31	10.15	10.92	11.65	12.33	12.98	13.61	14.22	14.80	15.35	15.90	16.45	16.98	17.50	18.00			18
19	6.11	7.55	8.67	9.63	10.49	11.28	12.02	12.72	13.38	14.02	14.64	15.24	15.80	16.38	16.90	17.46	17.99	18.48	19.00		19
20	6.34	7.82	8.96	9.94	10.82	11.63	12.38	13.10	13.78	14.43	15.06	15.65	16.26	16.81	17.40	17.93	18.46	18.98	19.51	20.00	20
21	6.57	8.08	9.25	10.25	11.15	11.97	12.74	13.47	14.17	14.83	15.47	16.08	16.69	17.26	17.83	18.36	18.95	19.46	19.98	20.50	21
22	6.79	8.34	9.54	10.56	11.47	12.31	13.10	13.84	14.55	15.22	15.87	16.50	17.11	17.71	18.27	18.84	19.40	19.95	20.46	20.98	22
23	7.02	8.60	9.82	10.86	11.80	12.65	13.45	14.20	14.92	15.61	16.27	16.90	17.53	18.13	18.73	19.27	19.85	20.38	20.89	21.47	23
24	7.24	8.85	10.10	11.16	12.11	12.98	13.80	14.56	15.29	15.99	16.66	17.31	17.94	18.56	19.15	19.73	20.29	20.87	21.41	21.94	24
25	7.46	9.11	10.38	11.46	12.43	13.31	14.14	14.92	15.66	16.37	17.05	17.72	18.35	18.97	19.57	20.15	20.73	21.30	21.85	22.41	25
26	7.67	9.36	10.65	11.75	12.74	13.64	14.48	15.27	16.02	16.74	17.43	18.10	18.75	19.39	20.01	20.60	21.20	21.77	22.34	22.87	26
27	7.89	9.61	10.92	12.04	13.05	13.96	14.82	15.62	16.38	17.12	17.81	18.50	19.15	19.79	20.42	21.01	21.62	22.20	22.77	23.33	27
28	8.10	9.85	11.19	12.33	13.35	14.28	15.15	15.96	16.74	17.48	18.20	18.89	19.55	20.20	20.82	21.44	22.05	22.62	23.20	23.78	28
29	8.32	10.10	11.46	12.62	13.65	14.60	15.48	16.31	17.09	17.85	18.57	19.26	19.94	20.60	21.24	21.85	22.47	23.07	23.66	24.22	29
30	8.53	10.34	11.73	12.90	13.95	14.91	15.80	16.64	17.44	18.20	18.94	19.64	20.33	20.99	21.64	22.27	22.88	23.48	24.08	24.66	30
31	8.74	10.58	11.99	13.19	14.25	15.22	16.13	16.98	17.79	18.56	19.30	20.02	20.71	21.39	22.04	22.67	23.29	23.92	24.50	25.10	31
32	8.94	10.82	12.25	13.47	14.55	15.54	16.45	17.31	18.13	18.91	19.67	20.39	21.09	21.77	22.44	23.08	23.72	24.33	24.94	25.53	32
33	9.15	11.06	12.51	13.74	14.84	15.84	16.77	17.64	18.47	19.26	20.03	20.76	21.47	22.16	22.83	23.48	24.13	24.75	25.35	25.96	33
34	9.36	11.29	12.77	14.02	15.13	16.14	17.08	17.97	18.81	19.61	20.38	21.13	21.85	22.54	23.21	23.88	24.53	25.16	25.79	26.38	34

* Courtesy, "Trane Air Conditioning Manual," pp. 348–349, copyright, 1941, by The Trane Company, La Crosse, Wisconsin. Reproduced by permission.

35	9.56	11.53	13.02	14.29	15.42	16.45	17.40	18.29	19.14	19.96	20.73	21.48	22.22	22.92	23.61	24.27	24.92	25.57	26.19	26.80	35
36	9.77	11.76	13.28	14.56	15.70	16.74	17.71	18.62	19.48	20.30	21.08	21.85	22.58	23.30	23.99	24.66	25.32	25.97	26.59	27.22	36
37	9.97	12.00	13.53	14.83	15.99	17.04	18.01	18.94	19.81	20.64	21.43	22.20	22.95	23.66	24.36	25.04	25.72	26.36	27.02	27.63	37
38	10.17	12.23	13.78	15.10	16.27	17.34	18.32	19.25	20.13	20.97	21.78	22.55	23.31	24.04	24.75	25.43	26.11	26.77	27.41	28.04	38
39	10.37	12.46	14.04	15.37	16.55	17.63	18.63	19.57	20.46	21.31	22.12	22.91	23.67	24.40	25.12	25.81	26.50	27.15	27.80	28.45	39
40	10.57	12.68	14.29	15.63	16.83	17.92	18.94	19.88	20.78	21.64	22.47	23.26	24.02	24.77	25.49	26.19	26.88	27.55	28.21	28.85	40
41	10.77	12.91	14.53	15.90	17.11	18.22	19.23	20.18	21.10	21.97	22.80	23.60	24.38	25.12	25.86	26.56	27.26	27.94	28.60	29.25	41
42	10.97	13.14	14.78	16.16	17.39	18.50	19.53	20.50	21.42	22.30	23.14	23.95	24.73	25.49	26.22	26.94	27.64	28.33	28.99	29.65	42
43	11.17	13.36	15.02	16.42	17.66	18.78	19.84	20.80	21.74	22.62	23.47	24.27	25.08	25.85	26.58	27.31	28.02	28.71	29.38	30.05	43
44	11.36	13.59	15.27	16.68	17.93	19.08	20.12	21.12	22.05	22.95	23.80	24.63	25.42	26.20	26.95	27.68	28.39	29.09	29.77	30.44	44
45	11.56	13.81	15.51	16.94	18.20	19.36	20.42	21.41	22.37	23.27	24.13	24.97	25.77	26.55	27.31	28.04	28.76	29.47	30.16	30.83	45
46	11.75	14.03	15.75	17.20	18.48	19.63	20.72	21.72	22.68	23.59	24.46	25.30	26.12	26.90	27.66	28.41	29.13	29.84	30.54	31.22	46
47	11.95	14.25	15.99	17.45	18.74	19.92	21.01	22.01	23.00	23.91	24.79	25.63	26.46	27.25	28.02	28.76	29.50	30.22	30.91	31.60	47
48	12.14	14.47	16.23	17.71	19.01	20.20	21.29	22.32	23.31	24.23	25.11	25.97	26.80	27.59	28.37	29.13	29.86	30.58	31.30	31.98	48
49	12.33	14.69	16.47	17.96	19.28	20.47	21.58	22.61	23.62	24.54	25.43	26.30	27.13	27.94	28.72	29.48	30.23	30.96	31.67	32.36	49
50	12.53	14.91	16.70	18.21	19.54	20.76	21.88	22.92	23.90	24.85	25.76	26.63	27.47	28.28	29.07	29.84	30.59	31.32	32.03	32.74	50
51	12.72	15.13	16.94	18.46	19.81	21.03	22.15	23.20	24.20	25.17	26.08	26.95	27.80	28.62	29.42	30.19	30.95	31.69	32.41	33.12	51
52	12.91	15.35	17.18	18.71	20.07	21.30	22.44	23.51	24.51	25.48	26.39	27.28	28.13	28.96	29.76	30.54	31.30	32.05	32.78	33.49	52
53	13.10	15.56	17.41	18.96	20.33	21.58	22.73	23.79	24.81	25.78	26.71	27.60	28.46	29.29	30.11	30.89	31.66	32.41	33.14	33.86	53
54	13.29	15.78	17.64	19.21	20.59	21.85	23.01	24.09	25.11	26.09	27.03	27.92	28.79	29.63	30.45	31.24	32.02	32.77	33.51	34.23	54
55	13.48	15.99	17.88	19.46	20.85	22.11	23.28	24.37	25.42	26.40	27.34	28.25	29.12	29.96	30.78	31.58	32.37	33.12	33.87	34.60	55
56	13.66	16.21	18.11	19.70	21.11	22.39	23.56	24.67	25.71	26.70	27.65	28.56	29.44	30.30	31.13	31.93	32.72	33.48	34.23	34.96	56
57	13.85	16.42	18.34	19.95	21.37	22.65	23.85	24.95	26.01	27.00	27.97	28.88	29.77	30.63	31.46	32.27	33.06	33.83	34.59	35.33	57
58	14.04	16.63	18.57	20.19	21.62	22.92	24.11	25.24	26.31	27.31	28.28	29.20	30.09	30.95	31.79	32.61	33.41	34.19	34.94	35.69	58
59	14.22	16.84	18.80	20.44	21.88	23.19	24.39	25.52	26.58	27.61	28.59	29.51	30.41	31.28	32.13	32.95	33.75	34.53	35.30	36.05	59
60	14.41	17.05	19.03	20.68	22.13	23.45	24.67	25.81	26.68	27.91	28.87	29.82	30.73	31.61	32.46	33.29	34.10	34.89	35.65	36.41	60
61	14.60	17.26	19.26	20.92	22.39	23.71	24.95	26.08	27.17	28.20	29.17	30.15	31.05	31.93	32.79	33.62	34.44	35.23	36.00	36.77	61
62	14.78	17.47	19.48	21.16	22.64	23.98	25.21	26.37	27.46	28.50	29.48	30.43	31.37	32.25	33.12	33.96	34.78	35.58	36.36	37.12	62
63	14.96	17.68	19.71	21.40	22.89	24.24	25.49	26.64	27.75	28.80	29.79	30.76	31.68	32.58	33.45	34.29	35.12	35.92	36.70	37.48	63
64	15.15	17.89	19.93	21.64	23.14	24.50	25.76	26.93	28.04	29.09	30.09	31.08	32.00	32.90	33.77	34.62	35.45	36.26	37.06	37.83	64
65	15.33	18.10	20.16	21.88	23.39	24.77	26.02	27.20	28.33	29.38	30.39	31.36	32.31	33.22	34.10	34.95	35.79	36.60	37.40	38.18	65
66	15.51	18.30	20.38	22.12	23.64	25.02	26.29	27.49	28.61	29.68	30.70	31.68	32.61	33.54	34.42	35.28	36.13	36.94	37.74	38.53	66
67	15.70	18.51	20.61	22.35	23.89	25.28	26.56	27.75	28.90	29.97	31.00	31.99	32.95	33.85	34.74	35.61	36.46	37.28	38.09	38.88	67
68	15.88	18.72	20.83	22.59	24.14	25.54	26.83	28.04	29.17	30.26	31.30	32.27	33.24	34.17	35.07	35.94	36.79	37.62	38.43	39.22	68
69	16.06	18.92	21.05	22.82	24.38	25.79	27.09	28.30	29.45	30.55	31.59	32.59	33.54	34.48	35.39	36.26	37.12	37.96	38.77	39.57	69
70	16.24	19.13	21.27	23.06	24.63	26.05	27.36	28.58	29.73	30.83	31.89	32.90	33.84	34.79	35.70	36.59	37.45	38.29	39.11	39.91	70
71	16.42	19.33	21.49	23.29	24.88	26.31	27.63	28.85	30.02	31.12	32.16	33.18	34.17	35.11	36.02	36.91	37.78	38.62	39.45	40.25	71
72	16.60	19.53	21.71	23.53	25.12	26.56	27.88	29.13	30.30	31.41	32.46	33.49	34.46	35.43	36.34	37.23	38.10	38.95	39.79	40.60	72
73	16.78	19.74	21.93	23.76	25.36	26.81	28.15	29.39	30.58	31.69	32.75	33.79	34.76	35.75	36.65	37.55	38.43	39.28	40.12	40.94	73
74	16.96	19.94	22.15	23.99	25.61	27.07	28.41	29.67	30.86	31.98	33.04	34.07	35.08	36.02	36.97	37.87	38.75	39.61	40.45	41.27	74
75	17.14	20.14	22.37	24.22	25.85	27.32	28.68	29.93	31.13	32.26	33.34	34.38	35.37	36.33	37.28	38.19	39.07	39.94	40.79	41.61	75
76	17.32	20.34	22.59	24.45	26.09	27.57	28.93	30.20	31.41	32.54	33.63	34.68	35.66	36.64	37.58	38.51	39.40	40.27	41.12	41.95	76
77	17.48	20.54	22.80	24.68	26.33	27.82	29.19	30.47	31.67	32.82	33.92	34.96	35.99	36.96	37.92	38.82	39.72	40.59	41.44	42.28	77
78	17.67	20.74	23.02	24.91	26.57	28.07	29.46	30.74	31.95	33.10	34.21	35.26	36.28	37.27	38.21	39.14	40.04	40.92	41.78	42.62	78
79	17.85	20.94	23.24	25.14	26.81	28.32	29.70	31.00	32.22	33.38	34.50	35.56	36.57	37.57	38.51	39.45	40.36	41.24	42.10	42.95	79

TABLE 51. MEAN TEMPERATURE DIFFERENCES—(*Continued*)

LARGE TEMP. DIFF.	SMALL TEMPERATURE DIFFERENCE																				LARGE TEMP. DIFF.
	21	22	23	24	25	26	27	28	29	30	31	32	33	34	35	36	37	38	39	40	
1																					1
2																					2
3																					3
4																					4
5																					5
6																					6
7																					7
8																					8
9																					9
10																					10
11																					11
12																					12
13																					13
14																					14
15																					15
16																					16
17																					17
18																					18
19																					19
20																					20
21	21.00																				21
22	21.51	22.00																			22
23	21.99	22.47	23.00																		23
24	22.46	22.99	23.49	24.00																	24
25	23.00	23.46	23.98	24.48	25.00																25
26	23.42	23.92	24.48	25.00	25.50	26.00															26
27	23.85	24.44	24.93	25.47	25.99	26.47	27.00														27
28	24.35	24.86	25.46	25.90	26.47	26.98	27.53	28.00													28
29	24.78	25.35	25.87	26.46	26.95	27.47	27.98	28.50	29.00												29
30	25.21	25.77	26.37	26.89	27.42	27.93	28.48	29.00	29.48	30.00											30
31	25.68	26.25	26.79	27.32	27.89	28.47	28.98	29.49	29.58	30.53	31.00										31
32	26.11	26.67	27.27	27.83	28.36	28.87	29.46	29.93	30.49	30.97	31.46	32.00									32
33	26.57	27.13	27.69	28.26	28.82	29.38	29.93	30.36	30.94	31.48	31.99	32.45	33.00								33
34	26.98	27.58	28.16	28.69	29.27	29.80	30.39	30.94	31.50	32.03	32.47	32.99	33.50	34.00							34
35	27.40	28.00	28.57	29.17	29.72	30.29	30.85	31.37	31.89	32.38	32.97	33.46	34.00	34.51	35.00						35
36	27.84	28.44	29.02	29.60	30.17	30.70	31.31	31.80	32.42	32.91	33.49	33.96	34.48	35.00	35.46	36.00					36
37	28.25	28.85	29.44	30.02	30.61	31.18	31.77	32.33	32.82	33.42	33.84	34.49	35.02	35.49	36.02	36.47	37.00				37
38	28.65	29.28	29.88	30.48	31.05	31.59	32.21	32.76	33.33	33.81	34.35	34.83	35.34	35.86	36.49	36.96	37.54	38.00			38
39	29.08	29.69	30.29	30.90	31.48	32.06	32.66	33.19	33.74	34.30	34.86	35.35	35.88	36.46	37.05	37.50	37.96	38.52	39.00		39

40	29.48	30.11	30.72	31.31	31.91	32.52	33.10	33.62	34.23	34.79	35.34	35.85	36.41	37.01	37.41	38.00	38.47	39.02	39.56	40.00	40
41	29.90	30.51	31.13	31.76	32.34	32.93	33.49	34.11	34.64	35.19	35.73	36.34	36.91	37.37	38.01	38.42	39.00	39.50	39.98	40.50	41
42	30.30	30.93	31.55	32.16	32.77	33.38	33.93	34.53	35.12	35.66	36.21	36.72	37.29	37.90	38.39	38.85	39.49	40.06	40.49	40.99	42
43	30.69	31.33	31.95	32.57	33.19	33.78	34.36	34.95	35.53	36.13	36.68	37.21	37.78	38.29	38.80	39.48	39.96	40.33	40.80	41.48	43
44	31.10	31.74	32.37	33.01	33.61	34.23	34.79	35.42	35.99	36.53	37.15	37.68	38.27	38.80	39.35	39.90	40.44	40.90	41.51	41.97	44
45	31.49	32.15	32.77	33.41	34.03	34.63	35.22	35.84	36.40	36.99	37.54	38.15	38.66	39.19	39.76	40.33	40.91	41.44	41.89	42.45	45
46	31.89	32.54	33.18	33.81	34.44	35.06	35.65	36.25	36.86	37.45	38.00	38.53	39.14	39.69	40.28	40.77	41.37	41.79	42.51	42.93	46
47	32.27	32.94	33.59	34.23	34.85	35.46	36.07	36.67	37.26	37.85	38.45	39.00	39.61	40.18	40.69	41.20	41.84	42.32	42.90	43.41	47
48	32.66	33.32	33.98	34.62	35.26	35.89	36.49	37.12	37.71	38.30	38.91	39.46	40.00	40.58	41.20	41.75	42.30	42.83	43.30	43.88	48
49	33.05	33.72	34.39	35.02	35.66	36.28	36.91	37.53	38.11	38.74	39.30	39.91	40.46	41.06	41.61	42.18	42.76	43.33	43.87	44.35	49
50	33.43	34.10	34.77	35.43	36.07	36.70	37.32	37.93	38.56	39.14	39.74	40.30	40.92	41.46	42.02	42.61	43.21	43.70	44.28	44.81	50
51	33.80	34.50	35.17	35.82	36.47	37.08	37.73	36.37	38.96	39.58	40.18	40.75	41.38	41.93	42.51	43.04	43.66	44.19	44.69	45.28	51
52	34.19	34.87	35.55	36.21	36.87	37.51	38.14	38.77	39.39	40.01	40.62	41.19	41.77	42.39	42.92	43.55	44.11	44.68	45.23	45.74	52
53	34.56	35.26	35.94	36.61	37.26	37.92	38.55	39.18	39.79	40.40	41.01	41.63	42.22	42.79	43.40	43.97	44.56	45.06	45.64	46.20	53
54	34.95	35.63	36.32	36.99	37.66	38.31	38.95	39.57	40.22	40.83	41.44	42.02	42.66	43.25	43.81†	44.39	45.00	45.54	46.05	46.65	54
55	35.31	36.01	36.71	37.38	38.05	38.72	39.35	40.00	40.61	41.26	41.87	42.46	43.05	43.64	44.28	44.82	45.44	46.01	46.57	47.10	55
56	35.68	36.40	37.08	37.77	38.44	39.10	39.75	40.40	41.03	41.64	42.29	42.89	43.49	44.09	44.68	45.24	45.81	46.39	46.98	47.55	56
57	36.06	36.76	37.47	38.15	38.83	39.50	40.15	40.79	41.42	42.07	42.68	43.31	43.92	44.54	45.08	45.72	46.25	46.86	47.39	48.00	57
58	36.42	37.14	37.83	38.53	39.21	39.88	40.55	41.21	41.84	42.48	43.10	43.70	44.31	44.93	45.54	46.14	46.69	47.32	47.89	48.44	58
59	36.78	37.50	38.22	38.92	39.60	40.28	40.94	41.60	42.25	42.87	43.52	44.12	44.74	45.37	45.95	46.55	47.12	47.71	48.30	48.89	59
60	37.15	37.88	38.58	39.29	39.98	40.65	41.33	41.98	42.64	43.28	43.93	44.54	45.17	45.76	46.40	46.96	47.55	48.16	48.78	49.33	60
61	37.51	38.24	38.96	39.66	40.36	41.05	41.72	42.37	43.05	43.69	44.31	44.96	45.60	46.20	46.80	47.43	47.98	48.61	49.19	49.76	61
62	37.88	38.61	39.32	40.04	40.74	41.42	42.11	42.78	43.43	44.07	44.72	45.34	46.02	46.59	47.24	47.84	48.41	49.00	49.60	50.20	62
63	38.23	38.97	39.70	40.41	41.11	41.81	42.50	43.16	43.83	44.48	45.13	45.75	46.40	47.02	47.64	48.25	48.89	49.45	50.07	50.63	63
64	38.58	39.33	40.06	40.78	41.49	42.18	42.88	43.54	44.21	44.88	45.51	46.17	46.82	47.44	48.03	48.65	49.26	49.89	50.47	51.06	64
65	38.95	39.69	40.43	41.16	41.86	42.56	43.26	43.94	44.61	45.26	45.91	46.58	47.20	47.83	48.47	49.06	49.68	50.27	50.88	51.49	65
66	39.30	40.05	40.78	41.52	42.23	42.95	43.64	44.32	44.99	45.66	46.32	46.95	47.61	48.25	48.86	49.51	50.10	50.71	51.34	51.92	66
67	39.65	40.41	41.15	41.88	42.60	43.31	44.02	44.70	45.39	46.06	46.72	47.36	48.02	48.63	49.29	49.91	50.52	51.14	51.74	52.34	67
68	40.00	40.76	41.51	42.25	42.97	43.69	44.37	45.07	45.76	46.43	47.09	47.76	48.40	49.05	49.68	50.31	50.93	51.58	52.14	52.77	68
69	40.35	41.12	41.87	42.61	43.34	44.05	44.76	45.47	46.15	46.82	47.49	48.16	48.80	49.47	50.11	50.71	51.34	51.96	52.59	53.19	69
70	40.70	41.47	42.23	42.97	43.71	44.43	45.13	45.84	46.52	47.22	47.88	48.53	49.21	49.85	50.49	51.15	51.75	52.39	52.99	53.61	70
71	41.05	41.82	42.58	43.34	44.07	44.79	45.50	46.21	46.91	47.58	48.28	48.93	49.58	50.26	50.88	51.54	52.16	52.81	53.39	54.03	71
72	41.39	42.17	42.94	43.69	44.43	45.16	45.87	46.60	47.28	47.97	48.64	49.33	49.98	50.63	51.30	51.94	52.57	53.19	53.83	54.44	72
73	41.74	42.52	43.29	44.04	44.79	45.52	46.24	46.96	47.67	48.36	49.03	49.72	50.38	51.04	51.68	52.33	52.98	53.61	54.23	54.86	73
74	42.08	42.86	43.65	44.41	45.15	45.89	46.61	47.33	48.03	48.73	49.42	50.09	50.78	51.45	52.10	52.72	53.38	54.03	54.66	55.27	74
75	42.42	43.22	43.99	44.76	45.51	46.25	47.06	47.69	48.42	49.11	49.81	50.48	51.15	51.82	52.48	53.15	53.78	54.41	55.06	55.68	75
76	42.76	43.56	44.35	45.11	45.87	46.62	47.34	48.08	48.78	49.49	50.17	50.87	51.55	52.22	52.89	53.54	54.18	54.82	55.45	56.09	76
77	43.10	43.90	44.69	45.47	46.23	46.97	47.71	48.44	49.16	49.85	50.58	51.25	51.94	52.59	53.27	53.92	54.48	55.24	55.88	56.49	77
78	43.44	44.25	45.04	45.82	46.58	47.33	48.07	48.80	49.52	50.23	50.94	51.62	52.30	52.99	53.65	54.31	54.98	55.61	56.26	56.90	78
79	43.78	44.59	45.38	46.16	46.93	47.70	48.43	49.18	49.90	50.61	51.32	52.00	52.69	53.36	54.05	54.73	55.37	56.02	56.65	57.30	79

To determine the area of heat transfer surface required by a heat exchange element through or over which fluids flow—i.e., fluids whose temperatures continually change and tend to become equal—apply the mtd to the equation

$$A = \frac{H}{U \times \text{mtd}}$$

where A = area, sq ft of cooling surface
U = over-all heat transfer, Btu per sq ft per hr per F
H = heat load, Btu
mtd = log mean temperature difference

Example: Determine the area of cooling surface required to extract 240,000 Btu per hr in order to maintain a condensing temperature of 86 F. Assume an mtd of 10.2 F and U equals 200 Btu per sq ft per hr per F.

$$A = \frac{H}{U \times \text{mtd}} = \frac{240{,}000}{200 \times 10.2}$$
$$= 117.64 \text{ sq ft}$$

EVAPORATOR PRESSURE DETERMINATIONS

The temperature of the refrigerant within the evaporator is determined by the service demands of the refrigeration system. After deciding the temperature necessary to maintain proper levels within the space or the substance being cooled, it is essential that the proper pressure corresponding to the selected temperature be controlled closely. Actually, what occurs in making such selection is that the boiling point of the liquid refrigerant is controlled at that temperature corresponding to the pressure.

The other variable to be considered here is the kind of refrigerant to be employed. Several kinds will be considered and the suction pressure determined which is necessary to maintain a saturated vapor condition within the evaporator at a temperature of 5 F.

To get this information, refer to the tables of properties of saturated vapor for the following refrigerants, and find the pressure of each corresponding to a temperature of 5 F.

Table 52. Evaporator Pressure, psia, Corresponding to 5 F for Some Refrigerants

Refrigerant @ 5 F Saturated Vapor	*Evaporator Pressure**
Ammonia	34.27 psia
Sulfur dioxide	11.81 psia
Freon-12	26.51 psia
Carbon dioxide	334.4 psia
Methyl chloride	20.89 psia

* To determine gage pressure, subtract 14.7 psi.

SPECIFIC VOLUME DETERMINATIONS

To estimate the amount of gas to be handled by the compressor, it is necessary to determine the volume involved as a result of vaporization within the evaporator. This value is based on the number of cubic feet of gas that is produced by 1 lb of liquid refrigerant upon vaporizing at 5 F and the corresponding boiling-point pressure.

If the density alone were known, the specific volume would be found by the formula described previously in Chapter 4, "Gases," viz.:

$$\text{Specific volume} = \frac{1}{\text{density}}$$

Referring to the refrigerant tables, it is found that 5 F ammonia has a density of 0.1227 lb per cu ft at a pressure of 34.27 psia. The specific volume for ammonia at this temperature and pressure, then, is

$$\text{Specific volume} = \frac{1}{0.1227} = 8.150 \text{ cu ft per lb}$$

The specific volume of 8.150 cu ft per lb, as calculated above, agrees with that shown in the table of saturated ammonia vapor at 5 F and 34.27 psia.

The specific volume of other refrigerants, as shown in their respective property tables at the standard temperature condition of 5 F, is as follows:

TABLE 53. SPECIFIC VOLUME, CU FT PER LB, AT 5 F FOR SOME REFRIGERANTS

Refrigerant	*Specific Volume @ 5 F (cu ft per lb)*
Ammonia	8.150
Sulfur dioxide	6.421
R-12	1.485
Carbon dioxide	0.2673
Methyl chloride	4.530

CONDENSER PRESSURE DETERMINATIONS

While the acutal temperature and subsequent pressure in the condenser of a simple refrigeration system depend on the temperature and the amount of condensing medium prevailing in relation to the surrounding temperature conditions, some standard basis must be adopted in order to estimate equipment sizes for given load conditions. For the purpose of demonstration here, the standard condenser temperature of 86 F will be used. This is the temperature to which the compressed vapor leaving the compressor is cooled in order to remove the heat of vaporization and compression and convert it to a liquid phase. Assuming that sufficient water at a temperature below 86 F is available, it is necessary that a pressure great enough to control the boiling point of the liquid at 86 F be exerted within the condensing side of the system.

Again referring to the Thermodynamic Property Tables, the pressure corresponding to the boiling-point temperature for the several kinds of refrigerants is as follows:

TABLE 54. CONDENSER PRESSURE, PSIA, CORRESPONDING TO 86 F FOR SOME REFRIGERANTS

Refrigerant	*Condenser Pressure 86 F*
Ammonia	169.2 psia
Sulfur dioxide	66.45 psia
R-12	107.9 psia
Carbon dioxide	1039.0 psia
Methyl chloride	95.53 psia

COMPRESSOR DISCHARGE TEMPERATURE

In the study of Boyle's, Charles', and Gay Lussac's laws, it was shown that a definite relationship exists between pressure, volume, and temperature of gases, and that under those laws this relationship is expressed as follows:

$$\frac{P_1}{P_2} = \frac{V_2}{V_1} \quad \text{or} \quad P_1V_1 = P_2V_2 \quad \text{where } T = K \text{ (Boyle's Law)}$$

$$\frac{P_1}{P_2} = \frac{T_1}{T_2} \quad \text{or} \quad P_1T_2 = P_2T_1 \quad \text{where } V = K \text{ (Charles' Law)}$$

$$\frac{V_1}{V_2} = \frac{T_1}{T_2} \quad \text{or} \quad V_1T_2 = V_2T_1 \quad \text{where } P = K \text{ (Gay Lussac's Law)}$$

$(K = \text{constant})$

Under adiabatic conditions, where pressure and volume are affected by temperature changes, due to the work of compression or expansion, a direct relationship according to the above laws no longer exists. Instead, their relationship is expressed as

$$\frac{P_2}{P_1} = \left(\frac{V_1}{V_2}\right)^y \quad \text{or} \quad P_1V_1{}^y = K \quad \text{where } y = \frac{C_p}{C_v}$$

To express any function of pressure, volume, and temperature adiabatically, the following are derived from $PV^y = K$:

Where P_1, V_1, and V_2 are known, and $P_2 = X$ (unknown),

(1) $$\frac{P_2}{P_1} = \left(\frac{V_1}{V_2}\right)^y \quad \text{then } P_2 = P_1\left(\frac{V_1}{V_2}\right)^y$$

Where P_2, V_1, and V_2 are known, and $P_1 = X$,

$$\frac{P_2}{P_1} = \left(\frac{V_1}{V_2}\right)^y \quad \text{then} \quad P_1 = P_2\left(\frac{V_2}{V_1}\right)^y$$

Where V_1, P_1 and P_2 are known, and $V_2 = X$,

(2) $$\frac{V_1}{V_2} = \left(\frac{P_2}{P_1}\right)^{\frac{1}{y}} \quad \text{then} \quad V_2 = V_1\left(\frac{P_1}{P_2}\right)^{\frac{1}{y}}$$

Where V_2, P_1, and P_2 are known, and $V_1 = X$,

$$\frac{V_1}{V_2} = \left(\frac{P_2}{P_1}\right)^{\frac{1}{y}} \quad \text{then} \quad V_1 = V_2\left(\frac{P_2}{P_1}\right)^{\frac{1}{y}}$$

Where V_1, T_2, and T_1 are known, and $V_2 = X$,

(3) $$\frac{V_1}{V_2} = \left(\frac{T_2}{T_1}\right)^{\frac{1}{y-1}} \quad \text{then} \quad V_2 = V_1\left(\frac{T_1}{T_2}\right)^{\frac{1}{y-1}}$$

Where V_2, T_1, and T_2 are known, and $V_1 = X$,

$$\frac{V_1}{V_2} = \left(\frac{T_2}{T_1}\right)^{\frac{1}{y-1}} \quad \text{then} \quad V_1 = V_2\left(\frac{T_2}{T_1}\right)^{\frac{1}{y-1}}$$

Where T_1, V_1, and V_2 are known, and $T_2 = X$,

(4) $$\frac{T_2}{T_1} = \left(\frac{V_1}{V_2}\right)^{y-1} \quad \text{then} \quad T_2 = T_1\left(\frac{V_1}{V_2}\right)^{y-1}$$

Where T_2, V_1, and V_2 are known, and $T_1 = X$,

$$\frac{T_2}{T_1} = \left(\frac{V_1}{V_2}\right)^{y-1} \quad \text{then} \quad T_1 = T_2\left(\frac{V_2}{V_1}\right)^{y-1}$$

Where P_2, T_2, and T_1 are known and $P_1 = X$,

(5) $$\frac{P_2}{P_1} = \left(\frac{T_2}{T_1}\right)^{\frac{y}{y-1}} \quad \text{then} \quad P_1 = P_2\left(\frac{T_1}{T_2}\right)^{\frac{y}{y-1}}$$

Where P_1, T_2, and T_1 are known and $P_2 = X$,

$$\frac{P_2}{P_1} = \left(\frac{T_2}{T_1}\right)^{\frac{y}{y-1}} \quad \text{then} \quad P_2 = P_1\left(\frac{T_2}{T_1}\right)^{\frac{y}{y-1}}$$

Where T_1, P_1, and P_2 are known and $T_2 = X$,

(6) $$\frac{T_2}{T_1} = \left(\frac{P_2}{P_1}\right)^{\frac{y-1}{y}} \quad \text{then} \quad T_2 = T_1\left(\frac{P_2}{P_1}\right)^{\frac{y-1}{y}}$$

Where T_2, P_1, and P_2 are known and $T_1 = X$,

$$\frac{T_2}{T_1} = \left(\frac{P_2}{P_1}\right)^{\frac{y-1}{y}} \quad \text{then} \quad T_1 = T_2\left(\frac{P_1}{P_2}\right)^{\frac{y-1}{y}}$$

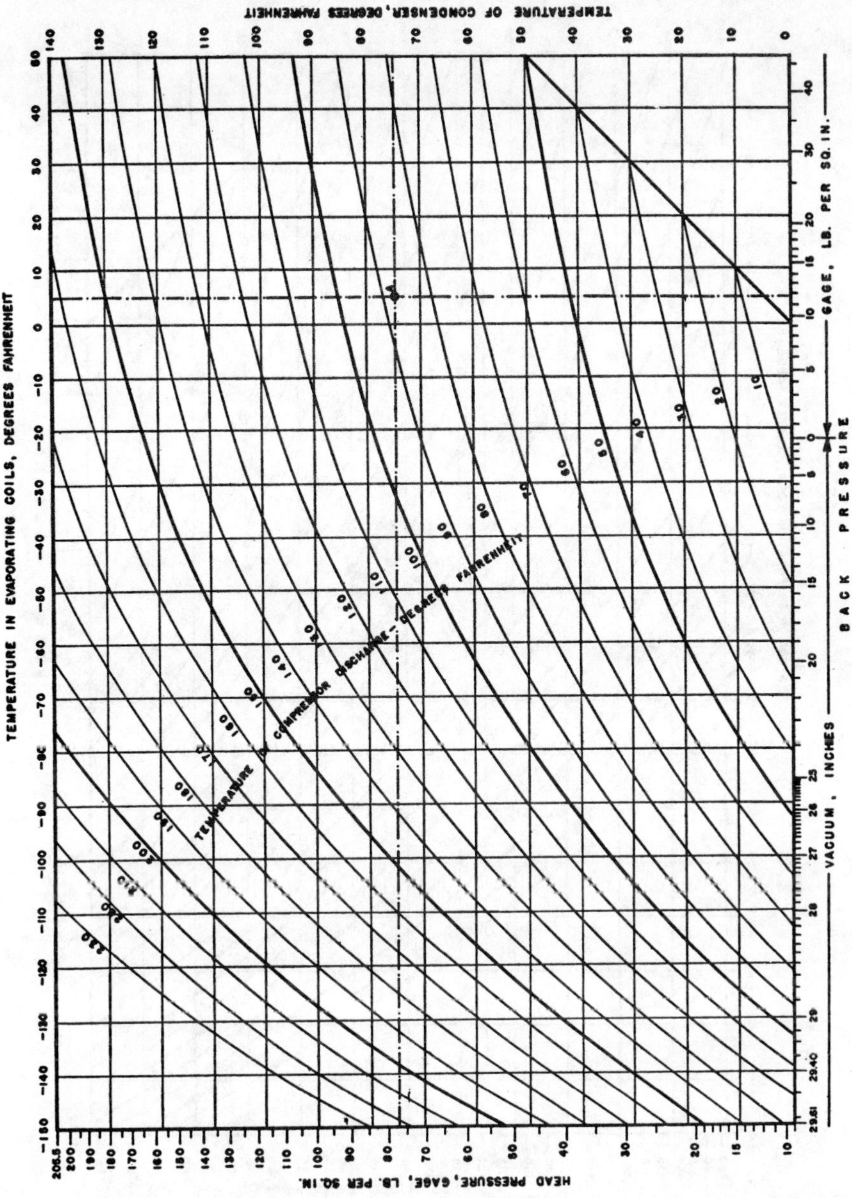

Fig. 356. This chart shows the pressures corresponding to the temperature conditions in condensers and evaporators of R-12 refrigeration systems. It also shows compressor-discharge temperature when the suction gas is saturated and compression is ideal.

Example: If the desired evaporator temperature is 5 F (top scale), the corresponding evaporator pressure is shown on the bottom scale as 11.7 psig. If the condenser temperature is 75 F (right-hand scale), the corresponding head pressure, shown on the left-hand scale, is 77.5 psig. The theoretical compressor-discharge temperature is shown at the intersection *A* to be about 88 F. (Courtesy, Mobil Oil Corporation.)

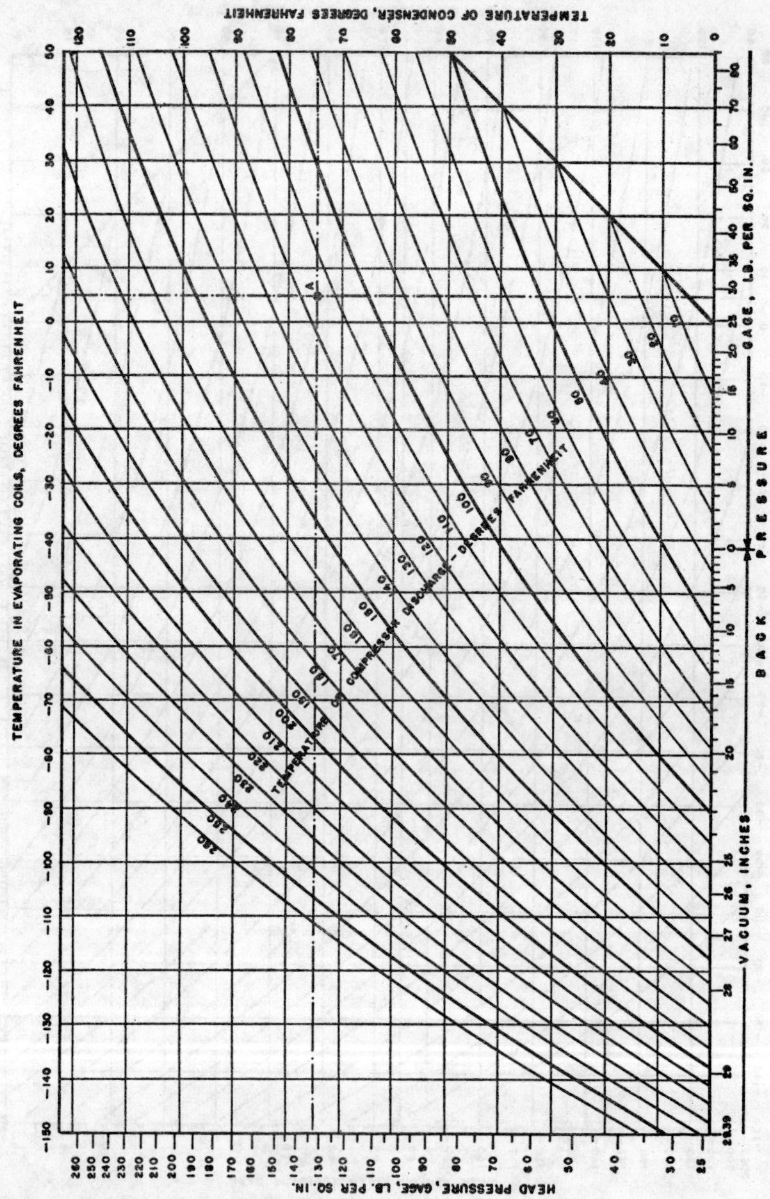

Fig. 357. This chart shows the pressures corresponding to the temperature conditions in condensers and evaporators of R-22 systems. It also shows the compressor-discharge temperature when the suction gas is saturated and compression is ideal.

Example: If the desired evaporator temperature is 5 F (top scale), the corresponding pressure in the evaporator is shown on the bottom scale as 28.5 psig. If the condenser temperature is 75 F (right-hand scale), the corresponding head pressure, shown on the left-hand scale, is 134 psig. The theoretical compressor-discharge temperature is shown at the intersection *A* to be about 115 F. (Courtesy, Mobil Oil Corporation.)

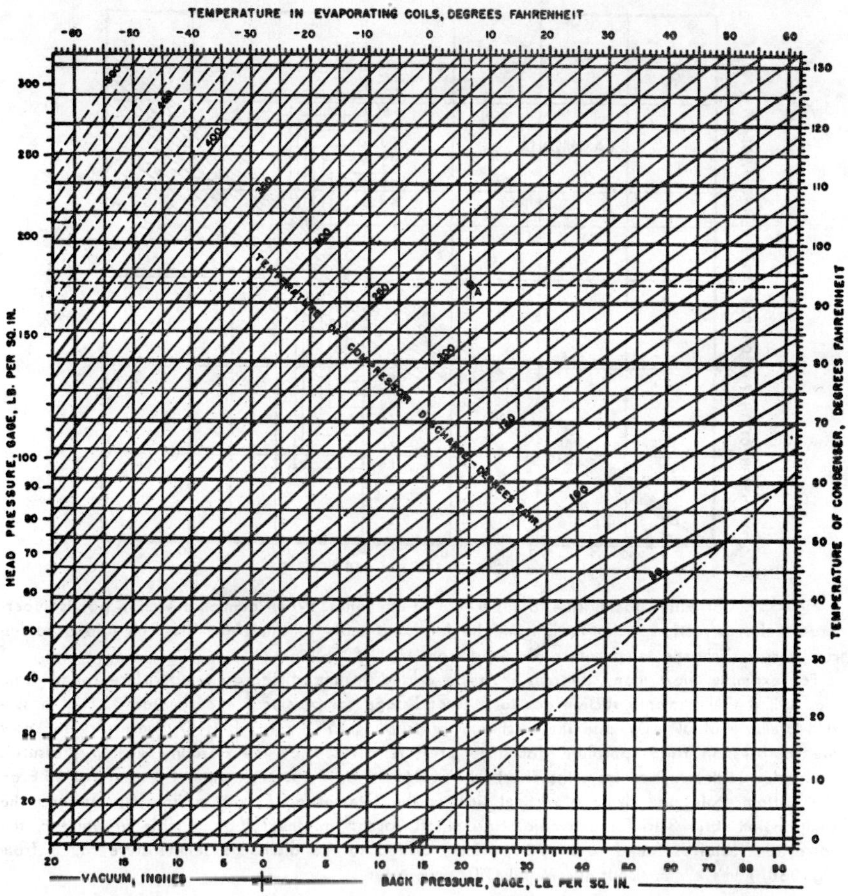

Fig. 358. This chart shows the pressures corresponding to the temperature conditions in condensers and evaporators of ammonia refrigeration systems. It also shows compressor-discharge temperature when the suction gas is saturated and compression is ideal.

Example: If the desired evaporator temperature is 6.8 F (top scale), the corresponding evaporator pressure is shown on the bottom scale as 21 psig. If the condenser temperature is 93.2 F (right-hand scale), the corresponding head pressure, shown on the left-hand scale, is 175 psig. The theoretical compressor-discharge temperature is shown at the intersection *A* to be about 222 F. (Courtesy, Mobil Oil Corporation.)

In determining the temperature of gases issuing from the discharge side of the refrigerant compressor, consideration must be given to the effect on final temperature of the heat of compression that further raises the compression pressure above that of pure compression, as expressed by Boyle's law.

In the study of gases (Chapter 4), it was pointed out that the additional

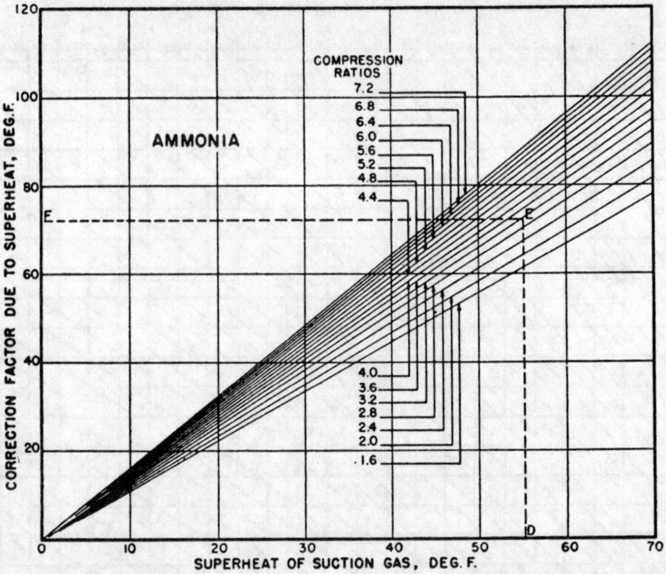

Fig. 359. Correction factor due to superheat in ammonia. When ammonia suction gas is superheated the correction factor read from the left-hand scale of this chart must be added to the compressor-discharge temperature obtained from Fig. 358.

For example, how many degrees must be added to the discharge temperature found from Fig. 358 if the ammonia suction gas to a second-stage compressor is superheated 55 F at a suction pressure of 39 psig, and the discharge pressure is 154.5 psig? It is first necessary to know the compression ratio (absolute discharge pressure divided by the absolute suction pressure). This value is determined from the chart in Fig. 360 to be 3.15. Then enter the chart at 55 F on the bottom scale and draw a vertical line to the intersection *E* with a line representing the compression ratio 3.15 for ammonia. Horizontally across on the left-hand scale at point *F*, the correction factor is found to be 72 F, which, added to the discharge temperature found from Fig. 358, gives 294 F. (Courtesy, Mobil Oil Corporation.)

pressure brought about by the inhibited tendency of the gas to expand, due to the transfer of mechanical energy in compressing the gas to heat energy gained by the gas, under adiabatic conditions, was a function of the ratio of C_p/C_v.

While adiabatic conditions are never obtained in practice because of the conduction characteristics of cylinder metals, the deliberate attempt to reduce temperatures via cooling water jackets, and because actual discharge temperatures cannot be exactly calculated due to mechanical variations, the theoretical adiabatic temperatures act as a useful guide in estimating actual results that may be expected. In operation, any measured discharge temperature that departs greatly from theoretical adiabatic results is reason enough to surmise compressor defects, such as broken or stuck valves, worn rings, carbon deposits, etc.

One method of determining compressor discharge temperatures is by applying the thermodynamic equation shown above, viz.:

$$\frac{T_2}{T_1} = \left(\frac{P_2}{P_1}\right)^{\frac{y-1}{y}}$$

For ammonia,

$$\frac{y-1}{y} = \frac{1.3-1}{1.3} = 0.230$$

Therefore, the thermodynamic equation for ammonia is

$$\frac{T_2}{T_1} = \left(\frac{P_2}{P_1}\right)^{0.23} \quad \text{and} \quad T_2 = T_1\left(\frac{P_2}{P_1}\right)^{0.23}$$

And for standard conditions,

$$T_1 = 459.6 + 5 \text{ F} = 464.6 \text{ F abs}$$

$$P_2 = 169.2 \text{ psia}$$

$$P_1 = 34.27 \text{ psia}$$

Substituting above values in the equation,

$$T_2 = 464.6 \times \left(\frac{169.2}{34.27}\right)^{0.23}$$

$$T_2 = 464.6 \times (4.9372)^{0.23}$$

$$\text{Log } 4.9372 = 0.693481$$

$$0.693481 \times 0.23 = 0.159500$$

$$\text{Antilog } 0.159500 = 1.443$$

$$T_2 = 464.6 \times 1.443 = 670.42 \text{ F abs}$$

$$t_2 = 670.42 - 459.6 = 210.8 \text{ F}$$

The compressor discharge temperature may be obtained also from (1) the Mollier Chart or (2) the Tables of Thermodynamic Properties. While the use of either of these two methods excludes knowledge of logarithms, they do require an understanding of entropy.

1. To obtain the discharge temperatures from the Mollier Chart for standard conditions for ammonia, proceed as shown on page following 240.

TABLE 55. COEFFICIENTS, DISCHARGE TEMPERATURES, FOR VARIOUS REFRIGERANTS HAVING VALUES OF $y = \frac{C_p}{C_v}$ AS GIVEN

Compression Ratio	$y = C_p/C_v$								
	1.3	**1.25**	**1.2**	**1.18**	**1.17**	**1.14**	**1.13**	**1.08**	**1.07**
1......	1.00	1.00	1.00	1.00	1.00	1.00	1.00	1.00	1.00
1.2....	1.043	1.038	1.031	1.028	1.026	1.022	1.021	1.014	1.012
1.4....	1.081	1.072	1.058	1.053	1.049	1.042	1.039	1.025	1.022
1.5....	1.098	1.087	1.070	1.064	1.060	1.050	1.048	1.030	1.027
1.6....	1.115	1.102	1.082	1.075 −	1.070	1.059	1.056	1.035	1.031
1.8....	1.145	1.129	1.103	1.094	1.088	1.074	1.070	1.044	1.039
2.0....	1.174	1.153	1.123	1.112	1.104	1.087	1.083	1.053	1.047
2.2....	1.200	1.176	1.141	1.128	1.119	1.100	1.095 −	1.060	1.053
2.4....	1.224	1.198	1.157	1.143	1.133	1.112	1.106	1.067	1.059
2.6....	1.247	1.218	1.173	1.157	1.146	1.123	1.116	1.073	1.064
2.8....	1.268	1.236	1.188	1.171	1.159	1.133	1.126	1.079	1.069
3.0....	1.289	1.254	1.201	1.183	1.170	1.142	1.135	1.085 −	1.074
3.2....	1.308	1.271	1.214	1.195 −	1.181	1.151	1.143	1.090	1.079
3.4....	1.327	1.287	1.227	1.206	1.191	1.160	1.151	1.095 −	1.083
3.6....	1.344	1.302	1.239	1.217	1.201	1.168	1.159	1.099	1.087
3.8....	1.361	1.317	1.250	1.227	1.210	1.175	1.166	1.104	1.091
4.0....	1.377	1.331	1.260	1.236	1.219	1.183	1.173	1.108	1.094
4.2....	1.393	1.344	1.271	1.246	1.228	1.190	1.179	1.112	1.098
4.4....	1.408	1.357	1.281	1.254	1.236	1.196	1.186	1.116	1.101
4.6....	1.423	1.369	1.290	1.263	1.244	1.203	1.192	1.120	1.104
4.8....	1.437	1.381	1.299	1.271	1.251	1.209	1.198	1.123	1.107
4.9....	1.444	1.387	1.304	1.275	1.255	1.212	1.201	1.125 −	1.109
5.0....	1.450	1.393	1.308	1.279	1.259	1.215 −	1.203	1.126	1.110
5.2....	1.464	1.404	1.317	1.287	1.266	1.221	1.209	1.130	1.113
5.4....	1.476	1.415	1.325	1.295 −	1.273	1.226	1.214	1.133	1.116
5.6....	1.489	1.426	1.333	1.302	1.279	1.232	1.219	1.136	1.118
5.8....	1.501	1.436	1.341	1.309	1.286	1.237	1.224	1.139	1.121
6.0....	1.513	1.446	1.349	1.315 −	1.292	1.242	1.229	1.142	1.124
6.2....	1.524	1.456	1.356	1.322	1.298	1.247	1.233	1.145 −	1.126
6.4....	1.534	1.466	1.363	1.328	1.304	1.252	1.238	1.147	1.128
6.6....	1.546	1.475	1.370	1.335 −	1.310	1.256	1.242	1.150	1.131
6.8....	1.557	1.484	1.377	1.341	1.315	1.261	1.247	1.152	1.133
7.0....	1.568	1.493	1.384	1.347	1.321	1.265	1.251	1.155 −	1.135 −
7.2....	1.578	1.502	1.391	1.353	1.326	1.270	1.255 −	1.157	1.137
7.4....	1.588	1.510	1.397	1.358	1.331	1.274	1.259	1.160	1.139
7.6....	1.598	1.519	1.403	1.364	1.336	1.278	1.263	1.162	1.141
7.8....	1.607	1.527	1.409	1.369	1.341	1.282	1.266	1.164	1.143
8.0....	1.617	1.535 −	1.415	1.375 −	1.346	1.286	1.270	1.166	1.145
8.2....	1.626	1.543	1.421	1.380	1.351	1.290	1.274	1.168	1.147
8.4....	1.635	1.550	1.427	1.385 −	1.356	1.294	1.277	1.171	1.148
8.6....	1.644	1.558	1.432	1.390	1.360	1.297	1.281	1.173	1.150
8.8....	1.653	1.565	1.438	1.395 −	1.365 −	1.301	1.284	1.175 −	1.152
9.0....	1.662	1.572	1.443	1.400	1.369	1.305 −	1.287	1.177	1.154
9.2....	1.670	1.580	1.449	1.404	1.373	1.308	1.291	1.178	1.155
9.4....	1.678	1.587	1.454	1.409	1.378	1.311	1.294	1.180	1.157
9.6....	1.686	1.593	1.459	1.413	1.382	1.315 −	1.297	1.182	1.158
9.8....	1.694	1.600	1.464	1.418	1.385	1.318	1.300	1.184	1.160
10.0....	1.702	1.607	1.469	1.423	1.390	1.321	1.303	1.186	1.161

1st Step: Determine the entropy value of saturated ammonia vapor at 5 F from the Properties of Saturated Vapor Tables. Entropy is 1.3253.

2nd Step: Locate the horizontal line on the Mollier Chart denoting suction pressure of 34.27 psia for ammonia under standard ton conditions, with intersection of 1.3253 constant entropy line. (Interpolation)

3rd Step: Follow the constant entropy line to the intersection of the horizontal line denoting discharge pressure of 169.2 psia.

4th Step: The position at the point of intersection of 1.3253 entropy and 169.2 psia in relation to the temperatures indicated by the constant-temperature lines on the chart denotes the discharge temperature at 169.2 psia.

By interpolation, read 210 F

2. Using the tables, proceed as follows:

1st Step: Determine the entropy value of the suction temperature of saturated vapor. For 5 F ammonia, this value is 1.3253.

2nd Step: Refer to the Superheated Ammonia Vapor tables under discharge pressure of 169.2 psi and locate entropy line S, denoting a value of 1.3253. (The 170 psi column is used.)

3rd Step: By interpolation, the temperature is found to be approximately 210 F+, reading from the nearest entropy value of 1.3249.

Another method for determining theoretical compressor discharge temperatures is through the use of the coefficients listed for various values of ratio $\frac{C_p}{C_v} = y$ (see Table 57) corresponding to some refrigerants in Table 55. Each of the several refrigerants shows coefficients of discharge temperature corresponding to compression ratios from 1 to 10 at intervals of 0.2. The theoretical discharge temperature for any system is determined by the equation

$$t_2 = (T_1 \times \text{coefficient}) - 460 \text{ F abs}$$

where t_2 = discharge temperature F
T_1 = initial temperature F abs

and the coefficient corresponds to the compression ratio for a given refrigerant.

Example: What is the theoretical discharge temperature of an ammonia system in which the compression ratio $\frac{P_2}{P_1} = 5.8$ and the suction temperature is -10 F?

$$T_1 = 460 + (-10) = 450 \text{ F abs}$$

from Table 55, the coefficient of discharge temperature corresponding to a compression ratio of 5.8 for ammonia is 1.501.

$$t_2 = (450 \times 1.501) - 460$$

$$t_2 = 215 \text{ F discharge temperature}$$

SPECIFIC VOLUME—COMPRESSOR DISCHARGE

As in the case of discharge temperatures, the specific volume of discharge vapor also may be established by employing similar methods.

As shown previously,

$$V_2 = V_1 \left(\frac{P_1}{P_2}\right)^{\frac{1}{\nu}}$$

Substituting standard conditions,

$$V_2 = 8.15 \times \left(\frac{34.27}{169.2}\right)^{\frac{1}{1.3}}$$

$$V_2 = 8.15 \times (0.202)^{0.769}$$

$$\text{Log } 0.202 = \bar{1}.30535$$

$$0.769 \times 0.30535 = 0.234815$$

$$0.769 \times \bar{1} = \frac{\bar{0}.769}{\bar{1}.465815}$$

$$\text{Antilog} = 0.292$$

$$V_2 = 8.15 \times 0.292 = 2.379 \text{ cu ft per lb}$$

Employing the equation

$$V_2 = V_1 \left(\frac{T_1}{T_2}\right)^{\frac{1}{\nu - 1}}$$

Substituting standard conditions,

$$V_2 = 8.15 \times \left(\frac{459.6 + 5}{459.6 + 210}\right)^{\frac{1}{1.3 - 1}}$$

$$V_2 = 8.15 \times (0.693)^{3.33}$$

$$\text{Log } 0.693 = \bar{1}.84037$$

$$0.84037 \times 3.33 = 2.79843$$

$$3.33 \times 0.\bar{1} = \frac{-1.33}{\bar{1}.46843}$$

$$\text{Antilog } 0.46843 = 0.2941$$

$$V_2 = 8.15 \times 0.2941$$

$$V_2 = 2.39 \text{ cu ft per lb}$$

The specific volume may be determined from the Mollier Chart by following a procedure similar to that used in determining discharge temperatures above. Instead of obtaining the reading from the temperature lines, as was done in the former case, the value is taken from the constant-volume lines. For standard conditions, 34.27 psi suction pressure and 169.2 psi discharge pressure, the specific volume for ammonia is 2.36 cu ft per lb.

From the refrigerant tables, the specific volume of discharge vapor is found in the same line as entropy and discharge temperature for superheated vapor at the discharge pressure. This would correspond to the 1.3253 entropy line of the superheat column for 169.2 psia discharge vapor. The nearest pressure value being 170 psia, it is necessary to interpolate, and for vapor at 1.3253 entropy and a temperature of 210.8 F, the specific volume would be approximately 2.36 cu ft per lb.

COMPRESSION RATIO

The refrigeration cycle in terms of thermodynamic laws is considered to be a *reversed heat system* and the compressor a heat pump. On the other hand, *true heat engines* are represented by steam, gasoline, and diesel engines. Simply defined, the difference between the two lies in the direction of heat flow, which becomes obvious when the principles of both types are considered. In the reversed heat engine, mechanical energy is converted to heat energy during compression of refrigerant vapors, whereas, in true heat engines, heat energy of the fuel is converted to mechanical energy. As will be observed in studying efficiencies of both engines, the effect of compression pressures is a highly important factor, and the reasons why high compression proves beneficial in one case is also the reason why it proves inefficient in the other. Therefore, in designing any heat engine, the amount of pressure allowed to develop within the compression chamber must be considered and weighed against both practicality or economy or both.

The final pressure in any compression chamber depends on the compression ratio, which may be defined simply as the ratio of initial volume within the cylinder when the piston is located at bottom-dead-center to the remaining or final volume when the piston is located at top-dead-center, or

$$\text{Compression ratio} = \frac{\text{initial volume}}{\text{final volume}}$$

Because volume varies inversely as the pressure under constant-temperature conditions, it follows that the compression ratio may be determined also by dividing the final pressure by initial pressure, or

$$\text{Compression ratio} = \frac{\text{final pressure}}{\text{initial pressure}}$$

Therefore, if the operating conditions of any refrigeration system are known or desired, by applying the above formula the compression ratio may be determined. Selecting standard conditions of 5 F evaporator temperature and 86 F condenser temperature, and assuming saturated vapor entering the compressor, the compression ratio for an ammonia system would be

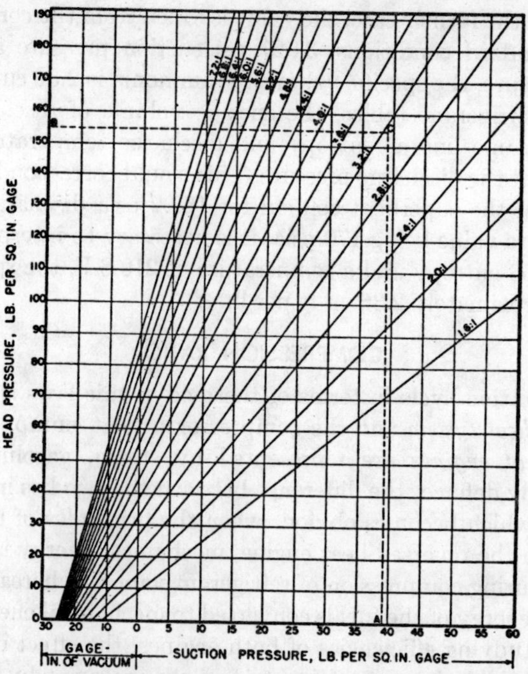

Fig. 360. Chart for finding compression ratio. For example, what is the compression ratio of a second-stage compressor if its suction pressure is 39 psig and its discharge pressure is 154.5 psig? Enter the chart at 39 on the bottom scale and draw a vertical line to intersect a horizontal line through the discharge pressure 154.5 on the left-hand scale. At the intersection C read the compression ratio, 3.15. (Courtesy, Mobil Oil Corporation.)

$$\text{Absolute pressure of 5 F saturated ammonia vapor} = 34.27 \text{ psi}$$

$$\text{Absolute pressure of 86 F liquid ammonia saturated} = 169.2 \text{ psi}$$

$$\text{Compression ratio} = \frac{\text{final pressure}}{\text{initial pressure}} = \frac{169.2}{34.27} = 4.94$$

Usually this value would be expressed as 4.94 to 1.

Under adiabatic or actual conditions, it is known that temperature increases as pressure. Therefore, because of cooling jacket limitations, any appreciable pressure increase is accompanied by a correspondingly higher compression temperature. This is borne out by the formula employed in computing discharge temperatures, viz.:

$$\frac{T_2}{T_1} = \left(\frac{P_2}{P_1}\right)^{\frac{\gamma-1}{\gamma}} \quad \text{or} \quad T_2 = T_1 \times \left(\frac{P_2}{P_1}\right)^{\frac{\gamma-1}{\gamma}}$$

where $\left(\frac{P_2}{P_1}\right)$ is equal to the compression ratio

The initial pressure P_1 being fixed by reason of desired evaporator temperature T_1, any increase in P_2 will also increase T_2. By referring to the tables

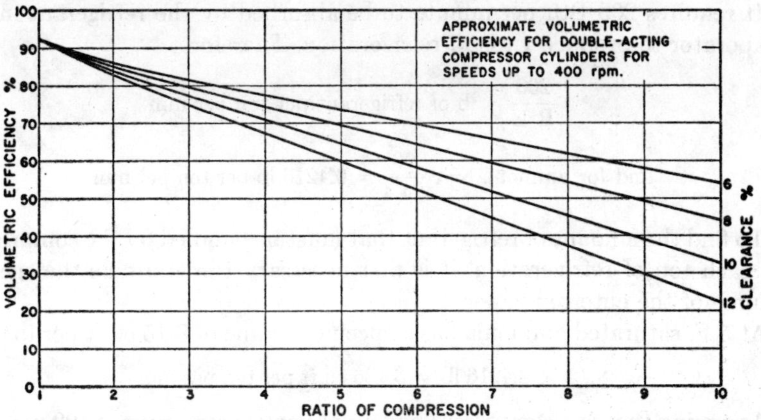

Fig. 361. Effect of compression ratio on volumetric efficiency of double-acting compressor cylinders at different clearance allowances. Speeds up to 400 rpm. (Courtesy, Worthington Corporation.)

of Superheated Ammonia Vapor, it can be observed that any increase in T_2 also increases specific volume, thereby reducing the density of vapors handled by the compressor. This, in turn, reduces the weight of refrigerant, which is the same as saying that it reduces volumetric efficiency, thereby decreasing the capacity of the compressor.

Economy of operation in any refrigeration system, therefore, is effected by compression ratio, the rule being that, *with every increase in spread between initial pressure and final pressure, a corresponding decrease in operating efficiency occurs.* It is for this reason that every precaution should be taken to prevent condenser temperatures from rising excessively during summer months due to higher cooling water temperatures. And where low and ultra-low evaporator temperatures are maintained (−20 F to −125 F), it becomes necessary to employ multi-stage (2 or 3) compressors with inter- and after-coolers in order to avoid excessive discharge temperatures. For temperatures from −50 F to −125 F, three compression stages are essential with any refrigerant. The employment of ammonia is impractical when evaporator temperatures below −90 F are required.

AMOUNT OF REFRIGERANT REQUIRED PER TON OF REFRIGERATION

The amount of refrigerant required to handle the load demand is calculated on the basis of pounds of refrigerant circulated per unit of time, usually per minute. This is accomplished by first determining the amount required for each ton of refrigeration capacity and multiplying the results by the total number of tons involved in the total load.

Referring to a previous section relating to refrigerating effect, it is found that, for ammonia under standard conditions of 5 F suction temperature and 86 F condenser temperature, R E = 474.4 Btu per lb.

It requires 200 Btu per minute to be absorbed by the refrigerant in the evaporator to produce 1 ton of refrigeration. Therefore,

$$\frac{200}{\text{R E}} = \text{lb of refrigerant per ton per min}$$

$$\text{and for ammonia} = \frac{200}{474.4} = 0.4216 \text{ lb per ton per min}$$

To find the amount of refrigerant that must be handled by the compressor for each ton of refrigeration, it is first necessary to determine the specific volume of the entering vapor.

At 5 F, saturated ammonia has a specific volume of 8.15 cu ft per lb

$$8.15 \times 0.4216 \text{ lb} = 3.436 \text{ cu ft per ton per min}$$

Assuming that the compressor had a volumetric efficiency of 100 percent, it would be required to handle 3.436 cu ft per min for each ton of refrigeration load. While a V E of 100 percent is not possible, and the results here are only theoretical, they do represent the basis on which equipment size and requirements are calculated.

WORK DONE BY COMPRESSOR

In order to compute the amount of power required to produce refrigeration for any given load, first it is necessary to determine the thermal work done by the compressor for each pound of refrigerant circulated. This is done by subtracting the enthalpy of the vapor just prior to entering the compressor at the suction valve from that contained in the superheated gas upon leaving the compressor discharge.

Assuming standard conditions of 5 F and 34.27 psia for the vapor at the compressor intake and a condenser pressure of 169.2 psia, the enthalpy values for ammonia are as follows:

Referring to the tables of saturated ammonia, the enthalpy of the vapor corresponding to a 5 F temperature and 34.27 psia pressure is 613.3 Btu per lb.

For superheated ammonia gas at 169.2 psia and constant entropy of 1.3253,* an interpolated value of 713.0 Btu per lb is found in the tables of superheated ammonia.

The thermal work done by the compressor is 713.0 − 613.3 = 99.7 Btu per lb of refrigerant circulated.

COEFFICIENT OF PERFORMANCE OR PERFORMANCE FACTOR

The ratio of the amount of energy of useful work to the energy expended in producing that work is expressed as the coefficient of performance of

* The entropy value is determined from saturated vapor tables corresponding to the suction temperature and pressure.

reversed heat engines. This ratio is the reverse of a true heat engine which in terms of input to output usually is expressed as efficiency.

Thus, for true heat engines.

$$\text{Efficiency} = \frac{\text{output}}{\text{input}} \times 100 \quad \text{or} \quad \frac{T_1 - T_2}{T_1}$$

And for reverse heat engines,

$$\text{Coefficient of performance} = \frac{\text{energy expended}}{\text{useful energy}} \quad \text{or} \quad \frac{T_1}{T_2 - T_1}$$

For a refrigeration system operating under standard conditions of 5 F and 86 F, the coefficient of performance would be found as follows:

$$\text{Coefficient of performance} = \frac{T_1}{T_2 - T_1} = \frac{460 + 5}{(460 + 86) - (460 + 5)}$$

$$= \frac{465}{81} = 5.74 \text{ performance factor}$$

In any system of refrigeration, the performance factor is found by dividing the heat absorbed in the evaporator by the heat equivalent of the work done by the compressor, or

$$\text{cop} = \frac{H_1}{H_2 - H_1}$$

At a 5 F evaporator temperature, the heat absorbed per pound of ammonia was found to be 474.4 Btu.

The work of the compressor against a 169.2 psia condenser pressure was found to be 713.0 − 613.3, or 99.7 Btu per lb. Therefore, the

$$\text{Performance factor} = \frac{474.4}{99.7} = 4.76$$

It readily can be observed from the above that best results are obtained when the difference between condenser and evaporator temperatures is held at a minimum.

HORSEPOWER REQUIRED PER TON OF REFRIGERATION

In order to select proper sized prime movers and auxiliary equipment, it is necessary that power requirements for total refrigeration loads be determined accurately. In the foregoing paragraphs, it was shown how the amount of thermal work done by the compressor per ton of refrigeration is computed. The method of determining the amount of refrigerant per ton per minute was also described. Both of these factors must necessarily be known in computing theoretical power requirements. And when such calculations are based on thermal loads, as represented in the formula for horsepower requirements, it becomes essential to convert Btu to their mechanical equivalent, or foot-pounds. From the study of thermal energy (Chapter 2), it is known that 778 ft-lb is the mechanical equivalent of 1 Btu, and that 33,000 ft-lb per minute defines 1 hp.

To determine the horsepower required per ton of refrigeration, the above factors are incorporated in the following formula:

$$\text{hp per ton} = \frac{\text{thermal work done by compressor} \times \text{lb refrigerant per ton per minute} \times 778\ \text{ft-lb}}{33{,}000}$$

Referring to results obtained in the foregoing paragraphs, the horsepower per ton under standard conditions is

$$\text{hp per ton} = \frac{99.6 \times 0.4216 \times 778}{33{,}000}$$
$$= 0.989\ \text{hp per ton of refrigeration}$$

0.989 hp is the theoretical indicated horsepower required per ton of refrigeration under standard conditions of 5 F evaporator and 86 F condenser temperatures.

MEAN EFFECTIVE PRESSURE

The mean effective pressure (mep) in a compressor cylinder under any set of operating conditions may be determined from an indicator card taken when equipment is already installed and operating. However, to determine this factor theoretically, the following formula may be used:

$$\text{mep} = \frac{y}{y-1} \times P_1 \left[\left(\frac{P_2}{P_1} \right)^{\frac{y-1}{y}} - 1 \right]$$

where mep = mean effective pressure
y = ratio of specific heat of gas at constant volume and constant pressure
P_1 = initial or evaporator pressure, psia
P_2 = final or condenser pressure, psia

For an evaporator pressure of 34.27 psia and a condenser pressure of 169.2 psia, which represents standard conditions, the mean effective pressure for an ammonia compressor is as follows:

$$\text{mep} = \frac{1.3}{1.3-1} \times 34.27 \times \left[\left(\frac{169.2}{34.27} \right)^{\frac{1.3-1}{1.3}} - 1 \right]$$
$$= 4.33 \times 34.27 \times [(4.9)^{0.23} - 1]$$
$$= 4.33 \times 34.27 \times (1.443 - 1)$$
$$= 65.7\ \text{psi mep}$$

To eliminate the necessity of employing logarithms in determining mep, as shown above, a factor corresponding to each compression ratio may be established. This factor is called the coefficient of mean effective pressure, and is computed for the various compression ratios as follows:

The ratio $\frac{y}{y-1}$ is constant for any specific refrigerant whose values are shown in Table 57. For ammonia, the ratio of $\frac{y}{y-1}$ is equal to $\frac{1.3}{1.3-1}$ = 4.33. Under standard conditions, the compression ratio is equal to $\frac{169.2}{34.2}$ = 4.9, and when raised to the $\frac{y-1}{y}$ power is equal to 1.443. Subtracting 1 from this figure, the expression $\left(\frac{P_2}{P_1}\right)^{\frac{y-1}{y}} - 1 = 0.443$. This latter value is constant for any ammonia system whose compression ratio is 4.9. And the coefficient of mep for such a system is found by multiplying 0.443 by the ratio $\frac{y}{y-1}$, or 4.33, to obtain 1.92.

Mean effective pressure (mep) may now be computed for an ammonia system operating under standard conditions, viz.:

$$\text{Initial pressure} \times \text{coefficient of mep} = \text{mep}$$

$$34.27 \times 1.92 = 65.8 \text{ psi mep}$$

Applying this method of computation, the coefficient of mep has been established for the various compression ratios of the several refrigerants, as listed in Table 56. To use Table 56, first establish the value of ratio $\frac{C_p}{C_v} = y$ from Table 57. Then follow the column for the required y value to the coefficient corresponding to the existing compression ratio.

For a compressor in operation, the theoretical indicated horsepower may be determined if the mep is known by applying the conventional power formula:

$$\text{Indicated hp} = \frac{\text{mep} \times \text{area of piston} \times \text{rpm} \times S}{33{,}000}$$

where Area of piston is in square inches
S is length of stroke in feet

A less difficult means of establishing mean effective pressure for any compressor when the compression ratio is known may be developed by establishing a constant based on the relationship of certain variables. If the constant thus established is used in an empirical formula, the use of logarithms may be eliminated and the work necessary to derive an answer shortened. While the results of such empirical formulas are not always exact, they are as near practically correct as any calculation can be, considering the many

Table 56. Coefficients, Mean Effective Pressures, for Various Refrigerants Having Values of $y = \frac{C_p}{C_v}$ as Given

Compression Ratio	$y = C_p/C_v$								
	1.3	**1.25**	**1.2**	**1.18**	**1.17**	**1.14**	**1.13**	**1.08**	**1.07**
1.2	0.186	0.184	0.186	0.184	0.182	0.182	0.182	0.189	0.183
1.4	0.351	0.349	0.348	0.348	0.343	0.347	0.339	0.338	0.336
1.5	0.424	0.422	0.420	0.420	0.419	0.413	0.417	0.405	0.413
1.6	0.498	0.495	0.492	0.492	0.489	0.487	0.487	0.473	0.474
1.8	0.628	0.626	0.618	0.617	0.615	0.611	0.608	0.594	0.596
2.0	0.753	0.742	0.738	0.735	0.727	0.718	0.721	0.716	0.719
2.2	0.866	0.854	0.846	0.840	0.832	0.825	0.826	0.810	0.810
2.4	0.970	0.960	0.942	0.938	0.930	0.924	0.921	0.905	0.902
2.6	1.070	1.057	1.038	1.030	1.021	1.015	1.008	0.986	0.979
2.8	1.160	1.145	1.128	1.122	1.111	1.097	1.095	1.067	1.055
3.0	1.251	1.232	1.206	1.200	1.188	1.172	1.173	1.148	1.131
3.2	1.334	1.314	1.284	1.279	1.265	1.246	1.243	1.215	1.208
3.4	1.416	1.392	1.362	1.351	1.335	1.320	1.312	1.282	1.269
3.6	1.490	1.465	1.434	1.423	1.405	1.386	1.382	1.337	1.330
3.8	1.563	1.537	1.500	1.489	1.468	1.444	1.443	1.404	1.391
4.0	1.632	1.605	1.560	1.548	1.531	1.510	1.503	1.458	1.437
4.2	1.702	1.668	1.626	1.614	1.594	1.568	1.556	1.512	1.498
4.4	1.767	1.731	1.686	1.666	1.650	1.617	1.616	1.566	1.544
4.6	1.832	1.790	1.740	1.725	1.706	1.675	1.668	1.620	1.590
4.8	1.892	1.848	1.794	1.778	1.754	1.724	1.721	1.661	1.636
4.9	1.923	1.877	1.824	1.804	1.782	1.749	1.747	1.687	1.667
5.0	1.949	1.906	1.848	1.830	1.810	1.774	1.764	1.701	1.682
5.2 . . .	2.009	1.959	1.902	1.883	1.859	1.823	1.816	1.755	1.728
5.4	2.061	2.013	1.950	1.935	1.908	1.865	1.860	1.796	1.774
5.6	2.117	2.066	1.998	1.981	1.950	1.914	1.903	1.836	1.804
5.8	2.169	2.115	2.046	2.027	1.999	1.955	1.947	1.877	1.850
6.0	2.221	2.163	2.094	2.066	2.041	1.997	1.990	1.917	1.896
6.2	2.269	2.212	2.136	2.112	2.083	2.038	2.025	1.958	1.927
6.4	2.312	2.260	2.178	2.152	2.125	2.079	2.068	1.985	1.957
6.6	2.364	2.304	2.220	2.198	2.167	2.112	2.103	2.025	2.003
6.8	2.412	2.347	2.262	2.237	2.202	2.153	2.146	2.052	2.034
7.0	2.459	2.391	2.304	2.276	2.244	2.186	2.181	2.092	2.064
7.2	2.503	2.435	2.346	2.316	2.279	2.228	2.216	2.120	2.095
7.4	2.546	2.474	2.382	2.348	2.314	2.261	2.251	2.160	2.125
7.6	2.589	2.517	2.418	2.388	2.349	2.294	2.285	2.187	2.156
7.8	2.628	2.556	2.454	2.421	2.384	2.327	2.312	2.214	2.186
8.0	2.672	2.595–	2.490	2.460	2.419	2.360	2.346	2.241	2.217
8.2	2.711	2.634	2.526	2.493	2.453	2.393	2.381	2.268	2.248
8.4	2.750	2.668	2.562	2.526	2.488	2.426	2.407	2.309	2.263
8.6	2.789	2.706	2.592	2.558	2.516	2.450	2.442	2.336	2.294
8.8	2.827	2.740	2.628	2.591	2.551	2.483	2.468	2.363	2.324
9.0	2.866	2.774	2.658	2.624	2.579	2.516	2.494	2.390	2.355
9.2	2.901	2.813	2.694	2.650	2.607	2.541	2.529	2.403	2.370
9.4	2.936	2.847	2.724	2.683	2.642	2.566	2.555–	2.430	2.401
9.6	2.970	2.876	2.754	2.709	2.670	2.599	2.581	2.457	2.416
9.8	3.005	2.910	2.784	2.742	2.691	2.624	2.607	2.484	2.446
10.0	3.040	2.944	2.814	2.775	2.726	2.648	2.633	2.511	2.462

variables to be considered in tempering a theoretical estimate to actual performance.

By employing the mep formula, viz.:

$$\text{mep} = \frac{y}{y-1} \times P_1 \left[\left(\frac{P_2}{P_1} \right)^{\frac{y-1}{y}} - 1 \right]$$

the mep can be calculated for any compression ratio, as represented by $\frac{P_2}{P_1}$.

Where P_1 is the same and P_2 a variable, then mep becomes a variable. It is known from the study of discharge temperatures that $T_2 = T_1 \left(\frac{P_2}{P_1} \right)^{\frac{y-1}{y}}$

Therefore $T_2 = T_1$ (compression ratio)$^{\frac{y-1}{y}}$. Signifying compression ratio as R, then

$$T_2 = T_1(R)^{\frac{y-1}{y}}$$

During adiabatic compression, entropy remains constant so that $\frac{H}{T} = K$. Therefore, as T_1 increases to T_2, then H increases proportionately to H_2, resulting in a definite relationship between enthalpy and pressure at any point along the line of piston travel.

If a list of thermal energy values is compiled for each increment of change in compression ratio based on a definite initial pressure in each case according to adiabatic law, a constant relationship between total enthalpy H and final pressure will be found. This would mean that when the same initial pressure P_1 was involved, then R would bear a constant relationship to H_2.

For a cylinder the capacity of which is 1 cu ft of vapor at the suction temperature and pressure corresponding to saturated conditions, the relationship between total enthalpy and pressure at any point along the line of piston travel would be

$$\frac{\frac{y}{y-1} \times P_1 \left[\left(\frac{P_2}{P_1} \right)^{\frac{y-1}{y}} - 1 \right]}{\text{density} \times (H_2 - H_1)} = \text{constant or } K$$

But

$$\frac{y}{y-1} \times P_1 \left[\left(\frac{P_2}{P_1} \right)^{\frac{y-1}{y}} - 1 \right] = \text{mep}$$

Therefore

$$\frac{\text{mep}}{\text{density} \times (H_2 - H_1)} = K$$

and

$$\text{mep} = K \times \text{density} \times (H_2 - H_1)$$

For standard conditions of 5 F evaporator temperature and 86 F condenser temperature, the constant for ammonia is

$$K = \frac{65.8}{0.1227 \times (713 - 613.7)} = 5.4$$

For conditions of 0 F and 75 F, the constant is approximately the same, or

$$K = \frac{55.64}{0.1097 \times (706.5 - 612.1)} = 5.37, \text{ or } 5.4 \text{ approx}$$

By similar computation for any conditions, $\frac{\text{mep}}{\text{density} \times (H_2 - H_1)}$ is constant and equal to 5.4 approx for ammonia. Therefore, in order to determine the mep for an ammonia compressor, the following formula may be applied for any set of conditions:

$$\text{mep} = 5.4 \times D \times (H_2 - H_1)$$

where D = density of vapor at P_1
H_1 = enthalpy of vapor at P_1 in Btu per lb
H_2 = total enthalpy of superheated vapor at P_2 in Btu per lb

TABLE 57. SPECIFIC HEAT OF GASES*

(C_p, C_v, Ratio C_p/C_v, Molecular Wt and Gas Constant)

Gas	C_p	C_v	(C_p/C_v) y	Mol Wt	Constant $\frac{1544}{\text{mol wt}}$	Approx y
Air (dry)	0.2375	0.1689	1.405	28.95	53.3	1.40
Ammonia	0.523	0.406	1.30	17.03	90.6	1.30
Carbon dioxide	0.205	0.160	1.28	44.00	35.1	1.30
Ethane	0.386		1.22	30.05	51.3	1.20
Ethyl chloride	0.274	0.23	1.20			1.20
Ethylene	0.399		1.18	28.03	55.08	1.20
Freon-11 @ 86 F			1.136			
@ 110 F	0.1374	0.1213	1.133	137.4	11.24	1.13
Freon-12 @ 86 F	0.1476	0.1297	1.138	120.9	12.77	1.14
Freon-13 @ −22 F			1.172	104.47	14.77	
Freon-14 @ −112 F			1.220	88.01	17.54	1.22
Freon-21 @ 86 F	0.1364	0.1516	1.185	102.93		1.185
Freon-22 @ 118 F			1.180	86.48	17.85	1.18
@ 212 F			1.157			1.16
Freon-113 @ 160 F	0.1633	0.1516	1.077	187.4		1.08
@ 210 F	0.1688	0.1573	1.073			1.07
Freon-114 @ 110 F	0.1629	0.1502	1.085	170.9		1.085
@ 160 F	0.1696	0.1571	1.080			
Freon-115 @ ...			1.091	154.48		1.10
Methane	0.593		1.32	16.03	96.32	1.30
Methyl chloride	0.24	0.20	1.20	50.48	30.58	1.20 to 1.28
Methyl formate	0.20	0.18	1.11			1.10
Nitrous oxide	0.22	0.176	1.25	44.02	35.07	
Steam @ 212 F	0.46	0.36	1.28			1.30
Sulfur dioxide	0.154	0.123	1.25	64.07	24.10	1.25

* Authorities differ as to the specific heat values for many gases. Approximate values as shown are accurate enough for practical problems involving C_p/C_v. Specific heat values for most substances vary with temperature either slightly or substantially according to the item. Values at room temperature (70 F) are given unless otherwise noted.

TABLE 58. APPROXIMATE EFFICIENCIES OF VERTICAL FREON COMPRESSORS*

Compression Ratio	*Volumetric Efficiency*	*Over-all Efficiency*
2.0	77.5%	78.0%
2.2	76.9	77.5
2.4	76.2	77.0
2.6	75.3	76.5
2.8	74.5	76.0
3.0	73.7	75.5
3.2	72.9	75.0
3.4	72.0	74.5
3.6	71.3	74.0
3.8	70.4	73.5
4.0	69.6	73.0
4.2	68.8	72.5
4.4	68.0	72.0
4.6	67.2	71.5
4.8	66.4	71.0
5.0	65.5	70.5
5.2	64.8	70.0
5.4	64.0	69.5
5.6	63.3	69.0
5.8	62.5	68.5
6.0	61.9	68.0
6.2	61.2	67.5
6.4	60.5	67.0
6.6	59.9	66.5
6.8	59.2	66.0

* Courtesy, "Trane Air Conditioning Manual," copyright, 1941, by The Trane Company, La Crosse, Wisconsin. Reproduced by permission.

PRESSURE DROP DUE TO LIQUID RISERS, OR LIFT

The pressure at the base of an open liquid column is equal to the height multiplied by the density of the fluid. This means that the pressure decreases as the distance from the top becomes less. If liquid is supplied to a closed liquid column at the bottom, the pressure at that point corresponds to the source pressure. Assuming that the closed liquid column is a vertical liquid line of a refrigeration system rising from the receiver level, the pressure at the base is equal to the head pressure. At the top of the liquid riser the pressure is equal to the head pressure less the pressure exerted by the weight of the fluid in the column. Thus, a loss of head pressure occurs in any system where the liquid must be carried to a level above that of the receiver.

For example, the loss in pressure within a riser carrying 86 F liquid R-12 would be

$$\frac{1 \text{ ft} \times \text{wt per cu ft}}{144 \text{ sq in.}} = \frac{1 \times 80.6}{144} = 0.56 \text{ psi per ft}$$

In a riser 30 ft high, the loss in pressure would amount to $30 \times 0.56 = 16.8$ psi. If the expansion valve were located on a level with the top of the riser, the entering pressure for an R-12 system operating under standard ton conditions would be $93.2 - 16.8 = 76.4$ psig. This latter pressure corresponds to a saturated temperature of approximately 75 F, or a drop of 11 F below the saturated temperature of 86 F in the condenser. This means that vaporization would occur in the upper portion of the riser if the surrounding temperature were above 75 F.

The loss of pressure for any reason in the high side of a refrigeration system results in decreased capacity. The reason lies first in the decrease in available refrigerating effect brought about by possible vaporization of the liquid before it reaches the evaporator. Secondly, the expansion valve performs at maximum capacity for a specific pressure drop across its orifice when the entering fluid is 100 percent liquid. The capacity of an expansion valve diminishes as the ratio of vapor to liquid of the entering fluid increases. Thirdly, the capacity of an expansion valve diminishes as the pressure differential across its orifice approaches equalization.

To compensate for losses due to lift in risers, friction in pipes, fittings, and distributor heads, three courses of action are open:

1. Increase head pressure by reducing the flow of condenser cooling water.
2. Subcool the liquid.
3. Use a larger expansion valve than would be necessary for a system in which the valve and the receiver are on the same level and in close proximity.

If none of the above courses is taken, the evaporator will become starved of liquid refrigerant under maximum design load demands. However, the first step—that of raising the head pressure—acts to reduce the over-all capacity of the system by reason of higher compression ratios. The second step tends to increase the cop but does not satisfy entirely the law of pressure differential in relation to flow capacity of a fixed orifice, which the expansion valve becomes when fully open. To obtain the necessary evaporator capacity, a larger expansion valve should be used; i.e., one that will pass sufficient liquid under the reduced pressure differential as will provide required tonnage.

The pressure drop in risers of various heights is shown in Table 59 for some commonly used refrigerants. The values given are determined by the equation $P_d = \frac{\text{ht} \times \text{density}}{144}$, or height × pressure drop per foot.

TABLE 59. PRESSURE DROP IN LIQUID RISERS, VARIOUS REFRIGERANTS

Height, ft	1	10	20	30	40	50	60	70	80	90	100
Refrigerant 12	0.56	5.6	11.2	16.8	22.4	28.0	33.6	39.2	44.8	50.4	56.0
Refrigerant 21	0.51	5.1	10.2	15.3	20.4	25.5	30.6	35.7	40.8	45.9	51.0
Ammonia	0.258	2.58	5.2	7.74	10.32	12.9	15.48	18.1	20.64	23.22	25.8
Sulfur dioxide	0.51	5.1	10.2	15.3	20.4	25.5	30.6	35.7	40.8	45.9	51.0
Methyl chloride	0.39	3.9	7.8	11.7	15.6	19.5	23.4	27.3	31.2	35.1	39.0
Refrigerant 22	0.73	7.3	14.6	21.9	29.2	36.5	43.8	51.1	58.4	65.7	73.0

PRESSURE DROP DUE TO FRICTION IN PIPES

In addition to the loss in head due to the presence of risers in the liquid line, the effect of friction in pipes and fittings during fluid flow must be considered. The loss of head (which may be termed variously "friction head," "drop in pressure," or "pressure drop," due to pipe friction) is a function of several factors related to the flow process; namely, (1) viscosity (fluidity), (2) fluid velocity, (3) specific gravity of the fluid, (4) size of pipe, (5) degree of internal surface roughness, (6) distance between points of measurement or length of pipe.

The loss due to friction in a conduit may be expressed in terms of feet-head of the fluid under consideration or pressure drop (psi). The relationship of the various factors involved is shown in the following equations for liquids:

For any conduit:

$$h_f = \frac{fLV^2}{4m2g}$$

For circular pipes:

$$h_f = \frac{fLV^2}{D2g} \qquad \text{(Fanning equation)}$$

The difference in the two equations lies in the relationship of m to the area-perimeter ratio in circular pipes. The factor m in the first equation is the hydraulic radius, found by dividing the area of the pipe by the internal perimeter. Thus $m = \frac{\pi d^2}{4\pi d}$ and $4m = \frac{\pi d^2}{\pi d} = d$ in inches or D in feet, internal diameter of the pipe. Substituting D for $4m$ in the first equation results in that shown for the circular pipe equation.

For gases, the equation becomes

$$h_p = \frac{fLV^2p}{24gd}$$

where, for all above equations

h_f = loss of head in feet due to friction
f = friction factor or coefficient of friction
L = length of pipe in feet, including equivalent length of fittings
V = fluid velocity, fps
D = inside diameter of circular pipe, ft
d = inside diameter of circular pipe, in.
m = hydraulic radius = $\frac{\text{area of pipe, sq ft}}{\text{wetted perimeter, ft}}$
g = gravitational acceleration in fps per sec, 32.2 (usually written 32.2 fps per sec)
h_p = loss due to friction, psi
p = density of gas at temperature and pressure of flow, lb per cu ft

It will be noted that all factors in the above equations may be known or safely assumed with the exception of the friction factor f. The value of f depends on the roughness of the internal pipe surface and the characteristics of flow.

TABLE 60. ROUGHNESS PARAMETER FOR PIPES

Pipe	*e, Feet*
Steel	0.00015
Wrought iron	0.00015
Cast iron, asphalt dipped	0.0004
Galvanized iron	0.0005
Cast iron	0.0004

$$\text{Roughness factor} = \frac{e}{D} = \frac{e\text{, feet as shown above}}{\text{internal diameter in feet}}$$

EXAMPLE: Roughness factor for schedule 40 steel pipe 18 in. OD.

$$\text{ID} = 1.4063 \text{ ft} \qquad e = 0.00015 \text{ ft}$$

$$\frac{e}{D} = \frac{0.00015}{1.4063} = 0.000107$$

Roughness Factor

The degree of internal roughness of any conduit is related to the material used in its construction. This means that, for any specific material, the degree of roughness may be assumed to be constant and is called the roughness factor. Scale, age, and internal treatment of pipe all affect the relative roughness. Certain methods of calculating pressure drop in a conduit require a numerical value for the roughness factor while others are

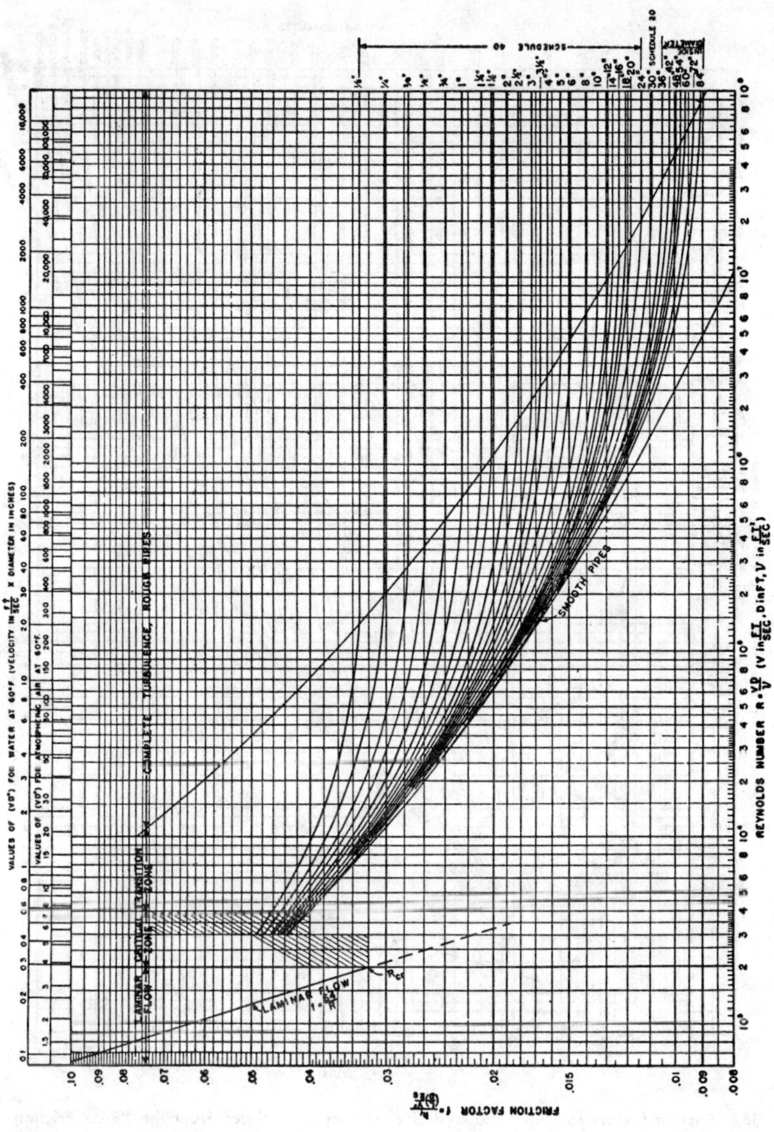

Fig. 362. Friction factors for steel or wrought iron pipe. (Reprinted from the "Pipe Friction Manual," pp. 59–60. Copyright, 1954, by the Hydraulic Institute, 122 E. 42nd St., New York 17, N. Y.)

based on curves constructed according to individual materials, such as drawn copper tubing, wrought iron or steel pipe, etc. It is obvious that the natural roughness of any specific material would have diminishing frictional effect with any increase in radius of the conduit. Therefore, the ratio of

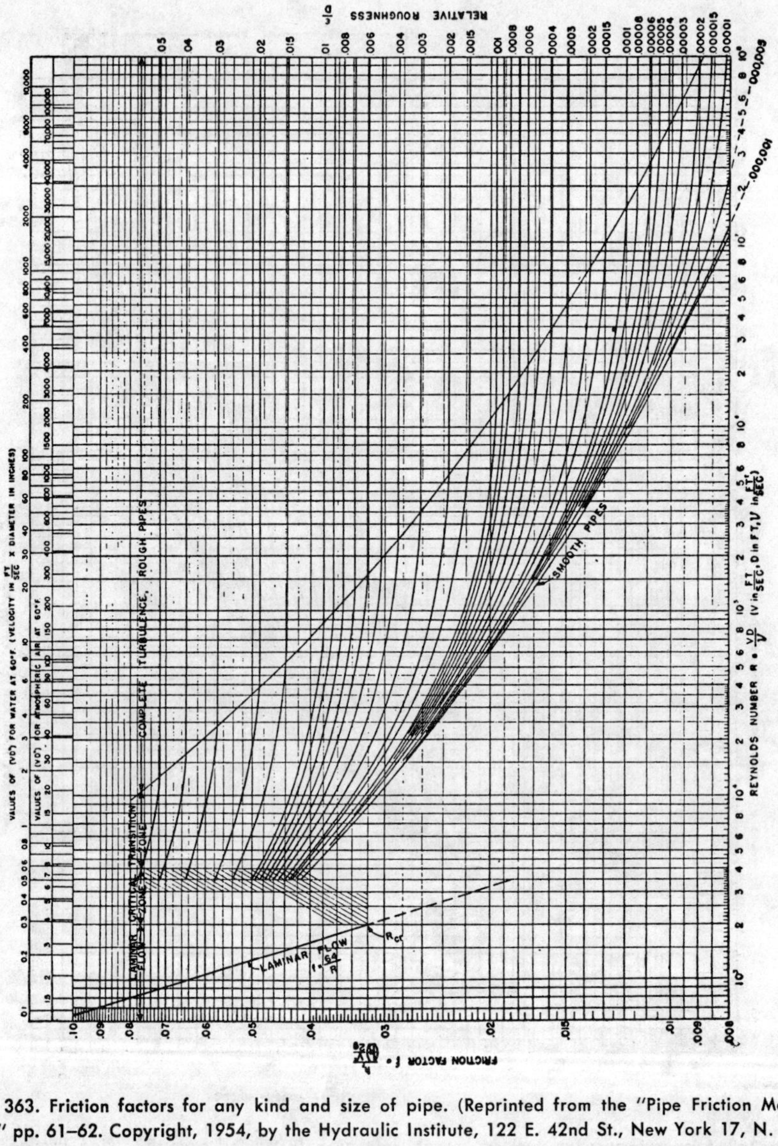

Fig. 363. Friction factors for any kind and size of pipe. (Reprinted from the "Pipe Friction Manual," pp. 61–62. Copyright, 1954, by the Hydraulic Institute, 122 E. 42nd St., New York 17, N. Y.)

roughness to diameter becomes more important than the degree of roughness alone. The ratio $\frac{\text{roughness parameter}}{\text{diameter}}$ is the relative roughness of the conduit surface. Some curves are based on the relative roughness values which range from 0.00001 for smooth pipes to 0.05 for rough pipes.

Flow Characteristics

The pattern of flow assumed by any fluid in moving through a conduit is governed by its velocity, viscosity, size of conduit, and internal surface roughness. Changes in direction of flow and restricting orifices also influence flow characteristics. Basically, there are two types of fluid flow: laminar flow and turbulent flow. Between the two are zones called the critical flow zone and the transition zone.

Laminar flow is variously called smooth pipe, stream line, or viscous flow. It assumes a zero velocity of the fluid layer immediately in contact or adjacent to the internal surface or *wetted* perimeter of the conduit. It also assumes a progressive velocity increase from the outermost layer until it reaches maximum velocity at the center or axis of the conduit. Laminar flow portrays fluid movement similar to that of a telescope in the act of being projected wherein neighboring concentric rings or layers of fluid proceed in a straight line. This type of flow is smooth, and layer mixing does not exist in any extensive proportion. Friction in fluids flowing under laminar conditions is due mainly to the resistance of the molecules of adjacent layers to slide over each other. This type of friction is called fluid friction and is a function of the nature of the individual fluid referred to as *viscosity*.

Turbulent flow, as its name implies, is unpredictable flow. Its flow pattern is not constant either in direction or in relative velocity. Turbulent flow is portrayed as a series of eddy currents in which the molecules are haphazard as to axial direction and take on motion which is both transverse and circulatory. The relative velocity of outer to axial layers prevalent in laminar flow is not present in turbulent flow. As a result, surface friction brought about by continual displacement of the static perimeter layer is an important consideration here.

To determine the influence of turbulent flow on friction, the relationship of pipe diameter, fluid velocity, fluid density, and viscosity must be established numerically. This relationship was first shown by Osborne Reynolds in 1883 and expressed as a number having a dimensionless value called the Reynolds number. The Reynolds number is established for any fluid by the general equation:

$$Re = \frac{DVp}{u}$$

where Re = Reynolds number
D = diameter of pipe (ft)
V = velocity of fluid
p = fluid density at flow conditions
u = absolute viscosity

When solving for the Reynolds number, all data must be based on the same system of measurement. If D is based on feet, then V must be con-

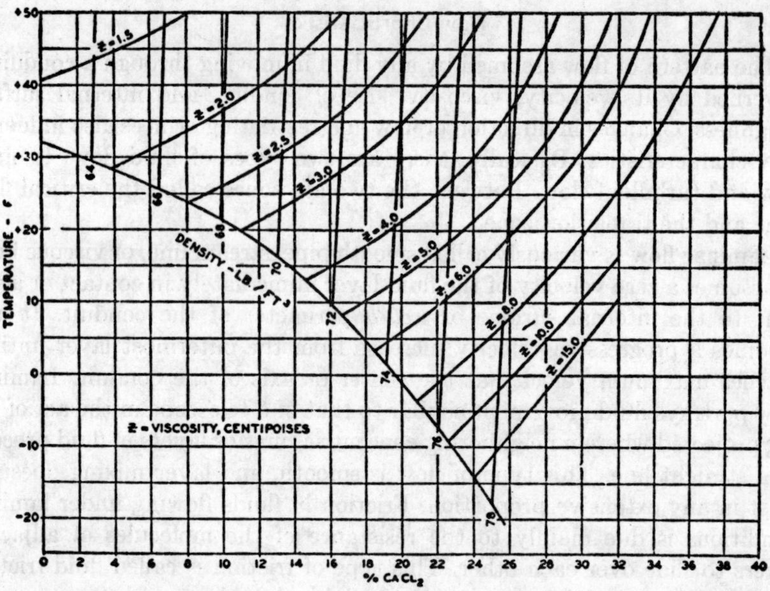

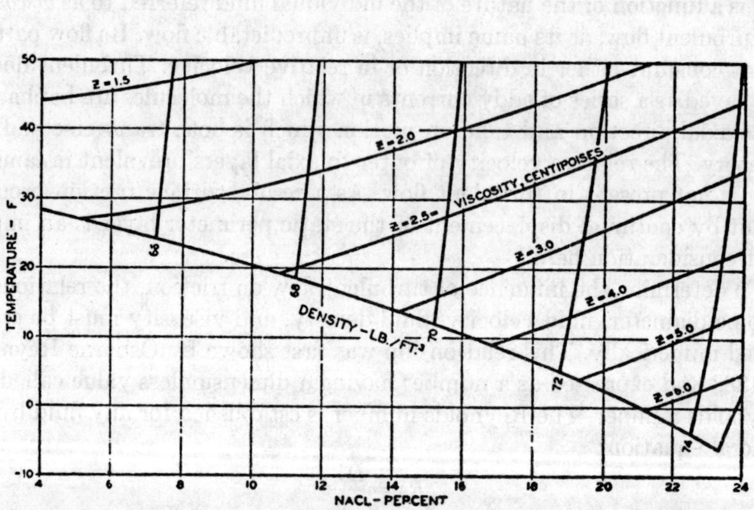

Fig. 364. (*Top*) Viscosity and density of calcium chloride brine. (Reprinted from the "Air Conditioning-Refrigerating Data Book 9th Edition—1955–56 Design Volume," by courtesy of The American Society of Refrigerating Engineers.)

Fig. 365. (*Bottom*) Viscosity and density of sodium chloride brine. (Reprinted from the "Air Conditioning-Refrigerating Data Book 9th Edition—1955–56 Design Volume," by courtesy of The American Society of Refrigerating Engineers.)

sidered in fps, p in lb per cu ft, and u in lb per ft per sec. To obtain u in lb per ft per sec, divide centipoises (z) by 1488, which is the factor for converting viscosity, as in this case, from the metric (centimeter-gram-second) system to the English (lb-ft-sec) system. The reciprocal of 1488 is $\frac{1}{1488} =$ 0.000672 so that the Reynolds number equation may be expressed as

$$Re = \frac{DVp}{0.000672z}$$

or

$$Re = \frac{124dVp}{z}$$

where d = diameter in inches
and z = viscosity in centipoises

Where fluid quantities are given in gallons per minute, as would be the case with water, then

$$Re = \frac{d/12 \times (0.408\ G/d^2 \times p)}{0.000672z}$$

$$Re = \frac{50.7Gp}{dz}$$

where G = gal per min

The Reynolds number may be established similarly when any quantity measurement is given, provided that the correct conversion factor and viscosity measurement are employed. The conversion factors and the viscosity relationship are shown in the following equations for various quantity expressions:

Key: B = bbl per hr
G = gal per min
w = lb per hr
c = cfs
C = cfm
D = diameter in ft
d = diameter in in.
z = centipoises
u = 0.000672z
k = centistokes
V = velocity, fps
p = weight, lb per cu ft

$$Re = \frac{DVp}{u} = \frac{124dVp}{z} = \frac{50.7Gp}{dz} = \frac{35.5Bp}{dz} = \frac{2214B}{dk} = \frac{22735cp}{dz} = \frac{3162G}{dk} =$$

$$\frac{7742dV}{k} = \frac{6.32w}{dz} = \frac{378.9Cp}{dz}$$

If the Reynolds number is below 2000 for a given flow process, it may be assumed that laminar flow occurs. In such a process, the friction factor is determined by the equation

$$f = \frac{64}{Re}$$

Many practical flow processes fall within the zone above the point of critical velocity so that turbulent flow prevails. To determine f in pipes subject to the latter, the following steps are taken:

(*a*) From the type of material used in construction of the conduit, determine the value of the roughness factor e.

(*b*) Determine $\frac{e}{D} = \frac{\text{roughness parameter}}{\text{diameter}}$.

(*c*) The values e and $\frac{e}{D}$ are necessary for some methods of calculations; i.e., when using Fig. 363. For the method shown here, only the type and the size of conduit are required; i.e., using Fig. 362.

(*d*) Determine Re from data given in the problem by converting to usable values of ft, lb per cu ft, etc., where necessary.

(*e*) Solve for f using the equation

$$f = \frac{64}{Re} \text{ for laminar flow; i.e., } Re \text{ is 2000 or less}$$

(*f*) Determine f from curves shown in Fig. 362 if the Re number indicates turbulent flow (above 2000).

(*g*) Compute h_f by substituting values found for a, b, c, d, e, and/or f above in the circular pipe equation $h_f = \frac{fLV^2}{D2g}$ or

(*h*) Compute h_f by similar substitutions in $h_f = \frac{fLV^2}{4m2g}$ for any conduit and using numerical values for m according to the ratio $\frac{\text{area}}{\text{perimete}}$.

EXAMPLE: Determine the pressure drop in 100 ft of 2-in. standard weight (Schedule 40) steel pipe carrying 60 F water at a velocity of 7.17 fps (75 gpm).

(*a*) A 2-in. nominal size standard weight steel pipe has an inside diameter of 2.067 in., according to Table 61.

TABLE 61. PROPERTIES OF STEEL AND WROUGHT IRON PIPE*

Nominal Diameter	Schedule	Outside Diameter	Wall Thickness	Internal Diameter	Internal Area	Internal Diameter	Internal Area	ε/D ε=0.00015 ft
Inches		Inches	Inches	Inches	Square Inches	Feet	Square Feet	
⅛	40 (S)	0.405	0.068	0.269	0.0568	0.0224	0.00039	0.00669
	80 (X)		0.095	0.215	0.0363	0.0179	0.00025	0.00837
¼	40 (S)	0.540	0.088	0.364	0.1041	0.0303	0.00072	0.00495
	80 (X)		0.119	0.302	0.0716	0.0252	0.00050	0.00596
⅜	40 (S)	0.675	0.091	0.493	0.1909	0.0411	0.00133	0.00365
	80 (X)		0.126	0.423	0.1405	0.0353	0.00098	0.00426
½	40 (S)	0.840	0.109	0.622	0.3039	0.0518	0.00211	0.00289
	80 (X)		0.147	0.546	0.2341	0.0455	0.00163	0.00330
	160		0.187	0.466	0.1706	0.0388	0.00118	0.00386
	(XX)		0.294	0.252	0.0499	0.0210	0.00035	0.00714
¾	40 (S)	1.050	0.113	0.824	0.5333	0.0687	0.00370	0.00218
	80 (X)		0.154	0.742	0.4324	0.0618	0.00300	0.00243
	160		0.218	0.614	0.2961	0.0512	0.00206	0.00293
	(XX)		0.308	0.434	0.1479	0.0362	0.00103	0.00415
1	40 (S)	1.315	0.133	1.049	0.8643	0.0874	0.00600	0.00172
	80 (X)		0.179	0.957	0.7193	0.0798	0.00500	0.00188
	160		0.250	0.815	0.5217	0.0679	0.00362	0.00221
	(XX)		0.358	0.599	0.2818	0.0499	0.00196	0.00301
1¼	40 (S)	1.660	0.140	1.380	1.496	0.1150	0.01039	0.00130
	80 (X)		0.191	1.278	1.283	0.1065	0.00890	0.00141
	160		0.250	1.160	1.057	0.0967	0.00734	0.00155
	(XX)		0.382	0.896	0.6305	0.0747	0.00438	0.00201
1½	40 (S)	1.900	0.145	1.610	2.036	0.1342	0.01414	0.00112
	80 (X)		0.200	1.500	1.767	0.1250	0.01227	0.00120
	160		0.281	1.338	1.406	0.1115	0.00976	0.00135
	(XX)		0.400	1.100	0.9503	0.0917	0.00660	0.00164
2	40 (S)	2.375	0.154	2.067	3.356	0.1723	0.02330	0.00087
	80 (X)		0.218	1.939	2.953	0.1616	0.02051	0.00093
	160		0.343	1.689	2.241	0.1408	0.01556	0.00107
	(XX)		0.436	1.503	1.774	0.1253	0.01232	0.00120
2½	40 (S)	2.875	0.203	2.469	4.788	0.2058	0.03325	0.000729
	80 (X)		0.276	2.323	4.238	0.1936	0.02943	0.000775
	160		0.375	2.125	3.547	0.1771	0.02463	0.000847
	(XX)		0.552	1.771	2.464	0.1476	0.01711	0.00102
3	40 (S)	3.500	0.216	3.068	7.393	0.2557	0.05134	0.000587
	80 (X)		0.300	2.900	6.605	0.2417	0.04587	0.000621
	160		0.437	2.626	5.416	0.2188	0.03761	0.000685
	(XX)		0.600	2.300	4.155	0.1917	0.02885	0.000783
3½	40 (S)	4.000	0.226	3.548	9.887	0.2957	0.06866	0.000507
	80 (X)		0.318	3.364	8.888	0.2803	0.06172	0.000535
	(XX)		0.636	2.728	5.845	0.2273	0.04059	0.000660
4	40 (S)	4.500	0.237	4.020	12.73	0.3355	0.08841	0.000447
	80 (X)		0.337	3.826	11.50	0.3188	0.07984	0.000470
	120		0.437	3.626	10.33	0.3022	0.07171	0.000496
	160		0.531	3.438	9.283	0.2865	0.06447	0.000524
	(XX)		0.674	3.152	7.803	0.2627	0.05419	0.000571

S = Wall thickness formerly designated, "standard weight".
X = Wall thickness formerly designated, "extra heavy".
XX = Wall thickness formerly designated, "double extra heavy".

* Courtesy, "Standards of the Hydraulic Institute," ed. 10. Copyright, 1955, by the Hydraulic Institute, 122 East 42nd St., New York 17, N. Y.

(*b*) Water at 60 F weighs 62.37 lb per cu ft, according to Table 62.

(*c*) The viscosity of 60 F water = 1.12 centipoises, from Table 64 (interpolation).

1st Step: Solve for *Re* (using appropriate equation).

$$Re = \frac{124dVp}{z} = \frac{124 \times 2.067 \times 7.17 \times 62.37}{1.12}$$

$$Re = 102{,}338$$

TABLE 61. PROPERTIES OF STEEL AND WROUGHT IRON PIPE—(*Continued*)

Nominal Diameter	Schedule	Outside Diameter	Wall Thickness	Internal Diameter	Internal Area	Internal Diameter	Internal Area	ϵ/D $\epsilon = 0.00015$ ft
Inches		Inches	Inches	Inches	Square Inches	Feet	Square Feet	
5	40 (S)	5.563	0.258	5.047	20.01	0.4206	0.1389	0.000357
	80 (X)		0.375	4.813	18.19	0.4011	0.1263	0.000374
	120		0.500	4.563	16.35	0.3803	0.1136	0.000394
	160		0.625	4.313	14.61	0.3594	0.1015	0.000417
	(XX)		0.750	4.063	12.97	0.3386	0.09004	0.000443
6	40 (S)	6.625	0.280	6.065	28.89	0.5054	0.2006	0.000293
	80 (X)		0.432	5.761	26.07	0.4801	0.1810	0.000312
	120		0.562	5.501	23.77	0.4584	0.1650	0.000327
	160		0.718	5.189	21.15	0.4324	0.1469	0.000347
	(XX)		0.864	4.897	18.83	0.4081	0.1308	0.000368
8	20	8.625	0.250	8.125	51.85	0.6771	0.3601	0.000222
	30 (S)		0.277	8.071	51.16	0.6726	0.3553	0.000223
	40 (S)		0.322	7.981	50.03	0.6651	0.3474	0.000226
	60		0.406	7.813	47.94	0.6511	0.3329	0.000230
	80 (X)		0.500	7.625	45.66	0.6354	0.3171	0.000236
	100		0.593	7.439	43.46	0.6199	0.3018	0.000242
	120		0.718	7.189	40.59	0.5991	0.2819	0.000250
	140		0.812	7.001	38.50	0.5834	0.2673	0.000257
	(XX)		0.875	6.875	37.12	0.5729	0.2578	0.000262
	160		0.906	6.813	36.46	0.5678	0.2532	0.000264
10	20	10.75	0.250	10.250	82.52	0.85417	0.5730	0.000176
	(S)		0.279	10.192	81.58	0.84933	0.5666	0.000177
	30 (S)		0.307	10.136	80.69	0.84467	0.5604	0.000178
	40 (S)		0.365	10.020	78.85	0.83500	0.5476	0.000180
	60 (X)		0.500	9.750	74.66	0.8125	0.5185	0.000185
	80		0.593	9.564	71.84	0.7970	0.4989	0.000188
	100		0.718	9.314	68.13	0.7762	0.4732	0.000193
	120		0.843	9.064	64.53	0.7553	0.4481	0.000199
	140		1.000	8.750	60.13	0.7292	0.4176	0.000206
	160		1.125	8.500	56.75	0.7083	0.3941	0.000212
12	20	12.75	0.250	12.250	117.86	1.0208	0.8185	0.000147
	30		0.330	12.090	114.80	1.0075	0.7972	0.000149
	(S)		0.375	12.000	113.10	1.0000	0.7854	0.000150
	40		0.406	11.938	111.93	0.99483	0.7773	0.000151
	(X)		0.500	11.750	108.43	0.97917	0.7530	0.000153
	60		0.562	11.626	106.16	0.96883	0.7372	0.000155
	80		0.687	11.376	101.64	0.94800	0.7058	0.000158
	100		0.843	11.064	96.14	0.92200	0.6677	0.000163
	120		1.000	10.750	90.76	0.89583	0.6303	0.000167
	140		1.125	10.500	86.59	0.87500	0.6013	0.000171
	160		1.312	10.126	80.53	0.84383	0.5592	0.000178
14 OD	10	14.00	0.250	13.500	143.14	1.1250	0.9940	0.000133
	20		0.312	13.376	140.52	1.1147	0.9758	0.000135
	30		0.375	13.250	137.89	1.1042	0.9575	0.000136
	40		0.437	13.126	135.32	1.0938	0.9397	0.000137
	60		0.593	12.814	128.96	1.0678	0.8956	0.000140
	80		0.750	12.500	122.72	1.0417	0.8522	0.000144
	100		0.937	12.126	115.49	1.0105	0.8020	0.000148
	120		1.062	11.876	110.77	0.98967	0.7692	0.000152
	140		1.250	11.500	103.87	0.95833	0.7213	0.000157
	160		1.406	11.188	98.31	0.93233	0.6827	0.000161

S = Wall thickness formerly designated, "standard weight".
X = Wall thickness formerly designated, "extra heavy".
XX = Wall thickness formerly designated, "double extra heavy".

(*Continued on next page.*)

2nd Step: Establish the friction head value f from Fig. 362 using curve corresponding to (Schedule 40) 2-in. ID pipe.

$$f = 0.0218$$

3rd Step: Solve for h_f in ft-head loss.

$$h_f = \frac{LV^2}{D2g} = 0.0218 \times \frac{100 \times 7.17^2}{0.172 \times 2 \times 32.2}$$

$$h_f = 10.1 \text{ ft}$$

TABLE 61. PROPERTIES OF STEEL AND WROUGHT IRON PIPE—(*Continued*)

Nominal Diameter	Schedule	Outside Diameter	Wall Thickness	Internal Diameter	Internal Area	Internal Diameter	Internal Area	ϵ/D ϵ=0.00015 ft
Inches		Inches	Inches	Inches	Square Inches	Feet	Square Feet	
16 OD	10	16.00	0.250	15.500	188.69	1.2917	1.3104	0.000116
	20		0.312	15.376	185.69	1.2813	1.2895	0.000117
	30		0.375	15.250	182.65	1.2708	1.2684	0.000118
	40		0.500	15.000	176.72	1.2500	1.2272	0.000120
	60		0.656	14.688	169.44	1.2240	1.1767	0.000121
	80		0.843	14.314	160.92	1.1928	1.1175	0.000126
	100		1.031	13.938	152.58	1.1615	1.0596	0.000129
	120		1.218	13.564	144.50	1.1303	1.0035	0.000133
	140		1.437	13.126	135.32	1.0938	0.9397	0.000137
	160		1.562	12.876	130.21	1.0730	0.9043	0.000140
18 OD	10	18.00	0.250	17.500	240.53	1.4583	1.6703	0.000103
	20		0.312	17.376	237.13	1.4480	1.6467	0.000104
	(S)		0.375	17.250	233.71	1.4375	1.6230	0.000104
	30		0.437	17.126	230.36	1.4272	1.5997	0.000105
	(X)		0.500	17.000	226.98	1.4167	1.5762	0.000106
	40		0.562	16.876	223.68	1.4063	1.5533	0.000107
	60		0.718	16.564	215.49	1.3803	1.4964	0.000109
	80		0.937	16.126	204.24	1.3438	1.4183	0.000112
	100		1.156	15.688	193.30	1.3073	1.3423	0.000115
	120		1.343	15.314	184.19	1.2762	1.2791	0.000118
	140		1.562	14.876	173.81	1.2397	1.2070	0.000121
	160		1.750	14.500	165.13	1.2083	1.1467	0.000124
20 OD	10	20.00	0.250	19.500	298.65	1.6250	2.0739	0.0000923
	20		0.375	19.250	291.04	1.6042	2.0211	0.0000935
	30		0.500	19.000	283.53	1.5833	1.9689	0.0000947
	40		0.593	18.814	278.01	1.5678	1.9306	0.0000957
	60		0.812	18.376	265.21	1.5313	1.8417	0.0000980
	80		1.031	17.938	252.72	1.4948	1.7550	0.000100
	100		1.250	17.500	240.53	1.4583	1.6703	0.000103
	120		1.500	17.000	226.98	1.4167	1.5762	0.000106
	140		1.750	16.500	213.83	1.3750	1.4849	0.000109
	160		1.937	16.126	204.24	1.3438	1.4183	0.000112
24 OD	10	24.00	0.250	23.500	433.74	1.9583	3.0121	0.0000766
	20		0.375	23.250	424.56	1.9375	2.9483	0.0000774
	(S)		0.500	23.000	415.48	1.9167	2.8852	0.0000783
	30		0.562	22.876	411.01	1.9063	2.8542	0.0000787
	40		0.687	22.626	402.07	1.8855	2.7922	0.0000796
	60		0.937	22.126	384.50	1.8438	2.6701	0.0000814
	80		1.218	21.564	365.22	1.7970	2.5362	0.0000835
	100		1.500	21.000	346.36	1.7500	2.4053	0.0000857
	120		1.750	20.500	330.06	1.7083	2.2921	0.0000878
	140		2.062	19.876	310.28	1.6563	2.1547	0.0000906
	160		2.312	19.376	294.86	1.6147	2.0476	0.0000929
30 OD	10	30.00	0.312	29.376	677.76	2.4480	4.7067	0.0000613
	20		0.500	29.000	660.52	2.4167	4.5869	0.0000621
	30		0.625	28.750	649.18	2.3958	4.5082	0.0000626

S = Wall thickness formerly designated, "standard weight".
X = Wall thickness formerly designated, "extra heavy".
XX = Wall thickness formerly designated, "double extra heavy".

4th Step: Convert ft-head loss to pressure drop, psi.

$$P_d = \text{ft} \times \text{psi per ft for fluid} = \frac{62.37}{144} = 0.433 \text{ for water @ 60 F}$$

$$P_d = 10.1 \times 0.433$$

$$P_d = 4.37 \text{ psi per 100 ft}$$

A practical value for h_f may be determined by using the Williams and Hazen empirical formula, viz.:

$$h_f = 0.2083 \times \left(\frac{100}{C}\right)^{1.85} \times \frac{q^{1.85}}{d^{4.8655}}$$

where h_f = friction head in feet of liquid per 100 ft of pipe
d = inside diameter, in.
q = flow, gpm
C = constant based on surface roughness
(Viscosity is not considered in the above formula)

For clean new wrought iron pipe, the average theoretical value for C is 130. For practical design purposes, C for the same pipe is given a value of 100. Equal values apply to pipes possessing similar surface qualities, such as those made of steel and cast iron. The difference in results using $C = 100$

TABLE 62. SPECIFIC GRAVITY AND DENSITY OF WATER

Temperature C	Temperature F	Specific Gravity Ratio $\frac{\text{wt } H_2O \text{ at given } t \text{ F}}{\text{wt } H_2O \text{ at } 39.2 \text{ F}}$	*Density, lb per cu ft*
0	32	0.9998	62.420
4	39.2	1.0000	62.4266
10	50	0.9997	62.410
20	68	0.9982	62.316
30	86	0.9957	62.157
40	104	0.9924	61.943
50	122	0.9881	61.680
60	140	0.9832	61.370
70	158	0.9778	61.040

as compared with $C = 130$ assumes a ratio of practical to theoretical values in the order of 1.62 to 1.00. Applying this ratio (1.62 as the multiplier) to the h_f value obtained by the theoretical method above, then

$$h_f = 9.83 \times 1.62$$

$$h_f = 15.92 \text{ ft-head loss per 100 ft}$$

and

$$P_d = 6.89 \text{ psi per 100 ft}$$

EXAMPLE: A 3-in. steel pipe delivers 260 gal of calcium chloride brine per min at 32 F through a system having an equivalent length of 600 ft of smooth steel pipe. Assuming a specific gravity of 1.143 (i.e., 16 percent salt) for the brine, determine the total pressure drop.

Data:

Sp gr = 1.143

Temp = 32 F

Length = 600 ft

Flow = 260 gpm

Pipe ID = 3.068 (internal diameter, ft)

TABLE 63. SPECIFIC GRAVITY AND DENSITY OF LIQUID REFRIGERANTS

Refrigerant	*Temperature °F*	*Specific Gravity at 1 atm*	*Density Lb Per Cu Ft at Pressure Corresponding to Temperature Shown*
Ammonia		0.684	
	70		38.0
	75		37.74
	80		37.48
	86		37.16
Refrigerant 12		1.48	
	70		82.717
	75		82.089
	80		81.450
	86		80.671
Refrigerant 22	0		83.90
	30		80.45
	70		75.46
	80		74.15
	86		73.36
Refrigerant 11		1.468	
	70		92.71
	80		91.88
	86		91.38
Refrigerant 113		1.559	
	50		99.83
	70		98.26
	86		06.06
Refrigerant 114	50		93.71
	70		91.63
	86		89.91
Butane	80		35.65
	85		35.45
Refrigerant 500	70		73.14
	80		71.96
	86		71.23
Methyl chloride		1.002	
	80		56.68
	86		56.21
Sulfur dioxide		1.357	
	80		85.03
	86		84.46

TABLE 64. ABSOLUTE VISCOSITY OF WATER, CENTIPOISES*

(°C)	*Temperature* °F	*Viscosity in Centipoises*	(°C)	*Temperature* °F	*Viscosity in Centipoises*
(0)	32.0	1.7921	(50)	122.0	0.5494
(2)	35.6	1.6728	(52)	125.6	0.5315
(4)	39.2	1.5674	(54)	129.2	0.5146
(6)	42.8	1.4728	(56)	132.8	0.4985
(8)	46.4	1.3860	(58)	136.4	0.4832
(10)	50.0	1.3077	(60)	140.0	0.4688
(12)	53.6	1.2363	(62)	143.6	0.4550
(14)	57.2	1.1709	(64)	147.2	0.4418
(16)	60.8	1.1111	(66)	158.8	0.4293
(18)	64.4	1.0559	(68)	154.4	0.4174
(20)	68.0	1.0050	(70)	158.0	0.4061
(20.2)	68.4	1.0000	(72)	161.6	0.3952
(22)	71.6	0.9579	(74)	165.2	0.3849
(24)	75.2	0.9142	(76)	168.8	0.3750
(26)	78.8	0.8737	(78)	172.4	0.3655
(28)	82.4	0.8360	(80)	176.0	0.3565
(30)	86.0	0.8007	(82)	179.6	0.3478
(32)	89.6	0.7679	(84)	183.2	0.3395
(34)	93.2	0.7371	(86)	186.8	0.3315
(36)	96.8	0.7085	(88)	190.4	0.3239
(38)	100.4	0.6814	(90)	194.0	0.3165
(40)	104.0	0.6560	(92)	197.6	0.3095
(42)	107.7	0.6321	(94)	201.2	0.3027
(44)	111.2	0.6097	(96)	204.8	0.2962
(46)	114.8	0.5883	(98)	208.4	0.2899
(48)	118.4	0.5683	(100)	212.0	0.2838

* Courtesy, Bureau of Standards.

Viscosity of CaCl brine from Fig. 364 is 2.677 centipoises. Brine, having a sp gr of 1.143, weighs 71.26 lb per cu ft.

$$Re = \frac{50.7GP}{dz} = \frac{50.7 \times 260 \times 71.26}{3.068 \times 2.677} = 114{,}373$$

$$f = 0.0207$$

$$h_f = f\frac{LV^2}{D2g} = 0.0207 \times \frac{600 \times 11.8^2}{0.2556 \times 64.4}$$

$$h_f = 105.06 \text{ ft-head loss}$$

$$P_d = 105.06 \times \frac{71.26}{144} = 105.06 \times 0.495$$

$$P_d = 52.0 \text{ psi pressure drop}$$

TABLE 65. VISCOSITY OF REFRIGERANTS*

Viscosity of Refrigerant Liquids at Saturation Pressure

(μ in centipoises)

t F	Liquids at saturation pressure									
	F-11*	F-12*	F-21*	F-22*	F-113*	F-114*	Methyl chloride*	NH_3†	SO_2†	CO_2†
−40	.980	.423	.629	.351		.879	.349			
−20	.801	.371	.547	.316	1.566	.711	.321			
0	.677	.335	.484	.291	1.263	.598	.298			
+ 5								.250	.503	.115
20	.586	.308	.436	.271	1.043	.516	.279	.240	.390	.110
40	.517	.286	.397	.256	.876	.454	.263	.230	.350	.095
60	.461	.269	.364	.243	.747	.405	.249			
80	.417	.255	.337	.232	.646	.366	.237	.210	.290	.064
100	.380	.242	.314	.223	.564	.334	.226	.200	.260	
120	.349	.232	.294	.214	.497	.307	.217			
140	.323	.222	.277	.207	.442	.284	.208			
160	.300	.214	.262	.201	.395		.200			
180	.281	.207	.248	.195	.356		.193			
200	.263	.200	.236		.322		.186			
220	.248		.225		.293		.180			
240	.232		.215		.268		.175			

* Data from Benning and Markwood, *Refrig. Eng.*, April 1939.
† Data from Thompson, *Refrig. Eng.*, Nov. 1942.

Viscosity of Refrigerant Vapors (μ in centipoises)

t F	Vapors at 1 atm‡									
	F-11*	F-12*	F-21*	F-22*	F-113*	F-114*	Methyl chloride*	NH_3†	SO_2†	CO_2†
−40	.0088	.0106	.0094	.0105		.0095	.0086			
−20	.0092	.0109	.0098	.0109	.0090	.0098	.0090			
0	.0096	.0113	.0101	.0113	.0093	.0102	.0094			
+ 5								.0085	.0111	.0133
20	.0099	.0116	.0105	.0118	.0095	.0106	.0098	.0088	.0115	.0135
40	.0103	.0119	.0108	.0122	.0098	.0109	.0101	.0093	.0120	.0140
60	.0106	.0123	.0111	.0126	.0101	.0112	.0105			
80	.0110	.0126	.0115	.0130	.0103	.0116	.0108			
100	.0113	.0129	.0118	.0133	.0106	.0119	.0111	.0105	.0134	.0155
120	.0116	.0132	.0121	.0137	.0108	.0122	.0115			
140	.0120	.0135	.0124	.0141	.0111	.0125	.0118			
150								.0116	.0144	.0165
160	.0123	.0138	.0127	.0145	.0113		.0122			
180	.0126	.0140	.0130	.0148	.0116		.0125			
200	.0129	.0143	.0133	.0152	.0118		.0128			
220	.0132	.0146	.0136	.0156	.0120		.0131			
240	.0135	.0149	.0139	.0159	.0123		.0134			

* Data from Benning and Markwood, *Refrig. Eng.*, April 1939.
† Data from Thompson, *Refrig. Eng.*, Nov. 1942.
‡ 0.1 atm. for F-113.

* Reprinted from the "Air Conditioning-Refrigerating Data Book 9th Edition—1955–56 Design Volume," by courtesy of The American Society of Refrigerating Engineers.

Table 66. Equivalent Length of New Straight Pipe for Valves and Fittings (For Turbulent Flow Only)*

Fittings			Pipe size																				
			1/4	3/8	1/2	3/4	1	1 1/4	1 1/2	2	2 1/2	3	4	5	6	8	10	12	14	16	18	20	24
Regular 90° ell	Screwed	Steel	2.3	3.1	3.6	4.4	5.2	6.6	7.4	8.5	9.3	11	13	...	...	...	...	...	...	...	...	...	...
		C.I.	...	...	...	...	...	...	...	...	...	9.0	11	...	...	...	...	...	...	...	...	...	...
	Flanged	Steel	...	...	.92	1.2	1.6	2.1	2.4	3.1	3.6	4.4	5.9	7.3	8.9	12	14	17	18	21	23	25	30
		C.I.	...	...	...	...	...	...	...	...	...	3.6	4.8	...	7.2	9.8	12	15	17	19	22	24	28
Long radius 90° ell	Screwed	Steel	1.5	2.0	2.2	2.3	2.7	3.2	3.4	3.6	3.6	4.0	4.6	...	...	...	...	...	...	...	...	...	...
		C.I.	...	...	...	...	...	...	...	...	...	3.3	3.7	...	...	...	...	...	...	...	...	...	...
	Flanged	Steel	...	...	1.1	1.3	1.6	2.0	2.3	2.7	2.9	3.4	4.2	5.0	5.7	7.0	8.0	9.0	9.4	10	11	12	14
		C.I.	...	...	...	...	...	...	...	...	...	2.8	3.4	...	4.7	5.7	6.8	7.8	8.6	9.6	11	11	13
Regular 45° ell	Screwed	Steel	.34	.52	.71	.92	1.3	1.7	2.1	2.7	3.2	4.0	5.5	...	...	...	...	...	...	...	...	...	...
		C.I.	...	...	...	...	...	...	...	...	...	3.3	4.5	...	...	...	...	...	...	...	...	...	...
	Flanged	Steel	...	...	.45	.59	.81	1.1	1.3	1.7	2.0	2.6	3.5	4.5	5.6	7.7	9.0	11	13	15	16	18	22
		C.I.	...	...	...	...	...	...	...	...	...	2.1	2.9	...	4.5	6.3	8.1	9.7	12	13	15	17	20
Tee-line flow	Screwed	Steel	.79	1.2	1.7	2.4	3.2	4.6	5.6	7.7	9.3	12	17	...	...	...	...	...	...	...	...	...	...
		C.I.	...	...	...	...	...	...	...	...	...	9.9	14	...	...	...	...	...	...	...	...	...	...
	Flanged	Steel	...	...	.69	.82	1.0	1.3	1.5	1.8	1.9	2.2	2.8	3.3	3.8	4.7	5.2	6.0	6.4	7.2	7.6	8.2	9.6
		C.I.	...	...	...	...	...	...	...	...	...	1.9	2.2	...	3.1	3.9	4.6	5.2	5.9	6.5	7.2	7.7	8.8
Tee-branch flow	Screwed	Steel	2.4	3.5	4.2	5.3	6.6	8.7	9.9	12	13	17	21	...	...	...	...	...	...	...	...	...	...
		C.I.	...	...	...	...	...	...	...	...	...	14	17	...	...	...	...	...	...	...	...	...	...
	Flanged	Steel	...	...	2.0	2.6	3.3	4.4	5.2	6.6	7.5	9.4	12	15	18	24	30	34	37	43	47	52	62
		C.I.	...	...	...	...	...	...	...	...	...	7.7	10	...	15	20	25	30	35	39	44	49	57
180° return bend	Screwed	Steel	2.3	3.1	3.6	4.4	5.2	6.6	7.4	8.5	9.3	11	13	...	...	...	...	...	...	...	...	...	...
		C.I.	...	...	...	...	...	...	...	...	...	9.0	11	...	...	...	...	...	...	...	...	...	...
	Reg. flanged	Steel	...	...	.92	1.2	1.6	2.1	2.4	3.1	3.6	4.4	5.9	7.3	8.9	12	14	17	18	21	23	25	30
		C.I.	...	...	...	...	...	...	...	...	...	3.6	4.8	...	7.2	9.8	12	15	17	19	22	24	28
	Long rad. flanged	Steel	...	...	1.1	1.3	1.6	2.0	2.3	2.7	2.9	3.4	4.2	5.0	5.7	7.0	8.0	9.0	9.4	10	11	12	14
		C.I.	...	...	...	...	...	...	...	...	...	2.8	3.4	...	4.7	5.7	6.8	7.8	8.6	9.6	11	11	13
Globe valve	Screwed	Steel	21	22	22	24	29	37	42	54	62	79	110	...	...	...	...	...	...	...	...	...	...
		C.I.	...	...	...	...	...	...	...	...	...	65	86	...	...	...	...	...	...	...	...	...	...
	Flanged	Steel	...	...	38	40	45	54	59	70	77	94	120	150	190	260	310	390	...	...	...	...	...
		C.I.	...	...	...	...	...	...	...	...	...	77	99	...	150	210	270	330	...	...	...	...	...
Gate valve	Screwed	Steel	.32	.45	.56	.67	.84	1.1	1.2	1.5	1.7	1.9	2.5	...	...	...	...	...	...	...	...	...	...
		C.I.	...	...	...	...	...	...	...	...	...	1.6	2.0	...	...	...	...	...	...	...	...	...	...
	Flanged	Steel	...	...	...	...	...	...	...	2.6	2.7	2.8	2.9	3.1	3.2	3.2	3.2	3.2	3.2	3.2	3.2	3.2	3.2
		C.I.	...	...	...	...	...	...	...	...	...	2.3	2.4	...	2.6	2.7	2.8	2.9	2.9	3.0	3.0	3.0	3.0
Angle valve	Screwed	Steel	12.8	15	15	15	17	18	18	18	18	18	18	...	...	...	...	...	...	...	...	...	...
		C.I.	...	...	...	...	...	...	...	...	...	15	15	...	...	...	...	...	...	...	...	...	...
	Flanged	Steel	...	...	15	15	17	18	18	21	22	28	38	50	63	90	120	140	160	190	210	240	300
		C.I.	...	...	...	...	...	...	...	...	...	23	31	...	52	74	98	120	150	170	200	230	280
Swing check valve	Screwed	Steel	7.2	7.3	8.0	8.8	11	13	15	19	22	27	38	...	...	...	...	...	...	...	...	...	...
		C.I.	...	...	...	...	...	...	...	...	...	22	31	...	...	...	...	...	...	...	...	...	...
	Flanged	Steel	...	...	3.8	5.3	7.2	10	12	17	21	27	38	50	63	90	120	140	...	...	...	...	...
		C.I.	...	...	...	...	...	...	...	...	...	22	31	...	52	74	98	120	...	...	...	...	...
Coupling or union	Screwed	Steel	.14	.18	.21	.24	.29	.36	.39	.45	.47	.53	.65	...	...	...	...	...	...	...	...	...	...
		C.I.	...	...	...	...	...	...	...	...	...	.44	.52	...	...	...	...	...	...	...	...	...	...
	Bell mouth inlet	Steel	.04	.07	.10	.13	.18	.26	.31	.43	.52	.67	.95	1.3	1.6	2.3	2.9	3.5	4.0	4.7	5.3	6.1	7.6
		C.I.	...	...	...	...	...	...	...	...	...	.55	.77	...	1.3	1.9	2.4	3.0	3.6	4.3	5.0	5.7	7.0
	Square mouth inlet	Steel	.44	.68	.96	1.3	1.8	2.6	3.1	4.3	5.2	6.7	9.5	13	16	23	29	35	40	47	53	61	76
		C.I.	...	...	...	...	...	...	...	...	...	5.5	7.7	...	13	19	24	30	36	43	50	57	70
	Re-entrant pipe	Steel	.88	1.4	1.9	2.6	3.6	5.1	6.2	8.5	10	13	19	25	32	45	58	70	80	95	110	120	150
		C.I.	...	...	...	...	...	...	...	...	...	11	15	...	26	37	49	61	73	86	100	110	140
	Sudden enlargement	$h = \frac{(V_1 - V_2)^2}{2g}$ FEET OF LIQUID; IF $V_2 = 0$ $h = \frac{V_1^2}{2g}$ FEET OF LIQUID																					

* Courtesy, "Standards of the Hydraulic Institute," ed. 10. Copyright, 1955, by the Hydraulic Institute, 122 East 42nd St., New York 17, N. Y.

EXAMPLE: Determine the pressure drop in a liquid ammonia line per 100 ft of length when standard ton conditions prevail and average velocity is 10 fps through a $1\frac{1}{4}$-in. pipe.

Data:

Pressure = 169.2 psia

Temp = 86 F

Density = 37.16 lb per cu ft (from Saturated Liquid Ammonia Tables)

Pipe size = 1.278 (from tables of extra strong steel pipe)

Viscosity − z = 0.207 NH_2 @ 86 F sat.

$$Re = \frac{124 \times 1.278 \times 10 \times 37.16}{0.207}$$

$$Re = 284{,}444$$

$$f = 0.0218 \text{ from curve Fig. 322.}$$

$$h_f = \frac{fLV^2}{D2g} = \frac{0.0218 \times 100 \times 100}{0.106 \times 64} = \frac{218}{6.8}$$

$$h_f = 32.06 \text{ ft-head of 86 F ammonia}$$

$$P_d = 32.06 \times \frac{37.16}{144}$$

$$P_d = 8.27 \text{ psi per 100 ft}$$

DROP IN PRESSURE DUE TO BENDS, FITTINGS, AND VALVES

Bends, fittings, and valves add to the resistance of fluids to flow because they effect change of direction and/or present what amounts to an increase in the over-all relative roughness of a system. As a result of their presence, the pressure drop is greater than it would be in any straight run pipe of equal length. The pressure drop increase depends on the type and the number of such components employed.

The resistance of each fitting may be determined and expressed in terms of friction head in the same manner described for straight pipe. The total friction head for any system, then, would be the sum h_{f1}, h_{f2}, h_{f3}, in which h_{f1}, h_{f2}, h_{f3}, etc., represent the friction head presented by the individual bend, fitting, and valve. The friction head of each type of component would be found to equal a specific length of straight pipe of the same diameter and construction material. The length representing the h_f of the individual component would then be established as the equivalent length of straight pipe. The total length of any flow system would be the actual length of straight pipe plus the sum of the equivalent lengths of each component. The friction head factors h_{f1}, h_{f2}, h_{f3}, etc., are replaced by $l_{e1} + l_{e2} + l_{e3}$, etc., in determining P_d where l_{e1}, l_{e2}, l_{e3}, etc., represent the equivalent length of each component.

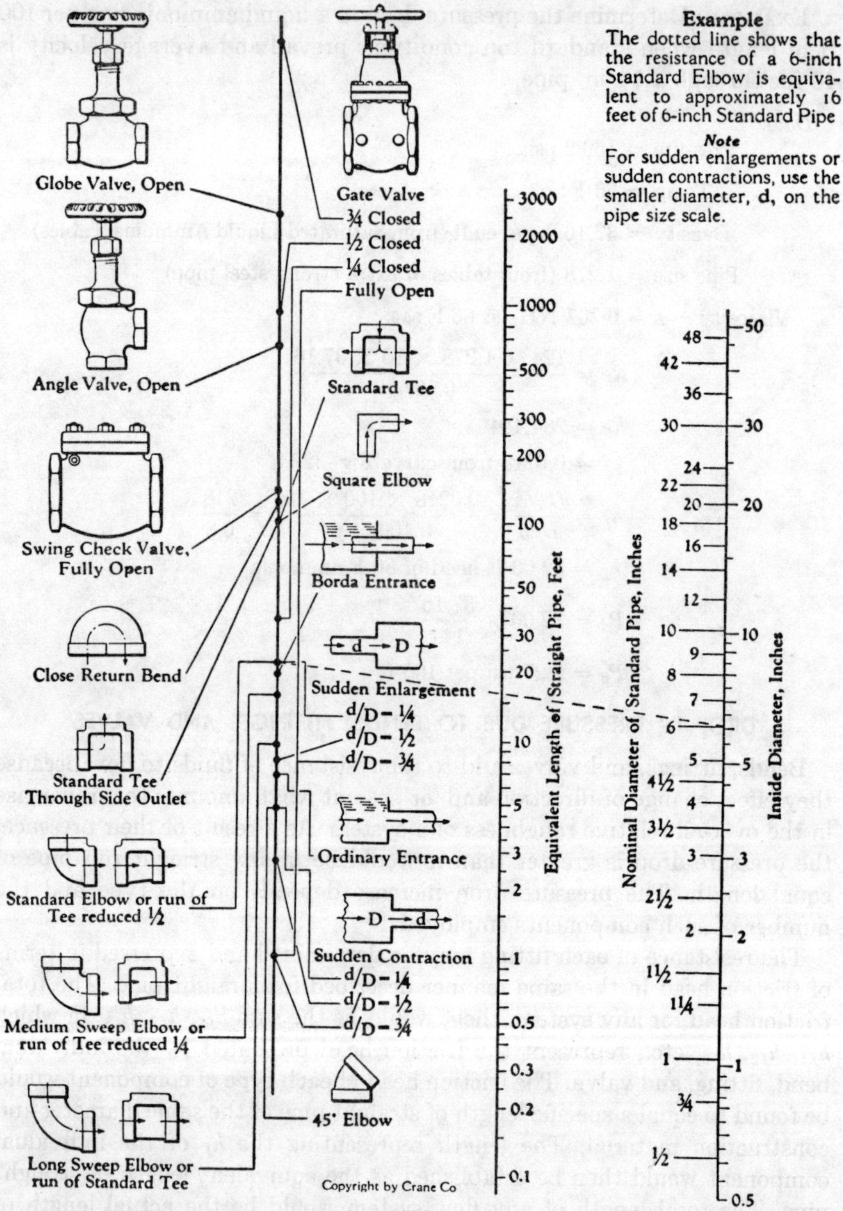

Fig. 366. Chart showing equivalent length of pipe for drop in pressure due to fittings and valves. (Courtesy, Crane Co.)

Equivalent lengths for the various type bends, fittings, and valves have been established and are shown in Table 66. They may also be determined from the chart in Fig. 366.

By connecting a point on the diameter line to a point on the line representing various valves and fittings, the equivalent length in straight pipe for the component selected may be determined. The broken line shown on the chart is for a 6-in. standard elbow and intersects the middle line to show an equivalent length of 16 ft for that fitting.

PRESSURE DROP ACROSS AN EXPANSION VALVE

The available pressure drop across an expansion valve is measured from the point of inlet to the point of outlet, or

$$P_d = p_2 - p_1$$

If the liquid receiver is within a short distance and on the same level as the expansion valve, the pressure drop will be very nearly the difference between head pressure and suction pressure. However, if the two components are separated by (1) any appreciable length of straight pipe, (2) one or more bends, (3) one or more valves (say a king valve), and (4) possibly a riser, then the inlet pressure is no longer the head pressure as found in the receiver or the condenser. Instead, it becomes equal to the receiver pressure minus the sum of the pressure losses sustained by 1, 2, 3, and 4.

EXAMPLE: Consider an ammonia system operating under standard ton conditions; i.e., 154.5 psig head pressure and 19.6 psig suction pressure. The expansion valve is separated from the receiver by a considerable run of straight pipe, several bends, a king valve, and a riser. What is the pressure at the expansion valve and the available pressure drop across its orifice if the individual restrictions are as follows:

$$P_d \text{ in riser} = 12.9 \text{ psi}$$
$$P_d \text{ in pipe} = 6.4 \text{ psi}$$

the latter of which is determined from the actual length of pipe plus the equivalent length of the valve plus the equivalent length of each bend, etc.

Expansion valve inlet pressure is

$$\begin{aligned} P_d \text{ in riser} &= 12.9 \\ P_d \text{ in pipe} &= 6.4 \\ \hline \text{Total } P_d &= 19.3 \text{ psi} \end{aligned}$$

$$154.5 - 19.3 = 135.2 \text{ psig, pressure at inlet}$$
$$135.2 - 19.6 = 115.6 \text{ psi, available pressure drop across expansion valve orifice}$$

TABLE 67. REDUCTION IN EXPANSION VALVE CAPACITIES DUE TO GAS IN LIQUID LINES*†

Refrigerant	*Condenser Pressure*	*Based on Solid Liquid at Condenser Outlet, and 5° Evaporator Temperature*									
		Pressure Drop from Condenser to Valve									
		5 lb	**10 lb**	**15 lb**	**20 lb**	**25 lb**	**30 lb**	**35 lb**	**40 lb**	**45 lb**	**50 lb**
	170 lb	5.9%	11.5%	17.1%	22.4%	27.5%	32.1%	36.4%	40.6%	44.3%	47.5%
Ammonia	200 lb	5.2%	10.2%	14.9%	19.7%	24.2%	28.2%	32.0%	35.4%	39.0%	42.2%
	230 lb	4.4%	8.7%	13.0%	17.0%	20.9%	24.7%	28.3%	31.7%	34.9%	38.3%

* In addition to creating capacity loss, the presence of gas in liquid refrigerant lines will quickly erode valve pins and seats, cause the valve to "hunt," and reduce the efficiency of the evaporator. It is urgently recommended that gas be eliminated by subcooling, or other means, rather than to make allowance for it by oversizing.

† Courtesy, Alco Valve Company.

HORSEPOWER AND HEAD-CENTRIFUGAL COMPRESSORS AND FANS

Of interest here is the work done by a compressor in delivering gas at a higher pressure than found at the inlet to the impeller or cylinder. Any one of three conditions may be considered in computing work of delivery, viz.:

(*a*) Isothermal (constant temperature)
(*b*) Adiabatic (constant entropy)
(*c*) Polytropic (actual conditions under which the work process occurs)

The relation of pressure P, volume V, and temperature T is shown in the pressure-volume diagram (Fig. 24). The polytropic process* is represented by the curve indicative of actual operating conditions. The curves show

$$PV = \text{constant for isothermal processes}$$

$$PV^2 = \text{constant for adiabatic processes}$$

where $y = \text{ratio}\ \dfrac{c_p \text{ (sp ht at constant pressure)}}{c_v \text{ (sp ht at constant volume)}}$

For polytropic conditions—i.e., conditions of actual work process—the exponent may be any value from zero (0) to infinity (∞), depending on the efficiency of the work system (compressor). Represented by the letter n, the value of the exponent may be expressed by the equation

$$\frac{n}{n-1} = \%\ \text{efficiency}\ (E) \times \frac{y}{y-1}$$

* When heat is transferred to or from the gas during compression at the rate expressed by PV^n = constant, the process is known as polytropic compression. The exponent n expresses the actual compression path of the co-ordinates P and V.

and

$$E = \frac{\frac{n}{n-1}}{\frac{y}{y-1}}$$

For an air compression system having an efficiency of 80 percent, then

$$\frac{n}{n-1} = 80\% \times \frac{y}{y-1}$$

$$\frac{n}{n-1} = 80\% \times \frac{1.405}{1.405-1} = 2.77$$

$$n = 1.56$$

The value of n is determined for a given system by experiment. Theoretically, n may have a value less or greater than y; i.e., $n = 0$ to $n =$ infinity (∞). For an isothermal condition, $n = 1$; for an adiabatic condition, $n = y$. For a given work system, n remains constant, as does y, along their own respective pressure-volume paths, so that

$$PV^n = \text{constant}$$

is the equation for a polytropic line. Therefore, the value of n can be determined for a polytropic curve when two points representing (any) co-ordinates of P_1V_1 and P_2V_2 are known. Considering the polytropic curve representing actual conditions (Fig. 24), the value of n may be found by the equation:

$$\frac{P_2}{P_1} = \frac{V_1{}^n}{V_2{}^n} = \left(\frac{V_1}{V_2}\right)^n$$

Since the power of a number is equal to its logarithm times its exponent, then

$$\log \frac{P_2}{P_1} = n \quad \log \frac{V_1}{V_2}$$

Transposing,

$$n = \frac{\log \frac{P_2}{P_1}}{\log \frac{V_1}{V_2}}$$

EXAMPLE: Determine the value of n from a given pressure-volume curve of a system having the co-ordinates

$$P_1 = 40 \text{ psia at } V_1 = 90\% \text{ vol}$$
$$P_2 = 230 \text{ psia at } V_2 = 25\% \text{ vol}$$

$$n = \frac{\log \frac{P_2}{P_1}}{\log \frac{V_1}{V_2}} = \frac{\log \frac{230}{40}}{\log \frac{90}{25}} = \frac{\log 5.75}{\log 3.6} = \frac{0.7596}{0.5563}$$

$$n = 1.365$$

When work is performed adiabatically, the relation of P, V, and T as developed from $P_1V_1^y = P_2V_2^y$ may be expressed

$$\frac{P_2}{P_1} = \left(\frac{V_1}{V_2}\right)^y = \left(\frac{T_2}{T_1}\right)^{\frac{y}{y-1}}$$

$$\frac{V_2}{V_1} = \left(\frac{P_1}{V_2}\right)^{\frac{1}{y}} = \left(\frac{T_1}{T_2}\right)^{\frac{1}{y-1}}$$

$$\frac{T_2}{T_1} = \left(\frac{P_2}{P_1}\right)^{\frac{y-1}{y}} = \left(\frac{V_1}{V_2}\right)^{y-1}$$

Transposing,

$$P_2 = P_1\left(\frac{V_1}{V_2}\right)^y \quad \text{or} \quad P_1\left(\frac{T_2}{T_1}\right)^{\frac{y}{y-1}}$$

$$P_1 = \frac{P_2}{\left(\frac{V_1}{V_2}\right)^y} \quad \text{or} \quad \frac{P_2}{\left(\frac{T_2}{T_1}\right)^{\frac{y}{y-1}}}$$

$$V_2 = V_1\left(\frac{P_1}{P_2}\right)^{\frac{1}{y}} \quad \text{or} \quad V_1\left(\frac{T_1}{T_2}\right)^{\frac{1}{y-1}}$$

$$V_1 = \frac{V_2}{\left(\frac{P_1}{P_2}\right)^{\frac{1}{y}}} \quad \text{or} \quad \frac{V_2}{\left(\frac{T_1}{T_2}\right)^{\frac{1}{y-1}}}$$

$$T_2 = T_1\left(\frac{P_2}{P_1}\right)^{\frac{y-1}{y}} \quad \text{or} \quad T_1\left(\frac{V_1}{V_2}\right)^{y-1}$$

$$T_1 = \frac{T_2}{\left(\frac{P_2}{P_1}\right)^{\frac{y-1}{y}}} \quad \text{or} \quad \frac{T_2}{\left(\frac{V_1}{V_2}\right)^{y-1}}$$

and for polytropic conditions the above expressions may take the form

$$T_2 = \left(\frac{P_2}{P_1}\right)^{\frac{n-1}{n}}$$

where T_2 = temperature, absolute, brought about by polytropic conditions.

There are two expressions related to work done on a fluid; namely, work and head. Usually work is expressed in terms of foot-pounds, or horsepower, while head is expressed in feet of fluid. The relation between work and head may be demonstrated, viz.:

$$\text{Head} = \frac{\text{work performed}}{\text{weight of fluid handled}} = \frac{\text{ft-lb}}{\text{lb}} = \text{ft}$$

Head in feet of gas is the work done per pound of gas.

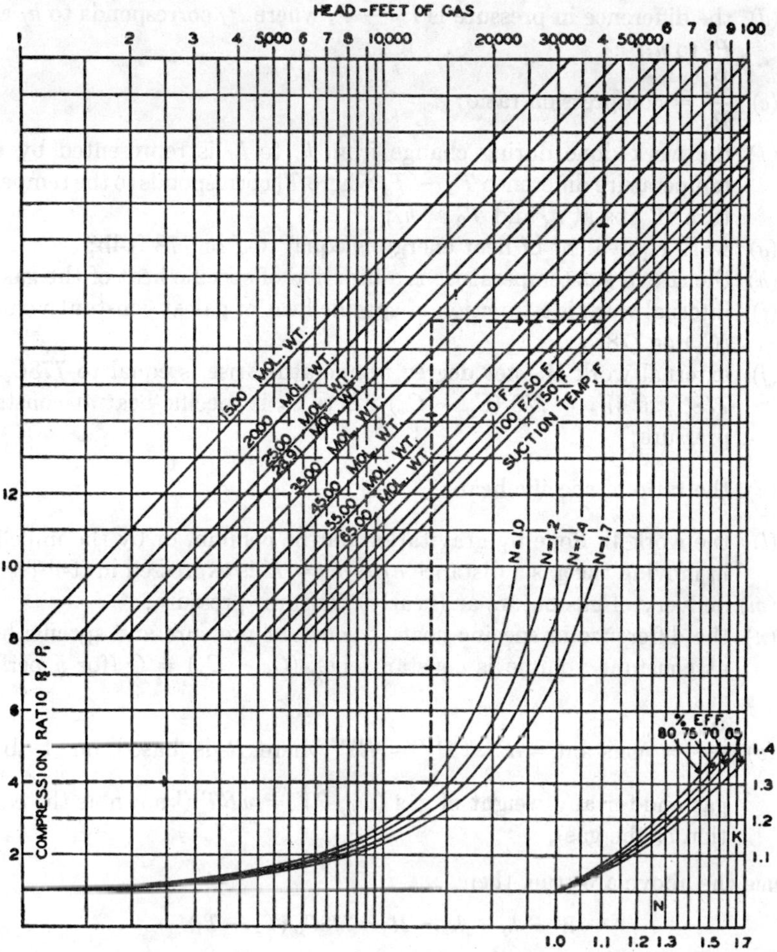

Fig. 367. Approximate head selection chart for compression of gases. (Reprinted from the "Air Conditioning-Refrigerating Data Book 9th Edition—1955–56 Design Volume," by courtesy of The American Society of Heating, Refrigeration and Air Conditioning Engineers.)

The equations for compressor or fan work and head-feet of gas are developed in consideration of the several thermodynamic and physical laws already discussed in this text. Consider, then, that

(*a*) an adiabatic condition exists in doing work on the gas;

(*b*) because an adiabatic condition exists, pressure (P), volume (V), and temperature (T) change during the work process;

(*c*) the head in head-feet of gas (H) is the distance between the base h_f and the top of the column h_i;

(*d*) the difference in pressure is $P_f - P_i$ where P_f corresponds to h_f and P_i to h_i;

(*e*) $\frac{P_2}{P_1}$ = compression ratio;

(*f*) the work done during change from h_i to h_f is represented by the temperature difference $T_2 - T_1$ where T_1 corresponds to the temperature °F abs at h_i and T_2 at h_f;

(*g*) work equivalent of heat energy is equal to J or 778 ft-lb;

(*h*) the rise in gas temperature is a function of specific heat of the gas;

(*i*) internal work is a function of specific heat of gas at constant volume (C_v) or $778C_v$;

(*j*) external work W_e per degree temperature rise is equal to $778C_p - 778C_v$ or $W_e = 778\,(C_p - C_v)$ where C_p is specific heat at constant pressure;

(*k*) the ratio of specific heats $\frac{C_p}{C_v} = y$;

(*l*) the work W done by gravitation on the column, or by the impeller, in moving the gas a distance $h_f - h_i = H$ is expressed in ft-lb;

(*m*) delivery of gas occurs at (nearly) constant pressure;

(*n*) the difference in specific heat at constant pressure and specific heat at constant volume is constant; i.e., $(C_p - C_v) = C$ (for a perfect gas);

(*o*) $\frac{PV}{T}$ = constant $= R \therefore PV = RT$ where R is based on 1 lb of gas and for any weight of gas (w); $PV = wRT$ (known as the equation of the gas).

Since the above are true, then

$$W = h_f - h_i = H = 778C_p\,(T_2 - T_1)$$

work per lb of gas corresponding to the rise in temperature.

External work is equal to the gas constant R for a specific gas so that

$$R = 778(C_p - C_v)$$

and

$$\frac{778C_p}{R} = \frac{C_p}{C_p - C_v} = \frac{y}{y - 1}$$

and

$$\frac{P_2}{P_1} = \left(\frac{T_2}{T_1}\right)^{\frac{778C_p}{R}} \quad \text{or} \quad \frac{P_2}{P_1} = \left(\frac{T_2}{T_1}\right)^{\frac{y}{y-1}}$$

so that

$$\frac{T_2}{T_1} = \left(\frac{P_2}{P_1}\right)^{\frac{y-1}{y}} \quad \text{or} \quad T_2 = T_1\left(\frac{P_2}{P_1}\right)^{\frac{y-1}{y}}$$

Combining the above to determine (1) work W = ft-lb and (2) head H = head-feet of gas, then

$$(1) \qquad W = \frac{778C_p}{778(C_p - C_v)} \times 144P_1 \times V_1 \left[\left(\frac{P_2}{P_1}\right) \times \frac{778(C_p - C_v)}{778C_p} - 1 \right]$$

where W = ft-lb of work performed *on* the gas.

Expressed in the conventional manner,

$$W = \frac{y}{y - 1} \times 144P_1V_1 \left[\left(\frac{P_2}{P_1}\right)^{\frac{y-1}{y}} - 1 \right]$$

To convert to horsepower, hp

$$\text{hp} = \frac{y}{y - 1} \times \frac{144P_1V_1}{33{,}000} \left[\left(\frac{P_2}{P_1}\right)^{\frac{y-1}{y}} - 1 \right]$$

in which P_1 = initial pressure
P_2 = delivery pressure, psia
V = cu ft of gas at P_1

$$(2) \quad H = \frac{778C_p}{778(C_p - C_v)} \times 778(C_p - C_v) \times T_1 \left[\left(\frac{P_2}{P_1}\right) \times \frac{778(C_p - C_v)}{778C_p} - 1 \right]$$

where H = head-feet of gas

Expressed in the conventional manner,

$$H = \frac{y}{y - 1} RT_1 \left[\left(\frac{P_2}{P_1}\right)^{\frac{y-1}{y}} - 1 \right]$$

or for actual (polytropic) work systems where the value of y becomes n as determined by experiment, then

$$H = \frac{n}{n - 1} RT_1 \left[\left(\frac{P_2}{P_1}\right)^{\frac{n-1}{n}} - 1 \right]$$

and where conditions require consideration of the factor of compressibility, Z, which is a function of the ratio of actual pressure and temperature to critical pressure and temperature of the gas, then

$$H = \frac{n}{n - 1} ZRT_1 \left[\left(\frac{P_2}{P_1}\right)^{\frac{n-1}{n}} - 1 \right]$$

where H = head-feet of gas
Z = compressibility factor (varies from 0.90 to 1.0)
R = gas constant
T_1 = initial or inlet temperature °F abs
P_1 = initial or inlet pressure psi
P_2 = final or discharge pressure psi
n = ratio $\frac{C_p}{C_v}$ modified for polytropic conditions

EXAMPLE: What horsepower is required to raise the pressure of 1000 cu ft of air per minute from 14.7 psia to 58.8 psia adiabatically?

(y = 1.405 for air)

$$hp = \frac{y}{y-1} \times \frac{P_1V_1}{33{,}000}\left[\left(\frac{P_2}{P_1}\right)^{\frac{y-1}{y}} - 1\right]$$

$$= \frac{1.405 \times 144 \times 14.7 \times 1000}{0.405 \times 33{,}000}\left[\left(\frac{58.8}{14.7}\right)^{\frac{1.405-1}{1.405}} - 1\right]$$

$$= 222.58\,(4^{0.288} - 1)$$

$$\log 4 \times 0.288 = 0.602060 \times 0.288 = 0.1734$$

Number corresponding to 0.1734 = 1.491

$$= 222.58 \times 0.491$$

$$= 109.29 \text{ hp}$$

EXAMPLE: Determine the head-feet of gas for the above compressor when the inlet temperature of air is 100 F. Assume n = 1.3 (polytropic conditions) and Z = 1.0.

$$H = \frac{n}{n-1} ZRT\left[\left(\frac{P_2}{P_1}\right)^{\frac{n-1}{n}} - 1\right]$$

Data:

$$n = 1.3$$

$$Z = 1.0$$

$$T = 100 \text{ F} + 460 \text{ F} = 560 \text{ F abs}$$

$$R = 53.3 \text{ for air*}$$

$$H = \frac{1.3}{1.3-1} \times 1.0 \times 53.3 \times 560\left[\left(\frac{58.8}{14.7}\right)^{\frac{1.3-1}{1.3}} - 1\right]$$

$$= 128{,}999\,[(4)^{0.23} - 1]$$

$$\log 4 \times 0.23 = 602060 \times 0.23 = 0.1384$$

Number corresponding to 0.1384 = 1.375

$$= 128{,}999 \times 0.375$$

$$= 48{,}374 \text{ ft-hd of gas}$$

TORQUE-HORSEPOWER RELATIONSHIP

The torque of an electric motor is proportional to motor horsepower.

Torque or twisting effort of a rotative body is expressed by the equation $f \times d$, where f is the acting force and d is the radius or distance between the point of applied force and the axis about which the force acts. The force f expressed in pounds and the radius in inches or feet creates the expression for torque as being lb-ft or lb-in.

* Gas constant R can be determined by the following methods:
(1) As shown in Chapter 4, "Gases"
(2) $R = 778(C_p - C_v)$
(3) $R = \dfrac{1544}{\text{molecular weight of the gas}}$

for air $R = \dfrac{1544}{28.95} = 53.3$

If the value of f is large enough relative to the rotative load, the body will turn. However, the value of f must become greater as the value of d diminishes in order to create a constant twisting effort. Assuming f to be great enough at the point of application, which may be at any radius d, revolving motion occurs. Work will now be performed. The rate at which work is performed depends on the speed of rotation or the distance moved in feet per minute (velocity). If the applied force and the rotating velocity are known, horsepower may be determined.

For example, an electric motor has a speed of 1750 rpm when the effective force on the rotor windings located 6 in. from the center of the shaft is 6 lb. The torque of the motor would be

$$\begin{aligned} \text{Torque} &= f \times d \\ &= 6 \text{ lb} \times 6 \text{ in.} = 6 \text{ lb} \times 0.5 \text{ ft} \\ &= 36 \text{ lb-in.} = 3 \text{ lb-ft} \end{aligned}$$

The horsepower rating of an electric motor having a torque of 36 in.-lb may be determined by the equation

$$\text{np} = \frac{\text{force in lb} \times \text{radius in ft} \times \text{rpm} \times \pi \times 2}{33{,}000}$$

$$\text{hp} = \frac{\text{torque (lb-ft)} \times \text{rpm} \times \pi \times 2}{33{,}000}$$

$$\text{hp} = \frac{3 \times 1750 \times 3.1416 \times 2}{33{,}000}$$

$$\text{hp} = 1 \text{ (approx)}$$

It may be said, then, that a 1-hp, 1750-rpm electric motor has an equivalent running torque of 36 in.-lb. Motor torque being proportional to motor hp, the torque of variously rated motors would be as shown in Table 64 when the speed is 1750 rpm.

TABLE 68. TORQUE-HORSEPOWER EQUIVALENCY

Motor Horsepower at 1750 rpm	*(Proportion)*	*Torque in.-lb*
$\frac{1}{4}$	$(\frac{1}{4} \times 36)$	9
$\frac{1}{3}$	$(\frac{1}{3} \times 36)$	12
$\frac{1}{2}$	$(\frac{1}{2} \times 36)$	18
$\frac{3}{4}$	$(\frac{3}{4} \times 36)$	27
1	(1×36)	36
2	(2×36)	72
3	(3×36)	108
5	(5×36)	180
	etc.	

To obtain the same horsepower rating at lower or higher speeds, the torque value must be increased or decreased, respectively, in the same proportion to such reduction or increase in speed. This applies whether speed changes are made within the motor or by some external component such as chains, belts, or gears. Also, by reducing the speed of a motor by the latter external component, an increase in torque becomes available. Thus, at a reduction ratio of 5 to 1 (from 1750 rpm to $\frac{1750}{5} = 350$ rpm), the available torque of a 1-hp motor increases from 36-in. lb to $5 \times 36 = 180$ in.-lb.

Again, say it is desirable to obtain a final speed of 350 rpm and 900 in.-lb torque is required. This means that a reduction of $\frac{1750}{350} = 5$ to 1 is necessary. Dividing 900 in.-lb by 5, it is found that a motor having a torque of 180 in.-lb at 1750 rpm is necessary to fill the requirement. From Table 68 the requirement corresponds to a 5-hp motor.

For any combination of conditions, the torque, the speed, and the horsepower may be determined as follows:

$$\text{hp} = \frac{\text{torque} \times \text{rpm}}{36 \text{ in.-lb} \times 1750 \text{ rpm}} = \frac{\text{torque} \times \text{rpm}}{63{,}000}$$

$$\text{torque} = \frac{36 \text{ in.-lb} \times 1750 \text{ rpm} \times \text{hp}}{\text{rpm}} = \frac{63{,}000 \times \text{hp}}{\text{rpm}}$$

$$\text{rpm} = \frac{36 \text{ in.-lb} \times 1750 \text{ rpm} \times \text{hp}}{\text{torque}} = \frac{63{,}000 \times \text{hp}}{\text{torque}}$$

where torque is expressed as in.-lb.

EXAMPLE: A compressor having a crank throw of 4 in. works against a maximum piston pressure of 200 psi. The total load using a 4-in. diameter piston would be

$$200 \times \frac{\pi d^2}{4} = 200 \times \frac{3.1416 \times 4^2}{4} = 2513.28 \text{ lb}$$

Crank torque = 2513.28 lb × 4 in. = 10,053 in.-lb. Disregarding friction, the hp requirements at 300 rpm would be:

$$\text{hp} = \frac{\text{torque} \times \text{rpm}}{63{,}000} = \frac{10{,}053 \times 300}{63{,}000}$$

$$\text{hp} = 47.87$$

SUMMARY OF CHAPTER 11

Heat Lost Equals Heat Gained

Standard Measurement of Refrigeration Load:

1 standard ton equals 200 Btu per min

Temperature Conversions:

$$°F = (9/5 \times °C) + 32, \qquad °C = 5/9 \times (°F - 32)$$

$$°R = 4/5 \times °C \qquad \text{or} \qquad °R = 4/9 \times (°F - 32)$$

Heat-Work Equivalents:

$$1 \text{ hp} = 42.4 \text{ Btu per min}$$

$$1 \text{ kw} = 56.8 \text{ Btu per min}$$

$$1 \text{ Watt} = 0.0568 \text{ Btu per min}$$

Specific Heat Factors:

Thermal kinetic energy (Btu) = weight of substance × specific heat × (final temperature − initial temperature)

Resultant Temperatures:

$$\text{Heat lost} = \text{heat gained}$$

$$\text{Weight} \times \text{sp ht} \times (t_1 - t_2) = \text{weight} \times \text{sp ht} \times (t_2 - t_1)$$

Refrigerating Effect:

Latent heat of refrigerant at evaporator pressure minus Btu loss in thermal kinetic energy of liquid in cooling from liquid temperature to evaporator temperature.

Heat Capacity of Refrigerants:

$$\frac{\text{Heat load per hour}}{\text{Refrigerating effect} \times 60} = \text{lb of refrigerant per min}$$

Condenser Load Calculations:

Btu in superheated gas − enthalpy of liquid at condenser temperature = heat to be extracted by condenser

$$\text{Amount of coolant required} = \frac{\text{heat to be extracted by condenser}}{\text{sp ht of coolant} \times \text{temperature differential}}$$

Latent Heat of Fusion—Food Products:

$$\frac{\%\ \text{water content by weight} \times 144}{100} = \text{latent heat of fusion}$$

Infiltration:

Heat load due to leakage of air

Heat Load—Humans:

Heat emitted by people adds to the total heat load of any compartment under refrigeration. People at work emit a greater amount of heat than do people at rest.

Heat Transfer Calculations:

Evaporators, condensers, heat exchangers, etc.

Coefficient of heat transfer:

$$U = \text{Btu per hr per sq ft per °F}$$

Area of heat transfer element for constant fluid temperatures:

$$A = \frac{\text{total heat transmitted}}{U \times t_d}$$

$$t_d = \text{temperature of warm fluid} - \text{temperature of cold fluid}$$

Area of heat transfer elements for variable temperatures:

$$A = \frac{\text{total heat transmitted}}{U \times \text{mtd}}$$

mtd = mean temperature differential

$$= \text{highest temperature} - \frac{\text{lowest temperature difference}}{2.3 \log_{10} \times \dfrac{\text{highest temperature difference}}{\text{lowest temperature difference}}}$$

Evaporator Pressure Determination:

Evaporator pressure depends on the temperature to be maintained in the evaporator. The pressure-temperature relationship varies with each kind of refrigerant.

Specific Volume Determinations:

$$\text{Specific volume} = \frac{1}{\text{density}}$$

Expressed in cubic feet per lb of refrigerant
Varies inversely as pressure

Condenser Pressure Determinations:

Condenser pressure depends on condensing temperature, which, in turn, depends on the temperature and the amount of cooling medium available. As temperatures rise, the condenser pressure must also rise correspondingly in order to effect condensation.

Compressor Discharge Temperatures:

$$T_2 = T_1\left(\frac{P_2}{P_1}\right)^{\frac{\nu-1}{\nu}}$$

Compressor discharge temperatures increase as compressor ratio increases; also rises as suction temperature becomes higher, for similar compression ratio.

Compression Ratio:

$$\text{Compression ratio} = \frac{\text{final or discharge pressure, psia}}{\text{initial or suction pressure, psia}}$$

Efficiency of compressors decreases as compression ratio increases.

Amount of Refrigerant Required Per Ton of Refrigeration:

$$\text{Lb of refrigerant per ton} = \frac{200}{\text{refrigerating effect per lb}}$$

Cubic feet of refrigerant per ton = lb × specific volume at evaporator temperature

Thermal Work Done by Compressor:

Work done by compressor per lb of refrigerant circulated = Btu per lb of superheated vapor at discharge pressure minus Btu per lb saturated vapor at suction temperature and pressure.

Coefficient of Performance:

$$\text{cop} = \frac{\text{energy expended}}{\text{useful energy}}$$

$$= \frac{T_1}{T_2 - T_1}$$

$$= \frac{H_1}{H_2 - H_1}$$

Horsepower Per Ton of Refrigeration:

$$\text{hp per ton} = \frac{\text{thermal work of compressor} \times \text{lb refrigerant per ton per minute} \times 778 \text{ ft-lb}}{33{,}000}$$

Mean Effective Pressure:

$$\text{mep} = \frac{y}{y-1} \times P_1 \left[\left(\frac{P_2}{P_1} \right)^{\frac{y-1}{y}} - 1 \right]$$

or

$$\text{mep} = P_1 \times \text{coef or mep}$$

where P_1 = suction pressure

$$\text{mep} = \text{constant} \times \text{density} \times (H_2 - H_1)$$

or

$$\text{mep} = \text{initial pressure} \times \text{coefficient of mep}$$

Indicated Horsepower for One End of Cylinder:

$$\text{ihp} = \frac{\text{mep} \times A \times \text{rpm} \times s}{33{,}000}$$

where A = area of piston, square inches
s = length of stroke in feet

$$\text{Total ihp} = \text{ihp (head end)} + \text{ihp (crank end)}$$

Volumetric Efficiency of Compressor:

$$\text{Volumetric efficiency} = \frac{\text{actual capacity}}{\text{theoretical capacity}} \times 100$$

$$= \frac{\text{amount of gas delivered}}{\text{piston displacement}}$$

$$\text{Total volumetric efficiency} = E_c \times E_f \times E_s$$

where E_c = vol. eff. due to clearance
E_f = vol. eff. due to friction
E_s = vol. eff. due to slippage

$$\text{Piston displacement per min} = \frac{\pi d^2}{4} \times s \times \text{rpm}$$

where d = diameter of piston
s = length of stroke
rpm = speed in revolutions per minute

Compressor Capacity:

$$\text{Actual compressor capacity} = \text{piston displacement} \times \frac{\text{vol. eff.}}{100}$$

$$\text{Size of compressor} = \frac{\text{cu ft per min required}}{\text{total volumetric efficiency}} \times 100$$

Pressure Drop in Liquid Risers:

$$\text{Pressure drop psi} = \frac{\text{height, ft} \times \text{wt, lb per cu ft}}{144}$$

Pressure Drop Due to Pipe Friction:

$$h_f = \frac{fLV^2}{D2g}$$

where f = friction factor
L = pipe length, ft
V = fluid velocity, fps
D = inside pipe diameter, in.
h_f = loss of head, ft
g = gravity factor = 32.2 fps per sec

Friction factor based on Reynolds number

$$R_e = \frac{DVp}{u}$$

where R_e = Reynolds number
p = fluid density at flow conditions
u = absolute viscosity

Knowing R_e and internal pipe condition, friction factor f may be determined from friction factor curve, Fig. 362 and Fig. 363.

Also

$$h_f = 0.2083 \left(\frac{100}{C}\right)^{1.85} \times \frac{q^{1.85}}{d^{4.8655}}$$

where C = 100 for new clean pipe

Drop in Pressure in Pipes Due to Bends, Fittings, and Valves:
Is expressed in equivalent length in feet of pipe. Determined from table or chart based on tests.

Drop in Pressure Across Expansion Valve:
Is equal to the difference in pressure between expansion valve inlet and outlet. The inlet pressure is equal to head pressure minus losses due to friction, risers, etc.

Horsepower-Gas Compression Machines:

$$\text{hp} = \frac{y}{y-1} \times \frac{144P_1V_1}{33{,}000}\left[\left(\frac{P_2}{P_1}\right)^{\frac{y-1}{y}} - 1\right]$$

where P_1 = initial pressure psia
P_2 = delivery pressure psia
V = cu ft of gas at P_1
$y = \frac{C_p}{C_v}$ (ratio of specific heats)

Head in Feet of Gas Corresponding to Compression Ratio $\frac{P_2}{P_1}$:

$$H = \frac{y}{y-1} \times RT_1\left[\left(\frac{P_2}{P_1}\right)^{\frac{y-1}{y}} - 1\right]$$

where H = head in feet of gas
R = gas constant
T_1 = initial temperature F abs
P_2 = delivery pressure psia
P_1 = initial pressure psia
y = the exponent for adiabatic conditions.

or

$$H = \frac{n}{n-1} \times ZRT_1 \left[\left(\frac{P_2}{P_1} \right)^{\frac{n-1}{n}} - 1 \right]$$

where n = the exponent for polytropic conditions
Z = the compressibility factor

Torque-Horsepower Relationship:

$$\text{hp} = \frac{f \times r \times \text{rpm} \times \pi \times 2}{33{,}000}$$

$$\text{hp} = \frac{\text{torque} \times \text{rpm} \times \pi \times 2}{33{,}000}$$

$$\text{hp} = \frac{3\ (\text{lb-ft}) \times 1750 \times 3.1416 \times 2}{33{,}000}$$

$$\text{hp} = 1\ (\text{approx})$$

With a torque of 36 in.-lb and a speed of 1750 rpm the horsepower equivalency equals 1.0.

$$\text{hp} = \frac{\text{torque} \times \text{rpm}}{36\ \text{in.-lb} \times 1750} = \frac{\text{torque} \times \text{rpm}}{63{,}000}$$

$$\text{torque} = \frac{36\ \text{in.-lb} \times 1750\ \text{rpm} \times \text{hp}}{\text{rpm}} = \frac{63{,}000 \times \text{hp}}{\text{rpm}}$$

$$\text{rpm} = \frac{36\ \text{in.-lb} \times 1750\ \text{rpm} \times \text{hp}}{\text{torque}} = \frac{63{,}000 \times \text{hp}}{\text{torque}}$$

STUDY OUTLINE FOR CHAPTER 12

Insulation versus conduction

Methods of heat transmission:
(a) Conduction
(b) Convection
(c) Radiation

Heat transmission through separating walls—controlling factors:
(a) Temperature differential
(b) Thickness of wall
(c) Wall material
(d) Velocity of fluids
(e) Character of wall surface
(f) State of fluids

Newton's law of radiation

Black body a perfect radiator:
(a) Absorptivity
(b) Emissivity
(c) Reflectivity

Kirchhoff's law:
Absorptivity = Emissivity

Rate of heat transmission—expressed as:
(a) Thermal coefficient of conductivity (k)
(b) Thermal coefficient of conductance (C)
(c) Thermal coefficient of surface conductance (f)
(d) Thermal coefficient of conductance of air spaces (a)
(e) Thermal coefficient of over-all transmission (U)

Resistivity (R):
Reciprocal of heat transmission
Resistivity or total resistance to heat flow through any wall equals the sum of the separate resistances of each layer or each material.

Mathematical determination of total heat transfer—building material:

$H = U \times \text{area} \times \text{temperature differential}$

when H = Btu per hr

Insulation materials:
Thermal properties
Structural and economic consideration

Types of insulation materials

Cork equivalents

Heat-transfer equipment (functional apparatus)—conductors of heat:
Evaporators
Condensers
Heat exchangers
After-coolers
Etc.

Conductors:
Metal tubes or piping:
Copper, iron, steel, aluminum, etc.

Heat-transfer factors—in equipment:
(a) Fluid factor
(b) Tube factor
(c) Temperature factor
(d) Surface factor
(e) Design factor

Mathematical determination of surface area requirements—in equipment:

$$A = \frac{\text{refrigeration load (Btu)}}{U \times \text{mean temperature difference}}$$

where A = sq ft of tube surface

Over-all coefficient of heat transfer—in equipment (U):
(a) Internal surface conductance
(b) Conductivity of tube material
(c) External surface conductance
(d) Scale factors (internal and external)

Heat transmission through compound cylinder wall

12

Heat Transfer

The tendency of heat to seek its own level by inducing flow from a region of high temperature to one of lower temperature gives rise to another important phase in the study of refrigeration. This phase involves the rate of heat transmission through walls of many different materials employed either in the functional equipment of the refrigeration system, such as condensers and evaporators, or for the construction of such compartments in which low temperatures are to be secured and maintained.

The *desired* rate at which heat is to be transmitted depends on the particular function to be performed; the rate at which it is *capable* of being transmitted or retarded is governed by a number of factors, the details of which comprise the subject matter of this chapter.

PRINCIPLES INVOLVED IN HEAT TRANSMISSION

The transmission of heat through a wall separating a region of high temperature from a region of lower temperature is influenced by conduction, radiation, and convection. Therefore, it is necessary that the combined effect of all three methods of heat transfer be considered when calculating heat loads due to this source. This rule applies to pipe or tubing used in evaporators and condensers as well as to walls, ceilings, and floors of cold-storage compartments or air-conditioned buildings.

While heat actually is transmitted by conduction through the homogeneous or heterogeneous mass that constitutes wall thickness, the amount transmitted depends on (1) the quantity conveyed to and from the wall surfaces by means of radiation and convection, (2) the absorptive and emissive properties of external and internal surfaces, respectively, (3) resistance to heat flow of the wall material, (4) wall thickness, (5) temperature difference between wall surfaces, and (6) kind and density of separated fluids.

If the effort is toward a reduced rate of heat transfer, as it would be in considering building materials, then only those materials possessing low

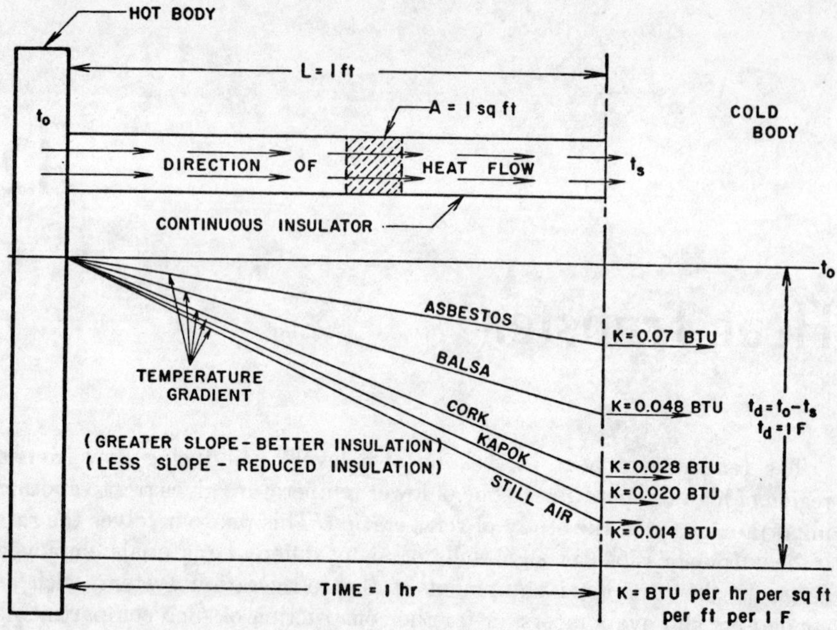

Fig. 368. Heat transfer by conduction. Relative insulating qualities of different materials based on Btu per hr per sq ft per ft per 1 F.

conduction capacity should be used. Structural components having such a property are known as insulation materials, the more common of which include cork, mineral wool, redwood bark, sawdust, etc. However, if rapid heat transfer is desired, the employment of high conduction materials becomes essential. Metals, being the best heat conductors, are used in functional equipment of mechanical refrigeration systems and include such excellent ones as copper, aluminum, steel, and iron.

METHODS OF HEAT TRANSFER

As stated previously, there are three methods of heat transfer; namely, conduction, radiation, and convection.

Conduction of heat involves molecular vibration within a mass or conductor. Being greatest at the point or area of contact with the heat source, molecular vibration diminishes as transmission progresses from one layer of molecules to another farther from the source. The highest rate of heat transfer by conduction is obtained in dense materials in which the molecules exist in closer contact with each other. For this reason, solids—particularly metals, such as silver, copper, and aluminum—prove to be the best conductors of heat energy as compared with liquids and gases, the latter being the least effective in this respect.

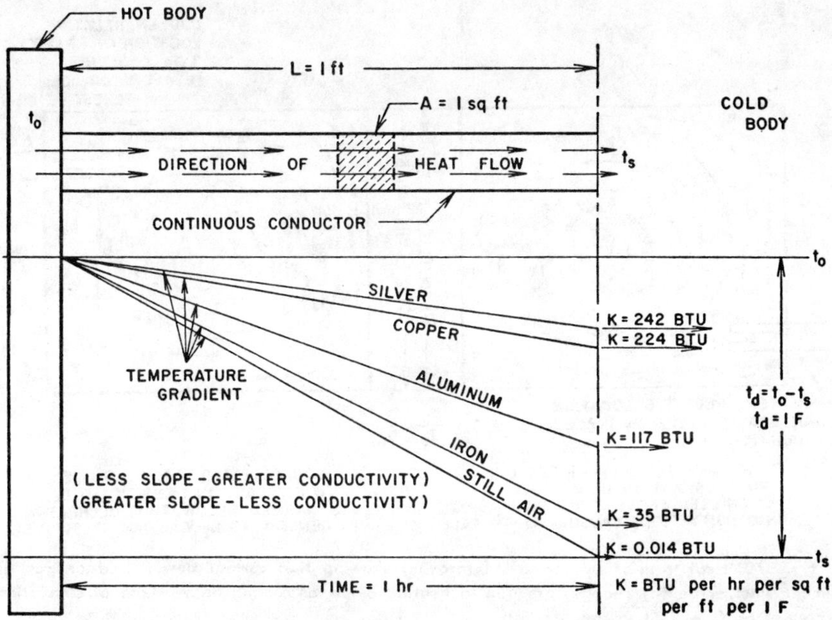

Fig. 369. Heat transfer by conduction. Relative heat-conducting qualities of different materials based on Btu per hr per sq ft per 1 F.

Convection is heat transfer within a fluid, either liquid or gaseous, promoted by density variations alone, as in *natural convection*, or by mechanical means, involving the use of a pump or blower, as in *forced convection*.

Applying heat to the bottom of a container of liquid establishes a vibratory motion throughout the liquid mass. Vibration reaches its greatest intensity in that portion of liquid immediately in contact with the area of heat concentration, tending to separate the molecules at that point to a greater degree than at any other location within the liquid mass. This results in expansion of the liquid in the heated area, causing it to rise within the mass to be replaced by the colder—and consequently heavier—liquid from above. As the new portion becomes heated, it, too, becomes lighter and in turn is replaced by still another portion from above. As heat constantly is applied, the rotation of the molecules from the heated area to the liquid mass above becomes continuous, so that a steady current is established over the heated area. The rate at which this current flows depends on the relative temperatures within the liquid mass becoming greater as the difference between top and bottom layers increases. This action would be the same if air in a room were substituted for the liquid in a container.

It is obvious that the highest rate of heat transfer by natural convection

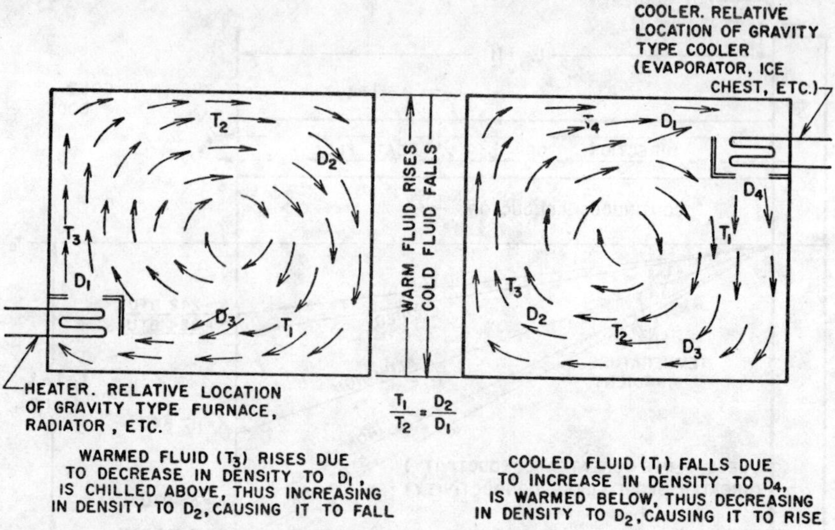

Fig. 370. Heat transfer by convection (gravity) showing that current flow (liquid or gas) is similar whether density changes are due to heating below or cooling above. Rate of circulation increases as $T_2 - T_1$ becomes greater.

is obtained when heat is applied to the lower region of the fluid mass. If, instead of deliberately adding heat to the bottom of the container, a cooling element were placed at the top, then the same rotational movement would be established for the reason that a higher density in the upper region is accomplished. In the design of thermal apparatus utilizing natural convection as the method of heat transfer, such as the domestic refrigerator, the cooling element or evaporator must always be placed in the top of the compartment. The same principle applies to a home heating system wherein hot water or air is circulated by gravity from the heater located in the basement.

If a stream of fluid is directed through a heating or cooling element by an external force, such as a fan or a pump, for the purpose of absorbing or dissipating heat, the action is considered one of forced convection. This method allows a greater latitude of heat distribution or concentration and promotes a more rapid heat transfer as a result of constant and wider temperature differentials as well as greater turbulence of the fluid at the thermal surfaces.

Radiation, the third method of heat transfer, does not require either a solid or a fluid medium for its transmission from source to receiver. The energy so transmitted is emitted by a body of highest intensity and absorbed by surrounding bodies of lower intensities. Energy so transmitted travels directly from the warm to the cooler substance without necessarily

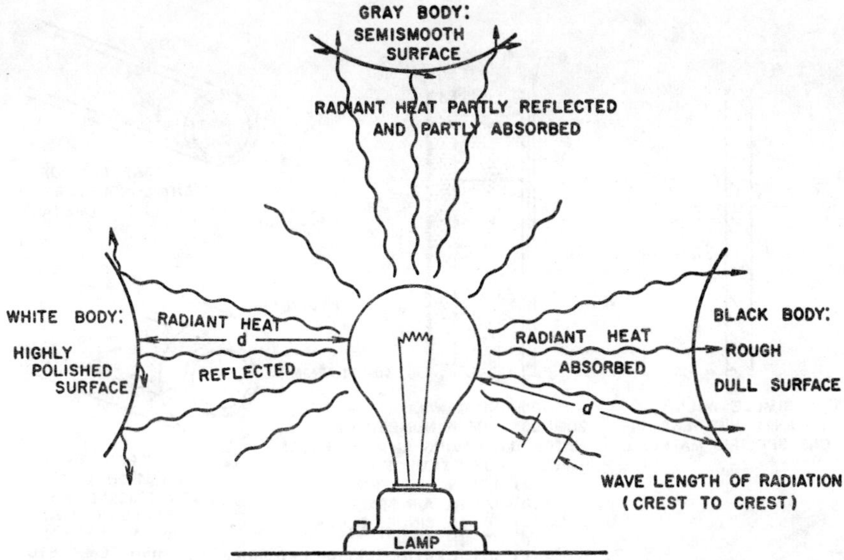

Fig. 371. Heat transfer by radiation. Radiant heat varies inversely as the square of the distance *d*. Wave length in magnitudes of 40 to 85 millionths inch depending on temperature of emanating body.

affecting the ether through which it passes. To prevent the transmission of radiant energy from its source to some specific object, a screen or block must be placed directly in the path of the two objects. However, in doing so only an absorber or reflector and not a destroyer of radiant energy is provided.

The amount of heat radiating from an object (radiator) in a unit time interval is governed by the temperature of that object modified somewhat by the character and the nature of its surface. The amount of radiant heat absorbed by an opposing colder body depends on its temperature and distance from the radiator. It is also modified by the character and the nature of its surface.

Newton's law states that *the rate of loss of heat by radiation is proportional to the difference between the temperature of the radiator and the surrounding substances*. While there is some question as to the accuracy of this law under any range of temperatures, it nevertheless holds true for practical refrigeration purposes.

It may also be stated in this connection that *the intensity of radiant heat is proportional to the temperature of the radiator and inversely proportional to the square of the distance from the radiator*. This means that more heat is emitted from a stove in which a fire is roaring than from one banked for the night. Also, a person standing 6 ft from a stove benefits only one quarter as much as a person standing 3 ft away.

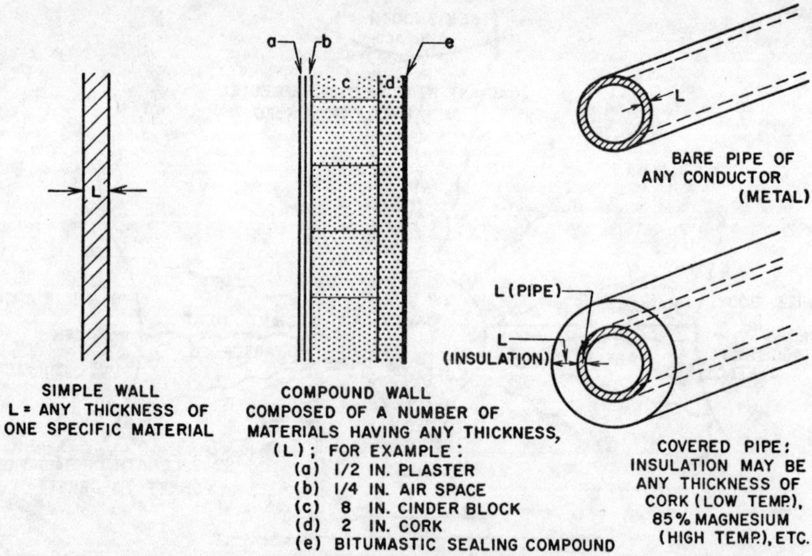

Fig. 372. Examples of separating walls.

HEAT-TRANSFER FACTORS—THROUGH A SEPARATING WALL

Six basic factors govern the rate of heat transmission through a wall separating regions of high and low temperature. They are discussed here in detail.

Temperature Differential. The greater the difference in temperature between two regions separated by a wall, the greater will be the rate at which heat will tend to flow. Comparing a heat exchanger with a water flow system, temperature differential in the former would be analogous to the effective pressure in the latter.

The amount of flow in any system, then, would depend on the effective head. Where H represents the rate of flow in a heat-transfer system, it is true that

$$H \text{ varies as } Td$$

where Td = temperature difference between thermal regions $(t_2 - t_1)$

Wall Thickness. Distance of movement in any flow system governs the rate at which flow occurs. Thus, the thickness of the separating wall is one of the determining factors governing the rate at which heat will flow from a higher to a lower temperature region. Under similar conditions and like materials, the resistance to heat flow increases with thickness. Conversely, a decrease in wall thickness increases the rate of heat transfer per unit area of exposed surface. In a water flow system, the length of pipe through which movement must occur would be analogous to wall

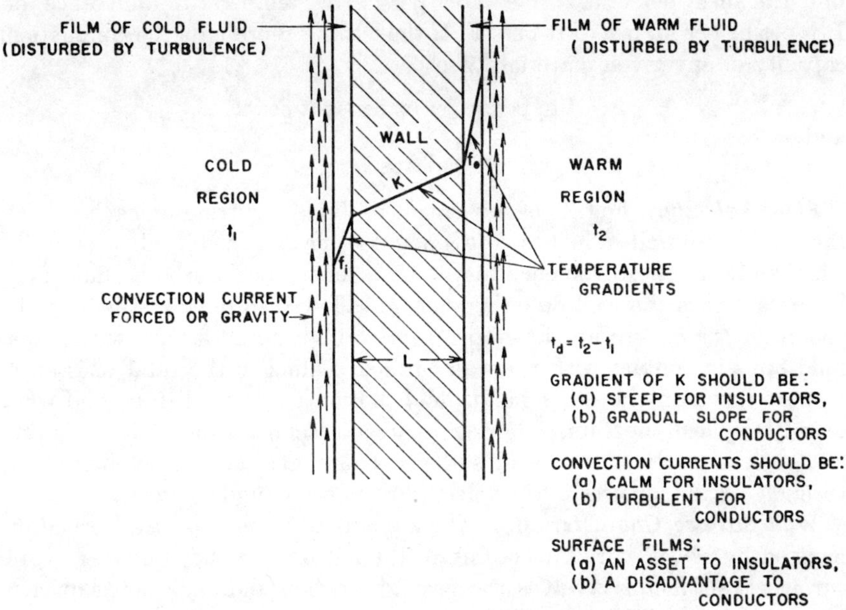

Fig. 373. Heat transfer through a wall separating cold and warm fluids. Represents wall of a room, building, pipe or tube thickness, etc.

thickness in the heat-transfer system. In this case, a long run presents greater total resistance to flow than a short run. Therefore, the rate of heat flow per unit area H varies inversely as the thickness of any given material, or

$$H \text{ varies as } \frac{1}{L}$$

where L = thickness of wall in inches (or feet*)

Wall Material. Materials used for separating walls vary in their ability to transmit heat. Therefore, the type of material employed in wall construction is another factor that determines the rate at which heat will pass through it, which in turn classifies a particular material as a conductor or an insulator.

Individual materials possess their own peculiar heat-conduction capacities which remain nearly constant under any specific set of conditions. Thus, for each degree of temperature difference, the rate at which heat may be conducted through a definite thickness of a given material will remain constant per unit area. Therefore, the rate of heat transfer through a wall varies as the conduction capacity of the material employed. This capacity is expressed as the number of Btu conducted per unit of time per

* As the inch is the most commonly used unit of measurement in refrigeration work, L will represent thickness in inches.

unit thickness per unit surface area per degree temperature difference, or Btu per hr per in. per sq ft per °F. If the letter k represents the conduction capacity of any given material, then

$$k = \text{Btu per hr per in. per sq ft per °F}$$

and

$$H \text{ varies as } k$$

Fluid Velocity. The rate of movement of fluids occurring on each side of the separating wall influences the amount of heat transmitted through it. This is due to the turbulence factor that tends to disturb the fluid film in contact with the wall on either side. If allowed to remain still, a static condition *tending* toward a temperature equilibrium is set up within the fluid films in contact with the wall surface, adding to the total resistance to heat flow. On the other hand, fluid motion tends to disturb and constantly replace the film, allowing a greater temperature differential to exist continuously at the wall surface. Therefore, the rate of heat flow through a wall increases as the velocity of opposing fluids increases.

Wall Surface Characteristics. The amount of radiant heat absorbed or emitted by a wall varies according to the nature and the character of its surface. This factor involves the heat absorption and radiation capacity of walls having different surface finish characteristics, including such elements as color, brightness, and degree of smoothness. Bright and polished finishes reflect a greater portion of radiant heat that falls on such surfaces, which therefore have less heat to radiate in turn than walls having dull or rough finishes. Likewise, a black surface absorbs and is capable of radiating a greater portion of radiant heat falling on it than does a light-colored surface.

A *perfect radiator* would be a black body that absorbs all radiant heat impinging on it without radiating or transmitting any of the heat absorbed. Such a body, of course, whose absorption capacity or *absorptivity* would have a value of 1.00 or 100 percent, is only theoretical.

Absorptivity of a body is the ratio of the amount of radiant heat it is capable of absorbing to the amount capable of being absorbed by a black body under similar thermal conditions.

Emissivity is the ratio of the amount of heat radiated by a body to that radiated by a black body under similar thermal conditions. An absolute black surface would have an emissivity of 1.00. Emissivity is expressed in terms of the ratio

$$\frac{\text{Radiating power of any surface}}{\text{Radiating power of a black surface}}$$

Reflectivity of a body is the ratio of radiant heat reflected by that body to that reflected by a black body under similar thermal conditions. A body whose absorptivity is 1.0 would have a reflectivity of 0.0.

From the above definitions of absorptivity, emissivity, and reflectivity, it follows that their desired values would depend on whether the surface involved is that of an insulator or a conductor of heat energy. If rapid transmission is required, then the conductor should possess high absorptivity and emissivity, but low reflectivity. On the other hand, insulators ought to possess high reflectivity, but low absorptivity and emissivity.

Kirchhoff's law states that the absorptivity of a substance is equal to its emissivity. This being true, then bodies having high absorption capacity are also good radiators. It must also be true that the absorption and the emissive capacity of a body diminish proportionately as its reflective ability increases.

Considering the influence of these various properties on heat-transfer wall surfaces, the following general laws may be stated:

(*a*) Bright, polished, or light-colored surfaces possess high reflectivity, and therefore act to resist heat absorptivity and emissivity.

(*b*) Dull, rough, or dark-colored surfaces are poor reflectors of heat and act as good absorbers and radiators.

State of Opposing Fluids. The amount of heat transmitted through a separating wall varies with the state of the fluids on either side of it. The most rapid movement of heat is accomplished when both the opposing fluids are liquid of high fluidity (low viscosity) or when one is in the process of changing state, assuming that all thermal conditions are similar when making comparisons. The least rapid heat movement occurs between two gases separated by a plate or wall. Between the two extremes lie liquid-to-gas, gas-to-boiling liquid, etc. In the case of a boiling liquid, the presence of bubbles on the transfer surfaces tends to reduce heat flow. In order to dispose of them, turbulence becomes an important factor.

The differences in the rate of heat transfer values for various combinations of fluid conditions are shown in Table 87.

Consolidating the variables that govern the transfer of heat from a region of high temperature to one of lower temperature, the basic equation may be formulated where H equals the total amount of heat transferred, or wall load in Btu.

$H \propto T$—Heat transfer varies as time
$H \propto (t_2 - t_1)$—Heat transfer varies directly as temperature differential
$H \propto \frac{1}{L}$—Heat transfer varies inversely as thickness
$H \propto k$—Heat transfer varies directly as conduction capacity
$H \propto A$—Heat transfer varies directly as area

Therefore,

$$H = \frac{kA(t_2 - t_1)}{L} \times T$$

Under fixed conditions of system design, the amount of heat transferred per unit area from one body to another is governed by the primary factors time, temperature, and turbulence, commonly referred to as the three T's of thermodynamics.

MEASUREMENT OF HEAT TRANSMISSION

The rate at which heat passes through any material can be determined only by actual test under laboratory control. While no method exists whereby the true heat-transfer value of a specific material may be established solely on the basis of mathematics, given a factor determined by testing under standardized physical and thermal conditions, the rate of transmission for the same materials under any combination of conditions may be calculated.

Considering this factor solely on the basis of *conducted* heat, those conditions that are otherwise variable—such as time, temperature, thickness, etc.,—must be assigned constant values for the benefit of standardization. Therefore, the results of laboratory tests are translated so as to represent similar conditions for all materials, thereby establishing a basis for comparison and computation.

The constants adopted as standard are as follows:

Variable Factor	*Constant Value*
Time	1 hr
Temperature difference	1 F
Thickness	1 in.
Wall area	1 sq ft

Conduction factors, therefore, are expressed and published in terms of Btu per hr, per sq ft, per in., per °F.

However, as practical conditions require that heat loads be known for walls having any thickness of a given material or combination of materials and for any existing surface condition, it becomes obvious that the terms employed in expressing heat transfer must differ in accordance with the existing conditions they represent.

There are five practical methods for expressing heat-transfer values or coefficients. They are: (1) thermal coefficient of conductivity, k; (2) thermal coefficient of conductance, C; (3) thermal coefficient of radiation and convection (surface conductance), f*; (4) thermal coefficient of conductance of air spaces, a*; and (5) thermal coefficient of over-all transmission, U.

* The values of f will be found in Table 70, Section A, in the column labeled C, and the values for a in Table 70, Section B, and also in Table 68, will be found in the column labeled C.

Thermal Coefficient of Conductivity

The thermal coefficient of conductivity is the amount of heat, expressed in Btu, that may be transmitted in 1 hr through 1 sq ft of a given material 1 in. thick when a temperature difference of 1 F exists between its two (inner and outer) surfaces. The numerical value (Btu) of this coefficient includes that heat transmitted by *conduction* only and when the distance traversed (thickness) is *no more* or *no less* than 1 in.

The coefficient of conductivity for any given material increases (1) as the density of the material increases and (2) as the level of operating temperatures increases. Because of such variations, density specifications (pounds per cubic foot), as well as applicable temperature ranges, should be included in tables listing conductivity values of all insulating materials.

Usually the thermal coefficient of conductivity is designated by the letter k where

$$k = \text{Btu per hr per sq ft per in. per °F temperature difference}$$

or

$$K = \text{Btu per hr per sq ft per ft per °F temperature difference}$$

Thermal Coefficient of Conductance

The amount of heat conducted through a wall varies inversely as the thickness. Therefore, the rate of heat transfer due to conduction decreases as the thickness of material increases, and, conversely, the rate of heat transfer due to conduction increases as the thickness decreases. Since k, the coefficient of conductivity, is based on a standard thickness of 1 in., the amount of heat that may be conducted through a wall of any thickness will be equal to k divided by its actual thickness. The quotient of $\frac{k}{L}$ would be termed the thermal coefficient of conductance, or

$$\text{Coefficient of conductance} = \frac{\text{coefficient of conductivity}}{\text{thickness in inches}}$$

Conductance may be defined as the amount of heat, expressed in Btu, that may be transmitted through a wall of any thickness by conduction per hour per square foot per 1 F temperature difference. While a conductance factor may be applied to any type of simple or compound wall, generally its use is confined to such materials whose thicknesses are standard, and which may be more or less than 1 in. Wall boards of various kinds, clay, roofing, hollow clay tile, and concrete block are some of the materials included in this category.

The thermal coefficient of conductance is designated by the letter C where

$$C = \frac{k}{L}$$

As with k, C also varies as density and temperature level.

Thermal Coefficient of Radiation and Convection (Surface Conductance)

The thermal coefficient of radiation and convection or surface conductance is the amount of heat, expressed in Btu, transmitted to or from 1 sq ft of wall surface by radiation and convection in 1 hr when a temperature difference of 1 F exists between the wall surface and the surrounding fluid.

Since it is only the over-all heat load that need be considered for purposes of estimating refrigeration requirements, it is not essential that the heat due to radiation and convection be determined individually. Therefore, it suffices to consider some factor that combines the effect of both. Such a factor is known as film or surface conductance designated generally by the letter and subscript f_i for internal and f_o for external surfaces.

f_i = Btu per sq ft per hr per °F transmitted by radiation and convection to or from internal surfaces (still fluid)

f_o = Btu per sq ft per hr per °F transmitted by radiation and convection to or from external surfaces (moving fluid)

The amount of heat transmitted by radiation and convection to or from a wall is determined by test like conductivity values. Because surface characteristics and fluid velocity are functions of radiation and convection, respectively, variations in either or both of these factors affect the surface conductance value and, therefore, must be considered in heat-load determinations.

As stated earlier in this chapter, the amount of radiated heat absorbed or emitted by a body decreases with surface smoothness and brightness. Therefore, the coefficient of radiation and convection must be considered in light of these specifications.

The heat transmitted to or from a body by convection increases as the turbulence of the fluid film in contact with its surface. Where the fluid is considered to be still, which means that any motion prevailing is due to natural convection alone, the surface film tends to approach a temperature equal to that of the wall surface, thereby acting as an insulator. While an equilibrium is never fully established where any degree of temperature difference exists between the surrounding fluid and the exterior wall surface, a nonturbulent condition does retard heat transmission to maintain minimum surface conductance. However, when a state of turbulence does exist, the surface film constantly is replaced by warmer fluid and the insulating effect thereby reduced.

Therefore, the surface conductance of any wall varies as the velocity of the surrounding fluid. For building materials, the velocity of the surrounding air is expressed in miles per hour, while in heat-exchange equipment it is expressed in feet per minute or feet per second. Because the surfaces of building materials commonly employed have nearly similar emissivity values, the surface conductance under still-air conditions may be accepted as being constant at 1.65. This value for surface conductance will not be true for walls covered with bright material such as aluminum foil.

Table 69 lists the surface conductances f_i and f_o for walls, ceilings, and roofs. The external surface conductance f_o is listed in relation to wind velocity during summer and winter. Where the external surface of a wall

TABLE 69. SURFACE CONDUCTANCE—BUILDING MATERIALS—BTU PER HR PER SQ FT PER °F*

Surface	f_i *Still Air*	f_o *8 mph Wind Velocity (Summer)*	f_o *15 mph Wind Velocity (Winter)*
Inside walls (conventional)	1.65	....	....
Ceilings	1.65	....	....
Outside walls (conventional)	1.65	4.00	6.00
Roofs	....	4.00	6.00

* Some authorities ignore the f_i and f_o factors in computing the coefficient of over-all heat transmission for heavily insulated compound walls, but they are included in all examples here for explanatory reasons. Their use is recommended by the author for all walls.

is located in the building, f_i and f_o are similar unless additional factors such as air blasts from unit heaters, open corridors, etc., are present to dictate otherwise.

Thermal Coefficient of Conductance of Air Spaces

Compound walls often include one or more built-in air spaces separating solid materials used in their construction. These air spaces affect the heat-transmission capacity of such walls and must be considered in determining refrigeration loads.

The rate at which heat is transferred from one surface to another of lower temperature across an air space is a function of conduction, radiation, and convection. Therefore, the amount transferred during any unit time interval depends on temperature range, temperature difference between opposing surfaces, width of air space, freedom of air motion, and character of both surfaces. It follows that the rate of heat transfer across such spaces may be reduced for any temperature condition by providing surfaces of high reflectivity such as aluminum foil and limiting convective currents. The advantages of the latter are displayed by certain materials whose excellent insulating properties result substantially from the presence

Table 70. Conductances of Air Space at Various Mean Temperatures*

(All conductance values expressed in Btu per hr per sq ft per °F temp diff)

Section A. Surface Conductances for Still Air[a]

Position of Surface	Direction of Heat Flow	Surface Emissivity: Non-Reflective $e = 0.90$ C	R	Reflective $e = 0.20$ C	R	Reflective $e = 0.05$ C	R
Horizontal	Upward	1.63	0.61	0.91	1.10	0.76	1.32
Sloping—45 deg	Upward	1.60	0.62	0.88	1.14	0.73	1.37
Vertical	Horizontal	1.46	0.68	0.74	1.35	0.59	1.70
Sloping—45 deg	Downward	1.32	0.76	0 60	1.67	0.45	2.22
Horizontal	Downward	1.08	0.92	0.37	2.70	0.22	4.55

[a] Conductances are for surfaces of the stated emissivity facing surroundings having an emissivity equal to 0.9 and at the same temperature as the ambient air. Values are based on a surface-air temperature difference of 10 F and for surface temperature of 70 F. (See Table 68 for surface conductances for moving air.)

Section B. Thermal Conductances and Resistances of a Plane[a] Air Space*

Position of Air Space	Heat Flow	Thickness[b] Inches	Mean Temp[c] F	Temp Diff[c] deg	Thermal Conductance-C, Value of $E^{c,d}$: 0.05	0.2	0.5	0.82	Thermal Resistance-R, Value of E^d: 0.05	0.2	0.5	0.82
Horiz.	Up	¾ to 4	50	10	0.41	0.55	0.82	1.11	2.44	1.83	1.22	0.90
			50	30	0.54	0.68	0.95	1.24	1.84	1.47	1.05	0.80
			90	10	0.41	0.58	0.92	1.28	2.47	1.73	1.09	0.78
45° Slope	Up	¾ to 4	50	10	0.35	0.49	0.76	1.05	2.84	2.05	1.31	0.95
			50	30	0.48	0.62	0.89	1.18	2.08	1.62	1.13	0.85
			90	10	0.35	0.52	0.86	1.23	2.85	1.92	1.16	0.81
Vert.	Horiz.	¾ to 4	50	10	0.28	0.42	0.69	0.99	3.52	2.37	1.44	1.02
			50	30	0.38	0.52	0.79	1.08	2.64	1.94	1.27	0.92
			90	10	0.29	0.46	0.80	1.16	3.49	2.19	1.25	0.86
45° Slope	Down	¾ to 4	50	10	0.24	0.38	0.65	0.94	4.14	2.65	1.54	1.06
			50	30	0.30	0.43	0.71	1.00	3.36	2.30	1.41	1.00
			90	10	0.25	0.42	0.76	1.12	4.07	2.40	1.32	0.89
Horiz.	Down[f]	¾	50	20	0.28	0.42	0.69	0.98	3.57	2.38	1.45	1.02
		1½	50	20	0.18	0.31	0.58	0.87	5.56	3.23	1.73	1.15
		4	50	20	0.11	0.25	0.52	0.81	8.94	4.03	1.92	1.23
		¾	90	20	0.31	0.48	0.82	1.18	3.23	2.08	1.22	0.85
		1½	90	20	0.20	0.37	0.71	1.07	5.00	2.70	1.41	0.93
		4	90	20	0.13	0.30	0.64	1.01	7.82	3.35	1.56	0.99

Notes: [a] Spaces of uniform thickness, bounded by moderately smooth surfaces.

[b] Where a range of thickness is given, the given conductance is the average over the range; extreme values within the range differ therefrom by less than 10 percent.

[c] Interpolation, and moderate extrapolation, of conductance values are permissible for other values of mean temperature, temperature difference and effective emissivity E.

[d] Effective emissivity of space, E, is given by $\frac{1}{E} = \frac{1}{e_1} + \frac{1}{e_2} - 1$ where e_1 and e_2 are the emissivities of the surfaces of the air space. (See Section C, below).

[f] The conductances of horizontal spaces with heat flow downward are substantially independent of temperature difference.

* Based on *National Bureau of Standards* data presented in Housing Research Paper No. 32, *Housing and Home Finance Agency*, 1954 (U. S. Government Printing Office, Washington, D. C.)

Section C. Reflectivity and Emissivity Values of Various Surfaces and Effective Emissivities of Air Spaces

Surface	Reflectivity in Percent	Average Emissivity e	Effective Emissivity E of Air Space: With one surface having emissivity e and other 0.90	With both surfaces of emissivity e
Aluminum foil, bright	92 to 97	0.05	0.05	0.03
Aluminum sheet	80 to 95	0.12	0.12	0.06
Aluminum coated paper, polished	75 to 84	0.20	0.20	0.11
Steel, galvanized, bright	70 to 80	0.25	0.24	0.15
Aluminum paint	30 to 70	0.50	0.47	0.35
Building materials: wood, paper, glass, masonry, non-metallic paints	5 to 15	0.90	0.82	0.82

* Reprinted by permission from "Heating Ventilating Air Conditioning Guide 1956," chapter 9.

of innumerable air cells distributed throughout their mass small enough to render convection a negligible factor. Sawdust, for example, is a better insulator than the solid wood from which it originated because of the air spaces between the particles when packed not too firmly.

Beyond a definite width of air space, the rate of heat transfer tends to become constant so that similar values may be established for distances of approximately 1 in. and above. However, the temperature level at which the system operates influences the rate of transfer, requiring tables to list average values according to their related temperature range.

Accurate values cannot be established for heat transfer across air spaces generally because of the many variations in wall construction. Therefore, those values listed in published tables, while only approximate (except for some specific set of physical and thermal conditions), provide data sufficiently accurate for most practical purposes.

To simplify calculations, heat-transfer values for air spaces usually are given as conductance coefficients, which is the number of Btu transmitted by conduction, radiation, and convection per square foot per hour per °F temperature difference. The coefficient of conductance for air spaces usually is designated by the letter *c*. However, it is represented in this text by the letter *a*.

Reasonable accuracy may be obtained in calculating over-all heat-transmission factors by adopting a value for *a* of 1.1 Btu per sq ft per hr per °F for most conditions found in compound walls of refrigerated compartments and air conditioned buildings. Where conditions are more definitely specified as to temperature and width of air space, the more selective conductance coefficient appearing in Table 70 may be employed.

Thermal Coefficient of Over-all Transmission

The thermal coefficient of over-all transmission is the amount of heat, expressed in Btu, transmitted per square foot through a wall of any thick-

TABLE 71. OVER-ALL HEAT TRANSFER COEFFICIENT U FOR GLASS*

Type	*Btu per hr per sq ft per 1 F* *U*
Single glass	1.10
Double glass	0.50
Triple glass	0.30

* *Note:* $H = A \times U \times t_d$

where A = area in sq ft of total window opening

ness and any material or combination of materials by the combined loads of conductance, radiation, and convection, in one hr, when a temperature difference of 1 F exists between the separated fluids.

Although the over-all coefficient of transmission, usually designated by the letter U, may be found in published tables for variously constructed walls, its value for any simple or compound wall may be computed when the factors of k, c, and a for component materials and air spaces, and f_i and f_o for given surface conditions, are known.

The thermal coefficient of over-all transmission is the number of Btu that will pass through any simple or built-up wall having an area of one sq ft in one hr when a temperature difference of 1 F exists, expressed simply as

$$U = \text{Btu per sq ft per hr per °F}$$

Therefore

$$H = U \times A \times \text{temperature difference } (t_2 - t_1)$$

where H = total amount of heat transmitted — Btu per hr
A = area of wall in square feet
t_2 = temperature of outside air
t_1 = temperature maintained in refrigerated compartment

U is employed also in connection with over-all heat transfer in condensers and evaporators to express Btu transmitted per hour per square foot per 1 F.

RESISTIVITY OR RESISTANCE TO HEAT TRANSMISSION

The amount of heat transmitted through a wall by conductivity, conductance, or over-all transmission is governed by the resistance to heat flow exerted by the specific materials employed and the prevailing surface conditions.

As the total resistance to heat flow is equal to the sum of the separate resistances exerted by each of the several materials in a compound wall, or of each 1 in. of thickness of the same material in a simple wall, then

$$R = r_1 + r_2 + r_3, \text{ etc.}$$

where R = total resistance
and r_1, r_2, r_3 = resistance of the separate materials in a compound wall or of each 1 in. of the same material in a simple wall

Resistivity or resistance to heat flow is equal to the reciprocal of transmission (U), or conductivity (k), or conductance (c), or film conductance (f_i, f_o), or conductance of air spaces (a). Therefore,

$$R = \frac{1}{U}$$

and

$$r = \frac{1}{k} \text{ or } \frac{1}{c} \text{ or } \frac{1}{f_i} \text{ or } \frac{1}{f_o} \text{ or } \frac{1}{a}$$

If, as above, $R = r_1 + r_2 + r_3 +$ etc., then, for a simple wall 1 in. thick,

$$R = \frac{1}{k} + \frac{1}{f_i} + \frac{1}{f_o}$$

or, for a simple wall L in. thick $\left(k = \frac{1}{L}\right)$,

$$R = \frac{L}{k} + \frac{1}{f_i} + \frac{1}{f_o}$$

and, for a compound wall built up of any number of materials of any thickness,

$$R = \frac{L_1}{k_1} + \frac{L_2}{k_2} + \frac{L_3}{k_3} + \text{etc.} + \frac{1}{f_i} + \frac{1}{f_o}$$

and for a compound wall built up of any number of materials of any thickness and having any number of built-in air spaces,

$$R = \frac{L_1}{k_1} + \frac{L_2}{k_2} + \frac{L_3}{k_3} + \text{etc.} + \frac{1}{f_i} + \frac{1}{f_o} + \frac{1}{a_1} + \frac{1}{a_2} + \text{etc.}$$

where k_1, k_2, k_3, etc. = coefficient of conductivity of each material
L_1, L_2, L_3, etc. = thickness in inches of each material, represented by k_1, k_2, k_3, etc.
f_i, f_o = internal and external film conductance
a_1, a_2, etc. = conductance of air spaces

However,

$$R - \frac{1}{U}$$

Therefore,

$$\frac{1}{U} = \frac{L_1}{k_1} + \frac{L_2}{k_2} + \frac{L_3}{k_3} + \text{etc.} + \frac{1}{f_i} + \frac{1}{f_o} + \frac{1}{a_1} + \frac{1}{a_2} + \text{etc.}$$

To solve for U, then

$$U = \frac{1}{\frac{L_1}{k_1} + \frac{L_2}{k_2} + \frac{L_3}{k_3} + \text{etc.} + \frac{1}{f_i} + \frac{1}{f_o} + \frac{1}{a_1} + \frac{1}{a_2} + \text{etc.}}$$

Where materials having conductance coefficients (c) normally, rather than conductivity coefficients (k) are included in the wall construction, they should be accounted for as a separate resistance $\frac{1}{c}$ and placed in the equation as a reciprocal to be added to all other resistances. For a wall built up of a homogeneous material of L thickness and a wall board having a coefficient of conductance of c, enclosing an air space between them, then

$$U = \frac{1}{\frac{L}{k} + \frac{1}{f_i} + \frac{1}{f_o} + \frac{1}{a_1} + \frac{1}{C}}$$

TABLE 72. CONDUCTIVITIES (k) AND CONDUCTANCES (C) OF BUILDING MATERIALS AND INSULATIONS*

(Design Values)[a]

These constants are expressed in Btu per (hour) (square foot) (Fahrenheit degree temperature difference). Conductivities (k) are per inch thickness and, conductances (C) are for thickness or construction stated, not per inch thickness.

Material	Description	Density (Lb. per Cu Ft)	Conductivity or Conductance (k)	Conductivity or Conductance (C)	Resistance (R) Per Inch Thickness $\left(\frac{1}{k}\right)$	Resistance (R) For Thickness Listed $\left(\frac{1}{C}\right)$
AIR SPACES[b]	See Table 66—Section B.					
	Position — Heat Flow — Thickness					
	Horizontal........Up..................¾–4 in.	—	—	1.18	—	0.85
	Sloping (45°)........Up..................¾–4 in.	—	—	1.11	—	0.90
	Vertical........Horizontal..........¾–4 in.	—	—	1.03	—	0.97
	Sloping (45°)........Down..............¾–4 in.	—	—	0.97	—	1.03
	Horizontal........Down..................¾ in.	—	—	0.98	—	1.02
	Horizontal........Down..................1½ in.	—	—	0.87	—	1.15
	Horizontal........Down..................4 in.	—	—	0.81	—	1.23
	Horizontal........Down..................8 in.	—	—	0.80	—	1.25
AIR SURFACES[c]	See Table 66—Section A					
Still Air	Position — Heat Flow					
	Horizontal..................Up	—	—	1.63	—	0.61
	Up Sloping (45°)..................Up	—	—	1.60	—	0.62
	Vertical..................Horizontal	—	—	1.46	—	0.68
	Sloping (45°)..................Down	—	—	1.32	—	0.76
	Horizontal..................Down	—	—	1.08	—	0.92
15 mph Wind	Any Position—Any Direction..........	—	—	6.00	—	0.17
7½ mph Wind	Any Position—Any Direction..........	—	—	4.00	—	0.25
BUILDING BOARD[d] Boards, Panels, Sheathing, Etc.	Asbestos-cement board..........	120	4.0	—	0.25	—
	Asbestos-cement board..........⅛ in.			33.		0.03
	Gypsum or plaster board..........⅜ in.	50	—	3.10	—	0.32
	Gypsum or plaster board..........½ in.	50	—	2.25	—	0.45
	Plywood..........	34	0.80	—	1.25	—
	Plywood..........¼ in.	34	—	3.20	—	0.31
	Plywood..........⅜ in.	34	—	2.12	—	0.47
	Plywood..........½ in.	34	—	1.60	—	0.63
	Plywood or Wood Panels..........¾ in.	—	—	1.07	—	0.94
	Wood fiber board, laminated or homogeneous..	26 31	0.42 0.50	—	2.38 2.00	—
	Wood fiber—hardboard type..........	65	1.40	—	0.72	—
	Wood fiber—hardboard type..........¼ in.	65	—	5.60	—	0.18
	Wood—fir or pine sheathing..........25/32 in.	—	—	1.02	—	0.98
BUILDING PAPER	Vapor—permeable felt..........	—	—	16.70	—	0.06
	Vapor—seal, 2 layers of mopped 15 lb felt..........	—	—	8.35	—	0.12
	Vapor—seal, plastic film..........	—	—	—	—	Negl
FLOORING MATERIALS	Asphalt tile..........⅛ in.	120	—	24.80	—	0.04
	Carpet and fibrous pad..........	—	—	0.48	—	2.08
	Carpet and rubber pad..........	—	—	0.81	—	1.23
	Ceramic tile..........1 in.	—	—	12.50	—	0.08
	Cork tile..........	25	0.45	—	2.22	—
	Cork tile..........⅛ in.	—	—	3.60	—	0.28
	Felt, flooring..........	—	—	16.70	—	0.06
	Floor tile or linoleum—av. value..........⅛ in.	—	—	20.00	—	0.05
	Linoleum..........⅛ in.	80	—	12.00	—	0.08
	Plywood subfloor..........⅝ in.	—	—	1.28	—	0.78
	Rubber or plastic tile..........⅛ in.	110	—	42.40	—	0.02
	Terrazzo..........1 in.	—	—	12.50	—	0.08
	Wood subfloor..........25/32 in.	—	—	1.02	—	0.98
	Wood, hardwood finish..........¾ in.	—	—	1.47	—	0.68
INSULATING MATERIALS Blanket and Batt	Cotton fiber[e]..........	0.8–2.0	0.26	—	3.85	—
	Mineral wool, fibrous form, processed from rock, slag, or glass[e]..........	1.5–4.0	0.27	—	3.70	—
	Wood fiber[e]..........	3.2–3.6	0.25	—	4.00	—
	Wood fiber, multilayer, stitched expanding[e]..........	1.5–2.0	0.27	—	3.70	—
Board	Glass fiber..........	9.5	0.25	—	4.00	—
	Wood or cane fiber					
	Acoustical tile[f]..........½ in.	—	—	0.84	—	1.19
	Acoustical tile[f]..........¾ in.	—	—	0.56	—	1.78
	Interior finish (plank, tile, lath)..........	15.0	0.35	—	2.86	—
	Interior finish (plank, tile, lath)..........½ in.	15.0	—	0.70	—	1.43
	Sheathing (impreg. or coated)..........	20.0	0.38	—	2.63	—
	Sheathing (impreg. or coated)..........½ in.	20.0	—	0.76	—	1.32
	Sheathing (impreg. or coated)..........25/32 in.	20.0	—	0.49	—	2.06

* Reprinted by permission from "Heating Ventilating Air Conditioning Guide 1956," chapter 9.

Table 72. Conductivities (k) and Conductances (C) of Building Materials and Insulations—(*Continued*)

(Design Values)[a]

Material	Description	Density (Lb per Cu Ft)	Conductivity or Conductance (k)	Conductivity or Conductance (C)	Resistance (R) Per Inch Thickness $\left(\frac{1}{k}\right)$	Resistance (R) For Thickness Listed $\left(\frac{1}{C}\right)$
INSULATING MATERIALS Board and Slabs	Cellular glass	9.0	0.40	—	2.50	—
	Corkboard (without added binder)	6.5–8.0	0.27	—	3.70	—
	Hog hair (with asphalt binder)	8.5	0.33	—	3.00	—
	Plastic (foamed)	1.62	0.29	—	3.45	—
	Wood shredded (cemented in preformed slabs)	22.0	0.55	—	1.82	—
Loose Fill	Macerated paper or pulp products	2.5–3.5	0.28	—	3.57	—
	Redwood bark shredded	2.5–3.5	0.30	—	3.33	—
	Mineral wool (glass, slag, or rock)	2.0–5.0	0.30	—	3.33	—
	Sawdust or shavings	8.0–15.0	0.45	—	2.22	—
	Vermiculite (expanded)	7.0	0.48	—	2.08	—
ROOF INSULATION	All types[g]					
	Preformed, for use above deck.					
	Approx. ½ in.	—	—	0.72	—	1.39
	Approx. 1 in.	—	—	0.36	—	2.78
	Approx. 1½ in.	—	—	0.24	—	4.17
	Approx. 2 in.	—	—	0.19	—	5.26
	Approx. 2½ in.	—	—	0.15	—	6.67
	Approx. 3 in.	—	—	0.12	—	8.33
MASONRY MATERIALS Concretes	Cement mortar	116	5.0	—	0.20	—
	Gypsum-fiber concrete 87½% gypsum, 12½% wood chips	51	1.66	—	0.60	—
	Lightweight aggregates including expanded shale, clay or slate; expanded slags; cinders; pumice; perlite; vermiculite; also cellular concretes	120	5.2	—	0.19	—
		100	3.6	—	0.28	—
		80	2.5	—	0.40	—
		60	1.7	—	0.59	—
		40	1.15	—	0.86	—
		30	0.90	—	1.11	—
		20	0.70	—	1.43	—
	Sand and gravel or stone aggregate (oven dried)	140	9.0	—	0.11	—
	Sand and gravel or stone aggregate (not dried)	140	12.0	—	0.08	—
	Stucco	116	5.0	—	0.20	—
MASONRY UNITS	Brick, common	120	5.0	—	0.20	—
	Brick, face	130	9.0	—	0.11	—
	Clay tile, hollow:					
	1 cell deep 3 in.	—	—	1.25	—	0.80
	1 cell deep 4 in.	—	—	0.90	—	1.11
	2 cells deep 6 in.	—	—	0.66	—	1.52
	2 cells deep 8 in.	—	—	0.54	—	1.85
	2 cells deep 10 in.	—	—	0.45	—	2.22
	3 cells deep 12 in.	—	—	0.40	—	2.50
	Concrete blocks, three oval core:					
	Sand & gravel aggregate 4 in.	—	—	1.40	—	0.71
	8 in.	—	—	0.90	—	1.11
	12 in.	—	—	0.78	—	1.28
	Cinder aggregate 3 in.	—	—	1.16	—	0.86
	4 in.	—	—	0.90	—	1.11
	8 in.	—	—	0.58	—	1.72
	12 in.	—	—	0.53	—	1.89
	Gypsum partition tile:					
	3 x 12 x 30 in. solid	—	—	0.79	—	1.26
	3 x 12 x 30 in. 4-cell	—	—	0.74	—	1.35
	4 x 12 x 30 in. 3-cell	—	—	0.60	—	1.67
	Lightweight aggregate (expanded shale, clay, slate or slag; pumice) 3 in.	—	—	0.79	—	1.27
	4 in.	—	—	0.67	—	1.50
	8 in.	—	—	0.50	—	2.00
	12 in.	—	—	0.44	—	2.27
	Stone, lime or sand	—	12.50	—	0.08	—
PLASTERING MATERIALS	Cement plaster, sand aggregate	116	5.0	—	0.20	—
	Sand aggregate ½ in.	—	—	10.00	—	0.10
	Sand aggregate ¾ in.	—	—	6.66	—	0.15
	Gypsum plaster:					
	Lightweight aggregate ½ in.	45	—	3.12	—	0.32
	Lightweight aggregate ⅝ in.	45	—	2.67	—	0.39
	Lightweight agg. on metal lath ¾ in.	—	—	2.13	—	0.47

TABLE 72. CONDUCTIVITIES (k) AND CONDUCTANCES (C) OF BUILDING MATERIALS AND INSULATIONS—(*Continued*)

(Design Values)[a]

Material	Description	Density (Lb per Cu Ft)	Conductivity or Conductance (k)	Conductivity or Conductance (C)	Resistance (R) Per Inch Thickness $\left(\frac{1}{k}\right)$	Resistance (R) For Thickness Listed $\left(\frac{1}{C}\right)$
PLASTERING MATERIALS	Perlite aggregate	45	1.5	—	0.67	—
	Sand aggregate	105	5.6	—	0.18	—
	Sand aggregate ½ in.	105	—	11.10	—	0.09
	Sand aggregate ⅝ in.	105	—	9.10	—	0.11
	Sand aggregate on metal lath ¾ in.	—	—	7.70	—	0.13
	Sand aggregate on wood lath	—	—	2.50	—	0.40
	Vermiculate aggregate	45	1.7	—	0.59	—
ROOFING	Asbestos-cement shingles	120	—	4.76	—	0.21
	Asphalt roll roofing	70	—	6.50	—	0.15
	Asphalt shingles	70	—	2.27	—	0.44
	Built-up roofing ⅜ in.	70	—	3.00	—	0.33
	Slate ½ in.	—	—	20.00	—	0.05
	Sheet metal	—	400+	—	Negl	
	Wood shingles	—	—	1.06	—	0.94
SIDING MATERIALS (ON FLAT SURFACE)	Shingles					
	Wood, 16-in. 7½-in. exposure	—	—	1.15	—	0.87
	Wood, double, 16-in., 12-in. exposure	—	—	0.84	—	1.19
	Wood, plus insul. backer board 5/16 in.	—	—	0.71	—	1.40
	Siding					
	Asbestos-cement, ¼ in., lapped	—	—	4.76	—	0.21
	Asphalt Roll Siding	—	—	6.50	—	0.15
	Asphalt Insulating Siding (½ in. bd.)	—	—	0.69	—	1.45
	Wood, Drop, 1 x 8 in.	—	—	1.27	—	0.79
	Wood, Bevel, ½ x 8 in., lapped	—	—	1.23	—	0.81
	Wood, Bevel, ¾ x 10 in., lapped	—	—	0.95	—	1.05
	Wood, Plywood, ⅜ in., lapped	—	—	1.59	—	0.59
	Structural glass			10.00		0.10
WOODS	Maple, oak, and similar hardwoods	45	1.10	—	0.91	—
	Fir, pine, and similar softwoods	32	0.80	—	1.25	—

[a] Representative values for dry materials at 75 F mean temperature, selected by the A.S.H.A.E. Technical Advisory Committee on Insulation. They are intended as design (not specification) values for materials of building construction in normal use. For conductivity of a particular product, the user may obtain the value supplied by the manufacturer or secure the results of unbiased tests.

[b] Air space resistance values shown here are based on a temperature difference of 20 F deg and a mean temperature of 50 F for spaces faced both sides with ordinary *non-reflective* materials ($e = 0.90$ and $E = 0.82$).

[c] Surface resistance values shown here are for ordinary *non-reflective* materials ($e = 0.90$).

[d] See also Insulating Materials, Board.

[e] Includes paper backing and facing if any. In cases where the insulation forms a boundary (highly reflective or otherwise) of an air space, refer to Table 66 Sections B and C, to obtain the insulating value of the air space for the appropriate effective emissivity and temperature conditions of the space.

[f] Insulating values of acoustical tile vary depending on density of the board and on the type, size and depth of the perforations. An average conductivity k value is 0.42.

[g] The *U. S. Department of Commerce*, Simplified Practice Recommendation for Thermal Conductance Factors for Preformed Above-Deck Roof Insulation, No. R 257-55, recognizes the specification of roof insulation on the basis of the C values shown. Roof insulation is made in thicknesses to meet these values. Therefore, thickness supplied by different manufacturers may vary depending on the conductivity k value of the particular material.

[h] Not oven dried—conductivity k prior to drying, approximately 1 to 3 years.

EXAMPLE 1: Determine the coefficient of conductance for a wall board $\frac{3}{8}$ in. thick when k for the same material is 2.7 Btu per sq ft per hr per in. per °F.

$$\text{Conductance} = \frac{\text{conductivity}}{\text{thickness}} = \frac{k}{L}$$

$$C = \frac{2.7}{0.375}$$

$$C = 7.2 \text{ Btu per sq ft per hr per °F}$$

EXAMPLE 2: What is the coefficient of over-all transmission of a simple wall constructed of 16 in. of concrete, both surfaces finished smooth? Assume still air conditions for both sides. Applying the fundamental formula

$$U = \frac{1}{\frac{L}{k} + \frac{1}{f_i} + \frac{1}{f_o}}$$

$$k = 12 \text{ for concrete}$$

$$f_i \,\&\, f_o = 1.65 \text{ for still air}$$

then

$$U = \frac{1}{\frac{16}{12} + \frac{1}{1.65} + \frac{1}{1.65}}$$

$$U = \frac{1}{1.33 + 0.606 + 0.606}$$

$$U = 0.393 \text{ Btu per sq ft per hr per °F}$$

EXAMPLE 3: Compute the coefficient of over-all heat transmission (U) for a compound wall constructed of 8-in. concrete block, ½-in. plaster, and 2-in. cork.

Assume the thermal coefficients of the various materials to be as follows:

8-in concrete block. . C = 0.84 Btu per sq ft per °F per hr

Plaster. . . . . k_1 = 2.32 Btu per sq ft per °F per in. per hr

Cork. . . . . k_2 = 0.30 Btu per sq ft per °F per in. per hr

and f_i = 1.65 Btu per sq ft per °F per hr

f_o = 6.0 Btu per sq ft per °F per hr

$$U = \frac{1}{\frac{1}{C} + \frac{L}{k_1} + \frac{L}{k_2} + \frac{1}{f_i} + \frac{1}{f_o}}$$

$$U = \frac{1}{\frac{1}{0.84} + \frac{0.5}{2.32} + \frac{2}{0.30} + \frac{1}{1.65} + \frac{1}{6.0}}$$

$$U = \frac{1}{1.19 + 0.215 + 6.6 + 0.606 + 0.166}$$

$$U = 0.113 \text{ Btu per sq ft per °F per hr}$$

EXAMPLE 4: Compute the coefficient of over-all heat transmission (U) for a compound wall constructed of 8-in. brick, 2-in. cork, and ½-in. plaster with a ¾-in. space between the brick and the cork.

Conductivities and conductance coefficients are:

Brick. . . . . $k_1 = 5$ Btu per sq ft per °F per hr per in.

Cork. . . . . $k_2 = 0.30$ Btu per sq ft per °F per hr per in.

Plaster. . . . . $k_3 = 2.32$ Btu per sq ft per °F per hr per in.

¾-in. air space. . . $a = 1.1$ Btu per sq ft per °F per hr

and $f_i = 1.65$ Btu per sq ft per °F per hr

$f_o = 6.00$ Btu per sq ft per °F per hr

$$U = \frac{1}{\frac{L}{k_1} + \frac{L}{k_2} + \frac{L}{k_3} + \frac{1}{a} + \frac{1}{f_i} + \frac{1}{f_o}}$$

$$U = \frac{1}{\frac{8}{5} + \frac{2}{0.30} + \frac{0.5}{2.32} + \frac{1}{1.1} + \frac{1}{1.65} + \frac{1}{6.0}}$$

$$U = \frac{1}{1.6 + 6.6 + 0.215 + 0.91 + 0.606 + 0.166}$$

$$U = 0.099 \text{ Btu per sq ft per °F per hr}$$

EXAMPLE 5: Determine the heat load per hour through the wall described in Example 4 if the inside and the outside temperatures were 20 F and 85 F, respectively. Total area of wall, 800 sq ft.

$$H = U \times A \times (t_2 - t_1)$$

$$H = 0.099 \times 800 \times (85 - 20)$$

$$H = 5148 \text{ Btu per hr}$$

FLOORS

Where floors separate two spaces, one of which may be a basement or another room, the coefficient of over-all heat transmission is computed in the same manner as shown for walls in the previous examples. However, if the floor is laid on the ground, then the outside surface conductance of f_o need not be included in the computation. For a simple floor laid directly on the ground,

$$U = \frac{1}{\frac{L}{k} + \frac{1}{f_i}}$$

and H for the floor equals $U \times$ floor area $\times$ the difference in temperature between the space above and below the ground.

TABLE 73. HEAT CONDUCTIVITY AND DENSITY OF VARIOUS INSULATING MATERIALS (U.S. BUREAU OF STANDARDS)*

Material	Thermal conductivity B.t.u. per square foot, per inch, per hour, per degree Fahrenheit	Density pounds per cubic foot	Description of material
Air	0.175	0.08	Ideal air space.
Air cell, ½ inch	0.458	8.80	Asbestos paper and air spaces.
Air cell, 1 inch	0.500	8.80	Asbestos paper and air spaces.
Asbestos mill board	0.830	61.0	Pressed asbestos.
Asbestos wood	3.700	123.0	Asbestos and cement.
Balsa wood	0.350	7.5	Light and soft across grain.
Calorax	0.221	4.0	Fluffy, finely divided mineral matter.
Cork	0.337	5.3	Granulated 1/8-3/16 inch.
Cork	0.330	10.0	Regranulated 1/16-1/8 inch.
Corkboard	0.279	6.9	No artificial binder—low density.
Corkboard	0.308	11.3	No artificial binder—low density.
Cotton wool	0.292	....	Loosely packed.
Fibrofelt	0.329	11.3	Felted vegetable fibers.
Fire felt wool	0.625	43.0	Asbestos sheet coated with cement.
Fire felt sheet	0.583	26.0	Soft, flexible asbestos sheet.
Flaxlinum	0.329	11.3	Felted vegetable fibers.
Hair felt	0.246	17.0	
Hard maple wood	1.125	44.0	Across grain.
Infusorial earth	0.583	43.0	Natural blocks.
Insulite	0.296	11.9	Pressed wool pulp—rigid.
Kapok	0.238	0.8	Vegetable fiber—loosely packed.
Keystone hair	0.271	19.0	Hair felt combined with building paper.
Linofelt	0.300	11.3	Vegetable fiber combined with paper.
Lithboard	0.379	12.5	Mineral wool and vegetable fibers.
Mineral wool	0.275	12.5	Medium packed.
Mineral wool	0.288	18.0	Felted in blocks.
Oak wood	1.000	38.0	Across grain.
Planer shavings	0.417	8.8	Various.
Pulp board	0.458	....	Stiff pasteboard.
Pure wool	0.263	5.0	
Rock cork	0.346	21.0	Mineral wool and binder—rigid.
Slag wool	0.750	15.0	
Tar roofing	0.707	55	
Virginia pine wood	0.958	34	Across grain.
White pine wood	0.791	32	Across grain.
Wool felt	0.363	21	Flexible paper stock.

* Courtesy, Calcium Chloride Institute, Washington, D. C.

Heat transmission through a floor laid on the ground usually is neglected for air-conditioning purposes. However, for refrigeration purposes, a 5 F temperature differential may be adopted as being more nearly correct. It allows for the initial load in freezing the ground to a considerable depth where sharp freezers or quick-freezing compartments are located on the ground floor.

INSULATION

Materials used in the construction of low-temperature chambers, such as domestic refrigerators, cabinets, and cold-storage rooms, or for pipe and duct covering, are selected primarily for their ability to retard the flow of heat through their homogeneous or heterogeneous mass. Those having extremely low heat-conduction capacity are called insulation materials. While there are numerous kinds of such materials, other factors that must

Table 74. Nominal Pipe Size Covering—Thickness* †

Nominal Pipe Size Inches	O.D. Pipe Inches	Light Duty Thickness		Medium Duty Thickness		Heavy Duty Thickness		Nominal 4" Wall Thickness		Nominal 5" Wall Thickness		Nominal 6" Wall Thickness		Nominal 7" Wall Thickness		Nominal 8" Wall Thickness	
		Thick-ness Inches	O.D. Covering Inches	Thick-ness Inches	O.D. Covering Inches	Thick-ness Inches	O.D. Covering Inches	Thick-ness Inches	O.D. Covering Inches	Thick-ness Inches	O.D. Covering Inches	Thick-ness Inches	O.D. Covering Inches	Thick-ness Inches	O.D. Covering Inches	Thick-ness Inches	O.D. Covering Inches
1/4	.540	1.35	3.25	1.85	4.25	2.88	6.31	4.16	8.87	5.29	11.12	5.82	12.19	7.03	14.63	8.41	17.38
3/8	.675	1.28	3.25	1.78	4.25	2.82	6.31	4.10	8.87	5.22	11.12	6.22	13.12	6.97	14.63	8.34	17.38
1/2	.840	1.20	3.25	1.70	4.25	2.73	6.31	4.01	8.87	5.14	11.12	6.14	13.12	6.88	14.63	8.25	17.38
3/4	1.050	1.35	3.75	1.85	4.69	2.63	6.31	3.91	8.87	5.03	11.12	6.03	13.12	6.78	14.63	8.16	17.38
1	1.315	1.47	4.25	2.00	5.31	2.97	7.25	4.16	9.62	4.90	11.12	5.90	13.12	7.47	16.25	8.03	17.38
1-1/4	1.660	1.48	4.69	2.32	6.31	3.10	7.87	3.98	9.62	5.26	12.19	6.48	14.62	7.28	16.25	7.84	17.38
1-1/2	1.900	1.42	4.69	2.50	6.75	2.99	7.87	3.86	9.62	5.14	12.19	6.36	14.62	7.16	16.25	7.72	17.38
2	2.375	1.47	5.31	2.44	7.25	3.25	8.87	3.94	10.25	4.90	12.19	6.12	14.62	6.94	16.25	8.38	19.13
2-1/2	2.875	1.37	5.62	2.50	7.87	3.00	8.87	4.12	11.12	5.12	13.12	5.87	14.62	7.25	17.38	8.13	19.13
3	3.500	1.56	6.75	2.68	8.87	3.06	9.62	4.34	12.19	4.81	13.12	6.37	16.25	6.94	17.38	7.81	19.13
3-1/2	4.000	1.62	7.25	2.81	9.62	3.56	11.12	4.09	12.19	5.31	14.62	6.12	16.25	7.53	19.13	8.47	21.00
4	4.500	1.68	7.87	2.56	9.62	3.31	11.12	4.31	13.12	5.06	14.62	6.43	17.37	7.28	19.13	8.22	21.00
5	5.563	1.65	8.87	2.78	11.12	3.78	13.12	4.53	14.62	5.34	16.25	5.90	17.37	6.75	19.13	7.69	21.00
6	6.625	1.81	10.25	2.78	12.19	4.00	14.62	4.00	14.62	5.37	17.37	6.31	19.25	7.13	21.00	8.00	22.63
8	8.625	1.78	12.19	3.00	14.62	3.81	16.25	4.37	17.37	5.31	19.25	6.19	21.00	7.00	22.63	8.00	24.63
10	10.750	1.83	14.62	3.31	17.37	4.19	19.12	----	----	5.12	21.00	6.00	22.75	7.00	24.75	8.00	26.75
12	12.750	1.75	16.25	3.18	19.12	4.12	21.00	----	----	5.00	22.75	6.00	24.75	7.00	26.75	8.00	28.75
14	14.00	1.50	17.00	3.50	21.00	4.00	22.00	----	----	5.00	24.00	6.00	26.00	7.00	28.00	8.00	30.00
16	16.00	1.50	19.00	3.00	23.00	4.00	24.00	----	----	5.00	26.00	6.00	28.00	7.00	30.00	8.00	32.00
18	18.00	1.50	21.00	3.00	25.00	4.00	26.00	----	----	5.00	28.00	6.00	30.00	7.00	32.00	8.00	34.00
20	20.00	1.50	23.00	3.00	27.00	4.00	28.00	----	----	5.00	30.00	6.00	32.00	7.00	34.00	8.00	36.00
22	22.00	1.50	25.00	3.00	29.00	4.00	30.00	----	----	5.00	32.00	6.00	34.00	7.00	36.00	8.00	38.00
24	24.00	1.50	27.00	3.00	31.00	4.00	32.00	----	----	5.00	34.00	6.00	36.00	7.00	38.00	8.00	40.00
26	26.00	1.50	29.00	3.00	33.00	4.00	34.00	----	----	5.00	36.00	6.00	38.00	7.00	40.00	8.00	42.00
28	28.00	1.50	31.00	3.00	35.00	4.00	36.00	----	----	5.00	38.00	6.00	40.00	7.00	42.00	8.00	44.00
30	30.00	1.50	33.00	3.00	37.00	4.00	38.00	----	----	5.00	40.00	6.00	42.00	7.00	44.00	8.00	46.00
32	32.00	1.50	35.00	3.00	39.00	4.00	40.00	----	----	5.00	42.00	6.00	44.00	7.00	46.00	8.00	48.00
34	34.00	1.50	37.00	3.00	41.00	4.00	42.00	----	----	5.00	44.00	6.00	46.00	7.00	48.00	8.00	50.00
36	36.00	1.50	39.00	3.00	43.00	4.00	44.00	----	----	5.00	46.00	6.00	48.00	7.00	50.00	8.00	52.00

* Courtesy, "Armstrong's Industrial Insulations," Lancaster, Pa., Armstrong Cork Company.

† For sizes *above* the heavy lines, molded cork covering is furnished, and for those sizes *below* the heavy lines, assembled cork lagging sections are furnished.

TABLE 75. TUBE COVERINGS—THICKNESS*

Tubing O. D. Pipe Inches	Light-Duty Thickness		Standard Thickness		Heavy-Duty Thickness	
	Thick-ness Inches	O. D. Covering Inches	Thick-ness Inches	O. D. Covering Inches	Thick-ness Inches	O. D. Covering Inches
3/8	1.43	3.25	1.93	4.25	2.96	6.31
1/2	1.37	3.25	1.87	4.25	2.90	6.31
5/8	1.31	3.25	1.81	4.25	2.84	6.31
7/8	1.43	3.75	1.90	4.69	2.71	6.31
1 1/8	1.56	4.25	2.09	5.31	3.06	7.25
1 3/8	1.65	4.69	2.46	6.31	3.24	7.87
1 5/8	1.53	4.69	2.56	6.75	3.12	7.87
2 1/8	1.59	5.31	2.56	7.25	3.37	8.87
2 5/8	1.34	5.31	2.62	7.87	3.12	8.87
3 1/8	1.81	6.75	2.87	8.87	3.24	9.62
3 5/8	1.81	7.25	2.99	9.62	3.74	11.12
4 1/8	1.87	7.87	2.74	9.62	3.49	11.12
5 1/8	1.87	8.87	3.00	11.12	3.99	13.12
6 1/8	2.06	10.25	3.02	12.19	4.24	14.62

* Courtesy, "Armstrong's Industrial Insulations," p. 51, Lancaster, Pa., Armstrong Cork Company.

TABLE 76. COMPARATIVE HEAT LOSS DATA FROM INSULATED (CORK) AND BARE PIPES*

Pipe size inches	Light-Duty Thickness Transmission in Btu per lin. ft. per deg. F. dif. in temp. for		Standard Thickness Transmission in Btu per lin. ft. per deg. F. dif. in temp. for		Heavy-Duty Thickness Transmission in Btu per lin. ft. per deg. F. dif. in temp. for		Bare Pipes Transmission in Btu per lin. ft. per deg. F. dif. in temp. for	
	1 hour	24 hours	1 hour	24 hours	1 hour	24 hours	1 hour	24 hours
1/4	.0802	1.92	.0713	1.71	.0611	1.47	0.452	10.8
3/8	.0906	2.17	.0976	1.91	.0670	1.60	0.494	11.9
1/2	.104	2.50	.0895	2.15	.0742	1.78	0.565	13.6
3/4	.111	2.66	.0973	2.34	.0829	1.98	0.670	16.0
1	.121	2.90	.105	2.52	.0873	2.09	0.817	19.5
1 1/4	.135	3.24	.110	2.64	.959	2.30	1.005	24.1
1 1/2	.155	3.73	.116	2.78	.104	2.50	1.130	27.2
2	.175	4.20	.131	3.14	.113	2.71	1.383	33.2
2 1/2	.207	4.98	.145	3.48	.132	3.17	1.650	39.6
3	.215	5.16	.157	3.77	.147	3.52	1.948	46.7
3 1/2	.236	5.67	.167	4.00	.146	3.51	2.20	52.8
4	.254	6.11	.192	4.61	.163	3.92	2.44	58.5
5	.300	7.20	.211	5.06	.173	4.15	2.97	71.3
6	.326	7.86	.238	5.70	.187	4.49	3.50	84.0
8	.400	9.60	.276	6.61	.227	5.45	4.45	107.0
10	.457	11.0	.329	7.90	.268	6.43	5.48	132.0
12	.658	15.8	.377	9.04	.303	7.27	6.48	155.0
14	.712	17.1	.415	9.96	.327	7.86	7.09	170.0
16	.819	19.6	.462	11.1	.363	8.71	8.15	196.0
18	.867	20.7	.509	12.2	.405	9.72	9.11	219.0
20	.972	23.3	.553	13.2	.438	10.5	10.1	242.0

* Courtesy, "Armstrong's Industrial Insulations," p. 60, Lancaster, Pa., Armstrong Cork Company.

be considered in their selection eliminate many as being of any practical importance in this field. While none of the remaining materials possesses all the necessary qualities to identify them as being ideal, some have proved to be highly suitable for the purpose.

In addition to their thermal resistant characteristics, insulation materials used for the purposes mentioned above must also be selected on the basis of economic and structural considerations. The ideal insulation material would be defined as having the following characteristics:

TABLE 77. THICKNESS OF CORK COVERING FOR PIPES*

Pipe Diameter Inches	4"	5"	6"	7"	8"
1/4	-200				
3/8	-200				
1/2	-200				
3/4	-120	-200			
1	-110	-180		-280	
1-1/4	-95	-150		-250	
1-1/2	-90	-145	-200	-245	-285
2	-80	-130	-185	-240	-270
2-1/2	-70	-120	-170	-225	-265
3	-65	-110	-155	-200	-240
3-1/2	-60	-100	-145	-195	-240
4	-55	-95	-140	-190	-235
5	-50	-85	-125	-170	-215
6	-40	-80	-120	-160	-200
8	-35	-70	-105	-140	-180
10	-30	-60	-95	-130	-165
12	-30	-60	-90	-120	-160
14	-30	-60	-85	-120	-155
16	-30	-55	-85	-115	-145
18	-30	-50	-80	-110	-140
20	-25	-50	-75	-105	-135
22	-25	-50	-75	-105	-135
24	-25	-50	-75	-105	-135
27	-25	-50	-75	-100	-130
30	-25	-50	-70	-100	-130
33	-25	-50	-70	-100	-130
36	-25	-45	-70	-100	-125

* Courtesy, "Armstrong's Industrial Insulations," p. 61, Lancaster, Pa., Armstrong Cork Company.

(*a*) ***Low heat-conducting properties.*** This means that the thermal coefficient of conductivity (k) should be low enough to avoid resorting to great thickness of material to increase thermal resistance.

(*b*) ***Permanence.*** Materials that may disintegrate as a result of internal chemical activity or as a result of exposure to surrounding conditions should not be considered.

(*c*) ***Strength.*** Especially refers to those members of a wall subject to structural stresses aside from such loose material as may act only as fill to provide greater thermal resistance.

(*d*) ***Fireproof.***

(*e*) ***Light weight.*** To avoid the use of heavy structural members and minimize outward pressure of heavy loose material.

(*f*) ***Water repellent.*** Moisture increases conductivity and materials that resist absorption either as free water or water vapor should be selected.

(*g*) ***Odorless.*** When either wet or dry.

(*h*) ***Sanitary.*** Materials that provide a medium for vermin infestation should be excluded as insulation.

(*i*) ***Nonsettling.*** Any degree of loose material (fill) leaves upper wall areas with reduced insulating effect and increases conductivity of lower areas as a result of increased density.

(*j*) ***Low cost.***

(*k*) ***Adaptability.*** For the use intended.

Insulating materials generally have four basic uses, all of which govern the selection of the kind and the type to be selected. They are: (1) to retard heat flow, (2) to prevent surface condensation, (3) to add structural rigidity, and (4) to control noise and vibration.

Fig. 374. Chest-type food freezer showing insulation. Note location of condensing unit and evaporator coils. (Courtesy, Frigidaire, Division of General Motors Corporation.)

While the first enters the selection of any material for purposes of insulation, the second, third, and fourth are considerations limited to the conditions involved. The prevention of surface condensation on walls automatically follows the selection of proper insulating materials. On the other hand, structural rigidity is not a factor where nonrigid or loose insulating materials must or will be selected. There are some instances, such as with water pipes or cold-air ducts, where the prevention of surface condensation is the only factor involved. There is a wide variety of insulating materials, some of which are adaptable to all requirements listed above, while others cannot be considered as structural members.

TYPES OF INSULATION

There are a number of types of insulating materials, the selection of which depends on the use for which it is intended. All types include proprietary as well as natural materials. The greatest insulating effect is obtained with materials having the lowest coefficient of conductivity. Where structural members do not provide the necessary resistance to heat

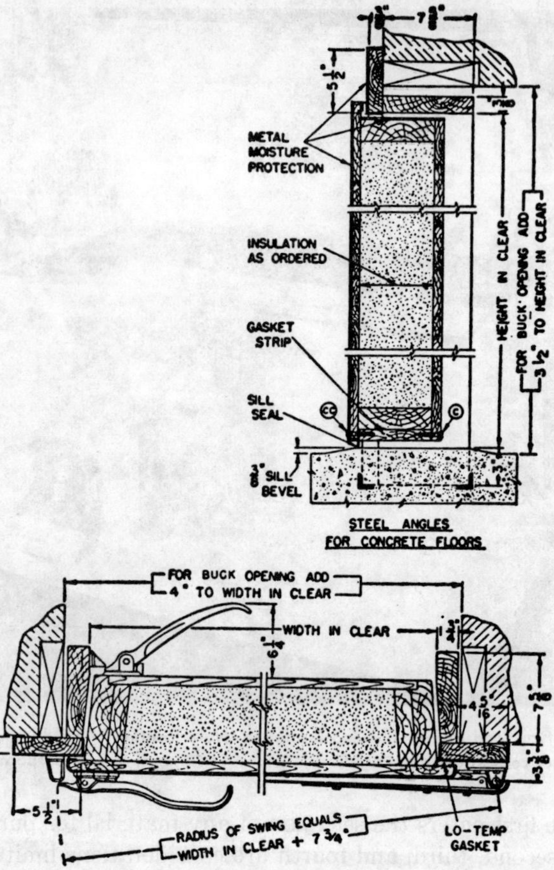

Fig. 375. Cross-section of low-temperature compartment door. The type shown provides for any kind of insulation material desired. (Courtesy, Jamison Cold Storage Door Co.)

flow, they must be supplemented with an appropriate insulation. For those conditions where the mean temperature difference is below 100 F, the k factor for such supplemental insulating materials should be below 0.5 Btu per hr per sq ft per in. per 1 F.

One method of expressing the relative insulating effect of a material is to compare its coefficient of conductance with that of cork ($k = 0.30$). Expressed in terms of *cork equivalent,* any material regardless of thickness having a coefficient of conductance of 0.30 would be equivalent to 1 in. of cork. Therefore, specifications calling for insulation having a conductance value equal to 1 in. of cork may be satisfied by using any material even though several inches of thickness are required to reduce its coefficient of conductance to equal 0.30 Btu per hr per sq ft per 1 F. *The thickness of any material having a coefficient of conductance of 0.30 Btu per hr per sq ft per 1 F is called the cork equivalent of that material.*

To obtain the cork equivalent for any thickness, divide the coefficient of conductivity, k, for cork by the desired thickness, L, in inches or

$$\text{Cork equivalent} = \frac{0.30}{L}$$

Fig. 376. Cold-storage door designed for low temperature. Note kick plate and method of sealing around door edges. (Courtesy, Jamison Cold Storage Door Co.)

For progressive thickness from 2 in. to 12 in., then, the cork equivalents are as follows (see Table 78).

TABLE 78. EQUIVALENT CONDUCTANCES FOR VARIOUS THICKNESSES OF CORK

Thickness of Cork, in.	*Btu per Hr per Sq Ft per 1 F*	*Thickness of Cork, in.*	*Btu per Hr per Sq Ft per 1 F*
2	0.15	8	0.0375
3	0.10	9	0.0333
4	0.075	10	0.030
5	0.060	11	0.0273
6	0.050	12	0.0250
7	0.043		

Such materials as wool, fiber, and hair are mentioned hereafter in Chapter 12. *Fiber* as associated with insulation signifies a cellulose or vegetable fiber, such as sugar cane, used with a binder to form wall boards. Celotex, a proprietary material, is an excellent example of a fire-

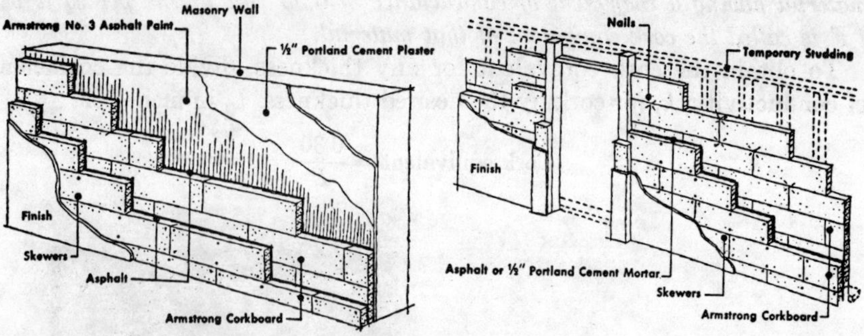

Fig. 377. (*Left*) Method of covering masonry wall with corkboard. (*Right*) Method of building solid corkboard wall. (Courtesy, Armstrong Cork Company.)

proof fiber wall board, in which bagasse (sugar cane after juice is removed) is a principal ingredient. Kapok is the silky fiber found within the seedpod of the ceiba tree, and usually is used in loose form.

The *wools*, such as rock wool, mineral wool, glass wool, and slag wool, are all names given to insulation materials made by melting the principal elements from which they are formed, such as rock, glass, slag, and certain type minerals, and blown by air into fibrous form, after which they are usually ground. They are used either as fill in their loose form or mixed with a binder to form wall boards.

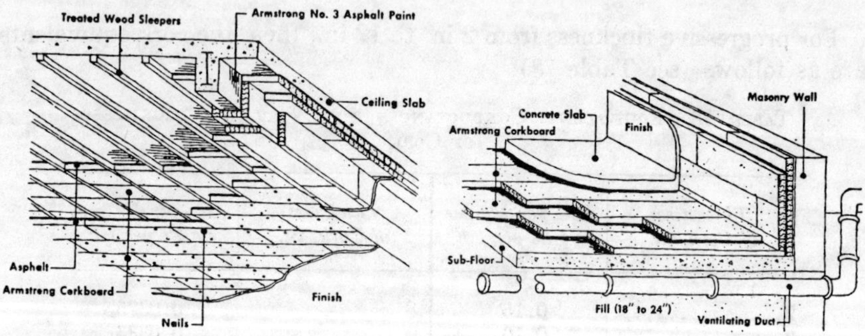

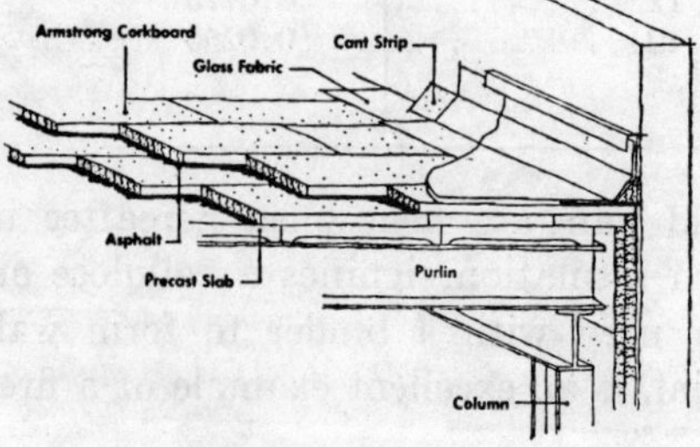

Fig. 378. Method of covering ceiling (*top, left*), floor (*top, right*), and roof (*bottom*) with corkboard. (Courtesy, Armstrong Cork Company.)

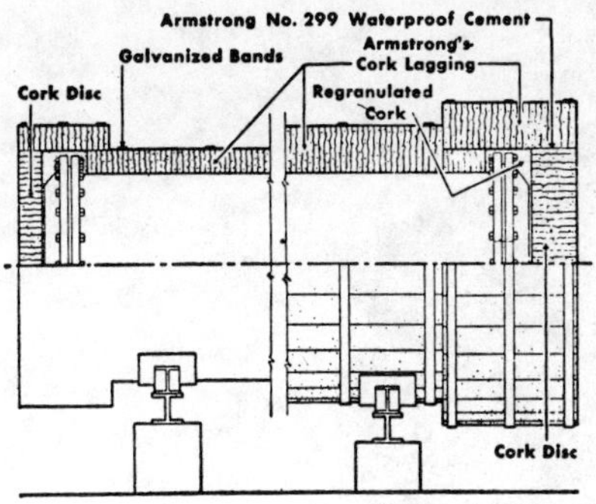

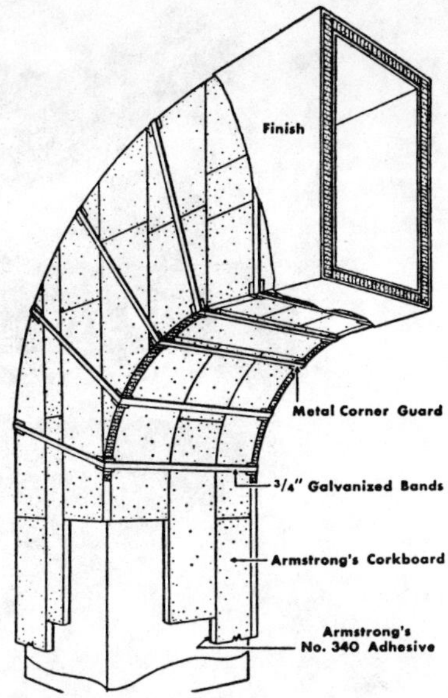

Fig. 379. Method of insulating (cork) tanks and vessels (*above*) and ducts (*right*). (Courtesy, Armstrong Cork Company.)

Types of insulating materials may be classified as follows:

Slab or block.	Cork, asbestos, glass, magnesia, etc.
Rigid board.	Cork board, balsa, mineral wool board, gypsum board, fiber board, asbestos, etc.
Batting.	Rock wool, cotton, glass wool, etc.
Blanket (flexible). . .	Hair felt, mineral wool, aluminum foil, eel grass, etc.
Loose or fill.	Granulated or regranulated cork, sawdust, wood shavings, slag wool, rock wool, diatomaceous earth, charcoal, kapok, etc.
Pipe covering.	Corrugated asbestos paper, cork, hair felt, magnesium, etc.
Sealers.	Asphalt, plaster, cement, paper, etc.
Reflective type.	Aluminum foil, sheet steel coated with a lead-tin alloy, aluminum powder coating, aluminum paint, etc.
Building materials (structural).	Hollow clay tile, cinder blocks, concrete blocks, shingles, wood siding or sheathing, brick, stone, etc.

Hair or *hair felt* designates such insulation made from animal hair and covered on both sides with paper or cloth to form a blanket.

Bats are insulating pads made of various wools (rock, slag, or glass) lined on one side to form a seal against condensation.

Natural insulation materials include cork, hair, eel grass, sawdust, wood shavings, and redwood bark. Proprietary materials incorporate a number of natural materials to form a heterogeneous mass with such binders and other processed substances as to produce wall boards, pipe covering, bats, blankets, and loose fill of low coefficient of conductivity. These proprietary types also include materials composed of various wools, fibers, and minerals

Fig. 380. Types of cork fitting coverings. (Courtesy, Armstrong Cork Company.)

such as magnesia, silica, and asbestos. The latter may be spun, woven, felted, or mixed with cement or other binders to form wall board. While asbestos products are excellent insulators at any temperature, the mineral itself holds a comparable position at higher intensities (above 100 F) as cork does at low temperature.

The principle of insulation is based primarily on the presence of dead air space within a material, first because of the low conduction property of air and, secondly because convected heat is held to a minimum under dead air (relatively still air) conditions. Because of the large number of air cells, the temperature difference across any one is so small as to exclude the radiation effect. The density of any given material is expected to become lower when the number of air cells within its mass increases. It follows that the heat conduction capacity of any material will increase with an increase in its density. The converse is also true, but with a limitation on the minimum density that may be obtained without incurring air leakage through the mass as a result of porosity.

The presence of moisture adversely affects the insulating ability of any material. They should be maintained in a dry condition even to the extent of sealing the joints of such cabinets or compartments under consideration. Where they are exposed to either the high or the low temperature region, there is always the danger of condensation occurring on the warm side.

Fig. 381. Fitted cork coverings. Made in various sizes and thicknesses to fit the service. LIGHT DUTY (1.20 to 1.93 in. thick) for temperatures of 30 F or above. MEDIUM DUTY (1.70 to 3.00 in. thick) for temperatures from 0 F to 25 F or where the temperature difference between refrigerant and surrounding air is 100 F or less. HEAVY DUTY (2.63 to 4.00 in. thick) for temperatures of 0 F to −25 F. Also made in special thicknesses above 4 in. (Courtesy, Armstrong Cork Company.)

Any sealing material used should be placed on the warm side to prevent condensation occurring inside the insulation.

HEAT-TRANSFER EQUIPMENT

In the foregoing section of this chapter, heat transfer as applied to insulating and building materials was discussed. In this connection emphasis lay in resisting heat flow through walls, floors, and ceilings of low-temperature compartments in order to minimize refrigeration loads while maintaining required low temperatures within the enclosed space. For refrigeration purposes, the resistance to heat flow must occur outwardly; i.e., from an external high-temperature region to an internal low-temperature region. The reverse is true in the case of compartment heating. Therefore, the same procedure for calculating heat losses may be used as is shown in the preceding examples for determining heat gain. The basic difference between a cooling system and a heating system is that in the one heat gained must be absorbed by the refrigerating element (evaporator) and in other heat lost must be replaced by a heating element (radiator). For economic reasons, both must be accomplished with all possible efficiency.

In this section dealing with heat-transfer equipment, principally evaporators and condensers, the objective becomes one of *rapid* heat transfer and not its retardation as previously desired for building materials. Consideration here involves heat conductors rather than heat insulators.

Heat-transfer equipment referred to here are those component parts of mechanical refrigeration systems wherein heat is absorbed from the space or substance being cooled or ejected by the refrigerant within the system or to an external source. Such activity involves evaporators, heat exchangers, intercoolers, after-coolers, and condensers. It also involves condenser water coolers or cooling towers where such equipment is employed.

Tubes and piping become the major heat-transfer elements in the equipment considered. As metals are the best conductors of heat and are also the materials from which tubing and piping are made, one of the major factors in rapid heat transfer becomes automatically congruent with standard manufacturing practice in promoting desirable conditions. It leaves only the kind of metal, state of opposing fluids, and design procedures to be considered.

Copper, brass, aluminum, iron, and steel are the metals most commonly used in refrigeration heat-transfer equipment. While the conduction properties of copper are greater than any of the others mentioned here, the thickness involved tends to offset the advantage that any one metal possesses over another in this respect. The rate of heat transfer per square foot of bare tube surface per hour per 1 F mean temperature difference for any one may be adopted as a constant for all under similar design and operating conditions. In fact, so great are the influences of other factors—such as turbulence, surface conditions, and fluid state—that the conduction property of any of the commonly used metals involved becomes negligible for most practical purposes in the case of plain (not finned) tubes.

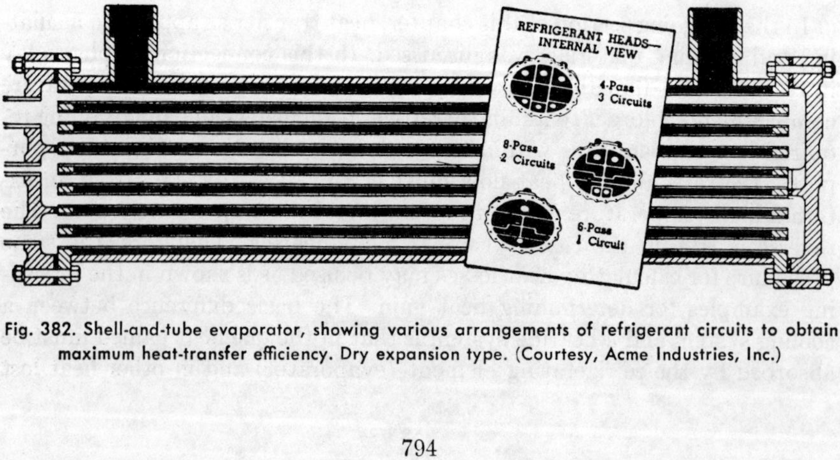

Fig. 382. Shell-and-tube evaporator, showing various arrangements of refrigerant circuits to obtain maximum heat-transfer efficiency. Dry expansion type. (Courtesy, Acme Industries, Inc.)

Heat-Transfer Factors in Equipment

The rate of heat transfer through a plate or tube wall is governed by the following factors, all of which individually affect the final result.

1. Fluid factor:
   (*a*) Kind
   (*b*) Velocity
   (*c*) State
   (*d*) Density
   (*e*) Viscosity
   (*f*) Thermal conductivity
   (*g*) Specific heat
2. Tube factor:
   (*a*) Kind of material
   (*b*) Thickness
   (*c*) Design—plain or finned
3. Temperature factor:
   (*a*) Mean temperature difference
   (*b*) Temperature level
4. Surface factor:
   (*a*) Character—rough, smooth, dull, or bright
   (*b*) Cleanliness
   (*c*) Turbulence
   (*d*) Surface temperature
   (*e*) Wetting characteristics
5. Design factor:
   (*a*) Tube diameter
   (*b*) Tube arrangement
   (*c*) Surface area

The influence exerted by the various factors listed above may be summarized by outlining the several laws and principles governing the rate of heat transfer through a conductor, viz.:

1. Fluid Factor. The coefficient of surface conductance (f) varies with the kind of fluid. Ammonia, for example, possesses a higher value in this respect than R-12.

The rate of heat transfer through a separating wall increases as the velocity, the density, the specific heat, and the thermal conductivity of a fluid increase. Fluid film conductance increases as fluid velocity increases.

The rate of heat transfer decreases as the viscosity increases. Liquid film conductance increases as liquid velocity increases.

The state of opposing fluids affects the rate of heat transmission. For example, the rate of transfer is greater between liquids than it is between gases, or liquid to gas.

TABLE 79. FILM COEFFICIENTS—EQUATIONS*

f = Btu per sq ft per hr per °F, heat transfer across a film between a surface (pipe, etc.) and a fluid

1. Moving fluid:

$$f = B\frac{k}{L}\left(\frac{LVp}{\mu}\right)^a\left(\frac{C_p}{k}\right)$$

2. Liquids heated or cooled outside single tubes, direction of flow normal to tubes:

$$f = 0.385\frac{k}{D}\left(\frac{DVp}{\mu}\right)^{0.56}\left(\frac{C_p\mu}{k}\right)^{0.3}$$

3. Gases heated or cooled inside tubes:

$$f = 0.0225\frac{k}{D}\left(\frac{DVp}{\mu}\right)^{0.8}\left(\frac{C_p\mu}{k}\right)^{0.4}$$

(provided that V_p is = or > than $1650P^{0.645}$ where P = pressure in atmospheres absolute

4. Any gas in turbulent flow within a tube:

$$f = 36C_pG^{0.8}\left(\frac{Z}{d}\right)^{0.2}$$

5. Turbulent flow of air across a single pipe:

$$f = 8.2G^{0.58}\frac{(1 + 0.000576t)}{d^{0.42}}$$

6. Turbulent flow of air across four or more rows deep of staggered tubes:

$$f = 1.7(t + 460)^{0.3} \times \frac{G^{2/3}}{d^{1/3}}$$

7. Low velocity (15 fps or less) air across smooth plane surface:

$$f = 0.8 + 0.22V$$

8. Air at higher velocities (than 7) across smooth plane surface:

$$f = \frac{(0.56V^{3/4})}{L^{1/4}}$$

9. Liquids heated inside tubes, turbulent flow, and $\frac{C_p\mu}{k} < 10$

$$f = 0.0225\left(\frac{DVp}{\mu}\right)^{0.8}\left(\frac{C_p\mu}{k}\right)^{0.4}$$

10. Liquids cooled inside tubes, turbulent flow:

$$f = 0.0225\left(\frac{DVp}{\mu}\right)^{0.8}\left(\frac{C_p\mu}{k}\right)^{0.}$$

11. Gases heated or cooled outside single tubes, direction of flow normal to tube:

$$= \text{constant}\ \frac{k}{D}\left(\frac{DVp}{\mu}\right)^{0.56}$$

where constant equals the value of $\frac{C_p\mu}{k}$ in equation (2) is as follows:

0.68 for gases having 1 atom per molecule
0.74 for gases having 2 atoms per molecule
0.75 for gases having 3 atoms per molecule
0.98 for gases having more than 3 atoms per molecule

and for air or other diatomic gases the constant is 0.35

* From the "Refrigerating Data Book 5th Edition," 1942, by courtesy of The American Society of Refrigerating Engineers.

TABLE 79. FILM COEFFICIENTS—EQUATIONS—(*Continued*)

12. For tube bundles in equation (11):

f should be raised 1.3 times for single tubes staggered

f should be raised 1.2 times for single tubes in line

13. Liquids flowing normally to tubes use equation (3) applying the factors 1.3 and 1.2 for reasons stated in equation (12).

14. Gravity flow of water in layer form over horizontal pipes:

$$= 65\left(\frac{w}{\frac{2L}{D_o}}\right)^{1/3}$$

15. Turbulent flow of water in layer form down inner walls of vertical pipe:

$$f = 120\left(\frac{w}{\pi D_o}\right)^{1/3}$$

16. Cold water flowing through straight tubes and warm water circulating through horizontal shell and directed back and forth across the tubes by baffles, f for the shell side of tube is

$$f = 0.73\,\frac{k}{D}\left(\frac{DVp}{\mu}\right)^{0.66}\left(\frac{C_p\mu}{k}\right)^{0.4}\left(\frac{d_s}{H}N_b\right)^{0.35}\frac{P^{0.95}}{D_o^{0.15}}$$

17. Turbulent flow of water in clean pipes:

$$f = 1.5(t + 100)\frac{V^{0.8}}{d^{0.2}}$$

18. Turbulent flow of oil in pipes, oil being heated:

$$f = \frac{70V}{Z^{0.63}}$$

19. Turbulent flow of oil in pipes, oil being cooled:

$$f = 0.75\,\frac{70V}{Z^{0.63}}$$

KEY:

- f = film coefficient, Btu per sq ft per hr per °F
- B = constant
- k = thermal conductivity of fluid, Btu per sq ft per hr per °F
- L = linear dimension of heat-transfer surface ft (pipe would be length)
- V = average linear velocity, ft per hr
- p = fluid density lb per cu ft
- μ = viscosity of fluid, lb per hr ft (centipoises $\times$ 2.42)
- C_p = specific heat of fluid at constant pressure, Btu per lb per °F
- a & b = exponents determined by experiment
- D = outside diameter of tube in feet
- G = mass velocity, lb per sq ft, sec ($V \times p$)
- d = tube diameter, in.
- t = gas or water temperature, F
- Z = viscosity in centipoises
- w = lb of water flowing over one tube per hr
- d_s = diameter of shell, ft
- N_b = number of baffles
- P = tube pitch, ft
- H = distance between inlet and outlet, ft

2. Tube Factor. The coefficient of conductivity varies with different materials. The relative k values of commonly used metals, as shown below, demonstrate this law:

TABLE 80. TUBE FACTOR—CONDUCTANCE (k) FOR SOME METALS USED IN HEAT-TRANSFER EQUIPMENT

Metal	k = *Btu per hr per sq ft per in. per °F*
1. Copper	2640
2. Aluminum	1400
3. Brass	750
4. Wrought iron	417
5. Steel—mild	312

The rate of heat transfer decreases as the thickness of the tube wall increases.

3. Temperature Factor. The rate of heat transfer increases as the mean temperature difference becomes greater. (In equipment design, the mtd is used, whereas the difference between inside and outside temperatures becomes a similar factor in calculating insulation requirements for buildings and low-temperature compartments.)

The coefficient of conductivity of metals usually decreases as the level of temperature increases.

4. Surface Factor. Rough or dull surfaces encourage a higher rate of heat transfer than do smooth or bright surfaces.

Scale, oil film, fluid film, ice, or snow deposited on surfaces decrease the rate of heat transfer.

Turbulence is an important factor in heat transfer, the latter increasing as the degree of turbulence increases. (While turbulence is a function of fluid velocity, it is also a function of design. However, it is considered here as a surface factor because of effect of film conductance in relation to surface temperature and condition.)

Wettable surfaces (oil or grease free) allow a greater liquid-metal contact, thus displacing bubbles that form an insulating film during evaporation processes. Bubbles having a small area of contact with the transfer surface break away more readily.

5. Design Factor. As the diameter of the tube increases, the ratio of volume to heat-transfer surface area becomes greater. The advantage lies with longer tubes of small diameter over short tubes of large diameter. Surface area, velocity, and conductivity of fluid are functions affected here.

Arrangement of tubes in relation to fluid flow is an important function

TABLE 81. THERMAL CONDUCTIVITIES OF LIQUIDS (K)

Liquid	*Temperature*	*Conductivity* K = Btu per hr per sq ft per °F per ft
Alcohols:		
Amyl (normal)	70	0.0938
Butyl	70	0.0886
Ethyl	70	0.0967
Methyl	70	0.1010
Ammonia	5–86	0.300
Calcium chloride brine	86	0.330 (av)
Carbon tetrachloride	70	0.0598
Ethylene glycol	70	0.1510
Freon-11	32	0.068
Freon-11	100	0.060
Freon-12	0	0.060
Freon-12	50	0.053
Freon-12	100	0.047
Freon-21	32	0.077
Freon-21	104	0.067
Freon-22	32	0.070
Freon-22	104	0.056
Freon-113	32	0.058
Freon-113	104	0.050
Glycerol	70	0.1701
Lubricating oil	80	0.0815
Methyl chloride	5–86	0.1112
Methylene chloride	32	0.0704
Propylene glycol	70	0.1161
Sodium chloride brine	86	0.335 (av)
Trichloroethylene	70	0.0672
Water	70	0.350

TABLE 82. THERMAL CONDUCTIVITIES OF REFRIGERANTS—VAPORS

Vapors	*Temperature °F*	*Conductivity* K = Btu per hr per sq ft per °F per ft
Ammonia	5	0.0120
	86	0.0144
	210	0.0185
R-11	86	0.0048
	194	0.0056
R-12	50	0.0051
	100	0.0058
	150	0.0064
R-22	86	0.0068
	194	0.0080
R-113	86	0.0045
	194	0.0059

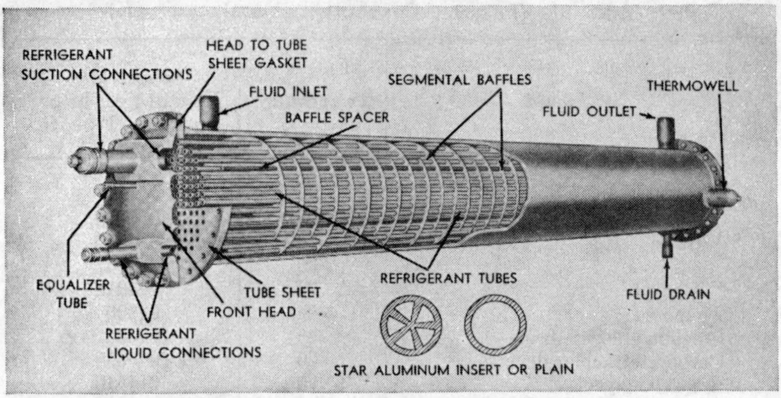

Fig. 383. Shell-and-tube dry expansion evaporator. Note baffle arrangement to obtain maximum heat transfer between refrigerant and fluid to be cooled. Star aluminum insert placed in refrigerant tubes increases rate of heat transfer. (Courtesy, Acme Industries, Inc.)

of heat transfer. Considerations include distance between tubes, relative location (staggering or uniform), and relative direction of fluid flow (counter or parallel flow).

The rate of heat transfer between opposing fluids increases as the area of transmission becomes greater. Forced circulation increases heat transfer by providing higher velocity and turbulence, eliminating liquid or vapor traps, and retarding oil and scale deposits.

Efficient design and/or maintenance of heat-transfer equipment is contingent on the adoption of certain rules based on the above laws and principles. Individual selection, however, depends on whether a change in state, i.e., evaporation or condensation, is occurring on one side of the tube wall or if heat transfer involves sensible heat only, as it would in cases of liquid-to-liquid, liquid-to-gas, etc. Such rules are as follows:

1. Use metals possessing high heat-conduction properties for tubing.
2. Where the internal film conductance is greater than the external film conductance for plain surfaces, external surfaces should be finned.
3. Maintain oil-free surfaces to impart wettability.
4. Employ oils that congeal only at temperatures below the surface temperature to prevent the formation of oil insulating films.
5. Use soft or treated water (when involved) to reduce scale formations on ferrous pipes.
6. Employ tubes of smaller diameter to minimize bubble size and to decrease volume-surface ratio. The limit to which diameters may be reduced is governed by internal fluid friction.
7. Incur turbulence on both sides by maintaining high velocity of fluid being cooled and/or forced circulation of refrigerant.

8. Employ dull surfaces.
9. Maintain effective temperature differentials.
10. Opposing fluids should flow in a direction counter to each other where practical.
11. Provide minimum thickness of tube wall commensurate with operating pressures and strength of material.
12. Tubes should be spaced closely to provide maximum turbulence.

TABLE 83. HEAT-TRANSFER COEFFICIENTS IN DOUBLE-PIPE CONDENSERS*

(10 F mtd)

Water Velocity (fpm)	*Heat Transfer (Btu per sq ft per hr per °F*
50	185
100	240
150	285
200	330

* Reprinted from the "Air Conditioning-Refrigerating Data Book 9th Edition—1955–56 Design Volume," by courtesy of The American Society of Refrigerating Engineers.

TABLE 84. HEAT-TRANSFER COEFFICIENTS IN HORIZONTAL SHELL-AND-TUBE AMMONIA CONDENSERS*

Velocity of Water (ft per min)	*Btu per sq ft per hr per 1 F (approx)*		
	Clean Tubes	*Dirty Condition*	*Practical Use Conditions*
100	245	160	120
200	375	220	180
300	440	240	215
400	460	250	225

* Reprinted from the "Air Conditioning-Refrigerating Data Book 9th Edition—1955–56 Design Volume," by courtesy of The American Society of Refrigerating Engineers.

TABLE 85. HEAT TRANSFER IN ATMOSPHERIC BLEEDER-TYPE CONDENSER*

Water Supply (gpm per ft of length)	*Btu per sq ft per hr per °F*
1	78
2	124
3	153
4	170
5	180

* Reprinted from the "Air Conditioning-Refrigerating Data Book 9th Edition—1955–56 Design Volume," by courtesy of The American Society of Refrigerating Engineers.

MATHEMATICAL DETERMINATIONS OF HEAT TRANSFER IN EQUIPMENT

Considering the number of dimensionless factors that affect the rate of heat transfer in equipment, outlined in the foregoing discussion, it is obvious that accurate determinations of an over-all coefficient of heat transmission must be augmented by some experimental data based on conditions similar to those immediately involved. The difficulty in deriving an accurate coefficient becomes more apparent when the number of variables, or combination of variables, present in the operation of such equipment are considered. Any changes in fluid velocity, fluid density, scale or ice formation, and oil film will render calculated results based on theoretical conditions such as clean tube surfaces, for example, somewhat impractical under operating conditions.

To include the necessary tables and graphs covering all the allowance factors of variables involved in heat-transfer equipment would require more space in this text than would be commensurate with their value for practical considerations. It is deemed sufficient to include here (1) the fundamental equation for calculating over-all coefficients which holds the key to the necessary allowance factors, and the value of U for different types of equipment, as shown in the various tables.

In its simplest form—i.e., where only clean tube surfaces are involved—U may be expressed by an equation that considers and combines the three basic heat-transfer factors. They are: (1) external film coefficient, (2) coefficient of conductivity of tube material, and (3) internal film coefficient.

A direct-expansion cooling coil exposed to air, as would be the case in some cold-storage compartments, serves as an excellent example assuming that the surfaces are free of frost. The over-all coefficient of heat transmission under such a condition is expressed by the equation

$$U = \frac{1}{\dfrac{A_o/A_i}{f_i} + \dfrac{L}{k} + \dfrac{1}{f_o}}$$

where U = coefficient of over-all heat transmission—Btu per hr per sq ft per °F
A_o = external surface area
A_i = internal surface area
A_o/A_i = ratio of external to internal surface area
f_o = external film conductance
f_i = internal film conductance
L = thickness of tube
k = coefficient of conductivity of tube material

The above equation expresses U for bare tubes free of any foreign surface film such as scale, oil, ice, etc. If tubes are finned, then f_o would necessarily be modified according to its fin efficiency, in which event $\frac{1}{f_o}$ would become $\frac{1}{ef_o}$ where e = fin efficiency. The value of f_o varies with external fluid velocity, reaching a minimum in still air (zero velocity). The value of f_i for a given refrigerant varies according to fluid conditions within the tube, reaching a maximum with wetted (liquid film) surfaces.

Where tube walls are relatively thin, the $\frac{L}{k}$ value for commonly used metals having high heat conduction properties becomes negligible and usually is omitted in the above equation. Neglecting $\frac{L}{k}$, the equation is reduced to its smallest expression, viz.:

$$U = \frac{1}{\dfrac{A_o/A_i}{f_i} + \dfrac{1}{f_o}}$$

In the practical application of heat-transfer equipment, surfaces free of foreign film cannot be assumed. Neither can it be assumed that such foreign films when present will continue to remain uniform in their thickness and so enable the use of constant allowance factors in determining over-all coefficients of heat transmission. While the equation for determining U is required to provide for the effect of foreign surface films, it should be recognized that their effective values vary according to the thickness and the type of film, which, in turn, are governed by operating conditions and such factors as pour and cloud point of oil, hardness of water, velocity of fluids, kind of tube material, and surface characteristics.

In equipment such as a shell-and-tube condenser, the coefficient of over-all heat transmission may be determined by the following equation, in which allowances for scale formation, both internally and externally, are made:

$$U = \frac{1}{\frac{1}{f_i} + \frac{1}{f_{sr}} + \frac{L}{k} + \frac{A_o/A_i}{f_{sw}} + \frac{A_o/A_i}{f_o}}$$

where f_{sr} = scale factor – refrigerant side – Btu per hr per sq ft per °F mtd
f_{sw} = scale factor – water side – Btu per hr per sq ft per °F mtd

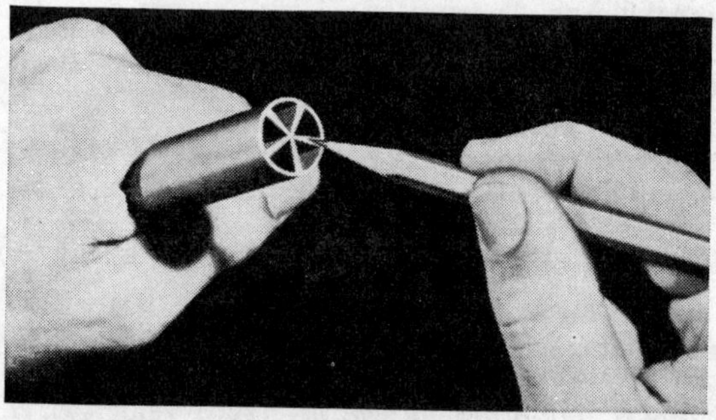

Fig. 384. Star aluminum insert placed in refrigerant tubes of dry expansion evaporators to increase heat-transfer rate. (Courtesy, Acme Industries, Inc.)

The value of f_{sw} decreases as the scale thickness increases and water velocity decreases. Water hardness (a measure of calcium and magnesium compounds present), water velocity, and tube temperatures govern the amount of scale that will build up on the surfaces. The presence of an oil film on tube surfaces as a result of carry-over from the compressor by the refrigerant decreases the value of f_{sr} to an extent commensurate with its thickness.

In order to maintain highest values of f_{sw} and f_{sr} for equipment in operation, it is necessary to provide against aggravation of scale and oil film deposits. This can be accomplished to a large extent by proper water treatment to minimize scale formation and employing oil having a pour point somewhat below the lowest temperature encountered within the equipment. Where already present, scale should be removed manually with wire brushes. Usually oil will leave the tube surfaces when the surface temperature is raised above the pour point of the oil, after which it can be drained from the lowest part of the unit.

From this discussion it readily can be deduced that considerable conjecture enters into any calculations relative to over-all coefficients of heat

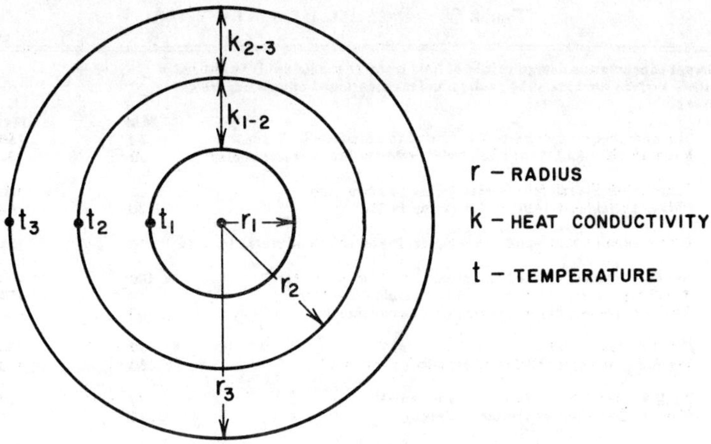

Fig. 385. Compound cylinder.

transfer in equipment when considered from an operating standpoint. Such conjectures may amount to considerable differences in computing heat transfer values solely on a theoretical basis with clean surfaces, optimum velocities, etc., as against practical considerations involving day-to-day variations in operating conditions and the degree of maintenance efficiency. It is the latter two factors that spell the difference between designing equipment to fulfill a load demand under conditions that are known and held within narrow limits of those conditions and the selection of standard equipment to perform satisfactorily under such adverse operating conditions as may reasonably be expected.

For the designer of heat-transfer equipment, the equations presented above are useful when correctly modified by factors established by experiment under conditions similar to those to which the designed unit will be exposed. As noted in the U equation for equipment, the variable factors include film conductances of the fluids involved and scale factors. Film conductances on the refrigerant side vary with the kind of refrigerant employed, the state of the fluid, temperature difference, and fluid turbulence. On the side being cooled (water, brine, oil, etc.), the film conductance varies as the velocity or turbulence and inversely as the viscosity. Scale factors vary according to the hardness of water and water velocity. On the refrigerant side, the scale factor is affected by the thickness of oil film or other foreign deposits.

OVER-ALL COEFFICIENT OF HEAT TRANSFER IN EQUIPMENT

Through experience and experiment, the amount of heat that will pass through a tube wall under most conditions found in practical operation

Table 86. Over-all Heat Transfer*

Current conservative design values of heat transfer coefficient Û based on outside surface for bare tube coolers, unless mentioned otherwise, are as follows:

	Min	Max
Flooded shell-and-tube cooler (water to ammonia or Freon)	50	150
Flooded shell-and-finned tube high velocity Freon water cooler	30	150
Flooded shell-and-tube cooler (brine to ammonia)	45	100
Flooded shell-and-tube cooler (brine to Freon)	30	90
Dry expansion shell-and-tube cooler, Freon in tubes, water in shell	50	115
Baudelot cooler, flooded (ammonia or Freon to water)	100	200
Baudelot cooler, dry expansion (ammonia to water)	60	150
Baudelot cooler, dry expansion (Freon to water)	60	120
Double-pipe cooler (water to ammonia)	50	150
Double-pipe cooler (brine to ammonia)	50	125
Shell-and-coil cooler (water to ammonia)	10	25
Shell-and-coil cooler (water to Freon)	10	25
Spray type shell-and-tube water coolers (ammonia or Freon)	150	250
Tank-and-agitator, coil type water cooler, ammonia, flooded	80	125
Tank-and-agitator, coil type water cooler, Freon flooded	60	100
Tank, ammonia, brine cooling, coils between can in ice tank	15	40
Tank, high velocity raceway type, brine to ammonia	80	110

* Reprinted from the "Air Conditioning-Refrigerating Data Book 9th Edition—1955–56 Design Volume," by courtesy of The American Society of Refrigerating Engineers.

Table 87. Average Over-all Coefficient of Heat Transfer

(Heat transferred between fluids separated by a metal wall, as from outside to inside of a clean pipe, tube, etc., or vice versa)

U = Btu per hr per sq ft per °F*

Fluids	*Free or Mildly Agitated Flow* U	*Forced Flow* U
Air to air	1	4
Water to water	40	200
Air to water	2	6
Water to air	2	7
Air to brine	3	12
Air to evaporating refrigerant	2	8
Water to evaporating refrigerant	12	100
Brine to evaporating refrigerant	12	150
Condensing refrigerant to air	2	7
Condensing refrigerant to water		200
Condensing refrigerant to evaporating refrigerant		500

* Values of U are given here only to show relative heat-transfer characteristics of various fluid combinations. Because the value of U varies with fluid velocity, pipe diameter, efficiency of the heat-transfer system, and kinds of fluids, equipment capacity is more accurately calculated using manufacturers' test data for given conditions.

TABLE 88. OVER-ALL COEFFICIENTS OF HEAT-TRANSFER MISCELLANEOUS EQUIPMENT

U = Btu per hr per sq ft per °F

Equipment	U
Vertical shell and tube ammonia condenser:	
2-in. tube 14 ft long—water flow 2 to 4 gpm per tube.	175 to 300
$2\frac{1}{2}$-in. tube, 10 ft long—water flow 2 to 4 gpm per tube	120 to 280
Double pipe condensers—water velocity 50 to 300 fpm..	175 to 300
Atmospheric condensers, bleeder type, 1 to 5 gpm water/ft length	100 to 200
Atmospheric condensers, ref. flow from top to bottom...	30 to 60
Horizontal shell and tube condensers—water velocity 50 to 400 ft per min	75 to 225
Shell and coil condenser—water velocity 50 to 250 ft per min	150 to 300
Air-cooled condenser—air velocity, still to 800 fpm	2 to 10

is well established. Such data have been compiled in various handbooks and manuals in terms of over-all coefficients of heat transfer, and it requires only the simple procedure of comparing the controlling factors of the immediate problem with similar factors in the table to determine applicable U values for estimating the required surface area for a given heat load. This procedure of determining U rather than the alternate computing method is recommended for most practical purposes.

The over-all coefficients of heat transfer in equipment are presented here in such a manner as will enable the student to select U on the basis of fluid conditions, as shown in Table 87, or according to the type of equipment, as shown in Table 88. The value of U in any case is given as a range rather than a constant because the rate of heat transfer increases as the velocity of the fluid on either or both sides of the separating wall increases.

HEAT-TRANSFER SURFACE AREA CALCULATIONS

When the over-all coefficient of heat transmission, U, is known, the surface area requirements in any thermal equipment for a given refrigeration load are determined by the equation

$$A = \frac{H}{U \times (t_2 - t_1)}$$

where A = heat transfer surface in square feet
H = refrigeration load, Btu per hour
U = coefficient of over-all heat transmission
$(t_2 - t_1)$ = mean temperature difference

EXAMPLE: How many square feet of $1\frac{1}{4}$-in. steel pipe surface are required for a shell-and-tube ammonia brine cooler operating under a refrigeration load of 20 tons and 5° mean temperature differential?

Solution:

1. From Table 86, a U valve of 100 Btu per hr per sq ft per °F for a shell-and-tube ammonia brine cooler is found.

2. The refrigeration load of 20 tons is equivalent to $20 \times 12{,}000$, or 240,000 Btu per hr.

3. Applying the formula

$$A = \frac{H}{U(t_2 - t_1)}$$

then

$$A = \frac{240{,}000}{100 \times 5}$$

$$A = 480 \text{ sq ft of pipe surface}$$

As it is an ammonia system, the steel pipe used will conform to Schedule 40 ASA B 36.10, as shown in Table 61, Standard Pipe Dimensions. For $1\frac{1}{4}$ in.-pipe 2.301 ft are required for each 1 sq ft of external surface.

To fulfill the problem requirements, it is necessary to multiply the lineal feet per square foot of external surface by the total area needed to maintain the load, or

Length of $1\frac{1}{4}$-in. pipe = total area (square feet) × length of pipe per square foot external surface
= 480 × 2.301
= 1104.48 lineal feet of $1\frac{1}{4}$-in. pipe

CONDUCTION, CYLINDRICAL WALLS

Heat transfer by conduction through cylindrical walls (pipes, tubes, silos, etc.), both simple and compound, i.e., having two or more insulation coverings or wall materials, may be determined as follows:

Simple wall

$$h_s = \frac{2\Pi k_1 (t_2 - t_1)}{\log_e \frac{r_i}{r_o}} \quad \text{Btu per sq ft per hr}$$

$$H_s = h_s \times L \quad \text{Total Btu transfer per hr}$$

Compound wall

$$h_c = \frac{2\Pi\ (t_3 - t_1)}{\dfrac{\log_e \frac{r_2}{r_1}}{k_1} + \dfrac{\log_e \frac{r_3}{r_2}}{k_2}} \quad \text{Btu per sq ft per hr}$$

$$H_c = h_c \times L \quad \text{Total Btu transfer per hr}$$

where h_s = Btu per sq ft per hr for simple wall
h_c = Btu per sq ft per hr for compound wall
k_1, k_2, etc. = thermal coefficient of conductivity of the respective insulations from r_2, r_3, etc., Btu per sq ft per hr per °F
t_1 = temperature °F at r_1 or r_i
t_2 = temperature °F at r_3 or r_o
r_i = internal radius, inches
r_o = external radius, inches
r_2, r_3 = external radius of separate materials
H_s, H_c = total heat transfer, Btu per hr per length L
L = length of cylinder considered

The above equations express values of h only when the materials are homogeneous (in which k_1 = constant, k_2 = constant, etc.) and the heat flows radially throughout the given length.

SUMMARY OF CHAPTER 12

Heat transfer occurs when a difference in temperature exists between bodies not perfectly insulated.

The direction of heat flow is from a region of higher temperature to one of lower temperature.

Materials that allow rapid heat transfer are known as conductors.

Materials that retard heat transfer are known as insulators or insulation materials.

Insulation materials are employed for purposes of maintaining a fixed temperature condition, such as in walls of cold-storage rooms through which heat transfer must be kept at a minimum.

Conductors are employed where rapid heat transfer is essential, as in evaporators and condensers of refrigeration systems.

Cork, mineral wool, balsa, and hair felt are materials possessing high insulating qualities and are employed as building materials.

Metals such as copper, aluminum, iron, and steel are excellent conductors and are used in making the heat-transfer elements of functional equipment.

Insulation materials owe their ability to resist heat flow to the presence of numerous "dead" air cells uniformly distributed throughout their mass. Cork, for example, is composed of 57 percent air and 43 percent wood fiber, by volume.

The rapid rate at which metals are capable of conducting heat is due to their compactness of grain structure or density.

Heat is transferred by conduction, convection, and radiation.

Conduction is heat transfer by molecular vibration and finds its best medium in solids and its poorest in gases, with liquids in between.

Convection is heat transfer by reason of density variations within a fluid mass due to expansion as a result of heat application causing motion throughout the mass.

Radiation is heat transfer by rays directly from a hotter body to a colder one. Radiant heat requires no transmitting medium.

Newton's law states that the rate of heat loss by radiation is proportional to the difference between the radiator temperature and the temperature of surrounding substances.

The intensity of radiant heat is proportional to the temperature of the radiator and inversely proportional to the square of the distance from the radiator.

The rate of heat transfer through a wall separating regions of different temperatures is governed by (1) degree of temperature, (2) wall thickness, (3) wall material, (4) velocity of fluid on either or both sides of wall, (5) surface character and condition, and (6) state of opposing fluids.

A perfect radiator would be a theoretical black body that absorbs all radiant heat impinging on it without radiating or transmitting any part of it.

Absorptivity is the ratio of the amount of heat absorbed by a body to the amount capable of being absorbed by a black body under similar thermal conditions. Absorptivity of a perfect radiator (black body) = 1.00.

Emissivity is the ratio of the amount of heat radiated (emitted) by a body to the amount radiated by a black body under similar thermal conditions.

Reflectivity is the ratio of radiant heat reflected by that body to the amount reflected by a black body under similar thermal conditions. Reflectivity of a perfect radiator = 0.0.

Bright, polished, and/or light-colored surfaces reflect heat and therefore are poor absorbers.

Dull, rough, and/or dark-colored surfaces are poor reflectors and good absorbers.

Heat transfer through a separating wall is expressed in Btu per hour, per square foot of surface, per 1 F temperature difference, which is called the coefficient for the method of transfer and wall thickness involved.

There are five methods for expressing coefficients of heat transfer:

k = thermal coefficient of conductivity = Btu that will pass through 1 sq ft of wall 1 in. thick by conduction only in 1 hr for each 1 F difference in temperature across its two surfaces.

c = thermal coefficient of conductance = Btu that will pass through 1 sq ft of wall of any thickness by conduction only in 1 hr per each 1 F difference in temperature across its two surfaces.

f = thermal coefficient of surface conductance = Btu transmitted to or from 1 sq ft of surface by radiation and convection in 1 hr for each 1 F difference in temperature between the surface and surrounding fluid.

a = thermal coefficient of conductance of air spaces = Btu transmitted from 1 sq ft of wall surface across an air space in 1 hr for each 1 F temperature difference between the external surfaces of walls enclosing the air space.

U = thermal coefficient of over-all transmission = Btu that will pass through 1 sq ft of wall having any thickness and any number of materials by the combined loads of conduction, radiation, and convection in 1 hr for each 1 F temperature difference between the regions separated.

Total heat transferred through any wall (H) is determined by multiplying the thermal coefficient of over-all transmission (U) by the total wall area in square feet (A) by the temperature difference ($t_2 - t_1$), or

$$H = U \times A \times (t_2 - t_1)$$

Resistivity means resistance to heat flow through a substance and varies inversely as U, k, c, f, or a, or

$$R = \frac{1}{U}, \frac{1}{k}, \text{ or } \frac{1}{c}, \text{ or } \frac{1}{f}, \text{ or } \frac{1}{a}$$

Therefore

$$U = \frac{1}{R}$$

and as R is the sum of the separate resistances, or

$$R = r_1 + r_2 + r_3 + r_4, \text{ etc.}$$

and

$$r = \frac{1}{k}, \text{ or } \frac{1}{c}, \text{ or } \frac{1}{a}, \text{ or } \frac{1}{f}, \text{ the separate resistances}$$

then

$$R = \frac{1}{k} + \frac{1}{f_i} + \frac{1}{f_o} \text{ for a simple wall}$$

or

$$R = \frac{L}{k_1} + \frac{L}{k_2} + \frac{L}{k_3} + \text{etc.}, + \frac{1}{c} + \frac{1}{a} + \frac{1}{f_i} + \frac{1}{f_o} \text{ for a compound wall}$$

Then

$$U = \frac{1}{\frac{L}{k} + \frac{1}{f_i} + \frac{1}{f_o}} \text{ for a simple wall}$$

and

$$U = \frac{1}{\frac{L}{k} + \frac{L}{k_2} + \text{etc.} + \frac{1}{c} + \frac{1}{a} + \frac{1}{f_i} + \frac{1}{f_o}}$$

$$= \text{coefficient of over-all transmission of a compound wall}$$

where L and k_1, k_2, etc. = the thickness and the coefficient of conductivity for the separate materials

STUDY OUTLINE FOR CHAPTER 13

Indirect systems of refrigeration
Secondary mediums:
- Water
- Air
- Brines:
  - Sodium chloride
  - Calcium chloride

Characteristics of brines:
- Density
- Specific heat
- Temperature of initial crystallization (freezing point)

Determination of salt content of brines:
- Specific gravity
- Hydrometry:
  - Archimedes' principle
- Salimeter
- Baumé gravity
- API gravity

Chemical balance of brines:
- General laws of chemistry:
  - Acids
  - Bases
  - Salts
  - Neutral solutions
- Ionization
- Degree of ionization
- Ionic definition of acids, bases, and salts

pH determinations of solutions

Measurement of pH:
- Litmus paper method
- Alkacid paper method
- Colorimetric method
- Glass electrode method

Chemical treatment of brines:
- Alkalinity control
- Corrosion retardation
- Neutralization of ammonia in brines due to leakage

Eutectic solutions

Anti-freeze solutions:
- Methanol
- Alcohols
- Glycerin
- Glycols

Endotherms and exotherms

The use of water in refrigeration and other processes

Major considerations in the use of water:
- Potability
- Impurities present that affect commercial processes
- Scale-forming impurities
- Conduit-fouling and -clogging impurities
- Corrosion due to:
  - Oxidation or rust
  - Scale
  - Acids
  - Alkalinity

Water-flow systems:
- Enclosed circulatory
- Open circulatory
- Enclosed noncirculatory
- Open noncirculatory

Water:
- Composition
- Characteristics, chemical and physical

Water treatment:
- Factors dictating type of treatment:
  - Ultimate use
  - Source of supply
- Impurities that affect use processes

Kinds of treatment required:
- Screening
- Sedimentation
- Coagulation
- Filtration:
  - Natural, coarse, slow, rapid (pressure), charcoal, miscellaneous
- Aeration (degasification)
- Deaeration
- Organic control:
  - Algae—copper sulfate
  - Bacteria, etc.—chlorination
- Inorganic control:
  - Precipitation of scale-forming compounds:
    - Phosphate as trisodium phosphate, etc., added to boiler feed waters
  - Distillation:
    - Neutralization
  - Demineralization:
    - Ion exchangers:
      - Cation (hydrogen zeolite, acid regenerant)
      - Anion (anion exchange resins, caustic soda regenerant)

Hardness reduction:
- Carbonate hardness:
  - Heating
  - Chemical treatment
- Noncarbonate hardness:
  - Lime soda-ash process (Precipitation)
  - Ion exchanger:
    - Sodium zeolite

Degree of water treatment necessary vs. ultimate use

Cost of water treatment vs. probable benefits

Cycles of concentration (of scale-forming solids in circulating cooling water)

Blow-down

13

Secondary or Indirect Refrigeration

Systems that employ a medium such as brine or water to transfer the heat from a body to be cooled to the evaporation coil of a refrigeration unit are known as secondary or indirect refrigeration systems. Such systems differ from direct expansion systems in which heat flow is accomplished directly from the body to be cooled to the evaporation coil.

Indirect systems require the use of suitable equipment in order to circulate the medium, such as pumps, fans, and agitators.

An excellent example of secondary refrigeration may be found in ice-manufacturing plants where brine, a salt water solution, is employed as the medium. Here cans containing pure water to be frozen are immersed in brine which, in turn, is cooled by direct contact with the evaporator coils. In order to facilitate freezing, an agitator is used to maintain the body of brine in motion around the ice cans to effect a greater rate of heat flow.

Another example of indirect refrigeration is the air-conditioning system where air is circulated from the space to be cooled through the cooling water or over the cooling coil and back to the cooled space. In this case, air is the medium. If spray water is used to cool the circulated air, as in the case of many large dehumidifying systems, a second medium is added. This type of system may be termed a compound secondary system of refrigeration.

Some installations such as cold-storage plants accomplish room cooling by the recirculation of brine through a pipe system leading from the evaporator coils (brine cooler) to the storage rooms where heat is absorbed and returned to the evaporator coils for recooling. Such a system is known as a simple indirect expansion system and has been found to have many advantages over the direct expansion type, particularly on larger installations.

Fig. 386 shows diagrammatic layout of three types of systems described

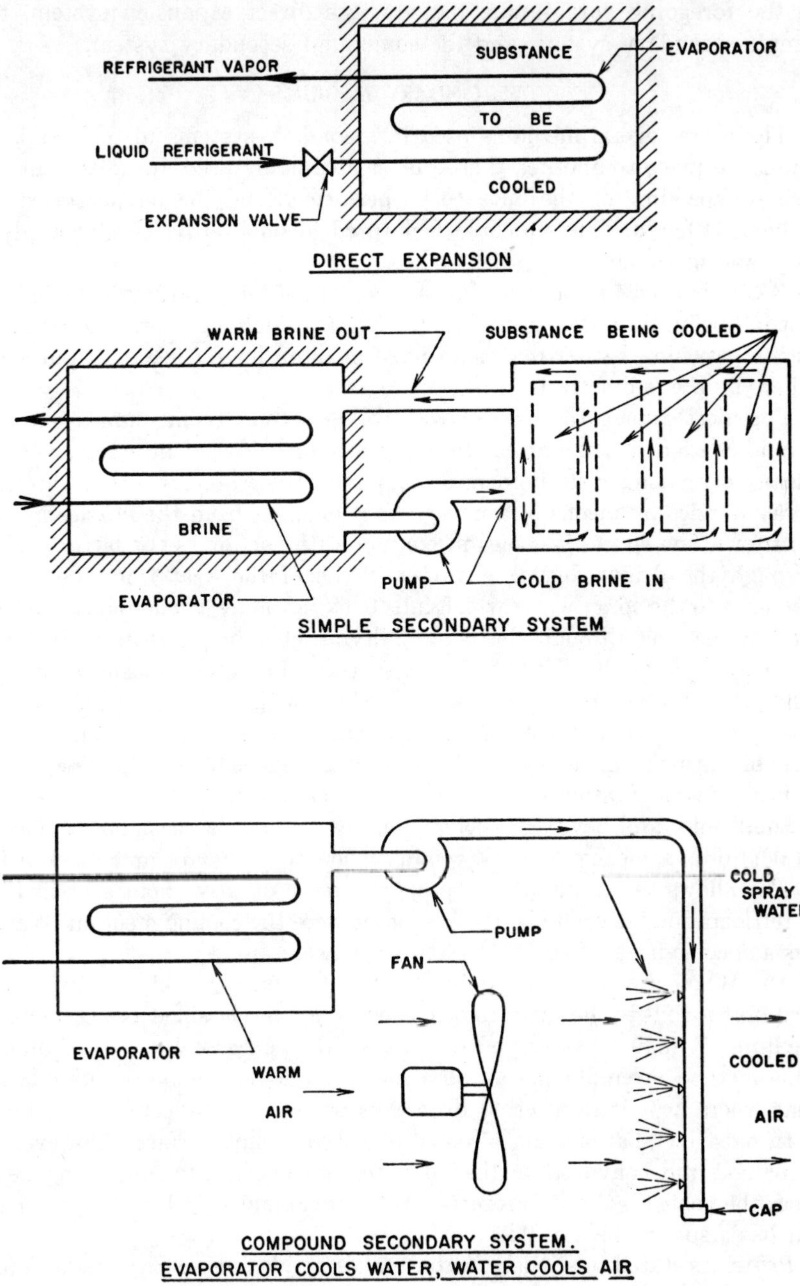

Fig. 386. Diagrammatic layout of refrigeration systems—direct, simple indirect, and compound indirect.

in the foregoing paragraphs; namely, the direct expansion system, the simple secondary system, and the compound secondary system.

SECONDARY MEDIUMS

There are several mediums used in secondary systems to transfer heat from one place to another. Those most commonly used are water, air, or brine, depending on the duty to be performed and the temperatures involved. Other mediums include denatured alcohol, ethylene glycol, glycerin, and methanol.

Water is used for the transfer of heat where temperatures involved do not fall below 36 F. Because of its fluidity and high heat-carrying capacity per unit volume, water makes an ideal medium for refrigeration purposes where its use is practical.

Foremost among the uses of water as a secondary medium is for dehumidification in true air-conditioning systems. Cooled in a reservoir by means of cooling coils immersed within it, the water is forced through spray nozzles by means of a pump. As it emerges from the nozzles in the form of a fine spray, it comes in contact with the air as the latter passes through the dehumidifying chamber. Some of the heat from the air is given up to the spray water which falls back in the reservoir to be recooled. The air passing through the dehumidifying chamber is reduced in temperature as a result of its contact with the cold water, causing a certain amount of its vapor to condense and return with the spray water to the reservoir. So actually the medium in this case not only transfers the sensible heat of the air to the cooling coils but also is responsible for the dissipation of heat of condensation of the water vapor of the air.

Small milk and beverage coolers employ water as a secondary medium as it affords a means of quick cooling when the cans are immersed in it. It also allows the container to be opened more often without a great loss in refrigeration, as would be the case if air were the cooling medium. Water has a specific heat of 1.0.

Air. When air is circulated by means of a fan or a blower from the evaporator coils to the space to be cooled, it can be classified as a secondary medium. It may be conveyed directly to the space or by way of ducts. This method of employing air as a medium should not be confused with that where heat is transferred by means of natural convection.

In most installations, air is forced over the cooling surface of an evaporator coil and conveyed to the body to be cooled. After absorbing heat from the cooled body, it is returned to the evaporator coils to be recooled. Air has a specific heat of 0.2415.

Brine. As stated previously, water makes an ideal secondary medium for the transfer of heat from one point to another. However, because of the limitations of this liquid as governed by its comparatively high freezing point (32 F), its use alone is not practical in low-temperature refrigeration systems. By adding an ingredient that would lower its freezing point to a considerable extent, water could be rendered suitable as a secondary medium. In doing so, some sacrifice of its desirable characteristics must be made, but this is far outweighed by the advantages remaining in the resulting solution.

If common table salt is dissolved in water, the resultant solution will have a lower "freezing temperature" than the water, and up to a certain point the more salt that is added a corresponding decrease in "freezing temperature" will be obtained. In the majority of secondary refrigeration installations, the solution of salt and water commonly called "brine" is employed. Brine is the name applied to any salt water solution.

BRINES

There are various types of salt used in commercial refrigeration for producing suitable brine solutions in order to provide for various temperature ranges. The two most commonly used are sodium chloride (common table salt) and calcium chloride.

Sodium Chloride Brine. Prepared by mixing common table salt (or rock salt) with water in ratios governed by the temperature at which it will be used, sodium chloride brine is used in secondary systems where brine temperatures are somewhat above 0 F. It may be employed for ice making where brine temperatures are carried within the range of 14 F to about 18 F. Sodium chloride brine finds an important place where meats, fish, and other commodities are chilled or frozen by means of brine spray or fog. It is used in place of calcium chloride where the brine may come in contact with food, as in the aforementioned cases, because of the dehydrating effect of the latter type salt.

What is often referred to as the "freezing point" of brine is actually the temperature at which crystallization starts. However, for convenience the term freezing point will be used in this text. The action of brines during salt crystallization is discussed more fully later in this chapter under the heading "Eutectic Solutions."

The temperature at which crystallization begins, or the freezing point, decreases as more salt is added or as the density of the brine increases up to a certain point. The lowest temperature to which sodium chloride brine can be carried without crystal formation is about −6 F. To obtain the latter temperature, the brine must contain 23 percent sodium chloride by weight. Its density at the latter concentration is 1.175 specific gravity, or its weight compared with an equal volume of water. The properties of sodium chloride brine are shown in Table 85.

Calcium Chloride Brine ($CaCl_2$). The salt calcium chloride is used in making a second important brine having a lower temperature of crystal-

Table 89. Complete Brine Table at 60 F ¶

Strength of brine					Salt				Water						Brine								Freezing point §	
					Lb salt per				Lb water per			Gal water per			Lb brine per				Gal brine per					
Salometer deg.	Baumé deg.	Spec. grav.	% Salt by wt in brine	% Salt by wt in water	1 lb water	1 gal water	1 lb brine	1 gal brine	1 lb salt	1 lb brine	1 gal brine	1 lb salt	1 lb brine	1 gal brine	1 lb salt	1 lb water	1 gal water	1 gal brine	1 lb salt	1 lb water	1 gal water	1 lb brine	°F	°C
0	0.0	1.000	0.0000	0.0000	.0000	0.0000	.0000	.0000	∞	1.0000	8.32823	∞	.12007	1.0000	∞	1.0000	8.328	8.32823	∞	.12007	1.0000	.12007	+32.0	0.0
1	0.3	1.002	0.2640	0.2646	.0026	0.0221	.0026	.0220	378.358	.9974	8.322	45.3356	.11976	.9992	384.6154	1.0026	8.350	8.344	45.3696	.12016	1.0008	.11985	+31.8	−0.1
2	0.6	1.004	0.5279	0.5307	.0053	0.0442	.0053	.0441	189.072	.9947	8.316	22.6254	.11944	.9985	188.6793	1.0053	8.372	8.360	22.6594	.12025	1.0015	.11962	+31.5	−0.3
3	0.9	1.006	0.7919	0.7982	.0079	0.0665	.0079	.0663	125.945	.9921	8.309	15.0435	.11912	.9977	126.5823	1.0080	8.395	8.376	15.0777	.12035	1.0023	.11939	+31.3	−0.4
4	1.1	1.008	1.0558	1.0671	.0107	0.0889	.0106	.0886	93.231	.9894	8.303	11.2526	.11881	.9970	94.3396	1.0107	8.417	8.392	11.2870	.12044	1.0031	.11917	+31.1	−0.5
5	1.3	1.010	1.3198	1.3374	.0134	0.1114	.0132	.1110	74.683	.9868	8.296	8.9781	.11849	.9962	75.7576	1.0134	8.440	8.407	9.0126	.12053	1.0038	.11894	+30.8	−0.7
6	1.6	1.011	1.5837	1.6092	.0160	0.1340	.0158	.1334	62.305	.9842	8.290	7.4617	.11817	.9954	63.2911	1.0161	8.462	8.423	7.4964	.12063	1.0046	.11872	+30.5	−0.8
7	1.9	1.013	1.8477	1.8824	.0188	0.1568	.0185	.1559	53.079	.9815	8.283	6.3786	.11785	.9946	54.0541	1.0188	8.485	8.439	6.4134	.12073	1.0054	.11850	+30.2	−1.0
8	2.1	1.015	2.1116	2.1572	.0225	0.1797	.0211	.1785	44.425	.9789	8.276	5.5663	.11754	.9938	47.3934	1.0216	8.508	8.455	5.6011	.12082	1.0063	.11827	+30.0	−1.1
9	2.4	1.017	2.3756	2.4334	.0243	0.2027	.0238	.2012	41.135	.9762	8.270	4.9345	.11722	.9930	42.0168	1.0244	8.531	8.471	4.9694	.12092	1.0071	.11805	+29.6	−1.3
10	2.7	1.019	2.6395	2.7111	.0271	0.2258	.0264	.2240	36.887	.9736	8.263	4.4290	.11690	.9921	37.8788	1.0271	8.554	8.487	4.4641	.12102	1.0079	.11783	+29.3	−1.5
11	3.0	1.021	2.9035	2.9903	.0299	0.2490	.0290	.2469	33.422	.9710	8.256	4.0155	.11659	.9913	34.4828	1.0299	8.577	8.503	4.0507	.12113	1.0088	.11761	+29.1	−1.6
12	3.3	1.023	3.1674	3.2710	.0327	0.2724	.0317	.2698	30.553	.9683	8.249	3.6708	.11627	.9905	31.5457	1.0327	8.601	8.519	3.7062	.12123	1.0096	.11739	+28.8	−1.8
13	3.5	1.025	3.4314	3.5533	.0355	0.2959	.0343	.2928	28.129	.9657	8.242	3.3792	.11595	.9896	29.1545	1.0355	8.624	8.535	3.4147	.12133	1.0105	.11717	+28.5	−1.9
14	3.7	1.027	3.6953	3.8371	.0384	0.3196	.0370	.3160	26.042	.9630	8.234	3.1293	.11564	.9887	27.0270	1.0384	8.648	8.550	3.1649	.12144	1.0114	.11695	+28.2	−2.1
15	4.0	1.029	3.9593	4.1225	.0412	0.3433	.0396	.3392	24.254	.9604	8.227	2.9126	.11532	.9879	25.2525	1.0412	8.672	8.566	2.9484	.12155	1.0123	.11673	+27.9	−2.3
16	4.2	1.031	4.2232	4.4094	.0441	0.3672	.0422	.3625	22.696	.9578	8.220	2.7231	.11500	.9870	23.6967	1.0441	8.696	8.582	2.7590	.12166	1.0132	.11652	+27.6	−2.4
17	4.5	1.032	4.4872	4.6980	.0470	0.3913	.0449	.3858	21.263	.9551	8.213	2.5559	.11469	.9861	22.2717	1.0470	8.720	8.598	2.5918	.12176	1.0141	.11630	+27.3	−2.6
18	4.8	1.034	4.7511	4.9881	.0499	0.4154	.0475	.4093	20.052	.9525	8.205	2.4072	.11437	.9852	21.0526	1.0499	8.744	8.615	2.4433	.12187	1.0150	.11608	+27.0	−2.8
19	5.0	1.036	5.0151	5.2798	.0528	0.4397	.0502	.4328	18.925	.9498	8.198	2.2742	.11405	.9843	19.9203	1.0529	8.768	8.631	2.3104	.12198	1.0159	.11587	+26.7	−2.9
20	5.3	1.038	5.2790	5.5732	.0557	0.4642	.0528	.4565	17.957	.9472	8.190	2.1545	.11373	.9834	18.9394	1.0557	8.792	8.647	2.1908	.12210	1.0169	.11565	+26.4	−3.1
21	5.6	1.040	5.5430	5.8682	.0587	0.4887	.0554	.4802	17.044	.9446	8.183	2.0462	.11342	.9825	18.0505	1.0586	8.817	8.663	2.0826	.12221	1.0178	.11543	+26.1	−3.3
22	5.8	1.042	5.8069	6.1649	.0615	0.5134	.0581	.5040	16.252	.9419	8.175	1.9477	.11310	.9816	17.2117	1.0617	8.842	8.679	1.9842	.12232	1.0187	.11522	+25.7	−3.5
23	6.1	1.044	6.0709	6.4632	.0647	0.5383	.0607	.5279	15.466	.9393	8.167	1.8578	.11278	.9807	16.4745	1.0646	8.867	8.695	1.8944	.12244	1.0197	.11501	+25.4	−3.7
24	6.4	1.046	6.3348	6.7632	.0677	0.5633	.0633	.5519	14.780	.9367	8.160	1.7754	.11247	.9797	15.7978	1.0676	8.892	8.711	1.8121	.12256	1.0207	.11479	+25.1	−3.8
25	6.7	1.048	6.5988	7.0649	.0707	0.5884	.0660	.5759	14.152	.9340	8.152	1.6996	.11215	.9788	15.1515	1.0707	8.917	8.728	1.7364	.12267	1.0217	.11458	+24.7	−4.1
26	6.9	1.050	6.8627	7.3684	.0737	0.6137	.0683	.6001	13.574	.9314	8.144	1.6296	.11183	.9779	14.5773	1.0737	8.942	8.744	1.6665	.12279	1.0226	.11437	+24.4	−4.2
27	7.2	1.052	7.1267	7.6735	.0767	0.6391	.0713	.6243	13.040	.9287	8.136	1.5648	.11152	.9769	14.0253	1.0768	8.967	8.760	1.6018	.12291	1.0237	.11415	+24.0	−4.4
28	7.4	1.054	7.3306	7.9804	.0798	0.6646	.0739	.6448	12.525	.9261	8.128	1.5046	.11120	.9759	13.5318	1.0798	8.993	8.776	1.5508	.12304	1.0247	.11394	+23.7	−4.6
29	7.7	1.056	7.6546	8.2890	.0829	0.6903	.0765	.6730	12.067	.9235	8.120	1.4486	.11088	.9750	13.0719	1.0828	9.019	8.793	1.4858	.12316	1.0257	.11373	+23.3	−4.8
30	7.9	1.058	7.9185	8.5994	.0860	0.7162	.0792	.6975	11.624	.9208	8.111	1.3963	.11056	.9740	12.6263	1.0860	9.044	8.809	1.4336	.12328	1.0267	.11352	+23.0	−5.0
31	8.2	1.060	8.1825	8.9117	.0891	0.7422	.0818	.7221	11.227	.9182	8.103	1.3474	.11025	.9730	12.2249	1.0891	9.070	8.825	1.3848	.12341	1.0278	.11331	+22.6	−5.2
32	8.5	1.062	8.4464	9.2256	.0923	0.7683	.0845	.7468	10.839	.9155	8.095	1.3015	.10993	.9720	11.8384	1.0923	9.097	8.842	1.3390	.12353	1.0288	.11310	+22.3	−5.4
33	8.7	1.064	8.7104	9.5414	.0954	0.7946	.0871	.7716	10.478	.9129	8.087	1.2584	.10961	.9710	11.4811	1.0954	9.123	8.858	1.2960	.12366	1.0299	.11289	+22.0	−5.6
34	9.0	1.066	8.9743	9.8591	.0986	0.8211	.0897	.7964	10.139	.9103	8.078	1.2179	.10930	.9700	11.1483	1.0985	9.149	8.875	1.2556	.12379	1.0310	.11268	+21.6	−5.8
35	9.2	1.068	9.2383	10.1786	.1018	0.8477	.0924	.8214	9.820	.9076	8.070	1.1797	.10898	.9690	10.8225	1.1018	9.176	8.891	1.2175	.12392	1.0320	.11247	+21.3	−5.9
36	9.5	1.070	9.5022	10.4999	.1050	0.8745	.0950	.8464	9.521	.9050	8.061	1.1436	.10866	.9679	10.5263	1.1050	9.203	8.908	1.1814	.12405	1.0331	.11226	+20.9	−6.2
37	9.7	1.072	9.7662	10.8232	.1082	0.9014	.0977	.8715	9.239	.9023	8.053	1.1094	.10835	.9669	10.2354	1.1083	9.230	8.924	1.1474	.12418	1.0342	.11206	+20.5	−6.4
38	10.0	1.074	10.0301	11.1483	.1115	0.9285	.1003	.8968	8.971	.8997	8.044	1.0771	.10803	.9658	9.9701	1.1115	9.257	8.941	1.1151	.12432	1.0353	.11185	+20.2	−6.6
39	10.2	1.076	10.2941	11.4753	.1147	0.9557	.1029	.9220	8.719	.8971	8.035	1.0464	.10771	.9648	9.7182	1.1147	9.284	8.957	1.0845	.12445	1.0365	.11164	+19.8	−6.8
40	10.5	1.078	10.5580	11.8043	.1181	0.9831	.1056	.9474	8.470	.8944	8.026	1.0172	.10740	.9637	9.4697	1.1181	9.311	8.974	1.0555	.12459	1.0376	.11144	+19.4	−7.0
41	10.7	1.080	10.8220	12.1352	.1213	1.0107	.1082	.9729	8.243	.8918	8.017	.9895	.10708	.9627	9.2421	1.1213	9.339	8.990	1.0278	.12473	1.0388	.11123	+19.1	−7.2
42	11.0	1.082	11.0859	12.4681	.1247	1.0384	.1109	.9985	8.020	.8891	8.008	.9630	.10676	.9616	9.0171	1.1247	9.367	9.007	1.0015	.12487	1.0399	.11103	+18.7	−7.4
43	11.2	1.084	11.3499	12.8030	.1281	1.0663	.1135	1.0242	7.806	.8865	7.999	.9379	.10644	.9605	8.8106	1.1280	9.395	9.024	.9764	.12501	1.0411	.11082	+18.3	−7.6
44	11.5	1.086	11.6138	13.1398	.1313	1.0943	.1161	1.0499	7.616	.8839	7.990	.9138	.10613	.9594	8.6133	1.1314	9.423	9.040	.9525	.12515	1.0423	.11062	+17.9	−7.8
45	11.7	1.088	11.8778	13.4787	.1348	1.1225	.1188	1.0758	7.418	.8812	7.981	.8908	.10581	.9583	8.4175	1.1348	9.451	9.057	.9296	.12530	1.0435	.11041	+17.5	−8.1
46	12.0	1.090	12.1417	13.8196	.1382	1.1509	.1214	1.1017	7.236	.8786	7.972	.8689	.10549	.9572	8.2372	1.1382	9.479	9.074	.9077	.12544	1.0447	.11021	+17.1	−8.3
47	12.2	1.092	12.4057	14.1626	.1416	1.1795	.1241	1.1277	7.062	.8759	7.963	.8478	.10518	.9561	8.0580	1.1417	9.508	9.090	.8868	.12559	1.0459	.11001	+16.7	−8.5
48	12.5	1.094	12.6696	14.5077	.1450	1.2082	.1267	1.1538	6.896	.8733	7.953	.8277	.10486	.9550	7.8927	1.1451	9.537	9.107	.8667	.12574	1.0472	.10981	+16.2	−8.8
49	12.7	1.096	12.9336	14.8548	.1486	1.2371	.1293	1.1800	6.729	.8707	7.944	.8083	.10454	.9538	7.7340	1.1485	9.565	9.124	.8474	.12589	1.0484	.10960	+15.8	−9.0
50	12.9	1.098	13.1975	15.2041	.1521	1.2662	.1320	1.2063	6.575	.8680	7.934	.7897	.10422	.9527	7.5758	1.1521	9.595	9.141	.8290	.12604	1.0497	.10940	+15.4	−9.2
51	13.2	1.100	13.4615	15.5554	.1555	1.2955	.1346	1.2327	6.431	.8654	7.925	.7719	.10391	.9515	7.4294	1.1555	9.624	9.157	.8112	.12619	1.0509	.10920	+15.0	−9.4
52	13.4	1.102	13.7254	15.9090	.1591	1.3249	.1373	1.2592	6.285	.8627	7.915	.7547	.10359	.9504	7.2833	1.1592	9.653	9.174	.7942	.12634	1.0522	.10900	+14.5	−9.7
53	13.7	1.104	13.9894	16.2647	.1626	1.3546	.1399	1.2858	6.150	.8601	7.905	.7382	.10328	.9492	7.1480	1.1627	9.683	9.191	.7777	.12650	1.0535	.10880	+14.1	−9.9
54	13.9	1.106	14.2533	16.6226	.1662	1.3844	.1425	1.3124	6.017	.8575	7.895	.7223	.10296	.9480	7.0175	1.1662	9.713	9.208	.7619	.12666	1.0548	.10860	+13.7	−10.2
55	14.1	1.108	14.5173	16.9827	.1699	1.4144	.1452	1.3392	5.886	.8548	7.886	.7070	.10264	.9468	6.8871	1.1699	9.743	9.225	.7467	.12681	1.0561	.10840	+13.3	−10.4
56	14.4	1.110	14.7812	17.3450	.1734	1.4445	.1478	1.3660	5.767	.8522	7.876	.6923	.10232	.9457	6.7659	1.1734	9.773	9.242	.7320	.12697	1.0575	.10821	+12.8	−10.7
57	14.6	1.112	15.0452	17.7096	.1771	1.4749	.1505	1.3930	5.646	.8495	7.866	.6780	.10201	.9445	6.6445	1.1772	9.803	9.259	.7179	.12713	1.0588	.10801	+12.3	−10.9
58	14.8	1.114	15.3091	18.0764	.1807	1.5055	.1531	1.4200	5.534	.8469	7.856	.6643	.10169	.9433	6.5317	1.1808	9.834	9.276	.7007	.12730	1.0602	.10781	+11.8	−11.2
59	15.1	1.116	15.5731	18.4456	.1844	1.5362	.1557	1.4472	5.423	.8443	7.846	.6510	.10137	.9420	6.4226	1.1844	9.864	9.293	.6910	.12746	1.0615	.10761	+11.4	−11.4

60	15.3	1.118	15.3370	18.8171	.1882	1.5671	.1584	1.4744	5.313	.8416	7.835	.6381	.10106	.9408	6.3131	1.1882	9.895	9.310	.6782	.12762	1.0629	.10741	+10.9	−11.7
61	15.5	1.120	16.1010	19.1909	.1919	1.5983	.1610	1.5017	5.211	.8390	7.825	.6257	.10074	.9396	6.2112	1.1919	9.927	9.327	.6659	.12779	1.0643	.10722	+10.4	−12.0
62	15.8	1.122	16.3649	19.5670	.1956	1.6296	.1636	1.5292	5.112	.8364	7.815	.6136	.10042	.9384	6.1125	1.1956	9.958	9.344	.6540	.12796	1.0657	.10702	+9.9	−12.3
63	16.0	1.124	16.6289	19.9456	.1995	1.6611	.1663	1.5567	5.013	.8337	7.805	.6020	.10011	.9371	6.0132	1.1995	9.989	9.361	.6424	.12813	1.0671	.10682	+9.4	−12.5
64	16.2	1.126	16.8928	20.3265	.2032	1.6928	.1689	1.5843	4.921	.8311	7.794	.5907	.09979	.9359	5.9207	1.2032	10.021	9.379	.6312	.12830	1.0685	.10663	+8.9	−12.8
65	16.5	1.128	17.1568	20.7099	.2071	1.7248	.1716	1.6120	4.828	.8284	7.784	.5798	.09947	.9346	5.8275	1.2071	10.053	9.396	.6203	.12847	1.0699	.10643	+8.4	−13.1
66	16.7	1.130	17.4207	21.0957	.2109	1.7569	.1742	1.6399	4.742	.8258	7.773	.5692	.09916	.9334	5.7405	1.2109	10.085	9.413	.6098	.12864	1.0714	.10623	+7.9	−13.4
67	17.0	1.132	17.6847	21.4840	.2149	1.7892	.1768	1.6678	4.653	.8232	7.763	.5589	.09884	.9321	5.6561	1.2148	10.118	9.431	.5996	.12882	1.0728	.10604	+7.3	−13.7
68	17.2	1.134	17.9486	21.8748	.2188	1.8218	.1795	1.6958	4.570	.8205	7.752	.5489	.09852	.9308	5.5710	1.2188	10.150	9.448	.5897	.12899	1.0743	.10584	+6.8	−14.0
69	17.5	1.137	18.2126	22.2681	.2228	1.8546	.1821	1.7239	4.488	.8179	7.742	.5392	.09820	.9296	5.4915	1.2226	10.183	9.466	.5801	.12917	1.0758	.10565	+6.3	−14.3
70	17.7	1.139	18.4765	22.6640	.2267	1.8875	.1848	1.7521	4.411	.8152	7.731	.5298	.09789	.9283	5.4113	1.2267	10.216	9.483	.5707	.12935	1.0773	.10545	+5.7	−14.6
71	17.9	1.141	18.7405	23.0625	.2307	1.9207	.1874	1.7805	4.335	.8126	7.720	.5206	.09757	.9270	5.3362	1.2306	10.249	9.501	.5616	.12953	1.0788	.10526	+5.2	−14.9
72	18.1	1.143	19.0044	23.4635	.2346	1.9541	.1900	1.8089	4.263	.8100	7.709	.5117	.09725	.9257	5.2632	1.2346	10.282	9.518	.5528	.12971	1.0803	.10506	+4.6	−15.2
73	18.4	1.145	19.2684	23.8672	.2396	1.9877	.1927	1.8374	4.174	.8073	7.699	.5031	.09694	.9244	5.1894	1.2387	10.316	9.536	.5442	.12989	1.0818	.10487	+4.0	−15.5
74	18.6	1.147	19.5323	24.2735	.2428	2.0216	1953	1.8661	4.119	.8047	7.688	.4947	.09662	.9231	5.1203	1.2427	10.350	9.554	.5359	.13008	1.0833	.10467	+3.4	−15.9
75	18.8	1.149	19.7963	24.6824	.2468	2.0556	1980	1.8948	4.052	.8020	7.677	.4865	.09630	.9218	5.0505	1.2469	10.384	9.571	.5278	.13026	1.0850	.10448	+2.8	−16.2
76	19.1	1.151	20.0602	25.0941	.2510	2.0899	2006	1.9236	3.984	.7994	7.666	.4785	.09599	.9204	4.9850	1.2509	10.418	9.589	.5199	.13045	1.0864	.10428	+2.2	−16.5
77	19.4	1.154	20.3242	25.5085	.2550	2.1244	2032	1.9526	3.922	.7968	7.655	.4707	.09567	.9191	4.9213	1.2550	10.453	9.607	.5121	.13064	1.0880	.10409	+1.6	−16.9
78	19.6	1.156	20.5881	25.9257	.2593	2.1592	2059	1.9816	3.857	.7941	7.643	.4631	.09535	9178	4.8567	1.2593	10.487	9.625	.5046	.13083	1.0896	.10390	+1.0	−17.2
79	19.8	1.158	20.8520	26.3457	.2635	2.1941	2085	2.0107	3.796	.7915	7.632	.4558	.09504	.9164	4.7962	1.2634	10.522	9.643	.4973	.13103	1.0912	.10370	+.4	−17.5
80	20.0	1.160	21.1160	26.7684	.2677	2.2293	.2112	2.0400	3.736	.7888	7.621	.4486	.09472	.9151	4.7348	1.2677	10.558	9.661	.4902	.13122	1.0928	.10351	−.4	−18.0
81	20.2	1.162	21.3800	27.1940	.2720	2.2648	.2138	2.0693	3.676	.7862	7.609	.4415	.09440	.9137	4.6773	1.2719	10.593	9.679	.4832	.13141	1.0945	.10332	−1.0	−18.3
82	20.4	1.164	21.6439	27.6225	.2762	2.3005	.2164	2.0988	3.621	.7836	7.598	.4347	.09408	.9123	4.6211	1.2761	10.629	9.697	.4765	.13161	1.0961	.10313	−1.6	−18.7
83	20.7	1.167	21.9079	28.0538	.2805	2.3364	.2191	2.1283	3.565	.7809	7.587	.4280	.09377	.9109	4.5641	1.2806	10.665	9.715	.4699	.13181	1.0978	.10293	−2.3	−19.0
84	21.0	1.169	22.1718	28.4881	.2848	2.3726	.2217	2.1580	3.511	.7783	7.575	.4215	.09345	.9096	4.5106	1.2849	10.701	9.733	.4634	.13201	1.0994	.10274	−3.0	−19.4
85	21.2	1.171	22.4358	28.9250	.2893	2.4090	.2244	2.1877	3.457	.7757	7.563	.4151	.09313	.9082	4.4563	1.2893	10.737	9.751	.4571	.13221	1.1011	.10255	−3.7	−19.8
86	21.4	1.173	22.6997	29.3656	.2937	2.4456	.2270	2.2176	3.405	.7730	7.552	.4089	.09282	.9067	4.4053	1.2937	10.774	9.769	.4509	.13242	1.1029	.10236	−4.4	−20.2
87	21.7	1.175	22.9637	29.8088	.2981	2.4826	.2296	2.2475	3.355	.7704	7.540	.4028	.09250	.9053	4.3554	1.2980	10.811	9.787	.4449	.13263	1.1046	.10217	−5.2	−20.7
88	21.9	1.177	23.2276	30.2551	.3026	2.5197	.2323	2.2776	3.305	.7677	7.528	.3969	.09218	.9039	4.3048	1.3026	10.848	9.805	.4391	.13284	1.1063	.10198	−5.8	−21.0
*88.3	22.0	1.178	23.3100	30.3951	.3039	2.5314	.2331	2.2870	3.291	.7669	7.524	.3950	.09208	.9034	4.2900	1.3040	10.860	9.811	.4373	.13290	1.1069	.10192	−6.0*	−21.1*
89	22.1	1.180	23.4916	30.7045	.3071	2.5572	.2349	2.3077	3.256	.7651	7.516	.3911	.09187	.9025	4.2571	1.3070	10.885	9.824	.4333	.13305	1.1081	.10180	−4.2	−20.1
90	22.3	1.182	23.7555	31.1570	.3115	2.5948	.2376	2.3380	3.210	.7624	7.504	.3854	.09155	.9010	4.2088	1.3116	10.923	9.842	.4277	.13327	1.1099	.10161	−1.1	−18.4
91	22.5	1.184	24.0195	31.6126	.3161	2.6328	.2402	2.3683	3.164	.7598	7.492	.3798	.09123	.8996	4.1632	1.3161	10.961	9.860	.4222	.13348	1.1117	.10142	+1.8	−16.8
92	22.8	1.186	24.2834	32.0714	.3206	2.6710	.2428	2.3988	3.119	.7572	7.479	.3744	.09092	.8981	4.1186	1.3207	10.999	9.878	.4169	.13370	1.1135	.10123	+4.8	−15.1
93	23.0	1.188	24.5474	32.5334	.3254	2.7095	.2455	2.4294	3.073	.7545	7.467	.3691	.09060	.8966	4.0733	1.3254	11.038	9.897	.4116	.13392	1.1153	.10104	+7.9	−13.4
94	23.2	1.190	24.8113	32.9987	.3296	2.7482	.2481	2.4600	3.034	.7519	7.455	.3639	.09028	.8951	4.0306	1.3300	11.076	9.915	.4065	.13414	1.1172	.10086	+11.1	−11.6
95	23.5	1.193	25.0753	33.4672	.3346	2.7872	.2508	2.4908	2.989	.7492	7.442	.3588	.08996	.8936	3.9872	1.3348	11.116	9.933	.4015	.13436	1.1190	.10067	+14.4	−9.8
96	23.7	1.195	25.3392	33.9391	.3394	2.8265	.2534	2.5217	2.946	.7466	7.430	.3538	.08965	.8921	3.9463	1.3394	11.155	9.952	.3966	.13459	1.1209	.10049	+18.0	−7.8
97	23.9	1.197	25.6032	34.4143	.3441	2.8661	.2560	2.5526	2.906	.7440	7.417	.3489	.08933	.8906	3.9063	1.3441	11.194	9.970	.3917	.13482	1.1228	.10030	+21.6	−5.8
98	24.2	1.199	25.8671	34.8929	.3489	2.9060	.2587	2.5837	2.866	.7413	7.405	.3441	.08901	.8891	3.8655	1.3490	11.234	9.988	.3870	.13505	1.1247	.10012	+25.5	−3.6
99	24.4	1.202	26.1311	35.3749	.3538	2.9462	.2613	2.6150	2.826	.7387	7.392	.3394	.08870	.8876	3.8270	1.3537	11.274	10.007	.3824	.13528	1.1267	.09993	+29.8	−1.2
99.6	24.5	1.203	26.2850	35.6576	.3566	2.9686	.2629	2.6323	2.804	.7372	7.385	.3369	.08851	.8867	3.8037	1.3567	11.298	10.018	.3799	.13541	1.1278	.09982	+32.3†	+0.2†
100	24.6	1.204	26.3950	35.8603	.3587	2.9865	.2640	2.6469	2.738	.7361	7.381	.3348	.08844	.8863	3.7879	1.3587	11.308	10.028	.3778	.13548	1.1283	.09978	+60.0‡	+15.56‡

*Eutectic point. For brines stronger than eutectic, the temperatures shown are the saturation temperatures for sodium chloride dihydrate. Brines stronger than eutectic deposit excess sodium chloride as dihydrate when cooled, and freeze at eutectic.

†Transition temperature from anhydrous salt to dihydrate.
‡Saturated brine at 60° F (15.56° C).
§Temperature at which freezing begins. Ice forms, brine concentrates, and freezing point lowers to eutectic.

¶ Courtesy, "Brine for Today's Industry," pp. 24–25, compiled and copyrighted by the International Salt Co., Inc., Industrial Division, Scranton, Pa., 1955.

TABLE 89—(*Continued*)

How to Use the Brine Table*

These data, as tabulated in Table 89, apply only to brines tested at 60 F. Preferably, all Salometer readings should be made at 60 F. For other brine temperatures, the observed Salometer readings must be converted before using them in the table.

For all practical purposes, the correction amounts to approximately one degree Salometer less for each ten degrees below 60 F, and one degree Salometer more for each 10 degrees above 60 F.

For more accurate conversion, subtract from the observed Salometer reading the following correction for each degree of temperature below 60 F; add for temperature between 60 and 100 F.

The data in the Complete Brine Table at 60 F are based on properties of chemically pure sodium chloride. For all practical purposes they may be applied without modification to solutions of salt mined or manufactured by the International Salt Company, Inc., Scranton, Pa.

Observed Salometer Reading	*Approximate Correction in Salometer Degrees*	
	Subtract per degree below 60 F	*Add per degree above 60 F*
0 to 10	0.049	0.060
11 to 20	0.064	0.082
21 to 30	0.077	0.094
31 to 40	0.087	0.103
41 to 50	0.095	0.112
51 to 60	0.102	0.118
61 to 70	0.107	0.123
71 to 80	0.112	0.128
81 to 90	0.116	0.131
91 to 100	0.120	0.134

Freezing Behavior of Brines*

Sometimes it is mistakenly assumed that the stronger the brine, the lower the temperature at which freezing begins. This is true only for brine strengths to and including 88.3 Salometer.

Pure water freezes at 32 F, and, if salt gradually is added, ice first appears at successively lower temperatures. At a strength of 88.3 S, brine has its lowest freezing point, −6.0 F, the eutectic temperature.

At the opposite extreme from pure water is fully saturated brine of 100 S strength. Any drop in temperature will deposit out a few grains of salt, but the quantity will be very small, because salt solubility changes but little with temperature.

If a saturated brine gradually is diluted, salt will appear at successively lower temperatures; even at the slight dilution to 99.6 S, the salt does not appear until a temperature of 32.3 F is reached. Below this particular strength and temperature, salt does not appear; but instead, sodium chloride dihydrate, which looks and behaves much like ice, and usually is mistaken for ice.

The temperature at which the dihydrate first appears becomes successively lower as the brine is diluted, until at 88.3 S strength the freezing temperature is −6.0 F. The solid material which appears in 88.3 S brine at −6.0 F is an intimate mixture of ice and dihydrate, although ice first forms in weaker brines, and dihydrate first forms in stronger brines.

Note that 1 S change in strength causes less than 1 F change in freezing point for brines slightly under 88.3 S strength, but approximately 3 F change for brines slightly over 88.3 S. Therefore, if freezing might cause trouble in refrigerating equipment, it is better to operate with brines somewhat under 88.3 S, rather than over, since slight variations in strength have less effect on the brine freezing point.

* Courtesy, "Brine for Today's Industry," pp. 24–25, compiled and copyrighted by the International Salt Co., Inc., Industrial Division, Scranton, Pa.

lization than is possible with sodium chloride. Therefore, calcium chloride brine can be utilized where subzero temperatures are required. It finds wide use in ice-making plants where ice blocks are formed by immersing cans of water and in circulation systems servicing sharp freezers where brine temperatures often reach 0 F and lower.

Like sodium chloride brine, the more calcium chloride added to water further reduces its temperature at which crystallization begins, up to a certain point. The temperature limit of reduction is −59.8 F when concentration reaches 29.6 percent salt by weight, corresponding to a specific gravity of 1.29.

TABLE 90. SPECIFIC HEAT OF SODIUM CHLORIDE BRINE

Btu per lb (Estimated)

Specific Gravity	*% NaCl by Weight*	*Freezing Point °F*	*Specific Heat, Btu per lb Range from Freezing Point to 50 F**
1.04	5.543	+26.1	0.925 to 0.935
1.06	8.183	+22.6	0.895 to 0.905
1.08	10.822	+19.1	0.875 to 0.885
1.10	13.462	+15.0	0.850 to 0.860
1.12	16.101	+10.4	0.830 to 0.840
1.14	18.611	+5.4	0.810 to 0.820
1.16	21.116	−0.4	0.775 to 0.807
1.178†	23.310	−6.0	0.760 to 0.795

* Approximate specific heat values may be obtained at any point between the freezing temperature and 50 F by interpolation.
† Eutectic point of NaCl.

One of the disadvantages of calcium chloride is its dehydrating effect on foods with which it may come in contact. Its affinity for moisture (hygroscopic property) causes whitish spots to appear on the surface and imparts a bitter taste to any food product so contaminated. However, care in the maintenance of equipment carrying calcium chloride brine will eliminate this hazard.

Brine Characteristics

It will be noted in the physical property tables for both sodium chloride and calcium chloride that the specific heat value of the solution decreases correspondingly with any increase in salt concentration. Because of this reduction in thermodynamic quality, the circulation of a greater volume of high gravity brine over a lower gravity brine is necessary in order to transfer an equal amount of heat.

TABLE 91. DENSITIES, FREEZING POINTS, AND CONCENTRATIONS OF CALCIUM CHLORIDE BRINES*

*Specific Gravity 60/60°F.	°Baumé at 60°F. (Am. Std.)	Salimeter Reading at 60°F.	Weight of Brine at 60°F.		% $CaCl_2$ Actual	Freezing Point °F.	Lbs. 73-75% Solid $CaCl_2$		Lbs. 77-80% Flake $CaCl_2$		**Lbs. Per Gal. of Water at 60°F.		***Final Volume in Gals. at 60°F.
			Lbs. Per Gallon	Lbs. Per Cu. Ft.			Lbs. Per Gallon of Brine at 60°F.	Lbs. Per Cu. Ft. of Brine at 60°F.	Lbs. Per Gallon of Brine at 60°F.	Lbs. Per Cu. Ft. of Brine at 60°F.	Solid 73-75 $CaCl_2$	Flake 77-80 $CaCl_2$	
1.000	0.0	0.0	8.34	62.37	0	+32.0	—	—	—	—	—	—	1.000
1.009	1.3	5.2	8.41	62.93	1.0	+31,1	0.11	0.86	0.10	0.81	0.11	0,10	1.004
1.010	1.4	5.6	8.42	62.99	1.2	+31.0	0.14	1.03	0.13	0.97	0.14	0.13	1.006
1.018	2.5	10.0	8.49	63.49	2.0	+30.4	0.23	1.73	0.22	1.63	0.23	0.22	1.008
1.020	2.8	11.2	8.50	63.61	2.3	+30.2	0.27	1.99	0.25	1.88	0.27	0.25	1.011
1.026	3.7	14.8	8.55	63.99	3.0	+29.5	0.35	2.61	0.33	2.46	0.36	0.33	1.014
1.030	4.2	16.8	8.59	64.24	3.5	+29.1	0.41	3.06	0.39	2.88	0.42	0.40	1.017
1.034	4.8	19.2	8.62	64.49	4.0	+28.6	0.47	3.51	0.44	3.31	0.48	0.45	1.020
1.040	5.6	22.4	8.67	64.86	4.6	+28.0	0.54	4.06	0.51	3.83	0.55	0.52	1.022
1.043	6.0	24.0	8.70	65.05	5.0	+27.7	0.59	4.43	0.56	4.17	0.61	0.57	1.024
1.050	6.9	27.6	8.75	65.48	5.9	+27.0	0.70	5.26	0.66	4.96	0.72	0.68	1.030
1.052	7.1	28.4	8.77	65.61	6.0	+26.8	0.72	5.36	0.68	5.05	0.75	0.70	1.031
1.060	8.2	32.8	8.84	66.11	7.0	+25.9	0.84	6.30	0.79	5.94	0.88	0.82	1.036
1.069	9.4	37.6	8.91	66.67	8.0	+24.6	0.97	7.26	0.91	6.84	1.02	0.95	1.043
1.070	9.5	38.0	8.92	66.73	8.1	+24.6	0.98	7.35	0.92	6.93	1.03	0.96	1.043
1.078	10.5	42.0	8.99	67.23	9.0	+23.5	1.10	8.23	1.04	7.76	1.16	1.09	1.049
1.080	10.7	42.8	9.00	67.36	9.2	+23.2	1.13	8.43	1.06	7.94	1.20	1.11	1.050
1.087	11.6	46.4	9.06	67.79	10.0	+22.3	1.23	9.22	1.16	8.69	1.31	1.22	1.055
1.090	12.0	48.0	9.09	67.98	10.3	+21.9	1.27	9.53	1.20	8.89	1.36	1.28	1.058
1.096	12.7	50.8	9.14	68.35	11.0	+20.8	1.37	10.23	1.29	9.64	1.47	1.37	1.062
1.100	13.2	52.8	9.17	68.60	11.4	+20.3	1.42	10.64	1.34	10.03	1.53	1.43	1.065
1.105	13.8	55.2	9.21	68.92	12.0	+19.3	1.50	11.25	1.41	10.60	1.62	1.51	1.069
1.110	14.3	57.2	9.25	69.23	12.4	+18.7	1.56	11.68	1.47	11.01	1.69	1.57	1.071
1.115	15.0	60.0	9.30	69.54	13.0	+17.6	1.64	12.30	1.55	11.59	1.79	1.67	1.076
1.120	15.5	62.0	9.34	69.85	13.5	+16.5	1.72	12.83	1.62	12.09	1.88	1.75	1.080
1.125	16.1	64.4	9.38	70.16	14.0	+15.5	1.79	13.36	1.69	12.59	1.97	1.83	1.084
1.130	16.7	66.8	9.42	70.47	14.5	+14.5	1.86	13.90	1.75	13.10	2.05	1.90	1.087
1.135	17.2	68.8	9.46	70.79	15.0	+13.5	1.93	14.45	1.82	13.62	2.14	1.99	1.092
1.140	17.8	71.2	9.50	71.10	15.6	+12.2	2.02	15.09	1.90	14.22	2.25	2.08	1.096
1.145	18.4	73.6	9.55	71.41	16.0	+11.2	2.08	15.54	1.96	14.64	2.32	2.16	1.099
1.150	18.9	75.6	9.59	71.72	16.6	+ 9.7	2.17	16.20	2.04	15.27	2.44	2.25	1.104
1.155	19.4	77.6	9.63	72.03	17.0	+ 8.6	2.23	16.66	2.10	15.70	2.51	2.33	1.109
1.160	20.0	80.0	9.67	72.35	17.6	+ 7.0	2.32	17.32	2.19	16.32	2.63	2.44	1.113
1.165	20.5	82.0	9.71	72.66	18.0	+ 5.9	2.38	17.79	2.24	16.76	2.71	2.50	1.118
1.170	21.1	84.4	9.75	72.97	18.6	+ 4.1	2.47	18.47	2.33	17.40	2.83	2.62	1.122
1.175	21.6	86.4	9.80	73.28	19.0	+ 2.8	2.53	18.94	2.38	17.85	2.90	2.68	1.127
1.180	22.1	88.4	9.84	73.59	19.5	+ 1.2	2.61	19.52	2.46	18.39	3.01	2.78	1.131
1.186	22.7	90.8	9.89	73.97	20.0	− 0.4	2.69	20.13	2.53	18.97	3.12	2.87	1.137
1.190	23.2	92.8	9.92	74.22	20.6	− 2.0	2.78	20.80	2.62	19.60	3.25	2.99	1.142
1.195	23.7	94.8	9.96	74.53	21.0	− 3.9	2.85	21.29	2.69	20.06	3.34	3.08	1.147

1.200	24.2	96.8	10.00	74.84	21.5	− 5.8	2.93	21.89	2.76	20.63	3.45	3.18	1.152
1.206	24.8	99.2	10.05	75.21	22.0	− 7.8	3.01	22.51	2.84	21.21	3.56	3.28	1.157
1.210	25.2	100.8	10.09	75.46	22.4	− 9.4	3.08	23.00	2.90	21.67	3.67	3.36	1.160
1.218	25.9	103.6	10.15	75.96	23.0	−11.9	3.18	23.77	3.00	22.40	3.80	3.50	1.168
1.220	26.2	104.8	10.17	76.09	23.3	−13.2	3.22	24.12	3.03	22.73	3.86	3.54	1.170
1.228	26.9	107.6	10.24	76.59	24.0	−16.2	3.34	25.01	3.15	23.57	4.04	3.71	1.178
1.230	27.1	108.4	10.25	76.71	24.2	−17.1	3.37	25.26	3.18	23.80	4.08	3.75	1.180
1.239	28.0	112.0	10.33	77.27	25.0	−21.0	3.51	26.28	3.31	24.76	4.29	3.93	1.189
1.240	28.1	112.4	10.34	77.33	25.1	−21.5	3.53	26.41	3.33	24.89	4.32	3.96	1.190
1.250	29.0	116.0	10.42	77.96	26.0	−25.8	3.69	27.58	3.48	25.99	4.57	4.18	1.202
1.260	29.9	119.6	10.50	78.58	27.0	−31.2	3.86	28.87	3.64	27.20	4.84	4.42	1.215
1.270	30.8	—	10.59	79.21	27.8	−37.1	4.01	29.96	3.79	28.23	5.08	4.65	1.227
1.272	31.0	—	10.60	79.33	28.0	−37.8	4.04	30.22	3.81	28.48	5.13	4.68	1.228
1.280	31.7	—	10.67	79.83	28.7	−44.3	4.17	31.17	3.93	29.37	5.35	4.86	1.237
1.283	32.0	—	10.70	80.02	29.0	−49.4	4.22	31.57	3.98	29.75	5.43	4.94	1.241
1.290	32.6	—	10.75	80.45	29.6	−59.8	4.33	32.40	4.08	30.53	5.62	5.10	1.250
1.295	33.0	—	10.80	80.76	30.0	−50.8	4.41	32.96	4.16	31.06	5.76	5.23	1.256
1.300	33.5	—	10.84	81.08	30.5	−41.8	4.50	33.65	4.24	31.71	5.92	5.36	1.264
1.306	34.0	—	10.89	81.45	31.0	−33.2	4.59	34.35	4.33	32.37	6.08	5.51	1.272
1.310	34.3	—	10.92	81.70	31.3	−29.2	4.65	34.79	4.38	32.78	6.18	5.58	1.278
1.317	34.9	—	10.98	82.14	32.0	−19.5	4.78	35.76	4.50	33.70	6.43	5.79	1.288
1.320	35.1	—	11.01	82.32	32.2	−17.0	4.82	36.06	4.54	33.98	6.50	5.86	1.293
1.328	35.8	—	11.07	82.82	33.0	− 6.9	4.97	37.18	4.68	35.03	6.79	6.11	1.305
1.330	36.0	—	11.09	82.95	33.2	− 4.7	5.01	37.47	4.72	35.31	6.87	6.18	1.309
1.340	36.8	—	11.17	83.57	34.0	+ 4.3	5.17	38.66	4.87	36.43	7.19	6.45	1.324
1.350	37.6	—	11.26	84.19	34.9	+14.3	5.35	39.98	5.04	37.67	7.56	6.76	1.341
1.351	37.7	—	11.26	84.26	35.0	+14.4	5.36	40.12	5.05	37.81	7.57	6.78	1.343
1.360	38.4	—	11.34	84.82	35.8	+21.7	5.52	41.31	5.20	38.93	7.91	7.07	1.359
1.363	38.6	—	11.36	85.01	36.0	+24.1	5.5[illegible]	41.64	5.24	39.24	8.00	7.14	1.363
1.370	39.0	—	11.42	85.44	36.4	+30.0	5.66	42.31	5.33	39.87	8.19	7.30	1.375
1.374	39.5	—	11.46	85.69	37.0	+33.4	5.77	43.14	5.44	40.65	8.47	7.54	1.383
1.380	39.9	—	11.51	86.07	37.4	+37.0	5.86	43.80	5.52	41.27	8.66	7.69	1.393
1.386	40.4	—	11.56	86.44	38.0	+42.1	5.98	44.69	5.63	42.11	8.95	7.93	1.407
1.390	40.7	—	11.59	86.69	38.3	+44.4	6.04	45.17	5.69	42.56	9.08	8.05	1.414
1.398	41.3	—	11.66	87.19	39.0	+49.6	6.19	46.26	5.83	43.59	9.45	8.35	1.431
1.400	41.4	—	11.67	87.31	39.2	+50.9	6.22	46.57	5.86	43.88	9.52	8.41	1.435
1.410	42.2	—	11.76	87.94	40.0	+55.9	6.40	47.86	6.03	45.10	9.97	8.79	1.457
					40.9	+61.0							
					41.0	+61.5							
					41.7	+64.8							
					42.0	+65.8							
					42.5	+68.0							
					43.0	+69.8							
					43.3	+70.9							
					44.0	+73.0							
					44.2	+73.6							
					45.0	+75.6							

* Hydrometers giving specific gravities and freezing points of calcium chloride brines can be obtained from any calcium chloride producer.

** Pounds of solid or flake calcium chloride per cubic foot if water at 60° F. may be obtained by multiplying the quantity in pounds per gallon by 7.481.

*** Final volume in gallons at 60° F. when quantity of flake calcium chloride shown in previous column is added to one gallon of water.

NOTE

At concentrations above 40.8% calcium chloride solutions contain suspended crystals at 60° F., and therefore hydrometer readings at 60° F. are of little value.

* Courtesy, "Calcium Chloride for Refrigeration Brine," pp. 18–19, Calcium Chloride Institute, Washington, D. C.

Table 92. Specific Heat of Calcium Chloride Brine

Btu per lb (Estimated)

Specific Gravity	*% CaCl by Weight*	*Freezing Point °F*	*Specific Heat, Btu per lb Range from Freezing Point to 50 F**
1.08.........	9.2	23.2	0.860 to 0.870
1.12.........	13.5	16.5	0.800 to 0.815
1.16.........	17.6	7.0	0.745 to 0.763
1.20.........	21.5	−5.8	0.700 to 0.720
1.24.........	25.1	−21.5	0.660 to 0.685
1.28.........	28.7	−44.3	0.640 to 0.655
1.29†........	29.6	−59.8	0.625 to 0.653

* Approximate specific heat values may be obtained at any point between the freezing temperature and 50 F by interpolation.
† Eutectic point of CaCl.

For example, as shown in Table 90, sodium chloride brine containing 18.611 percent salt content has an average specific heat value of 0.815, whereas a 5.543 percent solution possesses a specific heat value of 0.930, or a difference of 0.115 between the two solutions.

Similarly, calcium chloride brine as shown in Table 92 possesses a like characteristic except that the rate of specific heat loss is greater with each additional increment of salt concentration. Comparing the specific heat value of 0.865 for a 9.2 percent calcium chloride content with that of 0.670 as shown for a 25.1 percent solution, a difference in specific heat value of 0.195 is found which represents the loss of heat-transfer capacity for the higher specific gravity brine. In general, a reduction of about 2.5 percent in heat-transfer value (Btu per hr per sq ft per °F) follows each 1 percent increase in calcium chloride concentration.

In many cases, the importance of obtaining a lower point of crystallization warrants a sacrifice of thermodynamic qualities through an increase in specific gravity. However, the specific gravity should be maintained at a level to provide a minimum but safe temperature differential between the point of crystallization of the solution and the required operating temperature. Thus, by maintaining only a sufficient amount of salt concentration in the solution to meet a safe crystallization point for the work required, the highest heat-transfer value can be obtained.

Determination of Salt Content of Brines

In order to control the function of any brine, it is necessary that the freezing point be known at all times. The freezing point of a salt solution depends on the salt content, which, in turn, governs the specific gravity of the solution. This statement is borne out in the physical property tables, viz.: that with each increment of increase in salt concentration there is a corresponding increase in the specific gravity of a salt solution. Also, that for any given weight of salt in a salt water solution there is a definite specific gravity that is constant for the particular salt involved. If the specific gravity of a salt solution is known, the freezing point may be determined readily as the relationship between the two always remains constant for the same brine.

For example, a 10 percent solution of sodium chloride brine has a specific gravity of 1.074, which will always correspond to a freezing point of 20.2 F. For 20 percent sodium chloride brine the freezing point is always 2.2 F, etc.

Similarly, calcium chloride brine having a 10 percent salt content has a specific gravity of 1.087, the freezing point of which will always be 22.3 F. A 20 percent solution of the same type of brine will always correspond to a freezing point of −0.4 F, etc.

The following rule may therefore be formulated: *Because the specific gravity of any salt solution is a direct indication of its salt content and a specific relationship exists between the salt content and the freezing point of the solution, it is possible to determine the value of the remaining two properties if any one of three is known.*

Thus, if sodium chloride brine is found to have a specific gravity of 1.151 (determined when the brine is at a temperature of 60 F), by referring to the physical property tables for that brine, its salt content of 20.06 percent (by weight) and freezing point of +2.2 F are determined.

Operators of commercial installations utilizing brine as a secondary refrigeration medium employ a system whereby the specific gravity of the brine is measured periodically, thus enabling them to maintain a constant check on its freezing point. If for any reason a decrease in the original amount of salt concentration does take place, a decrease in specific gravity of the solution will be indicated and salt may be added to bring it back to the required operating level. In this way, danger of freezing brine will be eliminated.

Specific Gravity. The specific gravity of a substance is its weight as compared with the weight of an equal volume of water. It is determined by dividing the weight of a given volume of any substance with the weight of an equal volume of water. If the specific gravity of a substance is known, the weight of any volume of that substance may be calculated. The specific gravity of water is 1.00 as it is the basic substance to which all other substances are compared. Thus, if a substance such as sodium chloride brine having a salt content of 15 percent is found to have a specific gravity of 1.111, it means that it is 1.111 times heavier than water for a similar volume.

For example, water weighs 62.4 lb per cu ft. The weight of the 15 percent sodium chloride brine is determined by multiplying the weight of 1 cu ft of water, 62.4, by the specific gravity of the brine, 1.111, or 69.32 lb per cu ft. Conversely, if the weight of the brine or 69.32 lb per cu ft is divided by the weight of 1 cu ft of water, or 62.4 lb, the specific gravity of the brine is determined, or

$$\frac{69.32}{62.4} = 1.111 \text{ specific gravity of 15 percent NaCl brine.}$$

Principle of Hydrometry (Archimedes' Principle). Archimedes' principle of floating bodies states that *a floating body displaces an equal weight of the liquid in which it floats.* As an example of this, consider two pieces of wood having similar dimensions, one piece of oak and the other balsa. If placed in water, it will be found that, while both pieces float, the oak will have a greater portion of its mass submerged than the balsa. The reason is that the oak is heavier and exerts a greater pressure against the water than does the balsa. Because it submerges to a greater depth than the balsa by reason of its greater weight, it displaces a greater amount of water than the lighter wood. Also, if the oak were four times heavier than the balsa, it would displace four times more water by weight than the piece of balsa. It follows, too, that the oak would have four times more of its volume submerged.

If the same two pieces of wood were placed in a brine solution, neither piece would submerge as much as in water, the reason being that, although each piece exerts the same pressure in the brine as it did in the water, the brine because of its greater weight exerts a higher resistance to the floating bodies and therefore does not allow as much of its volume to be displaced. In other words, the brine provides more buoyancy than the water because of its greater density. The same relative proportion of volume of each piece of wood would be submerged in the brine as in the water even though the volumes submerged would be less.

This condition holds true with all floating materials. If a sealed glass tube is substituted for the wood, it will partially submerge to a depth depending on the density of the liquid in which it is placed. The lighter the liquid by weight, the deeper the glass tube will submerge, and, conversely, the heavier the liquid the less it will submerge. If a scale previously determined by test in liquids of known gravities is placed in the upper stem of the tube which is called a hydrometer, the gravity of any liquid in which it is placed may be determined directly. For liquids lighter than water, the scale starts at the bottom of the stem with a specific gravity value of 1.00 and the graduations decrease in value toward the top. The opposite is true with heavier-than-water liquids where the scale starting with 1.00 at the top of the stem increases in value toward the bottom.

Therefore, it may be stated that the hydrometer is governed by the principle that *a floating body displaces its own weight of the liquid in which it floats.*

The Hydrometer. There are two basic types of hydrometers in use, one to determine the density of liquids that are heavier than water and a second type for use with liquids lighter than water. The former embraces such liquids as acids, milk, brines, etc., while the latter includes petroleum products and liquid refrigerants among many others.

The hydrometer is an elongated glass tube sealed at both ends. One end is enlarged to control its buoyancy and is connected with another enlarged chamber in which sufficient weight in the form of lead shot or mercury is inserted to keep the instrument vertical while afloat in a liquid. In the narrow tube that extends upward from the enlarged chamber is placed a graduated scale that indicates in numerous ways the density or solute concentration, depending on the type of hydrometer, of the liquid in which it is placed.

As with salt solutions, so it is also true with many other solutions, that the gravity of a liquid varies proportionately with the amount of solute concentration. Therefore, the hydrometer may be designed to determine the solute content of any solution. Some of the systems of hydrometry in general use and their functions are described below.

Specific Gravity: a direct measure of density of liquids as compared with the density of pure water. The latter is given an arbitrary value of 1.00 at 60 F.

Salimeter: a hydrometer whose scale is designed to indicate the percentage of salt in solution. Used for either sodium chloride or calcium chloride brines.

Baumé Gravity: degree Baumé, a scale used for determining the relative densities of liquids either heavier or lighter than water.

(*a*) The scale for lighter-than-water liquids has a maximum value of 10.00 which corresponds to the specific gravity of pure water at 60 F of 1.00. By designating a zero value (0 Bé) on the scale at a point equal to the specific gravity of a 10 percent salt solution and dividing the range into 10 equal parts, the basis of the lighter-than-water Baumé scale is thus obtained. The graduations are so numbered that the °Bé designation increases as the specific gravity decreases.

Degrees Baumé for liquids having specific gravity values less than unity may be determined by the following formula:

$$\text{Degrees Bé} = \frac{140 \text{ (modulus)}}{\text{specific gravity}} - 130$$

Thus

$$70 \text{ Bé} = \frac{140}{0.70} - 130$$

(*b*) For liquids heavier-than-water or whose specific gravity is greater than unity, the Baumé scale has a minimum value of 0 which corresponds to a specific gravity of 1.00 or that of pure water at 60 F. The graduations for this scale are determined by dividing into 15 equal parts the range between 0 and the point on the scale corresponding to the specific gravity of a 15 percent brine solution which is designated at 15° Bé. Extension of equal divisions is made below and beyond the 15° Bé level to complete the scale. This permits a reading of degrees Baumé to be made of liquids heavier than 15 percent salt solutions. Differing from the lighter-than-water scale, the graduations are so designated that the degrees Baumé value increases correspondingly to an increase in specific gravity. This is shown in Table 94.

TABLE 93. RELATIONSHIP OF SPECIFIC GRAVITY AND °BÉ FOR LIGHTER-THAN-WATER LIQUIDS

Specific Gravity at 60 F	*Degrees Baumé*
1.00..........	10.00
0.90..........	25.56
0.80..........	45.00
0.70..........	70.00
0.60..........	103.33

TABLE 94. RELATIONSHIP OF SPECIFIC GRAVITY AND °BÉ FOR HEAVIER-THAN-WATER LIQUIDS

Specific Gravity at 60 F	*Degrees Baumé*
1.00..........	0.00
1.10..........	13.18
1.20..........	24.17
1.30..........	33.46
1.40..........	41.43
1.50..........	48.33
1.60..........	54.38
1.70..........	59.71
1.80..........	64.44

Degrees Baumé for liquids heavier-than-water may be determined, if the specific gravity is known, using the following formula:

$$\text{Degrees Baumé} = 145 - \frac{145\ (\text{modulus})}{\text{specific gravity}}$$

Thus

$$48.33° = 145 - \frac{145}{1.50}$$

A P I Scale: used by the petroleum industry and accepted by the American Petroleum Institute, the Bureau of Mines, and the Bureau of Standards for recording the relative gravity of gasolines, oils, etc. It differs from degrees Baumé in that the relative scale registers slightly higher for some liquids and a modulus of 141.5 is used in the conversion formula instead of 145 as in the Baumé scale. A P I Gravity is calculated from the following formula:

$$\text{Degrees API} = \frac{141.5}{\text{specific gravity (60 F)}} - 131.5$$

Thus

$$25.72° \text{ API} = \frac{141.5}{0.9} - 131.5$$

TABLE 95. RELATIONSHIP OF SPECIFIC GRAVITY AND °A P I FOR LIGHTER-THAN-WATER LIQUIDS

Specific Gravity at 60 F	*Degrees A P I*
1.00	10.00
0.90	25.72
0.80	45.38
0.70	70.64
0.60	104.33

Proof Scale: indicates the alcoholic content of spirits at 60 F.

Saccharometer: for the determination of the sugar content of sugar solutions.

There are other hydrometer scale systems employed throughout industry, but the above list indicates the possible uses to which such an instrument may be applied. Because of its convenience in determining salt concentration of brines, it is of great value to operators of refrigeration plants employing either sodium chloride or calcium chloride solutions.

Typical Brine Temperatures

For reasons of safety (from freezing) and efficiency, brine temperatures should be carried as high as practical in relation to the temperature required for the purpose intended. Table 96 shows the typical relationship between room, brine, and refrigerant temperatures.

CHEMICAL BALANCE OF BRINES

Brines employed in secondary refrigeration systems tend to undergo a gradual chemical change after short periods of operation. The causes of such a change are due, in general, to constant contact with metal surfaces and exposure to air. If not checked in its transformation, detrimental effects to metal parts result, causing an unsatisfactory operating condition. To insure chemical stability, some method of brine control becomes necessary.

To achieve this control, the modern operator is equipped to test brine solutions periodically to determine their chemical condition. These tests

TABLE 96. TYPICAL COLD-STORAGE ROOM, BRINE, AND REFRIGERANT TEMPERATURE RELATIONSHIPS*

Section (a)†								
Room temperature desired	−10	0	10	20	30	40	50	60
Brine temperature	−20	−12	−4	4	12	20	28	36
Refrigerant temperature	−28	−20	−13	−6	1	8	23	19
Temperature difference between brine and room	10	12	14	16	18	20	22	24
Temperature difference between refrigerant and brine	8	8	9	10	11	12	15	17
Temperature difference between refrigerant and room	18	20	23	26	29	32	37	41
Section (b)‡								
Room temperature desired	−10	0	10	20	30	40	50	60
Refrigerant temperature	−20	−15	−5	3	10	16	22	26
Temperature difference	10	15	15	17	20	24	28	34

* Reprinted from the "Air Conditioning—Refrigerating Data Book 5th Edition—1956–57 Applications Volume," by courtesy of The American Society of Heating, Refrigeration and Air Conditioning Engineers.

† (*a*) Relationship between room, brine, and refrigerant temperature, °F.

‡ (*b*) Relationship between room and refrigerant temperature with direct expansion, °F.

reveal two important factors, the determination of which are prerequisite to the application of control measures. These factors are: (*a*) the type of chemical change, if any and (*b*) the degree of change. This information enables the operator to prescribe with reasonable accuracy the necessary chemical treatment to apply in order to maintain a safe operating condition within the brine solution or restore it to that condition in the event that a chemical change has occurred.

Brine control can be possible only through an understanding of the chemical principles involved. To better comprehend the chemical changes that occur within the brine solution and the importance of its control, a résumé of these principles is given here.

Chemistry embraces the study of all matter contained throughout the universe, including that which is found beneath and on the surface of the earth and in the atmosphere above it. All matter is compounded from simpler substances called elements which cannot be chemically reduced further to lesser substances. Elements are basic in their structure, one or more being found in every object that comprises nature, either as a free agent or in combination with other elements. Elements can be divided physically into small particles called atoms. There are at present 101 known elements, many of which are familiar, such as iron, copper, lead, oxygen, nitrogen, mercury, etc. Table 97 lists all known elements, showing their chemical symbol, atomic number, and atomic weight.

TABLE 97. TABLE OF ELEMENTS

Atomic Number	*Name*	*Symbol*	*Atomic Weight*
0	neutron	n	
1	hydrogen	H	1.0080
2	helium	He	4.003
3	lithium	Li	6.940
4	beryllium	Be	9.013
5	boron	B	10.82
6	carbon	C	12.011
7	nitrogen	N	14.008
8	oxygen	O	16.000
9	fluorine	F	19.00
10	neon	Ne	20.183
11	sodium	Na	22.991
12	magnesium	Mg	24.32
13	aluminum	Al	26.98
14	silicon	Si	28.09
15	phosphorus	P	30.975
16	sulfur	S	32.066
17	chlorine	Cl	35.457
18	argon	A	39.944
19	potassium	K	39.100
20	calcium	Ca	40.08
21	scandium	Sc	44.96
22	titanium	Ti	47.90
23	vanadium	V	50.95
24	chromium	Cr	52.01
25	manganese	Mn	54.94
26	iron	Fe	55.85
27	cobalt	Co	58.94
28	nickel	Ni	58.71
29	copper	Cu	63.54
30	zinc	Zn	65.38
31	gallium	Ga	69.72
32	germanium	Ge	72.60
33	arsenic	As	74.91
34	selenium	Se	78.96
35	bromine	Br	79.916
36	krypton	Kr	83.80
37	rubidium	Rb	85.48
38	strontium	Sr	87.63
39	yttrium	Y	88.92
40	zirconium	Zr	91.22
41	niobium (*formerly* columbium, symbol Cb)	Nb	92.91
42	molybdenum	Mo	95.95
43	technetium	Tc	99.
44	ruthenium	Ru	101.1
45	rhodium	Rh	102.91
46	palladium	Pd	106.4
47	silver	Ag	107.880
48	cadmium	Cd	112.41
49	indium	In	114.82
50	tin	Sn	118.70

Atomic Number	*Name*	*Symbol*	*Atomic Weight*
51	antimony	Sb	121.76
52	tellurium	Te	127.61
53	iodine	I	126.91
54	xenon	Xe	131.30
55	cesium	Cs	132.91
56	barium	Ba	137.36
57	lanthanum	La	138.92
58	cerium	Ce	140.13
59	praseodymium	Pr	140.92
60	neodymium	Nd	144.27
61	promethium	Pm	(147)
62	samarium	Sm	150.35
63	europium	Eu	152.0
64	gadolinium	Gd	157.26
65	terbium	Tb	158.93
66	dysprosium	Dy	162.51
67	holmium	Ho	164.94
68	erbium	Er	167.27
69	thulium	Tm	168.94
70	ytterbium	Yb	173.04
71	lutetium	Lu	174.99
72	hafnium	Hf	178.50
73	tantalum	Ta	180.95
74	tungsten (*also called* wolfram)	W	183.86
75	rhenium	Re	186.22
76	osmium	Os	190.2
77	iridium	Ir	192.2
78	platinum	Pt	195.09
79	gold	Au	197.0
80	mercury	Hg	200.61
81	thallium	Tl	204.39
82	lead	Pb	207.21
83	bismuth	Bi	209.00
84	polonium	Po	210.0
85	astatine	At	(210)
86	emanation (*also called* radon, symbol Rn)	Em	222.0
87	francium	Fr	(223)
88	radium	Ra	226.05
89	actinium	Ac	227.0
90	thorium	Th	232.05
91	protactinium	Pa	231.0
92	uranium	U	238.07
93	neptunium	Np	237.0
94	plutonium	Pu	(239)
95	americium	Am	(241)
96	curium	Cm	(242)
97	berkelium	Bk	(243)
98	californium	Cf	(244)
99	einsteinium	E	254
100	fermium	Fm	255
101	mendelevium	Mv	256

When the atoms of two or more elements are chemically combined, the resulting substance is called a compound, the smallest particle of which is a molecule. Chemistry in general treats with compounds, the majority of which are divided into three important classes: (1) acids, (2) bases, and (3) salts. Each class has distinct characteristics differing from the other and indicative of their nature. Any given compound, or any molecule of a given compound, always contains the same elements and in the same proportions by weight. This is governed by the "law of fixed proportions." A molecule of water, for example, is a compound of one part oxygen and two parts hydrogen. It is always made up of these elements, never any other, and always in the same proportion. If hydrogen is given an arbitrary weight of 1, then oxygen must be given a weight of 16 because the weight of one atom of oxygen is 16 times heavier than a hydrogen atom. Water contains two atoms of hydrogen and one atom of oxygen, or a ratio by weight of 2:16. According to the law of fixed proportions, water is always found to contain these two elements in the same proportion by weight of 2:16 or 1:8. This principle holds true for all compounds.

The oxygen atom is 16 times heavier than the hydrogen atom, a fact borne out by the atomic theory which states that "atoms of different elements have different weights." All weights are relative in that they are based on the weight of the hydrogen atom, which is given a value of 1. This value does not denote 1 lb or 1 gram, or 1 of any specific weight unit, but is to be interpreted only as a relative value selected as the atomic weight of hydrogen because that element is the lightest of all known elements. Therefore, the atomic weight of any element is the ratio of its weight to the weight of the hydrogen atom. Furthermore, it is found that the atomic weight of iron is 55.84, which according to the above rule means that the iron atom is 55.84 times the weight of the hydrogen atom. This rule is true for all elements listed in Table 97.

Acids

Acids may be defined as a group of compounds having one or more nonmetallic elements in combination with the element hydrogen. They are usually sour to the taste. Blue litmus paper turns red in the presence of acids and compounds known as salts are formed in their reaction with metals.

The strength of acids is based on the vigorousness with which they react with metals or other compounds and is commonly expressed as being either strong or weak. Such terms, while denoting some difference in relative value, are totally inadequate for practical use in calculating chemical treatment of brines where more definite measurement is necessary.

At normal atmospheric temperatures, acids differ in their physical state in that some are liquid or solid while a few are gaseous. Sulfuric acid is an oily liquid prepared in one way by dissolving the gas, sulfur trioxide, in water. Boric acid occurs in nature in the form of white crystalline scales, while hydrogen chloride gas is prepared by heating a mixture of sulfuric acid and common table salt (sodium chloride). Each is an acid having different physical forms under atmospheric temperature conditions. When hydrogen chloride is dissolved in water, it forms hydrochloric acid. An acid concentrate is an acid that contains a small amount of water or in some instances none at all. Sulfuric acid, for example, is considered a concentrate when the water content is 70 percent by weight or below. Hydrochloric acid is considered a concentrated acid even though it contains 60 to 70 percent by weight of water. When an acid concentrate is mixed with water, it becomes a dilute acid solution and is identified under that name, such as "dilute sulfuric acid" or "dilute hydrochloric acid." The amount of dilution may be determined in any acid if the specific gravity of the concentrate of that acid is known. For commerical concentrated sulfuric acid the specific gravity is 1.84, which is almost twice that of water (1), while in dilute form it is lower depending on the amount of dilution. Some acids are not chemically active in their concentrated form as they are when diluted. For this reason, concentrated sulfuric acid (about 93.5 percent) can be shipped in iron containers or tank cars without serious effect whereas this would not be possible if it were mixed with water.

It may be stated, therefore, that any compound, liquid, solid, or gas having those characteristics pertinent to an acid *is* acid. If, when mixed with water—pure or impure—it continues to maintain those characteristics, the resulting dilution is also acid and is referred to as an acid solution. Thus, any liquid including brines that reacts with litmus paper to turn it red may be regarded as an acid solution and will react with other substances in the same way as a pure dilute acid. The strengths of such solutions are also measurable in their degree of acidity.

Bases

When one atom of oxygen combines with one atom of hydrogen, they characteristically act as a unit. Such a group of atoms is called hydroxyl. Hydroxyl forms the root or foundation of an important class of chemical compounds which are known as radicals, being represented by the chemical symbols OH.

Bases are a class of compounds that contain a metal element in combination with a hydroxyl or OH radical and are identified as hydroxides prefixed with the name of the particular metal element from which they are formed. Sodium hydroxide (NaOH), potassium hydroxide (KOH), and calcium hydroxide ($Ca(OH)_2$) are examples of typical bases. The OH radical bears the same relationship to a base as hydrogen bears to an acid.

Base solutions turn red litmus paper blue and have a bitter taste generally. Most of them are insoluble in water, but a few of the stronger ones are soluble. They react with acids to form salts and water, an important characteristic.

As in the case of acids, the strength of bases is predicated on their reaction with other substances and is commonly expressed as being either weak or strong. Highly active or strong bases are very soluble in water and are called alkalis, a group which includes the important hydroxides of sodium, potassium, and ammonium. Thus an alkali may be defined as a strong caustic (burns the flesh) soluble base.

Here, again, as with acids the terms strong and weak in expressing the strength of bases are not definite enough in their degree for close chemical calculations. The system of expressing base strength more precisely is associated closely with that relating to acids.

Bases occur as solids or liquids. Typical examples are (1) sodium hydroxide, commonly known as caustic soda, a white deliquescent solid formed by the interaction of water with the metal, sodium, and (2) ammonium hydroxide, a liquid formed by the reaction of ammonia gas and water. When bases are mixed with water, the resulting dilution is called a basic or alkaline solution. "Lime water" exemplifies such a solution and is made by dissolving the base, calcium hydroxide, in water. Like their pure basic hydroxides, the solutions react in the same way with indicators such as litmus paper. Again, particularly with liquid bases such as ammonium hydroxide, concentrated base compounds should not be confused with base solutions. Although water enters the reaction with ammonia gas to form the base itself, only enough water is required for complete atomic balance, leaving no free water molecules in the liquid base. In this case, all water becomes "saturated" with ammonia gas, thus forming a highly concentrated ammonium hydroxide.

Compounds, solid or liquid, having those characteristics described as base properties are bases, and their solutions, pure or impure, which maintain those characteristics can be considered as basic or alkalinic solutions. Thus, any liquid including brines that reacts with red litmus paper to turn it blue may be regarded as a basic solution and will have a tendency to react with other substances in a similar way as the pure dilute base.

To test for leaks in ammonia evaporator coils submerged in sodium chloride brine solutions, red litmus paper can be used. By dipping the paper directly in the brine, any ammonia present will turn the paper blue. The ammonia, being a base, causes the brine to become a basic solution.

Salts

Salts are formed when an acid and a base are mixed together. They combine a metal element with a nonmetal element, some of which also contain oxygen. Common table salt is a familiar member of the salt family, the chemical constituents of which are sodium and chlorine (sodium chloride, NaCl). Salts have a sour or a saline taste and vary in their water solubility. In most cases, salts do not affect either red or blue litmus paper and therefore are considered as neutral. As a general rule, they lack strong affinity for other substances, being unlike acids or bases in this respect, remaining unchanged under different types of environment. For this reason, many kinds of salts are found in the earth in a high state of purity. Sodium chloride and calcium chloride, both of which are used in brine solutions, are examples of salt compounds.

Chlorides, sulfates, and nitrates are major groups within the salt family each of which carries the name of the metal element as a prefix. Thus, sodium chloride, copper sulfate, and potassium nitrate are representative examples of the different groups.

In addition to their formation by the mixture of acids and bases, salts are also the product of combining an acid and a metal and/or an acid with another salt.

All salts are not necessarily neutral to litmus paper and may be classified as to their atomic composition as (1) normal salts, (2) acid salts, or (3) basic salts.

A **normal salt** is one in which all the hydrogen of its respective acid is replaced by a metal. Normal salts are generally neutral.

An **acid salt** is one in which hydrogen from its respective acid remains and is replaceable by a metal. It may be acid or neutral.

A **basic salt** is one that contains OH from its respective base and which is replaceable by a nonmetal element. Basic salts are usually neutral but in some cases basic.

Neutralization

When acids and bases are mixed together in proportions that assume atomic balance, each destroys the properties of the other. Thus, in eliminating the acid characteristics of the one and the basic characteristics of the other, the properties of the new substance become neutral and the chemical action one of neutralization. Salt is precipitated as one of the products of this neutralizing action and water the other.

The following transformation of elements occurs when the base, sodium hydroxide, is mixed with hydrochloric acid in their correct proportions:

Sodium hydroxide + hydrochloric acid = sodium chloride + water

$$NaOH + HCl = NaCl + H_2O$$

Base + acid = salt + water

Water is formed when the hydrogen atom of the acid combines with the OH radical of the base to make a total of two hydrogen atoms and one oxygen atom. As these two elements are the only ones required to form water, their ratio must be two to one of hydrogen, respectively. This leaves one atom of sodium and one atom of chlorine, which, having a natural affinity for each other, combine in that proportion to form the salt sodium chloride (NaCl). In separating the OH radical from the base and the hydrogen atom from the acid, the properties of both cease to exist and the resulting substance free from either of their properties is considered as being neutral.

Ionization

The theory of ionization proposed by the Swedish chemist Arrhenius states that, when acids, bases, and salts are dissolved in water, part of their molecules decompose into electrically charged particles called ions. The dissociation of molecules of these compounds upon ionization results in separate groups of atoms or radicals which carry either a positive (+) or a negative (−) electrical charge. However, ions must not be confused with atoms or radicals, but are to be considered as *electrically charged* atoms or radicals having entirely different characteristics.

When common salt (sodium chloride, NaCl) is dissolved in water, many of the molecules decompose into separate electrically charged atoms or ions of sodium and chloride. Because they are electrically charged, neither the sodium ion nor the chloride ion possesses the same characteristics as the atom of either. If this were not true, a violent reaction would take place upon the introduction of the sodium to the water and the solution would become discolored (yellowish) due to the effect of the chloride when these two elements separated.

What does happen, however, as a result of this dissociation is that the solution known as electrolyte becomes a conductor of electric current. In the case of sodium chloride, the sodium ion carries a positive charge while the chloride ion is negative. In the case of all acid, base, and salt compounds there are always an equal number of positive and negative charges. As a result, an electrical balance is established and the electrolyte appears static.

If an electric current were passed through the resulting electrolyte, a separation of the ions would take place, the positive ion traveling toward the negative terminal or cathode while the negative ion would be attracted by the positive terminal or anode. This action is one of *electrolysis*. The direction of travel of the ions as described here is governed by the law of magnetism, which states that *unlike poles attract while like poles repel.* Substituting silver nitrate for the sodium chloride and using a silver strip for the anode, silver deposits would be formed on the cathode as a result of passing an electric current through the electrolytic solution. Silver plating is accomplished commercially using this method of electrochemical reaction.

It was stated previously that molecules of acids, bases, and salts carry equal positive and negative charges upon ionization. If they were not balanced electrically, a bond between the atoms that constitute the molecules would not exist or at best would be highly unstable. It is this tendency toward electrical balance that determines the number of atoms of any element required to form a stable molecule or compound with one or more other elements or radicals.

It is readily seen that all elements characteristically carry their own type of electrical charge—that is, they must be either positive or negative by nature. It is also true that all atoms of the same element and all radicals made up of similar elements always carry the same amount of electrical charge with few exceptions. Therefore, an atom of sodium, no matter where it is found or with whatever other element it is combined, will carry the same amount of charge. These statements give rise to two important questions, viz.:

(*a*) How is the electrical nature of the different elements determined?
(*b*) What is the amount of charge carried by the various elements?

The first question can be satisfied by adapting several simple rules that determine the character of the various ions. These rules are as follows:

(*a*) Positive ions are formed from metals and hydrogen.
(*b*) Negative ions are formed from nonmetals except hydrogen.
(*c*) Negative ions are formed from any radical.

The answer to the second question can be determined best by reference to tables showing the electrical strength of various elements and radicals. Such information as given in these tables has been previously determined through calculations, observation, or experimentation. A partial list of important elements and radicals and their relative electrical strength is shown in Table 98.

TABLE 98. RELATIVE ELECTRICAL CHARGES OF SOME ELEMENTS AND RADICALS

Element or Radical	*Electrical Charges*	*Element or Radical*	*Electrical Charges*
Hydrogen	H^+	Copper	Cu^{++}
Potassium	K^+	Magnesium	Mg^{++}
Sodium	Na^+	Sulfate	SO_4^{--}
Chloride	Cl^-	Carbonate	CO_3^{--}
Bromide	Br^-	Chromate	CrO_4^{--}
Iodide	I^-	Aluminum	Al^{+++}
Calcium	Ca^{++}	Tin (stannic)	Sn^{++++}

It will be noted in Table 98 that calcium ions carry twice the charge of a hydrogen ion whereas the aluminum and tin ions carry three and four times, respectively, the amount of the hydrogen ion. These ratios will always be the same. By remembering the amount of charge carried by certain basic elements, it becomes easy for the student to calculate the electrical charge carried by many other elements and radicals through consideration of their chemical formulas. For example, consider the compound sulfuric acid, the symbol of which is H_2SO_4. The hydrogen ion always carries one electrical charge according to the table. However, sulfuric acid is made up of two hydrogen atoms as stated in the formulas, viz., H_2. This compound carries only one molecule of the radical (SO_4). An electrical balance must be formed in order to create a bond between the elements and/or radicals of any compound, so it is apparent that the radical (SO_4) must carry twice the charge of hydrogen. Conversely, because of the double electrical charge carried by the (SO_4) radical, it requires two atoms of hydrogen in combination to form the compound sulfuric acid. The ionic formula is written $2H^+(SO_4)^{--}$.

Table 99 lists compounds showing both their chemical and ionic formulas.

TABLE 99. CHEMICAL AND IONIC FORMULAS OF SOME COMPOUNDS

Compound	*Chemical Symbol*	*Ionic Symbol*	*Type*
Sodium chloride.......	$NaCl$	Na^+Cl^-	Metal + nonmetal
Calcium chloride......	$CaCl_2$	$Ca^{++}2Cl^-$	Metal + nonmetal
Silver nitrate..........	$Ag(NO_3)$	$Ag^+(NO_3)^-$	Metal + radical
Potassium hydroxide...	$K(OH)$	$K^+(OH)^-$	Metal + radical

Degree of Ionization. When a compound is dissolved in water, not all the molecules decompose. Some maintain their original form so that most electrolytes are composed of molecules and ions (positive and negative). The number of molecules of any acid, base, or salt that decompose to form ions depends on the concentration of the solution and the compound itself. In highly concentrated solutions, the number of molecules will be greater, whereas their number diminishes with greater dilution. However, with equal proportions of dilutions, the number of molecules varies with different compounds.

Because of this characteristic, it will be found that some acids and bases form stronger solutions than others, and are so classified. Salt solutions are not included in the category of strong or weak because most salts dissociate to approximately the same degree.

Hydrochloric acid (H^+Cl^-), for example, is considered to be a strong acid due to the tendency of its molecules to dissociate into ions when heavily diluted. Some authorities record 90 percent ionization of the hydrochloric acid molecules when diluted. On the other hand, acetic acid ($C_2H_4O_2$) does not ionize readily, 1.3 percent of its molecules dissociating in heavy dilutions. As a result of these differences in their tendency to dissociate, hydrochloric acid is considered to be a strong acid whereas acetic acid is a weak one.

Such differences in the degree to which various compounds will dissociate apply also to bases. Ninety percent of the molecules of sodium hydroxide will ionize when dissolved in water, and therefore it is considered a strong base. With ammonium hydroxide, a weak base, the molecules dissociate only to the extent of 1.4 percent.

Some understanding of the degree to which different compounds vary in their tendency to ionize when dissolved in water may be gained by reference to Table 100. Comparisons are made at the same temperatures and relative concentrations.

TABLE 100. RELATIVE IONIZATION OF VARIOUS TYPE COMPOUNDS

	Compounds	*Ions*	*% of Ionization*
Acids	Hydrochloric acid	HCl	90
	Nitric acid	HNO_3	90
	Sulfuric acid	H_2SO_4	60
	Acetic acid	$C_2H_4O_2$	1.3
Bases	Sodium hydroxide	NaOH	90
	Potassium hydroxide	KOH	90
	Ammonium hydroxide	NH_4OH	1.4
Salts	Sodium chloride	NaCl	90
	Potassium chloride	KCl	86
	Silver nitrate	$AgNO_3$	86

Ionic Definitions of Acids, Bases, and Salts. In view of the foregoing discussion of ionization, new definitions of acid, base, and salt compounds can now be formulated. Such definitions are based on their ionic development as compared with the conventional chemical definitions based on atomic compounding. Thus:

1. An acid is a substance whose water solution contains positive hydrogen ions.

2. A base is a substance whose water solution contains negative hydroxyl (OH) ions.

3. A salt is a substance resulting from the neutralizing effect of mixing an acid solution and a basic solution whereby the hydrogen atom of the acid combines with hydroxyl radical of the base to form water.

As was stated, high percentages of ionization are indicative of a strong acid or base. This means that strong acid solutions indicate high hydrogen ion concentration and a strong base solution is indicative of high hydroxyl ion concentration. The converse of this statement is also true. Therefore, the degree of strength of any acid solution would depend on the number of hydrogen ions it contained. Similarly, the degree of strength of any basic solution would depend on the number of hydroxyl ions present.

The number of ions present in any solution depends on the concentration of acid or base and/or the substance forming the solution, or both. This may be understood readily by comparing various concentrations of acetic acid (a weak acid). At a 0.1 concentration (60 grams per liter), acetic acid solutions contain 0.4 percent ionized acid. If it contained but $\frac{1}{10}$ of 0.1 concentration or 0.01 (6 grams per liter), or 10 times greater dilution, an increase in the number of ions or 1.3 percent ionization would result. This difference would be found true in varying degrees with all acids and bases. The conclusions drawn from this discussion may be summed up as follows: (*a*) the degree of acidity of any solution increases with an increase in the number of hydrogen ions, and (*b*) the degree of alkalinity of any solution increases with an increase in the number of hydroxyl (OH) ions.

Carrying these two laws farther, consider a substance such as pure water that ionizes to produce hydrogen ions (H^+) and hydroxyl ions (OH^-) in equal numbers. The ionic formula being

$$H_2O = H^+ + OH^-$$

it becomes obvious that pure water is neither acid nor alkaline and that it remains neutral of the equalizing effect of the hydrogen ions and the hydroxyl ions on each other.

Thus it is possible for substances to react in three principal ways, depending on their ion characteristics: acid, neutral, or alkaline.

A simple method of determining the reaction of any substance is by the use of litmus paper. The results are as follows:

Acid substances.......	turn blue litmus red turn white litmus red
Neutral substances....	have no effect on either red, white, or blue litmus (normally)
Alkaline substances....	turn red litmus blue

The use of litmus paper, although indicating the chemical reaction of substances, does not give an accurate clue as to the degree of acidity or alkalinity the substance may possess. In order to correct a chemical condition within a substance such as brine solutions, such information is highly important. In this connection, an analogy to temperature may be made. Stating that a substance is cold, warm, or hot provides only a meager amount of information as to its relative condition with some other substance of known heat intensity such as the human body. In order to calculate the size of a heating system to provide sufficient even warmth to a building of definite size, actual relative heat intensities called temperature must be known. To determine the amount of refrigeration required to freeze a definite amount of foods or water, it is important that initial and final temperatures be specifically stated.

In dealing with a substance, the degree of its acidity or alkalinity must be determined before it is possible to treat it in order to neutralize it or bring it to a desired chemical state.

THE pH SCALE

As in the measurement of heat intensity in which the degree of intensity is indicated by °F or °C as recorded by a thermometer specifically designed for the purpose, so it is possible to determine the degree of acidity or alkalinity of most substances. In place of the °F or °C reading, acid or alkaline intensities are stated in terms of pH values. Unlike temperature readings that are measured directly by a thermometer, pH values are not so simple to determine. Modern methods, however, have reduced the complications once prevalent in determining pH values to a comparatively simple procedure.

A pH value is a number denoting the degree of acidity or alkalinity of a substance. It does not in any way indicate the quantity of acids or alkalines in a solution as found by titration methods. It is derived by measuring the amount of hydrogen ions (H^+) in grams per liter liberated during ionization of acid and alkaline solutions at various stages of intensities. For example, pure water ionizes to produce 0.0000001 gram of hydrogen or acid ions per liter. Therefore, any substance producing 0.0000001 gram of acid ions (H^+) per liter is considered to be a neutral solution. It was also stated that, the greater the number of hydrogen ions present in a solution, the stronger becomes its acid reaction. Therefore, pure water being neutral and established as a relative point by which intensity comparisons can be made, it follows that (*a*) any solution producing more hydrogen ions than pure water will be acid and to the degree governed by the amount of difference, and (*b*) any solution producing less hydrogen ions than pure water will be alkaline (OH increases) to the degree governed by the amount of difference.

A substance producing 0.000001 gram of hydrogen ions per liter upon ionization will be slightly acid. On the other hand, a substance producing

0.00000001 gram of hydrogen ions per liter will be slightly alkaline. It will be noted that the fluctuations in measurements are increased or decreased by multiples of 10. This is no accident. For each whole graduation in the pH scale intensity of acidity and alkalinity is raised or lowered 10 times its preceding number. Because hydrogen ion concentration is measured in grams per liter, it requires the use of numerous decimal places for exact recording. In order to facilitate reading the pH scale, numbers from 0 to 14 have been selected to replace the otherwise awkward decimal system. The latter carried to its ultimate conclusion would require 14 decimal places if used in the scale, which obviously would cause many errors.

The selection of the numbers 0 through 14 is not arbitrary but is actually the number of decimal places required to record the measure of weight of hydrogen ions in grams per liter of solutions having ascending or descending values of acid and alkaline values. Thus, a solution having no decimal places represented by 1.0 or 1 gram of hydrogen ions per liter would be extremely acid as compared with pure water. Such a value on the pH scale would be represented as 0. A solution having 0.1 gram of hydrogen ions per liter would have a pH scale reading of 1. The acid intensity of a solution having a pH value of 14 is considered extremely alkaline or caustic, being 10 times more alkaline in intensity than a solution whose value measures pH 13.

The rule for reading the pH scale may be stated as follows:

pH 7 being the neutral point, a lesser value denotes acidity, the degree of which increases as the numbers decrease, whereas a higher value denotes alkalinity, the degree of which increases as the numbers increase.

Shown in Fig. 387 is a typical pH scale, and Table 101 shows the hydrogen ion concentration from which the pH scale numbers were derived.

In dealing with the correction of chemical conditions in solutions such as brine, interest is directed mainly to acid or alkaline intensities rather than the strength or the weakness of the individual compounds. For this reason, the pH scale of measuring such intensities becomes of great importance. The exact degree of acidity must be known before intelligent corrective measures can be applied.

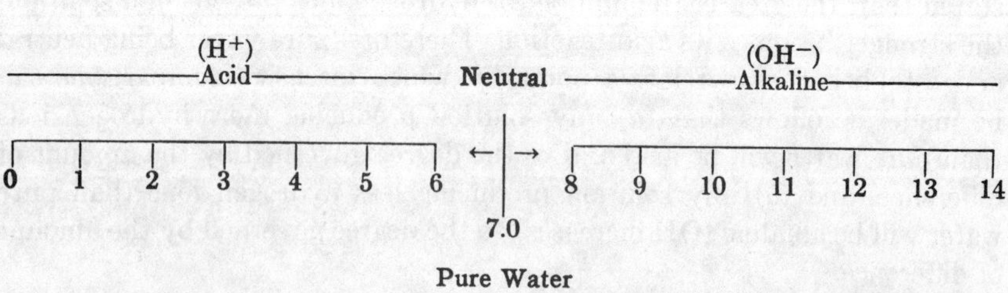

Fig. 387. The pH scale is divided into 14 parts, each part of which is again divided into 10 smaller graduations. Thus it is possible to have fractional pH values, such as 7.2, 9.5, 3.6, etc.

TABLE 101. THE pH SCALE*

pH Value	*Hydrogen Ion (H^+) Concentration Grams Ionizable H^+ per Liter*			*No. of Times Acidity or Alkalinity Exceeds that of Pure Water (pH 7)*	
0.0.....	1.0		Increasingly Acid ↑	10,000,000	10
1.0.....	0.1			1,000,000	10^{-1}
2.0.....	0.01			100,000	10^{-2}
3.0.....	0.001			10,000	10^{-3}
4.0.....	0.0001			1,000	10^{-4}
5.0.....	0.00001			100	10^{-5}
6.0.....	0.000001			10	10^{-6}
7.0.....	0.0000001	**Neutral**		1	10^{-7}
8.0.....	0.00000001		↓ Increasingly Alkaline	10	10^{-8}
9.0.....	0.000000001			100	10^{-9}
10.0.....	0.0000000001			1,000	10^{-10}
11.0.....	0.00000000001			10,000	10^{-11}
12.0.....	0.000000000001			100,000	10^{-12}
13.0.....	0.0000000000001			1,000,000	10^{-13}
14.0.....	0.00000000000001			10,000,000	10^{-14}

* *Note:* The pH scale value is equivalent to the number of decimal places required to denote the weight of H^+ in grams per liter.

In the discussion of basic chemistry, it was pointed out that neutralization of acids or acid solutions can be brought about by exposing them to the proper amount of alkali. Also, that if more alkali than the chemically equivalent amount is added, the acid is expended and an alkaline condition results. In order to determine the amount of alkali to be added to a brine system showing acid tendencies, some basis for calculation must be adopted to return it to the desired state. Determination by test of the pH value of the acid reactive brine presents such a basis.

MEASUREMENT OF pH

There are numerous methods for determining the pH value of brine solutions, some more accurate and precise in their measurements than others. While brine treatment should be controlled within certain specific limitations, it is not necessary to employ such elaborate pH testing equipment as would be practical only in highly specialized chemical laboratories. It is an asset toward economical brine plant operation, however, to employ fairly accurate test methods in an effort to organize a safe chemical control system.

Litmus Paper Method. Litmus is a coloring matter that is used to determine the presence of an acid or an alkali. It can be used either as a liquid or in the form of litmus paper. Blue litmus paper turns red in the presence

TABLE 102. pH VALUES OF VARIOUS ACIDS AND BASES*

Acids	*pH Value*	*Bases*	*pH Value*
Hydrochloric acid	1.0	Sodium bicarbonate	8.4
Sulfuric acid	1.2	Borax	9.2
Phosphoric acid	1.5	Ammonia	11.1
Sulfurous acid	1.5	Sodium carbonate	11.36
Acetic acid	2.9	Trisodium phosphate	12.0
Alum	3.2	Sodium metasilicate	12.2
Carbonic acid	3.8	Lime (saturated)	12.3
Boric acid	5.2	Sodium hydroxide	13.0

* Courtesy, "ABC of pH Control," Towson, Md., LaMotte Chemical Products Company.

TABLE 103. APPROXIMATE pH OF SOME COMMON SUBSTANCES*

Apples	2.9–3.3	Limes	1.8–2.0
Apricots (dried)	3.6–4.0	Magnesia, milk of	10.5
Asparagus	5.4–5.7	Milk, cows'	6.4–6.8
Beans	5.0–6.0	Molasses	5.0–5.4
Beer	4.0–5.0	Olives	3.6–3.8
Beets	4.9–5.6	Oranges	3.0–4.0
Blackberries	3.2–3.6	Peaches	3.4–3.6
Bread, white	5.0–6.0	Pears	3.6–4.0
Cabbage	5.2–5.4	Peas	5.8–6.4
Carrots	4.9–5.2	Pickles, dill	3.2–3.5
Cherries	3.2–4.1	Pickles, sour	3.0–3.5
Cider	2.9–3.3	Pimento	4.7–5.2
Corn	6.0–6.5	Plums	2.8–3.0
Crackers	7.0–8.5	Pumpkin	4.8–5.2
Dates	6.2–6.4	Raspberries	3.2–3.7
Flour, wheat	6.0–6.5	Rhubarb	3.1–3.2
Gingerale	2.0–4.0	Salmon	6.1–6.3
Gooseberries	2.8–3.1	Sauerkraut	3.4–3.6
Grapefruit	3.0–3.3	Shrimp	6.8–7.0
Grapes	3.5–4.5	Spinach	5.1–5.7
Hominy (lye)	6.9–7.9	Squash	5.0–5.3
Human blood plasma	7.3–7.5	Strawberries	3.1–3.5
Human duodenal contents	4.8–8.2	Sweet potatoes	5.3–5.6
Human feces	4.6–8.4	Tomatoes	4.1–4.4
Human gastric contents	1.0–3.0	Tuna	5.9–6.1
Human milk	6.6–7.6	Turnips	5.2–5.5
Human saliva	6.0–7.6	Vinegar	2.4–3.4
Human spinal fluid	7.3–7.5	Water, distilled (equil. with air)	5.8
Human urine	4.8–8.4	Water, distilled (CO_2 free)	6.8–7.0
Jams, fruit	3.5–4.0	Water, mineral	6.2–9.4
Jellies, fruit	3.0–3.5	Water, sea	8.0–8.4
Lemons	2.2–2.4	Wines	2.8–3.8

* Courtesy, "ABC of pH Control," Towson, Md., LaMotte Chemical Products Company.

of an acid while red litmus turns blue in an alkali solution. Litmus paper is one of the simplest indicators used to determine whether a substance is acid or alkali. The degree of color change indicates the relative acidity or alkalinity of the substance tested. Thus, if litmus changes from blue to a light red, the substance is considered a weak acid. If the change is to a dark red, a stronger acid is indicated. Any shade of red in between the two indicates an acidity the strength of which is related to either extremity. This also applies to red litmus in changing to various shades of blue when exposed to alkali solutions. The use of litmus paper for indicating relative acid or alkaline values is not an accurate method for the pH control of brines. It is difficult to judge slight variations in color closely enough to estimate pH differences within 1.0 or even 2.0 pH numbers. Litmus paper is recommended only when more accurate methods of determining pH are not available.

Alkacid Paper Method. This method embraces the principle of litmus paper except that the alkacid papers are capable of assuming a greater range of color between a deep red and a dark blue. The color the paper assumes upon immersion in solutions to be tested is compared with the standard colors printed on the label of the vial in which the papers are furnished. These standards are shown in Table 104.

TABLE 104. RELATIONSHIP OF pH TO COLOR CHANGES OF ALKACID PAPER

pH 2	Red
pH 4	Orange
pH 6	Yellow
pH 8	Green
pH 10	Blue

Measurements can be made within one pH unit by a careful operator through color interpolation. This is one of the simplest methods for measuring pH values.

Colorimetric Method. The colorimetric method of determining pH values is based on the use of vegetable dyes or indicators. Indicators are susceptible to color changes under definite pH conditions. A number of indicators are necessary for testing solutions over a wide range of pH. Tests for pH are conducted by the simple procedure of adding a few drops of the most suitable indicator solution into a small quantity of brine. Usually a beaker or a test tube is employed for the purpose. For more accurate determinations, an especially designed glass tube or cell is used. The latter is of uniform diameter, depth, and wall thickness, and usually is supplied by the test set manufacturer.

Note that the expression "most suitable indicator solution" was stated above. There are a number of indicators, each kind having its individual pH range within which it will give accurate readings. To save time and material, the brine system operator should acquaint himself with the approximate changes in pH over a given period. Then, when tests are to be made, he will be able to estimate accurately the one indicator to use.

Color changes in the test sample of brine may be estimated by simple visual inspection if the operator has sufficient experience. For more accurate determinations, the change in color is compared with standard color solutions. Color standards are prepared by the test set manufacturer and are reproduced to give the exact interval of color change for each indicator for a given pH range. The interval of range governs the accuracy which the particular test set is capable of producing. The closer the range interval, the greater the number of color standards required. The usual interval is 0.2 pH.

For most work, the portable block comparator may be satisfactory. In its simplest form, the block comparator consists of one suitable indicator, a set of corresponding color standards, a measuring pipet, and the necessary number of test tubes. These are mounted in a case the top of which also serves as a comparator.

A drop of indicator is added to a test sample of brine and the resultant color compared with that of the standard. The pH of the brine is then determined from the label of the color standard container.

For laboratory or general industrial plant requirements, large pH testing sets are necessary. A convenient arrangement is one in which several indicators are used whose corresponding color standards are mounted on a revolving table. The test sample is placed in holders located in a stationary rim. The color standards are rotated until identification is completed. A comparator window and a lamp located in the center of the rotating table aid in making color identification.

Six tubes are used in testing turbid solutions for pH. In this method, the tubes are placed in pairs, making three columns as shown:

$$\begin{array}{ccc} 1' & 2' & 3' \\ \uparrow & \uparrow & \uparrow \\ 1 & 2 & 3 \end{array}$$

Tubes 1 and 3 contain samples of liquid to be tested with no indicator. Tubes 1′ and 3′ are standard color indicators representing estimated pH values of neighboring intervals. Tube 2 contains pure water while the test sample containing the indicator solution is placed in 2′. The color of each column, looking in the direction of the arrows, is the combined color of the two liquids in line. When the color of 2 — 2′ is the same as either 1 — 1′ or 3 — 3′ then the pH of the sample 2′ is established.

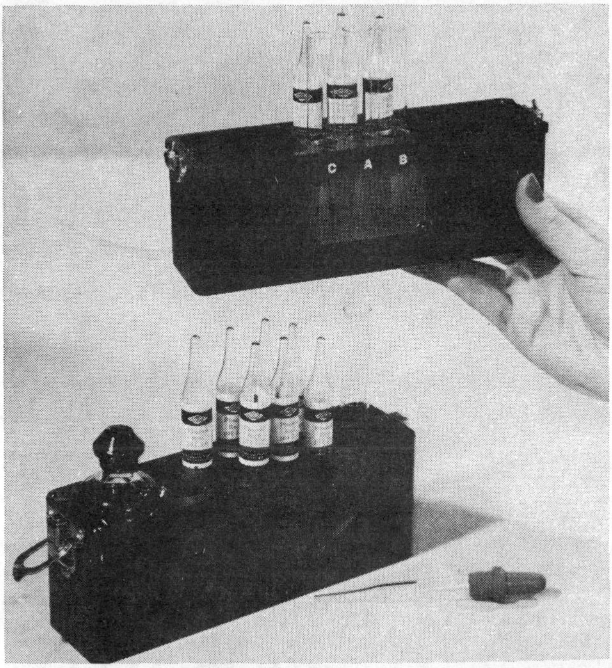

Fig. 388. (*Top*) Block comparator for pH determinations. (Courtesy, LaMotte Chemical Products Company.)

Fig. 389. (*Bottom*) Roulette comparator for pH determinations. (Courtesy, LaMotte Chemical Products Company.)

Colored glass slides are also used as indicators. This method involves the use of three tubes of test sample placed in a line, the middle one being treated with a measured amount of indicator solution. The pH is determined by moving the glass slide in front of the test samples until a color match is made. In performing this test, the sample holder and glass should be held up to some source of daylight.

A list of indicators is given in Table 105. Shown also are the color changes corresponding to the pH range of each indicator.

TABLE 105. pH INDICATORS FOR VARIOUS RANGES SHOWING RESULTING COLOR CHANGE

Indicator	*pH Range*	*Color Change*	
		From:	To:
Acid cresol red	0.2– 1.8	Red	Yellow
Acid meta cresol purple	1.2– 2.8	Red	Yellow
Methyl yellow	2.9– 4.0	Red	Yellow
Methyl orange	3.1– 4.4	Red	Orange
Bromphenol blue	3.0– 4.6	Yellow	Blue
Bromocresol green	3.8– 5.4	Yellow	Blue
Methyl red	4.4– 6.0	Red	Yellow
Chlorophenol red	4.8– 6.4	Yellow	Red
Bromcresol purple	5.2– 6.8	Yellow	Purple
Bromthymol blue	6.0– 7.6	Yellow	Blue
Phenol red	6.8– 8.4	Yellow	Red
Cresol red	7.2– 8.8	Yellow	Red
M-cresol purple	7.4– 9.0	Yellow	Purple
Thymol blue	8.0– 9.6	Yellow	Blue
Phenolphthalein	8.0– 9.8	Colorless	Red
Thymolphthalein	9.3–10.5	Yellow	Blue
Alizarin yellow	10.1–12.0	Yellow	Red
Parazo orange	11.0–12.6	Pale yellow	Deep orange
Acyl blue	12.0–13.6	Red	Blue

Glass Electrode Method. When two solutions of different pH values are separated by a thin membrane of soda-lime (or other) glass, a differential in electrical potential (voltage) is developed across the membrane. This principle is applied in the glass electrode method for determining pH. The potentiometer is composed of two electrodes connected to a voltmeter. One electrode called a calomel half-cell consists primarily of a glass tube containing a layer of pure mercury covered by a layer of mercurous chloride, called calomel, containing finely divided mercury over which is a definite concentration of potassium chloride (KCl) saturated with calomel. The voltage in the electrode containing the calomel half-cell is fixed. The second electrode acts to complete the electrical circuit when placed in the test solution with the calomel half-cell electrode. The voltage in the latter varies according to the pH of the test solution.

When the two electrodes are immersed in a solution having an unknown pH value, the electrical potential developed is indicated on the voltmeter. Because the voltage is small, due to high resistance of the circuit, it must be amplified (electronically) by an electron tube amplifier similar in principle to that used in radios. The meter scale is marked to indicate volts (actually millivolts) in equivalent pH values so that direct readings can be made. The scale is based on a zero millivolt reading at pH 7.0, or the neutral point. Because the polarity is reversed at about the neutral point, the potential is indicated as being negative (alkaline) or positive (acid) depending on the direction the needle moves—i.e., to either the right or the left, respectively.

Glass electrode pH meters vary in their method of test procedure according to different manufacturers, although their principle remains about the same as described above. Before testing a sample solution of unknown pH, the instrument is first standardized with a solution having a known pH value, usually pH 7.0. To standardize the instrument, the electrodes are placed in a beaker of the solution having a known pH. The latter is called a buffer solution. The needle is then adjusted to the scale by causing it to coincide with the pH of the buffer solution. This is done by means of an adjustment knob.

By means of a temperature correction dial, the instrument must be adjusted according to the temperature of the test sample. This is important because the millivolt readings vary according to temperature.

After standardizing and correcting the instrument for temperature, pH is determined by immersing the electrodes in the test sample. Other

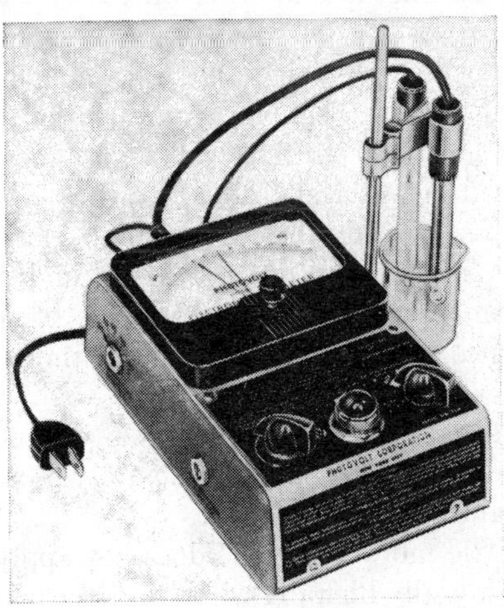

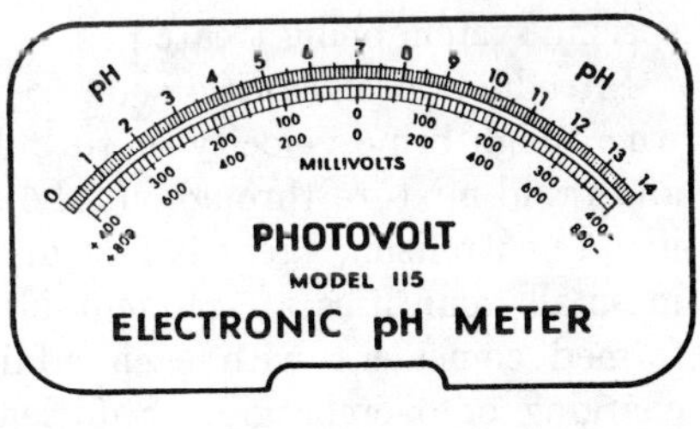

Fig. 390. Glass electrode pH meter. (Courtesy, Photovolt Corporation.)

electrodes used include the hydrogen half-cell (gold coated with palladium, platinum, or iridium black placed in a solution of hydrogen ions under a definite partial hydrogen pressure), the quinhydrone electrode, and the antimony electrode.

CHEMICAL TREATMENT OF BRINES

Both sodium chloride and calcium chloride brines are corrosive. The tendency to attack ferrous metals, such as iron and steel, becomes greater with any increase in their acidity.

One of the chief causes for increased acidity in brines is the carbon dioxide which enters the solution from the surrounding atmosphere. Carbon dioxide combines with other elements present in the brine solution to reduce alkalinity and bring about an acid condition. Chemical changes are often assisted by the catalytic action on the part of some types of metals used in constructing the system. Enclosed brine systems are less affected by carbon dioxide than are open systems having large areas of brine exposed to the atmosphere.

The purpose of chemically treating brines is to prevent corrosion of the metal parts of the brine system. Chemical treatment consists of two major steps, viz.: (1) maintain proper alkalinity of the brine; (2) establish a protective coating over the surface of submerged metal parts.

Control of Alkalinity

To increase and maintain the proper state of alkalinity, caustic soda (sodium hydroxide, NaOH) is added to brines. Treatment with caustic soda should be begun when the pH value has reached a low point within the range of allowable values. Only enough caustic soda should be added to raise the pH to the maximum value within the allowable range for the specific system being treated.

Introduction of caustic soda to the system should be made at a point where high brine velocity or turbulence occurs. This will assure thorough and rapid mixture throughout the system. Better results will be obtained if the neutralizing agent is first dissolved in water and added to the brine in small quantities at frequent intervals. Testing for pH should be performed coincident with each addition of caustic soda to avoid undercharging or overcharging. Sufficient time should be permitted the brine to circulate and allow complete distribution of the soda before test samples are taken.

Immediate reaction of the caustic soda with calcium chloride occurs, forming calcium hydroxide or hydrated lime. Being slightly soluble, more so in cold brine, the lime acts to neutralize any acid present.

The rule that "if a little is good a little more is better" does not apply when treating brines. Brines having an initially high pH will absorb carbon dioxide more readily than alkaline brines of lower pH. In addition to the rapid reduction in alkalinity of brines having excessive pH values, a precipitation problem also results. The caustic soda reacts with free carbon dioxide present to form insoluble calcium carbonate that drops out of the brine resulting in the formation of scale on the metal surfaces. Carbonate deposits can be offset to a large extent by allowing the pH to drop below 7.0 for short periods of time.

Systems that include some galvanized surfaces (zinc) such as ice-making cans should be maintained at a lower pH level than those constructed wholly of ferrous metals. Optimum pH levels for each of the two types of systems will be found within the following range:

Systems that include galvanized surfaces.........	pH 7.2—8.4
Systems constructed wholly of iron..............	pH 8.2—8.5

The minimum value for most brine systems is above the neutral point of pH 7.0.

Sometimes it is necessary to reduce the alkalinity of brine, which is usually the case with freshly made solutions. To lower the pH value, a mild acid is required. A slow but effective process is to dissolve carbon dioxide gas in the brine to form carbonic acid.

Sodium bichromate added to the brine in the same manner as that described for caustic soda is also used to decrease pH. It is first dissolved in water at a ratio of approximately 1 lb sodium bichromate per quart of water and then introduced to the system in gradual amounts. Test for pH should follow each addition of the chromate solution.

Metal Surface Protection

Providing a protective film over the surface of submerged metal parts of the system combats the corrosive effect of brine. This is accomplished by the use of certain corrosion retarders or inhibitors that are added to the brine solution. These agents "plate out" and form an invisible film over the metal surfaces and thus prevent contact between the metal and the corrosive element, oxygen, present in all brines. While such inhibitors possess an affinity for metals, they will not attack them.

(*a*) The chromates are the most effective corrosion inhibitors employed in brine systems, sodium dichromate being the compound most widely employed. While the normal chromate (sodium chromate) is best, the more available sodium dichromate ($Na_2Cr_2O_7 \cdot 2H_2O$) is more often adopted as the corrosion inhibitor in both sodium chloride and calcium chloride systems.

The quantities of sodium dichromate added as an initial charge to brine solutions should be equivalent to the following proportions:

Calcium chloride systems.................. or approximately 1.33 lb per 100 gal	100 lb per 1000 cu ft of brine
Sodium chloride systems.................... or approximately 2.66 lb per 100 gal	200 lb per 1000 cu ft of brine

Caustic soda in an amount equivalent to 2.7 lb for each 10 lb of sodium chromate should be added to counteract the reduction in pH that may result from the addition of the chromate. The addition of caustic soda also converts sodium dichromate to normal chromate.

(*b*) Disodium phosphate ($Na_2HPO_4 \cdot 12H_2O$) is also used to inhibit sodium chloride brines, particularly in open type systems. The quantity added as an initial charge should be equivalent to 100 lb disodium phosphate for each 1000 cu ft (7500 gal) of brine.

(*c*) Zinc dust may also be used to retard corrosion in open calcium chloride brines. The recommended quantity of this material is equivalent to 60 lb zinc dust for each 1000 cu ft of brine. Zinc dust should be added in gradual amounts.

Corrosion inhibitors should be added to brine systems in the same manner as that described for caustic soda; i.e., at a point of high brine velocity. A test for pH should follow each addition of the corrosion-retarding agent.

Chromates should not be used in brines that are to contact food. Because of the possibility of developing an irritating skin rash called "chrome itch," workmen handling brines treated with chromate should wear rubber gloves. The chromates give a clear, sparkling appearance to the brine.

To counteract ammonia in brine due to leakage, oxygen is used. Oxygen is introduced directly from the shipping cylinder into the brine at a point of maximum turbulence, usually in the agitator stream. Before treating the brine with oxygen, it should be certain that the leaks are eliminated.

EUTECTIC SOLUTIONS

When one or more substances acting as solutes are mixed with one or more solvents in such proportions as to obtain the lowest possible freezing point for the combination, the resulting solution is called a cryohydrate or eutectic mixture. The temperature at which a eutectic mixture freezes is called the eutectic temperature.

Brines become eutectic mixtures when the ratio of salt to water is such as to obtain the lowest possible freezing point. Any more or less salt present in the same quantity of solution results in a higher freezing point for the mixture.

If the salt content is in excess of a eutectic concentration, a continuous reduction in temperature will gradually force the surplus salt out of solution. As salt is forced out, it crystallizes. As the temperature is further reduced, a greater amount of excess salt becomes crystallized until, at the eutectic temperature, the brine becomes a combination of eutectic solution and free salt crystals. A further reduction in heat content—i.e., at the eutectic temperature of the liquid—results in freezing of the eutectic solution and the solid becomes a mechanical mixture of salt and ice.

When the salt content is less than that required to produce a eutectic solution, then the excess water will freeze out as the temperature reaches some point below 32 F. The temperature at which ice starts to form depends on the salt concentration. The formation of ice continues with further temperature reduction until the remaining solution reaches eutectic concentration. Through the latter stage the mixture becomes an ever-increasing slush as more ice is formed and less brine remains. At the eutectic temperature the mass becomes completely solid, forming a mechanical mixture of salt and ice. In crystallizing, the salt takes on six water molecules to form a hexahydrate such as calcium chloride hexahydrate ($CaCl_2 \cdot 6H_2O$).

The formation of salt crystals or ice during temperature reduction of solutions other than those of eutectic proportions is a function of the solubility-temperature relationship of the salt to water. Solubility decreases as temperatures drop. This relationship is shown in the solubility diagram (Fig. 391) for calcium chloride.

The state of the brine in the various areas shown on the diagram is as follows:

Area A. This area represents the state of the brine during temperature

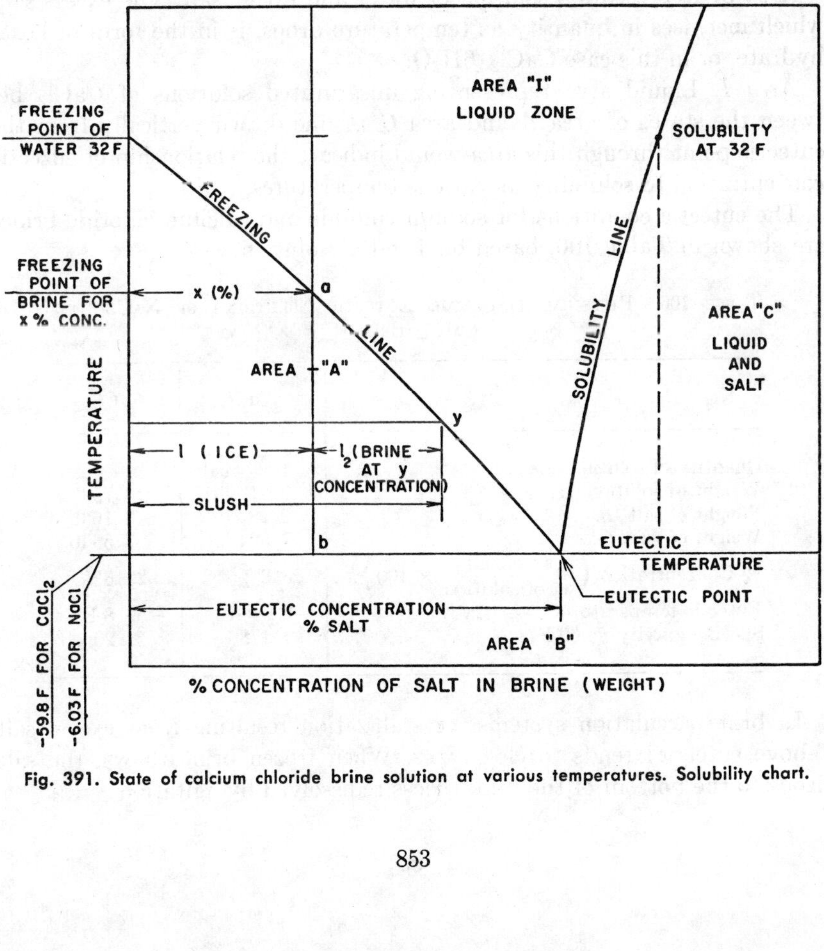

Fig. 391. State of calcium chloride brine solution at various temperatures. Solubility chart.

reduction. For eutectic solutions, the area represents liquid throughout. For solutions containing excess water, area A will represent a mixture of ice and brine, the proportions of which depend on the concentration (freezing point). In the case of, say, 20 percent $CaCl_2$ concentration, the proportion of ice to brine at any temperature below -0.4 F, the corresponding freezing point, may be determined by drawing a line from that point on the curve vertically downward to the eutectic temperature line. This is shown by the line ab on the curve. The proportion of ice to brine at, say, -40, will be according to the ratio of distance l to l_2, where l represents ice and l_2 represents brine. The mixture of ice and brine forms slush. Drawing a similar line for any concentration, the proportions of brine and ice may be determined in a like manner. It will be noted that the distance l_2 diminishes with higher salt concentration, disappearing entirely for brines of eutectic proportions.

Area B. The solid or frozen state of the brine below the eutectic temperature or -59.8 F for $CaCl_2$. The mixture in this area is ice and calcium chloride hexahydrate ($CaCl_2 \cdot 6H_2O$), a mechanical mixture.

Area C. Represents a solution of water and excess salt. The excess salt, which increases in quantity as temperature drops, is in the form of hexahydrate, or in this case $CaCl_2 \cdot 6H_2O$.

Area I. Liquid area representing unsaturated solutions of $CaCl_2$ between the states of area A and area C. A line drawn vertically from the eutectic point through this area would indicate the relationship of eutectic concentration to solubility at various temperatures.

The eutectic conditions for sodium chloride and calcium chloride brines are shown in Table 106, based on 1 gal of solution.

TABLE 106. PHYSICAL DATA FOR EUTECTIC SOLUTIONS OF NaCl AND $CaCl_2$ BRINES

	NaCl	*$CaCl_2$*
Quantity of solution	1 gal	1 gal
Weight of solution	9.81 lb	10.75 lb
Weight of salt	2.287 lb	3.16 lb
Weight of water	7.524 lb	7.59 lb
% Concentration $\left(\frac{\text{wt of salt}}{\text{wt of solution}} \times 100\right)$	23.3%	29.6%
Eutectic temperature	−6 F	−59.8 F
Specific gravity @ 60 F	1.178	1.29

In brine circulation systems, crystallization resulting from excess salt (above eutectic) tends to clog pipes. When frozen brine thaws, the salt drops to the bottom of the tank unless redissolved by agitation.

Eutectic solutions often replace ice, solid CO_2 (dry ice), and even small condensing units. Placed in suitable heat exchangers and frozen, eutectic solutions will maintain reduced temperatures in various type compartments for extended periods. Some types of eutectic solutions act as hold-over agents for refrigeration storage where advantage is taken of their latent heat of fusion. To prevent precipitation of the solute upon thawing, a jelling solution is employed in eutectic plates.

One of the important uses for eutectic solutions is in plate evaporators installed in insulated motor trucks. Called eutectic plates, the solution fills the space between the evaporator coils where it is frozen. Employed in this way, the eutectic acts as a hold-over medium providing refrigeration for perishable commodities during transport. In some instances, the eutectic is cooled by connecting a condensing unit permanently mounted on a stationary charging or loading platform. After charging, the condensing unit is disconnected and the eutectic plates maintain reduced temperature in the truck until the produce is unloaded. Where the condensing unit is mounted on the truck, the eutectic plates act to extend cycling periods, thus reducing the frequency of starting and stopping the compressor.

Eutectic solutions find wide use in small portable insulated containers in which food for camping and picnics is carried. Precooled in domestic refrigerators or freezers, small sealed cans of eutectic mixture placed in such containers will provide refrigeration for several hours. Larger containers of eutectic brines may be frozen in cold rooms and placed in storage compartments to maintain reduced temperatures.

The refrigeration capacity of frozen eutectic solutions is calculated on the basis of latent heat of fusion for the particular brine. The latent heat of fusion of sodium chloride and of calcium chloride brines is as follows:

$NaCl$..........	101 Btu per lb
$CaCl_2$.........	56 Btu per lb

OTHER BRINES OR ANTIFREEZE SOLUTIONS

Magnesium chloride brines have been used. While its eutectic temperature is higher than that of sodium chloride brine, it has no other particular advantage over the latter. Therefore, $MgCl_3$ brine is not used widely for refrigeration purposes.

Lithium chloride and water form an absorbent solution used in some continuous defrosting systems. A salt balance must be maintained for several reasons. First, condensate lowers the salt concentration so additional lithium chloride must be added to prevent freezing. Second, brines being hygroscopic, they absorb moisture from the air which becomes more aggravated with higher salt concentration. Therefore, a regenerator to maintain proper salt level must be incorporated in any continuous brine defrosting system.

Methanol or methyl alcohol, sometimes called "wood alcohol" (obtained from volatile matter of charcoal), is used to reduce the freezing point of water. Generally described as an antifreeze, it is soluble in water in all proportions. The aqueous solution decreases in freezing point as the amount of methanol increases. A 70 percent solution of methyl alcohol and water will result in a freezing point of -164 F.

Denatured alcohol is also used as an antifreeze agent for water with which it is soluble in all proportions. Denatured alcohol is a mixture of ethyl alcohol (grain alcohol) and methanol (about 10 percent) plus some unpalatable substance such as benzol.

Ethyl alcohol, obtained by fermentation of sugars, is also an antifreeze, but because of a high tax its use is generally prohibitive.

Glycerin, a sirupy liquid also called glycerol, is an antifreeze agent obtained from carbohydrates such as glucose, starch, etc. A freezing point of -38 F can be obtained with an aqueous solution containing 70 percent glycerin by weight. Beyond 70 percent the freezing point increases.

The glycols, derived from the petroleum by-products ethylene and propylene, are antifreeze agents gaining industrial importance. Both ethylene glycol and propylene glycol are miscible with water in all proportions, and act to lower the freezing point of solutions in proportion to the amount of glycol added, up to a certain point.

With ethylene glycol, the freezing point is depressed from 22.4 F corresponding to a 15 percent (by volume) glycol solution to -56 F for a 65 percent solution. Above 65 percent, the freezing point increases until it reaches 11.3 F for pure ethylene glycol.

With propylene glycol, the freezing point is depressed from 22.5 F corresponding to a 15 percent glycol solution to -57 F for a 60 percent solution. The use of glycols has several thermal and physical advantages. First, they are noncorrosive even when mixed with water. They are nonelectrolytic, therefore safe to use in systems made up of two or more metals. Glycols will not evaporate under normal operating temperatures. They are stable chemicals so will not deteriorate in service even when continuously circulated.

Propylene glycol offers strong possibilities as a lubricant for compressors handling highly solvent fluids such as propane, butane, ethane, etc. The latter does not increase the density of water solutions to any great extent, reaching a specific gravity of only 1.0651 for a mixture of 60 percent by weight of propylene glycol and 90 percent water at 5 F.

The specific heat of aqueous propylene glycol solutions decreases as the weight of the latter increases, for the same temperature. At 32 F the specific heat ranges from 1.0 (Btu per lb) for pure water to 0.8 for aqueous solutions containing 60 percent (by weight) propylene glycol, being higher or lower for temperatures above and below 32 F, respectively.

TABLE 107. CRYSTALLIZATION TEMPERATURE OF ANTIFREEZE-WATER SOLUTIONS

Ethyl Alcohol

% Alcohol by Weight	*Temperature °F of Crystallization*	*Sp. Gravity (68°F)*
2.5	30.2	.993
6.8	26.6	.987
16.4	18.5	.973
20.3	12.9	.968
26.7	3.2	.960
33.8	−10.5	.950
46.3	−29.0	.922
56.1	−41.8	.900
71.9	−60.3	.863

Glycerine

% Glycerol by Weight	*Temperature °F of Crystallization*	*Sp. Gravity (68°F)*
10	29.1	1.02
20	23.4	1.04
30	14.9	1.07
40	4.3	1.10
50	−9.4	1.13
60	−30.5	1.6
70	−38.0	1.8
80	−5.5	1.2
90	+29.1	1.24
100	+62.6	1.26

Ethylene Glycol

% Glycol by Volume	*Freezing Point °F*	*Sp. Gravity (60°F)*
12.5	25	1.02
17.0	20	1.03
25.0	10	1.04
32.5	0	1.05
38.5	−10	1.056
44.0	−20	1.06
49.0	−30	1.07
52.5	−40	1.073

ENDOTHERMS

There is a certain heat phenomenon related to chemical reactions when elements or compounds unite to form their specific compound. In some instances heat is liberated, while in others heat is absorbed by the mass. The thermal changes occurring during the reaction are called the heat of *formation*, the heat of *solution*, the heat of *dilution*, or the heat of ***hydration***, depending on the process involved. In some cases, thermal change involves the liberation of heat from the mass and is called an *exothermic* reaction. In other cases, heat is absorbed by the mass and is called an *endothermic* reaction.

When hydrogen and chlorine combine to form hydrogen chloride, an exothermic reaction occurs in which 22,000 calories per mol* are liberated. Combining hydrogen with iodine to form hydrogen iodide, an endothermic reaction occurs in which 6100 calories per mol are absorbed. During exothermic reaction, the heat liberated causes a temperature increase in surrounding substances, while endothermic reaction decreases surrounding temperatures.

Certain compounds—particularly salts—when mixed with water, ice, snow, or even other compounds form endothermic mixtures. Because they absorb heat when combined, endothermic mixtures are often employed to reduce the temperature of other substances. One of the common examples in the practical use of endothermic mixtures is familiar to makers of homemade ice cream. The container of ingredients placed in a mixture of crushed ice and salt is agitated until the ice cream hardens. The reaction here is one in which the latent heat of the ice cream is absorbed by the endothermic reaction of the salt and the ice. Similar results are obtained by using calcium chloride and snow, potash and snow, etc., each of which will produce a higher or lower final temperature in surrounding substances.

Other endothermic mixtures involve more than two compounds, such as ice, sodium chloride and ammonium nitrate, or sodium sulfate, ammonium chloride, potassium nitrate and dilute nitric acid. The resulting temperature drop will depend on the proportion of various compounds in each case.

WATER

Water and its physical and chemical condition are important to refrigeration and air-conditioning systems. It is used for condenser cooling, humidifying, dehumidifying, ice making, brine making, compressor cylinder jacket cooling, and intercooling, and in numerous industrial processes involving refrigeration. It is important that refrigeration personnel know some of the problems associated with the handling of water. The problems

* One mol of any substance is equal to its molecular weight multiplied by 1 gram. *Example:* 1 mol of 1 molecule of hydrogen having 2 atoms, or a molecular weight of 2, is equal to 2 × 1 gram or 2 grams.

of algae and mineral salt concentration have been discussed previously in connection with cooling towers and spray ponds.

To reduce or eliminate many water problems, various physical and chemical treatments are necessary. The major considerations may be summarized:

1. Potability of water
2. Freedom from impurities that interfere with commerical processes
3. Elimination of water-conduit fouling
4. Prevention of heat-transfer surface fouling
5. Retarding corrosion

Some of the specific benefits that are expected through the use of water treated in such a manner as to meet the need are given below:

Maintain health of the community
Render the water palatable for drinking and the preparation of food, drugs, etc.
Increase laundering quality and efficiency
Prevent discoloration of textiles during manufacturing
Assure positive dyeing of textiles
Extend boiler surface life and maintain their heat-transfer efficiency
Maintain maximum heat-transfer efficiency and life of water type heat exchangers
Provide continuous maximum flow through conduits

To gain expected benefits, both the organic and the inorganic matter found in all natural waters must be controlled or eliminated. Such control is accomplished by physical and chemical treatment of water. The degree of treatment depends on the economics associated with a given system. For example, it is unnecessary to thoroughly treat river water that is passed once through a condenser and immediately returned to its source. On the other hand, the degree of chemical treatment of water used for extended recirculation would be relatively high and commensurate with all factors involved. Those factors are numerous, and vary, either singly or in combination, according to the type of circuit and the ultimate use of the water. The following is one method of classifying water circuits showing a typical purpose of each type:

1. Enclosed types:
   (*a*) Purely circulatory—no appreciable evaporation. Heat-transfer systems
   (*b*) Circulatory—high rate of evaporation. Condensing steam boiler systems
   (*c*) Noncirculatory—high rate of evaporation. Noncondensing steam boiler systems. Some industrial processes
   (*d*) Noncirculatory—no appreciable evaporation. City water systems and industrial and food processes

2. Open types:
   (*a*) Circulatory—normal surface evaporation
   (*b*) Circulatory—rapid evaporation. Cooling towers, spray ponds, and atmospheric condensers
   (*c*) Noncirculatory—rapid evaporation. Evaporative condensers
   (*d*) Noncirculatory—no evaporation. Once through condensers, usually river or well water, although sometimes city water is used
   (*e*) Large bodies of water. Reservoirs
   (*f*) Water used to make ice

Water Characteristics

Water is a compound of hydrogen and oxygen the proportions of which are fixed at 2 to 1 by volume or 2 to 16 by weight. The gravitational composition of water is expressed as 11.18 percent hydrogen and 88.82 percent oxygen. Its chemical formula is H_2O. Pure water is odorless, colorless, and tasteless.

Water will dissolve many substances (solutes), including mineral salts, some liquids, and a number of gases. The resulting substance is called a *solution*. The degree of solubility is a function of temperature. Solubility of many solids becomes greater as temperature increases. The opposite is generally true in the case of soluble gases. Some substances are soluble in water in all proportions, such as alcohol. Others—oil, for example—are practically insoluble in water. Therefore, oil will settle out while alcohol will not; however, the latter can be removed by distillation. Some solids, of which sand is one, do not dissolve but will settle out of water, although silica in some forms is soluble to a limited degree.

When water is unable to take into solution any further quantity of a given soluble substance, it is said to be *saturated* with that substance. The amount it will hold of any normally solid soluble substance increases as the temperature rises. With gases, pressure is one of the important factors. A release of pressure on liquids containing a gas will cause a bubbling action known as *effervescence*. Carbonated water is one example of the latter. Refrigerant in oil is another.

Substances may be classified according to their nature of water solubility; i.e., *highly soluble or soluble in all proportions*, *slightly soluble*, and *insoluble*. All three types are of interest to the operator of water circuits.

A *concentrated solution* is one containing a large proportion of soluble substance; *dilute solutions* contain only a small proportion.

It was stated above that water is a compound containing a fixed ratio of hydrogen (H) to oxygen (O); i.e., 11.18 percent H to 88.82 percent O. This is in accordance with the law of definite proportions, sometimes called the law of constant composition. The law states that *a given chemical compound is always composed of the same elements in the same gravimetric proportion, i.e., by weight.*

Water boils to form steam at 212 F under an atmospheric pressure of 14.7 psi. It will freeze to form ice at 32 F. Between the temperatures of boiling and freezing water remains a liquid subject to slow evaporation, the rate of which depends on its vapor pressure relative to that of the surrounding atmosphere. Above atmospheric pressure, the boiling point of water increases as with other liquids. The physical properties of saturated water and steam may be found in the steam tables. (See Appendix, Table X.) Water forms the basis of comparing the physical and the thermal properties of other substances. In this connection, its specific (ratio) heat value is standardized at 1 Btu per lb per °F temperature difference and a specific (ratio) gravity of 1.00. The act of extracting water from a substance is called *dehydration*, and a substance from which all water has been extracted is said to be *anhydrous*.

Water supports life in many forms. Most of the known forms of bacteria cannot exist in the absence of moisture. Algae thrive in water, more particularly sun-lighted areas. Water interacts with some metals at ordinary temperatures to form hydrogen gas. Sodium, calcium, and potassium are included in the latter group. It also units with metallic oxides, called anhydrides, to form hydroxides or bases. Bases that are soluble are called alkalis. Alkalis that act readily on the skin are said to be caustic.

Some substances crystallize. Water that combines with such substances in forming the crystal is called *water of crystallization*. Sodium carbonate, sodium sulfate, and copper sulfate are crystals containing water of crystallization, and are called *hydrates*. Crystals such as borax and alum that lose their water crystallization when exposed to normal atmosphere are described as being *efflorescent*.

Substances having an affinity for moisture—calcium chloride, zinc chloride, and sodium chloride (common salt), to mention several common ones—are *hygroscopic*. The action is known as *deliquescence* and the substances are said to be deliquescent.

Hydrolysis is the decomposition of chemical compounds by water. It involves the action of ionization, a subject discussed fully in this chapter.

Water will combine with many oxides, called *anhydrides*, to form *hydroxides* and *acids*. Calcium hydroxide (hydrated lime) and sulfurous acid are examples of such compounds formed when water combines with calcium oxide (lime) and sulfur dioxide, respectively.

Many solid substances that do not dissolve to form a solution will become suspended in water. Those that do not float and rise will finally settle to the bottom of the water column. The time required to settle out depends on the nature and the size of the particle and the purity of the water. Much of the suspended matter may be either settled out or removed by filtering. When the particle of suspended matter is so minute that it cannot be removed readily by settling or filtering, it is called a *colloid*. Water containing any appreciable number of colloids is called a *colloidal suspension*.

Liquid substances such as oil will not form a solution when mixed with water. When shaken together, an intimate mixture called temporary emulsion is formed. Separation will occur after turbulence ceases, with the lighter oil forming an upper layer on the water surface. By adding a third ingredient or emulsifier, such as acid sludge or fatty acid, a more permanent emulsion occurs. This is due to the change in interfacial relationship between the two substances brought about by the wetting action of the emulsifier. Depending on the type of emulsifier, either a *water-in-oil* or an *oil-in-water* emulsion may be formed. The latter should not be confused with a true emulsion formed when a solid such as flour acts as the third ingredient dispersed in the water phase in a mixture of water and grease. Gravy is one of the more common substances in the latter group.

Pure water is chemically neutral, having a pH of 7.0, meaning that it is neither acid nor alkaline. Because of its solubility as related to some gases, liquids, and solids, its chemical property tends to change when exposed to natural environment or industrial processes. Even when water is comparatively free of all other influences, carbon dioxide from the atmosphere will enter it to reduce pH value due to the subsequent formation of carbonic acid. With all the changes that can be brought about, either by the influence of foreign matter on water or by water on other substances, it is understandable why an otherwise stable compound often proves to be a source of commercial difficulty.

Water Treatment

Water treatment may be simple or complex, depending on its ultimate use and natural or available chemical and physical condition. The impurities found in natural waters vary according to source and location. What may be a good water condition in some localities may be judged poor in others. While treatment of water is generally considered to involve the fresh supply, either from city lines or natural sources, flowing into a plant as make-up it is not confined to that category. Water that is continuously recirculated must also be attended against the impurities that develop during its useful life.

The factors to be dealt with in treating water include:

(*a*) Ultimate use of the water
(*b*) Source of supply (expected condition)
(*c*) Impurities that affect its use in a specific process
(*d*) Kind(s) of treatment required
(*e*) Degree of treatment necessary
(*f*) Cost of treating vs. probable benefits to the use process

Ultimate Use of Water

The important consideration is the ultimate use of the water. It raises the question as to what impurities will be detrimental to the processes involved. Health, operating efficiencies, corrosion, product quality, maintenance of flow capacity, etc., are some of the factors included in this category.

Source of Supply

The source of supply governs the kind and the magnitude of impurities to be found in water. Sources include shallow and deep wells, springs, lakes, ponds, rivers, reservoirs, mines, and quarries. They include surface and ground waters as well as rain water. In many cases, water from commercial mains having some definite pretreatment is the source of supply.

Commercial or city water is expected to be healthfully pure but may contain mineral salts from original source plus chemicals added for bacteria control. River water usually carries with it suspended substances such as sand, clay, and animal and vegetable matter, in addition to various mineral salts. While stream pollution is unlawful in many areas, trade wastes still pose a problem as a result of leaching and illegal practices. Deep wells, while comparatively free of suspended matter, contain a large percentage of mineral salts and carbon dioxide. The latter causes a reduction in pH. Mine waters are corrosive due to the presence of sulfuric acid. Algae and bacterial growth occur in circultation systems and continuous evaporation processes act to concentrate mineral impurities which are deposited and form scale on equipment surfaces.

Impurities that Affect Use Processes

Any undesirable substance contained in water is an impurity. Therefore, any dissolved or suspended matter that interferes with the specific use (or uses) of water is an impurity. To improve a specific process, any impurities present in water that interfere with that process must be removed or neutralized. Natural waters collect impurities in a number of ways, as follows:

1. Dissolving mineral salts from the earth
2. Dissolving gases occurring underground or in the atmosphere
3. Dissolving gases emanating from decayed vegetable and animal matter
4. Chemical reaction between
   (*a*) the various impurities,
   (*b*) the water and various impurities,
   (*c*) the water and products of the several reactions
5. Contamination resulting from vegetable, animal, and mineral matter dissolved or suspended in water
6. Growth of plant life and fungi
7. Growth of bacteria originating from trade waste and animal, vegetable, and mineral matter
8. Pollution

Table 108 lists the various impurities found in water and methods employed for their removal or neutralization. Impurities include those which are soluble, insoluble, suspended in either a colloidal or a noncolloidal state, and emulsified.

Kinds of Treatment Required

Ultimate use of water governs the type of treatment necessary. Treatment may be limited to the addition of a suitable chemical for algae control or subject to multiple processes for complete purification, as with drinking water. The various kinds of water treatment may be classified according to the purpose for which it will be used, as given below:

Chemical treatment (including electrolysis):
- Control boiler or water circuit scale
- Retard corrosion
- Reduce temporary hardness due to mineral salts
- Reduce permanent hardness due to mineral salts
- Eliminate undesirable minerals, such as silica, iron, lead, zinc, etc.

Physical treatment:
- Eliminate odors
- Improve color and taste
- Reduce turbidity
- Remove suspended contaminants

Biological treatment:
- Control algae, fungi, etc.

Bacteriological treatment:
- Destroy all harmful bacteria

The various processes employed to prepare water for its ultimate use are described below. In a number of instances, water may be subject to two or more of the processes shown here.

Screening. Grids or screens of various mesh, either stationary, revolving, or reciprocating, are installed to prevent entrance of leaves, branches, logs, and other floating material from rivers, ponds, and reservoirs.

Sedimentation. This is a settling process. Suspended matter heavier than water settles to the bottom of basins in which water is either held for a period or allowed to flow through slowly. Exits from such basins are built so that clear water flows off the top. Sedimentation is often supported by a coagulum.

Coagulation (Chemical and Physical Process). A coagulum is employed in conjunction with water sedimentation. A coagulant, usually aluminum sulfate (alum), introduced to the water reacts with natural bicarbonates or added carbonates (soda-ash). The resulting spongelike material of decomposition (aluminum hydroxide) precipitates and settles to the bottom,

Table 108. Common Impurities Found in Water*

CONSTITUENT	CHEMICAL FORMULA	DIFFICULTIES CAUSED	MEANS OF TREATMENT
Turbidity	None—expressed in analysis as SiO_2.	Imparts unsightly appearance to water. Deposits in water lines, process equipment, boilers, etc. Interferes with most process uses.	Coagulation, settling and filtration.
Color	None—expressed in analysis as "units" of color on arbitrary scale.	May cause foaming in boilers. Hinders precipitation methods such as iron removal, hot phosphate softening. Can stain product in process use.	Coagulation and filtration. Chlorination. Adsorption by activated carbon.
Hardness	Calcium and magnesium salts expressed as $CaCO_3$.	Chief source of scale in heat exchange equipment, boilers, pipe lines, etc. Forms curds with soap, interferes with dyeing, etc.	Softening. Distillation. Internal boiler water treatment. Surface active agents.
Alkalinity	Bicarbonate (HCO_3), carbonate (CO_3), and hydrate (OH), expressed as $CaCO_3$.	Foaming and carryover of solids with steam. Embrittlement of boiler steel. Bicarbonate and carbonate produce CO_2 in steam, a source of corrosion.	Lime and lime-soda softening. Acid treatment. Hydrogen zeolite softening. Demineralization. Dealkalization by anion exchange. Distillation.
Free Mineral Acid	H_2SO_4, HCl, etc. expressed as $CaCO_3$.	Corrosion	Neutralization with alkalies.
Carbon Dioxide	CO_2	Corrosion in water lines and particularly steam and condensate lines.	Aeration. Deaeration. Neutralization with alkalies. Filming and neutralizing amines.
pH	Hydrogen ion concentration defined as: $pH = \log \frac{1}{(H^+)}$	pH varies according to acidic or alkaline solids in water. Most natural waters have a pH of 6-8.	pH can be increased by alkalies and decreased by acids.
Sulfate	$(SO_4)^{--}$	Adds to solids content of water, but, in itself, is not usually significant. Combines with calcium to form calcium sulfate scale.	Demineralization. Distillation.
Chloride	Cl^-	Adds to solids content and increases corrosive character of water.	Demineralization. Distillation.
Nitrate	$(NO_3)^-$	Adds to solids content, but is not usually significant industrially. High concentrations cause methemoglobinemia in infants. Useful for control of boiler metal embrittlement.	Demineralization. Distillation.

Fluoride	F^-	Cause of mottled enamel in teeth. Also used for control of dental decay. Not usually significant industrially.	Adsorption with magnesium hydroxide, calcium phosphate, or bone black. Alum coagulation.
Silica	SiO_2	Scale in boilers and cooling water systems. Insoluble turbine blade deposits due to silica vaporization.	Hot process removal with magnesium salts. Adsorption by highly basic anion exchange resins, in conjunction with demineralization. Distillation.
Iron	Fe^{++} (ferrous) Fe^{+++} (ferric)	Discolors water on precipitation. Source of deposits in water lines, boilers, etc. Interferes with dyeing, tanning, paper mfr., etc.	Aeration. Coagulation and filtration. Lime softening. Cation exchange. Contact filtration. Surface active agents for iron retention.
Manganese	Mn^{++}	Same as iron.	Same as iron.
Oil	Expressed as oil or ether extractible matter.	Scale, sludge and foaming in boilers. Impedes heat exchange. Undesirable in most processes.	Baffle separators. Strainers. Coagulation and filtration. Diatomaceous earth filtration.
Oxygen	O_2	Corrosion of water lines, heat exchange equipment, boilers, return lines, etc.	Deaeration, Sodium sulfite. Corrosion inhibitors.
Hydrogen Sulfide	H_2S	Cause of "rotten egg" odor. Corrosion.	Aeration. Chlorination. Highly basic anion exchange.
Ammonia	NH_3	Corrosion of copper and zinc alloys by formation of complex soluble ion.	Cation exchange with hydrogen zeolite. Chlorination. Deaeration.
Conductivity	Expressed as micromhos, specific conductance.	Conductivity is the result of ionizable solids in solution. High conductivity can increase the corrosive characteristics of a water.	Any process which decreases dissolved solids content will decrease conductivity. Examples are demineralization, lime softening.
Dissolved Solids	None	"Dissolved solids" is measure of total amount of dissolved matter, determined by evaporation. High concentrations of dissolved solids are objectionable because of process interference and as a cause of foaming in boilers.	Various softening process, such as lime softening and cation exchange by hydrogen zeolite, will reduce dissolved solids. Demineralization, Distillation.
Suspended Solids	None	"Suspended Solids" is the measure of undissolved matter, determined gravimetrically. Suspended solids plug lines, cause deposits in heat exchange equipment, boilers, etc.	Subsidence. Filtration, usually preceded by coagulation and settling.
Total Solids	None	"Total Solids" is the sum of dissolved and suspended solids, determined gravimetrically.	See "Dissolved Solids" and "Suspended Solids."

* Courtesy, "Betz' Handbook of Industrial Water Conditioning," ed. 4, pp. 2–3. Copyright, 1953, by Betz Laboratories, Inc., Philadelphia, Pa.

carrying with it suspended matter, including much of the bacteria. Coagulation is usually followed by filtration.

Filtration (Mechanical Process). (*a*) NATURAL FILTRATION is accomplished by water flowing through ground soil. Very effective in pathogenically clean areas, as evidenced by the many potable springs and wells.

(*b*) COARSE FILTRATION utilizes coarse sand and fine crushed stone or gravel through which water gravitates to remove heavier suspended matter. A pretreatment to finer filtering is referred to as a preliminary or *roughing filtration.*

(*c*) SLOW FILTRATION utilizes a layer of fine sand that removes suspended matter, including colloids and bacteria, as the water gravitates through it. The sand is usually laid on a bed of gravel that covers the necessary drainpipes. Alternate beds are provided so that filtration may be continued while cleaning the used sand. Also called the continuous or English filter.

(*d*) RAPID FILTRATION, better known as mechanical filtration, is employed in conjunction with coagulation and utilizes a coarser sand than that used in slow filtration to increase flow rate. Hard coal graded to proper size is also used. Rectangular or cylindrical tanks containing a layer of sand over graduated size gravel are used. The water may be pretreated with coagulum (alum) which also can be added in the filter. Compressed air continuously blown through the sand prevents the fine and the coarse sand particles within the mass from becoming localized and forming a noneffective path. The filling medium is washed periodically by forcing clean water up through it. As the flow of water is reversed, it removes the filtered residue and carries it to the sewer. The operation is known as *backwashing.* The purifying results of rapid sand filtration equal that of the slower continuous process. However, the rate of rapid filtering is approximately 50 times greater than that of slow filtering for equal areas.

(*e*) PRESSURE FILTRATION is similar to the rapid filtration process except that the former is carried on in a fully enclosed shell under some pressure head other than gravity alone. When placed in regular water lines, a booster pump is usually unnecessary.

(*f*) CHARCOAL FILTRATION improves color and eliminates odors from water. Charcoal can be used alone or can form a layer in gravity sand filters.

(*g*) MISCELLANEOUS FILTERS include those that employ cellulose, diatomaceous earth, asbestos, Fiberglas, paper, or cloth disks arranged in layers through which water is forced. Some of these types are used only where a limited volume of water is required, such as laboratories.

Aeration. Aeration is accomplished by spraying water into the atmosphere or by cascading. Compressed air bubbling up through the water is another method of aeration. The aeration process removes odors and gases, oxidizes organic or inorganic matter, destroys some bacteria, and increases

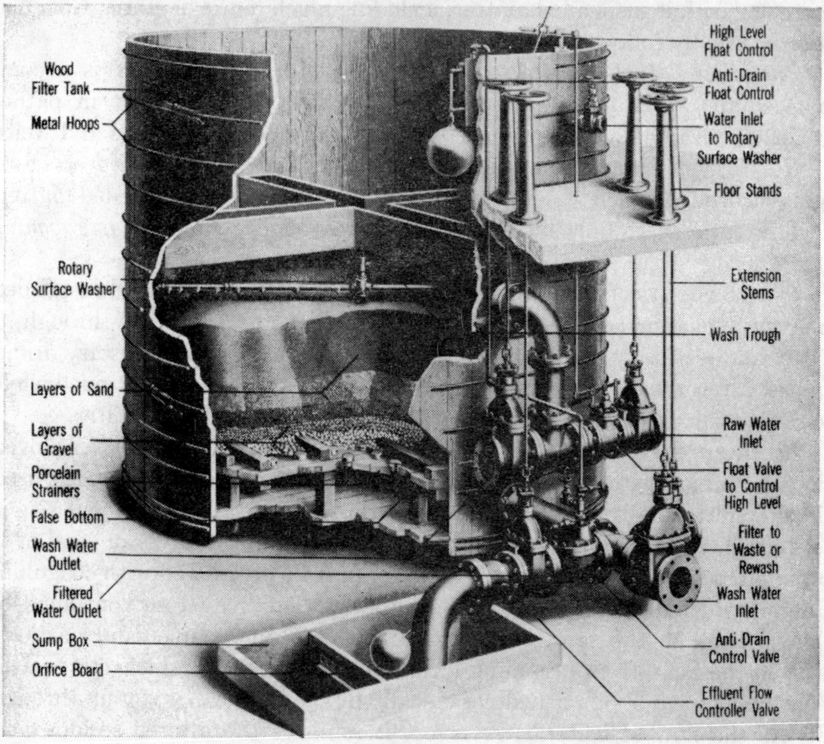

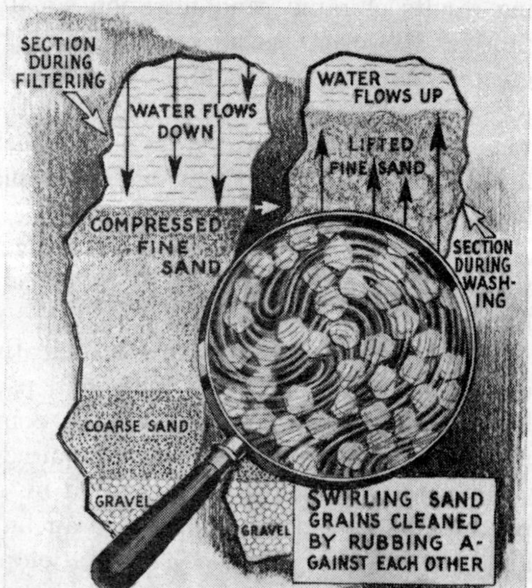

Fig. 392. (*Top*) Open type gravity water filter. (Courtesy, The Permutit Co.)

Fig. 393. (*Bottom*) Principle of backwashing filter media. (Courtesy, The Permutit Co.)

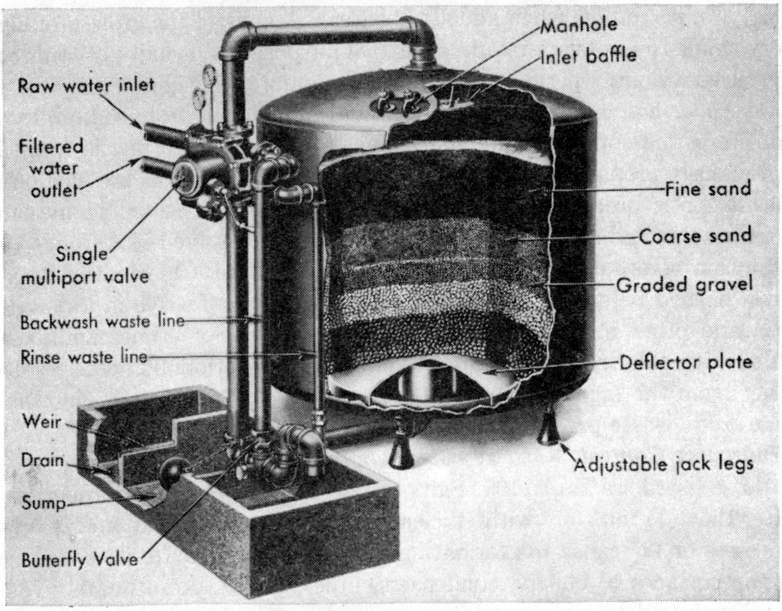

Fig. 394. Vertical pressure filter with multiport valve control. (Courtesy, The Permutit Co.)

pH. Eliminating gases such as free carbon dioxide and hydrogen sulfide reduces corrosive tendencies.

Deaeration. Deaeration is a process in which dissolved oxygen, carbon dioxide, and ammonia are removed. Such gases are corrosive and usually are removed by heating the water to saturation temperature corresponding to the pressure involved. Deaerating of boiler feed water prior to entrance to the boiler is carried on in special heaters designed for the purpose.

Distillation. Another method of removing the solids from water is by evaporation and subsequent condensation. This method is called distillation and is employed for some specific purpose, such as ice making, battery water where in conjunction with sulfuric acid it forms an electrolytic solution having a specific gravity of 1.3 when fully charged.

Neutralizing. Natural water may be found to have various pH values, depending on their source and type of impurities (solids). Acidic water may be treated with alkali to increase pH, whereas alkali water is treated with an acid to reduce pH. Treating acid water with trisodium phosphate not only increases alkalinity but also acts as a corrosion and scale inhibitor.

Organic Control. In addition to the above methods for obtaining more or less control of micro-organisms, some chemical treatment of water is also necessary. While *boiling*, *freezing*, ozonization, and exposure to ultraviolet rays are effective in destroying bacteria, algae, fungi, etc., they are not always practical for commercial purposes.

(*a*) *Copper sulfate* when added to water will control the growth of algae. Very small amounts are required for this purpose and usually it is added in measured amounts periodically through manual or automatic feeders.

(*b*) *Chlorination* is the process of liberating chlorine from sodium hyperchlorite or the application of chlorine gas to water.The former is applied as a previously prepared solution through an adjustable automatic feed pump. The latter is prepared as a liquid (pressurized) and is fed in measured amounts through a chlorinator in which it is predissolved in water to form a solution before entering the main body or pipeline. Chlorination may be accomplished before or after filtration. Chlorine is effective in killing bacteria and other micro-organisms, including algae. Sometimes ammonia is added with chlorine (chloramine) to offset the objectionable taste resulting either from the high chlorine content or from the reaction of chlorine on some trade waste products and decayed algae.

Inorganic Control. Natural waters contain in solution mineral salts such as those listed in Table 103. Salts impart certain chemical properties to water that (1) interfere with efficient industrial, commercial, and domestic processes or (2) cause the formation of a crustaceous material called scale on the surfaces of boilers, condensers, pipes, etc. Scale ultimately causes metal surface deterioration while reducing heat transfer during their formation.

Water containing certain mineral salts in solution is said to be hard. Hardness is evident by the resistance of the water to dissolve soap to produce suds. The process of removing the salts responsible for hardness is known as softening and results in soft water. Soft water will dissolve soap readily.

The salts that produce hardness in water include calcium bicarbonate, magnesium bicarbonate, calcium sulfate and magnesium sulfate, calcium chloride and magnesium chloride. Water is soft when it does not contain these salts. There are two types of hardness: (*a*) carbonate hardness, formerly known as temporary hardness, and (*b*) noncarbonate hardness, formerly known as permanent hardness.

Carbonate (temporary) hardness is caused by the presence of calcium and magnesium bicarbonates. Carbonate hardness can be removed somewhat by boiling the water.

Boiling the carbonate hard water breaks down the bicarbonates into carbonates, water, and carbon dioxide. The carbon dioxide is given off and the less soluble carbonates precipitate, viz.:

Heat + calcium bicarbonate → calcium carbonate + water + carbon dioxide

$$Ca(HCO_3)_2 = CaCO_3 + H_2O + CO_2$$

or

Heat + magnesium bicarbonate → magnesium carbonate + water + carbon dioxide

$$Mg(HCO_3)_2 = MgCO_3 + H_2O + CO_2$$

Carbonate hardness may be removed also by adding calcium oxide or calcium hydroxide to the water. In the latter case, the reaction changes the soluble bicarbonates to insoluble carbonates.

To remove the precipitates in either case, the water is filtered.

Noncarbonate or permanent hardness is due to the presence of calcium sulfate, magnesium sulfate, calcium chloride, and magnesium chloride. The breakdown of permanent hardness salts is not accomplished by boiling as the sulfates and the chlorides are soluble and will not precipitate. Therefore, chemical or some other type of treatment is necessary to eliminate them.

Water hardness is expressed in terms of calcium hardness and magnesium hardness. The sum of the two is called total hardness, expressed as calcium carbonate hardness. Usually hardness is reported in terms of parts per million (p.p.m.), or one part salt to one million parts of water, in which the representative part is the same for both salt and water—i.e., 1 lb salt per 1 million lb water, etc. P.p.m. may refer to the concentration of dissolved salts or solids, or to concentrations of undissolved or suspended solids in a water sample.

Hardness of natural water may vary from less than 15 p.p.m. for very soft water to over 200 p.p.m. for very hard water, with varying degrees of hardness in between.

There are a number of methods used to treat the hardness of water. The method or methods selected depend on the final required condition of the water. This latter statement infers a difference in degree with which any particular method removes hardness, and also the steps, if any, taken to rid the water of the remaining residues (sludge) after treatment is accomplished.

Some of the important mothods of reducing mineral salts in water include:

- Ion exchangers:
  - Demineralization
  - Sodium zeolite softening
- Lime and soda-ash process
- Cold lime process
- Acid treatment
- Phosphate softening

DEMINERALIZATION. The process of demineralization is employed to treat water for maximum reduction in solids (mineral salts). The principle of operation is based on an ion exchange. In the complete demineralizing system, both a cation (positive charged ions) exchanger and an anion (negative charged ions) exchanger are employed, augmented in some instances by an aerator or a degasifier. The full flow of raw water passes through both exchangers which are connected in series (shown in Fig. 395).

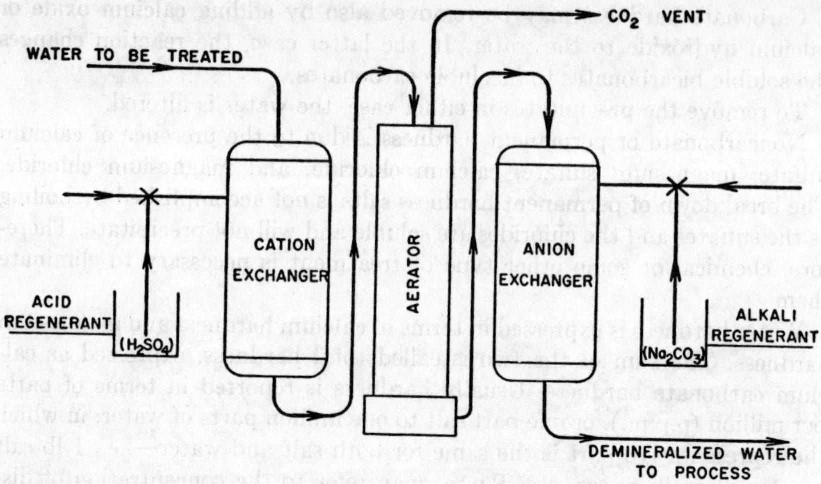

Fig. 395. Water demineralization system with optional aerator for Co2 removal.

The mineral salts dissolved in water are combinations of the elements calcium and magnesium, which produce positive charged ions Na^+, Ca^{++}, and Mg^{++} and their carbonates, sulfates, nitrates, and chlorides. The latter are negative charged and written CO_3^{--}, SO_4^{--}, NO_3^- and Cl^-, respectively.

The cation exchanger is charged with specially prepared resins, or hydrogen zeolite. Zeolite is the generic name given a group of hydrous aluminum sodium silicates found in the veins of basic igneous rocks. Sulfonated coal and coke are natural hydrogen zeolite softeners. The general formula for zeolite is $Na_2O \cdot Al_2 \cdot SiO_2$, but usually is expressed according to its type as Na_2Z for sodium zeolite, H_2Z for hydrogen zeolite, CaZ for calcium zeolite, etc.

When raw water containing sodium, calcium, and magnesium salts percolates through the cation exchange material, an ion exchange occurs. The positive ions (cations) Na^+, Ca^{++}, and Mg^{++} in the water are exchanged for the hydrogen ion in the cation exchange material. Thus the water leaves the exchanger free of sodium, calcium, and magnesium, but retains the negative ions (anions) of the nitrate NO_3^-, sulfate SO_4^{--}, and chloride Cl^-, plus carbon dioxide produced by the breakdown of carbonic acid (H_2CO_3) formed from the carbonate (CO_3^{--}). The cations of calcium Ca^{++} and magnesium Mg^{++} enter the bed of cation exchange material where they stay until washed out by an acid, usually sulfuric acid, regenerant. The sludge formed upon regeneration is disposed of upon being drained from the exchanger.

The ion exchange will be according to the following equations for salts, viz.:

$$Ca(HCO_3)_2 + H_2Z = CaZ + 2H_2CO_3$$

$$CaSO_4 \quad + H_2Z = CaZ + H_2SO_4$$

etc.

The softened water may enter a degasifier where the CO_2 is driven off or it may enter an anion exchanger where the sulfates, the nitrates, and the chlorides, which have reacted with the water to form sulfuric, nitric and hydrochloric acids, are removed. The anion exchanger is made up of a bed of basic organic resins, which will absorb any anions including sulfate, chloride, and nitrate. The anion exchange material, like the cation exchange material, must be regenerated. However, instead of acid, a caustic soda, for strongly basic anion exchanger, or sodium carbonate for weakly basic anion exchanger, is used.

Rinsing with water follows the regeneration process to remove excess regenerant in both the cation and the anion exchangers.

Strongly basic anion exchangers will remove silica, whereas weakly basic anion exchangers require fluoride to accomplish silica removal. The use of fluoride, however, makes the disposal of regenerated sludge a problem.

Proper design and sufficient volume of an effective ion exchange material relative to amount of water handled in gallons per minute will reduce water hardness to virtually zero.

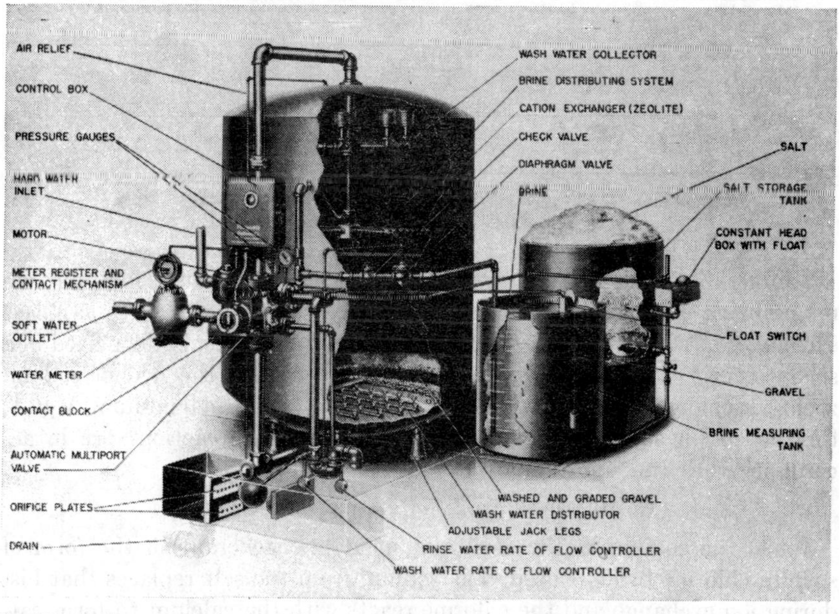

Fig. 396. Zeolite water softener, fully automatic. Embraces the principle of cation exchange, better known as base exchanger. (Courtesy, The Permutit Co.)

TABLE 109. METHODS OF FEEDING FOR COMMON TREATMENT CHEMICALS*

CHEMICAL	PROCESS USE	TYPE FEEDERS	POINTS OF APPLICATION
Alum	Coagulation Color Removal	Dry Pot Type Proportional Double Orifice Decanting Pumps	Coagulation and sedimentation systems; prior to pressure filters for removal of suspended matter and oil.
Sodium Aluminate	Coagulation	Dry Decanting Pumps Shot Feeders	Usually added with soda ash to softeners; used to some extent for internal boiler water treatment.
Ferric Salts	Coagulation Color Removal Oil Removal	Dry Decanting Double Orifice Pumps	Prior to coagulation and filtration systems.
Lime (Hydrated)	pH adjustment Softening	Dry Decanting Special Pumps (Slurry)	Prior to coagulation systems; to softeners; to treated water lines for adjustment of pH.
Soda Ash (Crystalline)	pH adjustment	Pot Type proportioning	Prior to pressure filters.
Soda Ash (Anhydrous)	pH and alkalinity adjustment Softening	All types of wet and dry feeders	To domestic systems, feedlines, softeners, coagulation and filtration systems; boilers.
Caustic Soda	pH adjustment Alkalinity adjustment Softening	All types of solution feeders	To softeners; oil removal systems; domestic water systems; boilers.
Acid Feed (H_2SO_4) (H_3PO_4) ($NaHSO_4$)	pH adjustment Reduction of alkalinity	Decanting Pumps	Treated water lines, prior to degassifiers or deaerating heaters; H_3PO_4 to phosphate softeners (for both softening and alkalinity reduction).
Surface Active Phosphates	Prevent calcium carbonate deposits Eliminate "red water"	All types of proportional solution feeders	Treated water lines.
Ortho-phosphates (Monosodium Phosphate) (Disodium Phosphate) (Trisodium Phosphate)	Prevent scale in boilers	Pumps Shot feeders	Added continuously to boiler drums; shot fed to drums or boiler feedline.
Sodium Sulfite	Prevent Corrosion due to oxygen in boilers, feedlines, economizers	All types continuous solution feeders	Storage section of deaerating heater; suction or pressure side of boiler feed pumps.
Sodium Nitrate	Inhibition of embrittlement	All types of solution feeders	Any point in boiler feedlines or direct to boilers.
Sodium and Potassium Chromates	Corrosion Inhibitor	All types of solution feeders	To brine systems and various circulating cooling and hot water systems.
Reactive Colloids (Sodium Manuronate) Protective Colloids (Starches) (Tannins)	Coagulation Particle Absorption and adsorption	All types of solution feeders	To boiler feedlines; circulating cooling systems.
Amines and Related Organic Compounds	Prevention of return line corrosion	Shot Feeders Pumps	Application depends upon material used. Some materials may be added to boilers and volatilize with steam; others are added to steam line direct, requiring pumps.

* Courtesy, "Betz' Handbook of Industrial Water Conditioning," by Betz Laboratories, Inc., Philadelphia, Pa.

SODIUM ZEOLITE SOFTENING. Sodium zeolite softening, like the demineralizing process, is based on ion exchange usually referred to as "base exchange." It is a cation exchanger in which sodium zeolite is placed in a shell to form a bed. Raw water percolating through the bed of sodium zeolite exchanges its cation ions, calcium, Ca^{++}, and magnesium, Mg^{++}, for the cation sodium, Na^{+}, from the zeolite. The reactions are in accordance with the equation:

$$Ca(HCO_3)_2 + Na_2Z \rightarrow CaZ + 2NaHCO_3, \text{ etc.}$$

When the sodium in the zeolite is spent, a regenerant in the form of sodium chloride brine is used. The sodium from the salt replaces that lost during ion exchange and the chlorine reacts with the calcium to form calcium chloride which is drained off. The zeolite is then rinsed, after which it is returned to service. The softened water contains sodium sulfate and sodium bicarbonate, however, which increase alkalinity.

Some softening arrangements include a sodium zeolite softener placed in parallel with a hydrogen zeolite unit to obtain a given alkalinity of the blended water.

Another arrangement places a chloride-anion exchanger after the sodium zeolite softener to remove the sodium salts carried over from the cation exchanger. This exchanger contains a highly basic resin that is also regenerated with sodium chloride brine. The anions, sulfate, SO_4^{--}, and bicarbonate, $(HCO_3)^-$, are exchanged for the anion chloride, Cl^-, in the exchange material to form sodium chloride, which remains in the water. This arrangement not only removes bicarbonates but also reduces alkalinity by a nonacid process.

The name zeolite is often used erroneously to identify ion exchange materials in general. However, there are a number of ion exchange materials on the market, both natural and synthetic, that are not zeolites. For example, certain carbonaceous (nonsiliceous) cation materials that can be regenerated with acid to act as a hydrogen ion exchanger, as does hydrogen zeolite, can also be regenerated with a brine solution to become a sodium ion exchanger in the same manner as sodium zeolite. Before selecting ion exchange material, it is advisable to seek recommendations from reliable sources such as the water-treatment experts who so courteously made available the various tables and pictures reproduced in this chapter.

LIME AND SODA-ASH PROCESS. The addition of lime (calcium hydroxide, $Ca(OH)_2$) and soda-ash (sodium carbonate, Na_2CO_3) to hard water brings about a reaction with soluble hardness salts to form insoluble salts that precipitate, leaving soft water. The chemical reaction occurring during this process is complex because of the subsequent reaction of new products formed by the initial reactions.

Assuming that sufficient quantities of both lime and soda-ash are added, and allowing enough time for all reactions to occur, the process may be divided into three phases:

1st—Lime breaks down the calcium and the magnesium bicarbonates into insoluble calcium carbonate and magnesium hydroxide. Water is also formed in the reaction.

2nd—Soda-ash breaks down the calcium sulfate and chloride into insoluble calcium carbonate and soluble sodium sulfate. Soluble sodium chloride is also produced.

3rd—Soda-ash and lime react in association with magnesium sulfate and chloride to produce insoluble calcium carbonate, magnesium hydroxide, and also the soluble sodium sulfate and chloride.

The lime also acts with carbon dioxide in the water to form insoluble calcium carbonate. The latter increases the alkalinity of the water. Filtering should follow the lime soda-ash treatment to remove the suspended insoluble materials. A coagulant is used to help the settling process, especially in instances of colloidal suspension.

Cold Lime Treatment. The formation of scale in open circulating water systems is due principally to carbonate hardness. The problem of scale formation becomes aggravated where a high rate of evaporation occurs, as it does in those systems employing cooling towers and evaporative condensers. Scale forms an insulating coat on the surfaces of heat-transfer equipment, reducing thermal efficiency.

Carbonate hardness is due to the presence of highly soluble bicarbonates in the water. When bicarbonates come in contact with heated surfaces, they are reduced to carbonates which form the scale. Surfaces do not have to be high in temperature to promote scaling. It becomes necessary to reduce the bicarbonates to insoluble carbonates which precipitate and eliminate them from the system by blow-down before the scale formation occurs. The major scale-forming ingredient is calcium bicarbonate, $Ca(HCO_3)_2$, also called acid calcium carbonate. Carbonates are salts of carbonic acid. Carbonic acid results from the reaction of carbon dioxide and water. By eliminating the carbon dioxide from water, bicarbonate will not form.

Where it is not possible to heat the water before it is fed to the system to break down calcium bicarbonate to calcium carbonate and to drive off carbon dioxide, other means must be employed. This is especially true where the water is subject to maximum contact with CO_2 in the atmosphere as it is in spray systems.

One method is to treat the make-up water of open circulating systems with slaked lime (calcium hydroxide, $Ca(OH)_2$). Calcium hydroxide reacts with the calcium bicarbonate to form calcium carbonate and water. The reaction is shown in the equation:

$$Ca(OH)_2 + Ca(HCO_3)_2 \rightarrow CaCO_3 + 2H_2O$$

The insoluble calcium carbonate precipitates and can be removed by periodic or continuous blow-down.

The calcium hydroxide also combines with the excess carbon dioxide to form calcium carbonate. Usually a coagulant is added with the lime.

It is obvious that the cold lime treatment gives only partial softening as the soluble sulfates and chlorides are not involved in the reaction. However, as the latter are not serious scale formers at the temperatures found in open cooling water systems, their removal except by normal blow-down is not generally required. In those cases where more complete softening is necessary, the lime soda-ash or zeolite treatment may be required.

Acid Treatment. To further reduce the carbonate content of recirculated cooling water following the cold lime treatment, a sulfuric acid treatment may be given. The sulfuric acid acts to convert calcium carbonate to soluble calcium sulfate. The reaction is as follows:

Sulfuric acid + calcium carbonate → calcium sulfate + water + carbon dioxide

$$H_2SO_4 + CaCO_3 \rightarrow CaSO_4 + H_2O + CO_2$$

The calcium sulfate will not form scale if kept below a maximum level (1700 p.p.m.), which is possible by proper blow-down. The addition of sulfuric acid to lime-treated water reduces alkalinity corresponding to the amount added. As slight alkalinity is required to prevent corrosion of ferrous metals, the amount of sulfuric acid added must be kept within the bounds of good chemical balance throughout the system. Hydrochloric and other acids are also used.

PHOSPHATE SOFTENING. The softened water from the general methods of treatment, some of which have already been discussed, retains various insoluble compounds either because they have not been removed or they remain as end products of chemical reactions with softening agents. While some of these compounds may not ordinarily be scale-forming, operating influences during the use process may convert them to a state whereby they will form deposits. One such process is in the making of high-pressure steam. To further soften make-up water and prevent scale from forming on

TABLE 110. CRYSTALLINE SCALE CONSTITUENTS, IDENTIFIED BY X-RAY DIFFRACTION*

Name	*Formula*
Acmite	$Na_2O \cdot Fe_2O_3 \cdot 4SiO_2$
Analcite	$Na_2O \cdot Al_2O_3 \cdot 4SiO_2 \cdot 2H_2O$
Anhydrite	$CaSO_4$
Aragonite	$CaCO_3$
Brucite	$Mg(OH)_2$
Calcite	$CaCO_3$
Cancrinite	$4Na_2O \cdot CaO \cdot 4Al_2O_3 \cdot 2CO_2 \cdot 9SiO_2 \cdot 3H_2O$
Hematite	Fe_2O_3
Hydroxyapatite	$Ca_{10}(OH)_2(PO_4)_6$
Magnetite	Fe_3O_4
Noselite	$4Na_2O \cdot 3Al_2O_3 \cdot 6SiO_2 \cdot SO_4$
Pectolite	$Na_2O \cdot 4CaO \cdot 6SiO_2 \cdot H_2O$
Quartz	SiO_2
Serpentine	$3MgO \cdot 2SiO_2 \cdot 2H_2O$
Thenardite	Na_2SO_4
Wollastonite	$CaSiO_3$
Xonotlite	$5CaO \cdot 5SiO_2 \cdot H_2O$

* Courtesy, "Betz' Handbook of Industrial Water Conditioning," ed. 4, p. 95. Betz Laboratories, Inc., Philadelphia, Pa.

TABLE 111. TYPICAL ANALYSIS OF SCALES FROM COOLING WATER SYSTEMS*

Source	*Condenser* %	*Pipe Cooler* %
Calcium as CaO	49.79	39.85
Magnesium as MgO	2.42	2.20
Iron and aluminum as R_2O_3	0.82	7.50
Organic matter and oil	6.05	15.00
Carbonate as CO_2	39.00	30.20
Sulfates as SO_3	1.29	0.50
Silica as SiO_2	0.15	3.85

* Courtesy, "Betz' Handbook of Industrial Water Conditioning," ed. 4, p. 160. Betz Laboratories, Inc., Philadelphia, Pa.

tubes and other steam boiler surfaces, the highly stable phosphate compounds are used as a secondary treatment.

There are three phosphate compounds used: trisodium, disodium, and monosodium phosphates. If the acid phosphates, disodium or monosodium, are used, caustic soda is added to convert them to normal or trisodium phosphate, Na_3PO_4.

Trisodium phosphate reacts with calcium compounds to form tricalcium phosphate, $Ca_3(PO_4)_2$, which precipitates. This and other precipitates are eliminated from the boiler by blow-down. If not removed periodically, the accumulation of precipitates forms a surface film on the boiler water surface and promotes foaming.

Trisodium phosphate is added to the boiler water by proportional or volume control pumps as it enters the boiler. This is called the internal treatment as the reactions occur inside the boiler. External phosphate treatment may also be given water already softened by lime-soda or zeolite methods.

SILICA. Soluble silica poses a problem to operators of steam boilers and turbines. Its removal from water is difficult, but certain types of strongly basic ion exchangers in which the silica is adsorbed appear to be the answer. Magnesium oxide is also used successfully. Another possibility lies in the development of better silica absorbers. Silica removal is facilitated under hot conditions. Silica, along with caustic soda—the latter formed by a breakdown of the natural compound, sodium bicarbonate—are believed to be the prime causes of certain types of boiler plate failures called caustic embrittlement. Silica combines with calcium and magnesium to form scale of extreme hardness.

IRON. Iron in water is a source of discoloration both of the water itself and materials in which high iron content water comes in contact. Usually present as soluble ferrous bicarbonate, it can be reduced to insoluble ferric hydroxide and carbon dioxide by aeration. During the aeration process oxidation occurs that causes the transformation.

Corrosion. Corrosion is one of the principal factors to be considered in open circulating water system maintenance. It is evident by the deterioration (pitting, rust, etc.) of metal surfaces that eventually leads to their destruction. Dissolved oxygen, highly prevalent in cooling tower and spray pond systems as a result of aeration, is the major corrosion factor. Dissolved oxygen is also the cause of corrosion in boilers. Its corrosive effect is considerably reduced by lower water temperatures (60 F or below) and a pH of 7.5 to 8.0. Dissolved sulfates and chlorides increase the corrosive effect of oxygen in water.

As in brine systems, corrosion in water circulating circuits results when pH values indicate either an acidic or a high alkaline condition. Carbon dioxide present in the water reduces pH value by reacting with the water to produce carbonic acid, which is corrosive.

To reduce the corrosion tendencies of water, it is necessary to eliminate the impurities that cause it. This means eliminating dissolved oxygen, carbon dioxide, and other corrosive gases (sulfur dioxide), and maintaining pH within a safe operating range commensurate with the water temperature.

The alternative is to employ inhibitors that act to stop or retard the electrochemical forces that cause corrosion by depositing an extremely thin protective coating of some material over the metal surfaces. Because of the electrolytic nature of the corrosion, activity inhibitors are considered as being anodic or cathodic in type. Corrosion inhibitors of the anodic type include sodium chromate, sodium nitrate, sodium phosphate, and similar potassium salts, among others. Cathodic types of inhibitors include the salts of zinc and nickel, such as zinc oxide (ZnO), zinc hydroxide ($Zn(OH)_2$), etc. However, no reasonable amount of inhibitor will fully protect iron and steel surfaces against corrosion in circuits having either a high acid content (low pH) or high alkalinity (high pH). Combining two or more anodic inhibitors to obtain better protection against corrosion than either one could accomplish separately is called the dianotic method of inhibitor treatment. This latter method is usually carried on under controlled pH conditions.

An ideal white protective coating that will not build up on itself may be established on the surfaces of heat exchangers of open circulation systems by a treatment of zinc, or sodium zincate, sodium dichromate, and phosphate. The proportionate amount of each compound should be such as to show by analysis zinc 0.5 p.p.m., sodium dichromate 16 p.p.m., and total phosphate 15 p.p.m. A pH of 7.5 to 8.0 is required to maintain an effective coating. The inhibitors should be fed separately to prevent precipitation resulting from their immediate reaction.

Certain amines are also used to inhibit corrosion in water circulating systems.

Solubility. In the foregoing discussion on water hardness, the various compounds were said to be soluble or insoluble. It is probable that no substance is completely insoluble. On the other hand, solubility is a matter of degree—i.e., substances range from slightly soluble to extremely soluble.

A soluble substance is one that dissolves readily in water while an insoluble one dissolves only slightly or (probably) not at all. The criterion of solubility lies in the amount of water required to dissolve a substance. If one thousand parts of water are required to dissolve one part of a substance, that substance would be considered practically insoluble. An extremely soluble substance would be one that would require only three to five parts of water to dissolve one part.

The important soluble mineral compounds found in water are:

Bicarbonates of calcium, magnesium, and sodium
Sulfates of calcium, magnesium, and sodium
Chlorides of calcium, magnesium, and sodium
Nitrates of calcium and magnesium
Silicon dioxide
Sodium silicate
Iron as ferrous bicarbonate

The important insolubles that occur in the natural water or result from chemical treatment which breaks down soluble compounds are:

Calcium carbonate
Magnesium hydroxide
Tricalcium phosphate
Ferric hydroxide

Important gases that dissolve in water, and are considered "free" in that they are not combined at the moment with another substance, are oxygen and carbon dioxide. They are both undesirable, usually, because oxygen is the major corrosion factor while carbon dioxide tends to form carbonic acid, lower pH, and otherwise assist oxygen in its corrosive activities. Both are reduced by degasification (mechanically) or by chemical treatment. For example, oxygen combines with sodium sulfite to form soluble sodium sulfate. Hydrogen sulfide is another gas present in some waters, and, because of its corrosive nature and obnoxious odor, is highly objectionable.

The solubility of gases diminishes as the temperature of the water increases. At 212 F and atmospheric pressure, the solubility of oxygen is nil, whereas at 70 F and similar pressure, oxygen content will approximate 8.5 p.p.m.

Gas solubility, according to Henry's law, is directly proportional to the pressure of the gas on the liquid when temperature remains constant. The amount of gas dissolved in water will depend on the difference between the vapor pressure exerted by the gas and the vapor pressure of the surrounding atmosphere. At 70 F and 10 psig, the solubility of oxygen will be about 13.0 p.p.m. of water. Both carbon dioxide and hydrogen sulfide are highly soluble in water, being about 1700 p.p.m. and 3900 p.p.m., respectively, at 70 F and atmospheric pressure. Unless overly saturated, such as is the case with carbonated water at the moment its container cap is removed, the partial pressure at normal atmospheric temperature of either gas is not sufficient to cause the water to expel it. Therefore, it is necessary to increase the partial pressure in the water to bring it in equilibrium with that of the atmosphere. This can be done by heating which raises the partial pressure in the water and lowers solubility, or by aeration which saturates the water with other gases, such as oxygen and hydrogen, and practically eliminates the lower vapor pressure gases CO_2 and H_2S. Another method is to lower the pressure of atmosphere by deaerating under vacuum.

Degree of Water Treatment Necessary

The degree and method of water treatment depends on the condition of raw water and its ultimate use. For example, in some instances it may be necessary to treat water for softening only while in others complete demineralization may be required. Again, it may be desirable to remove only calcium hardness and disregard magnesium hardness. A savings is thus made based on the cost of removing both. Considering the processes mentioned in the previous sections entitled "Ultimate Use of Water" and "Kinds of Treatment Required," the question of degree and method of treatment may be determined from Table 108.

Cost of Treating vs. Probable Benefits

Weighing the probable disadvantages to the use processes involved against the cost of treatment is essential. Only those treatments necessary to the specific requirement should be attempted, and then only to the degree that assures successful results.

Cycles of Concentration

In the discussion related to cooling towers it was stated that a concentration of dissolved mineral solids occurs in the circulating water as a result of evaporation. It was also stated that concentration of solids was offset to some extent by make-up water added to replace evaporation and windage losses. To further offset solids concentration, bleeding or blowdown must be practiced. The question arises as to the amount of bleeding that must be done to keep the system at a level of concentration wherein it will respond satisfactorily to scale treatment. Otherwise, the scale-forming salts reach such proportions that treatment in any reasonable amount will fail to produce necessary results.

The first step in determining the scale-forming characteristics of the circulating water is to ascertain the relative solids concentration. Because calcium, sodium, magnesium, and iron chlorides are highly soluble in water, they do not precipitate during the treatment process. Being proportional in parts-per-million to the solids concentration of circulating water, chlorides form the basis for determining the amount of bleeding necessary. The amount of chlorides formed in the circulating water taken from the outlet of the cooling tower, divided by the chloride content of the make-up water, results in a factor called *cycles of concentration*, or

$$\text{Cycles of concentration} = \frac{\text{p.p.m. chloride in circulating water}}{\text{p.p.m. chloride in make-up water}}$$

Cycles of concentration, then, is the criterion of scale-forming tendencies of the circulating water based on the relative chloride content. If the quotient in the above equation is, say, 4.0, it means that the solids concentration of the circulating water is four times that of the make-up water.

Windage loss in cooling towers and spray ponds tends to limit the level of solids concentration. This is due to the continual loss of relatively high concentration circulating water and its replacement by relatively low concentration make-up water. This leaves the process of evaporation as the major cause of solids build-up. It follows that the limit of solids concentration is a function of evaporation-windage loss relationship. This relationship is shown in Fig. 397.

For example, if the windage loss equals 1.4 percent and evaporation loss is 3.5 percent, the degree of solids concentration limit as expressed in terms of cycles of concentration would be 3.5, according to the curve in Fig. 397.

The safe level of cycles of concentration is based on the scale-forming tendencies of the water as indicated by pH value. Based on a temperature-solids concentration constant, the value of pH saturation may be determined from Langelier's Saturation Index. Subtracting pH saturation from actual pH results in a factor called the Saturation Index, which indicates the corrosive or scale-forming tendencies of the circulating water. By adding pH saturation to the Saturation Index as determined for maximum operating temperature of the circulating water, a factor called the Stability Index is determined. The Stability Index value conforming to levels of pH above

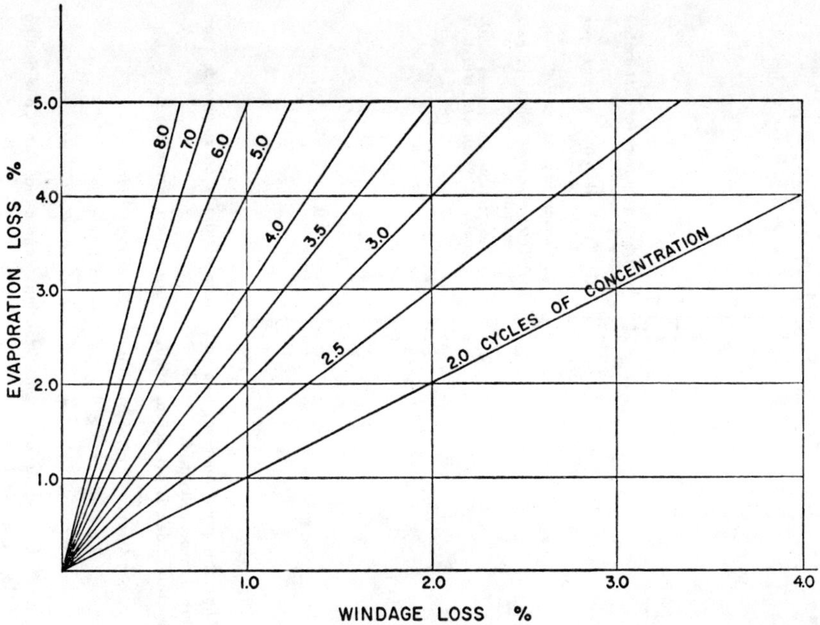

Fig. 397. Relationship between evaporation loss, windage loss, and cycles of concentration. (Redrawn from "Betz' Handbook of Industrial Water Conditioning," ed. 4, by Betz Laboratories, Inc., Philadelphia, Pa.)

or below the neutral point of 7.0 determines the scaling or nonscaling nature of the water as described previously in this chapter.

By actual testing, the correct cycles of concentration corresponding to a nonscaling stability index value can be ascertained. If tests show that cycles of concentration equal to 3.0 result in scale-forming conditions for a particular system, then that value is too high. Lowering the cycles of concentration to 2.0 by increased bleeding may prove to be the maximum level at which pH saturation can be maintained within the nonscaling range. If so, the system should be controlled continually at 2.0 cycles of concentration.

The amount of make-up water required to maintain a system at a given cycle of concentration may be determined as follows:

$$\text{Cycles of concentration} = \frac{\%\, e + \%\, bw}{bw}$$

where $\%\, e$ = percent of circulation rate lost by evaporation
$\%\, bw$ = percent of circulation rate lost by (b) bleeding and (w) windage

EXAMPLE: If the desired rate of circulation is 5000 gpm and 1.5 percent is lost by evaporation and 0.2 percent lost by windage, how many gallons

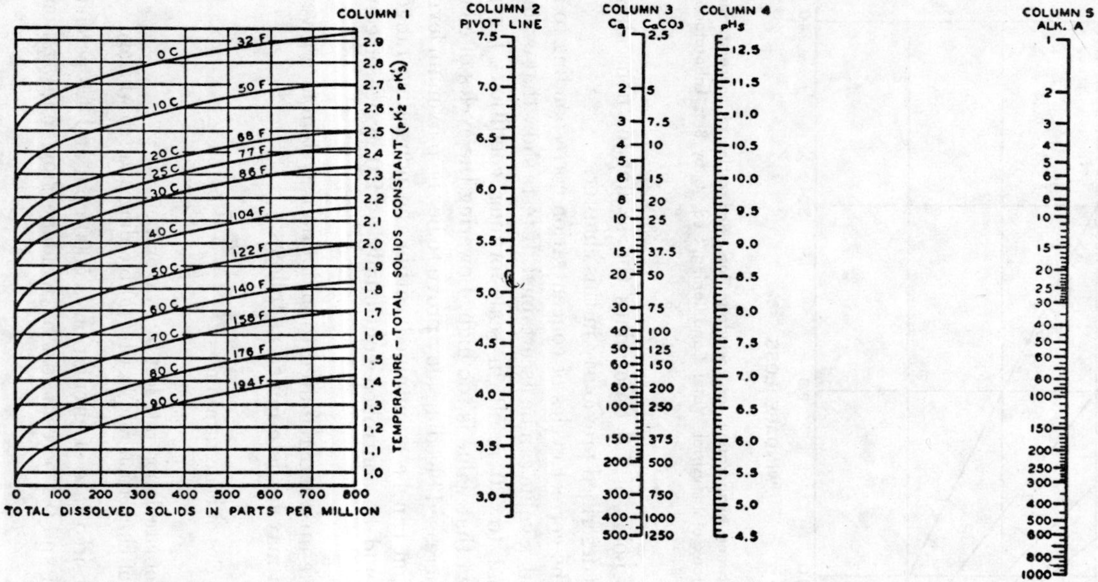

(Data required for determining *pH* saturation.

(1) Total alkalinity, as parts per million of $CaCO_3$.

(2) Calcium in parts per million.

(3) Total dissolved solids, in part per million.

(4) Temperature, in degrees centigrade, at which *pH* saturation desired.

Fig. 398. Graph and nomogram for determination of pH saturation by Langelier's formula. (Applicable within pH range 7.0–9.5.) (Based on an article in the Oct. 1936 issue of *American Water Works Association Journal* and later corrections for Tables 2 and 4 of the article. Prepared for C. P. Hoover of the Columbus (Ohio) Water Softening and Purification Plant by M. L. Riehl.)

Instructions for Using the Chart:

1. Knowing temperature and total dissolved solids, find temperature and total solids constant on column 1.

2. Align this constant with given value of calcium on column 3 of chart, then locate point on column 2 of chart (pivot line).

3. Align this point on pivot line with given alkalinity on column 5; read pH saturation on column 4.

Saturation index is pH actual minus pH saturation. For example:

pH Actual	*pH Saturation*	*Saturation Index*
7.6	8.1	−0.5 (corrosive)
8.4	7.8	+0.6 (scale-forming)

(Reprinted by permission from "Heating Ventilating Air Conditioning Guide 1956," chapter 43.)

per minute of water must be added as make-up to maintain the desired rate of circulation?

$$e = 1.5\% \times 5000 = 75 \text{ gpm}$$

$$w \doteq 0.2\% \times 5000 = 10 \text{ gpm}$$

$75 + 10 = 85$ gpm required as make-up to offset losses by evaporation and windage

The cycles of concentration for the above system would be

$$\text{cycles} = \frac{\%\,e + \%\,bw}{bw}$$

With no bleeding, then

$$\text{cycles} = \frac{1.5 + 0.2}{0.2}$$

$$= 8.5 \text{ cycles of concentration}$$

If it were found necessary to maintain 2.0 cycles of concentration for safe operation, then

$$bw = \frac{\%\,e}{\text{cycles} - 1}$$

$$bw = \frac{1.5}{2 - 1}$$

$$bw = 1.5\%$$

However,

$$b = 0.2\% \times 5000 = 10 \text{ gpm}$$

therefore

$$w = bw - b = 1.5 - 0.2 = 1.3\%$$

or

$$w = 1.3\% \times 5000 = 65 \text{ gpm}$$

and

$$e = 1.5\% \times 5000 = 75 \text{ gpm}$$

Total make-up would be the sum of all loses, or

$$\text{make-up} = 10 + 65 + 75$$

$$= 150 \text{ gpm}$$

Proof:

$$e + b + w = 1.5\% + 1.3\% + 0.2\%$$

$$= 3.0\% \text{ make-up}$$

$$3.0\% \times 5000 = 150 \text{ gpm}$$

General Summary of Water Problems

A general summarization of difficulties experienced with water and their causes may be made as follows:

(*a*) *Scale* in boilers is due principally to the presence of bicarbonates, sulfates, and silica and usually is formed by a combination of the three. Bicarbonates break down under heat and soluble carbonates, either forming a sludge that bakes on boiler surfaces or adhering directly on boiler surfaces, both of which form scale. Sulfates and silica precipitate directly on boiler surfaces. Scale in circulating-water cooling systems is due primarily to bicarbonate breakdown, leaving carbonates to form on heat-transfer surfaces.

(*b*) *Corrosion* in water systems is caused by dissolved gases such as oxygen, carbon dioxide, hydrogen sulfide, also natural acids or acids from trade wastes, and chlorides of sodium, calcium, and magnesium.

(*c*) *Fouling* is caused by scale, algae, and combinations of both accompanied by other contaminates such as soil, various organic materials, etc.

(*d*) Foaming is caused by impurities that form a surface layer and tend to restrict the escape of gases from a liquid mass such as water in a boiler. It is also caused by a sudden release of pressure on liquid surfaces during which entrapped gas expands and tends to expand the liquid mass also. The tendency of a liquid to foam varies with the tension required to separate the individual molecules of any given liquid. This latter is called the *interfacial tension* of a liquid.

(*e*) *Scale formed by silica* is the toughest of all scales and also makes other scales with which it forms tougher.

(*f*) *Carbonate scale* is porous except when formed with other denser scales from other substances such as silica.

(*g*) *Sulfate scale* is more dense than carbonate scale and therefore harder.

(*h*) *Scale is prevented* by eliminating the substances from which it is formed. This is done by chemical treatment such as cold lime, lime and soda-ash, phosphates, caustic soda, barium salts, and some types of coagulants. It is also accomplished by ion exchange methods using natural or synthetic zeolite (siliceous) materials containing silicates of sodium and aluminum called greensand, or carbonaceous zeolite materials using sulfonated coal or lignite. Both types of ion exchanger can be regenerated and used repeatedly. Siliceous zeolite is regenerated with sodium chloride (base exchange) while carbonaceous or synthetic resin zeolites may be regenerated either with sodium chloride or acids such as sulfuric or hydrochloric acid (hydrogen exchange).

(*i*) *Coagulants* are chemicals used to gather fine particles of other substances such as silt, sand, etc., and others formed by precipitation when, for example, soluble bicarbonates are broken down to form insoluble carbonates. This gathering together of such particles forms a coagulated material, commonly referred to as "floc" (from flocculent), that readily settles out of the water. Coagulants include aluminum sulfate (alum), sodium aluminate, and iron sulfate (copperas).

(*j*) *Gases* such as carbon dioxide and hydrogen sulfide are eliminated by aeration during which they are replaced by oxygen and hydrogen.

(*k*) *Oxygen* is eliminated either by deaeration or by chemically treating the water with sodium sulfite to form sodium sulfate. Deaeration is accomplished by heating or exposing water to a partial vacuum.

(*l*) *Corrosion* is retarded or eliminated by chemical treatment with phosphate, chromates, and zinc to form protective coating, or, by chemical treatment to prevent electrochemical action as with chromates, amines (inhibitors). Deaeration to eliminate oxygen, aeration to eliminate carbon dioxide and other gases also act to retard corrosion. Correct pH of water is important in corrosion control.

(*m*) Other considerations are *priming* (boilers) which is a sudden surge of water which is carried over with the steam. *Embrittlement* (caustic) is caused by the attack of caustic soda and silica on stressed boiler plate areas. *Blow-down* means the act of allowing water to escape from a system so that solid concentration may be reduced by subsequent replacement with fresh water.

CHELATING AGENTS

A relatively new method of treating water to reduce hardness caused by the presence of calcium and magnesium employs a select group of organic chemicals called *chelating agents* (pronounced ke-late-ing). The word chelate is derived from the Greek *chēlē,* meaning "claw," which fits their nature of developing a *claw structure* with metal ions This structure represents a very tenacious bond between the chelating agent and the ions of metal. The resulting reaction between the two is a complex chemical called a *chelate.*

Although there are several kinds of chelating agents, tetrasodium salt of ethylene diamine tetracidic acid (NaEDTA) is typical of the chelating agent family.

The use of chelating agents has been generally associated with boiler water treatment. Their discussion here is in the interest of completing the coverage of all types of water treatment that might be even remotely associated with the refrigeration plant. Like the more conventional types of organic-phosphate type treatments, the success of chelating agents lies in the selection, application and control of a proper treatment program. Their effectiveness depends on determining proper points of application, providing accurate control and assuring the correct treatment program based on the over-all boiler system conditions.

The product resulting in the reaction with hardness materials, i.e., calcium and magnesium, is a tight complex that is soluble. The chelated calcium and magnesium cannot be precipitated by carbonates or phosphates. Thus, in the concentrations commonly used in boilers, the resulting complex is sufficiently soluble that it effectively holds the calcium and magnesium in solution. Therefore, the most effective applications of chelating agents are in the control of calcium and magnesium deposits.

The tetrasodium salt of EDTA has the following structure:

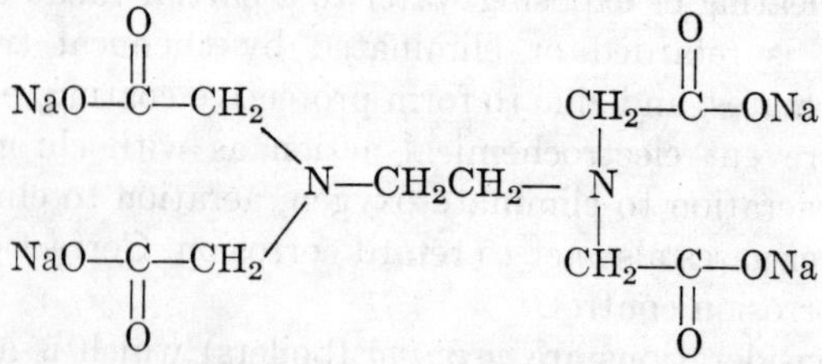

Reaction with Ca++ results in the following complex:

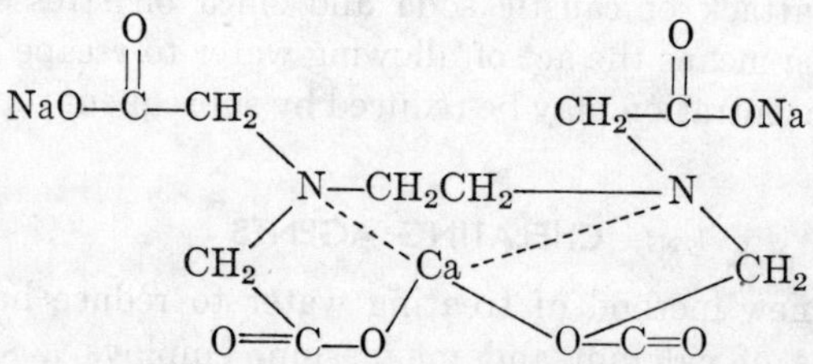

Chelation is a specialized part of the field of *coordination chemistry* of which the above is one of the complex structures. While the results of chelating iron and copper are not conclusive at this time, except under certain modifications, they are not advocated for use with these elements. While excellent results have been obtained with the more common hardness materials, such as calcium and magnesium, chelating agents should be applied only under a planned program recommended by the manufacturer.

Water Analysis

There are numerous tests that can be made on water. It is very seldom necessary to make a complete water analysis as such a term implies the performance of the entire gamut of tests involving physical, chemical, and bacteriological studies. For industrial purposes, it is necessary to determine periodically the condition of the water as it affects local processes. This means that specific tests are run for control purposes.

Hardness tests are essential in many industries. The operation of cooling water systems where evaporation is carried on at a high rate requires that chloride tests be made at close intervals in order to control salt concentrations through sufficient blow-down. Tests for pH are also necessary.

Methods of testing should be made as simple as possible, commensurate with the degree of accuracy required. Because of the number of tests possible and the complexities of the many industrial processes, no attempt is made here to discuss water analysis in detail. However, it should be said that plant operators faced with the problem of water conditioning and control should solicit the advice of water chemical and equipment firms concerning the most effective and efficient methods of procedures as to both treatment and analysis.

SUMMARY OF CHAPTER 13

Refrigeration systems that perform their cooling function through an intermediate substance are known as indirect or secondary systems.

Intermediate substances employed in secondary systems may be substances having a freezing point low enough to circulate at required temperatures.

Some of the substances employed as intermediaries are salt solutions called brines, water, air, ethylene glycol, denatured alcohol, methanol, and glycerol.

Brines are employed in commercial ice making, water in industrial processes, air in air-conditioning systems, and others in special cooling systems.

Sodium chloride brine is used where temperatures are maintained somewhat above 0 F; also used for brine spraying of meats, fish, etc., to be chilled or frozen.

Calcium chloride brine is used where temperatures are maintained somewhat below 10 F.

An increase in salt content of a brine up to the eutectic point decreases the freezing point.

An increase in salt content of brines increases the specific gravity.

An increase in specific gravity decreases the specific heat value of brines. It also increases the power required to pump it.

The physical characteristics of brines are interdependent. Knowing one of the three properties—i.e., specific gravity, freezing point, or specific heat—the remaining two may be established from tables.

The specific gravity of a brine, or its weight per unit volume as compared with an equal volume of water, is determined by a hydrometer.

The hydrometer is based on Archimedes' principle that a floating body displaces an equal weight of the liquid in which it floats. The specific gravity of water is assumed to be unity.

Baumé gravity is related to specific gravity according to the equations:

For liquids lighter in weight than water:

$$°\text{Bé} = \frac{140}{\text{specific gravity}} - 130$$

For liquids heavier than water:

$$°\text{Bé} = 145 - \frac{145}{\text{specific gravity}}$$

Brines change chemically during use. They tend to become acid.

Acidity of brines is expressed in terms of pH. Below pH 7.0 a substance is acid; above pH 7.0 it is alkaline. At pH 7.0 it is neutral.

pH is determined by the litmus, the colorimetric, or the glass electrode method. pH decreases with greater acidity and increases with greater alkalinity.

For general use, brines should be maintained at a pH of 8.2.

Caustic soda is used to raise the pH of brine.

A mild acid such as carbon dioxide gas is used to reduce alkalinity.

Brines are also treated to retard their corrosive effect. The chemical used depends on the kind of brine. Chromates, phosphates, and zinc dust used in correct proportions help retard corrosion.

A eutectic solution is one in which the mixture of two or more substances is of such proportions as to result in the lowest possible initial point of crystallization (freezing point) for the combination.

Hold-over brines are usually eutectic solutions frozen to supply refrigeration by utilizing latent heat of fusion during periods when condensing units are not available or to extend cycling periods.

Antifreeze agents, such as alcohols and glycerins, are used to reduce the freezing points of aqueous solutions.

An endothermic mixture is one in which heat is absorbed during the reaction of the substances constituting the mixture. A temperature drop results. *Example:*

Calcium chloride + crushed ice

An exothermic mixture is one in which heat develops during the reaction of the substances constituting the mixture. Temperature increases. *Example:*

Calcium oxide + water, having the reaction:

$$\text{Calcium oxide} + \text{water} \rightarrow \text{calcium hydroxide} + \text{heat}$$

$$CaO + H_2O \rightarrow Ca(OH)_2 + 15{,}540 \text{ calories per mol}$$

Water is an important substance in refrigerating process.

Natural water contains impurities that are detrimental to heat exchangers, conduits, and industrial processes. Mineral salts, algae, and gases are some of the impurities of water.

The detrimental effects of water impurities include corrosion of metal surfaces, scale formation, clogging of conduits, interference with processes such as laundering, ice making, etc.

Mineral salts cause water hardness, i.e., they resist the action of dissolving soap.

Water hardness is classified as:

Carbonate hardness

Noncarbonate hardness

Carbonate hardness is reduced by heating or by chemical treatment or ion exchange.

The two major water-softening processes are:

Lime soda-ash treatment

Zeolite softening

Zeolite softening is an ion exchange process. The following equations show the reaction of various water hardness compounds with sodium zeolite, Na_2Z, and hydrogen zeolite, H_2Z:

Sodium cycle:

$$\left.\begin{matrix} Ca \\ Mg \end{matrix}\right\} \text{salts} + Na_2Z \rightarrow \left\{\begin{matrix} Ca \\ Mg \end{matrix}\right\} Z + \text{Na salts}$$

Hydrogen cycle:

$$\begin{array}{lccl} \textit{Solubles} & \textit{Insoluble} & & \textit{Solubles} \\ \left.\begin{matrix} Ca \\ Mg \\ Na_2 \end{matrix}\right\} (HCO_3)_2 + & H_2Z & = & \left\{\begin{matrix} Ca \\ Mg \\ Na_2 \end{matrix}\right\} \begin{matrix} Z + 2H_2CO_3 \\ \text{or} \\ 2H_2O + 2CO_2 \end{matrix} \\ \left.\begin{matrix} Ca \\ Mg \\ Na_2 \end{matrix}\right\} SO_4 \quad + & H_2Z & = & \left\{\begin{matrix} Ca \\ Mg \\ Na_2 \end{matrix}\right\} Z + H_2SO_4 \\ \left.\begin{matrix} Ca \\ Mg \\ Na_2 \end{matrix}\right\} Cl_2 \quad + & H_2Z & = & \left\{\begin{matrix} Ca \\ Mg \\ Na_2 \end{matrix}\right\} Z + 2HCl \end{array}$$

Scale formation on heat-transfer surfaces is caused by the hardness compounds.

In cold circulating-water systems, scale is due chiefly to calcium bicarbonates which break down at the heating surfaces and cause calcium carbonate deposit.

Scale reduces the rate of heat transfer.

Some gases present in water are corrosive. They include oxygen, carbon dioxide, and hydrogen sulfide. Carbon dioxide and hydrogen sulfide can be removed by aeration. Oxygen can be removed by degasification (deaeration) or chemical treatment.

A concentration of mineral salts occurs in systems subject to evaporation. They may be controlled by softening the water and periodic or continuous blow-down.

Cycles of concentration is an expression indicating the degree of concentration of scale-forming compounds in circulating water. It is the ratio of chlorides in the circulating water to chlorides in the make-up water.

Chorides are highly soluble and remain in the water after general methods of treatment. Chlorides in water are proportional to the concentration of hardness compounds and therefore form the basis for determining amount of blow-down.

Chelating agents reduce water hardness due to calcium and magnesium salts by forming a tight complex chemical material that is soluble and cannot be precipitated by carbonates or phosphates. A typical chelating agent is NaEDTA (tetrasodium salt of ethylene diamine tetracidic acid).

STUDY OUTLINE FOR CHAPTER 14

The criterion of food spoilage and/or deterioration:
- Appearance
- Taste
- Odor
- Physical damage
- Chemical change:
  - Rancidity
  - Sourness
  - Putrefaction
- Loss of weight
- Overripeness (autolysis)

Causes of food spoilage:
- Enzymes
- Micro-organisms:
  - Bacteria
  - Yeasts
  - Molds
- Bruising
- Desiccation
- Freezing (improper)
- Absorption of odors
- Incorrect storage temperatures
- Excessive respiration

Measures for control of food spoilage:
- Establishment of conditions unfavorable to food spoilage agents by:
  - Canning
  - Fermentation or pickling
  - Dehydration
  - Smoking
  - Preservatives
  - Cold storage:
    - Chillers
    - Sharp freezing
    - Quick freezing
    - Freeze drying
    - Immersion freezing
    - Humidity control

Cold-storage practices:
- Chill-room conditions:
  - Temperature
  - Humidity
- Sharp-freezer conditions:
  - Temperature
- Quick-freezing conditions:
  - Methods
  - Temperatures
  - Preparation of foods
- Freeze drying methods
- Respiration:
  - Method of control

Rancidity in fatty foods (hydrolizing process)

Effect of low temperatures on milk

Causes of egg deterioration

Wine chilling

Use of dunnage

Cold-storage warehouses:
- Design factors
- Interior transportation facilities
- Construction details
- Utilities
- Control of operation

Frozen-food locker plants:
- Function
- Operation
- Temperature requirements
- Plant layouts

14

Principles of Cold Storage

Refrigeration plays a highly important role in the preservation of perishable commodities, the most important of which are foodstuffs. This is particularly true where the forces responsible for physical and chemical changes that cause food spoilage and ultimate decay may be rendered dormant or substantially retarded in their activity by their exposure to low-temperature conditions.

For this reason, the principles governing the deterioration of food substances naturally form an integral part of refrigeration study as a whole. The knowledge gained as a result of such study forms a basis on which to build a more complete understanding of what is required in the way of effective control measures for each of the individual food items to offset the forces of spoilage and decay.

That many items of food require stringent preparatory and storage handling to insure their safe preservation has been recognized universally since time immemorial. The science of food preservation has been highly developed by man in his effort to maintain a wide variety of perishable substances in an edible condition and as near the point of freshness as possible for extended periods of time. That the science was born of necessity can be appreciated when it is considered that food supply must be constant at all times while many important plants and trees produce their entire crop during one short season of the year only. Meats, fish, and dairy products, although produced continually, must be preserved against the time of distribution to outlying points and kept until required by the consumer, which may mean weeks or even months after processing.

In analyzing the complete subject of food preservation as a prerequisite toward its intelligent preparation for storage and control measures during storage periods, consideration should be given to the following basic questions: (1) What constitutes food spoilage? (2) What are the causes of food spoilage? (3) What control measures can be taken to allay or to prevent spoilage of foods?

FACTORS CONTRIBUTING TO FOOD SPOILAGE

The effect of spoilage varies in different types of perishable commodities, but the result is similar in all cases since it renders such items unfit for human or animal consumption. However, spoilage is generally apparent through noticeable changes in food substances—changes that render definite variations in original characteristics. Odor, color, taste, physical appearance, and chemical behavior are some of the traits by which the activity of spoilage agents can be detected once their presence becomes effective. Loss of weight, reduction in vitamin content, molding, wilting, softening, souring, and rotting—all are results of food deterioration and spoilage.

Fruit, as a result of continued chemical activity within the substance after harvesting, eventually becomes overripe and softens if not properly protected. Overripeness is closely followed by the initial stages of rotting, during which discoloration of the outer surface takes place. Starting with a rather spotty appearance, in many cases the defect soon spreads throughout the mass, causing spoilage of the fruit.

Vegetables, although similar in many respects to fruit in their appearance, when subjected to deteriorating influences also indicate such a condition by wilting.

Meat products such as beef, pork, and lamb change but little in physical appearance during the early stages of deterioration. However, the forces of spoilage carry on destructive activity known as putrefaction which produces gases that have offensive odors. When certain stages of putrefaction are reached, the meat surface takes on a slimy coating, a positive indication of spoilage.

Fatty foods, of which butter and cheese are examples, become progressively darker as deterioration advances. They also emit a rancid or sour odor resulting from chemical changes of the fats and the oils of which they are composed. The taste is also affected adversely.

Eggs lose weight due to evaporation and become strong-smelling upon spoiling. Broken eggs can be considered as spoiled unless they are destined to be frozen or powdered, when they again become a desirable product to the baker and the confectioner.

Milk sours gradually when exposed to conditions conducive to the activity of spoilage agents. The degree of sourness can be nearly estimated by the strength of its disagreeable odor or its bitterness of taste. The physical appearance of milk changes radically upon reaching its final states of deterioration, leaving a definite line of demarcation between the water and the more solid components as one separates from the other.

Considering that absolute spoilage is the final result of gradual but progressive chemical changes within the various food substances, it follows that deterioration can be identified according to its phase of development. Although this aspect is not so important when considered from a commercial point of view, nevertheless it does have some practical application. Overripe fruit, a good example, while still fit for human consumption, is impaired commercially. This is because the fruit in its overripeness contains no harmful ingredients and remains in a sweet-tasting condition, but because of its softness—a natural result of overripening—it can no longer be handled conveniently without damage and continued storage for long periods would be impractical.

Milk can be used for several purposes in a "sour" condition, one of which is in the making of certain types of butter. For this purpose, milk in such a condition cannot be considered spoiled even though it is bitter to the taste or disagreeable in odor as compared with the fresh, clean product. There are numerous other occasions where food products of different types are considered good even though they have deteriorated below the taste, odor, and other characteristic standards by which fresh products are judged. It must be noted, however, that even in those cases the phase to which deterioration has reached also has its limits.

Most of these substandard products are in a class by themselves, and usually remain within the province of the special processor which they serve. It still remains for the cold-storage plant and the student of food preservation practices to consider only those items which are received in their natural fresh and undamaged condition and which are expected to be maintained in that condition for reasonable time intervals.

CAUSES OF FOOD SPOILAGE

Food spoilage is brought about through a series of complicated chemical changes. These changes originate either within the food substances through the action of natural forces or by influences of external foreign contaminants. In many cases, a combination of both factors results in deterioration and final decay of food products.

Each of these factors is dependent on particular types of agents that activate the process of deterioration. These agents continue such process through progressive stages of chemical changes which finally conclude such action in absolute decay of the food substance. Generally, spoilage agents may be divided into two major groups:

I. Enzymes
II. Micro-organisms:
   (*a*) Bacteria
   (*b*) Yeasts
   (*c*) Molds

Enzymes

Enzymes are chemical substances of highly complex composition, about which little is known. Described generally as *catalysts* formed by living cells, they induce chemical changes within organic materials while remaining unchanged themselves. Enzymes occur naturally in all organic substances and are involved in the sprouting of seeds and in the life processes of fruits, vegetables, and animals, and they promote the chemical action necessary for the digestion of foods by man and animals.

There are various types of enzymes, each of which is specific in its individual action; i.e., each type brings about its own characteristic result. Because of their complex nature and the failure of science so far to determine their exact chemical composition, identification of enzymes depends largely on environment and the results of their chemical action. The names listed in Table 112, along with the substances on which they act and the product resulting from their chemical activity, have become a mark of identification for a few of the known enzymes.

TABLE 112. SOME COMMON ENZYMES AND THEIR HOSTS

Enzyme	*Substance on Which It Acts*	*Product*
Diastase	Starch	Malt sugar
Cytase	Cellulose	Sugar
Pepsin	Proteins	Peptones
Trypsin	Proteins	Amino acids
Lipase	Fats	Glycerin—fatty acids
Zymase	Sugars	Alcohol—carbon dioxide
Ptyalin	Starch	Sugar

Enzymes are responsible for the normal ripening of fruit and vegetables, and, if allowed to continue in their activity uncontrolled, will cause overripening, deterioration, and ultimate destruction. This life process continues even after harvesting, because enzymic action continues to induce such chemical changes as to bring about these conditions when surrounding temperatures are correct. They are proteins built up by the plant for the purpose of accelerating chemical processes that must be carried on in order to keep alive the living cells and tissues of the plant and its product.

Enzymes cannot be eliminated without changing the chemical structure of the organic substance in which they are contained. Cooking the food substance is the most effective way, as heat destroys these complex proteins. Freezing temperatures, however, will retard their activity without destroying the enzymes.

Disintegration of substances such as fruits and vegetables caused by enzymic action alone is known as *autolysis* (self-disintegration).

Food substances are made up of proteins (nitrogenous compounds), carbohydrates (sugar, starch, or cellulose), fats (a compound of organic acids and glycerin), and minerals (calcium, iron, phosphorus, etc.). With the exception of the mineral content all other compounds are subject to chemical change brought about through the influence of their own specific enzymes. The amount of enzymic substance contained in any organic material is surprisingly small in relation to the effective results they produce by their presence.

Free oxygen will greatly accelerate enzymic activity.

As shown in Table 112, enzymes induce chemical changes within such compounds as starch, cellulose, fats, and sugar, converting them into various types of sugars, peptones, fatty acids, and alcohols. Such activity is carried out under numerous environments, such as in the digestive system of living creatures, the growing and ripening processes of fruits and vegetables, and within the cellular tissues of meats, both living and dead.

However, enzymic action will take place only when surrounding conditions of temperature and acidity are conducive to its effectiveness. As was stated, heat above certain temperatures will destroy enzymes, while low temperatures—particularly below the freezing point of water—will retard their activity, although some types will continue their effectiveness very slowly even at such low temperature levels. Some enzymes thrive under acid conditions while others are destroyed when exposed to strong acid solutions. This is also true of alkali solutions.

One of the basic results of enzymic action on food substances is to transform insoluble materials to a soluble state. This action of conversion is also carried on within food substances of all kinds by their own enzymes, and as a result soluble materials are made available for digestion by *microorganisms* that are generally present in all matter.

It is important to note that, while certain types of enzymes exist as material substances of organic compounds, many other types are the result of bacterial secretion.

Micro-organisms

Bacteria

Bacteria are minute organisms, being the smallest and the simplest form of plant life known. They are made up of a single cell, some of which are larger than others, and reproduce through elongation of the cell and subsequent segmentation into two cells. Although reproduction is rapid, their life cycle is extremely short, being a matter of minutes in many types. Bacteria are classified as to type in respect to the chemical changes they bring about rather than to their physical characteristics. This is because of their minute size (1/30,000 in.). Some are definable under the microscope while others are so small they will pass through fine porcelain filters. Different species of bacteria are similar physically while the results of their activity vary widely.

Present in almost every known substance, bacteria will be found in large numbers in the air, in water, and in soil, as well as within animal bodies, on the surface of plants and their products, and in the secretive substances of animals, such as milk, eggs, butter, and cheese.

Bacteria in their relation to man's economy can be broken down into two classes: (*a*) those that cause spoilage and deterioration of foodstuffs and (*b*) those employed for food preservation.

The first type are those most subject to concern by the operators of cold-storage plants and, therefore, will receive most attention here. While the second type is important in the processing of certain foods (such as wine, vinegar, cheese, butter, and fermented milk products), and should not be overlooked in a thorough study of bacteriology, their control should be left to the processors of such items or to the student of such processes.

Bacteria are classed as plants rather than animals. Temperature conditions, light, moisture, oxygen, and food supply—all affect the growth and the reproduction of bacterial life. Bacteria vary according to their type in their susceptibility to changes under the above conditions, and for this reason, extreme care is required in the measures taken for their ultimate control. As an example, each type of bacteria has a maximum and a minimum temperature under which they will survive, and the optimum temperature (that is, the temperature at which they thrive best) varies with different species. As a result, some species thrive best when temperatures are near the freezing point of water while others multiply most rapidly at temperature levels approaching the boiling point. It is universally accepted, however, that steam under pressure at 257 F (125 C) will kill all known bacteria if held for at least 15 minutes. At the other extreme, most bacteria are rendered dormant or considerably retarded in their activity when subjected to temperatures below freezing. If held at levels of −10 to 0 F, all known types will become ineffective in their food destruction.

Under certain conditions, the temperature alone is not the governing factor in bacterial destruction, the length of time of exposure being of utmost importance.

Bacteria are the agents of fermentation and decay under favorable temperature conditions. Fermentation of carbohydrates, such as make up the cellular tissues of fruit, decomposition of nitrogenous compounds (proteins) of meat, known as putrefaction, and souring of milk are all results of bacterial activity that render such products unfit for human consumption.

The products of bacterial reproduction are heat, gases, pigments, light, and enzymes. As bacteria are entirely dependent on soluble substances as food for their existence, enzymic action becomes necessary in the life cycle of these organisms. While autolysis or self-deterioration may take place through enzymic action alone, decay of food substances generally requires the action of both enzymes and bacteria.

While dry heat of sufficient intensity will destroy bacteria, moist heat of similar intensity will kill them in a much shorter space of time. Therefore it has been found that steam is more effective in bacterial destruction than dry hot air.

Many species of bacteria pass through a spore phase when their cell wall becomes thickened. At this time they are capable of enduring extreme heat

TABLE 113. SOME VARIETIES OF BACTERIA FOUND IN FOODS, PLANTS, WATER, AND SOIL*

Name	*Habitat*
Bacillus bulgaricus	Fermented milk products
Bacillus pleofructi	Fermenting vegetables and plants
Acetobacter plicatum	Wine
Bacterium pasteurianum	Vinegar, beer, and wort
Bacillus punctatus	Water
Bacillus formosus	Soil
Bacterium aceti	Vinegar
Bacillus liquefaciens	Water
Bacillus anthracoides	Putrefying material
Bacillus (lactis) agglomeratus	Milk
Bacillus butyricus	Sour milk
Bacillus carotovorus	Carrots, cabbage, eggplant, etc. Causes soft rot
Bacillus citrimaculans	Citrus fruit. Spot disease
Bacillus tracheiphilus	Cucumbers, cantaloupe, muskmelons, pumpkins, squash. Causes wilt
Erwinia ananas	Pineapple. Causes rot
Bacterium apatatum	Sugar beets
Bacillus campestris	Cabbage, cauliflower, and related vegetables
Bacterium Jaggeri	Celery
Bacterium uliformica	Pears, beans
Bacillus proteus vulgaris	Putrifying substances. Causes food poisoning
Streptococcus thermophilus	Milk and milk products. Started for Swiss cheese
Bacterium Guntheri	Causes milk to sour
Bacillus botulinus	Produces a toxin in nonacid vegetables
Bacillus typhi	Causes typhoid fever. (Killed in frozen foods stored at 0 F). Some humans are natural carriers and must not handle foods

* Names of bacteria used are old terminology but easily identified. New terminology carries the genus of the various species as a prefix. Example: New terminology for *Bacterium aceti* is *Acetobacter aceti*; for *Bacillus liquefaciens* it is *Achromobacter liquefaciens*; for *Bacillus bulgaricus* it is *Lactobacillus bulgaricus*, etc. Bacteria are classified under 36 different genera.

or cold conditions without fatal injury. In this form they are very difficult to destroy unless subjected to high temperatures for prolonged periods.

As a general rule bacteria thrive best under nonacid conditions. In fact, a neutral or slightly alkaline environment is preferred by the majority of species. Nonacid substances as found in some vegetables, such as string beans, lima beans, peas, and squash, are highly susceptible to bacterial action. Salt solution, especially that of higher than 9 percent concentration, is very effective against bacterial life.

Darkness is desirable to most of the species of bacteria, while sunlight is usually fatal to all types.

Some understanding can be gained of the vast number of species of bacteria that are known to exist when some of the chemical changes for which they are responsible, and that take place in milk under ideal conditions, are considered, viz.:

1. Many species cause no change.
2. Some bacteria change flavor without changing the appearance of milk.
3. Others change both flavor and appearance.
4. Some types cause decomposition of the casein and the albumin, causing putrefaction accompanied by undesirable odors.
5. One species causes *"bloody" or red milk.*
6. Another species causes *blue milk.*
7. *Yellow milk* is the responsibility of still another species.
8. Eighteen species have been identified as causing *ropy milk.*

The fact is that *over 200 different species of bacteria have been found in milk alone.* Although bacteria are identified according to the chemical change they cause rather than their physical characteristics, they nevertheless can be classed in three groups as indicated by their shape. These are: (*a*) the bacillus (rod-shaped), (*b*) the coccus (ball-shaped), and (*c*) the spirillum (spiral-shaped).

When two cocci are found adhering to each other, they are called diplococci. If a number occur together to form a chain, they are called streptococci, etc. Others may divide in two planes at right angles, appearing not unlike a cluster of grapes, which are called staphylococci, etc.

Regardless of their shape or nomenclature, bacteria require food to live, which all too often is provided through the destruction of the human food supply or some other valuable commodity. It is also necessary that their food be provided under certain limited atmospheric conditions and within definite states of chemical environment. It is the responsibility of the cold-storage plant operator to see that foodstuffs are stored under conditions which prevent the growth of bacteria and to maintain those conditions as long as practicality and economic requirements dictate.

TABLE 114. SOME MILK BACTERIA AND THEIR MAJOR EFFECT*

Bacteria	*Effect*
Acid-forming types:	
Bacillus acidi lactici	Major cause of souring
Streptococcus mastitidis	Harmful to cows (mastitis) and humans (sore throat)
Bacillus bulgaricus	Souring
Bacillus acidophilus	Usually added as a culture. Beneficial for certain human ailments. Sour milk therapy
Alkali-forming types:	
Bacillus alkalescens	No major effect
Bacterium fluorescens	No major effect
Gas-forming types:	
Bacterium lactis aerogenes	Decomposes milk proteins (casein)
Bacillus coli	Decomposes milk proteins (casein)
Bacillus multifermentans	Forms gaseous curds
Bacillus perfringens	Forms gaseous curds
Peptonizing types:	
Bacillus proteus vulgaris (potato bacillus)	Putrefaction
Bacillus subtilis (hay bacillus)	Putrefaction
Color-changing types (pigment-forming):	
Bacillus syncyaneus	Causes blue milk
Bacillus prodigiosus	Causes red milk
Bacillus pyocyaneus	Causes blue-green milk
Sarcina flava	Causes yellow milk

* There are also a number of inert bacteria that cause no change in milk.

Yeasts

Of microscopic size (about 1/3000 in. in diameter), yeasts are one-celled plants of the fungus group. Like bacteria, they are present in many places, especially in the air. For this reason, exposed foods become hosts to these micro-organisms. They are round or oval in shape and multiply very rapidly by budding to form clusters under ideal conditions. Yeasts are one of the simplest *types* of fungi classified as Ascomycetes, a major division of the simplest *group* of plants called the Thallophyta. Ascomycetes are non-chlorophyll types of sac fungi that are unable to manufacture their own food, differing in this respect from the second major division of Thallophyta, the Algae.

While destructive to fresh foods, yeasts are highly important in the processing of certain food products such as wine, beer, and bread. The reason for their action is due to their ability to transform sugars into alcohol and carbon dioxide by the secretion of enzymes responsible for fermentation. Yeasts, therefore, are responsible for all fermented fruits and fruit products.

Yeasts require air, food, and moderately warm temperatures for growth, but are killed when exposed to high heat conditions such as boiling water or moist air at 212 F.

Yeast cells are hardy and will survive for long periods when deprived of food and moisture or exposed to low temperatures. Under such conditions, the micro-organisms pass into a dormant stage and growth ceases or is retarded greatly. Extremely low temperatures are injurious to commercial yeasts, however, and should be avoided wherever this product is handled for storage purposes.

Considering that 45 F is the ideal storage temperature for commercial (compressed) yeasts, it becomes an easy problem to control such micro-organisms when storage temperatures of foodstuffs are kept at 36 F or below.

There are numerous types of yeast, just as there are numerous types of bacteria. Some types of yeasts secrete zymose, which converts simple sugar such as glucose to carbon dioxide and ethyl alcohol through fermentation action. The type of fermentation is identified by the major product of the action. For example, fermentation that produces lactic acid is known as lactic fermentation, acetic acid as acetic fermentation, etc. *Saccharomyces cerevisiae* and *Saccharomyces ellipsoideus* are yeasts important in (*a*) brewing and baking and (*b*) wine making, respectively. However, the point to remember about all yeasts is the fact that all species excite fermentation, especially in fruits. Their actions in this respect are to be eliminated entirely, if possible, or at least retarded by the cold-storage plant operator through control measures.

Molds

Molds are a class of minute plants that thrive in a damp and dark environment. Encouraged by stagnant air, mold growth becomes apparent by the cottony, threadlike structure which appears furry or fuzzy to the naked eye. Molds are commonly familiar by both their fuzzy appearance and color, the latter ranging from white through red, brown, yellow, green, blue, pink, or black. Their growth is rapid under ideal conditions of temperature, moisture, acidity, and darkness (absence of bright light).

As mold grows, threadlike fuzz appears which develops a spore sack or "ball" at the end of each thread. The spore sack contains numerous spores that are microscopic in size and which float away into the atmosphere when the sacks are broken, usually by a slight movement of air. When these individual spores find their way to the surface of food substances, they thrive and reproduce a mold growth of their own. Mold enters fruit or vegetables through skin breaks, and if conditions are right rapid growth follows, rendering such products unfit for consumption. When they attach themselves to meat surfaces, a luxuriant growth of mold plant results, in time covering the entire surface. In many instances, the molded surface of meat products can be cut away and the meat may be consumed safely unless spoiled by other agents. As a matter of fact, the meat "ripening" or tenderizing process known as aging to produce aged steaks and "the roast beef of Old England" develops a surface mold over the entire exposed area of the beef.

While most common molds are nonpoisonous, some types are injurious to health. All molds are bitter in taste.

Molds thrive in damp locations. One location in cold-storage plants that is conducive to their rapid growth is the walls in the vicinity of doors leading directly from the corridors to the outside atmosphere. When these doors are open, they allow comparatively warm, moisture-laden air to enter the plant. When contact of this air is made with the cold walls, condensation occurs, leaving wet surfaces to act as an ideal incubator for molds. Mold-resistant paints, ultraviolet light, or constant cleaning is necessary to eliminate this condition.

Oxygen is essential to mold growth. Molds thrive in the presence of acids such as are found in fruit and fruit juices, and are the principal causes of spoilage in citrus fruits and apples.

Mold is prevalent on most surfaces where temperatures are above the freezing point of water. Where temperatures are below 32 F, the reproduction of mold life is retarded greatly or ceases altogether, due more to the absence of free moisture than to the low-temperature level. However, boiling temperatures or prolonged heat below boiling will kill all forms of molds.

TABLE 115. SOME MOLDS AND THEIR MAJOR HOSTS

Name	*Host and Effect*
Rhizopus nigricans	Bread mold
Penicillium expansum	Soft rot of apples in storage
Penicillium italicium	Spoilage of citrus fruit
Penicillium camemberti	Ripening of cheese (camembert)
Penicillium brevicaule	Spoilage of dairy products, including cheese
Aspergillus flavus	Grain mold
Aspergillus niger	Spoilage of general foods

Other Causes of Food Deterioration

Spoilage is a condition which renders foodstuffs unfit for human consumption. Deterioration, however, can be defined as either a definite phase in the journey toward spoilage or any phase in the change of state which is below a level of quality considered best for the particular item in question. Thus, on one hand stand food substances at the eminence of their quality, while on the other is the same substance that has undergone such chemical changes as to render it useless for any purpose. Between the two extremes at any point a phase of deterioration will be found. If deterioration is caused by micro-organisms, the journey between the two extremes will be short and chances are that the stage of unfitness will be reached quickly. However, if deterioration is brought about mechanically, for reasons other than enzymic or micro-organismic action, the result is in the lowering of the quality level. How far below depends on the nature of the mechanical abuse and the length of time the abuse is allowed to continue. A partial list of such abuses is shown below:

Bruising. The substance is broken by impact or squeezing. Causes are due generally to dropping or packing too tightly. Bruising is prevalent in most fruits and some vegetables.

Desiccation. The moisture content of a food substance is decreased below normal shrinkage allowances by excessive evaporation when exposed to dry atmospheric conditions. Loss of weight, shriveling, and wilting of fruits and vegetables are results of desiccation. Loss of weight in meats increases greatly when stored in air containing too little moisture. More important, however, in respect to quality in the case of meats, is the loss of firmness when its moisture content is reduced. Eggs lose considerable weight due to evaporation through the porous shell when the surrounding air is too dry.

Freezing. Except where proper steps are taken to follow out a planned program, such as found in the frozen-food industry, the freezing of fruits and vegetables means spoilage and total loss in many cases. The answer to this lies in the resultant difference in crystal size when such products are frozen slowly as compared with those subjected to rapid temperature reductions. Fruit and vegetable substances are made up of numerous cells that contain protoplasm, which, in turn, contains water. When frozen slowly, this water is drawn into the intercellular spaces and forms ice crystals that expand greatly during the freezing process. Sufficient expansion ruptures the membranes that act as a sack for each cell, allowing heavy losses of its moisture content upon thawing because of its inability to reabsorb such moisture in its damaged condition. The result is a dry, pithy substance made to appear even less appetizing by reason of its discoloration. Slight freezing does not definitely injure some products. Nevertheless, it should be avoided for the sake of good quality.

Although the cells of meats are affected similarly when frozen slowly, their membranes appear to be tougher and more elastic than those of fruits and vegetables. However, the fact that meat does not escape the effects of freezing entirely is borne out by the amount of bleeding that takes place upon thawing. Alternate freezing and thawing of meat greatly reduces its tender and tasting qualities and such practices should be avoided. Once frozen, all meats including poultry and fish should be kept in that condition until prepared for cooking.

Absorption of Odors. Butter, eggs, cheese, and meats, as well as some fruits, readily absorb foreign odors, such as are given off by other products or from gases that enter cold-storage plants from outside sources. Certain losses in eating qualities are the direct result of such contamination. Avoiding exposure of sensitive products during storage periods to others having distinctive odor characteristics will eliminate this factor to a large extent. The use of brines for cooling purposes rather than direct expansion ammonia systems will also insure protection against loss through gas contamination in the event of leaks. A thorough knowledge of all products in order to determine those that are companionable in respect to odors will prove an asset when decisions concerning storage locations for various items are to be made.

Incorrect Storage Temperatures. The effect of incorrect storage temperatures on food products—fruits and vegetables especially—should be considered from a physiological standpoint and not from the effects of micro-organisms or mechanical abuses. Deterioration is caused by chemical activity within the food substances, varying with each item and also with the origin and the pre-handling methods of any particular item. It is the unknown factors so prevalent in the handling of many such products that make the determination of correct storage conditions an uncertainty in many instances. It is known, for example, that apples suffer from low-temperature diseases when stored below their critical storage temperatures. Some varieties are susceptible to soggy breakdown at temperatures below 35 F, while others develop internal browning if stored below 37 F for extended periods. At 32 F, still other varieties will suffer from the disease called brown core if held too long.

Potatoes show harmful effects of low-temperature storage through the development of black heart symptoms after several weeks of exposure.

Citrus fruits such as grapefruit, on the other hand, are affected adversely when stored at relatively high temperatures. Rind pitting usually results under such conditions. Scald and watery breakdown in citrus fruits are chilling injuries and are prevalent when such products are exposed to temperatures below the critical point.

The after-affects of low-temperature storage must also be considered. Green tomatoes, if received in good condition, can be kept for 10 to 15 days without any apparent deterioration at 40 F. During this period ripening and coloring will be delayed but will be completed when again exposed to warm temperatures. However, if storage at 40 F or below is carried on for protracted periods beyond the 10 to 15 days' interval and they are then exposed to elevated temperatures, rapid breakdown of the tomatoes will take place before the ripening and coloring processes are satisfactorily completed.

Fruits or vegetables that are allowed to continue their ripening processes after harvesting for more than 24 or 48 hours will not keep as long in storage as the same product placed immediately under refrigeration. The keeping qualities may mean a difference of several weeks and possibly months of safe storage in some instances.

Product spoilage brought about by diseases as a result of natural influences when exposed to improper atmospheric conditions is often believed to be caused by freezing. This is a natural conclusion, considering the similarity in appearance of substances affected by either cause. However, constant vigilance will dispel such conclusions in most cases.

These examples are but few of the many harmful effects brought about by exposure of fruits and vegetables to incorrect temperatures. It must be remembered that each item has its own characteristics as related to the temperatures to which it is exposed, and that such characteristics must be considered in order to provide the correct condition for its safe storage for reasonable time intervals.

Each item, depending on the type, variety, physical condition, and purpose for which it is stored, demands a specific atmospheric environment measured in terms of temperature and moisture content to which it will respond for best storage results. Such conditions are called the *optimum* conditions for that individual item and the temperature at which it has been found to be the most effective in maintaining a high level of quality is known as the *optimum temperature* for that commodity. Temperatures denoted in storage tables are considered as indicative of the optimum temperatures for such products listed in Table 118.

Respiration. Fruits and vegetables continue to carry on their living functions even after they are separated from the plant, bush, or tree which produces them. This is true whether such items are picked green or fully ripened. Their life cycle embraces the function of absorbing oxygen and liberating carbon dioxide. What happens in the interim concerns the chemical activity necessary to produce the substances that convert raw materials taken from the soil and the air into palatable foods. If allowed to continue, such processes eventually disturb the chemical balance which is reached at the time of ideal ripeness and cause an increase in compounds that become less palatable as such changes proceed. Thus respiration continues in harvested products by promoting such changes to the substances stored within them during their growing periods.

Respiration consists of several separate but related processes, viz.: (*a*) oxygen is absorbed from the air, (*b*) carbohydrates become oxidized, (*c*) carbon dioxide, heat, and water are released.

Reducing the amount of oxygen available to the product diminishes the vigor of respiration. This can be accomplished by exposing fruits and vegetables to an accumulation of carbon dioxide within the surrounding atmosphere. Much work in this direction has been done through experimentation in the design of gas-tight compartments having controlled ventilation to bring about the proper proportions of oxygen and carbon dioxide. Carbon dioxide in such experiments has been supplied either artificially or by natural emanation from the products themselves.

Respiration is also retarded when fruits and vegetables are exposed to low temperatures and halted completely when they are frozen.

It may be noted here that the ripening process of fruit is hastened by exposure to ethylene gas.

CONTROL MEASURES TO BE TAKEN TO ALLAY OR TO PREVENT SPOILAGE OF FOODS

Obviously, the prerequisite for accurate control of any activity is a thorough understanding of fundamental principles involved. This rule was proved to be true in the operation of the refrigeration cycle and in brine-system maintenance, principles of which were discussed in previous chapters. The same rule applies to perishable-food preservation control to an even more important degree, considering that spoilage in food substances becomes final once it has taken place. Corrective measures do not exist for the rectification of damaged or spoiled foods and any errors in their processing or storage, through either carelessness or ignorance of the facts, may result in the total loss of a shipment or even an entire crop.

The principal reasons for the spoilage of food are now known to be the result of enzymic and micro-organismic activity. That these agents of destruction thrive best under certain physical, chemical, and temperature conditions has been proved by experiment and experience. It is also true that the requirements of various species of micro-organisms may or may not be similar in respect to their ideal environment. This latter is important and one which does not always present an advantage in the processing or storage of perishable commodities. Fortunately, however, all destructive agents are vulnerable to certain extreme physical, chemical, or temperature conditions in that they are greatly retarded in their activity or eliminated entirely when so exposed.

The methods employed in arriving at such extreme conditions detrimental to the life and/or the activity of these agents vary and depend on the means at hand or the final result desired. Before any procedure for food preservation can be adopted, first the characteristics of each type of spoilage agent must be determined. If both favorable and unfavorable conditions to the life processes of all spoilage agents are known, it remains a simple procedure of developing the mechanics of organizing methods of control by providing an unsatisfactory environment within and surrounding food substances.

Basically, the problem of food preservation narrows down to two important factors: (*a*) what are the conditions under which the forces of destruction may be retarded or eliminated? and (*b*) how may these conditions be brought about?

By establishing the known conditions, both favorable and unfavorable, to the growth and the activity of all spoilage agents, the answer to the first question may be determined. Table 116 lists all such conditions that are considered relevant to the problem of food preservation through the control of enzymes, bacteria, yeasts, and molds.

TABLE 116. CONDITIONS, FAVORABLE AND UNFAVORABLE, TO THE GROWTH AND ACTIVITY OF FOOD SPOILAGE AGENTS

Spoilage Agent	*Favorable Conditions for Growth and Activity*	*Unfavorable Conditions Resulting in Elimination or Retardation for Growth and Activity*
I. Enzymes.......	Optimum temperature Free oxygen (air) Moisture present Slight acidity	Low temperature Boiling temperature Strong acidity Strong alkalinity Absence of moisture
II. (*a*) Bacteria....	Slight alkalinity Moisture Optimum temperature High enzymic action	Low temperature Strong acidity Strong alkalinity High temperature Concentrated salt solution Absence of moisture Low enzymic action Concentrated sugar solution Sunlight or ultraviolet ray Antiseptics present
(*b*) Yeasts.....	Optimum temperature Free oxygen Moisture	Low temperature High temperature Sugar present
(*c*) Molds.....	Dampness Stagnant air Low light or darkness Slight acidity	Sunlight or ultraviolet ray High temperature Dryness

By careful analysis of conditions governing the processes of all spoilage agents as outlined in Table 116, it becomes apparent that their control may be affected by various methods. Bacteria, for example, are shown to be vulnerable to extreme temperature conditions of either a high or a low level. To provide the one, sufficient heat to boil the food substance must be applied, whereas the second method requires the use of refrigeration to reduce temperatures.

To afford a better understanding of the spoilage-control problem, it is appropriate to discuss briefly the most important and successful methods universally employed for the preservation of foods.

Canning. This process involves the application of heat to food substances, raising the temperature to a level fatal to all enzymes and micro-organisms and holding in that state for the period necessary to destroy all bacterial spores. Packing in sterilized containers while still in its heated condition and hermetically sealing assures a high state of preservation for indefinite periods. The process produces an unfavorable environment that proves fatal to all forms of spoilage agents by means of exposure to high temperatures. Secondly, by packing in sterilized containers which subsequently are sealed against the entrance of air, protection of the product from re-contamination by micro-organisms present in the air and on the surfaces of the container is provided.

The degree of temperature and the length of time necessary for its application depend on the type of product being processed. Most forms of bacteria life succumb when subjected to the temperatures of boiling water (212 F) for short periods of time, but because of the greater resistance to heat on the part of the spores, higher temperatures or prolonged heat application is necessary. Because quality is closely related to the length of the heating period, shorter processing time is accomplished through higher temperatures developed by means of pressure cooking. Thus bacteria requiring prolonged exposure to boiling temperatures are destroyed more quickly when subjected to higher temperatures obtained under pressure. Nonacid vegetables and meats are excellent examples of products that require extraordinary measures for their processing to assure safe preservation. As a result, such products are processed under pressure to kill all micro-organisms while maintaining a high state of quality.

Fermentation. This process, better known as pickling, is accomplished by submerging certain vegetables and meats in salt brine and keeping them in that solution during both the fermentation and the storage periods. The sugar contained within the food substances is gradually converted to lactic acid during the fermentation period while the salt, because of its preservative powers, controls bacterial action. When fermentation is complete, foods that ordinarily would support both bacteria and yeasts are exhausted.

Salt is used most generally for several reasons: because lactic acid bacteria are tolerant to salt and because it is an excellent curing agent for such products. Brines of sufficient salt concentration (9 percent and over) discourage the majority of spoilage agents, a characteristic which enhances the value of salt as a preservative.

During the process of fermentation, however, brines increase in acidity due to the formation of lactic acid, and in doing so become tolerant to mold development at the exposed surface. This can be avoided by spreading a layer of white mineral oil over the surface of the brine.

Solutions having a high concentration of either salt or sugar are not conducive to bacterial life. This is due to the difference in density between the solution and the cell protoplasm of the micro-organism resulting in extraction of water from the latter. The action is known as *osmosis*.

TABLE 117*†

Directions for Drying Fruit with Artificial Heat

Temperature: Start at 160 F to 175 F, decreasing gradually to 150 F

Fruit	*Preparation*	*Treatment*	*Appearance When Dried and Estimated Drying Time*
Apples (late variety best) and pears	Wash, pare, core. Remove blemishes. Cut into ¼-in. slices or rings	Put in salt solution (3 to 5 tsp salt to 1 gal water) for 10 min, or steam 5 to 7 min, or sulfur 30 min. Spread evenly in thin layers on drier trays	Dry, springy. Drying takes 5 to 6 hr
Apricots and peaches	Cut in half, remove stones. Leave skins on apricots	Steam 5 to 7 min or sulfur 30 min. Place stone side up on trays	Dry, leathery. Drying takes 10 to 20 hr
Cherries and berries	Pick fruit when ripe. Do not pit cherries	None. Or steam 2 min. Spread fruit in thin layers on trays. (Temperature should not rise above 150 F.) Do not sulfur	Dry and rattle on tray. Drying takes 8 to 12 hr

Directions for Drying Vegetables with Artificial Heat

Temperature: Start at 150 F to 160 F, decreasing gradually to 140 F to 150 F

Vegetable	*Preparation*	*Treatment*	*Appearance When Dried and Estimated Drying Time*
Beans (snap)	Prepare beans as for cooking, cutting them in such a way as to prevent beans from falling out of pod. Cut in ¾-in. lengths	Steam or blanch 5 min. Place on drying trays	Dry, tough, or brittle. Drying takes 8 to 12 hr
Beets (dark red, solid color best)	Slice or dice after precooking	Cook 20 to 30 min in steam before peeling	Dry and brittle Drying takes 8 to 12 hr
Carrots	Slice or dice after precooking	Steam or blanch 6 min. Place on drying trays	Dry and brittle Drying takes 8 to 12 hr
Corn (any sweet variety)	Select corn in the milk stage. Dry as soon as possible after picking	Steam or blanch 10 to 15 min or until milk is set. Cut corn from cob. Spread ½ to ¾ in. deep on trays to dry	Dry, brittle Drying takes 5 to 10 hr
Greens (spinach, kale, beet tops, and other greens)	Gather when ready for table use. Wash thoroughly. Remove stems, and strip out midrib where it is thick and heavy	Steam or blanch 2 to 4 min. Pile loosely on trays not more than 1 in. deep. Dry at temperature not to exceed 150 F	Dry, crisp Drying takes 6 to 10 hr
Peas and other legumes	Gather when mature, but when pods are not dry	Steam or blanch 10 min. Drain and spread on trays in layers	Dry, hard, wrinkled. Drying takes 8 to 12 hr

* While this table covers domestic processing, the same principles apply commercially. Drying time will vary with velocity of hot air when foods are exposed to forced draft.

† Courtesy, Agricultural Extension Service, Pennsylvania State University, Circular 247.

Dehydration. This process of drying or removal of a sufficient amount of moisture to insure preservation of foods is carried out by natural (sun and air) or artificial (ovens) means. Enzymes and micro-organisms of all types require moisture in order to continue their activities. By eliminating free moisture from food substances, an unfavorable environment exists that discourages such activity. Products subjected to the dehydration process can be maintained in an excellent state of preservation if stored in a cool, dry location. Greater osmotic pressure from within the food is exerted due to increased concentrations of normal solutes as the result of moisture removal.

While all minerals are preserved in dried fruits and vegetables, vitamin content is reduced considerably. Sulfuring some fruit preserves their natural flavor and color.

Smoking. During the process of exposing meat, meat products, and fish to the smoke of burning wood within a confined chamber, the food substance dries and becomes impregnated with an antiseptic present in wood smokes. Creosote is the antiseptic most generally found and one which is highly effective in micro-organismic control. The depth of penetration becomes greater with prolonged exposure, which, in turn, increases preserving qualities. Large meat packers regulate the number of days of exposure in accordance with the climatic conditions of the locality for which the product is destined. Shipments of meats to tropical climates are subjected to longer periods of exposure to smoke than those destined for more temperate or colder locations.

Preservatives. Salt, sugar, vinegar, and certain spices are all employed in the preservation of foods. Because such materials encourage unfavorable conditions for the growth of micro-organisms, their use in connection with such efforts has been universally accepted for centuries. Such unfavorable conditions are brought about through osmosis, acidity, or the antiseptic characteristics possessed by the materials employed. Sugar-cured hams and salt pork are examples of foods treated with preservatives. Jams and jellies are inhibited against fermentation because of the large percentage of sugar used in their processing. Commercial preservatives include such chemicals as benzoate of soda, saltpeter, and borax.

Cold Storage. This is the method of preserving perishable commodities in their fresh and wholesome state for extended periods by providing and controlling proper temperature and humidity conditions within the storage compartment. Cold storage exploits the effect of low temperature in retarding or arresting the activities of natural forces that cause food substances to deteriorate and decay. Because normal atmospheric conditions are such that temperature and humidity conditions are seldom at a level conducive to the safe and prolonged storage of perishable foods, it is necessary that artificial means be provided to produce such an environment. It is this method of preserving such products as meats, meat products, fish, dairy products, fruit, and vegetables that requires the employment of refrigeration equipment to provide the various degrees of low temperatures demanded.

The effort to maintain desired qualities in the numerous items adaptable to cold storage becomes more complex when it is realized that all products do not respond favorably to similar atmospheric conditions. Where economical results are expected, this factor must be recognized.

Actual cold-storage conditions are determined first by the type of food to be stored and second by the length of time such foods are to be held.

In general, types of foods are divided into two principal groups:

1. Products in whose substance living processes continue and on which low-temperature exposure has only a retarding effect, the degree of which depends on the temperature level. Fruits and vegetables are important examples of foods in this category in which life itself acts as a retarder to bacterial growth and activity.

2. Nonliving substances such as meats, meat products, fish, and dairy products that are highly susceptible to the activity of spoilage agents. Lacking any natural means of self-preservation, these products deteriorate rapidly unless drastic preventive measures are taken.

Whether products of either group are to be stored in a frozen or an unfrozen state depends chiefly on the length of time they are to be held, which may be for either short or prolonged periods. There are exceptions, of course, to this rule, as all uncooked products cannot be frozen without irreparable damage—potatoes being an excellent example.

Maintaining foods in temporary storage prescribes temperatures above their freezing point, which depends on the individual commodity. Items held in temporary storage will retain their fresh qualities for longer periods when favorable temperature and humidity conditions are provided. The definition of temporary storage varies with each item, as shown in Table 113, which lists vast differences in possible storage periods when products are exposed to their optimum atmospheric conditions.

Prolonged storage periods necessitate freezing of most commodities in which state they can be held indefinitely. Temperatures in this case are maintained considerably below freezing, usually at levels ranging from 10 F to −10 F or lower. Meats and other nonliving substances as a rule present few problems in their preparation for freezer storage, whereas fruits and vegetables require considerable processing before they can be frozen in order to safeguard them against deterioration while in storage.

It is obvious that cold-storage facilities must be designed to provide, as nearly as practicable, the various optimum conditions required by fresh foods for temporary storage, as well as the extremely low temperatures demanded for the prolonged storage of frozen products. To meet this situation, storage plants employ several systems for the safe preservation of the numerous items involved, each of which is separated from the other because of temperature differentials, but usually maintained by common equipment. The principal systems employed within general cold-storage plants are the *coolers* or *chill rooms* and *freezers.* As their names imply, coolers are maintained at temperatures above 32 F while freezers are held considerably below the freezing point. Here, again, a distinction must be made concerning the method of freezing employed, which is governed by the type of product to be stored. There are two widely employed methods in use today. The first, and the one most generally found in plants where storage is the main function, is called "*sharp freezing*" and the compartments identified as "*sharp freezers.*"

Quick freezing is the second method employed and one which has gained rapidly in importance during recent years. It accomplishes desired results in two hours or less at −50 F. This method of freezing is utilized largely by processors of fresh foods such as fish, fruits, and vegetables and also eggs, poultry, and meats. After the freezing process is completed, items are placed in compartments called *holding freezers* in which temperatures are maintained at a level above that required for freezing, usually 0 F. Sharp freezers usually are employed as holding freezers if the correct temperature is maintained constantly.

The principal difference between sharp freezing and quick freezing is in the size and the number of ice crystals formed within the food substances during solidification. Because of the rapid rate at which the crystalline structure is formed during quick freezing, cellular expansion is greatly reduced, decreasing the number of ruptured membranes and preventing the loss of valuable nutritious fluids. Ice crystals start forming within food substances at about 30 F and continue until the major portion of the moisture is solidified at approximately 25 F. Even at this temperature there remains a small quantity of highly concentrated fluid which contains dissolved quantities of phosphates, sulfates, chlorides, and other chemical compounds that form animal and plant substances. This fluid is the result of water detraction through crystallization by accretion, and continues to increase in concentration as temperature reduction takes place. Actually, some extremely concentrated fluid still remains unless food substances are exposed to temperature levels approaching −75 F, a point conducive to other harmful effects such as "bleaching." The range of temperature between 30 F and 25 F is referred to as the *zone of maximum water crystallization,* and the more rapidly heat reduction can be accomplished through this zone, the more desirable will be the result. Because the cellular tissues of nonliving substances are considerably tougher and more elastic than those that retain their life processes, the rapidity in accomplishing temperature reduction through the zone of water crystallization is not so important in meats and meat products as in the processing of fruits and vegetables. However, this should not be construed to mean that improved

TABLE 118. STORAGE REQUIREMENTS AND PROPERTIES OF PERISHABLE FOODS[1]

Commodity	Storage temperature, F	Relative humidity, Percent	Approximate storage life	Water content, Percent	Average freezing point, F	Specific heat above freezing[3]	Specific heat below freezing[3]	Latent heat (calculated)[4] Btu
Apples	30 to 32	85 to 90		84.1	28.4	.87	.45	121
Apricots	31 to 32	85 to 90	1 to 2 wks	85.4	28.1	.88	.46	122
Artichokes								
Globe	31 to 32	90 to 95	1 to 2 wks	83.7	29.1	.87	.45	120
Jerusalem	31 to 32	90 to 95	2 to 5 mos	79.5	27.5	.83	.44	114
Asparagus	32	90 to 95	3 to 4 wks	93.0	29.8	.94	.48	134
Avocados	45 to 55	85 to 90	4 wks	—	27.2	—	—	—
Bananas	—	—	—	74.8	—	.80	.42	108
Beans								
Green or snap	45	85 to 90	8 to 10 days	88.9	29.7	.91	.47	128
Lima	32 to 40	85 to 90	10 to 15 days	66.5	30.1	.73	.40	94
Beer, barrelled	32 to 40	—	several months	—	—	—	—	—
Beets								
Bunch	32	85 to 90	10 to 14 days	—	—	—	—	—
Topped	32	90 to 95	1 to 3 mos	87.6	26.9	.90	.46	126
Blackberries	31 to 32	85 to 90	7 to 10 days	85.3	28.9	.88	.46	122
Broccoli, sprouting	32 to 35	90 to 95	7 to 10 days	89.9	29.2	.92	.47	130
Brussels sprouts	32 to 35	90 to 95	3 to 4 wks	84.9	—	.88	.46	122
Cabbage, late	32	90 to 95	3 to 4 mos	92.4	31.2	.94	.47	132
Candy	0 to 33	40 to 65	—	—	—	—	—	—
Carrots								
Bunch	32	85 to 90	10 to 14 days	—	—	—	—	—
Prepackaged	32	85 to 90	3 to 4 wks	—	—	—	—	—
Topped	32	90 to 95	4 to 5 mos	88.2	29.6	.90	.46	126
Cauliflower	32	85 to 90	2 to 3 wks	91.7	30.1	.93	.47	132
Celeriac	32	90 to 95	3 to 4 mos	88.3	—	.91	.46	126
Celery	31 to 32	90 to 95	2 to 4 mos	93.7	29.7	.95	.48	135
Cherries	31 to 32	85 to 90	10 to 14 days	83.0	—[2]	.87	.45	120
Corn, sweet	31 to 32	85 to 90	—	73.9	28.9	.79	.42	106
Cranberries	36 to 40	85 to 90	1 to 3 mos	87.4	27.3	.90	.46	124
Cucumbers	45 to 50	90 to 95	10 to 14 days	96.1	30.5	.97	.49	137
Currants	32	80 to 85	10 to 14 days	84.7	—	.88	.45	120
Dairy products								
Cheese	35	65 to 70	—	37 to 38	Approx. 28.0	.50	.31	54
Butter	45	80 to 85	2 mos	15.5 to 16.5	—	.33	—	23
Butter	−10 to −5	80 to 85	1 yr	—	—	—	.25	—
Cream (sweetened)	−15	—	several months	—	—	—	—	—
Ice cream	−15	—	several months	—	—	—	—	—
Skim milk								
Dried	40	—	several months	—	—	—	—	—

Unsweetened	−15	—	short time	—	—	—	—	—
Sweetened	35	—	several months	—	—	—	—	—
Dates	—	—	—	20.0	−4.1	.36	.26	29
Dewberries	31 to 32	85 to 90	7 to 10 days	—	29.2	—	—	—
Dried fruits	32	50 to 75	9 to 12 mos	—	—	.30 to .32	—	17 to 21
Eggplant	45 to 50	85 to 90	2 to 3 weeks	92.7	30.4	.94	.48	132
Eggs								
Shell	29 to 31	85 to 90	9 mos	67.0	Approx. 31.0	.74	.40	96
Shell, farm cooler	40 to 55	75	—	67.0	Approx. 31.0	.74	.40	96
Frozen	−10 to 0	—	—	73.0	Approx. 31.0	—	.42	104
Dried, whole	35	Low as possible	6 mos	6.0	—	.25	.21	9
Dried, yolk	35	Low as possible	6 mos	—	—	—	—	—
Dried spray albumen	35	Low as possible	6 mos	Up to 6.0	—	.25	—	9
Fermented albumen	Room temp.	Low as possible	—	3 to 15	—	.22 to .32	—	4 to 21
Endive (escarole)	32	90 to 95	2 to 3 wks	93.3	30.9	.94	.48	132
Figs								
Dried	40 to 45	65 to 75	9 to 12 mos	24.0	—	.39	.27	34
Fresh	28 to 32	85 to 90	5 to 7 days	78.0	27.1[1]	.82	.43	112
Fish								
Fresh	33 to 40	90 to 95	5 to 20 days	—	—	—	—	—
Frozen	−10 to 0	90 to 95	8 to 10 mos	—	—	—	—	—
Smoked	40 to 50	50 to 60	6 to 8 mos	—	—	—	—	—
Brine salted	40 to 50	90 to 95	10 to 12 mos	—	—	—	—	—
Mild cured	28 to 35	75 to 90	4 to 8 mos	—	—	—	—	—
Frozen-pack fruits	−10 to 0	—	6 to 12 mos	—	—	—	—	—
Frozen-pack vegetables	−10 to 0	—	6 to 12 mos	—	—	—	—	—
Garlic, dry	32	70 to 75	6 to 8 mos	74.2	25.4	.79	.42	106
Gooseberries	31 to 32	80 to 85	3 to 4 wks	88.3	28.9	.90	.46	126
Grapefruit	32 to 50	85 to 90	4 to 12 wks	88.8	28.4	.91	.46	126
Grapes								
American type	31 to 32	85 to 90	3 to 8 wks	81.9	27.5	.86	.44	116
European type	30 to 31	85 to 90	3 to 6 mos	81.6	24.9	.86	.44	116
Honey	—	—	—	—	—	—	—	—
Hops (double compression)	38 to 40	50	several months	—	—	—	—	—
Horseradish	32	90 to 95	10 to 12 mos	73.4	26.4	.78	.42	104
Kohlrabi	32	90 to 95	2 to 4 wks	90.1	30.0	.92	.47	128
Lard (without antioxidant)	45	90 to 95	4 to 8 mos	—	—	—	—	—
"	0	90 to 95	12 to 14 mos	—	—	—	—	—
Leeks, green	32	85 to 90	1 to 3 mos	88.2	29.2	.90	.46	126
Lemons	32, 55 to 58	85 to 90	1 to 4 mos	89.3	28.1	.92	.46	127
Lettuce	32	90 to 95	3 to 4 wks	94.8	31.2	.96	.48	136
Limes	45 to 48	85 to 90	6 to 8 wks	86.0	29.3	.89	.46	122
Logan blackberries	31 to 32	85 to 90	7 to 10 days	82.9	29.5	.86	.45	118
Malt syrup	—	—	—	—	—	—	—	—
Maple syrup	—	—	—	—	—	—	—	—
Meat								
Bacon—Fresh	−10 to 0	90 to 95	4 to 6 mos	—	—	—	—	—
Smoked	60 to 65	85	4 to 6 mos	13 to 29	—	.30 to .43	.24 to .29	18 to 41

TABLE 118—(*Continued*)

Commodity	Storage temperature, F	Relative humidity, Percent	Approximate storage life	Water content, Percent	Average freezing point, F	Specific heat above freezing[5]	Specific heat below freezing[5]	Latent heat (calculated)[6] Btu
Beef—Fresh	32 to 34	88 to 92	1 to 6 wks	62 to 77	28 to 29	.70 to .84	.38 to .43	89 to 110
Frozen	−10 to 0	90 to 95	9 to 12 mos	—	—	—	—	—
Fat backs	38 to 40	85 to 90	0 to 3 mos	—	—	—	—	—
Hams and shoulders								
Fresh	32 to 34	85 to 90	7 to 12 days	47 to 54	28 to 29	.58 to .63	.34 to .36	67 to 77
Frozen	−10 to 0	90 to 95	6 to 8 mos	—	—	—	—	—
Cured	60 to 65	75 to 80	0 to 3 yrs	40 to 45	—	.52 to .56	.32 to .33	57 to 64
Lamb—Fresh	32 to 34	85 to 90	5 to 12 days	60 to 70	28 to 29	.68 to .76	.38 to .51	86 to 100
Frozen	−10 to 0	90 to 95	8 to 10 mos	—	—	—	—	—
Livers	−10 to 0	90 to 95	3 to 4 mos	70.0	—	—	—	—
Pork—Fresh	32 to 34	85 to 90	3 to 7 days	35 to 42	28 to 29	.48 to .54	.30 to .32	50 to 60
Frozen	−10 to 0	90 to 95	4 to 8 mos	—	—	—	—	—
Sausage								
Smoked	40 to 45	85 to 90	—	—	—	—	—	—
Casings	40 to 45	85 to 90	—	—	—	—	—	—
Veal	32 to 34	90 to 95	5 to 10 days	70 to 80	28 to 29	.76 to .84	.42 to .51	100 to 114
Melons								
Cantaloup and Persian	45 to 50	85 to 90	1 to 2 wks	92.7	29.0	.94	.48	132
Honey Dew and Honey Ball	45 to 50	85 to 90	2 to 4 wks	92.6	29.0	.94	.48	132
Casaba	45 to 50	85 to 90	4 to 6 wks	92.7	29.0	.94	.48	132
Watermelons	36 to 40	85 to 90	2 to 3 wks	92.1	29.2	.97	.48	132
Mushrooms	32 to 35	80 to 85	2 to 3 days	91.1	30.2	.93	.47	130
Nuts, dried	32 to 50	65 to 75	8 to 12 mos	3 to 6	—	.22 to .25	.21 to .22	4 to 8
Oil (vegetable salad)	55	—	6 mos	14.4 to 15.4	—	—	—	—
Oil (vegetable salad)	35	—	1 yr	—	—	—	—	—
Oleomargarine	55	60 to 70	6 mos	—	—	—	—	—
Oleomargarine	35	60 to 70	1 year	—	—	—	—	—
Olives, fresh	45 to 50	85 to 90	4 to 6 wks	75.2	28.5	.80	.42	108
Onions and onion sets	32	50 to 70	6 to 8 mos	87.5	30.1	.90	.46	124
Okra	45 to 50	85 to 90	7 to 10 days	89.8	30.1	.92	.46	128
Oranges	32 to 34	85 to 90	8 to 10 wks	87.2	28.0	.90	.46	124
Parsnips	32	90 to 95	2 to 4 mos	78.6	28.9	.84	.46	112
Peaches	31 to 32	85 to 90	2 to 4 wks	86.9	29.4	.90	.46	124
Pears	29 to 31	85 to 90	—	83.5 (Bartlett)	—	.86	.45	118
Peas, green	32	85 to 90	1 to 2 wks	74.3	30.0	.79	.42	106
Peppers								
Sweet	45 to 50	85 to 90	8 to 10 days	92.4	30.1	.94	.47	132
Chili (dry)	32 to 50	70 to 75	6 to 9 mos	—	—	—	—	—

[1] Reprinted from the "Air Conditioning-Refrigerating Data Book, ed. 5—1954–55 Applications Volume." by courtesy of The American Society of Refrigerating Engineers.

[2] Eastern Sour, 28.0; Eastern Sweet, 24.7; California Sweet, 24.2.

[3] Mission variety.

Persimmons	31 to 32	85 to 90	2 to 3 wks	78.2	28.3	.84	.43	112
Pineapples								
Mature green	50 to 60	85 to 90	3 to 4 wks	—	29.1	—	—	—
Ripe	40 to 45	85 to 90	2 to 4 wks	85.3	29.9	.88	.45	122
Plums, including fresh prunes	31 to 32	80 to 85	3 to 8 wks	85.7	28.0	.88	.45	123
Popcorn	32 to 40	85	—	—	—	—	—	—
Potatoes								
Early crop	50 to 55	85 to 90	—	—	—	—	—	—
Late crop	38 to 50	85 to 90	—	77.8	28.9	.82	.43	111
Poultry								
Fresh	32	—	1 wk.	—	—	—	—	—
Frozen, New York dressed	−10 to 0	—	6 to 9 mos	—	—	—	—	—
Frozen, eviscerated	−10 to 0	—	6 to 9 mos	—	—	—	—	—
Pumpkins	50 to 55	70 to 75	2 to 6 mos	90.5	30.2	.92	.47	130
Quinces	31 to 32	80 to 85	2 to 3 mos	85.3	28.1	.88	.45	122
Radishes—Spring, bunched or prepackaged	32	90 to 95	10 days	—	—	—	—	—
Winter	32	90 to 95	2 to 4 mos	93.6	—	.95	.48	134
Rabbits								
Fresh	32 to 34	90 to 95	1 to 5 days	—	—	—	—	—
Frozen	−10 to 0	90 to 95	0 to 6 mos	—	—	—	—	—
Raspberries								
Black	31 to 32	85 to 90	7 to 10 days	80.7	29.9	.84	.44	122
Red	31 to 32	85 to 90	7 to 10 days	83.4	30.4	.86	.45	122
Frozen (red or black)	−10 to 0		1 yr.	—	—	—	—	—
Rhubarb	32	90 to 95	2 to 3 wks	94.9	28.4	.96	.48	134
Rutabagas	32	90 to 95	2 to 4 mos	89.1	29.5	.91	.47	127
Salsify	32	90 to 95	2 to 4 mos	79.1	28.4	.83	.44	113
Spinach	32	90 to 95	10 to 14 days	92.7	30.3	.94	.48	132
Squash								
Summer	32 to 35	85 to 90	10 to 14 days	95	—	.96	—	135
Winter	50 to 55	75 to 80	4 to 6 mos.	88.6	29.3	.91	—	127
Strawberries								
Fresh	31 to 32	85 to 90	7 to 10 days	90.0	29.9	.92	.47	129
Frozen	−10 to 0	—	1 yr.	—	—	—	—	—
Sugar, granulated	45 to 50	below 60	1 to 3 years	0.5	—	.20	.20	72
Sweetpotatoes	55 to 60	90 to 95	4 to 6 mos	68.5	28.5	.75	.40	97
Tomatoes								
Mature green	55 to 70	85 to 90	3 to 5 wks	94.7	30.4	.95	.48	134
Ripe	40 to 50	85 to 90	7 to 10 days	94.1	30.4	.95	.48	134
Turnips	32	90 to 95	4 to 5 mos	90.9	30.5	.93	.47	130
Vegetable seed	32 to 50	50 to 65	—	—	—	—	—	—
Yeast	31 to 32	—	—	—	—	—	—	—

[5] Calculated by Siebel's formula. For values above freezing point $S = 0.008a + 0.20$. For values below freezing point $S = 0.003a + 0.20$. (Work on the calorimetric determination of specific heats is now in progress at the University of Texas, but values thus obtained are available for only a few products and therefore are not included in this tabulation.)

[6] Values for latent heat (latent heat of fusion) in Btu per lb, calculated by multiplying the percentage of water content by the latent heat of fusion of water, 143.4 Btu.

quality cannot be obtained when meats are subjected to quick freezing. It is believed that the freezing of animal and plant cells, if accomplished quickly enough, will not disturb the characteristics of their chemical constituents that are present in a colloidal state so as to radically affect the inherent qualities of the product.

Snap freezing is the latest method of preserving foods by refrigeration. It involves the use of cryogenic liquids, such as nitrogen, which has a boiling point of −325 F at atmospheric pressure. The process is fully described in Chapter 17.

Food spoilage—time temperature factor. Most foods can be maintained in a fairly fresh condition for short periods of time at temperatures above (32 F) freezing. The length of time depends on the kind of food and the temperature to which it is exposed. The time-temperature factor is most important in the operation of food vending machines that are in wide use in public gathering places, industrial plants, etc. A study of vending machine operations shows the following time-temperature relationship conducive to food storage without growth of harmful microorganisms:

Storage Temperature °F	*Number of Days*
40	5+
42	5
44	4
46	3
48	2
50	1

HUMIDITY—EFFECT ON COLD-STORAGE PRODUCTS. Shrinkage of nearly all food commodities is a factor to be considered in cold-storage handling. Loss in weight caused by a decrease in moisture content of food substances as a result of evaporation while in transit or storage is sometimes referred to as desiccation. Evaporation is the reverse action of condensation, both of which occur because of vapor-pressure differentials. Dalton's law states that *the total pressure of any gas is the sum of the pressures of all gases within a mixture.* The total pressure of the atmosphere is the sum of the pressures exerted by the air and water vapor, and any increase of the latter at the same temperature increases the total pressure. The converse is also true. If the water-vapor content is raised to the point where its pressure exerts a force equal to that of the air, the mixture is said to be saturated and will not tolerate further increases in water-vapor content at that temperature. At this point, the relative humidity of the atmosphere is 100 percent. Under such a condition, evaporation will cease. However, if the water-vapor content is reduced below the level required to saturate air at a given temperature, it becomes capable of accepting or "absorbing" additional moisture and evaporation is then possible at a rate proportional to the amount

TABLE 119. STORAGE LIFE AND TEMPERATURES OF FROZEN FOODS

Food	*Maximum Temperature °F*	*Approximate* Storage Life*
Vegetables	0 F	1 year or more
	10 F	4 months
Fruits	0 F	1 year or more
	10 F	4 months
Meats:		
Pork	0 F	6 months
	10 F	3 months
Beef	0 F	1 year or more
	10 F	4 months
Lamb	0 F	10 months
	10 F	3 to 4 months
Mildly cured ham and bacon	0 F	4 months
	10 F	2 months
Fish	0 F	4 months
Poultry	−10 F	1 year or more
	0 F	10 months
	10 F	4 months
Dairy products:		
Butter	0 F	1 year
	10 F	3 months

* Assuming that high-quality foods are placed in storage.

of water vapor required to cause saturation. Therefore, the rate at which evaporation occurs depends on the relative weight of water vapor contained in the atmosphere as compared with the amount necessary for saturation at the same temperature. If the amount of water vapor present is only half that required to cause saturation at a given temperature, the atmosphere is said to have a relative humidity of 50 percent, usually written 50 percent R.H. If it contains one quarter the amount, the relative humidity is then 25 percent, or 25 percent R.H., etc.

To reduce shrinkage in food products it is necessary that the moisture content of the storage atmosphere be maintained at a high level, or as nearly 100 percent as practicable for some items so as to discourage rapid evaporation. For this reason, relative humidity levels in Table 118 are as high as shown for the various products. That higher levels are not recommended for all products in order to further reduce evaporation is due to the encouragement of mold growth—and in some cases, as with meats, bacterial growth—when optimum conditions are exceeded.

Air motion within the storage compartment also governs the shrinkage rate, even though a constant relative humidity is maintained and increases in proportion to the air velocity over the food surfaces.

Chilling

Commodities destined for temporary storage are placed in cold-storage chambers in their freshly harvested condition after the necessary cleaning and sorting. Fruits and vegetables are the principal items subjected to

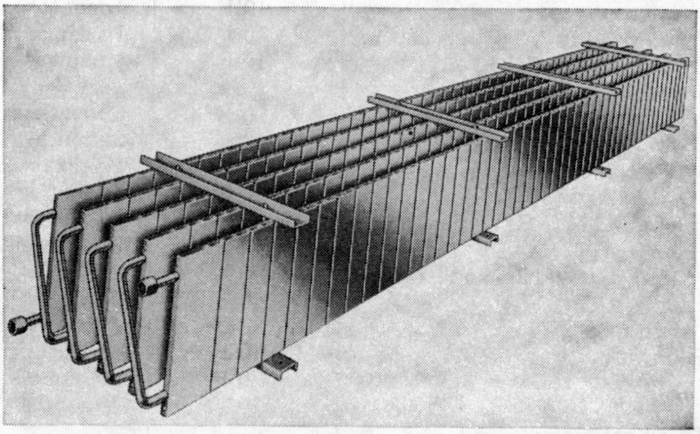

Fig. 399. Overhead plate evaporator of the type used in low-temperature compartments. (Courtesy, Peerless of America, Inc.)

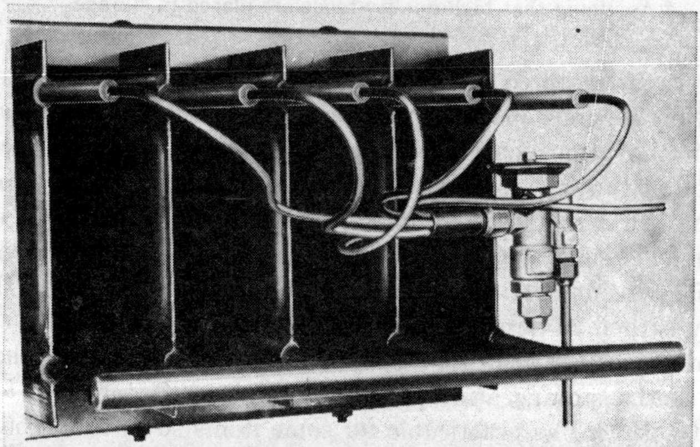

Fig. 400. Arrangement of thermostatic expansion valve and plate evaporator showing multi-outlet connections. (Courtesy, Sporlan Valve Company.)

Table 120. Condition in Individual Compartments of a General Cold-Storage Plant and Suggested Separation of Various Commodities

Compartment	*Optimum Temperature Range*	*Average Temperatures*	*Average Relative Humidity*
Chill Room #1:	**29 F–31 F**	**30 F**	**85 %**
Commodities Stored: Eggs,* cheese			
Chill Room #2:	**32 F–34 F**	**32 F**	**85–90 %**
Commodities Stored: Apples, cherries, quinces, strawberries, grapes, blackberries, raspberries, currants, plums, prunes, asparagus, carrots, mushrooms, radishes, turnips, onions, garlic			
Chill Room #3:	**32 F–34 F**	**32 F**	**85–95 %**
Commodities Stored: Cabbage, peas, lettuce, horseradish, broccoli, Brussels sprouts, oranges, peaches, pears, rhubarb, turnips, sweet corn, endive, leeks, lima beans			
Chill Room #4:	**40 F–45 F**	**42 F**	**90 %**
Commodities Stored: Potatoes, avocados, cucumbers, eggplants, limes, pineapples (ripe), peppers, melons, string beans, tomatoes (ripe)			
Chill Room #5:	**50 F–60 F**	**55 F**	**85 %**
Commodities Stored: Sweet potatoes, grapefruit, pumpkins, pineapples (mature green), squash (winter), tomatoes (green), lemons			

* Egg storage temperatures should not fall below 29.5 F.

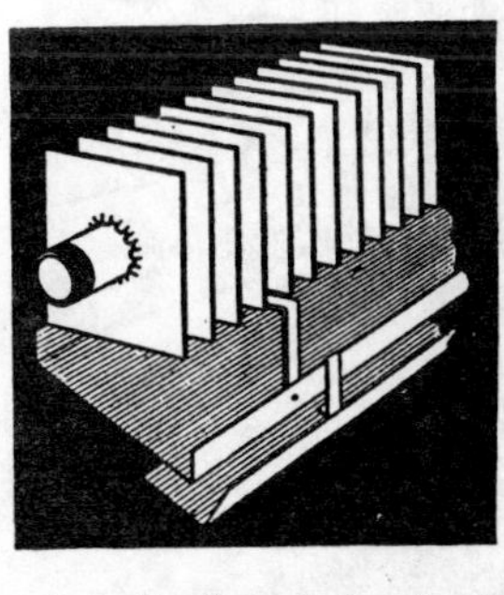

Fig. 401. (*Left*) Blower type cooling coil. (Courtesy, Peerless of America, Inc.)

Fig. 402. (*Right*) Finned pipe of the type used in cold-storage plants and freezers. Note ample distance between fins (1 to 1½ in.) and drip trough. Made of steel, the fins are clamped on the pipe under hydraulic pressure and then hot-dipped galvanized. (Courtesy, Frick Company.)

chill-room storage. If shipped long distances they are precooled and hauled in refrigerated cars or trucks to the plant to safeguard their fresh qualities. Only specimens in firm and healthy condition should be accepted for storage to assure their safe preservation through maximum holding periods. Because the optimum storage conditions for perishable commodities are not similar for all items, temperatures in the chill rooms are varied to meet required conditions. Chill rooms or coolers usually are provided with some means of increasing moisture content of the air contained within them, such as spray nozzles or humidifiers which can be controlled manually or automatically. As a practical measure, considering the variations in optimum temperatures of the individual items, an average condition should be selected for each compartment so as to include all commodities whose requirements fall within close range, preferably 3 F plus or minus. A study of Table 120 indicates that an ideal general storage capable of holding all commodities listed will have various chill compartments whose average temperatures will be approximately as shown.

Sharp Freezing

Sharp-freezer temperatures vary widely, depending on the type of storage operation carried on, ranging from 0 F to −20 F in the majority of commercial storage plants. True sharp freezers are utilized chiefly for the storage of meats received in an unfrozen state and held until temperatures are reduced to the holding temperature or lower, after which they are removed to holding freezers. During the freezing process, which begins immediately following exposure to the low temperatures maintained within

Fig. 403. Interior, cold-storage plant, showing method of storing apples. Note pallets at bottom and halfway up the tiers to facilitate handling with fork truck and to provide air circulation. Ceiling duct is for purpose of circulating air throughout room by drawing air from near floor level and discharging it near ceiling. A fan is used to move the air. (Courtesy, Quaker City Cold Storage Co., Inc.)

such compartments, heat transfer is comparatively slow (6 to 10 hours) and results in a decided time lag in passing through the zone of water crystallization. Some cellular damage occurs during the sharp-freezing process, and for this reason meats should be kept frozen until they are prepared for consumption. The subsequent loss of fluid upon each thawing considerably reduces the color, the flavor, and qualities of tenderness. All meats stored in sharp and holding freezers should be wrapped in suitable material, such as cloth or paper, to reduce shrinkage caused by excessive evaporation and to keep the surface clean.

Quick Freezing

This method of preserving fruits and vegetables has become increasingly important in recent years. Involving the exposure of freshly harvested and carefully prepared foods to subzero temperatures and holding them at 0 F during storage periods, this process provides a supply of high-quality, vitamin-rich fruits and vegetables at all times. Specimens selected for quick freezing must be fully ripened and of high quality for best results. Bruises and injured portions are removed in order to eliminate the micro-organisms that collect rapidly in impaired areas.

All items are washed thoroughly to remove micro-organisms adhering to their surface. Usually fruits are covered with sugar syrup when packed for quick freezing in an effort to prevent air from reaching the tissue during storage. Vegetables are blanched or precooked prior to freezing, during which much of their enzyme substance is destroyed. Blanching is accomplished by immersing the product in boiling water or steam for short periods, the length of which depends on the nature of the vegetable, ranging from 1 minute for hulled peas to 8 minutes for corn on the cob. The product should be chilled in cold water (50 F) immediately after blanching and then dried thoroughly before packing. Changes due to contact with oxygen in the air occur less rapidly when fruits and vegetables are kept at low temperatures. Blanching of fruit is not practiced, for to do so would destroy its fresh quality, a characteristic so highly valued.

All products prepared for quick freezing are packed in suitable moisture-proof and air-tight containers for further protection against enzymic action while in storage.

Temperatures employed commercially for quick freezing of food products range between −10 F and −50 F, after which they are stored continuously at 0 F. It is believed that fluctuations of 5 F, plus or minus, in holder storage temperatures are detrimental to color, flavor, and food value of most items and should be avoided. However, some doubt has been expressed by late investigators regarding the effects of temperature fluctuation storage. If products are selected, prepared, and stored, fruits packed in syrup will keep for a year or longer while vegetables stay comparatively fresh for six months.

Fig. 403(a). Continuous belt contact, in-line freezer. Unit shown has a capacity of 1200–3000 lbs. per hour of 0 F product and freezes many kinds of products by varying belt speed and/or raising or lowering brine temperature. No forced air movement prevents color change or dehydration of product. Underside of stainless steel belt carries product in direct contact with brine to assure rapid heat transfer. Feeding and delivery of product may be automatic. (Courtesy, Frick Company.)

Fig. 403(b). Freez-Pak. Continuous individual quick freezing unit that eliminates elaborate dewatering device. Product is washed and dewatered before entering freeze unit, then distributed over conveyor (1), by shaker spreader (2), and quickly glazed by updraft of cold air (3), to prevent dehydration and seal in flavors. Product is conveyed through tunnel under low temperature (−25 F to −30 F) by high velocity air from refrigerated coil (4), and discharged at unloading station (5). Individual coils are glycol defrosted (others continue in operation) by sequential program timer. Insulation is 4-in. foamed-in-place polyurethane (6). Entrance door and exit door (not shown) are (7) in illustration, axial flow fans (8), inner wall and epoxy coated steel plate (9), and aluminum outer covering (10). (Courtesy, Frick Company.)

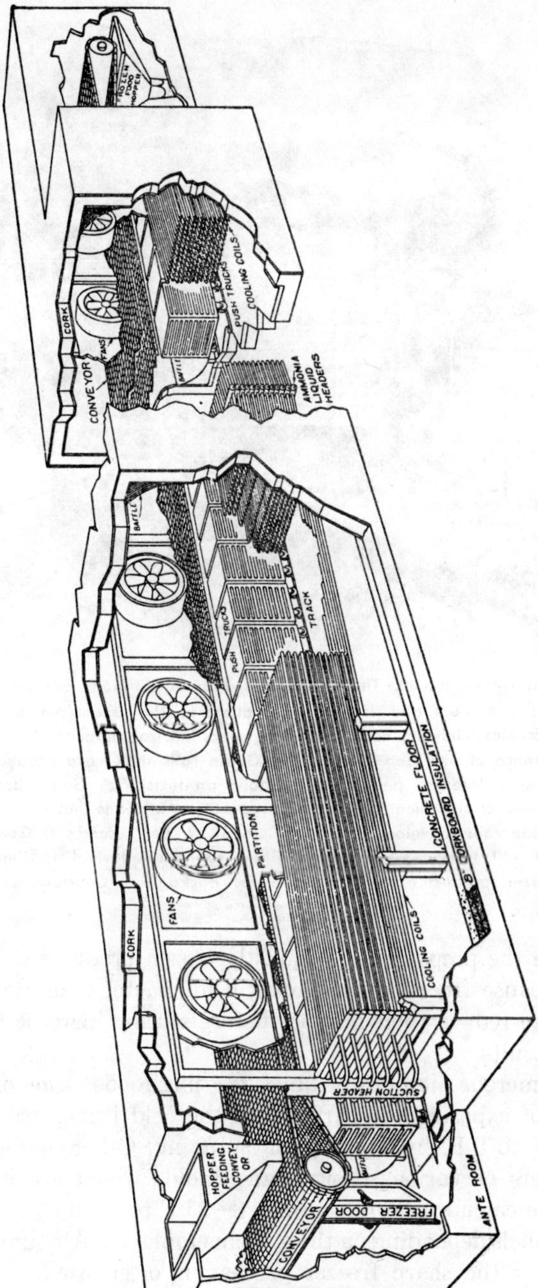

Fig. 404(a). Tunnel freezer for quick-freezing foods. (Courtesy, Frick Company.)

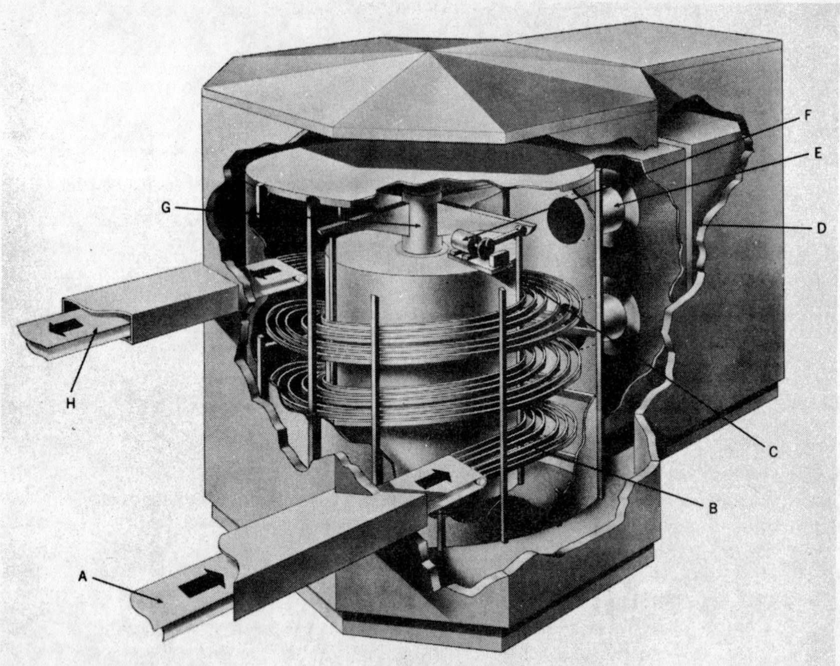

Fig. 404(b). Spiral freeze tunnel. The tunnel can process different products with freezing times from 30 minutes to 4 hours, and sizes from one ounce to 2½ gallons simultaneously. Having various shapes, it occupies minimum floor space due to cylindrical shape and unlimited vertical construction. (a) Entrance of unfrozen product. (b) Guide rails that move product (have timing control to allow various freezing periods for various products). (c) Guide bar and circular frame that controls product movement. (d) Finned evaporators. (e) Fans that circulate air through the tunnel along guide rails at velocities of 26 ft. to 33 ft. per second. (f) Gear motor which powers the movable rail track. (g) Hydraulic lifting cylinder attached to timer to regulate proper flow. (h) Frozen product exits here ready for packaging. (Courtesy, Gram Equipment of America, Inc.)

Precaution in the preparation of quickly frozen products is emphasized particularly because freezing does not kill all bacteria, as does canning, and all efforts to reduce their number before storing increase the chances for successful results.

There are numerous methods of quick freezing foods, some of which use the procedure of exposing the surfaces to the cold brine, usually sodium chloride, at 4 F to 6 F, by dipping or spraying. Other methods employ plates, containers of various designs and molds which are immersed in low-temperature calcium chloride brines (−25 F to −50 F) that circulate in tanks or channels depending on the system employed. Air-tunnel freezing, a modification of the sharp freezer process, is often used. As the name implies, cold air at −50 F is blown through or across a tunnel in which the product to be frozen is placed. It is usually wrapped to avoid excessive surface drying, referred to as freezer burn. In some instances, fish is dipped

Table 121*

Directions for Freezing Vegetables

Vegetable	*Preparation*	*Packing*
Asparagus: Martha Washington Mary Washington	Avoid small woody stalks. For tips, cut to fit in box, discarding woody ends. Wash thoroughly, removing the large scales after washing. Or cut asparagus in 1-in. lengths. Scald in boiling water 3 min or in steam $3\frac{1}{2}$ min. Cool at once. Drain well. Steam scalding is preferred to water scalding	Pack whole stalks parallel in box with heads in alternate directions. Pack cut asparagus compactly
Beans (lima): Fordhook 242 Fordhook (bush) King of the Garden (pole) Peerless (bush) Green-seeded Henderson (bush) Thorogreen (bush) Clarks Bush (bush) Challenger (pole)	Select medium-sized beans, not too meaty in texture. Shell the beans and sort according to size. Discard undesirable ones. Water-scald $2\frac{1}{2}$ min or steam-scald 3 min. Cool promptly. Drain	Package and freeze
Beans (snap): Kentucky Wonder Tendergreen Longgreen Blue Lake Refugee Stringless Greenpod Pencil Pod (wax)	Avoid overmature pods. Select freshly picked, crisp beans. Wash carefully, snip ends, and cut in $\frac{3}{4}$-in. lengths or lengthwise for French style. Water-scald 2 min or steam-scald $2\frac{1}{2}$ min. Varieties of green beans are more satisfactory than wax varieties	Package, excluding as much air as possible
Beets: Detroit Dark Red Crosby	Select young, tender beets. Water-scald small peeled beets $2\frac{1}{2}$ min or steam $3\frac{1}{2}$ min. Cook large beets in skins until tender. Cool and remove skins	Pack small beets whole; slice or quarter larger beets
Broccoli: Italian Green Sprouting	Select tender, compact heads of uniformly dark green color. Let stand in salt water $\frac{1}{2}$ hr to remove insects. Wash and look over carefully. Cut off woody stems. Skin stems if tough. Sep-	Package with heads in alternate directions

* Courtesy, Agricultural Extension Service, Pennsylvania State University, Circular 317.

TABLE 121—(*Continued*)

Vegetable	*Preparation*	*Packing*
	arate heads into convenient pieces for packaging. Scald 3 to 5 min depending on size of heads. Steam scalding is preferred. Cool and drain well	
Brussels Sprouts: Long Island Improved Half Dwarf Improved Catskill	Select firm, dark green compact heads, not wilted. Separate heads from stem. Wash carefully. Scald 3 to 4 min. Cool and drain well	Pack compactly
Cantaloupe	See Fruits, page 872	
Carrots: Red Core Chantenay Nantes or Coreless Touchon	Select only small, young carrots. Wash and scrape. If tiny, freeze whole. Or slice in ¼-in. slices or in lengthwise strips. Scald 3 min. Cool and drain.	Pack compactly
Cauliflower: Snowball Super Snowball "A" Holland Erfurt "X"	Select compact, tender white heads. Trim and cut into pieces about 1 in. in thickness. Wash carefully. Scald 3 to 4 min depending on size of pieces. Cool in ice water. Drain well	Pack compactly
Corn (on the cob): Goldencross Bantam Tendergold Lincoln Lee Seneca Chief Silvercross Bantam	Corn on the cob is not always worth the room it takes because the corn may acquire a cob taste. Select corn just right for table use. Husk, remove silk, and wash ears. Discard imperfect ears. Steam-scald 6 to 8 min, depending on size. Cool and drain. Yellow corn is preferred to white corn. Any available late variety of yellow corn is satisfactory	Wrap in cellophane or aluminum foil or place in moisture-vapor-proof bags
Corn (off the cob): Goldencross Bantam Lincoln Tendergold Lee Seneca Chief Silvercross Bantam Tendermost Goldengrain Bantam Evergreen	*Whole Kernel:* Scald corn on the cob 6 to 8 min. Cool promptly without letting ears stand in water too long. Cut from cob, being careful not to cut the cob. *Cream Style:* Scald corn on cob 8 min. Cool. Cut kernels from cob but not too deeply. Scrape the cob with back of knife. Place corn in freezer tray of refrigerator to cool. Or cut corn from freshly husked cobs, scraping cob. Cook in top of double boiler 12 min. Cool in refrigerator trays	Pack quickly after cutting from cob Pack when cool

TABLE 121—(*Continued*)

Vegetable	*Preparation*	*Packing*
Eggplant: New York Black Beauty New Hampshire Hybrid	Peel mature eggplant. Slice in $\frac{1}{3}$-in. slices. Scald 4 min. Rinse in citric and ascorbic acid solution ($4\frac{1}{2}$ tsp citric and ascorbic acid mixture to 1 qt water), then in cold water	Pack compactly
Greens: Swiss Chard Fordhook Giant Lucullus Spinach King of Denmark Long Standing Bloomsdale Kale Dwarf Scotch Beet tops	Curly leaf varieties more satisfactory for freezing. Wash and pick over greens carefully to remove all sand, dirt, and bruised, discolored, or decayed leaves. Cut off large stems. Scald small amount (to prevent leaves from matting) 2 to $2\frac{1}{2}$ min. Cool and drain. New Zealand spinach is not adapted to freezing, nor are salad greens, such as endive and dandelion	Slightly squeeze in hand before packing to take out water
Mixed Vegetables: Recommended varieties for the vegetables used	Peas, lima beans, corn, snap beans, etc., may be packaged together for soup mixture or succotash. Scald vegetables separately for time recommended for each one. Cool and drain	Mix different vegetables together and package at once
Mushrooms	Wash lightly. Slice large ones; leave small buttons whole. Brown 4 to 5 min in hot butter, allowing most of juice to evaporate. Or scald buttons 2 min; larger sizes and pieces, 3 to 4 min. Cool and drain	Package in small containers. A $1\frac{1}{2}\%$ salt solution may be used for the scalded mushrooms
Peas: Thomas Laxton (dark podded) Improved Gradus World's Record Hundredfold Alderman (telephone) Freezonia Little Marvel	Select young tender peas just right for eating. Shell, wash, and water-scald 1 min or steam-scald $1\frac{1}{2}$ min. Cool and drain well	Package at once and freeze
Peppers: Tomato or Cheese California Wonder Oak View Wonder Pennwonder	Wash, halve or slice, removing seeds and stem. Scald 2 min. Cool and drain. Or cut fine as for pepper cabbage and freeze without scalding	Package in small amounts for seasoning foods

TABLE 121—(*Continued*)

Vegetable	*Preparation*	*Packing*
Rhubarb: MacDonald Ruby Valentine	Pick early in season when color is best and stalks are not stringy. Cut off leaves and wash carefully. Cut in 1-in. pieces without peeling Scald pieces in boiling water 1½ min. Cool and package. Or scald, cool, and pack; cover with a 60% sirup. Or cook rhubarb in a 60% sirup, cool, and package. Or pack without scalding and without sugar.	If sirup is used, pack in waxed cartons. Dry pieces may be packed in cellophane-lined boxes or bags
Soybeans (green): Giant green Bansei Mendota	Boil in pods 5 min. Chill. Squeeze beans out of pods. Wash, drain, and pack	Pack compactly
Squash, summer: White Bush Scallop Straightneck Cocozelle	Select young, tender squash with tender rind. Wash, slice in ½-in. slices. Scald 3½ min. Cool and drain well	Package and freeze at once
Squash, winter: Delicious Hubbard Butternut Buttercup	Select well-ripened, fully mature squash with hard rind. Peel squash, cut in 1-in. cubes, and steam until fairly soft or bake squash slowly in oven. Mash and cool thoroughly in refrigerator trays	Pack when cold and freeze
Sweet Potatoes: (Freeze only those which cannot otherwise be stored) Yellow Jersey Maryland Golden Nancy Hall Porto Rico	Use only mature, cured potatoes. Wash, cook until tender. Remove skins. Cut in half and package whole, or mash and mix with a little orange juice	Package dry halves in cellophane-lined cartons. Allow 1-in. headspace in waxed cartons for mashed sweet potatoes
Tomatoes: Rutgers Marglobe Pritchard Stokesdale Oxheart Golden Jubilee (yellow)	Wash firm, perfect, ripe tomatoes. Wrap whole or slice and freeze on trays. More suitable for cooking than salad use	Wrap whole tomatoes in cellophane or aluminum foil before freezing. Pack frozen slices in cartons
Tomato Juice: Any of the previously mentioned varieties	Quarter or slice tomatoes, and heat only until they can be sieved. Work quickly. Do not reheat juice. Season as desired or freeze plain	Allow 1½-in. headspace in regular glass jars or tin cans; 1 in. in freezer jars or cartons. Seal and freeze

TABLE 121—(*Continued*)

Directions for Freezing Fruits

Fruit	*Preparation*	*Packing*
Apples: Cortland Rome Beauty Stayman Grimes Golden All firm winter varieties	Peel, slice into 50% sirup containing 2 tsp citric and ascorbic acid mixture to 2½ to 3 cups sirup. Or bake apples as for the table. Or precook halved or sliced apples 3 to 4 min in 50% sirup. When cooled, add 2 tsp citric and ascorbic acid mixture to 2½ to 3 cups sirup	Pack sliced or halved apples in heavily waxed cartons. Crumple cellophane to place in ½-in. headspace. Wrap each baked apple separately in cellophane and stack in waxed cylindrical carton
Applesauce	Prepare as for table use, cool. Do not add spices	Pack in heavily waxed cartons
Blackberries: Eldorado	Select fully ripened berries, discarding imperfect ones. Handle as little as possible. Wash and drain	For dessert pack in 40 to 50% sirup. For jam or pie use 1 part dry sugar to 4 parts fruit by weight. For muffins use ½ cup dry sugar to 1 qt berries
Blueberries: Any standard varieties or wild fruit	Select well-ripened berries. Wash; remove stems, immature berries, and leaves. Drain and crush slightly. Or steam-scald 1 min, cool, and drain well. Or immerse in boiling 75% sugar sirup 15 to 20 sec. Drain for 1 min. Cool	For dessert, pack in a 40 to 50% sugar sirup. For pies, pack with dry sugar, using 1 lb sugar to 4 lb berries
Cantaloupe: Delicious Sugar Rock Bender Hearts of Gold	Select only ripe, full-flavored melons. Slice, cube, or cut in balls with ball cutter	Pack dry or in 50% sirup in heavily waxed cartons
Cherries, sour: Montmorency	Select firm, fully ripe cherries, since immature fruit turns pale and overripe fruit darkens after freezing. Chill in ice water several hours. Two tbsp of vinegar per gallon of water will help remove spray residue. Drain quickly and pit	Mix carefully with 1 part dry sugar to 5 parts fruit by weight. More sugar can be added when fruit is to be used
Cherries, sweet: Bing Napoleon Schmidt Windsor	Select fully ripe but firm fruit. Stem and wash cherries. Pit or not as desired	Cover with 40 to 50% sugar sirup or mix with dry sugar

TABLE 121—(*Continued*)

Fruit	*Preparation*	*Packing*
Cranberries	Stem and wash carefully, remove all soft or poorly formed berries. (Cranberry sauce and relish also may be frozen)	Pack in a 50% sirup or freeze without any sugar. Or mix with 1 part dry sugar to 6 parts berries
Fruitcup	Any desired combinations of fruits may be used. Prepare as directed for each individual fruit. Choice of fruits will determine whether dry sugar or sirup is to be used. Consider color as well as flavor in making combinations	Pack in waxed cartons
Grapefruit	Peel ripe fruit, remove membrane and seeds, break into sections	Pack and cover with 60% sirup
Oranges	Peel, remove membrane and seeds, break into sections. Orange and grapefruit sections may be packed together	Pack and cover with 60% sirup
Peaches: Elberta Golden Jubilee Halehaven J. H. Hale South Haven Summer Crest	Select firm, evenly ripened fruit. Peel with a stainless steel knife, slice directly into cartons and with dry sugar. Use citric and ascorbic acid mixture if desired. Work quickly. Scalding to slip skins causes more discoloration	Leave ½-in. headspace in carton. Fill headspace with piece of crumpled cellophane. Freeze immediately
Pears: Bartlett(fair) (not suitable whole) Duchess	Select firm, ripe fruit. Peel and slice into 60% sirup or mix with dry sugar. Use citric and ascorbic acid mixture if desired	Pack the same as peaches
Pineapple	Peel, core, and slice or dice	Pack in containers and cover with a 60% sirup
Plums and fresh prunes: Stanley Italian Fellenberg	Select freshly picked tree-ripened fruit. Wash, sort, drain, halve, and pit	Pack in a 50 to 60% sirup
Raspberries, black: Cumberland Morrison New Logan	Frozen black raspberries are not too satisfactory as a dessert but can be used for pies or preserves. Select evenly ripened fruit, discarding any imperfect berries. Wash and drain, being careful not to bruise	Dry sugar pack is not recommended. Pack in a 40 to 50% sirup

TABLE 121—(*Continued*)

Fruit	*Preparation*	*Packing*
Raspberries, red: Latham Taylor Viking Milton	Clean freshly picked, firm ripe berries; handle them carefully. Wash in cold water	Mix berries with sugar, using 1 part sugar to 4 parts berries by weight
Strawberries: Catskill Premier	Select bright red, firm ripe berries. Avoid green or overripe fruit. Wash, hull, sort, and drain well. Work rapidly	*Sliced:* Slice in pieces ¼-in. thick or cut in halves lengthwise or chop. Use 1 part sugar to 4 parts fruit by weight. Mix carefully and thoroughly. Package and freeze promptly *Whole:* Only the most perfect berries should be frozen whole. Mix with enough dry sugar to coat them and pack into heavily waxed containers. Or pack in 50% sirup. Dry sugar is preferable

in water after completion of the freezing process to form a glaze of ice over its surface. This glaze greatly retards rancidity of oils and fats when stored for long periods.

Immersion Freezing

Immersing poultry, such as turkeys, in a brine solution at a temperature of 0 F, forms a relatively thick frozen crust over the entire outer surface of the carcass. This crust entraps the natural fluids within the body tissues, thus maintaining eating quality. The procedure is as follows:

(*a*) the carcasses are first chilled in ice,
(*b*) then floated breast down in the cold brine, having a velocity of about 250 fpm, while
(*c*) simultaneously being sprayed from an overhead brine sprayer to freeze the backs, after which they are
(*d*) frozen throughout in an air blast freezer.

The individual birds are wrapped in plastic bags before the above process is initiated. Sodium chloride brines have the advantage of more rapid freezing and allow reprocessing if plastic bags are punctured during the immersion process. However, calcium chloride and propylene glycol are also used.

Freeze Drying

Sublimation is the principle employed in the freeze drying of foods, beverages, pharmaceuticals, etc. Facility of transportation, preservation and lower cost of production are some of the advantages of the art. Solid carbon dioxide (dry ice), mothballs and camphor, which are converted from their solid state directly to their gaseous state without passing through a liquid state, are some examples of sublimation. (An example of sublimation in nature is the evaporation during sub-freezing weather, especially if a wind is blowing, of ice that forms in fields and on roads.)

Many substances are liquids that contain solids in *colloidal* form. The affinity of the colloid for its host liquid is called *lyophilization*. The fluid that forms the liquid of substances capable of being freeze-dried is water. One method of removing the water to leave a solid material is to evaporate it, either by raising its temperature above the boiling point at atmospheric pressure, i.e., 212 F or above, or boiling it at a reduced temperature under a vacuum. Some foods attain a "cooked" taste when prepared by the former method. With either method, the thermal energy involved is greater than that of freeze drying. For example, it requires 970 Btu per lb to convert water to steam at atmospheric pressure and 212 F, and between 300 and 350 Btu per lb to remove water under high vacuum and lower temperature at the corresponding state points in both cases. To remove water by sublimation, i.e., as ice, requires only 144 Btu per lb to freeze it and remove it as ice. Therefore, the thermal efficiency is higher with freeze drying than with heating processes.

The freezing-drying process then is one in which the substance to be reduced to a solid is frozen first and the water removed next under high vacuum, leaving behind desired residue, namely the colloids present in the original fluid. One of the latest substances to be freeze-dried is coffee.

One of the difficulties of freeze drying lies in removing the last remaining volume of water, which may amount to as much as 15–20% of the original volume. The reason for this is the loss of energy during the initial stages of evaporation. This is the thermal *energy of sublimation* utilized to convert the solid ice to vapor. As a result, the frozen product is further reduced in temperature, which retards further evaporation. As it is necessary to replace this heat loss from an external source in order to maintain and hasten the evaporation process, care must be taken to avoid burning the product, especially meat, by the concentration of applied heat in any water, thus formed by a substantial rise in temperature. Therefore, it is necessary to apply only the amount of heat equal to that lost through sublimation. The difficulty lies in getting the heat to the center of the product through the dried insulating mass surrounding it. The usual method of applying direct heat has some disadvantages.

The problem has been overcome by the use of *radio frequency* (*RF*) *waves* directed on the product which penetrates the dried mass without resistance to reach the ice remaining in the center. The rapid alternating effect of the radio frequency, known as *dielectric heating,* causes the polarity of the ice molecules to oscillate continually. Thus, the work of polarity oscillation causes a temperature increase and the reinforced energy enables evaporation to continue until all the water has been withdrawn from the product. The drying process requires about eight hours when accelerated by heat augmentation.

While radio frequency is in the range between that of audio sound (approximately 10,000 cycles per sec.) to infrared light portion of the spectrum (approximately 10,000,000 megacycles), there is a probability that a frequency level closer to the latter may prove equally or more successful. The removal of all water is necessary if the product is to achieve keeping qualities.

As long as thoroughly dried products are kept absolutely free of moisture, they do not require refrigerated storage. They are usually packed in moisture-proof packages and in some cases, the container is filled with an inert gas, such as nitrogen, to exclude all oxygen and water vapor present in the atmosphere.

Freeze Concentration

Whereas freeze-drying is the reduction of fluid materials to their dry state, freeze concentration is the removal of part of the water to form a slurry that is relatively more viscous than the original substance. Fruit juices are among the more popular freeze-concentrated substances in the food-beverage class. Beer is another. In the first instance, water is added to reconstitute it. Beer concentrate requires the addition of carbonated water to convert it to a fresh zesty beverage.

Freeze concentration of liquids is performed by first reducing their temperature until ice crystals of suitable size are formed. The slurry is then centrifuged to remove the liquid concentrate, which may again be reduced in temperature and the process repeated until the desired consistency is obtained. A slurry containing about 20 to 25% of the original amount of water is usual with fruit juice concentrates.

Respiration

Respiration is a process of oxidation which may be likened to that of combustion since oxygen is taken into fruit and vegetable substances and carbon dioxide is released. Meats, being made up of nonliving substances,

TABLE 122. APPROXIMATE RATE OF EVOLUTION OF HEAT (BTU PER TON PER 24 HR) BY CERTAIN FRESH FRUITS AND VEGETABLES WHEN STORED AT THE TEMPERATURES INDICATED*

Commodity	32 F	40 F	60 F
Apples	660 to 1000	1,110 to 1,760	4,400 to 6,600
Bananas[1]	—	—	—
Beans			
Green or snap	5,600 to 6,160	9,160 to 11,390	32,090 to 44,130
Lima	2,330 to 3,000	4,300 to 6,100	21,990 to 27,410
Beets, topped	2,650	4,060	7,240
Broccoli, sprouting	7,450	11,630	33,870
Cabbage	1,200	1,670	4,080
Carrots, topped	2,130	3,470	8,080
Celery	1,620	2,420	8,220
Cherries	1,320 to 1,760	—	11,000 to 13,200
Corn, sweet	6,560	9,390	38,410
Cranberries	600 to 720	870 to 970	1,650 to 1,800
Cucumbers	1,690	2,550	10,460
Grapefruit	460	1,070	2,770
Grapes			
American type	602	1,170	3,487
European type	300 to 430	—	—
Lemons	580	810	2,970
Lettuce	11,320	15,990	45,980
Melons, Cantaloupes	1,320	1,960	8,500
Mushrooms	6,160	22,000	58,000
Onions and onion sets	660 to 1,100	1,760 to 1,980	—
Oranges	690 to 900	1,400	5,000
Peaches	850 to 1,370	1,440 to 2,030	7,260 to 9,310
Pears	660 to 880		8,800 to 13,200
Peas, green	8,160	13,220	39,250
Peppers, sweet	2,720	4,700	8,470
Potatoes	440 to 880	1,100 to 1,760	2,200 to 3,520
Raspberries, red[2]	—	—	15,400 to 17,600
Spinach	4,240 to 4,860	7,850 to 11,210	36,920 to 38,000
Strawberries	2,730 to 3,800	5,130 to 6,600	15,640 to 19,140
Sweetpotatoes	1,190 to 2,440	1,710 to 3,350	38,400 to 42,806
Tomatoes			
Mature green	580	1,070	6,230
Ripe	1,020	1,260	5,640
Turnips	1,940	2,150	5,280

[1] Turning ripe at 68 F, 8,300.
[2] At 36 F, 4400 to 6600.

* Reprinted from the "Air Conditioning-Refrigerating Data Book, ed. 5—1954–55 Applications Volume," by courtesy of The American Society of Refrigerating Engineers.

do not respire, but are subject to other enzymic changes. The action of converting starch to sugar within living substances such as fruits and vegetables produces carbon dioxide and generates heat. Thus a chemical change takes place, induced by the presence of oxidizing enzymes. In that low temperature exposure acts as a retarder to enzymic activity, respiration diminishes. As respiratory retardation is important in the preservation of plant substances, it is necessary to control a function that can prove devastating in its effect if allowed to continue. The reason is simple. Uncontrolled respiration generates heat that in turn encourages greater enzymic activity due to a subsequent temperature increase within the mass. Thus, more heat is evolved as respiration is accelerated, resulting in rapid starch to sugar conversion and ultimate cessation of life due to starch exhaustion. Decay follows rapidly.

To prevent the acceleration of respiratory processes, freshly harvested fruits and vegetables are precooled before shipping any distance to cold storage or placed immediately therein if located close by. This procedure not only maintains quality but also minimizes the amount of heat evolved, consequently requiring less refrigeration load to bring respiration under control. The amount of heat generated through respiration is considerable and may be appreciated best by comparing the data compiled for several commodities at various temperature levels listed in Table 122.

RANCIDITY

Animal fats are chemical combinations of carbon, hydrogen, and oxygen,

(1) HIND SHANK
- 1 to 3—Soup bones
- 4—Hocks

(2) ROUND
- 1 to 14—Round steaks
- 15—Heel of round

(3) RUMP
- Steaks or roasts

(4) LOIN END
- 1 to 6—Sirloin steaks

(5) SHORT LOIN
- 1 to 3—Club or Delmonico steaks
- 4 to 11—Porterhouse steaks

(6) FLANK
- 1—Flank steak
- 2—Stews or hamburger

(7) RIB
- 1 to 4—Rib roasts
- 5—Short ribs

(8) TRIMMED CHUCK
- 1 and 2—Bottom chuck roasts
- 3 and 4—Top chuck Roasts
- 5 to 7—Chuck rib roasts

(9) NECK
- 1—Boneless roasts, stews, or hamburgers

(10) FORE SHANK
- 1 to 3—Soup bones
- 4—Shoulder clod

(11) BRISKET
- 1—Stews or boned and rolled roasts

(12) PLATE
- 1—Stews or boned and rolled roasts
- 2—Short ribs

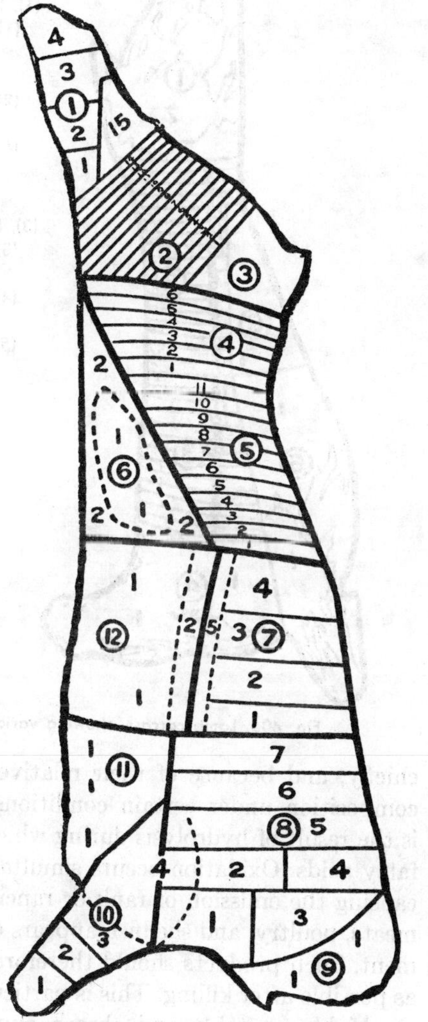

Fig. 405. Beef carcass. Grades (Federal grading) of beef include prime, choice, good, commercial, utility, cutter, and canner.

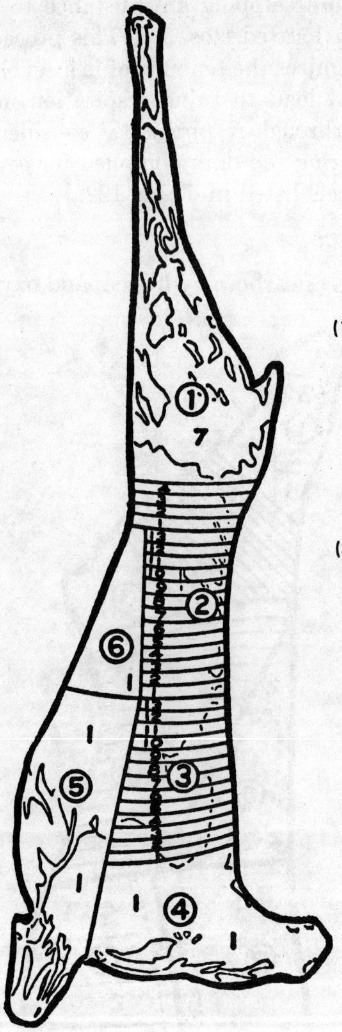

Fig. 406. Lamb carcass, showing various cuts and relation to the skeleton.

chiefly, and because of their relative instability are subject to rapid decomposition under certain conditions. Decomposition or spoilage of fats is the result of hydrolysis during which water reacts with the salts to form fatty acids. Oxidation occurs simultaneously with the *hydrolizing* process, causing the emission of rank or rancid odor. Rancidity in the fats of fish, meats, poultry, and shellfish appears quickly if exposed to a warm environment. Such products should therefore be placed in freezer storage as soon as possible after killing. This is particularly true with fish, the fats of which are highly unstable and absorb oxygen readily, causing a brown color. Careful packaging of fish before storage decreases this tendency and is widely practiced. Glazing with a thin coating of ice at 10 F is an excellent method of preserving fish. To obtain the glaze the fish are either dipped or sprayed after freezing.

Butter, lard, cream, bacon, and sausage contain high percentages of fat and are all products that tend to become rancid, some earlier than others under similar circumstances.

Rancidity is not checked in all products at similar temperatures. While beef fat will not be affected if exposed to temperatures below 14 F, some rancidity develops in pork fat at −5 F.

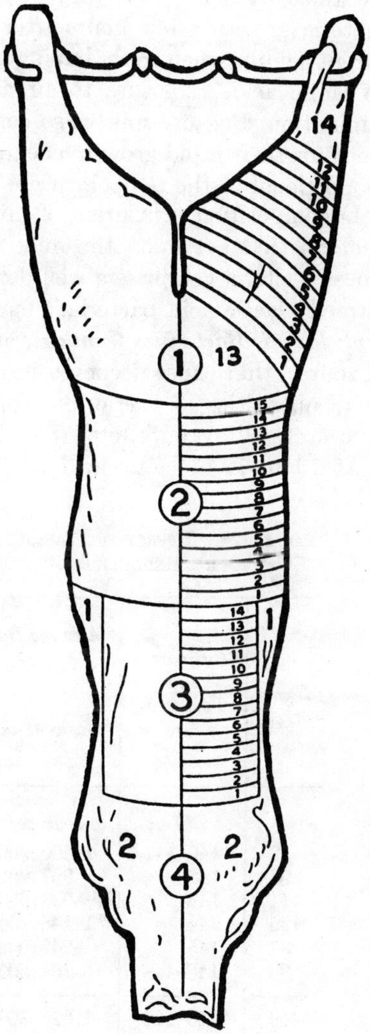

(1) and (2): HIND SADDLE
- **(1) Leg**
  - **1 to 12—Cutlets**
  - **13—Roast**
  - **14—Shank (stew)**
- **(2) Loin**
  - **1 to 15—Loin and kidney chops**

(3) and (4): FORE SADDLE
- **(3) Hotel rack**
  - **1 to 14—Rib chops**
- **(4) Chuck including shoulder, neck and breast**
  - **1—Stew**
  - **2—Roasts**

Fig. 407. Veal chart.

A discussion of the chemical changes that occur in fresh cow's milk when exposed to normal room temperatures may serve as the best means of describing bacterial activity that is carried on within a food product.

Milk is composed on the average of 87.1 percent water, 4 percent fat, 4.75 percent milk sugar (lactose), 3 percent casein, 0.4 percent albumin, and 0.75 percent minerals. When fresh, milk is not only free of harmful bacteria but also has a germicidal action believed due to the presence of antibody-like substances that destroy such organisms. These substances are killed by heat. That they are short-lived is borne out by the start of bacteria growth a few hours after the milk is produced. Milk has an apparent acidity when fresh, having a pH of 6.3 to 6.6 normally, and contains no lactic acid. Exposure to air and utensils causes early bacterial contamination. Because milk is so composed as to provide the ideal environment for their rapid growth, chemical decomposition soon follows at a rate proportional to the temperature.

Decomposition or souring of milk is brought about by certain types of bacteria that act upon the milk sugar or lactose to produce lactic acid. These bacteria comprise a well-defined group of related varieties generally termed "lactic acid bacteria," but more specifically identified as *Bacillus acidi lactici*, *Bacterium Guntheri*, and *Streptococcus lactis*. There are others, of course, that cause decomposition of additional milk ingredients to form small quantities of acetic and formic acids, and also to cause red milk, blue milk, slimy milk, etc. (See Table 114.)

At 70 to 72 F, lactic acid bacteria grow most rapidly and surpass in

TABLE 123. EFFECT OF VARIOUS TEMPERATURES OF PASTEURIZATION ON THE REDUCTION OF THE BACTERIAL COUNT OF MILK*

Runs	*Pasteurizing Temperature*	*Average Bacterial Count*		*Average Decrease in Bacteria per Cubic Centimeter*	
		Raw Milk	*After Pasteurization*		
	°F	*Number*	*Number*	*Number*	*Percent*
12	140–141	1,490,900	59,600	1,431,300	96.00
96	142	967,800	25,600	942,200	97.35
111	143	938,600	26,300	912,300	97.19
130	144	1,146,700	29,400	1,117,300	97.44
87	145	942,000	22,000	920,000	97.66
33	146–148	764,500	12,700	751,800	98.34
469	140–148	1,026,300	26,100	1,000,200	97.45

* Courtesy, U.S. Department of Agriculture, Circular 222.

numbers other types present. As bacteria growth progresses, so does the rate of lactose decomposition proportionately, resulting in early souring or what is sometimes referred to as normal fermentation of the milk. The lactic acid produced causes coagulation, the result of casein precipitation. Lactose decomposition continues at an accelerated rate until it becomes exhausted or until the acidity value reaches a point untenable to bacteria cells, at which time acid-producing cells become inactivated. Lactic acid content increases simultaneously as does coagulation during the decomposition periods. The final product shows a distinct separation between watery and coagulated substances. Natural milk fermentation limits acid development to 0.8 to 1.0 percent.

Bacteria content of milk is reduced considerably (97.3 percent) by heating to 145 F (varies from 142 F to 148 F) and holding it at that temperature for 30 minutes. The process is called *pasteurization*. Because it renders milk free of disease-producing bacteria, its practice has been widely adopted. (See Table 123.) Bacteria growth in milk is greatly retarded at temperatures below 50 F, although slight activity continues when held at or near the freezing point. The production of low bacteria count milk (under 30,000 per milliliter) rests upon adherence to several important rules, viz.: (*a*) sanitary conditions surrounding production, (*b*) immediate cooling below 50 F of fresh milk, (*c*) pasteurization followed by rapid cooling below 50 F, (*d*) bottling and storage at 40 F or slightly below.

EGGS

Bacterial invasion, dehydration, and fertilization are all factors that cause egg spoilage.

Bacteria enter eggs through pores of the shell to carry on their destructive activity. One step toward decreasing micro-organism penetration is early cleaning of the shell surface. If accompanied by immediate cooling, correct storage preparation is accomplished. Evaporation of the water content while in storage can be reduced greatly by maintaining a high relative humidity. However, moisture content of the air is limited, as with other commodities, because mold growth develops rapidly. For this reason, relative humidity levels of 85 to 90 percent are the rule. Dipping in mineral oil tends to clog the pores and retard evaporation. Candling, by holding eggs before a strong light, will determine the size of the air cell, and give an indication of water loss through evaporation. Accurate relative humidities can be determined by this method if periodic spot checking is carried out. Some idea of the value of candling can be obtained from the following excerpt taken from a bulletin of Illinois State Laws covering the handling of eggs:

> Grade A Eggs have an air cell $\frac{1}{8}''$ in depth or less; regular, localized. Shell clean, sound, normal. Yolk: well centered; outline indistinct; motion sluggish; free from visible germ development and other defects or blemishes. White: firm; clear.

Long storage periods of eggs demand nonfluctuating, low-temperature

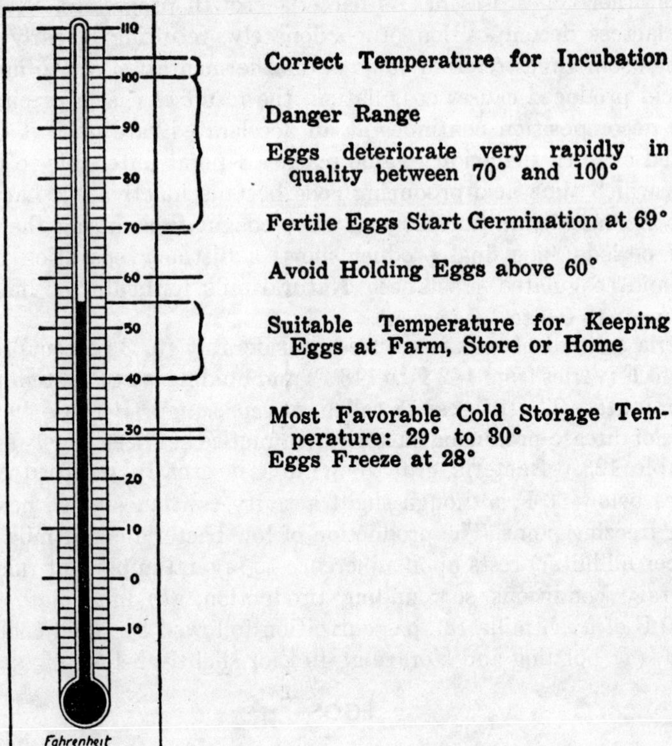

Fig. 408. Effect of temperature on egg quality. (From General Bulletin No. 545, Pennsylvania Department of Agriculture. Courtesy, Connecticut Department of Agriculture.)

TABLE 124. QUALITY OF EGGS VS. DEPTH OF AIR CELL*

Quality	*Depth*	*Movement*
AA..........	$\frac{1}{8}$ in.	$\frac{1}{4}$ in.
A...........	$\frac{1}{4}$ in.	$\frac{1}{4}$ in.
B...........	$\frac{3}{8}$ in.	May be free but not bubbly
C..........	No limit	No limit

* Courtesy, U.S. Department of Agriculture, Agricultural Handbook No. 75.

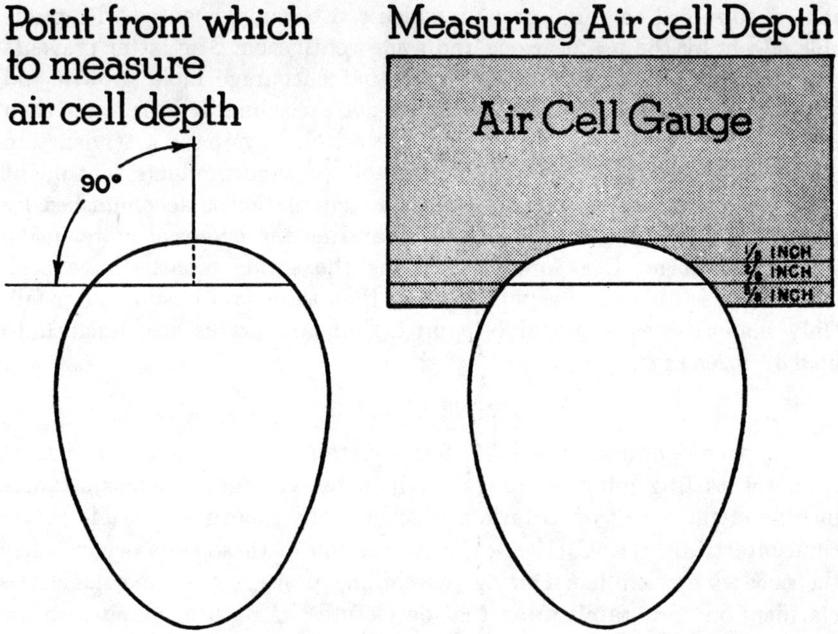

Fig. 409. Depth of air cell within an egg as shown by the gage. Any depth beyond the normal contraction level reached shortly after being laid is due to evaporation. (Courtesy, Agricultural Handbook No. 75, U.S. Department of Agriculture.)

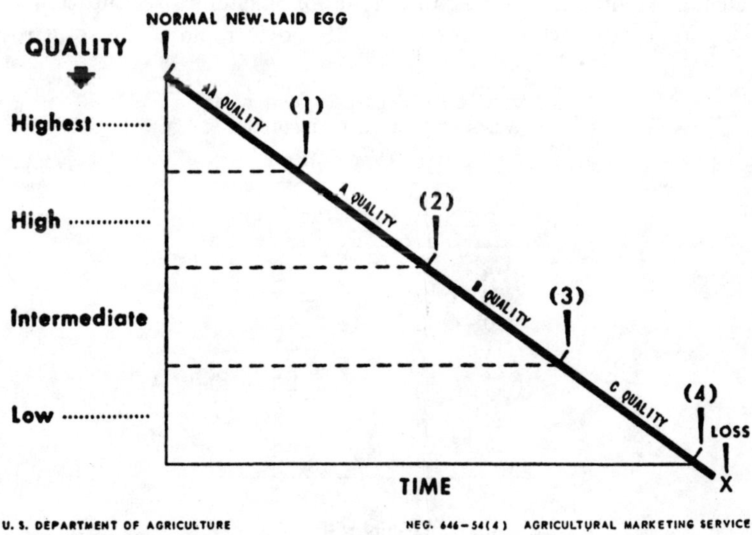

Fig. 410. Egg quality decline. Good cold-storage practice tends to decrease the slope of the quality decline, thereby maintaining high egg quality over longer periods. (Courtesy, Agricultural Handbook No. 75, U.S. Department of Agriculture.)

control to retard bacterial growth, which can be supplemented with desirable effects by the use of ozone and some ventilation. The latter prevents the formation of stagnant air pockets that encourage mold growth and absorption of odors. Temperature levels not exceeding 31 F or lower than 29.5 F have been found to be highly successful. Eggs have a tendency to absorb flavors and therefore should be isolated from products that might prove offensive in this respect. Proper air circulation is accomplished by providing aisles between stacks and separating the layers of crates using suitable dunnage. Eggs produced during the spring months have been found to possess better keeping qualities than those laid in summer or fall. Only nonfertile eggs should be stored. Standard crates are designed to hold 30 dozen eggs.

WINE CHILLING

New wine contains, in soluble form, certain substances which detract from the tasting qualities prevalent in older vintages. Such substances include tartar, a salt of tartaric acid, chemically known as a crude potassium bitartrate, $HK(C_4H_4O_6)$. Crystallization of these salts occurs when the new wine is chilled. During the chilling process, the wine undergoes clarification and stabilization that accelerates its natural aging to bring about a smoothness and mellowness nearer to that of more seasoned wines. After appropriate shock mixing of new wine to already chilled wine, causing a rapid temperature drop in the fresh product, it is agitated in refrigerated storage tanks for a period of four days or more, depending on its alcoholic content. The agitating action hastens precipitation of the salt crystals, after which the wine is filtered to remove the solid residue.

TABLE 125. FREEZING POINT OF WINE VS. NATURAL ALCOHOLIC CONTENT

% Alcohol	*Freezing Point °C*
8	−3.2
9	−3.7
10	−4.2
11	−4.7
12	−5.2
13	−5.7
14	−6.25
15	−6.85

DUNNAGE

Air circulation in chill rooms and coolers of cold-storage plants is highly important to prevent the formation of stagnant air pockets which provide an ideal environment for the growth of micro-organisms, particularly molds. Air motion is also required by perishable commodities to assure uniform cooling and to dissipate heat generated by respiration.

Owing to temperature differentials at various heights of the compartment, a natural air movement occurs during which air cooled by the brine pipes located near the ceiling is forced downward by its own weight and the pressure of the lighter warm air rising from the floor.

To assist the mass of air in its journey around the circuit, space between the layers of cases and/or sacks containing the stored products should be provided. Aisles, in addition to their convenience for handling and stacking, should be so arranged as to effect greater efficiency in air circulation.

Boards, preferably 2″ × 4″, are employed to provide sufficient space between the floor of the compartment and the bottom containers. Strips of wood, approximately 1″ × 2″, are placed between layers of containers. Such materials are called "dunnage" and form part of the required accessories maintained by well-managed cold-storage plants.

The larger dunnage recommended to separate bottom layers from the floor is selected because minimum air velocity occurs at this point and less resistance is afforded by the greater transit area provided. Modern plants employ pallets that act as dunnage and also allow handling by fork trucks.

Air space between meat carcasses when placed in sharp freezers in an unfrozen state is particularly required to assure a uniform temperature reduction throughout the mass. However, meats already frozen keep best when placed in close contact without air space as the combined cold mass is less susceptible to temperature change due to fluctuations in surrounding atmospheric temperature levels.

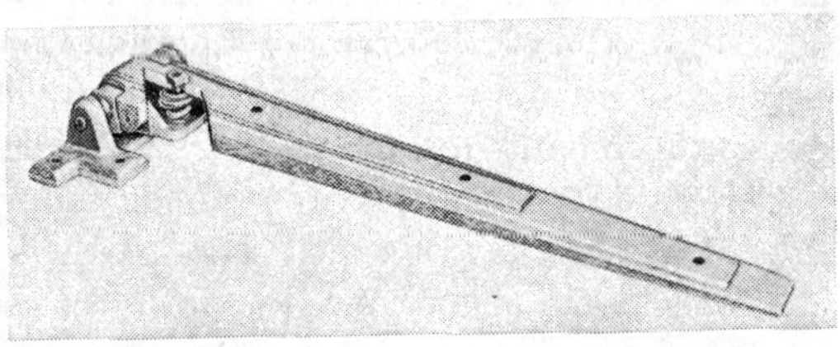

Fig. 411. Cold-storage door hardware. (*Left*) Handle and fastener. (*Right*) Hinge. (Courtesy, Jamison Cold Storage Door Co.)

COLD-STORAGE WAREHOUSES

The cold-storage warehouse may be any size, large or small. It may be operated for the purpose of holding food commodities for long periods, such

Fig. 412. One of the many types of cold-storage compartments. This one is used to hold cold dishes for a hotel kitchen. Note insulated doors and hardware. (Courtesy, Jamison Cold Storage Door Co.)

as one receiving a season crop to be shipped at some future date—possibly 30, 60, or 90 days or even six months from the time of harvesting. Or it may be a warehouse that holds produce for short periods, as do many commission merchants located in distributing centers of large cities.

Certain design features are common to all warehouses and must be considered at the time of construction. Others are peculiar to certain types and sizes. Some of the more important of these features are:

1. Space—sufficient cooler and freezer space:
   (*a*) For routine handling of scheduled deliveries to the plant relative to scheduled shipments from the plant
   (*b*) For heavier than usual demands
   (*c*) To allow for future expansion
2. Platforms—receiving and shipping:
   (*a*) Convenient for the approach of trucks and/or railroad reefers
   (*b*) Receiving platform above shipping dock to exploit gravity handling
   (*c*) Large enough to avoid congestion and to facilitate rapid loading and unloading
3. Interior transportation facilities:
   (*a*) Gravity conveyors for loading, unloading
   (*b*) Incline conveyors (powered)

(c) Elevators, where more than one story
(d) Chain conveyor for ice, if handled
(e) Fork and platform trucks, electric
(f) Hand trucks and dollies
(g) Overhead rail (trolley) for carcass meat with appropriate switches
(h) Convenient corridors

4. Construction details:
   (a) Firm footings and foundation to prevent floor and wall cracking
   (b) Adequate and properly applied insulation throughout
   (c) Expansion joints at wall and floor junctures, expecially if ground floor is not insulated
   (d) Curtained walls to eliminate exposure of cold room walls to outside temperature
   (e) Maximum temperature rooms located next to walls exposed to outside temperature
   (f) Adequate flashing at all outside structural junctures
   (g) Waterproof weather-exposed surfaces, especially roof
   (h) Mastic material between critical layers of structural materials, i.e., wall and insulation, roofing and insulation, etc.
   (i) Insulate roof rather than adjacent ceiling, if possible
   (j) Vestibules at doors and other openings to reduce entrance of outside air, since moisture from air condenses on walls and promotes mold growth
   (k) Mold-resistant paint on corridor and other walls
   (l) Installation of heavy insulated refrigerator doors
   (m) Fire walls included where size warrants

5. Refrigeration methods:
   (a) Determine the advantages of direct expansion or brine system. Provide large enough refrigeration plant
   (b) Isolate refrigeration plant from warehouse proper
   (c) Provide adequate standby equipment for emergencies
   (d) Employ forced convection wherever apropos
   (e) Install humidifiers such as spray heads in chill compartments for moisture control
   (f) Locate thermometers in all rooms
   (g) Provide gas-tight compartments with atmospheric washers for control of CO_2 where apples are to be stored in any quantity. (Sodium hydroxide solution—about 25 to 30 per cent—through which the entrapped atmosphere can be forced by circulating fan is usually employed in washers)
   (h) Use dunnage for air circulation between packages
   (i) Isolate ice-making tank and ice storage from warehouse proper
   (j) Minimize conversion of freezer rooms to chill rooms. Thawing of ice formed from leaks and condensation damages bonded insulation material such as cork board. Resulting drippage causes an untidy condition due to accompanying discoloration
   (k) Install temperature alarms (maximum or minimum in critical rooms)

6. Utilities:
   (a) Insure adequate and cheap supply of condenser cooling water or install appropriate cooling towers
   (b) Insure adequate source of power to operate compressor, fans, pumps, etc.
   (c) Provide good lighting in all parts of the plant and the warehouse
   (d) Utilize sufficient heat from condensers to heat offices where practical
   (e) Provide weighing scales

7. Control:
   (*a*) Ample office space for carrying on the normal business transactions of the warehouse and refrigeration plant. Should be isolated from warehouse proper
   (*b*) Checking points established at appropriate locations to control and record deliveries and shipments
   (*c*) Access to warehouse possible only at checking points to prevent losses
   (*d*) Fire extinguishers strategically located and ready for use

FROZEN-FOOD LOCKER PLANTS

Storage lockers for individual families are a source of extra income for operators of cold-storage plants supplying their own refrigeration. Because their refrigeration requirements are drawn from the large capacity source that also serves the warehouse locker plants, the cost per ton is relatively low. Another source of income for ice-making cold-storage plants is the sale of small blocks and packaged cracked ice through vending machines. However, plants operating either one or both of these enterprises should keep such activities remote of the main function of cold-storage operation.

The function of the frozen-food locker plant is to quick-freeze food commodities and reduce their temperature below the low point of the zone

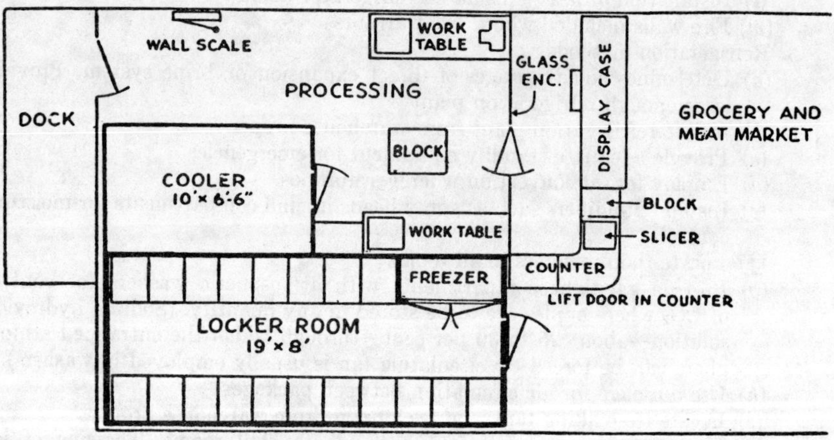

Fig. 413. Layout of small freezer and locker plant connected with retail grocery and meat market. (Courtesy, National Frozen Food Locker Institute.)

of crystallization and hold it at a temperature of preservation until needed. Its auxiliary function is to process such commodities in preparation for quick freezing.

A complete locker plant is designed and organized so that the following steps may be taken with full regard to sanitation, preservation, and efficiency throughout:

1. Receiving (not necessarily refrigerated). Proper facilities and space for receiving weighing commodities to be processed and stored.

2. Chill and aging (36 F). Meats must be prechilled before being cut into required size. This applies to fowl and fish as well. This requires a chill room. Beef is aged in the chill room for several days before processing.

3. Preparation (not necessarily refrigerated). (*a*) Meats, poultry, fish, etc. Involves cutting into roasts, steaks, chops, fillets, etc.; grinding, cleaning, and preparing poultry and fish. Must be accomplished before food becomes too warm. (*b*) Fruits and vegetables—involves washing and blanching.

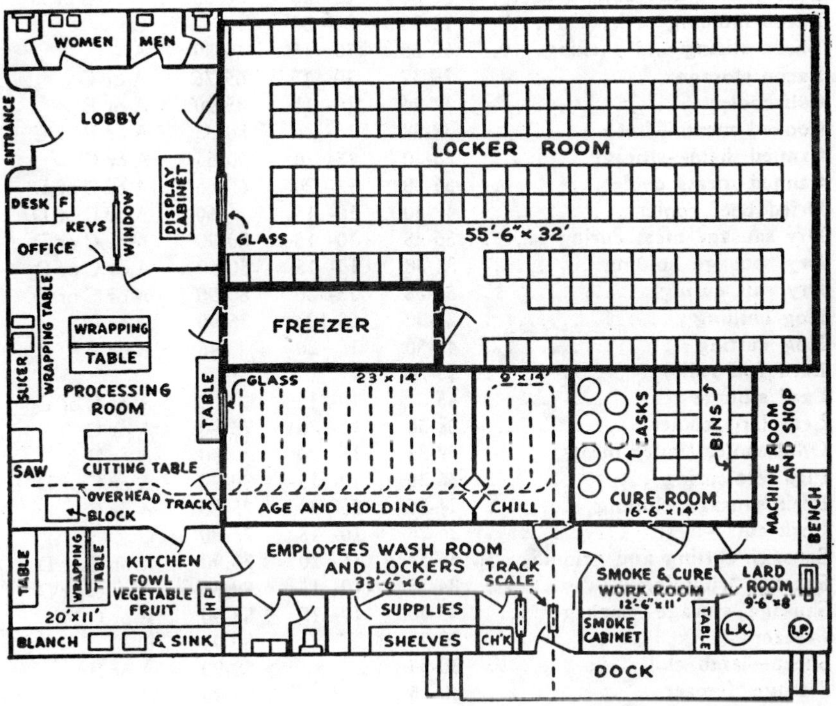

Fig. 414. Typical layout of complete frozen-food locker plant. (Courtesy, National Frozen Food Locker Institute.)

4. Wrapping or packaging (not necessarily refrigerated). All commodities must be properly packaged and/or wrapped before freezing. During this step the package is labeled to identify its content, weight, date, and owner.

5. Quick Freeze. Many small locker plants use the blast freeze method. This can be accomplished either by directing a blast of cold air into a walk-in freezing room or by placing the packages in the sharp freezer compartment of a blast freezer unit. Temperature in the blast freezer usually is set at −10 F. One type of installation obtains this temperature from the same unit that cools the locker storage room. The difference in temperature between the two rooms is due to the greater heat load of the

TABLE 126. RANGE OF TEMPERATURE AND HUMIDITY CONDITIONS IN VARIOUS ROOMS OF FROZEN-FOOD LOCKER PLANTS†

Department and Product Processed	Room temperature °F, DB	Frequency of air recirculation per hour (forced draft)	Relative humidity %	*Symbols for type of refrigerating system
Beef chill cooler	30-31	60-100	85-90	A or B
Beef carcass holding	33-34	12- 15	85-90	A or B
Beef curing (pickle)	36-38	.. ..	.. ..	A or C or D
Beef cutting and boning	36-38	15- 20	50-55	A or C or D
Bacon molding	12-15	20- 30	.. ..	C
Bacon slicing and packing	50-55	10- 15	45-50	A
Bacon storage	28-32	10- 15	65-75	A or C
Calf cooler	32-34	10- 15	85-90	A or B
Cooked meats—fresh	32-38	15- 20	80-85	A or B or C
Canned ham—storage	26-30	15- 20	60-65	A or C
Canned meats cooler	36-38	15- 20	60-65	A or C or D
Dried beef cooler	48-50	10- 15	60	A or C or D
Dry sausage meat curing	36-38	10- 15	80-90	A or C or D
Dry sausage holding	36-38	10- 15	50-60	A or C or D
Dry salt curing	34-38	15- 20	80-90	A or C or D
Hog chilling	28-30	60-100	85-90	A or B
Hog cutting	45-50	15- 20	45	A or C
Lard (off rolls)	65-70	.. ..	.. ..	
Lard storage	45	10- 15	.. ..	A or C or D
Leaf lard cooler	28-30	15- 25	75-80	A or C
Offal cooler (fancy meats)	32-34	15- 20	90	A or B
Pigs feet curing	36-38	10- 15	80-90	A or C or D
Pork sausage chilling	24-26	25- 40	75-80	A or C
Pork cuts—fresh	26-28	10- 15	75-80	C
Sausage cutting and grinding	36-38	15- 20	70-90	A or C or D
Sausage trimming	34-35	10- 15	80-85	A or C or D
Sausage storage—package	45-50	10- 15	88-90	A or C
Freezer	10-15	.. ..	80-90	C
Sheep—lamb chill	30-34	.. ..	88-90	A or B
Storage freezer	0- 5	.. ..	80-90	C
Smoked meats storage	28-32	10- 15	65-70	C
Smoked meats hanging	45-50	10- 15	65-75	A or C
S. P. curing	36-38	10- 15	80-90	A or C or D
S. P. retard cure	26-28	10- 15	80-90	C
Salted hides	50-55	10- 15	85-90	C or D

* A—Forced draft unit cooler either open brine spray or closed ammonia coil with brine defrosting.
B—Overhead deck with open brine spray or ammonia coils with brine defrosting.
C—Ammonia coils—prime surface or fin type or evaporator plates—not under forced draft.
D—Brine coils—not under forced draft.

† Courtesy, "Your Locker Plant," National Frozen Food Locker Institute, Elizabethtown, Pa.

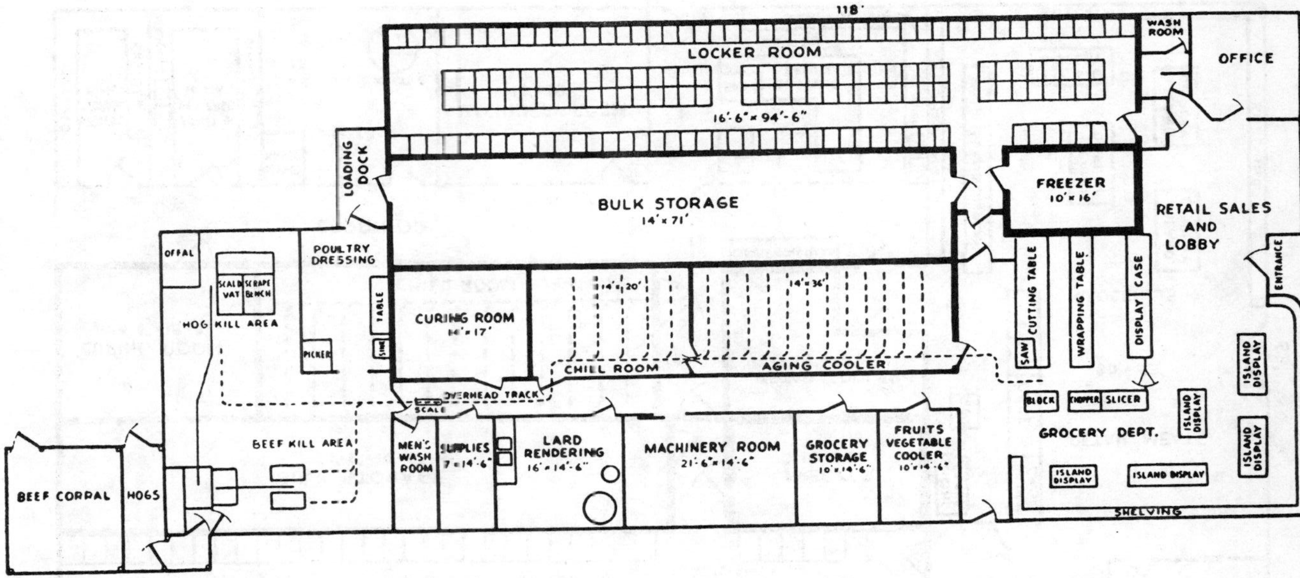

Fig. 415. Layout of frozen-food locker plant connected with beef and pork processing plant. (Courtesy, National Frozen Food Locker Institute.)

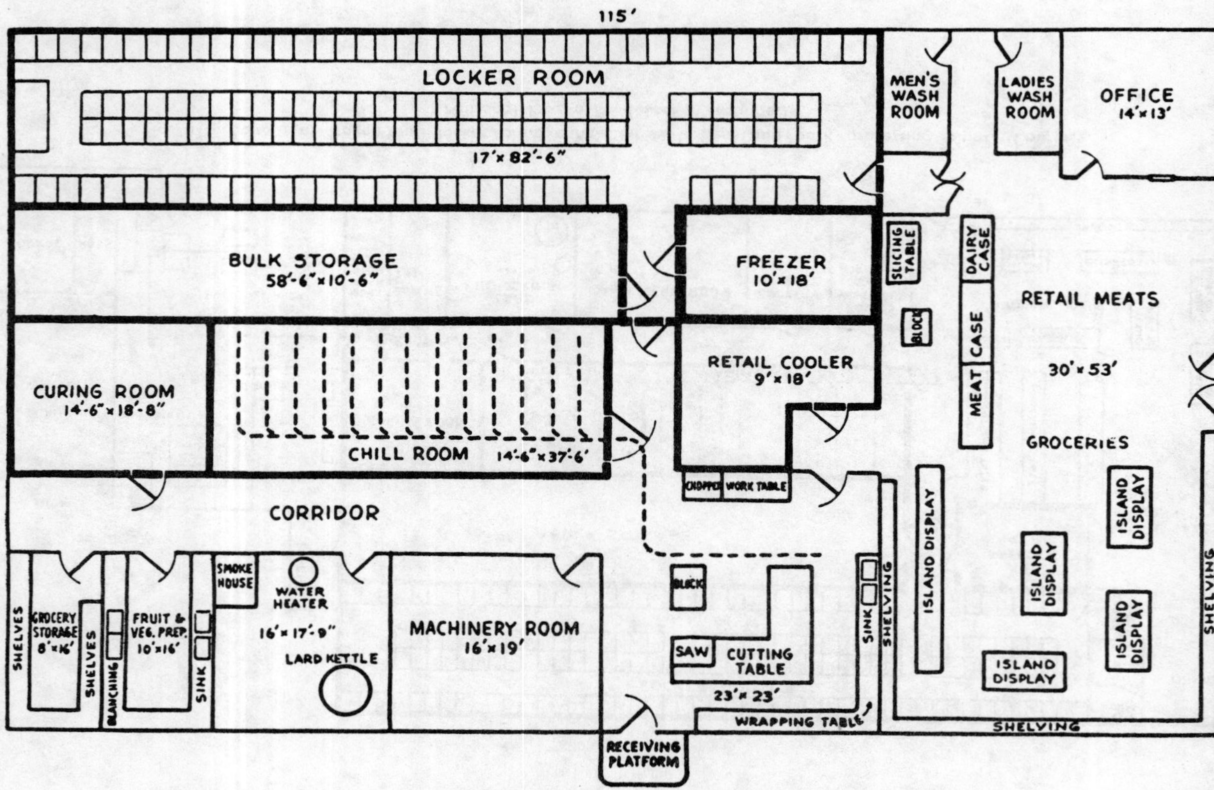

Fig. 416. Layout of large frozen-food locker plant connected with retail meat and grocery market. (Courtesy, National Frozen Food Locker Institute.)

TABLE 127. AIR HEAT GAIN DUE TO INFILTRATION IN FROZEN-FOOD LOCKER PLANTS AND OTHER STORAGE ROOMS. (ASSUMING RELATIVE HUMIDITY OF ENTERING AIR TO BE 50% AND DRY BULB TEMPERATURE OF 95 F)*

Storage Room Temperature °F	*Heat Gain (Btu per cu ft)*
45	2.12
40	2.31
35	2.49
30	2.26
20	2.62
10	2.93
5	3.12
0	3.28
−5	3.41
−10	3.56
−15	3.67
−20	3.88
−25	4.00

* Courtesy, National Frozen Food Locker Institute, Elizabethtown, Pa.

TABLE 128. INFILTRATION FOR VARIOUS SIZED COMPARTMENTS (BASED ON AIR CHANGES PER 24 HOURS)*

Size, cu ft	*Air Change per 24 Hours*
500	20.0
1,000	13.5
1,500	11.0
2,000	9.3
4,000	6.3
6,000	5.0
8,000	4.3
10,000	3.8
20,000	2.6
30,000	2.1

* Courtesy, National Frozen Food Locker Institute, Elizabethtown, Pa.

locker room. Contact or plate freezers are also used. In the latter type, packages are placed on the plate-type evaporator shelves. Evaporator temperatures are usually lower than that of blast freezers at about −20 F.

6. Storage. Lockers are located in the main storage room, held at 0 F, in which the quick-frozen foods are stored. Each locker is rented by individual families who call for one or more packages as required.

VAPOR BARRIERS

Cold storage walls act both as bulkheads to contain the refrigerated substances and to retard the inleak of heat. Assuming that an appropriate insulating material, such as cork, is used in the construction of the walls, there are other factors that tend to increase thermal conductivity. One of the most effective factors that reduce the natural insulating ability of materials is moisture. Condensation, hygroscopic processes, respiration and the presence of free water all tend to cause moisture to permeate the matrix of various insulators. Its presence is responsible for 10 to 30% increase in thermal conductivity depending on the material, the temperature difference and the volume of moisture to insulation relationship. The combination of moisture and frost within the cell structure forms a path for the conduction of heat from warm to colder areas.

To prevent the absorption of moisture from whatever external or internal (refrigerated compartment) sources, some material impervious to water should be stretched over the wall surfaces to act as a vapor barrier. Effective vapor barrier materials include polyethylene, aluminum foil and some types of waterproof fillers, such as those containing water glass (sodium or potassium silicate). Polyethylene is durable and should be lapped at the joints at least 9 in. A combination of aluminum foil and polyethylene is a highly effective vapor barrier.

The effect of substances to prevent the entrance of water is a measure of their permeance and is expressed in *perms*.

SUMMARY OF CHAPTER 14

Cold storage involves the employment of mechanical refrigeration to maintain temperatures at a level consistent with the keeping qualities of foods, drugs, furs, flowers, and other perishable items.

Perishable foods consist of fresh fruits and vegetables; dairy products such as eggs, milk, butter, and cheese; commercial yeasts; fish of all kinds, including skin, scale, and shell types; processed bakery items ready for the oven; fresh meats, including beef, lamb, mutton, pork, game, and all their parts; and poultry.

Fish spoil rapidly above 32 F. They should be iced or frozen immediately when caught. They are usually sharp frozen although quick freezing is carried on extensively.

Glazing of fish is performed to prevent dehydration and oxidation. Glazing is accomplished by repeatedly dipping or spraying frozen fish. Water treated with various chemicals to provide a crack-proof glaze that will not evaporate rapidly is used.

Furs are stored for the purpose of killing moths. The procedure is to expose furs at 20 F for three days, raise to 5 F for two days, lower to 20 F for three days, and finally raise and maintain at 36 F.

Flowers should be stored at a safe (nonfreezing) temperature and high relative humidity. With some exceptions, cut flowers stay fresh from 3 to 10 days at 36 F to 40 F and 85 percent relative humidity.

Food spoilage agents include enzymes, bacteria, yeasts, and molds.

Most spoilage agents may be controlled by reducing temperatures below freezing, 32 F.

All spoilage agents may be controlled indefinitely at temperatures below 0 F.

Storage of frozen commodities should be maintained at 0 F or below. Temperature fluctuation should be avoided.

The effects of food spoilage include off-taste, change in appearance, off-color, obnoxious odor, putrefaction, mold, wilt, souring, rancidity, softening, rotting, and various stages of decay. Loss of vitamin content also occurs.

Spoilage of food is caused by the activity of enzymes, bacteria, yeasts, and molds. Other causes include bruising, desiccation, improper freezing, absorption of foreign odors, and excessive respiration.

Enzymes and micro-organisms can be controlled by exposure of foods to optimum storage temperatures, maintaining sanitary storage and handling conditions, careful handling to avoid bruising and other damage, ozonization and other methods of mold control, eliminating sources of free moisture (condensation), proper preparation of frozen and other fruits and vegetables, quality control of incoming shipments, sorting and washing, preventing delays in placing food in storage, and inventory control (first in, first out).

Optimum temperature is the temperature at which an individual commodity maintains its quality longest during storage.

Respiration in fruits and vegetables can be reduced by precooling before or during shipment to storage, rapid cooling when placed in storage, maintaining tight storage compartments (with controlled ventilation) for natural CO_2 control, or by addition of CO_2 from external source.

Ripening of fruit may be hastened by exposure to gaseous ethylene.

Methods of reducing or preventing food spoilage include canning, fermenting or pickling, dehydrating, smoking, preserving, chilling and freezing (sharp or quick).

Wilting and shrinkage are some effects of desiccation and can be reduced by maintaining proper atmospheric moisture conditions (controlled relative humidity, usually between 85 and 95 percent).

Sharp freezing allows freezing to occur normally at usual storage temperatures. Applies to meat, although vegetables and fruits placed directly in home freezers or lockers are also sharp frozen.

Quick freezing causes freezing to occur rapidly at temperatures as low as −40 or −50 F. The reason for quick freezing is to pass through the zone of water crystallization as quickly as possible. Quick freezing is accomplished either by immersion in cold brine (on trays), in moisture-proof containers, on contact plate, or by air blast.

Freeze drying is the removal of water from a substance by first freezing it and then evaporating the ice under vacuum. The process is one of sublimation.

Freeze concentration of a liquid substance is the reduction of temperature to form ice crystals which are in turn removed by centrifuging to form a slurry.

Beef and lamb are chilled immediately after killing at temperatures of 32 F to 36 F. Aging for three days to one week often precedes the freezing process. Meat is chilled, cut, and wrapped before quick freezing. Carcasses are sharp frozen after aging.

Pork is chilled immediately after killing. Quick chilling is important for bacterial control. Use of hog chillers, cooled by air passing through an overhead brine spray chamber separated from the cooler by a deck, is the usual practice. Moisture control is necessary to prevent condensation on carcass.

Rancidity of fats, including butter, is due to a hydrolyzing action wherein fats and oils yield glycerin and fatty acids, carbohydrates are decomposed by the action of water, and proteins yield amino acids.

Milk sours as a result of the activity of certain species of bacteria (lacti) to break down milk sugar to form lactic acid. The action is retarded at temperatures below 40 F.

Pasteurization is the act of raising the temperature of milk (or other substances) to 145 F for 30 minutes and then reducing it immediately to 50 F or below. This is known as the "hold" method. Continuous pasteurizing (before homogenizing) involves raising the milk temperature to 161 F for 16 seconds and then cooling immediately to 36 F. Rapid cooling is obtained by passing milk over Baudelot coolers.

New wine is chilled to crystallize certain salts so that they may be precipitated and removed by filtration to give it a taste of that approaching older vintages.

Absorption of foreign odors and taste can be controlled by selective storage procedure.

Dunnage is wood strips (or other material) placed between layers of food containers to allow free passage of cold air.

Cold-storage warehouses should be designed to provide sufficient space, receiving and shipping platforms, interior (and exterior) transportation facilities, and convenient corridor system.

Cold-storage plants should be well constructed with firm footings, subfloor drains, well insulated, and waterproof. Curtain walls and vestibules reduce heat gain and condensation. Utilities should be adequate.

Methods of refrigeration must be considered on the basis of type of plant, commodities handled, economy of refrigerant distribution, etc. Circulating brine systems, central direct expansion systems, unit freezers, forced or gravity convection systems are some of the methods that can be selected. A combination of several types may prove more flexible and economical.

Frozen-food locker plants are a benefit to some communities. Their function ranges from storage of commodities prepared elsewhere to complete processing and storage of many different kinds of foods.

Temperatures maintained in various departments of frozen-food locker plants are: (*a*) freezer 0 F to −10 F; (*b*) locker room 0 F; (*c*) chill and/or aging room 36 F, processing of meats and produce 85 F or below to 36 F for cutting and boning beef (see Table 126).

STUDY OUTLINE FOR CHAPTER 15

Lubrication
- Definition
- Purpose

Friction
- Definition
- Friction vs. mechanical efficiency
- Frictional energy to heat energy
- Sliding friction
- Rolling friction
- Fluid friction
- General laws of friction
- Coefficient of friction
- Frictional components
  - Bearings
    - Radial plain bearings
      - Hard bearing metals
      - Soft bearing metals
      - Projected area
    - Guide bearings
    - Thrust bearings
    - Antifriction bearings—ball and roller types
      - Life factor
      - Metallurgy
        - Hardness
        - Steel alloy identification (SAE)
      - Radial load calculations
  - Cylinders:
    - Loads
    - Piston rings:
      - Types
  - Gears:
    - List of gear terms
    - Frictional characteristics of gears

Theory of lubrication:
- Boundary lubrication
- Viscosity:
  - Viscosity relation to SAE numbers
  - Viscosity index
- Hydrodynamic lubrication:
  - Flat surfaces, parallel
  - Flat surfaces, converging
  - Curved surfaces, full circular bearing
- Pressure distribution in full journal bearings
- Viscosity, speed and load relationship in journal bearings $\left(\frac{ZN}{P}\right)$
- Fluid film in gears
- Oiliness

Lubricants:
- Types

Hydrocarbon structures

Refining of lubricating oils:
- Distillation
- Refining methods:
  - Clay filtering
  - Acid treatment
  - Dewaxing

Classification of mineral lubricating oils

Operating problems associated with lubricants

Physical and chemical properties of mineral lubricating oils

Additives in lubricating oils

Fixed oils:
- Iodine number
- Saponification number
- Uses as lubricants

Plastic lubricants:
- Greases:
  - Consistency
  - Drop point
  - Classification

Solid lubricants

Application of lubricants:
- Factors
- Methods of application:
  - Oils
  - Greases

Refrigerator compressor lubrication:
- Parts to be lubricated—all types

Refrigerating compressor lubricating oils

Refrigerating system problems due to lubricating oils

Oil-refrigerant miscibility

Oil traps, chillers, and stills

Foam breakers

Copper plating

Gear lubrication and lubricants

Gear failures

15

Lubrication

When two bodies are in physical contact with each other, there is always an interposing force that resists motion of one body relative to the other. This reactionary force resisting motion is called friction; friction may be reduced but never eliminated. This fact alone makes perpetual motion a proverbial will-o'-the-wisp.

Lubrication is the science of reducing friction by the application of a suitable substance between the contiguous surfaces of bodies having relative motion. The friction-reducing substances are called lubricants.

Friction in some forms is destructive to the rubbing surfaces of engineering metals; in any form, friction causes energy to be wasted. Ultimately it is dissipated as heat, causing a temperature rise of the bodies involved. Because lubrication reduces the degree of destructive forms of friction, or converts the latter to less destructive forms, it is an essential phase of mechanical equipment maintenance.

Refrigeration equipment embraces all three basic types of frictional components; namely, bearings, gears, and cylinders. Protection of these components against (*a*) rapid wear and (*b*) excessive operating temperatures due to friction requires continual lubrication. The degree of protection afforded any frictional component depends on the quality of lubrication applied, including suitability of lubricant, method and interval of lubricant application, and the amount of preventive maintenance devoted to the lubrication system.

In mechanical refrigeration systems, only the compressor requires lubrication. If a reciprocating compressor is used, the lubricant enters the system. It enters between the piston rings and the cylinder walls, and/or with the refrigerant gas as it passes through the suction valves of trunk type pistons. This situation best demonstrates the importance of lubricant suitability. *Lubricants not excepted, any substance introduced into the refrigeration system (other than high-quality refrigerants) becomes at once a foreign material that tends to interfere with refrigerating processes and re-*

duces the efficiency of the system. This fact alone is reason enough to consider carefully the relative merits of various types and grades of lubricants available in order to select those that will, at best, prove least objectionable while performing their own function effectively.

While lubricant selection is based primarily on frictional conditions, consideration must be given also to such other processes carried on within a mechanical system whose degree of efficiency may depend on lubricant suitability. As for the three basic frictional components, the following rule will always apply, viz.: *Effective lubrication depends on the proper application of an adequate amount of correct lubricant at the time it is needed.*

FRICTION

As previously stated, friction is a force that opposes motion. It exists regardless of the shape, the size, or the nature of the bodies in contact. A journal rotating in a bearing, water flowing through a pipe, an airplane traveling through the air, or a wheel rolling over a road—all are examples of frictional conditions.

The principal causes of frictional development appear to be (*a*) molecular attraction existing between two bodies in physical contact, (*b*) interlocking of opposing surface asperities present on all surfaces, (*c*) surface waviness.

Friction has its advantages as well as its disadvantages. It serves a useful purpose in the case of vehicle brakes and clutches. It also furnishes the traction required between wheels of rolling equipment and their running surfaces. Friction can also be destructive, as already stated.

The amount of friction developed depends on the

(1) kind of frictional system involved,
(2) pressure exerted between contacting bodies,
(3) nature of rubbing surfaces,
(4) condition of rubbing surfaces.

The degree to which friction develops is apparent by the

(1) amount of heat generated,
(2) rate of surface wear of solid bodies,
(3) power required to produce motion,
(4) ratio of power output to power input of any mechanical system.

There are various kinds of friction; they include sliding friction, rolling friction, and fluid friction.

Sliding friction is resistance to relative motion set up between sliding solid bodies, such as iron, steel, wood, babbitt, etc., or any combination of such materials, in direct physical contact. Sliding friction develops only between dry (nonlubricated) surfaces.

Rolling friction is resistance to relative motion set up between two solid bodies when one or both roll over the surface of the other.

Fluid friction is a force that resists flow of liquids or gases.

Boundary friction is the force that resists motion of two bodies whose opposing surfaces are wetted by a lubricant but which surfaces are not fully separated by a lubricant film.

The force required to initiate motion of a body at rest is called *static friction*. The force required to continue motion of a moving body is called *kinetic friction*. Static friction is always greater than kinetic friction for the same bodies under similar conditions. For this reason, it takes less energy to keep a moving body in motion than it does to initiate motion of the same body at rest.

Friction vs. Mechanical Efficiency

A machine cannot produce an amount of work equivalent to the energy put into it because of friction. The relation of all three factors may be expressed as

$$\text{Work output} = \text{work input} - \text{friction}$$

From this expression, the mechanical efficiency of any machine may be derived, viz.:

$$\%\ \text{Efficiency} = \frac{\text{work input} - \text{friction}}{\text{work input}} \times 100$$

or

$$\%\ \text{Efficiency} = \frac{\text{work output}}{\text{work input}} \times 100$$

and

$$\text{Friction hp} = \text{hp input} - (\%\ \text{efficiency} \times \text{hp input})$$

Frictional Energy to Heat Energy

Frictional energy developed between any two rubbing surfaces is converted to heat energy. This conversion is in accordance with the *First Law of Thermodynamics* and the universal law governing the *Conservation of Energy*.

If 100 hp were applied to a simple gear train which delivered only 98 hp, the efficiency would be $\frac{98}{100} \times 100 = 98$ percent. The remaining 2 percent power input is converted to frictional energy which, in turn, is converted to heat energy. This latter is borne out by the subsequent temperature rise that occurs in all mechanical work systems.

The amount of heat developed in the above gear train would be determined by the equation:

$$\begin{aligned}\text{Btu per min} &= \text{frictional hp} \times \text{mechanical equivalent} \\ &= f \times J\end{aligned}$$

where f = frictional hp = hp input − (efficiency × hp input)

$f = 100 - (98\% \times 100) = 2$ hp

and $J = 42.5$ Btu per min per hp

Therefore

$$\text{Btu per min} = 2 \times 42.5$$
$$= 85 \text{ Btu per min, frictional heat}$$

Sliding Friction

A journal rotating in a fixed plain bearing, guides, ways, skids, thrusts, crossheads, and gears, are all machine parts that develop sliding friction. Under sliding (dry) conditions, friction reaches its maximum force. The reason for this is twofold:

First: Unwetted parallel surfaces permit maximum area contact when two solids are brought together, a condition under which molecular attraction attains its greatest influence. The bonding force of molecular attraction increases with better surface smoothness because the area of contact becomes proportionately greater. Johannsen blocks are difficult to pull apart or slide across each other because of their high degree of surface smoothness. If it were possible to obtain perfect smoothness, any two such surfaces would adhere to each other permanently and (probably) would require the same force to separate them as it would to tear apart the solid material of which they were created.

Second: The surfaces of all engineering metals, regardless of their "finish," consist of numerous asperities. The magnitude of such asperities governs surface smoothness, or roughness. It will be noted in Fig. 417 that surface irregularities are both jagged and wavy. The jagged phase is developed by innumerable projections and depressions that cover the entire surface area. The waviness phase is developed by a series of waves that present the surface with an irregular contour. When two such surfaces slide over each other, a confliction of asperities occurs tending to block the passage of one projection over another located on the opposing member.

If forced to slide on each other, the projections of the hardest member will shear their opposing projections. Shearing action becomes more aggravated as pressure between the two surfaces is increased. If the pressure is great enough, spot temperatures will reach a level high enough to cause the metal to soften. Plastic flow occurs at this point, permitting welding of the two surfaces. Seizure results, followed by wiping of the bearing metal.

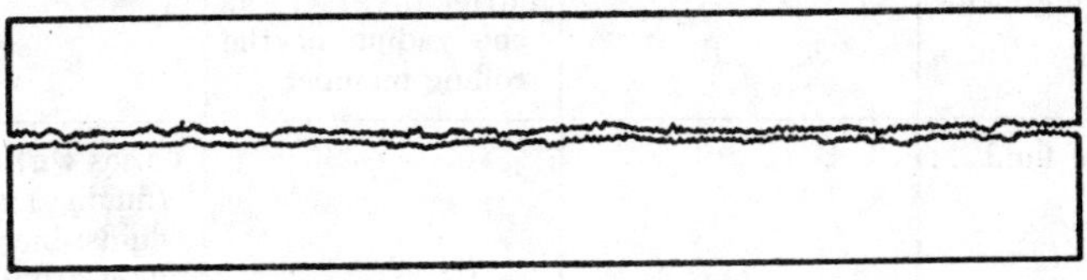

Fig. 417. How normally smooth surfaces appear when magnified.

Waves are relatively wide irregularities having a crest-to-crest dimension of $\frac{1}{32}$ in. or greater and a depth from crest-to-base of less than 0.002 in. Roughness (finely spaced minute projections) dimensions are on the order of ten-millionths of an inch for highly polished surfaces up to 200-millionths of an inch or more for relatively smooth surfaces.

The laws of sliding friction are given in Table 129.

TABLE 129. GENERAL LAW OF FRICTION

Condition	*Sliding Friction*	*Rolling Friction*	*Fluid Friction*
Load	Approximately proportional to the pressure between the sliding surfaces	Proportional to the load	Nearly independent of the pressure
Speed	Nearly independent of speed. Varies slightly with different ranges of velocity. Less at high velocity	Nearly independent of speed	Proportional to relative velocity at lower speeds and to the square of velocity at higher speeds (above 1600 ft. per min.)
Surface condition	Decreases with greater surface smoothness	Dependent on the condition of surfaces. Decreases with smoothness of surfaces	Dependent on the degree of smoothness, or roughness. Distortion of fluid layers occurs with pronounced surface roughness
Nature of materials	Varies with different materials and with different combinations of materials. Greater with soft materials	Dependent on the nature of materials. Decreases with surface hardness	Independent of the nature of materials
Area of contact	Nearly independent of area of contact		Directly proportional to the area of contact
Size of rolling member		Varies inversely as the radius of the rolling member	
Viscosity of fluid			Varies with viscosity (fluidity) of similar fluids, increasing as viscosity increases

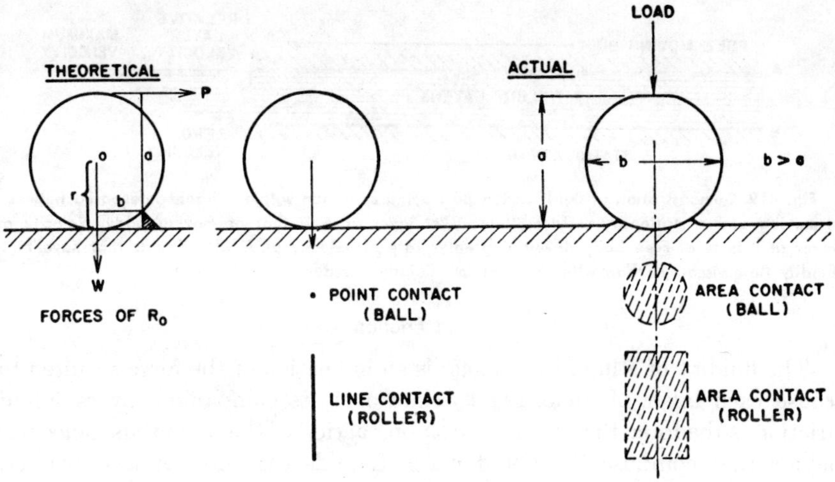

Fig. 418. Theoretical and actual (exaggerated) surface contact of ball and roller. Note distortion of rolling element and surface (stator) under load. Resistance due to friction (R_o) equals the coefficient of friction (f) times the load (w) divided by the radius (r), or $R_o = \frac{fw}{r}$.

Rolling Friction

Theoretically, a ball establishes point contact with its bearing surface. A roller establishes line contact. Under pure rolling conditions, friction would not exist since a point or a line has no area. However, applied loads tend to flatten a ball or a roller, and also indent the bearing surface or stator. Deformation of opposing members establishes area contact. The area of contact increases with greater loading and softer materials. Consequently, ball and roller bearings are made with hard steels.

As a ball or a roller rolls over a surface, it forms an indented band the width of which depends on the amount of flattening incurred on the ball or the length of roller. The bearing metal displaced as a result of indentation forms a temporary mound or "swell," immediately in front of the rolling member, the front being in the direction of rolling. An exaggeration of this condition is shown in Fig. 418. Rolling friction, then, is the sum of the resistances offered by the mound and the ever-present surface asperities. Some sliding action also occurs due to the slippage of the rolling member in its constant effort to lift itself over the obstructions.

The difference between static and kinetic friction for rolling bodies is slight, both of which are considerably less than comparable values for sliding friction.

Some general laws of rolling friction are given in Table 129.

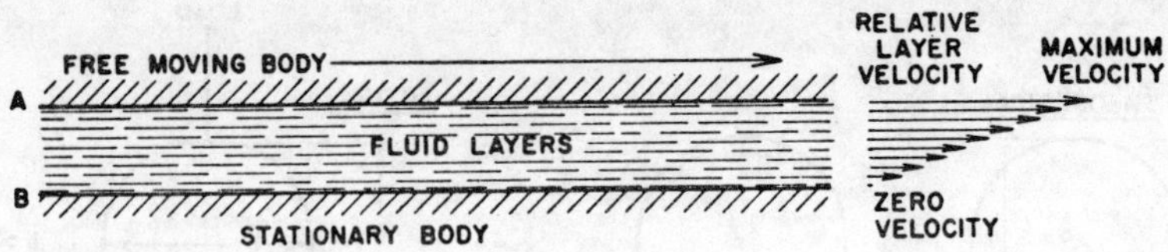

Fig. 419. Demonstration of fluid friction portraying condition within a liquid interposed between a moving and a stationary surface. Velocity of layer at A is that of moving body. Velocity of layer at B is zero. Free body develops greater rate of velocity at A with liquids that have high fluidity (low viscosity) than with those of low fluidity (viscous).

Fluid Friction

The fluidity of a liquid (or a gas) is an indication of the force required to cause each layer of molecules to slide over its companion layers. Fluid friction is the force that resists the sliding action of the numerous molecular layers that compose the liquid mass. Liquids that do not flow readily, such as castor oil, cylinder oil, and glycerin, have relatively high fluid friction. Rapid flowing liquids, including water, spindle oil, and alcohol, are fluids possessing relatively low fluid friction.

The action of the several molecular layers in sliding over each other is called *shear*. The speed with which they slide is known as *rate of shear*. The rate of shear within a fluid mass governs its fluidity. The fluidity of a liquid is expressed in terms of *viscosity*. Viscosity is the measure of liquid fluidity, the relationship of which is

$$\text{Viscosity} \propto \frac{1}{\text{fluidity}}$$

Fluidity decreases as viscosity increases. Viscosity, then, is the measure of fluid friction within a liquid mass. The laws governing fluid friction relative to bearings are given in Table 129.

Coefficient of Friction

Friction, being a force that resists motion, is measurable. Its value is equal to the counterforce required to initiate or maintain a body in motion. Because friction varies with any change in condition associated with moving bodies, the counterforce will also differ in each case. The value of frictional force remains constant only when all the conditions remain constant. Where all conditions remain the same except the load (i.e., the pressure between opposing surfaces), the force of friction will also vary, but in constant relationship to the load.

If a weight of 10 lb were placed on block *A* (Fig. 420), and a pull of 6 lb, as determined by a spring scale, were required to initiate motion, static friction being equal to the counterforce (pull) would also be 6 lb. If a drop in pulling force occurred, say to 4 lb, by continuing the block in

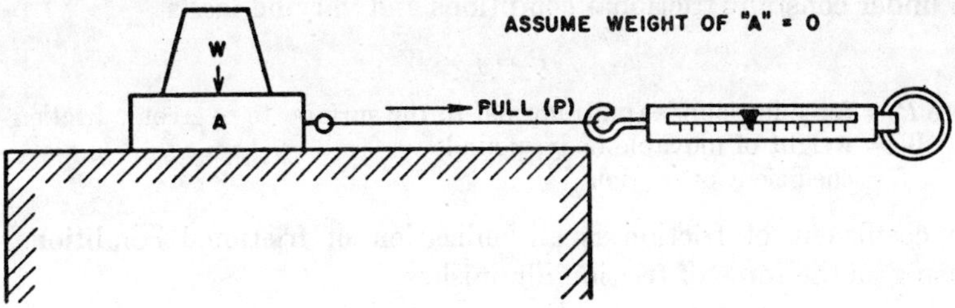

Fig. 420. Apparatus for measuring coefficient of friction, $f\left(f = \frac{P}{W}\right)$.

motion, it would indicate that kinetic friction is equal to the pull of 4 lb. Any increase in weight placed on block A would require a correspondingly greater pulling force provided that all other conditions remained the same.

By increasing the weight placed on block A to 20 lb, the values of static and kinetic friction would increase correspondingly to 12 and 8 lb, respectively. While the amount of pull varied with the load, their relationship remained constant as demonstrated below:

$$\text{Ratio } \frac{\text{Pull (lb)}}{\text{Load (lb)}} = \text{constant}$$

viz.:

$$\text{Kinetic friction} \begin{cases} \text{1st case: } \dfrac{\text{Pull}}{\text{Load}} = \dfrac{\text{friction}}{\text{weight}} = \dfrac{4}{10} = 0.4 \\ \text{2nd case: } \dfrac{\text{Pull}}{\text{Load}} = \dfrac{\text{friction}}{\text{weight}} = \dfrac{8}{20} = 0.4 \end{cases}$$

$$\text{Static friction} \begin{cases} \text{1st case: } \dfrac{\text{Pull}}{\text{Load}} = \dfrac{\text{friction}}{\text{weight}} = \dfrac{6}{10} = 0.6 \\ \text{2nd case: } \dfrac{\text{Pull}}{\text{Load}} = \dfrac{\text{friction}}{\text{weight}} = \dfrac{12}{20} = 0.6 \end{cases}$$

The ratio of the force required to overcome friction to the load (pressure) imposed between the surfaces of opposing bodies is called the *coefficient of friction*. The coefficient of friction may be determined for any frictional system, whether it be sliding, rolling, or fluid friction. It is constant for any specific set of conditions, but manifests the effect of any change in those conditions, as expressed by appropriate laws of friction.

The coefficient of friction may be determined in several ways, including the direct-pull method or by use of the inclined plane shown in Fig. 421.

The relationship of various forces present in the direct-pull method may be expressed by the equation

$$f = \frac{P}{W}$$

and under constant frictional conditions and varying loads

$$P = f \times W$$

where P = force in pounds (pull) normal to the surface, to overcome friction
W = weight of movable body, pounds
f = coefficient of friction

The coefficient of friction is an indication of frictional conditions, decreasing as the force of friction diminishes.

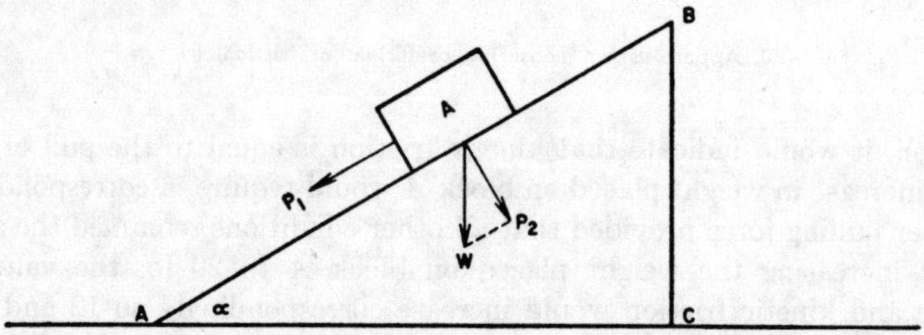

Fig. 421. Coefficient of friction, f determined by the inclined plane ($f = \tan \alpha$).

Gravity supplies the force P in the inclined plane method. Block A (Fig. 421), resting on the inclined plane BA, exerts force in two directions: P_1, parallel to BA, and P_2, perpendicular to the sliding surface BA. The sum of P_1 and P_2 equals W, the total force tending to move block A. The relative force exerted by either P_1 or P_2 depends on the slope represented by the angle α. The force P_1 increases while P_2 decreases correspondingly as the angle α becomes greater. The distance BC also increases as the angle α becomes greater. Therefore, the force P_1 varies as the distance BC. This is the same as saying that P_2 varies as the distance AC relative to BC. It follows that the forces P_1 and P_2 are functions of the relationship of the distances BC and AC expressed by the equation

$$\frac{P_1}{P_2} = \frac{BC}{AC}$$

But

$$\frac{BC}{AC} = \tan \alpha$$

Therefore

$$\frac{P_1}{P_2} = \frac{BC}{AC} = \tan \alpha$$

also

$$\frac{P_1}{P_2} =$$

and by substitution

$$\frac{BC}{AC} = f, \text{ which is also the tangent of } \alpha$$

Therefore, the coefficient of friction is equal to the tangent of the angle α, or

$$f = \tan \alpha$$

when the distance BC is great enough to cause A to move along AB in the direction of P_1.

The coefficient of friction varies with different materials or combination of materials. For example, steel on steel has a sliding friction (kinetic) coefficient of about 0.50, while f for cast iron is approximately 0.15. Steel on cast iron has a coefficient of friction of 0.25. The coefficient of friction also varies according to the frictional system, as shown in Table 130. Comparing f for the various types of frictional systems establishes their relative efficiency.

TABLE 130. COEFFICIENT OF FRICTION FOR VARIOUS FRICTIONAL SURFACES

Frictional Condition	*Coefficient of Friction f (Average)*
Fluid friction	0.001–0.005
Rolling friction:	
Ball	0.002
Roller	0.004
Sliding friction:	
Dry	0.15 –0.40
Boundary	0.080–0.10
Mixed film	0.020–0.08

Frictional Components

The three basic frictional components of mechanical equipment are bearings, cylinders, and gears.

Bearings

The chief function of bearings is to support moving elements, such as journals, crankshafts, piston pins, cross-heads, sliders, and riding components of various kinds. Bearings support motion that may be rectilinear, curvilinear, rotational, gyrational, vibratory, or oscillating. Many bearings are stationary while others form part of the moving element as in the case of wheels, connecting rods, and free pulleys. The three general types of bearings are radial, guide, and thrust. The radial types are plain, or anti-friction (ball or roller) bearings.

Radial Plain Bearings. Radial plain bearings support rotating motion, such as that of a shaft or a journal. Loads act perpendicular to the shaft axis. They are usually called journal bearings and include

(*a*) sleeve or bushing, which are solid full circle bearings;
(*b*) split type, also full circle but divided lengthwise;
(*c*) half bearing, which forms a semicircle and covers only half the journal circumference. Some types of partial bearings in use are less than semicircular;
(*d*) multi-part bearings, composed of more than two segments that together form a full circle.

While the solid full circle bearing of the sleeve type is used most widely, the others are required to meet certain design conditions, or to facilitate installation and removal of either the bearing or its journal. Of the various bearing materials (such as hard maple, lignum vitae, leather, plastics, and metals, the last are the most important. Metal bearings are classified as being hard or soft according to the materials that compose them.

HARD BEARING METALS include cast iron, copper base, cadmium base, leaded bronze base, and tin bronze base alloys.

SOFT BEARING METALS include tin base and lead base babbitt.

Other bearing materials include aluminum alloy, silver, silver alloys, and sintered carbide. Bronze and ferrous types are porous, making them capable of holding a lubricant for self-lubrication for definite periods. Bearings that hold lead in a copper matrix are known as grid bearings, in which copper lends strength and lead reduces the coefficient of friction. Another low frictional material is produced by impregnating some bearing metals (babbitt, bronze) with graphite.

Table 131 lists a number of bearing metals showing typical analysis and some of their characteristics.

The type of bearing material used depends on service conditions. Speeds, loads, temperature, and degree of shock are some of the service factors to be considered. Total load is not so important as the unit pressure exerted by that load. Therefore, bearing area serves as a compensating factor in bearing metal selection.

Unit pressures are measured and expressed in terms of *projected area.* Projected area is determined by multiplying the length of the bearing by its diameter. Unit pressure usually is expressed in psi of projected area. This means that the length and the diameter of the bearing are expressed in inches.

EXAMPLE: The projected area of a bearing 8 in. long and 4 in. in diameter will be $8 \times 4 = 32$ sq in. If a total load of 48,000 lb is exerted on the bearing, the unit pressure will be $\frac{48{,}000}{32} = 1500$ psi of projected area.

TABLE 131. TYPICAL ANALYSIS OF SOME BEARING METALS

Bearing Metals (Embeddability)‖	Brinell Hardness (Av)	% Tin	% Lead	% Copper	% Antimony	% Arsenic	% Silver	% Zinc	% Nickel	% Cadmium	% Phosphorus
Soft Metals:											
*Tin base babbitt (a)	25	75–93	0–18	2–6	5–15	0.1					
†Lead base babbitt (a)	17	1–11	74–89	0.5–0.6	9–16	0.1–1.10					
Hard Metals:											
Copper base alloys (copper-lead) (b)	25	0.25	24–40	60–75			0–1.0				
‡Cadmium base alloys (a)	35			0.4–0.75			0.5–1.0		1.0–1.5	98	
Bronze base alloys											
Tin bronze (Admiralty) (c)	70	10	0.5–4.0	88				2			
Lead bronze (c)	55	5	25	70							
Phosphorus bronze (c)	90+	10	10	80							0.3–0.4
§Aluminum base alloys (c)	50	6.5	1.0–2.0	(90% pure aluminum)					1.0		
Silver (overplated) (c)	25	(overplated on steel or copper-lead)									
Sintered bearings (c)		10	(2–4)¶	88	(1½% graphite)						
Grid bearings (c)	(lead in a copper matrix)										

* Heavy-duty material.
† Light-duty material.
‡ Indium plated for corrosion resistance.
§ One type contains 2% copper, 1% nickel, and about 1% magnesium.
‖ Embeddability: (a) good, (b) fair, (c) hard surface.
¶ 2 to 4% lead added to some types.

Unit pressures should not exceed those that have proved to be successful for a given service. Table 132 shows permissible mean bearing pressures for various types of service.

TABLE 132. CURRENT PRACTICE IN MEAN BEARING PRESSURES*

Type of bearing	Permissible pressure, psi, of projected area	Type of bearing	Permissible pressure, psi, of projected area
Diesel engines, main bearings	800–1,500	Automotive gasoline engines, main bearings	500–600
Crankpin	1,000–2,000	Crankpin	1,500–2,000
Wrist pin	1,800–2,000	Air compressors, main bearings	120–240
Electric motor bearings	100–200	Crankpin	240–400
Marine Diesel engines, main bearings	400–600	Crosshead pin	400–800
Crankpin	1,000–1,400	Aircraft engine crankpin	700–2,000
Marine line-shaft bearings	25–35	Centrifugal pumps	80–100
Steam engines, main bearings	150–500	Generators, low or medium speed	90–140
Crankpin	800–1,500	Roll-neck bearings	1,500–2,000
Crosshead pin	1,000–1,800	Locomotive crankpins	1,500–1,900
Flywheel bearings	200–250	Railway-car axle bearings	300–350
Marine steam engine, main bearings	275–500	Miscellaneous ordinary bearings	80–150
Crankpin	400–600	Light line shaft	15–25
Steam turbines and reduction gears	100–220	Heavy line shaft	100–150

* By permission from "Mechanical Engineers Handbook," edited by Lionel S. Marks. Copyright, 1951, by McGraw-Hill Book Company, Inc.

Another important factor in bearing design is the length-to-diameter ratio. Major considerations for $\frac{l}{d}$ values are end leakage of lubricant, vibration, and accuracy of shaft positioning. Ratios vary from $\frac{l}{d} = <1$ to $\frac{l}{d} = 4+$. Steam turbine shafts rotating at relatively high speeds (3600 rpm) are supported successfully by fixed bearings, constantly supplied with an abundance of lubricant, having a length-to-diameter ratio of 1.0 to 1.5. In contrast, line shaft bearings placed relatively far apart and subject to slow speeds (300 rpm or less), are designed with ratio $\frac{l}{d} = 3$ to 4. The usual practice is to maintain a ratio $\frac{l}{d}$ of between 1.0 and 2.0.

To allow for expansion and to provide lubricant space, some definite clearance between the journal and the bearing must be maintained, otherwise high frictional load and ultimate seizure may occur. For general purposes, clearances based on an allowance of 0.001 in. per in. diameter plus 0.001 in. is sufficient. For heated bearings (steam joints), greater

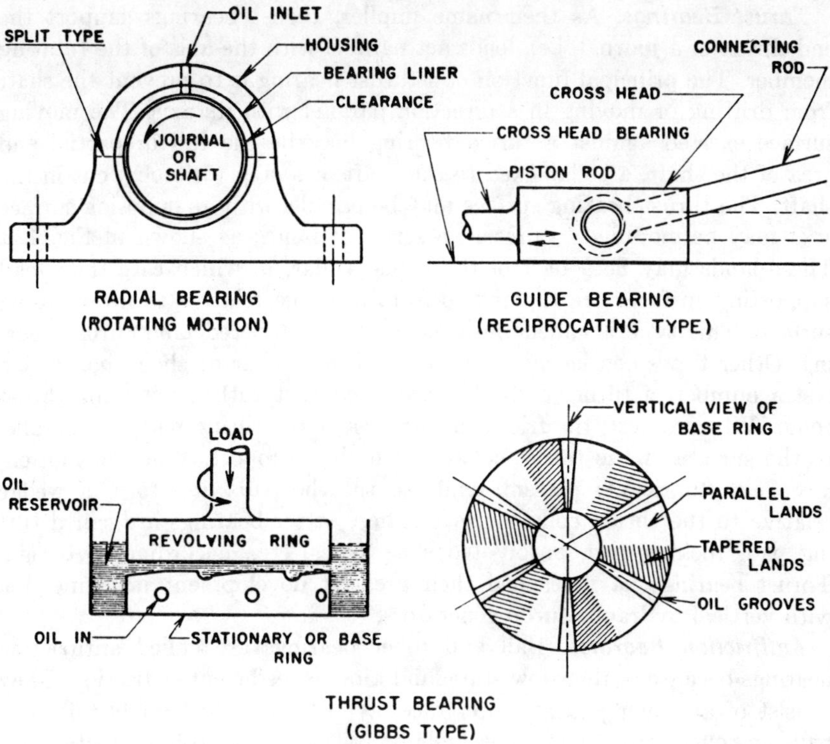

Fig. 422. Bearing types.

clearances should be allowed, while for precision machines bearing clearances may be reduced in relation to the effectiveness of the lubricant and the lubrication system.

Guide Bearings. Bearings that support rectilinear and curvilinear motion, such as crossheads, ways, and others of similar type, may be included under the heading of guide bearings. They are associated mostly with reciprocating motion, but not always. The cross-section of guide bearings may be flat, convex, concave, or V-shaped. They may be vertical as are elevator guides, they may be horizontal as in the case of machine tool ways, or they may assume any position between the two extremes. The riding member moves parallel with the guide bearing surface and the action is that of sliding. Contact pressure due to gravity reaches a maximum with horizontal guides and diminishes as they approach a vertical position.

Guide bearings are made from nearly all the materials listed in Table 131. However, the use of metals having a high coefficient of friction and/or improper lubricants will cause chattering of the guide bearing rider, especially at slow speeds.

Thrust Bearings. As their name implies, thrust bearings support the end thrust of a journal, i.e., loads act parallel with the axis of the rotating member. The principal function of a thrust bearing is to prevent the shaft from drifting or moving in a direction parallel with its axis. The moving surface exerted against a thrust bearing may be the full or partial end area of the shaft, a collar fixed to the shaft or a step (or steps) cut in the shaft. The thrust bearing surface may be parallel with its opposing surface or it may be modified to present a series of lands, as shown in Fig. 422. These lands may be a part of the thrust collar, in which case their load supporting surfaces are tapered to form a wedge relative to its opposing surface. This type is called the fixed wedge or tapered land thrust bearing. Other types consist of a stationary thrust collar or shoe upon which rest a number of tilting pads that are in contact with a revolving thrust collar. When at rest, the frictional surfaces of the tilting pads are parallel to the surface of the thrust collar which they support. When in motion, any fluid (lubricant) present tends to tilt the pads and form a wedge relative to the thrust collar surface. These latter bearings are called tilting pad, rocking pad, or pivoted shoe types, etc., according to design. Thrust bearings have reached their greatest development in connection with vertical hydraulic turbo-generators.

Antifriction Bearings. Ball and roller bearings are called antifriction bearings because of their low static and kinetic coefficient of friction. They consist of an inner and an outer raceway between which highly finished balls or rollers are placed. The balls or rollers impart rolling contact between the two races when either or both rotate about a common axis with relative motion. To maintain distance between the rolling members and to prevent contact with each other, separators or cages are used. These cages also prevent cylindrical rollers from skewing.

The size of antifriction bearings is indicated generally by the diameter of the inner race bore. For any given inner race bore, the relative overall diameter, width of race, and size of rolling members will vary according to their duty series, which is classified as extra light, light, medium, or heavy.

Shields are installed in some bearings to prevent the entrance of foreign materials. Seals are used also to retain the lubricant (grease) within the space containing the rolling members and to keep contaminants out.

Antifriction bearings are variously designed to meet different load conditions—i.e., radial, axial thrust, or a combination of radial and axial thrust. Deep grooving, or providing angular contact between race and ball, are methods of meeting combination radial and thrust loads in ball bearings. In roller bearings, combination loading is met by the use of tapered rollers and flanged races. As the ratio of thrust to radial load becomes higher, a greater cone angle must be provided the roller. Bearings

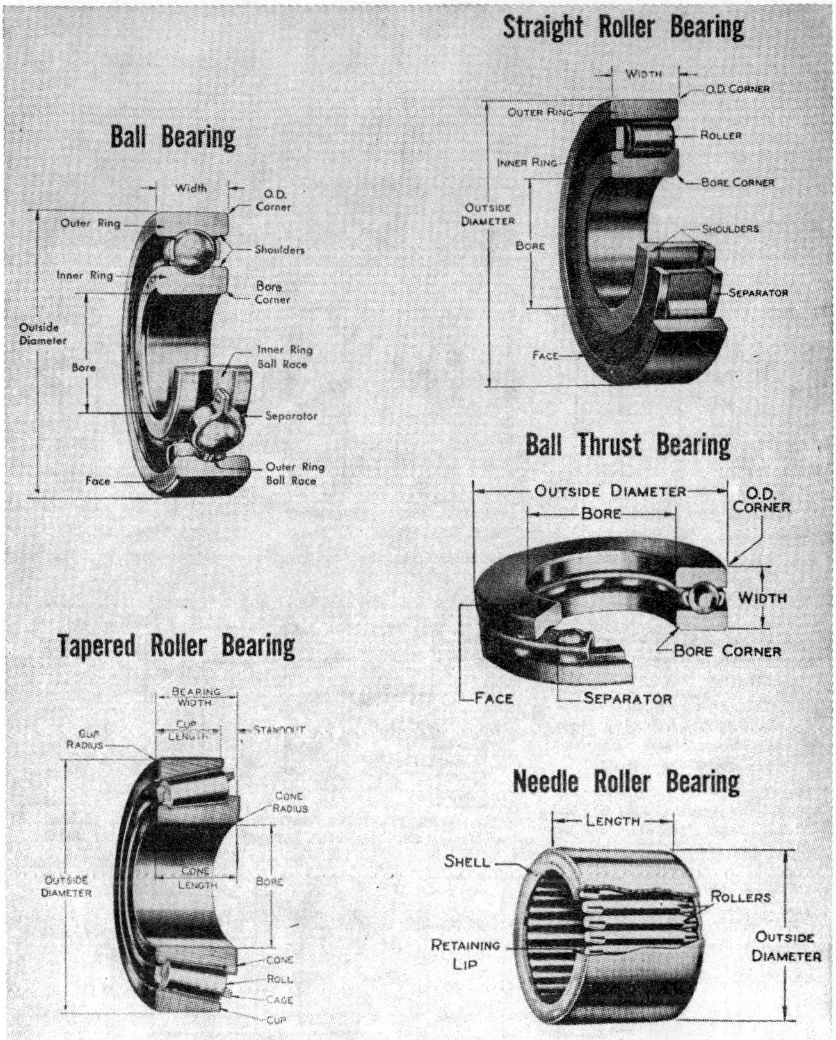

Fig. 423. Antifriction bearings showing parts and names. (Courtesy, Engineering Committees, Anti-Friction Bearing Manufacturers Association, Inc.)

designed to carry axial thrust only have their races located on an axial plane relative to each other, i.e., the races are side-by-side along the shaft with the rolling member in between adjacent races. Single-row angular contact ball and single-row tapered roller bearings will carry radial and thrust loads, the latter in one direction only. Double-row tapered roller and double row angular contact ball bearings will take thrust from both directions as well as radial load. Cylindrical rollers

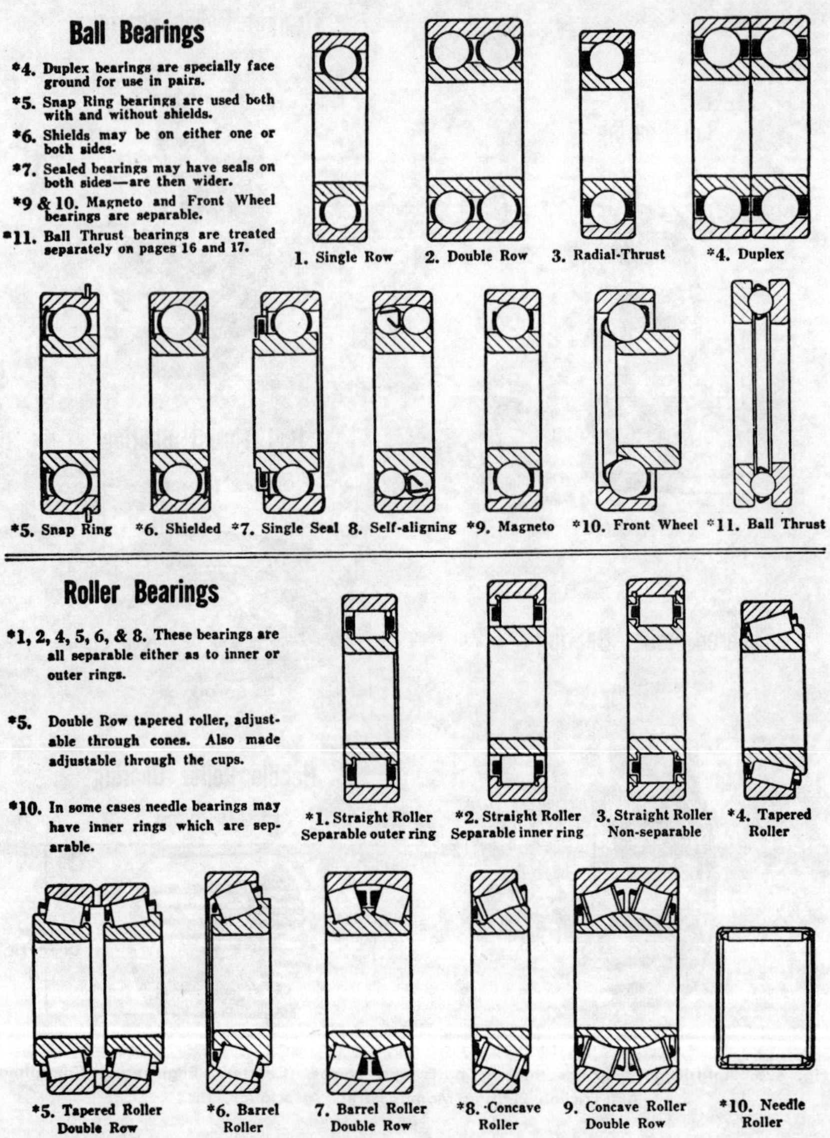

Fig. 424. Types of antifriction bearings. (Courtesy, Engineering Committees, Anti-Friction Bearing Manufacturers Association, Inc.)

with flanges will take heavy radial loads but only limited thrust. Using two single-row angular contact ball bearings placed back-to-back or face-to-face allows thrust to be carried in either direction. Where heavy thrusts are encountered along with high radial loads, both thrust and radial type bearings should be employed.

Spherical and concave rollers having their races cut with mating contours act to impart self-alignment to antifriction bearings. Self-aligning ball bearings are of two types, internal and external. The internal type has two rows of balls. The inner ring has two separate grooves while the outer ring has a single spherical raceway. The external type of self-aligning ball bearing is deep grooved and has two outer rings, one of which surrounds the other. Their spherical contacting surface allows the "internal" outer ring to swivel within the extreme outer ring and impart self-alignment.

TABLE 133. CHARACTERISTICS OF ANTIFRICTION BEARINGS

Type	*Radial Loads*	*Axial Loads*	*Combination Loads*	*Speed Characteristics*
BALL:				
Single-row, deep groove	M	H	M	High speed
Single-row, fitting slot or maximum ball	H	M	M	Intermediate
Angular contact	M	H	H	Intermediate
Double-row	H	H	H	Intermediate
Self-aligning (internal)	L	L	L	Intermediate
Self-aligning (external)	M	H	M	High speed
ROLLER:				
Single-row—cylindrical	H	—	—	High speed
Double-row—cylindrical	EH	—	—	Intermediate
Single-row—tapered	M	M	M	High speed
Double-row—tapered	H	M	H	Intermediate speed
Concave	M	L	M	Intermediate
Needle	EH	—	—	Intermediate

Key: L = light, M = moderate, H = heavy, EH = extra heavy.

Antifriction bearings are not subject to sudden deterioration. Their life span depends on the number of times their structural members are exposed to stresses of load, repeated exposure of such stresses finally causing metal fatigue and ultimate failure. The life expectancy of a given bearing, then, is based on the magnitude of the load and the frequency with which the stresses imposed by that load occur. The load magnitude or stress is expressed in pounds, while stress frequency is equivalent to the speed in revolutions per minute at which bearings operate. Life expectancy is rated in hours. Hence there are three factors, (1) load in pounds, (2) frequency of stress, rpm, and (3) bearing life, hours, all of which bear some relationship to the others. All factors may be variable.

Some basis of rating antifriction bearings that embraces the above three factors must be adopted. If, during a series of tests in which a number of bearings of a given group are used, it were found that the average life was 3800 hr at 1000 rpm under a load of 2200 lb, these values would form the basis of rating all bearings in that group. By the laws governing antifriction bearings, it would be expected that

(1) the life expectancy would decrease if the speed were increased and load remained the same,
(2) the life expectancy would decrease if the load were increased and speed remained the same,
(3) the life expectancy would decrease if both load and speed were increased,
(4) the life expectancy would increase if load were reduced and speed remained the same,
(5) the life expectancy would increase if speed were decreased and load remained the same,
(6) the life expectancy would increase if both load and speed were reduced,
(7) for a constant life expectancy the load must be reduced in some proportion to any increase in speed,
(8) for a constant life expectancy the speed must be reduced in some proportion to any increase in load,
(9) the general relation of life to load at constant speed is

$$\text{Life} \propto \frac{1}{\text{load}^3}$$

(10) the general relation of life to speed at constant load is

$$\text{Life} \propto \frac{1}{\text{speed}}$$

Life expectancy may be expressed in terms of minimum life or average life of bearings in hours. Minimum life is based on the number of hours that 90 percent of the bearings of a given group will operate at a given load and speed. Average life is the average number of hours all bearings in the group will operate under a given load and speed.

For ball bearings of the same group,

$$\text{Average life} = 5 \times \text{minimum life}$$

For roller bearings of the same group,

$$\text{Average life} = 3 \times \text{minimum life}$$

The relationship of the life expectancy, load, and speed of antifriction bearings may be expressed more definitely by the equation

$$C_r = \frac{C_s}{F_s}$$

where C_r = rated capacity in lb at a speed other than test, or listed speed
C_s = capacity in lb based on test speed as listed by manufacturer
F_s = speed factor

or $F_s = \sqrt[n]{\dfrac{\text{selected speed, rpm}}{\text{test or listed speed, rpm}}}$

or F_s = reciprocal of the nth root of the speed ratio
where n = 3 to 4, depending on the manufacturer

For example, the rated capacity of a ball bearing having an average life of 3800 hr at 1000 rpm and 2200 lb load when operated at 5000 rpm under the same load would be, assuming n to have a value of 3.0,

$$C_r = \frac{C_s}{F_s}$$

$$C_r = \frac{2200}{\sqrt[3]{\frac{5000}{1000}}} = \frac{2200}{\sqrt[3]{5}} = \frac{2200}{1.71}$$

$$C_r = 1286 \text{ lb}$$

If allowed to operate under a load of 1600 lb at 5000 rpm, the expected life of the same bearing would be determined as follows:

$$L_a = \left(\frac{C_r}{C_a}\right)^n \times L_r$$

where L_a = expected or actual life of the bearing in hours at operated speed and load
C_a = actual load
L_r = average life based on manufacturer's rating at standard speed and load

$$L_a = \left(\frac{1286}{1600}\right)^3 \times 3800 = 0.8 \times 3800$$

$$L_a = 3040$$

In determining loads, the effect of both radial and thrust forces must be considered. However, both forces combined are expressed as *equivalent radial load.* Combined loads produced by the simultaneous action of radial and thrust forces may be expressed as the square root of the sum of the square of the forces, or

$$\text{Combined load} = \sqrt{\text{radial load}^2 + \text{thrust load}^2}$$

$$L_c = \sqrt{L_r^2 + L_t^2}$$

Loads include the weight of structural members and the forces set up due to power transmission, torque, centrifugal action, etc. Loads are calculated on the basis of all factors involved in the operation of equipment whose forces impose either a radial or a thrust load. Bevel and hypoid gears, worm gears, propellers, and the weight of equipment associated with vertical or angular shafting are some of the components of thrust loads. Their influence can be determined by proper formulae. Radial forces rely on the location of the bearings in relation to the points of applied load.

For pure thrust (axial load), the combined load is equal to the thrust load, or $L_c = L_t$. For pure radial loads (L_r), $L_c = L_r$. For combined loads, however, the equivalent radial load is equal to radial load plus the thrust load, one or both of which may be modified, by appropriate coefficients, according to the type of bearing, magnitude of shock, rotating race, i.e., whether the inner, the outer, or both races turn, and other considerations. The relationship may be expressed as

$$L_c = L_r + L_t \text{ for some bearings}$$

and

$$L_c = xL_r + yL_t \text{ for others}$$

where x = coefficient of radial loads
y = coefficient of thrust loads

The values of x and y differ for different types of bearings and different manufacturers. For example, the equivalent radial load for a single-row radial ball bearing may be

$$L_c = L_r + L_t \text{ at low speeds}$$

and

$$L_c = L_r + 1.5\ L_t \text{ at high speeds}$$

and for a single-row, steep-angle taper roller bearing the equivalent radial load may be

$$L_c = 0.66\ L_r + 0.75\ L_t$$

etc.

Rotating Outer Race. When the inner race remains stationary while the outer race rotates, the effect is to direct the load toward the smaller area of the inner race. This combined with the higher rotating speed of the rolling members as compared with their speed if the inner race traveled at a comparable rate, increases the internal stresses and the stress frequency. Thus, by rotating, the outer race reduces capacity 10 to 30 percent, depending on the type of bearing and its manufacturer. This means that bearings will have a carrying capacity of 70 to 90 percent of their rated capacity when standard speeds apply to the outer race.

For example, a single-row light series ball bearing rated at 4500 lb at 1000 rpm of its inner race will have its carrying capacity reduced, assuming a rotating outer race coefficient of 0.87, to $0.87 \times 4500 = 3915$ lb. Because the coefficient varies with duty series classification, the probable value for a medium series bearing of the above type would be 0.84, and for the heavy series, 0.82.

Shock Loads. Antifriction bearing data are based on smooth, steady load. Therefore, some compensation must be made for shock and to a degree corresponding to shock magnitude. Shock factors range from 1.0 for smooth steady load to 3.0 for extreme shock loads, as shown in Table 134.

To meet a normal 4500-lb load having moderate shocks, the bearing selected would be required to have a carrying capacity of $2 \times 4500 = 9000$ lb. If a bearing having a normal rated capacity of 4500 lb were selected for the foregoing requirement, its life would be reduced to one third its rated average life.

LIFE FACTOR. The ratio of rated capacity at a given speed to actual load is called the *life factor* of an antifriction bearing. The rated capacity at the desired speed should be modified to allow for all deteriorating influences. Table 135 lists the various influences affecting bearing life, including those already discussed, along with their probable range of magnitude (f).

TABLE 134. CAPACITY MODIFYING FACTORS FOR ANTIFRICTION BEARINGS DUE TO SHOCK LOADS

Smooth, steady load	1.0
Light shock load	1.5
Moderate shock load	2.0
Heavy shock load	2.5
Extreme shock load	3.0

TABLE 135. ANTIFRICTION BEARING LIFE FACTORS

Condition	*Range of f*
Shock loads:	
Extreme	0.3–0.5
Heavy	0.4–0.6
Moderate	0.5–0.7
Light shock (or vibration)	0.85
Temperature of operation:	
High (above 250 F)	0.6–0.75
Moderate (180 to 250 F)	0.7 0.0
Normal (below 180 F)	0.9–1.0
Rotating outer race	0.7–0.9
Rotating load	0.7–0.9
Oscillating (high degree)	0.75 (av)

The life factor for a given bearing may be expressed by the equation

$$\text{Life factor} = \frac{\text{rated capacity at operating speed} \times f}{\text{actual load}}$$

$$= \frac{C_r \times f}{C_a}$$

and the number of hours of expected life under conditions expressed by the life factor may be determined by the equation

$$\text{Service hours} = (\text{life factor})^n \times K$$

where $n = 3.0$ to 4.0 (according to manufacturer)
K = number of hours on which manufacturer's ratings are based

METALLURGY. Metals selected for antifriction bearings must possess (*a*) high anti-wear qualities, (*b*) surface hardness capabilities, (*c*) good depth of hardness penetration characteristics. Certain alloy steels possess such characteristics. Alloying materials include chromium, nickel, manganese, molybdenum, silicon, sulfur, and phosphorus. Tungsten and vanadium are also used in some bearing alloys. Carbon content ranges from 0.10 to 0.25 percent in low carbon alloys and from 0.85 to 1.10 percent in high carbon steels.

Chromium is beneficial in giving deep and uniform hardness penetration as well as imparting good wear and fatigue resistance to steel. The effect of chromium is to obtain *through hardness*, i.e., penetration of hardness from surface to core center. The alternative is to obtain high surface hardness with tough centers by carburizing (case hardening). While it generally follows that the carrying capacity of a bearing increases with greater hardness, fineness of grain and depth of hardness are important qualities in preventing surface failures. Sulfur and phosphorus are present in very small amounts ranging from 0.020 to 0.05 percent maximum for nearly all bearing steels.

The popular alloy known as SAE 52100 possesses high antifriction bearing qualifications. The specifications of this alloy are in the order as shown below:

SAE 52100	
Carbon	0.95 to 1.10%
Chromium	1.30 to 1.60%
Manganese	0.25 to 0.45%
Silicon	0.20 to 0.35%
Phosphorus	0.025% (max)
Sulfur	0.025% (max)

The latter steel and other alloys having slight modifications in contents as given above can all be heat-treated to adequate hardness to produce antifriction bearings of high load-carrying capacity.

To provide corrosion-resistant bearings, stainless steels containing 13 and 17 percent chromium are used. Carbon content ranges from 0.25 to 0.35 percent and 0.95 to 1.05 percent, respectively. However, unless molybdenum is added in amounts approximating 0.50 percent, a hardness equal to that of properly heat-treated SAE 52100 (about Rockwell C-62) cannot be approached.

Hardness of Metals. The hardness of metals, including antifriction bearing steels, can be measured. The method of measuring involves the use of an indentor in the form of a sphere or a cone forced into the surface of a test specimen under some arbitrary load. The hardness of the material tested is based on (1) the depth of penetration or (2) the diameter of indentation, depending on the kind of test machine employed.

There are several indentation tests employed to determine metal hardness. They include the Brinell, Rockwell B, Rockwell C, Vickers, and Shore Scleroscope methods, of which the Brinell and the Rockwell test machines are the more widely used. Hard steels, including antifriction bearing materials, usually are expressed in terms of Rockwell C or Brinell hardness. For nonferrous metals and soft steels, the Rockwell B method is used. The hardness of nonferrous metals is expressed also in terms of Brinell number.

A spherical-tipped conical diamond called a "Brale" is used as the indentor in the *Rockwell C* method of hardness testing. An initial penetration is gained by the cone tip upon application of a minor load of 10 kilograms (kg). At this point in the test the depth-measuring scale is adjusted to an arbitrary value of 100. A major load of 150 kg is then applied under which greater penetration of the diamond cone into the test material is gained. The hardness number is read directly from the depth-measuring scale when the major load is removed and the force is returned to the original 10 kg. The scale is designed to register low penetration numbers for soft metals and progressively higher numbers with increased hardness. Rockwell C scale ranges from 15 to 70, the latter value representing the hardest material.

The *Rockwell B* method is similar to Rockwell C except that in the former a $\frac{1}{16}$-in. hardened steel ball is used under a major load of 100 kg. The minor load of 10 kg remains the same, but the depth-measuring scale is adjusted to an arbitrary value of 130 under Rockwell B procedure. Rockwell B numbers range from 0 to 100.

Brinell hardness is also determined by indentation. A hardened steel ball of 10 mm diameter is used. When steels are tested, a steady load of 3000 kg is applied to the ball. For softer metals, a steady load of 500 kg is applied, and 100 kg for very soft materials. The Brinell hardness number is determined by measuring the diameter of indentation and applying the equation

$$\text{Brinell number} = \frac{2P}{\pi D}\left(d - \sqrt{D^2 - d^2}\right)$$

where P = load in kg
D = diameter of ball, min
d = diameter of indentation, min

In the case of steels, the 3000-kg load should be held steady for at least 10 sec. For softer materials, the 500-kg load should be held for a longer period of 30 sec or more. Brinell numbers range up to 700. Above 500, however, an especially hardened steel ball should be used.

The relation of Brinell number (3000 kg) to Rockwell C hardness is about 10:1 up to about Rockwell C—60. Above the latter value the ratio increases from about 10.5:1 to about 11:1. The increase here is due to the flattening tendency of the Brinell ball by harder test steels. Therefore a material having a Brinell number of 500 would have a value comparable to Rockwell C—50, approximately.

Steel Identification. In order to identify the various steels, a numerical index has been adopted. Classified according to numbers, adopted by the Society of Automotive Engineers, each number in the index represents a different type steel, as shown in Table 136.

TABLE 136. NUMERICAL INDEX FOR IDENTIFYING STEELS (SAE)

Carbon steel	1
Nickel steel	2
Nickel-chromium steel	3
Molybdenum steel	4
Chromium steel	5
Chromium-vanadium steel	6
Tungsten steel	7
Silico-manganese steel	9

The numbers shown in Table 136 are the first figures of any SAE steel classification and represent the basic alloying element. The second number indicates the approximate amount of the basic element. The third, fourth, and fifth (if any) numbers indicate the approximate amount of carbon in the steel.

EXAMPLES: SAE 1020 steel is a carbon steel, as indicated by the first numeral 1. It contains 0.15 to 0.25 percent carbon, as indicated by the third and fourth numerals 2 and 0.

SAE 2115 is a nickel steel, as indicated by the first numeral 2. It contains 1.25 to 1.75 percent nickel (i.e., over 1.0 percent), as indicated by the second numeral 1. The third and fourth numerals 1 and 5 indicate a carbon content of 0.10 to 0.20 percent.

SAE 52100 is a chromium steel, according to the first numeral 5, of which there is approximately 1.5 percent (actually 1.3 to 1.6 percent) indicated by the second numeral 2. It contains about 1.0 percent carbon, denoted by the third and fourth numerals 1 and 0.

SAE 6140 is a chrome-vanadium steel containing about 0.35 to 1.10 percent chromium and 0.35 to 0.40 percent carbon. It is understood that vanadium is always present in chrome-vanadium steel, the minimum being 0.15 percent.

RADIAL LOADS. Radial loads are the forces acting on a bearing in a direction perpendicular to the axis of the journal. The weight of machine parts which a bearing supports plus the external work forces acting through the various parts represent the total bearing load. Usually a journal is

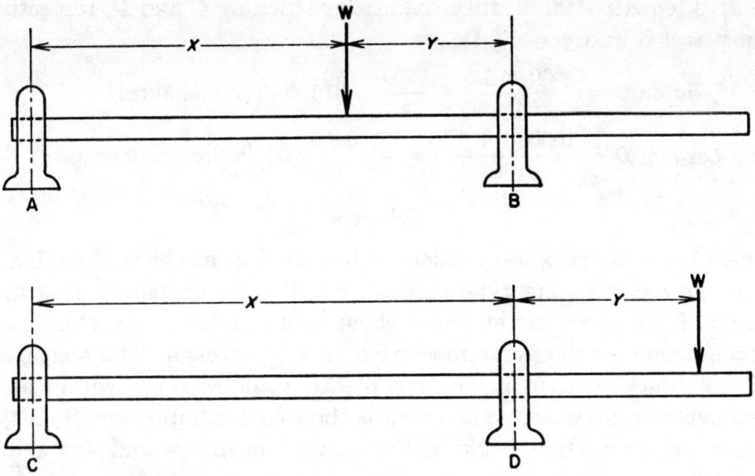

Fig. 425. Effect of bearing location on individual bearing load.

supported by two or more bearings. The amount of the total radial load carried by each bearing depends on their relative position in respect to the load.

If the load is exerted at a point halfway or is equally distributed on the journal between two bearings, the weight will be divided equally and each bearing will carry the same or one half the load. Figure 425 shows a shaft carrying loads that are not equally distributed or that are placed at any location relative to the bearings.

In Fig. 425 (*top*), the load on each of the bearings is determined by the equations

$$\text{Load, bearing } A = \frac{W \times y}{x + y}$$

$$\text{Load, bearing } B = \frac{W \times x}{x + y}$$

In Fig. 425 (*bottom*),

$$\text{Load, bearing } C = \frac{W \times y}{x}$$

$$\text{Load, bearing } D = \frac{W\,(x + y)}{x}$$

EXAMPLE:

1. If W equals 1000 lb, then the loads carried by A and B, respectively, when $x = 3$ ft and $y = 2$ ft, are

$$\text{Load on } A = \frac{1000 \times 2}{3 + 2} = \frac{2000}{5} = 400 \text{ lb}$$

$$\text{Load on } B = \frac{1000 \times 3}{3 + 2} = \frac{3000}{5} = 600 \text{ lb}$$

2. If W equals 1000 lb, then the loads carried by C and D, respectively, when $x = 4$ ft and $y = 1\frac{1}{2}$ ft, are

$$\text{Load on } C = \frac{1000 \times 1.5}{4} = \frac{1500}{4} = 375 \text{ lb (load is upward)}$$

$$\text{Load on } D = \frac{1000\,(4 + 1.5)}{4} = \frac{5500}{4} = 1375 \text{ lb (load is downward)}$$

Cylinders

There are two types of cylinders widely used in mechanical equipment. One is the reciprocating type through which a piston passes, lengthwise, back and forth to act as the power element of a steam or an internal combustion engine, or the pressure element of a compressor. The second type involves rotary motion in which a bladed rotor revolves within an enclosed cylinder, an example of which is the rotary compressor. Both types develop sliding friction between the moving members and the cylinder wall.

Reciprocating type cylinders assume either a horizontal, a vertical, or angular position. Frictional loads within a horizontal cylinder whose piston is connected directly with the crankshaft include the weight of the piston, angular thrust, and the pressure of gases acting on the piston rings to force them radially outward against the cylinder wall.

The weight of the piston exerts its greatest influence on horizontal cylinders, whereas in vertical cylinders the pressure on the cylinder wall due to this factor is nil. In radial type compressors the weight of the piston exerts a frictional force commensurate with the angularity of the cylinder, diminishing as the latter approaches a vertical position.

Angular thrust is always present in machines whose pistons are driven directly from the crankshaft through a connecting rod. The thrust occurs between the piston and the cylinder wall in the area adjacent to the piston pin bushings and on the right side of the cylinder if rotation is counterclockwise and on the left if clockwise. During operation, the force of connecting-rod thrust increases from zero to maximum as the crank moves from bottom-dead-center through 90 degrees of its revolution to maximum angularity at half-stroke. It then diminishes again to zero as it moves from half-stroke position to top-dead-center.

Where pistons are driven by a piston rod from a cross-head as in a double-acting compressor, the total angular thrust is taken by the cross-head guide bearing and therefore is not a factor in cylinders of this type machine.

In addition to the weight of the piston and angular thrust where either or both are present, there is a third force that creates considerable friction within a cylinder. This third force is the result of compression, combustion, or expansion, depending on whether a compressor, an internal combustion engine, or a steam engine is involved. The high-pressure gases fill the space within the piston grooves not taken up by the rings, causing the latter to expand outwardly and press against the cylinder wall. A downward pressure exerted by the same gas also causes the bottom of the ring to press against the groove surface (skirt side of groove) on which it bears. This force develops a high frictional load during the outward expanding motion of the ring.

The cylinder of a rotary vane type compressor is subject to the force imposed by the pressure of blades or vanes that are thrust radially outward from their rotor grooves. The force is developed by centrifugal action, becoming greater with increased shaft speed. In smaller compressors, centrifugal force is augmented by the tension of springs placed behind the blades to maintain a tight seal between blade and cylinder wall.

Piston Rings. Piston rings are placed in annular grooves cut in the piston for the purpose. The number and the type of rings used in any one piston vary according to the size of the latter and the service of the machine under consideration. The primary purpose of a piston ring is to form a seal between the piston and the cylinder wall. Because of the overall clearance necessary to prevent binding, a piston cannot be made sufficiently snug to provide an effective seal with its cylinder, especially under high compression ratios. In addition to the mechanical difficulties involved in obtaining a gas-tight fit, the temperature difference between the top and the bottom sections of the piston, and between the piston and the cylinder wall, make it necessary for the piston to be tapered. This means that the diameter at the head of the piston is less than at the skirt or open end. Tapering is necessary to allow for greater expansion of the piston area directly exposed to hot gases, which is at the top or crown. Subject only to conducted heat from the crown, the skirt of the piston has a lesser degree of expansion. It is also necessary to provide sufficient additional clearance to allow for the expansion variation between the piston and the cylinder walls because the latter is water-jacketed (cooled) and the former is not.

Piston rings are concentric, elastic, split metal rings capable of expanding and contracting in such a manner as to contact the wall of a cylinder as they ride with their piston along a reciprocating path. In doing so, they maintain a permanent seal between piston and cylinder. To maintain an effective seal, contact pressure must be uniform at all points around the circumference of the cylinder. The split makes expansion and contraction possible as the piston moves along the cylinder from the zone of minimum temperature to the zone of maximum temperature, or vice versa. When the ring reaches a condition of maximum expansion, which it does when the piston is at top-dead-center, the gap prevents crowding of the ring within the cylinder. The usual allowance for gap clearance is on the order of 0.004 in. per in. of cylinder bore, measured when the ring is fully compressed in its groove.

The split end of rings may be cut straight (butt), mitered, or in the form of a step. It is through the gap of an ill-fitted ring that appreciable leakage or blow-by occurs.

The two basic types of piston rings are (1) the compression ring and (2) the oil ring. The primary function of the compression ring is to provide a tight seal between the piston and the cylinder to prevent leakage of high-pressure gases from the compression chamber. A tight seal is also necessary in obtaining a vacuum within the compression chamber during the suction stroke.

The compression ring is designed with various cross-sectional shapes, including the rectangular and wedge or keystone types. Some are designed to allow full face (flat) contact with the cylinder wall. Others are tapered to provide minimum face contact. To decrease frictional resistance between ring face and cylinder wall, grooves are cut in the face of some type rings and filled with a material possessing low coefficient of friction, such as tin, bronze, etc. Rings having their surfaces plated with tin, chrome, cadmium, or zinc are also available and act to prevent scuffing during break-in periods. To allow oil to reach the frictional surfaces between ring face and cylinder wall, the bottom edge is beveled on some types of rings. Beveled edges also allow free passage of rings across suction ports present in certain types of compressor cylinders.

Oil rings usually are installed in the bottom grooves. Their function is to distribute lubricating oil evenly over the cylinder wall surface during the compression stroke, and scrape excess oil from the surface on the suction stroke. The cross-sections of oil rings are also variously designed, but the principle of riding over the oil on the up stroke and scraping it toward the crankcase on the down stroke prevails in nearly all types. To perform their function, it is necessary that the upper face edge be beveled and the lower face edge be relatively sharp. Double-facing by cutting an annular groove in the face completely around the ring and providing a bevel in the top of the upper face edge makes an efficient oil scraper ring. Usually such rings are ventilated by cutting slots (ventilators) in the bottom of the annular groove. The oil is scraped from the cylinder wall by the sharp bottom edge of the upper face, enters the ring groove from which it is forced through the ventilators to the back clearance space of the piston groove. The oil then enters the crankcase from the back clearance space through holes drilled through the piston skirt. A modification of the above is a single-faced ring made by cutting away the bottom land to leave an ell-shaped ring having a small width of face in contact with the cylinder wall. The upper edge of the face is also beveled in the latter type ring.

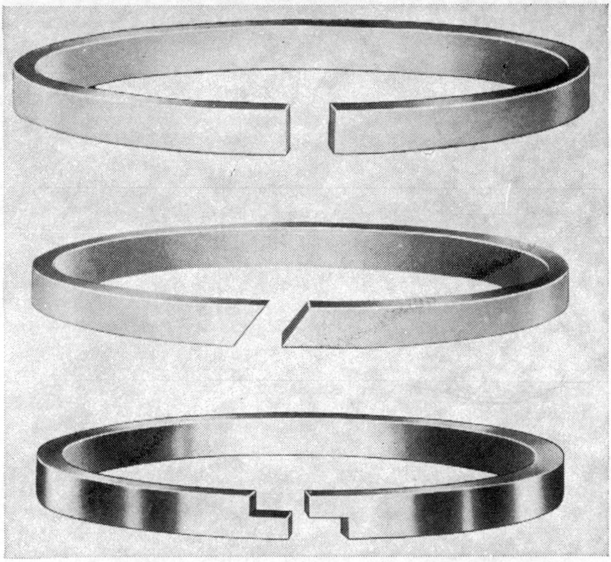

Fig. 426. Piston rings, compression type. (*Top to bottom*) Butt or straight, angle or miter cut, and step cut rings. (Courtesy, Double Seal Ring Company.)

Fig. 427. Two-piece piston ring. (Courtesy, Double Seal Ring Company.)

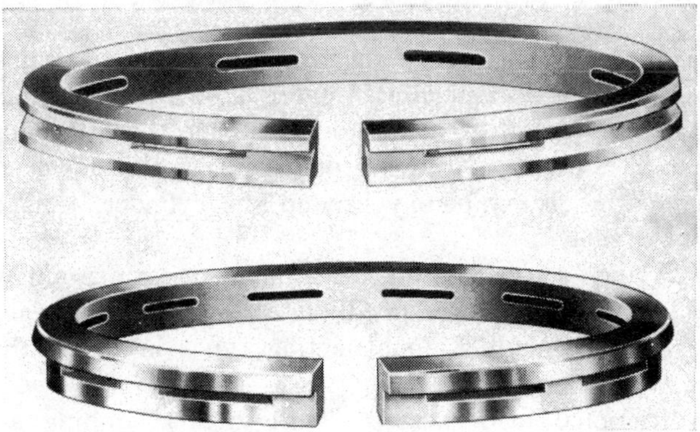

Fig. 428. Piston rings, oil type. (*Top*) Plain ventilated, (*bottom*) single scraper ventilated. (Courtesy, Double Seal Ring Company.)

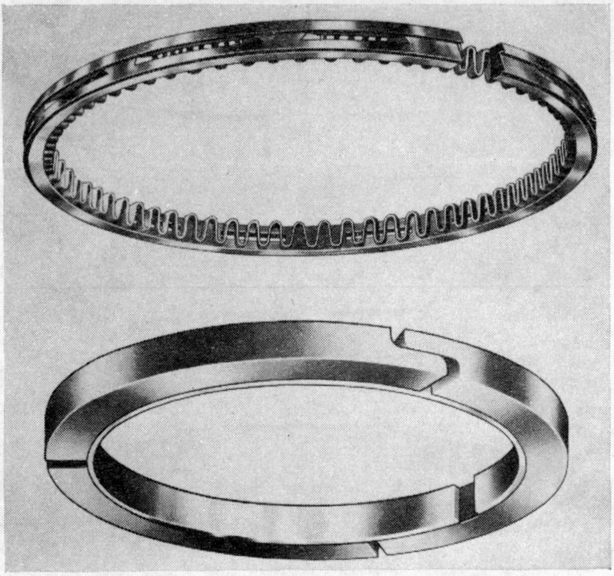

Fig. 429. Piston rings. (*Top*) Double scraper flexible ventilated oil ring, with expander, and (*bottom*) three-piece segmental rings. (Courtesy, Double Seal Ring Company.)

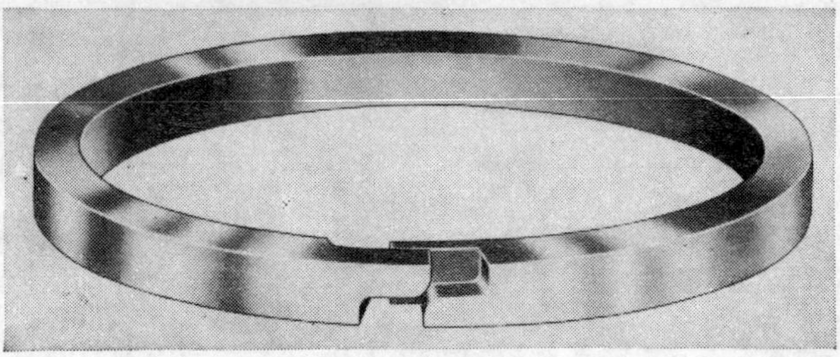

Fig. 430. One-piece double seal ring. Note double sealing effect of gap design. (Courtesy, Double Seal Ring Company.)

Where holes in the piston are not provided, excess oil drains down the cylinder wall into the crankcase. Annular grooves in oil rings, whether ventilated or not, permit excess oil to flow within them around the piston as they act in the role of moving dams during the down stroke. This action distributes oil completely around the cylinder, thus eliminating "dry" spots.

Segmental rings consist of two, three, or more segments mounted on a flat expanding hoop. Having stepped joints, they provide a seal for various type compressors, including double-acting types in which they combat pressure in two directions.

Because frictional load increases with greater rubbing area, piston rings should be designed to provide minimum face contact with the cylinder wall. However, face dimensions should be commensurate with efficient sealing and economical ring life.

Sufficient side clearance must be provided to allow rings to move freely in their grooves. To prevent excessive crowding of the ring which results in binding and ultimate damage to either the ring, the piston, or the cylinder (or all three), sufficient end clearance (gap) must be provided. The fixing of piston rings at one end to prevent their turning to line up the gaps is questionable procedure and not recommended.

Cast iron rings of uniform cross-section have proved to be highly successful in all types of reciprocating machines. Cast iron rings are often treated chemically to eliminate scuffing tendencies due to the presence of the soft pure iron (ferrite) contained in this metal. Chemical treatment converts ferrite to a depth of about 0.001 in. into iron salts, such as iron sulfide or iron phosphate. Steel rings are also used. Nonferrous metal rings employing bronze are satisfactory where operating conditions permit their use. Carbon rings have a low coefficient of friction and are employed in handling some types of gases where, because of fire hazard, the use of petroleum lubricating oil is prohibitive. Bakelite and other synthetic materials are employed for piston rings handling mild acid and alkaline gases.

To provide uniform pressure between the piston face and the cylinder wall around the entire circumference of the bore, stresses throughout the ring must also be uniform. Otherwise openings would occur in some areas to permit blow-by. Because of their design and their flexibility, rings have a tendency to become straight—i.e., to spread outward at the gap. This means that the greatest radial force would be exerted at the ends and at a point halfway around the ring from the gap. To balance the internal stresses and allow the ring to exert uniform radial pressure along its entire length of face, ring peening or hammering is practiced. This is done by peening the inside of the ring (as with a ball-peen hammer), increasing the peening force gradually as the action proceeds from the gap to a point halfway around the ring. From the halfway point the peening force diminishes in the same degree as it increased as the action continues toward the end opposite that of starting.

To increase the pressure between face and cylinder wall of oil rings, sometimes expanders are used. Expanders are abutment type springs designed to exert radial force. Placed at the back (around the inside circumference) of the ring, the radial force of the expander is transmitted to the ring, causing it to expand outwardly beyond its normal static limits. Compression type coils or ribbons made of spring steel make up the expander whose normal circumferential length is longer than the inside circumference of the ring when the latter is fully compressed to cylinder size.

Gears transmit power and motion from a revolving drive shaft to a driven shaft. The frictional forces involved depend on the type of gears employed. The type of gears selected is governed by the relative position of their shafts. Relative shaft positions may be (*a*) parallel to each other, their axes nonintersecting, noncrossing; (*b*) any angle to each other, their axes intersecting, noncrossing; (*c*) right angular to each other, their axes crossing, nonintersecting; (*d*) any angle to each other, their axes crossing, nonintersecting.

(*a*) Spur and herringbone gears are employed to transmit power and motion between parallel (nonintersecting, noncrossing) shafts.
(*b*) Bevel gears (straight tooth and spiral bevel types) transmit power and motion between intersecting shafts having any angle relative to each other. However, right-angle drives predominate.
(*c*) Worm and hypoid gears transmit power and motion between non-intersecting (their axes cross) shafts that form a right angle (90°) to each other.
(*d*) Helical gears may be designed to transmit motion (small loads) between shafts that cross (nonintersecting) each other at any angle.

Spur gears have their teeth cut straight across the face of the gear blank. The profile of spur gear teeth takes the form of an involute, scribed by a fixed point on a cord as it unwinds from around a given (base) circle. Spur gears have theoretical line contact straight across the face of meshing teeth. Deformation under load, however, develops area contact. As the teeth pass through mesh, the area of contact sweeps over the working surface of each tooth. This in effect develops a series of axial lines across the working surfaces of meshing teeth.

As the teeth start to mesh, radial sliding occurs at the initial line of contact, which is at the top of the driven tooth and at the bottom of the driving tooth. As meshing continues, the line sweeping upward on one tooth and downward on the other, sliding diminishes until a point immediately below and approaching the pitch line is reached. At this point, the meshing teeth start to roll over each other and continue the rolling action through the center of mesh (at the pitch line) to a point slightly beyond and above the pitch line. Sliding again commences, but in the reverse direction, increasing in magnitude until the rotation of the gear causes the teeth to break contact. Contact is broken at the tip of the driving tooth and at the root of the driven tooth.

Both sliding and rolling action occur when spur gears are in mesh and operating.

Table 137. List of Gear Terms

Addendum—That part of the tooth from the pitch circle (outside) to the outer diameter or the addendum circle.
Angle, helix—The angle of lead of a worm gear measured from a line perpendicular to the worm axis.
Back-lash—The difference between the space and the thickness of a tooth.
Circle, addendum—A circle formed by the outside surface of the teeth. The diameter of the addendum circle is the outside diameter of the gear.
Circle, base—A circle having a selected diameter on which the involute (profile of tooth face) is based. The base circle is always smaller in diameter than the pitch circle.
Circle, dedendum—A circle formed by the root of the teeth.
Circle, pitch—A circle scribed on the pitch diameter.
Contact, line of—A point, line, or area of contact is that surface of gear and pinion that touches simultaneously during mesh.
Dedendum—That part of the tooth from the pitch circle (inside) to the root of a tooth or the dedendum circle.
Depth, working—The distance from the top (outside surface) of the tooth to the root minus clearance.
Diameter, outside—The over-all diameter, scribed on a radius from the center of the gear to the outside surface of the teeth.
Diameter, pitch—The diameter of cylinders in contact when rotating about their axis. Located at a point about half of tooth depth.
Distance, center—The distance between the centers of gears in mesh.
Face—The working area of a tooth above the pitch circle.
Flank—The working area of a tooth below the pitch circle.
Lead—The distance each worm thread advances along its axis for each revolution of the worm.
Pitch, circular—The distance between the center of one tooth and the center of the next tooth measured on the pitch circle.
Pitch, diametral—The number of teeth per inch of pitch diameter.
Root—The point where a tooth meets the dedendum circle; the bottom depth of a tooth.
Space—The profile dimension between adjacent teeth measured at the pitch circle.
Tooth, thickness of—The profile dimension of a tooth at the pitch circle.
Tooth, width of—The dimension of a tooth across the gear blank width.

Herringbone gears (double helical gears) have similar frictional characteristics as described for spur gears; i.e., they have both sliding and rolling friction. However, instead of the straight contact line across the tooth face, as with spur gears, the helical design causes contact to be made along a diagonal line. In other words, the leading end of a helical tooth passes through a portion of mesh before the trailing end enters mesh. This diagonal or slanted line, which becomes steeper as the helix angle becomes greater, crosses the pitch line at the moment the gears reach the center of mesh. This means that the line of contact is shortest (infinitesimal) at the instant mesh commences, becoming increasingly longer as it sweeps upward and downward over the driving and driven teeth, respectively. The length of the diagonal contact line reaches maximum, i.e., across the entire face of the tooth, when the gears are in full mesh. As motion continues, the line

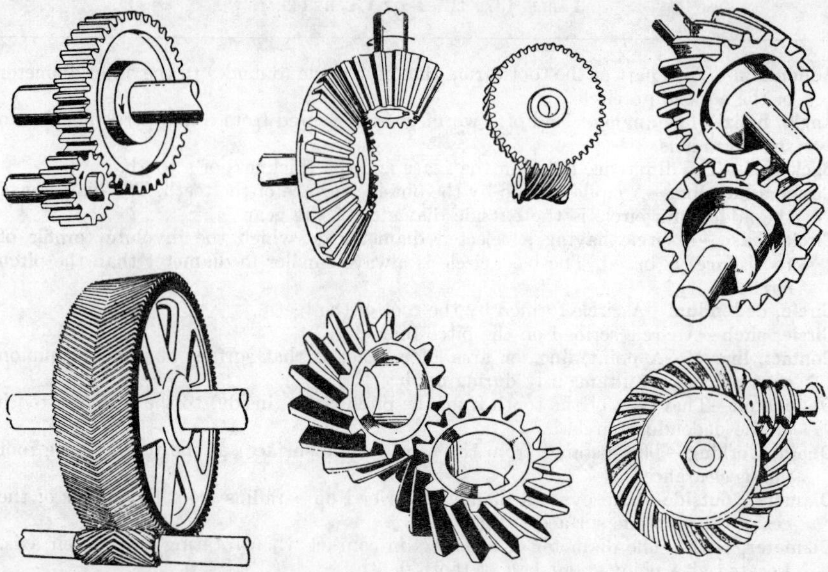

Fig. 431. Various types of gears. (*Left to right*) Spur, straight bevel, double-throated worm, helical, herringbone, spiral bevel, and hypoid.

of contact diminishes in length until it reaches zero at the instant contact is broken.

Because of the helix angle, herringbone gears are virtually continuous step gears. Step gears are made up of a number of gears placed side by side on the same shaft, each tooth of which is advanced slightly relative to its preceding neighbor to give a steplike appearance, rather than the straight-line effect of the spur gear. The effect of the helix angle allows several teeth to be in mesh simultaneously, which distributes load over a greater area and promotes smooth operation. Therefore, the small areas of contact developed during initial stages of approach to and final stages of departure from full mesh, are not subjected to any momentary pyramiding of unit pressure.

In the case of single helical gears, the helix angle, which may range from 20° to 45°, imposes an axial thrust along the shafts. This thrust load must be carried by the affected bearing. By using two opposing helical gears side by side on the same shaft, the thrust from one counteracts the thrust of the other. The herringbone gear is actually two opposing helical gears cut on the same gear blank, which gives the former a chevron-like or herringbone appearance.

Bevel gears have a cone shape and are designed with either straight or spiral teeth. Line contact prevails in both types and follows a similar pattern as described for spur and herringbone gears, respectively. Because the spiral angle is such as to allow one end of a spiral (of a spiral bevel gear) to engage its mate before the end of the preceding spiral has broken contact, smooth operation results.

The frictional load developed by straight bevel gears is due to rolling action. Rolling friction predominates in spiral bevel gears, although some sliding occurs.

Worm gears consist of a worm and a worm wheel or gear. The worm is formed by a thread having a definite lead, encircling a worm (blank) shaft, in which it is cut. The thread encircles the relatively small diameter blank one or more times. The lead is the distance that the thread advances for each complete turn of the worm. If the diameter of the pitch circle of the worm were increased indefinitely, it would tend to approach a straight line. Under the latter circumstance, the worm would become a *rack*, which makes the worm a form of endless rack. When the axis of the worm and the wheel are right angular to each other, the teeth of the wheel will mesh with the thread of the worm. Any advance of the thread, accomplished by rotating the worm, will cause the wheel to turn on its axis. The arc through which the wheel turns for each revolution of the worm depends on the lead or the helix angle. The helix angle is the angle made by the slant of the thread relative to a line perpendicular to the axis of the worm. The teeth of the worm gear are cut diagonally across the blank to conform with the helix angle of the worm.

Worm gears may be single, double, or multiple threaded. The ratio of the number of teeth on the gear to the number of threads on the worm is called the *velocity ratio*, or

$$\text{Velocity ratio} = \frac{\text{number of teeth on gear}}{\text{number of threads on worm}}$$

If the wheel has 60 teeth for a triple-threaded worm,

$$\text{Velocity ratio} = \tfrac{60}{3} = 20 \text{ to } 1$$

which means that the gear turns once for every 20 revolutions of the worm. For a double-threaded worm,

$$\text{Velocity ratio} = \tfrac{60}{2} = 30 \text{ to } 1$$

and for a single-threaded worm the velocity ratio would be 60 to 1. Thus, the worm gear is capable of large speed reductions. The number of teeth in mesh increase as the number of threads on the worm increase.

The length of the worm is usually six times the *pitch*, the latter being the center-to-center distance between threads. Worm gears may be nonthroated, single-throated, or double-throated. The nonthroated worm gear consists of (*a*) a straight worm, i.e., the outer surfaces of all threads are the same distance from the worm axis, and (*b*) the gear teeth are straight across the (width) outside surface of the gear blank. The single-throated worm gear also has a straight worm, but the outside surface of the gear teeth is curved (concave across the width) to conform with the curvature (circumference) of the worm at the root. The concave design permits the gear teeth to partially wrap around (envelop) the worm. Double-throated worm gears have, in addition to the throated gear teeth, a double-throated worm. Such a worm is designed to permit partial envelopment of the gear by the worm, accomplished by making the worm concave relative to its length. This means that the distance between the outside surface of the threads and the worm axis gradually becomes greater as the threads wind away from the center of the worm in both directions along the axis. Thus, the worm and the gear of a double-throated worm gear possess mutual envelopment. The concave form of the worm allows more teeth to be in mesh at the same time.

Modifying worm gears to obtain partial envelopment changes the frictional relationship between worm and gear teeth in the following manner:

(*a*) Nonthroated worm gears permit point contact only and loads transmitted by such gears act on a series of points as the rotation of the worm progresses through one revolution. For this reason, nonthroated worms are limited in their performance and seldom used for transmitting power of any appreciable magnitude.

(*b*) Single-throated worm gears attain *line* contact. As the worm revolves, contact is developed on a series of curved lines that run diagonally across the gear tooth. Starting at the tip of the gear tooth, the line of contact progresses radially to the limits of the working depth. The contact line of a worm gear is similar to the slanted line of a helical gear tooth in that it becomes progressively longer as full mesh is approached, after which it diminishes until contact is broken. Simultaneously, the contact line progresses outwardly on the worm threads. During rotation of the worm, considerable side slip occurs all during the period of contact. With worms of low helix angle the direction of side slip and contact line are nearly coincident. The line of contact becomes steeper, i.e., tends toward a radial direction as the helix angle becomes greater.

(*c*) It is probable that the contours of opposing surfaces of double-throated worm gears promote contact in the form of a radial band having an appreciable width rather than a line. As the band would extend simultaneously across the entire width of the gear tooth as in the case of line contact described for single-throated worms, the total bearing area would depend on the width of the band. Frictional relationship between gear and worm in a double-throated worm gear is one of sliding action only.

Because sliding action predominates in worm gear operation, the use of metals possessing a low coefficient of friction is desirable. A combination of bronze gear mated with a steel worm results in a worm gear set having good wearing characteristics, low coefficient of friction, and high load transmitting ability.

Hypoid gears may be considered as a compromise between a spiral bevel gear and a worm gear in many respects. In spiral bevel gears the axis of the pinion and the gear intersect, while in the worm gear the axis of the worm (pinion) is tangent to the gear. The axis of the hypoid pinion lies in a plane *anywhere* between the two extremes, but always right angular to the axis of its gear. It resembles the spiral bevel gear in appearance only.

As stated previously, rolling friction predominates in spiral bevel gears, although some sliding occurs. For all practical considerations, worm gears develop sliding friction only. Hypoid gears, being a combination of the two, develop both rolling and sliding friction, the latter due principally to side slip. The ratio of sliding to rolling friction, however, depends on whether the hypoid arrangement (offset of pinion relative to gear axis) approaches that of spiral bevel or worm gear. As the distance between gear and pinion shaft centerlines increases, the ratio of sliding to rolling friction becomes greater. Thus, when heavy loads are transmitted through hypoid gears having a pinion that is extremely offset, tremendous frictional force develops. Actually, the pressure developed between the contacting surfaces of hypoid gear and pinion under heavy load is great enough to promote welding and cause scoring of the teeth. To prevent scoring under heavy loading, special ingredients must be added to hypoid gear lubricants.

While hypoid gears may be designed for other than a 90-degree shaft angle, they are usually employed for right-angle drives.

Helical gears may be designed to transmit motion between shafts that cross each other at any angle. However, contact between driver and driven gear is made on a point. Both sliding and rolling occur between helical gears, the ratio of sliding to rolling friction increasing as the shaft angularity approaches nearer to 90 degrees. When helical gear shafts are at right angles, the action is one of side slip on a point, a condition not conducive to heavy loading. For this reason, helical gears are seldom used to transmit loads of any appreciable magnitude between nonparallel cross shafts.

Table 138 gives a summary of contact and frictional characteristics of various types of gears.

THEORY OF LUBRICATION

As previously stated, lubrication is the science of reducing friction by the application of a lubricant between sliding or rolling surfaces. At this point it will be assumed that a suitable lubricant is an oil made by refining petroleum. It will be assumed also that this oil will flow freely at room temperature. Because such an oil is made from petroleum, it is called mineral oil. Mineral oils are excellent lubricants.

TABLE 138. FRICTIONAL CHARACTERISTICS, MEANS OF CONTACT BETWEEN GEAR AND PINION, AND TOOTH DEVELOPMENT OF VARIOUS TYPES OF GEARS

Gear	*Means of Contact**	*Predominant Frictional Characteristic†*	*Pitch Surface on Which Teeth Are Developed*
Spur gears	Line	Sliding and rolling	Cylindrical
Herringbone gears	Line	Sliding and rolling	Cylindrical
Bevel gears:			Conical
Straight tooth	Line	Rolling and sliding	
Spiral bevel	Line	Rolling and sliding	
Worm gears:			Cylindrical
Nonthroated	Point	Sliding	
Single-throated	Line	Sliding	
Double-throated	Band	Sliding	
Hypoid	Line	Sliding	Hyperbolic
Helical (nonparallel shafts)	Point	Sliding	Cylindrical

* Deformation of tooth surface under load creates definite *area* contact in all gears. For teeth forming line contact the line becomes a band probably $\frac{1}{64}$ to $\frac{1}{4}$ in., depending on the size of gear.

† More or less rolling and sliding occur in nearly all gears; the table lists the predominant frictional characteristics.

Boundary Lubrication

Boundary lubrication is a condition in which a fluid film does not exist or is not thick enough to appreciably separate opposing frictional surfaces and prevent their intimate (metal-to-metal) contact. The lubricant probably fills the asperities and affords some reduction in coefficient of friction (as compared with solid sliding bodies) by whetting the surfaces in contact. Boundary lubrication may be considered as a compromise between solid sliding friction and fluid friction, although a rather poor one in most cases. Lubricants used to meet boundary conditions are not subject to the rules of hydrodynamic laws discussed in the section following. Instead, their selection is based chiefly on oiliness or slipperiness, and to a large extent on their strength and persistence of film. Where an oil film of sufficient thickness to maintain separation of two rubbing bodies cannot be established, the molecular layer development in fluids under hydrodynamic (fluid film) conditions must be replaced by oiliness and lubricant film strength factors. Greases of various kinds are often used to overcome the deficiencies of boundary conditions.

Where the quantity of oil between the rubbing surfaces is great enough to establish alternate periods of intermittent separation and metal-to-metal contact, a *mixed film* is said to exist.

Viscosity

The discussion of hydrodynamic lubrication following this section is based on a quality possessed by fluids that reflect their relative fluidity. That quality reflecting fluidity of fluids is called viscosity. Therefore, a study of viscosity must be made before the principle of hydrodynamics can be comprehended. The development of hydrodynamic forces follows the law of continuity of fluid flow. This law states, in effect, that for a fluid flowing in a smooth conduit of variable cross-section, the volume passing any point along the length of conduit must be constant. For any two points along the flow path xy (Fig. 434), then

$$P_1V_1A_1 = P_2V_2A_2$$

where P = pressure
V = velocity
A = area

In the case of liquids, which are incompressible:

$$V_1A_1 = V_2A_2$$

This means that the volume of liquid passing any point in the flow path per unit of time is constant. Therefore, VA = constant. Also, the pressure exerted by the liquid perpendicular to the axis of flow varies. Because of the variation of pressure in a converging flow path, it is possible to keep opposing nonparallel bearing surfaces separated during relative motion.

To abide by the law of continuity as stated above, fluids must adhere to definite characteristics based on their resistance to flow. The resistance of fluids to flow is called viscosity. When the flow characteristics are such that the shearing stress between companion molecular layers of fluid is proportional to the rate of shear (velocity) perpendicular to the direction of motion, it is said to act as a Newtonian fluid. In order to develop a hydrodynamic film, fluids must conform to viscosity as defined by Sir Isaac Newton.

Newton's definition of viscosity is based on experiments consisting of a cylinder placed concentrically within another of slightly larger diameter and both submerged in still water. This caused a thin film of water to exist between the outside area of the smaller cylinder and the inside area of the larger one. By rotating one cylinder, the other one also rotated at some relative speed.

The force required to establish relative motion between the cylinder was found to be proportional to the effective working area and to the speed of rotation. It was inversely proportional to the thickness of the water film separating the cylinders. Newton expressed the relationship of all factors in the following equation:

$$F = \mu A \frac{v}{h}$$

where F = force
μ = absolute viscosity
A = area of effective working surface
v = velocity
h = distance between cylinders or film thickness
$\frac{v}{h}$ = the velocity gradient associated with film thickness

The force necessary to establish relative motion varies with different fluids, but is constant for a given fluid at the same temperature. Force as applied to shearing stress of fluids, then, is a function of viscosity.

From the above, viscosity may be defined as the amount of force required to overcome friction between adjacent molecular layers of a fluid mass to initiate flow. Viscosity varies with different fluids having the same temperature. It generally varies for the same fluid at different temperatures. Viscosity is the measure of internal friction developed within a fluid mass during flow and, therefore, a measure of fluid friction. Fluid friction bears the same relation to adjacent fluid layers during flow as sliding friction does to rubbing solid surfaces.

Just as sliding friction between solids may be measured, so can fluid friction. Being a measure of force, viscosity is expressed as a force. To be intelligible, the measure of viscosity must have some basis of standardization. Being a force, viscosity standardization involves mass and velocity. Viscosity is expressed in various ways, all of which are interpretations of a basic measuring process. Values of measurement determined by the basic process (Newton's) are expressed in terms of *absolute viscosity.*

To summarize, then, viscosity is a quality of fluids that reflects their fluidity. As a means of interpretation, fluidity has been designated as a force. Force involves mass and velocity, and is, therefore, measurable. To be intelligible, force must be expressed in terms that include all pertinent conditions of measurement; i.e., distance, density, time, and temperature.

To establish a definite relationship between different fluids, temperature must be constant. In the case of Newtonian liquids, the variables include thickness of film (h), area of film (A), force required to support shear of fluids layers (F), distance of effort (d), and time of effort (t). If internal fluid friction is represented by μ, the relationship of the variables according to Newton's definition of viscosity is as follows:

$$\mu = \frac{F \times h}{A \times \frac{d}{t}} = \frac{F \times h \times t}{A \times d}$$

However,

$$\frac{d}{t} = \text{velocity } (v)$$

Therefore,

$$\mu = \frac{F \times h}{A \times v}$$

If the conditions of measurement are expressed in terms of the centimeter-gram-second system, then

$$\mu = \frac{F \times h}{A \times v} = \frac{\text{dynes} \times \text{cm} \times \text{sec}}{\text{cm}^2 \times \text{cm}} = \frac{\text{dynes} \times \text{sec}}{\text{cm}^2}$$

Where all variables are given a value of unity (1), then $\mu = 1.00$, a unit called a poise (after Poiseuille). 1 poise, then, is the viscosity of a fluid requiring a force of 1 dyne per sq cm to support a shear rate of 1 cm per sec when the film thickness is 1 cm. 1 centipoise = $\frac{1}{100}$ poise and is represented by z.

Where the values of all variables are expressed in terms of the English system—i.e., pounds-feet-seconds—then

$$\mu = \frac{F \times h}{A \times v} = \frac{\text{lb} \times \text{in.} \times \text{sec}}{\text{in.}^2 \times \text{in.}} = \frac{\text{lb} \times \text{sec}}{\text{in.}^2}$$

and as above when all variables are given a value of unity (1), then $\mu = 1.00$, a unit called a reyn (after Reynolds). 1 reyn, then, is the viscosity of a fluid requiring a force of 1 lb per sq in. to support a shear rate of 1 in. per sec when the film thickness is 1 in. 1 microreyn = $\frac{1}{1{,}000{,}000}$ reyn, a more practical value.

Consider, now, a liquid flowing through a capillary tube having a fixed internal diameter, instead of the mass resting on a plane surface having thickness and area. The volume of liquid that would flow in unit time is proportional to the fourth power of the tube radius (R) and to the pressure (P) on the liquid, and inversely proportional to the length of the tube (L) and the absolute viscosity (μ) of the liquid, or

$$V = \frac{P \times \pi \times R^4}{8\,\mu L} \qquad \text{(Poiseuille's formula)}$$

and

$$u = \frac{P \times \pi \times R^4}{8 \times V \times L}$$

If pressure depended on gravity, then

where $P = g \times h$
g = 32.2 ft per sec per sec
h = head

Substituting in Poiseuille's formula, then

$$\mu = \frac{g \times h \times \pi \times R}{8 \times V \times L}$$

In consideration of fluid density (p) as a factor governing the flow rate, the head becomes a function of $p \times g \times h$ in Poiseuille's formula, so that

$$\frac{\mu}{p} = \frac{g \times h \times \pi \times R^4}{8 \times V \times L}$$

The ratio of absolute viscosity (μ) of a liquid to its density (p) is called *kinematic viscosity*, and the unit of measurement is the *stoke* (after Stokes), represented by the Greek letter nu (ν). Or

$$\nu = \frac{\mu}{p}$$

and

$$1 \text{ stoke} = \frac{1 \text{ poise}}{\text{density}}$$

$$1 \text{ centistoke} = \tfrac{1}{100} \text{ stoke}$$

The specific gravity of water at 68 F is given an arbitrary value of 1.0. The specific gravity of other liquids, including oil, is based on the weight of water. Therefore, p is equal to the specific gravity of the test liquid. For 68 F water, than $\nu = \mu$.

The kinematic viscosity of 68 F distilled water is 1.007 centistokes. At 100 F and 130 F, it is 0.689 and 0.518 centistoke, respectively.

Kinematic viscosity of an oil is determined by an instrument called a kinematic viscosimeter. There are several kinds, but their principles are substantially the same. The test consists of allowing a given amount of oil at a given temperature to flow from one reservoir to another through a capillary tube having a specific length and internal diameter. The viscosimeter is immersed in a bath of water held at test temperature, usually 100 F, 130 F, or 210 F. The over-all length from the surface level of the oil when in the top reservoir to the same level in the bottom reservoir is an important factor, as are all other viscosimeter dimensions, oil quantity, and temperature. The length of time required for the given amount of oil (measured according to lines etched in the reservoir) to pass through the capillary tube is recorded in seconds. Conversion of time to kinematic viscosity in centistokes is accomplished by the use of tables based on design, and reference fluid against which it is calibrated.

Another method of reporting viscosity of lubricating oils is according to the results obtained by the use of the *Saybolt* viscosimeter. Saybolt viscosities have been used, and are still used by lubrication engineers as a general expression of oil fluidity. Saybolt viscosity denotes the time in seconds of a given amount of oil (60 milliliters) to flow by gravity from an immersed reservoir into a flask through a capillary tube of specified length (1.225 cm) and internal diameter (0.1765 cm). Tests are usually run at oil temperatures of 100 F, 130 F, and 210 F. The results are reported in Saybolt Seconds Universal at the desired temperature. Therefore, an oil at 100 F requiring 150 sec to complete the above test is said to have a viscosity of 150 SUS at 100 F.

When liquids are viscous enough to require more than 1000 sec to complete the Saybolt Universal viscosimeter test, they are tested more conveniently in a *Saybolt Furol* viscosimeter. The latter is similarly constructed and operated as the Universal machine, except that the internal diameter of the outlet tube is larger (0.313 cm). "Furol" is a contraction of the phrase "fuel and road oils." Furol time is approximately one tenth that of Universal time; i.e., an oil having a viscosity of 1000 SUS will have a Furol viscosity of 100 for the same temperature reading. Furol viscosity is often reported for 122 F temperature.

The Saybolt method of determining viscosities has certain disadvantages involving head variations and turbulence of flow. Therefore, the kinematic system of reporting viscosity in centistokes is gaining wide recognition as being closer to true values. For practical purposes, however, the Saybolt system of reporting viscosities that denote relative time-flow values is quite convenient.

Other viscosity determination systems include Redwood (England), Engler (Germany), and Redwood Admiralty (England), all of which are based on the same principle as the Saybolt Universal instrument. The volume of oil handled and test temperatures vary in each case, however. The viscosity values as determined by any one of the instruments mentioned have a corresponding value as reported for any of the remaining instruments based on the same temperature. Conversion of one to the other can be accomplished by mathematical formulae. Similarly, the values established by any of the above instruments can be converted to kinematic viscosity, and vice versa. To convert Saybolt, Redwood, and Engler viscosities to kinematic viscosity in centistokes, the following equations are used:

Saybolt Universal seconds to centistokes:

$$\nu = 0.22t - \frac{180}{t}$$

Redwood to centistokes:

$$\nu = 0.26t - \frac{171}{t}$$

Engler to centistokes:

$$\nu = 0.147t - \frac{374}{t}$$

where ν = kinematic viscosity, centistokes
t = Saybolt Universal viscosity, sec, at temperature under consideration

EXAMPLE: Determine the kinematic viscosity of an oil having a Saybolt Universal viscosity of 300 sec at 100 F.

$$\nu = 0.22t - \frac{180}{t}$$

$$\nu = 0.22 \times 300 - \frac{180}{300}$$

$$\nu = 65.4 \text{ centistokes}$$

To determine the absolute viscosity of the above oil if it had a specific gravity of 0.88 corresponding to an A.P.I. gravity of 29.3,

$$z = p \times \nu$$

$$z = 0.88 \times 65.4$$

$$z = 57.55 \text{ centipoises}$$

Viscosity Relation to SAE Numbers

A general designation of lubricating oil viscosities was devised by the Society of Automotive Engineers to simplify recommendations for automobile frictional components. SAE numbers classify lubricating oils on the basis of viscosity only and do not represent any other qualities or characteristics. Each number represents a range of viscosity at a given temperature. For the lower viscosity oils, the range represents Saybolt Universal seconds at 0 F. For higher viscosity oils, 210 F readings are designated. Table 139 shows the classification of lubricating oils according to SAE numbers. Engine oils are represented by the numbers SAE 5W to SAE 50, inclusive.

TABLE 139. SAE CLASSIFICATION OF CRANKCASE OILS

SAE No.	*0 F (Min.)*	*0 F (Max.)*	*210 F (Min.)*	*210 F (Max.)*
5W		4,000	—	—
10W	6,000*	Less than 12,000	—	—
20W	12,000†	48,000	—	—
20	—	—	45	Less than 58
30	—	—	58	Less than 70
40	—	—	70	Less than 85
50	—	—	85	110

* Minimum viscosity at 0 F can be waived provided that viscosity at 210 F is not below 40 Saybolt Universal Seconds.

† Minimum viscosity at 0 F can be waived provided that viscosity at 210 F is not below 45 Saybolt Universal Seconds.

Viscosity Index

Mineral lubricating oils decrease in viscosity as their temperature is raised. For example, a cylinder oil having a viscosity of, say, 3000 sec at 100 F (SUS) may be reduced to 1075 sec at 130 F and again to 175 sec at 210 F. The reduction from 3000 to 175 sec with a temperature increase of 100 F to 210 F, respectively, may be said to be the rate of viscosity change for that oil. Another cylinder oil of 3000 sec at 100 F (SUS) may have a lower viscosity at 210 F of, say, 160 sec (SUS). The second oil then would be considered to have a higher rate of viscosity change per degree temperature rise.

TABLE 140. VALUES OF H AND L IN DEAN AND DAVIS FORMULA FOR DETERMINING VISCOSITY INDEX*

Saybolt Universal Viscosity at 210 F	*H*	*L*	(*L* – *H*)
60	425	780	355
65	514	976	462
70	604	1182	578
75	697	1399	702
80	791	1627	836
85	888	1865	977
90	986	2115	1129
95	1087	2375	1288
100	1189	2646	1457
105	1294	2928	1634
110	1401	3220	1819
115	1510	3523	2013
120	1620	3838	2218
125	1733	4163	2430
130	1848	4498	2650
135	1965	4845	2880
140	2084	5202	3118
145	2205	5570	3365
150	2328	5949	3621
960	2580	6740	4160
170	2840	7573	4733
180	3109	8450	5341
190	3385	9370	5985
200	3670	10333	6663
225	4418	12930	8512
250	5217	15796	10579
300	6967	22340	15373

* Courtesy, ASTM D567: "Standard Method for Calculating Viscosity Index." American Society for Testing Materials.

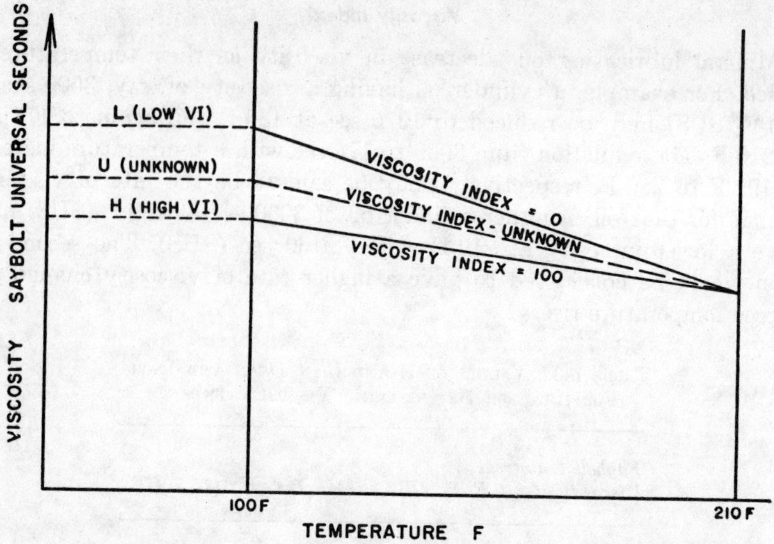

Fig. 432. Graphic demonstration of Viscosity Index determination.

The rate at which an oil changes viscosity with a rise (or drop) in temperature is designated by a comparative number called *viscosity index* (VI). A VI of 100 indicates that an oil possessing this value "thins out" less rapidly than oils having a VI of 0 (zero). Values in between indicate intermediate viscosity changes according to the comparative VI number.

VI for any oil having a viscosity change between the two extremes may be determined by the Dean and Davis formula, viz.:

$$\mathrm{VI} = \frac{L - U}{L - H} \times 100$$

where L = viscosity SUS of low VI (0) oil at 100 F
H = viscosity SUS of high VI (100) oil at 100 F
U = viscosity SUS of unknown VI oil at 100 F

Viscosity at 210 F of oils represented by H and U must be the same.

EXAMPLE: The viscosity of a paraffinic base oil has a viscosity at 100 F of 340 SUS and of 55 SUS at 210 F, and is rated at 100 VI. A naphthenic base oil having viscosities of 600 SUS at 100 F and 55 SUS at 210 F is rated at 0 VI. The VI of another oil having the same viscosity at 210, but between 340 and 600 sec at 100, say 440 sec, would have a VI as follows:

$$\mathrm{VI} = \frac{L - U}{L - H} \times 100$$

$$\mathrm{VI} = \frac{600 - 440}{600 - 340} \times 100$$

$$\mathrm{VI} = 61.5 \text{ or } 62 \text{ (approx)}$$

The values of H, L, and $L - H$ may be determined from Table 140. An example referring to Table 140 follows.

Determine the VI of an oil having a viscosity of 70 sec at 210 F and 640 sec at 100 F.

$$VI = \frac{L - U}{L - H}$$

$$VI = \frac{1182 - 640}{578} \times 100$$

$$VI = 94$$

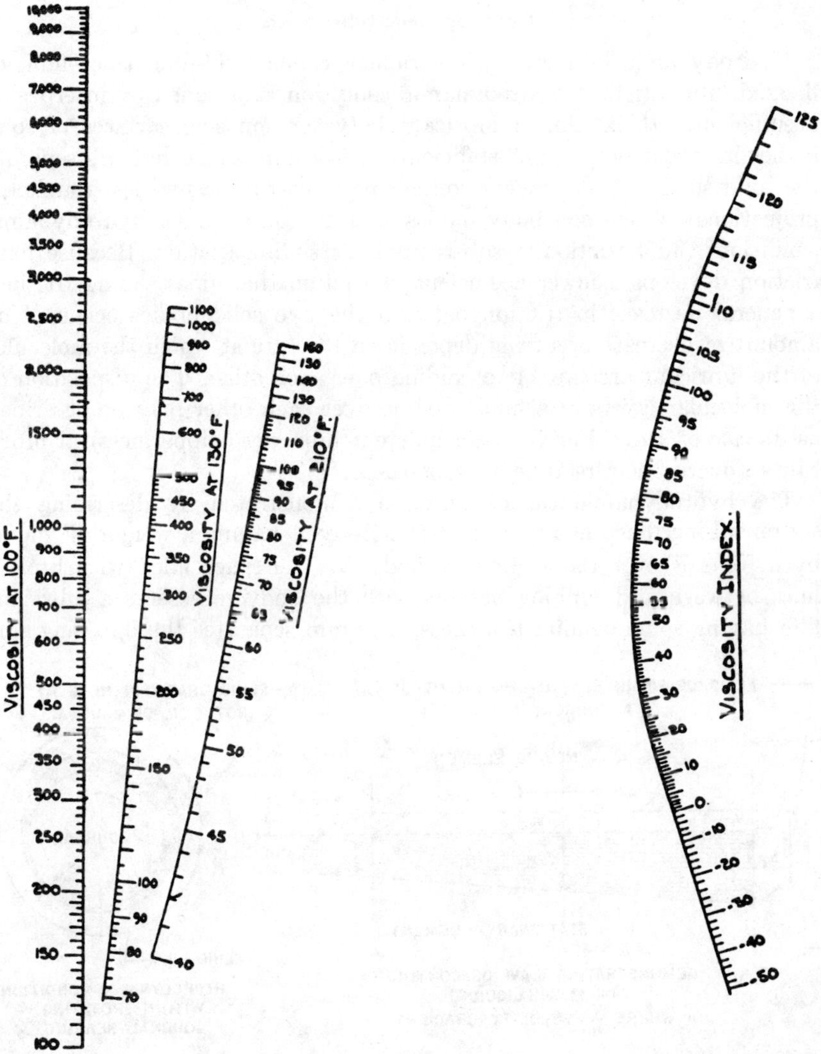

Fig. 433. Nomograph for determining Viscosity Index. VI found by lining up known viscosities (SUS) at 100 F (or 130 F), and 210 F with straight edge. Read VI where straight edge crosses VI line.

Paraffinic oils lose their viscosity less rapidly with increased temperature than naphthenic oils and subsequently have a higher VI generally. However, with newer methods of refining (solvent treating) and the use of certain type additives, the viscosity-temperature relationship of naphthenic oils has improved. The viscosity index of an oil is a valuable specification in predicting the lubricating-film-forming characteristics of an oil when exposed to elevated temperatures. It is extensively used in connection with internal combustion engine lubricants.

Hydrodynamic Lubrication

Hydrodynamic lubrication is variously called fluid-film, thick-film, or flooded lubrication. A hydrodynamic condition is obtained by interposing a sufficiently thick film of lubricant between opposing surfaces of solid bodies in relative motion. A sufficiently thick film is one that will separate the solid surfaces and prevent contact between their asperities (roughness projections). When one body moves over the other under hydrodynamic conditions, fluid friction is substituted for sliding friction. Because fluid friction develops a lower coefficient of friction than does sliding friction, a general decrease in friction between the two solid bodies occurs. The amount of decrease occurring depends on the rate at which the molecules of the lubricant are capable of sliding over each other. The disposition of the molecular layers of a fluid to slide over each other may be described as its *rate of shear*. Fluidity and inherent lubricity (slipperiness) of lubricants govern their relative rate of shear.

The hydrodynamic theory may be demonstrated by describing the action taking place in Fig. 434 (*left*). Body A having a weight W moves by a force F over the stationary body B. The space filled with broken lines between and running parallel with the body represents a lubricant film having some definite thickness. The film separates the opposing sur-

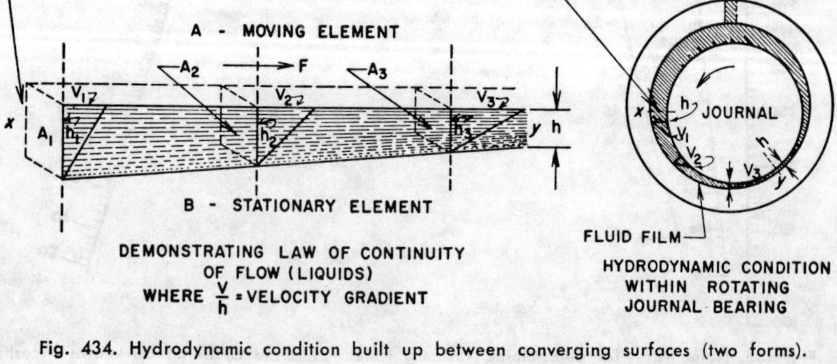

Fig. 434. Hydrodynamic condition built up between converging surfaces (two forms).

faces. Each broken line represents a molecular layer of fluid. Differences in length of the dash marks indicate the velocity of each layer relative to its companion layer. In the illustration the lines grow shorter with diminishing velocity. The velocity of the fluid layer contiguous to the moving surface of A is equal to the velocity of that surface. The velocity of each succeeding layer diminishes until a zero velocity is reached in that layer contiguous to the stationary surface of B. Thus, fluid layer velocity ranges from maximum, at the speed of the moving or fastest moving member, to minimum, corresponding to the speed of the slower moving member. In the case of one stationary member, as it is here, minimum velocity equals zero.

Because relative velocity exists between companion layers of fluid as shown, a multiplicity of sliding or shearing actions occur throughout the fluid mass. The rate of shear between each layer is a function of the speed of A, assuming that all other factors remain the same. The speed of A as a *free body** depends on the magnitude of fluid friction within the film mass. Stated differently, the speed of A as a free body is governed by the disposition of companion fluid layers to slide over each other. As a *captive* body, the speed of A will depend on the amount of power applied continuously parallel to motion. With fluids of high internal friction, a greater amount of power would be required by a captive body than would be necessary using fluids of low internal friction.

By interposing a fluid film between the surfaces of a bearing and its rotating journal, as shown in Fig. 134 (*right*), the same hydrodynamic condition may be established as described above.

Flat Surfaces, Parallel

A lubricant film of h thickness separating a moving body A having a force F across a stationary body B is shown in Fig. 435 (*left*). The surfaces of bodies A and B are parallel. Assume that both surfaces are of indefinite width (perpendicular to the page) so that no side leakage of lubricant occurs. Under such conditions, (a) all the lubricant entering the space, h, at the leading end, x, will leave at the trailing end, y, during motion of A; (b) the velocity of the lubricant layer contiguous to the stationary surface is zero and the layer contiguous to the moving member has a velocity equal to the latter (assuming no slippage). With the same quantity of oil flowing at all cross-sections of h throughout the length of the surfaces A and B, (x to y), and a constant relative velocity existing between each fluid layer at every point along xy, no increase in pressure is allowed to develop at any one point above that at any other point. Thickness h then becomes a function of load W. When W is great enough, h shrinks

* A free body is one that continues to move unaided after having received an initial push by some external force.

sufficiently in magnitude to allow the surfaces of A and B to make solid contact.

Thus, parallel surfaces do not sustain a hydrodynamic film where motion is accompanied by any appreciable loading. Guide bearings, crossheads, guides, and ways are frictional components whose surfaces are parallel to the opposing surfaces of their riding elements.

Flat Surfaces, Converging

When two opposing flat surfaces are not parallel, any lubricant interposed will develop a film having a wedge-shaped cross-section, as shown in Fig. 435 (*right*). Assuming that the direction of the moving body A is toward the point of smallest (separation) fluid film cross-section, a fluid lubricant will be drawn between the bodies at the leading end of the wedge, x. The same amount of lubricant must leave at the trailing end, y, as enters at x. Because the area at the outlet is smaller than the inlet area, an ever-increasing velocity occurs from, and in the direction of, x to y. Thus, the relative velocity between fluid layers changes at each increment of length between x and y—i.e., with diminishing film thickness. The change in velocity is called the velocity gradient and expressed by the ratio $\frac{V}{h}$ where V represents velocity and h is the film thickness at any point along xy. The load W attempts to force A more closely in contact with B and further restrict the outlet area at y. This tends to crowd the fluid, resulting in gradual pressure rise along the route xy. It is this rise in pressure that counteracts the load and maintains a fluid film thick enough to keep the bodies separated. Points x and y in Fig. 435 (*right*) are in the same relative positions as shown in Fig. 435 (*left*).

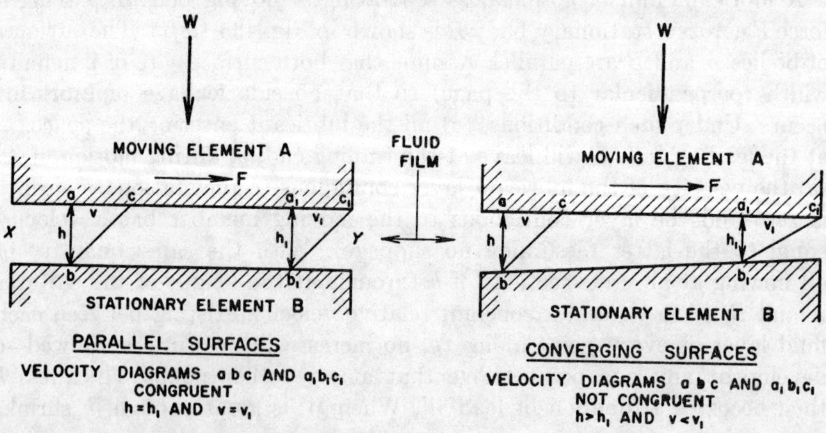

Fig. 435. Demonstrating the instability of a fluid film between parallel surfaces vs. fluid wedge formation with converging surfaces.

From the above it may be stated that for constant-speed, nonparallel surfaces employing the same fluid,

(*a*) Fluid film at the point of minimum clearance decreases in thickness as the load increases;
(*b*) Pressure within the fluid mass (hydrodynamic force) increases as the film thickness decreases due to load;
(*c*) Pressure within the fluid mass is greatest at some point approaching minimum clearance and lowest at the point of maximum clearance;
(*d*) Fluidity decreases, i.e., fluid becomes more viscous as applied pressure increases;
(*e*) Thickness of film increases at the point of minimum clearance as load decreases.

It is also true that

(*f*) Any increase in relative speed increases film thickness at the point of minimum clearance,

and where all other factors remain the same,

(*g*) Film thickness at the point of minimum clearance increases with the use of more viscous fluids.

Thrust bearings having tapered or beveled lands, pivoted thrust shoes, and tilting pads are all nonparallel, flat surface bearings. The surfaces mentioned may represent part of the stationary member, although the wedge developing component may also be the moving member. Kingsbury, Mitchell, and Gibbs thrust bearings are some of the types designed to produce an oil wedge and develop hydrodynamic lubrication.

Curved Surfaces—Full Circular Bearing

The clearance allowed between a journal and its bearing creates an eccentric relation between the two. Eccentricity is the distance between the center of the journal and the center of the bearing. The clearance space assumes the shape of a crescent, due to eccentricity, when the journal lies at rest, as shown (exaggerated) in Fig. 436 (*left*). At rest, the journal exerts its weight on the bottom surface of the bearing. If the journal is rotated, it would tend to climb up the side of the bearing much in the same manner as a wheel rolls uphill. At the moment rotation is initiated, a position would be assumed, as shown in Fig. 436 (*center*). Due to the weight imposed on it, however, the journal would continually slip back toward its original position.

If sufficient oil were poured continuously into the bearing through the hole in the top to keep the clearance space filled while the journal was

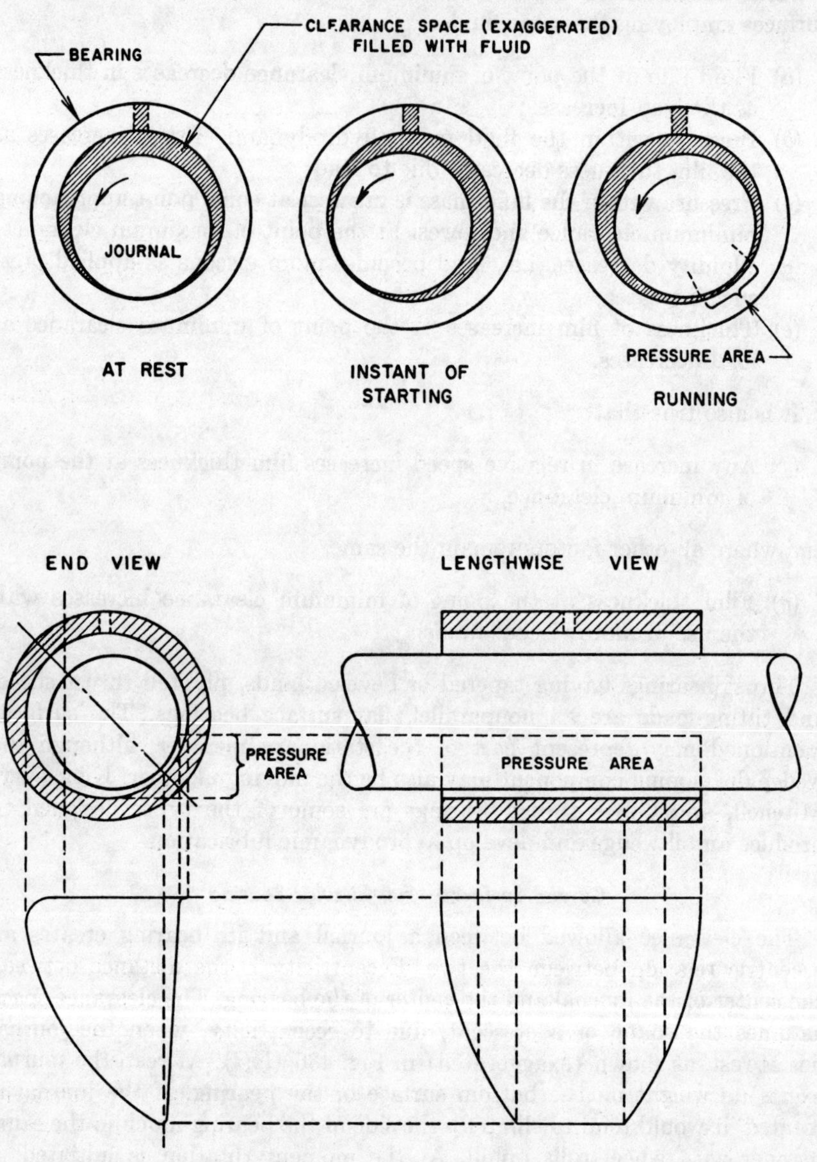

Fig. 436. (*Top*) Hydrodynamic film developed within a full journal bearing. (*Bottom*) Pressure distribution in a full journal bearing.

being rotated, a condition similar to that described for nonparallel flat surfaces would develop. The crescent-shaped clearance space would act to provide the oil wedge.

Upon initiating rotation, the journal attempts to climb the side of the bearing. As it starts to climb, the journal enters upon the thin oil film at the end of the crescent-shaped oil mass. In doing so, two actions occur. First, the journal loses traction rapidly by reason of the fluid film and slips back to its original position immediately. Second, oil adheres to the surface of the journal and is pulled between it and the bearing. As speed increases, the rotating journal acts as a rotary pump, pulling more and more oil beneath it, causing its separation with the bearing surface. The journal now rides on a fluid film. At this point, a similar condition exists between journal and bearing as described for nonparallel flat surfaces. The oil, entering between the journal and the bearing, continues to build up, during which it develops hydrodynamic pressure along the side of the bearing opposite the direction of rotation. The greatest pressure occurs at the tip of the crescent (converging wedge), which is the zone of thinnest oil film. During the initial stage of rotation, this zone is located slightly to the left of the extreme bottom of the bearing, assuming that the journal is turning counterclockwise and load is vertical.

Upon development of hydrodynamic pressure along the side opposite rotation, in the same manner it developed from x to y (Fig. 434), the journal is forced over to the other side of the bearing. The thinnest film is now located slightly to the right of the extreme bottom of the bearing and the crescent has shifted to its permanent running position, as shown in Fig. 436 (*right*). In this position, the journal continues rotation on a fluid film subject only to fluid friction, the amount of which depends on the rate of shear of the lubricant used. The actual relation of the rotating journal to its bearing, i.e., the position of the crescent, depends on load and speed. Higher speeds tend to shift the journal farther to the right (or to the left, depending on direction of rotation), while heavier loads attempt to keep it at or near the extreme bottom of the bearing. It will finally settle in a position of equilibrium governed by the resultant force of both speed and load. The zone of thinnest fluid film is the pressure area of a given journal bearing.

The dimension of the thinnest film when all other factors remain constant depends on the fluidity of the lubricant. Its relative thickness is based on the following rules:

(*a*) With any load the thickness of fluid film in the pressure area will increase as the viscousness (viscosity) of the fluid introduced into the bearing becomes greater.

(*b*) With a given fluid, the thickness of the film will decrease as the load becomes heavier.

(*c*) With a given load and fluid, the thickness of the film will increase as speed is increased.

(*d*) Fluid friction increases as the viscousness of the lubricant becomes greater. Conversely, fluids of high fluidity develop low fluid friction.

From the above rules, the following may be concluded for plain journal bearings.

1. For light loads and high speeds, use a lubricant of high relative fluidity.

2. For intermediate loads and reasonable speeds (within the normal range for the work system), use a lubricant of medium relative fluidity.

3. For heavy loads and slow to medium speeds, use a lubricant of low relative fluidity.

Pressure Distribution in Full Journal Bearings

The hydrodynamic forces developed between a journal and a bearing, as shown in Fig. 436 (*right*), follow nearly the same pattern as the forces developed between nonparallel flat surfaces. As the journal rotates, it tends to crowd oil between the two opposing surfaces in the direction of rotation. The clearance space being crescent-shaped, i.e., diminishing in a cross-sectional area, requires a higher flow velocity to develop in order to allow the same quantity of fluid to pass through it as is forced to enter it. A pressure increase develops coincident with velocity increase and reaches its highest level at the point of minimum film thickness. This rise in pressure tends to lift the journal and increase the thickness of this fluid film throughout the pressure zone or pressure area. The lubricant passes through the pressure area and escapes into a region beyond the point of minimum film thickness where the clearance space starts diverging. The pressure drops rapidly in the diverging region.

Viscosity, Speed, and Load Relationship in Journal Bearings

Considering the rules *a*, *b*, *c*, and *d* for journal bearings (above), it is obvious that conditions developed by the relation of viscosity, speed, and load affect the film thickness. Where all three factors are not balanced adequately to produce a fluid film, boundary lubrication results, causing metal-to-metal contact. A high coefficient of friction is produced under boundary conditions accompanied by wear of the rubbing surfaces, usually the bearing. Coefficient of friction of a journal bearing, then, is a function of viscosity, speed, and load.

The relation of viscosity, speed, and load may be constant for a given bearing, but varies for different bearings. The relationship may be expressed by a dimensionless number which value depends on combining the three factors in the following manner:

$$\frac{\text{Viscosity} \times \text{speed}}{\text{Load}} = \text{a number}$$

or

$$\frac{ZN}{P} = \text{a number}$$

where Z = absolute viscosity of lubricant, centipoises
N = speed, rpm
P = load, lb per sq in. (average based on projected area)

For example, where the viscosity of the oil film is, say, 19 centipoises, and the journal carries a load exerting a pressure of 350 psi at 1750 rpm, then

$$\frac{ZN}{P} = \frac{19 \times 1750}{350}$$

$$\frac{ZN}{P} = 95.0$$

If the viscosity of the oil were reduced to, say, 12 centipoises, then

$$\frac{ZN}{P} = \frac{12 \times 1750}{350} = 60.0$$

Reducing speed to 900 rpm and using the original oil, then

$$\frac{ZN}{P} = \frac{19 \times 900}{350} = 48.85$$

Increasing load to 500 psi and maintaining original oil and speed

$$\frac{ZN}{P} = \frac{19 \times 1750}{500} = 66.5$$

If the speed were reduced to 50 rpm in the original example, then

$$\frac{ZN}{P} = \frac{19 \times 50}{350} = 2.71$$

The above cases show the effect of changes in any of the three factors on the $\frac{ZN}{P}$ number. A significant change is that of speed. Using an oil of higher viscosity, i.e., less fluidity, a higher $\frac{ZN}{P}$ number than 2.71 may be established. Raising the viscosity to 90 centipoises, $\frac{ZN}{P} = \frac{90 \times 50}{350}$ = 12.85, a value that is more indicative of safe operating conditions through the establishment of a thicker fluid film. The reason for this may be explained as follows:

Because a definite relation exists between $\frac{ZN}{P}$ number and fluid film thickness, it follows that the coefficient of friction is affected also by any single or multiple change in viscosity, speed, and load. The coefficient of friction developed between a journal and a bearing is the ratio of tangential force required to turn the journal to the total load on the bearing. If a cord were wound around a loaded journal and pulled with sufficient force, the journal would rotate. By pulling the cord through a spring balance, the amount of force required to turn the journal could be measured. Under dry (nonlubricated) conditions, it might be established that the tangential pull would be 0.25 to 0.40 of the load, determined by the ratio $\frac{\text{tangential force in lb}}{\text{load in lb}}$. Therefore, the value 0.25 to 0.40 would be the coefficient of friction (f) for that bearing by definition.

Applying a small amount of lubricant to the bearing as to barely wet the surfaces (boundary condition), and applying force to the cord, the coefficient of friction would be approximately 0.1 to 0.15. Adding more lubricant to obtain a partial (mixed-film) fluid film would decrease f to about 0.02 to 0.10. Under hydrodynamic conditions (fluid film), f would be below 0.002. The amount that f falls below 0.002 depends on the hydrodynamic efficiency of the bearing system under the load, the viscosity of the fluid film, and the speed at which the journal is turned.

Operating the journal at constant speed and load with oils having different viscosities, a series of $\frac{ZN}{P}$ numbers could be established for that particular bearing. If the power required to turn the journal were measured during each change of lubricant, a relationship between lubricant viscosity and f would be established. By the same processes, $\frac{ZN}{P}$ and f could be determined for any variable combination of load, speed, and viscosity.

Using values of f as the ordinate and $\frac{ZN}{P}$ as the abscissa, a pattern of frictional behavior within the bearing could be plotted, having the general form shown in Fig. 437. Note that for low $\frac{ZN}{P}$ numbers boundary conditions exist. As $\frac{ZN}{P}$ values increase, a transition occurs to establish a partial or mixed-film conducive to a reduction in f. The lowest value of f is established at a point representing the initial phase of a fluid film. It is at this point where metal-to-metal contact ceases to exist and fluid friction is at minimum magnitude due to the thinness of film (minimum shear). It may be said that perfect lubrication exists slightly beyond or to the right of the low point in the curve and should be selected as the ideal running condition. However, any adverse change in Z, N, or P (shock, temperature rise, etc.) would cause the thin film to rupture and initiate

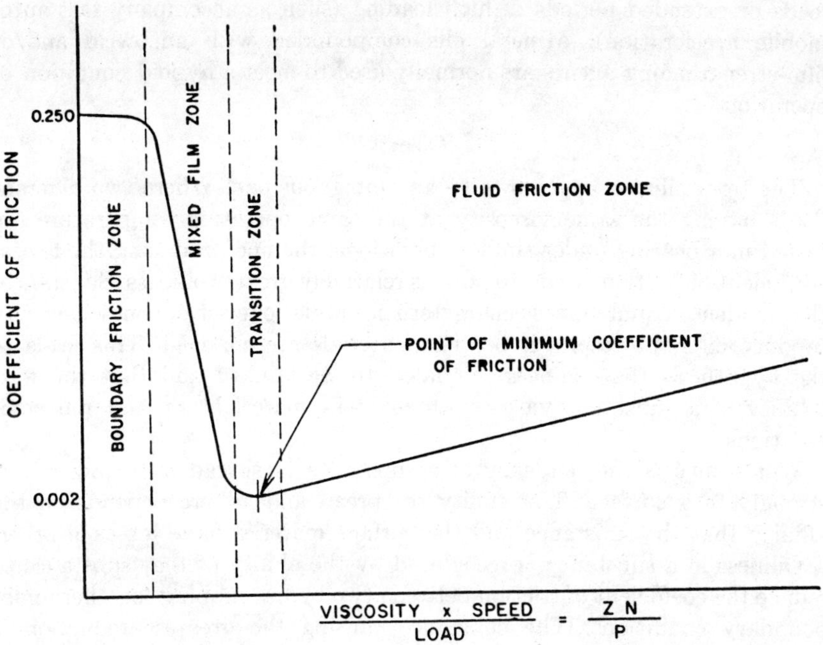

Fig. 437. Relation of coefficient of friction to $\frac{ZN}{P}$.

boundary conditions. For this reason, the safest condition lies in the zone of higher $\frac{ZN}{P}$ to the right of the low point where f is slightly higher due to increased fluid friction (thicker film).

Fluid Film in Gears

A study of tooth profiles and frictional characteristics of gears will determine the possibility of establishing a fluid film between meshing teeth. Spur, herringbone, and bevel gear teeth are so designed as to produce an oil wedge as they slide and roll over each other. Lubrication of such gears poses no problem if all other factors are considered properly. An important factor is one of lubricant viscosity.

Gears subject to a preponderance of side-slip during which a wiping action occurs simultaneously cannot establish a fluid film and must rely on the oiliness and the adhering qualities of the lubricant. Worm and hypoid gears fall in the latter category. Worm gears require a high oiliness condition accompanied by persistence of the film to combat wiping. Compounded mineral oils of relatively high viscosity appear to be the answer. Hypoid gears develop tremendous pressure when subject to shock loads or extended periods of high loading (such as accompany fast automobile acceleration). Mineral oils compounded with anti-weld and/or film-strengthening agents are normally used to meet a hypoid condition of operation.

Oiliness

The term oiliness appears to be an ambiguous one. Where two different fluids having the same viscosity at the same operating temperature are tested in a bearing under similar conditions, the one producing the lowest coefficient of friction is said to possess relatively greater oiliness. Substances that evidence equal oiliness characteristics under one set of conditions may produce different results if bearing materials are changed. This leads to the hypothesis that oiliness is related to wettability and that the wettability of a substance varies with different materials, or material combinations.

Wettability is the tendency of a substance to spread over and adhere strongly to a surface. The ability to spread and adhere depends on the affinity that the substance and the surface material have for each other.

Oiliness in a substance is recognized by the ability of that substance to reduce the coefficient of friction between two bodies rubbing together under boundary conditions. The substance causing the greatest reduction is said to possess better oiliness under the particular operating conditions. Animal, vegetable, and fish oils are substances that possess a high degree of oiliness, probably due to the presence of fatty acids alone or in combination with other compounds.

LUBRICANTS

Substances capable of reducing friction are called lubricants. While the definition of a lubricant includes any substance capable of counteracting the force of friction, practical applications demand the employment of only those possessing inherent characteristics that are compatible with sound economic practices. Consequently, comparatively few substances are selected for a majority of lubrication requirements.

The prime requisites of all lubricants are that they possess

(*a*) Proper fluidity, or plasticity, at the operating temperature,
(*b*) Adhesiveness,
(*c*) Film strength commensurate with load,
(*d*) Chemical stability,
(*e*) Purity,
(*f*) Lubricity.

Under certain operating conditions, water-separating ability and foam resistance should be included as primary qualities of lubricants.

Types of Lubricants

Lubricants may be gaseous, liquid, plastic, or solid. Their classification according to physical state is shown below:

Gaseous lubricant:
- Air

Liquid lubricants:
- Mineral oils:
  - Straight or unadulterated
  - Compounded with fixed oils
  - Compounded with special additives
- Fixed oils:
  - Animal
  - Vegetable
  - Fish
- Synthetic

Plastic lubricants:
- Petrolatums (petroleum jelly)
- Greases (soap-thickened oils)
- Greases (synthetic)

Solid lubricants:
- Graphite
- Molybdenum disulfide
- Mica
- Talc

A group of lubricants not shown above is comprised of those that provide lubrication through the process of melting when exposed directly to the conducted heat of the journal. This group includes block greases and various solid fats, such as suet. However, the latter is used but seldom and not at all in modern machinery.

Gaseous Lubricants. A gas, such as air, may be used under certain hydrodynamic conditions. However, where the only pressure available is that developed by hydrodynamic process, the load-carrying ability of air is low, about 10 psi, more or less. Where compressed air is used, hydrostatic conditions prevail and the carrying capacity of air-lubricated bearings will follow, nearly, the air pressure applied.

Liquid Lubricants. The most important of the liquid lubricants are the mineral oils produced from petroleum crudes by a series of refining processes. Basically, petroleum crudes are complex mixtures of the elements hydrogen and carbon, and, therefore, are classified as *hydrocarbons*. Crudes also contain other elements as impurities, such as sulfur, nitrogen, etc. Most of the impurities are removed from lubricating oils during the refining process. The ratio of carbon and hydrogen in lubricating oils produced from petroleum is approximately 85 percent and 15 percent by weight.

Petroleum crudes vary in chemical structure, depending on their source. Hence, they are often identified as being Pennsylvania, Mid-Continent, Gulf Coast, etc., crudes. They are classified, however, according to the hydrocarbon *series* to which their predominant constituents belong, or *base*. There are four major bases of petroleum crudes. These are listed in Table 141.

It must be recognized that crudes are not made up solely of paraffins or naphthenes or asphaltines, but may contain all three of those hydro-

TABLE 141

Base	Typical Source*
Paraffinic..........	Pennsylvania
Naphthenic........	California, Gulf Coast
Asphaltic..........	Mexico, California
Mixed.............	Mid-Continent

* The various types of crudes are not confined to the localities given. Mixed based crudes, for example, are distributed throughout the world.

carbons. However, their names do imply that they contain a predominant amount of the hydrocarbon with which they are identified. It is also true that oils produced from various based crudes possess individual and distinct physical and chemical characteristics. For example, paraffinic base oils have a higher API gravity than naphthenic base oils. The API gravity of mixed base oils is somewhere (about halfway) between the two.

HYDROCARBON STRUCTURAL FORMULAS

Paraffins have the general formula C_nH_{2n+2}, or the number of hydrogen atoms is *always* two more than twice the number of carbon atoms. The simplest member of the paraffin hydrocarbon family is methane, a compound described in Chapter 5. It has the formula CH_4 arranged as follows:

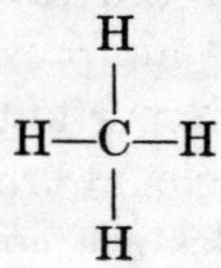

Ethane has the formula C_2H_6, or

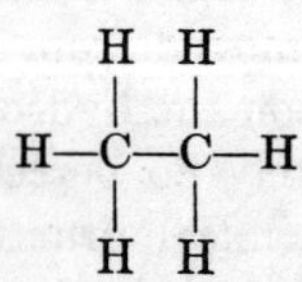

and propane, C_3H_8, which appears as

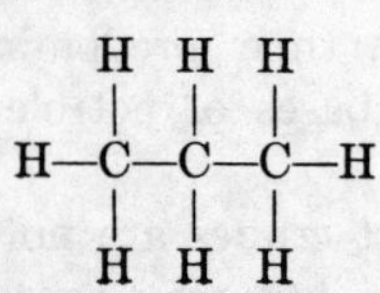

while the molecular formula for normal butane, C_4H_{10}, will be constructed

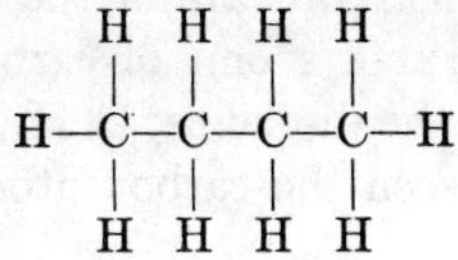

Note that all paraffinic molecules have their atoms linked chainlike together in a manner portraying saturation; i.e., all atoms are satisfied according to their valence. The indications are that paraffins are a chain series of hydrocarbon possessing chemical stability, which is true.

There is another series of paraffinic compounds that have the same general formula, but a different structural relationship between atoms. These are the iso-paraffins of which the isomer, iso-butane, is a member, having the formula $(CH_3)_3CH$ in contrast to C_4H_{10} for normal butane. The iso-butane molecule may be constructed as follows:

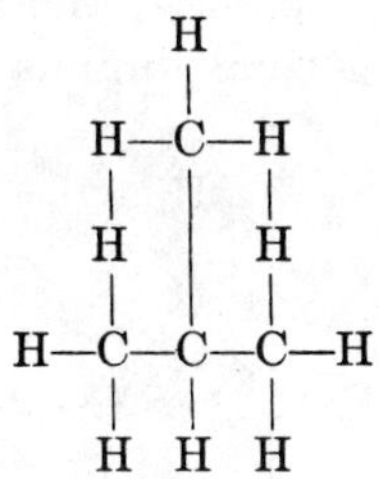

The basis for the structural development of hydrocarbon molecules, as shown above, is that hydrogen has a valence of one (univalent), while carbon has a valence of four (quadrivalent). This means that each carbon atom will combine with four hydrogen atoms. Each carbon atom must combine with four hydrogen atoms to enable the compound to remain stable. Otherwise, an unsaturated condition will exist since carbon will combine with other elements to satisfy its valence. The stable compound can be broken, but it takes outside influence in the form of heat, catalysts, stronger chemical elements, and pressure.

Another family of hydrocarbons having relatively good chemical stability are the naphthenes, the general formula of which is C_nH_{2n}. Thus, naphthenes always have twice as many hydrogen atoms as carbon atoms. Their general construction is the form of a ring of atoms, as shown below, for cyclobutane having the formula C_4H_8:

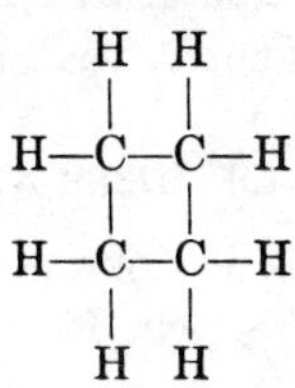

The paraffin hydrocarbons are more stable than the naphthenic compounds.

The more unstable hydrocarbons are of the olefin and aromatic series, both of which have unsaturated atoms of carbon. Mono-olefins have the general formula C_nH_{2n} and the di-olefins, C_nH_{2n-2}. They appear as follows (note the double bond between the carbon atoms, one in mono- and two in di-olefins):

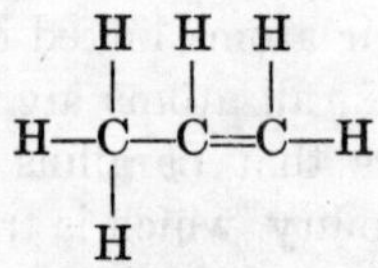

Mono-olefin C_3H_6

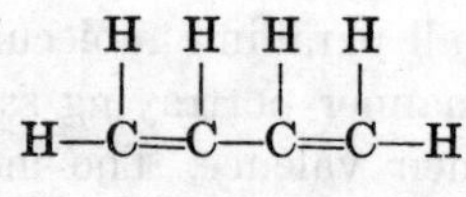

Di-olefin C_4H_6

Because of their unsaturated condition, olefins react readily with other elements, such as oxygen. It is probably this reason that causes their general absence from natural crudes. Another unsaturated group, aromatic hydrocarbons, of the general formula C_nH_n, has a ring structure as shown for benzene, C_6H_6

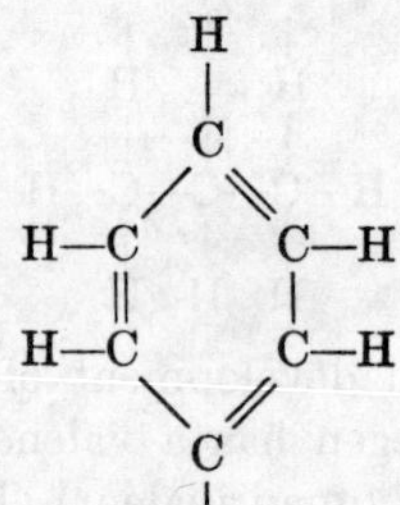

As produced from the well, petroleum crudes are not composed solely of simple molecules of paraffinic, napthenic, or aromatic structure, but may contain any number or all in combination. Lubricating oils produced under conditions that allow some unsaturated hydrocarbons to remain will be subject to the formation of oxy-materials, of which sludge is a familiar member. The tendency of poorly refined oils to break down and form sludges, gums, and other undesirable products is aggravated when subject to heat. Early reactions occur when such oils are subject to normal operating temperatures (110 F to 140 F) for extended periods or for much shorter periods at the higher temperatures prevalent in automotive engines. The above bears out the statement that good lubricants are made and not found.

REFINING OF LUBRICATING OILS

A complete discussion of the various processes for refining lubricating oils is not within the scope of this text. However, a brief résumé concerning the division of petroleum crudes into its various important components and some methods of oil refining may lead to a better understanding of the problems associated with lubrication of mechanical equipment.

Distillation

As shown in the discussion of saturated and unsaturated hydrocarbon molecules, crude petroleum is what its name implies insofar as lubricating oils are concerned. Crude oils are not made up of a given substance possessing a common boiling point. They are made up of numerous compounds having individual boiling points enabling them to be separated by heating. The process is called fractionating and is accomplished in a fractionating tower that has trays installed at various levels to catch the condensating fluids.

Initial fractionating is called crude reduction or topping. In the topping tower the light, low boiling point products are separated from the heavier, higher boiling point products. The remaining substance is made up of that material containing low viscosity lubricating oils down to heavy residuums. The crude is heated by steam before entering the tower. Vaporization occurs immediately. The vapor rises in the tower in which temperature and pressure from bottom to top are carefully controlled by live steam. As the heavier and higher boiling point vapors reach a certain level, they pass through bubble caps. The bubble caps that form part of the trays direct the vapors through a liquid body of the same fraction that has already condensed in the trays under ambient temperatures and pressure at that level. These, in turn, condense in the tray, the overflow of which flows to some source outside the tower.

The more volatile vapors continue upward in the tower passing through each tray level until they reach a point where the temperature corresponds to the individual boiling points of the various fractions. Thus, the least volatile vapor condenses at the bottom and the most volatile fraction condenses at the top of the tower. The least volatile fraction condensed in a topping tower may be gas oil, while gas or gasoline may represent the most volatile in the crude distillation unit. All the products resulting from boiling and consequent condensing produced in a distillation unit are called distillates. Thus, there are gasoline distillates, naphtha distillates, kerosene distillates, gas oil distillates, etc., named in the order of ascending boiling points. Light-bodied fuel oils and kerosene are representative intermediate boiling point fractions in the topping tower.

The residue from the crude still or topping tower, composed of fractions having higher boiling points than the temperatures carried for the topping process, is pumped to another tower after it is preheated. The distillation process is repeated at suitable temperatures with the light-bodied lubricating oils being drawn off at the top and the relatively heavier products

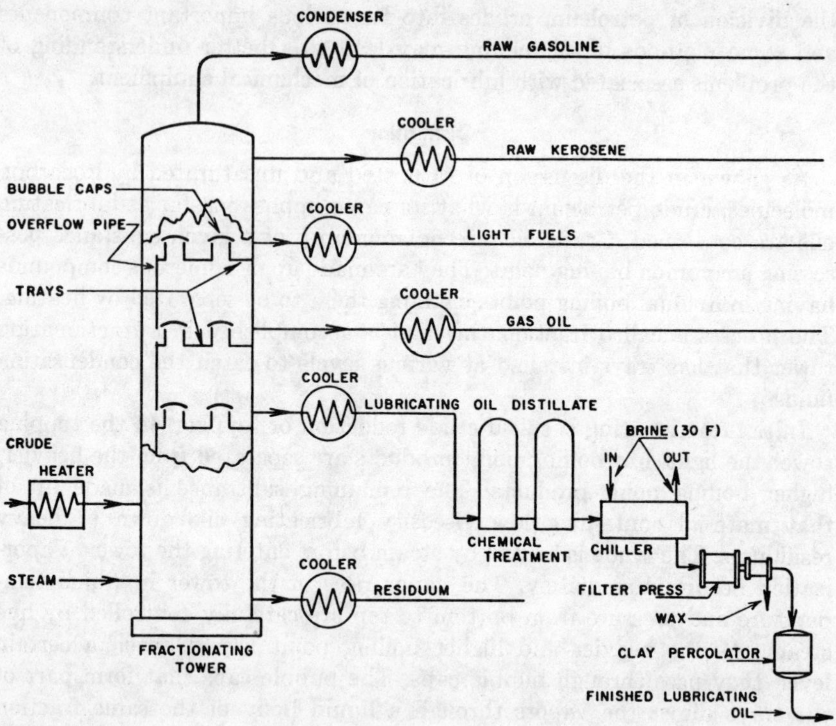

Fig. 438. Typical flow of petroleum from crude (well) to finished lubricating oil.

at the bottommost tray level. It is at this point in the oil preparation process that the flash point and the viscosity for any fixed combination of conditions carried on in subsequent refining operations are determined. The residue (residuums) left in the extreme bottom of the still may be heavy lubricating oils or asphalt, depending on the base of the crude. Intermediate levels produce lubricating oil stocks having relative viscosities between the two extremes.

The process of distillation is a continuous operation, and not one of intermittent charging.

High temperatures promote the decomposition of hydrocarbon molecules causing chemical changes that are undesirable. Such high temperatures are produced in the fire still, where heating the charge is accomplished by direct contact heating as carried on in the old batch process. To reduce the boiling point of the charge during distillation, several methods are employed.

One method is known as *steam distillation*. Steam is mixed with the crude to raise the vapor pressure of the mass above that of the surrounding atmosphere within the still. This lowers the boiling point below what it

Table 142. Petroleum Products*

- **NATURAL GAS**
  - **Natural Gas**
    - Industrial and Domestic Fuels
    - Crude-Oil Production Stimulus
    - Source of Synthetic Fuels and Chemicals
    - Carbon Black
  - **Natural Gasoline**
    - Components of Automotive and Aviation Fuels
    - Liquefied Gases, Industrial and Domestic Fuels, and Illuminants
    - Hydrocarbon Chemicals
    - Propane, Butane
    - Synthetic Bases
- **REFINERY GAS**
  - **Noncondensable Gases**
    - Hydrogen
    - Light Hydrocarbons
      - Chemical and Industrial Processing
      - Glass Making
      - Cutting and Welding
      - Carbon Black
      - Refinery Fuel
  - **Liquefied Gases**
    - Industrial and Domestic Fuels and Illuminants
    - Synthetic Motor Fuels
  - **Processed Derivatives**
    - Olefins, Di-olefins
      - Butadiene, Styrene — Synthetic Rubbers
      - Polymers — Lube-Oil Additives
      - Anti-knock Agents — Aviation-Gasoline Blending Agents
      - Alcohols, Esters, Ketones — Antifreeze, Fat, Lacquer, and Drug Solvents
      - Synthetic Plastic Waxes — Inks, Paints
      - Resins
- **LIGHT DISTILLATES**
  - **Gasoline**
    - Light Naphthas
      - Gas-Machine Gasoline
    - Intermediate Naphthas
      - Aviation Gasoline — Airplane Fuels
      - Motor Gasoline — Automobile Fuels
      - Commercial Solvents — Rubber and Fatty-Oil Solvents
      - Lacquer Diluent
      - Toluene, Xylene — Explosives
    - Heavy Naphthas
      - Varnish-Makers, Painters, Dyers and Cleaners Naphtha
      - Turpentine Substitutes
  - **Kerosene**
    - Refined Kerosene
      - Stove, Lamp, and Tractor Fuels; Railroad Signals, Lighthouse Oil, Ship Illuminants, Gas-Absorption Oils
- **MEDIUM DISTILLATES**
  - **Gas Oil**
    - Naphthenic Acids
      - Soaps
      - Lube-Oil Additives
      - Paint Driers
      - Fungicides
    - Water-Gas Carburetion Oil
    - Metallurgical Fuels; Heating Fuels
    - Diesel Fuels
  - **Absorber Oil**
    - Gasoline-Recovery Oils
    - Benzol-Recovery Oils

(*Continued*)

Table 142—(*Continued*)

HEAVY DISTILLATES

- **Lubricating Oils**
  - White Oils (Technical)
    - Insecticides, Tree Sprays, Bakers, Fruit-Packers, Candy-Makers, Egg-Packers, and Slab Oils, Recoil Oils, Hydraulic Oils
  - White Oils (Medicinal)
    - Medicines and Cosmetics
  - Saturating Oils, Emulsifying Oils, Flotation Oils
    - Textile, Leather, Paper, Wool, and Twine Oils, Coal-Spray, Dust-Laying, Cutting, Metal-Recovery, and Quenching Oils
  - Paraffin Wax
    - Candy-Makers, Chewing-Gum, Candle, Laundry, Sealing, Etchers, Saturating, Insulating, Match-Cardboard, Paper, Medical, and Canning Waxes
    - Synthetic Lubricants and Derivatives
    - Fatty Acids
      - Grease and Soap Lubricant
    - Fatty Alcohols and Sulfates
      - Rubber-Compounding, Detergent, and Wetting Agents
  - Light Lubricating Oils
    - Turbine, Light-Spindle, Transformer, Household-Lubricating, Compressor, Ice-Machine, Meter, Dust, and Tempering Oils
  - Medium Lubricating Oils
    - Journal, Motor, Diesel-Engine Aircraft, and Railroad Oils
  - Heavy Lubricating Oils
    - Steam-Cylinder, Valve, Transmission, and Printing Oils, Black Oils and Lubricating Grease, Tempering Oils
- **Lubricating-Oil Residues**
  - Medicinal Petrolatums
    - Salves, Creams, Ointments, and Petroleum Jelly
  - Technical Petrolatums
    - Rust-Preventing Compounds, Rubber Softeners, Lubricants and Cable-Coating Compounds

RESIDUES

- **Residual Fuel Oils**
  - Wood Preservative, Boiler Fuel, Gas-Manufacture Oils, Metallurgical Oils
- **Asphalts**
  - Liquid Asphalts, Binders, Fluxes
    - Roofing and Shoe Material, Shingle Saturants, Roofing Saturants, Road Oils, Emulsion Bases
  - Steam-Reduced Asphalts
    - Briquetting and Paving Asphalts, Paint Bases, Flooring Saturants
  - Oxidized Asphalts
    - Roof Coatings, Waterproofings, Rubber Substitutes, Insulating Asphalts
- **Coke**
  - Carbon Electrodes, Carbon Brush, Fuel Coke, Metallurgical Coke
- **Refinery Sludges**
  - Acid Coke
    - Fuel
  - Sulfonic Acid
    - Saponification Agents, Fat-Splitting Agents, Special Oils, Demulsifying Agents, Emulsifiers
  - Heavy Fuel Oils
    - Refinery Fuel

* Courtesy, American Petroleum Institute.

would be if all the pressure were to be developed within the charge by direct contact heating.

Another method used to reduce the boiling point of the charge is by direct contact heating, i.e., in a manner similar to boiling water over a gas flame.

Still another method used, particularly in distilling lubricating oils of of the ligher viscosities, is a modification of the above. The boiling point of distillation is reduced by vacuum, a process known as *vacuum distillation*. This method is similar to that described for steam distillation except that a partial vacuum is maintained within the tower during the fractionating process. Vacuum is developed by rapidly condensing light vapors as they emerge from the top of the tower. The action is much the same as that found in steam turbine condensers.

The description of the distillation processes given above is general and varies in many details for reasons dictated by types of crudes, end results, etc. It must also be recognized that fractionating between relative ranges of temperature results in individual products having definite minimum and maximum boiling points. This means that under closer control they, too, can be further divided (cut) into products having closer "end" points. The substances within a given mass of a distilled product having the lowest boiling point (most volatile) are referred to as light "ends." The less volatile substances in the same mass are called heavy "ends."

Blending

A finished lubricating oil made up by *blending* two oils of different fractionations (i.e., one of lower viscosity than the other, but of the same base crude) may have a range of volatility not compatible with some critical lubricating requirements. In other words, the "ends" may be too far apart for an express purpose. However, where such practices are kept within reasonable limits, satisfactory results can be obtained in most mechanical operations. Blending oils of different viscosities to obtain some desired compromise is widely practiced with excellent results. Some high-grade motor and industrial oils of certain intermediate viscosities are produced by this method.

Refining

During the refining process following distillation, unsaturates, asphalt color bodies, and waxes are removed from the lubricating oil distillates and residuums. This is accomplished by treating the unrefined oil with solvents, of which there are several kinds. A distillate of oat hulls treated with sulfuric acid, and known as furfural, sulfur dioxide, cresylic acid, and propane, are some of the solvents used. All act to dissolve certain impurities in the oil, after which the solvents are recovered and reused.

Mixed directly in the oil, solvents remove such impurities for which they are especially selected. The action is one of a solvent rather than a chemical, which accounts for solvent recoverability. Solvent refining may be accomplished variously by using one or two different solvents, processes known as single solvent and duosolvent refining systems, respectively. Sulfur dioxide in its liquid form removes unsaturates. Upon reduction of pressure on the solvent-unsaturate mixture, the sulfur dioxide evaporates, leaving the unsaturates behind. The sulfur dioxide is then recovered in a manner similar to that of any mechanical refrigeration system.

Selective solvent treatment is also carried out using two different cleaning agents, a refining method known as the duosolvent system. One method of double solvent treatment employs propane and cresylic acid, which act as a team to remove unsaturates, asphalts, and waxes. In the first stage, liquid propane is mixed with the oil under pressure. Asphalt is not solvent in propane and, therefore, precipitates. A reduction of pressure on the oil-propane mixture causes propane to evaporate, lowering the temperature which causes the formation of wax crystals in the oil. A further reduction of oil temperature followed by filtration through cloth filters removes the wax crystals. In the second stage, the oil is scrubbed by cresylic acid. Cresylic acid, being the heavier of the two, percolates down through the oil and removes the unsaturates.

Clay Filtering. Some lubricating oils are refined further by filtering through fuller's earth, a natural claylike substance. Artificial activated clays are also used to a large extent as a substitute for fuller's earth. Clay filtering improves the water-separating characteristics of oil and also acts as a decolorizing medium. The clay may be added directly to the oil in controlled quantities and then filtered out, or the process may be carried out by allowing the oil to percolate down through the clay. A petroleum solvent is added to heavier lubricating oils to increase flow rate during the clay filtering process. The process is one of adsorption.

Acid Treatment. An old method of refining lubricating oils and one still used in special cases is that of sulfuric acid treating. The reaction of sulfuric acid is both chemical and physical.

Acid is added to the oil and the mixture is agitated. Unsaturated hydrocarbon molecules react with the acid forming a sludge which precipitates and settles to the bottom of the lead-lined mixing tank. The acid sludge is then drawn from the tank. The process may be repeated several times, depending on the type of oil and the desired quality of the finished product.

Acid treating is followed by a neutralizing process accomplished by transferring the oil to another tank and treating with caustic soda. The caustic soda reacts with any acid remaining in the oil, the residue of which settles in the same manner as the acid sludge. After the caustic soda treatment, the oil is washed with water to remove all traces of the alkali.

Dewaxing. Wax from petroleum sources may be crystalline in form, resembling beeswax, or it may be formless, resembling jelly. The latter is called petrolatum. The amount of wax contained in a crude depends on the base, paraffins leading all others in this respect.

Waxes within a given crude congeal at different temperatures (melting point). For this reason, the temperature of the lubricating oil must be reduced to a level where a sufficient amount of wax may be removed to meet the conditions at which the oil will be used. Wax is the congealing agent in oils and will reduce or prevent flow when the temperature of the oil is reduced to a temperature commensurate with the amount of wax it contains. Therefore, wax content is an important consideration where the lubrication service requires oils to be exposed to low temperatures. Automotive engines and transmission components are affected during winter operation when the oil-congealing temperatures are not sufficiently low. Channeling results, leading to rapid wear. Hard starting is another disadvantage of using high wax content oils. Oils used to lubricate refrigeration compressors, especially in low-temperature systems, are particularly required to be of low wax content. Otherwise, wax separation will occur in the evaporators and cause restricted flow, reduce heat transfer, and create an undesirable condition generally.

To meet special service conditions, some lubricating oils must be dewaxed beyond the stage afforded during solvent treatment. This is accomplished by chilling the oil to some reduced temperature in a brine chiller. If drastic dewaxing is to be accomplished, chilling must be done in stages to facilitate filtration, which otherwise would be hampered by excessive wax precipitation.

The chiller is a long cylinder containing a spiral baffle through which the oil is pumped. Cold brine is circulated in a jacket surrounding the cylinder. After the oil is chilled to the desired temperature, say −40 F, it is forced through a filter press. The filter press is made up of a series of cloth filters backed by perforated metal plates. The wax is held by the filters and the dewaxed oil withdrawn from the press.

Summary of Lubricant Refining

Table 143 shows the order of the various processes through which light, medium, and residuum lubricating oils are run during refining. Actual process operations will vary from those described above for different crudes. They will also vary according to the product in greatest demand at the moment.

CLASSIFICATION OF MINERAL LUBRICATING OILS

A study of mineral oil distillation processes makes obvious the viscosity differences between those drawn off the still at different temperature levels

TABLE 143. FLOW CHART—LUBRICATING OIL REFINING

Light lubricating oils (distillates)	Pressing +	Acid treatment + decolorizing or Single solvent refining + decolorizing	
Purpose: To remove	Wax	low viscosity index and chemically unstable hydrocarbons	color bodies
Intermediate lubricating oils (distillates)	Acid treating or Single solvent refining	+ solvent dewaxing	+ decolorizing
Purpose: To remove	Low viscosity index and chemically unstable hydrocarbons	wax	color bodies
Heavier lubricating oils (residuum)	Acid treating or Propane deasphalting and Single solvent refining or Duosolvent refining	+ solvent dewaxing	+ decolorizing
Purpose: To remove	Asphalt, low viscosity index and unstable hydrocarbons	wax	color bodies

This is true because the lower viscosity oils are taken off the top, medium viscosity oils at intermediate points, and higher viscosity oils are taken from the lower levels of the still. The oils of greatest viscosity form the residuum left in the bottom of the still and are called cylinder stocks. Viscosities other than those represented by the various distillation cuts (fractions) may be obtained by blending oils from different cuts. Therefore, it is possible to obtain oils ranging in viscosity from that approaching the fluidity of water to one having the flow characteristics of molasses and having any viscosity between the two when all are measured at the same temperature.

One method of classifying mineral lubricating oils is according to viscosity. Employing the viscosity method of classifying lubricating oils would result in the following list, starting with the lighter bodied or low viscosity oils and progressing gradually to the heaviest bodied or high viscosity oils.

Low viscosity class (60 to 220″ @ 100 F, SUS)	Spindle oils Light machine oils Light turbine oils
Medium viscosity class (220 to 420″ @ 100 F, SUS)	Medium-bodied machine oils Compressor oils Medium turbine oils
Heavy viscosity class (420 to 1500″ @ 100 F, SUS)	Heavy machine oils Gear oils
Extra heavy viscosity class (1500 to 3000″ @ 100 F, SUS)	Extra heavy machine oils Heavy gear oils Cylinder oils

The above list represents the older method of classifying lubricating oils. Because such classification is based entirely on viscosity, it is not a sound method.

Various degrees of purity can be obtained, depending on the amount as well as the quality of refining to which an oil is subjected. For some purposes, as in refrigerating, turbine, and internal combustion engine systems where the lubricant may be affected by operating conditions and/or is reused for extended periods, a highly refined (chemically stable) oil is essential. Otherwise, chemical changes occur that result in the formation of gums, varnish-like materials, sludges, and carbon that deposit on the working surfaces of the lubricant system. When present on pistons, piston ring grooves, valves, cylinder heads, bearings, heat transfer surfaces, and the interior walls of oil lines, such deposits interfere with mechanical processes, restrict the flow of oil, and eventually cause machine failure. The life expectancy of a mineral lubricating oil is reduced by one half (50 percent) for each 18 F rise in its temperature. It follows that only chemically stable oils having inherent longevity should be selected for lubricating systems requiring their continual reuse, or in which conditions of operation may cause rapid deterioration of an oil of inferior quality.

For systems in which the oil is lost after serving its purpose for a short period, such as reservoirless bearings, chains, etc., any fairly good oil, having suitable characteristics otherwise, will suffice in many cases. In modern machinery, the latter type of lubrication system is diminishing, although some conditions of design will not permit its elimination entirely.

The foregoing bears out the fallacy of viscosity classification of lubricating oils. At least it implies that lubricant selection demands quality consideration in addition to the physical demands of a given requirement.

Table 144 shows that there are other important considerations. In addition to complicating the problem of lubrication, such considerations also serve as evidence that success in dealing with friction results only when lubricants have been selected intelligently.

Some of the problems associated with lubricants and lubrication systems are shown in Table 144.

TABLE 144. OPERATING PROBLEMS ASSOCIATED WITH LUBRICANTS AND LUBRICATION SYSTEMS

Problem	*Types of Lubrication System Where Problem Generally Exists*
Rust	Circulation systems—turbines, any system in which moisture is allowed to enter
Oil oxidation (forms gums, carbon, sludge, etc.)	Circulation systems, splash and bath reservoirs—turbines, gear cases; ring oil bearing reservoirs; internal combustion engine crankcases, valves, and combustion chambers; compressor crankcases, valves, compression chambers
Corrosion (acid)	Circulation systems, splash-internal combustion engine crankcases. Some types of bearing materials generally
Foaming	Any system where air may mix with oil during agitation, circulation, etc.
Emulsification	Any system in which water (condensate, etc.) may mix with oil. Inefficient steam seals, leaky oil coolers, moisture from atmosphere entering through breathers after machine or engine shutdown, etc.
Lubricant displacement by water (steam condensate)	Steam engine cylinders
Abrupt viscosity change due to change in temperature (low viscosity index)	Systems operating at relatively high temperature—internal combustion engine crankcases, etc.
Lubricating film rupture	Gears, bearings, etc., operating under heavy unit pressures with unfortified lubricants
Welding	Extreme pressure systems—hypoid gears at rapid automobile acceleration
Wax formation	Refrigeration systems operating with low evaporator temperatures
Precipitation of unsaturated hydrocarbons (tars, etc.)	Refrigerating systems using solvent type refrigerants. Precipitation occurs at the expansion valve usually
Chattering (vibration)	Ways, guides, etc., where conditions demand high film strength and oiliness characteristics
Insufficient adhesiveness (loss of lubricant by squeeze, or centrifugal action)	Shock loads, open gears, cams, guides, etc.
Moisture content of oil (due to hygroscopic nature of oils)	Refrigeration systems, transformers, and other electrical apparatus
Foreign deposits from products of combustion, oil oxidation, etc.	Internal combustion engine pistons, rings and ring grooves, combustion chamber, etc.

Some of the problems listed in Table 144 may be combated successfully through the use of properly refined straight mineral oils. For example, the water-separating ability of an oil depends on the absence of emulsifying agents. Such agents are not present in new highly refined, unadulterated lubricating oils. High-grade oils will serve indefinitely in well-maintained systems operating at temperatures up to about 140 F. They will also render excellent service in refrigeration systems if their wax content is low enough and of such nature as to be compatible with operating conditions. There are numerous other operating conditions where the only requirement demanded is that the oil employed be well refined and physically suitable.

Improved metallurgy, streamline design, and the demand for greater production, closer tolerances, and increased power per unit weight of machine have led to the development of specialized lubricants. Specialized lubricants are products that have been compounded or specially treated to meet certain conditions peculiar to one or more mechanical operations. Many of the problems listed in Table 144 stem from or have been aggravated by modern demands. For example, the public demand for faster acceleration of automobiles, plus low (in height) car styling brought about a wide use of hypoid gears. Under high acceleration of heavier automobiles, the hypoid gear is subject to loads that would ordinarily destroy them in a short operating period if special extreme pressure lubricants were not available. Hydraulic systems employed for the operation of controls, machine tools, and numerous other types of production machinery are subject to conditions that promote oil oxidation. The deposits resulting from oxidation cause valve sticking and leakage, loss of volumetric efficiency due to rapid pump wear, and other operating difficulties. To extend the life of oils used in hydraulic systems, they are treated to resist oxidation for longer periods than are attainable even with lubricants that have been refined to a maximum degree.

Steam engine cylinders subject to the washing effect of condensate, especially under "wet" steam conditions, require a lubricant that will emulsify and continue to provide protection to the frictional surfaces. Therefore, it is necessary to compound steam cylinder oils with other substances that provide the necessary emulsifying properties. Table 144 shows other operating conditions for which mineral oils must be especially prepared to meet the peculiarities of the system to which they are applied.

In view of the foregoing considerations, it is obvious that viscosity alone does not provide a sound basis for classifying lubricating oils. Classification should be made on the basis of service requirements, as shown in Table 148. However, before attempting to interpret the contents of Table 148, the properties of lubricants, both physical and chemical, as well as the additives required for various purposes, should be studied.

PHYSICAL AND CHEMICAL PROPERTIES OF MINERAL LUBRICATING OILS

Specifications of lubricating oils are developed by standard test procedures. The conditions and methods of testing lubricants follow the standards prescribed by the American Society for Testing Materials (A.S.T.M.) in their publication "A.S.T.M. Standards on Petroleum Products and Lubricants."

While there are a number of tests required by the producer for purposes of control and development of lubricating oils, the user need be acquainted with only a few. Lubricant specifications are necessary in selecting a suitable product for a given requirement, but they should not be adopted as the sole criterion for predicting final results. Buying lubricants on the basis of specification alone is extremely poor policy, especially where expensive machinery is concerned, or where oil deterioration will interfere with work processes. Quality and adaptability to a given lubrication requirement must also be considered. The latter factors have no numerical equivalency that can be adopted as a specification for the interpretation of service results. Quality lubricants cost more to produce than poor ones. The relative difference in unit price will usually result in lower maintenance cost, in fewer interruptions in machine operation, and in the long run reduced over-all lubrication cost.

Gravity

Gravity is associated with the weight per unit volume of oil. Specific gravity is the ratio of weight per unit volume of oil to the weight of an equal volume of water when both are at 60 F, or

$$\text{Specific gravity} = \frac{\text{weight per unit volume of oil at 60 F}}{\text{weight of equal volume of water at 60 F}}$$

To facilitate gravity readings, the method proposed by the American Petroleum Institute (A.P.I.) is used by the oil industry. Determined by an A.P.I. hydrometer (having a scale of some part of the range from 0 to 100 degrees A.P.I.), values are expressed in terms of A.P.I. gravity. An equivalent value of A.P.I. gravity may be determined if the specific gravity of an oil is known by the following equation, using 141.5 as a modulus:

$$\text{A.P.I. gravity} = \frac{141.5}{\text{specific gravity (60/60 F)}} - 131.5$$

A.P.I. gravity values correspond to specific gravity in the order of

0 degree A.P.I. = 1.076 sp gr

10 degrees A.P.I. = 1.00 sp gr

100 degrees A.P.I. = 0.6112 sp gr

EXAMPLE: Determine the A.P.I. gravities for lubricating oils having specific gravities of 0.94, 0.91, and 0.88.

1st case $$\text{A.P.I. gravity} = \frac{141.5}{0.94} - 131.5 = 19.0$$

2nd case $$\text{A.P.I. gravity} = \frac{141.5}{0.91} - 131.5 = 24.0$$

3rd case $$\text{A.P.I. gravity} = \frac{141.5}{0.88} - 131.5 = 29.3$$

Note that A.P.I. gravity values increase as specific gravity decreases. In other words, the A.P.I. gravity becomes correspondingly less with heavier (in weight) oils. Also note that gravity differences are more discernible when expressed in terms of A.P.I. gravity.

Gravity is an indication of the type of crude from which an oil is produced. Oils from different crudes will have gravities approximating values shown in Table 145 for oils of similar viscosity.

TABLE 145. A.P.I. GRAVITIES OF LUBRICATING OILS FROM VARIOUS TYPE CRUDES

Base of Lubricating Oil	*A.P.I. Gravity of Lubricating Oil (approximate)*
Paraffinic base oils (Pennsylvania)......	29.0–32.0
Mixed base oils (Mid-Continent).......	27.5–29.5
Asphaltic base oils (California).........	24.0–27.0
Naphthenic base oils (Gulf Coast)......	19.0–23.0

Most petroleum oils are lighter than water and, therefore, have A.P.I. gravities above 10, usually in the range of 20 to 30 for lubricating oils.

Flash Point

Flash point is the temperature at which an oil vaporizes sufficiently to sustain momentary ignition when exposed to a flame under atmospheric conditions and controlled test procedure. Placed in an open cup with a thermometer immersed, the oil is heated at the rate of 30 F per min temperature rise until it reaches 100 F below the probable flash point. The rate of heating is then reduced gradually until it occurs at about 10 F per min for the final 50 F. A flame is passed over the oil in the region immediately above the surface at intervals of 5 F temperature rise. The temperature at which a definite, self-extinguishing flash occurs on the surface of the oil is the flash point.

The flash point will vary with oils of similar viscosity made from different crudes. In this respect, paraffinic base oils have higher flash points than other types, with mixed, asphaltic, and naphthenic base oils following in the order named. The flash point will increase with higher viscosity oils from the same crude. For example, light spindle oils may have a flash point of 350 F, or less, while a cylinder oil from the same crude may not flash until temperatures above 600 F are reached. The flash point is a measure of lubricant volatility.

Fire Point

The fire point is the temperature at which an oil will sustain ignition continually when exposed to a flame under atmospheric conditions and controlled test procedure. The methods of determining both fire and flash points are similar, except that heating continues until the surface of the oil ignites and continues to burn for a period of at least 5 sec. The temperature at this point is called the fire point. The difference between flash and fire point temperatures varies with different oils by an amount ranging from 10 to 100 F, depending chiefly on viscosity. For general purposes, interpolation of fire points can be made on the basis of the flash point–fire point correlation, shown in Table 146.

TABLE 146. FLASH POINT–FIRE POINT CORRELATION

Flash Point, F	*Fire Point, F*
200	210
300	330
400	450
500	570
600	690
650	750

Pour Point

The lowest temperature at which an oil will flow, when tested (chilled) under certain specified conditions, is called the pour point. The oil to be tested is first placed in a testing jar tightly closed at the top by a cork, which also supports a thermometer that is immersed in the oil. The test involves, first, heating the oil in a bath to 115 F and then allowing it to cool in air (or in a water bath not over 77 F) to a temperature of 90 F. The jar of oil is then placed in a suitable cooling medium, such as crushed ice and calcium chloride crystals, and reduced in temperature. Beginning at a temperature of 20 F above the expected pour point, the jar is removed at each lower interval of 5 F and tilted to determine fluidity. The jar must be moved from one cooling medium to another having a lower temperature range at stipulated temperature levels above the pour point. The test continues under the above conditions until the oil will not show any movement when the jar is held in a horizontal position for 5 sec. The pour point of the oil tested will be 5 F above the temperature at which the sample solidified.

The pour point is a guide in predicting the temperature at which oil will flow by gravity under otherwise static conditions. Operational factors tend to contradict test results in many instances, but discrepancies occur for reasons not always apparent. Pressures and previous conditions to which oils are exposed in service affect the pour point and often in a direction of advantage rather than disadvantage. Treated with intelligence, laboratory results will serve as a safe guide in governing the use of oils, not only as they concern pour point, but many other physical tests as well.

Oils employed under low-temperature conditions or where they ultimately will find their way into cold operations must have low pour points. It is well to select oils having pour points substantially below the minimum (1) operating temperature of a system or (2) surrounding temperature to which the oil will be exposed. Refrigeration systems operating with low evaporator temperatures; worm gears of elevators having their drive mechanism located in unheated penthouses; fluid couplings, gears, pumps, chains, hydraulic systems, etc., operating out-of-doors during the winter; and many other systems operating under subnormal temperature conditions rely on the fluidity of the low pour point oils for maximum efficiency and wear protection.

Cloud Point

If the temperature of an oil is reduced sufficiently, wax will start to precipitate. The cloud point is the temperature at which wax precipitation starts. When precipitation occurs, the oil takes on a cloudy appearance, hence the name. Cloudiness is not always due entirely to wax crystallization, but may be caused partly by the presence of moisture. For this reason, it is well to use oils of low cloud point for the lubrication of refrigeration compressors, particularly in low-temperature systems.

Refrigerants that are miscible in oil, such as some of the Freons, tend to reduce the temperature at which wax precipitation occurs. Therefore, the cloud point is not a true criterion of oil behavior in refrigeration systems employing miscible refrigerants. In the case of ammonia and other nonmiscible fluids, the cloud point must be considered if wax deposits in low-temperature evaporators are to be avoided. For miscible refrigerant systems, the floc point is of greater importance than the cloud point.

Floc Point

The floc test is somewhat similar to the cloud test, except that a mixture of 90 percent R-12 and 10 percent oil serves as the test specimen instead of oil alone. The floc point is the temperature at which flocculation starts to occur in the mixture when chilled. While a more practical test would employ a mixture containing the miscible refrigerant of immediate concern, R-12 is adopted as a standard for reporting floc points of oils. The relative quantity of oil used in determining floc point appears to be reasonable considering that the amount of oil entering a refrigeration system does not usually exceed 10 percent of the total fluid circulated. In most cases, the percentage is less than 5 percent, depending on the design and the condition of the compressor.

The floc point is important in selecting lubricants for low-temperature systems employing miscible refrigerants. It is of little value where immiscible refrigerants are used and need not be considered in any case where evaporator temperatures are maintained above the freezing point of water.

Dielectric Strength

The principles of dielectric strength of materials were discussed in Chapter 10. The subject is of interest in the study of lubrication and refrigeration, because the insulating value of oils is directly associated with moisture content. Well-refined lubricating oils that are relatively free of moisture will resist the passage of an electric current through them. The less moisture they contain, the better insulators they become.

Corrosion caused directly by the presence of moisture, deterioration of refrigerants, copper plating within a refrigeration system, icing and clogging of orifices, oil breakdown, and loss of refrigerating performance are all related to moisture. One avenue by which moisture can enter a refrigeration system is via the lubricant.

Where dehydrators can be used, small amounts of moisture may be removed from a system after it enters by whatever means. In cases of hermetically sealed units where every precaution is taken to eliminate moisture, by drying the unit or other means, and on which a dehydrator cannot be used after final assembly, the moisture content of oil becomes an important consideration. Therefore, it is necessary to use oils that have been dehydrated to a condition of high dielectric strength.

Lubricating oils can be dehydrated to a condition of extremely low moisture content, resulting in a dielectric strength of 27,500 volts (v) or more. For the majority of refrigerating installations where some moisture enters by various means and subsequently is removed by cartridge dryers, it is poor economics to demand oils that are especially dehydrated. The reason for this is that the refiner normally prepares refrigerating oils having a dielectric strength of at least 25,000 v, which indicates a very low moisture content.

Because mineral oils are hygroscopic (i.e., absorb moisture from the air), they should be kept covered and otherwise handled carefully to prevent undue exposure. This applies particularly to oils in which any significant amount of moisture may prove deleterious.

Emulsification

The subject of emulsification involves the old belief that "water and oil will not mix." The statement has some semblance of truth, but only under extraordinary circumstances. New highly refined straight mineral oils possess qualities of good water separability. Such qualities will remain to a high degree just so long as conditions of handling oils are proper. They will not continue to demulsify under poor operating conditions where contaminants not only are permitted to enter the oil but also are allowed to remain indefinitely. Carbon dioxide from the air forms carbonic acid when dissolved in water that may be present in an oil in service. Acids form sludges that have more or less emulsification properties. Add to the latter numerous other contaminants that find their way into an oil and the conglomerate results in a chemical complexity that leads not only to emulsions, but also to other deteriorating influences.

Except in special cases, such as soluble oils used in metal cutting processes, emulsification is undesirable. Emulsification permits oil to retain water particles that (1) ultimately lead to lubricant deterioration, (2) displace oil in frictional areas, causing mechanical failure, and (3) cause rust and corrosion. Reservoir foaming also results in some cases.

The problem of oil emulsification is of great importance in circulation or other reuse systems. It cannot be tolerated to any extent in lubrication systems serving high-speed journals, such as in steam turbines.

Laboratory tests dealing with the emulsification tendencies of lubricants include the steam emulsion test.

The steam emulsion test involves passing steam (212 F) through a test tube of oil (20 ml at room temperature) for a period of 4 to 6.5 min. The test tube containing the oil-condensate mixture is then moved to a separating bath (200 to 203 F). The time in seconds required for the oil to separate from the water is called the *ASTM Steam Emulsion Number*, or simply the S.E. Number. If 20 ml of oil does not separate in 20 min, the sample will be reported as having an S.E. Number of 1200 plus.

Certain type additives used in turbine and other type oils tend to raise the S.E. Number. Therefore, the result of a steam emulsion test is not a true criterion on which to base conclusions relative to pertinent chemical changes in the oil due to operation.

Carbon Residue

The carbon residue test (Conradson) determines the percentage of carbon residue left by an oil upon evaporation under specified test conditions. The test sample placed in a crucible within a crucible is heated under atmospheric conditions until it is evaporated; i.e., its vapors no longer ignite. The residue is then cooled and weighed. The result is reported on the basis of weight ratio of residue to oil sample. The Ramsbottom test is another method of determining carbon residue by evaporation. Some discrepancy exists between the results obtained by the two methods.

The carbon residue test serves to permit some doubtful conclusions to be drawn relative to the carbonization properties of an oil. However, the nature of carbon may be deduced by residue inspection. Hard carbon is formed by paraffinic-base oils, while naphthenic-base oils produce soft fluffy carbon. The amount of carbon residue of an oil varies with different crudes, refining methods, and viscosity. Predicting results based on carbon residue without consideration of all other factors that might influence operation is an unsound practice.

Neutralization Number

The neutralization number is the number of milligrams of potassium hydroxide required to produce a pink color (by titration) in a mixture of test oil, distilled water, alcohol, and phenolphthalein. The result of the test for neutralization number must not be construed as a measure of absolute acidic or basic properties of an oil. It is intended to determine the pressure of acidic or basic constituents in an oil. The American Society for Testing Materials lists as constituents of new or used oils considered to have acidic characteristics, the following: organic and inorganic acids, esters, phenolic compounds, resins, salts of heavy metals, inhibitors, detergents, and others. Basic constituents include: organic and inorganic bases, amino compounds, salts of weak acids (soaps), salts of heavy metals, inhibitors, detergents, and others. Some of the constituents are the result of changes that may occur in an oil due to the deteriorating influence of oxidizing conditions. Such constituents may promote further deterioration either alone or in conjunction with some types of construction metals that act as catalysts.

The criterion for changing used oil in a lubrication system should not always be based on the level of the neutralization number as much as on the rate at which it reaches that level, assuming that the lubricant condition warrants such consideration. A decided rise in neutralization number accompanied by an increase in viscosity and other pertinent changes is an indication that a serious condition has been reached, or is being approached. Operations involving steam turbines, large internal combustion engines, hydraulic systems, and large refrigeration compressors are usually closely attended. The operational history of their oil charges should be followed and recorded. A correlation of an immediate oil condition with its historical background furnishes the soundest basis on which to decide the fate of an oil charge.

ADDITIVES

High speeds and pressures, as well as high and low temperatures, accompanying the development of modern machinery required that certain modifications be made in lubricants produced from petroleum. Straight mineral oils, even though highly refined, are no longer able satisfactorily to meet all the operating conditions brought about by the industrial change. Many of the problems that were brought about or intensified are shown in Table 144.

To meet the severe operating conditions prevalent in modern machines and processes, mineral lubricating oils are (1) specially treated during the refining process to remove deleterious materials and/or (2) compounded with chemical agents that improve (extend) their qualifications. (1) The lubricating oil flow chart (Table 143) shows some of the improvements made in refining, including those that remove low viscosity index and chemically unstable hydrocarbons, waxes, and color bodies. (2) The chemical agents are called *additives*, the most important types of which are described below. Some additives are shown in Table 147.

Anti-oxidants

Petroleum oils tend to oxidize and form sludge when exposed to elevated temperatures, oxygen, and materials (such as some metals) that act as oxidizing catalysts. *To retard* oxidation, chemical agents or additives known as oxidation inhibitors are added to some types of lubricating oils. Oxidation inhibitors combine with (1) oxygen and (2) oxidized substances to form innocuous materials. Anti-oxidants are used in oils employed in circulation systems (steam turbines, hydraulic machinery, paper machines, etc.), internal combustion engines, electrical transformers, and other systems in which the oil is continually reused or must perform for long periods without change. Regarding oxidation, the life expectancy of any petroleum lubricating oil is reduced approximately 50 percent with each 18 F rise in operating temperature.

Anticorrosive Agents

Certain organic acids are formed during the process of oil deterioration (oxidation). Acids attack metals. Metal bearings subject to corrosive activity are short-lived. Anticorrosive agents or corrosion inhibitors act to combat (1) the presence of acids by combining with them to form innocuous materials, or (2) acid activity on metal by establishing a protective film on their surfaces. In the first case, a chemical action is involved in which corrosive and catalytic substances are deactivated. Metals, such as copper, lead, and iron, act as catalysts that promote and increase the rate of oil oxidation. Metal deactivators act in much the same way as described for corrosive substances; i.e., they form insoluble or soluble compounds that respectively settle out or become inactive. In the second case, a physical action occurs whereby the additive material "plates out" to coat the metal surfaces with a thin film through which corrosive materials cannot permeate. The latter type additives are called metal passivators. Anticorrosion agents of the type discussed here are associated principally with internal combustion engine crankcase oils to protect hard alloy bearings employed in automotive service.

Rust Inhibitors

Rust is associated with ferrous materials of lubrication systems. Rust is a form of corrosion brought about by the oxidizing action of air and moisture. The action of rust inhibitors may be that of (1) an emulsifier or (2) polar material. The emulsifying action of the inhibitor causes water droplets to be taken up and surrounded by oil which holds it in suspension. The polar type material, as its name implies, sets up an electrolytic action whereby the electrical fields of the additive and the metal have mutual attraction. By attaching itself to the metal surfaces, the additive molecules form a strong moisture-resistant film. Only those iron or steel surfaces that are in contact with the oil are protected against rust by rust inhibitors. Other parts, such as cavities in bearing caps, must be protected otherwise. Rust-inhibited oils are used in all systems where the effect of moisture prevails, including compressors, circulating systems, and oil reuse systems in general. Many such oils are inhibited against both rust and oxidation.

Cleanliness Agents

The operation of internal combustion engines involves products of combustion, external contaminants (that enter with the fuel, air required for combustion, and through the breather), and products of oil oxidation. Such materials flocculate to form sludge or are deposited (baked) on heated surfaces of the engine (cylinder heads, rings, ring grooves, pistons, valves, etc.). To prevent deposit formations and sludge aggregation, additives known as "detergents" are used. Such additives act to surround particles of contamination and keep them in suspension (dispersed) within the oil. This action imparts a dirty appearance to the oil after a short time in service, but keeps the engine surfaces clean. While detergents do not act to dissolve (making the name a misnomer) deposits already formed, often they will loosen them, thereby cleaning an already fouled engine. An increase in oil consumption often occurs when worm and "stuck" piston rings are loosened by the introduction of an oil containing a substantial amount of detergent material.

Antifoam Agents

Foaming usually is caused by excessive aeration. Aeration occurs when oil is allowed to fall from a return line having its outlet above the oil surface in a reservoir. It is also caused by leaking pump suctions. Oil agitation in numerous ways causes aeration. Oils of high viscosity tend to foam more than lighter bodied oils. Foaming will also occur in oils containing soaplike substances present due to contamination, oil oxidation, or some types of additives. Where conditions that cause foaming cannot be eliminated, sometimes a defoamant is used. The action of a defoaming agent is to reduce the interfacial tension of the oil film surrounding the bubbles, causing them to join with others to form larger bubbles. In doing so, they become light enough in weight to be displaced more readily by the heavier oil and forced to the surface where they explode.

Where foaming can be eliminated or sufficiently reduced by overcoming the mechanical difficulties causing it, such action should be taken. Defoamants should be used only as a last resort.

Viscosity Index Improvers (VI)

The inherent characteristics of some types of crudes are such as to produce oils having a low viscosity index. Modern refining methods have improved this condition somewhat. Where refining influences are not sufficient to produce a desired VI in some type oils, viscosity index improvers are used. Such additives reduce the rate of viscosity change in an oil with a change in temperature.

Pour Point Depressants

Dewaxing of oils as conducted by refinery processes can produce oils having appreciably low pour points. When the temperature of oils is reduced sufficiently, a wax formation analogous to the lattice-like structure of a honeycomb forms, entrapping the liquid oil within the numerous cells. In this condition, the oil cannot flow. Cost of refining and the removal of desirable lubricating properties accompany the extreme dewaxing required to prepare some types of oils for low-temperature operation. To augment the dewaxing process, an additive that acts to (1) retard wax crystallization and (2) change the physical characteristics of wax crystals is used. The additive, known as a pour-point depressant, acts to relieve flow interference at lower temperature without any apparent change in other characteristics.

Antifriction Agents

Antifriction agents are referred to more commonly as oiliness agents. Their presence in an oil increases frictional surface adherence, provides slipperiness, and reduces the coefficient of friction. Antifriction agents are polar materials and act in the same manner described for the film-forming materials used to combat corrosion (see rust inhibitors). Their chief characteristics are adhesiveness, oiliness, and persistence of film. The three factors together result in strengthening the interposed film. Oiliness agents eliminate or appreciably reduce chattering and wear between parallel frictional surface (absence of oil wedge). They also help provide varying degrees of frictional protection to sparsely lubricated bearings. Usually antifriction agents are associated with oils employed for the lubrication of machine ways, guides, screws, etc., and bearings subject to boundary conditions. Fatty compounds of fixed oils, of which oleic acid is an important member, provide excellent antifriction agents.

Antiwear Agents

The principal purpose of any lubricant is to establish conditions between frictional surfaces that will prevent, or drastically reduce, the rate of wear. Where a fluid film can be established and maintained continually, a straight mineral oil will perform efficiently. Bearings operating at normal loads and speeds under boundary conditions require lubricants possessing oiliness properties to counteract an otherwise high coefficient of friction. Another condition precludes the possibility of either maintaining a fluid film or performing satisfactorily with lubricants that may possess oiliness properties, but lack other necessary film-forming characteristics to prevent rapid wear created by abnormally heavy loads. The most serious condition exists where loading is of such magnitude and disposition as to create extremes of film rupture and film wiping that resulting temperatures cause plastic flow and subsequent welding of the opposing surfaces. The four major conditions, as outlined above, may be identified as follows in the order of their lubrication facility: (1) ideal or full fluid film, (2) boundary or thin film, (3) heavy pressure type film, (4) extreme pressure type film.

1st Case. The *ideal or full fluid film* (developed hydrodynamically) depends on viscosity, the magnitude of which is a function of load and speed according to the relation $\frac{ZN}{P}$. Antiwear type lubricants are not required normally for service in this category.

2nd Case. Boundary or thin film conditions rely on viscosity and oiliness to minimize coefficient of friction and prevent rapid wear. Requires an oil of proper viscosity compounded with an oiliness agent (polar additive) of the type described in the preceding section.

3rd Case. Heavy pressure type films are established by polar materials, many of which increase their antiwear activity with higher rubbing temperatures. Materials required for this service react rapidly to form films on exposed surfaces with the development of frictional heat resulting from metal-to-metal contact. Frictional conditions represented in this category are those commonly classified as "mild extreme pressure" cases. The author's preference of "heavy pressure" to distinguish such cases is based on (1) the paradoxical significance of the words "mild" and "extreme," and (2) the fact that many bearings operating within the range of this category are constructed of materials incapable of withstanding loads of the magnitude inferred by the description as being "extreme." Additives used in heavy pressure lubricants include the metallic soaps, lead naphthenate, and lead oleate.

4th Case. Extreme pressure conditions depend on chemical action of the lubricant additive to prevent welding of opposing surfaces. Tremendous pressures accompanied by a drastic wiping action as one surface sweeps over the other precludes any cushioning effect of the lubricant. Plastic flow of surface metals (steel) results from the high localized temperature developed by solid friction. Thus is established an ideal welding condition. However, the antifluxing characteristic of extreme pressure lubricant additives prevents the occurrence of welding. There are few cases of extreme pressure other than that which is developed occasionally in hypoid gears. The latter can be categorized as developing heavy pressure conditions when operating under normal shaft loads. Antiflux materials include compounds of sulfur, phosphorus, and chlorine which react with hot metallic (iron or steel) surfaces to form films of sulfides, phosphides, and chlorides of iron.

Adhesive Agents

Cams, gears, and other machine components that depend on periodic hand lubrication require a lubricant that will not be thrown from their surfaces by centrifugal force. Other components subject to shock loading and vibratory motion depend on lubricants that will not be displaced by splash under conditions analogous to hammer blows. To meet such conditions, lubricants must possess high cohesive and adhesive properties. To impart cohesive properties to an oil (or grease), agents commonly referred to as "stringiness" additives are used. Lubricants compounded with additives of this type "string out" like a myriad of thin rubber bands when the surfaces to which they adhere are separated. Actually, such additives change the viscosity characteristics of the oil to which they are added.

Emulsifiers

Lubricants that must resist the washing effect of water are compounded with emulsifying agents. Steam cylinders operating under wet steam conditions, air tools that act as receptacles for moisture resulting from expansion and subsequent cooling of compressed air, and stern tubes of ships subject to washing by sea water are a few conditions requiring emulsifying agents. Water and emulsifiers act to form water-in-oil solutions. The resulting emulsion acts as a lubricant and also prevents contact of water with metallic surfaces. Because fixed oils, such as acidless tallow, lard, and blown rapeseed are used, lubricants treated with emulsifiers possess good lubricity.

RECLASSIFICATION OF LUBRICATING OILS

Earlier in the discussion of petroleum lubricating oils, a classification of lubricants based on viscosity was shown. Since then, a study of (1) the physical and chemical characteristics of lubricants and (2) the effects of additives as related to lubrication problems has been made. It may be deduced from such a study that it is a fallacy to base the classification of lubricants solely on viscosity. A more accurate method of classifying lubricating oils is shown in Table 148.

Table 147. Lubricating Oil Additives*

Additive Type	*Type Compounds Commonly Used*	*Reason for Use*	*Probable Mechanism*
Oxidation inhibitors	Zinc dithiophosphates, hindered phenols, aromatic amines, sulfurized phenols.	Retard oxidative decomposition of the oil which can result in varnish, sludge and corrosion.	Decompose peroxides, inhibit free radical formation and passivate metal surfaces.
Dispersants	High molecular weight alkyl succinimides, alkylthiophosphonates, organic boron compounds.	Maintain cleanliness by keeping oil insoluble material in suspension.	Primarily a physical process. Dispersant is attracted to sludge particle by polar forces. Oil solubility of dispersant keeps sludge suspended.
Metal deactivators	Zinc dithiophosphates, organic sulfides, certain organic nitrogen compounds.	Passivate catalytic metal surfaces to inhibit oxidation.	Form inactive protective film on metal surface. Form catalytically inactive complex with metal ions.
Viscosity index improvers	Methacrylate polymers, butylene polymers, polymerized olefins or isoolefins, alkylated styrene polymers, various selected copolymers.	To lower the rate of change of viscosity with temperature.	VI improvers are less affected by temperature than oil. They raise the viscosity at 210°F more in proportion than at 100°F due to changes in solubility.
Anti-wear agents	Zinc dithiophosphates, organic phosphates, and acid phosphates, organic sulfur and chlorine compounds, sulfurized fats, certain amines.	To reduce friction, prevent scoring and seizure. To reduce wear.	Film formed by chemical reaction on metal contacting surfaces which has lower shear strength than base metal thereby reducing friction and preventing welding and seizure of contacting surfaces when oil film is ruptured.
Rust inhibitors	Metal sulfonates, fatty acids and amines.	Prevent rusting of ferrous parts during storage and from acidic moisture accumulated during cold operation. This is a specific type of corrosion.	Preferential adsorption of polar type surface active material on metal surfaces. This film repels attack of water. Neutralizing corrosive acids.
Corrosion inhibitors	Zinc dithiophosphates, metal phenolates, basic metal sulfonates.	To prevent attack of corrosive oil contaminants on bearings and other parts.	Neutralization of acidic material and by the formation of a chemical film on metal surfaces.
Foam inhibitors	Silicone polymers.	To prevent the formation of stable foam.	Reduce surface tension which allows air bubbles to separate from the oil more readily.
Seal swellers	Organic phosphates, aromatics, halogenated hydrocarbons.	To swell seals slightly, reducing leakage.	Mild chemical modification of seal elastomer.
Friction modifiers	Organic fatty acids and amides, lard oil, sperm oil, high molecular weight organic phosphorus acids and esters.	To reduce the static coefficient of friction.	Preferential adsorption of surface active materials.

* Courtesy of Texaco Corp.

TABLE 148. GENERAL CLASSIFICATION OF LUBRICATING OILS ACCORDING TO SERVICE

Service	General* Viscosity Range SUS at 100 F	Principal Additive or Other Required Characteristic
I. Reuse Systems:		
Circulating lubricants:		
Spindle oils	60– 150	R, O
Turbine oils	150– 300	R, O
Hydraulic oils	150– 750	R, O
Bearings, general	300– 900	R, O, (AF)†
Gears, spur, bevel, herringbone	300– 900	R, O
heavy pressure	300–1500	AW, AF
hypoid	800–1500	EP
Splash or bath lubricants:		
Gears, spur, bevel, herringbone	600–1500	R, O
heavy loads	700–2500	AW
worm	1800–3000	AW or AF
hypoid	900–2500	EP
Bearings, air compressor	300– 600	R, O
internal combustion engine	150–1400	High VI R, O, AC, D
Chains	300– 750	R, O
Bearings, heavy duty, journal	900–1500	AW
medium duty, journal	300–1000	AW or AF
II. All-Loss Systems:		
Cylinder oils:		
Steam cylinders, dry	1800–3000	
wet	1800–3000	E
Air compressor cylinders		
Up to 150 psi, water cooled	300	R, O
Over 150 psi, water cooled	600	AF, E
Air cooled	300– 600	R, O
Air tools	150– 300	E
Refrigeration compressor cylinders		
Low temperature	100– 200‡	Low floc point
Above 20 F	300– 400	Low pour point
Inert gas compressor cylinders (same as air cylinders, generally)		
Boundary lubricants:		
Bearings—guide, ways, etc	150– 900	AF
journal		
High speed, light load	150– 300	A
Moderate speed, moderate load	300– 750	A or AW
Slow speed, heavy load	750–1500	AW

Key: (A) Adhesive agent, (AF) Antifriction agent, (AW) Antiwear agent, (AC) Anticorrosive agent, (E) Emulsifier, (EP) Extreme pressure agent, (D) Detergent, (O) Antioxidant, (R) Rust inhibitor, (VI) Viscosity index.

* Viscosity may be raised or lowered to meet other than normal conditions.

† Antifriction agent where loads require.

‡ Predicated on pour point.

FIXED OILS

Fixed oils are fatty substances that are extracted from animals, vegetable matter, and fish. They are called fixed oils because they will not volatilize without decomposing. When reduced in temperature sufficiently, fixed oils become fats. The temperature of solidification varies with the kind of oil.

Fixed oils will combine with oxygen of the air and become substances having a gumlike consistency. The process is known as *drying.* Some fixed oils dry readily, while others combine with oxygen at a much slower rate. The former types are called *drying oils* and are used in paint manufacturing, the most familiar one of which is linseed oil. The slower drying types are known as *nondrying oils* and include the types associated with lubrication. The tendency of a fixed oil to dry is determined by its *iodine number.* The iodine number is the amount (percent) of iodine absorbed by an oil under controlled test conditions. The higher the iodine number, the better the drying qualities of the oil.

Fixed oils are composed of fatty acids and alcohols, the radicals of which are combined to form fatty esters. A means of determining free and combined fatty acids in a fixed oil is the saponification test. The relative values of various fixed oils in respect to fatty acids can be determined by their *saponification number.* The saponification number is the number of milligrams of potassium hydroxide taken up by 1 gram of oil under controlled test conditions. The higher the saponification number, the greater the net amount of base-consuming constituent present in the oil.

Table 149 lists some of the more important fixed oils showing average iodine and saponification numbers, and their Saybolt viscosities at 100 F.

TABLE 149

Fixed Oil	*Saybolt Viscosity at 100 F*	*Saponification Number*	*Iodine Number*
Castor oil	1370	180	85
Corn oil	180	187	115
Lard oil	100–210	194	70
Linseed oil	130	150	180
Mineral oil	—	—	12
Neatsfoot oil	220	197	70
Olive oil	190	192	85
Peanut oil	330	191	101
Rapeseed oil	260	174	95
Rapeseed, blown	900–1000	206	60
Sperm oil	105	130	80
Tallow oil	225	194	65

Castor, lard, tallow, rapeseed (blown), and sperm oils represent the more important lubricating types of fixed oils. They are very seldom used alone, but usually are employed as an additive to increase oiliness or promote emulsibility of mineral oils. Lard oil, acidless tallow oil, and degras (lanolin) are used principally with steam cylinder oils where they lend emulsibility and heat-resistant qualities. Tallow is used in manufacturing greases. Lard and sperm oils are blended with mineral oils to produce metal cutting oils. Rapeseed when blown with air increases viscosity considerably. It is used with mineral oil for lubricating stern tubes and other water-washed bearings. Palm oil is an excellent metal drawing and roll oil. Porpoise jaw oil is used for lubricating watches and other small timepieces. Olive oil is employed in the combing of wool in the form of a soluble oil, known as wool oil. Corn oil is an excellent base for molybdenum disulfide, enabling the latter to adhere more readily to metal surfaces. Rosin oil is used in the manufacture of cold-set greases. Castor oil, like all other fixed oils, is an excellent lubricity agent. It has been used for lubricating racing cars, but it forms hard gummy deposits when exposed to heat, as will any other fixed oil.

Extracts of fixed oils, such as oleic acid, possess high oiliness properties and film strength characteristics. It is used to improve the lubricating qualities of mineral oils and is particularly valuable as an additive in oils employed for boundary lubrication. Oleic acid or elaine, as it is commonly called, or oils containing even small amounts of the fatty acid, should not be used in galvanized containers or applicators, with which it sets up a chemical action. About 1 percent of oleic acid is all that is required to be blended with mineral oils for maximum performance.

PLASTIC LUBRICANTS

All the members of this group of lubricants are not plastic at room temperature. However, those that are not plastic under normal ambient conditions can be made so with an appropriate change in temperature one way or the other. Petrolatums and greases comprise the plastic lubricant family, of which greases are the more important members. Plastic lubricants are employed where design conditions will not permit the use of oil, chiefly because of leakage.

Greases

Greases are made, generally, by saponifying a metal base with a fatty substance, such as tallow, and adding oil to the soap. The process, which includes agitation, is carried on under controlled temperature conditions, in either an open kettle or a pressure cooker. The bases used include calcium, sodium, aluminum, lithium, and lead. Barium and strontium are also used. Lubricants produced using various metal bases are identified as calcium- (lime-) soap, soda-soap, lithium-soap, aluminum-soap, etc., greases. To obtain the advantages of more than one soap, mixed-base greases are also available. The physical characteristics of a grease depend principally on the base used in their manufacture. The consistency of greases becomes more firm with greater soap content. The latter defines greases as soap-thickened oils. Oils of different viscosity are employed in manufacturing grease, the choice of which depends on service considerations. Oil viscosities range from 60 to over 1000 Saybolt seconds at 100 F.

Greases made with mineral oils contain varying amounts of water that act as a bonding agent. Water content of greases varies from a trace to approximately 3.0 percent. *Cold-set* greases, made with a mixture of lime, rosin, and mineral oil, contain over 8 percent water. These types of greases set or jell in their shipping containers in which they are poured after saponification is complete.

The property of water solubility varies with greases made with different bases. Lime-soap greases are water insoluble, while soda-soap greases are highly soluble. The structure of greases may be smooth, buttery, or fibrous. The fibers of the latter type may be long or short. The long fiber types commonly are referred to as sponge greases. The color of greases ranges from white to black with intermediate colors of amber, yellow, and various shades of brown and green.

Greases are treated with various additives for special purposes. Oxidation inhibitors extend the life of greases in the same manner with which they act on mineral oils. Solid materials, such as graphite, molybdenum disulfide, talc, mica, and other fillers, are added to some types of greases to assure protection to frictional surfaces and maintain low coefficient of friction in the event of grease film rupture under rough service. Heavy pressure greases are developed by combining the film-strengthening characteristics of lead and zinc soaps (neither of which is a thickening agent to any great extent) with the thickening qualities of lime-soap. Such products are called *mixed-base* greases, members of which include the successful leaded greases used extensively for certain steel mill requirements. Lead-soap is also used as a film strength additive for heavy pressure oils.

Grease Consistency. The consistency of greases ranges from that of a semifluid to hard blocks, with intermediate states of softness and firmness. Semifluid and soft greases are made so for several reasons, an important one being that of pumpability. Firmness of greases is dictated by service conditions, of which leakage and handling are prime requisites. Hard greases are called block greases and do not necessarily require a reservoir, except as a receptacle for holding them in a desired location on the journal.

The consistency of greases is measured in terms of penetration. Penetration values are no criterion of grease quality or kind, but merely indicate relative hardness, or softness, as measured by a penetrometer under

ASTM standard conditions. A penetrometer consists of a standard cone weighing 150 grams. The cone is connected by suitable linkage to a gage designed to measure millimeter changes in cone elevation. The cone is held in suspension with its tip just touching the surface of the test sample of grease contained in a standard cup. At this point in the test, the gage reads zero and the grease is at a standard temperature of 77 F. Penetration of the grease occurs when the cone is released. After a lapse of 5 sec, during which the cone exerts penetrating force, the gage is read to determine the distance of penetration; i.e., the number of millimeters the cone dropped. The depth of penetration made by the cone in tenths of a millimeter is called the penetration number. If the grease sample has been undisturbed before testing, the results are expressed as *unworked penetration.* If the sample has been subject to extrusion prior to testing, by reciprocating a perforated piston 60 times (60 double strokes) through a cylinder containing the grease, the results are expressed as *worked penetration.* The softer the grease, the greater will be the depth of cone penetration. Consequently, the softer the grease, the higher will be its penetration number and vice versa.

To identify the relative consistency of greases, the National Lubricating Grease Institute has devised a system in which a number represents the penetration range for a given grease. The system is devised so that zero represents the softest grease (semifluid) and 6 the hardest. The numbers given the various consistencies, as shown in Table 150, are not necessarily those used by all grease manufacturers to identify their products. It does, however, act as a universal method for expressing equivalent values.

TABLE 150. RELATIVE GREASE CONSISTENCY —N.L.G.I. NUMBERS VS. ASTM PENETRATION

N.L.G.I. Nos.	*Relative Consistency*	*ASTM Penetration*
0..............	Semifluid	355–385
1..............	Very soft	310–340
2..............	Soft	265–295
3..............	Semifirm	220–250
4..............	Firm	175–205
5..............	Very firm	130–160
6..............	Hard	85–115

Drop Point of Greases. The drop point of a grease is the temperature at which it liquefies when heated and drops from a cup through a bottom orifice. The cup and the thermometer are placed in a test tube and the tube immersed in a heated bath of oil. As the bath of oil is heated, the consistency of the grease decreases. When the consistency of the grease is reduced sufficiently, it will flow through the orifice in the bottom of the cup. The temperature at which the first drop occurs is the drop point. The drop point gives some indication of the consistency of a grease when exposed to a known operating temperature. It does not indicate the maximum temperature at which the grease will perform satisfactorily.

There is no standard rule for selecting a grease based on a relation of drop point to type of soap. Ingredients and manufacturing quality and processes control this property. This is not to infer, however, that certain type soaps do not tend to impart better high operating temperature characteristics than others. When operating temperatures are of such high level or duration to interfere with the water-stabilizing effect within a hydrous grease, a separation of oil and soap results. The oil runs away, leaving the soap to char and probably ruin the bearing.

Other Grease Tests. There are a number of other laboratory tests performed on greases. Such tests are made to determine special properties associated with specific service requirements. Such tests include those that determine chemical and physical stability, performance in wheel bearings, pumpability, wear resistance, and a number of others that may be performed in the laboratory or under field service conditions.

Grease Classification. The various soap-base greases are listed in Table 151, showing some general properties associated with service conditions.

SOLID LUBRICANTS

Solid lubricants include graphite, molybdenum disulfide, talc, and mica. Of the four mentioned, graphite and molybdenum disulfide are the more important. Talc and mica are typical fillers for axle greases. They are also used as lapping compounds for smoothing out frictional surfaces.

Graphite

Graphite is a crystalline form of the element carbon (C). It is a product of coke or hard coal. It is also mined. As a lubricant, it is in a powdered or colloidal form. It can be used alone or mixed with oil, grease, or other carriers. Graphite possesses an affinity for metal surfaces and imparts a high degree of lubricity to bearings. It has high chemical stability to withstand high temperatures, and will not dissolve in water. Graphite in the form used as a lubricant will enter minute clearances. Oil compounded with colloidal graphite has had some measure of success as a metal cutting fluid. It is mixed with lime-soap greases in varying quantities (2 to 20 percent) to produce graphite grease, an excellent lubricant especially suited for wet operation. Graphite depends on an adsorbed (moisture condensed on its surface) film of moisture for its lubricating action. If an adsorbed film of moisture is not present, graphite acts as an abrasive.

TABLE 151. GREASE CLASSIFICATION AND TYPICAL USES

Base (Soap)	*Temperature Limitations*	*Water Solubility*	*Lubricant Service (Typical)*
Calcium (lime)...	175 F	Insoluble	Cup grease. Plain bearings. Antifriction bearings of moderate speed and temperature. Water repellent. Pump glands. Presence of water. Buttery in structure. N.L.G.I. Nos. 1 to 6
Sodium (soda)....	200 F–350 F	Soluble	Antifriction bearings of high or moderate speeds. Will emulsify with water. Use under dry conditions generally. Will withstand higher operating temperatures than lime-soap greases. Do not require water as a stabilizer. Smooth, fibrous, or spongelike in structure. N.L.G.I. Nos. 1 to 6
Aluminum.......	180 F–190 F	Insoluble	Withstands shock loads without splashing. Cohesive and adhesive. Water repellent. Will not separate readily under centrifugal force. Auto chassis, flexible couplings, eccentrics, cams, small open gears. Smooth and stringy in structure
Lithium.........	Up to 300 F Down to −100 F	Slightly soluble	Practically water repellent. With low pour point oils can be used at extremely low temperature. With high viscosity oils can be used at higher temperatures. General purpose service. Buttery to stringy in structure
Calcium complex.	300 F+	Insoluble	Water repellent. Fairly high temperature lubricant. Good pumpability. Usually has high film strength when fortified with proper additives. Good stability. General purpose lubricant
Mixed base (lime-lead)	200 F	Insoluble	Water repellent. Heavy pressure service and severe operating conditions. Good pumpability for central grease systems with long feed lines. Smooth texture
Mixed base (lime-lithium)	250 F	Slightly soluble	Antifriction bearings subject to relatively high or low temperatures (depending on oil ingredient)

Such a condition may be brought about at high sliding velocities, during which a sufficient rise in temperature occurs to preclude the presence of moisture when dry graphite is used. Because of its resistance to heavy pressures and high temperatures, graphite can be used under extremely

TABLE 151 (*Continued*)

Base (*Soap*)	*Temperature Limitations*	*Water Solubility*	*Lubricant Service* (*Typical*)
Graphite grease . .	175 F	Insoluble	Water repellent. Usually a lime-soap grease to which has been added up to 20% colloidal or powdered graphite. For wet service conditions. Rotating water screen chains, pump plungers, hydraulic rams (water). Not good for high temperature or for anti-friction bearings unless colloidal graphite is used. Wet or dry plain bearings
Cold-set (lime-rosin)	180 F	Insoluble	Buttery in structure and consistency. Can be applied to open gears, axle bearings, rough machining in general. Water repellent
Block (sodium).	Up to 450 F	Soluble	Hard and brittle. High-temperature plain journal bearings where grease rides on rotating journal. Will form an emulsion with water which is sometimes desirable. Often wrapped in cloth before applying to journal for control of attrition. Large slow-turning journals. Some block greases contain other soaps in addition to sodium to impart high-temperature characteristic
Residuum	About 200 F	Will wash away unless specially treated	May or may not contain soaps. Heavy asphaltic type material of normally heavy body and high viscosity. Used principally for lubrication of large open gears. Often adulterated with a solvent to facilitate application. Solvent volatilizes upon application, leaving tough adhesive lubricating film. Lighter grades produced by blending with other petroleum stocks, or by distillation to desired consistency. Also called gear shields, flux, etc. Sometimes treated to improve film strength and water repellency
Non-soap	250 F–500 F	Insoluble	High temperature range. Insoluble in both hot and cold water. Acid resistant to some extent. Rust protection

severe operating conditions. Graphite has a low coefficient of friction. It has laminar structure.

Molybdenum Disulfide

Molybdenum disulfide, like graphite, has laminar structure, being made up of alternate layers of molybdenum and sulfur atoms. It is produced by purifying the mineral molybdenite (MoS_2). Molybdenum disulfide may be used (1) in powder form as a dry lubricant applied to clean metal surfaces or (2) mixed with oil, grease, or other carrier, such as silicone or polyalkylene glycol.

Molybdenum disulfide provides excellent lubrication under high or low temperature (−100 to 750 F) operating conditions. It will oxidize at temperatures above 750 F if not protected against the atmosphere. At ordinary temperatures, molybdenum is relatively inactive.

Some tests appear to show that an excellent lubricating film can be established with a mixture of molybdenum disulfide and corn oil. Because it resists heat, cold, water, and chemicals, molybdenum disulfide is finding wider use for extreme operating conditions. It is noncorrosive, possesses high film strength and low shear resistance, and has an affinity for metal surfaces. It is used as an additive in oils to provide protection when heavy pressures or wiping action destroy hydrodynamic films. It acts as an anti-flux to prevent welding when the oil film becomes ruptured under heavy loading.

Molybdenum disulfide has a melting point of 2165 F, is blue-gray to black in color, has a molecular weight of 160.08 and 4.8 to 5.0 specific gravity. It has a coefficient of friction of about 0.032.

COEFFICIENT OF FRICTION OF VARIOUS SUBSTANCES UNDER HIGH PRESSURE

The coefficient of friction of various lubricants and additives is shown in Table 152.

APPLICATION OF LUBRICANTS

Effective lubrication is dependent on the following basic conditions:

Type of lubricant:	Use correct lubricant.
Quantity of lubricant:	Sufficient lubricant must be supplied.
Interval of lubricant application:	Lubricant should be available when needed.
Method of lubricant application:	Apply lubricants in a manner and at point(s) that render maximum effectiveness.

The method of application depends on (*a*) design of frictional system, (*b*) type of lubricant required, (*c*) frictional heat developed, (*d*) functions

TABLE 152. AVERAGE COEFFICIENT OF FRICTION (f) OF SOME LUBRICATION SUBSTANCES UNDER HIGH PRESSURE AND ANGULAR DISPLACEMENT (TURNING ACTION OF OPPOSING SURFACES OF 10 TO 50°)

Substance	*Average (f)*
Stearic acid	0.025
Tungsten disulfide	0.035
Molybdenum disulfide	0.035
Lead iodide	0.066
Palm oil	0.069
Castor oil	0.074
Grease (zinc soap)	0.076
Grease (calcium soap)	0.077
Lard oil	0.078
Sperm oil	0.081
Turbine oil + graphite	0.088
Turbine oil	0.098
Oleic acid	0.116
Soapstone (powdered)	0.276
Mica	0.280

of the lubricant, and (*e*) required frequency of application. Some variables associated with each of the five major factors are:

(*a*) Either bearings, gears, or cylinders may be involved.
(*b*) The lubricant may be an oil or a grease or some other type of lubricant.
(*c*) An oil may function as a cooling agent, as well as a lubricant.
(*d*) The lubricant often acts as a hydraulic medium, as well as a lubricant. For example, turbine oils serve to lubricate the bearings and also to actuate the governor mechanism.
(*e*) Frequency of lubricant application depends on one or more of the following, some of which govern lubricant consumption in the case of all-loss lubricating systems:
Temperature
Speed
Load
Cleanliness
Leakage of lubricant from bearing (good or poor seals)
Clearance (leakage, blow-by, etc.)
Type of frictional part
Type of lubricant used
Method of lubricant application
Ambient conditions (moisture, dust, etc.)
Effect of other fluids on lubricant viscosity (i.e., mixture of oil and

refrigerant or other fluid being pumped)
Condition of piston rings
Other service conditions

Methods of applying lubricants are numerous, ranging from simple hand oiling to automatic central distributing systems serving many frictional points from a common reservoir. The methods commonly used are described below:

A. Oil application:

1. Manual or hand oiling:

(*a*) *Common oil can,* actuated by thumb pressure. Intermittent application permitting boundary lubrication only. Results in "feast or famine" type of lubricant supply.

(*b*) *Pistol type oil can,* actuated by trigger or plunger. Dispenses oil in measured amount according to plunger (pump) adjustment. Also permits only boundary lubrication, but is less wasteful due to control of feed.

2. Semiautomatic oil appliances:

(*a*) *Bottle oiler,* mounted on bearing cap of plain bearing, applies oil

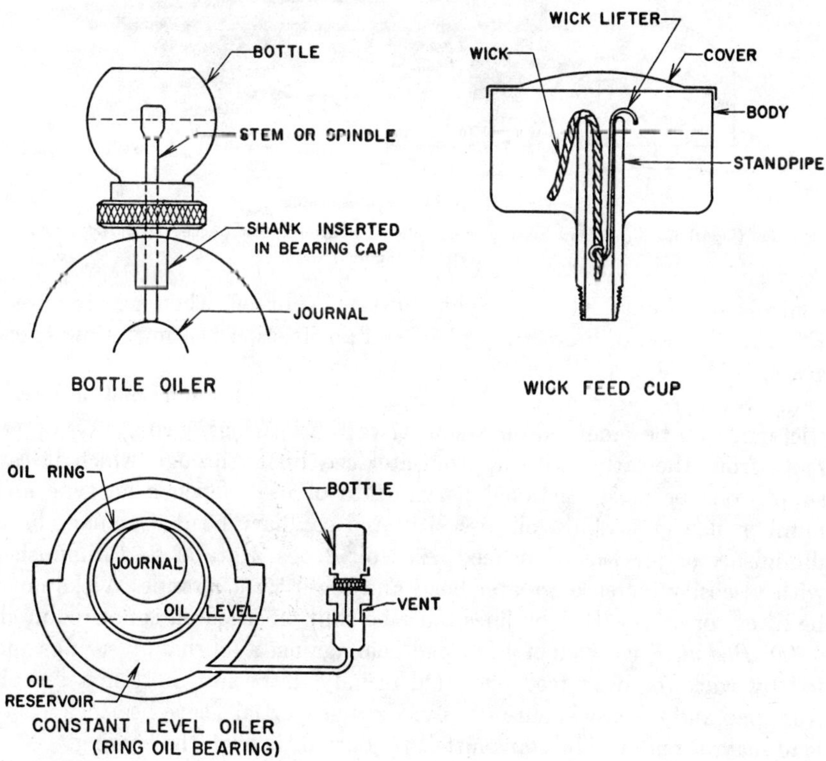

Fig. 439. Oil application devices.

direct to journal. Actuated by pumping action caused by vibration transmitted through stem or plunger from journal (on which stem rides) to oil in the bottle. Quantity of oil fed depends on journal speed and bearing temperature. One bottle per 6 in. of bearing length. Reduces consumption on all-loss system and eliminates "feast or famine" supply. Does not feed when journal is at rest.

(*b*) *Constant level oiler*, maintains oil level in main reservoir of ring oiled bearings and similar oil application methods. Actuated by drop in reservoir level to cause entering air to force more oil into reservoir to maintain given level. Assures constant, safe oil level in main reservoir.

(*c*) *Drop feed oil*, actuated by gravity or in some cases by pneumatic pressure, or bearing temperature. Oil container is mounted on bearing cap, gear case, etc., or remote from the latter with appropriate feed lines through which it can feed one or more frictional parts. Needle adjustment

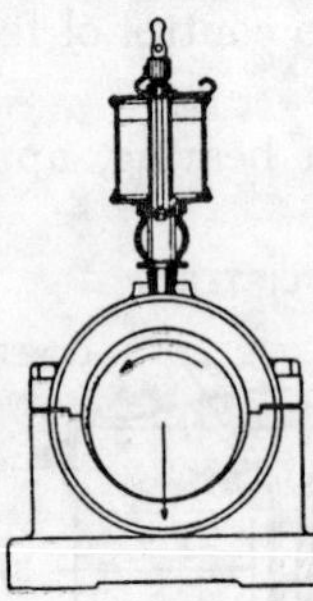

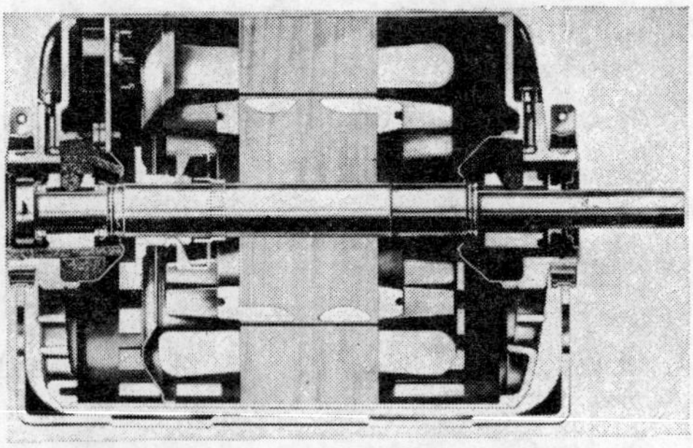

Fig. 440. (*Left*) Drop feed oiler.

Fig. 441. (*Right*) Oil lubricated electric motor sleeve bearings. Felt packed. (Courtesy, General Electric Company.)

controls oil supply. Gravity type must be shut off when not required. Efficient method of lubricating plain and antifriction bearings, slow speed gears, guides, etc.

(*d*) *Wick feed oiler*, actuated by capillary action through wool or cotton wicking. Can be mounted on casing or cap of bearings, gears, etc., or remote from the latter with appropriate feed lines, through which it can supply one or more frictional parts. Rate of feed depends on type and number of wick strands, oil viscosity, and capillary head. Capillary head diminishes as level of oil in feed reservoir drops. Rate of feed diminishes with viscosity increase, greater head and fewer wick strands. Wicks must be lifted, or crimped, or oil lines must be shut off when oil is not required.

(*e*) *Pad oiler*, consists of a felt pad held against a journal by springs and fed by wicks or drop feed oiler. Oil usually drops into bottom reservoir from pad and is reused continuously. Used on partial sleeve bearings where load may or may not be transmitted to journal through the bearing.

(*f*) *Waste packed oiler*, in which wool waste is held against journal by its own resiliency after proper packing. Oil feeds through waste by capillary action from own reservoir. Waste is soaked in oil and drained before installing. Horsehair mixed with waste imparts excellent resiliency. Oil in reservoir below waste is reused continuously. Used in electric motors and railway car journals. Glazing of waste or pads (*e*, above) should be avoided.

(*g*) *Central oiling system* (*manual*), consists of a hand pump, oil reservoir, feed lines, and measuring valves (ejectors). Measuring valves control oil feed to any number of frictional parts. Rate of feed controlled by pump adjustment and frequency of pump actuation by hand. Commonly referred to as "one shot" system of lubrication.

3. AUTOMATIC OIL APPLIANCES:

(*a*) *Oil mist* (*or fog*) *lubricator*, feeds oil to frictional parts by (1) vaporizing with compressed air and (2) condensing the oil at the point of motion of the journal, gear, etc. Vaporized oil is led from oil reservoir to one or more frictional parts through appropriate feed lines. Compressed air at about 10 to 15 psig required to vaporize oil in vaporizer, after which pressure is reduced to about 1 psig in feed line. Same oil generally used as normally would be employed by another method of application for a specific requirement. Efficient.

(*b*) *Air line lubricator*, acts on the principle of common atomizer, either by syphon or capillary action. Installed in compressed air line leading to pneumatic tools and machines. Usually coupled with an air filter. Oil feed is adjustable.

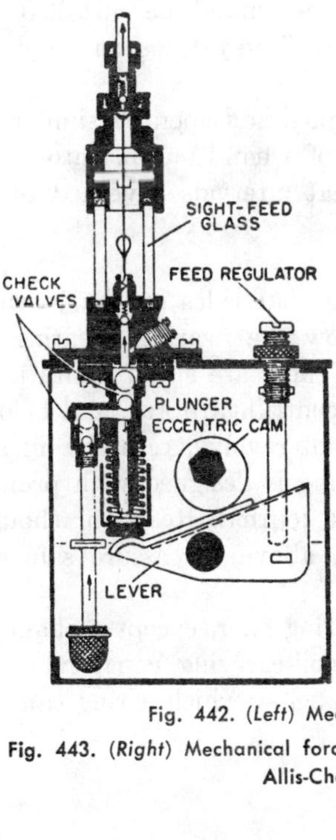

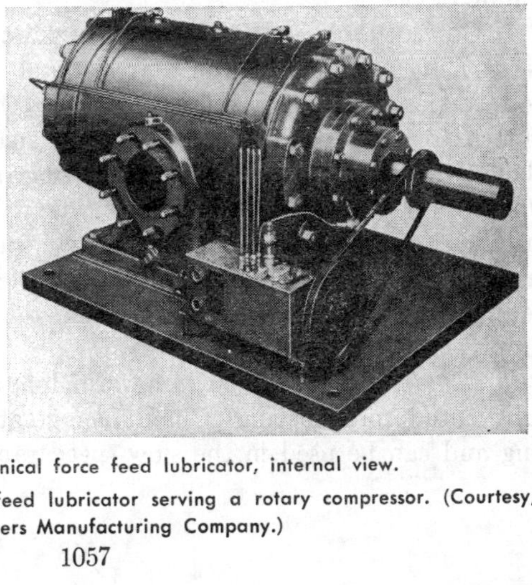

Fig. 442. (*Left*) Mechanical force feed lubricator, internal view.

Fig. 443. (*Right*) Mechanical force feed lubricator serving a rotary compressor. (Courtesy, Allis-Chalmers Manufacturing Company.)

(*c*) *Mechanical force feed lubricator*, is an adjustable oil feed pump. Usually, but not always, actuated by machine on which it is used through suitable linkage. Flexible as to number of pump plungers, all of which are driven from a common shaft through cams. Will serve one or more oil feed lines. Employed to lubricate all types of compressor and engine cylinders, as well as bearings, gears, etc. Adjustable as to rate of feed, usually in drops per minute determined by means of a sight glass in oil discharge lines.

(*d*) *Hydrostatic lubricator*, employs the principle of head differential acting on a fluid column. Used chiefly for steam cylinder lubrication by injecting oil into the high-pressure steam line leading to the steam chest. Lubricator is located at the bottom of a by-pass line, one end of which leads into the steam pipe an appropriate distance above the other end, which also opens into the same steam line. Steam enters the by-pass through the top connection, where it condenses and forms a column of water. The water column bears on the oil in the lubricator reservoir. The steam pressure bearing on the top of the water column, plus the weight of the water forming the column, adds up to a pressure slightly greater than the steam pressure at the entrance of the bottom by-pass connection. This pressure differential allows oil to be forced into the steam pipe through the bottom connection. As the oil enters the high-velocity downward flowing steam, it breaks up into fine particles (atomizes) and is carried into the cylinder. It then deposits on the valves and the cylinder walls to provide lubrication. The hydrostatic lubricator should be installed at least 6 ft from the steam chest, or any intermediate fittings, in order to obtain complete atomization.

(*e*) *Central oiling system* (automatic) is designed and operates similar to its manual counterpart (2, *g*, above). Instead of a hand pump, automatic types are motorized and operate to feed oil at intervals governed by a timing device.

4. Continuous (reuse) oiling systems:

(*a*) *Ring oiler*, consists of a free metal ring that rides on the journal and carries oil from a reservoir located below the bearing. Bearing is slotted to allow ring to ride on the journal. There are speed limitations above which oil will be thrown from ring by centrifugal force, and below which oil will drain from ring, or ring will not revolve, resulting in an insufficient supply of oil to the journal. Bearing is designed with proper grooves to distribute oil along the top of the journal. Reservoir should be equipped with constant level oilers (2, *b*, above) to assure safe oil supply.

(*b*) *Chain oiler* works on same principle as ring oiler, except a chain is substituted for the ring. Chain has greater oil-carrying capacity than ring and can be used in the slow-speed range below which a ring cannot function effectively. Chain will whip against sides of reservoir, lose contact with journal, and churn oil (foam) if subject to too high speed.

(*c*) *Fixed-collar oiler* also works on same principle of ring and chain oiler, except collar is fixed to the journal. Capable of delivering heavy oil to slow-speed journals, which ordinarily would interfere with the rotation of a chain or a ring.

(*d*) *Bath lubrication* provides lubricant to frictional parts by partially or wholly submerging the latter in a static bath (reservoir) of oil (or grease,

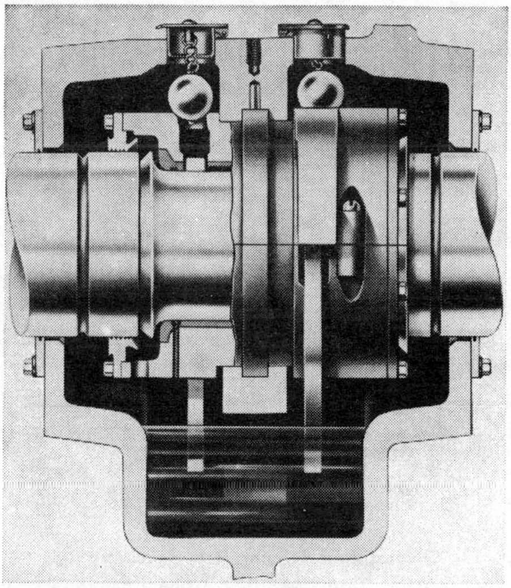

Fig. 444. Ring oiled bearing. (Courtesy, Allis-Chalmers Manufacturing Company.)

in some cases). Used to lubricate enclosed gears, chains, some types of bearings, etc. Oil level is predicated on the speeds of lubricated parts. Full or excessive submersion is usually accompanied by a rise in oil temperature and foaming, especially if speeds are appreciable.

(*e*) *Splash lubrication* is similar to that of the bath system, except that it includes the action of oil throw to other frictional parts within the same or connected housing, but remote of the main oil body. Gears, crankshafts, etc., if operated at sufficient speed, will throw or splash oil into cylinders, bearings, cross-heads, and also into troughs that distribute the lubricant to other frictional parts not accessible by splash. Scoops are often installed on the element in contact with the main oil body to increase splashing effect. This method of lubrication is often used in conjunction with a circulation system having an oil pump that distributes lubricant to other inaccessible bearings, such as piston pin bushings, connecting rod bearings, etc., of internal combustion engines. Small refrigeration compressors are lubricated by splash from their crankcases, while a combination of splash and circulation types, with the possible addition of a mechanical force-feed lubricator, to lubricate the cylinders, is employed on larger compressors.

(*f*) *Circulating oil systems* are composed of a suitable reservoir, pump, and appropriate distributing lines. They may also include an oil filter and/or a centrifuge, and an oil cooler. Sizes of circulating systems range

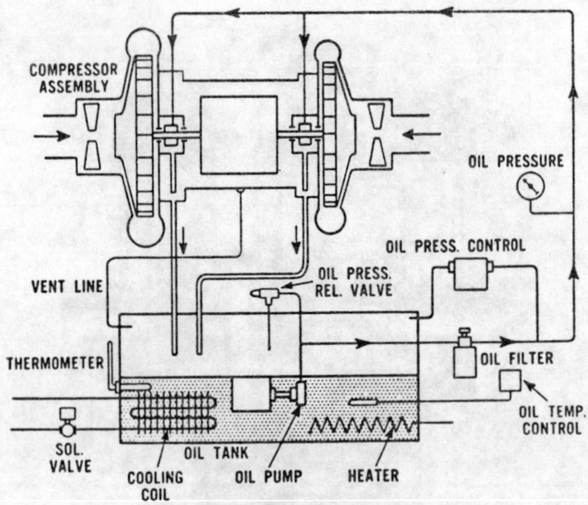

Fig. 445. Circulating oil system serving centrifugal refrigerant compressor system. (Copyright, The Trane Company. Reproduced by permission.)

from such small ones as required by machine tool spindles involving less than a pint of oil to large turbine installations handling several thousand gallons. The reservoir may be located below the frictional system, in which case the oil is pumped to the bearings under any required pressure, and returned by gravity. Or the frictional system may be fed by gravity from an oil reservoir located above, and to which the oil is returned by a pump. The circulation system is ideal for certain purposes as oil temperatures, cleanliness, and volume may all be controlled. The cycle of oil use may be shortened or extended, depending on the flow rate (pump capacity and size of conduits) and the quantity of oil contained in the system (size of reservoir).

(*g*) *Trip oilers* are stationary devices used to lubricate conveyor chains or wheels, while the latter are in motion. Oil under pressure (air or pump) is released through a nozzle directed into the friction areas by an automatic spring closing valve, which is opened when the moving part momentarily depresses, or otherwise acts on, a valve actuating lever. The oil may be released in the form of a stream or spray, depending on the pressure and nozzle design. Such devices may be modified to handle greases.

B. Grease application:

1. Hand application:

(*a*) Hand application of grease is accomplished by either *swabbing or packing*. Methods are crude, wasteful, and in many cases, as with packing, not too effective.

(*b*) *Pressure grease guns, hand operated,* are of the screw plunger, lever, or push types. Each bearing must be equipped with a fitting that fits the

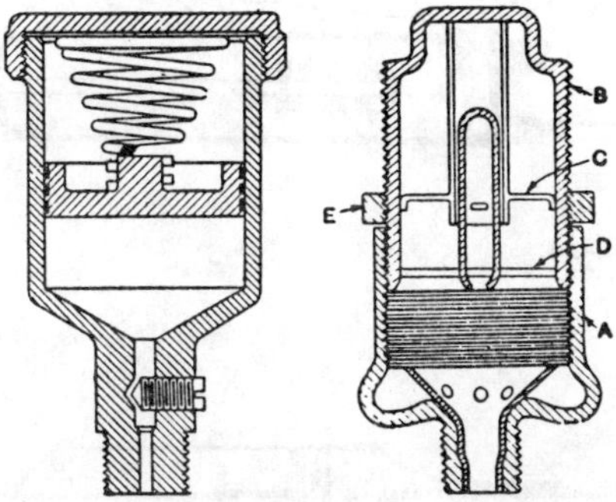

Fig. 446. Grease cups. (*Left*) Spring loaded, (*right*) compressed air type. Legend for the latter: (A) bottom grease retainer, (B) screw cap, (C) fixed disk, (D) movable disk, (E) lock nut. Air is trapped between C and D when B is turned down.

nozzle of the gun. They do not require an external source of power. Frequent filling of the gun cylinder is required where numerous bearings are serviced regularly. Push type is dangerous around moving machinery.

(*c*) *Pressure grease gun, power operated,* requires external pressure for actuation, such as compressed air or motor-driven pump. There are several types, the most convenient being one that is connected by a flexible hose with the grease container mounted on a conveyance (for mobility), or on the floor or platform where mobility is not a factor. A follower plate is installed in the container (in which the grease is shipped) that acts as a piston to force grease through the line leading to the pressure gun.

(*d*) *Screw-down* (*compression*) *grease cups* are cups fitted with a screw cap, or other arrangement, that, when turned on its thread, moves inward to force grease through an orifice in the floor of the cup. The cap must be turned slightly at periodic intervals until the grease supply is exhausted, at which time the cup is refilled. It is more positive than hand packing.

2. Semiautomatic:

(*a*) *Spring-loaded* (*compression*) *grease cups* act on the same principle as hand screw-down cups, except that a spring located in the cap assisted by a follower plate or piston maintains a continuous pressure on the grease to initiate flow as needed. This type of cup is filled by a pressure grease gun through a fitting located at the bottom. As grease is forced into the cup, it compresses the spring between the top of the cup and a follower plate or piston. For best results, the cup should be kept filled with grease of N.L.G.I. No. 2 consistency. Sometimes this type of cup is fitted with a grease level indicating stem.

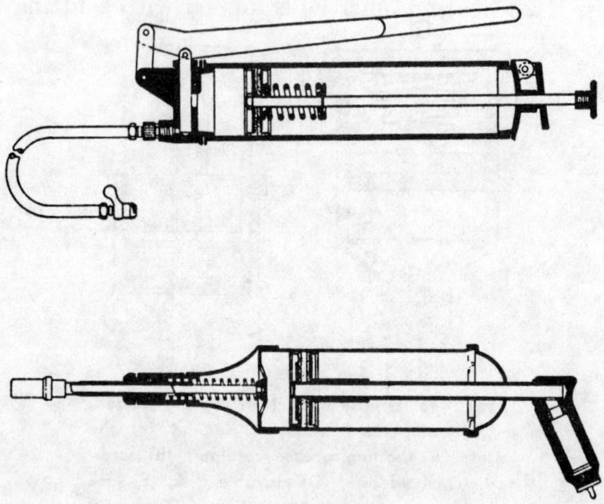

Fig. 447. Grease guns. (*Top*) Hand operated, (*bottom*) air operated.

(*b*) *Air-loaded* (*compression*) *grease cups* act on the same principle as spring-loaded cups. However, grease is forced into the bearing by air trapped in the cup when it expands during a bearing temperature rise. Positive application diminishes as air reaches maximum expansion relative to bearing temperature. However, some types are equipped with a compensating spiral to obtain more positive action as the effectiveness of either the spring (2, *a*, above), or air diminish.

3. Automatic grease systems:

Central grease systems are similar to those described for central oil systems. Higher pressures are required for greases than for oil. Also, a follower plate must be used in the container to assure the movement of the higher consistency grease to the pump suction. Central grease systems may be operated manually or automatically. If manual, a hand pump is used. A power-driven pump controlled by a timing device is employed with automatic systems. Frequency of application may be intermittent or continuous. Grease consistency is an important consideration in all systems involving relatively long lines. More frequent feeding of small

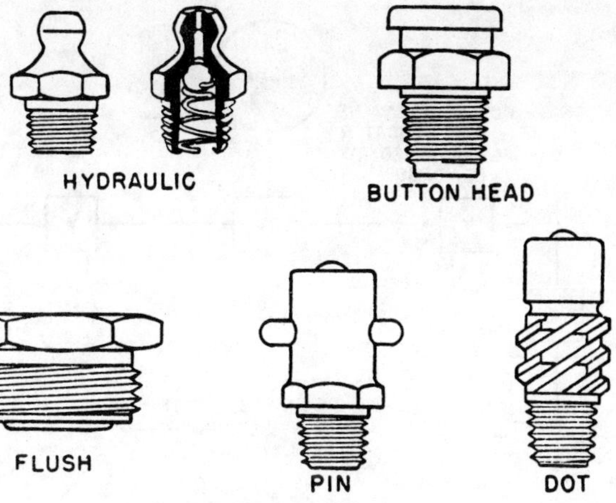

Fig. 448. Various types of pressure grease fittings.

amounts of grease is more advantageous than less frequent feeds of large amounts. Systems vary in their means of grease distribution. All are equipped with pumps, necessary feed lines, and feeder valves that control flow from main lines to branch or discharge lines. A control valve (line-selector) is used on two line systems, while a reversing valve is found on single-line, loop systems of the progressive type.

The various types of central grease systems include the following:

Two-line, nonprogressive, end-of-line system
Two-line, nonprogressive, loop system
Single-line, progressive, loop system
Single-line, nonprogressive, end-of-line system
Single-line, progressive, end-of-line system

Central grease systems should be designed and operated in consideration of the following:

(*a*) Eliminate dead-ends where grease may remain permanently static to finally harden in the line. Keep grease moving at frequent intervals.
(*b*) Bends should be designed with long sweeps, not sharp elbows.
(*c*) Avoid excessive length of run where possible.
(*d*) Insulate, or isolate feed lines from high-temperature areas.
(*e*) Use large enough tubing having sufficient strength to withstand maximum pressures.
(*f*) Avoid crimping or depressing the internal area of tubing.
(*g*) Install all parts in accessible locations for servicing.

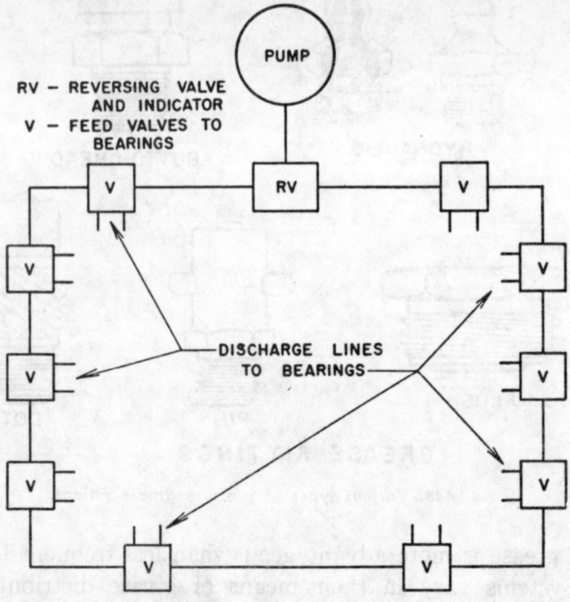

Fig. 449. Central grease system—single line, progressive, loop type.

(*h*) Use grease having good pumpability commensurate with size of tubing.

(*i*) Avoid air pockets. Introduction of grease should immediately follow a charge of oil pumped through the lines to remove air.

(*j*) Keep contaminants (dirt, chips, etc.) out of the system.

REFRIGERATION COMPRESSOR LUBRICATION

It was stated earlier in this chapter that the compressor is the only component in a refrigeration system requiring lubrication. Lubrication considerations depend on (1) the type of compressor, (2) the kind of refrigerant used, and (3) evaporator temperatures. In the case of centrif-

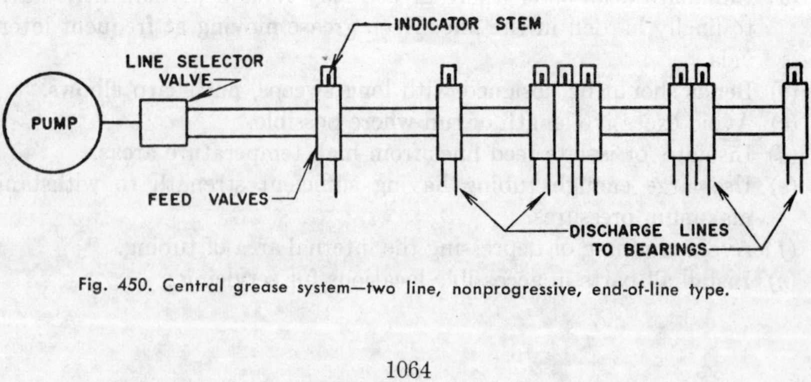

Fig. 450. Central grease system—two line, nonprogressive, end-of-line type.

ugal compressors, in which there are no internal components to lubricate, (2) and (3) are not important factors.

The frictional components of various type compressors are:

Reciprocating Types:	
Enclosed, vertical	Bearings:
	Main
	Crankshaft
	Connecting rod
	Piston pin bushings
	Cylinders:
	Walls
	Piston
	Piston rings
	Valves
	Shaft seals (some types)
Open, horizontal	Bearings:
	Main
	Crankshaft
	Cross-head guide
	Cylinders:
	Walls
	Piston
	Piston rings
	Valves
	Glands
Rotary Types:	
Vanes (or blades) in rotor	Bearings:
	Rotor shaft
	Cylinder:
	Walls
	Rotor:
	Blade slot surfaces
	Vanes, sliding areas within slots
Vanes in cylinder	Bearings:
	Rotor shaft
	Cylinder:
	Walls
	Rotor
	Vane slot surfaces
	Vanes, sliding area within slot
Lobe	Connecting (timing) gears
	Lobe shaft bearings
Screw	All parts
Centrifugal Types	Bearings
	Gears
	Shaft seal

All parts, including cylinders, of *small enclosed reciprocating compressors* are splash lubricated. The oil located in the crankcase is thrown to the various frictional parts by the rotating action of the crankshaft and the connecting rods as they dip into the main oil body. *Large enclosed recipro-*

cating compressors require a pump, usually located in the crankcase to circulate oil to all bearings and also to the shaft seal. The oil is pumped through appropriate tubing and holes are drilled through the crankshaft and the connecting rods to the main connecting rod and piston pin bearings. After performing its function, the oil drains back to the main body in the crankcase to be reused. Where the cylinders are so remote from the crankcase as to preclude their proper lubrication by splash, mechanical force feed lubricators are used. The latter feed oil through tubing leading to different points, in the same plane usually, around the cylinder. The number of oil entrances into the cylinder is predicated on proper distribution and varies with the size of the compressor.

Bearings and cross-heads of *horizontal reciprocating compressors* are usually splash lubricated. Splash-fed troughs carry oil from the crankcase to lubricate the cross-head. Cylinders of horizontal compressors invariably are lubricated by means of mechanical force feed lubricators, which also feed the piston rod packing gland.

Valves of all reciprocating compressors are lubricated with oil carried over by the compressed gas.

Rotary compressor cylinders are lubricated by mechanical force feed lubricators, as are the bearings and seals of some types. One type employs the oil drawn into the cylinder from the bearings to lubricate the vanes. To assure lubrication of the vanes during periods of starting, drop feed oilers are installed on the latter types. After starting, the drop feed oilers are shut off. Mechanical force-feed lubricators are also used to lubricate compressors, as shown in Fig. 451.

An oil circulation system is employed for the lubrication of *centrifugal compressor* bearings and to augment the shaft seal. Oil is fed to the bearings and the shaft seals under pressure developed by a gear pump usually driven from the impeller shaft. The bearings carrying the radial load are of the liner or sleeve type, while the axial load, caused by differences of pressure on the impellers, is taken up by a thrust bearing of the Kingsbury type. The thrust bearing consists of a collar that rotates with the shaft and a stationary base ring on which are mounted pivoted segments that tilt to allow the development of a hydrodynamic wedge with the rotating flat surface on which it bears. The thrust bearing is located on the suction end of the compressor outside the refrigerant gas zone. The radial bearings located on both sides of the impeller are also outside the gas zone. The oil pressure maintained in the bearings and seals is above that of the atmosphere, thus excluding the entrance of air into the compressor along the shaft. Oil slingers or collars fixed around the shaft deflect by centrifugal force any oil that tends to seep along the shaft, returning it to the reservoir located below the bearing on the suction end of the compressor. The lubricating oil of centrifugal compressors is usually water-cooled by a shell-and-tube heat exchanger located externally to the oil reservoir.

Because the oil circulates outside the refrigerant gas zone, it is not affected by the refrigerant. Any oil that may seep into the system is removed by an oil separator (see Chapter 8). Because normally the oil does not come in contact with the refrigerant, and, therefore, is not subjected to (1) the diluting action of that fluid or (2) heat of compression, lubricant

Fig. 451. Rotary compressor cylinder and bearings, lubricated with oil fed by mechanical force feed lubricator. Note seal and roller bearings. (Courtesy, Allis-Chalmers Manufacturing Company.)

selection for centrifugal machines depends chiefly on mechanical considerations.

The main oil pump being driven from the impeller shaft does not develop sufficient pressure to maintain safe operating conditions during periods of starting and stopping. To assure protection of bearings and maintain proper sealing at such periods, a motor-driven auxiliary pump is used that supplies oil at sufficient pressure and volume even before starting and after stopping. Controlled by means of a pressure switch, in conjunction with the main oil pump pressure, the auxiliary pump stops operating when the shaft reaches sufficient speed after starting. At this point, the oil pressure developed by the main pump driven from the shaft becomes greater than that of the constant-speed auxiliary pump. The same switch causes the auxiliary pump to start when the shaft speed is reduced to a predetermined level when stopping and causing a drop in main oil pump pressure. The auxiliary oil pump is put in operation manually before the com-

Fig. 452. Type VC reciprocating compressor showing oil reservoir. (Courtesy, Worthington Air Conditioning Company.)

pressor is started and continues in operation after the shaft comes to rest upon shut-down. After shut-down, the auxiliary pump continues to operate until automatically stopped by an adjustable timer. The importance of maintaining an oil film continuously to bearings of all types of high-speed machines, including centrifugal compressors, turbines, blowers, etc., cannot be overly emphasized. Any metal-to-metal contact between journal and bearing in such equipment, even for a fraction of a second, may mean seizure of frictional surfaces and subsequent bearing failure.

When not direct-connected, the speed change gears are lubricated from the same circulating system supplying oil to the bearings and seals. This is also true of other equipment, such as steam turbines where a common reservoir serves turbine and generator bearings, gears, and governor system. Where gears are included in the lubrication system, a higher viscosity oil than that required for direct-connected systems is used.

Screw compressors of the oil injection or oil flooded type use a unique method of supplying lubricant to their helically ground rotors. The driver (normally) or male rotor has lobes, while the driven or female rotor is

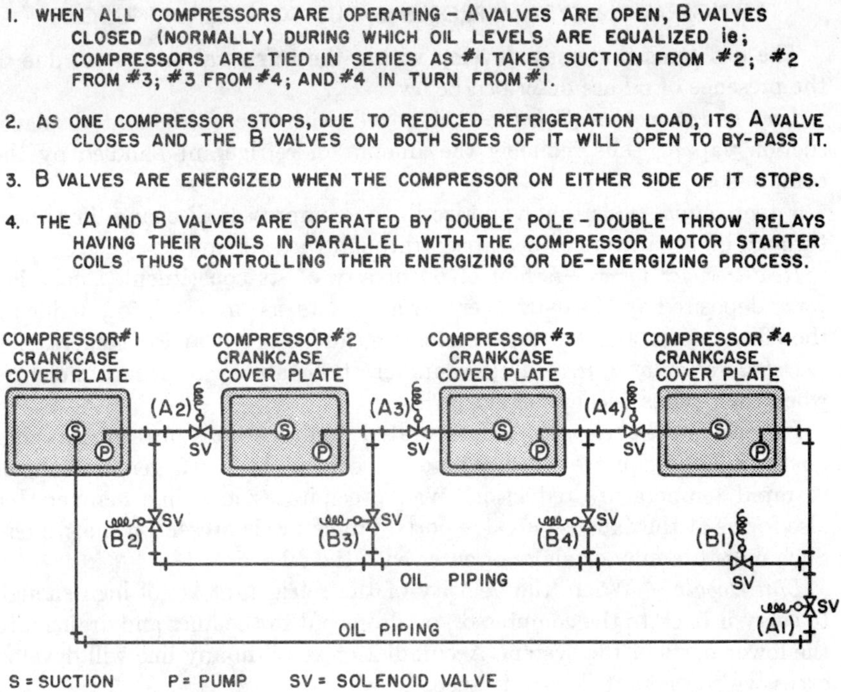

Fig. 453. Lubricating oil piping arrangement for maintaining proper level in multiple compressors in parallel.

designed with interlobes or gullies that mesh and develop metal-to-metal contact. The lubricating oil is injected into the rotor area beyond the suction zone through the slide valve cusp. The oil is dispersed by the high speed rotary motion of the rotors to lubricate the lobes and interlobes. In passing with the refrigerant gas through the enclosed casing, which also houses the electric motor that drives the rotors, the oil acts to substantially cool the motor windings. (See Chapter 8.)

REFRIGERATING COMPRESSOR LUBRICATING OILS

In the study of refrigerants (Chapter 5), the subject of oil-refrigerant relationship was discussed. The necessity for an oil to lubricate the moving parts of the compressor poses a number of problems that interfere with the operation and the thermal processes of the refrigeration system. Many of such problems exist in the lines and the equipment beyond the compressor. Some of them exist within the compressor itself. Such problems may develop because of improper oil selection. They often exist because of the relation of some refrigerants to oils. System design is also responsible for the existence (or absence) of a number of oil problems.

REFRIGERATION SYSTEM PROBLEMS DUE TO LUBRICATING OILS

The problems that may develop within the refrigeration system due to the presence of oil are described below:

Loss of volumetric efficiency—Oil mixed with the refrigerant displaces suction vapors, thus reducing the amount of refrigerant handled by the compressor.

Vaporization—Oils that vaporize in the compressor discharge lines pass through oil separators to be carried through the system.

Heat-transfer losses—A film of oil or any of its constituents, including wax, deposited on heat-transfer surfaces acts as an insulator, reducing the efficiency of evaporators, condensers, and other heat exchangers.

Deterioration of dryers—Desiccants lose their ability to absorb moisture when their pores become oil clogged.

Expansion-valve clogging—Unsaturated hydrocarbons present in poor quality oils precipitate when exposed to solvent type refrigerants subject to rapid temperature reduction. Wax precipitates in a like manner. Ice also forms at the expansion valve ports if moisture is present in the system. Such moisture may or may not enter with the oil.

Line clogging—Where the velocity of the refrigerant is not high enough to carry it back to the compressor, oil drops out in the lines and drains into the lower parts of the system. Accumulation of oil in any line will develop excessive refrigerant flow resistance.

Foaming—Foaming occurs in the crankcase when refrigerant absorbed in the oil is suddenly released with a rapid reduction of pressure caused when the compressor is started. Foam may reach such proportions as to allow oil carry-over to the discharge side of the system. Carried to extremes, loss of oil from the crankcase due to foaming may be great enough to leave splash-lubricated bearings without sufficient frictional protection.

Wax formation—Precipitation of wax in low-temperature systems results in the use of oils possessing relatively high pour or floc points. In this connection, pour point is an important factor when employing immiscible refrigerants, whereas floc point is a better wax-forming criterion for systems using miscible refrigerants.

Carbon deposits—Carbonaceous deposits that form on discharge valves, cylinder heads, and pistons and in discharge lines are caused by oil deterioration under high temperature. The condition is aggravated by the use of oils of poor chemical stability under conditions of high discharge temperatures and pressures. Discharge temperatures are reduced substantially by cooling the head of air-cooled compressors with liquid refrigerant. Jacket water cooling of the upper cylinder region is another method. Stage compression not only reduces discharge temperatures but also is more efficient both mechanically and volumetrically. Any reduction in discharge temperatures, accompanied by the use of a relatively lower viscosity oil of good quality, will reduce or eliminate the formation of carbon deposits.

Cylinder blow-by—Loss of compression is often due to the escape of gas back to the crankcase between the piston and the cylinder walls (blow-by) during the compression stroke. In well-fitted compressor cylinders, blow-by is caused by the failure of the lubricating oil to seal the space between the piston rings and their abutting surfaces, usually the cylinder wall. The reason for the lack of sealing may be caused by insufficient or improper lubricant. The insufficiency may be due to the method of lubricant application or to incorrect feeds. Low oil level in the crankcase, cylinders too remote from a splash oil reservoir, or improper adjustment of a mechanical force-feed lubricator are some of the causes of insufficient oil supply to the cylinders. An improper lubricant in this case is one having a viscosity too low for the operating condition. In the case of immiscible refrigerant systems, a miscalculation of mechanical requirements, including size of compressor, speeds, and pressures may lead to the selection of a lubricant not of sufficient viscosity. For systems employing miscible refrigerants, not only mechanical miscalculations are possible, but also the thinning effect on the oil due to dilution may not be estimated properly. While miscibility does decrease the viscosity of the lubricating oil, it has the advantage of appreciably reducing the temperature at which wax precipitation occurs, as indicated by the floc or cloud point.

In consideration of the above factors, lubricating oils selected for refrigeration compressors must possess the following qualifications:

1. *Chemical stability*—to resist oxidation at discharge temperatures. Free of unsaturated hydrocarbons to prevent precipitation at expansion valves.
2. *Low pour point*—having its wax content reduced sufficiently to meet the low-temperature conditions of the system.
3. *Low floc point*—type of wax and wax content should be such as to preclude excessive precipitation when diluted with refrigerant. May also indicate low moisture content.
4. *High dielectric strength*—especially for hermetically sealed units where moisture cannot be removed otherwise.
5. *Proper viscosity*—to meet the mechanical conditions of the compressor after considering the immiscibility or the miscibility of the refrigerant used.
6. *Low volatility*—to reduce oil vaporization at discharge temperatures and pressures.

OIL-REFRIGERANT MISCIBILITY

All refrigerants are miscible with oil to some extent, although some are more miscible than others. Miscibility, then, is a matter of degree of solubility with which two fluids mix together to form a phase. Refrigerants that are slightly miscible or possess a negligible miscible relationship with oil are considered immiscible for practical lubrication purposes. Some refrigerants are miscible with oil in all proportions and remain so throughout the operating range of refrigeration system temperatures. Others are completely miscible above a certain temperature, but become less so or partially miscible at temperatures below the critical point.

The oil miscibility relationships of the more important refrigerants are shown in Table 153.

TABLE 153. OIL-REFRIGERANT MISCIBILITY

Miscible in All Proportions at Refrigeration System Temperature Range	*Miscible at High Temperature. Limited Miscibility at Low Temperature*	*Slight or Immiscible*
R-11	Carrene-7	Ammonia
R-12	R-22	Carbon dioxide
R-21	R-114	R-13
R-113		R-14
Methane, ethane		Sulfur dioxide
Methyl chloride		
Propane, butane		

Because ammonia, carbon dioxide, sulfur dioxide, R-13, and R-14 are not miscible, they do not cause any viscosity change in lubricating oil. R-11, R-12, R-21, R-113, and methyl chloride are miscible, and, therefore, cause a decrease in lubricating oil viscosity. R-22 and R-114 are miscible under condenser pressure, but will separate from the oil when normal evaporator conditions are reached. Under evaporator pressures, the oil will rise and form a blanket over the surface of R-22. An oil separator in the discharge line is necessary to remove oil from R-22 before it reaches the evaporator.

Figure 454 shows the rate of viscosity change in an oil when mixed with various amounts of R-12.

Lubricant recommendations for reciprocating compressors follow a basic pattern with some modification necessary for low and ultra-low temperature operation to meet pour point requirements. For evaporator temperatures of 0 F and above, most reciprocating compressors handling ammonia, carbon dioxide, Freons, and methyl chloride can be lubricated properly with an oil having a viscosity of 300 sec SU at 100 F. When handling ammonia, the majority of the latter type compressors perform safely and efficiently with oils having a viscosity of 150 to 200 sec SU at 100 F. Some large open slow-speed ammonia compressors appear to improve cylinder wearing qualities when a 400 sec SU at 100 F oil is employed.

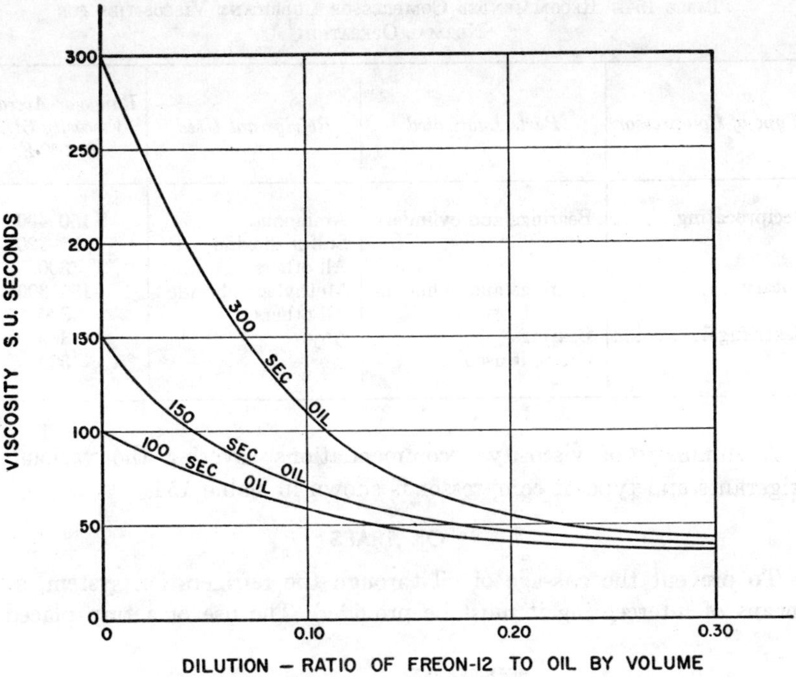

Fig. 454. Rate of viscosity change with various R-12 and oil dilutions.

Because of the dilution factor, compressors handling miscible refrigerants, such as Freons and methyl chloride, invariably require a 300 sec SU at 100 F oil, preferably of water white to extra lemon pale in color. For low-temperature operation, pour point considerations often force the use of oil having viscosities below 100 sec SU at 100 F. Such oils, however, while they have a pour point of −50 F or below, have a low flash point—a serious disadvantage, especially with the high discharge temperatures of single-stage compression.

The importance of pour point is often overemphasized. Where efficient oil separation is carried out through the use of separators, oil stills, and periodic blow-off, very little oil reaches the expansion valve. That which does is carried through the low pressure side by the refrigerant providing sufficient velocity is maintained.

The bearings of centrifugal compressors are lubricated with a 300 sec SU at 100 F oil. The oil is usually circulated and water cooled to maintain a viscosity of at least 80 sec SU at the operating temperature. The same oil lubricates the speed increaser or reducer if used.

A 300 sec SU at 100 F oil is normally recommended for rotary compressors handling any of the commonly used refrigerants.

TABLE 154. RECOMMENDED COMPRESSOR LUBRICANT VISCOSITIES FOR NORMAL OPERATION

Type of Compressor	*Parts Lubricated*	*Refrigerant Used*	*Range or Average Viscosity SUS at 100 F*
Reciprocating......	Bearings and cylinders	Ammonia	150–400
		Sulfur dioxide	75–220
		All others	300
Rotary............	Bearings and cylinders	Methylene choride	150–300
		All others	300
Centrifugal........	Bearings	Any	300
	Gears, if used		300

A summary of viscosity recommendations covering the various refrigerants and type of compressor is shown in Table 154.

OIL TRAPS

To prevent the passage of oil through the refrigeration system, some means of intercepting it must be provided. The use of a trap placed in

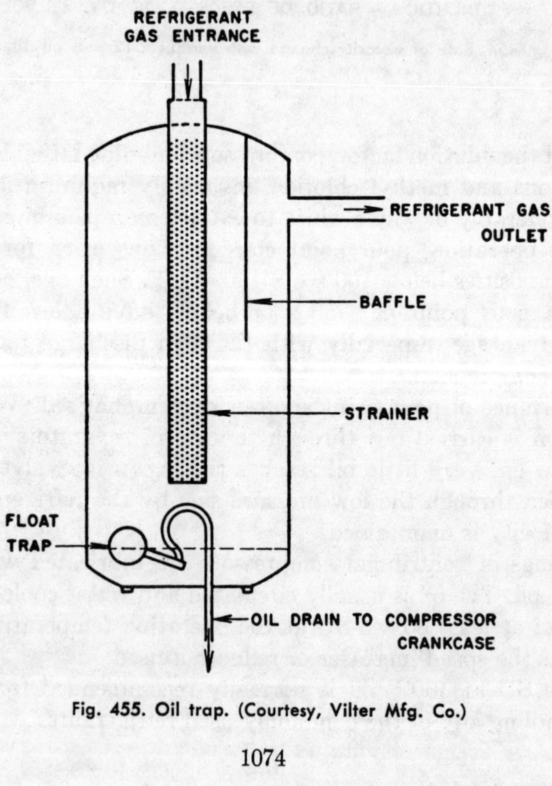

Fig. 455. Oil trap. (Courtesy, Vilter Mfg. Co.)

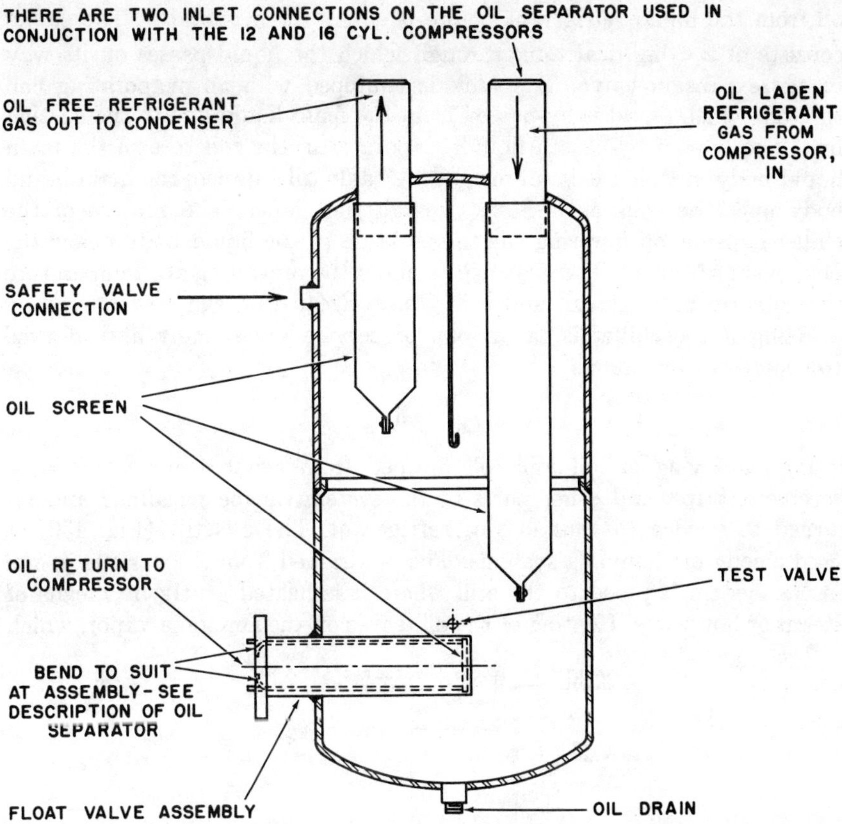

Fig. 455(b). Oil separator. (Courtesy, Vilter Manufacturing Corporation.)

the compressor discharge line is one method of removing oil from the hot gas. The gas and oil mixture enters the trap at the top and flows downward through a screen, as shown in Fig. 455. The gas must pass around a baffle before leaving through the trap outlet also located at the top. In doing so, it must perform an abrupt U-turn, during which action the oil is thrown from the gas by centrifugal force. The oil drains to the bottom of the trap from which it is returned to the crankcase through a float valve. This type of trap is effective for ammonia and R-22. It should be kept warmer than the condensing temperature of the refrigerant, otherwise liquid will return to the crankcase with the oil. Placed in the discharge line of miscible refrigerant systems, the trap catches the oil before it goes into solution with the liquid from which it cannot be separated.

OIL CHILLERS

To remove oil from low-temperature immiscible refrigerant systems, such as one employing ammonia, a chiller can be used. A chiller removes oil from the liquid refrigerant before it enters the evaporator. The chiller consists of a cylindrical tank through which the liquid passes on its way to the expansion valve. The tank is equipped with an evaporating coil through which liquid is by-passed from the main liquid line to the suction line. The flow of by-passed liquid evaporates in the coil to cool the main liquid body in which it is submerged. A baffle tube directs the main liquid body over the coils as it flows through the chiller. The success of the chiller depends on lowering the temperature of the liquid body below the pour point of the oil. The oil congeals under the prevailing low temperature and adheres to the baffle and coil. The congealed oil can be removed by draining if the chiller is taken out of service occasionally and allowed to become warm enough.

OIL STILLS

By employing an oil still, oil drained from evaporators, condensers, receivers, traps, and other parts of the system can be reclaimed and returned to service without loss of refrigerant. The oil still (Fig. 456) is used chiefly on ammonia systems. The oil drained from the various points in the system is piped to the still where it is heated gently by means of steam or hot water. Heating of the oil drives off the ammonia vapor, which

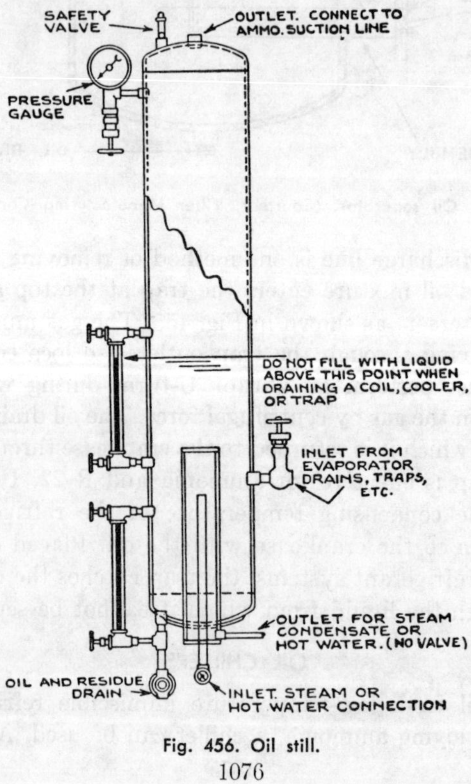

Fig. 456. Oil still.

is returned to the compressor suction through a line that connects it to the top of the still. The oil that remains in the bottom of the still can be drained, filtered, and returned to service.

FOAM BREAKERS

Compressors equipped with a suction manifold through which suction gas is drawn into the compression chamber through a suction valve in the head have passages between suction manifold and crankcase for the return of oil during operation. During shutdown, however, Freon drains into the crankcase and goes in solution with the oil. Upon starting, the rapid release of pressure causes foaming and the oil is carried over into the suction manifold with the gas. To prevent oil carry-over, a foam breaker is installed in vents leading to the suction manifold through the crank-

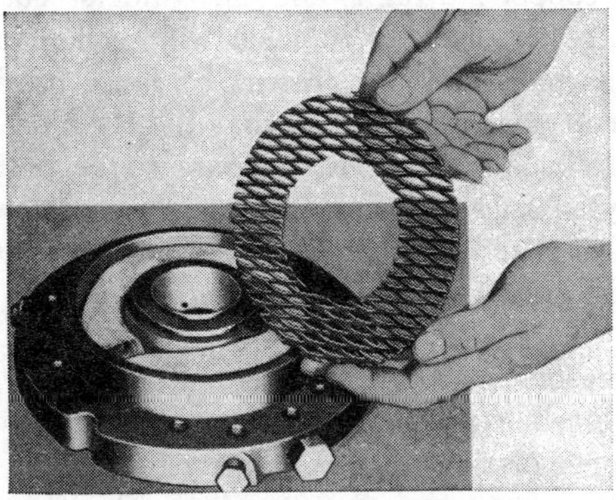

Fig. 457. Foam breaker. (Copyright, The Trane Company. Reproduced by permission.)

case housing. The pressure differential developed during foaming is equalized in an area provided by a rolled screen placed in the foam breaker cavity (Fig. 457). Gas flow becomes distributed over the entire inlet area of foam breaker, resulting in low velocity flow. The expanded metal shield (held by hand in the picture), assisted by the first layers of rolled screen, breaks the foam bubbles, and the oil adheres to the screen and shield surfaces. Because of the low gas velocity through the foam breaker pressure equalizing area, the oil drains back to the crankcase, while the gas continues on through the vent to the suction manifold. Check valves prevent flow from crankcase to manifold through any other oil return openings when the pressure in the crankcase is above that in the suction manifold.

COPPER PLATING

Under some internal operating conditions within refrigeration systems employing halogenated refrigerants, highly polished metal parts of the compressor are found to be plated with copper. Parts so plated include cylinder walls, pistons, valves, and bearings. Seals and other parts have also been found to be plated with copper. Plating is most common on parts that develop heat, such as frictional surfaces and discharge valves. If allowed to accumulate, thickness of plate may become of such magnitude as to cause binding of moving parts and cause compressor failure. The copper may emanate from tubing, motor windings (in sealed rotary compressor systems), or any other part of the system made of or containing this metal.

Copper is not used in ammonia systems so that plating is not a problem where that refrigerant is employed. Plating does not occur in sulfur dioxide systems, although copper is present. Halogenated refrigerant systems, such as those employing Freons, Genetrons and methyl chloride, are affected.

The reason for copper plating is not fully understood. Oil does dissolve copper, but tests prove that the solution will remain intact unless other influences, such as moisture, are present to cause precipitation. Copper plating will occur on glass surfaces so that the theory of precipitation by straight electromotive displacement appears to be unfounded. Theories involving hydrolysis, salts resulting from acid reaction with copper, ironing out of particles resulting from wear, stray electrical currents causing electroplating, and others have all been considered and more or less disproved.

Some day the reasons for copper plating may be found. In the meantime, it is fairly well established that systems containing moisture are more susceptible to copper plating than are dry systems. It is also evident that the problem is less serious in systems employing high quality oils than are those in which poor quality oils are used.

The answer to the problem of copper plating, then, appears to be one of maintaining a moisture-free system and the use of only highly refined lubricating oils.

GEAR LUBRICANTS

The lubricant recommendations for enclosed gears usually are made in accordance with the standards devised by the American Gear Manufacturing Association (A.G.M.A.) (see Table 155).

In general, the oils classified according to A.G.M.A. numbers meet the gear requirements as shown in Table 156.

Gear failure occurs when the teeth are broken for some reason or are worn so badly that they can no longer transmit power. In many instances, gears become noisy before failure occurs. Abrupt change in load or shock will cause excessively worn teeth to break or shear. Some of the difficulties encountered with gear teeth that may be caused by abrasives, overloading, surface roughness, metal fatigue, or improper lubrication are given below:

Pitting. Caused by repeated stresses on high spots of relatively rough tooth surfaces. Small particles broken out of the metal leave pits as the

TABLE 155. A.G.M.A. GEAR OIL CLASSIFICATION

A.G.M.A. No.	*Viscosity Range SSU*	
	100 F	210 F
1	180–240	
2	280–360	
3	490–700	
4	700–1000	
5		80–105
6		105–125
7		125–150
8		150–190

TABLE 156. GENERAL GEAR LUBRICANT RECOMMENDATIONS ACCORDING TO VISCOSITY REPRESENTED BY A.G.M.A. CLASSIFICATION NUMBERS*

Unit	*Ambient Temperature Range*		
	0–40 F	41–100 F	101–150 F
Gear motors (all sizes)	2	4	5
Parallel shafts (600–1800 rpm):			
Up to 20-in. centers	2 or 3†	4	5
Over 20-in. centers	3 or 4†	5	6
Bevel gears (spiral or straight):			
Cone distance up to 12 in	2 or 3†	4	5
Cone distance over 12 in	3 or 4†	5	6
Planetary gear units:			
OD of housing up to 16 in	2 or 3†	3	4
OD of housing over 16 in	3 or 4†	4	5
High-speed gears (above 1800 rpm)	1	2	3
Worm gears:			
Up to 600 rpm	3 or 4†	8	8
Over 600 rpm to 1800 rpm	3	7	8

* Compiled by the author; does not necessarily conform to other published tables.

† Heavier oil preferable if pour point is sufficiently low.

areas comprising the high spots become affected by uneven load distribution. The area most affected by pitting is that in close proximity to the pitch line.

Scratching. Broken tooth surface represented by a straight line running radially from root to tip. Usually caused by abrasion of a high spot on one tooth sliding across the surface of its meshing teeth.

Abrasion. Worn or scratched tooth surfaces caused by the presence of abrasive materials.

Galling. Usually starts at the tip of the tooth and proceeds progressively toward the root. Galling is the result of one tooth tearing metal from its meshing tooth, or from each other. Probably caused by metal-to-metal contact under heavy pressure accompanied by welding action.

Scuffing. Galling in milder form.

Spalling. Removal of large metal chips from the tooth, usually in or below the pitch line area. Due to metal fatigue caused by overloading or the inability of the metal to carry its designed load continuously.

Scoring. Similar to galling.

Feather-Edging. Caused by the squeezing action of one tooth against another during which action galled metal is dragged between them. The subsequent "ironing" effect causes the formation of a thin metal edge to project beyond the tip of the tooth, giving a featherlike appearance.

Broken Teeth. Entrance of large metal pieces (nuts, bolts, etc.), metal failure, excessive wear, and heavy shock loads are some of the reasons for broken teeth.

Gunther lubrication system. The writer has devised an effective system for the lubrication of frictional components, such as high speed gears, high temperature bearings, etc., by the use of an appropriate liquid refrigerant in conjunction with a lubricant selected for the operating conditions. It was originally designed to protect the teeth of gears having operating speeds of the magnitude of 100,000 rpm and over. At such high speeds, high fluid friction and rapid attritional wear occur when lubrication is attempted by any of the conventional methods of application, including that of the oil bath or splash systems. By injecting the proper lubricant into the gear housing by a high pressure liquid refrigerant, say, R-12, through an adjustable three-way valve, an oil mist develops that deposits on the moving surfaces to supply adequate lubrication. Simultaneously, the liquid refrigerant entering a lower pressure environment flashes into a vapor, causing a refrigerating effect, resulting in the dissipation of frictional heat generated by the meshing gears. At high speeds, the direction of the injected fluids should be in the same direction of rotation as the gear to which it is aimed. This will reduce the collision effect of the fluid stream into the faces of the gear teeth. The resulting refrigerant gas can be allowed to escape into the atmosphere through appropriate venting or reclaimed by conventional refrigeration system methods after it is properly filtered.

The amount of refrigerant required will depend on the power transmitted through the gears and the heat conduction capability of the gear housing.

SUMMARY OF CHAPTER 15

The subject of lubrication may be summarized best by listing typical recommendations for equipment common to an industrial manufacturing plant, and citing pertinent notes for maximum performance or general information. While each mechanical requirement portrays optimum performance under constant operating conditions with a lubricant having specific qualifications, including viscosity, oiliness characteristics, etc., the selection of such a lubricant is not always feasible for two important reasons. First, operating conditions, such as speed, temperatures, loads, etc., are not always constant, so that a lubricant that gives optimum results one moment may not do so continuously. Secondly, if lubricants were selected on the basis of maximum results for each individual machine, the number of different lubricants required would reach impractical, or at least ungainly, proportions. Therefore, plant equipment lubrication becomes a fair compromise; i.e., a compromise that will give maximum over-all results without detriment to any individual frictional part.

The lubricants recommended in the following chart are for general operating conditions, based on ambient temperatures of 40 to 100 F, and normal loads for the equipment listed. Where conditions vary one way or the other from normal, appropriate changes must be made in lubricant selection, depending on the effect of such variations on viscosity, pour point, flash points, etc. Design conditions will also alter the lubricant characteristics, in which case manufacturer's recommendations should be followed.

The lubricant recommendations listed below are set down in the following order: (1) *Name of major equipment*, (2) *Part to be lubricated*, (3) *Method of application*, (4) *Interval of application or oil change*, (5) *Type of lubricant*, (6) ***Rules where applicable.***

STEAM TURBINE, direct connected to generator, 3600 rpm

Turbine and generator bearings—oil circulation—indefinitely

Rust and oxidation inhibited oil—150 SUS at 100 F

Rules:—A clean, well-maintained system will enable a high-quality oil to give excellent service for years. Keep oil free of water and other contaminants. Use a centrifuge, filter, and reservoir screen. Reservoir to be equipped with a vapor extractor. Bottom of reservoir to be lower at one end and equipped with water-leg. Average oil temperature from bearings 130 F

PUMPS, miscellaneous, centrifugal (1800 to 3600 rpm)

Pump bearings, antifriction—Grease, pressure fittings—Quarterly

No. 2 soda-base grease

Glands—Grease, screw-down cup—Keep filled

No. 3 lime-base grease

Rules:—Antifriction bearings should never be lubricated more often than to keep the cavity about one third to one half filled with a proper grease. Otherwise heating will occur to rupture seals because of subsequent grease expansion. This rule applies to all antifriction bearings

STEAM-DRIVEN PUMP

Steam cylinder—oil-mechanical force-feed lubricator—Keep filled

Compounded cylinder oil—2500 SUS at 100 F

Linkage—hand oiled or device—Daily

Oil (no additive required usually)—300 SUS at 100 F

Open gears

Gear and pinion—Lubricant—swab—Weekly

Residuum, with or without antiwear material. Viscosity depends on ambient temperature and size

Steam turbine, geared, driving centrifugal pump

Bearings } oil—circulation—Indefinitely
Gears }

Rust and oxidation inhibited oil—300 SUS at 100 F

Note: Recommendation predicated on the use of an oil cooler

Electric motors, 1750 rpm

Bearings—Grease, pressure fittings—Quarterly

No. 2 soda base grease

Bearings—Oil, reservoir (ring, wick)—Keep filled to proper level

Rust and oxidation inhibited oil—300 SUS at 100 F

Rules:—Ring oil reservoir of large electric motors should be equipped with constant level oilers. Reservoirs should be inspected regularly and the oil changed semi-annually after thorough cleaning

Air compressors, horizontal (or vertical), average pressure 100 psi

Bearings—Oil—bath and splash—Semi-annually

Cylinders—Oil—mechanical force-feed—Keep filled lubricators

Rust and oxidation inhibited oil—300 SUS at 100 F

Rules:—Inspect discharge valves about every three months. Keep oil feeds to cylinders at required minimum for reduced carbon formation. Suggested rate of feed, to be used as a guide only

$$\text{Drops per minute} = RS \times 0.00128$$

where RS is the rubbing surface swept by piston in sq ft per min. Use an oil separator on discharge line and keep receiver free of oil

Speed reducers, herringbone, single reduction

Gears—Oil—bath } Annually
Bearings—Oil—splash }

Rust and oxidation inhibited oil or heavy pressure (antiwear) lubricating oil—600 to 750 SUS at 100 F

Speed reducers, herringbone, double reduction

Gears—Oil—bath } Annually
Bearings—Oil—splash }

Rust and oxidation inhibited oil, or heavy pressure (antiwear) lubricating oil—750 to 1000 SUS at 100 F

Flexible couplings, gear, chain, spring, and other lubricated types

Friction parts—Grease—pressure fitting—6 to 18 months

No. 2 or No. 3 aluminum base grease—(adhesive additive)

Rule:—Couplings should be inspected every six months

A substitute for grease in sound couplings is a cylinder oil of about 2500 SUS at 100 F

Gear motors

Gears—Oil—bath—Annually

Rust and oxidation inhibited oil—750 to 1000 SUS at 100 F

ENCLOSED CHAINS, silent or roller
Chain and sprocket—Oil—bath—Annually
Rust and oxidation inhibited oil—300 to 600 SUS at 100 F (depending on speed)

WORM GEARS
Gear, worm, and bearings—Oil—bath—Annually
Antifriction (compounded steam cylinder oil), or antiwear type lubricating oil—2500 SUS at 100 F
Note: Worm gears are subject to about 90 F rise in operating temperature

FANS, or blowers, large types, ventilating
Bearings, sleeve—Oil—ring—Semi-annually
Rust and oxidation inhibited oil—300 to 600 SUS at 100 F

FANS, or blowers, large types, forced draft
Bearings, sleeve—Oil—ring—Semi-annually
Rust and oxidation inhibited oil—600 to 750 SUS at 100 F

LINE SHAFT SYSTEM
Bearings—Oil—ring—Keep filled and clean semi-annually
Rust and oxidation inhibited oil—300 SUS at 100 F

Bearings, sleeve—grease—pressure fittings—Monthly
No. 2 lime-base grease
Bearings, antifriction—grease—pressure fittings—Quarterly
No. 2 soda-base grease

TRUNNION BEARINGS (pulverizers, kilns, etc.)
Bearings, plain—grease—block—Keep filled
Grease, block, melting point depending on temperature
Bearings, plain—grease—pack—Keep filled
No. 5 or No. 6 soda base (sponge) grease

CAMS, COGS (open types)
All frictional surfaces—Oil (or grease)—Daily or weekly
Oil—adhesive additive type—600 to 750 SUS at 100 F
No. 3 Grease—aluminum soap with adhesive additive

HIGH-SPEED SPINDLES—5000 rpm and over
Bearings—Oil—reservoir, circulation—Semi-annually
Oil—rust, oxidation and antifriction additives—60 to 90 SUS at 100 F

SPINDLES, below 5000 rpm
Bearings—Oil—reservoir—Semi-annually
Oil—rust, oxidation, and antifriction additives—90 to 125 SUS at 100 F

REFRIGERATING COMPRESSORS, reciprocating, air conditioning (any refrigerant)
Cylinders, Bearings } oil—any method of application—Indefinitely
Oil—Rust and oxidation inhibited or straight mineral oil—300 SUS at 100 F
Note: For low and ultra-low temperature conditions use an oil having appropriate pour or floc point

REFRIGERATING COMPRESSORS, centrifugal, air conditioning (any refrigerant)
Bearings—Oil—circulation—Indefinitely
Rust and oxidation inhibited oil—300 SUS at 100 F
REFRIGERATING COMPRESSORS, screw
All parts—300 SUS at 100 F

STUDY OUTLINE FOR CHAPTER 16

Simple refrigeration system

Multi-evaporator, single temperature system

Multi-temperature systems:
- Dual temperature system—single compressor
- Multi-temperature system—single compressor, three evaporators
- Multi-temperature system—single compressor, any number of evaporators
- Multi-temperature system—multi-compressors
- Multi-temperature system—nonthrottling back pressure, using individual and dual compression cylinders

Low temperature multi-stage system

Two-stage system with rotary booster

Brine cooler, two-stage (duplex) system

Pressure-enthalpy diagrams for two-stage system:
- External intercooler
- Heat exchange between liquid and vapor

Hermetic system

Cascade systems

High humidity systems

Defrosting:
- Internal methods
- External methods
- Controls

Ice manufacturing:
- Block ice—brine tanks
- Ice-making machines

Transportation equipment refrigeration:
- Ice-cooled vehicles
- Dry-ice-cooled vehicles
- Mechanical refrigeration of trucks and trailers
- Cryogenic fluid refrigerated trailers
- Trailer cooling tunnels

Self-refrigerated propane storage

Solid carbon dioxide (dry ice) manufacturing:
- Stage system
- Cascade system

Refrigeration capacity storage

Refrigeration in metallurgy

16

The Refrigeration System

In previous chapters, the principles of refrigeration and individual components, including various controls, were discussed in detail. This chapter deals with complete systems wherein those components and controls are shown in their general working relationship. Such a relationship concerns both function and position under different mechanical arrangements, according to the purpose of the system.

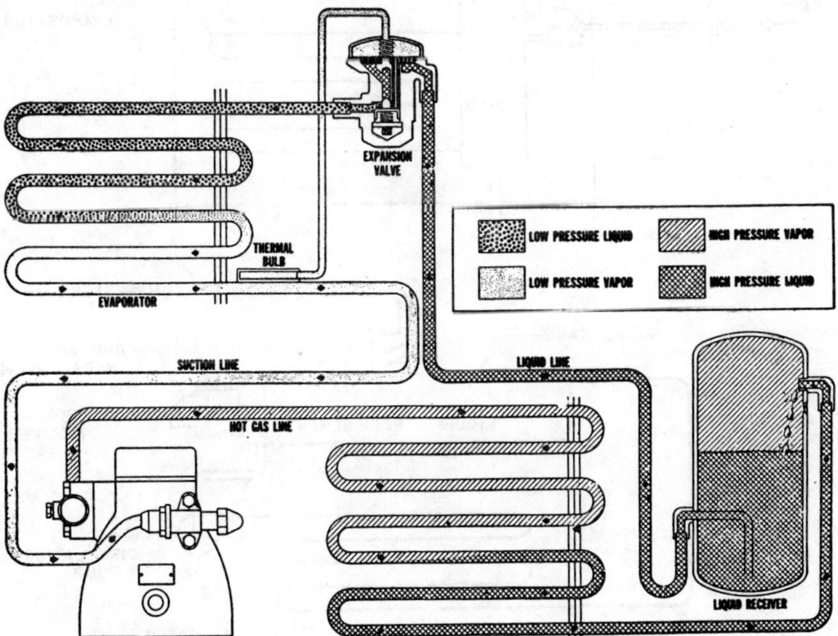

Fig. 458. Simple refrigeration system showing state of fluid as it progresses from the evaporator back to the evaporator. (Courtesy, General Electric Company.)

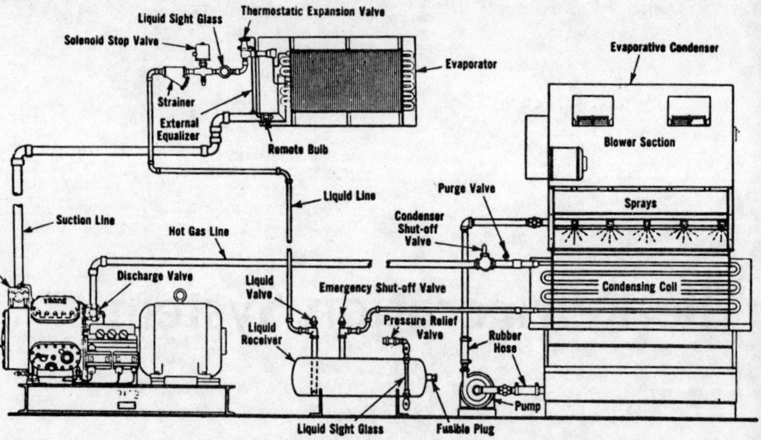

Fig. 459. Simple refrigeration system showing location of various controls and auxiliary equipment. Note use of evaporative condenser. (Copyright, The Trane Company. Reproduced by permission.)

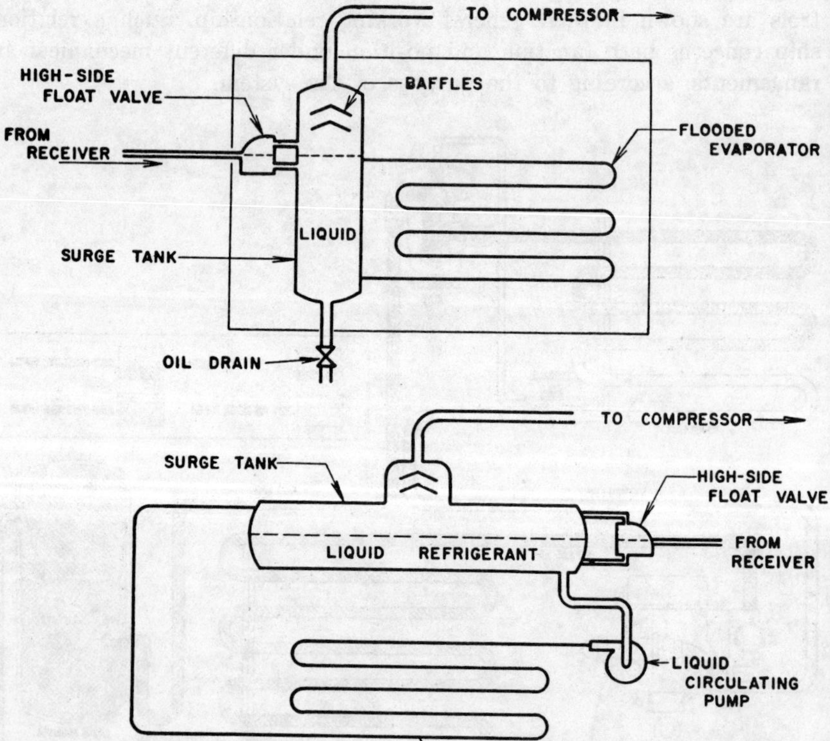

Fig. 460. Flooded evaporators. (*Top*) Gravity circulation; (*bottom*) forced circulation using liquid refrigerant pump. Evaporator can also be elevated above surge tank in latter arrangement.

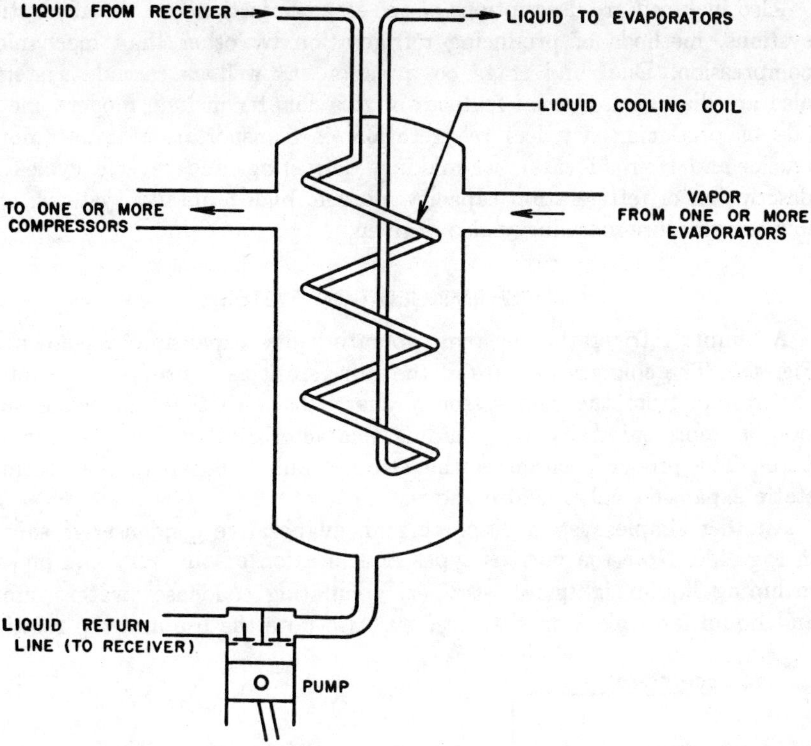

Fig. 461. Accumulator serving any number of compressors and evaporator suction lines. Where two or more accumulators serve the same system, they should be connected by an equalizing or by-pass line. Accumulator shown also acts as a liquid cooler.

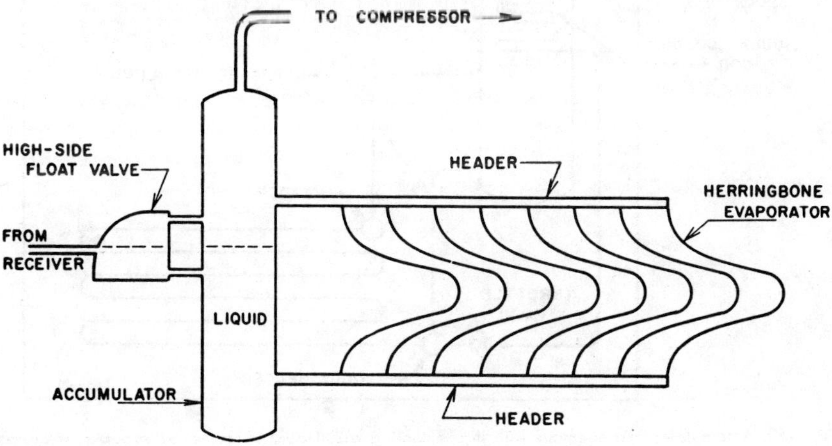

Fig. 462. Type of evaporator and accumulator arrangement employed for high-velocity brine (raceway) as in large ice-making tanks.

Also included are descriptions of the steam injection and the absorption systems, methods of producing refrigeration by other than mechanical compression. Dual and stage compression, as well as cascade systems, also are discussed. Special features of this chapter include modern methods of producing dry ice, refrigeration of transportation units (motor trucks and railroad cars), ice-making, defrosting, and reverse cycles. A description of refrigeration capacity storage, high humidity systems, and low-temperature metallurgy also is given.

SIMPLE REFRIGERATION SYSTEM

A simple refrigeration system, operating dry expansion, is shown in Fig. 458. The changes in state of the refrigerant as it progresses through the system from the compressor discharge back to the compressor suction is represented by the shading characteristics within the various parts. The proper position of the thermal bulb relative to the thermostatic expansion valve is also shown.

Another simple system employing an evaporative condenser is shown in Fig. 459. Note the various types and location of auxiliary equipment, including liquid sightglass, strainer, circulating condenser water pump, and liquid level glass on the receiver. Also note the use and the relative

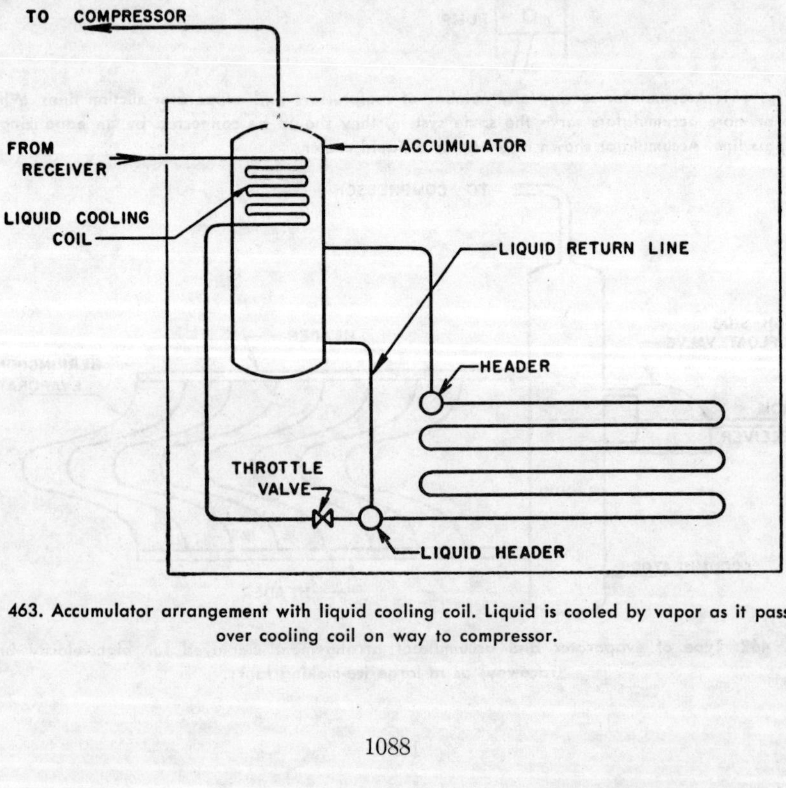

Fig. 463. Accumulator arrangement with liquid cooling coil. Liquid is cooled by vapor as it passes over cooling coil on way to compressor.

location of such controls as the solenoid stop valve, external equalizer, suction and discharge valves, purge valve, condenser shut-off valve, liquid valve, pressure-relief valve, and emergency shut-off valve.

Substituting the arrangement shown in Fig. 460 for the dry expansion evaporators in Figs. 458 and 459 forms the basis of a flooded system. In place of the expansion valves used in the dry expansion system, a typical flooded evaporator employs a float-operated expansion valve in connection with an accumulator, or surge tank.

Figure 462 shows other types of accumulator arrangements for flooded systems employing headers or serving multiple evaporators. The accumulator in Fig. 463 serves the double function of a surge tank and a liquid precooler.

MULTI-EVAPORATOR, SINGLE TEMPERATURE SYSTEMS

Numerous arrangements are possible where more than one evaporator, served from a single compressor and operating at a similar temperature, is required. Figure 464 shows three arrangements for this type of system, designated as evaporators in series, evaporators in parallel served by a common expansion valve, and evaporators in parallel served by individual expansion valves, all operating dry expansion.

Series. Placing evaporators in series has definite limitations and should not be used if runs are relatively long or where loads are not fairly constant in all compartments under refrigeration. This arrangement should be treated similarly to the continuous coil described in the evaporator section of Chapter 8, "Equipment." The coil area provided in each compartment having the same temperature requirement must be proportional to the respective loads. Thus, a compartment wherein the load demand is greater than another with which it is in series requires a proportionally greater coil area in order to maintain an equal temperature.

One successful but highly inefficient system of this kind, later changed on the recommendation of the author, was an abattoir in Central America wherein the carcasses were frozen in the first compartment and held in the second until shipped. Undisturbed for an extended period, both compartments would finally reach the same temperature. However, during periods when the freezer was being loaded with carcasses from the chill room (where it loses animal heat), and for a considerable time afterward, the heat load was great enough to starve the holding compartment coils and cause thawing. The condition resulted in excessive superheating and consequently a high loss of efficiency.

Parallel. This arrangement, as shown in Fig. 464(b), with one expansion valve feeding two or more coils, can be employed with good effect only under certain conditions. Because expansion occurs in the header,

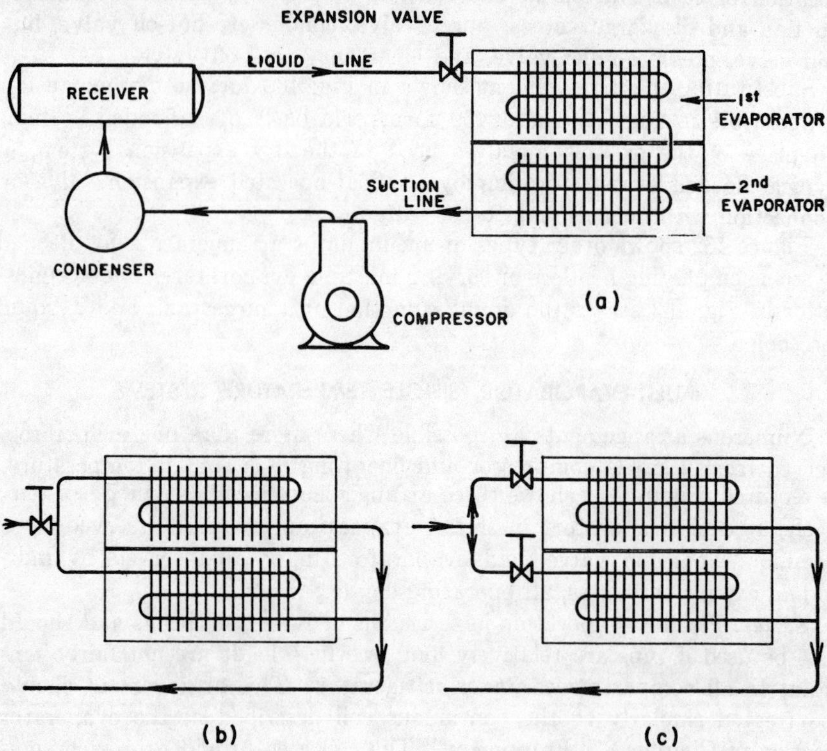

Fig. 464. Evaporator arrangements: (a) coils in series, (b) coils in parallel with common expansion valve, (c) coils in parallel with individual expansion valves.

it should be located within the cooled area. Satisfactory results can be obtained where the header feeds a number of branch lines submerged in a common bath or space and ends in the same suction header. However, it is not recommended for systems having their coils remote from each other or where load differences between coils may cause starving of the one meeting the greater demand.

The third arrangement includes an expansion valve for each coil. While this method of refrigerant distribution is included here for evaporators operating under similar temperature conditions, it can also be used for multi-temperature purposes with suitable controls, including expansion valves variously adjusted and evaporator pressure regulators. This arrangement forms the basis for all correctly designed multi-evaporator systems.

Multi-coil Distributor. The liquid distributor used in connection with a thermostatic expansion valve is an ideal arrangement for large-capacity evaporators of the type shown in Figs. 87 and 88. It can be used also with multi-plate systems (Fig. 400) or other similar type arrangements where the individual expansion elements are within reasonable proximity.

MULTI-TEMPERATURE SYSTEMS

In dealing with systems having two or more evaporators operating at different temperatures, the question of convenience versus economy becomes a principal consideration. In smaller systems, convenience often justifies the thermal losses that are incurred when one compressor is employed to serve multi-temperature evaporators. Another advantage of the single-compressor multi-temperature system is its lower initial cost. On large industrial type systems, however, the additional power requirements resulting from the sustained thermal losses of common compression are reason enough to adopt more efficient methods of refrigerant reclamation. Therefore, while convenience is sufficient reason for single compressors on domestic or small commercial systems, economy becomes progressively predominant as the tonnage requirements increase to favor a tendency toward separate compressors for each level of temperature. This is based on the law that:

Whenever two or more vapors of dissimilar pressures are throttled so as to permit their entrance to the compressor at a common pressure equal to that of the evaporator having the lowest temperature, an increase in superheat of the mixture results which consequently increases the work requirements of the compressor.

Classified in their simplest form, there are three basic types of multi-temperature systems:

1. Single compressor serving all evaporators.
2. Separate compressor for each temperature level.
3. Dual compression cylinder type compressor.

Figures 465 to 469, inclusive, show diagrammatic layouts of the various dual- and multi-temperature systems. Pertinent information regarding each one follows.

Dual-temperature System—Single Compressor (Fig. 465)

1. When high-temperature coil reaches low-temperature limit, thermostat de-energizes solenoid, closing individual suction line.
2. Compressor continues to operate until pressure in suction line reaches low-pressure limit of cut-off switch, due to closing of thermostatic expansion valve, to shut down compressor.
3. If high-temperature evaporator calls for refrigeration before low-temperature evaporator, solenoid valve will open as a result of thermostatic action, raising pressure in suction line to close low-pressure cut-off circuit, starting the compressor. Compressor operation will not affect pressure in low-temperature coil until thermal bulb allows thermostatic expansion valve to open.

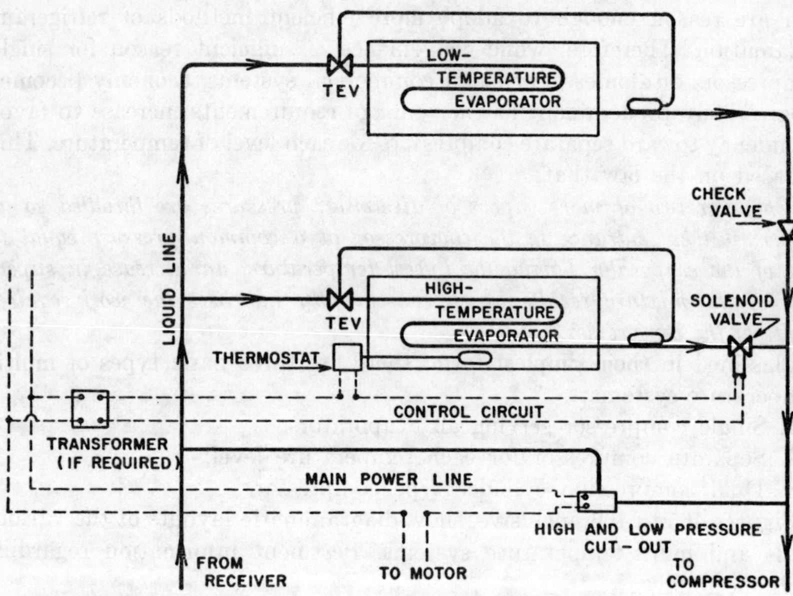

Fig. 465. Dual-temperature system, single compressor.

Multi-temperature System—Single Compressor, Three Evaporators (Fig. 466)

1. This system operates in a similar manner to that shown in Fig. 465, with one exception.
2. This exception is that the low-temperature evaporator is controlled by the setting of the low-pressure cut-out. Note that the suction line of the low-temperature evaporator is not equipped with a suction pressure regulator.

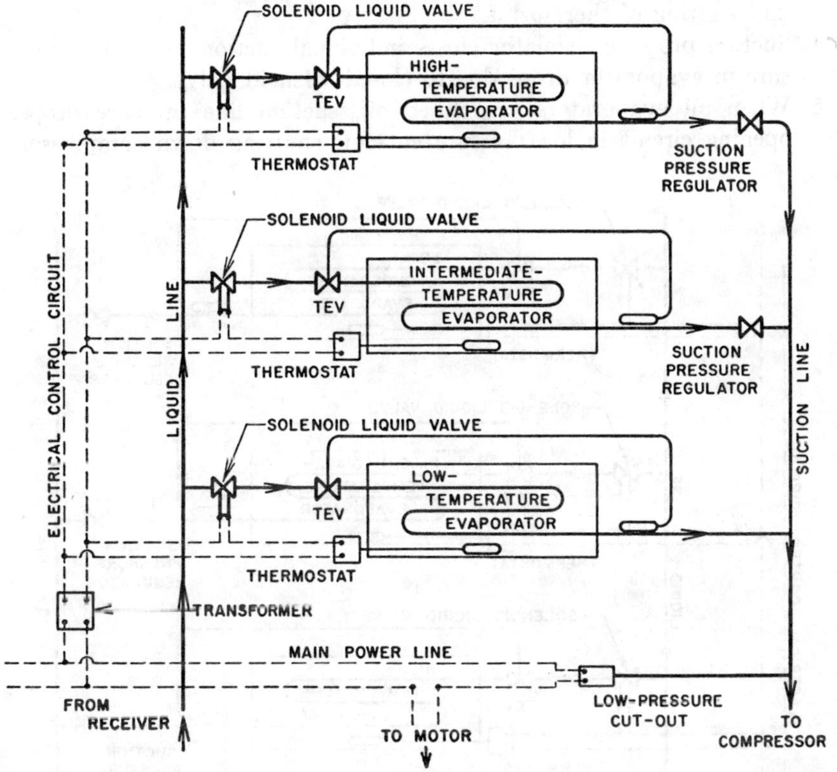

Fig. 466. Multi-temperature system, single compressor, three evaporators, in which low-temperature coil is controlled by low-pressure cut-out.

Multi-temperature System—Single Compressor, Any Number of Evaporators (Fig. 467)

1. Thermostatic expansion valves set for about 10 F superheat above desired evaporator temperature feed refrigerant when solenoid valve is open.
2. Thermostat set at desired compartment temperature controlling the flow of liquid by solenoid in the control circuit.
3. Solenoid valve closes when compartment temperature reaches low limit setting of thermostat.
4. Suction pressure regulator closes individual suction line when pressure in evaporator drops due to closed solenoid valve.
5. When all evaporators are closed off, suction line pressure drops, opening circuit in low-pressure cut-out, shutting down compressor.

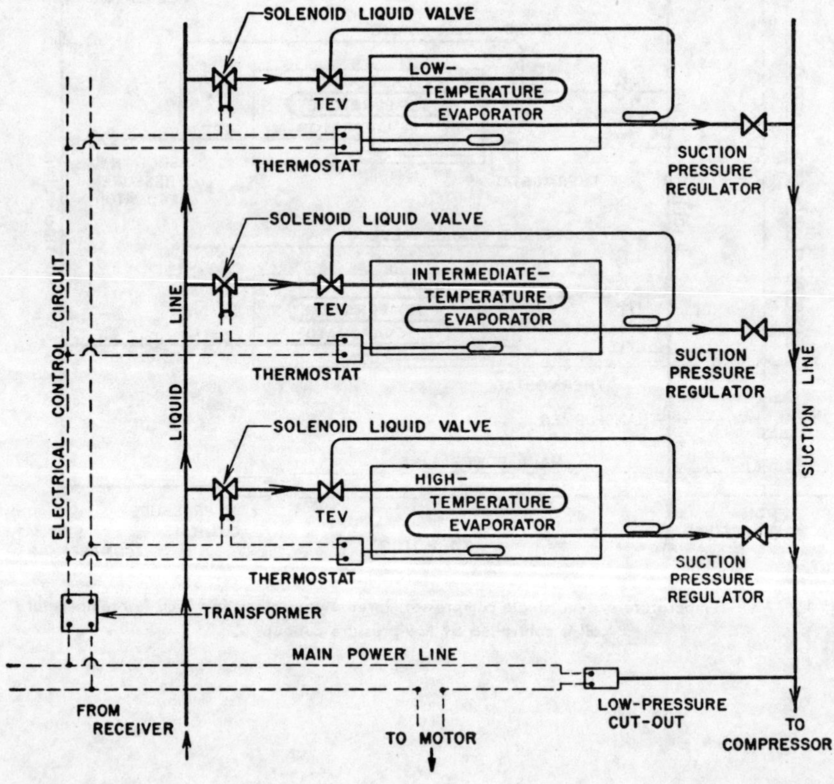

Fig. 467. Multi-temperature system, single compressor, any number of evaporators controlled by suction pressure regulators with low-pressue cut-out shutting down compressor operation when demand is zero.

Multi-temperature System—Multi-compressors (Fig. 468)

1. Evaporators having similar temperatures served by separate compressor.
2. Each compressor controlled by cut-out switch on own suction line to stop and start motor.

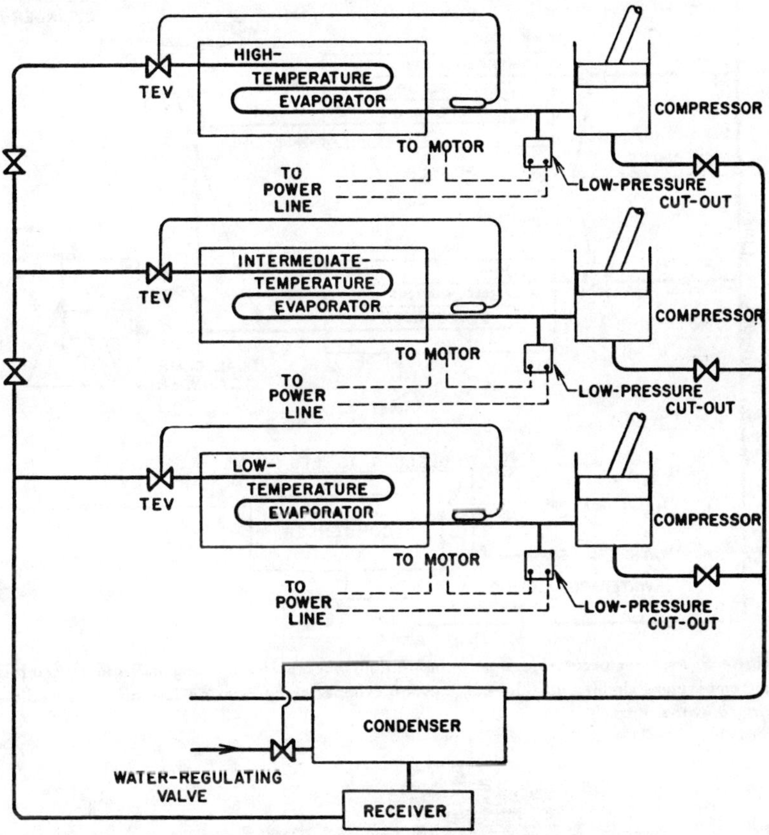

Fig. 468. Multi-temperature system, multi-compressors.

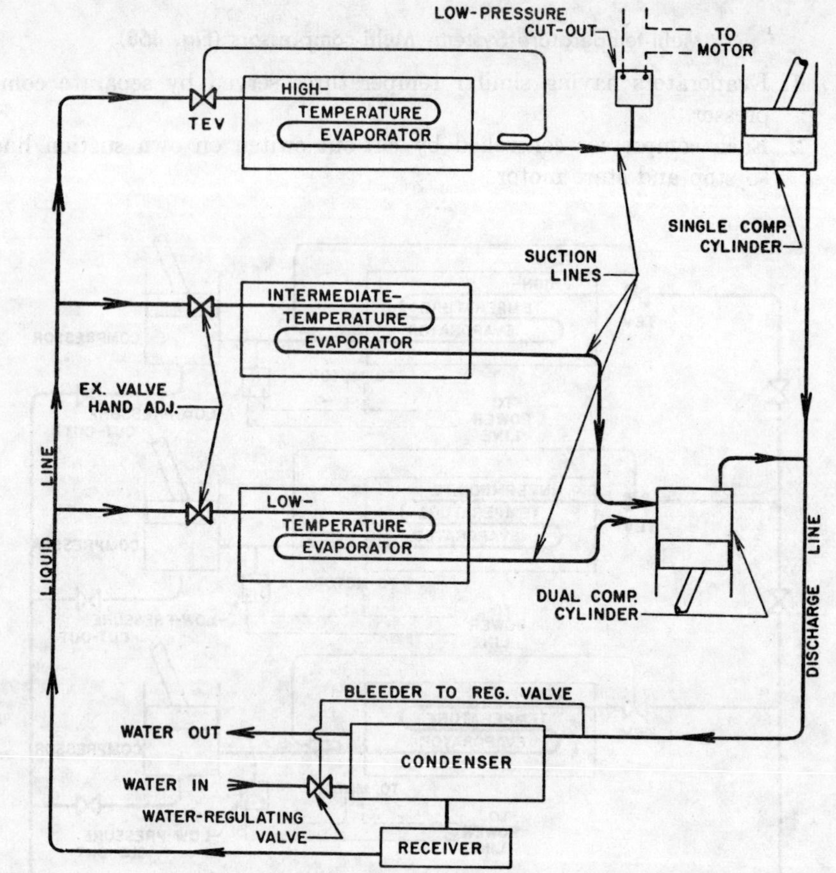

Fig. 469. Multi-temperature system, nonthrottling back pressure, using individual compressor on high temperature circuit and dual compression cylinder compressor on low and intermediate temperature evaporators.

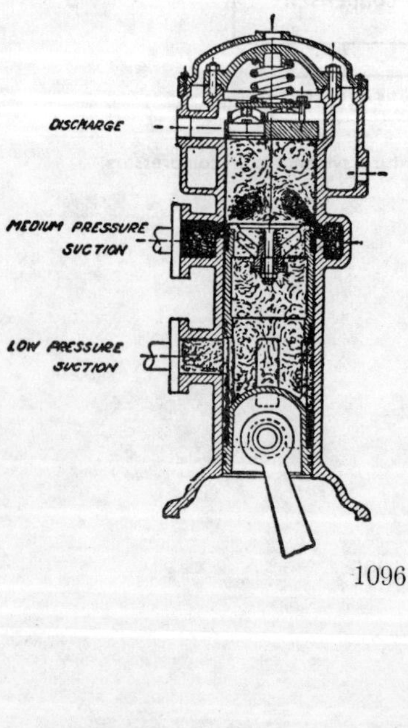

Fig. 470. Dual compression cylinder.
(Courtesy, Frick Company.)

Multi-temperature System—Nonthrottling Back Pressure, Using Individual and Dual Compression Cylinders (Fig. 469)

1. High-temperature evaporator served by its own compressor.
2. Intermediate- and low-temperature evaporators served by compressor having dual compression cylinders to eliminate throttling of intermediate evaporator pressure to low-temperature evaporator condition.
3. A common condenser is used for all compressors.
4. High-temperature evaporator controlled by a thermostatic expansion valve and compressor stops on low-pressure cut-out.
5. Intermediate- and low-pressure evaporators manually controlled. No automatic arrangement on dual compression machine.

LOW-TEMPERATURE MULTI-STAGE SYSTEM

A typical low-temperature system employing a flooded evaporator is shown in Fig. 471. Compression from low pressure required in the low-

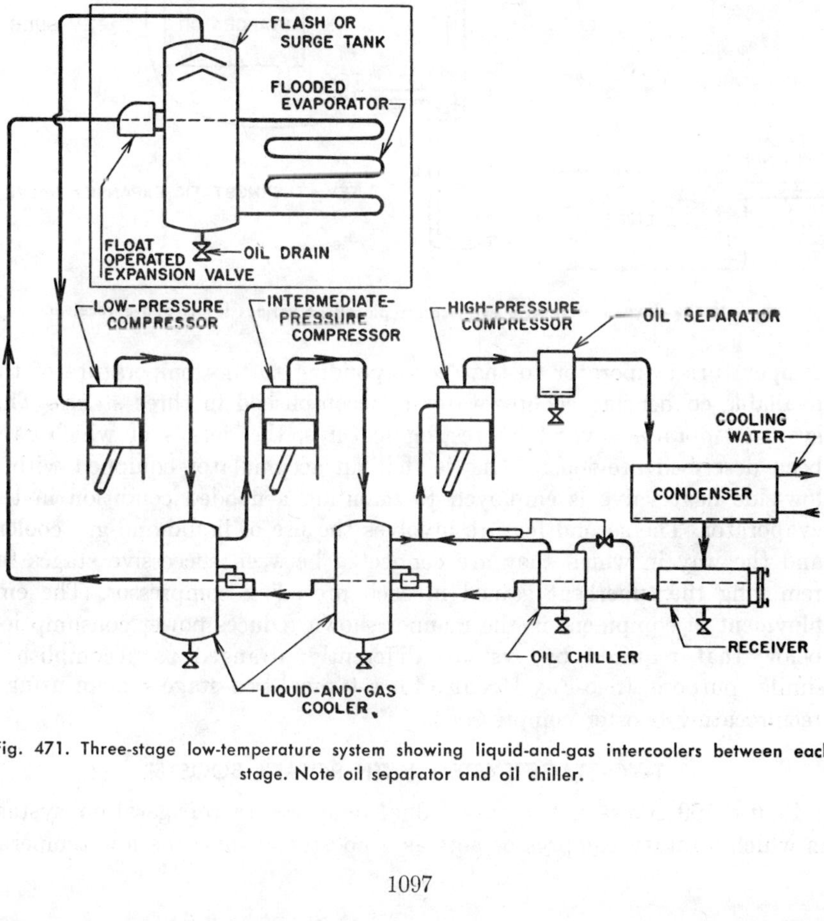

Fig. 471. Three-stage low-temperature system showing liquid-and-gas intercoolers between each stage. Note oil separator and oil chiller.

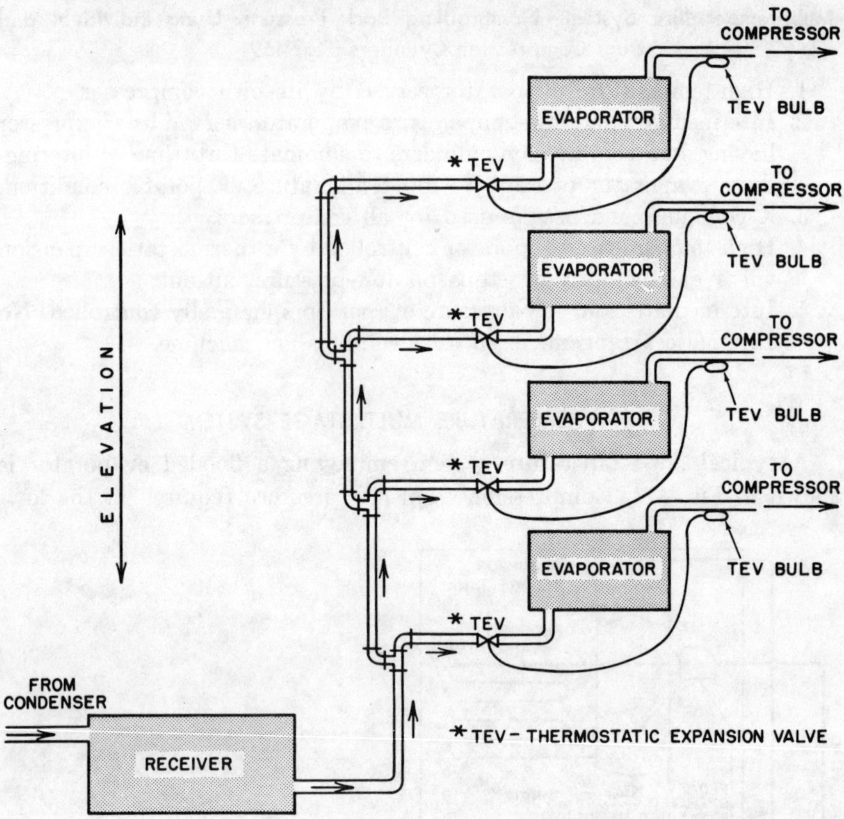

Fig. 472. Connection of several evaporators placed at various levels above receiver.

temperature evaporator to that corresponding to the temperature of the available condensing cooling water is accomplished in three stages. This system embraces several interesting features, the details of which have been described previously. One is that an accumulator equipped with a low-side float valve is employed to maintain a flooded condition in the evaporator. The second feature involves the use of liquid-and-gas coolers and the way in which they are connected between successive stages for removing the superheat gained in each preceding compressor. The employment of equipment in the manner shown reduces power consumption below that required by systems differently arranged to accomplish a similar purpose. (See Fig. 168 also for a typical two-stage system using a reciprocating booster compressor.)

TWO-STAGE SYSTEM WITH ROTARY BOOSTER

Figure 169 shows a two-stage, dual-temperature refrigeration system in which a rotary compressor acts as a booster to raise the low-temperature vapor through initial stage compression. An interesting point to note in this system is that the vapor from the high-temperature evaporator is drawn into the reciprocating or second-stage compressor without first passing through the booster. This action eliminates throttling that portion of the vapor representing the high-temperature load, thereby reducing the amount of work done by the compressor. In effect, the action is similar to that of a dual compression cylinder machine.

BRINE COOLER

A two-stage ammonia refrigeration system employing a duplex compressor is shown in Fig. 473, and represents the more common arrangement of equipment used to cool brine. From the cooler, a shell-and-tube evaporator, brine is circulated to provide refrigeration as a secondary medium to cold-storage rooms, ice-making tanks, etc. The illustration shows the position of a liquid cooler and an intercooler on the same system and the gravitational method of transferring excess liquid from the latter to the former. The liquid cooler is connected to the suction of the high-pressure stage compressor. It is through this connection that the vapor passes after extracting heat from the liquid before the latter enters the brine cooler.

Also included is an oil still into which the oil from separators, traps, and coils drains. The still is connected to the low-pressure suction line to draw off any liquid refrigerant that enters the still with the oil.

A pressure-enthalpy (Ph) diagram for a two-stage ammonia system provided with an external means of intercooling is shown in Fig. 474. (Also see Fig. 170 for a typical external cooling arrangement using water.) To attain a saturated condition of the interstage gas, 23.5 F at 52 psia, as shown, a fluid other than water must be employed for intercooling. The condition as represented in Fig. 474 would be ideal for increasing the coefficient of performance (cop), but would require the use of a cold brine or another refrigerant to bring about a saturated state between stages. The gain in system performance must be weighed against the cost of providing sufficient cooling from an external source to determine over-all results. By cooling the intermediate gas from 117 F to, say, 60 F with water, a savings in power over single-stage operation would still be obtained by reason of greater weight handled per minute of the more dense gas. The colder the water, the greater the reduction in intermediate gas temperature and consequently a higher power saving. In Fig. 170, a satisfactory condition is shown in which the gas is cooled by 65 F water from 128 F (103 F superheat) to 80 F (with 55 F superheat remaining).

The reduction in power requirements of a two-stage system over a single-stage system is proportional to the difference in total enthalpy of compression developed in both systems. This may be shown by comparing enthalpy values for the Ph diagram in Fig. 473, viz.:

For a single-stage system, enthalpy of compression for the conditions shown is 757 Btu per lb − 603.2 Btu per lb = 153.8 Btu per lb.

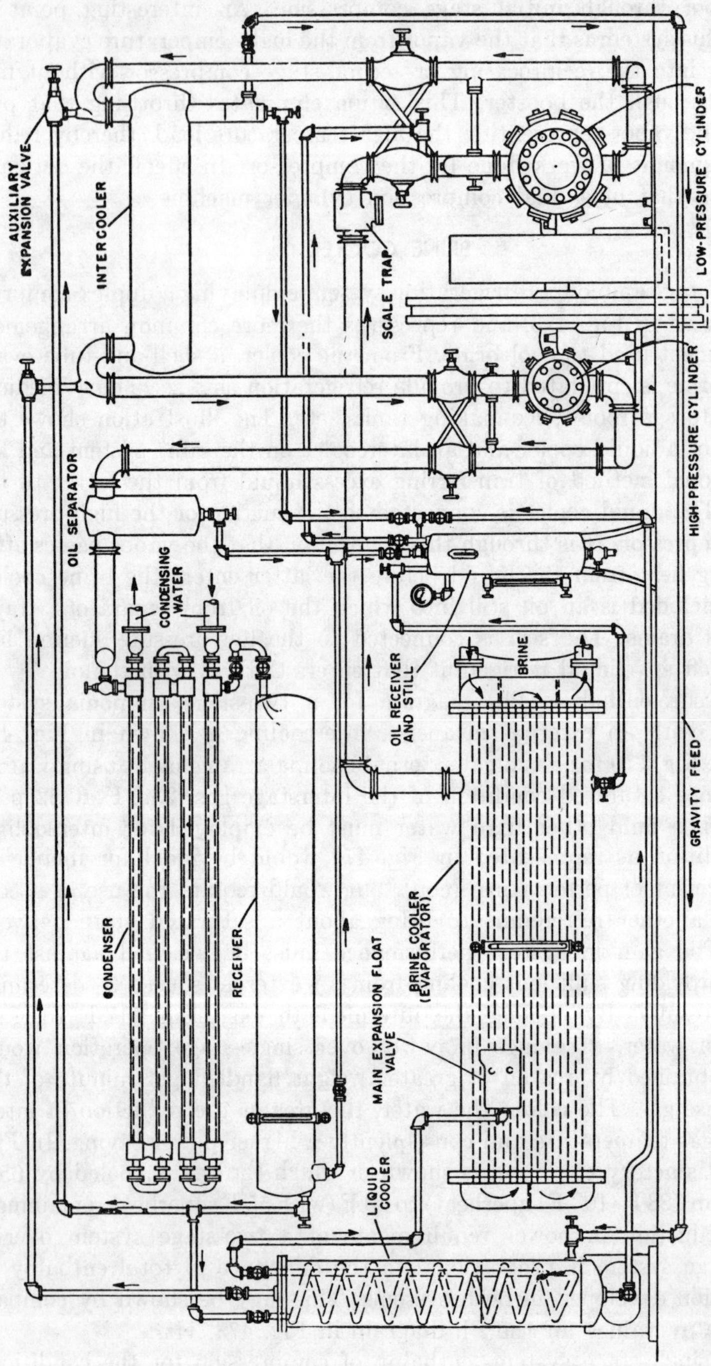

Fig. 473. Duplex compressor with high- and low-pressure cylinders (two-stage) with intercooler between stages. Note double-pipe condenser and shell-and-tube brine cooler. (Courtesy, Socony Mobil Oil Company, Inc.)

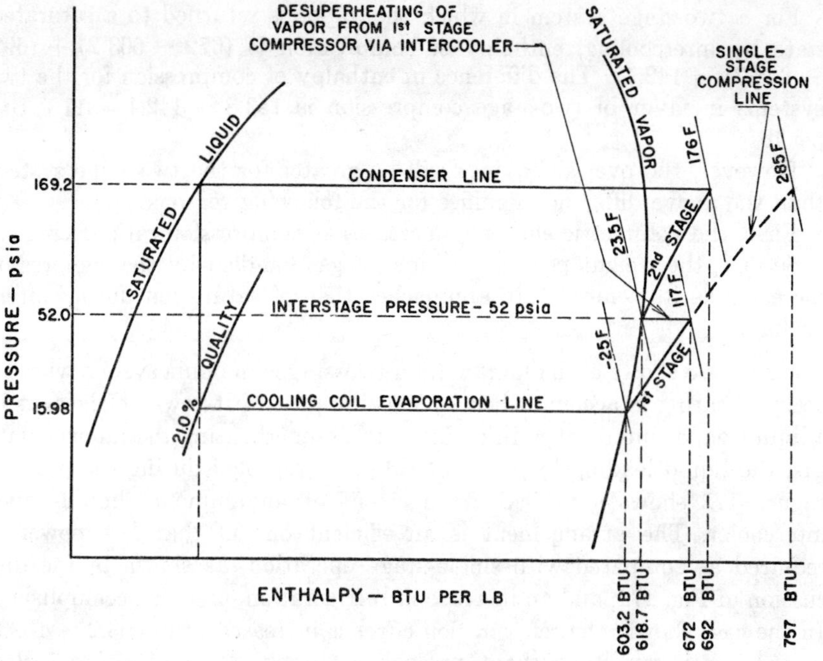

Fig. 474. Ph diagram for two-stage ammonia system with intercooling performed by external medium.

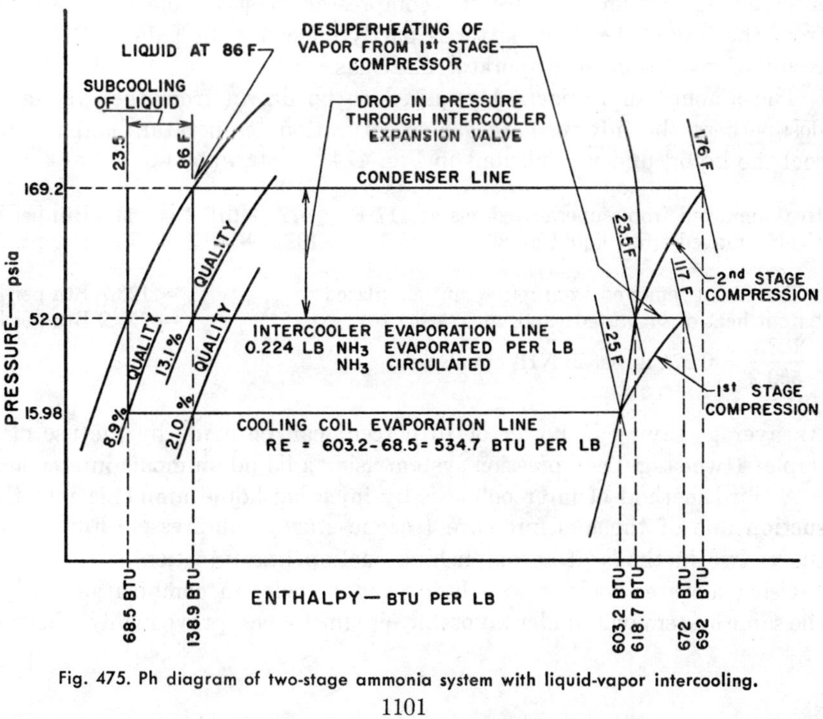

Fig. 475. Ph diagram of two-stage ammonia system with liquid-vapor intercooling.

For a two-stage system in which the vapor is returned to a saturated state by intercooling, enthalpy of compression is (672 − 603.2) + (692 − 618.7) = 142.1. The difference in enthalpy of compression for the two systems in favor of two-stage compression is 153.8 − 142.1 = 11.7 Btu per lb.

However, the over-all savings will be greater for the two-stage system than the above difference signifies for the following reasons:

First, the volumetric efficiency increases as compression ratio decreases.

Second, the weight per unit volume of gas handled by the compressor increases as its temperature approaches that of saturation during intercooling.

Figure 475 shows a Ph diagram for a two-stage ammonia system wherein desuperheating is accomplished by flashing part of its own refrigerant in a liquid-vapor intercooler. In addition to desuperheating the intermediate gas, the liquid leading to the evaporator is precooled during the process. Figure 172 shows a typical arrangement of an ammonia liquid-vapor intercooler. The arrangement is an efficient one in that less power is required as compared with single-stage operation, as shown in the discussion of Fig. 474, and an increase in refrigeration effect is accomplished. In the case shown, the refrigeration effect is increased from 603.2 − 138.9 = 464.3 Btu per lb, without precooling, to 603.2 − 68.5 = 534.7 Btu per lb with precooling. Some reduction in capacity occurs over external intercooling systems because the compressor must handle the flash gas from the intercooler. This latter is partly offset by the gain in R E as a result of reduction in evaporator flash gas.

The amount of refrigerant required to be drawn from the system to desuperheat the intermediate gas to saturation temperature and to precool the liquid under conditions in Fig. 474 is determined as follows:

Heat removed from superheated gas at 117 F = 672 − 618.7 =	53.3 Btu per lb
Heat removed from liquid at 86 F to 23.5 F = 138.9 − 68.5 =	70.4 Btu per lb
Total heat removed from refrigerant circulated	= 123.7 Btu per lb
Latent heat of saturated vapor at 52 psia	= 550.2 Btu per lb

$$\frac{123.7}{550.2} = 0.224 \text{ lb per lb } NH_3 \text{ circulated}$$

An average saving in power of 10 percent can be made by the use of a duplex (two-stage) compression system using a liquid ammonia intercooler.

A third method of intercooling is by injecting liquid ammonia into the suction line of the high-pressure (second-stage) compressor. Figure 171 shows two methods of accomplishing suction line injection.

Using a water cooler and a liquid-vapor cooler in combination within the same system will, under favorable circumstances, prove highly efficient.

The hermetically sealed refrigeration system is one in which the motor and the compressor are directly connected by a common shaft and encased in an air-tight housing. While the practice of hermetically sealing the compression unit was first applied to domestic refrigerators, it has now been extended to larger systems. The purpose of the hermetic unit is to eliminate the shaft seal necessary when the compressor is driven externally and thus prevent the entrance of air to the system and outward leakage of refrigerant. This results in a compact unit capable of operating unattended for long periods, an important factor in any domestic appliance. Their principles of operation in no way differ from other types of mechanical refrigeration systems.

Both reciprocating and rotary type compressors are employed in hermetic units driven by a squirrel-cage induction motor, of either the straight split phase or the capacitor type. A scotch yoke often replaces the con-

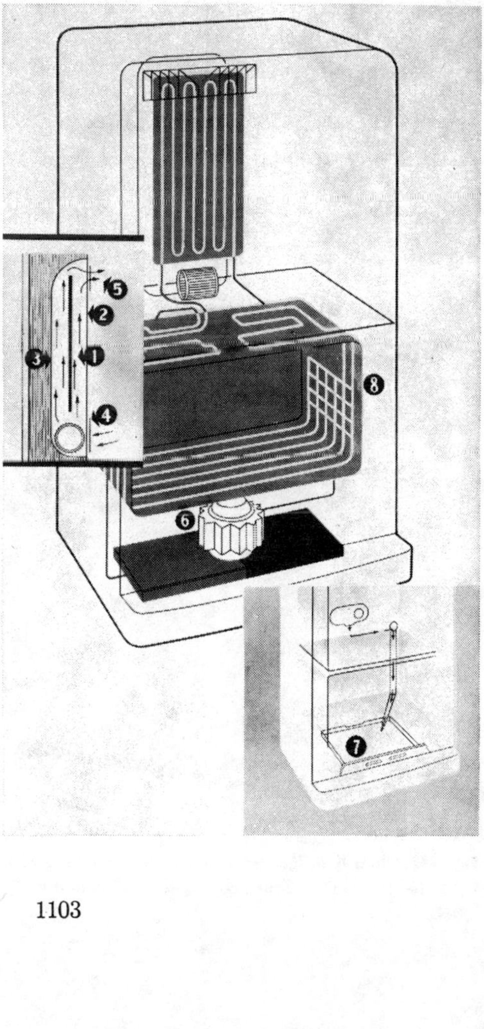

Fig. 476. Hermetically sealed unit as employed on domestic refrigeration (top compartment) and freezer (bottom compartment). Top insert shows details of cold air circulation within refrigerator. (1) Plate evaporator, (2) refrigerator liner, (3) cold air passage or duct. The fan (4) draws air from the refrigerator compartment and forces it over evaporator through the louvers (5) at top.

(*Bottom insert*) Shows defrost drain from refrigerator. When compressor stops, temperature in the top evaporator rises to 36 F, during which frost melts and moisture is directed down the back of liner to drain pan (7) and thence to a porcelain pan above the compressor (6). The heat of the compressor finally evaporates the moisture. The freezer (8) is not automatically defrosted. (Courtesy, Frigidaire, Division of General Motors Corporation.)

ventional connecting rod and wrist-pin arrangement in transmitting motion from the motor shaft to the piston of reciprocating compressors. The compressor and the motor are cooled by the suction gas which, upon returning from the evaporator, fills the remaining space not taken up by the lubricating oil within the housing before entering the compressor.

While many kinds of refrigerants have been employed in the hermetic unit, including sulfur dioxide, methyl chloride, R-12 and R-114, the latter two are favored by most manufacturers today. R-12 units employ either reciprocating or rotary compressors while R-114 systems require only the rotary type.

Capacity control of hermetic systems used to maintain a given temperature range is obtained by stopping and starting the compressor by means of a bi-metal or thermal bulb thermostat. Automatic defrosting is accomplished by various methods, some of which are described in another section of this chapter.

Condensers vary according to the manufacturer, so that many different types and shapes are found. Because of convenience in locating them on the back of the refrigerator, plate type condensers find wide popularity, although tube (both finned and prime) types are also used extensively. All are air cooled, however, when used in domestic service.

The capillary tube is favored as the throttling device on most hermetically sealed domestic units for several reasons. First, because it is simple in construction and has no moving parts to go out of order. Secondly, it allows the pressure in both sides of the system to equalize during off-cycle periods, thereby reducing the torque required by the motor

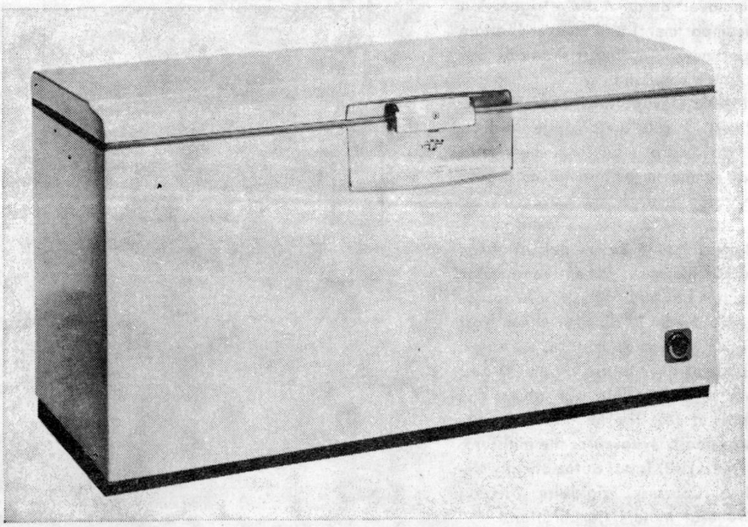

Fig. 477. Chest type food freezer. Another example of the progress made with hermetically sealed units. (Courtesy, Frigidaire, Division of General Motors Corporation.)

in starting. The amount of refrigerant required by the average hermetic household refrigerator and home freezer is relatively small, ranging as low as 1 lb on smaller models. Oil capacity varies from 8 oz to well over 32 oz on larger models. Motor ratings are low, ranging from $\frac{1}{8}$ to $\frac{1}{4}$ hp.

Older type domestic units employed evaporators of various shapes located in the top of the refrigerated compartment. Many of them were merely expansion coils shaped to form a convenient holder for ice-cube trays. Others of the same general shape were formed from two sheets of metal, one of which was embossed to produce the effect of a plate evaporator. These evaporators were placed in the top of the refrigerated compartment on the theory that cold air travels downward as a result of gravity. However, the same weakness prevailed in this type unit as it did in the earlier ice box: nonuniformity of temperatures within the compartment, causing the air at the bottom to be warmer than that at the top. This has been corrected by distributing a cooling coil around the sides and the bottom of the compartment between the inner wall and the insulation and providing a freezing space at the top of the unit to which access is permitted by a separate door in some instances.

One method of cooling the lower compartment of a domestic refrigerator is interesting and differs from any other method described here. Only the evaporator serving the freeze compartment is connected to the compressor and the condenser and that system is designed in the conventional manner.

The coils that cool the lower compartment are filled with refrigerant, but are not connected to a compressor or a condenser and depend entirely on the absorption of heat by the colder coils of the freezing unit with which a portion of their upper surface is in contact. Liquid in the lower sections of the coil absorbs heat from the storage compartment by conduction through the walls. This promotes boiling during which the vapor rises in the coil until it comes in contact with the surface whose temperature is below the boiling point due to heat absorption by the freezer coil. The vapor then condenses and flows back to the lower portion of the coil to again absorb heat from the storage compartment, vaporize, and repeat the process. The action is known as *thermosiphon circulation.*

It may be noted that the thermosiphon system described above for cooling the storage compartment of domestic refrigerators is also used to refrigerate motor trucks carrying perishable commodities. In place of the cooling coils which refrigerate the freezing space, dry ice is used to condense the vapor in preparation for its return as a liquid to the coils located in the truck body.

CASCADE SYSTEMS

There are three basic methods by which low and ultra-low temperatures may be obtained, using mechanical refrigeration as a means. They

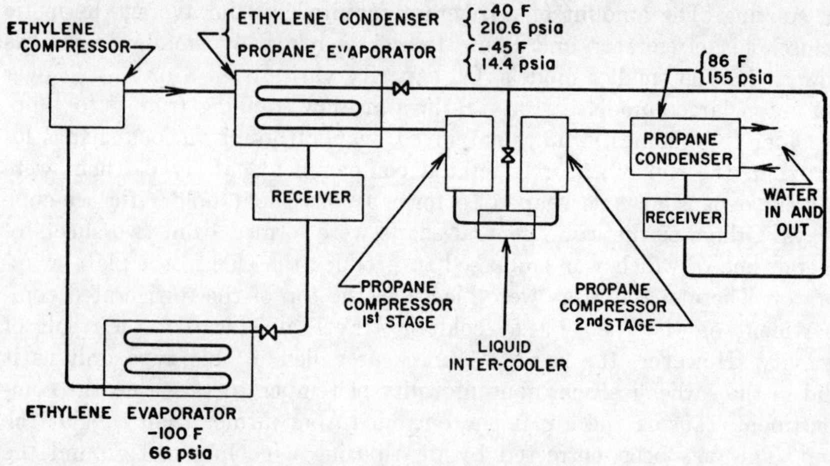

Fig. 478. Cascade system of refrigeration.

are: (*a*) multi-stage compression systems, (*b*) combination compression and expansion systems, and (*c*) cascade systems.

The first of these, (*a*), is used to produce low to ultra-low temperatures generally and was described previously in this text.

The second, (*b*), includes the *Linde* and *Claude* systems employed in liquefying atmospheric gases such as air, oxygen, and nitrogen. Air acts as its own refrigerant in this type of system, which includes expanding cylinders as precoolers to produce temperatures of −330 F and lower. Air compression and expansion systems that produce more moderate temperatures are included in this group.

Cascading, (*c*) the third method listed, involves the use of two or more refrigerants having progressively lower boiling points. The first of these expands in an evaporator that, in turn, acts as the condenser for the second fluid having a lower boiling point, whose evaporator is the condenser for the third fluid having a still lower boiling point, etc.

For example, a cascade system in which ethylene is employed as the low-temperature refrigerant and propane as the condenser control fluid would have an arrangement and fluid conditions on the order shown in Fig. 478. With a low-temperature evaporator requirement of −100 F, the pressure of ethylene in the low side will be 66 psia. Holding ethylene compression to a 3.0 to 3.5:1 ratio, the condenser pressure would be somewhere between 198 and 231 psia, say 210.6 psia, corresponding to a temperature of −40 F. This means that the condenser cooling medium, the second in the progressive order of ascending boiling point fluids, propane, must boil at a temperature below −40 F, or say −45 F, corresponding to a pressure of 14.4 psia. At a condensing temperature of 86 F, corresponding to a pressure of 155 psia for propane, the compression ratio for single-stage operation would be high at $\frac{155}{14.4} = 10.7$. Therefore, the condensing pressure of propane should be raised in two stages unless a third refrigerant having a higher boiling point is employed to condense propane at a lower pressure.

Cascade systems have been made up with various refrigerant com-

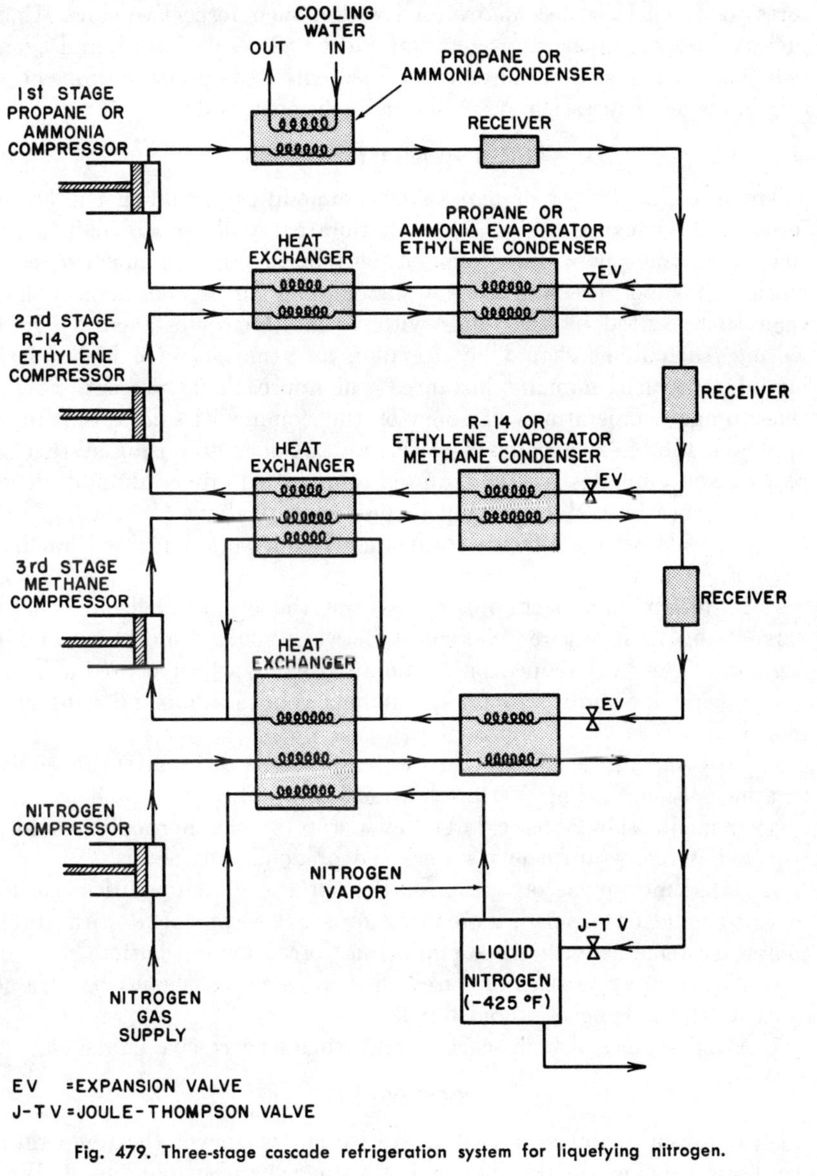

Fig. 479. Three-stage cascade refrigeration system for liquefying nitrogen.

binations, some of which included sulfur dioxide with carbon dioxide, and a more complex arrangement using ammonia, ethylene, and methane in series, all named in the order of descending boiling points. Other refrigerants used in the low temperature stages of cascade systems include R-13 and R-503.

The advantage of the cascade system lies in the possible selection of refrigerants having moderate pressures and volumes at the temperatures to which they become exposed within their respective stage. Under such conditions, greater refrigeration effect will result from liquid expansion (less flash gas) than may be obtained with such pressure drops occurring in stage compression systems. High efficiency follows.

HIGH HUMIDITY SYSTEMS

There are a number of perishable commodities, including cut flowers, ferns, and some vegetables, that deteriorate rapidly if stored in a relatively dry atmosphere. Such commodities tend to lose their moisture readily when the vapor pressure of the surrounding air is considerably lower than that exerted by the fluids within the stored substance. To retard wilting, such items should be stored in an atmosphere of high relative humidity, which in many instances will approach 95 percent. Because the storage temperatures of some of the commodities in this category approach the freezing point of water, an auxiliary atomizing system employing spray nozzles is often required to maintain proper humidity levels. With others whose storage temperatures are well above 32 F, a high relative humidity is not difficult to attain without spraying if several basic steps are taken.

1. Maintain high back pressure within the cooling coils in order to narrow the temperature differential between room atmosphere and refrigerant. This will reduce or eliminate frosting which causes a loss of atmospheric moisture through continuous condensation and subsequent freezing.

2. To compensate for the narrow temperature differential, an equivalent increase in cooling coil area is necessary. Therefore, high humidity compartments should have larger evaporators than normally would be required if temperature levels were the predominant factor.

3. Rapid movement of air across the surface of commodities sensitive to wilt should be avoided. This problem is not so prevalent with gravity cooling systems as with those employing forced air circulation.

4. Where spray nozzles are used, fine atomization should be attained to avoid the wetting of stored items.

5. Avoid sources of high sensible heat, such as excessive lights, etc.

DEFROSTING

Precipitation of moisture will always occur whenever the temperature of a body is below the dew point of the atmosphere surrounding it. When this occurs on the subfreezing surface of an evaporator or suction line, frost is formed. Because it acts as an insulator to reduce the heat-transfer capacity of cooling elements, frost should be removed periodically. This is true whether the system involved is a small domestic refrigerator or a large industrial installation. The frost occurring on evaporator coils is formed to a large extent from water vapor present in the atmosphere. When the change of state is from a vapor to a solid, a hard tenacious substance known as *rime frost* is formed. Rime frost is easily distinguished from the flaky structure of *crystalline* frost formed when the change of

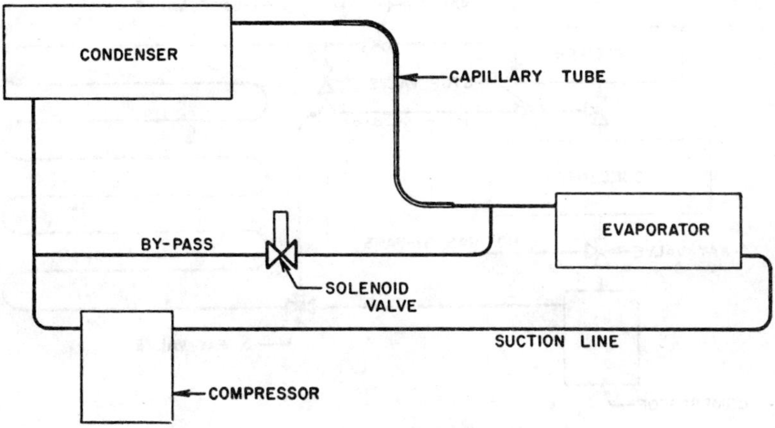

Fig. 480. Defrost system, domestic refrigerators (one method).

state is from the liquid such as entrained water to a solid. It is the flint-like nature of the rime frost that requires drastic measures in many instances for its removal.

There are numerous methods for accomplishing the bothersome chore of defrosting, but the one adopted must be that most practical for the type of system to which it will be applied. Basically, defrosting methods may be divided into two groups: *internal methods,* those that employ the refrigerant as the heating agent to remove frost, and *external methods,* whereby frost removal is accomplished by heat from an external source.

Internal Methods

This method lends itself well to automatic defrosting as practiced on small commercial and domestic units. While the thawing effect of hot refrigerant gas is the basic principle of this group, it is the methods used to provide for the introduction of such gas to the cooling coils that spell the difference in the various defrosting systems.

Method of Defrosting Domestic Refrigerators. One internal method of defrosting domestic units is shown in Fig. 480. The refrigeration system is equipped with a by-pass line whose cross-sectional area is larger than that of the capillary tube, running from the compressor discharge to the evaporator, circumventing the condenser and the throttling device. The by-pass line is equipped with a solenoid valve controlled by an electrically timed thermostatic switch set to open once every 24 hours by a dial on top of the refrigerator. With the solenoid valve open, hot discharge gas following the path of least resistance will flow through the by-pass directly into the evaporator to defrost it. The hot gas condenses in the evaporator and flows back to the compressor, where it picks up enough heat to vaporize it before being compressed and returned to the discharge line. This system is adaptable only where a small amount of refrigerant is handled, as in a domestic unit.

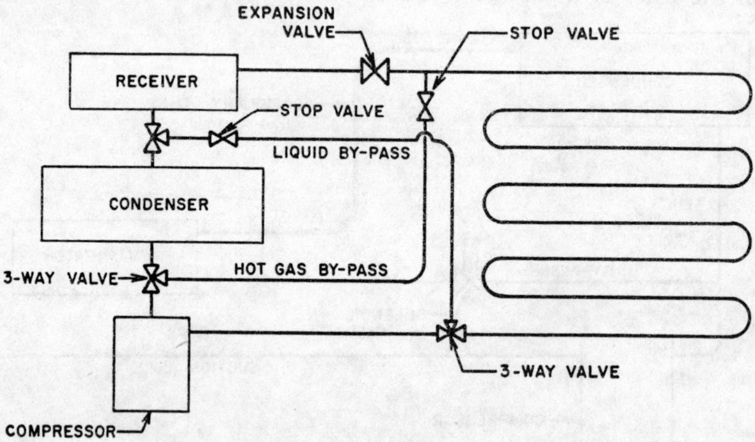

Fig. 481. Hot gas defrost system.

Hot Gas Defrosting—Industrial or Cold-Storage Evaporators. Because of the large amount of moisture given off by the product, open coils in cold-storage rooms are subject to heavy frosting. An effective method employed for its periodic removal involves the use of hot discharge gas directly into the coils from the compressor, by-passing the condenser and the expansion valve. (See Fig. 481.)

A line connecting the compressor discharge through a three-way valve is connected to the evaporator at a point beyond the expansion valve. Another line is run from the evaporator outlet, where they are joined by a three-way valve, to the receiver. With the two three-way valves open to allow flow through the by-pass lines and the expansion valve closed, the compressor forces hot gas through the coils, melting the frost. The hot gas condenses upon giving up heat in melting the frost, and returns to the receiver as a liquid. When all frost is melted, the system should be returned to normal operation as the condensing effect of the evaporator will drop and allow hot discharge gas to enter the receiver.

Reverse Cycle Defroster. A defrosting system employing the reverse cycle arrangement has been devised for use in commercial or transportation (truck) service. The defrosting cycle can be attended manually or automatically, the latter controlled by pressure, temperature, or a timer, or any combination of these three variables. Normally, the unit is controlled automatically with a combination timer and a pressure switch, the former starting the defrost at a set time and the latter ending the cycle when the pressure in the evaporator builds up after the frost has melted.

An expansion valve is placed between the receiver and the condenser and acts as the throttling device during the reverse cycle. However, there is a check valve built into this expansion valve to allow normal liquid flow when the system is operating as a cooling unit. The receiver is equipped with dip tubes placed at the inlet and the outlet to allow liquid to flow in either direction. A slide type, four-way valve, pilot operated by a solenoid valve, controls the direction of refrigerant flow so that during the defrost cycle the normal evaporator acts as a condenser and the normal condenser, fed by the auxiliary expansion valve, serves as an evaporator. The receiver continues to serve its intended purpose for either cycle.

The water dripping from the evaporator during the defrost cycle is caught in a drip pan and carried away through a drain tube which is bonded to the suction line, both of which are wrapped in a common insulation covering to prevent the drain water from freezing and also to maintain the heat within the suction line which is warmed during the defrosting periods.

Heat Bank System of Defrosting. A new system has been devised to defrost evaporators especially adaptable to the low-temperature field. It is called the Thermobank* (L type) system, the L representing the latent heat principle involved in its operation. The two basic types include one for large systems over 18,000 Btu per hr and one for smaller systems under 18,000 Btu per hr (Fig. 482). The major difference in the two is that suction gas from the evaporator *by-passes* the heat bank during normal operation of the larger type; in the smaller type, the suction gas from the evaporator *passes through* the heat bank during normal operation.

During the refrigeration cycle of larger systems, the refrigerant vapor is drawn from the evaporator directly back to the compressor. In smaller systems, the vapor from the evaporator is drawn through the hold-back valve "H" into the re-evaporation coil which is immersed in the heat bank water and thence to the compressor suction.

From the compressor of either system the discharge gas enters the heating coil, also immersed in the heat bank water, to which it transfers heat, and from there flows to the condenser. In order to control the temperature of the heat bank at a maximum level (about room temperature), a built-in by-pass is installed in the heating coil. The size of this

* Proprietary designation of the patent owners, Kramer Trenton Company.

by-pass is such as to force proportionately more discharge gas to flow directly to the condenser as the temperature of the heat bank increases.

The defrosting cycle, which lasts about 6 to 8 minutes, is initiated by a timer which opens the hot gas solenoid valve, closes the suction line solenoid valve present in larger types, and stops the evaporator fans. The hot gas enters the evaporator where it condenses and defrosts the coil.

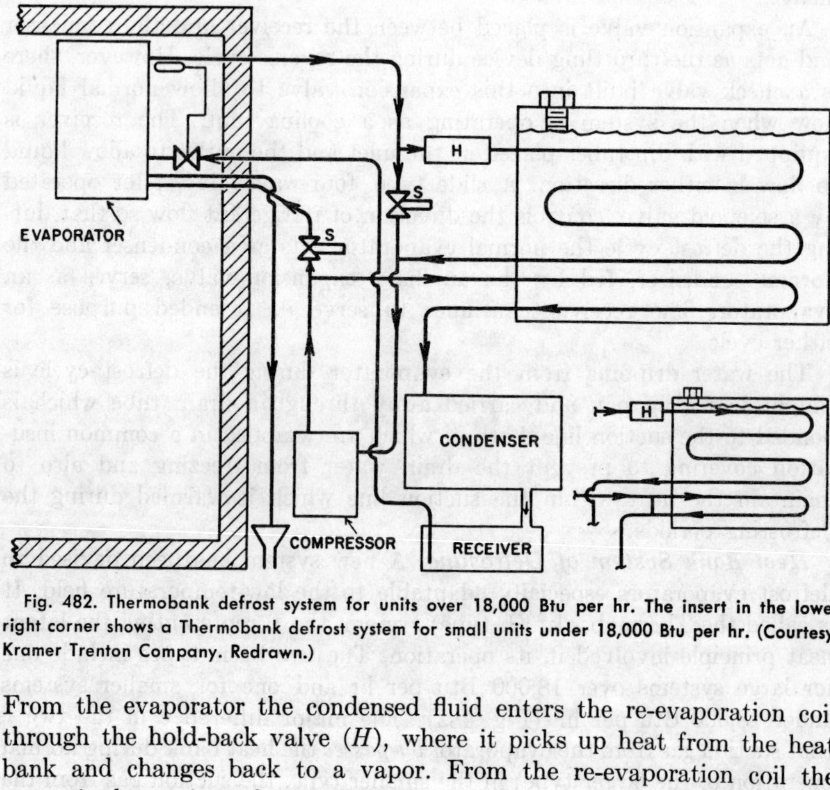

Fig. 482. Thermobank defrost system for units over 18,000 Btu per hr. The insert in the lower right corner shows a Thermobank defrost system for small units under 18,000 Btu per hr. (Courtesy, Kramer Trenton Company. Redrawn.)

From the evaporator the condensed fluid enters the re-evaporation coil through the hold-back valve (H), where it picks up heat from the heat bank and changes back to a vapor. From the re-evaporation coil the vapor is drawn into the suction side of the compressor. During the defrosting period, the hold-back valve acts as a constant-pressure expansion valve which feeds liquid from the evaporator to the re-evaporation coil. The rate of evaporation in the re-evaporation coil is controlled to maintain a suction pressure during defrosting about equal to that of the normal refrigeration cycle. By doing so, the compressor does not become overloaded during the defrosting cycle.

During re-evaporation of the liquid in the heat bank, the water surrounding the coils freezes. The abstraction of both sensible and latent heat from the heat bank water results in a greater heat storage potential at a reduced temperature level. After defrosting is complete, the timer closes the hot gas solenoid valve but does not start the fans (or open the suction line solenoid) until the evaporator pressure has been reduced to normal and the defrost water completely drained. During the period of delay between the closing of the hot gas line solenoid valve and return to normal operation, called the post-defrost, the remaining liquid in the low side is re-evaporated so that slugging does not occur.

The defrost water is drained from the evaporator as shown. Note that the hot gas line is in contact (soldered) with the drain line to keep it warm and prevent freezing.

Warm Liquid Refrigerant Defrosting. Where an accumulator is employed, such as in large ice-making plants, the warm liquid from the receiver can be used to defrost evaporator coils. A globe valve must be placed in the suction line leading from the coils to the accumulator.

By closing off the suction line by means of the globe valve and opening the expansion valve wide, the liquid fills the coils at the receiver pressure and temperature. Being warm under the latter conditions, the liquid transfers heat to the coils, causing the frost to loosen and break away or melt. The globe valve is then opened slightly (cracked) to allow the liquid to enter the line to the accumulator from which any resulting vapor is drawn off by the compressor, and the remaining liquid is returned to the coils after proper valve adjustments are made.

This method of defrosting should be practiced only where the compressor is protected against slugging, as with an accumulator, and where proper valves are installed. The evaporator must also be strong enough to withstand high-side pressure.

External Methods

Defrosting of cooling coils frequently is accomplished by supplying heat from sources other than the hot discharge gas from the compressor. Warm air, water, steam, and electricity are the principal sources of heat externally applied for the purpose. Some of the more important methods of applying heat to frosted cooling coils are described below.

Off-cycle. The simplest but slowest method of defrosting any type of system is to stop the compressor and allow the temperature within the refrigerated compartment to rise above the melting point (32 F). Placing open vessels of hot water in smaller units, such as household refrigerators and home freezers, or allowing warm air to enter through open doors of any cold compartment will hasten the melting process.

Control of the defrost period on domestic units varies with different models, all of which are equipped with a suitable switch for the purpose. One type requires that the defrost period be started and ended manually while another will automatically return to its cooling cycle after a given period of manual shutdown.

A range adjustment is the principle governing still another type of defrosting arrangement which allows the unit continually to provide refrigeration, but at a temperature high enough to permit defrosting, the adjustment being wholly manual.

One type of domestic off-cycle defroster depends on a time switch for full automatic operation. The defrosting cycle starts and ends according to the setting of a time (clock) mechanism which opens the power circuit at the desired time to defrost the unit and closes it after a set interval to return the system to normal operation.

Electrical Heating Elements. Employing electrical energy for defrosting purposes has wide possible application. As applied to unit coolers, heating elements are placed within the coil cabinet and energized after the suction line valve is shut off and the cabinet louvers closed. Defrosting is accomplished in about 30 minutes.

Heat applied to the liquid refrigerant after it leaves the capillary tube by an automatically controlled electric heating element is the basis of one external defrost system on domestic refrigerators. The defrost cycle starts upon opening of the refrigerator door a given number of times as counted mechanically.

Upon opening the door after a given number of times, an electrical heating element is automatically energized. This element heats the liquid refrigerant as it leaves the capillary tube to enter a storage tank located immediately ahead of the evaporator. Vaporized by the heat from the element, the refrigerant enters the storage tank from which it flows through a U-tube opening designed to carry only sufficient liquid to meet the cooling demand. Being too small to satisfy the entire vapor charge causes the pressure to build up in the storage tank and force some of the accumulated liquid into the evaporator. This spasmodic surge of liquid acts to boost the effectiveness of the warm vapor that follows each surge. Some of the liquid in the evaporator, which is of the flooded type, is forced over into the suction line by the incoming warm vapor, but vaporizes when it reaches the heat exchanger formed by the contact of a section of the suction line with an equivalent section of the discharge line. A pressure is built up within the suction line as a result of vaporization in the heat exchanger preventing further liquid carry-over from the evaporator. The vapor is then drawn from the heat exchanger into the compressor. During the process, all the refrigerating effect occurs in the heat exchanger, reducing the temperature of the discharge gas.

Water Bath Defrosters. Except in extremely low-temperature compartments, defrosting may be accomplished by spraying water over the surfaces of the cooling coils. Warm or cold water may be used in good effect. Catch pans must be placed under the coils and provided with suitable drain lines to prevent water from spilling onto the floor and freezing. Because of the presence of drip pans, water spray defrosting should not be employed in compartments depending on gravity cooling as they interfere with the descent of air currents.

Brine can be used in place of water for defrosting coils of low-temperature systems. However, corrosion of the coils must be guarded against wherever salt solutions are employed. Brines, being hygroscopic, must operate with a regenerator to keep the salt intact to a level that will prevent freezing.

Water circulated through tubes placed adjacent to cooling coils is another method of employing this medium for defrosting. Freeze-ups, however, should be prevented by draining the tubes after defrosting.

Defrost Cycling Control

Defrost cycling can be controlled manually or automatically. Time and trip frequency are the chief elements employed for automatic control.

The timer is a clock-like arrangement, equipped with appropriate linkage, which can be adjusted to open or close an electric circuit at desired intervals, such as every 24 hours. Electric circuit control may involve stopping the compressor and allowing the evaporator to defrost during the off-cycle. Or it may cause actuation of solenoids or energizing of electric heating elements, or initiate such other activities peculiar to the design of individual defrost systems.

Trip devices, or counters, may be designed to perform functions similar to the timer, except that their operation is based on the frequency or repetition of some appropriate action, such as refrigerator door openings.

The disadvantage of timers and trip devices is that they act regardless of the amount of frost accumulated during the period for which they are adjusted. This fault is not serious with small refrigeration systems, or where consistency of circumstances permits an accurate estimate of frost accumulation rate based on experience. On larger systems, such as railway car air-conditioning units, timers cause interruptions in cooling activity whether or not defrosting is necessary.

One method of controlling defrost cycling according to actual accumulation is based on the drop in air pressure across air-cooling coils. When the accumulation of frost reaches a magnitude as to cause a predetermined drop in air flow as reflected by increased pressure differential across the coil, the defrost cycle is initiated. Defrosting of coils on which the pressure drop principle is used is usually accomplished by means of encased electric heating elements that are built into and span the evaporator housing. While the pressure drop method appears ideal for stationary refrigeration systems when applied to railway car installations, some erratic defrost control occurs due to vibration.

HERMETIC ELECTRIC MOTORS

Motors that are so constructed and sealed as to be installed within the refrigerant flow circuit, *hermetic* types, are employed under appropriate conditions. The modern domestic refrigerator has, of course, employed such a motor for a number of years. Hermetic motors offer certain advantages. Because of the cooling effect of the refrigerant, usually R-12 or R-22, a motor smaller than that used for non-hermetic systems can be employed. Also, the shaft seals that separate the refrigerant from the atmosphere can be eliminated. See Figs. 96 and 97.

ICE-MAKING SYSTEMS

Ice made by freezing water in various forms and sizes is used widely for the storage and distribution of refrigeration. Natural ice is obtained from frozen bodies of water, cut to size on the lake or pond and stored in ice houses, insulated by a covering of sawdust until distributed to the consumer. Manufactured ice in the form of blocks weighing from 50 to 400 lb is made by immersing cans of distilled or raw water in cold brine until frozen. Ice in the form of small cubes, cylinders, and briquettes is made in specially designed refrigerated machines, or evaporators. Flake ice and shell ice also are made in ice-making machines.

Block Ice

Ice in block form is made in sizes weighing 50, 100, 200, 300, and 400 lb, depending on the size of the can in which the water is frozen. The ice cans are tapered to facilitate removal of the ice, being 1 in. less in width and breadth at the bottom than at the top. The top dimension of cans having a capacity of 200 lb or over is 11 in. × 22 in., although some measure $11\frac{1}{2}$ in. × $22\frac{1}{2}$ in. They vary in length according to capacity, being approximately 32 in., 46 in., and 60 in. for 200-, 300-, and 400-lb blocks, respectively. All are made of 16-gage galvanized sheet steel with the exception of the 400-lb cans, which are of 14-gage steel.

A favorable brine temperature is 16 F. If the brine temperature is too low, the ice will become hard and brittle, causing it to break readily into small pieces when subject to any appreciable shock. When too high, the freezing time will be excessive or freezing will not occur throughout the block.

Raw water is water that has not been distilled but is fit for human consumption. It contains varying amounts of impurities, including air, carbon dioxide, mineral substances, etc. When impurities are present, they become entrapped between the ice crystals that form during the freezing process and cause the block to be opaque or have a "milky" appearance. This condition is overcome by agitating the water with air during the freezing period. The agitating action prevents the impurities from resting and causes them finally to accumulate in the water remaining unfrozen in the core of the ice block. From there the water is removed by a suction nozzle, together with the accumulated impurities. The space left in the core is then filled with fresh water and allowed to freeze. The drop-pipes are removed from the can at the appropriate time to prevent their becoming frozen in the block.

The agitating air is supplied by a compressor or a blower through an appropriate arrangement consisting of a header and several branch lines to drop-pipes. The drop-pipes are immersed in the ice water at the center of the can to a depth of about 10 in. from the bottom. From there the air bubbles up through the water and prevents impurities from becoming entrapped. The volume of air fed to the cans depends on the amount of impurities contained in the water, and should be increased with any increase in the latter. The air volume also should be increased if the brine temperature is lowered. The air pressure depends on the depth of the can and should be increased with deeper cans.

The brine should be agitated or circulated within the tank to obtain maximum heat-transfer capacity. This is done by using a propeller that forces brine over a series of cooling coils and around the brine tank. The path of brine may be controlled by installing strategically placed baffles in the brine tank between which flow can be directed to obtain highest efficiency. Brines are discussed fully in Chapter 13.

Raceway coils are often used, over which brine flows at a velocity of 15 to 30 fpm. Open-end horizontal shell-and-tube evaporators also are used. Cooling by the latter methods is accomplished in a small area of the tank with the cold brine being circulated around the cans. Older systems employed evaporator coils through which the refrigerant was circulated through the tank between can rows and the brine agitated for maximum heat-transfer efficiency.

The latent heat of fusion for ice is 144 Btu per lb. However, this value is based on the conversion of 32 F water to 32 F ice or vice versa. In determining the heat load for ice-making, the water temperature at time of placing it in the brine must be considered. Assume an initial temperature of 80 F and a brine temperature of 16 F; the theoretical amount of heat to be removed per pound of water can be determined as follows:

To reduce water from 80 F to 32 F	$= 48 \times 1$	=	48 Btu
To convert 32 F water to 32 F ice	=		144 Btu
To reduce 32 F ice to 16 F ice	$= 16 \times 0.5$	=	8 Btu
Total heat to be removed from 1 lb of 80 F water		=	200 Btu

To compensate for refrigeration losses in the system, the theoretical value above should be increased 10 to 20 percent, depending on operating conditions. It will be noted that the load is affected by the temperatures of the incoming water and the brine. While it is not always possible to control the temperature of incoming water, some reduction in load can be obtained by maintaining the brine temperature as high as is economically feasible. Water precooling also helps to reduce load.

Under the conditions shown in the above example, the refrigeration requirements per ton of ice per day would be, assuming a 15 percent loss:

$$\frac{2000 \text{ lb of ice} \times 200 \text{ Btu per lb} \times 1.15}{288{,}000 \text{ Btu per ton per day}} = 1.597 \text{ tons}$$

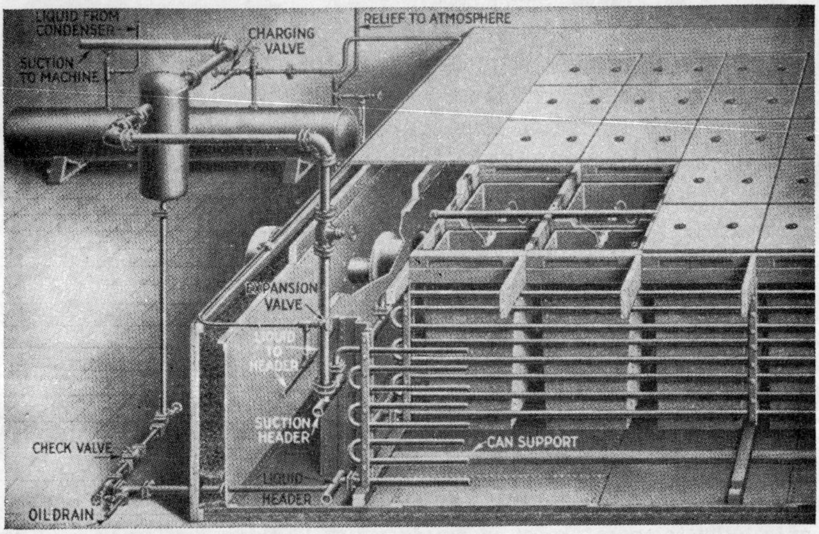

Fig. 483. (*Top*) Evaporator coils of the type used in brine tanks for making ice. (Courtesy, Frick Company.)

Fig. 484. (*Bottom*) Cross-sectional view through ice-making tank, showing arrangement of coils, headers, accumulator, agitator, insulation, cans, removable covers, etc. (Courtesy, Frick Company.)

or approximately 1.6 tons of refrigeration per ton of 16 F ice produced from 80 F water.

The time required to freeze ice varies as the square of the thickness (h^2) and inversely as the temperature difference between the freezing point of water and the brine temperature (t_d), or

$$\text{hours} = \frac{h^2}{t_d} \times K$$

where h = maximum thickness of block in inches (top of can)
t_d = difference between 32 F and brine temperatures
K = constant for a given operating and design condition, usually given a value of 6 to 7 for the average plant having a brine velocity of 20 to 30 fpm

The time in hours to freeze an $11\frac{1}{2}$-in. block of ice in 16 F brine using standard cans would be:

$$\frac{11.5 \times 11.5}{32 - 16} \times 6 = 49.59 \text{ hr}$$

For any given ice-making plant the freezing time can be increased by lowering the brine temperature.

The storage temperature for ice should be low enough to prevent melting and high enough to keep it from becoming too hard. Any melting and subsequent freezing causes the blocks to bond together. Hard ice tends to shatter when subject to shock. The most favorable temperature for storing ice appears to be 27 F, based on the author's experience in the tropics.

A modern development in ice distribution is the refrigerated vending machine. Block ice is cut into convenient sizes of 25- or 50-lb cakes and placed in the vending machine, wrapped or unwrapped, and dispensed when the proper coin for payment has been inserted. The ice must be maintained at the proper storage temperature during the entire handling process. Ice cubes are also dispensed from vending machines.

Ice-Cube Maker

Ice cubes may be produced from evaporators in which a number of trays are placed. Figure 485 shows the type of evaporator in which ice cubes are mass produced. The process is similar to that employed to produce ice cubes in domestic refrigerators. Cube makers are also available in fully automatic self-contained units that will produce up to 500 lb of cubes per day.

Shell-Ice Maker

Shell Ice is made by freezing water as it circulates over the outside surface of vertical stainless steel direct expansion evaporator tubes. When the shell of ice formed on the tubes reaches a desired thickness, the water

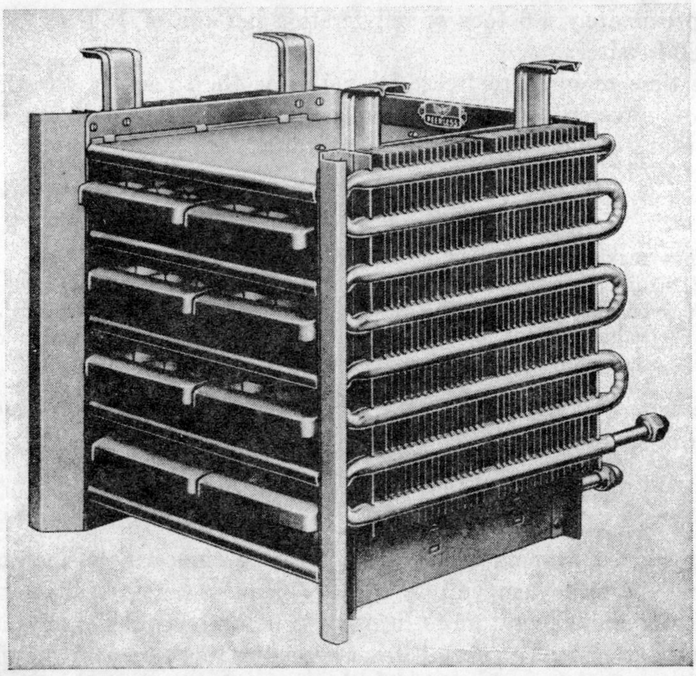

Fig. 485. Ice-cube maker. (Courtesy, Peerless of America, Inc.)

and the refrigerant automatically are turned off. Warm refrigerant is then admitted to the tubes to release the ice from the surface. The shell drops by its own weight to a storage bin. As it slips from the evaporator tube, the shell is broken into small pieces by an ice breaker. (See Fig. 486.)

Shell Ice weighs about 34 lb per cu ft. There are 9 sizes of Shell-Ice Makers, varying in capacity per 24 hr of from 2 to 4 tons for the smallest up to 20 to 46 tons for the largest. Shell Ice can be handled by blowing it through flexible tubes, which facilitates distribution within the food-storage compartment.

Tube-Ice Maker

The Vogt Tube-Ice Maker consists of a number of vertical tubes submerged in liquid ammonia. Water flows from the distributing header at the top of the tubes down through the tubes to a sump below. From the sump it is pumped back to the distributor so that it is constantly circulated during the freezing process. As the water flows through the tubes, it freezes on the surfaces, which are cooled by the surrounding refrigerant. Freezing proceeds inward, forming a tube of ice whose thickness depends on the time cycle. When the desired thickness is reached, an automatic timer shuts off the flow of liquid ammonia and admits a hot discharge gas. The hot, high-pressure gas forces the ammonia liquid into an auxiliary chamber and also releases the ice from the interior surface of the tubes.

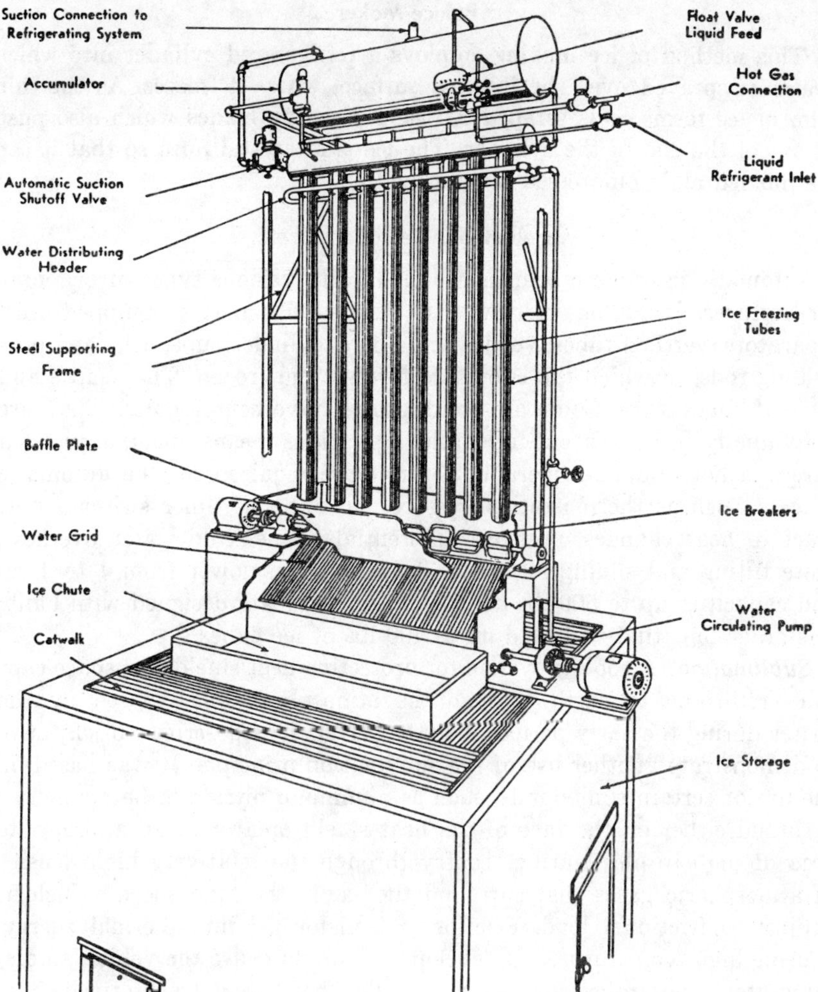

Fig. 486. Shell-Ice machine, showing names of principal parts. (Courtesy, Frick Company.)

The ice tubes drop by their own weight and are cut to length by a revolving cutter. The freezing process is then repeated.

Flakice Maker

Flakice is made by slowly rotating a brine-filled drum covered by a flexible cylinder in a bath of water. The thin film of ice forms on the external surface of the outer flexible cylinder. Because of its flexibility, the cylinder becomes distorted by a roller at the top and causes the ice to peel from its surface. As it peels from the cylinder surface, it breaks up into flakes and falls into a storage bin.

Pakice Maker

This method of ice making employs a refrigerated cylinder into which water is sprayed over the internal surface, where it freezes. As the thin film of ice forms, it is scraped off the surface by blades which also push it out of the end of the cylinder. The ice is in crystal form so that it can be molded and compressed into briquettes.

Miscellaneous Ice-cube Makers

Automatic ice-cube machines are made with various types of evaporator surfaces, including vertical and inclined flat plates equipped with separators, vertical tubes, vertical cylinders, cylinder cups, cells and protruding rods, in which the cubes are formed and frozen. The shapes and sizes of cubes vary. Some are solid, others have center holes. They are individually formed or cut in lengths by various means, including heated wire and hot liquid refrigerant. Controls are required for the automatic process, such as thermostatic, pressure, cut-out and timer switches that react to heat changes, pressure differentials, time, cold water overflow, plate tilting and sliding ice bath. They range in power from $\frac{1}{3}$ to 1 hp and capacities up to 500 lbs per 24 hours. Many are designed with built-in storage bins that will hold up to 400 lbs of ice cubes.

Sublimation. A cooling system for protecting heat shields for space capsules returning to earth through the atmosphere was devised by the writer during the early period of space travel. It is described briefly here to demonstrate another use of the sublimation principle. It was based on the use of certain compounds, such as aluminum nitride to be applied as a shroud either on the face of the heat shield or around an appropriate area of the capsule. During reentry through the relatively high density of atmospheric gases that surround the earth, the high speed vehicle is subject to frictional load. Friction is transformed into thermal energy, causing high temperatures to develop that would cause the vehicle to disintegrate if not otherwise protected. The heat thus transformed from frictional energy would tend to convert the sublimate or shroud directly to its vaporous form, thus dissipating the heat as in any other refrigerating system utilizing latent heat of vaporization.

The materials recommended for this process are solid under normal atmospheric conditions of temperature and pressure. They sublime only at the higher temperature prevalent under reentry conditions.

TRANSPORTATION EQUIPMENT

Perishable commodities must be refrigerated when being transported via trucks, trailers, and railway cars. The major problem associated with supplying mechanical refrigeration in any rolling equipment is that of obtaining power to operate the compressor and the extra weight imposed by the cooling system. One alternative is to use ice or dry ice to maintain the load in a safe condition during transportation. In cold weather, ventilating doors located at both the front and the rear of the vehicle will serve to cool the load during movement.

Ice-Cooled Vehicles

The ice (frozen water) usually is stored in bunkers through which air is circulated by either gravity or blower. Where a blower is used, it is driven by a small gasoline engine or a battery-powered electric motor. The bunker is located in the front of the vehicle with the blower at the top. The blower engine, or motor, is mounted on a platform outside the vehicle at the top and front, where it is a familiar sight to motorists. Hatches for replenishing the ice supply are located above the bunker. Floor drains allow the ice water to escape. Railway cars or "reefers" are equipped with two bunkers located at each end. Air circulation is obtained by gravity or fans in the latter case.

A grill or grate installed at the bottom of the bunker allows the warm return air from the load compartment to pass upward through the ice. The cold air issuing from the top may be blown out at random on to the load. Or it may be directed through a duct toward the opposite end of the vehicle, from where it flows back to the bunker intake, picking up heat en route. Installing racks along each side wall and on the floor permits proper air circulation through the vehicle.

Dry ice (solid carbon dioxide) is used also to refrigerate trucks and trailers. In the trucks the bunker for storing the dry ice is located at the top of the vehicle. A metal plate forming the floor of the bunker acts as a heat-transfer surface. The remaining surfaces are insulated. By placing a cooling coil filled with an anti-freeze solution in the dry-ice bunkers, and extending it along the sides or top of the vehicle proper to permit circulation, a more positive system is devised. The anti-freeze solution is circulated by means of a pump, which stops and starts by thermostatic control to maintain a constant temperature.

A finned dry-ice bunker is used to refrigerate large trailers carrying frozen foods. Placed in the nose of the trailer, air is circulated by a main fan over the fins and into the storage space much in the same manner described above for ice (water)—refrigerated vehicles. Controlled by a thermostat, the main fan starts and stops according to temperature demand. An auxiliary by-pass fan continues air circulation within the storage space when the main fan is shut off.

Mechanical Refrigeration of Trucks and Trailers

The mechanical refrigeration system employed on vehicles is similar in principle to stationary units. However, methods of supplying power to drive the compressor require special consideration due to their mobile

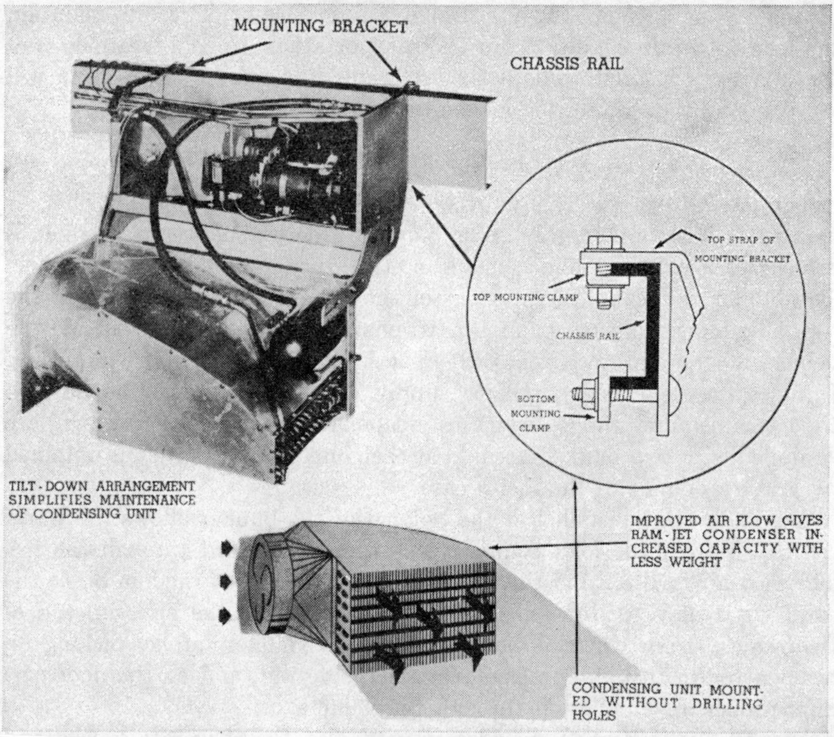

Fig. 487. Truck condensing unit mounted on chassis rail showing tilt-down arrangement to facilitate servicing of condenser. (Courtesy, Kold-Hold, Division of Tranter Mfg., Inc.)

characteristics. Compressors may be driven by individual gasoline engines, electric motors, or an engine-motor combination. The motor of the latter type is operated by current from a platform connection during loading or parking. Others are driven by utilizing the power plant of the vehicle, through some *power take-off* arrangement from the engine, transmission, or propeller shaft. The power take-off drive may be direct by belt, hydraulic system (Fig. 489), flexible shaft (Fig. 490), and/or some type of variable-speed clutch. An indirect method is to drive a generator which, in turn, supplies current to an electric motor driving the compressor. The condensing unit may be installed in a special compartment or under and on the side of the vehicle where it is usually mounted on the frame, or in front of the body at the top. The unit should be located to allow easy access for repairs. The condenser, the compressor, and the evaporator may be an integral unit, or they may be separate. The integral unit is mounted on the front of the vehicle body at the top with the evaporator projecting inside the storage space. Another type has a compressor and a condenser mounted under and at the side of the vehicle, serving an evaporator either of the blower (see Fig. 490) or the plate (see Fig. 489)

Fig. 488. Blower evaporator for installation in trucks and trailers. Note drain connection for defrost water, also connection for hot gas line defrosting. (*Left*) Front view; (*right*) rear view showing component parts. (Courtesy, Kold-Hold, Division of Tranter Mfg., Inc.)

type located in the vehicle body. In some cases, the condenser is mounted in front of the vehicle engine radiator. All are self-contained units; i.e., their source of power is located on the vehicle.

The hydraulic system shown in Fig. 489 derives its power from the engine crankshaft. A hydraulic pump located forward is driven from the front end of the engine crankshaft. This pump in turn transfers power through the hydraulic oil lines to a fluid motor that drives the compressor. The hydraulic pump delivers oil to the fluid motor at a constant rate regardless of engine speed. This results in a constant-speed compressor.

Power take-off drives should be augmented by an electric motor that can be connected to a power line at the loading platform. Thus the temperature in the vehicle body can be pulled down and/or maintained without running the vehicle engine. This also applies to compressors driven by individual gasoline engines.

Refrigerated vehicles must be sufficiently insulated. Because it is light in weight, Fiberglass commonly is used, the thickness of which depends on whether low (0 F) or high (32 F or above) temperatures are to be maintained.

Defrosting of evaporators is necessary for continual heat-transfer efficiency. This is done automatically, or manually in many cases, by the reverse cycle method described previously in this chapter.

Continuous hot gas defrosting is obtained automatically by means of a temperature-sensing device that is installed and set at the factory of the manufacturer.* The temperature-sensing device is a control bulb inserted among the cooling coil fins, but not in contact with the tubes.

* Kold-Hold, Division of Tranter Mfg., Inc.

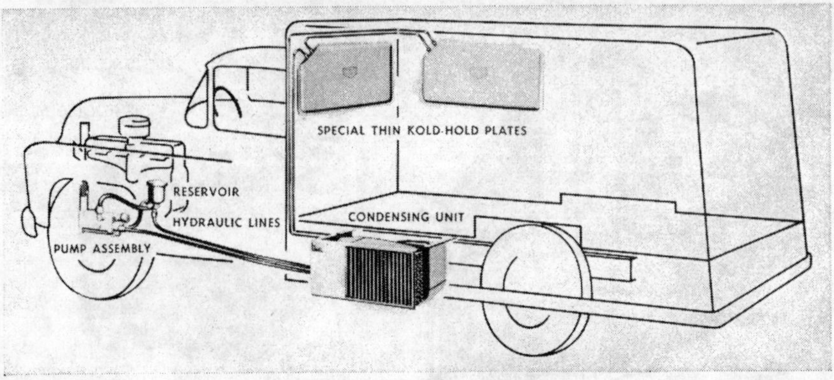

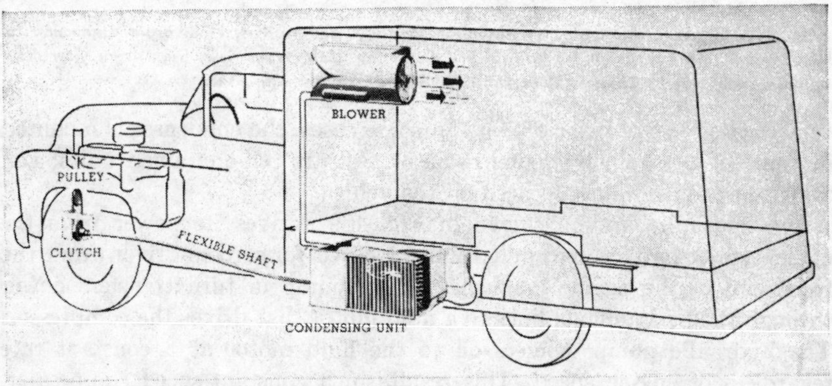

Fig. 489. (*Top*) Refrigerated truck installation. (Crown System) with eutectic plate evaporators and hydraulic system power take-off driving fluid motor. Actual temperature is determined by the eutectic solution employed, ranging from 26 F to −14 F. (Courtesy, Kold-Hold, Division of Tranter Mfg., Inc.)

Fig. 490. (*Bottom*) Refrigerated truck installation with blower evaporator and flexible shaft power take-off from engine crankshaft. (Courtesy, Kold-Hold, Division of Tranter Mfg., Inc.)

A hot gas line from the compressor discharge is connected to a gas inlet on the evaporator. The bulb is connected by a capillary tube to a valve on the hot gas line. When the fin temperature reaches 32.5 F, the fluid in the bulb expands to open the hot gas line valve, allowing hot gas to enter and defrost the evaporator. When the heat of the hot gas reaches the control bulb, the valve closes and refrigeration action is again initiated.

On high-temperature operation (above 35 F), defrosting will occur by allowing the air-circulating fan to run during the off-cycle period.

Eutectic Plates for Trucks and Trailers

Eutectic plates act as evaporators and refrigeration capacity storage (hold-over) units. They can be charged during night storage of the vehi-

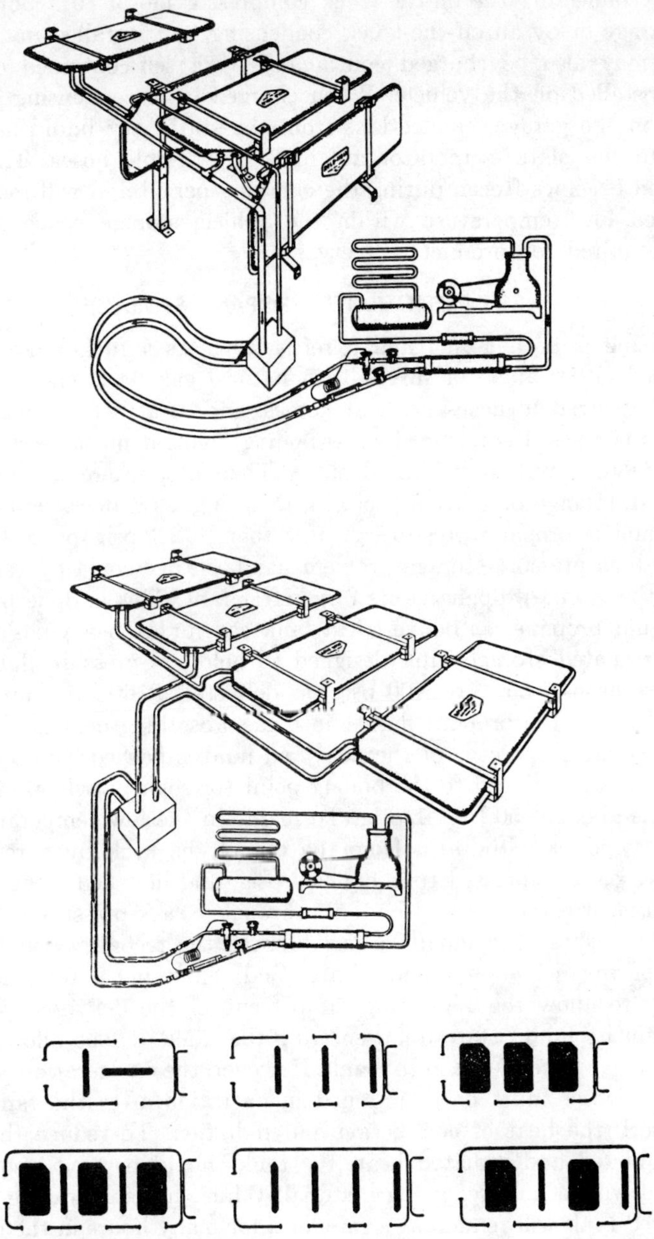

Fig. 491. Eutectic plate evaporator hook-up for vehicles showing various arrangements possible. (Courtesy, Kold-Hold, Division of Tranter Mfg., Inc.)

cle by connecting the on-the-truck compressor motor to a power line in the garage or by an off-the-truck condensing unit installed in the garage. They may also be charged continually by a self-contained condensing unit installed on the vehicle. When charged by a condensing system located in the garage, connections from the stationary liquid and suction lines to the plate evaporator are made by flexible hoses. The eutectic solution becomes frozen during the charging period and will maintain the required low temperature within the vehicle storage space during the time required for product delivery.

SELF-REFRIGERATED PROPANE STORAGE

Propane is both a fuel and a refrigerant. As a fuel, large quantities are used either alone or mixed with natural gas. It is employed to fire large industrial furnaces, to heat homes, and as a fuel for internal combustion engines. Because of its low boiling point, it must be stored under pressure to maintain its liquid state. Therefore, where large quantities are used, storage becomes a problem, requiring, as it does, tanks that will withstand a propane pressure greater than 172.3 psig on a 100 F day.

The high-pressure storage problem has been overcome by refrigeration using the stored propane as its own refrigerant. This is done by expanding liquid propane, as delivered in tank cars or from a pipe-line, into a large insulated storage tank designed to hold low-pressure fluid. As the pressure drops from, say, 100 psig (as delivered at 60 F) to about 1 psig (in the tank), the propane flashes into gas, absorbing heat in the process. When about 50 percent of the delivered fluid is flashed, the temperature in the tank is reduced to the boiling point (or slightly below) of propane, which is about -50 F at 1 psig. To maintain the low temperature, some of the vapor is withdrawn from the top of the tank by a refrigeration compressor. Compressed to about 200 psig, and liquefied in an air-cooled condenser, it is returned to the tank, through a back pressure valve, where it again flashes. Continual flashing of both the recompressed liquid and some of the delivered liquid creates and maintains a temperature low enough to allow the remaining 50 percent of the delivered propane to maintain its liquid state upon entering the tank. It also allows much of the flash gas to condense in the tank. However, the refrigeration compressor and condenser must be maintained in operation after the tank is filled to absorb the heat of conduction and radiation. To reduce the amount of conducted and radiated heat, the tank should be well insulated and covered with a heat-reflecting material. While the temperature in a well-insulated tank will remain at a safe level for many hours in the case of refrigeration plant failure, a safety valve should be installed as a safety precaution. The safety escape line may be exposed to a pilot light to burn the escaping gas.

TRAILER COOLING TUNNELS

A method of shipping perishable commodities over limited distances is to precool the cargo before leaving for its destination. Loading well insulated highway cargo trailers and running them into cooling tunnels for a period of time assure protection during shipment. The tunnels are precooled and the trailers are kept open during the cooling period. When the center of the mass reaches a predetermined (about 36 F) temperature, the trailer is closed, sealed and sent on its way. Lettuce, celery and various fruits are commodities shipped by this method.

CRYOGENIC COOLING

The cooling of vehicles carrying perishable products using liquid nitrogen and other cryogenes is fast being adopted. This method of transportation cooling is more fully described in Chapter 17.

DRY-ICE PRODUCTION

Solid carbon dioxide or dry ice, as it commonly is called, is made by converting the gas to a liquid at some pressure above 75.1 psia, the triple point, and then expanding the liquid in a chamber having a pressure below the triple point, where it flashes into loose snow. The snow is then pressed into blocks.

A three-stage compression system is employed to raise CO_2 gas from any low initial pressure to 900 to 1100 psia. The gas is filtered and dehydrated between the first and the second stages and intercooled between each stage. It is condensed at the discharge pressure of the third stage. The high-pressure liquid is then reduced from that of the third-stage discharge to some pressure below the triple point, usually atmospheric. Some of the liquid flashes into snow while the remainder converts to gas. The gas is drawn off by the first-stage compressor and recompressed. The expansion chamber into which the high-pressure liquid is expanded is also a press chamber in which the snow, remaining after the gas is drawn off, is compressed into a block of solid CO_2. The reduction in liquid pressure is accomplished in stages. The flash gas produced during each stage of pressure reduction returns to the compressor having a corresponding suction pressure. Thus, in a three-stage compression system having a three-stage pressure reduction, the flash gas from the first reduction stage will be drawn off by the third-stage compressor. The second-stage compressor will draw off the gas from the second reduction stage and the first-stage compressor will withdraw the gas from the final expansion chamber.

The carbon dioxide gas may also be liquefied by cooling it with another refrigerant, such as ammonia, thus eliminating the need of heavy equipment. The heat transfer is accomplished by a combination ammonia evaporator and carbon dioxide condenser, an arrangement similar to that described for cascade systems. One such system produces dry ice from carbon dioxide gas obtained from a well. A booster compressor draws the gas from the well, and discharges it into a mixing drum located at the producing plant five miles away at about 25 psig. It is cooled by the flash gas returning from the snow presses. Some moisture is removed during the cooling process. From the mixer the gas is drawn into two compressors in which the pressure is raised to 250 psig. It then passes through a water-cooled shell-and-tube vessel which serves as a precooler and oil separator, and thence to a bank of oil filters. The cleaned carbon dioxide gas is then liquefied in a shell-and-tube condenser in which ammonia is the cooling agent, evaporating at -28 F. The liquid CO_2 is then stored in a receiver. Upon leaving the receiver, the liquid CO_2 is cooled in a single-pass heat exchanger by the cold gas from the snow presses, which is -100 F or lower. The snow presses have chambers 20 in. square, fitted with a ram in the upper part and a power-operated cover plate at the bottom. The machine is hydraulically (oil at 2400 psi) operated. About 42.5 percent of the liquid CO_2 flashes into loose snow. About 55 percent of the CO_2 flashes into gas and is drawn back to the system and recompressed. The snow is compressed to a thickness of 10 in. The block is then cut by saws into smaller blocks 10 in. square, wrapped in paper, and shipped in heavily insulated boxes 5 ft long.

Dry ice weighs about 86 lb per cu ft, or 0.05 lb per cu in. It has a temperature of -110 F at atmospheric pressure, under which conditions its heat-absorbing capacity is 246.3 Btu per lb. Because it is noncorrosive and nontoxic, it is used widely as a cooling substance for foods, medicines, etc.

REFRIGERATION CAPACITY STORAGE

The refrigeration capacity of any system is based on the amount of heat it is capable of absorbing from a space or substance in a given period of time, as established previously in this text. It is possible, therefore, for a system temporarily to lag behind a heat load even though its capacity is normally equal to or greater than the load. This is especially true when an abrupt change in load occurs, as it often does in air-conditioned restaurants, stores, churches, and theaters. The abrupt increase in load in such places is caused by a heavy influx of people in a short space of time, such as is experienced by restaurants during mealtimes. The result is an increase in temperature for periods of 15 minutes or longer, during which time the system works continuously to finally meet the demand. If the same number of people entered the restaurant, say, over a period of one hour, then the load would be spread and so enable the system to maintain steady conditions.

In contrast with the concentrated loads made on the system during

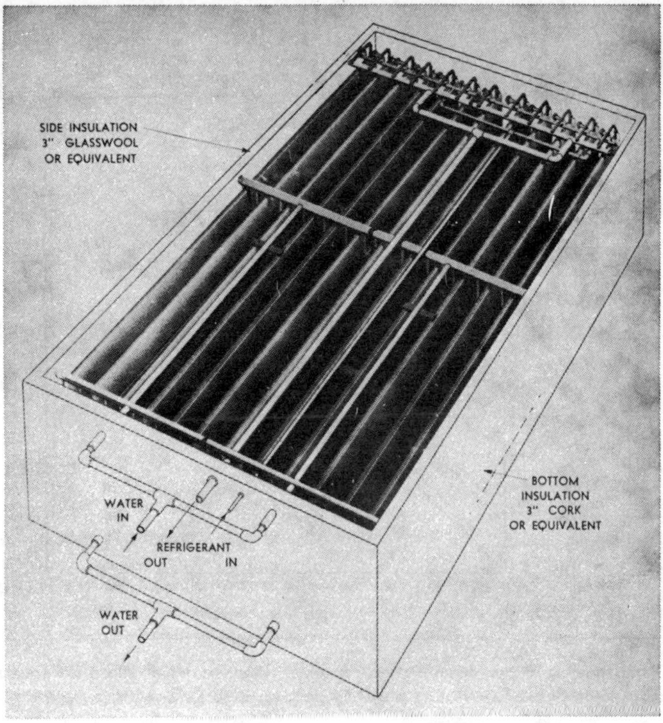

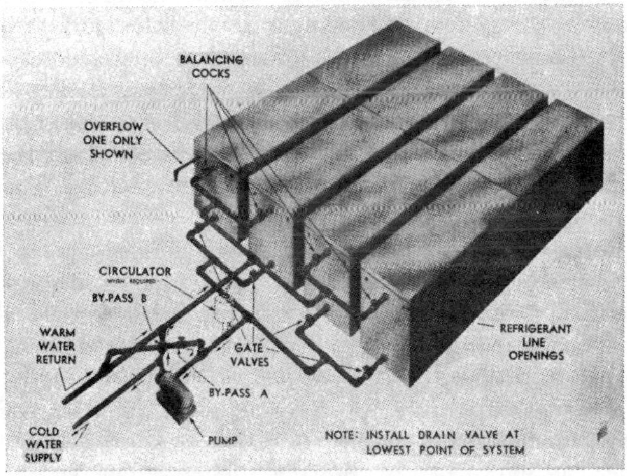

Fig. 492. (*Top*) Ice accumulator unit with plate evaporators showing refrigerant and circulating water piping. (Courtesy, Dole Refrigerating Company.)

Fig. 493. (*Bottom*) Arrangement for precooling and preheating circulating water for refrigeration accumulator. (Courtesy, Dole Refrigerating Company.)

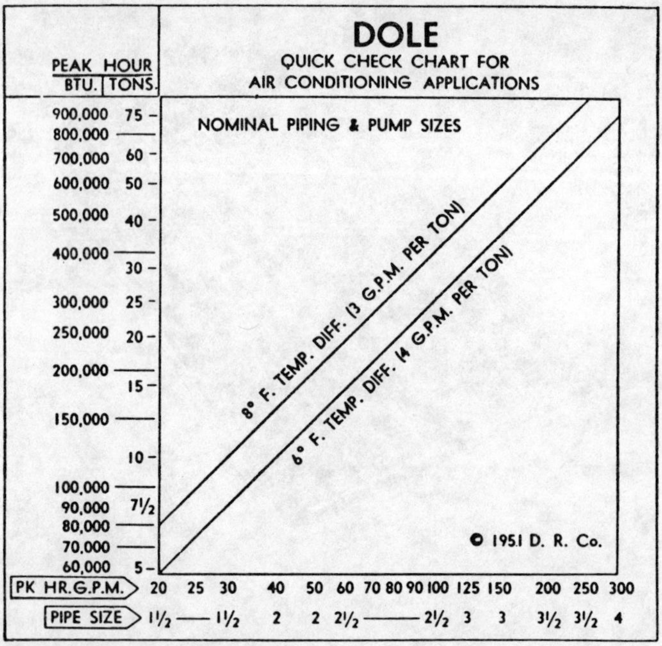

Fig. 494. Chart showing water requirements for given load vs. temperature difference between accumulator (Ice-Cel) inlet and outlet. (Courtesy, Dole Refrigerating Company.)

peak periods, only light demands prevail at other hours of the day. During these off-hours, the system does comparatively little work.

If, during the off-periods, the system could be operated to store refrigeration capacity, it would be the same as increasing its size to counterbalance the time element during peak loads. If the capacity so stored were released at the time when an abrupt increase in load occurred, desired temperatures could be maintained more readily. The situation brought about by abrupt load changes has necessitated refrigeration capacity storage systems whereby the system off-peak potential can be exploited.

One method of storing refrigeration is by producing ice during the off-peak periods and utilizing its latent heat to meet sudden heavy demands for short periods. Such a method is the ice accumulator system described below.

The ice accumulator unit consists of a tank in which plate evaporators are installed, as shown in Fig. 492. The plates, submerged in the water that fills the tank, are connected to the system condensing unit during off-peak hours. The system operates to build up a coating of ice on the plate surfaces to whatever thickness is required. When refrigeration is otherwise needed, the accumulator may be shunted or by-passed by installing appropriate valves in the refrigerant lines.

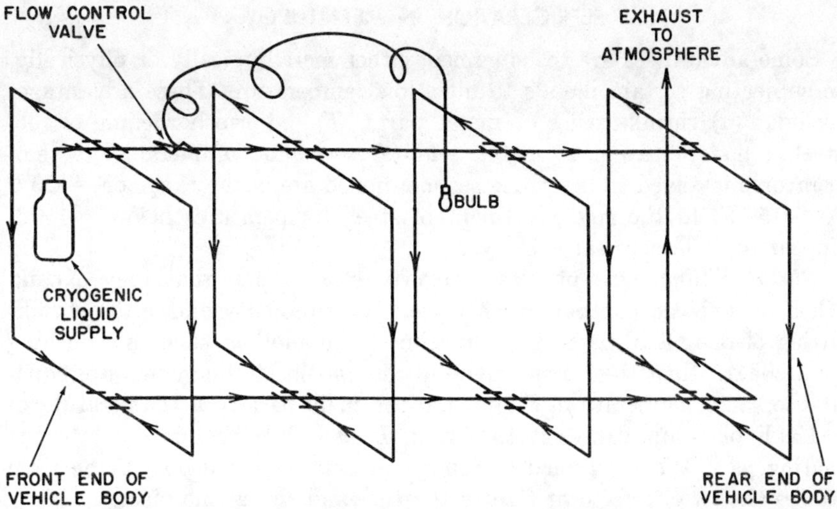

Fig. 495. Skeleton of cryogenic fluid distribution arrangement within a truck, trailer or railroad car. Liquid cryogenic vessel located outside and forward of the vehicle body. System is recharged by replacing empty vessel with a full one. Nitrogen liquid is commonly used as refrigerant.

Upon demand, the chilled water in the accumulator is circulated through a cooling coil placed in the conditioned air duct or stream. Here it gives up some heat in augmenting the main air-cooling element and returns to the accumulator. In passing over the ice previously formed on the plates, the water gives up heat. In giving up heat, it melts the ice so that accumulator capacity is based on the latent heat of fusion. For ice the latent heat of fusion equals 144 Btu per lb. Reduced in temperature, the water is returned to the cooling coil to repeat its function.

An arrangement for precooling or preheating the circulating water is shown in Fig. 493. If the water leaving the accumulator is too cold, it may be tempered by mixing with it some warm water from the coil. This is accomplished through by-pass *A*. Where the water from the cooling coil is too warm to attain a sufficient temperature reduction through the accumulator, it may be chilled by mixing it with water already cooled. This is accomplished through by-pass *B*. In many circumstances, the by-pass arrangement is not a necessity.

To insure uniform cooling—i.e., uniform ice melting—minimum water velocity through the accumulator must be maintained. The number of gallons of water per minute that must be circulated to obtain necessary velocities depends on the size of the tank and the number of plates in the tank.

To determine the quantity of water per minute required to handle a given load at a specific temperature difference between the inlet and the outlet of the accumulator, refer to the curve in Fig. 494.

REFRIGERATION IN METALLURGY

Some advantages are to be gained, either metallurgically or physically, in subjecting certain metals to ultra-low temperature. These advantages include: (*a*) shrink-fitting of metal parts, (*b*) uniform hardening of tool steel, (*c*) stabilization of metal, and (*d*) seasoning of metal. The temperatures involved in the processes mentioned are in the range of −120 F to −150 F. In the cold treatment of steel, temperatures below −150 F appear to be unnecessary.

Shrink-Fitting. One physical property of metals is that they expand when heated and contract when cooled. This property allows the shrink-fitting of one metal part—say, an arbor—into another, such as a bushing or a collar, when their respective outside and inside diameters are equal at the same temperature. Subjecting the male part to a temperature of −120 F or below causes it to shrink. It can then be inserted into the mating part. When warmed to room temperature, expansion of the male causes the two parts to fit tightly together and act as one piece.

Shrinkage varies with different metals. From room temperature to −120 F shrinkage per inch diameter is as follows:

Steel................	0.001 in. per in. diameter
Bronze..............	0.0015 in. per in. diameter
Aluminum...........	0.002 in. per in. diameter
Magnesium..........	0.002 in. per in. diameter

Uniform Hardening of Tool Steel. Steel is not a pure metal, but an aggregate of iron and carbon, the latter never exceeding about 1.75 percent. Iron itself will hold up to about 0.04 percent carbon in solid solution at ordinary temperatures. The remainder of the carbon exists as iron carbide. Iron and carbon in solid solution is a soft, ductile material called *ferrite*. The iron carbide is a hard, brittle material called *cementite*. Together they form a laminated structure of alternate layers of ferrite and cementite called *pearlite*. When steel is exclusively pearlite, it is called *eutectoid* steel and contains 0.85 percent carbon. When less than 0.85 percent carbon is present, free ferrite exists, resulting in *hypo-eutectoid* steel. Steel more highly carburized—i.e., over 0.85 percent carbon—contains an excess of ferrite and is known as *hyper-eutectoid* steel.

When eutectoid (0.85 percent carbon) steel is heated to a temperature of about 1350 F, the iron is capable of absorbing all carbon present, resulting in a solid solution called *austenite*. Austenite then is a solid solution in which a change in physical properties may occur without a change in substance. It does not have a constant composition, however, as the carbon content may vary upward of 1.7 percent. The same transformation occurs in hypo- and hyper-eutectoid steels at higher temperatures. The temperatures at which the above transformation begins and ends are known as the *lower* and the *upper* critical temperatures of a given specimen. When steel is heated above its upper critical temperature and then cooled rapidly, the carbon does not have time to separate from the iron phase. Rapid cooling by *quenching* in a liquid bath of the critically heated steel forms a needle-like structure called *martensite*, a hard, strong, but brittle material. Tempering the steel after quenching relieves the grain stresses developed as a result of rapid cooling.

The transformation of austenite to martensite during the quenching period proceeds approximately in the ratio shown in Table 157.

TABLE 157. RATIO OF AUSTENITE TO MARTENSITE IN STEEL COOLED FROM 400 F TO −150 F DURING HEAT AND COLD TREATMENT

Temperature °F	*% Austenite*	*% Martensite*
400	100	0
300	60	40
100	15	85
70	10	90
−150	3	97

Reducing the temperature to −150 F (see Table 157) reduces the austenite considerably below that obtainable when heat treatment ceases at room temperature. Because toughness decreases with larger percentages of retained austenite, it is advantageous to transform as much of it as possible to martensite.

If the specimen is allowed to remain at 70 F too long between heat and cold treatment, the austenite will become stabilized at the corresponding high level and cold treatment will be useless. However, if chilled too rapidly when the piece is under pre-tempering grain stress, some chance of cracking exists, much like that which occurs when heated glass is placed in cold water.

The following procedure of heat- and cold-treating steel should be modified to meet individual requirements:

1. Preheat to 750 F, or slightly above.
2. Preheat again to 1500 F.
3. Heat to final heat-treating temperature, depending on analysis of specimen.
4. Cool rapidly by quenching in liquid bath to above room temperature (say, 150 F).
5. Temper to some temperature below lower critical (550 to 700 F) for a given period—say, 3 hours—depending on desired hardness.
6. Remove from tempering bath and place in −150 F chill compartment for a period depending on mass (cross-sectional area)—usually 3 to 5 hours.

7. Remove from cold treatment and warm gradually to room temperature.

Stabilization of Metal. Steel tends to expand dimensionally and become distorted when aged at room temperature after heat-treating in the conventional manner. Cold treatment of steel reduces expanding tendencies during the aging or seasoning period.

Seasoning of Metal. Dimensional stability of metals may also be obtained by repeated and alternate exposure of specimens to high (slightly above room temperature) and ultra-low (−120 F) temperatures. The cycle of heating and chilling should be repeated from four to six times for best results.

The cold treatment of metal is carried on at temperatures between −120 F and −150 F in an insulated cabinet or compartment. Refrigeration for the cabinet is obtained from the low-temperature evaporator of a cascade system. Ethane may be used in the low-temperature phase while R-22 may be employed as the ethane condensing coolant.

SUMMARY OF CHAPTER 16

A simple refrigeration system consists of liquid receiver, expansion valve, evaporator, compressor, and condenser.

A single-stage compressor is used when the pressure difference between evaporator and condenser allows a compression ratio within the range of mechanical efficiency.

A multi-stage compressor arrangement is used when the pressure difference between evaporator and condenser is such as to develop excessive mechanical losses with any less number of stages.

The employment of cascade systems allows condensation of primary refrigerant at lower pressures and reduces power requirements per unit of refrigerant gas condensed as compared with stage operation.

Dual-compression cylinders reduce flash gas losses incurred when vapors from intermediate- or high-temperature evaporators are allowed to be reduced to low-evaporator pressure before compression.

A rotary booster compressor is used to precompress low density vapors (having relatively large volume per unit weight) prior to recompression in a higher stage compressor.

The hermetic system is one which is totally enclosed and sealed after all air and moisture have been excluded and charged with refrigerant and lubricating oil.

High humidity is maintained in refrigerated spaces by minimizing temperature differential between evaporator and surrounding air, increasing evaporator surface to compensate for small temperature differential, maintaining low velocity of convection currents, and keeping sensible heat load to a minimum.

Defrosting can be accomplished by external or internal heat sources. External heat sources may be warm air or water, or electrical heaters. Hot gases, variously fed to the evaporator either automatically (time cycling) or manually, comprise the internal defrosting medium.

Ice is manufactured in block form by immersing cans of water in cold brine.

The time required to freeze block ice depends on the thickness of the block, temperature of the brine, and efficiency of the systems. The time in hours is determined by the equation:

$$\text{hours} = \frac{\text{thickness}^2}{32\text{ F} - \text{brine temperature}} \times K$$

where $K = 6$ to 7, depending on the efficiency of the freezing system

Ice is also manufactured in cubes, cylinders, and other small forms on self-contained, automatic ice-making machines. The usual procedure is to expose water to evaporator surfaces in the shape of tubes, cylinders, etc., and loosen the ice by heating the surface with hot gas, or by scraping in some instances.

Trucks and trailers are refrigerated with ice, dry ice, or mechanical refrigeration systems to transport perishable commodities.

The cooling surface usually is placed in the upper part of the vehicle nose over which air is circulated before distribution over the load.

Power for mechanical refrigeration is supplied by an individual gasoline motor or from the power plant of the vehicle through a power take-off.

Power take-off may be from the engine crankshaft, transmission, or propeller shaft by various arrangements, including belts, hydraulic system, flexible shaft, etc.

Trucks may also be refrigerated by self-contained units placed in the truck body.

Eutectic plates are often used to act as hold-over units to maintain desired temperatures within the vehicle. Eutectic plates can be pre-charged at the garage or continually charged when required from the refrigeration system located and powered on the truck.

Trucks and truck trailers may be cooled with cryogenic fluids by allowing expansion to occur in cooling coils placed around the interior of the body.

Precooling perishable commodities in insulated cargo trailers to appropriate temperatures by placing loaded trailers in cooling tunnels assures safe shipment over limited distances.

Propane may be stored at low pressure by a method of self-refrigeration. By continually drawing off propane vapor from the holder, recompressing and condensing it and expanding it again in the holder reduces the mass to a liquefying temperature. Thus it is held as a liquid at about 1 psig and −50 F.

Dry ice is produced by compressing carbon dioxide gas to a high pressure in stages, condensing it to a liquid at high pressure, and then expanding the liquid in a compartment at a pressure below the critical point. The liquid is converted to a solid (snow) and gas in the lower compartment. The gas is returned to the system to be recompressed and the snow is pressed into blocks, having a density of about 86 lb per cu ft. Compression and expansion are usually accomplished in three stages.

Dry ice is also produced by liquefying the gas at lower temperature and pressure in a cascade system. This eliminates the need for the heavy equipment required for stage compression.

Refrigeration capacity storage is accomplished by producing ice during off-peak periods. Produced in cell-like compartments, the ice cools water, which in turn circulates through a booster cooling coil to augment the main coil during peak load periods of air-conditioning systems. The extra cooling afforded by the ice compensates for the time-lag that occurs with the surge of heat developed when a large number of people enter the air-conditioned space within a short period of time.

Subjecting metals, such as tool steel, to ultra-low temperatures (between −120 F and 150 F) after heat-treating succeeds in obtaining more uniform hardness, allows shrink-fitting of parts by expansion of the male part rather than heating of the female part, structural stabilization after treating, and controlled aging.

STUDY OUTLINE FOR CHAPTER 17

- Steam jet refrigeration system
  - Components
  - Construction
  - Primary booster or ejector
    - Theory
    - Function
  - Secondary ejectors
  - Multi-stage units
  - Condenser function
  - Capacity control
  - Operating factors
    - Steam consumption
- Absorption refrigeration system
  - Components
  - Construction
  - Principle of operation
  - Refrigerants and absorbers
  - Continuous operation, with pump
  - Continuous operation, without pump
  - Intermittent operation, Faraday's principle
  - Industrial type
- Temperature conversion factors
  - Fahrenheit vs. Centigrade vs. Rankine vs. Kelvin
  - Kelvin related to thermoelectric refrigeration
  - Rankine related to cryogenics
- Cryogenic processes
  - Definition
  - Cryogenes
    - Description of gases
    - Properties
  - Quasi-cryogenes
  - Liquefying processes
    - Cycle, basic
    - Cycles, various
      - Components
        - Cascade
        - Joule-Thompson refrigeration
        - Hampson cycle, liquefier
        - Linde cycle, liquefier
        - Claude cycle, liquefier
        - Linde dual pressure system
        - Kapitza cycle
        - Heylandt cycle
        - Claude dual-pressure system
- Separation of gases
  - Rectifiers
    - Construction
    - Operation
  - Compounded liquefiers
  - Deuterium separation
  - Regenerators
    - Function
  - Joule-Thompson valve
- Cryogenic engineering materials
  - Effect of ultra-low temperatures
    - On behavior of some substances
    - On structural properties of metals and other materials
- Cryogenic vessels
  - Simple containers
  - Sophisticated containers
  - Dewar flask
- Stirling cycle
  - Werkspoor cold-gas refrigerator
    - Process
    - Components
    - Application
    - Capabilities
- Cryogenic cooled transportation
  - Carriers
    - Truck, trailers
    - Railroad cars
    - Miscellaneous containers
- Nitrogen freezing tunnels
- Thermoelectric refrigeration
  - Seebeck effect (generator)
  - Peltier effect (refrigerator)
  - Thompson effect (Seebeck-Peltier relationship)
  - Parameters
  - Parameter relationships
    - Figure-of-merit
    - Refrigerating capacity
    - Coefficient of performance
  - Thermoelectric materials
  - Thermoelectric refrigerator
  - Semiconductors
    - Atomic structure, planetary bands
    - Conductors
    - Insulators
  - Principle of semiconductors
    - Intrinsic types
    - Extrinsic types
    - Doping agents
      - Donors and acceptors
      - n-type semiconductors (negative)
      - p-type semiconductors (positive)
    - The pn-junction
      - Diffusion
      - Drift
  - Practical thermoelectric refrigerator
  - The reverse cycle
    - Refrigerator
    - Heat pump
      - Basic cycle
      - Three- and four-way valve
      - Heat sources

17

Other Refrigeration Systems

In addition to the conventional type of refrigerator discussed in earlier chapters, there remain types that produce temperature reduction or state changes by other processes. Some of the latter types of cooling processes are of commercial importance and must be included in the overall study of refrigeration. They include the following systems:

1. Steam jet refrigeration
2. Absorption systems
3. Cryogenic processes
4. Thermoelectric cooling
5. Chemical refrigeration
6. Magnetic refrigeration
7. Vortex-tube cooling

The first four are in practical use and described in detail within this chapter. Also covered is the *heat pump or reverse cycle,* which is considered here because of its departure from the conventional process of cooling alone. The magnetic refrigeration process is important where temperatures below 3 R are involved.

STEAM JET REFRIGERATION

The steam jet water vapor system of refrigeration is now used widely by industry as a means of cooling water for various processes. It is applicable wherever sufficient steam and condensing water are available at a reasonable cost and where refrigeration requirements are limited to temperatures somewhat above 32 F, the freezing point of water. As the system will operate with low-pressure steam, an economic advantage is gained in those plants utilizing the exhaust from processes requiring steam at higher pressures.

Because refrigeration effect is obtained through the process of water evaporation under partial vacuum, no other refrigerant is required, thus establishing the steam jet system as a desirable one where safety is an essential factor. Other features that have led to its adoption, especially by the marine industry, include (*a*) freedom from vibration and noise, (*b*) minimum space requirements, and (*c*) no moving parts, except on circulating pumps. It can be and usually is located outdoors, thus saving valuable plant space.

Some typical applications of steam jet refrigeration are air conditioning, cooling or rubber mill rolls and mixers, cold water gas absorption in chemical plants, distillation as performed in petroleum refineries, chemical plants, etc., and beverage cooling in breweries, distilleries, etc., to name but a few. It has been adopted by the illuminating gas industry for lowering the dew point of gas below ground temperature to prevent corrosion due to precipitation of moisture in the mains, branch lines, appliances, and meters.

Construction

The principal parts of the steam jet refrigeration system are shown in Fig. 496. They consist of the (*a*) steam ejector assembly, (*b*) evaporator or flash chamber, (*c*) condenser, (*d*) chilled water circulating pump, and (*e*) condensate pump.

The ejector assembly is made up of nozzle plate, steam nozzles, steam dome, and booster. The steam dome is provided with an opening (*f*) to allow free passage of water vapor into the mixing chamber from the evaporator to which it is attached.

The object of the system is to create a partial vacuum within the evaporator to establish a lower boiling point of the circulating water, causing evaporation to occur at such a rapid rate as to reduce its temperature to a level corresponding to the absolute pressure created. Thus, the principle of the porous water jar used in the desert wherein water cools itself intrinsically by evaporation is applied artificially.

The partial vacuum in the evaporator, or flash chamber as it commonly is referred to, is created by directing a jet of high-velocity steam into the booster body (Fig. 497) at a point known as the combining chamber (*1*). From there it passes through the throat (*2*) into the diffuser (*3*) by its own momentum. As the high-velocity steam passes through the combining chamber, it combines with the water vapor from the flash chamber, carrying that fluid ultimately to the condenser. In the booster, the vapor mixture is compressed to a point at which it can be condensed in the condenser. The water vapor enters the booster through a manifold (*4*), connecting the latter with the flash chamber. Because the booster is open to the manifold at a point back of the nozzle openings, as related to direction of flow, the water vapor combines with the steam along a parallel path, thereby reducing impact losses to a minimum.

For each pound of water evaporated, the amount of heat given up by the main body of circulating water is equivalent to the latent heat of

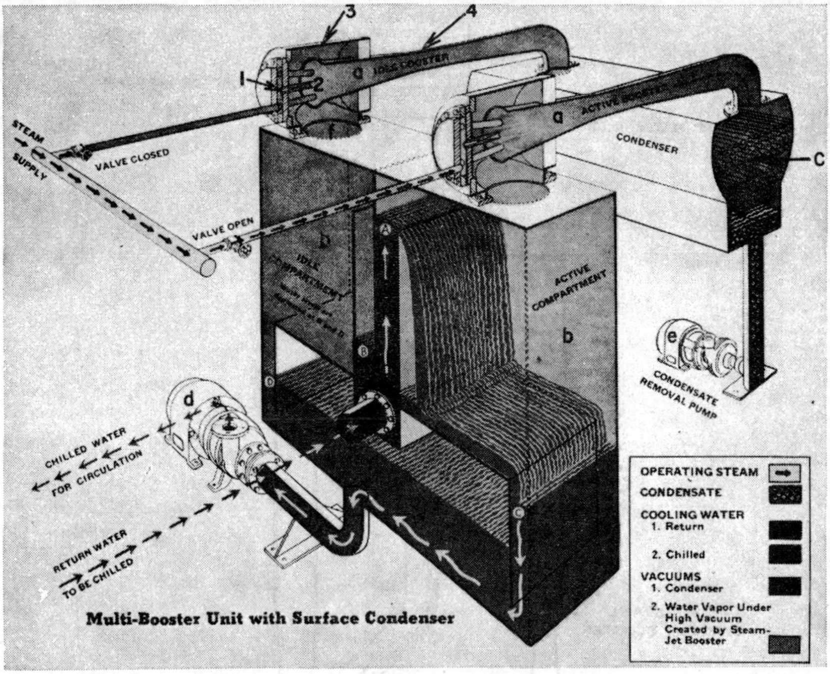

Fig. 496. Steam jet refrigeration system with two boosters. As shown, one booster is idle. The pressure in the idle compartment and condenser pressure are the same. The difference between the vacuum in the active compartment and the condenser vacuum causes a difference in water levels a and b, and also between c and d. The water column, which automatically adjusts itself when one booster is shut off, prevents short circuiting of condenser vapor back to the active compartment. Surface condenser is shown. (Coutresy, Ingersoll-Rand Company.)

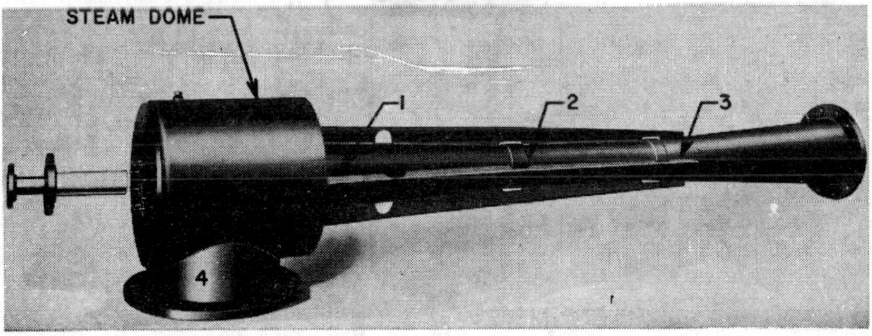

Fig. 497. Steam jet and booster assembly. (Courtesy, Ingersoll-Rand Company.)

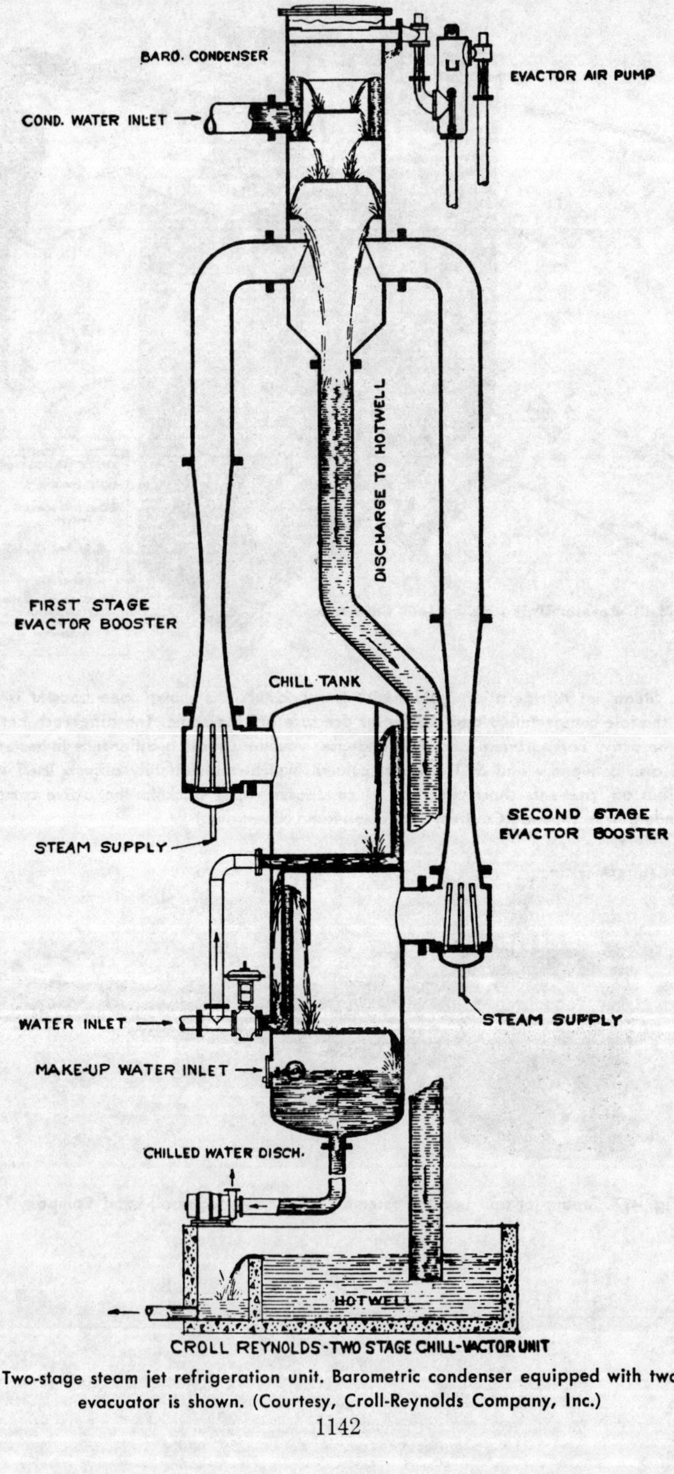

Fig. 498. Two-stage steam jet refrigeration unit. Barometric condenser equipped with two-stage air evacuator is shown. (Courtesy, Croll-Reynolds Company, Inc.)

TABLE 158. RELATIONSHIP OF PRESSURE, TEMPERATURE, HEAT, AND VOLUME OF WATER VAPOR

Water Temperature	*Absolute Pressure In. Hg*	*Volume of Vapor Cu Ft per Lb*	*Latent Heat Btu per Lb*
35	0.203	2940	1071.7
40	0.248	2440	1068.9
45	0.300	2030	1066.1
50	0.362	1700	1063.3
55	0.436	1430	1060.6
60	0.522	1210	1057.8
70	0.739	872	1052.3
80	1.029	637	1046.7
90	1.417	470	1041.2
100	1.926	351	1035.6
212	29.92	26.8	970.0

vaporization corresponding to its temperature. This in turn depends on the absolute pressure maintained in the flash chamber.

For example, by referring to the tables "Properties of Saturated Steam," it will be found that the pressure corresponding to a temperature of, say, 50 F, is 0.363 in. Hg abs. It will also be found that enthalpy of the vapor corresponding to the above conditions is 1065.6 Btu per lb. Therefore, for every pound of water evaporated in a flash chamber maintained at 0.363 in. Hg abs., 1065.6 Btu will be given up by the circulating water body. For every pound of water evaporated per minute under such conditions, the amount of refrigeration produced will be equivalent to $\frac{1065.6}{200} = 5.33$ tons.

EXAMPLE: How many pounds of water is it necessary to evaporate in a flash chamber maintained at 0.363 in. Hg abs., through which 300 gal of water circulates per minute, entering at 60 F and leaving at 50 F? What is the total tonnage of the system assuming no heat losses?

Solution:

Heat removed from circulating water:

$$300 \text{ gal} \times 8.33 \text{ lb per gal} \times (60 - 50) = 24{,}990 \text{ Btu}$$

Pounds of water evaporated:

$$\frac{24{,}990}{1065.6} = 23.45 \text{ lb of water evaporated per minute}$$

or

$$23.45 \times 60 = 1407 \text{ lb of water per hour}$$

Tonnage:

$$\frac{24{,}990}{200} = 124.95 \text{ tons of refrigeration}$$

The capacity of the steam jet refrigeration system usually is referred to in terms of *frame* tons. This is the capacity of a system operating under any set of temperature conditions as corrected to those conditions if the same system were operating to cool the circulating water to 50 F. When the leaving water temperature is below 50 F, the number of standard tons will be less than the number of frame tons. When the leaving water temperature is above 50 F, then the number of standard tons will be greater than its corresponding number of frame tons. The latter statements are true because of the wide variation in capacity, as expressed in standard refrigeration tons, for a given system operating at different leaving water temperatures. They follow the rule that the capacity of a given unit increases as the temperature of the leaving water increases.

The correction to frame tons for any steam jet system operating at leaving water temperatures other than 50 F can be made from the curve in Fig. 499, in which chilled water temperatures are plotted against percent refrigerating tons referred to a rating of 50 F. This curve shows that for a given amount of work performed by the booster, the refrigeration capacity of the unit increases approximately 15 percent for each 5 F rise in chilled water temperature.

The counterpart for the curve (Fig. 499) in mechanical refrigeration would be a curve showing the capacity decrease of a compressor operating at various suction pressures against a constant head. Because of the combined effect of higher compression ratio, diminishing heat of vaporization, and lower density of vapor handled, the refrigeration capacity per unit of work (horsepower) performed by the compressor would decrease with every increment of reduction in suction pressure.

The use of the curve is shown in the following example:

A system designed to chill 625 lb of water per minute from 61 F to 45 F would have a refrigeration capacity of $\dfrac{625 \times (61 - 45)}{200} = 50$ tons. To determine the capacity of the same unit if it were cooling through the same temperature differential, but at a higher level—say, from 66 F to 50 F—reference is made to the curve.

Starting from the point on the ordinate representing 45 F leaving water, draw a line parallel with the abscissa until it intersects the curve. Draw another line from the point of intersection downward and parallel with the ordinate to the abscissa. Where the vertical line meets the abscissa, read the percent refrigerating tonnage referred to a rating at 50 F, which in this case is 85 percent. The capacity of the unit if it were operating at 50 F leaving water temperature would be $\dfrac{50 \text{ standard tons at } 45\text{ F}}{0.85} = 58.82$ tons,

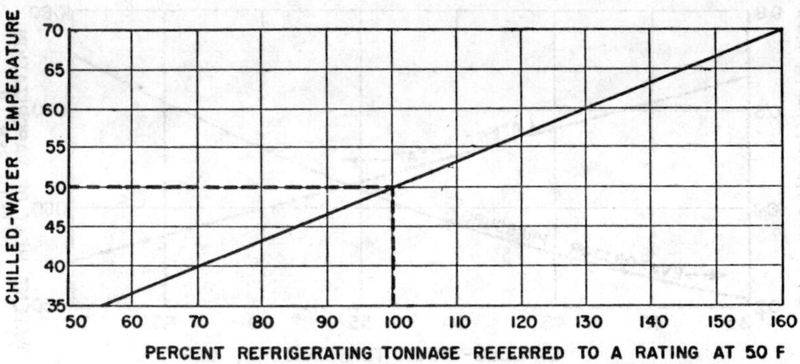

Fig. 499. Variation in tonnage according to water temperature. (Courtesy, Ingersoll-Rand Company.)

which would also be the number of frame tons because a correction to 50 F was made in this case. The same system operating through a similar temperature differential (16 F), but at a higher leaving water temperature—say, 60 F—would have a refrigeration capacity according to the chart of 30 percent above frame tonnage, or $58.82 \times 1.30 = 76.46$ tons.

It is obvious from the above that, when a system is operating with 50 F leaving water, frame tons equal standard refrigeration tons and are so designated on the curve by the reference value of 100 percent.

The Primary Booster or Ejector—Theory. The booster assembly is the prime component of the steam jet system to which it bears a relationship similar to that of the compressor to the mechanical refrigeration system. The operation of the booster is based on Bernoulli's principle, which in effect states that the pressure within a fluid in motion varies inversely as the velocity of that fluid. The application of that principle within the booster is accomplished by the passage of high-velocity steam in such a manner as to create a low absolute pressure within the combining chamber, the manifold, and, ultimately, the flash chamber. To do this necessitates the use of a series of pressure-velocity converters, which are represented here by the steam nozzle, the combining chamber, the throat, and the diffuser, the last three forming the booster body.

Expressing Bernoulli's principle in a different way bears out the relationship of potential and kinetic energy within a fluid subject to varying degrees of conditions, largely dynamic; viz.: the relative forces of potential and kinetic energy within a fluid particle vary according to the velocity of the particle, whose total energy remains constant. While it is assumed here that the total energy does remain constant, losses due to friction within the nozzle and the booster dispel that assumption in the actual machine. As kinetic energy is energy of motion, it follows that potential energy decreases as velocity increases, accompanied by a corresponding increase in kinetic energy. Thus, for purposes here, steam at high pressure and negligible velocity is converted to steam at lower pressure and high

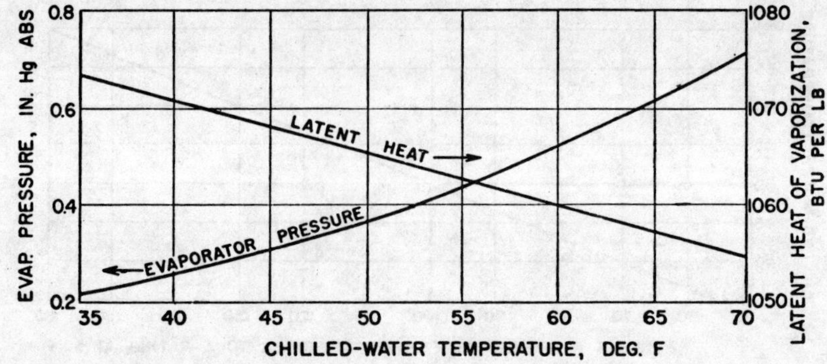

Fig. 500. Chart showing chilled water temperature corresponding to evaporator pressure and number of Btu removed per lb evaporated. (Courtesy, Ingersoll-Rand Company.)

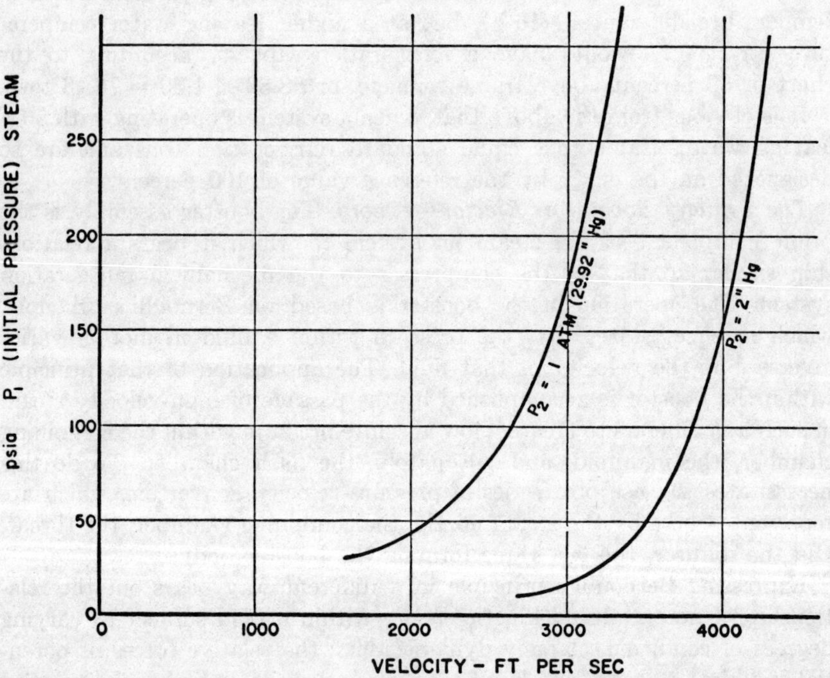

Fig. 501. Steam velocity, ft per sec, for various initial pressures against 1 atm (curve *A*), and 2 in. Hg (curve *B*).

velocity by means of the nozzle made to do work by impact with the water vapor in the combining chamber, and compressed to higher pressure by reconversion of velocity energy within the throat and the diffuser. In effect, then, the booster body acts as a compressor for the steam jet system.

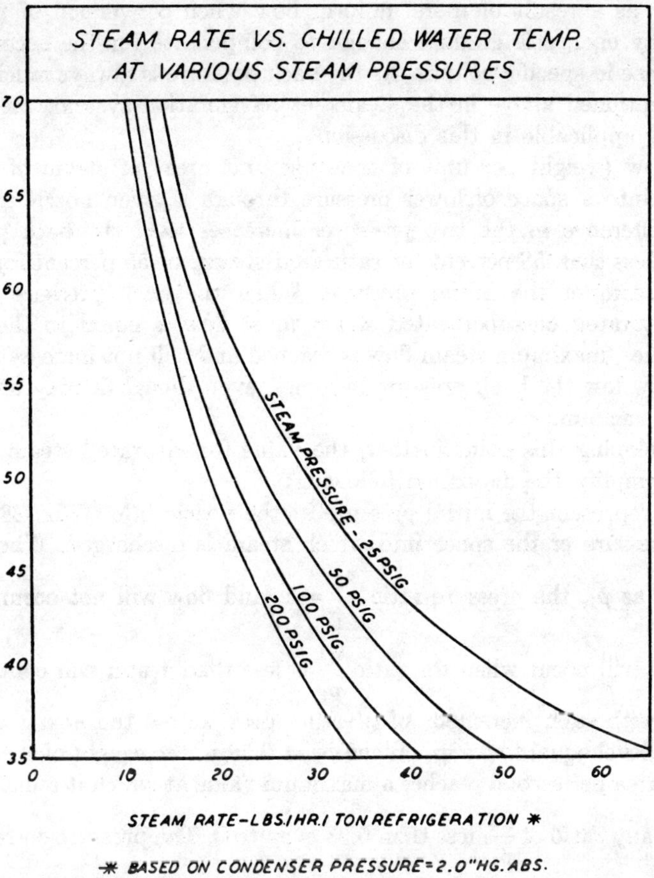

Fig. 502. Steam rate vs. chilled water temperature at various steam pressures. (Courtesy, Croll-Reynolds Company, Inc.)

Before demonstrating the effect of Bernoulli's principle within the various components of the booster assembly, it is well to note the conformation of those components. The steam nozzle leading from the steam dome to the combining chamber is of the *convergent-divergent* type; i.e., the cross-sectional area of the orifice diminishes from the point of entrance to a minimum at the throat and then gradually increases again to its maximum area at the exit or mouth. The booster body is of similar conformation, except that its cross-sectional area is considerably larger throughout its greater length.

The Steam Nozzle. The rule governing the design of nozzles is stated generally: when the back pressure or pressure against which steam must flow through an orifice is less than 58 percent for saturated steam or 55 percent for superheated steam of the initial pressure, the nozzle should be both convergent and divergent. The reason for this lies in greater efficiency as a result of more uniform flow when conversion of pressure to velocity energy is gradual enough to compensate for the accompanying increase in specific volume. As the back pressure is always much below those mentioned above in the steam jet refrigeration system, the rule is especially applicable in this discussion.

The flow (weight per unit of time per unit area) of steam of greater pressure into a space of lower pressure through a given nozzle increases as the difference in the two pressures increases *until* the back pressure becomes less than 58 percent for saturated steam, or 55 percent for superheated steam of the initial pressure. When the back pressure against which saturated or superheated steam must flow is equal to the stipulated values, maximum steam flow is reached and will not increase regardless of how low the back pressure becomes, even though it may approach a perfect vacuum.

In developing this point further, the value for saturated steam will be used to simplify the discussion following:

Let p_1 represent the initial pressure at the nozzle inlet (Fig. 281), and p_2 the pressure of the space into which steam is discharged. When p_2 is the same as p_1, the pressure ratio $\frac{p_2}{p_1} = 1$, and flow will not occur. However, flow will occur when the ratio $\frac{p_2}{p_1}$ is less than 1 and will continue to increase with each increment of pressure drop across the nozzle until p_2 reaches a level equal to $0.58p_1$. When $p_2 = 0.58p_1$, the weight of steam flow per unit area per second reaches a maximum value at which it remains constant for any ratio of $\frac{p_2}{p_1}$, less than 0.58 even to 0. The pressure corresponding to $0.58p_1$ is called the *critical pressure* of saturated steam and $\frac{p_2}{p_1} = 0.58$ is the *critical pressure ratio.* The critical pressure for saturated steam, then, is

$$p_c = 0.58p_1$$

The pressure in the throat is always $0.58p_1$ when p_2 is equal to or below the critical pressure.

The law governing the transmission of pressure waves and sound waves is the same. This law states that the velocity of pressure waves varies inversely as the square root of the density of the fluid through which transmission is to occur. When the velocity of pressure waves reaches a maximum for the given density, a state of equilibrium exists and cannot be increased regardless of how low the discharge pressure falls beyond critical level. A body cannot progress against a current moving at the same speed or faster than itself. As pressure waves move at the same speed as sound, which for steam at 212 F is 1328 fps, no effect can be made on the high-pressure side of the orifice through which steam passes at the pressure corresponding to that temperature by waves traveling upstream against flow at the same velocity.

The velocity of steam through a nozzle depends on the difference in pressure across the orifice and is determined by the fundamental equation $V^2 = gh$, in which h (head) for compressible fluids must be converted to energy released in foot-pounds as related to thermodynamic head. Therefore h is a function of total enthalpies corresponding to the pressures on either side of the orifice, so that H_1 and H_2, enthalpies of the vapor, are related adiabatically to p_1 and p_2, respectively. As 1 Btu is equivalent to 778 ft-lb of work, the head in feet corresponding to any pressure difference for compressible fluids becomes the product of 778 ft-lb times their enthalpy difference, or $J(H_1 - H_2)$. Substituting in the fundamental velocity equation,

$$V^2 = 2gJ(H_1 - H_2) = 2 \times 32.16 \times 778(H_1 - H_2)$$

or

$$V = 223.7\sqrt{H_1 - H_2}$$

where H_1 = enthalpy of vapor for steam at p_1
H_2 = enthalpy for steam at p_2 after isentropic (constant entropy) expansion

Example: Steam at an initial pressure of 75 psia, saturated, is expanded through a nozzle into a space maintained at 1 psia. What is its theoretical velocity upon leaving the nozzle? Assume initial velocity to be negligible.

Solution:

1st step: Determine H_1 from the Mollier chart (or saturated steam tables) corresponding to 75 psia steam pressure along the saturation line. Read 1181.2.

2nd step: Determine H_2 from the Mollier chart by following a path of constant entropy 1.6254 from the intersection of 75 psia line and saturation line to intersection of 1 psia line. Project this point of intersection to ordinate along constant enthalpy line. Read 908 Btu.

3rd step: Apply velocity equation for compressible fluids

$$\begin{aligned} V &= 223.7\sqrt{H_1 - H_2} \\ &= 223.7\sqrt{1181.2 - 908} \\ &= 3690 \text{ fps velocity of steam} \end{aligned}$$

Assuming an initial velocity of steam (entering the nozzle) of 200 ft per sec in the above, the discharge velocity would be

$$V = \sqrt{2g778(1181.2 - 908) + 200^2}$$

$$V = 3703 \text{ fps}$$

TABLE 159. NOZZLE DISCHARGE VELOCITY OF STEAM FOR DIFFERENT HEAD PRESSURES AGAINST A BACK PRESSURE OF 1 PSIA

Initial (Head) Pressure psia	*Entropy s*	*Initial Enthalpy H_1*	*Final Enthalpy H_2*	*Approximate Enthalpy Difference*	*Velocity fps*
2	1.919	1115.6	1072	43.6	1476
5	1.8435	1130.6	1030	100.6	2237
10	1.7874	1143.0	999	144.0	2684
15	1.7548	1150.6	982	169.0	2908
20	1.7317	1156.0	968	188.0	3065
25	1.7137	1160.2	957	203.2	3188
30	1.699	1163.7	949	214.7	3272
40	1.6759	1169.2	936	233.0	3416
50	1.6580	1173.5	926	247.0	3517
75	1.6254	1181.2	908	273.0	3690
100	1.6022	1186.6	893	293.0	3829

Table 159 lists the entropy, the initial and discharge enthalpies, and the theoretical velocities of steam for different initial pressures (p_1) when discharged through a nozzle against a back pressure (p_2) of 1 psia.

While Table 159 gives theoretical velocities, the actual velocity of steam through a nozzle is a function of nozzle efficiency, as determined by the ratio

$$\text{Nozzle efficiency} = \frac{\text{actual difference in enthalpies}}{\text{isentropic difference in enthalpies}}$$

in which the difference between actual and isentropic enthalpies is the amount of energy lost due to friction and eddying.

If y represents the percent of available energy lost due to friction and eddying, then the theoretical velocity values will be less than shown in Table 159, in accordance with the compensating equation

$$V = \sqrt{2g778[(H_1 - H_2)(1 - y)]}$$

or

$$V = 223.7\sqrt{(H_1 - H_2)(1 - y)}$$

The value of y is about 8 percent or 0.08 for a velocity of 1500 fps. It increases as the difference in enthalpies $(H_1 - H_2)$ increases at a rate approximating 2 percent for each velocity increase of 500 fps, which rate of increase rises slightly at velocities above 2500 fps, or $^*y = 15$ percent or 0.15 (approx) at 3000 fps.

In expanding steam to a pressure below the critical pressure, the nozzle should be convergent to the throat and then divergent, as previously stated. However, in determining the quantity of flow in pounds per sec-

* The loss due to y is called *reheat* because some of the available enthalpy is converted to heat.

ond, it is the throat pressure and the throat area that must be considered, and not the discharge pressure. A simple but approximate (within 3 percent) solution for determining steam flow through a nozzle according to the above rule is found in Napier's equation for *saturated* steam:

$$W = A_2 p_1 \div 70$$

where W = steam flow, lb per sec, when p_2 is less than p_c (critical pressure)
A_2 = area of throat in square feet
p_1 = initial pressure, psia

or Goudie's formula for *superheated* steam:

$$W = 0.3155 A_2 \sqrt{p_1/v_1}$$

where v_1 = specific volume at p_1, cu ft per lb

In the case of convergent-divergent nozzles, design must conform to the given steam conditions under which they are to operate; otherwise, underexpansion or overexpansion will occur when the discharge pressure is respectively below or above design value, decreasing the efficiency of the nozzle in either case. In designing a suitable nozzle, it is necessary to determine cross-sectional areas along its length in accordance with pressure changes occurring as flow proceeds from the high-pressure zone to the zone of low-pressure discharge.

To compute the cross-sectional areas of a nozzle at points representing descending pressures—say, of 125, 100, 75, 25, and 15 psia—along its length for expanding 2 lb per sec of steam at 150 psia to discharge pressure of 15 psia, the following steps are necessary:

1st step: Determine velocity of steam at the various points by the conventional formula

$$V = 223.7\sqrt{H_1 - H_{vp}}$$

where H_1 = enthalpy at 150 psia
H_{vp} = enthalpy at various points

2nd step: Determine specific volume for each pressure point from the Mollier chart for isentropic flow.

3rd step: Determine areas at the various points along the length of the nozzle according to the equation of continuity.

$$A_{vp} = \frac{W v_{vp}}{V_{vp}}$$

where A_{vp} = area of cross-section at the various points, sq ft
v_{vp} = specific volume at the various pressures for isentropic flow, cu ft
V_{vp} = velocity at the various points, ft per sec
W = steam flow, lb per sec

The Primary Booster or Ejector—Function. Steam expands in the nozzle to enter the combining chamber at velocities of 4000 fps and up-

Fig. 503. Single booster steam jet system installed inside of building. Installation shows compact arrangement possible with this type unit. (Courtesy, Croll-Reynolds Company, Inc.)

ward, during which potential energy is converted to kinetic energy within the fluid. This action establishes an extremely low absolute pressure in accordance with Bernouilli's principle, which influences a similar surrounding pressure that finally is reflected in the flash chamber through the manifold. The entire assembly, being hermetically sealed, allows the low pressure to be maintained continually on the surface of the water flowing through the flash chamber, causing a reduction in boiling point and a subsequent increase in evaporation rate commensurate with sustained pressures.

The water vapor following the path of diminishing pressure enters the combining chamber where it mixes by parallel flow with the steam. Some of the kinetic energy of the steam is imparted to the water vapor during the process. This action, while increasing the velocity of the water vapor, reduces the velocity of the mixture relative to the entering steam. The loss of kinetic energy of the steam is equal to the work performed during impact with the water vapor in the combining chamber.

Up to this point, steam under pressure was converted to a high-velocity fluid at a much reduced pressure by means of a nozzle, and, in accordance with Bernouilli's principle, created a partial vacuum within the combining chamber, which includes the space from the point of discharge to a point slightly ahead of the booster throat. The degree of partial vacuum created within the combining chamber is a function of velocity borne out by the law $p \propto \frac{1}{V}$. However, as the velocity of the steam leaving the nozzle is a function of the discharge pressure according to the equation $V = 223.7\sqrt{H_1 - H_2}$, it is essential that the pressure within the combining chamber be maintained as low as possible to attain maximum velocity. To do this it becomes necessary to evacuate both the steam and the water vapor at a rate commensurate with operational demand. This is accomplished by the mutual effect of a diffuser and a condenser.

Primary Condenser. As with all fluids, the pressure of the mixture of steam and water vapor must be raised to a level corresponding to the temperature and the quantity of the condenser cooling medium in order to convert it to liquid form. Therefore, the condensing pressure must be maintained above that corresponding to the temperature of the cooling water. The function of the booster is to deliver the high-velocity, low-

Fig. 504. Three booster steam jet system installed outside. (Courtesy, Ingersoll-Rand Company.)

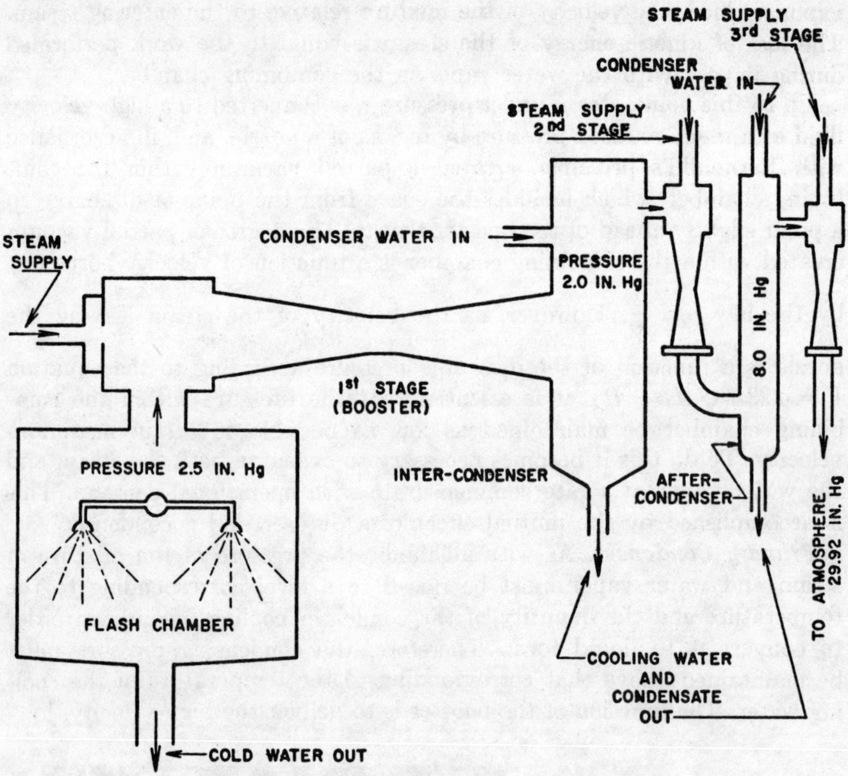

Fig. 505. Typical conditions in three-stage steam jet system.

pressure mixture to the condenser at a lower velocity and a higher pressure. In effect, the process is one of compression, so that the action within the diffuser is the reverse of that taking place in the nozzle.

The rate at which conversion of velocity energy to pressure energy occurs in a given diffuser depends on the rate at which the vapor is condensed, which, in turn, is dependent on condenser conditions. If the condensing temperature is high, the pressure will be correspondingly high, causing interference with steam jet velocity by reason of higher back pressures along the path of flow. This means that a greater amount of steam would be required to hold desired conditions within the flash chamber. If the condensing temperature is too low, the resulting pressure would approach so close to that corresponding to the cooling water temperature as to prevent rapid condensation. Under the latter circumstances, it appears that the only solution lies in providing greater quantities of cooling water to compensate for the small temperature difference. However, as decreased steam consumption follows any decrease in condenser pressure, the balance of heat exchange tends to remain somewhat constant so that very little additional water is required in the low-temperature condenser.

As in all compression systems, the pressure in the condenser should be so related to the pressure in the evaporator as to maintain the compression ratio within allowable limits of a single stage machine conducive to highest operating efficiency. For example, the evaporator pressure corresponding to a temperature of 50 F is 0.1780 psia. If the condenser temperature is maintained at 100 F, which corresponds to a pressure of 0.9487 psia, the compression ratio would be $\frac{0.9487}{0.1780} = 5.3$. This ratio compares favorably with single-stage mechanical systems employing other types of refrigerants and operating under moderate conditions.

Condenser Cooling Water Requirements. If 40 lb of 75 psia saturated steam per hour per ton of refrigeration were required to maintain 50 F evaporator temperature and the condenser pressure were held at 0.9487 psia (100 F), the cooling water demand if entering at 68 F would be determined as follows:

From the saturated steam tables find the enthalpy of the initial steam at 75 psia and condensate at 100 F, viz.:

H = enthalpy of vapor − steam at 75 psia	= 1181.9 Btu per lb
h = enthalpy of liquid, saturated, at 100 F	= 67.93 Btu per lb
Heat equivalent of 1 ton of refrigeration	= 12,000 Btu per hr
Steam consumption per hour per ton	= 40 lb
Entering water temperature	68 F

The total heat to be removed in the condenser per hour would be

$$40(1181.9 - 67.93) + 12{,}000 = 56\ 558 \text{ Btu}$$

and

$$\frac{56{,}558 \text{ Btu}}{60 \times 8.33 \times (100 - 68)} = 3.50 \text{ gal of water per min per ton}$$

Both barometric (see Fig. 498) and surface type (shell and tube) condensers are used. Where head room is a factor or it is desired to keep the condenser cooling water separated from the condensate, the surface type should be selected. However, less water is required by the barometric type because of the intimate association between water and steam.

Secondary Ejectors. Any air present in the condenser causes the pressure to rise considerably above required levels, and, therefore, must be evacuated by a continuing process. This is accomplished by means of a two-stage ejector operating in a similar manner as the primary ejector. Supplied by steam from the main header, the air is drawn from the condenser by the first-stage ejector and compressed to a higher pressure of approximately 8 in. Hg in an inter-condenser from which it is again drawn off by the second-stage ejector and compressed to atmospheric pressure in an after-condenser. From the after-condenser it is vented to the atmosphere.

Capacity Control. Very little is gained in the way of reducing the capacity of a steam jet system by throttling the flow of steam to the primary booster. Therefore, practical capacity control can be accomplished only by completely closing or opening the steam line to the nozzles according to the temperature of the chilled water. This is analogous to the cycling operation of a mechanical system and may be accomplished automatically by a thermostat located in the chilled water discharge line.

Multiple boosters are employed where greater flexibility of control is required. These are installed in parallel and discharge into a common condenser. By shutting down one or more boosters, the capacity of the system will be reduced by an amount proportional to the total number of boosters present. A system having three boosters will have a capacity variation from full load to no load in steps of one third full load capacity. This means that, when one booster is manually or automatically shut down, the remaining two will operate to maintain the system at two thirds full load capacity. If the second one shuts down, the system will then operate at one third full load capacity, etc. The reduction in motive steam consumption is proportional to the number of boosters rendered idle as a result of capacity control. In this respect, the steam jet system has an advantage of higher efficiency over mechanical systems operating at reduced capacity. This is apparent from the power requirements of a mechanical compressor operating 50 percent capacity with a cylinder by-pass arrangement, being 70 percent of full load demand, assuming similar conditions in both cases.

Practical industrial applications for steam jet refrigeration usually are limited to chilled water temperatures of 40 F to 45 F. They are built in sizes ranging from 20 to over 1200 tons and can operate with steam pressures as low as 2 psig.

Multi-stage Steam Jet Units. Steam jet units are of the single- or multi-stage type. The latter are built with two, three, four, five, or six stages, the selection of which depends primarily on the level of vacuum required. Single-stage ejectors usually discharge to atmospheric pressure or higher. Units of three or more stages are usually employed where industrial processes require extremely low absolute pressures.

The arrangement of the ejectors in multi-stage units varies. They may be placed in parallel and discharge into a common condenser. Or they may be placed in series with or without intercondensers. Where intercondensers are not used, each succeeding stage must carry both the original load, the motivating steam from the first stage, and its own motivating steam. In some instances, intercondensers are used only between certain stages while no intercooling occurs between the others.

Where intercooling is carried on between stages, the condenser acts as the vacuum chamber for the succeeding ejector. A two-stage series inter-

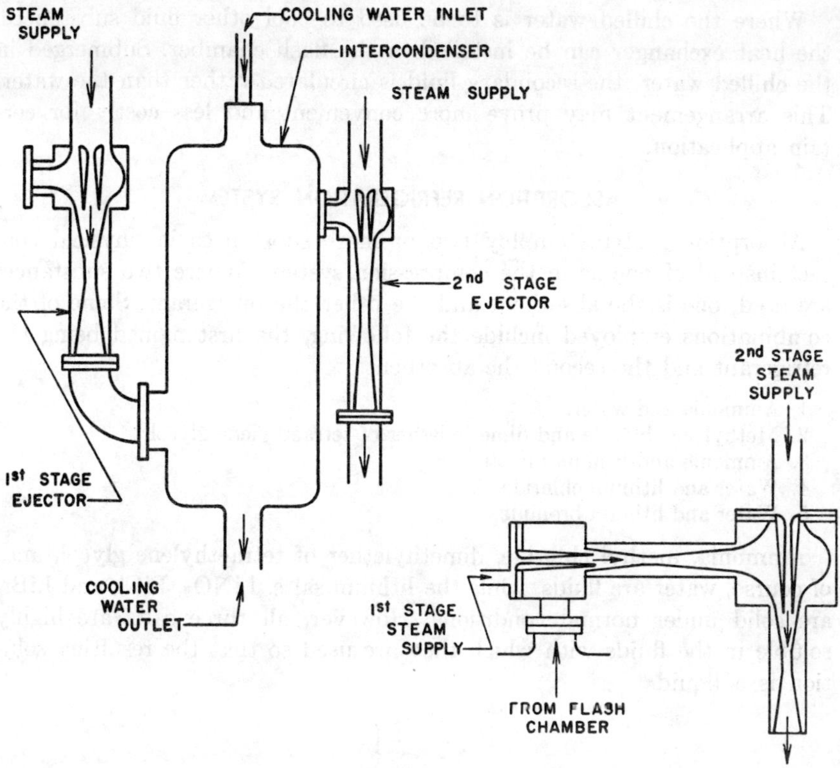

Fig. 506. Two-stage steam jet evacuator. (*Left*) Intercondensing, (*right*) nonintercondensing.

cooled unit, for example, may consist of a first-stage ejector or booster that discharges to a barometric intercooler, in which a partial vacuum has been drawn by a second-stage ejector that discharges to the atmosphere.

A five-stage unit may consist of three ejectors in series that discharge to a condenser, the latter of which is equipped with a two-stage evacuator or air pump.

Water may act as the motivating fluid in place of steam for creating vacuum. Steam and water may be used in combination with the water ejector employed in the second stage.

When arranged to develop maximum efficiency, i.e., with proper intercooling on multi-stage units, steam jet systems are capable of developing the following minimum absolute pressures: (1 in. = 25.4 mm)

Single-stage	100 mm Hg
Two-stage	12 mm Hg
Three-stage	3 mm Hg
Four-stage	0.15 mm Hg
Five-stage	0.005 mm Hg

Where the chilled water is to be used to cool other fluid substances the heat exchanger can be installed in the flash chamber. Submerged in the chilled water, the secondary fluid is circulated rather than the water. This arrangement may prove more convenient and less costly for certain application.

ABSORPTION REFRIGERATION SYSTEM

Absorption systems employ two or more substances in physical contact instead of one as in the compression system. Where two substances are used, one is the absorbent and the other the refrigerant. Some of the combinations employed include the following, the first named being the refrigerant and the second the absorbent.

1. Ammonia and water.
2. Methylene chloride and dimethylether of tetraethylene glycol.
3. Ammonia and lithium nitrate.
4. Water and lithium chloride.
5. Water and lithium bromide.

Ammonia, methyl chloride, dimethylether of tetraethylene glycol, and, of course, water are fluids, while the lithium salts, $LiNO_3$, LiCl, and LiBr, are solid under normal conditions. However, all three salts are highly soluble in the fluids with which they are used so that the resulting solution is a liquid.

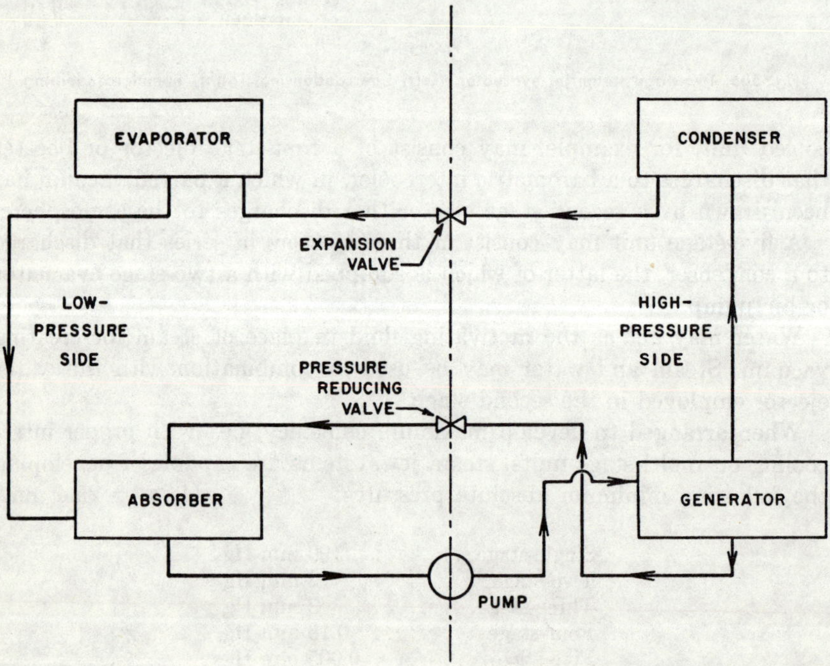

Fig. 507. Simple absorption refrigeration system.

Absorption systems may be continuous or intermittent in operation. They may be single- or multi-stage, the latter consisting of two or more stages in which the evaporators function under different pressures.

Construction. The simple absorption system consists of the following components that represent four major functional steps, as shown in Fig. 507.

Evaporator
Absorber
Generator
Condenser

For a greater efficiency and improved performance, a *heat exchanger*, an *analyzer*, and a *rectifier* are employed on more complex systems. The relative locations of the various components, including the *aqua pump*, *expansion* and *pressure reducing valves*, are shown in Fig. 508.

The evaporator and the absorber operate under low pressure and are analogous to the evaporator and suction side of the compressor in the compression system. The pressure in the low side corresponds to that of the evaporator temperature. The generator, the condenser, and the receiver, as well as the analyzer and the rectifier, operate under high

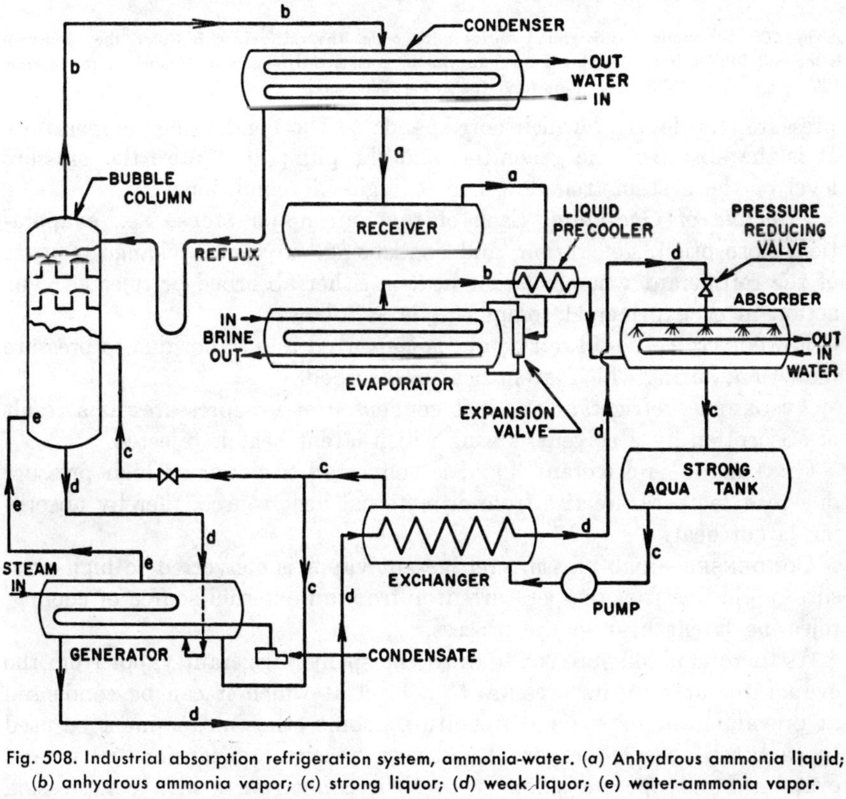

Fig. 508. Industrial absorption refrigeration system, ammonia-water. (a) Anhydrous ammonia liquid; (b) anhydrous ammonia vapor; (c) strong liquor; (d) weak liquor; (e) water-ammonia vapor.

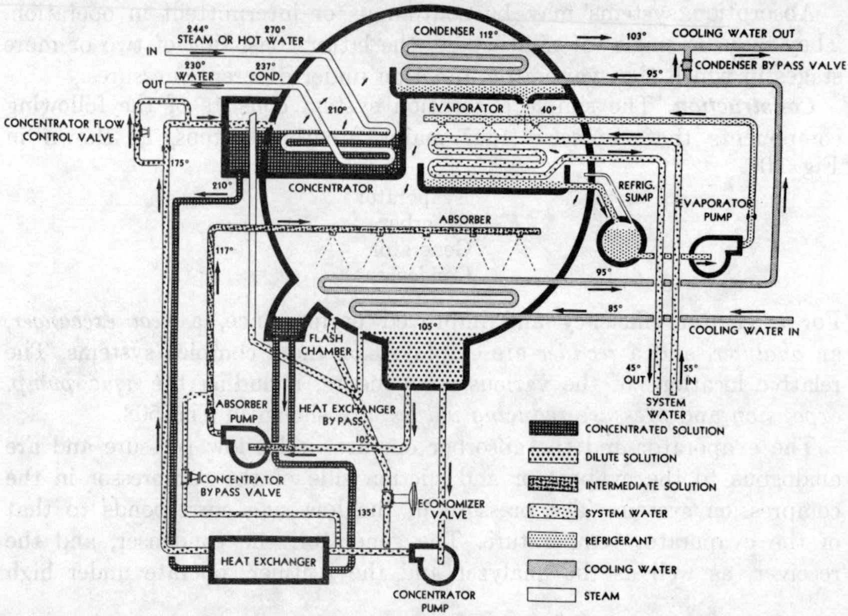

Fig. 509. Schematic of absorption refrigeration cycle. The refrigerant is water, the absorbent is the salt lithium bromide. This is an industrial or commercial type unit. Capacities range from 100 tons to over 1,000 tons. (Courtesy, The Trane Company.)

pressure, the level of which corresponds to the condensing temperature. It is the absorber, the generator, and the pump that raise the pressure level of the system from low-side to high-side conditions.

Principle of Operation. Each of the four major steps—i.e., evaporation, absorption, generation, and condensing—involve a change of state of the refrigerant wherein latent heat is either absorbed or rejected. The action in each principal component is as follows:

EVAPORATOR—liquid refrigerant is converted to a vapor due to pressure reduction, during which latent heat is absorbed.

ABSORBER—refrigerant vapor is condensed at low pressures as a result of absorption by a solvent, during which latent heat is rejected.

GENERATOR—refrigerant liquid is converted to vapor at high pressure due to a temperature rise from an external heat source, thereby absorbing latent heat.

CONDENSER—high-pressure refrigerant vapor is converted to high-pressure liquid due to a temperature drop from an external source of cooling, rejecting latent heat in the process.

As there is no compressor to draw the spent refrigerant vapor from the evaporator or raise its pressure to a level at which it can be condensed at prevailing air or water temperatures, some other means must be used to carry on these functions. Condenser, receiver, expansion valve, and evaporator all operate similarly to their counterparts in a compression

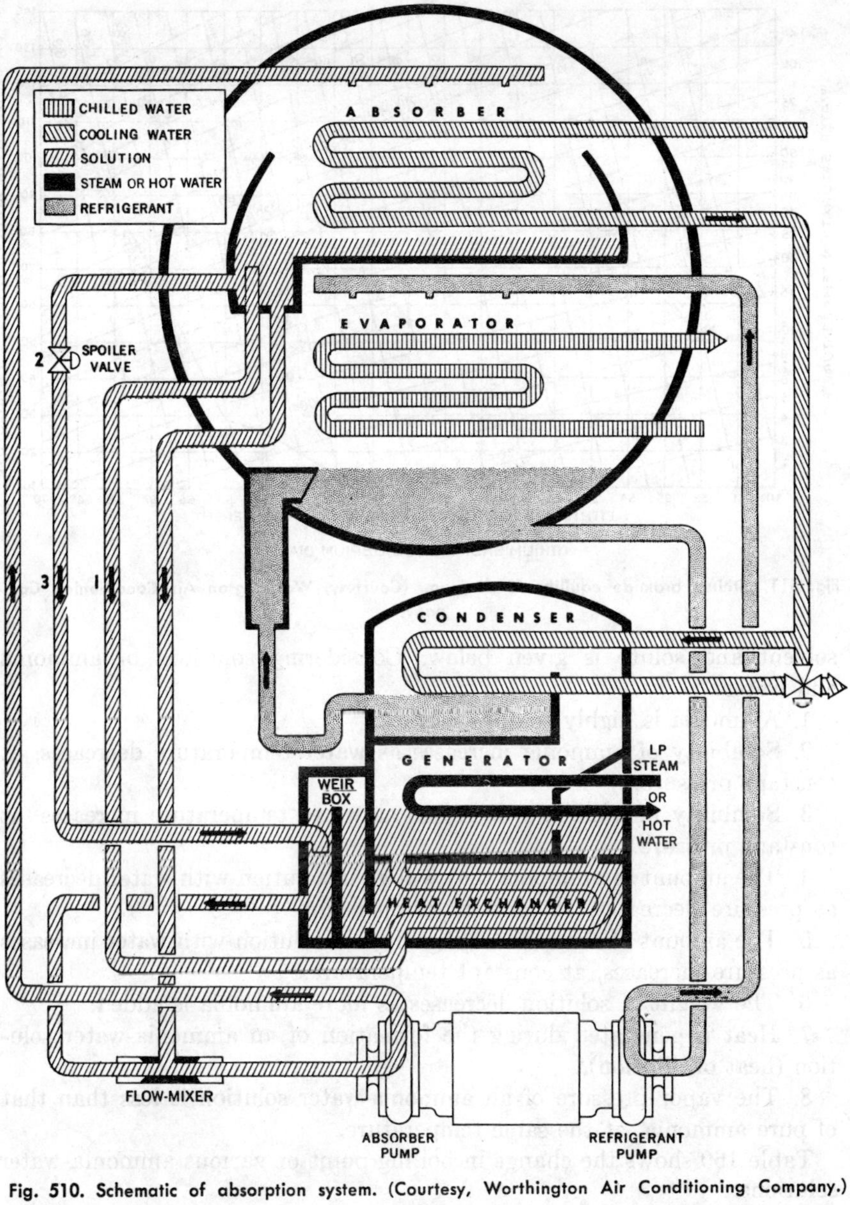

Fig. 510. Schematic of absorption system. (Courtesy, Worthington Air Conditioning Company.)

system. Such components theoretically handle only pure anhydrous refrigerant liquid or gas. However, the absorber and the generator handle the solution of refrigerant and solvent in its weak or strong form. Fluids are variously referred to as weak and strong liquors or weak and strong aqua.

Before describing the activity occurring in the absorber and the generator, a review of the pertinent factors governing the association of a

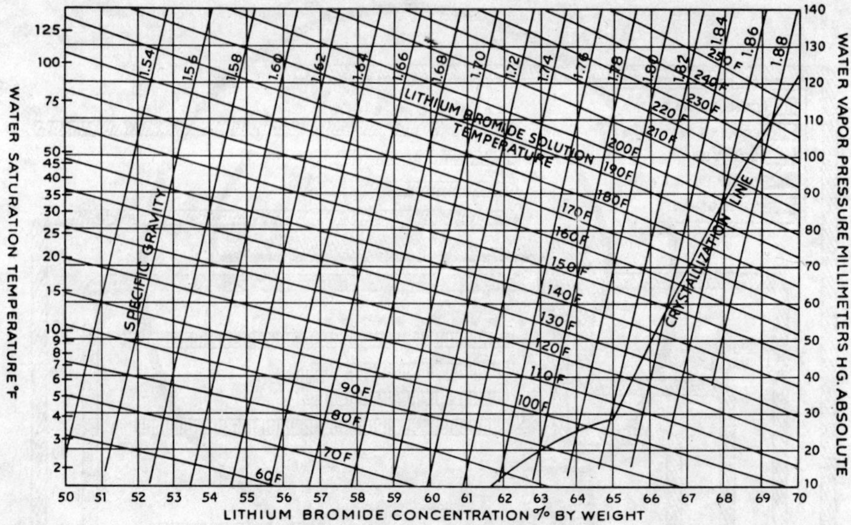

Fig. 511. Lithium bromide equilibrium diagram. (Courtesy, Worthington Air Conditioning Company.)

solvent and solute is given below. Considering solutions of ammonia and water:

1. Ammonia is highly soluble in water.

2. Solubility of ammonia increases as water temperature decreases, at constant pressure.

3. Solubility of ammonia decreases as water temperature increases, at constant pressure.

4. The amount of ammonia that will go in solution with water decreases as pressure decreases, at constant temperature.

5. The amount of ammonia that will go in solution with water increases as pressure increases, at constant temperature.

6. The weight of solution decreases as more ammonia is added.

7. Heat is generated during the formation of an ammonia-water solution (heat of dilution).

8. The vapor pressure of an ammonia-water solution is less than that of pure ammonia, at the same temperature.

Table 160 shows the change in boiling point of various ammonia-water solutions.

In chronological order, functions of the components of a typical industrial absorption system, as shown in Fig. 508, are described below:

Receiver—this component, like its compression system counterpart, holds the charge of anhydrous liquid ammonia under (high) pressure governed by condenser conditions.

Expansion valve—reduces the pressure of the refrigerant as it passes

TABLE 160. BOILING POINTS OF VARIOUS AMMONIA-WATER SOLUTIONS AT ATMOSPHERIC PRESSURE

Percent Water	*Boiling Point, F*	*Percent Water*	*Boiling Point, F*
0	−28	35	--4
5	−26	40	3
10	−24	50	21
15	−21	60	49
20	−17	75	90
25	−14	100	212
30	−10		

from the (high side) receiver to the (low side) evaporator. As in compression systems, the flash gas factor is present in absorption systems.

Evaporator—the cooling apparatus in which liquid ammonia evaporates to produce refrigeration.

Absorber—the component to which the vapor travels after producing its refrigerating effect in the evaporator. The absorber is a compartment in which a weak solution of ammonia and water is sprayed. As the vapor from the evaporator enters the absorber, it is absorbed by the weak solution. During the process of absorption, the following occurs:

1. The weak solution is converted to a strong solution.
2. Latent heat of condensation develops, raising the temperature of the solution.
3. Heat of dilution develops to help raise the temperature of solution.
4. The vapor pressure of ammonia is reduced to that of the solution.

In order to dissipate the heat of condensation and dilution, the absorber is cooled by water from an external source. Lowering the absorber temperature increases the amount of ammonia that will go in solution. The vapor pressure of the solution, being lower than that of the ammonia vapor in the evaporator, both of which are at the same ambient pressure, causes flow to occur from the evaporator to the absorber. Thus the absorbing action establishes the initial flow of refrigerant through the system. Dissipating the heat developed in the absorber as rapidly as possible, thereby maintaining minimum absorption temperatures, encourages a greater flow rate from the evaporator and increases refrigerating capacity. Therefore, it is advantageous to use the lowest temperature absorber cooling water available.

Strong Liquor Pump. This component, preferably of centrifugal type, serves to establish a pressure differential between the high side and the low side of the system. In doing so, it draws the strong solution from its region of origin, the absorber, and forces it *under pressure* into the generator.

Generator. The generator receives the strong solution from the absorber by means of the liquor pump. The latter increases the pressure to a level governed by condenser conditions, during transfer of the strong solution from the absorber to generator. The generator is equipped with some means of heating the solution, usually a steam coil. As the solution is heated, ammonia vapor is driven off, leaving behind a weak solution. The weak solution, being heavier in weight than the strong solution, gravitates to the bottom of the generator. The bottom of the generator is connected by a return pipe to the absorber. Thus the generator is the source of weak liquor used in the absorber to absorb the ammonia vapor as it comes from the evaporator. Because the generator is under high pressure and the absorber under low pressure, a *pressure reducing valve* is required in the weak liquor line to maintain a pressure differential between the two. The concentration of ammonia in the weak solution returned to the absorber is controlled by the amount of heat added in the generator. This is in accordance with the rule that the absorbing capacity of a solution varies as its temperature at constant pressure.

Analyzer. Ammonia and water vapor are driven off the strong solution during the heating process in the generator. As the vapors rise, they enter an analyzer attached to the top of the generator. The analyzer is primarily a fractionating (bubble) tower similar in principle to that described for distilling petroleum crude (see Chapter 15). As the vapors rise in the analyzer, they are cooled. The higher boiling point water vapor condenses and falls back to the generator. The ammonia vapor continues to rise until it leaves at the top of the analyzer to enter the condenser. The condensate absorbs some ammonia as it descends in the analyzer. It also becomes mixed with strong liquor bled from the high-pressure side of the main strong liquor line to provide additional cooling in the analyzer. Therefore, the condensate returns to the generator in the form of a weak solution. The heat exchange in the analyzer reduces both the generator heat load and the condenser cooling requirement.

Rectifier. In an effort further to reduce the amount of water carried over to the evaporator, a rectifier is used. The rectifier is in effect a water-cooled pre-condenser separated from the main condenser. It does not have the capacity to condense any appreciable amount of ammonia, but it does cause precipitation of higher boiling point water vapor remaining in the gas from the analyzer. This water is returned to the analyzer through a bleed-off line in the form of a weak solution called *reflux*. In most cases, a portion of the main condenser is used as a rectifier.

Condenser. The function of the condenser is to convert refrigerant vapor to liquid. It is similar in design and operation to those used on compression systems. The liquid refrigerant enters the receiver from the condenser where it is stored under high pressure until called for by the evaporator. In some systems, the condenser cooling water is also used to cool the absorber employing a series arrangement.

Heat Exchanger. Installed in the strong liquor line, the exchanger provides for an exchange of heat between the strong and the weak liquors. Thus the temperature of the weak liquor from the absorber is increased and the temperature of the strong liquor from the generator is decreased. A thermal advantage results.

Liquid Precooler. The liquid from the receiver may be reduced in temperature as it passes to the expansion valve by means of a liquid precooler. Cooling is provided by the refrigerant vapor as it passes from the evaporator to the absorber.

Capacity Control. When the load drops, less vapor enters the absorber from the evaporator. The absorber tends to continue the absorption process, thereby reducing the evaporator pressure. Lowering evaporator pressure reduces the boiling point of the liquid refrigerant contrary to the load demand. One method of preventing a decrease in evaporator pressure is to throttle the line leading to the absorber.

If the amount of weak liquor from the generator remains constant during the throttling period, the concentration of ammonia in the strong liquor returning to the generator will diminish, thus upsetting the balance of concentration throughout the system. When this occurs, the heat in the generator becomes excessive.

Changes in load demand are better met by controlling the quantity of weak liquor flowing to the absorber. Thus, only that amount of weak liquor necessary to absorb the lesser quantity of vapor produced and to maintain maximum strong liquor concentration would be fed to the absorber.

However, the degree of strong liquor concentration is also governed by the amount of heat imparted in the generator. Therefore, it is necessary to adjust heat input to the generator commensurate with the flow of weak liquor to the absorber to maintain the balance of concentration.

As in compression systems, the pressure differential between the low and the high side should be kept as low as possible. By maintaining absorber pressure as high as possible commensurate with refrigeration temperature, the heat demand in the generator will be minimized.

Absorption System, Domestic Refrigerator

With the use of an inert gas, such as hydrogen, to balance low-side and high-side pressures, an absorption system using a refrigerant and an absorbent will function without a pump or other moving parts. Such systems employ various refrigerant-absorbent combinations. It is necessary that the absorbing action be physical rather than chemical. Both solid and liquid absorbents may be used. With the use of a solid, the action is one of *adsorption* while *absorption* is the action occurring between two liquids. Water is usually the refrigerant when solid absorbents are used even though the final absorbing substance is a liquid solution or brine. Lithium chloride and lithium bromide solutions are hygroscopic brines.

A typical absorbent system for domestic service may use ammonia and water or lithium bromide and water. Air-conditioning systems require the use of a nontoxic substance, so lithium salts acting as the absorbent are used with water as the refrigerant.

Consider an absorption system having no moving parts using ammonia and water. It is adaptable to service in domestic refrigerators. Such a system consists of a generator, analyzer, rectifier, condenser, absorber, evaporator, and heat exchangers. It has no expansion valve, but employs a U-tube arrangement through which liquid refrigerant passes on its way to the evaporator. The low side of the system is charged with hydrogen which is not soluble and is lighter in weight than ammonia. The presence of hydrogen introduces the principle that follows Dalton's law of partial pressure; i.e., the total pressure of a gas mixture is equal to the sum of the pressures exerted by the individual gases. This means that, while the total pressures of high side and low side are equalized by the hydrogen, the ammonia pressure on the low side is below that of the high side. Therefore, evaporation of liquid ammonia in the evaporator will proceed as in the conventional manner.

The operation of the hydrogen-charged system is as follows:

The liquid ammonia flows from an air-cooled condenser to a U-tube, filling that component and establishing a trap between the high and the low side. The pressure at both ends of the U-tube being equalized, no surging occurs. Neither will there be any escape of hydrogen to the high side. As more liquid from the condenser enters the U-tube, the overflow passes into the evaporator, where refrigeration is produced in the usual manner. The ammonia vapor containing some hydrogen flows to the absorber, drawn there by the pressure reduction consequent to absorbing action. Because the hydrogen is not soluble in the weak liquor, it separates from the ammonia. Being lighter than ammonia, it rises from the absorber through a hydrogen return tube to the evaporator. Some of the hydrogen passes to a pressure vessel or reserve tank.

The ammonia enters the weak liquor to make strong liquor which flows back to a generator by gravity. The generator is heated by means of a gas burner which consequently boils the strong liquor. The boiling solution percolates upward through a percolator tube to a separator. The ammonia vapor leaving the solution passes through an analyzer and thence to the condenser. The separating chamber, having a liquid level higher than the absorber, causes the weak liquor to flow back to the absorber through a liquid heat exchanger. Entering at the top, the weak liquor drops downward through the absorber to meet the rising ammonia vapor and hydrogen in a counterflow action. The hydrogen passes through a gas heat exchanger on its way upward, where it is cooled by the ammonia vapor going to the absorber. The hydrogen, being light in weight, rises in the nonrestricted return line. Therefore, it will not filter back to the generator through the liquor lines which are at a low level in the system and filled with liquid.

Intermittent Absorption System

The tube shown in Fig. 512 is shaped to provide a reservoir at each end. In one end is placed ammonia chilled to a temperature at which it will exist in liquid form under the pressure existing within the tube. In

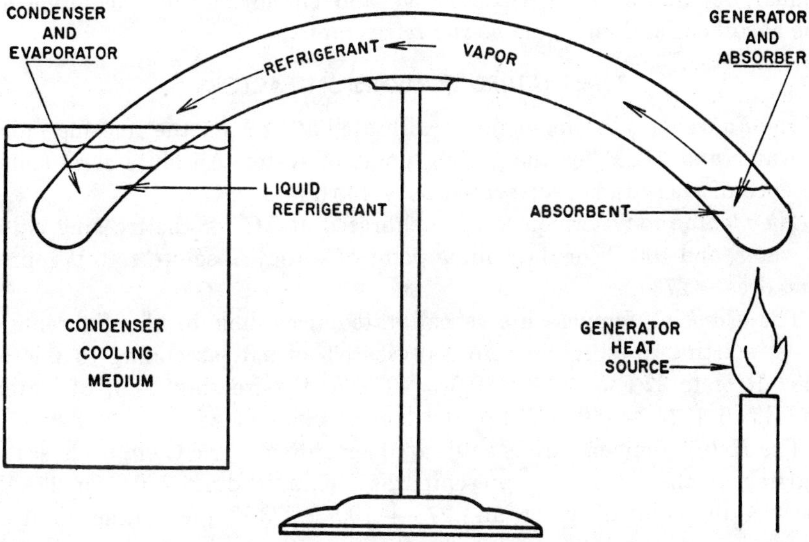

Fig. 512. Intermittent absorption apparatus. Regenerative cycle. Faraday's principle.

the other end is placed an absorbent whose temperature is reduced by external chilling. Under the above conditions very little activity will occur.

If the ammonia reservoir were immersed in a relatively warm substance, the ammonia would boil and start to evaporate. As in any evaporator, heat would be absorbed and the substance surrounding the ammonia reservoir would drop in temperature. The resulting vapor would rise and fill the vacant space within the tube. The vapor reaching the opposite end would be absorbed by the cooled absorbent material, causing a drop in pressure immediately above the absorption surface. Between the heat causing the ammonia to boil and the process of absorption, a continuous flow of vapor would be initiated. The action would continue until all the ammonia was evaporated. Thus the refrigeration cycle would be completed.

Heating the absorbent and chilling the ammonia reservoir would reverse the cycle. Under the latter conditions, ammonia would be driven off from the absorbent and condense in the chilled reservoir under the pressure developed by the heat applied to the absorber. Thus the regenerative cycle would be completed when all the ammonia was converted to liquid.

Constructed in a suitable manner, the apparatus described could be converted into an intermittent absorption refrigeration machine having (1) a refrigerating and (2) a regenerative cycle. The ammonia end would become intermittently a condenser and an evaporator while the opposite end would function successively as a generator and an absorber. The principle of absorption refrigeration was established by Faraday using a similar apparatus as described above and employing silver chloride as the absorbent and ammonia as the refrigerant.

TEMPERATURE CONVERSION FACTORS

The *Fahrenheit* thermometer is calibrated at 32 F for the freezing point of water and 212 F for the boiling point of water. Absolute zero (total absence of energy) is represented as −459.6 F.

The *Centigrade* thermometer is calibrated at 0 C for the freezing point of water, and 100 C for the boiling point of water. Absolute zero is represented as −273 C.

The *Rankine* temperature is calibrated according to the Fahrenheit scale, starting at absolute zero represented as 0 R, ascending to 0 F or 459.6 R or to 32 F or $32 + 459.6 = 491.6$ R, the freezing point of water, and $459.6 + 32 + 180 = 671.6$, the boiling point of water.

The *Kelvin* temperature is calibrated according to the Centigrade scale, starting at absolute zero represented as 0 K, ascending to 0 C or 273 K, the freezing point of water, and $273 + 100 = 373$ K, the boiling point of water.

All references to the state points of water are based on atmospheric pressure or 14.7 psia.

Conversion from one temperature scale to another is indicated as follows:

$$°F = 1.8\ (°C) + 32° \text{ or } °F = \frac{9}{5}\ (°C) + 32°$$

$$°C = 0.55555\ (°F - 32°) \text{ or } °C = \frac{5}{9}\ (°F - 32°)$$

$$°R = °F + 459.6°$$

$$°K = °C + 273°$$

$$°R = 1.8\ (°K)$$

$$°K = 0.55555\ (°R)$$

−40° represents the same temperature on both the Fahrenheit and Centigrade scale. Usually:

Ordinary refrigeration temperatures are expressed in °F.

Scientific or laboratory processes are expressed in °C.
Thermoelectric cooling processes are expressed in °K.
Cryogenic temperatures are expressed in °R.

There are exceptions in all cases given above. However, calculations must obey the same scale throughout, even to the extent of converting such property values based on a different scale.

TABLE 161. TEMPERATURE CONVERSION

°*K*	°*R*	°*C*	°*F*	°*K*	°*R*	°*C*	°*F*
0	0.0	-273.15	-459.67	140	252.0	-133.15	-207.6
1	1.8	-272.15	-457.8	150	270.0	-123.15	-189.6
2	3.6	-271.15	-456.0	160	288.0	-113.15	-171.6
3	5.4	-270.15	-454.2	170	306.0	-103.15	-153.6
4	7.2	-269.15	-452.4	180	324.0	- 93.15	-135.6
5	9.0	-268.15	-450.6	190	342.0	- 83.15	-117.6
6	10.8	-267.15	-448.8	200	360.0	- 73.15	- 99.6
7	12.6	-266.15	-447.0	210	378.0	- 63.15	- 81.6
8	14.4	-265.15	-445.2	220	396.0	- 53.15	- 63.6
9	16.8	-264.15	-443.4	230	414.0	- 43.15	- 45.6
10	18.0	-263.15	-441.6	240	432.0	- 33.15	- 27.6
11	19.8	-262.15	-439.8	250	450.0	- 23.15	- 9.6
12	21.6	-261.15	-438.0	260	468.0	- 13.15	+ 8.6
13	23.4	-260.15	-436.2	270	486.0	- 3.15	26.6
14	25.2	-259.15	-434.4	273	491.4	0.0	32.0
15	27.0	-258.15	-432.6	280	504.0	+ 7.15	44.6
16	28.8	-257.15	430.8	290	522.0	17.15	62.6
17	30.6	-256.15	-429.0	300	540.0	27.15	80.6
18	32.4	-255.15	-427.2	310	558.0	37.15	98.6
19	34.2	-254.15	-425.4	320	576.0	47.15	116.6
20	36.0	-253.15	-423.6	330	594.0	57.15	134.6
21	37.8	-252.15	-421.8	340	612.0	67.15	152.6
22	39.6	-251.15	-420.0	350	630.0	77.15	170.6
23	41.4	-250.15	-418.2	360	648.0	87.15	188.6
24	43.2	-249.15	-416.4	370	666.0	97.15	206.6
25	45.0	-248.15	-414.6	373	671.4	100.0	212.0
30	54.0	-243.15	-405.6	380	684.0	107.15	224.6
35	63.0	-238.15	-396.6	390	702.0	117.15	242.6
40	72.0	-233.15	-387.6	400	720.0	127.15	260.6
45	81.8	-228.15	-378.6	410	738.0	137.15	278.6
50	90.0	-223.15	-369.6	420	756.0	147.15	296.6
55	99.0	-218.15	-360.6	430	774.0	157.15	314.6
60	108.0	-213.15	-351.6	440	792.0	167.15	332.6
65	117.0	-208.15	-342.6	450	810.0	177.15	350.6
70	126.0	-203.15	-333.6	460	828.0	187.15	368.6
75	135.0	-198.15	-324.6	470	846.0	197.15	386.6
80	144.0	-193.15	-315.6	480	864.0	207.15	404.6
85	153.0	-188.15	-306.6	490	882.0	217.15	422.6
90	162.0	-183.15	-297.6	500	900.0	227.15	440.6
95	171.0	-178.15	-288.6	600	1080.0	327.15	620.6
100	180.0	-173.15	-279.6	700	1260.0	427.15	800.6
110	198.0	-163.15	-261.6	800	1440.0	527.15	980.6
120	216.0	-153.15	-243.6	900	1620.0	627.15	1160.6
130	234.0	-143.15	-225.6	1000	1800.0	727.15	1340.6

CRYOGENICS

Cryogenics is that branch of physics that deals with the production and use of temperatures that range from −240 F (−150 C) down to levels approaching absolute zero. The name of the art is derived from the Greek *kryos,* meaning "icy cold," and *genes,* "to become" or "produce."

The science of cryogenics is applied in a number of different fields involving low temperature substances both as the working mediums and the substances to which low temperatures are applied for various purposes. Cryogenic apparatus embrace the necessary mechanical equipment capable of converting extremely low boiling point fluids from a gas to liquid and, in some instances, to their solid state. The production of liquids from various gaseous elements, such as oxygen, hydrogen, helium, argon, nitrogen, neon, fluorine and methane is a cryogenic process, as is their separation from a host environment such as air. Some of these liquids are utilized as refrigerants to maintain proper temperatures for *preserving perishable commodities* and to treat metals for *metallurgical* purposes to mention only two of their numerous uses. Cryogenic effects lend advantages relative to the storage and transportation of liquefied gases, some of which are used as *fuels,* e.g., natural gas for domestic lighting and heating, hydrogen for propelling missiles, rockets, etc., or as *oxidizers* in rocket propulsion work, such as oxygen and fluorine. Although not considered a cryogenic fluid according to the limited temperature definition, propane is liquefied for storage purposes, a process (see Chapter 16) which is also applicable to natural gas.

While a degree of difficulty is related to the liquefaction and storage of fluids having boiling points at atmospheric pressure of −198 F (Refrigerant—R-14) up to −53.7 F (R-1270, propylene), they are not included in the family of cryogenic substances. Because such substances are considered as possessing an aura of cryogenes about them, they are widely regarded as such. To distinguish the latter group from

(1) the facile group of fluids represented within the range of higher boiling points at atmospheric pressure with R-13B1 (−72 F) at the low end, and R-11 (+74.7 F) at the high end of the range, and
(2) the already defined group of cryogenes,

they should be given some definite applicable nomenclature. While the various refrigerants are categorized as low-, intermediate-, high-, and extremely high-pressure fluids in Chapter 5, the classification is directed toward their temperature application rather than their relative storage and handling facility. For the latter purposes, the fluids represented within range between R-14 (−198.2 F) and R-744 (carbon dioxide) (−109.3 F) may be considered as *quasi-cryogenes* and are so identified herein.

Cryogenic Substances

Table 162 lists the important cryogenic substances and some of their physical properties.

CRYOGENES

Of the fluids listed in Table 162 neon, argon, krypton, xenon and helium are inert, colorless gases present in the atmosphere and valuable for their behavior when exposed to an electrical current, or in some cases to high and low temperature extremes.

Neon (*Ne*) is obtained from the atmosphere by liquefaction. Its chief use is in connection with lighting of airport beacons, advertising signs, etc. It emits a bright reddish-orange light when electrically energized within a sealed tube where it furnishes its own conduction between the positive and negative electrodes at each end.

Argon (*A*) is the most abundant member of the inert gases present in the atmosphere (0.935%) and is produced by careful fractionation of liquid air. Its most important use is in its application in incandescent light bulbs where it acts to retard evaporation of the filament. Argon is also employed for filling vapor-pressure thermometers for use at extremely low temperatures, and in ionization chambers for measuring X-ray intensity. It emits a purple glow when electrically energized.

Krypton (*Kr*) although considered inert will form a limited number of compounds whose bonds between the neighboring atoms are of low stability. Such bonds are said to possess a weak intermolecular character, described as *Van der Waals forces,* which is related to the behavior of gases. When electrically energized, it emits a pale-violet light. It is used in photographic lighting elements that produce a strong flash of extremely short duration, in decorative glow lamps and in some types of searchlights. Krypton is present in the atmosphere (dry) volumetrically 1 part in 910,000.

Xenon (*Xe*) is a quasi-cryogenic substance used in some vacuum tubes in the electronic field. It emits a light ranging from sky blue to green in color. It is present in the atmosphere in quantities approximately 1 part in 20,000,000 parts of dry air by volume. It is employed in mercury-xenon lamps having a quartz lens system, to simulate irradiation of the sun in space research.

Helium (*He*) is produced chiefly from natural gas occurring in a limited area of the Texas panhandle. It is also present in the atmosphere at the earth's surface in a ratio of 1 part to 185,000. It ranks second, below hydrogen, in lightness in weight of all other elements. Helium as a gas has a number of uses, including that of providing buoyancy (92% as buoyant as hydrogen) for lighter-than-air vehicles, such as dirigibles and balloons, and for the prevention of bends in humans working under

TABLE 162. SOME PROPERTIES OF CRYOGENIC FLUIDS

Cryogene	*Normal Boiling Point at 14.7 psia* °F	*Liquid Density* lbs/gal	*Liquid Density* lbs/cu ft	*Latent Heat of Vaporization* Btu/lb	*Normal Melting Point* °F	*Critical Temperatures* °F	*Shrinkage from Gas at Standard Condition to Liquid at Normal B.P.* cu ft*
Helium	−425	1.043	7.802	8.80		−450	755
Hydrogen	−423	.592	4.430	193.0	−435	−450	840
Neon	−411	9.917	74.18	31.1	−416	−380	1424
Nitrogen	−320	6.738	50.40	85.2	−346	−233	695
Fluorine	−307	12.59	94.18	74.1	−363	−200	959
Argon	−303	11.70	87.52	70.2	−309	−188	848
Oxygen	−297	9.527	71.27	91.7	−361	−182	862
Methane	−258	3.537	26.46	219.0	−296	−116	638
Krypton	−244	20.05	150.00	46.4	−251	− 83	692
†Xenon	−163	25.93	194.00	41.4	−169	+ 62	568
†Ethylene	−155	4.743	35.50	206.0	−272	+ 49	490
†Propane	− 44.2				−309.8	+202	
Air	−318					−220	

* Ratio of standard cubic feet of gas to cubic foot of liquid.
† Non-cryogenic.
From American Society of Heating, Refrigeration and Air Conditioning Journal.

high pressures related to deep-sea diving and caissons. It has high heat conductivity so is valuable for use in vacuum tubes and electric lamps. Having inert properties helium provides an environment for certain metallurgical operations, including the welding of light metals, the latter known as heliarc welding. In heliarc welding (blanket of helium gas) helium prevents burning of the metal. Helium emits a pinkish-violet light when electrically energized. The alpha rays of radioactive materials are of helium nuclei. It has a low refractive index, an advantage in the assembly of complex optical systems to fill the spaces between lenses when placed in series.

Helium will not solidify at atmospheric pressure even if the temperature is reduced to absolute zero. It is, therefore, one of the most difficult of all gases to liquefy. Liquid helium exists in two phases, *He I or normal liquid* and *He II or superfluid liquid.*

As a liquid, helium can be used to provide low temperatures for both commercial and laboratory purposes. Helium II has a heat conductivity value almost 200 times that of copper. Lead, mercury and other metals increase their electrical conductivity in the environment of an extremely low temperature liquid. Helium has the uncanny characteristic of creeping over cold surfaces seemingly unaccountable to turns, barriers or slopes. seeking a warm area. Helium is the remaining element when all other gases contained in natural gas are liquefied.

The rest of the cryogenic liquids listed are chemically active in various degrees, their characteristics being described below:

Hydrogen (H) is the lightest in weight of all known elements. It is made up of atoms (ordinary hydrogen) having one proton and one electron. A stable molecule is formed by a pair of hydrogen atoms that share their electrons to create a bond. The properties of hydrogen are determined by the nature of this molecule, chemically indicated as H_2. Hydrogen is a colorless, odorless and tasteless gas. Hydrogen, in appreciable quantities, occurs only in combination with other elements, such as carbon, to form the complex hydrocarbon molecules of natural gases and petroleum, or oxygen, to produce water (H_2O). At higher temperature, *normal hydrogen* is a mixture of two forms of hydrogen molecules, known as *ortho-hydrogen* (o-H_2) and *para-hydrogen* (p-H_2) in the proportion of 75% and 25% by volume, respectively. The difference between the two molecular forms lies in the relative direction of spin of their nuclear particles. Ortho-hydrogen is made up of molecules whose particles spin in the same direction, whereas the particles in para-hydrogen spin in opposite directions. The proportion of o-H_2 to p-H_2 varies according to temperature and defines an equilibrium mixture known as *equilibrium hydrogen* (e-H_2). As the temperature of the mixture decreases, the o-H_2 changes to p-H_2 until the equilibrium hydrogen becomes 99.8% p-H_2 at the boiling point. It might be added that only the atoms whose electrons spin in opposite directions will combine.

The behavior of hydrogen as related to its molecular conversion, i.e., from o-H_2 to p-H_2, is cryogenically important in that a release of energy occurs during the process, which is not accomplished rapidly. In fact, the process will continue after liquefaction is complete. The release of energy, related to spin reversal, is an *exothermic* action and the quantity of energy released is called *heat of conversion.* The amount of heat involved is sufficient to cause the liquefied hydrogen to boil away during storage. To overcome this problem, a catalyst is used to quicken the process and remove the heat of conversion during the liquefying stage.

Hydrogen either in liquid form or liquid hydrogen in combination with its solid state (to increase density) is a nearly ideal high energy fuel for providing thrust for space vehicles in conjunction with an oxidizer.

Hydrogen is flammable, but will not support combustion in the manner of oxygen. It is present in all acids.

Nitrogen (*N*) is a colorless gas and forms the major part (78% by volume) of the earth's atmosphere. It is obtained chiefly by distilling liquid air. Nitrogen does not combine readily with other elements, a characteristic that enables it to be used in direct contact with other materials. In its liquid form, nitrogen is useful in the field of space and missile development, and in the distillation of other gases at low temperatures. As a stable gas, it becomes an ideal purging agent for systems employing combustible or combustion supporting elements, such as hydrogen and oxygen. Under pressure, it furnishes mobility to liquids in their transfer through pipe-lines, and a separating agent between different kinds of liquids transported in tandem within the same conduit. Nitrogen is also used for the rapid freezing of foods, etc., as well as providing refrigeration in transportation vehicles and shipping containers.

Fluorine (*F*), a member of the halogen family of elements, is pale greenish-yellow in color, has an irritating odor and is one of the densest of cryogenic liquids. In its liquid state, it becomes clear yellow changing to a colorless solid at a temperature of about −422 F. Fluorine is the most electronegative and reactive of the known elements and will combine with all of them readily, excepting the rare gases and a few other compounds. It is chiefly obtained from the mineral fluorspar (calcium fluoride, C_aF_2). Because of its powerful oxidizing capability, it finds important use in rocket propulsion work. It is highly toxic, however, and must be employed under isolated conditions of use, storage and control.

Methane (CH_4) is a hydrocarbon compound that is the major constituent (79%+) of natural gas. It is a tasteless, odorless and colorless gas in its natural state. It is also called *marsh gas* because of its association with the bacterial decomposition of marsh plants under water and the resultant stagnation of swamp water. It is explosive when mixed with oxygen or air. Methane is the first of a series of hydrocarbons identified as *paraffins*. Each succeeding compound in order of its hydrocarbon composition, such as ethane (C_2H_6), propane (C_3H_8), butane (C_4H_{10}), etc., differs from its preceding one by the group, CH_2, called a *homolog*. The series is said to be *homologous*. Any homolog increase raises the boiling point of the resulting compound. Methane when combined with the various halogen elements forms an important group of fluids, referred to as the methane series of halogenated refrigerants. Ethane, the second hydrocarbon in the paraffinic series, having a boiling point of −128 F, also forms a series of halogenated refrigerants. (Both series are described in Chapter 5.) Its use has been found in carbonizing steel and as a starting compound for a number of chemical materials, such as that described above, relative to refrigerants. It solidifies in the form of needle-like crystals.

Oxygen (*O*) is tasteless, odorless, colorless and slightly heavier than air. Its atomic weight is 16, a number used as the standard on which to base the atomic weights of other elements. Oxygen is present in the atmosphere composing about 22% of the latter by volume. Water is about 88% oxygen. It is essential in sustaining life and in the support of combustion. Because of its influence in causing the burning of fuels and rusting of ferrous metals, such activities are defined as *oxidation*, respectively *rapid* or *slow*. Any element or compound that releases its oxygen faster than air is called an *oxidizer*. Therefore, pure oxygen is its own best representative in this connection, e.g., the metal iron will burn rapidly in an environment of pure oxygen gas. It is used for many industrial purposes, including oxy-acetylene welding and metal cutting. Oxygen is responsible for spontaneous combustion. Chemical *reduction* is the removal of oxygen from a compound. Oxygen may be described as both an essential and destructive element.

Ethylene (C_2H_4) is a hydrocarbon gas (normally) of the olefin series, in which it occupies the same relative position as methane does in the paraffin series, i.e., first and most volatile. It is colorless and has a sweetish taste. Like propane and xenon in the list of substances given in Table 162, ethylene is quasi-cryogenic having a boiling point of −154.7 F at 1 atmosphere. Ethylene is oxidized by a dilute solution of potassium-permanganate to produce ethylene glycol used for certain types of brine systems, and as an anti-freeze agent. (It is further described in Chapter 5, as a refrigerant suitable for use in the high pressure stages of cascade systems.)

Other quasi-cryogenes employed for refrigeration purposes are discussed in Chapter 5, "Refrigerants."

CRYOGENIC LIQUEFYING PROCESSES

The raw materials to produce cryogenic liquids and/or solids are those gases described above and listed in Table 162. They have one important physical characteristic in common, their boiling temperatures are below −240 F, except for the quasi-cryogenic fluids noted. The purpose of industrial cryogenic processes is to convert such raw materials to the state in which they become most useful to the specific application for which they are selected.

In the study of the refrigeration cycle, it was shown that to reclaim a spent refrigerant it must be first compressed and then cooled in order to return it to its liquid state. The process embraces a definite interchange of thermal energy and such variation in temperatures as required. The process is relatively simple in that the fluids used are selected on the basis of their nearly ideal behavior, both physically and chemically, under all the conditions to which they will be exposed.

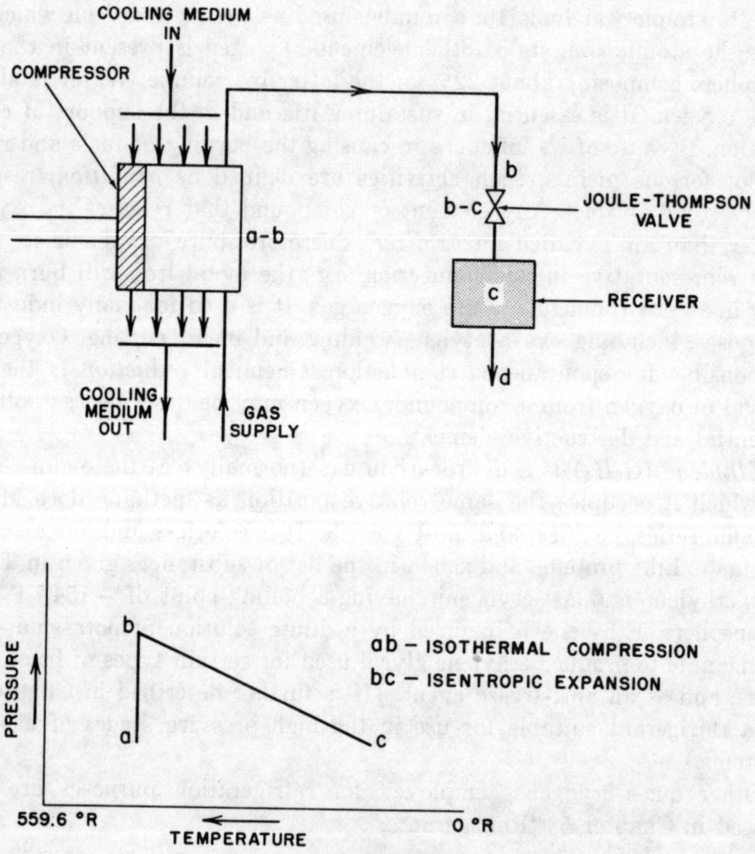

Fig. 513(a). Ideal cryogenic cycle (basic).

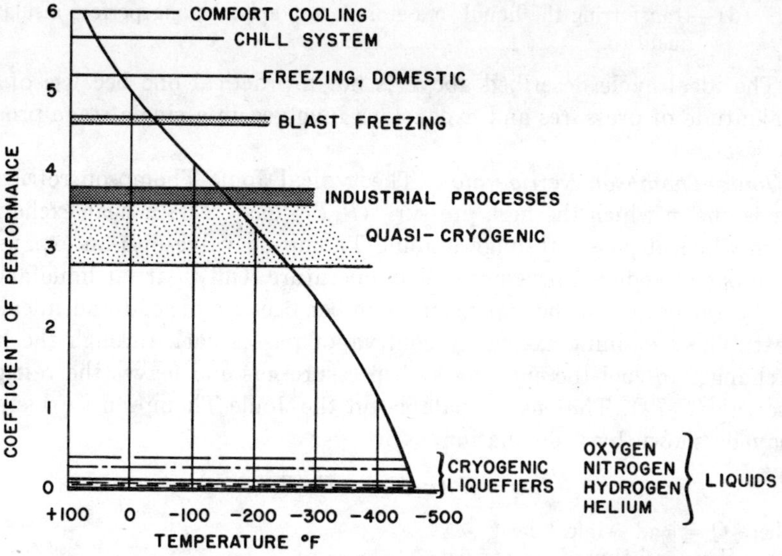

Fig. 513(*b*). Approximate COP values vs. temperature range of processes.

The situation is different in the field of cryogenics. The fluids subject to thermal changes in order to convert them to the state required are selected on the basis of their utility related to other processes not necessarily cryogenic. Therefore, cryogenic procedures must fit the peculiarities of the individual fluids, which vary greatly in many instances. Additionally, the magnitude of forces, both physical and thermal, is much greater as compared with the severest conditions found in pure refrigeration processes. The magnitude of such forces involves temperatures approaching total absence of internal energy and pressures 200 times greater than that of the atmosphere. An old colloquialism that "your work is cut out for you" fits the situation here.

Ideal Liquefication Cycle. The following describes an ideal cycle for liquefying a cryogenic gas. Referring to Fig. 513

(a–b)—compressing the gas isothermally to a high pressure during which heat is removed by transfer from the compressor cylinder wall to an external sink (cooling water).

(at b)—throttling the high pressure gas according to the *Joule-Thompson cooling effect for steady flow.* In this process, a drop in temperature follows a pressure reduction during which the kinetic energy generated within the fluid by the reduction in pressure is completely converted to internal enthalpy without a loss of heat. This is accomplished through a Joule-Thompson valve.

(b–c)—expansion of the gas *isentropically* (at constant entropy) to a pressure corresponding to its temperature of liquid conversion. This completes the cycle.

(d)—transferring the liquid to a storage container having perfect insulation qualities.

The ideal cycle described above is not a practical one because of the magnitude of pressures and expansions involved in a single stage process as described.

Joule-Thompson Refrigerator. The typical Joule-Thompson refrigerator is one in which the high pressure (P_1T_1) gas enters a heat exchanger from which it passes through a Joule-Thompson valve into an expansion chamber at reduced pressure and temperature. Only partial liquefaction of the gas occurs in the expansion chamber because of load and inleak of heat. The remaining extremely cold vapor passes back through the heat exchanger to cool the entering high pressure gas and leaves the refrigerator at (P_2T_2). The energy balance in the Joule-Thompson refrigerator may be shown by the equation

$$Q = H_2(P_2,T_2) - H_1(P_1T_1)$$

where Q = load + inleak heat
H_2 = enthalpy of gas at inlet
H_1 = enthalpy of gas at outlet

In such a system partial liquefied gas is obtained as a result of augmentation of the heat exchanger. It is strictly a refrigerator.

While the ideal cycle and the Joule-Thompson refrigerator serve to demonstrate some of the principles related to cryogenics, both must be substantially augmented in order to produce liquids from cryogenic gases. However, some gases, such as helium, hydrogen and neon require special considerations beyond those for air, for example, to bring about their liquefaction. The reason for this lies partly in their *maximum inversion temperature,* which is below ambient temperature for He, H and Ne. To understand the significance of the inversion properties of a gas during the expansion process at constant enthalpy, refer to Fig. 514, showing tem-

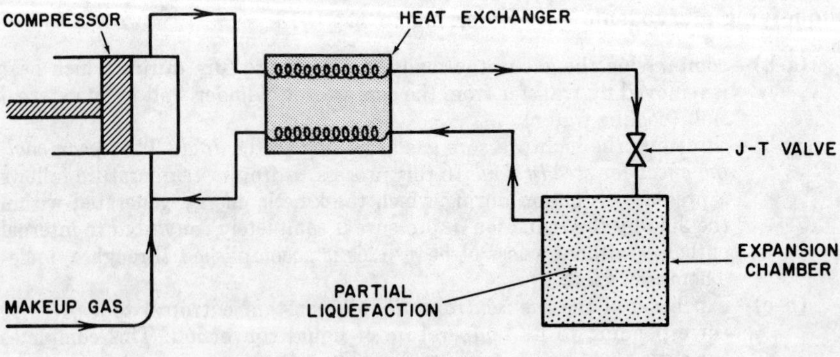

Fig. 523. Typical Joule-Thompson refrigerator.

perature behavior during a steady drop in pressure. From left to right, it shows a *temperature increase* of the gas during initial phases of diminishing pressure followed by a *decrease in temperature* as the pressure continues to drop. The highest point on the pressure-temperature curve indicates a reversal in temperature direction, which is the maximum inversion temperature for the gas under given state conditions represented by the individual curve being considered. Note also that the high point of each curve varies relative to the pressure for different initial state conditions. The inversion curve is established by connecting increment changes in inversion temperature. By comparing the conditions indicated by the various curves, it is obvious that the maximum inversion temperature decreases as the initial temperature of the gas being expanded is reduced. To reach a maximum inversion temperature below that of ambient conditions for some gases, it becomes necessary to reduce their temperature substantially by effective means of refrigeration before the expansion process. To demonstrate Joule-Thompson effects, gases must be below their maximum inversion temperature at the point where it starts initial expansion.

Cryogenic System Components

The preceding discussion showed the basic components of a gas liquefying system dealing with cryogenic fluids to include the following:

Precooling equipment involving the regular apparatus as discussed for refrigeration systems within this text. Refrigerants used for this purpose include carbon dioxide, ammonia, certain halogenated hydrocarbon fluids, and hydrocarbon fluids such as ethylene and methane. It also includes cryogenic fluids where extremely low temperatures are essential to liquefaction.

Heavy duty compressors for developing high cryogenic gas pressures up to 200 atmospheres (3000 psia). Usually multi-stage units.

Heat exchangers for the cooling and inter-cooling of stream gas. Regenerators that act both as heat exchangers and purifiers.

Expansion valves for pressure reduction, usually those having Joule-Thompson (J-T) effect.

Expanders for the reduction of pressure and simultaneous decrease in temperature; includes rotary (turbine) expanders and a piston-cylinder type expansion engine.

Safety devices, including pressure relief valves for the protection of human life and equipment.

Pumps necessary for the transfer of all fluids involved.

Instruments for the measuring and recording of state point properties of fluids.

Insulation and insulating systems for heat inleak.

Other equipment required for special purposes include:

Rectifiers for the separation of gases. Results obtained by flowing (cascading) liquid downward over several perforated plates counter to the rising vapor, so that the higher boiling point fluid exits at the bottom or at points approaching the bottom, with the lower boiling point fluid leaving the rectifier at the top or at points approaching the top.

Purifiers of gas to remove impurities. The purification of gas may be accomplished by *refrigeration, adsorption* (layer compilation on solid surface), *absorption* (penetration of a solid), chemical *adsorption* (strong reaction between the gas and solid) or *chemical processes* reacting with agents, such as desiccants to remove water vapor, or carbon dioxide reacting with sodium carbonate, etc.

Vaporizer for producing gas from liquid, usually oxygen or nitrogen.

Cryogenic Systems

The evolution of cryogenic systems for the production of liquid cryogenic gases depicts a pattern of increased capability and efficiency. Starting with the simple Joule-Thompson refrigerator the development points toward the conditions demonstrated for the ideal or most efficient cycle to produce a quality product safely, simply and at the lowest cost per unit quantity possible. The systems described below follow that evolution. The basic cycles are those employed for the liquefaction of oxygen, nitrogen, argon, krypton, xenon and fluorine, but do not include those extensions necessary for the liquefaction of helium, neon and hydrogen.

Cascade Systems. These are described more fully in Chapter 16 (Fig. 478) as multi-systems in which a lower pressure refrigerant acts as condenser cooling medium for a higher pressure refrigerant in a complex arrangement. It was the first system used to produce liquid air. Various refrigerants having descending boiling points corresponding to the number of ascending pressure stages are used. They include ammonia, ethylene and methane in the order of ascending pressures, as well as some of the halogenated methanes. Propane and ethylene are used in systems for liquefying natural gas.

Joule-Thompson (J-T) Refrigerator. The principles of this have already been discussed. It is not a practical system for the production of cryogenic liquids because of the high compression and expansion ratios. However, it is the natural cycle on which to develop more efficient systems.

Hampson or Linde-Hampson Liquefier. This is a simple method of liquefying gases. (See Fig. 515.) It improves on the J-T refrigerator by allowing the cold vapor from the liquid receiver to return to the compressor suction through the heat exchanger to cool the fresh gas entering the system. A four-stage compressor is used with air.

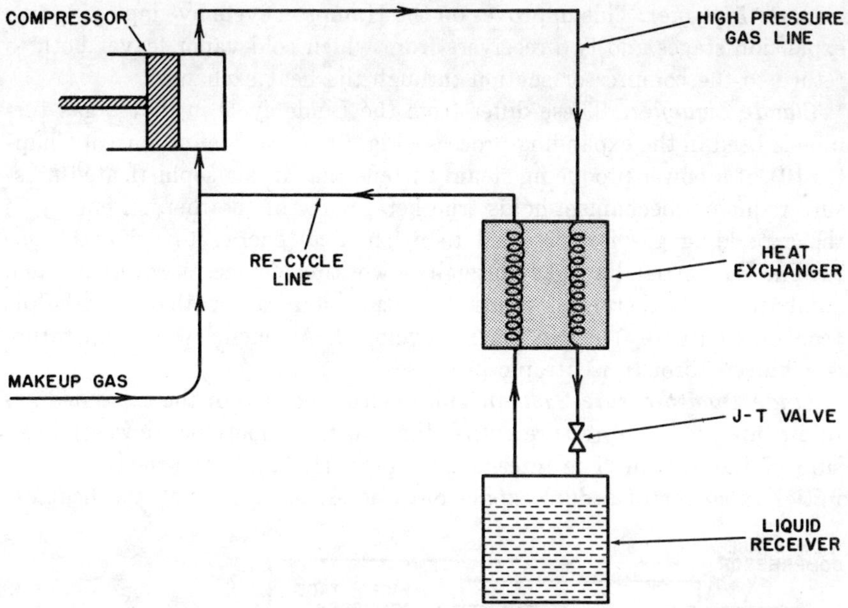

Fig. 515. Hampson liquefier.

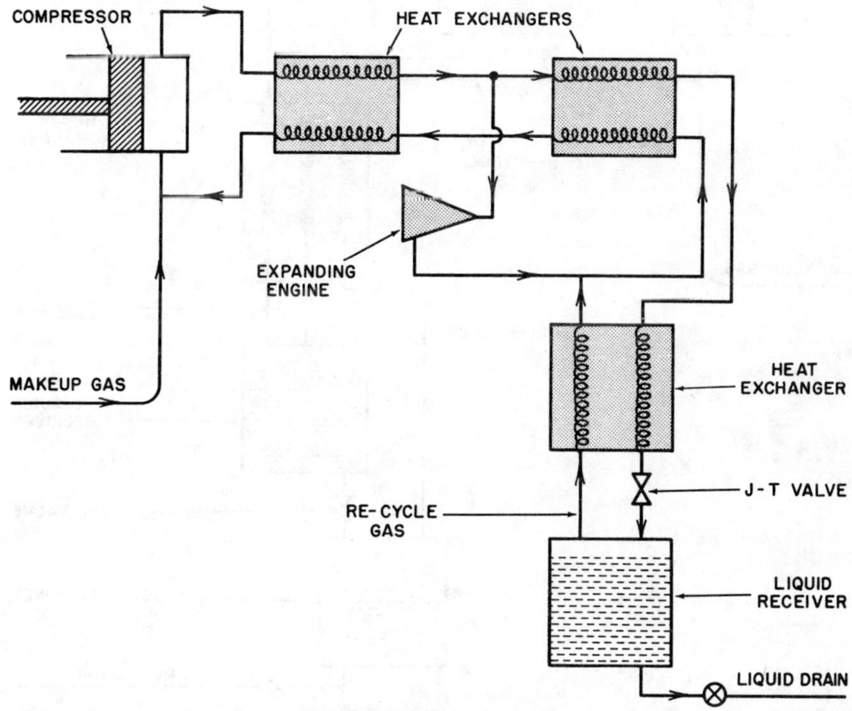

Fig. 516. Claude liquefier.

Linde Liquefier. This improves on the Hampson cycle by employing two expansion stages and two receivers from which cold vapor leaves both to return to the compressor suction through the heat exchanger.

Claude Liquefier. These differ from the Linde cycle in that a gas turbine is used in the expanding process (Fig. 516). Our definition (in Chapter 10) of a power producing steam turbine that it is a sophisticated pressure reducing mechanism holds true here. Some of the thermal energy in the expanding gas is converted to mechanical energy to drive the gas turbine. By expanding gas through a working device, a reversible and adiabatic and, therefore, isentropic expansion rather than isenthalpic (constant enthalpy) expansion is developed. A much lower temperature is achieved through isentropic processes.

Linde Dual-pressure System. This cycles a portion of the expanded gas to an intermediate pressure rather than to the liquefying (lowest) pressure of the system. The intermediate pressure gas is returned from the initial expander of a dual expansion process, back through the heat ex-

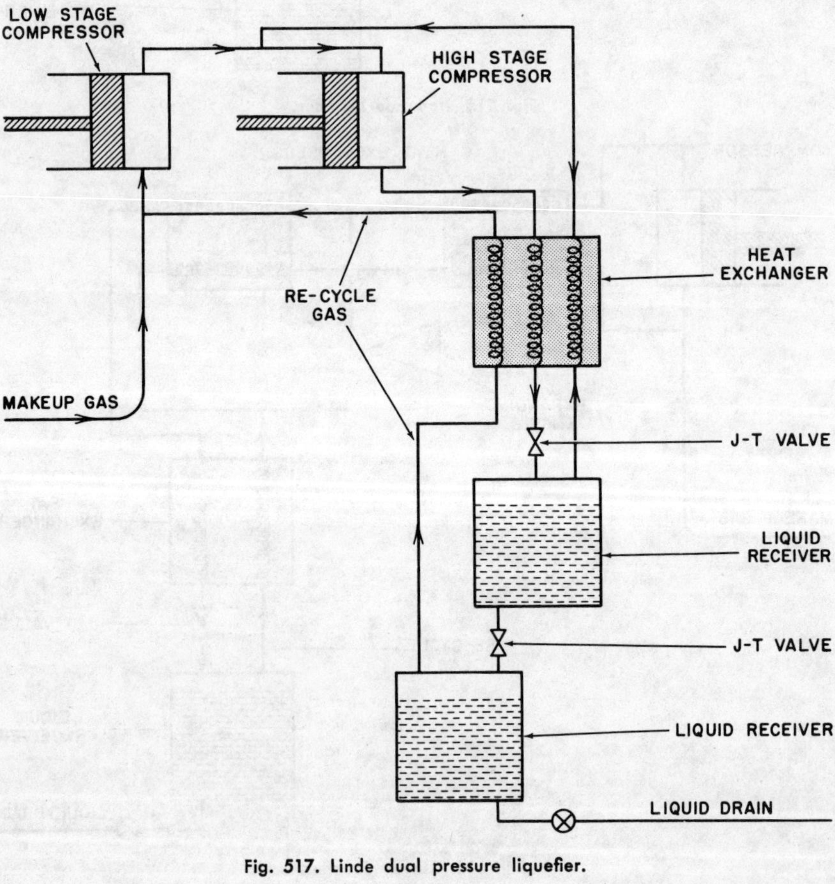

Fig. 517. Linde dual pressure liquefier.

changer to the suction line of the high pressure stage compressor. Because all the gas is not put through the low stage expander some production of liquid is sacrificed for a reduction in work load. As shown in Fig. 517, there are usually three heat exchangers involved in this type of cycle.

Kapitza Cycle. This is a modified Claude cycle by an arrangement which substitutes parallel *regenerators* (see p. 1191) for one of the heat exchangers and eliminating the third or low temperature heat exchanger of the Claude cycle, as shown in Fig. 518. The cycle operates at a relatively lower pressure than is usual for other types of liquefying systems.

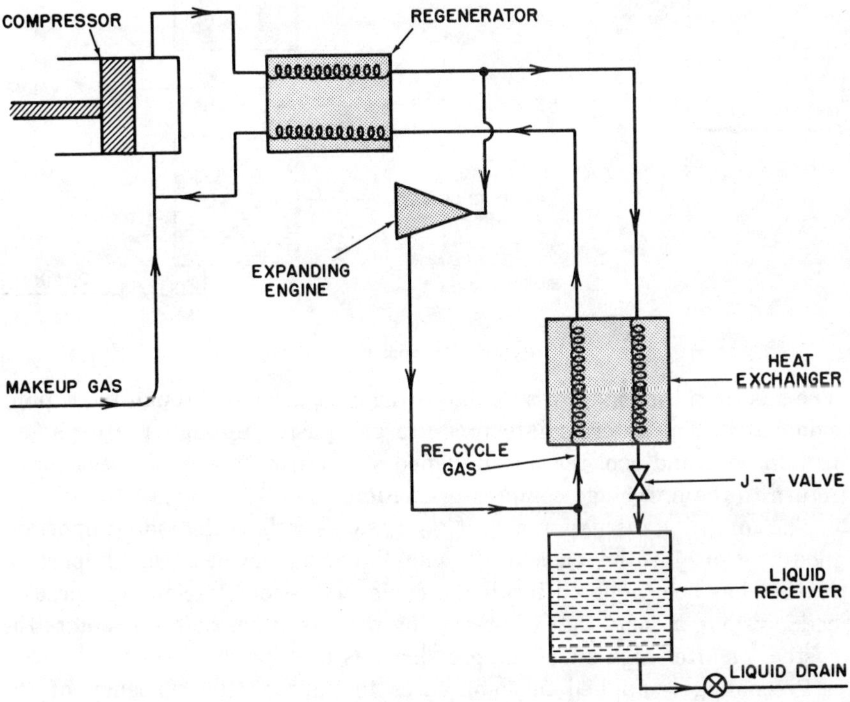

Fig. 518. Kapitza liquefier (low pressure).

Heylandt Cycle. This is a high pressure system for liquefying cryogenic gases. By employing a gas pressure approaching 3500 psia, the first Claude cycle heat exchanger is eliminated and the process of reducing to liquefying temperatures is divided almost equally between the expansion valve and the expansion component of the piston-cylinder type. Figure 519 for the Claude system can be considered for the Heylandt cycle by eliminating the first (nearest the compressor) heat exchanger.

Claude Dual-pressure System. As its name implies, this system is a modification of the basic Claude system in nearly the same manner as the dual-pressure Linde system is a modification of the basic Linde cycle.

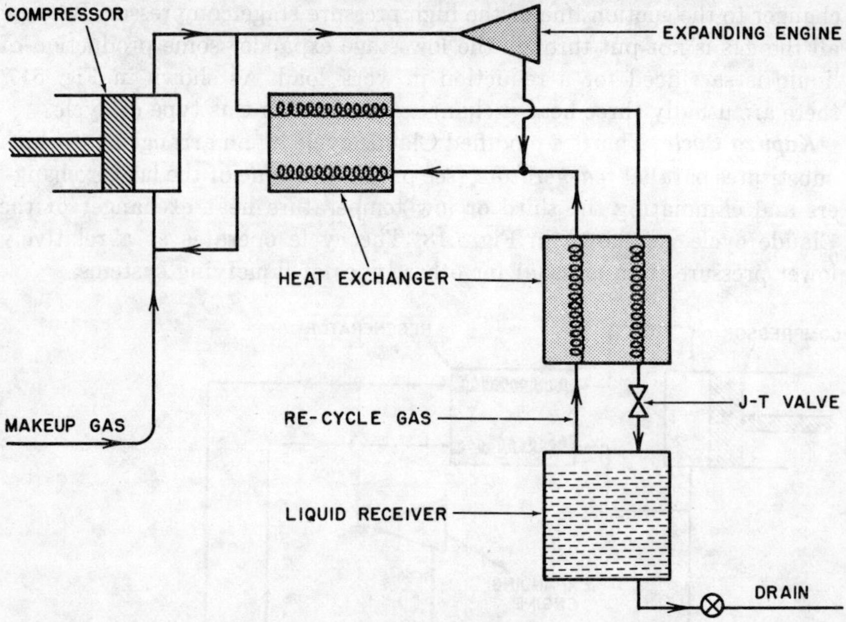

Fig. 519. Heylandt liquefier.

The gas from the high pressure line is the only gas put through the expansion valve. The intermediate pressure gas passes through the expander to join with and cool the non-liquefied vapor from the liquid receiver to return to the first stage compressor suction.

The foregoing discussion covers the basic as well as the more important modifications of the basic cryogenic liquefying cycles. One important feature has been ignored in all the cycle discussions, except for the cascade system, namely that of precooling the gas either before it enters the system or after high stage compression has been accomplished.

Precooling is applied in some cases to improve the efficiency of the process by reducing the load factor within the cycle. In others, it is essential if liquefaction is to be accomplished at all. Also, it depends on the nature of the gas being liquefied what temperature level must be provided before expansion occurs. Therefore, it is necessary that some gases, such as helium, be reduced below its relatively low maximum inversion temperature before liquefaction can be accomplished. Such factors define a cycle as to its utility nomenclature whether it serves as a simple air liquefier, helium liquefier, etc., and also if it is a batch, or continuous flow producer.

Most industrial liquefiers are continuous flow producers, while batch procedures are followed more often in laboratory work. Industrial steady flow apparatus may be stationary or portable.

Separation of Gases

Many normal gases are mixed with other fluids, such as nitrogen, oxygen and the rare gases argon, neon, krypton and xenon that compose the atmosphere. Helium is produced from natural gas and in much lesser quantities from the air. Their production by other than cryogenic processes, however, is beyond the scope of this text.

In the case of air products, then, separation and liquefaction are accomplished cryogenically. Helium from any other source is cryogenically important only in its liquefaction or use as a refrigerant within the limits of the process.

The above spheres of activity incorporate other processes within the art, including that of separation of the gases from their mix environment. Such separators are known as fractionating columns or *rectifiers*. Where gases are separated from a mixture of several gases, such components may be likened to the fractionating or bubble towers used in the petroleum industry for producing fluids according to their boiling points.

If mixtures exist by reason of mutual solubility in all proportions, the individual constituents may exist either in a gaseous or liquid state. In petroleum crudes the various constituents are separated according to their boiling points by heating the mass and allowing each component to condense at appropriate levels in the fractionating tower. This results from the separation of the gases methane, propanes, etc., and liquids, such as light to intermediate fuel oils, lubricating oils, etc., at their own fraction levels, leaving a residual in the forms of heavy cylinder oil stocks and tar.

The action in the rectifiers separating cryogenic components of air is based on the same general principle. Gas separating apparatus are made an integral part of the liquefying process in some cases, such as oxygen and nitrogen from air.

The *simple rectifier* (Fig. 520) is a single column appropriately insulated whose base forms a reservoir and whose dome-shaped top acts as a condenser, between which is located tube condensers and perforated levels within the column. It provides an inlet for the mixture of fluids to be separated, and outlets for taking off the various separated fluids at appropriate levels consistent with their individual boiling points. The fluid enters the column as a saturated vapor at its expanded pressure and temperature. The vapor flows upward through the perforations in the trays, where the lower boiling point component condenses to form a liquid through which the vapor that follows percolates. As the higher boiling point component of the vapor condenses as it bubbles up through the already formed liquid, the heat of condensation causes a gradual evaporation of the lower boiling point component. This results in an upward flow of vapor and a downward flow of liquid as the latter continues to form in the tray until it reaches a spill-over level controlled by a weir

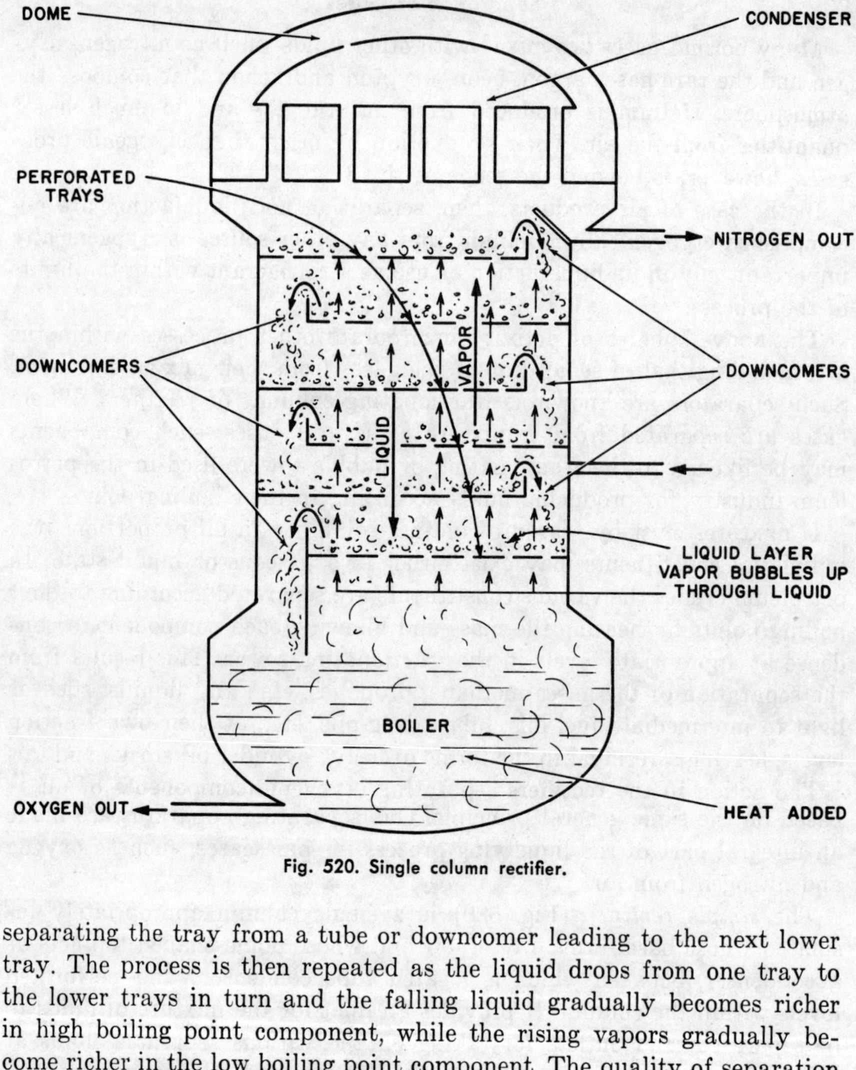

Fig. 520. Single column rectifier.

separating the tray from a tube or downcomer leading to the next lower tray. The process is then repeated as the liquid drops from one tray to the lower trays in turn and the falling liquid gradually becomes richer in high boiling point component, while the rising vapors gradually become richer in the low boiling point component. The quality of separation is a function of the number of plates when the thermal conditions are otherwise satisfied. Heat must be added in the base of the rectifier to supply vapor to the trays below the inlet, while heat must be removed from the top in order to produce some liquid for the uppermost trays. The method of adding and removing heat in each phase of the separating process is shown in the description of the dual-column rectifier that follows.

Rectifier Operation. Consider air as the raw material in the following description of the rectifier (gas separation) principle of operation. Figure

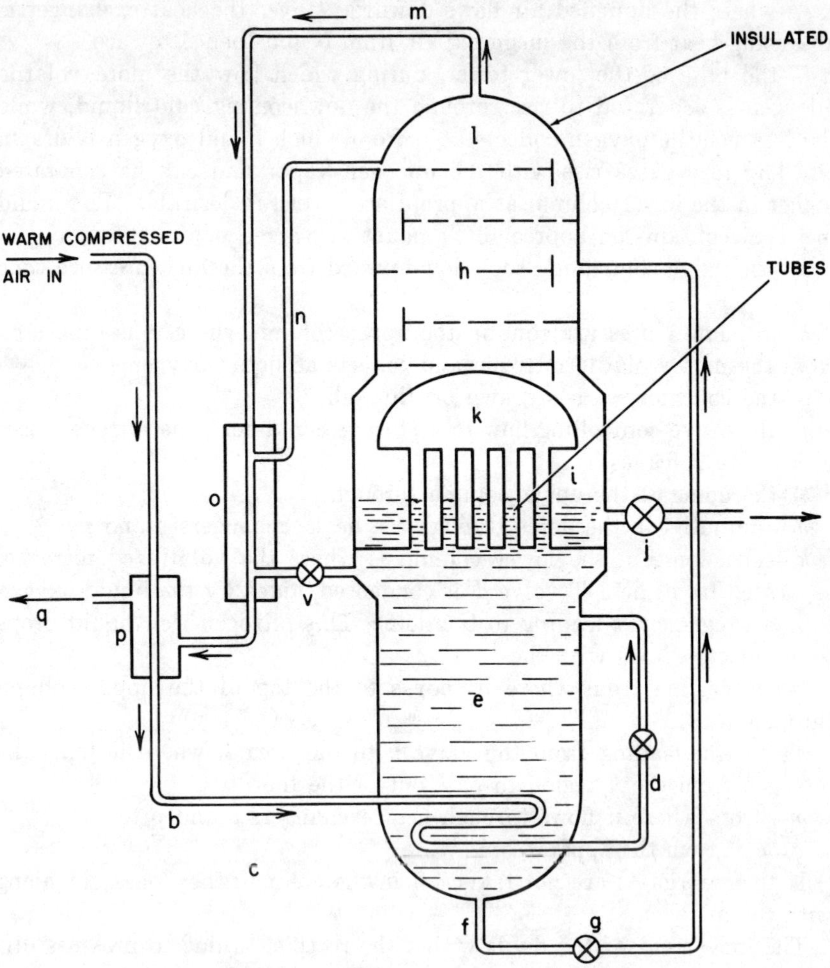

Fig. 521. Linde liquefier and double column rectifier.

521 shows a schematic double column rectifier containing its own liquefier. The process from *a* to *p* as indicated on the drawing is as follows:

a—relatively warm compressed air (about 2950 psia) enters and is cooled in the heat exchanger "p" to a lower temperature at

b—where precooled air leaves the heat exchanger "o," entering

c—another heat exchanger in the bottom of the column to furnish heat in that area before passing through

d—expansion valve (J-T), reducing the air to a final pressure of about 75 psia, and liquefaction temperature, from which it enters the lower tower *e*—where the liquefied air flows downward over the heat exchanger *c*, absorbing heat from the incoming air from *b* and thence toward

f—the base of the lower tower, during which flow the more volatile nitrogen is separated to rise through the downcoming cold liquid, while the less volatile oxygen collects at *f*, from which liquid oxygen is drawn off. The rare gases rise with the nitrogen vapor and can be separated higher in the lower column at appropriate levels, if desirable. The liquid in *f* that contains an appreciable amount of oxygen passes through

g—the valve controlling such flow upward from the high pressure zone *f* to

h—the lower pressure zone in the upper column where it is separated from the more volatile nitrogen and collects as liquid oxygen in

i—the collector, and is drawn off through

j—the valve controlling flow to a storage container. The nitrogen rises in *h* where it leaves

l—the dome of the upper column through

m—an outlet in the dome through the heat exchangers *o* and *p*.

k—The dome in the lower column is where the volatilized nitrogen, separated from the J-T valve *d*, is condensed, aided by the liquid oxygen tube arrangement *t* leading to the dome. This nitrogen rich liquid drops to a collector, level with the

v—valve line from where it moves to the top of the upper column through

n—a riser leading from the valve *v* to the area *l*, where it joins the nitrogen from the *h* region to pass out of the top at

m—from where it flows through heat exchangers *o* and *p* to

q—exit from the apparatus.

If the rare gases are not taken off in the column, they pass out along with the nitrogen.

The movement of the fluids within the rectifier is due to pressure differentials generated by the expansion valve *d* and the condensing action occurring in the lower and upper columns. At the tube and dome area of the lower column, the level of pressure is such as to allow both oxygen and nitrogen to condense at the same temperature.

Oxygen having a higher boiling point liquefies first, causing it to flow downward while the lower boiling point nitrogen rises as a vapor.

Helium (−425 F boiling point) and neon (−411 F bp) will go off with the nitrogen and must be liquefied by further refrigeration to attain separation. Argon (−303 F bp) may be taken off at the bottom of the upper column. Krypton (−244 F bp) may be removed just above the xenon (−163 F) at a low level of the lower column. From the above, it is obvious that the cryogenic components condense and may be taken off the column at ascending levels as individual boiling points decrease, i.e., the higher boiling point constituents toward the bottom and those of lower boiling points toward the top of the column.

Compounded Liquefying Systems

The basic liquefying systems, based on the Hampson or Linde-Hampson, Claude, Kapitsa, Heylandt, the dual column systems and even the cascade system, as previously described, are highly capable cycles for the production of liquid oxygen, nitrogen, fluorine, as well as for the higher boiling point rare gases argon, krypton and xenon. Considerations beyond those shown for the above processes, however, are essential in the gas-to-liquid conversion of the lowest boiling point cryogenic fluids, such as helium, hydrogen and neon, named in their order of ascending boiling points. We have identified the latter cycles as *compounded systems* because certain extensions or modification to the basic cycles are essential to either their separation and/or liquefaction. The basic simple Hampson cycle has already been shown to be extended by the innovation of the expanding engine, dual-pressure and dual-expansion processes, regeneration, and various combinations of such extensions. Operating efficiency is the prime objective of the latter extension processes.

The basic cycles may also be extended by the addition of external sources of refrigeration to provide essential extreme precooling beyond the capacity of intrinsic heat exchangers, for some cryogenic fluids. Catalytic processes are also extensions in some instances.

Helium Separation. Helium is one of the more difficult gases to liquefy because of its extremely low maximum inversion temperature, about −387 F. Therefore, it will not liquefy by simple expansion and intrinsic heat transfer processes. Nor will it liquefy at such temperatures as can be attained with most substances. Hence, a refrigerant, such as hydrogen, is required for precooling helium before it will liquefy. Hydrogen is more available and lower in cost than neon, the only other substance capable of meeting helium-liquefying conditions. An especially designed heat exchanger (to overcome variable specific heat values of hydrogen under different state conditions), using a liquid hydrogen bath placed in the helium gas stream prior to the J-T valve of a Hampson system, is one method of liquefying helium. Another is an extension of the Claude system by a second strategically placed expanding engine. (See Fig. 522.)

Hydrogen Liquefaction. The hydrogen molecules, as described previously, exist in two forms at higher temperatures, i.e., ortho- and para-hydrogen. In addition to the use of an external refrigeration system, having conventional inter-system heat exchangers using liquid nitrogen as the refrigerant, a catalyst accelerates the conversion of ortho-hydrogen to para-hydrogen within the liquefier, so that the heat of inversion is rejected in that component. Otherwise post-evaporation would occur dur-

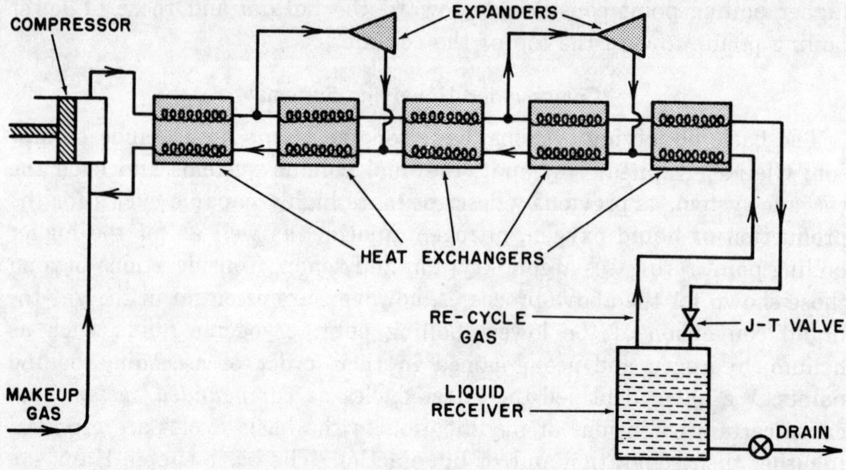

Fig. 522. Collins helium liquefier.

ing subsequent storage, resulting in a loss of liquid. One of the most active catalysts is small pebble-sized hydrous ferric oxide.

The distillation behavior of liquid hydrogen differs from most other cryogenic liquids, chiefly due to low density and high fluidity (low viscosity). The partial solution to this difficulty has been found to lie in the design of the rectifier relative to the area ratio of down-spouts to cross-sectional area of the rectifier.

Hydrogen gas separation from its source material, natural gas or fuel oil, is accomplished by (1) partial oxidation at high pressure, combined with (2) carbon dioxide and steam to form water gas, compressed to (3) a high pressure (up to 6000 psia) and subject to (4) a cascading system of refrigeration, the final stage being with nitrogen, as described above.

Neon is another cryogenic gas requiring drastic precooling before it can be liquefied. The process of liquefying neon is similar to that described for hydrogen.

Deuterium Separation. Hydrogen gas is diatomic, composed of the molecules H_2 (atomic mass = 1), described previously as ortho- and parahydrogen deuteride, HD. Hydrogen deuteride is made up of hydrogen and deuterium, a heavy isotope having an atomic mass of 2 compared to 1 for ordinary hydrogen. It is used to produce heavy water. Deuterium is contained in natural hydrogen in a very small volume, 0.015%, and has a higher boiling point of -417.4 F. Like hydrogen, deuterium exists in ortho- and para-forms. Unlike hydrogen, however, the para-form (p-D_2) converts to the ortho-form (o-D_2) when its temperature decreases. Because of its higher boiling point, deuterium can be separated from hydrogen by distillation. The mixture of H_2 and HD is liquefied in a primary rectifier column to concentrate the hydrogen deuteride (HD). The concentrated HD liquid then flows to the upper column of a dual-column rectifier where it is further concentrated. It then flows to a catalyzing process where the HD undergoes conversion to an equilibrium state consisting of HD, H_2 and D_2. The composite mixture is returned to the lower column of the dual-column rectifier where it is further separated, during which process the deuterium is liquefied and drawn off to storage or ultimate destination. The resulting hydrogen vapors (H_2) rise through the upper column and primary rectifier in turn and thence to storage.

Regenerators and Joule-Thompson Expansion Valve

Regenerators. As shown for the Kapitsa system, regenerators are basically heat exchangers that act also as purifiers, consisting of two parallel channels through which the hot gas from the compressor alternates periodically with the cold waste gas from the liquid receiver. A regenerator is located in each channel. Alternations are made at minute intervals by a valve arrangement designed for the purpose. The reason for the flow alternations is to reverse directions of both gases to facilitate purification. The impurities contained in the hot gas drop out in one regenerator due to cooling effected by the heat exchange and are subsequently removed by the cold gas upon directional alternation. Thus, purification becomes an inherent process associated with an effective heat transfer system and rotary type gas expander to develop a relatively efficient cryogenic liquid producer.

Joule-Thompson Valve. The Joule-Thompson valve is designed and constructed to develop expansion of a gas at constant enthalpy, i.e., no change in either kinetic or potential energy is assumed. In an ideal situation the Joule-Thompson effect would be positive or along the diminishing line shown on the isenthalpic curves in Fig. 523. The J-T effect is shown by the changes in temperature during the expansion process through the point of inversion. This change or effect is described as the J-T coefficient usually written uJT. The coefficient is *positive* in the zone of diminishing temperature, *negative* in the zone of increasing temperature, and *zero* at the point of inversion. Therefore, expansion through the J-T valve theoretically produces a positive effect where there is no work expended (as there is in the expanding engine), or no inleak of heat. Since the energy contained in the gas remains constant, the incoming enthalpy is equal to the outgoing enthalpy or $H_1 = H_2$ and the only change that occurs is a decrease in temperature and pressure of the fluid.

The valve consists of a body, stem, seat and appropriate seals. The seat opens in the direction of the fluid to assure continuous scouring by the high pressure gas stream and to prevent "hang-up" upon closing. The stem is usually made in the form of a light-weight hollow tube to reduce heat inleak and of such material as will provide self-lubrication

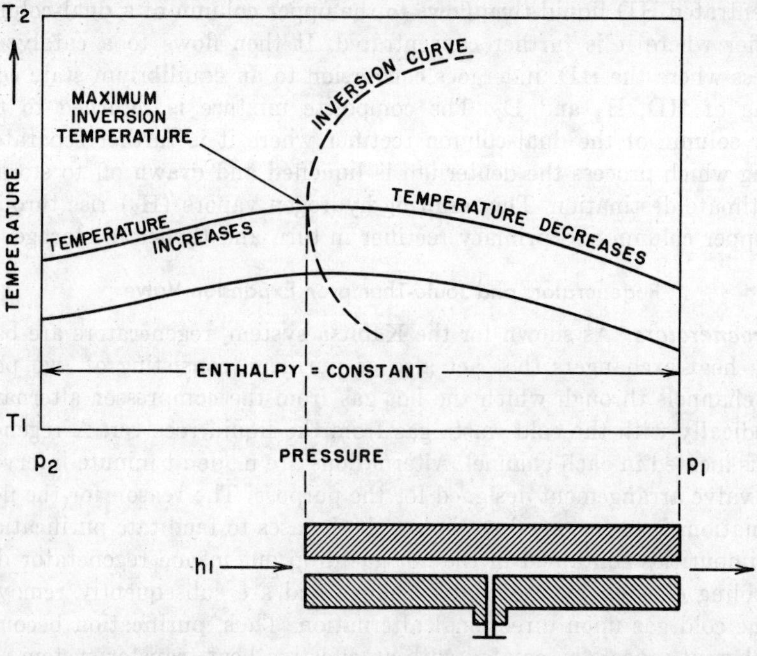

Fig. 523. Relationship of Joule-Thompson expansion to behavior of a real gas.

to the stem threads, usually brass, which is also a good cryogenic metal. The valve is variously designed to be operated by hand, pneumatically or hydraulically.

Behavior of Substance at Cryogenic Temperatures

At cryogenic temperatures, many substances behave very differently than at higher temperatures. Actually, their behavior in many fields is such as to warrant the addition of the prefix "super-" to the property being described. In many instances, the change in specific property magnitude offers significant benefits in some directions. They may also prove a liability in other directions. Some of the changes that occur at cryogenic temperatures are given below:

Some materials become superconductors of electricity.

Others become superconductors of thermal energy.

Cryogenic liquids are liquefied gases that remain liquid only under extremely low temperatures. Therefore, cryogenic fluids are compressible, whereas customary fluids are not.

Some metals become brittle at low temperatures.

Others remain normally ductile.

A few metals increase in tensile strength at cryogenic temperatures.

Specific heat increases when materials become superconductors.

Thermoelectric effects diminish to zero when semiconductors become superconductors.

Thermocouples consisting of superconducting materials are rendered to zero effect at cryogenic temperatures.

Liquid helium acts in an exotic manner, as described previously.

Fluorine is highly corrosive to many engineering metals.

The above comments relative to the various material behaviors are generally correct. However, there are a number of exceptions under certain conditions. One example of this is the loss of superconductivity by some materials that repulse a magnetic field of definite intensity, loses its superconductivity when a more intense magnetic field capable of magnetizing it is externally applied.

Engineering materials applicable to cryogenic environments must possess special characteristics to meet the stresses and thermal demands encountered at extremely low temperatures. Engineering materials must include those employed in the structural, heat conduction, insulating, electrical conduction and fluid property control phases of the industry. Their selection in some instances must be based on both the type of the material and its metallurgical treatment.

In general, favorable materials for structural purposes include aluminum, austenitic stainless steel, nickel-steel alloys, monel metals, copper and some forms of iron. Pyrex and some plastic materials also possess cryogenic merit, although the latter should be used only under restrained compression conditions. Monel, aluminum-copper alloys, nickel copper, and some steels are compatible with fluorine.

Teflon serves as a compatible lubricant where applicable.

Copper and other good electrical conductors are represented in the above list, selected on the basis of their feasibility for the purpose.

Appropriate materials for instrumentation may also be selected from the above list of structural metals.

Insulation processes include the application of low heat conducting structural metals, appropriate covering materials, high vacuum chambers, vacuum powders (powders reduce emissivity or heat radiation), heat deflection surfaces and cryogenic liquids. Powders include expanded perlite.

Coverings include certain expanded foams and fibrous materials, some of which are proprietary. Foams are cellular materials of polystyrene, glass, silicone, rubber and others. At temperature differences experienced in cryogenic processes, some foam materials have an extremely low coefficient of conduction as compared with cork, kapok and expanded rubber, to mention only a few common insulating materials.

Cryogenic Vessels

Vessels for storing and transporting cryogenic liquids are specially constructed to maintain minimum heat transfer between their content and ambient temperature conditions. Such liquids are highly volatile as their boiling points are considerably below ambient temperature at atmospheric pressure. Some of the characteristics associated with cryogenic storage are:

(*a*) The vessels are usually low pressure containers.

(*b*) They must be properly vented to prevent pressure build-up. Because fluorine is poisonous, it cannot be vented.

(*c*) Highly effective insulation design is important.

(*d*) Cryogenic materials, suitable for exposure to the liquid, must be used.

The simpler cryogenic vessels consist of an inner cylinder or sphere for containing the liquid (Fig. 524). Such containers may be constructed of

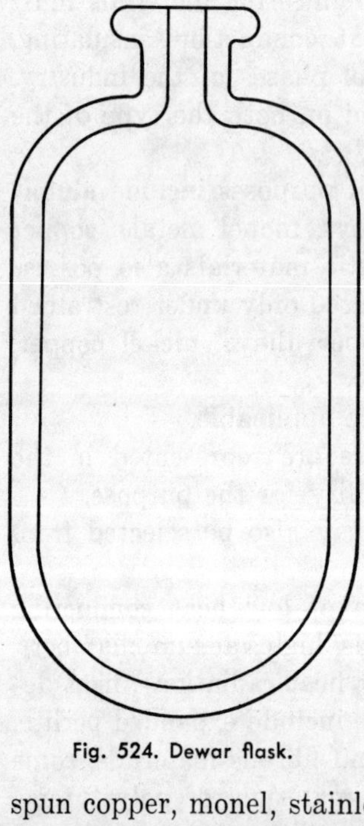

Fig. 524. Dewar flask.

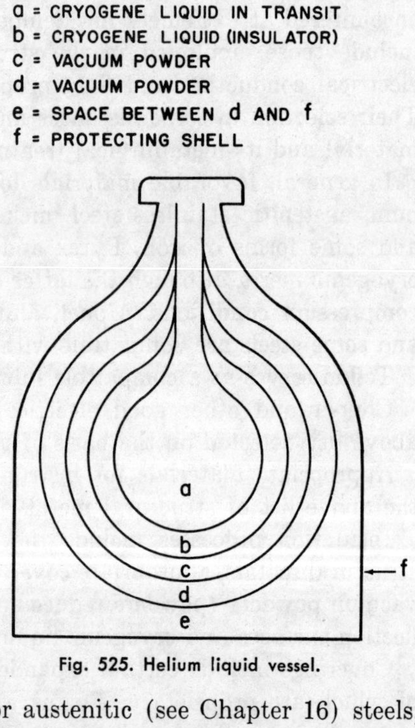

Fig. 525. Helium liquid vessel.

spun copper, monel, stainless steel, or austenitic (see Chapter 16) steels, depending on the product involved. The inner container is protected by a concentric outer shell, ruggedly designed, and mutually separated by a high vacuum chamber that entirely surrounds the inner liquid container. Both containers are connected rigidly together in a manner to minimize heat conduction between the surroundings ambient to the liquid. The vacuum chamber is first filled with an insulating powder, such as perlite or other suitable materials before evacuation of air.

Sophisticated containers may have as many as four concentric spheres placed one inside the other, all encased in an outer housing or casing for external protection. Such spheres are separated by two high vacuum spaces and the remaining space or reservoir is filled with a cryogenic liquid. This latter vessel is designed to hold liquid helium and the insulating liquid is usually hydrogen (Fig. 525).

The simple vessel described above is essentially a Dewar flask, the principle employed to produce "thermos" bottles. In practice, the Dewar vessel is equipped with a fill and drain line and an internal diffuser or vapor collector connected to a vent having a pressure relief valve.

TABLE 163. CRYOGENIC INSULATION

Insulating systems employed for ultra-low temperature vessels in the storage or transportation of cryogenic liquids.

Type of Insulator	*Materials Used*
1. Expanded foam	Foam rubber Silicone Glass foam Polyurethane foam Polystyrene foam
Foaming material—carbon dioxide	
2. Gas filled powders, or fibers	Powdered cork Fiber glass Rock wool Perlite
Filling material—helium	
3. Opacified powder (radiation reduction)	Aluminum Copper Various other materials
Usually added to evacuated powders	
4. Layering	Various foils Aluminum foil Proprietary foils
Alternating layers (multi-layer) of heat reflecting foils separated by a low-conducting spacer or mats, such as paper, fiber glass, etc. Used in conjunction with evacuated separating chambers.	
5. Vacuum	
Evacuation of air to minimum pressure (abs) levels within separating chambers. Dewar flask.	
6. Vacuum-powder or fibers	
Insulation more effective by evacuating all gas from insulating materials, such as the powders and fibers mentioned above.	
7. Cryogenic liquids and vapors	
Liquid nitrogen, etc., used as shields when contained in an intermediate chamber between the cryogenic liquid to be protected, such as helium, and a vacuum chamber.	

THE STIRLING CYCLE

A reversible, regenerative thermodynamic cycle embracing two constant volume and two isothermal processes is known as the *Stirling cycle.* It differs from the Ericsson cycle in that the two constant volume processes replace the two constant pressure processes of the latter. The Stirling cycle, developed by an English preacher early in the nineteenth century, has been adopted to produce cryogenic refrigeration after effecting higher efficiencies through modern design.

Basically, the Stirling apparatus consists of two cylinders separated by an insulated passage that allows a gas to flow in either direction from one cylinder to another motivated by a piston and a displacer located in their respective cylinders. The connecting passage is filled with a wire mesh or coil that acts as a thermal regenerator by absorbing heat from the gas as it passes through it in one direction and rejects that heat upon its return to the original cylinder. An external heat source, from which heat is absorbed to perform the cooling process, and a heat sink, to which the absorbed heat is rejected, are also required to complete the cycle.

The p-v and T-s diagrams representing the Stirling refrigerating cycle are shown in Fig. 526, as performed in the apparatus (Figs. 527(*a*) and (*b*)). The following action occurs within the apparatus, as described for the p-v diagram with simultaneous cyclic changes in temperature and entropy depicted on the T-s diagram. References to the components identified in Figs. 527(*a*) and (*b*) are included in the description of the Stirling processes.

Assume a gas, such as air, that maintains its gaseous state throughout the complete cycle. Starting at *a* on the p-v diagram, a given volume of

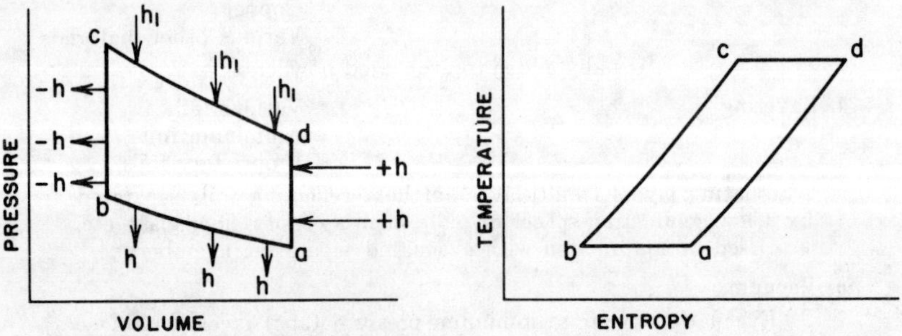

a-b = ISOTHERMAL COMPRESSION
b-c = CONSTANT VOLUME COOLING
c-d = ISOTHERMAL EXPANSION
d-a = CONSTANT VOLUME HEATING
h = HEAT OF COMPRESSION, REMOVED BY EXTERNAL COOLING (SINK)
h_1 = HEAT ABSORBED DURING EXPANSION FROM EXTERNAL SOURCE
-h = HEAT TO REGENERATOR REMOVED TEMPORARILY
+h = HEAT FROM REGENERATOR -h REABSORBED

Fig. 526. Stirling cycle, refrigerator and Ts diagrams.

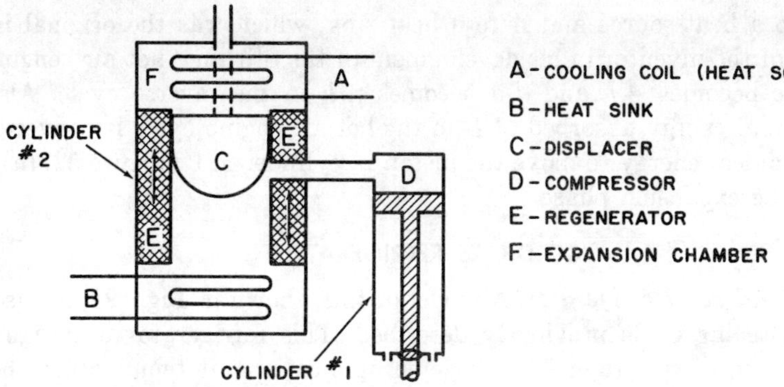

Fig. 527(a). Stirling cycle, two cylinders.

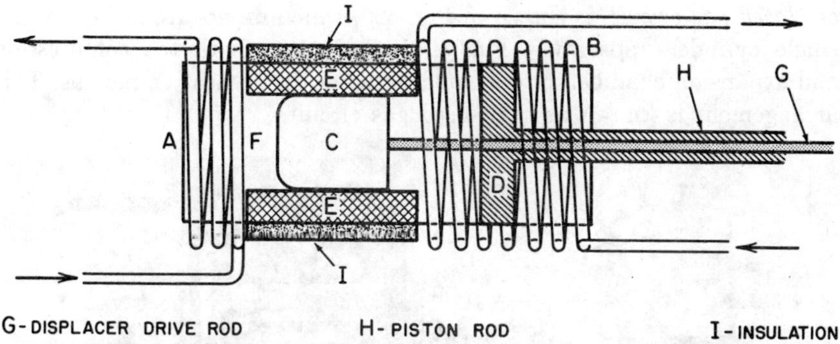

Fig. 527(b). Stirling cycle, (basic) single cylinder.

gas at atmospheric pressure and temperature, as measured by the compressor (D) displacement, is compressed to a higher pressure and temperature as represented along line a-b. As the hot, high pressure gas enters the second cylinder, it is cooled isothermally by the cooling coil (A), thus rejecting the heat of compression, as shown by h on the p-v diagram. The cooled gas then passes at constant volume through the regenerator in which it gives up additional heat to the wire mesh, $-h$, as represented by line b-c, to enter the expansion chamber (F) at point c of the p-v diagram. Also at point c, the compression piston (D) starts on its backward stroke allowing the gas to expand isothermally, during which it absorbs heat, h_1, from the refrigerating coil (B), as represented by the line c-d. The gas is then displaced by the displacer piston (C) and directed back to the compressor chamber (D) at constant volume, through the regenerator (in reverse direction) absorbing the heat (ideal), $+h$, deposited there on its way to the expansion chamber, as shown along line d-a. Upon reaching a the gas is again compressed to repeat the cycle. In the process $+h = -h$. The process can be reversed by converting B to a heat source and A to a heat sink, which was the original intention of the inventor in his development of the Stirling "hot air" engine. Line b-c becomes $+h$ and d-a becomes $-h$ in the reverse cycle. Also, the thermal energy absorbed at B in the hot air engine cycle is converted to mechanical energy to move the piston in cylinder #1 (Figure 527(a)), during the expansion phase.

COLD-GAS REFRIGERATOR

CVI Cold-gas Refrigerator. A model of this, shown in Fig. 528, is based on the Stirling cycle previously described. This refrigerator is designed to deal with a wide range of cryogenic applications at temperatures between 75° and 180° Kelvin. It operates with the compression and expansion processes occurring in separate chambers of the same cylinder, as shown in Fig. 527(a). A schematic cycle of the cold-gas refrigerator, or *closed gas circuit,* is shown in Fig. 529, which differs from the Stirling single cylinder apparatus (Fig. 527(a)) in that both the compression and expansion chambers are located between the respective pistons. This arrangement is known as the closed gas circuit.

Fig. 528. CVI cold-gas refrigerator. Skid mounted with 150 hp electric motor and explosion-proof controls. This particular model is in use as a natural gas liquefier. (Courtesy, CVI Corp., affiliate of Pennsalt Chemicals Corp.)

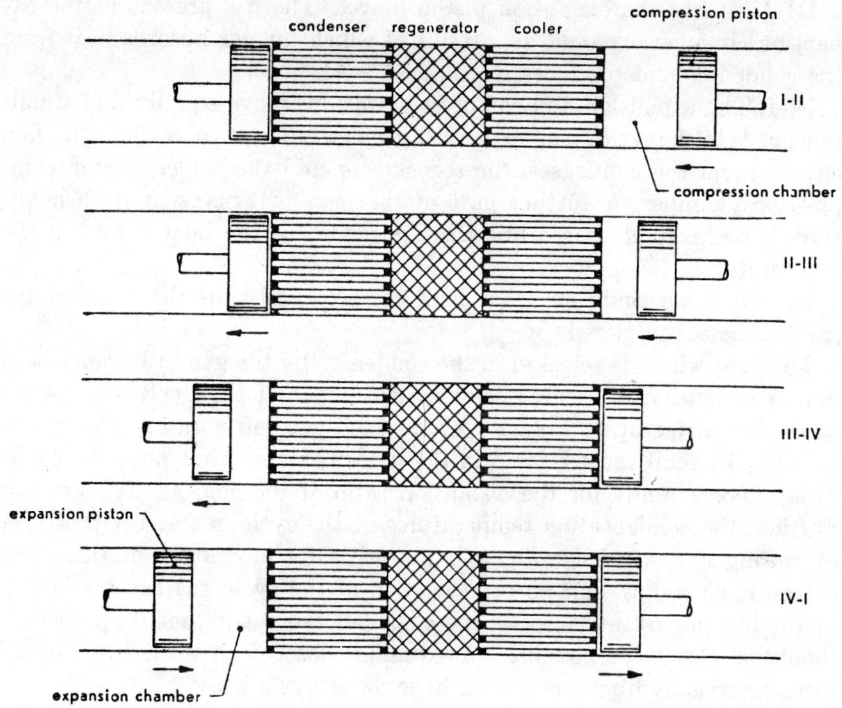

Fig. 529. Schematic cycle, cold-gas refrigerator (four views). (Courtesy, CVI Corp., affiliate of Pennsalt Chemicals Corp.)

As shown in the schematic cycle of the cold-gas refrigerator (Fig. 529), the gas circuit consists of a compression chamber, an externally water-cooled cooler, regenerator, condenser, which is externally heated by the substance being refrigerated (cooled and condensed) and an expansion chamber. The schematic cycle involves the mechanical and thermodynamic processes described below. (Refer to Fig. 529 and consider the use of hydrogen as the ideal gas for the purpose.)

I-II. Only the compression piston moves. The gas present in the compression chamber is compressed and the heat of compression is removed in the cooler.

II-III. The compression and expansion pistons move equally and simultaneously to the left. The compressed gas is displaced from the compression chamber via the cooler, the regenerator and the condenser, to the expansion chamber. At the same time a further part of the heat of compression is removed in the cooler. Furthermore, the gas releases heat to the regenerator, as a result of which this regenerator serves to store heat. If the temperature distribution in the cold-gas refrigerator is in equilibrium, the gas leaves the regenerator at a low temperature.

III-IV. Only the expansion piston moves. The gas present in the expansion chamber expands, as a result of which the gas absorbs heat from the condenser and produces mechanical work.

IV-I. The expansion and compression pistons move equally and simultaneously to the right. The expanded gas is displaced from the expansion chamber via the condenser, the regenerator and the cooler, to the compression chamber. A further part of the heat of expansion is then absorbed in the condenser, while the gas reabsorbs the heat stored in the regenerator.

By this time condition I has been restored and thus the cycle of the gas has been completed.

The heat which is released in the condenser by the gas to be condensed during expansion and subsequent displacement of the cycle gas constitutes the refrigerating capacity of the cold-gas refrigerator. The gas in the closed circuit must remain in the gaseous state throughout the cycle. This imposes limits for the expansion ratio of the cold-gas refrigerator. Further, the condensation temperatures of the cycle gas at the pressures prevailing in the refrigerator and the critical temperature of this cycle gas must be so low that no condensation of the cycle gas can occur.

Another important factor in the choice of the gas is that its properties should as closely as possible approximate those of an ideal gas. For all these reasons hydrogen is chosen to serve as cycle gas.

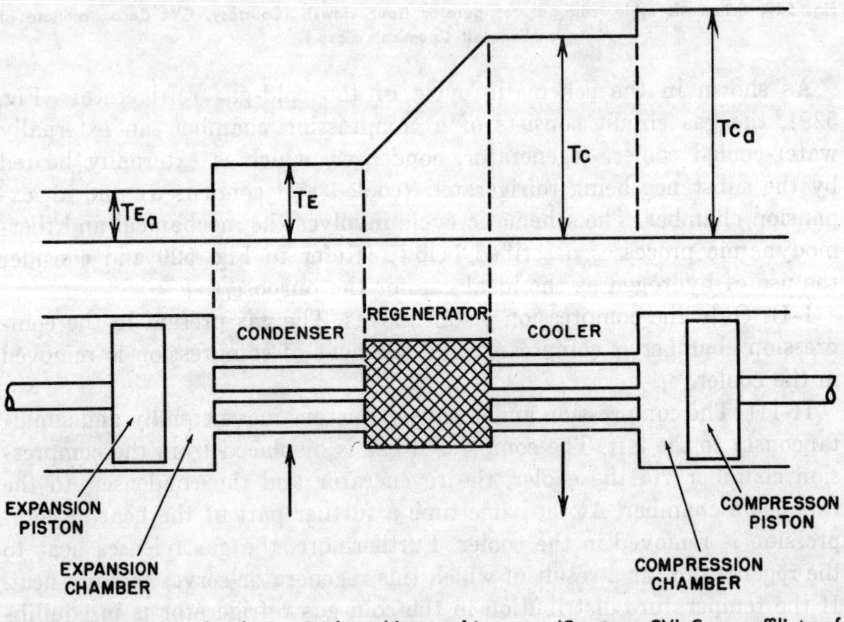

Fig. 530. Temperature distribution in the cold-gas refrigerator. (Courtesy, CVI Corp., affiliate of Pennsalt Chemicals Corp.)

Figure 530 shows a diagrammatic representation of the temperature equilibrium conditions in the cold-gas refrigerator. The cooler is at room temperature, T_C, the condenser at the low temperature, T_E. As a result of adiabatic losses, the mean temperature of the compression chamber T_{Ca} is above the temperature T_C, while the mean temperature of expansion chamber, T_{Ea}, is lower than the temperature T_E.

The cycle described above is similar to the basic Stirling cycle, in which the refrigeration process is developed in two separate cylinders, as shown. In the schematic cold-gas refrigeration cycle, the separate compression and expansion chambers within the one cylinder are a close analogy and the thermodynamic processes are similar.

Another arrangement having similar thermodynamic processes as already described, but one in which the construction involves only one cylinder in which the compression piston (power element) and the expansion piston or displacer move, is called the Werkspoor cold-gas refrigerator and is an improved modification of the Stirling cycle refrigerator previously discussed.

Werkspoor Cold-gas Refrigerator. The construction of the Werkspoor cold-gas refrigerator differs from the schematic apparatus (Fig. 529) in that the compression chamber is located between the power piston and the displacer, and the expansion chamber above the displacer, as shown in the basic single cylinder model shown in Fig. 527(*b*).

A schematic arrangement of the Werkspoor apparatus (CVI) is represented in Fig. 531, showing the relative locations of both moving elements, as well as the position of the cooler, regenerator and condenser in relation to the cylinder. The latter three thermal elements are of annular shape and constructed circumferentially around the displacer.

In the Werkspoor apparatus the changes in volume are not obtained with two equivalent pistons, but with the aid of a main compression

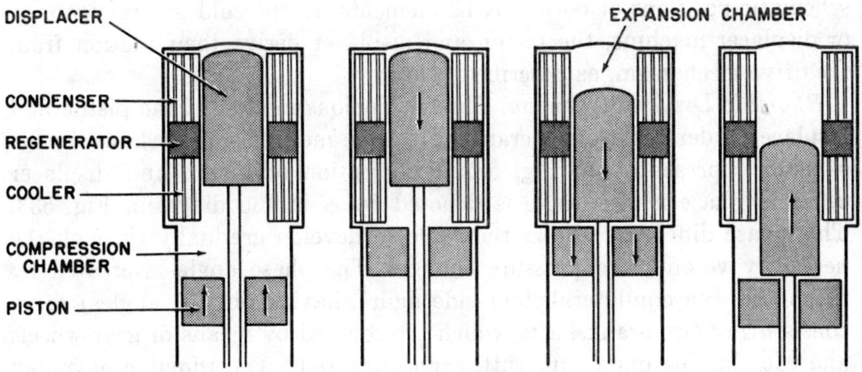

Fig. 531. Schematic cold-gas process in "displacer machine." (Courtesy, CVI Corp., affiliate of Pennsalt Chemicals Corp.)

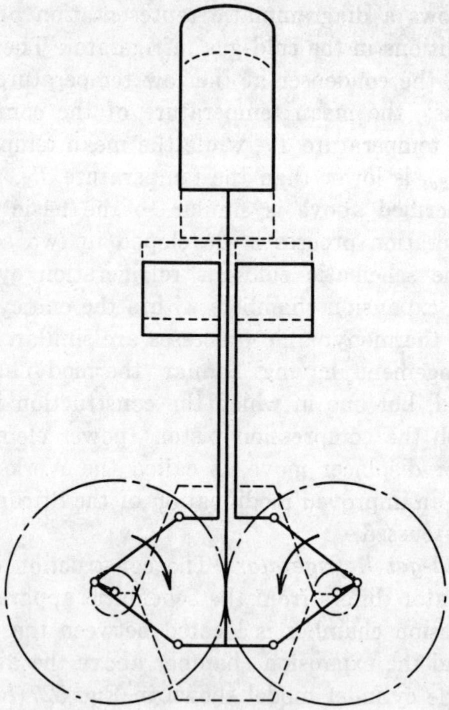

Fig. 532. Rhombic drive mechanism of a cold-gas refrigerator of the "displacer" type. (Courtesy, CVI Corp., affiliate of Pennsalt Chemicals Corp.)

piston and an auxiliary piston, previously described as the displacer. The cycle sequence of the Werkspoor apparatus is the same as described for the schematic arrangement (Fig. 529), and the various compression piston and displacer positions, as shown in Fig. 531, correspond to the schematic positions of both moving elements. In the cold-gas refrigerator or displacer machine, the piston and displacer derive their motion from the drive mechanism, as described below:

Rhombic Drive Mechanism. The harmonious motion of the piston and displacer is derived from a crankshaft drive mechanism, as shown in the skeleton representation, Fig. 532. The motion of piston and displacer differ in phase whose angle is denoted by ϕ in the diagram, Fig. 533. This phase difference allows the cycle to develop gradually through the necessary volume and pressure changes. The phase angle is created by a rhombic (an equilateral, four-side figure, having oblique angles) drive consisting of two crankshafts, which are coupled by means of gear wheels and each having one crank with connecting rods. The rhombic assembly imparts relative motion to the respective yokes of piston and displacer,

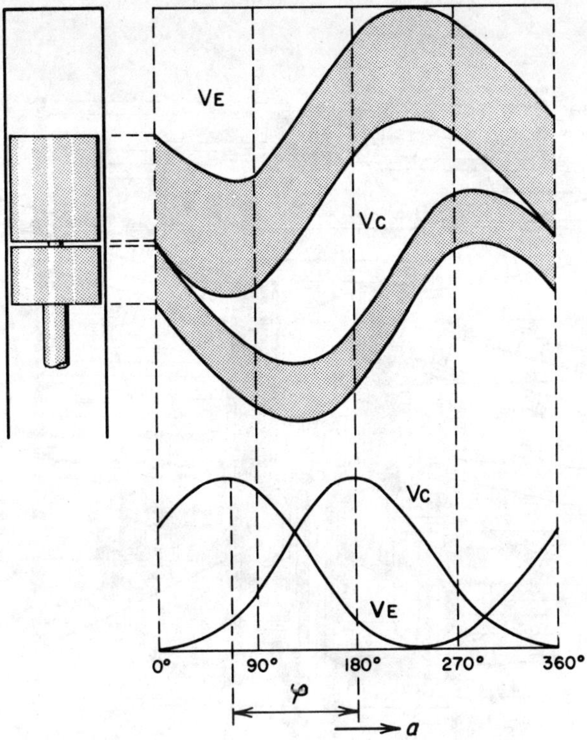

Fig. 533. Changes in volume of the compression and the expansion chambers as a function of the crank angle α. V_E is the change in volume of the expansion chamber and V_C that of the compression chamber. The angle ϕ denotes the phase difference which exists between the expansion and compression volumes. (Courtesy, CVI Corp., affiliate of Pennsalt Chemicals Corp.)

having a phase angular lead of approximately 90°. Note that the displacer driving rod moves through the hollow piston rod.

The changes in volume of expansion chamber V_E and of the compression chamber V_C originating from the harmonious motions of the power piston and displacer are represented in Fig. 533.

Components of the Cold-Gas Refrigerator. A cross-sectional drawing of the CVI Cold-Gas Refrigerator is portrayed in Fig. 534. The principal parts are identified below, corresponding to the letters on the drawing:

a—inlet gas to be refrigerated (condensed)
b—discharge, refrigerated (condensed) gas
c—displacer
d—condenser
e—regenerator

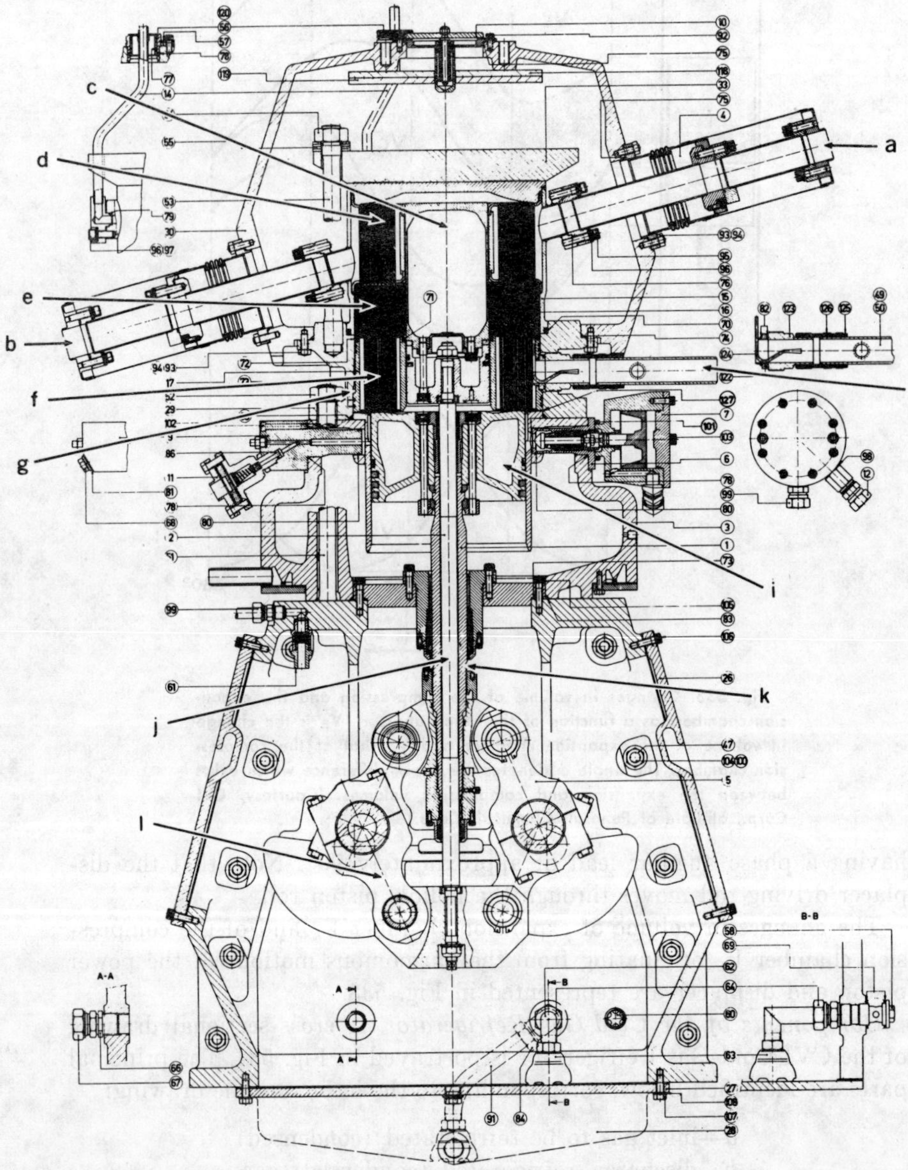

Fig. 534. Cross-section showing Rhombic drive mechanism and piston-displacer arrangement surrounded by a condenser regenerator and water cooler. (Courtesy, CVI Corp., affiliate of Pennsalt Chemicals Corp.)

f—cooler
g—cooling water inlet
h—cooling water outlet
i—power (compression) piston
j—displacer rod
k—piston rod
l—rhombic drive mechanism

Operation. The cold-gas refrigerator requires a temperature pull-down period of about 15 minutes after starting to reach a temperature of 77° Kelvin, upon which refrigeration can be initiated. Adjustment of the desired refrigerating capacity is effected by supplying cycle gas (hydrogen) from a regulating vessel to the apparatus, or by returning such gas to the vessel as thermal conditions dictate. The gas is supplied or returned through connections constructed into the apparatus, known as P_{min} and P_{max} connections, respectively.

The method of adjusting refrigerating capacity by employing the P_{min} and P_{max} arrangement depends on the size (volume) of the regulating vessels. In order to control the refrigerating capacity over a limited range and enable the gas content of the machine and regulating vessel to be maintained at a proper level, i.e., so that there remains a sufficient quantity to meet maximum requirements, but not a superabundance to cause the discharge of cyclic gas to the atmcsphere, a regulating vessel of proper size is essential. A 40 liter (10.567 gallons) hydrogen bottle is suggested by the builder.

The refrigerating capacity is usually controlled manually, but can be made automatic. Other controls include cooling water temperature and lubricating oil pressure cut-out switches to make the cold-gas refrigerator operate unattended. An electric motor overload trip device that may or may not be made dependent on refrigerating capacity is also a component of the control system.

Gages to report the lubricating oil, regulating vessel, minimum (P_{min}) and maximum (P_{max}) mean cycle and mean buffer pressures are also part of the complete operating machine.

The capacity of the cold-gas refrigerator varies and is produced with one, two, three or four cylinders. It will generate cryogenic refrigeration down to −320 F. In multi-cylinder units the compression-expansion process is complete within each cylinder in a single stage. One of the advantages of the machine described is that it can be skid-mounted for easy portability and installation. The rhombic drive feature is fully balanced and does not require heavy or special foundations. Another advantage is that the refrigerated product does not contact the cooling medium, i.e., the cycle gas, in any way.

The operation of the refrigerator can be equipped with temperature and/or capacity controls.

Refrigeration Capabilities. The cold-gas refrigeration unit has a refrigeration capability and input power requirement, when operated at 1750 rpm, of

37,000 Btu per hour at −320 F and 160 hp
92,000 Btu per hour at −100 F and 75 hp

It has a cool down time of about 10-15 minutes. The portable unit requires only a source of power and cooling water.

In addition to the electric motor shown in the cross-sectional drawing, other drives can be used, including internal combustion engines of either gasoline, diesel or natural gas types.

Application. The cold-gas refrigerator has many applications, some of which are listed below:

Production of liquid air, both at atmospheric pressure and at higher pressures up to 575 psia.

Production of low temperatures required in air separation plants in which gaseous or liquid oxygen and nitrogen are produced.

Recondensation of evaporated gases from storage tanks for liquefied gases.

Production of low process temperatures in food, dairy and chemical industries.

Metallurgical and plastic deflashing.

Frozen food transportation.

Portable cryogenic refrigeration.

Closed-cycle cryogenic refrigeration.

Cryobiology.

Ethylene reliquefier.

Natural gas liquefier.

Air liquefier for food freezing.

Etc.

CRYOGENICALLY COOLED CARRIERS

Cryogenic liquids are employed to refrigerate highway trucks and trailers, railroad cars, ship and plane space as well as shipping containers for the transportation of perishable products.

It involves a container of cryogenic liquid, such as nitrogen, which has a boiling point of −320 F at atmospheric pressure, control valve, vaporizer, cooling channels and control system. The liquid nitrogen flows from its storage cylinder through a flow control valve into the vaporizing channels. The cryogenic channels are usually placed so as to form a header lengthwise relative to the truck or car body which feeds a number of branch lines leading from the header to and down the sides to the floor and thence to a channel in the center of the body leading from the front to the back of the body. A riser from the center floor channel carries the spent gas to the top and rear of the body where it is exhausted into the atmosphere. A sensor is provided which controls the temperature by automatically regulating the control valve that governs the flow-rate of cryogenic liquid. The cooling channels are imbedded in, but not covered by, the insulating material. This latter eliminates interference with cargo movement within the vehicle and provides direct exposure to the cargo. This type of system can maintain temperatures from −20 F to 60 F.

Containerization is an expression recently adopted to describe a sophisticated method of enwrapping appropriate merchandise for the purpose of transportation either domestically or to other countries. While the procedures are not new, the methods of securing both stable and perishable commodities are undergoing changes in techniques that assure greater protection of such items against loss by deterioration, damage separation and pilfering. Container construction and design involve techniques that will provide sufficient strength, size, shape and any one or more of the following, where appropriate: shock absorbing elements, compactness, ventilation, handling components (for slinging, palleting, etc.), hermetic sealings, temperature, humidity, rust protection, mildew protection, and protection against theft.

It is not within the scope of this text to describe all the techniques involved in containerization, but only the role of refrigeration. This role is in the protection of items vulnerable to either rapid or gradual deterioration during transit from origin to destination. Because the situation is one of maximum mobility, the technique itself must be both mobile and flexible.

The best process for the protection of perishable commodities in the field of containerization is drawn from the science of cryogenics. In this, the technique employed for transportation vehicles discussed above is similar to that applied to containers.

A bottle of cryogenic liquid, usually nitrogen, is placed in the container and connected, through a flow control valve, to flexible conduits forming the cooling coil and capable of being appropriately placed to absorb heat uniformly. The size of the components depends on the magnitude of the container, temperature required in each case and the length of time in transit.

NITROGEN FREEZING TUNNEL—SNAP FREEZING

The latest method of freezing substances, particularly foods, e.g., meat products, is called *snap freezing* and has the important advantage of

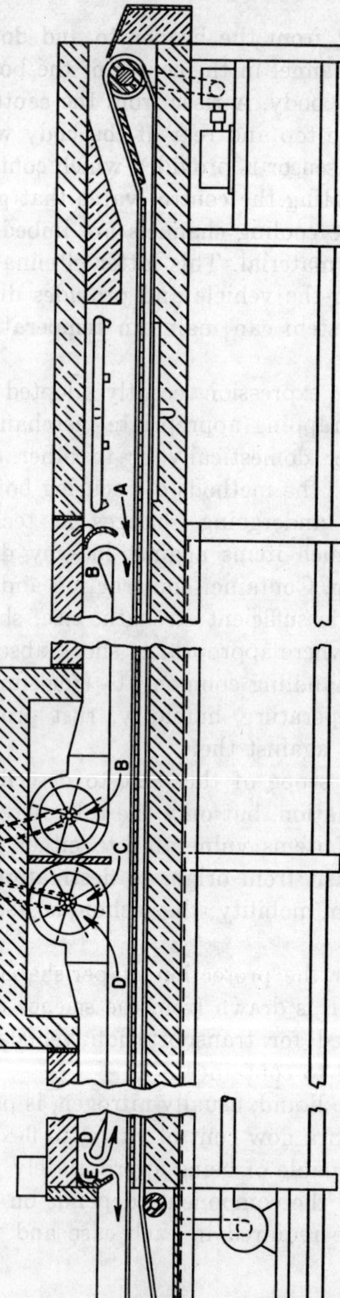

Fig. 535. Nitrogen cooling tunnel. (Courtesy, Air Products.)

solidifying substances more rapidly than is possible with any other freezing system, including (air) blast freezing as now practiced. In the case of, e.g., beef steak freezing, the time period, i.e., from the time the fresh steak at chill room temperature enters the processing tunnel until it leaves completely solidified, is seven to fourteen minutes, depending on thickness.

The snap freezing method utilizes the cryogenic temperature levels supplied by liquid nitrogen at −320 F. The process develops as follows:

(*a*) Fresh meat portions (steaks) are placed on an endless moving stainless-steel-mesh belt that transports them into the tunnel.

(*b*) They enter a chamber where they are swiftly cooled by cold circulating nitrogen gas.

(*c*) They continue to be carred into a pre-freeze chamber where they are exposed to yet colder nitrogen gas and where their surfaces are reduced to below freezing temperature.

(*d*) Continuing their travel through the tunnel, the steaks next enter the snap freezing chamber where they are directly sprayed with liquid nitrogen, which reduces their surface temperature to −125 F in a few seconds.

(*e*) They enter the final chamber where they are surrounded by a swirling super-cool nitrogen gas directly from the snap-freezer, during which their interior temperatures and surface temperature approach equilibrium from which

(*f*) they continue on to leave the tunnel thoroughly frozen and are discharged to a table where they are wrapped and packed; thence to storage.

The final or equilibrating, pre-freezing and pre-cooling chambers are supplied in the order named by the cold nitrogen gas from the snap freezing chamber where the nitrogen gas is produced from the boiling liquid nitrogen as it is sprayed on to the product being frozen.

The nitrogen gas increases in temperature as it proceeds chronologically through the above named chambers until it is spent and finally discharged into the atmosphere. Because nitrogen is inert, there is no danger in eliminating it in that manner.

Rapid freezing substantially reduces dehydration and sublimation experienced to some extent in blast freezing.

Another advantage of the snap-freezing method is the rapidity of temperature reduction through the thermal range of water crystallization, as described under "Principles of Cold Storage" in Chapter 14. The process greatly retards crystal growth and results in a frozen product that retains its fresh appearance and natural taste upon being thawed and properly cooked.

While the above description covers beef steaks, other items, such as scallops, shrimp, various kinds of fish, ground meat, etc., can be frozen by the snap-freeze method.

MAGNETIC REFRIGERATION

Magnetic refrigeration is a process that allows temperatures in the zone below 1°R to be obtained. The process is known as *adiabatic demagnetization* and involves certain salts that have *paramagnetic* properties. Paramagnetic materials have the property of being attracted by a magnetic pole and, when freely suspended in a magnetic field, of taking a position parallel (an alignment) to the lines of force. Therefore, they differ from *diamagnetic* materials that are repelled by a magnetic pole and take a position transverse to the axis of a magnetic field in which it may be freely suspended.

The process of magnetic cooling may be described best by certain analogies using, first, the position of iron filings when exposed to a magnetic field and, secondly, the behavior of gas molecules when subject to isothermal compression and adiabatic expansion.

First. Consider then that each of the individual segments of iron filings are dipolar, i.e., they have both a north and south pole. When freely suspended in a magnetic field all segments take a position parallel to the field, as shown in Fig. 536(*b*), as compared to their random position under non-magnetic conditions as in Fig. 536(*a*). It is true that the atoms of paramagnetic materials behave in a similar manner under the above circumstances. To proceed a step further, they do so regardless of the temperature.

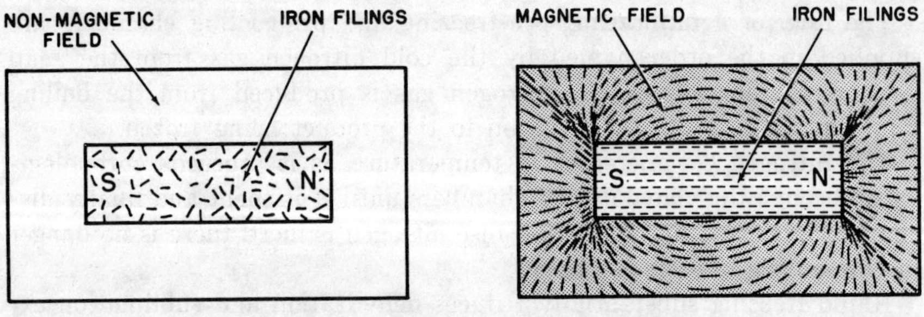

Fig. 536. Random position of dipolar iron filings in non-magnetic field (left). Polarized position of iron filings exposed to magnetic field (right).

Second. The molecular theory of fluids states that when subject to a change in temperature, the random velocity of the molecules also changes. It is also true that, when compressed, the molecules of a gas are forced closer to each other, causing an increase in density. If compression occurs without external cooling, the temperature rises, causing an increase in molecular velocity.

However, if compression is accomplished isothermally (constant temperature), a decrease in entropy occurs, but the random velocity of the molecules remains constant. Furthermore, by expanding the compressed gas adiabatically (heat is neither gained nor lost during the process), the state is created making it necessary for the molecules to separate in order to fill the increased volume, at which time the temperature drops causing a decrease in their random velocity. Entropy remains constant during the above conditions of expansion. Actually, it is the above compression-expansion process that occurs in cryogenic systems if the initial heat exchanger is considered to replace simultaneous external compression cooling in the above process and the J-T valve is the sole expansion device.

When a paramagnetic material is exposed to a magnetic field, the temperature of the material rises. The thermal energy so generated is called *heat of magnetization.* If magnetization occurs at constant temperature, i.e., heat of magnetization is removed by external means, an analogy to isothermal gas compression is demonstrated so that entropy decreases.

By removing the magnetic field during which the paramagnetic material is highly insulated to prevent a gain or loss of heat, an analogy to adiabatic gas expansion is demonstrated so that entropy remains constant. Also, as with molecular separation and diminishing velocity during gas expansion, some misalignment of the dipoles occurs, developing a temperature reduction. It is during this latter process that adiabatic demagnetization occurs as a result of work done against the magnetic field during its removal. The analogy here is the work done by a gas against the reduction in pressure during the expansion process, which effort is increased when expansion is accomplished in an expansion turbine.

Magnetic cooling, or more specifically adiabatic demagnetization, is possible only at extremely low temperatures. This is true because any appreciable temperature decrease can be accomplished only when the thermal effects of the crystalline structure (lattice effects) of the salts are less than the effects of dipolar energy during demagnetization. Such a condition exists only when the temperature of the salts is below approximately 3 R. The approximate analogy of this limitation is shown in the lattice effects occurring in thermoelectric semiconductors when the rate of electrical energy to thermal energy transformation is such as to overcome Peltier effects and nullify the temperature gradient.

To prepare the paramagnetic salts for adiabatic demagnetization processes requires initial cooling to the proper temperature in several stages. A typical procedure would be first to subject the substance to helium evaporation within a highly effective insulated chamber and then to apply a magnetic field around the precooled substance, during which the helium gas is extracted and the heat of magnetization removed with it.

The third stage of demagnetization is then applied to reduce the surroundings to a temperature approaching absolute zero, or somewhat below 1 R, which is about the limit of ordinary paramagnetic salts.

The paramagnetic materials involved in magnetic refrigeration include the following salts:

- Chromium potassium alum
- Iron ammonium alum
- Cerium magnesium nitrate
- Copper potassium sulfate
- Manganese ammonium sulfate
- Titanium cesium alum

The chronological procedure for reducing cryogenic fluids to extremely low temperatures or converting their gases to liquids is to apply conventional practices using low boiling point refrigerants, such as R-13, methane, etc., by staging, or cascading processes, followed by employment of the Joule-Thompson effect of isothermal compression and adiabatic expansion (or work-expansion engines) to further reduce temperatures and, finally, in certain cases subject the refrigerated substance to magnetic cooling to reach the ultra-cryogenic temperature levels below 3 R and in some instances down to the region of .01 R. The latter appears to be about the lowest temperature limit for ordinary paramagnetic salts.

Temperatures in the region of .004 R have been attained employing two stages of paramagnetic cooling. One stage (upper) is made up of ordinary paramagnetic salts, such as chromium potassium alum, and a second stage (lower) of copper as the paramagnetic material. Both stages are subject to large magnetic fields by using superconducting magnets. While both stages are in thermal contact, the first stage magnetic field is removed, after which the temperature of both reach a common level of about .01 R, a temperature necessary for the lower stage material to function as an adiabatic demagnetization element. Upon reaching the temperature of equalization, the second stage magnet is removed and a subsequent further reduction in temperature occurs in that stage.

The low temperature capabilities of any paramagnetic material are limited by its point of magnetic saturation (see Chapter 10). At saturation, the flux density is increased to the point where any further application of magnetomotive force has little or no additional magnetic effect on the material. This relationship is shown on the *BH* curve, Fig. 295.

The salts are usually pressed into pellet form along with a binder and a heat conducting material, such as fine copper wire, to facilitate the transfer of heat of magnetization from the salt to the cooling medium during the demagnetizing process. Magnetic fields having reluctance values over 25,000 oersteds are required to attain temperatures appreciably below 0.01 R.

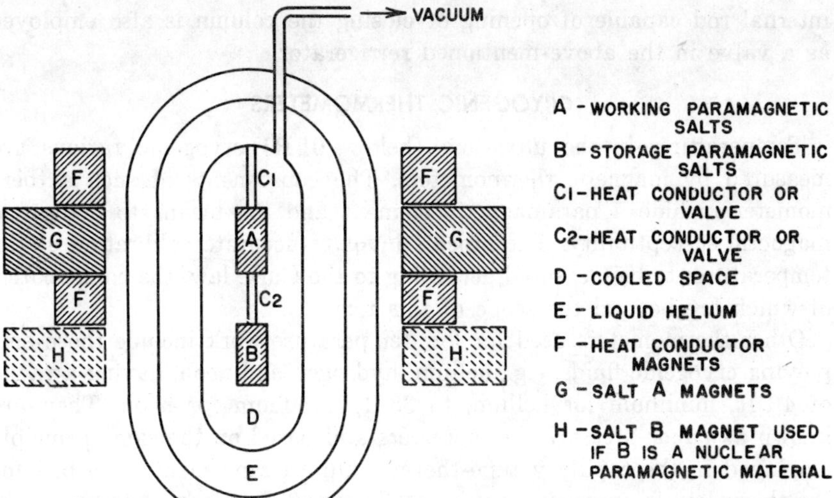

Fig. 537. Magnetic refrigerator.

A *magnetic refrigerator* is shown in Fig. 537. Its principle of operation is briefly described as follows:

Situation 1	Valve C_1 is open and Valve C_2 is closed. Magnet G field is fully applied about Salt A.
Process 1	Heat of magnetization is transferred from Salt A to liquid helium E through C_1 (open) to maintain constant temperature of A.
Situation 2	Valve C_1 is closed while Valve C_2 remains closed. Magnetic field from magnet G is reduced adiabatically.
Process 2	Temperature of Salt A decreases during partial demagnetization process.
Situation 3	Valve C_2 is open, Valve C_1 remains closed. Magnetic field about A is reduced to zero.
Process 3	Heat from Salt B is transferred isothermally to Salt A, reducing the temperature of B to required level.
Situation 4	Valve C_1 and C_2 are closed. Magnet G again energized adiabatically to full strength.
Process 4	Repetition of cycle initiated.

Magnets F are energized to maintain normal heat conductivity in such metals that may be used for valves or conductors. Otherwise, the heat conducting capabilities of certain metals are reduced when they enter a condition of superconductivity at the ultra-low temperatures. Therefore, to open or close a metal (such as a lead strip) valve, the magnetic field is applied or removed, respectively. A liquid helium column with an internal rod capable of opening or closing the column is also employed as a valve in the above-mentioned refrigerator.

CRYOGENIC THERMOMETERS

Temperatures in the ultra-low (below .01 R) cryogenic regions are measured by magnetic thermometers. The components of such a thermometer include a paramagnetic element and the means to determine magnetic susceptibility. The process involves accurate calibration at the temperatures to be measured according to the Curie law, the complexities of which are beyond the scope of this text.

Other thermometers used for low temperature work include those employing cryogenic fluids, e.g., helium, hydrogen and neon, having a range of 4.1 R, minimum for helium, to 38 R, minimum for neon. They are known as *vapor-pressure* thermometers and based on the same principle employed in the capillary tube-thermal bulb arrangement described for the thermostatic expansion valve in Chapter 9. In place of the valve in the latter case, the capillary tube is connected with a pressure indicator whose resulting measurement is related to the bulb temperature. The bulb fluid used depends on, and is usually the same as, the fluid whose temperature is to be measured.

Other thermometers include the thermocouples, resistance thermometers and the others described in Chapter 2.

AIR-CYCLE REFRIGERATION

The process of producing refrigeration by using air as the fluid medium is not efficient and, therefore, not economical for general purposes. For some special services the air-cycle system is convenient, but limited to small dimensional equipment design. In air-cycle systems fresh air is continuously introduced into the system and rejected back to the atmosphere when it has served its cooling purpose. Thus, it is a once-through arrangement.

There are a number of cooling requirements to which air-cycle refrigeration systems may be applied, the most important of which include airplane space cooling, instruments and instrument panels that tend to become excessively warm or overheat. They have been convenient when used in conjunction with jet engine compressors of airplanes to furnish compressed air, and with the cooling capacity of ram air (ram air is that which is scooped into the system from outside the airplane motivated by the speed of the plane), both of which are required in the air-cycle process.

The air-cycle refrigeration process is similar to that described for the cryogenic liquefaction systems, i.e., the basic cycle consists of

(1) A compressor to supply a sufficient quantity of high pressure, high temperature air,

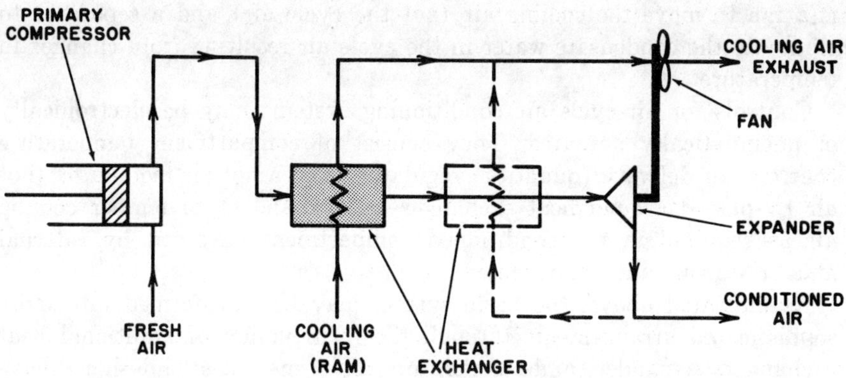

Fig. 538. Basic air cycle refrigeration system showing (dotted lines) modified regenerative arrangement. Secondary compression placed between heat exchangers forms "bootstrap" arrangement.

(2) A heat exchanger to dissipate the heat absorbed by the air during compression, and

(3) An expander through which the air, now relatively cooled in the heat exchanger, is expanded while simultaneously performing work as in an air turbine.

A schematic drawing of the basic air-cycle process is shown in Fig. 538.

The compressed air may be furnished from any available source, such as a stationary or portable air compressor, capable of delivering the required amount of air according to the magnitude of the cooling system, which may range in capacity from 300 to 8500 cfm. Or it may be the jet engine of an airplane as already mentioned.

The heat exchangers may be a simple shell-and-coil arrangement, or an evaporative type. In some instances, depending on the resultant temperature desired, both types of heat exchangers may be used and arranged in series, i.e., the evaporator is located upstream relative to the simple heat exchanger. A regenerative type heat exchanger is also used in some systems. In this arrangement, the regenerator is placed in series with the primary heat exchanger and uses cold air bled from the discharge line of the expanding device as the cooling medium.

The expander is usually a turbine harnessed to a work system, such as an electrical generator or fan. As in all gas expansion-cooling devices, a greater temperature reduction of the expanded fluid is accomplished when work is performed during the process as compared to simple expansion through an orifice, such as the throttling device or expansion valves discussed in refrigerant compression systems. The difference in temperature reduction here lies in the thermal energy expanded by the air (cooling medium) in performing work during expansion through a turbine.

Other auxiliary equipment that may be required in an air-cycle system is a fan to move the cooling air (not the cycle air) and a separator to eliminate the condensate water in the cycle air resulting from changes in temperature.

Controls for air-cycle air conditioning systems may be electronically or pneumatically actuated. They consist of compartment temperature controls, air delivery (quantity) regulators, and a hot air modulator (hot air by-passed around heat exchanger and expander) to temper cooling air as required in the conditioned compartment governed by internal and/or surrounding conditions.

As indicated above, the basic system may be transformed into more sophisticated arrangements through the introduction of additional heat exchangers, expanders and regenerators. They may also be designed having dual equipment served by the same compressor.

Some of the more important types include the (1) *basic* (described above), (2) *evaporative,* having an additional heat exchanger of the evaporator type, (3) *regenerative,* having an additional heat exchanger whose cooling medium is cooled air bled from the discharge line of the expander, (4) *bootstrap,* having a secondary compressor located between two air-cooled heat exchangers, and (5) others that may consist of combinations of the appropriate components described above, such as a primary expanding turbine located in the fresh air line prior to the primary heat exchanger.

VORTEX TUBE REFRIGERATOR

Some of the fruits of victory in World War II included scientific data and experimental information compiled by German engineers and physicists. One such project was the cooling device employing a *vortex tube* that converted compressed air at high velocity (sonic speeds) into a stratum of hot and cold air.

As its name implies, the high velocity air entering the tube tangentially is subject to a *vortical* or swirling motion generated by an element that is fitted circumferentially wihin the tube. This circumferential element develops a spiral as well as a rotary motion of the air mass and carries it toward the center of the tube to create the vortex effect. An analogy of the vortex tube design would be a water wheel or turbine in which the water enters tangentially and follows a series of spiral troughs to be discharged at the center. The air expands during the swirling process within the tube, having a much lower pressure environment, during which it is reduced in temperature. The centrifugal force generated by the vortical pattern as the air leaves the influence of the spiral element transforms its energy to thermal energy as the air is thrown toward the periphery of the tube. Thus, a temperature gradient develops radially represented by $\Delta T = T_c - T_h$ where $T_c =$ temperature at the center and $T_h =$ temperature at the periphery of the tube.

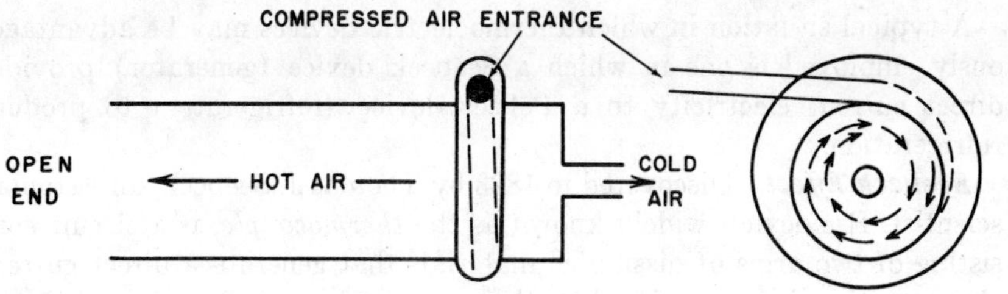

Fig. 539. Vortex-tube refrigerator.

The cold air leaves the tube through a centered orifice near the vortex element, while the hot air leaves at the open and opposite end, remote from the vortex. (See Fig. 539.)

The vortex tube system is applicable only to small refrigerating requirements, such as cooling wearing apparel for work in hot or humid environments, concentrated areas of electrical or electronic instruments, etc.

THERMOELECTRIC REFRIGERATION

A relatively new method of producing a practical refrigerator, although the principles involved were discovered in the nineteenth century, is based on the *thermoelectric* phenomena. While the potentialities of the art were recognized for a number of years, it was not until suitable materials were developed that its present state of progress was made possible.

Thermoelectric phenomena effect the conversion of electrical energy to thermal energy to produce a heat pump or *refrigerator*. It may also be modified to effect the conversion of thermal energy to electrical energy to produce a *generator*. The operation of thermoelectric devices are thermodynamically reversible. The phenomenon of an *electrical to thermal* energy conversion is called the *Peltier effect*. The phenomenon of *thermal to electrical* energy conversion is known as the *Seebeck effect*.

Thermoelectric *devices* are the practical solutions to situations where energy conversion may be accomplished with low power requirements under operating conditions that are free of moving parts. The art has found a high utility value in situations that for several reasons cannot employ such apparatus required by other conventional processes for the production of refrigeration, or electrical power generation. Modern and highly advanced vehicles, such as those related to space travel development, submarines, aircraft and ballistic missiles, all require energy conversion sources that are self-contained, reasonably efficient, easily maintained and/or repaired, light weight, dimensionally small and reliable for extended periods of time. While these factors are in some instances in opposition, it is necessary that every effort be made to approach the ideal.

A typical situation in which thermoelectric devices may be advantageously employed is one in which a Seebeck device (generator) provides direct current electricity to a Peltier device (refrigerator) to produce refrigeration.

Seebeck Effect. Discovered in 1823 by Thomas J. Seebeck, an Estonian scientist. His device, widely known as the *thermocouple,* is a circuit consisting of two arms of dissimilar materials that generate a direct current electromotive force (emf) when their junctions are exposed to different temperatures simultaneously. The electromotive force is known as the *Seebeck emf* and the effect of the device as the *Seebeck effect.* The magnitude of the emf so generated is *proportional to the temperature difference* between the junctions and depends on the *nature of the conductor materials.*

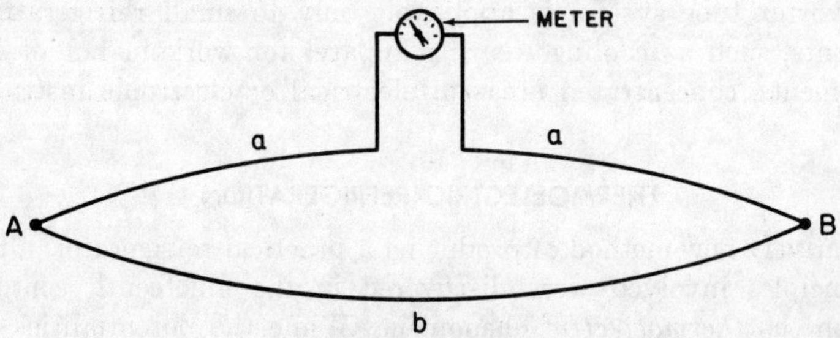

Fig. 540. Seebeck generator.

The Seebeck device, a sketch of which is shown in Fig. 540, is the basic element of the *thermoelectric generator* capable of converting thermal energy to electrical energy. It is a closed circuit in which a galvanometer may be placed in series to register an emf. One of its earliest uses is the thermocouple or thermoelectric thermometer. One type of pyrometer, a temperature measuring instrument, previously described in detail, is an example of the practical use of the thermoelectric Seebeck device.

Peltier Effect. Discovered in 1834 by Jean C. A. Peltier, a French physicist. He found that if an electric current is impressed across the junctions of the dissimilar materials of a Seebeck device heat is absorbed at one junction and rejected at the other. (See Fig. 541.) The particular junctions that act as heat source and heat sink depend on the direction of current flow through the circuit. The resulting energy conversion is called the *Peltier effect* and the thermal energy transferred from one junction to the other is known as the *Peltier heat.* The rate at which heat is transferred is a function of the nature of the dissimilar materials and is proportional to the magnitude of electrical current.

As heat is removed from the hot junction, the temperature of the cold junction is reduced correspondingly. This initiates a heat flow from the

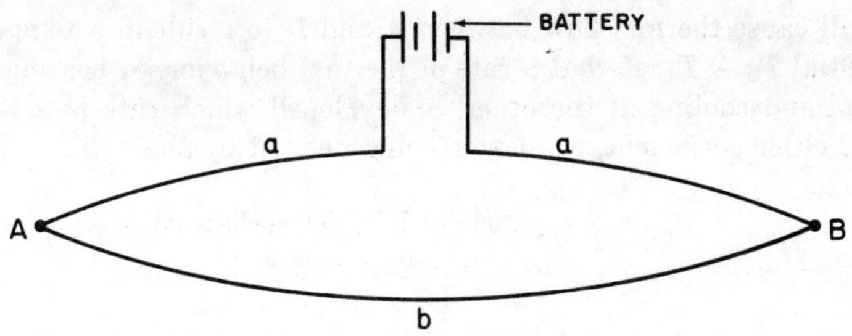

Fig. 541. Peltier cooler.

substance to be cooled (source) to the substance provided for the dissipation of transferred heat (sink). Thus, a refrigerator is established.

Thompson Effect. In 1857, William Thompson, an English physicist better known as Lord Kelvin, discovered a relationship between the Seebeck effect and the Peltier effect. It involved the effect of the surroundings on the complete circuit during electrical current flow when a constant and uniform temperature differential between increments of circuit dimension exists along the entire length of the conductor, the latter better described as the *temperature gradient.* Heat flow developed by the Thompson effect is a function of electrical current and temperature gradient.

Basic Thermoelectric Parameters

The factors identified with thermoelectric processes include Seebeck electromotive force (S), impressed electromotive force (E), electrical resistivity (ρ), current (I), thermal conductivity (k), thermal flow rate (q) and temperature gradient $\left(\dfrac{dT}{dx}\right)$. The relationship of these parameters may be shown by the Seebeck coefficient (α), the Peltier coefficient (Π) and the Thompson coefficient (τ).

In Fig. 540, where a and b represent the conductors, and A and B conductor terminals of a Seebeck device in which an emf is generated when a temperature differential $T_1 - T_2$ exists between A and B, the Seebeck coefficient or thermoelectric power may be expressed

$$\alpha_{ab} = \frac{S}{\Delta T} = \text{(differential Seebeck coefficient)}$$

where α_{ab} = thermoelectric power developed in the circuit ab, or specific Seebeck coefficient

S = electromotive force, circuit voltage

ΔT = temperature differential between A and B

Assuming that T_1 and T_2 are equal initially and that an electrical current (I) is applied across the terminal DC, a Peltier heat will develop.

This will cause thermal flow between A and B to result in a temperature differential $T_2 - T_1$, so that a rate of thermal behavior, of heating at one terminal and cooling at the other, is developed which rate is a function of the Peltier coefficient, the latter being defined by the ratio

$$\Pi_{ab} = \frac{q}{I} = \text{(differential Peltier coefficient)}$$

or

$$\Pi = \alpha T$$

Considering that, according to Thompson, in addition to the current flow I, an incremental temperature difference or temperature gradient exists, then the rate of cooling or heating may be expressed $\frac{dq}{dx}$ per unit length of conductor. So that combining the increments of change in electrical current, rate of heat flow, temperature differentials (gradient) for each increment of conductor length, the relationship may be expressed

$$\tau = \frac{\frac{dq}{dx}}{\frac{IdT}{dx}} = \text{(Thompson coefficient)}$$

where dx = incremental distances between A and B

Summarizing the parameters shown in the above development of pertinent coefficients, and adding parameters and appropriate thermoelectric terms and expressions, the following list is given along with the terms of measurement each represents:

Z—Figure of merit	$(\text{degree})^{-1}$
S—Relative Seebeck coefficient	Volts/degree T
ρ—Specific electrical resistivity	Ohm-cm
K—Total thermal conductance of an element	Watts/degree
k—Specific thermal conductivity	Watts/cm-degree
R—Total electrical resistance	Ohms
Π—Relative Peltier coefficient	Volts
A—Cross sectional area of an element	Sq cm
L—Length of an element (contact to contact)	Cm
$\frac{L}{A}$—Ratio of length to cross sectional area of an element	Ratio $\frac{\text{cm}}{\text{sq cm}}$
D—Form factor $= \frac{A}{L}$	Ratio $\frac{\text{sq cm}}{\text{cm}}$
T_c—Temperature at cold junction	°Kelvin
T_h—Temperature at hot junction	°Kelvin
ΔT—Temperature difference, $T_h - T_c$	°Kelvin
I, i—Circuit, total and circuit, couple, current	Amperes
E—Circuit voltage, applied	Volts
$\dot{Q}$—Rate of heat transfer to cold junction	Watts
Q—Heat absorbed at hot junction	Watts
P—Electrical power input (EI)	Watts

ϕ—Coefficient of performance | Dimensionless
η—Efficiency | %
T_m—Mean temperature | °Kelvin
q—Rate of cooling per unit length | Watts

Subscripts n and p refer to n-type and p-type arms.

The refrigerating capacity of a thermoelectric device (Peltier) depends on the *nature of the semiconductors, electric current* and the *ratio of the length to the cross sectional area of the arms.* Of importance, then, are the emf, and the amperes consumed by the system (the latter known as the heat pumping current), the heat transfer capacity, temperature difference ($T_h - T_c = \Delta T$) and the coefficient of performance of the system.

The above list of terms includes the foregoing parameters, as well as other parameters, the association of which assists in determining pertinent values related to cooling capabilities of a thermoelectric system.

Figure-of-Merit. The figure-of-merit is a value that indicates the performance or merit of a material as a thermoelectric semiconductor. It is numerically defined by the ratio (Z)

$$Z = \frac{\text{Seebeck coefficient}^2}{\text{electrical resistivity} \times \text{thermal conductivity}}$$

or $$Z = \frac{S^2}{\rho K}$$ for a single element

and $$Z_{np} = \frac{(S_p - S_n)^2}{(\rho_p K_p + \rho_n K_n)^2}$$ for a thermocouple having both an n-type and p-type element

also $$Z = \frac{2(\Delta T)}{T^2}$$ for a given couple when ΔT is maximum

Seebeck Coefficient. The thermoelectric power of an arm or a couple is called the Seebeck coefficient (S), expressed as volts per °Kelvin.

Peltier Coefficient. The Peltier coefficient (Π) is an absolute value derived by the equation Seebeck coefficient (S) times temperature (T)

$$\Pi = ST \text{ volts (Peltier heat)}$$

Power Input. This is the quantity of electrical power (P) required by a device (or devices) to produce a given refrigeration effect. It is the product of the amperes times the voltage applied

$$P = IE \text{ watts}$$

in which I for the system is equal to the i for the individual devices $\times$ the number of couples (N) or, $I = Ni$.

Ratio of Cross-Sectional Area to Length of an Element. This is expressed as the ratio, sometimes referred to as the *form factor* (D), as follows:

$$D = \frac{A}{L}$$

Ratio of Length to Cross-Sectional Area of an Element. This is expressed $\frac{L}{A}$.

Refrigeration capacity decreases as the $\frac{L}{A}$ increases. This means that the shape of the element should depart the form of a long cylinder.

Coefficient of Performance. The ratio of the power absorbed at the cold junction to the electrical power required by the device represents the coefficient of performance (cop) sometimes represented by ϕ.

$$\text{cop} = \frac{\dot{Q}}{P}$$

cop is independent of the $\frac{L}{A}$ ratio.

Mean Temperature. The mean temperature (T_m) of the couple is determined by the universal method establishing mean values, i.e.,

$$T_m = \frac{T_h + T_c}{2}$$

Refrigerating Capacity. The important factor for any thermoelectric cooling system is the effective transformation of electrical energy to thermal energy available at the junctions. It is a function of thermal conductance, temperature differential across the couple, form factor, and the figure of merit. The relationship of the foregoing parameters is shown in the equation

$$Q_{max} = 2\,\frac{A}{L}\,K(\tfrac{1}{2}ZT^2{}_c - \Delta T)$$

Maximum Coefficient of Performance. The value of the maximum coefficient of performance of a thermoelectric device relates to the Carnot cycle for the compression refrigerator, as described under "Thermodynamic Functions" in Chapter 7, in which $\frac{T_2}{T_1 - T_2}$. In terms of the figure of merit

$$\phi_{max} = \text{cop}_{max} = \frac{T_c}{T_h - T_c} \times \frac{1 + ZT_m - T_h/T_c}{\sqrt{1 + ZT_m} + 1}$$

or in terms of actual temperatures at the hot and cold junctions, then

$$\phi_{max} = \frac{T_c}{T_h - T_c} \times \frac{\sqrt{1 + \tfrac{1}{2}Z_{np}(T_h - T_c)} - T_h/T_c}{1 + \tfrac{1}{2}Z_{np}(T_h - T_c) + 1}$$

where Z_{np} = figure of merit for the couple having both an n- and p-type element.

Note that the cop is independent of ratio $\frac{L}{A}$ and, therefore, does not appear in the equation.

Efficiency. A parameter that does not influence the selection of element materials, as priority to other parameter considerations, is the efficiency (η) of the thermoelectric system.

$$\eta = \text{efficiency} = \frac{\text{cop of any system}}{\text{cop of Carnot cycle}}$$

Given below are average practical values representing the various parameters associated with thermoelectric materials. Bismuth telluride (Bi_2Te_3) is the semiconductor used as the guide here where appropriate.

(Z) Figure of merit $(°K)^{-1}$—.003
(S) Seebeck coefficient—220 microvolts/°C
(ρ) Resistivity—1.10^{-3} ohm-cm
(K) Thermal conductance—15.4 mw/cm°C
(ϕ) Coefficient of performance—.25
(I) Current, input—up to 30 ampere
(RE) Btu/hr/couple—<1.5 Btu
$\left(\frac{L}{A}\right)$ Length to area ratio of elements—2 to 1, to, 4 to 1

The negative sign used in the exponent of temperature indicates direction of decreasing temperature.

Thermoelectric Refrigerator

The practical Peltier or thermoelectric refrigerator device consists of two arms of dissimilar materials having a common junction connecting the ends of each material and of appropriate terminals for the application of an external electrical current to form a series or electrical circuit, as shown in Fig. 542. The device, better described as a thermocouple, is

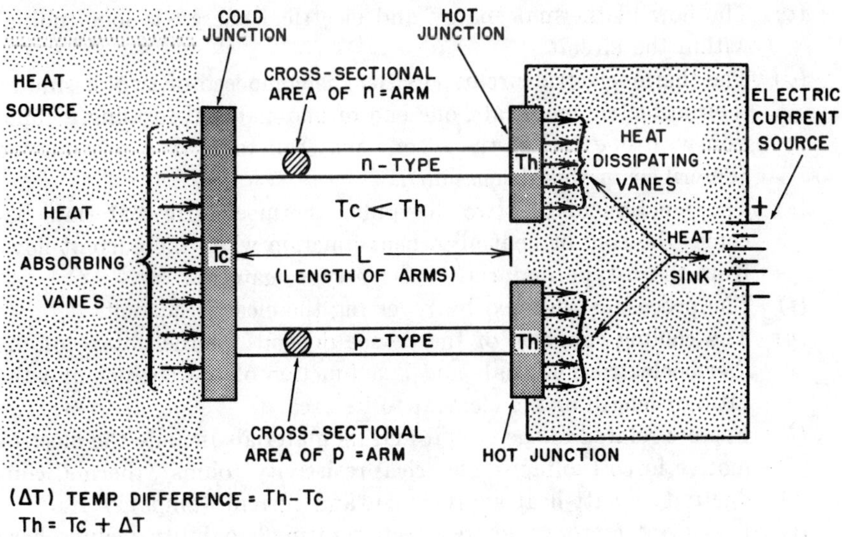

Fig. 542. Thermo-electric system of refrigeration (Peltier effect) device.

so arranged as to allow for the exposure of one junction to a zone of lower temperature and the second junction to a zone of higher temperature so that dT (temperature differential) is equal to $T_{hot} - T_{cold}$.

The thermoelectric refrigerator may be designed to include any number of Peltier devices or *elements*, depending on cooling load requirements. The elements also called *couples* are electrically connected in series, a convenient number of which form an assembly, called a *module*. Each module has its own positive and negative terminals to permit flow of d-c current in and out of the assembly. The necessary number of modules based on load capacity requirements and/or refrigerator design may be connected in parallel to form a *panel*. The panel may or may not constitute the wall construction of the refrigerator cabinet, or may be installed in the existing wall(s) of a larger refrigerated space. Couples may be composed of a single element or arranged in a pile to form a cascade system (Fig. 546).

For reasons related to electric current demands in producing refrigeration, it is not feasible to replace the system of multiple Peltier couples of small physical dimensions with a single but larger couple, having equivalent dimensional characteristics, in order to meet heat load demands.

Peltier Effects, Parameters and Conditions

The following are related to the development, application and limitations of the Peltier thermoelectric refrigerator:

(*a*) Converts electrical energy to thermal energy.

(*b*) Electrical energy is supplied by an external source of direct current.

(*c*) The flow of thermal energy and electrical energy is longitudinal within the circuit.

(*d*) The thermoelectric circuit (Peltier) is composed of two dissimilar thermoelectric materials, one end of one material forming a junction with one end of the second material to form a pn junction (described in this discussion).

(*e*) The junction of the two dissimilar thermoelectric materials becomes the cold or heat absorbent junction when the current flows from the *n*-type material to the *p*-type material.

(*f*) The process is reversed by reversing the electric current.

(*g*) The cooling capacity of the device depends on the nature of the thermoelectric materials and is a function of the current and the ratio of length of the element to its area.

(*h*) Factors related to the circuit and its materials include the electromotive force (voltage), electrical resistivity (ohms), thermal conductivity (watt-heat equivalent), and current (amps).

(*i*) Important factors related to refrigerating capability include heat pumping capacity, temperature difference between the hot and cold junctions, and coefficient of performance.

(j) Factors related to the physical conditions of the device are length of the element (L), the cross-sectional area of the element, and the ratio of $\frac{L}{A}$.

(k) Design requirements dictate that the elements be connected in such a manner as to be electrically in series and thermally in parallel.

(l) The capabilities of the system depend also on the suitability of the materials that compose the elements.

(m) The dissimilar materials are in series circuit.

(n) Elements that compose a module are in series circuit.

(o) Modules that compose a panel are in parallel circuit.

(p) Where necessary defrosting is accomplished by reversing the current by cycling.

(q) The thermoelectric materials are better described as semiconductors for reasons given later in this discussion.

Semiconductors

The dissimilar materials, previously referred to as composing the *arms* of thermocouples or thermoelectric devices especially of the Peltier type, are *semiconductors.* Semiconductors are materials that have neither the insulating effect of true insulators, such as rubber, or the conducting capability of true conductors, e.g., copper, platinum, etc. Their characteristics in relation to the materials mentioned may be roughly stated as being in between regarding their conduction capacity.

The factors governing the nature of any material to conduct energy through a distance barrier are a function of their atomic structure and the inherent force with which their atom particles are mutually bonded. Some materials that are non-conducting at certain energy levels (thermal or electrical) become conductors at higher energy levels. In some cases, they may be made conductors by incorporating certain impurities into their basic structure during preparation, such as molecules of another compound into the crystalline structure of an otherwise pure compound.

Atomic Structure. All matter is made up of *atoms* and combinations of atoms which unite chemically to form *molecules* of *compounds* (atomic theory). An atom is the smallest unit that retains the properties of an element (iron, copper, bismuth, etc.), or that can enter a chemical reaction with another atom of a different element. All atoms, except those of ordinary hydrogen, contain three primary particles of minute size, namely, *neutrons, protons* and *electrons.* A definite number of neutrons and protons make up the *nucleus* or core (with which the weight of the atom is related) of the atom. While neutrons and protons have about the same weight of 1 Atomic Mass Unit (amu), they differ in that the neutron is uncharged electrically, i.e., it is neutral, while the proton has a positive charge, making the nucleus positively charged. The electron is a minute particle, negatively charged, that revolves about the nucleus in the same manner in which the planets in the solar system travel around the sun. Thus, the atom is normally electrically neutral.

The simplest of all atoms is that of hydrogen that has no neutron particle and only one electron in the orbital path about its nucleus.

Actually the neutron has a mass slightly larger than that of a proton and is usually reported as 1+. Amu is the weight of an atom of other elements compared to the weight of an atom of oxygen, which is arbitrarily set at 16 amu. Hydrogen has an amu of approximately 1 (one) according to this scale. The nucleus contains almost all of the mass of the atom, yet the diameter of the atom is about 10,000 times the diameter of the nucleus. Neutrons differ from protons in their electrical charge, the former (neutron) being neutral, i.e., having no electrical charge, while the proton has on it a positive (+) electrical charge of +1. Therefore, the nucleus of the atom has a positive electrical charge. All the atoms of a *given* element have the same number of protons. However, the number of protons of any element differ from the number of protons in the atoms of any other element. This fact points to the essential difference between atoms of different elements, the number of which is called the *atomic number* of the element. For example, the hydrogen atom has 1 proton, or an atomic number of 1, the helium atom has 2 protons, or an atomic number of 2, the uranium atom has 92 protons, or an atomic number of 92, etc. The atomic number of an element is fundamentally more important than its atomic weight. Neutrons outside the nucleus are unstable and tend to form protons. It may be noted here that the forces between protons and neutrons within the nucleus of the atom are much greater than those between the atoms common to any element. For this reason, and where equal masses are involved, nuclear energy is tremendously greater in magnitude than any known conventional energy, such as the chemical energy released by high explosives (e.g., TNT) upon detonation. Hence, the equivalency indication of nuclear blasts to TNT explosions, such as the megaton power of fission and fusion interactions during nuclear explosions; hence also the detonation of high explosives, which only involves rearrangement of the atoms that constitute the molecules of the explosive materials.

The number of neutrons in the atom of a given element may vary, which means that the weight of the atom also varies. If atoms of the *same* element differ only in their ordinary weight, they are known as *isotopes.*

The electron is a minute particle having on it a negative (−) electrical charge of the smallest known quantity of electrical energy. Electrons have a charge of 1 electronic charge in magnitude, which is defined as equal to 4.774×10^{-10} *electrostatic units*. Electrons that revolve in an orbital path about and outside the nucleus are referred to as *planetary electrons*. They are extremely light in weight having practically no mass with an amu 1/1845th of that of the proton. Neutral atoms of any given element have a definite number of electrons *equal to the number of protons in the same atom*. Therefore, the number of planetary electrons have the same range as the atomic number associated with the element, i.e., from 1 as in hydrogen to 92 for uranium, for natural elements. The number of electrons in an element determines its chemical characteristics and behavior. The atoms of some elements readily lose and/or gain one or more electrons either under conditions of (1) radioactivity, or (2) by exposure to externally applied energy, such as from electrical and thermal sources. Materials, such as radium, are radioactive and fit into the first (1) category. Conductors are of the second (2) category.

Electron Shell. *Bohr's* theory of the atom offers the probability that electrons of an atom are arranged into a series of planetary paths about the nucleus. These planetary paths or orbits form definite patterns called *shells*, each of which expands to a greater circumference relative to the nucleus (Fig. 543). The number of shells vary according to the element and are identified as energy levels, each shell being identified by letters, starting with the innermost shell as *K*, *L*, *M*, *N*, *O*, *P*, *Q* depending on the element considered. Each shell of an atom of a given element varies somewhat in the number of electrons it contains. For example, the element hydrogen (H) has the simplest atomic structure of all elements having one proton in the nucleus and one electron in its planetary system. Therefore, hydrogen is composed of one shell. Bismuth (Bi) on the other hand

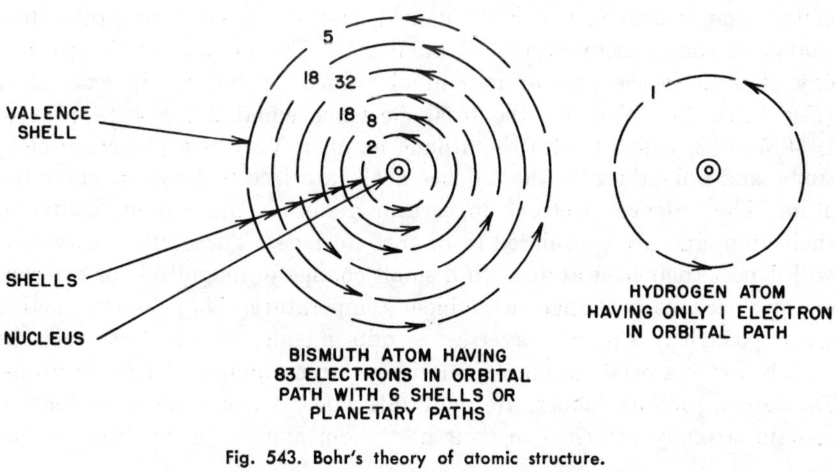

Fig. 543. Bohr's theory of atomic structure.

is a complex element having a planetary system containing six shells. The first shell, or the one nearest the nucleus, always contains 2 electrons; the second shell always 8 electrons; the third, 18 electrons; the fourth, 32 electrons; the fifth, 18 electrons; and the sixth, or outermost shell, 5 electrons. The sequence of electrons relative to their respective shells is shown for certain elements in the Periodic Table of Elements (see Table 164). The outer orbit is called the *valence shell*. The shell

TABLE 164. ATOMIC SHELL ARRANGEMENT IN SOME CONDUCTOR AND SEMI-CONDUCTOR ELEMENTS SHOWING NUMBER OF ELECTRONS IN EACH SUCCEEDING SHELL BEGINNING AT THE ONE NEAREST TO THE NUCLEUS

Element	*Symbol*	*Material*	*No. of Shells*	*Shell Arrangements Showing Number of Electrons from Nucleus*
Aluminum	Al	Metal	3	2, 8, 3
Silicon	Si	Non-metal	3	2, 8, 4
Iron	Fe	Metal	4	2, 8, 14, 2
Copper	Cu	Metal	4	2, 8, 18, 1
Germanium	Ge	Metal	4	2, 8, 18, 4
Selenium	Se	Non-metal	4	2, 8, 18, 6
Antimony	Sb	Metal	5	2, 8, 18, 18, 5
Tellurium	Te	Non-metal	5	2, 8, 18, 18, 6
Arsenic	As	Non-metal	4	2, 8, 18, 5
Indium	In	Metal	5	2, 8, 18, 18, 3
Bismuth	Bi	Metal	6	2, 8, 18, 22, 18, 5
Silver	Ag	Metal	5	2, 8, 18, 18, 1
Platinum	Pt	Metal	6	2, 8, 18, 32, 17, 1
Lead	Pb	Metal	6	2, 8, 18, 32, 18, 4
Phosphorus	P	Non-metal	3	2, 8, 5

sequence of bismuth is shown to be 2, 8, 18, 32, 18, 5. *Valence electrons* of the atom are those which are gained, lost or shared during the interchange of energy between reactive materials. The number of electrons in any element is the sum of individual values represented in each shell (Fig. 543). For bismuth, the total electrons equal $2 + 8 + 18 + 32 + 18 + 5 = 83$, which is also the atomic number. Very few planetary electrons, and only those in the valence shell, are free to leave or enter the atom. The valence electrons in certain materials are loosely bound to their atom and are considered to be *free electrons*. These "free" electrons will depart their host atom with a small change in magnitude of external energy, such as a change in ambient temperature. They become active when electrical energy traverses certain metals. *Conductors,* such as metals, are materials having a relatively large number of free electrons. *Insulators,* such as rubber, are materials whose valence electrons tend to remain strongly attached to their host atom and so do not become free to initiate the flow of energy from atom to atom within their parent material. Semiconductors, on the other hand, act as described previously relative to the availability of free electrons or other energy charged carriers.

Thermoelectric Materials. Semiconductor materials compose the energy transformation elements employed in thermoelectric and electronic devices, acting as *transistors, solar batteries, rectifiers* (converting a-c current to d-c current), *amplifiers* (strengthening electrical impulses) and other types of *solid-state* systems. Germanium and silicon are important elements that compose the semiconductor materials in electronic devices.

The problem of selecting a material that will serve as an efficient thermoelectric element goes further than possessing sufficient electrical conductor characteristics. Many metals have high conduction qualities, but lack the properties conducive to high thermoelectric effect. *The definition of a thermoelectric material is one that possesses low thermal conductivity and a relatively high electrical conductivity.* A metal having the proper balance ratio of the two properties has not been found.

In normal semiconductor materials, thermal conduction is compounded by two types of heat energy transport systems. There is the thermal energy conducted by electrons, as identified with metals, plus the thermal energy conducted by atoms of the *crystals* forming the semiconductor material. This condition substantially increases the thermal conductivity of the circuit, thus reducing the ratio of electrical to thermal conductivity, which in turn decreases thermoelectric capacity. This has been overcome by certain techniques employed in the preparation of thermoelectric materials, resulting in the reduction in thermal conductivity without any apparent effect on electrical conductivity.

As stated earlier, normal semiconductor materials lack the necessary free electrons to be efficient thermoelectric elements under certain energy level conditions. By the addition of small amounts of certain *discrete* impurities, free electrons are produced that can initiate and move an electric current.

Semiconductors have a crystalline structure and are, therefore, subject to *lattice* effects, a highly complex phenomenon.

Principle of Semiconductors. The more important thermoelectric materials are composed of elements represented in Groups IV, V and VI of the "Periodic Tables of the Elements" (Table 97). Such elements include bismuth, tellurium, antimony, selenium, germanium and silicon. The compounds are crystalline in structure and have *lattice* effects, particularly on thermal conductivity, an important advantage. Crystals are said to be either *intrinsic* or *extrinsic.*

Intrinsic crystals are those developed from the compounding of exacting amounts of pure elements composing the crystal, i.e., they are pure crystals which may have defects, however. Extrinsic crystals are those that, in addition to the molecules that form a pure crystal, contain an excess of atoms of either one of the elements composing the compound, or discrete atoms of another *selective* element. Any atom extraneous to the composition of the pure or intrinsic crystal is an *impurity,* and so designated. Impurities in controlled amounts are built into the crystals during their preparation as a selected type of semiconductors. In some instances an atom of impurity material replaces an atom of intrinsic material to become an extrinsic material. Intrinsic materials are also made extrinsic by causing an excess of one or the other element of a compound to remain present in the semiconductor material. Impurities purposely *added* to the crystal are called *doping agents.*

The kind of impurity selected governs the nature of the thermoelectric element and the electronic behavior of its current carriers as a semiconductor. It is the impurity from which, in some instances, current carriers originate.

It has already been explained that an electric current flowing through a conductor is carried by available free electrons. Materials having a greater number of free electrons available become more effective conductors. Free electrons are identified with the *conduction band* of the conducting material. Thermal energy (heat) from an external source tends to increase the supply of free electrons in the conduction-band, resulting in an increase in current flow.

When an electron carrying an electric current leaves the valence shell, it leaves a void called a *hole* or *vacancy.* Holes also become carriers of electric current within the valence band, which is at a lower energy level than the conduction band. With additional heat, more electrons are driven into the conduction band, allowing more holes to develop in the valence band. Thus, the activity within the crystal increases, allowing greater current flow, i.e., permitting the crystal to become more conductive. Upon extrication from its host atom, the free electron will travel about the crystal structure in the manner of a true conductor, such as copper. However, before becoming mobile, electrons whose energy level corresponds to that identified with the valence band are relatively stationary and, therefore, do not lend to the current carrying capacity of the semiconductor. To disassociate itself with the valence band and move into the conduction band, additional energy must be imparted to the electron. The measure of this energy required is indicated by the energy gap dividing the valence band and the conductor band. (See Fig. 544.)

In the case of intrinsic materials, an equal number of electrons and holes develop as a result of applying additional heat energy to the crystal. While the intrinsic material is not of major importance in simple thermoelectric devices, they are always used in conjunction with extrinsic materials to develop certain limiting effects on the operation of those devices.

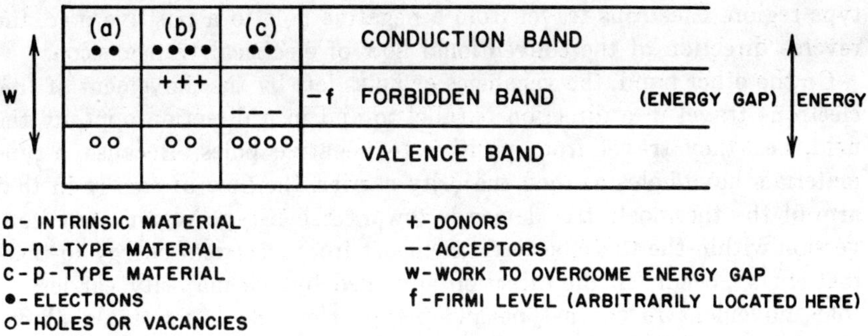

Fig. 544. Energy bands of atom electrons for intrinsic and extrinsic materials.

There are certain elements used to furnish impurities to intrinsic crystals to produce extrinsic crystals. The element selected governs the behavior of the current carriers within the semiconductor circuit.

Impurities are classified as either *donors* or *acceptors*. Donors are usually identified as n-type impurities and acceptors as p-type impurities, the semiconductor taking on the same identification as its impurity. The difference between donor and acceptor, simply stated, is as follows:

The atoms of donor or n-type materials produce one or more *free electrons* that can carry an electric current and initiate current flow within the crystals of the semiconductor. When an electron is in a donor state, it is neutral. A donor retains a positive charge when it becomes energized and is forced into the conduction band. In this state, it becomes ionized.

The atoms of acceptors or p-type materials react to cause electrons to break away from the atoms of the semiconductors and produce holes that also become current carriers. The activity of both electrons and holes is not unlike the behavior of gas within a conduit.

As stated previously, the displacement of an electron results in a vacancy or hole in its atom of origin. Holes also become free to move within the crystal and carry an electric current with them. In an n-type semiconductor, *the electron becomes the major current carrier and the hole, the minor current carrier*. The situation in the p-type semiconductor is reversed, with *the hole being the major and the electron, the minor carrier*. It may be said, then, that the movement of the electrons and holes results in the flow of current (conduction) through the thermoelectric materials.

When an electrical potential is applied across a thermoelectric element, it sets up an electrostatic field. Electrons move in a direction parallel to and in the direction of the electrostatic field. Because n-type materials have electrons as their *majority* carrier, the flow of energy within that arm of the thermoelectric device occurs in the direction of the p-type region. Electrons travel from a negative pole to a positive pole, the reverse direction of the conventional flow of electrical transmission.

On the other hand, the vacancies or holes left by the movement of free electrons travel in a direction parallel to and in a direction opposite the field, i.e., they travel from positive to negative poles. Because p-type materials have holes as their majority carrier, the flow of energy in that arm of the thermoelectric device is toward the n-type region. Any conversion within the thermoelectric elements from electrical energy to thermal energy occurs in the direction governed by the majority carriers so that movement in the n-type and p-type elements is as stated above. Both elements, then, tend to produce thermal energy flow from the cold junction T_c to the hot junction T_h across which a temperature differential of $T_h - T_c$ exists. Thus the phenomenon of thermoelectric effect as produced by the Peltier device is established. It should be noted here that the process is reversed by reversing the electric current and thus reversing the electrostatic field.

It is important to note here that the charged particles which are free to move are in constant random motion. This random motion is known as *diffusion,* an important phenomenon of lattice materials. The diffusion of charged particles or carriers is due to their density variations as they move from one point to another within a solid. Diffusion, then, is based on the *thermal* considerations affecting the motion of the particles.

When an electrical force is applied to the charged particles, they transport current within the semiconductor, depending on the direction of the field. The motion of particles dependent on an external electrical field is called *drift.* It is the drift that causes an electric current to flow.

The pn-junction allows a transition from a p-type to an n-type semiconductor. The transition is due to a change from acceptor density to a donor impurity density. A potential energy difference develops between the two regions as a result of the electrons and holes being swept away from the junction with the resulting space being taken by ionized donors and acceptors.

The random motion of the electrons is effected by an electrostatic field which results in a directional drift flow counter to the field. Drift flow of holes is in the direction of the field. This results in the majority carriers from the n-region enriching the minority carrier density in the p-region.

The interchange of energy occurs during the process of positional flow. The drift velocity of the charged particles, being directed against the field in Peltier devices, must overcome the (electrical) potential barrier. In doing so, it must absorb thermal energy from the crystal structure generating a heat flow toward the junction. The charged particles moving with the field gain thermal energy, which is transferred to the crystal structure and thence along the temperature gradient to the junction where it is rejected to the heat sink.

To form an effective junction of n-type and p-type materials, cleavages as would be present if two separate materials were joined mechanically must be avoided. Rather the junction should be formed by converting part of, say, the n-type material to a p-type material by applying a p-type dope to the n-type crystal. Upon heating that portion of the semiconductor and the dope to a melting temperature and allowing the alloy to cool, it will recrystallize and form a p-type material. Hence, we have a continuous structure one-half of which is n-type and the other half p-type material, whose line of demarcation is the pn-junction. The pn-junction may also be formed as described for bismuth telluride, in which an excess of one or the other elements results to form the dissimilar materials.

The solid continuance of the crystallized materials eliminates any barriers to the flow of energy carriers in either direction across the junction. The negatively charged electrons of the n-type material will scale the electrical potential by *absorbing* heat from the crystal (lattice) which in turn is *absorbed* at the junction. The positively charged holes of the p-type material flow with the potential, in the process gaining energy which is *rejected* at the junction. The potential decreases as the temperature difference diminishes.

Semiconductor Materials

There are a number of elements suitable for use as semiconductors. Their selection, however, must be based on the purpose for which they are intended. Semiconductors constitute the energy conversion elements of electronic devices, such as transistors, solar batteries, thermoelectric generators and thermoelectric refrigerators. Germanium (Gr) and silicon (Si) are the more important elements used for electronic purposes. Materials for thermoelectric conversion of energy as represented by the Seebeck and Peltier effects also require selective elements and a combination of elements to produce practical and effective results. The amount of electric current available or required, as the case may be, coupled with involved temperatures, is to be considered in the selection of suitable materials.

Thermoelectric materials involve both p-type and n-type semiconductors that form the arms of the respective devices. Because certain situations, such as isolated space vehicles and submarines, depend on self-contained energy conversion units, both thermoelectric generators and refrigerators should be of interest here. Because intrinsic crystals require an impurity to be made extrinsic for current carrying capabilities, doping agents are also included.

One of the more important materials employed for thermoelectric refrigeration (Peltier devices) is composed of the elements bismuth (Bi) and tellurium (Te) that form the compound bismuth telluride, having the chemical formula Bi_2Te_3. This material also has some importance for thermoelectric generators in the low temperature zone. Other materials that produce effective results for Peltier devices are lead telluride (PbTe), bismuth selenide (Bi_2Se_3) and antimony telluride (Sb_2Te_3).

Important semiconductor materials employed for thermoelectric generators (Seebeck devices) include bismuth telluride, lead telluride, zinc antimonide (ZnSb), manganese telluride (MnTe) and element combinations of cerium-sulphur (CeS) and germanium-tellurium. Thermoelectric materials of significant value are composed of bismuth telluride alloys containing antimony telluride (Sb_2Te_3) for p-type and bismuth selenide (Bi_2Se_3) for n-type elements. PbTe doped with sodium (Na) and PbTe doped with lead iodine (PbI_2) have also been used for thermoelectric devices.

Doping is the act of transforming an intrinsic crystal into an extrinsic crystal by the diffusion of an impurity in the form of another element or compound, or excess atoms of the same elements that compose the original intrinsic material.

The impurity selected depends on the type of material, i.e., n-type or p-type. Doping agents used include excess lead (n-type) or excess tellurium (p-type) dissolved into the intermetallic compound crystal PbTe. Or it may consist of excess bismuth to produce a p-type material or excess tellurium to produce an n-type material. Other doping agents include mercury (p-type), cuprous sulphide (n-type), cuprous bromide (n-type), sodium (Na) (p-type) and others of varying importance.

An excess of one or the other element composing the primary material of a semiconductor is produced during the process of crystallization. For example, if bismuth telluride is melted and then allowed to solidify, the initial substance to freeze out contains excess bismuth. Bismuth being a p-type material, it follows that the first material to freeze out produces a p-type material. As bismuth-rich Bi_2Te_3 continues to crystallize or freeze out, the melted material becomes progressively richer in tellurium to produce an n-type material.

Materials used at the junctions of thermocouples whose purpose is to transfer energy in or out of the thermoelectric device must be of such a nature as to impart minimum interference on Peltier or Seebeck processes. Therefore, both bonding and conduction materials should be highly efficient in performing their function. Copper strips serve successfully in carrying electrical current to and from the thermocouples and as series connectors between individual devices. While silver fins to conduct heat to and away from the thermocouple have been used effectively in experimental work, other good conducting materials, such as aluminum and copper, are practical heat conduits. It has been found that bonding materials at hot and cold junctions have little influence on optimum performance providing values approaching maximum efficiency are considered. However, bonds within the system must be carefully prepared.

Practical Thermoelectric System

While scientific research continues to develop improved thermoelectric materials, we have yet to perfect several physical areas through which system cooling capacity is lost. These areas include (*a*) resistance of energy flow at the junctions, and (*b*) heat transfer through surfaces other than those of primary concern. Better materials and procedures in both cases will help to diminish such losses, which often amount to 35–45% of theoretical capabilities. Moisture and electrostatic currents are responsible for the deterioration and corrosion of critical components of the system, both of which must be effectively restrained through construction techniques.

Cascade systems. Previous discussion dealt with single-stage thermoelectric devices in which heat from the source was directly rejected into the substance employed to dispose of it externally, such as water, air, etc. An arrangement whereby the cold junction (T_c) of a second stage couple becomes the sink for the hot junction of the first stage couple is known as *cascading.* This arrangement connects the couples of each succeeding stage thermally in series (Fig. 545). A cascade system consisting of more

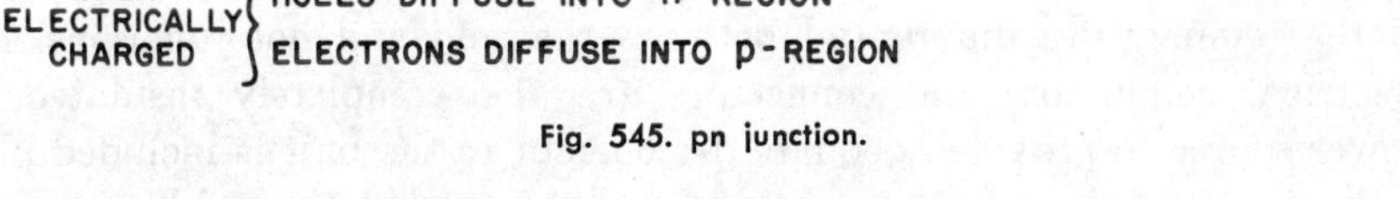

Fig. 545. pn junction.

than one stage is called a *pile.* It is usually designed to include a greater number of couples in the second stage than in the first stage of a two-stage system. It is important to design the first stage with sufficient capacity to satisfy the heat absorbing capability of the second stage. This rule applies to all subsequent stages that may be built into the system. Otherwise, a decrease in cop results.

A practical *single-stage* thermoelectric cooling system is shown in Fig. 546. It consists of a panel, say, of ten modules, each module having sixteen thermocouples or elements. Each element is connected electrically

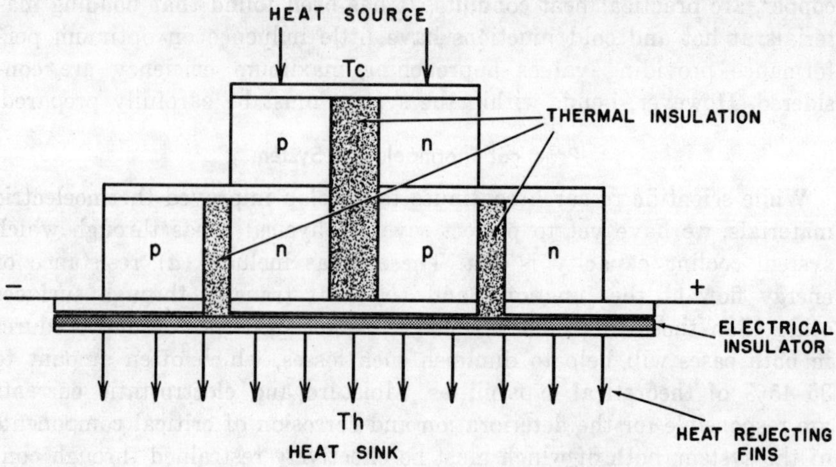

Fig. 546. Two-stage cascade thermo-electric refrigerator.

in series with all other elements of the same module, while the modules are connected electrically in parallel with all other modules. The cold junction of each element is exposed to the interior of the refrigerator where heat is absorbed, aided by forced convection by means of a fan blowing air over the surfaces of fins that in turn conduct heat to the thermocouples. The hot junctions are in contact with a header constructed of high heat conducting material that absorbs heat from thermocouples. Cooling water is circulated over the header surface to conduct the heat away to some predetermined sink. In some instances and where feasible, a fan and fin arrangement can be substituted for the cooling water to dissipate the heat from the hot junction in the same manner described for the cold junction.

Each element is completely insulated to maintain the temperature gradient within the element, as nearly as possible, as solely a function of the temperature differential between the cold and hot junctions. All electrical conductors and connectors are also completely insulated. A power source to provide adequate d-c current to the unit is included plus a thermostat for controlling the temperature within the refrigerator by modulation. Another control, not included in the figure, would provide protection against overheating of the elements by diminishing electrical current supply during periods when the temperature differential tends to approach too nearly a zero level. Also a defrosting arrangement, either by reversing electrical current which also reverses the function of the junctions, or by a heater arrangement in the path of the forced convection air flow around the fins. Either method of defrosting may be controlled manually or automatically with time cycling.

The refrigerating effect of the entire thermocouple panel, of course, depends on the capability of the semiconductors, both as to material and design. The capacity of the refrigerator depends on the refrigerating effect of the panel and the thermodynamic factors relative to its construction, including insulating materials, quality of joints, etc.

Thus, rejecting the heat at the hot junction depresses the temperature of the cold junction. Maximum temperature drop between the cold and hot junction is attained when the heat absorbed is equal to the heat rejected at the respective ends of the couple, which should cool itself by uniform heat transfer accomplished when the heat is dissipated at the same rate as it is absorbed. Any heat entering the couple from its surroundings other than the cold junction interferes with the basic function so that highly effective thermal insulation must be provided each couple. Any departure from the above condition adversely affects the figure of merit as shown in the equation

$$Z_{effective} = \frac{2(\Delta T)_{max}}{T^2_{cold}}$$

When assembled in a module which in turn is assembled in a panel, the individual couples will provide their own insulation, except at the extreme edges of the assembly.

Assume a refrigeration load of 6000 Btu per hour (½ ton) in Fig. 547, including fan, heat transfer inward through the floor and walls, inleak due to door openings and introduction of warm product. Also assume that each couple has a cooling capacity averaging 1.2 Btu per hour, net. The number of couples required would be

$$\frac{6000}{1.2} = 5000 \text{ couples}$$

This would constitute the number of couples contained in the complete panel. If the panel were designed to form part of one of the walls, the number of square modules required would be determined by first considering a convenient size and then calculating a feasible number of couples per module. Successful results have been obtained with module capacity ranging from less than 50 to more than 1500 couples and from about 3.5 inches square upwards.

If the panel is placed near the ceiling for the sake of isolation, it should be made rectangular rather than square and the dimensions governed by the internal size of the refrigerator. A convenient size then would be 50 couples per module and 100 modules per panel. The panel then can be made up of 20 horizontal modules × 5 tiers of vertically placed modules or any combination according to wall space available.

The above-mentioned hypothetical refrigerator has been included to convey to the student the versatility of the component parts of the thermoelectric refrigeration device.

By way of summarizing the thermoelectric cooling phenomenon, it may be said:

(*a*) By causing an electric current to flow, as from a battery, through a circuit comprised of semiconductor materials, heat will be absorbed at one junction and rejected at the other junction to cause a temperature differential across the circuit;

(*b*) Dissimilar materials consist of a p-type or positive arm and an n-type or negative arm whose junctions form a pn-junction;

(*c*) n-type semiconductors have electrons as their major current carriers that transport energy in the direction of the pn-junction and become energy donors;

(*d*) p-type semiconductors have holes or vacancies left by free electrons as their major current carriers that transport energy toward the pn-junction and become acceptors;

(*e*) The energy required to cause carriers to become mobile must be great enough to overcome the gap (forbidden) between the valence band and the conduction band;

(*f*) The temperature characteristics of conductivity in p-type materials are a function of the amount of donor electrons as governed by the ability of the impurities to make them available relative to the intrinsic electron flow through the forbidden band;

(*g*) A potential energy difference develops between the p- and n-regions as the electrons and holes are swept away to be replaced by their ionization to become donors and acceptors immediately at the junction as shown in Fig. 547;

(*h*) The movement of energy is not direct but subject to diffusion and drift causing a "slowing down" (through the element) process, which is an important reaction within the transport system;

(*i*) This slowing down process is a phenomenon helpful in thermoelectric elements in reducing heat flow between junctions and in maintaining electrical resistance at a sufficiently low level to preclude internal heat generation that would reduce cooling effect;

(*j*) Only certain crystalline structural materials allow the proper balance of electrical-thermal energy conversion and transport to occur;

(*k*) Thus the transformation of energies as progress against the electrical potential continues through the element from the cold junction to the hot junction;

(*l*) As this transformation occurs, the temperature increases gradually along each increment of distance between junctions (gradient) until it reaches the hot junction where it is rejected;

(*m*) As heat is rejected, i.e., absorbed by an external cooling medium, such as water or air, the temperature of the old junction is depressed, causing heat to flow into the element at that point to continue the process;

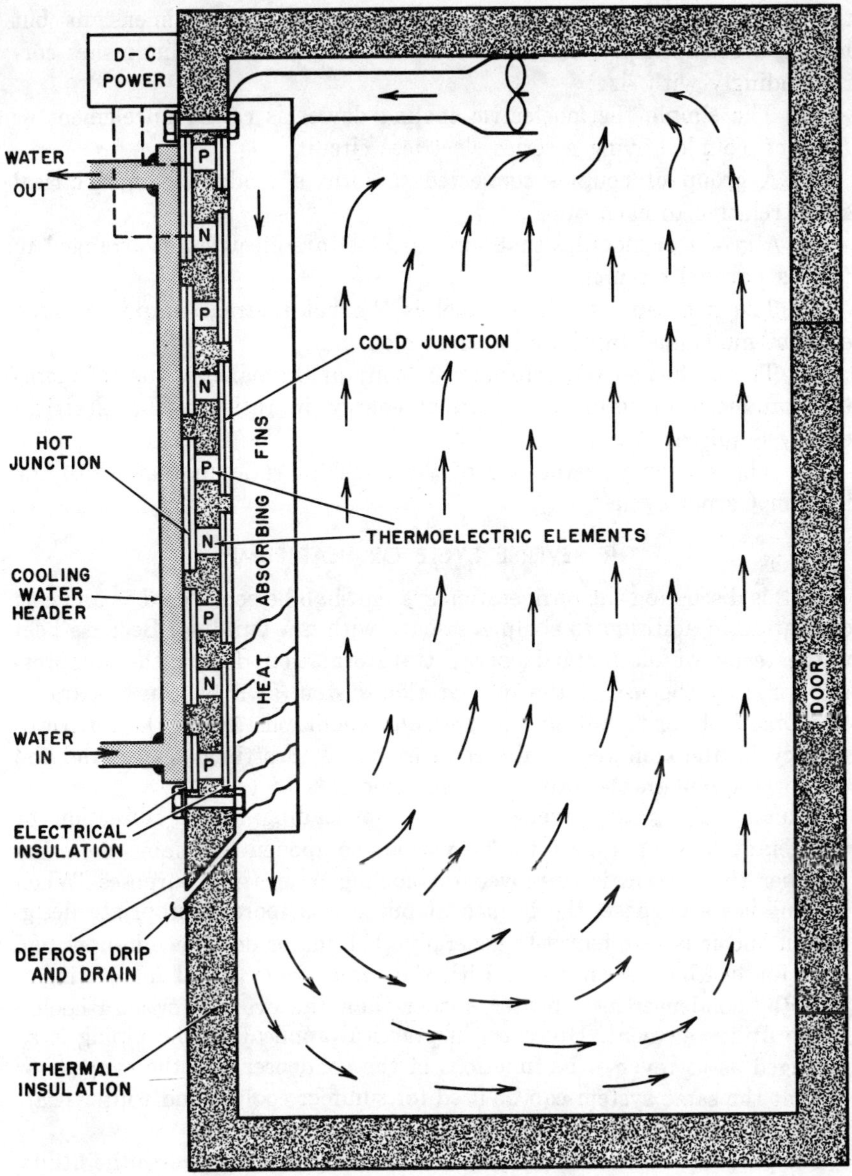

Fig. 547. Thermo-electric refrigerator.

(*n*) With greater impressed current, while maintaining balanced flow through the element, a greater temperature differential can be attained;

(*o*) The thermoelectric capabilities of the semiconductor material govern the thermal behavior (internal heat generation);

(*p*) Because of the compounded increase in electrical current, a system made up of a number of elements of small dimension is more practical than one made of an element of considerably larger dimensions but having a theoretically similar thermal capacity, i.e., cop diminishes correspondingly with size;

(*q*) The simple thermoelectric cooling device is called an element or (thermo)couple having a series electrical circuit;

(*r*) A group of couples connected to form a module is in electrical series relative to each other;

(*s*) A group of modules that form a panel are electrically arranged to form a parallel circuit;

(*t*) The net capacity of a panel is the net thermal output of each element multiplied by the number of elements;

(*u*) The coefficient of performance (cop) of the panel is the refrigeration produced in terms of electrical energy in ratio to the electrical energy required;

(*v*) The efficiency is the cop of the panel in ratio to the cop of the ideal or Carnot cycle.

THE REVERSE CYCLE OR HEAT PUMP

In this discussion the refrigeration system shall be considered as a means of heating in addition to cooling a space within a building. Because heat is the result of mechanical energy transformation during the compression process, the use of the refrigeration system as a heating apparatus is a practical consideration under certain conditions. As in the refrigeration cycle, the compressor, the condenser, the throttling device, and the evaporator remain the principal components.

Actually, the term reverse cycle is misleading as the direction of refrigerant flow in respect to the various components remains the same whether the system is employed for cooling or heating purposes. When heating is its purpose, the term heat pump is a more appropriate designation and it is so referred to generally. The major difference in a system used for heating as compared with the same system used for cooling is that the condenser as a heater, rather than the evaporator as a cooler, is the utility element. However, in practical application the piping is so arranged as to reverse the functions of the condenser and the evaporator so that the same system can be used for summer cooling and winter heating merely by closing certain valves and opening others.

For purposes of discussion, it will be assumed that to reverse the utility function of a system it is necessary merely to turn it around so that the condenser will always remain a condenser and the evaporator will continue to fulfill its primary purpose.

As a heat pump, the evaporator is located out-of-doors, or at least isolated from the heated space, while all other components, including the condenser, are installed indoors. The water or air used as the condenser cooling medium is also the heating medium. Therefore, the con-

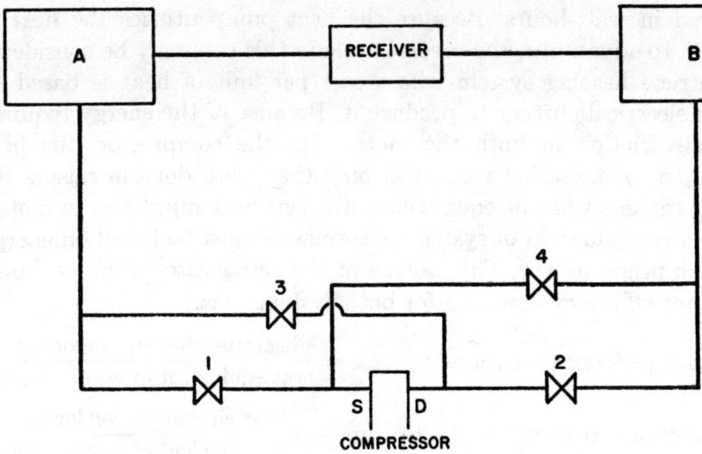

Fig. 548. Reverse cycle or heat pump. When *A* is evaporator, *B* is condenser (heater), so that valves 3 and 4 are closed and valves 1 and 2 are open. When *B* is evaporator, *A* is condenser, so that valves 1 and 2 are closed and valves 3 and 4 are open.

denser cooling medium is maintained in constant circulation, during which it absorbs heat from the refrigerant while passing through the condenser and ejects heat to the air within the building as it passes through suitable radiators.

The liquid refrigerant leaving the condenser is expanded in the evaporator where it absorbs heat from the surrounding outside atmosphere. This heat is conveyed by the refrigerant now in vaporous state to the compressor, where it receives additional heat as a result of the mechanical work done in raising its pressure to that in the condenser. It is the heat absorbed in the evaporator, plus heat of compression, that becomes available in the condenser for heating purposes.

A given system for the same range of temperatures has a higher coefficient of performance when operating as a heat pump than as a refrigerator. This is due to the availability of the heat of compression for heating purposes within the condenser. In the refrigerator, the heat of compression becomes a "necessary evil" to be disposed of together with the heat of vaporization. In the reverse cycle, it becomes an additional source of energy utilized to good advantage by assisting to heat the building and, therefore, compensating to a large extent for the cost to produce it.

In either system, it is the power required to operate the compressor that governs the cost of maintaining a desired condition. If driven by a motor, the cost is computed on the basis of electrical energy consumed as measured in watt-hours. Because the heat pump utilizes the heat of compression to advantage, the reverse cycle in this case may be considered as an electrical heating system whose cost per unit of heat is based on the cost of electrical current to produce it. Because of the energy required to overcome friction in both the motor and the compressor, the heat gained by the condenser as a result of only that work done in raising the pressure of the gas will not equal the equivalent heat input to the motor. Therefore, any evaluation of system performance must be based on energy input to the prime mover. This is true of the refrigerator also, as shown in the Carnot efficiency equation for both systems, viz.:

$$\text{Coefficient of performance for a refrigerator} = \frac{\text{heat absorbed in evaporator}}{\text{heat equivalent of energy input}}$$

$$\text{Coefficient of performance for a heat pump} = \frac{\text{heat ejected by condenser}}{\text{heat equivalent of energy input}}$$

As the coefficient of performance of both systems is based on energy input to the prime mover, and because the heat pump gains the advantage of that energy in *heating effect*, whereas the refrigerator does not gain in refrigerating effect because of it, it follows that the cop for a heat pump must always be larger than the cop of a refrigerator by 1.

EXAMPLE: An ammonia system operating under standard ton conditions, 5 F evaporator and 86 F condenser temperatures, according to the Mollier chart, has a refrigerating effect of 447.4 Btu per lb and 99.4 Btu heat of compression.

For the refrigeration cycle:

$$\text{Coefficient of performance} = \frac{\text{refrigerating effect}}{\text{heat of compression}} = \frac{474.4}{99.4} = 4.77$$

For the heating cycle:

$$\text{Coefficient of performance} = \frac{\text{refrigerating effect} + \text{heat of compression}}{\text{heat of compression}}$$

$$= \frac{474.4 + 99.4}{99.4} = 5.77$$

The R-12 system under standard ton conditions would have a cop for the refrigeration cycle of 4.69, and for the heating cycle 5.69, as determined by the same method used for ammonia.

It will be noted that the heat gained in the evaporator (which for ammonia under standard ton conditions is 474.4 Btu per lb) is 4.77 times the equivalent heat value of electrical energy required to produce it, or 99.4 Btu. Considering the power requirements for any auxiliary equipment, such as pumps and fans, the average ratio will be lower than given, depending on the size and the capacity of such units—usually between 3.5 and 4.0.

The employment of the refrigeration system to cool a building in summer and heat it in winter appears to warrant practical consideration, and in many instances has proved to be successful. However, the economy of the reverse cycle is predicated on several factors, including the source of heat for the evaporator and power costs.

Where outside winter temperatures are extremely low, the suction pressure and the temperature must be lower to cause evaporation. Low suction pressures are followed by relatively lower condenser load capacity, so that, at a time when a greater amount of heat is required, there is less available. This means that the utilization of outside air alone as a source of heat to the evaporator is not always practical. If available in sufficient quantities, well water provides an excellent source of heat. Other heat sources include city water, surface water, waste water, and the earth itself. The water from rivers, lakes, and ponds also provides a source of heat. However, there are limiting factors attached to all sources, the most important of which involve economics and/or continuous availability, as shown in Table 165.

A combination of those sources mentioned above may also be used, either simultaneously or individually when a thermal advantage can be exploited by such an arrangement. Some consideration should be given to waste heat from industrial processes that might otherwise be lost.

Heat from the condenser to be utilized for heating purposes may be transmitted to the space to be heated by water or air. In effect, an air-cooled or water-cooled condenser becomes a heater. Where the system receives heat from air and dissipates it by means of air, it is known as an air-to-air system. Other combinations include water-to-water, air-to-water, and water-to-air systems, named in the order of source-to-sink.

Solar systems, utilizing heat from the sun's rays, may be made more positive if some method of heat storage is provided. An effective arrangement is one in which a large body of suitable substance, such as water, is heated by means of sun rays concentrated through special glass to raise the temperature of the mass. The water may also be heated by other means (electricity, steam, etc.), and the stored heat used to supplement natural heat sources when the latter diminishes to a point below demand requirements. The storage of sensible heat requires a large volume of holding material, so that such systems are quite bulky. To overcome the problem of bulkiness, storage systems utilizing latent heat of fusion or vaporization of various substances have been tried with more or less success.

One of the major uses for the heat pump is in space air conditioning where heat is required during the winter and refrigeration is needed in the summer. Where the reverse cycle is to be employed, as in air conditioning, the system must be designed to allow for the convenient conversion of the evaporator to a heater and the condenser to an evaporator, or return both components to their original function, as circumstances demand. One arrangement for permitting conversion is the system shown in Fig. 548.

TABLE 165. HEAT PUMP HEAT SOURCES*

HEAT SOURCE	AIR	CITY WATER	WELL WATER	SURFACE WATER	WASTE WATER	EARTH	SOLAR
Source Classification	Primary	Primary or auxiliary	Primary	Primary	Primary or auxiliary	Primary or auxiliary	Auxiliary
Suitability as Heat Sink	Good	Good	Good	Good	Variable with source	Usually poor	None
Availability (Location)	Universal	Cities	Uncertain	Rare	Limited	Extensive	Universal
Availability (Time)	Continuous	Continuous—except local shortages	Continuous—unless well runs dry	Continuous	Variable	Continuous, temperature level drops from a maximum to a minimum as heat is removed, slowly rises when pump is inoperative	Intermittent. Unpredictable, except over extended time
Expense (Original)	Low, less than earth and water sources except city	Usually lowest	High, result of drilling well	Low	Variable	High, usually less than the cost of drilling a well	Unexplored
Expense (Operating)	Relatively low	High, usually prohibitive	Low to moderate	Relatively low	Relatively low	Relatively moderate	Unexplored. Promising as auxiliary for reducing operating cost
Temperature (Level)	Favorable 75-95% of time in most of U. S.	Usually satisfactory	Satisfactory	Satisfactory	Usually good	Initially good—drops with time and rate of heat withdrawal	Excellent
Temperature (Variation)	Extreme	Variable with location (10 to 25 F Deg)	Small	Moderate	Usually moderate	Large—less than for air, however	Extreme
Design Information	Inadequate	Usually adequate	Usually adequate	Usually adequate	Adequate if source is constant in supply and temperature	Inadequate	Practically available
Size of Equipment	Particularly bulky	Moderate	Moderate (except well)	Moderate	Variable (usually moderate)	Moderate (except ground coils)	Probably bulky
Adaptability to Mass Production	Excellent, can be factory assembled and tested	Excellent	Excellent, (except well)	Excellent	Poor	Fair	Fair
Sources it may Augment		Air, earth				Air, water	Air, earth, water
Special Problems	Least heat available when demand greatest. Coils may become frosted requiring extra capacity, alternate source, or standby heat. May require duct work. Variable air temperature makes control difficult.	Scale on coils. Local use restrictions during shortages. Disposal. Water temperature may become too low to permit further heat removal.	Corrosion, scale may form on heat transfer surface. Disposal. Water location, temperature, composition usually unknown until well drilled. Well may run dry.	Water may cause scale, corrosion, and algae fouling.	Usually scale forming or corrosive. Often insufficient supply. Very limited application, hence requires individual design.	Limited by local geology and climate. Installation costs difficult to estimate. Requires considerable ground area, may damage lawns, gardens.	Probably will require heat storage equipment at either evaporator or condenser side.

* Reprinted by permission from "Heating Ventilating Air Conditioning Guide 1956," Chapter 37.

In those localities having average winter temperatures between 40 F and 50 F, and where it provides summer cooling, the reverse cycle will succeed in meeting the occasional drop in temperature. Therefore, some advantage is gained in its use, either through utilization of natural heat or when supplemented by other heat sources.

HEAT PUMP SYSTEMS

Air and water are primary sources of heat for heat pump systems. They also provide a suitable sink for the rejection of heat. The conditioning of space involves one or both, depending on the type of air conditioning system used. During the summer, the system absorbs heat from the indoor air and rejects it to some external source. The process is reversed for heating during the winter when the inside air becomes the sink for heat generated by some external means.

The reverse cycle which produces the versatility of the heat pump furnishes both the source and sink during the cooling and heating process. The procedure follows the pattern described earlier that

1. The heat source during the winter is the prime mover-compressor-condenser package. The heat sink is the space to be heated. The condenser cooling medium and subsequently the space heating medium may be air or water. An additional major source of heat is necessary here; that required to vaporize the refrigerant. The latter must be provided from a source external to the space to be heated. An outdoor evaporator is necessary for the latter.

2. The heat source during the summer is the space to be cooled. The sink is external to the cooled space, usually the outdoor atmosphere or water supplied from various sources (see Table 165). The conveyance for the rejected heat is the condenser of the refrigeration system that serves the cooling process. Also, the heat from the cooled space may be absorbed by direct expansion or by means of a secondary medium, usually water (see Chapter 13).

Using the atmosphere as a source and sink for absorbing and rejecting heat requires an outdoor unit, of which there are two basic types: (1) air cooled condenser or *dry* type, and (2) evaporative condenser or *wet* type. Both types are used in the summer as heat sinks.

If water is the medium to be cooled or heated, the heat pump system requires only two components: (1) the outdoor condenser (dry or wet), and (2) the compressor package. The compressor package consists of a chiller and condenser to supply cold or hot water as required.

If air is the medium to be cooled or heated, the heat pump system requires three components, which include the outside unit (dry or wet), and the compressor package without the chiller and condenser mounted on it. The third component is a coil in the indoor air unit, which replaces the mounted chiller-condenser, and cools or heats the air as required.

When cooling and heating are required at the same time (simultaneously), additional components must necessarily be added to the compressor package or indoor air unit, which will cool or heat the conditioned air as required.

The *compressor package* consists of the compressor, drive motor and controls, accumulator (if used), refrigerant pump, oil return unit, controls, necessary structural supports, appropriate utilities (electrical), piping, valves, etc., and chiller-condenser *for air-to-water and air-to-water simultaneous systems only, or,* a condenser (without the chiller) *for air-to-water simultaneous systems only.*

Figure 554 shows the relationship of the various components of various heat pump packages, all of which are designed as split systems utilizing the atmosphere as their heat source and sink, whichever is required by the immediate process.

The various combinations that may be created with the split system, using air as the source in all instances are summarized in Table 166.

TABLE 166. AVAILABLE HEAT PUMP SYSTEMS

Type of Source	*Cooling Heating Medium*	*Type Outside Unit*	*Simultaneous Cooling and Heating S**	*System Type*
Air	Water	Dry	—	Air-to-water dry
Air	Water	Wet	—	Air-to-water wet
Air	Air	Dry	—	Air-to-air dry
Air	Air	Wet	—	Air-to-air wet
Air	Water	Dry	S	Air-to-water dry simultaneous
Air	Water	Wet	S	Air-to-water wet simultaneous
Air	Air	Dry	S	Air-to-air dry simultaneous
Air	Air	Wet	S	Air-to-air wet simultaneous

* Simultaneous systems provide cooling and heating at the same time.

The various types of heat pump arrangements are shown in Figs. 549–554.

SUMMARY OF CHAPTER 17

In addition to the conventional type of compression-condensing-evaporating refrigeration process, there are a number of other temperature reducing and/or state transforming systems, including

Steam jet chillers
Absorption process
Cryogenic systems
Cold-gas refrigerator (Stirling cycle)

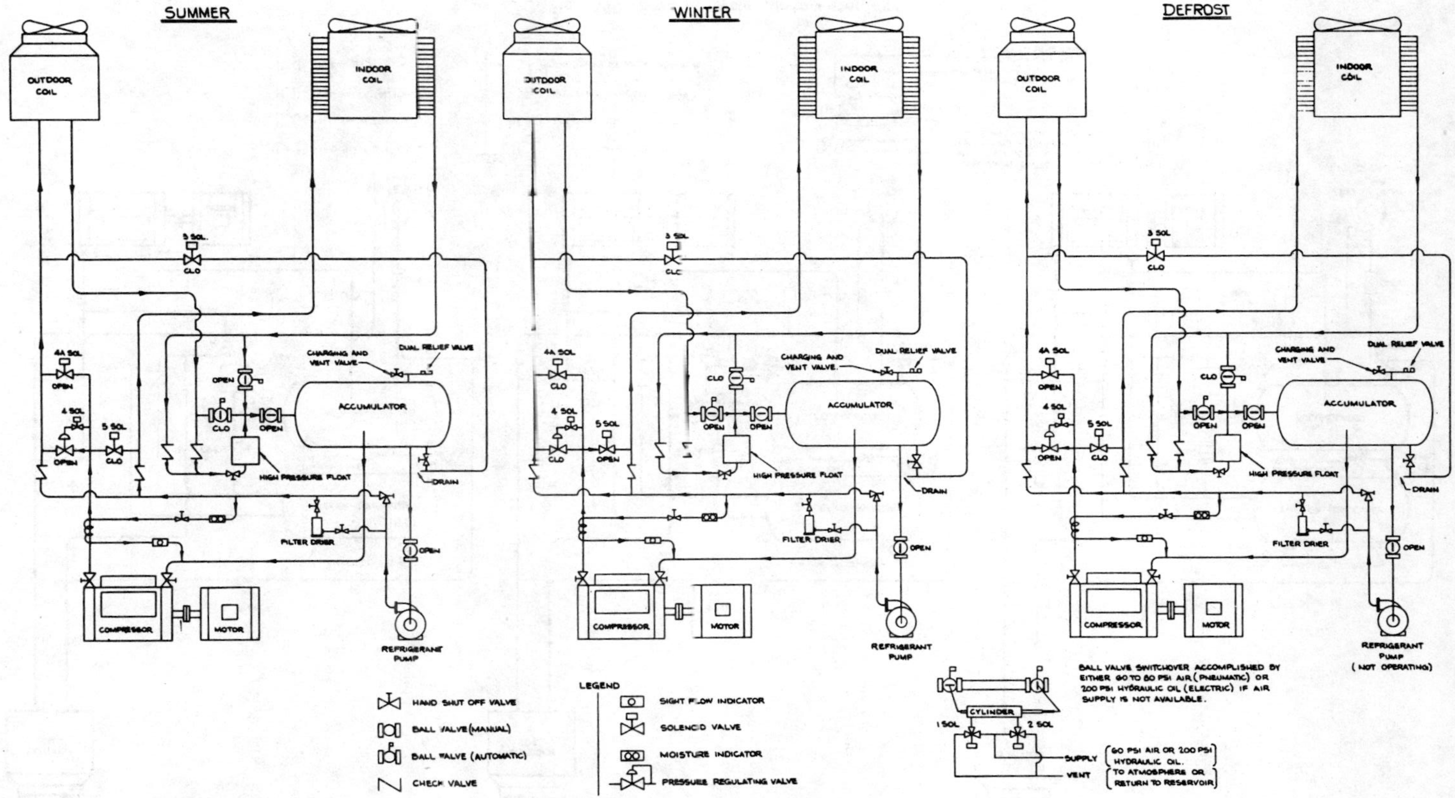

Fig. 549. Reverse cycle air-to-air schematic.

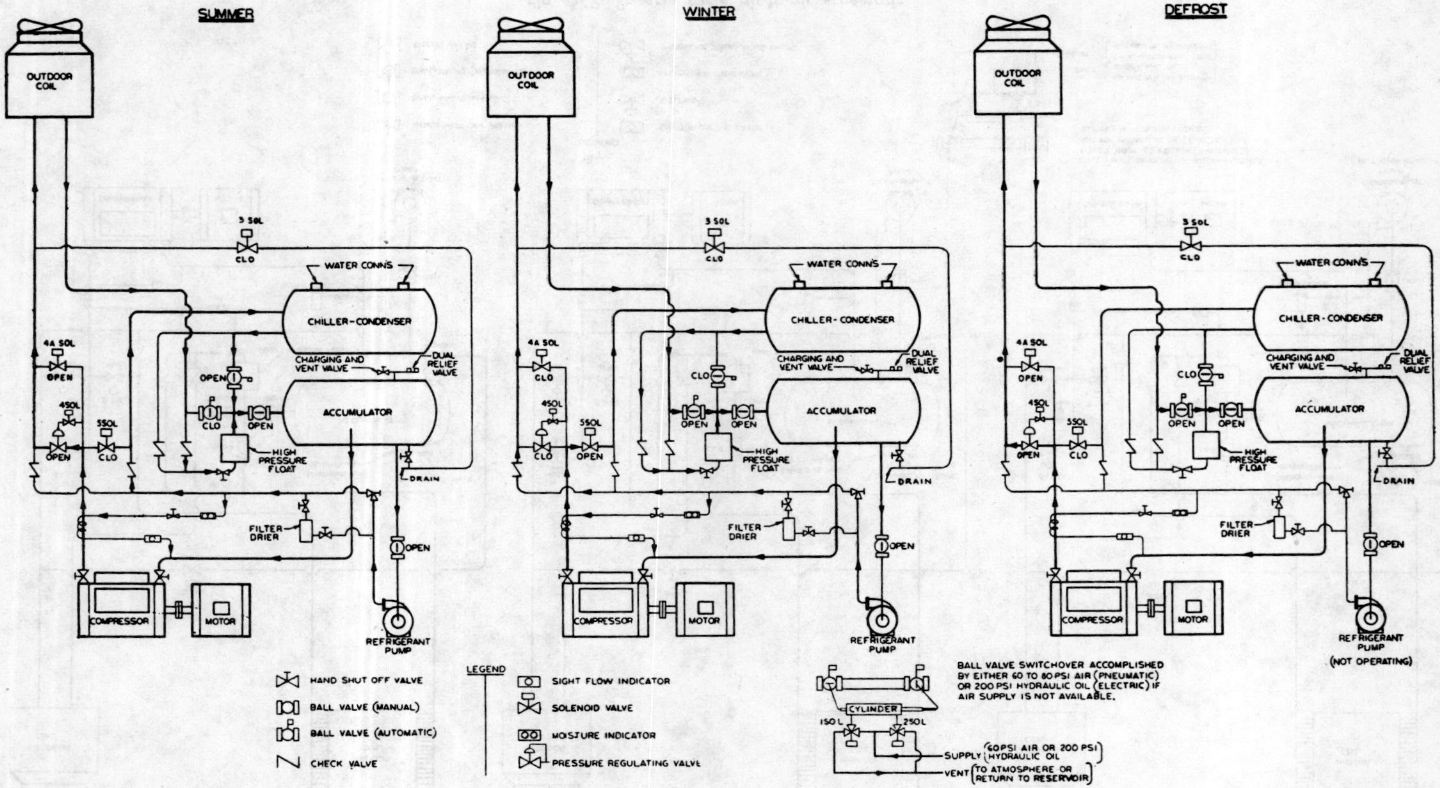

Fig. 550. Reverse cycle air-to-water dry.

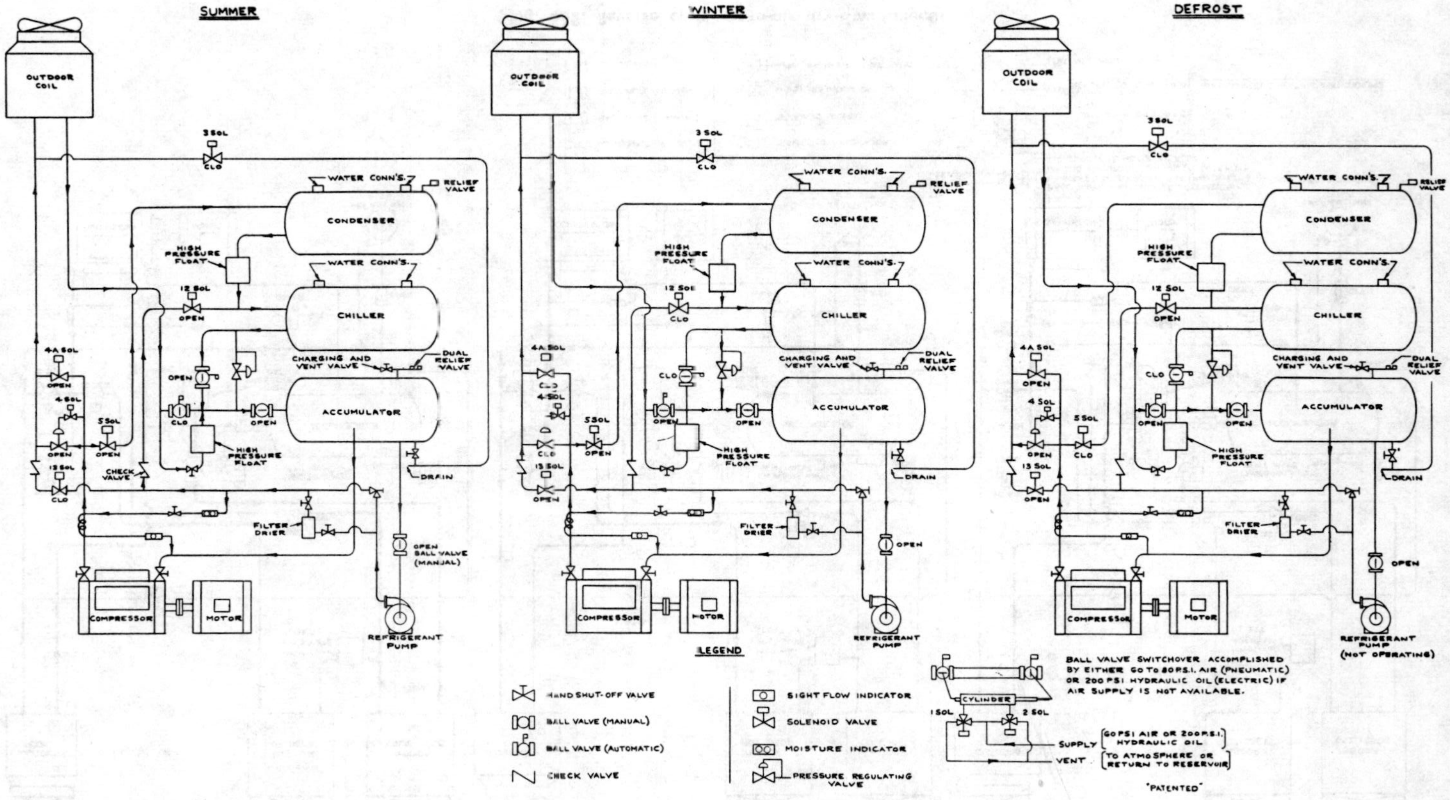

Fig. 551. Reverse cycle air-to-water wet-simultaneous.

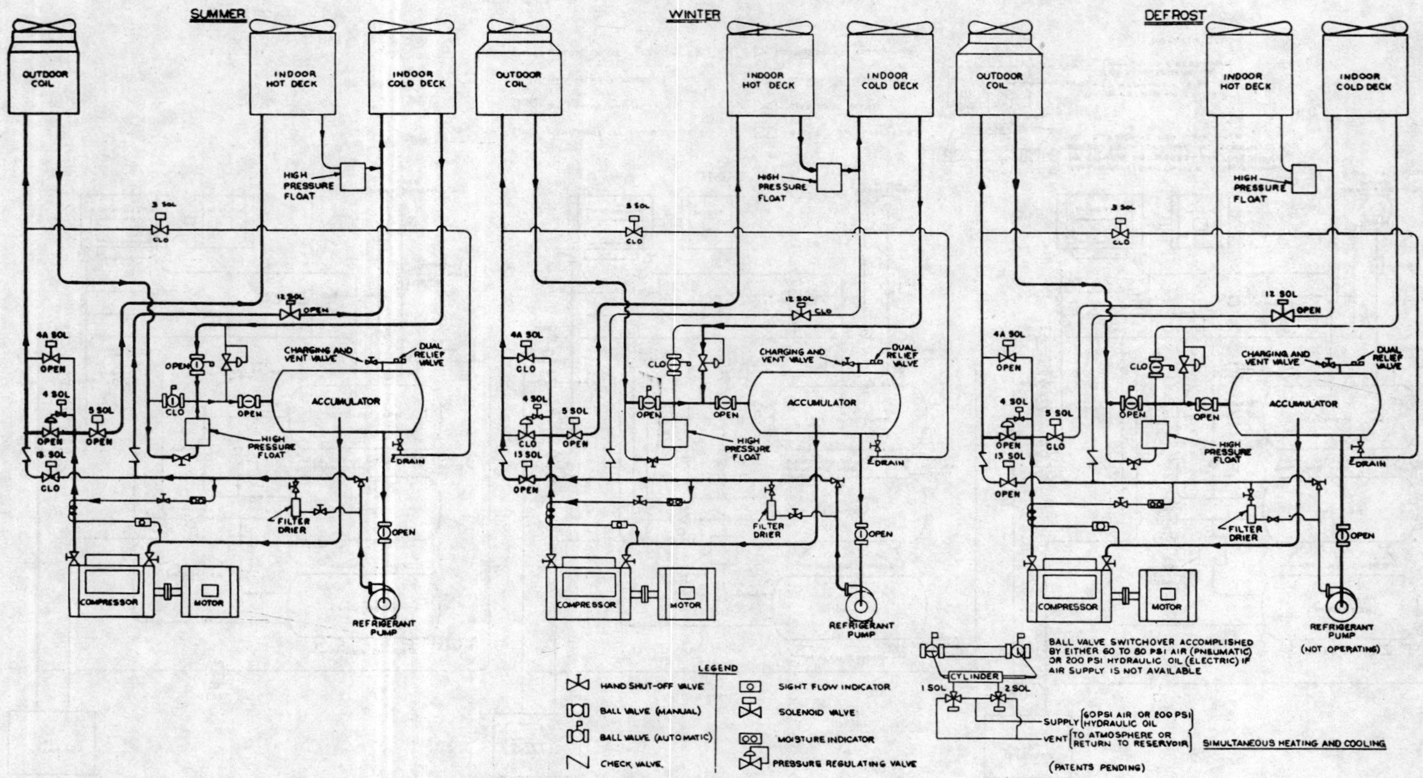

Fig. 552. Reverse cycle air-to-air dry-simultaneous.

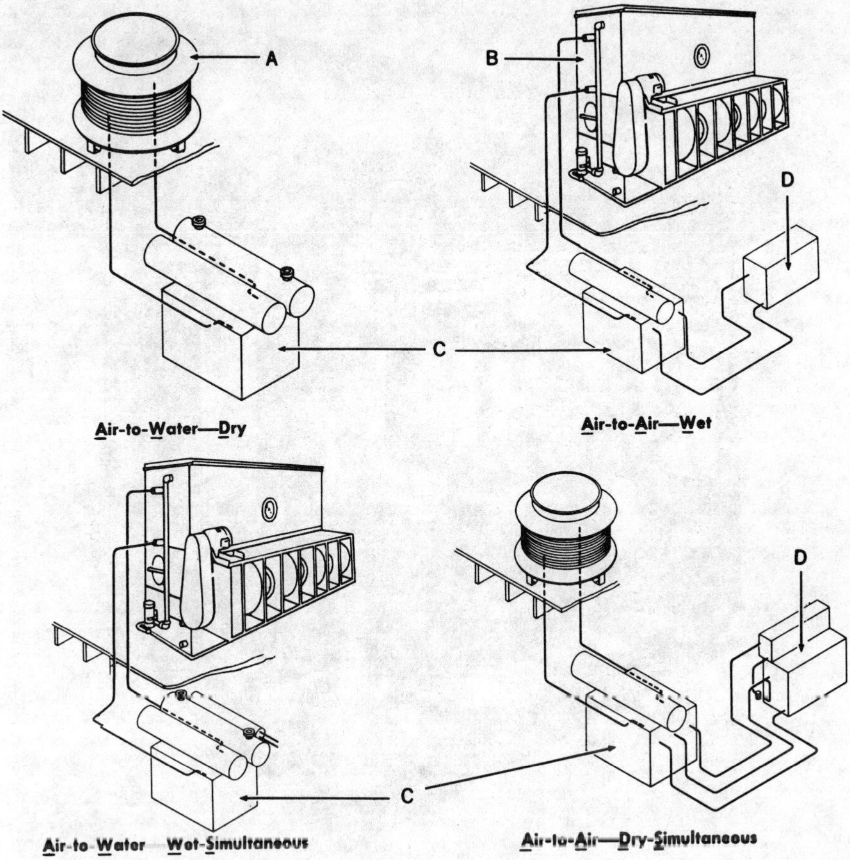

Fig. 553. Reverse cycle.

- Cryogenic liquid evaporators
- Magnetic refrigeration
- Air cycle cooling
- Vortex-tube refrigerator
- Thermoelectric cooling

Steam jet refrigeration is a method of chilling substances by their evaporation under vacuum. A partial vacuum is created when a high-velocity fluid such as steam, air, or water is directed across the mouth of a tight compartment. The vacuum created lowers the boiling point of any liquid contained within the compartment, causing evaporation to occur to reduce its temperature.

The steam jet system consists of steam nozzles, booster, condenser, evaporator or chill compartment, and necessary pumps. It is also equipped with an air injector on the condenser.

Capacity control of the steam jet refrigeration system is accomplished by reducing or increasing the number of boosters in operation.

Fig. 554. Compressor package for heat pump. (Courtesy, Worthington Air Conditioning Company.)

Absorption refrigeration systems employ at least two substances, one of which is the refrigerant, the other the absorbent. The refrigerant vapor is absorbed by an absorbent in an absorber, and driven from the mixture at a higher pressure by heating it in a generator. The pressure is raised either by a pump, as in large systems, or by the existence of an inert gas in the evaporator side, obeying Dalton's law of partial pressure.

The absorption system consists of liquid receiver, expansion valve, evaporator, absorber, strong liquor pump, generator, analyzer, rectifier, condenser, and pressure-reducing valve in the weak liquor line, between generator and absorber. Condensing, expansion, and evaporation are carried on as in any other refrigeration system.

The absorbent most generally used in industrial systems is lithium bromide.

The refrigerants most generally used are water (with lithium bromide as the absorber) and ammonia (with water as the absorber).

An inert gas, such as hydrogen, is used in some systems to equalize the pressure on both the high and low pressure sides without interfering with the evaporation of the refrigerant. The inert gas at high pressure eliminates the use of a pump or compressor.

Cryogenics is the branch of physics that deals with the production and use of temperatures in the range from −240 F down to levels approaching absolute zero.

Cryogenes are fluids whose normal boiling point temperatures at atmospheric pressure are below −240 F.

Fluids categorized as cryogenes are helium, hydrogen, neon, nitrogen, air, fluorine, argon, oxygen, methane and krypton.

Quasi-cryogenes are fluids whose boiling points at atmospheric pressure are above −240 F, but below −100 F.

Fluids categorized as quasi-cryogenes include carbon dioxide, Refrigerants 13, 14, Genetron 23/13 (R-503), nitrous oxide, ethane and ethylene.

Cryogenic processes produce low temperature liquids and gases for various purposes, ranging from facility of storage and/or transportation of cryogenes to the quick freezing of such substances as foods, chemicals, pharmaceuticals, etc. Liquid oxygen occupies approximately 1/800th the space of gaseous oxygen under standard atmospheric conditions.

Cryogenic processes include the separation of gases, such as oxygen and nitrogen contained in the air, and the liquefaction (condensing) of such gases.

Separation of gases cryogenically involves the employment of special refrigeration processes and rectifiers or bubble towers.

Liquefaction of gases cryogenically involves the employment of high pressure compressors (multi-stage), heat exchangers, regenerators (in some instances), expanding engines, isothermal expansion (J-T) valves, purifiers, and liquid receivers or storage vessels.

Helium will not solidify at atmospheric pressure regardless of the temperature.

Helium exists in two phases, He I and He II, and is one of the most difficult gases to liquefy. It contains the isotopes He^3 and He^4, the latter being the most common and the one ordinarily referred to as liquid helium.

Normal hydrogen is a mixture of two forms of molecules, ortho- and para-hydrogen (symbolized o-H_2 and p-H_2, respectively), whose proportions define equilibrium hydrogen, e-H_2.

Hydrogen tends toward maximum p-H_2 as its temperature approaches boiling point.

Heat of conversion is given off during the change from o-H_2 to p-H_2 and will cause boil-off if not completed before storage of liquid hydrogen.

A catalyst is employed to accelerate the molecular conversion within liquid hydrogen during the liquefying process.

Fluorine is highly toxic and corrosive to some engineering materials. It is the most chemically reactive of the known elements.

Methane is a hydrocarbon fluid and a member of a chemical group called homologs.

Homologs or homologous compounds are members of a particular series of compounds having corresponding constant gradient atomic structures, e.g., CH_4, C_2H_6, C_3H_8, C_4H_{10}, etc.

The typical Joule-Thompson refrigerator is considered an ideal but impractical process for liquefying cryogenic gases. Being impractical for liquefaction purposes, it is purely a refrigerator.

The basic Joule-Thompson process becomes practical for most cryogenic fluids when augmented by additional mechanical and thermal reinforcement.

Energy reinforcement of the basic cryogenic liquefaction process includes pre-cooling systems, intrinsic heat exchangers or regenerators, expansion engines, dual expansion, dual compression and inter-cooling.

The more difficult cryogenic fluids to liquefy, such as helium, hydrogen and neon, require inter-cooling in order to reduce their temperature below their maximum inversion temperature.

Inter-cooling and/or pre-cooling may be accomplished by using a separate refrigeration system, usually a cascade arrangement, or another cryogenic liquid, such as hydrogen.

Maximum inversion temperature is the point at which the temperature of a fluid is transformed from an ascending to a descending value during isenthalpic expansion.

Liquefaction of cryogenic gases is performed by several processes, all of which are modifications (compounded) of the basic (J-T) system. They include the cascade, Hampson, Linde, Claude, Linde dual-pressure, Kapitsa, Heylandt and Claude dual-pressure systems.

Compound liquefaction systems are defined as those systems employing energy processes in addition to those required in the basic system.

Deuterium is an isotope of hydrogen and is used to produce heavy water. It is removed as D_2 by liquefying the mixture of H_2 and hydrogen deuteride (HD), the concentrated fluid being subjected to catalytic action and finally separated and liquefied in a rectifier.

Regeneration is an alternating heat exchanger within which the direction of hot and cold fluids may be reversed to act simultaneously as a purifier.

The Joule-Thompson expansion valve is one in which expansion of the fluid occurs isothermally.

Cryogenic vessels are constructed to provide maximum insulation and are usually equipped with pressure-relief controls.

Liquid cryogenes, vacuum, vacuum powder using various heat reflecting materials are used as insulation media in cryogenic storage, etc.

The Dewar flask is represented by the simple thermos bottle.

Vehicles, including motor trucks, highway trailers, railroad cars (reefers) and shipping containers, as well as wharf lockers and ship space, are cooled with appropriately located coils in which cryogenic liquids are allowed to evaporate.

Containerization is a name given to a modern method of packing various kinds of commodities for both domestic and foreign shipment to reduce loss by damage, deterioration, pilfering, etc. Such containers carrying perishable items are refrigerated with cryogenic liquid evaporators supplied directly from a cryogenic bottle of suitable size.

Rapid freezing of foods and other commodities is accomplished by spraying them with a cryogenic liquid, such as nitrogen. Pre-cooling may be accomplished by exposure to the resultant vapor.

The cold-gas refrigerator produces cryogenic effects.

The Stirling cycle is a thermodynamic system involving two constant volume and two isothermal processes to produce refrigeration (or heating).

The cold-gas refrigerator is based on the Stirling cycle and basically consists of an expansion chamber, compression chamber, cooler, condenser and regenerator.

The regenerator connects the compression and expansion chamber and is fitted with a wire mesh for alternately absorbing and rejecting heat from cyclic gas as it passes back and forth between chambers.

Magnetic refrigeration or adiabatic demagnetization is a process in which paramagnetic salts undergo a temperature reduction upon being demagnetized, during which the heat of magnetization is externally rejected.

Temperatures in the range of 2 R (initial) to below .01 R (final) are involved in the magnetic process using conventional paramagnetic materials. About 0.7 R is the low limit of temperature that can be attained with liquid helium (He^3).

Ultra-low temperatures approaching zero degree absolute may be obtained through nuclear demagnetization, consisting of two stages using ordinary paramagnetic salts to reduce temperature in the first stage to .01 R, and a nuclear paramagnetic material, such as copper, to reach a final temperature of about 0.0004 R in the second stage.

Paramagnetic materials include chromium potassium alum, cerium magnesium nitrate, copper potassium sulfate, etc.

Air cycle refrigeration employs air as the refrigerant and is based largely on the compression-expansion process described for cryogenic liquefaction.

Air cycle refrigeration is employed in aircraft space cooling and for other purposes where the load demands are small.

Some of the more important types of air cycle systems are the basic, evaporative, regenerative, bootstrap, dual, and certain combinations of the foregoing. All are modifications of the basic type.

The air used in the air cycle system remains a gas throughout the process and is discharged to the atmosphere when the cycle is completed.

The vortex-tube refrigerator employs air at high pressure and high velocity which expands during vortical flow within a tube to undergo a simultaneous increase and decrease in temperature.

The air enters the vortex tube tangentially and moves circumferentially toward the center, during which a part of its energy is transformed to thermal energy.

The transformation of energy causes a part of the mass to become warmer, while the part from which the energy is derived becomes colder, developing a temperature gradient radially with T_{cold} at the center, T_{hot} at the periphery.

Cold air leaves the tube through a center orifice at the end close to the vortex, while warm air exits at the opposite open end remote from the vortex.

The vortex refrigerator is employed where the load demand is small.

Thermoelectric refrigeration is the cooling effect developed at one junction of a thermocouple, made up of two dissimilar materials, when the other junction is exposed to a higher temperature, and an electric current is applied to the circuit. It is called a Peltier device and the resultant cooling is known as the Peltier effect.

The thermoelectric generator develops an emf when the opposite junctions of two dissimilar materials are exposed to a temperature differential. It is called a Seebeck device or a thermocouple and the resultant emf is known as the Seebeck effect.

The dissimilar materials or arms are semiconductors, one of which is referred to as n-type (negative) and the other as p-type (positive) material. The junction of the two materials is called the pn-junction.

Semiconductors are crystalline materials that possess specific energy conducting characteristics whose conducting capabilities are not those of a true conductor (metals), and whose insulating capabilities are not those of a true insulator (rubber).

Semiconductors made up of pure crystals are called intrinsic materials and those crystals having combined impurities are called extrinsic materials. Extrinsic materials are richer in free electrons, and, therefore, are more effective conductors.

Electrons are energy carriers that move in a direction from a negative to a positive terminal when an electric current is applied to the lattice structure of crystalline materials.

The vacancies or holes left by mobile electrons are also energy carriers and move in the direction from a positive to a negative terminal when an electric current is applied.

The p-type semiconductors are those materials in which the holes are the major and the electrons the minor current carrier.

The n-type semiconductors are those materials in which the electrons are the major and the holes the minor current carrier.

The flow of both types of carriers establishes a condition in which the electron becomes a donor (n-type) of energy and the hole becomes an acceptor (p-type).

In moving against a potential within the lattice structure of a crystal, work is required that is transformed to thermal energy creating a temperature gradient across the structure. As the thermal energy finally reaches the pn-junction, it is rejected to an external source. This causes the temperature of the opposite junction to become depressed in order to continue its heat absorbing process.

The figure-of-merit is a value indicating the performance of a semiconductor as to its cooling capability.

Thermoelectric refrigerating capacity is a function of thermal conductance, temperature differential, form factor and the figure-of-merit.

The basic thermoelectric element is called a couple or device. A module is made up of any number of couples through which the electric current is in series. A panel is made up of a number of modules which are connected electrically in parallel.

The refrigerating capacity (theoretical) of a module or panel is found by multiplying the refrigerating effect of each couple (Btu/hour) by the number of couples. The actual capacity must account for any losses due to system defects which may run as high as 45%.

The heat pump or reverse cycle is a refrigeration system employed as a heater. The condenser acts as the heater while the evaporator acts as the heat source. Usually the system is designed to cool in summer and heat in winter. To reverse the function of the system from that of a heater to a refrigerator, the evaporator becomes a condenser and the condenser becomes an evaporator. The conversion is made chiefly by a series of stop valves so placed in the lines as to alter the path of the refrigerant to and from the compressor whenever desired.

The heat sources for heat pumps include the atmosphere, well water, waste water, earth, surface water, sun, and industrial processes.

The heat pump, in effect, is an electrical heating system measured according to the power required to drive the compressor and any auxiliary equipment such as fans, pumps, etc.

Heat pumps may be classified according to source and sink. Where the source is air and the sink is also air (inside or atmosphere) heat pumps may be classified as

Air-to-air—wet or dry
Air-to-water—wet or dry
Air-to-air—wet or dry simultaneous
Air-to-water—wet or dry simultaneous

Booster heat is employed to supply additional heat when required during the winter.

STUDY OUTLINE FOR CHAPTER 18

Air:
- Constituents
- Specific heat
- Humidity
- Weight
- Volume
- Volume, increased 1 F per 1 Btu

Humidity:
- Absolute
- Relative

Temperature of air:
- Dewpoint
- Dry-bulb
- Wet-bulb

Effect of additional moisture on dewpoint and dry-bulb temperature

Heat content:
- Sensible heat of air
- Latent heat of water vapor in air
- Total heat of air

Relationship of dewpoint, dry-bulb, and wet-bulb temperatures

Psychrometric laws

Psychrometric chart

Volume determination for any mixture of air and water vapor

Volume of air required to maintain relative humidity—humidifying

Volume of air required to absorb heat load—dehumidifying

Heat loads:
- Sources

Refrigeration load, dehumidifying or cooling

Air-conditioning apparatus

Ducts:
- Entrance losses
- Elbows
- Plenum chambers
- Shapes
- Sizes

Friction in ducts

Proportioning of ducts

Flow of air in ducts:
- Static pressure
- Velocity pressure
- Total pressure

Air purification

Automobile air conditioning

18

Principles of Air Conditioning

The air-conditioning system provides for the control of temperature, humidity, purity, and distribution of air within an enclosed space. The space may be a small compartment, such as an industrial test cabinet, or a large auditorium. The purpose of air conditioning is to furnish a sufficient volume of clean air containing a specific amount of water vapor and at a temperature capable of maintaining predetermined atmospheric conditions within a selected enclosure. Properly conditioned air results in greater comfort to occupants when applied to public gathering places, homes, offices, and factories. Industrially, it has many benefits, including better control of product quality, size, texture, and uniformity. Its beneficial effect on workers leads to increased productive efficiency.

AIR

The atmosphere surrounding the earth and to which all nature is exposed is a mixture of many gases, pure and impure. It also contains solid particles in the form of dust, the extent of which depends on the locality, and, in some instances, the circumstances. High in the mountains or over the oceans, it is expected that the air is relatively free of solid particles and impure gases, resulting from industrial processes. This is quite true except during periods of high winds, when such impurities are carried great distances from their place of origin, which might be from industrial centers, deserts, eroded agricultural areas, or volcanic formations. In city or suburban areas, impurities are always present in the air and in varying amounts, depending on the weather and the extent of industrial activity, the latter also determining the type.

Under normal outside conditions, air may be defined as a mechanical mixture of gases, water vapor, solid impurities, and small amounts of micro-organisms. A typical analysis of city air may be found to contain the following:

Natural gases
- Oxygen
- Nitrogen
- Hydrogen
- Water vapor
- Argon
- Helium
- Neon
- Krypton
- Xenon

Impurities
- Sulfur fumes
- Acid fumes
- Carbon dioxide
- Dust—Lime
  - Soil
  - Fly-ash
  - Other
    - Mineral
    - Vegetable
    - Animal
- Micro-organisms—Bacteria
  - Molds

The proportions of the above constituents will vary according to the location of the city, and, in many cases, with the degree of enforcement of laws governing industrial sanitation, smoke abatement, and fume control.

Pure dry air is composed of the following gases in approximately the amount shown:

	Volume %	*Weight %*
Oxygen........	20.9	23.1
Nitrogen......	78.1	76.0
Argon.........	1.0	0.9

Air will contain varying amounts of water vapor ranging from 0 percent to 4 percent under ordinary atmospheric conditions. Control of water vapor content is one of the functions of air conditioning. Helium, neon, krypton, and xenon are also present in the atmosphere. Their negligible quantity, however, excludes their individual consideration in air-conditioning calculations.

Specific Heat of Air at Constant Pressure

The specific heat of air at constant pressure is not constant at all temperatures. For air-conditioning purposes, an average value of 0.2415 or simply 0.24 Btu per lb per °F may be employed. For more accurate results, the specific heat of air at any given temperature may be determined by the formula:

$$\text{sp ht of air } (C_p) = 0.24112 + 0.000009t$$

where t = temperature of air under consideration

(For the specific heat of saturated air at various temperatures and constant pressure, see Table 168.)

Water Vapor Content of Air—Humidity

Humidity is the term used in reference to water vapor contained in or mixed with the atmosphere. One of the important properties of air is its ability to hold a given amount of water in its gaseous form, called water vapor. The amount of water vapor, usually referred to as *moisture content*, capable of being held by air depends on the temperature of the latter. Water vapor is a natural constituent of the atmosphere, originating as a result of evaporation from the surface of water bodies. The relative dryness of air is not so much dependent on the amount of water vapor it contains as on the relationship of water-vapor content to temperature. Air that is saturated—i.e., containing the highest possible amount of moisture at a given temperature—will become relatively dry when heated to a high temperature. The reverse is also true. The mixture of air and water vapor follows Dalton's law of partial pressures in that the total pressure of the mixture is the sum of the pressure exerted by all the gases that constitute dry air, plus the pressure exerted by the water vapor. (See Chapter 3.)

Weight of Air

The weight of a given volume of air varies with temperature, barometric pressure, and moisture content. Based on a temperature of 70 F and a barometric pressure of 29.92 in. Hg, the weights of dry air, of dry air in 1 cu ft of saturated mixture, and of saturated air are as follows:

Weight of dry air	0.07495 lb per cu ft
Weight of dry air in 1 cu ft of saturated air	0.07310 lb
Weight of saturated mixture	0.07425 lb

The weights of dry and saturated air for other temperatures and 29.92 in. Hg are shown in Tables 167 and 168.

The weight of air varies directly as the barometric pressure and inversely as the absolute temperature. To determine the weight of air for any pressure and temperature, apply the multiple factor equation:

$$W = W_1 \left(\frac{459.6 + t_1}{459.6 + t} \right) \frac{p}{p_1}$$

where W = weight of air at temperature t and pressure p
W_1 = weight of air at temperature t_1 and pressure p_1
p and p_1 = barometric pressure, in. Hg

Volume of Air

Air-conditioning calculations are based on air volume rather than weight. To convert weight to volume, apply the rule that volume in cu ft per lb is the reciprocal of weight (W) in lb per cu ft, or

$$V = \frac{1}{W}$$

Table 167*

PROPERTIES OF DRY AIR
Barometric Pressure 29.92 Inches

Temperature Degrees Fahr	Weight, Pounds per Cu Ft	Volume, Cu Ft per Pound	Btu Absorbed by One Cu Ft Dry Air per Degree Fahr	Cu Ft Dry Air Warmed One Degree per Btu	Temperature Degrees Fahr	Weight, Pounds per Cu Ft	Volume, Cu Ft per Pound	Btu Absorbed by One Cu Ft Dry Air per Degree Fahr	Cu Ft Dry Air Warmed One Degree per Btu
−25	.09134	10.95	.02189	45.68	56	.07698	12.99	.01848	54.13
−20	.09028	11.08	.02164	46.21	57	.07683	13.02	.01844	54.24
−15	.08926	11.20	.02139	46.74	58	.07668	13.04	.01840	54.35
−10	.08827	11.33	.02116	47.27	59	.07653	13.07	.01837	54.44
−5	.08732	11.45	.02093	47.80	60	.07640	13.09	.01834	54.54
0	.08635	11.58	.02070	48.31	61	.07624	13.12	.01830	54.64
5	.08542	11.71	.02048	48.82	62	.07610	13.14	.01826	54.76
10	.08452	11.83	.02026	49.35	63	.07595	13.17	.01823	54.85
15	.08363	11.96	.02005	49.87	64	.07580	13.19	.01819	54.97
20	.08275	12.09	.01984	50.40	65	.07566	13.22	.01816	55.06
21	.08260	12.11	.01980	50.50	66	.07552	13.24	.01812	55.18
22	.08241	12.13	.01976	50.60	67	.07538	13.27	.01809	55.27
23	.08226	12.16	.01972	50.70	68	.07524	13.29	.01806	55.37
24	.08207	12.19	.01968	50.81	69	.07509	13.32	.01802	55.48
25	.08192	12.21	.01964	50.91	70	.07495	13.34	.01799	55.58
26	.08173	12.24	.01959	51.04	71	.07481	13.37	.01795	55.69
27	.08155	12.26	.01955	51.14	72	.07468	13.39	.01792	55.79
28	.08140	12.28	.01951	51.24	73	.07454	13.42	.01789	55.88
29	.08125	12.31	.01948	51.34	74	.07440	13.44	.01786	55.97
30	.08107	12.34	.01944	51.44	75	.07424	13.47	.01783	56.09
31	.08089	12.36	.01940	51.54	76	.07412	13.49	.01780	56.19
32	.08074	12.38	.01937	51.64	77	.07398	13.52	.01776	56.30
33	.08056	12.41	.01933	51.75	78	.07384	13.54	.01773	56.40
34	.08042	12.44	.01929	51.86	79	.07370	13.57	.01770	56.49
35	.08025	12.46	.01925	51.96	80	.07357	13.59	.01767	56.59
36	.08009	12.49	.01921	52.07	81	.07343	13.62	.01764	56.70
37	.07992	12.51	.01917	52.19	82	.07330	13.64	.01761	56.80
38	.07978	12.54	.01913	52.30	83	.07316	13.67	.01758	56.90
39	.07960	12.56	.01909	52.41	84	.07302	13.70	.01755	57.01
40	.07945	12.59	.01906	52.51	85	.07290	13.72	.01752	57.11
41	.07928	12.61	.01902	52.61	86	.07275	13.74	.01748	57.22
42	.07914	12.64	.01898	52.72	87	.07261	13.77	.01744	57.35
43	.07897	12.66	.01894	52.83	88	.07248	13.80	.01741	57.44
44	.07880	12.69	.01890	52.94	89	.07234	13.82	.01738	57.54
45	.07866	12.71	.01886	53.04	90	.07223	13.85	.01735	57.63
46	.07850	12.74	.01882	53.14	91	.07209	13.87	.01732	57.73
47	.07835	12.76	.01878	53.25	92	.07196	13.90	.01729	57.83
48	.07819	12.79	.01874	53.36	93	.07182	13.92	.01726	57.92
49	.07804	12.81	.01871	53.46	94	.07170	13.95	.01723	58.02
50	.07788	12.84	.01867	53.56	95	.07156	13.97	.01720	58.15
51	.07773	12.86	.01864	53.64	96	.07144	14.00	.01717	58.25
52	.07758	12.89	.01860	53.73	97	.07131	14.02	.01714	58.35
53	.07743	12.92	.01857	53.84	98	07120	14.05	.01711	58.45
54	.07728	12.94	.01855	53.94	99	.07106	14.07	.01708	58.55
55	.07713	12.97	.01851	54.05	100	.07093	14.10	.01705	58.65

* Courtesy, Madison, Richard D., ed.: "Fan Engineering—An Engineer's Handbook," ed. 5, pp. 18–19, Buffalo, N. Y., Buffalo Forge Company, 1949. Based on data from "International Critical Tables," published for the National Research Council by the McGraw-Hill Book Co., New York; and on data from Keenan and Keyes' "Steam Tables," published by John Wiley & Sons, New York.

TABLE 167—(*Continued*)

PROPERTIES OF DRY AIR
Barometric Pressure 29.92 Inches

Temperature Degrees Fahr	Weight, Pounds per Cu Ft	Volume, Cu Ft per Pound	Btu Absorbed by One Cu Ft Dry Air per Degree Fahr	Cu Ft Dry Air Warmed One Degree per Btu	Temperature Degrees Fahr	Weight, Pounds per Cu Ft	Volume, Cu Ft per Pound	Btu Absorbed by One Cu Ft Dry Air per Degree Fahr	Cu Ft Dry Air Warmed One Degree per Btu
101	.07080	14.12	.01702	58.74	215	.05884	17.00	.01421	70.34
102	.07067	14.15	.01699	58.84	220	.05840	17.12	.01411	70.87
103	.07056	14.17	.01696	58.93	230	.05756	17.37	.01392	72.89
104	.07042	14.20	.01693	59.03	240	.05674	17.62	.01373	71.88
105	.07031	14.22	.01690	59.12	250	.05594	17.88	.01355	73.87
106	.07017	14.25	.01687	59.20	260	.05517	18.13	.01339	74.84
107	.07006	14.27	.01684	59.30	270	.05442	18.38	.01320	75.76
108	.06993	14.30	.01681	59.40	280	.05367	18.63	.01302	76.79
109	.06982	14.32	.01678	59.49	290	.05296	18.88	.01285	77.81
110	.06968	14.35	.01675	59.58	300	.05226	19.13	.01269	78.80
111	.06957	14.37	.01672	59.69	310	.05158	19.39	.01253	79.81
112	.06944	14.40	.01669	59.85	320	.05092	19.64	.01238	80.77
113	.06933	14.42	.01666	60.02	330	.05028	19.89	.01225	81.63
114	.06920	14.45	.01663	60.13	340	.04965	20.14	.01210	82.62
115	.06908	14.48	.01660	60.24	350	.04904	20.39	.01196	83.61
116	.06896	14.50	.01657	60.35	360	.04844	20.64	.01182	84.60
117	.06884	14.53	.01654	60.46	370	.04785	20.90	.01169	85.55
118	.06872	14.55	.01651	60.57	380	.04728	21.15	.01156	86.51
119	.06860	14.58	.01648	60.67	390	.04672	21.40	.01144	87.41
120	.06849	14.60	.01646	60.75	400	.04618	21.65	.01131	88.42
121	.06837	14.63	.01643	60.87	420	.04513	22.16	01107	90.33
122	.06826	14.65	.01640	60.98	440	.04413	22.66	.01084	92.25
123	.06814	14.68	.01637	61.08	460	.04317	23.16	.01066	93.62
124	.06802	14.70	.01635	61.18	480	.04225	23.67	.01044	95.78
125	.06790	14.72	.01632	61.28	500	.04138	24.17	.01025	97.56
130	.06732	14.86	.01619	61.76	520	.04053	24.68	.01006	99.40
135	.06676	14.98	.01604	62.34	540	.03974	25.18	.00988	101.2
140	.06614	15.11	.01591	62.85	560	.03894	25.68	.00972	102.8
145	.06565	15.23	.01580	63.29	580	.03819	26.19	.00954	104.8
150	.06512	15.36	.01568	63.77	600	.03751	26.69	.00940	106.4
155	.06459	15.48	.01555	64.31	650	.03578	27.95	.00903	110.7
160	.06406	15.61	.01543	64.81	700	.03424	29.21	.00868	115.2
165	.06356	15.74	.01531	65.31	750	.03282	30.47	.00836	119.6
170	.06304	15.86	.01519	65.83	800	.03152	31.73	.00809	123.6
175	.06256	15.99	.01507	66.35	850	.03032	32.99	.00779	128.3
180	.06207	16.11	.01496	66.81	900	.02920	34.24	.00754	132.6
185	.06158	16.24	.01484	67.33	950	.02816	35.51	.00739	135.3
190	.06111	16.37	.01473	67.85	1000	.02720	36.76	.00716	139.6
195	.06065	16.49	.01462	68.25	1050	.02630	38.02	.00689	145.1
200	.06018	16.62	.01452	68.85	1100	.02546	39.28	.00674	148.3
205	.05973	16.74	.01442	69.34	1150	.02467	40.54	.00659	151.8
210	.05928	16.87	.01432	69.83	1200	.02392	41.80	.00643	155.5

Table 168*

PROPERTIES OF SATURATED AIR

Weights of Air, Water Vapor and Saturated Mixture of Air and Water Vapor at Different Temperatures and at 29.92 Inches of Mercury

Temperature Degrees Fahr	Weight of the Vapor		Weight in a Cu Ft of Mixture			Volume-Cu Ft per pound of Dry Air	Btu Absorbed by one Cu Ft Sat Air per Deg F	Cu Ft Sat Air Warmed One Degree per Btu	Specific Heat Btu per Pound of Mixture
	Per Pound of Dry Air	Per Pound of Mixture	Dry Air, Pounds	Vapor, Pounds	Total Weight‡				
—25	.00020	.00020	.09134	.000018	.09136	10.95	.02192	45.60	.2400
—20	.00027	.00027	.09025	.000024	.09027	11.07	.02167	46.15	.2400
—15	.00035	.00035	.08922	.000031	.08925	11.21	.02142	46.69	.2400
—10	.00046	.00046	.08820	.000041	.08824	11.34	.02119	47.20	.2401
— 5	.00061	.00061	.08723	.000053	.08728	11.46	.02096	47.69	.2401
0	.00080	.00080	.08625	.000068	.08632	11.59	.02073	48.24	.2402
5	.00102	.00102	.08529	.000087	.08538	11.72	.02051	48.62	.2402
10	.00130	.00130	.08434	.000110	.08445	11.85	.02029	49.29	.2403
15	.00168	.00168	.08340	.000140	.08354	11.99	.02008	49.80	.2404
20	.00213	.00213	.08247	.000176	.08264	12.12	.01987	50.33	.2404
21	.00225	.00224	.08230	.000185	.08248	12.15	.01983	50.43	.2404
22	.00235	.00234	.08210	.000193	.08229	12.18	.01979	50.54	.2405
23	.00246	.00245	.08193	.000202	.08213	12.20	.01975	50.64	.2405
24	.00260	.00259	.08173	.000213	.08194	12.23	.01971	50.74	.2405
25	.00272	.00271	.08156	.000222	.08178	12.26	.01967	50.84	.2405
26	.00285	.00284	.08136	.000233	.08159	12.29	.01963	50.95	.2406
27	.00300	.00299	.08117	.000243	.08141	12.32	.01959	51.05	.2406
28	.00314	.00313	.08099	.000254	.08124	12.34	.01955	51.16	.2406
29	.00328	.00327	.08083	.000264	.08109	12.37	.01952	51.26	.2407
30	.00345	.00344	.08063	.000277	.08090	12.40	.01947	51.36	.2407
31	.00362	.00361	.08043	.000290	.08072	12.43	.01943	51.47	.2407
32	.00378	.00376	.08025	.000303	.08055	12.46	.01940	51.57	.2408
33	.00393	.00392	.08006	.000315	.08038	12.49	.01935	51.68	.2408
34	.00409	.00408	.07989	.000327	.08022	12.51	.01932	51.76	.2408
35	.00426	.00425	.07970	.000339	.00004	12.54	.01928	51.87	.2409
36	.00444	.00442	.07952	.000353	.07987	12.57	.01924	51.98	.2409
37	.00460	.00458	.07933	.000364	.07969	12.60	.01920	52.09	.2409
38	.00480	.00478	.07916	.000380	.07954	12.63	.01917	52.17	.2410
39	.00499	.00496	.07897	.000394	.07936	12.66	.01913	52.28	.2410
40	.00519	.00516	.07880	.000409	.07921	12.69	.01909	52.39	.2410
41	.00541	.00538	.07860	.000425	.07902	12.72	.01905	52.50	.2411
42	.00561	.00558	.07843	.000440	.07887	12.75	.01902	52.58	.2411
43	.00583	.00579	.07825	.000456	.07871	12.78	.01898	52.69	.2412
44	.00606	.00602	.07805	.000473	.07852	12.81	.01894	52.80	.2412
45	.00630	.00626	.07788	.000491	.07837	12.84	.01891	52.89	.2413
46	.00655	.00651	.07768	.000509	.07819	12.87	.01887	53.00	.2413
47	.00680	.00675	.07750	.000527	.07803	12.90	.01884	53.08	.2414
48	.00705	.00700	.07731	.000545	.07785	12.93	.01880	53.19	.2415
49	.00734	.00728	.07714	.000567	.07771	12.96	.01877	53.31	.2415
50	.00762	.00756	.07694	.000587	.07753	12.99	.01873	53.42	.2416
51	.00792	.00786	.07676	.000608	.07737	13.02	.01869	53.51	.2416
52	.00823	.00819	.07657	.000632	.07720	13.06	.01866	53.60	.2417
53	.00854	.00845	.07637	.000651	.07702	13.09	.01862	53.71	.2418
54	.00884	.00877	.07620	.000675	.07687	13.12	.01859	53.80	.2419
55	.00921	.00913	.07600	.000700	.07670	13.15	.01856	53.88	.2420

‡Reciprocal gives volume in cu ft per lb of mixture.

* Courtesy, Madison, Richard D., ed.: "Fan Engineering—An Engineer's Handbook," ed. 5, pp. 20–22, Buffalo, N. Y., Buffalo Forge Company, 1949. Based on data from "International Critical Tables," published for the National Research Council by the McGraw-Hill Book Co., New York; and on data from Keenan and Keyes' "Steam Tables," published by John Wiley & Sons, New York.

TABLE 168—(*Continued*)

PROPERTIES OF SATURATED AIR
Weights of Air, Water Vapor and Saturated Mixture of Air and Water Vapor at Different Temperatures and at 29.92 Inches of Mercury

Temperature Degrees Fahr	Weight of the Vapor		Weight in a Cu Ft of Mixture			Volume-Cu Ft per pound of Dry Air	Btu Absorbed by one Cu Ft Sat Air per Deg F	Cu Ft Sat Air Warmed One Degree per Btu	Specific Heat Btu per Pound of Mixture
	Per Pound of Dry Air	Per Pound of Mixture	Dry Air, Pounds	Vapor, Pounds	Total Weight‡				
56	.00952	.00943	.07582	.000723	.07654	13.19	.01853	53.97	.2421
57	.00989	.00980	.07562	.000749	.07637	13.22	.01849	54.09	.2421
58	.01026	.01016	.07544	.000775	.07622	13.25	.01845	54.20	.2421
59	.01063	.01052	.07524	.000801	.07604	13.29	.01842	54.30	.2422
60	.01103	.01091	.07506	.000829	.07589	13.32	.01839	54.38	.2423
61	.01143	.01130	.07486	.000857	.07572	13.35	.01836	54.47	.2425
62	.01185	.01171	.07468	.000886	.07557	13.39	.01833	54.56	.2426
63	.01229	.01214	.07447	.000916	.07539	13.42	.01830	54.65	.2427
64	.01273	.01257	.07429	.000947	.07524	13.46	.01826	54.77	.2427
65	.01320	.01303	.07408	.000979	.07506	13.49	.01822	54.89	.2428
66	.01368	.01349	.07390	.001012	.07491	13.53	.01819	54.98	.2429
67	.01417	.01397	.07369	.001045	.07473	13.57	.01816	55.07	.2430
68	.01468	.01447	.07350	.001080	.07458	13.60	.01813	55.16	.2431
69	.01520	.01497	.07330	.001115	.07441	13.64	.01810	55.25	.2432
70	.01576	.01551	.07310	.001152	.07425	13.68	.01807	55.34	.2434
71	.01630	.01604	.07290	.001189	.07409	13.71	.01804	55.44	.2435
72	.01691	.01662	.07270	.001229	.07393	13.75	.01801	55.53	.2436
73	.01748	.01717	.07250	.001268	.07377	13.79	.01798	55.62	.2437
74	.01812	.01780	.07229	.001310	.07360	13.83	.01795	55.71	.2439
75	.01876	.01841	.07208	.001352	.07343	13.87	.01792	55.80	.2440
76	.01941	.01904	.07188	.001395	.07328	13.91	.01789	55.90	.2442
77	.02008	.01968	.07166	.001439	.07310	13.95	.01786	56.00	.2443
78	.02079	.02036	.07144	.001485	.07293	13.99	.01783	56.09	.2445
79	.02150	.02106	.07124	.001532	.07277	14.03	.01780	56.18	.2446
80	.02223	.02174	.07104	.001579	.07262	14.08	.01777	56.28	.2447
81	.02301	.02249	.07081	.001629	.07244	14.12	.01773	56.40	.2448
82	.02380	.02325	.07059	.001680	.07227	14.16	.01770	56.50	.2449
83	.02462	.02403	.07038	.001733	.07211	14.21	.01768	56.57	.2452
84	.02545	.02482	.07015	.001785	.07193	14.26	.01765	56.66	.2454
85	.02631	.02566	.06993	.001840	.07177	14.30	.01762	56.76	.2455
86	.02723	.02651	.06970	.001898	.07160	14.34	.01759	56.85	.2457
87	.02813	.02736	.06947	.001954	.07142	14.39	.01756	56.95	.2459
88	.02908	.02826	.06925	.002014	.07126	14.44	.01754	57.02	.2461
89	.03002	.02915	.06902	.002072	.07109	14.48	.01751	57.12	.2463
90	.03109	.03015	.06880	.002139	.07094	14.53	.01748	57.21	.2464
91	.03211	.03111	.06855	.002201	.07075	14.58	.01745	57.31	.2466
92	.03318	.03212	.06832	.002267	.07058	14.63	.01743	57.38	.2469
93	.03428	.03314	.06809	.002334	.07042	14.69	.01740	57.48	.2471
94	.03543	.03422	.06785	.002404	.07025	14.73	.01738	57.54	.2474
95	.03660	.03531	.06760	.002474	.07007	14.79	.01735	57.64	.2476
96	.03780	.03642	.06736	.002546	.06991	14.84	.01732	57.74	.2478
97	.03904	.03757	.06711	.002620	.06973	14.90	.01730	57.81	.2481
98	.04025	.03870	.06688	.002692	.06957	14.95	.01728	57.88	.2484
99	.04159	.03993	.06660	.002770	.06931	15.01	.01725	57.98	.2487
100	.04300	.04124	.06634	.002853	.06919	15.07	.01723	58.04	.2489

‡ Reciprocal gives volume in cu ft per lb of mixture.

TABLE 168—(*Continued*)

PROPERTIES OF SATURATED AIR
Weights of Air, Water Vapor and Saturated Mixture of Air and Water Vapor at Different Temperatures and at 29.92 Inches of Mercury

Temperature Degrees Fahr	Weight of the Vapor		Weight in a Cu Ft of Mixture			Volume—Cu Ft per pound of Dry Air	Btu Absorbed by one Cu Ft Sat Air per Deg F	Cu Ft Sat Air Warmed One Degree per Btu	Specific Heat Btu per Pound of Mixture
	Per Pound of Dry Air	Per Pound of Mixture	Dry Air, Pounds	Vapor, Pounds	Total Weight‡				
101	.04443	.04255	.06610	.002937	.06904	15.12	.01720	58.14	.2491
102	.04586	.04385	.06583	.003019	.06885	15.18	.01717	58.24	.2494
103	.04737	.04523	.06557	.003106	.06868	15.25	.01715	58.31	.2497
104	.04890	.04662	.06530	.003193	.06849	15.31	.01713	58.38	.2501
105	.05048	.04806	.06504	.003283	.06832	15.37	.01711	58.45	.2504
106	.05212	.04953	.06477	.003375	.06814	15.44	.01709	58.52	.2508
107	.05379	.05105	.06451	.003470	.06798	15.50	.01707	58.58	.2511
108	.05556	.05264	.06421	.003568	.06778	15.57	.01705	58.65	.2515
109	.05734	.05422	.06394	.003666	.06761	15.64	.01703	58.72	.2519
110	.05917	.05587	.06364	.003766	.06741	15.71	.01700	58.83	.2522
111	.06111	.05760	.06336	.003872	.06723	15.78	.01698	58.90	.2526
112	.06308	.05934	.06306	.003978	.06704	15.85	.01696	58.96	.2530
113	.06507	.06110	.06278	.004085	.06686	15.93	.01694	59.03	.2534
114	.06722	.06299	.06247	.004199	.06667	16.00	.01692	59.10	.2538
115	.06935	.06486	.06216	.004311	.06647	16.08	.01690	59.18	.2542
116	.07157	.06678	.06186	.004427	.06629	16.16	.01688	59.25	.2547
117	.07390	.06882	.06154	.004548	.06609	16.24	.01686	59.32	.2552
118	.07625	.07084	.06124	.004669	.06591	16.32	.01685	59.35	.2557
119	.07869	.07296	.06092	.004794	.06571	16.41	.01683	59.42	.2561
120	.08121	.07511	.06060	.004921	.06552	16.50	.01681	59.49	.2566
121	.08376	.07729	.06027	.005049	.06532	16.58	.01679	59.56	.2571
122	.08646	.07958	.05995	.005183	.06513	16.68	.01678	59.60	.2576
123	.08925	.08194	.05960	.005319	.06492	16.77	.01676	59.67	.2582
124	.09204	.08428	.05927	.005456	.06473	16.87	.01674	59.74	.2587
125	.09502	.08677	.05892	.005598	.06452	16.96	.01673	59.78	.2593
130	.11125	.10010	.05713	.006355	.06349	17.49	.01666	60.03	.2625
135	.13026	.11523	.05524	.007195	.06244	18.10	.01660	60.24	.2659
140	.15280	.13255	.05319	.008128	.06132	18.79	.01656	60.39	.2701
145	.17966	.15230	.05100	.009162	.06016	19.60	.01653	60.54	.2748
150	.21178	.17478	.04865	.010303	.05895	20.55	.01652	60.64	.2802
155	.25038	.20022	.04612	.011547	.05767	21.67	.01652	60.68	.2865
160	.29810	.22962	.04340	.012937	.05634	23.03	.01653	60.61	.2934
165	.35660	.26285	.04048	.014436	.05492	24.69	.01657	60.54	.3017
170	.43168	.30150	.03734	.016118	.05346	26.77	.01665	60.32	.3115
175	.52750	.34530	.03398	.017926	.05191	29.43	.01673	59.99	.3223
180	.65580	.39525	.03035	.019905	.05036	32.94	.01685	59.70	.3346
185	.83410	.45425	.02645	.022062	.04851	37.78	.01703	59.28	.3501
190	1.0948	.52270	.02228	.024393	.04667	44.85	.01719	58.72	.3684
195	1.5153	.60240	.01779	.026957	.04475	56.20	.01744	59.04	.3898
200	2.2923	.69660	.01297	.029730	.04270	77.11	.01770	57.21	.4145
205	4.1838	.80500	.00782	.032715	.04064	127.9	.01798	56.47	.4425
210	15.493	.93700	.00232	.035942	.03836	431.0	.01836	55.47	.4787
212	Inf.	1.0000	.00000	.037298	.03730	—	.01850	55.07	.4960

‡ Reciprocal gives volume in cu ft per lb of mixture.

For the three conditions given above for 70 F air at 29.92 in. Hg, then

$$V = \frac{1}{0.07495} = 13.34 \text{ cu ft per lb, dry air}$$

$$= \frac{1}{0.07310} = 13.68 \text{ cu ft of dry air in 1 lb of saturated mixture}$$

$$= \frac{1}{0.07425} = 13.47 \text{ cu ft per lb saturated mixture}$$

For volume of air, cu ft per lb at any temperature, see Tables 167 and 168.

Volume of Air Increased 1 F per 1 Btu

The capacity of air-conditioning systems is based on the volume of air required to maintain desired conditions under predetermined heat loads. To simplify calculations, a factor representing the volume of air in cu ft and 29.92 in. Hg, capable of absorbing 1 Btu per 1 F temperature rise, is employed. The factor, K, is determined by the equation:

$$K = \frac{1}{c_p \times W}$$

or

$$K = \frac{V}{c_p}$$

where K = cu ft of air warmed 1 F per 1 Btu
c_p = specific heat of air mixture at temperature t
W = lb per cu ft of mixture at temperature t
V = cu ft per lb of air at temperature t

For 70 F air, $K = \frac{13.34}{0.24} = 55.58$ cu ft of dry air

For 70 F saturated air, $K = \frac{13.47}{0.2434} = 55.34$ cu ft of saturated air.

The value of K will range from 45.6 cu ft for -25 F saturated air to 60.68 cu ft for 155 F saturated air. For purposes of air-conditioning calculations where temperatures are maintained within the human comfort zone, the value $K = 56$ may be employed with reasonable accuracy. For exact values at various conditions and temperatures, see Tables 167 and 168.

ABSOLUTE HUMIDITY

The moisture or water vapor mixed with air is referred to as humidity. The actual weight of water vapor contained in a given volume of air is the absolute humidity. Weight of water vapor is expressed in grains or pounds per cu ft of air. The numerical value of absolute humidity alone does not indicate relative dryness or moistness of air. Temperature must also be known.

RELATIVE HUMIDITY (RH)

It was stated previously that air is capable of holding moisture in an amount depending on its temperature. If the moisture content of a definite volume of air is such that precipitation would occur if an additional amount were added at constant temperature and pressure, the air is said to be saturated. When air is saturated, it has reached its maximum limit of absolute humidity for the existing temperature. Air containing its maximum moisture capacity is said to be saturated and to have a relative humidity (RH) of 100 percent. Air that contains 50 percent of its maximum capacity to hold moisture at the existing temperature has an RH of 50 percent.

Relative humidity may be defined as the ratio of the absolute humidity at the existing temperature to the absolute humidity corresponding to its maximum moisture capacity at the same temperature. It is expressed as percent and determined by the formula

$$\text{Relative humidity (RH)} = \frac{\text{absolute humidity at given temperature}}{\text{saturation capacity at given temperature}} \times 100$$

EXAMPLE: Determine the relative humidity of 70 F air containing 4 grains of moisture per cu ft. (The maximum moisture capacity of 70 F air is 8 grains per cu ft.)

$$\text{RH} = \frac{\text{4 grains per cu ft}}{\text{8 grains per cu ft}} \times 100 = 50 \text{ percent}$$

Air at 80 F will hold 11 grains of moisture per cu ft. Therefore, if the air in the above example were raised to 80 F without an increase in moisture content, the final RH would be:

$$\text{RH} = \tfrac{4}{11} \times 100 = 36.36 \text{ percent, or about 36 percent}$$

By adding 2 grains of moisture to the air at 80 F, then the RH would be increased, viz.:

$$\text{RH} = \frac{4+2}{11} \times 100 = 54.54 \text{ percent, or about 55 percent}$$

If 5 more grains of moisture were added to each cu ft of the 80 F air, then it would become saturated as shown:

$$\text{RH} = \frac{6+5}{11} \times 100 = 100 \text{ percent}$$

If the air when it contained 4 grains of moisture per cu ft were reduced to 50 F, it would become saturated because the moisture capacity at 50 percent is 4 grains per cu ft. Therefore, it would have an RH of 100 percent at 50 F.

Relative humidity may be defined as the ratio of the weight of moisture in a given volume of air at a given temperature to the weight of moisture necessary to saturate it at the same temperature. The RH of a given volume of air may be increased by: (*a*) reducing temperature when moisture content remains constant; (*b*) increasing moisture content while temperature remains constant.

The RH may be decreased by: (*a*) raising temperature while moisture content remains constant; (*b*) decreasing moisture content while maintaining constant temperature (as with a desiccant).

A more accurate definition of relative humidity, that would encompass all variables, is based on partial pressure ratios existing between the air-water vapor mixture and water vapor for the two states; i.e., actual and saturated conditions. The equation based on partial pressures is:

$$\text{RH} = \frac{\text{vapor pressure of the mixture}}{\text{vapor pressure of the saturated mixture at the existing temperature}}$$

or

$$\text{RH} = \frac{P_m}{P_s}$$

DEWPOINT TEMPERATURE (DP)

The dewpoint temperature is the measure of moisture content of the air. It is the temperature at which air becomes saturated if cooled to that level without the addition or subtraction of moisture. The dewpoint of air cannot be changed without a change in moisture content regardless of the amount of heat added or subtracted above the dewpoint level. If air is cooled to a temperature below the existing dewpoint, condensation occurs and a new dewpoint is established. The level of the new dewpoint will depend on the amount of heat extracted from that which was contained in the saturated air at the original dewpoint temperature.

The dewpoint may be reduced by decreasing the water vapor content of a given weight of air. It may be increased by adding water vapor to air having a relative humidity below 100 percent; i.e., when in an unsaturated condition.

The saturation curve shown in Fig. 555 is constructed along increment state points of a mixture of air and water vapor. As the temperature changes, so does the ability of air to hold water vapor change. If 70 F saturated air is cooled, say, 5 F, precipitation occurs. The loss of water vapor will correspond to the 5 F drop, but the air will still be saturated at 65 F. The dewpoint is decreased from 70 F to 65 F during the cooling process. If cooled to 60 F, more water vapor will precipitate and the air will still be saturated at the lower temperature; i.e., its new dewpoint temperature will be 60 F. A state of saturation will continue as long as the temperature is reduced below the saturation temperature. Each new

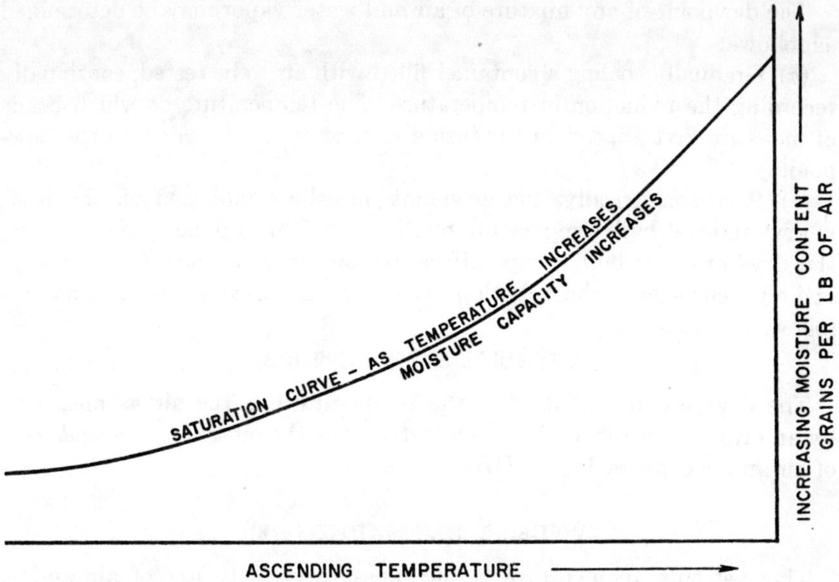

Fig. 555. Saturation curve, air and water vapor mixture.

saturation temperature will be the dewpoint of the mixture. This means that some water vapor will always remain in the air in an amount depending on the temperature to which it is finally cooled. If the final temperature is 40 F, then 40 F is the dewpoint of the mixture. The state points, represented by each 1 F reduction in temperature, would develop a path indicated by the saturation line (Fig. 555) in a downward direction. Adding heat to the mixture, as by an electric heater, will raise its temperature above 40 F as measured by an ordinary thermometer, but the dewpoint will remain constant provided that water vapor is not added during the heating process. When the mixture is finally returned to its original 70 F, the water vapor content will remain the same as it was at 40 F. Therefore, the dewpoint of the 70 F air is still 40 F and the water vapor content corresponds to that of the 40 F saturated mixture.

If, on the other hand, the 40 F saturated air were heated to 70 F by passing it through a spray of warm water, evaporation would occur and it would remain saturated. At 60 F, it would contain the same water vapor content as it did when it passed through the 60 F level during the cooling process. The same condition would be true for each corresponding increment temperature change. Therefore, the state points during the heating process would be coincident with the state points developed when the mixture was being cooled, but in an upward direction along the saturation line (Fig. 555).

The dewpoint of any mixture of air and water vapor may be determined as follows:

(*a*) Gradually cooling a container filled with air to be tested, continually recording the reduction in temperature. The temperature at which beads of moisture first appear on the inside surface of the container is the dewpoint.

(*b*) Psychrometrically, the dewpoint may be established on the basis of dry- and wet-bulb temperature relationship. This is done by determining the dry- and wet-bulb temperatures by means of a sling psychrometer and a psychrometric chart, both of which are discussed in the sections that follow.

DRY-BULB TEMPERATURE (DB)

The dry-bulb temperature is the temperature of the air as measured by an ordinary or dry-bulb thermometer. It is the measure of *sensible heat* of air and is expressed as F, DB.

WET-BULB TEMPERATURE (WB)

The wet-bulb temperature is the measure of *total heat* of air and is expressed as F, WB. It is determined by covering the bulb of a dry-bulb thermometer with a water-moistened wick and passing air rapidly (about 1500 fpm) over the wick. By whirling the wet-bulb thermometer in the air, the moisture contained in the wick tends to evaporate. The cooling effect of evaporation lowers the temperature as registered by the dry-bulb to a point corresponding to the rate at which evaporation occurs. If the surrounding air is dry (i.e., has a low RH), evaporation will be rapid and the drop in the temperature as registered by the wet-bulb thermometer will be appreciable. If the surrounding air is moisture laden (i.e., has a high RH), evaporation will be slow and consequently the difference between wet- and dry-bulb temperatures will be relatively small. Evaporation will not occur if the air is saturated; i.e., has a 100 percent RH. It will occur when the RH is below 100 percent. The difference between wet- and dry-bulb temperatures is called the wet-bulb depression.

The heat necessary to cause evaporation in the manner described above is present in the air as sensible heat, as measured by the dry-bulb thermometer. During the process of evaporation, the sensible heat is converted to latent heat of vaporization. However, the total heat of the air remains the same and, therefore, the wet-bulb temperature is constant. Because the conversion of sensible to latent heat is accomplished without a change in total heat, an adiabatic change occurs. If evaporation continues under the described circumstances until a state of saturation exists (i.e., the dry-bulb temperature is reduced to that of the existing wet-bulb temperature), the condition will be one approaching adiabatic saturation.

The wet-bulb temperature may be variously defined, then, as (*a*) the measure of total heat of the air and (*b*) the temperature at which air would become saturated if moisture were added without an increase or a decrease in heat content.

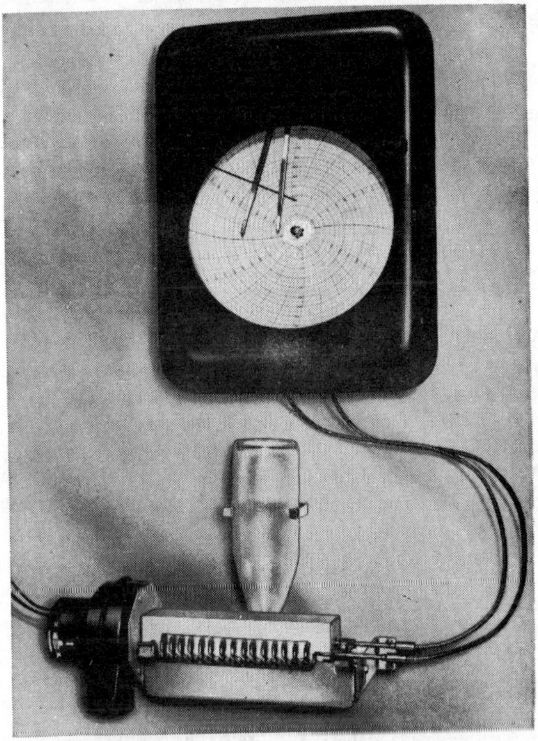

Fig. 556. Recording and controlling psychrometer. Air passes over wet- and dry-bulb in lower receptacle. Wet-bulb wick kept moist by bottle feed. Instrument shown records both temperatures. (Courtesy, United States Gauge, Division of American Machine and Metals, Inc.)

Under saturated conditions, dewpoint, dry-bulb, and wet-bulb temperatures of the air are the same. During the process of adding moisture to the air under adiabatic conditions (i.e., constant wet-bulb temperature), the dry-bulb temperature will decrease 8.5 F for each grain of moisture absorbed.

EFFECT OF ADDING MOISTURE TO AIR ON DEWPOINT AND DRY-BULB TEMPERATURE

When moisture is added to an unsaturated mixture of air and moisture, the dewpoint increases according to the amount added. For 70 F, DB air, the dewpoint corresponding to various amounts of moisture per cubic foot of air will be approximately as shown in Table 169.

TABLE 169. RELATIONSHIP OF DEWPOINT, MOISTURE CONTENT, AND RELATIVE HUMIDITY OF 70 F AIR

Dewpoint, F	*Grains of Moisture per cu ft of 70 F Air*	*Relative Humidity, %*
50	4	50
55	5	60
60	5.7	72
65	6.8	83
70	8	100

From Table 169, it will be noted that the DP and the RH increase with any increase in moisture content of air.

SENSIBLE HEAT OF AIR

The heat contained in 1 lb of dry air above 0 F is sensible heat and is determined by the dry-bulb thermometer and the application of the equation

$$h = C_p (t_2 - t_1)$$

where h = Btu per lb of sensible heat
C_p = specific heat of air at constant pressure
t_1 = initial dry-bulb temperature, 0 F
t_2 = final dry-bulb temperature

LATENT HEAT OF WATER VAPOR

The heat necessary to convert water from its liquid form to water vapor at constant pressure is the latent heat of steam or water vapor. The latent heat contained in the air corresponds to the weight of water vapor present. The latent heat of vaporization for water at 70 F is 1052.7 Btu per lb. For other temperatures between 40 F and 150 F, the latent heat may be determined by subtracting the product of the existing temperature × the average change in latent heat per 1 F from the latent heat of steam at 0 F or 1091.6 Btu per lb, or

$$L = 1091.6 - 0.55t$$

where L = latent heat of vaporization at the existing wet-bulb temperature, Btu per lb

Heat of vaporization required to evaporate water is taken from the air, causing a drop in dry-bulb temperature during the evaporation process.

TOTAL HEAT OF AIR

The total heat of air is the sum of the sensible heat and the latent heat of water vapor contained in the air.

$$H = h + L$$

where H = total heat of air, Btu per lb
h = sensible heat of air at dry-bulb temperature
L = latent heat of vaporization, Btu per lb of water vapor present

The total heat of the air is measured by the wet-bulb temperature. The wet-bulb temperature of the air remains constant if the total heat of the air is constant.

The total heat of the air is determined by the equation

$$H = C_p \times t + L \times W$$

where H = total heat of air above 0 F, Btu per lb
C_p = specific heat of air at temperature t
t = wet-bulb temperature of the air
L = latent heat of water vapor at wet-bulb temperature t
W = weight of water vapor, lb per lb of dry air in mixture

EXAMPLE: Compute the total heat above 0 F (t_1) contained in 1 lb of 70 F saturated air.

Data:
Weight of air, 1 lb
Volume of air at 70 F (V), 13.68 cu ft per lb (from psychrometric chart or Table 168)
Specific heat of 70 F air (c_p), 0.2415, average
Latent heat of water vapor, 1091.6 Btu per lb at 0 F (from Steam Tables)
Temperature of saturated air (t_2), 70 F
Weight of water vapor (w), 0.001152 lb per cu ft of moisture at 70 F (from Table 168)

1st step: Total weight, W, of water vapor:

$$W = V \times w$$

2nd step: Sensible heat, h:

$$h = c_p(t_2 - t_1)$$

3rd step: Latent heat, L:

$$L = W(1091.6 - 0.55t)$$

4th step: Total heat, H:

$$H = h + L$$
$$H = c_p(t_2 - t_1) + W(1091.6 - 0.55t)$$
$$H = 0.2415(70 - 0) + (13.68 \times 0.001152)\,[1091.6 - (0.55 \times 70)]$$
$$H = 33.5 \text{ Btu per lb, total heat in 1 lb of saturated 70 F air}$$

RELATIONSHIP OF DEWPOINT, DRY-BULB, AND WET-BULB TEMPERATURES

The curve in Fig. 557 shows the relation between the dewpoint, the dry-bulb, and the wet-bulb temperatures of air passing through a water spray during which the total heat remains constant. The curve portrays a

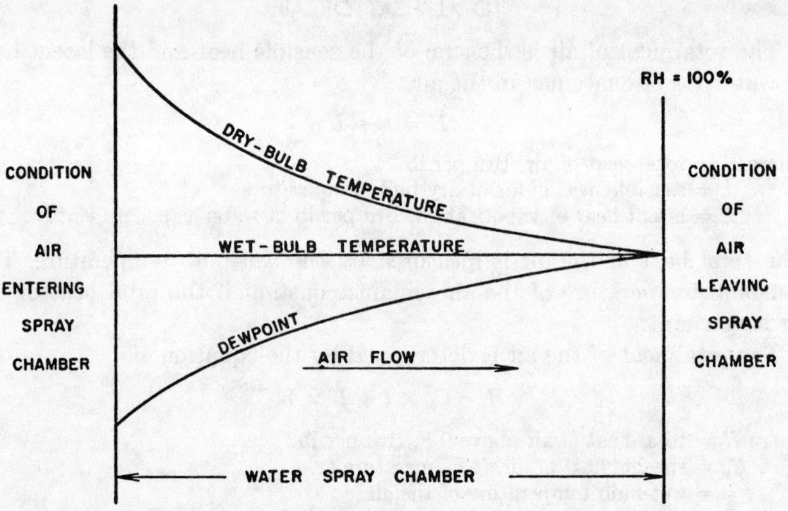

Fig. 557. Relation of dewpoint, dry-bulb and wet-bulb temperature of air passing through water spray chamber (saturation process) at constant wet-bulb temperature.

condition in which water constantly replenished from a constant temperature supply, such as a cooling tower, is employed in the spray chamber. The air enters the chamber in an unsaturated condition. The temperature of the water is the same as the wet-bulb temperature of the air so that no change in the latter occurs. As water is absorbed, the dewpoint rises. With each grain of moisture per cu ft of air absorbed, the dry-bulb temperature decreases 8.5 F. The air finally becomes saturated at the wet-bulb temperature, which is also the dewpoint and the dry-bulb temperature of the leaving air. This is called evaporative cooling. The process is also one of *humidifying* due to the moisture increase of the leaving air.

If, upon leaving the spray chamber, the air absorbed heat from the entering space, the dewpoint, dry-bulb, and wet-bulb temperature relationship will be as shown in Fig. 558. Assuming no moisture is absorbed, the dewpoint will remain the same as when leaving the spray chamber. The sensible heat will increase, causing a rise in the dry-bulb temperature, and also in wet-bulb temperature in consequence of an increase in total heat. As no moisture is absorbed, no conversion of sensible heat to latent heat occurs, causing the dry-bulb temperature to rise at a greater rate than the wet-bulb temperature.

PSYCHROMETRIC LAWS

The principles governing the relationship between pure air and water vapor, as outlined by Willis H. Carrier in his "Rational Psychrometric Formulae," are reproduced below. Close study of these laws reveals the

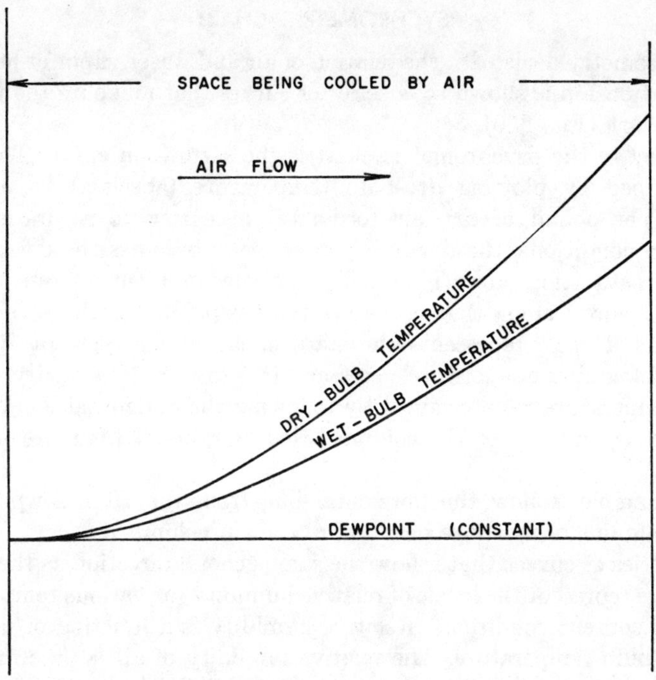

Fig. 558. Relation of dewpoint, dry-bulb and wet-bulb temperature of air passing through conditioned space. Constant dewpoint.

basis upon which the foregoing comments relative to humidity and temperature behavior are founded:

(*a*) When dry air is saturated adiabatically, the temperature is reduced as the absolute humidity is increased, and the decrease of sensible heat is exactly equal to the simultaneous increase in latent heat due to evaporation.

(*b*) As the moisture content of air is increased adiabatically, the temperature is reduced simultaneously until the vapor pressure corresponds to the temperature, when no further heat metamorphosis is possible. This ultimate temperature may be termed the temperature of adiabatic saturation.

(*c*) When an insulated body of water is permitted to evaporate freely in the air, it assumes the temperature of adiabatic saturation of that air and is unaffected by convection; i.e., the true wet-bulb temperature of air is identical with its temperature of adiabatic saturation.

(*d*) The true wet-bulb temperature of the air depends entirely on the total of the sensible and the latent heat in the air and is independent of their relative proportions. In other words, the wet-bulb temperature of the air is constant, provided that the total heat of the air is constant.

PSYCHROMETRIC CHART

Psychrometry deals with the relation of air and water vapor or humidity. This relationship is shown by a series of curves that make up the psychrometric chart (Fig. 559).

The key to the psychrometric chart is the saturation curve. This curve is developed by plotting dry-bulb temperature (abscissa) vs. moisture content per pound of dry air (ordinate) necessary to saturate. Under saturated conditions, the dry-bulb temperature becomes the dewpoint, so that the saturation curve is actually the dewpoint curve, and the temperatures shown along that curve are the dewpoints for the several state conditions. It also represents the state of air at 100 percent RH. The number of grains of moisture necessary to saturate 1 lb of dry air at a given temperature is determined by following the horizontal line from the desired dewpoint left to the column titled "Grains of Moisture per lb of Dry Air."

For example, follow the horizontal line from the 70 F dewpoint and read 110 grains of moisture per lb of dry air in column *A*.

The series of curves that follow the same general direction as the saturation curve represent the levels of relative humidity for various temperature-moisture content conditions. Relative humidity is a function of dewpoint and dry-bulb temperature. The relative humidity of air is determined by following the horizontal line from a given dewpoint right to the intersection of the vertical line representing a given dry-bulb temperature. For example, follow the horizontal line from 70 F dewpoint to where it intersects the vertical 80 F dry-bulb temperature line, and read 72 percent RH.

The diagonal lines running downward from and to the right of the saturation curve represent wet-bulb temperatures. The values of WB temperature for any combination of psychrometric conditions are found by following the diagonal lines to the saturation curve. Under saturated conditions, the DP, DB, and WB temperatures are the same and are represented by the temperatures shown along the saturation curve. To determine the WB temperature for a given DB temperature and DP, proceed as follows: from the point of intersection of DP and DB temperature follow the diagonal WB temperature line to the saturation curve.

For example, the WB temperature of air having a 70 F DP and 80 F DB temperature (or 72 percent RH) is 73 F.

Total heat content of air depends on the WB temperature. To determine this value, proceed vertically from the given WB temperature to the curve *C*, representing "total heat above 0 F contained in 1 lb of dry air saturated with moisture." From the point of intersection follow a horizontal line to the right and determine the value in the total heat column. For example, the total heat of air having a 70 F WB is 33.5 Btu per lb.

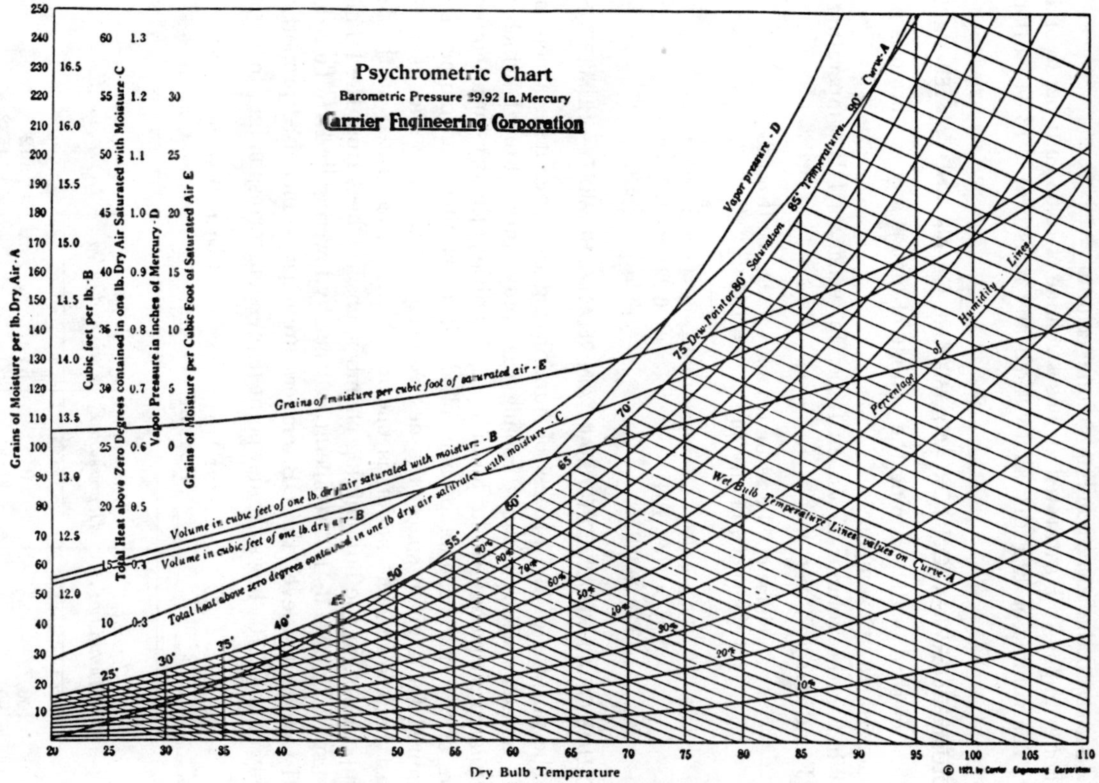

Fig. 559. Psychrometric chart. (Courtesy, Carrier Corporation.)

Volume in cu ft per lb of air depends on the DB temperature. To determine this value, proceed horizontally to the left from the intersection of DB temperature and the appropriate volume curve (dry or saturated air as desired) to the corresponding volume column.

For example, 70 F saturated air has a volume of 13.68 cu ft per lb of dry air. The volume of 70 F dry air (without any moisture) is 13.34 cu ft per lb.

VOLUME DETERMINATION FOR ANY MIXTURE OF AIR AND WATER VAPOR

The method of determining the volume of air for any condition is best demonstrated by example.

What is the volume of air, cu ft per lb, when its DB temperature is 70 F and RH is 70 percent?

Volume of 70 F saturated air	= 13.68	cu ft per lb
Volume of 70 F dry air	= 13.34	" " " "
Difference	= 0.34	" " " "
0.34 × 70% RH	= 0.238	" " " "
13.34 + 0.238	= 13.578	" " " "

VOLUME OF AIR REQUIRED TO MAINTAIN RELATIVE HUMIDITY—HUMIDIFYING

The purpose of the process of humidifying is to maintain a constant RH within a space without cooling the spray water through which recirculated air passes. Under humidifying conditions the dry-bulb temperature of the space cannot be controlled. To determine the volume of air required for a given space, the desired relative humidity and heat load must be known. The action is one in which air, saturated at the WB temperature, in the humidifying chamber, absorbs heat from the humidified space to raise its DB temperature, thereby lowering its RH. To reduce RH to a desired level, the DB temperature must be increased proportionately. Table 170 shows the dewpoint depression corresponding to various

TABLE 170. DEWPOINT DEPRESSIONS FOR VARIOUS RELATIVE HUMIDITY VALUES

Relative Humidity %	*Dewpoint Depression F*	*Relative Humidity %*	*Dewpoint Depression F*
100	0	65	12.9
95	1.8	60	15.2
90	3.2	55	17.6
85	4.9	50	20.4
80	6.8	45	23.3
75	8.7	40	26.5
70	10.8		

RH values. For example, the final DB temperature must be 10.8 F above the DP to maintain a 70 percent RH. The heat necessary to raise the DB temperature is absorbed in the humidified space.

The volume of humidified air necessary to maintain constant RH is determined by the equation:

$$\frac{\text{Load in Btu per min} \times 56}{\text{DP depression corresponding to RH}} = \text{cu ft per min}$$

EXAMPLE: Determine the number of cu ft per min of air required to maintain 60 percent RH in a space having a heat load of 3600 Btu per min.

1st step: Select DP depression corresponding to 60 percent RH:

From Table 170, DP depression = 15.2

2nd step: Determine cu ft of air warmed 1 F per 1 Btu:

Use standard factor of 56 cu ft

3rd step: Apply above equation:

$$\frac{3600 \text{ Btu} \times 56}{15.2} = 13{,}263.1 \text{ cu ft per min}$$

VOLUME OF AIR REQUIRED TO ABSORB HEAT LOAD—DEHUMIDIFYING

To determine the volume of air required to maintain desired temperature and humidity conditions within a cooled space, the following must be known:

Heat load...............	Btu per min
Relative humidity........	Percent
Dry-bulb temperature.....	F

EXAMPLE: Determine the number of cu ft of air required to maintain 70 F DB and 50 percent RH in a space having a heat load of 2000 Btu per min.

1st step: Determine DP from psychrometric chart:

For 70 F and 50 percent RH air, DP = 50 F

2nd step: Determine difference between DB and DP:

70 − 50 = 20 F DP depression

3rd step: Determine cu ft of air warmed 1 F per 1 Btu:

Use standard factor of 56 cu ft

4th step: Apply equation:

$$\frac{\text{load in Btu per min} \times 56}{\text{DP depression}} = \text{cu ft per min}$$

or

$$\frac{2000 \times 56}{20} = 5600 \text{ cu ft of saturated air at 50 F}$$

The spray water in the dehumidifier must be cooled to at least 50 F. To allow for the heating effect of the air passing through it, a 5 F spread should be allowed between required DP and water temperature, necessitating a 45 F spray temperature.

HEAT LOADS

The following factors, listed in the form of a work sheet, must be considered in calculating heat loads for dehumidifying or cooling systems. Estimates should be based on Btu per hr.

Sensible heat:

1. Conduction:
   - Exterior walls .... $A \times U \times t_d$ = Btu
   - Exterior glass .... $A \times U \times t_d$ = Btu
   - Floor .... $A \times U \times t_d$ = Btu
   - Ceiling (roof) .... $A \times U \times t_d$ = Btu
   - Partitions (interior walls) .... $A \times U \times t_d$ = Btu
2. Sunlight gain:
   - Walls, south and west .... $A \times f \times 60$ = Btu
   - Windows, south and west .... $A \times f \times 60$ = Btu
   - Roof, exposed to sun .... $A \times f \times 60$ = Btu
   - Skylight, exposed to sun .... $A \times f \times 60$ = Btu
3. Electrical, mechanical equipment, gas, etc.:
   - Electric lights .... watts $\times e$ = Btu
   - Electric motors .... hp $\times e$ = Btu
   - Gas appliances .... cu ft per hr $\times$ Btu per cu ft = Btu
4. Occupants:
   - People .... no. $\times h$ = Btu
5. Infiltration:
   - Room volume $\times$ no. of changes per hr $\div$ cu ft per lb .... lb $\times 0.24 \times t_d$ * = Btu

Latent heat:

1. People:
   - At rest .... no. $\times l$ = Btu
   - Active .... no. $\times l$ = Btu
2. Infiltration:

$$\frac{\left(\begin{array}{c}\text{Room volume cu ft} \times \text{air changes per hr} \times [\text{grains}\\ \text{per lb, outside air} - \text{grains per lb, inside DP}]\end{array}\right)}{\text{cu ft per lb} \times 7000} \times 1080.2 = \quad \text{Btu}$$

The letter symbols in the above work sheet represent the following:

A = area, sq ft
U = over-all coefficient of heat transmission, Btu per hr per sq ft per 1 F
t_d = temperature differential between conditioned space and outside, or other high-temperature zones

* Based on outside air conditions.

f = factor for sunlight gain. The values of f in Btu per sq ft per min for various surfaces are as follows:

Windows, no protection	3.0
Windows, with white shades	1.5
Windows, with awnings	0.5
Skylight (vertical)	1.0
Skylight (horizontal)	1.5
Average wall (12 in.)	0.1
Concrete wall (4 in.)	0.5
Roof (insulated)	0.15
Roof (poorly insulated)	0.5

e = electrical or mechanical equivalent, Btu per hr
e = 3.4 Btu per hr per watt
e = 2,545 Btu per hr per hp
h = sensible heat factor for people
h = 225 Btu per hr per person
l = latent heat factor for people
l = 175 Btu per hr per person, at rest
l = 450 Btu per hr per person, active

Conduction need not be considered when computing loads for humidifying requirements, except for RH of 75 percent or above. The following temperature differentials between inside and outside for conditions of 75 percent and above may be used in computing conducted heat loads:

75%	$8Ft_d$
80%	$10Ft_d$
85%	$12Ft_d$

Heat loads should be calculated only for wall areas not exposed to the sun (shaded areas).

REFRIGERATION LOAD, DEHUMIDIFYING OR COOLING

The heat-load factors that must be considered in calculating refrigeration requirements of cooling systems are as follows:

1. Total heat gains in space to be cooled. This is taken off work sheet and includes sensible and latent heat.
2. Heat gain from fresh air. Usual practice is to allow 5 to 10 percent of fresh air to be drawn in by the fan through fresh-air louvers. This not only provides periodic air changes, but also permits a positive pressure to develop within the conditioned space.

The heat gain due to fresh air is determined by the equation:

$$\frac{\text{Volume, cu ft per hr, fresh air}}{\text{cu ft per lb}} \times (H_{fa} - H_{dp}) = \text{Btu}$$

where H_{fa} = total heat contained in 1 lb of fresh air at outside temperature
H_{dp} = total heat contained in conditioned air at DP temperature

Other factors that may be included are loss of heat through apparatus (dehumidifier) due to conduction, and water circulating pump where used. The total heat gain is found by adding the individual gains. The refrigeration load is determined by dividing the total heat gain in Btu per hr by 12,000, or

$$\frac{\text{total heat gain, Btu per hr}}{12{,}000} = \text{tons, refrigeration}$$

INFILTRATION

The number of air changes per hour that occur due to leakage varies according to the tightness of construction, the number of windows, the degree of outside exposure, the physical location, etc., of the space to be air conditioned. One air change per hour means that the entire volume of a space will be replaced by air from an outside source once every hour. Well-constructed spaces with few openings may average $\frac{1}{2}$ to 1 air change per hour while spaces having a number of doors and windows will change air 2 to 3 times per hour. Usual practice is to allow for 1 air change per hour for a room with one outside wall, 1.5 for two outside walls, and 2 when three or more walls are exposed to the outside.

AIR-CONDITIONING APPARATUS

The air-conditioning system consists primarily of a fan to circulate air, ducts for distributing air (central systems) to the various parts of the space to be conditioned, a cooling element to cool and dehumidify the air before distribution, appropriate controls, and a refrigeration unit. (See Fig. 560.)

The cooling element or dehumidifier may be an air washer in which several banks of nozzles are installed to develop a water spray covering the entire cross-sectional area of the apparatus. The cold water entering the dehumidifier passes through an arrangement of pipes that act as headers for the nozzles. The bottom is a trough that catches the spray water. Baffles are placed at the entrance and the outlet of the dehumidifier. The baffles at the exit act to eliminate entrained water from the air and are called eliminators. An overflow pipe is located in the reservoir to carry away excess water formed by condensing moisture from the air.

Complete saturation is not always obtained in air washers. Their efficiency is rated on the actual decrease in DB temperature of air passing through it. Humidifying efficiency E_h is expressed by the equation:

$$E_h = \frac{DB_1 - DB_2}{DB_1 - WB} \times 100$$

where DB_1 = dry-bulb temperature of entering air
DB_2 = dry-bulb temperature of leaving air
WB = wet-bulb temperature of entering air

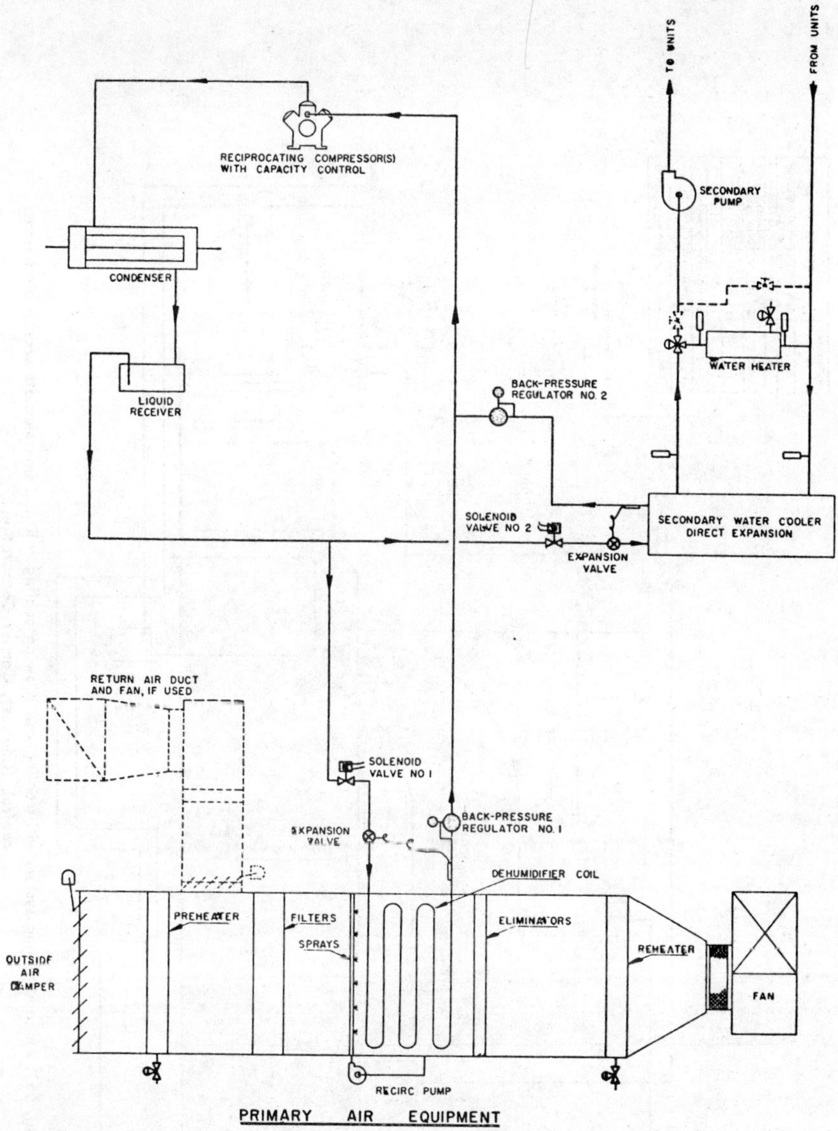

Fig. 560. Air-conditioning system with water spray and direct expansion dehumidifying coil. (Courtesy, Carrier Corporation.)

The cooling element may be a direct expansion coil, as shown in Fig. 561. The heat absorbed by the air is given up directly to the refrigerant. Direct expansion coils are used in both central systems and unit space coolers. In such an arrangement a drain must be provided to carry away

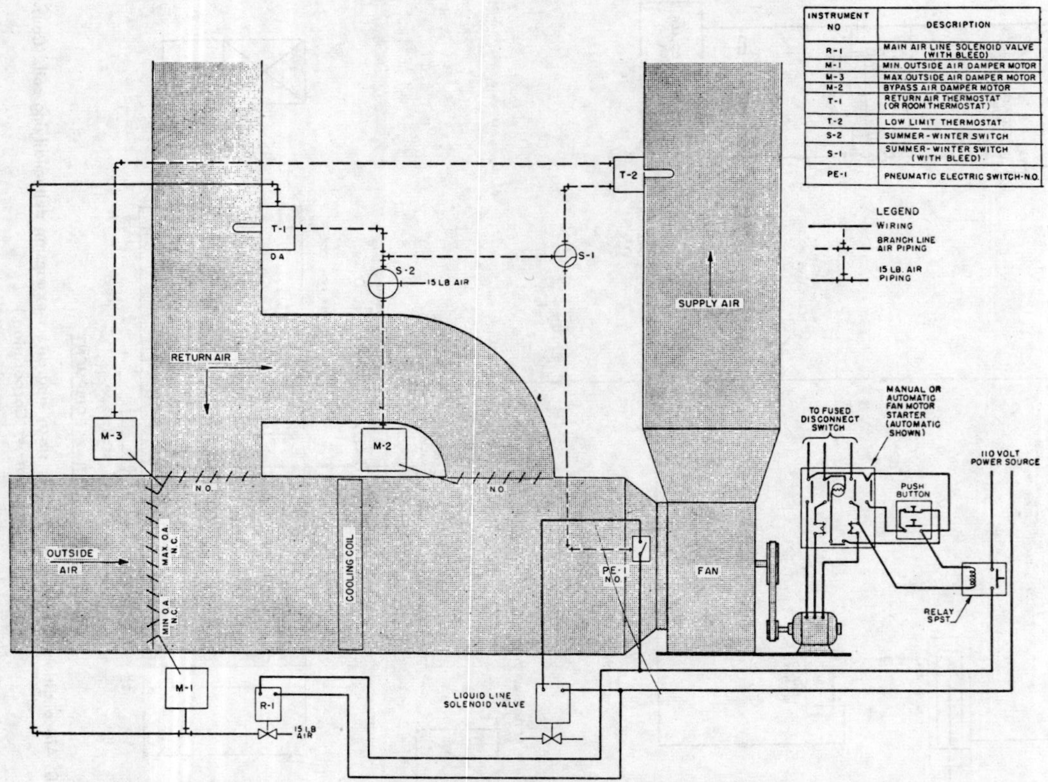

Fig. 561. Air-conditioning system with cooling coil showing cooling coil and intermediate season air by-pass control. (Courtesy, Carrier Corporation.)

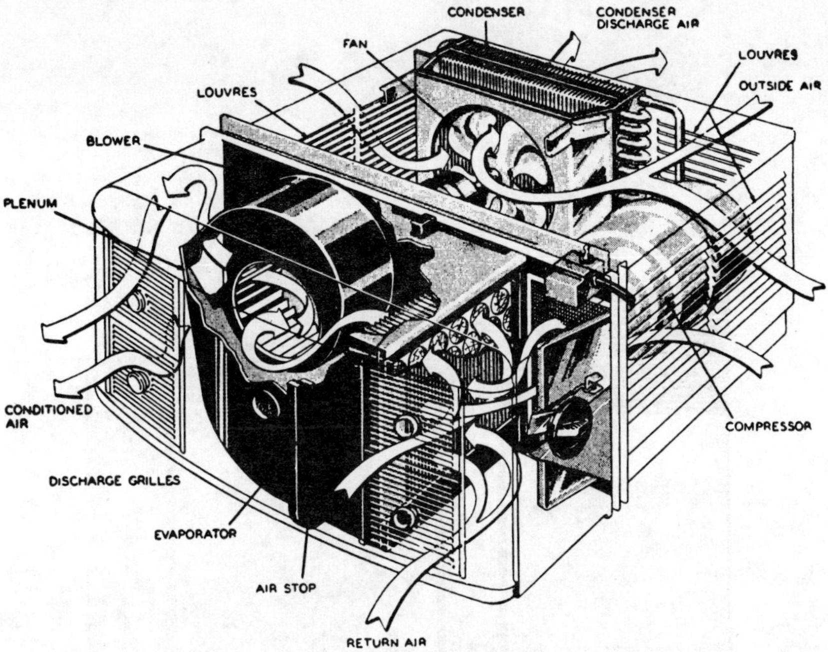

Fig. 562. Room air conditioner, window type. (Courtesy, York Corporation.)

the condensate emanating from the air. Unit coolers are self-contained, requiring only a source of electricity to operate their motor or motors. They are equipped with a condensing system, evaporator coils, filters, fan or fans, and appropriate controls. Window units for home and office cooling range from $\frac{1}{3}$ hp to larger, the 1-hp size proving to be adequate for an average size room. Window units are equipped with air-cooled condensers that must be placed out-of-doors to eliminate the heat absorbed by the refrigerant indoors. Condensers of larger type floor models are either water cooled or air cooled. The industry now identifies the size of small air-conditioning units according to their Btu capacity rather than horsepower.

DUCTS

Ducts are used to distribute air to various parts of the conditioned space and to return it to the cooling element of central systems. The cooling element of central systems may be located below, above, or on the same level as the conditioned space. Risers, elbows, headers, and laterals may be present on one system. Other types use a plenum chamber from which the air is distributed directly into the space or through individual ducts to reach remote areas.

Fig. 563. Unit air conditioner, floor type. (Worthington Air Conditioning Company.)

EXAMPLE – FIND SIZE OF UNIT TO REASONABLY COOL A ROOM 18' X 32' OR 576 SQ. FT. (NOTE LINE (a).

SOLUTION – LOCATE 576 (APPROX.) ON LINE A. EXTEND POINT 576 HORIZONTALLY TO LINE B. DROP VERTICALLY TO LINE C. READ 10,600 BTU.

SELECT (1 SOLUTION)
2-5500 BTU UNITS TO MEET LENGTH LIMITATIONS BY PLACING 1 UNIT AT EACH END OF ROOM.

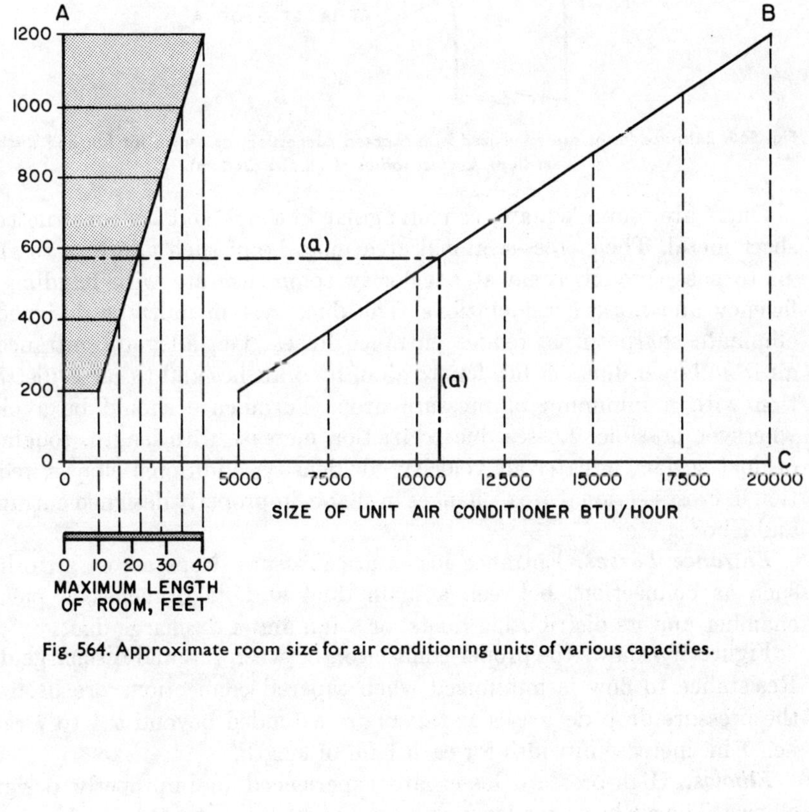

Fig. 564. Approximate room size for air conditioning units of various capacities.

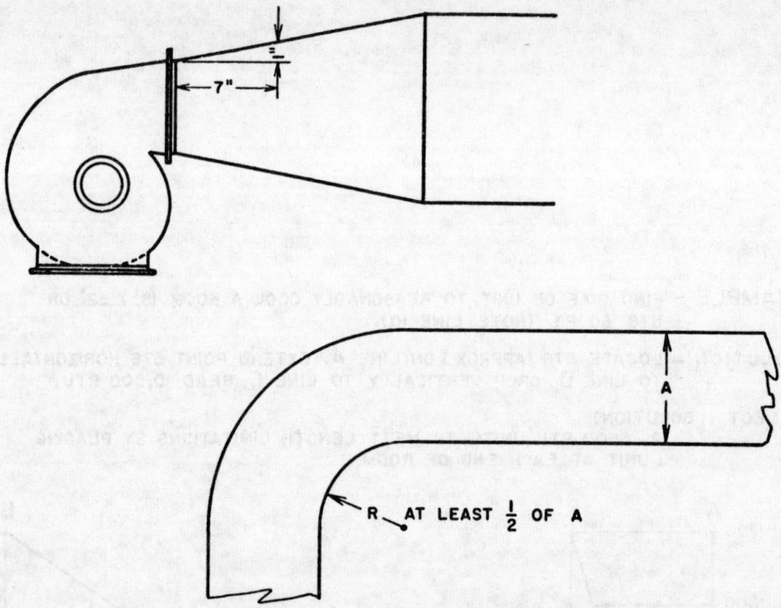

Fig. 565. Entrance losses are minimized with tapered connection, as shown for fan and discharge duct (*top*). Correct radius of elbows (*bottom*).

Ducts are round, square, or rectangular in shape and are constructed of sheet metal. Their cross-sectional area must be of such a size as to allow air to pass through them at a velocity commensurate with handling efficiency and noise considerations. The duct system must be designed to eliminate sharp turns, reduce entrance losses; i.e., alleviate entrance of air from larger ducts or headers to all of its branches and to cause distribution with a minimum of pressure drop. Turbulence should be avoided wherever possible. Losses due to friction increase with length, roughness of duct surface, greater air velocity, air density, additional elbows, reduction in cross-sectional area, changes in shape, improperly designed entrances and elbows, etc.

Entrance Losses. Entrance losses are pressure drops across an orifice, such as connections between a main duct and its branches, a plenum chamber and its distributing ducts, or a fan and a discharge duct.

Figure 565 shows the proper connection between fan and discharge duct. Resistance to flow is minimized when tapered connections are used and the pressure drop decreases as tapers are extended beyond a 1 to 7 ratio; i.e., 1 in. increase in width for each 7 in. of length.

Elbows. High-pressure losses are experienced in improperly designed elbows. The air has a tendency to crowd and become turbulent when bends are made too sharp. Long sweeps are best, but not always feasible. Right-angle turns should be made with elbows having their radius equal to or greater than one and one-half times the width or diameter of the duct. (See Fig. 565, *bottom.*)

Plenum Chambers. Large chambers in which the fan discharges and from which air passes through ducts to remote areas or through ejector nozzles directly into the conditioned space are called plenum chambers. Entrance losses between plenum chamber and ducts may be minimized by adopting the taper connection shown for fans in Fig. 565, using similar ratios.

Duct Shapes. Ducts may be round, square, or rectangular. Round ducts under equal operating conditions offer less resistance than other shapes. Rectangular ducts are often used to increase head room. Their width to depth ratio should be maintained as near to a square (1 to 1) as possible.

Duct Sizes. The cross-sectional area of ducts will depend on the volume and the velocity with which air passes through them. For a given volume and velocity, the area of the duct may be determined by the equation:

$$A = \frac{Q}{V}$$

where A = area in sq ft
Q = volume in cu ft per min
V = velocity in ft per min

EXAMPLE: Determine the area in sq in. for a duct carrying 15,000 cu ft of air per min at a velocity of 1200 fpm.

$$A = \frac{15{,}000}{1200} = 12.5 \text{ sq ft}$$

$$A = 12.5 \times 144 = 1800 \text{ sq in.}$$

To convert area into round-duct diameter use the conventional formula:

$$A = \frac{\pi d^2}{4}$$

$$d = \sqrt{\frac{4A}{\pi}} = \sqrt{\frac{4 \times 1800}{3.1416}}$$

$$d = 47.9 \text{ in., or 48 in. (approx)}$$

To find the dimension of rectangular or square ducts for equivalent round duct diameters, refer to Table 171. For the duct in the above example, follow the "Equivalent Diameter—Inches" across Table 171 to 48 in. Read down the 48 in. column to find desirable short dimension, say, 35.5 in. Proceed across the long dimension column at the left of the table and read 54 in. The dimensions of the rectangular duct selected will be 35.5 in. × 54 in. This means that a rectangular duct 35.5 in. × 54 in. will have

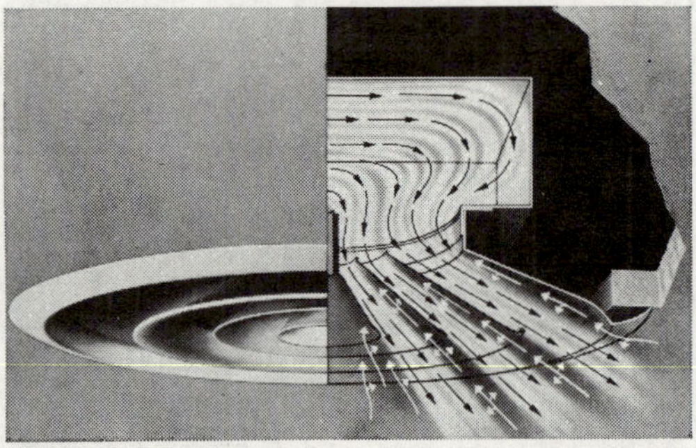

Fig. 566. Room air diffusers (*top*). Note aspiration principle (*bottom*) that causes room air to be siphoned into diffuser to become thoroughly mixed with supply air to equalize temperature and humidity, and avoid stagnant air pockets. (Courtesy, Anemostat Corporation of America.)

a loss in friction equal to that of a round duct having a diameter of 48 in. for the same length. Table 171 is of value for determining equivalent rectangular duct sizes from given diameters of round ducts when friction losses are calculated on the basis of the latter shape.

Friction in Ducts. Friction losses in ducts are based on the drop in air pressure between two points along the path of flow and expressed as inches of water. For low-pressure drops, under which conditions compressibility is not a factor, the formula demonstrated for flow of liquids in pipes in a previous chapter may be used. It is shown again here, viz.:

$$h_f = f\frac{Lv^2}{D2g}$$

where h_f = pressure loss in duct having uniform dimensions
f = dimensionless coefficient, determined by test
L = length of pipe
D = diameter of pipe in ft
v = velocity, fps
g = 32.2 fps per sec, acceleration due to gravity

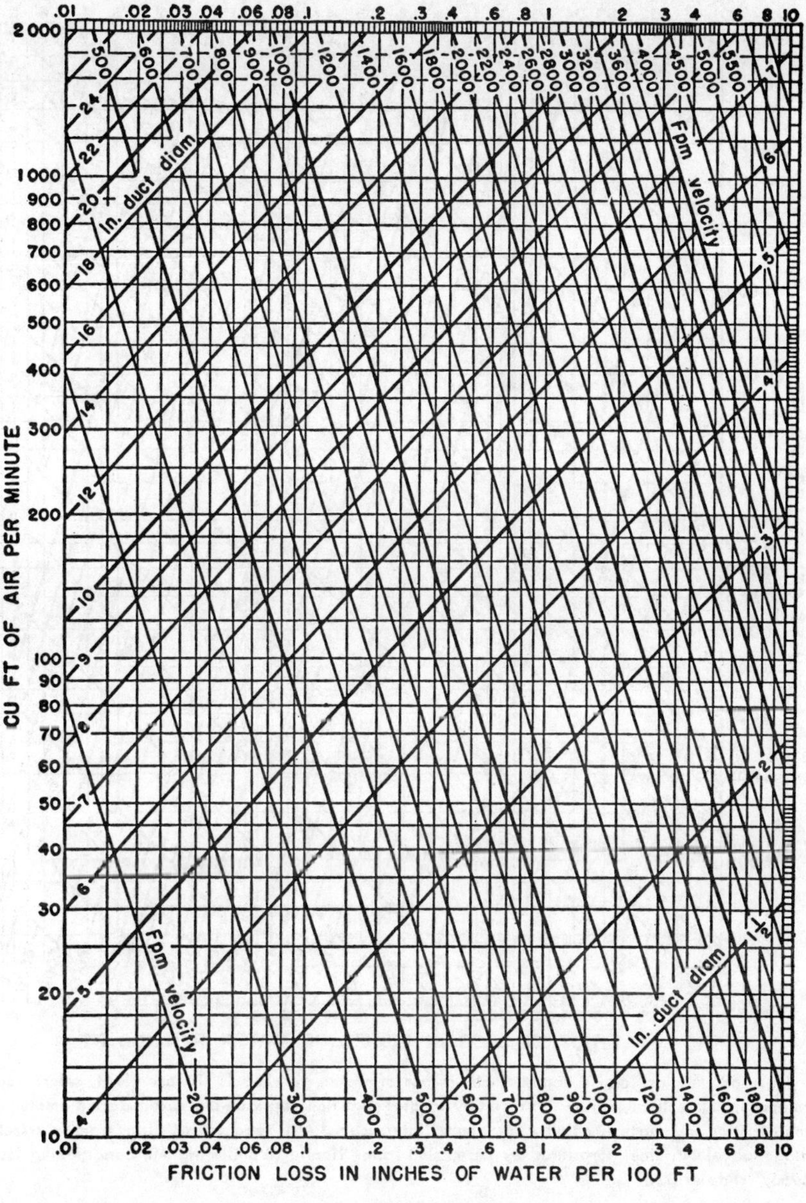

Fig. 567. Friction loss in round ducts, inches of water per 100 ft. To use chart, select known volume in left column, follow horizontally to diagonal line representing given duct diameter and read friction loss vertically below (or above) intersection. Also read velocity fpm from intersecting diagonal velocity line. (Reprinted by permission from "Heating Ventilating Air Conditioning Guide 1956," chapter 32.)

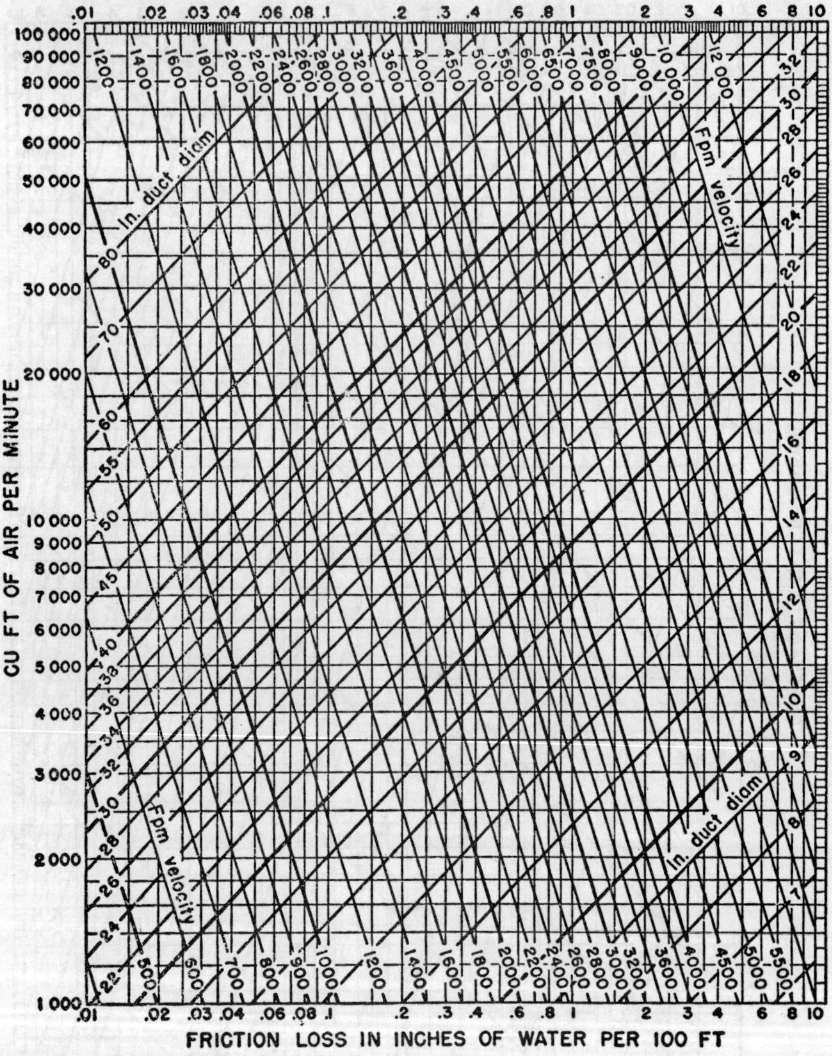

Fig. 568. Friction loss in round ducts, inches of water per 100 ft. To use chart, select known volume in left column, follow horizontally to diagonal line representing given duct diameter and read friction loss vertically below (or above) intersection. Also read velocity fpm from intersecting diagonal velocity line. (Reprinted by permission from "Heating Ventilating Air Conditioning Guide 1956," chapter 32.)

Table 171. Short Sides of Rectangular Ducts for Various Equivalent Diameters (Inches)*

Long Side of Rect-angular Duct Inches	Equivalent Diameter — Inches																						Long Side of Rect-angular Duct Inches
	4	5	6	7	8	9	10	11	12	13	14	15	16	17	18	19	20	21	22	23	24	25	
4	3.3																						4
5		4.1																					5
6		3.5	5.0																				6
7		3.0	4.3	5.8																			7
8			3.8	5.1	6.6																		8
9			3.4	4.6	5.9	7.4																	9
10			3.2	4.2	5.4	6.7	8.3	10.0															10
11				3.9	5.0	6.1	7.5	5.1	10.8														11
12				3.6	4.6	5.7	6.9	8.3	9.9	11.6													12
13				3.4	4.3	5.3	6.5	7.8	9.2	10.7	12.4												13
14				3.2	4.0	5.0	6.1	7.3	8.6	10.0	11.6	13.2											14
15				3.0	3.8	4.8	5.8	6.8	8.1	9.4	10.8	12.4	14.1										15
16					3.6	4.5	5.4	6.5	7.6	8.9	10.2	11.7	13.2	14.9									16
17					3.5	4.3	5.2	6.2	7.2	8.4	9.7	11.0	12.5	14.0	15.8								17
18					3.3	4.1	5.0	5.9	6.8	8.0	9.1	10.4	11.8	13.3	14.9	16.5							18
19					3.2	3.9	4.7	5.6	6.6	7.6	8.7	10.0	11.2	12.6	14.1	15.7	17.4						19
20					3.1	3.8	4.5	5.4	6.3	7.3	8.4	9.5	10.8	12.0	13.4	14.9	16.5	18.2	20.0				20
21						3.7	4.4	5.2	6.0	7.1	8.0	9.1	10.3	11.6	12.8	14.3	15.7	17.3	19.0	20.8			21
22						3.5	4.2	5.0	5.9	6.8	7.7	8.8	9.9	11.1	12.3	13.7	15.0	16.6	18.2	19.8	21.6		22
23						3.4	4.1	4.8	5.7	6.5	7.5	8.5	9.5	10.6	11.8	13.1	14.5	15.9	17.4	19.0	20.6	22.4	23
24						3.3	4.0	4.7	5.5	6.3	7.2	8.2	9.2	10.3	11.4	12.6	13.9	15.3	16.7	18.2	19.8	21.4	24
25						3.2	3.8	4.6	5.3	6.1	7.0	7.9	8.9	10.0	11.0	12.2	13.4	14.7	16.1	17.6	19.1	20.6	25
26						3.1	3.7	4.4	5.1	5.9	6.7	7.6	8.6	9.6	10.7	11.8	13.0	14.2	15.5	16.9	18.4	19.9	26
27						3.0	3.6	4.3	5.0	5.7	6.6	7.4	8.3	9.3	10.4	11.4	12.6	13.7	15.0	16.3	17.7	19.2	27
28							3.5	4.2	4.9	5.6	6.4	7.2	8.1	9.0	10.1	11.1	12.2	13.4	14.6	15.8	17.1	18.5	28
29							3.5	4.1	4.7	5.4	6.2	7.0	7.9	8.8	9.7	10.8	11.8	12.9	14.2	15.3	16.6	17.9	29

30							3.4	4.0	4.6	5.3	6.0	6.8	7.6	8.6	9.5	10.5	11.5	12.6	13.7	14.9	16.1	17.4	30
32							3.2	3.8	4.4	5.1	5.8	6.5	7.3	8.1	9.0	9.9	10.9	11.9	13.0	14.1	15.3	16.4	32
34							3.1	3.6	4.2	4.8	5.5	6.2	7.0	7.8	8.5	9.5	10.4	11.4	12.3	13 4	14.5	15.6	34
36								3.5	4.0	4.6	5.3	6.0	6.7	7.4	8.2	9.0	9.9	10.8	11.8	12.8	13.8	14.9	36
38								3.3	3.9	4.4	5.1	5.7	6.4	7.1	7.9	8.7	9.5	10.3	11.3	12.2	13.2	14.2	38
40								3.3	3.8	4.3	4.9	5.5	6.2	6.8	7.6	8.3	9.1	10.0	10.8	11.7	12.7	13.6	40
42								3.2	3.7	4.2	4.7	5.3	5.9	6.6	7.3	8.0	8.7	9.6	10.4	11.3	12.1	13.1	42
44								3.0	3.5	4.0	4.6	5.1	5.8	6.4	7.1	7.7	8.5	9.2	10.0	10.9	11.7	12.6	44
46									3.4	4.0	4.4	5.0	5.6	6.2	6.8	7.5	8.2	8.9	9.7	10.5	11.3	12.1	46
48									3.4	3.8	4.3	4.8	5.4	6.0	6.6	7.2	8.0	8.6	9.4	10.2	10.9	11.8	48
50									3.2	3.8	4.2	4.7	5.2	5.8	6.4	7.0	7.7	8.4	9.1	9.8	10.6	11.4	50
52									3.1	3.6	4.1	4.6	5.1	5.7	6.2	6.8	7.5	8.2	8.8	9.5	10.3	11.1	52
54									3.1	3.5	4.0	4.5	5.0	5.5	6.0	6.6	7.2	7.9	8.6	9.3	10.0	10.7	54
56									3.0	3.4	3.9	4.4	4.9	5.4	5.9	6.4	7.1	7.7	8.3	9.0	9.7	10.4	56
58										3.4	3.8	4.3	4.8	5.2	5.8	6.4	6.9	7.4	8.1	8.8	9.5	10.2	58
60										3.3	3.7	4.2	4.7	5.2	5.6	6.2	6.7	7.3	7.9	8.5	9.2	9.9	60
62										3.2	3.6	4.1	4.5	5.0	5.5	6.0	6.6	7.1	7.8	8.4	9.0	9.6	62
64										3.1	3.5	4.0	4.5	4.9	5.4	6.0	6.5	7.0	7.6	8.1	8.8	9.4	64
66										3.1	3.4	3.9	4.4	4.8	5.3	5.8	6.3	6.9	7.4	7.9	8.6	9.2	66
68										3.1	3.3	3.8	4.3	4.8	5.2	5.6	6.2	6.7	7.3	7.8	8.4	9.0	68
70										3.0	3.3	3.7	4.2	4.7	5.1	5.6	6.1	6.6	7.1	7 7	8.3	8.8	70
72											3.2	3.7	4.1	4.5	5.0	5.5	6.0	6.5	7.0	7.6	8.1	8.6	72
74											3.2	3.6	4.0	4.4	4.9	5.4	5.8	6.4	6.9	7.4	7.9	8.5	74
76											3.1	3.6	4.0	4.4	4.8	5.3	5.8	6.2	6.7	7.3	7.8	8.4	76
78											3.1	3.5	3.9	4.4	4.8	5.2	5.7	6.2	6.6	7.2	7.6	8.2	78
80											3.1	3.4	3.8	4 2	4.7	5.1	5.6	6.1	6.5	7.0	7.5	8.0	80
82											3.0	3.4	3.8	4.1	4.6	5.0	5.5	6.0	6.4	6.9	7.5	8.0	82
84											3.0	3.4	3.7	4.1	4.5	5.0	5.4	5.8	6.3	6.8	7.3	7 8	84
86											3.0	3.4	3.7	4.0	4.5	4.8	5.2	5.8	6.3	6 7	7.2	7.7	86
88												3.3	3.6	4.0	4.4	4.8	5.2	5.6	6.1	6.6	7.1	7.6	88
90												3.2	3.6	4.0	4.3	4.8	5.1	5.6	6.0	6.5	7.0	7.5	90
92												3.2	3.5	3.9	4.3	4.7	5.1	5.5	6.0	6.3	6.9	7.4	92
94												3.2	3.5	3.9	4.2	4.5	5.0	5.5	5.8	6.3	6.8	7.2	94
96												3.2	3.5	3.8	4.1	4 5	4.9	5.3	5.8	6.2	6.6	7.1	96
98												3.1	3.4	3.7	4.1	4.5	4.9	5.2	5.7	6 2	6.6	7.1	98
100												3.1	3.4	3.7	4.0	4.5	4.8	5.2	5.6	6.0	6.5	6 9	100

* Courtesy, "Trane Air Conditioning Manual," pp. 360–361, copyright, 1941, by The Trane Company, La Crosse, Wisconsin. Reproduced by permission.

Table 171. Short Sides of Rectangular Ducts for Various Equivalent Diameters (Inches)—(*Continued*)

Long Side of Rectangular Duct Inches	Equivalent Diameter — Inches																				Long Side of Rectangular Duct Inches
	26	27	28	29	30	32	34	36	38	40	42	44	46	48	50	52	54	56	58	60	
24	23.2																				24
25	22.4	24.1																			25
26	21.5	23.2	24.9																		26
27	20.7	22.3	23.9	25.7																	27
28	20.0	21.6	23.1	24.8	26.5																28
29	19.3	20.8	22.4	24.0	25.6																29
30	18.8	20.2	21.6	23.2	24.8	28.1															30
32	17.7	19.0	20.4	21.8	23.3	26.4	29.8														32
34	16.8	18.1	19.3	20.6	21.9	24.9	28.1	31.3													34
36	16.0	17.2	18.4	19.6	20.9	23.7	26.6	29.7	33.1												36
38	15.3	16.4	17.5	18.7	20.0	22.5	25.3	28.2	31.4	34.8											38
40	14.6	15.7	16.8	17.9	19.1	21.5	24.1	26.9	29.9	33.0	36.4	40.0									40
42	14.1	15.1	16.1	17.2	18.3	20.6	23.1	25.7	28.6	31.5	34.7	38.1	41.5								42
44	13.5	14.5	15.5	16.5	17.6	19.8	22.2	24.7	27.4	30.2	33.3	36.3	39.6	43.2							44
46	13.1	13.9	15.0	15.9	16.9	19.0	21.3	23.7	26.3	29.0	31.8	34.8	38.0	41.3	44.8						46
48	12.6	13.5	14.4	15.4	16.3	18.4	20.5	22.9	25.3	27.8	30.6	33.5	36.5	39.6	43.0	46.4					48
50	12.2	13.0	14.0	14.9	15.8	17.8	19.9	22.1	24.4	26.8	29.4	32.2	35.1	38.2	41.3	44.7	48.1				50
52	11.9	12.7	13.5	14.5	15.3	17.2	19.2	21.3	23.6	26.0	28.5	31.0	33.8	36.7	39.8	43.0	46.7	49.8			52
54	11.5	12.3	13.2	14.0	14.9	16.7	18.7	20.7	22.9	25.2	27.5	30.0	32.7	35.5	38.3	41.5	44.6	48.0	51.4		54
56	11.2	11.9	12.8	13.6	14.4	16.2	18.1	20.1	22.2	24.4	26.7	29.2	31.6	34.2	37.0	40.0	43.1	46.3	49.7	53.0	56
58	10.8	11.7	12.5	13.2	14.0	15.7	17.6	19.5	21.5	23.7	25.9	28.3	30.7	33.3	35.9	38.7	41.6	44.8	47.9	51.4	58
60	10.6	11.3	12.1	12.9	13.7	15.3	17.2	19.0	20.9	23.1	25.2	27.4	29.8	32.2	34.7	37.6	40.4	43.3	46.4	49.6	60
62	10.4	11.0	11.8	12.5	13.3	14.9	16.6	18.5	20.4	22.3	24.4	26.7	29.0	31.3	33.8	36.5	39.1	41.9	44.9	48.0	62
64	10.0	10.8	11.5	12.2	13.1	14.5	16.3	18.0	19.8	21.8	23.9	26.0	28.2	30.5	32.9	35.5	38.1	40.7	43.6	46.7	64
66	9.9	10.6	11.2	11.9	12.7	14.3	15.8	17.6	19.4	21.3	23.3	25.3	27.4	29.6	32.1	34.5	37.1	39.5	42.4	45.2	66
68	9.7	10.3	11.0	11.7	12.4	13.9	15.5	17.1	18.9	20.7	22.8	24.6	26.7	29.0	31.1	33.7	36.0	38.6	41.2	44.0	68
70	9.4	10.2	10.8	11.4	12.2	13.5	15.2	16.8	18.6	20.2	22.1	24.2	26.0	28.2	30.4	32.8	35.1	37.5	40.2	42.8	70
72	9.3	9.9	10.5	11.2	11.9	13.3	14.8	16.3	18.1	19.8	21.6	23.5	25.6	27.6	29.7	32.0	34.2	36.6	39.2	41.8	72
74	9.1	9.7	10.3	11.0	11.7	13.0	14.6	16.1	17.7	19.4	21.2	23.0	24.9	27.0	29.1	31.3	33.5	35.9	38.3	40.8	74
76	8.9	9.5	10.1	10.8	11.4	12.8	14.2	15.7	17.3	19.0	20.7	22.5	24.4	26.4	28.5	30.6	32.8	35.0	37.4	39.8	76
78	8.7	9.3	10.0	10.6	11.2	12.5	14.0	15.4	16.9	18.6	20.3	22.0	23.9	25.8	27.8	29.8	31.9	34.3	36.6	38.9	78

80	8.6	9.2	9.8	10.4	11.0	12.3	13.6	15 1	16.6	18.2	19.9	21.6	23.4	25.3	27.2	29.3	31.3	33.5	35.8	38.1	80
82	8.4	9.0	9.5	10.2	10.8	12.1	13.4	14.9	16.3	17.9	19.5	21.2	23.0	24.8	26.7	28.8	30.9	32.9	35.0	37.4	82
84	8.3	8.8	9.4	9.9	10.6	11.9	13.2	14.6	16.0	17.5	19.2	20.8	22.5	24.3	26.1	28.2	30.2	32.3	34.3	36.5	84
86	8.2	8.8	9.3	9.8	10 4	11.6	12.9	14 4	15.7	17.2	18.7	20.4	22.1	23.8	25.6	27.6	29.7	31.6	33.5	35.8	86
88	8.1	8.6	9.2	9.7	10.2	11.5	12.7	14.1	15.4	17.0	18.4	20.0	21.7	23.4	25.2	27.1	29.0	31.0	33.0	35.1	88
90	7.9	8.5	9.0	9.4	10.1	11.2	12.5	13.9	15.1	16.6	18.1	19.7	21.3	23.0	24.8	26.6	28.4	30.3	32.4	34 6	90
92	7.8	8.3	8.8	9.4	9.9	11.0	12.3	13.6	15.0	16.4	17.8	19.3	21.0	22.6	24.3	26.1	28.1	29.9	31.8	33.9	92
94	7.8	8.3	8.7	9.3	9.9	10.9	12.1	13.3	14.7	16.1	17.6	19.1	20.6	22.3	23.9	25.7	27.4	29.3	31.3	33 3	94
96	7.7	8.2	8.6	9.1	9.7	10.8	12.0	13.2	14.4	15.8	17.3	18.7	20.4	21.9	23.5	25.2	27.1	28.8	30.7	32.7	96
98	7.5	8.0	8.5	9.0	9.5	10 6	11.8	13.0	14.2	15.6	17.1	18.4	19.9	21.6	23.1	24 9	26.7	28.3	30.2	32 0	98
100	7.5	7.9	8.4	9.0	9.5	10.5	11.6	12.8	14.0	15.4	16.8	18.1	19 6	21.2	22.8	24.5	26.1	27.9	29.7	31.6	100

EQUIVALENT DIAMETER — INCHES

	62	64	66	68	70	72	74	76	78	80	82	84	86	88	90	92	94	96	98	100	
58	54.8																				58
60	52.9	56.4	60.0																		60
62	51.2	54.4	58.0	61.4																	62
64	49.9	52.9	56.2	59.6	63.1																64
66	48.2	51.3	54.4	57.8	61.2	64.7															66
68	46.9	49.8	53.0	56.1	59.4	62.9	66.3														68
70	45.6	48.4	51.6	54.5	57.8	61 2	64.5	68.0													70
72	44.4	47.2	50.1	53.1	56.2	59.4	62.7	66.2	69.7												72
74	43.4	46.1	48 8	51.8	54.8	57.9	61.0	64.4	67.9	71.3											74
76	42.3	44.8	47.6	50.5	53.4	56.3	59.5	62.7	66.0	69.5	73.0										76
78	41.3	44.0	46 5	49.3	52.0	55.1	58.1	61.2	64.4	67.8	71.1	74.6									78
80	40.5	42.9	45.6	48.1	50 8	53.7	56.7	59.8	62.8	66.0	69.4	72.8	76.3	80.0							80
82	39.7	42.1	44.6	47.2	49.9	52.6	55.4	58.4	61.4	64.6	67.7	71.0	74.5	77 9	81.5						82
84	38.8	41.2	43.7	46.2	48.6	51.3	54.2	57.1	60.1	63.1	66 2	69.4	72.7	76.0	79.5	83.0					84
86	38.0	40.3	42.8	45.3	47.7	50.3	52.9	55.9	58.7	61.6	64.8	67.9	71.0	74.4	77.8	81.3	84.7				86
88	37.4	39.6	41 8	44 4	46 7	49.2	51.9	54.6	57.5	60.4	63.2	66.4	69.4	72.7	75.9	79.4	82.7	86.3			88
90	36.6	38.7	41 0	43.5	45.9	48.3	50.8	53.6	56.3	59.0	61.9	64.7	68.1	71.1	74.3	77.7	81.1	84.5	88.0		90
92	36.0	38.1	40.4	42.6	45 1	47.5	50 0	52.6	55.1	57.9	60.6	63.6	66.5	69.7	72.7	76.0	79.3	82.6	86.1	89.6	92
94	35.3	37.5	39.6	41.8	44.2	46 6	49.2	51.6	54.1	56.8	59 4	62.4	65.2	68.2	70.7	74.5	77.5	80.9	84.4	87.9	94
96	34.8	36.9	38.9	41.1	43 4	45.7	48.1	50.6	53.2	55.7	58.3	61.1	64.0	66.9	69.9	72 9	76.1	79.3	82.7	86.0	96
98	34.2	36.3	38.2	40.5	42.6	44.9	47.2	49.7	52.1	54.6	57.2	59 9	62.7	65.5	68 5	71 6	74.6	77 7	80.9	84 2	98
100	33.6	35.6	37.6	39.7	41.9	44.1	46.5	48.8	51.2	53.6	56.3	58.9	61.5	64.3	67.2	70.1	73.1	76.2	79 4	82.6	100

TABLE 172. RELATION BETWEEN PERCENTAGE OF AIR VOLUME TO THE PERCENTAGE OF DUCT AREAS

% Vol.	% Cap.	% Vol.	% Cap.	% Vol.	% Cap.
1	2	34	42	67	73.5
2	3.5	35	43	68	74.5
3	5.5	36	44	69	75.5
4	7.0	37	45	70	76.5
5	9.0	38	46	71	77.5
6	10.5	39	47	72	78
7	11.5	40	48	73	79
8	13.0	41	49	74	80
9	14.5	42	50	75	80.5
10	16.5	43	51	76	81
11	17.5	44	52	77	82
12	18.5	45	53	78	83
13	19.5	46	54	79	84
14	20.5	47	55	80	84.5
15	21.5	48	56	81	85.5
16	23.0	49	57	82	86
17	24.0	50	58	83	87
18	25.0	51	59	84	87.5
19	26.0	52	60	85	88.5
20	27.0	53	61	86	89.5
21	28	54	62	87	90
22	29.5	55	63	88	90.5
23	30.5	56	64	89	91
24	31.5	57	65	90	92
25	32.5	58	65.6	91	93
26	33.5	59	66.5	92	94
27	34.5	60	67.5	93	94.5
28	35.5	61	68	94	95
29	36.5	62	69	95	96
30	37.5	63	70	96	96.5
31	39	64	71	97	97.5
32	40	65	71.5	98	98
33	41	66	72.5	99	99
				100	100

The friction loss in inches of water per 100 ft of duct may be determined from the charts shown in Figs. 567 and 568. An example as to how the charts are used is as follows:

EXAMPLE: Determine the friction loss in 90 ft of duct having a diameter of 40 in. carrying 30,000 cu ft of air per min.

1st step: Find 30,000 cu ft per min at right of chart (Fig. 568).

2nd step: Proceed to the right horizontally to intersection of diagonal line representing 40 in. duct diameter.

3rd step: From intersection proceed vertically, upward or downward, and read 0.4 in. of water per 100 ft, friction loss.

4th step: For 90 ft of duct the loss would be $\frac{90}{100} \times 0.4 = 0.36$ in. of water.

The fan should be selected on its ability to deliver the required cubic feet of air per minute against the total friction of the duct system expressed as inches of water.

Proportioning of Ducts. A satisfactory method of determining the size of distributing ducts is on the basis of equal friction per unit length of duct. Thus the drop in pressure throughout the system will be equivalent to the loss in an equal length of duct having a constant dimension. Table 160 lists the relation between percentage of air volume to percentage of duct area. For example, if a branch duct were to take off 25 percent of the total air volume flowing in a main duct, the area of the branch duct would be 32.5 percent of that of the main duct.

The use of the above method of proportioning main and branch ducts decreases the velocity in the mains as volume decreases. It has the advantage (1) of utilizing velocity to produce static pressure, (2) of uniform static pressure and distribution throughout the system, and (3) of reducing friction in the smaller branch ducts.

Flow of Air in Ducts. The Pitot tube is used to measure the static and the velocity pressures of air flowing through a duct.

STATIC PRESSURE is the measure of potential energy exerted by the air. It is exerted in all directions and is the cause of friction in ducts.

VELOCITY PRESSURE is the kinetic energy of the air and is the force exerted in its movement through the duct.

TOTAL PRESSURE is the sum of static and velocity pressures.

TABLE 173. GOOD OPERATING VELOCITIES AND TIP SPEEDS FOR VENTILATING FANS*

STATIC PRESSURE INCHES OF WATER	FORWARD CURVED BLADE FANS		BACKWARD TIPPED AND DOUBLE CURVED BLADE FANS		TUBEAXIAL AND VANEAXIAL FANS
	Outlet Velocity Feet per Minute	Tip Speed Feet per Minute	Outlet Velocity Feet per Minute	Top Speed Feet per Minute	Wheel Velocity[a] Feet per Minute
1/4	1000–1100	1520–1700	800–1100	2600–3100	1100–1500
3/8	1000–1100	1760–1900	800–1150	3000–3500	1250–1700
1/2	1000–1200	1970–2150	900–1300	3400–4000	1400–1900
5/8	1200–1400	2225–2450	1000–1500	3800–4500	1500–2100
3/4	1300–1500	2480–2700	1100–1650	4200–5000	1650–2350
7/8	1400–1700	2660–2910	1200–1750	4500–5300	1800–2500
1	1500–1800	2820–3120	1200–1900	4800–5750	1900–2700
1 1/4	1600–1900	3162–3450	1300–2100	5300–6350	2150–3000
1 1/2	1800–2100	3480–3810	1400–2300	5750–6950	2350–3300
1 3/4	1900–2200	3760–4205	1500–2500	6200–7550	2500–3550
2	2000–2400	4000–4500	1600–2700	6650–8050	2700–3800
2 1/4	2200–2600	4250–4740	1700–2800	7050–8550	
2 1/2	2300–2600	4475–4970	1800–2950	7450–9000	
3	2500–2800	4900–5365	2000–3200	8200–9850	

[a] **Wheel velocity is the axial mean air velocity through the inside diameter of the housing cylinder at the point of wheel location.**

* Reprinted by permission from "Heating Ventilating Air Conditioning Guide 1956," chapter 33.

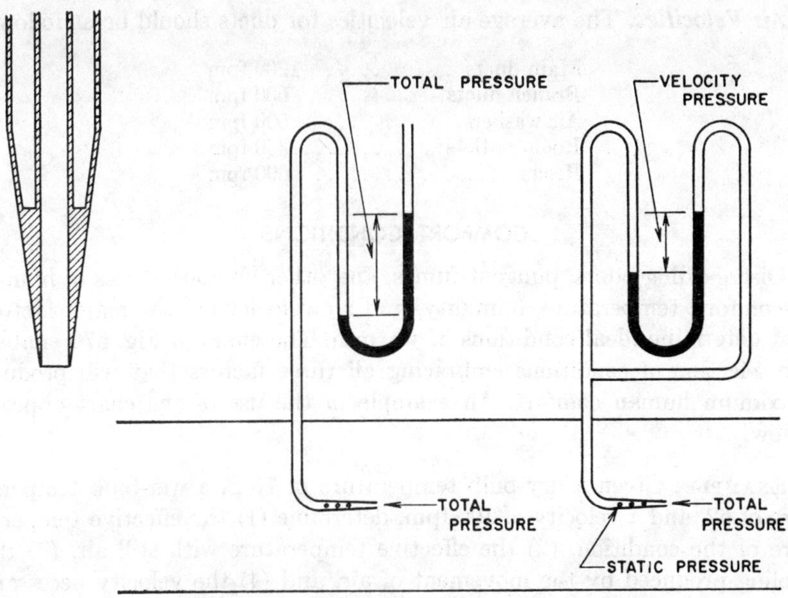

Fig. 569. Pilot tube.

The Pitot tube is a tube within a tube, the outer tube being closed at one end. The inner tube is open at the extreme end, while the outer tube is drilled at the side to permit entrance of air by static pressure only. When the bent portion is placed in the duct so that the inner tube opening faces the flow of air, the effects of both velocity and static pressures (or total pressure) are measured by the difference in water level in the U-tube, as shown in Fig. 569 (*left*). If the U-tube is hooked up, as shown in Fig. 569 (*right*), in which static pressure is neutralized by its own counterforce, the difference in water level will be that of velocity pressure only.

The readings should be made so as to obtain an average pressure reading for the duct area. This is done by traversing the duct at various levels and at regular intervals along the path of each level. Readings should be taken in a straight run of duct at a point whose distance is at least 50 diameters away from any bends or other turbulent elements.

From the readings obtained by the Pitot tube, the velocity of the air may be calculated by the equation:

$$V = 4005\sqrt{p}$$

where V = velocity in fpm of standard air (70 F and 29.92 in. Hg)
p = pressure, in. of water (velocity head)

and the value of 4005 is the velocity of standard air corresponding to a pressure of 1 in. of water.

Air Velocities. The average air velocities for ducts should be as follows:

Main ducts	1200 fpm
Branch ducts	600 fpm
Air washers	500 fpm
Room outlets	350 fpm
Risers	600 fpm

COMFORT CONDITIONS

Disregarding odors, pungent fumes, and other obvious causes of human discomfort, temperature, humidity, and air velocity are the major factors that determine ideal conditions of warmth. The chart in Fig. 570 enables the selection of conditions embracing all three factors that will produce maximum human comfort. An example in the use of the chart appears below.

EXAMPLE: Given a dry-bulb temperature of 76 F, a wet-bulb temperature of 62, and a velocity of 100 fpm, determine (1) the effective temperature of the condition, (2) the effective temperature with still air, (3) the cooling produced by the movement of air, and (4) the velocity necessary to reduce the condition to 66 F effective temperature.

1st step: Draw a line *AB* through given DB and WB temperatures. Where it intersects with the 100 fpm line, it also crosses the 69 F line. This is the effective temperature of the condition.

2nd step: Follow line *AB* to the right of its intersection with the 0 velocity line. Read 70.4 F. This is the effective temperature with still air.

3rd step: Subtract 69 F from 70.4 F, a difference of 1.4 F. This is the effective temperature difference produced by air movement.

4th step: Follow line *AB* to the left to the intersection of the 66 F temperature line. The velocity at this point reads 340 fpm. This is the velocity to which the air must be raised for maximum comfort.

AIR PURIFICATION

Impurities, in the form of either gases or solids, must be eliminated during the air-conditioning process. Solids may be separated from the air by filtering, using mechanical filters, baffles, or scrubbers (water washing). The majority of impure gases, including carbon dioxide, are soluble in water and may be removed by scrubbing. Filters and scrubbers are also effective in removing micro-organisms.

Complete air-conditioning systems include a water spray chamber through which the air passes on its way to the conditioned space. This spray chamber acts as air cooler, dehumidifier, humidifier, degasifier, and air washer, according to the immediate function. Solid particles that escape the water washer are removed by centrifugal force, impact, adhe-

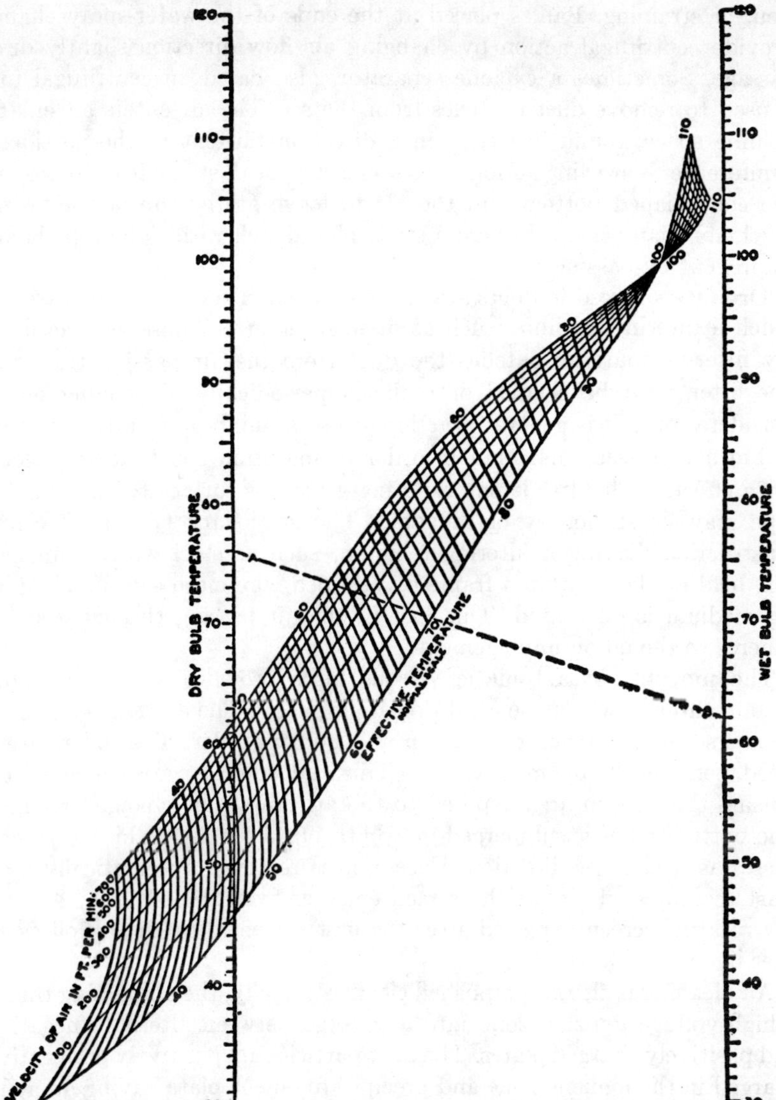

Fig. 570. Effective temperature chart showing normal scale of effective temperature, applicable to inhabitants of the United States under the following conditions (*A—Clothing*) Customary indoor clothing; (*B—Activity*) sedentary or light muscular work; (*C—Heating Methods*) convection type; i.e., warm air, direct steam, or hot water radiators, plenum system. (Reprinted by permission from "Heating Ventilating Air Conditioning Guide 1956," chapter 6.)

sion, or straining. Baffles placed at the ends of the water-spray chamber provide centrifugal action by changing air flow direction slightly during passage. Sometimes a cyclone separator, also based on centrifugal force, is used to remove dust particles from the air. The air enters an enlarged volume space, round in shape, in a direction tangent to the outside circumference. A swirling action occurs, causing the dust particles to drop into the cone-shaped bottom and the air to leave at the top of the swirling (cyclone) chamber up through a stack placed well within the top chamber to prevent by-passing.

Dry filters placed in ducts are also employed. They consist of a frame in which a filtering medium, such as fiber glass, or cellulose, is placed. The dry filtering material catches the dust from the air passing through it. The filter must be cleaned or replaced periodically, depending on cost considerations. It is general practice to use a number of individual filters within a cross-sectional area in order to maintain constant air velocity.

An effective air filter is one exposing a viscous surface to air flow. This type may be stationary or movable. The stationary type is of cellular construction, having a filtering medium, such as steel wool, spun glass, etc., held in place within a frame by crimped wire-mesh screen. The filtering medium is oil coated. The dust in the air passing through the filter adheres to the oil on impingement.

The movable or automatic viscous filter provides a vertical endless curtain through which the conditioned air passes. The curtain is an endless bronze screen stretched over a top and bottom roller. The rollers are rotated periodically by means of a chain and sprocket arrangement, thus causing the curtain area exposed to the air flow to be changed regularly. The bottom roller is submerged in a bath of oil through which the screen passes as it changes direction. Excess oil is removed automatically by air blast (compressed air) as the screen emerges from the oil bath. A second dry rotating screen is placed after the first screen to catch any oil carried over by the conditioned air.

Air cleaning is also accomplished electrostatically, the air passing through a high-voltage ionizing zone into a passage between alternate negatively and positively charged plates. The dust particles are positively or negatively charged in the ionizing zone and precipitate on the plate having an attractive charge; i.e., positively charged particles to negative plates and negatively charged particles to positive plates. The negative plates are movable and can be cleaned automatically. Wipers attached to moving chains clean the stationary positive plates.

Odors are removed from air by passing it through activated carbon produced from coconut shells. The action is one in which the activated carbon absorbs odors.

AUTOMOBILE AIR CONDITIONING

The automobile air-conditioning system consists primarily of a compressor, condenser, receiver, expansion valve, evaporator, dehydrator, filter, sight-glass, blower, and air-distributing system. It also includes some type of capacity control.

Typical System. A typical system would have its compressor (axial, in-line, and V-types) driven by belt from the engine, the condenser (finned coil type) located forward in front of the radiator, and the evaporator either under the dash or behind the rear seat. A filter, dehydrator, and sight-glass are installed in the liquid line. The expansion valve is of the thermostatic type with an equalizer. The circulating blower, usually of the centrifugal type, draws conditioned air from the interior of the automobile through return ducts and forces it over the evaporator coil back to the cooled space. Some fresh air is forced in with the return air under pressure provided by means of air scoops, or a fresh-air fan. Air scoops have external openings facing toward the front of the car which "scoop" up air during forward motion. The mixing of fresh warm air with return conditioned air has a tempering effect and acts to prevent evaporator icing. Both fresh and return air pass through a filter located on the suction side of the recirculat-

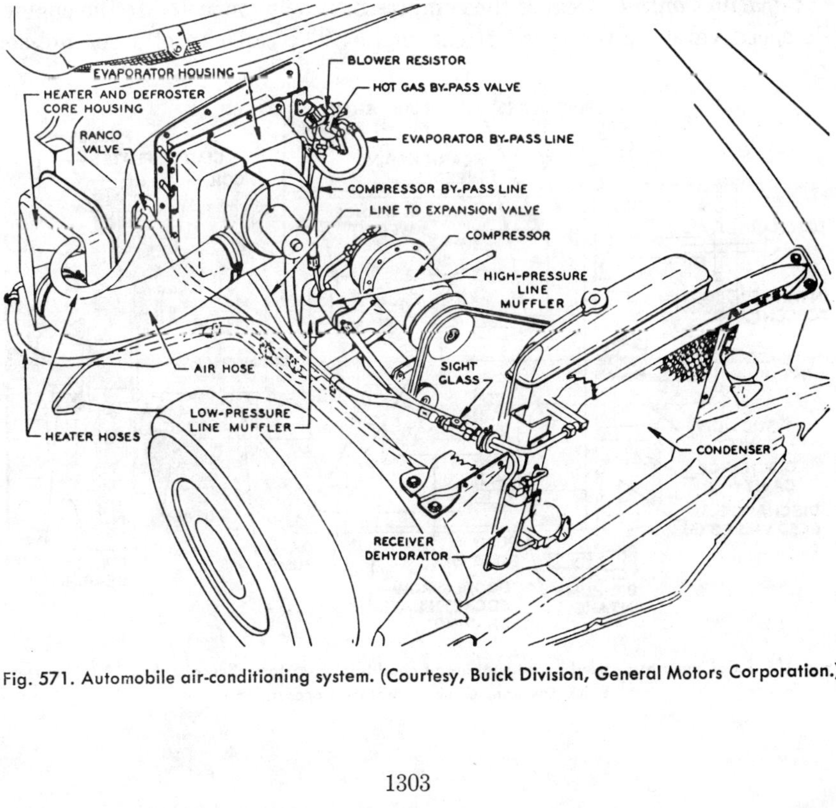

Fig. 571. Automobile air-conditioning system. (Courtesy, Buick Division, General Motors Corporation.)

ing blower to remove dust. The blower is driven by a direct-current motor receiving current from the car battery.

Conditioned air controls may include those that turn the system on and off, regulate the amount of fresh and recirculated air, vary the temperature, and adjust the speed of the blower. Some systems are not furnished with all the controls mentioned, the simplest being an on-or-off control.

The refrigerants used are the Freons because of their adaptability to air-conditioning requirements.

For the average automobile, the capacity of the blower should be about 150 cu ft per min. The refrigerating capacity varies, the minimum requirements being 1 ton, corresponding to a speed of 800 rpm (for axial compressors) for an inside temperature of 75 F for tightly constructed, well-insulated bodies. Capacity increases with engine speed reaching $2\frac{1}{2}$ or more tons at an engine speed of about 3000 rpm, using an axial compressor. The maximum capacity for rotary compressors is less than for axial compressors at all comparable speeds. At a belted speed of 400 rpm the capacity drops to approximately $\frac{1}{2}$ ton for axial compressors. The time necessary to pull the temperature down from 95 F to 75 F (pull-down time) should be about 25 min or less at a maintained speed of 30 mph. At higher speeds, the pull-down time is correspondingly less.

Capacity Control. Because the compressor is belt connected to the engine, its speed varies as the speed of the engine. The capacity of a compressor

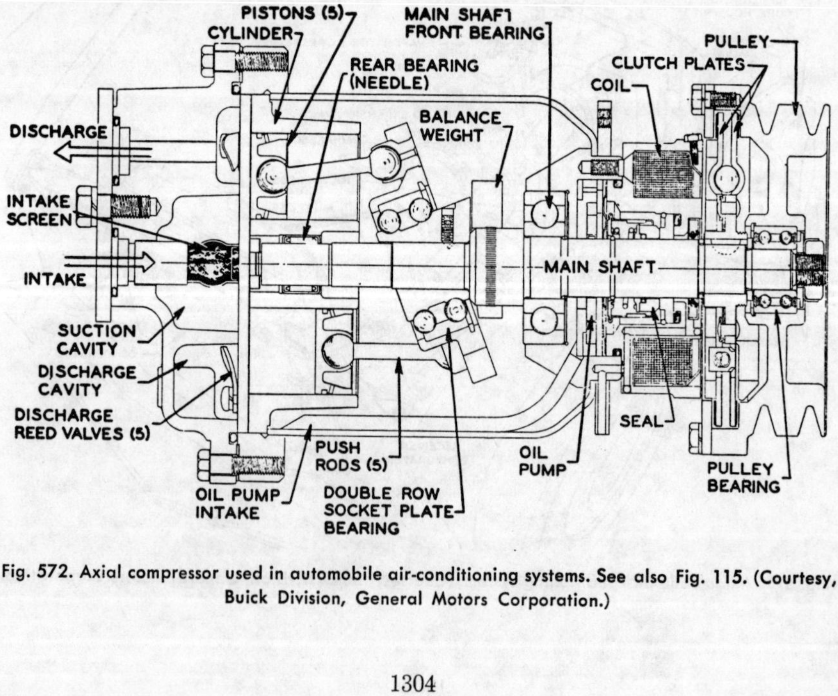

Fig. 572. Axial compressor used in automobile air-conditioning systems. See also Fig. 115. (Courtesy, Buick Division, General Motors Corporation.)

varies with speed. Variable compressor speed poses a problem in automobile air conditioning. To maintain desired conditions within the automobile during engine idling periods, the capacity of the compressor must be based on minimum crankshaft speed. This means that the refrigerating capacity is excessive during high-speed operation, especially during periods when the outside temperature is below the rated maximum level for the system. An alternative would be to rate the compressor capacity on the basis of the higher cruising speed and provide some means of increasing compressor speed during idling periods. Another alternative is to rate the compressor capacity at a lower speed, and provide a means of maintaining such a speed constantly.

Where constant speed cannot be attained by reason of design, it is necessary to affect some means of capacity control during high automobile speed operation, or when for any reasons the heat load is not sufficient to utilize the full refrigeration capacity of the system.

During normal operation, the thermostatic-expansion valve regulates the flow of refrigerant into the evaporator according to the suction temperature. To avoid icing (due to condensate from the air) and consequent fouling of the evaporator, the suction temperature must be controlled at a level above freezing or 32 F. If the compressor continues to operate at high speed without some measure of control, it would cause a reduced suction pressure, and, therefore, a lower evaporator temperature. It would also cause a higher condenser pressure.

One method of avoiding excessive capacity is by means of a *magnetic clutch* on the compressor drive pulley. Through an electric circuit the magnetic clutch is controlled by a thermostat located on the evaporator. At a given temperature level, the thermostat causes the circuit to open and break the magnetic field of the clutch. This in turn breaks the connection between the drive pulley and the compressor shaft and the compressor ceases to operate. When the temperature of the evaporator rises, the thermostat closes the electric circuit to energize the magnetic field to start the compressor operating again.

Another method of controlling capacity or governing evaporator temperatures is to employ some type of suction by-pass valve that will allow diversion of all or part of the discharge gas to the suction side of the compressor, with a reduction in load, or when there is a possibility of the evaporator temperature falling below 32 F. The by-pass valve acts to short-circuit (see Fig. 573) the discharge gas back to the compressor suction through a by-pass line when the evaporator pressure or temperature reaches a given level. Thus some or all of the gas merely circulates through and around the compressor when it is not required by the evaporator.

There are several types of by-pass valves used in automobile air-conditioning systems. Included are the pressure-actuated by-pass valve, the solenoid by-pass valve, and the thermostatically actuated by-pass valve.

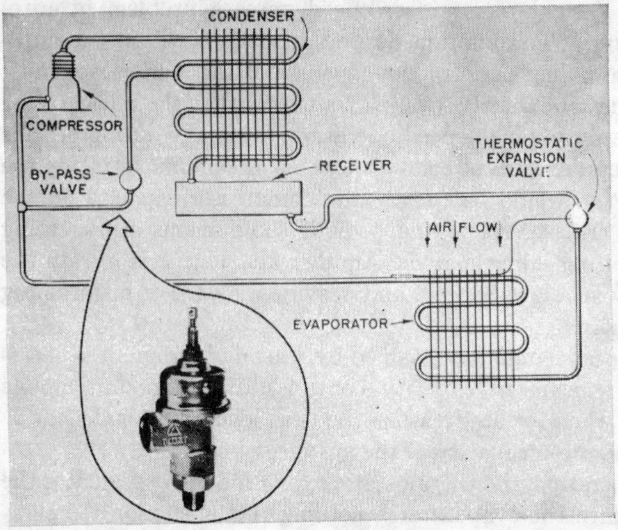

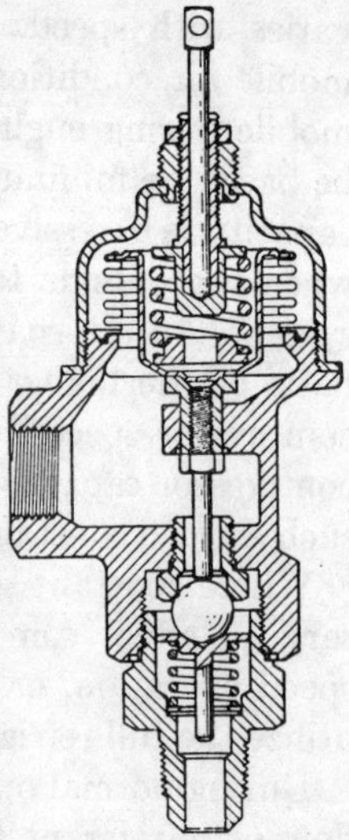

Fig. 573. (*Left*) Refrigeration system for automobile air-conditioning system showing method of capacity control using pressure-actuated by-pass valve. (Courtesy, Detroit Controls Corporation.)

Fig. 574. (*Right*) Pressure-actuated by-pass valve. (Courtesy, Detroit Controls Corporation.)

Pressure-Actuated By-Pass Valve. This type is actuated by pressure differential between the high- and the low-side of the system. As the suction pressure diminishes, the pressure differential increases. The valve is adjusted so that it will start to open when the difference between high- and low-side pressure reaches a predetermined value. It will continue to open until maximum flow through the suction by-pass line is accomplished when the valve has moved to the top limit of its pressure range. Thus the action of a pressure-actuated by-pass valve is one of modulation. (See Fig. 574.)

Solenoid By-Pass Valve. The solenoid by-pass valve is operated electrically by a thermostat located on the evaporator. As the thermostat acts to close or open the electric circuit, it energizes or de-energizes the solenoid. The solenoid either opens or closes the by-pass line completely, so does not have the modulating characteristics of the pressure-actuated valve. Another arrangement is to use a reverse-acting thermostat controlled by a thermal bulb placed in the return air stream. As the temperature of the return air drops, the bulb fluid contracts and acts on the thermostat to close the electrical control circuit. This energizes the solenoid to open the by-pass. The reverse type (or heating thermostat) is used in order to employ the conventional type solenoid; i.e., one that is held open when energized.

Thermostatically Actuated By-Pass Valve. This type of valve operates on the same principle as the thermostatic expansion valve. The valve stem is actuated by a diaphragm that flexes under pressure of the fluid in a thermal bulb and capillary tube. The bulb may be mounted on the evaporator or in the return air duct. The valve is designed to open and permit flow of hot compressor discharge gas through the suction by-pass with a decrease in temperature. Like the pressure-actuated by-pass valve, it permits modulated flow so that the amount of hot gas shunted to the suction line ranges from zero to maximum; i.e., the lower the temperature, the greater becomes the valve opening.

Another method of controlling capacity is to use a by-pass valve and an automatic expansion valve. The by-pass valve allows liquid to flow into the suction line when the evaporator pressure drops below the adjusted level. The automatic expansion valve is installed in the suction line between the by-pass connection and the compressor intake. When liquid enters the suction line through the by-pass, it is expanded in the automatic expansion valve before entering the compressor. Thus liquid is prevented from entering the compressor and the arrangement eliminates the condition of a compressor drawing a high vacuum in a completely closed off line.

AIR CURTAINS

An air curtain is virtually a sheet or blanket of warm or cooled air distributed over the entire area of a doorway. The curtain may be developed from conditioned air (warm or cool) delivered through nozzles from a plenum chamber located either on each side, or immediately above, the door. Or it may be directed from a fan or air conditioner having a plenum especially designed to adequately distribute the air over the entire area of the doorway. A simple air curtain for separating two compartments, such as a restaurant kitchen from the dining room to prevent the flow of cooking odors to the latter, may be provided by cascading conditioned air by means of a fan over a half dome shaped hood above the connecting doorway. The curtain is more effective if the dining room is air conditioned and in which a positive pressure exists.

Air curtains provide protection against odors, dust, insects, as well as providing an insulating blanket where it is necessary to maintain temperature differentials.

Vestibules of public buildings and department stores, as well as the doorways of warehouses, food packing and processing plants, cold storage plants, bakeries, etc., are potential users of air curtain facilities.

SUMMARY OF CHAPTER 18

Air is a mixture of gases, the major ones being oxygen and nitrogen in a ratio of 21 percent to 79 percent, respectively, by volume. The atmosphere also contains water vapor.

The specific heat of air at constant pressure is accepted as 0.24 for general purposes.

Humidity refers to the water vapor contained in the atmosphere.

Under standard conditions of 70 F and 29.92 in. Hg barometric pressure, dry air weighs 0.07495 lb per cu ft. Air saturated with moisture weighs 0.07425 lb per cu ft.

Volume of air is equal to the reciprocal of weight, lb per cu ft, based on temperature, moisture content, and pressure.

The volume of air that will rise 1 F per 1 Btu is

$$K = \frac{1}{c_p \times W} = \frac{v}{c_p}$$

$$K = 56 \text{ cu ft for general air-conditioning purposes}$$

The absolute humidity is the actual amount of water vapor contained in a given volume of air. It is expressed in grains, or lb per cu ft.

Relative humidity is the ratio of actual moisture content of air to the amount necessary to saturate at the same temperature.

Dewpoint is the measure of moisture in the air. It is the temperature at which saturation occurs when air is cooled to that level.

Dry-bulb temperature is the temperature of air as measured by an ordinary thermometer.

The dewpoint depression is dry-bulb temperature minus the dewpoint.

Wet-bulb temperature is the measure of total heat in the air.

Sensible heat is the heat content of dry air above 0 F.

The latent heat of water vapor is determined by the equation:

$$L = 1091.6 - 0.55t$$

Total heat of the air is the sum of the sensible heat and the latent heat of air for a given condition.

When air is saturated with moisture, dewpoint, dry-bulb, and wet-bulb temperatures are the same.

The psychrometric chart is used to determine the properties of air under various conditions of temperature and moisture content. It is based on 29.92 in. Hg barometric pressure.

The volume of air required to maintain constant RH is expressed as

$$\text{cu ft per min} = \frac{\text{Btu per min (load)} \times 56}{\text{dewpoint depression corresponding to RH}}$$

The volume of air required to maintain constant DB temperature and constant RH (dehumidify) is expressed as

$$\text{cu ft per min} = \frac{\text{Btu per min (load)} \times 56}{\text{Actual dewpoint depression}}$$

Refrigeration load for cooling is equal to the sum of all heat gains per hour divided by 12,000.

Infiltration is the natural leakage occurring in a conditioned space. It is expressed in number of room volume changes per hour.

Ducts are used to distribute air to remote spaces. Friction in ducts depends on length, number and types of elbows, design of entrances and exits, velocity of air, density of air, surface condition of duct, etc.

Static pressure is the measure of potential energy in the air.

Velocity pressure is the measure of kinetic energy in the air.

Total pressure is the sum of static and velocity pressures.

Air flow is determined by a Pitot tube that measures the static, velocity, and total pressures of air flowing through conduits.

Air may be purified by filtering, by separation, by centrifugal force, by washing, and electrostatically.

Study Reference

The following list of articles and publications that deal with various subjects pertaining to refrigeration and associated subjects is offered for two reasons.

First: It provides a handy reference for pursuing details of certain specific subjects which, because of space limitations, could not possibly be included in this or any single-volume text.

Secondly: It acquaints the student with reference sources that he will learn to employ not only in his pursuance of the listed subjects, but also for other associated material.

In some cases the list includes the titles of pertinent articles, together with the names of the author and the source. In other cases a general source is provided for the subject material. Many of the items listed may be obtained readily, either free or at a nominal charge. Others are out of print, but are available at the larger public libraries.

Sources of information not listed are those manufacturers credited with supplying the visual-aid material portrayed throughout this text. Other sources include the numerous specialty texts produced by the various publishers of technical books. It is suggested that the student become acquainted with the handbooks and the periodicals published by the various technical societies and manufacturers' associations, and the educational possibilities offered through such media.

The Industrial Index, available in public libraries, is an invaluable aid in locating sources of technical material.

In addition to the sources listed for various tables, others include the Smithsonian Physical Tables and the International Critical Tables. Many authoritative engineering handbooks are also available from various publishers.

The technical information compiled by the various agencies of the Federal Government is available from the Superintendent of Documents, U.S. Government Printing Office, Washington, D.C. 20402. Such publications are listed in catalogs of available material. Universities and colleges are also sources of specialized information.

AIR CONDITIONING

"Rational Psychrometric Formulae"—Willis H. Carrier.
American Society of Mechanical Engineers.

"Air Conditioning in Deep Mines"—R. W. Waterfill.
American Institute of Mining and Metallurgical Engineers.

"The Temperatures of Evaporation of Water into Air"—W. H. Carrier and Daniel C. Lindsay.
American Society of Refrigerating Engineers.

"Pipes, Ducts, and Fittings for Warm-Air Heating and Air Conditioning Systems."
U.S. Department of Commerce, R 207-56.
Superintendent of Documents, Washington 25, D. C.

"Theory of Evaporation with Special Reference to the Effect of Air Velocity and of Moisture Diffusion"—A. E. Stacey, Jr.
Journal of the American Ceramic Society, vol. 8, No. 7.

"Air Conditioning as Applied in Theatres and Film Laboratories"—D. C. Lindsay.
Transactions of the Society of Motion Picture Engineers, vol. xi.

"Comfort Air Conditioning."
"Air Conditioning Refrigerating Data Book—Applications, 1956–1957" (section VII).
American Society of Refrigerating Engineers.

"Industrial Air Conditioning."
"Air Conditioning and Refrigerating Data Book—Applications, 1956–1957" (section VIII).
American Society of Refrigerating Engineers.

"Automotive Air Conditioning"—Willis Stafford.
Detroit Controls Corporation, Detroit 8, Mich.

"Trane Air Conditioning Manual."
The Trane Company, La Crosse, Wis.

"Fan Engineering."
Buffalo Forge Company, Buffalo, N. Y.

"American Society of Heating, Ventilating and Air Conditioning Engineers Guide, 1957."
American Society of Heating and Air-Conditioning Engineers.

BRINES

"Air Conditioning Refrigerating Data Book—Design, 1955–1956" (chapter 8).
American Society of Refrigerating Engineers.

"Properties of Sodium Chloride Brines."
International Salt Company, Inc.

"Calcium Chloride for Refrigeration Brine."
Calcium Chloride Institute.

"Calcium Chloride in Refrigeration."
The Solvay Process Company, 40 Rector St., New York, N. Y.

COMPRESSORS

"Centrifugal Compression as Applied to Refrigeration"—W. H. Carrier.
Refrigerating Engineering, February, 1926.
American Society of Refrigerating Engineers.

"Centrifugal Compressors"—L. C. Loewenstein.
General Electric Review.
General Electric Company.

"Compressed Air Data."
Compressed Air Magazine.
Ingersoll-Rand Company, New York 4, N. Y.

"Small Compressors."
"Air Conditioning Refrigerating Data Book—Design, 1955–1956" (chapter 18).
American Society of Refrigerating Engineers.

"Large Compressors."
"Air Conditioning Refrigerating Data Book—Design, 1955–1956" (chapter 19).

American Society of Refrigerating Engineers.

"Large Screw Compressors for Refrigeration"—H. Soumerai.

Dunham Bush, Inc.

CONTROLS

"Solenoid Valves"—H. F. Spoehrer.

Service Engineer, June, 1948. RSES, *or*

Sporlan Valve Company, St. Louis, Mo.

"A Differential Temperature Expansion Valve for Low Temperature Applications"—F. Y. Carter.

American Society of Refrigerating Engineers, *or*

Detroit Lubricator Company, Detroit 8, Mich.

"Air Conditioning Refrigerating Data Book—Design, 1955–1956" (chapter 27).

American Society of Refrigerating Engineers.

"Operation and Testing Thermostatic Expansion Valves."

Refrigeration News, or

Detroit Controls Corporation, Detroit 8, Mich.

"Handbook of Automatic Refrigerant Controls."

Alco Valve Company, St. Louis 5, Mo.

"Thermal-electric Expansion Valve"—Alan A. Matthies.

Controls Co. of America.

ELECTRICITY

"Fundamentals of Electricity."

Westinghouse Electric and Manufacturing Company, and Science Service, Inc.

"Basic Electricity," Navy Training Courses, Navpers 10086.

Superintendent of Documents, Washington 25, D. C.

"Catechism of Electrical Machinery."

Fairbanks Morse & Company, Chicago, Ill.

FANS

"Fan Engineering."

Buffalo Forge Company, Buffalo, N. Y.

FOODS—COMPOSITION AND STORAGE REQUIREMENTS

U.S. Department of Agriculture.

Refer to U.S.D.A. list of available publications for specific items.

"Air Conditioning and Refrigerating Data Book—Applications, 1956–1957" (sections I, II, III, and IV).

HEAT TRANSFER—BUILDING MATERIALS

"American Society of Heating, Ventilating and Air Conditioning Engineers Guide, 1956."

American Society of Heating and Air-Conditioning Engineers.

"Armstrong Industrial Insulations."

Armstrong Cork Company, Lancaster, Pa.

HEAT TRANSFER—EQUIPMENT

"Air Conditioning Refrigerating Data Book—Design, 1955–1956."

Fundamentals (chapter 2).

Condensers (chapter 20).

Evaporators (chapter 22).

American Society of Refrigerating Engineers.

LUBRICANTS, ADDITIVES

"Lubricant Additives with a Purpose."

Power Engineering, April, 1957, courtesy of "Lubrication," The Texas Company, Beacon, N. Y.

"Lubricating Oil Additives" (5 parts).

Petroleum Refiner (Gulf Publishing Company).

Vol. 28, Nos. 6, 7, 8, 9, 10.

"Chemical Additives for Petroleum Lubricants."

American Cyanamid Company, 30 Rockefeller Plaza, New York 20, N. Y.

LUBRICANTS, TESTS

"A.S.T.M. Standards on Petroleum Products and Lubricants."
American Society for Testing Materials, Philadelphia 3, Pa.

LUBRICATION

"Refrigeration Compressors and Their Lubrication."
Socony Mobil Oil Company, Inc., New York, N. Y.

"Lubrication of Refrigeration and Air Conditioning Equipment," Sun Technical Bulletin B-3, Revised, January, 1950.
Sun Oil Company, Philadelphia, Pa.

"Plain Bearings, Fluid-Film Lubrication."
Socony Mobil Oil Company, Inc., New York, N. Y.

"Plain Bearings, Boundary Lubrication."
Socony Mobil Oil Company, Inc., New York, N. Y.

"Bearings and Their Lubrication."
Gulf Oil Corporation, Pittsburgh, Pa.

"Gears and Their Lubrication."
Socony Mobil Oil Company, Inc., New York, N. Y.

"A Study of Oil Separators in Refrigeration Systems"—James W. Archibald.
Temprite Products Corp.

pH CONTROL

"The ABC of pH Control."
LaMotte Chemical Products Company, Towson 4, Md.

"Betz Handbook of Industrial Water Conditioning"—Betz Laboratories, Inc.
Gillingham & Worth Sts., Philadelphia 24, Pa.

PIPING

"Pipe Friction Manual."
Hydraulic Institute, 122 E. 42nd St., New York 17, N. Y.

PUMPS

"Cameron Hydraulic Data."
Cameron Pump Division, Ingersoll-Rand Company, New York 14, N. Y.

"Standards of the Hydraulic Institute."
Hydraulic Institute, 122 E. 42nd St., New York 17, N. Y.

"Pump Handbook."
Worthington Corporation, Harrison, N. J.

REFRIGERANTS—THERMODYNAMIC PROPERTY TABLES AND MOLLIER CHARTS

"Ammonia Bulletin 142," Department of Commerce, Bureau of Standards.
Superintendent of Documents, Washington 25, D. C.

Carbon Dioxide:

"*Industrial and Engineering Chemistry*, vol. 38, No. 2.

"Air Conditioning and Refrigerating Data Book—Design, 1955–1956."

Carrene-7:

"Air Conditioning and Refrigerating Data Book—Design, 1955–1956."

Ethane:

The Linde Air Products Company.

"Air Conditioning and Refrigerating Data Book—Design, 1955–1956."

Ethylene:

"Air Conditioning and Refrigerating Data Book—Design, 1955–1956."

R-11, -12, -13, -14, -21, -22, -113, -114:

E. I. duPont de Nemours, Wilmington, Del. (Freons)

Industrial Chemicals Div., Allied Chemical Corp. (Genetrons)

"Air Conditioning and Refrigerating Data Book—Design, 1955–1956."

Air Conditioning and Refrigeration News, January 9 and January 16, 1950, issues.

Iso-butane:

Refrigerating Engineering, vol. 12, No. 12.

Industrial and Engineering Chemistry, June, 1938.

"Air Conditioning and Refrigerating Data Book—Design, 1955–1956."

Kulene-131:

Eston Chemicals, Inc., 3100 E. 26th St., Los Angeles 23, Calif.

Methyl Chloride:

"Air Conditioning and Refrigerating Data Book—Design, 1955–1956."

Propane:

Refrigerating Engineering, vol. 12, No. 12.

"Air Conditioning and Refrigerating Data Book—Design, 1955–1956."

Propylene:

Transactions of the American Institute of Chemical Engineers, vol. 41, No. 3.

Sulfur dioxide
Ethylamine
Methylamine
Methyl formate
Methylene chloride
Nitrous oxide
} "Air Conditioning and Refrigerating Data Book—Design, 1955–1956."

Refrigerants, Safety Classification:

National Board of Fire Underwriters.

Cyrogenics:

"The Werkspoor Cold-Gas Refrigerator."

CVI Corporation.

WATER

"Water Conditioning Handbook."

The Permutit Company, 330 W. 42nd St., New York 18, N. Y.

"Betz Handbook of Industrial Water Conditioning"—Betz Laboratories, Inc.

Gillingham & Worth Sts., Philadelphia 24, Pa.

"The Control of Fouling Organisms in Fresh and Salt Water Circuits"—John G. Dobson.

Transactions of the American Society of Mechanical Engineers, April, 1946, *or*

Wallace & Tiernan, Belleville 9, N. J.

"Questions and Answers on Boiler Feed Water Treatment."

Bulletin 30, National Aluminate Corporation, 6216 W. 66th Place, Chicago 38, Ill.

"Chlorination of Cooling Waters"—N. C. Estes.

Reprinted from *Refiner and Natural Gas Manufacturer*.

Wallace & Tiernan, Belleville 9, N. J.

"Identification and Characteristics of Typical Slime-Forming Organisms"—R. B. Martin and John G. Dobson.

Paper Trade Journal, October 11, 1945.

Wallace & Tiernan, Belleville 9, N. J.

"Bacterial Corrosion of Cooling Water Pipes"—C. L. Clark and W. J. Nungester.

Timken Roller Bearing Company.

Wallace & Tiernan, Belleville 9, N. J.

Appendix

Table I. Saturated Ammonia: Temperature Table†

Temp. °F.	Pressure. Absolute. lbs./in.²	Pressure. Gage. lbs./in.²	Volume vapor. ft³/lb.	Density vapor. lbs./ft.³	Heat content. Liquid. Btu./lb.	Heat content. Vapor. Btu./lb.	Latent heat. Btu./lb.	Entropy. Liquid. Btu./lb.°F.	Entropy. Vapor. Btu./lb.°F.	Temp. °F.
t	p	$g.p.$	V	$1/V$	h	H	L	s	S	t
−60	5.55	*18.6	44.73	0.02235	−21.2	589.6	610.8	−0.0517	1.4769	**−60**
−59	5.74	*18.2	43.37	.02306	−20.1	590.0	610.1	−.0490	.4741	−59
−58	5.93	*17.8	42.05	.02378	−19.1	590.4	609.5	−.0464	.4713	−58
−57	6.13	*17.4	40.79	.02452	−18.0	590.8	608.8	−.0438	.4686	−57
−56	6.33	*17.0	39.56	.02528	−17.0	591.2	608.2	−.0412	.4658	−56
−55	6.54	*16.6	38.38	0.02605	−15.9	591.6	607.5	−0.0386	1.4631	**−55**
−54	6.75	*16.2	37.24	.02685	−14.8	592.1	606.9	.0360	.4604	−54
−53	6.97	*15.7	36.15	.02766	−13.8	592.4	606.2	−.0334	.4577	−53
−52	7.20	*15.3	35.09	.02850	−12.7	592.9	605.6	−.0307	.4551	−52
−51	7.43	*14.8	34.06	.02936	−11.7	593.2	604.9	−.0281	.4524	−51
−50	7.67	*14.3	33.08	0.03023	−10.6	593.7	604.3	−0.0256	1.4497	**−50**
−49	7.91	*13.8	32.12	.03113	−9.6	594.0	603.6	−.0230	.4471	−49
−48	8.16	*13.3	31.20	.03205	−8.5	594.4	602.9	−.0204	.4445	−48
−47	8.42	*12.8	30.31	.03299	−7.4	594.9	602.3	−.0179	.4419	−47
−46	8.68	*12.2	29.45	.03395	−6.4	595.2	601.6	−.0153	.4393	−46
−45	8.95	*11.7	28.62	0.03494	−5.3	595.6	600.9	−0.0127	1.4368	**−45**
−44	9.23	*11.1	27.82	.03595	−4.3	596.0	600.3	−.0102	.4342	−44
−43	9.51	*10.6	27.04	.03698	−3.2	596.4	599.6	−.0076	.4317	−43
−42	9.81	*10.0	26.29	.03804	−2.1	596.8	598.9	−.0051	.4292	−42
−41	10.10	*9.3	25.56	.03912	−1.1	597.2	598.3	−.0025	.4267	−41
−40	10.41	*8.7	24.86	0.04022	0.0	597.6	597.6	0.0000	1.4242	**−40**
−39	10.72	*8.1	24.18	.04135	1.1	598.0	596.9	.0025	.4217	−39
−38	11.04	*7.4	23.53	.04251	2.1	598.3	596.2	.0051	.4193	−38
−37	11.37	*6.8	22.89	.04369	3.2	598.7	595.5	.0076	.4169	−37
−36	11.71	*6.1	22.27	.04489	4.3	599.1	594.8	.0101	.4144	−36
−35	12.05	*5.4	21.68	0.04613	5.3	599.5	594.2	0.0126	1.4120	**−35**
−34	12.41	*4.7	21.10	.04739	6.4	599.9	593.5	.0151	.4096	−34
−33	12.77	*3.9	20.54	.04868	7.4	600.2	592.8	.0176	.4072	−33
−32	13.14	*3.2	20.00	.04999	8.5	600.6	592.1	.0201	.4048	−32
−31	13.52	*2.4	19.48	.05134	9.6	601.0	591.4	.0226	.4025	−31
−30	13.90	*1.6	18.97	0.05271	10.7	601.4	590.7	0.0250	1.4001	**−30**
−29	14.30	*0.8	18.48	.05411	11.7	601.7	590.0	.0275	.3978	−29
−28	14.71	0.0	18.00	.05555	12.8	602.1	589.3	.0300	.3955	−28
−27	15.12	0.4	17.54	.05701	13.9	602.5	588.6	.0325	.3932	−27
−26	15.55	0.8	17.09	.05850	14.9	602.8	587.9	.0350	.3909	−26
−25	15.98	1.3	16.66	0.06003	16.0	603.2	587.2	0.0374	1.3886	**−25**
−24	16.42	1.7	16.24	.06158	17.1	603.6	586.5	.0399	.3863	−24
−23	16.88	2.2	15.83	.06317	18.1	603.9	585.8	.0423	.3840	−23
−22	17.34	2.6	15.43	.06479	19.2	604.3	585.1	.0448	.3818	−22
−21	17.81	3.1	15.05	.06644	20.3	604.6	584.3	.0472	.3796	−21
−20	18.30	3.6	14.68	0.06813	21.4	605.0	583.6	0.0497	1.3774	**−20**
−19	18.79	4.1	14.32	.06985	22.4	605.3	582.9	.0521	.3752	−19
−18	19.30	4.6	13.97	.07161	23.5	605.7	582.2	.0545	.3729	−18
−17	19.81	5.1	13.62	.07340	24.6	606.1	581.5	.0570	.3708	−17
−16	20.34	5.6	13.29	.07522	25.6	606.4	580.8	.0594	.3686	−16
−15	20.88	6.2	12.97	0.07709	26.7	606.7	580.0	0.0618	1.3664	**−15**
−14	21.43	6.7	12.66	.07898	27.8	607.1	579.3	.0642	.3643	−14
−13	21.99	7.3	12.36	.08092	28.9	607.5	578.6	.0666	.3621	−13
−12	22.56	7.9	12.06	.08289	30.0	607.8	577.8	.0690	.3600	−12
−11	23.15	8.5	11.78	.08490	31.0	608.1	577.1	.0714	.3579	−11
−10	23.74	9.0	11.50	0.08695	32.1	608.5	576.4	0.0738	1.3558	**−10**

*** Inches of mercury below one standard atmosphere (29.92 in.).**

† Courtesy, Circular of the Bureau of Standards, No. 142, pp. 18–21, Washington, U.S. Government Printing Office, 1945.

TABLE I. SATURATED AMMONIA: TEMPERATURE TABLE—(*Continued*)

Temp. °F.	Pressure. Absolute. lbs./in.²	Gage. lbs./in.²	Volume vapor. ft³/lb.	Density vapor. lbs./ft.³	Heat content. Liquid. Btu./lb.	Vapor. Btu./lb.	Latent heat. Btu./lb.	Entropy. Liquid. Btu./lb.°F.	Vapor. Btu./lb.°F.	Temp. °F.
t	p	$g.\ p.$	V	$1/V$	h	H	L	s	S	t
−10	23.74	9.0	11.50	0.08695	32.1	608.5	576.4	0.0738	1.3558	**−10**
−9	24 35	9.7	11.23	.08904	33.2	608.8	575.6	.0762	.3537	−9
−8	24.97	10.3	10.97	.09117	34.3	609.2	574.9	.0786	.3516	−8
−7	25.61	10.9	10.71	.09334	35.4	609.5	574.1	.0809	.3495	−7
−6	26.26	11.6	10.47	.09555	36.4	609.8	573.4	.0833	.3474	−6
−5	26.92	12.2	10.23	0.09780	37.5	610.1	572.6	0.0857	1.3454	**−5**
−4	27.59	12.9	9.991	.1001	38.6	610.5	571.9	.0880	.3433	−4
−3	28.28	13.6	9.763	.1024	39.7	610.8	571.1	.0904	.3413	−3
−2	28.98	14.3	9.541	.1048	40.7	611.1	570.4	.0928	.3393	−2
−1	29.69	15.0	9.326	.1072	41.8	611.4	569.6	.0951	.3372	−1
0	30.42	15.7	9.116	0.1097	42.9	611.8	568.9	0.0975	1.3352	**0**
1	31.16	16.5	8.912	.1122	44.0	612.1	568.1	.0998	.3332	1
2	31.92	17.2	8.714	.1148	45.1	612.4	567.3	.1022	.3312	2
3	32.69	18.0	8.521	.1174	46.2	612.7	566.5	.1045	.3292	3
4	33.47	18.8	8.333	.1200	47.2	613.0	565.8	.1069	.3273	4
5	34.27	19.6	8.150	0.1227	48.3	613.3	565.0	0.1092	1.3253	**5**
6	35.09	20.4	7.971	.1254	49.4	613.6	564.2	.1115	.3234	6
7	35.92	21.2	7.798	.1282	50.5	613.9	563.4	.1138	.3214	7
8	36.77	22.1	7.629	.1311	51.6	614.3	562.7	.1162	.3195	8
9	37.63	22.9	7.464	.1340	52.7	614.6	561.9	.1185	.3176	9
10	38.51	23.8	7.304	0.1369	53.8	614.9	561.1	0.1208	1.3157	**10**
11	39.40	24.7	7.148	.1399	54.9	615.2	560.3	.1231	.3137	11
12	40.31	25.6	6.996	.1429	56.0	615.5	559.5	.1254	.3118	12
13	41.24	26.5	6.847	.1460	57.1	615.8	558.7	.1277	.3099	13
14	42.18	27.5	6.703	.1492	58.2	616.1	557.9	.1300	.3081	14
15	43.14	28.4	6.562	0.1524	59.2	616.3	557.1	0.1323	1.3062	**15**
16	44.12	29.4	6.425	.1556	60.3	616.6	556.3	.1346	.3043	16
17	45.12	30.4	6.291	.1590	61.4	616.9	555 5	.1369	.3025	17
18	46.13	31.4	6.161	.1623	62.5	617.2	554.7	.1392	.3006	18
19	47.16	32.5	6.034	.1657	63.6	617.5	553.9	.1415	.2988	19
20	48.21	33.5	5.910	0.1692	64.7	617.8	553.1	0.1437	1.2969	**20**
21	49.28	34.6	5.789	.1728	65.8	618.0	552.2	.1460	.2951	21
22	50.36	35.7	5.671	.1763	66.9	618.3	551.4	.1483	.2933	22
23	51.47	36.8	5.556	.1800	68.0	618.6	550.6	.1505	.2915	23
24	52.59	37.9	5.443	.1837	69.1	618.9	549.8	.1528	.2897	24
25	53.73	39.0	5.334	0.1875	70.2	619.1	548.9	0.1551	1.2879	**25**
26	54.90	40.2	5.227	.1913	71.3	619.4	548.1	.1573	.2861	26
27	56.08	41.4	5.123	.1952	72.4	619.7	547.3	.1596	.2843	27
28	57.28	42.6	5.021	.1992	73.5	619.9	546.4	.1618	.2825	28
29	58.50	43.8	4.922	.2032	74.6	620.2	545.6	.1641	.2808	29
30	59.74	45.0	4.825	0.2073	75.7	620.5	544.8	0.1663	1.2790	**30**
31	61.00	46.3	4.730	.2114	76.8	620.7	543.9	.1686	.2773	31
32	62.29	47.6	4.637	.2156	77.9	621.0	543.1	.1708	.2755	32
33	63.59	48.9	4.547	.2199	79.0	621.2	542.2	.1730	.2738	33
34	64.91	50.2	4.459	.2243	80.1	621.5	541.4	.1753	.2721	34
35	66.26	51.6	4.373	0.2287	81.2	621.7	540.5	0.1775	1.2704	**35**
36	67.63	52.9	4.289	.2332	82.3	622.0	539.7	.1797	.2686	36
37	69.02	54.3	4.207	.2377	83.4	622.2	538.8	.1819	.2669	37
38	70.43	55.7	4.126	.2423	84.6	622.5	537.9	.1841	.2652	38
39	71.87	57.2	4.048	.2470	85.7	622.7	537.0	.1863	.2635	39
40	73.32	58.6	3.971	0.2518	86.8	623.0	536.2	0.1885	1,2618	**40**

TABLE I. SATURATED AMMONIA: TEMPERATURE TABLE—(*Continued*)

Temp. °F.	Pressure.		Volume vapor. ft.³/lb.	Density vapor. lbs./ft.³	Heat content.		Latent heat. Btu./lb.	Entropy.		Temp. °F.
	Absolute. lbs./in.²	Gage. lbs./in.²			Liquid. Btu./lb.	Vapor. Btu./lb.		Liquid. Btu./lb. °F.	Vapor. Btu./lb. °F.	
t	*p*	*g. p.*	*V*	*1/V*	*h*	*H*	*L*	*s*	*S*	*t*
40	73.32	58.6	3.971	0.2518	86.8	623.0	536.2	0.1885	1.2618	**40**
41	74.80	60.1	3.897	.2566	87.9	623.2	535.3	.1908	.2602	41
42	76.31	61.6	3.823	.2616	89.0	623.4	534.4	.1930	.2585	42
43	77.83	63.1	3.752	.2665	90.1	623.7	533.6	.1952	.2568	43
44	79.38	64.7	3.682	.2716	91.2	623.9	532.7	.1974	.2552	44
45	80.96	66.3	3.614	0.2767	92.3	624.1	531.8	0.1996	1.2535	**45**
46	82.55	67.9	3.547	.2819	93.5	624.4	530.9	.2018	.2519	46
47	84.18	69.5	3.481	.2872	94.6	624.6	530.0	.2040	.2502	47
48	85.82	71.1	3.418	.2926	95.7	624.8	529.1	.2062	.2486	48
49	87.49	72.8	3.355	.2981	96.8	625.0	528.2	.2083	.2469	49
50	89.19	74.5	3.294	0.3036	97.9	625.2	527.3	0.2105	1.2453	**50**
51	90.91	76.2	3.234	.3092	99.1	625.5	526.4	.2127	.2437	51
52	92.66	78.0	3.176	.3149	100.2	625.7	525.5	.2149	.2421	52
53	94.43	79.7	3.119	.3207	101.3	625.9	524.6	.2171	.2405	53
54	96.23	81.5	3.063	.3265	102.4	626.1	523.7	.2192	.2389	54
55	98.06	83.4	3.008	0.3325	103.5	626.3	522.8	0.2214	1.2373	**55**
56	99.91	85.2	2.954	.3385	104.7	626.5	521.8	.2236	.2357	56
57	101.8	87.1	2.902	.3446	105.8	626.7	520.9	.2257	.2341	57
58	103.7	89.0	2.851	.3508	106.9	626.9	520.0	.2279	.2325	58
59	105.6	90.9	2.800	.3571	108.1	627.1	519.0	.2301	.2310	59
60	107.6	92.9	2.751	0.3635	109.2	627.3	518.1	0.2322	1.2294	**60**
61	109.6	94.9	2.703	.3700	110.3	627.5	517.2	.2344	.2278	61
62	111.6	96.9	2.656	.3765	111.5	627.7	516.2	.2365	.2262	62
63	113.6	98.9	2.610	.3832	112.6	627.9	515.3	.2387	.2247	63
64	115.7	101.0	2.565	.3899	113.7	628.0	514.3	.2408	.2231	64
65	117.8	103.1	2.520	0.3968	114.8	628.2	513.4	0.2430	1.2216	**65**
66	120.0	105.3	2.477	.4037	116.0	628.4	512.4	.2451	.2201	66
67	122.1	107.4	2.435	.4108	117.1	628.6	511.5	.2473	.2186	67
68	124.3	109.6	2.393	.4179	118.3	628.8	510.5	.2494	.2170	68
69	126.5	111.8	2.352	.4251	119.4	628.9	509.5	.2515	.2155	69
70	128.8	114.1	2.312	0.4325	120.5	629.1	508.6	0.2537	1.2140	**70**
71	131.1	116.4	2.273	.4399	121.7	629.3	507.6	.2558	.2125	71
72	133.4	118.7	2.235	.4474	122.8	629.4	506.6	.2579	.2110	72
73	135.7	121.0	2.197	.4551	124.0	629.6	505.6	.2601	.2095	73
74	138.1	123.4	2.161	.4628	125.1	629.8	504.7	.2622	.2080	74
75	140.5	125.8	2.125	0.4707	126.2	629.9	503.7	0.2643	1.2065	**75**
76	143.0	128.3	2.089	.4786	127.4	630.1	502.7	.2664	.2050	76
77	145.4	130.7	2.055	.4867	128.5	630.2	501.7	.2685	.2035	77
78	147.9	133.2	2.021	.4949	129.7	630.4	500.7	.2706	.2020	78
79	150.5	135.8	1.988	.5031	130.8	630.5	499.7	.2728	.2006	79
80	153.0	138.3	1.955	0.5115	132.0	630.7	498.7	0.2749	1.1991	**80**
81	155.6	140.9	1.923	.5200	133.1	630.8	497.7	.2769	.1976	81
82	158.3	143.6	1.892	.5287	134.3	631.0	496.7	.2791	.1962	82
83	161.0	146.3	1.861	.5374	135.4	631.1	495.7	.2812	.1947	83
84	163.7	149.0	1.831	.5462	136.6	631.3	494.7	.2833	.1933	84
85	166.4	151.7	1.801	0.5552	137.8	631.4	493.6	0.2854	1.1918	**85**

Temp. °F.	Pressure.		Volume vapor. ft.³/lb.	Density vapor. lbs./ft.³	Heat content.		Latent heat. Btu./lb.	Entropy.		Temp. °F.
	Absolute. lbs./in.²	Gage. lbs./in.²			Liquid. Btu./lb.	Vapor. Btu./lb.		Liquid. Btu./lb.°F.	Vapor. Btu./lb.°F.	
t	*p*	*g. p.*	*V*	*1/V*	*h*	*H*	*L*	*s*	*S*	*t*
85	166.4	151.7	1.801	0.5552	137.8	631.4	493.6	0.2854	1.1918	**85**
86	169.2	154.5	1.772	.5643	138.9	631.5	492.6	.2875	.1904	86
87	172.0	157.3	1.744	.5735	140.1	631.7	491.6	.2895	.1889	87
88	174.8	160.1	1.716	.5828	141.2	631.8	490.6	.2917	.1875	88
89	177.7	163.0	1.688	.5923	142.4	631.9	489.5	.2937	.1860	89
90	180.6	165.9	1.661	0.6019	143.5	632.0	488.5	0.2958	1.1846	**90**
91	183.6	168.9	1.635	.6116	144.7	632.1	487.4	.2979	.1832	91
92	186.6	171.9	1.609	.6214	145.8	632.2	486.4	.3000	.1818	92
93	189.6	174.9	1.584	.6314	147.0	632.3	485.3	.3021	.1804	93
94	192.7	178.0	1.559	.6415	148.2	632.5	484.3	.3041	.1789	94
95	195.8	181.1	1.534	0.6517	149.4	632.6	483.2	0.3062	1.1775	**95**
96	198.9	184.2	1.510	.6620	150.5	632.6	482.1	.3083	.1761	96
97	202.1	187.4	1.487	.6725	151.7	632.8	481.1	.3104	.1747	97
98	205.3	190.6	1.464	.6832	152.9	632.9	480.0	.3125	.1733	98
99	208.6	193.9	1.441	.6939	154.0	632.9	478.9	.3145	.1719	99
100	211.9	197.2	1.419	0.7048	155.2	633.0	477.8	0.3166	1.1705	**100**
101	215.2	200.5	1.397	.7159	156.4	633.1	476.7	.3187	.1691	101
102	218.6	203.9	1.375	.7270	157.6	633.2	475.6	.3207	.1677	102
103	222.0	207.3	1.354	.7384	158.7	633.3	474.6	.3228	.1663	103
104	225.4	210.7	1.334	.7498	159.9	633.4	473.5	.3248	.1649	104
105	228.9	214.2	1.313	0.7615	161.1	633.4	472.3	0.3269	1.1635	**105**
106	232.5	217.8	1.293	.7732	162.3	633.5	471.2	.3289	.1621	106
107	236.0	221.3	1.274	.7852	163.5	633.6	470.1	.3310	.1607	107
108	239.7	225.0	1.254	.7972	164.6	633.6	469.0	.3330	.1593	108
109	243.3	228.6	1.235	.8095	165.8	633.7	467.9	.3351	.1580	109
110	247.0	232.3	1.217	0.8219	167.0	633.7	466.7	0.3372	1.1566	**110**
111	250.8	236.1	1.198	.8344	168.2	633.8	465.6	.3392	.1552	111
112	254.5	239.8	1.180	.8471	169.4	633.8	464.4	.3413	.1538	112
113	258.4	243.7	1.163	.8600	170.6	633.9	463.3	.3433	.1524	113
114	262.2	247.5	1.145	.8730	171.8	633.9	462.1	.3453	.1510	114
115	266.2	251.5	1.128	0.8862	173.0	633.9	460.9	0.3474	1.1497	**115**
116	270.1	255.4	1.112	.8996	174.2	634.0	459.8	.3495	.1483	116
117	274.1	259.4	1.095	.9132	175.4	634.0	458.6	.3515	.1469	117
118	278.2	263.5	1.079	.9269	176.6	634.0	457.4	.3535	.1455	118
119	282.3	267.6	1.063	.9408	177.8	634.0	456.2	.3556	.1441	119
120	286.4	271.7	1.047	0.9549	179.0	634.0	455.0	0.3576	1.1427	**120**
121	290.6	275.9	1.032	.9692	180.2	634.0	453.8	.3597	.1414	121
122	294.8	280.1	1.017	.9837	181.4	634.0	452.6	.3618	.1400	122
123	299.1	284.4	1.002	.9983	182.6	634.0	451.4	.3638	.1386	123
124	303.4	288.7	0.987	1.0132	183.9	634.0	450.1	.3659	.1372	124
125	307.8	293.1	0.973	1.028	185.1	634.0	448.9	0.3679	1.1358	**125**

TABLE II. SATURATED AMMONIA: ABSOLUTE PRESSURE TABLE*

Pressure (abs.). lbs./in.²	Temp. °F.	Volume vapor. ft.³/lb.	Density vapor. lbs./ft.³	Heat content. Liquid. Btu./lb.	Heat content. Vapor. Btu./lb.	Latent heat. Btu./lb.	Entropy. Liquid. Btu./lb. °F.	Entropy. Evap. Btu./lb.°F.	Entropy. Vapor. Btu./lb.°F.	Pressure (abs.). lbs./in.²
p	*t*	*V*	*1/V*	*h*	*H*	*L*	*s*	*L/T*	*S*	*p*
5.0	-63.11	49.31	0.02029	-24.5	588.3	612.8	-0.0599	1.5456	1.4857	**5.0**
5.5	-60.27	45.11	.02217	-21.5	589.5	611.0	- .0524	.5301	.4777	5.5
6.0	-57.64	41.59	.02405	-18.7	590.6	609.3	- .0455	.5158	.4703	6.0
6.5	-55.18	38.59	.02591	-16.1	591.6	607.7	- .0390	.5026	.4636	6.5
7.0	-52.88	36.01	.02777	-13.7	592.5	606.2	- .0330	.4904	.4574	7.0
7.5	-50.70	33.77	0.02962	-11.3	593.4	604.7	-0.0274	1.4790	1.4516	**7.5**
8.0	-48.64	31.79	.03146	- 9.2	594.2	603.4	- .0221	.4683	.4462	8.0
8.5	-46.69	30.04	.03329	- 7.1	595.0	602.1	- .0171	.4582	.4411	8.5
9.0	-44.83	28.48	.03511	- 5.1	595.7	600.8	- .0123	.4486	.4363	9.0
9.5	-43.05	27.08	.03693	- 3.2	596.4	599.6	- .0077	.4396	.4319	9.5
10.0	-41.34	25.81	0.03874	- 1.4	597.1	598.5	-0.0034	1.4310	1.4276	**10.0**
10.5	-39.71	24.66	.04055	+ 0.3	597.7	597.4	+ .0007	.4228	.4235	10.5
11.0	-38.14	23.61	.04235	2.0	598.3	596.3	.0047	.4149	.4196	11.0
11.5	-36.62	22.65	.04414	3.6	598.9	595.3	.0085	.4074	.4159	11.5
12.0	-35.16	21.77	.04593	5.1	599.4	594.3	.0122	.4002	.4124	12.0
12.5	-33.74	20.96	0.04772	6.7	600.0	593.3	0.0157	1.3933	1.4090	**12.5**
13.0	-32.37	20.20	.04950	8.1	600.5	592.4	.0191	.3866	.4057	13.0
13.5	-31.05	19.50	.05128	9.6	601.0	591.4	.0225	.3801	.4026	13.5
14.0	-29.76	18.85	.05305	10.9	601.4	590.5	.0257	.3739	.3996	14.0
14.5	-28.51	18.24	.05482	12.2	601.9	589.7	.0288	.3679	.3967	14.5
15.0	-27.29	17.67	0.05658	13.6	602.4	588.8	0.0318	1.3620	1.3938	**15.0**
15.5	-26.11	17.14	.05834	14.8	602.8	588.0	.0347	.3564	.3911	15.5
16.0	-24.95	16.64	.06010	16.0	603.2	587.2	.0375	.3510	.3885	16.0
16.5	-23.83	16.17	.06186	17.2	603.6	586.4	.0403	.3456	.3859	16.5
17.0	-22.73	15.72	.06361	18.4	604.0	585.6	.0430	.3405	.3835	17.0
17.5	-21.66	15.30	0.06535	19.6	604.4	584.8	0.0456	1.3354	1.3810	**17.5**
18.0	-20.61	14.90	.06710	20.7	604.8	584.1	.0482	.3305	.3787	18.0
18.5	-19.59	14.53	.06884	21.8	605.1	583.3	.0507	.3258	.3765	18.5
19.0	-18.58	14.17	.07058	22.9	605.5	582.6	.0531	.3211	.3742	19.0
19.5	-17.60	13.83	.07232	23.9	605.8	581.9	.0555	.3166	.3721	19.5
20.0	-16.64	13.50	0.07405	25.0	606.2	581.2	0.0578	1.3122	1.3700	**20.0**
20.5	-15.70	13.20	.07578	26.0	606.5	580.5	.0601	.3078	.3679	20.5
21.0	-14.78	12.90	.07751	27.0	606.8	579.8	.0623	.3036	.3659	21.0
21.5	-13.87	12.62	.07924	27.9	607.1	579.2	.0645	.2995	.3640	21.5
22.0	-12.98	12.35	.08096	28.9	607.4	578.5	.0666	.2955	.3621	22.0
22.5	-12.11	12.09	0.08268	29.8	607.7	577.9	0.0687	1.2915	1.3602	**22.5**
23.0	-11.25	11.85	.08440	30.8	608.1	577.3	.0708	.2876	.3584	23.0
23.5	-10.41	11.61	.08612	31.7	608.3	576.6	.0728	.2838	.3566	23.5
24.0	- 9.58	11.39	.08783	32.6	608.6	576.0	.0748	.2801	.3549	24.0
24.5	- 8.76	11.17	.08955	33.5	608.9	575.4	.0768	.2764	.3532	24.5
25.0	- 7.96	10.96	0.09126	34.3	609.1	574.8	0.0787	1.2728	1.3515	**25.0**
25.5	- 7.17	10.76	.09297	35.2	609.4	574.2	.0805	.2693	.3498	25.5
26.0	- 6.39	10.56	.09468	36.0	609.7	573.7	.0824	.2658	.3482	26.0
26.5	- 5.63	10.38	.09638	36.8	609.9	573.1	.0842	.2625	.3467	26.5
27.0	- 4.87	10.20	.09809	37.7	610.2	572.5	.0860	.2591	.3451	27.0
27.5	- 4.13	10.02	0.09979	38.4	610.4	572.0	0.0878	1.2558	1.3436	**27.5**
28.0	- 3.40	9.853	.1015	39.3	610.7	571.4	.0895	.2526	.3421	28.0
28.5	- 2.68	9.691	.1032	40.0	610.9	570.9	.0912	.2494	.3406	28.5
29.0	- 1.97	9.534	.1049	40.8	611.1	570.3	.0929	.2463	.3392	29.0
29.5	- 1.27	9.383	.1066	41.6	611.4	569.8	.0945	.2433	.3378	29.5
30.0	- 0.57	9.236	0.1083	42.3	611.6	569.3	0.0962	1.2402	1.3364	**30.0**

* Courtesy, Circular of the Bureau of Standards, No. 142, pp. 22–25, Washington, U.S. Government Printing Office, 1945.

TABLE II. SATURATED AMMONIA: ABSOLUTE PRESSURE TABLE—(*Continued*)

Pressure (abs.). lbs./in.²	Temp. °F.	Volume vapor. ft.³/lb.	Density vapor. lbs./ft.³	Heat content.		Latent heat. Btu./lb.	Entropy.			Pressure (abs.). lbs./in.²
				Liquid. Btu./lb.	Vapor. Btu./lb.		Liquid. Btu./lb. °F.	Evap. Btu./lb.°F.	Vapor. Btu./lb.°F.	
p	t	V	$1/V$	h	H	L	s	L/T	S	p
30	−0.57	9.236	0.1083	42.3	611.6	569.3	0.0962	1.2402	1.3364	**30**
31	+0.79	8.955	.1117	43.8	612.0	568.2	.0993	.2343	.3336	31
32	2.11	8.693	.1150	45.2	612.4	567.2	.1024	.2286	.3310	32
33	3.40	8.445	.1184	46.6	612.8	566.2	.1055	.2230	.3285	33
34	4.66	8.211	.1218	48.0	613.2	565.2	.1084	.2176	.3260	34
35	5.89	7.991	0.1251	49.3	613.6	564.3	0.1113	1.2123	1.3236	**35**
36	7.09	7.782	.1285	50.6	614.0	563.4	.1141	.2072	.3213	36
37	8.27	7.584	.1319	51.9	614.3	562.4	.1168	.2022	.3190	37
38	9.42	7.396	.1352	53.2	614.7	561.5	.1195	.1973	.3168	38
39	10.55	7.217	.1386	54.4	615.0	560.6	.1221	.1925	.3146	39
40	11.66	7.047	0.1419	55.6	615.4	559.8	0.1246	1.1879	1.3125	**40**
41	12.74	6.885	.1452	56.8	615.7	558.9	.1271	.1833	.3104	41
42	13.81	6.731	.1486	57.9	616.0	558.1	.1296	.1788	.3084	42
43	14.85	6.583	.1519	59.1	616.3	557.2	.1320	.1745	.3065	43
44	15.88	6.442	.1552	60.2	616.6	556.4	.1343	.1703	.3046	44
45	16.88	6.307	0.1586	61.3	616.9	555.6	0.1366	1.1661	1.3027	**45**
46	17.87	6.177	.1619	62.4	617.2	554.8	.1389	.1620	.3009	46
47	18.84	6.053	.1652	63.4	617.4	554.0	.1411	.1580	.2991	47
48	19.80	5.934	.1685	64.5	617.7	553.2	.1433	.1540	.2973	48
49	20.74	5.820	.1718	65.5	618.0	552.5	.1454	.1502	.2956	49
50	21.67	5.710	0.1751	66.5	618.2	551.7	0.1475	1.1464	1.2939	**50**
51	22.58	5.604	.1785	67.5	618.5	551.0	.1496	.1427	.2923	51
52	23.48	5.502	.1818	68.5	618.7	550.2	.1516	.1390	.2906	52
53	24.36	5.404	.1851	69.5	619.0	549.5	.1536	.1354	.2890	53
54	25.23	5.309	.1884	70.4	619.2	548.8	.1556	.1319	.2875	54
55	26.00	5.218	0.1917	71.4	619.4	548.0	0.1575	1.1284	1.2859	**55**
56	26.94	5.129	.1950	72.3	619.7	547.4	.1594	.1250	.2844	56
57	27.77	5.044	.1983	73.3	619.9	546.6	.1613	.1217	.2830	57
58	28.59	4.962	.2015	74.2	620.1	545.9	.1631	.1184	.2815	58
59	29.41	4.882	.2048	75.0	620.3	545.3	.1650	.1151	.2801	59
60	30.21	4.805	0.2081	75.9	620.5	544.6	0.1668	1.1119	1.2787	**60**
61	31.00	4.730	.2114	76.8	620.7	543.9	.1685	.1088	.2773	61
62	31.78	4.658	.2147	77.7	620.9	543.2	.1703	.1056	.2759	62
63	32.55	4.588	.2180	78.5	621.1	542.6	.1720	.1026	.2746	63
64	33.31	4.519	.2213	79.4	621.3	541.9	.1737	.0996	.2733	64
65	34.06	4.453	0.2245	80.2	621.5	541.3	0.1754	1.0966	1.2720	**65**
66	34.81	4.389	.2278	81.0	621.7	540.7	.1770	.0937	.2707	66
67	35.54	4.327	.2311	81.8	621.9	540.1	.1787	.0907	.2694	67
68	36.27	4.207	.2344	82.6	622.0	539.4	.1803	.0879	.2682	68
69	36.99	4.208	.2377	83.4	622.2	538.8	.1819	.0851	.2670	69
70	37.70	4.151	0.2409	84.2	622.4	538.2	0.1835	1.0823	1.2658	**70**
71	38.40	4.095	.2442	85.0	622.6	537.6	.1850	.0795	.2645	71
72	39.09	4.041	.2475	85.8	622.8	537.0	.1866	.0768	.2634	72
73	39.78	3.988	.2507	86.5	622.9	536.4	.1881	.0741	.2622	73
74	40.46	3.937	.2540	87.3	623.1	535.8	.1896	.0715	.2611	74
75	41.13	3.887	0.2573	88.0	623.2	535.2	0.1910	1.0689	1.2599	**75**
76	41.80	3.838	.2606	88.8	623.4	534.6	.1925	.0663	.2588	76
77	42.46	3.790	.2638	89.5	623.5	534.0	.1940	.0637	.2577	77
78	43.11	3.744	.2671	90.2	623.7	533.5	.1954	.0612	.2566	78
79	43.76	3.699	.2704	90.9	623.8	532.9	.1968	.0587	.2555	79
80	44.40	3.655	0.2736	91.7	624.0	532.3	0.1982	1.0563	1.2545	**80**

TABLE II. SATURATED AMMONIA: ABSOLUTE PRESSURE TABLE—(*Continued*)

Pressure (abs.). lbs./in.²	Temp. °F.	Volume vapor. ft.³/lb.	Density vapor. lbs./ft.³	Heat content.		Latent heat. Btu./lb.	Entropy.			Pressure (abs.). lbs./in.²
				Liquid. Btu./lb.	Vapor. Btu./lb.		Liquid. Btu./lb. °F.	Evap. Btu./lb.°F.	Vapor. Btu./lb.°F.	
p	*t*	*V*	*1/V*	*h*	*H*	*L*	*s*	*L/T*	*S*	*p*
80	44.40	3.655	0.2736	91.7	624.0	532.3	0.1982	1.0563	1.2545	**80**
81	45.03	3.612	.2769	92.4	624.1	531.7	.19 6	.0538	.2534	81
82	45.66	3.570	.2801	93.1	624.3	531.2	.2010	.0514	.2524	82
83	46.28	3.528	.2834	93.8	624.4	530.6	.2024	.0490	.2514	83
84	46.89	3.488	.2867	94.5	624.6	530.1	.2037	.0467	.2504	84
85	47.50	3.449	0.2899	95.1	624.7	529.6	0.2051	1.0443	1.2494	**85**
86	48.11	3.411	.2932	95.8	624.8	529.0	.2064	.0420	.2484	86
87	48.71	3.373	.2964	96.5	625.0	528.5	.2077	.0397	.2474	87
88	49.30	3 337	.2997	97.2	625.1	527.9	.2090	.0375	.2465	88
89	49.89	3.301	.3030	97.8	625.2	527.4	.2103	.0352	.2455	89
90	50.47	3.266	0.3062	98.4	625.3	526.9	0.2115	1.0330	1.2445	**90**
91	51.05	3.231	.3095	99.1	625.5	526.4	.2128	.0308	.2436	91
92	51.62	3.198	.3127	99.8	625.6	525.8	.2141	.0286	.2427	92
93	52.19	3.165	.3160	100.4	625.7	525.3	.2153	.0265	.2418	93
94	52.76	3.132	.3192	101.0	625.8	524.8	.2165	.0243	.2408	94
95	53.32	3.101	0.3225	101.6	625.9	524.3	0.2177	1.0222	1.2399	**95**
96	53.87	3.070	.3258	102.3	626.1	523.8	.2190	.0201	.2391	96
97	54.42	3.039	.3290	102.9	626.2	523.3	.2201	.0181	.2382	97
98	54.97	3.010	.3323	103.5	626.3	522.8	.2213	.0160	.2373	98
99	55.51	2.980	.3355	104.1	626.4	522.3	.2225	.0140	.2365	99
100	56.05	2.952	0.3388	104.7	626.5	521.8	0.2237	1.0119	1.2356	**100**
102	57.11	2.896	.3453	105.9	626.7	520.8	.2260	.0079	.2339	102
104	58.16	2.843	.3518	107.1	626.9	519.8	.2282	.0041	.2323	104
106	59.19	2.791	.3583	108.3	627.1	518.8	.2305	1.0002	.2307	106
108	60.21	2.741	.3648	109.4	627.3	517.9	.2327	0.9964	.2291	108
110	61.21	2.693	0.3713	110.5	627.5	517.0	0.2348	0.9927	1.2275	**110**
112	62.20	2.647	.3778	111.7	627.7	516.0	.2369	.9890	.2259	112
114	63.17	2.602	.3843	112.8	627.9	515.1	.2390	.9854	.2244	114
116	64.13	2.559	.3909	113.9	628.1	514.2	.2411	.9819	.2230	116
118	65.08	2.517	.3974	114.9	628.2	513.3	.2431	.9784	.2215	118
120	66.02	2.476	0.4039	116.0	628.4	512.4	0.2452	0.9749	1.2201	**120**
122	66.94	2.437	.4104	117.1	628.6	511.5	.2471	.9715	.2186	122
124	67.86	2.399	.4169	118.1	628.7	510.6	.2491	.9682	.2173	124
126	68.76	2.362	.4234	119.1	628.9	509.8	.2510	.9649	.2159	126
128	69.65	2.326	.4299	120.1	629.0	508.9	.2529	.9616	.2145	128
130	70.53	2.291	0.4364	121.1	629.2	508.1	0.2548	0.9584	1.2132	**130**
132	71.40	2.258	.4429	122.1	629.3	507.2	.2567	.9552	.2119	132
134	72.26	2.225	.4494	123.1	629.5	506.4	.2585	.9521	.2106	134
136	73.11	2.193	.4559	124.1	629.6	505.5	.2603	.9490	.2093	136
138	73.95	2.162	.4624	125.1	629.8	504.7	.2621	.9460	.2081	138
140	74.79	2.132	0.4690	126.0	629.9	503.9	0.2638	0.9430	1.2068	**140**
142	75.61	2.103	.4755	126.9	630.0	503.1	.2656	.9400	.2056	142
144	76.42	2.075	.4820	127.9	630.2	502.3	.2673	.9371	.2044	144
146	77.23	2.047	.4885	128.8	630.3	501.5	.2690	.9342	.2032	146
148	78.03	2.020	.4951	129.7	630.4	500.7	.2707	.9313	.2020	148
150	78.81	1.994	0.5016	130.6	630.5	499.9	0.2724	0.9285	1.2009	**150**

TABLE II. SATURATED AMMONIA: ABSOLUTE PRESSURE TABLE—(*Continued*)

Pressure (abs.). lbs./in.²	Temp. °F.	Volume vapor. ft.³/lb.	Density vapor. lbs./ft.³	Heat content. Liquid. Btu./lb.	Heat content. Vapor. Btu./lb.	Latent heat. Btu./lb.	Entropy. Liquid. Btu./lb. °F.	Entropy. Evap. Btu./lb.°F.	Entropy. Vapor. Btu./lb.°F.	Pressure (abs.). lbs./in.²
p	*t*	*V*	*1/V*	*h*	*H*	*L*	*s*	*L/T*	*S*	*p*
150	78.81	1.994	0.5016	130.6	630.5	499.9	0.2724	0.9285	1.2009	**150**
152	79.60	1.968	.5081	131.5	630.6	499.1	.2740	.9257	.1997	152
154	80.37	1.943	.5147	132.4	630.7	498.3	.2756	.9229	.1985	154
156	81.13	1.919	.5212	133.3	630.9	497.6	.2772	.9202	.1974	156
158	81.89	1.895	.5277	134.2	631.0	496.8	.2788	.9175	.1963	158
160	82.64	1.872	0.5343	135.0	631.1	496.1	0.2804	0.9148	1.1952	**160**
162	83.39	1.849	.5408	135.9	631.2	495.3	.2820	.9122	.1942	162
164	84.12	1.827	.5473	136.8	631.3	494.5	.2835	.9096	.1931	164
166	84.85	1.805	.5539	137.6	631.4	493.8	.2850	.9070	.1920	166
168	85.57	1.784	.5604	138.4	631.5	493.1	.2866	.9044	.1910	168
170	86.29	1.764	0.5670	139.3	631.6	492.3	0.2881	0.9019	1.1900	**170**
172	87.00	1.744	.5735	140.1	631.7	491.6	.2895	.8994	.1889	172
174	87.71	1.724	.5801	140.9	631.7	490.8	.2910	.8969	.1879	174
176	88.40	1.705	.5866	141.7	631.8	490.1	.2925	.8944	.1869	176
178	89.10	1.686	.5932	142.5	631.9	489.4	.2939	.8920	.1859	178
180	89.78	1.667	0.5998	143.3	632.0	488.7	0.2954	0.8896	1.1850	**180**
182	90.46	1.649	.6063	144.1	632.1	488.0	.2968	.8872	.1840	182
184	91.14	1.632	.6129	144.8	632.1	487.3	.2982	.8848	.1830	184
186	91.80	1.614	.6195	145.6	632.2	486.6	.2996	.8825	.1821	186
188	92.47	1.597	.6261	146.4	632.3	485.9	.3010	.8801	.1811	188
190	93.13	1.581	0.6326	147.2	632.4	485.2	0.3024	0.8778	1.1802	**190**
192	93.78	1.564	.6392	147.9	632.4	484.5	.3037	.8755	.1792	192
194	94.43	1.548	.6458	148.7	632.5	483.8	.3050	.8733	.1783	194
196	95.07	1.533	.6524	149.5	632.6	483.1	.3064	.8710	.1774	196
198	95.71	1.517	.6590	150.2	632.6	482.4	.3077	.8688	.1765	198
200	96.34	1.502	0.6656	150.9	632.7	481.8	0.3090	0.8666	1.1756	**200**
205	97.90	1.466	.6821	152.7	632.8	480.1	.3122	.8612	.1734	205
210	99.43	1.431	.6986	154.6	633.0	478.4	.3154	.8559	.1713	210
215	100.94	1.398	.7152	156.3	633.1	476.8	.3185	.8507	.1692	215
220	102.42	1.367	.7318	158.0	633.2	475.2	.3216	.8455	.1671	220
225	103.87	1.336	0.7484	159.7	633.3	473.6	0.3246	0.8405	1.1651	**225**
230	105.30	1.307	.7650	161.4	633.4	472.0	.3275	.8356	.1631	230
235	106.71	1.279	.7817	163.1	633.5	470.4	.3304	.8307	.1611	235
240	108.09	1.253	.7984	164.7	633.6	468.9	.3332	.8260	.1592	240
245	109.46	1.227	.8151	166.4	633.7	467.3	.3360	.8213	.1573	245
250	110.80	1.202	0.8319	168.0	633.8	465.8	0.3388	0.8167	1.1555	**250**
255	112.12	1.178	.8487	169.5	633.8	464.3	.3415	.8121	.1536	255
260	113.42	1.155	.8655	171.1	633.9	462.8	.3441	.8077	.1518	260
265	114.71	1.133	.8824	172.6	633.9	461.3	.3468	.8033	.1501	265
270	115.97	1.112	.8993	174.1	633.9	459.8	.3494	.7989	.1483	270
275	117.22	1.091	0.9162	175.6	634.0	458.4	0.3519	0.7947	1.1466	**275**
280	118.45	1.072	.9332	177.1	634.0	456.9	.3545	.7904	.1449	280
285	119.66	1.052	.9502	178.6	634.0	455.4	.3569	.7863	.1432	285
290	120.86	1.034	.9672	180.0	634.0	454.0	.3594	.7821	.1415	290
295	122.05	1.016	.9843	181.5	634.0	452.5	.3618	.7781	.1399	295
300	123.21	0.999	1.0015	182.9	634.0	451.1	0.3642	0.7741	1.1383	**300**

Table III. Properties of Liquid Ammonia†

Temp. °F.	Saturation.						Latent heat of pressure variation.	Variation of h with p (t constant).	Compressibility.	Temp. °F.
	Pressure (abs.). lbs./in.²	Volume. ft.³/lb.	Density. lbs./ft.³	Specific heat. Btu./lb. °F.	Heat content. Btu./lb.	Latent heat. Btu./lb.	$\frac{\text{Btu./lb.}}{\text{lb./in.}^2}$	$\frac{\text{Btu./lb.}}{\text{lb./in.}^2}$	per lb./in.²×10⁶	
t	p	v	$\frac{1}{v}$	c	h	L	l	$\left(\frac{\partial h}{\partial p}\right)_t$	$-\frac{1}{v}\left(\frac{\partial v}{\partial p}\right)_t$	t
Triple point. {	......	0.01961*	51.00*	......	......	......	......	......	......	
	0.88	.02182	45.83	......	......	......	......	......	......	−107.86
−100	1.24	0.02197	45.52	(1.040)	(−63.0)	(633)	......	......	......	**−100**
−95	1.52	.02207	45.32	(1.042)	(−57.8)	(631)	......	......	......	−95
−90	1.86	.02216	45.12	(1.043)	(−52.6)	(628)	......	......	......	−90
−85	2.27	.02226	44.92	(1.045)	(−47.4)	(625)	......	......	......	−85
−80	2.74	.02236	44.72	(1.046)	(−42.2)	(622)	......	......	......	−80
−75	3.29	0.02246	44.52	(1.048)	(−36.9)	(619)	......	......	......	**−75**
−70	3.94	.02256	44.32	(1.050)	(−31.7)	(616)	......	......	......	−70
−65	4.69	.02267	44.11	(1.052)	(−26.4)	(613)	......	......	......	−65
−60	5.55	.02278	43.91	1.054	−21.18	610.8	−0.0016	0.0026	4.4	−60
−55	6.54	.02288	43.70	1.056	−15.90	607.5	−.0016	.0026	4.5	−55
−50	7.67	0.02299	43.49	1.058	−10.61	604.3	−0.0017	0.0026	4.6	**−50**
−45	8.95	.02310	43.28	1.060	−5.31	600.9	−.0017	.0026	4.7	−45
−40	10.41	.02322	43.08	1.062	0.00	597.6	−.0018	.0025	4.8	−40
−35	12.05	.02333	42.86	1.064	+5.32	594.2	−.0018	.0025	5.0	−35
−30	13.90	.02345	42.65	1.066	10.66	590.7	−.0019	.0025	5.1	−30
−25	15.98	0.02357	42.44	1.068	16.00	587.2	−0.0019	0.0024	5.2	**−25**
−20	18.30	.02369	42.22	1.070	21.36	583.6	−.0020	.0024	5.4	−20
−15	20.88	.02381	42.00	1.073	26.73	580.0	−.0020	.0024	5.5	−15
−10	23.74	.02393	41.78	1.075	32.11	576.4	−.0021	.0023	5.7	−10
−5	26.92	.02406	41.56	1.078	37.51	572.6	−.0022	.0023	5.8	−5
0	30.42	0.02419	41.34	1.080	42.92	568.9	−0.0022	0.0022	6.0	**0**
5	34.27	.02432	41.11	1.083	48.35	565.0	−.0023	.0022	6.2	5
10	38.51	.02446	40.89	1.085	53.79	561.1	−.0024	.0021	6.4	10
15	43.14	.02460	40.66	1.088	59.24	557.1	−.0025	.0021	6.6	15
20	48.21	.02474	40.43	1.091	64.71	553.1	−.0025	.0020	6.8	20
25	53.73	0.02488	40.20	1.094	70.20	548.9	−0.0026	0.0020	7.0	**25**
30	59.74	.02503	39.96	1.097	75.71	544.8	−.0027	.0019	7.3	30
35	66.26	.02518	39.72	1.100	81.23	540.5	−.0028	.0019	7.5	35
40	73.32	.02533	39.49	1.104	86.77	536.2	−.0029	.0018	7.8	40
45	80.96	.02548	39.24	1.108	92.34	531.8	−.0030	.0017	8.1	45
50	89.19	0.02564	39.00	1.112	97.93	527.3	−0.0031	0.0017	8.4	**50**
55	98.06	.02581	38.75	1.116	103.54	522.8	−.0032	.0016	8.8	55
60	107.6	.02597	38.50	1.120	109.18	518.1	−.0033	.0015	9.1	60
65	117.8	.02614	38.25	1.125	114.85	513.4	−.0034	.0014	9.5	65
70	128.8	.02632	38.00	1.129	120.54	508.6	−.0035	.0013	10.0	70
75	140.5	0.02650	37.74	1.133	126.25	503.7	−0.0037	0.0012	10.4	**75**

***Properties of solid ammonia at the triple point (−107.86 °F.)**

† Courtesy, Circular of the Bureau of Standards, No. 142, pp. 30, 31, Washington, U.S. Government Printing Office, 1945.

Table III. Properties of Liquid Ammonia—(*Continued*)

Temp. °F.	Saturation. Pressure (abs.). lbs./in.²	Volume. ft.³/lb.	Density. lbs./ft.³	Specific heat. Btu./lb. °F	Heat content. Btu./lb.	Latent heat. Btu./lb.	Latent heat of pressure variation. Btu./lb. / lb./in.²	Variation of h with p (t constant). Btu./lb. / lb./in.²	Compressibility. per lb./in.² × 10⁶	Temp. °F.
t	p	v	$\frac{1}{v}$	c	h	L	l	$\left(\frac{\partial h}{\partial p}\right)_t$	$-\frac{1}{v}\left(\frac{\partial v}{\partial p}\right)_t$	t
75	140. 5	0. 02650	37. 74	1. 133	126. 25	503. 7	−0. 0037	0. 0012	10. 4	**75**
80	153. 0	. 02668	37. 48	1. 138	131. 99	498. 7	− . 0038	. 0011	10. 9	80
85	166. 4	. 02687	37. 21	1. 142	137. 75	493. 6	− . 0040	. 0010	11. 4	85
90	180. 6	. 02707	36. 95	1. 147	143. 54	488. 5	− . 0041	. 0009	12. 0	90
95	195. 8	. 02727	36. 67	1. 151	149. 36	483. 2	− . 0043	. 0008	12. 6	95
100	211. 9	0. 02747	36. 40	1. 156	155. 21	477. 8	−0. 0045	0. 0006	13. 3	**100**
105	228. 9	. 02769	36. 12	1. 162	161. 09	472. 3	− . 0047	. 0005	14. 1	105
110	247. 0	. 02790	35. 84	1. 168	167. 01	466. 7	− . 0049	. 0003	14. 9	110
115	266. 2	. 02813	35. 55	1. 176	172. 97	460. 9	− . 0051	. 0001	15. 8	115
120	286. 4	. 02836	35. 26	1. 183	178. 98	455. 0	− . 0053	. 0000	16. 7	120
125	307. 8	0. 02860	34. 96	(1. 189)	(185)	(449)	....	....	....	**125**
130	330. 3	. 02885	34. 66	(1. 197)	(191)	(443)	....	....	....	130
135	354. 1	. 02911	34. 35	(1. 205)	(197)	(436)	....	....	....	135
140	379. 1	. 02938	34. 04	(1. 213)	(203)	(430)	....	....	....	140
145	405. 5	. 02966	33. 72	(1. 222)	(210)	(423)	....	....	....	145
150	433. 2	0. 02995	33. 39	(1. 23)	(216)	(416)	....	....	....	**150**
155	462. 3	. 03025	33. 06	(1. 24)	(222)	(409)	....	....	....	155
160	492. 8	. 03056	32. 72	(1. 25)	(229)	(401)	....	....	....	160
165	524. 8	. 03089	32. 37	(1. 26)	(235)	(394)	....	....	....	165
170	558. 4	. 03124	32. 01	(1. 27)	(241)	(386)	....	....	....	170
175	593. 5	0. 03160	31. 65	(1. 29)	(248)	(377)	....	....	....	**175**
180	630. 3	. 03198	31. 27	(1. 30)	(255)	(369)	....	....	....	180
185	668. 7	. 03238	30. 88	(1. 32)	(262)	(360)	....	....	....	185
190	708. 9	. 03281	30. 48	(1. 34)	(269)	(351)	....	....	....	190
195	750. 9	. 03326	30. 06	(1. 36)	(276)	(342)	....	....	....	195
200	794. 7	0. 03375	29. 63	(1. 38)	(283)	(332)	....	....	....	**200**
210	888. 1	. 03482	28. 72	(1. 43)	(297)	(310)	....	....	....	210
220	989. 5	. 0361	27. 7	(1. 49)	(313)	(287)	....	....	....	220
230	1099. 5	. 0376	26. 6	(1. 57)	(329)	(260)	....	....	....	230
240	1218. 5	. 0395	25. 3	(1. 70)	(346)	(229)	....	....	....	240
250	1347	. 0422	23. 7	(1. 90)	(365)	(192)	....	....	....	**250**
260	1486	. 0463	21. 6	(2. 33)	(387)	(142)	....	....	....	260
270	1635	. 0577	17. 3	(5. 30)	(419)	(52)	....	....	....	270
Critical.	1657	. 0686	14. 6	∞	(433)	0	−∞	−∞	∞	271. 4

Note.—The figures in parentheses were calculated from empirical equations given in Bureau of Standards Scientific Papers Nos. 313 and 315 and represent values obtained by extrapolation beyond the range covered in the experimental work.

TABLE IV. PROPERTIES OF SUPERHEATED AMMONIA VAPOR*

Temp. °F.	Absolute pressure in lbs./in.² (Saturation temperature in italics.)									Temp. °F.
	5 *−63.11°*			**6** *−57.64°*			**7** *−52.88°*			
	V	*H*	*S*	*V*	*H*	*S*	*V*	*H*	*S*	
Sat.	*49.31*	*588.3*	*1.4857*	*41.59*	*590.6*	*1.4703*	*36.01*	*592.5*	*1.4574*	*Sat.*
−50	51.05	595.2	1.5025	42.44	594.6	1.4803	36.29	594.0	1.4611	**−50**
−40	52.36	600.3	.5149	43.55	599.8	.4928	37.25	599.3	.4739	−40
−30	53.67	605.4	.5269	44.64	604.9	.5049	38.19	604.5	.4861	−30
−20	54.97	610.4	.5385	45.73	610.0	.5166	39.13	609.6	.4979	−20
−10	56.26	615.4	.5498	46.82	615.1	.5280	40.07	614.7	.5094	−10
0	57.55	620.4	1.5608	47.90	620.1	1.5391	41.00	619.8	1.5206	**0**
10	58.84	625.4	.5716	48.98	625.2	.5499	41.93	624.9	.5314	10
20	60.12	630.4	.5821	50.05	630.2	.5605	42.85	629.9	.5421	20
30	61.41	635.4	.5925	51.12	635.2	.5708	43.77	635.0	.5525	30
40	62.69	640.4	.6026	52.19	640.2	.5810	44.69	640.0	.5627	40
50	63.96	645.5	1.6125	53.26	645.2	1.5910	45.61	645.0	1.5727	**50**
60	65.24	650.5	.6223	54.32	650.3	.6008	46.53	650.1	.5825	60
70	66.51	655.5	.6319	55.39	655.3	.6104	47.44	655.2	.5921	70
80	67.79	660.6	.6413	56.45	660.4	.6199	48.36	660.2	.6016	80
90	69.06	665.6	.6506	57.51	665.5	.6292	49.27	665.3	.6110	90
100	70.33	670.7	1.6598	58.58	670.6	1.6384	50.18	670.4	1.6202	**100**
110	71.60	675.8	.6689	59.64	675.7	.6474	51.09	675.5	.6292	110
120	72.87	680.9	.6778	60.70	680.8	.6563	52.00	680.7	.6382	120
130	74.14	686.1	.6865	61.76	685.9	.6651	52.91	685.8	.6470	130
140	75.41	691.2	.6952	62.82	691.1	.6738	53.82	691.0	.6557	140
150	76.68	696.4	1.7038	63.87	696.3	1.6824	54.73	696.2	1.6643	**150**
160	77.95	701.6	.7122	64.93	701.5	.6909	55.63	701.4	.6727	160
170	79.21	706.8	.7206	65.99	706.7	.6992	56.54	706.6	.6811	170
180	80.48	712.1	.7289	67.05	712.0	.7075	57.45	711.9	.6894	180
	10 *−41.34°*			**11** *−38.14°*			**12** *−35.16°*			
Sat.	*25.81*	*597.1*	*1.4276*	*23.61*	*598.3*	*1.4196*	*21.77*	*599.4*	*1.4124*	*Sat.*
−30	26.58	603.2	1.4420	24.12	602.7	1.4300	22.07	602.3	1.4190	−30
−20	27.26	608.5	.4542	24.74	608.1	.4423	22.64	607.7	.4314	−20
−10	27.92	613.7	.4659	25.35	613.3	.4542	23.20	613.0	.4434	−10
0	28.58	618.9	1.4773	25.95	618.5	1.4656	23.75	618.2	1.4549	**0**
10	29.24	624.0	.4884	26.55	623.7	.4768	24.31	623.4	.4661	10
20	29.90	629.1	.4992	27.15	628.9	.4876	24.86	628.6	.4770	20
30	30.55	634.2	.5097	27.74	634.0	.4982	25.41	633.7	.4877	30
40	31.20	639.3	.5200	28.34	639.1	.5085	25.95	638.9	.4980	40
50	31.85	644.4	1.5301	28.93	644.2	1.5187	26.49	644.0	1.5082	**50**
60	32.49	649.5	.5400	29.52	649.3	.5286	27.03	649.1	.5182	60
70	33.14	654.6	.5497	30.10	654.4	.5383	27.57	654.3	.5279	70
80	33.78	659.7	.5593	30.69	659.6	.5479	28.11	659.4	.5375	80
90	34.42	664.8	.5687	31.28	664.7	.5578	28.65	664.5	.5470	90
100	35.07	670.0	1.5779	31.86	669.8	1.5666	29.19	669.7	1.5562	**100**
110	35.71	675.1	.5870	32.44	675.0	.5757	29.72	674.8	.5654	110
120	36.35	680.3	.5960	33.03	680.1	.5847	30.26	680.0	.5744	120
130	36.99	685.4	.6049	33.61	685.3	.5936	30.79	685.2	.5833	130
140	37.62	690.6	.6136	34.19	690.5	6023	31.33	690.4	.5920	140
150	38.26	695.8	1.6222	34.77	695.7	1.6109	31.86	695.6	1.6006	**150**
160	38.90	701.1	.6307	35.35	700.9	.6194	32.39	700.8	.6092	160
170	39.54	706.3	.6391	35.93	706.2	.6278	32.92	706.1	.6176	170
180	40.17	711.6	.6474	36.51	711.5	.6362	33.46	711.4	.6259	180
190	40.81	716.9	.6556	37.09	716.8	.6444	33.99	716.7	.6341	190
200	41.45	722.2	1.6637	37.67	722.1	1.6525	34.52	722.0	1.6422	**200**

V, volume in ft³ per lb; *H*, heat content in Btu per lb; *S*, entropy in Btu per lb, °F.

* Courtesy, Circular of the Bureau of Standards, No. 142, pp. 32–47, Washington, U.S. Government Printing Office, 1945.

TABLE IV. PROPERTIES OF SUPERHEATED AMMONIA VAPOR—(*Continued*)

Temp. °F.	Absolute pressure in lbs./in.² (Saturation temperature in italics.)									Temp. °F.
	8 *−48.64°*			**9** *−44.38°*			**10** *−41.34°*			
	V	*H*	*S*	*V*	*H*	*S*	*V*	*H*	*S*	
Sat.	*31.79*	*594.2*	*1.4462*	*28.48*	*595.7*	*1.4363*	*25.81*	*597.1*	*1.4276*	*Sat.*
−50	.....	.....	.....	.....	.....	.....	.....	.....	.....	**−50**
−40	32.52	598.8	1.4573	28.85	598.3	1.4426	25.90	597.8	1.4293	−40
−30	33.36	604.1	.4697	29.59	603.6	.4551	26.58	603.2	.4420	−30
−20	34.19	609.3	.4816	30.34	608.9	.4672	27.26	608.5	.4542	−20
−10	35.01	614.4	.4932	31.07	614.0	.4788	27.92	613.7	.4659	−10
0	35.83	619.5	1.5044	31.80	619.2	1.4902	28.58	618.9	1.4773	**0**
10	36.64	624.6	.5154	32.53	624.3	.5012	29.24	624.0	.4884	10
20	37.45	629.7	.5261	33.26	629.4	.5119	29.90	629.1	.4992	20
30	38.26	634.7	.5365	33.98	634.5	.5224	30.55	634.2	.5097	30
40	39.07	639.8	.5467	34.70	639.5	.5327	31.20	639.3	.5200	40
50	39.88	644.8	1.5568	35.42	644.6	1.5427	31.85	644.4	1.5301	**50**
60	40.68	649.9	.5666	36.13	649.7	.5526	32.49	649.5	.5400	60
70	41.48	655.0	.5763	36.85	654.8	.5623	33.14	654.6	.5497	70
80	42.28	660.1	.5858	37.56	659.9	.5718	33.78	659.7	.5593	80
90	43.08	665.2	.5952	38.27	665.0	.5812	34.42	664.8	.5687	90
100	43.88	670.3	1.6044	38.98	670.1	1.5904	35.07	670.0	1.5779	**100**
110	44.68	675.4	.6135	39.70	675.3	.5995	35.71	675.1	.5870	110
120	45.48	680.5	.6224	40.40	680.4	.6085	36.35	680.3	.5960	120
130	46.27	685.7	.6312	41.11	685.6	.6173	36.99	685.4	.6049	130
140	47.07	690.9	.6399	41.82	690.7	.6260	37.62	690.6	.6136	140
150	47.87	696.1	1.6485	42.53	695.9	1.6346	38.26	695.8	1.6222	**150**
160	48.66	701.3	.6570	43.24	701.2	.6431	38.90	701.1	.6307	160
170	49.46	706.5	.6654	43.95	706.4	.6515	39.54	706.3	.6391	170
180	50.25	711.8	.6737	44.65	711.7	.6598	40.17	711.6	.6474	180
	13 *−32.37°*			**14** *−29.76°*			**15** *−27.29°*			
Sat.	*20.20*	*600.5*	*1.4057*	*18.85*	*601.4*	*1.3996*	*17.67*	*602.4*	*1.3938*	*Sat.*
−30	20.33	601.8	1.4088	.....	.....	.....	.....	.....	.....	−30
−20	20.86	607.2	.4213	19.33	606.8	1.4119	18.01	606.4	1.4031	−20
−10	21.38	612.6	.4334	19.82	612.2	.4241	18.47	611.9	.4154	−10
0	21.90	617.9	1.4450	20.30	617.6	1.4358	18.92	617.2	1.4272	**0**
10	22.41	623.1	.4563	20.78	622.8	.4472	19.37	622.5	.4386	10
20	22.92	628.3	.4672	21.26	628.0	.4582	19.82	627.8	.4497	20
30	23.43	633.5	.4779	21.73	633.2	.4688	20.26	633.0	.4604	30
40	23.93	638.6	.4883	22.20	638.4	.4793	20.70	638.2	.4709	40
50	24.43	643.8	1.4985	22.67	643.6	1.4896	21.14	643.4	1.4812	**50**
60	24.94	648.9	.5085	23.14	648.7	.4996	21.58	648.5	.4912	60
70	25.43	654.1	.5183	23.60	653.9	.5094	22.01	653.7	.5011	70
80	25.93	659.2	.5279	24.06	659.0	.5191	22.44	658.9	.5108	80
90	26.43	664.4	.5374	24.53	664.2	.5285	22.88	664.0	.5203	90
100	26.93	669.5	1.5467	24.99	669.4	1.5378	23.31	669.2	1.5296	**100**
110	27.42	674.7	.5558	25.45	674.5	.5470	23.74	674.4	.5388	110
120	27.92	679.9	.5649	25.91	679.7	.5560	24.17	679.6	.5478	120
130	28.41	685.1	.5737	26.37	884.9	.5649	24.60	684.8	.5567	130
140	28.90	690.3	.5825	26.83	690.1	.5737	25.03	690.0	.5655	140
150	29.40	695.5	1.5911	27.29	695.4	1.5824	25.46	695.3	1.5742	**150**
160	29.89	700.7	.5997	27.74	700.6	.5909	25.88	700.5	.5827	160
170	30.38	706.0	.6081	28.20	705.9	.5993	26.31	705.8	.5911	170
180	30.87	711.3	.6164	28.66	711.2	.6076	26.74	711.1	.5995	180
190	31.36	716.6	.6246	29.11	716.5	.6159	27.16	716.4	.6077	190
200	31.85	721.9	1.6328	29.57	721.8	1.6240	27.59	721.7	1.6158	**200**

V, volume in ft³ per lb; *H*, heat content in Btu per lb; *S*, entropy in Btu per lb, °F.

TABLE IV. PROPERTIES OF SUPERHEATED AMMONIA VAPOR—(*Continued*)

Temp. °F.	Absolute pressure in lbs./in.² (Saturation temperature in italics.)									Temp. °F.
	15 *−27.29°*			**16** *−24.95°*			**17** *−22.73°*			
	V	*H*	*S*	*V*	*H*	*S*	*V*	*H*	*S*	
Sat.	*17.67*	*602.4*	*1.3938*	*16.64*	*603.2*	*1.3885*	*15.72*	*604.0*	*1.3835*	*Sat.*
−20	18.01	606.4	1.4031	16.86	606.0	1.3948	15.83	605.6	1.3870	−20
−10	18.47	611.9	.4154	17.29	611.5	.4072	16.24	611.1	.3994	−10
0	18.92	617.2	1.4272	17.72	616.9	1.4191	16.65	616.6	1.4114	**0**
10	19.37	622.5	.4386	18.14	622.2	.4306	17.05	621.9	.4230	10
20	19.82	627.8	.4497	18.56	627.5	.4417	17.45	627.2	.4342	20
30	20.26	633.0	.4604	18.97	632.7	.4525	17.84	632.5	.4450	30
40	20.70	638.2	.4709	19.39	638.0	.4630	18.23	637.7	.4556	40
50	21.14	643.4	1.4812	19.80	643.2	1.4733	18.62	642.9	1.4659	**50**
60	21.58	648.5	.4912	20.21	648.3	.4834	19.01	648.1	.4761	60
70	22.01	653.7	.5011	20.62	653.5	.4933	19.39	653.3	.4860	70
80	22.44	658.9	.5108	21.03	658.7	.5030	19.78	658.5	.4957	80
90	22.88	664.0	.5203	21.43	663.9	.5125	20.16	663.7	.5052	90
100	23.31	669.2	1.5296	21.84	669.1	1.5218	20.54	668.9	1.5146	**100**
110	23.74	674.4	.5388	22.24	674.3	.5310	20.92	674.1	.5238	110
120	24.17	679.6	.5478	22.65	679.5	.5401	21.30	679.3	.5328	120
130	24.60	684.8	.5567	23.05	684.7	.5490	21.68	684.5	.5418	130
140	25.03	690.0	.5655	23.45	689.9	.5578	22.06	689.8	.5506	140
150	25.46	695.3	1.5742	23.86	695.1	1.5665	22.44	695.0	1.5593	**150**
160	25.88	700.5	.5827	24.26	700.4	.5750	22.82	700.3	.5678	160
170	26.31	705.8	.5911	24.66	705.7	.5835	23.20	705.6	.5763	170
180	26.74	711.1	.5995	25.06	711.0	.5918	23.58	710.9	.5846	180
190	27.16	716.4	.6077	25.46	716.3	.6001	23.95	716.2	.5929	190
200	27.59	721.7	1.6158	25.86	721.6	1.6082	24.33	721.5	1.6010	**200**
220	28.44	732.4	.6318	26.66	732.3	.6242	25.08	732.2	.6170	220
	20 *−16.64°*			**21** *−14.78°*			**22** *−12.98°*			
Sat.	*13.50*	*606.2*	*1.3700*	*12.90*	*606.8*	*1.3659*	*12.35*	*607.4*	*1.3621*	*Sat.*
−10	13.74	610.0	1.3784	13.06	609.6	1.3720	12.45	609.2	1.3659	**−10**
0	14.09	615.5	1.3907	13.40	615.2	1.3844	12.77	614.8	1.3784	**0**
10	14.44	621.0	.4025	13.73	620.7	.3962	13.09	620.4	.3903	10
20	14.78	626.4	.4138	14.06	626.1	.4077	13.40	625.8	.4018	20
30	15.11	631.7	.4248	14.38	631.5	.4187	13.71	631.2	.4129	30
40	15.45	637.0	.4356	14.70	636.8	.4295	14.02	636.6	.4237	40
50	15.78	642.3	1.4460	15.02	642.1	1.4400	14.32	641.9	1.4342	**50**
60	16.12	647.5	.4562	15.34	647.3	.4502	14.63	647.1	.4445	60
70	16.45	652.8	.4662	15.65	652.6	.4602	14.93	652.4	.4545	70
80	16.78	658.0	.4760	15.97	657.8	.4700	15.23	657.7	.4643	80
90	17.10	663.2	.4856	16.28	663.1	.4796	15.53	662.9	.4740	90
100	17.43	668.5	1.4950	16.59	668.3	1.4891	15.83	668.1	1.4834	**100**
110	17.76	673.7	.5042	16.90	673.5	.4983	16.12	673.4	.4927	110
120	18.08	678.9	.5133	17.21	678.8	.5075	16.42	678.6	.5019	120
130	18.41	684.2	.5223	17.52	684.0	.5165	16.72	683.9	.5109	130
140	18.73	689.4	.5312	17.83	689.3	.5253	17.01	689.2	.5197	140
150	19.05	694.7	1.5399	18.14	694.6	1.5340	17.31	694.4	1.5285	**150**
160	19.37	700.0	.5485	18.44	699.8	.5426	17.60	699.7	.5371	160
170	19.70	705.3	.5569	18.75	705.1	.5510	17.89	705.0	.5456	170
180	20.02	710.6	.5653	19.06	710.5	.5595	18.19	710.4	.5539	180
190	20.34	715.9	.5736	19.36	715.8	.5678	18.48	715.7	.5622	190
200	20.66	721.2	1.5817	19.67	721.1	1.5759	18.77	721.1	1.5704	**200**
220	21.30	732.0	.5978	20.28	731.9	.5920	19.35	731.8	.5865	220
240	21.94	742.8	.6135	20.89	742.7	.6077	19.94	742.7	.6022	240

V, volume in ft³ per lb; *H*, heat content in Btu per lb; *S*, entropy in Btu per lb, °F.

TABLE IV. PROPERTIES OF SUPERHEATED AMMONIA VAPOR—(*Continued*)

Temp. °F.	Absolute pressure in lbs./in.² (Saturation temperature in italics.)									Temp. °F.
	18 *−20.61°*			19 *−18.58°*			20 *−16.64°*			
	V	H	S	V	H	S	V	H	S	
Sat.	*14.90*	*604.8*	*1.3787*	*14.17*	*605.5*	*1.3742*	*13.50*	*606.2*	*1.3700*	*Sat.*
−20	14.93	605.1	1.3795	....	....	....	....	....	....	−20
−10	15.32	610.7	.3921	14.49	610.3	1.3851	13.74	610.0	1.3784	−10
0	15.70	616.2	1.4042	14.85	615.9	1.3973	14.09	615.5	1.3907	**0**
10	16.08	621.6	.4158	15.21	621.3	.4090	14.44	621.0	.4025	10
20	16.46	626.9	.4270	15.57	626.7	.4203	14.78	626.4	.4138	20
30	16.83	632.2	.4380	15.93	632.0	.4312	15.11	631.7	.4248	30
40	17.20	637.5	.4486	16.28	637.3	.4419	15.45	637.0	.4356	40
50	17.57	642.7	1.4590	16.63	642.5	1.4523	15.78	642.3	1.4460	**50**
60	17.94	647.9	.4691	16.98	647.7	.4625	16.12	647.5	.4562	60
70	18.30	653.1	.4790	17.33	653.0	.4724	16.45	652.8	.4662	70
80	18.67	658.4	.4887	17.67	658.2	.4822	16.78	858.0	.4760	80
90	19.03	663.6	.4983	18.02	663.4	.4918	17.10	663.2	.4856	90
100	19.39	668.8	1.5077	18.36	668.6	1.5012	17.43	668.5	1.4950	**100**
110	19.75	674.0	.5169	18.70	673.8	.5104	17.76	673.7	.5042	110
120	20.11	679.2	.5260	19.04	679.1	.5195	18.08	678.9	.5133	120
130	20.47	684.4	.5349	19.38	684.3	.5285	18.41	684.2	.5223	130
140	20.83	689.7	.5438	19.72	689.5	.5373	18.73	689.4	.5312	140
150	21.19	694.9	1.5525	20.06	694.8	1.5460	19.05	694.7	1.5399	**150**
160	21.54	700.2	.5610	20.40	700.1	.5546	19.37	700.0	.5485	160
170	21.90	705.5	.5695	20.74	705.4	.5631	19.70	705.3	.5569	170
180	22.26	710.8	.5778	21.08	710.7	.5714	20.02	710.6	.5653	180
190	22.61	716.1	.5861	21.42	716.0	.5797	20.34	715.9	.5736	190
200	22.97	721.4	1.5943	21.75	721.3	1.5878	20.66	721.2	1.5817	**200**
220	23.68	732.2	.6103	22.43	732.1	.6039	21.30	732.0	.5978	220
	23 *−11.25°*			24 *−9.58°*			25 *−7.96°*			
Sat.	*11.85*	*608.1*	*1.3584*	*11.39*	*608.6*	*1.3549*	*10.96*	*609.1*	*1.3516*	*Sat.*
−10	11.89	608.8	1.3600	....	....	....	....	....	....	−10
0	12.20	614.5	1.3726	11.67	614.1	1.3670	11.19	613.8	1.3616	**0**
10	12.50	620.0	.3846	11.96	619.7	.3791	11.47	619.4	.3738	10
20	12.80	625.5	.3961	12.25	625.2	.3907	11.75	625.0	.3855	20
30	13.10	630.9	.4073	12.54	630.7	.4019	12.03	630.4	.3967	30
40	13.40	636.3	.4181	12.82	636.1	.4128	12.30	635.8	.4077	40
50	13.69	641.6	1.4287	13.11	641.4	1.4234	12.57	641.2	1.4183	50
60	13.98	646.9	.4390	13.39	646.7	.4337	12.84	646.5	.4287	60
70	14.27	652.2	.4491	13.66	652.0	.4438	13.11	651.8	.4388	70
80	14.56	657.5	.4589	13.94	657.3	.4537	13.37	657.1	.4487	80
90	14.84	662.7	.4686	14.22	662.6	.4634	13.64	662.4	.4584	90
100	15.13	668.0	1.4780	14.49	667.8	1.4729	13.90	667.7	1.4679	**100**
110	15.41	673.2	.4873	14.76	673.1	.4822	14.17	673.0	.4772	110
120	15.70	678.5	.4965	15.04	678.4	.4914	14.43	678.2	.4864	120
130	15.98	683.8	.5055	15.31	683.6	.5004	14.69	683.5	.4954	130
140	16.26	689.0	.5144	15.58	688.9	.5093	14.95	688.8	.5043	140
150	16.55	694.3	1.5231	15.85	694.2	1.5180	15.21	694.1	1,5131	**150**
160	16.83	699.6	.5317	16.12	699.5	.5266	15.47	699.4	.5217	160
170	17.11	704.9	.5402	16.39	704.8	.5352	15.73	704.7	.5303	170
180	17.39	710.3	.5486	16.66	710.2	.5436	15.99	710.1	.5387	180
190	17.67	715.6	.5569	16.93	715.5	.5518	16.25	715.4	.5470	190
200	17.95	721.0	1.5651	17.20	720.9	1.5600	16.50	720.8	1.5552	**200**
220	18.51	731.7	.5812	17.73	731.7	.5761	17.02	731.6	.5713	220
240	19.07	742.6	.5969	18.27	742.6	.5919	17.53	742.5	.5870	240

V, volume in ft³ per lb; *H*, heat content in Btu per lb; *S*, entropy in Btu per lb, °F.

Temp. °F.	Absolute pressure in lbs./in.² (Saturation temperature in italics.) 25 *−7.96°*			26 *−6.39°*			27 *−4.87°*			Temp. °F.
	V	*H*	*S*	*V*	*H*	*S*	*V*	*H*	*S*	
Sat.	*10.96*	*609.1*	*1.3515*	*10.56*	*609.7*	*1.3482*	*10.20*	*610.2*	*1.3451*	*Sat.*
0	11.19	613.8	1.3616	10.74	613.4	1.3564	10.33	613.0	1.3513	**0**
10	11.47	619.4	.3738	11.01	619.1	.3686	10.59	618.8	.3637	10
20	11.75	625.0	.3855	11.28	624.7	.3804	10.85	624.4	.3755	20
30	12.03	630.4	.3967	11.55	630.2	.3917	11.11	629.9	.3869	30
40	12.30	635.8	.4077	11.81	635.6	.4027	11.37	635.4	.3979	40
50	12.57	641.2	1.4183	12.08	641.0	1.4134	11.62	640.8	1.4087	**50**
60	12.84	646.5	.4287	12.34	646.3	.4238	11.87	646.1	.4191	60
70	13.11	651.8	.4388	12.59	651.6	.4339	12.12	651.5	.4292	70
80	13.37	657.1	.4487	12.85	656.9	.4439	12.37	656.8	.4392	80
90	13.64	662.4	.4584	13.11	662.2	.4536	12.61	662.1	.4489	90
100	13.90	667.7	1.4679	13.36	667.5	1.4631	12.86	667.4	1.4585	**100**
110	14.17	673.0	.4772	13.61	672.8	.4725	13.10	672.7	.4679	110
120	14.43	678.2	.4864	13.87	678.1	.4817	13.34	678.0	.4771	120
130	14.69	683.5	.4954	14.12	683.4	.4907	13.59	683.3	.4861	130
140	14.95	688.8	.5043	14.37	688.7	.4996	13.83	688.6	.4950	140
150	15.21	694.1	1.5131	14.62	694.0	1.5084	14.07	693.9	1.5038	**150**
160	15.47	699.4	.5217	14.87	699.3	.5170	14.31	699.2	.5125	160
170	15.73	704.7	.5303	15.12	704.6	.5256	14.55	704.5	.5210	170
180	15.99	710.1	.5387	15.37	710.0	.5340	14.79	709.9	.5295	180
190	16.25	715.4	.5470	15.62	715.3	.5423	15.03	715.2	.5378	190
200	16.50	720.8	1.5552	15.86	720.7	1.5505	15.27	720.6	1.5460	**200**
220	17.02	731.6	.5713	16.36	731.5	.5666	15.75	731.4	.5621	220
240	17.53	742.5	.5870	16.85	742.4	.5824	16.23	742.3	.5779	240
260	18.04	753.4	.6025	17.35	753.3	.5978	16.70	753.2	.5933	260

Temp. °F.	30 *−0.57°*			31 *+0.79°*			32 *+2.11°*			Temp. °F.
Sat.	*9.236*	*611.6*	*1.3364*	*8.955*	*612.0*	*1.3336*	*8.693*	*612.4*	*1.3310*	*Sat.*
10	9.492	617.8	1.3497	9.173	617.4	1.3453	8.874	617.1	1.3411	10
20	9.731	623.5	.3618	9.405	623.2	.3574	9.099	622.9	.3532	20
30	9.966	629.1	.3733	9.633	628.8	.3691	9.321	628.5	.3649	30
40	10.20	634.6	.3845	9.858	634.4	.3803	9.540	634.1	.3762	40
50	10.43	640.1	1.3953	10.08	639.9	1.3912	9.757	639.6	1.3871	**50**
60	10.65	645.5	.4059	10.30	645.3	.4017	9.972	645.1	.3977	60
70	10.88	650.9	.4161	10.52	650.7	.4120	10.18	650.5	.4080	70
80	11.10	656.2	.4261	10.74	656.1	.4221	10.40	655.9	.4181	80
90	11.33	661.6	.4359	10.96	661.4	.4319	10.61	661.2	.4280	90
100	11.55	666.9	1.4456	11.17	666.7	1.4415	10.81	666.6	1.4376	**100**
110	11.77	672.2	.4550	11.38	672.1	.4510	11.02	671.9	.4470	110
120	11.99	677.5	.4642	11.60	677.4	.4602	11.23	677.3	.4563	120
130	12.21	682.9	.4733	11.81	682.7	.4693	11.44	682.6	.4655	130
140	12.43	688.2	.4823	12.02	688.1	.4783	11.64	687.9	.4744	140
150	12.65	693.5	1.4911	12.23	693.4	1.4871	11.85	693.3	1.4833	**150**
160	12.87	698.8	.4998	12.44	698.7	.4958	12.05	698.6	.4920	160
170	13.08	704.2	.5083	12.66	704.1	.5044	12.26	704.0	.5006	170
180	13.30	709.6	.5168	12.87	709.5	.5129	12.46	709.4	.5090	180
190	13.52	714.9	.5251	13.07	714.8	.5212	12.66	714.7	.5174	190
200	13.73	720.3	1.5334	13.28	720.2	1.5294	12.86	720.1	1.5256	**200**
220	14.16	731.1	.5495	13.70	731.1	.5456	13.27	731.0	.5418	220
240	14.59	742.0	.5653	14.12	742.0	.5614	13.67	741.9	.5576	240
260	15.02	753.0	.5808	14.53	752.9	.5769	14.08	752.9	.5731	260
280	15.45	764.1	.5960	14.95	764.0	.5921	14.48	763.9	.5883	280

V, volume in ft³ per lb; *H*, heat content in Btu per lb; *S*, entropy in Btu per lb, °F.

TABLE IV. PROPERTIES OF SUPERHEATED AMMONIA VAPOR—(*Continued*)

Temp. °F.	Absolute pressure in lbs./in.² (Saturation temperature in italics.)									Temp. °F.
	28 *−3.40°*			**29** *−1.97°*			**30** *−0.57°*			
	V	H	S	V	H	S	V	H	S	
Sat.	*9.853*	*610.7*	*1.3421*	*9.534*	*611.1*	*1.3392*	*9.236*	*611.6*	*1.3364*	*Sat.*
0	9.942	612.7	1.3465	9.584	612.3	1.3417	9.250	611.9	1.3371	**0**
10	10.20	618.4	.3589	9.834	618.1	.3542	9.492	617.8	.3497	10
20	10.45	624.1	.3708	10.08	623.8	.3662	9.731	623.5	.3618	20
30	10.70	629.6	.3822	10.32	629.4	.3777	9.966	629.1	.3733	30
40	10.95	635.1	.3933	10.56	634.9	.3888	10.20	634.6	.3845	40
50	11.19	640.5	1.4041	10.80	640.3	1.3996	10.43	640.1	1.3953	**50**
60	11.44	645.9	.4145	11.03	645.7	.4101	10.65	645.5	.4059	60
70	11.68	651.2	.4247	11.26	651.1	.4204	10.88	650.9	.4161	70
80	11.92	656.6	.4347	11.50	656.4	.4304	11.10	656.2	.4261	80
90	12.15	661.9	.4445	11.73	661.7	.4401	11.33	661.6	.4359	90
100	12.39	667.2	1.4540	11.96	667.1	1.4497	11.55	666.9	1.4456	**100**
110	12.63	672.5	.4634	12.18	672.4	.4591	11.77	672.2	.4550	110
120	12.86	677.8	.4726	12.41	677.7	.4684	11.99	677.5	.4642	120
130	13.10	683.1	.4817	12.64	683.0	.4775	12.21	682.9	.4733	130
140	13.33	688.4	.4906	12.86	688.3	.4864	12.43	688.2	.4823	140
150	13.56	693.7	1.4994	13.09	693.6	1.4952	12.65	693.5	1.4911	**150**
160	13.80	699.1	.5081	13.31	699.0	.5039	12.87	698.8	.4998	160
170	14.03	704.4	.5167	13.54	704.3	.5124	13.08	704.2	.5083	170
180	14.26	709.8	.5251	13.76	709.7	.5209	13.30	709.6	.5168	180
190	14.49	715.1	.5334	13.99	715.0	.5292	13.52	714.9	.5251	190
200	14.72	720.5	1.5416	14.21	720.4	.5374	13.73	720.3	1.5334	**200**
220	15.18	731.3	.5578	14.65	731.2	.5536	14.16	731.1	.5495	220
240	15.64	742.2	.5736	15.10	742.2	.5694	14.59	742.0	.5653	240
260	16.10	753.2	.5890	15.54	753.1	.5848	15.02	753.0	.5808	260
	33 *3.40°*			**34** *4.66°*			**35** *5.89°*			
Sat.	*8.445*	*612.8*	*1.3285*	*8.211*	*613.2*	*1.3260*	*7.991*	*613.6*	*1.3236*	*Sat.*
10	8.592	616.8	1.3369	8.328	616.4	1.3328	8.078	616.1	1.3289	10
20	8.812	622.6	.3492	8.542	622.3	.3452	8.287	622.0	.3413	20
30	9.028	628.3	.3609	8.753	628.0	.3570	8.493	627.7	.3532	30
40	9.242	633.9	.3722	8.960	633.6	.3684	8.695	633.4	.3646	40
50	9.452	639.4	1.3832	9.166	639.2	1.3793	8.895	638.9	1.3756	**50**
60	9.661	644.9	.3938	9.369	644.7	.3900	9.093	644.4	.3863	60
70	9.868	650.3	.4042	9.570	650.1	.4004	9.289	649.9	.3967	70
80	10.07	655.7	.4143	9.770	655.5	.4105	9.484	655.3	.4069	80
90	10.28	661.1	.4241	9.969	660.9	.4204	9.677	660.7	.4168	90
100	10.48	666.4	1.4338	10.17	666.3	1.4301	9.869	666.1	1.4265	**100**
110	10.68	671.8	.4433	10.36	671.6	.4396	10.06	671.5	.4360	110
120	10.88	677.1	.4526	10.56	677.0	.4489	10.25	676.8	.4453	120
130	11.08	682.5	.4617	10.75	682.3	.4581	10.44	682.2	.4545	130
140	11.28	687.8	.4707	10.95	687.7	.4671	10.63	687.6	.4635	140
150	11.48	693.2	1.4795	11.14	693.0	1.4759	10.82	692.9	1.4724	**150**
160	11.68	698.5	.4883	11.33	698.4	.4846	11.00	698.3	.4811	160
170	11.88	703.9	.4968	11.53	703.8	.4932	11.19	703.7	.4897	170
180	12.08	709.3	.5053	11.72	709.2	.5017	11.38	709.1	.4982	180
190	12.27	714.6	.5137	11.91	714.5	.5101	11.56	714.5	.5066	190
200	12.47	720.0	1.5219	12.10	720.0	1.5183	11.75	719.9	1.5148	**200**
220	12.86	730.9	.5381	12.48	730.8	.5346	12.12	730.7	.5311	220
240	13.26	741.8	.5540	12.86	741.7	.5504	12.49	741.7	.5469	240
260	13.65	752.8	.5695	13.24	752.7	.5659	12.86	752.7	.5624	260
280	14.04	763.9	.5846	13.62	763.8	.5811	13.23	763.7	.5776	280

V, volume in ft³ per lb; *H*, heat content in Btu per lb; *S*, entropy in Btu per lb, °F.

TABLE IV. PROPERTIES OF SUPERHEATED AMMONIA VAPOR—(*Continued*)

Temp. °F.	Absolute pressure in lbs./in.² (Saturation temperature in italics.) 35 (*5.89°*)			36 (*7.09°*)			37 (*8.27°*)			Temp. °F.
	V	*H*	*S*	*V*	*H*	*S*	*V*	*H*	*S*	
Sat.	*7.991*	*613.6*	*1.3236*	*7.782*	*614.0*	*1.3213*	*7.584*	*614.3*	*1.3190*	Sat.
10	8.078	616.1	1.3289	7.842	615.7	1.3250	7.619	615.4	1.3212	10
20	8.287	622.0	.3413	8.046	621.7	.3375	7.819	621.4	.3338	20
30	8.493	627.7	.3532	8.247	627.4	.3494	8.015	627.2	.3458	30
40	8.695	633.4	.3646	8.445	633.1	.3609	8.208	632.9	.3573	40
50	8.895	638.9	1.3756	8.640	638.7	1.3720	8.398	638.5	1.3684	**50**
60	9.093	644.4	.3863	8.833	644.2	.3827	8.587	644.0	.3792	60
70	9.289	649.9	.3967	9.024	649.7	.3932	8.773	649.5	.3897	70
80	9.484	655.3	.4069	9.214	655.2	.4033	8.958	655.0	.3999	80
90	9.677	660.7	.4168	9.402	660.6	.4133	9.142	660.4	.4098	90
100	9.869	666.1	1.4265	9.589	666.0	1.4230	9.324	665.8	1.4196	**100**
110	10.06	671.5	.4360	9.775	671.3	.4325	9.506	671.2	.4291	110
120	10.25	676.8	.4453	9.961	676.7	.4419	9.686	676.6	.4385	120
130	10.44	682.2	.4545	10.15	682.1	.4510	9.866	681.9	.4477	130
140	10.63	687.6	.4635	10.33	687.4	.4601	10.05	687.3	.4567	140
150	10.82	692.9	1.4724	10.51	692.8	1.4689	10.22	692.7	1.4656	**150**
160	11.00	698.3	.4811	10.69	698.2	.4777	10.40	698.1	.4744	160
170	11.19	703.7	.4897	10.88	703.6	.4863	10.58	703.5	.4830	170
180	11.38	709.1	.4982	11.06	709.0	.4948	10.76	708.9	.4915	180
190	11.56	714.5	.5066	11.24	714.4	.5032	10.93	714.3	.4999	190
200	11.75	719.9	1.5148	11.42	719.8	1.5115	11.11	719.7	1.5082	**200**
220	12.12	730.7	.5311	11.78	730.6	.5277	11.46	730.6	.5244	220
240	12.49	741.7	.5469	12.14	741.6	.5436	11.81	741.5	.5403	240
260	12.86	752.7	.5624	12.50	752.6	.5591	12.16	752.5	.5558	260
280	13.23	763.7	.5776	12.86	763.7	.5743	12.51	763.6	.5710	280
	40 (*11.66°*)			42 (*13.81°*)			44 (*15.88°*)			
Sat.	*7.047*	*615.4*	*1.3125*	*6.731*	*616.0*	*1.3084*	*6.442*	*616.6*	*1.3046*	Sat.
20	7.203	620.4	1.3231	6.842	619.8	1.3164	6.513	619.1	1.3099	20
30	7.387	626.3	.3353	7.019	625.8	.3287	6.683	625.2	.3224	30
40	7.568	632.1	.3470	7.192	631.6	.3405	6.850	631.1	.3343	40
50	7.746	637.8	1.3583	7.363	637.3	1.3519	7.014	636.8	1.3457	**50**
60	7.922	643.4	.3692	7.531	643.0	.3628	7.176	642.5	.3567	60
70	8.096	648.9	.3797	7.697	648.5	.3734	7.336	648.1	.3674	70
80	8.268	654.4	.3900	7.862	654.1	.3838	7.494	653.7	.3778	80
90	8.439	659.9	.4000	8.026	659.5	.3939	7.650	659.2	.3880	90
100	8.609	665.3	1.4098	8.188	665.0	1.4037	7.806	664.7	1.3978	**100**
110	8.777	670.7	.4194	8.349	670.4	.4133	7.960	670.1	.4075	110
120	8.945	676.1	.4288	8.510	675.9	.4228	8.114	675.6	.4170	120
130	9.112	681.5	.4381	8.669	681.3	.4320	8.267	681.0	.4263	130
140	9.278	686.9	.4471	8.828	686.7	.4411	8.419	686.4	.4354	140
150	9.444	692.3	1.4561	8.986	692.1	1.4501	8.570	691.9	1.4444	**150**
160	9.609	697.7	.4648	9.144	697.5	.4589	8.721	697.3	.4532	160
170	9.774	703.1	.4735	9.301	702.9	.4676	8.871	702.7	.4619	170
180	9.938	708.5	.4820	9.458	708.3	.4761	9.021	708.1	.4704	180
190	10.10	714.0	.4904	9.614	713.8	.4845	9.171	713.6	.4789	190
200	10.27	719.4	1.4987	9.770	719.2	1.4928	9.320	719.0	1.4872	**200**
220	10.59	730.3	.5150	10.08	730.1	.5091	9.617	730.0	.5035	220
240	10.92	741.3	.5309	10.39	741.1	.5251	9.913	741.0	.5195	240
260	11.24	752.3	.5465	10.70	752.2	.5406	10.21	752.0	.5350	260
280	11.56	763.4	.5617	11.01	763.3	.5559	10.50	763.1	.5503	280
300	11.88	774.6	.5766	11.31	774.5	1.5708	10.80	774.3	1.5652	**300**

V, volume in ft³ per lb; *H*, heat content in Btu per lb; *S*, entropy in Btu per lb, °F.

TABLE IV. PROPERTIES OF SUPERHEATED AMMONIA VAPOR—(*Continued*)

Temp. °F.	Absolute pressure in lbs./in.² (Saturation temperature in italics.)									Temp. °F.
	38 *9.42°*			**39** *10.55°*			**40** *11.66°*			
	V	H	S	V	H	S	V	H	S	
Sat.	*7.396*	*614.7*	*1.3168*	*7.217*	*615.0*	*1.3146*	*7.047*	*615.4*	*1.3125*	*Sat.*
10	7.407	615.0	1.3175	....	....	....	....	....	....	10
20	7.603	621.0	.3301	7.398	620.7	1.3266	7.203	620.4	1.3231	20
30	7.795	626.9	.3422	7.586	626.6	.3387	7.387	626.3	.3353	30
40	7.983	632.6	.3538	7.770	632.4	.3504	7.568	632.1	.3470	40
50	8.170	638.3	1.3650	7.952	638.0	1.3616	7.746	637.8	1.3583	**50**
60	8.353	643.8	.3758	8.132	643.6	.3724	7.922	643.4	.3692	60
70	8.535	649.3	.3863	8.310	649.1	.3830	8.096	648.9	.3797	70
80	8.716	654.8	.3965	8.486	654.6	.3932	8.268	654.4	.3900	80
90	8.895	660.2	.4065	8.661	660.1	.4032	8.439	659.9	.4000	90
100	9.073	665.6	1.4163	8.835	665.5	1.4130	8.609	665.3	1.4098	**100**
110	9.250	671.0	.4258	9.008	670.9	.4226	8.777	670.7	.4194	110
120	9.426	676.4	.4352	9.179	676.3	.4320	8.945	676.1	.4288	120
130	9.602	681.8	.4444	9.351	681.7	.4412	9.112	681.5	.4381	130
140	9.776	687.2	.4534	9.521	687.1	.4503	9.278	686.9	.4471	140
150	9.950	692.6	1.4623	9.691	692.5	1.4592	9.444	692.3	1.4561	**150**
160	10.12	698.0	.4711	9.860	697.8	.4679	9.609	697.7	.4648	160
170	10.30	703.3	.4797	10.03	703.2	.4766	9.774	703.1	.4735	170
180	10.47	708.7	.4883	10.20	708.6	.4851	9.938	708.5	.4820	180
190	10.64	714.2	.4966	10.36	714.1	.4935	10.10	714.0	.4904	190
200	10.81	719.6	1.5049	10.53	719.5	1.5018	10.27	719.4	1.4987	**200**
220	11.16	730.5	.5212	10.87	730.4	.5181	10.59	730.3	.5150	220
240	11.50	741.4	.5371	11.20	741.3	.5340	10.92	741.3	.5309	240
260	11.84	752.4	.5526	11.53	752.4	.5495	11.24	752.3	.5465	260
280	12.18	763.5	.5678	11.86	763.5	.5647	11.56	763.4	.5617	280
	46 *17.87°*			**48** *19.80°*			**50** *21.67°*			
Sat.	*6.177*	*617.2*	*1.3009*	*5.934*	*617.7*	*1.2973*	*5.710*	*618.2*	*1.2939*	*Sat.*
20	6.213	618.5	1.3036	5.937	617.8	1.2976	....	....	....	20
30	6.377	624.6	.3162	6.096	624.0	.3103	5.838	623.4	1.3046	30
40	6.538	630.5	.3283	6.251	630.0	.3225	5.988	629.5	.3169	40
50	6.696	636.4	1.3398	6.404	635.9	1.3341	6.135	635.4	1.3286	**50**
60	6.851	642.1	.3509	6.554	641.6	.3453	6.280	641.2	.3399	60
70	7.005	647.7	.3617	6.702	647.3	.3561	6.423	646.9	.3508	70
80	7.157	653.3	.3721	6.848	652.9	.3666	6.564	652.6	.3613	80
90	7.308	658.9	.3823	6.993	658.5	.3768	6.704	658.2	.3716	90
100	7.457	664.4	1.3922	7.137	664.0	1.3868	6.843	663.7	1.3816	**100**
110	7.605	669.8	.4019	7.280	669.5	.3965	6.980	669.2	.3914	110
120	7.753	675.3	.4114	7.421	675.0	.4061	7.117	674.7	.4009	120
130	7.899	680.7	.4207	7.562	680.5	.4154	7.252	680.2	.4103	130
140	8.045	686.2	.4299	7.702	685.9	.4246	7.387	685.7	.4195	140
150	8.190	691.6	1.4389	7.842	691.4	1.4336	7.521	691.1	1.4286	**150**
160	8.335	697.1	.4477	7.981	696.8	.4425	7.655	696.6	.4374	160
170	8.479	702.5	.4564	8.119	702.3	.4512	7.788	702.1	.4462	170
180	8.623	707.9	.4650	8.257	707.7	.4598	7.921	707.5	.4548	180
190	8.766	713.4	.4735	8.395	713.2	.4683	8.053	713.0	.4633	190
200	8.909	718.8	1.4818	8.532	718.7	1.4766	8.185	718.5	1.4716	**200**
220	9.194	729.8	.4981	8.805	729.6	.4930	8.448	729.4	.4880	220
240	9.477	740.8	.5141	9.077	740.6	.5090	8.710	740.5	.5040	240
260	9.760	751.9	.5297	9.348	751.7	.5246	8.970	751.6	.5197	260
280	10.04	763.0	.5450	9.619	762.9	.5399	9.230	762.7	.5350	280
300	10.32	774.2	1.5599	9.888	774.1	1.5548	9.489	774.0	1.5500	**300**

V, volume in ft³ per lb; H, heat content in Btu per lb; S, entropy in Btu per lb, °F.

TABLE IV. PROPERTIES OF SUPERHEATED AMMONIA VAPOR—(*Continued*)

Temp. °F.	Absolute pressure in lbs./in.² (Saturation temperatures in italics.)									Temp. °F.
	50 *21.67°*			**52** *23.48°*			**54** *25.23°*			
	V	*H*	*S*	*V*	*H*	*S*	*V*	*H*	*S*	
Sat.	*5.710*	*618.2*	*1.2939*	*5.502*	*618.7*	*1.2906*	*5.309*	*619.2*	*1.2875*	*Sat.*
30	5.838	623.4	1.3046	5.599	622.8	1.2991	5.378	622.2	1.2937	30
40	5.988	629.5	.3169	5.744	629.0	.3114	5.519	628.4	.3062	40
50	6.135	635.4	1.3286	5.887	634.9	1.3233	5.657	634.4	1.3181	**50**
60	6.280	641.2	.3399	6.027	640.8	.3346	5.793	640.3	.3295	60
70	6.423	646.9	.3508	6.165	646.5	.3456	5.927	646.1	.3406	70
80	6.564	652.6	.3613	6.302	652.2	.3562	6.059	651.8	.3513	80
90	6.704	658.2	.3716	6.437	657.8	.3665	6.190	657.5	.3616	90
100	6.843	663.7	1.3816	6.571	663.4	1.3766	6.319	663.1	1.3717	**100**
110	6.980	669.2	.3914	6.704	668.9	.3864	6.447	668.6	.3816	110
120	7.117	674.7	.4009	6.835	674.4	.3960	6.575	674.2	.3912	120
130	7.252	680.2	.4103	6.966	679.9	.4054	6.701	679.7	.4006	130
140	7.387	685.7	.4195	7.096	685.4	.4146	6.827	685.2	.4099	140
150	7.521	691.1	1.4286	7.225	690.9	1.4237	6.952	690.7	1.4190	**150**
160	7.655	696.6	.4374	7.354	696.4	.4326	7.076	696.1	.4279	160
170	7.788	702.1	.4462	7.483	701.8	.4413	7.200	701.6	.4367	170
180	7.921	707.5	.4548	7.611	707.3	.4500	7.323	707.1	.4453	180
190	8.053	713.0	.4633	7.738	712.8	.4585	7.446	712.6	.4538	190
200	8.185	718.5	1.4716	7.865	718.3	1.4668	7.569	718.1	1.4622	**200**
210	8.317	724.0	.4799	7.992	723.8	.4751	7.691	723.6	.4705	210
220	8.448	729.4	.4880	8.118	729.3	.4833	7.813	729.1	.4787	220
240	8.710	740.5	.5040	8.370	740.3	.4993	8.056	740.2	.4947	240
260	8.970	751.6	.5197	8.621	751.4	.5149	8.298	751.3	.5104	260
280	9.230	762.7	1.5350	8.871	762.6	1.5303	8.539	762.5	1.5257	**280**
300	9.489	774.0	.5500	9.120	773.8	.5453	8.779	773.7	.5407	300
	60 *30.21°*			**62** *31.78°*			**64** *33.31°*			
Sat.	*4.805*	*620.5*	*1.2787*	*4.658*	*620.9*	*1.2759*	*4.519*	*621.3*	*1.2733*	*Sat.*
40	4.933	626.8	1.2913	4.762	626.2	1.2866	4.602	625.6	1.2820	40
50	5.060	632.9	1.3035	4.886	632.4	1.2989	4.723	631.9	1.2944	**50**
60	5.184	639.0	.3152	5.007	638.5	.3107	4.842	638.0	.3063	60
70	5.307	644.9	.3265	5.127	644.4	.3220	4.958	644.0	.3177	70
80	5.428	650.7	.3373	5.244	650.3	.3330	5.072	649.9	.3287	80
90	5.547	656.4	.3479	5.360	656.0	.3435	5.185	655.7	.3393	90
100	5.665	662.1	1.3581	5.474	661.7	1.3538	5.296	661.4	1.3496	**100**
110	5.781	667.7	.3681	5.588	667.4	.3638	5.406	667.1	.3597	110
120	5.897	673.3	.3778	5.700	673.0	.3736	5.516	672.7	.3695	120
130	6.012	678.9	.3873	5.811	678.6	.3831	5.624	678.3	.3791	130
140	6.126	684.4	.3966	5.922	684.2	.3925	5.731	683.9	.3885	140
150	6.239	689.9	1.4058	6.032	689.7	1.4017	5.838	689.5	1.3977	**150**
160	6.352	695.5	.4148	6.142	695.2	.4107	5.944	695.0	.4067	160
170	6.464	701.0	.4236	6.250	700.8	.4195	6.050	700.5	.4156	170
180	6.576	706.5	.4323	6.359	706.3	.4282	6.155	706.1	.4243	180
190	6.687	712.0	.4409	6.467	711.8	.4368	6.260	711.6	.4329	190
200	6.798	717.5	1.4493	6.574	717.3	1.4453	6.364	717.2	1.4413	**200**
210	6.909	723.1	.4576	6.681	722.9	.4536	6.468	722.7	.4497	210
220	7.019	728.6	.4658	6.788	728.4	.4618	6.572	728.3	.4579	220
230	7.129	734.1	.4739	6.895	734.0	.4699	6.675	733.8	.4660	230
240	7.238	739.7	.4819	7.001	739.5	.4779	6.778	739.4	.4741	240
260	7.457	750.9	1.4976	7.213	750.7	1.4937	6.984	750.6	1.4898	**260**
280	7.675	762.1	.5130	7.424	761.9	.5091	7.188	761.8	.5052	280
300	7.892	773.3	.5281	7.634	773.2	.5241	7.392	773.1	.5203	300

V, volume in ft³ per lb; *H*, heat content in Btu per lb; *S*, entropy in Btu per lb, °F.

Table IV. Properties of Superheated Ammonia Vapor—(*Continued*)

Temp. °F.	Absolute pressure in lbs./in.² (Saturation temperature in italics.)									Temp. °F.
	56 *26.94°*			**58** *28.59°*			**60** *30.21°*			
	V	*H*	*S*	*V*	*H*	*S*	*V*	*H*	*S*	
Sat.	*5.129*	*619.7*	*1.2844*	*4.962*	*620.1*	*1.2815*	*4.805*	*620.5*	*1.2787*	*Sat.*
30	5. 172	621. 6	1. 2884	4. 981	621. 0	1. 2834	....	....	....	30
40	5. 310	627. 9	. 3011	5. 115	627. 3	. 2961	4. 933	626. 8	1. 2913	40
50	5. 444	633. 9	1. 3131	5. 245	633. 4	1. 3082	5. 060	632. 9	1. 3035	**50**
60	5. 576	639. 9	. 3246	5. 373	639. 4	. 3199	5. 184	639. 0	. 3152	60
70	5. 706	645. 7	. 3357	5. 499	645. 3	. 3310	5. 307	644. 9	. 3265	70
80	5. 834	651. 4	. 3465	5. 624	651. 1	. 3418	5. 428	650. 7	. 3373	80
90	5. 960	657. 1	. 3569	5. 746	656. 8	. 3523	5. 547	656. 4	. 3479	90
100	6. 085	662. 7	1. 3670	5. 868	662. 4	1. 3625	5. 665	662. 1	1. 3581	**100**
110	6. 209	668. 3	. 3769	5. 988	668. 0	. 3724	5. 781	667. 7	. 3681	110
120	6. 333	673. 9	. 3866	6. 107	673. 6	. 3821	5. 897	673. 3	. 3778	120
130	6. 455	679. 4	. 3961	6. 226	679. 1	. 3916	6. 012	678. 9	. 3873	130
140	6. 576	684. 9	. 4053	6. 343	684. 7	. 4009	6. 126	684. 4	. 3966	140
150	6. 697	690. 4	1. 4144	6. 460	690. 2	1. 4100	6. 239	689. 9	1. 4058	**150**
160	6. 817	695. 9	. 4234	6. 577	695. 7	. 4190	6. 352	695. 5	. 4148	160
170	6. 937	701. 4	. 4322	6. 692	701. 2	. 4278	6. 464	701. 0	. 4236	170
180	7. 056	706. 9	. 4408	6. 808	706. 7	. 4365	6. 576	706. 5	. 4323	180
190	7. 175	712. 4	. 4494	6. 923	712. 2	. 4450	6. 687	712. 0	. 4409	190
200	7. 294	717. 9	1. 4578	7. 037	717. 7	1. 4535	6. 798	717. 5	1. 4493	**200**
210	7. 412	723. 4	. 4661	7. 151	723. 2	. 4618	6. 909	723. 1	. 4576	210
220	7. 529	728. 9	. 4743	7. 265	728. 8	. 4700	7. 019	728. 6	. 4658	220
240	7. 764	740. 0	. 4903	7. 492	739. 9	. 4860	7. 238	739. 7	. 4819	240
260	7. 998	751. 1	. 5060	7. 718	751. 0	. 5017	7. 457	750. 9	. 4976	260
280	8. 230	762. 3	1. 5213	7. 943	762. 2	. 5171	7. 675	762. 1	1. 5130	**280**
300	8. 462	773. 6	. 5364	8. 167	773. 5	. 5321	7. 892	773. 3	. 5281	300
	66 *34.81°*			**68** *36.27°*			**70** *37.70°*			
Sat.	*4.389*	*621.7*	*1.2707*	*4.267*	*622.0*	*1.2682*	*4.151*	*622.4*	*1.2658*	*Sat.*
40	4. 452	625. 1	1. 2775	4. 310	624. 5	1. 2731	4. 177	623. 9	1. 2688	40
50	4. 570	631. 4	1. 2900	4. 426	630. 9	1. 2858	4. 290	630. 4	1. 2816	**50**
60	4. 686	637. 6	. 3020	4. 539	637. 1	. 2978	4. 401	636. 6	. 2937	60
70	4. 799	643. 6	. 3135	4. 650	643. 2	. 3094	4. 509	642. 7	. 3054	70
80	4. 910	649. 5	. 3245	4. 758	649. 1	. 3205	4. 615	648. 7	. 3166	80
90	5. 020	655. 3	. 3352	4. 865	655. 0	. 3312	4. 719	654. 6	. 3274	90
100	5. 129	661. 1	1. 3456	4. 971	660. 7	1. 3417	4. 822	660. 4	1. 3378	**100**
110	5. 236	666. 8	. 3557	5. 075	666. 5	. 3518	4. 924	666. 1	. 3480	110
120	5. 342	672. 4	. 3655	5. 179	672. 1	. 3617	5. 025	671. 8	. 3579	120
130	5. 447	678. 0	. 3751	5. 281	677. 8	. 3713	5. 125	677. 5	. 3676	130
140	5. 552	683. 6	. 3846	5. 383	683. 4	. 3807	5. 224	683. 1	. 3770	140
150	5. 656	689. 2	1. 3938	5. 484	689. 0	1. 3900	5. 323	688. 7	1. 3863	**150**
160	5. 759	694. 8	. 4028	5. 585	694. 5	. 3991	5. 420	694. 3	. 3954	160
170	5. 862	700. 3	. 4117	5. 685	700. 1	. 4080	5. 518	699. 9	. 4043	170
180	5. 964	705. 9	. 4205	5. 784	705. 7	. 4167	5. 615	705. 5	. 4131	180
190	6. 066	711. 4	. 4291	5. 883	711. 2	. 4254	5. 711	711. 0	. 4217	190
200	6. 167	717. 0	1. 4375	5. 982	716. 8	1. 4338	5. 807	716. 6	1. 4302	**200**
210	6. 268	722. 5	. 4459	6. 080	722. 3	. 4422	5. 902	722. 2	. 4386	210
220	6. 369	728. 1	. 4541	6. 179	727. 9	. 4505	5. 998	727. 7	. 4469	220
230	6. 470	733. 7	. 4623	6. 275	733. 5	. 4586	6. 093	733. 3	. 4550	230
240	6. 570	739. 2	. 4703	6. 373	739. 1	. 4666	6. 187	738. 9	. 4631	240
260	6. 769	750. 4	1. 4861	6. 567	750. 3	1. 4824	6. 376	750. 1	1. 4789	**260**
280	6. 968	761. 7	. 5015	6. 760	761. 5	. 4979	6. 563	761. 4	. 4943	280
300	7. 165	773. 0	. 5166	6. 952	772. 8	. 5130	6. 750	772. 7	. 5095	300

V, volume in ft³ per lb; *H*, heat content in Btu per lb; *S*, entropy in Btu per lb, °F.

Table IV. Properties of Superheated Ammonia Vapor—(*Continued*)

Temp. °F.	Absolute pressure in lbs./in.² (Saturation temperature in italics.)									Temp. °F.
	75 *41.13°*			80 *44.40°*			85 *47.50°*			
	V	*H*	*S*	*V*	*H*	*S*	*V*	*H*	*S*	
Sat.	*3.887*	*623.2*	*1.2599*	*3.655*	*624.0*	*1.2545*	*3.449*	*624.7*	*1.2494*	*Sat.*
50	3. 982	629. 1	1. 2715	3. 712	627. 7	1. 2619	3. 473	626. 4	1. 2527	**50**
60	4. 087	635. 5	. 2839	3. 812	634. 3	. 2745	3. 569	633. 0	. 2656	60
70	4. 189	641. 7	. 2957	3. 909	640. 6	. 2866	3. 662	639. 5	. 2779	70
80	4. 289	647. 7	. 3071	4. 005	646. 7	. 2981	3. 753	645. 7	. 2896	80
90	4. 388	653. 7	. 3180	4. 098	652. 8	. 3092	3. 842	651. 8	. 3008	90
100	4. 485	659. 6	1. 3286	4. 190	658. 7	1. 3199	3. 930	657. 8	1. 3117	**100**
110	4. 581	665. 4	. 3389	4. 281	664. 6	. 3303	4. 016	663. 8	. 3221	110
120	4. 676	671. 1	. 3489	4. 371	670. 4	. 3404	4. 101	669. 6	. 3323	120
130	4. 770	676. 8	. 3586	4. 460	676. 1	. 3502	4. 186	675. 4	. 3422	130
140	4. 863	682. 5	. 3682	4. 548	681. 8	. 3598	4. 269	681. 2	. 3519	140
150	4. 956	688. 1	1. 3775	4. 635	687. 5	1. 3692	4. 352	686. 9	1. 3614	**150**
160	5. 048	693. 7	. 3866	4. 722	693. 2	. 3784	4. 434	692. 6	. 3706	160
170	5. 139	699. 3	. 3956	4. 808	698. 8	. 3874	4. 515	698. 2	. 3797	170
180	5. 230	704. 9	. 4044	4. 893	704. 4	. 3963	4. 596	703. 9	. 3886	180
190	5. 320	710. 5	. 4131	4. 978	710. 0	. 4050	4. 677	709. 5	. 3974	190
200	5. 410	716. 1	1. 4217	5. 063	715. 6	1. 4136	4. 757	715. 2	1. 4060	**200**
210	5. 500	721. 7	. 4301	5. 147	721. 3	. 4220	4. 836	720. 8	. 4145	210
220	5. 589	727. 3	. 4384	5. 231	726. 9	. 4304	4. 916	726. 4	. 4228	220
230	5. 678	732. 9	. 4466	5. 315	732. 5	. 4386	4. 995	732. 1	. 4311	230
240	5. 767	738. 5	. 4546	5. 398	738. 1	. 4467	5. 074	737. 7	. 4392	240
250	5. 855	744. 1	1. 4625	5. 482	743. 8	1. 4547	5. 152	743. 4	1. 4472	**250**
260	5. 943	749. 8	. 4705	5. 565	749. 4	. 4626	5. 230	749. 0	. 4551	260
280	6. 119	761. 1	. 4860	5. 730	760. 7	. 4781	5. 386	760. 4	. 4707	280
300	6. 294	772. 4	. 5011	5. 894	772. 1	. 4933	5. 541	771. 8	. 4859	300
	100 *56.05°*			105 *58.67°*			110 *61 21°*			
Sat.	*2.952*	*626.5*	*1.2356*	*2.817*	*627.0*	*1.2314*	*2.693*	*627.5*	*1.2275*	*Sat.*
70	3. 068	636. 0	1. 2539	2. 907	634. 9	1. 2464	2. 761	633. 7	1. 2392	70
80	3. 149	642. 6	. 2661	2. 985	641. 5	. 2589	2. 837	640. 5	. 2519	80
90	3. 227	649. 0	. 2778	3. 061	648. 0	. 2708	2. 910	647. 0	. 2640	90
100	3. 304	655. 2	1. 2891	3. 135	654. 3	1. 2822	2. 981	653. 4	1. 2755	**100**
110	3. 380	661. 3	. 2999	3. 208	660. 5	. 2931	3. 051	659. 7	. 2866	110
120	3. 454	667. 3	. 3104	3. 279	666. 6	. 3037	3. 120	665. 8	. 2972	120
130	3. 527	673. 3	. 3206	3. 350	672. 6	. 3139	3. 188	671. 9	. 3076	130
140	3. 600	679. 2	. 3305	3. 419	678. 5	. 3239	3. 255	677. 8	. 3176	140
150	3. 672	685. 0	1. 3401	3. 488	684. 4	1. 3336	3. 321	683. 7	1. 3274	**150**
160	3. 743	690. 8	. 3495	3. 556	690. 2	. 3431	3. 386	689. 6	. 3370	160
170	3. 813	696. 6	. 3588	3. 623	696. 0	. 3524	3. 451	695. 4	. 3463	170
180	3. 883	702. 3	. 3678	3. 690	701. 8	. 3615	3. 515	701. 2	. 3555	180
190	3. 952	708. 0	. 3767	3. 757	707. 5	. 3704	3. 579	707. 0	. 3644	190
200	4. 021	713. 7	1. 3854	3. 823	713. 3	1. 3792	3. 642	712. 8	1. 3732	**200**
210	4. 090	719 4	. 3940	3. 888	719. 0	. 3878	3. 705	718. 5	. 3819	210
220	4. 158	725. 1	. 4024	3. 954	724. 7	. 3963	3. 768	724. 3	. 3904	220
230	4. 226	730. 8	. 4108	4. 019	730. 4	. 4046	3. 830	730. 0	. 3988	230
240	4. 294	736. 5	. 4190	4. 083	736. 1	. 4129	3. 892	735. 7	. 4070	240
250	4. 361	742. 2	1. 4271	4. 148	741. 9	1. 4210	3. 954	741. 5	1. 4151	**250**
260	4. 428	747. 9	. 4350	4. 212	747. 6	. 4290	4. 015	747. 2	. 4232	260
270	4. 495	753. 6	. 4429	4. 276	753. 3	. 4369	4. 076	752. 9	. 4311	270
280	4. 562	759. 4	. 4507	4. 340	759. 0	. 4447	4. 137	758. 7	. 4389	280
290	4. 629	765. 1	. 4584	4. 403	764. 8	. 4524	4. 198	764. 5	. 4466	290
300	4. 695	770. 8	1. 4660	4. 466	770. 5	1. 4600	4. 259	770. 2	1. 4543	**300**

V, volume in ft³ per lb; *H*, heat content in Btu per lb; *S*, entropy in Btu per lb, °F.

TABLE IV. PROPERTIES OF SUPERHEATED AMMONIA VAPOR—(*Continued*)

Temp. °F.	Absolute pressure in lbs./in.² (Saturation temperature in italics.) 90 *50.47°*			95 *53.32°*			100 *56.05°*			Temp. °F.
	V	*H*	*S*	*V*	*H*	*S*	*V*	*H*	*S*	
Sat.	*3.266*	*625.3*	*1.2445*	*3.101*	*625.9*	*1.2399*	*2.952*	*626.5*	*1.2356*	*Sat.*
50	......	......	......	......	......	......	......	......	......	**50**
60	3.353	631.8	1.2571	3.160	630.5	1.2489	2.985	629.3	1.2409	60
70	3.442	638.3	.2695	3.245	637.2	.2616	3.068	636.0	.2539	70
80	3.529	644.7	.2814	3.329	643.6	.2736	3.149	642.6	.2661	80
90	3.614	650.9	.2928	3.411	649.9	.2852	3.227	649.0	.2778	90
100	3.698	657.0	1.3038	3.491	656.1	1.2963	3.304	655.2	1.2891	**100**
110	3.780	663.0	.3144	3.570	662.1	.3070	3.380	661.3	.2999	110
120	3.862	668.9	.3247	3.647	668.1	.3174	3.454	667.3	.3104	120
130	3.942	674.7	.3347	3.724	674.0	.3275	3.527	673.3	.3206	130
140	4.021	680.5	.3444	3.799	679.8	.3373	3.600	679.2	.3305	140
150	4.100	686.3	1.3539	3.874	685.6	1.3469	3.672	685.0	1.3401	**150**
160	4.178	692.0	.3633	3.949	691.4	.3562	3.743	690.8	.3495	160
170	4.255	697.7	.3724	4.022	697.1	.3654	3.813	696.6	.3588	170
180	4.332	703.4	.3813	4.096	702.8	.3744	3.883	702.3	.3678	180
190	4.408	709.0	.3901	4.168	708.5	.3833	3.952	708.0	.3767	190
200	4.484	714.7	1.3988	4.241	714.2	1.3919	4.021	713.7	1.3854	**200**
210	4.560	720.4	.4073	4.313	719.9	.4005	4.090	719.4	.3940	210
220	4.635	726.0	.4157	4.384	725.6	.4089	4.158	725.1	.4024	220
230	4.710	731.7	.4239	4.455	731.3	.4172	4.226	730.8	.4108	230
240	4.785	737.3	.4321	4.526	736.9	.4254	4.294	736.5	.4190	240
250	4.859	743.0	1.4401	4.597	742.6	1.4334	4.361	742.2	1.4271	**250**
260	4.933	748.7	.4481	4.668	748.3	.4414	4.428	747.9	.4350	260
280	5.081	760.0	.4637	4.808	759.7	.4570	4.562	759.4	.4507	280
300	5.228	771.5	.4789	4.947	771.2	.4723	4.695	770.8	.4660	300

Temp. °F.	115 *63.65°*			120 *66.02°*			125 *68.31°*			Temp. °F.
Sat.	*2.580*	*628.0*	*1.2237*	*2.476*	*628.4*	*1.2201*	*2.380*	*628.8*	*1.2166*	*Sat.*
70	2.628	632.5	1.2323	2.505	631.3	1.2255	2.392	630.0	1.2189	70
80	2.701	639.4	.2451	2.576	638.3	.2386	2.461	637.2	.2322	80
90	2.772	646.0	.2574	2.645	645.0	.2510	2.528	644.0	.2448	90
100	2.841	652.5	1.2690	2.712	651.0	1.2628	2.593	650.7	1.2568	**100**
110	2.909	658.8	.2802	2.778	658.0	.2741	2.657	657.1	.2682	110
120	2.975	665.0	.2910	2.842	664.2	.2850	2.719	663.5	.2792	120
130	3.040	671.1	.3015	2.905	670.4	.2956	2.780	669.7	.2899	130
140	3.105	677.2	.3116	2.967	676.5	.3058	2.840	675.8	.3002	140
150	3.168	683.1	1.3215	3.029	682.5	1.3157	2.900	681.8	1.3102	**150**
160	3.231	689.0	.3311	3.089	688.4	.3254	2.958	687.8	.3199	160
170	3.294	694.9	.3405	3.149	694.3	.3348	3.016	693.7	.3294	170
180	3.355	700.7	.3497	3.209	700.2	.3441	3.074	699.6	.3387	180
190	3.417	706.5	.3587	3.268	706.0	.3531	3.131	705.5	.3478	190
200	3.477	712.3	1.3675	3.326	711.8	1.3620	3.187	711.3	1.3567	**200**
210	3.538	718.1	.3762	3.385	717.6	.3707	3.243	717.2	.3654	210
220	3.598	723.8	.3847	3.442	723.4	.3793	3.299	723.0	.3740	220
230	3.658	729.6	.3931	3.500	729.2	.3877	3.354	728.8	.3825	230
240	3.717	735.3	.4014	3.557	734.9	.3960	3.409	734.5	.3908	240
250	3.776	741.1	1.4096	3.614	740.7	1.4042	3.464	740.3	1.3990	**250**
260	3.835	746.8	.4176	3.671	746.5	.4123	3.519	746.1	.4071	260
270	3.894	752.6	.4256	3.727	752.2	.4202	3.573	751.9	.4151	270
280	3.952	758.4	.4334	3.783	758.0	.4281	3.627	757.7	.4230	280
290	4.011	764.1	.4411	3.839	763.8	.4359	3.681	763.5	.4308	290
300	4.069	769.9	1.4488	3.895	769.6	1.4435	3.735	769.3	1.4385	**300**

V, volume in ft³ per lb; *H*, heat content in Btu per lb; *S*, entropy in Btu per lb, °F.

TABLE IV. PROPERTIES OF SUPERHEATED AMMONIA VAPOR—(*Continued*)

Temp. °F.	Absolute pressure in lbs./in.² (Saturation temperature in italics.)									Temp. °F.
	125 *68.31°*			**130** *70.53°*			**135** *72.69°*			
	V	*H*	*S*	*V*	*H*	*S*	*V*	*H*	*S*	
Sat.	*2.380*	*628.8*	*1.2166*	*2.291*	*629.2*	*1.2132*	*2.209*	*629.6*	*1.2100*	*Sat.*
80	2.461	637.2	1.2322	2.355	636.0	1.2260	2.2[illegible]7	634.9	1.2199	80
90	2.528	644.0	.2448	2.421	643.0	.2388	2.321	642.0	.2329	90
100	2.593	650.7	1.2568	2.484	649.7	1.2509	2.382	648.8	1.2452	**100**
110	2.657	657.1	.2682	2.546	656.3	.2625	2.442	655.4	.2569	110
120	2.719	663.5	.2792	2.606	662.7	.2736	2.501	661.9	.2681	120
130	2.780	669.7	.2899	2.665	668.9	.2843	2.559	668.2	.2790	130
140	2.840	675.8	.3002	2.724	675.1	.2947	2.615	674.4	.2894	140
150	2.900	681.8	1.3102	2.781	681.2	1.3048	2.671	680.5	1.2996	**150**
160	2.958	687.8	.3199	2.838	687.2	.3146	2.726	686.6	.3094	160
170	3.016	693.7	.3294	2.894	693.2	.3241	2.780	692.6	.3191	170
180	3.074	699.6	.3387	2.949	699.1	.3335	2.834	698.6	.3284	180
190	3.131	705.5	.3478	3.004	705.0	.3426	2.887	704.5	.3376	190
200	3.187	711.3	1.3567	3.059	710.9	1.3516	2.940	710.4	1.3466	**200**
210	3.243	717.2	.3654	3.113	716.7	.3604	2.992	716.2	.3554	210
220	3.299	723.0	.3740	3.167	722.5	.3690	3.044	722.1	.3641	220
230	3.354	728.8	.3825	3.220	728.3	.3775	3.096	727.9	.3726	230
240	3.409	734.5	.3908	3.273	734.1	.3858	3.147	733.7	.3810	240
250	3.464	740.3	1.3990	3.326	739.9	1.3941	3.198	739.6	1.3893	**250**
260	3.519	746.1	.4071	3.379	745.7	.4022	3.249	745.4	.3974	260
270	3.573	751.9	.4151	3.431	751.5	.4102	3.300	751.2	.4054	270
280	3.627	757.7	.4230	3.483	757.3	.4181	3.350	757.0	.4133	280
290	3.681	763.5	.4308	3.535	763.1	.4259	3.400	762.8	.4212	290
300	3.735	769.3	1.4385	3.587	769.0	1.4336	3.450	768.6	1.4289	**300**
320	3.842	780.9	.4536	3.690	780.6	.4487	3.550	780.3	.4441	320
	150 *78.81°*			**160** *82.64°*			**170** *86.29°*			
Sat.	*1.994*	*630.5*	*1.2009*	*1.872*	*631.1*	*1.1952*	*1.764*	*631.6*	*1.1900*	*Sat.*
90	2.061	638.8	1.2161	1.914	636.6	1.2055	1.784	634.4	1.1952	90
100	2.118	645.9	1.2289	1.969	643.9	1.2186	1.837	641.9	1.2087	**100**
110	2.174	652.8	.2410	2.023	651.0	.2311	1.889	649.1	.2215	110
120	2.228	659.4	.2526	2.075	657.8	.2429	1.939	656.1	.2336	120
130	2.281	665.9	.2638	2.125	664.4	.2542	1.988	662.8	.2452	130
140	2.334	672.3	.2745	2.175	670.9	.2652	2.035	669.4	.2563	140
150	2.385	678.6	1.2849	2.224	677.2	1.2757	2.081	675.9	1.2669	**150**
160	2.435	684.8	.2949	2.272	683.5	.2859	2.127	682.3	.2773	160
170	2.485	690.9	.3047	2.319	689.7	.2958	2.172	688.5	.2873	170
180	2.534	696.9	.3142	2.365	695.8	.3054	2.216	694.7	.2971	180
190	2.583	702.9	.3236	2.411	701.9	.3148	2.260	700.8	.3066	190
200	2.631	708.9	1.3327	2.457	707.9	1.3240	2.303	706.9	1.3159	**200**
210	2.679	714.8	.3416	2.502	713.9	.3331	2.346	713.0	.3249	210
220	2.726	720.7	.3504	2.547	719.9	.3419	2.389	719.0	.3338	220
230	2.773	726.6	.3590	2.591	725.8	.3506	2.431	724.9	.3426	230
240	2.820	732.5	.3675	2.635	731.7	.3591	2.473	730.9	.3512	240
250	2.866	738.4	1.3758	2.679	737.6	1.3675	2.514	736.8	1.3596	**250**
260	2.912	744.3	.3840	2.723	743.5	.3757	2.555	742.8	.3679	260
270	2.958	750.1	.3921	2.766	749.4	.3838	2.596	748.7	.3761	270
280	3.004	756.0	.4001	2.809	755.3	.3919	2.637	754.6	.3841	280
290	3.049	761.8	.4079	2.852	761.2	.3998	2.678	760.5	.3921	290
300	3.095	767.7	1.4157	2.895	767.1	1.4076	2.718	766.4	1.3999	**300**
320	3.185	779.4	.4310	2.980	778.9	.4229	2.798	778.3	.4153	320
340	3.274	791.2	.4459	3.064	790.7	.4379	2.878	790.1	.4303	340

V, volume in ft^3 per lb; *H*, heat content in Btu per lb; *S*, entropy in Btu per lb, °F.

TABLE IV. PROPERTIES OF SUPERHEATED AMMONIA VAPOR—(*Continued*)

Temp. °F.	Absolute pressure in lbs./in.² (Saturation temperature in italics.)									Temp. °F.
	140 *74.79°*			145 *76.83°*			150 *78.81°*			
	V	H	S	V	H	S	V	H	S	
Sat.	*2.132*	*629.9*	*1.2068*	*2.061*	*630.2*	*1.2038*	*1.994*	*630.5*	*1.2009*	Sat.
80	2.166	633.8	1.2140	2.080	632.6	1.2082	2.001	631.4	1.2025	80
90	2.228	640.9	.2272	2.141	639.9	.2216	2.061	638.8	.2161	90
100	2.288	647.8	1.2396	2.200	646.9	1.2342	2.118	645.9	1.2289	**100**
110	2.347	654.5	.2515	2.257	653.6	.2462	2.174	652.8	.2410	110
120	2.404	661.1	.2628	2.313	660.2	.2577	2.228	659.4	.2526	120
130	2.460	667.4	.2738	2.368	666.7	.2687	2.281	665.9	.2638	130
140	2.515	673.7	.2843	2.421	673.0	.2793	2.334	672.3	.2745	140
150	2.569	679.9	1.2945	2.474	679.2	1.2896	2.385	678.6	1.2849	**150**
160	2.622	686.0	.3045	2.526	685.4	.2996	2.435	684.8	.2949	160
170	2.675	692.0	.3141	2.577	691.4	.3093	2.485	690.9	.3047	170
180	2.727	698.0	.3236	2.627	697.5	.3188	2.534	696.9	.3142	180
190	2.779	704.0	.3328	2.677	703.4	.3281	2.583	702.9	.3236	190
200	2.830	709.9	1.3418	2.727	709.4	1.3372	2.631	708.9	1.3327	**200**
210	2.880	715.8	.3507	2.776	715.3	.3461	2.679	714.8	.3416	210
220	2.931	721.6	.3594	2.825	721.2	.3548	2.726	720.7	.3504	220
230	2.981	727.5	.3679	2.873	727.1	.3634	2.773	726.6	.3590	230
240	3.030	733.3	.3763	2.921	732.9	.3718	2.820	732.5	.3675	240
250	3.080	739.2	1.3846	2.969	738.8	1.3801	2.866	738.4	1.3758	**250**
260	3.129	745.0	.3928	3.017	744.6	.3883	2.912	744.3	.3840	260
270	3.179	750.8	.4008	3.064	750.5	.3964	2.958	750.1	.3921	270
280	3.227	756.7	.4088	3.111	756.3	.4043	3.004	756.0	.4001	280
290	3.275	762.5	.4166	3.158	762.2	.4122	3.049	761.8	.4079	290
300	3.323	768.3	1.4243	3.205	768.0	1.4199	3.095	767.7	1.4157	**300**
320	3.420	780.0	.4395	3.298	779.7	.4352	3.185	779.4	.4310	320
	180 *89.78°*			190 *93.13°*			200 *96.34°*			
Sat.	*1.667*	*632.0*	*1.1850*	*1.581*	*632.4*	*1.1802*	*1.502*	*632.7*	*1.1756*	Sat.
90	1.668	632.2	1.1853	....	....	....	....	....	....	90
100	1.720	639.9	1.1992	1.615	637.8	1.1899	1.520	635.6	1.1809	**100**
110	1.770	647.3	.2123	1.663	645.4	.2034	1.567	643.4	.1947	110
120	1.818	654.4	.2247	1.710	652.6	.2160	1.612	650.9	.2077	120
130	1.865	661.3	.2364	1.755	659.7	.2281	1.656	658.1	.2200	130
140	1.910	668.0	.2477	1.799	666.5	.2396	1.698	665.0	.2317	140
150	1.955	674.6	1.2586	1.842	673.2	1.2506	1.740	671.8	1.2429	**150**
160	1.999	681.0	.2691	1.884	679.7	.2612	1.780	678.4	.2537	160
170	2.042	687.3	.2792	1.925	686.1	.2715	1.820	684.9	.2641	170
180	2.084	693.6	.2891	1.966	692.5	.2815	1.859	691.3	.2742	180
190	2.126	699.8	.2987	2.005	698.7	.2912	1.897	697.7	.2840	190
200	2.167	705.9	1.3081	2.045	704.9	1.3007	1.935	703.9	1.2935	**200**
210	2.208	712.0	.3172	2.084	711.1	.3099	1.972	710.1	.3029	210
220	2.248	718.1	.3262	2.123	717.2	.3189	2.009	716.3	.3120	220
230	2.288	724.1	.3350	2.161	723.2	.3278	2.046	722.4	.3209	230
240	2.328	730.1	.3436	2.199	729.3	.3365	2.082	728.4	.3296	240
250	2.367	736.1	1.3521	2.236	735.3	1.3450	2.118	734.5	1.3382	**250**
260	2.407	742.0	.3605	2.274	741.3	.3534	2.154	740.5	.3467	260
270	2.446	748.0	.3687	2.311	747.3	.3617	2.189	746.5	.3550	270
280	2.484	753.9	.3768	2.348	753.2	.3698	2.225	752.5	.3631	280
290	2.523	759.9	.3847	2.384	759.2	.3778	2.260	758.5	.3712	290
300	2.561	765.8	1.3926	2.421	765.2	1.3857	2.295	764.5	1.3791	**300**
320	2.637	777.7	.4081	2.493	777.1	.4012	2.364	776.5	.3947	320
340	2.713	789.6	.4231	2.565	789.0	.4163	2.432	788.5	.4099	340

V, volume in ft³ per lb; *H*, heat content in Btu per lb; *S*, entropy in Btu per lb, °F.

TABLE IV. PROPERTIES OF SUPERHEATED AMMONIA VAPOR—(*Continued*)

Temp. °F.	Absolute pressure in lbs./in.² (Saturation temperature in italics.) 200 *96.34°*			210 *99.43°*			220 *102.42°*			Temp. °F.
	V	*H*	*S*	*V*	*H*	*S*	*V*	*H*	*S*	
Sat.	*1.502*	*632.7*	*1.1756*	*1.431*	*633.0*	*1.1713*	*1.367*	*633.2*	*1.1671*	*Sat.*
110	1.567	643.4	1.1947	1.480	641.5	1.1863	1.400	639.4	1.1781	110
120	1.612	650.9	.2077	1.524	649.1	.1996	1.443	647.3	.1917	120
130	1.656	658.1	.2200	1.566	656.4	.2121	1.485	654.8	.2045	130
140	1.698	665.0	.2317	1.608	663.5	.2240	1.525	662.0	.2167	140
150	1.740	671.8	1.2429	1.648	670.4	1.2354	1.564	669.0	1.2281	**150**
160	1.780	678.4	.2537	1.687	677.1	.2464	1.601	675.8	.2394	160
170	1.820	684.9	.2641	1.725	683.7	.2569	1.638	682.5	.2501	170
180	1.859	691.3	.2742	1.762	690.2	.2672	1.675	689.1	.2604	180
190	1.897	697.7	.2840	1.799	696.6	.2771	1.710	695.5	.2704	190
200	1.935	703.9	1.2935	1.836	702.9	1.2867	1.745	701.9	1.2801	**200**
210	1.972	710.1	.3029	1.872	709.2	.2961	1.780	708.2	.2896	210
220	2.009	716.3	.3120	1.907	715.3	.3053	1.814	714.4	.2989	220
230	2.046	722.4	.3209	1.942	721.5	.3143	1.848	720.6	.3079	230
240	2.082	728.4	.3296	1.977	727.6	.3231	1.881	726.8	.3168	240
250	2.118	734.5	1.3382	2.011	733.7	1.3317	1.914	732.9	1.3255	**250**
260	2.154	740.5	.3467	2.046	739.8	.3402	1.947	739.0	.3340	260
270	2.189	746.5	.3550	2.080	745.8	.3486	1.980	745.1	.3424	270
280	2.225	752.5	.3631	2.113	751.8	.3568	2.012	751.1	.3507	280
290	2.260	758.5	.3712	2.147	757.9	.3649	2.044	757.2	.3588	290
300	2.295	764.5	1.3791	2.180	763.9	1.3728	2.076	763.2	1.3668	**300**
320	2.364	776.5	.3947	2.246	775.9	.3884	2.140	775.3	.3825	320
340	2.432	788.5	.4099	2.312	787.9	.4037	2.203	787.4	.3978	340
360	2.500	800.5	.4247	2.377	800.0	.4186	2.265	799.5	.4127	360
380	2.568	812.5	.4392	2.442	812.0	.4331	2.327	811.6	.4273	380
	250 *110.80°*			260 *113.42°*			270 *115.97°*			
Sat.	*1.202*	*633.8*	*1.1555*	*1.155*	*633.9*	*1.1518*	*1.112*	*633.9*	*1.1483*	*Sat.*
120	1.240	641.5	1.1690	1.182	639.5	1.1617	1.128	637.5	1.1544	120
130	1.278	649.6	.1827	1.220	647.8	.1757	1.166	645.9	.1689	130
140	1.316	657.2	.1956	1.257	655.6	.1889	1.202	653.9	.1823	140
150	1.352	664.6	1.2078	1.292	663.1	1.2014	1.236	661.6	1.1950	**150**
160	1.386	671.8	.2195	1.326	670.4	.2132	1.269	669.0	.2071	160
170	1.420	678.7	.2306	1.359	677.5	.2245	1.302	676.2	.2185	170
180	1.453	685.5	.2414	1.391	684.4	.2354	1.333	683.2	.2296	180
190	1.486	692.2	.2517	1.422	691.1	.2458	1.364	690.0	.2401	190
200	1.518	698.8	1.2617	1.453	697.7	1.2560	1.394	696.7	1.2504	**200**
210	1.549	705.3	.2715	1.484	704.3	.2658	1.423	703.3	.2603	210
220	1.580	711.7	.2810	1.514	710.7	.2754	1.452	709.8	.2700	220
230	1.610	718.0	.2902	1.543	717.1	.2847	1.481	716.2	.2794	230
240	1.640	724.3	.2993	1.572	723.4	.2938	1.509	722.6	.2885	240
250	1.670	730.5	1.3081	1.601	729.7	1.3027	1.537	728.9	1.2975	**250**
260	1.699	736.7	.3168	1.630	736.0	.3115	1.565	735.2	.3063	260
270	1.729	742.9	.3253	1.658	742.2	.3200	1.592	741.4	.3149	270
280	1.758	749.1	.3337	1.686	748.4	.3285	1.620	747.7	.3234	280
290	1.786	755.2	.3420	1.714	754.5	.3367	1.646	753.9	.3317	290
300	1.815	761.3	1.3501	1.741	760.7	1.3449	1.673	760.0	1.3399	**300**
320	1.872	773.5	.3659	1.796	772.9	.3608	1.726	772.3	.3559	320
340	1.928	785.7	.3814	1.850	785.2	.3763	1.778	784.6	.3714	340
360	1.983	797.9	.3964	1.904	797.4	.3914	1.830	796.9	.3866	360
380	2.038	810.1	.4111	1.957	809.6	.4062	1.881	809.1	.4014	380
400	2.093	822.3	1.4255	2.009	821.9	1.4206	1.932	821.4	1.4158	**400**

V, volume in ft³ per lb; *H*, heat content in Btu per lb; *S*, entropy in Btu per lb, °F.

TABLE IV. PROPERTIES OF SUPERHEATED AMMONIA VAPOR—(*Continued*)

Temp. °F.	Absolute pressure in lbs./in.² (Saturation temperature in italics.)									Temp. °F.
	230 *105.30°*			240 *108.09°*			250 *110.80°*			
	V	H	S	V	H	S	V	H	S	
Sat.	*1.307*	*633.4*	*1.1631*	*1.253*	*633.6*	*1.1592*	*1.202*	*633.8*	*1.1555*	**Sat.**
110	1. 328	637. 4	1. 1700	1. 261	635. 3	1. 1621	....	....	....	110
120	1. 370	645. 4	. 1840	1. 302	643. 5	. 1764	1. 240	641. 5	1. 1690	120
130	1. 410	653. 1	. 1971	1. 342	651. 3	. 1898	1,278	649. 6	. 1827	130
140	1. 449	660. 4	. 2095	1. 380	658. 8	. 2025	1. 316	657. 2	. 1956	140
150	1. 487	667. 6	1. 2213	1. 416	666. 1	1. 2145	1. 352	664. 6	1, 2078	**150**
160	1. 524	674. 5	. 2325	1. 452	673. 1	. 2259	1. 386	671. 8	. 2195	160
170	1. 559	681. 3	. 2434	1. 487	680. 0	. 2369	1. 420	678. 7	. 2306	170
180	1. 594	687. 9	. 2538	1. 521	686. 7	. 2475	1. 453	685. 5	. 2414	180
190	1. 629	694. 4	. 2640	1. 554	693. 3	. 2577	1. 486	692. 2	. 2517	190
200	1. 663	700. 9	1. 2738	1. 587	699. 8	1. 2677	1. 518	698. 8	1. 2617	**200**
210	1. 696	707. 2	. 2834	1. 619	706. 2	. 2773	1. 549	705. 3	. 2715	210
220	1. 729	713. 5	. 2927	1. 651	712. 6	. 2867	1. 580	711. 7	. 2810	220
230	1. 762	719. 8	. 3018	1. 683	718. 9	. 2959	1. 610	718. 0	. 2902	230
240	1. 794	726. 0	. 3107	1. 714	725. 1	. 3049	1. 640	724. 3	. 2993	240
250	1. 826	732. 1	1. 3195	1. 745	731. 3	1. 3137	1. 670	730. 5	1. 3081	**250**
260	1. 857	738. 3	. 3281	1. 775	737. 5	. 3224	1. 699	736. 7	. 3168	260
270	1. 889	744. 4	. 3365	1. 805	743. 6	. 3308	1. 729	742. 9	. 3253	270
280	1. 920	750. 5	. 3448	1. 835	749. 8	. 3392	1. 758	749. 1	. 3337	280
290	1. 951	756. 5	. 3530	1. 865	755. 9	. 3474	1. 786	755. 2	. 3420	290
300	1. 982	762. 6	1. 3610	1. 895	762. 0	1. 3554	1. 815	761. 3	1. 3501	**300**
320	2. 043	774. 7	. 3767	1. 954	774. 1	. 3712	1. 872	773. 5	. 3659	320
340	2. 103	786. 8	. 3921	2. 012	786. 3	. 3866	1. 928	785. 7	. 3814	340
360	2. 163	798. 9	. 4070	2. 069	798. 4	. 4016	1. 983	797. 9	. 3964	360
380	2. 222	811. 1	. 4217	2. 126	810. 6	. 4163	2. 038	810. 1	. 4111	380
	280 *118.45°*			290 *120.86°*			300 *123.21°*			
Sat.	*1.072*	*634.0*	*1.1449*	*1.034*	*634.0*	*1.1415*	*0.999*	*634.0*	*1.1383*	**Sat.**
120	1. 078	635. 4	1. 1473	....	....	....	....	....	....	120
130	1. 115	644. 0	. 1621	1. 068	642. 1	1. 1554	1. 023	640. 1	1. 1487	130
140	1. 151	652. 2	. 1759	1. 103	650. 5	. 1695	1. 058	648. 7	. 1632	140
150	1. 184	660. 1	1. 1888	1. 136	658. 5	1. 1827	1. 091	656. 9	1. 1767	**150**
160	1. 217	667. 6	. 2011	1. 168	666. 1	. 1952	1. 123	664. 7	. 1894	160
170	1. 249	674. 9	. 2127	1. 199	673. 5	. 2070	1. 153	672. 2	. 2014	170
180	1. 279	681. 9	. 2239	1. 229	680. 7	. 2183	1. 183	679. 5	. 2129	180
190	1. 309	688. 9	. 2346	1. 259	687. 7	. 2292	1. 211	686. 5	. 2239	190
200	1. 339	695. 6	1, 2449	1. 287	694. 6	1. 2396	1. 239	693. 5	1. 2344	**200**
210	1. 367	702. 3	. 2550	1. 315	701. 3	. 2497	1. 267	700. 3	. 2447	210
220	1. 396	708. 8	. 2647	1. 343	707. 9	. 2596	1. 294	706. 9	. 2546	220
230	1. 424	715. 3	. 2742	1. 370	714. 4	. 2691	1. 320	713. 5	. 2642	230
240	1. 451	721. 8	. 2834	1. 397	720. 9	. 2784	1. 346	720. 0	. 2736	240
250	1. 478	728. 1	1. 2924	1. 423	727. 3	1. 2875	1. 372	726. 5	1. 2827	**250**
260	1. 505	734. 4	. 3013	1. 449	733. 7	. 2964	1. 397	732. 9	. 2917	260
270	1. 532	740. 7	. 3099	1. 475	740. 0	. 3051	1. 422	739. 2	. 3004	270
280	1. 558	747. 0	. 3184	1. 501	746. 3	. 3137	1. 447	745. 5	. 3090	280
290	1. 584	753. 2	. 3268	1. 526	752. 5	. 3221	1. 472	751. 8	. 3175	290
300	1. 610	759. 4	1. 3350	1. 551	758. 7	1. 3303	1. 496	758. 1	1. 3257	**300**
320	1. 661	771. 7	. 3511	1. 601	771. 1	. 3464	1. 544	770. 5	. 3419	320
340	1. 712	784. 0	. 3667	1. 650	783. 5	. 3621	1. 592	782. 9	. 3576	340
360	1. 762	796. 3	. 3819	1. 698	795. 8	. 3773	1. 639	795. 3	. 3729	360
380	1. 811	808. 7	. 3967	1. 747	808. 2	. 3922	1. 686	807. 7	. 3878	380
400	1. 861	821. 0	1. 4112	1. 794	820. 5	1. 4067	1. 732	820. 1	1. 4024	**400**

V, volume in ft³ per lb; *H*, heat content in Btu per lb; *S*, entropy in Btu per lb, °F.

Table V. Freon-11—Properties of Liquid and Saturated Vapor‡

Temp F t	Pressure		Liquid, sp vol	Vapor, sp vol	Enthalpy, datum − 40 F Btu per lb		Entropy, datum − 40 F Btu per lb F	
	psia	psig	cu ft/lb v_f	cu ft/lb v_g	Liquid h_f	Vapor h_g	Liquid s_f	Vapor s_g
−90	0.100	29.72*	0.00964	289.0	−9.7	81.7	−0.0247	0.2226
−80	.157	29.60*	.00961	189.0	−7.7	82.9	− .0195	.2192
−70	.240	29.43*	.00967	127.0	−5.8	84.0	− .0144	.2161
−60	.358	29.19*	.00974	87.0	−3.8	85.2	− .0095	.2132
−50	.520	28.86*	.00981	61.3	−1.9	86.3	− .0047	.2106
−40	0.739	28.42*	0.00988	44.21	0.00	87.48	0.0000	0.2085
−36	0.847	28.20*	.00991	38.93	0.79	87.96	.0019	.2076
−32	0.968	27.95*	.00993	34.37	1.58	88.44	.0037	.2068
−28	1.103	27.67*	.00996	30.44	2.36	88.91	.0055	.2060
−24	1.253	27.37*	.00999	27.03	3.15	89.39	.0073	.2053
−20	1.420	27.03*	0.01002	24.06	3.94	89.87	0.0091	0.2046
−16	1.605	26.65*	.01005	21.47	4.73	90.35	.0109	.2040
−12	1.810	26.24*	.01008	19.20	5.52	90.83	.0127	.2033
− 8	2.035	25.78*	.01011	17.21	6.31	91.31	.0145	.2027
− 4	2.283	25.27*	.01015	15.47	7.10	91.79	.0162	.2021
0	2.555	24.72*	0.01018	13.94	7.89	92.27	0.0179	0.2015
4	2.852	24.11*	.01021	12.58	8.68	92.75	.0197	.2010
5†	2.931	23.95*	.01022	12.27	8.88	92.88	.0201	.2009
8	3.179	23.45*	.01024	11.38	9.48	93.24	.0213	.2005
12	3.534	22.73*	.01027	10.31	10.28	93.72	.0231	.2000
16	3.923	21.94*	.01031	9.359	11.07	94.21	.0248	.1996
20	4.342	21.08*	0.01034	8.519	11.87	94.69	0.0264	0.1991
24	4.801	20.15*	.01037	7.760	12.68	95.18	.0281	.1987
28	5.294	19.14*	.01041	7.087	13.48	95.66	.0297	.1983
32	5.830	18.05*	.01044	6.481	14.28	96.15	.0314	.1979
36	6.411	16.87*	.01048	5.934	15.08	96.63	.0330	.1976
40	7.032	15.61*	0.01051	5.447	15.89	97.11	0.0346	0.1972
44	7.702	14.24*	.01055	5.006	16.70	97.60	.0362	.1969
48	8.422	12.78*	.01058	4.607	17.52	98.08	.0378	.1966
52	9.199	11.20*	.01062	4.245	18.33	98.56	.0394	.1963
56	10.02	9.53*	.01066	3.921	19.15	99.05	.0410	.1960
60	10.90	7.73*	0.01069	3.636	19.96	99.53	0.0426	0.1958
64	11.85	5.80*	.01073	3.356	20.78	100.01	.0442	.1955
68	12.87	3.72*	.01077	3.107	21.61	100.49	.0457	.1953
72	13.95	1.53*	.01081	2.883	22.43	100.97	.0473	.1950
76	15.09	0.39	.01085	2.679	23.26	101.45	.0489	.1948
80	16.31	1.61	0.01088	2.492	24.09	101.93	0.0504	0.1947
84	17.60	2.90	.01092	2.322	24.93	102.41	.0519	.1945
86†	18.28	3.58	.01094	2.242	25.34	102.65	.0527	.1944
88	18.97	4.27	.01096	2.165	25.76	102.89	.0535	.1943
92	20.43	5.73	.01101	2.020	26.60	103.36	.0550	.1941
96	21.97	7.27	.01105	1.887	27.43	103.83	.0565	.1940
100	23.60	8.90	0.01109	1.765	28.27	104.30	0.0580	0.1938
104	25.33	10.63	.01113	1.652	29.12	104.77	.0595	.1937
108	27.15	12.45	.01117	1.548	29.97	105.24	.0610	.1936
112	29.05	14.35	.01122	1.452	30.82	105.71	.0625	.1935
116	31.07	16.37	.01126	1.363	31.67	106.17	.0639	.1934
120	33.20	18.50	0.01130	1.281	32.53	106.63	0.0654	0.1933
124	35.42	20.72	.01135	1.206	33.38	107.09	.0669	.1932
128	37.74	23.04	.01139	1.135	34.24	107.55	.0683	.1931
132	40.23	25.53	.01144	1.068	35.10	108.00	.0698	.1930
136	42.80	28.10	.01149	1.007	35.97	108.46	.0712	.1929
140	45.50	30.80	0.01154	0.9505	36.84	108.91	0.0727	0.1929
144	48.35	33.65	.01159	.8970	37.71	109.35	.0741	.1928
148	51.31	36.61	.01163	.8476	38.59	109.80	.0755	.1927
152	54.41	39.71	.01168	.8014	39.46	110.24	.0770	.1927
156	57.65	42.95	.01173	.7581	40.35	110.69	.0784	.1927
160	61.04	46.34	0.01179	0.7176	41.23	111.12	0.0798	0.1926

* Inches of mercury below one standard atmosphere.
† Standard cycle temperatures.
Taken from ASRE DATA BOOK, 1951 Edition, and reprinted by permission of the American Society of Refrigerating Engineers.

‡ Courtesy, "Handbook of Automatic Refrigerant Controls," p. 122, copyright 1955 by Alco Valve Company, St. Louis 5, Mo.

TABLE VI. FREON-12—PROPERTIES OF LIQUID AND SATURATED VAPOR‡

Temp F t	Pressure		Liquid sp vol	Vapor, sp vol	Enthalpy, datum −40 F Btu per lb		Entropy, datum −40 F Btu per lb F	
	psia	psig	cu ft/lb v_f	cu ft/lb v_g	iquid h_f	Vapor h_g	Liquid s_f	Vapor s_g
−155	0.1163	29 68*	0.00954	232.29	−24.61	60.00	−0.0686	0.2092
−150	0.1527	29 61*	0.00957	179.79	−23.50	60.57	−0.0650	0.2065
−145	.1985	29 52*	.00961	140.52	−22.39	61.14	− .0615	.2040
−140	.2554	29.40*	.00965	110.92	−21.29	61.72	− .0580	.2017
−135	.3256	29.26*	.00969	88.34	−20.19	62.30	− .0546	.1995
−130	.4116	29.08*	.00973	70.94	−19.10	62.88	− .0512	.1975
−125	0.5160	28.87*	0.00977	57.42	−18.02	63.46	−0.0480	0.1955
−120	.6417	28.61*	.00981	46.84	−16.94	64.04	− .0448	.1937
−115	.7921	28 31*	.00985	38.49	−15.85	64.63	− .0416	.1919
−110	.9709	27.94*	.00989	31.84	−14.78	65.22	− .0385	.1903
−105	1.182	27.51*	.00994	26.51	−13.71	65.81	− .0355	.1888
−100	1.430	27.01*	0.00998	22.20	−12.64	66.40	−0.0325	0.1873
− 95	1.719	26.42*	.01003	18.71	−11.58	66.99	− .0295	.1860
− 90	2.054	25.74*	.01007	15.86	−10.51	67.59	− .0266	.1847
− 85	2.441	24.95*	.01012	13.51	− 9.46	68.18	− .0238	.1835
− 80	2.885	24.05*	.01016	11.57	− 8.40	68.77	− .0210	.1823
− 75	3.393	23.01*	0.01021	9.958	− 7.35	69.36	−0.0182	0.1813
− 70	3.971	21.84*	.01026	8.608	− 6.30	69.95	− .0155	.1802
− 65	4.626	20.50*	.01031	7.474	− 5.25	70.54	− .0128	.1793
− 60	5.365	19.00*	.01036	6.516	− 4.20	71.13	− .0102	.1783
− 55	6.195	17.31*	.01041	5.704	− 3.15	71.72	− .0076	.1774
− 50	7.125	15.42*	0.01047	5.012	− 2.11	72.31	−0.0050	0.1767
− 45	8.163	13.31*	.01052	4.420	− 1.06	72.91	− .0025	.1759
− 40	9.317	10.96*	.01057	3.911	0.00	73.50	.0000	.1752
− 35	10.60	8.34*	.01063	3.471	1.01	74.10	+ .0024	.1745
− 30	12.02	5.45*	0.0107	3.088	2.03	74.70	0.00471	0.17387
− 28	12.62	4.23*	.0107	2.950	2.44	74.94	.00565	.17364
− 26	13.26	2.93*	.0107	2.820	2.85	75.18	.00659	.17340
− 24	13.90	1.63*	.0108	2.690	3.25	75.41	.00753	.17317
− 22	14.58	0.24*	.0108	2.583	3.66	75.64	.00846	.17296
− 20	15.28	0.58	0.0108	2.474	4.07	75.87	0.00940	0.17275
− 18	16.01	1.31	.0108	2.370	4.48	76.11	.01033	.17253
− 16	16.77	2.07	.0108	2.271	4.89	76.34	.01126	.17232
− 14	17.55	2.85	.0109	2.177	5.30	76.57	.01218	.17212
− 12	18.37	3.67	.0109	2.088	5.72	76.81	.01310	17194
− 10	19.20	4.50	0.0109	2.003	6.14	77.05	0.01403	0.17175
− 8	20.08	5.38	.0109	1.922	6.57	77.29	.01496	.17158
− 6	20.98	6.28	.0110	1.845	6.99	77.52	.01589	.17140
− 4	21.91	7.21	.0110	1.772	7.41	77.75	.01682	.17123
− 2	22.87	8.17	.0110	1.703	7.83	77.98	.01775	.17107
0	23.87	9.17	0.0110	1.637	8.25	78.21	0.01869	0.17091
2	24.89	10.19	.0110	1.574	8.67	78.44	.01961	.17075
4	25.96	11.26	.0111	1.514	9.10	78.67	.02052	.17060
5†	26.51	11.81	.0111	1.485	9.32	78.79	.02097	.17052
6	27.05	12.35	.0111	1.457	9.53	78.90	.02143	.17045
8	28.18	13.48	.0111	1.403	9.96	79.13	.02235	.17030
10	29.35	14.65	0.0112	1.354	10.39	79.36	0.02328	0.17015
12	30.56	15.86	.0112	1.301	10.82	79.59	.02419	.17001
14	31.80	17.10	.0112	1.253	11.26	79.82	.02510	.16987
16	33.08	18.38	.0112	1.207	11.70	80.05	.02601	.16974
18	34.40	19.70	.0113	1.163	12.12	80.27	.02692	.16961
20	35.75	21.05	0.0113	1.121	12.55	80.49	0.02783	0.16949
22	37.15	22.45	.0113	1.081	13.00	80.72	.02873	.16938
24	38.58	23.88	.0113	1.043	13.44	80.95	.02963	.16926
26	40.07	25.37	.0114	1.007	13.88	81.17	.03053	.16913
28	41.59	26.89	.0114	0.973	14.32	81.39	.03143	.16900
30	43.16	28.46	0.0115	0.939	14.76	81.61	0.03233	0.16887
32	44.77	30.07	.0115	.908	15.21	81.83	.03323	.16876
34	46.42	31.72	.0115	.877	15.65	82.05	.03413	.16865
36	48.13	33.43	.0116	.848	16.10	82.27	.03502	.16854
38	49.88	35.18	.0116	.819	16.55	82.49	.03591	.16843

*** Inches of mercury below one atmosphere.**
† Standard cycle temperature.

‡ Courtesy, "Handbook of Automatic Refrigerant Controls," pp. 114–115, copyright 1955 by Alco Valve Company, St. Louis 5, Mo.

TABLE VI. FREON-12—PROPERTIES OF LIQUID AND SATURATED VAPOR—(*Continued*)

Temp F t	Pressure		Liquid, sp vol	Vapor, sp vol	Enthalpy, datum −40 F Btu per lb		Entropy, datum −40 F Btu per lb F	
	psia	psig	cu ft/lb v_f	cu ft/lb v_g	Liquid h_f	Vapor h_g	Liquid s_f	Vapor s_g
40	51.68	36.98	0.0116	0.792	17.00	82.71	0.03680	0.16833
42	53.51	38.81	.0116	.767	17.46	82.93	.03770	.16823
44	55.40	40.70	.0117	.742	17.91	83.15	.03859	.16813
46	57.35	42.65	.0117	.718	18.36	83.36	.03948	.16803
48	59.35	44.65	.0117	.695	18.82	83.57	.04037	.16794
50	61.39	46.69	0.0118	0.673	19.27	83.78	0.04126	0.16785
52	63.49	48.79	.0118	.652	19.72	83.99	.04215	.16776
54	65.63	50.93	.0118	.632	20.18	84.20	.04304	.16767
56	67.84	53.14	.0119	.612	20.64	84.41	.04392	.16758
58	70.10	55.40	.0119	.593	21.11	84.62	.04480	.16749
60	72.41	57.71	0.0119	0.575	21.57	84.82	0.04568	0.16741
62	74.77	60.07	.0120	.557	22.03	85.02	.04657	.16733
64	77.20	62.50	.0120	.540	22.49	85.22	.04745	.16725
66	79.67	64.97	.0120	.524	22.95	85.42	.04833	.16717
68	82.24	67.54	.0121	.508	23.42	85.62	.04921	.16709
70	84.82	70.12	0.0121	0.493	23.90	85.82	0.05009	0.16701
72	87.50	72.80	.0121	.479	24.37	86.02	.05097	.16693
74	90.20	75.50	.0122	.464	24.84	86.22	.05185	.16685
76	93.00	78.30	.0122	.451	25.32	86.42	.05272	.16677
78	95.85	81.15	.0123	.438	25.80	86.61	.05359	.16669
80	98.76	84.06	0.0123	0.425	26.28	86.80	0.05446	0.16662
82	101.7	87.00	.0123	.413	26.76	86.90	.05534	.16655
84	104.8	90.1	.0124	.401	27.24	87.18	.05621	.16648
86†	107.9	93.2	.0124	.389	27.72	87.37	.05708	.16640
88	111.1	96.4	.0124	.378	28.21	87.56	.05795	.16632
90	114.3	99.6	0.0125	0.368	28.70	87.74	0.05882	0.16624
92	117.7	103.0	.0125	.357	29.19	87.92	.05969	.16616
94	121.0	106.3	.0126	.347	29.68	88.10	.06056	.16608
96	124.5	109.8	.0126	.338	30.18	88.28	.06143	.16600
98	128.0	113.3	.0126	.328	30.67	88.45	.06230	.16592
100	131.6	116.9	0.0127	0.319	31.16	88.62	0.06316	0.16584
102	135.3	120.6	.0127	.310	31.65	88.79	.06403	.16576
104	139.0	124.3	.0128	.302	32.15	88.95	.06490	.16568
106	142.8	128.1	.0128	.293	32.65	89.11	.06577	.16560
108	146.8	132.1	.0129	.285	33.15	89.27	.06663	.16551
110	150.7	136.0	0.0129	0.277	33.65	89.43	0.06749	0.16542
112	154.8	140.1	.0130	.269	34.15	89.58	.06836	.16533
114	158.9	144.2	.0130	.262	34.65	89.73	.06922	.16524
116	163.1	148.4	.0131	.254	35.15	89.87	.07008	.16515
118	167.4	152.7	.0131	.247	35.65	90.01	.07094	.16505
120	171.8	157.1	0.0132	0.240	36.16	90.15	0.07180	.16495
122	176.2	161.5	.0132	.233	36.66	90.28	.07266	.16484
124	180.8	166.1	.0133	.277	37.16	90.40	.07352	.16473
126	185.4	170.7	.0133	.220	37.67	90.52	.07437	.16462
128	190.1	175.4	.0134	.214	38.18	90.64	.07522	.16450
130	194.9	180.2	0.0134	0.208	38.69	90.76	0.07607	0.16438
132	199.8	185.1	.0135	.202	39.19	90.86	.07691	.16425
134	204.8	190.1	.0135	.196	39.70	90.96	.07775	.16411
136	209.9	195.2	.0136	.191	40.21	91.06	.07858	.16396
138	215.0	200.3	.0137	.185	40.72	91.15	.07941	.16380
140	220.2	205.5	0.0138	0.180	41.24	91.24	0.08024	0.16363

† Standard cycle temperature.

Taken from ASRE DATA BOOK, 1951 Edition, and reprinted by permission of the American Society of Refrigerating Engineers.

TABLE VII. CARRENE-7—PROPERTIES OF LIQUID AND SATURATED VAPOR†

Temp F	Pressure		Liquid, density	Vapor, sp vol	Enthalpy, datum −40 F Btu per lb		Entropy, datum −40 F Btu per lb F	
	psia	psig	lb/cu ft $1/v_f$	cu ft/lb v_g	Liquid h_f	Vapor h_g	Liquid s_f	Vapor s_g
−40	10.84	7.86*	84.20	4.0764	0.00	89.92	0.00000	0.21424
−38	11.42	6.68*	84.03	3.8825	0.51	90.19	.00122	.21388
−36	12.03	5.44*	83.86	3.7008	1.02	90.47	.00243	.21353
−34	12.66	4.15*	83.69	3.5279	1.54	90.75	.00365	.21319
−32	13.31	2.83*	83.51	3.3663	2.06	91.03	.00485	.21286
−30	14.00	1.40*	83.34	3.2133	2.57	91.30	0.00606	0.21254
−28	14.71	0.01	83.16	3.0679	3.09	91.58	.00725	.21222
−26	15.45	0.75	82.99	2.9311	3.61	91.86	.00845	.21191
−24	16.21	1.51	82.81	2.8015	4.13	92.13	.00963	.21159
−22	17.01	2.31	82.63	2.6789	4.66	92.41	.01083	.21131
−20	17.84	3.14	82.45	2.5627	5.18	92.68	0.01202	0.21102
−18	18.70	4.00	82.27	2.4525	5.70	92.95	.01322	.21074
−16	19.59	4.89	82.09	2.3482	6.23	93.23	.01438	.21045
−14	20.51	5.81	81.91	2.2489	6.76	93.50	.01557	.21018
−12	21.47	6.77	81.72	2.1552	7.28	93.77	.01674	.20991
−10	22.46	7.76	81.54	2.0657	7.81	94.04	0.01793	0.20966
− 8	23.49	8.79	81.35	1.9811	8.34	94.31	.01909	.20940
− 6	24.55	9.85	81.16	1.9006	8.88	94.58	.02027	.20916
− 4	25.65	10.95	80.97	1.8242	9.41	94.85	.02144	.20892
− 2	26.79	12.09	80.78	1.7512	9.94	95.11	.02259	.20867
0	27.96	13.26	80.59	1.6817	10.48	95.38	0.02376	0.20844
2	29.18	14.48	80.40	1.6160	11.01	95.64	.02491	.20821
4	30.43	15.73	80.21	1.5532	11.56	95.91	.02608	.20799
6	31.73	17.03	80.01	1.4930	12.09	96.17	.02722	.20776
8	33.06	18.36	79.81	1.4361	12.63	96.43	.02839	.20755
10	34.45	19.75	79.62	1.3816	13.17	96.69	0.02954	0.20734
12	35.88	21.18	79.42	1.3295	13.71	96.95	.03067	.20712
14	37.35	22.65	79.22	1.2799	14.26	97.21	.03182	.20693
16	38.86	24.16	79.02	1.2326	14.81	97.47	.03296	.20672
18	40.42	25.72	78.81	1.1873	15.36	97.73	.03411	.20653
20	42.03	27.33	78.61	1.1441	15.91	97.98	0.03526	0.20635
22	43.69	28.99	78.41	1.1027	16.46	98.24	.03638	.20615
24	45.40	30.70	78.20	1.0632	17.01	98.49	.03752	.20597
26	47.15	32.45	77.99	1.0255	17.56	98.74	.03865	.20578
28	48.96	34.26	77.79	0.9892	18.12	98.99	.03978	.20560
30	50.82	36.12	77.58	0.9546	18.67	99.24	0.04092	0.20544
32	52.74	38.04	77.37	.9214	19.23	99.49	.04203	.20526
34	54.71	40.01	77.15	.8895	19.80	99.74	.04316	.20508
36	56.72	42.02	76.94	.8591	20.35	99.98	.04430	.20493
38	58.80	44.10	76.73	.8298	20.91	100.22	.04542	.20476
40	60.94	46.24	76.51	0.8018	21.47	100.46	0.04654	0.20460
42	63.14	48.44	76.30	.7748	22.03	100.70	.04764	.20443
44	65.39	50.69	76.08	.7490	22.60	100.94	.04875	.20427
46	67.71	53.01	75.86	.7242	23.17	101.18	.04986	.20411
48	70.09	55.39	75.64	.7002	23.75	101.41	.05099	.20395
50	72.52	57.82	75.42	0.6774	24.32	101.65	0.05212	0.20382
52	75.02	60.32	75.20	.6554	24.89	101.88	.05321	.20366
54	77.57	62.87	74.97	.6343	25.47	102.11	.05431	.20350
56	80.22	65.52	74.75	.6138	26.04	102.33	.05543	.20335
58	82.91	68.21	74.52	.5944	26.60	102.56	.05649	.20321
60	85.66	70.96	74.29	0.5756	27.19	102.78	0.05763	0.20307
62	88.49	73.79	74.07	.5575	27.76	103.00	.05872	.20293
64	91.39	76.69	73.84	.5401	28.33	103.22	.05979	.20278
66	94.37	79.67	73.61	.5232	28.93	103.44	.06091	.20264
68	97.41	82.71	73.37	.5069	29.52	103.65	.06203	.20250
70	100.51	85.81	73.14	0.4914	30.11	103.87	0.06311	0.20235
72	103.71	89.01	72.91	.4763	30.69	104.08	.06419	.20221
74	106.97	92.27	72.67	.4619	31.28	104.29	.06527	.20207
76	110.29	95.59	72.43	.4479	31.87	104.49	.06637	.20193
78	113.70	99.00	72.20	.4344	32.47	104.69	.06749	.20180
80	117.20	102.5	71.96	0.4213	33.06	104.89	0.06855	0.20164
82	120.78	106.1	71.72	.4088	33.63	105.09	.06959	.20150
84	124.40	109.7	71.48	.3967	34.25	105.29	.07071	.20136
86	128.14	113.4	71.23	.3849	34.85	105.48	.07179	.20122
88	131.96	117.3	70.99	.3736	35.45	105.67	.07288	.20109

*** Inches of Mercury below one standard atmosphere (29.92 in.).**
W. A. Pennington, Compiler.

† Courtesy, "Handbook of Automatic Refrigerant Controls," pp. 119–120, copyright 1955 by Alco Valve Company, St. Louis 5, Mo.

TABLE VII. CARRENE-7—PROPERTIES OF LIQUID AND SATURATED VAPOR—(*Continued*)

Temp F t	Pressure		Liquid, density	Vapor, sp vol	Enthalpy, datum −40 F Btu per lb		Entropy, datum −40 F Btu per lb F	
	psia	psig	lb/cu ft $1/v_f$	cu ft/lb v_g	Liquid h_f	Vapor h_g	Liquid s_f	Vapor s_g
90	135.86	121.2	70.74	0.3626	36.05	105.86	0.07395	0.20093
92	139.83	125.1	70.50	.3520	36.67	106.04	.07505	.20078
94	143.90	129.2	70.25	.3418	37.26	106.22	.07611	.20065
96	148.03	133.3	70.00	.3319	37.89	106.40	.07722	.20050
98	152.27	137.6	69.75	.3224	38.48	106.58	.07826	.20036
100	156.61	141.9	69.50	0.3131	39.10	106.75	0.07935	0.20021
102	161.02	146.3	69.25	.3041	39.71	106.91	.08042	.20005
104	165.55	150.9	69.00	.2954	40.34	107.08	.08151	.19990
106	170.14	155.4	68.74	.2870	40.97	107.24	.08261	.19975
108	174.84	160.1	68.49	.2788	41.59	107.39	.08367	.19958
110	179.62	164.9	68.23	0.2709	42.23	107.55	0.08479	0.19943
112	184.51	169.8	67.97	.2633	42.83	107.70	.08582	.19928
114	189.47	174.8	67.71	.2559	43.48	107.85	.08691	.19911
116	194.55	179.9	67.45	.2486	44.11	107.98	.08800	.19893
118	199.71	185.0	67.19	.2418	44.73	108.13	.08905	.19879
120	204.99	190.3	66.93	0.2350	45.36	108.26	0.09012	0.19862
122	210.40	195.7	66.66	.2283	46.03	108.38	.09124	.19842
124	215.88	201.2	66.40	.2219	46.69	108.50	.09236	.19825
126	221.44	206.7	66.13	.2157	47.33	108.62	.09342	.19806
128	227.13	212.4	65.86	.2097	47.99	108.74	.09451	.19787
130	232.89	218.2	65.60	0.2038	48.65	108.84	0.09560	0.19767
132	238.80	224.1	65.33	.1982	49.29	108.95	.09666	.19748
134	244.84	230.1	65.05	.1928	49.92	109.05	.09771	.19730
136	250.96	236.3	64.78	.1873	50.60	109.14	.09883	.19709
138	257.17	242.5	64.51	.1820	51.31	109.22	.09996	.19684
140	263.52	248.8	64.23	0.1768	52.02	109.29	0.10113	0.19662

W. A. Pennington, Compiler.

Taken from ASRE DATA BOOK, 1951 Edition, and reprinted by permission of the American Society of Refrigerating Engineers.

TABLE VIII. SULFUR DIOXIDE—PROPERTIES OF LIQUID AND SATURATED VAPOR‡

Temp F	Pressure		Liquid, sp vol	Vapor, sp vol	Enthalpy, datum—40F Btu per lb		Entropy, datum —40F Btu per lb F		
t	psia	psig	cu ft/lb v_f	cu ft/lb v_g	Liquid h_f	Vapor h_g	Liquid s_f	Evap	Vapor s_g
−40	3.136	23.53*	0.01044	22.42	0.00	178.61	0.00000	0.42562	0.42562
−30	4.331	21.10*	.01053	16.56	2.93	179.90	.00674	.41190	.41864
−20	5.883	17.94*	.01063	12.42	5.98	181.07	.01366	.39826	.41192
−10	7.863	13.91*	.01072	9.44	9.16	182.13	.02075	.38469	.40544
0	10.35	8.85*	0.01082	7.280	12.44	183.07	0.02795	0.37122	0.39917
2	10.91	7.71*	.01084	6.923	13.12	183.25	.02941	.36853	.39794
4	11.50	6.51*	.01086	6.584	13.78	183.41	.03084	.36586	.39670
5†	11.81	5.88*	0.01087	6.421	14.11	183.49	0.03155	0.36454	0.39609
6	12.12	5.25*	.01088	6.266	14.45	183.57	.03228	.36319	.39547
8	12.75	3.97*	.01090	5.967	15.13	183.73	.03373	.36053	.39426
10	13.42	2.61*	0.01092	5.682	15.80	183.87	0.03519	0.35787	0.39306
11	13.77	1.89*	.01093	5.548	16.14	183.94	.03592	.35654	.39246
12	14.12	1.18*	.01094	5.417	16.48	184.01	.03664	.35521	.39185
13	14.48	0.45*	.01095	5.289	16.81	184.07	.03737	.35388	.39125
14	14.84	0.14	.01096	5.164	17.15	184.14	.03808	.35257	.39065
15	15.21	0.51	0.01097	5.042	17.49	184.21	0.03880	0.35125	0.39005
16	15.59	0.89	.01098	4.926	17.84	184.28	.03953	.34993	.38946
17	15.98	1.28	.01099	4.812	18.18	184.34	.04026	.34861	.38887
18	16.37	1.67	.01100	4.701	18.52	184.40	.04098	.34729	.38827
19	16.77	2.07	.01101	4.593	18.86	184.46	.04169	.34598	.38767
20	17.18	2.48	0.01102	4.487	19.20	184.52	0.04241	0.34466	0.38707
21	17.60	2.90	.01104	4.386	19.55	184.58	.04313	.34335	.38648
22	18.03	3.33	.01105	4.287	19.90	184.64	.04385	.34204	.38589
23	18.46	3.76	.01106	4.190	20.24	184.69	.04457	.34073	.38530
24	18.89	4.19	.01107	4.096	20.58	184.74	.04528	.33943	.38471
25	19.34	4.64	0.01108	3.994	20.92	184.79	0.04600	0.33812	0.38412
26	19.80	5.10	.01109	3.915	21.26	184.84	.04671	.33683	.38354
27	20.26	5.56	.01110	3.829	21.61	184.89	.04743	.33553	.38296
28	20.73	6.03	.01112	3.744	21.96	184.94	.04814	.33422	.38236
29	21.21	6.51	.01113	3.662	22.30	184.98	.04886	.33292	.38178
30	21.70	7.00	0.01114	3.581	22.64	185.02	0.04956	0.33163	0.38119
31	22.20	7.50	.01115	3.503	22.98	185.06	.05027	.33034	.38061
32	22.71	8.01	.01116	3.437	23.33	185.10	.05099	.32904	.38003
33	23.23	8.53	.01118	3.355	23.68	185.14	.05171	.32774	.37945
34	23.75	9.05	.01119	3.283	24.03	185.18	.05242	.32645	.37887
35	24.28	9.58	0.01120	3.212	24.38	185.22	0.05312	0.32517	0.37829
40	27.10	12.40	.01126	2.887	26.12	185.37	.05668	.31873	.37541
45	30.15	15.45	.01132	2.601	27.86	185.48	.06020	.31234	.37254
50	33.45	18.75	.01138	2.348	29.61	185.56	.06370	.30599	.36969
55	37.05	22.35	0.01144	2.124	31.36	185.60	0.06715	0.29971	0.36686
60	40.93	26.23	.01150	1.926	33.10	185.59	.07060	.29345	.36405
65	45.13	30.43	.01156	1.749	34.84	185.54	.07401	.28724	.36125
70	49.62	34.92	.01163	1.590	36.58	185.46	.07736	.28110	.35846
75	54.47	39.77	.01169	1.448	38.32	185.34	.08070	.27498	.35568
80	59.68	44.98	0.01176	1.321	40.05	185.17	0.08399	0.26897	0.35296
81	60.77	46.07	.01177	1.297	40.39	185.13	.08462	.26772	.35234
82	61.88	47.18	.01179	1.274	40.73	185.09	.08525	.26652	.35177
83	63.01	48.31	.01180	1.253	41.08	185.05	.08589	.26532	.35121
84	64.14	49.44	.01181	1.229	41.43	185.01	.08653	.26412	.35065
85	65.28	50.58	0.01183	1.207	41.78	184.97	0.08718	0.26291	0.35009
86†	66.45	51.75	.01184	1.185	42.12	184.92	.08783	.26171	.34954
87	67.64	52.94	.01185	1.164	42.46	184.87	.08847	.26052	.34899
88	68.84	54.14	.01187	1.144	42.80	184.82	.08910	.25993	.34843
89	70.04	55.34	.01188	1.124	43.15	184.77	.08974	.25813	.34787
90	71.25	56.55	0.01190	1.104	43.50	184.72	0.09038	0.25693	0.34731
95	77.60	62.90	.01196	1.011	45.20	184.43	.09349	.25103	.34452
100	84.52	69.82	.01204	0.9262	46.90	184.10	.09657	.24516	.34173
110	99.76	85.06	.01219	.7804	50.26	183.31	.10254	.23357	.33611
120	120.93	106.23	.01236	.6598	53.58	182.36	.10829	.22217	.33046
140	158.61	143.91	.01272	.4758	60.04	179.94	.11893	.19990	.31883

* Inches of mercury below one standard atmosphere.

† Standard cycle temperatures.

Note—For values used in making chart No. 18 see Transactions of American Institute of Chemical Engineers, vol. 41, no. 3, June 25, 1945.

Taken from ASRE DATA BOOK, 1951 Edition, and reprinted by permission of the American Society of Refrigerating Engineers.

‡ Courtesy, "Handbook of Automatic Refrigerant Controls," p. 127, copyright 1955 by Alco Valve Company, St. Louis 5, Mo.

Table IX. Vapor Pressures* at Various Fahrenheit Temperatures

(For Genetrons 113, 11, 114, 12, 22 and 13, classified respectively as R-113, R-11, R-114, R-12, R-22 and R-13)

Temp. °F	*Genetron 113*	*Genetron 11*	*Genetron 114*	*Genetron 12*	*Genetron 22*	*Genetron 13*
−150	—	—	—	—	*29.4*	*20.8*
−140	—	—	—	—	*29.0*	*16.8*
−130	—	—	—	*29.1*	*28.5*	*11.4*
−120	—	—	—	*28.6*	*27.7*	*4.5*
−110	—	—	—	*27.9*	*26.5*	2.12
−100	—	—	—	*27.0*	*25.0*	7.53
−90	—	—	—	*25.7*	*22.9*	14.2
−80	—	—	*29.0*	*24.0*	*20.1*	22.3
−70	—	—	*28.6*	*21.8*	*16.5*	32.0
−60	—	—	*28.0*	*19.0*	*11.9*	43.5
−50	—	*28.9*	*27.2*	*15.4*	*6.1*	57.0
−40	—	*28.4*	*26.1*	*10.9*	0.6	72.7
−30	*29.3*	*27.8*	*24.7*	*5.5*	4.9	90.9
−20	*29.0*	*27.0*	*22.9*	0.6	10.2	111.7
−15	*28.9*	*26.5*	*21.8*	2.5	13.2	123.1
−10	*28.7*	*26.0*	*20.6*	4.5	16.6	135.4
−5	*28.4*	*25.4*	*19.3*	6.8	20.2	148.4
0	*28.2*	*24.7*	*17.8*	9.2	24.1	162.1
5	*27.9*	*24.0*	*16.1*	11.8	28.3	176.7
10	*27.5*	*23.1*	*14.3*	14.7	32.9	192.1
15	*27.2*	*22.1*	*12.3*	17.8	37.9	208.5
20	*26.7*	*21.1*	*10.1*	21.1	43.3	225.7
25	*26.3*	*19.9*	*7.6*	24.7	49.0	244.0
30	*25.7*	*18.6*	*5.0*	28.5	55.2	263.2
35	*25.1*	*17.2*	*2.1*	32.6	61.9	283.5
40	*24.4*	*15.6*	0.5	37.0	69.0	304.9
45	*23.7*	*13.9*	2.2	41.7	76.6	327.5
50	*22.9*	*12.0*	4.0	46.8	84.7	351.2
55	*22.0*	*9.9*	6.0	52.1	93.3	376.1
60	*21.0*	*7.7*	8.1	57.8	102.4	402.3
65	*19.8*	*5.3*	10.4	63.9	112.2	429.8
70	*18.6*	*2.6*	12.9	70.3	122.5	458.7
75	*17.3*	0.1	15.5	77.1	133.4	489.0
80	*15.8*	1.6	18.3	84.3	145.0	520.8
85	*14.2*	3.2	21.4	91.9	157.2	—
90	*12.5*	5.0	24.6	100.0	170.0	—
95	*10.7*	6.9	28.0	108.4	183.6	—
100	*8.6*	8.9	31.7	117.4	197.9	—
105	*6.4*	11.1	35.6	126.8	212.9	—
110	*4.1*	13.4	39.7	136.7	228.6	—
115	*1.5*	15.9	44.1	147.1	245.2	—
120	0.6	18.5	48.7	158.0	262.5	—
125	2.1	21.3	53.7	169.5	280.7	—
130	3.6	24.3	58.8	181.5	299.7	—
135	5.3	27.5	64.3	194.1	319.6	—
140	7.0	30.8	70.1	207.2	340.3	—
145	8.9	34.4	76.2	221.0	362.0	—
150	11.0	38.1	82.6	235.4	384.6	—

* Vapor pressures are shown as psig.
Italic figures are shown as inches of mercury vacuum.
Courtesy of Allied Chemical Corporation, General Chemical Division.

Table X. Properties of Saturated Steam*

Abs. Press. psi (*p*)	*Temp.* °F (*t*)	*Specific Volume* (*V*)		*Enthalpy*			*Entropy* (*s*)			*Internal Energy*
		Liquid	Vapor	Liquid (*h*)	Evap. (*L*)	Vapor (*H*)	Liquid	Evap.	Vapor	Evap. (h_1)
1.0	101.74	0.01614	333.6	69.70	1036.3	1106.0	0.1326	1.8456	1.9782	974.6
2.0	126.08	0.01623	173.73	93.99	1022.2	1116.2	0.1749	1.7451	1.9200	957.9
3.0	141.48	0.01630	118.71	109.37	1013.2	1122.6	0.2008	1.6855	1.8863	947.3
4.0	152.97	0.01636	90.63	120.86	1006.4	1127.3	0.2198	1.6427	1.8625	939.3
5.0	162.24	0.01640	73.52	130.13	1001.0	1131.1	0.2347	1.6094	1.8441	933.0
6.0	170.06	0.01645	61.98	137.96	996.2	1134.2	0.2472	1.5820	1.8292	927.5
7.0	176.85	0.01649	53.64	144.76	992.1	1136.9	0.2581	1.5586	1.8167	922.7
8.0	182.86	0.01653	47.34	150.79	988.5	1139.3	0.2674	1.5383	1.8057	918.4
9.0	188.28	0.01656	42.40	156.22	985.2	1141.4	0.2759	1.5203	1.7962	914.6
10	193.21	0.01659	38.42	161.17	982.1	1143.3	0.2835	1.5041	1.7876	911.1
14.696	212.00	0.01672	26.80	180.07	970.3	1150.4	0.3120	1.4446	1.7566	897.5
15	213.03	0.01672	26.29	181.11	969.7	1150.8	0.3135	1.4415	1.7549	896.7
20	227.96	0.01683	20.089	196.16	960.1	1156.3	0.3356	1.3962	1.7319	885.8
25	240.07	0.01692	16.303	208.42	952.1	1160.6	0.3533	1.3606	1.7139	876.8
30	250.33	0.01701	13.746	218.82	945.3	1164.1	0.3680	1.3313	1.6993	869.1
35	259.28	0.01708	11.898	227.91	939.2	1167.1	0.3807	1.3063	1.6870	862.3
40	267.25	0.01715	10.498	236.03	933.7	1169.7	0.3919	1.2844	1.6763	856.1
45	274.44	0.01721	9.401	243.36	928.6	1172.0	0.4019	1.2650	1.6669	850.5
50	281.01	0.01727	8.515	250.09	924.0	1174.1	0.4110	1.2474	1.6585	845.4
55	287.07	0.01732	7.787	256.30	919.6	1175.9	0.4193	1.2316	1.6509	840.6
60	292.71	0.01738	7.175	262.09	915.5	1177.6	0.4270	1.2168	1.6438	836.0
65	297.97	0.01743	6.655	267.50	911.6	1179.1	0.4342	1.2032	1.6374	831.8
70	302.92	0.01748	6.206	272.61	907.9	1180.6	0.4409	1.1906	1.6315	827.8
75	307.60	0.01753	5.816	277.43	904.5	1181.9	0.4472	1.1787	1.6259	824.0
80	312.03	0.01757	5.472	282.02	901.1	1183.1	0.4531	1.1676	1.6207	820.3
85	316.25	0.01761	5.168	286.39	897.8	1184.2	0.4587	1.1571	1.6158	816.8
90	320.27	0.01766	4.896	290.56	894.7	1185.3	0.4641	1.1471	1.6112	813.4
95	324.12	0.01770	4.652	294.56	891.7	1186.2	0.4692	1.1376	1.6068	810.2
100	327.81	0.01774	4.432	298.40	888.8	1187.2	0.4740	1.1286	1.6026	807.1
110	334.77	0.01782	4.049	305.66	883.2	1188.9	0.4832	1.1117	1.5948	801.2
120	341.25	0.01789	3.728	312.44	877.9	1190.4	0.4916	1.0062	1.5878	795.6
130	347.32	0.01706	3.455	318.81	872.9	1191.7	0.4995	1.0817	1.5812	790.2
140	353.02	0.01802	3.220	324.82	868.2	1193.0	0.5069	1.0682	1.5751	785.2
150	358.42	0.01809	3.015	330.51	863.6	1194.1	0.5138	1.0556	1.5694	780.5
160	363.53	0.01815	2.834	335.93	859.2	1195.1	0.5204	1.0436	1.5640	775.8
170	368.41	0.01822	2.675	341.09	854.9	1196.0	0.5266	1.0324	1.5590	771.4
180	373.06	0.01827	2.532	346.03	850.8	1196.9	0.5325	1.0217	1.5542	767.1
190	377.51	0.01833	2.404	350.79	846.8	1197.6	0.5381	1.0116	1.5497	763.0
200	381.79	0.01839	2.288	355.36	843.0	1198.4	0.5435	1.0018	1.5453	759.0
210	385.90	0.01844	2.183	359.77	839.2	1199.0	0.5487	0.9925	1.5412	755.2
220	389.86	0.01850	2.087	364.02	835.6	1199.6	0.5537	0.9835	1.5372	751.3
230	393.68	0.01854	1.9992	368.13	832.0	1200.1	0.5585	0.9750	1.5334	747.7
240	397.37	0.01860	1.9183	372.12	828.5	1200.6	0.5631	0.9667	1.5298	744.1
250	400.95	0.01865	1.8438	376.00	825.1	1201.1	0.5675	0.9588	1.5263	740.7

* Extracted from "Thermodynamic Properties of Steam," by Keenan and Keyes, published by John Wiley & Sons, Inc., New York.

TABLE XI. DIMENSIONS OF COPPER TUBING*

Type L Copper Tube

Diameter			Wall Thickness	Circumference		Transverse Area			Length of Pipe Ft./Sq. Ft.		Length of Pipe Ft/Cu. Ft.	Wt./Ft. Length in Lbs.
Nom.	O.D.	I.D.		Ext.	Int.	Ext.	Int.	Metal	Ext.	Int.		
⅛	.250	.200	0.025	.785	.628	.049	.031	.018	15.290	19.120	4645.00	0.68
¼	.375	.315	0.030	1.178	.990	.110	.078	.032	10.200	12.130	1847.00	.126
⅜	.500	.430	0.035	1.570	1.350	.196	.146	.050	7.650	8.900	987.00	.198
½	.625	.545	0.040	1.964	1.712	.307	.233	.074	6.110	7.010	618.00	.285
⅝	750	.666	0.042	2.356	2.090	.442	.348	.094	5.100	5.740	414.00	.362
¾	.875	.785	0.045	2.749	2.465	.601	.484	.117	4.370	4.870	297.50	.455
1	1.125	1.025	0.050	3.534	3.220	.994	.825	.169	3.390	3.730	174.60	.655
1¼	1 375	1.265	0.055	4.320	3.975	1.485	1.256	.229	2.775	3.020	114.70	.884
1½	1.625	1.505	0.060	5.105	4.730	2.074	1.780	.294	2.350	2.540	80.90	1.140
2	2.125	1.985	0.070	6.676	6.240	3.547	3.094	.453	1.798	1.925	46.60	1.750
2½	2.625	2.465	0.080	8.247	7.750	5.412	4.770	.642	1.455	1.550	30.20	2.480
3	3.125	2.945	0.090	9.818	9.250	7.670	6.812	.858	1.222	1.298	21.14	3.330
3½	3.625	3.425	0.100	11.388	10.760	10.321	9.213	1.108	1.054	1.115	15.64	4.290
4	4.125	3.905	0.110	12.959	12.270	13.360	11.970	1.390	.927	.979	12.03	5.390
5	5.125	4.875	0.125	16.100	15.300	20.630	18.670	1.960	.745	.785	7.72	7.610
6	6.125	5.845	0.140	19.242	18.370	29.460	26.830	2.630	.624	.654	5.36	10.200
8	8.125	7.725	0.200	25.526	24.270	51.850	46.870	4.980	.471	.494	3.07	19.300
10	10.125	9.625	0.250	31.809	30.250	80.520	72.760	7.760	.377	.397	1.98	30.060
12	12.125	11.565	0.280	38.092	36.300	115.470	107.050	8.420	.315	.331	1.34	40.390

Type K Copper Tube

Nom.	O.D.	I.D.	Wall Thickness	Circ. Ext.	Circ. Int.	Area Ext.	Area Int.	Area Metal	Ft./Sq. Ft. Ext.	Ft./Sq. Ft. Int.	Ft/Cu. Ft.	Wt./Ft.
⅛	.250	.186	0.032	.785	.585	.049	.027	022	15.290	20.500	5330.00	.085
¼	.375	.311	0.032	1.178	.977	.110	.076	.034	10.200	12.290	1895.00	.134
⅜	.500	.402	0.049	1.570	1.263	.196	.127	.069	7.650	9.500	1133.00	.269
½	625	.527	0.049	1.964	1.655	.307	.218	.089	6.110	7.250	660.00	.344
⅝	750	.652	0.049	2.356	2.047	.442	.334	.108	5.100	5.870	431.00	.418
¾	.875	.745	0.065	2.749	2.340	.601	.436	.165	4.370	5.130	330.00	.641
1	1.125	.995	0.065	3.534	3.126	.994	.778	.216	3.390	3.840	185.00	.839
1¼	1.375	1.245	0.065	4.320	3.910	1.485	1.217	.268	2.775	3.070	118.30	1.040
1½	1.625	1.481	0.072	5.105	4.650	2.074	1.722	.352	2 350	2.580	83.60	1.360
2	2.125	1.959	0.083	6.676	6.150	3.547	3.014	.533	1.798	1.950	47.80	2.060
2½	2.625	2.435	0.095	8.247	7.650	5.412	4.656	.756	1.455	1.570	30.90	2.920
3	3.125	2.907	0.109	9.818	9.130	7.670	6.637	1.033	1.222	1.315	21.70	4.000
3½	3.625	3.385	0.120	11.388	10.630	10.321	8.999	1.322	1.054	1.128	16.00	5.120
4	4.125	3.857	0.134	12.959	12.100	13.360	11.680	1.680	.927	.992	12.33	6.510
5	5.125	4.805	0.160	16.100	15.100	20.630	18.130	2.500	.745	.795	6.81	9.670
6	6.125	5.741	0.192	19.242	18.030	29.460	25.880	3.580	.624	.665	5.56	13.870
8	8.125	7.583	0.271	25.526	23.800	51.850	45.160	6.690	.471	.504	3.19	25.900
10	10.125	9.449	0.338	31.809	29.680	80.520	70.120	10.400	.377	.404	2.05	40.280
12	12.125	11.315	0.405	38.092	35.540	115.470	100.550	14.920	.315	.337	1.43	57.800

* Courtesy, "Handbook of Automatic Refrigerant Controls," p. 149, copyright 1955 by Alco Valve Company, St. Louis 5, Mo.

TABLE XII. ENGINEERING DETAILS, 2-STAGE RECIPROCATING COMPRESSOR†

No. of Cylinders Speed and Drive	*6 Cyl. Std. Speed Belt Driven*	*6 Cyl. High Speed Belt Driven*	*12 Cyl. Std. Speed Belt Driven*	*12 Cyl. High Speed Belt Driven*	*6 Cyl. Std. Speed Direct Conn.*	*6 Cyl. High Speed Direct Conn.*	*12 Cyl. Std. Speed Direct Conn.*	*12 Cyl. High Speed Direct Conn.*
Standard speeds rpm	1000* 900 810 720 650	1200* 1100	1000* 900 810 720 650	1200* 1100	900	1200*	900	1200*
Number of booster cylinders	4	4	8	8	4	4	8	8
Booster displacement at max rpm cu ft/min	128.9	154.6	257.8	309.3	116.0	154.6	232.0	309.3
Booster displacement cu in./rev	222.7	222.7	445.4	445.4	222.7	222.7	445.4	445.4
Number of high stage cylinders	2	2	4	4	2	2	4	4
High stage displacement @ max rpm cu ft/min	64.45	77.3	128.9	154.6	58.0	77.3	116.0	154.6
High stage displacement cu in./rev	111.35	111.35	222.7	222.7	111.35	111.35	222.7	222.7

Above information applies to *all* VMC two-stage compressors regardless of refrigerant used.
* Maximum.
† Courtesy, Vilter Manufacturing Co.

TABLE XIII. SOLID AND SATURATED CARBON DIOXIDE (CO_2)

Temp., °F.	Pressure		Volume		Enthalpy from —40°		Entropy from 32° + 1.0	
	Abs., lb./sq. in.	Gage, lb./sq. in.	Solid, cu. ft./lb.	Vapor, cu. ft./lb.	Solid, Btu./lb.	Vapor, Btu./lb.	Solid or liquid, Btu./lb. °F.	Vapor Btu./lb. °F.
t	p	p_d	v_f	v_g	h_f	h_g	s_f	s_g
—140	3.18	23.44*	0.01008	24.320	—121.5	129.2	.6065	1.3909
—130	5.39	18.84*	0.01012	14.740	—118.8	130.6	.6150	1.3717
—120	8.90	11.82*	0.01018	9.179	—116.0	132.0	.6232	1.3535
—110	14.22	1.00*	0.01024	5.848	—113.2	133.2	.6314	1.3362
—109.3	14.70	0.00	0.01025	5.665	—113.0	133.3	.6318	1.3358
—105	17.82	3.12	0.01028	4.703	—116.6	133.8	.6359	1.3279
—100	22.22	7.52	0.01032	3.804	—110.1	134.3	.6403	1.3199
— 95	27.54	12.84	0.01035	3.093	—108.5	134.7	.6449	1.3119
— 90.	33.98	19.28	0.01040	2.525	—106.7	135.1	.6499	1.3033
— 85	41.67	26.97	0.01044	2.070	—104.6	135.5	.6551	1.2960
— 80	50.85	36.15	0.01048	1.700	—102.5	135.7	.6607	1.2881
— 79	52.89	38.19	0.01049	1.636	—102.0	135.7	.6618	1.2866
— 78	55.00	40.30	0.01050	1.575	—101.5	135.8	.6630	1.2850
— 77	57.19	42.49	0.01051	1.516	—101.1	135.8	.6642	1.2836
— 76	59.45	44.75	0.01052	1.460	—100.5	135.8	.6655	1.2819
— 75	61.79	47.09	0.01053	1.407	—100.2	135.8	.6665	1.2802
— 74	64.21	49.51	0.01054	1.356	— 99.8	135.8	.6677	1.2788
— 73	66.72	52.02	0.01055	1.306	— 99.3	135.8	.6688	1.2772
— 72	69.33	54.63	0.01057	1.257	— 98.9	135.9	.6699	1.2757
— 71	72.03	57.33	0.01058	1.209	— 98.4	135.9	.6712	1.2741
— 70	74.82	60.12	0.01059	1.162	— 98.0	135.9	.6724	1.2726
— 69.9	75.10	60.40	0.01059	1.157	— 97.9	135.9	.6725	1.2724
— 69.9	75.10	60.40	0.01360	1.1570	— 13.7	135.9	.8885	1.2724
— 62	90.5	75.80	0.01379	0.9686	— 10.1	136.5	.8975	1.2662
— 60	94.7	80.0	0.01384	0.9270	— 9.2	136.6	.8997	1.2647
— 58	99.1	84.4	0.01389	0.8875	— 8.3	136.7	.9020	1.2632
— 56	103.7	89.0	0.01393	0.8502	— 7.4	136.8	.9042	1.2617
— 52	113.2	98.5	0.01404	0.7812	— 5.6	137.1	.9087	1.2587
— 50	118.2	103.5	0.01409	0.7492	— 4.7	137.2	.9110	1.2572
— 48	123.4	108.7	0.01414	0.7188	— 3.8	137.3	.9131	1.2558
— 46	128.7	114.0	0.01420	0.6899	— 2.9	137.4	.9153	1.2544
— 42	139.9	125.2	0.01431	0.6362	— 1.0	137.7	.9196	1.2516
— 40	145.8	131.1	0.01437	0.6113	.00	137.8	.9218	1.2503
— 38	151.8	137.1	0.01442	0.5876	+ .9	137.8	.9239	1.2489
— 36	158.0	143.3	0.01448	0.5649	1.8	138.0	.9261	1.2475
— 32	171.0	156.3	0.01460	0.5227	3.6	138.1	.9303	1.2449
— 30	177.8	163.1	0.01466	0.5029	4.5	138.2	.9325	1.2436
— 28	184.8	170.1	0.01472	0.4841	5.4	138.3	.9345	1.2424
— 26	192.0	177.3	0.01479	0.4661	6.3	138.3	.9366	1.2411
— 24	199.4	184.7	0.01485	0.4489	7.2	138.4	.9387	1.2398
— 22	207.0	192.3	0.01491	0.4325	8.1	138.4	.9408	1.2385
— 20	214.9	200.2	0.01498	0.4168	9.1	138.5	.9430	1.2372
— 18	223.0	208.3	0.01504	0.4092	10.1	138.6	.9449	1.2358
— 16	231.2	216.5	0.01511	0.3872	11.1	138.7	.9470	1.2345
— 14	239.6	224.9	0.01518	0.3666	12.0	138.7	.9491	1.2333
— 12	248.3	233.6	0.01525	0.3600	12.9	138.7	.9511	1.2321

*Inches of mercury below one atmosphere.
Courtesy American Society of Refrigerating, Air Conditioning and Ventilating Eng.

TABLE XIII. SATURATED CARBON DIOXIDE (CO_2) (CONCLUDED)

Temp., °F.	Pressure		Volume		Enthalpy from −40°		Entropy from 32° + 1.0	
	Abs., lb./sq. in.	Gage lb./sq. in.	Liquid, cu. ft./lb.	Vapour, cu. ft./lb.	Liquid, Btu./lb.	Vapor, Btu./lb.	Liquid, Btu./lb. °F.	Vapour, Btu./lb. °F.
t	p	p_d	v_f	v_g	h_f	h_g	s_f	s_g
−10	257.3	242.6	0.01532	0.3472	13.9	138.7	0.9532	1.2303
− 8	266.5	251.8	.01540	.3349	14.9	138.8	.9552	1.2297
− 6	275.9	261.2	.01547	.3231	15.9	138.8	.9573	1.2284
− 4	285.4	270.7	.01555	.3118	16.9	138.9	.9594	1.2272
− 2	295.3	280.6	.01563	.3009	17.9	138.9	.9614	1.2259
0	305.5	290.8	0.01570	0.2904	18.8	138.9	0.9636	1.2247
2	315.9	305.2	.01579	.2803	19.8	138.8	.9657	1.2235
4	326.5	311.8	.01588	.2707	20.8	138.8	.9679	1.2224
6	337.4	322.7	.01596	.2614	21.8	138.7	.9701	1.2212
8	348.7	334.0	.01605	.2526	22.9	138.7	.9722	1.2199
10	360.2	345.5	0.01614	0.2437	24.0	138.7	0.9744	1.2188
12	371.9	357.2	.01623	.2354	25.0	138.6	.9765	1.2175
14	383.9	369.2	.01632	.2274	26.1	138.6	.9787	1.2163
16	396.2	381.5	.01642	.2197	27.2	138.5	.9810	1.2151
18	408.9	394.2	.01652	.2121	28.3	138.4	.9833	1.2139
20	421.8	407.1	0.01663	0.2049	29.4	138.3	0.9856	1.2127
22	434.0	419.3	.01673	.1979	30.5	138.2	.9879	1.2115
24	448.4	433.7	.01684	.1912	31.7	138.1	.9902	1.2103
26	462.2	444.5	.01695	.1846	32.9	138.0	.9927	1.2091
28	476.3	461.6	.01707	.1783	34.1	137.9	.9951	1.2079
30	490.8	476.1	0.01719	0.1722	35.4	137.8	0.9976	1.2067
32	505.5	491.8	.01731	.1663	36.7	137.7	1.0000	1.2055
34	522.6	507.9	.01744	.1603	37.9	137.4	1.0023	1.2038
36	536.0	521.3	.01759	.1550	39.1	137.2	1.0046	1.2024
38	551.7	537.0	.01773	.1496	40.4	136.9	1.0071	1.2009
40	567.8	553.1	0.01787	0.1444	41.7	136.7	1.0092	1.1994
42	584.3	560.6	.01801	.1393	42.9	136.3	1.0115	1.1979
44	601.1	586.4	.01817	.1344	44.3	136.1	1.0140	1.1964
46	618.2	603.5	.01834	.1297	45.6	135.7	1.0166	1.1949
48	635.7	621.0	.01851	.1250	47.0	135.4	1.0194	1.1933
50	653.6	638.9	0.01868	0.1205	48.4	135.0	1.0218	1.1917
52	671.9	657.2	.01887	.1161	49.8	134.5	1.0254	1.1901
54	690.6	675.9	.01906	.1117	51.2	133.9	1.0272	1.1882
56	709.5	694.8	.01927	.1075	52.6	133.4	1.0283	1.1865
58	728.8	714.1	.01948	.1034	54.0	132.7	1.0312	1.1846
60	748.6	733.9	0.01970	0.0994	55.5	132.1	1.0353	1.1826
64	789.4	774.7	.02020	.0918	58.6	130.6	1.0410	1.1803
66	810.3	795.6	.02048	.0880	60.2	129.7	1.0438	1.1780
68	831.6	816.9	.02079	.08422	61.9	128.7	1.0468	1.1754
70	853.4	838.7	0.02112	0.08040	63.7	127.5	1.0500	1.1724
74	898.2	883.5	.02192	.07269	67.3	124.5	1.0568	1.1690
76	921.3	906.6	.02242	.06875	69.4	122.8	1.0607	1.1655
78	944.8	930.1	.02300	.06473	71.6	120.9	1.0649	1.1605
80	968.7	954.0	0.02370	0.06064	73.9	118.7	1.0694	1.1555
82	993.0	978.3	.02456	.05648	76.4	116.6	1.0740	1.1505
84	1017.7	1003.0	.02553	.05223	79.4	113.9	1.0790	1.1455
86	1043.0	1028.3	.02686	.04789	83.3	110.4	1.0854	1.1381
88	1069.4	1054.7	.03454	.03454	97.0	97.0	1.1098	1.1098

Index

RAYMOND C. GUNTHER

is a graduate of Pratt Institute in mechanical engineering and is now retired from the United States Army with the rank of Colonel. He has served in the Ordnance, Quartermaster and Civil Affairs branches of the Army. Colonel Gunther served as Officer-in-Charge of the Refrigeration School, Army Service Forces Training Center, at Fort Lee, Virginia, and as Refrigeration and Cold Storage Officer of the Caribbean Defense Command. Headquartered at Corozal in the Panama Canal Zone, his responsibilities were in Central and South America and the Pacific during World War II.

Colonel Gunther was a Field Engineer for Carrier Engineering Corp.; an Engineering Consultant for Mobil Oil Corp. to industry and machinery builders; a contributor to *World Book Encyclopedia*, "Refrigeration" Section; a Consulting Engineer, Refrigeration and Lubrication; a Registered Professional Engineer in New Jersey and Pennsylvania; a Qualified Fall-out Shelter Analyst for the Department of Defense, Office of Civil Defense; and Director of Civil Defense in East Vincent Township in Chester County, Pa.